# Cutting Edge Microbiology R
## for *Today's* Learners

The 13th Edition of Tortora, Funke, and Case's *Microbiology: An Introduction* brings a 21st-century lens to this trusted market-leading introductory textbook. New and updated features, such as **Exploring the Microbiome** boxes and **Big Picture** spreads, emphasize how our understanding of microbiology is constantly expanding. New **In the Clinic Video Tutors** in **Mastering**™ **Microbiology** illustrate how students can apply their learning to their future careers. Mastering Microbiology also includes new Ready-to-Go Teaching Modules that guide you through the most effective teaching tools available.

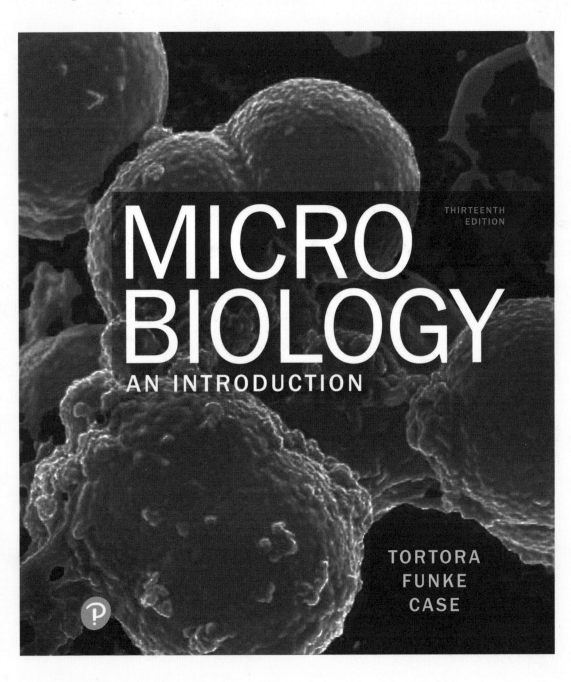

THIRTEENTH EDITION

# MICRO BIOLOGY
## AN INTRODUCTION

TORTORA
FUNKE
CASE

# Do your students struggle to make connections between course

## EXPLORING THE MICROBIOME  Do Artificial Sweeteners (and the Intestinal Microbiota That Love Them) Promote Diabetes?

For years, beverages made with artificial sweeteners were embraced by diabetics and weight watchers because, unlike sugar, artificial sweeteners don't impact blood glucose levels and don't provide calories. However, recent research indicates artificial sweeteners may actually increase the risk of nondiabetics developing the disease. One study published in 2009 by the American Diabetes Association found that daily consumption of diet soda was associated with a 67% greater relative risk of developing type 2 diabetes.

Undigestible by humans, artificial sweeteners provide zero calories to us when we consume them. But they are a great source of nutrients for *Bacteroides* bacteria living in the colon. As *Bacteroides* break down the sweeteners and increase in numbers, other types of microbiota simultaneously decline. Among these are *Lactobacillus* bacteria. Studies indicate that high *Lactobacillus* levels in the intestine are associated with decreased blood sugar levels. The exact mechanism remains unclear, but it is hypothesized that decreases in the population of *Lactobacillus* bacteria lead to higher blood glucose levels, thereby forcing the body to produce more insulin to control the rising blood glucose. Prolonged high insulin levels may lead to insulin resistance, a condition where the body stops responding correctly to the hormone. Insulin resistance is the hallmark sign of type 2 diabetes.

Recent and current research are exploring whether ingesting probiotics with *Lactobacillus acidophilus* and *Bifidobacterium animalis* may be a useful treatment for type 2 diabetes. Initial studies were promising, showing that these species might lower blood glucose levels. If proven effective, one day bacteria could be key weapons in preventing a deadly disease.

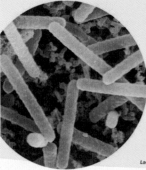

*Lactobacillus acidophilus.*

## EXPLORING THE MICROBIOME  Antimicrobial Soaps: Doing More Harm Than Good?

*Staphylococcus aureus* is a normal member of the human microbiome, found on the skin and in the nose. *S. aureus* is also a significant cause of healthcare-associated infections in patients. The bacterium can switch from benign member of the skin community to a disease-causing pathogen if it gains entry to the body through a wound.

Since most hospital-acquired *S. aureus* infections are endogenous—that is, caused by bacteria that have colonized in or on the body before someone became a patient—hospitals have long used a disinfectant called triclosan in clinical soaps and skin lotions to prevent staphylococcal infections. Over the years, triclosan was also added to many household products, such as dishwashing detergent, toothpastes, and body washes. However, using these antimicrobial products daily seems to be a case of "too much of a good thing."

Triclosan enters the blood and is excreted in urine. Therefore, triclosan can be found in many areas of the body, including the nasal mucosa, of people who use it. The nose is the primary habitat of *S. aureus*. In an example of unintended consequences, presence of triclosan in blood is also associated with nasal colonization of the *S. aureus*. *S. aureus* is more likely to bind to host-cell-membrane proteins in the presence of triclosan. Moreover, constant exposure to triclosan selects for triclosan-resistant mutants over generations of bacterial growth. Triclosan-resistant bacteria avoid death by removing the chemical from their cells using transporter proteins. These transporters can also remove some antibiotics from the bacterial cells. Moreover, methicillin-resistant *S. aureus* (MRSA) is more resistant to triclosan than methicillin-sensitive staphylococci.

Starting in late 2016, the Federal Drug Association banned triclosan from over-the-counter consumer washing products. The American Medical Association recommends using plain soap and water and proper handwashing techniques instead— these products and techniques remove microbes without the harmful unintended consequences associated with widespread triclosan use.

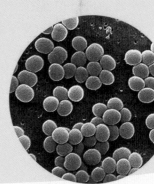

*Staphylococcus aureus.*

# content and their future careers?

**New! In the Clinic Video Tutors** bring to life the scenarios in the chapter-opening In the Clinic features. Concepts related to infection control, principles of disease, and antimicrobial therapies are integrated throughout the chapters, providing a platform for instructors to introduce clinically relevant topics throughout the term. Each Video Tutor has a series of assessments assignable in Mastering Microbiology that are tied to learning outcomes.

**NEW! Ready-to-Go Teaching Modules** in the Instructor Resources of Mastering Microbiology help instructors efficiently make use of the available teaching tools for the toughest topics in microbiology. Pre-class assignments, in-class activities, and post-class assessments are provided for ease of use.

Within the Ready-to-Go Teaching Modules, **Adopt a Microbe** modules enable instructors to select specific pathogens for additional focus throughout the text.

# Do your students need help understanding the toughest

**Interactive Microbiology** is a dynamic suite of interactive tutorials and animations that teach key microbiology concepts. Students actively engage with each topic and learn from manipulating variables, predicting outcomes, and answering assessment questions that test their understanding of basic concepts and their ability to integrate and build on these concepts. These are available in Mastering Microbiology.

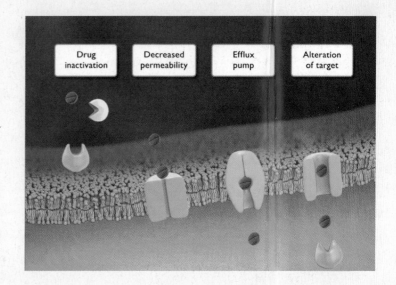

**NEW! Even more Interactive Microbiology** modules are available for Fall 2018. Additional titles include:

- Antimicrobial Resistance: Mechanisms
- Antimicrobial Resistance: Selection
- Aerobic Respiration in Prokaryotes
- The Human Microbiome

Antimicrobial

# concepts **in microbiology?**

**MicroBoosters** are a suite of brief video tutorials that cover key concepts some students may need to review or relearn. Titles include Study Skills, Math, Scientific Terminology, Basic Chemistry, Cell Biology, and Basic Biology.

**Energy** is the capacity or ability to cause change.

Types of Energy:

1. **Potential energy** — stored energy based on location or structure

*Lowest potential energy state at bottom of slide*

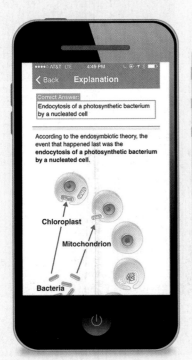

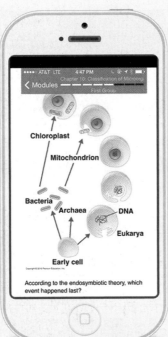

**Dynamic Study Modules** help students acquire, retain, and recall information faster and more efficiently than ever before. The flashcard-style modules are available as a self-study tool or can be assigned by the instructor.

**NEW!** Instructors can now remove questions from **Dynamic Study Modules** to better fit their course.

# Do your students have trouble organizing and synthesizing

**Big Picture** spreads integrate text and illustrations to help students gain a broad, "big picture" understanding of important course topics.

**Each Big Picture spread** includes an overview that **breaks down important concepts** into manageable steps and gives students a clear learning framework for related chapters. Each spread includes Key Concepts that **help students make the connection** between the presented topic and previously learned microbiology principles. Each spread is paired with a coaching activity and assessment questions in Mastering Microbiology.

---

**BIG PICTURE** | **Bioterrorism**

*Biological agents were first tapped by armies, and now by terrorists. Today, technology and ease of travel increase the potential damage.*

## History of Bioweapons

Biological weapons (bioweapons)—pathogens intentionally used for hostile purposes—are not new. The "ideal" bioweapon is one that disseminates by aerosol, spreads efficiently from human to human, causes debilitating disease, and has no readily available treatment.

The earliest recorded use of a bioweapon occurred in 1346 during the Siege of Kaffa, in what is now known as Feodosia, Ukraine. There the Tartar army catapulted their own dead soldiers' plague-ridden bodies over city walls to infect opposing troops. Survivors from that attack went on to introduce the "Black Death" to the rest of Europe, sparking the plague pandemic of 1348–1350.

In the eighteenth century, blankets contaminated with smallpox were intentionally introduced into Native American populations by the British during the French and Indian War. And during the Sino-Japanese War (1937–1945), Japanese planes dropped canisters of fleas carrying *Yersinia pestis* bacteria, the causative agent of plague, on China. In 1975, *Bacillus anthracis* endospores were accidentally released from a bioweapon production facility in Sverdlovsk.

A citadel in Ukraine, location of the first known biowarfare attack in history.

### Selected Diseases Identified as Potential Bioweapons

| Bacterial | Viral |
|---|---|
| Anthrax (*Bacillus anthracis*) | Nonbacterial meningitis (Arenaviruses) |
| Psittacosis (*Chlamydophila psittaci*) | Hantavirus disease |
| Botulism (*Clostridium botulinum* toxin) | Hemorrhagic fevers (Ebola, Marburg, Lassa) |
| Tularemia (*Francisella tularensis*) | Monkeypox |
| Cholera (*Vibrio cholerae*) | Nipah virus infection |
| Plague (*Yersinia pestis*) | Smallpox |

## Biological Weapons Banned in the Twentieth Century

The Geneva Conventions are internationally agreed upon standards for conducting war. Written in the 1920s, they prohibited deploying bioweapons—but did not specify that possessing or creating them was illegal. As such, most powerful nations in the twentieth century continued to create bioweapons, and the growing stockpiles posed an ever-growing threat. In 1975, the Biological Weapons Convention banned both possession and development of biological weapons. The majority of the world's nations ratified the treaty, which stipulated that any existing bioweapons be destroyed and related research halted.

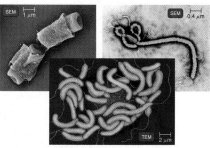

*(Clockwise from top left): Bacillus anthracis, Ebolavirus, and Vibrio cholerae are just a few microbes identified as potential bioterrorism agents.*

## Emergence of Bioterrorism

Unfortunately, the history of biowarfare doesn't end with the ratification of the Biological Weapons Convention. Since then, the main actors engaging in biowarfare have not been nations but rather radical groups and individuals. One of the most publicized bioterrorism incidents occurred in 2001, when five people died from, and many more were infected with, anthrax that an army researcher sent through the mail in letters.

**Kansas City, Mo.** Spores found in postal facility

**Indianapolis, Ind.** Spores found in postal facility

**New York City** Four confirmed cases; one death

**Princeton, Hamilton, N.J.** Five confirmed cases

**Washington, D.C., area** Three confirmed cases and two deaths

**Boca Raton, Lake Worth, Fla.** One confirmed case; one death

© 2008 MCT
Source: U.S. Centers for Disease Control and Prevention, AP
Graphic: Staff

Map showing location of 2001 bioterrorism anthrax attacks.

# visual information?

Play MicroFlix 3D Animation
@MasteringMicrobiology

## Public Health Authorities Try to Meet the Threat of Bioterrorism

One of the problems with bioweapons is that they contain living organisms, so their impact is difficult to control or even predict. However, public health authorities have created some protocols to deal with potential bioterrorism incidents.

CAUTION

BIOHAZARD

Biological hazard symbol.

### New Technologies and Techniques to Identify Bioweapons

Monitoring public health, and reporting incidence of diseases of note, is the first step in any bioterrorism defense plan. The faster a potential incident is uncovered, the greater the chance for containment. Rapid tests are being investigated to detect genetic changes in hosts due to bioweapons even before symptoms develop. Early-warning systems, such as DNA chips or recombinant cells that fluoresce in the presence of a bioweapon, are also being developed.

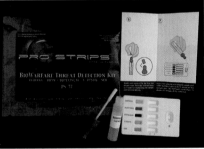

Pro Strips Rapid Screening System, developed by ADVNT Biotechnologies LLC, is the first advanced multi-agent biowarfare detection kit that tests for anthrax, ricin toxin, botulinum toxin, plague, and SEB (staphylococcal enterotoxin B).

### Vaccination: A Key Defense

When the use of biological agents is considered a possibility, military personnel and first-responders (health care personnel and others) are vaccinated—if a vaccine for the suspected agent exists. New vaccines are being developed, and existing vaccines are being stockpiled for use where needed.

The current plan to protect civilians in the event of an attack with a microbe is illustrated by the smallpox preparedness plan. This killer disease has been eradicated from the population, but unfortunately, a cache of the virus remains preserved in research facilities, meaning that it might one day be weaponized. It's not practical to vaccinate all people against the disease. Instead, the U.S. government's strategy following a confirmed smallpox outbreak includes "ring containment and voluntary vaccination." A "ring" of vaccinated/protected individuals is built around the bioterrorism infection case and their contacts to prevent further transmission.

Examining mail for *B. anthracis.*

### KEY CONCEPTS

- Vaccination is critical to preventing spread of infectious diseases, especially those that can be weaponized. **(See Chapter 18, "Principles and Effects of Vaccinations," pages 500–501.)**
- Many organisms that could be used for weapons require BSL-3 facilities. **(See Chapter 6, "Special Culture Techniques," pages 161–162.)**
- Tracking pathogen genomics provides information on its source. **(See Chapter 9, "Forensic Microbiology," pages 258–260.)**

697

Three Big Picture spreads focus on **important fundamental topics in microbiology:**

- Metabolism
- Genetics
- Immunity

Eight Big Picture spreads focus on **diseases and related public health issues that present complex real-world challenges:**

- Vaccine-Preventable Diseases
- The Hygiene Hypothesis
- Neglected Tropical Diseases
- Vertical Transmission: Mother to Child
- Climate Change and Disease
- Bioterrorism
- Cholera After Natural Disasters
- STI Home Test Kits

# Additional Instructor and Student Resources

**Learning Catalytics** is a "bring your own device" (laptop, smartphone, or tablet) student engagement, assessment, and classroom intelligence system. With **Learning Catalytics**, instructors can assess students in real time using open-ended tasks to probe student understanding. Mastering Microbiology users may select from Pearson's library of questions designed especially for use with **Learning Catalytics**.

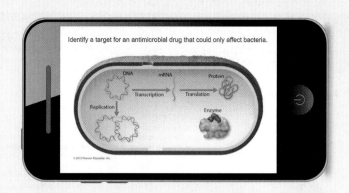

## Instructor Resource Materials for *Microbiology: An Introduction*

The Instructor Resource Materials organize all instructor media resources by chapter into one convenient and easy-to-use package containing:

- All figures, photos, and tables from the textbook in both labeled and unlabeled formats
- TestGen Test Bank
- MicroFlix animations
- Instructor's Guide

A wealth of additional classroom resources can be downloaded from the Instructor Resources area of Mastering Microbiology.

## *Laboratory Experiments in Microbiology*, 12th Edition by Johnson/Case

0-134-60520-9 / 978-0-134-60520-3

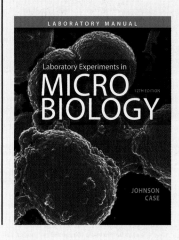

Engaging, comprehensive and customizable, *Laboratory Experiments in Microbiology* is the perfect companion lab manual for *Microbiology: An Introduction*, 13th Edition.

# MICRO BIOLOGY

## AN INTRODUCTION

### THIRTEENTH EDITION

Gerard J. Tortora
BERGEN COMMUNITY COLLEGE

Berdell R. Funke
NORTH DAKOTA STATE UNIVERSITY

Christine L. Case
SKYLINE COLLEGE

Editor-in-Chief: Serina Beauparlant
Courseware Portfolio Manager: Jennifer McGill Walker
Managing Producer: Nancy Tabor
Content & Design Manager: Michele Mangelli, Mangelli Productions, LLC
Courseware Director, Content Development: Barbara Yien
Courseware Sr. Analyst: Erin Strathmann
Courseware Editorial Assistant: Dapinder Dosanjh
Rich Media Content Producer: Lucinda Bingham and Tod Regan
Production Supervisor: Karen Gulliver
Copyeditor: Sally Peyrefitte
Proofreaders: Betsy Dietrich and Martha Ghent

Compositor: iEnergizer Aptara®, Ltd
Art Coordinator: Jean Lake
Interior & Cover Designer: Hespenheide Design
Illustrators: Imagineering STA Media Services, Inc.
Rights & Permissions Management: Ben Ferrini
Rights & Permissions Project Manager: Cenveo © Publishing Services, Matt Perry
Photo Researcher: Kristin Piljay
Manufacturing Buyer: Stacey Weinberger
Director of Product Marketing, Science: Allison Rona

Cover photo: Science Source

**Library of Congress Cataloging-in-Publication Data**
Names: Tortora, Gerard J., author. | Funke, Berdell R., author. | Case, Christine L., 1948- , author.
Title: Microbiology : an introduction / Gerard J. Tortora, Bergen Community College, Berdell R. Funke, North Dakota State University, Christine L. Case, Skyline College.
Description: Thirteenth edition. | Boston : Pearson, [2019] | Includes bibliographical references and index.
Identifiers: LCCN 2017044147| ISBN 9780134605180 (student edition) | ISBN 0134605187 (student edition) | ISBN 9780134709260 (instructor's review copy) | ISBN 0134709268 (instructor's review copy)
Subjects: LCSH: Microbiology.
Classification: LCC QR41.2 .T67 2019 | DDC 579--dc23 LC record available at https://lccn.loc.gov/2017044147

1   17

 Pearson

ISBN 10: 0-13-460518-7; ISBN 13: 978-0-13-460518-0 (Student edition)
ISBN 10: 0-13-470926-8; ISBN 13: 978-0-13-470926-0 (Instructor's Review Copy)

www.pearson.com

# About the Authors

**Gerard J. Tortora** Jerry Tortora is professor of biology and former biology coordinator at Bergen Community College in Paramus, New Jersey. He received his bachelor's degree in biology from Fairleigh Dickinson University and his master's degree in science education from Montclair State College. He has been a member of many professional organizations, including the American Society of Microbiology (ASM), the Human Anatomy and Physiology Society (HAPS), the American Association for the Advancement of Science (AAAS), the National Education Association (NEA), and the Metropolitan Association of College and University Biologists (MACUB).

Above all, Jerry is devoted to his students and their aspirations. In recognition of this commitment, MACUB presented Jerry with the organization's 1992 President's Memorial Award. In 1995, he was selected as one of the finest faculty scholars of Bergen Community College and was named Distinguished Faculty Scholar. In 1996, he received a National Institute for Staff and Organizational Development (NISOD) excellence award from the University of Texas and was selected to represent Bergen Community College in a campaign to increase awareness of the contributions of community colleges to higher education.

Jerry is the author of several best-selling science textbooks and laboratory manuals, a calling that often requires an additional 40 hours per week beyond his full-time teaching responsibilities. Nevertheless, he still makes time for four or five weekly aerobic workouts. He also enjoys attending opera performances at the Metropolitan Opera House, Broadway plays, and concerts. He spends his quiet time at his beach home on the New Jersey Shore.

To all my children, the most important gift I have: Lynne, Gerard Jr., Kenneth, Anthony, and Drew, whose love and support have been such an important part of my personal life and professional career.

**Berdell R. Funke** Bert Funke received his Ph.D., M.S., and B.S. in microbiology from Kansas State University. He has spent his professional years as a professor of microbiology at North Dakota State University. He taught introductory microbiology, including laboratory sections, general microbiology, food microbiology, soil microbiology, clinical parasitology, and pathogenic microbiology. As a research scientist in the Experiment Station at North Dakota State, he has published numerous papers in soil microbiology and food microbiology.

**Christine L. Case** Chris Case is a professor of microbiology at Skyline College in San Bruno, California, where she has taught for the past 46 years. She received her Ed.D. in curriculum and instruction from Nova Southeastern University and her M.A. in microbiology from San Francisco State University. She was Director for the Society for Industrial Microbiology and is an active member of the ASM. She received the ASM and California Hayward outstanding educator awards. Chris received the SACNAS Distinguished Community College Mentor Award for her commitment to her students, several of whom have presented at undergraduate research conferences and won awards. In addition to teaching, Chris contributes regularly to the professional literature, develops innovative educational methodologies, and maintains a personal and professional commitment to conservation and the importance of science in society. Chris is also an avid photographer, and many of her photographs appear in this book.

I owe my deepest gratitude to Don Biederman and our three children, Daniel, Jonathan, and Andrea, for their unconditional love and unwavering support.

# Digital Authors

**Warner B. Bair III** Warner Bair is a professor of biology at Lone Star College–CyFair in Cypress, Texas. He has a bachelor of science in general biology and a Ph.D. in cancer biology, both from the University of Arizona. He has over 10 years of higher education teaching experience, teaching both general biology and microbiology classes. Warner is the recipient of multiple educational awards, including the National Institute for Staff and Organizational Development (NISOD) excellence award from the University of Texas and the League for Innovation in the Community College John and Suanne Roueche Excellence Award. Warner has previously authored Interactive Microbiology® videos and activities for the MasteringMicrobiology website and is a member of the American Society for Microbiology (ASM). He is also a certified Instructional Skill Workshop (ISW) facilitator, where he assists other professors in the development of engaging and active classroom instruction. When not working, Warner enjoys outdoor activities and travel. Warner would like to thank his wife, Meaghan, and daughter, Aisling, for their support and understanding of the many late nights and long weekends he spends pursuing his writing.

**Derek Weber** Derek Weber is a professor of biology and microbiology at Raritan Valley Community College in Somerville, New Jersey. He received his B.S. in chemistry from Moravian College and his Ph.D. in biomolecular chemistry from the University of Wisconsin–Madison. His current scholarly work focuses on the use of instructional technology in a flipped classroom to create a more active and engaging learning environment. Derek has received multiple awards for these efforts, including the Award for Innovative Excellence in Teaching, Learning and Technology at the International Teaching and Learning Conference. As part of his commitment to foster learning communities, Derek shares his work at state and national conferences and is a regular attendee at the annual American Society for Microbiology Conference for Undergraduate Educators (ASMCUE). He has previously authored MicroBooster Video Tutorials, available in MasteringMicrobiology, which remediate students on basic concepts in biology and chemistry as they apply to microbiology. Derek acknowledges the support of his patient wife, Lara, and his children, Andrew, James, and Lilly.

# Preface

Since the publication of the first edition nearly 30 years ago, well over 1 million students have used *Microbiology: An Introduction* at colleges and universities around the world, making it the leading microbiology textbook for non-majors. The thirteenth edition continues to be a comprehensive beginning text, assuming no previous study of biology or chemistry. The text is appropriate for students in a wide variety of programs, including the allied health sciences, biological sciences, environmental science, animal science, forestry, agriculture, nutrition science, and the liberal arts.

The thirteenth edition has retained the features that have made this book so popular:

- **An appropriate balance between microbiological fundamentals and applications, and between medical applications and other applied areas of microbiology.** Basic microbiological principles are given greater emphasis, and health-related applications are featured.
- **Straightforward presentation of complex topics**. Each section of the text is written with the student in mind.
- **Clear, accurate, and pedagogically effective illustrations and photos.** Step-by-step diagrams that closely coordinate with narrative descriptions aid student comprehension of concepts.
- **Flexible organization.** We have organized the book in what we think is a useful fashion while recognizing that the material might be effectively presented in other sequences. For instructors who wish to use a different order, we have made each chapter as independent as possible and have included numerous cross-references. The Instructor's Guide provides detailed guidelines for organizing the material in several other ways.
- **Clear presentation of data regarding disease incidence.** Graphs and other disease statistics include the most current data available.
- **Big Picture core topic features.** These two-page spreads focus on the most challenging topics for students to master: metabolism (Chapter 5), genetics (Chapter 8), and immunology (Chapter 16). Each spread breaks down these important concepts into manageable steps and gives students a clear learning framework for the related chapters. Each refers the student to a related MicroFlix video accessible through MasteringMicrobiology.
- **Big Picture disease features.** These two-page spreads appear within each chapter in Part Four, Microorganisms and Human Disease (Chapters 21–26), as well as Chapters 18 (Practical Applications of Immunology) and 19 (Disorders of the Immune System). Each spread focuses on one significant public health aspect of microbiology.

- **ASM guidelines.** The American Society for Microbiology has released six underlying concepts and 27 related topics to provide a framework for key microbiological topics deemed to be of lasting importance beyond the classroom. The thirteenth edition explains the themes and competencies at the beginning of the book and incorporates callouts when chapter content matches one of these 27 topics. Doing so addresses two key challenges: it helps students and instructors focus on the enduring principles of the course, and it provides another pedagogical tool for instructors to assess students' understanding and encourage critical thinking.
- **Cutting-edge media integration.** MasteringMicrobiology (www.masteringmicrobiology.com) provides unprecedented, cutting-edge assessment resources for instructors as well as self-study tools for students. Big Picture Coaching Activities are paired with the book's Core Topics and Clinical Features. Interactive Microbiology is a dynamic suite of interactive tutorials and animations that teach key concepts in microbiology; and MicroBoosters are brief video tutorials that cover key concepts that some students may need to review or relearn.

## New to the Thirteenth Edition

The thirteenth edition focuses on big-picture concepts and themes in microbiology, encouraging students to visualize and synthesize more difficult topics such as microbial metabolism, immunology, and microbial genetics.

The thirteenth edition meets all students at their respective levels of skill and understanding while addressing the biggest challenges that instructors face. Updates to the thirteenth edition enhance the book's consistent pedagogy and clear explanations. Some of the highlights follow.

- **Exploring the Microbiome.** Each chapter has a new box featuring an aspect of microbiome study related to the chapter. Most feature the human microbiome. The boxes are designed to show the importance of microorganisms in health, their importance to life on Earth, and how research on the microbiome is being done.
- **In the Clinic videos accompanying each chapter opener.** In the Clinic scenarios that appear at the start of every chapter include critical-thinking questions that encourage students to think as health care professionals would in various clinical scenarios and spark student interest in the forthcoming chapter content. For the thirteenth edition, videos have been produced for the In the Clinic features for Chapters 1 through 20 and are accessible through MasteringMicrobiology.

- **New Big Picture disease features.** New Big Picture features include Vaccine-Preventable Diseases (Chapter 18), Vertical Transmission: Mother to Child (Chapter 22), and Bioterrorism (Chapter 24).
- **Reworked immunology coverage in Chapters 17, 18, and 19.** New art and more straightforward discussions make this challenging and critical material easier for students to understand and retain.

## Chapter-by-Chapter Revisions

Data in text, tables, and figures have been updated. Other key changes to each chapter are summarized below.

### Chapter 1
- The resurgence in microbiology is highlighted in sections on the Second and Third Golden Ages of Microbiology.
- The Emerging Infectious Diseases section has been updated.
- A discussion of normal microbiota and the human microbiome has been added.

### Chapter 2
- A discussion of the relationship between starch and normal microbiota has been added.

### Chapter 3
- Coverage of super-resolution light microscopy has been added.

### Chapter 4
- The description of the Gram stain method of action has been revised.
- Archaella are now covered.

### Chapter 5
- The potential for probiotic therapy using lactic acid bacteria is introduced.
- Reoxidation of NADH in fermentation is now shown in Figure 5.18.

### Chapter 6
- Discussion has been added regarding the influence of carrying capacity on the stationary phase of microbial growth.
- Discussion of quorum sensing in biofilms is included.
- The plate-streaking figure is revised.

### Chapter 7
- A new section on plant essential oils has been added.

### Chapter 8
- The discussion of operons, induction, and repression has been revised.

- Riboswitches are defined.
- A new box about tracking Zika virus is included.

### Chapter 9
- Discussion of gene editing using CRISPR technology has been added.

### Chapter 10
- Rapid identification using mass spectrophotometry is included.

### Chapter 11
- The genus *Prochlorococcus* is now included.
- The phylum Tenericutes has been added.

### Chapter 12
- The classification of algae and protozoa is updated.

### Chapter 13
- Baltimore classification is included.
- Virusoids are defined.

### Chapter 14
- Discussions of herd immunity and the control of healthcare-associated infections are expanded.
- Clinical trials are defined.
- Congenital transmission of infection is included.
- Discussion of the emerging HAI pathogen *Elizabethkingia* is now included.
- Epidemiological data have been updated.

### Chapter 15
- Genotoxin information is updated.

### Chapter 16
- The discussion of the role of normal microbiota in innate immunity is expanded.
- A table of chemical mediators of inflammation is included.

### Chapter 17
- A new table listing cytokines and their functions has been added.
- Cells involved in cell-mediated immunity are summarized in a table.

### Chapter 18
- Vaccine-preventable diseases are discussed in a new Big Picture.
- Coverage of recombinant vector vaccines has been added.

### Chapter 19
- The discussion of autoimmune diseases has been updated.
- The discussion of HIV/AIDS has been updated.
- The Big Picture box has been revised to expand discussion of dysbiosis-linked disorders.

## Chapter 20

- Tables have been reorganized.
- Coverage regarding the mechanisms of action of antimicrobial drugs has been updated.
- In the Clinical Focus box, data on antibiotics in animal feed have been updated.

## Chapter 21

- All data are updated.
- The Big Picture on Neglected Tropical Diseases has been revised to include river blindness.

## Chapter 22

- All data are updated.
- Coverage of Zika virus disease has been added.
- Discussion of Bell's palsy has been added.
- A new Big Picture covering vertical transmission of congenital infections has been added.

## Chapter 23

- All data are updated.
- The new species of *Borrelia* are included.
- Maps showing local transmission of vector-borne diseases have been updated.

## Chapter 24

- All data, laboratory tests, and drug treatments have been updated.
- The emerging pathogen *Enterovirus* D68 is included.
- A new Big Picture covering bioterrorism has been added.

## Chapter 25

- All data, laboratory tests, and drug treatments are updated.
- *Salmonella* nomenclature has been revised to reflect CDC usage.
- Images of protozoan oocysts and helminth eggs have been added to illustrate laboratory identification.

## Chapter 26

- All data, laboratory tests, and drug treatments have been updated.
- STIs that do not affect the genitourinary system are cross-referenced to the organ system affected.
- Discussion of ocular syphilis is now included.

## Chapter 27

- The concept of the Earth microbiome is introduced.
- Discussion of hydrothermal vent communities has been added.
- The discussions of bioremediation of oil and wastewater have been updated.

## Chapter 28

- The discussion of industrial fermentation has been updated.
- The definition of *biotechnology* is included.
- A discussion of the iChip has been added.
- A table listing fermented foods has been added.
- Discussion of microbial fuels cells is now included.

# Acknowledgments

In preparing this textbook, we have benefited from the guidance and advice of a large number of microbiology instructors across the country. These reviewers have provided constructive criticism and valuable suggestions at various stages of the revision. We gratefully acknowledge our debt to these individuals. Special thanks to retired epidemiologist Joel A. Harrison, Ph.D., M.P.H. for his thorough review and editorial suggestions.

## Contributor

Special thanks to Janette Gomos Klein, *CUNY Hunter College*, for her work on Chapters 17, 18, and 19.

## Reviewers

Jason Adams, *College of DuPage*
D. Sue Katz Amburn, *Rogers State University*
Ana Maria Barral, *National University*
Anar Brahmbhatt, *San Diego Mesa College*
Carron Bryant, *East Mississippi Community College*
Luti Erbeznik, *Oakland Community College*
Tod R. Fairbanks, *Palm Beach State College*
Myriam Alhadeff Feldman, *North Seattle College*
Kathleen Finan, *College of DuPage*
Annissa Furr, *Kaplan University*
Pattie S. Green, *Tacoma Community College*
Julianne Grose, *Brigham Young University*
Amy Jo M. Hammett, *Texas Woman's College*
Justin Hoshaw, *Waubonsee Community College*
Huey-Jane Liao, *Northern Virginia Community College*
Anne Montgomery, *Pikes Peak Community College*
Jessica Parilla, *Georgia State University*
Taylor Robertson, *Snead State Community College*
Michelle Scanavino, *Moberly Area Community College*
John P. Seabolt, *University of Kentucky*
Ginny Webb, *University of South Carolina Upstate*

We also thank the staff at Pearson Education for their dedication to excellence. Kelsey Churchman guided the early stages of this revision, and Jennifer McGill Walker brought it across the finish line. Erin Strathmann edited the new Exploring the Microbiome boxes, Chapters 17–19, and four new Big Picture spreads. Margot Otway edited the new In the Clinic videos. Serina Beauparlant and Barbara Yien kept the project moving during a period of staff transitioning.

Michele Mangelli, Mangelli Productions, LLC, managed the book from beginning to end. She expertly guided the team through the editorial phase, managed the new design, and then oversaw the production team and process. Karen Gulliver expertly guided the text through the production process and managed the day-to-day workflow. Sally Peyrefitte's careful attention to continuity and detail in her copyedit of both text and art served to keep concepts and information clear throughout. The talented staff at Imagineering gracefully managed the high volume and complex updates of our art and photo program. Jean Lake coordinated the many complex stages of the art and photo processing and kept the entire art team organized and on-track. Our photo researcher, Kristin Piljay, made sure we had clear and striking images throughout the book. Gary Hespenheide created the elegant interior design and cover. The skilled team at iEnergizer Aptara®, Ltd moved this book through the composition process. Maureen Johnson prepared the index, Betsy Dietrich carefully proofread the art, while Martha Ghent proofread pages. Stacey Weinberger guided the book through the manufacturing process. A special thanks goes to Amy Siegesmund for her detailed review of the pages. Lucinda Bingham, Amanda Kaufmann, and Tod Regan managed this book's robust media program. Courtney Towson managed the print ancillaries through the complex production stages.

Allison Rona, Kelly Galli, and the entire Pearson sales force did a stellar job presenting this book to instructors and students and ensuring its unwavering status as the best-selling microbiology textbook.

We would like to acknowledge our spouses and families, who have provided invaluable support throughout the writing process.

Finally, we have an enduring appreciation for our students, whose comments and suggestions provide insight and remind us of their needs. This text is for them.

*Gerard J. Tortora      Berdell R. Funke      Christine Case*

# Contents

# Features

## LIFE CYCLE FIGURES

## CLINICAL FOCUS

## DISEASES IN FOCUS

# ASM Recommended Curriculum Guidelines for Undergraduate Microbiology

The American Society for Microbiology (ASM) endorses a concept-based curriculum for introductory microbiology, emphasizing skills and concepts that remain important long after students exit the course. The ASM *Curriculum Guidelines for Undergraduate Microbiology Education* provide a framework for key microbiological topics and agree with scientific literacy reports from the American Association for the Advancement of Science and Howard Hughes Medical Institute. This textbook references part one of curriculum guidelines throughout chapters. When a discussion touches on one of the concepts, readers will see the ASM icon, along with a summary of the relevant statement.

ASM:

## ASM Guideline Concepts and Statements

### Evolution

- Cells, organelles (e.g., mitochondria and chloroplasts), and all major metabolic pathways evolved from early prokaryotic cells.
- Mutations and horizontal gene transfer, with the immense variety of microenvironments, have selected for a huge diversity of microorganisms.
- Human impact on the environment influences the evolution of microorganisms (e.g., emerging diseases and the selection of antibiotic resistance).
- The traditional concept of species is not readily applicable to microbes due to asexual reproduction and the frequent occurrence of horizontal gene transfer.
- The evolutionary relatedness of organisms is best reflected in phylogenetic trees.

### Cell Structure and Function

- The structure and function of microorganisms have been revealed by the use of microscopy (including brightfield, phase contrast, fluorescent, and electron).
- Bacteria have unique cell structures that can be targets for antibiotics, immunity, and phage infection.
- Bacteria and Archaea have specialized structures (e.g. flagella, endospores, and pili) that often confer critical capabilities.
- While microscopic eukaryotes (for example, fungi, protozoa, and algae) carry out some of the same processes as bacteria, many of the cellular properties are fundamentally different.
- The replication cycles of viruses (lytic and lysogenic) differ among viruses and are determined by their unique structures and genomes.

### Metabolic Pathways

- Bacteria and Archaea exhibit extensive, and often unique, metabolic diversity (e.g., nitrogen fixation, methane production, anoxygenic photosynthesis).
- The interactions of microorganisms among themselves and with their environment are determined by their metabolic abilities (e.g., quorum sensing, oxygen consumption, nitrogen transformations).
- The survival and growth of any microorganism in a given environment depend on its metabolic characteristics.
- The growth of microorganisms can be controlled by physical, chemical, mechanical, or biological means.

### Information Flow and Genetics

- Genetic variations can impact microbial functions (e.g., in biofilm formation, pathogenicity, and drug resistance).
- Although the central dogma is universal in all cells, the processes of replication, transcription, and translation differ in Bacteria, Archaea, and Eukaryotes.
- The regulation of gene expression is influenced by external and internal molecular cues and/or signals.
- The synthesis of viral genetic material and proteins is dependent on host cells.
- Cell genomes can be manipulated to alter cell function.

### Microbial Systems

- Microorganisms are ubiquitous and live in diverse and dynamic ecosystems.
- Most bacteria in nature live in biofilm communities.
- Microorganisms and their environment interact with and modify each other.
- Microorganisms, cellular and viral, can interact with both human and nonhuman hosts in beneficial, neutral, or detrimental ways.

### Impact of Microorganisms

- Microbes are essential for life as we know it and the processes that support life (e.g., in biogeochemical cycles and plant and/or animal microbiota).
- Microorganisms provide essential models that give us fundamental knowledge about life processes.
- Humans utilize and harness microorganisms and their products.
- Because the true diversity of microbial life is largely unknown, its effects and potential benefits have not been fully explored.

# The Microbial World and You 1

The overall theme of this textbook is the relationship between microbes—very small organisms that usually require a microscope to be seen—and our lives. We've all heard of epidemics of infectious diseases such as plague or smallpox that wiped out populations. However, there are many positive examples of human-microbe interactions. For example, we use microbial fermentation to ensure safe food supplies, and the human microbiome, a group of microbes that lives in and on our bodies, helps keep us healthy. We begin this chapter by discussing how organisms are named and classified and then follow with a short history of microbiology. Next, we discuss the incredible diversity of microorganisms and their ecological importance, noting how they recycle chemical elements such as carbon and nitrogen among the soil, organisms, and the atmosphere. We also examine how microbes are used to treat sewage, clean pollutants,

> ASM: Microorganisms provide essential models that give us fundamental knowledge about life processes.

control pests, and produce foods, chemicals, and drugs. Finally, we will discuss microbes as the cause of diseases such as Zika virus disease, avian (bird) flu, Ebola virus disease, and diarrhea, and we examine the growing public health problem of antibiotic-resistant bacteria.

Shown in the photograph are *Staphylococcus aureus* (STAF-i-lō-kok'kus OR-ē-us) bacteria on human nasal epithelial cells. These bacteria generally live harmlessly on skin or inside the nose. Misuse of antibiotics, however, allows the survival of bacteria with antibiotic-resistance genes, such as methicillin-resistant *S. aureus* (MRSA). As illustrated in the Clinical Case, an infection caused by these bacteria is resistant to antibiotic treatment.

◀ *Staphylococcus aureus* bacteria on skin cell culture.

## In the Clinic

As the nurse practitioner in a rural hospital, you are reviewing a microscope slide of a skin scraping from a 12-year-old girl. The slide shows branched, intertwined nucleated hyphae. The girl has dry, scaly, itchy patches on her arms. **What is causing her skin problem?**

*Hint: Read about types of microorganisms (pages 4–6).*

Play **In the Clinic** Video @ MasteringMicrobiology

# Microbes in Our Lives

## LEARNING OBJECTIVES

**1-1** List several ways in which microbes affect our lives.

**1-2** Define *microbiome*, *normal microbiota*, and *transient microbiota*.

For many people, the words *germ* and *microbe* bring to mind a group of tiny creatures that do not quite fit into any of the categories in that old question, "Is it animal, vegetable, or mineral?" *Germ* actually comes from the Latin word *germen*, meaning to spout from, or germinate. Think of wheat germ, the plant embryo from which the plant grows. It was first used in relation to microbes in the nineteenth century to explain the rapidly growing cells that caused disease. **Microbes**, also called **microorganisms**, are minute living things that individually are usually too small to be seen with the unaided eye. The group includes bacteria, fungi (yeasts and molds), protozoa, and microscopic algae. It also includes viruses, those noncellular entities sometimes regarded as straddling the border between life and nonlife (Chapters 11, 12, and 13, respectively).

We tend to associate these small organisms only with infections and inconveniences such as spoiled food. However, the majority of microorganisms actually help maintain the balance of life in our environment. Marine and freshwater microorganisms form the basis of the food chain in oceans, lakes, and rivers. Soil microbes break down wastes and incorporate nitrogen gas from the air into organic compounds, thereby recycling chemical elements among soil, water, living organisms, and air. Certain microbes play important roles in *photosynthesis*, a food- and oxygen-generating process that is critical to life on Earth.

Microorganisms also have many commercial applications. They are used in the synthesis of such chemical products as vitamins, organic acids, enzymes, alcohols, and many drugs. For example, microbes are used to produce acetone and butanol, and the vitamins $B_2$ (riboflavin) and $B_{12}$ (cobalamin) are made biochemically. The process by which microbes produce acetone and butanol was discovered in 1914 by Chaim Weizmann, a Russian-born chemist working in England. With the outbreak of World War I in August of that year, the production of acetone became very important for making cordite (a smokeless form of gunpowder used in munitions). Weizmann's discovery played a significant role in determining the outcome of the war.

The food industry also uses microbes in producing, for example, vinegar, sauerkraut, pickles, soy sauce, cheese, yogurt, bread, and alcoholic beverages. In addition, enzymes from microbes can now be manipulated to cause the microbes to produce substances they normally don't synthesize, including cellulose, human insulin, and proteins for vaccines.

## The Microbiome

An adult human is composed of about 30 trillion body cells and harbors another 40 trillion bacterial cells. Microbes that live stably in and on the human body are called the human **microbiome**, or **microbiota**. Humans and many other animals depend on these microbes to maintain good health. Bacteria in our intestines, including *E. coli*, aid digestion (see Exploring the Microbiome on page 3) and even synthesize some vitamins that our bodies require, including B vitamins for metabolism and vitamin K for blood clotting. They also prevent growth of **pathogenic** (disease-causing) species that might otherwise take up residence, and they seem to have a role in training our immune system to know which foreign invaders to attack and which to leave alone. (See Chapter 14 for more details on relationships between normal microbiota and the host.)

Even before birth, our bodies begin to be populated with bacteria. As newborns, we acquire viruses, fungi, and bacteria (**Figure 1.1**). For example, *E. coli* and other bacteria acquired from foods take residence in the large intestine. Many factors influence where and whether a microbe can indefinitely colonize the body as benign **normal microbiota** or be only a fleeting member of its community (known as **transient microbiota**). Microbes can colonize only those body sites that can supply the appropriate nutrients. Temperature, pH, and the presence or absence of chemical compounds are some factors that influence what types of microbes can flourish.

To determine the makeup of typical microbiota of various areas of the body, and to understand the relationship between changes in the microbiome and human diseases, is the goal of the **Human Microbiome Project**, which began in 2007. Likewise, the **National Microbiome Initiative (NMI)** launched in 2016 to expand our understanding of the role microbes play in different ecosystems, including soil, plants, aquatic environments, and the human body. Throughout the book, look for

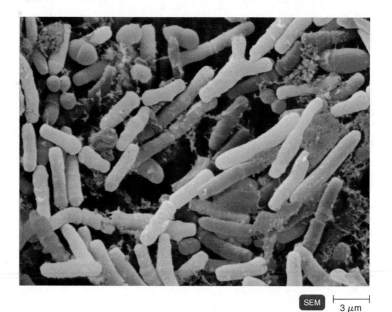

SEM | ⊢———⊣ 3 μm

**Figure 1.1** Several types of bacteria found as part of the normal microbiota in an infant's intestine.

**Q** How do we benefit from the production of vitamin K by microbes?

# How Does Your Microbiome Grow?

The specific traits of microbes that reside in human intestines can vary greatly—even within the same microbial species. Take *Bacteroides*, a bacterium commonly found in gastrointestinal tracts of humans worldwide. The strain residing in Japanese people has specialized enzymes that break down nori, the red algae used as the wrap component of sushi. These enzymes are absent from *Bacteroides* found in the gastrointestinal tracts of North Americans.

How did the Japanese *Bacteroides* acquire the ability to digest algae? It's thought the skill hails from *Zobellia galactanivorans*, a marine bacterium that lives on this alga. Not surprisingly, *Zobellia* readily breaks down the alga's main carbohydrate with enzymes. Since people living in Japan consumed algae regularly, *Zobellia* routinely met up with *Bacteroides* that lived in the human intestine. Bacteria can swap genes with other species—a process called *horizontal gene transfer*—and at some point, *Zobellia* must have given *Bacteroides* the genes to produce algae-digesting enzymes. (For more on horizontal gene transfer, see Chapter 8).

In an island nation where algae are an important diet component, the ability to extract more nutrition from algal carbohydrates would give an intestinal microbe a competitive advantage over others that couldn't use it as a food source. Over time, this *Bacteroides* strain became the dominant one found within the gastrointestinal tracts of people living in Japan.

You may be wondering whether North American sushi eaters can expect their own *Bacteroides* to shift to the algae-eating variety, too. Researchers say this is unlikely. Traditional Japanese food included raw algae, which allowed for living *Zobellia* to reach the large intestine. By contrast, the algae used in foods today is usually roasted or dried; these processes kill any bacteria that may be present on the surface.

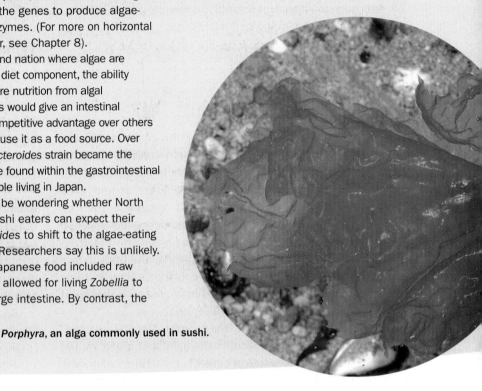

***Porphyra*, an alga commonly used in sushi.**

stories related to the human microbiome, highlighted in the Exploring the Microbiome feature boxes.

Our realization that some microbes are not only harmless to humans, but also are actually essential, represents a large shift from the traditional view that the only good microbe was a dead one. In fact, only a minority of microorganisms are pathogenic to humans. Although anyone planning to enter a health care profession needs to know how to prevent the transmission and spread of pathogenic microbes, it's also important to know that pathogens are just one aspect of our full relationship with microbes.

Today we understand that microorganisms are found almost everywhere. Yet not long ago, before the invention of the microscope, microbes were unknown to scientists. Next we'll look at the major groups of microbes and how they are named and classified. After that, we'll examine a few historic milestones in microbiology that have changed our lives.

## CHECK YOUR UNDERSTANDING

✔ 1-1* Describe some of the destructive and beneficial actions of microbes.

✔ 1-2 What percentage of all the cells in the human body are bacterial cells?

## CLINICAL CASE  A Simple Spider Bite?

Andrea is a normally healthy 22-year-old college student who lives at home with her mother and younger sister, a high school gymnast. She is trying to work on a paper for her psychology class but is having a hard time because a red, swollen sore on her right wrist is making typing difficult. "Why won't this spider bite heal?" she wonders. "It's been there for days!" She makes an appointment with her doctor so she can show him the painful lesion. Although Andrea does not have a fever, she does have an elevated white blood cell count that indicates a bacterial infection. Andrea's doctor suspects that this isn't a spider bite at all, but a staph infection. He prescribes a β-lactam antibiotic, cephalosporin. Learn more about the development of Andrea's illness on the following pages.

**What is staph? Read on to find out.**

| 3 | 16 | 18 | 19 | |
|---|----|----|----|--|

---

* The numbers preceding Check Your Understanding questions refer to the corresponding Learning Objectives.

# Naming and Classifying Microorganisms

## LEARNING OBJECTIVES

**1-3** Recognize the system of scientific nomenclature that uses two names: a genus and a specific epithet.

**1-4** Differentiate the major characteristics of each group of microorganisms.

**1-5** List the three domains.

## Nomenclature

The system of nomenclature (naming) for organisms in use today was established in 1735 by Carolus Linnaeus. Scientific names are latinized because Latin was the language traditionally used by scholars. Scientific nomenclature assigns each organism two names—the **genus** (plural: **genera**) is the first name and is always capitalized; the **specific epithet (species name)** follows and is not capitalized. The organism is referred to by both the genus and the specific epithet, and both names are underlined or italicized. By custom, after a scientific name has been mentioned once, it can be abbreviated with the initial of the genus followed by the specific epithet.

Scientific names can, among other things, describe an organism, honor a researcher, or identify the habitat of a species. For example, consider *Staphylococcus aureus*, a bacterium commonly found on human skin. *Staphylo-* describes the clustered arrangement of the cells; *-coccus* indicates that they are shaped like spheres. The specific epithet, *aureus*, is Latin for golden, the color of many colonies of this bacterium. The genus of the bacterium *Escherichia coli* (esh'er-IK-ē-ah KŌ-lī, or KŌ-lē) is named for a physician, Theodor Escherich, whereas its specific epithet,

*coli*, reminds us that *E. coli* live in the colon, or large intestine. Table 1.1 contains more examples.

**CHECK YOUR UNDERSTANDING**

✔ **1-3** Distinguish a genus from a specific epithet.

## Types of Microorganisms

In health care, it is very important to know the different types of microorganisms in order to treat infections. For example, antibiotics can be used to treat bacterial infections but have no effect on viruses or other microbes. Here is an overview of the main types of microorganisms. (The classification and identification of microorganisms are discussed in Chapter 10.)

### Bacteria

**Bacteria** (singular: **bacterium**) are relatively simple, single-celled (unicellular) organisms. Because their genetic material is not enclosed in a special nuclear membrane, bacterial cells are called **prokaryotes** (prō-KAR-e-ōts), from Greek words meaning prenucleus. Prokaryotes include both bacteria and archaea.

Bacterial cells generally appear in one of several shapes. *Bacillus* (bah-SIL-lus) (rodlike), illustrated in **Figure 1.2a**, *coccus* (KOK-kus) (spherical or ovoid), and *spiral* (corkscrew or curved) are among the most common shapes, but some bacteria are star-shaped or square (see Figures 4.1 through 4.5, pages 74–75). Individual bacteria may form pairs, chains, clusters, or other groupings; such formations are usually characteristic of a particular genus or species of bacteria.

Bacteria are enclosed in cell walls that are largely composed of a carbohydrate and protein complex called *peptidoglycan*.

---

**TABLE 1.1** Making Scientific Names Familiar

Use the word roots guide to find out what the name means. The name will not seem so strange if you translate it. When you encounter a new name, practice saying it out loud (guidelines for pronunciation are given in Appendix D). The exact pronunciation is not as important as the familiarity you will gain.

Following are some examples of microbial names you may encounter in the popular press as well as in the lab.

| | Pronunciation | Source of Genus Name | Source of Specific Epithet |
|---|---|---|---|
| *Salmonella enterica* (bacterium) | sal'mō-NEL-lah en-TER-i-kah | Honors public health microbiologist Daniel Salmon | Found in the intestines (*entero-*) |
| *Streptococcus pyogenes* (bacterium) | strep'tō-KOK-kus pī-AH-jen-ēz | Appearance of cells in chains (*strepto-*) | Forms pus (*pyo-*) |
| *Saccharomyces cerevisiae* (yeast) | sak'kar-ō-MĪ-sēz se-ri-VIS-ē-ī | Fungus (*-myces*) that uses sugar (*saccharo-*) | Makes beer (*cerevisia*) |
| *Penicillium chrysogenum* (fungus) | pen'i-SIL-lē-um krī-SO-jen-um | Tuftlike or paintbrush (*penicill-*) appearance microscopically | Produces a yellow (*chryso-*) pigment |
| *Trypanosoma cruzi* (protozoan) | tri'pa-nō-SŌ-mah KROOZ-ē | Corkscrew- (*trypano-*, borer; *soma-*, body) | Honors epidemiologist Oswaldo Cruz |

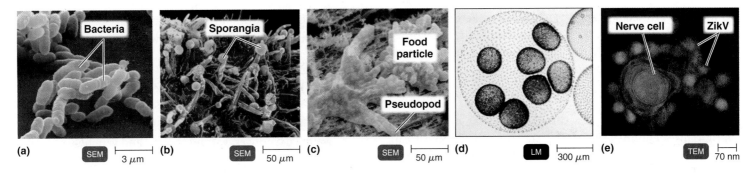

**Figure 1.2 Types of microorganisms.**
**(a)** The rod-shaped bacterium *Haemophilus influenzae*, one of the bacterial causes of pneumonia. **(b)** *Mucor*, a common bread mold, is a type of fungus. When released from sporangia, spores that land on a favorable surface germinate into a network of hyphae (filaments) that absorb nutrients. **(c)** An ameba, a type of protozoan, approaching a food particle. **(d)** The pond alga *Volvox*. **(e)** Zika virus (ZikV). *NOTE:* Throughout the book, a red icon under a micrograph indicates that the micrograph has been artificially colored. SEM (scanning electron microscope) and LM (light microscope) are discussed in detail in Chapter 3.

**Q** How are bacteria, archaea, fungi, protozoa, algae, and viruses distinguished on the basis of structure?

(By contrast, cellulose is the main substance of plant and algal cell walls.) Bacteria generally reproduce by dividing into two equal cells; this process is called *binary fission*. For nutrition, most bacteria use organic chemicals, which in nature can be derived from either dead or living organisms. Some bacteria can manufacture their own food by photosynthesis, and some can derive nutrition from inorganic substances. Many bacteria can "swim" by using moving appendages called *flagella*. (For a complete discussion of bacteria, see Chapter 11.)

### Archaea

Like bacteria, **archaea** (ar-KĒ-ah) consist of prokaryotic cells, but if they have cell walls, the walls lack peptidoglycan. Archaea, often found in extreme environments, are divided into three main groups. The *methanogens* produce methane as a waste product from respiration. The *extreme halophiles* (*halo* = salt; *philic* = loving) live in extremely salty environments such as the Great Salt Lake and the Dead Sea. The *extreme thermophiles* (*therm* = heat) live in hot sulfurous water, such as hot springs at Yellowstone National Park. Archaea are not known to cause disease in humans.

### Fungi

**Fungi** (singular: **fungus**) are **eukaryotes** (ū-KAR-ē-ōts), organisms whose cells have a distinct nucleus containing the cell's genetic material (DNA), surrounded by a special envelope called the *nuclear membrane*. Organisms in the Kingdom Fungi may be unicellular or multicellular (see Chapter 12, page 324). Large multicellular fungi, such as mushrooms, may look somewhat like plants, but unlike most plants, fungi cannot carry out photosynthesis. True fungi have cell walls composed primarily of a substance called *chitin*. The unicellular forms of fungi, *yeasts*, are oval microorganisms that are larger than bacteria. The most typical fungi are *molds* (Figure 1.2b). Molds form visible masses called *mycelia*, which are composed of long filaments (*hyphae*) that branch and intertwine. The cottony growths sometimes found on bread and fruit are mold mycelia. Fungi can reproduce sexually or asexually. They obtain nourishment by absorbing organic material from their environment—whether soil, seawater, freshwater, or an animal or plant host. Organisms called *slime molds* are actually ameba-like protozoa (see Chapter 12).

### Protozoa

**Protozoa** (singular: **protozoan**) are unicellular eukaryotic microbes (see Chapter 12, page 341). Protozoa move by pseudopods, flagella, or cilia. Amebae (Figure 1.2c) move by using extensions of their cytoplasm called *pseudopods* (false feet). Other protozoa have long *flagella* or numerous shorter appendages for locomotion called *cilia*. Protozoa have a variety of shapes and live either as free entities or as *parasites* (organisms that derive nutrients from living hosts) that absorb or ingest organic compounds from their environment. Some protozoa, such as *Euglena* (ū-GLĒ-nah), are photosynthetic. They use light as a source of energy and carbon dioxide as their chief source of carbon to produce sugars. Protozoa can reproduce sexually or asexually.

### Algae

**Algae** (singular: **alga**) are photosynthetic eukaryotes with a wide variety of shapes and both sexual and asexual reproductive forms (Figure 1.2d). The algae of interest to microbiologists are usually unicellular (see Chapter 12, page 337). The cell walls of many algae are composed of a carbohydrate called *cellulose*. Algae are abundant in freshwater and saltwater, in soil, and in association with plants. As photosynthesizers, algae need light, water, and carbon dioxide for food production and growth, but they do not generally require organic compounds from the environment. As a result of photosynthesis, algae produce oxygen and carbohydrates that are then utilized by other organisms, including animals. Thus, they play an important role in the balance of nature.

### Viruses

**Viruses** (Figure 1.2e) are very different from the other microbial groups mentioned here. They are so small that most can be seen only with an electron microscope, and they are acellular (that is, they are not cells). Structurally very simple, a virus particle contains a core made of only one type of nucleic acid, either DNA or RNA. This core is surrounded by a protein coat, which is sometimes encased by a lipid membrane called an *envelope*. All living cells have RNA *and* DNA, can carry out chemical reactions, and can reproduce as self-sufficient units. Viruses can reproduce only by using the cellular machinery of other organisms. Thus, on the one hand, viruses are considered to be living only when they multiply within host cells they infect. In this sense, viruses are parasites of other forms of life. On the other hand, viruses are not considered to be living because they are inert outside living hosts. (Viruses will be discussed in detail in Chapter 13.)

### Multicellular Animal Parasites

Although multicellular animal parasites are not strictly microorganisms, they are of medical importance and therefore will be discussed in this text. Animal parasites are eukaryotes. The two major groups of parasitic worms are the flatworms and the roundworms, collectively called **helminths** (see Chapter 12, page 347). During some stages of their life cycle, helminths are microscopic in size. Laboratory identification of these organisms includes many of the same techniques used for identifying microbes.

> **CHECK YOUR UNDERSTANDING**
>
> ✔ **1-4** Which groups of microbes are prokaryotes? Which are eukaryotes?

### Classification of Microorganisms

Before the existence of microbes was known, all organisms were grouped into either the animal kingdom or the plant kingdom. When microscopic organisms with characteristics of animals and plants were discovered late in the seventeenth century, a new system of classification was needed. Still, biologists couldn't agree on the criteria for classifying these new organisms until the late 1970s.

In 1978, Carl Woese devised a system of classification based on the cellular organization of organisms. It groups all organisms in three domains as follows:

1. Bacteria (cell walls contain a protein–carbohydrate complex called peptidoglycan)

2. Archaea (cell walls, if present, lack peptidoglycan)

3. Eukarya, which includes the following:

   • Protists (slime molds, protozoa, and algae)
   • Fungi (unicellular yeasts, multicellular molds, and mushrooms)
   • Plants (mosses, ferns, conifers, and flowering plants)
   • Animals (sponges, worms, insects, and vertebrates)

Classification will be discussed in more detail in Chapters 10 through 12.

> **CHECK YOUR UNDERSTANDING**
>
> ✔ **1-5** What are the three domains?

# A Brief History of Microbiology

> **LEARNING OBJECTIVES**
>
> **1-6** Explain the importance of observations made by Hooke and van Leeuwenhoek.
>
> **1-7** Compare spontaneous generation and biogenesis.
>
> **1-8** Identify the contributions to microbiology made by Needham, Spallanzani, Virchow, and Pasteur.
>
> **1-9** Explain how Pasteur's work influenced Lister and Koch.
>
> **1-10** Identify the importance of Koch's postulates.
>
> **1-11** Identify the importance of Jenner's work.
>
> **1-12** Identify the contributions to microbiology made by Ehrlich and Fleming.
>
> **1-13** Define *bacteriology, mycology, parasitology, immunology,* and *virology.*
>
> **1-14** Explain the importance of microbial genetics, molecular biology, and genomics.

Bacterial ancestors were the first living cells to appear on Earth. For most of human history, people knew little about the true causes, transmission, and effective treatment of disease. Let's look now at some key developments in microbiology that have spurred the field to its current technological state.

### The First Observations

In 1665, after observing a thin slice of cork through a crude microscope, Englishman Robert Hooke reported that life's smallest structural units were "little boxes," or "cells." Using his improved microscope, Hooke later saw individual cells. Hooke's discovery marked the beginning of the **cell theory**—the theory that *all living things are composed of cells.*

Though Hooke's microscope was capable of showing large cells, it lacked the resolution that would have allowed him to see microbes clearly. Dutch merchant and amateur scientist Anton van Leeuwenhoek was probably the first to observe live microorganisms through the magnifying lenses of the more than

400 microscopes he constructed. Between 1673 and 1723, he wrote about the "animalcules" he saw through his simple, single-lens microscopes. Van Leeuwenhoek made detailed drawings of organisms he found in rainwater, feces, and material scraped from teeth. These drawings have since been identified as representations of bacteria and protozoa (**Figure 1.3**).

**CHECK YOUR UNDERSTANDING**

✔ **1-6** What is the cell theory?

## The Debate over Spontaneous Generation

After van Leeuwenhoek discovered the previously "invisible" world of microorganisms, the scientific community became interested in the origins of these tiny living things. Until the second half of the nineteenth century, many scientists and philosophers believed that some forms of life could arise spontaneously from nonliving matter; they called this hypothetical process **spontaneous generation**. Not much more than 100 years ago, people commonly believed that toads, snakes, and mice could be born of moist soil; that flies could emerge from manure; and that maggots (which we now know are the larvae of flies) could arise from decaying corpses.

Physician Francesco Redi set out in 1668 to demonstrate that maggots did not arise spontaneously. Redi filled two jars with decaying meat. The first was left unsealed, allowing flies to lay eggs on the meat, which developed into larvae. The second jar was sealed, and because the flies could not get inside, no maggots appeared. Still, Redi's antagonists were not convinced; they claimed that fresh air was needed for spontaneous generation. So Redi set up a second experiment, in which he covered a jar with a fine net instead of sealing it. No larvae appeared in the gauze-covered jar, even though air was present.

Redi's results were a serious blow to the long-held belief that large forms of life could arise from nonlife. However, many scientists still believed that small organisms, such as van Leeuwenhoek's "animalcules," were simple enough to generate from nonliving materials.

The case for spontaneous generation of microorganisms seemed to be strengthened in 1745, when John Needham found that even after he heated chicken broth and corn broth before pouring them into covered flasks, the cooled solutions were soon teeming with microorganisms. Needham claimed that microbes developed spontaneously from the fluids. Twenty years later, Lazzaro Spallanzani suggested that microorganisms from the air probably entered Needham's solutions after they were boiled. Spallanzani showed that nutrient fluids heated after being sealed in a flask did not develop microbial growth. Needham responded by claiming the "vital force" necessary for spontaneous generation had been destroyed by the heat and was kept out of the flasks by the seals.

**(a)** Van Leeuwenhoek using his microscope

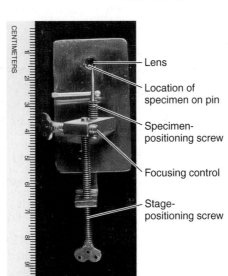

**(b)** Microscope replica

CENTIMETERS

Lens

Location of specimen on pin

Specimen-positioning screw

Focusing control

Stage-positioning screw

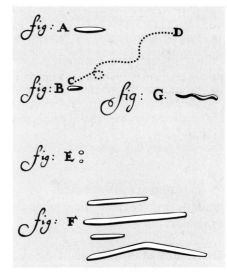

**(c)** Drawings of bacteria

*fig: A*     **D**
*fig: B*     *fig: G.*
*fig: E*
*fig: F*

**Figure 1.3 Anton van Leeuwenhoek's microscopic observations. (a)** By holding his brass microscope toward a source of light, van Leeuwenhoek was able to observe living organisms too small to be seen with the unaided eye. **(b)** The specimen was placed on the tip of the adjustable point and viewed from the other side through the tiny, nearly spherical lens. The highest magnification possible with his microscopes was about 300× (times). **(c)** Some of van Leeuwenhoek's drawings of bacteria, made in 1683. The letters represent various shapes of bacteria. C–D represents a path of motion he observed.

Needham = S.G
Lazzaro = Non-SG

**Q** **Why was van Leeuwenhoek's discovery so important?**

# Disproving Spontaneous Generation

According to the hypothesis of spontaneous generation, life can arise spontaneously from nonliving matter, such as dead corpses and soil. Pasteur's experiment, described below, demonstrated that microbes are present in nonliving matter—air, liquids, and solids.

**1** Pasteur first poured beef broth into a long-necked flask.

**2** Next he heated the neck of the flask and bent it into an S-shape; then he boiled the broth for several minutes.

**3** Microorganisms did not appear in the cooled solution, even after long periods.

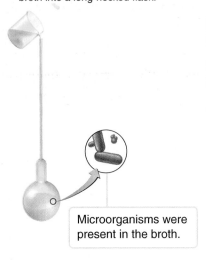

Microorganisms were present in the broth.

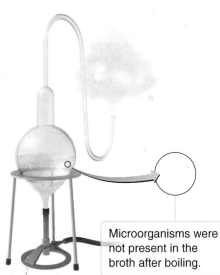

Microorganisms were not present in the broth after boiling.

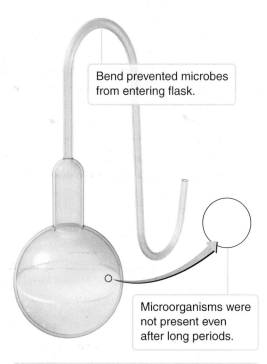

Bend prevented microbes from entering flask.

Microorganisms were not present even after long periods.

## KEY CONCEPTS

- Pasteur demonstrated that microbes are responsible for food spoilage, leading researchers to the connection between microbes and disease.

- His experiments and observations provided the basis of aseptic techniques, which are used to prevent microbial contamination, as shown in the photo at right.

Some of these original vessels are still on display at the Pasteur Institute in Paris. They have been sealed but show no sign of contamination more than 100 years later.

---

Spallanzani's observations were also criticized on the grounds that there was not enough oxygen in the sealed flasks to support microbial life.

### The Theory of Biogenesis

In 1858 Rudolf Virchow challenged the case for spontaneous generation with the concept of **biogenesis,** hypothesizing that living cells arise only from preexisting living cells. Because he could offer no scientific proof, arguments about spontaneous generation continued until 1861, when the issue was finally resolved by the French scientist Louis Pasteur.

Pasteur demonstrated that microorganisms are present in the air and can contaminate sterile solutions, but that air itself does not create microbes. He filled several short-necked flasks with beef broth and then boiled their contents. Some were then left open and allowed to cool. In a few days, these flasks were found to be contaminated with microbes. The other flasks, sealed after boiling, were free of microorganisms. From these results, Pasteur reasoned that microbes in the air were the agents responsible for contaminating nonliving matter.

Pasteur next placed broth in open-ended, long-necked flasks and bent the necks into S-shaped curves (**Figure 1.4**). The contents of these flasks were then boiled and cooled. The broth in the flasks did not decay and showed no signs of life, even after months. Pasteur's unique design allowed air to pass into the flask, but the curved neck trapped any airborne microorganisms that might contaminate the broth. (Some of these original vessels are still on display at the Pasteur Institute in

Paris. They have been sealed but, like the flask in Figure 1.4, show no sign of contamination more than 100 years later.)

Pasteur showed that microorganisms can be present in non-living matter—on solids, in liquids, and in the air. Furthermore, he demonstrated conclusively that microbial life can be destroyed by heat and that methods can be devised to block the access of airborne microorganisms to nutrient environments. These discoveries form the basis of **aseptic techniques**, procedures that prevent contamination by unwanted microorganisms, which are now the standard practice in laboratory and many medical procedures. Modern aseptic techniques are among the first and most important concepts that a beginning microbiologist learns.

Pasteur's work provided evidence that microorganisms cannot originate from mystical forces present in nonliving materials. Rather, any appearance of "spontaneous" life in nonliving solutions can be attributed to microorganisms that were already present in the air or in the fluids themselves. Scientists now believe that a form of spontaneous generation probably did occur on the primitive Earth when life first began, but they agree that this does not happen under today's environmental conditions.

CHECK YOUR UNDERSTANDING

1-7 What evidence supported spontaneous generation?

1-8 How was spontaneous generation disproved?

## The First Golden Age of Microbiology

The period from 1857 to 1914 has been appropriately named the First Golden Age of Microbiology. Rapid advances, spearheaded mainly by Pasteur and Robert Koch, led to the establishment of microbiology. Discoveries included both the agents of many diseases and the role of immunity in preventing and curing disease. During this productive period, microbiologists studied the chemical activities of microorganisms, improved the techniques for performing microscopy and culturing microorganisms, and developed vaccines and surgical techniques. Some of the major events that occurred during the First Golden Age of Microbiology are listed in **Figure 1.5**.

**First Golden Age of MICROBIOLOGY**

| | |
|---|---|
| 1857 | **Pasteur**—Fermentation |
| 1861 | **Pasteur**—Disproved spontaneous generation |
| 1864 | **Pasteur**—Pasteurization |
| 1867 | **Lister**—Aseptic surgery |
| 1876 | **Koch***—Germ theory of disease |
| 1879 | **Neisser**—*Neisseria gonorrhoeae* |
| 1881 | **Koch***—Pure cultures |
| | **Finlay**—Yellow fever |
| 1882 | **Koch***—*Mycobacterium tuberculosis* |
| | **Hess**—Agar (solid) media |
| 1883 | **Koch***—*Vibrio cholerae* |
| 1884 | **Metchnikoff***—Phagocytosis |
| | **Gram**—Gram-staining procedure |
| | **Escherich**—*Escherichia coli* |
| 1887 | **Petri**—Petri dish |
| 1889 | **Kitasato**—*Clostridium tetani* |
| 1890 | **von Bering***—Diphtheria antitoxin |
| | **Ehrlich***—Theory of immunity |
| 1892 | **Winogradsky**—Sulfur cycle |
| 1898 | **Shiga**—*Shigella dysenteriae* |
| 1908 | **Ehrlich***—Syphilis treatment |
| 1910 | **Chagas**—*Trypanosoma cruzi* |
| 1911 | **Rous***—Tumor-causing virus (1966 Nobel Prize) |

**Louis Pasteur (1822–1895)**
Demonstrated that life did not arise spontaneously from nonliving matter.

**Joseph Lister (1827–1912)**
Performed surgery under aseptic conditions using phenol. Proved that microbes caused surgical wound infections.

**Robert Koch (1843–1910)**
Established experimental steps for directly linking a specific microbe to a specific disease.

**Figure 1.5 Milestones in the First Golden Age of Microbiology.** An asterisk (*) indicates a Nobel laureate.

**Q** Why do you think the First Golden Age of Microbiology occurred when it did?

## Fermentation and Pasteurization

One of the key steps that established the relationship between microorganisms and disease occurred when a group of French merchants asked Pasteur to find out why wine and beer soured. They hoped to develop a method that would prevent spoilage when those beverages were shipped long distances. At the time, many scientists believed that air converted the sugars in these fluids into alcohol. Pasteur found instead that microorganisms called yeasts convert the sugars to alcohol in the absence of air. This process, called **fermentation** (see Chapter 5, page 128), is used to make wine and beer. Souring and spoilage are caused by different microorganisms, called bacteria. In the presence of air, bacteria change the alcohol into vinegar (acetic acid).

Pasteur's solution to the spoilage problem was to heat the beer and wine just enough to kill most of the bacteria that caused the spoilage. The process, called **pasteurization**, is now commonly used to reduce spoilage and kill potentially harmful bacteria in milk and other beverages as well as in some alcoholic beverages.

## The Germ Theory of Disease

Before the time of Pasteur, effective treatments for many diseases were discovered by trial and error, but the causes of the diseases were unknown. The realization that yeasts play a crucial role in fermentation was the first link between the activity of a microorganism and physical and chemical changes in organic materials. This discovery alerted scientists to the possibility that microorganisms might have similar relationships with plants and animals—specifically, that microorganisms might cause disease. This idea was known as the **germ theory of disease**.

The germ theory met great resistance at first—for centuries, disease was believed to be punishment for an individual's crimes or misdeeds. When the inhabitants of an entire village became ill, people often blamed the disease on demons appearing as foul odors from sewage or on poisonous vapors from swamps. Most people born in Pasteur's time found it inconceivable that "invisible" microbes could travel through the air to infect plants and animals or remain on clothing and bedding to be transmitted from one person to another. Despite these doubts, scientists gradually accumulated the information needed to support the new germ theory.

In 1865, Pasteur was called upon to help fight silkworm disease, which was ruining the silk industry in Europe. Decades earlier, amateur microscopist Agostino Bassi had proved that another silkworm disease was caused by a fungus. Using data provided by Bassi, Pasteur found that the more recent infection was caused by a protozoan, and he developed a method for recognizing afflicted silkworm moths.

In the 1860s, Joseph Lister, an English surgeon, applied the germ theory to medical procedures. Lister was aware that in the 1840s, the Hungarian physician Ignaz Semmelweis had demonstrated that physicians, who at the time did not disinfect their hands, routinely transmitted infections (puerperal, or childbirth, fever) from one obstetrical patient to another. Lister had also heard of Pasteur's work connecting microbes to animal diseases. Disinfectants were not used at the time, but Lister knew that phenol (carbolic acid) kills bacteria, so he began treating surgical wounds with a phenol solution. The practice so reduced the incidence of infections and deaths that other surgeons quickly adopted it. His findings proved that microorganisms cause surgical wound infections.

The first proof that bacteria actually cause disease came from Robert Koch (kōk) in 1876. Koch, a German physician, was Pasteur's rival in the race to discover the cause of anthrax, a disease that was destroying cattle and sheep in Europe. Koch discovered rod-shaped bacteria now known as *Bacillus anthracis* (bah-SIL-lus an-THRĀ-sis) in the blood of cattle that had died of anthrax. He cultured the bacteria on nutrients and then injected samples of the culture into healthy animals. When these animals became sick and died, Koch isolated the bacteria in their blood and compared them with the originally isolated bacteria. He found that the two sets of blood cultures contained the same bacteria.

Koch thus established **Koch's postulates**, a sequence of experimental steps for directly relating a specific microbe to a specific disease (see Figure 14.3, page 339). During the past 100 years, these same criteria have been invaluable in investigations proving that specific microorganisms cause many diseases. Koch's postulates, their limitations, and their application to disease will be discussed in greater detail in Chapter 14.

## Vaccination

Often a treatment or preventive procedure is developed before scientists know why it works. The smallpox vaccine is an example. Almost 70 years before Koch established that a specific microorganism causes anthrax, Edward Jenner, a young British physician, embarked on an experiment to find a way to protect people from smallpox. The disease periodically swept through Europe, killing thousands, and it wiped out 90% of the Native Americans on the East Coast when European settlers first brought the infection to the New World.

When a young milkmaid informed Jenner that she couldn't get smallpox because she already had been sick from cowpox—a much milder disease—he decided to put the girl's story to the test. First Jenner collected scrapings from cowpox blisters. Then he inoculated a healthy 8-year-old volunteer with the cowpox material by scratching the child's arm with a pox-contaminated needle. The scratch turned into a raised bump. In a few days, the volunteer became mildly sick but recovered and never again contracted either cowpox or smallpox. The protection from disease provided by vaccination (or by recovery from the disease itself) is called **immunity**. (We will discuss the mechanisms of immunity in Chapter 17.)

Years after Jenner's experiment, Pasteur discovered why vaccinations work. He found that the bacterium that causes fowl cholera lost its ability to cause disease (lost its *virulence*, or became *avirulent*) after it was grown in the laboratory for long periods. However, it—and other microorganisms with decreased virulence—was able to induce immunity against subsequent infections by its virulent counterparts. The discovery of this phenomenon provided a clue to Jenner's successful experiment with cowpox. Both cowpox and smallpox are caused by viruses. Even though cowpox virus is not a laboratory-produced derivative of smallpox virus, it is so closely related to the smallpox virus that it can induce immunity to both viruses. Pasteur used the term *vaccine* for cultures of avirulent microorganisms used for preventive inoculation. (The Latin word *vacca* means cow—thus, the term *vaccine* honored Jenner's earlier cowpox inoculation work.)

Jenner's experiment was actually not the first time a living viral agent—in this case, the cowpox virus—was used to produce immunity. Starting in the 1500s, physicians in China had immunized patients from smallpox by removing scales from drying pustules of a person suffering from a mild case of smallpox, grinding the scales to a fine powder, and inserting the powder into the nose of the person to be protected.

Some vaccines are still produced from avirulent microbial strains that stimulate immunity to the related virulent strain. Other vaccines are made from killed virulent microbes, from isolated components of virulent microorganisms, or by genetic engineering techniques.

> **CHECK YOUR UNDERSTANDING**
>
> ✔ **1-9** Summarize in your own words the germ theory of disease.
>
> ✔ **1-10** What is the importance of Koch's postulates?
>
> ✔ **1-11** What is the significance of Jenner's discovery?

## The Second Golden Age of Microbiology

After the relationship between microorganisms and disease was established, medical microbiologists next focused on the search for substances that could destroy pathogenic microorganisms without damaging the infected animal or human.

Treatment of disease by using chemical substances is called **chemotherapy**. (The term also commonly refers to chemical treatment of noninfectious diseases, such as cancer.) Chemicals produced naturally by bacteria and fungi that act against other microorganisms are called **antibiotics**. Chemotherapeutic agents prepared from chemicals in the laboratory are called **synthetic drugs**. The success of chemotherapy is based on the fact that some chemicals are more poisonous to microorganisms than to the hosts infected by the microbes. Antimicrobial therapy will be discussed in further detail in Chapter 20.

### The First Synthetic Drugs

Paul Ehrlich was the imaginative thinker who fired the first shot in the chemotherapy revolution. As a medical student, Ehrlich speculated about a "magic bullet" that could hunt down and destroy a pathogen without harming the infected host. In 1910, after testing hundreds of substances, he found a chemotherapeutic agent called *salvarsan*, an arsenic derivative effective against syphilis. The agent was named *salvarsan* because it was considered to offer salvation from syphilis and it contained arsenic. Before this discovery, the only known chemical in Europe's medical arsenal was an extract from the bark of a South American tree, *quinine*, which had been used by Spanish conquistadors to treat malaria.

By the late 1930s, researchers had developed several other synthetic drugs that could destroy microorganisms. Most of these drugs were derivatives of dyes. This came about because the dyes synthesized and manufactured for fabrics were routinely tested for antimicrobial qualities by microbiologists looking for a "magic bullet." In addition, *sulfonamides* (sulfa drugs) were synthesized at about the same time.

### A Fortunate Accident—Antibiotics

The first antibiotic was discovered by accident. Alexander Fleming, a Scottish physician and bacteriologist, almost tossed out some culture plates that had been contaminated by mold. Fortunately, he noticed the curious pattern of growth on the plates—a clear area where bacterial growth had been inhibited encircled the mold (**Figure 1.6**). Fleming was looking at a mold that inhibited growth of a bacterium. The mold became known as *Penicillium chrysogenum* (pen'i-SIL-lē-um krī-SO-jen-um), and the mold's active inhibitor was called *penicillin*. Thus, penicillin

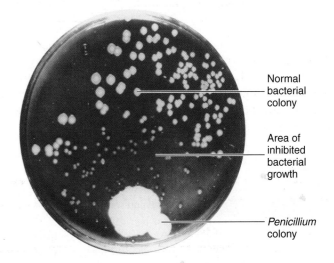

Normal bacterial colony

Area of inhibited bacterial growth

*Penicillium* colony

**Figure 1.6 The discovery of penicillin.** Alexander Fleming took this photograph in 1928. The colony of *Penicillium* mold accidentally contaminated the plate and inhibited nearby bacterial growth.

**Q** Why do you think penicillin is no longer as effective as it once was?

is an antibiotic produced by a fungus. The Second Golden Age of Microbiology began in the 1940s, when the enormous usefulness of penicillin became apparent and the drug came into common use.

Since these early discoveries, thousands of other antibiotics have been discovered. Unfortunately, use of antibiotics and other chemotherapeutic drugs is not without problems. Many antimicrobial chemicals kill pathogenic microbes but also damage the infected host. For reasons we will discuss later, toxicity to humans is a particular problem in the development of drugs for treating viral diseases. Viral growth depends on life processes of normal host cells. Thus, there are very few successful antiviral drugs, because a drug that would interfere with viral reproduction would also likely affect uninfected cells of the body.

Over the years, more and more microbes also developed resistance to antibiotics that were once very effective against them. Drug resistance results from genetic changes in microbes that enable them to tolerate a certain amount of an antibiotic that would normally inhibit them (see the box in Chapter 26, page 771). For example, a microbe might produce enzymes that inactivate antibiotics, or a microbe might undergo changes to its surface that prevent an antibiotic from attaching to it or entering it.

The recent appearance of vancomycin-resistant *Staphylococcus aureus* and *Enterococcus faecalis* (en'ter-ō-KOK-kus fē-KĀ-lis) has alarmed health care professionals because it indicates that some previously treatable bacterial infections may soon be impossible to treat with antibiotics.

The quest to solve drug resistance, identify viruses, and develop vaccines requires sophisticated research techniques and correlated studies that were never dreamed of in the days of Koch and Pasteur. Other microbiologists also used these

| | | |
|---|---|---|
| **Second Golden Age of MICROBIOLOGY** | 1940s 1950s | **Fleming, Chain, and Florey**—Penicillin |
| | | **Waksman**—Streptomycin |
| | | **H. Krebs**—Chemical steps of the Krebs cycle |
| | | **Enders, Weller, and Robbins**—Poliovirus cultured in cell cultures |
| | | **Beadle and Tatum**—Genetic control of biochemical reactions |
| | 1960s | **Medawar**—Acquired immune tolerance |
| | 1980s | **Sanger and Gilbert**—Techniques for sequencing DNA |
| | | **Jerne, Köhler, and Milstein**—Technique for producing monoclonal (single pure) antibodies |
| | | **Tonegawa**—Genetics of antibody production |
| **Third Golden Age of MICROBIOLOGY** | | **Bishop and Varmus**—Cancer-causing genes (oncogenes) |
| | 1990s | **Murray and Thomas**—First successful transplants using immunosuppressive drugs |
| | | **Fischer and E. Krebs**—Enzymes that regulate cell growth (protein kinases) |
| | | **Roberts and Sharp**—Genes can be present in separated segments of DNA |
| | | **Mullis**—Polymerase chain reactions that amplify (make multiple copies of) DNA |
| | | **Doherty and Zinkernagel**—Cell-mediated immunity |
| | 2000s | **Agre and MacKinnon**—Water and ion channels in plasma membranes |
| | | **Marshall and Warren**—*Helicobacter pylori* as the cause of peptic ulcers |
| | | **Barré-Sinoussi and Montagnier**—Discovery of HIV |
| | 2010s | **Ramakrishnan, Steitz, and Yonath**—Detailed structure and function of ribosomes |
| | | **Beutler, Hoffmann, and Steinman**—Innate immunity; dendritic cells in adaptive immunity |
| | | **Tu**—Treatment for malaria |

**César Milstein (1927–)**
Fused cancerous cells with antibody-producing cells to produce a hybrid cell that grows continuously and produces therapeutic antibodies.

**Françoise Barré-Sinoussi (1947–)**
Discovered a virus in a patient with swollen lymph nodes; the virus was human immunodeficiency virus.

**Youyou Tu (1930–)**
Extracted artemisinin from a Chinese sage plant. Artemisinin inhibits the malaria parasite.

**Figure 1.7 Second and Third Golden Ages of Microbiology.** All researchers listed are Nobel laureates.

**Q** What advances occurred during the Second Golden Age of Microbiology?

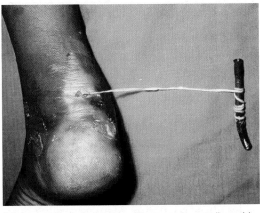

**(a)** A parasitic guinea worm (*Dracunculus medinensis*) is removed from the subcutaneous tissue of a patient by winding it onto a stick. This procedure may have been used for the design of the symbol in part (b).

**(b)** Rod of Asclepius, symbol of the medical profession

**Figure 1.8  Parasitology: the study of protozoa and parasitic worms.**

**Q** How do you think parasitic worms survive and live off a human host?

techniques to investigate industrial applications and roles of microorganisms in the environment.

### Bacteriology, Mycology, and Parasitology

The groundwork laid during the First Golden Age of Microbiology provided the basis for several monumental achievements in the years following that era (**Figure 1.7**). New branches of microbiology were developed, including immunology and virology.

**Bacteriology**, the study of bacteria, began with van Leeuwenhoek's first examination of tooth scrapings. New pathogenic bacteria are still discovered regularly. Many bacteriologists, like Pasteur, look at the roles of bacteria in food and the environment. One intriguing discovery came in 1997, when Heide Schulz discovered a bacterium large enough to be seen with the unaided eye (0.2 mm wide). This bacterium, named *Thiomargarita namibiensis* (THĪ-ō-mar-gar′ē-tah nah′mib-ē-EN-sis), lives in mud on the African coast. *Thiomargarita* is unusual because of its size and its ecological niche. The bacterium consumes hydrogen sulfide, which would be toxic to mud-dwelling animals (Figure 11.28, page 320).

**Mycology**, the study of fungi, includes medical, agricultural, and ecological branches. Fungal infection rates have been rising during the past decade, accounting for 10% of hospital-acquired infections. Climatic and environmental changes (severe drought) are thought to account for the tenfold increase in *Coccidioides immitis* (kok′sid-ē-OID-ēz IM-mi-tis) infections in California. New techniques for diagnosing and treating fungal infections are currently being investigated.

**Parasitology** is the study of protozoa and parasitic worms. Because many parasitic worms are large enough to be seen with the unaided eye, they have been known for thousands of years. It has been speculated that the medical symbol, the rod of Asclepius, represents the removal of parasitic guinea worms (**Figure 1.8**). Asclepius was a Greek physician who practiced about 1200 B.C.E. and was deified as the god of medicine.

The clearing of rain forests has exposed laborers to previously undiscovered parasites. Parasitic diseases unknown until recently are also being found in patients whose immune systems have been suppressed by organ transplants, cancer chemotherapy, or AIDS.

### Immunology

**Immunology** is the study of immunity. Knowledge about the immune system has accumulated steadily and expanded rapidly. Vaccines are now available for numerous diseases, including measles, rubella (German measles), mumps, chickenpox, pneumococcal pneumonia, tetanus, tuberculosis, influenza, whooping cough, polio, and hepatitis B. The smallpox vaccine was so effective that the disease has been eliminated. Public health officials estimate that polio will be eradicated within a few years because of the polio vaccine.

A major advance in immunology occurred in 1933, when Rebecca Lancefield (**Figure 1.9**) proposed that streptococci be classified according to serotypes (variants within a species) based on certain components in the cell walls of the bacteria. Streptococci are responsible for a variety of diseases, such as sore throat (strep throat), streptococcal toxic shock, and septicemia (blood poisoning).

In 1960, interferons, substances generated by the body's own immune system, were discovered. Interferons inhibit replication of viruses and have triggered considerable research related

**Figure 1.9  Rebecca Lancefield (1895–1981), who discovered differences in the chemical composition of a polysaccharide in the cell walls of many pathogenic streptococci.** Rapid laboratory tests using immunologic techniques now identify and classify streptococci into Lancefield groups based on this carbohydrate.

**Q** Why is it important to identify streptococci quickly?

to the treatment of viral diseases and cancer. One of today's biggest challenges for immunologists is learning how the immune system might be stimulated to ward off the virus responsible for AIDS, a disease that destroys the immune system.

### Virology

The study of viruses, **virology**, originated during the First Golden Age of Microbiology. In 1892, Dmitri Iwanowski reported that the organism that caused mosaic disease of tobacco was so small that it passed through filters fine enough to stop all known bacteria. At the time, Iwanowski was not aware that the organism in question was a virus. In 1935, Wendell Stanley demonstrated that the organism, called tobacco mosaic virus (TMV), was fundamentally different from other microbes and so simple and homogeneous that it could be crystallized like a chemical compound. Stanley's work facilitated the study of viral structure and chemistry. Since the development of the electron microscope in the 1930s, microbiologists have been able to observe the structure of viruses in detail, and today much is known about their structure and activity.

### Molecular Genetics

Once science turned to the study of unicellular life, rapid progress was made in genetics. **Microbial genetics** studies the mechanisms by which microorganisms inherit traits, and **molecular biology** looks at how genetic information is carried in molecules of DNA.

In the 1940s, George W. Beadle and Edward L. Tatum demonstrated the relationship between genes and enzymes; DNA was established as the hereditary material by Oswald Avery, Colin MacLeod, and Maclyn McCarty. Joshua Lederberg and Edward L. Tatum discovered that genetic material could be transferred from one bacterium to another by a process called conjugation. Then in the 1950s, James Watson and Francis Crick proposed a model for the structure and replication of DNA. In the early 1960s, François Jacob and Jacques Monod discovered messenger RNA (ribonucleic acid), a chemical involved in protein synthesis, and later they made the first major discoveries about the regulation of gene function in bacteria. During the same period, scientists were able to break the genetic code and thus understand how the information for protein synthesis in messenger RNA is translated into the amino acid sequence for making proteins.

Although molecular genetics encompasses all organisms, much of our knowledge of how genes determine specific traits has been revealed through experiments with bacteria. Unicellular organisms, primarily bacteria, have several advantages for genetic and biochemical research. Bacteria are less complex than plants and animals, and the life cycles of many bacteria last less than an hour, so scientists can cultivate very large numbers of bacteria for study in a relatively short time.

## The Third Golden Age of Microbiology

Stephen Jay Gould said we now live in the "age of bacteria." The bacteria aren't new, but our understanding of their importance to the Earth and to our health is. New DNA-sequencing tools and computers allow investigators to study all the DNA in an organism, helping them to identify genes and their functions. Moreover, through **genomics**, the study of all of an organism's genes, scientists are able to classify bacteria and fungi according to their genetic relationships with other bacteria, fungi, and protozoa. These microorganisms were originally classified according to a limited number of visible characteristics. The tools of genomics are being used to identify microbes in the ocean, on leaves, and on humans, many of which are newly discovered and haven't been grown in laboratories. After microbes are discovered, the next step is to find out what they are doing. The Exploring the Microbiome boxes throughout this textbook give examples of this research.

Microorganisms can now be genetically modified to manufacture large amounts of human hormones and other urgently needed medical substances. This development had its origins in the late 1960s, when Paul Berg showed that fragments of human or animal DNA (genes) that code for important proteins can be attached to bacterial DNA. The resulting hybrid was the first example of **recombinant DNA**. **Recombinant DNA (rDNA) technology** inserts recombinant DNA into bacteria (or other microbes) to make large quantities of a desired protein. The development of recombinant DNA technology has revolutionized research and practical applications in all areas of microbiology.

## Microbes and Human Welfare

### LEARNING OBJECTIVES

**1-15** List at least four beneficial activities of microorganisms.

**1-16** Name two examples of biotechnology that use recombinant DNA technology and two examples that do not.

As mentioned earlier, only a minority of all microorganisms are pathogenic. Microbes that cause food spoilage—such as soft spots on fruits and vegetables, decomposition of meats, and rancidity of fats and oils—are also a minority. The vast majority of microbes benefit humans, other animals, and plants in many ways. For example, microbes produce methane and ethanol that can be used as alternative fuels to generate electricity and power vehicles. Biotechnology companies are using bacterial enzymes to break down plant cellulose so that yeast can metabolize the resulting simple sugars and produce ethanol. The following sections outline some of these beneficial activities. In later chapters, we will discuss these activities in greater detail.

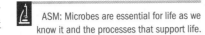

ASM: Microbes are essential for life as we know it and the processes that support life.

ASM: Humans utilize and harness microorganisms and their products.

## Recycling Vital Elements

Discoveries made by two microbiologists in the 1880s have formed the basis for today's understanding of the biogeochemical cycles that support life on Earth. Martinus Beijerinck and Sergei Winogradsky were the first to show how bacteria help recycle vital elements between the soil and the atmosphere. **Microbial ecology**, the study of the relationship between microorganisms and their environment, originated with the work of these scientists. Today, microbial ecology has branched out and includes the study of how microbial populations interact with plants and animals in various environments. Among the concerns of microbial ecologists are water pollution and toxic chemicals in the environment.

The chemical elements carbon, nitrogen, oxygen, sulfur, and phosphorus are essential for life and abundant, but not necessarily in forms that organisms can use. Microorganisms are primarily responsible for converting these elements into forms that plants and animals can use. Microorganisms, especially bacteria and fungi, return carbon dioxide to the atmosphere when they decompose organic wastes and dead plants and animals. Algae, cyanobacteria, and higher plants use the carbon dioxide during photosynthesis to produce carbohydrates for animals, fungi, and bacteria. Nitrogen is abundant in the atmosphere but in that form is not usable by plants and animals. Only bacteria can naturally convert atmospheric nitrogen to a form available to plants and animals.

## Sewage Treatment: Using Microbes to Recycle Water

Our society's growing awareness of the need to preserve the environment has made people more conscious of the responsibility to recycle precious water and prevent pollution of rivers and oceans. One major pollutant is sewage, which consists of human excrement, wastewater, industrial wastes, and surface runoff. Sewage is about 99.9% water, with a few hundredths of 1% suspended solids. The remainder is a variety of dissolved materials.

Sewage treatment plants remove the undesirable materials and harmful microorganisms. Treatments combine various physical processes with the action of beneficial microbes. Large solids such as paper, wood, glass, gravel, and plastic are removed from sewage; left behind are liquid and organic materials that bacteria convert into such by-products as carbon dioxide, nitrates, phosphates, sulfates, ammonia, hydrogen sulfide, and methane. (We will discuss sewage treatment in detail in Chapter 27.)

## Bioremediation: Using Microbes to Clean Up Pollutants

In 1988, scientists began using microbes to clean up pollutants and toxic wastes produced by various industrial processes. For example, some bacteria can actually use pollutants as energy sources; others produce enzymes that break down toxins into less harmful substances. By using bacteria in these ways—a process known as **bioremediation**—toxins can be removed from underground wells, chemical spills, toxic waste sites, and oil spills, such as the massive oil spill from a British Petroleum offshore drilling rig in the Gulf of Mexico in 2010. In addition, bacterial enzymes are used in drain cleaners to remove clogs without adding harmful chemicals to the environment. In some cases, microorganisms indigenous to the environment are used; in others, genetically modified microbes are used. Among the most commonly used microbes are certain species of bacteria of the genera *Pseudomonas* and *Bacillus*. *Bacillus* enzymes are also used in household detergents to remove spots from clothing.

## Insect Pest Control by Microorganisms

Besides spreading diseases, insects can cause devastating crop damage. Insect pest control is therefore important for both agriculture and the prevention of human disease.

The bacterium *Bacillus thuringiensis* (ther-IN-jē-en-sis) has been used extensively in the United States to control such pests as alfalfa caterpillars, bollworms, corn borers, cabbageworms, tobacco budworms, and fruit tree leaf rollers. It is incorporated into a dusting powder that is applied to the crops these insects eat. The bacteria produce protein crystals that are toxic to the digestive systems of the insects. The toxin gene also has been inserted into some plants to make them insect resistant.

By using microbial rather than chemical insect control, farmers can avoid harming the environment. In contrast, many chemical insecticides, such as DDT, remain in the soil as toxic pollutants and are eventually incorporated into the food chain.

## Biotechnology and Recombinant DNA Technology

Earlier we touched on the commercial use of microorganisms to produce some common foods and chemicals. Such practical applications of microbiology are called **biotechnology**. Although biotechnology has been used in some form for centuries, techniques have become much more sophisticated in the past few decades. In the last several years, biotechnology has undergone a revolution through the advent of recombinant DNA technology to expand the potential of bacteria, viruses, and yeast and other fungi as miniature biochemical factories. Cultured plant and animal cells, as well as intact plants and animals, are also used as recombinant cells and organisms.

The applications of recombinant DNA technology are increasing with each passing year. Recombinant DNA techniques have been used thus far to produce a number of natural proteins, vaccines, and enzymes. Such substances have great potential for medical use; some of them are described in Table 9.2 on page 256.

A very exciting and important outcome of recombinant DNA techniques is **gene therapy**—inserting a missing gene or replacing a defective one in human cells. This technique uses a harmless virus to carry the missing or new gene into certain host cells, where the gene is picked up and inserted into the appropriate chromosome. Since 1990, gene therapy has been used to treat patients with adenosine deaminase (ADA) deficiency, a cause of severe combined immunodeficiency disease (SCID), in which cells of the immune system are inactive or missing; Duchenne's muscular dystrophy, a muscle-destroying disease; cystic fibrosis, a disease of the secreting portions of the respiratory passages, pancreas, salivary glands, and sweat glands; and LDL-receptor deficiency, a condition in which low-density lipoprotein (LDL) receptors are defective and LDL cannot enter cells. The LDL remains in the blood in high concentrations and leads to fatty plaque formation in blood vessels, increasing the risk of atherosclerosis and coronary heart disease. Results of gene therapy are still being evaluated. Other genetic diseases may also be treatable by gene therapy in the future, including hemophilia, an inability of the blood to clot normally; diabetes, elevated blood sugar levels; and sickle cell disease, caused by an abnormal kind of hemoglobin.

Beyond medical applications, recombinant DNA techniques have also been applied to agriculture. For example, genetically altered strains of bacteria have been developed to protect fruit against frost damage, and bacteria are being modified to control insects that damage crops. Recombinant DNA has also been used to improve the appearance, flavor, and shelf life of fruits and vegetables. Potential agricultural uses of recombinant DNA include drought resistance, resistance to insects and microbial diseases, and increased temperature tolerance in crops.

**CHECK YOUR UNDERSTANDING**

✔ **1-15** Name two beneficial uses of bacteria.

✔ **1-16** Differentiate biotechnology from recombinant DNA technology.

# Microbes and Human Disease

**LEARNING OBJECTIVES**

**1-17** Define *resistance*.

**1-18** Define *biofilm*.

**1-19** Define *emerging infectious disease*.

When is a microbe a welcome part of a healthy human, and when is it a harbinger of disease? The distinction between health and disease is in large part a balance between the natural defenses of the body and the disease-producing properties of microorganisms. Whether our bodies overcome the offensive tactics of a particular microbe depends on our **resistance**—the ability to ward off diseases. Important resistance is provided by the barrier of the skin, mucous membranes, cilia, stomach acid, and antimicrobial chemicals such as interferons. Microbes can be destroyed by white blood cells, by the inflammatory response, by fever, and by specific responses of our immune system. Sometimes, when our natural defenses are not strong enough to overcome an invader, they have to be supplemented by antibiotics or other drugs.

## Biofilms

In nature, microorganisms may exist as single cells that float or swim independently in a liquid, or they may attach to each other and/or some usually solid surface. This latter mode of behavior is called a **biofilm**, a complex aggregation of microbes. The slime covering a rock in a lake is a biofilm. Use your tongue to feel the biofilm on your teeth. Biofilms can be beneficial. They protect your mucous membranes from harmful microbes, and biofilms in lakes are an important food for aquatic animals. Biofilms can also be harmful. They can clog water pipes, and on medical implants such as joint prostheses and catheters (**Figure 1.10**), they can cause such infections as endocarditis (inflammation of the heart). Bacteria in biofilms are often resistant to antibiotics because the biofilm offers a protective barrier. Biofilms will be discussed in Chapter 6.

**CLINICAL CASE**

"Staph" is the common name for *Staphylococcus aureus* bacteria, which are carried on the skin of about 30% of the human population. Although Andrea is diligent about taking her antibiotic as prescribed, she doesn't seem to be improving. After 3 days, the lesion on her wrist is even larger than before and is now draining yellow pus. Andrea also develops a fever. Her mother insists that she call her doctor to tell him about the latest developments.

**Why does Andrea's infection persist after treatment?**

3   16   18   19

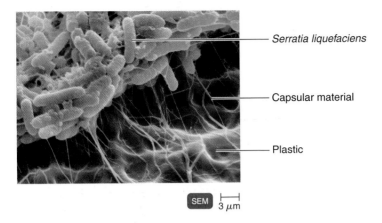

Figure 1.10 **Biofilm on a piece of plastic.** Bacteria stick to solid surfaces, forming a slimy layer. The filaments in the photo may be capsular material. Bacteria that break away from biofilms on medical implants can cause infections.

**Q** How does a biofilm's protective barrier make it resistant to antibiotics?

## Infectious Diseases

An **infectious disease** is a disease in which pathogens invade a susceptible host, such as a human or an animal. In the process, the pathogen carries out at least part of its life cycle inside the host, and disease frequently results. By the end of World War II, many people believed that infectious diseases were under control. They thought malaria would be eradicated through the use of the insecticide DDT to kill mosquitoes, that a vaccine would prevent pertussis, and that improved sanitation measures would help prevent cholera transmission. Expectations did not match reality: Malaria is far from eliminated; since 1986, local outbreaks have been identified in New Jersey, California, Florida, New York, and Texas, and the disease infects over 200 million people worldwide. Pertussis is not eliminated but vaccination has decreased the incidence from 200,000 cases to 12,000 cases annually. And cholera outbreaks still occur in less-developed parts of the world.

## Emerging Infectious Diseases

These recent outbreaks point to the fact that infectious diseases are not disappearing, but rather seem to be reemerging and increasing. In addition, a number of new diseases—**emerging infectious diseases (EIDs)**—have cropped up in recent years. These are diseases that are new or changing and are increasing or have the potential to increase in incidence in the near future. Some of the factors that have contributed to the development of EIDs are evolutionary changes in existing organisms (e.g., *Vibrio cholerae*; VIB-rē-ō KOL-er-ī) and the spread of known diseases to new geographic regions or populations by modern transportation. Some EIDs are the result of increased human exposure to new, unusual infectious agents in areas that are undergoing ecologic changes such as deforestation and construction (e.g., Venezuelan hemorrhagic virus). Some EIDs are due to changes in the pathogen's ecology. For example, Powassan virus (POWV) was transmitted by ticks that don't usually bite humans. However, the virus recently became established in the same deer ticks that transmit Lyme disease. An increasing number of incidents in recent years highlights the extent of the problem.

### Zika Virus Disease

In 2015, the world became aware of Zika virus disease. Zika virus is spread by the bite of an infected *Aedes* mosquito; sexual transmission has also occurred. Zika is a mild disease usually presenting with fever, rash, and joint pain. However, Zika infection during pregnancy can cause severe birth defects in a fetus. The virus was discovered in 1947 in the Zika Forest of Uganda, but until 2007, only 14 cases of Zika virus disease were known. The first Zika epidemic occurred on the island of Yap in Micronesia in 2007, when 73% of the people became infected. Between 2013 and 2015, Zika epidemics occurred in French Polynesia and Brazil. Over 1600 cases of Zika have occurred in the United States. Until mid-2016, they were all acquired during travel to endemic areas (except one laboratory-acquired infection). However, the first U.S. cases of transmission by mosquitoes occurred in Florida during the summer of 2016.

### Middle East Respiratory Syndrome (MERS)

Since 2014, there have been 1800 confirmed human cases and 630 deaths caused by a new virus called **Middle East respiratory syndrome coronavirus (MERS-CoV)**. The virus belongs to the same family that causes illnesses from the common cold to severe acute respiratory syndrome (SARS). Because the first reported cases were linked to the Middle East, this latest emerging infectious disease is called **Middle East respiratory syndrome (MERS)**. MERS has spread to Europe and Asia, and two travel-associated cases occurred in the United States in 2014.

### Influenza

**H1N1 influenza (flu)**, also known as *swine flu*, is a type of influenza caused by a new virus called *influenza H1N1*. H1N1 was first detected in the United States in 2009, and that same year WHO declared H1N1 flu to be a *pandemic disease* (a disease that affects large numbers of individuals in a short period of time and occurs worldwide).

Avian influenza A (H5N1), or **bird flu**, caught the attention of the public in 2003, when it killed millions of poultry and 24 people in southeast Asia. Avian influenza viruses occur in birds worldwide. In 2013, a different avian influenza, H7N9, sickened 131 people in China. In 2015, two cases of H7N9 were reported in Canada.

Influenza A viruses are found in many different animals, including ducks, chickens, pigs, whales, horses, and seals. Normally, each subtype of influenza A virus is specific to certain species. However, influenza A viruses normally seen in one species sometimes can cross over and cause illness in another species, and all subtypes of influenza A virus can infect pigs. Although it is unusual for people to get influenza infections directly from animals, sporadic human infections and outbreaks caused by certain avian influenza A viruses and pig influenza viruses have been reported. Fortunately, the virus has not yet evolved to be transmitted successfully among humans.

Human infections with avian influenza viruses detected since 1997 have not resulted in sustained human-to-human transmission. However, because influenza viruses have the potential to change and gain the ability to spread easily between people, monitoring for human infection and person-to-person transmission is important (see the box in Chapter 13 on page 367).

## Antibiotic-Resistant Infections

Antibiotics are critical in treating bacterial infections. However, years of overuse and misuse of these drugs have created environments in which antibiotic-resistant bacteria thrive. Random mutations in bacterial genes can make a bacterium resistant to an antibiotic. In the presence of that antibiotic, this bacterium has an advantage over other, susceptible bacteria and is able to proliferate. Antibiotic-resistant bacteria have become a global health crisis.

*Staphylococcus aureus* causes a wide range of human infections from pimples and boils to pneumonia, food poisoning, and surgical wound infections, and it is a significant cause of hospital-associated infections. After penicillin's initial success in treating *S. aureus* infection, penicillin-resistant *S. aureus* became a major threat in hospitals in the 1950s, requiring the use of methicillin. In the 1980s, **methicillin-resistant S. aureus**, called **MRSA**, emerged and became endemic in many hospitals, leading to increasing use of vancomycin. In the late 1990s, *S. aureus* infections that were less sensitive to vancomycin (**vancomycin-intermediate S. aureus**, or **VISA**) were reported. In 2002, the first infection caused by **vancomycin-resistant S. aureus (VRSA)** in a patient in the United States was reported.

In 2010, the World Health Organization (WHO) reported that in Asia and eastern Europe, about 35% of all individuals with tuberculosis (TB) had the multidrug-resistant form of the disease (MDR-TB). Multidrug-resistant TB is caused by bacteria that are resistant to at least the antibiotics isoniazid and rifampicin, the most effective drugs against tuberculosis.

The antibacterial substances added to various household cleaning products inhibit bacterial growth when used

**CLINICAL CASE**

The *S. aureus* bacterium responsible for Andrea's infection is resistant to the β-lactam antibiotic prescribed by Andrea's doctor. Concerned about what his patient is telling him, Andrea's doctor calls the local hospital to let them know he is sending a patient over. In the emergency department, a nurse swabs Andrea's wound and sends it to the hospital lab for culturing. The culture shows that Andrea's infection is caused by methicillin-resistant *Staphylococcus aureus* (MRSA). MRSA produces β-lactamase, an enzyme that destroys β-lactam antibiotics. The attending physician surgically drains the pus from the sore on Andrea's wrist.

**How does antibiotic resistance develop?**

3    16    **18**    19

correctly. However, wiping every household surface with these antibacterial agents creates an environment in which the resistant bacteria survive. Unfortunately, when you really need to disinfect your homes and hands—for example, when a family member comes home from a hospital and is still vulnerable to infection—you may encounter mainly resistant bacteria.

Routine housecleaning and handwashing are necessary, but standard soaps and detergents (without added antibacterials) are fine for these tasks. In addition, quickly evaporating chemicals, such as chlorine bleach, alcohol, ammonia, and hydrogen peroxide, remove potentially pathogenic bacteria but do not leave residues that encourage the growth of resistant bacteria.

In 2004, emergence of a new epidemic strain of *Clostridium difficile* (klo-STRID-ē-um DIF-fi-sē-il) was reported. The epidemic strain produces more toxins than others and is more resistant to antibiotics. In the United States, *C. difficile* infections kill nearly 29,000 people a year. Nearly all of the *C. difficile* infections occur in health care settings, where the infection is frequently transmitted between patients via health care personnel whose hands are contaminated after contact with infected patients or their surrounding environment.

## Ebola Virus Disease

First detected in 1995, **Ebola virus disease** causes fever, hemorrhaging, and blood clotting in vessels. In the first outbreak, 315 people in the Democratic Republic of Congo contracted the disease, and over 75% of them died. The epidemic was controlled through use of protective equipment and educational measures in the community. Close personal contact with infectious blood or other body fluids or tissue (see Chapter 23) leads to human-to-human transmission.

In 2014, a new outbreak in West Africa occurred. The countries Sierra Leone, Guinea, and Liberia experienced the worst

(See the Clinical Focus box in Chapter 26 on page 771.) The breakdown of public health measures for previously controlled infections has resulted in unexpected cases of tuberculosis, whooping cough, and measles (see Chapter 24).

## CLINICAL CASE

Mutations develop randomly in bacteria; some mutations are lethal, some have no effect, and some may be beneficial. Once these mutations develop, the offspring of the mutated parent cell also carry the same mutation. Because they have an advantage in the presence of the antibiotic, bacteria that are resistant to antibiotics soon outnumber those that are susceptible to antibiotic therapy. The widespread use of antibiotics selectively allows the resistant bacteria to grow, whereas the susceptible bacteria are killed. Eventually, almost the entire population of bacteria is resistant to the antibiotic.

The emergency department physician prescribes a different antibiotic, vancomycin, which will kill the MRSA in Andrea's wrist. She also explains to Andrea what MRSA is and why it's important they find out where Andrea acquired the potentially lethal bacteria.

**What can the emergency department physician tell Andrea about MRSA?**

3   16   18   **19**

impacts, with over 28,000 people infected over the next two years. Over one-third of those infected died. This time, a small number of health care workers from the United States and Europe who had been working with Ebola patients in Africa brought the disease back with them to their home countries, sparking fears that the disease would gain a foothold elsewhere in the world.

### Marburg Virus

Recorded cases of **Marburg virus**, another hemorrhagic fever virus, are rare. The first cases were laboratory workers in Europe who handled African green monkeys from Uganda. Thirteen outbreaks were identified in Africa between 1975 and 2016, involving 1 to 252 people, with 57% mortality. African fruit bats are the natural reservoir for the Marburg virus, and microbiologists suspect that bats are also the reservoir for Ebola.

Just as microbiological techniques helped researchers in the fight against syphilis and smallpox, they will help scientists discover the causes of new emerging infectious diseases in the twenty-first century. Undoubtedly there will be new diseases. *Ebolavirus* and *Influenzavirus* are examples of viruses that may be changing their abilities to infect different host species. Emerging infectious diseases will be discussed further in Chapter 14 on page 411.

Infectious diseases may reemerge because of antibiotic resistance and through the use of microorganisms as weapons.

### CHECK YOUR UNDERSTANDING

✔ 1-17 Differentiate normal microbiota and infectious disease.
✔ 1-18 Why are biofilms important?
✔ 1-19 What factors contribute to the emergence of an infectious disease?

* * *

The diseases we have mentioned are caused by viruses, bacteria, and protozoa—types of microorganisms. This book introduces you to the enormous variety of microscopic organisms. It shows you how microbiologists use specific techniques and procedures to study the microbes that cause such diseases as AIDS and diarrhea—and diseases that have yet to be discovered. You will also learn how the body responds to microbial infection and how certain drugs combat microbial diseases. Finally, you will learn about the many beneficial roles that microbes play in the world around us.

## CLINICAL CASE  Resolved

The first MRSA was healthcare–associated MRSA (HA-MRSA), transmitted between staff and patients in health care settings. In the 1990s, infections by a genetically different strain, community-associated MRSA (CA-MRSA), emerged as a major cause of skin disease in the United States. CA-MRSA enters skin abrasions from environmental surfaces or other people. Andrea has never been hospitalized before now, so they are able to rule out the hospital as the source of infection. Her college courses are all online, so she didn't contract MRSA at the university, either. The local health department sends someone to her family home to swab for the bacteria there.

MRSA is isolated from Andrea's living room sofa, but how did it get there? A representative from the health department, knowing that clusters of CA-MRSA infections have been seen among athletes, suggests swabbing the mats used by the gymnasts at the school Andrea's sister attends. The cultures come back positive for MRSA. Andrea's sister, although not infected, transferred the bacteria from her skin to the sofa, where Andrea laid her arm. (A person can carry MRSA on the skin without becoming infected.) The bacteria entered through a scratch on Andrea's wrist.

3   16   18   **19**

## Study Outline

 Go to @MasteringMicrobiology for **Interactive Microbiology**, *In the Clinic videos, MicroFlix, MicroBoosters, 3D animations, practice quizzes, and more.*

### Microbes in Our Lives (pp. 2–3)

1. Living things too small to be seen with the unaided eye are called microorganisms.
2. Microorganisms are important in maintaining Earth's ecological balance.
3. Everyone has microorganisms in and on the body; these make up the normal microbiota or human microbiome. The normal microbiota are needed to maintain good health.
4. Some microorganisms are used to produce foods and chemicals.
5. Some microorganisms cause disease.

### Naming and Classifying Microorganisms (pp. 4–6)

#### Nomenclature (p. 4)

1. In a nomenclature system designed by Carolus Linnaeus (1735), each living organism is assigned two names.
2. The two names consist of a genus and a specific epithet, both of which are underlined or italicized.

#### Types of Microorganisms (pp. 4–6)

3. Bacteria are unicellular organisms. Because they have no nucleus, the cells are described as prokaryotic.
4. Most bacteria have a peptidoglycan cell wall; they divide by binary fission, and they may possess flagella.
5. Bacteria can use a wide range of chemical substances for their nutrition.
6. Archaea consist of prokaryotic cells; they lack peptidoglycan in their cell walls.
7. Archaea include methanogens, extreme halophiles, and extreme thermophiles.
8. Fungi (mushrooms, molds, and yeasts) have eukaryotic cells (cells with a true nucleus). Most fungi are multicellular.
9. Fungi obtain nutrients by absorbing organic material from their environment.
10. Protozoa are unicellular eukaryotes.
11. Protozoa obtain nourishment by absorption or ingestion through specialized structures.
12. Algae are unicellular or multicellular eukaryotes that obtain nourishment by photosynthesis.
13. Algae produce oxygen and carbohydrates that are used by other organisms.
14. Viruses are noncellular entities that are parasites of cells.
15. Viruses consist of a nucleic acid core (DNA or RNA) surrounded by a protein coat. An envelope may surround the coat.
16. The principal groups of multicellular animal parasites are flatworms and roundworms, collectively called helminths.
17. The microscopic stages in the life cycle of helminths are identified by traditional microbiological procedures.

#### Classification of Microorganisms (p. 6)

18. All organisms are classified into one of three domains: Bacteria, Archaea, and Eukarya. Eukarya include protists, fungi, plants, and animals.

### A Brief History of Microbiology (pp. 6–14)

#### The First Observations (pp. 6–7)

1. Hooke's observations laid the groundwork for development of the cell theory, the concept that all living things are composed of cells.
2. Anton van Leeuwenhoek, using a simple microscope, was the first to observe microorganisms (1673).

#### The Debate over Spontaneous Generation (pp. 7–9)

3. Until the mid-1880s, many people believed in spontaneous generation, the idea that living organisms could arise from nonliving matter.
4. Francesco Redi demonstrated that maggots appear on decaying meat only when flies are able to lay eggs on the meat (1668).
5. John Needham claimed that microorganisms could arise spontaneously from heated nutrient broth (1745).
6. Lazzaro Spallanzani repeated Needham's experiments and suggested that Needham's results were due to microorganisms in the air entering his broth (1765).
7. Rudolf Virchow introduced the concept of biogenesis: living cells can arise only from preexisting cells (1858).
8. Louis Pasteur demonstrated that microorganisms are in the air everywhere and offered proof of biogenesis (1861).
9. Pasteur's discoveries led to the development of aseptic techniques used in laboratory and medical procedures to prevent contamination by microorganisms.

#### The First Golden Age of Microbiology (pp. 9–11)

10. The science of microbiology advanced rapidly between 1857 and 1914.
11. Pasteur found that yeasts ferment sugars to alcohol and that bacteria can oxidize the alcohol to acetic acid.
12. A heating process called pasteurization is used to kill bacteria in some alcoholic beverages and milk.
13. Agostino Bassi (1835) and Pasteur (1865) showed a causal relationship between microorganisms and disease.
14. Joseph Lister introduced the use of a disinfectant to clean surgical wounds in order to control infections in humans (1860s).
15. Robert Koch proved that microorganisms cause disease. He used a sequence of procedures, now called Koch's postulates (1876), that are used today to prove that a particular microorganism causes a particular disease.
16. In 1798, Edward Jenner demonstrated that inoculation with cowpox material provides humans with immunity to smallpox.
17. About 1880, Pasteur discovered that avirulent bacteria could be used as a vaccine for fowl cholera.
18. Modern vaccines are prepared from living avirulent microorganisms or killed pathogens, from isolated components of pathogens, and by recombinant DNA techniques.

#### The Second Golden Age of Microbiology (pp. 11–14)

19. The Second Golden Age began with the discovery of penicillin's effectiveness against infections.

20. Two types of chemotherapeutic agents are synthetic drugs (chemically prepared in the laboratory) and antibiotics (substances produced naturally by bacteria and fungi that inhibit the growth of bacteria).

21. Paul Ehrlich introduced an arsenic-containing chemical called salvarsan to treat syphilis (1910).

22. Alexander Fleming observed that the *Penicillium* fungus inhibited the growth of a bacterial culture. He named the active ingredient penicillin (1928).

23. Researchers are tackling the problem of drug-resistant microbes.

24. Bacteriology is the study of bacteria, mycology is the study of fungi, and parasitology is the study of parasitic protozoa and worms.

25. The study of AIDS and analysis of the action of interferons are among the current research interests in immunology.

26. New techniques in molecular biology and electron microscopy have provided tools for advancing our knowledge of virology.

27. The development of recombinant DNA technology has helped advance all areas of microbiology.

### The Third Golden Age of Microbiology  (p. 14)

28. Microbiologists are using genomics, the study of all of an organism's genes, to study microbiomes in different environments.

## Microbes and Human Welfare  (pp. 14–16)

1. Microorganisms degrade dead plants and animals and recycle chemical elements to be used by living plants and animals.

2. Bacteria are used to decompose organic matter in sewage.

3. Bioremediation processes use bacteria to clean up toxic wastes.

4. Bacteria that cause diseases in insects are being used as biological controls of insect pests. Biological controls are specific for the pest and do not harm the environment.

5. Using microbes to make products such as foods and chemicals is called biotechnology.

6. Using recombinant DNA, bacteria can produce important substances such as proteins, vaccines, and enzymes.

7. In gene therapy, viruses are used to carry replacements for defective or missing genes into human cells.

8. Genetically modified bacteria are used in agriculture to protect plants from frost and insects and to improve the shelf life of produce.

## Microbes and Human Disease  (pp. 16–19)

1. The disease-producing properties of a species of microbe and the host's resistance are important factors in determining whether a person will contract a disease.

2. Bacterial communities that form slimy layers on surfaces are called biofilms.

3. An infectious disease is one in which pathogens invade a susceptible host.

4. An emerging infectious disease (EID) is a new or changing disease showing an increase in incidence in the recent past or a potential to increase in the near future.

# Study Questions

For answers to the Knowledge and Comprehension questions, turn to the Answers tab at the back of the textbook.

## Knowledge and Comprehension

### Review

1. How did the idea of spontaneous generation come about?

2. Briefly state the role microorganisms play in each of the following:
   a. biological control of pests
   b. recycling of elements
   c. normal microbiota
   d. sewage treatment
   e. human insulin production
   f. vaccine production
   g. biofilms

3. Into which field of microbiology would the following scientists best fit?

| Researcher Who | Field |
| --- | --- |
| _____ **a.** Studies biodegradation of toxic wastes | **1.** Biotechnology |
| _____ **b.** Studies the causative agent of Ebola virus disease | **2.** Immunology |
|  | **3.** Microbial ecology |
|  | **4.** Microbial genetics |

| | |
| --- | --- |
| _____ **c.** Studies the production of human proteins by bacteria | **5.** Microbial physiology |
| _____ **d.** Studies the symptoms of AIDS | **6.** Molecular biology |
| _____ **e.** Studies the production of toxin by *E. coli* | **7.** Mycology |
| _____ **f.** Studies biodegradation of pollutants | **8.** Virology |
| _____ **g.** Develops gene therapy for a disease | |
| _____ **h.** Studies the fungus *Candida albicans* | |

4. Match the microorganisms in column A to their descriptions in column B.

| Column A | Column B |
| --- | --- |
| _____ **a.** Archaea | **1.** Not composed of cells |
| _____ **b.** Algae | **2.** Cell wall made of chitin |
| _____ **c.** Bacteria | **3.** Cell wall made of peptidoglycan |
| _____ **d.** Fungi | **4.** Cell wall made of cellulose; photosynthetic |
| _____ **e.** Helminths | |
| _____ **f.** Protozoa | **5.** Unicellular, complex cell structure lacking a cell wall |
| _____ **g.** Viruses | **6.** Multicellular animals |
| | **7.** Prokaryote without peptidoglycan cell wall |

5. Match the people in column A to their contribution toward the advancement of microbiology, in column B.

| Column A | Column B |
|---|---|
| _____ **a.** Avery, MacLeod, and McCarty | **1.** Developed vaccine against smallpox |
| | **2.** Discovered how DNA controls protein synthesis in a cell |
| _____ **b.** Beadle and Tatum | **3.** Discovered penicillin |
| _____ **c.** Berg | **4.** Discovered that DNA can be transferred from one bacterium to another |
| _____ **d.** Ehrlich | **5.** Disproved spontaneous generation |
| _____ **e.** Fleming | **6.** First to characterize a virus |
| _____ **f.** Hooke | **7.** First to use disinfectants in surgical procedures |
| _____ **g.** Iwanowski | |
| _____ **h.** Jacob and Monod | **8.** First to observe bacteria |
| | **9.** First to observe cells in plant material and name them |
| _____ **i.** Jenner | |
| _____ **j.** Koch | **10.** Observed that viruses are filterable material |
| _____ **k.** Lancefield | **11.** Proved that DNA is the hereditary material |
| _____ **l.** Lederberg and Tatum | **12.** Proved that microorganisms can cause disease |
| _____ **m.** Lister | **13.** Said living cells arise from preexisting living cells |
| _____ **n.** Pasteur | **14.** Showed that genes code for enzymes |
| _____ **o.** Stanley | **15.** Spliced animal DNA to bacterial DNA |
| _____ **p.** van Leeuwenhoek | **16.** Used bacteria to produce acetone |
| _____ **q.** Virchow | **17.** Used the first synthetic chemotherapeutic agent |
| _____ **r.** Weizmann | **18.** Proposed a classification system for streptococci based on antigens in their cell walls |

6. It is possible to purchase the following microorganisms in a retail store. Provide a reason for buying each.
   a. *Bacillus thuringiensis*
   b. *Saccharomyces*

7. NAME IT What type of microorganism has a peptidoglycan cell wall, has DNA that is not contained in a nucleus, and has flagella?

8. DRAW IT Show where airborne microbes ended up in Pasteur's experiment.

## Multiple Choice

1. Which of the following is a scientific name?
   a. *Mycobacterium tuberculosis*
   b. Tubercle bacillus

2. Which of the following is *not* a characteristic of bacteria?
   a. are prokaryotic
   b. have peptidoglycan cell walls
   c. have the same shape
   d. grow by binary fission
   e. have the ability to move

3. Which of the following is the most important element of Koch's germ theory of disease? The animal shows disease symptoms when
   a. the animal has been in contact with a sick animal.
   b. the animal has a lowered resistance.
   c. a microorganism is observed in the animal.
   d. a microorganism is inoculated into the animal.
   e. microorganisms can be cultured from the animal.

4. Recombinant DNA is
   a. DNA in bacteria.
   b. the study of how genes work.
   c. the DNA resulting when genes of two different organisms are mixed.
   d. the use of bacteria in the production of foods.
   e. the production of proteins by genes.

5. Which of the following statements is the best definition of *biogenesis*?
   a. Nonliving matter gives rise to living organisms.
   b. Living cells can only arise from preexisting cells.
   c. A vital force is necessary for life.
   d. Air is necessary for living organisms.
   e. Microorganisms can be generated from nonliving matter.

6. Which of the following is a beneficial activity of microorganisms?
   a. Some microorganisms are used as food for humans.
   b. Some microorganisms use carbon dioxide.
   c. Some microorganisms provide nitrogen for plant growth.
   d. Some microorganisms are used in sewage treatment processes.
   e. all of the above

7. It has been said that bacteria are essential for the existence of life on Earth. Which of the following is the essential function performed by bacteria?
   a. control insect populations
   b. directly provide food for humans
   c. decompose organic material and recycle elements
   d. cause disease
   e. produce human hormones such as insulin

8. Which of the following is an example of bioremediation?
   a. application of oil-degrading bacteria to an oil spill
   b. application of bacteria to a crop to prevent frost damage
   c. fixation of gaseous nitrogen into usable nitrogen
   d. production by bacteria of a human protein such as interferon
   e. all of the above

9. Spallanzani's conclusion about spontaneous generation was challenged because Antoine Lavoisier had just shown that oxygen was the vital component of air. Which of the following statements is true?
   a. All life requires air.
   b. Only disease-causing organisms require air.
   c. Some microbes do not require air.
   d. Pasteur kept air out of his biogenesis experiments.
   e. Lavoisier was mistaken.

10. Which of the following statements about *E. coli* is *false*?
    a. *E. coli* was the first disease-causing bacterium identified by Koch.
    b. *E. coli* is part of the normal microbiome of humans.
    c. *E. coli* is beneficial in human intestines.
    d. *E. coli* gets nutrients from intestinal contents.
    e. None of the above; all the statements are true.

## Analysis

1. How did the theory of biogenesis lead the way for the germ theory of disease?

2. Even though the germ theory of disease was not demonstrated until 1876, why did Semmelweis (1840) and Lister (1867) argue for the use of aseptic techniques?

3. The genus name of a bacterium is "erwinia," and the specific epithet is "amylovora." Write the scientific name of this organism correctly. Using this name as an example, explain how scientific names are chosen.

4. Find at least three supermarket products made by microorganisms. (*Hint:* The label will state the scientific name of the organism or include the word *culture, fermented,* or *brewed.*)

5. In the 1960s, many physicians and the public believed that infectious diseases were retreating and would be fully conquered. Discuss why this didn't happen. Is it possible?

## Clinical Applications and Evaluation

1. The prevalence of arthritis in the United States is 1 in 100,000 children. However, 1 in 10 children in Lyme, Connecticut, developed arthritis between June and September 1973. Allen Steere, a rheumatologist at Yale University, investigated the cases in Lyme and found that 25% of the patients remembered having a skin rash during their arthritic episode and that the disease was treatable with penicillin. Steere concluded that this was a new infectious disease and did not have an environmental, genetic, or immunologic cause.
    a. What was the factor that caused Steere to reach his conclusion?
    b. What is the disease?
    c. Why was the disease more prevalent between June and September?

2. In 1864, Lister observed that patients recovered completely from simple fractures but that compound fractures had "disastrous consequences." He knew that the application of phenol (carbolic acid) to fields in the town of Carlisle prevented cattle disease. Lister treated compound fractures with phenol, and his patients recovered without complications. How was Lister influenced by Pasteur's work? Why was Koch's work still needed?

3. Discuss whether antibacterial soaps and detergents should be used in the home.

# 2 Chemical Principles

Like all organisms, microorganisms use nutrients to make chemical building blocks for growth and other functions essential to life. For most microorganisms, synthesizing these building blocks requires them to break down nutrient substances and use the energy released to assemble the resulting molecular fragments into new substances.

We see evidence of these microbial chemical reactions in the world routinely, from a fallen tree rotting on the forest floor to milk going sour in the refrigerator. Although most people give little thought to the causes of such things, the chemistry of microbes is one of the most important concerns of microbiologists. Knowledge of chemistry is essential to understanding what roles microorganisms play in nature, how they cause disease, how methods for diagnosing disease are developed, how the body's defenses combat infection, and how antibiotics and vaccines are produced to combat the harmful effects of microbes.

The *Bacillus anthracis* (bah-SIL-lus an-THRĀ-sis) bacteria in the photograph make a capsule that is not readily digested by animal cells. As discussed in the Clinical Case, these bacteria can grow in mammals by avoiding host defenses. Researchers are investigating ways to identify unique chemicals made by *B. anthracis* and other potential biological weapons in order to detect bioterrorism. To understand the changes that occur in microorganisms and the changes microbes make in the world around us, we need to know how molecules are formed and how they interact.

▶ *Bacillus anthracis* bacteria produce heat-resistant endospores (red).

## In the Clinic

As the health advisory nurse at a health service company, you receive a call from a man who is concerned that his blood sugar levels have not decreased, even though he has switched to using organic sugar. **How would you respond to the man?**

*Hint: Read about important biological molecules later in this chapter on pages 31–47.*

Play **In the Clinic** Video
@Mastering**Microbiology**

# The Structure of Atoms

## LEARNING OBJECTIVE

**2-1** Describe the structure of an atom and its relation to the physical properties of elements.

All matter—whether air, rock, or a living organism—is made up of small units called atoms. An **atom** is the smallest component of a substance, and it cannot be subdivided into smaller substances without losing its properties. Atoms combine to form **molecules**. Living cells are made up of molecules, some of which are very complex. The science of the interaction between atoms and molecules is called **chemistry**.

Atoms are the smallest units of matter that enter into chemical reactions. Every atom has a centrally located **nucleus** and negatively (−) charged particles called **electrons** that move around the nucleus in regions called *electron shells* (Figure 2.1). The nucleus is made up of positively (+) charged particles called **protons** and uncharged (neutral) particles called **neutrons**. The nucleus, therefore, bears a net positive charge. All atoms contain an equal number of electrons and protons. Because the total positive charge of the nucleus equals the total negative charge of the electrons, each atom is electrically neutral.

The nuclei of most atoms are stable—that is, they do not change spontaneously—and nuclei do not participate in chemical reactions. The number of protons in an atomic nucleus ranges from one (in a hydrogen atom) to more than 100 (in the largest atoms known). Atoms are often listed by their **atomic number**, the number of protons in the nucleus. Protons and neutrons are approximately the same weight, which is about 1840 times that of an electron, and the total number of protons and neutrons in an atom is its approximate **atomic mass**.

## Chemical Elements

All atoms with the same number of protons behave the same way chemically and are classified as the same **chemical element**. Each element has its own name and a one- or two-letter symbol, usually derived from the English or Latin name for the element. For example, the symbol for the element hydrogen is H, and the symbol for carbon is C. The symbol for sodium is Na—the first two letters of its Latin name, *natrium*—distinguish it from nitrogen, N, and from sulfur, S. There are 92 naturally occurring elements. However, only about 26 elements are commonly found in living things. Table 2.1 lists some of the chemical elements found in living organisms.

Most elements have several **isotopes**—atoms with different numbers of neutrons in their nuclei. All isotopes of an element have the same number of protons in their nuclei, but their atomic masses differ because of the difference in the number of neutrons. For example, in a natural sample of oxygen,

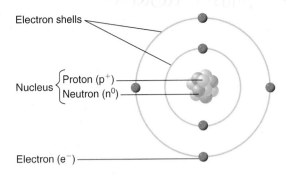

Figure 2.1 **The structure of an atom.** In this simplified diagram of a carbon atom, note the central location of the nucleus. The nucleus contains six neutrons and six protons, although not all the neutrons are visible in this view. The six electrons move about the nucleus in regions called electron shells, shown here as circles.

**Q** What is the atomic number of this atom?

all the atoms contain eight protons. However, 99.76% of the atoms have eight neutrons, 0.04% contain nine neutrons, and the remaining 0.2% contain ten neutrons. Therefore, the three isotopes composing a natural sample of oxygen have atomic masses of 16, 17, and 18, although all will have the atomic number 8. Atomic numbers are written as a subscript to the left of an element's chemical symbol. Atomic masses are written as a superscript above the atomic number. Thus, natural oxygen isotopes are represented as $^{16}_{8}O$, $^{17}_{8}O$, and $^{18}_{8}O$. Isotopes of certain elements are extremely useful in biological research, medical diagnosis, the treatment of some disorders, and some forms of sterilization.

**CLINICAL CASE**  Drumming Up Dust

Jonathan, a 52-year-old drummer, is doing his best to ignore the cold sweat that is breaking out all over his body. He and his bandmates are performing in a local Philadelphia nightclub, and they are just about finished with the second set of the evening. Jonathan hasn't been feeling well for a while, actually; he has been feeling weak and short of breath for the last 3 days or so. Jonathan makes it to the end of the song, but the noise from the clapping and cheering audience seems to come from far away. He stands up to bow and collapses. Jonathan is admitted to a local emergency department with a mild fever and severe shaking. He is able to tell the admitting nurse that he also has had a dry cough for the last few days. The attending physician orders a chest X-ray exam and sputum culture. Jonathan is diagnosed with bilateral pneumonia caused by *Bacillus anthracis*. The attending physician is astonished by this diagnosis.

**How did Jonathan become infected by *B. anthracis*? Read on to find out.**

25    42    44    46

*most abundant chemical elements in living organisms*

| TABLE 2.1 The Elements of Life* | | | |
|---|---|---|---|
| **Element** | **Symbol** | **Atomic Number** | **Approximate Atomic Mass** |
| Hydrogen | H | 1 | 1 |
| Carbon | C | 6 | 12 |
| Nitrogen | N | 7 | 14 |
| Oxygen | O | 8 | 16 |
| Sodium | Na | 11 | 23 |
| Magnesium | Mg | 12 | 24 |
| Phosphorus | P | 15 | 31 |
| Sulfur | S | 16 | 32 |
| Chlorine | Cl | 17 | 35 |
| Potassium | K | 19 | 39 |
| Calcium | Ca | 20 | 40 |
| Iron | Fe | 26 | 56 |
| Iodine | I | 53 | 127 |

*Hydrogen, carbon, nitrogen, and oxygen are the most abundant chemical elements in living organisms.

## Electronic Configurations

In an atom, electrons are arranged in **electron shells**, which are regions corresponding to different **energy levels**. The arrangement is called an **electronic configuration**. Shells are layered outward from the nucleus, and each shell can hold a characteristic maximum number of electrons—two electrons in the innermost shell (lowest energy level), eight electrons in the second shell, and eight electrons in the third shell, if it is the atom's outermost (valence) shell. The fourth, fifth, and sixth electron shells can each accommodate 18 electrons, although there are some exceptions to this generalization. Table 2.2 shows the electronic configurations for atoms of some elements found in living organisms.

The number of electrons in the outermost shell determines an atom's tendency to react with other atoms. An atom can give up, accept, or share electrons with other atoms to fill the outermost shell. When its outer shell is filled, the atom is chemically stable, or inert: it does not tend to react with other atoms. Helium (atomic number 2) and neon (atomic number 10) are examples of atoms of inert gases whose outer shells are filled.

When an atom's outer electron shell is only partially filled, the atom is chemically unstable. These unstable atoms react with other atoms, depending, in part, on the degree to which the outer energy levels are filled. Notice the number of electrons in the outer energy levels of the atoms in Table 2.2. We will see later how the number correlates with the chemical reactivity of the elements.

| TABLE 2.2 Electronic Configurations for the Atoms of Some Elements Found in Living Organisms | | | | |
|---|---|---|---|---|
| **Element** | **Diagram** | **Number of Valence (Outermost) Shell Electrons** | **Number of Unfilled Spaces** | **Maximum Number of Bonds Formed** |
| Hydrogen | | 1 | 1 | 1 |
| Carbon | | 4 | 4 | 4 |
| Nitrogen | | 5 | 3 | 5 |
| Oxygen | | 6 | 2 | 2 |
| Magnesium | | 2 | 6 | 2 |
| Phosphorus | | 5 | 3 | 5 |
| Sulfur | | 6 | 2 | 6 |

**CHECK YOUR UNDERSTANDING**

✔ **2-1** How does $^{14}_{6}C$ differ from $^{12}_{6}C$? What is the atomic number of each carbon atom? The atomic mass?

# How Atoms Form Molecules: Chemical Bonds

## LEARNING OBJECTIVE

2-2 Define *ionic bond*, *covalent bond*, *hydrogen bond*, *molecular weight*, and *mole*.

When the outermost energy level of an atom is not completely filled by electrons, you can think of it as having either unfilled spaces or extra electrons in that energy level, depending on whether it is easier for the atom to gain or lose electrons. For example, an atom of oxygen, with two electrons in the first energy level and six in the second, has two unfilled spaces in the second electron shell; an atom of magnesium has two extra electrons in its outermost shell. The most chemically stable configuration for any atom is to have its outermost shell filled. Therefore, for these two atoms to attain that state, oxygen must gain two electrons, and magnesium must lose two electrons. Because all atoms tend to combine so that the extra electrons in the outermost shell of one atom fill the spaces of the outermost shell of the other atom, oxygen and magnesium combine so that the outermost shell of each atom has the full complement of eight electrons.

The **valence**, or combining capacity, of an atom is the number of extra or missing electrons in its outermost electron shell. For example, hydrogen has a valence of 1 (one unfilled space, or one extra electron), oxygen has a valence of 2 (two unfilled spaces), carbon has a valence of 4 (four unfilled spaces, or four extra electrons), and magnesium has a valence of 2 (two extra electrons).

Basically, atoms achieve the full complement of electrons in their outermost energy shells by combining to form molecules, which are made up of atoms of one or more elements. A molecule that contains at least two different kinds of atoms, such as $H_2O$ (the water molecule), is called a **compound**. In $H_2O$, the subscript 2 indicates that there are two atoms of hydrogen; the absence of a subscript indicates that there is only one atom of oxygen. Molecules hold together because the valence electrons of the combining atoms form attractive forces, called **chemical bonds**, between the atomic nuclei. Therefore, valence may also be viewed as the bonding capacity of an element.

In general, atoms form bonds in one of two ways: by either gaining or losing electrons from their outer electron shell, or by sharing outer electrons. When atoms have gained or lost outer electrons, the chemical bond is called an *ionic bond*. When outer electrons are shared, the bond is called a *covalent bond*. Although we will discuss ionic and covalent bonds separately, the kinds of bonds actually found in molecules do not belong entirely to either category. Instead, bonds range from the highly ionic to the highly covalent.

## Ionic Bonds

Atoms are electrically neutral when the number of positive charges (protons) equals the number of negative charges (electrons). But when an isolated atom gains or loses electrons, this balance is upset. If the atom gains electrons, it acquires an overall negative charge; if the atom loses electrons, it acquires an overall positive charge. Such a negatively or positively charged atom (or group of atoms) is called an **ion**.

Consider the following examples. Sodium (Na) has 11 protons and 11 electrons, with one electron in its outer electron shell. Sodium tends to lose the single outer electron; it is an *electron donor* (**Figure 2.2a**). When sodium donates an electron to another atom, it is left with 11 protons and only 10 electrons and so has an overall charge of +1. This positively charged sodium atom is called a sodium ion and is written as $Na^+$. Chlorine (Cl) has a total of 17 electrons, seven of them in the outer electron shell. Because this outer shell can hold eight electrons, chlorine tends to pick up an electron that has been lost by another atom; it is an *electron acceptor* (see Figure 2.2a). By accepting an electron, chlorine totals 18 electrons. However, it still has only 17 protons in its nucleus. The chloride ion therefore has a charge of −1 and is written as $Cl^-$.

The opposite charges of the sodium ion ($Na^+$) and chloride ion ($Cl^-$) attract each other. The attraction, an ionic bond, holds the two atoms together, and a molecule is formed (Figure 2.2b). The formation of this molecule, called sodium chloride (NaCl) or table salt, is a common example of ionic bonding. Thus, an **ionic bond** is an attraction between ions of opposite charge that holds them together to form a stable molecule. Put another way, an ionic bond is an attraction between atoms in which one atom loses electrons and another atom gains electrons. Strong ionic bonds, such as those that hold $Na^+$ and $Cl^-$ together in salt crystals, have limited importance in living cells. But the weaker ionic bonds formed in aqueous (water) solutions are important in biochemical reactions in microbes and other organisms. For example, weaker ionic bonds assume a role in certain antigen–antibody reactions—that is, reactions in which molecules produced by the immune system (antibodies) combine with foreign substances (antigens) to combat infection.

In general, an atom whose outer electron shell is less than half-filled will lose electrons and form positively charged ions, called **cations**. Examples of cations are the potassium ion ($K^+$), calcium ion ($Ca^{2+}$), and sodium ion ($Na^+$). When an atom's outer electron shell is more than half-filled, the atom will gain electrons and form negatively charged ions, called **anions**. Examples are the iodide ion ($I^-$), chloride ion ($Cl^-$), and sulfide ion ($S^{2-}$).

## Covalent Bonds

A **covalent bond** is a chemical bond formed by two atoms sharing one or more pairs of electrons. Covalent bonds are stronger

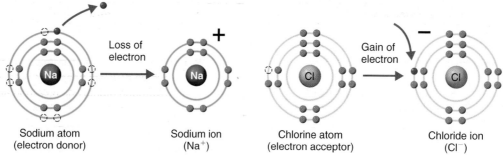

**(a)** A sodium atom (Na) loses one electron to an electron acceptor and forms a sodium ion (Na⁺). A chlorine atom (Cl) accepts one electron from an electron donor to become a chloride ion (Cl⁻).

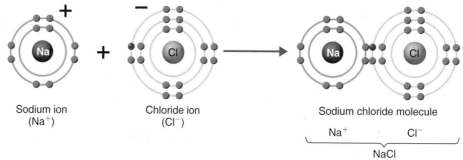

**(b)** The sodium and chloride ions are attracted because of their opposite charges and are held together by an ionic bond to form a molecule of sodium chloride.

**Figure 2.2 Ionic bond formation.**

**Q** What is an ionic bond?

and far more common in organisms than are true ionic bonds. In the hydrogen molecule, $H_2$, two hydrogen atoms share a pair of electrons. Each hydrogen atom has its own electron plus one electron from the other atom (**Figure 2.3a**). The shared pair of electrons actually orbits the nuclei of both atoms. Therefore, the outer electron shells of both atoms are filled. Atoms that share only one pair of electrons form *a single covalent bond.* For simplicity, a single covalent bond is expressed as a single line between the atoms (H—H). Atoms that share two pairs of electrons form a *double covalent bond,* expressed as two single lines (═). A *triple covalent bond,* expressed as three single lines (≡), occurs when atoms share three pairs of electrons.

The principles of covalent bonding that apply to atoms of the same element also apply to atoms of different elements. Methane ($CH_4$) is an example of covalent bonding between atoms of different elements (Figure 2.3b). The outer electron shell of the carbon atom can hold eight electrons but has only four; each hydrogen atom can hold two electrons but has only one. Consequently, in the methane molecule the carbon atom gains four hydrogen electrons to complete its outer shell, and each hydrogen atom completes its pair by sharing one electron from the carbon atom. Each outer electron of the carbon atom orbits both the carbon nucleus and a hydrogen nucleus. Each hydrogen electron orbits both its own nucleus and the carbon nucleus.

Elements such as hydrogen and carbon, whose outer electron shells are half-filled, form covalent bonds quite easily. In fact, in living organisms, carbon almost always forms covalent bonds; it almost never becomes an ion. *Remember:* Covalent bonds are formed by the *sharing* of electrons between atoms. Ionic bonds are formed by *attraction* between atoms that have lost or gained electrons and are therefore positively or negatively charged.

## Hydrogen Bonds

Another chemical bond of special importance to all organisms is the **hydrogen bond,** in which a hydrogen atom that is covalently bonded to one oxygen or nitrogen atom is attracted to another oxygen or nitrogen atom. Such bonds are weak and do not bind atoms into molecules. However, they do serve as bridges between different molecules or between various portions of the same molecule.

When hydrogen combines with atoms of oxygen or nitrogen, the relatively large nucleus of these larger oxygen or nitrogen atoms has more protons and attracts the hydrogen electron more strongly than does the small hydrogen nucleus. Thus, in a molecule of water ($H_2O$), all the electrons tend to be closer to the oxygen nucleus than to the hydrogen nuclei. As a result, the oxygen portion of the molecule has a slightly negative charge, and the hydrogen portion of the molecule has a slightly positive charge (**Figure 2.4a**). When the positively charged end of one

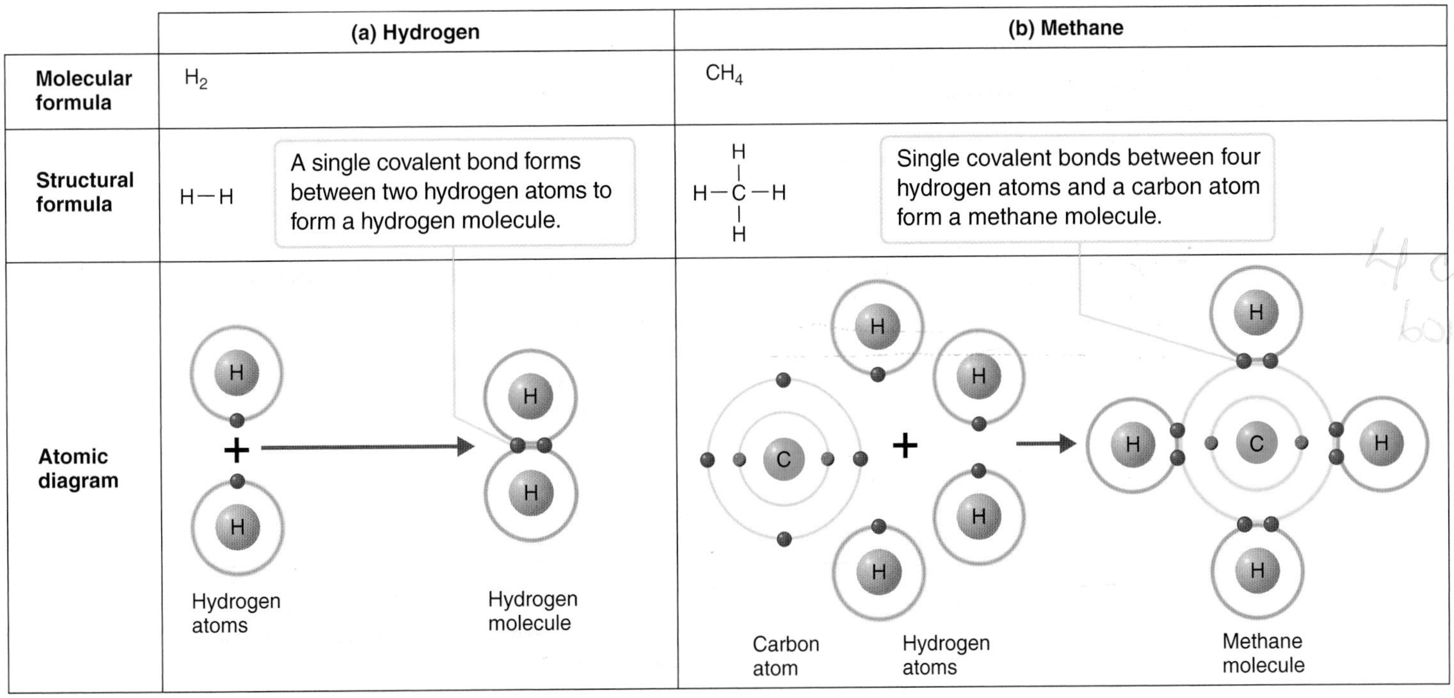

| | (a) Hydrogen | (b) Methane |
|---|---|---|
| **Molecular formula** | $H_2$ | $CH_4$ |
| **Structural formula** | H—H  A single covalent bond forms between two hydrogen atoms to form a hydrogen molecule. | H—C—H (with H above and below)  Single covalent bonds between four hydrogen atoms and a carbon atom form a methane molecule. |
| **Atomic diagram** | 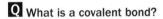 Hydrogen atoms → Hydrogen molecule | Carbon atom + Hydrogen atoms → Methane molecule |

*4 covalent bonds* (handwritten)

**Figure 2.3 Covalent bond formation.** The molecular formula shows the number and types of atoms in a molecule. In structural formulas, each covalent bond is written as a straight line between the symbols for two atoms. In molecular formulas, the number of atoms in each molecule is noted by subscripts.

*On Slides Examples* (handwritten)

**Q** What is a covalent bond?

*On Slides* (handwritten)

molecule is attracted to the negatively charged end of another molecule, a hydrogen bond is formed (Figure 2.4b). This attraction can also occur between hydrogen and other atoms of the same molecule, especially in large molecules, but oxygen and nitrogen are the elements most frequently involved in hydrogen bonding.

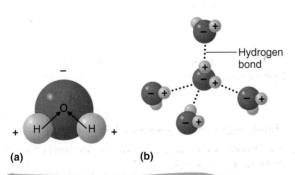

(a)　　　　　(b)

Hydrogen bond

**Figure 2.4 Hydrogen bond formation in water. (a)** In a water molecule, the electrons of the hydrogen atoms are strongly attracted to the oxygen atom. Therefore, the part of the water molecule containing the oxygen atom has a slightly negative charge, and the part containing hydrogen atoms has a slightly positive charge. **(b)** In a hydrogen bond between water molecules, the hydrogen of one water molecule is attracted to the oxygen of another water molecule. Many water molecules may be attracted to each other by hydrogen bonds (black dots).

**Q** Which chemical elements are usually involved in hydrogen bonding?

Hydrogen bonds are considerably weaker than either ionic or covalent bonds; they have only about 5% of the strength of covalent bonds. Consequently, hydrogen bonds are formed and broken relatively easily. This property accounts for the temporary bonding that occurs between certain atoms of large and complex molecules, such as proteins and nucleic acids. Even though hydrogen bonds are relatively weak, large molecules containing several hundred of these bonds have considerable strength and stability. A summary of ionic, covalent, and hydrogen bonds is shown in Table 2.3.

## Molecular Mass and Moles

You have seen that bond formation usually results in the creation of molecules. Molecules are often discussed in terms of units of measure called molecular mass and moles. The **molecular mass** of a molecule is the sum of the atomic masses of all its atoms. To relate the molecular level to the laboratory level, we use a unit called the mole. One **mole** of a substance is its molecular mass expressed in grams. The unit of molecular mass is a **dalton (da)**. For example, 1 mole of water weighs 18 grams because the molecular mass of $H_2O$ is 18 da, or $[(2 \times 1) + 16]$.

*Is slide* (handwritten)

**CHECK YOUR UNDERSTANDING**

✔ 2-2 Differentiate an ionic bond from a covalent bond.

| TABLE 2.3 | Comparison among Ionic, Covalent, and Hydrogen Bonds |
|---|---|
| **Type of Bond** | **Definition and Importance** |
| Ionic | An *attraction* between ions of opposite charge that holds them together to form a stable molecule. Weaker ionic bonds are important in biochemical reactions such as antigen–antibody reactions. |
| Covalent | A bond formed by two atoms that *share* one or more pairs of electrons. Covalent bonds are the most common type of chemical bond in organisms and are responsible for holding together the atoms of most molecules in organisms. |
| Hydrogen | A relatively weak bond in which a hydrogen atom that is covalently bonded to one oxygen or nitrogen atom is attracted to another oxygen or nitrogen atom. Hydrogen bonds do not bind atoms into molecules, but serve as *bridges between different molecules* or different portions of the same molecule, for example, within proteins and nucleic acids. |

# Chemical Reactions

## LEARNING OBJECTIVE

2-3  Diagram three basic types of chemical reactions.

As we said earlier, chemical reactions involve the making or breaking of bonds between atoms. After a chemical reaction, the total number of atoms remains the same, but there are new molecules with new properties because the atoms have been rearranged.

## Energy in Chemical Reactions

All chemical bonds require energy to form or break. It is important to note that initially, *activation energy* is needed to break a bond (see page 111). In the chemical reactions of metabolism, energy is released when new bonds are formed after the original bonds break; this is the energy cells use to do work. A chemical reaction that absorbs more energy than it releases is called an **endergonic reaction** (*endo* = within), meaning that energy is directed inward. A chemical reaction that releases more energy than it absorbs is called an **exergonic reaction** (*exo* = out), meaning that energy is directed outward.

In this section we will look at three basic types of chemical reactions common to all living cells. By becoming familiar with these reactions, you will be able to understand the specific chemical reactions we will discuss later (particularly in Chapter 5).

## Synthesis Reactions

When two or more atoms, ions, or molecules combine to form new and larger molecules, the reaction is called a **synthesis reaction**. To synthesize means to put together, and a synthesis reaction *forms new bonds*. Synthesis reactions can be expressed in the following way:

$$
\begin{array}{ccc}
\text{reactants} & \text{combine to form} & \text{product} \\
A \quad + \quad B & \longrightarrow & AB \\
\text{Atom, ion,} \quad \text{Atom, ion,} & & \text{New molecule AB} \\
\text{or molecule A} \quad \text{or molecule B} & &
\end{array}
$$

The combining substances, A and B, are called the *reactants*; the substance formed by the combination, AB, is the *product*. The *arrow* indicates the direction in which the reaction proceeds.

Pathways of synthesis reactions in living organisms are collectively called anabolic reactions, or simply **anabolism** (an-AB-ō-liz-um). The combining of sugar molecules to form starch and of amino acids to form proteins are two examples of anabolism.

## Decomposition Reactions

The reverse of a synthesis reaction is a **decomposition reaction**. To decompose means to break down into smaller parts, and in a decomposition reaction *bonds are broken*. Typically, decomposition reactions split large molecules into smaller molecules, ions, or atoms. A decomposition reaction occurs in the following way:

$$
\begin{array}{ccc}
\text{reactant} \quad \text{breaks down into} & & \text{products} \\
AB & \longrightarrow & A \quad + \quad B \\
\text{Molecule AB} & & \text{Atom, ion,} \quad \text{Atom, ion,} \\
& & \text{or molecule A} \quad \text{or molecule B}
\end{array}
$$

Decomposition reactions that occur in living organisms are collectively called catabolic reactions, or simply **catabolism** (ka-TAB-ō-liz-um). An example of catabolism is the breakdown of sucrose (table sugar) into simpler sugars, glucose and fructose, during digestion.

## Exchange Reactions

All chemical reactions are based on synthesis and decomposition. Many reactions, such as **exchange reactions**, are actually part synthesis and part decomposition. An exchange reaction works in the following way:

$$
\begin{array}{ccc}
\text{reactants} \quad \text{recombine to form} & & \text{products} \\
AB + CD & \longrightarrow & AD + BC
\end{array}
$$

First, the bonds between A and B and between C and D are broken in a decomposition process. New bonds are then formed between A and D and between B and C in a synthesis process. For example, an exchange reaction occurs when sodium hydroxide (NaOH) and hydrochloric acid (HCl) react to form table salt (NaCl) and water ($H_2O$), as follows:

$$ NaOH + HCl \longrightarrow NaCl + H_2O $$

## The Reversibility of Chemical Reactions

All chemical reactions are, in theory, reversible; that is, they can occur in either direction. In practice, however, some reactions do this more easily than others. A chemical reaction that is readily reversible (when the end product can revert to the original molecules) is termed a **reversible reaction** and is indicated by two arrows, as shown here:

$$\underset{A + B}{\overset{reactants}{}} \quad \underset{break\ down\ into}{\overset{combine\ to\ form}{\rightleftharpoons}} \quad \underset{AB}{\overset{product}{}}$$

Some reversible reactions occur because neither the reactants nor the end products are very stable. Other reactions reverse only under special conditions:

$$\underset{A + B}{\overset{reactants}{}} \quad \underset{water}{\overset{heat}{\rightleftharpoons}} \quad \underset{AB}{\overset{product}{}}$$

Whatever is written above or below the arrows indicates the special condition under which the reaction in that direction occurs. In this case, A and B react to produce AB only when heat is applied, and AB breaks down into A and B only in the presence of water. See Figure 2.8 on page 35 for another example.

In Chapter 5 we will examine the various factors that affect chemical reactions.

---

**CHECK YOUR UNDERSTANDING**

✔ **2-3** The chemical reaction below is used to remove chlorine from water. What type of reaction is it?

$$HClO + Na_2SO_3 \longrightarrow Na_2SO_4 + HCl$$

---

# Important Biological Molecules

Biologists and chemists divide compounds into two principal classes: inorganic and organic. **Inorganic compounds** are defined as molecules, usually small and structurally simple, which typically lack carbon and in which ionic bonds may play an important role. Inorganic compounds include water, molecular oxygen ($O_2$), carbon dioxide, and many salts, acids, and bases.

**Organic compounds** always contain carbon and hydrogen and typically are structurally complex. Carbon is a unique element because it has four electrons in its outer shell and four unfilled spaces. It can combine with a variety of atoms, including other carbon atoms, to form straight or branched chains and rings. Carbon chains form the basis of many organic compounds in living cells, including sugars, amino acids, and vitamins. Organic compounds are held together mostly or entirely by covalent bonds. Some organic molecules, such as polysaccharides, proteins, and nucleic acids, are very large and usually contain thousands of atoms. Such giant molecules are called *macromolecules*. In the following section we will discuss inorganic and organic compounds that are essential for cells.

## Inorganic Compounds

### LEARNING OBJECTIVES

**2-4** List several properties of water that are important to living systems.

**2-5** Define *acid*, *base*, *salt*, and *pH*.

### Water

All living organisms require a wide variety of inorganic compounds for growth, repair, maintenance, and reproduction. Water is one of the most important, as well as one of the most abundant, of these compounds, and it is particularly vital to microorganisms. Outside the cell, nutrients are dissolved in water, which facilitates their passage through cell membranes.

Inside the cell, water is the medium for most chemical reactions. In fact, water is by far the most abundant component of almost all living cells. Water makes up between 65% and 75% of every cell on average. Simply stated, no organism can survive without water.

Water has structural and chemical properties that make it particularly suitable for its role in living cells. As we discussed, the total charge on the water molecule is neutral, but the oxygen region of the molecule has a slightly negative charge, and the hydrogen region has a slightly positive charge (see Figure 2.4a). Any molecule having such an unequal distribution of charges is called a **polar molecule**. The polar nature of water gives it four characteristics that make it a useful medium for living cells.

First, every water molecule is capable of forming four hydrogen bonds with nearby water molecules (see Figure 2.4b). This property results in a strong attraction between water molecules and makes water an excellent temperature buffer. Because of this strong attraction, a great deal of heat is required to separate water molecules from each other to form water vapor; thus, water has a relatively high boiling point (100°C, 212°F). Because water has such a high boiling point, it exists in the liquid state on most of the Earth's surface. Conversely, water temperatures must drop significantly in order for it to freeze. Secondly, the hydrogen bonding between water molecules affects the density of water, depending on whether it occurs as ice or a liquid. For example, the hydrogen bonds in the crystalline structure of water (ice) make ice take up more space. As a result, ice has fewer molecules than an equal volume of liquid water. This makes its crystalline structure less dense than liquid water. For this reason, ice floats and can serve as an insulating layer on the surfaces of lakes and streams that harbor living organisms.

Third, the polarity of water makes it an excellent dissolving medium, or **solvent**. Many polar substances undergo **dissociation**, or separation, into individual molecules in water—that is, they dissolve. The negative part of the water molecules

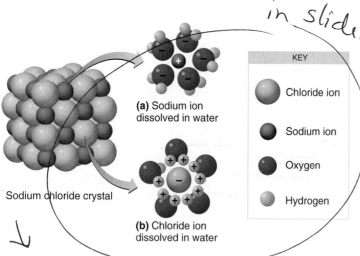

*in slides*

**(a)** Sodium ion dissolved in water

**(b)** Chloride ion dissolved in water

Sodium chloride crystal

| KEY | |
|---|---|
| ● | Chloride ion |
| ● | Sodium ion |
| ● | Oxygen |
| ● | Hydrogen |

**Figure 2.5   How water acts as a solvent for sodium chloride (NaCl). (a)** The positively charged sodium ion (Na⁺) is attracted to the negative part of the water molecule. **(b)** The negatively charged chloride ion (Cl⁻) is attracted to the positive part of the water molecule. In the presence of water molecules, the bonds between the Na⁺ and Cl⁻ are disrupted, and the NaCl dissolves in the water.

**Q** What happens during ionization?

is attracted to the positive part of the molecules in the **solute**, or dissolving substance, and the positive part of the water molecules is attracted to the negative part of the solute molecules. Substances (such as salts) that are composed of atoms (or groups of atoms) held together by ionic bonds tend to dissociate into separate cations and anions in water. Thus, the polarity of water allows molecules of many different substances to separate and become surrounded by water molecules (**Figure 2.5**).

Fourth, polarity accounts for water's characteristic role as a reactant or product in many chemical reactions. Its polarity facilitates the splitting and rejoining of hydrogen ions (H⁺) and hydroxide ions (OH⁻). Water is a key reactant in the digestive processes of organisms, whereby larger molecules are broken down into smaller ones. Water molecules are also involved in synthetic reactions; water is an important source of the hydrogen and oxygen that are incorporated into numerous organic compounds in living cells.

## Acids, Bases, and Salts

As we saw in Figure 2.5, when inorganic salts such as sodium chloride (NaCl) are dissolved in water, they undergo **ionization** or *dissociation*; that is, they break apart into ions. Substances called acids and bases show similar behavior.

An **acid** can be defined as a substance that dissociates into one or more hydrogen ions (H⁺) and one or more negative ions (anions). Thus, an acid can also be defined as a proton (H⁺) donor. A **base** dissociates into one or more negatively charged hydroxide ions (OH⁻) that can accept, or combine with, protons, and one or more positive ions (cations). Thus, sodium hydroxide (NaOH) is a base because it dissociates to release

OH⁻, which has a strong attraction for protons and is among the most important proton acceptors. A **salt** is a substance that dissociates in water into cations and anions, neither of which is H⁺ or OH⁻. **Figure 2.6** shows common examples of each type of compound and how they dissociate in water.

## Acid–Base Balance: The Concept of pH

An organism must maintain a fairly constant balance of acids and bases to remain healthy. For example, if a particular acid or base concentration is too high or too low, enzymes change in shape and no longer effectively promote chemical reactions in a cell. In the aqueous environment within organisms, acids dissociate into hydrogen ions (H⁺) and anions. Bases, in contrast, dissociate into hydroxide ions (OH⁻) and cations. The more hydrogen ions that are free in a solution, the more acidic the solution is. Conversely, the more hydroxide ions that are free in a solution, the more basic, or alkaline, it is.

Biochemical reactions—that is, chemical reactions in living systems—are extremely sensitive to even small changes in the acidity or alkalinity of the environments in which they occur. In fact, H⁺ and OH⁻ are involved in almost all biochemical processes, and any deviation from a cell's narrow band of normal H⁺ and OH⁻ concentrations can dramatically modify the cell's functions. For this reason, the acids and bases that are continually formed in an organism must be kept in balance.

It is convenient to express the amount of H⁺ in a solution by a logarithmic **pH scale**, which ranges from 0 to 14 (**Figure 2.7**). The term *pH* means potential of hydrogen. On a logarithmic scale, a change of one whole number represents a *tenfold* change from the previous concentration. Thus, a solution of pH 1 has ten times more hydrogen ions than a solution of pH 2 and has 100 times more hydrogen ions than a solution of pH 3.

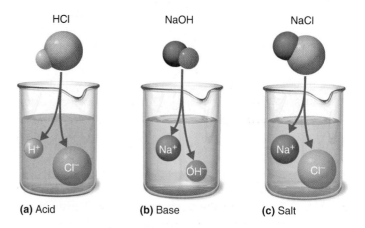

**(a)** Acid    **(b)** Base    **(c)** Salt

**Figure 2.6   Acids, bases, and salts. (a)** In water, hydrochloric acid (HCl) dissociates into H⁺ and Cl⁻. **(b)** Sodium hydroxide (NaOH), a base, dissociates into OH⁻ and Na⁺ in water. **(c)** In water, table salt (NaCl) dissociates into positive ions (Na⁺) and negative ions (Cl⁻), neither of which are H⁺ or OH⁻.

**Q** How do acids and bases differ?

*On slides*

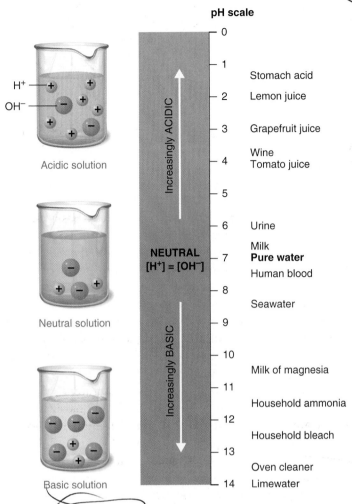

**pH scale**

| pH | Substance |
|----|-----------|
| 0 | |
| 1 | Stomach acid |
| 2 | Lemon juice |
| 3 | Grapefruit juice |
| 4 | Wine / Tomato juice |
| 5 | |
| 6 | Urine |
| 7 | Milk / **Pure water** / Human blood |
| 8 | Seawater |
| 9 | |
| 10 | |
| 11 | Milk of magnesia |
| 12 | Household ammonia |
| 13 | Household bleach |
| 14 | Oven cleaner / Limewater |

Increasingly ACIDIC

**NEUTRAL** $[H^+] = [OH^-]$

Increasingly BASIC

$H^+$ +

$OH^-$ –

Acidic solution

Neutral solution

Basic solution

**Figure 2.7 The pH scale.** As pH values decrease from 14 to 0, the $H^+$ concentration increases. Thus, the lower the pH, the more acidic the solution; the higher the pH, the more basic the solution. If the pH value of a solution is below 7, the solution is acidic; if the pH is above 7, the solution is basic (alkaline). The approximate pH values of some human body fluids and common substances are shown next to the pH scale.

**Q** At what pH are the concentrations of $H^+$ and $OH^-$ equal?

A solution's pH is calculated as $-\log_{10}[H^+]$, the negative logarithm to the base 10 of the hydrogen ion concentration (denoted by brackets), determined in moles per liter $[H^+]$. For example, if the $H^+$ concentration of a solution is $1.0 \times 10^{-4}$ moles/liter, or $10^{-4}$, its pH equals $-\log_{10}10^{-4} = -(-4) = 4$; this is about the pH value of wine (see Appendix B). The pH values of some human body fluids and other common substances are also shown in Figure 2.7. In the laboratory, you will usually measure the pH of a solution with a pH meter or with chemical test papers.

Acidic solutions contain more $H^+$ than $OH^-$ and have a pH lower than 7. If a solution has more $OH^-$ than $H^+$, it is a basic, or alkaline, solution. In pure water, a small percentage of the molecules are dissociated into $H^+$ and $OH^-$, so it has a pH of 7. Because the concentrations of $H^+$ and $OH^-$ are equal, this pH is said to be the pH of a neutral solution.

Keep in mind that the pH of a solution can be changed. We can increase its acidity by adding substances that will increase the concentration of hydrogen ions. As a living organism takes up nutrients, carries out chemical reactions, and excretes wastes, its balance of acids and bases tends to change, and the pH fluctuates. Fortunately, organisms possess natural pH **buffers**, compounds that help keep the pH from changing drastically. But the pH in our environment's water and soil can be altered by waste products from organisms, pollutants from industry, or fertilizers used in agricultural fields or gardens. When bacteria are grown in a laboratory medium, they excrete waste products such as acids that can alter the pH of the medium. If this effect were to continue, the medium would become acidic enough to inhibit bacterial enzymes and kill the bacteria. To prevent this problem, pH buffers are added to the culture medium. One very effective pH buffer for some culture media uses a mixture of $K_2HPO_4$ and $KH_2PO_4$ (see Table 6.2, page 159).

Different microbes function best within different pH ranges, but most organisms grow best in environments with a pH value between 6.5 and 8.5. Among microbes, fungi are best able to tolerate acidic conditions, whereas the prokaryotes called cyanobacteria tend to do well in alkaline habitats. *Propionibacterium acnes* (prō-pē'on-ē-bak-TI-rē-um AK-nēz), a bacterium that causes acne, has as its natural environment human skin, which tends to be slightly acidic, with a pH of about 4. *Acidithiobacillus ferrooxidans* (a'sid-ē-thī'ō-bah-SIL-lus fer'rō-OKS-i-danz) is a bacterium that metabolizes elemental sulfur and produces sulfuric acid ($H_2SO_4$). Its pH range for optimum growth is from 1 to 3.5. The sulfuric acid produced by this bacterium in mine water is important in dissolving uranium and copper from low-grade ore (see Chapter 28).

**CHECK YOUR UNDERSTANDING**

✔ 2-4 Why is the polarity of a water molecule important?

✔ 2-5 Antacids neutralize acid by the following reaction.

$$Mg(OH)_2 + 2HCl \rightarrow MgCl_2 + H_2O$$

Identify the acid, base, and salt.

# Organic Compounds

**LEARNING OBJECTIVES**

2-6 Distinguish organic and inorganic compounds.

2-7 Define *functional group*.

2-8 Identify the building blocks of carbohydrates.

2-9 Differentiate simple lipids, complex lipids, and steroids.

2-10 Identify the building blocks and structure of proteins.

2-11 Identify the building blocks of nucleic acids.

2-12 Describe the role of ATP in cellular activities.

Inorganic compounds, excluding water, constitute about 1–1.5% of living cells. These relatively simple components, whose molecules have only a few atoms, cannot be used by cells to perform complex biological functions. Organic molecules, whose carbon atoms can combine in an enormous variety of ways with other carbon atoms and with atoms of other elements, are relatively complex and thus are capable of more complex biological functions.

## Structure and Chemistry

In the formation of organic molecules, carbon's four outer electrons can participate in up to four covalent bonds, and carbon atoms can bond to each other to form straight-chain, branched-chain, or ring structures.

In addition to carbon, the most common elements in organic compounds are hydrogen (which can form one bond), oxygen (two bonds), and nitrogen (three bonds). Sulfur (two bonds) and phosphorus (five bonds) appear less often. Other elements are found, but only in relatively few organic compounds. The elements that are most abundant in living organisms are the same as those that are most abundant in organic compounds (see Table 2.1).

The chain of carbon atoms in an organic molecule is called the **carbon skeleton**; a huge number of combinations is possible for carbon skeletons. Most of these carbons are bonded to hydrogen atoms. The bonding of other elements with carbon and hydrogen forms characteristic **functional groups**, specific groups of atoms that are most commonly involved in chemical reactions and are responsible for most of the characteristic chemical properties and many of the physical properties of a particular organic compound (Table 2.4).

Different functional groups confer different properties on organic molecules. For example, the hydroxyl group of alcohols is hydrophilic (water-loving) and thus attracts water molecules to it. This attraction helps dissolve organic molecules containing hydroxyl groups. Because the carboxyl group is a source of hydrogen ions, molecules containing it have acidic properties. Amino groups, by contrast, function as bases because they readily accept hydrogen ions. The sulfhydryl group helps stabilize the intricate structure of many proteins.

Functional groups help us classify organic compounds. For example, the —OH group is present in each of the following molecules:

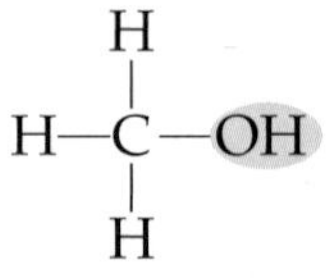

Methanol

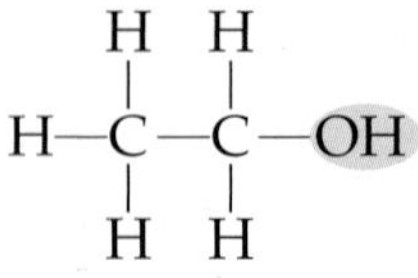

Ethanol

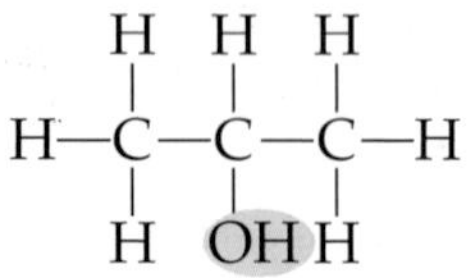

Isopropanol

**TABLE 2.4** Representative Functional Groups and the Compounds in Which They Are Found

| Structure | Name of Group | Biological Importance |
|---|---|---|
| R—O—H | Alcohol | Lipids; carbohydates |
| R—C(=O)—H | Aldehyde* | Reducing sugars such as glucose; polysaccharides |
| R—C(=O)—R | Ketone* | Metabolic intermediates |
| R—C(H)(H)—H | Methyl | DNA; energy metabolism |
| R—C(H)(H)—NH₂ | Amino | Proteins |
| R—C(=O)—O—R' | Ester | Bacterial and eukaryotic plasma membranes |
| R—C(H)(H)—O—C(H)(H)—R' | Ether | Archaeal plasma membranes |
| R—C(H)(H)—SH | Sulfhydryl | Energy metabolism; protein structure |
| R—C(=O)—OH | Carboxyl | Organic acids; lipids; proteins |
| R—O—P(=O)(O⁻)—O⁻ | Phosphate | ATP; DNA |

*In an aldehyde, a C=O is at the end of a molecule, in contrast to the internal C=O in a ketone.

Because the characteristic reactivity of the molecules is based on the —OH group, they are grouped together in a class called alcohols. The —OH group is called the *hydroxyl group* and is not to be confused with the *hydroxide ion* (OH⁻) of bases. The hydroxyl group of alcohols does not ionize at neutral pH; it is covalently bonded to a carbon atom.

When a class of compounds is characterized by a certain functional group, the letter $R$ can be used to stand for the remainder of the molecule. For example, alcohols in general may be written $R—OH$.

Frequently, more than one functional group is found in a single molecule. For example, an amino acid molecule contains both amino and carboxyl groups. The amino acid glycine has the following structure: *in slides!*

Amino group — Carboxyl group

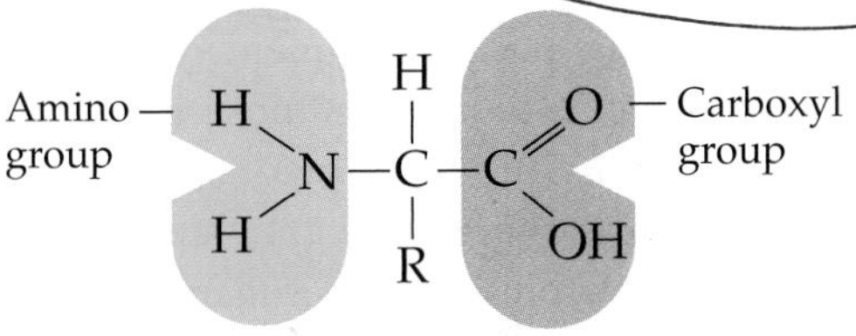

Most of the organic compounds found in living organisms are quite complex; a large number of carbon atoms form the skeleton, and many functional groups are attached. In organic molecules, it is important that each of the four bonds of carbon be satisfied (attached to another atom) and that each of the attaching atoms have its characteristic number of bonds satisfied. Because of this, such molecules are chemically stable.

*4 types*

Small organic molecules can be combined into very large molecules called **macromolecules** (*macro* = large). Macromolecules are usually **polymers** (*poly* = many; *mers* = parts): polymers are formed by covalent bonding of many repeating small molecules called **monomers** (*mono* = one). When two monomers join together, the reaction usually involves the elimination of a hydrogen atom from one monomer and a hydroxyl group from the other; the hydrogen atom and the hydroxyl group combine to produce water:

$$R— OH+OH — R' \longrightarrow R—R' + H_2O$$

This type of exchange reaction is called **dehydration synthesis** (*de* = from; *hydro* = water), or a **condensation reaction**, because a molecule of water is released (**Figure 2.8a**). Such macromolecules as carbohydrates, lipids, proteins, and nucleic acids are assembled in the cell, essentially by dehydration synthesis. However, other molecules must also participate to provide energy for bond formation. ATP, the cell's chief energy provider, is discussed at the end of this chapter.

**CHECK YOUR UNDERSTANDING**

2-6 Define *organic*.

2-7 Add the appropriate functional group(s) to the ethyl group below to produce each of the following compounds: ethanol, acetic acid, acetaldehyde, ethanolamine, diethyl ether.

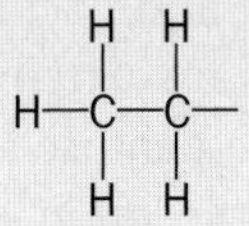

## Carbohydrates

The **carbohydrates** are a large and diverse group of organic compounds that includes sugars and starches. Carbohydrates perform a number of major functions in living systems. For instance, one type of sugar (deoxyribose) is a building block of deoxyribonucleic acid (DNA), the molecule that carries hereditary information. Other sugars are needed for the cell walls. Simple carbohydrates are used in the synthesis of amino acids and fats or fatlike substances, which are used to build cell membranes and other structures. Macromolecular carbohydrates function as food reserves. The principal function of carbohydrates, however, is to fuel cell activities with a ready source of energy.

Carbohydrates are made up of carbon, hydrogen, and oxygen atoms. The ratio of hydrogen to oxygen atoms is always 2:1 in simple carbohydrates. This ratio can be seen in the formulas for the carbohydrates ribose ($C_5H_{10}O_5$), glucose ($C_6H_{12}O_6$), and sucrose ($C_{12}H_{22}O_{11}$). Although there are exceptions, the

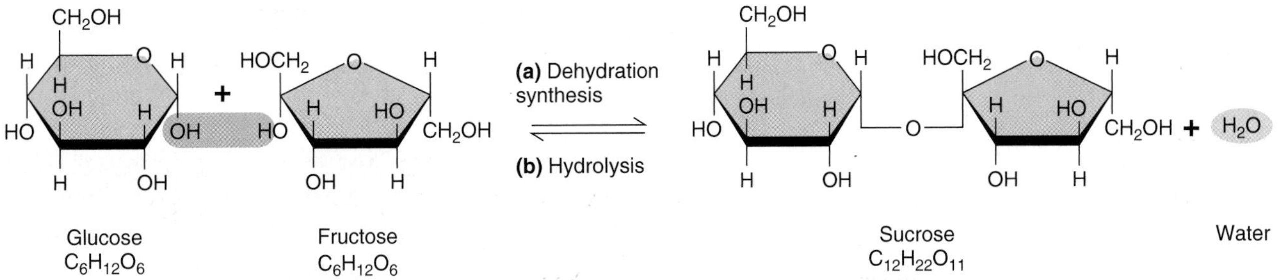

Glucose $C_6H_{12}O_6$    Fructose $C_6H_{12}O_6$    (a) Dehydration synthesis    (b) Hydrolysis    Sucrose $C_{12}H_{22}O_{11}$    Water

**Figure 2.8 Dehydration synthesis and hydrolysis.** **(a)** In dehydration synthesis (left to right), the monosaccharides glucose and fructose combine to form a molecule of the disaccharide sucrose. A molecule of water is released in the reaction. **(b)** In hydrolysis (right to left), the sucrose molecule breaks down into the smaller molecules glucose and fructose. For the hydrolysis reaction to proceed, water must be added to the sucrose.

**Q** What is the difference between a polymer and a monomer?

general formula for carbohydrates is $(CH_2O)_n$, where *n* indicates that there are three or more $CH_2O$ units. Carbohydrates can be classified into three major groups on the basis of size: monosaccharides, disaccharides, and polysaccharides.

## Monosaccharides

Simple sugars are called **monosaccharides** (*sacchar* = sugar); each molecule contains three to seven carbon atoms. The number of carbon atoms in the molecule of a simple sugar is indicated by the prefix in its name. For example, simple sugars with three carbons are called trioses. There are also tetroses (four-carbon sugars), pentoses (five-carbon sugars), hexoses (six-carbon sugars), and heptoses (seven-carbon sugars). Pentoses and hexoses are extremely important to living organisms. Deoxyribose is a pentose found in DNA. Glucose, a very common hexose, is the main energy-supplying molecule of living cells.

## Disaccharides

**Disaccharides** (*di* = two) are formed when two monosaccharides bond in a dehydration synthesis reaction.* For example, molecules of two monosaccharides, glucose and fructose, combine to form a molecule of the disaccharide sucrose (table sugar) and a molecule of water (see Figure 2.8a). Similarly, the dehydration synthesis of the monosaccharides glucose and galactose forms the disaccharide lactose (milk sugar).

It may seem odd that glucose and fructose have the same chemical formula (see Figure 2.8), even though they are different monosaccharides. The positions of the oxygens and carbons differ in the two different molecules; consequently, the molecules have different physical and chemical properties. Two molecules with the same chemical formula but different structures and properties are called **isomers** (*iso* = same).

Disaccharides can be broken down into smaller, simpler molecules when water is added. This chemical reaction, the reverse of dehydration synthesis, is called **hydrolysis** (*hydro* = water; *lysis* = to loosen) (Figure 2.8b). A molecule of sucrose, for example, may be hydrolyzed (digested) into its components of glucose and fructose by reacting with the $H^+$ and $OH^-$ of water.

As you will see in Chapter 4, the cell walls of bacterial cells are composed of disaccharides and proteins, which together are called peptidoglycan.

## Polysaccharides

Carbohydrates in the third major group, the **polysaccharides**, consist of tens or hundreds of monosaccharides joined through dehydration synthesis. Polysaccharides often have side chains branching off the main structure and are classified as macromolecules. Like disaccharides, polysaccharides can be split apart into their constituent sugars through hydrolysis. Unlike monosaccharides and disaccharides, however, they usually lack the characteristic sweetness of sugars such as fructose and sucrose and usually are not soluble in water.

One important polysaccharide is *glycogen*, which is composed of glucose subunits and is synthesized as a storage material by animals and some bacteria. *Cellulose*, another important glucose polymer, is the main component of the cell walls of plants and most algae. Although cellulose is the most abundant carbohydrate on Earth, it can be digested by only a few organisms that have the appropriate enzyme. The polysaccharide *dextran*, which is produced as a sugary slime by certain bacteria, is used in a blood plasma substitute. *Chitin* is a polysaccharide that makes up part of the cell wall of most fungi and the exoskeletons of lobsters, crabs, and insects. *Starch* is a polymer of glucose produced by plants and used as food by humans. Digestion of starch by intestinal bacteria is important to human health (see Exploring the Microbiome, page 37).

Many animals, including humans, produce enzymes called *amylases* that can break the bonds between the glucose molecules in glycogen. However, this enzyme cannot break the bonds in cellulose. Bacteria and fungi that produce enzymes called *cellulases* can digest cellulose. Cellulases from the fungus *Trichoderma* (TRIK-ō-der-mah) are used for a variety of industrial purposes. One of the more unusual uses is producing stone-washed or distressed denim. Because washing the fabric with rocks would damage washing machines, cellulase is used to digest, and therefore soften, the cotton.

---

**CHECK YOUR UNDERSTANDING**

✔ 2-8 Give an example of a monosaccharide, a disaccharide, and a polysaccharide.

---

## Lipids

If lipids were suddenly to disappear from the Earth, all living cells would collapse in a pool of fluid, because lipids are essential to the structure and function of membranes that separate living cells from their environment. **Lipids** (*lip* = fat) are a second major group of organic compounds found in living matter. Like carbohydrates, they are composed of atoms of carbon, hydrogen, and oxygen, but lipids lack the 2:1 ratio between hydrogen and oxygen atoms. Even though lipids are a very diverse group of compounds, they share one common characteristic: they are *nonpolar* molecules so, unlike water, do not have a positive and a negative end (pole). Therefore, most lipids are insoluble in water but dissolve readily in nonpolar solvents, such as ether and chloroform. Lipids provide the structure of membranes and some cell walls and function in energy storage.

---

*Carbohydrates composed of 2 to about 20 monosaccharides are called **oligosaccharides** (*oligo* = few). Disaccharides are the most common oligosaccharides.

# Feed Our Intestinal Bacteria, Feed Ourselves: A Tale of Two Starches

Structurally speaking, starch found in plants appears as either a many-branched chain called amylopectin or as a straight-chained variety called amylose. Amylopectin-rich foods common to our diet include sticky rice and waxy varieties of corn and potatoes. Foods containing high levels of amylose include beans and other legumes and whole grains. Although both are starches, they produce markedly different effects when we eat them.

In the small intestine, enzymes rapidly convert amylopectin into glucose, which is the preferred carbohydrate for many essential metabolic reactions our cells conduct. By contrast, amylose's structure has less surface area for enzymes to react with, making it more resistant to digestion. Since it isn't easily broken down in our small intestine, the amylose starch continues through the gastrointestinal tract to the colon, where it's available for bacteria living there to ferment.

You might assume that amylopectin, the starch that is easily broken down, must be the best type for us to eat. However, feeding our microbiome amylose seems to provide excellent health benefits. Amylose-fermenting bacteria, including members of the *Prevotella* and *Lachnospira* genera, produce short-chain fatty acids. Several types of these molecules are important players in how our intestinal cells absorb electrolytes (ions).

Research also suggests that butyrate, one type of short-chain fatty acid linked to *Prevotella* metabolism, may protect us against colorectal cancer. In another study, mice were treated with antibiotics and then inoculated with pathogenic *Clostridium difficile* bacteria. What happened to the animals next was stark: some rapidly developed lethal infections, while others were colonized by the bacteria and experienced only mild disease. The fate of the study mice seemed to come down to their microbiome composition before

*C. difficile* was introduced. Mice with large numbers of *Lachnospira* bacteria were more likely to survive, whereas those with intestinal microbiomes dominated by *E. coli* were much more likely to die.

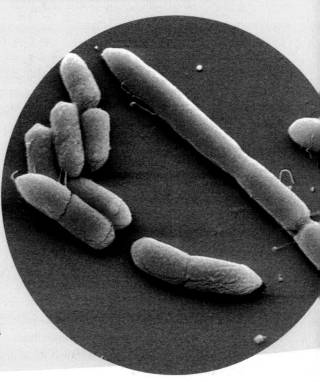

**Helpful *Prevotella* species thrive in the intestines by fermenting amylose we can't digest.**

## Simple Lipids

*Simple lipids*, called *fats* or *triglycerides*, contain an alcohol called *glycerol* and a group of compounds known as *fatty acids*. Glycerol molecules have three carbon atoms to which are attached three hydroxyl ($-$OH) groups (**Figure 2.9a**). Fatty acids consist of long hydrocarbon chains (composed only of carbon and hydrogen atoms) ending in a carboxyl ($-$COOH, organic acid) group (Figure 2.9b). Most common fatty acids contain an even number of carbon atoms.

A molecule of fat is formed when a molecule of glycerol combines with one to three fatty acid molecules. The number of fatty acid molecules determines whether the fat molecule is a monoglyceride, diglyceride, or triglyceride (Figure 2.9c). In the reaction, one to three molecules of water are formed (dehydration), depending on the number of fatty acid molecules reacting. The chemical bond formed where the water molecule is removed is called an *ester linkage*. In the reverse reaction, hydrolysis, a fat molecule is broken down into its component fatty acid and glycerol molecules.

Because the fatty acids that form lipids have different structures, there is a wide variety of lipids. For example, three molecules of fatty acid A might combine with a glycerol molecule. Or one molecule each of fatty acids A, B, and C might unite with a glycerol molecule (see Figure 2.9c).

The primary function of lipids is to form plasma membranes that enclose cells. A plasma membrane supports the cell and allows nutrients and wastes to pass in and out; therefore, the lipids must maintain the same viscosity, regardless of the surrounding temperature. The membrane must be about as viscous as olive oil, without getting too fluid when warmed or too thick when cooled. As everyone who has ever cooked a meal knows, animal fats (such as butter) are usually solid at room temperature, whereas vegetable oils are usually liquid at room temperature. The difference in their respective melting points is due to the degrees of saturation of the fatty acid chains. A fatty acid is said to be *saturated* when it has no double bonds, in which case the carbon skeleton contains the maximum

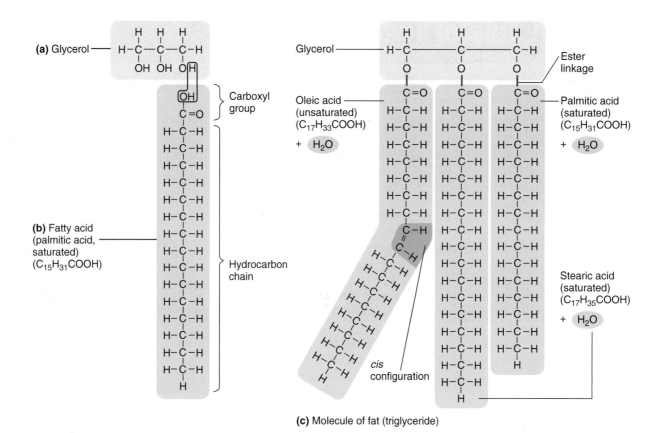

(c) Molecule of fat (triglyceride)

**Figure 2.9 Structural formulas of simple lipids. (a)** Glycerol. **(b)** Palmitic acid, a saturated fatty acid. **(c)** The chemical combination of a molecule of glycerol and three fatty acid molecules (palmitic, stearic, and oleic in this example) forms one molecule of fat (triglyceride) and three molecules of water in a dehydration synthesis reaction. Oleic acid is a *cis* fatty acid. The bond between glycerol and each fatty acid is called an ester linkage.

The addition of three water molecules to a fat forms glycerol and three fatty acid molecules in a hydrolysis reaction.

**Q** How do saturated and unsaturated fatty acids differ?

number of hydrogen atoms (see Figure 2.9c and **Figure 2.10a**). Saturated chains become solid more easily because they are relatively straight and are thus able to pack together more closely than unsaturated chains. The double bonds of *unsaturated* chains create kinks in the chain, which keep the chains apart from one another (Figure 2.10b). Note in Figure 2.9c that the H atoms on either side of the double bond in oleic acid are on the same side of the unsaturated fatty acid. Such an unsaturated fatty acid is called a *cis* fatty acid. If, instead, the H atoms are on opposite sides of the double bond, the unsaturated acid is called a *trans* fatty acid.

### Complex Lipids   — in slides.

*Complex lipids* contain such elements as phosphorus, nitrogen, and sulfur, in addition to the carbon, hydrogen, and oxygen found in simple lipids. The complex lipids called *phospholipids* are made up of glycerol, two fatty acids, and, in place of a third fatty acid, a phosphate group bonded to one of several organic groups (see Figure 2.10a). Phospholipids are the lipids that build membranes; they are essential to a cell's survival. Phospholipids have polar as well as nonpolar regions (Figure 2.10a

and b; see also Figure 4.14, page 86). When placed in water, phospholipid molecules twist themselves in such a way that all polar (hydrophilic) portions orient themselves toward the polar water molecules, with which they then form hydrogen bonds. (Recall that *hydrophilic* means water-loving.) This forms the basic structure of a plasma membrane (Figure 2.10c). Polar portions consist of a phosphate group and glycerol. In contrast to the polar regions, all nonpolar (hydrophobic) parts of the phospholipid make contact only with the nonpolar portions of neighboring molecules. (*Hydrophobic* means water-fearing.) Nonpolar portions consist of fatty acids. This characteristic behavior makes phospholipids particularly suitable for their role as a major component of the membranes that enclose cells. Phospholipids enable the membrane to act as a barrier that separates the contents of the cell from the water-based environment in which it lives.

Some complex lipids are useful in identifying certain bacteria. For example, the cell wall of *Mycobacterium tuberculosis* (mī'kō-bak-TI-rē-um too'ber-kū-LŌ-sis), the bacterium that causes tuberculosis, is distinguished by its lipid-rich content. The cell wall contains complex lipids such as waxes and glycolipids (lipids

*Know*

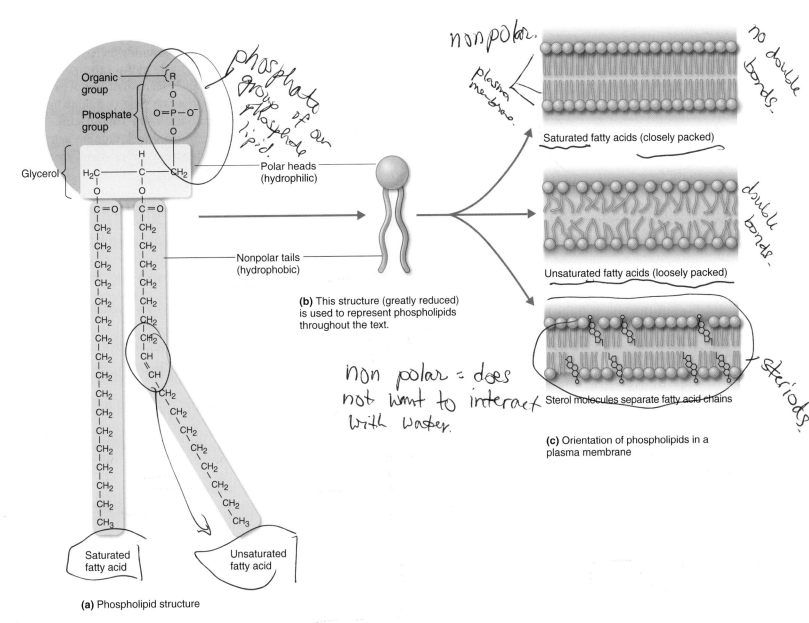

Handwritten annotations: "phosphate group of our lipid", "nonpolar", "plasma membrane", "no double bonds.", "double bonds.", "steroids.", "non polar = does not want to interact with water."

(b) This structure (greatly reduced) is used to represent phospholipids throughout the text.

Saturated fatty acids (closely packed)

Unsaturated fatty acids (loosely packed)

Sterol molecules separate fatty acid chains

(c) Orientation of phospholipids in a plasma membrane

Saturated fatty acid

Unsaturated fatty acid

(a) Phospholipid structure

**Figure 2.10 Phospholipid and orientation, showing saturated and unsaturated fatty acids and the molecules' polarity.**

**Q** Where are phospholipids found in cells?

with carbohydrates attached) that give the bacterium distinctive staining characteristics. Cell walls rich in such complex lipids are characteristic of all members of the genus *Mycobacterium*.

### Steroids

**Steroids** are structurally very different from lipids. **Figure 2.11** shows the structure of the steroid cholesterol, with the four interconnected carbon rings that are characteristic of steroids. When an —OH group is attached to one of the rings, the steroid is called a *sterol* (an alcohol). Sterols are important constituents of the plasma membranes of animal cells and of one group of bacteria (mycoplasmas), and they are also found in fungi and plants. The sterols separate the fatty acid chains and

**Figure 2.11 Cholesterol, a steroid.** Note the four "fused" carbon rings (labeled A–D), which are characteristic of steroid molecules. The hydrogen atoms attached to the carbons at the corners of the rings have been omitted. The —OH group (colored red) makes this molecule a sterol.

**Q** Where are sterols found in cells?

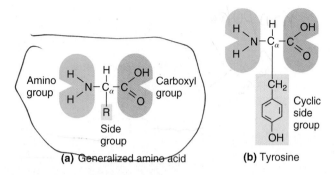

**Figure 2.12 Amino acid structure. (a)** The general structural formula for an amino acid. The alpha-carbon ($C_\alpha$) is shown in the center. Different amino acids have different R groups, also called side groups. **(b)** Structural formula for the amino acid tyrosine, which has a cyclic side group.

**Q** **What distinguishes one amino acid from another?**

thus prevent the packing that would harden the plasma membrane at low temperatures (see Figure 2.10c).

**CHECK YOUR UNDERSTANDING**

✔ 2-9 How do simple lipids differ from complex lipids?

## Proteins

**Proteins** are organic molecules that contain carbon, hydrogen, oxygen, and nitrogen. Some also contain sulfur. If you were to separate and weigh all the groups of organic compounds in a living cell, the proteins would tip the scale. Hundreds of different proteins can be found in any single cell, and together they make up 50% or more of a cell's dry weight.

Proteins are essential ingredients in all aspects of cell structure and function. *Enzymes* are the proteins that speed up biochemical reactions. But proteins have other functions as well. *Transporter proteins* help transport certain chemicals into and out of cells. Other proteins, such as the *bacteriocins* produced by many bacteria, kill other bacteria. Certain *toxins*, called exotoxins, produced by some disease-causing microorganisms are also proteins. Some proteins play a role in the *contraction* of animal muscle cells and the *movement* of microbial and other types of cells. Other proteins are integral parts of *cell structures* such as walls, membranes, and cytoplasmic components. Still others, such as the *hormones* of certain organisms, have regulatory functions. As we will see in Chapter 17, proteins called *antibodies* play a role in vertebrate immune systems.

### Amino Acids

Just as monosaccharides are the building blocks of larger carbohydrate molecules, and just as fatty acids and glycerol are the building blocks of fats, **amino acids** are the building blocks of proteins. Amino acids contain at least one carboxyl (—COOH) group and one amino (—$NH_2$) group attached to the same carbon atom, called an alpha-carbon (written $C_\alpha$) **(Figure 2.12a)**.

Such amino acids are called *alpha-amino acids*. Also attached to the alpha-carbon is a side group (R group), which is the amino acid's distinguishing feature. The side group can be a hydrogen atom, an unbranched or branched chain of atoms, or a ring structure that is cyclic (all carbon) or heterocyclic (when an atom other than carbon is included in the ring). Figure 2.12b shows the structural formula of tyrosine, an amino acid that has a cyclic side group. The side group can contain functional groups, such as the sulfhydryl group (—SH), the hydroxyl group (—OH), or additional carboxyl or amino groups. These side groups and the carboxyl and alpha-amino groups affect the total structure of a protein, described later. The structures and standard abbreviations of the 20 amino acids found in proteins are shown in Table 2.5.

Most amino acids exist in either of two configurations called **stereoisomers**, designated by D and L. These configurations are mirror images, corresponding to "right-handed" (D) and "left-handed" (L) three-dimensional shapes **(Figure 2.13)**. The amino acids found in proteins are always the L-isomers (except for glycine, the simplest amino acid, which does not

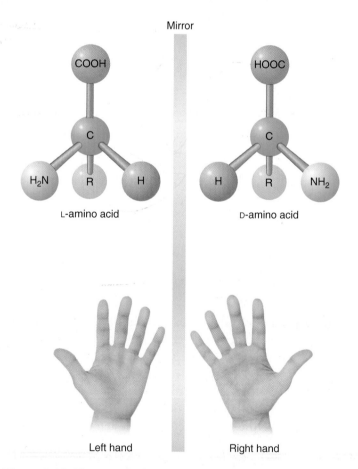

**Figure 2.13 The L- and D-isomers of an amino acid, shown with ball-and-stick models.** The two isomers, like left and right hands, are mirror images of each other and cannot be superimposed on one another. (Try it!)

**Q** **Which isomer is always found in proteins?**

Subunits of proteins.     in slide

**TABLE 2.5** The 20 Amino Acids Found in Proteins*

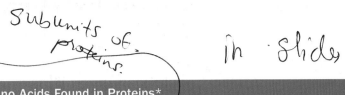

| Glycine (Gly, G) | Alanine (Ala, A) | Valine (Val, V) | Leucine (Leu, L) | Isoleucine (Ile, I) |
|---|---|---|---|---|
| Hydrogen atom | Unbranched chain | Branched chain | Branched chain | Branched chain |

| Serine (Ser, S) | Threonine (Thr, T) | Cysteine (Cys, C) | Methionine (Met, M) | Glutamic acid (Glu, E) |
|---|---|---|---|---|
| Hydroxyl (—OH) group | Hydroxyl (—OH) group | Sulphur-containing (—SH) group | Thioether (SC) group | Additional carboxyl (—COOH) group, acidic |

| Aspartic acid (Asp, D) | Lysine (Lys, K) | Arginine (Arg, R) | Asparagine (Asn, N) | Glutamine (Gln, Q) |
|---|---|---|---|---|
| Additional Carboxyl (—COOH) group, acidic | Additional amino (—NH₂) group, basic | Additional amino (—NH₂) group, basic | Additional amino (—NH₂) group, basic | Additional amino (—NH₂) group, basic |

| Phenylalanine (Phe, F) | Tyrosine (Tyr, Y) | Histidine (His, H) | Tryptophan (Trp, W) | Proline (Pro, P) |
|---|---|---|---|---|
| Cyclic | Cyclic | Heterocyclic | Heterocyclic | Heterocyclic |

*Shown are the amino acid names, including the three-letter and one-letter abbreviations in parentheses (above), their structural formulas (center), and characteristic R group (in green). Note that cysteine and methionine are the only amino acids that contain sulfur.

While Jonathan is in intensive care, his wife, DeeAnn, and adult daughter talk with his physician and an investigator from the Centers for Disease Control and Prevention (CDC) to find the source of Jonathan's *B. anthracis* infection. Environmental investigations uncover *B. anthracis* at Jonathan's home, in his van, and in his workplace, but neither his wife nor children show signs of infection. His bandmates are also tested; they are all negative for *B. anthracis*. The CDC investigator explains to Jonathan's family that *B. anthracis* forms endospores that can survive in soil for up to 60 years. It is rare in humans; however, grazing animals and people who handle their hides or other by-products can become infected. *B. anthracis* cells have capsules that are composed of poly-D-glutamic acid.

**Why are the capsules resistant to digestion by phagocytes? (Phagocytes are white blood cells that engulf and destroy bacteria.)**

25    42    44    46

have stereoisomers). However, D-amino acids occasionally occur in nature—for example, in certain bacterial cell walls and antibiotics.

Although only 20 different amino acids occur naturally in proteins, a single protein molecule can contain from 50 to hundreds of amino acid molecules, which can be arranged in an almost infinite number of ways to make proteins of different lengths, compositions, and structures. The number of proteins is practically endless, and every living cell produces many different proteins.

### Peptide Bonds

Amino acids bond between the carbon atom of the carboxyl (—COOH) group of one amino acid and the nitrogen atom of the amino (—NH$_2$) group of another (**Figure 2.14**). The bonds between amino acids are called **peptide bonds.** For every peptide bond formed between two amino acids, one water molecule is released; thus, peptide bonds are formed by dehydration synthesis. The resulting compound in Figure 2.14 is called a *dipeptide* because it consists of two amino acids joined by a peptide bond. Adding another amino acid to a dipeptide would form a *tripeptide.* Further additions of amino acids would produce a long, chainlike molecule called

a *peptide* (4–9 amino acids) or *polypeptide* (10–2000 or more amino acids).

### Levels of Protein Structure

Proteins vary tremendously in structure. Different proteins have different architectures and different three-dimensional shapes. This variation in structure is directly related to their diverse functions.

When a cell makes a protein, the polypeptide chain folds spontaneously to assume a certain shape. One reason for folding of the polypeptide is that some parts of a protein are attracted to water and other parts are repelled by it. In practically every case, the function of a protein depends on its ability to recognize and bind to some other molecule. For example, an enzyme binds specifically with its substrate. A hormonal protein binds to a receptor on a cell whose function it will alter. An antibody binds to an antigen (foreign substance) that has invaded the body. The unique shape of each protein permits it to interact with specific other molecules in order to carry out specific functions.

Proteins are described in terms of four levels of organization: primary, secondary, tertiary, and quaternary. The *primary structure* is the unique sequence in which the amino acids are linked together to form a polypeptide chain (**Figure 2.15a**). This sequence is genetically determined. Alterations in sequence can have profound metabolic effects. For example, a single incorrect amino acid in a blood protein can produce the deformed hemoglobin molecule characteristic of sickle cell disease. But proteins do not exist as long, straight chains. Each polypeptide chain folds and coils in specific ways into a relatively compact structure with a characteristic three-dimensional shape.

A protein's *secondary structure* is the localized, repetitious twisting or folding of the polypeptide chain. This aspect of a protein's shape results from hydrogen bonds joining the atoms of peptide bonds at different locations along the polypeptide chain. The two types of secondary protein structures are clockwise spirals called *helices* (singular: *helix*) and pleated sheets, which form from roughly parallel portions of the chain (Figure 2.15b). Both structures are held together by hydrogen bonds between oxygen or nitrogen atoms that are part of the polypeptide's backbone.

*Tertiary structure* refers to the overall three-dimensional structure of a polypeptide chain (Figure 2.15c). The folding is not repetitive or predictable, as in secondary structure. Whereas secondary structure involves hydrogen bonding between atoms

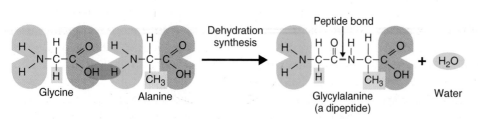

Glycine    Alanine

Dehydration synthesis

Peptide bond

Glycylalanine (a dipeptide)    + H$_2$O    Water

**Figure 2.14 Peptide bond formation by dehydration synthesis.** The amino acids glycine and alanine combine to form a dipeptide. The newly formed bond between the carbon atom of glycine and the nitrogen atom of alanine is called a peptide bond.

**Q** How are amino acids related to proteins?

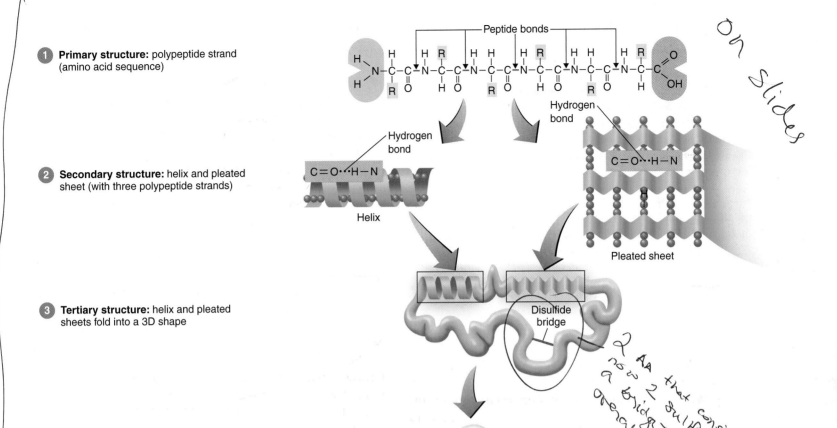

*On slides*

1. **Primary structure:** polypeptide strand (amino acid sequence)

2. **Secondary structure:** helix and pleated sheet (with three polypeptide strands)

3. **Tertiary structure:** helix and pleated sheets fold into a 3D shape

4. **Quaternary structure:** the relationship of several folded polypeptide chains, forming a protein

Peptide bonds

Hydrogen bond

Hydrogen bond

C=O···H—N

C=O···H—N

Helix

Pleated sheet

Disulfide bridge

*2 AA that consist of a sulfur atom now 2 sulfur atoms are creating a bridge—for structure & overall shape of protein*

**Figure 2.15 Protein structure.** 1 Primary structure, the amino acid sequence. 2 Secondary structures: helix and pleated sheet. 3 Tertiary structure, the overall three-dimensional folding of a polypeptide chain. 4 Quaternary structure, the relationship between several polypeptide chains that make up a protein. Shown here is the quaternary structure of a hypothetical protein composed of two polypeptide chains.

**Q** **What property of a protein enables it to carry out specific functions?**

of the amino and carboxyl groups involved in the peptide bonds, tertiary structure involves several interactions between various amino acid side groups in the polypeptide chain. For example, amino acids with nonpolar (hydrophobic) side groups usually interact at the core of the protein, out of contact with water. This *hydrophobic interaction* helps contribute to tertiary structure. Hydrogen bonds between side groups, and ionic bonds between oppositely charged side groups, also contribute to tertiary structure. Proteins that contain the amino acid cysteine form strong covalent bonds called *disulfide bridges*. These bridges form when two cysteine molecules are brought close together by the folding of the protein. Cysteine molecules contain sulfhydryl groups (—SH), and the sulfur of one cysteine molecule bonds to the sulfur on another, forming (by

the removal of hydrogen atoms) a disulfide bridge (S—S) that holds parts of the protein together.

Some proteins have a *quaternary structure,* which consists of an aggregation of two or more individual polypeptide chains (subunits) that operate as a single functional unit. Figure 2.15d shows a hypothetical protein consisting of two polypeptide chains. More commonly, proteins have two or more kinds of polypeptide subunits. The bonds that hold a quaternary structure together are basically the same as those that maintain tertiary structure. The overall shape of a protein may be globular (compact and roughly spherical) or fibrous (threadlike).

If a protein encounters a hostile environment in terms of temperature, pH, or salt concentrations, it may unravel and lose its characteristic shape. This process is called **denaturation**

The host's phagocytes cannot easily digest D-forms of amino acids, such as D-glutamic acid found in the capsules of *B. anthracis*. Therefore, infection can develop. The CDC investigator's mention of animal hides gives DeeAnn an idea. Jonathan plays West African drums called *djembe*; the drum skins are made from dried imported goat hides from West Africa. Although most of these hides are legally imported, some slip through the cracks. It's possible that the hides on Jonathan's drums have been illegally imported and therefore have not been inspected by the U.S. Department of Agriculture. To create *djembe* drums, the hides are soaked in water, stretched over the drum body, and then scraped and sanded. The scraping and sanding generates a large amount of aerosolized dust as the hides dry. Sometimes this dust contains *B. anthracis* endospores, which contain dipicolinic acid.

**What is the functional group in dipicolinic acid? See the figure above.**

| 25 | 42 | **44** | 46 | |
|---|---|---|---|---|

(see Figure 5.6, page 115). As a result of denaturation, the protein is no longer functional. This process will be discussed in more detail in Chapter 5 with regard to denaturation of enzymes.

The proteins we have been discussing are *simple proteins*, which contain only amino acids. *Conjugated proteins* are combinations of amino acids with other organic or inorganic components. Conjugated proteins are named by their non–amino acid component. Thus, glycoproteins contain sugars, nucleoproteins contain nucleic acids, metalloproteins contain metal atoms, lipoproteins contain lipids, and phosphoproteins contain phosphate groups. Phosphoproteins are important regulators of activity in eukaryotic cells. Bacterial synthesis of phosphoproteins may be important for the survival of bacteria such as *Legionella pneumophila* that grow inside host cells.

**CHECK YOUR UNDERSTANDING**

✔ 2-10 What two functional groups are in all amino acids?

# Nucleic Acids

In 1944, three American microbiologists—Oswald Avery, Colin MacLeod, and Maclyn McCarty—discovered that a substance called **deoxyribonucleic acid (DNA)** is the substance of which genes are made. Nine years later, James Watson and Francis Crick, working with molecular models and X-ray information supplied by Maurice Wilkins and Rosalind Franklin, identified the physical structure of DNA. In addition, Crick suggested a mechanism for DNA replication and how it works as the hereditary material. DNA and another substance called **ribonucleic acid (RNA)** are together referred to as **nucleic acids** because they were first discovered in the nuclei of cells. Just as amino acids are the structural units of proteins, nucleotides are the structural units of nucleic acids.

Each **nucleotide** has three parts: a nitrogen-containing base, a pentose (five-carbon) sugar (either **deoxyribose** or **ribose**), and a phosphate group (phosphoric acid). The nitrogen-containing bases are cyclic compounds made up of carbon, hydrogen, oxygen, and nitrogen atoms. The bases are named adenine (A), thymine (T), cytosine (C), guanine (G), and uracil (U). A and G are double-ring structures called **purines**, whereas T, C, and U are single-ring structures referred to as **pyrimidines**.

Nucleotides are named according to their nitrogen-containing base. Thus, a nucleotide containing thymine is a *thymine nucleotide*, one containing adenine is an *adenine nucleotide*, and so on. The term **nucleoside** refers to the combination of a purine or pyrimidine plus a pentose sugar; it does not contain a phosphate group.

## DNA

According to the model proposed by Watson and Crick, a DNA molecule consists of two long strands wrapped around each other to form a **double helix** (Figure 2.16). The double helix looks like a twisted ladder, and each strand is composed of many nucleotides.

Every strand of DNA composing the double helix has a "backbone" consisting of alternating deoxyribose sugar and phosphate groups. The deoxyribose of one nucleotide is joined to the phosphate group of the next. The nitrogen-containing bases make up the rungs of the ladder. Note that the purine A is always paired with the pyrimidine T and that the purine G is always paired with the pyrimidine C. The bases are held together by hydrogen bonds; A and T are held by two hydrogen bonds, and G and C by three. DNA does not contain uracil (U).

The order in which the nitrogen base pairs occur along the backbone is extremely specific and in fact contains the genetic instructions for the organism. Nucleotides form genes, and a single DNA molecule may contain thousands of genes. Genes determine all hereditary traits, and they control all the activities that take place within cells.

One very important consequence of nitrogen-containing base pairing is that if the sequence of bases of one strand is known, then the sequence of the other strand is also known. For example, if one strand has the sequence . . . ATGC . . . , then the other strand has the sequence . . . TACG . . . . Because the sequence of

# The Structure of DNA

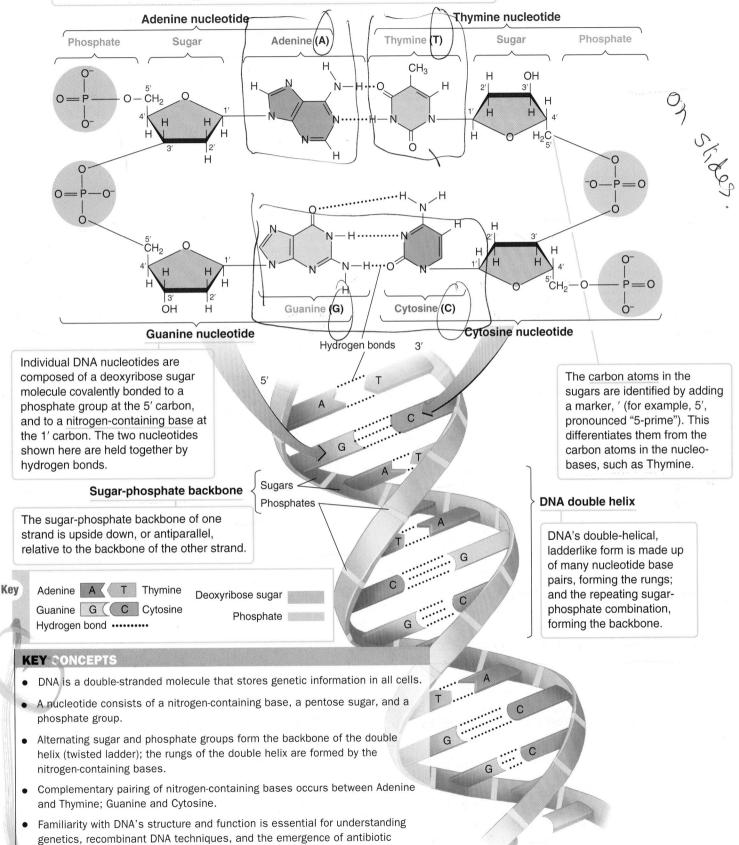

Adenine, Thymine, Cytosine, and Guanine are nitrogenous bases, or nucleobases.

**Adenine nucleotide**

Phosphate | Sugar | Adenine (A)

**Thymine nucleotide**

Thymine (T) | Sugar | Phosphate

**Guanine nucleotide**

Guanine (G) | Cytosine (C) | **Cytosine nucleotide**

Hydrogen bonds     3'

on slides.

Individual DNA nucleotides are composed of a deoxyribose sugar molecule covalently bonded to a phosphate group at the 5' carbon, and to a nitrogen-containing base at the 1' carbon. The two nucleotides shown here are held together by hydrogen bonds.

The carbon atoms in the sugars are identified by adding a marker, ' (for example, 5', pronounced "5-prime"). This differentiates them from the carbon atoms in the nucleo-bases, such as Thymine.

**Sugar-phosphate backbone**

The sugar-phosphate backbone of one strand is upside down, or antiparallel, relative to the backbone of the other strand.

Sugars
Phosphates

**DNA double helix**

DNA's double-helical, ladderlike form is made up of many nucleotide base pairs, forming the rungs; and the repeating sugar-phosphate combination, forming the backbone.

**Key**

Adenine [A] ⟨ [T] Thymine
Guanine [G] ⟨ [C] Cytosine
Hydrogen bond ••••••••

Deoxyribose sugar
Phosphate

## KEY CONCEPTS

- DNA is a double-stranded molecule that stores genetic information in all cells.

- A nucleotide consists of a nitrogen-containing base, a pentose sugar, and a phosphate group.

- Alternating sugar and phosphate groups form the backbone of the double helix (twisted ladder); the rungs of the double helix are formed by the nitrogen-containing bases.

- Complementary pairing of nitrogen-containing bases occurs between Adenine and Thymine; Guanine and Cytosine.

- Familiarity with DNA's structure and function is essential for understanding genetics, recombinant DNA techniques, and the emergence of antibiotic resistance and new diseases.

*on slide*

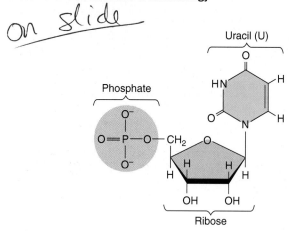

Uracil (U)

Phosphate

Ribose

**Figure 2.17 A uracil nucleotide of RNA.**

**Q** How are DNA and RNA similar in structure?

bases of one strand is determined by the sequence of bases of the other, the bases are said to be *complementary*. The actual transfer of information becomes possible because of DNA's unique structure and will be discussed further in Chapter 8.

### RNA

RNA, the second principal kind of nucleic acid, differs from DNA in several respects. Whereas DNA is double-stranded, RNA is usually single-stranded. The five-carbon sugar in the RNA nucleotide is ribose, which has one more oxygen atom than deoxyribose. Also, one of RNA's bases is uracil (U) instead of thymine (**Figure 2.17**). The other three bases (A, G, C) are the same as DNA. Three major kinds of RNA have been identified in cells. They are **messenger RNA (mRNA)**, **ribosomal RNA (rRNA)**, and **transfer RNA (tRNA)**, each of which has a specific role in protein synthesis (see Chapter 8).

DNA and RNA are compared in Table 2.6.

**CHECK YOUR UNDERSTANDING**

✔ 2-11 How do DNA and RNA differ?

### Adenosine Triphosphate (ATP)

**Adenosine triphosphate (ATP)** is the principal energy-carrying molecule of all cells and is indispensable to the life of the cell. It stores the chemical energy released by some chemical reactions, and it provides the energy for reactions that require energy. ATP consists of an adenosine unit, composed of adenine and ribose, with three phosphate groups (**P**) attached (**Figure 2.18**). In other words, it is an adenine nucleotide (also called adenosine monophosphate, or AMP) with two extra phosphate groups. ATP is called a high-energy molecule because it releases a large amount of usable energy when the third phosphate group is hydrolyzed to become **adenosine diphosphate (ADP)**. This reaction can be represented as follows:

The functional group in dipicolinic acid is carboxyl. *B. anthracis* infection is contracted by contact, ingestion, or inhalation of the endospores. In Jonathan's case, the process of stretching, scraping, and sanding the goat hides had created dust that settled on the drum skin and any surrounding crevices. *B. anthracis* endospores became airborne, or aerosolized, whenever Jonathan beat on the drum. He makes a full recovery, and from now on he makes certain that all parts of any drum he purchases have been legally imported.

25    42    44    **46**

*on slides.*

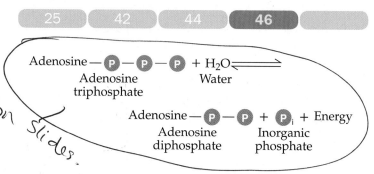

Adenosine — P — P — P + $H_2O$ ⇌

Adenosine triphosphate      Water

Adenosine — P — P + P$_i$ + Energy

Adenosine diphosphate      Inorganic phosphate

A cell's supply of ATP at any particular time is limited. Whenever the supply needs replenishing, the reaction goes in the reverse direction; the addition of a phosphate group to ADP and the input of energy produces more ATP. The energy required to attach the terminal phosphate group to ADP is supplied by the cell's various oxidation reactions, particularly the oxidation of glucose. ATP can be made in every cell, where its potential energy is released when needed.

**CHECK YOUR UNDERSTANDING**

✔ 2-12 Which can provide more energy for a cell and why: ATP or ADP?

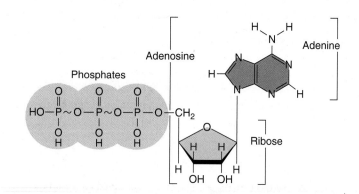

Phosphates     Adenosine     Adenine

Ribose

**Figure 2.18 The structure of ATP.** High-energy phosphate bonds are indicated by wavy lines. When ATP breaks down to ADP and inorganic phosphate, a large amount of chemical energy is released for use in other chemical reactions.

**Q** How is ATP similar to a nucleotide in RNA? In DNA?

**TABLE 2.6 Comparison between DNA and RNA**

| Backbone | DNA | RNA |
|---|---|---|
| Strands | Double-stranded in cells and most DNA viruses to form a double helix; single-stranded in some viruses (parvoviruses). | Single-stranded in cells and most RNA viruses; double-stranded in some viruses (reoviruses). |
| Composition | The sugar is deoxyribose. | The sugar is ribose. |
|  | The nitrogen-containing bases are cytosine (C), guanine (G), adenine (A), and thymine (T). | The nitrogen-containing bases are cytosine (C), guanine (G), adenine (A), and uracil (U). |
| Function | Determines all hereditary traits. | Protein synthesis. |

# Study Outline

 Go to @MasteringMicrobiology for **Interactive Microbiology**, *In the Clinic* videos, MicroFlix, MicroBoosters, 3D animations, practice quizzes, and more.

## Introduction (p. 24)

1. The science of the interaction between atoms and molecules is called chemistry.
2. The metabolic activities of microorganisms involve complex chemical reactions.
3. Microbes break down nutrients to obtain energy and to make new cells.

## The Structure of Atoms (pp. 25–27)

1. An atom is the smallest unit of a chemical element that exhibits the properties of that element.
2. Atoms consist of a nucleus, which contains protons and neutrons, and electrons, which move around the nucleus.
3. The atomic number is the number of protons in the nucleus; the total number of protons and neutrons is the atomic mass.

## Chemical Elements (p. 25)

4. Atoms with the same number of protons and the same chemical behavior are classified as the same chemical element.
5. Chemical elements are designated by abbreviations called chemical symbols.
6. About 26 elements are commonly found in living cells.
7. Atoms that have the same atomic number (are of the same element) but different atomic masses are called isotopes.

## Electronic Configurations (p. 26)

8. In an atom, electrons are arranged around the nucleus in electron shells.
9. Each shell can hold a characteristic maximum number of electrons.
10. The chemical properties of an atom are due largely to the number of electrons in its outermost shell.

## How Atoms Form Molecules: Chemical Bonds (pp. 27–29)

1. Molecules are made up of two or more atoms; molecules consisting of at least two different kinds of atoms are called compounds.
2. Atoms form molecules in order to fill their outermost electron shells.
3. Attractive forces that bind two atoms together are called chemical bonds.
4. The combining capacity of an atom—the number of chemical bonds the atom can form with other atoms—is its valence.

## Ionic Bonds (p. 27)

5. A positively or negatively charged atom or group of atoms is called an ion.
6. A chemical attraction between ions of opposite charge is called an ionic bond.
7. To form an ionic bond, one ion is an electron donor, and the other ion is an electron acceptor.

## Covalent Bonds (pp. 27–28)

8. In a covalent bond, atoms share pairs of electrons.
9. Covalent bonds are stronger than ionic bonds and are far more common in organic molecules.

## Hydrogen Bonds (pp. 28–29)

10. A hydrogen bond exists when a hydrogen atom covalently bonded to one oxygen or nitrogen atom is attracted to another oxygen or nitrogen atom.
11. Hydrogen bonds form weak links between different molecules or between parts of the same large molecule.

## Molecular Mass and Moles (p. 29)

12. The molecular mass is the sum of the atomic masses of all the atoms in a molecule.
13. A mole of an atom, ion, or molecule is equal to its atomic or molecular mass expressed in grams.

## Chemical Reactions (pp. 30–31)

1. Chemical reactions are the making or breaking of chemical bonds between atoms.
2. A change of energy occurs during chemical reactions.
3. Endergonic reactions require more energy than they release; exergonic reactions release more energy.
4. In a synthesis reaction, atoms, ions, or molecules are combined to form a larger molecule.
5. In a decomposition reaction, a larger molecule is broken down into its component molecules, ions, or atoms.
6. In an exchange reaction, two molecules are decomposed, and their subunits are used to synthesize two new molecules.
7. The products of reversible reactions can readily revert to form the original reactants.

## Important Biological Molecules (pp. 31–47)

### Inorganic Compounds (pp. 31–33)

1. Inorganic compounds are usually small, ionically bonded molecules.

### Water (pp. 31–32)

2. Water is the most abundant substance in cells.
3. Because water is a polar molecule, it is an excellent solvent.
4. Water is a reactant in many of the decomposition reactions of digestion.
5. Water is an excellent temperature buffer.

### Acids, Bases, and Salts (p. 32)

6. An acid dissociates into $H^+$ and anions.
7. A base dissociates into $OH^-$ and cations.
8. A salt dissociates into negative and positive ions, neither of which is $H^+$ or $OH^-$.

### Acid–Base Balance: The Concept of pH (pp. 32–33)

9. The term *pH* refers to the concentration of $H^+$ in a solution.
10. A solution of pH 7 is neutral; a pH value below 7 indicates acidity; pH above 7 indicates alkalinity.
11. The pH inside a cell and in culture media is stabilized with pH buffers.

### Organic Compounds (pp. 33–47)

1. Organic compounds always contain carbon and hydrogen.
2. Carbon atoms form up to four bonds with other atoms.
3. Organic compounds are mostly or entirely covalently bonded.

### Structure and Chemistry (pp. 34–35)

4. A chain of carbon atoms forms a carbon skeleton.
5. Functional groups of atoms are responsible for most of the properties of organic molecules.
6. The letter *R* may be used to denote the remainder of an organic molecule.
7. Frequently encountered classes of molecules are R—OH (alcohols) and R—COOH (organic acids).
8. Small organic molecules may combine into very large molecules called macromolecules.
9. Monomers usually bond together by dehydration synthesis, or condensation reactions, that form water and a polymer.
10. Organic molecules may be broken down by hydrolysis, a reaction involving the splitting of water molecules.

### Carbohydrates (pp. 35–36)

11. Carbohydrates are compounds consisting of atoms of carbon, hydrogen, and oxygen, with hydrogen and oxygen in a 2:1 ratio.
12. Monosaccharides contain from three to seven carbon atoms.
13. Isomers are two molecules with the same chemical formula but different structures and properties—for example, glucose ($C_6H_{12}O_6$) and fructose ($C_6H_{12}O_6$).
14. Monosaccharides may form disaccharides and polysaccharides by dehydration synthesis.

### Lipids (pp. 36–40)

15. Lipids are a diverse group of compounds distinguished by their insolubility in water.
16. Simple lipids (fats) consist of a molecule of glycerol and three molecules of fatty acids.
17. A saturated lipid has no double bonds between carbon atoms in the fatty acids; an unsaturated lipid has one or more double bonds. Saturated lipids have higher melting points than unsaturated lipids.
18. Phospholipids are complex lipids consisting of glycerol, two fatty acids, and a phosphate group.
19. Steroids have carbon ring structures; sterols have a functional hydroxyl group.

### Proteins (pp. 40–44)

20. Amino acids are the building blocks of proteins.
21. Amino acids consist of carbon, hydrogen, oxygen, nitrogen, and sometimes sulfur.
22. Twenty amino acids occur naturally in proteins.
23. By linking amino acids, peptide bonds (formed by dehydration synthesis) allow the formation of polypeptide chains.
24. Proteins have four levels of structure: primary (sequence of amino acids), secondary (helices or pleated), tertiary (overall three-dimensional structure of a polypeptide), and quaternary (two or more polypeptide chains).
25. Conjugated proteins consist of amino acids combined with inorganic or other organic compounds.

### Nucleic Acids (pp. 44–46)

26. Nucleic acids—DNA and RNA—are macromolecules consisting of repeating nucleotides.
27. A nucleotide is composed of a pentose, a phosphate group, and a nitrogen-containing base. A nucleoside is composed of a pentose and a nitrogen-containing base.
28. A DNA nucleotide consists of deoxyribose (a pentose) and one of the following nitrogen-containing bases: thymine or cytosine (pyrimidines) or adenine or guanine (purines).
29. DNA consists of two strands of nucleotides wound in a double helix. The strands are held together by hydrogen bonds between purine and pyrimidine nucleotides: AT and GC.
30. Genes consist of sequences of nucleotides.
31. An RNA nucleotide consists of ribose (a pentose) and one of the following nitrogen-containing bases: cytosine, guanine, adenine, or uracil.

### Adenosine Triphosphate (ATP) (p. 46)

32. ATP stores chemical energy for various cellular activities.
33. When the bond to ATP's terminal phosphate group is hydrolyzed, energy is released.
34. The energy from oxidation reactions is used to regenerate ATP from ADP and inorganic phosphate.

# Study Questions

For answers to the Knowledge and Comprehension questions, turn to the Answers tab at the back of the textbook.

## Knowledge and Comprehension

### Review

1. What is a chemical element?
2. **DRAW IT** Diagram the electronic configuration of a carbon atom.
3. What type of bond holds the following atoms together?
   a. $Li^+$ and $Cl^-$ in LiCl
   b. carbon and oxygen atoms in methanol
   c. oxygen atoms in $O_2$
   d. a hydrogen atom of one nucleotide to a nitrogen or oxygen atom of another nucleotide in:

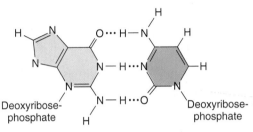

   Deoxyribose-phosphate    Deoxyribose-phosphate

   Guanine    Cytosine

4. Classify the following types of chemical reactions.
   a. glucose + fructose → sucrose + $H_2O$
   b. lactose → glucose + galactose
   c. $NH_4Cl + H_2O \rightarrow NH_4OH + HCl$
   d. ATP $\rightleftharpoons$ ADP+ Ⓟ$_i$
5. Bacteria use the enzyme urease to obtain nitrogen in a form they can use from urea in the following reaction:

$$CO(NH_2)_2 \quad + \quad H_2O \quad \rightarrow \quad 2NH_3 \quad + \quad CO_2$$
   Urea                              Ammonia      Carbon
                                                  dioxide

   What purpose does the enzyme serve in this reaction? What type of reaction is this?
6. Classify the following as subunits of either a carbohydrate, lipid, protein, or nucleic acid.
   a. $CH3-(CH_2)_7-CH=CH-(CH_2)_7-COOH$
      Oleic acid
   b.

   NH$_2$
   |
   H—C—COOH
   |
   CH$_2$
   |
   OH
   Serine

   c. $C_6H_{12}O_6$
   d. Thymine nucleotide

7. **DRAW IT** The artificial sweetener aspartame, or NutraSweet®, is made by joining aspartic acid to methylated phenylalanine, as shown below.

   a. What types of molecules are aspartic acid and phenylalanine?
   b. What direction is the hydrolysis reaction (left to right or right to left)?
   c. What direction is the dehydration synthesis reaction?
   d. Circle the atoms involved in the formation of water.
   e. Identify the peptide bond.

8. **DRAW IT** The following diagram shows the bacteriorhodopsin protein. Indicate the regions of primary, secondary, and tertiary structure. Does this protein have quaternary structure?

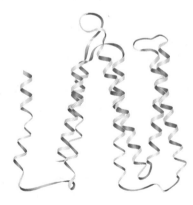

9. **DRAW IT** Draw a simple lipid, and show how it could be modified to a phospholipid.
10. **NAME IT** What type of microorganism has a chitin cell wall, has DNA that is contained in a nucleus, and has ergosterol in its plasma membrane?

## Multiple Choice

Radioisotopes are frequently used to label molecules in a cell. The fate of atoms and molecules in a cell can then be followed. This process is the basis for questions 1–3.

1. Assume *E. coli* bacteria are grown in a nutrient medium containing the radioisotope $^{16}$N. After a 48-hour incubation period, the $^{16}$N would most likely be found in the *E. coli's*
   a. carbohydrates.
   b. lipids.
   c. proteins.
   d. water.
   e. none of the above

2. If *Pseudomonas* bacteria are supplied with radioactively labeled cytosine, after a 24-hour incubation period this cytosine would most likely be found in the cells'
   a. carbohydrates.
   b. DNA.
   c. lipids.
   d. water.
   e. proteins.

3. If *E. coli* were grown in a medium containing the radioactive isotope $^{32}$P, the $^{32}$P would be found in all of the following molecules of the cell *except*
   a. ATP.
   b. carbohydrates.
   c. DNA.
   d. plasma membrane.
   e. none of the above

4. The optimum pH of *Acidithiobacillus* bacteria (pH 3,) is _____ times more acid than blood (pH 7).
   a. 4
   b. 10
   c. 100
   d. 1000
   e. 10,000

5. The best definition of ATP is that it is
   a. a molecule stored for food use.
   b. a molecule that supplies energy to do work.
   c. a molecule stored for an energy reserve.
   d. a molecule used as a source of phosphate.

6. Which of the following is an organic molecule?
   a. $H_2O$ (water)
   b. $O_2$ (oxygen)
   c. $C_{18}H_{29}SO_3$ (Styrofoam)
   d. FeO (iron oxide)
   e. $F_2C$=$CF_2$ (Teflon)

Classify each of the molecules on the left as an acid, base, or salt. The dissociation products of the molecules are shown to help you.

7. $HNO_3 \rightarrow H^+ + NO_3^-$      a. acid
8. $H_2SO_4 \rightarrow 2H^+ + SO_4^{2-}$   b. base
9. $NaOH \rightarrow Na^+ + OH^-$       c. salt
10. $MgSO_4 \rightarrow Mg^{2+} + SO_4^{2-}$

## Analysis

1. When you blow bubbles into a glass of water, the following reactions take place:

$$H_2O + CO_2 \xrightarrow{A} H_2CO_3 \xrightarrow{B} H^+ + HCO_3^-$$

   a. What type of reaction is *A*?
   b. What does reaction *B* tell you about the type of molecule $H_2CO_3$ is?

2. What are the common structural characteristics of ATP and DNA molecules?

3. What happens to the relative amount of unsaturated lipids in the plasma membrane when *E. coli* bacteria grown at 25°C are then grown at 37°C?

4. Giraffes, termites, and koalas eat only plant matter. Because animals cannot digest cellulose, how do you suppose these animals get nutrition from the leaves and wood they eat?

## Clinical Applications and Evaluation

1. *Ralstonia* bacteria make poly-β-hydroxybutyrate (PHB), which is used to make a biodegradable plastic. PHB consists of many of the monomers shown below. What type of molecule is PHB? What is the most likely reason a cell would store this molecule?

2. *Acidithiobacillus ferrooxidans* was responsible for destroying buildings in the Midwest by causing changes in the earth. The original rock, which contained lime ($CaCO_3$) and pyrite ($FeS_2$), expanded as bacterial metabolism caused gypsum ($CaSO_4$) crystals to form. How did *A. ferrooxidans* bring about the change from lime to gypsum?

3. Newborn babies are tested for phenylketonuria (PKU), an inherited disease. Individuals with this disease are missing an enzyme to convert phenylalanine (phe) to tyrosine; the resulting accumulation of phe can cause mental retardation, brain damage, and seizures. The Guthrie test for PKU involves culturing *Bacillus subtilis*, which requires phe to grow. The bacteria are grown on media with a drop of the baby's blood.
   a. What type of chemical is phenylalanine?
   b. What does "no growth" in the Guthrie test mean?
   c. Why must individuals with PKU avoid the sweetener aspartame?

4. The antibiotic amphotericin B causes leaks into cells by combining with sterols in the plasma membrane. Would you expect to use amphotericin B against a bacterial infection? A fungal infection? Offer a reason why amphotericin B has severe side effects in humans.

5. You can smell sulfur when boiling eggs. What amino acids do you expect in the egg?

# Observing Microorganisms Through a Microscope 3

Microorganisms are much too small to be seen with the unaided eye; they must be observed with a microscope. The word *microscope* is derived from the Latin word *micro* (small) and the Greek word *skopos* (to look at). Modern microbiologists use microscopes that produce, with great clarity, magnifications that range from ten to thousands of times greater than those of van Leeuwenhoek's single lens (see Figure 1.3c on page 7). This chapter describes how different types of microscopes function and why one type might be used in preference to another. *Helicobacter pylori* (HĒ-lik-ō-bak'ter PĪ-lor-ē), shown in the photograph, is a spiral-shaped bacterium that was first seen in cadaver stomachs in 1886. The bacterium was largely ignored until the resolving ability of microscopes was improved. Microscopic examination of these bacteria is described in the Clinical Case.

Some microbes are more readily visible than others because of their larger size or more easily observable features. Many microbes, however, must undergo several staining procedures before their cell walls, capsules, and other structures lose their colorless natural state. The last part of this chapter explains some of the more commonly used methods of preparing specimens for examination through a light microscope.

You may wonder how we are going to sort, count, and measure the specimens we will study. To answer these questions, this chapter opens with a discussion of how to use the metric system for measuring microbes.

◀ *Helicobacter pylori* bacteria live in the human stomach and can cause ulcers.

## In the Clinic

Mike is one of your regular patients at the homeless clinic where you are a volunteer nurse. He has a severe cough and is quite thin. Last week you sent Mike's sputum sample to the laboratory and requested a Gram stain and acid-fast stain. The stain results are now posted in his chart: It says "acid-fast +." **What infection does Mike have?**

*Hint: Read the acid-fast stain section on page 66.*

Play In the Clinic Video
@MasteringMicrobiology

# Units of Measurement

### LEARNING OBJECTIVE

**3-1** List the units used to measure microorganisms.

When measuring microorganisms, we use the metric system. A major advantage of the metric system is that units relate to each other by factors of 10. Thus, 1 meter (m) equals 10 decimeters (dm) or 100 centimeters (cm) or 1000 millimeters (mm). Units in the U.S. system of measure do not have the advantage of easy conversion by a single factor of 10. For example, 3 feet, or 36 inches, equals 1 yard.

Microorganisms are measured in even smaller units, such as micrometers and nanometers. A **micrometer (μm)** equals 0.000001 m ($10^{-6}$ m). The prefix *micro* indicates the unit following it should be divided by 1 million, or $10^6$ (see the "Exponential Notation" section in Appendix B). A **nanometer (nm)** equals 0.000000001 m ($10^{-9}$ m). Angstrom (Å) was previously used for $10^{-10}$ m, or 0.1 nm.

**Table 3.1** presents the basic metric units of length and some U.S. equivalents.

> **CHECK YOUR UNDERSTANDING**
> ✔ **3-1** How many nanometers is 10 μm?

# Microscopy: The Instruments

### LEARNING OBJECTIVES

**3-2** Diagram the path of light through a compound microscope.

**3-3** Define *total magnification* and *resolution*.

**3-4** Identify a use for darkfield, phase-contrast, differential interference contrast, fluorescence, confocal, two-photon, and scanning acoustic microscopy, and compare each with brightfield illumination.

**3-5** Explain how electron microscopy differs from light microscopy.

**3-6** Identify uses for the transmission electron microscope (TEM), scanning electron microscope (SEM), and scanned-probe microscopes.

The simple microscope used by van Leeuwenhoek in the seventeenth century had only one lens and was similar to a magnifying glass. However, van Leeuwenhoek was the best lens grinder in the world in his day. His lenses were ground with such precision that a single lens could magnify a microbe 300×. His simple microscopes enabled him to be the first person to see bacteria (see Figure 1.3, page 7).

Contemporaries of van Leeuwenhoek, such as Robert Hooke, built compound microscopes, which have multiple lenses. In fact, a Dutch spectacle maker, Zaccharias Janssen, is credited with making the first compound microscope around 1600. However, these early compound microscopes were of poor quality and could not be used to see bacteria. It was not until about 1830 that a significantly better microscope was developed by Joseph Jackson Lister (the father of Joseph Lister). Various improvements to Lister's microscope resulted in the development of the modern compound microscope, the kind

Play Microscopy and Staining: *Overview* @MasteringMicrobiology

used in microbiology laboratories today. Microscopic studies of live specimens have revealed dramatic interactions between microbes.

## Light Microscopy

**Light microscopy** refers to the use of any kind of microscope that uses visible light to observe specimens. Here we examine several types of light microscopy.

### Compound Light Microscopy

A modern **compound light microscope (LM)** has a series of lenses and uses visible light as its source of illumination (**Figure 3.1a**). With a compound light microscope, we can examine very small specimens as well as some of their fine detail. A series of finely ground lenses (Figure 3.1b) forms a clearly focused image

> 🔬 ASM: The structure and function of microorganisms have been revealed by the use of microscopy (including brightfield phase-contrast, fluorescent, and electron).

| **TABLE 3.1** Metric Units of Length and U.S. Equivalents | | | |
|---|---|---|---|
| **Metric Unit** | **Meaning of Prefix** | **Metric Equivalent** | **U.S. Equivalent** |
| 1 kilometer (km) | *kilo* = 1000 | 1000 m = $10^3$ m | 3280.84 ft or 0.62 mi; 1 mi = 1.61 km |
| 1 meter (m) | | Standard unit of length | 39.37 in or 3.28 ft or 1.09 yd |
| 1 decimeter (dm) | *deci* = 1/10 | 0.1 m = $10^{-1}$ m | 3.94 in |
| 1 centimeter (cm) | *centi* = 1/100 | 0.01 m = $10^{-2}$ m | 0.394 in; 1 in = 2.54 cm |
| 1 millimeter (mm) | *milli* = 1/1000 | 0.001 m = $10^{-3}$ m | |
| 1 micrometer (μm) | *micro* = 1/1,000,000 | 0.000001 m = $10^{-6}$ m | |
| 1 nanometer (nm) | *nano* = 1/1,000,000,000 | 0.000000001 m = $10^{-9}$ m | |
| 1 picometer (pm) | *pico* = 1/1,000,000,000,000 | 0.000000000001 m = $10^{-12}$ m | |

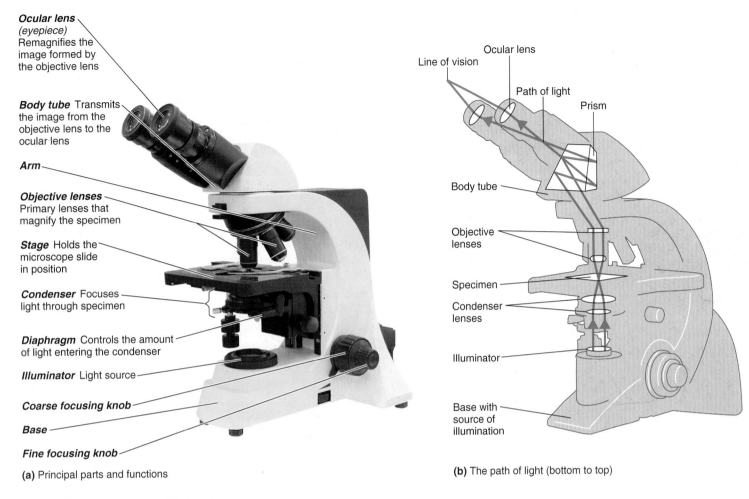

**Ocular lens** *(eyepiece)* Remagnifies the image formed by the objective lens

**Body tube** Transmits the image from the objective lens to the ocular lens

**Arm**

**Objective lenses** Primary lenses that magnify the specimen

**Stage** Holds the microscope slide in position

**Condenser** Focuses light through specimen

**Diaphragm** Controls the amount of light entering the condenser

**Illuminator** Light source

**Coarse focusing knob**

**Base**

**Fine focusing knob**

**(a)** Principal parts and functions

Line of vision

Ocular lens

Path of light

Prism

Body tube

Objective lenses

Specimen

Condenser lenses

Illuminator

Base with source of illumination

**(b)** The path of light (bottom to top)

**Figure 3.1 The compound light microscope.**

**Q** What is the total magnification of a compound light microscope with objective lens magnification of 40× and ocular lens of 10×?

that is many times larger than the specimen itself. This magnification is achieved when light rays from an **illuminator**, the light source, pass through a condenser, which has lenses that direct the light rays through the specimen. From here, light rays pass into the **objective lenses**, the lenses closest to the specimen. The image of the specimen is magnified again by the **ocular lens**, or *eyepiece*.

We can calculate the **total magnification** of a specimen by multiplying the objective lens magnification (power) by the ocular lens magnification (power). Most microscopes used in microbiology have several objective lenses, including 10× (low power), 40× (high power), and 100× (oil immersion). Most ocular lenses magnify specimens by a factor of 10. Multiplying the magnification of a specific objective lens with that of the ocular, we see that the total magnifications would be 100× for low power, 400× for high power, and 1000× for oil immersion.

**CLINICAL CASE** Microscopic Mayhem

Maryanne, a 42-year-old marketing executive and mother of three, has been experiencing recurrent stomach pain, which seems to be getting worse. She jokes her husband should buy stock in Pepto-Bismol, because she buys so much of it. At her husband's urging, she makes an appointment to see her primary care physician. After hearing that Maryanne feels better immediately after taking Pepto-Bismol, the doctor suspects Maryanne may have a peptic ulcer associated with *Helicobacter pylori*.

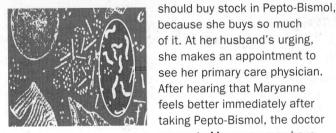

5 µm

**What is *Helicobacter pylori*? Read on to find out.**

53    60    66    67

# Microscopes and Magnification

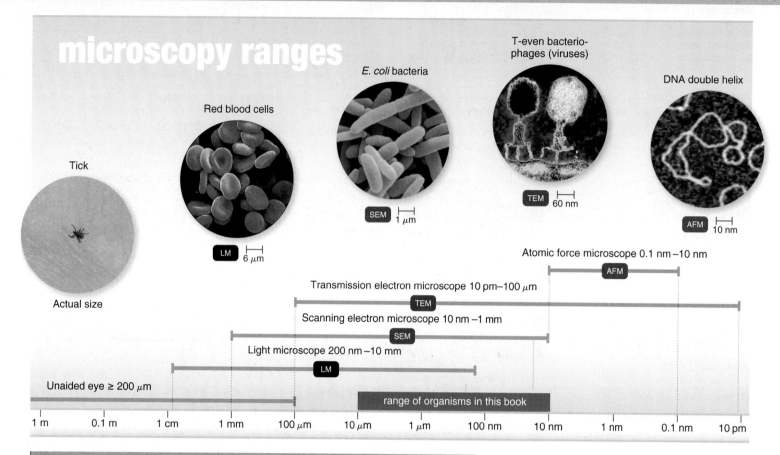

## microscopy ranges

Tick

Actual size

Red blood cells

E. coli bacteria

T-even bacterio-phages (viruses)

DNA double helix

LM 6 μm

SEM 1 μm

TEM 60 nm

AFM 10 nm

Atomic force microscope 0.1 nm –10 nm

AFM

Transmission electron microscope 10 pm–100 μm

TEM

Scanning electron microscope 10 nm –1 mm

SEM

Light microscope 200 nm –10 mm

LM

Unaided eye ≥ 200 μm

range of organisms in this book

| 1 m | 0.1 m | 1 cm | 1 mm | 100 μm | 10 μm | 1 μm | 100 nm | 10 nm | 1 nm | 0.1 nm | 10 pm |

## KEY CONCEPTS

- Microscopes are used to magnify small objects.

- Because different microscopes have different resolution ranges, the size of a specimen determines which microscopes can be used to view the specimen effectively.

- Most micrographs shown in this textbook (like the ones above) have size bars and symbols to help you identify the actual size of the specimen and the type of microscope used for that image.

- A red icon indicates that a micrograph has been artificially colorized.

- Resolution increases with decreasing wavelength.

micro tip

If a bacterium is 1 micrometer long and your index finger is 6.5 cm long, how many of the bacteria can you place end-to-end on your finger? Answer: 65,000.

**Resolution** (also called *resolving power*) is the ability of the lenses to distinguish fine detail and structure. Specifically, it refers to the ability of the lenses to distinguish two points that are a specified distance apart. For example, if a microscope has a resolving power of 0.4 nm, it can distinguish two points if they are at least 0.4 nm apart. A general principle of microscopy is that the shorter the wavelength of light used in the instrument, the greater the resolution. The white light used in a compound light microscope has a relatively long wavelength and cannot resolve structures smaller than about 0.2 μm. This and other considerations limit the magnification achieved by the best compound light microscopes to about 1500×. By comparison, van Leeuwenhoek's tiny spherical lenses had a resolution of 1 μm. **Figure 3.2** shows various specimens that can be resolved by the human eye and microscopes.

To obtain a clear, finely detailed image under a compound light microscope, specimens must contrast sharply with their *medium* (the substance in which they are suspended). To attain such contrast, we must change the refractive index of specimens from that of their medium. The **refractive index** is a measure of the light-bending ability of a medium. We change the refractive index of specimens by staining them, a procedure we will discuss shortly. Light rays move in a straight line through a single medium. After the specimen is stained, the specimen and its medium have different refractive indexes. When light rays pass through the two materials (the specimen and its medium), the rays change direction (refract) from a straight path by bending or changing angle at the boundary between the materials. This increases the image's contrast between the specimen and the medium. As the light rays travel away from the specimen, they spread out and enter the objective lens, and the image is thereby magnified.

To achieve high magnification (1000×) with good resolution, the objective lens must be small. Although we want light traveling through the specimen and medium to refract differently, we do not want to lose light rays after they have passed through the stained specimen. To preserve the direction of light rays at the highest magnification, immersion oil is placed between the glass slide and the oil immersion objective lens (**Figure 3.3**). The immersion oil has the same refractive index as glass, so the oil becomes part of the optics of the glass of the microscope. Unless immersion oil is used, light rays are refracted as they enter the air from the slide,

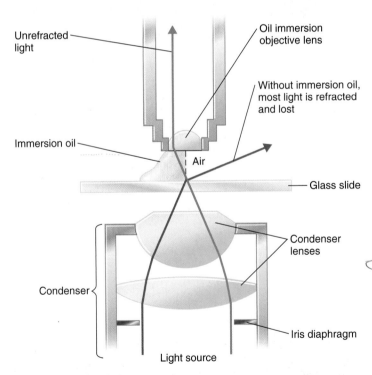

**Figure 3.3 Refraction in the compound microscope using an oil immersion objective lens.** Because the refractive indexes of the glass microscope slide and immersion oil are the same, the light rays do not refract when passing from one to the other when an oil immersion objective lens is used. Use of immersion oil is necessary at magnifications greater than 900×.

**Q** Why is immersion oil necessary at 1000× but not with the lower power objectives?

and the objective lens would have to be increased in diameter to capture most of them. The oil has the same effect as increasing the objective lens diameter; therefore, it improves the resolving power of the lenses. If oil is not used with an oil immersion objective lens, the image has poor resolution and becomes fuzzy.

Under usual operating conditions, the field of vision in a compound light microscope is brightly illuminated. By focusing the light, the condenser produces a **brightfield illumination** (**Figure 3.4a**).

It is not always desirable to stain a specimen, but an unstained cell has little contrast with its surroundings and is therefore difficult to see. Unstained cells are more easily observed with the modified compound microscopes described in the next section.

### Darkfield Microscopy

A **darkfield microscope** is used to examine live microorganisms that either are invisible in the ordinary light microscope, cannot be stained by standard methods, or are so distorted by staining that their characteristics are obscured. A darkfield microscope uses a darkfield condenser that contains an opaque disk. The disk blocks light that would enter the objective lens directly. Only light that is reflected off (turned away from) the specimen enters the objective lens. Because there is no direct background light, the specimen appears light against a black background—the dark field (Figure 3.4b). This technique is frequently used to examine unstained microorganisms suspended in liquid. One use for darkfield microscopy is the examination of very thin spirochetes, such as *Treponema pallidum* (trep-ō-NĒ-mah PAL-li-dum), the causative agent of syphilis.

### Phase-Contrast Microscopy

Another way to observe microorganisms is with a **phase-contrast microscope**. Phase-contrast microscopy is especially useful because the internal structures of a cell become more sharply defined, permitting detailed examination of *living* microorganisms. In addition, it isn't necessary to fix (attach the microbes to the microscope slide) or stain the specimen—procedures that could distort or kill the microorganisms.

In a phase-contrast microscope, one set of light rays comes directly from the light source. The other set comes from light that is reflected or diffracted from a particular structure in the specimen. (*Diffraction* is the scattering of light rays as they

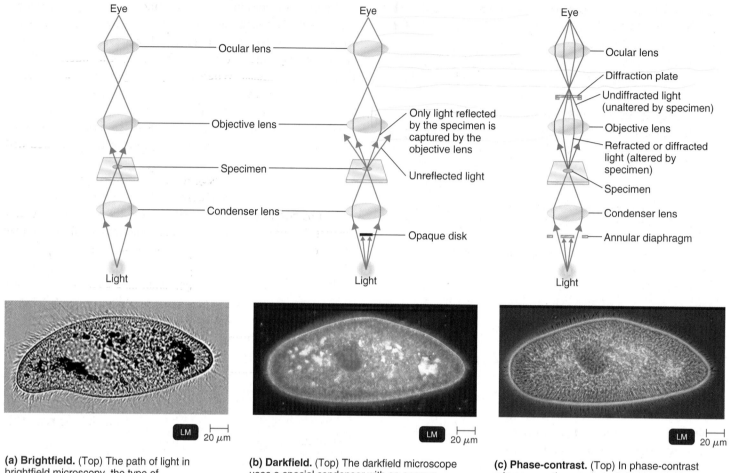

**(a) Brightfield.** (Top) The path of light in brightfield microscopy, the type of illumination produced by regular compound light microscopes. (Bottom) Brightfield illumination shows internal structures and the outline of the transparent pellicle (external covering).

**(b) Darkfield.** (Top) The darkfield microscope uses a special condenser with an opaque disk that eliminates all light in the center of the beam. The only light that reaches the specimen comes in at an angle; thus, only light reflected by the specimen (blue lines) reaches the objective lens. (Bottom) Against the black background seen with darkfield microscopy, edges of the cell are bright, some internal structures seem to sparkle, and the pellicle is almost visible.

**(c) Phase-contrast.** (Top) In phase-contrast microscopy, the specimen is illuminated by light passing through an annular (ring-shaped) diaphragm. Direct light rays (unaltered by the specimen) travel a different path from light rays that are reflected or diffracted as they pass through the specimen. These two sets of rays are combined at the eye. Reflected or diffracted light rays are indicated in blue; direct rays are red. (Bottom) Phase-contrast microscopy shows greater differentiation of internal structures and clearly shows the pellicle.

**Figure 3.4  Brightfield, darkfield, and phase-contrast microscopy.** The photographs compare the protozoan *Paramecium* using these three different microscopy techniques.

**Q** What are the advantages of brightfield, darkfield, and phase-contrast microscopy?

"touch" a specimen's edge. The diffracted rays are bent away from the parallel light rays that pass farther from the specimen.) When the two sets of light rays—direct rays and reflected or diffracted rays—are brought together, they form an image of the specimen on the ocular lens, containing areas that are relatively light (in phase), through shades of gray, to black (out of phase; Figure 3.4c).

### Differential Interference Contrast (DIC) Microscopy

**Differential interference contrast (DIC) microscopy** is similar to phase-contrast microscopy in that it uses differences in refractive indexes. However, a DIC microscope uses two beams of light instead of one. In addition, prisms split each light beam, adding contrasting colors to the specimen. Therefore, the resolution of a DIC microscope is higher than that of a standard phase-contrast microscope. Also, the image is brightly colored and appears nearly three-dimensional (**Figure 3.5**).

### Fluorescence Microscopy

**Fluorescence microscopy** takes advantage of **fluorescence**, the ability of substances to absorb short wavelengths of light

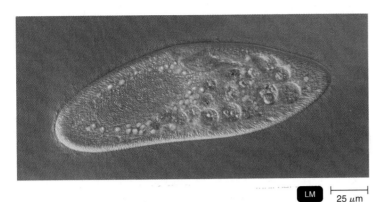

**Figure 3.5  Differential interference contrast (DIC) microscopy.**
Like phase-contrast, DIC uses differences in refractive indexes to produce
an image, in this case a *Paramecium*. The colors in the image are
produced by prisms that split the two light beams used in this process.

**Q** Why is a DIC micrograph brightly colored?

(ultraviolet) and give off light at a longer wavelength (visible).
Some organisms fluoresce naturally under ultraviolet light; if
the specimen to be viewed does not naturally fluoresce, it is
stained with one of a group of fluorescent dyes called *fluorochromes.* When microorganisms stained with a fluorochrome
are examined under a fluorescence microscope with an ultraviolet or near-ultraviolet light source, they appear as luminescent, bright objects against a dark background.

Fluorochromes have special attractions for different microorganisms. For example, the fluorochrome auramine O, which
glows yellow when exposed to ultraviolet light, is strongly
absorbed by *Mycobacterium tuberculosis*, the bacterium that
causes tuberculosis. When the dye is applied to a sample of
material suspected of containing the bacterium, the bacterium
can be detected by the appearance of bright yellow organisms against a dark background. *Bacillus anthracis*, the causative agent of anthrax, appears apple green when stained with
another fluorochrome, fluorescein isothiocyanate (FITC).

The principal use of fluorescence microscopy is a diagnostic technique called the **fluorescent-antibody (FA) technique**,
or **immunofluorescence. Antibodies** are natural defense molecules that are produced by humans and many animals in reaction to a foreign substance, or **antigen**. Fluorescent antibodies
for a particular antigen are obtained as follows: an animal is
injected with a specific antigen, such as a bacterium, and the
animal then begins to produce antibodies against that antigen. After a sufficient time, the antibodies are removed from
the serum of the animal. Next, as shown in **Figure 3.6a**, a fluorochrome is chemically combined with the antibodies. These
fluorescent antibodies are then added to a microscope slide
containing an unknown bacterium. If this unknown bacterium
is the same bacterium that was injected into the animal, the
fluorescent antibodies bind to antigens on the surface of the
bacterium, causing it to fluoresce.

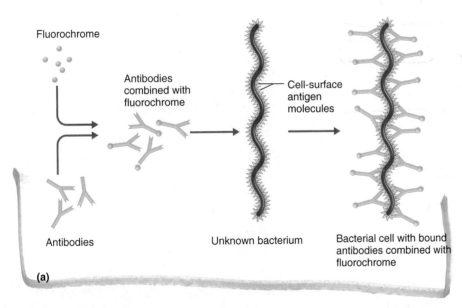

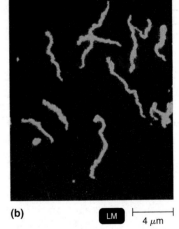

**Figure 3.6  The principle of immunofluorescence. (a)** A type of fluorochrome is combined
with antibodies against a specific type of bacterium. When the preparation is added to bacterial
cells on a microscope slide, the antibodies attach to the bacterial cells, and the cells fluoresce
when illuminated with ultraviolet light. **(b)** In the fluorescent treponemal antibody absorption
(FTA-ABS) test for syphilis shown here, *Treponema pallidum* shows up as green cells against a
darker background.

**Q** Why won't other bacteria fluoresce in the FTA-ABS test?

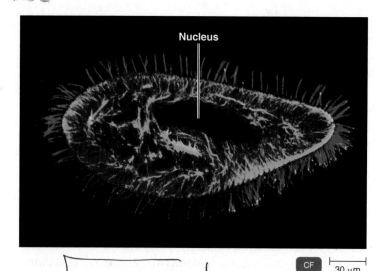

Nucleus

CF ⊢ 30 μm

**Figure 3.7 Confocal microscopy.** Confocal microscopy produces three-dimensional images and can be used to look inside cells. Shown here is the nucleus in *Paramecium tetraurelia*.

**Q** What are the advantages of confocal microscopy?

This technique can detect bacteria or other pathogenic microorganisms, even within cells, tissues, or other clinical specimens (Figure 3.6b). Of paramount importance, it can be used to identify a microbe in minutes. Immunofluorescence is especially useful in diagnosing syphilis and rabies. We will say more about antigen–antibody reactions and immunofluorescence in Chapter 18.

### Confocal Microscopy

Confocal microscopy is a technique in light microscopy used to reconstruct three-dimensional images. Like fluorescent microscopy, specimens are stained with fluorochromes so they will emit, or return, light. But instead of illuminating the entire field, one plane of a small region of a specimen is illuminated with a short-wavelength (blue) light which passes the returned light through an aperture aligned with the illuminated region. Each plane corresponds to an image of a fine slice that has been physically cut from a specimen. Successive planes and regions are illuminated until the entire specimen has been scanned. Because confocal microscopy uses a pinhole aperture, it eliminates blurring that occurs with other microscopes. As a result, exceptionally clear two-dimensional images can be obtained, with improved resolution of up to 40% over that of other microscopes.

Most confocal microscopes are used in conjunction with computers to construct three-dimensional images. The scanned planes of a specimen, which resemble a stack of images, are converted to a digital form that can be used by a computer to construct a three-dimensional representation. The reconstructed images can be rotated and viewed in any orientation. This technique has been used to obtain three-dimensional images

of entire cells and cellular components (Figure 3.7). In addition, confocal microscopy can be used to evaluate cellular physiology by monitoring the distributions and concentrations of substances such as ATP and calcium ions.

Play Light Microscopy
@MasteringMicrobiology

### Two-Photon Microscopy

As in confocal microscopy, specimens are stained with a fluorochrome for **two-photon microscopy** (TPM). Two-photon microscopy uses long-wavelength (red) light, and therefore two photons, instead of one, are needed to excite the fluorochrome to emit light. The longer wavelength allows imaging of living cells in tissues up to 1 mm (1000 μm) deep (**Figure 3.8**). Confocal microscopy can image cells in detail only to a depth of less than 100 μm. Additionally, the longer wavelength is less likely to generate singlet oxygen, which damages cells (see page 156). Another advantage of TPM is that it can track the activity of cells in real time. For example, cells of the immune system have been observed responding to an antigen.

### Super-Resolution Light Microscopy

Until recently, the maximum resolution for light microscopes was 0.2 μm. However, in 2014, the Nobel Prize in Chemistry was awarded to Eric Betzig, Stefan Hell, and William Moerner for the development of a microscope that uses two laser beams. In **super-resolution light microscopy**, one wavelength stimulates fluorescent molecules to glow, and another wavelength

Nucleus

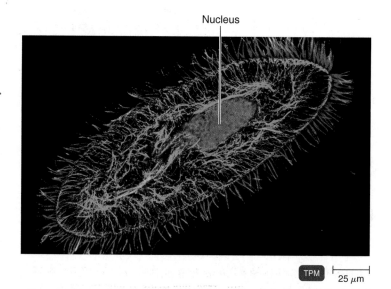

TPM ⊢ 25 μm

**Figure 3.8 Two-photon microscopy (TPM).** This procedure makes it possible to image cells up to 1 mm deep in detail. This image shows a *Paramecium*. Immunofluorescence is used to show microtubules and the nucleus.

**Q** What are the differences between TPM and confocal microscopy?

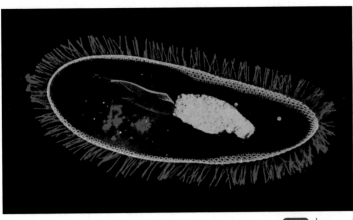

SRM ⊢ 25 μm

**Figure 3.9 Super-resolution light microscopy.** Super-resolution light microscopy scans cells one nanometer at a time to provide resolution of less than 0.1 μm.

**Q** What is the advantage of super-resolution light microscopy?

cancels out all fluorescence except for that in one nanometer. Cells can be stained with fluorescent dyes that are specific for certain molecules such as DNA or protein, allowing even a single molecule to be tracked in a cell. A computer tells the microscope to scan the specimen nanometer by nanometer and then puts the images together (**Figure 3.9**).

picometers

## Scanning Acoustic Microscopy

**Scanning acoustic microscopy (SAM)** basically consists of interpreting the action of a sound wave sent through a specimen. A sound wave of a specific frequency travels through the specimen, and a portion of it is reflected back every time it hits an interface within the material. The resolution is about 1 μm. SAM is used to study living cells attached to another surface, such as cancer cells, artery plaque, and bacterial biofilms that foul equipment (**Figure 3.10**).

> **CHECK YOUR UNDERSTANDING**
>
> ✔ 3-4 How are brightfield, darkfield, phase-contrast, and fluorescence microscopy similar?

## Electron Microscopy

Objects smaller than about 0.2 μm, such as viruses or the internal structures of cells, must be examined with an **electron microscope**. In electron microscopy, a beam of electrons is used instead of light. Like light, free electrons travel in waves. The resolving power of the electron microscope is far greater than that of the other microscopes described here so far. The better resolution of electron microscopes is due to the shorter wavelengths of electrons; the wavelengths of electrons are about 100,000 times smaller than the wavelengths of visible light. Thus, electron

microscopes are used to examine structures too small to be resolved with light microscopes. Images produced by electron microscopes are always black and white, but they may be colored artificially to accentuate certain details.

Instead of using glass lenses, an electron microscope uses electromagnetic lenses to focus a beam of electrons onto a specimen. There are two types: the transmission electron microscope and the scanning electron microscope.

### Transmission Electron Microscopy

In the **transmission electron microscope (TEM)**, a finely focused beam of electrons from an electron gun passes through a specially prepared, ultrathin section of the specimen (**Figure 3.11a**). The beam is focused on a small area of the specimen by an electromagnetic condenser lens that performs roughly the same function as the condenser of a light microscope—directing the beam of electrons in a straight line to illuminate the specimen.

Instead of being placed on a glass slide, as in light microscopes, the specimen is usually placed on a copper mesh grid. The beam of electrons passes through the specimen and then through an electromagnetic objective lens, which magnifies the image. Finally, the electrons are focused by an electromagnetic projector lens (rather than by an ocular lens as in a light microscope) onto a viewing screen and saved as a digital image. The final image, called a *transmission electron micrograph*, appears as many light and dark areas, depending on the number of electrons absorbed by different areas of the specimen.

The transmission electron microscope can resolve objects as close together as 10 pm, and objects are generally magnified 10,000 to 10,000,000×. Because most microscopic specimens are so thin, the contrast between their ultrastructures and the background is weak. Contrast can be greatly enhanced by using

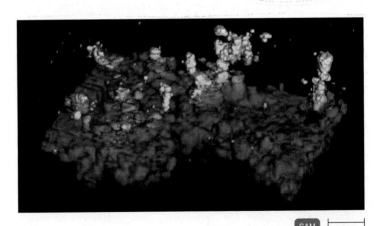

SAM ⊢ 170 μm

**Figure 3.10 Scanning acoustic microscopy (SAM) of a bacterial biofilm on glass.** Scanning acoustic microscopy essentially consists of interpreting the action of sound waves through a specimen. © 2006 IEEE.

**Q** What is the principal use of SAM?

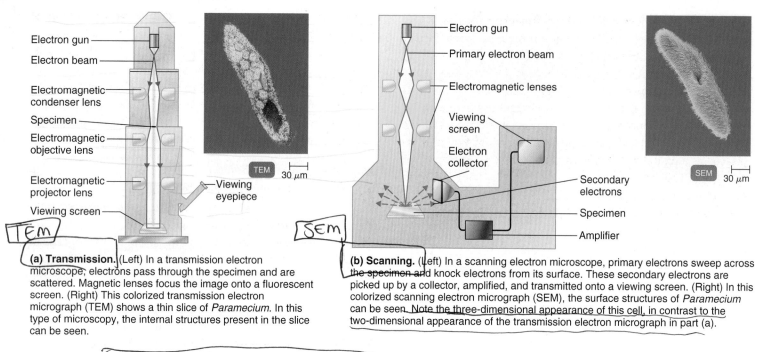

**TEM**

Electron gun

Electron beam

Electromagnetic condenser lens

Specimen

Electromagnetic objective lens

Electromagnetic projector lens

Viewing screen

Viewing eyepiece

TEM  30 μm

**SEM**

Electron gun

Primary electron beam

Electromagnetic lenses

Viewing screen

Electron collector

Secondary electrons

Specimen

Amplifier

SEM  30 μm

**(a) Transmission.** (Left) In a transmission electron microscope, electrons pass through the specimen and are scattered. Magnetic lenses focus the image onto a fluorescent screen. (Right) This colorized transmission electron micrograph (TEM) shows a thin slice of *Paramecium*. In this type of microscopy, the internal structures present in the slice can be seen.

**(b) Scanning.** (Left) In a scanning electron microscope, primary electrons sweep across the specimen and knock electrons from its surface. These secondary electrons are picked up by a collector, amplified, and transmitted onto a viewing screen. (Right) In this colorized scanning electron micrograph (SEM), the surface structures of *Paramecium* can be seen. Note the three-dimensional appearance of this cell, in contrast to the two-dimensional appearance of the transmission electron micrograph in part (a).

**Figure 3.11** **Transmission and scanning electron microscopy.** The illustrations show a *Paramecium* viewed with both of these types of microscopes. Although electron micrographs are normally black and white, these and other electron micrographs in this book have been artificially colorized for emphasis.

**Q** How do TEM and SEM images of the same organism differ?

a "dye" that absorbs electrons and produces a darker image in the stained region. Salts of various heavy metals, such as lead, osmium, tungsten, and uranium, are commonly used as stains. These metals can be fixed onto the specimen (*positive staining*) or used to increase the electron opacity of the surrounding field (*negative staining*). Negative staining is useful for the study of the very smallest specimens, such as virus particles, bacterial flagella, and protein molecules.

In addition to positive and negative staining, a microbe can be viewed by a technique called *shadow casting*. In this procedure, a heavy metal such as platinum or gold is sprayed at an angle of about 45° so that it strikes the microbe from only one side. The metal piles up on one side of the specimen, and the uncoated area on the opposite side of the specimen leaves a clear area behind it as a shadow. This gives a three-dimensional effect to the specimen and provides a general idea of the size and shape of the specimen (see the TEM in Figure 11.9a, page 304).

Transmission electron microscopy has high resolution and is extremely valuable for examining different layers of specimens. However, it does have certain disadvantages. Because electrons have limited penetrating power, only a very thin section of a specimen (about 100 nm) can be studied effectively. Thus, the specimen has no three-dimensional aspect. In addition, specimens must be fixed, dehydrated, and viewed under a high vacuum to prevent electron scattering. These treatments not only kill the specimen, but also cause shrinkage and

distortion, sometimes to the extent that there may appear to be additional structures in a prepared cell. Structures that appear as a result of the preparation method are called *artifacts*.

### Scanning Electron Microscopy

The **scanning electron microscope (SEM)** overcomes the sectioning problems associated with a transmission electron

**CLINICAL CASE**

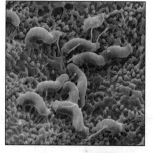

*H*elicobacter pylori is a spiral-shaped, gram-negative bacterium with multiple flagella. It is the most common cause of peptic ulcers in humans and can also cause stomach cancer. The first electron micrograph of *H. pylori* was viewed in the 1980s, when Australian physician Robin Warren used an electron microscope to see *H. pylori* in stomach tissue.

TEM  2 μm

**Why was an electron microscope necessary to see the *H. pylori* bacteria?**

microscope. It provides striking three-dimensional views of specimens (Figure 3.11b). An electron gun produces a finely focused beam of electrons called the primary electron beam. These electrons pass through electromagnetic lenses and are directed over the surface of the specimen. The primary electron beam knocks electrons out of the surface of the specimen, and the secondary electrons thus produced are transmitted to an electron collector, amplified, and used to produce an image on a viewing screen that is saved as a digital image. The image is called a *scanning electron micrograph*. This microscope is especially useful in studying the surface structures of intact cells and viruses. In practice, it can resolve objects as close together as 10 nm, and objects are generally magnified 1000 to 500,000×.

> ▶ **Play Electron Microscopy**
> @Mastering**Microbiology**

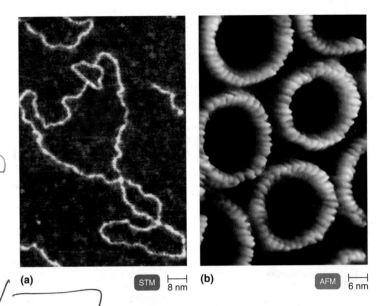

(a)  STM ⊢ 8 nm    (b)  AFM ⊢ 6 nm

**Figure 3.12 Scanned-probe microscopy. (a)** Scanning tunneling microscopy (STM) image of a double-stranded DNA molecule. **(b)** Atomic force microscopy (AFM) image of perfringolysin O toxin from *Clostridium perfringens*. This protein makes holes in human plasma membranes.

**Q** What is the principle employed in scanned-probe microscopy?

**CHECK YOUR UNDERSTANDING**

✔ 3-5 Why do electron microscopes have greater resolution than light microscopes?

## Scanned-Probe Microscopy

Since the early 1980s, several new types of microscopes, called **scanned-probe microscopes**, have been developed. They use various kinds of probes to examine the surface of a specimen using electric current, which does not modify the specimen or expose it to damaging, high-energy radiation. Such microscopes can be used to map atomic and molecular shapes, to characterize magnetic and chemical properties, and to determine temperature variations inside cells. Among the new scanned-probe microscopes are the scanning tunneling microscope and the atomic force microscope, discussed next.

### Scanning Tunneling Microscopy

Scanning tunneling microscopy (STM) uses a thin tungsten probe that scans a specimen and produces an image that reveals the bumps and depressions of the atoms on the surface of the specimen (Figure 3.12a). The resolving power of an STM is much greater than that of an electron microscope; it can resolve features that are only about 1/100 the size of an atom. Moreover, special preparation of the specimen for observation is not needed. STMs are used to provide incredibly detailed views of molecules such as DNA.

### Atomic Force Microscopy

In **atomic force microscopy** (AFM), a metal-and-diamond probe is gently forced down onto a specimen. As the probe moves along the surface of the specimen, its movements are recorded, and a three-dimensional image is produced (Figure 3.12b). As with STM, AFM does not require special specimen preparation. AFM is used to image both biological substances (in nearly atomic detail) (see also Figure 17.4c on page 480) and

molecular processes (such as the assembly of fibrin, a component of a blood clot).

The various types of microscopy just described are summarized in Table 3.2.

**CHECK YOUR UNDERSTANDING**

✔ 3-6 For what is TEM used? SEM? Scanned-probe microscopy?

## Preparation of Specimens for Light Microscopy

**LEARNING OBJECTIVES**

3-7 Differentiate an acidic dye from a basic dye.

3-8 Explain the purpose of simple staining.

3-9 List Gram stain steps, and describe the appearance of gram-positive and gram-negative cells after each step.

3-10 Compare and contrast the Gram stain and the acid-fast stain.

3-11 Explain why each of the following is used: capsule stain, endospore stain, flagella stain.

Most microorganisms appear almost colorless when viewed through brightfield microscopy, so we must prepare them for observation. One way is to stain (color) the specimen. Next we discuss several different staining procedures.

### Preparing Smears for Staining

Most initial observations of microorganisms are made with stained preparations. **Staining** simply means coloring the

## TABLE 3.2  A Summary of Various Types of Microscopes

| Microscope Type | Distinguishing Features | Typical Image | Principal Uses |
|---|---|---|---|
| **Light** | | | |
| Brightfield | Uses visible light as a source of illumination; cannot resolve structures smaller than about 0.2 μm; specimen appears against a bright background. Inexpensive and easy to use. | *Paramecium*  LM  25 μm | To observe various stained specimens and to count microbes; does not resolve very small specimens, such as viruses. |
| Darkfield | Uses a special condenser with an opaque disk that blocks light from entering the objective lens directly; light reflected by specimen enters the objective lens, and the specimen appears light against a black background. | *Paramecium*  LM  25 μm | To examine living microorganisms that are invisible in brightfield microscopy, do not stain easily, or are distorted by staining; frequently used to detect *Treponema pallidum* in the diagnosis of syphilis. |
| Phase-contrast | Uses a special condenser containing an annular (ring-shaped) diaphragm. The diaphragm allows direct light to pass through the condenser, focusing light on the specimen and a diffraction plate in the objective lens. Direct and reflected or diffracted light rays are brought together to produce the image. No staining required. | *Paramecium*  LM  25 μm | To facilitate detailed examination of the internal structures of living specimens. |
| Differential interference contrast (DIC) | Like phase-contrast, uses differences in refractive indexes to produce images. Uses two beams of light separated by prisms; the specimen appears colored as a result of the prism effect. No staining required. | *Paramecium*  LM  23 μm | To provide three-dimensional images. |
| Fluorescence | Uses an ultraviolet or near-ultraviolet source of illumination that causes fluorescent compounds in a specimen to emit light. | *Treponema pallidum*  LM  2 μm | For fluorescent-antibody techniques (immunofluorescence) to rapidly detect and identify microbes in tissues or clinical specimens. |

## TABLE 3.2 (continued)

| Microscope Type | Distinguishing Features | Typical Image | Principal Uses |
|---|---|---|---|
| **Confocal** | Uses a single photon to illuminate one plane of a specimen at a time. | <br>*Paramecium*  CF ⊢ 25 μm | To obtain two- and three-dimensional images of cells for biomedical applications. |
| **Two-Photon** | Uses two photons to illuminate a specimen. | <br>*Paramecium*  TPM ⊢ 22 μm | To image living cells, up to depth of 1 mm, reduce phototoxicity, and observe cell activity in real time. |
| **Super-Resolution Light Microscopy** | Uses two laser lights to illuminate one nanometer at a time. | <br>*Paramecium*  SRM ⊢ 25 μm | To observe the locations of molecules in cells. |
| **Scanning Acoustic** | Uses a sound wave of specific frequency that travels through the specimen with a portion being reflected when it hits an interface within the material. | <br>Biofilm  SAM ⊢ 180 μm | To examine living cells attached to another surface, such as cancer cells, artery plaque, and biofilms. |
| **Electron** Transmission | Uses a beam of electrons instead of light; electrons pass through the specimen; because of the shorter wavelength of electrons, structures smaller than 0.2 μm can be resolved. The image produced is two-dimensional. | <br>*Paramecium*  TEM ⊢ 25 μm | To examine viruses or the internal ultrastructure in thin sections of cells (usually magnified 10,000–10,000,000×). |

*(continued)*

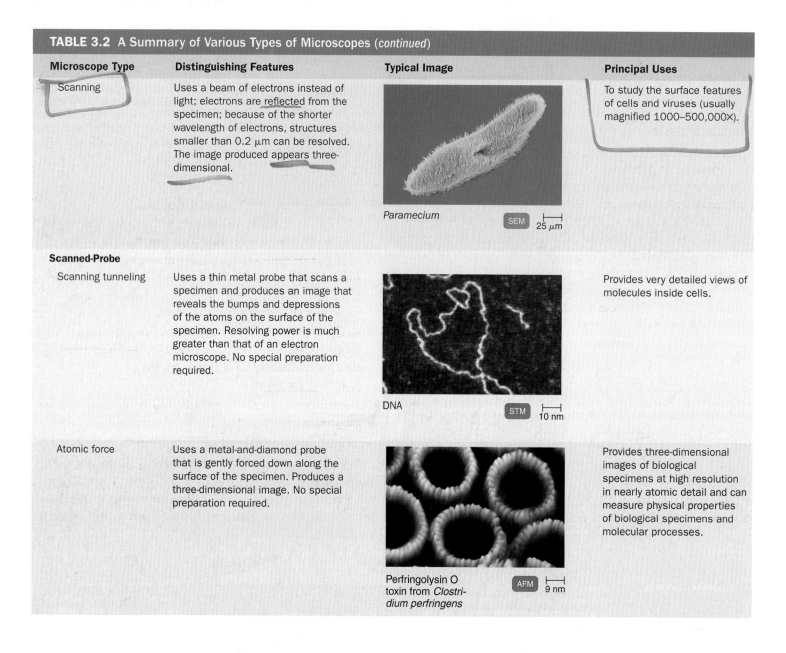

**TABLE 3.2** A Summary of Various Types of Microscopes (*continued*)

| Microscope Type | Distinguishing Features | Typical Image | Principal Uses |
|---|---|---|---|
| Scanning | Uses a beam of electrons instead of light; electrons are reflected from the specimen; because of the shorter wavelength of electrons, structures smaller than 0.2 μm can be resolved. The image produced appears three-dimensional. | *Paramecium*　SEM　⊢ 25 μm | To study the surface features of cells and viruses (usually magnified 1000–500,000×). |
| **Scanned-Probe** | | | |
| Scanning tunneling | Uses a thin metal probe that scans a specimen and produces an image that reveals the bumps and depressions of the atoms on the surface of the specimen. Resolving power is much greater than that of an electron microscope. No special preparation required. | DNA　STM　⊢ 10 nm | Provides very detailed views of molecules inside cells. |
| Atomic force | Uses a metal-and-diamond probe that is gently forced down along the surface of the specimen. Produces a three-dimensional image. No special preparation required. | Perfringolysin O toxin from *Clostridium perfringens*　AFM　⊢ 9 nm | Provides three-dimensional images of biological specimens at high resolution in nearly atomic detail and can measure physical properties of biological specimens and molecular processes. |

microorganisms with a dye that emphasizes certain structures. Before the microorganisms can be stained, however, they must be **fixed** (attached) to the microscope slide. Fixing simultaneously kills the microorganisms and fixes them to the slide. It also preserves various parts of microbes in their natural state with only minimal distortion.

When a specimen is to be fixed, a thin film of material containing the microorganisms is spread over the surface of the slide. This film, called a **smear**, is allowed to air dry. In most staining procedures the slide is then fixed by passing it through the flame of a Bunsen burner several times, smear side up, or by covering the slide with methanol for 1 minute. Stain is applied and then washed off with water; then the slide is blotted with absorbent paper. Without fixing, the stain might wash the microbes off the slide. The stained microorganisms are now ready for microscopic examination.

Stains are salts composed of a positive and a negative ion, one of which is colored and is known as the *chromophore*. The color of so-called **basic dyes** is in the cation; in **acidic dyes**, it is in the anion. Bacteria are slightly negatively charged at pH 7. Thus, the colored cation in a basic dye is attracted to the negatively charged bacterial cell. Basic dyes, which include crystal violet, methylene blue, malachite green, and safranin, are more commonly used than acidic dyes. Acidic dyes are not attracted to most types of bacteria because the dye's negative ions are repelled by the negatively charged bacterial surface, so the stain colors the background instead. Preparing colorless bacteria against a colored background is called **negative staining**. It is valuable for observing overall cell shapes, sizes, and capsules because the cells are made highly visible against a contrasting dark background (see Figure 3.15a on page 68). Distortions of cell size and shape are minimized because fixing isn't necessary and the cells don't pick up the stain. Examples of acidic dyes are eosin, acid fuchsin, and nigrosin.

To apply acidic or basic dyes, microbiologists use three kinds of staining techniques: simple, differential, and special.

## Simple Stains

A **simple stain** is an aqueous or alcohol solution of a single basic dye. Although different dyes bind specifically to different parts of cells, the primary purpose of a simple stain is to highlight the entire microorganism so that cellular shapes and basic structures are visible. The stain is applied to the fixed smear for a certain length of time and then washed off. The slide is dried and examined. Occasionally, a chemical is added to the solution to intensify the stain; such an additive is called a **mordant**. One function of a mordant is to increase the affinity of a stain for a biological specimen; another is to coat a structure (such as a flagellum) to make it thicker and easier to see after it is stained with a dye. Some of the simple stains commonly used in the laboratory are methylene blue, carbolfuchsin, crystal violet, and safranin.

> ## CHECK YOUR UNDERSTANDING
>
> ✔ **3-7** Why doesn't a negative stain color a cell?
>
> ✔ **3-8** Why is fixing necessary for most staining procedures?

## Differential Stains

Unlike simple stains, **differential stains** react differently with different kinds of bacteria and thus can be used to distinguish them. The differential stains most frequently used for bacteria are the Gram stain and the acid-fast stain.

### Gram Stain

The **Gram stain** was developed in 1884 by the Danish bacteriologist Hans Christian Gram. It is one of the most useful staining procedures because it classifies bacteria into two large groups: gram-positive and gram-negative.

In this procedure (**Figure 3.13a**),

❶ A heat-fixed smear is covered with a basic purple dye, usually crystal violet. Because the purple stain imparts its color to all cells, it is referred to as a **primary stain**.

❷ After a short time, the purple dye is washed off, and the smear is covered with iodine, a mordant. When the iodine is washed off, both gram-positive and gram-negative bacteria appear dark violet or purple.

❸ Next, the slide is washed with alcohol or an alcohol-acetone solution. This solution is a **decolorizing agent**, which removes the purple from the cells of some species but not from others.

❹ The alcohol is rinsed off, and the slide is then stained with safranin, a basic red dye. The smear is washed again, blotted dry, and examined microscopically.

The purple dye and the iodine combine in the cell wall of each bacterium and color it dark violet or purple. Bacteria that retain this color after the alcohol has attempted to decolorize them are classified as **gram-positive bacteria**; bacteria that lose

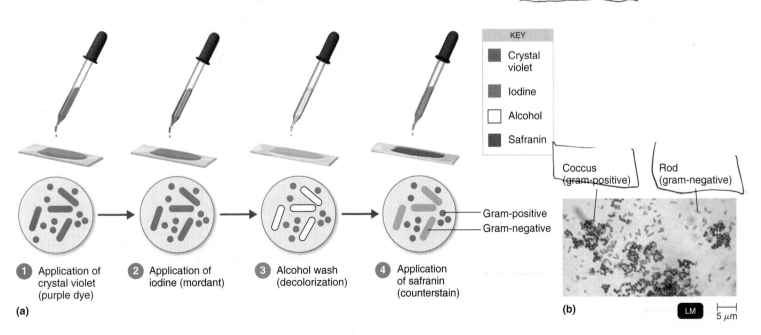

KEY
- Crystal violet
- Iodine
- Alcohol
- Safranin

Coccus (gram-positive)   Rod (gram-negative)

Gram-positive
Gram-negative

❶ Application of crystal violet (purple dye)
❷ Application of iodine (mordant)
❸ Alcohol wash (decolorization)
❹ Application of safranin (counterstain)

(a)

(b)    LM    5 μm

**Figure 3.13  Gram staining. (a)** Procedure. **(b)** Micrograph of Gram-stained bacteria. The cocci (purple) are gram-positive, and the rods (pink) are gram-negative.

**Q** How can the Gram reaction be useful in prescribing antibiotic treatment?

the dark violet or purple color after decolorization are classified as **gram-negative bacteria** (Figure 3.13b). Because gram-negative bacteria are colorless after the alcohol wash, they are no longer visible. This is why the basic dye safranin is applied; it turns the gram-negative bacteria pink. Stains such as safranin that have a contrasting color to the primary stain are called **counterstains**. Because gram-positive bacteria retain the original purple stain, they are not affected by the safranin counterstain.

As you will see in Chapter 4, different kinds of bacteria react differently to the Gram stain because structural differences in their cell walls affect the retention or escape of a combination of crystal violet and iodine, called the crystal violet–iodine (CV–I) complex. Among other differences, gram-positive bacteria have a thicker peptidoglycan cell wall (disaccharides and amino acids) than gram-negative bacteria. In addition, gram-negative bacteria contain a layer of lipopolysaccharide (lipids and polysaccharides) as part of their cell wall (see Figure 4.13, page 82). When applied to both gram-positive and gram-negative cells, crystal violet and then iodine readily enter the cell walls. Inside the cell walls, the crystal violet and iodine combine to form CV–I. Crystal violet or iodine alone is water soluble and will drain out of the cell. However, CV–I is insoluble in water, so it doesn't easily leave the cell wall. Consequently, gram-positive cells retain the color of the crystal violet dye. In gram-negative cells, however, the alcohol wash disrupts the outer lipopolysaccharide layer, and the CV–I complex is washed out of the thin layer of peptidoglycan. As a result, gram-negative cells are colorless until counterstained with safranin, after which they are pink.

In summary, gram-positive cells retain the dye and remain purple. Gram-negative cells do not retain the dye; they are colorless until counterstained with a red dye.

The Gram method is one of the most important staining techniques in medical microbiology. But Gram staining results are not universally applicable, because some bacterial cells stain poorly or not at all. The Gram reaction is most consistent when it is used on young, growing bacteria. Because of these limitations, Gram staining may overestimate the number of gram-negative bacteria in the human intestine (see Exploring the Microbiome on page 67).

The Gram reaction of a bacterium can provide valuable information for the treatment of disease. Gram-positive bacteria tend to be killed easily by penicillins and cephalosporins. Gram-negative bacteria are generally more resistant because the antibiotics cannot penetrate the lipopolysaccharide layer. Some resistance to these antibiotics among both gram-positive and gram-negative bacteria is due to bacterial inactivation of the antibiotics.

### Acid-Fast Stain

Another important differential stain (one that differentiates bacteria into distinctive groups) is the **acid-fast stain**, which binds strongly only to bacteria that have a waxy material in their cell walls. Microbiologists use this stain to identify all bacteria

The resolving power of the electron microscope is much greater than that of a light microscope. The higher resolution provided unequivocal proof of the presence of spiral bacteria.

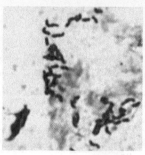

LM | 5 μm

Although bismuth (the key ingredient in Pepto-Bismol) kills *H. pylori*, it is not a cure. Maryanne's physician prescribes the antibiotic clarithromycin. However, one week after finishing treatment, Maryanne's symptoms continue. To see whether *H. pylori* is still present, her physician orders a stomach biopsy to obtain a sample of the mucous lining of Maryanne's stomach. The lab uses light microscopy and a Gram stain to view the sample.

**What does the Gram stain above show?**

| 53 | 60 | **66** | 67 | |

in the genus *Mycobacterium* (mī′kō-bak-TI-rē-um), including the two important pathogens *Mycobacterium tuberculosis*, the causative agent of tuberculosis, and *Mycobacterium leprae* (LEP-rī), the causative agent of leprosy. This stain is also used to identify the pathogenic strains of the genus *Nocardia* (nō-KAR-dē-ah). Bacteria in the genera *Mycobacterium* and *Nocardia* are acid-fast.

In the acid-fast staining procedure, the red dye carbolfuchsin is applied to a fixed smear, and the slide is gently heated for several minutes. (Heating enhances penetration and retention of the dye.) Then the slide is cooled and washed with water. The smear is next treated with acid-alcohol, a decolorizer, which removes the red stain from bacteria that are not acid-fast. The acid-fast microorganisms retain the pink or red color because the carbolfuchsin is more soluble in the cell wall lipids than in the acid-alcohol (**Figure 3.14**). In non–acid-fast bacteria, whose cell walls lack the lipid components, the carbolfuchsin is rapidly removed during decolorization, leaving the cells colorless. The smear is then stained with a methylene blue counterstain. Non–acid-fast cells appear blue after the counterstain is applied.

**CHECK YOUR UNDERSTANDING**

✔ 3-9 Why is the Gram stain so useful?

✔ 3-10 Which stain would be used to identify microbes in the genera *Mycobacterium* and *Nocardia*?

## Special Stains

**Special stains** are used to color parts of microorganisms, such as endospores, flagella, or capsules.

**Obtaining a More Accurate Picture of Our Microbiota**

Studying the relationship between intestinal microbiota, human health, and disease is a challenge. First, it's difficult to come up with growth media outside the body that can support the array of species found in the body. Additionally, many species residing in the human gut are obligate anaerobes, meaning that exposure to the oxygen in air kills them. So even though many bacteria pass out of the body routinely in feces, one cannot simply inoculate nutrient agar with a fecal sample and expect cells to begin growing in the lab. Culture techniques for anaerobic samples do exist, but they are difficult and expensive. So, some researchers have tried tackling this problem by using microscopy in new ways.

In one study, fecal samples were diluted, spread on a slide, heat fixed, Gram stained, and examined with oil immersion light microscopy. Samples were also examined via transmission electron microscopy (TEM). Scientists then categorized the microbes they viewed.

The work showed that the different microscopy techniques show a different variety of organisms in the samples. Light microscopy revealed many gram-negative prokaryotes, which comprised 58 to 68% of the organisms categorized. However, under TEM, gram-positive bacteria predominated, making up 51 to 52% of the organisms seen. The scientists concluded that Gram staining overestimated the gram-negative bacteria because some gram-positive cells, called gram-variable, are easily decolorized.

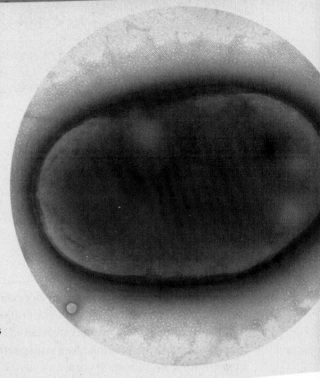

***Microvirga* was discovered in human intestines in 2016. It is nearly twice as long as other bacteria.**

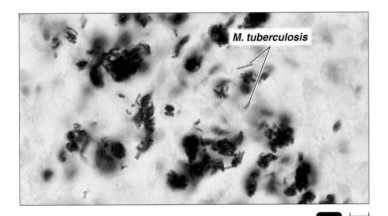

**Figure 3.14 Acid-fast bacteria.** The *Mycobacterium tuberculosis* bacteria in a lung have been stained pink or red with an acid-fast stain.

**Q** Why is *Mycobacterium tuberculosis* easily identified by the acid-fast stain?

### Negative Staining for Capsules

Many microorganisms contain a gelatinous covering called a **capsule**. (We will discuss capsules in our examination of the prokaryotic cell in Chapter 4.) In medical microbiology, demonstrating the presence of a capsule is a means of determining the organism's **virulence**, the degree to which a pathogen can cause disease.

**CLINICAL CASE** Resolved

Because it's gram-negative, *H. pylori* turns pink from the counterstain. Lab results indicate the *H. pylori* is still present in Maryanne's stomach lining. Suspecting the bacteria are resistant to clarithromycin, Maryanne's physician now prescribes tetracycline and metronidazole. This time, Maryanne's symptoms do not return. Soon, she feels like her old self again and is back in the office full time.

| 53 | 60 | 66 | **67** | |

Capsule staining is more difficult than other types of staining procedures because capsular materials are water soluble and may be dislodged or removed during rigorous washing. To demonstrate the presence of capsules, a microbiologist can mix the bacteria in a solution containing a fine colloidal suspension of colored particles (usually india ink or nigrosin) to provide a contrasting background and then stain the bacteria with a simple stain, such as safranin (**Figure 3.15a**). Because of their chemical composition, capsules do not accept most biological dyes, such as safranin, and thus appear as halos surrounding each stained bacterial cell.

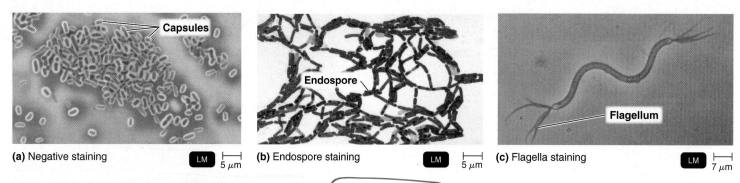

**(a)** Negative staining   LM   5 µm

**(b)** Endospore staining   LM   5 µm

**(c)** Flagella staining   LM   7 µm

**Figure 3.15 Special staining. (a)** Capsule staining provides a contrasting background, so the capsules of these bacteria, *Klebsiella pneumoniae*, show up as light areas surrounding the stained cells. **(b)** Endospores are seen as green ovals in these rod-shaped cells of the bacterium *Bacillus anthracis*, using the Schaeffer-Fulton endospore stain. **(c)** Flagella appear as wavy extensions from the ends of these cells of the bacterium *Spirillum volutans*. In relation to the body of the cell, the flagella are much thicker than normal because layers of the stain have accumulated from treatment of the specimen with a mordant.

← Red

**Q** Of what value are capsules, endospores, and flagella to bacteria?

## Endospore (Spore) Staining

An **endospore** is a special resistant, dormant structure formed within a cell that protects a bacterium from adverse environmental conditions. Although endospores are relatively uncommon in bacterial cells, they can be formed by a few genera of bacteria. Endospores cannot be stained by ordinary methods, such as simple staining and Gram staining, because the dyes don't penetrate the endospore's wall.

The most commonly used endospore stain is the *Schaeffer-Fulton endospore stain* (Figure 3.15b). Malachite green, the primary stain, is applied to a heat-fixed smear and heated to steaming for about 5 minutes. The heat helps the stain penetrate the endospore wall. Then the preparation is washed for about 30 seconds with water to remove the malachite green from all of the cells' parts except the endospores. Next, safranin, a counterstain, is applied to the smear to stain portions of the cell other than

| Stain | Principal Uses |
|---|---|
| **TABLE 3.3** A Summary of Various Stains and Their Uses | |
| **Stain** | **Principal Uses** |
| **Simple** (methylene blue, carbolfuchsin, crystal violet, safranin) | Used to highlight microorganisms to determine cellular shapes and arrangements. Aqueous or alcohol solution of a single basic dye stains cells. (Sometimes a mordant is added to intensify the stain.) |
| **Differential** <br> Gram | Used to distinguish different kinds of bacteria. <br> Classifies bacteria into two large groups: gram-positive and gram-negative. Gram-positive bacteria retain the crystal violet stain and appear purple. Gram-negative bacteria do not retain the crystal violet stain; they remain colorless until counterstained with safranin and then appear pink. |
| Acid-fast | Used to identify *Mycobacterium* species and some species of *Nocardia*. Acid-fast bacteria, once stained with carbolfuchsin and treated with acid-alcohol, remain pink or red because they retain the carbolfuchsin stain. Non–acid-fast bacteria, when stained and treated the same way and then stained with methylene blue, appear blue because they lose the carbolfuchsin stain and are then able to accept the methylene blue stain. |
| **Special** | Used to color and isolate various structures, such as capsules, endospores, and flagella; sometimes used as a diagnostic aid. |
| Negative | Used to demonstrate the presence of capsules. Because capsules do not accept most stains, the capsules appear as unstained halos around bacterial cells and stand out against a contrasting background. |
| Endospore | Used to detect the presence of endospores in bacteria. When malachite green is applied to a heat-fixed smear of bacterial cells, the stain penetrates the endospores and stains them green. When safranin (red) is then applied, it stains the remainder of the cells red or pink. |
| Flagella | Used to demonstrate the presence of flagella. A mordant is used to build up the diameters of flagella until they become visible microscopically when stained with carbolfuchsin. |

endospores. In a properly prepared smear, the endospores appear green within red or pink cells. Because endospores are highly refractive, they can be detected under the light microscope when unstained, but without a special stain they cannot be differentiated from inclusions of stored material.

### Flagella Staining

Bacterial **flagella** (singular: **flagellum**) are structures of locomotion too small to be seen with a light microscope without staining. A tedious and delicate staining procedure uses a mordant and the stain

Play Staining
@MasteringMicrobiology

carbolfuchsin to build up the diameters of the flagella until they become visible under the light microscope (Figure 3.15c). Microbiologists use the number and arrangement of flagella as diagnostic aids.

A summary of stains is presented in Table 3.3. In the next chapters we will take a closer look at the structure of microbes and how they protect, nourish, and reproduce themselves.

**CHECK YOUR UNDERSTANDING**

✓ **3-11** How do unstained endospores appear? Stained endospores?

## Study Outline

Go to @MasteringMicrobiology for **Interactive Microbiology**, *In the Clinic videos, MicroFlix, MicroBoosters, 3D animations, practice quizzes, and more.*

### Units of Measurement (p. 52)

1. Microorganisms are measured in micrometers, μm ($10^{-6}$ m), and in nanometers, nm ($10^{-9}$ m).

### Microscopy: The Instruments (pp. 52–61)

1. A simple microscope consists of one lens; a compound microscope has multiple lenses.

### Light Microscopy (pp. 52–58)

2. The most common microscope used in microbiology is the compound light microscope ( LM ).
3. The total magnification of an object is calculated by multiplying the magnification of the objective lens by the magnification of the ocular lens.
4. The compound light microscope uses visible light.
5. The maximum resolution, or resolving power (the ability to distinguish two points) of a compound light microscope is 0.2 μm; maximum magnification is 1500×.
6. Specimens are stained to increase the difference between the refractive indexes of the specimen and the medium.
7. Immersion oil is used with the oil immersion lens to reduce light loss between the slide and the lens.
8. Brightfield illumination is used for stained smears.
9. Unstained cells are more productively observed using darkfield, phase-contrast, or DIC microscopy.
10. The darkfield microscope shows a light silhouette of an organism against a dark background. It is most useful for detecting the presence of extremely small organisms.
11. A phase-contrast microscope brings direct and reflected or diffracted light rays together (in phase) to form an image of the specimen on the ocular lens. It allows the detailed observation of living organisms.
12. The DIC microscope provides a colored, three-dimensional image of living cells.
13. In fluorescence microscopy, specimens are first stained with fluorochromes and then viewed through a compound microscope

by using an ultraviolet light source. The microorganisms appear as bright objects against a dark background.
14. Fluorescence microscopy is used primarily in a diagnostic procedure called fluorescent-antibody (FA) technique, or immunofluorescence.
15. In confocal microscopy, a specimen is stained with a fluorescent dye and illuminated with short-wavelength light.

### Two-Photon Microscopy (p. 58)

16. In TPM, a live specimen is stained with a fluorescent dye and illuminated with long-wavelength light.

### Super-Resolution Light Microscopy (pp. 58–59)

17. Super-resolution light microscopy uses two lasers to excite fluorescent molecules.
18. When a computer is used to process the images, two-dimensional and three-dimensional images of cells can be produced.

### Scanning Acoustic Microscopy (p. 59)

19. Scanning acoustic microscopy (SAM) is based on the interpretation of sound waves through a specimen.
20. It is used to study living cells attached to surfaces such as biofilms.

### Electron Microscopy (pp. 59–61)

21. Instead of light, a beam of electrons is used with an electron microscope.
22. Instead of glass lenses, electromagnets control focus, illumination, and magnification.
23. Thin sections of organisms can be seen in an electron micrograph produced using a transmission electron microscope ( TEM ). Magnification: 10,000–10,000,000×. Resolving power: 10 pm.
24. Three-dimensional views of the surfaces of whole microorganisms can be obtained with a scanning electron microscope ( SEM ). Magnification: 1000–500,000×. Resolution: 10 nm.

### Scanned-Probe Microscopy (p. 61)

25. Scanning tunneling microscopy (STM) and atomic force microscopy (AFM) produce three-dimensional images of the surface of a molecule.

## Preparation of Specimens for Light Microscopy

(pp. 61–69)

### Preparing Smears for Staining (pp. 61–65)

1. Staining means coloring a microorganism with a dye to make some structures more visible.

2. Fixing uses heat or alcohol to kill and attach microorganisms to a slide.

3. A smear is a thin film of material used for microscopic examination.

4. Bacteria are negatively charged, and the colored positive ion of a basic dye will stain bacterial cells.

5. The colored negative ion of an acidic dye will stain the background of a bacterial smear; a negative stain is produced.

### Simple Stains (p. 65)

6. A simple stain is an aqueous or alcohol solution of a single basic dye.

7. A mordant may be used to improve bonding between the stain and the specimen.

### Differential Stains (pp. 65–66)

8. Differential stains, such as the Gram stain and acid-fast stain, differentiate bacteria according to their reactions to the stains.

9. The Gram stain procedure uses a purple stain, iodine as a mordant, an alcohol decolorizer, and a red counterstain.

10. Gram-positive bacteria remain purple after decolorization; gram-negative bacteria do not, and appear pink from the counterstain.

11. Acid-fast microbes, such as members of the genera *Mycobacterium* and *Nocardia*, retain carbolfuchsin after acid-alcohol decolorization and appear red; non–acid-fast microbes take up the methylene blue counterstain and appear blue.

### Special Stains (pp. 66–69)

12. Negative staining is used to make microbial capsules visible.

13. The endospore stain and flagella stain are special stains that are used to visualize specific structures in bacterial cells.

# Study Questions

For answers to the Knowledge and Comprehension questions, turn to the Answers tab at the back of the textbook.

## Knowledge and Comprehension

### Review

1. Fill in the following blanks.
   a. $1 \ \mu m = $ _____ m
   b. $1$ _____ $= 10^{-9}$ m
   c. $1 \ \mu m = $ _____ nm

2. Which type of microscope would be best to use to observe each of the following?
   a. a stained bacterial smear
   b. unstained bacterial cells: the cells are small, and no detail is needed
   c. unstained live tissue when it is desirable to see some intracellular detail
   d. a sample that emits light when illuminated with ultraviolet light
   e. intracellular detail of a cell that is 1 μm long
   f. unstained live cells in which intracellular structures are shown in color

3. DRAW IT Label the parts of the compound light microscope in the figure below, and then draw the path of light from the illuminator to your eye.

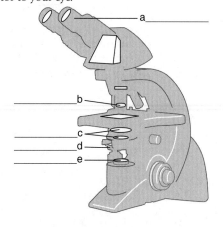

4. Calculate the total magnification of the nucleus of a cell being observed through a compound light microscope with a 10× ocular lens and an oil immersion lens.

5. The maximum magnification of a compound microscope is (a) _____; that of an electron microscope, (b) _____. The maximum resolution of a compound microscope is (c) _____; that of an electron microscope, (d) _____. One advantage of a scanning electron microscope over a transmission electron microscope is (e) _____.

6. Why is a mordant used in the Gram stain? In the flagella stain?

7. What is the purpose of a counterstain in the acid-fast stain?

8. What is the purpose of a decolorizer in the Gram stain? In the acid-fast stain?

9. Fill in the following table regarding the Gram stain:

| | Appearance After This Step | |
| Steps | Gram-Positive Cells | Gram-Negative Cells |
| --- | --- | --- |
| Crystal violet | a. _____ | e. _____ |
| Iodine | b. _____ | f. _____ |
| Alcohol-acetone | c. _____ | g. _____ |
| Safranin | d. _____ | h. _____ |

10. DRAW IT A sputum sample from Calle, a 30-year-old Asian elephant, was smeared onto a slide and air dried. The smear was fixed, covered with carbolfuchsin, and heated for 5 minutes. After washing with water, acid-alcohol was placed on the smear for 30 seconds. Finally, the smear was stained with methylene blue for 30 seconds, washed with water, and dried. On examination at 1000×, the zoo veterinarian saw red rods on the slide. What microbe do these results suggest?

## Multiple Choice

1. Assume you stain *Bacillus* by applying malachite green with heat and then counterstain with safranin. Through the microscope, the green structures are
   a. cell walls.
   b. capsules.
   c. endospores.
   d. flagella.
   e. impossible to identify.

2. Three-dimensional images of live cells can be produced with
   a. darkfield microscopy.
   b. fluorescence microscopy.
   c. transmission electron microscopy.
   d. confocal microscopy.
   e. phase-contrast microscopy.

3. Carbolfuchsin can be used as a simple stain and a negative stain. As a simple stain, the pH is
   a. 2.
   b. higher than the negative stain.
   c. lower than the negative stain.
   d. the same as the negative stain.

4. Looking at the cell of a photosynthetic microorganism, you observe the chloroplasts are green in brightfield microscopy and red in fluorescence microscopy. You conclude:
   a. chlorophyll is fluorescent.
   b. the magnification has distorted the image.
   c. you're not looking at the same structure in both microscopes.
   d. the stain masked the green color.
   e. none of the above

5. Which of the following is *not* a functionally analogous pair of stains?
   a. nigrosin and malachite green
   b. crystal violet and carbolfuchsin
   c. safranin and methylene blue
   d. ethanol-acetone and acid-alcohol
   e. All of the above pairs are functionally analogous.

6. Which of the following pairs is *mismatched*?
   a. capsule—negative stain
   b. cell arrangement—simple stain
   c. cell size—negative stain
   d. Gram stain—bacterial identification
   e. none of the above

7. Assume you stain *Clostridium* by applying a basic stain, carbolfuchsin, with heat, decolorizing with acid-alcohol, and counterstaining with an acidic stain, nigrosin. Through the microscope, the endospores are ___1___, and the cells are stained ___2___.
   a. 1—red; 2—black
   b. 1—black; 2—colorless
   c. 1—colorless; 2—black
   d. 1—red; 2—colorless
   e. 1—black; 2—red

8. Assume that you are viewing a Gram-stained field of red cocci and blue rods through the microscope. You can safely conclude that you have
   a. made a mistake in staining.
   b. two different species.
   c. old bacterial cells.
   d. young bacterial cells.
   e. none of the above

9. In 1996, scientists described a new tapeworm parasite that had killed at least one person. The initial examination of the patient's abdominal mass was most likely made using
   a. brightfield microscopy.
   b. darkfield microscopy.
   c. electron microscopy.
   d. phase-contrast microscopy.
   e. fluorescence microscopy.

10. Which of the following is *not* a modification of a compound light microscope?
    a. brightfield microscopy
    b. darkfield microscopy
    c. electron microscopy
    d. phase-contrast microscopy
    e. fluorescence microscopy

## Analysis

1. In a Gram stain, a step could be omitted and still allow differentiation between gram-positive and gram-negative cells. What is it?

2. Using a good compound light microscope with a resolving power of 0.3 µm, a 10× ocular lens, and a 100× oil immersion lens, would you be able to discern two objects separated by 3 µm? 0.3 µm? 300 nm?

3. Why isn't the Gram stain used on acid-fast bacteria? If you did Gram stain acid-fast bacteria, what would their Gram reaction be? What is the Gram reaction of non–acid-fast bacteria?

4. Endospores can be seen as refractile structures in unstained cells and as colorless areas in Gram-stained cells. Why is it necessary to do an endospore stain to verify the presence of endospores?

## Clinical Applications and Evaluation

1. In 1882, German bacteriologist Paul Erhlich described a method for staining *Mycobacterium* and noted, "It may be that all disinfecting agents which are acidic will be without effect on this [tubercle] bacillus, and one will have to be limited to alkaline agents." How did he reach this conclusion without testing disinfectants?

2. Laboratory diagnosis of *Neisseria gonorrhoeae* infection is based on microscopic examination of Gram-stained pus. Locate the bacteria in this light micrograph. What is the disease?

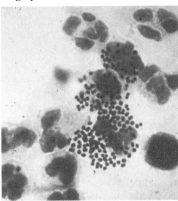

 LM  5 µm

3. Assume that you are viewing a Gram-stained sample of vaginal discharge. Large (10 µm) nucleated red cells are coated with small (0.5 µm wide by 1.5 µm long) blue cells on their surfaces. What is the most likely explanation for the red and blue cells?

# 4 Functional Anatomy of Prokaryotic and Eukaryotic Cells

Despite their complexity and variety, all living cells can be classified into two groups, prokaryotes and eukaryotes, based on certain structural and functional characteristics. In general, prokaryotes are structurally simpler and smaller than eukaryotes. The DNA (genetic material) of prokaryotes is usually a single, circularly arranged chromosome and is not surrounded by a membrane; the DNA of eukaryotes is found in multiple chromosomes in a membrane-enclosed nucleus.

Plants and animals are entirely composed of eukaryotic cells. In the microbial world, bacteria and archaea are prokaryotes. Other cellular microbes—fungi (yeasts and molds), protozoa, and algae—are eukaryotes. Both eukaryotic and prokaryotic cells can have a sticky glycocalyx surrounding them. In nature, most bacteria are found sticking to solid surfaces, including other cells, rather than free-floating. The glycocalyx is the glue that holds the cells in place. *E. coli* bacteria use their fimbriae to attach to the urinary bladder, resulting in infection, as described in the Clinical Case.

▶ *Serratia* bacteria adhere to plastic using strands of glycocalyx (yellow).

## In the Clinic

As a pediatric nurse, you see an 8-year-old patient, Sophia, who has just been diagnosed with a urinary tract infection (UTI). You explain to Sophia's mother that UTIs are quite common in children, especially girls. When you give the mother a prescription for *nitrofurantoin*, she asks why Sophia can't just have penicillin again—that was the drug she received last winter to treat a chest infection. **How would you answer the mother's question?**

*Hint: Read about the cell wall on pages 80–85.*

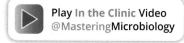

Play **In the Clinic** Video
@Mastering**Microbiology**

# Comparing Prokaryotic and Eukaryotic Cells: An Overview

**LEARNING OBJECTIVE**

4-1 Compare the cell structure of prokaryotes and eukaryotes.

Prokaryotes and eukaryotes both contain nucleic acids, proteins, lipids, and carbohydrates. They use the same kinds of chemical reactions to metabolize food, build proteins, and store energy. It is primarily the structure of cell walls and ribosomes, and the absence of *organelles* (specialized cellular structures that have specific functions), that distinguish prokaryotes from eukaryotes. The chief distinguishing characteristics of **prokaryotes** (from the Greek words meaning prenucleus) are as follows:

1. Typically their DNA is not enclosed within a membrane and is usually a singular, circularly arranged chromosome. *Gemma obscuriglobus* has a double membrane around its nucleus. (Some bacteria, such as *Vibrio cholerae*, have two chromosomes, and some bacteria have a linearly arranged chromosome.)
2. Their DNA is not associated with histones (special chromosomal proteins found in eukaryotes); other proteins are associated with the DNA.
3. They generally lack organelles. Advances in microscopy reveal a few membrane-enclosed organelles (for example, some inclusions). However, prokaryotes lack other membrane-enclosed organelles such as nuclei, mitochondria, and chloroplasts.

4. Their cell walls almost always contain the complex polysaccharide peptidoglycan.
5. They usually divide by **binary fission**, where DNA is copied, and the cell splits into two cells. This involves fewer structures and processes than eukaryotic cell division.

**Eukaryotes** (from the Greek words meaning true nucleus) have the following distinguishing characteristics:

1. Their DNA is found in the cell's nucleus, which is separated from the cytoplasm by a nuclear membrane, and the DNA is found in multiple chromosomes.
2. Their DNA is consistently associated with chromosomal proteins called histones and with nonhistones.
3. They have a number of membrane-enclosed organelles, including mitochondria, endoplasmic reticulum, Golgi complex, lysosomes, and sometimes chloroplasts.
4. Their cell walls, when present, are chemically simple.
5. Cell division usually involves mitosis, in which chromosomes replicate and an identical set is distributed into each of two nuclei. Division of the cytoplasm and other organelles follows so that the two cells produced are identical to each other.

**CHECK YOUR UNDERSTANDING**

✔ 4-1 What is the main feature that distinguishes prokaryotes from eukaryotes?

# The Prokaryotic Cell

Prokaryotes make up a vast group of very small unicellular organisms that include bacteria and archaea. The majority are bacteria. Although bacteria and archaea look similar, their chemical composition is different. The thousands of species of bacteria are differentiated by many factors, including morphology (shape), chemical composition, nutritional requirements, biochemical activities, and sources of energy. It is estimated that 99% of the bacteria in nature exist in biofilms (see pages 157–159).

## The Size, Shape, and Arrangement of Bacterial Cells

**LEARNING OBJECTIVE**

4-2 Identify the three basic shapes of bacteria.

Most bacteria range from 0.2 to 2.0 μm in diameter and from 2 to 8 μm in length. They may be spherical-shaped **coccus** (plural: **cocci**, meaning berries), rod-shaped **bacillus** (plural: **bacilli**, meaning little rods or walking sticks), and **spiral**.

**CLINICAL CASE** Infection Detection

Irene Matthews, an infection control nurse, is in a quandary. Three patients in her hospital have contracted postprocedure bacterial septicemia. All have a fever and dangerously low blood pressure. The patients are in different units and had different procedures. The first patient, Joe, a 32-year-old construction worker, is recovering from rotator cuff surgery. He is in relatively good health, otherwise. The second patient, Jessie, a 16-year-old student in intensive care, is in critical condition following an automobile accident. She is on a ventilator and cannot breathe on her own. The third patient, Maureen, a 57-year-old grandmother, is recovering from coronary artery bypass surgery. As far as Irene can tell, the only thing these patients have in common is the infectious agent—*Klebsiella pneumoniae*.

**How can three patients in different parts of a hospital contract *Klebsiella pneumoniae*? Read on to find out.**

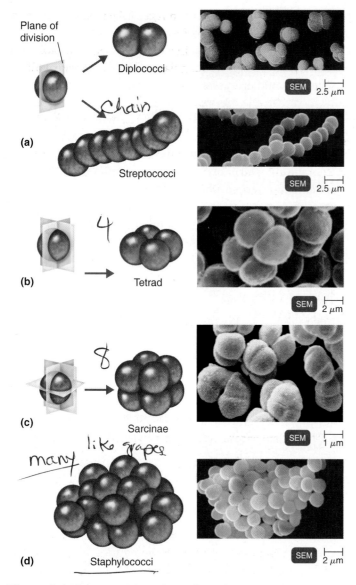

*[handwritten annotations: "chain", "4", "8", "many like grapes"]*

**Figure 4.1 Arrangements of cocci. (a)** Division in one plane produces diplococci and streptococci. **(b)** Division in two planes produces tetrads. **(c)** Division in three planes produces sarcinae, and **(d)** division in multiple planes produces staphylococci.

**Q** How do planes of division determine cell arrangement?

Cocci are usually round but can be oval, elongated, or flattened on one side. When cocci divide to reproduce, the cells can remain attached to one another. Cocci that remain in pairs after dividing are called **diplococci**; those that divide and remain attached in chainlike patterns are called **streptococci** (**Figure 4.1a**). Those that divide in two planes and remain in groups of four are known as **tetrads** (Figure 4.1b). Those that divide in three planes and remain attached in cubelike groups of eight are called **sarcinae** (Figure 4.1c). Those that divide in multiple planes and form grapelike clusters or broad sheets are called **staphylococci** (Figure 4.1d). These group characteristics are frequently helpful in identifying certain cocci.

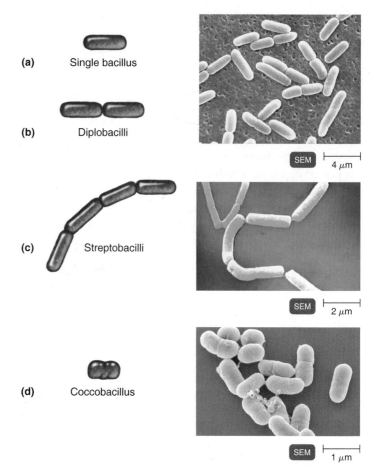

**Figure 4.2 Bacilli. (a)** Single bacilli. **(b)** Diplobacilli. In the top micrograph, a few joined pairs of bacilli could serve as examples of diplobacilli. **(c)** Streptobacilli. **(d)** Coccobacilli.

**Q** Why don't bacilli form tetrads or clusters?

Bacilli divide only across their short axis, so there are fewer groupings of bacilli than of cocci. Most bacilli appear as single rods, called **single bacilli** (**Figure 4.2a**). **Diplobacilli** appear in pairs after division (Figure 4.2b), and **streptobacilli** occur in chains (Figure 4.2c). Some bacilli look like straws. Others have tapered ends, like cigars. Still others are oval and look so much like cocci that they are called **coccobacilli** (Figure 4.2d).

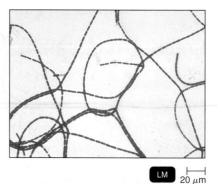

**Figure 4.3 Gram-stained *Bacillus anthracis*.** As *Bacillus* cells age, their walls thin, and their Gram reaction is variable.

**Q** What is the difference between the term bacillus and *Bacillus*?

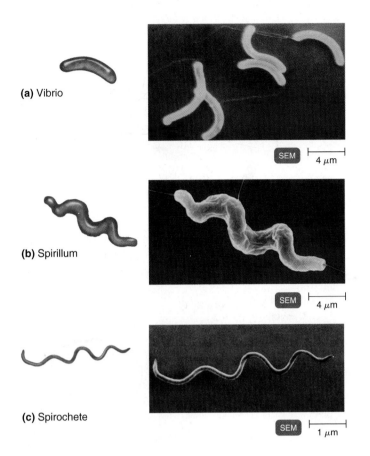

**(a)** Vibrio

SEM | 4 μm

**(b)** Spirillum

SEM | 4 μm

**(c)** Spirochete

SEM | 1 μm

**Figure 4.4 Spiral bacteria.**

**Q** **What is the distinguishing feature of spirochete bacteria?**

"Bacillus" has two meanings in microbiology. As we have just used it, bacillus refers to a bacterial shape. When capitalized and italicized, it refers to a specific genus. For example, the bacterium *Bacillus anthracis* is the causative agent of anthrax. Bacillus cells often form long, twisted chains of cells (**Figure 4.3**).

Spiral bacteria have one or more twists; they are never straight. Bacteria that look like curved rods are called **vibrios** (**Figure 4.4a**). Others, called **spirilla** (singular: **spirillum**), have a helical shape, like a corkscrew, and fairly rigid bodies (Figure 4.4b). Yet another group of spirals are helical and flexible; they are called **spirochetes** (Figure 4.4c). Unlike the spirilla, which

use propeller-like external appendages called flagella to move, spirochetes move by means of axial filaments, which resemble flagella but are contained within a flexible external sheath. There are also star-shaped and rectangular prokaryotes (**Figure 4.5**).

The shape of a bacterium is determined by heredity. Genetically, most bacteria are **monomorphic**; that is, they maintain a single shape. However, a number of environmental conditions can alter that shape. If the shape is altered, identification becomes difficult. Moreover, some bacteria, such as *Rhizobium* (rī-ZŌ-bē-um) and *Corynebacterium* (kor'ī-nē-bak-TI-rē-um), are genetically **pleomorphic**, which means they can have many shapes, not just one.

The structure of a typical prokaryotic cell is shown in **Figure 4.6**. We will discuss structures external to the cell wall, the cell wall itself, and structures internal to the cell wall.

**CHECK YOUR UNDERSTANDING**

✔ **4-2** How can you identify streptococci with a microscope?

# Structures External to the Cell Wall

**LEARNING OBJECTIVES**

**4-3** Describe the structure and function of the glycocalyx.

**4-4** Differentiate flagella, axial filaments, fimbriae, and pili.

Among the possible structures external to the prokaryotic cell wall are the glycocalyx, flagella, axial filaments, fimbriae, and pili.

## Glycocalyx

Many prokaryotes secrete on their surface a substance called glycocalyx. **Glycocalyx** (meaning sugar coat) is the general term used for substances that surround cells. The bacterial glycocalyx is a viscous (sticky), gelatinous polymer that is external to the cell wall and composed of polysaccharide, polypeptide, or both. Its chemical composition varies widely with the species. For the most part, it is made inside the cell and secreted to the cell surface. If the substance is organized and is firmly attached to the

ASM: Bacteria have unique cell structures that are targets for antibiotics, immunity, and phage infection.

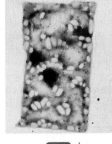

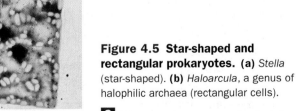

**(a)** Star-shaped bacteria

TEM | 0.5 μm

**(b)** Rectangular bacteria

TEM | 1 μm

**Figure 4.5 Star-shaped and rectangular prokaryotes. (a)** *Stella* (star-shaped). **(b)** *Haloarcula*, a genus of halophilic archaea (rectangular cells).

**Q** **What are the common bacterial shapes?**

# The Structure of a Prokaryotic Cell

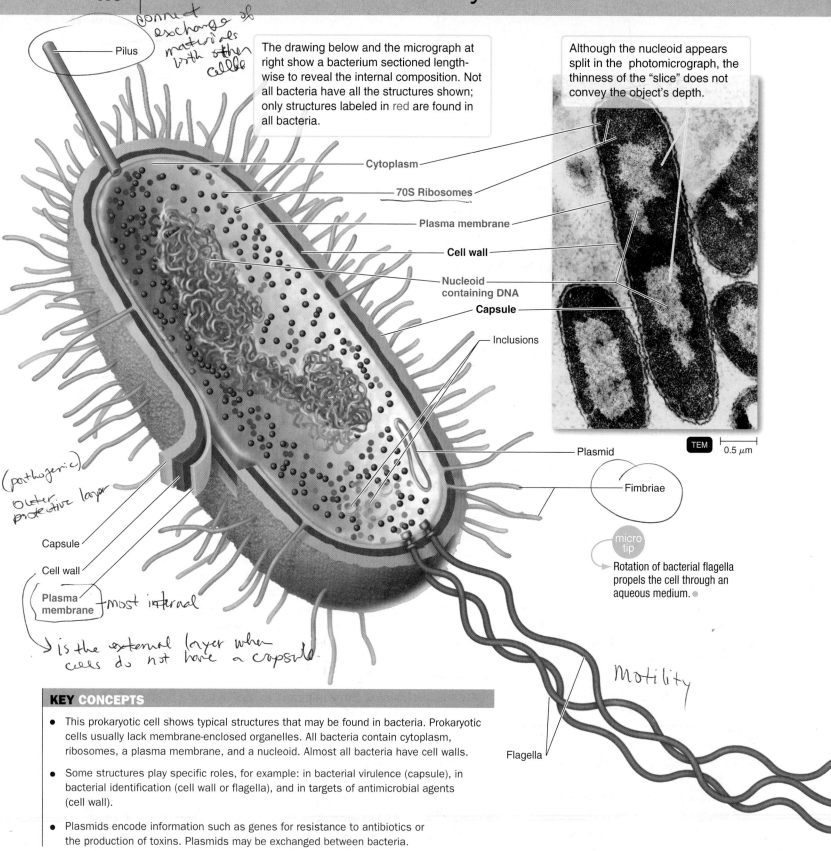

Pilus

*connect exchange of materials with other cells* (handwritten)

The drawing below and the micrograph at right show a bacterium sectioned length-wise to reveal the internal composition. Not all bacteria have all the structures shown; only structures labeled in red are found in all bacteria.

Although the nucleoid appears split in the photomicrograph, the thinness of the "slice" does not convey the object's depth.

Cytoplasm

70S Ribosomes

Plasma membrane

Cell wall

Nucleoid containing DNA

Capsule

Inclusions

Plasmid

TEM    0.5 μm

Fimbriae

*(pathogenic) outer protective layer* (handwritten)

Capsule

Cell wall

Plasma membrane — *most internal* (handwritten)

*is the external layer when cells do not have a capsule.* (handwritten)

micro tip

Rotation of bacterial flagella propels the cell through an aqueous medium. ●

Flagella

*Motility* (handwritten)

## KEY CONCEPTS

- This prokaryotic cell shows typical structures that may be found in bacteria. Prokaryotic cells usually lack membrane-enclosed organelles. All bacteria contain cytoplasm, ribosomes, a plasma membrane, and a nucleoid. Almost all bacteria have cell walls.

- Some structures play specific roles, for example: in bacterial virulence (capsule), in bacterial identification (cell wall or flagella), and in targets of antimicrobial agents (cell wall).

- Plasmids encode information such as genes for resistance to antibiotics or the production of toxins. Plasmids may be exchanged between bacteria.

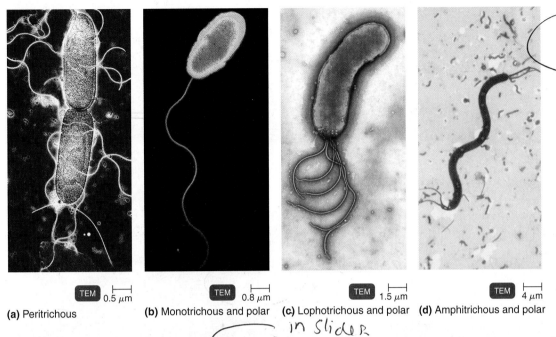

**Figure 4.7 Arrangements of bacterial flagella.**
**(a)** Peritrichous. **(b)–(d)** Polar.

**Q** Not all bacteria have flagella. What are bacteria without flagella called?

TEM 0.5 μm  TEM 0.8 μm  TEM 1.5 μm  TEM 4 μm

**(a)** Peritrichous  **(b)** Monotrichous and polar  **(c)** Lophotrichous and polar  **(d)** Amphitrichous and polar

in slider

cell wall, the glycocalyx is described as a **capsule**. The presence of a capsule can be determined by using negative staining, described in Chapter 3 (see Figure 3.15a, page 68). If the substance is unorganized and only loosely attached to the cell wall, the glycocalyx is described as a **slime layer**.

In certain species, capsules are important in contributing to bacterial virulence (the degree to which a pathogen causes disease). Capsules often protect pathogenic bacteria from phagocytosis by the cells of the host. (As you will see later, phagocytosis is the ingestion and digestion of microorganisms and other solid particles.) For example, *Bacillus anthracis* produces a capsule of D-glutamic acid. (Recall from Chapter 2 that the D forms of amino acids are unusual.) Because only encapsulated *B. anthracis* causes anthrax, it is speculated that the capsule may prevent it from being destroyed by phagocytosis.

Another example involves *Streptococcus pneumoniae* (strep′tō-KOK-kus noo-MŌ-nē-ī), which causes pneumonia only when the cells are protected by a polysaccharide capsule. Unencapsulated *S. pneumoniae* cells cannot cause pneumonia and are readily phagocytized. The polysaccharide capsule of *Klebsiella* also prevents phagocytosis and allows the bacterium to adhere to and colonize the respiratory tract.

The glycocalyx is a very important component of biofilms (see pages 157–159). A glycocalyx that helps cells in a biofilm attach to their target environment and to each other is called an **extracellular polymeric substance (EPS)**. The EPS protects the cells within it, facilitates communication among them, and enables the cells to survive by attaching to various surfaces in their natural environment.

Through attachment, bacteria can grow on diverse surfaces such as rocks in fast-moving streams, plant roots, human teeth, medical implants, water pipes, and even other bacteria. *Streptococcus mutans* (MŪ-tanz), an important cause of dental caries,

attaches itself to the surface of teeth by a glycocalyx. *S. mutans* may use its capsule as a source of nutrition by breaking it down and utilizing the sugars when energy stores are low. *Vibrio cholerae* (VIB-rē-ō KOL-er-ī), the cause of cholera, produces a glycocalyx that helps it attach to the cells of the small intestine. A glycocalyx also can protect a cell against dehydration, and its viscosity may inhibit the movement of nutrients out of the cell.

## Flagella and Archaella

Some bacterial cells have **flagella** (singular: **flagellum**), which are long filamentous appendages that propel bacteria. Bacteria that lack flagella are referred to as **atrichous** (without projections). Flagella may be **peritrichous** (distributed over the entire cell; **Figure 4.7a**) or **polar** (at one or both poles or ends of the cell). If polar, flagella may be **monotrichous** (a single flagellum at one pole; Figure 4.7b), **lophotrichous** (a tuft of flagella coming from one pole; Figure 4.7c), or **amphitrichous** (flagella at both poles of the cell; Figure 4.7d).

A flagellum has three basic parts (**Figure 4.8**). The long outermost region, the *filament*, is constant in diameter and contains the globular (roughly spherical) protein *flagellin* arranged in several chains that intertwine and form a helix around a hollow core. In most bacteria, filaments are not covered by a membrane or sheath, as in eukaryotic cells. The filament is attached to a slightly wider *hook*, consisting of a different protein. The third portion of a flagellum is the *basal body*, which anchors the flagellum to the cell wall and plasma membrane.

The basal body is composed of a small central rod inserted into a series of rings. Gram-negative bacteria contain two pairs of rings; the outer pair of rings is anchored to various portions of the cell wall, and the inner pair of rings is anchored to the plasma membrane. In gram-positive bacteria, only the inner

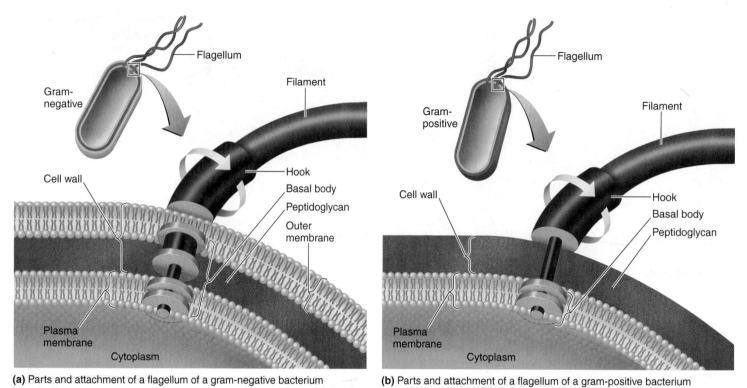

(a) Parts and attachment of a flagellum of a gram-negative bacterium

(b) Parts and attachment of a flagellum of a gram-positive bacterium

**Figure 4.8 The structure of a bacterial flagellum.** The parts and attachment of a flagellum of a gram-negative bacterium and gram-positive bacterium are shown in these highly schematic diagrams.

**Q** How do the basal bodies of gram-negative and gram-positive bacteria differ?

pair is present. As you will see later, the flagella (and cilia) of eukaryotic cells are more complex than those of bacteria.

Each bacterial flagellum is a semirigid, helical structure that moves the cell by rotating from the basal body. The rotation of a flagellum is either clockwise or counterclockwise around its long axis. (Eukaryotic flagella, by contrast, undulate in a wavelike motion.) The movement of a bacterial flagellum results from rotation of its basal body and is similar to the movement of the shaft of an electric motor. As the flagella rotate, they form a bundle that pushes against the surrounding liquid and propels the bacterium. Flagellar rotation depends on the cell's continuous generation of energy.

Bacterial cells can alter the speed and direction of rotation of flagella and thus are capable of various patterns of motility, the ability of an organism to move by itself. When a bacterium moves in one direction for a length of time, the movement is called a "run" or "swim." "Runs" are interrupted by periodic, abrupt, random changes in direction called "tumbles." Then, a "run" resumes. "Tumbles" are caused by a reversal of flagellar rotation (**Figure 4.9a**). Some species of bacteria endowed with many flagella—*Proteus* (PRŌ-tē-us), for example (Figure 4.9b)—can "swarm," or show rapid wavelike movement across a solid culture medium.

One advantage of motility is that it enables a bacterium to move toward a favorable environment or away from an adverse

one. The movement of a bacterium toward or away from a particular stimulus is called **taxis**. Such stimuli include chemicals (**chemotaxis**) and light (**phototaxis**). Motile bacteria contain receptors in various locations, such as in or just under the cell wall. These receptors pick up chemical stimuli, such as oxygen, ribose, and galactose. In response to the stimuli, information is passed to the flagella. If the chemotactic signal is positive, called an *attractant*, the bacteria move toward the stimulus with many runs and few tumbles. If the chemotactic signal is negative, called a *repellent*, the frequency of tumbles increases as the bacteria move away from the stimulus.

The flagellar protein called **H antigen** is useful for distinguishing among **serovars**, or variations within a species, of gram-negative bacteria (see page 304). For example, there are at least 50 different H antigens for *E. coli*. Those serovars identified as *E. coli* O157:H7 are associated with foodborne epidemics.

Play Motility; Flagella: *Structure, Movement, Arrangement* @MasteringMicrobiology

### Archaella

Motile archaeal cells have **archaella** (singular: **archaellum**). Archaella share similarities with bacterial flagella and pili (discussed on page 79). A knoblike structure anchors archaella to the cell. No basal-body type anchor has been found for pili.

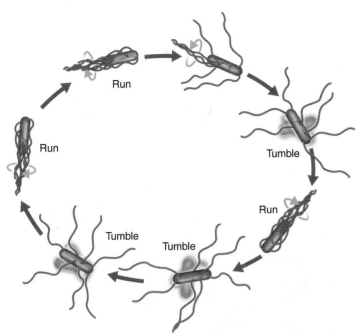

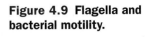

**Figure 4.9 Flagella and bacterial motility.**

**Q** Do bacterial flagella push or pull a cell?

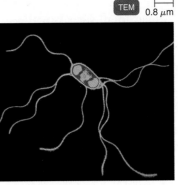

**(b)** A *Proteus* cell in the swarming stage may have more than 1000 peritrichous flagella.

**(a)** A bacterium running and tumbling. Notice that the direction of flagellar rotation (blue arrows) determines which of these movements occurs. Gray arrows indicate direction of movement of the microbe.

Archaella rotate like flagella, an action that pushes the cell through water, and, like pili, archaella use ATP for energy and lack a cytoplasmic core. Archaella consist of glycoproteins called archaellins.

## Axial Filaments

Spirochetes are a group of bacteria that have unique structure and motility. One of the best-known spirochetes is *Treponema pallidum* (trep-ō-NĒ-mah PAL-li-dum), the causative agent of syphilis. Another spirochete is *Borrelia burgdorferi* (bor-RĒ-lē-ah burg-DOR-fer-ē), the causative agent of Lyme disease. Spirochetes move by means of **axial filaments**, or **endoflagella**, bundles of fibrils that arise at the ends of the cell beneath an outer sheath and spiral around the cell (**Figure 4.10**).

Axial filaments, which are anchored at one end of the spirochete, have a structure similar to that of flagella. The rotation of the filaments produces a movement of the outer sheath that propels the spirochetes in a spiral motion. This type of movement is similar to the way a corkscrew moves through a cork. This corkscrew motion probably enables a bacterium such as *T. pallidum* to move effectively through bodily fluids.

Play Spirochetes
@MasteringMicrobiology

## Fimbriae and Pili

Many gram-negative bacteria contain hairlike appendages that are shorter, straighter, and thinner than flagella. These structures, which consist of a protein called *pilin* arranged helically around a central core, are divided into two types, fimbriae and

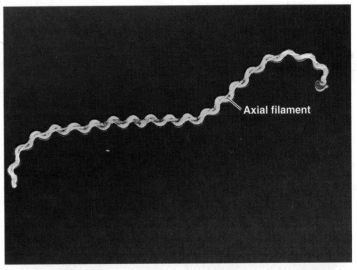

**(a)** A photomicrograph of the spirochete *Leptospira*, showing an axial filament

Axial filament

SEM 5 μm

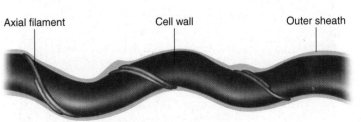

Axial filament          Cell wall          Outer sheath

**(b)** A diagram of axial filaments wrapping around part of a spirochete

**Figure 4.10 Axial filaments.**

**Q** How are endoflagella different from flagella?

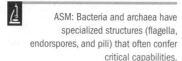

pili, having very different functions. (Some microbiologists use the two terms interchangeably to refer to all such structures, but we distinguish between them.)

**Fimbriae** (singular: **fimbria**) can occur at the poles of the bacterial cell or can be evenly distributed over the entire surface of the cell. They can number anywhere from a few to several hundred per cell (**Figure 4.11**). Fimbriae have a tendency to adhere to each other and to surfaces. As a result, they are involved in forming biofilms and other aggregations on the surfaces of liquids, glass, and rocks. Fimbriae can also help bacteria adhere to epithelial surfaces in the body. For example, fimbriae on the bacterium *Neisseria gonorrhoeae* (nī-SE-rē-ah go-nōr-RĒ-ī), the causative agent of gonorrhea, help the microbe colonize mucous membranes. Once colonization occurs, the bacteria can cause disease. The fimbriae of *E. coli* O157 enable this bacterium to adhere to the lining of the small intestine, where it causes a severe watery diarrhea. When fimbriae are absent (because of genetic mutation), colonization cannot happen, and no disease ensues.

**Pili** (singular: **pilus**) are usually longer than fimbriae and number only one or two per cell. Pili are involved in motility and DNA transfer. In one type of motility, called **twitching motility**, a pilus extends by the addition of subunits of pilin, makes contact with a surface or another cell, and then retracts (powerstroke) as the pilin subunits are disassembled. This is called the *grappling hook model* of twitching motility and results in short, jerky, intermittent movements. Twitching motility has been observed in *Pseudomonas aeruginosa*, *Neisseria gonorrhoeae*, and some strains of *E. coli*. The other type of motility associated with pili is **gliding motility**, the smooth gliding movement of myxobacteria. Although the exact mechanism is unknown

> ASM: Bacteria and archaea have specialized structures (flagella, endospores, and pili) that often confer critical capabilities.

for most myxobacteria, some utilize pilus retraction. Gliding motility provides a means for microbes to travel in environments with a low water content, such as biofilms and soil.

Some pili are used to bring bacteria together, allowing the transfer of DNA from one cell to another, a process called conjugation. Such pili are called **conjugation (sex) pili** (see pages 232–233 or Figure 8.29a, page 233). In this process, the conjugation pilus of one bacterium called an F+ cell connects to receptors on the surface of another bacterium of its own species or a different species. The two cells make physical contact, and DNA from the F+ cell is transferred to the other cell. The exchanged DNA can add a new function to the recipient cell, such as antibiotic resistance or the ability to digest its medium more efficiently.

**CHECK YOUR UNDERSTANDING**

✔ **4-3** Why are bacterial capsules medically important?

✔ **4-4** How do bacteria move?

# The Cell Wall

**LEARNING OBJECTIVES**

**4-5** Compare and contrast the cell walls of gram-positive bacteria, gram-negative bacteria, acid-fast bacteria, archaea, and mycoplasmas.

**4-6** Compare and contrast archaea and mycoplasmas.

**4-7** Differentiate *protoplast*, *spheroplast*, and *L form*.

The **cell wall** of the bacterial cell is a complex, semirigid structure responsible for the shape of the cell. Almost all prokaryotes have a cell wall that surrounds the underlying, fragile plasma (cytoplasmic) membrane and protects it and the interior of the cell from adverse changes in the outside environment (see Figure 4.6).

The major function of the cell wall is to prevent bacterial cells from rupturing when the water pressure inside the cell is greater than that outside the cell (see Figure 4.18d, page 89). It also helps maintain the shape of a bacterium and serves as a point of anchorage for flagella. As the volume of a bacterial cell increases, its plasma membrane and cell wall extend as needed. Clinically, the cell wall is important because it contributes to the ability of some species to cause disease and is the site of action of some antibiotics. In addition, the chemical composition of the cell wall is used to differentiate major types of bacteria.

Although the cells of some eukaryotes, including plants, algae, and fungi, have cell walls, their walls differ chemically from those of prokaryotes, are simpler in structure, and are less rigid.

## Composition and Characteristics

The bacterial cell wall is composed of a macromolecular network called **peptidoglycan** (also known as *murein*), which is present either alone or in combination with other substances.

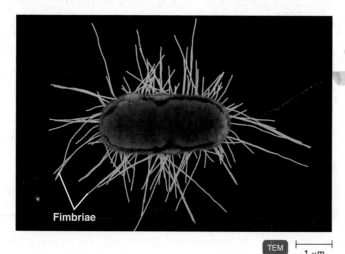

Fimbriae

TEM   ⊢——⊣ 1 μm

**Figure 4.11 Fimbriae.** The fimbriae seem to bristle from this *E. coli* cell, which is beginning to divide.

**Q** Why are fimbriae necessary for colonization?

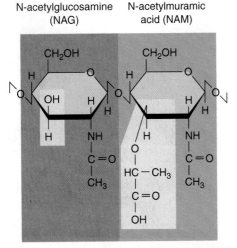

N-acetylglucosamine
(NAG)          N-acetylmuramic
acid (NAM)

**Figure 4.12 N-acetylglucosamine (NAG) and N-acetylmuramic acid (NAM) joined as in a peptidoglycan.** The gold areas show the differences between the two molecules. The linkage between them is called a β-1,4 linkage.

**Q** What kind of molecules are these: carbohydrates, lipids, or proteins?

*on Slides.*

Peptidoglycan consists of a repeating disaccharide connected by polypeptides to form a lattice that surrounds and protects the entire cell. The disaccharide portion is made up of monosaccharides called N-acetylglucosamine (NAG) and N-acetylmuramic acid (NAM) (from *murus*, meaning wall), which are related to glucose. The structural formulas for NAG and NAM are shown in **Figure 4.12**.

The various components of peptidoglycan are assembled in the cell wall (**Figure 4.13a**). Alternating NAM and NAG molecules are linked in rows of 10 to 65 sugars to form a carbohydrate "backbone" (the glycan portion of peptidoglycan). Adjacent rows are linked by **polypeptides** (the peptide portion of peptidoglycan). Although the structure of the polypeptide link varies, it always includes *tetrapeptide side chains,* which consist of four amino acids attached to NAMs in the backbone. The amino acids occur in an alternating pattern of D and L forms (see Figure 2.13, page 40). This is unique because the amino acids found in other proteins are L forms. Parallel tetrapeptide side chains may be directly bonded to each other or linked by a *peptide cross-bridge,* consisting of a short chain of amino acids.

Penicillin interferes with the final linking of the peptidoglycan rows by peptide cross-bridges (see Figure 4.13a). As a result, the cell wall is greatly weakened and the cell undergoes lysis, destruction caused by rupture of the plasma membrane and the loss of cytoplasm.

## Gram-Positive Cell Walls

In most gram-positive bacteria, the cell wall consists of many layers of peptidoglycan, forming a thick, rigid structure (Figure 4.13b). By contrast, gram-negative cell walls contain only a thin layer of peptidoglycan (Figure 4.13c). The space between the cell wall and plasma membrane of gram-positive bacteria is the periplasmic space. It contains the granular layer, which is composed of lipoteichoic acid. In addition, the cell walls of gram-positive bacteria contain *teichoic acids,* which consist primarily of an alcohol (such as glycerol or ribitol) and phosphate. There are two classes of teichoic acids: *lipoteichoic acid,* which spans the peptidoglycan layer and is linked to the plasma membrane, and *wall teichoic acid,* which is linked to the peptidoglycan layer. Because of their negative charge (from the phosphate groups), teichoic acids may bind and regulate the movement of cations (positive ions) into and out of the cell. They may also assume a role in cell growth, preventing extensive wall breakdown and possible cell lysis. Finally, teichoic acids provide much of the wall's antigenic specificity and thus make it possible to identify gram-positive bacteria by certain laboratory tests (see Chapter 10). Similarly, the cell walls of gram-positive streptococci are covered with various polysaccharides that allow them to be grouped into medically significant types.

## Gram-Negative Cell Walls

The cell walls of gram-negative bacteria consist of one or a very few layers of peptidoglycan and an outer membrane (see Figure 4.13c). The peptidoglycan is bonded to lipoproteins in the outer membrane and is in the *periplasm,* a gel-like fluid in the periplasmic space of gram negative bacteria, the region between the outer membrane and the plasma membrane. The periplasm contains a high concentration of degradative enzymes and transport proteins. Gram-negative cell walls do not contain teichoic acids. Because the cell walls of gram-negative bacteria contain only a small amount of peptidoglycan, they are more susceptible to mechanical breakage.

The *outer membrane* of the gram-negative cell consists of lipopolysaccharides (LPS), lipoproteins, and phospholipids (see Figure 4.13c). The outer membrane has several specialized functions. Its strong negative charge is an important factor in evading phagocytosis and the actions of complement (lyses cells and promotes phagocytosis), two components of the defenses of the host (discussed in detail in Chapter 16). The outer membrane also provides a barrier to detergents, heavy metals, bile salts, certain dyes, antibiotics (for example, penicillin), and digestive enzymes such as lysozyme.

However, the outer membrane doesn't block all environmental substances because nutrients must enter to sustain the metabolism of the cell. Part of the permeability of the outer membrane is due to proteins in the membrane, called **porins**, that form channels. Porins permit the passage of molecules such as nucleotides, disaccharides, peptides, amino acids, vitamin $B_{12}$, and iron.

The **lipopolysaccharide (LPS)** of the outer membrane is a large, complex molecule that contains lipids and carbohydrates and consists of three components: (1) lipid A, (2) a core polysaccharide, and (3) an O polysaccharide. **Lipid A** is the lipid portion of the LPS and is embedded in the top layer of the outer membrane. When gram-negative bacteria die, they release lipid A,

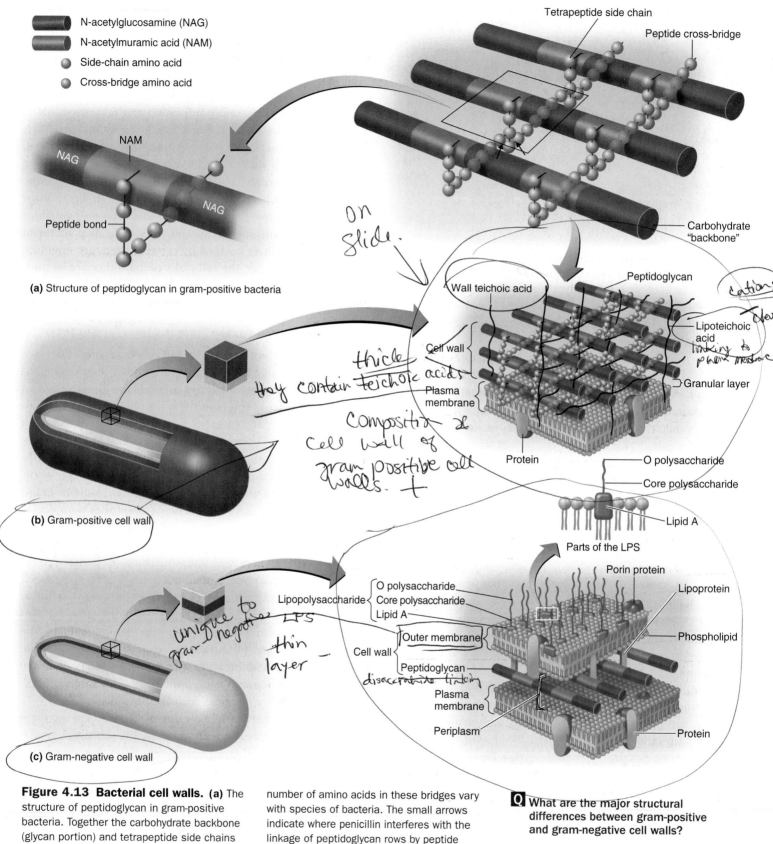

**Figure 4.13 Bacterial cell walls. (a)** The structure of peptidoglycan in gram-positive bacteria. Together the carbohydrate backbone (glycan portion) and tetrapeptide side chains (peptide portion) make up peptidoglycan. The frequency of peptide cross-bridges and the number of amino acids in these bridges vary with species of bacteria. The small arrows indicate where penicillin interferes with the linkage of peptidoglycan rows by peptide cross-bridges. **(b)** A gram-positive cell wall. **(c)** A gram-negative cell wall.

**Q** What are the major structural differences between gram-positive and gram-negative cell walls?

Irene reviews what she knows about gram-negative *K. pneumoniae* bacteria. Although this bacterium is part of normal intestinal microbiota, outside its typical environment it can cause serious infection. *K. pneumoniae* accounts for about 8% of all healthcare–associated infections. Irene surmises that the bacteria had to come from the hospital somewhere.

**What is causing the patients' fever and low blood pressure?**

73    83    85    91    94

which functions as an endotoxin (Chapter 15). Lipid A is responsible for the symptoms associated with infections by gram-negative bacteria, such as fever, dilation of blood vessels, shock, and blood clotting. The core polysaccharide is attached to lipid A and contains unusual sugars. Its role is structural—to provide stability. The O polysaccharide extends outward from the core polysaccharide and is composed of sugar molecules. The O polysaccharide functions as an antigen and is useful for distinguishing serovars of gram-negative bacteria. For example, the foodborne pathogen *E. coli* O157:H7 is distinguished from other serovars by certain laboratory tests that test for these specific antigens. This role is comparable to that of teichoic acids in gram-positive cells.

## Cell Walls and the Gram Stain Mechanism

Now that you've studied the Gram stain (in Chapter 3, page 65) and the chemistry of the bacterial cell wall (in the previous section), it's easier to understand how the Gram stain works. The mechanism of the Gram stain is based on differences in the cell wall structure of gram-positive and gram-negative bacteria and how each reacts to various reagents (substances used for producing a chemical reaction). Crystal violet, the primary stain, stains both gram-positive and gram-negative cells purple because the dye combines with the peptidoglycan. When iodine (the mordant) is applied, it forms large crystals with the dye that are not soluble in water. The application of alcohol dissolves the outer membrane of gram-negative cells, and the crystal-violet iodine is washed out of the thin peptidoglycan layer. Because gram-negative bacteria are colorless after the alcohol wash, the addition of safranin (the counterstain) turns the cells pink or red. Safranin provides a contrasting color to the primary stain (crystal violet). Although gram-positive and gram-negative cells both absorb safranin, the pink or red color of safranin is masked by the darker purple dye previously absorbed by gram-positive cells.

In any population of cells, some gram-positive cells will give a gram-negative response. These cells are usually dead. However, there are a few gram-positive genera that show an increasing number of gram-negative cells as the culture ages.

*Bacillus* and *Clostridium* are examples and are often described as *gram-variable* (see Figure 4.3).

Table 4.1 compares some of the characteristics of gram-positive and gram-negative bacteria.

## Atypical Cell Walls

Among prokaryotes, certain types of cells have no walls or have very little wall material. These include members of the genus *Mycoplasma* (mī-kō-PLAZ-mah) and related organisms (see Figure 11.24, page 316). Mycoplasmas are the smallest known bacteria that can grow and reproduce outside living host cells. Because of their size and because they have no cell walls, they pass through most bacterial filters and were first mistaken for viruses. Their plasma membranes are unique among bacteria in having lipids called *sterols*, which are thought to help protect them from lysis (rupture).

Archaea may lack walls or may have unusual walls composed of polysaccharides and proteins but not peptidoglycan. These walls do, however, contain a substance similar to peptidoglycan called *pseudomurein*. Pseudomurein contains N-acetyltalosaminuronic acid instead of NAM and lacks the D-amino acids found in bacterial cell walls. Archaea generally cannot be Gram-stained but appear gram-negative because they do not contain peptidoglycan.

## Acid-Fast Cell Walls

Recall from Chapter 3 that the acid-fast stain is used to identify all bacteria of the genus *Mycobacterium* and pathogenic species of *Nocardia*. These bacteria contain high concentrations (60%) of a hydrophobic waxy lipid (mycolic acid) in their cell wall that prevents the uptake of dyes, including those used in the Gram stain. The mycolic acid forms a layer outside of a thin layer of peptidoglycan. The mycolic acid and peptidoglycan are held together by a polysaccharide. The hydrophobic waxy cell wall causes broth cultures of *Mycobacterium* to clump and stick to the walls of the flask. Acid-fast bacteria can be stained with carbolfuchsin, which penetrates bacteria more effectively when heated. The carbolfuchsin penetrates the cell wall, binds to the cytoplasm, and resists removal by washing with acid-alcohol. Acid-fast bacteria retain the red color of carbolfuchsin because it's more soluble in the cell wall's mycolic acid than in the acid-alcohol. If the mycolic acid layer is removed from the cell wall of acid-fast bacteria, they will stain gram-positive with the Gram stain.

## Damage to the Cell Wall

Chemicals that damage bacterial cell walls, or interfere with their synthesis, often do not harm the cells of an animal host because the bacterial cell wall is made of chemicals unlike those in eukaryotic cells. Thus, cell wall synthesis is the target for some antimicrobial drugs. One way the cell wall can be damaged is by exposure to the digestive enzyme *lysozyme*. This enzyme occurs naturally in some eukaryotic cells and is a constituent of

**TABLE 4.1** Some Comparative Characteristics of Gram-Positive and Gram-Negative Bacteria

| Characteristic | Gram-Positive | Gram-Negative |
|---|---|---|
| | LM ⊢ 12 μm | LM ⊢ 15 μm |
| Gram Reaction | Retain crystal violet dye and stain blue or purple | Can be decolorized to accept counterstain (safranin) and stain pink or red |
| Peptidoglycan Layer | Thick (multilayered) | Thin (single-layered) |
| Teichoic Acids | Present in many | Absent |
| Periplasmic Space | Granular layer | Periplasm |
| Outer Membrane | Absent | Present |
| Lipopolysaccharide (LPS) Content | Virtually none | High |
| Lipid and Lipoprotein Content | Low (acid-fast bacteria have lipids linked to peptidoglycan) | High (because of presence of outer membrane) |
| Flagellar Structure | 2 rings in basal body | 4 rings in basal body |
| Toxins Produced | Exotoxins | Endotoxin and exotoxins |
| Resistance to Physical Disruption | High | Low |
| Cell Wall Disruption by Lysozyme | High | Low (requires pretreatment to destabilize outer membrane) |
| Susceptibility to Penicillin and Sulfonamide | High | Low |
| Susceptibility to Streptomycin, Chloramphenicol, and Tetracycline | Low | High |
| Inhibition by Basic Dyes | High | Low |
| Susceptibility to Anionic Detergents | High | Low |
| Resistance to Sodium Azide | High | Low |
| Resistance to Drying | High | Low |

perspiration, tears, mucus, and saliva. Lysozyme is particularly active on the major cell wall components of most gram-positive bacteria, making them vulnerable to lysis. Lysozyme catalyzes hydrolysis of the bonds between the sugars in the repeating disaccharide "backbone" of peptidoglycan. This act is analogous to cutting the steel supports of a bridge with a cutting torch: the gram-positive cell wall is almost completely destroyed by lysozyme. The cellular contents that remain surrounded by the plasma membrane may remain intact if lysis does not occur; this wall-less cell is termed a **protoplast**. Typically, a protoplast is spherical and is still capable of carrying on metabolism.

Some members of the genus *Proteus*, as well as other genera, can lose their cell walls and swell into irregularly shaped cells called **L forms**, named for the Lister Institute, where they were discovered. They may form spontaneously or develop in response to penicillin (which inhibits cell wall formation) or lysozyme (which removes the cell wall). L forms can live and divide repeatedly or return to the walled state.

When lysozyme is applied to gram-negative cells, usually the wall is not destroyed to the same extent as in gram-positive cells; some of the outer membrane also remains. In this case, the cellular contents, plasma membrane, and remaining outer wall layer are called a **spheroplast**, also a spherical structure. For lysozyme to exert its effect on gram-negative cells, the cells are first treated with EDTA (ethylenediaminetetraacetic acid). EDTA weakens ionic bonds in the outer membrane and thereby damages it, giving the lysozyme access to the peptidoglycan layer.

Protoplasts and spheroplasts burst in pure water or very dilute salt or sugar solutions because the water molecules from the surrounding fluid rapidly move into and enlarge the cell, which has a much lower internal concentration of water. This rupturing, called **osmotic lysis**, will be discussed in detail shortly.

cell and discuss the structures and functions of the plasma membrane and components within the cytoplasm of the cell.

## The Plasma (Cytoplasmic) Membrane

The **plasma (cytoplasmic) membrane** (or *inner membrane*) is a thin structure lying inside the cell wall and enclosing the cytoplasm of the cell (see Figure 4.6). The plasma membrane of prokaryotes consists primarily of phospholipids (see Figure 2.10, page 39), which are the most abundant chemicals in the membrane, and proteins. Eukaryotic plasma membranes also contain carbohydrates and sterols, such as cholesterol. Because they lack sterols, prokaryotic plasma membranes are less rigid than eukaryotic membranes. One exception is the wall-less prokaryote *Mycoplasma*, which contains membrane sterols.

### Structure

In electron micrographs, prokaryotic and eukaryotic plasma membranes (and the outer membranes of gram-negative bacteria) look like two-layered structures; there are two dark lines with a light space between the lines (**Figure 4.14a**). The phospholipid molecules are arranged in two parallel rows, called a *lipid bilayer* (Figure 4.14b). Each phospholipid molecule (see Chapter 2) contains a polar head, composed of a phosphate group and glycerol that is hydrophilic (water-loving) and soluble in water, and nonpolar tails, composed of fatty acids that are hydrophobic (water-fearing) and insoluble in water (Figure 4.14c). The polar heads are on the two surfaces of the lipid bilayer, and the nonpolar tails are in the interior of the bilayer.

The protein molecules in the membrane can be arranged in a variety of ways. Some, called *peripheral proteins*, are easily removed from the membrane by mild treatments and lie at the inner or outer surface of the membrane. They may function as enzymes that catalyze chemical reactions, as a "scaffold" for support, and as mediators of changes in membrane shape during movement. Other proteins, called *integral proteins*, can be removed from the membrane only after disrupting the lipid bilayer (by using detergents, for example). Most integral proteins penetrate the membrane completely and are called *transmembrane proteins*. Some integral proteins are channels that have a pore, or hole, through which substances enter and exit the cell.

Many of the proteins and some of the lipids on the outer surface of the plasma membrane have carbohydrates attached to them. Proteins attached to carbohydrates are called **glycoproteins**, and lipids attached to carbohydrates are called **glycolipids**. Both glycoproteins and glycolipids help protect and lubricate the cell and are involved in cell-to-cell interactions. For example, glycoproteins play a role in certain infectious diseases. The influenza virus and the toxins that cause cholera and botulism enter their target cells by first binding to glycoproteins on their plasma membranes.

---

The outer membrane of *K. pneumoniae*'s gram-negative cell wall contains the endotoxin, lipid A, which causes fever and capillary dilation.

Irene works with Joe's, Jessie's, and Maureen's physicians to combat this potentially deadly infection. Irene is particularly concerned about Jessie because of her already weakened respiratory condition. All three patients are treated with a β-lactam antibiotic, imipenem. *Klebsiella* bacteria are resistant to many antibiotics, but imipenem seems to be working for Joe and Maureen. Jessie, however, is getting worse.

**Why are Jessie's symptoms worsening if the bacteria are being killed?**

| 73 | 83 | **85** | 91 | 94 |

---

As noted earlier, certain antibiotics, such as penicillin, destroy bacteria by interfering with the formation of the peptide cross-bridges of peptidoglycan, thus preventing the formation of a functional cell wall. Most gram-negative bacteria are not as susceptible to penicillin as gram-positive bacteria are because the outer membrane of gram-negative bacteria forms a barrier that inhibits the entry of this and other substances, and gram-negative bacteria have fewer peptide cross-bridges. However, gram-negative bacteria are quite susceptible to some β-lactam antibiotics that penetrate the outer membrane better than penicillin. Antibiotics will be discussed in more detail in Chapter 20.

**CHECK YOUR UNDERSTANDING**

✓ 4-5 Why are drugs that target cell wall synthesis useful?

✓ 4-6 Why are mycoplasmas resistant to antibiotics that interfere with cell wall synthesis?

✓ 4-7 How do protoplasts differ from L forms?

## Structures Internal to the Cell Wall

### LEARNING OBJECTIVES

4-8 Describe the structure, chemistry, and functions of the prokaryotic plasma membrane.

4-9 Define *simple diffusion, facilitated diffusion, osmosis, active transport*, and *group translocation*.

4-10 Identify the functions of the nucleoid and ribosomes.

4-11 Identify the functions of four inclusions.

4-12 Describe the functions of endospores, sporulation, and endospore germination.

Thus far, we have discussed the prokaryotic cell wall and structures external to it. We will now look inside the prokaryotic

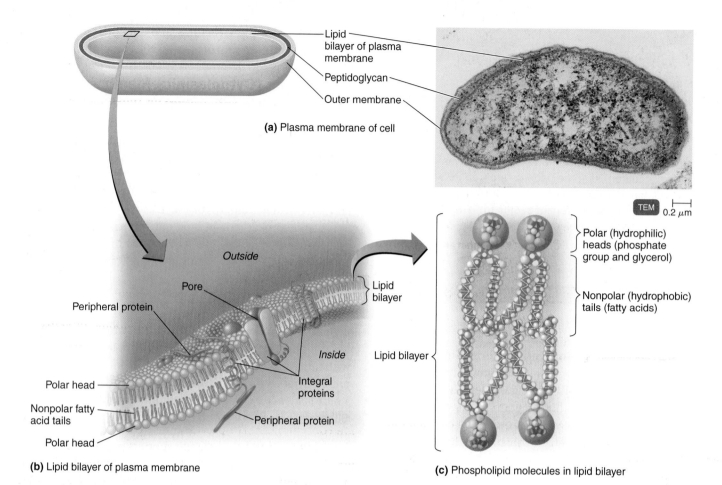

**(a)** Plasma membrane of cell

Lipid bilayer of plasma membrane

Peptidoglycan

Outer membrane

TEM 0.2 μm

Outside

Pore

Peripheral protein

Lipid bilayer

Inside

Lipid bilayer

Polar head

Integral proteins

Nonpolar fatty acid tails

Peripheral protein

Polar head

Polar (hydrophilic) heads (phosphate group and glycerol)

Nonpolar (hydrophobic) tails (fatty acids)

**(b)** Lipid bilayer of plasma membrane

**(c)** Phospholipid molecules in lipid bilayer

**Figure 4.14 Plasma membrane.**
**(a)** A diagram and micrograph showing the lipid bilayer forming the inner plasma membrane of the gram-negative bacterium *Vibrio cholerae.* Layers of the cell wall, including the outer membrane, can be seen outside the inner membrane. **(b)** A portion of the inner membrane showing the lipid bilayer and proteins. The outer membrane of gram-negative bacteria is also a lipid bilayer. **(c)** Space-filling models of several phospholipid molecules as they are arranged in the lipid bilayer.

**Q** What is the difference between a peripheral and an integral protein?

Studies have demonstrated that the phospholipid and protein molecules in membranes are not static but move quite freely within the membrane surface. This movement is most probably associated with the many functions performed by the plasma membrane. Because the fatty acid tails cling together, phospholipids in the presence of water form a self-sealing bilayer; as a result, breaks and tears in the membrane heal themselves. The membrane must be about as viscous as olive oil, which allows membrane proteins to move freely enough to perform their functions without destroying the structure of the membrane. This dynamic arrangement of phospholipids and proteins is referred to as the **fluid mosaic model**.

### Functions

The most important function of the plasma membrane is to serve as a selective barrier through which materials enter and exit the cell. In this function, plasma membranes have **selective permeability** (sometimes called *semipermeability*). This term indicates that certain molecules and ions are allowed to pass through the membrane but others are stopped. The permeability of the membrane depends on several factors. Large molecules (such as proteins) cannot pass through the plasma membrane, possibly because these molecules are larger than the pores in integral proteins that function as channels. But smaller molecules (such as water, oxygen, carbon dioxide, and some simple sugars) usually pass through easily. Ions penetrate the membrane very slowly. Substances that dissolve easily in lipids (such as oxygen, carbon dioxide, and nonpolar organic molecules) enter and exit more easily than other substances because the membrane consists mostly of phospholipids. The movement of materials across plasma membranes also depends on transporter molecules, which will be described shortly.

Plasma membranes are also important to the breakdown of nutrients and the production of energy. The plasma membranes of bacteria contain enzymes capable of catalyzing the chemical reactions that break down nutrients and produce ATP. In some bacteria, pigments and enzymes involved in photosynthesis are found in infoldings of the plasma membrane that extend into the cytoplasm. These membranous structures are called **chromatophores** (**Figure 4.15**).

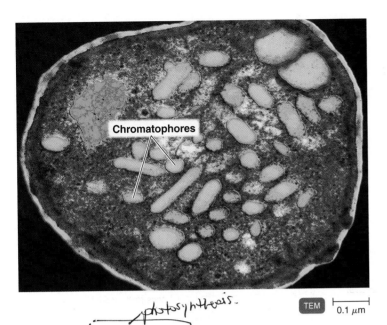

Chromatophores

*photosynthesis.*

TEM  0.1 μm

**Figure 4.15 Chromatophores.** In this micrograph of a purple sulfur bacterium, the chromatophores are clearly visible.

**Q** What is the function of chromatophores?

When viewed with an electron microscope, bacterial plasma membranes often appear to contain one or more large, irregular folds called **mesosomes**. Many functions have been proposed for mesosomes. However, it is now known that they are artifacts, not true cell structures. Mesosomes are believed to be folds in the plasma membrane that develop by the process used for preparing specimens for electron microscopy.

> ▶ **Play Membrane Structure; Membrane Permeability**
> @Mastering**Microbiology**

### Destruction of the Plasma Membrane by Antimicrobial Agents

Because the plasma membrane is vital to the bacterial cell, it is not surprising that several antimicrobial agents exert their effects at this site. In addition to the chemicals that damage the cell wall and thereby indirectly expose the membrane to injury, many compounds specifically damage plasma membranes. These compounds include certain alcohols and quaternary ammonium compounds, which are used as disinfectants. By disrupting the membrane's phospholipids, a group of antibiotics known as the *polymyxins* cause leakage of intracellular contents and subsequent cell death. This mechanism will be discussed in Chapter 20.

## The Movement of Materials across Membranes

Materials move across plasma membranes of both prokaryotic and eukaryotic cells by two kinds of processes: passive and active. In *passive processes*, substances cross the membrane from an area

of high concentration to an area of low concentration (move with the concentration gradient, or difference), without any expenditure of energy by the cell. In *active processes*, the cell must use energy to move substances from areas of low concentration to areas of high concentration (against the concentration gradient).

### Passive Processes

Passive processes include simple diffusion, facilitated diffusion, and osmosis.

**Simple diffusion** is the net (overall) movement of molecules or ions from an area of high concentration to an area of low concentration (**Figure 4.16** and **Figure 4.17a**). The movement continues until the molecules or ions are evenly distributed. The point of even distribution is called *equilibrium*. Cells rely on simple diffusion to transport certain small molecules, such as oxygen and carbon dioxide, across their cell membranes.

In **facilitated diffusion**, integral membrane proteins function as channels or carriers that facilitate the movement of ions or large molecules across the plasma membrane. Such integral proteins are called *transporter proteins* or *permeases*. Facilitated diffusion is similar to simple diffusion in that the cell *does not* expend energy, because the substance moves from a high to a low concentration. The process differs from simple diffusion in its use of transporters. Some transporters permit the passage of mostly small, inorganic ions that are too hydrophilic to penetrate the nonpolar interior of the lipid bilayer (Figure 4.17b). These transporters, which are common in prokaryotes, are nonspecific and allow a wide variety of ions or small molecules to pass through channels in integrated membrane proteins. Other

(a)                          (b)

**Figure 4.16 The principle of simple diffusion. (a)** After a dye pellet is put into a beaker of water, the molecules of dye in the pellet diffuse into the water from an area of high dye concentration to areas of low dye concentration. **(b)** The dye potassium permanganate in the process of diffusing.

**Q** Why are passive processes important to a cell?

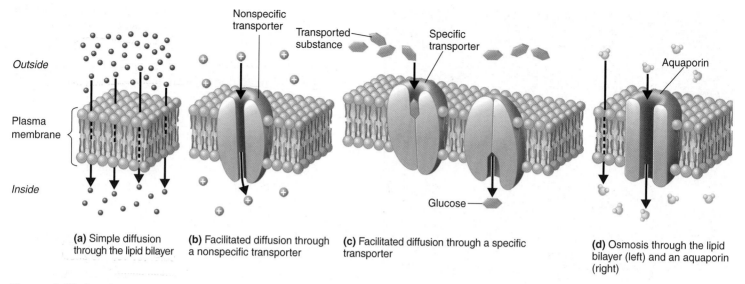

**(a)** Simple diffusion through the lipid bilayer

**(b)** Facilitated diffusion through a nonspecific transporter

**(c)** Facilitated diffusion through a specific transporter

**(d)** Osmosis through the lipid bilayer (left) and an aquaporin (right)

**Figure 4.17 Passive processes.**

**Q** How does simple diffusion differ from facilitated diffusion?

transporters, which are common in eukaryotes, are specific and transport only specific, usually larger, molecules, such as simple sugars (glucose, fructose, and galactose) and vitamins. In this process, the transported substance binds to a specific transporter (integrated membrane protein) on the outer surface of the plasma membrane, which undergoes a change of shape; then the transporter releases the substance on the other side of the membrane (Figure 4.17c).

In some cases, molecules that bacteria need are too large to be transported into the cells by these methods. Most bacteria, however, produce enzymes that can break down large molecules into simpler ones (such as proteins into amino acids, or polysaccharides into simple sugars). Such enzymes, which are released by the bacteria into the surrounding medium, are appropriately called *extracellular enzymes*. Once the enzymes degrade the large molecules, the subunits move into the cell with the help of transporters. For example, specific carriers retrieve DNA bases, such as the purine guanine, from extracellular media (substances outside the cell) and bring them into the cell's cytoplasm.

**Osmosis** is the net movement of water molecules across a selectively permeable membrane from an area with a high concentration of water molecules (low concentration of solute molecules) to an area of low concentration of water molecules (high concentration of solute molecules). Water molecules may pass through plasma membranes by moving through the lipid bilayer by simple diffusion or through integral membrane proteins, called *aquaporins*, that function as water channels (Figure 4.17d).

Osmosis may be demonstrated with the apparatus shown in **Figure 4.18a**. A sack constructed from cellophane, which is a selectively permeable membrane, is filled with a solution of 20% sucrose (table sugar). The cellophane sack is placed into a

beaker containing distilled water. Initially, the concentrations of water on either side of the membrane are different. Because of the sucrose molecules, the concentration of water is lower inside the cellophane sack. Therefore, water moves from the beaker (where its concentration is higher) into the cellophane sack (where its concentration is lower).

Sugar does not move out of the cellophane sack into the beaker, however, because the cellophane is impermeable to molecules of sugar—the sugar molecules are too large to go through the pores of the membrane. As water moves into the cellophane sack, the sugar solution becomes increasingly dilute, and, because the cellophane sack has expanded to its limit as a result of an increased volume of water, water begins to move up the glass tube. In time, the water that has accumulated in the cellophane sack and the glass tube exerts a downward pressure that forces water molecules out of the cellophane sack and back into the beaker. This movement of water through a selectively permeable membrane produces osmotic pressure. **Osmotic pressure** is the pressure required to prevent the movement of pure water (water with no solutes) into a solution containing some solutes. In other words, osmotic pressure is the pressure needed to stop the flow of water across the selectively permeable membrane (cellophane). When water molecules leave and enter the cellophane sack at the same rate, equilibrium is reached (Figure 4.18b).

A bacterial cell may be subjected to any of three kinds of osmotic solutions: isotonic, hypotonic, or hypertonic. An **isotonic solution** is a medium in which the overall concentration of solutes equals that found inside a cell (*iso* means equal). Water leaves and enters the cell at the same rate (no net change); the cell's contents are in equilibrium with the solution outside the cytoplasmic membrane (Figure 4.18c).

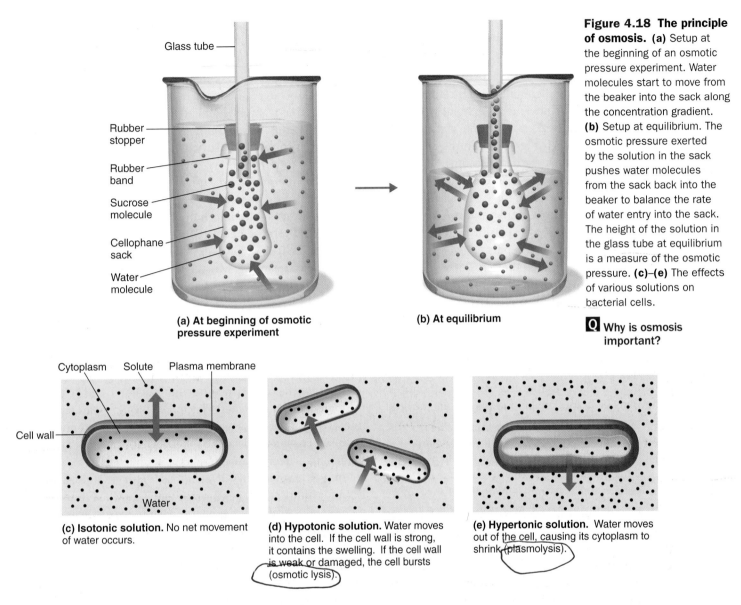

Glass tube

Rubber stopper

Rubber band

Sucrose molecule

Cellophane sack

Water molecule

**(a) At beginning of osmotic pressure experiment**

**(b) At equilibrium**

**Figure 4.18 The principle of osmosis. (a)** Setup at the beginning of an osmotic pressure experiment. Water molecules start to move from the beaker into the sack along the concentration gradient. **(b)** Setup at equilibrium. The osmotic pressure exerted by the solution in the sack pushes water molecules from the sack back into the beaker to balance the rate of water entry into the sack. The height of the solution in the glass tube at equilibrium is a measure of the osmotic pressure. **(c)–(e)** The effects of various solutions on bacterial cells.

**Q** Why is osmosis important?

Cytoplasm   Solute   Plasma membrane

Cell wall

Water

**(c) Isotonic solution.** No net movement of water occurs.

**(d) Hypotonic solution.** Water moves into the cell. If the cell wall is strong, it contains the swelling. If the cell wall is weak or damaged, the cell bursts (osmotic lysis).

**(e) Hypertonic solution.** Water moves out of the cell, causing its cytoplasm to shrink (plasmolysis).

Earlier we mentioned that lysozyme and certain antibiotics (such as penicillin) damage bacterial cell walls, causing the cells to rupture, or lyse. Such rupturing occurs because bacterial cytoplasm usually contains such a high concentration of solutes that additional water enters the cell by osmosis when the wall is weakened or removed. The damaged (or removed) cell wall cannot constrain the swelling of the cytoplasmic membrane, and the membrane bursts. This is an example of osmotic lysis caused by immersion in a hypotonic solution. A **hypotonic solution** outside the cell is a medium whose concentration of solutes is lower than that inside the cell (*hypo* means under or less). Most bacteria live in hypotonic solutions, and the cell wall resists further osmosis and protects cells from lysis. Cells with weak cell walls, such as gram-negative bacteria, may burst or undergo osmotic lysis as a result of excessive water intake (Figure 4.18d).

A **hypertonic solution** is a medium having a higher concentration of solutes than that inside the cell (*hyper* means above or

more). Most bacterial cells placed in a hypertonic solution shrink and collapse or *plasmolyze* because water leaves the cells by osmosis (Figure 4.18e). Keep in mind that the terms *isotonic, hypotonic,* and *hypertonic* describe the concentration of solutions outside the cell *relative to* the concentration inside the cell.

 **Play Passive Transport:** *Principles of Diffusion, Special Types of Diffusion* @Mastering**Microbiology**

### Active Processes

Simple diffusion and facilitated diffusion are useful mechanisms for transporting substances into cells when the concentrations of the substances are greater outside the cell. However, when a bacterial cell is in an environment in which nutrients are in low concentration, the cell must use active processes, such as active transport and group translocation, to accumulate the needed substances.

In performing **active transport,** the cell *uses energy* in the form of ATP to move substances across the plasma membrane.

Among the substances actively transported are ions (for example Na$^+$, K$^+$, H$^+$, Ca$^{2+}$, and Cl$^-$), amino acids, and simple sugars. Although these substances can also be moved into cells by passive processes, their movement by active processes can go against the concentration gradient, allowing a cell to accumulate needed materials. The movement of a substance in active transport is usually from outside to inside, even though the concentration might be much higher inside the cell. Like facilitated diffusion, active transport depends on transporter proteins in the plasma membrane (see Figure 4.17b, c). There appears to be a different transporter for each substance or group of closely related substances. Active transport enables microbes to move substances across the plasma membrane at a constant rate, even if they are in short supply.

*active process*

In active transport, the substance that crosses the plasma membrane is not altered by transport across the membrane. In **group translocation**, a special form of active transport that occurs exclusively in prokaryotes, the substance is chemically altered during transport across the membrane. Once the substance is altered and inside the cell, the plasma membrane is impermeable to it, so it remains inside the cell. This important mechanism enables a cell to accumulate various substances even though they may be in low concentrations outside the cell. Group translocation requires energy supplied by high-energy phosphate compounds, such as phosphoenolpyruvic acid (PEP).

One example of group translocation is the transport of the sugar glucose, which is often used in growth media for bacteria. While a specific carrier protein is transporting the glucose molecule across the membrane, a phosphate group is added to the sugar. This phosphorylated form of glucose, which cannot be transported out, can then be used in the cell's metabolic pathways.

Some eukaryotic cells (those without cell walls) can use two additional active transport processes called phagocytosis and pinocytosis. These processes, which do not occur in bacteria, are explained on page 97.

Play Active Transport:
*Types, Overview*
@Mastering**Microbiology**

**CHECK YOUR UNDERSTANDING**

✔ **4-8** Which agents can cause injury to the bacterial plasma membrane?

✔ **4-9** How are simple diffusion and facilitated diffusion similar? How are they different?

## Cytoplasm

For a prokaryotic cell, the term **cytoplasm** refers to the substance of the cell inside the plasma membrane (see Figure 4.6). Cytoplasm is about 80% water and contains primarily proteins (enzymes), carbohydrates, lipids, inorganic ions, and many low-molecular-mass compounds. Inorganic ions are present in much higher concentrations in cytoplasm than in most media. Cytoplasm is thick, aqueous, semitransparent, and elastic. The major structures in the cytoplasm of prokaryotes are a nucleoid (containing DNA), particles called ribosomes, and reserve deposits called inclusions.

The term **cytoskeleton** is a collective term for a series of fibers (small rods and cylinders) in the cytoplasm. Not long ago, it was believed that the absence of a cytoskeleton was a distinguishing feature of prokaryotes. However, use of atomic force microscopy shows that prokaryotic cells have a cytoskeleton similar to that of eukaryotes. Components include MreB and ParM, crescetin, and FtsZ, which correspond to the microfilaments, intermediate filaments, and microtubules of the eukaryotic cytoskeleton, respectively. The prokaryotic cytoskeleton assumes roles in cell division, maintaining cell shape, growth, DNA movement, protein targeting, and alignment of organelles. The cytoplasm of prokaryotes is not capable of cytoplasmic streaming, which will be described later.

## The Nucleoid

The **nucleoid** of a bacterial cell (see Figure 4.6) usually contains a single long, continuous, and frequently circularly arranged thread of double-stranded DNA called the **bacterial chromosome**. This is the cell's genetic information, which carries all the information required for the cell's structures and functions. Unlike the chromosomes of eukaryotic cells, bacterial chromosomes are not surrounded by a nuclear envelope (membrane) and do not include histones. The nucleoid can be spherical, elongated, or dumbbell-shaped. In actively growing bacteria, as much as 20% of the cell volume is occupied by DNA because such cells presynthesize DNA for future cells. The chromosome is attached to the plasma membrane. Proteins in the plasma membrane are believed to be responsible for replication of the DNA and segregation of the new chromosomes to daughter cells during cell division.

In addition to the bacterial chromosome, bacteria often contain small usually circular, double-stranded DNA molecules called **plasmids** (see Figure 8.25, page 229). These molecules are extrachromosomal genetic elements; that is, they are not connected to the main bacterial chromosome, and they replicate independently of chromosomal DNA. Research indicates that plasmids are associated with plasma membrane proteins. Plasmids usually contain from 5 to 100 genes that are generally not crucial for the survival of the bacterium under normal environmental conditions; plasmids may be gained or lost without harming the cell. Under certain conditions, however, plasmids are an advantage to cells. Plasmids may carry genes for such activities as antibiotic resistance, tolerance to toxic metals, the production of toxins, and the synthesis of enzymes. Plasmids can be transferred from one bacterium to another. In fact, plasmid DNA is used for gene manipulation in biotechnology.

## Ribosomes

All eukaryotic and prokaryotic cells contain **ribosomes**, where protein synthesis takes place. Cells that have high rates of protein synthesis, such as those that are actively growing, have a large number of ribosomes. The cytoplasm of a prokaryotic cell

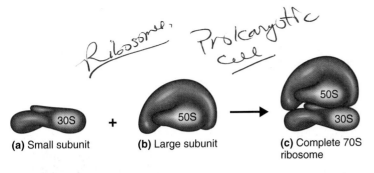

*Ribosome, Prokaryotic cell* (handwritten)

**(a)** Small subunit  +  **(b)** Large subunit  →  **(c)** Complete 70S ribosome

30S + 50S → 50S / 30S

**Figure 4.19  The prokaryotic ribosome. (a)** A small 30S subunit and **(b)** a large 50S subunit make up **(c)** the complete 70S prokaryotic ribosome.

**Q** What is the importance of the differences between prokaryotic and eukaryotic ribosomes with regard to antibiotic therapy?

contains tens of thousands of ribosomes, giving the cytoplasm a granular appearance (see Figure 4.6).

Ribosomes are composed of two subunits, each of which consists of protein and a type of RNA called *ribosomal RNA (rRNA)*. Prokaryotic ribosomes differ from eukaryotic ribosomes in the number of proteins and rRNA molecules they contain; they are also somewhat smaller and less dense than ribosomes of eukaryotic cells. Accordingly, prokaryotic ribosomes are called 70S ribosomes (**Figure 4.19**), and those of eukaryotic cells are known as 80S ribosomes. The letter S refers to Svedberg units, which indicate the relative rate of sedimentation during ultra-high-speed centrifugation. Sedimentation rate is a function of the size, weight, and shape of a particle. The subunits of a 70S ribosome are a small 30S subunit containing one molecule of rRNA and a larger 50S subunit containing two molecules of rRNA. (Note that the 70S "value" is not the sum of 30S plus 50S. The apparent arithmetic error here has often puzzled students. However, you can think of a Svedberg unit as a unit of size rather than weight. Therefore, the combination here of 50S and 30S is not the same as combining 50 grams and 30 grams.)

Several antibiotics work by inhibiting protein synthesis at prokaryotic ribosomes. Antibiotics such as streptomycin and gentamicin attach to the 30S subunit and interfere with protein synthesis. Other antibiotics, such as erythromycin and chloramphenicol, interfere with protein synthesis by attaching to the 50S subunit. Because of differences in prokaryotic and eukaryotic ribosomes, the microbial cell can be killed by the antibiotic while the eukaryotic host cell remains unaffected.

## Inclusions

Within the cytoplasm of prokaryotic cells are several kinds of reserve deposits, known as **inclusions**. Cells may accumulate certain nutrients when they are plentiful and use them when the environment is deficient. Evidence suggests that macromolecules concentrated in inclusions avoid the increase in osmotic pressure that would result if the molecules were dispersed in the cytoplasm. Some inclusions are common to a wide variety of bacteria, whereas others are limited to a small number of species and therefore serve as a basis for identification. Some inclusions, such as magnetosomes, are membrane-enclosed

organelles, while other inclusions, such as carboxysomes, are enclosed in protein complexes.

### Metachromatic Granules

**Metachromatic granules** are large inclusions that take their name from the fact that they sometimes stain red with certain blue dyes such as methylene blue. Collectively they are known as **volutin**. Volutin represents a reserve of inorganic phosphate (polyphosphate) that can be used in the synthesis of ATP. It is generally formed by cells that grow in phosphate-rich environments. Metachromatic granules are found in algae, fungi, and protozoa, as well as in bacteria. These granules are characteristic of *Corynebacterium diphtheriae*, the causative agent of diphtheria; thus, they have diagnostic significance.

### Polysaccharide Granules

Inclusions known as **polysaccharide granules** typically consist of glycogen and starch, and their presence can be demonstrated when iodine is applied to the cells. In the presence of iodine, glycogen granules appear reddish brown and starch granules appear blue.

### Lipid Inclusions

**Lipid inclusions** appear in various species of *Mycobacterium, Bacillus, Azotobacter* (ah-ZŌ-tō-bak-ter), *Spirillum* (spī-RIL-lum), and other genera. A common lipid-storage material, one unique to bacteria, is the polymer *poly-β-hydroxybutyric acid*. Lipid inclusions are revealed by staining cells with fat-soluble dyes, such as Sudan dyes.

### Sulfur Granules

Certain bacteria—for example, the "sulfur bacteria" that belong to the genus *Acidithiobacillus*—derive energy by oxidizing sulfur and sulfur-containing compounds. These bacteria

**CLINICAL CASE**

The antibiotic killed the bacteria, but endotoxin is released when the cells die, causing Jessie's condition to worsen. Jessie's physician prescribes polymyxin, an antibiotic that does not cause release of endotoxin, and to which Jessie responds favorably.

As Irene sits with Jessie, she notices another patient being fed ice chips by a relative. On a hunch, Irene hurries back to her office to find out whether the ice machines had been swabbed. They have not. She immediately orders the machines to be swabbed and cultured. Her hunch turns out to be correct: the samples are positive for *K. pneumoniae*. Bacteria growing in the hospital's water pipes entered the ice machine with incoming water.

**How can *K. pneumoniae* grow in water pipes?**

73 | 83 | 85 | **91** | 94

may deposit **sulfur granules** in the cell, where they serve as an energy reserve.

## Carboxysomes

**Carboxysomes** are inclusions that contain the enzyme ribulose 1,5-bisphosphate carboxylase. Photosynthetic bacteria use carbon dioxide as their sole source of carbon and require this enzyme for carbon dioxide fixation. Among the bacteria containing carboxysomes are nitrifying bacteria, cyanobacteria, and acidithiobacilli.

## Gas Vacuoles

Hollow cavities found in many aquatic prokaryotes, including cyanobacteria, anoxygenic photosynthetic bacteria, and halobacteria are called **gas vacuoles**. Each vacuole consists of rows of several individual *gas vesicles*, which are hollow cylinders covered by protein. Gas vacuoles maintain buoyancy so that the cells can remain at the depth in the water appropriate for them to receive sufficient amounts of oxygen, light, and nutrients.

## Magnetosomes

**Magnetosomes** are inclusions of iron oxide ($Fe_3O_4$) surrounded by invaginations of the plasma membrane. Magnetosomes are formed by several gram-negative bacteria such as *Magnetospirillum magnetotacticum* and act like magnets (**Figure 4.20**). Bacteria may use magnetosomes to move downward until they reach a suitable attachment site. In vitro, magnetosomes can decompose hydrogen peroxide, which forms in cells in the presence of oxygen. Researchers speculate that magnetosomes may protect the cell against hydrogen peroxide accumulation.

*help remove $H_2O_2$ (hydrogen peroxide)*

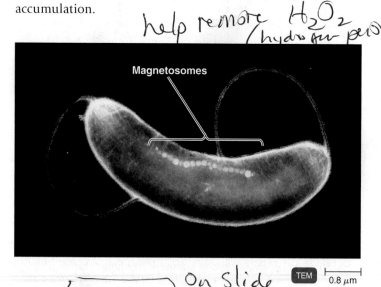

Magnetosomes

On Slide  TEM  0.8 μm

**Figure 4.20 Magnetosomes.** This micrograph of *Magnetospirillum magnetotacticum* shows a chain of magnetosomes. This bacterium is usually found in shallow freshwater mud.

**Q** How do magnetosomes behave like magnets?

## Endospores

When essential nutrients are depleted, certain gram-positive bacteria, such as those of the genera *Clostridium* and *Bacillus*, form specialized "resting" cells called **endospores** (**Figure 4.21**). As you will see later, some members of the genus *Clostridium* cause diseases such as gangrene, tetanus, botulism, and food poisoning. Some members of the genus *Bacillus* cause anthrax and food poisoning. Unique to bacteria, endospores are highly durable dehydrated cells with thick walls and additional layers. They are formed internal to the bacterial cell membrane. When released into the environment, they can survive extreme heat, lack of water, and exposure to many toxic chemicals and radiation. For example, 7500-year-old endospores of *Thermoactinomyces vulgaris* (ther′mō-ak-tin-ō-MĪ-sēz vul-GAR-is) from the freezing muds of Elk Lake in Minnesota have germinated when rewarmed and placed in a nutrient medium, and 25- to 40-million-year-old endospores found in the gut of a stingless bee entombed in amber (hardened tree resin) in the Dominican Republic are reported to have germinated when placed in nutrient media. Although true endospores are found in gram-positive bacteria, one gram-negative species, *Coxiella burnetii* (KOKS-ē-el-lah ber-NE-tē-ē), the cause of Q fever (usually a mild disease with flulike symptoms), forms endospore-like structures that resist heat and chemicals and can be stained with endospore stains (see Figure 24.13, page 706).

The process of endospore formation within a vegetative cell takes several hours and is known as **sporulation** or **sporogenesis** (Figure 4.21a). Vegetative cells of endospore-forming bacteria begin sporulation when a key nutrient, such as the carbon or nitrogen source, becomes scarce or unavailable. In the first observable stage of sporulation, a newly replicated bacterial chromosome and a small portion of cytoplasm are isolated by an ingrowth of the plasma membrane called a *spore septum*. The spore septum becomes a double-layered membrane that surrounds the chromosome and cytoplasm. This structure, entirely enclosed within the original cell, is called a *forespore*. Thick layers of peptidoglycan are laid down between the two membrane layers. Then a thick *spore coat* of protein forms around the outside membrane; this coat is responsible for the resistance of endospores to many harsh chemicals. The original cell is degraded, and the endospore is released.

The diameter of the endospore may be the same as, smaller than, or larger than the diameter of the vegetative cell. Depending on the species, the endospore might be located *terminally* (at one end), *subterminally* (near one end), or *centrally* (Figure 4.21b) inside the vegetative cell. When the endospore matures, the vegetative cell wall ruptures (lyses), killing the cell, and the endospore is freed.

Most of the water present in the forespore cytoplasm is eliminated by the time sporulation is complete, and endospores do not carry out metabolic reactions. The endospore contains a large amount of an organic acid called *dipicolinic acid*

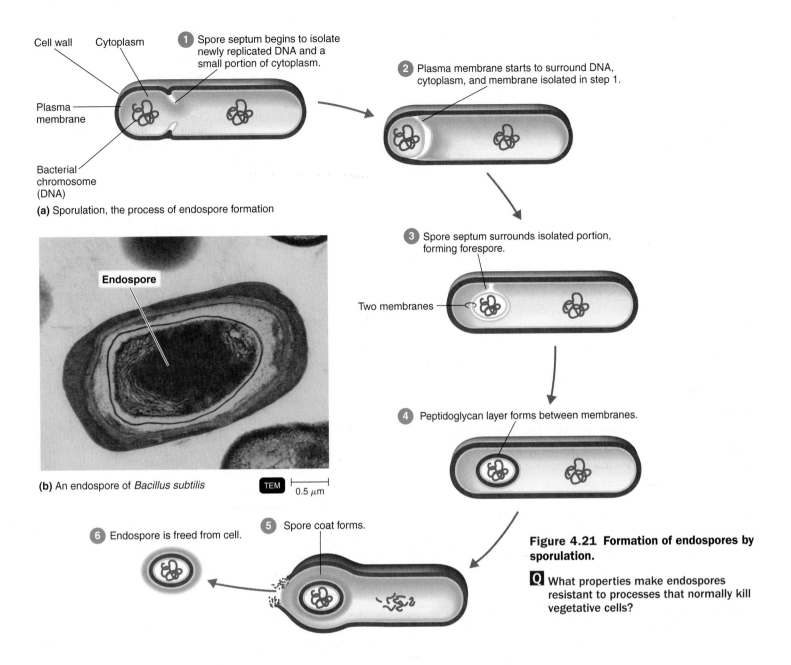

Cell wall   Cytoplasm

**1** Spore septum begins to isolate newly replicated DNA and a small portion of cytoplasm.

Plasma membrane

Bacterial chromosome (DNA)

**(a)** Sporulation, the process of endospore formation

Endospore

**(b)** An endospore of *Bacillus subtilis*   TEM   0.5 μm

**2** Plasma membrane starts to surround DNA, cytoplasm, and membrane isolated in step 1.

**3** Spore septum surrounds isolated portion, forming forespore.

Two membranes

**4** Peptidoglycan layer forms between membranes.

**6** Endospore is freed from cell.   **5** Spore coat forms.

**Figure 4.21  Formation of endospores by sporulation.**

**Q** What properties make endospores resistant to processes that normally kill vegetative cells?

(DPA), which is accompanied by a large number of calcium ions. Evidence indicates that DPA protects the endospore DNA against damage. The highly dehydrated endospore core contains only DNA, small amounts of RNA, ribosomes, enzymes, and a few important small molecules. These cellular components are essential for resuming metabolism later.

Endospores can remain dormant for thousands of years. An endospore returns to its vegetative state by a process called **germination**. Germination is triggered by high heat, such as is used in canning, or small triggering molecules called *germinants.* Germinants identified to date are alanine and inosine (a nucleotide). The endospore's enzymes then break down the extra layers surrounding the endospore, water enters, and metabolism resumes. Because one vegetative cell forms a single endospore,

which, after germination, remains one cell, sporulation in bacteria is *not* a means of reproduction. This process does not increase the number of cells. Bacterial endospores differ from spores formed by (prokaryotic) actinomycetes and the eukaryotic fungi and algae, which detach from the parent and develop into another organism and, therefore, represent reproduction.

Endospores are important from a clinical viewpoint and in the food industry because they're resistant to processes that normally kill vegetative cells. Such processes include heating, desiccation, use of chemicals, and radiation. Whereas most vegetative cells are killed by temperatures above 70°C, endospores can survive in boiling water for several hours or more. Endospores of thermophilic (heat-loving) bacteria can survive in boiling water for 19 hours. Endospore-forming bacteria are a

# Eukaryotes Are Microbiota, Too

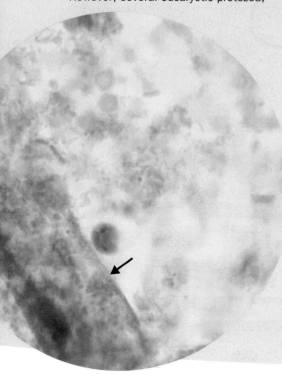

The vast majority of research being conducted on the microbiome revolves around prokaryotes. However, several eucaryotic protozoa, collectively called the *eukaryome*, are also part of the human microbiome.

*Retortamonas intestinalis*, *Pentatrichomonas hominis*, and *Enteromonas hominis* have all been found living in the large intestines of healthy individuals without causing disease. The exact location within the large intestine may be important to the protozoa's ability to cause disease. Parasitic protozoa, such as *Giardia*, attach to the intestinal lining, whereas harmless protozoa are more likely to stay in the inside space of the intestine without attaching to cells.

Diversity of organisms in the eukaryome is lower in highly developed countries and higher in less industrialized areas with less rigorous sanitation and hygiene practices. No doubt, the decreased diversity of the eukaryome among people living in developed nations is due to improved hygiene and the availability of medications.

At the same time, the incidence of autoimmune diseases such as Crohn's disease and irritable bowel syndrome is higher in industrialized countries than in countries with greater incidence of eukaryotic infections (pathogenic and otherwise). Scientists now believe that the eukaryome in particular may be important in setting the sensitivity of the human immune response and that removal of certain parts of our microbiota during childhood may lead the body to "overrespond," and initiate autoimmune disorders.

More work needs to be done to characterize the eukaryome and to learn about the full range of effects, both positive and negative, that occur in the body when certain microbes are present or absent. Completely avoiding contact with microbes in childhood may be counterproductive for our long-term health.

*Pentatrichomonas hominis* (at the arrow) **is a nonpathogenic protozoan in the human intestines.**

---

## CLINICAL CASE   Resolved

It is the glycocalyx that enables bacteria in water to stick inside a pipe. The bacteria grow slowly in the nutrient-poor tap water but do not get washed away by the flowing water. A slimy layer of bacteria can accumulate in a pipe. Irene discovers that the disinfectant in the hospital's water supply was inadequate to prevent bacterial growth. Some bacteria can get dislodged by flowing water, and even normally harmless bacteria can infect a surgical incision or weakened host.

| 73 | 83 | 85 | 91 | 94 |

problem in the food industry because they are likely to survive underprocessing, and, if conditions for growth occur, some species produce toxins and disease. Special methods for controlling organisms that produce endospores are discussed in Chapter 7.

### CHECK YOUR UNDERSTANDING

✔ 4-10 Where is the DNA located in a prokaryotic cell?

✔ 4-11 What is the general function of inclusions?

✔ 4-12 Under what conditions do endospores form?

* * *

Having examined the functional anatomy of the prokaryotic cell, we will now look at the functional anatomy of the eukaryotic cell.

# The Eukaryotic Cell

As mentioned earlier, eukaryotic organisms include algae, protozoa, fungi, plants, and animals. Some eukaryotes cause disease, but others are part of the normal human microbiome (see Exploring the Microbiome). The eukaryotic cell is typically larger and structurally more complex than the prokaryotic cell (**Figure 4.22**). When the structure of the prokaryotic cell in Figure 4.6 is compared with that of the eukaryotic cell, the differences between the two types of cells become apparent.

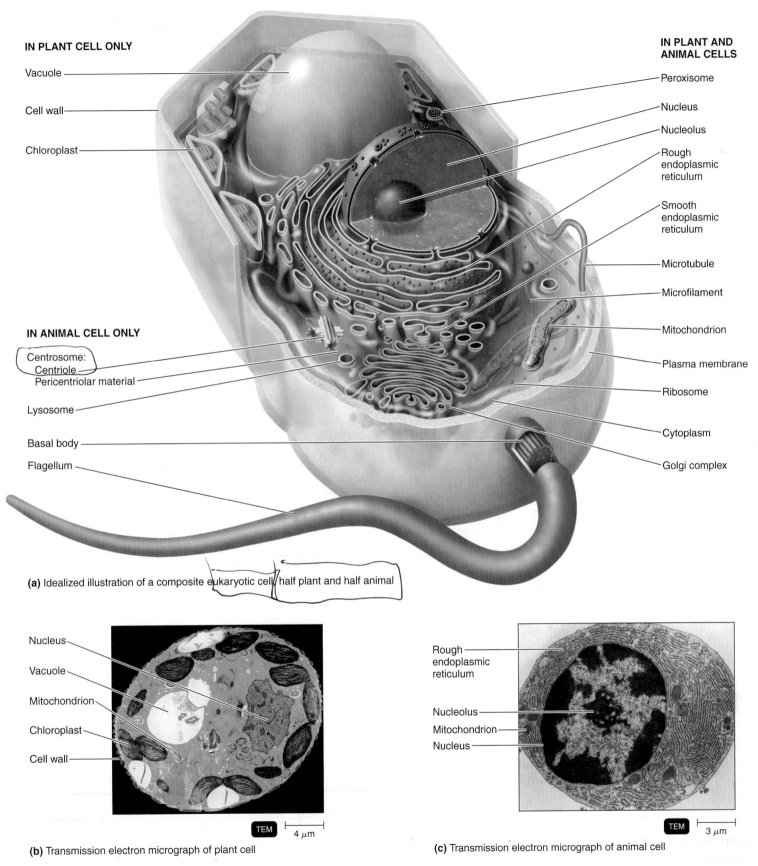

**IN PLANT CELL ONLY**

Vacuole

Cell wall

Chloroplast

**IN ANIMAL CELL ONLY**

Centrosome:
   Centriole
   Pericentriolar material

Lysosome

Basal body

Flagellum

**IN PLANT AND
ANIMAL CELLS**

Peroxisome

Nucleus

Nucleolus

Rough
endoplasmic
reticulum

Smooth
endoplasmic
reticulum

Microtubule

Microfilament

Mitochondrion

Plasma membrane

Ribosome

Cytoplasm

Golgi complex

**(a)** Idealized illustration of a composite eukaryotic cell, half plant and half animal

Nucleus

Vacuole

Mitochondrion

Chloroplast

Cell wall

TEM | 4 μm

**(b)** Transmission electron micrograph of plant cell

Rough
endoplasmic
reticulum

Nucleolus

Mitochondrion

Nucleus

TEM | 3 μm

**(c)** Transmission electron micrograph of animal cell

**Figure 4.22 Eukaryotic cells showing typical structures.**

**Q** What kingdoms contain eukaryotic organisms?

**TABLE 4.2** Principal Differences between Prokaryotic and Eukaryotic Cells

| Characteristic | Prokaryotic | Eukaryotic |
|---|---|---|
| Size of Cell | Typically 0.2–2.0 μm in diameter | Typically 10–100 μm in diameter |
| Nucleus | Typically no nuclear membrane or nucleoli except *Gemmata* (see Figure 11.23) | True nucleus, consisting of nuclear membrane and nucleoli |
| Membrane-Enclosed Organelles | Relatively few | Present; examples include nuclei, lysosomes, Golgi complex, endoplasmic reticulum, mitochondria, and chloroplasts |
| Flagella | Consist of two protein building blocks | Complex; consist of multiple microtubules |
| Glycocalyx | Present as a capsule or slime layer | Present in some cells that lack a cell wall |
| Cell Wall | Usually present; chemically complex (typical bacterial cell wall includes peptidoglycan) | When present, chemically simple (includes cellulose and chitin) |
| Plasma Membrane | Carbohydrates and generally lacks sterols | Sterols and carbohydrates that serve as receptors |
| Cytoplasm | Cytoskeleton (MreB and ParM, cresetin, and FtsZ proteins); no cytoplasmic streaming | Cytoskeleton (microfilaments, intermediate filaments, and microtubules); cytoplasmic streaming |
| Ribosomes | Smaller size (70S) | Larger size (80S); smaller size (70S) in organelles |
| Chromosome (DNA) | Usually single circular chromosome; typically lacks histones | Multiple linear chromosomes with histones |
| Cell Division | Binary fission | Involves mitosis |
| Sexual Recombination | None; transfer of DNA only | Involves meiosis |

The principal differences between prokaryotic and eukaryotic cells are summarized in Table 4.2.

The following discussion of eukaryotic cells will parallel our discussion of prokaryotic cells by starting with structures that extend to the outside of the cell.

> ASM: While microscopic eukaryotes (for example, fungi, protozoa, and algae) carry out some of the same processes as bacteria, many of the cellular properties are fundamentally different.

# Flagella and Cilia

**LEARNING OBJECTIVE**

4-13 Differentiate prokaryotic and eukaryotic flagella.

Many types of eukaryotic cells have projections that are used for cellular locomotion or for moving substances along the surface of the cell. These projections contain cytoplasm and are enclosed by the plasma membrane. If the projections are few and are long in relation to the size of the cell, they are called **flagella**. If the projections are numerous and short, they are called **cilia** (singular: **cilium**).

Algae of the genus *Euglena* (ū-GLĒ-nah) use a flagellum for locomotion, whereas protozoa, such as *Tetrahymena* (tet'rah-HĪ-me-nah), use cilia for locomotion (**Figure 4.23**a and Figure 4.23b). Both flagella and cilia are anchored to the plasma membrane by a basal body, and both consist of nine pairs of microtubules (doublets) arranged in a ring, plus another two microtubules in the center of the ring, an arrangement called a *9 + 2 array* (Figure 4.23c). **Microtubules** are long, hollow tubes made up of a protein called *tubulin*. A prokaryotic flagellum rotates, but a eukaryotic flagellum moves in a wavelike manner (Figure 4.23d). To help keep foreign material out of the lungs, ciliated cells of the human respiratory system move the material along the surface of the cells in the bronchial tubes and trachea toward the throat and mouth (see Figure 16.3, page 444).

# The Cell Wall and Glycocalyx

**LEARNING OBJECTIVE**

4-14 Compare and contrast prokaryotic and eukaryotic cell walls and glycocalyxes.

Most eukaryotic cells have cell walls, although they are generally much simpler than those of prokaryotic cells. Many algae have cell walls consisting of the polysaccharide *cellulose* (as do all plants); other chemicals may be present as well. Cell walls of some fungi also contain cellulose, but in most fungi the principal structural component of the cell wall is the polysaccharide *chitin*, a polymer of N-acetylglucosamine (NAG) units. (Chitin is also the main structural component of the exoskeleton of crustaceans and insects.) The cell walls of yeasts contain the polysaccharides *glucan* and *mannan*. In eukaryotes that lack a cell wall, the plasma

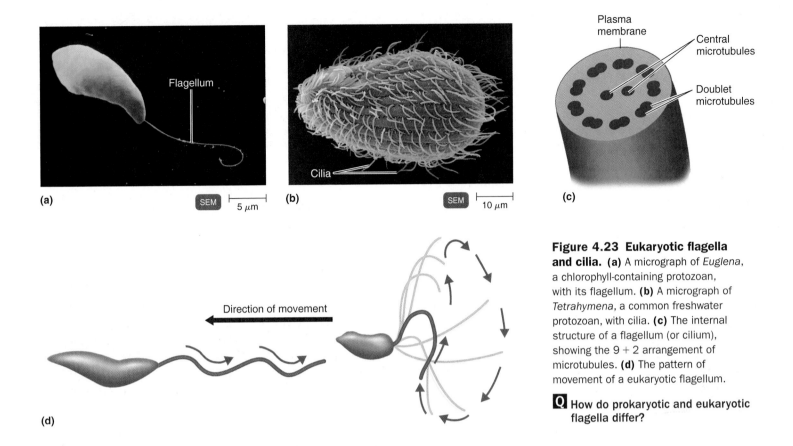

**Figure 4.23 Eukaryotic flagella and cilia.** **(a)** A micrograph of *Euglena*, a chlorophyll-containing protozoan, with its flagellum. **(b)** A micrograph of *Tetrahymena*, a common freshwater protozoan, with cilia. **(c)** The internal structure of a flagellum (or cilium), showing the 9 + 2 arrangement of microtubules. **(d)** The pattern of movement of a eukaryotic flagellum.

**Q** How do prokaryotic and eukaryotic flagella differ?

membrane may be the outer covering; however, cells that have direct contact with the environment may have coatings outside the plasma membrane. Protozoa do not have a typical cell wall; instead, they have a flexible outer protein covering called a *pellicle*.

In other eukaryotic cells, including animal cells, the plasma membrane is covered by a **glycocalyx**, a layer of material containing substantial amounts of sticky carbohydrates. Some of these carbohydrates are covalently bonded to proteins and lipids in the plasma membrane, forming glycoproteins and glycolipids that anchor the glycocalyx to the cell. The glycocalyx strengthens the cell surface, helps attach cells together, and may contribute to cell–cell recognition.

Eukaryotic cells do not contain peptidoglycan, the framework of the prokaryotic cell wall. This is significant medically because antibiotics, such as penicillins and cephalosporins, act against peptidoglycan and therefore do not affect human eukaryotic cells.

# The Plasma (Cytoplasmic) Membrane

### LEARNING OBJECTIVE

4-15 Compare and contrast prokaryotic and eukaryotic plasma membranes.

The **plasma (cytoplasmic) membrane** of eukaryotic and prokaryotic cells is very similar in function and basic structure. There are, however, differences in the types of proteins found in the membranes. Eukaryotic membranes also contain carbohydrates, which serve as attachment sites for bacteria and as receptor sites that assume a role in such functions as cell–cell recognition. Eukaryotic plasma membranes also contain *sterols*, complex lipids not found in prokaryotic plasma membranes (with the exception of *Mycoplasma* cells). Sterols seem to be associated with the ability of the membranes to resist lysis resulting from increased osmotic pressure.

Substances can cross eukaryotic and prokaryotic plasma membranes by simple diffusion, facilitated diffusion, osmosis, or active transport. Group translocation does not occur in eukaryotic cells. However, eukaryotic cells can use a mechanism called **endocytosis**. This occurs when a segment of the plasma membrane surrounds a particle or large molecule, encloses it, and brings it into the cell.

The three types of endocytosis are phagocytosis, pinocytosis, and receptor-mediated endocytosis. During *phagocytosis*, cellular projections called pseudopods engulf particles and bring them into the cell. Phagocytosis is used by white blood cells to destroy bacteria and foreign substances (see Figure 16.8, page 452, and further discussion in Chapter 16). In *pinocytosis*, the plasma membrane folds inward, bringing extracellular fluid into the cell, along with whatever substances are dissolved in the fluid. In *receptor-mediated endocytosis*, substances (ligands) bind to receptors in the membrane. When binding occurs, the membrane folds inward. Receptor-mediated endocytosis is one of the ways viruses can enter animal cells (see Figure 13.14a, page 377).

# Cytoplasm

## LEARNING OBJECTIVE

4-16   Compare and contrast prokaryotic and eukaryotic cytoplasms.

The **cytoplasm** of eukaryotic cells encompasses the substance inside the plasma membrane and outside the nucleus (see Figure 4.22). The cytoplasm is the substance in which various cellular components are found. (The term **cytosol** refers to the fluid portion of cytoplasm.) The **cytoskeleton** of eukaryotes consists of small rods (*microfilaments* and *intermediate filaments*) and cylinders (*microtubules*). Recall these correspond to the MreB and ParM, crescetin, and FtsZ of the prokaryotic cytoskeleton, respectively. The cytoskeleton of eukaryotes provides support, shape, and assistance in transporting substances through the cell (and even in moving the entire cell, as in phagocytosis). The movement of eukaryotic cytoplasm from one part of the cell to another, which helps distribute nutrients and move the cell over a surface, is called **cytoplasmic streaming**. Another difference between prokaryotic and eukaryotic cytoplasm is that many of the important enzymes found in the cytoplasmic fluid of prokaryotes are sequestered in the organelles of eukaryotes.

# Ribosomes    *eukaryotic*

## LEARNING OBJECTIVE

4-17   Compare the structure and function of eukaryotic and prokaryotic ribosomes.

Attached to the outer surface of the rough endoplasmic reticulum (discussed on page 99) are **ribosomes** (see Figure 4.19), which are also found free in the cytoplasm. As in prokaryotes, ribosomes are the sites of protein synthesis in the cell.

The ribosomes of cells are somewhat larger and denser than those of prokaryotic cells. These eukaryotic ribosomes are 80S ribosomes, each of which consists of a large 60S subunit containing three molecules of rRNA and a smaller 40S subunit with one molecule of rRNA. The subunits are made separately in the nucleolus and, once produced, exit the nucleus and join together in the cytosol. Chloroplasts and mitochondria contain 70S ribosomes, which indicates their evolution from prokaryotes. (This theory is discussed on page 102.) The role of ribosomes in protein synthesis will be discussed in more detail in Chapter 8.

Some ribosomes, called *free ribosomes*, are unattached to any structure in the cytoplasm. Primarily, free ribosomes synthesize proteins used *inside* the cell. Other ribosomes, called *membrane-bound ribosomes*, attach to the nuclear membrane and the endoplasmic reticulum. These ribosomes synthesize proteins destined for insertion in the plasma membrane or for export from the cell. Ribosomes located within mitochondria synthesize mitochondrial proteins.

# Organelles

## LEARNING OBJECTIVES

4-18   Define *organelle*.

4-19   Describe the functions of the nucleus, endoplasmic reticulum, Golgi complex, lysosomes, vacuoles, mitochondria, chloroplasts, peroxisomes, and centrosomes.

**Organelles** are structures with specific shapes and specialized functions and are characteristic of eukaryotic cells. They include the nucleus, endoplasmic reticulum, Golgi complex, lysosomes, vacuoles, mitochondria, chloroplasts, peroxisomes, and centrosomes. Not all of the organelles described are found in all cells, and certain cells have their own type and distribution of organelles based on specialization, age, and level of activity.

## The Nucleus

The most characteristic eukaryotic organelle is the nucleus (see Figure 4.22). The nucleus (Figure 4.24) is usually spherical or oval, is frequently the largest structure in the cell, and contains almost all of the cell's hereditary information (DNA). Some DNA is also found in mitochondria and in the chloroplasts of photosynthetic organisms.

The nucleus is surrounded by a double membrane called the **nuclear envelope**. Both membranes resemble the plasma membrane in structure. Tiny channels in the membrane called **nuclear pores** allow the nucleus to communicate with the cytoplasm (Figure 4.24b). Nuclear pores control the movement of substances between the nucleus and cytoplasm. Within the nuclear envelope are one or more spherical bodies called **nucleoli** (singular: **nucleolus**). Nucleoli are actually condensed regions of chromosomes where ribosomal RNA is being synthesized. Ribosomal RNA is an essential component of ribosomes.

The nucleus also contains most of the cell's DNA, which is combined with several proteins, including some basic proteins called histones and nonhistones. The combination of about 165 base pairs of DNA and 9 molecules of histones is referred to as a *nucleosome*. When the cell is not reproducing, the DNA and its associated proteins appear as a threadlike mass called **chromatin**. During nuclear division, the chromatin coils into shorter and thicker rodlike bodies called **chromosomes**. Prokaryotic chromosomes do not undergo

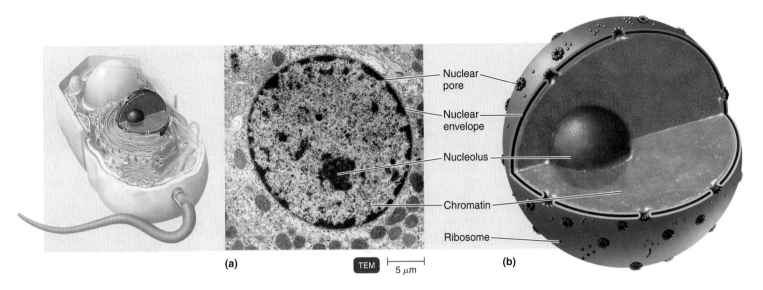

(a)

TEM ⊢ 5 μm

(b)

**Figure 4.24 The eukaryotic nucleus.** **(a)** A micrograph of a nucleus. **(b)** Drawing of details of a nucleus.

**Q** What keeps the nucleus suspended in the cell?

this process, do not have histones, and are not enclosed in a nuclear envelope.

Eukaryotic cells require two elaborate mechanisms: mitosis and meiosis to segregate chromosomes prior to cell division. Neither process occurs in prokaryotic cells.

## Endoplasmic Reticulum

Within the cytoplasm of eukaryotic cells is the **endoplasmic reticulum**, or **ER**, an extensive network of flattened membranous sacs or tubules called **cisternae** (**Figure 4.25**). The ER network is continuous with the nuclear envelope (see Figure 4.22a).

Most eukaryotic cells contain two distinct, but interrelated, forms of ER that differ in structure and function. The membrane of **rough ER is continuous with the nuclear membrane and usually unfolds into a series of flattened sacs.** The outer surface of rough ER is studded with ribosomes, the sites of protein synthesis. Proteins synthesized by ribosomes that are attached to rough ER enter cisternae within the ER for processing and sorting. In some cases, enzymes within the cisternae attach the proteins to carbohydrates to form glycoproteins. In other cases, enzymes attach the proteins to phospholipids, also synthesized by rough ER. These molecules may be incorporated into

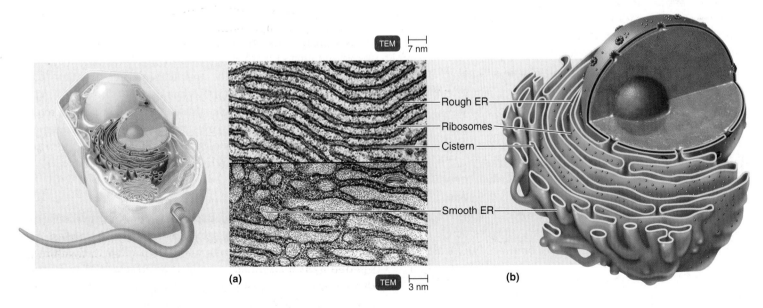

TEM ⊢ 7 nm

(a)

TEM ⊢ 3 nm

(b)

Rough ER
Ribosomes
Cistern
Smooth ER

**Figure 4.25 Endoplasmic reticulum.** **(a)** A micrograph of the rough and smooth endoplasmic reticulum and ribosomes. **(b)** A drawing of details of the endoplasmic reticulum.

**Q** What functions of the smooth ER and rough ER are similar?

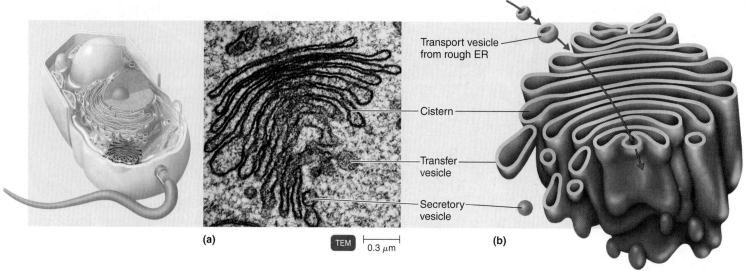

**Figure 4.26 Golgi complex. (a)** A micrograph of a Golgi complex. **(b)** A drawing of details of a Golgi complex.

**Q** What is the function of the Golgi complex?

organelle membranes or the plasma membrane. Thus, rough ER is a factory for synthesizing secretory proteins and membrane molecules.

**Smooth ER** extends from the rough ER to form a network of membrane tubules (see Figure 4.25). Unlike rough ER, smooth ER does not have ribosomes on the outer surface of its membrane. However, smooth ER contains unique enzymes that make it functionally more diverse than rough ER. Although it does not synthesize proteins, smooth ER does synthesize phospholipids, as does rough ER. Smooth ER also synthesizes fats and steroids, such as estrogens and testosterone. In liver cells, enzymes of the smooth ER help release glucose into the bloodstream and inactivate or detoxify drugs and other potentially harmful substances (for example, alcohol). In muscle cells, calcium ions released from the sarcoplasmic reticulum, a form of smooth ER, trigger the contraction process.

## Golgi Complex

Most of the proteins synthesized by ribosomes attached to rough ER are ultimately transported to other regions of the cell. The first step in the transport pathway is through an organelle called the **Golgi complex**. It consists of 3 to 20 cisternae that resemble a stack of pita bread (**Figure 4.26**). The cisternae are often curved, giving the Golgi complex a cuplike shape.

Proteins synthesized by ribosomes on the rough ER are surrounded by a portion of the ER membrane, which eventually buds from the membrane surface to form a **transport vesicle**. The transport vesicle fuses with a cistern of the Golgi complex, releasing proteins into the cistern. The proteins are modified and move from one cistern to another via **transfer vesicles** that bud from the edges of the cisternae. Enzymes in the cisternae

modify the proteins to form glycoproteins, glycolipids, and lipoproteins. Some of the processed proteins leave the cisternae in **secretory vesicles**, which detach from the cistern and deliver the proteins to the plasma membrane, where they are discharged by exocytosis. Other processed proteins leave the cisternae in vesicles that deliver their contents to the plasma membrane for incorporation into the membrane. Finally, some processed proteins leave the cisternae in vesicles that are called **storage vesicles**. The major storage vesicle is a lysosome, whose structure and functions are discussed next.

## Lysosomes

**Lysosomes** are formed from Golgi complexes and look like membrane-enclosed spheres. Unlike mitochondria, lysosomes have only a single membrane and lack internal structure (see Figure 4.22). But they contain as many as 40 different kinds of digestive enzymes capable of breaking down various molecules. Moreover, these enzymes can also digest bacteria that enter the cell. Human white blood cells, which use phagocytosis to ingest bacteria, contain large numbers of lysosomes.

## Vacuoles

A **vacuole** (see Figure 4.22) is a space or cavity in the cytoplasm of a cell that is enclosed by a membrane called a *tonoplast*. In plant cells, vacuoles may occupy 5 to 90% of the cell volume, depending on the type of cell. Vacuoles are derived from the Golgi complex and have several diverse functions. Some vacuoles serve as temporary storage organelles for substances such as proteins, sugars, organic acids, and inorganic ions. Other vacuoles form during endocytosis to help bring food into the cell. Many plant cells also store metabolic wastes and poisons

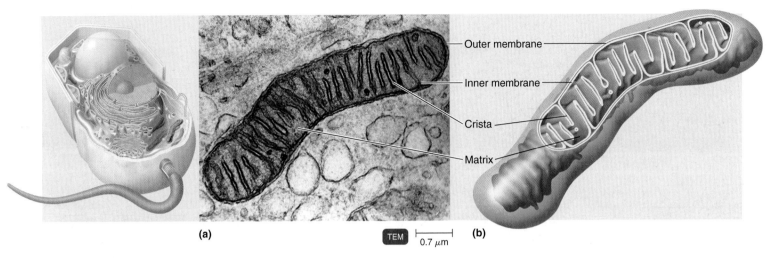

**(a)** TEM $\vdash$ 0.7 $\mu$m $\dashv$ **(b)**

**Figure 4.27 Mitochondria. (a)** A micrograph of a mitochondrion from a rat pancreas cell.
**(b)** A drawing of details of a mitochondrion.

**Q** How are mitochondria similar to prokaryotic cells?

that would otherwise be injurious if they accumulated in the cytoplasm. Finally, vacuoles may take up water, enabling plant cells to increase in size and also providing rigidity to leaves and stems.

## Mitochondria

Elongated, irregularly shaped organelles called **mitochondria** (singular: **mitochondrion**) appear throughout the cytoplasm of most eukaryotic cells (see Figure 4.22). The number of mitochondria per cell varies greatly among different types of cells. For example, the protozoan *Giardia* has no mitochondria, whereas liver cells contain 1000 to 2000 per cell. A mitochondrion has two membranes similar in structure to the plasma membrane **(Figure 4.27)**. The outer mitochondrial membrane is smooth, but the inner mitochondrial membrane is arranged in a series of folds called **cristae** (singular: **crista**). The center of the mitochondrion is a semifluid substance called the **matrix**. Because of the nature and arrangement of the cristae, the inner membrane provides an enormous surface area on which chemical reactions can occur. Some proteins that function in cellular respiration, including the enzyme that makes ATP, are located on the cristae of the inner mitochondrial membrane, and many of the metabolic steps involved in cellular respiration are concentrated in the matrix (see Chapter 5). Mitochondria are often called the "powerhouses of the cell" because of their central role in ATP production.

Mitochondria contain 70S ribosomes and some DNA of their own, as well as the machinery necessary to replicate, transcribe, and translate the information encoded by their DNA. In addition, mitochondria can reproduce more or less on their own by growing and dividing in two.

## Chloroplasts

Algae and green plants contain a unique organelle called a chloroplast **(Figure 4.28)**, a double membrane-enclosed structure that contains both the pigment chlorophyll and the enzymes required for the light-gathering phases of photosynthesis (see Chapter 5). The chlorophyll is contained in flattened membrane sacs called **thylakoids;** stacks of thylakoids are called **grana** (singular: **granum**) (see Figure 4.28).

Like mitochondria, chloroplasts contain 70S ribosomes, DNA, and enzymes involved in protein synthesis. They are capable of multiplying on their own within the cell. The way both chloroplasts and mitochondria multiply—by increasing in size and then dividing in two—is strikingly reminiscent of bacterial multiplication.

## Peroxisomes

Organelles similar in structure to lysosomes, but smaller, are called **peroxisomes** (see Figure 4.22). Although peroxisomes were once thought to form by budding off the ER, it is now generally agreed that they form by the division of preexisting peroxisomes.

Peroxisomes contain one or more enzymes that can oxidize various organic substances. For example, substances such as amino acids and fatty acids are oxidized in peroxisomes as part of normal metabolism. In addition, enzymes in peroxisomes oxidize toxic substances, such as alcohol. A by-product of the oxidation reactions is hydrogen peroxide ($H_2O_2$), a potentially toxic compound. However, peroxisomes also contain the enzyme *catalase*, which decomposes $H_2O_2$ (see Chapter 6, page 157). Because the generation and degradation of $H_2O_2$ occurs within the same organelle, peroxisomes protect other parts of the cell from the toxic effects of $H_2O_2$.

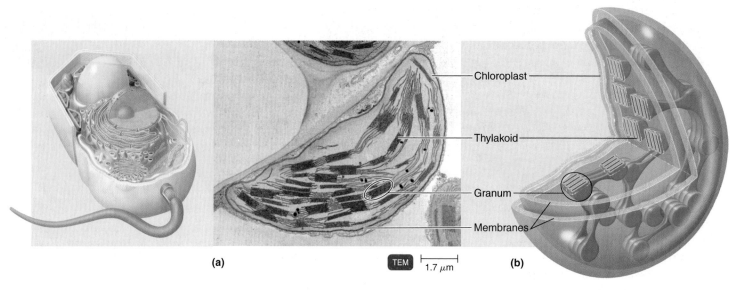

(a)

TEM ├─── 1.7 μm

(b)

**Figure 4.28 Chloroplasts.** Photosynthesis occurs in chloroplasts; the light-trapping pigments are located on the thylakoids. **(a)** A micrograph of chloroplasts in a plant cell. **(b)** A drawing of details of a chloroplast, showing grana.

**Q** What are the similarities between chloroplasts and prokaryotic cells?

## Centrosome

The **centrosome**, located near the nucleus, consists of two components: the pericentriolar area and centrioles (see Figure 4.22). The *pericentriolar material* is a region of the cytosol composed of a dense network of small protein fibers. This area is the organizing center for the mitotic spindle, which plays a critical role in cell division, and for microtubule formation in nondividing cells. Within the pericentriolar material is a pair of cylindrical structures called *centrioles*, each of which is composed of nine clusters of three microtubules (triplets) arranged in a circular pattern, an arrangement called a *9 + 0 array*. The 9 refers to the nine clusters of microtubules, and the 0 refers to the absence of microtubules in the center. The long axis of one centriole is at a right angle to the long axis of the other.

> **CHECK YOUR UNDERSTANDING**
>
> ✔ 4-18 Compare the structure of the nucleus of a eukaryote and the nucleoid of a prokaryote.
>
> ✔ 4-19 How do rough and smooth ER compare structurally and functionally?

## The Evolution of Eukaryotes

### LEARNING OBJECTIVE

4-20 Discuss evidence that supports the endosymbiotic theory of eukaryotic evolution.

Biologists generally believe that life arose on Earth in the form of very simple organisms, similar to prokaryotic cells, about 3.5 to 4 billion years ago. About 2.5 billion years ago, the first

eukaryotic cells evolved from prokaryotic cells. Recall that prokaryotes and eukaryotes differ mainly in that eukaryotes contain highly specialized organelles. The theory explaining the origin of eukaryotes from prokaryotes, pioneered by Lynn Margulis, is the **endosymbiotic theory**. According to this theory, larger bacterial cells lost their cell walls and engulfed smaller bacterial cells. This relationship, in which one organism lives within another, is called *endosymbiosis (symbiosis = living together)*.

> ASM: Cells, organelles (e.g., mitochondria and chloroplasts) and all major metabolic pathways evolved from early prokaryotic cells.

According to the endosymbiotic theory, the ancestral eukaryote developed a rudimentary nucleus when the plasma membrane folded around the chromosome (see Figure 10.2, page 273). This cell, called a nucleoplasm, may have ingested aerobic bacteria. Some ingested bacteria lived inside the host nucleoplasm. This arrangement evolved into a symbiotic relationship in which the host nucleoplasm supplied nutrients and the endosymbiotic bacterium produced energy that could be used by the nucleoplasm. Similarly, chloroplasts may be descendants of photosynthetic prokaryotes ingested by this early nucleoplasm.

Studies comparing prokaryotic and eukaryotic cells provide evidence for the endosymbiotic theory. For example, both mitochondria and chloroplasts resemble bacteria in size and shape. These organelles contain circular DNA, which is typical of prokaryotes, and the organelles can reproduce independently of their host cell. Moreover, mitochondrial and chloroplast ribosomes resemble those of prokaryotes, and their mechanism of protein synthesis is more similar to that found in bacteria than eukaryotes. Finally, the same antibiotics that inhibit protein synthesis on ribosomes in bacteria also inhibit protein synthesis on ribosomes in mitochondria and chloroplasts.

**CHECK YOUR UNDERSTANDING**

✔ **4-20** Which three organelles are not associated with the Golgi complex? What does this suggest about their origin?

# Study Outline

 **Go to** @MasteringMicrobiology for **Interactive Microbiology**, *In the Clinic* videos, *MicroFlix, MicroBoosters, 3D animations, practice quizzes, and more.*

## Comparing Prokaryotic and Eukaryotic Cells: An Overview (p. 73)

1. Prokaryotic and eukaryotic cells are similar in their chemical composition and chemical reactions.
2. Prokaryotic cells typically lack membrane-enclosed organelles (including a nucleus).
3. Peptidoglycan is found in prokaryotic cell walls but not in eukaryotic cell walls.
4. Eukaryotic cells have a membrane-bound nucleus and other organelles.

## The Prokaryotic Cell (pp. 73–94)

1. Bacteria are unicellular, and most of them multiply by binary fission.
2. Bacterial species are differentiated by morphology, chemical composition, nutritional requirements, biochemical activities, and source of energy.

## The Size, Shape, and Arrangement of Bacterial Cells (pp. 73–75)

1. Most bacteria are 0.2 to 2.0 μm in diameter and 2 to 8 μm in length.
2. The three basic bacterial shapes are coccus (spherical), bacillus (rod-shaped), and spiral (twisted).
3. Pleomorphic bacteria can assume several shapes.

## Structures External to the Cell Wall (pp. 75–80)

### Glycocalyx (pp. 75–77)

1. The glycocalyx (capsule, slime layer, or extracellular polysaccharide) is a gelatinous polysaccharide and/or polypeptide covering.
2. Capsules may protect pathogens from phagocytosis.
3. Capsules enable adherence to surfaces, prevent desiccation, and may provide nutrients.

### Flagella and Archaella (pp. 77–79)

4. Bacterial flagella and archael archaella rotate to push the cell.
5. Flagella are relatively long filamentous appendages consisting of a filament, hook, and basal body.
6. Motile bacteria exhibit taxis; positive taxis is movement toward an attractant, and negative taxis is movement away from a repellent.
7. Flagellar (H) protein is an antigen.

### Axial Filaments (pp. 79)

8. Spiral cells that move by means of an axial filament (endoflagellum) are called spirochetes.

9. Axial filaments are similar to flagella, except that they wrap around the cell.

### Fimbriae and Pili (pp. 79–80)

10. Fimbriae help cells adhere to surfaces.
11. Pili are involved in twitching motility and DNA transfer.

## The Cell Wall (pp. 80–85)

### Composition and Characteristics (pp. 80–83)

1. The cell wall surrounds the plasma membrane and protects the cell from changes in water pressure.
2. The bacterial cell wall consists of peptidoglycan, a polymer consisting of NAG and NAM and short chains of amino acids.
3. Gram-positive cell walls consist of many layers of peptidoglycan and also contain teichoic acids.
4. Gram-negative bacteria have a lipopolysaccharide-lipoprotein-phospholipid outer membrane surrounding a thin peptidoglycan layer.
5. The outer membrane protects the cell from phagocytosis and from penicillin, lysozyme, and other chemicals.
6. Porins are proteins that permit small molecules to pass through the outer membrane; specific channel proteins allow other molecules to move through the outer membrane.
7. The lipopolysaccharide component of the outer membrane consists of sugars (O polysaccharides), which function as antigens, and lipid A, which is an endotoxin.

### Cell Walls and the Gram Stain Mechanism (p. 83)

8. The crystal violet–iodine complex combines with peptidoglycan.
9. The decolorizer removes the lipid outer membrane of gram-negative bacteria and washes out the crystal violet.

### Atypical Cell Walls (p. 83)

10. *Mycoplasma* is a bacterial genus that naturally lacks cell walls.
11. Archaea have pseudomurein; they lack peptidoglycan.
12. Acid-fast cell walls have a layer of mycolic acid outside a thin peptidoglycan layer.

### Damage to the Cell Wall (pp. 83–85)

13. In the presence of lysozyme, gram-positive cell walls are destroyed, and the remaining cellular contents are referred to as a protoplast.
14. In the presence of lysozyme, gram-negative cell walls are not completely destroyed, and the remaining cellular contents are referred to as a spheroplast.
15. L forms are gram-positive or gram-negative bacteria that do not make a cell wall.
16. Antibiotics such as penicillin interfere with cell wall synthesis.

## Structures Internal to the Cell Wall (pp. 85–94)

### The Plasma (Cytoplasmic) Membrane (pp. 85–87)

1. The plasma membrane encloses the cytoplasm and is a lipid bilayer with peripheral and integral proteins (the fluid mosaic model).
2. The plasma membrane is selectively permeable.
3. Plasma membranes contain enzymes for metabolic reactions, such as nutrient breakdown, energy production, and photosynthesis.
4. Mesosomes, irregular infoldings of the plasma membrane, are artifacts, not true cell structures.
5. Plasma membranes can be destroyed by alcohols and polymyxins.

### The Movement of Materials across Membranes (pp. 87–90)

6. Movement across the membrane may be by passive processes, in which materials move from areas of higher to lower concentration and no energy is expended by the cell.
7. In simple diffusion, molecules and ions move until equilibrium is reached.
8. In facilitated diffusion, substances are transported by transporter proteins across membranes from areas of high to low concentration.
9. Osmosis is the movement of water from areas of high to low concentration across a selectively permeable membrane until equilibrium is reached.
10. In active transport, materials move from areas of low to high concentration by transporter proteins, and the cell must expend energy.
11. In group translocation, energy is expended to modify chemicals and transport them across the membrane.

### Cytoplasm (p. 90)

12. Cytoplasm is the fluid component inside the plasma membrane.
13. The cytoplasm is mostly water, with inorganic and organic molecules, DNA, ribosomes, inclusions, and cytoskeleton proteins.
14. A cytoskeleton is present, but cytoplasmic streaming does not occur.

### The Nucleoid (p. 90)

15. The nucleoid contains the DNA of the bacterial chromosome.
16. Bacteria can also contain plasmids, which are circular, extrachromosomal DNA molecules.

### Ribosomes (pp. 90–91)

17. The cytoplasm of a prokaryote contains numerous 70S ribosomes; ribosomes consist of rRNA and protein.
18. Protein synthesis occurs at ribosomes; it can be inhibited by certain antibiotics.

### Inclusions (pp. 91–92)

19. Inclusions are reserve deposits found in prokaryotic and eukaryotic cells.
20. Among the inclusions found in bacteria are metachromatic granules (inorganic phosphate), polysaccharide granules (usually glycogen or starch), lipid inclusions, sulfur granules, carboxysomes (ribulose 1,5-diphosphate carboxylase), magnetosomes ($Fe_3O_4$), and gas vacuoles.

### Endospores (pp. 92–94)

21. Endospores are resting structures formed by some bacteria; they allow survival during adverse environmental conditions.

## The Eukaryotic Cell (pp. 94–102)

### Flagella and Cilia (p. 96)

1. Flagella are few and long in relation to cell size; cilia are numerous and short.
2. Flagella and cilia are used for motility, and cilia also move substances along the surface of the cells.
3. Both flagella and cilia consist of an arrangement of nine pairs and two single microtubules.

### The Cell Wall and Glycocalyx (pp. 96–97)

1. The cell walls of many algae and some fungi contain cellulose.
2. The main material of fungal cell walls is chitin.
3. Yeast cell walls consist of glucan and mannan.
4. Animal cells are surrounded by a glycocalyx, which strengthens the cell and provides a means of attachment to other cells.

### The Plasma (Cytoplasmic) Membrane (p. 97)

1. Like the prokaryotic plasma membrane, the eukaryotic plasma membrane is a phospholipid bilayer containing proteins.
2. Eukaryotic plasma membranes contain carbohydrates attached to the proteins and sterols not found in prokaryotic cells (except *Mycoplasma* bacteria).
3. Eukaryotic cells can move materials across the plasma membrane by the passive processes and by active transport used by prokaryotes and endocytosis (phagocytosis, pinocytosis, and receptor-mediated endocytosis).

### Cytoplasm (p. 98)

1. The cytoplasm of eukaryotic cells includes everything inside the plasma membrane and external to the nucleus.
2. The chemical characteristics of the cytoplasm of eukaryotic cells resemble those of the cytoplasm of prokaryotic cells.
3. Eukaryotic cytoplasm has a cytoskeleton and exhibits cytoplasmic streaming.

### Ribosomes (p. 98)

1. 80S ribosomes are found in the cytoplasm or attached to the rough endoplasmic reticulum.

### Organelles (pp. 98–102)

1. Organelles are specialized membrane-enclosed structures in the cytoplasm of eukaryotic cells.
2. The nucleus, which contains DNA in the form of chromosomes, is the most characteristic eukaryotic organelle.
3. The nuclear envelope is connected to a system of membranes in the cytoplasm called the endoplasmic reticulum (ER).
4. The ER provides a surface for chemical reactions and serves as a transport network. Protein synthesis and transport occur on the rough ER; lipid synthesis occurs on the smooth ER.
5. The Golgi complex consists of flattened sacs called cisternae. It functions in membrane formation and protein secretion.
6. Lysosomes are formed from Golgi complexes. They store digestive enzymes.

7. Vacuoles are membrane-enclosed cavities derived from the Golgi complex or endocytosis. They are usually found in plant cells that store various substances and provide rigidity to leaves and stems.

8. Mitochondria are the primary sites of ATP production. They contain 70S ribosomes and DNA, and they multiply by binary fission.

9. Chloroplasts contain chlorophyll and enzymes for photosynthesis. Like mitochondria, they contain 70S ribosomes and DNA and multiply by binary fission.

10. A variety of organic compounds are oxidized in peroxisomes. Catalase in peroxisomes destroys $H_2O_2$.

11. The centrosome consists of the pericentriolar material and centrioles. Centrioles are 9 triplet microtubules involved in formation of the mitotic spindle and microtubules.

## The Evolution of Eukaryotes (pp. 102–103)

1. According to the endosymbiotic theory, eukaryotic cells evolved from symbiotic prokaryotes living inside other prokaryotic cells.

# Study Questions

For answers to the Knowledge and Comprehension questions, turn to the Answers tab at the back of the textbook.

## Knowledge and Comprehension

### Review

1. **DRAW IT** Diagram each of the following flagellar arrangements:
   a. lophotrichous
   b. monotrichous
   c. peritrichous
   d. amphitrichous
   e. polar

2. Endospore formation is called (a) _____. It is initiated by (b) _____. Formation of a new cell from an endospore is called (c) _____. This process is triggered by (d) _____.

3. **DRAW IT** Draw the bacterial shapes listed in (a), (b), and (c). Then draw the shapes in (d), (e), and (f), showing how they are special conditions of a, b, and c, respectively.
   a. spiral
   b. bacillus
   c. coccus
   d. spirochetes
   e. staphylococci
   f. streptobacilli

4. Match the structures in column A to their functions in column B.

| Column A | Column B |
| --- | --- |
| _____ a. Cell wall | 1. Attachment to surfaces |
| _____ b. Endospore | 2. Cell wall formation |
| _____ c. Fimbriae | 3. Motility |
| _____ d. Flagella | 4. Protection from osmotic lysis |
| _____ e. Glycocalyx | 5. Protection from phagocytes |
| _____ f. Pili | 6. Resting |
| _____ g. Plasma membrane | 7. Protein synthesis |
| _____ h. Ribosomes | 8. Selective permeability |
| | 9. Transfer of genetic material |

5. Why is an endospore called a resting structure? Of what advantage is an endospore to a bacterial cell?

6. Compare and contrast the following:
   a. simple diffusion and facilitated diffusion
   b. active transport and facilitated diffusion
   c. active transport and group translocation

7. Answer the following questions using the diagrams provided, which represent cross sections of bacterial cell walls.
   a. Which diagram represents a gram-positive bacterium? How can you tell?

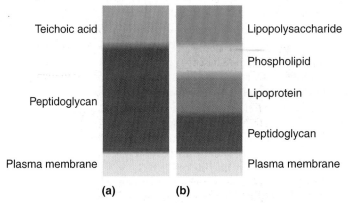

(a)          (b)

   b. Explain how the Gram stain works to distinguish these two types of cell walls.
   c. Why does penicillin have no effect on most gram-negative cells?
   d. How do essential molecules enter cells through each wall?
   e. Which cell wall is toxic to humans?

8. Starch is readily metabolized by many cells, but a starch molecule is too large to cross the plasma membrane. How does a cell obtain the glucose molecules from a starch polymer? How does the cell transport these glucose molecules across the plasma membrane?

9. Match the characteristics of eukaryotic cells in column A with their functions in column B.

| Column A | Column B |
| --- | --- |
| _____ a. Pericentriolar material | 1. Digestive enzyme storage |
| _____ b. Chloroplasts | 2. Oxidation of fatty acids |
| _____ c. Golgi complex | 3. Microtubule formation |
| _____ d. Lysosomes | 4. Photosynthesis |
| _____ e. Mitochondria | 5. Protein synthesis |
| _____ f. Peroxisomes | 6. Respiration |
| _____ g. Rough ER | 7. Secretion |

10. **NAME IT** What group of microbes is characterized by cells that form filaments, reproduce by spores, and have peptidoglycan in their cell walls?

## Multiple Choice

1. Which of the following is *not* a distinguishing characteristic of prokaryotic cells?
   a. They usually have a single, circular chromosome.
   b. They have 70S ribosomes.
   c. They have cell walls containing peptidoglycan.
   d. Their DNA is not associated with histones.
   e. They lack a plasma membrane.

**Use the following choices to answer questions 2–4.**
   a. No change will result; the solution is isotonic.
   b. Water will move into the cell.
   c. Water will move out of the cell.
   d. The cell will undergo osmotic lysis.
   e. Sucrose will move into the cell from an area of higher concentration to one of lower concentration.

2. Which statement best describes what happens when a gram-positive bacterium is placed in distilled water and penicillin?

3. Which statement best describes what happens when a gram-negative bacterium is placed in distilled water and penicillin?

4. Which statement best describes what happens when a gram-positive bacterium is placed in an aqueous solution of lysozyme and 10% sucrose?

5. Which of the following statements best describes what happens to a cell exposed to polymyxins that destroy phospholipids?
   a. In an isotonic solution, nothing will happen.
   b. In a hypotonic solution, the cell will lyse.
   c. Water will move into the cell.
   d. Intracellular contents will leak from the cell.
   e. Any of the above might happen.

6. Which of the following is *false* about fimbriae?
   a. They are composed of protein.
   b. They may be used for attachment.
   c. They are found on gram-negative cells.
   d. They are composed of pilin.
   e. They may be used for motility.

7. Which of the following pairs is *mismatched*?
   a. glycocalyx—adherence
   b. pili—reproduction
   c. cell wall—toxin
   d. cell wall—protection
   e. plasma membrane—transport

8. Which of the following pairs is *mismatched*?
   a. metachromatic granules—stored phosphates
   b. polysaccharide granules—stored starch
   c. lipid inclusions—poly-β-hydroxybutyric acid
   d. sulfur granules—energy reserve
   e. ribosomes—protein storage

9. You have isolated a motile, gram-positive cell with no visible nucleus. You can assume this cell has
   a. ribosomes.
   b. mitochondria.
   c. an endoplasmic reticulum.
   d. a Golgi complex.
   e. all of the above

10. The antibiotic amphotericin B disrupts plasma membranes by combining with sterols; it will affect all of the following cells *except*
   a. animal cells.
   b. gram-negative bacterial cells.
   c. fungal cells.
   d. *Mycoplasma* cells.
   e. plant cells.

## Analysis

1. How can prokaryotic cells be smaller than eukaryotic cells and still carry on all the functions of life?

2. The smallest eukaryotic cell is the motile alga *Ostreococcus*. What is the minimum number of organelles this alga must have?

3. Two types of prokaryotic cells have been distinguished: bacteria and archaea. How do these cells differ from each other? How are they similar?

4. In 1985, a 0.5-mm cell was discovered in surgeonfish and named *Epulopiscium fishelsoni* (see Figure 11.20, page 313). It was presumed to be a protozoan. In 1993, researchers determined that *Epulopiscium* is actually a gram-positive bacterium. Why do you suppose this organism was initially identified as a protozoan? What evidence would change the classification to bacterium?

5. When *E. coli* cells are exposed to a hypertonic solution, the bacteria produce a transporter protein that can move $K^+$ (potassium ions) into the cell. Of what value is the active transport of $K^+$, which requires ATP?

## Clinical Applications and Evaluation

1. *Clostridium botulinum* is a strict anaerobe; that is, it is killed by the molecular oxygen ($O_2$) present in air. Humans can die of botulism from eating foods in which *C. botulinum* is growing. How does this bacterium survive on plants picked for human consumption? Why are home-canned foods most often the source of botulism?

2. A South San Francisco child enjoyed bath time at his home because of the colorful orange and red water. The water did not have this rusty color at its source, and the water department could not culture the *Acidithiobacillus* bacteria responsible for the rusty color from the source. How were the bacteria getting into the household water? What bacterial structures make this possible?

3. Live cultures of *Bacillus thuringiensis* (Dipel®) and *B. subtilis* (Kodiak®) are sold as pesticides. What bacterial structures make it possible to package and sell these bacteria? For what purpose is each product used? (*Hint*: Refer to Chapter 11.)

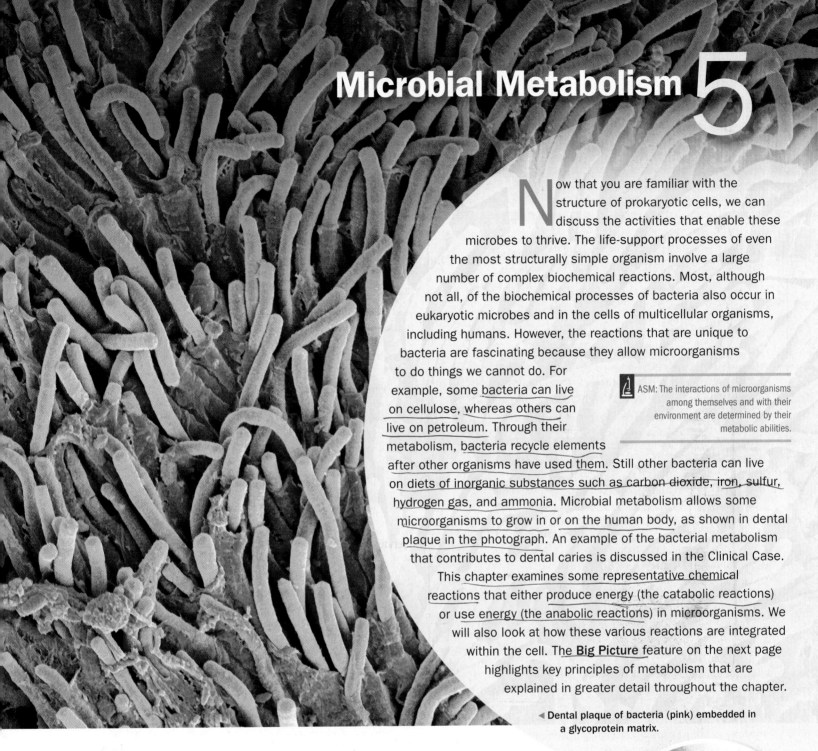

# Microbial Metabolism 5

Now that you are familiar with the structure of prokaryotic cells, we can discuss the activities that enable these microbes to thrive. The life-support processes of even the most structurally simple organism involve a large number of complex biochemical reactions. Most, although not all, of the biochemical processes of bacteria also occur in eukaryotic microbes and in the cells of multicellular organisms, including humans. However, the reactions that are unique to bacteria are fascinating because they allow microorganisms to do things we cannot do. For example, some bacteria can live on cellulose, whereas others can live on petroleum. Through their metabolism, bacteria recycle elements

ASM: The interactions of microorganisms among themselves and with their environment are determined by their metabolic abilities.

after other organisms have used them. Still other bacteria can live on diets of inorganic substances such as carbon dioxide, iron, sulfur, hydrogen gas, and ammonia. Microbial metabolism allows some microorganisms to grow in or on the human body, as shown in dental plaque in the photograph. An example of the bacterial metabolism that contributes to dental caries is discussed in the Clinical Case. This chapter examines some representative chemical reactions that either produce energy (the catabolic reactions) or use energy (the anabolic reactions) in microorganisms. We will also look at how these various reactions are integrated within the cell. The **Big Picture** feature on the next page highlights key principles of metabolism that are explained in greater detail throughout the chapter.

◀ Dental plaque of bacteria (pink) embedded in a glycoprotein matrix.

## In the Clinic

As the nurse researcher in a large medical center, you are working with the gastroenterology physicians on a project to study the effect of diet on intestinal gas. People in the test group who developed the most gas ate broccoli and beans, which are high in raffinose and stachyose, and eggs, which are high in methionine and cysteine. Intestinal gas is composed of $CO_2$, $CH_4$, $H_2S$, and $H_2$. **How are these gases produced?**

*Hint: Read about carbohydrates (pages 35–36), amino acids (pages 40–42), carbohydrate catabolism (pages 119–132), and protein catabolism (page 133).*

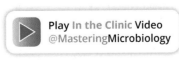
Play **In the Clinic** Video
@Mastering**Microbiology**

## Metabolism is the buildup and breakdown of nutrients within a cell. These chemical reactions provide energy and create substances that sustain life.

Two key players in metabolism are enzymes and the molecule adenosine triphosphate (ATP).

Enzymes catalyze reactions for specific molecules called **substrates**. During enzymatic reactions, substrates are transformed into new substances called **products**.

Substrate

Products

Enzyme

Enzymes, which are generally proteins, may need other nonprotein molecules called cofactors to work. Inorganic cofactors include metal ions. Organic cofactors, or coenzymes, include the electron carriers FAD, NAD$^+$ and NADP$^+$.

Substrate
Cofactor
Enzyme

However, without energy, certain reactions will never occur, even if enzymes are present. Adenosine triphosphate (ATP) is a molecule that cells use to manage energy needs.

If a reaction results in excess energy, some can be captured in the form of ATP's bonds. A cell can then break those same bonds and use the released energy to fuel other reactions.

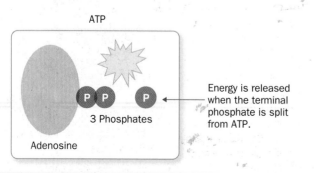

ATP

3 Phosphates

Adenosine

Energy is released when the terminal phosphate is split from ATP.

The chemistry of metabolism can seem overwhelming at first, with *pathways*, or sets of many coordinated reactions, working together toward common goals. But the basic rules of metabolism are actually quite simple. Pathways can be categorized into two general types—**catabolic** and **anabolic**.

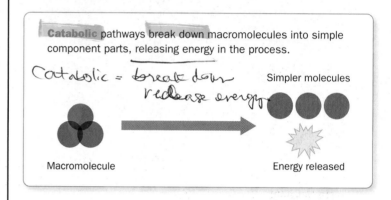

**Catabolic** pathways break down macromolecules into simple component parts, releasing energy in the process.

Catabolic = break down
release energy

Simpler molecules

Macromolecule

Energy released

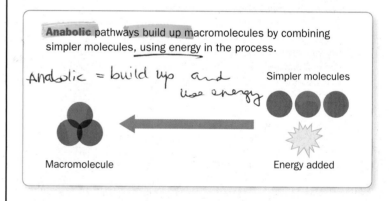

**Anabolic** pathways build up macromolecules by combining simpler molecules, using energy in the process.

Anabolic = build up and
use energy

Simpler molecules

Macromolecule

Energy added

In other words, catabolic and anabolic pathways are linked by **energy**. Catabolic reactions provide the energy needed for anabolic reactions.

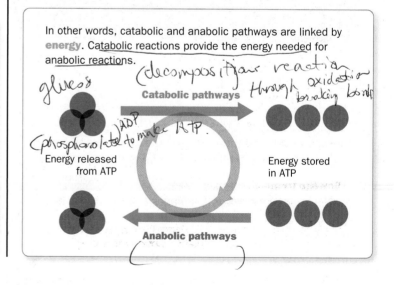

glucose

(decomposition reaction through oxidation breaking bonds

Catabolic pathways

(phosphorylate to make ATP

ADP

Energy released from ATP

Energy stored in ATP

**Anabolic pathways**

## Although microbial metabolism can cause disease and food spoilage, many pathways are beneficial rather than pathogenic.

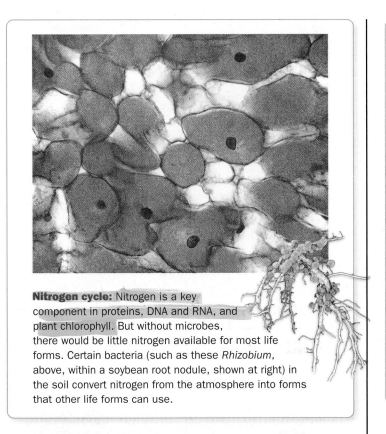

**Nitrogen cycle:** Nitrogen is a key component in proteins, DNA and RNA, and plant chlorophyll. But without microbes, there would be little nitrogen available for most life forms. Certain bacteria (such as these *Rhizobium*, above, within a soybean root nodule, shown at right) in the soil convert nitrogen from the atmosphere into forms that other life forms can use.

**Beverages and food:** Various bacteria and yeasts (such as *Saccharomyces cerevisiae,* shown at right) carry out catabolic reactions called fermentation. Beer, wine, and foods such as cheeses, yogurt, pickles, sauerkraut, and soy sauce tap microbial metabolism as a crucial part of production.

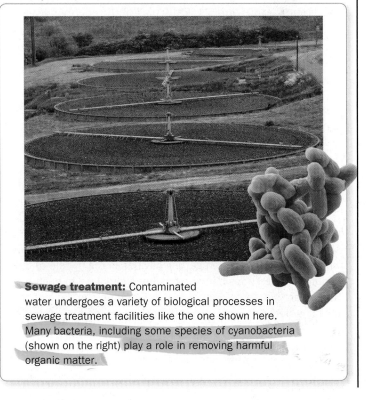

**Sewage treatment:** Contaminated water undergoes a variety of biological processes in sewage treatment facilities like the one shown here. Many bacteria, including some species of cyanobacteria (shown on the right) play a role in removing harmful organic matter.

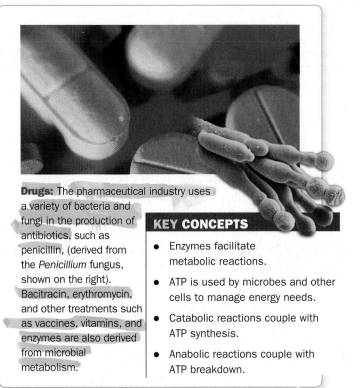

**Drugs:** The pharmaceutical industry uses a variety of bacteria and fungi in the production of antibiotics, such as penicillin, (derived from the *Penicillium* fungus, shown on the right). Bacitracin, erythromycin, and other treatments such as vaccines, vitamins, and enzymes are also derived from microbial metabolism.

### KEY CONCEPTS

- Enzymes facilitate metabolic reactions.
- ATP is used by microbes and other cells to manage energy needs.
- Catabolic reactions couple with ATP synthesis.
- Anabolic reactions couple with ATP breakdown.

*on slides,*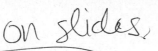

# Catabolic and Anabolic Reactions

## LEARNING OBJECTIVES

**5-1** Define *metabolism*, and describe the fundamental differences between anabolism and catabolism.

**5-2** Identify the role of ATP as an intermediate between catabolism and anabolism.

We use the term **metabolism** to refer to the sum of all chemical reactions within a living organism. Because chemical reactions either release or require energy, metabolism can be viewed as an energy-balancing act. Accordingly, metabolism can be divided into two classes of chemical reactions: those that release energy and those that require energy.

In living cells, the enzyme-regulated chemical reactions that release energy are generally the ones involved in **catabolism**, the breakdown of complex organic compounds into simpler ones. These reactions are called *catabolic*, or *degradative*, reactions. Catabolic reactions are generally *hydrolytic* reactions (reactions which use water and in which chemical bonds are broken), and they are *exergonic* (produce more energy than they consume). An example of catabolism occurs when cells break down sugars into carbon dioxide and water.

The enzyme-regulated energy-requiring reactions are mostly involved in **anabolism**, the building of complex organic molecules from simpler ones. These reactions are called *anabolic*, or *biosynthetic*, reactions. Anabolic processes often involve *dehydration synthesis* reactions (reactions that release water), and they are *endergonic* (consume more energy than they produce). Examples of anabolic processes are the formation of proteins from amino acids, nucleic acids from nucleotides, and polysaccharides from simple sugars. These biosynthetic reactions generate the materials for cell growth.

Catabolic reactions provide building blocks for anabolic reactions and furnish the energy needed to drive anabolic reactions. This coupling of energy-requiring and energy-releasing reactions is made possible through the molecule adenosine triphosphate (ATP). (You can review its structure in Figure 2.18, page 46.) ATP stores energy derived from catabolic reactions and releases it later to drive anabolic reactions and perform other cellular work. Remember that a molecule of ATP consists of an adenine, a ribose, and three phosphate groups. When the terminal phosphate group is split from ATP, adenosine diphosphate (ADP) is formed, and energy is released to drive anabolic reactions. Using ⓟ to represent a phosphate group (ⓟ$_i$ represents inorganic phosphate, which is not bound to any other molecule), we write this reaction as follows:

$$ATP \rightarrow ADP + ⓟ_i + energy$$

Then, the energy from catabolic reactions is used to combine ADP and a ⓟ$_i$ to resynthesize ATP:

$$ADP + ⓟ_i + energy \rightarrow ATP$$

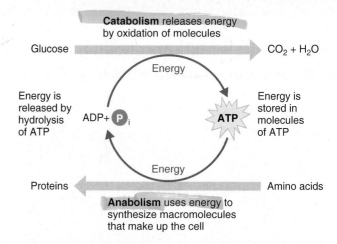

**Figure 5.1 The role of ATP in coupling anabolic and catabolic reactions.** When complex molecules are split apart (catabolism), some of the energy is transferred to and trapped in ATP, and the rest is given off as heat. When simple molecules are combined to form complex molecules (anabolism), ATP provides the energy for synthesis, and again some energy is given off as heat.

**Q** How does ATP provide the energy for synthesis?

Thus, anabolic reactions are coupled to ATP breakdown, and catabolic reactions are coupled to ATP synthesis. This concept of coupled reactions is very important; you will see why by the end of this chapter. For now, you should know that the chemical composition of a living cell is constantly changing: some molecules are broken down while others are being synthesized. This balanced flow of chemicals and energy maintains the life of a cell.

The role of ATP in coupling anabolic and catabolic reactions is shown in **Figure 5.1**. Only part of the energy released in catabolism is actually available for cellular functions because part of the energy is lost to the environment as heat. Because the cell must use energy to maintain life, it has a continuous need for new external sources of energy.

### CLINICAL CASE More Than a Sweet Tooth

Dr. Antonia Rivera is a pediatric dentist in St. Louis, Missouri. Her latest patient, 7-year-old Micah Thompson, has just left the office with strict instructions about brushing and flossing regularly. What most worries Dr. Rivera, however, is that Micah is her seventh patient this week to present with multiple dental caries, or cavities. Dr. Rivera is used to seeing some increase in tooth decay after Halloween and Easter, but why are all these children getting cavities in the middle of the summer? When possible, she has been speaking to each of the patient's parents or grandparents, but no one has noticed anything out of the ordinary in the children's diets.

**Why do so many of Dr. Rivera's patients have multiple dental caries? Read on to find out.**

Before we discuss how cells produce energy, let's first consider the principal properties of a group of proteins involved in almost all biologically important chemical reactions: enzymes. A cell's **metabolic pathways** (sequences of chemical reactions) are determined by its enzymes, which are in turn determined by the cell's genetic makeup.

> **Play Metabolism:** *Overview*
> @Mastering**Microbiology**

# Enzymes

## LEARNING OBJECTIVES

**5-3** Identify the components of an enzyme.

**5-4** Describe the mechanism of enzymatic action.

**5-5** List the factors that influence enzymatic activity.

**5-6** Distinguish competitive and noncompetitive inhibition.

**5-7** Define *ribozyme*.

## Collision Theory

Chemical reactions occur when chemical bonds are formed or broken. For reactions to take place, atoms, ions, or molecules must collide. The **collision theory** explains how chemical reactions occur and how certain factors affect the rates of those reactions. The basis of the collision theory is that all atoms, ions, and molecules are continuously moving and colliding with one another. The energy transferred by the particles in the collision can disrupt their electron structures enough to break chemical bonds or form new bonds.

Several factors determine whether a collision will cause a chemical reaction: the velocities of the colliding particles, their energy, and their specific chemical configurations. Up to a point, the higher the particles' velocities, the more probable that their collision will cause a reaction. Also, each chemical reaction requires a specific level of energy. But even if colliding particles possess the minimum energy needed for reaction, no reaction will take place unless the particles are properly oriented toward each other.

Let's assume that molecules of substance AB (the reactant) are to be converted to molecules of substances A and B (the products). In a given population of molecules of substance AB, at a specific temperature, some molecules possess relatively little energy; the majority of the population possesses an average amount of energy; and a small portion of the population has high energy. If only the high-energy AB molecules are able to react and be converted to A and B molecules, then only relatively few molecules at any one time possess enough energy to react in a collision. The collision energy required for a chemical

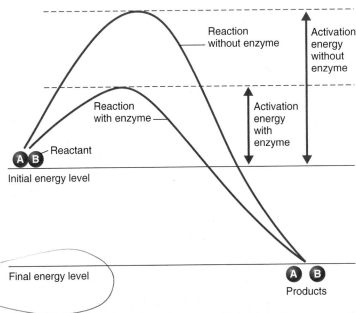

**Figure 5.2 Energy requirements of a chemical reaction.** This graph shows the progress of the reaction AB → A + B both without (blue line) and with (red line) an enzyme. The presence of an enzyme lowers the activation energy of the reaction (see arrows). Thus, more molecules of reactant AB are converted to products A and B because more molecules of reactant AB possess the activation energy needed for the reaction.

**Q** Why does a chemical reaction require increased activation energy without an enzyme as a biological catalyst?

reaction is its **activation energy**, which is the amount of energy needed to disrupt the stable electronic configuration of any specific molecule so that the electrons can be rearranged.

The **reaction rate**—the frequency of collisions containing sufficient energy to bring about a reaction—depends on the number of reactant molecules at or above the activation energy level. One way to increase the reaction rate of a substance is to raise its temperature. By causing the molecules to move faster, heat increases both the frequency of collisions and the number of molecules that attain activation energy. The number of collisions also increases when pressure is increased or when the reactants are more concentrated (because the distance between molecules is thereby decreased). In living systems, enzymes increase the reaction rate without raising the temperature.

## Enzymes and Chemical Reactions

Substances that can speed up a chemical reaction without being permanently altered themselves are called **catalysts**. In living cells, **enzymes** serve as biological catalysts. As catalysts, each enzyme acts on a specific substance, called the enzyme's **substrate** (or substrates, when there are two or more reactants), and each catalyzes only one reaction. For example, sucrose (table sugar) is the substrate of the enzyme sucrase, which catalyzes the hydrolysis of sucrose to glucose and fructose.

As catalysts, enzymes typically accelerate chemical reactions by lowering their activation energy (**Figure 5.2**). The

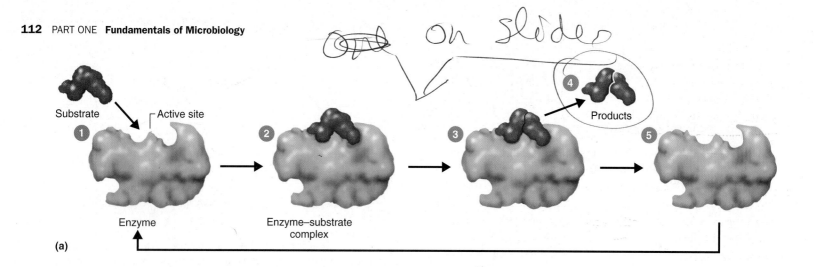

*on slides*

#2

**Figure 5.3 The mechanism of enzymatic action. (a)** ❶ The substrate contacts the active site on the enzyme to form ❷ an enzyme–substrate complex. ❸ The substrate is then transformed into products, ❹ the products are released, and ❺ the enzyme is recovered unchanged. In the example shown, the transformation into products involves a breakdown of the substrate into two products. Other transformations may occur, however. **(b)** Left: A molecular model of the enzyme in step ❶ of part (a). The active site of the enzyme can be seen here as a groove on the surface of the protein. Right: As the enzyme and substrate meet in step ❷ of part (a), the enzyme changes shape slightly to fit together more tightly with the substrate.

**Q** **Give an example of enzymatic specifity.**

enzyme therefore speeds up the reaction by increasing the number of AB molecules that attain sufficient activation energy to react. The general sequence of events in enzyme action is as follows (**Figure 5.3a**).

❶ The surface of the substrate contacts a specific region of the surface of the enzyme molecule, called the *active site*.

❷ A temporary intermediate compound forms, called an **enzyme–substrate complex**. The enzyme orients the substrate into a position that increases the probability of reaction, which enables the collisions to be more effective.

❸ The substrate molecule is transformed by the rearrangement of existing atoms, the breakdown of the substrate molecule, or in combination with another substrate molecule.

❹ The transformed substrate molecules—the products of the reaction—are released from the enzyme molecule because they no longer fit in the active site of the enzyme.

❺ The unchanged enzyme is now free to react with other substrate molecules.

An enzyme's ability to accelerate a reaction without the need for an increase in temperature is crucial to living systems because a significant temperature increase would destroy cellular proteins. The crucial function of enzymes, therefore, is to speed up biochemical reactions at a temperature that is compatible with the normal functioning of the cell.

## Enzyme Specificity and Efficiency

Enzymes have specificity for particular substrates. For example, a specific enzyme may be able to hydrolyze a peptide bond only between two specific amino acids. Other enzymes can hydrolyze starch but not cellulose; even though both starch and cellulose are polysaccharides composed of glucose subunits, the orientations of the subunits in the two polysaccharides differ. Each of the thousands of known enzymes have this specificity because the three-dimensional shape of the specific amino acids of the active site fits the substrate somewhat as a lock fits with its key (Figure 5.3b). The unique configuration of each enzyme enables it to "find" the correct substrate from among the diverse molecules in a cell. However, the active site and substrate are flexible, and they change shape somewhat as they meet to fit together more tightly. The substrate is usually much smaller than the enzyme, and relatively few of the enzyme's amino acids make up the active site. A certain compound can be a substrate for several different enzymes that catalyze different reactions, so the fate of a compound depends on the enzyme that acts on it. At least four different enzymes can act on glucose 6-phosphate, a molecule important in cell metabolism, and each reaction will yield a different product.

> **Play Enzymes:** *Overview, Steps in a Reaction*
> @MasteringMicrobiology

Enzymes are extremely efficient. Under optimum conditions, they can catalyze reactions at rates $10^8$ to $10^{10}$ times (up to 10 billion times) higher than those of comparable reactions without enzymes. The **turnover number** (the maximum number of substrate molecules an enzyme molecule converts to product each second) is generally between 1 and 10,000 and can be as high as 500,000. For example, the enzyme DNA polymerase I, which participates in the synthesis of DNA, has a turnover number of 15, whereas the enzyme lactate dehydrogenase, which removes hydrogen atoms from lactic acid, has a turnover number of 1000.

Many enzymes exist in the cell in both active and inactive forms. The rate at which enzymes switch between these two forms is determined by the cellular environment.

## Naming Enzymes

The names of enzymes usually end in -ase. All enzymes can be grouped into six classes, according to the type of chemical reaction they catalyze (Table 5.1). Enzymes within each of the major classes are named according to the more specific types of reactions they assist. For example, the class called *oxidoreductases* is

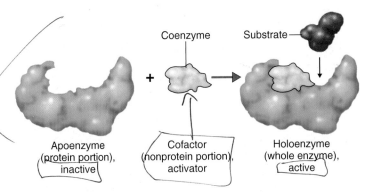

**Figure 5.4 Components of a holoenzyme.** Many enzymes require both an apoenzyme (protein portion) and a cofactor (nonprotein portion) to become active. The cofactor can be a metal ion, or if it is an organic molecule, it is called a coenzyme (as shown here). The apoenzyme and cofactor together make up the holoenzyme, or whole enzyme. The substrate is the reactant acted upon by the enzyme.

**Q** How does the enzyme–substrate complex lower the activation energy of the reaction?

involved with oxidation-reduction reactions (described shortly). Enzymes in the oxidoreductase class that remove hydrogen (H) from a substrate are called *dehydrogenases;* those that add electrons to molecular oxygen ($O_2$) are called *oxidases*. As you will see later, dehydrogenase and oxidase enzymes have even more specific names, such as lactate dehydrogenase and cytochrome oxidase, depending on the specific substrates on which they act.

### Enzyme Components

Although some enzymes consist entirely of proteins, most consist of both a protein portion, called an **apoenzyme**, and a nonprotein component, called a **cofactor**. Ions of iron, zinc, magnesium, or calcium are examples of cofactors. If the cofactor is an organic molecule, it is called a **coenzyme**. Apoenzymes are inactive by themselves; they must be activated by cofactors. Together, the apoenzyme and cofactor form a **holoenzyme**, or whole, active enzyme (**Figure 5.4**). If the cofactor is removed, the apoenzyme will not function.

| Class | Type of Chemical Reaction Catalyzed | Examples |
|---|---|---|
| Oxidoreductase | Oxidation-reduction, in which oxygen and hydrogen are gained or lost | Cytochrome oxidase, lactate dehydrogenase |
| Transferase | Transfer of functional groups, such as an amino group, acetyl group, or phosphate group | Acetate kinase, alanine deaminase |
| Hydrolase | Hydrolysis (addition of water) | Lipase, sucrase |
| Lyase | Removal of groups of atoms without hydrolysis | Oxalate decarboxylase, isocitrate lyase |
| Isomerase | Rearrangement of atoms within a molecule | Glucose-phosphate isomerase, alanine racemase |
| Ligase | Joining of two molecules (using energy usually derived from the breakdown of ATP) | Acetyl-CoA synthetase, DNA ligase |

**TABLE 5.1 Enzyme Classification Based on Type of Chemical Reaction Catalyzed**

**TABLE 5.2** Selected Vitamins and Their Coenzymatic Functions

| Vitamin | Function |
|---|---|
| Vitamin B$_1$ (Thiamine) | Part of coenzyme cocarboxylase; has many functions, including the metabolism of pyruvic acid |
| Vitamin B$_2$ (Riboflavin) | Coenzyme in flavoproteins; active in electron transfers |
| Niacin (Nicotinic Acid) | Part of NAD molecule*; active in electron transfers |
| Vitamin B$_6$ (Pyridoxine) | Coenzyme in amino acid metabolism |
| Vitamin B$_{12}$ (Cyanocobalamin) | Coenzyme (methyl cyanocobalamide) involved in the transfer of methyl groups; active in amino acid metabolism |
| Pantothenic Acid | Part of coenzyme A molecule; involved in the metabolism of pyruvic acid and lipids |
| Biotin | Involved in carbon dioxide fixation reactions and fatty acid synthesis |
| Folic Acid | Coenzyme used in the synthesis of purines and pyrimidines |
| Vitamin E | Needed for cellular and macromolecular syntheses |
| Vitamin K | Coenzyme used in electron transport |

*NAD = nicotinamide adenine dinucleotide

Cofactors may help catalyze a reaction by forming a bridge between an enzyme and its substrate. For example, magnesium ($Mg^{2+}$) is required by many phosphorylating enzymes (enzymes that transfer a phosphate group from ATP to another substrate). The $Mg^{2+}$ can form a link between the enzyme and the ATP molecule. Most trace elements required by living cells are probably used in some such way to activate cellular enzymes.

Coenzymes may assist the enzyme by accepting atoms removed from the substrate or by donating atoms required by the substrate. Some coenzymes act as electron carriers, removing electrons from the substrate and donating them to other molecules in subsequent reactions. Many coenzymes are derived from vitamins (Table 5.2). Two of the most important coenzymes in cellular metabolism are **nicotinamide adenine dinucleotide (NAD$^+$)** and **nicotinamide adenine dinucleotide phosphate (NADP$^+$)**. Both compounds contain derivatives of the B vitamin niacin (nicotinic acid), and both function as electron carriers. Whereas NAD$^+$ is primarily involved in catabolic (energy-yielding) reactions, NADP$^+$ is primarily involved in anabolic (energy-requiring) reactions. The flavin coenzymes, such as **flavin mononucleotide (FMN)** and **flavin adenine dinucleotide (FAD)**, contain derivatives of the B vitamin riboflavin and are also electron carriers. Another important coenzyme, **coenzyme A (CoA)**, contains a derivative of pantothenic acid, another B vitamin. This coenzyme plays an important role in the synthesis and breakdown of fats and in a series of oxidizing reactions called the Krebs cycle. We will come across all of these coenzymes in our discussion of metabolism later in the chapter.

### Factors Influencing Enzymatic Activity

Enzymes are subject to various cellular controls. Two primary types are the control of enzyme *synthesis* (see Chapter 8) and the control of enzyme *activity* (how much enzyme is present versus how active it is).

Several factors influence the activity of an enzyme. Among the more important are temperature, pH, substrate concentration, and the presence or absence of inhibitors.

### Temperature

The rate of most chemical reactions increases as the temperature increases. Molecules move more slowly at lower temperatures than at higher temperatures and so may not have enough energy to cause a chemical reaction. For enzymatic reactions, however, elevation beyond a certain temperature (the optimal temperature) drastically reduces the rate of reaction (**Figure 5.5a**). The optimal temperature for most disease-producing bacteria in the human body is between 35°C and 40°C. The rate of reaction declines beyond the optimal temperature because of the enzyme's **denaturation**, the loss of its characteristic three-dimensional structure (tertiary configuration) (**Figure 5.6**). Denaturation of a protein involves the breakage of hydrogen bonds and other noncovalent bonds; a common example is the transformation of uncooked egg white (a protein called albumin) to a hardened state by heat.

Denaturation of an enzyme changes the arrangement of the amino acids in the active site, altering its shape and causing the enzyme to lose its catalytic ability. In some cases, denaturation is partially or fully reversible. However, if denaturation continues until the enzyme has lost its solubility and coagulates, the enzyme cannot regain its original properties. Enzymes can also be denatured by concentrated acids, bases, heavy-metal ions (such as lead, arsenic, or mercury), alcohol, and ultraviolet radiation.

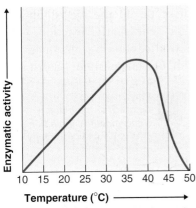

(a) **Temperature.** The enzymatic activity (rate of reaction catalyzed by the enzyme) increases with increasing temperature until the enzyme, a protein, is denatured by heat and inactivated. At this point, the reaction rate falls steeply.

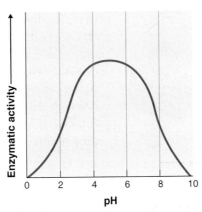

(b) **pH.** The enzyme illustrated is most active at about pH 5.0.

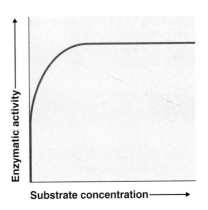

(c) **Substrate concentration.** With increasing concentration of substrate molecules, the rate of reaction increases until the active sites on all the enzyme molecules are filled, at which point the maximum rate of reaction is reached.

**Figure 5.5 Factors that influence enzymatic activity, plotted for a hypothetical enzyme.**

**Q** How will this enzyme act at 25°C? At 45°C? At pH 7?

## pH

Typically, enzymes have an optimum pH at which they are most active. Above or below this pH value, enzyme activity, and therefore the reaction rate, decline (Figure 5.5b). When the $H^+$ concentration (pH) in the medium is changed drastically, the protein's three-dimensional structure is altered. Extreme changes in pH can cause denaturation. Acids (and bases) alter a protein's three-dimensional structure because the $H^+$ (and $OH^-$) compete with hydrogen and ionic bonds in an enzyme, resulting in the enzyme's denaturation.

### Substrate Concentration

Under conditions of high substrate concentration, an enzyme is said to be in **saturation**; that is, its active site is always occupied by substrate or product molecules, and it's catalyzing a

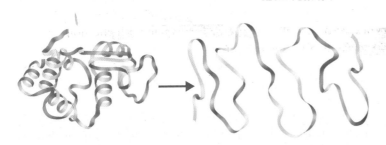

Active (functional) protein                    Denatured protein

**Figure 5.6 Denaturation of a protein.** Breakage of the noncovalent bonds (such as hydrogen bonds) that hold the active protein in its three-dimensional shape renders the denatured protein nonfunctional.

**Q** When is denaturation irreversible?

specific reaction at its maximum rate. This maximum rate can be attained only when the concentration of substrate(s) is extremely high. In this condition, a further increase in substrate concentration will not affect the reaction rate because all active sites are already in use (Figure 5.5c). Under normal cellular conditions, enzymes are not saturated with substrate(s). At any given time, many of the enzyme molecules are inactive for lack of substrate; thus, the substrate concentration is likely to influence the rate of reaction.

### Inhibitors

An effective way to control the growth of bacteria is to control, or inhibit, their enzymes. Certain poisons, such as cyanide, arsenic, and mercury, combine with enzymes and prevent the bacteria from functioning. As a result, the cells stop functioning and die.

Enzyme inhibitors are classified as either competitive or noncompetitive inhibitors (**Figure 5.7**). **Competitive inhibitors** fill the active site of an enzyme and compete with the normal substrate for the active site. A competitive inhibitor can do this because its shape and chemical structure are similar to those of the normal substrate (Figure 5.7b). However, unlike the substrate, it does not undergo any reaction to form products. Some competitive inhibitors bind irreversibly to amino acids in the active site, preventing any further interactions with the substrate. Others bind reversibly, alternately occupying and leaving the active site; these slow the enzyme's interaction with the substrate. Increasing the substrate concentration can overcome reversible competitive inhibition. As active sites become available, more substrate molecules than competitive inhibitor molecules are available to attach to the active sites of enzymes.

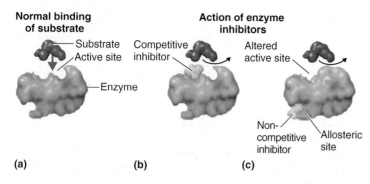

Normal binding of substrate

Action of enzyme inhibitors

(a)  (b)  (c)

**Figure 5.7 Enzyme inhibitors. (a)** An uninhibited enzyme and its normal substrate. **(b)** A competitive inhibitor. **(c)** One type of noncompetitive inhibitor, causing allosteric inhibition.

**Q** How do competitive inhibitors operate in comparison to noncompetitive inhibitors?

One good example of a competitive inhibitor is sulfanilamide (a potent antibacterial drug), which inhibits the enzyme whose normal substrate is *para*-aminobenzoic acid (PABA):

Sulfanilamide    PABA

PABA is an essential nutrient used by many bacteria in the synthesis of folic acid, a vitamin that functions as a coenzyme. When sulfanilamide is administered to bacteria, the enzyme that normally converts PABA to folic acid combines instead with the sulfanilamide. Folic acid is not synthesized, and the bacteria cannot grow. Because human cells do not use PABA to make their folic acid, sulfanilamide can kill bacteria but does not harm human cells.

**Noncompetitive inhibitors** do not compete with the substrate for the enzyme's active site; instead, they interact with another part of the enzyme (Figure 5.7c). In this process, called **allosteric** ("other space") **inhibition**, the inhibitor binds to a site on the enzyme other than the substrate's binding site, called the **allosteric site**. This binding causes the active site to change its shape, making it nonfunctional. As a result, the enzyme's activity is reduced. This effect can be either reversible or irreversible, depending on whether the active site can return to its original shape. In some cases, allosteric interactions can activate an enzyme rather than inhibit it. Another type of noncompetitive inhibition can operate on enzymes that require metal ions for their activity.

Certain chemicals can bind or tie up the metal ion activators and thus prevent an enzymatic reaction. Cyanide can bind the iron in iron-containing enzymes, and fluoride can bind calcium or magnesium. Substances such as cyanide and fluoride are sometimes called *enzyme poisons* because they permanently inactivate enzymes.

Play Enzymes: *Competitive Inhibition, Noncompetitive Inhibition* @MasteringMicrobiology

### Feedback Inhibition   #5

Noncompetitive, or allosteric, inhibitors play a role in a type of biochemical control called **feedback inhibition**, or **end-product inhibition**. This control mechanism stops the cell from making more of a substance than it needs and thereby wasting chemical resources. In some metabolic reactions, several steps are required for the synthesis of a particular chemical compound, called the **end-product**. The process is similar to an assembly line, with each step catalyzed by a separate enzyme (**Figure 5.8**). In many metabolic pathways, the end-product can allosterically inhibit the activity of one of the enzymes earlier in the pathway.

Feedback inhibition generally acts on the first enzyme in a metabolic pathway (similar to shutting down an assembly line by stopping the first worker). Because the enzyme is inhibited, the product of the first enzymatic reaction in the pathway is not synthesized. Because that unsynthesized product would normally be the substrate for the second enzyme in the pathway, the second reaction stops immediately as well. Thus, even though only the first enzyme in the pathway is inhibited, the entire pathway shuts down, and no new end-product is formed. By inhibiting the first enzyme in the pathway, the cell also keeps metabolic intermediates from accumulating. As the cell uses up the existing end-product, the first enzyme's allosteric site remains unbound more frequently, and the pathway resumes activity.

The bacterium *E. coli* can be used to demonstrate feedback inhibition in the synthesis of the amino acid isoleucine, which is required for the cell growth. In this metabolic pathway, the amino acid threonine is enzymatically converted to isoleucine in five steps. If isoleucine is added to the growth medium for *E. coli*, it inhibits the first enzyme in the pathway, and the bacteria stop synthesizing isoleucine. This condition is maintained until the supply of isoleucine is depleted. This type of feedback inhibition is also involved in regulating the cells' production of other amino acids, as well as vitamins, purines, and pyrimidines.

### Ribozymes

Prior to 1982, it was believed that only protein molecules had enzymatic activity. Then researchers working on microbes

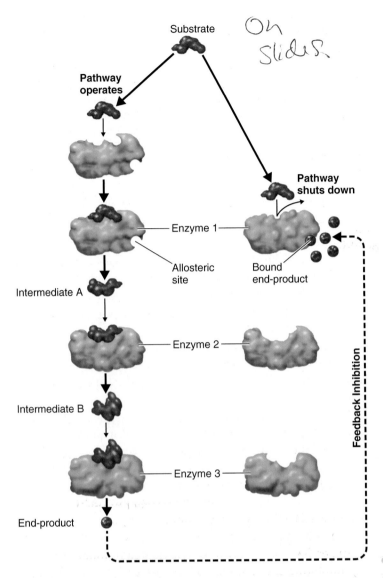

*On slides*

**Figure 5.8 Feedback inhibition.**

**Q** Explain the differences between competitive inhibition and feedback inhibition.

discovered a unique type of RNA called a **ribozyme**. Like protein enzymes, ribozymes function as catalysts, have active sites that bind to substrates, and are not used up in a chemical reaction. Ribozymes cut and splice RNA and are involved in protein synthesis at ribosomes.

**CHECK YOUR UNDERSTANDING**

✔ **5-3** What is a coenzyme?

✔ **5-4** Why is enzyme specificity important?

✔ **5-5** What happens to an enzyme below its optimal temperature? Above its optimal temperature?

✔ **5-6** Why is feedback inhibition noncompetitive inhibition?

✔ **5-7** What is a ribozyme?

# Energy Production

**LEARNING OBJECTIVES**

**5-8** Explain the term *oxidation-reduction*.

**5-9** List and provide examples of three types of phosphorylation reactions that generate ATP.

**5-10** Explain the overall function of metabolic pathways.

Nutrient molecules, like all molecules, have energy associated with the electrons that form bonds between their atoms. When it's spread throughout the molecule, this energy is difficult for the cell to use. Various reactions in catabolic pathways, however, concentrate the energy into the bonds of ATP, which serves as a convenient energy carrier. ATP is generally referred to as having "high-energy" bonds. Actually, a better term is probably *unstable bonds*. Although the amount of energy in these bonds is not exceptionally large, it can be released quickly and easily. In a sense, ATP is similar to a highly flammable liquid such as kerosene. Although a large log might eventually burn to produce more heat than a cup of kerosene, the kerosene is easier to ignite and provides heat more quickly and conveniently. In a similar way, the "high-energy" unstable bonds of ATP provide the cell with readily available energy for anabolic reactions.

Before discussing the catabolic pathways, we will consider two general aspects of energy production: the concept of oxidation-reduction and the mechanisms of ATP generation.

## Oxidation-Reduction Reactions

**Oxidation** is the removal of electrons ($e-$) from an atom or molecule, a reaction that often produces energy. **Figure 5.9** shows an example of an oxidation in which molecule A loses an electron to molecule B. Molecule A has undergone oxidation (meaning that it has lost one or more electrons), whereas molecule B has undergone **reduction** (meaning that it has gained one or more electrons).* Oxidation and reduction reactions are always coupled: each time one substance is oxidized, another is simultaneously reduced. The pairing of these reactions is called **oxidation-reduction** or a **redox reaction**.

In many cellular oxidations, electrons and protons (hydrogen ions, $H^+$) are removed at the same time; this is equivalent to the removal of hydrogen atoms, because a hydrogen atom is made up of one proton and one electron (see Table 2.2, page 26). Because most biological oxidations involve the loss

---

* The terms do not seem logical until one considers the history of the discovery of these reactions. When mercury is heated, it gains weight as mercuric oxide is formed; this was called *oxidation*. Later it was determined that the mercury actually *lost* electrons, and the observed *gain* in oxygen was a direct result of this. Oxidation, therefore, is a *loss* of electrons, and reduction is a *gain* of electrons, but the gain and loss of electrons is not usually apparent as chemical-reaction equations are usually written. For example, in this chapter's equations for aerobic respiration on page 128, notice that each carbon in glucose had only one oxygen originally, and later, as carbon dioxide, each carbon now has two oxygens. However, the gain or loss of electrons is not apparent in the equations.

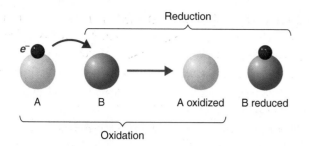

**Figure 5.9 Oxidation-reduction.** An electron is transferred from molecule A to molecule B. In the process, molecule A is oxidized, and molecule B is reduced.

**Q** How do oxidation and reduction differ?

*[handwritten: involves removing H+ atoms it contains an electron and one proton]*

of hydrogen atoms, they are also called **dehydrogenation reactions.** Figure 5.10 shows an example of a biological oxidation. An organic molecule is oxidized by the loss of two hydrogen atoms, and a molecule of NAD$^+$ is reduced. Remember that NAD$^+$ assists enzymes by accepting hydrogen atoms that have been removed from the substrate, in this case the organic molecule. As shown in Figure 5.10, NAD$^+$ accepts two electrons and one proton. One proton (H$^+$) is left over and is released into the surrounding medium. The reduced coenzyme, NADH, contains more energy than NAD$^+$. This energy can be used to generate ATP in later reactions.

It's important to remember that cells use biological oxidation-reduction reactions in catabolism to extract energy from nutrient molecules. Cells take nutrients, some of which serve as energy sources, and degrade them from highly reduced compounds (with many hydrogen atoms) to highly oxidized compounds. For example, when a cell oxidizes a molecule of glucose (C$_6$H$_{12}$O$_6$) to CO$_2$ and H$_2$O, the energy in the glucose molecule is removed in a stepwise manner and ultimately is trapped by ATP, which can then serve as an energy source for energy-requiring reactions. Compounds such as glucose that have many hydrogen atoms are highly

> ▶ Play Oxidation-Reduction Reactions @MasteringMicrobiology

reduced compounds, containing a large amount of potential energy. Thus, glucose is a valuable nutrient for organisms.

**CHECK YOUR UNDERSTANDING**

✔ 5-8 Why is glucose such an important molecule for organisms?

## The Generation of ATP

Much of the energy released during oxidation-reduction reactions is trapped within the cell by the formation of ATP. Specifically, an inorganic phosphate group, **P**$_i$, is added to ADP with the input of energy to form ATP:

$$\overbrace{\text{Adenosine}-\textcircled{P}\sim\textcircled{P}}^{\text{ADP}} + \text{Energy} + \textcircled{P}_i \longrightarrow$$

$$\underbrace{\text{Adenosine}-\textcircled{P}\sim\textcircled{P}\sim\textcircled{P}}_{\text{ATP}}$$

The symbol ~ designates a "high-energy" bond that can readily be broken to release usable energy. The high-energy bond that attaches the third ⊙ contains the energy stored in this reaction, in a sense. When this ⊙ is removed, usable energy is released. The addition of ⊙ to a chemical compound is called **phosphorylation**. Organisms use three mechanisms of phosphorylation to generate ATP from ADP. *[handwritten: on slides]*

### Substrate-Level Phosphorylation

In **substrate-level phosphorylation**, ATP is usually generated when a high-energy ⊙ is directly transferred from a phosphorylated compound (a substrate) to ADP. Generally, the ⊙ has acquired its energy during an earlier reaction in which the substrate itself was oxidized. The following example shows only the carbon skeleton and the ⊙ of a typical substrate:

$$\text{C}-\text{C}-\text{C}\sim\textcircled{P} + \text{ADP} \rightarrow \text{C}-\text{C}-\text{C} + \text{ATP}$$

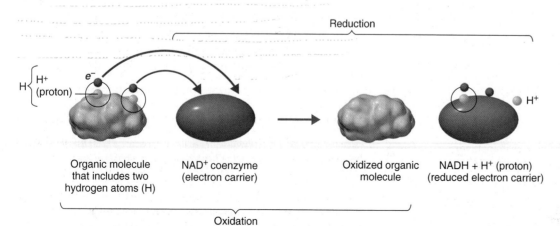

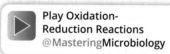

**Figure 5.10 Representative biological oxidation.** Two electrons and two protons (altogether equivalent to two hydrogen atoms) are transferred from an organic substrate molecule to a coenzyme, NAD$^+$. NAD$^+$ actually receives one hydrogen atom and two electrons, and one proton is released into the medium. NAD$^+$ is reduced to NADH, which is a more energy-rich molecule.

**Q** How do organisms use oxidation-reduction reactions?

## Oxidative Phosphorylation

In oxidative phosphorylation, electrons are transferred from organic compounds to one group of electron carriers (usually to $NAD^+$ and FAD). Then the electrons are passed through a series of different electron carriers to molecules of oxygen ($O_2$) or other oxidized inorganic and organic molecules. This process occurs in the plasma membrane of prokaryotes and in the inner mitochondrial membrane of eukaryotes. The sequence of electron carriers used in oxidative phosphorylation is called an **electron transport chain (system)** (see Figure 5.14). The transfer of electrons from one electron carrier to the next releases energy, some of which is used to generate ATP from ADP through a process called **chemiosmosis**, to be described on pages 126–127.

#7

## Photophosphorylation

The third mechanism of phosphorylation, **photophosphorylation**, occurs only in photosynthetic cells, which contain light-trapping pigments such as chlorophylls. In photosynthesis, organic molecules, especially sugars, are synthesized with the energy of light from the energy-poor building blocks, carbon dioxide and water. Photophosphorylation starts this process by converting light energy to the chemical energy of ATP and NADPH, which, in turn, are used to synthesize organic molecules. As in oxidative phosphorylation, an electron transport chain is involved.

**CHECK YOUR UNDERSTANDING**

✔ 5-9 Outline the three ways that ATP is generated.

## Metabolic Pathways of Energy Production

Organisms release and store energy from organic molecules by a series of controlled reactions rather than in a single burst. If the energy were released all at once as a large amount of heat, it could not be readily used to drive chemical reactions and would, in fact, damage the cell. To extract energy from organic compounds and store it in chemical form, organisms pass electrons from one compound to another through a series of oxidation-reduction reactions.

As noted earlier, a sequence of enzymatically catalyzed chemical reactions occurring in a cell is called a metabolic pathway. Below is a hypothetical metabolic pathway that converts starting material A to end-product E in a series of five steps.

NAD$^+$ NADH + H$^+$

A ———①———→ B
Starting material

— on slides

ADP + P$_i$  ATP

B ——②——→ C ⇌③ D  $O_2$  →④→ E
CO$_2$ H$_2$O   End-product

① Molecule A converts to molecule B. The curved arrow indicates that the reduction of coenzyme $NAD^+$ to NADH is coupled to that reaction; the electrons and protons come from molecule A.

② Similarly, the arrow shows a coupling of two reactions. As B is converted to C, ADP is converted to ATP; the energy needed comes from B as it transforms into C.

③ The reaction converting C to D is readily reversible, as indicated by the double arrow.

④ The arrow leading from $O_2$ indicates that $O_2$ is a reactant. The arrows leading to $CO_2$ and $H_2O$ indicate that these substances are secondary products produced in the reaction, in addition to E, the end-product that (presumably) interests us the most.

Secondary products, such as $CO_2$ and $H_2O$, are sometimes called *by-products* or *waste products*.

Keep in mind that almost every reaction in a metabolic pathway is catalyzed by a specific enzyme; sometimes the name of the enzyme is printed near the arrow.

**CHECK YOUR UNDERSTANDING**

✔ 5-10 What is the purpose of metabolic pathways?

## Carbohydrate Catabolism

**LEARNING OBJECTIVES**

5-11 Describe the chemical reactions of glycolysis.

5-12 Identify the functions of the pentose phosphate and Entner-Doudoroff pathways.

5-13 Explain the products of the Krebs cycle.

5-14 Describe the chemiosmotic model for ATP generation.

5-15 Compare and contrast aerobic and anaerobic respiration.

5-16 Describe the chemical reactions of, and list some products of, fermentation.

**Play Interactive Microbiology**
@MasteringMicrobiology
*See how prokaryotic aerobic respiration affects a patient's health*

Most microorganisms oxidize carbohydrates as their primary source of cellular energy. **Carbohydrate catabolism**, the breakdown of carbohydrate molecules to produce energy, is therefore of great importance in cell metabolism. Glucose is the most common carbohydrate energy source used by cells. Microorganisms can also catabolize various lipids and proteins for energy production (page 133).

To produce energy from glucose, microorganisms use two general processes: *cellular respiration and fermentation*. (In discussing cellular respiration, we frequently refer to the process simply as respiration, but it should not be confused with breathing.) Both cellular respiration and fermentation usually start with the same first step, glycolysis, but follow different subsequent pathways (**Figure 5.11**). Before examining the details of glycolysis, respiration, and fermentation, we will first look at a general overview of the processes.

# An Overview of Respiration and Fermentation

*On slides*

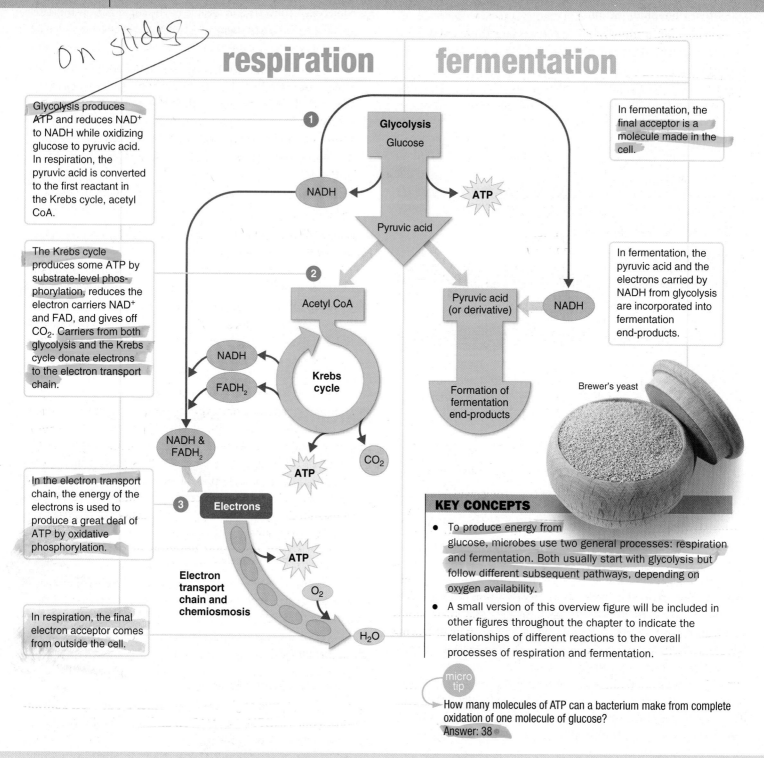

## respiration

## fermentation

Glycolysis produces ATP and reduces NAD$^+$ to NADH while oxidizing glucose to pyruvic acid. In respiration, the pyruvic acid is converted to the first reactant in the Krebs cycle, acetyl CoA.

The Krebs cycle produces some ATP by substrate-level phosphorylation, reduces the electron carriers NAD$^+$ and FAD, and gives off $CO_2$. Carriers from both glycolysis and the Krebs cycle donate electrons to the electron transport chain.

In the electron transport chain, the energy of the electrons is used to produce a great deal of ATP by oxidative phosphorylation.

In respiration, the final electron acceptor comes from outside the cell.

In fermentation, the final acceptor is a molecule made in the cell.

In fermentation, the pyruvic acid and the electrons carried by NADH from glycolysis are incorporated into fermentation end-products.

**1** **Glycolysis**
Glucose

NADH

ATP

Pyruvic acid

**2** Acetyl CoA

Pyruvic acid (or derivative)

NADH

NADH

FADH$_2$

**Krebs cycle**

NADH & FADH$_2$

ATP

$CO_2$

Formation of fermentation end-products

Brewer's yeast

**3** Electrons

ATP

**Electron transport chain and chemiosmosis**

$O_2$

$H_2O$

### KEY CONCEPTS

- To produce energy from glucose, microbes use two general processes: respiration and fermentation. Both usually start with glycolysis but follow different subsequent pathways, depending on oxygen availability.

- A small version of this overview figure will be included in other figures throughout the chapter to indicate the relationships of different reactions to the overall processes of respiration and fermentation.

micro tip

How many molecules of ATP can a bacterium make from complete oxidation of one molecule of glucose?
Answer: 38

---

As shown in Figure 5.11, the respiration of glucose typically occurs in three principal stages: glycolysis, the Krebs cycle, and the electron transport chain (system).

**1** Glycolysis is the oxidation of glucose to pyruvic acid with the production of some ATP and energy-containing NADH.

**2** The Krebs cycle is the oxidation of acetyl CoA (a derivative of pyruvic acid) to carbon dioxide, with the production of

some ATP, energy-containing NADH, and another reduced electron carrier, FADH$_2$ (the reduced form of flavin adenine dinucleotide).

**3** In the electron transport chain (system), NADH and FADH$_2$ are oxidized, contributing the electrons they have carried from the substrates to a "cascade" of oxidation-reduction reactions involving a series of additional electron carriers. Energy from these reactions is used to generate a considerable amount of ATP. In respiration, most of the ATP is generated in the third step.

Because respiration involves a long series of oxidation-reduction reactions, the entire process can be thought of as involving a flow of electrons from the energy-rich glucose molecule to the relatively energy-poor $CO_2$ and $H_2O$ molecules. The coupling of ATP production to this flow is somewhat analogous to producing electrical power by using energy from a flowing stream. Carrying the analogy further, you could imagine a stream flowing down a gentle slope during glycolysis and the Krebs cycle, supplying energy to turn two old-fashioned waterwheels. Then the stream rushes down a steep slope in the electron transport chain, supplying energy for a large modern power plant. In a similar way, glycolysis and the Krebs cycle generate a small amount of ATP and also supply the electrons that generate a great deal of ATP at the electron transport chain stage.

Typically, the initial stage of fermentation is also glycolysis (Figure 5.11). However, once glycolysis has taken place, the pyruvic acid is converted into one or more different products, depending on the type of cell. These products might include alcohol (ethanol) and lactic acid. Unlike respiration, there is no Krebs cycle or electron transport chain in fermentation. Accordingly, the ATP yield, which comes only from glycolysis, is much lower.

*# 9*

## Glycolysis

**Glycolysis**, the oxidation of glucose to pyruvic acid, is usually the first stage in carbohydrate catabolism. Most microorganisms use this pathway; in fact, it occurs in most living cells.

Glycolysis is also called the *Embden-Meyerhof pathway*. The word *glycolysis* means splitting of sugar, and this is exactly what happens. The enzymes of glycolysis catalyze the splitting of glucose, a six-carbon sugar, into two three-carbon sugars. These sugars are then oxidized, releasing energy, and their atoms are rearranged to form two molecules of pyruvic acid. During glycolysis $NAD^+$ is reduced to NADH, and there is a net production of two ATP molecules by substrate-level phosphorylation. Glycolysis does not require oxygen; it can occur whether oxygen is present or not. This pathway is a series of ten chemical reactions, each catalyzed by a different enzyme. The steps are outlined in **Figure 5.12**; see also Figure A.2 in Appendix A for a more detailed representation of glycolysis.

To summarize the process, glycolysis consists of two basic stages, a preparatory stage and an energy-conserving stage:

1 First, in the preparatory stage (steps 1 – 4 in Figure 5.12), two molecules of ATP are used as a six-carbon glucose molecule is phosphorylated, restructured, and split into two three-carbon compounds: glyceraldehyde 3-phosphate (GP) and dihydroxyacetone phosphate (DHAP). 5 DHAP is readily converted to GP. (The reverse reaction may also occur.)

The conversion of DHAP into GP means that from this point on in glycolysis, two molecules of GP are fed into the remaining chemical reactions.

2 In the energy-conserving stage (steps 6 – 10), the two three-carbon molecules are oxidized in several steps to two molecules of pyruvic acid. In these reactions, two molecules of $NAD^+$ are reduced to NADH, and four molecules of ATP are formed by substrate-level phosphorylation.

Because two molecules of ATP were needed to get glycolysis started and four molecules of ATP are generated by the process, *there is a net gain of two molecules of ATP for each molecule of glucose that is oxidized.*

Play Glycolysis: *Overview, Steps* @**Mastering**Microbiology

## Additional Pathways to Glycolysis

Many bacteria have another pathway in addition to glycolysis for the oxidation of glucose. The most common alternative is the *pentose phosphate pathway;* another alternative is the *Entner-Doudoroff pathway.*  *#10 On slides*

### The Pentose Phosphate Pathway

The **pentose phosphate pathway** (or *hexose monophosphate shunt*) operates simultaneously with glycolysis and provides a means for the breakdown of five-carbon sugars (pentoses) as well as glucose (see Figure A.3 in Appendix A for a more detailed representation of the pentose phosphate pathway). A key feature of this pathway is that it produces important intermediate pentoses used in the synthesis of (1) nucleic acids, (2) glucose from carbon dioxide in photosynthesis, and (3) certain amino acids. The pathway is an important producer of the reduced coenzyme NADPH from $NADP^+$. The pentose phosphate pathway yields a net gain of only one molecule of ATP for each molecule of glucose oxidized. Bacteria that use the pentose phosphate pathway include *Bacillus subtilis* (SU-til-us), *E. coli*, *Leuconostoc mesenteroides* (loo-kō-NOS-tok mes'en-TER-oi-dēz), and *Enterococcus faecalis* (fē-KĀ-lis).

*They do not undergo Glycolysis.*

### The Entner-Doudoroff Pathway  *gram-negative bacteria*

From each molecule of glucose, the **Entner-Doudoroff pathway** produces one NADPH, one NADH, and one ATP for use in cellular biosynthetic reactions (see Figure A.4 in Appendix A for a more detailed representation). Bacteria that have the enzymes for the Entner-Doudoroff pathway can metabolize glucose without either glycolysis or the pentose phosphate pathway. The Entner-Doudoroff pathway is found in some gram-negative bacteria, including *Rhizobium* (rī-ZŌ-bē-um), *Pseudomonas* (soo'dō-MŌ-nas), and *Agrobacterium* (A-grō-bak-ti'rē-um); it is generally not found among gram-positive bacteria. Tests for the ability to oxidize glucose by this pathway are sometimes used to identify *Pseudomonas* in the clinical laboratory.

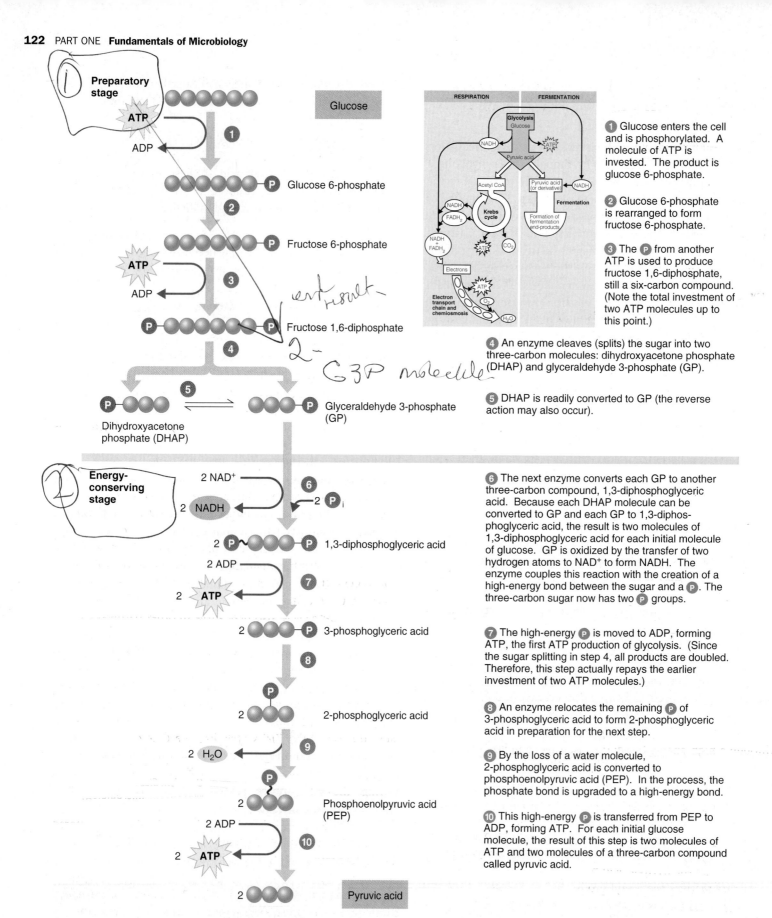

**Figure 5.12   An outline of the reactions of glycolysis (Embden-Meyerhof pathway).** The inset indicates the relationship of glycolysis to the overall processes of respiration and fermentation. A more detailed version of glycolysis is presented in Figure A.2 in Appendix A.

**Q** What is glycolysis?

## Cellular Respiration

After glucose has been broken down to pyruvic acid, the pyruvic acid can be channeled into the next step of either fermentation (page 128) or cellular respiration (see Figure 5.11). **Cellular respiration**, or simply **respiration**, is defined as an ATP-generating process in which molecules are oxidized and the final electron acceptor comes from outside the cell and is (almost always) an inorganic molecule. An essential feature of respiration is the operation of an electron transport chain.

There are two types of respiration, depending on whether an organism is an **aerobe**, which uses oxygen, or an **anaerobe**, which does not use oxygen and may even be killed by it. In **aerobic respiration**, the final electron acceptor is $O_2$; in **anaerobic respiration**, the final electron acceptor is an inorganic molecule other than $O_2$ or, rarely, an organic molecule. First we will describe respiration as it typically occurs in an aerobic cell.

Citric acid cycle #13

### Aerobic Respiration

**The Krebs Cycle** The Krebs cycle, also called the *tricarboxylic acid (TCA) cycle* or *citric acid cycle*, is a series of biochemical reactions in which the large amount of potential chemical energy stored in acetyl CoA is released step by step (see Figure 5.11). In this cycle, a series of oxidations and reductions transfer that potential energy, in the form of electrons, to electron carrier coenzymes, chiefly $NAD^+$ and $FADH_2$. The pyruvic acid derivatives are oxidized; the coenzymes are reduced.

Pyruvic acid, the product of glycolysis, cannot enter the Krebs cycle directly. In a preparatory step, it must lose one molecule of $CO_2$ and become a two-carbon compound (**Figure 5.13**, at top). This process is called **decarboxylation**. The two-carbon compound, called an *acetyl group*, attaches to coenzyme A through a high-energy bond; the resulting complex is known as *acetyl coenzyme A (acetyl CoA)*. During this reaction, pyruvic acid is also oxidized, and $NAD^+$ is reduced to NADH.

Remember that the oxidation of one glucose molecule produces two molecules of pyruvic acid, so for each molecule of glucose, two molecules of $CO_2$ are released, two molecules of NADH are produced, and two molecules of acetyl CoA are formed. Once the pyruvic acid has undergone decarboxylation and its derivative (the acetyl group) has attached to CoA, the resulting acetyl CoA is ready to enter the Krebs cycle. 2 Carbs.

As acetyl CoA enters the Krebs cycle, CoA detaches from the acetyl group. The acetyl group combines with oxaloacetic acid to form citric acid. This synthesis reaction requires energy, which is

provided by the cleavage of the high-energy bond between the acetyl group and CoA. The formation of citric acid is thus the first step in the Krebs cycle. The major chemical reactions of this cycle are outlined in Figure 5.13 (also see Figure A.5 in Appendix A for a more detailed representation of the Krebs cycle). Keep in mind that each reaction is catalyzed by a specific enzyme.

The chemical reactions of the Krebs cycle fall into several general categories; one of these is decarboxylation. For example, in step ❸, isocitric acid is decarboxylated to a compound called α-ketoglutaric acid. Another decarboxylation takes place in step ❹. All three carbon atoms in pyruvic acid are eventually released as $CO_2$ by the Krebs cycle. The conversion to $CO_2$ of all six carbon atoms contained in the original glucose molecule is completed in two turns of the Krebs cycle.

Another general category of Krebs cycle chemical reactions is oxidation-reduction. For example, in step ❸, isocitric acid is oxidized. Hydrogen atoms are also released in the Krebs cycle in steps ❹, ❻, and ❽ and are picked up by the coenzymes $NAD^+$ and FAD. Because $NAD^+$ picks up two electrons but only one additional proton, its reduced form is represented as NADH; however, FAD picks up two complete hydrogen atoms and is reduced to $FADH_2$.

If we look at the Krebs cycle as a whole, we see that for every two molecules of acetyl CoA that enter the cycle, four molecules of $CO_2$ are liberated by decarboxylation, six molecules of NADH and two molecules of $FADH_2$ are produced by oxidation-reduction reactions, and two molecules of ATP are generated by substrate-level phosphorylation. A molecule of guanosine triphosphate (GTP), formed from guanosine diphosphate (GDP + Ⓟi), is similar to ATP and serves as an intermediary at this point in the cycle. Many of the intermediates in the Krebs cycle also play a role in other pathways, especially in amino acid biosynthesis (page 143).

The $CO_2$ produced in the Krebs cycle is ultimately liberated into the atmosphere as a gaseous by-product of aerobic respiration. (Humans produce $CO_2$ from the Krebs cycle in most cells of the body and discharge it through the lungs during exhalation.) The reduced coenzymes NADH and $FADH_2$ are the most important products of the Krebs cycle because they contain most of the energy that was originally stored in glucose. During the next phase of respiration, a series of oxidation-reduction reactions indirectly transfer the energy stored in those coenzymes to ATP. These reactions are collectively called the electron transport chain. #14

Play Krebs Cycle:
*Overview, Steps*
@MasteringMicrobiology

**The Electron Transport Chain (System)** An **electron transport chain (electron transport system)** consists of a sequence of carrier molecules that are capable of oxidation and reduction. As electrons are passed through the chain, there occurs a stepwise release of energy, which is used to drive the

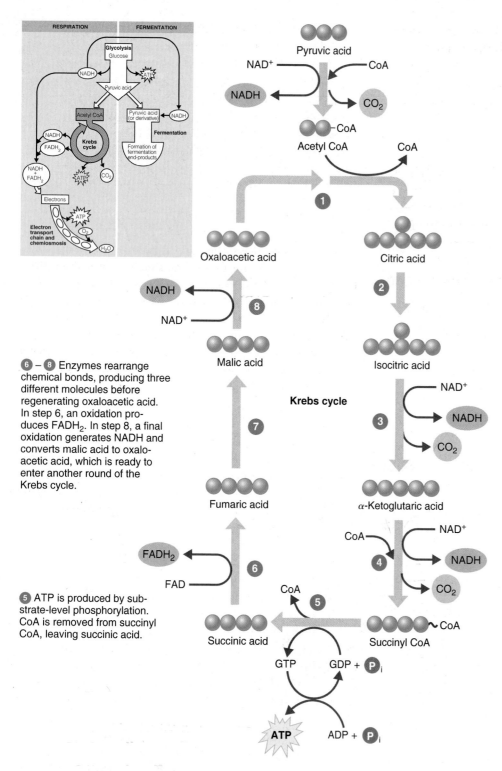

**Figure 5.13 The Krebs cycle.** The inset indicates the relationship of the Krebs cycle to the overall process of respiration. A more detailed version of the Krebs cycle is presented in Figure A.5 in Appendix A.

**Q** What are the products of the Krebs cycle?

**1** A turn of the cycle begins as enzymes strip off the CoA portion from acetyl CoA and combine the remaining two-carbon acetyl group with oxaloacetic acid. Adding the acetyl group produces the six-carbon molecule citric acid.

**2 – 4** Oxidations generate NADH. Step 2 is a rearrangement. Steps 3 and 4 combine oxidations and decarboxylations to dispose of two carbon atoms that came from oxaloacetic acid. The carbons are released as $CO_2$, and the oxidations generate NADH from $NAD^+$. During the second oxidation (step 4), CoA is added into the cycle, forming the compound succinyl CoA.

**6 – 8** Enzymes rearrange chemical bonds, producing three different molecules before regenerating oxaloacetic acid. In step 6, an oxidation produces $FADH_2$. In step 8, a final oxidation generates NADH and converts malic acid to oxaloacetic acid, which is ready to enter another round of the Krebs cycle.

**5** ATP is produced by substrate-level phosphorylation. CoA is removed from succinyl CoA, leaving succinic acid.

chemiosmotic generation of ATP, to be described shortly. The final oxidation is irreversible. In eukaryotic cells, the electron transport chain is contained in the inner membrane of mitochondria; in prokaryotic cells, it is found in the plasma membrane.

There are three classes of carrier molecules in electron transport chains:

**Flavoproteins** contain flavin, a coenzyme derived from riboflavin (vitamin $B_2$), and are capable of performing alternating oxidations and reductions. One important flavin coenzyme is flavin mononucleotide (FMN). **Cytochromes** are proteins with an iron-containing group (heme) capable of existing alternately as a reduced form ($Fe^{2+}$) and an oxidized form ($Fe^{3+}$). The cytochromes involved in electron transport chains include cytochrome $b$ (cyt $b$), cytochrome $c_1$ (cyt $c_1$), cytochrome *c* (cyt *c*), cytochrome *a* (cyt *a*), and cytochrome *a₃* (cyt *a₃*). **Ubiquinones**, or **coenzyme Q (Q)**, are small non-protein carriers.

The electron transport chains of bacteria are somewhat diverse, in that the particular carriers used by a bacterium and the order in which they function may differ from those of other bacteria and from those of eukaryotic mitochondrial systems. Even a single bacterium may have several types of electron transport chains. However, keep in mind that all electron transport chains achieve the same basic goal: to release energy while electrons are transferred from higher-energy compounds to lower-energy compounds. Much is known about the electron transport chain in the mitochondria of eukaryotic cells, so we'll describe its steps here.

①  High-energy electrons transfer from NADH to FMN, the first carrier in the chain (**Figure 5.14**). A hydrogen atom with two electrons passes to FMN, which picks up an additional $H^+$ from the surrounding aqueous medium. As a result NADH is oxidized to $NAD^+$, and FMN is reduced to $FMNH_2$.

②  $FMNH_2$ passes $2H^+$ to the other side of the mitochondrial membrane (see Figure 5.16) and passes two electrons to Q.

As a result, $FMNH_2$ is oxidized to FMN. Q also picks up an additional $2H^+$ from the surrounding aqueous medium and releases it on the other side of the membrane.

③  Electrons are passed successively from Q to cyt *b*, cyt *c₁*, cyt *c*, cyt *a*, and cyt *a₃*. Each cytochrome in the chain is reduced as it picks up electrons and is oxidized as it gives up electrons. The last cytochrome, cyt *a₃*, passes its electrons to molecular oxygen ($O_2$), which becomes negatively charged and picks up protons from the surrounding medium to form $H_2O$.

Notice that Figure 5.14 shows $FADH_2$, which is derived from the Krebs cycle, as another source of electrons. However, $FADH_2$ adds its electrons to the electron transport chain at a lower level than NADH. Because of this, the electron transport chain produces about one-third less energy for ATP generation when $FADH_2$ donates electrons than when NADH is involved.

An important feature of the electron transport chain is the presence of some carriers, such as FMN and Q, that accept and release protons and electrons, and other carriers, such as cytochromes, that transfer electrons only. Electron flow down

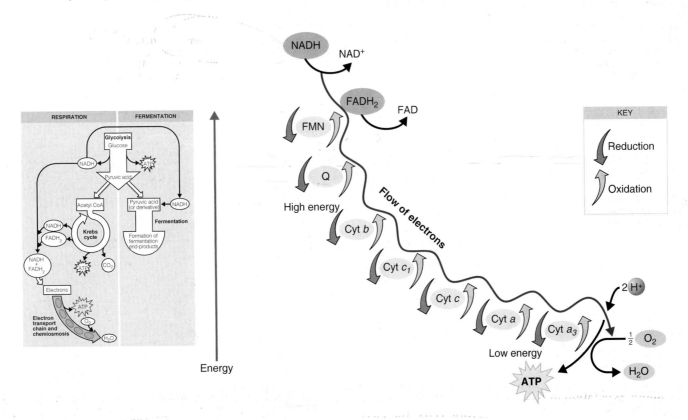

**Figure 5.14  An electron transport chain (system).** The inset indicates the relationship of the electron transport chain to the overall process of respiration. In the mitochondrial electron transport chain shown, the electrons pass along the chain in a gradual and stepwise fashion, so energy is released in manageable quantities (see Figure 5.16 to learn where ATP is formed).

**Q**  How many ATPs can be made from the oxidation of one NADH in the electron transport chain?

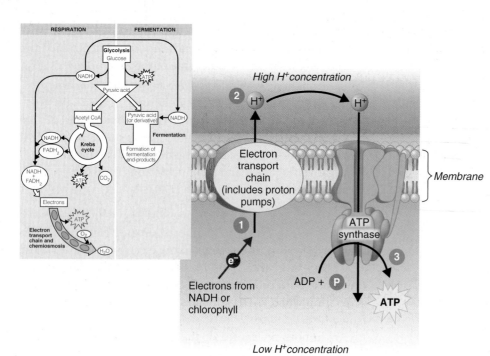

**Q** What is the proton motive force?

the chain is accompanied at several points by the active transport (pumping) of protons from the matrix side of the inner mitochondrial membrane to the opposite side of the membrane. The result is a buildup of protons on one side of the membrane. Just as water behind a dam stores energy that can be used to generate electricity, this buildup of protons provides energy that the chemiosmotic mechanism uses to generate ATP.

*Aerobic Respiration* *# 14 B*

### The Chemiosmotic Mechanism of ATP Generation ATP synthesis using the electron transport chain is called **chemiosmosis**, and it involves **oxidative phosphorylation**. To understand chemiosmosis, we need to review several concepts related to the movement of materials across membranes. (See Chapter 4, page 87.) Substances diffuse passively across membranes from areas of high concentration to areas of low concentration; this diffusion along a concentration gradient yields energy. The movement of substances *against* a concentration gradient *requires* energy that is usually provided by ATP. In chemiosmosis, the energy released when a substance moves along a gradient is used to *synthesize* ATP. The "substance" in this case refers to protons. In respiration, chemiosmosis is responsible for most of the ATP that is generated. The steps of chemiosmosis are as follows (**Figure 5.15**):

**1** As energetic electrons from NADH (or chlorophyll) pass down the electron transport chain, some of the carriers in the chain pump actively transport protons across the membrane. Such carrier molecules are called *proton pumps*.

**2** The phospholipid membrane is normally impermeable to protons, so this one-directional pumping establishes a proton gradient (a difference in the concentrations of protons on the two sides of the membrane). In addition to a concentration gradient, there is an electrical charge gradient. The excess $H^+$ on one side of the membrane makes that side positively charged compared with the other side. The resulting electrochemical gradient has potential energy, called the *proton motive force*.

**3** The protons on the side of the membrane with the higher proton concentration can diffuse across the membrane only through special protein channels that contain an enzyme called *ATP synthase*. When this flow occurs, energy is released and is used by the enzyme to synthesize ATP from ADP and **P**$_i$.

The detailed steps that show how the electron transport chain operates in eukaryotes to drive the chemiosmotic mechanism are as follows (**Figure 5.16**):

**1** Energetic electrons from NADH pass down the electron transport chains. Within the inner mitochondrial membrane, the carriers of the electron transport chain are organized into three complexes, with Q transporting electrons between the first and second complexes, and cyt *c* transporting them between the second and third complexes.

**2** Three components of the system pump protons. At the end of the chain, electrons join with protons and oxygen ($O_2$) in the matrix fluid to form water ($H_2O$). Thus, $O_2$ is the final electron acceptor.

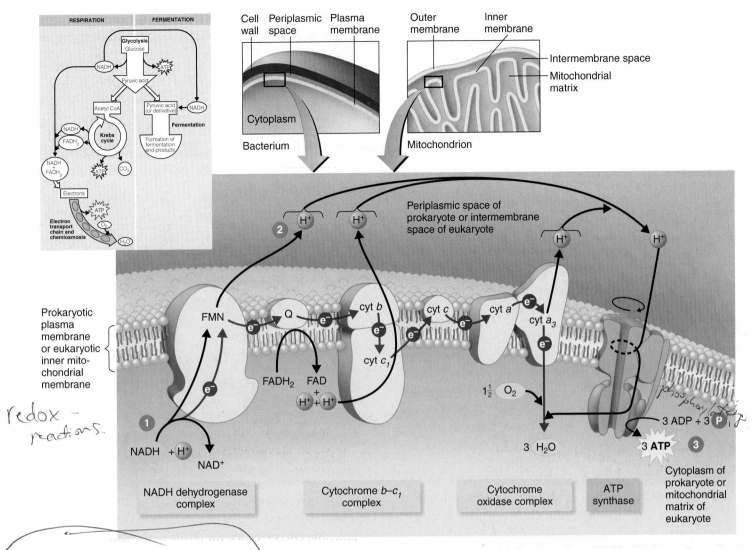

Figure 5.16 **Electron transport and the chemiosmotic generation of ATP.** Electron carriers are organized into three complexes, and protons (H$^+$) are pumped across the membrane at three points. In a prokaryotic cell, protons are pumped across the plasma membrane from the cytoplasmic side. In a eukaryotic cell, they are pumped from the matrix side of the mitochondrial membrane to the opposite side. The flow of electrons is indicated with red arrows.

**Q** Where does chemiosmosis occur in eukaryotes? In prokaryotes?

❸ Both prokaryotic and eukaryotic cells use the chemiosmotic mechanism to generate energy for ATP production. However, in eukaryotic cells, the inner mitochondrial membrane contains the electron transport carriers and ATP synthase. In most prokaryotic cells, the plasma membrane does so. An electron transport chain also operates in photophosphorylation and is located in the thylakoid membrane of cyanobacteria and eukaryotic chloroplasts.

**A Summary of Aerobic Respiration** The electron transport chain regenerates NAD$^+$ and FAD, which can be used again in glycolysis and the Krebs cycle. The various electron transfers in the electron transport chain generate about 34 molecules of ATP from each molecule of glucose oxidized: approximately three from each of the ten molecules of NADH (a total of 30), and approximately two from each of the two molecules of FADH$_2$ (a total of four). In aerobic respiration among prokaryotes, each molecule of glucose generates 38 ATP molecules: 34 from chemiosmosis plus 4 generated by oxidation in glycolysis and the Krebs cycle. Table 5.3 provides a detailed accounting of the ATP yield during prokaryotic aerobic respiration, and **Figure 5.17** presents a summary of the stages of aerobic respiration in prokaryotes.

Aerobic respiration among eukaryotes produces a total of only 36 molecules of ATP. There are fewer ATPs than in prokaryotes because some energy is lost when pyruvate is shuttled across the mitochondrial membranes that separate glycolysis

**TABLE 5.3** ATP Yield during Prokaryotic Aerobic Respiration of One Glucose Molecule

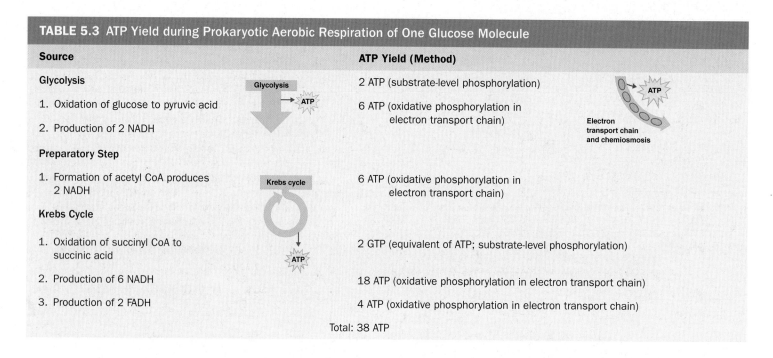

| Source | ATP Yield (Method) |
|---|---|
| **Glycolysis** | |
| 1. Oxidation of glucose to pyruvic acid | 2 ATP (substrate-level phosphorylation) |
| | 6 ATP (oxidative phosphorylation in electron transport chain) |
| 2. Production of 2 NADH | |
| **Preparatory Step** | |
| 1. Formation of acetyl CoA produces 2 NADH | 6 ATP (oxidative phosphorylation in electron transport chain) |
| **Krebs Cycle** | |
| 1. Oxidation of succinyl CoA to succinic acid | 2 GTP (equivalent of ATP; substrate-level phosphorylation) |
| 2. Production of 6 NADH | 18 ATP (oxidative phosphorylation in electron transport chain) |
| 3. Production of 2 FADH | 4 ATP (oxidative phosphorylation in electron transport chain) |
| | Total: 38 ATP |

(in the cytoplasm) from the electron transport chain. No such separation exists in prokaryotes. We can now summarize the overall reaction for aerobic respiration in prokaryotes as follows:

$$C_6H_{12}O_6 + 6\,O_2 + 38\,ADP + 38\,P_i \longrightarrow$$

Glucose    Oxygen

$$6\,CO_2 + 6\,H_2O + 38\,ATP$$

Carbon    Water
dioxide

#15

### Anaerobic Respiration

In anaerobic respiration, the final electron acceptor is an inorganic substance other than oxygen ($O_2$). Some bacteria, such as *Pseudomonas* and *Bacillus*, can use a nitrate ion ($NO_3^-$) as a final electron acceptor; the nitrate ion is reduced to a nitrite ion ($NO_2^-$), nitrous oxide ($N_2O$), or nitrogen gas ($N_2$). Other bacteria, such as *Desulfovibrio* (dē-sul-fō-VIB-rē-ō), use sulfate ($SO_4^{2-}$) as the final electron acceptor to form hydrogen sulfide ($H_2S$). Some archaea use carbon dioxide to form methane ($CH_4$). Anaerobic respiration by bacteria using nitrate and sulfate as final acceptors is essential for the nitrogen and sulfur cycles that occur in nature. The amount of ATP generated in anaerobic respiration varies with the organism and the pathway. Because only part of the Krebs cycle operates under anaerobic conditions, and because only some of the carriers in the electron transport chain participate in anaerobic respiration, the ATP yield is never as high as in aerobic respiration. Accordingly, anaerobes tend to grow more slowly than aerobes.

> **Play Electron Transport Chain:** *Overview, The Process, Factors Affecting ATP Yield*
> @Mastering**Microbiology**

**CHECK YOUR UNDERSTANDING**

✔ **5-13** What are the principal products of the Krebs cycle?

✔ **5-14** How do carrier molecules function in the electron transport chain?

✔ **5-15** Compare the energy yield (ATP) of aerobic and anaerobic respiration.

## Fermentation

After glucose has been oxidized into pyruvic acid, the pyruvic acid can be completely broken down in respiration, as previously described, or it can be converted to an organic product in fermentation, whereupon $NAD^+$ and $NADP^+$ are regenerated and can enter another round of glycolysis (see Figure 5.11). Fermentation is defined as a process that

1. releases energy from sugars or other organic molecules;

2. does not require oxygen (but can occur in its presence);

3. does not require the use of the Krebs cycle or an electron transport chain;

4. uses an organic molecule synthesized in the cell as the final electron acceptor.

Fermentation produces only small amounts of ATP (only one or two ATP molecules for each molecule of starting material) because much of the original energy in glucose remains in the chemical bonds of the organic end-products, such as lactic acid or ethanol. However, the advantage of fermentation for a cell is that it produces ATP quickly.

During fermentation, electrons are transferred (along with protons) from reduced coenzymes (NADH, NADPH) to pyruvic

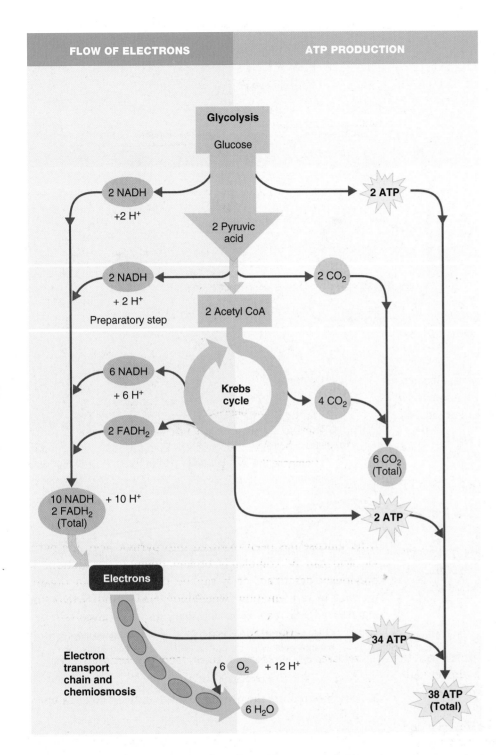

| FLOW OF ELECTRONS | ATP PRODUCTION |

**Glycolysis**

Glucose

2 NADH +2 H$^+$

**2 ATP**

2 Pyruvic acid

2 NADH + 2 H$^+$

**Preparatory step**

2 CO$_2$

2 Acetyl CoA

6 NADH + 6 H$^+$

**Krebs cycle**

4 CO$_2$

2 FADH$_2$

6 CO$_2$ (Total)

10 NADH + 10 H$^+$
2 FADH$_2$
(Total)

**2 ATP**

**Electrons**

**Electron transport chain and chemiosmosis**

6 O$_2$ + 12 H$^+$

**34 ATP**

6 H$_2$O

**38 ATP (Total)**

**Figure 5.17  A summary of aerobic respiration in prokaryotes.** Glucose is broken down completely to carbon dioxide and water, and ATP is generated. This process has three major phases: glycolysis, the Krebs cycle, and the electron transport chain. The preparatory step is between glycolysis and the Krebs cycle. The key event in aerobic respiration is that electrons are picked up from intermediates of glycolysis and the Krebs cycle by NAD$^+$ or FAD and are carried by NADH or FADH$_2$ to the electron transport chain. NADH is also produced during the conversion of pyruvic acid to acetyl CoA. Most of the ATP generated by aerobic respiration is made by the chemiosmotic mechanism during the electron transport chain phase; this is called oxidative phosphorylation.

**Q** How do aerobic and anaerobic respiration differ?

acid or its derivatives (**Figure 5.18a**). Those final electron acceptors are reduced to the end-products shown in Figure 5.18b. An essential function of fermentation is to ensure a steady supply of NAD$^+$ and NADP$^+$ so that glycolysis can continue. In fermentation, ATP is generated only during glycolysis.

The aforementioned lactic acid is the same substance associated with muscle fatigue in your body. During strenuous exercise, the cardiovascular system cannot supply enough oxygen to skeletal muscles and the heart for them to generate sufficient

energy. In such cases, muscles shift from aerobic respiration to fermentation. In the absence of oxygen, pyruvic acid is oxidized to lactic acid.

Microorganisms can ferment various substrates; the end-products depend on the particular microorganism, the substrate, and the enzymes that are present and active. Chemical analyses of these end-products are useful in identifying microorganisms. Two of the more important processes are lactic acid fermentation and alcohol fermentation.

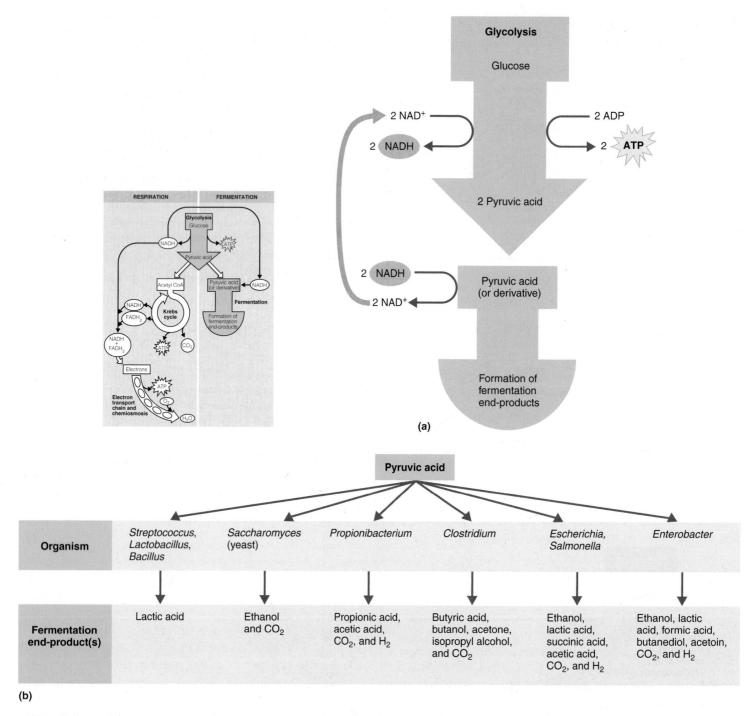

**Figure 5.18 Fermentation.** The inset indicates the relationship of fermentation to the overall energy-producing processes. **(a)** An overview of fermentation. The first step is glycolysis, the conversion of glucose to pyruvic acid. In the second step, the reduced coenzymes from glycolysis (NADH) or its alternative (NADPH) donate their electrons and hydrogen ions to pyruvic acid or a derivative to form a fermentation end-product and reoxidize the NADH to be available for glycolysis. **(b)** End-products of various microbial fermentations.

**Q** During which phase of fermentation is ATP generated?

## Lactic Acid Fermentation

During glycolysis, which is the first phase of **lactic acid fermentation,** a molecule of glucose is oxidized to two molecules of pyruvic acid (**Figure 5.19**; see also Figure 5.10). This oxidation generates the energy that is used to form the two molecules of ATP. In the next step, the two molecules of pyruvic acid are reduced by two molecules of NADH to form two molecules of lactic acid (Figure 5.19a). Because lactic acid is the end-product of the reaction, it undergoes no further oxidation, and most of the energy produced by the reaction remains

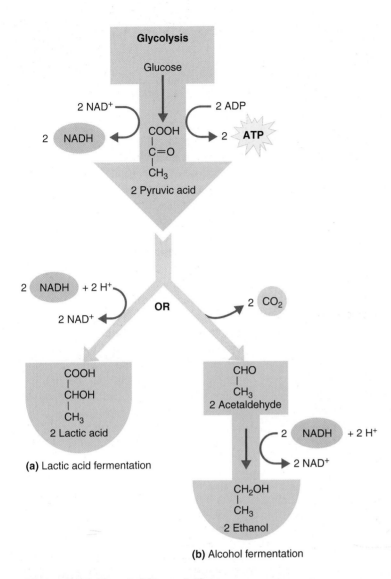

**(a)** Lactic acid fermentation

**(b)** Alcohol fermentation

**Figure 5.19  Types of fermentation.**

**Q** What is the difference between homolactic and heterolactic fermentation?

stored in the lactic acid. Thus, this fermentation yields only a small amount of energy.

Two important genera of lactic acid bacteria are *Streptococcus* and *Lactobacillus* (lak′tō-bah-SIL-lus). Because these microbes produce only lactic acid, they are referred to as **homolactic** (or *homofermentative*). Lactic acid fermentation can result in food spoilage. However, the process can also produce yogurt from milk, sauerkraut from fresh cabbage, and pickles from cucumbers. The homolactic fermenters in the human intestine are important for health. (See Exploring the Microbiome on page 132.)

## Alcohol Fermentation

**Alcohol fermentation** also begins with the glycolysis of a molecule of glucose to yield two molecules of pyruvic acid and two

molecules of ATP. In the next reaction, the two molecules of pyruvic acid are converted to two molecules of acetaldehyde and two molecules of $CO_2$ (Figure 5.19b). The two molecules of acetaldehyde are reduced by two molecules of NADH to form two molecules of ethanol. Again, alcohol fermentation is a low-energy-yield process because most of the energy contained in the initial glucose molecule remains in the ethanol, the end-product.

Alcohol fermentation is carried out by a number of bacteria and yeasts. The ethanol and carbon dioxide produced by the yeast *Saccharomyces* (sak′kar-ō-MĪ-sēz) are waste products for yeast cells but are useful to humans. Ethanol made by yeasts is the alcohol in alcoholic beverages, and carbon dioxide made by yeasts causes bread dough to rise.

Organisms that produce lactic acid as well as other acids or alcohols are known as **heterolactic** (or *heterofermentative*) and often use the pentose phosphate pathway.

Table 5.4 lists some of the various microbial fermentations used by industry to convert inexpensive raw materials into useful end-products. Table 5.5 provides a summary comparison of aerobic respiration, anaerobic respiration, and fermentation.

▶ **Play Fermentation**
@MasteringMicrobiology

### CHECK YOUR UNDERSTANDING

✔ 5-16 List four compounds that can be made from pyruvic acid by an organism that uses fermentation.

# EXPLORING THE MICROBIOME  Do Artificial Sweeteners (and the Intestinal Microbiota That Love Them) Promote Diabetes?

For years, beverages made with artificial sweeteners were embraced by diabetics and weight watchers because, unlike sugar, artificial sweeteners don't impact blood glucose levels and don't provide calories. However, recent research indicates artificial sweeteners may actually increase the risk of nondiabetics developing the disease. One study published in 2009 by the American Diabetes Association found that daily consumption of diet soda was associated with a 67% greater relative risk of developing type 2 diabetes.

Undigestible by humans, artificial sweeteners provide zero calories to us when we consume them. But they are a great source of nutrients for *Bacteroides* bacteria living in the colon. As *Bacterioides* break down the sweeteners and increase in numbers, other types of microbiota simultaneously decline. Among these are *Lactobacillus* bacteria. Studies indicate that high *Lactobacillus* levels in the intestine are associated with decreased blood sugar levels. The exact mechanism remains unclear, but it is hypothesized that decreases in the population of *Lactobacillus* bacteria lead to higher blood glucose levels, thereby forcing the body to produce more insulin to control the rising blood glucose. Prolonged high insulin levels may lead to insulin resistance, a condition where the body stops responding correctly to the hormone. Insulin resistance is the hallmark sign of type 2 diabetes.

Recent and current research are exploring whether ingesting probiotics with *Lactobacillus acidophilus* and *Bifidobacterium animalis* may be a useful treatment for type 2 diabetes. Initial studies were promising, showing that these species might lower blood glucose levels. If proven effective, one day bacteria could be key weapons in preventing a deadly disease.

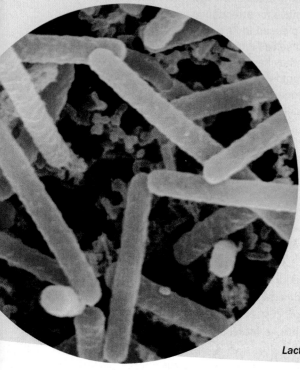

*Lactobacillus acidophilus.*

## TABLE 5.4  Some Industrial Uses for Different Types of Fermentations*

| Fermentation End-Product(s) | Industrial or Commercial Use | Starting Material | Microorganism |
|---|---|---|---|
| Ethanol | Beer, wine | Starch, sugar | *Saccharomyces cerevisiae* (yeast, a fungus) |
| | Fuel | Agricultural wastes | *Saccharomyces cerevisiae* (yeast) |
| Acetic Acid | Vinegar | Ethanol | *Acetobacter* |
| Lactic Acid | Cheese, yogurt | Milk | *Lactobacillus, Streptococcus* |
| | Rye bread | Grain, sugar | *Lactobacillus delbrueckii* |
| | Sauerkraut | Cabbage | *Lactobacillus plantarum* |
| | Summer sausage | Meat | *Pediococcus* |
| Propionic Acid and Carbon Dioxide | Swiss cheese | Lactic acid | *Propionibacterium freudenreichii* |
| Acetone and Butanol | Pharmaceutical, industrial uses | Molasses | *Clostridium acetobutylicum* |
| Citric Acid | Flavoring | Molasses | *Aspergillus* (fungus) |
| Methane | Fuel | Acetic acid | *Methanosarcina* (archaeon) |
| Sorbose | Vitamin C (ascorbic acid) | Sorbitol | *Gluconobacter* |

*Unless otherwise noted, the microorganisms listed are bacteria.

**TABLE 5.5** Aerobic Respiration, Anaerobic Respiration, and Fermentation

| Energy-Producing Process | Growth Conditions | Final Hydrogen (Electron) Acceptor | Type of Phosphorylation Used to Generate ATP | ATP Molecules Produced per Glucose Molecule |
|---|---|---|---|---|
| Aerobic Respiration | Aerobic | Molecular oxygen ($O_2$) | Substrate-level and oxidative | 36 (eukaryotes)<br>38 (prokaryotes) |
| Anaerobic Respiration | Anaerobic | Usually an inorganic substance (such as $NO_3^-$, $SO_4^{2-}$, or $CO_3^{2-}$) but not molecular oxygen ($O_2$) | Substrate-level and oxidative | Variable (fewer than 38 but more than 2) |
| Fermentation | Aerobic or anaerobic | An organic molecule | Substrate-level | 2 |

# Lipid and Protein Catabolism

### LEARNING OBJECTIVE

**5-17** Describe how lipids and proteins undergo catabolism.

Our discussion of energy production has emphasized the oxidation of glucose, the main energy-supplying carbohydrate. However, microbes also oxidize lipids and proteins, and the oxidations of all these nutrients are related.

Recall that fats are lipids consisting of fatty acids and glycerol. Microbes produce extracellular enzymes called *lipases* that break fats down into their fatty acid and glycerol components. Each component is then metabolized separately (**Figure 5.20**). The Krebs cycle functions in the oxidation of glycerol and fatty acids. Many bacteria that hydrolyze fatty acids can use the same enzymes to degrade petroleum products. Whereas beta-oxidation (fatty acid oxidation) of petroleum is a nuisance when these bacteria grow in a fuel storage tank, it is beneficial when they grow in oil spills.

Proteins are too large to pass unaided through plasma membranes. Microbes produce extracellular *proteases* and *peptidases*, enzymes that break down proteins into their component amino acids, which can cross the membranes. However, before amino acids can be catabolized, they must be enzymatically converted to other substances that can enter the Krebs cycle. In one such conversion, called **deamination**, the amino group of an amino acid is removed and converted to an ammonium ion ($NH_4^+$), which can be excreted from the cell. The remaining organic acid can enter the Krebs cycle. Other conversions involve **decarboxylation** (the removal of —COOH) and **desulfurization** (removal of —SH).

A summary of the interrelationships of carbohydrate, lipid, and protein catabolism is shown in **Figure 5.21**.

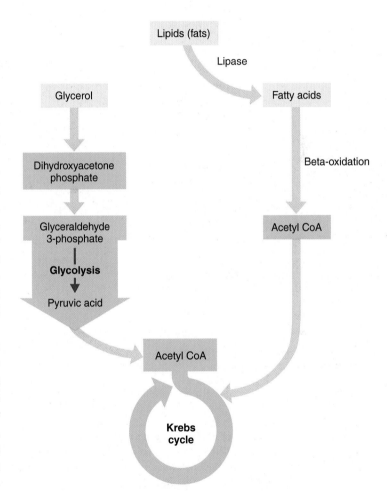

**Figure 5.20 Lipid catabolism.** Glycerol is converted into dihydroxyacetone phosphate (DHAP) and catabolized via glycolysis and the Krebs cycle. Fatty acids undergo beta-oxidation, in which carbon fragments are split off two at a time to form acetyl CoA, which is catabolized via the Krebs cycle.

**Q** What is the role of lipases?

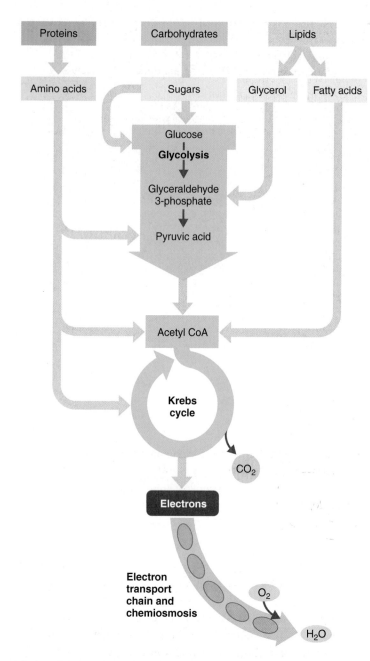

**Figure 5.21 Catabolism of various organic molecules.**
Proteins, carbohydrates, and lipids can all be sources of electrons and protons for respiration. These food molecules enter glycolysis or the Krebs cycle at various points.

**Q** What are the catabolic pathways through which high-energy electrons from all kinds of organic molecules flow on their energy-releasing pathways?

# Biochemical Tests and Bacterial Identification

### LEARNING OBJECTIVE

5-18 Provide two examples of the use of biochemical tests to identify bacteria in the laboratory.

Biochemical testing is frequently used to identify bacteria and yeasts because different species produce different enzymes. Such biochemical tests are designed to detect the presence of enzymes. One type of biochemical test detects amino acid catabolizing enzymes involved in decarboxylation and dehydrogenation (discussed on pages 123 and 118; **Figure 5.22**).

Another biochemical test is a **fermentation test** (**Figure 5.23**). The test medium contains protein, a single carbohydrate, a pH indicator, and an inverted Durham tube, which is used to capture gas. Bacteria inoculated into the tube can use the protein or carbohydrate as a carbon and energy source. If they catabolize the carbohydrate and produce acid, the pH indicator changes color. Some organisms produce gas as well as acid from carbohydrate catabolism. The presence of a bubble in the Durham tube indicates gas formation.

*E. coli* ferments the carbohydrate sorbitol. The pathogenic *E. coli* O157 strain, however, does not ferment sorbitol, a characteristic that differentiates it from nonpathogenic, commensal *E. coli*.

Another example of the use of biochemical tests is shown in Figure 10.8 on page 281.

In some instances, the waste products of one microorganism can be used as a carbon and energy source by another species. *Acetobacter* (ah-sē'tō-BAK-ter) bacteria oxidize ethanol made by yeast. *Propionibacterium* (prō'-pē-on-ē-bak-TI-rē-um) can use lactic acid produced by other bacteria. Propionibacteria convert lactic acid to pyruvic acid in preparation for the Krebs cycle. During the Krebs cycle, propionic acid and $CO_2$ are made. The holes in Swiss cheese are formed by the accumulation of $CO_2$ gas.

**Figure 5.22 Detecting amino acid catabolizing enzymes in the lab.** Bacteria are inoculated into tubes containing glucose, a pH indicator, and a specific amino acid. **(a)** The pH indicator turns to yellow when bacteria produce acid from glucose. **(b)** Alkaline products from decarboxylation turn the indicator to purple.

**Q** What is decarboxylation?

**Figure 5.23 A fermentation test. (a)** An uninoculated fermentation tube containing the carbohydrate mannitol. **(b)** *Staphylococcus epidermidis* grew on the protein but did not use the carbohydrate. This organism is described as mannitol −. **(c)** *Staphylococcus aureus* produced acid but not gas. This species is mannitol +. **(d)** *Escherichia coli* is also mannitol + and produced acid and gas from mannitol. The gas is trapped in the inverted Durham tube.

**Q** What is the *S. epidermidis* using as its energy source?

Biochemical tests are used to identify bacteria that cause disease. All aerobic bacteria use the electron transport chain (ETC), but their ETCs are not all identical. Some bacteria have cytochrome *c*, but others do not. In the former, *cytochrome c oxidase* is the last enzyme, which transfers electrons to oxygen. The oxidase test is routinely used to quickly identify *Neisseria gonorrhoeae*. *Neisseria* is positive for cytochrome oxidase. The oxidase test can also be used to distinguish some gram-negative rods: *Pseudomonas* is oxidase-positive, and *Escherichia* is oxidase-negative.

*Shigella* causes dysentery and is differentiated from *E. coli* by biochemical tests. Unlike *E. coli*, *Shigella* does not produce gas from lactose.

*Salmonella* bacteria are readily distinguishable from *E. coli* by the production of hydrogen sulfide ($H_2S$). Hydrogen sulfide is released when the bacteria remove sulfur from amino acids (**Figure 5.24**).

The Clinical Focus box on page 141 describes how biochemical tests were used to determine the cause of disease in a young child in Dallas, Texas.

**Figure 5.24 Use of peptone iron agar to detect the production of $H_2S$.** $H_2S$ produced in the tube precipitates with iron in the medium as ferrous sulfide.

**Q** What chemical reaction causes the release of $H_2S$?

Dental caries are caused by oral streptococci, including *S. mutans*, *S. salivarius*, and *S. sobrinus*, that attach to tooth surfaces. Oral streptococci ferment sucrose and produce lactic acid, which lowers the salivary pH. Dr. Rivera decides to ask the camp counselors to substitute the bubblegum with a sugarless gum made with xylitol. A study has shown that chewing gum sweetened with xylitol, a naturally occuring sugar alcohol, can significantly lower the number of dental caries in children because it lowers the number of *S. mutans* in the mouth.

**Why might xylitol reduce the number of *S. mutans*?**

110    131    **135**    137

**CHECK YOUR UNDERSTANDING**

✔ 5-18 On what biochemical basis are *Pseudomonas* and *Escherichia* differentiated?

# Photosynthesis

#19

**LEARNING OBJECTIVES**

5-19 Compare and contrast cyclic and noncyclic photophosphorylation.

5-20 Compare and contrast the light-dependent and light-independent reactions of photosynthesis.

5-21 Compare and contrast oxidative phosphorylation and photophosphorylation.

In all of the metabolic pathways just discussed, organisms obtain energy for cellular work by oxidizing organic compounds. But where do organisms obtain these organic compounds? Some, including animals and many microbes, feed on matter produced by other organisms. For example, bacteria may catabolize compounds from dead plants and animals, or they may obtain nourishment from a living host.

Other organisms synthesize complex organic compounds from simple inorganic substances. The major mechanism for such synthesis is a process called **photosynthesis**, which is carried out by plants and many microbes. Essentially, photosynthesis is the conversion of light energy from the sun into chemical energy. The chemical energy is then used to convert $CO_2$ from the atmosphere to more reduced carbon compounds, primarily sugars. The word *photosynthesis* summarizes the process: *photo* means light, and *synthesis* refers to the assembly of organic compounds. This synthesis of sugars by using carbon atoms from $CO_2$ gas is also called **carbon fixation**. Continuation of life as we know it on Earth depends on the recycling of carbon in this way (see Figure 27.3 on page 790). Cyanobacteria, algae, and green plants all contribute to this vital recycling with photosynthesis.

Photosynthesis can be summarized with the following equations:

1. Plants, algae, and cyanobacteria use water as a hydrogen donor, releasing $O_2$.

$$6\,CO_2 + 12\,H_2O + \text{Light energy} \rightarrow$$
$$C_6H_{12}O_6 + 6\,H_2O + 6\,O_2$$

2. Purple sulfur and green sulfur bacteria use $H_2S$ as a hydrogen donor, producing sulfur granules.

$$6\,CO_2 + 12\,H_2S + \text{Light energy} \rightarrow$$
$$C_6H_{12}O_6 + 6\,H_2O + 12\,S$$

In the course of photosynthesis, electrons are taken from hydrogen atoms, an energy-poor molecule, and incorporated into sugar, an energy-rich molecule. The energy boost is supplied by light energy, although indirectly.

Photosynthesis takes place in two stages. In the first stage, called the **light-dependent (light) reactions**, light energy is used to convert ADP and ⓟ to ATP. In addition, in the predominant form of the light-dependent reactions, the electron carrier $NADP^+$ is reduced to NADPH. The coenzyme NADPH, like NADH, is an energy-rich carrier of electrons. In the second stage, the **light-independent (dark) reactions**, these electrons are used along with energy from ATP to reduce $CO_2$ to sugar.

> ▶ Play Photosynthesis: *Overview*
> @Mastering**Microbiology**

## The Light-Dependent Reactions: Photophosphorylation

**Photophosphorylation** is one of the three ways ATP is formed, and it occurs only in photosynthetic cells. In this mechanism, light energy is absorbed by chlorophyll molecules in the photosynthetic cell, exciting some of the molecules' electrons. The chlorophyll principally used by green plants, algae, and cyanobacteria is *chlorophyll a*. It is located in the membranous thylakoids of chloroplasts in algae and green plants (see Figure 4.28, page 102) and in the thylakoids found in the photosynthetic structures of cyanobacteria. Other bacteria use *bacteriochlorophylls*.

The excited electrons jump from the chlorophyll to the first of a series of carrier molecules, an electron transport chain similar to that used in respiration. As electrons are passed along the series of carriers, protons are pumped across the membrane, and ADP is converted to ATP by chemiosmosis. Chlorophyll and other pigments are packed into thylakoids of chloroplasts and are called **photosystems**. *Photosystem II* is so numbered because even though it was most likely the first photosystem to evolve, it was the second one discovered. It contains chlorophyll that is sensitive to wavelengths of light of 680 nm. The chlorophyll in *photosystem I* is sensitive to wavelengths of

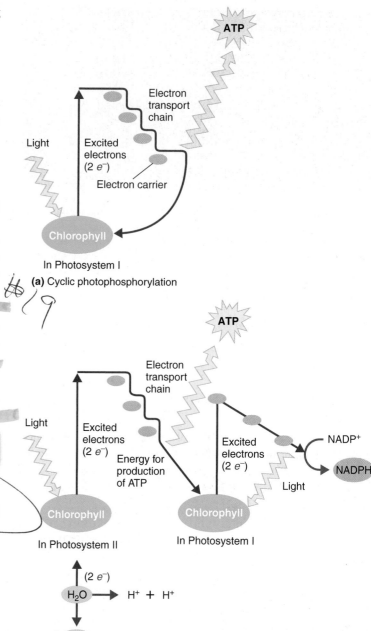

**(a)** Cyclic photophosphorylation

**(b)** Noncyclic photophosphorylation

**Figure 5.25 Photophosphorylation. (a)** In cyclic photophosphorylation, electrons released from chlorophyll by light in photosystem I return to chlorophyll after passage along the electron transport chain. The energy from electron transfer is used to synthesize ATP. **(b)** In noncyclic photophosphorylation, electrons released from chlorophyll in photosystem II are replaced by electrons from the hydrogen atoms in water. This process also releases hydrogen ions. Electrons from chlorophyll in photosystem I are passed along the electron transport chain to the electron acceptor $NADP^+$. $NADP^+$ combines with electrons and with hydrogen ions from water, forming NADPH.

**Q** How are oxidative phosphorylation and photophosphorylation similar?

S. *mutans* cannot ferment xylitol; consequently, it doesn't grow and can't produce acid in the mouth. The camp counselors agree to switch to sugarless gum made with xylitol, and Dr. Rivera is pleased. She understands that there will be other sources of sucrose in the children's diets, but at least her patients are no longer going to be adversely affected by the camp's well-intentioned incentives. Researchers are still investigating ways that antimicrobials and vaccines can be used to reduce bacterial colonization. However, reducing consumption of sucrose-containing gum and candy may be an effective preventive measure.

110    131    135    **137**

light of 700 nm. In **cyclic photophosphorylation**, the electrons released from chlorophyll in photosystem I eventually return to chlorophyll (**Figure 5.25a**). That is, the electrons in photosystem I remain in photosystem I. In **noncyclic photophosphorylation**, which is used in oxygenic organisms, both photosystems are required. The electrons released from the chlorophyll in photosystem II and photosystem I do not return to chlorophyll but become incorporated into NADPH (Figure 5.25b). The electrons lost from chlorophyll are replaced by electrons from $H_2O$. To summarize: the products of noncyclic photophosphorylation are ATP (formed by chemiosmosis using energy released in an electron transport chain), $O_2$ (from water molecules), and NADPH (carrying electrons from chlorophyll and protons derived from water).

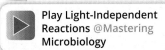
▶ Play Light Reaction: Cyclic Photophosphorylation; Light Reaction: Noncyclic Photophosphorylation @MasteringMicrobiology

## The Light-Independent Reactions: The Calvin-Benson Cycle

The light-independent reactions are so named because they don't require light directly. They include a complex cyclic pathway called the **Calvin-Benson cycle**, in which $CO_2$ is "fixed"—that is, it's used to synthesize sugars (**Figure 5.26**; see also Figure A.1 in Appendix A).

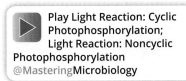
▶ Play Light-Independent Reactions @Mastering Microbiology

### CHECK YOUR UNDERSTANDING

✔ **5-19** How is photosynthesis important to catabolism?

✔ **5-20** What is made during the light-dependent reactions?

✔ **5-21** How are oxidative phosphorylation and photophosphorylation similar?

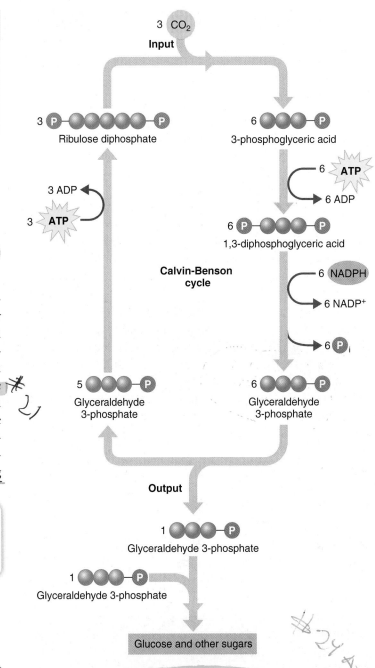

**Figure 5.26  A simplified version of the Calvin-Benson cycle.** This diagram shows three turns of the cycle, in which three molecules of $CO_2$ are fixed and one molecule of glyceraldehyde 3-phosphate is produced and leaves the cycle. Two molecules of glyceraldehyde 3-phosphate are needed to make one molecule of glucose. Therefore, the cycle must turn six times for each glucose molecule produced, requiring a total investment of 6 molecules of $CO_2$, 18 molecules of ATP, and 12 molecules of NADPH. A more detailed version of this cycle is presented in Figure A.1 in Appendix A.

**Q** In the Calvin-Benson cycle, which molecule is used to synthesize sugars?

# A Summary of Energy Production Mechanisms

**LEARNING OBJECTIVE**

5-22 Write a sentence to summarize energy production in cells.

In the living world, energy passes from one organism to another in the potential energy contained in the bonds of chemical compounds. Organisms obtain the energy from oxidation reactions. To obtain energy in a usable form, a cell must have an electron (or hydrogen) donor, which serves as an initial energy source within the cell. Electron donors are diverse and can include photosynthetic pigments, glucose or other organic compounds, elemental sulfur, ammonia, or hydrogen gas (**Figure 5.27**). Next, electrons removed from the chemical energy sources are transferred to electron carriers, such as the coenzymes $NAD^+$, $NADP^+$, and FAD. This transfer is an oxidation-reduction reaction; the initial energy source is oxidized as this first electron carrier is reduced. During this phase, some ATP is produced. In the third stage, electrons are transferred from electron carriers to their final electron acceptors in further oxidation-reduction reactions, producing more ATP.

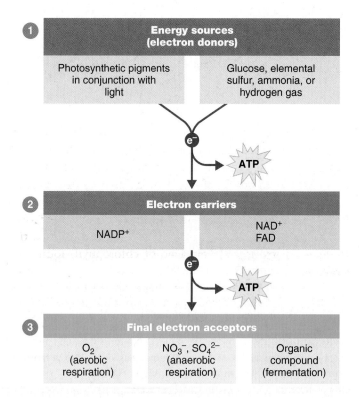

**Figure 5.27 Requirements of ATP production.** The production of ATP requires **1** an energy source (electron donor), **2** the transfer of electrons to an electron carrier during an oxidation-reduction reaction, and **3** the transfer of electrons to a final electron acceptor.

**Q** Are energy-generating reactions oxidations or reductions?

In aerobic respiration, oxygen ($O_2$) serves as the final electron acceptor. In anaerobic respiration, substances from the environment other than oxygen, such as nitrate ions ($NO_3^-$) or sulfate ions ($SO_4^{2-}$), serve as the final electron acceptors. In fermentation, compounds in the cytoplasm serve as the final electron acceptors. In aerobic and anaerobic respiration, a series of electron carriers called an electron transport chain releases energy that is used by the mechanism of chemiosmosis to synthesize ATP. Regardless of their energy sources, all organisms use similar oxidation-reduction reactions to transfer electrons and similar mechanisms to use the energy released to produce ATP.

**CHECK YOUR UNDERSTANDING**

✔ 5-22 Summarize how oxidation enables organisms to get energy from glucose, sulfur, or sunlight.

# Metabolic Diversity among Organisms

**LEARNING OBJECTIVE**

5-23 Categorize the various nutritional patterns among organisms according to carbon source and mechanisms of carbohydrate catabolism and ATP generation.

We have looked in detail at some of the energy-generating metabolic pathways that are used by animals and plants, as well as by many microbes. Some microbes can sustain themselves on inorganic substances by using pathways that are unavailable to either plants or animals. All organisms, including microbes, can be classified metabolically according to their *nutritional pattern*—their source of energy and their source of carbon.

ASM: Bacteria and Archaea exhibit extensive, and often unique, metabolic diversity.

First considering the energy source, we can generally classify organisms as phototrophs or chemotrophs. **Phototrophs** use light as their primary energy source, whereas **chemotrophs** depend on oxidation-reduction reactions of inorganic or organic compounds for energy. For their principal carbon source, **autotrophs** (self-feeders) use carbon dioxide, and **heterotrophs** (feeders on others) require an organic carbon source. Autotrophs are also referred to as *lithotrophs* (rock eating), and heterotrophs are also referred to as *organotrophs*.

If we combine the energy and carbon sources, we derive the following nutritional classifications for organisms: *photoautotrophs*, *photoheterotrophs*, *chemoautotrophs*, and *chemoheterotrophs* (**Figure 5.28**). Almost all of the medically important microorganisms discussed in this book are chemoheterotrophs. Typically, infectious organisms catabolize substances obtained from the host.

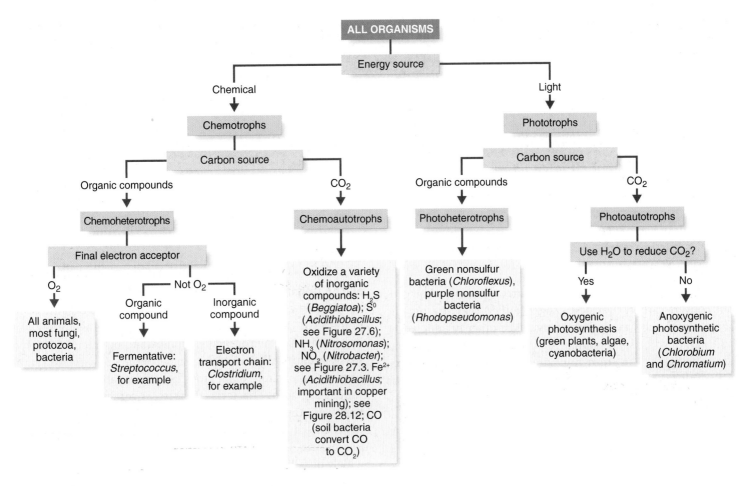

**Figure 5.28  A nutritional classification of organisms.**

**Q** What is the basic difference between chemotrophs and phototrophs?

## Photoautotrophs

**Photoautotrophs** use light as a source of energy and carbon dioxide as their chief source of carbon. They include photosynthetic bacteria (green and purple bacteria and cyanobacteria), algae, and green plants. In the photosynthetic reactions of cyanobacteria, algae, and green plants, the hydrogen atoms of water are used to reduce carbon dioxide, and oxygen gas is given off. Because this photosynthetic process produces $O_2$ it is sometimes called **oxygenic.**

In addition to the cyanobacteria (see Figure 11.13, page 307), there are several other families of photosynthetic prokaryotes. Each is classified according to the way it reduces $CO_2$. These bacteria cannot use $H_2O$ to reduce $CO_2$ and cannot carry on photosynthesis when oxygen is present (they must have an anaerobic environment). Consequently, their photosynthetic process does not produce $O_2$ and is called **anoxygenic**. The anoxygenic photoautotrophs are the green and purple bacteria. The **green sulfur bacteria**, such as *Chlorobium* (klor-Ō-bē-um), use sulfur (S), sulfur compounds (such as hydrogen sulfide, $H_2S$),

or hydrogen gas ($H_2$) to reduce carbon dioxide and form organic compounds. Applying the energy from light and the appropriate enzymes, these bacteria oxidize sulfide ($S^{2-}$) or sulfur (S) to sulfate ($SO_4^{2-}$) or oxidize hydrogen gas to water ($H_2O$). The **purple sulfur bacteria**, such as *Chromatium* (krō-MĀ-shum), also use sulfur, sulfur compounds, or hydrogen gas to reduce carbon dioxide. They are distinguished from the green bacteria by their type of chlorophyll, location of stored sulfur, and ribosomal RNA.

The chlorophylls used by these photosynthetic bacteria are called *bacteriochlorophylls*, and they absorb light at longer wavelengths than that absorbed by chlorophyll *a*. Bacteriochlorophylls of green sulfur bacteria are found in vesicles called *chlorosomes* (or *chlorobium vesicles*) underlying and attached to the plasma membrane. In the purple sulfur bacteria, the bacteriochlorophylls are located in invaginations of the plasma membrane (*chromatophores*).

Table 5.6 summarizes several characteristics that distinguish eukaryotic photosynthesis from prokaryotic photosynthesis.

**TABLE 5.6** Photosynthesis Compared in Selected Eukaryotes and Prokaryotes

| Characteristic | Eukaryotes | Prokaryotes | | |
|---|---|---|---|---|
| | Algae, Plants | Cyanobacteria | Green Bacteria | Purple Bacteria |
| Substance That Reduces $CO_2$ | H atoms of $H_2O$ | H atoms of $H_2O$ | Sulfur, sulfur compounds, $H_2$ gas | Sulfur, sulfur compounds, $H_2$ gas |
| Oxygen Production | Oxygenic | Oxygenic (and anoxygenic) | Anoxygenic | Anoxygenic |
| Type of Chlorophyll | Chlorophyll *a* | Chlorophyll *a* | Bacteriochlorophyll *a* | Bacteriochlorophyll *a* or *b* |
| Site of Photosynthesis | Thylakoids in chloroplasts | Thylakoids | Chlorosomes | Chromatophores |
| Environment | Aerobic | Aerobic (and anaerobic) | Anaerobic | Anaerobic |

## Photoheterotrophs

**Photoheterotrophs** use light as a source of energy but cannot convert carbon dioxide to sugar; rather, they use organic compounds, such as alcohols, fatty acids, other organic acids, and carbohydrates, as sources of carbon. Photoheterotrophs are anoxygenic. The **green nonsulfur bacteria,** such as *Chloroflexus* (klor-ō-FLEX-us), and **purple nonsulfur bacteria,** such as *Rhodopseudomonas* (rō'dō-soo'dō-MŌ-nas), are photoheterotrophs (see also page 308).

## Chemoautotrophs

**Chemoautotrophs** use the electrons from reduced inorganic compounds as a source of energy, and they use $CO_2$ as their principal source of carbon. They fix $CO_2$ in the Calvin-Benson Cycle (see Figure 5.26). Inorganic sources of energy for these organisms include hydrogen sulfide ($H_2S$) for *Beggiatoa* (BEJ-jē-ah-tō-ah); elemental sulfur (S) for *Acidithiobacillus thiooxidans;* ammonia ($NH_3$) for *Nitrosomonas* (NĪ-trō-sō-mō-nas); nitrite ions ($NO_2^-$) for *Nitrobacter* (ni'trō-BAK-ter); hydrogen gas ($H_2$) for *Cupriavidus* (koo-prē-AH-vid-us); ferrous iron ($Fe^{2+}$) for *Acidithiobacillus ferrooxidans;* and carbon monoxide (CO) for *Pseudomonas carboxydohydrogena* (kar-box'id-ō-hī-DRO-je-nah). The energy derived from the oxidation of these inorganic compounds is eventually stored in ATP, which is produced by oxidative phosphorylation.

## Chemoheterotrophs

When we discuss photoautotrophs, photoheterotrophs, and chemoautotrophs, it's easy to categorize the energy and carbon sources because they occur as separate entities. However, in chemoheterotrophs, the distinction isn't as clear because the energy and carbon sources are usually the same organic compound—glucose, for example. **Chemoheterotrophs** specifically use the electrons from hydrogen atoms in organic compounds as their energy source.

Heterotrophs are further classified according to their source of organic molecules. **Saprophytes** live on dead organic matter, and **parasites** derive nutrients from a living host. Most bacteria, and all fungi, protozoa, and animals, are chemoheterotrophs.

Bacteria and fungi can use a wide variety of organic compounds for carbon and energy sources. This is why they can live in diverse environments. Understanding microbial diversity is scientifically interesting and economically important. In some situations microbial growth is undesirable, such as when rubber-degrading bacteria destroy a gasket or shoe sole. However, these same bacteria might be beneficial if they decomposed discarded rubber products, such as used tires. *Rhodococcus erythropolis* (rō-dō-KOK-kus er'i-THROP-ō-lis) is widely distributed in soil and can cause disease in humans and other animals. However, this species is able to replace sulfur atoms in petroleum with atoms of oxygen. Removing sulfur from crude oil is an important step in the oil-refining process. Sulfur corrodes equipment and pipelines, and contributes to acid precipitation and pollution-related respiratory problems in people. A Texas company is currently using *R. erythropolis* to produce desulfurized oil.

**CHECK YOUR UNDERSTANDING**

✔ **5-23** Almost all medically important microbes belong to which of the four aforementioned groups?

\* \* \*

We will next consider how cells use ATP pathways for the synthesis of organic compounds such as carbohydrates, lipids, proteins, and nucleic acids.

# Metabolic Pathways of Energy Use

**LEARNING OBJECTIVE**

**5-24** Describe the major types of anabolism and their relationship to catabolism.

Up to now we've been considering energy production. Through the oxidation of organic molecules, organisms produce energy by aerobic respiration, anaerobic respiration, and

# Human Tuberculosis—Dallas, Texas

As you read through this box, you will encounter a series of questions that laboratory technicians ask themselves as they identify bacteria. Try to answer each question before going on to the next one.

1. Julia, a 12-month-old Latina, is brought by her parents to the emergency department of a Dallas, Texas, hospital. She has a fever of 39°C, a distended abdomen, some abdominal pain, and watery diarrhea. Julia is admitted to the pediatric wing of the hospital, pending results of laboratory and radiologic tests. Test results suggest peritoneal tuberculosis (TB). Caused by one of several closely related species in the *Mycobacterium tuberculosis* complex, TB is a reportable condition in the United States. Peritoneal TB is a disease of the intestines and abdominal cavity.
What organ is usually associated with tuberculosis? How might someone get peritoneal TB?

2. Pulmonary TB is contracted by inhaling the bacteria; ingesting the bacteria can result in peritoneal TB. A laparoscopy reveals that nodules are present in Julia's abdominal cavity. A portion of a nodule is removed for biopsy so that it can be observed for the presence of acid-fast bacteria. Based on the presence of the abdominal nodules, Julia's physician begins conventional antituberculosis treatment. This long-term treatment can last up to 12 months.
What is the next step?

3. The lab results confirm that acid-fast bacteria are indeed present in Julia's abdominal cavity. The laboratory now needs to identify the *Mycobacterium* species. Speciation of the *M. tuberculosis* complex is done

**Figure 5.A An identification scheme for selected species of slow-growing mycobacteria.**

by biochemical testing in reference laboratories (Figure A). The bacteria need to be grown in culture media. Slow-growing mycobacteria may take up to 6 weeks to form colonies.
After colonies have been isolated, what is the next step?

4. Two weeks later, the laboratory results show that the bacteria are slow-growing. According to the identification scheme, the urease test (Figure B) should be performed.
What is the result shown in Figure B?

5. Because the urease test is positive, the nitrate reduction test is performed. It shows that the bacteria do not produce the enzyme nitrate reductase. Julia's physician lets her parents know that they are very close to identifying the pathogen that is causing Julia's illness.
What is the bacterium?

6. *M. bovis* is a pathogen that primarily infects cattle. However, humans can become infected by consuming unpasteurized dairy products or inhaling infectious droplets from cattle. Human-to-human

Test     Control

**Figure 5.B The urease test. In a positive test, bacterial urease hydrolyzes urea, producing ammonia. The ammonia raises the pH, and the indicator in the medium turns to fuchsia.**

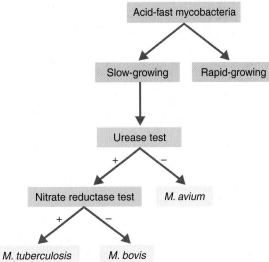

transmission occurs only rarely. The clinical and pathologic characteristics of *M. bovis* TB are indistinguishable from *M. tuberculosis* TB, but identification of the bacterium is important for prevention and treatment. Children may be at higher risk. In one study, almost half of the culture-positive pediatric TB cases were caused by *M. bovis*.

Unfortunately, Julia does not recover from her illness. Her cardiovascular system collapses, and she dies. The official cause of death is peritoneal tuberculosis caused by *M. bovis*. Everyone should avoid consuming products from unpasteurized cow's milk, which carry the risk of transmitting *M. bovis* if imported from countries where the bacterium is common in cattle.

*Source:* Adapted from *MMWR* 65(8): 197–201, March 4, 2016.

---

fermentation. Much of this energy is given off as heat. The complete metabolic oxidation of glucose to carbon dioxide and water is considered a very efficient process, but about 45% of the energy of glucose is lost as heat. Cells use the remaining energy, which is trapped in the bonds of ATP, in a variety of ways. Microbes use ATP to provide energy for the transport of substances across plasma membranes by active transport.

(See Chapter 4, page 89.) Microbes also use some of their energy for flagellar motion (also discussed in Chapter 4). Most of the ATP, however, is used in the production of new cellular components. This production is a continuous process in cells and, in general, is faster in prokaryotic cells than in eukaryotic cells.

Autotrophs build their organic compounds by fixing carbon dioxide in the Calvin-Benson cycle (see Figure 5.26).

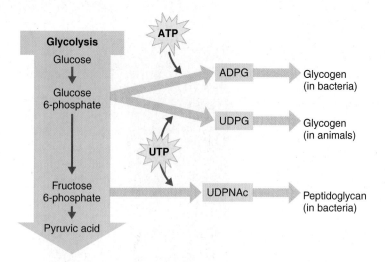

**Figure 5.29 The biosynthesis of polysaccharides.**

**Q** How are polysaccharides used in cells?

This requires both energy (ATP) and electrons (from the oxidation of NADPH). Heterotrophs, by contrast, must have a ready source of organic compounds for biosynthesis—the production of needed cellular components, usually from simpler molecules. The cells use these compounds as both the carbon source and the energy source. We will next consider the biosynthesis of a few representative classes of biological molecules: carbohydrates, lipids, amino acids, purines, and pyrimidines. As we do so, keep in mind that synthesis reactions require a net input of energy.

## Polysaccharide Biosynthesis

Microorganisms synthesize sugars and polysaccharides. The carbon atoms required to synthesize glucose are derived from the intermediates produced during processes such as glycolysis and the Krebs cycle, and from lipids or amino acids. After synthesizing glucose (or other simple sugars), bacteria may assemble it into more complex polysaccharides, such as glycogen. For bacteria to build glucose into glycogen, glucose units must be phosphorylated and linked. The product of glucose phosphorylation is glucose 6-phosphate. Such a process involves the expenditure of energy, usually in the form of ATP. In order for bacteria to synthesize glycogen, a molecule of ATP is added to glucose 6-phosphate to form *adenosine diphosphoglucose (ADPG)* (**Figure 5.29**). Once ADPG is synthesized, it is linked with similar units to form glycogen.

Using a nucleotide called uridine triphosphate (UTP) as a source of energy and glucose 6-phosphate, animals synthesize glycogen (and many other carbohydrates) from *uridine diphosphoglucose, UDPG* (see Figure 5.29). A compound related to UDPG, called *UDP-N-acetylglucosamine (UDPNAc)*, is a key starting material in the biosynthesis of peptidoglycan, the substance that forms bacterial cell walls. UDPNAc is formed from fructose 6-phosphate, and the reaction also uses UTP.

## Lipid Biosynthesis

Because lipids vary considerably in chemical composition, they are synthesized by a variety of routes. Cells synthesize fats by joining glycerol and fatty acids. The glycerol portion of the fat is derived from dihydroxyacetone phosphate, an intermediate formed during glycolysis. Fatty acids, which are long-chain hydrocarbons (hydrogen linked to carbon), are built up when two-carbon fragments of acetyl CoA are successively added to each other (**Figure 5.30**). As with polysaccharide synthesis, the building units of fats and other lipids are linked via dehydration synthesis reactions that require energy, not always in the form of ATP.

The most important role of lipids is to serve as structural components of biological membranes, and most membrane lipids are phospholipids. A lipid of a very different structure, cholesterol, is also found in plasma membranes of eukaryotic cells. Waxes are lipids that are important components of the cell wall of acid-fast bacteria. Other lipids, such as carotenoids,

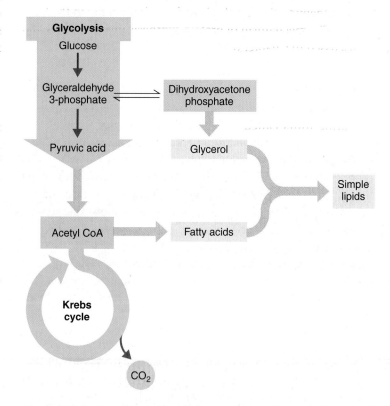

**Figure 5.30 The biosynthesis of simple lipids.**

**Q** What is the primary use of lipids in cells?

provide the red, orange, and yellow pigments of some microorganisms. Some lipids form portions of chlorophyll molecules. Lipids also function in energy storage. Recall that the breakdown products of lipids after biological oxidation feed into the Krebs cycle.

## Amino Acid and Protein Biosynthesis

Amino acids are required for protein biosynthesis. Some microbes, such as *E. coli*, contain the enzymes necessary to use starting materials, such as glucose and inorganic salts, for the synthesis of all the amino acids they need. Organisms with the necessary enzymes can synthesize all amino acids directly or indirectly from intermediates of carbohydrate metabolism (**Figure 5.31a**). Other microbes require that the environment provide some preformed amino acids.

One important source of the precursors used in amino acid synthesis is the Krebs cycle. Adding an amine group to pyruvic acid or to an appropriate organic acid of the Krebs cycle converts the acid into an amino acid. This process is called **amination**. If the amine group comes from a preexisting amino acid, the process is called **transamination** (Figure 5.31b).

Most amino acids within cells are destined to be building blocks for protein synthesis. Proteins play major roles in the cell as enzymes, structural components, and toxins, to name just a few uses. The joining of amino acids to form proteins involves dehydration synthesis and requires energy in the form of ATP. The mechanism of protein synthesis involves genes and is discussed in Chapter 8.

## Purine and Pyrimidine Biosynthesis

The informational molecules DNA and RNA consist of repeating units called *nucleotides*, each of which consists of a purine or pyrimidine, a pentose (five-carbon sugar), and a phosphate group. (See Chapter 2.) The five-carbon sugars of nucleotides are derived from either the pentose phosphate pathway or the Entner-Doudoroff pathway. Certain amino acids—aspartic acid, glycine, and glutamine—made from intermediates produced during glycolysis and in the Krebs cycle participate in the biosyntheses of purines and pyrimidines (**Figure 5.32**). The carbon and nitrogen atoms derived from these amino acids form the purine and pyrimidine rings, and the energy for synthesis is provided by ATP. DNA contains all the information necessary to determine the specific structures and functions of cells. Both RNA and DNA are required for protein synthesis. In addition, nucleotides such as ATP, $NAD^+$, and $NADP^+$ assume roles in stimulating and inhibiting the rate of cellular metabolism. The synthesis of DNA and RNA from nucleotides will be discussed in Chapter 8.

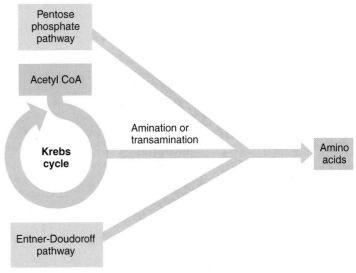

**(a)** Amino acid biosynthesis

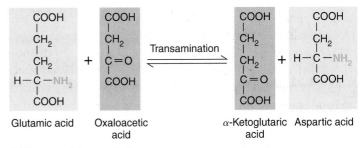

**(b)** Process of transamination

**Figure 5.31 The biosynthesis of amino acids. (a)** Pathways of amino acid biosynthesis through amination or transamination of intermediates of carbohydrate metabolism from the Krebs cycle, pentose phosphate pathway, and Entner-Doudoroff pathway. **(b)** Transamination, a process by which new amino acids are made with the amine groups from old amino acids. Glutamic acid and aspartic acid are both amino acids; the other two compounds are intermediates in the Krebs cycle.

**Q** What is the function of amino acids in cells?

**CHECK YOUR UNDERSTANDING**

5-24 Where do amino acids required for protein synthesis come from?

# The Integration of Metabolism

**LEARNING OBJECTIVE**

5-25 Define *amphibolic pathways*.

We have seen thus far that the metabolic processes of microbes produce energy from light, inorganic compounds, and organic compounds. Reactions also occur in which

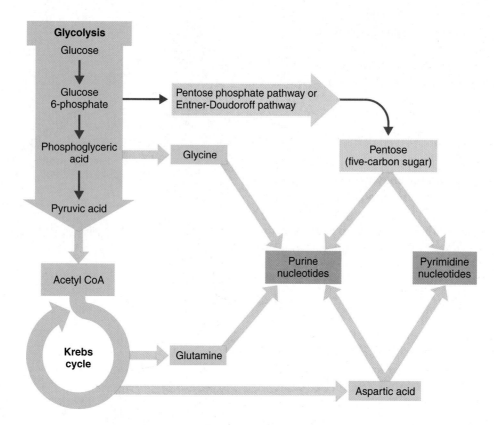

**Figure 5.32 The biosynthesis of purine and pyrimidine nucleotides.**

**Q** What are the functions of nucleotides in a cell?

energy is used for biosynthesis. With such a variety of activity, you might imagine that anabolic and catabolic reactions occur independently of each other in space and time. Actually, anabolic and catabolic reactions are joined through a group of common intermediates (identified as key intermediates in **Figure 5.33**). Both anabolic and catabolic reactions also share some metabolic pathways, such as the Krebs cycle. For example, reactions in the Krebs cycle not only participate in the oxidation of glucose but also produce intermediates that can be converted to amino acids. Metabolic pathways that function in both anabolism and catabolism are called **amphibolic pathways**, meaning that they are dual-purpose.

Amphibolic pathways bridge the reactions that lead to the breakdown and synthesis of carbohydrates, lipids, proteins, and nucleotides. Such pathways enable simultaneous reactions to occur in which the breakdown product formed in one reaction is used in another reaction to synthesize a different compound, and vice versa. Because various intermediates are common to both anabolic and catabolic reactions, mechanisms exist that regulate synthesis and breakdown pathways and allow these reactions to occur simultaneously.

One such mechanism involves the use of different coenzymes for opposite pathways. For example, $NAD^+$ is involved in catabolic reactions, whereas $NADP^+$ is involved in anabolic reactions. Enzymes can also coordinate anabolic and catabolic reactions by accelerating or inhibiting the rates of biochemical reactions.

The energy stores of a cell can also affect the rates of biochemical reactions. For example, if ATP begins to accumulate, feedback inhibition to an enzyme shuts down glycolysis; this control helps to synchronize the rates of glycolysis and the Krebs cycle. Thus, if citric acid consumption increases, either because of a demand for more ATP or because anabolic pathways are draining off intermediates of the Krebs cycle, glycolysis accelerates and meets the demand.

Play Metabolism: *The Big Picture* @Mastering Microbiology

**CHECK YOUR UNDERSTANDING**

✔ **5-25** Summarize the integration of metabolic pathways using peptidoglycan synthesis as an example.

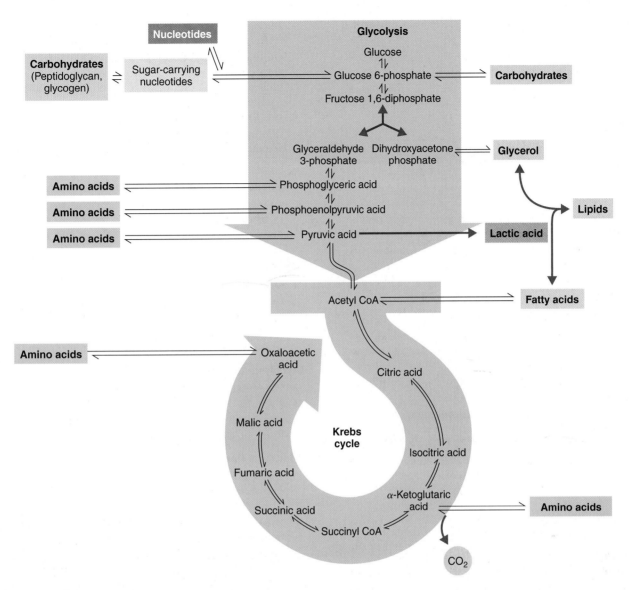

**Figure 5.33 The integration of metabolism.** Key intermediates are shown. Although not indicated in the figure, amino acids and ribose are used in the synthesis of purine and pyrimidine nucleotides (see Figure 5.32). The double arrows indicate amphibolic pathways.

 What is the purpose of an amphibolic pathway?

## Study Outline

> Go to @MasteringMicrobiology for Interactive Microbiology, *In the Clinic* videos, MicroFlix, MicroBoosters, 3D animations, practice quizzes, and more.

### Catabolic and Anabolic Reactions (pp. 110–111)

1. The sum of all chemical reactions within a living organism is known as metabolism.

2. Catabolism refers to chemical reactions that result in the breakdown of more complex organic molecules into simpler substances. Catabolic reactions usually release energy.

3. Anabolism refers to chemical reactions in which simpler substances are combined to form more complex molecules. Anabolic reactions usually require energy.

4. The energy of catabolic reactions is used to drive anabolic reactions.

5. The energy for chemical reactions is stored in ATP.

### Enzymes (pp. 111–116)

1. Enzymes are proteins, produced by living cells, that catalyze chemical reactions by lowering the activation energy.

2. Enzymes are generally globular proteins with characteristic three-dimensional shapes.

3. Enzymes are efficient, can operate at relatively low temperatures, and are subject to various cellular controls.

4. When an enzyme and substrate combine, the substrate is transformed, and the enzyme is recovered.

5. Enzymes are characterized by specificity, which is a function of their active sites.

### Naming Enzymes (p. 113)

6. Enzyme names usually end in -ase.

7. The six classes of enzymes are defined on the basis of the types of reactions they catalyze.

### Enzyme Components (pp. 113–114)

8. Most enzymes are holoenzymes, consisting of a protein portion (apoenzyme) and a nonprotein portion (cofactor).

9. The cofactor can be a metal ion (iron, copper, magnesium, manganese, zinc, calcium, or cobalt) or a complex organic molecule known as a coenzyme ($NAD^+$, $NADP^+$, FMN, FAD, or coenzyme A).

### Factors Influencing Enzymatic Activity (pp. 114–116)

10. At high temperatures, enzymes undergo denaturation and lose their catalytic properties; at low temperatures, the reaction rate decreases.

11. The pH at which enzymatic activity is maximal is known as the optimum pH.

12. Enzymatic activity increases as substrate concentration increases until the enzymes are saturated.

13. Competitive inhibitors compete with the normal substrate for the active site of the enzyme. Noncompetitive inhibitors act on other parts of the apoenzyme or on the cofactor and decrease the enzyme's ability to combine with the normal substrate.

### Feedback Inhibition (p. 116)

14. Feedback inhibition occurs when the end-product of a metabolic pathway inhibits an enzyme's activity near the start of the pathway.

### Ribozymes (pp. 116–117)

15. Ribozymes are enzymatic RNA molecules involved in protein synthesis.

## Energy Production (pp. 117–119)

### Oxidation-Reduction Reactions (pp. 117–118)

1. Oxidation is the removal of one or more electrons from a substrate. Protons ($H^+$) are often removed with the electrons.

2. Reduction of a substrate refers to its gain of one or more electrons.

3. Each time a substance is oxidized, another is simultaneously reduced.

4. $NAD^+$ is the oxidized form; NADH is the reduced form.

5. Glucose is a reduced molecule; energy is released during a cell's oxidation of glucose.

### The Generation of ATP (pp. 118–119)

6. Energy released during certain metabolic reactions can be trapped to form ATP from ADP and ⓟ$_i$ (phosphate). Addition of a ⓟ$_i$ molecule is called phosphorylation.

7. During substrate-level phosphorylation, a high-energy ⓟ from an intermediate in catabolism is added to ADP.

8. During oxidative phosphorylation, energy is released as electrons are passed to a series of electron acceptors (an electron transport chain) and finally to $O_2$ or another inorganic compound.

9. During photophosphorylation, energy from light is trapped by chlorophyll, and electrons are passed through a series of electron acceptors. The electron transfer releases energy used for the synthesis of ATP.

### Metabolic Pathways of Energy Production (p. 119)

10. A series of enzymatically catalyzed chemical reactions called metabolic pathways store energy in and release energy from organic molecules.

## Carbohydrate Catabolism (p. 119–132)

1. Most of a cell's energy is produced from the oxidation of carbohydrates.

2. The two major types of carbohydrate catabolism are respiration, in which a sugar is completely broken down, and fermentation, in which the sugar is partially broken down.

### Glycolysis (p. 121)

3. The most common pathway for the oxidation of glucose is glycolysis. Pyruvic acid is the end-product.

4. Glycolysis yields two ATP and two NADH molecules are produced from one glucose molecule.

### Additional Pathways to Glycolysis (p. 121)

5. The pentose phosphate pathway is used to oxidize five-carbon sugars; one ATP and 12 NADPH molecules are produced from one glucose molecule.

6. The Entner-Doudoroff pathway yields one ATP and two NADPH molecules from oxidation of one glucose molecule.

### Cellular Respiration (pp. 123–128)

7. During respiration, organic molecules are oxidized. Energy is generated from oxidations in the electron transport chain.

8. In aerobic respiration, $O_2$ functions as the final electron acceptor.

9. In anaerobic respiration, the final electron acceptor is not $O_2$; the electron acceptors in anaerobic respiration include $NO_3^-$, $SO_4^{2-}$, and $CO_3^{2-}$.

10. Decarboxylation of pyruvic acid produces one $CO_2$ molecule and one acetyl group.

11. Two-carbon acetyl groups are oxidized in the Krebs cycle. Electrons are picked up by $NAD^+$ and FAD for the electron transport chain.

12. Oxidation of one molecule of glucose, oxidation produces six molecules of NADH, two molecules of $FADH_2$, and two molecules of ATP.

13. Decarboxylation produces six molecules of $CO_2$ in the Krebs Cycle.

14. NADH and $FADH_2$ carry electrons to the electron transport chain by NADH.

15. The electron transport chain consists of carriers, including flavoproteins, cytochromes, and ubiquinones.

16. Protons being pumped across the membrane generate a proton motive force as electrons move through a series of acceptors or carriers.

17. Energy produced from movement of the protons back across the membrane is used by ATP synthase to make ATP from ADP and ⓟi.

18. In eukaryotes, electron carriers are located in the inner mitochondrial membrane; in prokaryotes, electron carriers are in the plasma membrane.

19. In aerobic prokaryotes, 38 ATP molecules can be produced from complete oxidation of a glucose molecule in glycolysis, the Krebs cycle, and the electron transport chain.

20. In eukaryotes, 36 ATP molecules are produced from complete oxidation of a glucose molecule.

21. The total ATP yield in anaerobic respiration is less than in aerobic respiration because only part of the Krebs cycle operates under anaerobic conditions.

### Fermentation (pp. 128–132)

22. Fermentation releases energy from sugars or other organic molecules by oxidation.

23. $O_2$ is not required in fermentation.

24. Two ATP molecules are produced by substrate-level phosphorylation.

25. Electrons removed from the substrate reduce $NAD^+$.

26. The final electron acceptor is a substance from inside the cell.

27. In lactic acid fermentation, pyruvic acid is reduced by NADH to lactic acid.

28. In alcohol fermentation, acetaldehyde is reduced by NADH to produce ethanol.

29. Heterolactic fermenters can use the pentose phosphate pathway to produce lactic acid and ethanol.

### Lipid and Protein Catabolism (p. 133)

1. Lipases hydrolyze lipids into glycerol and fatty acids.

2. Fatty acids and other hydrocarbons are catabolized by beta-oxidation.

3. Catabolic products can be further broken down in glycolysis and the Krebs cycle.

4. Before amino acids can be catabolized, they must be converted to various substances that enter the Krebs cycle.

5. Transamination, decarboxylation, and desulfurization reactions convert the amino acids to be catabolized.

### Biochemical Tests and Bacterial Identification (pp. 134–135)

1. Bacteria and yeast can be identified by detecting action of their enzymes.

2. Fermentation tests are used to determine whether an organism can ferment a carbohydrate to produce acid and gas.

### Photosynthesis (pp. 135–137)

1. Photosynthesis is the conversion of light energy from the sun into chemical energy; the chemical energy is used for carbon fixation.

### The Light-Dependent Reactions:
### Photophosphorylation (pp. 136–137)

2. Chlorophyll *a* is used by green plants, algae, and cyanobacteria.

3. Electrons from chlorophyll pass through an electron transport chain, from which ATP is produced by chemiosmosis.

4. Photosystems are made up of chlorophyll and other pigments packed into thylakoid membranes.

5. In cyclic photophosphorylation, the electrons return to the chlorophyll.

6. In noncyclic photophosphorylation, the electrons are used to reduce $NADP^+$. The electrons from $H_2O$ or $H_2S$ replace those lost from chlorophyll.

7. When $H_2O$ is oxidized by green plants, algae, and cyanobacteria, $O_2$ is produced; when $H_2S$ is oxidized by the sulfur bacteria, $S^0$ granules are produced.

### The Light-Independent Reactions:
### The Calvin-Benson Cycle (p. 137)

8. $CO_2$ is used to synthesize sugars in the Calvin-Benson cycle.

### A Summary of Energy Production Mechanisms (p. 138)

1. Sunlight is converted to chemical energy in oxidation reactions carried on by phototrophs. Chemotrophs can use this chemical energy.

2. In oxidation delete reactions, energy is derived from the transfer of electrons.

3. To produce energy, a cell needs an electron donor (organic or inorganic), a system of electron carriers, and a final electron acceptor (organic or inorganic).

### Metabolic Diversity among Organisms (pp. 138–140)

1. Photoautotrophs obtain energy by photophosphorylation and fix carbon from $CO_2$ via the Calvin-Benson cycle to synthesize organic compounds.

2. Cyanobacteria are oxygenic phototrophs. Green bacteria and purple bacteria are anoxygenic phototrophs.

3. Photoheterotrophs use light as an energy source and an organic compound for their carbon source and electron donor.

4. Chemoautotrophs use inorganic compounds as their energy source and carbon dioxide as their carbon source.

5. Chemoheterotrophs use complex organic molecules as their carbon and energy sources.

### Metabolic Pathways of Energy Use (pp. 140–143)

### Polysaccharide Biosynthesis (p. 142)

1. Glycogen is formed from ADPG.

2. UDPNAc is the starting material for the biosynthesis of peptidoglycan.

### Lipid Biosynthesis (pp. 142–143)

3. Lipids are synthesized from fatty acids and glycerol.

4. Glycerol is derived from dihydroxyacetone phosphate, and fatty acids are built from acetyl CoA.

### Amino Acid and Protein Biosynthesis (p. 143)

5. Amino acids are required for protein biosynthesis.

6. All amino acids can be synthesized either directly or indirectly from intermediates of carbohydrate metabolism, particularly from the Krebs cycle.

### Purine and Pyrimidine Biosynthesis (p. 143)

7. The sugars composing nucleotides are derived from either the pentose phosphate pathway or the Entner-Doudoroff pathway.

8. Carbon and nitrogen atoms from certain amino acids form the backbones of the purines and pyrimidines.

### The Integration of Metabolism (pp. 143–145)

1. Anabolic and catabolic reactions are integrated through a group of common intermediates.

2. Such integrated metabolic pathways are referred to as amphibolic pathways.

# Study Questions

For answers to the Knowledge and Comprehension questions, turn to the Answers tab at the back of the textbook.

## Knowledge and Comprehension

### Review

Use the following diagrams (a), (b), and (c) for question 1.

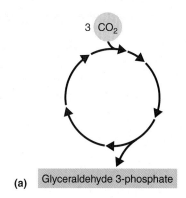

(a)

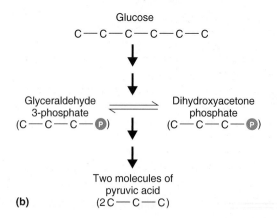

(b)

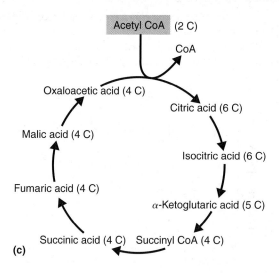

(c)

1. Name pathways diagrammed in parts (a), (b), and (c) of the figure.
   a. Show where glycerol is catabolized and where fatty acids are catabolized.
   b. Show where glutamic acid (an amino acid) is catabolized:

$$HOOC-CH_2-CH_2-\overset{\overset{\displaystyle H}{|}}{\underset{\underset{\displaystyle NH_2}{|}}{C}}-COOH$$

   c. Show how these pathways are related.
   d. Where is ATP required in pathways (a) and (b)?
   e. Where is $CO_2$ released in pathways (b) and (c)?
   f. Show where a long-chain hydrocarbon such as petroleum is catabolized.
   g. Where is NADH (or $FADH_2$ or NADPH) used and produced in these pathways?
   h. Identify four places where anabolic and catabolic pathways are integrated.

2. **DRAW IT** Using the diagrams below, show each of the following:
   a. where the substrate will bind
   b. where the competitive inhibitor will bind
   c. where the noncompetitive inhibitor will bind
   d. which of the four elements could be the inhibitor in feedback inhibition
   e. What effect will the reactions in (a), (b), and (c) have?

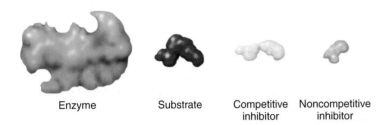

| Enzyme | Substrate | Competitive inhibitor | Noncompetitive inhibitor |

3. **DRAW IT** An enzyme and substrate are combined. The rate of reaction begins as shown in the following graph. To complete the graph, show the effect of increasing substrate concentration on a constant enzyme concentration. Show the effect of increasing temperature.

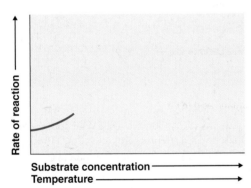

4. Define *oxidation-reduction*, and differentiate the following terms:
   a. aerobic and anaerobic respiration
   b. respiration and fermentation
   c. cyclic and noncyclic photophosphorylation

5. There are three mechanisms for the phosphorylation of ADP to produce ATP. Write the name of the mechanism that describes each of the reactions in the following table.

| ATP Generated by | Reaction |
|---|---|
| a. _____ | An electron, liberated from chlorophyll by light, is passed down an electron transport chain. |
| b. _____ | Cytochrome *c* passes two electrons to cytochrome *a*. |
| c. _____ | $CH_2$ ‖ $C$—O~Ⓟ → $CH_3$ ‖ $C$=O COOH COOH Phosphoenolpyruvic acid  Pyruvic acid |

6. All of the energy-producing biochemical reactions that occur in cells, such as photophosphorylation and glycolysis, are _____ reactions.

7. Fill in the following table with the carbon source and energy source of each type of organism.

| Organism | Carbon Source | Energy Source |
|---|---|---|
| Photoautotroph | a. _____ | b. _____ |
| Photoheterotroph | c. _____ | d. _____ |
| Chemoautotroph | e. _____ | f. _____ |
| Chemoheterotroph | g. _____ | h. _____ |

8. Write your own definition of the chemiosmotic mechanism of ATP generation. On Figure 5.16, mark the following using the appropriate letter:
   a. the acidic side of the membrane
   b. the side with a positive electrical charge
   c. potential energy
   d. kinetic energy

9. Why must NADH be reoxidized? How does this happen in an organism that uses respiration? Fermentation?

10. **NAME IT** What nutritional type is a colorless microbe that uses the Calvin-Benson cycle, uses $H_2$ as the electron donor to its ETC, and uses elemental S as the final electron acceptor in the ETC?

## Multiple Choice

1. Which substance in the following reaction is being reduced?

$$\begin{array}{c} H \\ | \\ C=O \\ | \\ CH_3 \end{array} + NADH + H^+ \longrightarrow \begin{array}{c} H \\ | \\ H-C-OH \\ | \\ CH_3 \end{array} + NAD^+$$

Acetaldehyde          Ethanol

   a. acetaldehyde
   b. NADH
   c. ethanol
   d. $NAD^+$

2. Which of the following reactions produces the most molecules of ATP during aerobic metabolism?
   a. glucose → glucose 6-phosphate
   b. phosphoenolpyruvic acid → pyruvic acid
   c. glucose → pyruvic acid
   d. acetyl CoA → $CO_2 + H_2O$
   e. succinic acid → fumaric acid

3. Which of the following processes does *not* generate ATP?
   a. photophosphorylation
   b. the Calvin-Benson cycle
   c. oxidative phosphorylation
   d. substrate-level phosphorylation
   e. All of the above generate ATP

4. Which of the following compounds has the greatest amount of energy for a cell?
   a. $CO_2$
   b. ATP
   c. glucose
   d. $O_2$
   e. lactic acid

5. Which of the following is the best definition of the Krebs cycle?
   a. the oxidation of pyruvic acid
   b. the way cells produce $CO_2$
   c. a series of chemical reactions in which NADH is produced from the oxidation of pyruvic acid
   d. a method of producing ATP by phosphorylating ADP
   e. a series of chemical reactions in which ATP is produced from the oxidation of pyruvic acid

6. Which of the following is the best definition of *respiration*?
   a. a sequence of carrier molecules with $O_2$ as the final electron acceptor
   b. a sequence of carrier molecules with the final electron acceptor from the environment
   c. a method of generating ATP
   d. the complete oxidation of glucose to $CO_2$ and $H_2O$
   e. a series of reactions in which pyruvic acid is oxidized to $CO_2$ and $H_2O$

Use the following choices to answer questions 7–10.
   a. *E. coli* growing in glucose broth at 35°C with $O_2$ for 5 days
   b. *E. coli* growing in glucose broth at 35°C without $O_2$ for 5 days
   c. both a and b
   d. neither a nor b

7. Which culture produces the most lactic acid?
8. Which culture produces the most ATP?
9. Which culture uses $NAD^+$?
10. Which culture uses the most glucose?

## Analysis

1. Explain why, even under ideal conditions, *Streptococcus* grows slowly.
2. The following graph shows the normal rate of reaction of an enzyme and its substrate (blue) and the rate when an excess of competitive inhibitor is present (red). Explain why the graph appears as it does.

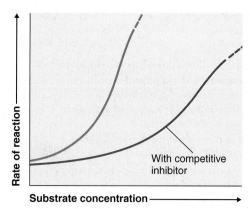

3. Compare and contrast carbohydrate catabolism and energy production in the following bacteria:
   a. *Pseudomonas*, an aerobic chemoheterotroph
   b. *Spirulina*, an oxygenic photoautotroph
   c. *Ectothiorhodospira*, an anoxygenic photoautotroph

4. How much ATP could be obtained from the complete oxidation of one molecule of glucose? From one molecule of butterfat containing one glycerol and three 12-carbon chains?

5. The chemoautotroph *Acidithiobacillus* can obtain energy from the oxidation of arsenic ($As^{3+} \rightarrow As^{5+}$). How does this reaction provide energy? How can humans put this bacterium to use?

## Clinical Applications and Evaluation

1. *Haemophilus influenzae* requires hemin (X factor) to synthesize cytochromes and $NAD^+$ (V factor) from other cells. For what does it use these two growth factors? What diseases does *H. influenzae* cause?

2. The drug Hivid, also called ddC, inhibits DNA synthesis. It is used to treat HIV infection and AIDS. Compare the following illustration of ddC to the structure of DNA nucleotides in Figure 2.16 on page 45. How does this drug work?

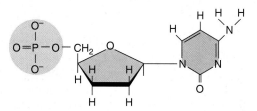

3. The bacterial enzyme streptokinase is used to digest fibrin (blood clots) in patients with atherosclerosis. Why doesn't injection of streptokinase cause a streptococcal infection? How do we know the streptokinase will digest fibrin only and not good tissues?

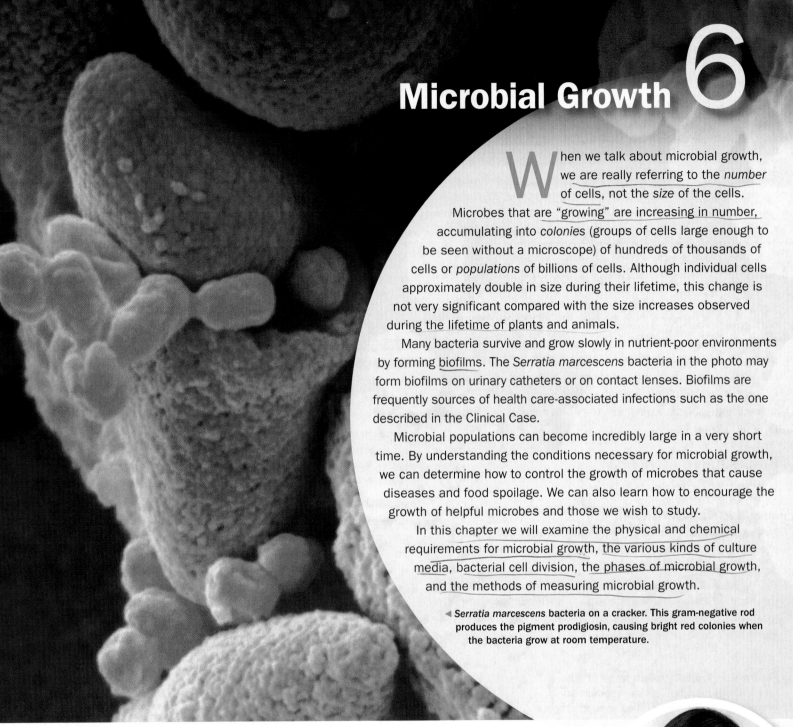

# Microbial Growth 6

When we talk about microbial growth, we are really referring to the *number* of cells, not the *size* of the cells. Microbes that are "growing" are increasing in number, accumulating into *colonies* (groups of cells large enough to be seen without a microscope) of hundreds of thousands of cells or *populations* of billions of cells. Although individual cells approximately double in size during their lifetime, this change is not very significant compared with the size increases observed during the lifetime of plants and animals.

Many bacteria survive and grow slowly in nutrient-poor environments by forming biofilms. The *Serratia marcescens* bacteria in the photo may form biofilms on urinary catheters or on contact lenses. Biofilms are frequently sources of health care-associated infections such as the one described in the Clinical Case.

Microbial populations can become incredibly large in a very short time. By understanding the conditions necessary for microbial growth, we can determine how to control the growth of microbes that cause diseases and food spoilage. We can also learn how to encourage the growth of helpful microbes and those we wish to study.

In this chapter we will examine the physical and chemical requirements for microbial growth, the various kinds of culture media, bacterial cell division, the phases of microbial growth, and the methods of measuring microbial growth.

◀ *Serratia marcescens* bacteria on a cracker. This gram-negative rod produces the pigment prodigiosin, causing bright red colonies when the bacteria grow at room temperature.

## In the Clinic

As a nurse in a plastic surgery clinic, you instruct patients on postsurgical care of their sutures. You tell patients to wash hands before removing bandages, to wash gently around the surgical site with soap and water, and to swab the wound with hydrogen peroxide. **One day a patient calls, alarmed that the hydrogen peroxide caused her wound to bubble. What would you tell the patient?**

*Hint: Read about catalase on page 157.*

 Play **In the Clinic** Video
@Mastering**Microbiology**

# The Requirements for Growth

### LEARNING OBJECTIVES

6-1 Classify microbes into five groups on the basis of preferred temperature range.

6-2 Identify how and why the pH of culture media is controlled.

6-3 Explain the importance of osmotic pressure to microbial growth.

6-4 Name a use for each of the four elements (carbon, nitrogen, sulfur, and phosphorus) needed in large amounts for microbial growth.

6-5 Explain how microbes are classified on the basis of oxygen requirements.

6-6 Identify ways in which aerobes avoid damage by toxic forms of oxygen.

The requirements for microbial growth can be divided into two main categories: physical and chemical. Physi-

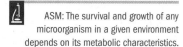 ASM: The survival and growth of any microorganism in a given environment depends on its metabolic characteristics.

cal aspects include temperature, pH, and osmotic pressure. Chemical requirements include sources of carbon, nitrogen, sulfur, phosphorus, oxygen, trace elements, and organic growth factors.

## Physical Requirements

### Temperature

Most microorganisms grow well at the temperatures that humans favor. However, certain bacteria are capable of growing at extremes of temperature that would certainly hinder the survival of almost all eukaryotic organisms.

Microorganisms are classified into three primary groups on the basis of their preferred range of temperature: **psychrophiles** (cold-loving microbes), **mesophiles** (moderate-temperature-loving microbes), and **thermophiles** (heat-loving microbes). Most

bacteria grow only within a limited range of temperatures, and their maximum and minimum growth temperatures are only about 30°C apart. They grow poorly at the high and low temperature extremes within their range.

Each bacterial species grows at particular minimum, optimum, and maximum temperatures. The **minimum growth temperature** is the lowest temperature at which the species will grow. The **optimum growth temperature** is the temperature at which the species grows best. The **maximum growth temperature** is the highest temperature at which growth is possible. By graphing the growth response over a temperature range, we can see that the optimum growth temperature is usually near the top of the range; above that temperature the rate of growth drops off rapidly (**Figure 6.1**). This happens presumably because the high temperature has inactivated necessary enzymatic systems of the cell.

The ranges and maximum growth temperatures that define bacteria as psychrophiles, mesophiles, or thermophiles are not rigidly defined. Psychrophiles, for example, were originally considered simply to be organisms capable of growing at 0°C. However, there seem to be two fairly distinct groups capable of growth at that temperature. One group, composed of psychrophiles in the strictest sense, can grow at 0°C but has an optimum growth temperature of about 15°C. Most of these organisms are so sensitive to higher temperatures that they will not even grow in a reasonably warm room (25°C). Found mostly in the oceans' depths or in certain polar regions, such organisms seldom cause problems in food preservation. The other group that can grow at 0°C has higher optimum temperatures, usually 20–30°C and cannot grow above about 40°C. Organisms of this type are much more common than psychrophiles and are the most likely to be encountered in low-temperature food spoilage because they grow fairly well at refrigerator temperatures. We will use the term **psychrotrophs**, which food microbiologists favor, for this group of spoilage microorganisms.

**Figure 6.1 Typical growth rates of different types of microorganisms in response to temperature.** The peak of the curve represents optimum growth (fastest reproduction). Notice that the reproductive rate drops off very quickly at temperatures only a little above the optimum. At either extreme of the temperature range, the reproductive rate is much lower than the rate at the optimum temperature.

**Q** Why is it difficult to define *psychrophile, mesophile,* and *thermophile*?

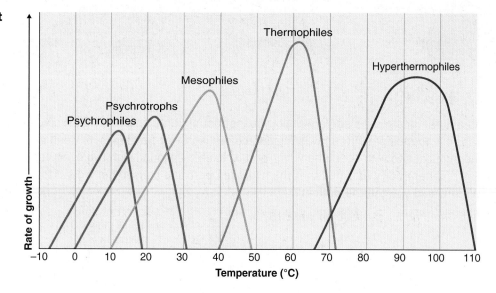

Refrigeration is the most common method of preserving household food supplies. It is based on the principle that microbial reproductive rates decrease at low temperatures. Although microbes usually survive even subfreezing temperatures (they might become entirely dormant), they gradually decline in number. Some species decline faster than others. Psychrotrophs do not grow well at low temperatures, except in comparison with other organisms; given time, however, they are able to slowly degrade food. Such spoilage might take the form of mold mycelium, slime on food surfaces, or off-tastes or off-colors in foods. The temperature inside a properly set refrigerator will greatly slow the growth of most spoilage organisms and will entirely prevent the growth of all but a few pathogenic bacteria. **Figure 6.2** illustrates the importance of low temperatures for preventing the growth of spoilage and disease organisms. When large amounts of food must be refrigerated, it is important to remember that a large quantity of warm food cools at a relatively slow rate (**Figure 6.3**).

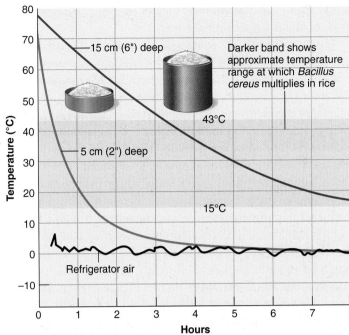

**Figure 6.3 The effect of the amount of food on its cooling rate in a refrigerator and its chance of spoilage.** Notice that in this example, the pan of rice with a depth of 5 cm (2 in) cooled through the incubation temperature range of the *Bacillus cereus* in about 1 hour, whereas the pan of rice with a depth of 15 cm (6 in) remained in this temperature range for about 5 hours.

**Q** Given a shallow pan and a deep pot with the same volume, which would cool faster? Why?

Mesophiles, with an optimum growth temperature of 25–40°C, are the most common type of microbe. Organisms that have adapted to live in the bodies of animals usually have an optimum temperature close to that of their hosts. The optimum temperature for many pathogenic bacteria is about 37°C, and incubators for clinical cultures are usually set at about this

**CLINICAL CASE Glowing in the Dark**

Reginald MacGruder, an investigator at the Centers for Disease Control and Prevention (CDC) in Atlanta, Georgia, has a mystery on his hands. Earlier this year, he was involved in the recall of an intravenous heparin solution that was blamed for causing *Pseudomonas fluorescens* bloodstream infections in patients in four different states. It seemed that everything was under control, but now, three months after the recall, 19 patients in two other states develop the same *P. fluorescens* bloodstream infections. It makes no sense to Dr. MacGruder; how could this infection be popping up again so soon after the recall? Could another heparin batch be tainted?

**What is *P. fluorescens*? Read on to find out.**

153    164    172    174

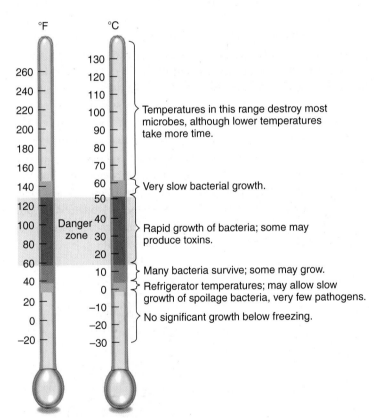

**Figure 6.2 Food preservation temperatures.** Low temperatures decrease microbial reproduction rates, which is the basic principle of refrigeration. There are always some exceptions to the temperature responses shown here; for example, certain bacteria grow well at high temperatures that would kill most bacteria, and a few bacteria can actually grow at temperatures well below freezing.

**Q** Which bacterium would theoretically be more likely to grow at refrigerator temperatures: a human intestinal pathogen or a soilborne plant pathogen?

temperature. The mesophiles include most of the common spoilage and disease organisms.

Thermophiles are microorganisms capable of growth at high temperatures. Many of these organisms have an optimum growth temperature of 50–60°C, about the temperature of water from a hot water tap. Such temperatures can also be reached in sunlit soil and in thermal waters such as hot springs. Remarkably, many thermophiles cannot grow at temperatures below about 45°C. Endospores formed by thermophilic bacteria are unusually heat resistant and may survive the usual heat treatment given canned goods. Although elevated storage temperatures may cause surviving endospores to germinate and grow, thereby spoiling the food, these thermophilic bacteria are not considered a public health problem. Thermophiles are important in organic compost piles (see Figure 27.8 on page 795), in which the temperature can rise rapidly to 50–60°C.

Some microbes, members of the Archaea (page 5), have an optimum growth temperature of 80°C or higher. These organisms are called **hyperthermophiles** or, sometimes, **extreme thermophiles**. Most of these organisms live in hot springs associated with volcanic activity, and sulfur is usually important in their metabolic activity. The known record for bacterial growth and replication at high temperatures is about 121°C near deep-sea hydrothermal vents. The immense pressure in the ocean depths prevents water from boiling even at temperatures well above 100°C.

## pH

Recall from Chapter 2 (pages 32–33) that the pH refers to the acidity or alkalinity of a solution. Most bacteria grow best in a narrow pH range near neutrality, between pH 6.5 and 7.5. Very few bacteria grow at an acidic pH below about pH 4. This is why a number of foods, such as sauerkraut, pickles, and many cheeses, are preserved from spoilage by acids produced by bacterial fermentation. Nonetheless, some bacteria, called **acidophiles**, are remarkably tolerant of acidity. One type of chemoautotrophic bacteria, which is found in the drainage water from coal mines and oxidizes sulfur to form sulfuric acid, can survive at a pH 1. Molds and yeasts will grow over a greater pH range than bacteria will, but the optimum pH of molds and yeasts is generally below that of bacteria, usually about pH 5 to 6. Alkalinity also inhibits microbial growth but is rarely used to preserve foods.

When bacteria are cultured in the laboratory, they often produce acids that eventually interfere with their own growth. To neutralize the acids and maintain the proper pH, chemical buffers are included in the growth medium. The peptones and amino acids in some media act as buffers, and many media also contain phosphate salts. Phosphate salts have the advantage of exhibiting their buffering effect in the pH growth range of most bacteria. They are also nontoxic; in fact, they provide phosphorus, an essential nutrient.

## Osmotic Pressure

Microorganisms obtain almost all their nutrients in solution from the surrounding water. Thus, they require water for growth, and their composition is 80–90% water. High osmotic pressures have the effect of removing necessary water from a cell. When a microbial cell is in a solution whose concentration of solutes is higher than in the cell (the environment is *hypertonic* to the cell), the cellular water passes out through the plasma membrane to the high solute concentration. (See the discussion of osmosis in Chapter 4, pages 88–89, and review Figure 4.18 for the three types of solution environments a cell may encounter.) This osmotic loss of water causes **plasmolysis**, or shrinkage of the cell's cytoplasm (**Figure 6.4**).

**Figure 6.4 Plasmolysis.**

**Q** Why is sugar added to fruit to make jams and jellies?

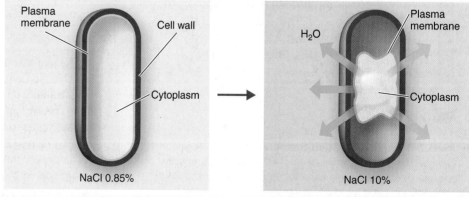

(a) **Cell in isotonic solution.** Under these conditions, the solute concentration in the cell is equivalent to a solute concentration of 0.85% sodium chloride (NaCl).

(b) **Plasmolyzed cell in hypertonic solution.** If the concentration of solutes such as NaCl is higher in the surrounding medium than in the cell (the environment is hypertonic), water tends to leave the cell. Growth of the cell is inhibited.

The growth of the cell is inhibited as the plasma membrane pulls away from the cell wall. Thus, the addition of salts (or other solutes) to a solution, and the resulting increase in osmotic pressure, can be used to preserve foods. Salted fish, honey, and sweetened condensed milk are preserved largely by this mechanism; the high salt or sugar concentrations draw water out of any microbial cells that are present and thus prevent their growth.

Some organisms, called **extreme halophiles**, have adapted so well to high salt concentrations that they actually require them for growth. In this case, they may be termed **obligate halophiles**. Organisms from such saline waters as the Dead Sea often require nearly 30% salt, and the inoculating loop (a device for handling bacteria in the laboratory) used to transfer them must first be dipped into a saturated salt solution. More common are **facultative halophiles**, which do not require high salt concentrations but are able to grow at salt concentrations up to 2%, a concentration that inhibits the growth of many other organisms. A few species of facultative halophiles can tolerate even 15% salt.

Most microorganisms, however, must be grown in a medium that is nearly all water. For example, the concentration of agar (a complex polysaccharide isolated from marine algae) used to solidify microbial growth media is usually about 1.5%. If markedly higher concentrations are used, the increased osmotic pressure can inhibit the growth of some bacteria.

If the osmotic pressure is unusually low (the environment is *hypotonic*)—such as in distilled water, for example—water tends to enter the cell rather than leave it. Some microbes that have a relatively weak cell wall may be lysed by such treatment.

### CHECK YOUR UNDERSTANDING

✔ 6-1 Why are hyperthermophiles that grow at temperatures above 100°C seemingly limited to oceanic depths?

✔ 6-2 Other than controlling acidity, what is an advantage of using phosphate salts as buffers in growth media?

✔ 6-3 Why might primitive civilizations have used food preservation techniques that rely on osmotic pressure?

## Chemical Requirements

### Carbon

Besides water, one of the most important requirements for microbial growth is carbon. Carbon is the structural backbone of living matter; it is needed for all the organic compounds that make up a living cell. Half the dry weight of a typical bacterial cell is carbon. Chemoheterotrophs get most of their carbon from the source of their energy—organic materials such as proteins, carbohydrates, and lipids. Chemoautotrophs and photoautotrophs derive their carbon from carbon dioxide.

### Nitrogen, Sulfur, and Phosphorus

In addition to carbon, microorganisms need other elements to synthesize cellular material. For example, protein synthesis requires considerable amounts of nitrogen as well as some sulfur. The syntheses of DNA and RNA also require nitrogen and some phosphorus, as does the synthesis of ATP, the molecule so important for the storage and transfer of chemical energy within the cell. Nitrogen makes up about 14% of the dry weight of a bacterial cell, and sulfur and phosphorus together constitute about another 4%.

Organisms use nitrogen primarily to form the amino group of the amino acids of proteins. Many bacteria meet this requirement by decomposing protein-containing material and reincorporating the amino acids into newly synthesized proteins and other nitrogen-containing compounds. Other bacteria use nitrogen from ammonium ions ($NH_4^+$), which are already in the reduced form and are usually found in organic cellular material. Still other bacteria are able to derive nitrogen from nitrates (compounds that dissociate to give the nitrate ion, $NO_3^-$, in solution).

Some important bacteria, including many of the photosynthesizing cyanobacteria (page 307), use gaseous nitrogen ($N_2$) directly from the atmosphere. This process is called **nitrogen fixation**. Some organisms that can use this method are free-living, mostly in the soil, but others live cooperatively in symbiosis with the roots of legumes such as clover, soybeans, alfalfa, beans, and peas. The nitrogen fixed in the symbiosis is used by both the plant and the bacterium (see Chapter 27).

Sulfur is used to synthesize sulfur-containing amino acids and vitamins such as thiamine and biotin. Important natural sources of sulfur include the sulfate ion ($SO_4^{2-}$), hydrogen sulfide, and the sulfur-containing amino acids.

Phosphorus is essential for the synthesis of nucleic acids and the phospholipids of cell membranes. Among other places, it is also found in the energy bonds of ATP. A source of phosphorus is the phosphate ion ($PO_4^{3-}$). Potassium, magnesium, and calcium are also elements that microorganisms require, often as cofactors for enzymes (see Chapter 5, page 114).

### Trace Elements

Microbes require very small amounts of other mineral elements, such as iron, copper, molybdenum, and zinc; these are referred to as **trace elements**. Most are essential for the functions of certain enzymes, usually as cofactors. Although these elements are sometimes added to a laboratory medium, they are usually assumed to be naturally present in tap water and other components of media. Even most distilled waters contain adequate amounts, but tap water is sometimes specified to ensure that these trace minerals will be present in culture media.

### Oxygen

We are accustomed to thinking of molecular oxygen ($O_2$) as a necessity of life, but it is actually in a sense a poisonous gas.

**TABLE 6.1** The Effect of Oxygen on the Growth of Various Types of Bacteria

| | a. Obligate Aerobes | b. Facultative Anaerobes | c. Obligate Anaerobes | d. Aerotolerant Anaerobes | e. Microaerophiles |
|---|---|---|---|---|---|
| **Effect of Oxygen on Growth** | Only aerobic growth; oxygen required. | Both aerobic and anaerobic growth; greater growth in presence of oxygen. | Only anaerobic growth; growth ceases in presence of oxygen. | Only anaerobic growth; but growth continues in presence of oxygen. | Only aerobic growth; oxygen required in low concentration. |
| **Bacterial Growth in Tube of Solid Growth Medium** | | | | | |
| **Explanation of Growth Patterns** | Growth occurs only where high concentrations of oxygen have diffused into the medium. | Growth is best where most oxygen is present, but occurs throughout tube. | Growth occurs only where there is no oxygen. | Growth occurs evenly; oxygen has no effect. | Growth occurs only where a low concentration of oxygen has diffused into medium. |
| **Explanation of Oxygen's Effects** | Presence of enzymes catalase and superoxide dismutase (SOD) allows toxic forms of oxygen to be neutralized; can use oxygen. | Presence of enzymes catalase and SOD allows toxic forms of oxygen to be neutralized; can use oxygen. | Lacks enzymes to neutralize harmful forms of oxygen; cannot tolerate oxygen. | Presence of one enzyme, SOD, allows harmful forms of oxygen to be partially neutralized; tolerates oxygen. | Produce lethal amounts of toxic forms of oxygen if exposed to normal atmospheric oxygen. |

Very little molecular oxygen existed in the atmosphere during most of Earth's history—in fact, it is possible that life could not have arisen had oxygen been present. However, many current forms of life have metabolic systems that require oxygen for aerobic respiration. Hydrogen atoms that have been stripped from organic compounds combine with oxygen to form water, as shown in Figure 5.14 (page 125). This process yields a great deal of energy while neutralizing a potentially toxic gas—a very neat solution, all in all.

Microbes that use molecular oxygen (aerobes) extract more energy from nutrients than microbes that do not use oxygen (anaerobes). Organisms that require oxygen to live are called **obligate aerobes** (Table 6.1a).

Obligate aerobes are at a disadvantage because oxygen is poorly soluble in the water of their environment. Therefore, many of the aerobic bacteria have developed, or retained, the ability to continue growing in the absence of oxygen. Such organisms are called **facultative anaerobes** (Table 6.1b). In other words, facultative anaerobes can use oxygen when it is present but are able to continue growth by using fermentation or anaerobic respiration when oxygen is not available. However, their efficiency in producing energy decreases in the absence of oxygen. An example of facultative anaerobes is the familiar *Escherichia coli* that are found in the human intestinal tract. Many yeasts are also facultative anaerobes. When growing

anaerobically, yeast use fermentation to produce energy, whereas many bacteria use anaerobic respiration. (See the discussion of anaerobic respiration in Chapter 5, page 128).

**Anaerobes** (Table 6.1c) are bacteria that are unable to use molecular oxygen for energy-yielding reactions. In fact, most are harmed by it. The genus *Clostridium* (klo-STRID-ē-um), which contains the species that cause tetanus and botulism, is the most familiar example. *Clostridium* obtains energy by anaerobic respiration.

Understanding how organisms can be harmed by oxygen requires a brief discussion of the toxic forms of oxygen:

1. **Singlet oxygen** ($^1O_2^-$) is normal molecular oxygen ($O_2$) that has been boosted into a higher-energy state and is extremely reactive.

2. **Superoxide radicals** ($O_2^-$), or **superoxide anions**, are formed in small amounts during the normal respiration of organisms that use oxygen as a final electron acceptor, forming water. In the presence of oxygen, obligate anaerobes also appear to form some superoxide radicals, which are so toxic to cellular components that all organisms attempting to grow in atmospheric oxygen must produce an enzyme, **superoxide dismutase (SOD)**, to neutralize them. Their toxicity is caused by their great instability, which leads them to steal an electron from a

neighboring molecule, which in turn becomes a radical and steals an electron, and so on. Aerobic bacteria, facultative anaerobes growing aerobically, and aerotolerant anaerobes (discussed shortly) produce SOD, with which they convert the superoxide radical into molecular oxygen ($O_2$) and hydrogen peroxide ($H_2O_2$):

$$O_2^- + O_2^- + 2\,H^+ \longrightarrow H_2O_2 + O_2$$

3. The hydrogen peroxide produced in this reaction contains the **peroxide anion** ($O_2^{2-}$) and is also toxic. It is the active principle in the antimicrobial agents hydrogen peroxide and benzoyl peroxide. (See Chapter 7, page 196.) Because the hydrogen peroxide produced during normal aerobic respiration is toxic, microbes have developed enzymes to neutralize it. The most familiar of these is **catalase**, which converts it into water and oxygen:

$$2\,H_2O_2 \longrightarrow 2H_2O + O_2$$

Catalase is easily detected by its action on hydrogen peroxide. When a drop of hydrogen peroxide is added to a colony of bacterial cells producing catalase, oxygen bubbles are released. Anyone who has put hydrogen peroxide on a wound will recognize that cells in human tissue also contain catalase. The other enzyme that breaks down hydrogen peroxide is **peroxidase**, which differs from catalase in that its reaction does not produce oxygen:

$$H_2O_2 + 2\,H^+ \longrightarrow 2\,H_2O$$

Another important form of reactive oxygen is **ozone ($O_3$)** (discussed on page 197).

4. The **hydroxyl radical** (OH·) is another intermediate form of oxygen and probably the most reactive. It is formed in the cellular cytoplasm by ionizing radiation. Most aerobic respiration produces traces of hydroxyl radicals, but they are transient.

These toxic forms of oxygen are an essential component of one of the body's most important defenses against pathogens, phagocytosis (see page 451 and Figure 16.7). In the phagolysosome of the phagocytic cell, ingested pathogens are killed by exposure to singlet oxygen, superoxide radicals, peroxide anions of hydrogen peroxide, hydroxyl radicals, and other oxidative compounds.

Obligate anaerobes usually produce neither superoxide dismutase nor catalase. Because aerobic conditions probably lead to an accumulation of superoxide radicals in their cytoplasm, obligate anaerobes are extremely sensitive to oxygen.

**Aerotolerant anaerobes** (Table 6.1d) are fermentative and cannot use oxygen for growth, but they tolerate it fairly well. On the surface of a solid medium, they will grow without the use of special techniques (discussed later) required for obligate anaerobes. Common examples of lactic acid–producing aerotolerant anaerobes are the lactobacilli used in the production of many acidic fermented foods, such as pickles and cheese.

In the laboratory, they are handled and grown much like any other bacteria, but they make no use of the oxygen in the air. These bacteria can tolerate oxygen because they possess SOD or an equivalent system that neutralizes the toxic forms of oxygen previously discussed.

A few bacteria are **microaerophiles** (Table 6.1e). They are aerobic; they do require oxygen. However, they grow only in oxygen concentrations lower than those in air. In a test tube of solid nutrient medium, they grow only at a depth where small amounts of oxygen have diffused into the medium; they do not grow near the oxygen-rich surface or below the narrow zone of adequate oxygen. This limited tolerance is probably due to their sensitivity to superoxide radicals and peroxides, which they produce in lethal concentrations under oxygen-rich conditions.

## Organic Growth Factors

Essential organic compounds an organism is unable to synthesize are known as **organic growth factors**; they must be directly obtained from the environment. One group of organic growth factors for humans is vitamins. Most vitamins function as coenzymes, the organic cofactors required by certain enzymes in order to function. Many bacteria can synthesize all their own vitamins and do not depend on outside sources. However, some bacteria lack the enzymes needed for the synthesis of certain vitamins, and for them those vitamins are organic growth factors. Other organic growth factors required by some bacteria are amino acids, purines, and pyrimidines.

### CHECK YOUR UNDERSTANDING

✔ 6-4 If bacterial cells were given a sulfur source containing radioactive sulfur ($^{35}$S) in their culture media, in what molecules would the $^{35}$S be found in the cells?

✔ 6-5 How would one determine whether a microbe is a strict anaerobe?

✔ 6-6 Oxygen is so pervasive in the environment that it would be very difficult for a microbe to always avoid physical contact with it. What, therefore, is the most obvious way for a microbe to avoid damage?

# Biofilms

**LEARNING OBJECTIVE**

6-7 Describe the formation of biofilms and their potential for causing infection.

Play Interactive Microbiology @MasteringMicrobiology *See how biofilms affect a patient's health*

In nature, microorganisms seldom live in the isolated single-species colonies that we see on laboratory plates. They more typically live in communities called **biofilms**, which are a thin, slimy layer encasing bacteria that adheres to a surface. This fact was not well appreciated until the development of confocal microscopy (see page 58) made the three-dimensional structure of biofilms more visible. A biofilm also can be considered a *hydrogel*, which is a complex polymer containing many times its dry weight in water. Cell-to-cell chemical communication, or *quorum sensing*, allows bacteria to coordinate their activity and group together into communities that provide benefits not unlike those of multicellular organisms. Therefore, biofilms are not just bacterial slime layers but biological systems; the bacteria are organized into a coordinated, functional community. Biofilms are usually attached to a surface, such as a rock in a pond, a human tooth (plaque; see Figure 25.3 on page 724), or a mucous membrane. This community might be of a single species or of a diverse group of microorganisms. Trillions of pieces of plastic, about 5 mm in diameter, float on the world's oceans. Biofilms consisting of hundreds of species of bacteria and algae have been found in biofilms on the these pieces of plastic. Interestingly, the different species are found on different types of plastic (e.g., polypropylene or polyethylene). Biofilms also might take other, more varied forms. In fast-flowing streams, the biofilm might be in the form of filamentous streamers. Within a biofilm community, the bacteria are able to share nutrients and are sheltered from harmful factors in the environment, such as desiccation, antibiotics, and the body's immune system. The close proximity of microorganisms within a biofilm might also have the advantage of facilitating the transfer of genetic information by, for example, conjugation.

 ASM: Most bacteria in nature live in biofilm communities.

A biofilm usually begins to form when a free-swimming (*planktonic*) bacterium attaches to a surface. (See Figure 1.10 on page 17.) In bacterial cells, cell density alters gene expression in a process called **quorum sensing**. In law, a quorum is the minimum number of members necessary to conduct business. Quorum sensing is the ability of bacteria to communicate and coordinate behavior. Bacteria that use quorum sensing produce and secrete a signaling chemical called an *inducer*. As the inducer diffuses into the surrounding medium, other bacterial cells move toward the source and begin producing inducer. The concentration of inducer increases as cell numbers increase. This, in turn, attracts more cells and initiates synthesis of more inducer.

If these bacteria grew in a uniformly thick monolayer, they would become overcrowded, nutrients would not be available in lower depths, and toxic wastes could accumulate. Microorganisms in biofilm communities sometimes avoid these problems by forming pillar-like structures (**Figure 6.5**) with channels between them, through which water can carry incoming nutrients and outgoing wastes. This constitutes a primitive circulatory system. Individual microbes and clumps of slime occasionally leave the established biofilm and move to a new location where the biofilm becomes extended. Such a biofilm is generally composed of a surface layer about 10 μm thick, with pillars that extend up to 200 μm above it.

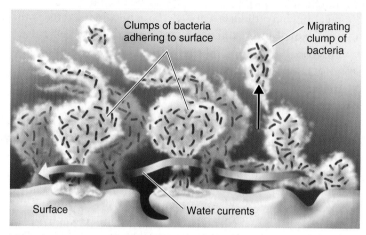

Clumps of bacteria adhering to surface

Migrating clump of bacteria

Surface

Water currents

10 μm

Water currents move, as shown by the blue arrow, among pillars of slime formed by the growth of bacteria attached to solid surfaces. This allows efficient access to nutrients and removal of bacterial waste products. Individual slime-forming bacteria or bacteria in clumps of slime detach and move to new locations.

**Figure 6.5 Biofilms.**

**Q** Why is the prevention of biofilms important in a health care environment?

The microorganisms in biofilms can work cooperatively to carry out complex tasks. An example is myxobacteria, which are found in decaying organic material and freshwater throughout the world. Although they are bacteria, many myxobacteria never exist as individual cells. *Myxococcus xanthus* cells appear to hunt in packs. In their natural aqueous habitat, *M. xanthus* cells form spherical colonies that surround prey bacteria, where they can secrete digestive enzymes and absorb the nutrients. On solid substrates, other myxobacterial cells glide over a solid surface, leaving slime trails that are followed by other cells. When food is scarce, the cells aggregate to form a mass. Cells within the mass differentiate into a fruiting body that consists of a slime stalk and clusters of spores (see Figure 11.11, page 306).

Biofilms are an important factor in human health. For example, microbes in biofilms are probably 1000 times more resistant to microbicides. Experts at the Centers for Disease Control and Prevention (CDC) estimate that 70% of human bacterial infections involve biofilms. Most healthcare-associated infections are probably related to biofilms on medical catheters (see Figure 1.10 on page 17 and Figure 21.3 on page 593). In fact, biofilms form on almost all indwelling medical devices, including mechanical heart valves. Biofilms, which also can be formed by fungi such as *Candida*, are encountered in many disease conditions, such as infections related to the use of contact lenses, dental caries (see page 724), and infections by pseudomonad bacteria (see page 301).

One approach to preventing biofilm formation is to incorporate antimicrobials into surfaces on which biofilms might form. Because the inducers that allow quorum sensing are essential to biofilm formation, research is under way to determine the makeup of these inducers and perhaps block them. Another approach involves the discovery that lactoferrin (see page 462), which is abundant in many human secretions, can inhibit biofilm formation. Lactoferrin binds iron, making it unavailable to bacteria. The lack of iron inhibits the surface motility essential for the aggregation of the bacteria into biofilms. Loss of lactoferrin in cystic fibrosis patents allows pseudomonad biofilms and recurring lung infections in these patients.

Most laboratory methods in microbiology today use organisms being cultured in their planktonic mode. However, microbiologists now predict that there will be an increasing focus on how microorganisms actually live in relation to one another and that this will be considered in industrial and medical research.

### CHECK YOUR UNDERSTANDING

✔ 6-7 Identify a way in which pathogens find it advantageous to form biofilms.

# Culture Media

### LEARNING OBJECTIVES

6-8  Distinguish chemically defined and complex media.

6-9  Justify the use of each of the following: anaerobic techniques, living host cells, candle jars, selective and differential media, enrichment medium.

6-10  Differentiate biosafety levels 1, 2, 3, and 4.

A nutrient material prepared for the growth of microorganisms in a laboratory is called a **culture medium**. Some bacteria can grow well on just about any culture medium; others require special media, and still others cannot grow on any nonliving medium yet developed. Microbes that are introduced into a culture medium to initiate growth are called an **inoculum**. The microbes that grow and multiply in or on a culture medium are referred to as a **culture**.

Suppose we want to grow a culture of a certain microorganism, perhaps the microbes from a particular clinical specimen. What criteria must the culture medium meet? First, it must contain the right nutrients for the specific microorganism we want to grow. It should also contain sufficient moisture, a properly adjusted pH, and a suitable level of oxygen, perhaps none at all. The medium must initially be **sterile**—that is, it must initially contain no living microorganisms—so that the culture will contain only the microbes (and their offspring) we add to the medium. Finally, the growing culture should be incubated at the proper temperature.

A wide variety of media are available for the growth of microorganisms in the laboratory. Most of these media, which are available from commercial sources, have premixed components and require only the addition of water and then sterilization. Media are constantly being developed or revised for use in the isolation

and identification of bacteria that are of interest to researchers in such fields as food, water, and clinical microbiology.

When it is desirable to grow bacteria on a solid medium, a solidifying agent such as agar is added to the medium. A complex polysaccharide derived from a marine alga, **agar** has long been used as a thickener in foods such as jellies and ice cream.

Agar has some very important properties that make it valuable to microbiology, and no satisfactory substitute has yet been found. Few microbes can degrade agar, so it remains solid. Also, agar liquefies at about 100°C (the boiling point of water) and at sea level remains liquid until the temperature drops to about 40°C. For laboratory use, agar is held in water baths at about 50°C. At this temperature, it does not injure most bacteria when it is poured over them (as shown in Figure 6.17a, page 170). Once the agar has solidified, it can be incubated at temperatures approaching 100°C before it again liquefies; this property is particularly useful when thermophilic bacteria are being grown.

Agar media are usually contained in test tubes or *Petri dishes*. The test tubes are called *slants* when their contents are allowed to solidify with the tube held at an angle so that a large surface area for growth is available. When the agar solidifies in a vertical tube, it is called a *deep*. Petri dishes, named for their inventor, are shallow dishes with a lid that nests over the bottom to prevent contamination; when filled, they are called *Petri* (or *culture*) *plates*.

## Chemically Defined Media

To support microbial growth, a medium must provide an energy source, as well as sources of carbon, nitrogen, sulfur, phosphorus, and any organic growth factors the organism is unable to synthesize. A **chemically defined medium** is one whose exact chemical composition is known. For a chemoheterotroph, the chemically defined medium must contain organic compounds that serve as a source of carbon and energy. For example, as shown in Table 6.2, glucose is included in the medium for growing the chemoheterotroph *E. coli*.

| TABLE 6.2 | A Chemically Defined Medium for Growing a Typical Chemoheterotroph, Such as *Escherichia coli* | |
|---|---|
| **Constituent** | **Amount** |
| Glucose | 5.0 g |
| Ammonium phosphate, monobasic ($NH_4H_2PO_4$) | 1.0 g |
| Sodium chloride (NaCl) | 5.0 g |
| Magnesium sulfate ($MgSO_4 \cdot 7H_2O$) | 0.2 g |
| Potassium phosphate, dibasic ($K_2HPO_4$) | 1.0 g |
| Water | 1 liter |

| Defined Culture Medium for | |
|:---|:---|
| **TABLE 6.3** *Leuconostoc mesenteroides* | |

**Carbon and Energy**

Glucose, 25 g

**Salts**

NH$_4$Cl, 3.0 g

K$_2$HPO$_4$*, 0.6 g

KH$_2$PO$_4$*, 0.6 g

MgSO$_4$, 0.1 g

**Amino Acids, 100–200 μg each**

Alanine, arginine, asparagine, aspartate, cysteine, glutamate, glutamine, glycine, histidine, isoleucine, leucine, lysine, methionine, phenylalanine, proline, serine, threonine, tryptophan, tyrosine, valine

**Purines and Pyrimidines, 10 mg of each**

Adenine, guanine, uracil, xanthine

**Vitamins, 0.01–1 mg each**

Biotin, folate, nicotinic acid, pyridoxal, pyridoxamine, pyridoxine, riboflavin, thiamine, pantothenate, p-aminobenzoic acid

**Trace Elements, 2–10 μg each**

Fe, Co, Mn, Zn, Cu, Ni, Mo

**Buffer, pH 7**

Sodium acetate, 25 g

**Distilled Water, 1000 ml**

*Also serves as buffer.

As Table 6.3 shows, many organic growth factors must be provided in the chemically defined medium used to cultivate a species of *Leuconostoc*. Organisms that require many growth factors are described as *fastidious*. Organisms of this type, such as *Lactobacillus* (page 314), are sometimes used in tests that determine the concentration of a particular vitamin in a substance. To perform such a *microbiological assay*, a growth medium is prepared that contains all the growth requirements of the bacterium except the vitamin being assayed. Then the medium, test substance, and bacterium are combined, and the growth of bacteria is measured. Bacterial growth, which is reflected by the amount of lactic acid produced, will be proportional to the amount of vitamin in the test substance. The more lactic acid, the more the *Lactobacillus* cells have been able to grow, so the more vitamin is present.

## Complex Media

Chemically defined media are usually reserved for laboratory experimental work or for the growth of autotrophic bacteria. Most heterotrophic bacteria and fungi, such as you would work with in an introductory lab course, are routinely grown on complex media made up of nutrients including extracts from yeasts, meat, or plants, or digests of proteins from these and other sources. The exact chemical composition varies slightly from batch to batch. Table 6.4 shows one widely used recipe.

In complex media, the energy, carbon, nitrogen, and sulfur requirements of the growing microorganisms are provided primarily by protein. Proteins are large, relatively insoluble molecules that only a minority of microorganisms can utilize directly. Partial digestion by acids or enzymes reduces proteins to shorter chains of amino acids called *peptones*. These small, soluble fragments can be digested by most bacteria.

Vitamins and other organic growth factors are provided by meat extracts or yeast extracts. The soluble vitamins and minerals from the meats or yeasts are dissolved in the extracting water, which is then evaporated, so these factors are concentrated. (These extracts also supplement the organic nitrogen and carbon compounds.) Yeast extracts are particularly rich in the B vitamins. If a complex medium is in liquid form, it is called **nutrient broth**. When agar is added, it is called **nutrient agar**. (This terminology can be confusing; just remember that agar itself is not a nutrient.)

## Anaerobic Growth Media and Methods

The cultivation of anaerobic bacteria poses a special problem. Because anaerobes might be killed by exposure to oxygen, special media called **reducing media** must be used. These media contain ingredients, such as sodium thioglycolate, that chemically combine with dissolved oxygen and deplete the oxygen in the culture medium. To routinely grow and maintain pure cultures of obligate anaerobes, microbiologists use reducing media stored in ordinary, tightly capped test tubes. These media are heated shortly before use to drive off absorbed oxygen.

When the culture must be grown in Petri plates to observe individual colonies, several methods are available. Laboratories that work with relatively few culture plates at a time can use systems that can incubate the microorganisms in sealed boxes and jars in which the oxygen is chemically removed after the culture

| Composition of Nutrient Agar, a Complex Medium for the Growth | |
|:---|:---|
| **TABLE 6.4** of Heterotrophic Bacteria | |

| Constituent | Amount |
|:---|:---|
| Peptone (partially digested protein) | 5.0 g |
| Beef extract | 3.0 g |
| Sodium chloride | 8.0 g |
| Agar | 15.0 g |
| Water | 1 liter |

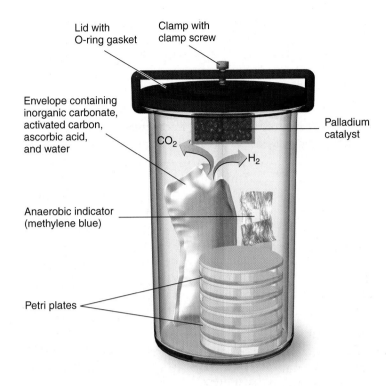

Lid with
O-ring gasket

Clamp with
clamp screw

Envelope containing
inorganic carbonate,
activated carbon,
ascorbic acid,
and water

$CO_2$

$H_2$

Palladium
catalyst

Anaerobic indicator
(methylene blue)

Petri plates

**Figure 6.6  A jar for cultivating anaerobic bacteria on Petri plates.** When water is mixed with the chemical packet containing sodium bicarbonate and sodium borohydride, hydrogen and carbon dioxide are generated. Reacting on the surface of a palladium catalyst in a screened reaction chamber, which may also be incorporated into the chemical packet, the hydrogen and atmospheric oxygen in the jar combine to form water. The oxygen is thus removed. Also in the jar is an anaerobic indicator containing methylene blue, which is blue when oxidized and turns colorless when the oxygen is removed (as shown here).

**Q** What is the technical name for bacteria that require a higher-than-atmospheric-concentration of $CO_2$ for growth?

plates have been introduced and the container sealed as shown in **Figure 6.6**. In one system, the envelope of chemicals (the active ingredient is ascorbic acid) is simply opened to expose it to oxygen in the container's atmosphere. The atmosphere in such containers usually has less than 1% oxygen, about 18% $CO_2$, and no hydrogen. In a recently introduced system, each individual Petri plate (OxyPlate™) becomes an anaerobic chamber. The medium in the plate contains an enzyme, oxyrase, which combines oxygen with hydrogen, removing oxygen as water is formed.

Laboratories that have a large volume of work with anaerobes often use an anaerobic chamber, such as that shown in **Figure 6.7**. The chamber is filled with inert gases (typically about 85% $N_2$, 10% $H_2$, and 5% $CO_2$) and is equipped with air locks to introduce cultures and materials.

## Special Culture Techniques

Many bacteria have never been successfully grown on artificial laboratory media. *Mycobacterium leprae*, the leprosy bacillus, is now usually grown in armadillos, which have a relatively

low body temperature that matches the requirements of the microbe. Another example is the syphilis spirochete, although certain nonpathogenic strains of this microbe have been grown on laboratory media. With few exceptions, the obligate intracellular bacteria, such as the rickettsias and the chlamydias, do not grow on artificial media. Like viruses, they can reproduce only in a living host cell. See the discussion of cell culture, page 371.

Many clinical laboratories have special *carbon dioxide incubators* in which to grow aerobic bacteria that require concentrations of $CO_2$ higher or lower than that found in the atmosphere. Desired $CO_2$ levels are maintained by electronic controls. High $CO_2$ levels are also obtained with simple *candle jars*. Cultures are placed in a large sealed jar containing a lighted candle, which consumes oxygen. The candle stops burning when the air in the jar has a lowered concentration of oxygen (at about 17% $O_2$, still adequate for the growth of aerobic bacteria). An elevated concentration of $CO_2$ (about 3%) is also present. Microbes that grow better at high $CO_2$ concentrations are called **capnophiles**. The low-oxygen, high-$CO_2$ conditions resemble those found in the intestinal tract, respiratory tract, and other body tissues where pathogenic bacteria grow.

Candle jars are still used occasionally, but more often commercially available chemical packets are used to generate carbon dioxide atmospheres in containers. When only one or two Petri plates of cultures are to be incubated, clinical laboratory investigators often use small plastic bags with self-contained chemical gas generators that are activated by crushing the packet or

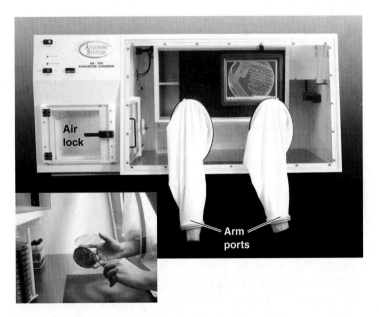

Air lock

Arm ports

**Figure 6.7  An anaerobic chamber.** Materials are introduced through the small doors in the air-lock chamber at the left. The operator works through arm ports in airtight sleeves. The airtight sleeves extend into the cabinet when it is in use. This unit also features an internal camera and monitor.

**Q** In what way would an anaerobic chamber resemble the Space Station Laboratory orbiting in the vacuum of space?

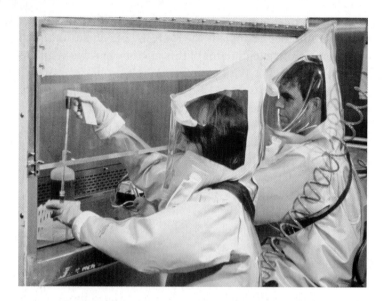

**Figure 6.8 Technicians in a biosafety level 4 (BSL-4) laboratory.** Personnel working in a BSL-4 facility wear a "space suit" that is connected to an outside air supply. Air pressure in the suit is higher than the atmosphere, preventing microbes from entering the suit.

**Q** If a technician were working with pathogenic prions, how would material leaving the lab be rendered noninfectious? (*Hint*: See Chapter 7.)

moistening it with a few milliliters of water. These packets are sometimes specially designed to provide precise concentrations of carbon dioxide (usually higher than can be obtained in candle jars) and oxygen for culturing organisms such as the microaerophilic *Campylobacter* bacteria (page 307).

Some microorganisms, such as *Ebolavirus*, are so dangerous that they can be handled only under extraordinary systems of containment called *biosafety level 4 (BSL-4)*. BSL-4 labs are popularly known as "the hot zone." Only a handful of such labs exists in the United States. The lab is a sealed environment within a larger building and has an atmosphere under negative pressure, so that aerosols containing pathogens will not escape. Both intake and exhaust air is filtered through high-efficiency particulate air filters (see HEPA filters, page 185); the exhaust air is filtered twice. All waste materials leaving the lab are rendered noninfectious. The personnel wear "space suits" that are connected to an air supply (**Figure 6.8**).

Less dangerous organisms are handled at lower levels of biosafety. For example, a basic microbiology teaching laboratory might be BSL-1. Organisms that present a moderate risk of infection can be handled at BSL-2 levels, that is, on open laboratory benchtops with appropriate gloves, lab coats, or possibly face and eye protection. BSL-3 labs are intended for highly infectious airborne pathogens such as the tuberculosis agent. Biological safety cabinets similar in appearance to the anaerobic chamber shown in Figure 6.7 are used. The laboratory itself should be negatively pressurized and equipped with air filters to prevent release of the pathogen from the laboratory.

## Selective and Differential Media

In clinical and public health microbiology, it is frequently necessary to detect the presence of specific microorganisms associated with disease or poor sanitation. For this task, selective and differential media are used. **Selective media** are designed to suppress the growth of unwanted bacteria and encourage the growth of the desired microbes. For example, bismuth sulfite agar is one medium used to isolate the typhoid bacterium, the gram-negative *Salmonella* Typhi (TĪ-fē), from feces. Bismuth sulfite inhibits gram-positive bacteria and most gram-negative intestinal bacteria (other than *Salmonella* Typhi), as well. Sabouraud's dextrose agar, which has a pH of 5.6, is used to isolate fungi that outgrow most bacteria at this pH.

**Differential media** make it easier to distinguish colonies of the desired organism from other colonies growing on the same plate. Similarly, pure cultures of microorganisms have identifiable reactions with differential media in tubes or plates. Blood agar (which contains red blood cells) is a medium that microbiologists often use to identify bacterial species that destroy red blood cells. These species, such as *Streptococcus pyogenes* (pī-AH-jen-ēz), the bacterium that causes strep throat, show a clear ring around their colonies where they have lysed the surrounding blood cells (**Figure 6.9**).

Sometimes, selective and differential characteristics are combined in a single medium. Suppose we want to isolate the

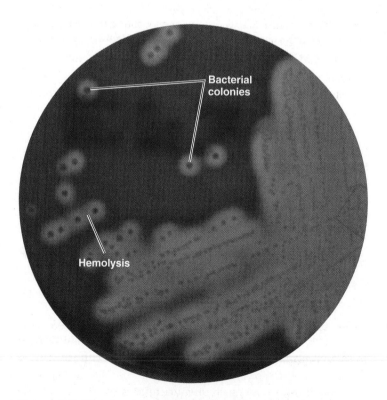

**Figure 6.9 Blood agar, a differential medium containing red blood cells.** The bacteria have lysed the red blood cells (beta-hemolysis), causing the clear areas around the colonies.

**Q** Of what value are hemolysins to pathogens?

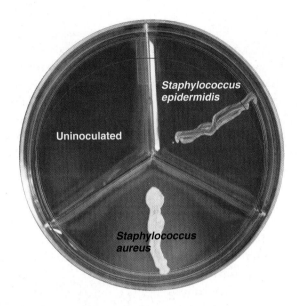

**Figure 6.10 Differential medium.** This medium is mannitol salt agar, and bacteria capable of fermenting the mannitol in the medium to acid (*Staphylococcus aureus*) cause the medium to change color to yellow. This **differentiates** between bacteria that can ferment mannitol and those that cannot. Actually, this medium is also **selective** because the high salt concentration prevents the growth of most bacteria but not *Staphlylococcus* spp.

**Q** Are bacteria capable of growing at a high osmotic pressure likely to be capable of growing in the mucus found in nostrils?

common bacterium *Staphylococcus aureus*, found in the nasal passages. This organism has a tolerance for high concentrations of sodium chloride; it can also ferment the carbohydrate mannitol to form acid. Mannitol salt agar contains 7.5% sodium chloride, which will discourage the growth of competing organisms and thus *select for* (favor the growth of) *S. aureus*. This salty medium also contains a pH indicator that changes color if the mannitol in the medium is fermented to acid; the mannitol-fermenting colonies of *S. aureus* are thus *differentiated from* colonies of bacteria that do not ferment mannitol. Bacteria that grow at the high salt concentration *and* ferment mannitol to acid can be readily identified by the color change (**Figure 6.10**). These are probably colonies of *S. aureus*, and their identification can be confirmed by additional tests. The use of differential media to identify toxin-producing *E. coli* is discussed in Chapter 5, page 134.

## Enrichment Culture

Because bacteria present in small numbers can be missed, especially if other bacteria are present in much larger numbers, it is sometimes necessary to use an **enrichment culture**. This is often the case for soil or fecal samples. The medium (enrichment medium) for an enrichment culture is usually liquid and provides nutrients and environmental conditions that favor the growth of a particular microbe but not others. In this sense, it is also a selective medium, but it is designed to increase very small numbers of the desired type of organism to detectable levels.

**TABLE 6.5** Culture Media

| Type | Purpose |
| --- | --- |
| Chemically Defined | Growth of chemoautotrophs and photoautotrophs; microbiological assays |
| Complex | Growth of most chemoheterotrophic organisms |
| Reducing | Growth of obligate anaerobes |
| Selective | Suppression of unwanted microbes; encouraging desired microbes |
| Differential | Differentiation of colonies of desired microbes from others |
| Enrichment | Similar to selective media but designed to increase numbers of desired microbes to detectable levels |

Suppose we want to isolate from a soil sample a microbe that can grow on phenol and is present in much smaller numbers than other species. If the soil sample is placed in a liquid enrichment medium in which phenol is the only source of carbon and energy, microbes unable to metabolize phenol will not grow. The culture medium is allowed to incubate for a few days, and then a small amount of it is transferred into another flask of the same medium. After a series of such transfers, the surviving population will consist of bacteria capable of metabolizing phenol. The bacteria are given time to grow in the medium between transfers; this is the enrichment stage. Any nutrients in the original inoculum are rapidly diluted out with the successive transfers. When the last dilution is streaked onto a solid medium of the same composition, only those colonies of organisms capable of using phenol should grow. A remarkable aspect of this particular technique is that phenol is normally lethal to most bacteria.

Table 6.5 summarizes the purposes of the main types of culture media.

**CHECK YOUR UNDERSTANDING**

✔ 6-8 Could humans exist on chemically defined media, at least under laboratory conditions?

✔ 6-9 Could Louis Pasteur, in the 1800s, have grown rabies viruses in cell culture instead of in living animals?

✔ 6-10 What BSL is your laboratory?

# Obtaining Pure Cultures

**LEARNING OBJECTIVES**

6-11 Define *colony*.

6-12 Describe how pure cultures can be isolated by using the streak plate method.

Most infectious materials, such as pus, sputum, and urine, contain several different kinds of bacteria; so do samples of soil, water, or food. If these materials are plated out onto the surface of a solid medium, colonies will form that are exact copies of the original organism. A visible **colony** theoretically arises from a single spore or vegetative cell or from a group of the same microorganisms attached to one another in clumps or chains. Estimates are that only about 1% of bacteria in ecosystems produce colonies by conventional culture methods. Microbial colonies often have a distinctive appearance that distinguishes one microbe from another (see Figure 6.10). The bacteria must be distributed widely enough so that the colonies are visibly separated from each other.

Most bacteriological work requires pure cultures, or clones, of bacteria. The isolation method most commonly used to get pure cultures is the **streak plate method** (Figure 6.11). A sterile inoculating loop is dipped into a mixed culture that contains more than one type of microbe and is streaked in a pattern over

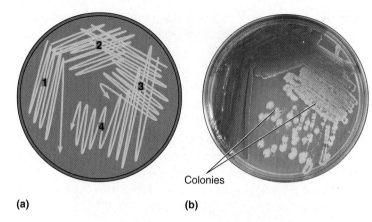

(a)                          (b)

Colonies

**Figure 6.11 The streak plate method for isolating pure bacterial cultures.** **(a)** Arrows indicate the direction of streaking. Streak series 1 is made from the original bacterial culture. The inoculating loop is sterilized following each streak series. In series 2, 3, and 4 the loop picks up bacteria from the previous series, diluting the number of cells each time. There are numerous variants of such patterns. **(b)** In series 4 of this example, notice that well-isolated colonies of bacteria of two different types, red and white, have been obtained.

**Q** Is a colony formed as a result of streaking a plate always derived from a single bacterium? Why or why not?

the surface of the nutrient medium. As the pattern is traced, bacteria are rubbed off the loop onto the medium. The last cells to be rubbed off the loop are far enough apart to grow into isolated colonies. These colonies can be picked up with an inoculating loop and transferred to a test tube of nutrient medium to form a pure culture containing only one type of bacterium.

The streak plate method works well when the organism to be isolated is present in large numbers relative to the total population. However, when the microbe to be isolated is present only in very small numbers, its numbers must be greatly increased by selective enrichment before it can be isolated with the streak plate method.

**CHECK YOUR UNDERSTANDING**

✔ **6-11** Can you think of any reason why a colony does not grow to an infinite size, or at least fill the confines of the Petri plate?

✔ **6-12** Could a pure culture of bacteria be obtained by the streak plate method if there were only one desired microbe in a bacterial suspension of billions?

# Preserving Bacterial Cultures

**LEARNING OBJECTIVE**

**6-13** Explain how microorganisms are preserved by deep-freezing and lyophilization (freeze-drying).

Refrigeration can be used for the short-term storage of bacterial cultures. Two common methods of preserving microbial cultures for long periods are deep-freezing and lyophilization.

**CLINICAL CASE**

*P. fluorescens* is an aerobic, gram-negative rod that grows best between 25°C and 30°C and grows poorly at the standard hospital microbiology incubation temperatures (35°C to 37°C). The bacteria are so named because they produce a pigment that fluoresces under ultraviolet light. While reviewing the facts of the latest outbreak, Dr. MacGruder learns that most recent patients were last exposed to the contaminated heparin 84 to 421 days before onset of their infections. On-site investigations confirmed that the patients' clinics are no longer using the recalled heparin and had, in fact, returned all unused inventory. Concluding that these patients did not develop infections during the previous outbreak, Dr. MacGruder must look for a new source of infection. The patients all have indwelling venous catheters: tubes that are inserted into a vein for long-term delivery of concentrated solutions, such as anticancer drugs.

Dr. MacGruder orders cultures of the new heparin being used, but the results do not recover any organisms. He then orders blood and catheter cultures from each of the patients.

**Illuminated with**      **Illuminated with**
**white light**            **ultraviolet light**

**The organism cultured from both the patients' blood and their catheters is shown in the figure. What organism is it?**

**Deep-freezing** is a process in which a pure culture of microbes is placed in a suspending liquid and quick-frozen at temperatures ranging from −50°C to −95°C. The culture can usually be thawed and cultured even several years later. During **lyophilization (freeze-drying)**, a suspension of microbes is quickly frozen at temperatures ranging from −54°C to −72°C, and the water is removed by a high vacuum (sublimation). While under vacuum, the container is sealed by melting the glass with a high-temperature torch. The remaining powderlike residue that contains the surviving microbes can be stored for years. The organisms can be revived at any time by hydration with a suitable liquid nutrient medium.

> **CHECK YOUR UNDERSTANDING**
>
> ✔ 6-13 If the Space Station in Earth orbit suddenly ruptured, the humans on board would die instantly from cold and the vacuum of space. Would all the bacteria in the capsule also be killed?

# The Growth of Bacterial Cultures

**LEARNING OBJECTIVES**

6-14 Define *bacterial growth*, including *binary fission*.

6-15 Compare the phases of microbial growth, and describe their relation to generation time.

6-16 Explain four direct methods of measuring cell growth.

6-17 Differentiate direct and indirect methods of measuring cell growth.

6-18 Explain three indirect methods of measuring cell growth.

Being able to represent graphically the enormous populations resulting from the growth of bacterial cultures is an essential part of microbiology. It is also necessary to be able to determine microbial numbers, either directly, by counting, or indirectly, by measuring their metabolic activity.

## Bacterial Division

As we mentioned at the beginning of the chapter, bacterial growth refers to an increase in bacterial numbers, not an increase in the size of the individual cells. Bacteria normally reproduce by **binary fission (Figure 6.12)**.

A few bacterial species reproduce by **budding**; they form a small initial outgrowth (a bud) that enlarges until its size approaches that of the parent cell, and then it separates. Some filamentous bacteria (certain actinomycetes) reproduce by producing chains of conidiospores (see Figure 11.25, page 317) (an asexual spore) carried externally at the tips of the filaments. A few filamentous species simply fragment, and the fragments initiate the growth of new cells.

▶ **Play Binary Fission;**
**Bacterial Growth:** *Overview*
@MasteringMicrobiology

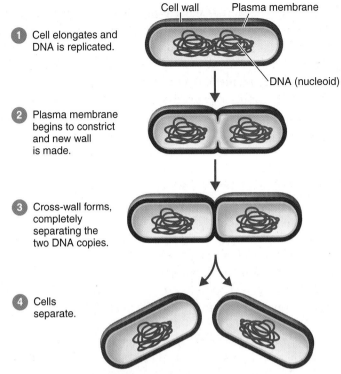

① Cell elongates and DNA is replicated.

Cell wall    Plasma membrane

DNA (nucleoid)

② Plasma membrane begins to constrict and new wall is made.

③ Cross-wall forms, completely separating the two DNA copies.

④ Cells separate.

**(a)** A diagram of the sequence of cell division

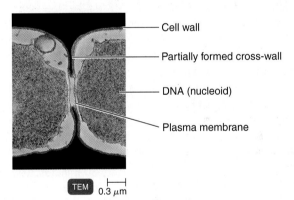

Cell wall

Partially formed cross-wall

DNA (nucleoid)

Plasma membrane

TEM  |—| 0.3 μm

**(b)** A thin section of an *E.coli* cell starting to divide

**Figure 6.12  Binary fission in bacteria.**

**Q** **In what way is budding different from binary fission?**

## Generation Time

For purposes of calculating the generation time of bacteria, we will consider only reproduction by binary fission, which is by far the most common method. As you can see in **Figure 6.13**, one cell's division produces two cells, two cells' divisions produce four cells, and so on. When the number of cells in each generation is expressed as a power of 2, the exponent tells the number of doublings (generations) that have occurred.

The time required for a cell to divide (and its population to double) is called the **generation time**. It varies considerably among organisms and with environmental conditions, such

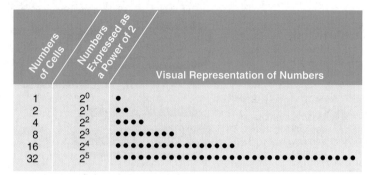

**(a)** Visual representation of increase in bacterial number over five generations. The number of bacteria doubles in each generation. The superscript indicates the generation; that is, $2^5$ = 5 generations.

| Generation Number | Number of Cells | | Log$_{10}$ of Number of Cells |
|---|---|---|---|
| 0 | $2^0$ = | 1 | 0 |
| 5 | $2^5$ = | 32 | 1.51 |
| 10 | $2^{10}$ = | 1,024 | 3.01 |
| 15 | $2^{15}$ = | 32,768 | 4.52 |
| 16 | $2^{16}$ = | 65,536 | 4.82 |
| 17 | $2^{17}$ = | 131,072 | 5.12 |
| 18 | $2^{18}$ = | 262,144 | 5.42 |
| 19 | $2^{19}$ = | 524,288 | 5.72 |
| 20 | $2^{20}$ = | 1,048,576 | 6.02 |

**(b)** Conversion of the number of cells in a population into the logarithmic expression of this number. To arrive at the numbers in the center column, use the $y^x$ key on your calculator. Enter 2 on the calculator; press $y^x$; enter 5; then press the = sign. The calculator will show the number 32. Thus, the fifth-generation population of bacteria will total 32 cells. To arrive at the numbers in the right-hand column, use the log key on your calculator. Enter the number 32; then press the log key. The calculator will show, rounded off, that the log$_{10}$ of 32 is 1.51.

**Figure 6.13  Cell division.**

**Q** If a single bacterium reproduced every 30 minutes, how many would there be in 2 hours?

as temperature. Most bacteria have a generation time of 1 to 3 hours; others require more than 24 hours per generation. (The math required to calculate generation times is presented in Appendix B.) If binary fission continues unchecked, an enormous number of cells will be produced. If a doubling occurred every 20 minutes—which is the case for *E. coli* under favorable conditions—after 20 generations a single initial cell would increase to over 1 million cells. This would require a little less than 7 hours. In 30 generations, or 10 hours, the population would be 1 billion, and in 24 hours it would be a number trailed by 21 zeros. It is difficult to graph population changes of such enormous magnitude by using arithmetic numbers. This is why logarithmic scales are generally used to graph bacterial growth. Understanding logarithmic representations of bacterial populations requires some use of mathematics and is necessary for anyone studying microbiology. (See Appendix B.)

## Logarithmic Representation of Bacterial Populations

To illustrate the difference between logarithmic and arithmetic graphing of bacterial populations, let's express 20 bacterial generations both logarithmically and arithmetically. In five generations ($2^5$), there would be 32 cells; in ten generations ($2^{10}$), there would be 1024 cells, and so on. (If your calculator has a $y^x$ key and a log key, you can duplicate the numbers in the third column of Figure 6.13.)

In **Figure 6.14**, notice that the arithmetically plotted line (solid) does not clearly show the population changes in the early stages of the growth curve at this scale. In fact, the first ten generations do not even appear to leave the baseline, whereas the logarithmic plot point for the tenth generation (3.01) is halfway up the graph. Furthermore, another one or two arithmetic generations graphed to the same scale would greatly increase the height of the graph and take the line off the page.

The dashed line in Figure 6.14 shows how these plotting problems can be avoided by graphing the log$_{10}$ of the population numbers. The log$_{10}$ of the population is plotted at 5, 10,

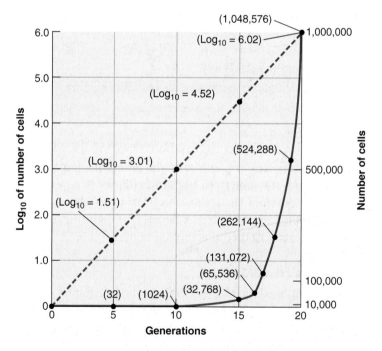

**Figure 6.14  A growth curve for an exponentially increasing population, plotted logarithmically (dashed line) and arithmetically (solid line).** For demonstration purposes, this graph has been drawn so that the arithmetic and logarithmic curves intersect at 1 million cells. This figure demonstrates why it is necessary to graph changes in the immense numbers of bacterial populations by logarithmic plots rather than by arithmetic numbers. For example, note that at ten generations the line representing arithmetic numbers has not even perceptibly left the baseline, whereas the logarithmic plot point for the tenth generation (3.01) is halfway up the graph.

**Q** If the arithmetic numbers (solid line) were plotted for two more generations, would the line still be on the page?

# Understanding the Bacterial Growth Curve

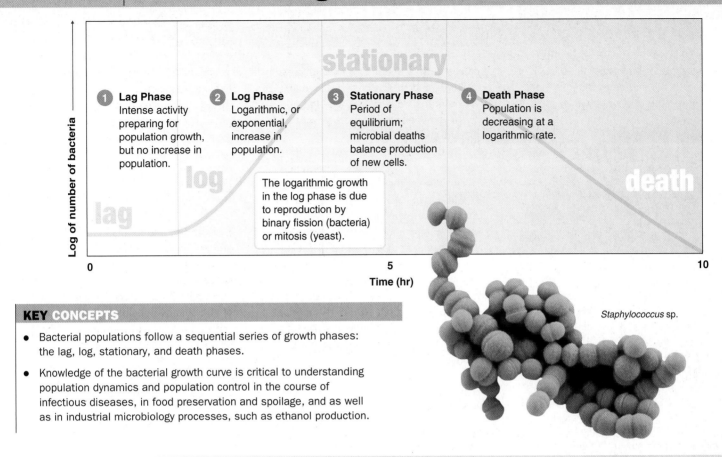

**1** **Lag Phase**
Intense activity preparing for population growth, but no increase in population.

**2** **Log Phase**
Logarithmic, or exponential, increase in population.

**3** **Stationary Phase**
Period of equilibrium; microbial deaths balance production of new cells.

**4** **Death Phase**
Population is decreasing at a logarithmic rate.

The logarithmic growth in the log phase is due to reproduction by binary fission (bacteria) or mitosis (yeast).

*Staphylococcus* sp.

**KEY CONCEPTS**

- Bacterial populations follow a sequential series of growth phases: the lag, log, stationary, and death phases.

- Knowledge of the bacterial growth curve is critical to understanding population dynamics and population control in the course of infectious diseases, in food preservation and spoilage, and as well as in industrial microbiology processes, such as ethanol production.

---

15, and 20 generations. Notice that a straight line is formed and that a thousand times this population (1,000,000,000, or $\log_{10}$ 9.0) could be accommodated in relatively little extra space. However, this advantage is obtained at the cost of distorting our "common sense" perception of the actual situation. We are not accustomed to thinking in logarithmic relationships, but it is necessary for a proper understanding of graphs of microbial populations.

**CHECK YOUR UNDERSTANDING**

✔ 6-14 Can a complex organism, such as a beetle, divide by binary fission?

## Phases of Growth

When a few bacteria are inoculated into an environment such as the large intestine (see Exploring the Microbiome on the next page) or a liquid growth medium and the population is counted at intervals, it is possible to plot a **bacterial growth curve** that shows the growth of cells over time (**Figure 6.15**). There are four basic phases of growth: the lag, log, stationary, and death phases.

**Play Bacterial Growth Curve**
@Mastering**Microbiology**

### The Lag Phase

For a while, the number of cells changes very little because the cells do not immediately reproduce in a new medium. This period of little or no cell division is called the **lag phase**, and it can last for 1 hour or several days. During this time, however, the cells are not dormant. The microbial population is undergoing a period of intense metabolic activity involving, in particular, synthesis of enzymes and various molecules. (The situation is analogous to a factory being equipped to produce automobiles; there is considerable tooling-up activity but no immediate increase in the automobile population.)

### The Log Phase

Eventually, the cells begin to divide and enter a period of growth, or logarithmic increase, called the **log phase**, or **exponential growth phase**. Cellular reproduction is most active during this period, and generation time (intervals during which the population doubles) reaches a constant minimum. Because the generation time is constant, a logarithmic plot of growth during the log phase is a straight line. The log phase is the time when cells are most active metabolically and is preferred for industrial purposes where, for example, a product needs to be produced efficiently.

# Circadian Rhythms and Microbiota Growth Cycles

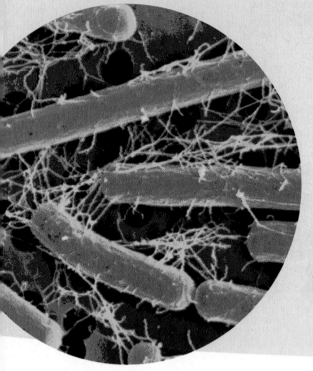

It's strange to think that microbes, especially those deep within us that never see the light of day, may grow at different rates depending on what time it is. But circadian rhythms—cyclical changes in a host that roughly follow a 24-hour cycle—do impact microbiota growth and, therefore, human health.

Studies show that introducing bacteria to a germ-free animal results in colonization of the host with the expected growth curve. For instance, when germ-free zebrafish were inoculated with intestinal bacteria, populations grew from the few starting cells to many thousands, following the timing and stages expected for those particular species: first came a lag phase with no increase in cell numbers, followed by a log phase with exponential growth, and then a stationary phase once the environmental carrying capacity was reached.

However, the hosts' activities lead to fascinating changes in the growth curve. Results of studies on mice and humans show that the stationary phase is altered by sleep changes such as those caused by jet lag. In a normal cycle, bacteria in the order Clostridiales dominated the intestinal microbiota during the active time of hosts. During resting time, *Lactobacillus* was more prevalent. But when the host's clock is disrupted by jet lag, the change in eating and activity causes dysbiosis, or a change in the microbiota. These disruptions seem capable of causing problems with the host over time. Surprisingly, the combination of microbiome species found in mice and humans with dysbiosis appeared to cause obesity when transferred and grown in germ-free mice.

*Lactobacillus* species like the one shown here seem to grow the best when their host is resting.

## The Stationary Phase

If exponential growth continued unchecked, startlingly large numbers of cells could arise. For example, a single bacterium (at a weight of $9.5 \times 10^{-13}$ g per cell) dividing every 20 minutes for only 25.5 hours can theoretically produce a population equivalent in weight to that of an 80,000-ton aircraft carrier. In reality, this does not happen. Eventually, the growth rate slows, the number of microbial deaths balances the number of new cells, and the population stabilizes. This period of equilibrium is called the **stationary phase**.

Exponential growth stops because the bacteria approach the **carrying capacity**, the number of organisms that an environment can support. Carrying capacity is controlled by available nutrients, accumulation of wastes, and space. When a population exceeds the carrying capacity, it will run out of nutrients and space.

## The Death Phase

The number of deaths eventually exceeds the number of new cells formed, and the population enters the **death phase**, or **logarithmic decline phase**. This phase continues until the population is diminished to a tiny fraction of the number of cells in the previous phase or until the population dies out entirely.

Some species pass through the entire series of phases in only a few days; others retain some surviving cells almost indefinitely. Microbial death will be discussed further in Chapter 7.

### CHECK YOUR UNDERSTANDING

✔ 6-15 If two mice started a family within a fixed enclosure, with a fixed food supply, would the population curve be the same as a bacterial growth curve?

## Direct Measurement of Microbial Growth

The growth of microbial populations can be measured in a number of ways. Some methods measure cell numbers; other methods measure the population's total mass, which is often directly proportional to cell numbers. Population numbers are usually recorded as the number of cells in a milliliter of liquid or in a gram of solid material. Because bacterial populations are usually very large, most methods of counting them are based on direct or indirect counts of very small samples; calculations then determine the size of the total population. Assume, for example, that a millionth of a milliliter ($10^{-6}$ ml) of sour milk is found to contain 70 bacterial cells. Then there must be 70 times 1 million, or 70 million, cells per milliliter.

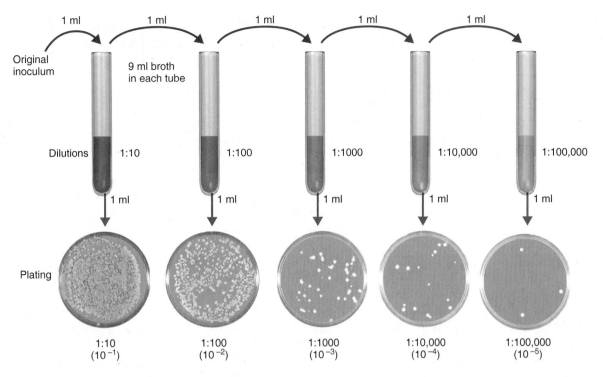

Calculation: Number of colonies on plate × reciprocal of dilution of sample = number of bacteria/ml
(For example, if 54 colonies are on a plate of 1:1000 dilution, then the count is 54 × 1000 = 54,000 bacteria/ml in sample.)

**Figure 6.16  Serial dilutions and plate counts.** In serial dilutions, the original inoculum is diluted in a series of dilution tubes. In our example, each succeeding dilution tube will have only one-tenth the number of microbial cells as the preceding tube. Then, samples of the dilution are used to inoculate Petri plates, on which colonies grow and can be counted. This count is then used to estimate the number of bacteria in the original sample.

**Q** Why were the dilutions of 1:10,000 and 1:100,000 not counted? Theoretically, how many colonies should appear on the 1:100 plate?

However, it is not practical to measure out a millionth of a milliliter of liquid or a millionth of a gram of food. Therefore, the procedure is done indirectly, in a series of dilutions. For example, if we add 1 ml of milk to 99 ml of water, each milliliter of this dilution now has one-hundredth as many bacteria as each milliliter of the original sample had. By making a series of such dilutions, we can readily estimate the number of bacteria in our original sample. To count microbial populations in solid foods (such as hamburger), an homogenate of one part food to nine parts water is finely ground in a food blender. Samples of this initial one-tenth dilution can then be transferred with a pipette for further dilutions or cell counts.

### Plate Counts

The most frequently used method of measuring bacterial populations is the **plate count**. An important advantage of this method is that it measures the number of viable cells. One disadvantage may be that it takes some time, usually 24 hours or more, for visible colonies to form. This can be a serious problem in some applications, such as quality control of milk, when it is not possible to hold a particular lot for this length of time.

Plate counts assume that each live bacterium grows and divides to produce a single colony. This is not always true, because bacteria frequently grow linked in chains or as clumps (see Figure 4.1, page 74). Therefore, a colony often results, not from a single bacterium, but from short segments of a chain or from a bacterial clump. To reflect this reality, plate counts are often reported as **colony-forming units (CFU)**.

When a plate count is performed, it is important that only a limited number of colonies develop in the plate. When too many colonies are present, some cells are overcrowded and do not develop; these conditions cause inaccuracies in the count. The U.S. Food and Drug Administration convention is to count only plates with 25 to 250 colonies, but many microbiologists prefer plates with 30 to 300 colonies. To ensure that some colony counts will be within this range, the original inoculum is diluted several times in a process called **serial dilution (Figure 6.16)**.

**Serial Dilutions** Let's say, for example, that a milk sample has 10,000 bacteria per milliliter. If 1 ml of this sample were plated out, there would theoretically be 10,000 colonies formed in the Petri plate of medium. Obviously, this would not produce a

**Figure 6.17 Methods of preparing plates for plate counts. (a)** The pour plate method. **(b)** The spread plate method.

**Q** In what instances would the pour plate method be more appropriate than the spread plate method?

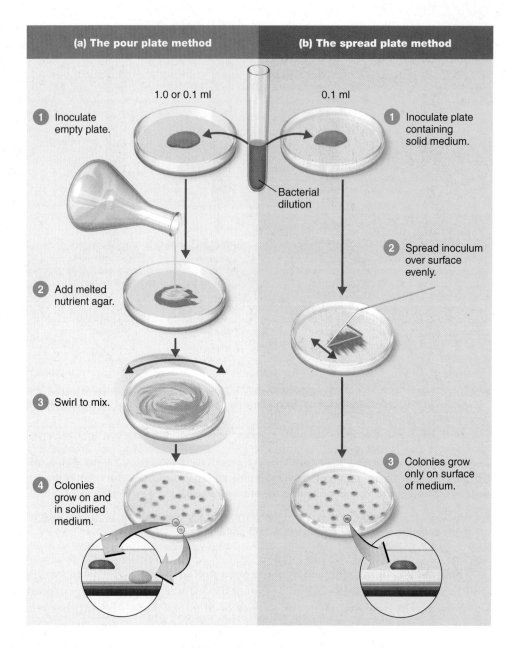

countable plate. If 1 ml of this sample were transferred to a tube containing 9 ml of sterile water, each milliliter of fluid in this tube would now contain 1000 bacteria. If 1 ml of this sample were inoculated into a Petri plate, there would still be too many potential colonies to count on a plate. Therefore, another serial dilution could be made. One milliliter containing 1000 bacteria would be transferred to a second tube of 9 ml of water. Each milliliter of this tube would now contain only 100 bacteria, and if 1 ml of the contents of this tube were plated out, potentially 100 colonies would be formed—an easily countable number.

**Pour Plates and Spread Plates** A plate count is done by either the pour plate method or the spread plate method. The **pour plate method** follows the procedure shown in **Figure 6.17a**. Either 1 ml or 0.1 ml of dilutions of the bacterial suspension

is introduced into a Petri dish. The nutrient medium, in which the agar is kept liquid by holding it in a water bath at about 50°C, is poured over the sample, which is then mixed into the medium by gentle agitation of the plate. When the agar solidifies, the plate is incubated. With the pour plate technique, colonies will grow within the nutrient agar (from cells suspended in the nutrient medium as the agar solidifies) as well as on the surface of the agar plate.

This technique has some drawbacks because some relatively heat-sensitive microorganisms may be damaged by the melted agar and will therefore be unable to form colonies. Also, when certain differential media are used, the distinctive appearance of the colony on the surface is essential for diagnostic purposes. Colonies that form beneath the surface of a pour plate are not satisfactory for such tests. To avoid these problems, the **spread plate method**

**Figure 6.18  Counting bacteria by filtration.**

**Q** Could you make a pour plate in the usual Petri dish with a 10-ml inoculum? Why or why not?

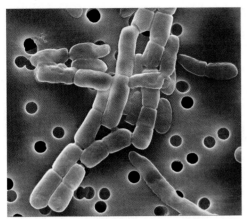

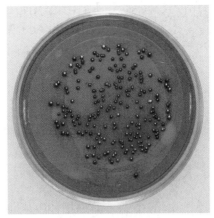

**(a)** The bacterial populations in bodies of water can be determined by passing a sample through a membrane filter. Here, the bacteria in a 100-ml water sample have been sieved out onto the surface of a membrane filter. These bacteria form visible colonies when placed on the surface of a suitable medium.

SEM | 1.5 μm

**(b)** A membrane filter with bacteria on its surface, as described in (a), has been placed on Endo agar. This medium is selective for gram-negative bacteria; lactose fermenters, such as the coliforms, form distinctive colonies. There are 214 colonies visible, so we would record 214 bacteria per 100 ml in the water sample.

is frequently used instead (Figure 6.17b). A 0.1-ml inoculum is added to the surface of a prepoured, solidified agar medium. The inoculum is then spread uniformly over the surface of the medium with a specially shaped, sterilized glass or metal rod. This method positions all the colonies on the surface and avoids contact between the cells and melted agar.

### Filtration

When the quantity of bacteria is very small, as in lakes or relatively pure streams, bacteria can be counted by **filtration** methods (**Figure 6.18**). In this technique, at least 100 ml of water are passed through a thin membrane filter whose pores are too small to allow bacteria to pass. Thus, the bacteria are filtered out and retained on the surface of the filter. This filter is then transferred to a Petri dish containing nutrient medium, where colonies arise from the bacteria on the filter's surface. This method is applied frequently to detection and enumeration of coliform bacteria, which are indicators of fecal contamination of food or water (see Chapter 27). The colonies formed by these bacteria are distinctive when a differential nutrient medium is used. (The colonies shown in Figure 6.18b are examples of coliforms.)

### The Most Probable Number (MPN) Method

Another method for determining the number of bacteria in a sample is the **most probable number (MPN) method**, illustrated in **Figure 6.19**. This statistical estimating technique is based on the fact that the greater the number of bacteria in a sample, the more dilution is needed to reduce the density to the point at which no bacteria are left to grow in the tubes in a dilution series. The MPN method is most useful when the microbes being counted will not grow on solid media (such as the chemoautotrophic nitrifying bacteria). It is also useful

when the growth of bacteria in a liquid differential medium is used to identify the microbes (such as coliform bacteria, which selectively ferment lactose to acid, in water testing). The MPN is only a statement that there is a 95% chance that the bacterial population falls within a certain range and that the MPN is statistically the most probable number.

### Direct Microscopic Count

In the method known as the **direct microscopic count**, a measured volume of a bacterial suspension is placed within a defined area on a microscope slide. Because of time considerations, this method is often used to count the number of bacteria in milk. A 0.01-ml sample is spread over a marked square centimeter of slide, stain is added so that the bacteria can be seen, and the sample is viewed under the oil immersion objective lens. The area of the viewing field of this objective can be determined. Once the number of bacteria has been counted in several different fields, the average number of bacteria per viewing field can be calculated. From these data, the number of bacteria in the square centimeter over which the sample was spread can also be calculated. Because this area on the slide contained 0.01 ml of sample, the number of bacteria in each milliliter of the suspension is the number of bacteria in the sample times 100.

A specially designed slide called a *Petroff-Hausser cell counter* is used in direct microscopic counts (**Figure 6.20**).

Motile bacteria are difficult to count by this method, and, as happens with other microscopic methods, dead cells are about as likely to be counted as live ones. In addition to these disadvantages, a rather high concentration of cells is required to be countable—about 10 million bacteria per milliliter. The chief advantage of microscopic counts is that no incubation

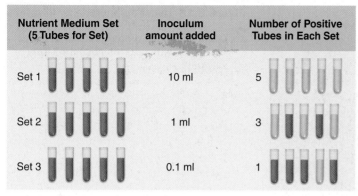

(a) Most probable number (MPN) dilution series

| Combination of Positives | MPN Index/ 100 ml | 95% Confidence Limits | |
| --- | --- | --- | --- |
| | | Lower | Upper |
| 4-2-0 | 22 | 6.8 | 50 |
| 4-2-1 | 26 | 9.8 | 70 |
| 4-3-0 | 27 | 9.9 | 70 |
| 4-3-1 | 33 | 10 | 70 |
| 4-4-0 | 34 | 14 | 100 |
| 5-0-0 | 23 | | 70 |
| 5-0-1 | 31 | 10 | 70 |
| 5-0-2 | 43 | 14 | 100 |
| 5-1-0 | 33 | 10 | 100 |
| 5-1-1 | 46 | 14 | 120 |
| 5-1-2 | 63 | 22 | 150 |
| 5-2-0 | 49 | 15 | 150 |
| 5-2-1 | 70 | 22 | 170 |
| 5-2-2 | 94 | 34 | 230 |
| 5-3-0 | 79 | 22 | 220 |
| 5-3-1 | 110 | 34 | 250 |
| 5-3-2 | 140 | 52 | 400 |

**(b) MPN table.** MPN tables enable us to calculate for a sample the microbial numbers that are statistically likely to lead to such a result. The number of positive (yellow) tubes is recorded for each set: in the shaded example, 5, 3, and 1. If we look up this combination in an MPN table, we find that the MPN index per 100 ml is 110. Statistically, this means that 95% of the water samples that give this result contain 34–250 bacteria, with 110 being the most probable number.

**Figure 6.19 The most probable number (MPN) method.**

**Q** Under what circumstances is the MPN method used to determine the number of bacteria in a sample?

time is required, and they are usually reserved for applications in which time is the primary consideration. This advantage also holds for *electronic cell counters*, sometimes known as *Coulter counters*, which automatically count the number of cells in a measured volume of liquid. These instruments are used in some research laboratories and hospitals.

**CHECK YOUR UNDERSTANDING**

✔ **6-16** Why is it difficult to measure realistically the growth of a filamentous mold isolate by the plate count method?

## Estimating Bacterial Numbers by Indirect Methods

It is not always necessary to count microbial cells to estimate their numbers. In science and industry, microbial numbers and activity are determined by some of the following indirect means as well.

### Turbidity

For some types of experimental work, estimating **turbidity** is a practical way of monitoring bacterial growth. As bacteria multiply in a liquid medium, the medium becomes turbid, or cloudy with cells.

The instrument used to measure turbidity is a *spectrophotometer* (or colorimeter). In the spectrophotometer, a beam of light is transmitted through a bacterial suspension to a light-sensitive detector (**Figure 6.21**). As bacterial numbers increase, less light will reach the detector. This change of light will register on the instrument's scale as the *percentage of transmission* (%T). Also printed on the instrument's scale is a logarithmic expression called the *absorbance* (sometimes called *optical density*, or *OD*). The absorbance is used to plot bacterial growth. When the bacteria are in logarithmic growth or decline, a graph of absorbance versus time will form an approximately straight line. If absorbance readings are matched with plate counts of the same culture, this correlation can be used in future estimations of bacterial numbers obtained by measuring turbidity.

More than a million cells per milliliter must be present for the first traces of turbidity to be visible. About 10 million to 100 million cells per milliliter are needed to make a suspension turbid enough to be read on a spectrophotometer. Therefore, turbidity is not a useful measure of contamination of liquids by relatively small numbers of bacteria.

**CLINICAL CASE**

The bacteria in the blood and catheter cultures fluoresce under ultraviolet light. The results from the culture show that *P. fluorescens* is present in the blood of 15 patients, in 17 catheters, and in the blood and catheters of 4 patients. The bacteria survived even after the heparin recall. Dr. MacGruder would like to have some idea how many bacteria are colonizing a patient's catheter. Because the amount of nutrients in a patient's catheter is minimal, he concludes that the bacteria grow slowly. He does some calculations based on the assumption that five *Pseudomonas* cells, with a generation time of 35 hours, may have been originally introduced into the catheters.

**Approximately how many cells would there be after a month?**

153  164  **172**  174

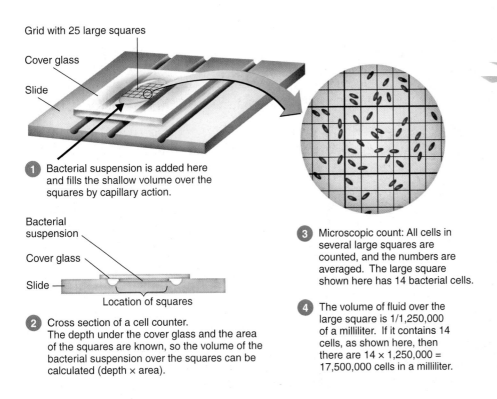

Grid with 25 large squares

Cover glass

Slide

**1** Bacterial suspension is added here and fills the shallow volume over the squares by capillary action.

Bacterial suspension

Cover glass

Slide

Location of squares

**2** Cross section of a cell counter. The depth under the cover glass and the area of the squares are known, so the volume of the bacterial suspension over the squares can be calculated (depth × area).

**Figure 6.20  Direct microscopic count of bacteria with a Petroff-Hausser cell counter.** The average number of cells within a large square multiplied by a factor of 1,250,000 gives the number of bacteria per milliliter.

**Q** This type of counting, despite its obvious disadvantages, is often used in estimating the bacterial population in dairy products. Why?

**3** Microscopic count: All cells in several large squares are counted, and the numbers are averaged. The large square shown here has 14 bacterial cells.

**4** The volume of fluid over the large square is 1/1,250,000 of a milliliter. If it contains 14 cells, as shown here, then there are 14 × 1,250,000 = 17,500,000 cells in a milliliter.

**Figure 6.21  Turbidity estimation of bacterial numbers.** The amount of light striking the light-sensitive detector on the spectrophotometer is inversely proportional to the number of bacteria under standardized conditions. The less light transmitted, the more bacteria in the sample. The turbidity of the sample could be reported as either 20% transmittance or 0.7 absorbance. Readings in absorbance are a logarithmic function and are sometimes useful in plotting data.

**Q** Why is turbidity more useful in measuring contamination of liquids by large numbers, rather than small numbers, of bacteria?

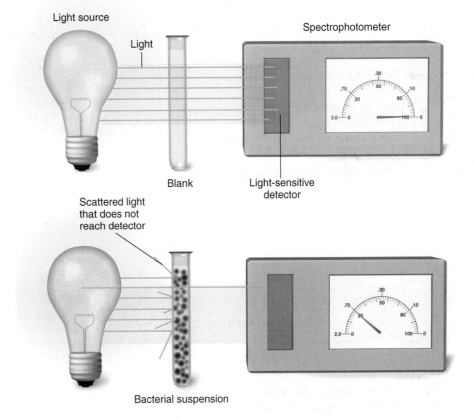

Light source

Light

Spectrophotometer

Blank

Light-sensitive detector

Scattered light that does not reach detector

Bacterial suspension

## Metabolic Activity

Another indirect way to estimate bacterial numbers is to measure a population's *metabolic activity*. This method assumes that the amount of a certain metabolic product, such as acid, $CO_2$, ATP, or DNA, is in direct proportion to the number of bacteria present. An example of a practical application of a metabolic test is the microbiological assay (described on page 160), in which acid production is used to determine amounts of vitamins.

## Dry Weight

For filamentous bacteria and molds, the usual measuring methods are less satisfactory. A plate count would not measure this increase in filamentous mass. In plate counts of actinomycetes (see Figure 11.26, page 317) and molds, it is mostly the number of asexual spores that is counted instead. This is not a good measure of growth. One of the better ways to measure the growth of filamentous organisms is by *dry weight*. In this procedure, the fungus is removed from the growth medium, filtered to remove extraneous material, and dried in a desiccator. It is then weighed. For bacteria, the same basic procedure is followed.

> **CHECK YOUR UNDERSTANDING**
>
> ✔ **6-17** Direct methods usually require an incubation time for a colony. Why is this not always feasible for analyzing foods?
>
> ✔ **6-18** If there is no good method for analyzing a product for its vitamin content, what is a feasible method of determining the vitamin content?

* * *

You now have a basic understanding of the requirements for, and measurements of, microbial growth. In Chapter 7, we will look at how this growth is controlled in laboratories, hospitals, industry, and our homes.

### CLINICAL CASE Resolved

Biofilms are dense accumulations of cells. Five cells might go through 20 generations in a month, producing $7.79 \times 10^6$ cells. Now Dr. MacGruder knows that the *P. fluorescens* bacteria are present in the patients' indwelling catheters. He orders the catheters to be replaced and has the CDC examine the used catheters with scanning electron microscopy. They discover that the *P. fluorescens* colonized the inside of the catheters by forming biofilms. In his report to the CDC, Dr. MacGruder explains that the *P. fluorescens* bacteria may have entered the bloodstreams of these patients at the same time as the first outbreak, but not in sufficient quantities to cause symptoms at that time. Biofilm formation enabled the bacteria to persist in the patients' catheters. He notes that previous electron microscopy studies indicate that nearly all indwelling vascular catheters become colonized by microorganisms that are embedded in a biofilm layer and that heparin has been reported to stimulate biofilm formation. Dr. MacGruder concludes that the bacteria in the biofilm were dislodged by subsequent uncontaminated intravenous solutions and released into the bloodstream, finally causing infections months after initial colonization.

153    164    172    **174**

# Study Outline

 Go to @MasteringMicrobiology for **Interactive Microbiology**, *In the Clinic videos, MicroFlix, MicroBoosters, 3D animations, practice quizzes, and more.*

## The Requirements for Growth (pp. 152–157)

1. The growth of a population is an increase in the number of cells.
2. The requirements for microbial growth are both physical and chemical.

### Physical Requirements (pp. 152–155)

3. On the basis of preferred temperature ranges, microbes are classified as psychrophiles (cold-loving), mesophiles (moderate-temperature–loving), and thermophiles (heat-loving).
4. The minimum growth temperature is the lowest temperature at which a species will grow, the optimum growth temperature is the temperature at which it grows best, and the maximum growth temperature is the highest temperature at which growth is possible.
5. Most bacteria grow best at a pH value between 6.5 and 7.5.
6. In a hypertonic solution, most microbes undergo plasmolysis; halophiles can tolerate high salt concentrations.

### Chemical Requirements (pp. 155–157)

7. All organisms require a carbon source; chemoheterotrophs use an organic molecule, and autotrophs typically use carbon dioxide.
8. Nitrogen is needed for protein and nucleic acid synthesis. Nitrogen can be obtained from the decomposition of proteins or from $NH_4^+$ or $NO_3^-$; a few bacteria are capable of nitrogen ($N_2$) fixation.
9. On the basis of oxygen requirements, organisms are classified as obligate aerobes, facultative anaerobes, obligate anaerobes, aerotolerant anaerobes, and microaerophiles.
10. Aerobes, facultative anaerobes, and aerotolerant anaerobes must have the enzymes superoxide dismutase ($2\,O_2^- + 2\,H^+ \longrightarrow O_2 + H_2O_2$) and either catalase ($2\,H_2O_2 \longrightarrow 2\,H_2O + O_2$) or peroxidase ($H_2O_2 + 2\,H^+ \longrightarrow 2\,H_2O$).
11. Other chemicals required for microbial growth include sulfur, phosphorus, trace elements, and, for some microorganisms, organic growth factors.

## Biofilms (pp. 157–159)

1. Microbes adhere to surfaces and accumulate as biofilms on solid surfaces in contact with water.

Play Interactive Microbiology @MasteringMicrobiology *See how biofilms affect a patient's health*

2. Most bacteria live in biofilms.

3. Microbes in biofilms are more resistant to antibiotics than are free-swimming microbes.

## Culture Media (pp. 159–163)

1. A culture medium is any material prepared for the growth of bacteria in a laboratory.

2. Microbes that grow and multiply in or on a culture medium are known as a culture.

3. Agar is a common solidifying agent for a culture medium.

### Chemically Defined Media (pp. 159–160)

4. A chemically defined medium is one in which the exact chemical composition is known.

### Complex Media (p. 160)

5. A complex medium is one in which the exact chemical composition varies slightly from batch to batch.

### Anaerobic Growth Media and Methods (pp. 160–161)

6. Reducing media chemically remove molecular oxygen ($O_2$) that might interfere with the growth of anaerobes.

7. Petri plates can be incubated in an anaerobic jar, anaerobic chamber, or OxyPlate™.

### Special Culture Techniques (pp. 161–162)

8. Some parasitic and fastidious bacteria must be cultured in living animals or in cell cultures.

9. $CO_2$ incubators or candle jars are used to grow bacteria that require an increased $CO_2$ concentration.

10. Procedures and equipment to minimize exposure to pathogenic microorganisms are designated as biosafety levels 1 through 4.

### Selective and Differential Media (pp. 162–163)

11. By inhibiting unwanted organisms with salts, dyes, or other chemicals, selective media allow growth of only the desired microbes.

12. Differential media are used to distinguish different organisms.

### Enrichment Culture (p. 163)

13. An enrichment culture is used to encourage the growth of a particular microorganism in a mixed culture.

## Obtaining Pure Cultures (pp. 163–164)

1. A colony is a visible mass of microbial cells that theoretically arose from one cell.

2. Pure cultures are usually obtained by the streak plate method.

## Preserving Bacterial Cultures (pp. 164–165)

1. Microbes can be preserved for long periods of time by deep-freezing or lyophilization (freeze-drying).

## The Growth of Bacterial Cultures (pp. 165–174)

### Bacterial Division (p. 165)

1. The normal reproductive method of bacteria is binary fission, in which a single cell divides into two identical cells.

2. Some bacteria reproduce by budding, aerial spore formation, or fragmentation.

### Generation Time (pp. 165–166)

3. The time required for a cell to divide or a population to double is known as the generation time.

### Logarithmic Representation of Bacterial Populations (pp. 166–167)

4. Bacterial division occurs according to a logarithmic progression (two cells, four cells, eight cells, and so on).

### Phases of Growth (pp. 167–168)

5. During the lag phase, there is little or no change in the number of cells, but metabolic activity is high.

6. During the log phase, the bacteria multiply at the fastest rate possible under the conditions provided.

7. During the stationary phase, there is an equilibrium between cell division and death.

8. During the death phase, the number of deaths exceeds the number of new cells formed.

### Direct Measurement of Microbial Growth (pp. 168–172)

9. A heterotrophic plate count reflects the number of viable microbes and assumes that each bacterium grows into a single colony; plate counts are reported as number of colony-forming units (CFU).

10. A plate count may be done by either the pour plate method or the spread plate method.

11. In filtration, bacteria are retained on the surface of a membrane filter and then transferred to a culture medium to grow and subsequently be counted.

12. The most probable number (MPN) method can be used for microbes that will grow in a liquid medium; it is a statistical estimation.

13. In a direct microscopic count, the microbes in a measured volume of a bacterial suspension are counted with the use of a specially designed slide.

### Estimating Bacterial Numbers by Indirect Methods (pp. 172–174)

14. A spectrophotometer is used to determine turbidity by measuring the amount of light that passes through a suspension of cells.

15. An indirect way of estimating bacterial numbers is measuring the metabolic activity of the population (for example, acid production).

16. For filamentous organisms such as fungi, measuring dry weight is a convenient method of growth measurement.

## Study Questions

For answers to the Knowledge and Comprehension questions, turn to the Answers tab at the back of the textbook.

### Knowledge and Comprehension

#### Review

1. Describe binary fission.
2. Macronutrients (needed in relatively large amounts) are often listed as CHONPS. What does each of these letters indicate, and why are they needed by the cell?
3. Define and explain the importance of each of the following:
   a. catalase
   b. hydrogen peroxide
   c. peroxidase
   d. superoxide radical
   e. superoxide dismutase
4. Seven methods of measuring microbial growth were explained in this chapter. Categorize each as either a direct or an indirect method.
5. By deep-freezing, bacteria can be stored without harm for extended periods. Why do refrigeration and freezing preserve foods?
6. A pastry chef accidentally inoculated a cream pie with six *S. aureus* cells. If *S. aureus* has a generation time of 60 minutes, how many cells would be in the cream pie after 7 hours?
7. Nitrogen and phosphorus added to beaches following an oil spill encourage the growth of natural oil-degrading bacteria. Explain why the bacteria do not grow if nitrogen and phosphorus are not added.
8. Differentiate complex and chemically defined media.
9. DRAW IT  Draw the following growth curves for *E. coli*, starting with 100 cells with a generation time of 30 minutes at 35°C, 60 minutes at 20°C, and 3 hours at 5°C.
   a. The cells are incubated for 5 hours at 35°C.
   b. After 5 hours, the temperature is changed to 20°C for 2 hours.
   c. After 5 hours at 35°C, the temperature is changed to 5°C for 2 hours followed by 35°C for 5 hours.

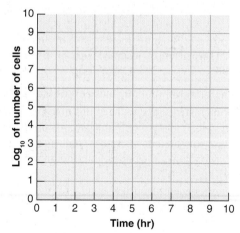

10. NAME IT  A prokaryotic cell hitched a ride to Earth on a space shuttle from some unknown planet. The organism is a psychrophile, an obligate halophile, and an obligate aerobe. Based on the characteristics of the microbe, describe the planet.

### Multiple Choice

Use the following information to answer questions 1 and 2. Two culture media were inoculated with four different bacteria. After incubation, the following results were obtained:

| Organism | Medium 1 | Medium 2 |
|---|---|---|
| *Escherichia coli* | Red colonies | No growth |
| *Staphylococcus aureus* | No growth | Growth |
| *Staphylococcus epidermidis* | No growth | Growth |
| *Salmonella enterica* | Colorless colonies | No growth |

1. Medium 1 is
   a. selective.
   b. differential.
   c. both selective and differential.
2. Medium 2 is
   a. selective.
   b. differential.
   c. both selective and differential.

Use the following graph to answer questions 3 and 4.

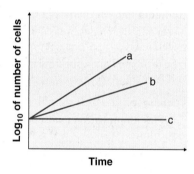

3. Which of the lines best depicts the log phase of a thermophile incubated at room temperature?
4. Which of the lines best depicts the log phase of *Listeria monocytogenes* growing in a human?
5. Assume you inoculated 100 facultatively anaerobic cells onto nutrient agar and incubated the plate aerobically. You then inoculated 100 cells of the same species onto nutrient agar and incubated the second plate anaerobically. After incubation for 24 hours, you should have
   a. more colonies on the aerobic plate.
   b. more colonies on the anaerobic plate.
   c. the same number of colonies on both plates.
6. The term *trace elements* refers to
   a. the elements CHONPS.
   b. vitamins.
   c. nitrogen, phosphorus, and sulfur.
   d. small mineral requirements.
   e. toxic substances.
7. Which one of the following temperatures would most likely kill a mesophile?
   a. −50°C
   b. 0°C
   c. 9°C
   d. 37°C
   e. 60°C

8. Which of the following is *not* a characteristic of biofilms?
   a. antibiotic resistance
   b. hydrogel
   c. iron deficiency
   d. quorum sensing
9. Which of the following types of media would *not* be used to culture aerobes?
   a. selective media
   b. reducing media
   c. enrichment media
   d. differential media
   e. complex media
10. An organism that has peroxidase and superoxide dismutase but lacks catalase is most likely an
    a. aerobe.
    b. aerotolerant anaerobe.
    c. obligate anaerobe.

## Analysis

1. *E. coli* was incubated with aeration in a nutrient medium containing two carbon sources, and the following growth curve was made from this culture.
   a. Explain what happened at the time marked *x*.
   b. Which substrate provided "better" growth conditions for the bacteria? How can you tell?

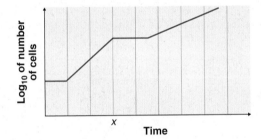

2. *Clostridium* and *Streptococcus* are both catalase-negative. *Streptococcus* grows by fermentation. Why is *Clostridium* killed by oxygen, whereas *Streptococcus* is not?
3. Most laboratory media contain a fermentable carbohydrate and peptone because the majority of bacteria require carbon, nitrogen, and energy sources in these forms. How are these three needs met by glucose–minimal salts medium? (*Hint*: See Table 6.2.)
4. Flask A contains yeast cells in glucose–minimal salts broth incubated at 30°C with aeration. Flask B contains yeast cells in glucose–minimal salts broth incubated at 30°C in an anaerobic jar. The yeasts are facultative anaerobes.
   a. Which culture produced more ATP?
   b. Which culture produced more alcohol?
   c. Which culture had the shorter generation time?
   d. Which culture had the greater cell mass?
   e. Which culture had the higher absorbance?

## Clinical Applications and Evaluation

1. Assume that after washing your hands, you leave ten bacterial cells on a new bar of soap. You then decide to do a plate count of the soap after it was left in the soap dish for 24 hours. You dilute 1 g of the soap $1:10^6$ and plate it on heterotrophic plate count agar. After 24 hours of incubation, there are 168 colonies. How many bacteria were on the soap? How did they get there?
2. Heat lamps are commonly used to maintain foods at about 50°C for as long as 12 hours in cafeteria serving lines. The following experiment was conducted to determine whether this practice poses a potential health hazard.

   Beef cubes were surface-inoculated with 500,000 bacterial cells and incubated at 43–53°C to establish temperature limits for bacterial growth. The following results were obtained from heterotrophic plate counts performed on beef cubes at 6 and 12 hours after inoculation:

| | Temp. (°C) | Bacteria per Gram of Beef After 6 hr | 12 hr |
|---|---|---|---|
| Staphylococcus aureus | 43 | 140,000,000 | 740,000,000 |
| | 51 | 810,000 | 59,000 |
| | 53 | 650 | 300 |
| Salmonella Typhimurium | 43 | 3,200,000 | 10,000,000 |
| | 51 | 950,000 | 83,000 |
| | 53 | 1,200 | 300 |
| Clostridium perfringens | 43 | 1,200,000 | 3,600,000 |
| | 51 | 120,000 | 3,800 |
| | 53 | 300 | 300 |

   Draw the growth curves for each organism. What holding temperature would you recommend? Assuming that cooking kills bacteria in foods, how could these bacteria contaminate the cooked foods? What disease does each organism cause? (*Hint*: See Chapter 25.)
3. The number of bacteria in saliva samples was determined by collecting the saliva, making serial dilutions, and inoculating nutrient agar by the pour plate method. The plates were incubated aerobically for 48 hours at 37°C.

| | Bacteria per ml Saliva Before Using Mouthwash | After Using Mouthwash |
|---|---|---|
| Mouthwash 1 | $13.1 \times 10^6$ | $10.9 \times 10^6$ |
| Mouthwash 2 | $11.7 \times 10^6$ | $14.2 \times 10^5$ |
| Mouthwash 3 | $9.3 \times 10^5$ | $7.7 \times 10^5$ |

What can you conclude from these data? Did all the bacteria present in each saliva sample grow?

# 7 The Control of Microbial Growth

The scientific control of microbial growth began only about 100 years ago. Recall from Chapter 1 that Pasteur's work on microorganisms led scientists to believe that microbes were a possible cause of disease. In the mid-1800s, the Hungarian physician Ignaz Semmelweis and English physician Joseph Lister used this thinking to develop some of the first microbial control practices for medical procedures. These practices included washing hands with microbe-killing chloride of lime [$Ca(OCl)_2$] and using the techniques of **aseptic surgery** to prevent microbial contamination of surgical wounds. Until that time, hospital-acquired infections, or *nosocomial infections*, were the cause of death in at least 10% of surgical cases and as many as 25% of deaths among delivering mothers. Ignorance of microbes was such that during the American Civil War, a surgeon might have cleaned his scalpel on his boot sole between incisions. We now know that handwashing is the best way to prevent transmission of pathogens such as the norovirus in the photo. Controlling noroviruses on environmental surfaces is the topic of the Clinical Case.

Over the last century, scientists have continued to develop a variety of physical methods and chemical agents to control microbial growth. In Chapter 20, we will discuss methods for controlling microbes once infection has occurred, focusing mainly on antibiotic chemotherapy.

▶ Norovirus can be spread on environmental surfaces.

## In the Clinic

As a hospital's infection control nurse, you become aware that 15 patients developed *Clostridium difficile* infections within a month. This rate of infection is 10 per 1000 patients—nearly 300% higher than the average of 2.7 cases per 1000 patients seen in previous months. You request that housekeeping clean rooms and equipment with a hypochlorite-based disinfectant rather than the standard hospital (quat) disinfectant normally used. The next month, the rate of infection is 3 cases per 1000. **Did your cleaning program work? How did you reach your conclusion?**

*Hint: Read about chlorines (pages 191–192) and quats (page 194); see Table 7.7.*

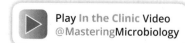

Play **In the Clinic** Video
@MasteringMicrobiology

# The Terminology of Microbial Control

### LEARNING OBJECTIVE

**7-1** Define the following key terms related to microbial control: *sterilization, disinfection, antisepsis, degerming, sanitization, biocide, germicide, bacteriostasis,* and *asepsis*.

A word frequently used, and misused, in discussions about the control of microbial growth is *sterilization*. **Sterilization** is the removal or destruction of *all* living microorganisms. Heating is the most common method used for killing microbes, including the most resistant forms, such as endospores. A sterilizing agent is called a **sterilant**. Liquids or gases can be sterilized by filtration.

 ASM: The growth of microorganisms can be controlled by physical, chemical, mechanical, or biological means.

One would think that canned food in the supermarket is completely sterile. In reality, the heat treatment required to ensure absolute sterility would unnecessarily degrade the quality of the food. Instead, food is subjected only to enough heat to destroy the endospores of *Clostridium botulinum* (BO-tū-lī′num), which can produce a deadly toxin. This limited heat treatment is termed **commercial sterilization**. The endospores of a number of thermophilic bacteria, capable of causing food spoilage but not human disease, are considerably more resistant to heat than *C. botulinum*. If present, they will survive, but their survival is usually of no practical consequence; they will not grow at normal food storage temperatures. If canned foods in a supermarket were incubated at temperatures in the growth range of these thermophiles (above about 45°C), significant food spoilage would occur.

Complete sterilization is often not required in other settings. For example, the body's normal defenses can cope with a few microbes entering a surgical wound. A drinking glass or a fork in a restaurant requires only enough microbial control to prevent the transmission of possibly pathogenic microbes from one person to another.

Control directed at destroying harmful microorganisms is called **disinfection**. The term usually refers to the destruction of vegetative (non–endospore-forming) pathogens, which is not the same thing as complete sterility. Disinfection might make use of chemicals, ultraviolet radiation, boiling water, or steam. In practice, the term is most commonly applied to the use of a chemical (a *disinfectant*) to treat an inert surface or substance. When this treatment is directed at living tissue, it is called **antisepsis**, and the chemical is then called an *antiseptic*. Therefore, in practice the same chemical might be called a disinfectant for one use and an antiseptic for another. Of course, many chemicals suitable for wiping a tabletop would be too harsh to use on living tissue.

There are modifications of disinfection and antisepsis. For example, when someone is about to receive an injection, the skin is swabbed with alcohol—the process of **degerming** (or *degermation*), which mostly results in the mechanical removal, rather than the killing, of most of the microbes in a limited area. Restaurant glassware, china, and tableware are subjected to **sanitization**, which is intended to lower microbial counts to safe public health levels and minimize the chances of disease transmission from one user to another. This is usually accomplished by high-temperature washing or, in the case of glassware in a bar, washing in a sink followed by a dip in a chemical disinfectant.

Table 7.1 summarizes the terminology relating to the control of microbial growth.

Names of treatments that cause the outright death of microbes have the suffix *-cide*, meaning kill. A **biocide**, or **germicide**, kills microorganisms (usually with certain exceptions, such as endospores); a *fungicide* kills fungi; a *virucide* inactivates viruses; and so on. Other treatments only inhibit the growth and multiplication of bacteria; their names have the suffix *-stat* or *-stasis*, meaning to stop or to steady, as in **bacteriostasis**. Once a bacteriostatic agent is removed, growth might resume.

**Sepsis**, from the Greek for decay or putrid, indicates bacterial contamination, as in septic tanks for sewage treatment. (The term is also used to describe a disease condition; see Chapter 23, page 652.) *Aseptic* means that an object or area is free of pathogens. **Asepsis** is the absence of significant contamination (see Chapter 1). Aseptic techniques are important in surgery to minimize contamination from the instruments, operating personnel, and the patient.

### CHECK YOUR UNDERSTANDING

✔ **7-1** The usual definition of *sterilization* is the removal or destruction of all forms of microbial life; how could there be practical exceptions to this simple definition?

### CLINICAL CASE  A School Epidemic

It is 9:00 A.M. on a Wednesday morning, and Amy Garza, the school nurse at Westview Elementary School in Rockville, Maryland, has been on the phone since she came in to work at 7:00 A.M. So far this morning, she has received reports of students unable to attend school today because of some sort of gastrointestinal ailment. They all have the same symptoms: nausea and vomiting, diarrhea, and a low-grade fever. As Amy picks up the phone to call the principal to give her an update, she receives her eighth call of the day. Keith Jackson, a first-grade teacher who has been out sick since Monday, calls to tell Amy that his physician sent his stool sample to the laboratory for testing. The results came back positive for norovirus.

**What is norovirus? Read on to find out.**

**TABLE 7.1** Terminology Relating to the Control of Microbial Growth

| | Definition | Comments |
|---|---|---|
| Sterilization | Destruction or removal of all forms of microbial life, including endospores but with the possible exception of prions. | Usually done by steam under pressure or a sterilizing gas, such as ethylene oxide. |
| Commercial Sterilization | Sufficient heat treatment to kill endospores of *Clostridium botulinum* in canned food. | More-resistant endospores of thermophilic bacteria may survive, but they will not germinate and grow under normal storage conditions. |
| Disinfection | Destruction of vegetative pathogens on inanimate objects. | May make use of physical or chemical methods. |
| Antisepsis | Destruction of vegetative pathogens on living tissue. | Treatment is almost always by chemical antimicrobials. |
| Degerming | Removal of microbes from a limited area, such as the skin around an injection site. | Mostly a mechanical removal by soap and water or an alcohol-soaked swab. |
| Sanitization | Treatment is intended to lower microbial counts on eating and drinking utensils to safe public health levels. | May be done with high-temperature washing or by dipping into a chemical disinfectant. |

# The Rate of Microbial Death

**LEARNING OBJECTIVE**

7-2 Describe the patterns of microbial death caused by treatments with microbial control agents.

When bacterial populations are heated or treated with antimicrobial chemicals, they usually die at a constant rate. For example, suppose a population of 1 million microbes has been treated for 1 minute, and 90% of the population has died. We are now left with 100,000 microbes. If the population is treated for another minute, 90% of *those* microbes die, and we are left with 10,000 survivors. In other words, for each minute the treatment is applied, 90% of the remaining population is killed (Table 7.2). If the death curve is plotted logarithmically, the death rate is constant, as shown by the straight line in **Figure 7.1a**.

Several factors influence the effectiveness of antimicrobial treatments:

- *The number of microbes.* The more microbes there are to begin with, the longer it takes to eliminate the entire population (Figure 7.1b).
- *Environmental influences.* Most disinfectants work somewhat better in warm solutions.

The presence of organic matter often inhibits the action of chemical antimicrobials. In hospitals, the presence of organic matter in blood, vomitus, or feces influences the selection of disinfectants. Microbes in surface biofilms, when they are encased in the mucoid matrix (see page 157), are difficult for biocides to reach effectively. Because their activity is due to temperature-dependent chemical reactions, disinfectants work somewhat better under warm conditions.

The nature of the suspending medium is also a factor in heat treatment. Fats and proteins are especially protective, and a medium rich in these substances protects microbes, which will then have a higher survival rate. Heat is also measurably more effective under acidic conditions.

- *Time of exposure.* Chemical antimicrobials often require extended exposure to affect more-resistant microbes or endospores. See the discussion of equivalent treatments on page 185.
- *Microbial characteristics.* The concluding section of this chapter discusses how microbial characteristics affect the choice of chemical and physical control methods.

**CHECK YOUR UNDERSTANDING**

7-2 How is it possible that a solution containing a million bacteria would take longer to sterilize than one containing a half-million bacteria?

# Actions of Microbial Control Agents

**LEARNING OBJECTIVE**

7-3 Describe the effects of microbial control agents on cellular structures.

In this section, we examine the ways various agents actually kill or inhibit microbes.

**TABLE 7.2** Microbial Exponential Death Rate: An Example

| Time (min) | Deaths per Minute | Number of Survivors |
|---|---|---|
| 0 | 0 | 1,000,000 |
| 1 | 900,000 | 100,000 |
| 2 | 90,000 | 10,000 |
| 3 | 9000 | 1000 |
| 4 | 900 | 100 |
| 5 | 90 | 10 |
| 6 | 9 | 1 |

# Understanding the Microbial Death Curve

Plotting the typical microbial death curve **logarithmically** (**red line**) results in a straight line.

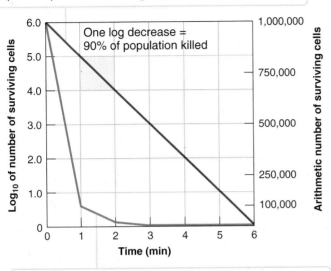

One log decrease = 90% of population killed

Log₁₀ of number of surviving cells (left axis): 0, 1.0, 2.0, 3.0, 4.0, 5.0, 6.0
Arithmetic number of surviving cells (right axis): 100,000, 250,000, 500,000, 750,000, 1,000,000
Time (min): 0 1 2 3 4 5 6

**(a)** Plotting the typical microbial death curve **arithmetically** (**blue line**) is impractical: at 3 minutes the population of 1000 cells would only be a hundredth of the graphed distance between 100,000 and the baseline.

Sterile surgical equipment

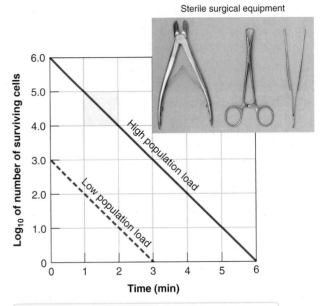

High population load

Low population load

Log₁₀ of number of surviving cells: 0, 1.0, 2.0, 3.0, 4.0, 5.0, 6.0
Time (min): 0 1 2 3 4 5 6

**(b)** Logarithmic plotting (**red**) reveals that if the rate of killing is the same, it will take longer to kill all members of a larger population than a smaller one, whether using heat or chemical treatments.

## KEY CONCEPTS

- Bacterial populations usually die at a constant rate when heated or when treated with antimicrobial chemicals.

- It is necessary to use logarithmic numbers to graph bacterial populations effectively.

- Understanding logarithmic death curves for microbial populations, including the elements of time and the size of the initial population, is especially useful in food preservation and in the sterilization of media or medical supplies.

Foods preserved by heat

## Alteration of Membrane Permeability

A microorganism's plasma membrane (see Figure 4.14, page 86), located just inside the cell wall, is the target of many microbial control agents. This membrane actively regulates the passage of nutrients into the cell and the elimination of wastes from the cell. Damage to the lipids or proteins of the plasma membrane by antimicrobial agents causes cellular contents to leak into the surrounding medium and interferes with the growth of the cell.

## Damage to Proteins and Nucleic Acids

Bacteria are sometimes thought of as "little bags of enzymes." Enzymes, which are primarily protein, are vital to all cellular activities. Recall that the functional properties of proteins are the result of their three-dimensional shape (see Figure 2.15, page 43). This shape is maintained by chemical bonds that link adjoining portions of the amino acid chain as it folds back and forth upon itself. Some of those bonds are hydrogen bonds, which are susceptible to breakage by heat or certain chemicals; breakage results in denaturation of the protein. Covalent bonds are stronger but are also subject to attack. For example, disulfide bridges, which play an important role in protein structure by joining amino acids with exposed sulfhydryl (—SH) groups, can be broken by certain chemicals or by sufficient heat.

The nucleic acids DNA and RNA are the carriers of the cell's genetic information. Damage to these nucleic acids by heat, radiation, or chemicals is frequently lethal to the cell; the cell can no longer replicate, nor can it carry out normal metabolic functions such as the synthesis of enzymes.

### CHECK YOUR UNDERSTANDING

✔ 7-3 Would a chemical microbial control agent that affects plasma membranes affect humans?

181

# Physical Methods of Microbial Control

## LEARNING OBJECTIVES

7-4 Compare the effectiveness of moist heat (boiling, autoclaving, pasteurization) and dry heat.

7-5 Describe how filtration, low temperatures, high pressure, desiccation, and osmotic pressure suppress microbial growth.

7-6 Explain how radiation kills cells.

As early as the Stone Age, humans likely were already using some physical methods of microbial control to preserve foods. Drying (desiccation) and salting (osmotic pressure) were probably among the earliest techniques.

When selecting methods of microbial control, one must consider what else, besides the microbes, a particular method will affect. For example, heat might inactivate certain vitamins or antibiotics in a solution. Repeated heating damages many laboratory and hospital materials, such as rubber and latex tubing. There are also economic considerations; for example, it may be less expensive to use presterilized, disposable plasticware than to repeatedly wash and resterilize glassware.

## Heat

A visit to any supermarket will demonstrate that heat-preserved canned goods represent one of the most common methods of food preservation. Heat is also usually used to sterilize laboratory media and glassware and hospital instruments. Heat appears to kill microorganisms by denaturing their enzymes; the resultant changes to the three-dimensional shapes of these proteins inactivate them (see Figure 5.6, page 115).

Heat resistance varies among different microbes; these differences can be expressed through the concept of thermal death point. **Thermal death point (TDP)** is the lowest temperature at which all the microorganisms in a particular liquid suspension will be killed in 10 minutes.

Another factor to be considered in sterilization is the length of time required. This is expressed as **thermal death time (TDT)**, the minimal length of time for all bacteria in a particular liquid culture to be killed at a given temperature. Both TDP and TDT are useful guidelines that indicate the severity of treatment required to kill a given population of bacteria.

**Decimal reduction time (DRT)**, or *D value*, is a third concept related to bacterial heat resistance. DRT is the time, in minutes, in which 90% of a population of bacteria at a given temperature will be killed (in Table 7.2 and Figure 7.1a, DRT is 1 minute). Chapter 28 describes an important application of DRT in the canning industry.

## Moist Heat Sterilization

Moist heat kills microorganisms primarily by coagulating proteins (denaturation), which is caused by breakage of the hydrogen bonds that hold the proteins in their three-dimensional structure. This coagulation process is familiar to anyone who has watched an egg white frying.

One type of moist heat "sterilization" is boiling, which kills vegetative forms of bacterial pathogens, many viruses (papovaviruses, however, are resistant to boiling), and fungi and their spores within about 10 minutes, usually much faster. Free-flowing (unpressurized) steam is essentially the same temperature as boiling water. Endospores and some viruses, however, are not destroyed this quickly. For example, some bacterial endospores can resist boiling for more than 20 hours. Boiling is therefore not always a reliable sterilization procedure. However, brief boiling, even at high altitudes, will kill most pathogens. The use of boiling to sanitize glass baby bottles is a familiar example.

Reliable sterilization with moist heat requires temperatures above that of boiling water. These high temperatures are most commonly achieved by steam under pressure in an **autoclave (Figure 7.2)**. Autoclaving is the preferred method of sterilization in health care environments, unless the material to be sterilized can be damaged by heat or moisture. The higher the pressure in the autoclave, the higher the temperature. For example, when free-flowing steam at a temperature of 100°C is placed under a pressure of 1 atmosphere above sea level pressure—that is, about 15 pounds of pressure per square inch (psi)—the temperature rises to 121°C. Increasing the pressure to 20 psi raises the temperature to 126°C. The relationship between temperature and pressure is shown in Table 7.3.

Sterilization in an autoclave is most effective when the organisms either are contacted by the steam directly or are contained in a small volume of aqueous (primarily water) liquid. Under these conditions, steam at a pressure of about 15 psi (121°C) will kill *all* organisms (but not prions; see page 198) and their endospores in about 15 minutes. Sterilizing the surface of a solid requires that steam actually contact it. To sterilize dry glassware, bandages, and the like, care must be taken to ensure that steam contacts all surfaces. For example, aluminum foil is impervious to steam and should not be used to wrap dry materials that are to be sterilized; paper should be used instead. Care should also be taken to avoid trapping air in the bottom of a dry container: trapped air will not be replaced by steam, because steam is lighter than air. The trapped air is the equivalent of a small hot-air oven, which, as we will see shortly, requires a higher temperature and longer time to sterilize materials. Containers that can trap air should be placed in a tipped position so that the steam will force out the air. Products that do not permit penetration by moisture, such as mineral oil or petroleum jelly, are not sterilized by the same methods used to sterilize aqueous solutions. Large industrial autoclaves are called *retorts*, but the same principle applies for the common household pressure cooker used in the home canning of foods.

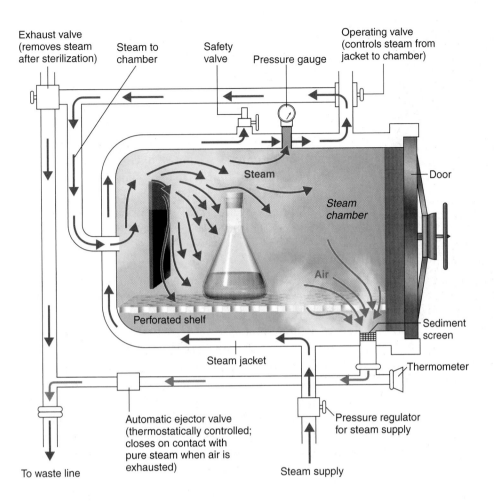

**Figure 7.2  An autoclave.** The entering steam forces the air out of the bottom (blue arrows). The automatic ejector valve remains open as long as an air-steam mixture is passing out of the waste line. When all the air has been ejected, the higher temperature of the pure steam closes the valve, and the pressure in the chamber increases.

**Q** How would an empty, uncapped flask be positioned for sterilization in an autoclave?

Exhaust valve (removes steam after sterilization)

Steam to chamber

Safety valve

Pressure gauge

Operating valve (controls steam from jacket to chamber)

Steam

Door

Steam chamber

Air

Perforated shelf

Sediment screen

Steam jacket

Thermometer

Automatic ejector valve (thermostatically controlled; closes on contact with pure steam when air is exhausted)

To waste line

Pressure regulator for steam supply

Steam supply

Autoclaving is used to sterilize culture media, instruments, dressings, intravenous equipment, applicators, solutions, syringes, transfusion equipment, and numerous other items that can withstand high temperatures and pressures.

Heat requires extra time to reach the center of solid materials, such as canned meats, because such materials do not develop the efficient heat-distributing convection currents that occur in liquids. Heating large containers also requires extra time. Table 7.4 shows the different time requirements for sterilizing liquids in various container sizes.

Several commercially available methods can indicate whether heat treatment has achieved sterilization. Some of these are chemical reactions in which an indicator changes color when the proper times and temperatures have been reached (**Figure 7.3**). In some designs, the word *sterile* or *autoclaved* appears on wrappings or tapes. This test has a practical advantage; it does not require an incubation time for growth of a test organism, which is important if the sterilized product is intended for immediate consumption. A widely used test consists of preparations of specified species of bacterial endospores impregnated into paper strips. After the strips are autoclaved, they can then be aseptically inoculated into culture media. Growth in the culture media indicates survival of the endospores and therefore inadequate processing. Other designs use endospore suspensions that can be released, after heating, into a surrounding culture medium within the same vial.

Steam under pressure fails to sterilize when the air is not completely exhausted. This can happen with the premature closing of the autoclave's automatic ejector valve (see Figure 7.2). The principles of heat sterilization have a direct bearing on home canning.

| TABLE 7.3 | The Relationship between the Pressure and Temperature of Steam at Sea Level* | |
|---|---|---|
| **Pressure (psi in Excess of Atmospheric Pressure)** | | **Temperature (°C)** |
| 0 | | 100 |
| 5 | | 110 |
| 10 | | 116 |
| 15 | | 121 |
| 20 | | 126 |
| 30 | | 135 |

*At higher altitudes, the atmospheric pressure is less, a phenomenon that must be taken into account in operating an autoclave. For example, to reach sterilizing temperatures (121°C) in Denver, Colorado, whose altitude is 5280 feet (1600 meters), the pressure shown on the autoclave gauge would need to be higher than the 15 psi shown in the table.

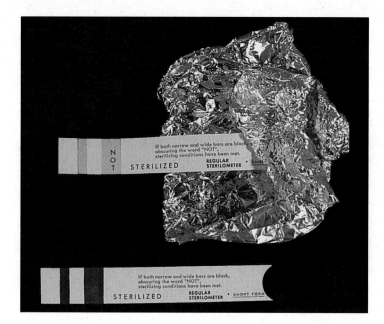

**Figure 7.3 Examples of sterilization indicators.** The strips indicate whether the item has been properly sterilized. The word *NOT* appears if heating has been inadequate. In the illustration, the indicator that was wrapped with aluminum foil was not sterilized because steam couldn't penetrate the foil.

**Q** **What should have been used to wrap the items instead of aluminum foil?**

As anyone familiar with home canning knows, the steam must flow vigorously out of the valve in the lid for several minutes to carry with it all the air before the pressure cooker is sealed. If the air is not completely exhausted, the container will not reach the temperature expected for a given pressure. Because of the possibility of botulism, a kind of food poisoning resulting from improper canning methods (see Chapter 22, page 626), anyone doing home canning should obtain reliable directions and follow them exactly.

### Pasteurization

In the early days of microbiology, Louis Pasteur found a practical method of preventing the spoilage of beer and wine (see Chapter 1). Pasteur used mild heating, which was sufficient to kill the organisms that caused the particular spoilage problem without seriously damaging the taste of the product. The same principle was later applied to milk to produce what we now call pasteurized milk. The intent of **pasteurization** of milk was to eliminate pathogenic microbes. It also lowers microbial numbers, which prolongs milk's good quality under refrigeration. Many relatively heat-resistant (**thermoduric**) bacteria survive pasteurization, but these are unlikely to cause disease or cause refrigerated milk to spoil.

Products other than milk, such as ice cream, yogurt, and beer, all have their own pasteurization times and temperatures, which often differ considerably. There are several reasons for these variations. For example, heating is less efficient in foods that are more viscous, and fats in food can have a protective effect on microorganisms. The dairy industry routinely uses a test to determine whether products have been pasteurized: the *phosphatase test* (phosphatase is an enzyme naturally present in milk). If the product has been pasteurized, phosphatase will have been inactivated.

Most milk pasteurization today uses temperatures of at least 72°C, but for only 15 seconds. This treatment, known as **high-temperature short-time (HTST) pasteurization**, is applied as the milk flows continuously past a heat exchanger. In addition to killing pathogens, HTST pasteurization lowers total bacterial counts, so the milk keeps well under refrigeration.

### Sterilization

Milk can also be sterilized—something quite different from pasteurization—by **ultra-high-temperature (UHT) treatments**. It can then be stored for several months without refrigeration (also see *commercial sterilization*, page 811). UHT-treated milk is widely sold in Europe and is especially necessary in less developed parts of the world where refrigeration facilities are not always available. In the United States, UHT is sometimes used on the small containers of coffee creamers found in restaurants. To avoid giving the milk a cooked taste, the process prevents the milk from touching a surface hotter than the milk itself. Usually, the liquid milk (or juice) is sprayed through a nozzle into a chamber filled with high-temperature steam under pressure. A small volume of fluid sprayed into an atmosphere of high-temperature steam exposes a relatively large surface area on the fluid droplets to heating by the steam; sterilizing temperatures are reached almost instantaneously. After reaching a temperature of 140°C for 4 seconds, the fluid is rapidly cooled in a vacuum chamber. The milk or juice is then packaged in a presterilized, airtight container.

| TABLE 7.4 | The Effect of Container Size on Autoclave Sterilization Times for Liquid Solutions* | |
|---|---|---|
| **Container Size** | **Liquid Volume** | **Sterilization Time (min)** |
| Test tube: 18 × 150 mm | 10 ml | 15 |
| Erlenmeyer flask: 125 ml | 95 ml | 15 |
| Erlenmeyer flask: 2000 ml | 1500 ml | 30 |
| Fermentation bottle: 9000 ml | 6750 ml | 70 |

*Sterilization times in the autoclave include the time for the contents of the containers to reach sterilization temperatures. For smaller containers, this is only 5 min or less, but for a 9000-ml bottle it might be as much as 70 min. Liquids in an autoclave boil vigorously, so their containers usually are filled only up to 75% of capacity.

The heat treatments we have just discussed illustrate the concept of **equivalent treatments**: as the temperature is increased, much less time is needed to kill the same number of microbes. For example, suppose that the destruction of highly resistant endospores might take 70 minutes at 115°C, whereas, in this hypothetical example, only 7 minutes might be needed at 125°C. Both treatments yield the same result.

### Dry Heat Sterilization

Dry heat kills by oxidation effects. A simple analogy is the slow charring of paper in a heated oven, even when the temperature remains below the ignition point of paper. One of the simplest methods of dry heat sterilization is direct **flaming**. You will use this procedure many times in the microbiology laboratory when you sterilize inoculating loops. To effectively sterilize the inoculating loop, you heat the wire to a red glow. A similar principle is used in *incineration*, an effective way to sterilize and dispose of contaminated paper cups, bags, and dressings.

Another form of dry heat sterilization is **hot-air sterilization**. Items to be sterilized by this procedure are placed in an oven. Generally, a temperature of about 170°C maintained for nearly 2 hours ensures sterilization. The longer period and higher temperature (relative to moist heat) are required because the heat in water is more readily transferred to a cool body than is the heat in air. For example, imagine the different effects of immersing your hand in boiling water at 100°C (212°F) and of holding it in a hot-air oven at the same temperature for the same amount of time.

## Filtration

*Filtration* is the passage of a liquid or gas through a screenlike material with pores small enough to retain microorganisms (often the same apparatus used for counting; see Figure 6.18, page 171). A vacuum is created in the receiving flask; air pressure then forces the liquid through the filter. Filtration is used to sterilize heat-sensitive materials, such as some culture media, enzymes, vaccines, and antibiotic solutions.

Some operating theaters and rooms occupied by burn patients receive filtered air to lower the numbers of airborne microbes. **High-efficiency particulate air (HEPA) filters** remove almost all microorganisms larger than about 0.3 μm in diameter.

In the early days of microbiology, hollow candle-shaped filters of unglazed porcelain were used to filter liquids. The long and indirect passageways through the walls of the filter adsorbed the bacteria. Unseen pathogens that passed through the filters (causing such diseases as rabies) were called *filterable viruses*. (See the discussion of filtration in modern water treatment on page 799.)

In recent years, **membrane filters**, composed of such substances as cellulose esters or plastic polymers, have become popular for industrial and laboratory use (**Figure 7.4**). These

**Figure 7.4 Filter sterilization with a disposable, presterilized plastic unit.** The sample is placed into the upper chamber and forced through the membrane filter by a vacuum in the lower chamber. Pores in the membrane filter are smaller than the bacteria, so bacteria are retained on the filter. The sterilized sample can then be decanted from the lower chamber. Similar equipment with removable filter disks is used to count bacteria in samples (see Figure 6.18).

**Q** How is a plastic filtration apparatus presterilized? (Assume the plastic cannot be heat sterilized.)

filters are only 0.1 mm thick. The pores of membrane filters, for example, 0.22-μm and 0.45-μm sizes, are intended for bacteria. Some very flexible bacteria, such as spirochetes, or the wall-less mycoplasma will sometimes pass through such filters, however. Filters are available with pores as small as 0.01 μm, a size that will retain viruses and even some large protein molecules.

## Low Temperatures

The effect of low temperatures on microorganisms depends on the particular microbe and the intensity of the application. For example, at temperatures of ordinary refrigerators (0–7°C), the metabolic rate of most microbes is so reduced that they cannot reproduce or synthesize toxins. In other words, ordinary refrigeration has a bacteriostatic effect. Yet psychrotrophs do grow slowly at refrigerator temperatures and will alter the appearance and taste of foods over time. For example, a single microbe reproducing only three times a day would reach a population of more than 2 million within a week. Pathogenic bacteria generally will not grow at refrigerator temperatures, but *Listeria* is one important exception (see the discussion of listeriosis in Chapter 22, page 623).

Surprisingly, some bacteria can grow at temperatures several degrees below freezing. Most foods remain unfrozen until –2°C or lower. Rapidly attained subfreezing temperatures tend to render microbes dormant but do not necessarily kill them. Slow freezing is more harmful to bacteria; the ice crystals that form and grow disrupt the cellular and molecular structure of the bacteria. Thawing, being inherently slower, is actually the more damaging part of a freeze-thaw cycle. Once frozen, one-third of the population of some vegetative bacteria might survive a year, whereas other species might have very few survivors after this time. Many eukaryotic parasites, such as the roundworms that cause human trichinellosis, are killed by several days of freezing temperatures. Some important temperatures associated with microorganisms and food spoilage are shown in Figure 6.2 (page 153).

## High Pressure

High pressure applied to liquid suspensions is transferred instantly and evenly throughout the sample. If the pressure is high enough, it alters the molecular structures of proteins and carbohydrates, resulting in the rapid inactivation of vegetative bacterial cells. Endospores are relatively resistant to high pressure. Fruit juices preserved by high-pressure treatments are sold in Japan and the United States. An advantage is that these treatments preserve the flavors, colors, and nutrient values of the products.

## Desiccation

In the absence of water, known as **desiccation**, microorganisms cannot grow or reproduce but can remain viable for years. Then, when water is made available to them, they can resume their growth and division. This is the principle that underlies lyophilization, or freeze-drying, a laboratory process for preserving microbes described in Chapter 6 (page 165). Certain foods are also freeze-dried (for example, coffee and some fruit additives for dry cereals).

The resistance of vegetative cells to desiccation varies with the species and the organism's environment. For example, the gonorrhea bacterium can withstand desiccation for only about an hour, but the tuberculosis bacterium can remain viable for months. Viruses are generally resistant to desiccation, but they are not as resistant as bacterial endospores, some of which have survived for centuries. This ability of certain dried microbes and endospores to remain viable is important in a hospital setting. Dust, clothing, bedding, and dressings might contain infectious microbes in dried mucus, urine, pus, and feces.

## Osmotic Pressure

The use of high concentrations of salts and sugars to preserve food is based on the effects of *osmotic pressure*. High concentrations of these substances create a hypertonic environment that causes water to leave the microbial cell (see Figure 6.4, page 154). This process resembles preservation by desiccation, in that both methods deny the cell the moisture it needs for growth. The principle of osmotic pressure is used in the preservation of foods. For example, concentrated salt solutions are used to cure meats, and thick sugar solutions are used to preserve fruits.

As a general rule, molds and yeasts are much more capable than bacteria of growing in materials with low moisture or high osmotic pressures. This property of molds, sometimes combined with their ability to grow under acidic conditions, is the reason that molds, rather than bacteria, cause spoilage of fruits and grains. It is also part of the reason molds are able to form mildew on a damp wall or a shower curtain.

## Radiation

Radiation has various effects on cells, depending on its wavelength, intensity, and duration. Radiation that kills microorganisms (sterilizing radiation) is of two types: ionizing and nonionizing.

**Ionizing radiation**—gamma rays, X rays, and high-energy electron beams—has a wavelength shorter than that of nonionizing radiation, less than about 1 nm. Therefore, it carries much more energy (**Figure 7.5**). *Gamma rays* are emitted by certain radioactive elements such as cobalt. Electron beams are produced by accelerating electrons to high energies in special machines, and *X rays*, which are produced by machines in a manner similar to the production of electron beams, are similar to gamma rays. Gamma rays penetrate deeply but may require hours to sterilize large masses; *high-energy electron beams* have much lower penetrating power but usually require only a few seconds of exposure. The principal effect of ionizing radiation is the ionization of water, which forms highly reactive hydroxyl radicals (see the discussion of toxic forms of oxygen in Chapter 6, pages 156–157). These radicals kill organisms by reacting with organic cellular components, especially DNA, and damaging them.

The so-called target theory of damage by radiation supposes that ionizing particles, or packets of energy, pass through or close to vital portions of the cell; these constitute "hits." One, or a few, hits may only cause nonlethal mutations, some of them conceivably useful. More hits are likely to cause sufficient mutations to kill the microbe.

The food industry is expanding the use of radiation for food preservation (discussed more fully in Chapter 28). Low-level ionizing radiation, used for years in many countries, has been approved in the United States for processing spices and certain meats and vegetables. Ionizing radiation, especially high-energy electron beams, is used to sterilize pharmaceuticals and disposable dental and medical supplies, such as plastic syringes, surgical gloves, suturing materials, and catheters. As a protection against bioterrorism, the postal service often uses electron beam radiation to sterilize certain classes of mail.

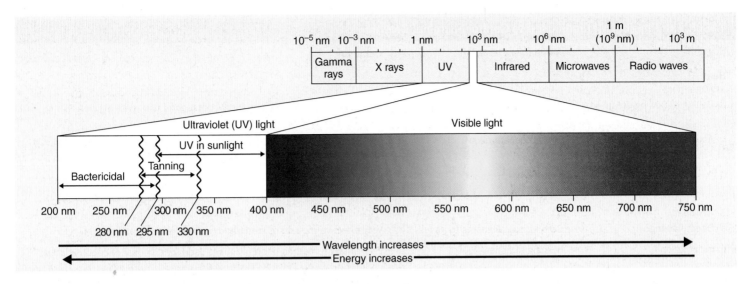

**Figure 7.5  The radiant energy spectrum.** Visible light and other forms of radiant energy radiate through space as waves of various lengths. Ionizing radiation, such as gamma rays and X rays, has a wavelength shorter than 1 nm. Nonionizing radiation, such as ultraviolet (UV) light, has a wavelength between 1 nm and about 380 nm, where the visible spectrum begins.

**Q** How might increased UV radiation (due to decrease in the ozone layer) affect the Earth's ecosystems?

**Nonionizing radiation** has a wavelength longer than that of ionizing radiation, usually greater than about 1 nm. The best example of nonionizing radiation is ultraviolet (UV) light. UV light damages the DNA of exposed cells by causing bonds to form between adjacent pyrimidine bases, usually thymines, in DNA chains (see Figure 8.21, page 223). These *thymine dimers* inhibit correct replication of the DNA during cell reproduction. The UV wavelengths most effective for killing microorganisms are about 260 nm; these specific wavelengths are absorbed by cellular DNA. UV radiation is also used to control microbes in the air. A UV, or "germicidal," lamp is commonly found in hospital rooms, nurseries, operating rooms, and cafeterias. UV light is also used to disinfect vaccines and other medical products. A major disadvantage of UV light as a disinfectant is that the radiation is not very penetrating, so the organisms to be killed must be directly exposed to the rays. Organisms protected by solids and such coverings as paper, glass, and textiles are not affected. Another potential problem is that UV light can damage human eyes, and prolonged exposure can cause burns and skin cancer in humans.

Sunlight contains some UV radiation, but the shorter wavelengths—those most effective against bacteria—are screened out by the ozone layer of the atmosphere. The antimicrobial effect of sunlight is due almost entirely to the formation of singlet oxygen in the cytoplasm (see Chapter 6, page 156). Many pigments produced by bacteria provide protection from sunlight.

Visible blue light (470 nm) kills methicillin-resistant *Staphylococcus aureus* (MRSA) in laboratory cultures and mice. It may be a practical treatment for skin infections.

**Microwaves** do not have much direct effect on microorganisms, and bacteria can readily be isolated from the interior of recently operated microwave ovens. Moisture-containing foods are heated by microwave action, and the heat will kill most vegetative pathogens. Solid foods heat unevenly because of the uneven distribution of moisture. For this reason, pork cooked in a microwave oven has been responsible for outbreaks of trichinellosis.

Table 7.5 summarizes the physical methods of microbial control.

### CHECK YOUR UNDERSTANDING

✔ 7-4 How is microbial growth in canned foods prevented?

✔ 7-5 Why would a can of pork take longer to sterilize at a given temperature than a can of soup that also contained pieces of pork?

✔ 7-6 What is the connection between the killing effect of radiation and hydroxyl radical forms of oxygen?

# Chemical Methods of Microbial Control

### LEARNING OBJECTIVES

7-7  List the factors related to effective disinfection.

7-8  Interpret the results of use-dilution tests and the disk-diffusion method.

7-9  Identify the methods of action and preferred uses of chemical disinfectants.

**TABLE 7.5 Physical Methods Used to Control Microbial Growth**

| Methods | Mechanism of Action | Comment |
|---|---|---|
| **Heat** | | |
| 1. Moist heat | | |
| a. Boiling or flowing steam | Protein denaturation | Kills vegetative bacterial and fungal pathogens and many viruses within 10 min; less effective on endospores. |
| b. Autoclaving | Protein denaturation | Very effective method of sterilization; at about 15 psi of pressure (121°C), all vegetative cells and their endospores are killed in about 15 min. |
| 2. Pasteurization | Protein denaturation | Heat treatment for milk (72°C for about 15 sec) that kills all pathogens and most nonpathogens. |
| 3. Dry heat | | |
| a. Direct flaming | Burning contaminants to ashes | Very effective method of sterilization. Used for inoculating loops. |
| b. Incineration | Burning to ashes | Very effective method of sterilization. Used for disposal of contaminated dressings, animal carcasses, and paper. |
| c. Hot-air sterilization | Oxidation | Very effective method of sterilization but requires temperature of 170°C for about 2 hr. Used for empty glassware. |
| **Filtration** | Separation of bacteria from suspending liquid | Removes microbes by passage of a liquid or gas through a screenlike material; most filters in use consist of cellulose acetate or nitrocellulose. Useful for sterilizing liquids (e.g., enzymes, vaccines) that are destroyed by heat. |
| **Cold** | | |
| 1. Refrigeration | Decreased chemical reactions and possible changes in proteins | Has a bacteriostatic effect. |
| 2. Deep-freezing (see Chapter 6, page 165) | Decreased chemical reactions and possible changes in proteins | An effective method for preserving microbial cultures, food, and drugs. |
| 3. Lyophilization (see Chapter 6, page 165) | Decreased chemical reactions and possible changes in proteins | Most effective method for long-term preservation of microbial cultures, food, and drugs. |
| **High Pressure** | Alteration of molecular structure of proteins and carbohydrates | Preserves colors, flavors, nutrient values of fruit juices. |
| **Desiccation** | Disruption of metabolism | Involves removing water from microbes; primarily bacteriostatic. |
| **Osmotic Pressure** | Plasmolysis | Results in loss of water from microbial cells. |
| **Radiation** | | |
| 1. Ionizing | Destruction of DNA | Used for sterilizing pharmaceuticals and medical and dental supplies. |
| 2. Nonionizing | Damage to DNA | Radiation is not very penetrating. |

7-10 Differentiate halogens used as antiseptics from halogens used as disinfectants.

7-11 Identify the appropriate uses for surface-active agents.

7-12 List the advantages of glutaraldehyde over other chemical disinfectants.

7-13 Identify chemical sterilizers.

Chemical agents are used to control the growth of microbes on both living tissue and inanimate objects. Unfortunately, few chemical agents achieve sterility; most of them merely reduce microbial populations to safe levels or remove vegetative forms of pathogens from objects. A common problem in disinfection is the selection of an agent. No single disinfectant is appropriate for all circumstances.

## Principles of Effective Disinfection

By reading the label, we can learn a great deal about a disinfectant's properties. Usually the label indicates what groups of organisms the disinfectant is effective against. Remember that the concentration of a disinfectant affects its action, so it should always be diluted exactly as specified by the manufacturer.

Also consider the nature of the material being disinfected. For example, are organic materials present that might interfere with the action of the disinfectant? Similarly, the pH of the medium often has a great effect on a disinfectant's activity.

Another very important consideration is whether the disinfectant will easily make contact with the microbes. An area might need to be scrubbed and rinsed before the disinfectant is applied. In general, disinfection is a gradual process. Thus, to be effective, a disinfectant might need to be left on a surface for several hours.

## Evaluating a Disinfectant

### Use-Dilution Tests

There is a need to evaluate the effectiveness of disinfectants and antiseptics. The current standard is the American Official Analytical Chemist's **use-dilution test**. Metal or glass cylinders (8 mm × 10 mm) are dipped into standardized cultures of the test bacteria grown in liquid media, removed, and dried at 37°C for a short time. The dried cultures are then placed into a solution of the disinfectant at the concentration recommended by the manufacturer and left there for 10 minutes at 20°C. Following this exposure, the cylinders are transferred to a medium that permits the growth of any surviving bacteria. The number of cultures that grow indicates the effectiveness of the disinfectant.

Variations of this method are used for testing the effectiveness of antimicrobial agents against endospores, viruses, fungi, and mycobacteria that cause tuberculosis, because they are difficult to control with chemicals. Also, tests of antimicrobials intended for special purposes, such as dairy utensil disinfection, can substitute for other test bacteria.

### The Disk-Diffusion Method

The **disk-diffusion method** is used in teaching laboratories to evaluate the efficacy of a chemical agent. A disk of filter paper is soaked with a chemical and placed on an agar plate that has been previously inoculated and incubated with the test organism. After incubation, if the chemical is effective, a clear zone representing inhibition of growth can be seen around the disk (**Figure 7.6**).

Disks containing antibiotics are commercially available and used to determine microbial susceptibility to antibiotics (see Figure 20.17, page 578).

## Types of Disinfectants

### Phenol and Phenolics

Lister was the first to use **phenol** (carbolic acid) to control surgical infections in the operating room. Its use had been suggested by its effectiveness in controlling odor in sewage. It is now rarely used as an antiseptic or disinfectant because it irritates the skin and has a disagreeable odor. It is often used in throat lozenges for its local anesthetic effect but has little antimicrobial effect at the low concentrations used. At concentrations above 1% (such as in some throat sprays), however,

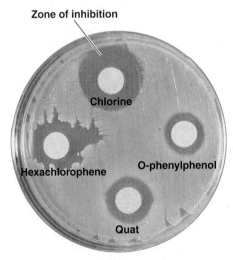

*Staphylococcus aureus*
(gram-positive)

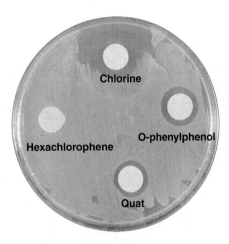

*Escherichia coli*
(gram-negative)

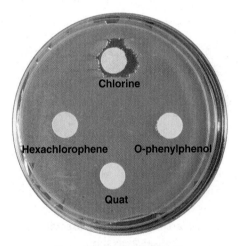

*Pseudomonas aeruginosa*
(gram-negative)

**Figure 7.6 Evaluation of disinfectants by the disk-diffusion method.** In this experiment, paper disks are soaked in a solution of disinfectant and placed on the surface of a nutrient medium on which a culture of test bacteria has been spread to produce uniform growth.

At the top of each plate, the tests show that chlorine (as sodium hypochlorite) was effective against all the test bacteria but was more effective against gram-positive bacteria.

At the bottom row of each plate, the tests show that the quaternary ammonium compound ("quat") was also more effective against the gram-positive bacteria, but it did not affect the pseudomonads at all.

At the left side of each plate, the tests show that hexachlorophene was effective against gram-positive bacteria only.

At the right sides, O-phenylphenol was ineffective against pseudomonads but was almost equally effective against the

gram-positive bacteria and the gram-negative bacteria.

All four chemicals worked against the gram-positive test bacteria, but only one of the four chemicals affected pseudomonads.

**Q** Why are the pseudomonads less affected by the four chemicals shown in the figure?

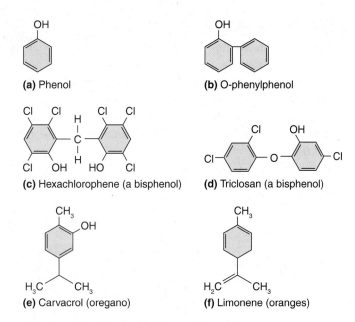

**Figure 7.7  The structure of phenol (a), phenolics (b and e), bisphenols (c and d), and terpenes (f).** Terpenes lack the hydroxyl group of phenol.

**Q** Spice traders became wealthy in the fifteenth century because spices were more valuable than gold. Why were spices so important?

phenol has a significant antibacterial effect. The structure of a phenol molecule is shown in **Figure 7.7a**.

Derivatives of phenol, called **phenolics**, contain a molecule of phenol that has been chemically altered to reduce its irritating qualities or increase its antibacterial activity in combination with a soap or detergent. Phenolics exert antimicrobial activity by injuring lipid-containing plasma membranes, which results in leakage of cellular contents. The cell wall of mycobacteria, the causes of tuberculosis and leprosy, are rich in lipids, which make them susceptible to phenol derivatives. A useful property of phenolics as disinfectants is that they remain active in the presence of organic compounds, are stable, and persist for long periods after application. For these reasons, phenolics are suitable agents for disinfecting pus, saliva, and feces.

One of the most frequently used phenolics is derived from coal tar, a group of chemicals called *cresols*. A very important cresol is *O-phenylphenol* (see Figure 7.6 and Figure 7.7b), the main ingredient in most formulations of Lysol®. Cresols are very good surface disinfectants.

## Bisphenols

**Bisphenols** are derivatives of phenol that contain two phenolic groups connected by a bridge (*bis* indicates *two*). One bisphenol, *hexachlorophene* (Figure 7.6 and Figure 7.7c), is an ingredient of a prescription lotion, pHisoHex®, used for surgical and

hospital microbial control procedures. Gram-positive staphylococci and streptococci, which can cause skin infections in newborns, are particularly susceptible to hexachlorophene, so it is often used to control such infections in nurseries.

Another widely used bisphenol is *triclosan* (Figure 7.7d), an ingredient in antibacterial soaps, toothpastes, and mouthwashes. Triclosan has even been incorporated into kitchen cutting boards and the handles of knives and other plastic kitchenware. Its use is now so widespread that resistant bacteria have been reported, and concerns have been raised that bacterial resistance to triclosan may also lead to resistance to certain antibiotics. This, along with concerns regarding triclosan's effects on the human microbiome, led the U.S. Food and Drug Administration to ban its use in most products for home consumption (see Exploring the Microbiome). The one exception is toothpaste; triclosan has been shown to reduce plaque and gingivitis.

Triclosan inhibits an enzyme needed for the biosynthesis of fatty acids (lipids), which mainly affects the integrity of the plasma membrane. It is especially effective against gram-positive bacteria but there is no evidence that triclosan washes are any better than soap and water.

## Biguanides

Biguanides have a broad spectrum of activity, with a mode of action primarily affecting bacterial cell membranes. They are especially effective against gram-positive bacteria. Biguanides are also effective against gram-negative bacteria, with the significant exception of most pseudomonads. Biguanides are not sporicidal but have some activity against enveloped viruses. The best-known biguanide is *chlorhexidine*, which is frequently used for microbial control on skin and mucous membranes. Combined with a detergent or alcohol, chlorhexidine is very often used for surgical hand scrubs and preoperative skin preparation in patients. *Alexidine* is a similar biguanide and is more rapid in its action than chlorhexidine. Eventually, alexidine is expected to replace povodone-iodine in many applications (see below).

## Essential Oils

**Essential oils (EOs)** are a mixture of hydrocarbons extracted from plants. You may be familiar with many EOs—peppermint oil, pine oil, and orange oil are examples. EOs were used for centuries in traditional medicine and for preserving food. They are now undergoing renewed interest because they are not toxic in the concentrations used, have pleasant odors, and are biodegradable. Their antimicrobial action is primarily due to phenolics (Figure 7.7e) and terpenes (Figure 7.7f). The method of action of EOs is similar to that of phenolics. There is a wide range of antimicrobial activity in these plant oils, corresponding to the types of plants from which they are derived.

# Antimicrobial Soaps: Doing More Harm Than Good?

Staphylococcus aureus is a normal member of the human microbiome, found on the skin and in the nose. S. aureus is also a significant cause of healthcare-associated infections in patients. The bacterium can switch from benign member of the skin community to a disease-causing pathogen if it gains entry to the body through a wound.

Since most hospital-acquired S. aureus infections are endogenous—that is, caused by bacteria that have colonized in or on the body before someone became a patient—hospitals have long used a disinfectant called triclosan in clinical soaps and skin lotions to prevent staphylococcal infections. Over the years, triclosan was also added to many household products, such as dishwashing detergent, toothpastes, and body washes. However, using these antimicrobial products daily seems to be a case of "too much of a good thing."

Triclosan enters the blood and is excreted in urine. Therefore, triclosan can be found in many areas of the body, including the nasal mucosa, of people who use it. The nose is the primary habitat of S. aureus. In an example of unintended consequences, presence of triclosan in blood is also associated with nasal colonization by S. aureus. S. aureus is more likely to bind to host-cell-membrane proteins in the presence of triclosan. Moreover, constant exposure to triclosan selects for triclosan-resistant mutants over generations of bacterial growth. Triclosan-resistant bacteria avoid death by removing the chemical from their cells using transporter proteins. These transporters can also remove some antibiotics from the bacterial cells. Morever, methicillin-resistant S. aureus (MRSA) is more resistant to triclosan than methicillin-sensitive staphylococci.

Starting in late 2016, the Federal Drug Association banned triclosan from over-the-counter consumer washing products. The American Medical Association recommends using plain soap and water and proper handwashing techniques instead—

these products and techniques remove microbes without the harmful unintended consequences associated with widespread triclosan use.

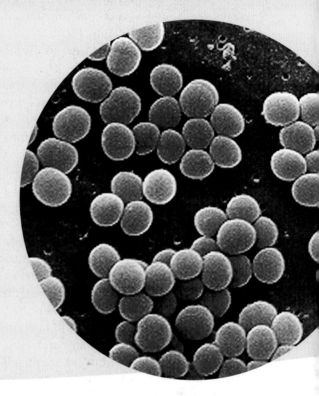

**Staphylococcus aureus.**

Generally, EOs have greater activity against gram-positive bacteria than against gram-negative bacteria. Pine oil and tea tree oil have a broad spectrum of activity, including killing gram-negative bacteria and fungi. Some EOs are used to disinfect hard surfaces such as countertops, and some can be used on skin. Their effectiveness against viruses has not been studied.

## Halogens

The **halogens**, particularly iodine and chlorine, are effective antimicrobial agents, both alone and as constituents of inorganic or organic compounds. *Iodine* ($I_2$) is one of the oldest and most effective antiseptics. It is active against all kinds of bacteria, many endospores, various fungi, and some viruses. Iodine impairs protein synthesis and alters cell membranes, apparently by forming complexes with amino acids and unsaturated fatty acids.

Iodine is available as a **tincture**—that is, in solution in aqueous alcohol—and as an iodophor. An **iodophor** is a combination of iodine and an organic molecule, from which the iodine is released slowly. Iodophors have the antimicrobial activity of iodine, but they do not stain and are less irritating.

The most common commercial preparation is Betadine®, which is a *povidone-iodine*. Povidone is a surface-active iodophor that improves the wetting action and serves as a reservoir of free iodine. Iodines are used mainly for skin disinfection and wound treatment. Many campers are familiar with using iodine for water treatment; it is even used to disinfect drinking water on the International Space Station.

*Chlorine* ($Cl_2$), as a gas or in combination with other chemicals, is another widely used disinfectant. Its germicidal action is caused by the hypochlorous acid (HOCl) that forms when chlorine is added to water:

(1)
$$Cl_2 + H_2O \rightleftharpoons H^+ + Cl^- + HOCl$$

| Chlorine | Water | Hydrogen ion | Chloride ion | Hypochlorous acid |

(2)
$$HOCl \rightleftharpoons H^+ + OCl^-$$

| Hypochlorous acid | Hydrogen ion | Hypochlorite ion |

Hypochlorite is a strong oxidizing agent that prevents much of the cellular enzyme system from functioning. Hypochlorous acid is the most effective form of chlorine because it is neutral in electrical charge and diffuses as rapidly as water through the cell wall. Because of its negative charge, the hypochlorite ion (OCl⁻) cannot enter the cell freely.

A liquid form of compressed chlorine gas is used extensively for disinfecting municipal drinking water, water in swimming pools, and sewage. Several compounds of chlorine are also effective disinfectants. For example, solutions of *calcium hypochlorite* $[Ca(OCl)_2]$ are used to disinfect dairy equipment and restaurant eating utensils. This compound, once called chloride of lime, was used as early as 1825—long before the concept of a germ theory for disease—to soak hospital dressings in Paris hospitals. It was also the disinfectant Semmelweis used in the 1840s to control hospital infections during childbirth, as mentioned in Chapter 1, page 10. Another chlorine compound, *sodium hypochlorite* (NaOCl; see Figure 7.6), is used as a household disinfectant and bleach (Clorox®), and as a disinfectant in dairies, food-processing establishments, and hemodialysis systems. When the quality of drinking water is in question, household bleach can provide a rough equivalent of municipal chlorination. After two drops of bleach are added to a liter of water (four drops if the water is cloudy) and the mixture has sat for 30 minutes, the water is considered safe for drinking under emergency conditions.

*Chloramine* $(NH_2Cl)$ is used to treat drinking water in several cities in the United States and Europe. (Chloramine is toxic to aquarium fish, but pet shops sell chemicals to neutralize it.) U.S. military forces in the field are issued tablets (Chlor-Floc®) that contain a chloramine combined with an agent that flocculates (coagulates) suspended materials in a water sample, causing them to settle out, clarifying the water. Chloramine is also used to sanitize eating utensils and food-manufacturing equipment. Chloramine is a relatively stable compound that releases chlorine over long periods. It is more stable and less irritating than hypochlorite or chlorine gas. Chloramine is less likely to react with organic compounds in the water than hypochlorite.

## Alcohols

**Alcohols** effectively kill bacteria and fungi but not endospores and nonenveloped viruses. Alcohol usually denatures protein, but it can also disrupt membranes and dissolve many lipids, including the lipid component of enveloped viruses. One advantage of alcohols is that they act and then evaporate rapidly, leaving no residue. When the skin is swabbed (degermed) before an injection, most of the microbial control activity comes from simply wiping away dirt and microorganisms, along with skin oils. However, alcohols are unsatisfactory antiseptics when applied to wounds. They cause coagulation of a layer of protein under which bacteria continue to grow.

**TABLE 7.6** Biocidal Action of Various Concentrations of Ethanol in Aqueous Solution against *Streptococcus pyogenes*

| Concentration of Ethanol (%) | Time of Exposure (sec) | | | | |
|---|---|---|---|---|---|
| | 10 | 20 | 30 | 40 | 50 |
| 100 | G | G | G | G | G |
| 95 | NG | NG | NG | NG | NG |
| 90 | NG | NG | NG | NG | NG |
| 80 | NG | NG | NG | NG | NG |
| 70 | NG | NG | NG | NG | NG |
| 60 | NG | NG | NG | NG | NG |
| 50 | G | G | NG | NG | NG |
| 40 | G | G | G | G | G |

Note:
G = growth
NG = no growth

Two of the most commonly used alcohols are ethanol and isopropanol. The recommended optimum concentration of *ethanol* is 70%, but concentrations between 60% and 95% seem to kill as well (Table 7.6). Pure ethanol is less effective than aqueous solutions (ethanol mixed with water) because denaturation requires water. *Isopropanol*, often sold as rubbing alcohol, is slightly superior to ethanol as an antiseptic and disinfectant. Moreover, it is less volatile, less expensive, and more easily obtained than ethanol.

Alcohol-based (about 62% alcohol) hand sanitizers such as Purell® and Germ-X® are very popular for use on hands that are not visibly soiled; the product should be rubbed over the surfaces of the hands and fingers until they are dry. Claims that products will kill 99.9% of germs should be viewed with caution; such effectiveness is seldom reached under typical user's conditions. Also, certain pathogens, such as the spore-forming *Clostridium difficile* and viruses that lack a lipid envelope, are comparatively resistant to alcohol-based hand sanitizers.

Ethanol and isopropanol are often used to enhance the effectiveness of other chemical agents. For example, an aqueous solution of Zephiran® (described on page 194) kills about 40% of the population of a test organism in 2 minutes, whereas a tincture of Zephiran® kills about 85% in the same period. To compare the effectiveness of tinctures and aqueous solutions, see Figure 7.10.

## Heavy Metals and Their Compounds

Several heavy metals, including silver, mercury, and copper, can be biocidal or antiseptic. The ability of very small amounts of heavy metals, especially silver and copper, to exert antimicrobial

**Figure 7.8 Oligodynamic action of heavy metals.** Clear zones where bacterial growth has been inhibited are seen around the sombrero charm (pushed aside), the dime, and the penny. The charm and the dime contain silver; the penny contains copper.

**Q** The coins used in this demonstration were minted many years ago; why were more contemporary coins not used?

activity is referred to as **oligodynamic action** (*oligo* means few). Centuries ago, Egyptians found that putting silver coins in water barrels served to keep the water clean of unwanted organic growths. This action can be demonstrated by placing a coin or other clean piece of metal containing silver or copper on a culture on an inoculated Petri plate. Extremely small amounts of metal diffuse from the coin and inhibit the growth of bacteria for some distance around the coin (**Figure 7.8**). This effect is produced by the action of heavy-metal ions on microbes. When the metal ions combine with the sulfhydryl groups on cellular proteins, denaturation results.

Silver is used as an antiseptic in a 1% *silver nitrate* solution. At one time, many states required that the eyes of newborns be treated with a few drops of silver nitrate to guard against an infection of the eyes called ophthalmia neonatorum, which the infants might have contracted as they passed through the birth canal. In recent years, antibiotics have replaced silver nitrate for this purpose.

Silver-impregnated dressings that slowly release silver ions have proven especially useful against antibiotic-resistant bacteria. The enthusiasm for incorporating silver in all manner of consumer products is increasing. Among the newer products being sold are plastic food containers infused with silver nanoparticles, which are intended to keep food fresher, and silver-infused athletic shirts and socks, which are claimed to minimize odors.

A combination of silver and the drug sulfadiazine, *silver-sulfadiazine*, is the most common formulation. It is available as a topical cream for use on burns. Silver can also be incorporated into indwelling catheters, which are a common source

of hospital infections, and in wound dressings. Surfacine® is a relatively new antimicrobial for application to surfaces, either animate or inanimate. It contains water-insoluble silver iodide in a polymer carrier and is very persistent, lasting at least 13 days. When a bacterium contacts the surface, the cell's outer membrane is recognized, and a lethal amount of silver ions is released.

Inorganic mercury compounds, such as *mercuric chloride*, have a long history of use as disinfectants. They have a very broad spectrum of activity; their effect is primarily bacteriostatic. However, their use is now limited because of their toxicity and ineffectiveness in organic matter. At present, the primary use of mercurials is to control mildew in exterior latex paints.

Copper in the form of *copper sulfate* or other copper-containing additives is used chiefly to destroy green algae (algicide) that grow in reservoirs, stock ponds, swimming pools, and fish tanks. If the water does not contain excessive organic matter, copper compounds are effective in concentrations of one part per million of water. To prevent mildew, copper compounds such as *copper 8-hydroxyquinoline* are sometimes included in paint. Mixtures based on copper ions (known as Bordeaux mixture) have long been used to control fungal diseases of plants.

Copper and silver ions are used to disinfect drinking water and swimming pools and to control *Legionella* in hospital water supplies. Copper and silver electrodes release $Ag^+$ and $Cu^{2+}$ ions when an electric current is applied. Silver-impregnated dressings that slowly release silver ions have proven especially useful against antibiotic-resistant bacteria. These have been used for wound dressings and to prevent growth of biofilms on indwelling devices such as catheters.

Long-term use of alcohol-based hand sanitizers often causes problems with skin dryness. A relatively new hand sanitizer, Xgel, does not contain alcohol but uses copper contained in a skin lotion formulation. Xgel may be more effective as an antimicrobial than alcohol-based hand sanitizers.

Another metal used as an antimicrobial is zinc. The effect of trace amounts of zinc can be seen on weathered roofs of buildings downslope from galvanized (zinc-coated) fittings. The color of the roof is lighter where biological growth, mostly algae, is impeded. Copper- and zinc-treated roofing shingles are available. *Zinc chloride* is a common ingredient in mouthwashes, and *zinc pyrithione* is an ingredient in antidandruff shampoos.

## Surface-Active Agents

**Surface-active agents**, or **surfactants**, can decrease surface tension among molecules of a liquid. Such agents include soaps and detergents.

**Soaps and Detergents** Soap has little value as an antiseptic, but it does have an important function in the mechanical removal of microbes through scrubbing. The skin normally contains

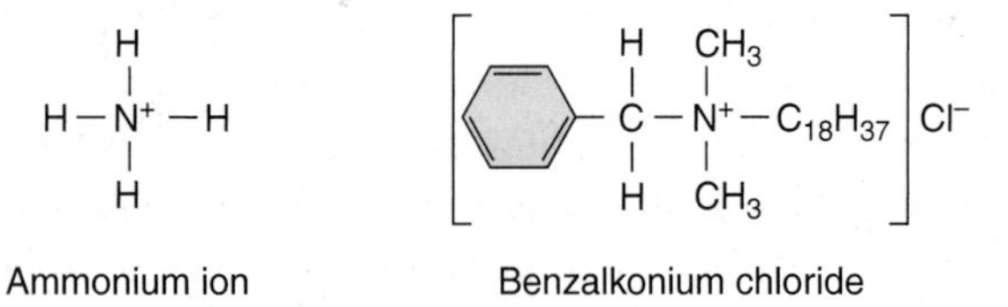

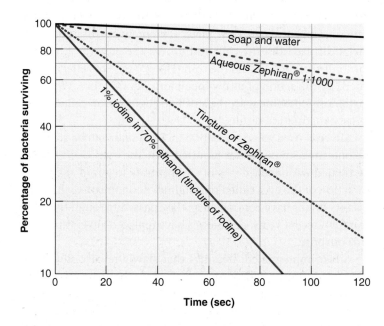

Ammonium ion          Benzalkonium chloride

**Figure 7.9 The ammonium ion and a quaternary ammonium compound, benzalkonium chloride (Zephiran®).** Notice how other groups replace the hydrogens of the ammonium ion.

**Q** Are quats more effective against gram-positive or gram-negative bacteria?

dead cells, dust, dried sweat, microbes, and oily secretions from oil glands. Soap breaks the oily film into tiny droplets, a process called *emulsification*, and the water and soap together lift up the emulsified oil and debris and float them away as the lather is washed off. In this sense, soaps are good degerming agents. Washing hands with soap and water is an effective sanitation method. Use soap and *warm* water (if possible), and rub hands together for 20 seconds (imagine singing "Happy Birthday" twice through). Then rinse, dry with a paper towel or air dryer, and try to use a paper towel to turn off the faucet.

**Acid-Anionic Sanitizers** *Acid-anionic* sanitizers are very important in cleaning food-processing facilities, especially dairy utensils and equipment. They are usually combinations of phosphoric acid with a surface-active agent. Their sanitizing ability is related to the negatively charged portion (anion) of the molecule, which reacts with the plasma membrane. These sanitizers, which act on a wide spectrum of microbes, including troublesome thermoduric bacteria, are odorless, nontoxic, noncorrosive, and fast acting.

**Quaternary Ammonium Compounds (Quats)** The most widely used surface-active agents are the cationic detergents, especially the **quaternary ammonium compounds (quats)**. Their cleansing ability is related to the positively charged portion— the cation—of the molecule. Their name is derived from the fact that they are modifications of the four-valence ammonium ion, $NH_4^+$ (**Figure 7.9**). Quaternary ammonium compounds are strongly bactericidal against gram-positive bacteria and less active against gram-negative bacteria (see Figure 7.6).

Quats are also fungicidal, amebicidal, and virucidal against enveloped viruses. They do not kill endospores or mycobacteria. (See the Clinical Focus box on page 197.) Their chemical mode of action is unknown, but they probably affect the plasma membrane. They change the cell's permeability and cause the loss of essential cytoplasmic constituents, such as potassium.

Two popular quats are Zephiran®, a brand name of *benzalkonium chloride* (see Figure 7.9), and Cēpacol®, a brand name of *cetylpyridinium chloride*. They are strongly antimicrobial,

colorless, odorless, tasteless, stable, easily diluted, and nontoxic, except at high concentrations. If your mouthwash bottle fills with foam when shaken, the mouthwash probably contains a quat. However, organic matter interferes with their activity, and quats are rapidly neutralized by soaps and anionic detergents.

Anyone involved in medical applications of quats should remember that certain bacteria, such as some species of *Pseudomonas*, not only survive in quaternary ammonium compounds but actively grow in them. These microbes are resistant not only to the disinfectant solution but also to gauze and bandages moistened with it, because the fibers tend to neutralize the quats.

Before we move on to the next group of chemical agents, refer to **Figure 7.10**, which compares the effectiveness of some of the antiseptics we have discussed so far.

### Chemical Food Preservatives

Chemical preservatives are frequently added to foods to retard spoilage. *Sulfur dioxide* ($SO_2$) has long been used as a disinfectant, especially in winemaking. Homer's *Odyssey*, written nearly 2800 years ago, mentions its use. Among the more common additives are sodium benzoate, sorbic acid, and calcium propionate. These chemicals are simple organic acids, or salts of organic acids, which the body readily metabolizes and

**Figure 7.10 A comparison of the effectiveness of various antiseptics.** The steeper the downward slope of the killing curve, the more effective the antiseptic. A 1% iodine in 70% ethanol solution is the most effective; soap and water are the least effective. Notice that a tincture of Zephiran® is more effective than an aqueous solution of the same antiseptic.

**Q** Why is the tincture of Zephiran® more effective than the aqueous solution?

which are generally judged to be safe in foods. *Sorbic acid*—or its more soluble salt, *potassium sorbate*—and *sodium benzoate* prevent molds from growing in certain acidic foods, such as cheese and soft drinks. Such foods, usually with a pH of 5.5 or lower, are most susceptible to spoilage by molds. *Calcium propionate*, an effective fungistat used in bread, prevents the growth of surface molds and the *Bacillus* bacteria that produce the mucuslike secretion that cause ropy bread. These organic acids inhibit mold growth, not by affecting the pH but by interfering with the mold's metabolism or the integrity of the plasma membrane.

*Sodium nitrate* and *sodium nitrite* are added to many meat products, such as ham, bacon, hot dogs, and sausage. The active ingredient is sodium nitrite, which certain bacteria in the meats can also produce from sodium nitrate by anaerobic respiration. The nitrite has two main functions: to preserve the pleasing red color of the meat by reacting with blood components in the meat, and to prevent the germination and growth of any botulism endospores that might be present. Nitrite selectively inhibits certain iron-containing enzymes of *Clostridium botulinum*. There has been some concern that the reaction of nitrites with amino acids can form certain carcinogenic products known as **nitrosamines**, and the amount of nitrites added to foods has generally been reduced in recent years for this reason. However, the use of nitrites continues because of their established value in preventing botulism. Because nitrosamines are formed in the body from other sources, the added risk posed by a limited use of nitrates and nitrites in meats is lower than was once thought.

## Antibiotics

The antimicrobials discussed in this chapter are not useful for ingestion or injection to treat disease. Antibiotics are used for this purpose. At least two antibiotics have considerable use in food preservation. Neither is of value for clinical purposes. *Nisin* is often added to cheese to inhibit the growth of certain endospore-forming spoilage bacteria. It is an example of a bacteriocin, a protein that is produced by one bacterium and inhibits another (see Chapter 8, page 228). Nisin is present naturally in small amounts in many dairy products. It is tasteless, readily digested, and nontoxic. *Natamycin* (pimaricin) is an antifungal antibiotic approved for use in foods, mostly cheese.

## Aldehydes

**Aldehydes** are among the most effective antimicrobials. Two examples are formaldehyde and glutaraldehyde. They inactivate proteins by forming covalent cross-links with several organic functional groups on proteins (—NH$_2$ —OH, —COOH, and —SH). *Formaldehyde gas* is an excellent disinfectant. However, it is more commonly available as *formalin*, a 37% aqueous solution of formaldehyde gas. Formalin was once used extensively to preserve biological specimens and inactivate bacteria and viruses in vaccines.

*Glutaraldehyde* is a chemical relative of formaldehyde that is less irritating and more effective than formaldehyde. Glutaraldehyde is used to disinfect hospital instruments, including endoscopes and respiratory therapy equipment, but they must be carefully cleaned first. When used in a 2% solution (Cidex®), glutaraldehyde is bactericidal, tuberculocidal, and virucidal in 10 minutes and sporicidal in 3 to 10 hours. Glutaraldehyde is one of the few liquid chemical disinfectants that can be considered a sterilizing agent. For practical purposes, 30 minutes is often considered the maximum time allowed for a sporicide to act, which is a criterion glutaraldehyde cannot meet. Both glutaraldehyde and formalin are used by morticians for embalming.

A possible replacement for glutaraldehyde for many uses is *ortho-phthalaldehyde* (OPA), which is more effective against many microbes and has fewer irritating properties.

## Chemical Sterilization

Sterilization with liquid chemicals is possible, but even sporicidal chemicals such as glutaraldehyde are usually not considered to be practical sterilants. However, the gaseous chemosterilants are frequently used as substitutes for physical sterilization processes. Their application requires a closed chamber similar to a steam autoclave. Probably the most familiar example is *ethylene oxide*:

$$H_2C—CH_2$$
$$\diagdown\diagup$$
$$O$$

Norovirus, a nonenveloped virus, is one cause of acute gastroenteritis. It can be spread by consuming fecally contaminated food or water, coming in direct contact with an infected person, or touching a contaminated surface. Amy is able to rule out foodborne transmission immediately because the small private school does not have a school lunch program; all students and staff bring their lunches from home. After meeting with the principal, Amy speaks to the custodial staff and directs them to use a quat to clean the school. She asks them to pay special attention to areas with high potential for fecal contamination, especially toilet seats, flush handles, toilet stall inner door handles, and restroom door inner handles. Amy is sure she has avoided a major outbreak, but by Friday, 42 students and six more staff members call in to report similar symptoms.

**Why didn't the quat work to kill the virus?**

179  195  197  200

Its activity depends on *alkylation*, that is, replacing the proteins' labile hydrogen atoms in a chemical group (such as —SH, —COOH, or —CH₂CH₂OH) with a chemical radical. This leads to cross-linking of nucleic acids and proteins and inhibits vital cellular functions. Ethylene oxide kills all microbes and endospores but requires a lengthy exposure period of several hours. It is toxic and explosive in its pure form, so it is usually mixed with a nonflammable gas, such as carbon dioxide. Among its advantages is that it carries out sterilization at ambient temperatures and is highly penetrating. Larger hospitals often are able to sterilize even mattresses in special ethylene oxide sterilizers.

*Chlorine dioxide* is a short-lived gas that is usually manufactured at the place of use. Notably, it has been used to fumigate enclosed building areas contaminated with endospores of anthrax. It is much more stable in aqueous solution. Its most common use is in water treatment prior to chlorination, where its purpose is to remove, or reduce the formation of, certain carcinogenic compounds sometimes formed in the chlorination of water.

The food-processing industry makes wide use of chlorine dioxide solution as a surface disinfectant because it does not leave residual tastes or odors. As a disinfectant, it has a broad spectrum of activity against bacteria and viruses and at high concentrations is even effective against cysts and endospores. At low concentrations, chlorine dioxide can be used as an antiseptic.

### Plasmas

In addition to the traditional three states of matter—liquid, gas, and solid—a fourth state of matter exists, called plasma. **Plasma** is a state of matter in which a gas is excited, in this case by an electromagnetic field, to make a mixture of nuclei with assorted electrical charges and free electrons. Health care facilities are increasingly facing the challenge of sterilizing metal or plastic surgical instruments used for many newer procedures in arthroscopic or laparoscopic surgery. Such devices have long, hollow tubes, many with an interior diameter of only a few millimeters, and are difficult to sterilize. *Plasma sterilization* is a reliable method for this. The instruments are placed in a container in which a combination of a vacuum, electromagnetic field, and chemicals such as hydrogen peroxide (sometimes with peracetic acid, as well) form the plasma. Such plasmas have many free radicals that quickly destroy even endospore-forming microbes. The advantage of plasma sterilization, which has elements of both physical and chemical sterilization, is that it requires only low temperatures, but it is relatively expensive.

### Supercritical Fluids

The use of supercritical fluids in sterilization combines chemical and physical methods. When carbon dioxide is compressed into a "supercritical" state, it has properties of both a liquid (with increased solubility) and a gas (with a lowered surface tension). Organisms exposed to *supercritical carbon dioxide* are inactivated, including most vegetative organisms that cause spoilage and foodborne pathogens. Even endospore inactivation requires a temperature of only about 45°C. Used for a number of years in treating certain foods, supercritical carbon dioxide has more recently been used to decontaminate medical implants, such as bone, tendons, or ligaments taken from donor patients.

### Peroxygens and Other Forms of Oxygen

**Peroxygens** are a group of oxidizing agents that includes hydrogen peroxide and peracetic acid.

Hydrogen peroxide is an antiseptic found in many household medicine cabinets and in hospital supply rooms. It is not a good antiseptic for open wounds. It is quickly broken down to water and gaseous oxygen by the action of the enzyme catalase, which is present in human cells (see Chapter 6, page 157). However, hydrogen peroxide does effectively disinfect inanimate objects; in such applications, it is even sporicidal at high concentrations. On a nonliving surface, the normally protective enzymes of aerobic bacteria and facultative anaerobes are overwhelmed by high concentrations of peroxide.

Because of these factors, and its rapid degradation into harmless water and oxygen, the food industry is increasing its use of hydrogen peroxide for aseptic packaging (see page 812). The packaging material passes through a hot solution of the chemical before being assembled into a container. In addition, many wearers of contact lenses are familiar with hydrogen peroxide's use as a disinfectant. After the lens is disinfected, a platinum catalyst in the lens-disinfecting kit destroys residual hydrogen peroxide so that it does not persist on the lens, where it might cause eye irritation. Hydrogen peroxide is used to sterilize components of spacecraft before flight to prevent contamination of landing sites.

Heated, gaseous hydrogen peroxide can be used as a sterilant of atmosphere and surfaces. Hospital rooms, for example, can be decontaminated quickly and routinely with equipment available under the brand name Bioquell. The room is sealed with the generating apparatus inside and the controls on the outside. Once the sealed room has undergone a decontamination cycle, the hydrogen peroxide vapor is catalytically converted into water vapor and oxygen.

*Peracetic acid* (*peroxyacetic acid*, or *PAA*) is one of the most effective liquid chemical sporicides available and can be used as a sterilant. Its mode of action is similar to that of hydrogen peroxide. It is generally effective on endospores and viruses within 30 minutes and kills vegetative bacteria and fungi in less than 5 minutes. PAA has many applications in the disinfection of food-processing and medical equipment, especially endoscopes, because it leaves no toxic residues (only water and

# Infection Following Cosmetic Surgery

As you read through this box, you will encounter a series of questions that infection control officers ask themselves as they track the source of infection. Try to answer each question before going on to the next one.

1. Dr. Priya Agarwal, an infectious disease physician, called the department of health to report that in the last 3 months, she had seen 12 patients with *Mycobacterium abscessus* soft-tissue infections. Slow-growing mycobacteria, including *M. tuberculosis and M. leprae,* are common human pathogens, but Dr. Agarwal was concerned because these infections were caused by rapidly growing mycobacteria (RGM).
   Where are these RGM normally found? (Hint: Read page 316.)

2. RGM are usually found in soil and water. In Dr. Agarwal's report, she noted that all 12 patients had undergone liposuction at the same clinic. The injection procedure consisted of cleaning the skin with cotton balls soaked in diluted (1:10) solution of benzalkonium chloride (Zephiran®), painting the skin with commercially prepared iodine swabs, and anesthetizing the area with 0.5 ml of 1% solution of lidocaine in a sterile 22-gauge needle and syringe.
   What did Dr. Agarwal need to do to determine the source of the infection?

3. Dr. Agarwal ordered cultures from the inside surface of an open metal container of forceps and the inside surface of a metal container of cotton balls. She also requested cultures of the iodine prep swabs, Zephiran®-soaked cotton balls, lidocaine, diluted and

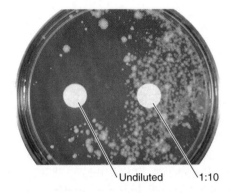

**Disk-diffusion test of Zephiran® against M. abscessus.**

Undiluted     1:10

undiluted Zephiran®, and a sealed bottle of distilled water used to dilute the Zephiran®.
   What type of disinfectant is Zephiran®?

4. Zephiran® is a quat. Results of cultures from the laboratory showed that only the cotton balls soaked in Zephiran® had grown *M. abscessus*. Dr. Agarwal then performed a disk-diffusion assay (see the figure) of both diluted and undiluted Zephiran®.
   What did the assay reveal? What should Dr. Agarwal tell the physician about preventing future infections?

5. Recent research has shown that soaking cotton balls in disinfectant selects for resistant bacteria. Additionally, the presence of organic material, such as cotton balls or towels, reduces the disinfecting ability of Zephiran® and other quats. Dr. Agarwal informed the physician that cotton balls should not be stored in the disinfectant; doing so can cause patients to be inoculated with bacteria.

*Source: Adapted from MMWR 63(9): 201–202, March 7, 2014.*

---

small amounts of acetic acid) and is minimally affected by the presence of organic matter. The FDA has approved use of PAA for the washing of fruits and vegetables.

## CLINICAL CASE

Quats are virucidal against enveloped viruses. By Monday, a total of 103 out of 266 staff and students call in sick with vomiting and diarrhea. With almost half the school either out sick or returning to school after being sick, Amy decides to call the Maryland State Health Department. After going over her records with a health department statistician, she finds out that the most significant risk factors for infection are contact with an ill person or being in the first grade. All but five first-graders have reported sick with diarrheal illness. Because the school is so small, the first-grade classroom also houses the computer lab for the school. Both students and staff share these computers. The health department sends someone to swab the first-grade classroom, and norovirus is cultured from a computer mouse.

**How did the virus get from a computer mouse in the first grade classroom to all of the other grades and staff?**

Other oxidizing agents include *benzoyl peroxide*, which is probably most familiar as the main ingredient in over-the-counter medications for acne. *Ozone ($O_3$)* is a highly reactive form of oxygen that is generated by passing oxygen through high-voltage electrical discharges. It is responsible for the air's rather fresh odor after a lightning storm, in the vicinity of electrical sparking, or around an ultraviolet light. Ozone is often used to supplement chlorine in the disinfection of water because it helps neutralize tastes and odors. Although ozone is a more effective killing agent than chlorine, its residual activity is difficult to maintain in water.

### CHECK YOUR UNDERSTANDING

✔ 7-7 If you wanted to disinfect a surface contaminated by vomit and a surface contaminated by a sneeze, why would your choice of disinfectant make a difference?

✔ 7-8 Which is more likely to be used in a medical clinic laboratory, a use-dilution test or a disk-diffusion test?

✔ 7-9 Why is alcohol effective against some viruses and not others?

✔ 7-10 Is Betadine® an antiseptic or a disinfectant when it is used on skin?

179    195    197    200

7-11 What characteristics make surface-active agents attractive to the dairy industry?

7-12 What chemical disinfectants can be considered sporicides?

7-13 What chemicals are used to sterilize?

# Microbial Characteristics and Microbial Control

## LEARNING OBJECTIVE

7-14 Explain how the type of microbe affects the control of microbial growth.

Many biocides tend to be more effective against gram-positive bacteria, as a group, than against gram-negative bacteria. A principal factor in this relative resistance to biocides is the external lipopolysaccharide layer of gram-negative bacteria. Within gram-negative bacteria, members of the genera *Pseudomonas* and *Burkholderia* are of special interest. These closely related bacteria are unusually resistant to biocides (see Figure 7.6) and will even grow actively in some disinfectants and antiseptics, most notably the quaternary ammonium compounds. These bacteria are also resistant to many antibiotics (see Chapter 20). This resistance to chemical antimicrobials is related mostly to the characteristics of their *porins* (structural openings in the wall of gram-negative bacteria; see Figure 4.13c, page 82). Porins are highly selective of molecules that they permit to enter the cell.

The mycobacteria are another group of non–endospore-forming bacteria that exhibit greater-than-normal resistance to chemical biocides. (See the Clinical Focus box on page 197.) This group includes *Mycobacterium tuberculosis*, the pathogen that causes tuberculosis. The cell wall of this organism and other members of this genus have a waxy, lipid-rich component. Instruction labels on disinfectants often state whether they are tuberculocidal, indicating that they are effective against mycobacteria. Special tuberculocidal tests have been developed to evaluate the effectiveness of biocides against this bacterial group.

Bacterial endospores are affected by relatively few biocides. (The activity of the major chemical antimicrobial groups against mycobacteria and endospores is summarized in Table 7.7.) The cysts and oocysts of protozoa are also relatively resistant to chemical disinfection.

The resistance of viruses to biocides largely depends on the presence or absence of an envelope. Antimicrobials that are lipid-soluble are more likely to be effective against enveloped viruses. The label of such an agent will indicate that it is effective against lipophilic viruses. Nonenveloped viruses, which

| | Effectiveness of Chemical Antimicrobials | |
|---|---|---|
| **TABLE 7.7** | against Endospores and Mycobacteria | |
| **Chemical Agent** | **Effect against Endospores** | **Effect against Mycobacteria** |
| Glutaraldehyde | Fair | Good |
| Chlorines | Fair | Fair |
| Alcohols | Poor | Good |
| Iodine | Poor | Good |
| Phenolics | Poor | Good |
| Chlorhexidine | None | Fair |
| Bisphenols | None | None |
| Quats | None | None |
| Silver | None | None |

have only a protein coat, are more resistant—fewer biocides are active against them.

A special problem, not yet completely solved, is the reliable killing of prions. Prions are infectious proteins that are the cause of neurological diseases known as transmissible spongiform encephalopathies, such as the popularly named mad cow disease (see Chapter 22, page 643). To destroy prions, infected animal carcasses are incinerated. A major problem is the disinfection of surgical instruments exposed to prion contamination. Normal autoclaving has proven to be inadequate. The World Health Organization (WHO) and the U.S. Centers for Disease Control and Prevention (CDC) have recommended the combined use of a solution of sodium hydroxide and autoclaving at 121°C for 1 hour or 134°C for 18 minutes. Disposable instruments should be incinerated.

In summary, it is important to remember that microbial control methods, especially biocides, are not uniformly effective against all microbes.

Table 7.8 summarizes chemical agents used to control microbial growth.

## CHECK YOUR UNDERSTANDING

7-14 The presence or absence of endospores has an obvious effect on microbial control, but why are gram-negative bacteria more resistant to chemical biocides than gram-positive bacteria?

The compounds discussed in this chapter are not generally useful in the treatment of diseases. Antibiotics and the pathogens against which they are active will be discussed in Chapter 20.

## TABLE 7.8  Chemical Agents Used to Control Microbial Growth

| Chemical Agent | Mechanism of Action | Preferred Use |
|---|---|---|
| Phenol | Disruption of plasma membrane, denaturation of enzymes. | Rarely used, except as a standard of comparison. Seldom used as a disinfectant or antiseptic because of its irritating qualities and disagreeable odor. |
| Phenolics | Disruption of plasma membrane, denaturation of enzymes. | Environmental surfaces, instruments, skin surfaces, and mucous membranes. |
| Bisphenols | Probably, disruption of plasma membrane. | Disinfectant hand soaps and skin lotions. |
| Biguanides (Chlorhexidine) | Disruption of plasma membrane. | Skin disinfection, especially for surgical hand scrubbing. Bactericidal. |
| Terpenes | Disruption of plasma membrane | Plant essential oils; used in foods and disinfecting hard surfaces. |
| Halogens | Iodine inhibits protein function and is a strong oxidizing agent; chlorine forms the strong oxidizing agent hypochlorous acid, which alters cellular components. | Iodine is an effective antiseptic available as a tincture and an iodophor; chlorine gas is used to disinfect water; chlorine compounds are used to disinfect dairy equipment, eating utensils, household items, and glassware. |
| Alcohols | Protein denaturation and lipid dissolution. | Bactericidal and fungicidal, but not effective against endospores or nonenveloped viruses. When the skin is swabbed with alcohol before an injection, most of the disinfecting action probably comes from a simple wiping away (degerming) of dirt and some microbes. |
| Heavy Metals and Their Compounds | Denaturation of enzymes and other essential proteins. | Silver nitrate may be used to prevent ophthalmia neonatorum; silver-sulfadiazine is used as a topical cream on burns; copper sulfate is an algicide. |
| Soaps and Detergents | Mechanical removal of microbes through scrubbing. | Skin degerming and removal of debris. |
| Acid-Anionic Sanitizers | Not certain; may involve enzyme inactivation or disruption. | Sanitizers in dairy and food-processing industries. |
| Quaternary Ammonium Compounds (Cationic Detergents) | Enzyme inhibition, protein denaturation, and disruption of plasma membranes. | Antiseptic for skin. Bactericidal, bacteriostatic, fungicidal, and virucidal against enveloped viruses. Used on instruments, utensils, rubber goods. |
| Organic Acids | Metabolic inhibition, mostly affecting molds; action not related to their acidity. | Control mold and bacterial growth in foods and cosmetics. Sorbic acid and benzoic acid effective at low pH; parabens used in cosmetics, shampoos; calcium propionate used in bread. |
| Nitrates/Nitrites | Active ingredient is nitrite, which is produced by bacterial action on nitrate. Nitrite inhibits certain iron-containing enzymes of anaerobes. | Meat products such as ham, bacon, hot dogs, sausage. Prevents growth of *Clostridium botulinum* in food; also imparts a red color. |
| Aldehydes | Protein denaturation. | Glutaraldehyde (Cidex®) is less irritating than formaldehyde and is used for disinfecting medical equipment. |
| Ethylene Oxide and Other Gaseous Sterilants | Inhibits vital cellular functions. | Mainly for sterilization of materials that would be damaged by heat. |
| Plasma Sterilization | Inhibits vital cellular functions. | Especially useful for tubular medical instruments. |
| Supercritical Fluids | Inhibits vital cellular functions. | Especially useful for sterilizing organic medical implants. |
| Peroxygens and Other Forms of Oxygen | Oxidation. | Contaminated water and surfaces; some deep wounds, in which they are very effective against oxygen-sensitive anaerobes. |

Norovirus is an extremely contagious virus and can spread quickly from person to person. It is also nonenveloped, so it cannot be easily destroyed by a biocide. Amy asks the principal if, once the school is back to full capacity, she can hold an assembly to discuss the importance of handwashing with the students and staff. Proper washing with soap and water can eliminate the transmission of norovirus to other people or surfaces. Amy also meets again with the custodial staff to discuss the health department's recommendations.

According to the health department, when cleaning environmental surfaces that are visibly soiled with feces or vomitus, the staff should wear masks and gloves, use a disposable towel that has been soaked in dilute detergent to wipe the surface for at least 10 seconds, and then apply a 1:10 household bleach solution for at least 1 minute. Although Amy knows this won't be the last time her school is affected by a virus, she is certain she has taken a positive step toward protecting her students and staff from this particular virus.

179    195    197    **200**

## Study Outline

 Go to @MasteringMicrobiology for **Interactive Microbiology**, *In the Clinic* *videos, MicroFlix, MicroBoosters, 3D animations, practice quizzes, and more.*

### The Terminology of Microbial Control (p. 179)

1. The control of microbial growth can prevent infections and food spoilage.
2. Sterilization is the process of removing or destroying all microbial life on an object.
3. Commercial sterilization is heat treatment of canned foods to destroy *C. botulinum* endospores.
4. Disinfection is the process of reducing or inhibiting microbial growth on a nonliving surface.
5. Antisepsis is the process of reducing or inhibiting microorganisms on living tissue.
6. The suffix *-cide* means to kill; the suffix *-stat* means to inhibit.
7. Sepsis is bacterial contamination.

### The Rate of Microbial Death (p. 180)

1. Bacterial populations subjected to heat or antimicrobial chemicals usually die at a constant rate.
2. Such a death curve, when plotted logarithmically, shows this constant death rate as a straight line.
3. The time it takes to kill a microbial population is proportional to the number of microbes.
4. Microbial species and life cycle phases (e.g., endospores) have different susceptibilities to physical and chemical controls.
5. Organic matter may interfere with heat treatments and chemical control agents.
6. Longer exposure to lower heat can produce the same effect as shorter time at higher heat.

### Actions of Microbial Control Agents (pp. 180–181)

#### Alteration of Membrane Permeability (p. 181)

1. The susceptibility of the plasma membrane is due to its lipid and protein components.
2. Certain chemical control agents damage the plasma membrane by altering its permeability.

#### Damage to Proteins and Nucleic Acids (p. 181)

3. Some microbial control agents damage cellular proteins by breaking hydrogen bonds and covalent bonds.
4. Other agents interfere with DNA and RNA and protein synthesis.

### Physical Methods of Microbial Control (pp. 182–187)

#### Heat (pp. 182–185)

1. Heat is frequently used to kill microorganisms.
2. Moist heat kills microbes by denaturing enzymes.
3. Thermal death point (TDP) is the lowest temperature at which all the microbes in a liquid culture will be killed in 10 minutes.
4. Thermal death time (TDT) is the length of time required to kill all bacteria in a liquid culture at a given temperature.
5. Decimal reduction time (DRT) is the length of time in which 90% of a bacterial population will be killed at a given temperature.
6. Boiling (100°C) kills many vegetative cells and viruses within 10 minutes.
7. Autoclaving (steam under pressure) is the most effective method of moist heat sterilization. The steam must directly contact the material to be sterilized.
8. In HTST pasteurization, a high temperature is used for a short time (72°C for 15 seconds) to destroy pathogens without altering the flavor of the food. Ultra-high-temperature (UHT) treatment (140°C for 4 seconds) is used to sterilize dairy products.
9. Methods of dry heat sterilization include direct flaming, incineration, and hot-air sterilization. Dry heat kills by oxidation.
10. Different methods that produce the same effect (reduction in microbial growth) are called equivalent treatments.

#### Filtration (p. 185)

11. Filtration is the passage of a liquid or gas through a filter with pores small enough to retain microbes.
12. Microbes can be removed from air by high-efficiency particulate air (HEPA) filters.
13. Membrane filters composed of cellulose esters are commonly used to filter out bacteria, viruses, and even large proteins.

**Low Temperatures** (pp. 185–186)

14. The effectiveness of low temperatures depends on the particular microorganism and the intensity of the application.

15. Most microorganisms do not reproduce at ordinary refrigerator temperatures (0–7°C).

16. Many microbes survive (but do not grow) at the subzero temperatures used to store foods.

**High Pressure** (p. 186)

17. High pressure denatures proteins in vegetative cells.

**Desiccation** (p. 186)

18. In the absence of water, microorganisms cannot grow but can remain viable.

19. Viruses and endospores can resist desiccation.

**Osmotic Pressure** (p. 186)

20. Microorganisms in high concentrations of salts and sugars undergo plasmolysis.

21. Molds and yeasts are more capable than bacteria of growing in materials with low moisture or high osmotic pressure.

**Radiation** (pp. 186–187)

22. The effects of radiation depend on its wavelength, intensity, and duration.

23. Ionizing radiation (gamma rays, X rays, and high-energy electron beams) has a high degree of penetration and exerts its effect primarily by ionizing water and forming highly reactive hydroxyl radicals.

24. Ultraviolet (UV) radiation, a form of nonionizing radiation, has a low degree of penetration and causes cell damage by making thymine dimers in DNA that interfere with DNA replication; the most effective germicidal wavelength is 260 nm.

25. Microwaves can kill microbes indirectly as materials get hot.

## Chemical Methods of Microbial Control (pp. 187–198)

1. Chemical agents are used on living tissue (as antiseptics) and on inanimate objects (as disinfectants).

2. Few chemical agents achieve sterility.

**Principles of Effective Disinfection** (p. 188)

3. Careful attention should be paid to the properties and concentration of the disinfectant to be used.

4. The presence of organic matter, degree of contact with microorganisms, and temperature should also be considered.

**Evaluating a Disinfectant** (p. 189)

5. The use-dilution test is used to determine bacterial survival in the manufacturer's recommended dilution of a disinfectant.

6. The use-dilution test can also be used to evaluate the effectiveness of agents against viruses, endospore-forming bacteria, mycobacteria, and fungi.

7. In the disk-diffusion method, a disk of filter paper is soaked with a chemical and placed on an inoculated agar plate; a zone of inhibition indicates effectiveness.

**Types of Disinfectants** (pp. 189–197)

8. Phenolics exert their action by injuring plasma membranes.

9. The bisphenol hexachlorophene is used as a skin disinfectant.

10. Biguanides damage plasma membranes of vegetative cells.

11. Terpenes and phenolics in essential oils of plants have antimicrobial activity.

12. Iodine may combine with certain amino acids to inactivate enzymes and other cellular proteins.

13. The germicidal action of chlorine is based on the formation of hypochlorous acid when chlorine is added to water.

14. Alcohols exert their action by denaturing proteins and dissolving lipids.

15. In tinctures, alcohols enhance the effectiveness of other antimicrobial chemicals.

16. Silver, mercury, copper, and zinc exert their antimicrobial action through oligodynamic action. When heavy metal ions combine with sulfhydryl (—SH) groups, proteins are denatured.

17. Soaps have limited germicidal action but assist in removing microorganisms.

18. Acid-anionic detergents are used to clean dairy equipment.

19. Quats are cationic detergents attached to $NH_4^+$ that disrupt plasma membranes.

20. $SO_2$, sorbic acid, benzoic acid, and propionic acid inhibit fungal metabolism and are used as food preservatives.

21. Nitrate and nitrite salts prevent germination of *C. botulinum* endospores in meats.

22. Nisin and natamycin are antibiotics used to preserve foods, especially cheese.

23. Aldehydes such as formaldehyde and glutaraldehyde are among the most effective chemical disinfectants. They exert their antimicrobial effect by inactivating proteins.

24. Ethylene oxide is the gas most frequently used for sterilization. It penetrates most materials and kills all microorganisms by protein denaturation.

25. Free radicals in plasma gases are used to sterilize plastic instruments.

26. Supercritical fluids, which have properties of liquid and gas, can sterilize at low temperatures.

27. Hydrogen peroxide, peracetic acid, benzoyl peroxide, and ozone exert their antimicrobial effect by oxidizing molecules inside cells.

## Microbial Characteristics and Microbial Control
(p. 198)

1. Gram-negative bacteria are generally more resistant than gram-positive bacteria to disinfectants and antiseptics.

2. Mycobacteria, endospores, and protozoan cysts and oocysts are very resistant to disinfectants and antiseptics.

3. Nonenveloped viruses are generally more resistant than enveloped viruses to disinfectants and antiseptics.

4. Prions are resistant to disinfection and autoclaving.

# Study Questions

For answers to the Knowledge and Comprehension questions, turn to the Answers tab at the back of the textbook.

## Knowledge and Comprehension

### Review

1. The thermal death time for a suspension of *Bacillus subtilis* endospores is 30 minutes in dry heat and less than 10 minutes in an autoclave. Which type of heat is more effective? Why?

2. If pasteurization does not achieve sterilization, why is pasteurization used to treat food?

3. Thermal death point is not considered an accurate measure of the effectiveness of heat sterilization. List three factors that can alter thermal death point.

4. The antimicrobial effect of gamma radiation is due to (a) _____. The antimicrobial effect of ultraviolet radiation is due to (b) _____.

5. ☐ DRAW IT ☐ A bacterial culture was in log phase in the following figure. At time *x*, an antibacterial compound was added to the culture. Draw the lines indicating addition of a bactericidal compound and a bacteriostatic compound. Explain why the viable count does not immediately drop to zero at *x*.

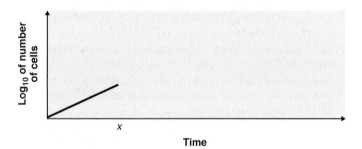

6. How do autoclaving, hot air, and pasteurization illustrate the concept of equivalent treatments?

7. How do salts and sugars preserve foods? Why are these considered physical rather than chemical methods of microbial control? Name one food that is preserved with sugar and one preserved with salt. How do you account for the occasional growth of *Penicillium* mold in jelly, which is 50% sucrose?

8. The use-dilution values for two disinfectants tested under the same conditions are as follows: Disinfectant A—1:2; Disinfectant B—1:10,000. If both disinfectants are designed for the same purpose, which would you select?

9. A large hospital washes burn patients in a stainless steel tub. After each patient, the tub is cleaned with a quat. It was noticed that 14 of 20 burn patients acquired *Pseudomonas* infections after being bathed. Provide an explanation for this high rate of infection.

10. ☐ NAME IT ☐ What bacteria have porins, are resistant to triclosan, and survive and may grow in quats?

## Multiple Choice

1. Which of the following does *not* kill endospores?
   a. autoclaving
   b. incineration
   c. hot-air sterilization
   d. pasteurization
   e. All of the above kill endospores.

2. Which of the following is most effective for sterilizing mattresses and plastic Petri dishes?
   a. chlorine
   b. ethylene oxide
   c. glutaraldehyde
   d. autoclaving
   e. nonionizing radiation

3. Which of these disinfectants does *not* act by disrupting the plasma membrane?
   a. phenolics
   b. phenol
   c. quats
   d. halogens
   e. biguanides

4. Which of the following *cannot* be used to sterilize a heat-labile solution stored in a plastic container?
   a. gamma radiation
   b. ethylene oxide
   c. supercritical fluids
   d. autoclaving
   e. short-wavelength radiation

5. Which of the following is used to control microbial growth in foods?
   a. organic acids
   b. alcohols
   c. aldehydes
   d. heavy metals
   e. all of the above

Use the following information to answer questions 6 and 7. The data were obtained from a use-dilution test comparing four disinfectants against *Salmonella* Choleraesuis. *G* = growth, *NG* = no growth

| | Bacterial Growth after Exposure to | | | |
|---|---|---|---|---|
| Dilution | Disinfectant A | Disinfectant B | Disinfectant C | Disinfectant D |
| 1:2 | NG | G | NG | NG |
| 1:4 | NG | G | NG | G |
| 1:8 | NG | G | G | G |
| 1:16 | G | G | G | G |

6. Which disinfectant is the most effective?

7. Which disinfectant(s) is (are) bactericidal?
   a. A, B, C, and D
   b. A, C, and D
   c. A only
   d. B only
   e. none of the above

8. Which of the following is *not* a characteristic of quaternary ammonium compounds?
   a. bactericidal against gram-positive bacteria
   b. sporicidal
   c. amebicidal
   d. fungicidal
   e. kills enveloped viruses

9. A classmate is trying to determine how a disinfectant might kill cells. You observed that when he spilled the disinfectant in your reduced litmus milk, the litmus turned blue again. You suggest to your classmate that
   a. the disinfectant might inhibit cell wall synthesis.
   b. the disinfectant might oxidize molecules.
   c. the disinfectant might inhibit protein synthesis.
   d. the disinfectant might denature proteins.
   e. he take his work away from yours.

10. Which of the following is most likely to be bactericidal?
   a. membrane filtration
   b. ionizing radiation
   c. lyophilization (freeze-drying)
   d. deep-freezing
   e. all of the above

## Analysis

1. The disk-diffusion method was used to evaluate three disinfectants. The results were as follows:

| Disinfectant | Zone of Inhibition |
|---|---|
| X | 0 mm |
| Y | 5 mm |
| Z | 10 mm |

   a. Which disinfectant was the most effective against the organism?
   b. Can you determine whether compound Y was bactericidal or bacteriostatic?

2. For each of the following bacteria, explain why it is often resistant to disinfectants.
   a. *Mycobacterium*
   b. *Pseudomonas*
   c. *Bacillus*

3. A use-dilution test was used to evaluate two disinfectants against *Salmonella* Choleraesuis. The results were as follows:

| | Bacterial Growth after Exposures | | |
|---|---|---|---|
| Time of Exposure (min) | Disinfectant A | Disinfectant B Diluted with Distilled Water | Disinfectant B Diluted with Tap Water |
| 10 | G | NG | G |
| 20 | G | NG | NG |
| 30 | NG | NG | NG |

   a. Which disinfectant was the most effective?
   b. Which disinfectant should be used against *Staphylococcus*?

4. To determine the lethal action of microwave radiation, two $10^5$ suspensions of *E. coli* were prepared. One cell suspension was exposed to microwave radiation while wet, whereas the other was lyophilized (freeze-dried) and then exposed to radiation. The results are shown in the following figure. Dashed lines indicate the temperature of the samples. What is the most likely method of lethal action of microwave radiation? How do you suppose these data might differ for *Clostridium*?

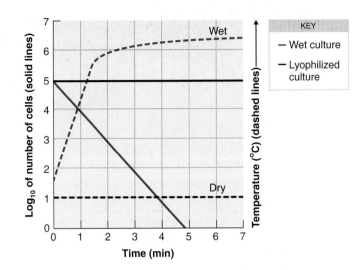

## Clinical Applications and Evaluation

1. *Entamoeba histolytica* and *Giardia intestinalis* were isolated from the stool sample of a 45-year-old man, and *Shigella sonnei* was isolated from the stool sample of an 18-year-old woman. Both patients experienced diarrhea and severe abdominal cramps, and prior to onset of digestive symptoms both had been treated by the same chiropractor. The chiropractor had administered colonic irrigations (enemas) to these patients. The device used for this treatment was a gravity-dependent apparatus using 12 liters of tap water. There were no check valves to prevent backflow, so all parts of the apparatus could have become contaminated with feces during each colonic treatment. The chiropractor provided colonic treatment to four or five patients per day. Between patients, the adaptor piece that is inserted into the rectum was placed in a "hot-water sterilizer."

   What two errors did the chiropractor make?

2. Between March 9 and April 12, five chronic peritoneal dialysis patients at one hospital became infected with *Pseudomonas aeruginosa*. Four patients developed peritonitis (inflammation of the abdominal cavity), and one developed a skin infection at the catheter insertion site. All patients with peritonitis had low-grade fever, cloudy peritoneal fluid, and abdominal pain. All patients had permanent indwelling peritoneal catheters, which the nurse wiped with gauze that had been soaked with an iodophor solution each time the catheter was connected to or disconnected from the machine tubing. Aliquots of the iodophor were transferred from stock bottles to small in-use bottles. Cultures from the dialysate concentrate and the internal areas of the dialysis machines were negative; iodophor from a small in-use plastic container yielded a pure culture of *P. aeruginosa*.

   What improper technique led to this infection?

3. You are investigating a national outbreak of *Ralstonia mannitolilytica* associated with use of a contaminated oxygen-delivery device among pediatric patients. The device adds moisture to and warms the oxygen. Each hospital followed the manufacturer's recommendation to use a detergent to clean the reusable components of the device between patients. Tap water is permitted in the device because the device uses a reusable 0.01-μm filter as a biological barrier between the air and water compartments. *Ralstonia* is a gram-negative rod commonly found in water.

   Why did disinfection fail?

   What do you recommend for disinfecting? The device cannot be autoclaved.

# 8 Microbial Genetics

Virtually all the microbial traits you have read about in earlier chapters are controlled or influenced by heredity. The inherited characteristics of microbes include shape, structural features, metabolism, ability to move, and interactions with other organisms. Individual organisms transmit these characteristics to their offspring through genes.

The development of antibiotic resistance in microorganisms is often carried on plasmids such as those in the photo, which are readily transferred between bacterial cells. They are responsible for the emergence of methicillin-resistant *Staphylococcus aureus* and the recent emergence of carbapenem-resistant *Klebsiella pneumoniae*. The emergence of vancomycin-resistant *S. aureus* (VRSA) poses a serious threat to patient care. In this chapter you will see how VRSA acquired this characteristic.

Emerging diseases provide another reason why it is important to understand genetics. New diseases are the results of genetic changes in some existing organism; for example, *E. coli* O157:H7 acquired the genes for Shiga toxin from *Shigella*.

Currently, microbiologists are using genetics to study unculturable microbes and the relationship between hosts and microbes.

The **Big Picture** on pages 206–207 highlights key principles of genetics that are explained in greater detail throughout the chapter.

▶ Plasmids exist in cells separate from chromosomes.

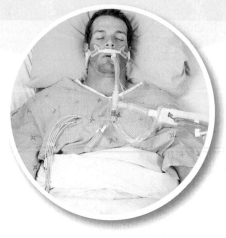

## In the Clinic

As a nurse at a U.S. military hospital, you treat service members injured in the recent Middle East conflicts. You notice that wounds infected by *Acinetobacter baumannii* are not responding to antibiotics. The Centers for Disease Control and Prevention reports that the antibiotic-resistance genes found in *A. baumannii* are the same as those in *Pseudomonas*, *Salmonella*, and *Escherichia*. Cephalosporin-resistance genes are on the chromosome, tetracycline resistance is encoded by a plasmid, and streptomycin resistance is associated with a transposon. **Can you suggest mechanisms by which *Acinetobacter* acquired this resistance?**

*Hint: Read about genetic recombination on pages 229–235.*

Play **In the Clinic** Video
@Mastering**Microbiology**

# Structure and Function of the Genetic Material

## LEARNING OBJECTIVES

8-1 Define *genetics, genome, chromosome, gene, genetic code, genotype, phenotype,* and *genomics.*

8-2 Describe how DNA serves as genetic information.

8-3 Describe the process of DNA replication.

8-4 Describe protein synthesis, including transcription, RNA processing, and translation.

8-5 Compare protein synthesis in prokaryotes and eukaryotes.

**Genetics** is the science of heredity. It includes the study of genes: how they carry information, how they replicate and pass to subsequent generations of cells or between organisms, and how the expression of their information within an organism determines its characteristics. The genetic information in a cell is called the **genome**. A cell's genome includes its chromosomes and plasmids. **Chromosomes** are structures containing DNA that physically carry hereditary information; the chromosomes contain the genes. **Genes** are segments of DNA (except in some viruses, in which they are made of RNA) that code for functional products. Usually these products are proteins, but they can also be RNAs (ribosomal RNA, transfer RNA, or microRNA).

We saw in Chapter 2 that DNA is a macromolecule composed of repeating units called *nucleotides.* Each nucleotide consists of a nucleobase (adenine, thymine, cytosine, or guanine), deoxyribose (a pentose sugar), and a phosphate group (see Figure 2.16, page 45). The DNA within a cell exists as long strands of nucleotides twisted together in pairs to form a double helix. Each strand has a string of alternating sugar and phosphate groups (its *sugar-phosphate backbone*), and a nitrogenous base is attached to each sugar in the backbone. The two strands are held together by hydrogen bonds between their nitrogenous bases. The **base pairs** always occur in a specific way: adenine always pairs with thymine, and cytosine always pairs with guanine. Because of this specific base pairing, the base sequence of one DNA strand determines the base sequence of the other strand. The two strands of DNA are thus *complementary.*

The structure of DNA helps explain two primary features of biological information storage. First, the linear sequence of bases provides the actual information. Genetic information is encoded by the sequence of bases along a strand of DNA, in much the same way as our written language uses a linear sequence of letters to form words and sentences. The genetic language, however, uses an alphabet with only four letters—the four kinds of nucleobases in DNA (or RNA). But 1000 of these four bases, the number contained in an average-sized gene, can be arranged in $4^{1000}$ different ways. This astronomically large number explains how genes can be varied enough to provide all the information a cell needs to grow and perform its functions. The **genetic code**, the set of rules that determines how a

nucleotide sequence is converted into the amino acid sequence of a protein, is discussed in more detail later in this chapter.

Second, the complementary structure allows for the precise duplication of DNA during cell division. Each offspring cell receives one of the original strands from the parent, thus ensuring one strand that functions correctly.

Much of cellular metabolism is concerned with translating the genetic message of genes into specific proteins. A gene is usually copied to make a messenger RNA (mRNA) molecule, which ultimately results in the formation of a protein. When the ultimate molecule for which a gene codes (a protein, for example) has been produced, we say that the gene has been **expressed**. The flow of genetic information can be shown as flowing from DNA to RNA to proteins, as follows:

This theory was called the **central dogma** by Francis Crick in 1956, when he first proposed that the sequence of nucleotides in DNA determines the sequence of amino acids in a protein.

ASM: Although the central dogma is universal in all cells, the processes differ in prokaryotes and eukaryotes, as we shall see in this chapter.

## Genotype and Phenotype

The **genotype** of an organism is its genetic makeup—all its DNA—the information that codes for all the particular characteristics of the organism. The genotype represents *potential*

### CLINICAL CASE  Where There's Smoke

Marcel DuBois, a 70-year-old grandfather of 12, quietly hangs up the phone. His doctor has just called him with the results of his stool DNA test that he undertook at the Mayo Clinic last week. Marcel's doctor suggested this new, noninvasive screening tool for colorectal cancer because Marcel is not comfortable with the colonoscopy procedure and usually tries to postpone getting one. The stool DNA test, however, uses stool samples, which contain cells that have been shed from the colon lining. The DNA from these cells is tested for DNA markers that may indicate the presence of precancerous polyps or cancerous tumors. Marcel makes an appointment to come in to see his doctor the next afternoon.

Once in the office, the doctor explains to Marcel and his wife, Janice, that the stool DNA test detected the presence of serrated colorectal polyps. This type of polyp is usually difficult to see with a colonoscopy because it is not raised and can be the same color as the colon wall.

**How can DNA show whether a person has cancer? Read on to find out.**

205  224  227  234

# Genetics is the science of heredity. It includes the study of genes: how they are replicated, expressed, and passed on from one generation to another.

The central dogma of molecular biology describes how, typically, DNA is transcribed to messenger RNA, which, in turn, is translated into proteins that carry out vital cellular functions. Mutations introduce change into this process—ultimately leading to new or lost functions.

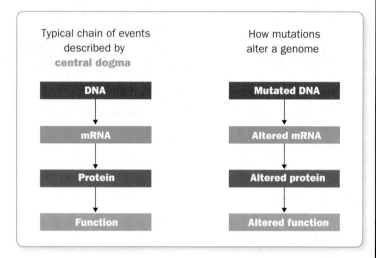

Mutations can be caused by base substitutions or frameshift mutations.

In base substitution mutations, a single DNA base pair is altered.

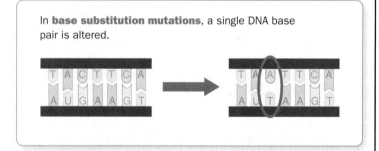

In frameshift mutations, DNA base pairs are added or removed from the sequence, causing a shift in the sequence reading.

Groups of genes in operons can be inducible or repressible.

An inducible operon includes genes that are in the "off" mode, with the repressor bound to the DNA, and is turned "on" by the environmental inducer.

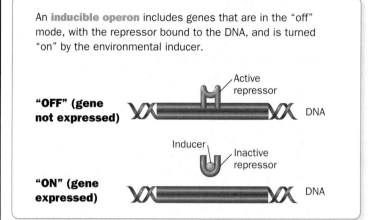

A repressible operon includes genes that are in the "on" mode, without the repressor bound to the DNA, and is turned "off" by the environmental corepressor and repressor.

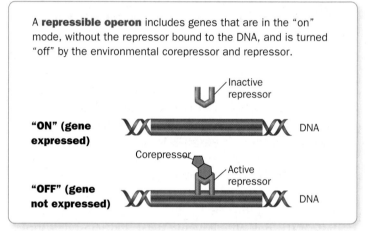

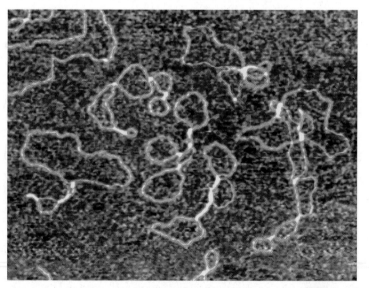

Atomic force micrograph showing DNA molecules.  AFM  ├───┤ 7 nm

# Alteration of bacterial genes and/or gene expression may cause disease, prevent disease treatment, or be manipulated for human benefit.

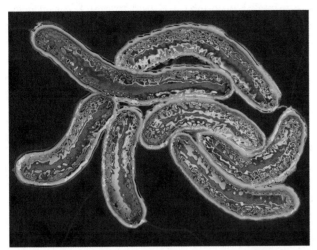

TEM   0.4 μm

**Diseases:** Many bacterial diseases are caused by the presence of toxic proteins that damage human tissue. These toxic proteins are coded for by genes. *Vibrio cholerae*, shown above, produces an enterotoxin that causes diarrhea and severe dehydration, which can be fatal if left untreated.

SEM   5 μm

**Biofilms:** Biofilms, such as the one seen here growing on a toothbrush bristle, are produced by altered bacterial gene expression when populations are large enough. Various *Streptococcus* species, including *S. mutans,* form biofilms on teeth and gums, contributing to the development of dental plaque and dental caries.

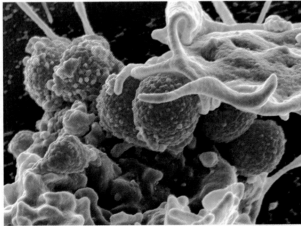

SEM   0.3 μm

**Antibiotic resistance:** Mutations in the bacterial genome are one of the first steps toward the development of antibiotic resistance. This process has occurred with *Staphylococcus aureus,* which is currently resistant to beta-lactam antibiotics such as penicillin. Methicillin was introduced to treat penicillin-resistant *S. aureus.* Methicillin-resistant *S. aureus* (MRSA), shown in purple above, is now a leading cause of healthcare-associated infections.

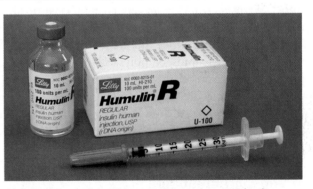

**Biotechnology:** Scientists can alter a microorganism's genome, adding genes that will produce human proteins used in treating disease. Insulin, used for treatment of diabetes, is produced in this manner.

## KEY CONCEPTS

- DNA expression leads to cell function via the production of proteins.

- Genes in operons are turned on or off together.

- Mutations alter DNA sequences.

- DNA mutations can change bacterial function.

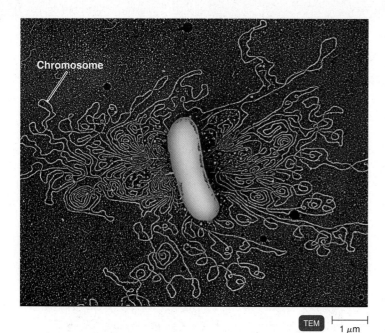

**Figure 8.1 A prokaryotic chromosome.**

**Q** How many times longer than the 2-μm cell is the chromosome?

properties, but not the properties themselves. **Phenotype** refers to *actual, expressed* properties, such as the organism's ability to perform a particular chemical reaction. Phenotype, then, is the manifestation of genotype. For example, *E. coli* with the *stx* gene can produce the stx (Shiga toxin) protein.*

In a sense, an organism's phenotype is its collection of proteins, because most of a cell's properties derive from the structures and functions of proteins. In microbes, most proteins are either *enzymatic* (catalyze particular reactions) or *structural* (participate in large functional complexes such as membranes or flagella). Even phenotypes that depend on structural macromolecules such as lipids or polysaccharides rely indirectly on proteins. For instance, the structure of a complex lipid or polysaccharide molecule results from catalytic activities of enzymes that synthesize, process, and degrade those molecules. Thus, saying that phenotypes are due to proteins is a useful simplification.

## DNA and Chromosomes

Bacteria typically have a single circular chromosome consisting of a single circular molecule of DNA with associated proteins. The chromosome is looped and folded and attached at one or several points to the plasma membrane. The DNA of *E. coli* has about 4.6 million base pairs and is about 1 mm long—1000 times longer than the entire cell (**Figure 8.1**). However, the chromosome takes up only about 10% of the cell's volume because the DNA is twisted, or *supercoiled*.

_____

*Gene names are italicized, but the protein name is not italicized.

The entire genome does not consist of back-to-back genes. Noncoding regions called **short tandem repeats (STRs)** occur in most genomes, including that of *E. coli*. STRs are repeating sequences of two- to five-base sequences. These are used in DNA fingerprinting (discussed on page 258).

Now, the complete base sequences of chromosomes can be determined. Computers are used to search for *open reading frames*, that is, regions of DNA that are likely to encode a protein. As you will see later, these are base sequences between start and stop codons. The sequencing and molecular characterization of genomes is called **genomics**. The use of genomics to track Zika virus is described in the Clinical Focus box on page 218.

## The Flow of Genetic Information

DNA replication makes possible the flow of genetic information from one generation to the next. This is called **vertical gene transfer**. As shown in **Figure 8.2**, the DNA of a cell replicates before cell division so that each offspring cell receives a chromosome identical to the parent's. Within each metabolizing cell, the genetic information contained in DNA also flows in another way: it is transcribed into mRNA and then translated into protein. We describe the processes of transcription and translation later in this chapter.

## DNA Replication

In DNA replication, one "parental" double-stranded DNA molecule is converted to two identical offspring molecules. The complementary structure of the nitrogenous base sequences in the DNA molecule is the key to understanding DNA replication. Because the bases along the two strands of double-helical DNA are complementary, one strand can act as a template for the production of the other strand (**Figure 8.3a**).

DNA replication requires the presence of several cellular proteins that direct a particular sequence of events. Enzymes involved in DNA replication and other processes are listed in **Table 8.1**. When replication begins, the supercoiling is relaxed by *topoisomerase* or *gyrase*. The two strands of parental DNA are unwound by *helicase* and separated from each other in one small DNA segment after another. Free nucleotides present in the cell cytoplasm are matched up to the exposed bases of the single-stranded parental DNA. Where thymine is present on the original strand, only adenine can fit into place on the new strand; where guanine is present on the original strand, only cytosine can fit into place, and so on. Any bases that are improperly base-paired are removed and

# The Flow of Genetic Information

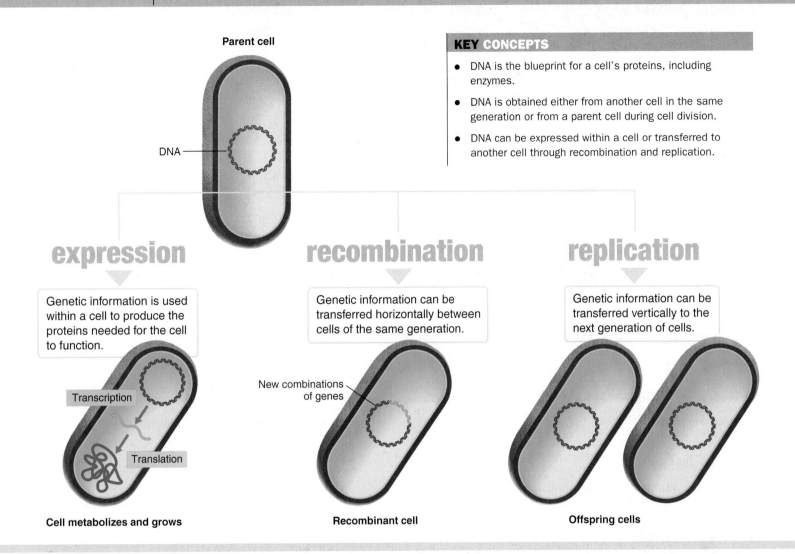

**Parent cell**

DNA

## expression

Genetic information is used within a cell to produce the proteins needed for the cell to function.

Transcription

Translation

**Cell metabolizes and grows**

## recombination

Genetic information can be transferred horizontally between cells of the same generation.

New combinations of genes

**Recombinant cell**

## replication

Genetic information can be transferred vertically to the next generation of cells.

**Offspring cells**

---

replaced by replication enzymes. Once aligned, the newly added nucleotide is joined to the growing DNA strand by an enzyme called **DNA polymerase**. Then the parental DNA is unwound a bit further to allow the addition of the next nucleotides. The point at which replication occurs is called the *replication fork*.

As the replication fork moves along the parental DNA, each of the unwound single strands combines with new nucleotides. The original strand and this newly synthesized daughter strand then rewind. Because each new double-stranded DNA molecule contains one original (conserved) strand and one new strand, the process of replication is referred to as **semiconservative replication**.

Before looking at DNA replication in more detail, let's discuss the structure of DNA (see Figure 2.16 on page 45 for an overview). It is important to understand that the paired DNA strands are oriented in opposite directions (antiparallel) relative to each other. The carbon atoms of the sugar component of each nucleotide are numbered 1′ (pronounced "one prime")

to 5′. For the paired bases to be next to each other, the sugar components in one strand are upside down relative to the other. The end with the hydroxyl attached to the 3′ carbon is called the 3′ end of the DNA strand; the end having a phosphate attached to the 5′ carbon is called the 5′ end. The way in which the two strands fit together dictates that the $5′ \rightarrow 3′$ direction of one strand runs counter to the $5′ \rightarrow 3′$ direction of the other strand (Figure 8.3b). This structure of DNA affects the replication process because DNA polymerases can add new nucleotides to the 3′ end only. Therefore, as the replication fork moves along the parental DNA, the two new strands must grow in different directions.

One new strand, called the *leading strand*, is synthesized continuously in the $5′ \rightarrow 3′$ direction (from a template parental strand running $3′ \rightarrow 5′$). In contrast, the *lagging strand* of the new DNA is synthesized discontinuously in fragments of about 1000 nucleotides, called **Okazaki fragments**. These must be joined later to make the continuous strand.

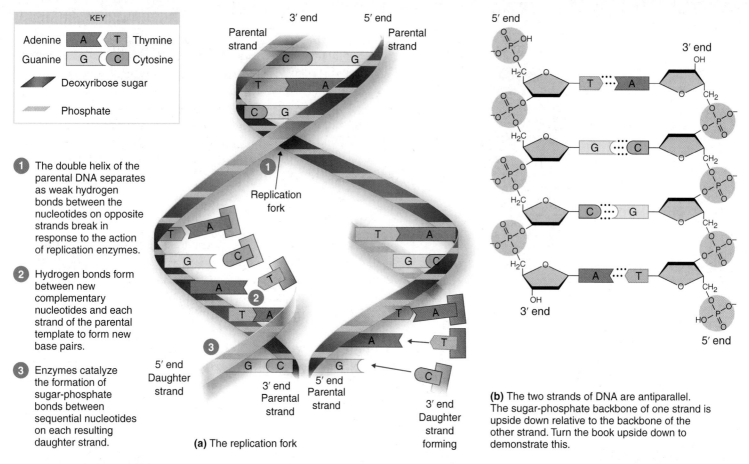

**KEY**

| | |
|---|---|
| Adenine A T Thymine | |
| Guanine G C Cytosine | |
| Deoxyribose sugar | |
| Phosphate | |

**①** The double helix of the parental DNA separates as weak hydrogen bonds between the nucleotides on opposite strands break in response to the action of replication enzymes.

**②** Hydrogen bonds form between new complementary nucleotides and each strand of the parental template to form new base pairs.

**③** Enzymes catalyze the formation of sugar-phosphate bonds between sequential nucleotides on each resulting daughter strand.

**(a)** The replication fork

**(b)** The two strands of DNA are antiparallel. The sugar-phosphate backbone of one strand is upside down relative to the backbone of the other strand. Turn the book upside down to demonstrate this.

**Figure 8.3 DNA replication.**

**Q** What is the advantage of semiconservative replication?

**TABLE 8.1  Important Enzymes in DNA Replication, Expression, and Repair**

| | |
|---|---|
| **DNA Gyrase** | Relaxes supercoiling ahead of the replication fork |
| **DNA Ligase** | Makes covalent bonds to join DNA strands; Okazaki fragments, and new segments in excision repair |
| **DNA Polymerases** | Synthesize DNA; proofread and facilitate repair of DNA |
| **Endonucleases** | Cut DNA backbone in a strand of DNA; facilitate repair and insertions |
| **Exonucleases** | Cut DNA from an exposed end of DNA; facilitate repair |
| **Helicase** | Unwinds double-stranded DNA |
| **Methylase** | Adds methyl group to selected bases in newly made DNA |
| **Photolyase** | Uses visible light energy to separate UV-induced pyrimidine dimers |
| **Primase** | An RNA polymerase that makes RNA primers from a DNA template |
| **Ribozyme** | RNA enzyme that removes introns and splices exons together |
| **RNA Polymerase** | Copies RNA from a DNA template |
| **snRNP** | RNA-protein complex that removes introns and splices exons together |
| **Topoisomerase or Gyrase** | Relaxes supercoiling ahead of the replication fork; separates DNA circles at the end of DNA replication |
| **Transposase** | Cuts DNA backbone, leaving single-stranded "sticky ends" |

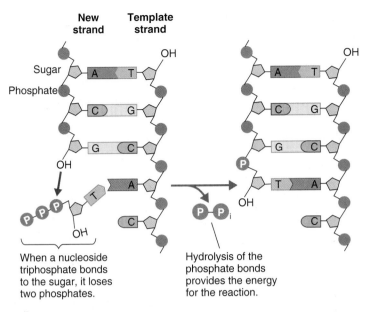

New strand   Template strand

Sugar

Phosphate

When a nucleoside triphosphate bonds to the sugar, it loses two phosphates.

Hydrolysis of the phosphate bonds provides the energy for the reaction.

**Figure 8.4  Adding a nucleotide to DNA.**

**Q** Why is one strand "upside down" relative to the other strand? Why can't both strands "face" the same way?

## Energy Needs

DNA replication requires a great deal of energy. The energy is supplied from the nucleotides, which are actually nucleoside triphosphates. You already know about ATP; the only difference between ATP and the adenine nucleotide in DNA is the sugar component. Deoxyribose is the sugar in the nucleosides used to synthesize DNA, and nucleoside triphosphates with ribose are used to synthesize RNA. Two phosphate groups are removed to add the nucleotide to a growing strand of DNA; hydrolysis of the nucleoside is exergonic and provides energy to make the new bonds in the DNA strand (**Figure 8.4**).

**Figure 8.5** provides more detail about the many steps that go into this complex process.

DNA replication by some bacteria, such as *E. coli*, goes *bidirectionally* around the chromosome (**Figure 8.6**). Two replication forks move in opposite directions away from the origin of replication. Because the bacterial chromosome is a closed loop, the replication forks eventually meet when replication is completed. The two loops must be separated by a topoisomerase. Much evidence shows an association between the bacterial plasma membrane and the origin of replication. After duplication, if each copy of the origin binds to the membrane at

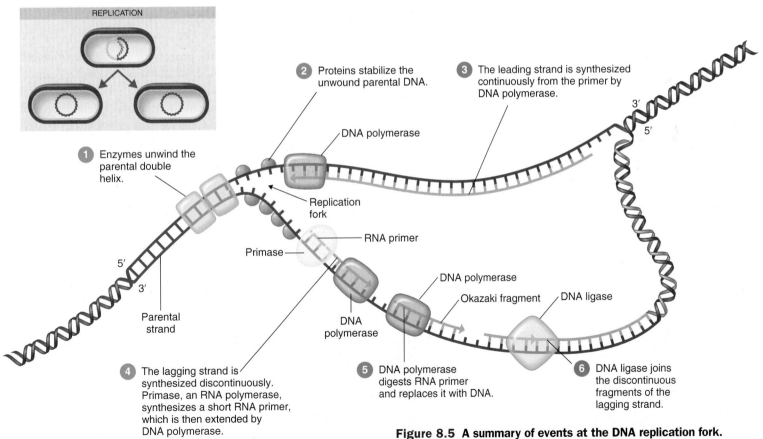

REPLICATION

1 Enzymes unwind the parental double helix.

2 Proteins stabilize the unwound parental DNA.

3 The leading strand is synthesized continuously from the primer by DNA polymerase.

DNA polymerase

Replication fork

RNA primer

Primase

DNA polymerase

Okazaki fragment

DNA ligase

5′

3′

Parental strand

DNA polymerase

4 The lagging strand is synthesized discontinuously. Primase, an RNA polymerase, synthesizes a short RNA primer, which is then extended by DNA polymerase.

5 DNA polymerase digests RNA primer and replaces it with DNA.

6 DNA ligase joins the discontinuous fragments of the lagging strand.

**Figure 8.5  A summary of events at the DNA replication fork.**

**Q** Why is one strand of DNA synthesized discontinuously?

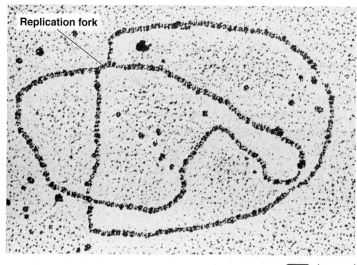

Replication fork

**(a)** An *E. coli* chromosome in the process of replicating

SEM    20 nm

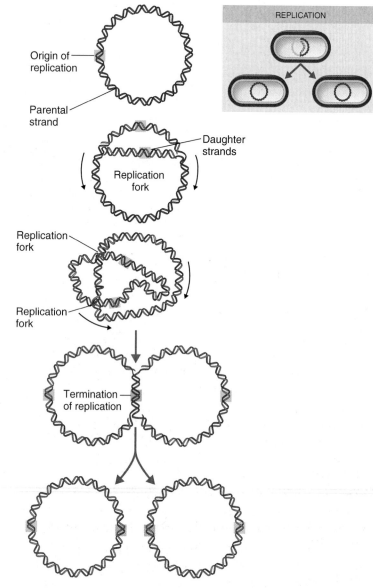

Origin of
replication

Parental
strand

Daughter
strands

Replication
fork

REPLICATION

Replication
fork

Replication
fork

Termination
of replication

**(b)** Bidirectional replication of a circular bacterial DNA molecule

**Figure 8.6 Replication of bacterial DNA.**

Q **What is the origin of replication?**

opposite poles, then each offspring cell receives one copy of the DNA molecule—that is, one complete chromosome.

DNA replication is an amazingly accurate process. Typically, mistakes are made at a rate of only 1 in every 10 billion bases incorporated. Such accuracy is largely due to the *proofreading* capability of DNA polymerase. As each new base is added, the enzyme evaluates whether it forms the proper complementary base-pairing structure. If not, the enzyme excises the improper base and replaces it with the correct one. In this way, DNA can be replicated very accurately, allowing each daughter chromosome to be virtually identical to the parental DNA.

> **Play DNA Replication:**
> *Overview, Forming the Replication Fork, Replication Proteins, Synthesis*
> @MasteringMicrobiology

**CHECK YOUR UNDERSTANDING**

✔ **8-3** Describe DNA replication, including the functions of DNA gyrase, DNA ligase, and DNA polymerase.

## RNA and Protein Synthesis

How is the information in DNA used to make the proteins that control cell activities? In the process of *transcription*, genetic information in DNA is copied, or transcribed, into a complementary base sequence of RNA. The cell then uses the information encoded in this RNA to synthesize specific proteins through the process of *translation*. We now take a closer look at these two processes as they occur in a bacterial cell.

### Transcription in Prokaryotes

**Transcription** is the synthesis of a complementary strand of RNA from a DNA template. We will discuss transcription in prokaryotic cells here. Transcription in eukaryotes is discussed on page 215.

**Ribosomal RNA (rRNA)** forms an integral part of ribosomes, the cellular machinery for protein synthesis. Transfer RNA is also involved in protein synthesis, as we will see. **Messenger RNA (mRNA)** carries the coded information for making specific proteins from DNA to ribosomes, where proteins are synthesized.

During transcription, a strand of mRNA is synthesized using a specific portion of the cell's DNA as a template. In other words, the genetic information stored in the sequence of nucleobases of DNA is rewritten so that the same information appears in the base sequence of mRNA.

As in DNA replication, a guanine (G) in the DNA template dictates a cytosine (C) in the mRNA being made, and a C in the DNA template dictates a G in the mRNA. Likewise, a thymine (T) in the DNA template dictates an adenine (A) in the mRNA.

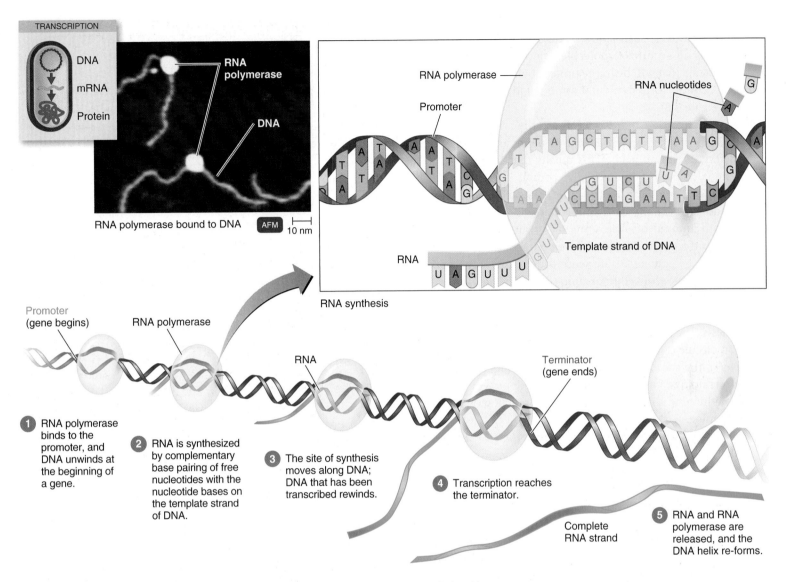

**TRANSCRIPTION**

DNA → mRNA → Protein

RNA polymerase

DNA

RNA polymerase bound to DNA — AFM — 10 nm

RNA polymerase

Promoter

RNA nucleotides

G

A

Template strand of DNA

RNA

U A G U U U

RNA synthesis

Promoter (gene begins)

RNA polymerase

RNA

Terminator (gene ends)

1 RNA polymerase binds to the promoter, and DNA unwinds at the beginning of a gene.

2 RNA is synthesized by complementary base pairing of free nucleotides with the nucleotide bases on the template strand of DNA.

3 The site of synthesis moves along DNA; DNA that has been transcribed rewinds.

4 Transcription reaches the terminator.

Complete RNA strand

5 RNA and RNA polymerase are released, and the DNA helix re-forms.

**Figure 8.7 The process of transcription.** The orienting diagram indicates the relationship of transcription to the overall flow of genetic information within a cell.

**Q** When does transcription stop?

However, an adenine in the DNA template dictates a uracil (U) in the mRNA, because RNA contains uracil instead of thymine. (Uracil has a chemical structure slightly different from thymine, but it base-pairs in the same way.) If, for example, the template portion of DNA has the base sequence 3′-ATGCAT, the newly synthesized mRNA strand will have the complementary base sequence 5′-UACGUA.

The process of transcription requires both an enzyme called *RNA polymerase* and a supply of RNA nucleotides (**Figure 8.7**). Transcription begins when RNA polymerase binds to the DNA at a site called the **promoter**. Only one of the two DNA strands serves as the template for RNA synthesis for a given gene. Like DNA, RNA is synthesized in the 5′ → 3′ direction. RNA synthesis continues until RNA polymerase reaches a site on the DNA called the **terminator**.

Transcription allows the cell to produce short-term copies of genes that can be used as the direct source of information for protein synthesis. Messenger RNA acts as an intermediate between the permanent storage form, DNA, and the process that uses the information, translation.

Play Transcription: *Overview, Process* @Mastering**Microbiology**

**Translation**

We have seen how the genetic information in DNA transfers to mRNA during transcription. Now we will see how mRNA serves as the source of information for the synthesis of proteins. Protein synthesis is called **translation** because it involves decoding the "language" of nucleic acids and converting it into the "language" of proteins.

The language of mRNA is in the form of **codons**, groups of three nucleotides, such as AUG, GGC, or AAA. The sequence of codons on an mRNA molecule determines the sequence of amino acids that will be in the protein being synthesized. Each codon "codes" for a particular amino acid. This is the genetic code (**Figure 8.8**).

Codons are written in terms of their base sequence in mRNA. Notice in Figure 8.8 that there are 64 possible codons but only 20 amino acids. This means that most amino acids are signaled by several alternative codons, a situation referred to as the **degeneracy** of the code. For example, leucine has six codons, and alanine has four codons. Degeneracy allows for a certain amount of misreading of, or mutation in, the DNA without affecting the protein ultimately produced.

Of the 64 codons, 61 are sense codons, and 3 are nonsense codons. **Sense codons** code for amino acids, and **nonsense codons** (also called *stop codons*) do not. Rather, the nonsense codons—UAA, UAG, and UGA—signal the end of the protein molecule's synthesis. The start codon that initiates the synthesis of the protein molecule is AUG, which is also the codon for methionine. In bacteria, the start AUG codes for formylmethionine rather than the methionine found in other parts of the protein. The initiating methionine is often removed later, so not all proteins contain methionine.

During translation, codons of an mRNA are "read" sequentially; and, in response to each codon, the appropriate amino acid is assembled into a growing chain. The site of translation is the ribosome, and **transfer RNA (tRNA)** molecules both recognize the specific codons and transport the required amino acids.

Each tRNA molecule has an **anticodon**, a sequence of three bases that is complementary to a codon. In this way, a tRNA molecule can base-pair with its associated codon. Each tRNA can also carry on its other end the amino acid encoded by the codon that the tRNA recognizes. The functions of the ribosome are to direct the orderly binding of tRNAs to codons and to assemble the amino acids brought there into a chain, ultimately producing a protein.

**Figure 8.9** shows the details of translation. The two ribosomal subunits, a tRNA with the anticodon UAC, and the mRNA molecule to be translated, along with several additional protein factors, all assemble. This sets up the start codon (AUG) in the proper position to allow translation to begin. After the ribosome joins the first two amino acids with a peptide bond, the first tRNA molecule leaves the ribosome. The ribosome then moves along the mRNA to the next codon. As the proper amino acids are brought into line one by one, peptide bonds are formed between them, and a polypeptide chain results. (Also see Figure 2.14, page 42.) Translation ends when one of the three nonsense codons in the mRNA is reached. The ribosome then comes apart into its two subunits, and the mRNA and newly synthesized polypeptide chain are released. The ribosome, the mRNA, and the tRNAs are then available to be used again.

The ribosome moves along the mRNA in the $5' \rightarrow 3'$ direction. As a ribosome moves along the mRNA, it will soon allow the start codon to be exposed. Additional ribosomes can then assemble and begin synthesizing protein. In this way, there are usually a number of ribosomes attached to a single mRNA, all at various stages of protein synthesis. In prokaryotic cells, the translation of mRNA into protein can begin even before

| First position | Second position: U | Second position: C | Second position: A | Second position: G | Third position |
|---|---|---|---|---|---|
| U | UUU Phe | UCU Ser | UAU Tyr | UGU Cys | U |
| U | UUC Phe | UCC Ser | UAC Tyr | UGC Cys | C |
| U | UUA Leu | UCA Ser | UAA Stop | UGA Stop | A |
| U | UUG Leu | UCG Ser | UAG Stop | UGG Trp | G |
| C | CUU Leu | CCU Pro | CAU His | CGU Arg | U |
| C | CUC Leu | CCC Pro | CAC His | CGC Arg | C |
| C | CUA Leu | CCA Pro | CAA Gln | CGA Arg | A |
| C | CUG Leu | CCG Pro | CAG Gln | CGG Arg | G |
| A | AUU Ile | ACU Thr | AAU Asn | AGU Ser | U |
| A | AUC Ile | ACC Thr | AAC Asn | AGC Ser | C |
| A | AUA Ile | ACA Thr | AAA Lys | AGA Arg | A |
| A | AUG Met/start | ACG Thr | AAG Lys | AGG Arg | G |
| G | GUU Val | GCU Ala | GAU Asp | GGU Gly | U |
| G | GUC Val | GCC Ala | GAC Asp | GGC Gly | C |
| G | GUA Val | GCA Ala | GAA Glu | GGA Gly | A |
| G | GUG Val | GCG Ala | GAG Glu | GGG Gly | G |

**Figure 8.8 The genetic code.** The three nucleotides in an mRNA codon are designated, respectively, as the first position, second position, and third position of the codon on the mRNA. Each set of three nucleotides specifies a particular amino acid, represented by a three-letter abbreviation (see Table 2.5, page 41). The codon AUG, which specifies the amino acid methionine, is also the start of protein synthesis. The word *Stop* identifies the nonsense codons that signal the termination of protein synthesis.

**Q What is the advantage of the degeneracy of the genetic code?**

transcription is complete (**Figure 8.10**). Because mRNA is produced in the cytoplasm in prokaryotes, the start codons of an mRNA being transcribed are available to ribosomes before the entire mRNA molecule is even made.

### Transcription in Eukaryotes

In eukaryotic cells, transcription takes place in the nucleus. The mRNA must be completely synthesized and moved through the nuclear membrane to the cytoplasm before translation can begin. In addition, the RNA undergoes processing before it leaves the nucleus. In eukaryotic cells, the regions of genes that code for proteins are often interrupted by noncoding DNA. Thus, eukaryotic genes are composed of **exons**, the regions of DNA *expressed*, and **introns**, the *intervening* regions of DNA that do not encode protein. In the nucleus, RNA polymerase synthesizes a molecule called an RNA transcript that contains copies of the introns. Particles called **small nuclear ribonucleoproteins**, abbreviated **snRNPs** and pronounced "snurps," remove the introns and splice the exons together. In some organisms, the introns act as ribozymes to catalyze their own removal (**Figure 8.11**).

\* \* \*

To summarize, genes are the units of biological information encoded by the sequence of nucleotide bases in DNA. A gene is expressed, or turned into a product within the cell, through the processes of transcription and translation. The genetic information carried in DNA is transferred to a temporary mRNA molecule by transcription. Then, during translation, the mRNA directs the assembly of amino acids into a polypeptide chain: a ribosome attaches to mRNA, tRNAs deliver the amino acids to the ribosome as directed by the mRNA codon sequence, and the ribosome assembles the amino acids into the chain that will be the newly synthesized protein.

 **Play Translation:** *Overview, Genetic Code, Process* @Mastering**Microbiology**

---

**CHECK YOUR UNDERSTANDING**

✔ **8-4** What is the role of the promoter, terminator, and mRNA in transcription?

✔ **8-5** How does mRNA production in eukaryotes differ from the process in prokaryotes?

# The Regulation of Bacterial Gene Expression

### LEARNING OBJECTIVES

**8-6** Define *operon*.

**8-7** Explain pre-transcriptional regulation of gene expression in bacteria.

 **Play Interactive Microbiology** @Mastering**Microbiology** *See how operons affect a patient's health*

**8-8** Explain post-transcriptional regulation of gene expression.

A cell's genetic and metabolic machineries are integrated and interdependent. The bacterial cell carries out an enormous number of metabolic reactions (see Chapter 5). The common feature of all metabolic reactions is that they are catalyzed by enzymes that are proteins synthesized via transcription and translation. Feedback inhibition stops a cell from performing unneeded chemical reactions (Chapter 5, page 116) by stopping enzymes that have already been synthesized. We will now look at mechanisms to prevent synthesis of enzymes that are not needed.

 ASM: The regulation of gene expression is influenced by external and internal molecular cues and/or signals.

Because protein synthesis requires a huge amount of energy, cells save energy by making only those proteins needed at a particular time. Next we look at how chemical reactions are regulated by controlling gene expression.

Many genes, perhaps 60–80%, are not regulated but are instead *constitutive*, meaning that their products are constantly produced at a fixed rate. Usually these genes, which are effectively turned on all the time, code for enzymes that the cell needs in fairly large amounts for its major life processes. Glycolysis enzymes are examples. The production of other enzymes is regulated so that they are present only when needed. *Trypanosoma*, the protozoan parasite that causes African sleeping sickness, has hundreds of genes coding for surface glycoproteins. Each protozoan cell turns on only one glycoprotein gene at a time. As the host's immune system kills parasites with one type of surface molecule, parasites expressing a different surface glycoprotein can continue to grow.

## Pre-transcriptional Control

Two genetic control mechanisms known as repression and induction regulate the transcription of mRNA and, consequently, the synthesis of enzymes from them. These mechanisms control the formation and amounts of enzymes in the cell, not the activities of the enzymes.

### The Operon Model of Gene Expression

Details of the control of gene expression by induction and repression are described by the operon theory formulated in the 1960s by François Jacob and Jacques Monod. An **operon** is a group of genes that are transcribed together and controlled by one promoter. We'll look first at an inducible operon, in which transcription must be turned on. In *E. coli*, the enzymes of the *lac* operon are needed to metabolize lactose. In addition to β-galactosidase, these enzymes include lac permease, which is involved in the transport of lactose into the cell, and transacetylase, which metabolizes certain disaccharides other than lactose.

The genes for the three enzymes involved in lactose uptake and utilization are next to each other on the bacterial

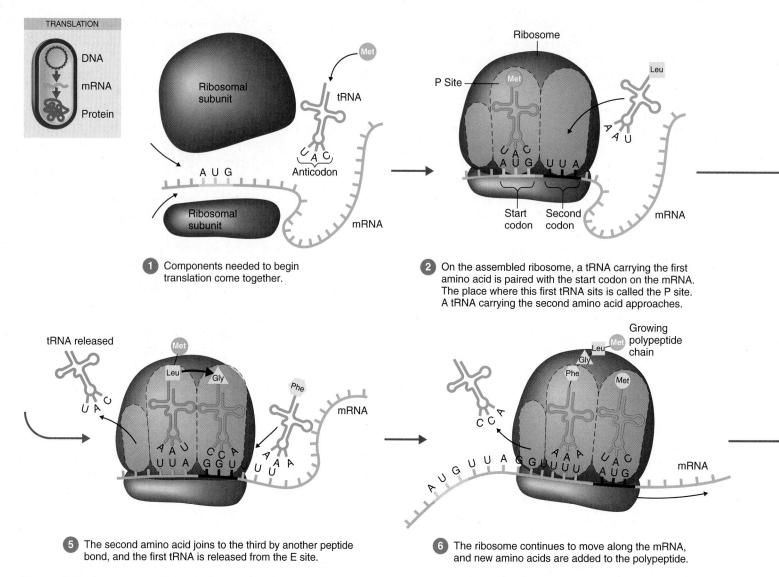

**TRANSLATION**

1 Components needed to begin translation come together.

2 On the assembled ribosome, a tRNA carrying the first amino acid is paired with the start codon on the mRNA. The place where this first tRNA sits is called the P site. A tRNA carrying the second amino acid approaches.

5 The second amino acid joins to the third by another peptide bond, and the first tRNA is released from the E site.

6 The ribosome continues to move along the mRNA, and new amino acids are added to the polypeptide.

**Figure 8.9 The process of translation.** The overall goal of translation is to produce proteins using mRNAs as the source of biological information. The complex cycle of events illustrated here shows the primary role of tRNA and ribosomes in the decoding of this information. The ribosome acts as the site where the mRNA-encoded information is decoded, as well as the site where individual amino acids are connected into polypeptide chains. The tRNA molecules act as the actual "translators"—one end of each tRNA recognizes a specific mRNA codon, while the other end carries the amino acid encoded by that codon.

**Q When does translation stop?**

chromosome and are regulated together (**Figure 8.12**). These genes, which determine the structures of proteins, are called *structural genes* to distinguish them from an adjoining control region on the DNA. When lactose is introduced into the culture medium, the *lac* structural genes are all transcribed and translated rapidly and simultaneously. We will now see how this regulation occurs.

In the control region of the *lac* operon are two relatively short segments of DNA. One, the promoter, is the segment where RNA polymerase initiates transcription. The other is the **operator**, which is like a traffic light that acts as a go or stop signal for transcription of the structural genes. A set of operator and promoter sites and the structural genes they control define an **operon**; thus, the combination of the three *lac* structural genes and the adjoining control regions is called the *lac* operon.

A regulatory gene called the *I gene* encodes a repressor protein that switches inducible and repressible operons on or off. The *lac* operon is an **inducible operon** (see Figure 8.12). In the absence of lactose, the repressor binds to the operator site, thus preventing transcription. If lactose is present, the repressor binds

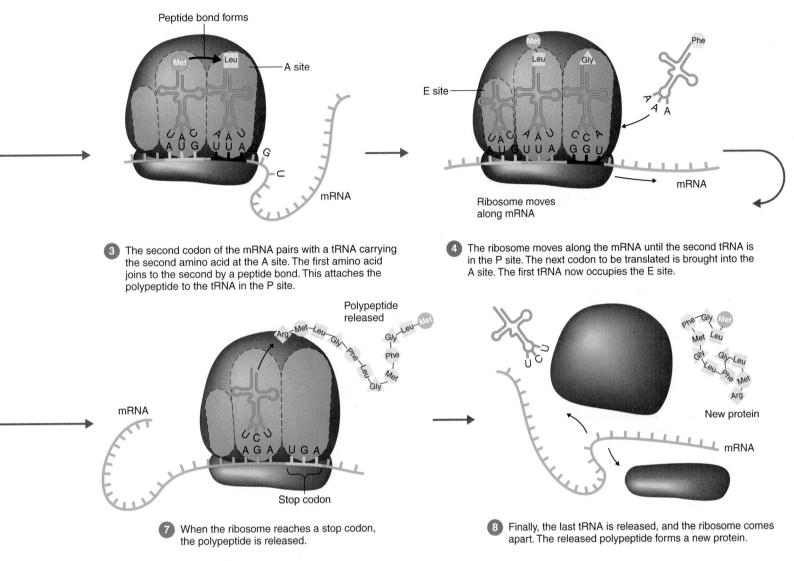

**3** The second codon of the mRNA pairs with a tRNA carrying the second amino acid at the A site. The first amino acid joins to the second by a peptide bond. This attaches the polypeptide to the tRNA in the P site.

**4** The ribosome moves along the mRNA until the second tRNA is in the P site. The next codon to be translated is brought into the A site. The first tRNA now occupies the E site.

**7** When the ribosome reaches a stop codon, the polypeptide is released.

**8** Finally, the last tRNA is released, and the ribosome comes apart. The released polypeptide forms a new protein.

**Figure 8.9 The process of translation.** (*continued*)

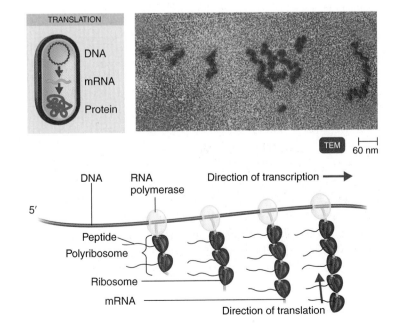

**Figure 8.10 Simultaneous transcription and translation in bacteria.** Many molecules of mRNA are being synthesized simultaneously. The longest mRNA molecules were the first to be transcribed at the promoter. Note the ribosomes attached to the newly forming mRNA. The micrograph shows a polyribosome (many ribosomes) in a single bacterial gene.

**Q** Why can translation begin before transcription is complete in prokaryotes but not in eukaryotes?

# Tracking Zika Virus

In 2014, Brazilian physicians reported clusters of patients with fever and rash. Reverse transcription polymerase chain reaction (RT-PCR) was used to detect dengue, chikungunya, West Nile, and Zika viruses. Public health officials were relieved when the cause was identified as Zika virus (ZIKV) because ZIKV had never made anyone sick enough to go to the hospital. ZIKV is an arbovirus (*arthropod-borne* virus) that is spread between susceptible vertebrate hosts by blood-feeding arthropods, such as mosquitoes. At the same time, local health officials saw a fourfold increase in microcephaly—fetal brains were not developing at the same rate as the body. Some mothers reported having rashes and achy joints, but these symptoms weren't long-lasting, and Zika virus disease was common.

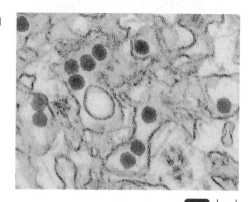

**Zika virus.**

TEM 40 μm

Adriana Melo, an obstetrician, sent samples of amniontic fluid from two patients to be tested. RT-PCR confirmed the presence of ZIKV. By 2016, nearly 5000 cases of microcephaly had been reported in Brazil.

This Old World flavivirus was first identified in 1947 in monkeys in the Zika Forest of Uganda. Prior to 2000, only 14 human cases had been documented in the world. In 2007, an outbreak occurred on the island of Yap in Micronesia. Over 70% of Yap residents were infected with ZIKV. However, no deaths or neurological complications were reported.

The ZIKV genome consists of a positive, single-stranded RNA consisting of 10,794 base pairs. (Positive RNA can act as mRNA and be translated.) The polyprotein encoded by the genome is cut to produce the proteins that make up the virus. The virus has acquired several mutations, and researchers are looking for clues in these mutations to determine the virus's journey around the world.

1. Using the portions of the genomes (shown below) that encode viral proteins, can you determine how similar these viruses are? Can you figure out its movement around the world?

   *Determine the amino acids encoded, and group the viruses based on percentage of similarity to the Uganda strain.*

2. Based on amino acids, there are two groups called clades.

   *Can you identify the two groups?*

3. The two clades are the African and Asian.

   *Calculate the percentage of difference between nucleotides to see how the viruses are related within their clade.*

4. The virus in the Americas is most closely related to the Asian strain that circulated in French Polynesia.

*Source: GenBank genome sequences.*

| | | | | | | | | | | | | | | | | | | |
|---|---|---|---|---|---|---|---|---|---|---|---|---|---|---|---|---|---|---|
| **Brazil** | K | K | R | R | S | A | E | T | S | G | L | L | L | T | A | M | A | V | S | K |
| **Colombia** | K | K | R | R | S | A | E | T | S | G | L | L | L | T | A | M | A | V | N | K |
| **French Polynesia** | K | K | R | R | G | A | D | T | S | L | L | L | L | T | A | M | A | V | S | K |
| **Haiti** | K | K | R | R | G | A | D | T | S | L | L | L | L | T | A | M | A | I | S | K |
| **Mexico** | K | K | R | R | S | A | E | T | S | L | L | L | L | T | A | M | A | V | N | E |
| **Micronesia** | K | K | R | R | G | A | D | T | S | L | L | L | L | T | A | M | A | I | S | K |
| **Nigeria** | R | K | R | R | G | A | D | T | S | L | L | L | L | T | V | M | A | I | S | K |
| **Uganda 1947** | R | K | R | R | G | A | D | A | S | L | L | L | L | T | V | M | A | I | S | K |
| **United States** | K | K | R | R | G | A | E | T | S | L | L | L | L | T | A | M | A | V | S | K |

to a metabolite of lactose instead of to the operator, and lactose-digesting enzymes are transcribed.

**In repressible operons**, the structural genes are transcribed until they are turned off (**Figure 8.13**). The genes for the enzymes involved in the synthesis of tryptophan are regulated in this manner. The structural genes are transcribed and translated, leading to tryptophan synthesis. When excess tryptophan is present, the tryptophan acts as a **corepressor** binding to the repressor protein. The repressor protein can now bind to the operator, stopping further tryptophan synthesis.

Play Operons: *Overview, Induction, Repression*
@MasteringMicrobiology

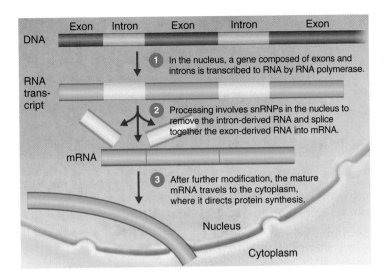

**Figure 8.11 RNA processing in eukaryotic cells.**

**Q** Why can't the RNA transcript be used for translation?

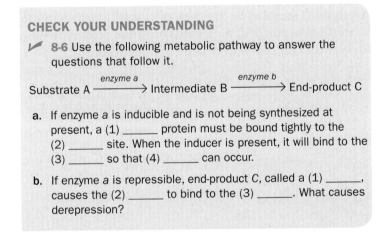

**CHECK YOUR UNDERSTANDING**

✔ 8-6 Use the following metabolic pathway to answer the questions that follow it.

Substrate A $\xrightarrow{enzyme\ a}$ Intermediate B $\xrightarrow{enzyme\ b}$ End-product C

a. If enzyme *a* is inducible and is not being synthesized at present, a (1) _____ protein must be bound tightly to the (2) _____ site. When the inducer is present, it will bind to the (3) _____ so that (4) _____ can occur.

b. If enzyme *a* is repressible, end-product *C*, called a (1) _____, causes the (2) _____ to bind to the (3) _____. What causes derepression?

## Positive Regulation

Regulation of the lactose operon also depends on the level of glucose in the medium, which in turn controls the intracellular level of the small molecule **cyclic AMP (cAMP)**, a substance derived from ATP that serves as a cellular alarm signal. Enzymes that metabolize glucose are constitutive, and cells grow at their maximal rate with glucose as their carbon source because they can use it most efficiently (**Figure 8.14**). When glucose is no longer available, cAMP accumulates in the cell. The cAMP binds to the allosteric site of *catabolic activator protein (CAP)*. CAP then binds to the *lac* promoter, which initiates transcription by making it easier for RNA polymerase to bind to the promoter. Thus transcription of the *lac* operon requires both the presence of lactose and the absence of glucose (**Figure 8.15**).

Cyclic AMP is an example of an *alarmone*, a chemical alarm signal that promotes a cell's response to environmental or nutritional stress. (In this case, the stress is the lack of glucose.)

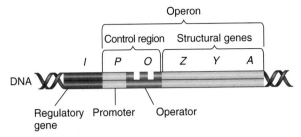

**1** **Structure of the operon.** The operon consists of the promoter (*P*) and operator (*O*) sites and structural genes that code for the protein. The operon is regulated by the product of the regulatory gene (*I*).

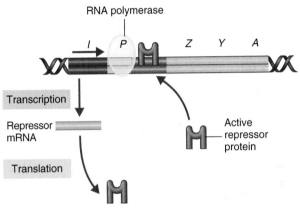

**2** **Repressor active, operon off.** The repressor protein binds with the operator, preventing transcription from the operon.

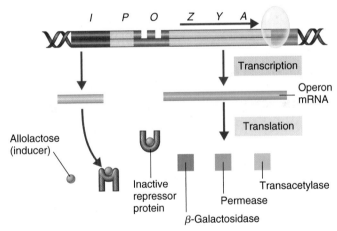

**3** **Repressor inactive, operon on.** When the inducer allolactose binds to the repressor protein, the inactivated repressor can no longer block transcription. The structural genes are transcribed, ultimately resulting in the production of the enzymes needed for lactose catabolism.

**Figure 8.12 An inducible operon.** Lactose-digesting enzymes are produced in the presence of lactose. In *E. coli*, the genes for the three enzymes are in the *lac* operon. β-galactosidase is encoded by *lacZ*. The *lacY* gene encodes the lac permease, and *lacA* encodes transacetylase, whose function in lactose metabolism is still unclear.

**Q** What causes transcription of an inducible enzyme?

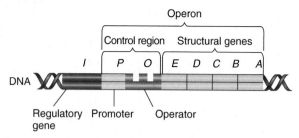

**1** **Structure of the operon.** The operon consists of the promoter (*P*) and operator (*O*) sites and structural genes that code for the protein. The operon is regulated by the product of the regulatory gene (*I*).

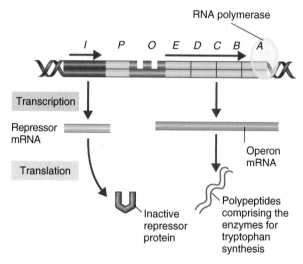

**2** **Repressor inactive, operon on.** The repressor is inactive, and transcription and translation proceed, leading to the synthesis of tryptophan.

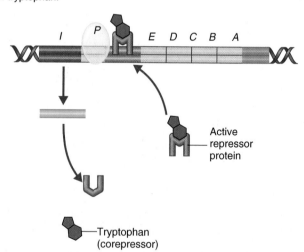

**3** **Repressor active, operon off.** When the corepressor tryptophan binds to the repressor protein, the activated repressor binds with the operator, preventing transcription from the operon.

**Figure 8.13  A repressible operon.** Tryptophan, an amino acid, is produced by anabolic enzymes encoded by five structural genes. Accumulation of tryptophan represses transcription of these genes, preventing further synthesis of tryptophan. The *E. coli trp* operon is shown here.

**Q** What causes transcription of a repressible enzyme?

The same mechanism involving cAMP allows the cell to use other sugars. Inhibition of the metabolism of alternative carbon sources by glucose is termed **catabolite repression** (or the *glucose effect*). When glucose is available, the level of cAMP in the cell is low, and consequently CAP is not bound.

### Epigenetic Control

Eukaryotic and bacterial cells can turn genes off by methylating certain nucleotides—that is, by adding a methyl group ($-CH_3$). The methylated (off) genes are passed to offspring cells. Unlike mutations, this isn't permanent, and the genes can be turned on in a later generation. This is called *epigenetic inheritance* (*epigenetic* = on genes). Epigenetics may explain why bacteria behave differently in a biofilm.

### Post-transcriptional Control

Some regulatory mechanisms stop protein synthesis after transcription has occurred. A part of an mRNA molecule, called a **riboswitch**, that binds to a substrate can change the mRNA structure. Depending on the type of change, translation can be initiated or stopped. Both eukaryotes and prokaryotes use riboswitches to control expression of some genes.

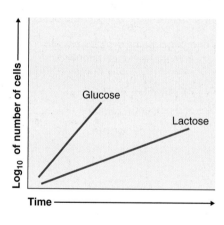

**(a)** Bacteria growing on glucose as the sole carbon source grow faster than on lactose.

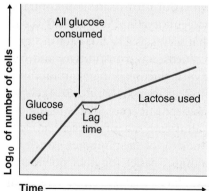

**(b)** Bacteria growing in a medium containing glucose and lactose first consume the glucose and then, after a short lag time, the lactose. During the lag time, intracellular cAMP increases, the *lac* operon is transcribed, lactose is transported into the cell, and β-galactosidase is synthesized to break down lactose.

**Figure 8.14  The growth rate of *E. coli* on glucose and lactose.**

**Q** When both glucose and lactose are present, why will cells use glucose first?

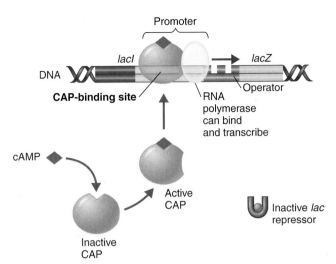

**(a) Lactose present, glucose scarce (cAMP level high).** If glucose is scarce, the high level of cAMP activates CAP, and the *lac* operon produces large amounts of mRNA for lactose digestion.

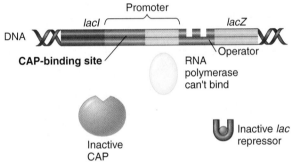

**(b) Lactose present, glucose present (cAMP level low).** When glucose is present, cAMP is scarce, and CAP is unable to stimulate transcription.

**Figure 8.15 Positive regulation of the *lac* operon.**

**Q** Will transcription of the *lac* operon occur in the presence of lactose and glucose? In the presence of lactose and the absence of glucose? In the presence of glucose and the absence of lactose?

Single-stranded RNA molecules of approximately 22 nucleotides, called **microRNAs (miRNAs)**, inhibit protein production in eukaryotic cells. In humans, miRNAs produced during development allow different cells to produce different proteins. Heart cells and skin cells have the same genes, but the cells in each organ produce different proteins because of miRNAs produced in each cell type during development. Similar short RNAs in bacteria enable the cell to cope with environmental stresses, such as low temperature or oxidative damage. An miRNA base-pairs with a complementary mRNA, forming a double-stranded RNA. This double-stranded RNA is enzymatically destroyed so that the mRNA-encoded protein is not made (**Figure 8.16**). The action of another type of RNA, siRNA, is similar and is discussed on page 256.

is discussed on page 256.

# Changes in Genetic Material

**LEARNING OBJECTIVES**

8-9 Classify mutations by type.

8-10 Describe two ways mutations can be repaired.

8-11 Describe the effect of mutagens on the mutation rate.

8-12 Outline the methods of direct and indirect selection of mutants.

8-13 Identify the purpose of and outline the procedure for the Ames test.

A cell's DNA can be changed by mutations and horizontal gene transfer. Changes in DNA result in genetic variations that can impact microbial function (e.g., biofilm formation, pathogenicity, and antibiotic resistance). Survival and reproduction of

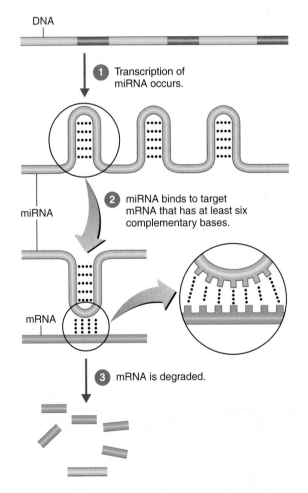

**Figure 8.16 MicroRNAs control a wide range of activities in cells.**

**Q** In mammals, some miRNAs hybridize with viral RNA. What would happen if a mutation occurred in the miRNA gene?

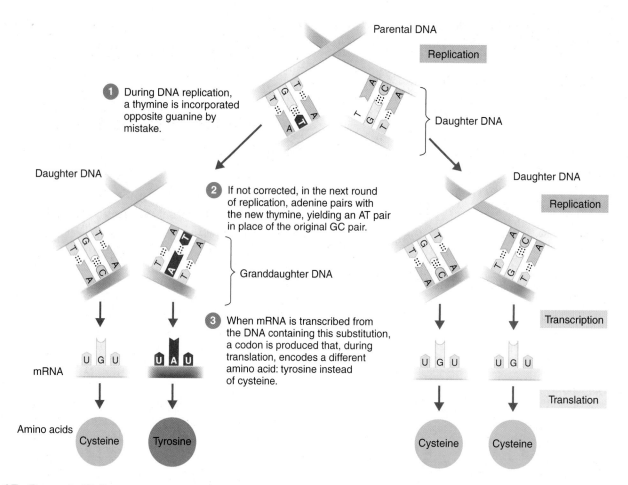

**Figure 8.17 Base substitutions.** This mutation leads to an altered protein in a grandchild cell.

**Q** Does a base substitution always result in a different amino acid?

the bacteria with a new genotype can be favored by natural and human-influenced environments and result in a huge diversity of microorganisms. The survival of new genotypes is called **natural selection**.

## Mutation

A **mutation** is a permanent change in the base sequence of DNA. Such a change will sometimes cause a change in the product encoded by that gene. For example, when the gene for an enzyme mutates, the enzyme encoded by the gene may become inactive or less active because its amino acid sequence has changed. Such a change in genotype may be disadvantageous, or even lethal, if the cell loses a phenotypic trait it needs. However, a mutation can be beneficial if, for instance, the altered enzyme encoded by the mutant gene has a new or enhanced activity that benefits the cell. See the Clinical Focus box in Chapter 26, page 771.

## Types of Mutations

Many simple mutations are silent (neutral); the change in DNA base sequence causes no change in the activity of the product encoded by the gene. Silent mutations commonly occur when one nucleotide is substituted for another in the DNA, especially at a location corresponding to the third position of the mRNA codon. Because of the degeneracy of the genetic code, the resulting new codon might still code for the same amino acid. Even if the amino acid is changed, the function of the protein may not change if the amino acid is in a nonvital portion of the protein, or is chemically very similar to the original amino acid.

The most common type of mutation involving single base pairs is **base substitution** (or *point mutation*), in which a single base at one point in the DNA sequence is replaced with a different base. When the DNA replicates, the result is a substituted base pair (**Figure 8.17**). For example, AT might be substituted for GC, or CG for GC. If a base substitution occurs within a gene that codes for a protein, the mRNA transcribed from the gene will carry an incorrect base at that position. When the mRNA is translated into protein, the incorrect base may cause the insertion of an incorrect amino acid in the protein. If the base substitution results in an amino acid substitution in the synthesized

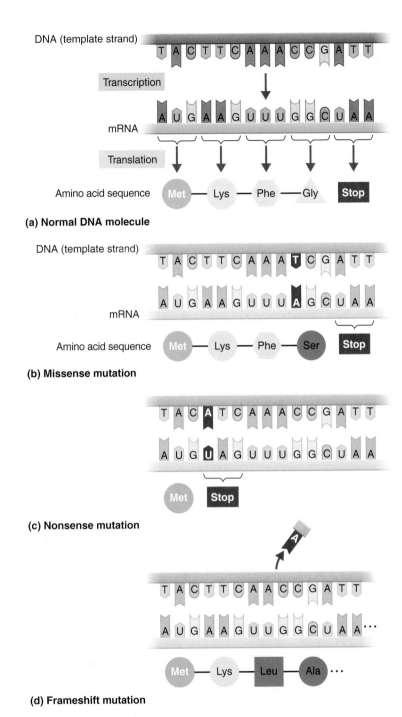

DNA (template strand)

Transcription

mRNA

Translation

Amino acid sequence

**(a) Normal DNA molecule**

DNA (template strand)

mRNA

Amino acid sequence

**(b) Missense mutation**

**(c) Nonsense mutation**

**(d) Frameshift mutation**

**Figure 8.18 Types of mutations and their effects on the amino acid sequences of proteins.**

**Q** What happens if base 9 in (a) is changed to a C?

protein, this change in the DNA is known as a **missense mutation** (**Figure 8.18a** and Figure 8.18b).

The effects of such mutations can be dramatic. For example, sickle cell disease is caused by a single change in the gene for globin, the protein component of hemoglobin. Hemoglobin is primarily responsible for transporting oxygen from the lungs to the tissues. A single change from an A to a T at a specific site results in the change from glutamic acid to valine in the protein. This causes the shape of the hemoglobin molecule to change under conditions of low oxygen, which, in turn, alters the shape of the red blood cells.

By creating a nonsense (stop) codon in the middle of an mRNA molecule, some base substitutions effectively prevent the synthesis of a complete functional protein; only a fragment is synthesized. A base substitution resulting in a nonsense codon is thus called a **nonsense mutation** (Figure 8.18c).

Besides base-pair mutations, there are also changes in DNA called **frameshift mutations**, in which one or a few nucleotide pairs are deleted or inserted in the DNA (Figure 8.18d). This mutation can shift the "translational reading frame"—that is, the three-by-three grouping of nucleotides recognized as codons by the tRNAs during translation. For example, deleting one nucleotide pair in the middle of a gene causes changes in many amino acids downstream from the site of the original mutation. Frameshift mutations almost always result in a long stretch of altered amino acids and the production of an inactive protein from the mutated gene. In most cases, a nonsense codon will eventually be encountered and thereby terminate translation.

Base substitutions and frameshift mutations may occur spontaneously because of occasional mistakes made during DNA replication. These **spontaneous mutations** apparently occur in the absence of any mutation-causing agents.

> **CHECK YOUR UNDERSTANDING**
>
> ✔ 8-9 How can a mutation be beneficial?

## Mutagens

### Chemical Mutagens

Agents in the environment, such as certain chemicals and radiation, that directly or indirectly bring about mutations are called **mutagens**.

One of the many chemicals known to be a mutagen is nitrous acid. **Figure 8.19** shows how exposing DNA to nitrous acid can convert the base adenine to a form that pairs with cytosine instead of the usual thymine. When DNA containing such modified adenines replicates, one daughter DNA molecule will have a base-pair sequence different from that of the parent DNA. Eventually, some AT base pairs of the parent will have been changed to GC base pairs in a granddaughter cell. Nitrous acid makes a specific base-pair change in DNA. Like all mutagens, it alters DNA at random locations.

Another type of chemical mutagen is the **nucleoside analog**. These molecules are structurally similar to normal nitrogenous bases, but they have slightly altered base-pairing

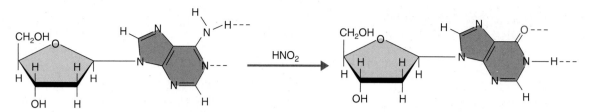

**(a)** Adenosine nucleoside normally base-pairs by hydrogen bonds with an oxygen and a hydrogen of a thymine or uracil nucleotide.

Altered adenine will hydrogen bond with a hydrogen and a nitrogen of a cytosine nucleotide.

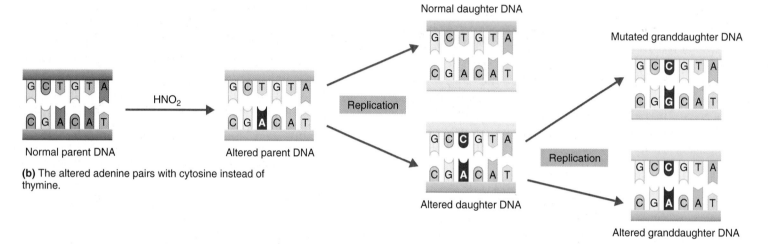

Normal daughter DNA

Mutated granddaughter DNA

Replication

Replication

Normal parent DNA

Altered parent DNA

Altered daughter DNA

Altered granddaughter DNA

**(b)** The altered adenine pairs with cytosine instead of thymine.

**Figure 8.19 Oxidation of nucleotides makes a mutagen.** The nitrous acid emitted into the air by burning fossil fuels oxidizes adenine.

**Play MicroFlix 3D Animation** @MasteringMicrobiology

**Q** What is a mutagen?

properties. Examples, 2-aminopurine and 5-bromouracil, are shown in **Figure 8.20**. When nucleoside analogs are given to growing cells, the analogs are randomly incorporated into cellular DNA in place of the normal bases. Then, during DNA replication, the analogs cause mistakes in base pairing. The

**CLINICAL CASE**

A person's DNA can undergo mutations. One improper nucleotide in DNA creates a mutation, which could alter the function of the gene. Cancer is abnormal cell growth caused by mutations. These mutations can be inherited.

As Marcel and his wife, Janice, drive home from the doctor's office, they review Marcel's family history. Marcel's brother, Robert, passed away from colon cancer 10 years ago, but Marcel has always been the picture of health. Even at 70, he hasn't given a thought to retiring from his Memphis barbeque restaurant that he once co-owned with his brother until Robert's death.

**What factors may have contributed to Marcel's colon cancer?**

**Normal nitrogenous base**            **Analog**

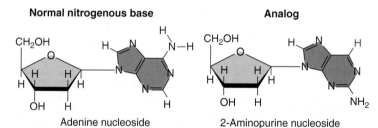

Adenine nucleoside            2-Aminopurine nucleoside

**(a)** The 2-aminopurine is incorporated into DNA in place of adenine but can pair with cytosine, so an AT pair becomes a CG pair.

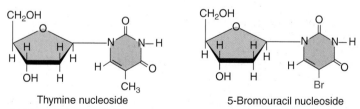

Thymine nucleoside            5-Bromouracil nucleoside

**(b)** The 5-bromouracil is used as an anticancer drug because it is mistaken for thymine by cellular enzymes but pairs with cytosine. In the next DNA replication, an AT pair becomes a GC pair.

**Figure 8.20 Nucleoside analogs and the nitrogenous bases they replace.** A nucleoside is phosphorylated, and the resulting nucleotide used to synthesize DNA.

**Q** Why do these drugs kill cells?

incorrectly paired bases will be copied during subsequent replication of the DNA, resulting in base-pair substitutions in the progeny cells. Some antiviral and antitumor drugs are nucleoside analogs, including AZT (azidothymidine), used to treat HIV infection.

Still other chemical mutagens cause small deletions or insertions, which can result in frameshifts. For instance, under certain conditions, benzopyrene, which is present in smoke and soot, is an effective *frameshift mutagen*. Aflatoxin—produced by *Aspergillus flavus* (a-sper-JIL-lus FLĀ-vus), a mold that grows on peanuts and grain—is a frameshift mutagen. Frameshift mutagens usually have the right size and chemical properties to slip between the stacked base pairs of the DNA double helix. They may work by slightly offsetting the two strands of DNA, leaving a gap or bulge in one strand or the other. When the staggered DNA strands are copied during DNA synthesis, one or more base pairs can be inserted or deleted in the new double-stranded DNA. Interestingly, frameshift mutagens are often potent carcinogens.

### Radiation

X rays and gamma rays are forms of radiation that are potent mutagens because of their ability to ionize atoms and molecules. The penetrating rays of ionizing radiation cause electrons to pop out of their usual shells (see Chapter 2). These electrons bombard other molecules and cause more damage, and many of the resulting ions and free radicals (molecular fragments with unpaired electrons) are very reactive. Some of these ions oxidize bases in DNA, resulting in errors in DNA replication and repair that produce mutations (see Figure 8.19). An even more serious outcome is the breakage of covalent bonds in the sugar-phosphate backbone of DNA, which causes physical breaks in chromosomes.

Another form of mutagenic radiation is ultraviolet (UV) light, a nonionizing component of ordinary sunlight. However, the most mutagenic component of UV light (wavelength 260 nm) is screened out by the ozone layer of the atmosphere. The most important effect of direct UV light on DNA is the formation of harmful covalent bonds between pyrimidine bases. Adjacent thymines in a DNA strand can cross-link to form thymine dimers. Such dimers, unless repaired, may cause serious damage or death to the cell because it cannot properly transcribe or replicate such DNA.

Bacteria and other organisms have enzymes that can repair UV-induced damage. **Photolyases**, also known as *light-repair enzymes*, use visible light energy to separate the dimer back to the original two thymines. **Nucleotide excision repair**, shown in **Figure 8.21**, is not restricted to UV-induced damage; it can repair mutations from other causes as well. Enzymes cut out the incorrect base and fill in the gap with newly synthesized

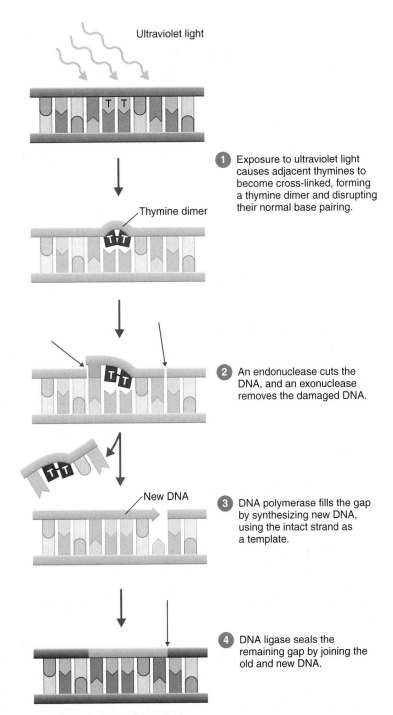

Ultraviolet light

Thymine dimer

New DNA

1 Exposure to ultraviolet light causes adjacent thymines to become cross-linked, forming a thymine dimer and disrupting their normal base pairing.

2 An endonuclease cuts the DNA, and an exonuclease removes the damaged DNA.

3 DNA polymerase fills the gap by synthesizing new DNA, using the intact strand as a template.

4 DNA ligase seals the remaining gap by joining the old and new DNA.

**Figure 8.21  The creation and repair of a thymine dimer caused by ultraviolet light.** After exposure to UV light, adjacent thymines can become cross-linked, forming a thymine dimer. In the absence of visible light, the nucleotide excision repair mechanism is used in a cell to repair the damage.

**Q** How do excision repair enzymes "know" which strand is incorrect?

DNA that is complementary to the correct strand. For many years biologists questioned how the incorrect base could be distinguished from the correct base if it was not physically distorted like a thymine dimer. In 1970, Hamilton Smith provided the answer with the discovery of **methylases.** These enzymes add a methyl group to selected bases soon after a DNA strand is made. A repair endonuclease then cuts the nonmethylated strand.

## The Frequency of Mutation

The **mutation rate** is the probability that a gene will mutate when a cell divides. The rate is usually stated as a power of 10, and because mutations are very rare, the exponent is always a negative number. For example, if there is one chance in a million that a gene will mutate when the cell divides, the mutation rate is 1/1,000,000, which is expressed as $10^{-6}$. Spontaneous mistakes in DNA replication occur at a very low rate, perhaps only once in $10^9$ replicated base pairs (a mutation rate of one in a billion). Because the average gene has about $10^3$ base pairs, the spontaneous rate of mutation is about one in $10^6$ (a million) replicated genes.

Mutations usually occur more or less randomly along a chromosome. The occurrence of random mutations at low frequency is an essential aspect of the adaptation of species to their environment, for evolution requires that genetic diversity be generated randomly and at a low rate. For example, in a bacterial population of significant size—say, greater than $10^7$ cells—a few new mutant cells will always be produced in every generation. Most mutations either are harmful and likely to be removed from the gene pool when the individual cell dies or are neutral. However, a few mutations may be beneficial. For example, a mutation that confers antibiotic resistance is beneficial to a population of bacteria that is regularly exposed to antibiotics. Once such a trait has appeared through mutation, cells carrying the mutated gene are more likely than other cells to survive and reproduce as long as the environment stays the same. Soon most of the cells in the population will have the gene; an evolutionary change will have occurred, although on a small scale.

A mutagen usually increases the spontaneous rate of mutation, which is about one in $10^6$ replicated genes, by a factor of 10 to 1000 times. In other words, in the presence of a mutagen, the normal rate of $10^{-6}$ mutations per replicated gene becomes a rate of $10^{-5}$ to $10^{-3}$ per replicated gene. Mutagens are used experimentally to enhance the production of mutant cells for research on the genetic properties of microorganisms and for commercial purposes.

 Play Mutations: *Types, Repair* @Mastering**Microbiology**

### CHECK YOUR UNDERSTANDING

✔ 8-10 How can mutations be repaired?

✔ 8-11 How do mutagens affect the mutation rate?

## Identifying Mutants

Mutants can be detected by selecting or testing for an altered phenotype. Whether or not a mutagen is used, mutant cells with specific mutations are always rare compared with other cells in the population. The problem is detecting such a rare event.

Experiments are usually performed with bacteria because they reproduce rapidly, so large numbers of organisms (more than $10^9$ per milliliter of nutrient broth) can easily be used. Furthermore, because bacteria generally have only one copy of each gene per cell, the effects of a mutated gene are not masked by the presence of a normal version of the gene, as in many eukaryotic organisms.

**Positive (direct) selection** involves the detection of mutant cells by rejection of the unmutated parent cells. For example, suppose we were trying to find mutant bacteria that are resistant to penicillin. When the bacterial cells are plated on a medium containing penicillin, the mutant can be identified directly. The few cells in the population that are resistant (mutants) will grow and form colonies, whereas the normal, penicillin-sensitive parental cells cannot grow.

To identify mutations in other kinds of genes, **negative (indirect) selection** can be used. This process selects a cell that cannot perform a certain function, using the technique of **replica plating.** For example, suppose we wanted to use replica plating to identify a bacterial cell that has lost the ability to synthesize the amino acid histidine (**Figure 8.22**). First, about 100 bacterial cells are inoculated onto an agar plate. This plate, called the master plate, contains a medium with histidine on which all cells will grow. After 18 to 24 hours of incubation, each cell reproduces to form a colony. Then a pad of sterile material, such as latex, filter paper, or velvet, is pressed over the master plate, and some of the cells from each colony adhere to the velvet. Next, the velvet is pressed down onto two (or more) sterile plates. One plate contains a medium without histidine, and one contains a medium with histidine on which the original, nonmutant bacteria can grow. Any colony that grows on the medium with histidine on the master plate but that cannot synthesize its own histidine will not be able to grow on the medium without histidine. The mutant colony can then be identified on the master plate. Of course, because mutants are so rare (even those induced by mutagens), many plates must be screened with this technique to isolate a specific mutant.

Replica plating is a very effective means of isolating mutants that require one or more new growth factors. Any mutant microorganism having a nutritional requirement that is absent in the parent is known as an **auxotroph.** For example, an auxotroph may lack an enzyme needed to synthesize a particular amino acid and will therefore require that amino acid as a growth factor in its nutrient medium.

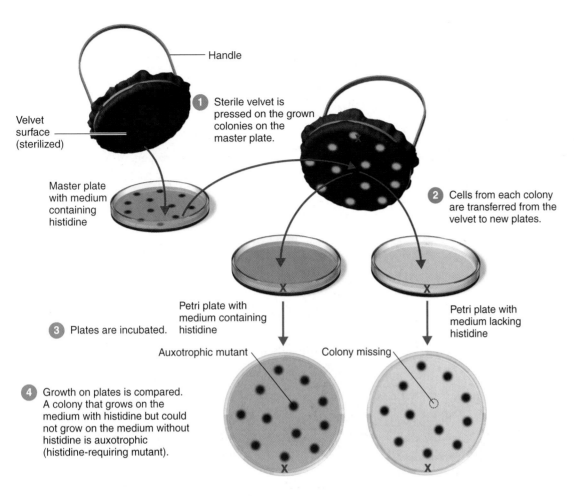

**Figure 8.22 Replica plating.** In this example, the auxotrophic mutant cannot synthesize histidine. The plates must be carefully marked (with an X here) to maintain orientation so that colony positions are known in relation to the original master plate.

**Q** What is an auxotroph?

Handle

Velvet surface (sterilized)

**1** Sterile velvet is pressed on the grown colonies on the master plate.

Master plate with medium containing histidine

**2** Cells from each colony are transferred from the velvet to new plates.

**3** Plates are incubated.

Petri plate with medium containing histidine

Petri plate with medium lacking histidine

**4** Growth on plates is compared. A colony that grows on the medium with histidine but could not grow on the medium without histidine is auxotrophic (histidine-requiring mutant).

Auxotrophic mutant

Colony missing

## Identifying Chemical Carcinogens

Many known mutagens have been found to be **carcinogens**, substances that cause cancer in animals, including humans. In recent years, chemicals in the environment, the workplace, and the diet have been implicated as causes of cancer in humans. Animal testing procedures are time-consuming and expensive, so some faster and less expensive procedures for preliminary screening of potential carcinogens that do not use animals have been developed. One of these, called the **Ames test**, uses bacteria as carcinogen indicators.

The Ames test is based on the observation that exposure of mutant bacteria to mutagenic substances may cause new mutations that reverse the effect (the change in phenotype) of the original mutation. These are called *reversions*. Specifically, the test measures the reversion of histidine auxotrophs of *Salmonella* (so-called his⁻ cells, mutants that have lost the ability to synthesize histidine) to histidine-synthesizing cells (his⁺) after treatment with a mutagen (**Figure 8.23**). Bacteria are incubated in both the presence and absence of the substance being tested. Because animal enzymes must activate many chemicals into forms that are chemically reactive for mutagenic or carcinogenic activity to appear, the chemical to be tested

**CLINICAL CASE**

Not all mutations are inherited; some are induced by genotoxins, that is, chemicals that damage a cell's genetic material. Marcel is not overweight and has never smoked. Researchers have known since the 1970s that people who consume cooked meat and meat products are more likely to develop colon cancer. The suspect cancer-causing chemicals are aromatic amines that form during high-heat cooking.

Marcel has owned his Memphis barbeque restaurant for over 50 years. He is a hands-on type of employer and is always in the kitchen overseeing the cooking process. All of his barbequed meat is seared over high heat and then slow-cooked for hours. Marcel is considered the expert in this technique, but now it seems as if his profession could be a factor in his disease.

**What test can be used to determine whether a chemical is genotoxic?**

205  224  **227**  234

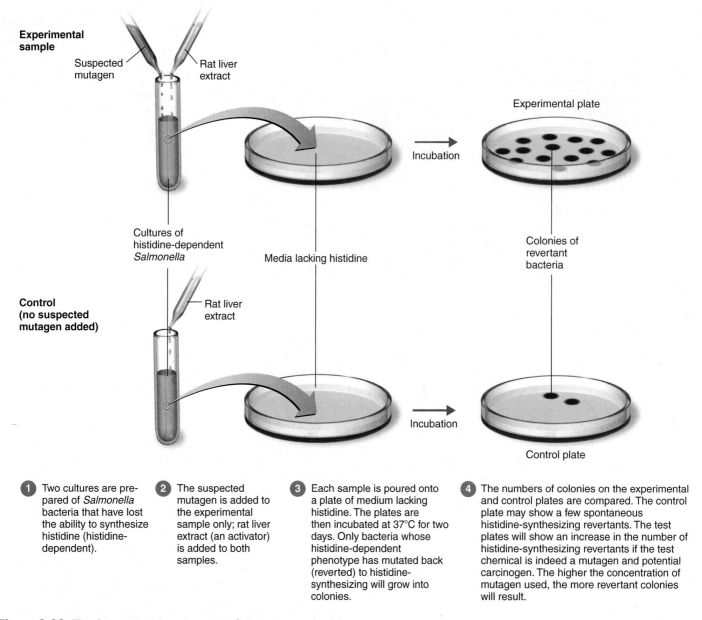

Experimental
sample

Suspected
mutagen

Rat liver
extract

Cultures of
histidine-dependent
*Salmonella*

Media lacking histidine

Control
(no suspected
mutagen added)

Rat liver
extract

Incubation

Experimental plate

Colonies of
revertant
bacteria

Incubation

Control plate

1 Two cultures are prepared of *Salmonella* bacteria that have lost the ability to synthesize histidine (histidine-dependent).

2 The suspected mutagen is added to the experimental sample only; rat liver extract (an activator) is added to both samples.

3 Each sample is poured onto a plate of medium lacking histidine. The plates are then incubated at 37°C for two days. Only bacteria whose histidine-dependent phenotype has mutated back (reverted) to histidine-synthesizing will grow into colonies.

4 The numbers of colonies on the experimental and control plates are compared. The control plate may show a few spontaneous histidine-synthesizing revertants. The test plates will show an increase in the number of histidine-synthesizing revertants if the test chemical is indeed a mutagen and potential carcinogen. The higher the concentration of mutagen used, the more revertant colonies will result.

**Figure 8.23  The Ames reverse gene mutation test.**

**Q** Do all mutagens cause cancer?

and the mutant bacteria are incubated together with rat liver extract, a rich source of activation enzymes. If the substance being tested is mutagenic, it will cause the reversion of his⁻ bacteria to his⁺ bacteria at a rate higher than the spontaneous reversion rate. The number of observed revertants indicates the degree to which a substance is mutagenic and therefore possibly carcinogenic.

The Ames test can be performed in liquid media with a pH indicator in a 96-well plate. Several potential mutagens or different concentrations of mutagens can be qualitatively tested in different wells. Bacterial growth is determined by a color change of the pH indicator. The Ames test is routinely used to evaluate new chemicals and air and water pollutants.

About 90% of the substances found by the Ames test to be mutagenic have also been shown to be carcinogenic in animals. By the same token, the more mutagenic substances have generally been found to be more carcinogenic.

**CHECK YOUR UNDERSTANDING**

✔ 8-12 How would you isolate an antibiotic-resistant bacterium? An antibiotic-sensitive bacterium?

✔ 8-13 What is the principle behind the Ames test?

# Genetic Transfer and Recombination

## LEARNING OBJECTIVES

**8-14** Describe the functions of plasmids and transposons.

**8-15** Differentiate horizontal and vertical gene transfer.

**8-16** Compare the mechanisms of genetic recombination in bacteria.

**Genetic recombination** refers to the exchange of genes between two DNA molecules to form new combinations of genes on a chromosome. **Figure 8.24** shows one mechanism for genetic recombination. If a cell picks up foreign DNA (called donor DNA in the figure), some of it could insert into the cell's chromosome—a process called **crossing over**—and some of the genes carried by the chromosomes are shuffled. The DNA has recombined, so that the chromosome now carries a portion of the donor's DNA.

If A and B represent DNA from different individuals, how are they brought close enough together to recombine? In eukaryotes, genetic recombination is an ordered process that usually occurs as part of the sexual cycle of the organism. Crossing over generally takes place during the formation of reproductive cells, such that these cells contain recombinant DNA. In bacteria, genetic recombination can happen in a number of ways, which we will discuss in the following sections.

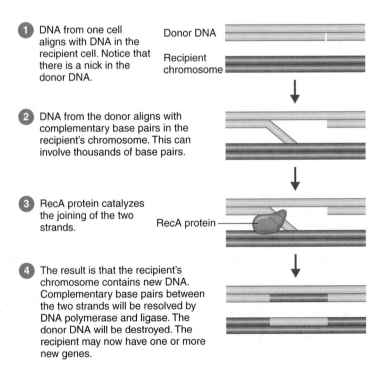

① DNA from one cell aligns with DNA in the recipient cell. Notice that there is a nick in the donor DNA.

Donor DNA

Recipient chromosome

② DNA from the donor aligns with complementary base pairs in the recipient's chromosome. This can involve thousands of base pairs.

③ RecA protein catalyzes the joining of the two strands.

RecA protein

④ The result is that the recipient's chromosome contains new DNA. Complementary base pairs between the two strands will be resolved by DNA polymerase and ligase. The donor DNA will be destroyed. The recipient may now have one or more new genes.

**Figure 8.24 Genetic recombination by crossing over.** Foreign DNA can be inserted into a chromosome by breaking and rejoining the chromosome. This can insert one or more new genes into the chromosome. A photograph of RecA protein is shown in Figure 3.11a, page 60.

**Q** What type of enzyme breaks the DNA?

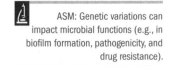

ASM: Genetic variations can impact microbial functions (e.g., in biofilm formation, pathogenicity, and drug resistance).

Like mutation, genetic recombination contributes to a population's genetic diversity, which is the source of variation in evolution. In highly evolved organisms such as present-day microbes, recombination is more likely to be beneficial than mutation because recombination will less likely destroy a gene's function and may bring together combinations of genes that enable the organism to carry out a valuable new function.

The major protein that constitutes the flagella of *Salmonella* is also one of the primary proteins that causes our immune systems to respond. However, these bacteria have the capability of producing two different flagellar proteins. As our immune system mounts a response against those cells containing one form of the flagellar protein, those organisms producing the second are not affected. Which flagellar protein is produced is determined by a recombination event that apparently occurs somewhat randomly within the chromosomal DNA. Thus, by altering the flagellar protein produced, *Salmonella* can better avoid the defenses of the host.

**Vertical gene transfer** occurs when genes are passed from an organism to its offspring. Plants and animals transmit their genes by vertical transmission. Bacteria can pass their genes not only to their offspring, but also laterally, to other microbes  of the same generation. This is known as **horizontal gene transfer** (see Figure 8.2). Horizontal gene transfer between normal microbiota and pathogens may be important in the spread of antibiotic resistance. Horizontal gene transfer between bacteria occurs in several ways. In all of the mechanisms, the transfer involves a **donor cell** that gives a portion of its total DNA to a **recipient cell**. Once transferred, part of the donor's DNA can be incorporated into the recipient's DNA; the remainder is degraded by cellular enzymes. The recipient cell that incorporates donor DNA into its own DNA is called a *recombinant*. The transfer of genetic material between bacteria is by no means a frequent event; it may occur in only 1% or less of an entire population. Let's examine in detail the specific types of genetic transfer.

**Play Horizontal Gene Transfer:** Overview @Mastering**Microbiology**

## Plasmids and Transposons

Plasmids and transposons are genetic elements that exist outside chromosomes. They occur in both prokaryotic and eukaryotic organisms, but this discussion focuses on their role in genetic change in prokaryotes. Plasmids and transposons are called **mobile genetic elements** because they can move from one chromosome to another or from one cell to another.

### Plasmids

Recall from Chapter 4 (page 90) that plasmids are self-replicating, gene-containing, circular pieces of DNA about 1–5% the size of

**Horizontal Gene Transfer and the Unintended Consequences of Antibiotic Usage**

The number of antibiotic-resistant bacteria in our intestinal microbiome increases with age. The reason: exposure to antibiotics. In the presence of a bacteria-killing drug, a resistant mutant will grow while the nonresistant, or susceptible, bacteria die off. So over the human life span, which includes many episodes of illnesses and treatments, we end up populated with more and more antibiotic-resistant microbes.

At first, this seems like a desirable effect. For instance, if beneficial intestinal microbes survive a course of drugs meant to treat your pneumonia, then you may not experience medication side effects such as GI discomfort or diarrhea. Unfortunately, recent evidence shows that a drug-resistant microbiome may actually threaten us in ways we previously didn't understand.

Scientists suspect that drug resistance in pathogenic bacteria often originates from drug-resistant normal microbiota. Nearly half of the resistance genes identified in intestinal bacteria are identical to resistance genes found in pathogens. Swapping of genes between species that come in contact with each other (horizontal gene transfer) happens easily in the intestines, where large numbers of different microbes mingle. In one study, *Escherichia coli* bacteria that were resistant to the drugs sulfonamide and ampicillin were found residing in volunteers who ingested *E. coli* bacteria that were susceptible to these antibiotics. How could this be? The researchers traced the drug-resistance genes to a plasmid found in *E. coli* that had resided in the volunteers before the study— the resistant bacteria had transferred the plasmid to the drug-susceptible bacteria once the different strains met up in the intestine. Likewise, resistance to the drug vancomycin is believed to have transferred from the commensal bacterium *Enterococcus faecalis* to pathogenic strains of *Staphylococcus aureus*. The resistance gene was found on a conjugative plasmid in both species.

Antibiotics remain an essential part of modern health care. However, these days, weighing whether an antibiotic is truly needed is all the more important.

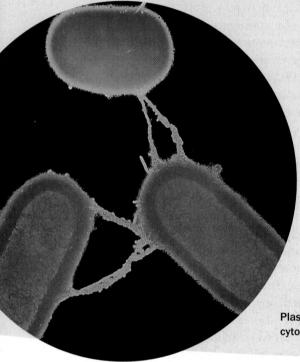

**Plasmids can be transferred between unrelated bacteria through cytoplasmic bridges between cells.**

the bacterial chromosome (**Figure 8.25**). They are found mainly in bacteria but also in some eukaryotic microorganisms, such as *Saccharomyces cerevisiae*. The F factor is a **conjugative plasmid** that carries genes for sex pili and for the transfer of the plasmid to another cell. Although plasmids are usually dispensable, under certain conditions genes carried by plasmids can be crucial to the survival and growth of the cell. For example, **dissimilation plasmids** code for enzymes that trigger the catabolism of certain unusual sugars and hydrocarbons. Some species of *Pseudomonas* can actually use such exotic substances as toluene, camphor, and petroleum as primary carbon and energy sources because they have catabolic enzymes encoded by genes carried on plasmids. Such specialized capabilities permit the survival of those microorganisms in very diverse and challenging environments. Because of their ability to degrade and detoxify a variety of unusual compounds, many of them are being investigated for possible use in the cleanup of environmental wastes.

Other plasmids code for proteins that enhance the pathogenicity of a bacterium. The strain of *E. coli* that causes infant diarrhea and traveler's diarrhea carries plasmids that code for toxin production and for bacterial attachment to intestinal cells. Without these plasmids, *E. coli* is a harmless resident of the large intestine; with them, it is pathogenic. Other plasmid-encoded toxins include the exfoliative toxin of *Staphylococcus aureus*, *Clostridium tetani* neurotoxin, and toxins of *Bacillus anthracis*. Still other plasmids contain genes for the synthesis of **bacteriocins**, toxic proteins that kill other bacteria. These plasmids have been found in many bacterial genera, and they are useful markers for the identification of certain bacteria in clinical laboratories.

**Resistance factors (R factors)** are plasmids that have significant medical importance. They were first discovered in Japan in the late 1950s after several dysentery epidemics. In some of these epidemics, the infectious agent was resistant to the usual antibiotic. Following isolation, the pathogen was also found to be resistant to a number of different antibiotics. In addition, other normal bacteria from the patients (such as *E. coli*) proved to be resistant as well. Researchers soon discovered that these bacteria acquired resistance through the spread of genes from one organism to another. The plasmids that mediated this transfer are R factors.

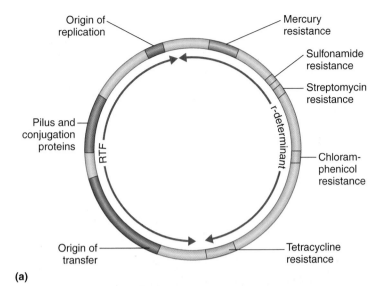

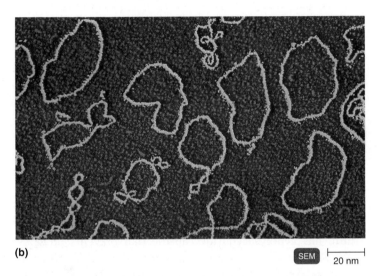

**Figure 8.25 R factor, a type of plasmid. (a)** A diagram of an R factor, which has two parts: the RTF contains genes needed for plasmid replication and transfer of the plasmid by conjugation, and the r-determinant carries genes for resistance to four different antibiotics and mercury (*sul* = sulfonamide resistance, *str* = streptomycin resistance, *cml* = chloramphenicol resistance, *tet* = tetracycline resistance, *mer* = mercury resistance); numbers are base pairs × 1000. **(b)** Plasmids from *E. coli* bacteria.

**Q** Why are R factors important in the treatment of infectious diseases?

R factors carry genes that confer upon their host cell resistance to antibiotics, heavy metals, or cellular toxins. Many R factors contain two groups of genes. One group is called the **resistance transfer factor (RTF)** and includes genes for plasmid replication and conjugation. The other group, the **r-determinant**, has the resistance genes; it codes for the production of enzymes that inactivate certain drugs or toxic substances (Figure 8.25a). Different R factors, when present in the same cell, can recombine to produce R factors with new combinations of genes in their r-determinants.

In some cases, the accumulation of resistance genes on a single plasmid is quite remarkable. For example, Figure 8.25a shows a genetic map of resistance plasmid R100. This particular plasmid can be transferred between a number of enteric genera, including *Escherichia*, *Klebsiella*, and *Salmonella*.

R factors present very serious problems for treating infectious diseases with antibiotics. The widespread use of antibiotics in medicine and agriculture (see the box in Chapter 20 on page 583) has led to the preferential survival (selection) of bacteria that have R factors, so populations of resistant bacteria grow larger and larger. The transfer of resistance between bacterial cells of a population, and even between bacteria of different genera, also contributes to the problem. The ability to reproduce sexually with members of its own species defines a eukaryotic species. However, a bacterial species can conjugate and transfer plasmids to other species. *Neisseria* may have acquired its penicillinase-producing plasmid from *Streptococcus*, and *Agrobacterium* can transfer plasmids to plant cells (see Figure 9.20, page 262). Nonconjugative plasmids may be transferred from one cell to another by inserting themselves into a conjugative plasmid or a chromosome or by transformation when released from a dead cell. Insertion is made possible by an insertion sequence, which will be discussed shortly.

Plasmids are an important tool for genetic engineering, discussed in Chapter 9 (pages 243–247).

### Transposons

**Transposons** are small segments of DNA that can move (be "transposed") from one region of a DNA molecule to another. These pieces of DNA are 700 to 40,000 base pairs long.

In the 1950s, American geneticist Barbara McClintock discovered transposons in corn, but they occur in all organisms and have been studied most thoroughly in microorganisms. They may move from one site to another site on the same chromosome or to another chromosome or plasmid. As you might imagine, the frequent movement of transposons could wreak havoc inside a cell. For example, as transposons move about on chromosomes, they may insert themselves *within* genes, inactivating them. Fortunately, transposition occurs relatively rarely. The frequency of transposition is comparable to the spontaneous mutation rate that occurs in bacteria—that is, from $10^{-5}$ to $10^{-7}$ per generation.

All transposons contain the information for their own transposition. As shown in **Figure 8.26a**, the simplest transposons, also called **insertion sequences (IS)**, contain only a gene that codes for an enzyme (*transposase*, which catalyzes the cutting and resealing of DNA that occurs in transposition) and recognition sites. *Recognition sites* are short inverted repeat sequences of DNA that the enzyme recognizes as recombination sites between the transposon and the chromosome.

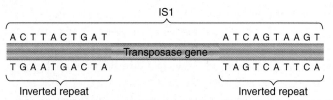

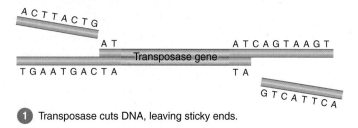

IS1

ACTTACTGAT        ATCAGTAAGT
— Transposase gene —
TGAATGACTA        TAGTCATTCA

Inverted repeat        Inverted repeat

**(a)** An insertion sequence (IS), the simplest transposon, contains a gene for transposase, the enzyme that catalyzes transposition. The tranposase gene is bounded at each end by inverted repeat sequences that function as recognition sites for the transposon. IS1 is one example of an insertion sequence, shown here with simplified IR sequences.

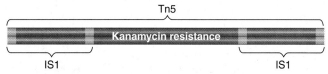

ACTTACTG

AT                ATCAGTAAGT
— Transposase gene —
TGAATGACTA        TA

GTCATTCA

**1** Transposase cuts DNA, leaving sticky ends.

Tn5

Kanamycin resistance

IS1        IS1

**(b)** Complex transposons carry other genetic material in addition to transposase genes. The example shown here, Tn5, carries the gene for kanamycin resistance and has complete copies of the insertion sequence IS1 at each end.

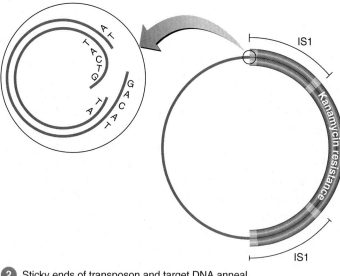

IS1

Kanamycin resistance

IS1

**2** Sticky ends of transposon and target DNA anneal.

**(c)** Insertion of the transposon Tn5 into R100 plasmid

**Figure 8.26  Transposons and insertion.**

**Q** Why are transposons sometimes referred to as "jumping genes"?

Complex transposons also carry other genes not connected with the transposition process. For example, bacterial transposons may contain genes for enterotoxin or for antibiotic resistance (Figure 8.26b). Plasmids such as R factors are frequently made up of a collection of transposons (Figure 8.26c).

Transposons with antibiotic resistance genes are of practical interest, but there is no limitation on the kinds of genes that transposons can have. Thus, transposons provide a natural mechanism for the movement of genes from one chromosome to another. Furthermore, because they may be carried between cells on plasmids or viruses, they can also spread from one organism—or even species—to another. For example, vancomycin resistance was transferred from *Enterococcus faecalis* to *Staphylococcus aureus* via a transposon called Tn1546. Transposons are thus a potentially powerful mediator of evolution in organisms.

**Play Transposons:** *Overview, Insertion Sequences, Complex Transposons* @Mastering**Microbiology**

**CHECK YOUR UNDERSTANDING**

✔ **8-14** What types of genes do plasmids carry?

## Transformation in Bacteria

During the process of **transformation**, genes are transferred from one bacterium to another as "naked" DNA in solution. This process was first demonstrated over 70 years ago, although it was not understood at the time. Not only did transformation show that genetic material could be transferred from one bacterial cell to another, but study of this phenomenon eventually led to the conclusion that DNA is the genetic material. The initial experiment on transformation was performed by Frederick Griffith in England in 1928 while he was working with two strains of *Streptococcus pneumoniae*. One, a virulent strain, has a polysaccharide capsule that prevents phagocytosis. The bacteria grow and cause pneumonia. The other, an avirulent strain, lacks the capsule and does not cause disease.

Griffith was interested in determining whether injections of heat-killed bacteria of the encapsulated strain could be used to vaccinate mice against pneumonia. As he expected, injections of living encapsulated bacteria killed the mouse (**Figure 8.27a**); injections of live nonencapsulated bacteria (Figure 8.27b) or dead encapsulated bacteria (Figure 8.27c) did not kill the mouse. However, when the dead encapsulated bacteria were mixed with live nonencapsulated bacteria and injected into the mice, many of the mice died. In the blood of the dead mice, Griffith found living, encapsulated bacteria. Hereditary material (genes) from the dead bacteria had entered the live cells and changed them genetically so that their progeny were encapsulated and therefore virulent (Figure 8.27d).

Subsequent investigations based on Griffith's research revealed that bacterial transformation could be carried out without mice. A broth was inoculated with live nonencapsulated bacteria. Dead encapsulated bacteria were then added to the broth. After incubation, the culture was found to contain living bacteria that were encapsulated and virulent. The nonencapsulated bacteria had been transformed; they had acquired

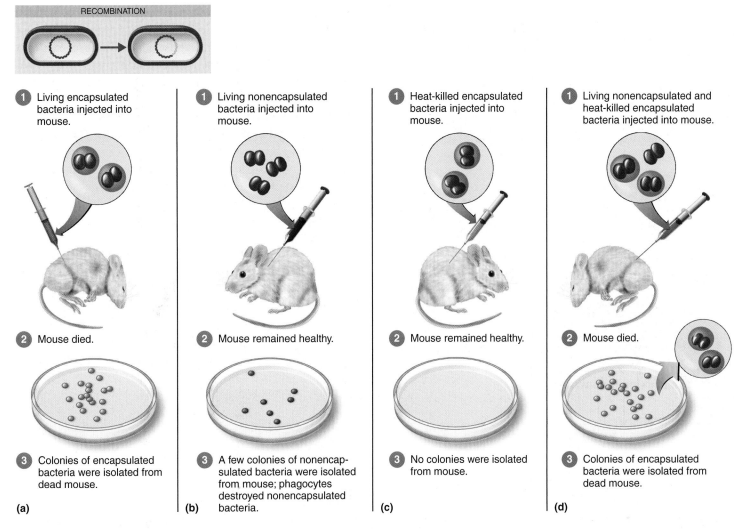

RECOMBINATION

**(a)**

**1** Living encapsulated bacteria injected into mouse.

**2** Mouse died.

**3** Colonies of encapsulated bacteria were isolated from dead mouse.

**(b)**

**1** Living nonencapsulated bacteria injected into mouse.

**2** Mouse remained healthy.

**3** A few colonies of nonencapsulated bacteria were isolated from mouse; phagocytes destroyed nonencapsulated bacteria.

**(c)**

**1** Heat-killed encapsulated bacteria injected into mouse.

**2** Mouse remained healthy.

**3** No colonies were isolated from mouse.

**(d)**

**1** Living nonencapsulated and heat-killed encapsulated bacteria injected into mouse.

**2** Mouse died.

**3** Colonies of encapsulated bacteria were isolated from dead mouse.

**Figure 8.27 Griffith's experiment demonstrating genetic transformation.**
**(a)** Living encapsulated bacteria caused disease and death when injected into a mouse. **(b)** Living nonencapsulated bacteria are readily destroyed by the phagocytic defenses of the host, so the mouse remained healthy after injection. **(c)** After being killed by heat, encapsulated bacteria lost the ability to cause disease. **(d)** However, the combination of living nonencapsulated bacteria and heat-killed encapsulated bacteria (neither of which alone causes disease) did cause disease. Somehow, the live nonencapsulated bacteria were transformed by the dead encapsulated bacteria so that they acquired the ability to form capsules and therefore cause disease. Subsequent experiments proved the transforming factor to be DNA.

**Q** Why did encapsulated bacteria kill the mouse while nonencapsulated bacteria did not? What killed the mouse in (d)?

a new hereditary trait by incorporating genes from the killed encapsulated bacteria.

The next step was to extract various chemical components from the killed cells to determine which component caused the transformation. These crucial experiments were performed in the United States by Oswald T. Avery and his associates Colin M. MacLeod and Maclyn McCarty. After years of research, they announced in 1944 that the component responsible for transforming harmless *S. pneumoniae* into virulent strains was DNA. Their results provided one of the conclusive indications that DNA was indeed the carrier of genetic information.

Since the time of Griffith's experiment, considerable information has been gathered about transformation. In nature, some bacteria, perhaps after death and cell lysis, release their DNA into the environment. Other bacteria can then encounter the DNA and, depending on the particular species and growth conditions, take up fragments of DNA and integrate them into their own chromosomes by recombination. A protein called RecA binds to the cell's DNA and then to donor DNA causing the exchange of strands. A recipient cell with this new combination of genes is a kind of hybrid, or recombinant cell (**Figure 8.28**). All the descendants of such a recombinant cell will be identical to it. Transformation occurs naturally among very few genera of bacteria, including *Bacillus*, *Haemophilus* (hē-MAH-fil-us), *Neisseria*, *Acinetobacter* (a-sin-E-tō-bak'ter), and certain strains of the genera *Streptococcus* and *Staphylococcus*.

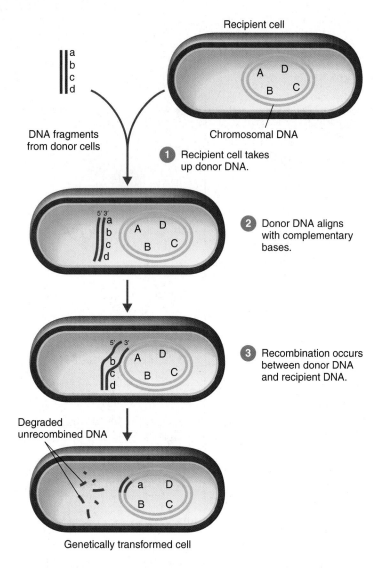

Figure 8.28 **The mechanism of genetic transformation in bacteria.** Some similarity is needed for the donor and recipient to align. Genes *a, b, c,* and *d* may be mutations of genes *A, B, C,* and *D*.

**Q** What type of enzyme cuts the donor DNA?

Even though only a small portion of a cell's DNA is transferred to the recipient, the molecule that must pass through the recipient cell wall and membrane is still very large. When a recipient cell is in a physiological state in which it can take up the donor DNA, it is said to be competent. **Competence** results from alterations in the cell wall that make it permeable to large DNA molecules.

Play Transformation
@MasteringMicrobiology

## Conjugation in Bacteria

Another mechanism by which genetic material is transferred from one bacterium to another is known as **conjugation.** Conjugation is mediated by a *conjugative plasmid,* (discussed on page 229).

Conjugation differs from transformation in two major ways. First, conjugation requires direct cell-to-cell contact. Second, the conjugating cells must generally be of opposite mating type; donor cells must carry the plasmid, and recipient cells usually do not. In gram-negative bacteria, the plasmid carries genes that code for the synthesis of *sex pili,* projections from the donor's cell surface that contact the recipient and help bring the two cells into direct contact (**Figure 8.29a**). Gram-positive bacterial cells produce sticky surface molecules that cause cells to come into direct contact with each other. In the process of conjugation, the plasmid is replicated during the transfer of a single-stranded copy of the plasmid DNA to the recipient, where the complementary strand is synthesized (Figure 8.29b).

Because most experimental work on conjugation has been done with *E. coli,* we will describe the process in this organism. In *E. coli,* the **F factor (fertility factor)** was the first plasmid observed to be transferred between cells during conjugation. Donors carrying F factors (F$^+$ cells) transfer the plasmid to recipients (F$^-$ cells), which become F$^+$ cells as a result (**Figure 8.30a**). In some cells carrying F factors, the factor

---

**CLINICAL CASE** Resolved

The Ames test allows rapid screening of chemicals for genotoxicity. The his$^-$ mutant *Salmonella* bacteria used in the Ames test are spread over glucose–minimal salts agar plates. A paper disk saturated with 2-aminofluorene (2-AF), an aromatic amine, is placed on the culture. The figure, for example, shows that reversion of the his$^-$ mutation allowed the *Salmonella* to grow. This indicates that the chemical is mutagenic and is therefore potentially carcinogenic.

There are studies indicating that 2-AF activated by enzymes is more damaging than 2-AF alone, suggesting that the interaction between diet and intestinal microbiota is more likely to cause cancer than just diet. Variations in diet produce little change in the kinds of bacteria in the intestine, but they produce dramatic changes in the metabolic activity of the bacteria.

The detection of serrated colorectal polyps from Marcel's stool DNA test led to a diagnosis of an early, rather than late, stage of colorectal cancer. The offending polyps are found and removed, and Marcel undergoes chemotherapy to kill any missed cancer cells in his colon.

205   224   227   **234**

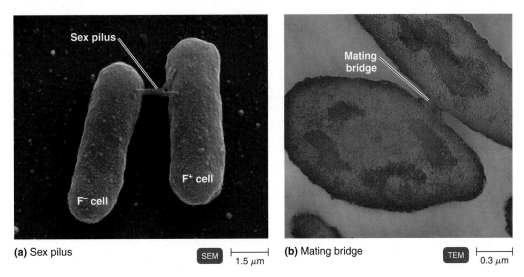

(a) Sex pilus

SEM | 1.5 μm

(b) Mating bridge

TEM | 0.3 μm

**Figure 8.29 Bacterial conjugation.**

**Q** What is an F$^+$ cell?

integrates into the chromosome, converting the F$^+$ cell to an **Hfr cell** (high frequency of recombination) (Figure 8.30b). When conjugation occurs between an Hfr cell and an F$^-$ cell, the Hfr cell's chromosome (with its integrated F factor) replicates, and a parental strand of the chromosome is transferred to the recipient cell (Figure 8.30c). Replication of the Hfr chromosome begins in the middle of the integrated F factor, and a small piece of the F factor leads the chromosomal genes into the F$^-$ cell. Usually, the chromosome breaks before it is completely transferred. Once within the recipient cell, donor DNA can recombine with the recipient's DNA. (Donor DNA that is not integrated is degraded.) Therefore, by conjugation with an Hfr cell, an F$^-$ cell may acquire new versions of chromosomal genes (just as in transformation). However, it remains an F$^-$ cell because it did not receive a complete F factor during conjugation.

Conjugation is used to map the location of genes on a bacterial chromosome (**Figure 8.31**). The genes for the synthesis of threonine (*thr*) and leucine (*leu*) are first, reading clockwise from 0. Their locations were determined by conjugation experiments. Assume that conjugation is allowed for only 1 minute between an Hfr strain that is his$^+$, pro$^+$, thr$^+$, and leu$^+$, and an F$^-$ strain that is his$^-$, pro$^-$, thr$^-$, and leu$^-$. If the F$^-$ acquired the ability to synthesize threonine, then the *thr* gene is located early in the chromosome, between 0 and 1 minute. If after 2 minutes the F$^-$ cell now becomes thr$^+$ and leu$^+$, the order of these two genes on the chromosome must be *thr*, *leu*.

**Play Conjugation:** *Overview, F Factor, Hfr Conjugation, Chromosome Mapping* @Mastering**Microbiology**

## Transduction in Bacteria

A third mechanism of genetic transfer between bacteria is **transduction.** In this process, bacterial DNA is transferred from a donor cell to a recipient cell inside a virus that infects bacteria, called a **bacteriophage,** or **phage.** (Phages will be discussed further in Chapter 13.)

To understand how transduction works, we will consider the life cycle of one type of transducing phage of *E. coli*; this phage carries out **generalized transduction** (**Figure 8.32**).

During phage reproduction, phage DNA and proteins are synthesized by the host bacterial cell. The phage DNA should be packaged inside the phage protein coat. However, bacterial DNA, plasmid DNA, or even DNA of another virus may be packaged inside a phage protein coat.

**Play Transduction:** *Generalized Transduction* @Mastering**Microbiology**

All genes contained within a bacterium infected by a generalized transducing phage are equally likely to be packaged in a phage coat and transferred. In another type of transduction, called **specialized transduction**, only certain bacterial genes are transferred (see page 375). In one type of specialized transduction, the phage codes for certain toxins produced by their bacterial hosts, such as diphtheria toxin for *Corynebacterium diphtheriae* (kor′ĭ-nē-bak-TI-rē-um dif-THI-rē-ī), erythrogenic toxin for *Streptococcus pyogenes*, and Shiga toxin for *E. coli* O157:H7.

**CHECK YOUR UNDERSTANDING**

✔ **8-15** Differentiate horizontal and vertical gene transfer.

✔ **8-16** Compare conjugation between the following pairs: F$^+$ × F$^-$, Hfr × F$^-$.

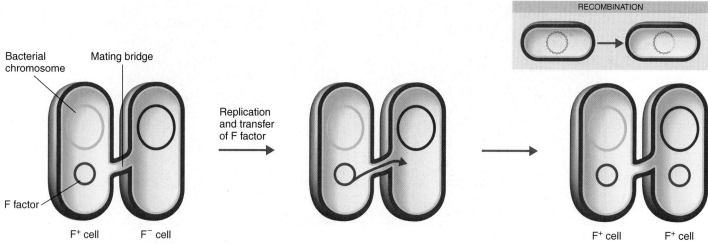

**(a)** When an F factor (a plasmid) is transferred from a donor (F⁺) to a recipient (F⁻), the F⁻ cell is converted to an F⁺ cell.

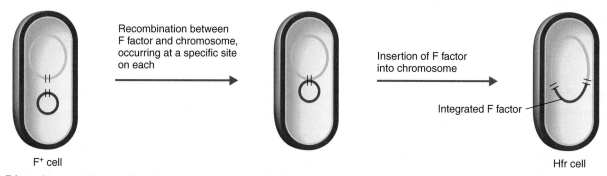

**(b)** When an F factor becomes integrated into the chromosome of an F⁺ cell, it makes the cell a high frequency of recombination (Hfr) cell.

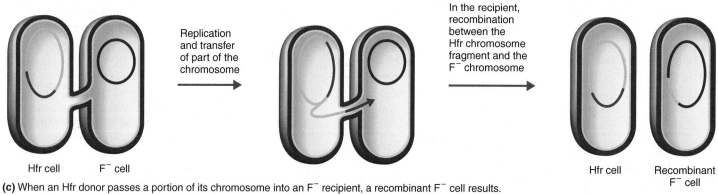

**(c)** When an Hfr donor passes a portion of its chromosome into an F⁻ recipient, a recombinant F⁻ cell results.

**Figure 8.30 Conjugation in *E. coli*.**

**Q** Do bacteria reproduce during conjugation?

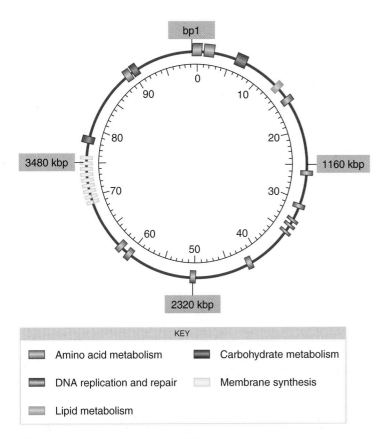

**Figure 8.31  A genetic map of the chromosome of *E. coli*.** This map is made by observing recombinant cells after conjugation. The numbers inside the circle indicate the number of minutes it takes to transfer the genes during mating between two cells; the numbers in colored boxes indicate the number of base pairs. 1 kbp = 1000 base pairs.

**Q** How many minutes of conjugation would be needed to transfer genes for membrane synthesis on this chromosome?

# Genes and Evolution

### LEARNING OBJECTIVE

**8-17** Discuss how genetic mutation and recombination provide material for natural selection to act upon.

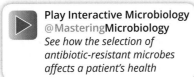

**Play Interactive Microbiology** @Mastering**Microbiology** *See how the selection of antibiotic-resistant microbes affects a patient's health*

We have now seen how gene activity can be controlled by the cell's internal regulatory mechanisms and how genes themselves can be altered or rearranged by mutation, transposition, and recombination. All these processes provide diversity in the descendants of cells. Diversity provides the raw material for evolution, and natural selection provides its driving force. Natural selection will act on diverse populations to ensure the survival of those fit for that particular environment. The different kinds of microorganisms that exist today are the result of a long history of evolution. Microorganisms have continually changed by alterations in their genetic properties and acquisition of adaptations

to many different habitats. See Exploring the Microbiome on page 230 and the box on antibiotic resistance in Chapter 26, page 771, for examples of natural selection.

### CHECK YOUR UNDERSTANDING

✔ **8-17** Natural selection means that the environment favors survival of some genotypes. From where does diversity in genotypes come?

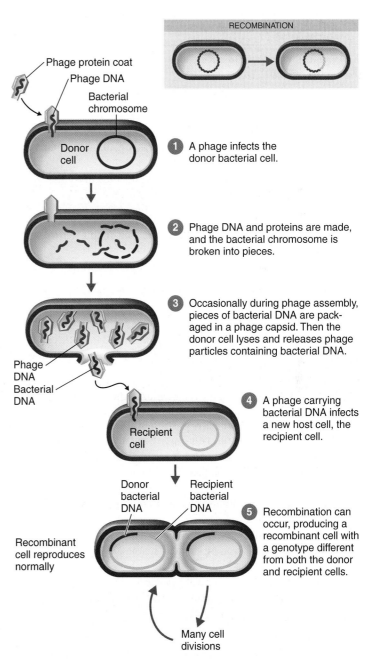

RECOMBINATION

**Phage protein coat**
**Phage DNA**
**Bacterial chromosome**
**Donor cell**

1 A phage infects the donor bacterial cell.

2 Phage DNA and proteins are made, and the bacterial chromosome is broken into pieces.

3 Occasionally during phage assembly, pieces of bacterial DNA are packaged in a phage capsid. Then the donor cell lyses and releases phage particles containing bacterial DNA.

**Phage DNA**
**Bacterial DNA**

**Recipient cell**

4 A phage carrying bacterial DNA infects a new host cell, the recipient cell.

**Donor bacterial DNA**  **Recipient bacterial DNA**

**Recombinant cell reproduces normally**

5 Recombination can occur, producing a recombinant cell with a genotype different from both the donor and recipient cells.

**Many cell divisions**

**Figure 8.32  Transduction by a bacteriophage.** Shown here is generalized transduction, in which any bacterial DNA can be transferred from one cell to another.

**Q** How could *E. coli* acquire the Shiga toxin gene?

# Study Outline

Go to @MasteringMicrobiology for **Interactive Microbiology**, *In the Clinic videos, MicroFlix, MicroBoosters, 3D animations, practice quizzes, and more.*

## Structure and Function of the Genetic Material (pp. 205–217)

1. Genetics is the study of what genes are, how they carry information, how their information is expressed, and how they are replicated and passed to subsequent generations or other organisms.

2. DNA in cells exists as a double-stranded helix; the two strands are held together by hydrogen bonds between specific nitrogenous base pairs: AT and CG.

3. A gene is a sequence of nucleotides, that encodes a functional product, usually a protein.

4. The DNA in a cell is duplicated before the cell divides, so each offspring cell receives the same genetic information.

### Genotype and Phenotype (pp. 205, 208)

5. Genotype is the genetic composition of an organism, its entire complement of DNA.

6. Phenotype is the expression of the genes: the proteins of the cell and the properties they confer on the organism.

### DNA and Chromosomes (p. 208)

7. The DNA in a chromosome exists as one long double helix associated with various proteins that regulate genetic activity.

8. Genomics is the molecular characterization of genomes.

### The Flow of Genetic Information (p. 208)

9. Following cell division, each offspring cell receives a chromosome that is virtually identical to the parent's.

10. Information contained in the DNA is transcribed into RNA and translated into proteins.

### DNA Replication (pp. 208–212)

11. During DNA replication, the two strands of the double helix separate at the replication fork, and each strand is used as a template by DNA polymerases to synthesize two new strands of DNA according to the rules of complementary base pairing.

12. The result of DNA replication is two new strands of DNA, each having a base sequence complementary to one of the original strands.

13. Because each double-stranded DNA molecule contains one original and one new strand, the replication process is called semiconservative.

14. DNA is synthesized in one direction designated $5' \rightarrow 3'$. At the replication fork, the leading strand is synthesized continuously and the lagging strand discontinuously.

15. DNA polymerase proofreads new molecules of DNA and removes mismatched bases before continuing DNA synthesis.

### RNA and Protein Synthesis (pp. 212–217)

16. During transcription, the enzyme RNA polymerase synthesizes a strand of RNA from one strand of double-stranded DNA, which serves as a template.

17. RNA is synthesized from nucleotides containing the bases A, C, G, and U, which pair with the bases of the DNA strand being transcribed.

18. RNA polymerase binds the promoter; transcription begins at AUG; the region of DNA that is the end point of transcription is the terminator; RNA is synthesized in the $5' \rightarrow 3'$ direction.

19. Translation is the process in which the information in the nucleotide base sequence of mRNA is used to dictate the amino acid sequence of a protein.

20. The mRNA associates with ribosomes, which consist of rRNA and protein.

21. Three-base codons of mRNA specify amino acids.

22. The genetic code refers to the relationship among the nucleotide base sequence of DNA, the corresponding codons of mRNA, and the amino acids for which the codons code.

23. Specific amino acids are attached to molecules of tRNA Another portion of the tRNA has a base triplet called an anticodon.

24. The base pairing of codon and anticodon at the ribosome results in specific amino acids being brought to the site of protein synthesis.

25. The ribosome moves along the mRNA strand as amino acids are joined to form a growing polypeptide; mRNA is read in the $5' \rightarrow 3'$ direction.

26. Translation ends when the ribosome reaches a stop codon on the mRNA.

## The Regulation of Bacterial Gene Expression (pp. 217–221)

1. Regulating protein synthesis at the gene level is energy-efficient because proteins are synthesized only as they are needed.

2. Constitutive genes are expressed at a fixed rate. Examples are genes for the enzymes in glycolysis.

### Pre-transcriptional Control (pp. 217–220)

3. In bacteria, a group of coordinately regulated structural genes with related metabolic functions, plus the promoter and operator sites that control their transcription, is called an operon.

4. In the operon model for an inducible system, a regulatory gene codes for the repressor protein.

5. When the inducer is absent, the repressor binds to the operator, and no mRNA is synthesized.

6. When the inducer is present, it binds to the repressor so that it cannot bind to the operator; thus, mRNA is made, and enzyme synthesis is induced.

7. In repressible systems, the repressor requires a corepressor in order to bind to the operator site; thus, the corepressor controls enzyme synthesis.

8. Transcription of structural genes for catabolic enzymes (such as β-galactosidase) is induced by the absence of glucose. Cyclic AMP and CRP must bind to a promoter in the presence of an alternative carbohydrate.

9. Methylated nucleotides are not transcribed in epigenetic control.

### Post-transcriptional Control (pp. 220–221)

10. mRNA as a riboswitch regulates translation.

11. MicroRNAs combine with mRNA; the resulting double-stranded RNA is destroyed.

## Changes in Genetic Material (pp. 221–228)

1. Mutations and horizontal gene transfer can change a bacterium's genotype.

### Mutation (p. 222)

2. A mutation is a change in the nitrogenous base sequence of DNA; that change causes a change in the product coded for by the mutated gene.

3. Many mutations are neutral, some are disadvantageous, and others are beneficial.

### Types of Mutations (pp. 222–223)

4. A base substitution occurs when one base pair in DNA is replaced with a different base pair.

5. Alterations in DNA can result in missense mutations, frameshift, or nonsense mutations.

6. Spontaneous mutations occur without the presence of any mutagen.

### Mutagens (pp. 223–226)

7. Mutagens are agents in the environment that cause permanent changes in DNA.

8. Ionizing radiation causes the formation of ions and free radicals that react with DNA; base substitutions or breakage of the sugar-phosphate backbone results.

9. Ultraviolet (UV) radiation is nonionizing; it causes bonding between adjacent thymines.

### The Frequency of Mutation (p. 226)

10. Mutation rate is the probability that a gene will mutate when a cell divides; the rate is expressed as 10 to a negative power.

11. A low rate of spontaneous mutations is beneficial in providing the genetic diversity needed for evolution.

### Identifying Mutants (p. 226)

12. Mutants can be detected by selecting or testing for an altered phenotype.

13. Positive selection involves the selection of mutant cells and the rejection of nonmutated cells.

14. Replica plating is used for negative selection—to detect, for example, auxotrophs that have nutritional requirements not possessed by the parent (nonmutated) cell.

### Identifying Chemical Carcinogens (pp. 227–228)

15. The Ames test is a relatively inexpensive and rapid test for identifying possible chemical carcinogens.

16. The test assumes that a mutant cell can revert to a normal cell in the presence of a mutagen and that many mutagens are carcinogens.

## Genetic Transfer and Recombination (pp. 229–237)

1. Genetic recombination, the rearrangement of genes from separate groups of genes, usually involves DNA from different organisms; it contributes to genetic diversity.

2. In crossing over, genes from two chromosomes are recombined into one chromosome containing some genes from each original chromosome.

3. Vertical gene transfer occurs during reproduction when genes are passed from an organism to its offspring.

4. Horizontal gene transfer in bacteria involves a portion of the cell's DNA being transferred from donor to recipient.

5. When some of the donor's DNA has been integrated into the recipient's DNA, the resultant cell is called a recombinant.

### Plasmids and Transposons (pp. 229–232)

6. Plasmids are self-replicating circular molecules of DNA carrying genes that are not usually essential for the cell's survival.

7. There are several types of plasmids, including conjugative plasmids, dissimilation plasmids, plasmids carrying genes for toxins or bacteriocins, and resistance factors.

8. Transposons are small segments of DNA that can move from one region to another region of the same chromosome or to a different chromosome or a plasmid.

9. Complex transposons can carry any type of gene, including antibiotic-resistance genes, and are thus a natural mechanism for moving genes from one chromosome to another.

### Transformation in Bacteria (pp. 232–234)

10. During this process, genes are transferred from one bacterium to another as "naked" DNA in solution.

### Conjugation in Bacteria (pp. 234–235)

11. This process requires contact between living cells.

12. One type of genetic donor cell is an $F^+$; recipient cells are $F^-$. F cells contain plasmids called F factors; these are transferred to the $F^-$ cells during conjugation.

### Transduction in Bacteria (pp. 235–237)

13. In this process, DNA is passed from one bacterium to another in a bacteriophage and is then incorporated into the recipient's DNA.

14. In generalized transduction, any bacterial genes can be transferred.

## Genes and Evolution (p. 237)

1. Diversity is the precondition for evolution.

2. Genetic mutation and recombination provide diversity of organisms, and the process of natural selection allows the growth of those best adapted to a given environment.

 **Play Interactive Microbiology**
**@MasteringMicrobiology**
*See how the selection of antibiotic-resistant microbes affects a patient's health*

# Study Questions

For answers to the Knowledge and Comprehension questions, turn to the Answers tab at the back of the textbook.

## Knowledge and Comprehension

### Review

1. Briefly describe the components of DNA, and explain its functional relationship to RNA and protein.

2. DRAW IT  Identify and mark each of the following on the portion of DNA undergoing replication: replication fork, DNA polymerase, RNA primer, parent strands, leading strand, lagging strand, the direction of replication on each strand, and the 5′ end of each strand.

5′
3′

3. Match the following examples of mutagens.

| Column A | Column B |
|---|---|
| _____ **a.** A mutagen that is incorporated into DNA in place of a normal base | **1.** Frameshift mutagen |
| _____ **b.** A mutagen that causes the formation of highly reactive ions | **2.** Nucleoside analog |
| _____ **c.** A mutagen that alters adenine so that it base-pairs with cytosine | **3.** Base-pair mutagen |
| _____ **d.** A mutagen that causes insertions | **4.** Ionizing radiation |
| _____ **e.** A mutagen that causes the formation of pyrimidine dimers | **5.** Nonionizing radiation |

4. The following is a code for a strand of DNA.

| DNA | 3′ A T A T _ _ _ T T T _ _ _ _ _ _ _ _ _ _ |
|---|---|
|  | 1 2 3 4 5 6 7 8 9 10 11 12 13 14 15 16 17 18 19 |
| mRNA | C G U      U G A |
| tRNA | U G G |
| Amino Acid | Met _____ |

ATAT = Promoter sequence

   a. Using the genetic code provided in Figure 8.8, fill in the blanks to complete the segment of DNA shown.
   b. Fill in the blanks to complete the sequence of amino acids coded for by this strand of DNA.

   c. Write the code for the complementary strand of DNA completed in part (a).
   d. What would be the effect if C were substituted for T at base 10?
   e. What would be the effect if A were substituted for G at base 11?
   f. What would be the effect if G were substituted for T at base 14?
   g. What would be the effect if C were inserted between bases 9 and 10?
   h. How would UV radiation affect this strand of DNA?
   i. Identify a nonsense sequence in this strand of DNA.

5. When iron is not available, *E. coli* can stop synthesis of all proteins, such as superoxide dismutase and succinate dehydrogenase, that require iron. Describe a mechanism for this regulation.

6. Identify when (before transcription, after transcription but before translation, after translation) each of the following regulatory mechanisms functions.
   a. ATP combines with an enzyme, altering its shape.
   b. A short RNA is synthesized that is complementary to mRNA.
   c. Methylation of DNA occurs.
   d. An inducer combines with a repressor.

7. Which sequence is the best target for damage by UV radiation: AGGCAA, CTTTGA, or GUAAAU? Why aren't all bacteria killed when they are exposed to sunlight?

8. You are provided with cultures with the following characteristics:
   Culture 1: F⁺, genotype $A^+ B^+ C^+$
   Culture 2: F⁻, genotype $A^- B^- C^-$
   a. Indicate the possible genotypes of a recombinant cell resulting from the conjugation of cultures 1 and 2.
   b. Indicate the possible genotypes of a recombinant cell resulting from conjugation of the two cultures after the F⁺ has become an Hfr cell.

9. Why are mutation and recombination important in the process of natural selection and the evolution of organisms?

10. NAME IT  Normally a commensal in the human intestine, this bacterium became pathogenic after acquiring a toxin gene from a *Shigella* bacterium.

## Multiple Choice

Match the following terms to the definitions in questions 1 and 2.
   a. conjugation
   b. transcription
   c. transduction
   d. transformation
   e. translation

1. Transfer of DNA from a donor to a recipient cell by a bacteriophage.

2. Transfer of DNA from a donor to a recipient as naked DNA in solution.

3. Feedback inhibition differs from repression because feedback inhibition
   a. is less precise.
   b. is slower acting.
   c. stops the action of preexisting enzymes.
   d. stops the synthesis of new enzymes.
   e. all of the above

4. Bacteria can acquire antibiotic resistance by all of the following *except*
   a. mutation.
   b. insertion of transposons.
   c. conjugation.
   d. snRNPs.
   e. transformation.

5. Suppose you inoculate three flasks of minimal salts broth with *E. coli*. Flask A contains glucose. Flask B contains glucose and lactose. Flask C contains lactose. After a few hours of incubation, you test the flasks for the presence of β-galactosidase. Which flask(s) do you predict will have this enzyme?
   a. A
   b. B
   c. C
   d. A and B
   e. B and C

6. Plasmids differ from transposons in that plasmids
   a. become inserted into chromosomes.
   b. are self-replicated outside the chromosome.
   c. move from chromosome to chromosome.
   d. carry genes for antibiotic resistance.
   e. none of the above

Use the following choices to answer questions 7 and 8:
   a. catabolite repression
   b. DNA polymerase
   c. induction
   d. repression
   e. translation

7. Mechanism by which the presence of glucose inhibits the *lac* operon.

8. The mechanism by which lactose controls the *lac* operon.

9. Two offspring cells are most likely to inherit which one of the following from the parent cell?
   a. a change in a nucleotide in mRNA
   b. a change in a nucleotide in tRNA
   c. a change in a nucleotide in rRNA
   d. a change in a nucleotide in DNA
   e. a change in a protein

10. Which of the following is *not* a method of horizontal gene transfer?
    a. binary fission
    b. conjugation
    c. integration of a transposon
    d. transduction
    e. transformation

## Analysis

1. Nucleoside analogs and ionizing radiation are used in treating cancer. These mutagens can cause cancer, so why do you suppose they are used to treat the disease?

2. Replication of the *E. coli* chromosome takes 40 to 45 minutes, but the organism has a generation time of 26 minutes. How does the cell have time to make complete chromosomes for each offspring cell?

3. *Pseudomonas* has a plasmid containing the *mer* operon, which includes the gene for mercuric reductase. This enzyme catalyzes the reduction of the mercuric ion $Hg^{2+}$ to the uncharged form of mercury, $Hg^0$. $Hg^{2+}$ is quite toxic to cells; $Hg^0$ is not.
   a. What do you suppose is the inducer for this operon?
   b. The protein encoded by one of the *mer* genes binds $Hg^{2+}$ in the periplasm and brings it into the cell. Why would a cell bring in a toxin?

## Clinical Applications and Evaluation

1. Ciprofloxacin, erythromycin, and acyclovir are used to treat microbial infections. Ciprofloxacin inhibits DNA gyrase. Erythromycin binds in front of the A site on the 50S subunit of a ribosome. Acyclovir is a guanine analog.
   a. What steps in protein synthesis are inhibited by each drug?
   b. Which drug is more effective against bacteria? Why?
   c. Which drugs will have effects on the host's cells? Why?
   d. Use the index to identify the disease for which acyclovir is primarily used. Why is it more effective than erythromycin for treating this disease?

2. HIV, the virus that causes AIDS, was isolated from three individuals, and the amino acid sequences for the viral coat were determined. Of the amino acid sequences shown below, which two of the viruses are most closely related? How can these amino acid sequences be used to identify the source of a virus?

| Patient | Viral Amino Acid Sequence | | | | | | | | | | | |
|---|---|---|---|---|---|---|---|---|---|---|---|---|
| A | Asn | Gln | Thr | Ala | Ala | Ser | Lys | Asn | Ile | Asp | Ala | Leu |
| B | Asn | Leu | His | Ser | Asp | Lys | Ile | Asn | Ile | Ile | Leu | Leu |
| C | Asn | Gln | Thr | Ala | Asp | Ser | Ile | Val | Ile | Asp | Ala | Leu |

3. Human herpesvirus-8 (HHV-8) is common in parts of Africa, the Middle East, and the Mediterranean, but is rare elsewhere except in AIDS patients. Genetic analyses indicate that the African strain is not changing, whereas the Western strain is accumulating changes. Using the portions of the HHV-8 genomes (shown below) that encode one of the viral proteins, how similar are these two viruses? What mechanism can account for the changes? What disease does HHV-8 cause?

**Western**      3'-ATGGAGTTCTTCTGGACAAGA

**African**      3'-ATAAACTTTTTCTTGACAACG

# 9 Biotechnology and DNA Technology

For thousands of years, people have consumed foods produced by the action of microorganisms. Bread, chocolate, and soy sauce are some of the best-known examples. But it was only just over 100 years ago that scientists showed that microorganisms are responsible for these products. This knowledge opened the way for using microorganisms to produce other important products. Since World War I, microbes have been used to produce a variety of chemicals, such as ethanol, acetone, and citric acid. Since World War II, microorganisms have been grown to produce antibiotics. More recently, microbes and their enzymes are replacing a variety of chemical processes involved in manufacturing such products as paper, textiles, and fructose. Using microbes or their enzymes instead of chemical syntheses offers several advantages: microbes may use inexpensive, abundant raw materials; microbes work at normal temperatures and pressure, thereby avoiding the need for expensive and dangerous systems; and microbes don't produce toxic, hard-to-treat wastes. In the past 30 years, DNA technology has been added to the tools used to make products.

In this chapter you will learn the tools and techniques that are used to research and develop a product. You will also learn how DNA technology is used to track outbreaks of infectious disease and to provide evidence for courts of law in forensic microbiology. The Clinical Case illustrates the use of DNA technology to track HIV (see the photo).

▶ **Human immunodeficiency virus (HIV) (yellow) budding from a host cell.**

## In the Clinic

A crime suspect claims he is innocent. His clothing became bloodstained, he says, when he tried to resuscitate the victim. The pattern of blood on the suspect may have resulted from striking the victim, but it is also consistent with spatter from the victim's nose and mouth during CPR. As a forensic nurse for a police department, you collect bloodstained fabric from the suspect and a blood sample from the crime scene. You request PCR for streptococci on both samples. The test is positive for the fabric, but negative for the blood at the scene. **How can a PCR test detect evidence of *Streptococcus* bacteria in such small samples? Do these results help or hurt the suspect?**

Play **In the Clinic** Video
@MasteringMicrobiology

*Hint: Read about the polymerase chain reaction technique on page 247.*

# Introduction to Biotechnology

## LEARNING OBJECTIVES

**9-1** Compare and contrast biotechnology, genetic modification, and recombinant DNA technology.

**9-2** Identify the roles of a clone and a vector in making recombinant DNA.

**Biotechnology** is the use of microorganisms, cells, or cell components to make a product. Microbes have been used in the commercial production of foods, vaccines, antibiotics, and vitamins for years. Bacteria are also used in mining to extract valuable elements from ore (see Figure 28.12, page 822). Additionally, animal cells have been used to produce viral vaccines since the 1950s. Until the 1980s, products made by living cells were all made by naturally occurring cells; the role of scientists was to find the appropriate cell and develop a method for their large-scale cultivation.

Now, microorganisms and plants are being used as "factories" to produce chemicals that the organisms don't naturally make. This is accomplished by inserting, deleting, or modifying genes with **recombinant DNA (rDNA) technology,** which is sometimes called *genetic engineering.* The development of rDNA technology is expanding the practical applications of biotechnology almost beyond imagination.

> ASM: Cell genomes can be manipulated to alter cell function.

## Recombinant DNA Technology

Recombination of DNA occurs naturally in microbes (see Chapter 8). In the 1970s and 1980s, scientists developed artificial techniques for making rDNA.

A gene from one organism can be inserted into the DNA of a bacterium or a yeast. In many cases, the recipient can then be made to express the gene, which may code for a commercially useful product. Thus, bacteria with genes for human insulin are now being used to produce insulin for treating diabetes, and a vaccine for hepatitis B is being made by yeast carrying a gene for part of the hepatitis virus (the yeast produces a viral coat protein). Scientists hope that such an approach may prove useful in producing vaccines against other infectious agents, thus eliminating the need to use whole organisms, as in conventional vaccines.

The rDNA techniques can also be used to make thousands of copies of the same DNA molecule—to *amplify* DNA—thus generating sufficient DNA for various kinds of experimentation and analysis. This technique has practical application for identifying microbes, such as viruses, that can't be cultured.

## An Overview of Recombinant DNA Procedures

An overview of some of the procedures typically used for making rDNA, along with some promising applications, is shown in **Figure 9.1**. A **vector** is a DNA molecule that transports foreign DNA into a cell. (See more on vectors on page 246.) The gene of interest is inserted into the vector DNA in vitro. In Figure 9.1, the vector is a plasmid. The DNA molecule chosen as a vector must be self-replicating, such as a plasmid or a viral genome. This recombinant vector DNA is taken up by a cell such as a bacterium, where it can multiply. The cell containing the recombinant vector is then grown in culture to form a **clone** of many genetically identical cells, each of which carries copies of the vector, and therefore many copies of the gene of interest. This is why DNA vectors are often called *gene-cloning vectors,* or simply *cloning vectors.* (In addition to referring to a culture of identical cells, the word *clone* is also routinely used as a verb, to describe the entire process, as in "to clone a gene.")

The final step varies according to whether the gene itself or the product of the gene is of interest. From the cell clone, the researcher may isolate ("harvest") large quantities of the gene of interest, which may then be used for a variety of purposes. The gene may even be inserted into another vector for introduction into another kind of cell (such as a plant or animal cell). Alternatively, if the gene of interest is expressed (transcribed and translated) in the cell clone, its protein product can be harvested and used for a variety of purposes.

The advantages of using rDNA for obtaining such proteins is illustrated by one of its early successes, the production of human growth hormone (hGH) in *E. coli* bacteria. Some individuals don't produce adequate amounts of hGH, so their growth is stunted. In the past, hGH had to be obtained from human pituitary glands at autopsy. (Growth hormone

---

Dr. B. is closing his dental practice after 20 years. Four years ago, he went to his family doctor because of debilitating exhaustion. He thought he had a flu virus that he could not shake, and he was also having night sweats. His doctor ordered a myriad of blood tests, but only one came back positive. Dr. B. had HIV. Although he immediately began an HIV treatment regimen, one year later he was diagnosed with AIDS. Now, two years later, Dr. B. is very ill and can no longer work.

Dr. B. lets his employees know the situation and suggests that they all get tested for HIV. All of Dr. B.'s employees, including the hygienists, test negative. Dr. B. also writes an open letter to his patients informing them of his decision to close his practice and why he is doing so. This letter prompts 400 former patients to be tested for HIV, seven of whom test positive for antibodies against HIV.

**What type of test can determine whether these patients contracted HIV from Dr. B.? Read on to find out.**

243   249   252   254   257

# A Typical Genetic Modification Procedure

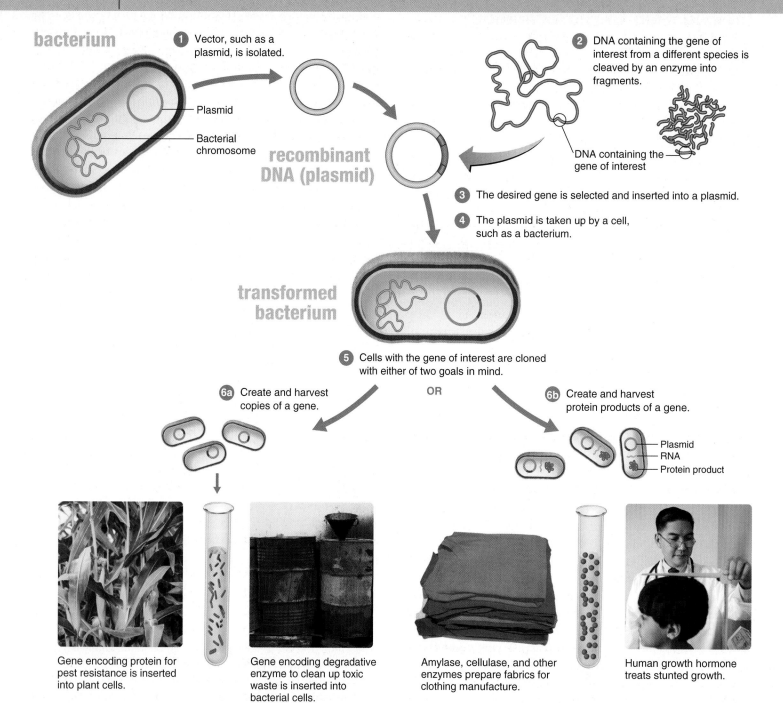

**bacterium**

**① Vector, such as a plasmid, is isolated.**

Plasmid

Bacterial chromosome

**recombinant DNA (plasmid)**

**② DNA containing the gene of interest from a different species is cleaved by an enzyme into fragments.**

DNA containing the gene of interest

**③ The desired gene is selected and inserted into a plasmid.**

**④ The plasmid is taken up by a cell, such as a bacterium.**

**transformed bacterium**

**⑤ Cells with the gene of interest are cloned with either of two goals in mind.**

**⑥a Create and harvest copies of a gene.**

**OR**

**⑥b Create and harvest protein products of a gene.**

Plasmid
RNA
Protein product

Gene encoding protein for pest resistance is inserted into plant cells.

Gene encoding degradative enzyme to clean up toxic waste is inserted into bacterial cells.

Amylase, cellulase, and other enzymes prepare fabrics for clothing manufacture.

Human growth hormone treats stunted growth.

## KEY CONCEPTS

- Genes from one organism's cells can be inserted and expressed in another organism's cells.

- Genetically modified cells can be used to create a wide variety of useful products and applications.

from other animals is not effective in humans.) This practice was not only expensive but also dangerous because on several occasions neurological diseases were transmitted with the hormone. Human growth hormone produced by genetically modified *E. coli* is a pure and cost-effective product. Recombinant DNA techniques also result in faster production of the hormone than traditional methods might allow.

**CHECK YOUR UNDERSTANDING**

✔ 9-1 Differentiate biotechnology and rDNA technology.

✔ 9-2 In one sentence, describe how a vector and clone are used.

# Tools of Biotechnology

**LEARNING OBJECTIVES**

9-3 Compare selection and mutation.

9-4 Define *restriction enzymes*, and outline how they are used to make rDNA.

9-5 List the four properties of vectors.

9-6 Describe the use of plasmid and viral vectors.

9-7 Outline the steps in PCR, and provide an example of its use.

Research scientists and technicians isolate bacteria and fungi from natural environments such as soil and water to find, or *select*, the organisms that produce a desired product. The selected organism can be mutated to make more product or to make a better product.

## Selection

In nature, organisms with characteristics that enhance survival are more likely to survive and reproduce than are variants that lack the desirable traits. This is called *natural selection*. Humans use **artificial selection** to select desirable breeds of animals or strains of plants to cultivate. As microbiologists learned how to isolate and grow microorganisms in pure culture, they were able to select the ones that could accomplish a desired objective, such as brewing beer more efficiently or producing a new antibiotic. Over 2000 strains of antibiotic-producing bacteria have been discovered by testing soil bacteria and selecting the strains that produce an antibiotic.

## Mutation

Mutations are responsible for much of the diversity of life (see Chapter 8). A bacterium with a mutation that confers resistance to an antibiotic will survive and reproduce in the presence of that antibiotic. Biologists working with antibiotic-producing microbes discovered that they could create new strains by exposing microbes to mutagens. After random mutations were created in penicillin-producing *Penicillium* by exposing fungal

cultures to radiation, the highest-yielding variant among the survivors was selected for another exposure to a mutagen. Using mutations, biologists increased the amount of penicillin the fungus produced by over 1000 times.

Screening each mutant for penicillin production is a tedious process. **Site-directed mutagenesis** is more targeted and can be used to make a specific change in a gene. Suppose you determine that changing one amino acid will make a laundry enzyme work better in cold water. Using the genetic code (see Figure 8.8, page 214), you could, using the techniques described next, produce the sequence of DNA that encodes that amino acid and insert it into that enzyme's gene.

The science of molecular genetics has advanced to such a degree that many routine cloning procedures are performed using prepackaged materials and procedures that are very much like cookbook recipes. Scientists have a grab bag of methods from which to choose, depending on the ultimate application of their experiments. Next we describe some of the most important tools and techniques, and later we will consider some applications.

## Restriction Enzymes

Recombinant DNA technology has its technical roots in the discovery of **restriction enzymes,** a special class of DNA-cutting enzymes that exist in many bacteria. First isolated in 1970, restriction enzymes in nature had actually been observed earlier, when certain bacteriophages were found to have a restricted host range. If these phages were used to infect bacteria other than their usual hosts, restriction enzymes in the new host destroyed almost all the phage DNA. Restriction enzymes protect a bacterial cell by hydrolyzing phage DNA. The bacterial DNA is protected from digestion because the cell **methylates** (adds methyl groups to) some of the cytosines in its DNA. The purified forms of these bacterial enzymes are used in today's laboratories.

What is important for rDNA techniques is that a restriction enzyme recognizes and cuts, or *digests*, only one particular sequence of nucleotide bases in DNA, and it cuts this sequence the same way each time. Typical restriction enzymes used in cloning experiments recognize four-, six-, or eight-base sequences. Hundreds of restriction enzymes are known, each producing DNA fragments with characteristic ends. A few restriction enzymes are listed in Table 9.1. You can see they are named for their bacterial source. Some of these enzymes (e.g., *Hae*III) cut both strands of DNA in the same place, producing **blunt ends,** and others make staggered cuts in the two strands—cuts that are not directly opposite each other (**Figure 9.2**). These staggered ends, or **sticky ends,** are most useful in rDNA because they can be used to join two different pieces of DNA that were cut by the same restriction enzyme. The sticky ends "stick" to stretches of single-stranded DNA by complementary base pairing.

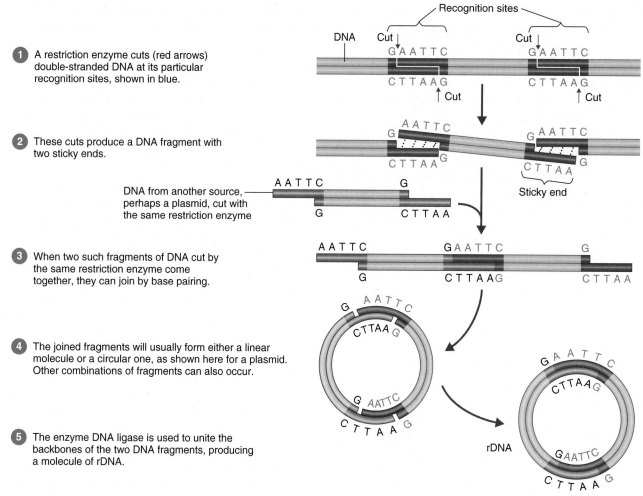

1. A restriction enzyme cuts (red arrows) double-stranded DNA at its particular recognition sites, shown in blue.

2. These cuts produce a DNA fragment with two sticky ends.

DNA from another source, perhaps a plasmid, cut with the same restriction enzyme

3. When two such fragments of DNA cut by the same restriction enzyme come together, they can join by base pairing.

4. The joined fragments will usually form either a linear molecule or a circular one, as shown here for a plasmid. Other combinations of fragments can also occur.

5. The enzyme DNA ligase is used to unite the backbones of the two DNA fragments, producing a molecule of rDNA.

**Figure 9.2  A restriction enzyme's role in making rDNA.**

**Q** Why are restriction enzymes used to make rDNA?

| **TABLE 9.1** | Selected Restriction Enzymes Used in rDNA Technology | |
|---|---|---|
| **Enzyme** | **Bacterial Source** | **Recognition Sequence** |
| BamHI | *Bacillus amyloliquefaciens* | G↓G A T C C<br>C C T A G↑G |
| EcoRI | *Escherichia coli* | G↓A A T T C<br>C T T A A↑G |
| HaeIII | *Haemophilus aegyptius* | G G↓C C<br>C C↑G G |
| HindIII | *Haemophilus influenzae* | A↓A G C T T<br>T T C G A↑A |

Notice in Figure 9.2 that the darker base sequences on the two strands are the same but run in opposite directions. Staggered cuts leave stretches of single-stranded DNA at the ends of the DNA fragments. If two fragments of DNA from different sources have been produced by the action of the same restriction enzyme, the two pieces will have identical sets of sticky ends and can be spliced (recombined) in vitro. The sticky ends join spontaneously by hydrogen bonding (base pairing). The enzyme DNA ligase is used to covalently link the backbones of the DNA pieces, producing an rDNA molecule.

> ▶ Play Recombinant DNA Technology
> @MasteringMicrobiology

## Vectors

Many different types of DNA molecules can serve as vectors, provided they have certain properties. The most important property is self-replication; once in a cell, a vector must be capable of replicating. Any DNA that is inserted in the vector will be replicated in the process. Thus, vectors serve as vehicles for the replication of desired DNA sequences.

Vectors also need to be large enough to be manipulated outside the cell during rDNA procedures. Smaller vectors are more

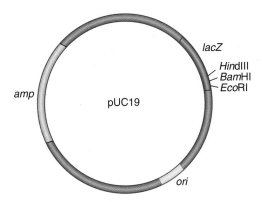

**Figure 9.3 A plasmid used for cloning.** A plasmid vector used for cloning in the bacterium *E. coli* is pUC19. An origin of replication (*ori*) allows the plasmid to be self-replicating. Two genes, one encoding resistance to the antibiotic ampicillin (*amp*) and one encoding the enzyme β-galactosidase (*lacZ*), serve as marker genes. Foreign DNA can be inserted at the restriction enzyme sites.

**Q** What is a vector in rDNA technology?

easily manipulated than larger DNA molecules, which tend to be more fragile. Preservation is another important property of vectors. The DNA molecule's circular form protects the vector's DNA from being destroyed by its recipient. Notice in **Figure 9.3** that the DNA of a plasmid is circular. Another preservation mechanism occurs when a virus's DNA inserts itself quickly into the chromosome of the host.

When it is necessary to retrieve cells that contain the vector, a marker gene in the vector often helps make selection easy. Common selectable marker genes are for antibiotic resistance or for an enzyme that carries out an easily identified reaction.

Plasmids are one of the primary vectors in use, particularly variants of R factor plasmids. Plasmid DNA can be cut with the same restriction enzymes as the DNA that will be cloned, so that all pieces of the DNA will have the same sticky ends. When the pieces are mixed, the DNA to be cloned will be inserted into the plasmid (Figure 9.2). Note that other fragment combinations can occur as well, including the plasmid reforming a circle with no DNA inserted.

Some plasmids are capable of existing in several different species. They are called **shuttle vectors** and can be used to move cloned DNA sequences among organisms, such as among bacterial, yeast, and mammalian cells, or among bacterial, fungal, and plant cells. Shuttle vectors can be very useful in the process of genetically modifying multicellular organisms—for example, when herbicide resistance genes are inserted into plants.

A different kind of vector is viral DNA. This type of vector can usually accept much larger pieces of foreign DNA than plasmids can. After the DNA has been inserted into the viral vector, it can be cloned in the virus's host cells. The choice of a suitable vector depends on many factors, including the organism that will receive the new gene and the size of the DNA to be cloned. Retroviruses, adenoviruses, and herpesviruses are

being used to insert corrective genes into human cells that have defective genes. Gene therapy is discussed on page 255.

Gene therapy is discussed on page 255.

### CHECK YOUR UNDERSTANDING

☑ 9-3 How are selection and mutation used in biotechnology?

☑ 9-4 What is the value of restriction enzymes in rDNA technology?

☑ 9-5 What criteria must a vector meet?

☑ 9-6 Why is a vector used in rDNA technology?

## Polymerase Chain Reaction

The **polymerase chain reaction (PCR)** is a technique by which small samples of DNA can be quickly amplified, that is, increased to quantities that are large enough for analysis.

Starting with just one gene-sized piece of DNA, PCR can be used to make billions of copies in only a few hours. The PCR process is shown in **Figure 9.4**.

Each strand of the target DNA will serve as a template for DNA synthesis. Added to this DNA are a supply of the four nucleotides (for assembly into new DNA) and the enzyme for catalyzing the synthesis, DNA polymerase (see Chapter 8, page 209). Short pieces of nucleic acid called primers are also added to help start the reaction. The primers are complementary to the ends of the target DNA and will hybridize to the fragments to be amplified. Then, the polymerase synthesizes new complementary strands. After each cycle of synthesis, the DNA is heated to convert all the new DNA into single strands. Each newly synthesized DNA strand serves in turn as a template for more new DNA.

As a result, the process proceeds exponentially. All of the necessary reagents are added to a tube, which is placed in a *thermal cycler*. The thermal cycler can be set for the desired temperatures, times, and number of cycles. Use of an automated thermal cycler is made possible by the use of DNA polymerase taken from a thermophilic bacterium such as *Thermus aquaticus*; the enzyme from such organisms can survive the heating phase without being destroyed. Thirty cycles, completed in just a few hours, will increase the amount of target DNA by more than a billion times.

The amplified DNA can be seen by gel electrophoresis. In *real-time PCR*, or *quantitative PCR* (*qPCR*), the newly made DNA is tagged with a fluorescent dye, so that the levels of fluorescence can be measured after every PCR cycle (that's the *real time* aspect). Another PCR procedure called *reverse-transcription (RT-PCR)* uses viral RNA or a cell's mRNA as the template. The enzyme, reverse transcriptase, makes DNA from the RNA template, and the DNA is then amplified.

Note that PCR can only be used to amplify relatively small, specific sequences of DNA as determined by the choice of primers. It cannot be used to amplify an entire genome.

PCR can be applied to any situation that requires the amplification of DNA. Especially noteworthy are diagnostic tests that use PCR to detect the presence of infectious agents in situations in

## PREPARATION

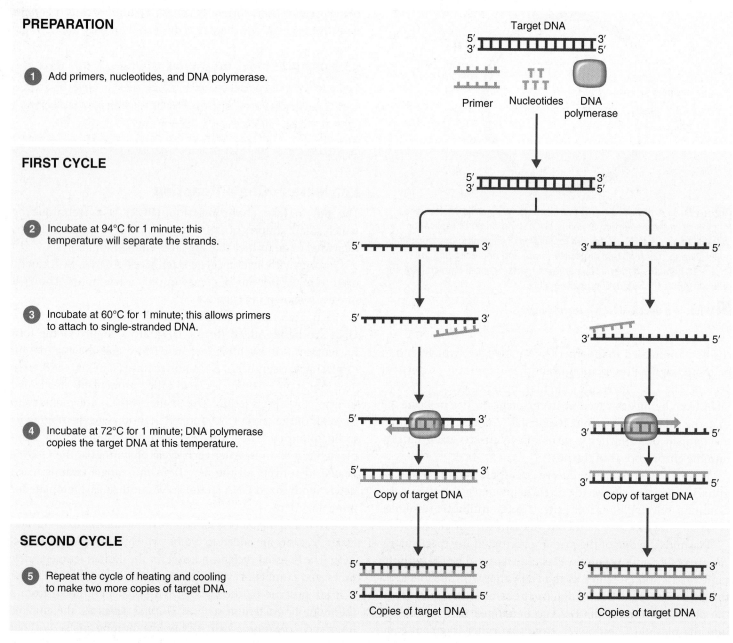

① Add primers, nucleotides, and DNA polymerase.

Target DNA

Primer    Nucleotides    DNA polymerase

## FIRST CYCLE

② Incubate at 94°C for 1 minute; this temperature will separate the strands.

③ Incubate at 60°C for 1 minute; this allows primers to attach to single-stranded DNA.

④ Incubate at 72°C for 1 minute; DNA polymerase copies the target DNA at this temperature.

Copy of target DNA          Copy of target DNA

## SECOND CYCLE

⑤ Repeat the cycle of heating and cooling to make two more copies of target DNA.

Copies of target DNA          Copies of target DNA

**Figure 9.4 The polymerase chain reaction.** Deoxynucleotides (dNTPs) base-pair with the target DNA: adenine pairs with thymine, and cytosine pairs with guanine.

**Q** How does *reverse-transcription PCR* differ from this figure?

which they would otherwise be undetectable. A qPCR test provides rapid identification of drug-resistant *Mycobacterium tuberculosis.* Otherwise, this bacterium can take up to 6 weeks to culture, leaving patients untreated for a significant period of time.

 **Play PCR:** *Overview, Components, Process* @Mastering**Microbiology**

**CHECK YOUR UNDERSTANDING**

✔ **9-7** For what is each of the following used in PCR: primer, DNA polymerase, 94°C?

# Techniques of Genetic Modification

**LEARNING OBJECTIVES**

**9-8** Describe five ways of getting DNA into a cell.

**9-9** Describe how a genomic library is made.

**9-10** Differentiate cDNA from synthetic DNA.

**9-11** Explain how each of the following is used to locate a clone: antibiotic-resistance genes, DNA probes, gene products.

**9-12** List one advantage of modifying each of the following: *E. coli, Saccharomyces cerevisiae,* mammalian cells, plant cells.

Reverse-transcription PCR using a primer for an HIV gene can be used to amplify DNA for analysis. The Centers for Disease Control and Prevention (CDC) interviews the seven former patients to determine whether their histories show any additional risk factors for contracting HIV. Five out of the seven have no identified risk factors for HIV other than having had invasive procedures performed on them by Dr. B. The CDC then performs reverse-transcription PCR on DNA from white blood cells in Dr. B.'s peripheral blood and the seven HIV-positive patients (see the figure).

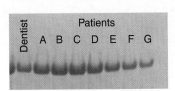

**What can be concluded from the PCR amplification in the figure?**

243    249    252    254    257

## Inserting Foreign DNA into Cells

Recombinant DNA procedures require that DNA molecules be manipulated outside the cell and then returned to living cells. There are several ways to introduce DNA into cells. The choice of method is usually determined by the type of vector and host cell being used.

In nature, plasmids are usually transferred between closely related microbes by cell-to-cell contact, such as in conjugation. To modify a cell, a plasmid must be inserted into a cell by **transformation**, a procedure during which cells can take up DNA from the surrounding environment (see Chapter 8, page 232). Many cell types, including *E. coli*, yeast, and mammalian cells, do not naturally transform; however, simple chemical treatments can make all of these cell types *competent*, or able to take up external DNA. For *E. coli*, the procedure for making cells competent is to soak them in a solution of calcium chloride for a brief period. Following this treatment, the now-competent cells are mixed with the cloned DNA and given a mild heat shock. Some of these cells will then take up the DNA.

There are other ways to transfer DNA to cells. A process called **electroporation** uses an electrical current to form microscopic pores in the membranes of cells; the DNA then enters the cells through the pores. Electroporation is generally applicable to all cells; those with cell walls often must be converted to protoplasts first. **Protoplasts** are produced by enzymatically removing the cell wall, thereby allowing more direct access to the plasma membrane.

The process of **protoplast fusion** also takes advantage of the properties of protoplasts. Protoplasts in solution fuse at a low but significant rate; the addition of polyethylene glycol increases the frequency of fusion (**Figure 9.5**). In the new hybrid cell, the DNA derived from the two "parent" cells may undergo natural recombination. This method is especially valuable in the genetic manipulation of plant and algal cells.

A remarkable way of introducing foreign DNA into plant cells is to literally shoot it directly through the thick cellulose walls using a gene gun (**Figure 9.6**). Microscopic particles of tungsten or gold are coated with DNA and propelled by a burst of helium through the plant cell walls. Some of the cells express the introduced DNA as though it were their own.

DNA can be introduced directly into an animal cell by **microinjection**. This technique requires the use of a glass micropipette with a diameter that is much smaller than the cell. The micropipette punctures the plasma membrane, and DNA can be injected through it (**Figure 9.7**).

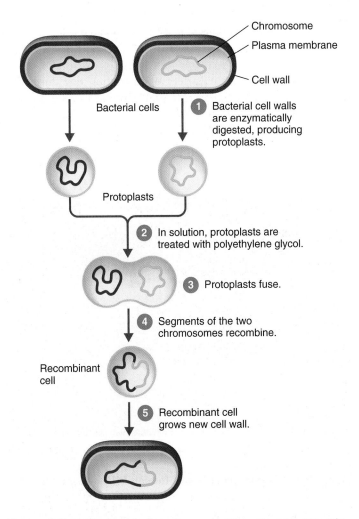

**Figure 9.5 Protoplast fusion.** Removal of the cell wall leaves only the delicate plasma membranes, which will fuse together, allowing the exchange of DNA.

**Q** What is a protoplast?

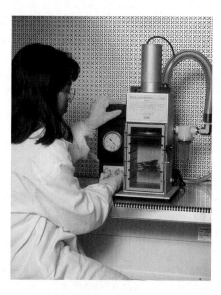

**Figure 9.6  A gene gun, which can be used to insert DNA-coated "bullets" into a cell.**

**Q** Name four other methods of inserting DNA into a cell.

Thus, there is a great variety of restriction enzymes, vectors, and methods of inserting DNA into cells. But foreign DNA will survive only if it's either present on a self-replicating vector or incorporated into one of the cell's chromosomes by recombination.

## Obtaining DNA

We have seen how genes can be cloned into vectors by using restriction enzymes and how genes can be transformed or transferred into a variety of cell types. But how do biologists obtain the genes they are interested in? There are two main sources of genes: (1) genomic libraries containing either natural copies of genes or cDNA copies of genes made from mRNA, and (2) synthetic DNA.

### Genomic Libraries

Isolating specific genes as individual pieces of DNA is seldom practical. Therefore, researchers interested in genes from a particular organism start by extracting the organism's DNA, which can be obtained from cells of any organism, whether plant, animal, or microbe, by lysing the cells and precipitating the DNA. This process results in a DNA mass that includes the organism's entire genome. After the DNA is digested by restriction enzymes, the restriction fragments are then spliced into plasmid or phage vectors, and the recombinant vectors are introduced into bacterial cells. The goal is to make a collection of clones large enough to ensure that at least one clone exists for every gene in the organism. This collection of clones containing different DNA fragments is called a **genomic library**; each "book" is a bacterial or phage strain that contains a fragment of the genome (**Figure 9.8**). Such libraries are essential for

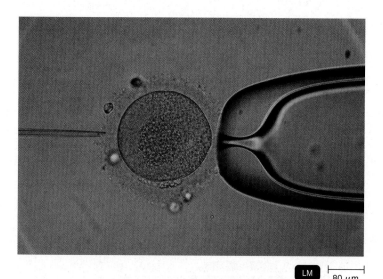

**Figure 9.7  The microinjection of foreign DNA into an egg.** The egg is first immobilized by applying mild suction to the large, blunt, holding pipette (right). Several hundred copies of the gene of interest are then injected into the nucleus of the cell through the tiny end of the micropipette (left).

**Q** Why is microinjection impractical for bacterial and fungal cells?

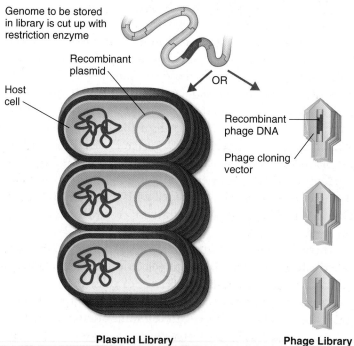

Genome to be stored in library is cut up with restriction enzyme

Recombinant plasmid

Host cell

OR

Recombinant phage DNA

Phage cloning vector

**Plasmid Library**

**Phage Library**

**Figure 9.8  Genomic libraries.** Each fragment of DNA, containing about one gene, is carried by a vector, either a plasmid within a bacterial cell or a phage.

**Q** Differentiate a restriction fragment from a gene.

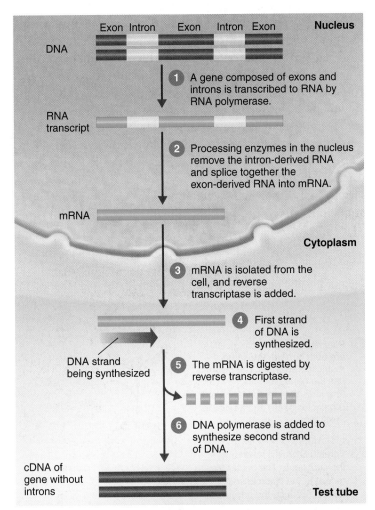

**Figure 9.9 Making complementary DNA (cDNA) for a eukaryotic gene.** Reverse transcriptase catalyzes the synthesis of double-stranded DNA from an RNA template.

**Q** How does reverse transcriptase differ from DNA polymerase?

maintaining and retrieving DNA clones; they can even be purchased commercially.

Cloning genes from eukaryotic organisms presents a specific problem. Genes of eukaryotic cells generally contain both **exons**, stretches of DNA that code for protein, and **introns**, intervening stretches of DNA that do not code for protein. When the RNA transcript of such a gene is converted to mRNA, the introns are removed (see Figure 8.11 on page 219). To clone genes from eukaryotic cells, it's desirable to use a version of the gene that lacks introns because a gene that includes introns may be too large to work with easily. In addition, if such a gene is put into a bacterial cell, the bacterium usually won't be able to remove the introns from the RNA transcript. Therefore, it won't be able to make the correct protein product. However, an artificial gene that contains only exons can be produced by using an enzyme called **reverse transcriptase** to synthesize **complementary DNA (cDNA)** from an mRNA template

(Figure 9.9). This synthesis is the reverse of the normal DNA-to-RNA transcription process. A DNA copy of mRNA is produced by reverse transcriptase. Following this, the mRNA is enzymatically digested away. DNA polymerase then synthesizes a complementary strand of DNA, creating a double-stranded piece of DNA containing the information from the mRNA. Molecules of cDNA produced from a mixture of all the mRNAs from a tissue or cell type can then be cloned to form a cDNA library.

The cDNA method is the most common method of obtaining eukaryotic genes. A difficulty with this method is that long molecules of mRNA may not be completely reverse-transcribed into DNA; the reverse transcription often aborts, forming only parts of the desired gene.

### Synthetic DNA

Under certain circumstances, genes can be made in vitro with the help of DNA synthesis machines (**Figure 9.10**). A keyboard on the machine is used to enter the desired sequence of nucleotides, much as letters are entered into a word processor to compose a sentence. A microprocessor controls the synthesis of the DNA from stored supplies of nucleotides and the other necessary reagents. A short chain of about 200 nucleotides, called an *oligonucleotide*, can be synthesized by this method. Unless the gene is very small, at least several chains must be synthesized separately and linked together to form an entire gene.

The difficulty of this approach, of course, is that the sequence of the gene must be known before it can be synthesized. If the gene hasn't already been isolated, then the only

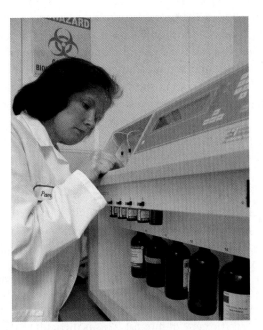

**Figure 9.10 A DNA synthesis machine.** Short sequences of DNA can be synthesized by instruments such as this one.

**Q** What four reagents (in the brown bottles) are necessary to synthesize DNA?

## CLINICAL CASE

The primer amplifies all eight samples and confirms that Dr. B. and seven of his former patients are all infected with HIV. The CDC then sequences the amplified DNA and compares the sequencing to an HIV isolate from Cleveland (local control) and an isolate from Haiti (outlier). A portion of the coding (5' to 3') is shown below.

| Patient A | GCTTG | GGCTG | GCGCT | GAAGT | GAGA |
|---|---|---|---|---|---|
| Patient B | GCTAT | TGCTG | GCGCT | GAATT | GCAC |
| Patient C | GCCAT | AGCTG | GCGCA | GAAGT | GCAC |
| Patient D | GCTAT | TGGCG | TGGCT | GACAG | AGAA |
| Patient E | GCACC | TGCTG | GCGCT | GAAGT | GAAA |
| Patient F | CAGAT | TGTGT | TGATT | GAACC | TCAC |
| Patient G | GCTAT | TGCTG | GCGCT | GAAGT | GAAA |
| Dentist | GCTAT | TGCTG | GCGCT | GAAGT | GCAC |
| Local control | CAGAC | TACTG | CTAGG | AAAAA | TATT |
| Outlier | GAAGA | CGAAA | GGACT | GCTAT | TCAG |

**What is the percent similarity among the viruses?**

243  249  **252**  254  257

way to predict the DNA sequence is by knowing the amino acid sequence of the gene's protein product. If this amino acid sequence is known, in principle you can work backward through the genetic code to obtain the DNA sequence. Unfortunately, the degeneracy of the code prevents definitive determination; thus, if the protein contains a leucine, for example, which of the six codons for leucine is the one in the gene?

For these reasons, it's rare to clone a gene by synthesizing it directly, although some commercial products such as insulin, interferon, and somatostatin are produced from chemically synthesized genes. Desired restriction sites are added to the synthetic genes so the genes can be inserted into plasmid vectors and cloned in *E. coli*. Synthetic DNA plays a much more useful role in selection procedures, as we will see.

### CHECK YOUR UNDERSTANDING

✔ **9-8** Contrast the five ways of putting DNA into a cell.

✔ **9-9** What is the purpose of a genomic library?

✔ **9-10** Why isn't cDNA synthetic?

## Selecting a Clone

In cloning, it's necessary to select the particular cell that contains the specific gene of interest. This is difficult to do because out of millions of cells, only a very few might contain the desired gene. Here we'll examine a typical screening procedure known as *blue-white screening*, from the color of the bacterial colonies formed at the end of the screening process.

The plasmid vector used contains a gene (*amp*) encoding resistance to the antibiotic ampicillin. The host bacterium won't be able to grow on the test medium, which contains ampicillin, unless the vector has transferred the ampicillin-resistance gene. The plasmid vector also contains a second gene, this one for the enzyme β-galactosidase (*lacZ*). Notice in Figure 9.3 that there are several sites in *lacZ* that can be cut by restriction enzymes.

In the blue-white screening procedure shown in **Figure 9.11**, a library of bacteria is cultured in a medium called X-gal. X-gal contains two essential components other than those necessary to support normal bacterial growth. One is the antibiotic ampicillin, which prevents the growth of any bacterium that

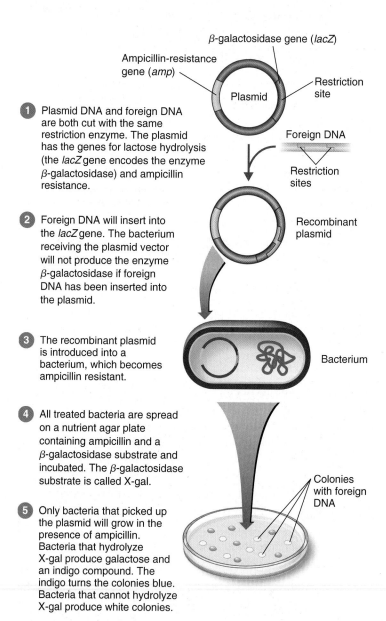

1. Plasmid DNA and foreign DNA are both cut with the same restriction enzyme. The plasmid has the genes for lactose hydrolysis (the *lacZ* gene encodes the enzyme β-galactosidase) and ampicillin resistance.

2. Foreign DNA will insert into the *lacZ* gene. The bacterium receiving the plasmid vector will not produce the enzyme β-galactosidase if foreign DNA has been inserted into the plasmid.

3. The recombinant plasmid is introduced into a bacterium, which becomes ampicillin resistant.

4. All treated bacteria are spread on a nutrient agar plate containing ampicillin and a β-galactosidase substrate and incubated. The β-galactosidase substrate is called X-gal.

5. Only bacteria that picked up the plasmid will grow in the presence of ampicillin. Bacteria that hydrolyze X-gal produce galactose and an indigo compound. The indigo turns the colonies blue. Bacteria that cannot hydrolyze X-gal produce white colonies.

**Figure 9.11 Blue-white screening, one method of selecting recombinant bacteria.**

**Q** Why are some colonies blue and others white?

has not successfully received the ampicillin-resistance gene from the plasmid. The other, called X-gal, is a substrate for β-galactosidase.

Only bacteria that picked up the plasmid will grow, because they are now ampicillin resistant. Bacteria that picked up the recombinant plasmid—in which the new gene was inserted into the *lacZ* gene—will not hydrolyze lactose and will produce white colonies. If a bacterium received the original plasmid containing the intact *lacZ* gene, the cells will hydrolyze X-gal to produce a blue-colored compound; the colony will be blue.

What remains to be done can still be difficult. The above procedure has isolated white colonies known to contain foreign DNA, but it is still not known whether it's the desired fragment of foreign DNA. A second procedure is needed to identify these bacteria. If the foreign DNA in the plasmid encodes the production of an identifiable product, the bacterial isolate only needs to be grown in culture and tested. However, in some cases the gene itself must be identified in the host bacterium.

**Colony hybridization** is a common method of identifying cells that carry a specific cloned gene. **DNA probes**, short segments of single-stranded DNA that are complementary to the desired gene, are synthesized. If the DNA probe finds a match, it will adhere to the target gene. The DNA probe is labeled with an enzyme or fluorescent dye so its presence can be detected. A typical colony hybridization experiment is shown in **Figure 9.12**. An array of DNA probes arranged in a DNA chip can be used to identify pathogens (see Figure 10.17, page 287).

## Making a Gene Product

We have just seen how to identify cells carrying a particular gene. Gene products are frequently the reason for genetic modification. Most of the earliest work in genetic modification used *E. coli* to synthesize the gene products. *E. coli* is easily grown, and researchers are very familiar with the bacterium and its genetics. For example, some inducible promoters, such as that of the *lac* operon, have been cloned, and cloned genes can be attached to such promoters. The synthesis of great amounts of the cloned gene product can then be directed by the addition of an inducer. Such a method has been used to produce gamma interferon in *E. coli* (**Figure 9.13**). However, *E. coli* also has several disadvantages. Like other gram-negative bacteria, it produces endotoxins as part of the outer layer of its cell wall. Because endotoxins cause fever and shock in mammals, their accidental presence in products intended for human use would be a serious problem.

Another disadvantage of *E. coli* is it doesn't usually secrete protein products. To obtain a product, cells must usually be broken open and the product purified from the resulting "soup" of cell components. Recovering the product from such a mixture is expensive when done on an industrial scale. It's more economical to have an organism secrete the product so it

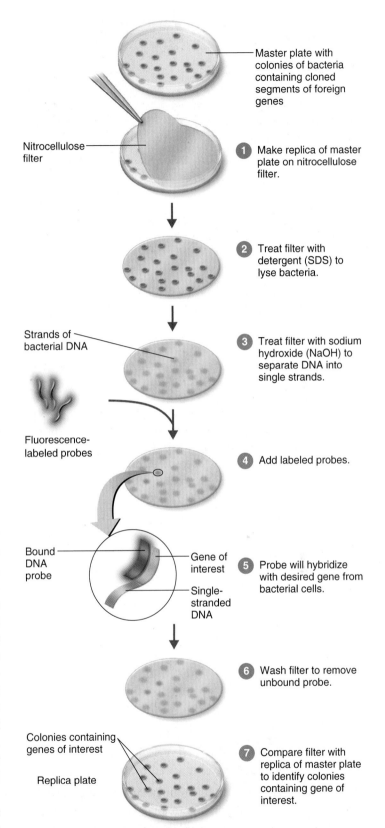

**Figure 9.12  Colony hybridization: using a DNA probe to identify a cloned gene of interest.**

**Q** What is a DNA probe?

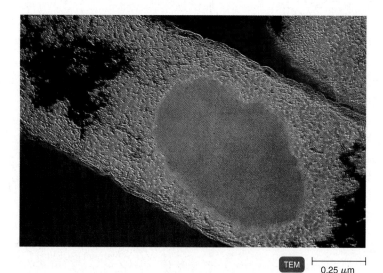

**Figure 9.13 *E. coli* genetically modified to produce gamma interferon, a human protein that promotes an immune response.** The product, visible here as a red mass in a violet ring, can be released by lysis of the cell.

**Q** What is one advantage of using *E. coli* for genetic engineering? One disadvantage?

can be recovered continuously from the growth medium. One approach has been to link the product to a natural *E. coli* protein that the bacterium does secrete. However, gram-positive bacteria, such as *Bacillus subtilis*, are more likely to secrete their products and are often preferred industrially for that reason.

Another microbe being used as a vehicle for expressing rDNA is baker's yeast, *Saccharomyces cerevisiae*. Its genome is only about four times larger than that of *E. coli* and is probably the best understood eukaryotic genome. Yeasts may carry plasmids, which are easily transferred into yeast cells whose cell walls have been removed. As eukaryotic cells, yeasts may be more successful in expressing foreign eukaryotic genes than bacteria. Furthermore, yeasts are likely to continuously secrete the product. Because of all these factors, yeasts have become the eukaryotic workhorse of biotechnology.

Mammalian cells in culture, even human cells, can be genetically modified much like bacteria to produce various products. Scientists have developed effective methods of growing certain mammalian cells in culture as hosts for growing viruses (see Chapter 13, page 371). Mammalian cells are often the best suited to making protein products for medical use because the cells secrete their products and there's a low risk of toxins or allergens. Using mammalian cells to make foreign gene products on an industrial scale often requires a preliminary step of cloning the gene in bacteria. Consider the example of colony-stimulating factor (CSF). A protein produced naturally in tiny amounts by white blood cells, CSF is valuable because it stimulates the growth of certain cells that protect

against infection. To produce huge amounts of CSF industrially, the gene is first inserted into a plasmid. Bacteria are used to make multiple copies of the plasmid (see Figure 9.1), and the resulting recombinant plasmids are then inserted into mammalian cells that are grown in bottles.

Plant cells can also be grown in culture, altered by rDNA techniques, and then used to generate genetically modified plants. Such plants may prove useful as sources of valuable products, such as plant alkaloids (the painkiller codeine, for example), the isoprenoids that are the basis of synthetic rubber, and melanin (the animal skin pigment) for use in sunscreens. Genetically modified plants have many advantages for the production of human therapeutic agents, including vaccines and antibodies. The advantages include large-scale, low-cost agricultural production and a low risk of product contamination by mammalian pathogens or cancer-causing genes. Genetically modifying plants often requires use of a bacterium. We'll return to the topic of genetically modified plants later in the chapter (page 260).

**CHECK YOUR UNDERSTANDING**

✔ 9-11 How are recombinant clones identified?

✔ 9-12 What types of cells are used for cloning rDNA?

# Applications of DNA Technology

**LEARNING OBJECTIVES**

**9-13** List at least five applications of DNA technology.

**9-14** Define RNAi.

**9-15** Discuss the value of genome projects.

**9-16** Define the following terms: *random shotgun sequencing, bioinformatics, proteomics*.

**9-17** Diagram the Southern blotting procedure, and provide an example of its use.

**9-18** Diagram DNA fingerprinting, and provide an example of its use.

**9-19** Outline genetic engineering with *Agrobacterium*.

**CLINICAL CASE**

The sequences from Dr. B. and patients A, B, C, E, and G share 87.5% of the nucleotide sequence, which is comparable to reported similarities for known linked infections.

**Identify the amino acids encoded by the viral DNA. Did this change the percent similarity? (Hint: Refer to Figure 8.8 on page 214).**

243   249   252   **254**   257

We have now described the entire sequence of events in cloning a gene. As indicated earlier, such cloned genes can be applied in a variety of ways. One is to produce useful substances more efficiently and less expensively. Another is to obtain information from the cloned DNA that is useful for either basic research, medicine, or forensics. A third is to use cloned genes to alter the characteristics of cells or organisms.

## Therapeutic Applications

An extremely valuable pharmaceutical product is the hormone insulin, a small protein produced by the pancreas that controls the body's uptake of glucose from blood. For many years, people with insulin-dependent diabetes controlled their disease by injecting insulin obtained from the pancreases of slaughtered animals. Obtaining this insulin is an expensive process, and the insulin from animals is not as effective as human insulin.

Because of the value of human insulin and the protein's small size, producing human insulin by rDNA techniques was an early goal for the pharmaceutical industry. To produce the hormone, synthetic genes were first constructed for each of the two short polypeptide chains that make up the insulin molecule. The small size of these chains—only 21 and 30 amino acids long—made it possible to use synthetic genes. Following the procedure described earlier (page 249), each of the two synthetic genes was inserted into a plasmid vector and linked to the end of a gene coding for the bacterial enzyme β-galactosidase, so that the insulin polypeptide was coproduced with the enzyme. Two different E. coli bacterial cultures were used, one to produce each of the insulin polypeptide chains. The polypeptides were then recovered from the bacteria, separated from the β-galactosidase, and chemically joined to make human insulin. This accomplishment was one of the early commercial successes of DNA technology, and it illustrates a number of the principles and procedures discussed in this chapter.

Another human hormone that is now being produced commercially by genetic modification of E. coli is somatostatin. At one time 500,000 sheep brains were needed to produce 5 mg of animal somatostatin for experimental purposes. By contrast, only 8 liters of a genetically modified bacterial culture are now required to obtain the equivalent amount of the human hormone.

**Subunit vaccines**, consisting only of a protein portion of a pathogen, are being made by genetically modifying yeasts. Subunit vaccines have been produced for a number of diseases, notably hepatitis B. One of the advantages of a subunit vaccine is that there is no chance that the vaccine will cause an infection. The protein is harvested from genetically modified cells and purified for use as a vaccine. Animal viruses such as vaccinia virus can be genetically modified to carry a gene for another microbe's surface protein. When injected, the virus acts as a vaccine against the other microbe.

**DNA vaccines** are usually circular plasmids that include a gene encoding a viral protein that's under the transcriptional control of a promoter region active in human cells. The plasmids are then cloned in bacteria. A DNA vaccine to protect against Zika virus disease is currently in clinical trials. Vaccines are discussed in further detail in Chapter 18 (page 503). Table 9.2 lists some other important rDNA products used in medical therapy.

The importance of rDNA technology to medical research cannot be emphasized enough. Artificial blood for use in transfusions can now be prepared with human hemoglobin produced in genetically modified pigs. Sheep have also been genetically modified to produce a number of drugs in their milk. This procedure has no apparent effect upon the sheep, and they provide a ready source of raw material for the product that does not require sacrificing animals.

**Gene therapy** may eventually provide cures for some genetic diseases. It is possible to imagine removing some cells from a person and transforming them with a normal gene to replace a defective or mutated gene. When these cells are returned to the person, they should function normally. For example, gene therapy has been used to treat hemophilia B and severe combined immunodeficiency. Adenoviruses and retroviruses are used most often to deliver genes; however, some researchers are working with plasmid vectors. An attenuated retrovirus was used as the vector when the first gene therapy to treat hemophilia in humans was performed in 1990. Glybera® is a gene therapy drug licensed in Europe to treat lipoprotein lipase deficiency. It uses an adenovirus to deliver the lipase gene to cells. Antisense DNA (page 262) introduced into cells is also being explored. Fomivirsen is an antisense DNA drug used in the treatment of cytomegalovirus retinitis.

Thus far, gene therapy results have not been impressive; there have even been a few deaths attributed to the viral vectors. A great deal of preliminary work remains to be done, but cures may not be possible for all genetic diseases.

**Gene editing** is a promising new technology to correct genetic mutations at precise locations. Gene editing uses CRISPR (pronounced "crisper"), which stands for clustered regularly interspaced short palindromic repeats. CRISPR enzymes are found in archaea and bacteria, where they destroy foreign DNA. A small RNA molecule, complementary to the desired target, binds DNA, and then the Cas9 enzyme cuts the DNA like molecular scissors. The cell's DNA polymerase and DNA ligase reattach the ends. A researcher can add template DNA for the correct gene, which can be attached by the DNA ligase. If may be possible to correct mutations in the human genome to treat genetic causes of disease. Gene editing was used to repair a defective muscle protein gene in mice with Duchenne muscular dystrophy. A parvovirus was used to deliver the gene-editing system into mice. In 2016, the

## TABLE 9.2 Some Pharmaceutical Products of rDNA

| Product | Comments |
|---|---|
| Cervical Cancer Vaccine | Consists of viral proteins; produced by *Saccharomyces cerevisiae* or by insect cells |
| Epidermal Growth Factor (EGF) | Heals wounds, burns, ulcers; produced by *E. coli* |
| Erythropoietin (EPO) | Treatment of anemia; produced by mammalian cell culture |
| Interferon | |
|   IFN–α | Therapy for leukemia, melanoma, and hepatitis; produced by *E. coli* and *S. cerevisiae* (yeast) |
|   IFN–β | Treatment for multiple sclerosis; produced by mammalian cell culture |
|   IFN–γ | Treatment of chronic granulomatous disease; produced by *E. coli* |
| Hepatitis B Vaccine | Produced by *S. cerevisiae* that carries hepatitis-virus gene on a plasmid |
| Human Growth Hormone (hGH) | Corrects growth deficiencies in children; produced by *E. coli* |
| Human Insulin | Therapy for diabetes; better tolerated than insulin extracted from animals; produced by *E. coli* |
| Influenza Vaccine | Vaccine made from *E. coli* or *S. cerevisiae* carrying virus genes |
| Interleukins | Regulate the immune system; possible treatment for cancer; produced by *E. coli* |
| Orthoclone OKT3 Muromonab-CD3 | Monoclonal antibody used in transplant patients to help suppress the immune system, reducing the chance of tissue rejection; produced by mouse cells |
| Pulmozyme (rhDNase) | Enzyme used to break down mucous secretions in cystic fibrosis patients; produced by mammalian cell culture |
| Relaxin | Used to ease childbirth; produced by *E. coli* |
| Superoxide Dismutase (SOD) | Minimizes damage caused by oxygen free radicals when blood is resupplied to oxygen-deprived tissues; produced by *S. cerevisiae* and *Komagataella pastoris* (yeast) |
| Taxol | Plant product used for treating ovarian cancer; produced in *E. coli* |
| Tissue Plasminogen Activator | Dissolves the fibrin of blood clots; therapy for heart attacks; produced by mammalian cell culture |
| Tumor Necrosis Factor (TNF) | Causes disintegration of tumor cells; produced by *E. coli* |
| Veterinary Use | |
|   Canine Distemper Vaccine | Canarypox virus carrying canine distemper virus genes |
|   Feline Leukemia Vaccine | Canarypox virus carrying feline leukemia virus genes |

first clinical trials were approved to modify a patient's T cells (see page 476) to fight cancer.

**Gene silencing** is a natural process that occurs in a wide variety of eukaryotes and is apparently a defense against viruses and transposons. Gene silencing is similar to miRNA (page 219) in that a gene encoding a small piece of RNA is transcribed. Following transcription, RNAs called **small interfering RNAs (siRNAs)** are formed after processing by an enzyme called *Dicer*. The siRNA molecules bind to mRNA, which is then destroyed by proteins called the **RNA-induced silencing complex (RISC)**, thus *silencing* the expression of a gene (**Figure 9.14**).

New technology called **RNA interference (RNAi)** holds promise for gene therapy for treating genetic diseases. A small DNA insert encoding siRNA against the gene of interest could be cloned into a plasmid. When transferred into a cell, the cell would produce the desired siRNA. Clinical trials are currently being conducted to test RNAi to prevent Ebola and respiratory syncytial virus infections.

### CHECK YOUR UNDERSTANDING

✔ 9-13 Explain how DNA technology can be used to treat disease and to prevent disease.

✔ 9-14 What is gene silencing?

## Genome Projects

The first genome to be sequenced was from a bacteriophage in 1977. In 1995, the genome of a free-living cell—*Haemophilus influenzae*—was sequenced. Since then, 1000 prokaryotic genomes and over 400 eukaryotic genomes have been sequenced.

In **shotgun sequencing**, small pieces of a genome of a free-living cell are sequenced, and the sequences are then assembled using a computer. Any gaps between the pieces then have to be found and sequenced (**Figure 9.15**). This technique can be used on environmental samples to study the genomes of microorganisms that haven't been cultured. The study of

The amino acid sequence reflects the nucleotide sequence. Analysis of the amino acid signature pattern confirms that the viruses from the dentist and patients are closely related. HIV has a high mutation rate, so HIVs from different individuals are genetically distinct. Dr. B.'s HIV is different from the local control and from the outlier. Dr. B.'s amino acid sequences and those of patients A, B, C, E, and G are distinct from those in the control and in the outlier and from two dental patients with known behavioral risks for HIV infection.

PCR and RFLP analyses have made it possible to track transmission of disease between individuals, communities, and countries. This tracking works best with pathogens that have enough genetic variation to identify different strains.

\* \* \*

Dr. B. died before the mode of transmission could be established. But in the era when he practiced dentistry, it was not always the norm to wear gloves when performing procedures. Patient interviews indicated that Dr. B. didn't like to wear gloves. It is likely that HIV was transmitted when a cut on the doctor's bare hands allowed the virus to enter patients' gums. Today the CDC and state health departments ask dental care providers to use universal precautions, including wearing gloves and masks and sterilizing equipment that is to be reused. Had Dr. B. used standard precautions, it is extremely unlikely he would have infected patients.

243  249  252  254  **257**

genetic material taken directly from environmental samples is called **metagenomics**.

The Human Genome Project was an international 13-year effort, formally begun in October 1990 and completed in 2003. The goal of the project was to sequence the entire human genome, approximately 3 billion nucleotide pairs, comprising 20,000 to 25,000 genes. Thousands of people in 18 countries participated in this project. Researchers collected blood (female) or sperm (male) samples from a large number of donors. Only a few samples were processed as DNA resources, and the source names are protected so that neither donors nor scientists knew whose samples were used. Development of shotgun sequencing greatly speeded the process, and 99% of the genome has been sequenced.

One surprising finding was that less than 2% of the genome encodes a functional product—the other 98% includes miRNA genes, viral remnants, repetitive sequences (called *short tandem repeats*), introns, the chromosome ends (called *telomeres*), and transposons (page 231).

The next goal of researchers is the Human Proteome Project, which will map all the proteins expressed in human cells. Even before it is completed, however, it's yielding data that are of immense value to our understanding of biology. It will also

eventually be of great medical benefit, especially for the diagnosis and treatment of genetic diseases.

## Scientific Applications

Recombinant DNA technology can be used to make products, but this isn't its only important application. Because of its ability to produce many copies of DNA, it can serve as a sort of DNA "printing press." Once a large amount of a particular piece of DNA is available, various analytic techniques, discussed in this section, can be used to "read" the information contained in the DNA.

In 2010, researchers synthesized the smallest known cellular genome during the Minimal Genome Project. A copy of the *Mycoplasma mycoides* genome was synthesized and transplanted into an *M. capricolum* cell that had had its own DNA removed. The modified cell produced *M. mycoides* proteins. This experiment showed that large-scale changes to a genome can be made and that an existing cell will accept this DNA.

DNA sequencing has produced an enormous amount of information that has spawned the new field of **bioinformatics**, the science of understanding the function of genes through computer-assisted analysis. DNA sequences are stored in

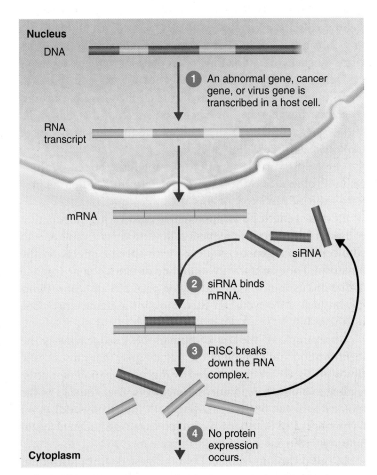

**Figure 9.14 Gene silencing could provide treatments for a wide range of diseases.**

**Q** Does RNAi act during or after transcription?

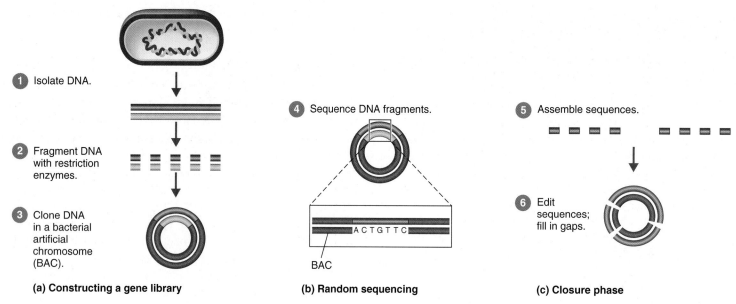

1. Isolate DNA.

2. Fragment DNA with restriction enzymes.

3. Clone DNA in a bacterial artificial chromosome (BAC).

4. Sequence DNA fragments.

ACTGTTC

BAC

5. Assemble sequences.

6. Edit sequences; fill in gaps.

**(a) Constructing a gene library**

**(b) Random sequencing**

**(c) Closure phase**

**Figure 9.15 Shotgun sequencing.** In this technique, a genome is cut into pieces, and each piece is sequenced. Then the pieces are fit together. There may be gaps if a specific DNA fragment was not sequenced.

**Q** Does this technique identify genes and their locations?

web-based databases referred to as GenBank. Genomic information can be searched with computer programs to find specific sequences or to look for similar patterns in the genomes of different organisms. Microbial genes are now being searched to identify molecules that are the virulence factors of pathogens. By comparing genomes, researchers discovered that *Chlamydia trachomatis* (tra-KŌ-ma-tis) produces a toxin similar to that of *Clostridium difficile* (DIF-fi-sē-il).

The next goal is to identify the proteins encoded by these genes. **Proteomics** is the science of determining all of the proteins expressed in a cell.

**Reverse genetics** is an approach to discovering the function of a gene from a genetic sequence. Reverse genetics attempts to connect a given genetic sequence with specific effects on the organism. For example, if you mutate or block a gene (see the earlier discussions of gene editing on page 255 and gene silencing on page 256), you can then look for a characteristic the organism lost.

An example of the use of human DNA sequencing is the identification and cloning of the mutant gene that causes cystic fibrosis (CF). CF is characterized by the oversecretion of mucus, leading to blocked respiratory passageways. The sequence of the mutated gene can be used as a diagnostic tool in a hybridization technique called **Southern blotting (Figure 9.16)**, named for Ed Southern, who developed the technique in 1975.

In this technique, subject DNA is digested with a restriction enzyme, yielding thousands of fragments of various sizes. The fragments are called **RFLPs** (pronounced "rif-lip"), for *restriction fragment length polymorphisms*. The different fragments are then separated by **gel electrophoresis**. The fragments are put

in a well at one end of a layer of agarose gel. Then an electrical current is passed through the gel. While the charge is applied, the different-sized RFLPs migrate through the gel at different rates. The RFLPs are transferred onto a filter by blotting and are exposed to a labeled probe made from the cloned gene of interest, in this case the CF gene. The probe will hybridize to this mutant gene but not to the normal gene. Fragments to which the probe binds are identified by a colored dye. With this method, any person's DNA can be tested for the presence of the mutated gene.

**Genetic testing** can now be used to screen for several hundred genetic diseases. Such screening procedures can be performed on prospective parents and also on fetal tissue. Two of the more commonly screened genes are those associated with inherited forms of breast cancer and the gene responsible for Huntington's disease. Genetic testing can help a physician prescribe the correct medication for a patient. The drug herceptin, for example, is effective only in breast cancer patients with a specific nucleotide sequence in the HER2 gene.

### Forensic Microbiology

For several years, microbiologists have used RFLPs in a method of identification known as **DNA fingerprinting** to identify bacterial or viral pathogens (**Figure 9.17**).

DNA chips (see Figure 10.18, page 288) or *PCR microarrays* that can screen a sample for multiple pathogens at once are now being used. In a DNA chip, up to 22 primers from different microorganisms can be used to initiate the PCR. A suspect microorganism is identified if DNA is copied from one of the primers. At the Centers for Disease Control and Prevention

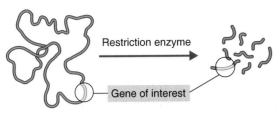

**1** DNA containing the gene of interest is extracted from human cells and cut into fragments by restriction enzymes. Fragments are called restriction fragment length polymorphisms, or RFLPs (pronounced "rif-lips").

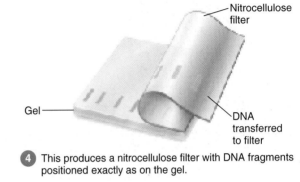

**2** The fragments are separated according to size by gel electrophoresis. Each band contains many copies of a particular DNA fragment. The bands are invisible but can be made visible by staining.

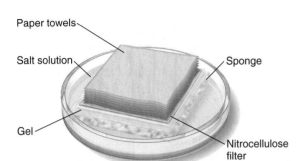

**3** The DNA bands are transferred to a nitrocellulose filter by blotting. The solution passes through the gel and filter to the paper towels by capillary action.

**4** This produces a nitrocellulose filter with DNA fragments positioned exactly as on the gel.

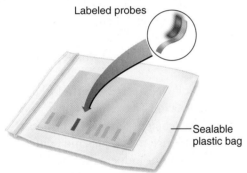

**5** The filter is exposed to a labeled probe for a specific gene. The probe will base-pair (hybridize) with a short sequence present on the gene.

**6** The fragment containing the gene of interest is identified by a band on the filter.

**Figure 9.16 Southern blotting.**

**Q** What is the purpose of Southern blotting?

(CDC), PulseNet uses RFLPs to track outbreaks of foodborne disease. In some cases, PCR using specific primers can be used to track a bacterial strain to locate the source of an outbreak.

The genomics of pathogens has become a mainstay of monitoring, preventing, and controlling infectious disease. The use of genomics to trace a disease outbreak is described in the Clinical Focus box on page 264. The new field of **forensic microbiology** developed because hospitals, food manufacturers, and individuals can be sued in courts of law and because microorganisms can be used as weapons. In the 2001 anthrax attacks in the United States, DNA fingerprints of *Bacillus anthracis* were used to track the source and then the alleged attacker.

Northern Arizona University researchers determined that the *B. anthracis* endospores used in a 1993 attack by a cult in Japan were actually a nonpathogenic vaccine strain. No one was hurt when those endospores were released. Currently, a DNA database is being developed for microorganisms that could be used in biological crimes.

Microbial forensics has been used in court. In the 1990s, DNA fingerprints of HIV were used for the first time to obtain a rape conviction. Since then, a physician was convicted of injecting his former lover with HIV from one of his patients, based on the DNA fingerprint of the HIV.

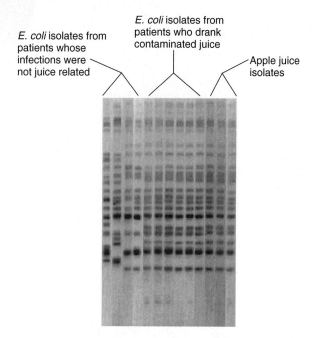

*E. coli* isolates from patients whose infections were not juice related

*E. coli* isolates from patients who drank contaminated juice

Apple juice isolates

**Figure 9.17 DNA fingerprints used to track an infectious disease.** This figure shows the RFLP patterns of bacterial isolates from an outbreak of *Escherichia coli* O157:H7. The isolates from apple juice are identical to the patterns of isolates from patients who drank the contaminated juice but different from those from patients whose infections were not juice related.

**Q What is forensic microbiology?**

The requirements to prove in a court of law the source of a microbe are stricter than for the medical community. For example, to prove intent to commit harm requires collecting evidence properly and establishing a chain of custody of that evidence. Microbial properties that are unimportant in public health may be important clues in forensic investigations. The genetic fingerprint of sexually transmitted pathogens, for instance, has been used as evidence in sexual abuse and rape cases. In the Clinic (page 242) offers another example of the use of bacterial genomics in a criminal investigation. These developments suggest that the human microbiome may become an important law enforcement tool. The American Academy of Microbiology recently proposed professional certification in forensic microbiology.

### Nanotechnology

**Nanotechnology** deals with the design and manufacture of extremely small electronic circuits and mechanical devices built at the molecular level of matter. Molecule-sized robots or computers can be used to detect contamination in food, diseases in plants, or biological weapons. However, the small machines require small (a nanometer is $10^{-9}$ meters; 1000 nm fit in 1 $\mu$m) wires and components. Bacteria may provide the needed small metals without producing the toxic waste associated with chemical manufacture. Bacteria have been isolated

that make nanoparticles from a variety of elements, including gold, silver, selenium, and cadmium. (**Figure 9.18**). Nanotechnology research is growing, with researchers developing innovative ways of using bacteria to produce nanospheres for potential drug targeting and delivery. Researchers with the U.S. Department of Energy are using bacteria in nanoscale electrical circuits to make hydrogen gas. Swedish researchers are using *Acetobacter xylinum* to build cellulose nanofibers for artificial blood vessels.

**CHECK YOUR UNDERSTANDING**

✔ **9-15, 9-16** How are shotgun sequencing, bioinformatics, and proteomics related to genome projects?

✔ **9-17** What is Southern blotting?

✔ **9-18** Why do RFLPs result in a DNA fingerprint?

### Agricultural Applications

The process of selecting for genetically desirable plants has always been time-consuming. Conventional plant cross-breeding is laborious and involves waiting for the planted seed to germinate and the resulting plant to mature in order to learn whether the plant has the desired traits. Plant breeding has been revolutionized by the use of plant cells grown in culture. Clones of plant cells, including cells that have been genetically altered by rDNA techniques, can be grown in large numbers. These cells can then be induced to regenerate whole plants, from which seeds can be harvested.

Recombinant DNA can be introduced into plant cells in several ways. Previously we mentioned protoplast fusion and the use of DNA-coated "bullets." The most elegant method, however, makes use of a plasmid called the **Ti plasmid** (*Ti* stands for tumor-inducing), which occurs naturally in the bacterium *Agrobacterium tumefaciens* (TOO-mah-fas'ē-enz). This bacterium infects certain plants, in which the Ti plasmid

**Figure 9.18 *Bacillus* cells growing on selenium form chains of elemental selenium.**

**Q What might bacteria provide for nanotechnology?**

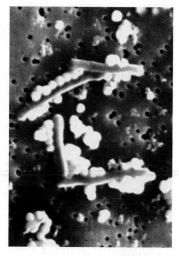

SEM | 1 $\mu$m

# Crime Scene Investigation and Your Microbiome

Fingerprints, blood types, and DNA were once new to crime scene investigations (CSI). Each technique uses unique profiles from the human body to draw conclusions about a person's actions or whereabouts. Now the microbiome might be the next CSI tool.

Even after we wash our hands, certain bacteria persist. These microbes can also be transferred to objects in the home or office or to other people we live with. But which microbes commonly live on the body also varies greatly throughout the population as a whole—meaning that the microbiome can become a unique identifier in certain situations.

A research project called The Home Microbiome Project followed seven families and their pets over 6 weeks. Researchers discovered distinct microbial communities in each house. Couples and their young children shared most of their microbial community. When three of the families moved, it took less than a day for the new house to have the same microbial population as the old one.

In another study, it was shown that a person's "microbiome fingerprint" remains fairly consistent over time. All this research suggests that microbiome composition may be the basis for a reliable forensic tool. Microbiome profiles could be used to track whether a person lived somewhere, used a particular cell phone, or walked over a surface. Humans also exchange microbes during intercourse, so microbes on pubic hair might also provide evidence of sexual assault.

**Microbiota, like this skin biofilm, may one day be another crime scene "fingerprint."**

---

causes the formation of a tumorlike growth called a crown gall (**Figure 9.19**). A part of the Ti plasmid, called T-DNA, integrates into the genome of the infected plant. The T-DNA stimulates local cellular growth (the crown gall) and simultaneously causes the production of certain products used by the bacteria as a source of nutritional carbon and nitrogen.

For plant scientists, the attraction of the Ti plasmid is that it provides a vehicle for introducing rDNA into a plant (**Figure 9.20**). A scientist can insert foreign genes into the T-DNA, put the recombinant plasmid back into the *Agrobacterium* cell, and use the bacterium to insert the recombinant Ti plasmid into a plant cell. The plant cell with the foreign gene can then be used to generate a new plant. With luck, the new plant will express the foreign gene. Unfortunately, *Agrobacterium* does not naturally infect grasses, so it cannot be used to improve grains such as wheat, rice, or corn.

Noteworthy accomplishments of this approach are the introduction into plants of resistance to the herbicide glyphosate. Normally, the herbicide kills both weeds and useful plants by inhibiting an enzyme necessary for making certain essential amino acids. Some *Salmonella* bacteria happen to have this enzyme, but are resistant to the herbicide. When the DNA for this enzyme is introduced into a crop plant, the crop

Crown gall

**Figure 9.19 Crown gall disease on a rose plant.** The tumorlike growth is stimulated by a gene on the Ti plasmid that *Agrobacterium tumefaciens* inserted into a plant cell.

**Q** What are some of the agricultural applications of rDNA technology?

261

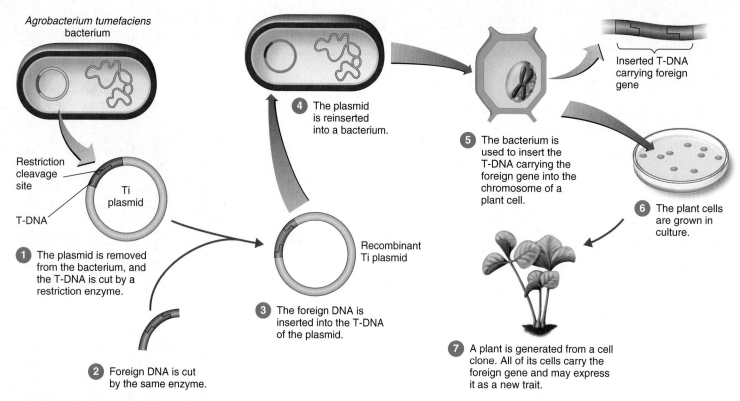

**Figure 9.20 Using the Ti plasmid as a vector for genetic modification in plants.**

**Q** Why is the Ti plasmid important to biotechnology?

becomes resistant to the herbicide, which then kills only the weeds. The Bt gene from *Bacillus thuringiensis* has been inserted into a variety of crop plants, including cotton and potatoes, so insects that eat the plants will be killed. Resistance to drought, viral infection, and several other environmental stresses has also been engineered into crop plants.

Another example involves FlavrSavr™ tomatoes, which stay firm after harvest because the gene for polygalacturonase (PG), the enzyme that breaks down pectin, is suppressed. The suppression was accomplished by **antisense DNA** technology. First, a length of DNA complementary to the PG mRNA is synthesized. This antisense DNA is taken up by the cell and binds to the mRNA to inhibit translation. The DNA-RNA hybrid is broken down by the cell's enzymes, freeing the antisense DNA to disable another mRNA.

An example of a genetically modified bacterium now in agricultural use is *Pseudomonas fluorescens* that has been engineered to produce Bt toxin, normally produced by *Bacillus thuringiensis*. The genetically altered *Pseudomonas*, which produces much more toxin than *B. thuringiensis*, can be added to plant seeds and in time will enter the vascular system of the growing plant. Its toxin is ingested by the feeding insect larvae and kills them (but is harmless to humans and other warm-blooded animals).

Animal husbandry has also benefited from rDNA technology to develop disease-resistant food animals. Techniques for making cattle resistant to bovine spongiform encephalopathy and chickens and pigs resistant to avian influenza are currently being researched.

Table 9.3 lists several rDNA products used in agriculture and animal husbandry.

**CHECK YOUR UNDERSTANDING**

✔ 9-19 Of what value is the plant pathogen *Agrobacterium*?

# Safety Issues and the Ethics of Using DNA Technology

### LEARNING OBJECTIVE

9-20 List the advantages of, and problems associated with, the use of genetic modification techniques.

There will always be concern about the safety of any new technology, and genetic modification and biotechnology are certainly no exceptions. One reason for this concern is it's nearly impossible to prove that something is entirely safe under all conceivable conditions. People worry that the same techniques that can alter a microbe or plant to make them useful to humans could also inadvertently make them pathogenic to

**TABLE 9.3** Some Agriculturally Important Products of rDNA Technology

| Product | Comments |
|---|---|
| **AGRICULTURAL PRODUCTS** | |
| Button mushroom (*Agaricus bisporus*) | Gene for polyphenyl oxidase, which causes browning, is deleted. |
| Bt cotton and Bt corn | Plants have toxin-producing gene from *Bacillus thuringiensis*; toxin kills insects that eat plants. |
| Genetically modified tomatoes, raspberries | Antisense gene blocks pectin degradation, so fruits have longer shelf life. |
| *Pseudomonas syringae*, ice-minus bacterium | Lacks normal protein product that initiates undesirable ice formation on plants. |
| RoundUp (glyphosate)-resistant crops | Plants have bacterial gene; allows use of herbicide on weeds without damaging crops. |
| **ANIMAL PRODUCTS** | |
| *Aedes aegypti* | Male mosquito with a gene that causes larvae to die; used to control spread of Zika virus. |
| Atlantic salmon | Salmon grow faster with a gene from Chinook salmon and promoter from another fish (pout). |
| GloFish® | Brightly colored fluorescent aquarium fish with the color-protein genes from marine invertebrates. |

humans or otherwise dangerous to living organisms or could create an ecological nightmare. Therefore, laboratories engaged in rDNA research must meet rigorous standards of control to avoid either accidental release of genetically modified organisms into the environment or exposure of humans to any risk of infection. To reduce risk further, microbiologists engaged in genetic modification often delete from the microbes' genomes certain genes that are essential for growth in environments outside the laboratory. Genetically modified organisms intended for use in the environment (in agriculture, for example) may be engineered to contain "suicide genes"—genes that eventually turn on to produce a toxin that kills the microbes, thus ensuring that they will not survive in the environment for very long after they have accomplished their task.

The safety issues in agricultural biotechnology are similar to those concerning chemical pesticides: toxicity to humans and to nonpest species. Although not shown to be harmful, genetically modified foods have not been popular with consumers. In 1999, researchers in Ohio noticed that humans may develop allergies to *Bacillus thuringiensis* (Bt) toxin after working in fields sprayed with the insecticide. And an Iowa study showed that the caterpillar stage of Monarch butterflies could be killed by ingesting windblown Bt-carrying pollen that landed on milkweed, the caterpillars' normal food. Crop plants can be genetically modified for herbicide resistance so that fields can be sprayed to eliminate weeds without killing the desired crop. However, if the modified plants pollinate related weed species,

weeds could become resistant to herbicides, making it more difficult to control unwanted plants. An unanswered question is whether releasing genetically modified organisms will alter evolution as genes move to wild species.

These developing technologies also raise a variety of ethical issues. Genetic testing for diseases is becoming routine. Who should have access to this information? Should employers have the right to know the results of such tests? How can we be assured that such information will not be used to discriminate against certain groups? Should individuals be told they will get an incurable disease? If so, when?

Genetic counseling, which provides advice and counseling to prospective parents with family histories of genetic disease, is becoming more important in considerations about whether to have children.

There are probably just as many harmful applications of a new technology as there are helpful ones. It is particularly easy to imagine DNA technology being used to develop new and powerful biological weapons. In addition, because such research efforts are performed under top-secret conditions, it is virtually impossible for the general public to learn of them.

Perhaps more than most new technologies, molecular genetics holds the promise of affecting human life in previously unimaginable ways. It is important that society and individuals be given every opportunity to understand the potential impact of these new developments.

Norovirus—Who Is Responsible for the Outbreak?

As you read through this box, you will encounter a series of questions that microbiologists ask themselves as they trace a disease outbreak. Whether the microbiologist is called as an expert witness in court will depend on whether a lawsuit is filed. Try to answer each question before going on to the next one.

1. On May 7, Nadia Koehler, a microbiologist at a county health department, is notified of a gastroenteritis outbreak among 115 people. The case is defined as vomiting and diarrhea and fever, cramps, or nausea.
   What information does Nadia need?

2. Nadia needs to find out where the ill people have been in the past 48 hours. After several interviews, Nadia finds out that the ill people include 23 school employees, 55 publishing company employees, 9 employees of a social service organization, and 28 other people (see Figure A).
   Now what does Nadia need to know?

3. Next, Nadia finds out what these 115 people have in common. In her investigation, Nadia discovers that on May 2, the school staff had been served a party-sized sandwich catered by a national franchise restaurant. On May 3, the publishing company and social service staff luncheons were catered by the same restaurant. The remaining 28 people ate sandwiches at the same restaurant, at varying times between these two days.
   What does Nadia do next?

4. Nadia analyzes exposures to 16 food items; the results show that eating lettuce is significantly associated with illness.
   What is Nadia's next step?

5. Nadia then requests a reverse-transcription PCR (RT-PCR) using a norovirus primer to be done on stool samples (Figure B).
   What did Nadia conclude?

6. RT-PCR confirmed norovirus infection. Nadia's next request is for a sequence analysis to be performed on 21 stool specimens. The results demonstrated 100% sequence homology for the 21 specimens.
   What should Nadia do next?

7. Nadia learns that a food handler employed by the restaurant had experienced vomiting and diarrhea on May 1. The food handler believes he had acquired the illness from his child. The child's illness was traced to an ill cousin who had been exposed to norovirus at a child-care center. The food handler's vomiting ended by the early morning of May 2, and he returned to work at the restaurant later that morning.
   What should Nadia look for now?

8. Now Nadia compares the virus strains from the food handler to the ones from the ill customers. She requests a sequence analysis on viruses from the food handler and eight ill customers. They are identical to the strains identified in step 6.
   Where does Nadia look next?

9. Nadia looks for any areas in the restaurant that still may be contaminated by the norovirus. She finds out that the lettuce was sliced each morning by the food handler who had been sick. Nadia's inspection reveals that the food preparation sink is also used for handwashing. The sink was not sanitized before and after the lettuce was washed. The health department closes the restaurant until it can be cleaned with the proper sanitizers.

Noroviruses are the most common cause of outbreaks of acute gastroenteritis worldwide. Annually, norovirus causes 20 million cases of gastroenteritis. During 2015, 316 norovirus outbreaks in the United States were reported.

*Source:* Adapted from CDC, Foodborne Outbreak Online Database (FOOD).

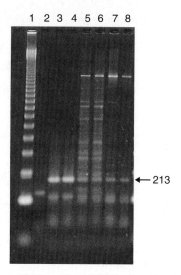

**Figure B Results of PCR of patient samples. Lane 1, 123-bp size ladders. Lane 2, negative RT-PCR control; Lanes 3–8, patient samples. Norovirus is identified by the 213-bp band of DNA.**

← 213

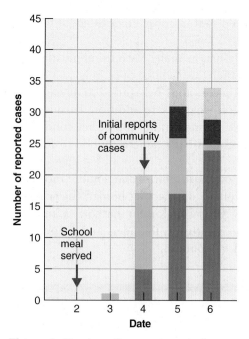

KEY

- Reported community cases
- Social service group
- School employees
- Publishing company employees

**Figure A Number of cases reported.**

Like the invention of the microscope, the development of DNA techniques is causing profound changes in science, agriculture, and human health care. With this technology not quite 50 years old, it is difficult to predict exactly what changes will occur. However, it is likely that within another 30 years, many of the treatments and diagnostic methods discussed in this book will have been replaced by far more powerful techniques based on the unprecedented ability to manipulate DNA precisely.

### CHECK YOUR UNDERSTANDING

✔ 9-20 Identify two advantages and two problems associated with genetically modified organisms.

## Study Outline

 Go to @MasteringMicrobiology for **Interactive Microbiology**, *In the Clinic* videos, *MicroFlix, MicroBoosters, 3D animations, practice quizzes, and more.*

### Introduction to Biotechnology (pp. 243–245)

1. Biotechnology is the use of microorganisms, cells, or cell components to make a product.

### Recombinant DNA Technology (p. 243)

2. Closely related organisms can exchange genes in natural recombination.
3. Genes can be transferred among unrelated species via laboratory manipulation, called rDNA technology.
4. Recombinant DNA is DNA that has been artificially manipulated to combine genes from two different sources.

### An Overview of Recombinant DNA Procedures (pp. 243–245)

5. A desired gene is inserted into a DNA vector, such as a plasmid or a viral genome.
6. The vector inserts the DNA into a new cell, which is grown to form a clone.
7. Large quantities of the gene product can be harvested from the clone.

### Tools of Biotechnology (pp. 245–248)

#### Selection (p. 245)

1. Microbes with desirable traits are selected for culturing by artificial selection.

#### Mutation (p. 245)

2. Mutagens are used to cause mutations that might result in a microbe with desirable traits.
3. Site-directed mutagenesis is used to change a specific codon in a gene.

#### Restriction Enzymes (pp. 245–246)

4. Prepackaged kits are available for rDNA techniques.
5. A restriction enzyme recognizes and cuts only one particular nucleotide sequence in DNA.
6. Some restriction enzymes produce sticky ends, short stretches of single-stranded DNA at the ends of the DNA fragments.
7. Fragments of DNA produced by the same restriction enzyme will spontaneously join by base pairing. DNA ligase can covalently link the DNA backbones.

#### Vectors (pp. 246–247)

8. Vectors are DNA used to transfer other DNA between cells.
9. A plasmid containing a new gene can be inserted into a cell by transformation.

10. A virus containing a new gene can insert the gene into a cell.

### Polymerase Chain Reaction (pp. 247–248)

11. The polymerase chain reaction (PCR) is used to make multiple copies of a desired piece of DNA enzymatically.
12. PCR can be used to increase the amounts of DNA in samples to detectable levels. This may allow sequencing of genes, the diagnosis of genetic diseases, or the detection of viruses.

### Techniques of Genetic Modification (pp. 248–254)

#### Inserting Foreign DNA into Cells (pp. 249–250)

1. Cells can take up naked DNA by transformation. Chemical treatments are used to make cells that are not naturally competent take up DNA.
2. Pores made in protoplasts and animal cells by electric current in the process of electroporation can provide entrance for new pieces of DNA.
3. Protoplast fusion is the joining of cells whose cell walls have been removed.
4. Foreign DNA can be introduced into plant cells by shooting DNA-coated particles into the cells or by using a thin micropipette.

#### Obtaining DNA (pp. 250–252)

5. Genomic libraries can be made by cutting up an entire genome with restriction enzymes and inserting the fragments into bacterial plasmids or phages.
6. Complementary DNA (cDNA) made from mRNA by reverse transcription can be cloned in genomic libraries.
7. Synthetic DNA can be made in vitro by a DNA synthesis machine.

#### Selecting a Clone (pp. 252–253)

8. Antibiotic-resistance markers on plasmid vectors are used to identify cells containing the engineered vector by direct selection.
9. In blue-white screening, the vector contains the genes for *amp* and β-galactosidase.
10. The desired gene is inserted into the β-galactosidase gene site, destroying the gene.
11. Clones containing the recombinant vector will be resistant to ampicillin and unable to hydrolyze X-gal (white colonies).
12. Clones containing foreign DNA can be tested for the desired gene product.
13. A short piece of labeled DNA called a DNA probe can be used to identify clones carrying the desired gene.

### Making a Gene Product (pp. 253–254)

14. *E. coli* is used to produce proteins using rDNA because *E. coli* is easily grown and its genomics are well understood.

15. Efforts must be made to ensure that *E. coli*'s endotoxin does not contaminate a product intended for human use.

16. To recover the product, *E. coli* must be lysed, or the gene must be linked to a gene that produces a naturally secreted protein.

17. Yeasts can be genetically modified and are likely to secrete a gene product continuously.

18. Genetically modified mammalian cells can be grown to produce proteins such as hormones for medical use.

19. Genetically modified plant cells can be grown and used to produce plants with new properties.

## Applications of DNA Technology (pp. 254–262)

1. Cloned DNA is used to produce products, study the cloned DNA, and alter the phenotype of an organism.

### Therapeutic Applications (pp. 255–256)

2. Synthetic genes linked to the β-galactosidase gene (*lacZ*) in a plasmid vector were inserted into *E. coli*, allowing *E. coli* to produce and secrete the two polypeptides used to make human insulin.

3. Cells and viruses can be modified to produce a pathogen's surface protein, which can be used as a vaccine.

4. DNA vaccines consist of rDNA cloned in bacteria.

5. Gene therapy can be used to cure genetic diseases by replacing the defective or missing gene.

6. RNAi may be useful to prevent expression of abnormal proteins.

### Genome Projects (pp. 256–257)

7. Nucleotide sequences of genomes from more than 1000 organisms, including humans, have been completed.

8. This leads to determining the proteins produced in a cell.

### Scientific Applications (pp. 257–260)

9. DNA can be used to increase understanding of DNA, for genetic fingerprinting, and for gene therapy.

10. DNA sequencing machines are used to determine the nucleotide base sequence of restriction fragments in shotgun sequencing.

11. Bioinformatics is the use of computer applications to study genetic data; proteomics is the study of a cell's proteins.

12. Southern blotting can be used to locate a gene in a cell.

13. DNA probes can be used to quickly identify a pathogen in body tissue or food.

14. Forensic microbiologists use DNA fingerprinting to identify the source of bacterial or viral pathogens.

15. Bacteria may be used to make nano-sized materials for nanotechnology machines.

### Agricultural Applications (pp. 260–262)

16. Cells from plants with desirable characteristics can be cloned to produce many identical cells. These cells can then be used to produce whole plants from which seeds can be harvested.

17. Plant cells can be modified by using the Ti plasmid vector. The tumor-producing T genes are replaced with desired genes, and the rDNA is inserted into *Agrobacterium*. The bacterium naturally transforms its plant hosts.

18. Antisense DNA can prevent expression of unwanted proteins.

## Safety Issues and the Ethics of Using DNA Technology (pp. 262–265)

1. Strict safety standards are used to avoid the accidental release of genetically modified microorganisms.

2. Some microbes used in rDNA cloning have been altered so that they cannot survive outside the laboratory.

3. Microorganisms intended for use in the environment may be modified to contain suicide genes so that the organisms do not persist in the environment.

4. Genetic testing raises a number of ethical questions: Should employers have access to a person's genetic records? Will genetic information be used to discriminate against people? Will genetic counseling be available to everyone?

5. Genetically modified crops must be safe for consumption and for release in the environment.

# Study Questions

For answers to Knowledge and Comprehension questions, turn to the Answers tab at the back of the textbook.

## Knowledge and Comprehension

### Review

1. Compare and contrast the following terms:
   a. *cDNA* and *gene*
   b. *RFLP* and *gene*
   c. *DNA probe* and *gene*
   d. *DNA polymerase* and *DNA ligase*
   e. *rDNA* and *cDNA*
   f. *genome* and *proteome*

2. Differentiate the following terms. Which one is "hit and miss"—that is, does *not* add a specific gene to a cell?
   a. protoplast fusion
   b. gene gun
   c. microinjection
   d. electroporation

3. Some commonly used restriction enzymes are listed in Table 9.1 on page 246.
   a. Indicate which enzymes produce sticky ends.
   b. Of what value are sticky ends in making rDNA?
4. Suppose you want multiple copies of a gene you have synthesized. How would you obtain the necessary copies by cloning? By PCR?
5. **DRAW IT** Using the following map of plasmid pMICRO, diagram the locations of the restriction fragments that result from digesting pMICRO with *Eco*RI, *Hin*dIII, and both enzymes together following electrophoresis. Which enzyme makes the smallest fragment containing the tetracycline resistance gene?

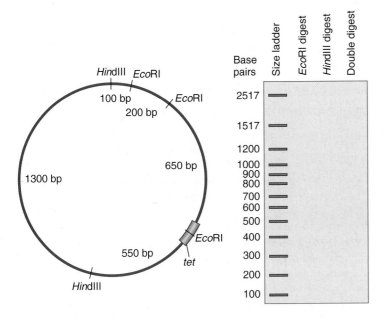

6. Describe an rDNA experiment in two or three sentences. Use the following terms: intron, exon, DNA, mRNA, cDNA, RNA polymerase, reverse transcriptase.
7. List at least two examples of the use of rDNA in medicine and in agriculture.
8. You are attempting to insert a gene for saltwater tolerance into a plant by using the Ti plasmid. In addition to the desired gene, you add a gene for tetracycline resistance (*tet*) to the plasmid. What is the purpose of the *tet* gene?
9. How does RNAi "silence" a gene?
10. **NAME IT** This virus family, normally associated with AIDS, may be useful for gene therapy.

## Multiple Choice

1. Restriction enzymes were first discovered with the observation that
   a. DNA is restricted to the nucleus.
   b. phage DNA is destroyed in a host cell.
   c. foreign DNA is kept out of a cell.
   d. foreign DNA is restricted to the cytoplasm.
   e. all of the above
2. The DNA probe, 3'-GGCTTA, will hybridize with which of the following?
   a. 5'-CCGUUA                    d. 3'-CCGAAT
   b. 5'-CCGAAT                    e. 3'-GGCAAU
   c. 5'-GGCTTA
3. Which of the following is the fourth basic step to genetically modify a cell?
   a. transformation
   b. ligation
   c. plasmid cleavage
   d. restriction-enzyme digestion of gene
   e. isolation of gene
4. The following enzymes are used to make cDNA. What is the second enzyme used to make cDNA?
   a. reverse transcriptase
   b. ribozyme
   c. RNA polymerase
   d. DNA polymerase
5. If you put a gene in a virus, the next step in genetic modification would be
   a. insertion of a plasmid.        d. PCR.
   b. transformation.                e. Southern blotting.
   c. transduction.
6. You have a small gene that you want replicated by PCR. You add radioactively labeled nucleotides to the PCR thermal cycler. After three replication cycles, what percentage of the DNA single strands are radioactively labeled?
   a. 0%                            d. 87.5%
   b. 12.5%                         e. 100%
   c. 50%

Match the following choices to the statements in questions 7 through 10.
   a. antisense                      d. Southern blot
   b. clone                          e. vector
   c. library

7. Pieces of human DNA stored in yeast cells.
8. A population of cells carrying a desired plasmid.
9. Self-replicating DNA for transmitting a gene from one organism to another.
10. DNA that hybridizes with mRNA.

## Analysis

1. Design an experiment using vaccinia virus to make a vaccine against the AIDS virus (HIV).

2. Why did the use of DNA polymerase from the bacterium *Thermus aquaticus* allow researchers to add the necessary reagents to tubes in a preprogrammed heating block?

3. The following picture shows bacterial colonies growing on X-gal plus ampicillin in a blue-white screening test. Which colonies have the recombinant plasmid? The small satellite colonies do not have the plasmid. Why did they start growing on the medium 48 hours after the larger colonies?

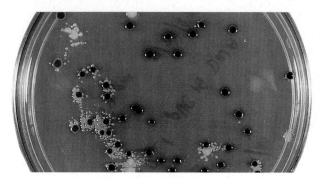

## Clinical Applications and Evaluation

1. PCR has been used to examine oysters for the presence of *Vibrio cholerae*. Oysters from different areas were homogenized, and DNA was extracted from the homogenates. The DNA was digested by the restriction enzyme *Hinc*II. A primer for the hemolysin gene of *V. cholerae* was used for the PCR reaction. After PCR, each sample was electrophoresed and stained with a probe for the hemolysin gene. Which of the oyster samples were (was) positive for *V. cholerae*? How can you tell? Why look for *V. cholerae* in oysters? What is the advantage of PCR over conventional biochemical tests to identify the bacteria?

2. Using the restriction enzyme *Eco*RI, the following gel electrophoresis patterns were obtained from digests of various DNA molecules from a transformation experiment. Can you conclude from these data that transformation occurred? Explain why or why not.

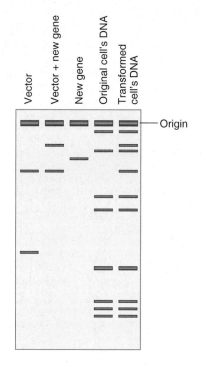

# Classification of Microorganisms 10

The science of classification, especially the classification of living forms, is called *taxonomy* (from the Greek for orderly arrangement). The objective of taxonomy is to classify living organisms—that is, to establish the relationships between one group of organisms and another and to differentiate them. There may be as many as 100 million different living organisms, but fewer than 10% have been discovered, much less identified and classified.

Taxonomy also provides a common reference for identifying organisms already classified. For example, when a bacterium suspected of causing a specific disease is isolated from a patient, characteristics of that isolate are matched to lists of characteristics of previously classified bacteria to identify the isolate (see the Clinical Focus box on page 280). Finally, taxonomy is a basic and necessary tool for scientists, providing a universal language of communication.

Modern taxonomy is an exciting and dynamic field. The ability to rapidly sequence DNA, even entire genomes, has led to new insights into classification and evolution and has given rise to the current Third Golden Age of Microbiology (Chapter 1, page 14). In this chapter, you will learn the various classification systems, the different criteria used for classification, and tests that are used to identify microorganisms that have already been classified. The contribution of taxonomy in shedding new light on previously discovered organisms, such as the *Salmonella enterica* shown in the photograph, will be discussed in this chapter.

◄ *Salmonella enterica* bacteria. *S. enterica* includes over 2500 serovars, many of which cause gastroenteritis.

## In the Clinic

As a hospice nurse, you are caring for a 75-year-old patient undergoing chemotherapy for cancer who recently developed pneumonia. The patient is homebound, and none of his visitors have been sick. While at the patient's house to collect a sputum sample for the laboratory, you notice the man's dog also has a cough. You end up swabbing the pet's nose and sending that sample in, too. **The cultures for both man and dog grow gram-negative, oxidase-positive, urease- and $H_2S$ positive bacteria. What is the causative agent of these infections?**

*Hint: Look at the figure on page 280 to narrow the possibilities.*

 Play **In the Clinic** Video @MasteringMicrobiology

# The Study of Phylogenetic Relationships

## LEARNING OBJECTIVES

**10-1** Define *taxonomy*, *taxon*, and *phylogeny*.

**10-2** Discuss the limitations of a two-kingdom classification system.

**10-3** Identify the contributions of Linnaeus, Whittaker, and Woese.

**10-4** Discuss the advantages of the three-domain system.

**10-5** List the characteristics of the Bacteria, Archaea, and Eukarya domains.

In 2001, an international project called the All Species Inventory was launched. The project's purpose is to identify and record every species of life on Earth in the next 25 years. These researchers have undertaken a challenging goal: whereas biologists have identified more than 1.7 million different organisms thus far, it is estimated that the number of living species ranges from 10 to 100 million.

Among these many and diverse organisms, however, are many similarities. For example, all organisms are composed of cells surrounded by a plasma membrane, use ATP for energy, and store their genetic information in DNA. These similarities are the result of evolution, or descent from a common ancestor. In 1859, the English naturalist Charles Darwin proposed that natural selection was responsible for the similarities as well as the differences among organisms. The differences can be attributed to the survival of organisms with traits best suited to a particular environment.

To facilitate research, scholarship, and communication, we use **taxonomy**—that is, we put organisms into categories, or **taxa** (singular: *taxon*), to show degrees of similarities among organisms. These similarities are due to relatedness—all organisms are related through evolution. **Systematics**, or **phylogeny**, is the study of the evolutionary history of organisms, and the hierarchy of taxa reflects their evolutionary, or *phylogenetic*, relationships.

The way we have classified organisms has changed greatly over the centuries. From the time of Aristotle, living organisms were categorized in just two ways, as either plants or animals. In 1735 Carolus Linnaeus introduced a formal system of classification with two kingdoms—Plantae and Animalia. As the biological sciences developed, a *natural classification* system—one that groups organisms based on ancestral relationships and allows us to see the order in life—was sought. In the 1800s, Carl von Nägeli proposed that bacteria and fungi be placed in the plant kingdom while Ernst Haeckel proposed the Kingdom Protista, to include bacteria, protozoa, algae, and fungi. For 100 years, biologists continued to follow von Nägeli's placement of bacteria and fungi in the plant kingdom—ironic, given that DNA sequencing shows that fungi are closer to animals than plants. Fungi were placed in their own kingdom in 1959.

The term *prokaryote* was introduced in 1937 to distinguish cells having no nucleus from the nucleated cells of plants and animals. In 1968, Robert G. E. Murray proposed the Kingdom Prokaryotae.

In 1969, Robert H. Whittaker founded the five-kingdom system in which prokaryotes were placed in the Kingdom Prokaryotae, or Monera, and eukaryotes comprised the other four kingdoms. The Kingdom Prokaryotae had been based on microscopic observations. Subsequent advances in molecular biology revealed that there are actually two types of prokaryotic cells and one type of eukaryotic cell.

> **CHECK YOUR UNDERSTANDING**
>
> ✔ **10-1** Of what value are taxonomy and systematics?
>
> ✔ **10-2, 10-3** Why shouldn't bacteria be placed in the plant kingdom?

## The Three Domains

The discovery of three cell types was based on the observations that ribosomes are not the same in all cells (see Chapter 4, page 90). Ribosomes are present in all cells. Comparing the sequences of nucleotides in ribosomal RNA (see page 288) from different kinds of cells shows that there are three distinctly different cell groups: the eukaryotes and two different types of prokaryotes—the bacteria and the archaea.

In 1978, Carl R. Woese proposed elevating the three cell types to a level above kingdom, called domain. Woese believed that the archaea and the bacteria, although similar in appearance, should form their own separate domains on the evolutionary tree (**Figure 10.1**). In addition to differences in rRNA, the three domains differ in membrane lipid structure, transfer RNA molecules, and sensitivity to antibiotics (Table 10.1).

In this widely accepted scheme, animals, plants, and fungi are kingdoms in the Domain **Eukarya**. The Domain **Bacteria** includes all of the pathogenic prokaryotes as well as many of the nonpathogenic prokaryotes found in soil and water. The photoautotrophic prokaryotes are also in this domain. The Domain **Archaea** includes prokaryotes that do not have peptidoglycan in their cell walls. They often live in extreme environments and carry out unusual metabolic processes. Archaea include three major groups:

1. Methanogens, strict anaerobes that produce methane ($CH_4$) from carbon dioxide and hydrogen

2. Extreme halophiles, which require high concentrations of salt for survival

3. Hyperthermophiles, which normally grow in extremely hot environments

The evolutionary relationship of the three domains is the subject of current research by biologists. Based on rRNA analysis,

# Three-Domain System

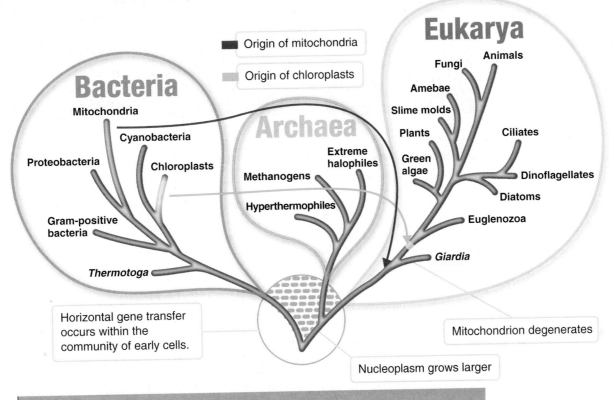

## KEY CONCEPTS

- All organisms evolved from cells that formed over 3.5 billion years ago.

- The DNA passed on from ancestors is described as *conserved*.

- The Domain Eukarya includes the Kingdoms Fungi, Plantae, and Animalia, as well as protists. The Domains Bacteria and Archaea are prokaryotes.

three cell lineages clearly emerged as cells were forming 3.5 billion years ago. That led to the Archaea, the Bacteria, and what eventually became the nucleoplasm of the eukaryotes. However, the three cell lines were not isolated from each other; horizontal gene transfer (page 229) appears to have occurred among them. Analysis of complete genomes shows that each domain shares genes with other domains. One-quarter of the genes of the bacterium *Thermotoga* were probably acquired from an archaeon. Gene transfer also has been seen between eukaryotic hosts and their prokaryote symbionts.

The oldest known fossils are the remains of prokaryotes that lived more than 3.5 billion years ago. Eukaryotic cells evolved more recently, about 2.5 billion years ago. According to the endosymbiotic theory, eukaryotic cells evolved from prokaryotic cells living inside one another, as endosymbionts (see Chapter 4, page 102). In fact, the similarities between prokaryotic cells and eukaryotic organelles provide striking evidence for this endosymbiotic relationship (Table 10.2).

## CLINICAL CASE  Full-Flavor Outbreak

Monica Jackson, a 32-year-old production assistant at a Reno, Nevada, television station, has made an appointment with the nurse practitioner at her physician's office. Monica tells the nurse practitioner that she has had diarrhea, nausea, and abdominal cramping for almost 12 hours. She also feels tired and has a low-grade fever. Monica felt fine one minute, and the next she was violently ill. Monica informs the nurse practitioner that she and her good friend, who is also sick, had been at the same luncheon the day before. The nurse practitioner takes a stool sample and sends it to the hospital laboratory for analysis.

**What will the laboratory do first to look for a bacterial pathogen? Read on to find out.**

271  281  283  287  289  290

## TABLE 10.1 Some Characteristics of Archaea, Bacteria, and Eukarya

|  | **Archaea** | **Bacteria** | **Eukarya** |
|---|---|---|---|
|  | *Sulfolobus* sp.  SEM 1 μm | *E. coli*  SEM 1 μm | *Amoeba* sp.  SEM 5 μm |
| **Cell Type** | Prokaryotic | Prokaryotic | Eukaryotic |
| **Cell Wall** | Varies in composition; contains no peptidoglycan | Contains peptidoglycan | Varies in composition; contains carbohydrates |
| **Membrane Lipids** | Composed of branched carbon chains attached to glycerol by ether linkage | Composed of straight carbon chains attached to glycerol by ester linkage | Composed of straight carbon chains attached to glycerol by ester linkage |
| **First Amino Acid in Protein Synthesis** | Methionine | Formylmethionine | Methionine |
| **Antibiotic Sensitivity** | No | Yes | No |
| **rRNA Loop*** | Lacking | Present | Lacking |
| **Common Arm of tRNA†** | Lacking | Present | Present |

*Binds to ribosomal protein; found in all bacteria.

†A sequence of bases in tRNA found in all eukaryotes and bacteria: guanine-thymine-pseudouridine-cytosine-guanine.

## TABLE 10.2 Prokaryotic Cells and Eukaryotic Organelles Compared

|  | **Prokaryotic Cell** | **Eukaryotic Cell** | **Eukaryotic Organelles (Mitochondria and Chloroplasts)** |
|---|---|---|---|
| **DNA** | One circular; some two circular; some linear | Linear | Circular |
| **Histones** | In archaea | Yes | No |
| **First Amino Acid in Protein Synthesis** | Formylmethionine (bacteria) Methionine (archaea) | Methionine | Formylmethionine |
| **Ribosomes** | 70S | 80S | 70S |
| **Growth** | Binary fission | Mitosis | Binary fission |

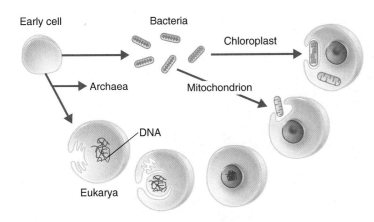

**Figure 10.2 A model of the origin of eukaryotes.** Invagination of the plasma membrane may have formed the nuclear envelope and endoplasmic reticulum. Similarities, including rRNA sequences, indicate that endosymbiotic prokaryotes gave rise to mitochondria and chloroplasts.

**Q** How many membranes make up the nuclear envelope of a eukaryotic cell?

The original nucleoplasmic cell was prokaryotic. However, infoldings in its plasma membrane may have surrounded the nuclear region to produce a true nucleus (**Figure 10.2**). French researchers provided support for this hypothesis with their observations of a true nucleus in *Gemmata* bacteria (see Figure 11.16 on page 311). Over time, the nucleoplasm's chromosome may have acquired pieces such as transposons (page 231). In some cells, this large chromosome may have fragmented into smaller linear chromosomes. Perhaps cells with linear chromosomes had an advantage in cell division over those with a large, unwieldy circular chromosome.

That nucleoplasmic cell provided the original host in which endosymbiotic bacteria developed into organelles (see page 102). An example of a modern prokaryote living in a eukaryotic cell is shown in **Figure 10.3**. The cyanobacterium-like cell and its eukaryotic host require each other for survival.

Taxonomy provides tools for clarifying the evolution of organisms, as well as their interrelationships. New organisms are being discovered every day, and taxonomists continue to search for a natural classification system that reflects phylogenetic relationships.

## A Phylogenetic Tree

In a phylogenetic tree, grouping organisms according to common properties implies that a group of organisms evolved from a common ancestor; each species retains some of the characteristics of the ancestor. Some of the information used to classify and determine phylogenetic relationships in higher organisms comes from fossils. Bones, shells, or stems that contain mineral matter or

ASM: The evolutionary relatedness of organisms is best reflected in phylogenetic trees.

have left imprints in rock that was once mud are examples of fossils.

The structures of most microorganisms are not readily fossilized. Some exceptions are the following:

- A marine protist whose fossilized colonies form the White Cliffs of Dover, England.
- Stromatolites, the fossilized remains of filamentous bacteria and sediments that flourished between 0.5 and 2 billion years ago (**Figure 10.4a** and Figure 10.4b).
- Cyanobacteria-like fossils found in rocks that are 3.0 to 3.5 billion years old. These are widely believed to be the oldest known fossils (Figure 10.4c).

Because fossil evidence isn't available for most prokaryotes, their phylogeny must be based on other evidence. In one notable exception, scientists may have isolated living bacteria and yeast that are 25 to 40 million years old. In 1995, the American microbiologist Raul Cano and his colleagues reported growing *Bacillus sphaericus* and other as yet unidentified microorganisms that had survived embedded in amber (fossilized plant resin) for millions of years. If confirmed, this discovery should provide more information about the evolution of microorganisms.

Similarities in genomes can be used to group organisms into taxa and to provide a timeline for the emergence of taxa. This is especially important for microorganisms that usually don't leave fossil evidence. This concept of a molecular clock based on the differences in amino acids in hemoglobin among different animals was first proposed in the 1960s. A **molecular clock** for evolution is based on nucleotide sequences in the

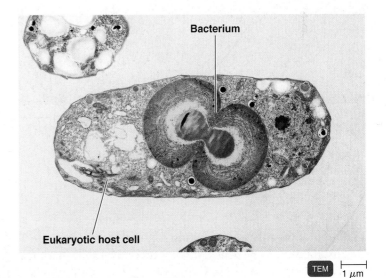

**Figure 10.3 *Cyanophora paradoxa*.** This organism, in which the eukaryotic host and the bacterium require each other for survival, provides a modern example of how eukaryotic cells might have evolved.

**Q** What features do chloroplasts, mitochondria, and bacteria have in common?

**(a)** Bacterial communities form rocklike pillars called stromatolites. These began growing about 3000 years ago.

30 cm

**(b)** Cut section through a fossilized stomatolite that flourished 2 billion years ago

2 cm

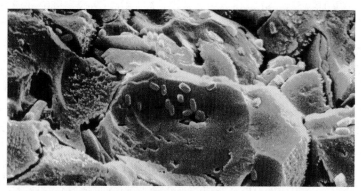

**(c)** Rod-shaped prokaryotes from the Early Precambrian (3.5 billion years ago) of South Africa

SEM  5 μm

**Figure 10.4 Fossilized prokaryotes.**

**Q** What evidence is used to determine the phylogeny of prokaryotes?

genomes of organisms. Mutations accumulate in a genome at a constant rate. Some genes, such as those for rRNA, have few mutations—these are highly conserved genes. Other regions of a genome change with no apparent effect on the organism. Comparing the number of mutations between two organisms with the expected rate of change provides an estimate of when the two diverged from a common ancestor. This technique was used to track the path of Zika virus to the United States. (See the box on page 218.)

Conclusions from rRNA sequencing and DNA hybridization studies (discussed on page 286) of selected orders and families of eukaryotes are in agreement with the fossil records. This has encouraged scientists to use DNA hybridization and rRNA sequencing to gain a better understanding of the evolutionary relationships among prokaryotic groups.

**CHECK YOUR UNDERSTANDING**

✔ **10-4** What evidence supports classifying organisms into three domains?

✔ **10-5** Compare archaea and bacteria; bacteria and eukarya; and archaea and eukarya.

# Classification of Organisms

**LEARNING OBJECTIVES**

**10-6** Explain why scientific names are used.

**10-7** List the major taxa.

**10-8** Differentiate *culture*, *clone*, and *strain*.

**10-9** List the major characteristics used to differentiate the three kingdoms of multicellular Eukarya.

**10-10** Define *protist*.

**10-11** Differentiate eukaryotic, prokaryotic, and viral species.

Living organisms are grouped according to similar characteristics (classification), and each organism is assigned a unique scientific name. The rules for classifying and naming, which are used by biologists worldwide, are discussed next.

## Scientific Nomenclature

In a world inhabited by millions of living organisms, biologists must be sure they know exactly which organism is being discussed. We cannot use common names, because the same name is often used for many different organisms in different locales. For example, there are two different organisms with the common name Spanish moss, and neither one is actually a moss. Plus, common names can be misleading and are in different languages. A system of scientific names was developed to solve this problem.

Every organism is assigned two names, or a binomial (Chapter 1, page 4). These names are the **genus** name and **specific epithet (species)**, and both names are printed underlined or italicized. The genus name is always capitalized and is always a noun. The species name is lowercase and is usually an adjective. Because this system gives two names to each organism, the system is called **binomial nomenclature**.

Let's consider some examples. Our own genus and specific epithet are *Homo sapiens* (HŌ-mō SĀ-pē-enz). The genus means man; the specific epithet means wise. A mold that contaminates bread is called *Rhizopus stolonifer* (RĪ-zō-pus stō-LON-i-fer). *Rhizo-* (root) describes rootlike structures on the fungus; *stolo-* (a shoot) describes the long hyphae. Table 1.1 on page 4 contains more examples.

Binomials are used by scientists worldwide, regardless of their native language. This nomenclature enables them to share knowledge efficiently and accurately. Several scientific entities are responsible for establishing rules governing the naming of organisms. Scientific names are taken from Latin (a genus name can be taken from Greek) or latinized by the addition of the appropriate suffix. Suffixes for order and family are *-ales* and *-aceae*, respectively.

As new laboratory techniques make more detailed characterizations of microbes possible, two genera may be reclassified as a single genus, or a genus may be divided into two or more genera. For example, rRNA analysis (see page 288) indicated that "Streptococcus faecalis" was only distantly related to the other streptococcal species; consequently, a new genus called *Enterococcus* was created, and this species was renamed *E. faecalis*. Making the transition to a new name can be confusing, so the old name is often written in parentheses. For example, a physician looking for information on the cause of a patient's pneumonia-like symptoms (melioidosis) would find the bacterial name *Burkholderia (Pseudomonas) pseudomallei* (berk-HŌL-der′ē-ah soo-dō-MAL-lē-ē).

*# 4*

## The Taxonomic Hierarchy

All organisms can be grouped into a series of subdivisions that make up the taxonomic hierarchy. Linnaeus developed this hierarchy for his classification of plants and animals. A **eukaryotic species** is a group of closely related organisms that breed among themselves. (Bacterial species will be discussed shortly.) A genus consists of species that differ from each other in certain ways but are related by descent. For example, *Quercus* (KWER-kus), the genus name for oak, consists of all types of oak trees (white oak, red oak, bur oak, velvet oak, and so on). Even though each species of oak differs from every other species, they are all related genetically. Just as a number of species make up a genus, related genera make up a **family**. A group of similar families constitutes an **order**, and a group of similar orders makes up a **class**. Related classes, in turn, make up a **phylum**. Thus, a particular organism (or species) has a genus name and specific epithet and belongs to a family, order, class, and phylum.

All phyla that are related to each other make up a **kingdom**, and related kingdoms are grouped into a **domain** (Figure 10.5).

**CHECK YOUR UNDERSTANDING**

✔ **10-6** Using *Escherichia coli* and *Entamoeba coli* as examples, explain why the genus name must be written out on first use. Why is binomial nomenclature preferable to common names?

✔ **10-7** Find the gram-positive bacteria *Staphylococcus* in Appendix E. To which bacteria is this genus more closely related: *Bacillus* or *Streptococcus*?

## Classification of Prokaryotes

The taxonomic classification scheme for prokaryotes is found in *Bergey's Manual of Systematic Bacteriology*, 2nd edition (see Appendix E). In *Bergey's Manual*, prokaryotes are divided into two domains: Bacteria and Archaea. Each domain is divided into phyla. Remember, the classification is based on similarities in rRNA nucleotide sequences. Classes are divided into orders; orders, into families; families, into genera; and genera, into species.

A prokaryotic species is defined somewhat differently from a eukaryotic species, which is a group of closely related organisms that can interbreed. Unlike reproduction in eukaryotic organisms, cell division in bacteria is not directly tied to sexual conjugation, which is infrequent and does not always need to be species-specific. A **prokaryotic species**, therefore, is defined simply as a population of cells with similar characteristics. (The types of characteristics will be discussed later in this chapter.) The members of a bacterial species are nearly indistinguishable from each other but are distinguishable from members of other species, usually on the basis of several features. As you know, bacteria grown in media are called a culture. A pure culture is often a **clone**, a population of cells derived from a single parent cell. All cells in the clone should be identical, but in some cases, pure cultures of the same species are not identical in all

 ASM: The traditional concept of species is not readily applicable to microbes due to asexual reproduction and the frequent occurrence of horizontal gene transfer.

ways. Each such group is called a **strain**. Strains are identified by numbers, letters, or names that follow the specific epithet.

*Bergey's Manual* provides a reference for identifying bacteria in the laboratory, as well as a classification scheme for bacteria. One scheme for the evolutionary relationships of bacteria is shown in **Figure 10.6**. Characteristics used to classify and identify bacteria are discussed in Chapter 11.

## Classification of Eukaryotes   *# 2*

Some kingdoms in the domain Eukarya are shown in Figure 10.1.

In 1969, simple eukaryotic organisms, mostly unicellular, were grouped as the kingdom *Protista*, a catchall kingdom for a variety of organisms. Historically, eukaryotic organisms that didn't fit into other kingdoms were placed in the Protista. Approximately 200,000 species of protists have been identified thus far, and these organisms are nutritionally quite diverse—from photosynthetic to obligate intracellular parasite. Ribosomal RNA sequencing is making it possible to divide protists into groups based on their descent from common ancestors. Consequently, the organisms once classified as protists are being divided into **clades**, or genetically related groups. For convenience, we'll continue to use the term *protist* to refer to unicellular eukaryotes and their close relatives. These organisms will be discussed in Chapter 12.

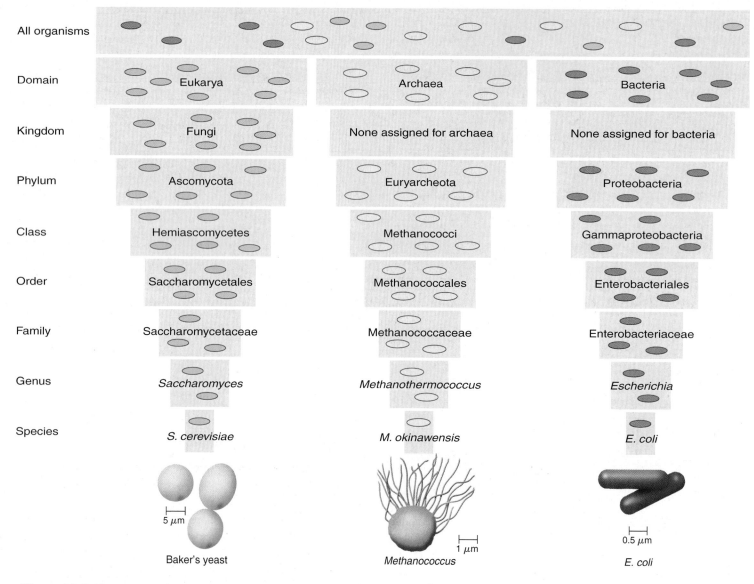

**Figure 10.5 The taxonomic hierarchy.** Organisms are grouped according to relatedness. Species that are closely related are grouped into a genus. For example, the baker's yeast, *Saccharomyces cerevisiae*, belongs to the genus that includes sourdough yeast (*S. exiguus*). Related genera, such as *Saccharomyces* and *Candida*, are placed in a family, and so on. Each group is more comprehensive. The domain Eukarya includes all organisms with eukaryotic cells.

**Q** What is the biological definition of *family*?

Fungi, plants, and animals make up the three kingdoms of more complex eukaryotic organisms, most of which are multicellular.

The Kingdom **Fungi** includes the unicellular yeasts, multicellular molds, and macroscopic species such as mushrooms. A fungus absorbs dissolved organic matter through its plasma membrane to obtain raw materials for vital functions. The cells of a multicellular fungus are commonly joined to form thin tubes called *hyphae*. Fungi develop from spores or from fragments of hyphae. (See Figure 12.2, page 326.)

The Kingdom **Plantae** (plants) includes mosses, ferns, conifers, and flowering plants. All members of this kingdom are multicellular. To obtain energy, a plant uses photosynthesis, a process that converts carbon dioxide and water into organic molecules used by the cell.

The kingdom of multicellular organisms called **Animalia** (animals) includes sponges, various worms, insects, and animals with backbones (vertebrates). Animals obtain nutrients and energy by ingesting organic matter through a mouth of some kind.

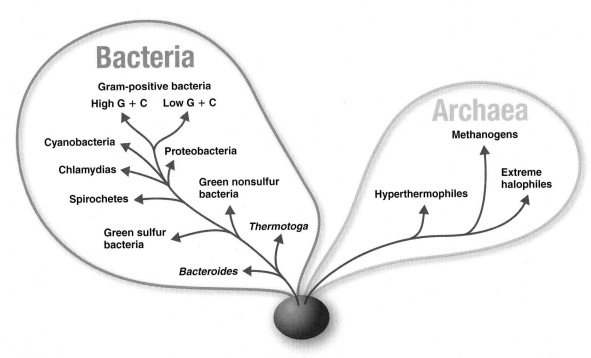

**Figure 10.6 Phylogenetic relationships of prokaryotes.** Arrows indicate major lines of descent of bacterial groups. Selected phyla are indicated.

**Q** Members of which phylum can be identified by Gram staining?

## Classification of Viruses

Viruses aren't classified as part of any of the three domains. They aren't composed of cells, and they use the anabolic machinery within living host cells to multiply. A viral genome can direct biosynthesis inside a host cell, and some viral genomes can be incorporated into the host genome. The ecological niche of a virus is its specific host cell, so viruses may be more closely related to their hosts than to other viruses. The International Committee on Taxonomy of Viruses defines a **viral species** as a population of viruses with similar characteristics (including morphology, genes, and enzymes) that occupies a particular ecological niche.

Viruses are obligatory intracellular parasites. Viral genes carried in the genomes of other organisms provide a record of viral evolution. Recent analysis shows that bornavirus and retrovirus genes integrated into mammals, including humans, at least 40 million years ago. There are three hypotheses on the origin of viruses: (1) They arose from independently replicating strands of nucleic acids (such as plasmids). (2) They developed from degenerative cells that, through many generations, gradually lost the ability to survive independently but could survive when associated with another cell. (3) They coevolved with host cells. Viruses will be discussed in Chapter 13.

### CHECK YOUR UNDERSTANDING

✔ **10-8** Use the terms *species*, *culture*, *clone*, and *strain* in one sentence to describe growing methicillin-resistant *Staphylococcus aureus* (MRSA).

✔ **10-9** You discover a new multicellular, nucleated, heterotrophic organism with cell walls. To what kingdom does it belong?

✔ **10-10** Write your own definition of *protist*.

✔ **10-11** Why doesn't the definition of *viral species* work for bacteria?

## Methods of Classifying and Identifying Microorganisms

### LEARNING OBJECTIVES

**10-12** Compare and contrast classification and identification.

**10-13** Explain the purpose of *Bergey's Manual*.

**10-14** Describe how staining and biochemical tests are used to identify bacteria.

**10-15** Differentiate Western blotting from Southern blotting.

**10-16** Explain how serological tests and phage typing can be used to identify an unknown bacterium.

**10-17** Describe how a newly discovered microbe can be classified by DNA sequencing, DNA fingerprinting, and PCR.

**10-18**  Describe how microorganisms can be identified by nucleic acid hybridization, Southern blotting, DNA chips, ribotyping, and FISH.

**10-19**  Differentiate a dichotomous key from a cladogram.

A classification scheme provides a list of characteristics and a means for comparison to aid in the identification of an organism. Once an organism is identified, it can be placed into a previously devised classification scheme. Microorganisms are *identified* for practical purposes—for example, to determine an appropriate treatment for an infection. They are not necessarily identified by the same techniques by which they are *classified*. Most identification procedures are easily performed in a laboratory and use as few procedures or tests as possible. Protozoa, parasitic worms, and fungi can usually be identified microscopically. Most prokaryotic organisms don't have distinguishing morphological features or even much variation in size and shape. Consequently, microbiologists have developed a variety of methods to test metabolic reactions and other characteristics to identify prokaryotes.

*Bergey's Manual of Determinative Bacteriology* has been a widely used reference since the first edition was published in 1923. *Bergey's Manual* does not classify bacteria according to evolutionary relatedness but instead provides identification (determinative) schemes based on such criteria as cell wall composition, morphology, differential staining, oxygen requirements, and biochemical testing.* The majority of Bacteria and Archaea haven't been cultured, and scientists estimate that only 1% of these microbes have been discovered.

Medical microbiology (the branch of microbiology dealing with human pathogens) has dominated the interest in microbes, and this interest is reflected in many identification schemes. However, to put the pathogenic properties of bacteria in perspective, of the more than 11,500 species listed in the *Approved Lists of Bacterial Names*, fewer than 5% are human pathogens.

We next discuss several criteria and methods for the classification and routine identification of microorganisms. In addition to properties of the organism itself, the source and habitat of a bacterial isolate are part of the identification processes. In clinical microbiology, a physician will swab a patient's pus or tissue surface. The swab is inserted into a tube of transport medium. **Transport media** are usually not nutritive and are designed to prolong viability of fastidious pathogens. The physician will note the type of specimen and test(s) requested on a lab requisition form (**Figure 10.7**). The lab technician's results will help the physician begin treatment (see the box in Chapter 5, page 141).

## Morphological Characteristics

Morphological (structural) characteristics have helped taxonomists classify organisms for 200 years. Higher organisms are frequently classified according to observed anatomical detail. But many microorganisms look too similar to be classified by their structures alone. Organisms that might differ in metabolic or physiological properties may look alike under a microscope. Literally hundreds of bacterial species are small rods or small cocci.

Larger size and the presence of intracellular structures don't always mean easy classification, however. *Pneumocystis* (noo-mō-SIS-tis) pneumonia is the most common opportunistic infection in AIDS and other immunocompromised patients. Until the AIDS epidemic, the causative agent of this infection, *P. jirovecii* (ye-rō-VET-zē-ē) [formerly "*P. carinii*" (kar-I-nē-ē)] was rarely seen in humans. *Pneumocystis* lacks structures that can be easily used for identification (see Figure 24.19, page 714), and its taxonomic position has been uncertain since its discovery in 1909. It was originally classified as a protozoan; however, in 1988 rRNA sequencing showed that *Pneumocystis* is actually a member of the Kingdom Fungi. New treatments are being investigated as researchers take this organism's relatedness to fungi into account.

Cell morphology tells us little about phylogenetic relationships. However, morphological characteristics are still useful in identifying bacteria. For example, differences in such structures as endospores or flagella can be helpful.

## Differential Staining

One of the first steps in identifying bacteria is differential staining (see Chapter 3). Most bacteria are either gram-positive or gram-negative. Other differential stains, such as the acid-fast stain, can be useful for a more limited group of microorganisms. Recall that these stains are based on the chemical composition of cell walls and therefore are not useful in identifying either the wall-less bacteria or the archaea with unusual walls. Microscopic examination of a Gram stain or an acid-fast stain is used to obtain information quickly in the clinical environment.

## Biochemical Tests

Enzymatic activities are widely used to differentiate bacteria. Even closely related bacteria can usually be separated into distinct species by subjecting them to biochemical tests. For example, biochemical tests are used to identify bacteria in humans and in marine mammals (see the boxes on pages 141 and 280). Moreover, biochemical tests can provide insight into a species' niche in the ecosystem. For example, a bacterium that can fix nitrogen gas or oxidize elemental sulfur will provide important nutrients for plants and animals. This will be discussed in Chapter 27.

Enteric gram-negative bacteria are a large heterogeneous group of microbes whose natural habitat is the intestinal tract of humans and other animals. This family contains several

---

*Both *Bergey's Manual of Systematic Bacteriology* (see page 275) and *Bergey's Manual of Determinative Bacteriology* are referred to simply as *Bergey's Manual*; the complete titles are used when the information under discussion is found in one but not the other, for example, an identification table.

**MICROBIOLOGY REQUISITION**

Date:

Time:

Slip prepared by:

Lab:

Date, time received:

Physician name:

Collected by:

Patient ID#:

→ DO NOT WRITE BELOW THIS LINE ←

**USE SEPARATE SLIP FOR EACH REQUEST**

**GRAM STAIN REPORT**

- ☐ GRAM POS. COCCI, GROUPS
- ☐ GRAM POS. COCCI, PAIRS/CHAIN
- ☐ GRAM POS. RODS
- ☒ GRAM NEG. COCCI
- ☐ GRAM NEG. RODS
- ☐ GRAM NEG. COCCOBACILLI
- ☐ YEAST
- ☐ OTHER

- ☐ NO GROWTH
- ☐ NO GROWTH IN ___ DAYS
- ☐ MIXED MICROBIOTA
- ☐ SPECIMEN IMPROPERLY COLLECTED OR TRANSPORTED
- ☐ ___ DIFFERENT TYPES OF ORGANISMS
- ☐ NEGATIVE FOR SALMONELLA, SHIGELLA, AND CAMPYLOBACTER
- ☐ NO OVA, CYSTS, OR PARASITES SEEN
- ☒ OXIDASE-POSITIVE GRAM-NEGATIVE DIPLOCOCCI
- ☐ PRESUMPTIVE BETA STREP GROUP A BY BACITRACIN

**SOURCE OF SPECIMEN**

- ☐ BLOOD
- ☐ CEREBROSPINAL FLUID
- ☐ FLUID (Specify Source) _____
- ☐ THROAT
- ☐ SPUTUM, expectorated
- ☐ OTHER Respiratory (Describe) _____
- ☐ URINE, Clean Catch Midstream
- ☐ URINE, Indwelling Catheter
- ☐ URINE, Straight Catheter
- ☐ URINE, Entire First Morning
- ☐ URINE, Other (Describe) _____
- ☐ STOOL
- ☒ GU (Specify Source) _vag._
- ☐ ABSCESS (Specify Source) _____
- ☐ TISSUE (Specify Source) _____
- ☐ ULCER (Specify Source) _____
- ☐ WOUND (Specify Source) _____
- ☐ STERILIZER TEST

**TEST(S) REQUESTED**

**Bacterial**
- ☐ **Routine culture;** Gram stain, anaerobic culture, susceptibility testing. Throats done for Gp A Strep.
- ☐ *Legionella* culture
- ☐ *Bartonella*
- ☐ Blood Culture

**Other Non-Routine Cultures**
- ☐ *E. coli* O157:H7
- ☐ *Vibrio*
- ☐ *Yersinia*
- ☒ *H. ducreyi*
- ☐ *B. pertussis*
- ☐ Other _____

**Screening Cultures**
- ☒ Gonococci
- ☐ Group B Strep
- ☐ Group A Strep
- ☐ Other _____

- ☐ **ACID-FAST BACILLI**

- ☐ **FUNGAL**

**VIRAL**
- ☐ Routine culture
- ☐ Herpes simplex
- ☐ Direct FA for _____

**PARASITOLOGY**
- ☐ Exam for intestinal ova and parasites
- ☐ *Giardia* immunoassay
- ☐ *Cryptosporidium*
- ☐ Pinworm prep
- ☐ Blood parasites
- ☐ Filaria concentration
- ☐ *Trichomonas*
- ☐ Other _____

**TOXIN ASSAY**
- ☐ *Clostridium difficile*

**DIRECT (Antigen Detection)**
- ☐ Cryptococcal antigen-CSF only
- ☐ Bacterial antigens (Specify) _____

**SPECIAL**
- ☒ Antimicrobial tests (MIC)

Filled out by one person | Filled out by different person

**Figure 10.7 A clinical microbiology lab report form.** In health care, morphology and differential staining are important in determining the proper treatment for microbial diseases. A clinician completes the form to identify the sample and specific tests. In this case, a genitourinary sample will be examined for sexually transmitted infections. The red notations are the lab technician's report of the Gram stain and culture results. (Minimal inhibitory concentration [MIC] of antibiotics will be discussed in Chapter 20, page 578.)

**Q** What diseases are suspected if the "acid-fast bacilli" box is checked?

pathogens that cause diarrheal illness. A number of tests have been developed so technicians can quickly identify the pathogens, a clinician can then provide appropriate treatment, and epidemiologists can locate the source of an illness. All members of the family Enterobacteriaceae are oxidase-negative. Among the enteric bacteria are members of the genera *Escherichia, Enterobacter, Shigella, Citrobacter,* and *Salmonella. Escherichia, Enterobacter,* and *Citrobacter,* which ferment lactose to produce acid and gas, can be distinguished from *Salmonella*

and *Shigella,* which do not. Further biochemical testing, as represented in **Figure 10.8**, can differentiate among the genera.

The time needed to identify bacteria can be reduced considerably by the use of selective and differential media or by rapid identification methods. Selective media contain ingredients that suppress the growth of competing organisms and encourage the growth of desired ones, and differential media allow the desired organism to form a colony that is somehow distinctive (see Chapter 6, page 162).

# Mass Deaths of Marine Mammals Spur Veterinary Microbiology

Over the past 20 years, thousands of marine mammals have died unexpectedly all over the world from a variety of infectious diseases. Notable outbreaks and problems include:

- The deaths of over 500 bottlenose dolphins along the mid-Atlantic coast due to *Brucella* spp. during 2010–2013

- The decline of the California sea otter population in recent years, with toxoplasmosis and other bacterial species responsible for a 40% mortality rate

- The deaths of over 100 harbor seals along the New England coast in 2011 due to influenza A H3N8

- The 2013 deaths of hundreds of bottlenose dolphins in the Atlantic

Ocean from cetacean morbillivirus, likely transmitted to dolphins from pilot whales

- Large numbers of pathogens, including *Nocardia* and *Arcanobacterium,* have been identified in stranded dolphins, harbor seals, and sea otters

## Information Is Scarce

Such issues are the concern of veterinary microbiology, which until recently has been a neglected branch of medical microbiology. Although the diseases of cattle, chickens, and mink have been studied, partly because of their availability to researchers, the microbiology of wild animals, especially marine mammals, is a relatively new field. Gathering samples from animals that live in the open ocean and performing bacteriological analyses on them is

very difficult. The animals being studied are those that have been stranded and those that come onto the shore to breed, such as the northern fur sea lion.

Microbiologists are identifying bacteria in marine mammals by using conventional test batteries (see figure) and genomic data of known species. The FISH technique is being used to find new species of bacteria in marine mammals (see page 288).

Veterinary microbiologists hope that studying the microbiology of wild animals, including marine mammals, will improve wildlife management and also provide models for studying human diseases.

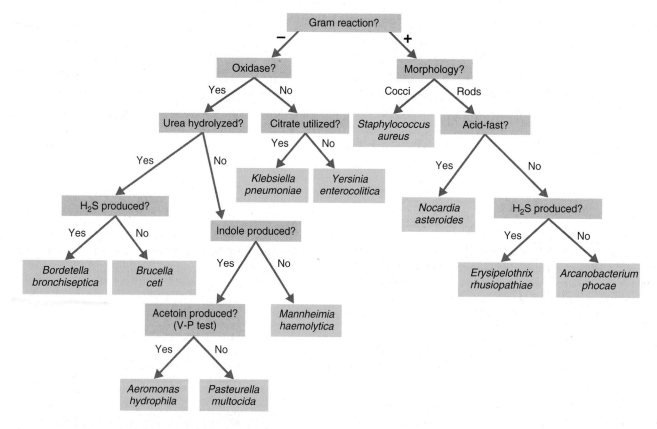

**Biochemical tests used to identify selected species of human pathogens isolated from marine mammals.**

**Q** Assume you isolated a gram-negative rod that is oxidase-positive, is indole-negative, and does not produce urease or acetoin. What is the bacterium?

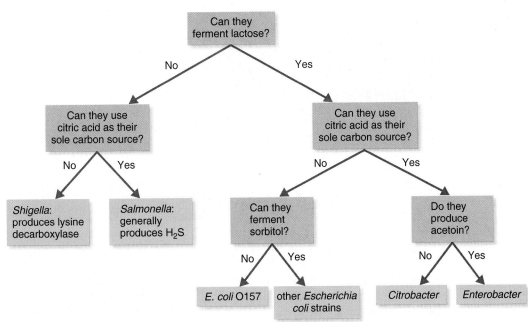

**Figure 10.8 The use of metabolic characteristics to identify selected genera of enteric bacteria.**

**Q** Assume you have a gram-negative bacterium that produces acid from lactose and cannot use citric acid as its sole carbon source. What is the bacterium?

*Bergey's Manual* doesn't evaluate the relative importance of each biochemical test or always describe strains. In diagnosing an infection, clinicians must identify a particular species and even a particular strain to proceed with proper treatment. A limitation of biochemical testing is that mutations and plasmid acquisition can result in strains with different characteristics. Unless a large number of tests is used, an organism could be incorrectly identified. To this end, specific series of biochemical tests have been developed for fast identification in hospital laboratories. Rapid test systems have been developed for yeasts and other fungi, as well as bacteria.

**Rapid identification methods** are manufactured for groups of medically important bacteria, such as the enterics. Such tools are designed to perform several biochemical tests simultaneously and can identify bacteria within 4 to 24 hours. This is sometimes called **numerical identification** because the results of each test are assigned a number. In the simplest form, a positive test would be assigned a value of 1, and a negative is assigned a value of 0. In most commercial testing kits, test results are assigned numbers ranging from 1 to 4 that are based on the relative reliability and importance of each test, and the resulting total is compared to a database of known organisms.

In the example shown in **Figure 10.9**, an unknown enteric bacterium is inoculated into a tube designed to perform 15 biochemical tests. After incubation, results in each compartment are recorded. Notice that each test is assigned a value; a code number is derived from scoring all the tests. Fermentation of glucose is important, and a positive reaction is valued at 4, compared with the production of indole, which has a value of 1. A computerized interpretation of the simultaneous test results is essential and is provided by the manufacturer.

Automated rapid identification is now available for some medically important bacteria and yeasts. Cells from a single colony are lysed, and their proteins are extracted in acetonitrile. Cellular proteins in the extract are read using a mass spectrophotometer that measures the molecular mass of proteins in the sample (**Figure 10.10**). The data obtained are then compared to commercial databases.

## Serology

**Serology** is the science that studies serum and immune responses that are evident in serum (see Chapter 18). Microorganisms are antigenic; that is, microorganisms that enter an animal's body stimulate it to form antibodies. Antibodies are proteins that circulate in the blood and combine in a highly specific way with the bacteria that caused their production. For example, the immune system of a rabbit injected with killed

**CLINICAL CASE**

The laboratory cannot just Gram stain a stool sample to look for a bacterial pathogen. The large number of gram-negative rods would be indistinguishable in a Gram stain made directly from feces. The stool sample should be cultured on selective and differential media to distinguish among bacteria in the stool. Monica's stool sample is cultured on bismuth sulfite agar. Black colonies are present on the agar after 24 hours.

**Can gram-positive bacteria grow on this medium? Refer to Chapter 6 if you need a hint.**

**1** One tube containing media for 15 biochemical tests is inoculated with an unknown enteric bacterium.

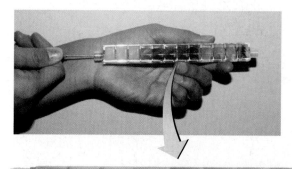

**2** After incubation, the tube is observed for results.

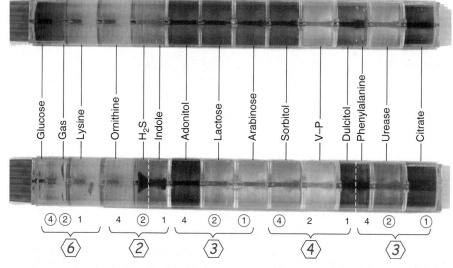

**3** The value for each positive test is circled, and the numbers from each group of tests are added to give the code number.

**4** Comparing the resultant code number with a computerized listing shows that the organism in the tube is *Citrobacter freundii*.

| Code Number | Microorganism | Atypical Test Results |
|---|---|---|
| 62342 | *Citrobacter freundii* | Citrate |
| 62343 | *Citrobacter freundii* | None |

**Figure 10.9  One type of rapid identification method for bacteria: EnteroPluri test from BD Diagnostics.** This example shows results for a typical strain of *C. freundii*; however, other strains may produce different test results, which are listed in the Atypical Test Results column.

**Q** How can one species have two different code numbers?

typhoid bacteria (antigens) responds by producing antibodies against typhoid bacteria. Solutions of such antibodies used in the identification of many medically important microorganisms are commercially available; such a solution is called an **antiserum** (plural: *antisera*). If an unknown bacterium is isolated from a patient, it can be tested against known antisera and often is identified quickly.

In a procedure called a **slide agglutination test**, samples of an unknown bacterium are placed in a drop of saline on each of several slides. Then a different known antiserum is added to each sample. The bacteria agglutinate (clump) when mixed with antibodies that were produced in response to that species or strain of bacterium; a positive test is indicated by the presence of agglutination. Positive and negative slide agglutination tests are shown in **Figure 10.11**.

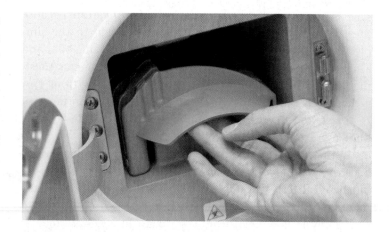

**Figure 10.10  Cellular proteins detected by mass spectrophotometry create a spectrum that can be compared to a database.**

**Q** Identify one advantage and one disadvantage of automated systems.

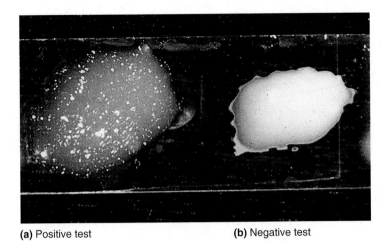

**(a)** Positive test  **(b)** Negative test

**Figure 10.11 A slide agglutination test. (a)** In a positive test, the grainy appearance is due to the clumping (agglutination) of the bacteria. **(b)** In a negative test, the bacteria are still evenly distributed in the saline and antiserum.

**Q** Agglutination results when the bacteria are mixed with _____.

Serological testing can differentiate not only among microbial species, but also among strains within species. Strains with different antigens are called serotypes, serovars, or biovars. See the discussion of *Escherichia* and *Salmonella* serovars on page 303. As mentioned in Chapter 1, Rebecca Lancefield was able to classify streptococcal serotypes by studying serological reactions. She found that the different antigens in the cell walls of various serotypes of streptococci stimulate the formation of different antibodies. In contrast, because closely related bacteria also produce some of the same antigens, serological testing can be used to screen bacterial isolates for possible similarities. If an antiserum reacts with proteins from different bacterial species or strains, these bacteria can be tested further for relatedness.

Serological testing was used to determine whether the increased number of necrotizing fasciitis cases in the United States and England since 1987 was due to a common source of the infections. No common source was located, but there has been an increase in two serotypes of *Streptococcus pyogenes* that have been dubbed the "flesh-eating" bacteria.

A test called the enzyme-linked immunosorbent assay (ELISA) is widely used because it is fast and can be read by a computer scanner (Figure 10.12; see also Figure 18.14, page 517). In a direct ELISA, known antibodies are placed in (and adhere to) the wells of a microplate, and an unknown type of bacterium is added to each well. A reaction between the known antibodies and the bacteria provides identification of the bacteria. An ELISA is used in AIDS testing to detect the presence of antibodies against human immunodeficiency virus (HIV), the virus that causes AIDS (see Figure 19.14, page 546).

Another serological test, Western blotting, is also used to identify antibodies in a patient's serum (Figure 10.13). HIV infection is confirmed by Western blotting, and Lyme disease,

**Figure 10.12 An ELISA test.**

**Q** What are the similarities between the slide agglutination test and the ELISA test?

**(a)** A technician uses a micropipette to add samples to a microplate for an ELISA.

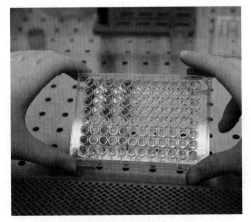

**(b)** ELISA results are then read using a spectrophotometer.

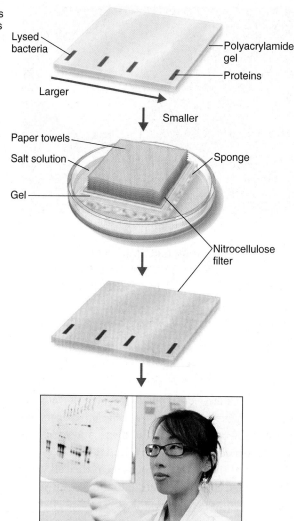

1 If Lyme disease is suspected in a patient: Electrophoresis is used to separate *Borrelia burgdorferi* proteins. Proteins move at different rates based on their charge and size when the gel is exposed to an electric current.

2 The bands are transferred to a nitrocellulose filter by blotting. Each band consists of many molecules of a particular protein (antigen). The bands are not visible at this point.

3 The proteins (antigens) are positioned on the filter exactly as they were on the gel. The filter is then washed with patient's serum followed by anti-human antibodies tagged with an enzyme. The patient antibodies that combine with their specific antigen are visible (shown here in red) when the enzyme's substrate is added.

4 The test is read. If the tagged antibodies stick to the filter, evidence of the presence of the microorganism in question—in this case, *B. burgdorferi*—has been found in the patient's serum.

**Figure 10.13 The Western blot.** Proteins separated by electrophoresis can be detected by their reactions with antibodies.

**Q** Name two diseases that may be diagnosed by Western blotting.

caused by *Borrelia burgdorferi*, is often diagnosed by the Western blot.

## Phage Typing

Like serological testing, phage typing looks for similarities among bacteria. Both techniques are useful in tracing the origin and course of a disease outbreak. **Phage typing** is a test for determining to which phages a bacterium is susceptible. Bacteriophages (phages) are bacterial viruses that usually cause lysis of the bacterial cells they infect (Chapter 13, page 373). They are highly specialized, in that they usually infect only members of a particular species or even particular strains within a species. One bacterial strain might be susceptible to two different phages, whereas another strain of the same species might be susceptible to those two phages plus a third phage. Bacteriophages will be discussed further in Chapter 13.

The sources of food-associated infections can be traced by phage typing. One version of this procedure starts with a plate totally covered with bacteria growing on agar. A drop of each different phage type to be used in the test is then placed on the bacteria. Wherever the phages are able to infect and lyse the bacterial cells, clearings in the bacterial growth (called plaques) appear (**Figure 10.14**). Such a test might show, for instance, that bacteria isolated from a surgical wound have the same pattern of phage sensitivity as those isolated from the operating surgeon or surgical nurses. This result establishes that the surgeon or a nurse is the source of infection.

## Fatty Acid Profiles

Bacteria synthesize a wide variety of fatty acids, and in general, these fatty acids are constant for a particular species. Commercial systems have been designed to separate cellular fatty acids

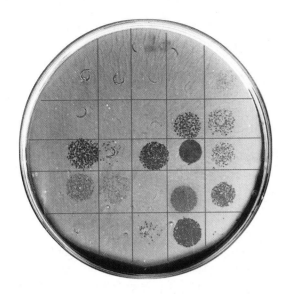

**Figure 10.14 Phage typing of a strain of *Salmonella enterica*.** The tested strain was grown over the entire plate. Plaques, or areas of lysis, were produced by bacteriophages, indicating that the strain was sensitive to infection by these phages. Phage typing is used to distinguish *S. enterica* serotypes and *Staphylococcus aureus* types.

**Q** **What is being identified in phage typing?**

to compare them to fatty acid profiles of known organisms. Fatty acid profiles, called **FAME** (*f*atty *a*cid *m*ethyl *e*ster), are widely used in clinical and public health laboratories.

## Flow Cytometry

**Flow cytometry** can be used to identify bacteria in a sample without culturing the bacteria. In a *flow cytometer*, a moving fluid containing bacteria is forced through a small opening (see Figure 18.12, page 516). The simplest method detects the presence of bacteria by detecting the difference in electrical conductivity between cells and the surrounding medium. If the fluid passing through the opening is illuminated by a laser, the scattering of light provides information about the cell size, shape, density, and surface, which is analyzed by a computer. Fluorescence can be used to detect naturally fluorescent cells, such as *Pseudomonas*, or cells tagged with fluorescent dyes.

A proposed test that uses flow cytometry to detect *Listeria* in milk could save time because the bacteria wouldn't need to be cultured for identification. Antibodies against *Listeria* can be labeled with a fluorescent dye and added to the milk to be tested. The milk is passed through the flow cytometer, which records the fluorescence of the antibody-labeled cells.

## DNA Sequencing

Taxonomists can use an organism's **DNA base composition** to draw conclusions about relatedness. This base composition is usually expressed as the percentage of guanine plus cytosine (G + C). The base composition of a single species is theoretically a fixed property; thus, a comparison of the G + C content in different species can reveal the degree of species relatedness. Each guanine (G) in DNA has a complementary cytosine (C) (see Chapter 8). Similarly, each adenine (A) in the DNA has a complementary thymine (T). Therefore, the percentage of DNA bases that are GC pairs also tells us the percentage that are AT pairs (GC+AT= 100%). Two organisms that are closely related and have many identical or similar genes will have similar amounts of the various bases in their DNA. However, if there is a difference of more than 10% in their percentage of GC pairs (for example, if one bacterium's DNA contains 40% GC and another bacterium has 60% GC), then these two organisms probably aren't related. Of course, two organisms that have the same percentage of GC aren't necessarily closely related; other supporting data are needed to draw conclusions about their phylogenetic relationship.

During the last decade, comparison of DNA sequences has led to great strides in reclassifying known species and identifying new species. Genetic sequences of hundreds of organisms are compiled in databases that can be used online through the NCBI Genome Database. In 2016, researchers discovered a new Lyme disease pathogen, *Borrelia mayonii*, using DNA sequencing.

## DNA Fingerprinting

Determining the entire sequence of bases in an organism's DNA is impractical for laboratory identification because it requires a great amount of time. However, the use of restriction enzymes enables researchers to compare the base sequences of different organisms. Restriction enzymes cut a molecule of DNA wherever a specific base sequence occurs, producing restriction fragments (as discussed in Chapter 9, page 245). For example, the enzyme *Eco*RI cuts DNA at the arrows in the sequence

$$...\ G^{\downarrow}A\ A\ T\ T\ C\ ...$$
$$...\ C\ T\ T\ A\ A_{\uparrow}G\ ...$$

In this technique, the DNA from two microorganisms is treated with the same restriction enzyme, and the restriction fragments (RFLPs) produced are separated by electrophoresis, producing a **DNA fingerprint** (see Figure 9.17, page 260). Comparing the number and sizes of restriction fragments that are produced from different organisms provides information about their genetic similarities and differences; the more similar the patterns, or *DNA fingerprints*, the more closely related the organisms are expected to be (**Figure 10.15**).

DNA fingerprinting is used to determine the source of hospital-acquired infections. In one hospital, patients undergoing coronary bypass surgery developed infections caused by *Gordonia bronchialis* (gor-DŌ-nē-ah bron-kē-AL-is). The DNA fingerprints of the patients' bacteria and the bacteria of one nurse were identical. The hospital was thus able to break the infection's chain of transmission by encouraging the nurse to use aseptic technique.

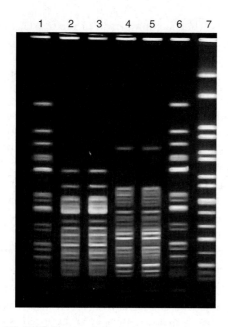

**Figure 10.15 DNA fingerprints.** DNA from seven different bacteria was digested with the same restriction enzyme. Each digest was put in a different well (origin) in the agarose gel. An electrical current was then applied to the gel to separate the fragments by size and electrical charge. The DNA was made visible by staining with a dye that fluoresces under ultraviolet light. Comparison of the lanes shows that DNA samples (and therefore the bacteria) in lanes 2 and 3; 4 and 5; and 1 and 6 are identical.

**Q** Differentiate between a gene and an RFLP.

This has led to interest in finding a few genes that are present in all species and that provide a large variation between species. Primers for these genes would be used for PCR to produce a *DNA bar code* for each species. This was first proposed in 2003 for eukaryotic species, but the necessary six to nine genes for bacterial identification have not been found yet.

## Nucleic Acid Hybridization

If a double-stranded molecule of DNA is subjected to heat, the complementary strands will separate as the hydrogen bonds between the bases break. If the single strands are then cooled slowly, they will reunite to form a double-stranded molecule identical to the original double strand. (This reunion occurs because the single strands have complementary sequences.) When this technique is applied to separated DNA strands from two different organisms, it is possible to determine the extent of similarity between the base sequences of the two organisms. This method is known as **nucleic acid hybridization**. The procedure assumes that if two species are similar or related, a major portion of their nucleic acid sequences will also be similar. The procedure measures the ability of DNA strands from one organism to hybridize (bind through complementary base pairing) with the DNA strands of another organism (**Figure 10.16**). Hybridization of 70% or more indicates the two organisms belong to the same species.

Similar hybridization reactions can occur between any single-stranded nucleic acid chain: DNA-DNA, RNA-RNA, DNA-RNA. An RNA transcript will hybridize with the separated template DNA to form a DNA-RNA hybrid molecule. Nucleic acid hybridization reactions are the basis of several techniques (described below) that are used to detect the presence of microorganisms and to identify unknown organisms.

### Nucleic Acid Amplification Tests (NAATs)

When a microorganism cannot be cultured by conventional methods, the causative agent of an infectious disease might not be recognized. However, **nucleic acid amplification tests (NAATs)** can be used to increase the amount of microbial DNA to levels that can be tested by gel electrophoresis. NAATs use PCR, reverse-transcription PCR, and real-time PCR (see Chapter 9, page 247). If a primer for a specific microorganism is used, the presence of amplified DNA indicates that the microorganism is present.

In 1992, researchers used PCR to determine the causative agent of Whipple's disease, which was previously an unknown bacterium now named *Tropheryma whipplei* (trō-fer-Ē-mah WHIP-plē-ī). Whipple's disease was first described in 1907 by George Whipple as a gastrointestinal and nervous system disorder caused by an unknown bacillus. No one has been able to

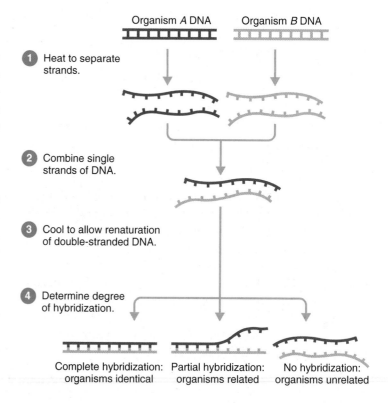

**Figure 10.16 DNA-DNA hybridization.** The greater the amount of pairing between DNA strands from different organisms (hybridization), the more closely the organisms are related.

**Q** What is the principle involved in DNA probes?

culture the bacterium to identify it; thus PCR provides the only reliable methods of diagnosing and treating the disease.

In recent years, PCR made possible several discoveries. For example, in 1992, Raul Cano used PCR to amplify DNA from *Bacillus* bacteria in amber that was 25 to 40 million years old. These primers were made from rRNA sequences in living *B. circulans* to amplify DNA coding for rRNA in the amber. These primers will cause amplification of DNA from other *Bacillus* species but do not cause amplification of DNA from other bacteria that might be present, such as *Escherichia* or *Pseudomonas*. The DNA was sequenced after amplification. This information was used to determine the relationships between the ancient bacteria and modern bacteria.

PCR is used to diagnosis Zika virus in pregnant women who may have been exposed to the virus. PCR is used to identify the source of rabies viruses, as well (see the Clinical Focus box in Chapter 22, page 636).

In 2013, public health scientists used real-time PCR to identify a new strain of H7N9 influenza virus.

### Southern Blotting

Nucleic acid hybridization can be used to identify unknown microorganisms by **Southern blotting** (see Figure 9.16, page 259).

In addition, rapid identification methods using **DNA probes** are being developed. One method involves breaking DNA extracted from *Salmonella* into fragments with a restriction enzyme, then selecting a specific fragment as the probe for *Salmonella* (**Figure 10.17**). This fragment must be able to hybridize with the DNA of all *Salmonella* strains, but not with the DNA of closely related enteric bacteria.

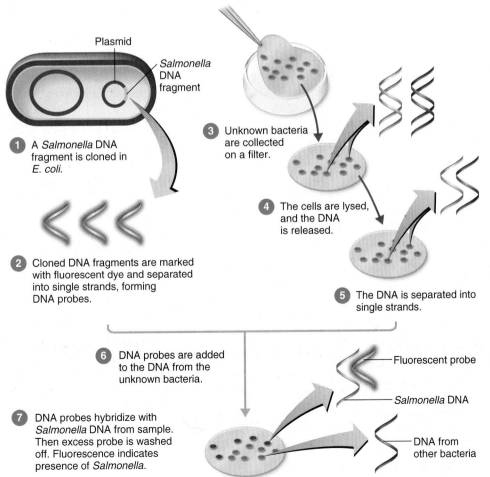

**Figure 10.17  A DNA probe used to identify bacteria.** Southern blotting is used to detect specific DNA. This modification of the Southern blot is used to detect *Salmonella*.

**Q** Why do the DNA probe and cellular DNA hybridize?

Plasmid

*Salmonella* DNA fragment

**1** A *Salmonella* DNA fragment is cloned in *E. coli*.

**2** Cloned DNA fragments are marked with fluorescent dye and separated into single strands, forming DNA probes.

**3** Unknown bacteria are collected on a filter.

**4** The cells are lysed, and the DNA is released.

**5** The DNA is separated into single strands.

**6** DNA probes are added to the DNA from the unknown bacteria.

**7** DNA probes hybridize with *Salmonella* DNA from sample. Then excess probe is washed off. Fluorescence indicates presence of *Salmonella*.

Fluorescent probe

*Salmonella* DNA

DNA from other bacteria

## DNA Chips

An exciting new technology is the **DNA chip**, or **microarray**, which can quickly detect a pathogen in a host or the environment by identifying a gene that is unique to that pathogen (**Figure 10.18**). The DNA chip is composed of DNA probes.

**(a)** A DNA chip can be manufactured to contain hundreds of thousands of synthetic single-stranded DNA sequences. Assume that each DNA sequence was unique to a different gene.

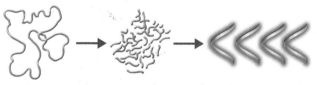

**(b)** Unknown DNA from a sample is separated into single strands, enzymatically cut, and labeled with a fluorescent dye.

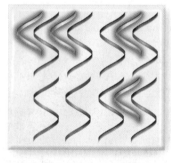

**(c)** The unknown DNA is inserted into the chip and allowed to hybridize with the DNA on the chip.

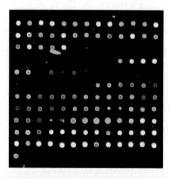

**(d)** The tagged DNA will bind only to the complementary DNA on the chip. The bound DNA will be detected by its dye and analyzed by a computer. In this *Salmonella* antimicrobial-resistance gene microarray, *Salmonella* Typhimurium–specific antibiotic-resistance gene probes are green, *Salmonella* Typhi–specific resistance gene probes are red, and antibiotic-resistance genes found in both serovars appear yellow. Blue indicates absence of the gene.

**Figure 10.18 DNA chip.** This DNA chip contains probes for antibiotic-resistance genes. It is used to detect antibiotic-resistant bacteria in samples collected from animals on a farm or in slaughter facilities.

**Q** What is on the chip to make it specific for a particular microorganism?

A sample containing DNA from an unknown organism is labeled with a fluorescent dye and added to the chip. Hybridization between the probe DNA and DNA in the sample is detected by fluorescence.

**Ribotyping and Ribosomal RNA Sequencing** Ribotyping is currently being used to determine the phylogenetic relationships among organisms. There are several advantages to using rRNA. First, all cells contain ribosomes. Second, RNA genes have undergone few changes over time, so all members of a domain, phylum—and, in some cases, a genus—have the same "signature" sequences in their rRNA. The rRNA used most often is a component of the smaller portion of ribosomes. A third advantage of rRNA sequencing is that cells don't have to be cultured in the laboratory.

DNA can be amplified by PCR using an rRNA primer for specific signature sequences. The amplified fragments are subsequently cut with one or more restriction enzymes and separated by electrophoresis. The resulting band patterns can then be compared. The rRNA genes in the amplified fragments can be sequenced to determine evolutionary relationships between organisms. This technique is useful for classifying a newly discovered organism to domain or phylum or to determine the general types of organisms present in one environment. More specific probes (see page 253) are needed to identify individual species, however.

**Fluorescent In Situ Hybridization (FISH)** Fluorescent dye–labeled RNA or DNA probes are used to stain microorganisms in place, or in situ. This technique is called **fluorescent in situ hybridization**, or **FISH**. Cells are treated so the probe enters the cells and reacts with target DNA in the cell (in situ). FISH is used to determine the identity, abundance, and relative activity of microorganisms in an environment and can be used to detect bacteria that have not yet been cultured. Using FISH, researchers discovered a tiny bacterium called *Pelagibacter* (pel-AJ-ē-bak-ter) in the ocean and determined that it's related to the rickettsias (page 297). As probes are developed, FISH can be used to detect bacteria in drinking water or bacteria in a patient without the normal 24-hour or longer wait required to culture the bacteria (**Figure 10.19**).

## Putting Classification Methods Together

Morphological characteristics, differential staining, and biochemical testing were the only identification tools available just a few years ago. Technological advancements are making it possible to use nucleic acid analysis techniques, once reserved for classification, for routine identification. Information obtained about microbes is used to identify and classify the organisms (see Exploring the Microbiome, page 291). Two methods of using the information are described next.

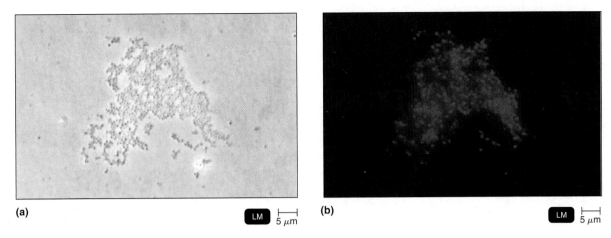

(a)  LM 5 μm    (b)  LM 5 μm

**Figure 10.19 FISH, or fluorescent in situ hybridization.** A DNA or RNA probe attached to fluorescent dyes is used to identify chromosomes. Bacteria seen with phase-contrast microscopy **(a)** are identified with a fluorescent-labeled probe that hybridizes with a specific sequence of DNA in *Staphylococcus aureus* **(b)**.

**Q** What is stained using the FISH technique?

## Dichotomous Keys

**Dichotomous keys** are widely used for identification. In a dichotomous key, identification is based on successive questions, and each question has two possible answers (*dichotomous* means cut in two). After answering one question, the investigator is directed to another question until an organism is identified. Although these keys often have little to do with phylogenetic relationships, they're invaluable for identification.

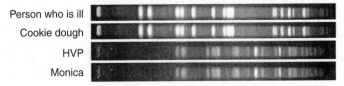

### CLINICAL CASE

*Salmonella* isolates from each of the 29 infected people are sent to the state's public health laboratory for DNA fingerprinting. The DNA fingerprints are then sent to the Centers for Disease Control and Prevention (CDC). At the CDC, computer software compares each of the *Salmonella* DNA fingerprints to determine whether all 29 cases of *Salmonella* Tennessee are identical. At this point, the CDC has received over 400 samples from 20 states, indicating a potential nationwide outbreak. Below is a figure of Monica's *Salmonella* DNA fingerprint along with other DNA fingerprint samples.

Person who is ill
Cookie dough
HVP
Monica

**What can the CDC conclude about the outbreak based on these DNA fingerprints?**

271  281  283  287  **289**  290

For example, a dichotomous key for bacteria could begin with an easily determined characteristic, such as cell shape, and move on to the ability to ferment a sugar. Dichotomous keys are shown in Figure 10.8 and in the boxes on pages 141 and 280.

**Play Dichotomous Keys:**
*Overview, Sample with Flowchart, Practice*
@MasteringMicrobiology

## Cladograms

**Cladograms** are maps that show evolutionary relationships among organisms (*clado-* means branch). Cladograms are shown in Figures 10.1 and 10.6. Each branch point on the cladogram is defined by a feature shared by various species on that branch. Historically, cladograms for vertebrates were made using fossil evidence; however, rRNA sequences are now being used to confirm fossil-based assumptions. As we said earlier, most microorganisms don't leave fossils; therefore, rRNA sequencing is primarily used to make cladograms for microorganisms. The small rRNA subunit used has 1500 bases, and computer programs do the calculations. The steps for constructing a cladogram are shown in **Figure 10.20**.

❶ Two rRNA sequences are aligned.

❷ The percentage of similarity between the sequences is calculated.

❸ Then the horizontal branches are drawn in a length proportional to the calculated percent similarity. All species beyond a node (branch point) have similar rRNA sequences, suggesting that they arose from an ancestor at that node.

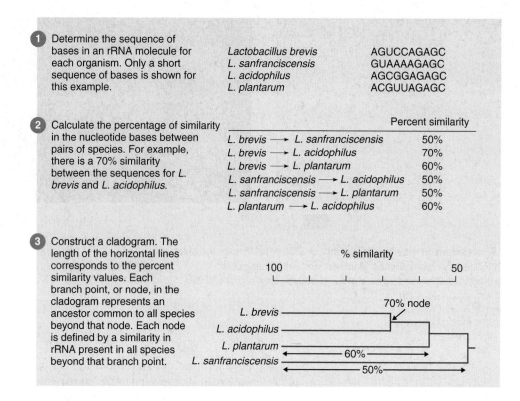

**Figure 10.20 Building a cladogram.**

**Q** Why do *L. brevis* and *L. acidophilus* branch from the same node?

---

**CLINICAL CASE** Resolved

Early in this outbreak, there was a cluster of *Salmonella* Tennessee illness due to consumption of raw eggs. Ill people and randomly chosen uninfected people completed questionnaires about foods they ate. The ill people were significantly more likely than well persons to report eating raw cookie dough, which contains uncooked eggs. However, the CDC soon determines that the cookie dough cluster involves a different strain of *Salmonella* Tennessee from the strain involved in the current outbreak. This strain is associated with hydrolyzed vegetable protein (HVP), a flavor enhancer used in a variety of foods, including a vegetable dip that Monica and her friend ate the day before they became ill. In conjunction with the CDC and the U.S. Food and Drug Administration, the manufacturer recalls that batch of HVP. Monica and her friend fully recover.

Tracing *Salmonella* infections to their source is essential because *Salmonella* can be transmitted through a variety of foods. It causes an estimated 1.4 million illnesses and 400 deaths annually in the United States.

DNA fingerprinting is used in laboratories to distinguish among *Salmonella* strains because strain-specific primers aren't needed. RFLPs can also detect strains because they are made from the entire genome rather than from amplification of a few nucleotide sequences.

---

**CHECK YOUR UNDERSTANDING**

✔ **10-13** What is in *Bergey's Manual*?

✔ **10-14** Design a rapid test for a *Staphylococcus aureus*. (*Hint:* See Figure 6.10, page 163.)

✔ **10-15** What is tested in Western blotting and Southern blotting?

✔ **10-16** What is identified by phage typing?

✔ **10-17** Why does PCR identify a microbe?

✔ **10-18** Which techniques involve nucleic acid hybridization?

✔ **10-12, 10-19** Is a cladogram used for identification or classification?

# Techniques for Identifying Members of Your Microbiome

In the past, the main way to identify a microbe from your microbiome—say, a bacterium in your intestines—would be to take a sample, isolate a species, grow it as a pure culture on the appropriate media, and then examine it under a microscope. But not every microbe can be grown in the lab. DNA analysis expands what members of the microbial community we can see and study.

In prokaryotes, studying ribosomal DNA allows us to identify and compare relatedness of microbial species. The 16S gene, which makes up part of the 30S ribosomal subunit, is highly conserved from an evolutionary standpoint. (In other words, it has not changed much over time in a given species.) Looking at a microbe's 16S gene is therefore useful. Although complete genome sequences, not just 16S genes, are needed to identify a species, the more similar the 16S gene is to another known microbe's, the more closely related the two

species are. Automated systems using genetic analysis have been developed. These systems can rapidly recognize individual organisms from a mixed sample and are thus useful for identifying members of the intestinal microbiome, where a wide variety of bacteria coexist.

Genetic identification techniques have revolutionized the study of microbes. However, that doesn't mean Petri plates and microscopes are now obsolete. Genetic analysis isn't appropriate for all situations and won't tell us all the information we may want to know about a microbe in a sample. For example, genetic analysis may not detect species if they are low in number in the sample. Additionally, differences within a species, such as a new antibiotic-resistance gene, can be identified only by culturing. Growing bacteria is also essential to understanding their metabolism and their relationship to their host—both

crucial factors when it comes to studying microbiota. So in the end, we need both new and old techniques to properly study the microbiome.

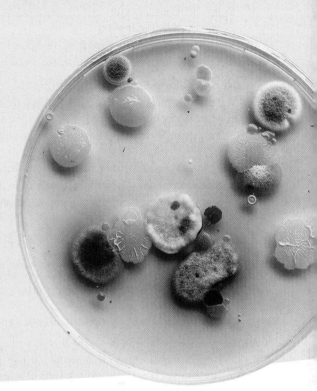

**Microbes grown from a 5-year-old's handprint.**

---

## Study Outline

Go to @MasteringMicrobiology for **Interactive Microbiology**, *In the Clinic videos, MicroFlix, MicroBoosters, 3D animations, practice quizzes, and more.*

### Introduction (p. 269)

1. Taxonomy is the science of the classification of organisms. Its goal is to show relationships among organisms.
2. Taxonomy also provides a means of identifying organisms.

### The Study of Phylogenetic Relationships
(pp. 270–274)

1. Phylogeny is the evolutionary history of a group of organisms.
2. The taxonomic hierarchy shows evolutionary, or phylogenetic, relationships among organisms.
3. Bacteria were separated into the Kingdom Prokaryotae in 1968.
4. Living organisms were divided into five kingdoms in 1969.

### The Three Domains (pp. 270–273)

5. Living organisms are currently classified into three domains. A domain can be divided into kingdoms.

6. In this system, plants, animals, and fungi belong to the Domain Eukarya.
7. Bacteria (with peptidoglycan) form a second domain.
8. Archaea (with unusual cell walls) are placed in the Domain Archaea.

### A Phylogenetic Tree (pp. 273–274)

9. Organisms are grouped into taxa according to phylogenetic relationships (from a common ancestor).
10. Some of the information for eukaryotic relationships is obtained from the fossil record.
11. Prokaryotic relationships are determined by rRNA sequencing.

### Classification of Organisms (pp. 274–277)

#### Scientific Nomenclature (pp. 274–275)

1. According to scientific nomenclature, each organism is assigned two names, or a binomial: a genus and a specific epithet, or species.

### The Taxonomic Hierarchy (p. 275)

2. A eukaryotic species is a group of organisms that interbreed with each other but do not breed with individuals of another species.

3. Similar species are grouped into a genus; similar genera are grouped into a family; families, into an order; orders, into a class; classes, into a phylum; phyla, into a kingdom; and kingdoms, into a domain.

### Classification of Prokaryotes (p. 275)

4. *Bergey's Manual of Systematic Bacteriology* is the standard reference on bacterial classification.

5. A group of bacteria derived from a single cell is called a strain.

6. Closely related strains constitute a bacterial species.

### Classification of Eukaryotes (pp. 275–276)

7. Eukaryotic organisms may be classified into the Kingdom Fungi, Plantae, or Animalia.

8. Protists are mostly unicellular organisms; these organisms are currently being assigned to kingdoms.

9. Fungi are absorptive chemoheterotrophs that develop from spores.

10. Multicellular photoautotrophs are placed in the Kingdom Plantae.

11. Multicellular ingestive heterotrophs are classified as Animalia.

### Classification of Viruses (p. 277)

12. Viruses are not placed in a kingdom. They are not composed of cells and do not have ribosomes.

13. A viral species is a population of viruses with similar characteristics that occupies a particular ecological niche.

### Methods of Classifying and Identifying Microorganisms (pp. 277–290)

1. *Bergey's Manual of Determinative Bacteriology* is the standard reference for laboratory identification of bacteria.

2. Morphological characteristics are useful in identifying microorganisms, especially when aided by differential staining techniques.

3. The presence of various enzymes, as determined by biochemical tests, is used in identifying bacteria and yeasts.

4. Serological tests, involving the reactions of microorganisms with specific antibodies, are useful in determining the identity of strains and species, as well as relationships among organisms. ELISA and Western blotting are examples of serological tests.

5. Phage typing is the identification of bacterial species and strains by determining their susceptibility to various phages.

6. Fatty acid profiles can be used to identify some organisms.

7. Flow cytometry measures physical and chemical characteristics of cells.

8. The percentage of GC base pairs in the nucleic acid of cells can be used in the classification of organisms.

9. The number and sizes of DNA fragments, or DNA fingerprints, produced by restriction enzymes are used to determine genetic similarities.

10. Single strands of DNA, or of DNA and RNA, from related organisms will hydrogen-bond to form a double-stranded molecule; this bonding is called nucleic acid hybridization.

11. NAATs can be used to amplify a small amount of microbial DNA in a sample. The presence or identification of an organism is indicated by amplified DNA.

12. PCR, Southern blotting, DNA chips, and FISH are examples of nucleic acid hybridization techniques.

13. The sequence of bases in ribosomal RNA can be used in the classification of organisms.

14. Dichotomous keys are used for identifying organisms. Cladograms show phylogenetic relationships among organisms.

## Study Questions

For answers to the Knowledge and Comprehension questions, turn to the Answers tab at the back of the textbook.

## Knowledge and Comprehension

### Review

1. Which of the following organisms are most closely related? Are any two the same species? On what did you base your answer?

| Characteristic | A | B | C | D |
|---|---|---|---|---|
| Morphology | Rod | Coccus | Rod | Rod |
| Gram Reaction | + | − | − | + |
| Glucose Utilization | Fermentative | Oxidative | Fermentative | Fermentative |
| Cytochrome Oxidase | Present | Present | Absent | Absent |
| GC Moles % | 48–52 | 23–40 | 50–54 | 49–53 |

2. Here is some additional information on the organisms in question 1:

| Organism | % DNA Hybridization |
|---|---|
| A and B | 5–15 |
| A and C | 5–15 |
| A and D | 70–90 |
| B and C | 10–20 |
| B and D | 2–5 |

Which of these organisms are most closely related? Compare this answer with your response to review question 1.

3. **DRAW IT** Use the additional information on the next page to construct a cladogram for some of the organisms used in question 4. What is the purpose of a cladogram? How does your cladogram differ from a dichotomous key for these organisms?

| | Similarity in rRNA Bases |
|---|---|
| P. aeruginosa—M. pneumoniae | 52% |
| P. aeruginosa—C. botulinum | 52% |
| P. aeruginosa—E. coli | 79% |
| M. pneumoniae—C. botulinum | 65% |
| M. pneumoniae—E. coli | 52% |
| E. coli—C. botulinum | 52% |

% similarity

100          50

4. DRAW IT Use the information in the table below to complete the dichotomous key to these organisms. What is the purpose of a dichotomous key? Look up each genus in Chapter 11, and provide an example of why this organism is of interest to humans.

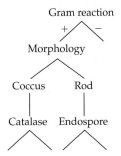

| | Morphology | Gram Reaction | Acid from Glucose | Growth in Air (21% $O_2$) | Motile by Peritrichous Flagella | Presence of Cytochrome Oxidase | Produce Catalase |
|---|---|---|---|---|---|---|---|
| Staphylococcus aureus | Coccus | + | + | + | − | − | + |
| Streptococcus pyogenes | Coccus | + | + | + | − | − | − |
| Mycoplasma pneumoniae | Coccus | − | + | + (Colonies < 1 mm) | − | − | + |
| Clostridium botulinum | Rod | + | + | − | + | − | − |
| Escherichia coli | Rod | − | + | + | + | − | + |
| Pseudomonas aeruginosa | Rod | − | + | + | − | + | + |
| Campylobacter fetus | Vibrio | − | − | − | − | + | + |
| Listeria monocytogenes | Rod | + | + | + | + | − | + |

5. NAME IT Use the key in the Clinical Focus box on page 280 to identify the gram-negative rod causing pneumonia in a sea otter. It is $H_2S$-positive, indole-negative, and urease-positive.

## Multiple Choice

1. *Bergey's Manual of Systematic Bacteriology* differs from *Bergey's Manual of Determinative Bacteriology* in that the former
   a. groups bacteria into species.
   b. groups bacteria according to phylogenetic relationships.
   c. groups bacteria according to pathogenic properties.
   d. groups bacteria into 19 species.
   e. all of the above

2. *Bacillus* and *Lactobacillus* are not in the same order. This indicates that which one of the following is *not* sufficient to assign an organism to a taxon?
   a. biochemical characteristics
   b. amino acid sequencing
   c. phage typing
   d. serology
   e. morphological characteristics

3. Which of the following is used to classify organisms into the Kingdom Fungi?
   a. ability to photosynthesize; possess a cell wall
   b. unicellular; possess cell wall; prokaryotic
   c. unicellular; lacking cell wall; eukaryotic
   d. absorptive; possess cell wall; eukaryotic
   e. ingestive; lacking cell wall; multicellular; prokaryotic

4. Which of the following is *false* about scientific nomenclature?
   a. Each name is specific.
   b. Names vary with geographical location.
   c. The names are standardized.
   d. Each name consists of a genus and specific epithet.
   e. It was first designed by Linnaeus.

5. You could identify an unknown bacterium by all of the following *except*
   a. hybridizing a DNA probe from a known bacterium with the unknown's DNA.
   b. making a fatty acid profile of the unknown.
   c. specific antiserum agglutinating the unknown.
   d. ribosomal RNA sequencing.
   e. percentage of guanine + cytosine.

6. The wall-less mycoplasmas are considered to be related to gram-positive bacteria. Which of the following would provide the most compelling evidence for this?
   a. They share common rRNA sequences.
   b. Some gram-positive bacteria and some mycoplasmas produce catalase.
   c. Both groups are prokaryotic.
   d. Some gram-positive bacteria and some mycoplasmas have coccus-shaped cells.
   e. Both groups contain human pathogens.

Use the following choices to answer questions 7 and 8.
   a. Animalia
   b. Fungi
   c. Plantae
   d. Firmicutes (gram-positive bacteria)
   e. Proteobacteria (gram-negative bacteria)

7. Into which group would you place a multicellular organism that has a mouth and lives inside the human liver?

8. Into which group would you place a photosynthetic organism that lacks a nucleus and has a thin peptidoglycan wall surrounded by an outer membrane?

Use the following choices to answer questions 9 and 10.

1. 9 + 2 flagella
2. 70S ribosome
3. fimbria
4. nucleus
5. peptidoglycan
6. plasma membrane

9. Which is (are) found in all three domains?
   a. 2, 6
   b. 5
   c. 2, 4, 6
   d. 1, 3, 5
   e. all six

10. Which is (are) found *only* in prokaryotes?
   a. 1, 4, 6
   b. 3, 5
   c. 1, 2
   d. 4
   e. 2, 4, 5

## Analysis

1. The GC content of *Micrococcus* is 66–75 moles %, and of *Staphylococcus*, 30–40 moles %. According to this information, would you conclude that these two genera are closely related?

2. Describe the use of a DNA probe and PCR for:
   a. rapid identification of an unknown bacterium.
   b. determining which of a group of bacteria are most closely related.

3. SF medium is a selective medium, developed in the 1940s, to test for fecal contamination of milk and water. Only certain gram-positive cocci can grow in this medium. Why is it named SF? Using this medium, which genus will you culture? (*Hint*: Refer to page 275.)

## Clinical Applications and Evaluation

1. A 55-year-old veterinarian was admitted to a hospital with a 2-day history of fever, chest pain, and cough. Gram-positive cocci were detected in his sputum, and he was treated for lobar pneumonia with penicillin. The next day, another Gram stain of his sputum revealed gram-negative rods, and he was switched to ampicillin and gentamicin. A sputum culture showed biochemically inactive gram-negative rods identified as *Pantoea (Enterobacter) agglomerans*.

After fluorescent-antibody staining and phage typing, *Yersinia pestis* was identified in the patient's sputum and blood, and chloramphenicol and tetracycline were administered. The patient died 3 days after admission to the hospital. Tetracycline was given to his 220 contacts (hospital personnel, family, and co-workers). What disease did the patient have? Discuss what went wrong in the diagnosis and how his death might have been prevented. Why were the 220 other people treated? (*Hint*: Refer to Chapter 23.)

2. A 6-year-old girl was admitted to a hospital with endocarditis. Blood cultures showed a gram-positive, aerobic rod identified by the hospital laboratory as *Corynebacterium xerosis*. The girl died after 6 weeks of treatment with intravenous penicillin and chloramphenicol. The bacterium was tested by another laboratory and identified as *C. diphtheriae*. The following test results were obtained by each laboratory:

| | Hospital Lab | Other Lab |
|---|---|---|
| Catalase | + | + |
| Nitrate reduction | + | + |
| Urea hydrolysis | – | – |
| Esculin hydrolysis | – | – |
| Glucose fermentation | + | + |
| Sucrose fermentation | – | + |
| Serological test for toxin production | Not done | + |

Provide a possible explanation for the incorrect identification. What are the potential public health consequences of misidentifying *C. diphtheriae*? (*Hint*: Refer to Chapter 24.)

3. Using the following information, create a dichotomous key for distinguishing these unicellular organisms. Which cause human disease?

| | Mitochondria? | Chlorophyll? | Nutritional Type? | Motile? |
|---|---|---|---|---|
| *Euglena* | + | + | Both | + |
| *Giardia* | – | – | Heterotroph | + |
| *Nosema* | – | – | Heterotroph | – |
| *Pfiesteria* | + | + | Autotroph | + |
| *Trichomonas* | – | – | Heterotroph | + |
| *Trypanosoma* | + | – | Heterotroph | + |

Using the additional information shown below, create a cladogram for these organisms. Do your two keys differ? Explain why. Which key is more useful for laboratory identification? For classification?

**rRNA base #**

| | 1 | 2 | 3 | 4 | 5 | 6 | 7 | 8 | 9 | 10 | 11 | 12 | 13 | 14 | 15 | 16 | 17 | 18 | 19 | 20 |
|---|---|---|---|---|---|---|---|---|---|---|---|---|---|---|---|---|---|---|---|---|
| *Euglena* | C | C | A | G | G | U | U | G | U | U | C | C | A | G | U | U | U | U | A | A |
| *Giardia* | C | C | A | U | A | U | U | U | U | U | G | A | C | G | A | A | G | G | U | C |
| *Nosema* | C | C | A | U | A | U | U | U | U | U | A | A | C | G | A | A | G | G | C | C |
| *Pfiesteria* | C | C | A | A | C | U | U | A | U | U | C | C | A | G | U | U | U | C | A | G |
| *Trichomonas* | C | C | A | U | A | U | U | U | U | U | G | A | C | G | A | A | G | G | G | C |
| *Trypanosoma* | C | C | A | C | G | U | U | G | U | U | C | C | A | G | U | U | U | A | A | A |

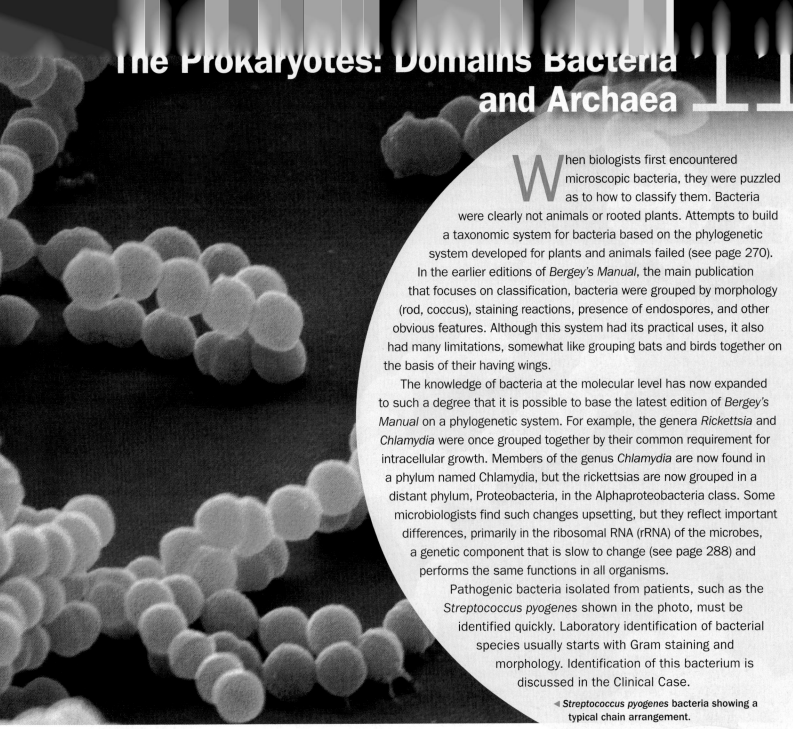

# The Prokaryotes: Domains Bacteria and Archaea 11

When biologists first encountered microscopic bacteria, they were puzzled as to how to classify them. Bacteria were clearly not animals or rooted plants. Attempts to build a taxonomic system for bacteria based on the phylogenetic system developed for plants and animals failed (see page 270). In the earlier editions of *Bergey's Manual*, the main publication that focuses on classification, bacteria were grouped by morphology (rod, coccus), staining reactions, presence of endospores, and other obvious features. Although this system had its practical uses, it also had many limitations, somewhat like grouping bats and birds together on the basis of their having wings.

The knowledge of bacteria at the molecular level has now expanded to such a degree that it is possible to base the latest edition of *Bergey's Manual* on a phylogenetic system. For example, the genera *Rickettsia* and *Chlamydia* were once grouped together by their common requirement for intracellular growth. Members of the genus *Chlamydia* are now found in a phylum named Chlamydia, but the rickettsias are now grouped in a distant phylum, Proteobacteria, in the Alphaproteobacteria class. Some microbiologists find such changes upsetting, but they reflect important differences, primarily in the ribosomal RNA (rRNA) of the microbes, a genetic component that is slow to change (see page 288) and performs the same functions in all organisms.

Pathogenic bacteria isolated from patients, such as the *Streptococcus pyogenes* shown in the photo, must be identified quickly. Laboratory identification of bacterial species usually starts with Gram staining and morphology. Identification of this bacterium is discussed in the Clinical Case.

◄ *Streptococcus pyogenes* bacteria showing a typical chain arrangement.

## In the Clinic

As a perioperative clinical nurse specialist, you need to identify the source of infection in seven cardiovascular surgery patients. Nutrient-agar cultures from the patients grow red colonies consisting of gram-negative bacteria. You take cultures of selected hospital staff, and the same bacterium is cultured from a scrub nurse who wears artificial fingernails. **Removal of the fingernails ends the outbreak. What is the causative agent of the outbreak?**

*Hint: The red pigment produced by this bacterium is distinctive. (See page 304.)*

**Play In the Clinic Video**
@Mastering**Microbiology**

# The Prokaryotic Groups

In the second (current) edition of *Bergey's Manual*, the prokaryotes are grouped into two **domains**, the **Archaea** and the **Bacteria**. Spelled without capitalization, archaea and bacteria denote organisms

 ASM: Mutations and horizontal gene transfer, with the immense variety of microenvironments, have selected for a huge diversity of microorganisms.

that fit into these domains. Each domain is divided into phyla, each phylum into classes, each class into orders, each order into families, each family into genera, and finally, each genus into species. Note that bacteria are also commonly distinguished by whether they are gram-negative or gram-positive.

The phyla discussed in this chapter are summarized in Table 11.1 (see also Appendix E).

## Domain Bacteria

Most of us think of bacteria as invisible, potentially harmful little creatures. Actually, relatively few species of bacteria cause disease in humans, animals, plants, or any other organisms. Once you have completed a course in microbiology, you'll realize that without bacteria, much of life as we know it would be impossible. In this chapter, you'll learn how bacterial groups are differentiated from each other and how important bacteria

are in the world. Our discussion in this chapter emphasizes bacteria considered to be of practical importance, those that are important in medicine, or those that illustrate biologically unusual or interesting principles.

The Learning Objectives and Check Your Understanding questions throughout this chapter will help you to become familiar with these organisms and to look for similarities and

**TABLE 11.1 Classification of Selected Prokaryotes***

| Domain | Phyla | Selected Classes | Notes |
|---|---|---|---|
| **BACTERIA** | | | |
| (Gram-Negative) | Proteobacteria | · Alphaproteobacteria | Includes *Ehrlichia* and *Rickettsia* |
| | | · Betaproteobacteria | Includes *Bordetella* and *Burkholderia* |
| | | · Gammaproteobacteria | Includes *Vibrio, Salmonella, Helicobacter,* and *Escherichia* |
| | | · Deltaproteobacteria | Includes *Bdellovibrio* |
| | | · Epsilonproteobacteria | *Campylobacter* and *Helicobacter* |
| | Cyanobacteria | · Cyanobacteria | Oxygenic photosynthetic bacteria |
| | Chlorobi | · Chlorobia | Photosynthetic; anoxygenic; green sulfur bacteria |
| | Chloroflexi | · Chloroflexi | Include anoxygenic, photosynthetic, filamentous green nonsulfur bacteria |
| | Chlamydiae | · Chlamydiae | Grow only in eukaryotic host cells |
| | Planctomycetes | · Planctomycetacia | Aquatic bacteria; some are stalked |
| | Bacteroidetes | · Bacteroidetes | Phylum members include opportunistic pathogens |
| | Fusobacteria | · Fusobacteria | Anaerobic; some cause tissue necrosis and septicemia in humans |
| | Spirochaetes | · Spirochaetes | Classes include pathogens that cause syphilis and Lyme disease |
| | Deinococcus-Thermus | · Deinococci | *Deinococcus* and *Thermus* |
| **BACTERIA** | | | |
| (Gram-Positive) | Firmicutes | · Bacilli | Low G + C gram-positive rods and cocci |
| | | · Clostridia | |
| | Tenericutes | · Mycoplasmas | Low G + C wall-less bacteria |
| | Actinobacteria | · Actinobacteria | High G + C gram-positive bacteria |
| **ARCHAEA** | | | |
| | Crenarchaeota | · Thermoprotei | Thermophiles and hyperthermophiles |
| | Euryarchaeota | · Methanobacteria | Methanobacteria are important sources of methane |
| | | · Halobacteria | |
| | Nanoarchaeota | | Obligate symbionts with other archaea |
| | Korarchaeota | | Hyerthermophiles |

*For a complete list of genera discussed in this text, see Appendix E.

differences between organisms. You'll also learn to draw dichotomous keys to differentiate the bacteria described in each group.

We've drawn the first of these dichotomous keys (for alphaproteobacteria) for you as an example (on the right).

# Gram-Negative Bacteria

### LEARNING OBJECTIVES

**11-1** Differentiate the alphaproteobacteria described in this chapter by drawing a dichotomous key.

**11-2** Differentiate the betaproteobacteria described in this chapter by drawing a dichotomous key.

**11-3** Differentiate the gammaproteobacteria described in this chapter by drawing a dichotomous key.

**11-4** Differentiate the deltaproteobacteria described in this chapter by drawing a dichotomous key.

**11-5** Differentiate the epsilonproteobacteria described in this chapter by drawing a dichotomous key.

## Proteobacteria

The **proteobacteria**, which include most of the gram-negative, chemoheterotrophic bacteria, are presumed to have arisen from a common photosynthetic ancestor. They are now the largest taxonomic group of bacteria. However, few are now photosynthetic; other metabolic and nutritional capacities have arisen to replace this characteristic. The phylogenetic relationship in these groups is based upon rRNA studies. The name *Proteobacteria* was taken from the mythological Greek god Proteus, who could assume many shapes. The proteobacteria are separated into five classes designated by Greek letters: alphaproteobacteria, betaproteobacteria, gammaproteobacteria, deltaproteobacteria, and epsilonproteobacteria.

### The Alphaproteobacteria

As a group, the alphaproteobacteria includes most of the proteobacteria that are capable of growth with very low levels of nutrients. Some have unusual morphology, including protrusions such as stalks or buds known as **prosthecae**. The alphaproteobacteria also include agriculturally important bacteria capable of nitrogen fixation in symbiosis with plants, and several plant and human pathogens.

**Pelagibacter** One of the most abundant microorganisms on Earth, certainly in the ocean environment, is *Pelagibacter ubique* (pel-AJ-ē-bak-ter OO-bēk). It is a member of a group of marine microbes named SAR 11 because of their original discovery in the Sargasso Sea. *P. ubique* is the first member of this group to be successfully cultivated. Its genome has been sequenced and found to have only 1354 genes. This number is very low for a free-living organism, although several mycoplasmas (see page 315) have even fewer genes. Bacteria in a symbiotic relationship

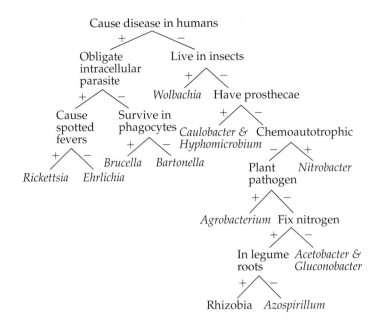

have lower metabolic requirements and have the smallest genomes (see page 319). The bacterium is extremely small, a little over 0.3 μm diameter. Its small size and minimal genome probably give it a competitive advantage for survival in low-nutrient environments. In fact, it seems to be the most abundant living organism in the oceans on the basis of weight. (Part of its name, *ubique*, is derived from *ubiquitous*.) Its sheer numbers must give it an important role in the Earth's carbon cycle.

**Azospirillum** Agricultural microbiologists have been interested in members of the genus *Azospirillum* (ā-zō-spī-RIL-lum), a soil bacterium that grows in close association with the roots of many plants, especially tropical grasses. It uses nutrients

---

**CLINICAL CASE**  **Mercy**

Sheree Walker, a neonatologist at a local hospital, is checking on Mercy, a 48-hour-old infant girl. Mercy was delivered normally at 39 weeks and appeared to be a healthy baby. In the last 2 days, however, she has taken a bad turn and is admitted to the neonatal intensive care unit (NICU). Mercy is limp, has difficulty breathing, and has a body temperature of 35°C; but her lungs are clear, and her heart exam is normal. Dr. Walker speaks with Mercy's mother, who confirms that she received appropriate prenatal care and has no other medical problems. Dr. Walker orders a lumbar puncture for Mercy to check her cerebrospinal fluid (CSF) for possible infection. The report back from the lab shows blood in Mercy's CSF. Dr. Walker diagnoses Mercy with meningitis and then orders a venous blood culture to check for bacteria.

**What bacteria could be the cause of Mercy's meningitis? Read on to find out.**

297    315    317    318    319

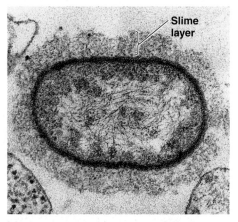

Slime layer

**(a)** A rickettsial cell that has just been released from a host cell

TEM 0.4 μm

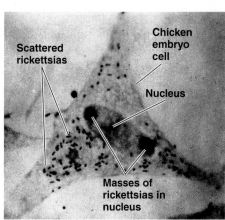

Scattered rickettsias

Chicken embryo cell

Nucleus

Masses of rickettsias in nucleus

LM 5 μm

**Figure 11.1 Rickettsias.**

**Q** How are rickettsias transmitted from one host to another?

**(b)** Rickettsias grow only within a host cell, such as the chicken embryo cell shown here. Note the scattered rickettsias within the cell and the compact masses of rickettsias in the cell nucleus.

excreted by the plants and in return fixes nitrogen from the atmosphere. This form of nitrogen fixation is most significant in tropical grasses and in sugarcane, although the organism can be isolated from the root system of many temperate-climate plants, such as corn. The prefix *azo-* is frequently encountered in nitrogen-fixing genera of bacteria. It's derived from *a* (without) and *zo* (life), in reference to the early days of chemistry, when oxygen was removed, by a burning candle, from an experimental atmosphere. Presumably, mammalian life wasn't possible in this nitrogen-rich atmosphere. Hence, nitrogen came to be associated with absence of life.

**Acetobacteraceae** *Acetobacter* (ah-SĒ-tō-bak-ter) and *Gluconobacter* (glū-kon'ō-BAK-ter) are industrially important aerobic organisms that convert ethanol into acetic acid (vinegar). The recently identified *Granulibacter* (GRAN-ū-li-bak-ter) is an emerging pathogen found in patients with chronic granulomatous disease.

**Rickettsia** In the first edition of *Bergey's Manual*, the genera *Rickettsia*, *Coxiella*, and *Chlamydia* were grouped closely because they are all obligate intracellular parasites—that is, they reproduce only within a mammalian cell. In the second edition they are now widely separated. Rickettsias, chlamydias, and viruses are compared in Table 13.1, page 362.

The rickettsias are gram-negative rod-shaped bacteria, or coccobacilli (**Figure 11.1a**). One distinguishing feature of most rickettsias is that they are transmitted to humans by insect and tick bites, unlike the *Coxiella* (discussed later with gammaproteobacteria). Rickettsia enter their host cell by inducing phagocytosis. They quickly enter the cytoplasm of the cell and begin reproducing by binary fission (Figure 11.1b). They can usually be cultivated artificially in cell culture or chick embryos (Chapter 13, pages 370–372).

The rickettsias are responsible for a number of diseases known as the spotted fever group. These include epidemic typhus, caused by *Rickettsia prowazekii* (ri-KET-sē-ah

prou-wah-ZE-kē-ē) and transmitted by lice (page 355); endemic murine typhus, caused by *R. typhi* (TĪ-fē) and transmitted by rat fleas; and Rocky Mountain spotted fever, caused by *R. rickettsii* (ri-KET-sē-ē) and transmitted by ticks (page 355). In humans, rickettsial infections damage the permeability of blood capillaries, which results in a characteristic spotted rash.

**Ehrlichia** Ehrlichiae are gram-negative, rickettsia-like bacteria that live obligately within white blood cells. *Ehrlichia* (er-LIK-ē-ah) species are transmitted by ticks to humans and cause ehrlichiosis, a sometimes fatal disease (page 666).

**Caulobacter and Hyphomicrobium** Members of the genus *Caulobacter* (kah-lō-BAK-ter) are found in low-nutrient aquatic environments, such as lakes. They feature stalks that anchor the organisms to surfaces (**Figure 11.2**). This arrangement increases their nutrient uptake because they are exposed to a continuously changing flow of water and because the stalk increases the surface-to-volume ratio of the cell. Also, if the surface to which they anchor is a living host, these bacteria can use the host's excretions as nutrients. When the nutrient concentration is exceptionally low, the size of the stalk increases, evidently to provide an even greater surface area for nutrient absorption.

Budding bacteria don't divide by binary fission into two nearly identical cells. The budding process resembles the asexual reproductive processes of many yeasts (see Figure 12.4, page 327). The parent cell retains its identity while the bud increases in size until it separates as a complete new cell. An example is the genus *Hyphomicrobium* (hī'fō-mī-KRŌ-bē-um), as shown in **Figure 11.3**. These bacteria, like the caulobacteria, are found in low-nutrient aquatic environments and have even been found growing in laboratory water baths. Both *Caulobacter* and *Hyphomicrobium* produce prominent prosthecae.

**Rhizobium, Bradyrhizobium, and Agrobacterium** The *Rhizobium* (rī-ZŌ-bē-um) and *Bradyrhizobium* (brā'dē-rī-ZŌ-bē-um) are

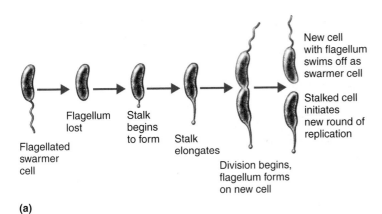

**(a)**

Flagellated swarmer cell → Flagellum lost → Stalk begins to form → Stalk elongates → Division begins, flagellum forms on new cell → New cell with flagellum swims off as swarmer cell / Stalked cell initiates new round of replication

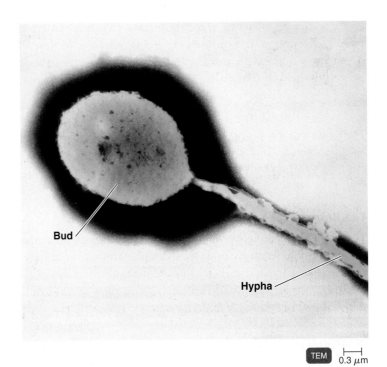

**(b)**

TEM    2 μm

**Figure 11.2  *Caulobacter*.**

**Q** What is the competitive advantage provided by attaching to a surface?

two of the more important genera of a group of agriculturally important bacteria that specifically infect the roots of leguminous plants, such as beans, peas, or clover. For simplicity these bacteria are known by the common name of **rhizobia**. The presence of rhizobia in the roots leads to formation of nodules in which the rhizobia and plant form a symbiotic relationship, resulting in the fixation of nitrogen from the air for use by the plant (see Figure 27.4 on page 792).

Like rhizobia, the genus *Agrobacterium* (A-grō-bak-ti′rē-um) has the ability to invade plants. However, these bacteria do not induce root nodules or fix nitrogen. Of particular interest is *Agrobacterium tumefaciens*. This is a plant pathogen that causes a disease called crown gall; the crown is the area of the plant where the roots and stem merge. The tumorlike gall is induced when *A. tumefaciens* inserts a plasmid containing bacterial genetic information into the plant's chromosomal DNA (see Figure 9.19, page 261). For this reason, microbial geneticists are very interested in this organism. Plasmids are the most common vector that scientists use to carry new genes into a plant cell because the thick wall of plants is especially difficult to penetrate (see Figure 9.20, page 262).

***Bartonella*** The genus *Bartonella* (BAR-tō-nel-lah) contains several members that are human pathogens. The best-known is *Bartonella henselae*, a gram-negative bacillus that causes cat-scratch disease (page 660).

***Brucella*** *Brucella* (broo-SEL-lah) bacteria are small nonmotile coccobacilli. All species of *Brucella* are obligate parasites of mammals and cause the disease brucellosis (page 656). Of medical interest is the ability of *Brucella* to survive phagocytosis, an important element of the body's defense against bacteria (see Chapter 16, page 456).

***Nitrobacter* and *Nitrosomonas*** *Nitrobacter* (ni′trō-BAK-ter) and *Nitrosomonas* (NĪ-trō-sō-mō-nas) are genera of nitrifying bacteria that are of great importance to the environment and to agriculture. They are chemoautotrophs capable of using inorganic chemicals as energy sources and carbon dioxide as the only source of carbon, from which they synthesize all of their complex chemical makeup. The energy sources of the genera *Nitrobacter* and *Nitrosomonas* (the latter is a member of the betaproteobacteria) are reduced nitrogenous compounds. *Nitrosomonas* species oxidize ammonium ($NH_4^+$) to nitrite ($NO_2^-$), which is in turn oxidized by *Nitrobacter* species to nitrates ($NO_3^-$) in the process of *nitrification*. Nitrate is important to agriculture; it's a nitrogen form that's highly mobile in soil and therefore likely to be encountered and used by plants.

Bud

Hypha

TEM    0.3 μm

**Figure 11.3  *Hyphomicrobium*, a type of budding bacterium.**

**Q** Most bacteria do not reproduce by budding; what method do they use?

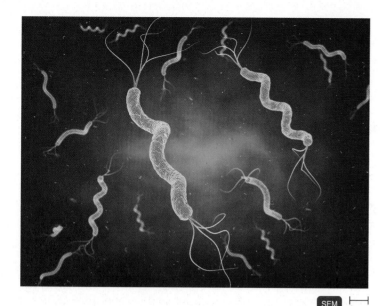

SEM | 4 μm

**Figure 11.4 *Spirillum volutans*.** These large helical bacteria are found in aquatic environments. Note the polar flagella.

**Q** Is this bacterium motile? How can you tell?

***Wolbachia*** *Wolbachia* (wol-BAK-ē-ah) is probably the most common infectious bacterial genus in the world; they live only inside the cells of their hosts, usually insects (a relationship known as *endosymbiosis*). Therefore, *Wolbachia* escapes detection by the usual culture methods. *Wolbachia* interferes with reproduction and egg development in infected insects. *Wolbachia*-infected male *Aedes aegypti* mosquitoes are being released in several places including Brazil, Florida, California, and southeast Asia to prevent the spread of Zika, chikungunya, and dengue viruses.

---

**CHECK YOUR UNDERSTANDING**

✔ 11-1 Make a dichotomous key to distinguish the alphaproteobacteria described in this chapter. (*Hint:* See page 297 for a completed example.)

---

## The Betaproteobacteria

There is considerable overlap between the betaproteobacteria and the alphaproteobacteria, for example, among the nitrifying bacteria discussed earlier. The betaproteobacteria often use nutrient substances that diffuse away from areas of anaerobic decomposition of organic matter, such as hydrogen gas, ammonia, and methane. Several important pathogenic bacteria are found in this group.

***Acidithiobacillus*** *Acidithiobacillus* (a′sid-ē-thi′ō-bah-SIL-lus) species and other sulfur-oxidizing bacteria are important in the sulfur cycle (see Figure 27.6, page 793). These chemoautotrophic bacteria are capable of obtaining energy by oxidizing the reduced forms of sulfur, such as hydrogen sulfide ($H_2S$), or elemental sulfur ($S^0$), into sulfates ($SO_4^{2-}$).

***Spirillum*** The genus *Spirillum* (spī-RIL-lum) is found mainly in freshwater. *Spirillum* bacteria are motile by conventional polar flagella, an important morphological distinction from the helical spirochetes (page 310), which use axial filaments. The spirilla are relatively large, gram-negative, aerobic bacteria. *Spirillum volutans* (VOL-ū-tanz) is often used as a demonstration slide when microbiology students are first introduced to the operation of the microscope (**Figure 11.4**).

***Sphaerotilus*** Sheathed bacteria, which include *Sphaerotilus natans* (sfer-O-ti-lus NĀ-tanz), are found in freshwater and in sewage. These gram-negative bacteria with polar flagella form a hollow, filamentous sheath in which to live (**Figure 11.5**). Sheaths are protective and also aid in nutrient accumulation. *Sphaerotilus* probably contributes to bulking, an important problem in sewage treatment (see Chapter 27).

***Burkholderia*** The genus *Burkholderia* was formerly grouped with the genus *Pseudomonas*, which is now classified under the gammaproteobacteria. Like the pseudomonads, almost all *Burkholderia* species are motile by a single polar flagellum or tuft of flagella. The best-known species is the aerobic, gram-negative rod *Burkholderia cepacia* (berk-HŌL-der′ē-ah se-PĀ-sē-ah). It has an extraordinary nutritional spectrum and is capable of degrading more than 100 different organic molecules. This capability is often a factor in the contamination of equipment and

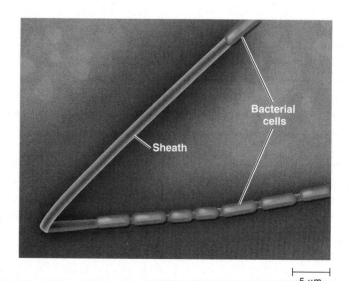

Bacterial cells

Sheath

5 μm

**Figure 11.5 *Sphaerotilus natans*.** These sheathed bacteria are found in dilute sewage and aquatic environments. They form elongated sheaths in which the bacteria live. The bacteria have flagella (not visible here) and can eventually swim free of the sheath.

**Q** How does the sheath help the cell?

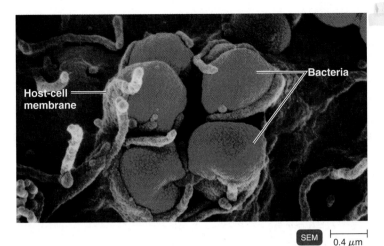

**Figure 11.6 The gram-negative coccus *Neisseria gonorrhoeae*.**
This bacterium uses fimbriae and an outer membrane protein called Opa to attach to host cells. After the (red) bacterium attaches, the (green) host cell membrane surrounds it.

**Q How do fimbriae contribute to pathogenicity?**

drugs in hospitals; these bacteria may actually grow in disinfectant solutions (see the Clinical Case in Chapter 15). This bacterium is also a problem for people with the genetic lung disease cystic fibrosis, in whom it metabolizes accumulated respiratory secretions.

*Burkholderia pseudomallei* (soo'dō-MAL-lē-ē) is a resident in moist soils and is the cause of a severe disease (melioidosis) endemic in southeast Asia and northern Australia (page 707).

***Bordetella*** Of special importance is the nonmotile, aerobic, gram-negative rod *Bordetella pertussis* (bor'de-TEL-lah per-TUS-sis). This serious pathogen is the cause of pertussis, or whooping cough (page 695).

***Neisseria*** Bacteria of the genus *Neisseria* (nī-SE-rē-ah) are aerobic, gram-negative cocci that usually inhabit the mucous membranes of mammals. Pathogenic species include the gonococcus bacterium *Neisseria gonorrhoeae* (go-nōr-RĒ-ī), the causative agent of gonorrhoea (page 766, **Figure 11.6**, and the box in Chapter 26, page 771), and *N. meningitidis* (ME-nin-ji-ti-dis), the agent of meningococcal meningitis (page 622).

***Zoogloea*** The genus *Zoogloea* (zō-ō-GLĒ-ah) is important in the context of aerobic sewage-treatment processes, such as the activated sludge system. As they grow, *Zoogloea* bacteria form fluffy, slimy masses that are essential to the proper operation of such systems.

**CHECK YOUR UNDERSTANDING**

✔ **11-2** Make a dichotomous key to distinguish the betaproteobacteria described in this chapter.

## The Gammaproteobacteria

The gammaproteobacteria constitute the largest subgroup of the proteobacteria and include a great variety of physiological types.

**Thiotrichales** Members of the order Thiotrichales include *Thiomargarita namibiensis*, which not only is the largest known bacterium but also exhibits several unusual characteristics. (See the discussion of microbial diversity, page 319). Other members of this order include the nutritionally distinctive genus *Beggiatoa* and *Francisella tularensis*, the pathogen causing tularemia.

***Beggiatoa*** *Beggiatoa* species (BEJ-jē-ah-tō-ah), are an unusual genus that grows the only in aquatic sediments at the interface between the aerobic and anaerobic layers. Morphologically, the genus resembles certain filamentous cyanobacteria (page 307), but it is not photosynthetic. Its motility is enabled by the production of slime, which attaches to the surface on which movement occurs and provides lubrication, allowing the organism to glide.

Nutritionally, *Beggiatoa* use hydrogen sulfide ($H_2S$) as an energy source and accumulates internal granules of sulfur. The ability of this organism to obtain energy from an inorganic compound was an important factor in the discovery of autotrophic metabolism.

***Francisella*** *Francisella* (FRAN-sis-el-lah) is a genus of small, pleomorphic bacteria that grow only on complex media enriched with blood or tissue extracts. *Francisella tularensis* (TOO-lar-en-sis) causes the disease tularemia. (See the Clinical Focus box in Chapter 23, page 659.)

**Pseudomonadales** Members of the order Pseudomonadales are gram-negative aerobic rods or cocci. The most important genus in this group is *Pseudomonas*. The order also includes *Azotobacter*, *Azomonas*, *Moraxella*, and *Acinetobacter*.

***Pseudomonas*** A very important genus, *Pseudomonas* (soo'dō-MŌ-nas) consists of aerobic, gram-negative rods that are motile by polar flagella, either single or in tufts (**Figure 11.7**). Pseudomonads are very common in soil and other natural environments.

Many species of pseudomonads excrete extracellular, water-soluble pigments that diffuse into their media. One species, *Pseudomonas aeruginosa* (Ā-roo-ji'nō-sah), produces a soluble, blue-green pigmentation. Under certain conditions, particularly in weakened hosts, this organism can infect the urinary tract, burns, and wounds, and can cause blood infections (sepsis; page 652), abscesses, and meningitis. Other pseudomonads produce soluble fluorescent pigments that glow when illuminated by ultraviolet light. One species, *P. syringae* (ser-IN-gī), is an occasional plant pathogen. (Some species of *Pseudomonas* have been transferred, based upon rRNA studies, to the genus *Burkholderia*, which was discussed previously with the betaproteobacteria.)

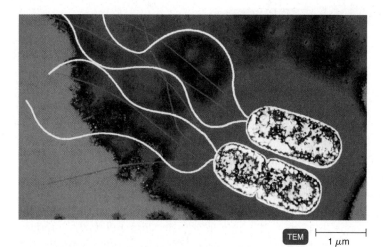

**Figure 11.7  *Pseudomonas.*** This photo of a pair of *Pseudomonas* bacteria shows polar flagella that are a characteristic of the genus. In some species, only a single flagellum is present (see Figure 4.7b, page 77). Note that one cell (on the bottom) is beginning to divide.

**Q** How does the nutritional diversity of these bacteria make them a problem in hospitals?

Pseudomonads have almost as much genetic capacity as the eukaryotic yeasts and almost half as much as a fruit fly. Although these bacteria are less efficient than some other heterotrophic bacteria in utilizing many of the more common nutrients, they compensate for this by making use of their genetic capabilities. For example, pseudomonads synthesize an unusually large number of enzymes and can metabolize a wide variety of substrates. Therefore, they probably contribute significantly to the decomposition of uncommon chemicals, such as pesticides, that are added to soil.

In hospitals and other places where pharmaceutical agents are prepared, the ability of pseudomonads to grow on minute traces of unusual carbon sources, such as soap residues or cap-liner adhesives found in a solution, has been unexpectedly troublesome. Pseudomonads are even capable of growth in some antiseptics, such as quaternary ammonium compounds. Their resistance to most antibiotics has also been a source of medical concern. This resistance is probably related to the characteristics of the cell wall porins, which control the entrance of molecules through the cell wall (see Chapter 4, page 81). The large genome of pseudomonads also codes for several very efficient efflux pump systems (page 581) that eject antibiotics from the cell before they can function. Pseudomonads are responsible for about one in ten healthcare-associated infections; see page 408), especially among infections in burn units. People with cystic fibrosis are also especially prone to infections by *Pseudomonas* and the closely related *Burkholderia.*

Although pseudomonads are classified as aerobic, some are capable of substituting nitrate for oxygen as a terminal electron in anaerobic respiration (see page 128). In this way,

pseudomonads cause important losses of valuable nitrogen in fertilizer and soil. Nitrate ($NO_3^-$) is the form of nitrogen fertilizer most easily used by plants. Under anaerobic conditions, as in water-logged soil, pseudomonads eventually convert this valuable nitrate into nitrogen gas ($N_2$), which is lost to the atmosphere (see Figure 27.3, page 790).

Many pseudomonads can grow at refrigerator temperatures. This characteristic, combined with their ability to utilize proteins and lipids, makes them an important contributor to food spoilage.

***Azotobacter* and *Azomonas***   Some nitrogen-fixing bacteria, such as *Azotobacter* (a-ZŌ-tō-bak-ter) and *Azomonas* (ā-zō-MŌ-nas), are free-living in soil. These large, ovoid, heavily capsulated bacteria are frequently used in laboratory demonstrations of nitrogen fixation. However, to fix agriculturally significant amounts of nitrogen, they would require energy sources, such as carbohydrates, that are in limited supply in soil.

***Moraxella***   Members of the genus *Moraxella* (MOR-ax-el-lah) are strictly aerobic coccobacilli—that is, intermediate in shape between cocci and rods. *Moraxella lacunata* (LAH-koo-nah-tah) is implicated in conjunctivitis, an inflammation of the conjunctiva, the membrane that covers the eye and lines the eyelids (page 612).

***Acinetobacter***   The genus *Acinetobacter* (ah-sin-Ē-tō-bak′ter) is aerobic and typically forms pairs. The bacteria occur naturally in soil and water. A member of this genus, *Acinetobacter baumanii* (BŌ-man-nē-ē), is an increasing concern to the medical community because of the rapidity with which it becomes resistant to antibiotics. Some strains are resistant to most available antibiotics. Not yet widespread in the United States, *A. baumanii* is an opportunistic pathogen primarily found in health care settings. The antibiotic resistance of the pathogen, combined with the weakened health of infected patients, has resulted in an unusually high mortality rate. *A. baumanii* is primarily a respiratory pathogen, but it also infects skin, soft tissues, and wounds and occasionally invades the bloodstream. It is more environmentally hardy than most gram-negative bacteria, and once established in a health care setting, it becomes difficult to eliminate.

**Legionellales**   The genera *Legionella* and *Coxiella* are closely associated in the second edition of *Bergey's Manual*, where both are placed in the same order, Legionellales. Because the *Coxiella* share an intracellular lifestyle with the rickettsial bacteria, they were previously considered rickettsial in nature and grouped with them. *Legionella* bacteria grow readily on suitable artificial media.

***Legionella***   *Legionella* (lē-JEN-el-lah) bacteria were originally isolated during a search for the cause of an outbreak of pneumonia now known as legionellosis (page 704). The search was difficult because these bacteria didn't grow on the usual laboratory isolation media then available. After intensive effort, special media were developed that enabled researchers to isolate and culture the first *Legionella*. Microbes of this genus are now known to be relatively common in streams, and they colonize such habitats as warm-water supply lines in hospitals and water in the cooling towers of air conditioning systems. (See the box in Chapter 24, page 708.) An ability to survive and reproduce within aquatic amebae often makes them difficult to eradicate in water systems.

**Coxiella** *Coxiella burnetii* (KOKS-ē-el-lah ber-NE-tē-ē), which causes Q fever (page 706) *Coxiella* bacteria require a mammalian host cell to reproduce. *Coxiella* is most commonly transmitted by contaminated milk. A sporelike body is present in *C. burnetii* (see Figure 24.13, page 706). This might explain the bacterium's relatively high resistance to the stresses of airborne transmission and heat treatment.

**Vibrionales** Members of the order Vibrionales are facultatively anaerobic gram-negative rods. They are found mostly in aquatic habitats. *Vibrio* (VIB-rē-ō) species are rods that are often slightly curved (**Figure 11.8**). One important pathogen is *Vibrio cholerae* (KOL-er-ī), the causative agent of cholera (page 732). The disease is characterized by a profuse and watery diarrhea. *V. parahaemolyticus* (par'ah-hē-mō-LI-ti-kus) causes a less serious form of gastroenteritis. Usually inhabiting coastal salt waters, it's transmitted to humans mostly by raw or undercooked shellfish.

**Enterobacteriales** The members of the order Enterobacteriales are facultatively anaerobic, gram-negative rods that are, if motile, peritrichously flagellated. Morphologically, the rods are straight. This is an important bacterial group, often commonly called **enterics**. This reflects the fact that they inhabit the intestinal tracts of humans and other animals. Most enterics are active fermenters of glucose and other carbohydrates.

Because of the clinical importance of enterics, there are many techniques to isolate and identify them. An identification method for some enterics is shown in Figure 10.9 (page 282), which incorporates a modern tool using 15 biochemical tests. Biochemical tests are especially important in clinical laboratory work and in food and water microbiology.

Enterics have fimbriae that help them adhere to surfaces or mucous membranes. Specialized sex pili aid in the exchange of genetic information between cells, which often includes antibiotic resistance (see Figures 8.29 and 8.30, pages 235 and 236).

Enterics, like many bacteria, produce proteins called bacteriocins that cause the lysis of closely related species of bacteria. Bacteriocins may help maintain the ecological balance of various enterics in the intestines. Important genera of the order include *Escherichia*, *Salmonella*, *Shigella*, *Klebsiella*, *Serratia*, *Proteus*, *Yersinia*, *Erwinia*, *Enterobacter*, and *Cronobacter*—all discussed next.

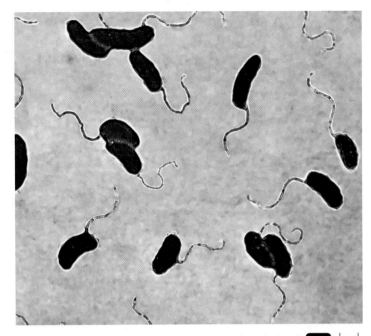

LM  |⊢——⊣| 0.7 μm

**Figure 11.8 *Vibrio cholerae*.** Notice the curvature of these rods, which is a characteristic of the genus.

**Q** **What is the flagellar arrangement of these cells?**

**Escherichia** The bacterial species *Escherichia coli* is a common inhabitant of the human intestinal tract and is probably the most familiar organism in microbiology. A great deal is known about the biochemistry and genetics of *E. coli*, and it continues to be an important tool for basic biological research—many researchers consider it almost a laboratory pet. Its presence in water or food is an indication of fecal contamination (see Chapter 27, page 798). *E. coli* is not usually pathogenic. However, it can be a cause of urinary tract infections, and certain strains produce enterotoxins that cause traveler's diarrhea (page 736) and occasionally cause very serious foodborne disease (see *E. coli* O157:H7, page 736). A newly identified emerging pathogenic species of *E. albertii* has been associated with sporadic infections in humans, birds, and calves.

**Salmonella** Almost all members of the genus *Salmonella* (sal'mōn-EL-lah) are potentially pathogenic. Accordingly, there are extensive biochemical and serological tests to clinically isolate and identify salmonellae. Salmonellae are common inhabitants of the intestinal tracts of many animals, especially poultry and cattle. Under unsanitary conditions, they can contaminate food.

There are only two species of *Salmonella*: *S. enterica* and *S. bongori*. *S. bongori* (BON-gor-ē) is a resident of "cold-blooded" animals—it was originally isolated from a lizard in the town of Bongor in the African desert nation of Chad—and is rarely found in humans. *S. enterica* (en-TER-i-kah) bacteria are infectious

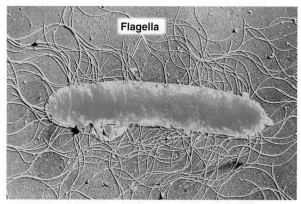

(a) *Proteus mirabilis* with peritrichous flagella | TEM | 0.4 μm

(b) A swarming colony of *Proteus mirabilis*, showing concentric rings of growth

**Figure 11.9 *Proteus mirabilis*.** Chemical communication between bacterial cells causes changes from cells adapted to swimming in fluid (few flagella) to cells that are able to move on surfaces (numerous flagella). The concentric growth **(b)** results from periodic synchronized conversion to the highly flagellated form capable of movement on surfaces.

**Q** The photo of the *Proteus* cell is probably a swarmer cell. How would you confirm swarming capability?

to warm-blooded animals. *S. enterica* includes more than 2500 **serovars**, that is, *serological varieties*. The term **serotype** is often used to mean the same thing. By way of explanation of these terms, when salmonellae are injected into appropriate animals, their flagella, capsules, and cell walls serve as *antigens* that cause the animals to form *antibodies* in their blood that are specific for each of these structures. Thus, *serological* means are used to differentiate the microorganisms. Serology is discussed more fully in Chapter 18, but for now it's sufficient to state that it can be used to differentiate and identify bacteria.

A serovar such as *Salmonella* Typhimurium (tī-fi-MYER-ē-um) is not a species and should be more properly written as "*Salmonella enterica* serovar Typhimurium." The convention now used by the Centers for Disease Control and Prevention (CDC) is to spell out the entire name at the first mention and then abbreviate it as, for example, *Salmonella* Typhimurium.

Specific antibodies, which are available commercially, can be used to differentiate *Salmonella* serovars by a system known as the Kauffmann-White scheme. This scheme designates an organism by numbers and letters that correspond to specific antigens on the organism's capsule, cell wall, and flagella, which are identified by the letters K, O, and H, respectively. For example, the antigenic formula for the bacterium *S.* Typhimurium is O1,4,[5],12:H,i,1,2.* Many salmonellae are named only by their antigenic formulas.

*The letters derive from the original German usage. Colonies that spread in a thin film over the agar surface were described by the German word for film, *Hauch.* The motility needed to form a film implied the presence of flagella, and the letter H came to be assigned to the antigens of flagella. clear Nonmotile bacteria were described as *ohne Hauch*, without film, and the O came to be assigned to the cell surface or body antigens. This terminology is also used in the naming of *E. coli* O157:H7, *Vibrio cholerae* O:1, and others.

Serovars can be further differentiated by special biochemical or physiological properties into **biovars**, or **biotypes.**

Typhoid fever, caused by *Salmonella* Typhi (TĪ-fē), is the most severe illness caused by any member of the genus *Salmonella* (page 732). A less severe gastrointestinal disease caused by other *S. enterica* serovars is called salmonellosis. Salmonellosis is one of the most common forms of foodborne illness. (See the box in Chapter 25 on page 731.)

***Shigella*** Species of *Shigella* (shi-GEL-lah) are responsible for a disease called bacillary dysentery, or shigellosis (page 728). Unlike salmonellae, they are found only in humans. Some strains of *Shigella* can cause life-threatening dysentery.

***Klebsiella*** Members of the genus *Klebsiella* (kleb-SĒ-el-lah) are commonly found in soil or water. Many isolates are capable of fixing nitrogen from the atmosphere, which has been proposed as being a nutritional advantage in isolated populations with little protein nitrogen in their diet. The species *Klebsiella pneumoniae* (noo-MŌ-nē-ī) occasionally causes a serious form of pneumonia in humans.

***Serratia*** *Serratia marcescens* (ser-RĀ-shah mar-SE-sens) is a bacterial species distinguished by its production of red pigment. In hospital situations, the organism can be found on catheters, in saline irrigation solutions, and in other supposedly sterile solutions. Such contamination is probably the cause of many urinary and respiratory tract infections in hospitals.

***Proteus*** Colonies of *Proteus* (PRŌ-tē-us) bacteria growing on agar exhibit a swarming type of growth. Swarmer cells with many flagella (**Figure 11.9a**) move outward on the edges of the colony and then revert to normal cells with only a few flagella and reduced motility. Periodically, new generations of highly motile swarmer cells develop, and the process is repeated. As a result, a *Proteus* colony has the distinctive appearance of a series of concentric rings (Figure 11.9b). This genus of bacteria is implicated in many infections of the urinary tract and in wounds.

**Yersinia** *Yersinia pestis* (yer-SIN-ē-ah PES-tis) causes plague, the Black Death of medieval Europe (page 661). Urban rats in some parts of the world and ground squirrels in the American Southwest carry these bacteria. Fleas usually transmit the organisms among animals and to humans, although contact with respiratory droplets from infected animals and people can be involved in transmission.

**Erwinia** *Erwinia* (er-WI-nē-ah) species are primarily plant pathogens; some cause plant soft-rot diseases. These species produce enzymes that hydrolyze the pectin between individual plant cells. This causes the plant cells to separate from each other, a disease that plant pathologists term *soft rot*.

**Enterobacter** Two *Enterobacter* (en'ter-ō-BAK-ter) species, *E. cloacae* (KLŌ-ā-kī), and *E. aerogenes* (ar-O-jen-ēz), can cause urinary tract infections and healthcare-associated infections. They are widely distributed in humans and animals, as well as in water, sewage, and soil.

**Cronobacter** *Cronobacter* (krō-nō-BAK-ter) is a genus of gram-negative, rod-shaped bacteria of the family Enterobacteriaceae. This genus was introduced in 2007, and there are now seven named species. These bacteria are facultatively anaerobic and generally motile. The type species is *Cronobacter sakazakii* (sa-kah-ZAH-kē-ē), which was previously known as *Enterobacter sakazakii*. This organism can cause meningitis and necrotizing enterocolitis in infants. It is widespread in a range of environments and foods. Most cases occur in adults, although the most publicized outbreaks have been associated with contaminated infant formulas.

**Pasteurellales** The bacteria in the order Pasteurellales are nonmotile; they are best known as human and animal pathogens.

**Pasteurella** The genus *Pasteurella* (pas-tyer-EL-lah) is primarily known as a pathogen of domestic animals. It causes sepsis in cattle, fowl cholera in chickens and other fowl, and pneumonia in several types of animals. The best-known species is *Pasteurella multocida* (mul-tō-SID-ah), which can be transmitted to humans by dog and cat bites.

**Haemophilus** *Haemophilus* (hē-MAH-fil-us) is a very important genus of pathogenic bacteria. These organisms inhabit the mucous membranes of the upper respiratory tract, mouth, vagina, and intestinal tract. The best-known species that affects humans is *Haemophilus influenzae* (IN-flū-en-zī), named long ago because of the erroneous belief that it was responsible for influenza.

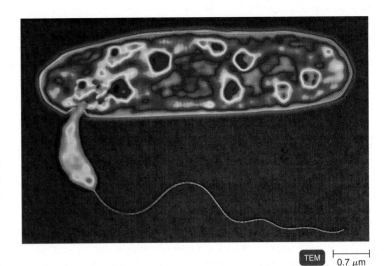

TEM | 0.7 μm

**Figure 11.10 *Bdellovibrio bacteriovorus*.** The yellow bacterium is *B. bacteriovorus*. It is attacking a bacterial cell shown in blue.

**Q** **Would this bacterium attack *Staphylococcus aureus*?**

The name *Haemophilus* is derived from the bacteria's requirement for blood in their culture medium (*hemo* = blood). They are unable to synthesize important parts of the cytochrome system needed for respiration, and they obtain these substances from the heme fraction, known as the **X factor**, of blood hemoglobin. The culture medium must also supply the cofactor nicotinamide adenine dinucleotide (from either $NAD^+$ or $NADP^+$), which is known as **V factor**. Clinical laboratories use tests for the requirement of X and V factors to identify isolates as *Haemophilus* species.

*Haemophilus influenzae* is responsible for several important diseases. It has been a common cause of meningitis in young children and is a frequent cause of earaches. Other clinical conditions caused by *H. influenzae* include epiglottitis (a life-threatening condition in which the epiglottis becomes infected and inflamed), septic arthritis in children, bronchitis, and pneumonia. *Haemophilus ducreyi* (doo-KRĀ-ē) is the cause of the sexually transmitted disease chancroid (page 775).

**CHECK YOUR UNDERSTANDING**

✔ **11-3** Make a dichotomous key to distinguish the orders of gammaproteobacteria described in this chapter.

## The Deltaproteobacteria

The deltaproteobacteria are distinctive in that they include some bacteria that are predators on other bacteria. Bacteria in this group are also important contributors to the sulfur cycle.

**Bdellovibrio** *Bdellovibrio* (del-lō-VIB-rē-ō) is a particularly interesting genus. It attacks other gram-negative bacteria. It attaches tightly (*bdella* = leech; **Figure 11.10**), and after penetrating the outer layer of gram-negative bacteria, it reproduces within the periplasm. There, the cell elongates into a tight spiral, which then fragments almost simultaneously into several individual flagellated cells. The host cell then lyses, releasing the *Bdellovibrio* cells.

**Desulfovibrionales** Members of the order Desulfovibrionales are sulfur-reducing bacteria. They are obligately anaerobic bacteria that use oxidized forms of sulfur, such as sulfates ($SO_4^{2-}$) or elemental sulfur ($S^0$) rather than oxygen as electron acceptors. The product of this reduction is hydrogen sulfide ($H_2S$). (Because the $H_2S$ is not assimilated as a nutrient, this type of metabolism is termed *dissimilatory*.) The activity of these bacteria releases millions of tons of $H_2S$ into the atmosphere every year and plays a key part in the sulfur cycle (see Figure 27.6 on page 793). Sulfur-oxidizing bacteria such as *Beggiatoa* are able to use $H_2S$ either as part of photosynthesis or as an autotrophic energy source.

*Desulfovibrio* (de′sul-fō-VIB-rē-ō), the best-studied sulfur-reducing genus, is found in anaerobic sediments and in the intestinal tracts of humans and animals. Sulfur-reducing and sulfate-reducing bacteria use organic compounds such as lactate, ethanol, or fatty acids as electron donors. This reduces sulfur or sulfate to $H_2S$. When $H_2S$ reacts with iron it forms insoluble FeS, which is responsible for the black color of many sediments.

**Myxococcales** In earlier editions of *Bergey's Manual*, the Myxococcales were classified among the fruiting and gliding bacteria. They illustrate the most complex life cycle of all bacteria, part of which is predatory upon other bacteria.

Vegetative cells of the myxobacteria (*myxo* = mucus) move by gliding and leave behind a slime trail. *Myxococcus xanthus* (mix′ō-KOK-us ZAN-thus) and *M. fulvus* (FUL-vus) are well-studied representatives of the genus *Myxococcus*. As they move, their source of nutrition is the bacteria they encounter, enzymatically lyse, and digest. Large numbers of these gram-negative microbes eventually aggregate (**Figure 11.11**). Where the moving cells aggregate, they differentiate and form a macroscopic stalked fruiting body that contains large numbers of resting cells called *myxospores*. Differentiation is usually triggered by low nutrients. Under proper conditions, usually a change in nutrients, the myxospores germinate and form new vegetative gliding cells. You might note the resemblance to the life cycle of the eukaryotic cellular slime molds in Figure 12.22 (page 348).

**CHECK YOUR UNDERSTANDING**

✔ 11-4 Make a dichotomous key to distinguish the deltaproteobacteria described in this chapter.

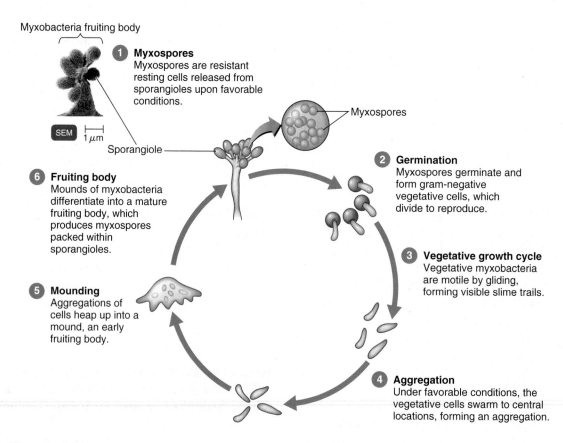

Myxobacteria fruiting body

SEM | 1 μm

Sporangiole

**1 Myxospores**
Myxospores are resistant resting cells released from sporangioles upon favorable conditions.

Myxospores

**2 Germination**
Myxospores germinate and form gram-negative vegetative cells, which divide to reproduce.

**3 Vegetative growth cycle**
Vegetative myxobacteria are motile by gliding, forming visible slime trails.

**4 Aggregation**
Under favorable conditions, the vegetative cells swarm to central locations, forming an aggregation.

**5 Mounding**
Aggregations of cells heap up into a mound, an early fruiting body.

**6 Fruiting body**
Mounds of myxobacteria differentiate into a mature fruiting body, which produces myxospores packed within sporangioles.

**Figure 11.11 Myxococcales.**

**Q** What is the feeding stage of this organism?

## The Epsilonproteobacteria

The epsilonproteobacteria are slender gram-negative rods that are helical or curved. We will discuss the two important genera, both of which are motile by means of flagella and are microaerophilic.

***Campylobacter*** *Campylobacter* (KAM-pi-lō-bak-ter) bacteria are microaerophilic vibrios; each cell has one polar flagellum. One species, *C. fetus* (FĒ-tus), causes spontaneous abortion in domestic animals. Another species, *C. jejuni* (je-JU-nē), is a leading cause of outbreaks of foodborne intestinal disease.

***Helicobacter*** *Helicobacter* bacteria are microaerophilic curved rods with multiple flagella. The species *Helicobacter pylori* (HĒ-lik-ō-bak'ter pī-LOR-ē) has been identified as the most common cause of peptic ulcers in humans and a cause of stomach cancer (**Figure 11.12**; see also Figure 25.12 on page 738).

> ### CHECK YOUR UNDERSTANDING
>
> ✔ **11-5** Make a dichotomous key to distinguish the epsilonproteobacteria described in this chapter.

## The Nonproteobacteria Gram-Negative Bacteria

### LEARNING OBJECTIVES

**11-6** Differentiate planctomycetes, chlamydias, Bacteroidetes, *Cytophaga*, and Fusobacteria by drawing a dichotomous key.

**11-7** Compare and contrast purple and green photosynthetic bacteria with the cyanobacteria.

**11-8** Describe the features of spirochetes and *Deinococcus*.

There are a number of important gram-negative bacteria that are not closely related to the gram-negative proteobacteria.

### Cyanobacteria (The Oxygenic Photosynthetic Bacteria)

The cyanobacteria, named for their characteristic blue-green (*cyan*) pigmentation, were once called blue-green algae. Although they resemble the eukaryotic algae and often occupy the same environmental niches, this is a misnomer because they are bacteria, algae are not. However, cyanobacteria do carry out oxygenic photosynthesis, as do the eukaryotic plants and algae (see Chapter 12). Many of the cyanobacteria are capable of fixing nitrogen from the atmosphere. In most cases, this activity is located in specialized cells called **heterocysts**, which contain enzymes that fix nitrogen gas ($N_2$) into ammonium ($NH_4^+$) that can be used by the growing cell (**Figure 11.13a**). Species that grow in water usually have gas vacuoles that provide buoyancy, helping the cell float at a favorable environment. Cyanobacteria that move about on solid surfaces use gliding motility.

Cyanobacteria are morphologically varied. They range from unicellular forms that divide by simple binary fission

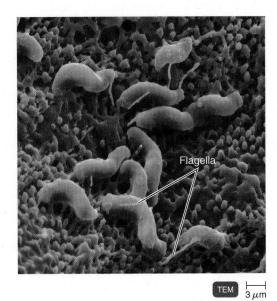

**Figure 11.12** ***Helicobacter pylori* on stomach cells.** *H. pylori*, a curved rod, is an example of a helical bacterium that does not make a complete twist.

**Q** How do helical bacteria differ from spirochetes?

(Figure 11.13b), to colonial forms that divide by multiple fission, to filamentous forms (see Figure 11.13a) that reproduce by fragmentation of the filaments. The filamentous forms usually exhibit some differentiation of cells that are often bound together within an envelope or sheath.

At less than 1 μm, *Prochlorococcus* (prō-KLŌR-ō-kok-kus) is the smallest known photosynthesizer. It is one of the most abundant organisms on Earth, comprising most of the photosynthetic population in tropical and subtropical oceans. *Prochlorococcus* possesses a variety of pigments that enable it to grow as deep as 200 m. Evidence indicates that oxygenic cyanobacteria played an important part in the development of life on Earth, which originally had very little free oxygen that would support life as we are familiar with it. Fossil evidence indicates that when cyanobacteria first appeared, the atmosphere contained only about 0.1% free oxygen. When oxygen-producing eukaryotic plants appeared millions of years later, the concentration of oxygen was more than 10%. The increase presumably was a result of photosynthetic activity by cyanobacteria. The atmosphere we breathe today contains about 20% oxygen. Cyanobacteria occupy environmental niches similar to those occupied by the eukaryotic algae (see Figure 12.12, page 338). The environmental role of cyanobacteria is presented more fully in Chapter 27, in the discussion of eutrophication (the nutritional over-enrichment of bodies of water). Several cyanobacteria produce toxins that sicken humans and other animals swimming in water containing large numbers of cyanobacteria (Table 15.3, page 435).

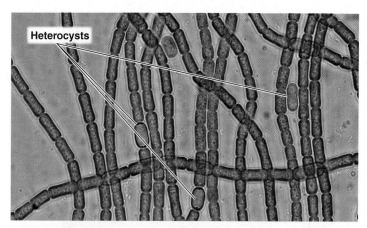

(a) Filamentous cyanobacterium showing heterocysts, in which nitrogen-fixing activity is located

LM 10 μm

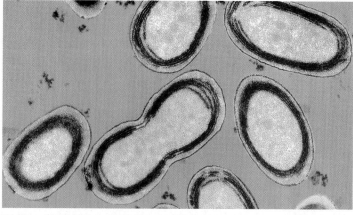

(b) The unicellular, nonfilamentous cyanobacterium *Prochlorococcus* may be the world's most abundant photosynthetic organism. (*Electron micrograph courtesy of Claire Ting, Williams College*)

TEM 0.15 μm

**Figure 11.13 Cyanobacteria.**

**Q** What does *anoxygenic* mean?

### The Phyla Chlorobi and Chloroflexi (The Anoxygenic Photosynthetic Bacteria)

The photosynthetic bacteria are taxonomically confusing, but they represent some interesting ecological niches. The phyla Cyanobacteria, Chlorobi, and Chloroflexi are gram-negative, but they are not included in the proteobacteria. Members of the photosynthetic phylum Chlorobi (representative genus: *Chlorobium*) are called **green sulfur bacteria**. Members of the phylum Chloroflexi (representative genus: *Chloroflexus*) are called **green nonsulfur bacteria**. Both these varieties of bacteria produce no oxygen during photosynthesis. The photosynthesizing bacteria are summarized in Table 11.2.

However, there are photosynthetic gram-negative bacteria that *are* genetically included in the proteobacteria. These are the **purple sulfur bacteria** (gammaproteobacteria) and the **purple nonsulfur bacteria** (alphaproteobacteria and betaproteobacteria).

In these bacterial groups, the term *sulfur bacteria* indicates that the microbes can use $H_2S$ as an electron donor (see the equations that follow). If classified as *nonsulfur bacteria*, the microbes have at least a limited capability of phototrophic growth, but without the production of oxygen. Cyanobacteria, as well as eukaryotic plants and algae, produce oxygen ($O_2$) from water ($H_2O$) as they carry out photosynthesis:

$$(1)\ 2H_2O + CO_2 \xrightarrow{\text{light}} (CH_2O) + H_2O + O_2$$

The *purple sulfur bacteria* and *green sulfur bacteria* use reduced sulfur compounds, such as hydrogen sulfide ($H_2S$), instead of water, and they produce granules of sulfur ($S^0$) rather than oxygen, as follows:

$$(2)\ 2H_2S + CO_2 \xrightarrow{\text{light}} (CH_2O) + H_2O + 2S^0$$

*Chromatium* (krō-MĀ-shum), shown in (**Figure 11.14**), is a representative genus. At one time, an important question in

**TABLE 11.2 Selected Characteristics of Photosynthesizing Bacteria**

| Common Name | Example | Phylum | Comments | Electron Donor for $CO_2$ Reduction | Oxygenic or Anoxygenic |
|---|---|---|---|---|---|
| Cyanobacteria | *Anabaena* | Cyanobacteria | Plantlike photosynthesis; some use bacterial photosynthesis under anaerobic conditions | Usually $H_2O$ | Usually oxygenic |
| Green nonsulfur bacteria | *Chloroflexus* | Chloroflexi | Grow chemoheterotrophically in aerobic environments | Organic compounds | Anoxygenic |
| Green sulfur bacteria | *Chlorobium* | Chlorobi | Deposit sulfur granules inside cells | Usually $H_2S$ | Anoxygenic |
| Purple nonsulfur bacteria | *Rhodospirillum* | Proteobacteria | Can grow chemoheterotrophically as well | Organic compounds | Anoxygenic |
| Purple sulfur bacteria | *Chromatium* | Proteobacteria | Deposit sulfur granules inside cells | Usually $H_2S$ | Anoxygenic |

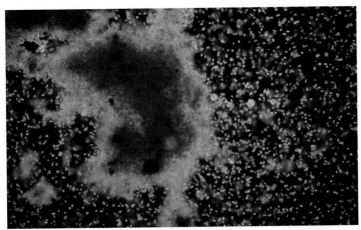

LM ⊢——⊣ 10 μm

**Figure 11.14 Purple sulfur bacteria.** The purple color of these *Chromatium* cells is due to carotenoid pigments. These pigments capture light energy and transfer electrons to bacteriochlorophyll.

**Q** How does the photosynthesis of cyanobacteria differ from that of the purple sulfur bacteria?

biology concerned the source of the oxygen produced by plant photosynthesis: was it from $CO_2$ or from $H_2O$? Until the introduction of radioisotope tracers, which traced the oxygen in water and carbon dioxide and settled the question, comparison of equations 1 and 2 was the best evidence that the oxygen source was $H_2O$. It is also important to compare these two equations to understand how reduced sulfur compounds, such as $H_2S$, can substitute for $H_2O$ in photosynthesis. See "Life Without Sunshine" on page 792.

Other photoautotrophs, the *purple nonsulfur bacteria* and *green nonsulfur bacteria*, use organic compounds, such as acids and carbohydrates, for the photosynthetic reduction of carbon dioxide. Morphologically, the photosynthetic bacteria are very diverse, with spirals, rods, cocci, and even budding forms.

## Chlamydiae

Members of the phylum Chlamydiae are grouped with other genetically similar bacteria that do not contain peptidoglycan in their cell walls. We will discuss only the genera *Chlamydia* and *Chlamydophila*. Earlier editions of *Bergey's Manual* grouped these bacteria with the rickettsial bacteria because they all grow intracellularly within host cells. The rickettsias are now classified according to their genetic content with the alphaproteobacteria.

***Chlamydia* and *Chlamydophila*** *Chlamydia* and *Chlamydophila*, which we will call by the common name of the chlamydias, have a unique developmental cycle that is perhaps their most distinguishing characteristic (**Figure 11.15a**). They are gram-negative coccoid bacteria (Figure 11.15b). The **elementary body** shown in Figure 11.15 is the infective agent. Chlamydias are

transmitted to humans by interpersonal contact or by airborne respiratory routes. The chlamydias can be cultivated in laboratory animals, in cell cultures, or in the yolk sac of embryonated chicken eggs.

There are three species of the chlamydias that are significant pathogens for humans. *Chlamydia trachomatis* (kla-MI-dē-ah tra-KŌ-ma-tis) is the best-known pathogen of the group and responsible for more than one major disease. These include trachoma, one of the most common causes of blindness in humans in the less-developed countries (page 612). It is also considered to be the primary causative agent of both nongonococcal urethritis, which may be the most common sexually transmitted disease in the United States, and lymphogranuloma venereum, another sexually transmitted disease (page 775).

Two members of the genus *Chlamydophila* (KLA-mi-dah'fil-ah) are well-known pathogens. *Chlamydophila psittaci* (SIT-tah-sē) is the causative agent of the respiratory disease psittacosis (ornithosis) (page 705). *Chlamydophila pneumoniae* (noo-MŌ-nē-ī) is the cause of a mild form of pneumonia that is especially prevalent in young adults.

## Planctomycetes

The planctomycetes, a group of gram-negative, budding bacteria, are said to "blur the definition of what bacteria are." Although their DNA places them among the bacteria, they resemble the archaea in the makeup of their cell walls, and some even have organelles that resemble the nucleus of a eukaryotic cell. The members of the genus *Planctomyces* are aquatic bacteria that produce stalks resembling *Caulobacter* (page 298) and have cell walls similar to those of the archaea, that is, without peptidoglycan.

One species of planctomycetes, *Gemmata obscuriglobus* (JEM-mah-tah OB-scyer-i-glō'bus), has a double internal membrane around its DNA, resembling a eukaryotic nucleus (**Figure 11.16**). Biologists wonder whether this might make *Gemmata* a model for the origin of the eukaryotic nucleus.

## Bacteroidetes

The phylum Bacteroidetes includes several genera of aerobic or anaerobic bacteria. Bacteroidetes are common members of the human microbiome, especially the gastrointesintal tract. The genus *Prevotella* is found in the human mouth, and the genus *Elizabethkingia* is an emerging cause of healthcare-associated infections.

***Bacteroides*** Bacteria of the genus *Bacteroides* (bak-ter-OI-dēz) live in the human intestinal tract in numbers approaching 1 billion per gram of feces. Some *Bacteroides* species also reside in anaerobic habitats such as the gingival crevice (see Figure 25.2 on page 724) and are also frequently recovered from deep-tissue infections. *Bacteroides* bacteria are gram-negative, are nonmotile, and do not

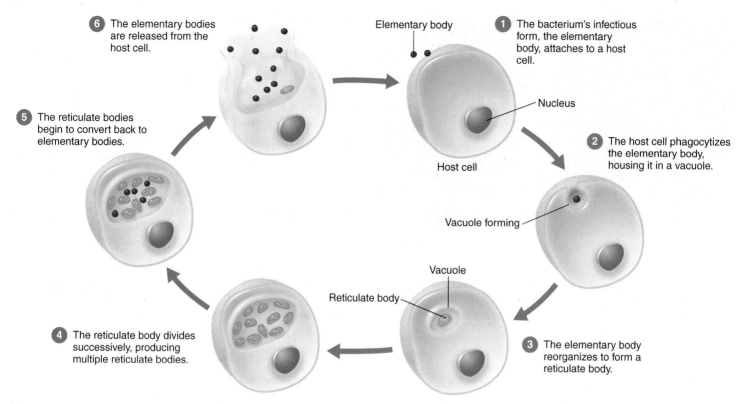

**6** The elementary bodies are released from the host cell.

**5** The reticulate bodies begin to convert back to elementary bodies.

**4** The reticulate body divides successively, producing multiple reticulate bodies.

Elementary body

**1** The bacterium's infectious form, the elementary body, attaches to a host cell.

Nucleus

Host cell

**2** The host cell phagocytizes the elementary body, housing it in a vacuole.

Vacuole forming

Vacuole

Reticulate body

**3** The elementary body reorganizes to form a reticulate body.

**(a)** Life cycle of the chlamydias, which takes about 48 hours to complete

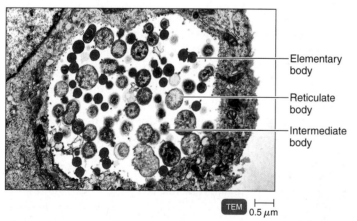

Elementary body

Reticulate body

Intermediate body

TEM ⊢──┤ 0.5 μm

**(b)** Micrograph of *Chlamydophila psittaci* in the cytoplasm of a host cell. The **elementary bodies** are the infectious stage; they are dense, dark, and relatively small. **Reticulate bodies,** the form in which chlamydias reproduce within the host cell, are larger with a speckled appearance. **Intermediate bodies,** a stage between the two, have a dark center.

**Figure 11.15 Chlamydias.**

**Q** Which stage of the life cycle is infectious to humans?

form endospores. Infections caused by *Bacteroides* often result from puncture wounds or surgery. *Bacteroides* are a frequent cause of peritonitis, an inflammation resulting from a perforated bowel.

**Cytophaga** Members of the genus *Cytophaga* (sī-TO-fa-gah) are important in the degradation of cellulose and chitin, which are both abundant in soil. Gliding motility places the microbe in close contact with these substrates, resulting in very efficient enzymatic action.

### Fusobacteria

The fusiform bacteria comprise another phylum of anaerobes. These bacteria are often pleomorphic but, as their name suggests, may be spindle-shaped (*fuso* = spindle).

**Fusobacterium** Members of the genus *Fusobacterium* (fu'-so-bak-TI-rē-um) are long, slender, gram-negative rods with pointed rather than blunt ends (**Figure 11.17**). In humans, they are found most often in the gingival crevice of the gums and may be responsible for some dental abscesses.

### Spirochaetes

The spirochetes have a coiled morphology, resembling a metal spring; some are more tightly coiled than others. The most distinctive characteristic of this phylum, however, is the cells' method of motility, which makes use of two or more **axial filaments** (or *endoflagella*) enclosed in the space between an outer sheath and the body of the cell. One end of each axial filament is attached near one of the cell's poles (see Figure 4.10,

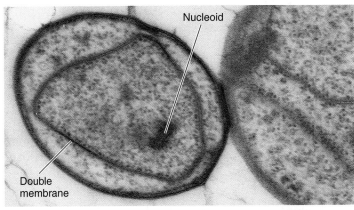

**Figure 11.16 *Gemmata obscuriglobus.*** This planctomycete exhibits a double membrane surrounding its nucleoid, which resembles a eukaryotic nucleus.

**Q** Can you see a resemblance between the double membrane around the nucleoid in this photo and the membrane around the nuclear envelope shown in Figure 4.24?

page 79, and **Figure 11.18**). By rotating its axial filament, the cell rotates in the opposite direction, like a corkscrew, which is very efficient in moving the organism through liquids. For bacteria, this is more difficult than it might seem. At the scale of a bacterium, water is as viscous as molasses is to a human. However, a bacterium can typically move about 100 times its body length in a second (or about 50 μm/sec); by comparison, a large, fast fish, such as a tuna, can move only about 10 times its body length in this time.

Many spirochetes are found in the human oral cavity and are probably among the first microorganisms described by van Leeuwenhoek in the 1600s. An extraordinary location for spirochetes is on the surfaces of some of the cellulose-digesting protozoa found in termites, where they may function as substitutes for flagella.

**Treponema** The spirochetes include a number of important pathogenic bacteria. The best known is the genus *Treponema* (trep-ō-NĒ-mah), which includes *Treponema pallidum* (PAL-li-dum), the cause of syphilis (page 772).

**Borrelia** Members of the genus *Borrelia* (bor-RĒ-lē-ah) cause relapsing fever (page 664) and Lyme disease (page 664), serious diseases that are usually transmitted by ticks or lice.

**Leptospira** Leptospirosis is a disease usually spread to humans by water contaminated by *Leptospira* (lep-tō-SPĪ-rah) species (page 764). The bacteria are excreted in the urine of dogs, rats, and swine, so domestic dogs and cats are routinely immunized against leptospirosis. The tightly coiled cells of *Leptospira* are shown in Figure 26.4 on page 765.

## Deinococcus-Thermus

The deinococci include two species of bacteria that have been widely studied because of their resistance to extremes in the environment. They stain gram-positive but have a cell wall that differs slightly in chemical structure from those of other gram-positives. *Deinococcus radiodurans* (dī-NŌ-kok-kus RĀ-dē-ō-der-anz) is exceptionally resistant to radiation, even more so than endospores. It can survive exposure to radiation doses as high as 15,000 Grays (see page 813), which is 1500 times the radiation dosage that would kill a human. The mechanism for this extraordinary resistance lies in a unique arrangement of its DNA that facilitates rapid repair of radiation damage. It is similarly resistant to many mutagenic chemicals.

*Thermus aquaticus* (THER-mus ah-KWA-ti-kus), another unique member of this group, is a bacterium that is unusually heat stable. It was isolated from a hot spring in Yellowstone National Park and is the source of the heat-resistant enzyme *Taq polymerase*, which is essential to the polymerase chain reaction (PCR). This is the method by which traces of DNA are amplified and used for identification (see page 247).

### CHECK YOUR UNDERSTANDING

✔ **11-6** Which gram-negative group has a life cycle that includes different stages?

✔ **11-7** Both the purple and green photosynthetic bacteria and the photosynthetic cyanobacteria use plantlike $CO_2$ fixation to make carbohydrates. In what way does the photosynthesis carried out by these two groups differ from plant photosynthesis?

✔ **11-8** The axial filament distinguishes what genera of bacteria?

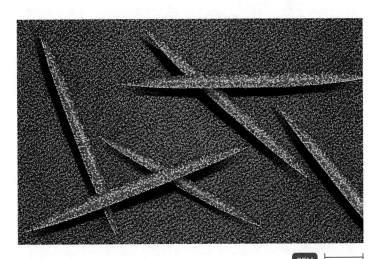

**Figure 11.17 *Fusobacterium.*** This is a common anaerobic rod found in the human intestine. Notice the characteristic pointed ends.

**Q** In what other place in the human body do you often find *Fusobacterium*?

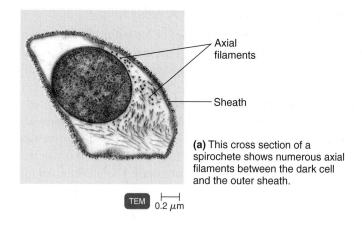

**(a)** This cross section of a spirochete shows numerous axial filaments between the dark cell and the outer sheath.

TEM $\vdash\!\!\dashv$ 0.2 μm

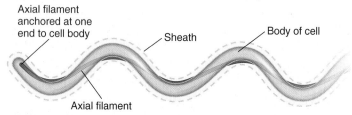

Axial filament anchored at one end to cell body

Sheath

Body of cell

Axial filament

**(b)** The end of an axial filament (endoflagellum) is attached to the cell and extends through most of the length of the cell. Another axial filament is attached to the opposite end of the cell. These axial filaments do not extend away from the cell but remain between the body of the cell and the external sheath. Their contractions and relaxations cause the helical cell to rotate in a corkscrew fashion.

**Figure 11.18 Spirochetes.** Spirochetes are helical and have axial filaments under an outer sheath that enables them to move by a corkscrewlike rotation.

**Q** How does a spirochete's motility differ from that of *Spirillum* (see Figure 11.4)?

# The Gram-Positive Bacteria

## LEARNING OBJECTIVES

**11-9** Differentiate the genera of firmicutes and tenericutes described in this chapter by drawing a dichotomous key.

**11-10** Differentiate the actinobacteria described in this chapter by drawing a dichotomous key.

The gram-positive bacteria can be divided into two groups: those that have a high G + C ratio, and those that have a low G + C ratio (see "Nucleic Acids," page 44). To illustrate the variations in G + C ratio, the genus *Streptococcus* has a low G + C content of 33 to 44%; and the genus *Clostridium* has a low content of 21 to 54%. Included with the gram-positive, low G + C bacteria are the mycoplasmas, even though they lack a cell wall and therefore do not have a Gram reaction. Their G + C ratio is 23 to 40%.

By contrast, filamentous actinomycetes of the genus *Streptomyces* have a high G + C content of 69 to 73%. Gram-positive bacteria of a more conventional morphology, such as the genera *Corynebacterium* and *Mycobacterium*, have a G + C content of 51 to 63% and 62 to 70%, respectively.

These bacterial groups are placed into separate phyla, the **Firmicutes (low G + C ratios)** and **Actinobacteria (high G + C ratios)**.

## Firmicutes (Low G + C Gram-Positive Bacteria)

Low G + C gram-positive bacteria are assigned to the phylum Firmicutes. This group includes important endospore-forming bacteria such as the genera *Clostridium* and *Bacillus*. Also of extreme importance in medical microbiology are the genera *Staphylococcus*, *Enterococcus*, and *Streptococcus*. In industrial microbiology, the genus *Lactobacillus*, which produces lactic acid, is well known. The mycoplasmas, which do not possess a cell wall, are also found in this phylum.

### Clostridiales

***Clostridium*** Members of the genus *Clostridium* (klo-STRID-ē-um) are obligate anaerobes. The rod-shaped cells contain endospores that usually distend the cell (**Figure 11.19**). The formation of endospores by bacteria is important to both medicine and the food industry because of the endospore's resistance to heat and many chemicals. Diseases associated with clostridia include tetanus (page 625), caused by *C. tetani* (TE-tan-ē); botulism (page 626), caused by *C. botulinum* (bo-tū-LI-num); and gas gangrene (page 659), caused by

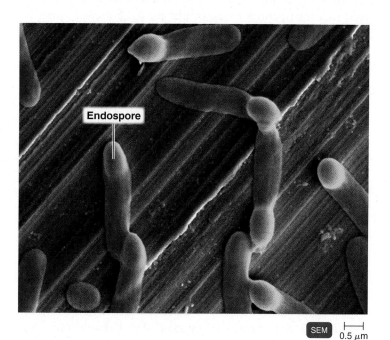

Endospore

SEM $\vdash\!\!\dashv$ 0.5 μm

**Figure 11.19 *Clostridium difficile*.** The endospores of clostridia usually distend the cell wall, as seen in some of these cells.

**Q** What physiological characteristic of *Clostridium* makes it a problem in contamination of deep wounds?

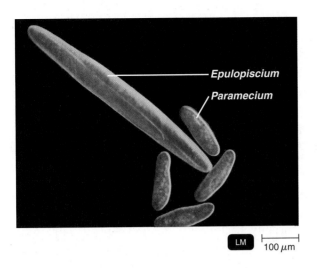

LM ⊢———⊣ 100 μm

**Figure 11.20  A giant prokaryote, *Epulopiscium fishelsoni.***
*Paramecium* is a protozoan, a group whose members are normally larger than bacterial cells.

**Q** Why is *Epulopiscium* not in the same domain as *Paramecium*?

*C. perfringens* (per-FRIN-jenz) and other clostridia. *C. perfringens* is also the cause of a common form of foodborne diarrhea. *C. difficile* (DIF-fi-sē-il) is an inhabitant of the intestinal tract that may cause a serious diarrhea (page 738). This occurs only when antibiotic therapy alters the normal intestinal microbiota, allowing overgrowth by toxin-producing *C. difficile*.

***Epulopiscium*** Biologists have long considered bacteria to be small by necessity because they lack the nutrient transport systems used by higher, eukaryotic organisms and because they depend on simple diffusion to obtain nutrients. These characteristics would seem to critically limit size. So, when a cigar-shaped organism living symbiotically in the gut of the Red Sea surgeonfish was first observed in 1985, it was considered to be a protozoan. Certainly, its size suggested this: the organism was as large as 80 μm × 600 μm—over half a millimeter in length—large enough to be seen with the unaided eye (**Figure 11.20**). Compared to the familiar bacterium *E. coli*, which is about 1 μm × 2 μm, this organism would be about a million times larger in volume.

Further investigation of the new organism showed that certain external structures thought to resemble the cilia of protozoa were actually similar to bacterial flagella, and it didn't have a membrane-enclosed nucleus. Ribosomal RNA analysis conclusively placed *Epulopiscium* (ep′ū-lō-PIS-sē-um) with the prokaryotes. (The name means "guest at the banquet of a fish." It is literally bathed in semidigested food.) It most closely resembles gram-positive bacteria of the genus *Clostridium*. Strangely, the species *Epulopiscium fishelsoni* (fish-el-SŌ-ne) doesn't reproduce by binary fission. Daughter cells formed within the cell are released through a slit opening in the parent cell. This may be related to the evolutionary development of sporulation.

Recently it was discovered that this bacterium does not rely on diffusion to distribute nutrients. Instead, it makes use of its larger genetic capacity—it has 25 times as much DNA as a human cell and as many as 85,000 copies of at least one gene—to manufacture proteins at internal sites where they are needed. (We describe another, more recently discovered, giant bacterium, *Thiomargarita*, on page 319.)

### Bacillales

The order Bacillales includes several important genera of gram-positive rods and cocci.

***Bacillus*** Bacteria of the genus *Bacillus* are typically rods that produce endospores. They are common in soil, and only a few are pathogenic to humans. Several species produce antibiotics.

*Bacillus anthracis* (bah-SIL-lus an-THRĀ-sis) causes anthrax, a disease of cattle, sheep, and horses that can be transmitted to humans (page 656). It's often mentioned as a possible agent of biological warfare. The anthrax bacillus is a nonmotile facultative anaerobe, often forming chains in culture. The centrally located endospore does not distend the walls. *Bacillus thuringiensis* (ther-IN-jē-en-sis) is probably the best-known microbial insect pathogen (**Figure 11.21**). It produces intracellular crystals

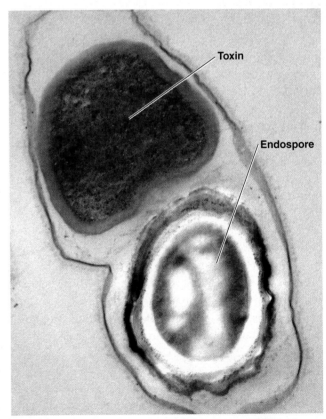

Toxin

Endospore

*Bacillus thuringiensis.* The protein shown next to the endospore, called a parasporal body, is toxic to an insect that ingests it.

TEM ⊢———⊣ 0.2 μm

**Figure 11.21  *Bacillus.***

**Q** What structure is made by both *Clostridium* and *Bacillus*?

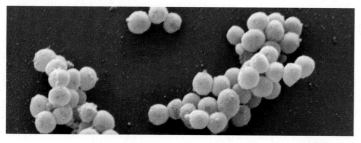

**Figure 11.22 *Staphylococcus aureus.*** Notice the grapelike clusters of these gram-positive cocci.

**Q** What is an environmental advantage of a pigment?

when it sporulates. Commercial preparations containing endospores and crystalline toxin (Bt) of this bacterium are sold in gardening supply shops to be sprayed on plants. *Bacillus cereus* (SEER-ē-us) is a common bacterium in the environment and occasionally is identified as a cause of food poisoning, especially in starchy foods such as rice (page 739).

The three species of the genus *Bacillus* that we have just described are dramatically different in important ways, especially their disease-causing properties. However, they are so closely related that taxonomists consider them to be variants of a single species.

*Staphylococcus* Staphylococci typically occur in grapelike clusters (**Figure 11.22**). The most important staphylococcal species is *Staphylococcus aureus* (STAF-i-lō-kok′kus OR-ē-us), which is named for its yellow-pigmented colonies (*aureus* = golden). Members of this species are facultative anaerobes.

Some characteristics of the staphylococci account for their pathogenicity, which takes many forms. They grow comparatively well under conditions of high osmotic pressure and low moisture, which partially explains why they can grow and survive in nasal secretions (many of us carry the bacteria in our nostrils) and on the skin. This also explains how *S. aureus* can grow in some foods with high osmotic pressure (such as ham and other cured meats) or in low-moisture foods that tend to inhibit the growth of other organisms. The yellow pigment probably confers some protection from the antimicrobial effects of sunlight. The wide variety of diseases caused by *S. aureus* are described in Part Four of this textbook.

## Lactobacillales

Several important genera are found in the order Lactobacillales. The genus *Lactobacillus* is a representative of the industrially important lactic acid–producing bacteria. Most lack a cytochrome system and are unable to use oxygen as an electron acceptor. Unlike most obligate anaerobes, though, they are aerotolerant and capable of growth in the presence of oxygen. They grow poorly compared to oxygen-utilizing microbes.

However, the production of lactic acid from simple carbohydrates inhibits the growth of competing organisms and allows them to grow competitively in spite of their inefficient metabolism. The genus *Streptococcus* shares the metabolic characteristics of the genus *Lactobacillus*. There are several industrially important species, but the streptococci are best known for their pathogenicity. The genera *Enterococcus* and *Listeria* are more conventional metabolically. Both are facultative anaerobes, and several species are important pathogens.

*Lactobacillus* In humans, bacteria of the genus *Lactobacillus* (lak′tō-bah-SIL-lus) are located in the vagina, intestinal tract, and oral cavity. Lactobacilli are used commercially in the production of sauerkraut, pickles, buttermilk, and yogurt. Typically, a succession of lactobacilli, each more acid tolerant than its predecessor, participates in these lactic acid fermentations.

*Streptococcus* Members of the genus *Streptococcus* (strep′tō-KOK-kus) are spherical, gram-positive bacteria that typically appear in chains (**Figure 11.23**). They are a taxonomically complex group, probably responsible for more illnesses and causing a greater variety of diseases than any other group of bacteria.

Pathogenic streptococci produce several extracellular substances that contribute to their pathogenicity. Among them are products that destroy phagocytic cells that ingest them. Enzymes produced by some streptococci spread infections by digesting the host's connective tissue, which may also result in extensive tissue destruction. (See the discussion of necrotizing

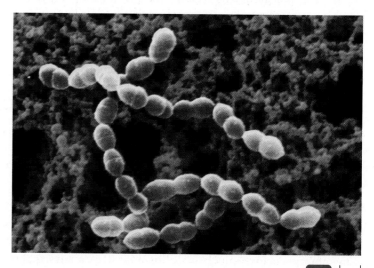

**Figure 11.23 *Streptococcus.*** Notice the chains of cells characteristic of most streptococci. Many of the spherical cells are dividing and are somewhat oval in appearance—especially when viewed with a light microscope, which has lower magnification than this electron micrograph.

**Q** How does the arrangement of *Streptococcus* differ from *Staphylococcus*?

fasciitis on page 597). Also, bacterial enzymes digest the fibrin (a threadlike protein) of blood clots, allowing infections to spread from the sites of injury.

A few nonpathogenic species of streptococci are important in the production of dairy products (see Chapter 28, page 815).

**Beta-hemolytic streptococci** A useful basis for classifying some streptococci is their colonial appearance when grown on blood agar. The *beta-hemolytic* species produce a hemolysin that forms a clear zone of hemolysis on blood agar (see Figure 6.9 on page 162). This group includes the principal pathogen of the streptococci, *Streptococcus pyogenes* (pī-AH-jen-ēz), also known as the beta-hemolytic group A streptococcus. Group A represents one of an antigenic group (A through G) within the hemolytic streptococci. Among the diseases caused by *S. pyogenes* (page 595) are scarlet fever, pharyngitis (sore throat), erysipelas, impetigo, and rheumatic fever. The most important virulence factor is the M protein on the bacterial surface (see Figure 21.6, page 597) by which the bacteria avoid phagocytosis. Another member of the beta-hemolytic streptococci is *Streptococcus agalactiae* (ā-gal-AK-tē-ī), in the beta-hemolytic group B. It is the only species with the group B antigen and is the cause of an important disease of the newborn, neonatal sepsis (page 654).

**Non-beta-hemolytic streptococci** Certain streptococci are not beta-hemolytic, but when grown on blood agar, their colonies are surrounded by a distinctive greening. These are the *alpha-hemolytic* streptococci. The greening represents a partial destruction of the red blood cells caused mostly by the action of bacteria-produced hydrogen peroxide, but it appears only when the bacteria grow in the presence of oxygen. The most

important pathogen in this group is *Streptococcus pneumoniae*, the cause of pneumococcal pneumonia (page 703). Also included among the alpha-hemolytic streptococci are species of streptococci called *viridans streptococci*. However, not all species form the alpha-hemolytic greening (*virescent* = green), so this isn't really a satisfactory group name. Probably the most significant pathogen of the group is *Streptococcus mutans* (MŪ-tanz), the primary cause of dental caries (page 724).

***Enterococcus*** The enterococci are adapted to areas of the body that are rich in nutrients but low in oxygen, such as the gastrointestinal tract, vagina, and oral cavity. They are also found in large numbers in human stool. Because they are relatively hardy microbes, they persist as contaminants in a hospital environment, on hands, bedding, and even as a fecal aerosol. In recent years they have become a leading cause of healthcare-associated infections, especially because of their high resistance to most antibiotics. Two species, *Enterococcus faecalis* (en'ter-ō-KOK-kus fē-KĀ-lis) and *Enterococcus faecium* (FĒ-sē-um), are responsible for much of the infections of surgical wounds and the urinary tract. In medical settings they frequently enter the bloodstream through invasive procedures, such as indwelling catheters (see page 408).

***Listeria*** The pathogenic species of the genus *Listeria*, *Listeria monocytogenes* (lis-TEER-ē-ah mon'ō-sī-TO-je-nēz), can contaminate food, especially dairy products. Important characteristics of *L. monocytogenes* are that it survives within phagocytic cells and is capable of growth at refrigeration temperatures. If it infects a pregnant woman, the organism poses the threat of stillbirth or serious damage to the fetus.

## Tenericutes

The Tenericutes phylum includes wall-less bacteria called mycoplasmas. Once included in the Firmicutes because of their low G + C content, mycoplasmas are now in their own phylum, Tenericutes. The mycoplasmas are highly pleomorphic because they lack a cell wall (**Figure 11.24**) and can produce filaments that resemble fungi, hence their name (*mykes* = fungus, and *plasma* = formed).

Cells of the genus *Mycoplasma* (mī-kō-PLAZ-mah) are very small, ranging in size from 0.1 to 0.25 μm, with a cell volume that is only about 5% of that of a typical bacillus. Because their size and plasticity allowed them to pass through filters that retained bacteria, they were originally considered to be viruses. Mycoplasmas may represent the smallest self-replicating organisms that are capable of a free-living existence. One species has only 517 genes; the minimum necessary is between 265 and 350. Studies of their DNA suggest that they are genetically related to the gram-positive Lactobacillales but have gradually lost genetic material. The term *degenerative evolution* has been used to describe this process.

Two species of bacteria that can cause bacterial meningitis are *Neisseria meningitidis* and *Streptococcus pneumoniae*. Dr. Walker then requests that the lab perform a Gram stain of Mercy's CSF and venous blood. Below is the Gram stain of Mercy's venous blood.

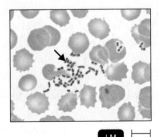

LM 7 μm

**What do you see that would cause you to modify your list of possible causes?**

297 **315** 317 318 319

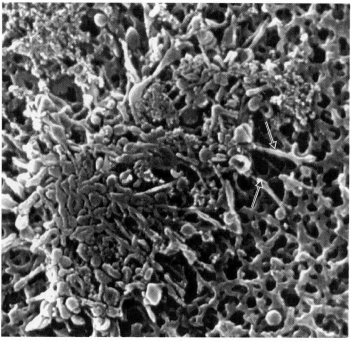

SEM    0.75 µm

**Figure 11.24 *Mycoplasma pneumoniae*.** This micrograph shows the filamentous growth of *M. pneumoniae*. This bacterium does not have a cell wall; the cell membrane is the outermost layer. The cells are so small that they cannot be examined by light microscopy. Individual cells (arrows) have extensions at each end that probably aid in gliding motility and in attachment to the host cells. They depend on the host for survival and do not survive as free-living organisms.

**Q** **How can the cell structure of mycoplasmas account for their pleomorphism?**

The most significant human pathogen among the mycoplasmas is *M. pneumoniae* (noo-MŌ-nē-ī), which is the cause of a common form of mild pneumonia.

Mycoplasmas can be grown on artificial media that provide them with sterols (if necessary) and other special nutritional or physical requirements. Colonies are less than 1 mm in diameter and have a characteristic "fried egg" appearance when viewed under magnification (see Figure 24.12, page 705). For many purposes, cell culture methods are often more satisfactory. In fact, mycoplasmas grow so well by this method that they are a frequent contamination problem in cell culture laboratories.

**CHECK YOUR UNDERSTANDING**

✔ **11-9** To which genus is *Enterococcus* more closely related: *Staphylococcus* or *Lactobacillus*?

## Actinobacteria (High G + C Gram-Positive Bacteria)

High G + C gram-positive bacteria are in the phylum Actinobacteria. Many bacteria in this phylum are highly pleomorphic in their morphology; the genera *Corynebacterium* and *Gardnerella*,

for example, and several genera such as *Streptomyces* grow only as extended, often-branching filaments. Several important pathogenic genera are found in the Actinobacteria, such as the *Mycobacterium* species causing tuberculosis and leprosy. The genera *Streptomyces*, *Frankia*, *Actinomyces*, and *Nocardia* are often informally called actinomycetes (from the Greek *actino* = ray) because they have a radiate, or starlike, form of growth by reason of their often-branching filaments. Superficially, their morphology resembles that of filamentous fungi; however, the actinomycetes are prokaryotic cells, and their filaments have a diameter much smaller than that of the eukaryotic molds. Some actinomycetes further resemble molds by their possession of externally carried asexual spores that are used for reproduction. Filamentous bacteria, like filamentous fungi, are very common inhabitants in soil, where a filamentous pattern of growth has advantages. The filamentous organism can bridge water-free gaps between soil particles to move to a new nutritional site. This morphology also gives the organism a much higher surface-to-volume ratio and improves its ability to absorb nutrients in the highly competitive soil environment.

***Mycobacterium***   The mycobacteria are aerobic, non–endospore-forming rods. The name *myco*, meaning funguslike, was derived from their occasional exhibition of filamentous growth (see Figure 24.7, page 698). Many of the characteristics of mycobacteria, such as acid-fast staining, drug resistance, and pathogenicity, are related to their distinctive cell wall, which is structurally similar to gram-negative bacteria (see Figure 4.13c, page 82). However, the outermost lipopolysaccharide layer in mycobacteria is replaced by mycolic acids, which form a waxy, water-resistant layer. This makes the bacteria resistant to stresses such as drying. Also, few antimicrobial drugs are able to enter the cell. (See the box in Chapter 7 on page 197.) Nutrients enter the cell through this layer very slowly, which is a factor in the slow growth rate of mycobacteria; it sometimes takes weeks for visible colonies to appear. The mycobacteria include the important pathogens *Mycobacterium tuberculosis* (mī′kō-bak-TI-rē-um too′ber-kū-LŌ-sis), which causes tuberculosis (page 698), and *M. leprae* (LEP-rī), which causes leprosy (page 629).

The mycobacteria are generally separated into two groups: (1) the slow growers, such as *M. tuberculosis*, and (2) the fast, or rapid, growers, which form visible colonies on appropriate media within 7 days. The slow-growing mycobacteria are more likely to be pathogenic to humans. The rapidly growing group also contains a number of occasional, nontuberculous human pathogens, which most commonly infect wounds. However, these mycobacteria are more likely to be nonpathogenic soil and water microbes.

***Corynebacterium***   The corynebacteria (*coryne* = club-shaped) tend to be pleomorphic, and their morphology often varies with the age of the cells. The best-known species is *Corynebacterium diphtheriae* (kor′ī-nē-bak-TI-rē-um dif-THI-rē-ī), the causative agent of diphtheria (page 692).

The Gram stain of Mercy's venous blood shows gram-positive cocci among the red blood cells. Needing to identify the species of bacteria that is causing Mercy's meningitis, the lab then cultures the cocci on blood agar. See below for the results.

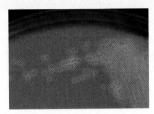

**Based on this new information, what bacteria are responsible for Mercy's meningitis? (Hint: See Figure 6.9 on page 162.)**

297    315    **317**    318    319

**Propionibacterium** The name of the genus *Propionibacterium* (pro-pē′on-ē-bak-TI-rē-um) is derived from the organism's ability to form propionic acid; some species are important in the fermentation of Swiss cheese. *Propionibacterium acnes* (AK-nēz) are bacteria that are commonly found on human skin and are implicated as the primary bacterial cause of acne.

**Gardnerella** *Gardnerella vaginalis* (gard-ne-REL-lah vaj-i-NA-lis) is a bacterium that causes one of the most common forms of vaginitis (page 776). There has always been some difficulty in assigning a taxonomic position in this species, which is gram-variable and exhibits a highly pleomorphic morphology.

**Frankia** The genus *Frankia* (FRANK-ē-ah) causes nitrogen-fixing nodules to form in alder tree roots, much as rhizobia cause nodules on the roots of legumes (see Figure 27.4, page 792).

**Streptomyces** The genus *Streptomyces* (strep′tō-MĪ-sēs) is the best known of the actinomycetes and is one of the bacteria most commonly isolated from soil (**Figure 11.25**). The reproductive asexual spores of *Streptomyces* are formed at the ends of aerial filaments. If each spore lands on a suitable substrate, it is capable of germinating into a new colony. These organisms are strict aerobes. They often produce extracellular enzymes that enable them to utilize proteins, polysaccharides (such as starch and cellulose), and many other organic materials found in soil. *Streptomyces* bacteria characteristically produce a gaseous compound called *geosmin*, which gives fresh soil its typical musty odor. Species of *Streptomyces* are valuable because they produce most of our commercial antibiotics (see Table 20.1, page 560). This has led to intensive study of the genus—there are nearly 500 described species.

**Actinomyces** The genus *Actinomyces* (ak-tin-ō-MĪ-sēs) consists of facultative anaerobes that are found in the mouth and throat of humans and animals. They occasionally form filaments called *hyphae* that can fragment (**Figure 11.26**). One species, *Actinomyces israelii* (is-RĀ-lē-ē), causes actinomycosis, a tissue-destroying disease usually affecting the head, neck, or lungs.

**Nocardia** The genus *Nocardia* (nō-KAR-dē-ah) morphologically resembles *Actinomyces*; however, these bacteria are aerobic. To reproduce, they form rudimentary filaments, which fragment into short rods. The structure of their cell wall resembles that of the mycobacteria; therefore, they are often acid-fast. *Nocardia* species are common in soil. Some species, such as *Nocardia asteroides* (AS-ter-oid-ēz), occasionally cause a chronic, difficult-to-treat pulmonary infection. *N. asteroides* is also one of the causative agents of mycetoma, a localized destructive infection of the feet or hands.

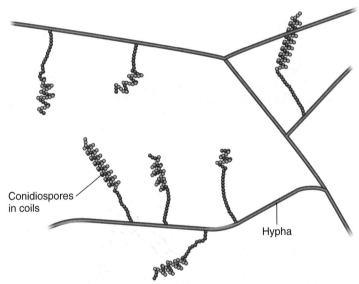

Conidiospores in coils

Hypha

**(a)** Drawing of a typical streptomycete showing filamentous, branching hyphae with asexual reproductive conidiospores at the filament tips

Hypha          Conidiospores in coils

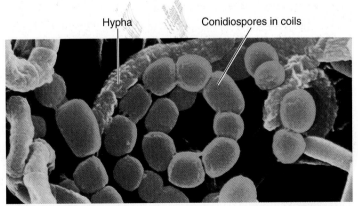

**(b)** Coils of conidiospores supported by hyphae of the streptomycete

SEM    0.75 μm

**Figure 11.25  *Streptomyces*.**

**Q** Why is *Streptomyces* not classified with fungi?

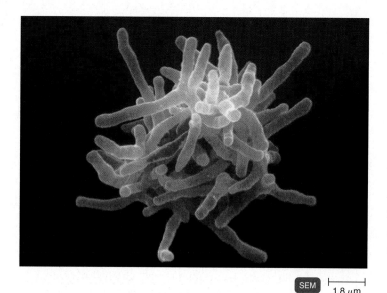

SEM ⊢———⊣ 1.8 μm

**Figure 11.26 *Actinomyces*.** Notice the branched filamentous hyphae.

Ⓠ Why are these bacteria not classified as fungi?

**CHECK YOUR UNDERSTANDING**

✔ **11-10** What group of bacteria makes most of the commercially important antibiotics?

# Domain Archaea

In the late 1970s, a distinctive type of prokaryotic cell was discovered. Most strikingly, the cell walls of these prokaryotes lacked the peptidoglycan common to most bacteria. It soon became clear that they also shared many rRNA sequences, and the sequences were different from both those of the Domain Bacteria and the Domain Eukarya. These differences were so significant that these organisms now constitute a new taxonomic grouping, the Domain Archaea.

## Diversity within the Archaea

**LEARNING OBJECTIVE**

**11-11** Name a habitat for each group of archaea.

This exceptionally interesting group of prokaryotes is highly diverse. Most archaea are of conventional morphology, that is, rods, cocci, and helices, but some are of very unusual morphology, as illustrated in **Figure 11.27**. Some are gram-positive, others gram-negative; some may divide by binary fission, others by fragmentation or budding; a few lack cell walls. Cultivated members of the archaea (singular: *archaeon*) can be placed into five physiological or nutritional groups.

Physiologically, archaea are found under extreme environmental conditions. **Extremophiles**, as they are known, include halophiles, thermophiles, and acidophiles. There are no known pathogenic archaea.

Halophiles thrive in salt concentrations of more than 25%, such as found in the Great Salt Lake and solar evaporating ponds. Examples of these are found in the genus *Halobacterium* (hā'lo-bak-TI-rē-um), some of which may even require such salt concentrations in order to grow. The optimal growth temperatures of extremely thermophilic archaea is 80°C or higher. The present record growth temperature is 121°C, established by archaea growing near a hydrothermal vent at 2000 meters

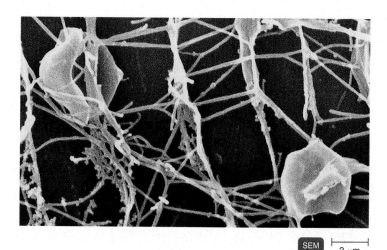

SEM ⊢———⊣ 3 μm

**Figure 11.27 Archaea.** *Pyrodictium abyssi*, an unusual member of the archaea found growing in deep-ocean sediment at a temperature of 110°C. The cells are disk-shaped with a network of tubules (cannulae). Most archaea are more conventional in their morphology.

Ⓠ Do the terms included in the name, *pyro* and *abyssi*, suggest a basis for the naming of this bacterium?

deep in the ocean. See Exploring the Microbiome on page 794. Acidophilic archaea can be found growing at pH values below zero and frequently at elevated temperatures, as well. An example is *Sulfolobus* (SUL-fō-lō-bus), whose optimal pH is about 2 and optimal temperature is more than 70°C.

Nutritionally, the ocean contains numerous *nitrifying* archaea that oxidize ammonia for energy. Some might also be found in soils. Methanogens are strictly anaerobic archaea that produce methane as an end-product by combining hydrogen ($H_2$) with carbon dioxide ($CO_2$). There are no known bacterial methanogens. These archaea are of considerable economic importance when they are used in sewage treatment (see the discussion of sludge digestion in Chapter 27 on pages 803–804). Methanogens are also part of the microbiota of the human colon, vagina, and mouth.

> **CHECK YOUR UNDERSTANDING**
> ✔ **11-11** What kind of archaea would populate solar evaporating ponds?

**CLINICAL CASE**  Resolved

GBS is often part of the normal intestinal or genitourinary microbiota, but it can cause disease in immunocompromised individuals. GBS emerged as a major cause of neonatal bacterial sepsis in the 1970s and is a leading infectious cause of neonatal morbidity in the United States. The bacterium, a common colonizer of the maternal genital tract, can infect the fetus during gestation, causing fetal death. GBS can also be acquired by the fetus during passage through the birth canal during delivery. Prevention includes screening all pregnant women for GBS at 35 to 37 weeks' gestation and administering antibiotics to carriers during labor. Although Mercy's mother tested negative during her pregnancy, her results were a rare false negative. Mercy is put on intravenous antibiotics and remains in the hospital for 10 days until the infection clears. She is sent home after 2 weeks and is now a healthy, happy 2-month-old girl.

297   315   317   318   **319**

# Microbial Diversity

The Earth provides a seemingly infinite number of environmental niches, and novel life forms have evolved to fill them. Many of the microbes that exist in these niches cannot be cultivated by conventional methods on conventional growth media and have remained unknown. In recent years, however, isolation and identification methods have become much more sophisticated, and microbes that fill these niches are being identified—many without being cultivated. For example, see the discussion of *Pelagibacter* on page 297. The effects of space travel on bacteria are being studied as the human microbiome is now going on spaceflights; see Exploring the Microbiome (page 320).

> ASM: Microorganisms are ubiquitous and live in diverse and dynamic ecosystems. Because the true diversity of microbial life is largely unknown, its effects and potential benefits have not been fully explored.

## Discoveries Illustrating the Range of Diversity

### LEARNING OBJECTIVE

**11-12** List two factors that contribute to the limits of our knowledge of microbial diversity.

Earlier in this chapter, we described the giant bacterium *Epulopiscium*. In 1999, another, even larger, giant bacterium was discovered 100 meters deep in the sediments of the coastal waters off Namibia, on the southwestern coast of Africa. Named *Thiomargarita namibiensis* (THĪ-ō-mar-gar'ē-tah nah'mib-ē-EN-sis), meaning "sulfur pearl of Namibia," these spherical organisms, classified with the gammaproteobacteria, are as large as 750 μm in diameter (**Figure 11.28**). This is a bit larger than the size of a period at the end of this sentence.

As we have mentioned, a factor that limits the size of prokaryotic cells is that nutrients must enter the cytoplasm by simple diffusion. *T. namibiensis* minimizes this problem by resembling a fluid-filled balloon, the vacuole in the interior being surrounded by a relatively thin outer layer of cytoplasm. This cytoplasm is equal in volume to that of most other prokaryotes. Its energy source is essentially hydrogen sulfide, which is plentiful in the sediments in which it is normally found, and nitrate, which it must extract intermittently from nitrate-rich seawaters when storms stir the loose sediment. The cell's interior vacuole, which makes up about 98% of the bacterium's volume, serves as a storage space to hold the nitrate between recharging of its supply. The cell's energy is derived from the oxidation of hydrogen sulfide; the nitrate, although a source of nutritional nitrogen, primarily serves as an electron acceptor in the absence of oxygen.

The discovery of uniquely large bacteria has raised the question of how large a prokaryotic cell can be and still absorb nutrients. At the other extreme, is there a lower limit for the size of microorganisms—especially their genome? There are reports of bacteria as small as 0.02 to 0.08 μm (nanobacteria) found in deep rock formations and even blood vessels and kidney stones. Most microbiologists have concluded that these are nonliving particles that have crystallized from minerals. Theoretical considerations have been used to calculate that a cell with a significant metabolism would have to have

# Microbiome in Space

Since space exploration began, scientists have worried about potential human-microbe interactions beyond Earth. In the 1960s, caution over "space germ" contamination led to 30-day quarantines of astronauts, equipment, and samples upon return from lunar missions. Today, scientists are equally concerned that spacecraft such as the Mars landers might inadvertently contaminate other planets we visit. But although equipment can often be sterilized before entering space, the human microbiome travels with us wherever we go and spreads to the surfaces we frequently touch.

The physical environment within a space craft is different from Earth's—there's less gravity and more ionizing radiation. Being in space depresses cells in the immune system, and antibiotics seem to lose potency. Meanwhile, *Salmonella* Typhimurium and *Pseudomonas aeruginosa* grown on the Atlantis space shuttle were more virulent than the same strains grown on Earth. Studies have shown that bacteria grown in space often take on different biofilm structures from those seen on Earth.

To date, only one astronaut has developed a serious microbial infection while in space, with the cause identified as *P. aeruginosa*. But could the microbiome that is so essential to us on Earth become a traitor in space? Astronauts stationed on the International Space Station are conducting the Astronaut Microbiome Project to find out. Since air within the space station is filtered, it's possible to examine the debris that is strained out and identify microbes in it. Actinobacteria are the most abundant bacteria on the International Space Station. *Ralstonia eutropha* and *P. aeruginosa* were also detected. Samples taken from various body sites, as well as fecal samples of astronauts before, during, and after a space mission, will also be analyzed during the Astronaut Microbiome Project, so we can identify other microbial changes that may take place in space.

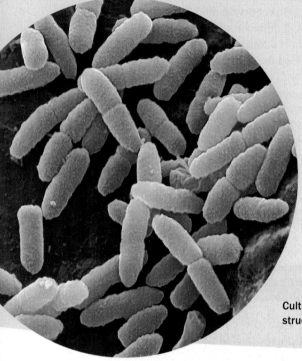

Cultures of *Pseudonomas aeruginosa* grown in space showed a "column and canopy" structure never seen before in cultures grown on Earth.

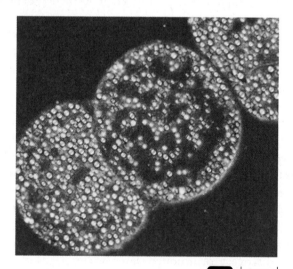

**Figure 11.28 *Thiomargarita namibiensis*.** This micrograph shows the intracellular sulfur granules in the thin layer of cytoplasm near the cell membrane. Sulfur accumulates when the bacteriium oxidizes $H_2S$.

**Q** Would a bacterium of this size be theoretically possible if the interior were cytoplasm rather than a fluid-filled vacuole?

a diameter of at least 0.1 μm. Certain bacteria have extraordinarily small genomes. For example, *Carsonella ruddii* (KAR-son-el-lah RUD-dē-ē) is a bacterium that lives in a *symbiotic relationship* with its insect host, a sap-eating psyllid (plant louse), and requires less genetic capability than would a free-living microbe. It has only 182 genes, which is close to the 151 genes that is the calculated theoretical minimum even for a microbe in such a symbiotic relationship. (Compare this with the minimal genetic requirements of the *free-living* mycoplasmas, on page 315). *C. ruddii* is not completely parasitic in its relationship with its host insect because it supplies the host with some essential amino acids. It is therefore probably in the evolutionary process of becoming an organelle, like the mitochondria of mammalian cells (see page 273).

Until now, microbiologists have described only about 5000 bacterial species, of which about 3000 are listed in *Bergey's Manual*. The true number may be in the millions. Many bacteria in soil or water, or elsewhere in nature, cannot be cultivated with the media and conditions normally used for bacterial growth. Moreover, some bacteria are part of complex food chains and can grow only in the presence of other microbes that supply specific growth requirements. Recently, researchers have been using the polymerase chain reaction (PCR) to make millions of copies of genes found at random in a soil sample. By comparing the genes found in many repetitions of this process, researchers can estimate the different bacterial species in such a sample. One report indicates that a single gram of soil may contain 10,000 or so bacterial types—about twice as many as have ever been described.

**CHECK YOUR UNDERSTANDING**

✔ **11-12** How can you detect the presence of a bacterium that cannot be cultured?

# Study Outline

 Go to @MasteringMicrobiology for Interactive Microbiology, *In the Clinic* videos, *MicroFlix, MicroBoosters, 3D animations, practice quizzes, and more.*

## Introduction (p. 295)

1. *Bergey's Manual* categorizes bacteria into taxa based on rRNA sequences.
2. *Bergey's Manual* lists identifying characteristics such as Gram stain reaction, cellular morphology, oxygen requirements, and nutritional properties.

## The Prokaryotic Groups (p. 296)

1. Prokaryotic organisms are classified into two domains: Archaea and Bacteria.

## Domain Bacteria (pp. 296–318)

1. Bacteria are essential to life on Earth.

## The Gram-Negative Bacteria (pp. 297–311)

### The Proteobacteria (pp. 297–307)

1. Members of the phylum Proteobacteria are gram-negative.
2. Alphaproteobacteria include nitrogen-fixing bacteria, chemoautotrophs, and chemoheterotrophs.
3. Betaproteobacteria include chemoautotrophs and chemoheterotrophs.
4. Pseudomonadales, Legionellales, Vibrionales, Enterobacteriales, and Pasteurellales are classified as gammaproteobacteria.
5. *Bdellovibrio* and *Myxococcus* in the deltaproteobacteria prey on other bacteria.
6. Epsilonproteobacteria include *Campylobacter* and *Helicobacter.*

## The Nonproteobacteria Gram-Negative Bacteria (pp. 307–312)

7. Cyanobacteria are photoautotrophs that use light energy and $CO_2$ and do produce $O_2$.
8. Purple and green photosynthetic bacteria are photoautotrophs that use light energy and $CO_2$ and do not produce $O_2$.
9. *Deinococcus* and *Thermus* are resistant to environmental extremes.
10. Planctomycetes, Chlamydiae, Spirochetes, Bacteroidetes, and Fusobacteria are phyla of gram-negative, chemoheterotrophic bacteria.

## The Gram-Positive Bacteria (pp. 312–318)

1. In *Bergey's Manual*, gram-positive bacteria are divided into those that have low G + C ratio and those that have high G + C ratio.
2. Low G + C gram-positive bacteria include common soil bacteria, the lactic acid bacteria, and several human pathogens.
3. High G + C gram-positive bacteria include mycobacteria, corynebacteria, and actinomycetes.

## Domain Archaea (pp. 318–319)

1. Extreme halophiles, extreme thermophiles, and methanogens are included in the Archaea.

## Microbial Diversity (pp. 319–321)

1. Few of the total number of different prokaryotes have been isolated and identified.
2. PCR can be used to uncover the presence of bacteria that can't be cultured in the laboratory.

# Study Questions

For answers to the Knowledge and Comprehension questions, turn to the Answers tab at the back of the textbook.

## Knowledge and Comprehension

### Review

1. The following outline can be used to identify important bacteria. Fill in a representative genus in the space provided.

**Representative Genus**

I. Gram-positive
   A. Endospore-forming rod
      1. Obligate anaerobe          (a) _____
      2. Not obligate anaerobe      (b) _____
   B. Non–endospore-forming
      1. Cells are rods
         a. Produce conidiospores   (c) _____
         b. Acid-fast               (d) _____
      2. Cells are cocci
         a. Lack cytochrome system  (e) _____
         b. Use aerobic respiration (f) _____
II. Gram-negative
   A. Cells are helical or curved
      1. Axial filament             (g) _____
      2. No axial filament          (h) _____
   B. Cells are rods
      1. Aerobic, nonfermenting     (i) _____
      2. Facultatively anaerobic    (j) _____

III. Lack cell walls (k) _____

IV. Obligate intracellular parasites

A. Transmitted by ticks (l) _____

B. Reticulate bodies in host cells (m) _____

2. Compare and contrast each of the following:
   a. Cyanobacteria and algae
   b. Actinomycetes and fungi
   c. *Bacillus* and *Lactobacillus*
   d. *Pseudomonas* and *Escherichia*
   e. *Leptospira* and *Spirillum*
   f. *Escherichia* and *Bacteroides*
   g. *Rickettsia* and *Chlamydia*
   h. *Mycobacterium* and *Mycoplasma*

3. **DRAW IT** Draw a key to differentiate the following bacteria: cyanobacteria, *Cytophaga*, *Desulfovibrio*, *Frankia*, *Hyphomicrobium*, methanogens, myxobacteria, *Nitrobacter*, purple bacteria, *Sphaerotilus*, and *Sulfolobus*.

4. **NAME IT** These organisms are important in sewage treatment and can produce a fuel used for home heating and for generating electricity.

## Multiple Choice

1. If you Gram-stained the bacteria that live in the human intestine, you would expect to find mostly
   a. gram-positive cocci.
   b. gram-negative rods.
   c. gram-positive, endospore-forming rods.
   d. gram-negative, nitrogen-fixing bacteria.
   e. all of the above.

2. Which of the following does *not* belong with the others?
   a. Enterobacteriales
   b. Lactobacillales
   c. Legionellales
   d. Pasteurellales
   e. Vibrionales

3. Pathogenic bacteria can be
   a. motile.
   b. rods.
   c. cocci.
   d. anaerobic.
   e. all of the above

4. Which of the following is an intracellular parasite?
   a. *Rickettsia*
   b. *Mycobacterium*
   c. *Bacillus*
   d. *Staphylococcus*
   e. *Streptococcus*

5. Which of the following terms is the most specific?
   a. bacillus
   b. *Bacillus*
   c. gram-positive
   d. endospore-forming rods and cocci
   e. anaerobic

6. Which one of the following does *not* belong with the others?
   a. *Enterococcus*
   b. *Lactobacillus*
   c. *Staphylococcus*
   d. *Streptococcus*
   e. All are grouped together.

7. Which of the following pairs is *mismatched*?
   a. anaerobic endospore-forming gram-positive rods—*Clostridium*
   b. facultatively anaerobic gram-negative rods—*Escherichia*
   c. facultatively anaerobic gram-negative rods—*Shigella*
   d. pleomorphic gram-positive rods—*Corynebacterium*
   e. spirochete—*Helicobacter*

8. *Spirillum* is *not* classified as a spirochete because spirochetes
   a. do not cause disease.
   b. possess axial filaments.
   c. possess flagella.
   d. are prokaryotes.
   e. none of the above

9. When *Legionella* was newly discovered, why was it classified with the pseudomonads?
   a. It is a pathogen.
   b. It is an aerobic gram-negative rod.
   c. It is difficult to culture.
   d. It is found in water.
   e. none of the above

10. Unlike purple and green phototrophic bacteria, cyanobacteria
    a. produce oxygen during photosynthesis.
    b. do not require light.
    c. use $H_2S$ as an electron donor.
    d. have a membrane-enclosed nucleus.
    e. all of the above

## Analysis

1. Use of culture-independent techniques such as rRNA sequencing and FISH have increased our understanding of microbial diversity without cultivation. Do microbiologists still need to attempt to grow new species? Briefly explain.

2. To which of the following is the photosynthetic bacterium *Chromatium* most closely related? Briefly explain why.
   a. cyanobacteria
   b. *Chloroflexus*
   c. *Escherichia*

3. Bacteria are single-celled organisms that must absorb their nutrients by simple diffusion. The dimensions of *Thiomargarita namibiensis* are several hundred times larger than those of most bacteria, much too large for simple diffusion to operate. How does the bacterium solve this problem?

## Clinical Applications and Evaluation

1. After contact with a patient's spinal fluid, a lab technician developed fever, nausea, and purple lesions on her neck and extremities. A throat culture grew gram-negative diplococci. What is the genus of the bacteria?

2. Between April 1 and May 15 of one year, 22 children in three states developed diarrhea, fever, and vomiting. The children had each received pet ducklings. Gram-negative, facultatively anaerobic bacteria were isolated from both the patients' and the ducks' feces; the bacteria were identified as serovar C2. What is the genus of these bacteria?

3. A woman complaining of lower abdominal pain with a temperature of 39°C gave birth soon after to a stillborn baby. Blood cultures from the infant revealed gram-positive rods. The woman had a history of eating unheated hot dogs during her pregnancy. Which organism is most likely involved?

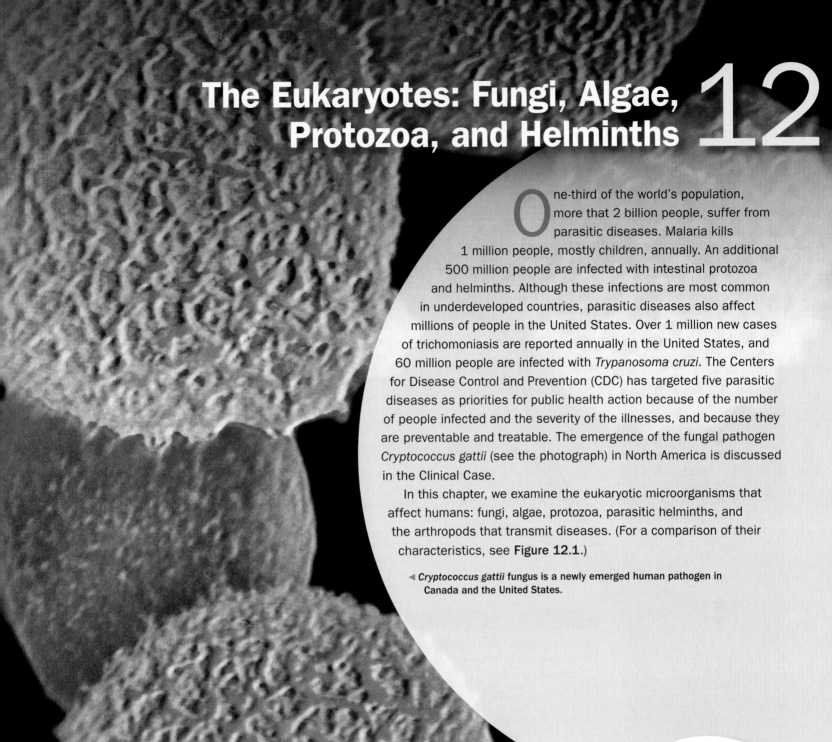

# The Eukaryotes: Fungi, Algae, Protozoa, and Helminths 12

One-third of the world's population, more that 2 billion people, suffer from parasitic diseases. Malaria kills 1 million people, mostly children, annually. An additional 500 million people are infected with intestinal protozoa and helminths. Although these infections are most common in underdeveloped countries, parasitic diseases also affect millions of people in the United States. Over 1 million new cases of trichomoniasis are reported annually in the United States, and 60 million people are infected with *Trypanosoma cruzi*. The Centers for Disease Control and Prevention (CDC) has targeted five parasitic diseases as priorities for public health action because of the number of people infected and the severity of the illnesses, and because they are preventable and treatable. The emergence of the fungal pathogen *Cryptococcus gattii* (see the photograph) in North America is discussed in the Clinical Case.

In this chapter, we examine the eukaryotic microorganisms that affect humans: fungi, algae, protozoa, parasitic helminths, and the arthropods that transmit diseases. (For a comparison of their characteristics, see **Figure 12.1**.)

◄ *Cryptococcus gattii* **fungus is a newly emerged human pathogen in Canada and the United States.**

## In the Clinic

As a Peace Corps nurse in West Africa, you meet a 4-year-old girl with a particularly swollen stomach. The girl's mother shows you a large (10-cm) white worm that the girl coughed up. The worm is cylindrical with tapered ends. **What is the worm? How did she contract it?**

*Hint: Read about helminths (pages 347–355).*

 Play **In the Clinic** Video @MasteringMicrobiology

# Exploring Pathogenic Eukaryotes

**Arthropods** are animals with jointed legs. The arthropods that transmit diseases are important in microbiology. These include ticks and some insects; most often, members of the mosquito family are responsible for transmitting disease.

**Helminths** are multicellular animals. They are chemoheterotrophs. Most obtain nutrients by ingestion through a mouth; some are absorptive. Parasitic helminths often have elaborate life cycles including egg, larva, and adult.

animals

Arthropods

Helminths

fungi

**Fungi** are in the Fungi kingdom. They are chemoheterotrophs and acquire food by absorption. With the exception of yeasts, fungi are multicellular. Most reproduce with sexual and asexual spores.

algae

**Algae** belong to several super clades and can reproduce both sexually and asexually. They are photoautotrophs and produce several different photosynthetic pigments. They obtain nutrients by diffusion. Some are multicellular, forming colonies, filaments, or even tissues. A few produce toxins.

protozoa

**Protozoa** belong to several super clades. Most are chemoheterotrophic, but a few are photoautotrophic. They obtain nutrients by absorption or ingestion. All are unicellular, and many are motile. Parasitic protozoans often form resistant cysts.

## KEY CONCEPTS

- Fungi, protozoa, and helminths cause diseases in humans. Most of these diseases are diagnosed by microscopic examination. Like bacteria, fungi are cultured on laboratory media.

- Infections caused by eukaryotes are difficult to treat because humans have eukaryotic cells.

- Algal diseases of humans are not infectious; they are intoxications because the symptoms result from ingesting algal toxins.

- Arthropods that transmit infectious diseases are called vectors. Arthropod-borne diseases such as West Nile encephalitis are best controlled by limiting exposure to the arthropod.

# Fungi

## LEARNING OBJECTIVES

12-1 List the defining characteristics of fungi.

12-2 Differentiate asexual from sexual reproduction, and describe each of these processes in fungi.

12-3 List defining characteristics of the four phyla of fungi.

12-4 Identify two beneficial and two harmful effects of fungi.

Of the more than 100,000 species of fungi, only about 200 are pathogenic to humans and animals. However, over the last 10 years, the incidence of serious fungal infections has been increasing. These infections are occurring in health care settings and in people with compromised immune systems. These infections are often caused by fungi normally found in and on the human body. In addition, thousands of fungal diseases afflict economically important plants, costing more than $1 billion annually.

**TABLE 12.1  Selected Features of Fungi and Bacteria Compared**

|  | Fungi | Bacteria |
|---|---|---|
| **Cell Type** | Eukaryotic | Prokaryotic |
| **Cell Membrane** | Sterols present | Sterols absent, except in *Mycoplasma* |
| **Cell Wall** | Glucans; mannans; chitin (no peptidoglycan) | Peptidoglycan |
| **Spores** | Sexual and asexual reproductive spores | Endospores (not for reproduction); some asexual reproductive spores |
| **Metabolism** | Limited to heterotrophic; aerobic, facultatively anaerobic | Heterotrophic, autotrophic; aerobic, facultatively anaerobic, anaerobic |

Fungi are also beneficial. They're important in the food chain because they decompose dead plant matter, thereby recycling vital elements. The hard parts of plants, which animals can't digest, are decomposed primarily by fungi through the use of extracellular enzymes, such as cellulases. Nearly all plants depend on symbiotic fungi, known as **mycorrhizae**, which help their roots absorb minerals and water from the soil (see Chapter 27). Fungi are also valuable to animals. Fungi-farming ants cultivate fungi that break down cellulose and lignin from plants, providing glucose that the ants can then digest. Humans use fungi for food (mushrooms) and to produce foods (bread and citric acid) and drugs (alcohol and penicillin).

The study of fungi is called **mycology**. A pathogen must be accurately identified if the disease is to be properly treated and its spread prevented. We will first look at the structures that are the basis of fungal identification in a clinical laboratory. Then we will explore their life cycles and nutritional needs. All fungi are chemoheterotrophs, requiring organic compounds for energy and carbon. Fungi are aerobic or facultatively anaerobic; only a few anaerobic fungi are known.

Table 12.1 lists the basic differences between fungi and bacteria.

## Characteristics of Fungi

Identifying yeasts and bacteria requires biochemical tests. However, multicellular fungi are identified on the basis of physical appearance, including colony characteristics and reproductive spores.

### Vegetative Structures

Fungal colonies are described as **vegetative** structures because they're composed of the cells involved in catabolism and growth.

**Molds and Fleshy Fungi**  The **thallus** (body) of a mold or fleshy fungus consists of long filaments of cells joined together; these filaments are called **hyphae** (singular: **hypha**). Hyphae can grow to immense proportions. Using DNA fingerprinting,

scientists mapped the hyphae of a single fungus in Oregon (a mushroom) that extended over 4 square miles.

In most molds, the hyphae contain cross-walls called **septa** (singular: **septum**), which divide them into distinct, uninucleate (one-nucleus) cell-like units. These hyphae are called **septate hyphae** (**Figure 12.2a**). In a few classes of fungi, the hyphae contain no septa and appear as long, continuous cells with many nuclei. These are called **coenocytic hyphae** (Figure 12.2b). Even in fungi with septate hyphae, there are usually openings in the septa that make the cytoplasm of adjacent "cells" continuous; these fungi are actually coenocytic organisms, too.

Hyphae grow by elongating at the tips (Figure 12.2c). Each part of a hypha is capable of growth, and when a fragment breaks off, it can elongate to form a new hypha. In the laboratory, fungi are usually grown from fragments obtained from a fungal thallus.

The portion of a hypha that obtains nutrients is called the *vegetative hypha*; the portion concerned with reproduction

### CLINICAL CASE  Man's Best Friend

Ethan, a 26-year-old computer programmer, is coaxing his dog, Waldo, into his truck. Waldo is very ill, and Ethan is taking him to the veterinary clinic in Bellingham, Washington, to get Waldo checked out. Waldo not only has nasal discharge, noisy breathing, coughing, and sneezing, but also is losing weight and having difficulty walking. Ethan already had to look all over the property to find Waldo, and by the time Ethan locates him in the barn, carries him to the driveway, and lifts the 60-pound Labrador into the bed of his truck, Ethan has to stop to rest. As a matter of fact, he thinks to himself, he doesn't look much better than Waldo these days! Ethan has been fighting what he thinks is some sort of virus as well.

The veterinarian examines Waldo and prescribes fluconazole, an antibiotic. Ethan, now very tired by this point, takes Waldo home. They both lie down to get some rest.

**What type of infection could Waldo have? Read on to find out.**

325   332   334

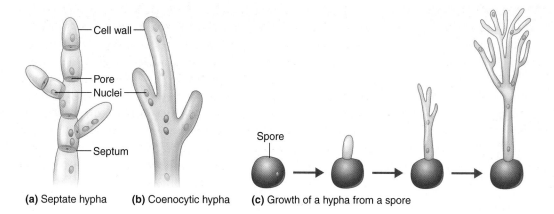

**Figure 12.2 Characteristics of fungal hyphae. (a)** Septate hyphae have cross-walls, or septa, dividing the hyphae into cell-like units. **(b)** Coenocytic hyphae lack septa. **(c)** Hyphae grow by elongating at the tips.

**Q** What is a hypha? A mycelium?

(a) Septate hypha  (b) Coenocytic hypha  (c) Growth of a hypha from a spore

is the *reproductive* or *aerial hypha*, so named because it projects above the surface of the medium on which the fungus is growing. Aerial hyphae often bear reproductive spores (**Figure 12.3a**), discussed later. When environmental conditions are suitable, the hyphae grow to form a filamentous mass called a **mycelium**, which is visible to the unaided eye (Figure 12.3b).

**Yeasts** Yeasts are nonfilamentous, unicellular fungi that are typically spherical or oval. Like molds, yeasts are widely distributed in nature; they are frequently found as a white powdery coating on fruits and leaves. **Budding yeasts**, such as *Saccharomyces* (sak-kar-ō'MĬ'sēz), divide unevenly.

In budding (**Figure 12.4**), the parent cell forms a protuberance (bud) on its outer surface. As the bud elongates, the parent cell's nucleus divides, and one nucleus migrates into the bud. Cell wall material is then laid down between the bud and parent cell, and the bud eventually breaks away.

One yeast cell can in time produce up to 24 daughter cells by budding. Some yeasts produce buds that fail to detach themselves; these buds form a short chain of cells called a **pseudohypha**. *Candida albicans* (KAN-dē-dah AL-bi-kanz) attaches to human epithelial cells as a yeast but usually requires pseudohyphae to invade deeper tissues (see Figure 21.17a, page 609).

**Fission yeasts**, such as *Schizosaccharomyces* (SKI-zō-sak'kar-ō'mī-sēz), divide evenly to produce two new cells. During fission, the parent cell elongates, its nucleus divides, and two offspring cells are produced. Increases in the number of yeast cells on a solid medium produce a colony similar to a bacterial colony.

Yeasts are capable of facultative anaerobic growth, which allows these fungi to survive in various environments. If given access to oxygen, yeasts perform aerobic respiration to metabolize carbohydrates into carbon dioxide and water; denied oxygen, they ferment carbohydrates and produce ethanol and carbon dioxide. This fermentation is used in the brewing,

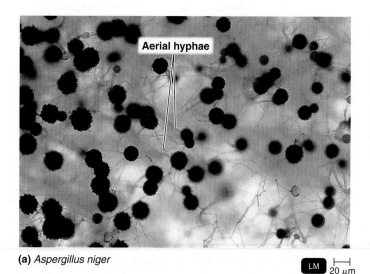

(a) *Aspergillus niger*

LM ⊢—⊣ 20 μm

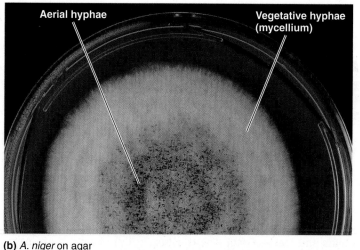

(b) *A. niger* on agar

**Figure 12.3 Aerial and vegetative hyphae. (a)** A photomicrograph of aerial hyphae, showing reproductive spores. **(b)** A colony of *Aspergillus niger* grown on a glucose agar plate, showing both vegetative and aerial hyphae.

**Q** How do fungal colonies differ from bacterial colonies?

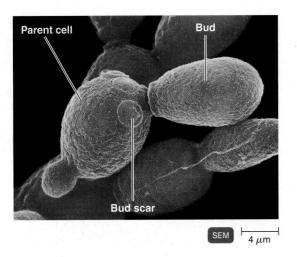

**Figure 12.4 A budding yeast.** A micrograph of *Saccharomyces cerevisiae* in various stages of budding.

**Q** How does a bud differ from a spore?

wine-making, and baking industries. *Saccharomyces* species produce ethanol in brewed beverages and carbon dioxide for leavening bread dough.

**Dimorphic Fungi** Some fungi, most notably the pathogenic species, exhibit **dimorphism**—two forms of growth. Such fungi can grow either as a mold or as a yeast. The moldlike forms produce vegetative and aerial hyphae; the yeastlike forms reproduce by budding. Dimorphism in pathogenic fungi is temperature-dependent: at 37°C, the fungus is yeastlike, and at 25°C, it is moldlike. (See Figure 24.15, page 712.) However, the appearance of the dimorphic (in this instance, nonpathogenic) fungus shown in **Figure 12.5** changes with $CO_2$ concentration.

## Life Cycle

Filamentous fungi can reproduce asexually by fragmentation of their hyphae. In addition, both sexual and asexual reproduction in fungi occurs by the formation of **spores**. In fact, fungi are usually identified by spore type.

Fungal spores, however, are quite different from bacterial endospores. Bacterial endospores allow a bacterial cell to survive adverse environmental conditions (see pages 92–94). A single vegetative bacterial cell forms one endospore, which eventually germinates to produce a single vegetative bacterial cell. This process isn't reproduction because it doesn't increase the total number of bacterial cells. But after a mold forms a spore, the spore detaches from the parent and germinates into a new mold (see Figure 12.2c). Unlike the bacterial endospore, this is a true reproductive spore; a second organism grows from the spore. Although fungal spores can survive for extended periods in dry or hot environments, most do not exhibit the extreme tolerance and longevity of bacterial endospores.

Spores are formed from aerial hyphae in a number of different ways, depending on the species. Fungal spores can be either asexual or sexual. **Asexual spores** are formed by the hyphae of one organism. When these spores germinate, they become organisms that are genetically identical to the parent. **Sexual spores** result from the fusion of nuclei from two opposite mating strains of the same species of fungus. Sexual spores require two different mating strains and so are made less frequently than asexual spores. Organisms that grow from sexual spores will have genetic characteristics of both parental strains. Because spores are of considerable importance in identifying fungi, next we'll look at some of the various types of asexual and sexual spores.

**Asexual Spores** Asexual spores are produced by an individual fungus through mitosis and subsequent cell division; there is no fusion of the nuclei of cells. Two types of asexual spores are produced by fungi. One type is a **conidiospore**, or **conidium** (plural: *conidia*), a unicellular or multicellular spore that is not enclosed in a sac (**Figure 12.6a**). Conidia are produced in a chain at the end of a **conidiophore**. Such spores are produced by *Penicillium* and *Aspergillus* (as-per-JIL-lus). Conidia formed by the fragmentation of a septate hypha into single, slightly thickened cells are called **arthroconidia** (Figure 12.6b). One species that produces such spores is *Coccidioides immitis* (KOK-sid-ē-oi-dēz IM-mi-tis) (see Figure 24.17, page 613). Another type of conidium, **blastoconidia**, are formed from the buds of its parent cell (Figure 12.6c). Such spores are found in some yeasts, such as *Candida albicans* and *Cryptococcus*. A **chlamydoconidium** is a thick-walled spore formed by rounding and enlargement within a hyphal segment (Figure 12.6d). A fungus that produces chlamydoconidia is the yeast *Candida albicans*.

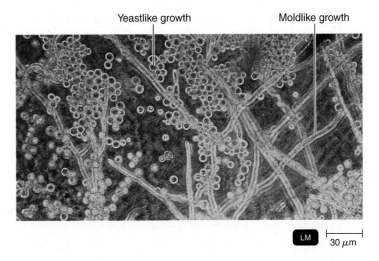

**Figure 12.5 Fungal dimorphism.** Dimorphism in the fungus *Mucor indicus* depends on $CO_2$ concentration. On the agar surface, *Mucor* exhibits yeastlike growth, but in the agar where $CO_2$ from metabolism has accumulated, it is moldlike.

**Q** What is fungal dimorphism?

**(a)** Conidia are arranged in chains at the end of an *Aspergillus niger* conidiophore.  12 μm

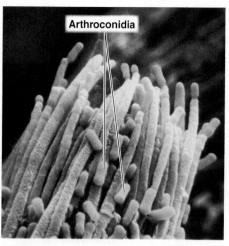

**(b)** Fragmentation of hyphae results in the formation of arthroconidia in *Ceratocystis ulmi*. SEM 2.5 μm

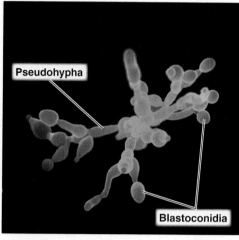

**(c)** Blastoconidia are formed from the buds of a parent cell of *Candida albicans*. SEM 13 μm

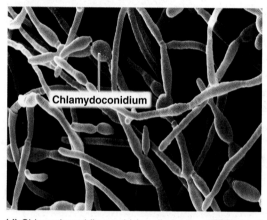

**(d)** Chlamydoconidia are thick-walled cells within hyphae of this *Candida albicans*. SEM 5 μm

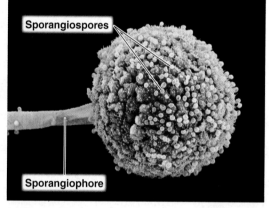

**(e)** Sporangiospores are formed within a sporangium of *Rhizopus stolonifer*. SEM 5 μm

**Figure 12.6 Representative asexual spores.**

 What are the green powdery structures on moldy food?

The other type of asexual spore is a **sporangiospore**, formed within a **sporangium**, or sac, at the end of an aerial hypha called a **sporangiophore**. The sporangium can contain hundreds of sporangiospores (Figure 12.6e). Such spores are produced by *Rhizopus*.

**Sexual Spores** A fungal sexual spore results from sexual reproduction, which consists of three phases:

1. **Plasmogamy**. A haploid nucleus of a donor cell (+) penetrates the cytoplasm of a recipient cell (−).

2. **Karyogamy**. The (+) and (−) nuclei fuse to form a diploid zygote nucleus.

3. **Meiosis**. The diploid nucleus gives rise to haploid nuclei (sexual spores), some of which may be genetic recombinants.

The sexual spores produced by fungi characterize the phyla. Clinical identification is based on microscopic examination of asexual spores, because most fungi exhibit only asexual spores in laboratory settings.

**Nutritional Adaptations**

Fungi are generally adapted to environments that would be hostile to bacteria. Fungi are chemoheterotrophs, and, like bacteria, they absorb nutrients rather than ingesting them as animals do. However, fungi differ from bacteria in certain environmental requirements and in the following nutritional characteristics:

- Fungi usually grow better in an environment with a pH of about 5, which is too acidic for the growth of most common bacteria.

- Almost all molds are aerobic. Most yeasts are facultative anaerobes.

- Most fungi are more resistant to osmotic pressure than bacteria; most can therefore grow in relatively high sugar or salt concentrations.

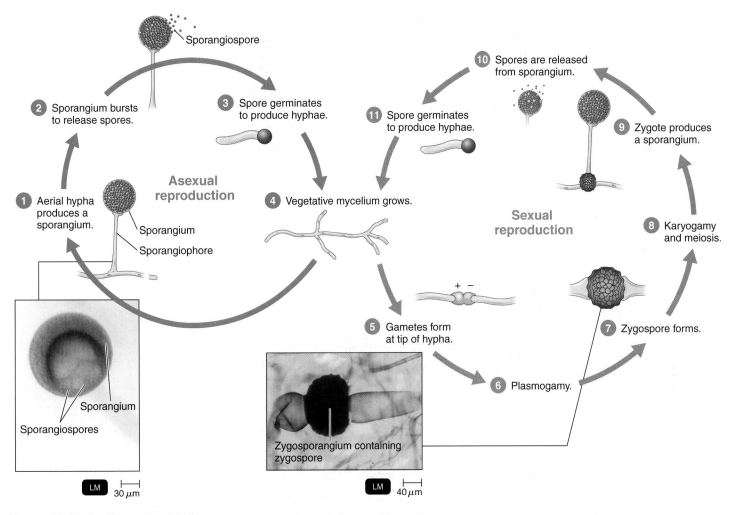

**Figure 12.7  The life cycle of *Rhizopus*, a zygomycete.** This fungus will reproduce asexually most of the time. Two opposite mating strains (designated + and −) are necessary for sexual reproduction.

**Q** What is an opportunistic mycosis?

- Fungi can grow on substances with a very low moisture content, generally too low to support the growth of bacteria.
- Fungi require somewhat less nitrogen than bacteria for an equivalent amount of growth.
- Fungi are often capable of metabolizing complex carbohydrates, such as lignin (a component of wood), that most bacteria can't use for nutrients.

These characteristics enable fungi to grow on such unlikely substrates as bathroom walls, shoe leather, and discarded newspapers.

**CHECK YOUR UNDERSTANDING**

✔ **12-1** Assume you isolated a single-celled organism that has a cell wall. How would you determine that it is a fungus and not a bacterium?

✔ **12-2** Contrast the mechanism of sexual and asexual spore formation.

## Medically Important Fungi

This section provides an overview of medically important phyla of fungi. The actual diseases they cause will be studied in Chapters 21 through 26. Note that only a relatively small number of fungi cause disease and some fungi protect us against infection (see Exploring the Microbiome, page 335).

The genera named in the following phyla include many that are readily found as contaminants in foods and in laboratory bacterial cultures. Although these genera are not all of primary medical importance, they are typical examples of their respective groups.

### Zygomycota

The Zygomycota, or conjugation fungi, are saprophytic molds that have coenocytic hyphae. An example is *Rhizopus stolonifer*, the common black bread mold. The asexual spores of *Rhizopus* are sporangiospores (**Figure 12.7**). The dark sporangiospores inside the sporangium give *Rhizopus* its descriptive common

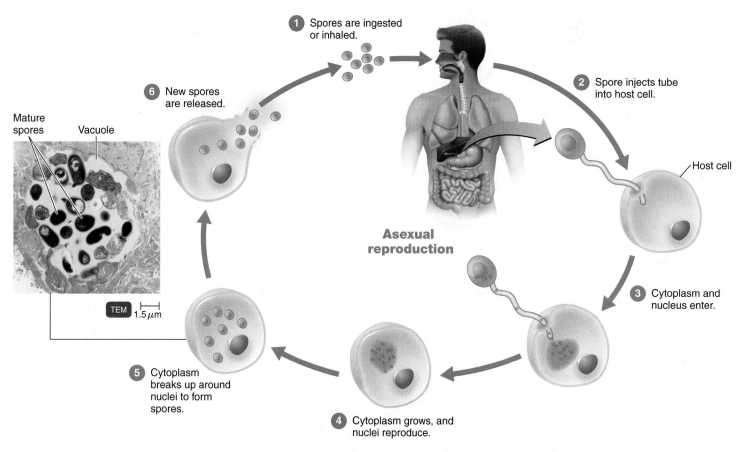

**Figure 12.8 The life cycle of _Encephalitozoon_, a microsporidian.** Microsporidiosis is an emerging opportunistic infection in immunocompromised patients and the elderly. _E. intestinalis_ causes diarrhea. Sexual reproduction has not been observed.

**Q** Why were microsporidia so difficult to classify?

name. When the sporangium breaks open, the sporangiospores are dispersed. If they fall on a suitable medium, they will germinate into a new mold thallus.

The sexual spores are zygospores. A **zygospore** is a large spore enclosed in a thick wall (Figure 12.7, step 7). This type of spore forms when the nuclei of two cells that are morphologically similar to each other fuse.

### Microsporidia

**Microsporidia** are unusual eukaryotes because they lack mitochondria. Microsporidia don't have microtubules (see Chapter 4, page 98), and they're obligate intracellular parasites. In 1857, when they were discovered, microsporidians were classified as fungi. They were reclassified as protists in 1983 because they lack mitochondria. Recent genome sequencing, however, reveals that the microsporidians are fungi. Sexual reproduction has not been observed but probably occurs within the host (**Figure 12.8**). Microsporidia have been reported to be the cause of a number of human diseases, including chronic diarrhea and keratoconjunctivitis

(inflammation of the conjunctiva near the cornea), most notably in AIDS patients.

### Ascomycota

The Ascomycota, or sac fungi, include molds with septate hyphae and some yeasts. Their asexual spores are usually conidia produced in long chains from the conidiophore. The term _conidia_ means dust, and these spores freely detach from the chain at the slightest disturbance and float in the air like dust.

An **ascospore** forms when the nuclei of two cells that can be either morphologically similar or dissimilar fuse. These spores are produced in a saclike structure called an **ascus** (**Figure 12.9**, lower right). The members of this phylum are called sac fungi because of the ascus.

### Basidiomycota

The Basidiomycota, or club fungi, also possess septate hyphae. This phylum includes fungi that produce mushrooms. **Basidiospores** are formed externally on a base pedestal called a

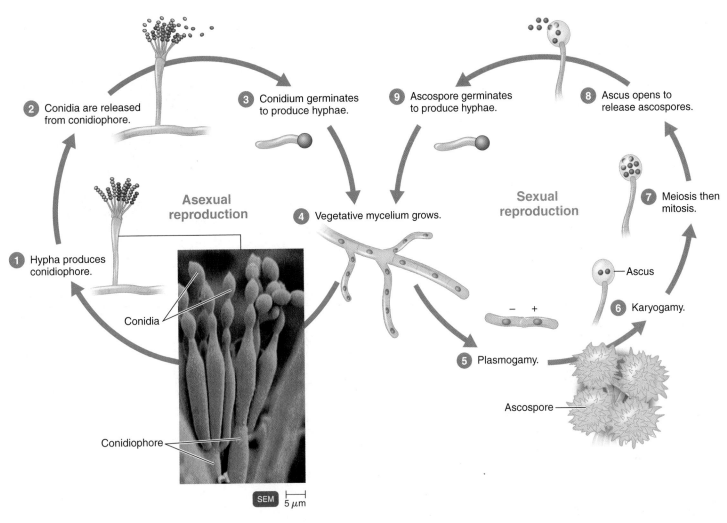

**2** Conidia are released from conidiophore.

**3** Conidium germinates to produce hyphae.

**9** Ascospore germinates to produce hyphae.

**8** Ascus opens to release ascospores.

**Asexual reproduction**

**1** Hypha produces conidiophore.

Conidia

**4** Vegetative mycelium grows.

**Sexual reproduction**

**7** Meiosis then mitosis.

Ascus

Conidiophore

− +

**6** Karyogamy.

**5** Plasmogamy.

Ascospore

SEM  5 µm

**Figure 12.9  The life cycle of *Talaromyces*, an ascomycete.** Occasionally, when two opposite mating cells from two different strains (+ and −) fuse, sexual reproduction occurs.

**Q** Name one ascomycete that can infect humans.

**basidium** (Figure 12.10). (The common name of the fungus is derived from the club shape of the basidium.) There are usually four basidiospores per basidium. Some of the basidiomycota produce asexual conidiospores.

\* \* \*

The fungi we have looked at thus far are **teleomorphs**; that is, they produce both sexual and asexual spores. Some ascomycetes have lost the ability to reproduce sexually. These asexual fungi are called **anamorphs**. *Penicillium* is an example of an anamorph that arose from a mutation in a teleomorph. Historically, fungi whose sexual cycle had not been observed were put in a "holding category" called *Deuteromycota*. Now, mycologists are using rRNA sequencing to classify these organisms. Most of these previously unclassified deuteromycetes are anamorphic phases of Ascomycota, and a few are basidiomycetes.

Table 12.2 lists some fungi that cause human diseases. Both teleomorphic and anamorphic names are given for some of the fungi because some medically important fungi are known by their anamorph, or asexual, names.

## Fungal Diseases

Any fungal infection is called a **mycosis**. Mycoses are generally chronic (long-lasting) infections because fungi grow slowly. Mycoses are classified into five groups according to the degree of tissue involvement and mode of entry into the host: systemic, subcutaneous, cutaneous, superficial, or opportunistic. Fungi are related to animals (as we saw in Chapter 10). Consequently, drugs that affect fungal cells may also affect animal cells, making fungal infections of humans and other animals often difficult to treat. Some fungi cause disease by producing toxins, and those are discussed in Chapter 15.

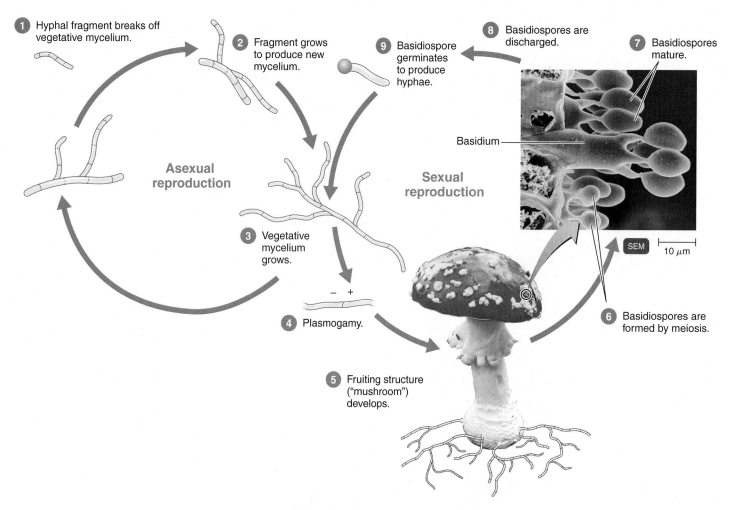

1 Hyphal fragment breaks off vegetative mycelium.

2 Fragment grows to produce new mycelium.

3 Vegetative mycelium grows.

4 Plasmogamy.

5 Fruiting structure ("mushroom") develops.

6 Basidiospores are formed by meiosis.

7 Basidiospores mature.

8 Basidiospores are discharged.

9 Basidiospore germinates to produce hyphae.

**Asexual reproduction**

**Sexual reproduction**

Basidium

SEM    10 μm

**Figure 12.10 A generalized life cycle of a basidiomycete.** Mushrooms appear after cells from two mating strains (+ and −) have fused.

**Q** On what basis are fungi classified into phyla?

Five days later, Ethan's doctor sends him to the hospital. For the past week Ethan has had shortness of breath, fever, chills, headache, night sweats, loss of appetite, nausea, and muscle pain. He has no other symptoms and is treated with amoxicillin for a presumed lower respiratory tract infection. Three days later, Ethan's condition deteriorates; his respiratory rate increases, but no other systemic symptoms are apparent. Based on Ethan's respiratory symptoms, his doctor orders a chest X-ray exam. The X-ray film shows that Ethan has a mass in his lungs.

**Both Ethan and his dog seem to have acquired the same infection. Make a short list of possible pathogens based on this new information.**

**Systemic mycoses** are fungal infections deep within the body. They are not restricted to any particular region of the body but can affect a number of tissues and organs. Systemic mycoses are usually caused by fungi that live in the soil. The spores are transmitted by inhalation; these infections typically begin in the lungs and then spread to other body tissues. They are not contagious from animal to human or from human to human. Two systemic mycoses, histoplasmosis and coccidioidomycosis, are discussed in Chapter 24.

**Subcutaneous mycoses** are fungal infections beneath the skin caused by saprophytic fungi that live in soil and on vegetation. Sporotrichosis is a subcutaneous infection acquired by gardeners and farmers (Chapter 21, page 608). Infection occurs by direct implantation of spores or mycelial fragments into a puncture wound in the skin.

Fungi that infect only the epidermis, hair, and nails are called **dermatophytes**, and their infections are called *dermatomycoses* or **cutaneous mycoses** (see Figure 21.16, page 608).

**TABLE 12.2** Characteristics of Some Pathogenic Fungi

| Phylum | Growth Characteristics | Asexual Spore Types | Human Pathogens | Habitat | Type of Mycosis | Page |
|---|---|---|---|---|---|---|
| **Zygomycota** | Nonseptate hyphae | Sporangiospores | *Rhizopus* | Ubiquitous | Systemic | 715 |
| | | | *Mucor* | Ubiquitous | Systemic | 715 |
| **Microsporidia** | No hyphae | Nonmotile spores | *Encephalitozoon*, *Nosema* | Humans, other animals | Diarrhea, keratoconjunctivitis | — — |
| **Ascomycota** | | Conidia | *Aspergillus* | Ubiquitous | Systemic | 715 |
| | | | *Claviceps purpurea* | Grasses | Toxin ingestion | 438 |
| | Dimorphic | | *Blastomyces** (or *Ajellomyces*†) *dermatitidis* | Unknown | Systemic | 714 |
| | | | *Histoplasma** (or *Ajellomyces*†) *capsulatum* | Soil | Systemic | 711 |
| | Septate hyphae, strong affinity for keratin | Conidia | *Microsporum* | Soil, animals | Cutaneous | 607 |
| | | Arthroconidia Chlamydoconidia | *Trichophyton** (or *Arthroderma*†) | Soil, animals | Cutaneous | 607 |
| Anamorphs | Septate hyphae Dimorphic | Conidia | *Epidermophyton* | Soil, humans | Cutaneous | 607 |
| | | | *Sporothrix schenckii,* | Soil | Subcutaneous | 608 |
| | | | *Stachybotrys* | Soil | | |
| | | Arthroconidia | *Coccidioides immitis* | Soil | Systemic | 712 |
| | Yeastlike, pseudohyphae | Chlamydoconidia | *Candida albicans* | Human normal microbiota | Cutaneous, systemic, mucocutaneous | 608, 779–780 |
| | Unicellular | None | *Pneumocystis* | Human lungs | Systemic | 713 |
| **Basidiomycota** | Septate hyphae; includes rusts and smuts, and plant pathogens; yeastlike encapsulated cells | Conidia | *Cryptococcus** (or *Filobasidiella*†) | Soil, bird feces | Systemic | 639 |
| | | | *Malassezia* | Human skin | Cutaneous | 592 |
| | | | *Amanita* spp. | Soil | Toxin ingestion | 439 |

*Anamorph name.
†Teleomorph name.

Dermatophytes secrete keratinase, an enzyme that degrades **keratin**, a protein found in hair, skin, and nails. Infection is transmitted from human to human or from animal to human by direct contact or by contact with infected hairs and epidermal cells (as from barber shop clippers or shower room floors).

The fungi that cause **superficial mycoses** are localized along hair shafts and in superficial (surface) epidermal cells. These infections are prevalent in tropical climates.

An **opportunistic pathogen** is generally harmless in its normal habitat but can become pathogenic in a host who is seriously debilitated or traumatized, who is under treatment with broad-spectrum antibiotics, whose immune system is suppressed by drugs or by an immune disorder, or who has

a lung disease. A number of fungi, including opportunistic pathogens, are found in and on the human body.

*Pneumocystis* is an opportunistic pathogen in individuals with compromised immune systems and is the most common life-threatening infection in AIDS patients (see Figure 24.19, page 714). It was first classified as a protozoan, but recent studies of its RNA reveal it's a unicellular anamorphic fungus. Another example of an opportunistic pathogen is the fungus *Stachybotrys* (STA-kē-bah-tris), which normally grows on cellulose found in dead plants but in recent years has been found growing on water-damaged walls of homes.

Mucormycosis is an opportunistic mycosis caused by *Rhizopus* and *Mucor* (MŪ-kor); the infection occurs mostly in patients who have diabetes mellitus, have leukemia, or

Ethan's physician, suspecting that Ethan has a fungal infection, orders a biopsy of the lung mass. **Figure A** and **Figure B** show the microscopic examination and a culture from the biopsied tissue.

**Figure A** Microscopic examination from lung mass.

**Figure B** Microscopic appearance of culture.

**Based on the figures, what is the most likely pathogen?**

325    332    **334**

---

are undergoing treatment with immunosuppressive drugs. Aspergillosis is also an opportunistic mycosis; it is caused by *Aspergillus* (see Figure 12.3). This disease occurs in people who have debilitating lung diseases or cancer and have inhaled *Aspergillus* spores. Opportunistic infections by *Cryptococcus* and *Penicillium* can cause fatal diseases in AIDS patients. These opportunistic fungi may be transmitted from one person to an uninfected person but do not usually infect immunocompetent people. **Yeast infection**, or candidiasis, is most frequently caused by *Candida albicans* and may occur as vulvovaginal candidiasis or thrush, a mucocutaneous candidiasis. Candidiasis frequently occurs in newborns, in people with AIDS, and in people being treated with broad-spectrum antibiotics (see Figure 21.17, page 609).

Some fungi cause disease by producing toxins. These toxins are discussed in Chapter 15, pages 438–439.

## Economic Effects of Fungi

Fungi have been used in biotechnology for many years. *Aspergillus niger* (NĪ-jer), for example, has been used to produce citric acid for foods and beverages since 1914. The yeast *Saccharomyces cerevisiae* is used to make bread and wine. It is also genetically modified to produce a variety of proteins, including hepatitis B vaccine. *Trichoderma* is used commercially to produce the enzyme cellulase, which is used to remove plant cell walls to produce a clear fruit juice. When the anticancer drug taxol, which is produced by yew trees, was discovered, there was concern that the yew forests of the U.S. Northwest coast would be decimated to harvest the drug. However, the fungus *Taxomyces* (TAKS-ō-mī-sēz) also produces taxol.

Fungi are used as biological controls of pests. In 1990, the fungus *Entomophaga* (en-tō-MŌ-fah-gah) unexpectedly proliferated and killed gypsy moths that were destroying trees in the eastern United States. Scientists are investigating the use of several fungi to kill pests:

- The fungus *Coniothyrium minitans* (kon'ē-ō-THER-ē-um MIN-i-tanz) feeds on fungi that destroy soybeans and other bean crops.
- A foam filled with *Paecilomyces fumosoroseus* (pī sil-ō-MĪ-sēz foo'mō-sō-RŌ-sē-us) is being used as a biological alternative to chemicals to kill termites hiding inside tree trunks and other hard-to-reach places.

In contrast to these beneficial effects, fungi can have undesirable effects for agriculture because of their nutritional adaptations. As most of us have observed, mold spoilage of fruits, grains, and vegetables is relatively common, but bacterial spoilage of such foods is not. There is little moisture on the unbroken surfaces of such foods, and the interiors of fruits are too acidic for many bacteria to grow there. Jams and jellies also tend to be acidic, and they have a high osmotic pressure from the sugars they contain. These factors all discourage bacterial growth but readily support the growth of molds.

*Cryptococcus gattii*, an emerging fungal infection in the United States, is a dimorphic fungus found in the soil. It grows as a yeast at 37°C and produces hyphae at 25°C. Based on the yeastlike appearance and presence of hyphae, the lab confirms the presence of *C. gattii* in Ethan's lung mass. Ethan regularly takes Waldo hiking in the Douglas fir forest of the Northwest, so it isn't possible to know exactly where or when they contracted their infections. Since the first reported case in 1999, over 200 cases have been reported in British Columbia alone. In the United States Pacific Northwest, 96 people and 100 companion animal cases have been confirmed since 2004.

Ethan is placed on intravenous therapy with the antifungal agents amphotericin B and flucytosine. After a 6-week hospital stay, Ethan and Waldo are back at home and almost ready to go hiking again.

325    332    **334**

# The Mycobiome

Just as viruses and bacteria are members of a healthy human microbiome, so are fungi. But unlike the bacterial portion of the microbiome, the so-called mycobiome is just beginning to be studied.

Members of the yeast genus *Candida* are the most common fungi that live as normal microbiota in the mouth, intestine, and vagina. The most diverse population of fungi is found in the mouth, where 101 species have been identified. In addition to *Candida*, the oral mycobiome includes *Saccharomyces, Cladosporium,* and *Pichia*. *Saccharomyces* and *Cladosporium* are common in the intestines, and *Malassezia* is the most abundant genus living on the skin.

Certain members of the mycobiome are well-known opportunistic pathogens. Yeast can proliferate and cause vaginal infections

after other members of the microbiome die off from a course of antimicrobial drugs. *Candida albicans* in particular causes invasive infections in organ transplant recipients and other immune-compromised individuals, such as people with AIDS, or cancer patients undergoing chemotherapy.

Understanding how fungi among the mycobiome interact with each other is important. For instance, a decrease in oral *Pichia* abundance coincides with an increase in *Candida* growth in AIDS patients—this observation led to the discovery that *Pichia* secretes several proteins that inhibit *Candida*. Likewise, noting how fungi and bacteria within the microbiome compete with each other is also a key area of research. The yeast *Saccharomyces boulardii,* for example, makes a protease (enzyme) that digests *Clostridium difficile* toxin. *Clostridium*

*difficile* is a tough-to-treat bacterium that causes diarrhea after antibiotics kill off other members of the intestinal microbiome.

**Micrograph of *Candida albicans* yeast**

---

The spreading chestnut tree, of which Longfellow wrote, no longer grows in the United States except in a few widely isolated locations; a fungal blight killed virtually all of them. This blight was caused by the ascomycete *Cryphonectria parasitica* (kri-fō-NEK-trē-ah par-ah-SI-ti-kah), which was introduced from China around 1904. The fungus allows the tree roots to live and put forth shoots regularly, but then it kills the shoots just as regularly. *Cryphonectria*-resistant chestnuts are being developed. Another imported fungal plant disease is Dutch elm disease, caused by *Ceratocystis ulmi* (ser-AH-tō-sis-tis UL-mē). Carried from tree to tree by a bark beetle, the fungus blocks the afflicted tree's circulation. The disease has devastated the American elm population.

### CHECK YOUR UNDERSTANDING

✔ **12-3** List the asexual and sexual spores made by zygomycetes, ascomycetes, and basidiomycetes.

✔ **12-4** Why are microsporidia classified as fungi?

✔ **12-4** Are yeasts beneficial or harmful?

## Lichens

### LEARNING OBJECTIVES

**12-5** List the distinguishing characteristics of lichens, and describe their nutritional needs.

**12-6** Describe the roles of the fungus and the alga in a lichen.

A **lichen** is a combination of a green alga (or a cyanobacterium) and a fungus. Lichens are placed in the Kingdom Fungi and are classified according to the fungal partner, most often an ascomycete. The two organisms exist in a *mutualistic* relationship, in which each partner benefits. The lichen is very different from either the alga or fungus growing alone, and if the partners are separated, the lichen no longer exists. Approximately 13,500 species of lichens occupy quite diverse habitats. Because they can inhabit areas in which neither fungi nor algae could survive alone, lichens are often the first life forms to colonize newly exposed soil or rock. Lichens secrete organic acids that chemically weather rock, and they accumulate nutrients needed for plant growth. Also found on trees, concrete structures, and

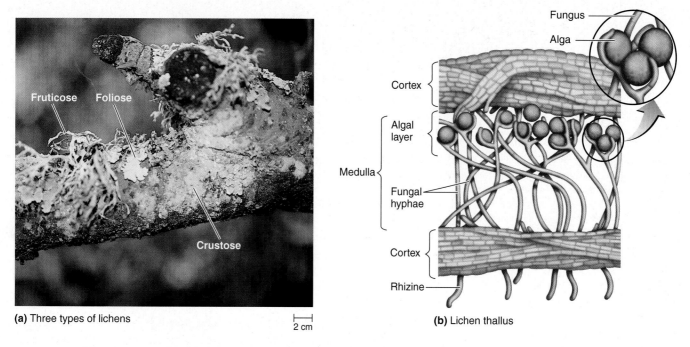

**(a)** Three types of lichens

2 cm

**(b)** Lichen thallus

**Figure 12.11 Lichens.** The lichen medulla is composed of fungal hyphae surrounding the algal layer. The protective cortex is a layer of fungal hyphae that covers the surface and sometimes the bottom of the lichen.

**Q** In what ways are lichens unique?

rooftops, lichens are some of the slowest-growing organisms on Earth.

Lichens can be grouped into three morphologic categories (**Figure 12.11a**). *Crustose lichens* grow flush or encrusted onto the substratum, *foliose lichens* are more leaflike, and *fruticose lichens* have fingerlike projections. The lichen's thallus, or body, forms when fungal hyphae grow around algal cells to become the **medulla** (Figure 12.11b). Fungal hyphae project below the lichen body to form **rhizines**, or holdfasts. Fungal hyphae also form a **cortex**, or protective covering, over the algal layer and sometimes under it as well. After incorporation into a lichen thallus, the alga continues to grow, and the growing hyphae can incorporate new algal cells.

When the algal partner is cultured separately in vitro, about 1% of the carbohydrates produced during photosynthesis are released into the culture medium; however, when the alga is associated with a fungus, the algal plasma membrane is more permeable, and up to 60% of the products of photosynthesis are released to the fungus or are found as end-products of fungal metabolism. The fungus clearly benefits from this association. The alga, while giving up valuable nutrients, is in turn compensated; it receives from the fungus both attachment (rhizines) and protection from desiccation (cortex).

Lichens had considerable economic importance in ancient Greece and other parts of Europe as dyes for clothing. Usnic acid from *Usnea* is used as an antimicrobial agent in China.

Erythrolitmin, the dye used in litmus paper to indicate changes in pH, is extracted from a variety of lichens. Some lichens or their acids can cause allergic contact dermatitis in humans.

Populations of lichens readily incorporate cations (positively charged ions) into their thalli. Therefore, the concentrations and types of cations in the atmosphere can be determined by chemical analyses of lichen thalli. In addition, the presence or absence of species that are quite sensitive to pollutants can be used to ascertain air quality. A 1985 study in the Cuyahoga Valley in Ohio revealed that 81% of the 172 lichen species that were present in 1917 were gone. Because this area is severely affected by air pollution, the inference is that air pollutants, primarily sulfur dioxide (the major contributor to acid precipitation), caused the death of sensitive species.

Lichens are the major food for tundra herbivores such as caribou and reindeer. After the 1986 Chernobyl nuclear disaster, 70,000 reindeer in Lapland that had been raised for food had to be destroyed because of high levels of radiation. The lichens on which the reindeer fed had absorbed radioactive cesium-137, which had spread in the air.

**CHECK YOUR UNDERSTANDING**

✔ **12-5** What is the role of lichens in nature?

✔ **12-6** What is the role of the fungus in a lichen?

# Algae

## LEARNING OBJECTIVES

**12-7** List the defining characteristics of algae.

**12-8** List the outstanding characteristics of the five phyla of algae discussed in this chapter.

**12-9** Identify two beneficial and two harmful effects of algae.

Algae are familiar as the large brown kelp in coastal waters, the green scum in a puddle, and the green stains on soil or on rocks. A few algae are responsible for food poisonings. Some algae are unicellular; others form chains of cells (are filamentous); and a few have thalli.

"Algae" is not a taxonomic group; it is a way to describe photoautotrophs that lack the roots and stems of plants. Historically they were considered plants, but they lack the embryos of true plants. The classification of algae is currently in flux as DNA and structural data are collected. Algae are currently grouped into super clades—see Table 12.3. Algae are mostly aquatic, although some are found in soil or on trees when sufficient moisture is available there. Unusual algal habitats include the hair of both the sedentary South American sloth and the polar bear. Water is necessary for physical support, reproduction, and the diffusion of nutrients. Generally, algae are found in cool temperate waters, although the large floating mats of the brown alga *Sargassum* (sar-GAS-sum) are found in the subtropical Sargasso Sea, and some species of brown algae grow in antarctic waters.

## Characteristics of Algae

Algae are relatively simple eukaryotic photoautotrophs that lack the tissues (roots, stem, and leaves) of plants. The identification of unicellular and filamentous algae requires microscopic examination. Most algae are found in the ocean. Their locations depend on the availability of appropriate nutrients, wavelengths of light, and surfaces on which they can grow. Probable locations for representative algae are shown in **Figure 12.12a**.

### Vegetative Structures

The body of a multicellular alga is called a thallus. Thalli of the larger multicellular algae, those commonly called seaweeds, consist of branched **holdfasts** (which anchor the alga to a rock), stemlike and often hollow **stipes**, and leaflike **blades** (Figure 12.12b). The cells covering the thallus can carry out photosynthesis. The thallus lacks the conductive tissue (xylem and phloem) characteristic of vascular plants; algae absorb nutrients from the water over their entire surface. The stipe is not lignified or woody, so it does not offer the support of a plant's stem; instead, the surrounding water supports the algal thallus; some algae are also buoyed by a floating, gas-filled bladder called a *pneumatocyst*.

### Life Cycle

All algae can reproduce asexually. Multicellular algae with thalli and filamentous forms can fragment; each piece is capable of forming a new thallus or filament. When a unicellular alga divides, its nucleus divides (mitosis), and the two nuclei move to opposite parts of the cell. The cell then divides into two complete cells (cytokinesis).

## TABLE 12.3 Characteristics of Selected Algae

| | Brown Algae | Diatoms | Dinoflagellates | Water Molds | Red Algae | Green Algae |
|---|---|---|---|---|---|---|
| Super Clade | SAR* | SAR | SAR | SAR | Archaeplastida | Archaeplastida |
| Phylum | Phaeophyta | Bacillariophyta | Dinoflagellata | Oomycota | Rhodophyta | Chlorophyta |
| Color | Brownish | Brownish | Brownish | Colorless, white | Reddish | Green |
| Cell Wall | Cellulose and alginic acid | Pectin and silica | Cellulose in membrane | Cellulose | Cellulose | Cellulose |
| Cell Arrangement | Multicellular | Unicellular | Unicellular | Multicellular | Multicellular (most) | Unicellular and multicellular |
| Photosynthetic Pigments | Chlorophyll *a* and *c*, xanthophylls | Chlorophyll *a* and *c*, carotene, xanthophylls | Chlorophyll *a* and *c*, carotene, xanthins | None | Chlorophyll *a* and *d*, phycobiliproteins | Chlorophyll *a* and *b* |
| Sexual Reproduction | Yes | Yes | In a few | Yes (similar to Zygomycota) | Yes | Yes |
| Storage Material | Carbohydrate | Oil | Starch | None | Glucose polymer | Starch |
| Pathogenicity | None | Toxins | Toxins | Parasitic | A few produce toxins | None |

*SAR stands for Stramenopiles, Alveolates, Rhizaria.

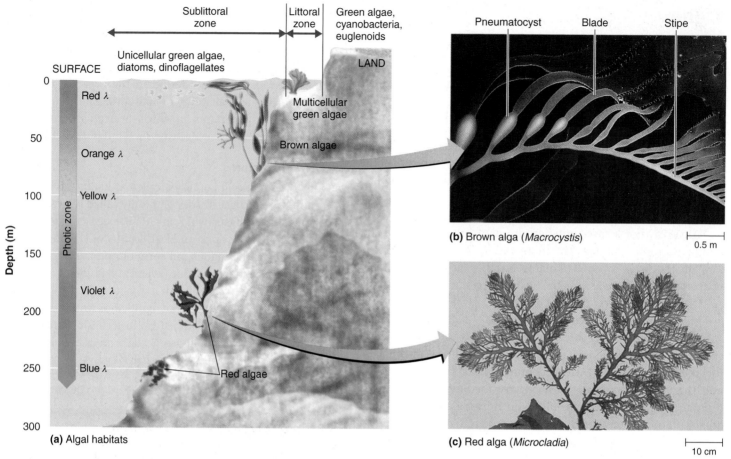

**(a) Algal habitats**

**(b) Brown alga (*Macrocystis*)** ⊢ 0.5 m ⊣

**(c) Red alga (*Microcladia*)** ⊢ 10 cm ⊣

**Figure 12.12 Algae and their habitats. (a)** Although unicellular and filamentous algae can be found on land, they frequently exist in marine and freshwater environments as plankton. Multicellular green, brown, and red algae require a suitable attachment site, adequate water for support, and light of the appropriate wavelengths. **(b)** *Macrocystis porifera,* a brown alga. The hollow stipe and gas-filled pneumatocysts hold the thallus upright, ensuring that sufficient sunlight is received for growth. **(c)** *Microcladia,* a red alga. The delicately branched red algae get their color from phycobiliprotein accessory pigments.

**Q** What red alga is toxic for humans?

Algae can also reproduce sexually (**Figure 12.13**). In some species, asexual reproduction may occur for several generations and then, under different conditions, the same species reproduce sexually. Other species alternate generations so that the offspring resulting from sexual reproduction reproduce asexually, and the next generation then reproduces sexually.

**Nutrition**

*Algae* is a common name that includes several phyla (Table 12.3). Most algae are photosynthetic; however, the oomycotes, or fungal-like algae, are chemoheterotrophs. Photosynthetic algae are found throughout the photic (light) zone of bodies of water. Chlorophyll *a* (a light-trapping pigment) and accessory pigments involved in photosynthesis are responsible for the distinctive colors of many algae.

Algae are classified according to their rRNA sequences, structures, pigments, and other qualities (see Table 12.3). Following are descriptions of some phyla of algae.

## Selected Phyla of Algae

The *brown algae,* or kelp, are macroscopic; some reach lengths of 50 m (see Figure 12.12b). Most brown algae are found in coastal waters. Brown algae have a phenomenal growth rate. Some grow at rates exceeding 20 cm per day and therefore can be harvested regularly. **Algin**, a thickener used in many foods (such as ice cream and cake decorations), is extracted from their cell walls. Algin is also used in the production of a wide variety of nonfood goods, including rubber tires and hand lotion. The brown alga *Laminaria japonica* is used to induce vaginal dilation before surgical entry into the uterus through the vagina.

Most *red algae* have delicately branched thalli and can live at greater ocean depths than other algae (see Figure 12.12c). The thalli of a few red algae form crustlike coatings on rocks and shells. The red pigments enable red algae to absorb the blue light that penetrates deepest into the ocean. The agar used in microbiological media is extracted from many red algae. Another gelatinous material, carrageenan, comes from a

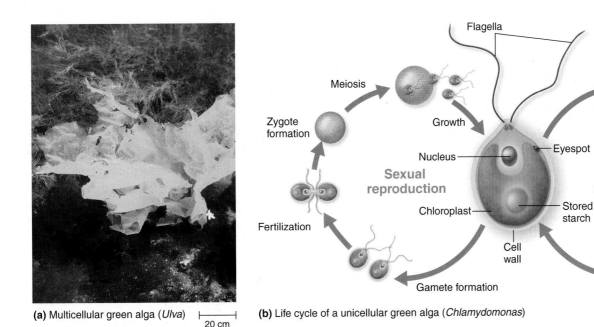

(a) Multicellular green alga (*Ulva*) |——————| 20 cm

(b) Life cycle of a unicellular green alga (*Chlamydomonas*)

**Figure 12.13 Green algae.** (a) The multicellular green alga *Ulva*. (b) The life cycle of the unicellular green alga *Chlamydomonas*. Two whiplike flagella propel this cell.

**Q** What is the primary role of algae in the ecosystem?

species of red algae commonly called Irish moss. Carrageenan and agar can be a thickening ingredient in evaporated milk, ice cream, and pharmaceutical agents. *Gracilaria* species, which grow in the Pacific Ocean, are used by humans for food. However, some members of this genus can produce a lethal toxin.

*Green algae* have cellulose cell walls, contain chlorophyll *a* and *b*, and store starch, as plants do (see Figure 12.13a). Green algae are believed to have given rise to terrestrial plants. Most green algae are microscopic, although they may be either unicellular or multicellular. Some filamentous kinds form grass-green scum in ponds.

Diatoms, dinoflagellates, and water molds are grouped into the Kingdom Stramenopila. *Diatoms* (**Figure 12.14**) are unicellular or filamentous algae with complex cell walls that consist of pectin and a layer of silica. The two parts of the wall fit together like the halves of a Petri dish. The distinctive patterns of the walls are a useful tool in diatom identification. Diatoms store energy captured through photosynthesis in the form of oil.

The first reported outbreak of a neurological disease caused by diatoms was reported in 1987 in Canada. Affected people ate mussels that had been feeding on diatoms. The diatoms produced *domoic acid*, a toxin that was concentrated in the mussels. Symptoms occurred within 24 hours of eating and included diarrhea and memory loss. The fatality rate was less than 4%. Subsequent outbreaks showed that in some cases, the brain damage may be permanent. Since 1991, hundreds of marine birds and sea lions have died from the same **domoic acid toxicosis** in California.

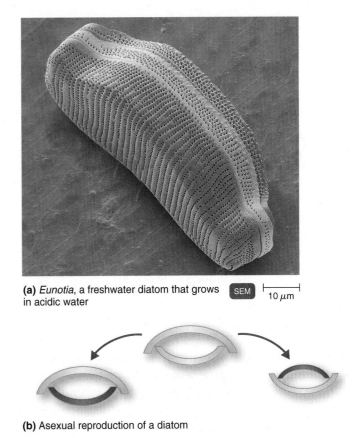

(a) *Eunotia*, a freshwater diatom that grows in acidic water [SEM] |——| 10 μm

(b) Asexual reproduction of a diatom

**Figure 12.14 Diatoms.** (a) In this micrograph of *Eunotia serra*, notice how the two parts of the cell wall fit together. (b) Asexual reproduction in a diatom. During mitosis, each daughter cell retains one-half of the cell wall from the parent (yellow) and must synthesize the remaining half (pink).

**Q** What human disease is caused by diatoms?

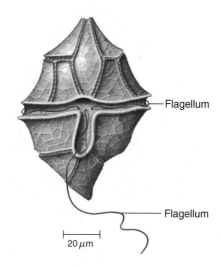

Figure 12.15 *Peridinium*, a dinoflagellate. Like all dinoflagellates, *Peridinium* has two flagella in perpendicular, opposing grooves. When the two flagella beat simultaneously, they cause the cell to spin.

**Q** What human diseases are caused by dinoflagellates?

*Dinoflagellates* are unicellular algae collectively called **plankton**, or free-floating organisms (**Figure 12.15**). Their rigid structure is due to cellulose embedded in the plasma membrane. Some dinoflagellates produce neurotoxins. In the last 20 years, a worldwide increase in toxic marine algae has killed millions of fish, hundreds of marine mammals, and even some humans. When fish swim through large numbers of the dinoflagellate *Karenia brevis* (kar-en-Ē-ah BREV-is), the algae trapped in the gills of the fish release a neurotoxin that stops the fish from breathing. Dinoflagellates in the genus *Alexandrium* (a'leks-AN-drē-um) produce neurotoxins (called **saxitoxins**) that cause **paralytic shellfish poisoning (PSP)**. The toxin is concentrated when large numbers of dinoflagellates are eaten by mollusks, such as mussels or clams. Humans who eat these mollusks develop PSP. Large concentrations of *Alexandrium* give the ocean a deep red color, from which the name **red tide** originates (Figure 27.10, page 798). Mollusks should not be harvested for consumption during a red tide. A disease called **ciguatera** occurs when the dinoflagellate *Gambierdiscus toxicus* (GAM-bē-er-dis-kus TOKS-i-kus) passes up the food chain and is concentrated in large fish. Ciguatera is endemic (constantly present) in the south Pacific Ocean and the Caribbean Sea. An emerging disease associated with heterotrophic *Pfiesteria* (fē-STEER-ē-ah) is responsible for periodic massive fish deaths along the Atlantic Coast.

Most *water molds*, or *Oomycota*, are decomposers. They form the cottony masses on dead algae and animals, usually in fresh water (**Figure 12.16**). Asexually, the oomycotes resemble the zygomycete fungi in that they produce spores in a sporangium (spore sac). However, oomycote spores, called **zoospores** (Figure 12.16, top right), have two flagella; fungi do not have flagella. Because of their superficial similarity to fungi, oomycotes were previously classified with the fungi. Their cellulose cell walls always raised the question about their relationship to algae, and recent DNA analyses have confirmed that oomycotes are more closely related to diatoms and dinoflagellates than to fungi. Many of the terrestrial oomycotes are plant parasites. The USDA inspects imported plants for white rust and other parasites. Often travelers, or even commercial plant importers, do not realize that one little blossom or seedling could carry a pest that is capable of causing millions of dollars' worth of damage to U.S. agriculture.

In Ireland during the mid-1800s, 1 million people died when the country's potato crop failed. The alga that caused the great potato blight, *Phytophthora infestans* (fī-TOF-thor-ah in-FES-tans), was one of the first microorganisms to be associated with a disease. Today, *Phytophthora* infects soybeans, potatoes, and cocoa worldwide. Vegetative hyphae produce motile zoospores as well as specialized sex hyphae (see Figure 12.16). All of the U.S. strains were one mating type ("sex") called A1. In the 1990s, the other mating type, A2, was identified in the United States. When in close proximity, A1 and A2 will differentiate to produce haploid gametes that can mate to form a zygote. When the zygote germinates, the resulting alga will have genes from both parents.

In Australia, *P. cinnamoni* has infected about 20% of one species of *Eucalyptus* tree. *Phytophthora* was introduced into the United States in the 1990s and caused widespread damage to fruit and vegetable crops. When California oak trees suddenly started dying in 1995, University of California scientists identified the cause of this "sudden oak death" to be a new species, *P. ramorum*. *P. ramorum* also infects redwood trees.

## Roles of Algae in Nature

Algae are an important part of any aquatic food chain because they fix carbon dioxide into organic molecules that can be consumed by chemoheterotrophs. Using the energy produced in photophosphorylation, algae convert carbon dioxide in the atmosphere into carbohydrates. Molecular oxygen ($O_2$) is a by-product of their photosynthesis. The top few meters of any body of water contain planktonic algae. As 75% of the Earth is covered with water, it is estimated that 80% of the Earth's $O_2$ is produced by planktonic algae.

Seasonal changes in nutrients, light, and temperature cause fluctuations in algal populations; periodic increases in numbers of planktonic algae are called **algal blooms**. Blooms of dinoflagellates are responsible for seasonal red tides. Blooms of a certain few species indicate that the water in which they grow is polluted because these algae thrive in high concentrations of organic materials that exist in sewage or industrial wastes. When algae die, the decomposition of the large numbers of cells associated with an algal bloom depletes the level of dissolved oxygen in the water. (This phenomenon is discussed in Chapter 27.)

Much of the world's petroleum was formed from diatoms and other planktonic organisms that lived several million years

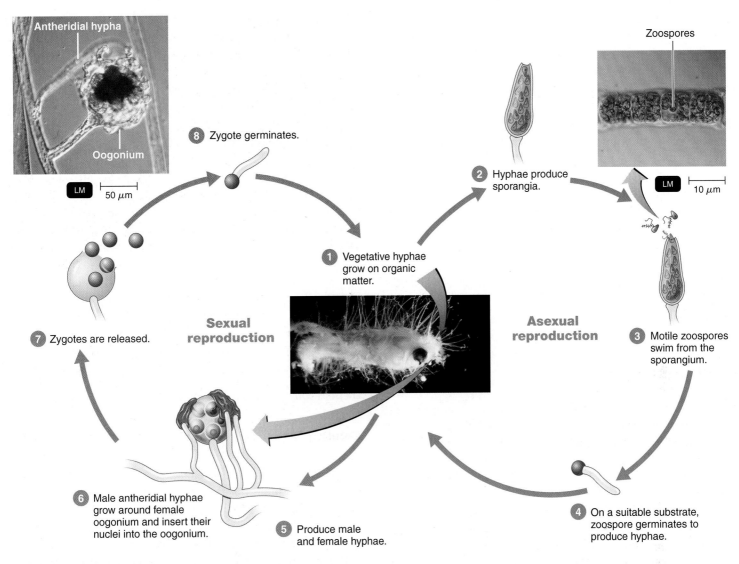

**Figure 12.16 Oomycotes.** These fungus-like algae are common decomposers in freshwater. A few cause diseases of fish and terrestrial plants. Note the fuzzy mycelium of *Saprolegnia ferax* on the fish.

**Q** Is this oomycote more closely related to *Penicillium* or to diatoms?

ago. When such organisms died and were buried by sediments, the organic molecules they contained did not decompose to be returned to the carbon cycle as $CO_2$. Heat and pressure resulting from the Earth's geologic movements altered the oil stored in the cells, as well as the cell membranes. Oxygen and other elements were eliminated, leaving a residue of hydrocarbons in the form of petroleum and natural gas deposits.

Many unicellular algae are symbionts in animals. The giant clam *Tridacna* (trī-DAK-nah) has evolved special organs that host dinoflagellates. As the clam sits in shallow water, the algae proliferate in these organs when they are exposed to the sun. The algae release glycerol into the clam's bloodstream, thus supplying the clam's carbohydrate requirement. In addition, evidence suggests that the clam gets essential proteins by phagocytizing old algae.

**CHECK YOUR UNDERSTANDING**

✔ **12-7** How do algae differ from bacteria? From fungi?

✔ **12-8, 12-9** List the cell wall composition and diseases caused by the following algae: diatoms, dinoflagellates, oomycotes.

# Protozoa

**LEARNING OBJECTIVES**

**12-10** List the defining characteristics of protozoa.

**12-11** Describe the outstanding characteristics of the seven phyla of protozoa discussed in this chapter, and give an example of each.

**12-12** Differentiate an intermediate host from a definitive host.

Protozoa are unicellular, eukaryotic organisms. Among the protozoa are many variations on this cell structure, as we shall see. Protozoa inhabit water and soil. The feeding and growing stage, or **trophozoite**, feeds upon bacteria and small particulate nutrients. Some protozoa are part of the normal microbiota of animals. See Exploring the Microbiome in Chapter 4, page 94. Researchers at the U.S. Department of Agriculture are studying an apicomplexan protozoan that reduces egg production by fire ants. Fire ants cause millions of dollars in agricultural damage each year and can cause painful stings. Of the nearly 20,000 species of protozoa, relatively few cause human disease. Those few, however, have significant health and economic impact. Malaria is the fourth leading cause of death in children in Africa.

## Characteristics of Protozoa

The term *protozoa* means "first animals," meant to describe animal-like nutrition. However, protozoa are quite different from animals—a few are photosynthetic, and many have complex life cycles that enable them to get from one host to the next. Protozoa are now classified into the same super clades as algae, based on DNA analyses (Table 12.4).

### Life Cycle

Protozoa reproduce asexually by fission, budding, or schizogony. **Schizogony** is multiple fission; the nucleus undergoes multiple divisions before the cell divides. After many nuclei are formed, a small portion of cytoplasm concentrates around each nucleus, and then the single cell separates into daughter cells.

Sexual reproduction has been observed in some protozoa. The ciliates, such as *Paramecium*, reproduce sexually by **conjugation (Figure 12.17)**, which is very different from the bacterial process of the same name (see Figure 8.30, page 236). During protozoan conjugation, two cells fuse, and a haploid nucleus (the micronucleus) from each cell migrates to the other cell. This haploid micronucleus fuses with the haploid micronucleus within the cell. The parent cells separate, each now a fertilized cell. When the cells later divide, they produce daughter cells with recombined DNA. Some protozoa produce **gametes (gametocytes)**, which are haploid sex cells. During reproduction, two gametes fuse to form a diploid zygote.

**Encystment** Under certain adverse conditions, some protozoa produce a protective capsule called a **cyst**. A cyst permits the organism to survive when food, moisture, or oxygen are lacking, when temperatures are not suitable, or when toxic chemicals are present. A cyst also enables a parasitic species to survive outside a host. This is important because parasitic protozoa may have to be excreted from one host in order to get to a new host. The cyst formed by members of the phylum Apicomplexa is called an **oocyst**. It is a reproductive structure in which new cells are produced asexually.

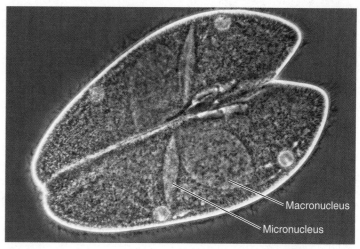

**Figure 12.17 Conjugation in the ciliate protozoan *Paramecium.*** Sexual reproduction in ciliates is by conjugation. Each cell has two nuclei: a micronucleus and a macronucleus. The micronucleus is haploid and is specialized for conjugation. One micronucleus from each cell will migrate to the other cell during conjugation. Both cells will then go on to produce two daughter cells.

**Q** Does conjugation result in more cells?

### Nutrition

Protozoa are mostly aerobic heterotrophs, although many intestinal protozoa are capable of anaerobic growth. Two chlorophyll-containing groups, dinoflagellates and euglenoids, are often studied with algae.

All protozoa live in areas with a large supply of water. Some protozoa transport food across the plasma membrane. However, some have a protective covering, or *pellicle*, and thus require specialized structures to take in food. Ciliates take in food by waving their cilia toward a mouthlike opening called a **cytostome**. Amebae engulf food by surrounding it with pseudopods and phagocytizing it. In all protozoa, digestion takes place in membrane-enclosed **vacuoles**, and waste may be eliminated through the plasma membrane or through a specialized **anal pore**.

## Medically Important Protozoa

The biology of protozoa is discussed in this chapter. Diseases caused by protozoa are described in Part Four, Microorganisms and Human Disease.

Protozoa are a large and diverse group. Current schemes of classifying protozoan species into phyla are based on DNA data and morphology. As more information is obtained, some of the phyla discussed here may be grouped to form kingdoms.

### Feeding Grooves

Single-celled eukaryotes with a feeding groove in the cytoskeleton have been placed in the Excavata. Most are spindle-shaped

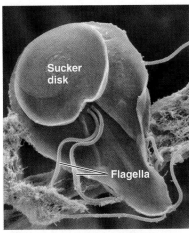

**(a)** *Giardia intestinalis.* This parasite has eight flagella and a ventral sucker disk that the parasite uses to attach itself to the intestine.

SEM ⊢——⊣ 3 μm

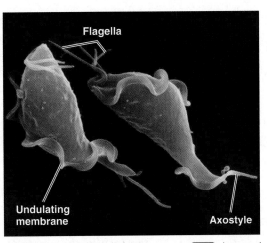

**(b)** *Trichomonas vaginalis.* This flagellate causes urinary and genital tract infections. Notice the small undulating membrane. This flagellate does not have a cyst stage.

SEM ⊢——⊣ 7.5 μm

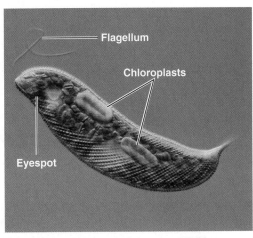

**(c)** *Euglena.* Euglenoids are autotrophs. Semirigid rings supporting the pellicle allow *Euglena* to change shape.

SEM ⊢——⊣ 10 μm

**Figure 12.18 Members of the superkingdom Excavata are spindle-shaped and have flagella.**

**Q** How does *Giardia* obtain energy without mitochondria?

and possess flagella (**Figure 12.18**). This superkingdom includes two phyla that lack mitochondria and the phylum Euglenozoa.

A parasite without mitochondria is *Giardia intestinalis* (JAR-dē-ah in'tes-tin-AL-is), sometimes called *G. lamblia* or *G. duodenalis.* The parasite (Figure 12.18a and Figure 25.16a, page 747) is found in the small intestine of humans and other mammals. It is excreted in the feces as a cyst (Figure 25.16b, page 747) and survives in the environment before being ingested by the next host. Diagnosis of giardiasis, the disease caused by *G. intestinalis,* is often based on the identification of cysts in feces.

Another human parasite that lacks mitochondria is *Trichomonas vaginalis* (TRIK-ō-mō-nas vaj-i-NA-lis), shown in Figure 12.18b and Figure 26.16 on page 780. Like some other flagellates, *T. vaginalis* has an **undulating membrane**, which consists of a membrane bordered by a flagellum. *T. vaginalis* does not have a cyst stage and must be transferred from host to host quickly before desiccation occurs. *T. vaginalis* is found in the vagina and in the male urinary tract. It is usually transmitted by sexual intercourse but can also be transmitted by toilet facilities or towels.

### Euglenozoa

Two groups of flagellated cells are included in the **Euglenozoa** based on common rRNA sequences, disk-shaped mitochondria, and absence of sexual reproduction.

**Euglenoids** are photoautotrophs (Figure 12.18c). Euglenoids have a semirigid plasma membrane called a pellicle, and they move by means of a flagellum at the anterior end. Most euglenoids also have a red *eyespot* at the anterior end.

This carotenoid-containing organelle senses light and directs the cell in the appropriate direction by using a *preemergent flagellum.* Some euglenoids are facultative chemoheterotrophs. In the dark, they ingest organic matter through a cytostome. Euglenoids are frequently studied with algae because they can photosynthesize.

The **hemoflagellates** (blood parasites) are transmitted by the bites of blood-feeding insects and are found in the circulatory system of the bitten host. To survive in this viscous fluid, hemoflagellates usually have long, slender bodies and an undulating membrane. The genus *Trypanosoma* (TRI-pa-nō-sō-mah) includes the species that causes African sleeping sickness, *T. brucei* (BROOS-e-ē), which is transmitted by the tsetse fly. *T. cruzi* (KROOZ-ē; see Figure 23.22, page 675), the causative agent of Chagas disease, is transmitted by the "kissing bug," so named because it bites on the face (see Figure 12.32d on page 356). After entering the insect, the trypanosome rapidly multiplies by schizogony. If the insect then defecates while biting a human, it can release trypanosomes that can contaminate the bite wound.

### Amebae

The **amebae** move by extending blunt, lobelike projections of the cytoplasm called **pseudopods** (Figure 12.19a). Any number of pseudopods can flow from one side of the ameba, and the rest of the cell will flow toward the pseudopods.

*Entamoeba histolytica* (en-tah-MĒ-bah his-tō-LI-ti-kah) is the only pathogenic ameba found in the human intestine. As many as 10% of the human population may be colonized by this ameba. New techniques, including DNA analyses and

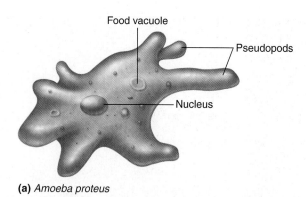

**(a)** *Amoeba proteus*

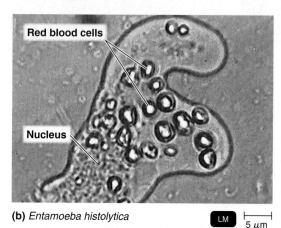

**(b)** *Entamoeba histolytica*    LM   5 μm

**Figure 12.19 Amebae. (a)** To move and to engulf food, amebae (such as this *Amoeba proteus*) extend cytoplasmic structures called pseudopods. Food vacuoles are created when pseudopods surround food and bring it into the cell. **(b)** *Entamoeba histolytica.* The presence of ingested red blood cells is diagnostic for *Entamoeba.*

**Q** How do amebic dysentery and bacillary dysentery differ?

lectin binding, have revealed that the amebae thought to be *E. histolytica* are actually two distinct species. The nonpathogenic species, *E. dispar* (DIS-par) is most common. The invasive *E. histolytica* (Figure 12.19b) causes amebic dysentery. In the human intestine, *E. histolytica* uses proteins called lectins to attach to the galactose of the plasma membrane and causes cell lysis. *E. dispar* does not have galactose-binding lectins. *Entamoeba* is transmitted between humans through ingestion of the cysts that are excreted in the feces of the infected person. *Acanthamoeba* (ah-kan-thah-MĒ-bah) growing in water, including tap water, can infect the cornea and cause blindness.

Since 1990, *Balamuthia* (bal-ah-MOO-thē-ah) has been reported as the cause of brain abscesses called granulomatous amebic encephalitis in the United States and other countries. The ameba most often infects immunocompromised people. Like *Acanthamoeba*, *Balamuthia* is a free-living ameba found in water and is not transmitted from human to human.

### Apicomplexa

The **Apicomplexa** are not motile in their mature forms and are obligate intracellular parasites. Apicomplexans are characterized by the presence of a complex of special organelles at the apexes (tips) of their cells (hence the phylum name). The organelles in these apical complexes contain enzymes that penetrate the host's tissues.

Apicomplexans have a complex life cycle that involves transmission between several hosts. An example of an apicomplexan is *Plasmodium* (plaz-MŌ-dē-um), the causative agent of malaria. Malaria affects 10% of the world's population, with 300 to 500 million new cases each year. The complex life cycle makes it difficult to develop a vaccine against malaria.

*Plasmodium* grows by sexual reproduction in the *Anopheles* (an-AH-fe-lēz) mosquito (**Figure 12.20**). When an *Anopheles* carrying the infective stage of *Plasmodium*, called a **sporozoite**, bites a human, sporozoites can be injected into the human. The sporozoites undergo schizogony in liver cells and produce thousands of trophozoites called **merozoites**, which infect red blood cells. The young merozoite looks like a ring in which the nucleus and cytoplasm are visible. This is called a **ring stage** (see Figure 23.25b, page 678). The red blood cells eventually rupture and release more merozoites. Upon release of the merozoites, their waste products, which cause fever and chills, are also released. Most of the merozoites infect new red blood cells and perpetuate their cycle of asexual reproduction. However, some develop into male and female sexual forms (gametocytes). Even though the gametocytes themselves cause no further damage, they can be picked up by the bite of another *Anopheles* mosquito; they then enter the mosquito's intestine and begin their sexual cycle. Their progeny can then be injected into a new human host by the biting mosquito.

The mosquito is the **definitive host** because it harbors the sexually reproducing stage of *Plasmodium*. The host in which the parasite undergoes asexual reproduction (in this case, the human) is the **intermediate host**.

Malaria is diagnosed in the laboratory by microscopic observation of thick blood smears for the presence of *Plasmodium* (see Figure 23.25, page 678). A peculiar characteristic of malaria is that the interval between periods of fever caused by the release of merozoites is always the same for a given species of *Plasmodium* and is always a multiple of 24 hours. The reason and mechanism for such precision have intrigued scientists. After all, why should a parasite need a biological clock? *Plasmodium*'s development is regulated by the host's body temperature, which normally fluctuates over a 24-hour period. The parasite's careful timing ensures that gametocytes are mature at night, when *Anopheles* mosquitoes are feeding, and thereby facilitates transmission of the parasite to a new host.

Another apicomplexan parasite of red blood cells is *Babesia microti* (bah-BĒ-sē-ah mi-KRŌ-tē). *Babesia* causes fever and anemia in immunosuppressed individuals. In the United States, it is transmitted by the tick *Ixodes scapularis* (IKS-ō-des skap-ū-LAR-is).

*Toxoplasma gondii* (toks-ō-PLAZ-mah GON-dē-ē) is another apicomplexan intracellular parasite of humans. The life cycle

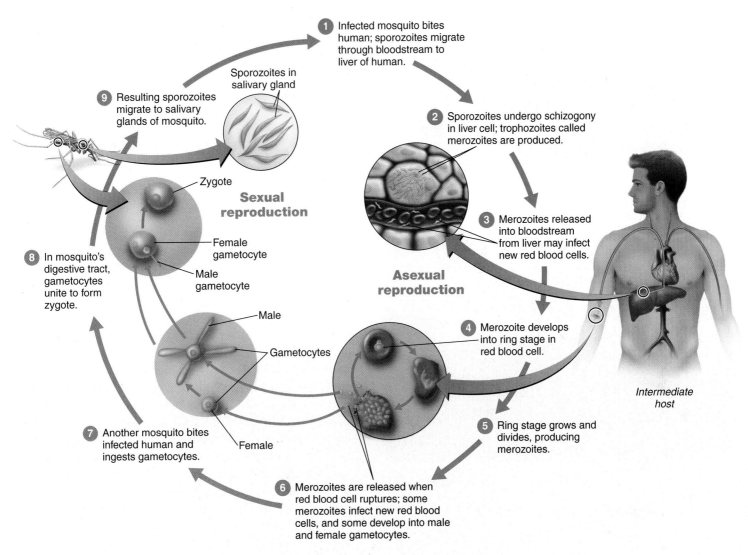

**Figure 12.20 The life cycle of *Plasmodium vivax*, the apicomplexan that causes malaria.** Asexual reproduction (schizogony) of the parasite takes place in the liver and in the red blood cells of a human host. Sexual reproduction occurs in the intestine of an *Anopheles* mosquito after the mosquito has ingested gametocytes.

**Q** What is the definitive host for *Plasmodium*?

of this parasite involves domestic cats. The trophozoites, called **tachyzoites**, reproduce sexually and asexually in an infected cat, and **oocysts**, each containing eight sporozoites, are excreted with feces. If the oocysts are ingested by humans or other animals, the sporozoites emerge as trophozoites, which can reproduce in the tissues of the new host (see Figure 23.23, page 678). *T. gondii* is dangerous to pregnant women because it can cause congenital infections in utero. Tissue examination and observation of *T. gondii* are used for diagnosis. Antibodies may be detected by ELISA and by indirect fluorescent-antibody tests (see Chapter 18).

*Cryptosporidium* (KRIP-tō-spor-i′dē-um) lives inside the cells lining the small intestine and can be transmitted to humans through the feces of cows, rodents, dogs, and cats. Inside the host cell, each *Cryptosporidium* organism forms four oocysts (see Figure 25.17, page 749), each containing four sporozoites. When the oocyst ruptures, sporozoites may infect new cells in the host or be released with the feces. See the Clinical Focus box on page 351.

During the 1980s, epidemics of waterborne diarrhea were identified on every continent except Antarctica. The causative agent was misidentified as a cyanobacterium because the outbreaks occurred during warm months and the disease agent looked like a prokaryotic cell. In 1993, the organism was identified as an apicomplexan similar to *Cryptosporidium*. In 2013, the new parasite, *Cyclospora cayetanensis* (sī-klō-SPOR-ah kī′ ah-tan-EN-sis), was responsible for 600 cases of diarrhea associated with cilantro in the United States and Canada.

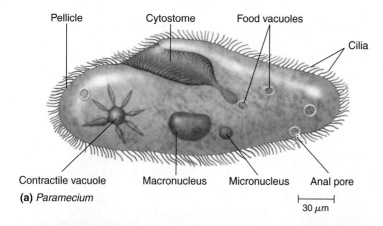

Pellicle    Cytostome    Food vacuoles

Cilia

Contractile vacuole    Macronucleus    Micronucleus    Anal pore

**(a)** *Paramecium*

30 μm

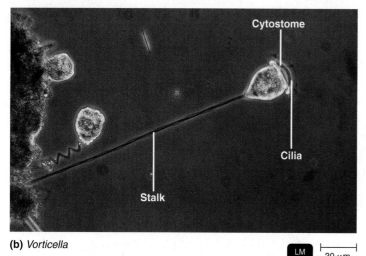

Cytostome

Cilia

Stalk

**(b)** *Vorticella*

LM    30 μm

**Figure 12.21 Ciliates. (a)** *Paramecium* is covered with rows of cilia. It has specialized structures for ingestion (cytostome), elimination of wastes (anal pore), and the regulation of osmotic pressure (contractile vacuoles). The macronucleus is involved with protein synthesis and other ongoing cellular activities. The micronucleus functions in sexual reproduction. **(b)** *Vorticella* attaches to objects in water by the base of its stalk. The springlike stalk can expand, allowing *Vorticella* to feed in different areas. Cilia surround its cytostome.

**Q  What ciliate can cause disease in humans?**

### Ciliates

Ciliates have cilia that are similar to but shorter than flagella. The cilia are arranged in precise rows on the cell (**Figure 12.21**). They are moved in unison to propel the cell through its environment and to bring food particles to the mouth.

The only ciliate that is a human parasite is *Balantidium coli* (bal′an-TID-ē-um KŌ-lē), the causative agent of a severe, though rare, type of dysentery. When the host ingests cysts, they enter the large intestine, into which the trophozoites are released. The trophozoites produce proteases and other substances that destroy host cells. The trophozoite feeds on host cells and tissue fragments. Its cysts are excreted with feces.

Table 12.4 lists some typical parasitic protozoa and the diseases they cause.

## Slime Molds

### LEARNING OBJECTIVE

**12-13** Compare and contrast cellular slime molds and plasmodial slime molds.

**Slime molds** are closely related to amebae and are placed in the phylum Amoebozoa. There are two taxa of slime molds: cellular and plasmodial. **Cellular slime molds** are typical eukaryotic cells that resemble amebae. In the life cycle of cellular slime molds (**Figure 12.22**), the ameboid cells live and grow by ingesting fungi and bacteria by phagocytosis. Cellular slime molds are of interest to biologists who study cellular migration and aggregation, because when conditions are unfavorable, large numbers of ameboid cells aggregate to form a single structure. This aggregation occurs because some individual amebae produce the chemical cyclic AMP (cAMP), toward which the other amebae migrate. Some of the ameboid cells form a stalk; others swarm up the stalk to form a spore cap, and most of these differentiate into spores. When spores are released under favorable conditions, they germinate to form single amebae.

In 1973, a Dallas resident discovered a pulsating red blob in his backyard. The news media claimed that a "new life form" had been found. For some people, the "creature" evoked spine-chilling recollections of an old science fiction movie. Before imaginations got carried away too far, biologists calmed everyone's worst fears (or dashed their highest hopes). The amorphous mass was merely a plasmodial slime mold, they explained. But its unusually large size—46 cm in diameter—startled even scientists.

**Plasmodial slime molds** were first scientifically reported in 1729. A plasmodial slime mold exists as a mass of protoplasm with many nuclei (it is multinucleated). This mass of protoplasm is called a **plasmodium** (**Figure 12.23**). The entire plasmodium moves as a giant ameba; it engulfs organic debris and bacteria. Biologists have found that musclelike proteins forming microfilaments account for the movement of the plasmodium.

When plasmodial slime molds are grown in laboratories, a phenomenon called **cytoplasmic streaming** is observed, during which the protoplasm within the plasmodium moves and changes both its speed and direction so that the oxygen and nutrients are evenly distributed. The plasmodium continues to grow as long as there is enough food and moisture for it to thrive.

When either is in short supply, the plasmodium separates into many groups of protoplasm; each of these groups forms a stalked sporangium, in which haploid spores (a resistant,

**TABLE 12.4** Some Representative Pathogenic Protozoa

| Super Clade | Phylum | Human Pathogens | Distinguishing Features | Disease | Source of Human Infections | Page |
|---|---|---|---|---|---|---|
| Excavata | Diplomonads | Giardia intestinalis | Two nuclei, eight flagella | Giardial enteritis | Fecal contamination of drinking water | 747–749 |
| | Parabasalids | Trichomonas vaginalis | No encysting stage | Urethritis, vaginitis | Contact with vaginal-urethral discharge | 780 |
| | Euglenozoa | Leishmania | Flagellated form in sand fly; ovoid form in vertebrate host | Leishmaniasis | Sand fly bite (Phlebotomus) | 675 |
| | | Naegleria fowleri | Flagellated and ameboid forms | Meningoencephalitis | Water (during swimming) | 640 |
| | | Trypanosoma cruzi | Undulating membrane | Chagas disease | Triatoma bite (kissing bug) | 675 |
| | | T. brucei gambiense, T. b. rhodesiense | | African trypanosomiasis | Tsetse fly bite | 639 |
| Amorphea | Amoebozoa | Acanthamoeba | Pseudopods | Keratitis | Water | 613, 642 |
| | | Entamoeba histolytica, E. dispar | | Amebic dysentery | Fecal contamination of drinking water | 750 |
| | | Balamuthia | | Encephalitis | Water | — |
| SAR* | Apicomplexa | Babesia microti | Complex life cycles may require multiple hosts | Babesiosis | Domestic animals, ticks | 680 |
| | | Cryptosporidium | | Diarrhea | Water, humans, other animals | 749 |
| | | Cyclospora | | Diarrhea | Water | 750 |
| | | Plasmodium | | Malaria | Anopheles mosquito bite | 677 |
| | | Toxoplasma gondii | | Toxoplasmosis | Cats; beef; congenital | 676 |
| | Dinoflagellates | Alexandrium, Pfiesteria | Photosynthetic (most) | Paralytic shellfish poisoning; ciguatera | Ingestion of dinoflagellates in mollusks or fish | 439 |
| | Ciliates | Balantidium coli | Only parasitic ciliate of humans | Balantidial dysentery | Fecal contamination of drinking water | — |

*SAR stands for Stramenopiles, Alveolates, Rhizaria.

resting form of the slime mold) develop. When conditions improve, these spores germinate, fuse to form diploid cells, and develop into a multinucleated plasmodium.

# Helminths

## LEARNING OBJECTIVES

12-14  List the distinguishing characteristics of parasitic helminths.

12-15  Provide a rationale for the elaborate life cycle of parasitic worms.

12-16  List the characteristics of the two classes of parasitic platyhelminths, and give an example of each.

12-17  Describe a parasitic infection in which humans serve as a definitive host, as an intermediate host, and as both.

12-18  List the characteristics of parasitic nematodes, and give an example of infective eggs and infective larvae.

12-19  Compare and contrast platyhelminths and nematodes.

A number of parasitic animals spend part or all of their lives in humans. Most of these animals belong to two phyla: Platyhelminthes (flatworms) and Nematoda (roundworms). These worms are commonly called **helminths**. There are also free-living species in these phyla, but we'll limit our discussion to the parasitic species. Diseases caused by parasitic worms are discussed in Part Four.

## Characteristics of Helminths

Helminths are multicellular eukaryotic animals that generally possess digestive, circulatory, nervous, excretory, and

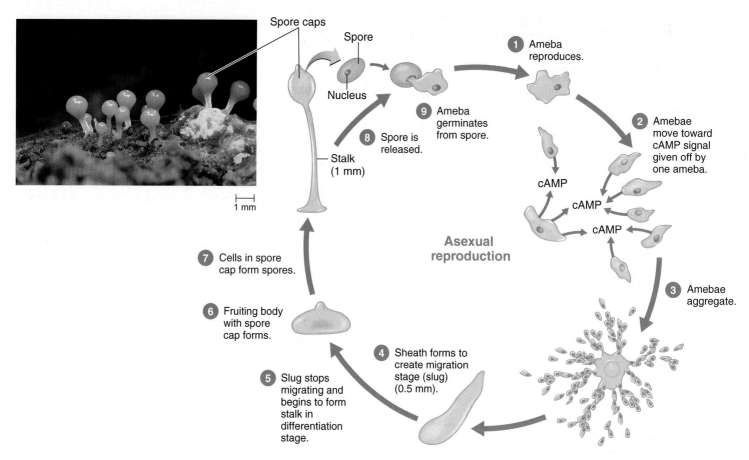

**Figure 12.22 The generalized life cycle of a cellular slime mold.** The micrograph shows a spore cap of *Hemitrichia*.

**Q** What characteristics do slime molds share with protozoa? With fungi?

reproductive systems. Parasitic helminths must be highly specialized to live inside their hosts. The following generalizations distinguish parasitic helminths from their free-living relatives:

1. *They may lack a digestive system.* They can absorb nutrients from the host's food, body fluids, and tissues.

2. *Their nervous system is reduced.* They do not need an extensive nervous system because they do not have to search for food or respond much to their environment. The environment within a host is fairly constant.

3. *Their means of locomotion is occasionally reduced or completely lacking.* Because they are transferred from host to host, they don't need to search actively for a suitable habitat.

4. *The reproductive system is often complex.* Individuals produce large numbers of eggs, by which a suitable host is infected.

### Life Cycle

The life cycle of parasitic helminths can be extremely complex, involving a succession of intermediate hosts for completion of each **larval** (developmental) stage of the parasite and a definitive host for the adult parasite.

Adult helminths may be **dioecious**; male reproductive organs are in one individual, and female reproductive organs are in another. In those species, reproduction occurs only when two adults of the opposite sex are in the same host.

Adult helminths may also be **monoecious**, or **hermaphroditic**—one animal has both male and female reproductive organs. Two hermaphrodites may copulate and simultaneously fertilize each other. A few types of hermaphrodites fertilize themselves.

**CHECK YOUR UNDERSTANDING**

✔ **12-14** Why are drugs for parasitic helminths often toxic to the host?

✔ **12-15** Of what value is the complicated life cycle of parasitic helminths?

### Platyhelminths

Members of the phylum Platyhelminthes, the **flatworms**, are dorsoventrally flattened. The classes of parasitic flatworms include the trematodes and cestodes. These parasites cause disease or developmental disturbances in a wide variety of animals (**Figure 12.24**).

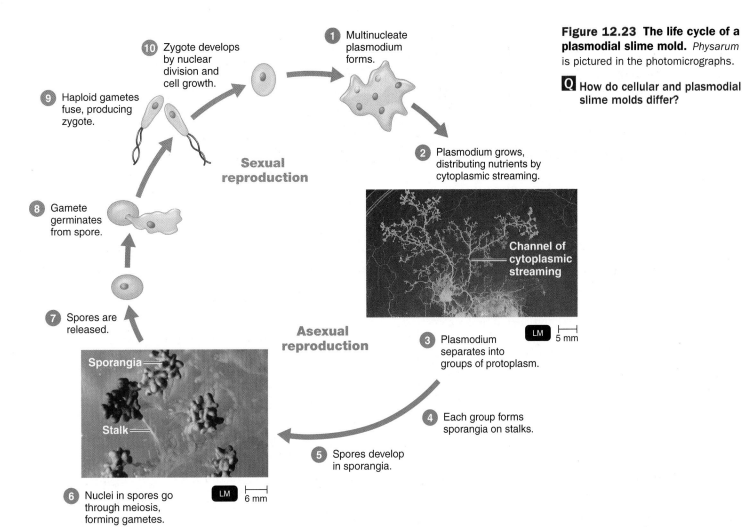

**Figure 12.23 The life cycle of a plasmodial slime mold.** *Physarum* is pictured in the photomicrographs.

**Q** **How do cellular and plasmodial slime molds differ?**

(10) Zygote develops by nuclear division and cell growth.

(9) Haploid gametes fuse, producing zygote.

(1) Multinucleate plasmodium forms.

**Sexual reproduction**

(8) Gamete germinates from spore.

(2) Plasmodium grows, distributing nutrients by cytoplasmic streaming.

Channel of cytoplasmic streaming

LM  5 mm

(7) Spores are released.

**Asexual reproduction**

(3) Plasmodium separates into groups of protoplasm.

Sporangia

Stalk

(4) Each group forms sporangia on stalks.

(5) Spores develop in sporangia.

(6) Nuclei in spores go through meiosis, forming gametes.

LM  6 mm

## Trematodes

Trematodes, or **flukes**, often have flat, leaf-shaped bodies with a ventral sucker and an oral sucker (**Figure 12.25**). The suckers hold the organism in place. Flukes obtain food by absorbing it through their nonliving outer covering, called the **cuticle.** Flukes are given common names according to the tissue of the definitive host in which the adults live (for example, lung fluke, liver fluke, blood fluke). The Asian liver fluke *Clonorchis sinensis* (klon-OR-kis sin-EN-sis) is occasionally seen in immigrants in the United States, but it cannot be transmitted because its intermediate hosts are not in the United States.

To exemplify a fluke's life cycle, let's look at the lung fluke, *Paragonimus* spp. (par-ah-GŌ-nĕ-mus). *Paragonimus* species occur throughout the world. *P. kellicotti* (kel-li-KOT-tē), endemic in the United States, has been associated with eating raw crayfish on river raft trips. The adult lung fluke lives in the bronchioles of humans and other mammals and is approximately 6 mm wide and 12 mm long. The hermaphroditic adults liberate eggs into the bronchi. Because sputum that contains eggs is frequently swallowed, the eggs are usually excreted in feces of the definitive host. If the life cycle is to continue, the eggs must reach a body of water. A series of steps occurs that

**Figure 12.24 Infection by a parasitic platyhelminth.** An increase in the trematode *Ribeiroia* in recent years has caused deformities in frogs. Frogs with multiple limbs have been found from Minnesota to California. Cercaria of the trematode infect tadpoles. The encysted metacercariae displace developing limb buds, causing abnormal limb development. The increase in the parasite may be due to fertilizer runoff that increases algae, which are food for the parasite's intermediate host snail.

**Q** **What tailed stage of the parasite lives in a snail?**

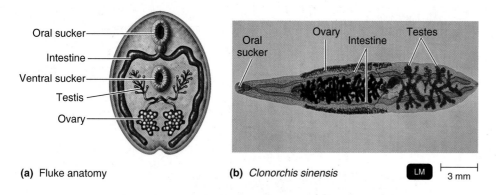

**(a)** Fluke anatomy

**(b)** *Clonorchis sinensis*

LM    3 mm

**Figure 12.25 Flukes. (a)** General anatomy of an adult fluke, shown in cross section. The oral and ventral suckers attach the fluke to the host. The mouth is located in the center of the oral sucker. Flukes are hermaphroditic; each animal contains both testes and ovaries. **(b)** The Asian liver fluke *Clonorchis sinensis*. Notice the incomplete digestive system. Heavy infestations may block bile ducts from the liver.

**Q** Why is the flatworm digestive system called "incomplete"?

ensure adult flukes can mature in the lungs of a new host. The life cycle is shown in **Figure 12.26**.

In a laboratory diagnosis, sputum and feces are examined microscopically for fluke eggs. Infection results from eating undercooked freshwater crustaceans, and the disease can be prevented by thoroughly cooking crayfish and freshwater crabs.

The cercariae of the blood fluke *Schistosoma* (shis-tō-SŌ-mah) are not ingested. Instead, they burrow through the skin of the human host and enter the circulatory system. The adults are found in certain abdominal and pelvic veins. The disease schistosomiasis is a major world health problem; it will be discussed further in Chapter 23 (page 681).

## Cestodes

Cestodes, or **tapeworms**, are intestinal parasites. Their structure is shown in **Figure 12.27**. The head, or **scolex** (plural: *scoleces*),

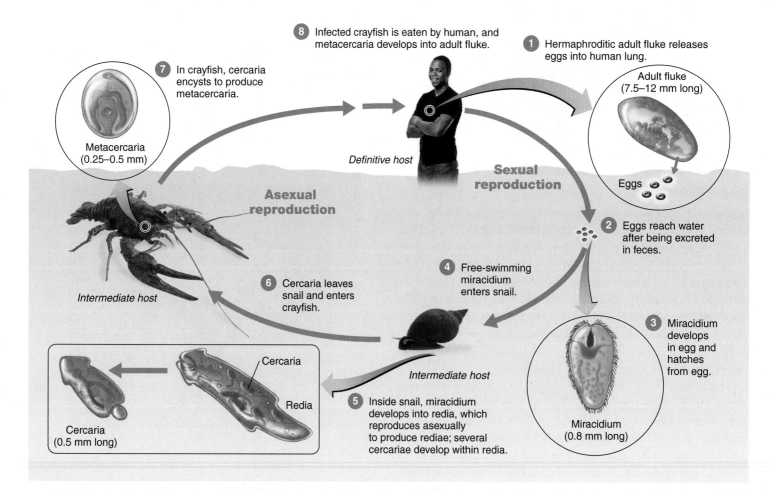

⑧ Infected crayfish is eaten by human, and metacercaria develops into adult fluke.

① Hermaphroditic adult fluke releases eggs into human lung.

⑦ In crayfish, cercaria encysts to produce metacercaria.

Adult fluke (7.5–12 mm long)

Metacercaria (0.25–0.5 mm)

*Definitive host*

**Asexual reproduction**

**Sexual reproduction**

Eggs

② Eggs reach water after being excreted in feces.

⑥ Cercaria leaves snail and enters crayfish.

④ Free-swimming miracidium enters snail.

*Intermediate host*

③ Miracidium develops in egg and hatches from egg.

Cercaria

Redia

*Intermediate host*

⑤ Inside snail, miracidium develops into redia, which reproduces asexually to produce rediae; several cercariae develop within redia.

Cercaria (0.5 mm long)

Miracidium (0.8 mm long)

**Figure 12.26 The life cycle of the lung fluke, *Paragonimus* spp.** The trematode reproduces sexually in a human and asexually in a snail, its first intermediate host. Larvae encysted in the second intermediate host, freshwater crabs and crayfish, infect humans and other mammals when ingested. Also see the *Schistosoma* life cycle in Figure 23.27 (page 682).

**Q** Of what value is this complex life cycle to *Paragonimus*?

# The Most Frequent Cause of Recreational Waterborne Diarrhea

As you read through this box, you'll encounter a series of questions that microbiologists ask themselves as they try to diagnose a disease. Try to answer each question before going on to the next one.

1. One week after her birthday party, 8-year-old Chloe had watery diarrhea, vomiting, and abdominal cramping. Her mother took her to the pediatrician because Chloe's symptoms weren't going away on their own.

   What diseases are possible? (*Hint:* See Diseases in Focus 25.2 and 25.5 on pages 740 and 748).

2. Possible diseases included giardiasis, cryptosporidiosis, cyclosporiasis, and amebic dysentery. Chloe's pediatrician took a stool sample from Chloe and sent it to the lab for testing. The result of acid-fast staining of her stool is shown in Figure A.

   What is the disease?

3. The acid-fast staining stains the *Cryptosporidium* oocysts red, therefore making them easy to identify. In this case, the sporozoites are made visible inside the oocyst at the arrow. The oocysts are infectious immediately upon being excreted in feces.

   What else do you need to know?

4. Chloe's birthday party was held at a community water park. Chloe's mother immediately followed up with the parents of the other children who attended the party. She found out that the other 20 children also had watery diarrhea, vomiting, or abdominal cramps. All the children recovered from the infection 2 to 10 days after becoming ill.

   How is this disease transmitted?

5. *Cryptosporidium* infection is transmitted by the fecal–oral route. It results from ingesting *Cryptosporidium* oocysts through the consumption of fecally

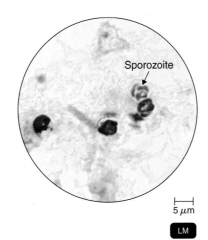

Sporozoite

5 μm

LM

**Figure A  Acid-fast stain of Chloe's feces.**

contaminated food or water or through direct person-to-person or animal-to-person contact. The infectious dose is low; feeding studies have demonstrated that ingesting as few as 10 to 30 oocysts can cause infection in healthy persons. Infected persons

have been reported to shed $10^8$ to $10^9$ oocysts in a single bowel movement and to excrete oocysts for up to 50 days after cessation of diarrhea.

6. *Cryptosporidium* has emerged as the most frequently recognized cause of recreational water-associated outbreaks of gastroenteritis, even in venues with disinfected water. It became a reportable disease in 1994 (Figure B).

   How can *Cryptosporidium* outbreaks be prevented?

   *Cryptosporidium* species are known to be resistant to most chemical disinfectants, such as chlorine. Recommendations to reduce the risk of infection include the following:

   • Do not swim during and for 2 weeks after diarrheal illness.

   • Avoid swallowing pool water.

   • Wash hands after using the restroom or changing diapers.

   *Source:* Adapted from *MMWR* 65(5), 1–26, May 20, 2016.

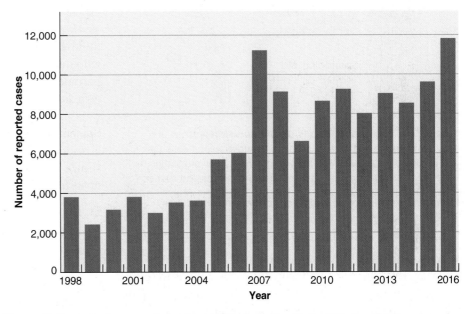

**Figure B  Reported cases of cryptosporidiosis in the United States. CDC.**

has suckers for attaching to the intestinal mucosa of the definitive host; some species also have small hooks for attachment. Tapeworms do not ingest the tissues of their hosts; in fact, they completely lack a digestive system. To obtain nutrients from the small intestine, they absorb food through their cuticle. The body consists of segments called **proglottids**. Proglottids are continually produced by the neck region of the scolex, as long as the scolex is attached and alive. Each mature proglottid contains

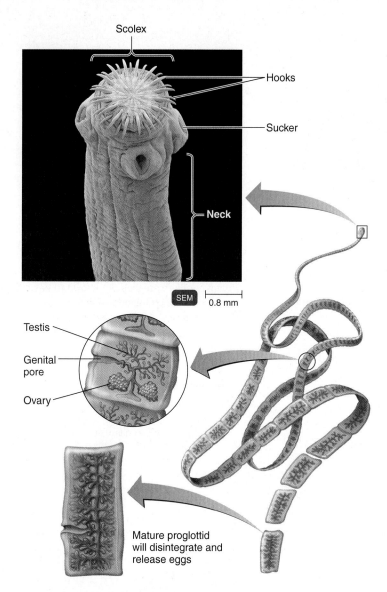

**Figure 12.27 General anatomy of an adult tapeworm.** The scolex, shown in the micrograph, consists of suckers and hooks that attach to the host's tissues. The body lengthens as new proglottids form at the neck. Each mature proglottid contains both testes and ovaries.

**Q What are the similarities between tapeworms and flukes?**

both male and female reproductive organs. The proglottids farthest away from the scolex are the mature ones containing fertilized eggs. Mature proglottids are essentially bags of eggs, each of which is infective to the proper intermediate host.

**Humans as Definitive Hosts** The adults of *Taenia saginata* (TE-nē-ah sa-jin-AH-tah), the beef tapeworm, live in humans and can reach a length of 6 m. The scolex is about 2 mm long and is followed by a thousand or more proglottids. The feces of an infected human contain mature proglottids, each of which contains thousands of eggs. As the proglottids wriggle away

from the fecal material, they increase their chances of being ingested by an animal that is grazing. Upon ingestion by cattle, the larvae hatch from the eggs and bore through the intestinal wall. The larvae migrate to muscle (meat), in which they encyst as **cysticerci**. When the cysticerci are ingested by humans, all but the scolex is digested. The scolex anchors itself in the small intestine and begins producing proglottids.

Diagnosis of tapeworm infection in humans is based on the presence of mature proglottids and eggs in feces (Figure 25.22, page 752). Cysticerci can be seen macroscopically in meat; their presence is referred to as "measly beef." Inspecting beef that is intended for human consumption for "measly" appearance is one way to prevent infections by beef tapeworm. Another method of prevention is to avoid the use of untreated human sewage as fertilizer in grazing pastures.

Humans are the only known definitive host of the pork tapeworm, *Taenia solium*. Adult worms living in the human intestine produce eggs, which are passed out in feces. When eggs are eaten by pigs, the larval helminth encysts in the pig's muscles; humans become infected when they eat undercooked pork. The human-pig-human cycle of *T. solium* is common in Latin America, Asia, and Africa. In the United States, however, *T. solium* is virtually nonexistent in pigs; the parasite is transmitted from human to human. Eggs shed by one person and ingested by another person hatch, and the larvae encyst in the brain and other parts of the body, causing cysticercosis (see Figure 25.21, page 751). The human hosting *T. solium*'s larvae is serving as an intermediate host. Approximately 7% of the few hundred cases reported in recent years were acquired by people who had never been outside the United States. They may have become infected through household contact with people who were born in or had traveled in other countries.

**Humans as Intermediate Hosts** Humans are the intermediate hosts for *Echinococcus granulosus* (ē-KĪ-nō-kok′kus gran-ū-LŌ-sus), shown in **Figure 12.28**. Dogs and coyotes are the definitive hosts for this minute (2–8 mm) tapeworm.

1 Eggs are excreted with feces.
2 Eggs are ingested by deer, sheep, or humans. Humans can also become infected by contaminating their hands with dog feces or saliva from a dog that has licked itself.
3 The eggs hatch in the human's small intestine, and the larvae migrate to the liver or lungs.
4 The larva develops into a **hydatid cyst**. The cyst contains "brood capsules," from which thousands of scoleces might be produced.
5 Humans are a dead end for the parasite, but in the wild, the cysts might be in a deer that is eaten by a wolf.
6 The scoleces would be able to attach themselves in the wolf's intestine and produce proglottids.

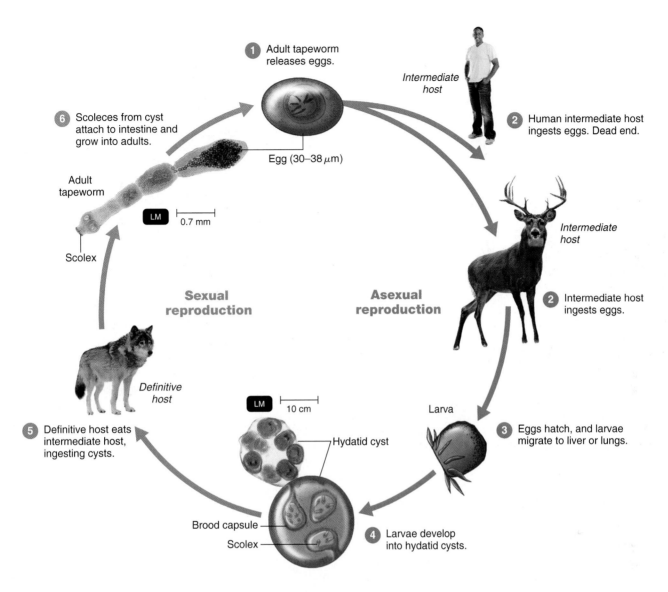

**1** Adult tapeworm releases eggs.

*Intermediate host*

**2** Human intermediate host ingests eggs. Dead end.

**6** Scoleces from cyst attach to intestine and grow into adults.

Egg (30–38 μm)

Adult tapeworm

LM 0.7 mm

Scolex

*Intermediate host*

**Sexual reproduction**

**Asexual reproduction**

**2** Intermediate host ingests eggs.

*Definitive host*

LM 10 cm

Larva

**5** Definitive host eats intermediate host, ingesting cysts.

**3** Eggs hatch, and larvae migrate to liver or lungs.

Hydatid cyst

Brood capsule

Scolex

**4** Larvae develop into hydatid cysts.

**Figure 12.28  The life cycle of the tapeworm, *Echinococcus* spp.** Dogs are the most common definitive host of *E. granulosus. E. multilocularis* infections in humans are rare. The parasite can complete its life cycle only if the cysts are ingested by a definitive host that eats the intermediate host.

**Q** Why isn't being in a human of benefit to *Echinococcus*?

Diagnosis of hydatid cysts is frequently made only on autopsy, although X rays can detect the cysts (see Figure 25.23, page 752).

**CHECK YOUR UNDERSTANDING**

✔ **12-16** Differentiate *Paragonimus* and *Taenia*.

## Nematodes

Members of the phylum Nematoda, the **roundworms**, are cylindrical and tapered at each end. Roundworms have a *complete* digestive system, consisting of a mouth, an intestine, and an anus. Most species are dioecious. Males are smaller than

females and have one or two hardened **spicules** on their posterior ends. Spicules are used to guide sperm to the female's genital pore.

Some species of nematodes are free-living in soil and water, and others are parasites of plants and animals. Some nematodes pass their entire life cycle, from egg to mature adult, in a single host.

Intestinal roundworms are the most common causes of chronic infectious diseases. The most common are *Ascaris*, hookworms, and whipworms, infecting more than 2 billion people worldwide. Nematode infections of humans can be divided into two categories: those in which the egg is infective, and those in which the larva is infective.

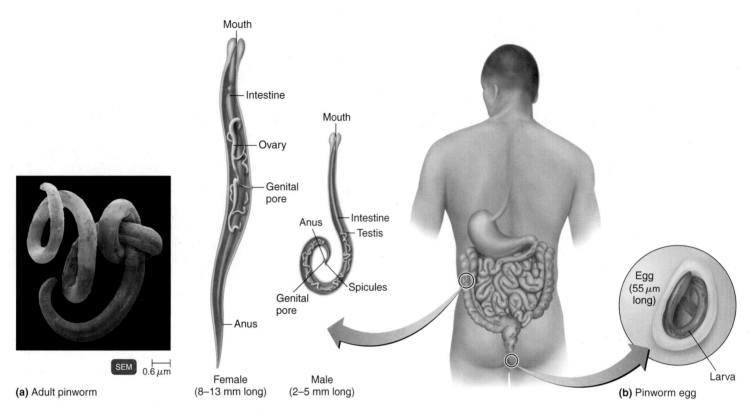

(a) Adult pinworm

SEM 0.6 μm

Female (8–13 mm long)

Male (2–5 mm long)

(b) Pinworm egg

Egg (55 μm long)

Larva

**Figure 12.29 The pinworm, *Enterobius vermicularis*.** **(a)** Adult pinworms live in the large intestine of humans. Most roundworms are dioecious, and the female (left and photomicrograph) is often distinctly larger than the male (right). **(b)** Pinworm eggs are deposited by the female on the perianal skin at night.

**Q** Are humans the definitive or intermediate host for pinworms?

### Eggs Infective for Humans

*Ascaris lumbricoides* (AS-kar-is lum-bri-KOY-dēz) is a large nematode (30 cm in length) that infects over 1 billion people worldwide (Figure 25.25, page 753). It is dioecious with **sexual dimorphism**; that is, the male and female worms look distinctly different, the male being smaller with a curled tail. The adult *Ascaris* lives in the small intestines of humans exclusively; it feeds primarily on semidigested food. Eggs, excreted with feces, can survive in the soil for long periods until accidentally ingested by another host. The eggs hatch in the small intestine of the host. The larvae then burrow out of the intestine and enter the blood. They are carried to the lungs, where they grow. The larvae will then be coughed up, swallowed, and returned to the small intestine, where they mature into adults.

Raccoon roundworm, *Baylisascaris procyonis*, is an emerging roundworm in North America. Raccoons are the definitive host, although the adult roundworm can also live in domestic dogs. Eggs are shed with feces and ingested by an intermediate host, usually a rabbit. The ingested eggs hatch in the intestines of rabbits and humans. The larvae migrate through a variety of tissues, causing a condition called *larva migrans*. Infection often results in severe neurological symptoms or death. Larva migrans can also be caused by *Toxocara canis* (from dogs) and

*T. cati* (from cats). These companion animals are the intermediate and definitive hosts, but humans can become infected by ingesting *Toxocara* eggs shed in the animals' feces. It is estimated that 14% of the U.S. population has been infected. Children are most likely to be infected probably because they play in soil and sandboxes, where animal feces can be found.

One billion people worldwide are infected with *Trichuris trichiura*, or whipworm. The worms are spread from person to person by fecal–oral transmission or through feces-contaminated food. The disease occurs most often in areas with tropical weather and poor sanitation practices and among children.

The pinworm *Enterobius vermicularis* (en-te-RŌ-bē-us ver-mi-kū-LAR-is) spends its entire life in a human host (**Figure 12.29**). Adult pinworms are found in the large intestine. From there, the female pinworm migrates to the anus to deposit her eggs on the perianal skin. The eggs can be ingested by the host or by another person exposed through contaminated clothing or bedding.

### Larvae Infective for Humans

A few nematode larvae live in soil and can enter a human host directly through the skin. *Strongyloides* (stron-gel-OI-dēz) nematodes infect 30 to 100 million people worldwide. Most

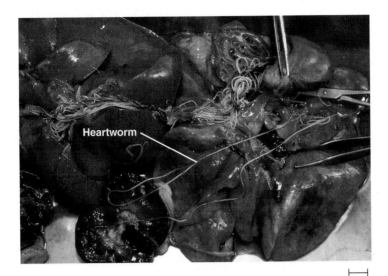

Heartworm

├──┤
20 mm

**Figure 12.30 The heartworm, *Dirofilaria immitis*.** Four adult
*D. immitis* in the right ventricle of a dog's heart. Each worm is
12–30 cm long.

**Q** How do roundworms and flatworms differ?

infections are limited to a rash where the nematode entered,
but the larvae can migrate to the intestine, causing abdominal
pain, or to the lungs, causing a cough.

Adult hookworms, *Necator americanus* (ne-KĂ-tor ah-mer-
i-KĂ-nus) and *Ancylostoma duodenale* (an'sil-ō-STŌ-mah DOO-
ah-den-al-ē), live in the small intestine of humans (Figure
25.24, page 753); the eggs are excreted in feces. The larvae
hatch in the soil, where they feed on bacteria. A larva enters
its host by penetrating the host's skin. It then enters a blood or
lymph vessel, which carries it to the lungs. It is coughed up in
sputum, swallowed, and finally carried to the small intestine.

Trichinellosis is caused by a nematode that the host acquires
by eating encysted larvae in undercooked meat of infected
animals (see page 754). The nematode, *Dirofilaria immitis*
(dir-ō -fil-AIR-ē-ah IM-mi-tis), is spread from host to host
through the bites of *Aedes* mosquitoes. It primarily affects dogs
and cats, but it can infest human skin, conjunctiva, or lungs.
Larvae injected by the mosquito migrate to various organs,
where they mature into adults. The parasitic worm is called
a **heartworm** because the adult stage is often in the animal
host's heart, where it can kill its host through congestive heart
failure (**Figure 12.30**). The disease occurs on every continent
except Antarctica. *Wolbachia* bacteria appear to be essential to
development of the worm embryos.

Four genera of roundworms called *anisakines,* or wriggly
worms, can be transmitted to humans from infected fish and
squid. Anisakine larvae are in the fish's intestinal mesenteries
and migrate to the muscle when the fish dies. Freezing or thorough cooking will kill the larvae.

Table 12.5 on page 356 lists representative parasitic helminths of each phylum and class and the diseases they cause.

**CHECK YOUR UNDERSTANDING**

✔ **12-17** What is the definitive host for *Enterobius*?

✔ **12-18** What stage of *Dirofilaria immitis* is infectious for dogs
and cats?

✔ **12-19** You find a parasitic worm in a baby's diapers. How
would you know whether it's a *Taenia* or a *Necator*?

# Arthropods as Vectors

**LEARNING OBJECTIVES**

**12-20** Define *arthropod vector*.

**12-21** Differentiate a tick from a mosquito, and name a disease
transmitted by each.

Arthropods are animals characterized by segmented bodies, hard
external skeletons, and jointed legs. With nearly 1 million species, this is the largest phylum in the animal kingdom. Although
arthropods are not microbes themselves, we will briefly describe
them here because a few suck the blood of humans and other
animals and can transmit microbial diseases while doing so.
Arthropods that carry pathogenic microorganisms are called
**vectors**. Scabies and pediculosis are diseases that are caused by
arthropods themselves (see Chapter 21, pages 609–611).

Representative classes of arthropods include the following:

- Arachnida (eight legs): spiders, mites, ticks
- Crustacea (four antennae): crabs, crayfish
- Insecta (six legs): bees, flies, lice

Table 12.6 lists those arthropods that are important vectors,
and **Figures 12.31** and **12.32** illustrate some of them. These
insects and ticks reside on an animal only when they are feeding. An exception to this is the louse, which spends its entire
life on its host and cannot survive for long away from a host.

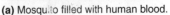

**(a)** Mosquito filled with human blood.    **(b)** Tick engorged with blood.

**Figure 12.31 Blood-sucking mosquitoes and ticks.** Mosquitoes
are the vector for several pathogens to humans, including Zika virus,
malaria, and dengue virus. Ticks are the vector for Lyme disease.

**Q** When is a vector also a definitive host?

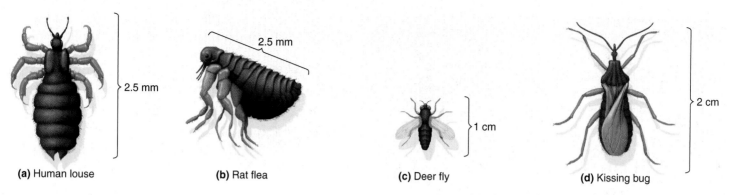

**(a)** Human louse     **(b)** Rat flea     **(c)** Deer fly     **(d)** Kissing bug

**Figure 12.32 Arthropod vectors. (a)** The human louse, *Pediculus*. **(b)** The rat flea, *Xenopsylla*. **(c)** The deer fly, *Chrysops*. **(d)** The kissing bug, *Triatoma*.

**Q** Name one pathogen carried by each of these vectors.

**TABLE 12.5** Representative Parasitic Helminths

| Phylum | Class | Human Parasites | Intermediate Host | Definitive Host Site | Stage Passed to Humans; Method | Disease | Figure Reference |
|---|---|---|---|---|---|---|---|
| Platyhelminthes | Trematodes | *Paragonimus* | Freshwater snails and crayfish | Humans; lungs | Metacercaria in crustaceans; ingested | Paragonimiasis (lung fluke) | 12.26 |
| | | *Schistosoma* | Freshwater snails | Humans | Cercariae; through skin | Schistosomiasis | 23.27 23.28 |
| | Cestodes | *Echinococcus granulosus* | Humans | Dogs and other animals; intestines | Eggs from other animals; ingested | Hydatidosis | 12.28 25.23 |
| | | *Taenia saginata* | Cattle | Humans; small intestine | Cysticerci in beef; ingested | Tapeworm | 25.21, 25.22b |
| | | *Taenia solium* | Humans; pigs | Humans | Eggs; ingested | Neurocysticercosis | — |
| Nematoda | | *Ancylostoma duodenale* | — | Humans; small intestine | Larvae; through skin | Hookworm | 25.24 |
| | | Anisakines | Marine fish and squid | Marine mammals | Larvae in fish; ingested | Anisakiasis (sashimi worms) | — |
| | | *Ascaris lumbricoides* | — | Humans; small intestine | Eggs; ingested | Ascariasis | 25.25 |
| | | *Baylisascaris procyonis* | Rabbits | Raccoons; large intestine | Eggs; ingested | Raccoon roundworm | — |
| | | *Enterobius vermicularis* | — | Humans; large intestine | Eggs; ingested | Pinworm | 12.29 |
| | | *Necator americanus* | — | Humans; small intestine | Larvae; through skin | Hookworm | — |
| | | *Strongyloides stercoralis* | Soil | Humans | Larvae; through skin | Strongyloidiasis | — |
| | | *Trichinella spiralis* | Humans and other mammals | Humans; small intestine | Eggs; ingested | Trichinellosis | 25.26 |
| | | *Trichuris trichiura* | — | Humans, pigs, and other mammals; small intestine | Larvae; ingested | Whipworm | 25.22f |
| | | *Toxocara canis, T. cati* | Dogs, cats | Dogs, cats; small intestine | Eggs; ingested | Toxocariasis | — |

**TABLE 12.6 Important Arthropod Vectors of Human Diseases**

| Class | Order | Vector | Disease | Figure Reference |
|-------|-------|--------|---------|------------------|
| Arachnida | Mites and ticks | *Dermacentor* (tick) | Rocky Mountain spotted fever | — |
| | | *Ixodes* (tick) | Lyme disease, babesiosis, ehrlichiosis | 12.31b |
| | | *Ornithodorus* (tick) | Relapsing fever | — |
| Insecta | Sucking lice | *Pediculus* (human louse) | Epidemic typhus, relapsing fever | 12.32a |
| | Fleas | *Xenopsylla* (rat flea) | Endemic murine typhus, plague | 12.32b |
| | True flies | *Chrysops* (deer fly) | Tularemia | 12.32c |
| | | *Aedes* (mosquito) | Dengue Zika virus disease, heartworm | 12.31a |
| | | *Anopheles* (mosquito) | Malaria | — |
| | | *Culex* (mosquito) | Arboviral encephalitis | — |
| | | *Glossina* (tsetse fly) | African trypanosomiasis | — |
| | True bugs | *Triatoma* (kissing bug) | Chagas disease | 12.32d |

Some vectors are just a mechanical means of transport for a pathogen. For example, houseflies lay their eggs on decaying organic matter, such as feces. While doing so, a housefly can pick up a pathogen on its feet or body and transport the pathogen to our food.

Some parasites multiply in their vectors. When this happens, the parasites can accumulate in the vector's feces or saliva. Large numbers of parasites can then be deposited on or in the host while the vector is feeding there. The spirochete that causes Lyme disease is transmitted by ticks in this manner (see Chapter 23, page 664), and the West Nile virus is transmitted in the same way by mosquitoes (see Chapter 22, page 637).

As discussed earlier, *Plasmodium* is an example of a parasite that requires that its vector also be the definitive host.

*Plasmodium* can sexually reproduce only in the gut of an *Anopheles* mosquito. *Plasmodium* is introduced into a human host with the mosquito's saliva, which acts as an anticoagulant that keeps blood flowing.

To eliminate vectorborne diseases, health workers focus on eradicating the vectors.

**CHECK YOUR UNDERSTANDING**

✓ 12-20 Vectors can be divided into three major types, according to the roles they play for the parasite. List the three types of vectors and a disease transmitted by each.

✓ 12-21 Assume you see an arthropod on your arm. How will you determine whether it is a tick or a flea?

# Study Outline

 Go to @MasteringMicrobiology for **Interactive Microbiology**, *In the Clinic videos, MicroFlix, MicroBoosters, 3D animations, practice quizzes, and more.*

## Fungi (pp. 324–335)

1. Mycology is the study of fungi.
2. The number of serious fungal infections is increasing.
3. Fungi are aerobic or facultatively anaerobic chemoheterotrophs.
4. Most fungi are decomposers, and a few are parasites of plants and animals.

## Characteristics of Fungi (pp. 325–329)

5. A fungal thallus consists of filaments of cells called hyphae; a mass of hyphae is called a mycelium.
6. Yeasts are unicellular fungi. To reproduce, fission yeasts divide symmetrically, whereas budding yeasts divide asymmetrically.
7. Buds that do not separate from the parent cell form pseudohyphae.
8. Pathogenic dimorphic fungi are yeastlike at 37°C and moldlike at 25°C.
9. Fungi are classified according to rRNA.
10. Sporangiospores and conidiospores are produced asexually.
11. Sexual spores are usually produced in response to special circumstances, often changes in the environment.

12. Fungi can grow in acidic, low-moisture, aerobic environments.
13. They are able to metabolize complex carbohydrates.

## Medically Important Fungi (pp. 329–331)

14. The Zygomycota have coenocytic hyphae and produce sporangiospores and zygospores.
15. Microsporidia lack mitochondria and microtubules; they cause diarrhea in AIDS patients.
16. The Ascomycota have septate hyphae and produce ascospores and frequently conidiospores.
17. Basidiomycota have septate hyphae and produce basidiospores; some produce conidiospores.
18. Teleomorphic fungi produce sexual and asexual spores; anamorphic fungi produce asexual spores only.

## Fungal Diseases (pp. 331–334)

19. Systemic mycoses are fungal infections deep within the body that affect many tissues and organs.
20. Subcutaneous mycoses are fungal infections beneath the skin.
21. Cutaneous mycoses affect keratin-containing tissues such as hair, nails, and skin.

22. Superficial mycoses are localized on hair shafts and superficial skin cells.

23. Opportunistic mycoses are caused by fungi that are not usually pathogenic.

24. Opportunistic mycoses can infect any tissues. However, they are usually systemic.

### Economic Effects of Fungi (pp. 334–335)

25. *Saccharomyces* and *Trichoderma* are used in the production of foods.

26. Fungi are used for the biological control of pests.

27. Mold spoilage of fruits, grains, and vegetables is more common than bacterial spoilage of these products.

28. Many fungi cause diseases in plants.

## Lichens (pp. 335–336)

1. A lichen is a mutualistic combination of an alga (or a cyanobacterium) and a fungus.

2. The alga photosynthesizes, providing carbohydrates for the lichen; the fungus provides a holdfast.

3. Lichens colonize habitats that are unsuitable for either the alga or the fungus alone.

4. Lichens may be classified on the basis of morphology as crustose, foliose, or fruticose.

## Algae (pp. 337–341)

1. Algae are unicellular, filamentous, or multicellular (thallic).

2. Most algae live in aquatic environments.

### Characteristics of Algae (pp. 337–338)

3. Algae are eukaryotic; most are photoautotrophs.

4. The thallus of multicellular algae usually consists of a stipe, a holdfast, and blades.

5. Algae reproduce asexually by cell division and fragmentation.

6. Many algae reproduce sexually.

7. Photoautotrophic algae produce oxygen.

8. Algae are classified according to their structures and pigments.

### Selected Phyla of Algae (pp. 338–340)

9. Brown algae (kelp) may be harvested for algin.

10. Red algae grow deeper in the ocean than other algae.

11. Green algae have cellulose and chlorophyll *a* and *b* and store starch.

12. Diatoms are unicellular and have pectin and silica cell walls; some produce a neurotoxin.

13. Dinoflagellates produce neurotoxins that cause paralytic shellfish poisoning and ciguatera.

14. The oomycotes are heterotrophic; they include decomposers and pathogens.

### Roles of Algae in Nature (pp. 340–341)

15. Algae are the primary producers in aquatic food chains.

16. Planktonic algae produce most of the molecular oxygen in the Earth's atmosphere.

17. Petroleum is the fossil remains of planktonic algae.

18. Unicellular algae are symbionts in such animals as *Tridacna*.

## Protozoa (pp. 341–346)

1. Protozoa are unicellular, eukaryotic chemoheterotrophs.

2. Protozoa are found in soil and water and as normal microbiota in animals.

### Characteristics of Protozoa (p. 342)

3. The vegetative form is called a trophozoite.

4. Asexual reproduction is by fission, budding, or schizogony.

5. Sexual reproduction is by conjugation.

6. During ciliate conjugation, two haploid nuclei fuse to produce a zygote.

7. Some protozoa can produce a cyst that provides protection during adverse environmental conditions.

8. Protozoa have complex cells with a pellicle, a cytostome, and an anal pore.

### Medically Important Protozoa (pp. 342–346)

9. *Trichomonas* and *Giardia* lack mitochondria and have flagella.

10. Euglenozoa move by means of flagella and lack sexual reproduction; they include *Trypanosoma*.

11. Amebae include *Entamoeba* and *Acanthamoeba*.

12. Apicomplexa have apical organelles for penetrating host tissue; they include *Plasmodium* and *Cryptosporidium*.

13. Ciliates move by means of cilia; *Balantidium coli* is the only human parasitic ciliate.

## Slime Molds (pp. 346–347)

1. Cellular slime molds resemble amebae and ingest bacteria by phagocytosis.

2. Plasmodial slime molds consist of a multinucleated mass of protoplasm that engulfs organic debris and bacteria as it moves.

## Helminths (pp. 347–335)

1. Parasitic flatworms belong to the phylum Platyhelminthes.

2. Parasitic roundworms belong to the phylum Nematoda.

### Characteristics of Helminths (pp. 347–348)

3. Helminths are multicellular animals; a few are parasites of humans.

4. The anatomy and life cycle of parasitic helminths are modified for parasitism.

5. The adult stage of a parasitic helminth is found in the definitive host.

6. Each larval stage of a parasitic helminth requires an intermediate host.

7. Helminths can be monoecious or dioecious.

### Platyhelminths (pp. 348–353)

8. Flatworms are dorsoventrally flattened animals; parasitic flatworms may lack a digestive system.

9. Adult trematodes, or flukes, have an oral and ventral sucker with which they attach to host tissue.

10. Eggs of trematodes hatch into free-swimming miracidia that enter the first intermediate host; two generations of rediae develop; the rediae become cercariae that bore out of the first intermediate host and penetrate the second intermediate host; cercariae encyst as metacercariae; the metacercariae develop into adults in the definitive host.

11. A cestode, or tapeworm, consists of a scolex (head) and proglottids.

12. Humans serve as the definitive host for the beef tapeworm, and cattle are the intermediate host.

13. Humans serve as the definitive host and can be an intermediate host for the pork tapeworm.

14. Humans serve as the intermediate host for *Echinococcus granulosus*; the definitive hosts are dogs, wolves, and foxes.

### Nematodes   (pp. 353–355)

15. Roundworms have a complete digestive system.

16. The nematodes that infect humans with their eggs include *Ascaris, Trichuris,* and *Enterobius.*

17. The nematodes that infect humans with their larvae include hookworms and *Trichinella.*

### Arthropods as Vectors   (pp. 355–357)

1. Jointed-legged animals, including ticks and insects, belong to the phylum Arthropoda.

2. Arthropods that carry diseases are called vectors.

3. Vectorborne diseases are most effectively eliminated by controlling or eradicating the vectors.

# Study Questions

For answers to the Knowledge and Comprehension questions, turn to the Answers tab at the back of the textbook.

## Knowledge and Comprehension

### Review

1. Following is a list of fungi, their methods of entry into the body, and sites of infections they cause. Categorize each type of mycosis as cutaneous, opportunistic, subcutaneous, superficial, or systemic.

| Genus | Method of Entry | Site of Infection | Mycosis |
|---|---|---|---|
| *Blastomyces* | Inhalation | Lungs | (a) _____ |
| *Sporothrix* | Puncture | Ulcerative lesions | (b) _____ |
| *Microsporum* | Contact | Fingernails | (c) _____ |
| *Trichosporon* | Contact | Hair shafts | (d) _____ |
| *Aspergillus* | Inhalation | Lungs | (e) _____ |

2. A mixed culture of *Escherichia coli* and *Penicillium chrysogenum* is inoculated onto the following culture media. On which medium would you expect each to grow? Why?
   a. 0.5% peptone in tap water
   b. 10% glucose in tap water

3. NAME IT   Identify the structures of this eukaryote, which has an affinity for keratin.

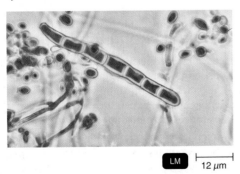

LM  ⊢────┤  12 μm

4. Briefly discuss the importance of lichens in nature. Briefly discuss the importance of algae in nature.

5. Differentiate cellular and plasmodial slime molds. How does each survive adverse environmental conditions?

6. Complete the following table.

| Phylum | Method of Motility | One Human Parasite |
|---|---|---|
| Diplomonads | (a) _____ | (b) _____ |
| Microsporidia | (c) _____ | (d) _____ |
| Amebae | (e) _____ | (f) _____ |
| Apicomplexa | (g) _____ | (h) _____ |
| Ciliates | (i) _____ | (j) _____ |
| Euglenozoa | (k) _____ | (l) _____ |
| Parabasalids | (m) _____ | (n) _____ |

7. Why is it significant that *Trichomonas* does not have a cyst stage? Name a protozoan parasite that does have a cyst stage.

8. By what means are helminthic parasites transmitted to humans?

9. Most roundworms are dioecious. What does this term mean? To what phylum do roundworms belong?

10. DRAW IT   A generalized life cycle of the liver fluke *Clonorchis sinensis* is shown below. Label the fluke's stages. Identify the intermediate host(s). Identify the definitive host(s). To what phylum and class does this animal belong?

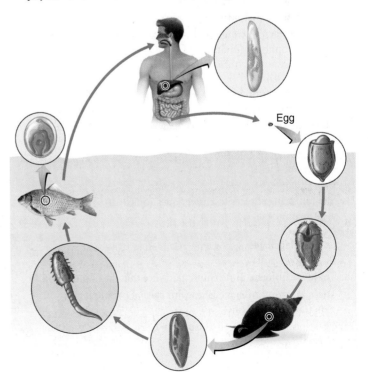

Egg

## Multiple Choice

1. How many phyla are represented in the following list of organisms: *Echinococcus, Cyclospora, Aspergillus, Taenia, Toxoplasma, Trichinella?*
   a. 1        d. 4
   b. 2        e. 5
   c. 3

Use the following choices to answer questions 2 and 3:
   1. metacercaria        4. miracidium
   2. redia               5. cercaria
   3. adult

2. Put the above stages in order of development, beginning with the egg.
   a. 5, 4, 1, 2, 3
   b. 4, 2, 5, 1, 3
   c. 2, 5, 4, 3, 1
   d. 3, 4, 5, 1, 2
   e. 2, 4, 5, 1, 3

3. If a snail is the first intermediate host of a parasite with these stages, which stage would be found in the snail?
   a. 1        d. 4
   b. 2        e. 5
   c. 3

4. Fleas are the intermediate host for *Dipylidium caninum* tapeworm, and dogs are the definitive host. Which stage of the parasite could be found in the flea?
   a. cysticercus larva
   b. proglottids
   c. scolex
   d. adult

5. Which of the following statements about yeasts are true?
   1. Yeasts are fungi.              a. 1, 2, 3, 4
   2. Yeasts can form pseudohyphae.  b. 3, 4, 5, 6
   3. Yeasts reproduce asexually by budding.  c. 2, 3, 4, 5
   4. Yeasts are facultatively anaerobic.  d. 1, 3, 5, 6
   5. All yeasts are pathogenic.     e. 2, 3, 4
   6. All yeasts are dimorphic.

6. Which of the following events follows cell fusion in an ascomycete?
   a. conidiophore formation
   b. conidiospore germination
   c. ascus opening
   d. ascospore formation
   e. conidiospore release

7. The definitive host for *Plasmodium vivax* is
   a. human.
   b. *Anopheles.*
   c. a sporocyte.
   d. a gametocyte.

Use the following choices to answer questions 8–10:
   a. Apicomplexa
   b. ciliates
   c. dinoflagellates
   d. Microsporidia

8. These are obligate intracellular parasites that lack mitochondria.

9. These are nonmotile parasites with special organelles for penetrating host tissue.

10. These photosynthetic organisms can cause paralytic shellfish poisoning.

## Analysis

1. *Alexandrium* (red tide) has been called a plant, protist, protozoan, and alga in the past. Now it's in the SAR clade, along with *Plasmodium* and *Paramecium.* Are all SAR members photosynthetic? Are any? Explain why it has been so difficult to accurately classify *Alexandrium.*

2. The life cycle of the fish tapeworm *Diphyllobothrium* is similar to that of *Taenia saginata,* except that the intermediate host is fish. Describe the life cycle and method of transmission to humans. Why are freshwater fish more likely to be a source of tapeworm infection than marine fish?

3. *Trypanosoma brucei gambiense*—part (a) in the figure—is the causative agent of African sleeping sickness. To what phylum does it belong? Part (b) shows a simplified life cycle for *T. b. gambiense.* Identify the host and vector of this parasite.

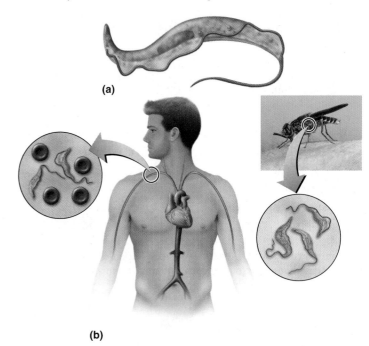

(a)

(b)

## Clinical Applications and Evaluation

1. A girl developed generalized seizures. A CT scan revealed a single brain lesion consistent with a tumor. Biopsy of the lesion showed a cysticercus. The patient lived in South Carolina and had never traveled outside the state. What parasite caused her disease? How is this disease transmitted? How might it be prevented?

2. A California farmer developed a low-grade fever, myalgia, and cough. A chest X-ray exam revealed an infiltrate in the lung. Microscopic examination of the sputum revealed round, budding cells. A sputum culture grew mycelia and arthroconidia. What organism is most likely the cause of the symptoms? How is this disease transmitted? How might it be prevented?

3. A teenaged male in California complained of remittent fever, chills, and headaches. A blood smear revealed ring-shaped cells in his red blood cells. He was successfully treated with primaquine and chloroquine. The patient lives near the San Luis Rey River and has no history of foreign travel, blood transfusion, or intravenous drug use. What is the disease? How was it acquired?

# Viruses, Viroids, and Prions 13

Viruses are too small to be seen with a light microscope and can't be cultured outside their hosts. Therefore, although viral diseases aren't new, the viruses themselves couldn't be studied until the twentieth century. In 1886, the Dutch chemist Adolf Mayer showed that tobacco mosaic disease (TMD) was transmissible from a diseased plant to a healthy plant. In 1892, in an attempt to isolate the cause of TMD, the Russian bacteriologist Dimitri Iwanowski filtered the sap of diseased plants through a porcelain filter that was designed to retain bacteria. He expected to find the microbe trapped in the filter; instead, he found that the infectious agent had passed through the minute pores of the filter. When he infected healthy plants with the filtered fluid, they contracted TMD. The first human disease associated with a filterable agent was yellow fever. Advances in molecular biological techniques in the 1980s and 1990s led to the recognition of several new viruses, including human immunodeficiency virus (HIV) and SARS-associated coronavirus. Viral hepatitis is one of the most common infectious diseases in the world. Several different hepatitis viruses have been identified, including the bloodborne hepatitis B virus and hepatitis C virus, and the foodborne hepatitis A virus (see the photo) discussed in the Clinical Case. The World Health Organization's list of the top 10 emerging pathogens likely to cause severe outbreaks in the near future are all viral diseases. These include highly pathogenic human corornaviruses, chikungunya virus, Zika virus, and thrombocytopenia syndrome virus. Human diseases caused by viruses will be discussed in Part Four. In this chapter, we'll study the biology of viruses.

◀ Hepatitis A virus is transmitted by ingestion.

## In the Clinic

A woman brings her 8-month-old daughter into the urgent care clinic where you work as a nurse practitioner. The baby has a runny nose and a fever of 39°C. The child's ears and lungs are free of infection. The woman is annoyed because she has already taken her daughter to see her pediatrician. She requested an antibiotic, but the doctor refused to prescribe it. She once again requests that the baby receive antibiotics. **Will you prescribe the medication the mother wants? Explain why or why not—and how you would discuss this with the mother.**

*Hint: Read about virus structure on page 363 and multiplication of animal viruses on page 376.*

Play **In the Clinic** Video
@MasteringMicrobiology

# General Characteristics of Viruses

## LEARNING OBJECTIVE

**13-1** Differentiate a virus from a bacterium.

One hundred years ago, researchers couldn't imagine sub-microscopic particles, so they described the infectious agent as *contagium vivum fluidum*—a contagious fluid. By the 1930s, scientists had begun using the word *virus*, the Latin word for poison, to describe these filterable agents. The nature of viruses, however, remained elusive until 1935, when Wendell Stanley, an American chemist, isolated tobacco mosaic virus, making it possible for the first time to carry out chemical and structural studies on a purified virus. At about the same time, the invention of the electron microscope made it possible to see viruses.

The question of whether viruses are living organisms has an ambiguous answer. Life can be defined as a complex set of processes resulting from the actions of proteins specified by nucleic acids. The nucleic acids of living cells are in action all the time. Because viruses are inert outside living host cells, in this sense they aren't considered to be living organisms. However, once viruses enter a host cell, the viral nucleic acids become active, and viral multiplication results. In this sense, viruses are alive when they multiply in the host cells they infect. From a clinical point of view, viruses are alive because they cause infection and disease, just as pathogenic bacteria, fungi, and protozoa do. Depending on one's viewpoint, a virus may be regarded as an exceptionally complex aggregation of nonliving chemicals, or as an exceptionally simple living microorganism.

How, then, do we define *virus*? Viruses were originally distinguished from other infectious agents because they are especially small (filterable) and because they are **obligatory intracellular parasites**—that is, they absolutely require living host cells in order to multiply. However, both of these properties are shared by certain small bacteria, such as some rickettsias. Viruses and bacteria are compared in Table 13.1.

The truly distinctive features of viruses are now known to relate to their simple structural organization and their mechanism of multiplication. Accordingly, **viruses** are entities that

- Contain a single type of nucleic acid, either DNA or RNA.
- Contain a protein coat (sometimes itself enclosed by an envelope of lipids, proteins, and carbohydrates) that surrounds the nucleic acid.
- Multiply inside living cells by using the synthesizing machinery of the cell.
- Cause the synthesis of specialized structures that can transfer the viral nucleic acid to other cells.

Viruses have few or no enzymes of their own for metabolism; for example, they lack enzymes for protein synthesis and ATP generation. To multiply, viruses must take over the metabolic machinery of the host cell. This fact has considerable medical significance for the development of antiviral drugs,

**TABLE 13.1**  Viruses and Bacteria Compared

| | Bacteria | | |
|---|---|---|---|
| | Typical Bacteria | Rickettsias/ Chlamydias | Viruses |
| **Intracellular Parasite** | No | Yes | Yes |
| **Plasma Membrane** | Yes | Yes | No |
| **Binary Fission** | Yes | Yes | No |
| **Pass through Bacteriological Filters** | No | No/Yes | Yes |
| **Possess Both DNA and RNA** | Yes | Yes | No |
| **ATP-Generating Metabolism** | Yes | Yes/No | No |
| **Ribosomes** | Yes | Yes | No |
| **Sensitive to Antibiotics** | Yes | Yes | No |
| **Sensitive to Interferon** | No | No | Yes |

because most drugs that would interfere with viral multiplication would also interfere with the functioning of the host cell and therefore are too toxic for clinical use. (Antiviral drugs are discussed in Chapter 20.)

## Host Range

The **host range** of a virus is the spectrum of host cells the virus can infect. There are viruses that infect invertebrates, vertebrates, plants, protists, fungi, and bacteria. However, most viruses are able to infect specific types of cells of only one host species. In rare cases, viruses cross the host-species barrier, thus expanding their host range. An example is described in the Clinical Focus box on page 367. In this chapter, we're concerned mainly with viruses that infect either humans or bacteria. Viruses that infect bacteria are called **bacteriophages**, or **phages**.

The particular host range of a virus is determined by the virus's requirements for its specific attachment to the host cell and the availability within the potential host of cellular factors required for viral multiplication. For the virus to infect the host cell, the outer surface of the virus must chemically interact with specific receptor sites on the surface of the cell. The two complementary components are held together by weak bonds, such as hydrogen bonds. The combination of many attachment and receptor sites leads to a strong association between host cell and virus. For some bacteriophages, the receptor site is part of the cell wall of the host; in other cases, it is part of the fimbriae or flagella. For animal viruses, the receptor sites are on the plasma membranes of the host cells.

The potential to use viruses to treat diseases is intriguing because of their narrow host range and their ability to kill their host cells. The idea of *phage therapy*—using bacteriophages to treat bacterial infections—was developed in France in 1919

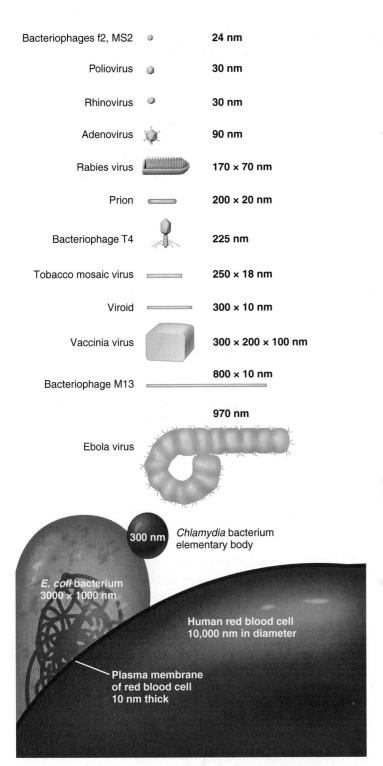

| | | |
|---|---|---|
| Bacteriophages f2, MS2 | | **24 nm** |
| Poliovirus | | **30 nm** |
| Rhinovirus | | **30 nm** |
| Adenovirus | | **90 nm** |
| Rabies virus | | **170 × 70 nm** |
| Prion | | **200 × 20 nm** |
| Bacteriophage T4 | | **225 nm** |
| Tobacco mosaic virus | | **250 × 18 nm** |
| Viroid | | **300 × 10 nm** |
| Vaccinia virus | | **300 × 200 × 100 nm** |
| Bacteriophage M13 | | **800 × 10 nm** |
| Ebola virus | | **970 nm** |

**300 nm** *Chlamydia* bacterium elementary body

*E. coli* bacterium 3000 × 1000 nm

Human red blood cell 10,000 nm in diameter

Plasma membrane of red blood cell 10 nm thick

**Figure 13.1 Virus sizes.** The sizes of several viruses (teal blue) and bacteria (brown) are compared with a human red blood cell, shown below the microbes. Dimensions are given in nanometers (nm) and are either diameters or length by width.

**Q** **How do viruses differ from bacteria?**

and used until 1979. Phage therapy is used currently in Russia and in neighboring Georgia. There is renewed interest in phage therapy in the United States and western Europe to treat infections caused by antibiotic-resistant bacteria. Furthermore, new

evidence suggests that bacteriophages in the human microbiome play a role in maintaining health. See Exploring the Microbiome.

## Viral Size

Viral sizes are determined with the aid of electron microscopy. Different viruses vary considerably in size. Although most are quite a bit smaller than bacteria, some of the larger viruses (such as the vaccinia virus) are about the same size as some very small bacteria (such as the mycoplasmas, rickettsias, and chlamydias). Viruses range from 20 to 1000 nm in length. The comparative sizes of several viruses and bacteria are shown in **Figure 13.1**.

---

**CHECK YOUR UNDERSTANDING**

✔ **13-1** How could the small size of viruses have helped researchers detect viruses before the invention of the electron microscope?

---

# Viral Structure

**LEARNING OBJECTIVE**

**13-2** Describe the chemical and physical structure of both an enveloped and a nonenveloped virus.

A **virion** is a complete, fully developed, infectious viral particle composed of nucleic acid and surrounded by a protein coat outside a host cell. Viruses are classified by their nucleic acid and by differences in the structures of their coats.

## Nucleic Acid

In contrast to prokaryotic and eukaryotic cells, in which DNA is always the primary genetic material (and RNA plays an auxiliary role), viral genes are encoded by either DNA or RNA—but never both. The genome of a virus can be single-stranded

---

**CLINICAL CASE** An Inconvenient Outbreak

Tina Markham, a 42-year-old pharmaceutical sales representative, has been home from work because of a very high, persistent fever (40°C). She is taking medications to reduce the fever, but they work only for a few hours. Tina makes an appointment to see her physician; he notices right away that Tina's skin is jaundiced. When he palpates her abdomen, she winces in pain; it is very tender. Sensing an issue with her liver, Tina's physician sends a blood sample to the local laboratory for a liver function test (LFT). The results show abnormal findings.

**What disease could be causing Tina's symptoms? Read on to find out.**

363  380  384  385

A healthy human harbors up to 10 permanent infectious viruses. The *human virome* is the viral portion of the microbiome. These viruses are found in the same sites as the majority of bacterial microbiome—the mouth, nose, skin, vagina, and intestines—and include persistent and latent viruses.

Retrovirus genetic material integrated into human chromosomes makes up about 8% of the human genome. However, the vast majority of the human virome consists of bacteriophages. Bacteriophages affect human health and disease by controlling growth of normal and pathogenic bacteria.

For example, large numbers of bacteriophages are present in mucus of the mouth and intestine. Mucus is normally considered a physical barrier to infection that protects underlying epithelial tissue. The exact ways phages impact their human environments is not fully understood yet, but it's likely that these phage interactions fall under two basic scenarios within the virome.

In the first scenario, "kill the winner," bacteriophages kill bacterial colonizers in the body. The capsids of some bacteriophages adhere to specific mucus glycoproteins. This puts the bacteriophage where it can encounter the bacterial cells that are its ultimate host. The bacteriophage benefits by having a host for reproduction, and the human host benefits by having the bacteriophage prevent colonization by pathogens.

In the second scenario, "kill the competition," some bacteriophages may protect the bacterial microbiome from invasion by other bacteria vying to gain a foothold in the area. For example, in the intestines, *Enterococcus* bacteria release lytic bacteriophages, from prophages (discussed later in the chapter), when competing enterococci are present—thus killing the competition.

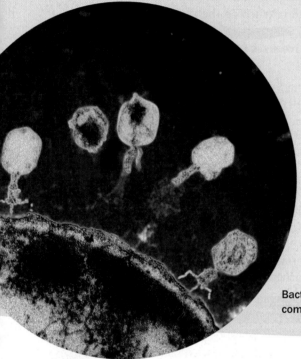

**Bacteriophages infecting an *E. coli* bacteria cell—a common member of our intestinal microbiota.**

or double-stranded. Thus, there are viruses with the familiar double-stranded DNA, with single-stranded DNA, with double-stranded RNA, and with single-stranded RNA. Depending on the virus, the nucleic acid can be linear or circular. In some viruses (such as the influenza virus), the nucleic acid is in several separate segments.

The percentage of nucleic acid in relation to protein is about 1% for the influenza virus and about 50% for certain bacteriophages. The total amount of nucleic acid varies from a few thousand nucleotides (or pairs) to as many as 250,000 nucleotides. (*E. coli*'s chromosome consists of approximately 4 million nucleotide pairs.)

## Capsid and Envelope

The nucleic acid of a virus is protected by a protein coat called the **capsid** (Figure 13.2a). The structure of the capsid is ultimately determined by the viral nucleic acid and accounts for most of the mass of a virus, especially of small ones. Each capsid is composed of protein subunits called **capsomeres**. In some viruses, the proteins composing the capsomeres are of a single type; in other viruses, several types of protein may be present. Individual capsomeres are often visible in electron micrographs

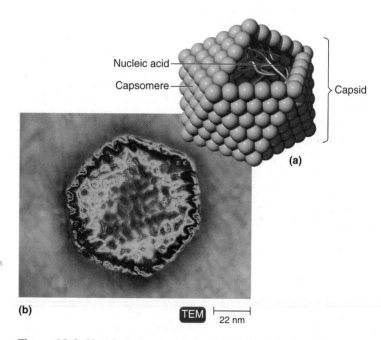

(a)

(b)

TEM ⊢ 22 nm

**Figure 13.2 Morphology of a nonenveloped polyhedral virus.**
(a) A diagram of a polyhedral (icosahedral) virus. (b) A micrograph of the adenovirus *Mastadenovirus*. Individual capsomeres are visible.

**Q** What is the chemical composition of a capsid?

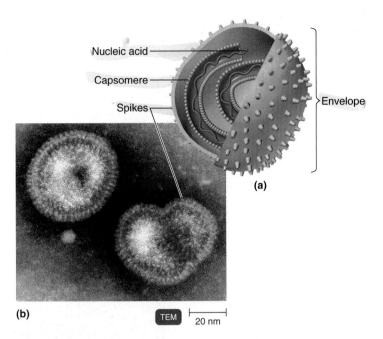

**Figure 13.3 Morphology of an enveloped helical virus. (a)** A diagram of an enveloped helical virus. **(b)** A micrograph of *Influenzavirus* A2. Notice the halo of spikes projecting from the outer surface of each envelope (see Chapter 24).

**Q** What is the nucleic acid in a virus?

(see Figure 13.2b for an example). The arrangement of capsomeres is characteristic of a particular type of virus.

In some viruses, the capsid is covered by an **envelope** (**Figure 13.3a**), which usually consists of some combination of lipids, proteins, and carbohydrates. Some animal viruses are released from the host cell by an extrusion process that coats the virus with a layer of the host cell's plasma membrane; that layer becomes the viral envelope. In many cases, the envelope contains proteins determined by the viral nucleic acid and materials derived from normal host cell components.

Depending on the virus, envelopes may or may not be covered by **spikes,** which are carbohydrate-protein complexes that project from the surface of the envelope. Some viruses attach to host cells by means of spikes. Spikes are such a reliable characteristic of some viruses that they can be used as a means of identification. The ability of certain viruses, such as the influenza virus (Figure 13.3b), to clump red blood cells is associated with spikes. Such viruses bind to red blood cells and form bridges between them. The resulting clumping is called *hemagglutination* and is the basis for several useful laboratory tests. (See Figure 18.8, page 512.)

Viruses whose capsids aren't covered by an envelope are known as **nonenveloped viruses** (see Figure 13.2). The capsid of a nonenveloped virus protects the nucleic acid from nuclease enzymes in biological fluids and promotes the virus's attachment to susceptible host cells.

When the host has been infected by a virus, the host's immune system is stimulated to produce antibodies (proteins that react with the surface proteins of the virus). This interaction between host antibodies and virus proteins should inactivate the virus and stop the infection. However, some viruses can escape antibodies because regions of the genes that code for these viruses' surface proteins are susceptible to mutations. The progeny of mutant viruses have altered surface proteins, such that the antibodies aren't able to react with them. Influenza virus frequently undergoes such changes in its spikes. This is why you can get influenza more than once. Although you may have produced antibodies to one influenza virus, the virus can mutate and infect you again.

## General Morphology

Viruses may be classified into several different morphological types on the basis of their capsid architecture. The structure of these capsids has been revealed by electron microscopy and a technique called X-ray crystallography.

### Helical Viruses

Helical viruses resemble long rods that may be rigid or flexible. The viral nucleic acid is found within a hollow, cylindrical capsid that has a helical structure (**Figure 13.4**). The viruses that cause rabies and Ebola are helical viruses.

### Polyhedral Viruses

Many animal, plant, and bacterial viruses are polyhedral, or many-sided, viruses. The capsid of most polyhedral viruses is in the shape of an *icosahedron*, a regular polyhedron with 20 triangular

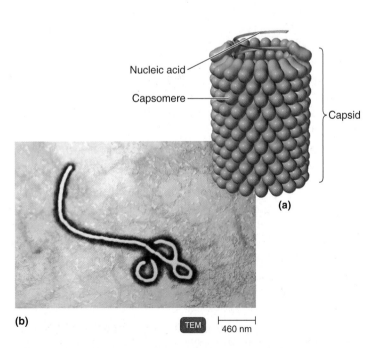

**Figure 13.4 Morphology of a helical virus. (a)** A diagram of a portion of a helical virus. A row of capsomeres has been removed to reveal the nucleic acid. **(b)** A micrograph of Ebola virus, a filovirus, showing a helical rodlike shape.

**Q** What is the chemical composition of a capsomere?

faces and 12 corners (see Figure 13.2a). The capsomeres of each face form an equilateral triangle. An example of a polyhedral virus in the shape of an icosahedron is the adenovirus (shown in Figure 13.2b). Another icosahedral virus is the poliovirus.

### Enveloped Viruses

As noted earlier, the capsid of some viruses is covered by an envelope. Enveloped viruses are roughly spherical. When helical or polyhedral viruses are enclosed by envelopes, they are called *enveloped helical* or *enveloped polyhedral viruses*. An example of an enveloped helical virus is the influenza virus (see Figure 13.3b). An example of an enveloped polyhedral (icosahedral) virus is the human herpes virus (see Figure 13.16b).

### Complex Viruses

Some viruses, particularly bacterial viruses, have complicated structures and are called **complex viruses**. One example of a complex virus is a bacteriophage. Some bacteriophages have capsids to which additional structures are attached. In **Figure 13.5a**, notice that the capsid (head) is polyhedral and that the tail sheath is helical. The head contains the nucleic acid. Later in the chapter, we'll discuss the functions of the other structures, such as the tail sheath, tail fibers, baseplate, and pin. Another example of complex viruses are poxviruses, which don't contain clearly identifiable capsids but do have several coats around the nucleic acid (Figure 13.5b).

> **CHECK YOUR UNDERSTANDING**
>
> ✔ 13-2 Diagram a nonenveloped polyhedral virus that has spikes.

# Taxonomy of Viruses

### LEARNING OBJECTIVES

**13-3** Define *viral species*.

**13-4** Give an example of a family, genus, and common name for a virus.

Just as we need taxonomic categories of plants, animals, and bacteria, we need viral taxonomy to help us organize and understand newly discovered organisms. The oldest classification of viruses is based on symptomatology, such as for diseases that affect the respiratory system. This system was convenient but not scientifically acceptable because the same virus may cause more than one disease, depending on the tissue affected. In addition, this system artificially grouped viruses that don't infect humans.

Viruses are now grouped according to how their mRNA is produced. New, fast DNA sequencing allows further classification of viruses into families based on genomics and structure. The suffix *-virus* is used for genus names; family names end in *-viridae*; and order names end in *-ales*. In formal usage, the family and genus names are used in the following manner: Family Herpesviridae, genus *Simplexvirus*, human herpesvirus-2.

A **viral species** is a group of viruses sharing the same genetic information and ecological niche (host range). Specific epithets for viruses aren't used. Thus, viral species are designated by descriptive common names, such as human immunodeficiency virus (HIV), with subspecies (if any) designated by a number (HIV-1). Table 13.2 presents a summary of the classification of viruses that infect humans.

> **CHECK YOUR UNDERSTANDING**
>
> ✔ 13-3 How does a virus species differ from a bacterial species?
>
> ✔ 13-4 Attach the proper endings to *Papilloma-* to show the family and genus that includes HPV, the cause of cervical cancer.

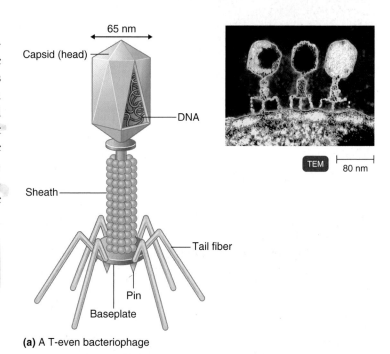

**(a)** A T-even bacteriophage

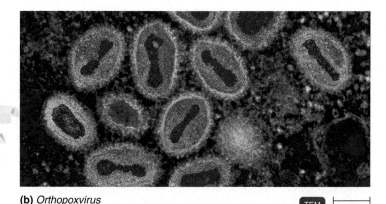

**(b)** *Orthopoxvirus*

**Figure 13.5 Morphology of complex viruses. (a)** A diagram and micrograph of a T-even bacteriophage. **(b)** A micrograph of variola virus, a species in the genus *Orthopoxvirus*, which causes smallpox.

**Q** What is the value of a capsid to a virus?

Influenza A viruses are found in many different animals, including birds, pigs, whales, horses, and seals. Sometimes influenza A viruses seen in one species can cross over and cause illness in another species. For example, up until 1998, only H1N1 viruses circulated widely in the U.S. pig population. In 1998, H3N2 viruses from humans were introduced into the pig population and caused widespread disease among pigs. The subtypes differ because of certain proteins on the surface of the virus (hemagglutinin [HA] and neuraminidase [NA] proteins). There are 16 different HA subtypes and 9 different NA subtypes of influenza A viruses.

How many different combinations of H and N proteins are possible?

Each combination is a different subtype. When we talk about "human flu viruses," we are referring to those subtypes that occur widely in humans. There are only three known subtypes of human influenza viruses (H1N1, H1N2, and H3N2).

What's different about bird flu?

H5 and H7 subtypes occur mainly in birds. Avian influenza (bird flu) viruses don't usually infect humans. All human cases of avian flu can be attributed to outbreaks in poultry, except one noteworthy probable transmission from a daughter to her mother. Avian influenza viruses may be transmitted to humans (1) directly from birds or from avian-virus-contaminated environments or (2) through an intermediate host, such as a pig.

Why are pigs important?

Pigs can be infected with both human and avian flu. The influenza virus genome is composed of eight separate segments. A segmented genome allows virus genes to mix and create a new influenza A virus if viruses from two different species infect the same person or animal (see the figure). This is known as *antigenic shift.*

The 2009 H1N1 virus was originally referred to as "swine flu" because laboratory testing showed that many of the genes in the virus were very similar to influenza viruses that normally occur in North American pigs. But further study has shown that the 2009 H1N1 is very different from that which normally circulates in North American pigs. It has two genes from flu viruses that normally circulate in pigs in Europe and Asia, avian influenza genes, and human genes. This is called a *quadruple reassortant* virus. The figure shows an example of *triple reassortment* between three viruses.

Pandemics

During the last 100 years, the emergence of new influenza A virus subtypes caused three pandemics, all of which spread around the world within one year of being detected (see the table). Some genetic parts of all of these influenza A strains originally came from birds.

*Source:* Adapted from *MMWR* sources.

**Model for antigenic shift in influenza virus. If a pig were infected with a human influenza virus and an avian influenza virus at the same time, the viruses could reassort and produce a new virus that had most of the genes from the human virus but a hemagglutinin and/or neuraminidase from the avian virus. The resulting new virus might then be able to infect humans and spread from person to person, but it would have surface proteins (hemagglutinin and/or neuraminidase) not previously seen in influenza viruses that infect humans.**

Human virus gene pool

Avian virus gene pool

Swine virus gene pool

Triple reassortment

2009 H1N1 pandemic

1918 H1N1 pandemic

## Influenza A Pandemics During the Past 100 Years

| | |
|---|---|
| 1918–19 | H1N1 caused up to 50 million deaths worldwide. Virus has avian flu–like genes. |
| 1957–58 | H2N2 caused about 70,000 deaths in the United States. First identified in China in late February 1957. Viruses contained a combination of genes from a human influenza virus and an avian influenza virus. |
| 1968–69 | H3N2 caused about 34,000 deaths in the United States. This virus contained genes from a human influenza virus and an avian influenza virus. |
| 2009–10 | H1N1 caused at least 14,000 deaths worldwide. A vaccine was available in developed and developing countries 3 months after the first cases. |

## TABLE 13.2 Families of Viruses That Affect Humans

| Class* | Characteristics/ Dimensions | Viral Family | Important Genera | Clinical or Special Features |
|---|---|---|---|---|
| I | **DOUBLE-STRANDED DNA** | | | |
| | Nonenveloped | | | |
| | 70–90 nm | Adenoviridae | *Mastadenovirus* | Medium-sized viruses that cause various respiratory infections in humans; some cause tumors in animals. |
| | 40–57 nm | Papovaviridae | *Papillomavirus* (human wart virus) *Polyomavirus* | Small viruses that cause warts and cervical and anal cancer in humans. Refer to Chapters 21 and 26. |
| | Enveloped | | | |
| | 200–350 nm | Poxviridae | *Orthopoxvirus* (vaccinia and smallpox viruses) *Molluscipoxvirus* | Very large, complex, brick-shaped viruses that cause smallpox (variola), molluscum contagiosum (wartlike skin lesion), and cowpox. Refer to Chapter 21. |
| | 150–200 nm | Herpesviridae | *Simplexvirus* (HHV-1 and -2) *Varicellovirus* (HHV-3) *Lymphocryptovirus* (HHV-4) *Cytomegalovirus* (HHV-5) *Roseolovirus* (HHV-6 and HHV-7) *Rhadinovirus* (HHV-8) | Medium-sized viruses that cause various human diseases: fever blisters, chickenpox, shingles, and infectious mononucleosis; cause a type of human cancer called Burkitt's lymphoma. Refer to Chapters 21, 23, and 26. |
| II | **SINGLE-STRANDED DNA** | | | |
| | Nonenveloped | | | |
| | 18–25 nm | Parvoviridae | Human parvovirus B19 | Fifth disease; anemia in immunocompromised patients. Refer to Chapter 21. |
| III | **DOUBLE-STRANDED RNA** | | | |
| | Nonenveloped 60–80 nm | Reoviridae | *Reovirus* *Rotavirus* | Generally mild respiratory infections transmitted by arthropods; Colorado tick fever is the best-known. Refer to Chapter 25. |
| IV | **SINGLE-STRANDED RNA, + STRAND** | | | |
| | Nonenveloped | | | |
| | 28–30 nm | Picornaviridae | *Enterovirus* *Rhinovirus* (common cold virus) Hepatitis A virus | Includes the polio-, coxsackie-, and echoviruses; hand-foot-mouth virus; more than 100 rhinoviruses exist and are the most common cause of colds. Refer to Chapters 22, 24, and 25. |
| | 35–40 nm | Caliciviridae | Hepatitis E virus *Norovirus* | Includes causes of gastroenteritis and hepatitis E. Refer to Chapter 25. |
| | Enveloped | | | |
| | 60–70 nm | Togaviridae | *Alphavirus* *Rubivirus* (rubella virus) | Includes many viruses transmitted by arthropods (*Alphavirus*); diseases include eastern equine encephalitis (EEE), western equine encephalitis (WEE), and chikungunya. Rubella virus is transmitted by the respiratory route. Refer to Chapters 21, 22, and 23. |

**TABLE 13.2** (continued)

| Class* | Characteristics/ Dimensions | Viral Family | Important Genera | Clinical or Special Features |
|---|---|---|---|---|
| | 40–50 nm | Flaviviridae  | *Flavivirus* <br> *Pestivirus* <br> Hepatitis C virus | Can replicate in arthropods that transmit them; diseases include yellow fever, dengue, Zika, and West Nile encephalitis. Refer to Chapters 22, 23, and 25. |
| | 80–160 nm | Coronaviridae  | *Coronavirus* | Associated with upper respiratory tract infections and the common cold; SARS virus, MERS-CoV. Refer to Chapter 24. |
| V | **SINGLE-STRANDED RNA, − STRAND** | | | |
| | **One strand of RNA, Enveloped** | | | |
| | 70–180 nm | Rhabdoviridae 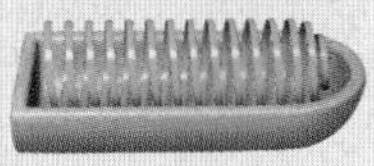 | *Vesiculovirus* (vesicular stomatatis virus) <br> *Lyssavirus* (rabies virus) | Bullet-shaped viruses with a spiked envelope; cause rabies and numerous animal diseases. Refer to Chapter 22. |
| | 80–14,000 nm | Filoviridae 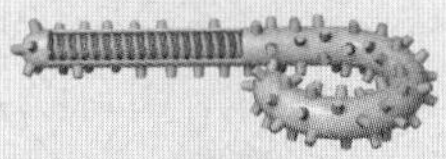 | *Filovirus* | Enveloped, helical viruses; Ebola and Marburg viruses are filoviruses. Refer to Chapter 23. |
| | 150–300 nm | Paramyxoviridae  | *Paramyxovirus Morbillivirus* (measles virus) | Paramyxoviruses cause parainfluenza, mumps, and Newcastle disease in chickens. Refer to Chapters 21, 24, and 25. |
| | **Virusoid or Satellite RNA** | | | |
| | 32 nm | Deltaviridae  | Hepatitis D | Depends on coinfection with hepadnavirus. Refer to Chapter 25. |
| | **Multiple Strands of RNA, Enveloped** | | | |
| | 80–200 nm | Orthomyxoviridae  | Influenza virus A, B, and C | Envelope spikes can agglutinate red blood cells. Refer to Chapter 24. |
| | 90–120 nm | Bunyaviridae  | *Bunyavirus* (California encephalitis virus) <br><br> *Hantavirus* | Hantaviruses cause hemorrhagic fevers such as Korean hemorrhagic fever and *Hantavirus* pulmonary syndrome; associated with rodents. Refer to Chapters 22 and 23. |
| | 110–130 nm | Arenaviridae  | *Arenavirus* | Helical capsids contain RNA-containing granules; cause lymphocytic choriomeningitis, Venezuelan hemorrhagic fever, and Lassa fever. Refer to Chapter 23. |
| VI | **SINGLE-STRANDED RNA, PRODUCE DNA** | | | |
| | **Enveloped** | | | |
| | 100–120 nm | Retroviridae  | Oncoviruses <br> *Lentivirus* (HIV) | Includes all RNA tumor viruses. Oncoviruses cause leukemia and tumors in animals; *Lentivirus* causes AIDS. Refer to Chapter 19. |

*(continued)*

| Class* | Characteristics/ Dimensions | Viral Family | Important Genera | Clinical or Special Features |
|---|---|---|---|---|
| VII | DOUBLE-STRANDED DNA, USE REVERSE TRANSCRIPTASE ENVELOPED | | | |
| | 42 nm | Hepadnaviridae | *Hepadnavirus* (hepatitis B virus) | After protein synthesis, hepatitis B virus uses reverse transcriptase to produce its DNA from mRNA; causes hepatitis B and liver tumors. Refer to Chapter 25. |

**TABLE 13.2** Families of Viruses That Affect Humans *(continued)*

*The Baltimore classification scheme was developed by David Baltimore, the discoverer of retroviruses.

# Isolation, Cultivation, and Identification of Viruses

## LEARNING OBJECTIVES

13-5 Describe how bacteriophages are cultured.

13-6 Describe how animal viruses are cultured.

13-7 List three techniques used to identify viruses.

The fact that viruses can't multiply outside a living host cell complicates their detection, enumeration, and identification. Viruses must be provided with living cells instead of a fairly simple chemical medium. Living plants and animals are difficult and expensive to maintain, and pathogenic viruses that grow only in higher primates and human hosts cause additional complications. However, viruses that use bacterial cells as a host (bacteriophages) are rather easily grown on bacterial cultures. This is one reason so much of our understanding of viral multiplication has come from bacteriophages.

## Growing Bacteriophages in the Laboratory

Bacteriophages can be grown either in suspensions of bacteria in liquid media or in bacterial cultures on solid media. The use of solid media makes it possible to use the *plaque method* to detect and count viruses. A bacteriophage sample is mixed with host bacteria and melted agar. The agar containing the bacteriophages and host bacteria is then poured into a Petri plate containing a hardened layer of agar growth medium. The virus-bacteria mixture solidifies into a thin top layer that contains a layer of bacteria approximately one cell thick. Each virus infects a bacterium, multiplies, and releases several hundred new viruses. These newly produced viruses infect other bacteria in the immediate vicinity, and more new viruses are produced. Following several viral multiplication cycles, all the bacteria in the area surrounding the original virus are destroyed. This produces a number of clearings, or **plaques**, visible against a lawn of bacterial growth on the surface of the agar (**Figure 13.6**). While the plaques form, uninfected bacteria elsewhere in the Petri plate multiply rapidly and produce a turbid background.

Theoretically each plaque corresponds to a single virus in the initial suspension. Therefore, the concentrations of viral suspensions measured by the number of plaques are usually expressed in terms of **plaque-forming units (PFU)**.

## Growing Animal Viruses in the Laboratory

In the laboratory, three methods are commonly used for culturing animal viruses. These methods involve using living animals, embryonated eggs, or cell cultures.

### In Living Animals

Some animal viruses can be cultured only in living animals, such as mice, rabbits, and guinea pigs. Most experiments to study the immune system's response to viral infections must also be performed in virally infected live animals. Animal inoculation may be used as a diagnostic procedure for identifying and isolating a virus from a clinical specimen. After the animal is inoculated with the specimen, the animal is observed for signs of disease or is killed so that infected tissues can be examined for the virus.

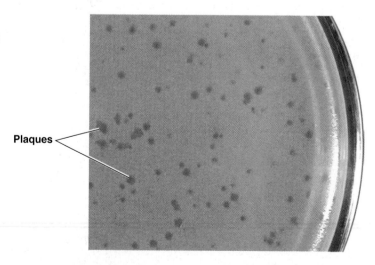

Plaques

**Figure 13.6 Viral plaques formed by bacteriophages.** Clear viral plaques of various sizes have been formed by bacteriophage λ (lambda) on a lawn of *E. coli*.

**Q** What is a plaque-forming unit?

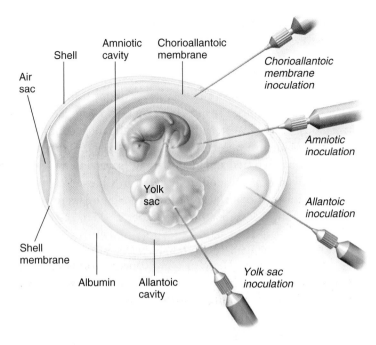

**Figure 13.7 Inoculation of an embryonated egg.** The viruses will grow on the membrane at the inoculation site.

**Q** Why are viruses grown in eggs and not in culture media?

Some human viruses can't be grown in animals or can be grown but don't cause disease. The lack of natural animal models for AIDS has slowed our understanding of its disease process and prevented experimentation with drugs that inhibit growth of the virus in vivo. Chimpanzees can be infected with one subspecies of human immunodeficiency virus (HIV-1, genus *Lentivirus*), but because they don't show symptoms of the disease, they can't be used to study the effects of viral growth and disease treatments. AIDS vaccines are presently being tested in humans, but the disease progresses so slowly in humans that it can take years to determine the effectiveness of these vaccines. In 1986, simian AIDS (an immunodeficiency disease of green monkeys) was reported, followed in 1987 by feline AIDS (an immunodeficiency

disease of domestic cats). These diseases are caused by lentiviruses, which are closely related to HIV, and the diseases develop within a few months, thus providing a model for studying viral growth in different tissues. In 1990, a way to infect mice with HIV was found when immunodeficient mice were grafted to produce human T cells and human gamma globulin. The mice provide a reliable model for studying viral replication, although they don't provide models for vaccine development.

### In Embryonated Eggs

If the virus will grow in an *embryonated egg*, this can be a fairly convenient and inexpensive form of host for many animal viruses. A hole is drilled in the shell of the embryonated egg, and a viral suspension or suspected virus-containing tissue is injected into the egg's fluid. There are several membranes in an egg, and the virus is injected near the one most appropriate for its growth (**Figure 13.7**). Viral growth is signaled by the death of the embryo, by embryo cell damage, or by the formation of typical pocks or lesions on the egg membranes. This method was once the most widely used method of isolating and growing viruses, and it's still used to grow viruses for some vaccines. For this reason, you may be asked if you're allergic to eggs before receiving a vaccination, because egg proteins may be present in the viral vaccine preparations. (Allergic reactions will be discussed in Chapter 19.)

### In Cell Cultures

**Cell cultures** have replaced embryonated eggs as the preferred type of growth medium for many viruses. Cell cultures consist of cells grown in culture media in the laboratory. Because these cultures are generally rather homogeneous collections of cells and can be propagated and handled much like bacterial cultures, they are more convenient to work with than whole animals or embryonated eggs.

Cell culture lines are started by treating a slice of animal tissue with enzymes that separate the individual cells (**Figure 13.8**). These cells are suspended in a solution that

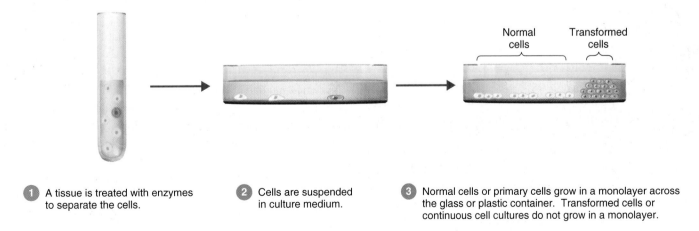

**1** A tissue is treated with enzymes to separate the cells.

**2** Cells are suspended in culture medium.

**3** Normal cells or primary cells grow in a monolayer across the glass or plastic container. Transformed cells or continuous cell cultures do not grow in a monolayer.

**Figure 13.8 Cell cultures.** Transformed cells can be grown indefinitely in laboratory culture.

**Q** Why are transformed cells referred to as "immortal"?

provides the osmotic pressure, nutrients, and growth factors needed for the cells to grow. Normal cells tend to adhere to the glass or plastic container and reproduce to form a monolayer. Viruses infecting such a monolayer sometimes cause the cells of the monolayer to deteriorate as they multiply. This cell deterioration, called **cytopathic effect (CPE)**, is illustrated in **Figure 13.9**. CPE can be detected and counted in much the same way as plaques caused by bacteriophages on a lawn of bacteria and reported as PFU/ml.

Viruses may be grown in primary or continuous cell lines. **Primary cell lines**, derived from tissue slices, tend to die out after only a few generations. Certain cell lines, called **diploid cell lines**, developed from human embryos can be maintained for about 100 generations and are widely used for culturing viruses that require a human host. Cell lines developed from embryonic human cells are used to culture rabies virus for a rabies vaccine called human diploid culture vaccine (see Chapter 22).

When viruses are routinely grown in a laboratory, **continuous cell lines** are used. These are transformed (cancerous) cells that can be maintained through an indefinite number of generations, and they're sometimes called immortal cell lines (see the discussion of transformation on page 385). One of these, the HeLa cell line, was isolated from the cancer of a woman (*Henrietta Lacks*) who died in 1951. In spite of the success of cell culture in viral isolation and growth, there are still some viruses that have never been successfully cultivated in cell culture.

The idea of cell culture dates back to the end of the nineteenth century, but it wasn't a practical laboratory technique until the development of antibiotics in the years following World War II. A major problem with cell culture is that the cell lines must be kept free of microbial contamination. The maintenance of cell culture lines requires trained technicians with considerable experience working on a full-time basis. Because of these difficulties, most hospital laboratories and many state health laboratories don't isolate and identify viruses in clinical work. Instead, the tissue or serum samples are sent to central laboratories that specialize in such work.

## Viral Identification

Identifying viral isolates isn't an easy task. For one thing, viruses can't be seen without the use of an electron microscope. Serological methods, such as Western blotting, are the most commonly used means of identification (see Figure 10.13, page 284). In these tests, the virus is detected and identified by its reaction with antibodies. We'll discuss antibodies in detail in Chapter 17 and a number of immunological tests for identifying viruses in Chapter 18. Observation of cytopathic effects, described in Chapter 15 (pages 436–438), is also useful for identifying a virus.

Virologists can identify and characterize viruses by using such modern molecular methods as use of restriction fragment length polymorphisms (RFLPs) (Chapter 9, page 258) and the polymerase chain reaction (PCR) (Chapter 9, page 247).

---

**CHECK YOUR UNDERSTANDING**

✔ **13-5** What is the plaque method?

✔ **13-6** Why are continuous cell lines of more practical use than primary cell lines for culturing viruses?

✔ **13-7** What tests could you use to identify influenza virus in a patient?

---

# Viral Multiplication

**LEARNING OBJECTIVES**

**13-8**   Describe the lytic cycle of T-even bacteriophages.

**13-9**   Describe the lysogenic cycle of bacteriophage lambda.

**13-10** Compare and contrast the multiplication cycle of DNA and RNA-containing animal viruses.

The nucleic acid in a virion contains only a few of the genes needed for the synthesis of new viruses. These include genes for the virion's structural components, such as the capsid proteins, and genes for a few of the enzymes used in the viral life cycle. These enzymes are synthesized and functional only when the virus is within the host cell. Viral enzymes are almost entirely concerned with replicating or processing viral nucleic acid. Enzymes needed for protein synthesis, ribosomes, tRNA,

> ASM: The synthesis of viral genetic material and proteins is dependent on host cells.

and energy production are supplied by the host cell and are used for synthesizing viral proteins, including viral enzymes. Although the smallest nonenveloped virions don't contain preformed enzymes, the larger virions may contain one or a few enzymes or mRNA, which usually function in helping the virus penetrate the host cell; replicate its own nucleic acid; or begin protein synthesis.

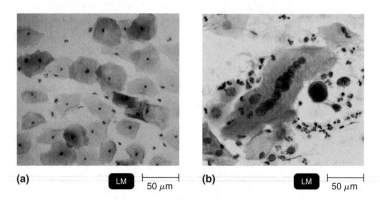

**(a)** ⬛ LM ⊢ 50 μm      **(b)** ⬛ LM ⊢ 50 μm

**Figure 13.9 The cytopathic effect of viruses. (a)** Uninfected human cervical cells; each has one nucleus. **(b)** After infection with HHV-2, the red cervical cell has many nuclei that are filled with viruses.

**Q** How did HHV-2 infection affect the cells?

Thus, for a virus to multiply, it must invade a host cell and take over the host's metabolic machinery. A single virion can give rise to several or even thousands of similar viruses in a single host cell. This process can drastically change the host cell and usually causes its death. In a few viral infections, cells survive and continue to produce viruses indefinitely.

The multiplication of viruses can be demonstrated with a **one-step growth curve** (**Figure 13.10**). The data are obtained by infecting every cell in a culture and then testing the culture medium and cells for virions and viral proteins and nucleic acids.

## Multiplication of Bacteriophages

Although the means by which a virus enters and exits a host cell may vary, the basic mechanism of viral multiplication is similar for all viruses. Bacteriophages can multiply by two alternative mechanisms: the lytic cycle or the lysogenic cycle. The **lytic cycle** ends with the lysis and death of the host cell, whereas the host cell remains alive in the **lysogenic cycle**. Because the *T-even bacteriophages* (T2, T4, and T6) have been studied most extensively, we'll describe the multiplication of T-even bacteriophages in their host, *E. coli*, as an example of the lytic cycle.

### T-Even Bacteriophages: The Lytic Cycle

The virions of T-even bacteriophages are large, complex, and nonenveloped, with a characteristic head-and-tail structure shown in Figure 13.5a and **Figure 13.11**. The length of DNA contained in these bacteriophages is only about 6% of that contained in *E. coli*, yet the phage has enough DNA for over 100 genes. The multiplication cycle of these phages, like that of

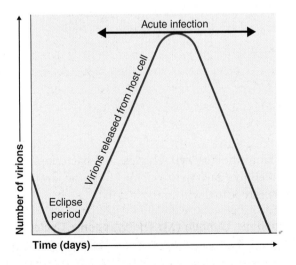

**Figure 13.10 A viral one-step growth curve.** No new infective virions are found in a culture until after biosynthesis and maturation have taken place. Most infected cells die as a result of infection; consequently, new virions won't be produced.

**Q** What can be found in the cell during biosynthesis and maturation?

all viruses, occurs in five distinct stages: attachment, penetration, biosynthesis, maturation, and release.

**Attachment** ① After a chance collision between phage particles and bacteria, *attachment*, or *adsorption*, occurs. During this process, an attachment site on the virus attaches to a complementary receptor site on the bacterial cell. This attachment is a chemical interaction in which weak bonds are formed between the attachment and receptor sites. T-even bacteriophages use fibers at the end of the tail as attachment sites. The complementary receptor sites are on the bacterial cell wall.

**Penetration** ② After attachment, the T-even bacteriophage injects its DNA (nucleic acid) into the bacterium. To do this, the bacteriophage's tail releases an enzyme, **phage lysozyme**, which breaks down a portion of the bacterial cell wall. During the process of *penetration*, the tail sheath of the phage contracts, and the tail core is driven through the cell wall. When the tip of the core reaches the plasma membrane, the DNA from the bacteriophage's head passes through the tail core, through the plasma membrane, and enters the bacterial cell. The capsid remains outside the bacterial cell. Therefore, the phage particle functions like a hypodermic syringe to inject its DNA into the bacterial cell.

**Biosynthesis** ③ Once the bacteriophage DNA has reached the cytoplasm of the host cell, the biosynthesis of viral nucleic acid and protein occurs. Host protein synthesis is stopped by virus-induced degradation of the host DNA, viral proteins that interfere with transcription, or the repression of translation.

Initially, the phage uses the host cell's nucleotides and several of its enzymes to synthesize many copies of phage DNA. Soon after, the biosynthesis of viral proteins begins. Any RNA transcribed in the cell is mRNA transcribed from phage DNA for the biosynthesis of phage enzymes and capsid proteins. The host cell's ribosomes, enzymes, and amino acids are used for translation. Genetic controls regulate when different regions of phage DNA are transcribed into mRNA during the multiplication cycle. For example, early messages are translated into early phage proteins, the enzymes used in the synthesis of phage DNA. Also, late messages are translated into late phage proteins for the synthesis of capsid proteins.

For several minutes following infection, complete phages can't be found in the host cell. Only separate components—DNA and protein—can be detected. The period during viral multiplication when complete, infective virions aren't yet present is called the **eclipse period**.

**Maturation** ④ In the next sequence of events, *maturation* occurs. In this process, bacteriophage DNA and capsids are assembled into complete virions. The viral components essentially assemble into a viral particle spontaneously, eliminating

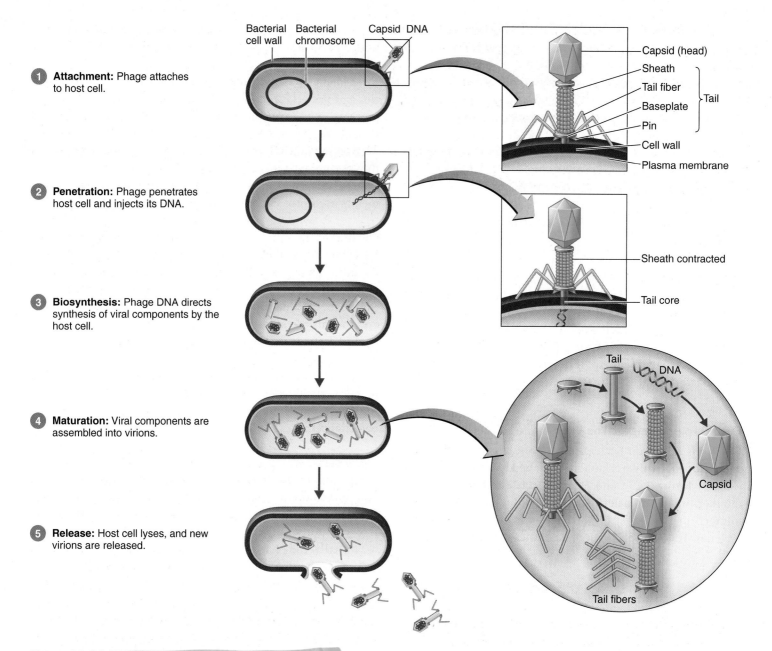

1. **Attachment:** Phage attaches to host cell.

2. **Penetration:** Phage penetrates host cell and injects its DNA.

3. **Biosynthesis:** Phage DNA directs synthesis of viral components by the host cell.

4. **Maturation:** Viral components are assembled into virions.

5. **Release:** Host cell lyses, and new virions are released.

Bacterial cell wall   Bacterial chromosome   Capsid   DNA

Capsid (head)
Sheath
Tail fiber
Baseplate
Pin
Cell wall
Plasma membrane
Tail

Sheath contracted
Tail core

Tail
DNA
Capsid
Tail fibers

**Figure 13.11 The lytic cycle of a T-even bacteriophage.**

**Q** What is the result of the lytic cycle?

the need for many nonstructural genes and gene products. The phage heads and tails are separately assembled from protein subunits, and the head is filled with phage DNA and attached to the tail.

**Release** 5 The final stage of viral multiplication is the *release* of virions from the host cell. The term **lysis** is generally used for this stage in the multiplication of T-even phages because in this case, the plasma membrane actually breaks open (lyses). Lysozyme, which is encoded by a phage gene, is synthesized within the cell. This enzyme causes the bacterial cell wall to break down, and the newly produced bacteriophages are

released from the host cell. The released bacteriophages infect other susceptible cells in the vicinity, and the viral multiplication cycle is repeated within those cells.

### Bacteriophage Lambda (λ): The Lysogenic Cycle

In contrast to T-even bacteriophages, some viruses don't cause lysis and death of the host cell when they multiply. These *lysogenic phages* (also called *temperate phages*) may indeed proceed through a lytic cycle, but they are also capable of incorporating their DNA into the host cell's DNA to begin a lysogenic cycle. In **lysogeny**, the phage remains latent (inactive). The participating bacterial host cells are known as *lysogenic cells*.

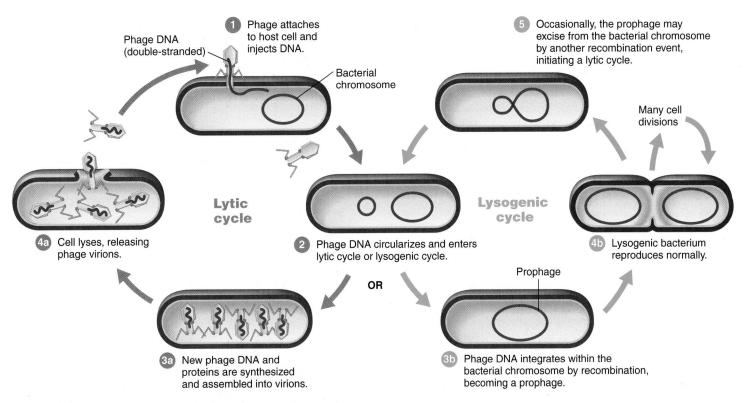

**Figure 13.12 The lysogenic cycle of bacteriophage λ in *E. coli*.**

**Q** How does lysogeny differ from the lytic cycle?

We will use the bacteriophage λ (lambda), a well-studied lysogenic phage, as an example of the lysogenic cycle (**Figure 13.12**).

❶ Upon penetration into an *E. coli* cell,

❷ the originally linear phage DNA forms a circle.

❸ₐ This circle can multiply and be transcribed,

❹ᵦ leading to the production of new phage and to cell lysis (the lytic cycle).

❸ᵦ Alternatively, the circle can recombine with and become part of the circular bacterial DNA (the lysogenic cycle). The inserted phage DNA is now called a **prophage**. Most of the prophage genes are repressed by two repressor proteins that are the products of phage genes. These repressors stop transcription of all the other phage genes by binding to operators. Thus, the phage genes that would otherwise direct the synthesis and release of new virions are turned off, in much the same way that the genes of the *E. coli lac* operon are turned off by the *lac* repressor (Figure 8.12, page 219).

Every time the host cell's machinery replicates the bacterial chromosome,

❹ᵦ it also replicates the prophage DNA. The prophage remains latent within the progeny cells.

❺ However, a rare spontaneous event, or the action of UV light or certain chemicals, can lead to the excision (popping-out) of the phage DNA, and to initiation of the lytic cycle.

There are three important results of lysogeny. First, the lysogenic cells are immune to reinfection by the same phage. (However, the host cell isn't immune to infection by other phage types.) The second result of lysogeny is **phage conversion**; that is, the host cell may exhibit new properties. For example, the bacterium *Corynebacterium diphtheriae*, which causes diphtheria, is a pathogen whose disease-producing properties are related to the synthesis of a toxin. The organism can produce toxin only when it carries a lysogenic phage, because the prophage carries the gene coding for the toxin. As another example, only streptococci carrying a lysogenic phage are capable of causing toxic shock syndrome. The toxin produced by *Clostridium botulinum*, which causes botulism, is encoded by a prophage gene.

The third result of lysogeny is that it makes **specialized transduction** possible. Bacterial genes can be picked up in a phage coat and transferred to another bacterium in a process called *generalized transduction* (see Figure 8.32, page 237). Any bacterial genes can be transferred by generalized transduction because the host chromosome is broken down into fragments, any of which can be packaged into a phage coat. In specialized transduction, however, only certain bacterial genes can be transferred.

Specialized transduction is mediated by a lysogenic phage, which packages bacterial DNA *along with* its own DNA in the same capsid. When a prophage is excised from the host chromosome, adjacent genes from either side may remain attached to the phage DNA. In **Figure 13.13**, bacteriophage λ has picked up the *gal* gene for galactose fermentation from its galactose-positive host. The phage carries this gene to a galactose-negative cell, which then becomes galactose-positive.

Certain animal viruses can undergo processes very similar to lysogeny. Animal viruses that can remain latent in cells for long periods without multiplying or causing disease may become inserted into a host chromosome or remain separate from host DNA in a repressed state (as some lysogenic phages). Cancer-causing viruses may also be latent, as we'll discuss later in the chapter.

Play Viral Replication: *Virulent Bacteriophages, Temperate Bacteriophages;* **Transduction:** *Specialized Transduction* @**Mastering**Microbiology

### CHECK YOUR UNDERSTANDING

✔ **13-8** How do bacteriophages get nucleotides and amino acids if they don't have any metabolic enzymes?

✔ **13-9** *Vibrio cholerae* produces toxin and is capable of causing cholera only when it is lysogenic. What does this mean?

## Multiplication of Animal Viruses

The multiplication of animal viruses follows the basic pattern of bacteriophage multiplication but has several differences, summarized in Table 13.3. Animal viruses differ from phages in their mechanism of entering the host cell. Also, once the virus is inside, the synthesis and assembly of the new viral components are somewhat different, partly because of the differences between prokaryotic cells and eukaryotic cells. Animal viruses may have certain types of enzymes not found in phages. Finally, the mechanisms of maturation and release, and the effects on the host cell, differ in animal viruses and phages.

A virus needs live host cells but must stop synthesis of host proteins, so the viral genes are translated. Research currently indicates that viruses use several mechanisms to inhibit expression of host cell genes. Early proteins translated from the viral genome may block transcription, existing mRNA, or in-progress translation. In the following discussion of the multiplication of animal viruses, we'll consider the processes that are shared by both DNA- and RNA-containing animal viruses. These processes are attachment, entry, uncoating, and release. We'll also examine how DNA- and RNA-containing viruses differ with respect to their processes of biosynthesis.

### Attachment

Like bacteriophages, animal viruses have attachment sites that attach to complementary receptor sites on the host cell's surface. However, the receptor sites of animal cells are proteins

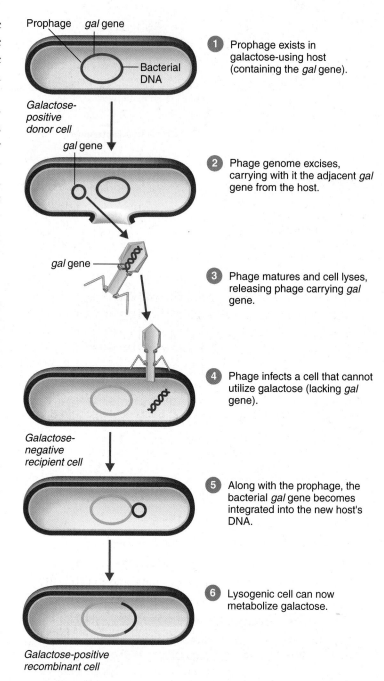

**Figure 13.13 Specialized transduction.** When a prophage is excised from its host chromosome, it can take with it a bit of the adjacent DNA from the bacterial chromosome.

**Q** How does specialized transduction differ from the lytic cycle?

and glycoproteins of the plasma membrane. Moreover, animal viruses don't possess appendages like the tail fibers of some bacteriophages. The attachment sites of animal viruses are distributed over the surface of the virus, and the sites themselves vary from one group of viruses to another. In adenoviruses, which are icosahedral viruses, the attachment sites are small fibers at the corners of the icosahedron (see Figure 13.2b). In

**TABLE 13.3** Bacteriophage and Animal Viral Multiplication Compared

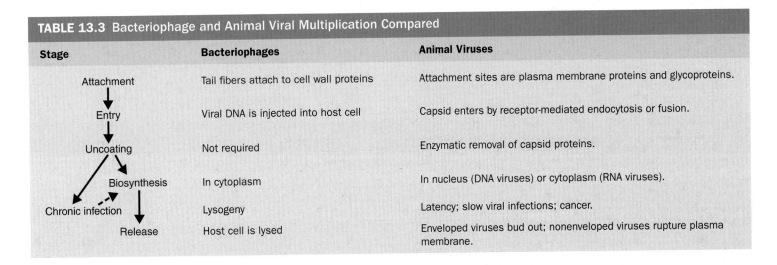

| Stage | Bacteriophages | Animal Viruses |
|---|---|---|
| Attachment | Tail fibers attach to cell wall proteins | Attachment sites are plasma membrane proteins and glycoproteins. |
| Entry | Viral DNA is injected into host cell | Capsid enters by receptor-mediated endocytosis or fusion. |
| Uncoating | Not required | Enzymatic removal of capsid proteins. |
| Biosynthesis | In cytoplasm | In nucleus (DNA viruses) or cytoplasm (RNA viruses). |
| Chronic infection | Lysogeny | Latency; slow viral infections; cancer. |
| Release | Host cell is lysed | Enveloped viruses bud out; nonenveloped viruses rupture plasma membrane. |

many of the enveloped viruses, such as influenza virus, the attachment sites are spikes located on the surface of the envelope (see Figure 13.3b). As soon as one spike attaches to a host receptor, additional receptor sites on the same cell migrate to the virus. Attachment is completed when many sites are bound.

Receptor sites are proteins of the host cell. The proteins have normal functions for the host and are hijacked by the virus. This could account for the individual differences in susceptibility to a particular virus. For example, people who lack the cellular receptor (called P antigen) for parvovirus B19 are naturally resistant to infection and don't get fifth disease (see page 607). Understanding the nature of attachment can lead to the development of drugs that prevent viral infections. Monoclonal antibodies (discussed in Chapter 18) that combine with a virus's attachment site or the cell's receptor site may soon be used to treat some viral infections.

### Entry

Following attachment, entry occurs. Many viruses enter into eukaryotic cells by **receptor-mediated endocytosis** (Chapter 4, page 97). A cell's plasma membrane continuously folds inward to form vesicles. These vesicles contain elements that originate outside the cell and are brought into the interior of the cell to be digested. If a virion attaches to the plasma membrane of a potential host cell, the host cell will enfold the virion and form a vesicle (**Figure 13.14a**).

Enveloped viruses can enter by an alternative method called **fusion**, in which the viral envelope fuses with the plasma membrane and releases the capsid into the cell's cytoplasm (Figure 13.14b).

### Uncoating

Viruses disappear during the eclipse period of an infection because they are taken apart inside the cell. **Uncoating** is the separation of the viral nucleic acid from its protein coat. This process varies with the type of virus. Some animal viruses

accomplish uncoating by the action of lysosomal enzymes of the host cell. These enzymes degrade the proteins of the viral capsid. The uncoating of poxviruses is completed by a specific enzyme encoded by the viral DNA and synthesized soon after infection. Uncoating of influenza virus occurs at the lower pH in a vesicle. Uncoating of togaviruses occurs at ribosomes in the host cytoplasm.

The biggest differences between viruses occur during biosynthesis of viral components. We'll discuss biosynthesis of DNA viruses and then biosynthesis of RNA viruses.

ASM: The replication cycles of viruses (lytic and lysogenic) differ among viruses and are determined by their unique structures and genomes.

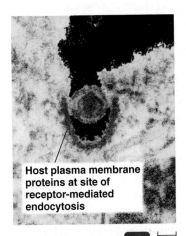

Host plasma membrane proteins at site of receptor-mediated endocytosis

TEM |— 40 nm

**(a)** Entry of pig retrovirus by receptor-mediated endocytosis

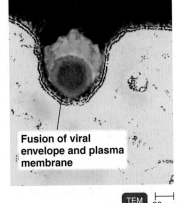

Fusion of viral envelope and plasma membrane

TEM |— 60 nm

**(b)** Entry of herpesvirus by fusion

**Figure 13.14 The entry of viruses into host cells.** After attachment, viruses enter host cells by **(a)** receptor-mediated endocytosis or **(b)** fusion of the viral envelope and cell membrane.

**Q** In which process is the cell actively taking in the virus?

**TABLE 13.4** The Biosynthesis of DNA and RNA Viruses Compared

| Viral Nucleic Acid | Virus Family | Special Features of Biosynthesis |
|---|---|---|
| DNA, single-stranded | Parvoviridae | Cellular enzyme transcribes viral DNA in nucleus. |
| DNA, double-stranded | Herpesviridae | Cellular enzyme transcribes viral DNA in nucleus. |
| | Papovaviridae | Cellular enzyme transcribes viral DNA in nucleus. |
| | Poxviridae | Viral enzyme transcribes viral DNA in cytoplasm. |
| DNA, reverse transcriptase | Hepadnaviridae | Cellular enzyme transcribes viral DNA in nucleus; reverse transcriptase copies mRNA to make viral DNA. |
| RNA, + strand | Picornaviridae | Viral RNA functions as a template for synthesis of RNA polymerase, which copies − strand RNA to make mRNA in cytoplasm. |
| | Togaviridae | |
| RNA, − strand | Rhabdoviridae | Viral enzyme copies viral RNA to make mRNA in cytoplasm. |
| RNA, double-stranded | Reoviridae | Viral enzyme copies − strand RNA to make mRNA in virion, in cytoplasm. |
| RNA, reverse transcriptase | Retroviridae | Viral enzyme copies viral RNA to make DNA in cytoplasm; DNA moves to nucleus. |

## The Biosynthesis of DNA Viruses

Generally, DNA-containing viruses replicate their DNA in the nucleus of the host cell by using viral enzymes, and they synthesize their capsid and other proteins in the cytoplasm by using host cell enzymes. Then the proteins migrate into the nucleus and are joined with the newly synthesized DNA to form virions. These virions are transported along the endoplasmic reticulum to the host cell's membrane for release. Herpesviruses, papovaviruses, adenoviruses, and hepadnaviruses all follow this pattern of biosynthesis (Table 13.4). Poxviruses are an exception because all of their components are synthesized in the cytoplasm.

As an example of the multiplication of a DNA virus, the sequence of events in papovavirus is shown in Figure 13.15.

**❶-❷** Following attachment, entry, and uncoating, the viral DNA is released into the nucleus of the host cell.

**❸** Transcription of a portion of the viral DNA—the "early" genes—occurs next. Translation follows. The products of these genes are enzymes that are required for the multiplication of viral DNA. In most DNA viruses, early transcription is carried out with the host's transcriptase (RNA polymerase); poxviruses, however, contain their own transcriptase.

**❹** Sometime after the initiation of DNA replication, transcription and translation of the remaining "late" viral genes occur. Late proteins include capsid and other structural proteins.

**❺** This leads to the synthesis of capsid proteins, which occurs in the cytoplasm of the host cell.

**❻** After the capsid proteins migrate into the nucleus of the host cell, maturation occurs; the viral DNA and capsid proteins assemble to form complete viruses.

**❼** Complete viruses are then released from the host cell.

Some DNA viruses are described below.

**Adenoviridae** Named after adenoids, from which they were first isolated, adenoviruses cause acute respiratory diseases—the common cold (**Figure 13.16a**).

**Poxviridae** All diseases caused by poxviruses, including smallpox and cowpox, include skin lesions (see Figure 21.10, page 603). *Pox* refers to pus-filled lesions. Viral multiplication is started by viral transcriptase; the viral components are synthesized and assembled in the cytoplasm of the host cell.

**Herpesviridae** Nearly 100 herpesviruses are known (Figure 13.16b). They are named after the spreading (*herpetic*) appearance of cold sores. Species of human herpesviruses (HHV) include HHV-1 and HHV-2, both in the genus *Simplexvirus*, which cause cold sores; HHV-3, *Varicellovirus*, which causes chickenpox; HHV-4, *Lymphocryptovirus*, which causes infectious mononucleosis; HHV-5, *Cytomegalovirus*, which causes CMV inclusion disease; HHV-6, *Roseolovirus*, which causes roseola; HHV-7, *Roseolovirus*, which infects most infants, causing measleslike rashes; and HHV-8, *Rhadinovirus*, which causes Kaposi's sarcoma, primarily in AIDS patients.

**Papovaviridae** Papovaviruses are named for *pa*pillomas (warts), *po*lyomas (tumors), and *va*cuolation (cytoplasmic vacuoles produced by some of these viruses). Warts are caused by members of the genus *Papillomavirus*. Some *Papillomavirus* species are capable of transforming cells and causing cancer. Viral DNA is replicated in the host cell's nucleus along with host cell chromosomes. Host cells may proliferate, resulting in a tumor.

# Replication of a DNA-Containing Animal Virus

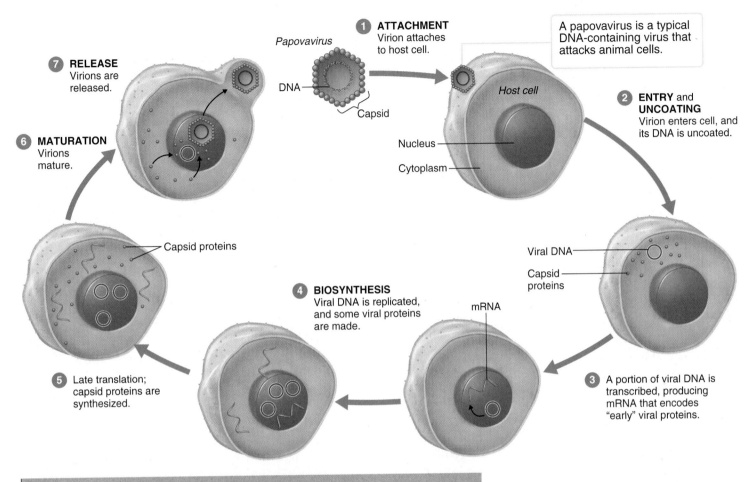

**1 ATTACHMENT**
Virion attaches
to host cell.

*Papovavirus*

DNA

Capsid

A papovavirus is a typical
DNA-containing virus that
attacks animal cells.

Host cell

Nucleus

Cytoplasm

**2 ENTRY and UNCOATING**
Virion enters cell, and
its DNA is uncoated.

**7 RELEASE**
Virions are
released.

**6 MATURATION**
Virions
mature.

Capsid proteins

Viral DNA

Capsid
proteins

**4 BIOSYNTHESIS**
Viral DNA is replicated,
and some viral proteins
are made.

mRNA

**5** Late translation;
capsid proteins are
synthesized.

**3** A portion of viral DNA is
transcribed, producing
mRNA that encodes
"early" viral proteins.

## KEY CONCEPTS

- Viral replication in animals generally follows these steps: attachment, entry, uncoating, biosynthesis of nucleic acids and proteins, maturation, and release.

- Knowledge of viral replication phases is important for drug development strategies and for understanding disease pathology.

Capsomeres

**(a)** *Mastadenovirus*

SEM
100 nm

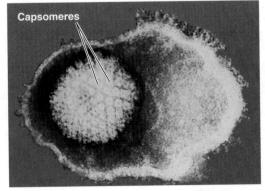

Capsomeres

**(b)** *Simplexvirus*

TEM
50 nm

**Figure 13.16 DNA-containing animal viruses. (a)** Negatively stained adenoviruses that have been concentrated in a centrifuge. The individual capsomeres are clearly visible. **(b)** The envelope around this human herpes virus capsid has broken, giving a "fried egg" appearance.

**Q** What is the morphology of these viruses?

**Hepadnaviridae** Hepadnaviridae are so named because they cause *hepatitis* and contain *DNA* (Figure 25.14, page 742). The only genus in this family causes hepatitis B. (Hepatitis A, C, D, E, F, and G viruses, although not related to each other, are RNA viruses. Hepatitis is discussed in Chapter 25.) Hepadnaviruses differ from other DNA viruses because they synthesize DNA by copying RNA, using viral reverse transcriptase. This DNA is the template for viral mRNA and the virus's DNA genome. This enzyme is discussed later with the retroviruses, the only other family with reverse transcriptase.

### The Biosynthesis of RNA Viruses

The multiplication of RNA viruses is essentially the same as that of DNA viruses, except RNA viruses multiply in the host cell's cytoplasm. Several different mRNA formation mechanisms occur among different groups of RNA viruses (see Table 13.4). Although the details of these mechanisms are beyond the scope of this text, for comparative purposes we'll trace the multiplication cycles of the four nucleic acid types of RNA viruses (three of which are shown in **Figure 13.17**). The major differences among the multiplication processes lie in how mRNA and viral RNA are produced. These viruses have **RNA-dependent RNA polymerase**. This enzyme isn't encoded in any cell's genome. Viral genes causes the enzyme to be made by a host cell. This enzyme catalyzes the synthesis of another strand of RNA, which is complementary in base sequence to the original infecting strand. Once viral RNA and viral proteins are synthesized, maturation occurs by similar means among all animal viruses, as we'll discuss shortly.

**Picornaviridae** Picornaviruses, such as enteroviruses and polio virus (see Chapter 22, page 630), are single-stranded RNA viruses. They are the smallest viruses; and the prefix *pico-* (small) plus *RNA* gives these viruses their name. The RNA within the virion is called a **sense strand** (+ **strand**), because it can act as mRNA. After attachment, penetration, and uncoating are completed, the single-stranded viral RNA is translated into two principal proteins. One inhibits the host cell's synthesis of RNA, and the other is RNA-dependent RNA polymerase. This enzyme copies the virus sense strand and makes an **antisense strand** (− **strand**), which serves as a template to produce additional + strands. The + strands may serve as mRNA for the translation of capsid proteins, may become incorporated into capsid proteins to form a new virus, or may serve as a template for continued RNA multiplication. Once viral RNA and viral protein are synthesized, maturation occurs.

**Togaviridae** Togaviruses, which include *arthropod-borne arbo*-viruses in the genus *Alphavirus* (see Chapter 22, page 637),

also contain a single + strand of RNA. Togaviruses are enveloped viruses; their name is from the Latin word for covering, *toga*. Keep in mind that these aren't the only enveloped viruses. After a − strand is made from the + strand, two types of mRNA are transcribed from the − strand. One type of mRNA is a short strand that codes for envelope proteins; the other, longer strand serves as mRNA for capsid proteins and can become incorporated into a capsid.

**Rhabdoviridae** Rhabdoviruses, such as rabies virus (genus *Lyssavirus*; see Chapter 22, page 632), are usually bullet-shaped (**Figure 13.18a**). *Rhabdo-* is from the Greek word for rod, which isn't really an accurate description of their morphology. They contain a single − strand of RNA. They also contain an RNA-dependent RNA polymerase that uses the − strand as a template from which to produce a + strand. The + strand serves as mRNA and as a template for synthesis of new viral RNA.

**Reoviridae** Reoviruses were named for their habitats: the respiratory and enteric (digestive) systems of humans. They weren't associated with any diseases when first discovered, so they were considered orphan viruses. Their name comes from the first letters of *r*espiratory, *e*nteric, and *o*rphan. Three serotypes are now known to cause respiratory tract and intestinal tract infections.

The capsid containing the double-stranded RNA is digested upon entering a host cell. Viral mRNA is produced in the cytoplasm, where it's used to synthesize more viral proteins. One of the newly synthesized viral proteins acts as RNA-dependent RNA polymerase to produce more − strands of RNA.

### CLINICAL CASE

Based on the abnormal LFT, Tina's physician diagnoses her with infectious hepatitis. This isn't the first case he's seen this month. As a matter of fact, the local health department has received reports of 31 other people with hepatitis. For a town of 4000 people, this is a large number. The health department needs to know what type of hepatitis they are dealing with because the term *hepatitis* describes any inflammation of the liver. Infectious hepatitis can be caused by a member of the Picornaviridae, Hepadnaviridae, or Flaviviridae.

**The health department will need to differentiate among these viral families. List the method of transmission, morphology, nucleic acid, and type of replication for these three viral families. (*Hint:* See Diseases in Focus 25.3 on page 743 for a complete list of the hepatitis viruses.)**

363 **380** 384 385

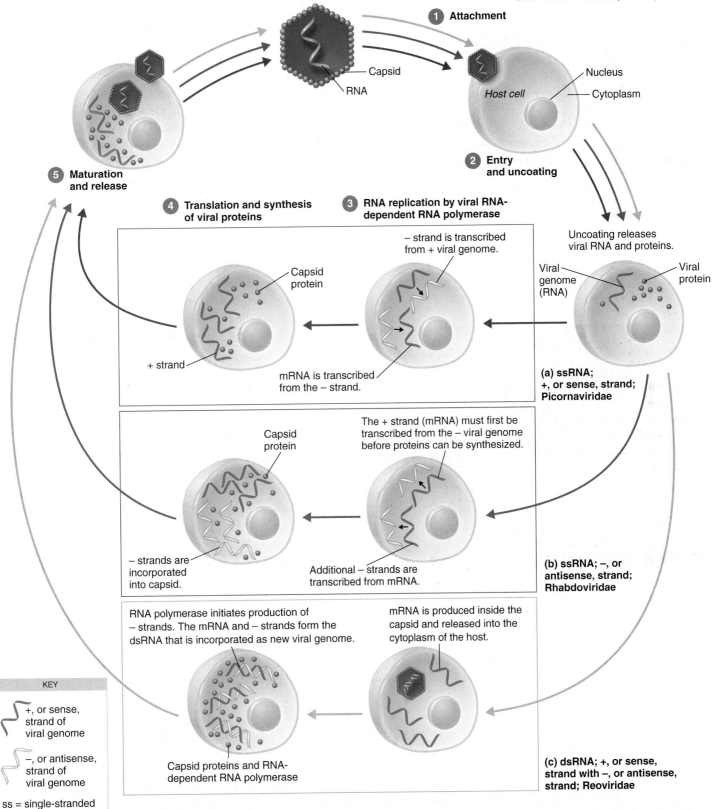

**Figure 13.17 Pathways of multiplication used by various RNA-containing viruses. (a)** After uncoating, single-stranded RNA (ssRNA) viruses with a + strand genome are able to synthesize proteins directly from their + strand. Using the + strand as a template, they transcribe − strands to produce additional + strands to serve as mRNA and be incorporated into capsid proteins as the viral genome. **(b)** The ssRNA viruses with a − strand genome must transcribe a + strand to serve as mRNA before they begin synthesizing proteins. The mRNA transcribes additional − strands for incorporation into capsid protein. Both ssRNA and **(c)** dsRNA viruses must use mRNA (+ strand) to code for proteins, including capsid proteins.

**Q** Why is − strand RNA made by picornaviruses and reoviruses? By rhabdoviruses?

KEY

+, or sense, strand of viral genome

−, or antisense, strand of viral genome

ss = single-stranded
ds = double-stranded

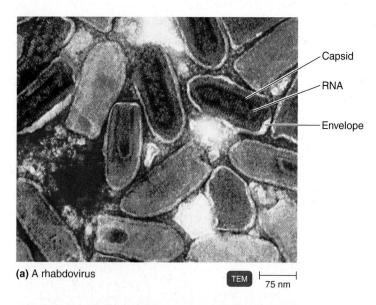

**(a)** A rhabdovirus

TEM | 75 nm

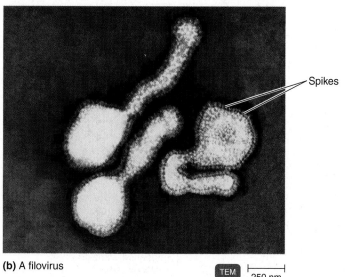

**(b)** A filovirus

TEM | 250 nm

**Figure 13.18 RNA-containing animal viruses. (a)** Vesicular stomatitis viruses, a member of the family Rhabdoviridae. **(b)** Marburg virus, found in African cave bats, causes hemorrhagic fever in humans.

**Q** Why do viruses with a + strand of RNA make a − strand of RNA?

The mRNA and − strand form the double-stranded RNA that is then surrounded by capsid proteins.

### Biosynthesis of RNA Viruses That Use DNA

This group includes retroviruses and oncogenic RNA viruses.

**Retroviridae** Many retroviruses infect vertebrates (see Figure 13.20b). One genus of retrovirus, *Lentivirus*, includes the subspecies HIV-1 and HIV-2, which cause AIDS (see Chapter 19, pages 544–554). The retroviruses that cause cancer will be discussed later in this chapter.

The formation of mRNA and RNA for new retrovirus virions is shown in **Figure 13.19**. These viruses carry **reverse transcriptase**, which uses the viral RNA as a template to produce complementary double-stranded DNA. This enzyme also degrades the original viral RNA. The name *retrovirus* is derived from the first letters of *reverse transcriptase*. The viral DNA is then integrated into a host cell chromosome as a **provirus**. Unlike a prophage, the provirus never comes out of the chromosome. As a provirus, HIV is protected from the host's immune system and antiviral drugs.

Sometimes the provirus simply remains in a latent state and replicates when the DNA of the host cell replicates. In other cases, the provirus is expressed and produces new viruses, which may infect adjacent cells. Mutagens such as gamma radiation can induce expression of a provirus. In oncogenic retroviruses, the provirus can also convert the host cell into a tumor cell; possible mechanisms for this phenomenon will be discussed later.

### Maturation and Release

The first step in viral maturation is the assembly of the protein capsid; this assembly is usually a spontaneous process. The capsids of many animal viruses are enclosed by an envelope consisting of protein, lipid, and carbohydrate, as noted earlier. Examples of such viruses include orthomyxoviruses and paramyxoviruses. The envelope protein is encoded by the viral genes and is incorporated into the plasma membrane of the host cell. The envelope lipid and carbohydrate are encoded by host cell genes and are present in the plasma membrane. The envelope actually develops around the capsid by a process called **budding** (Figure 13.20).

After the sequence of attachment, entry, uncoating, and biosynthesis of viral nucleic acid and protein, the assembled capsid containing nucleic acid pushes through the plasma membrane. As a result, a portion of the plasma membrane, now the envelope, adheres to the virus. This extrusion of a virus from a host cell is one method of release. Budding doesn't kill the host cell immediately, and in some cases the host cell survives.

Nonenveloped viruses are released through ruptures in the host cell plasma membrane. In contrast to budding, this type of release usually results in the death of the host cell.

Play Viral Replication:
*Overview, Animal Viruses*
@Mastering**Microbiology**

**CHECK YOUR UNDERSTANDING**

✔ **13-10** Describe the principal events of attachment, entry, uncoating, biosynthesis, maturation, and release of an enveloped DNA-containing virus.

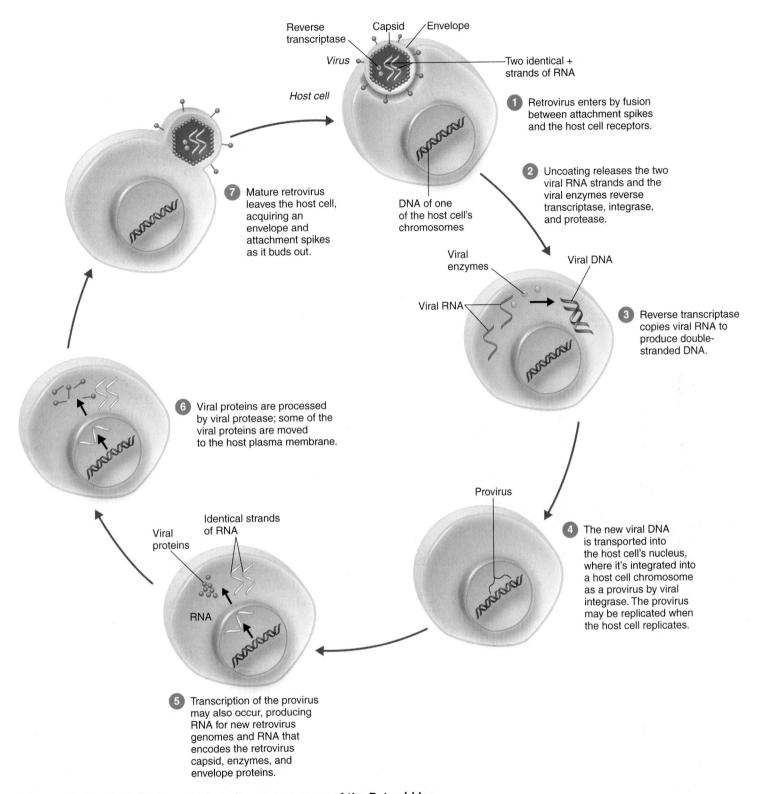

**Figure 13.19 Multiplication and inheritance processes of the Retroviridae.**
A retrovirus may become a provirus that replicates in a latent state, and it may produce new retroviruses.

**Q**  How does the biosynthesis of a retrovirus differ from that of other RNA viruses?

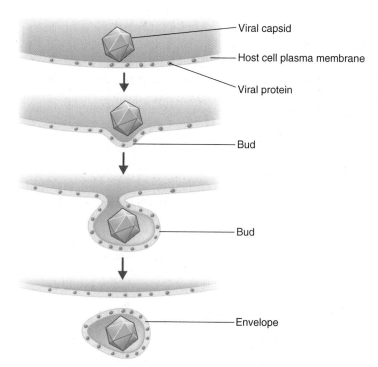

Viral capsid

Host cell plasma membrane

Viral protein

Bud

Bud

Envelope

**(a)** Release by budding

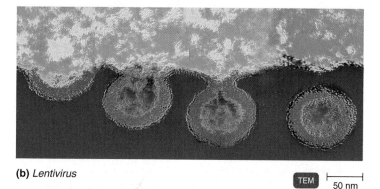

**(b)** *Lentivirus*

TEM  |— 50 nm —|

**Figure 13.20  Budding of an enveloped virus. (a)** A diagram of the budding process. **(b)** HIV budding from a T cell. Notice that the four budding viruses acquire their coats from the host plasma membrane.

**Q** Of what is a viral envelope composed?

# Viruses and Cancer

### LEARNING OBJECTIVES

**13-11**  Define *oncogene* and *transformed cell*.

**13-12**  Discuss the relationship between DNA- and RNA-containing viruses and cancer.

Several types of cancer are now known to be caused by viruses. Molecular biological research shows that the mechanisms of the diseases are similar, even when a virus doesn't cause the cancer.

The relationship between cancers and viruses was first demonstrated in 1908, when virologists Wilhelm Ellerman and Olaf Bang, working in Denmark, were trying to isolate the causative agent of chicken leukemia. They found that leukemia could be transferred to healthy chickens by cell-free filtrates that contained viruses. Three years later, F. Peyton Rous, working at the Rockefeller Institute in New York, found that a chicken **sarcoma** (cancer of connective tissue) can be similarly transmitted. Virus-induced **adenocarcinomas** (cancers of glandular epithelial tissue) in mice were discovered in 1936. At that time, it was clearly shown that mouse mammary gland tumors are transmitted from mother to offspring through the mother's milk. The cancer-causing SE polyoma virus was discovered and isolated in 1953 by American scientists Bernice Eddy and Sarah Stewart.

The viral cause of cancer can often go unrecognized for several reasons. First, most of the particles of some viruses infect cells but don't induce cancer. Second, cancer might not develop until long after viral infection. Third, cancers, even those caused by viruses, don't seem to be contagious, as viral diseases usually are.

## The Transformation of Normal Cells into Tumor Cells

Almost anything that can alter the genetic material of a eukaryotic cell has the potential to make a normal cell cancerous. These cancer-causing alterations to cellular DNA affect parts of the genome called **oncogenes**. Oncogenes were first identified in cancer-causing viruses and were thought to be a part of the normal viral genome. However, American microbiologists J. Michael Bishop and Harold E. Varmus received the 1989 Nobel Prize in Medicine for proving that the cancer-inducing genes carried by viruses are actually derived from animal cells. Bishop and Varmus showed that the cancer-causing *src* gene in avian sarcoma viruses is derived from a normal part of chicken genes.

It is improbable that over 30 people of different ages and backgrounds are all IV drug users, so the most likely virus is hepatitis A. To investigate the source of the viral infection, the health department compares foods eaten by the 32 ill people with asymptomatic household members. All 32, including Tina, had eaten an ice-slush beverage purchased from a local convenience store. The health department determines that the convenience store clerk, unknowingly infected with hepatitis A virus, transferred the virus to the machine that makes the icy beverage. Over the next several months, Tina's symptoms subside, and her liver function returns to normal.

**How can knowing the identity of the virus affect the health department's recommendation for treatment and for prevention of future outbreaks?**

363   380   384   385

Oncogenes can be activated to abnormal functioning by a variety of agents, including mutagenic chemicals, high-energy radiation, and viruses. Viruses capable of inducing tumors in animals are called **oncogenic viruses**, or *oncoviruses*. Approximately 10% of cancers are known to be virus-induced. A defining feature of all oncogenic viruses is that their genetic material integrates into the host cell's DNA and replicates along with the host cell's chromosome. This mechanism is similar to the phenomenon of lysogeny in bacteria, and it can alter the host cell's characteristics in the same way.

Tumor cells undergo **transformation**; that is, they acquire properties that are distinct from the properties of uninfected cells or from infected cells that don't form tumors. After being transformed by viruses, many tumor cells contain a virus-specific antigen on their cell surface, called **tumor-specific transplantation antigen (TSTA)**, and transformed cells tend to be irregularly shaped, compared to normal cells.

## DNA Oncogenic Viruses

Oncogenic viruses are found within several families of DNA-containing viruses. These groups include the Adenoviridae, Herpesviridae, Poxviridae, Papovaviridae, and Hepadnaviridae.

Virtually all cervical and anal cancers are caused by human papillomavirus (HPV). A vaccine against four HPVs is recommended for 11- to 12-year-old girls and boys.

Epstein-Barr (EB) virus was isolated from Burkitt's lymphoma cells in 1964 by Michael Epstein and Yvonne Barr. The proof that EB virus can cause cancer was accidentally demonstrated in 1985 when a 12-year-old boy known only as David received a bone marrow transplant. Several months after the transplant, he died of cancer. An autopsy revealed that the virus had been unwittingly introduced into the boy with the bone marrow transplant.

Another DNA virus that causes cancer is hepatitis B virus (HBV). Many animal studies have been performed that have clearly indicated the causal role of HBV in liver cancer. In one human study, virtually all people with liver cancer had previous HBV infections.

## RNA Oncogenic Viruses

Among the RNA viruses, only the oncoviruses in the family Retroviridae cause cancer. The human T cell leukemia viruses (HTLV-1 and HTLV-2) are retroviruses that cause adult T cell leukemia and lymphoma in humans. (T cells are a type of white blood cell involved in the immune response.)

Sarcoma viruses of cats, chickens, and rodents, and the mammary tumor viruses of mice, are also retroviruses. Another retrovirus, feline leukemia virus (FeLV), causes leukemia in cats and is transmissible among cats. There is a test to detect the virus in cat serum.

The ability of retroviruses to induce tumors is related to their production of a reverse transcriptase by the mechanism described earlier (see Figure 13.19). The provirus, which is the double-stranded DNA molecule synthesized from the viral RNA, becomes integrated into the host cell's DNA; new genetic material is thereby introduced into the host's genome and is the key reason retroviruses can contribute to cancer. Some retroviruses contain oncogenes; others contain promoters that turn on oncogenes or other cancer-causing factors.

Drugs and vaccines work against specific viruses. There are no special treatments for hepatitis, but the preventive measures are different. For example, in this case, the health department recommends that everyone who ate at this convenience store in the previous 2 weeks receive hepatitis A vaccine and hepatitis A immunoglobulin.

Viruses were originally named for the symptoms they caused, hence the name "hepatitis virus" for a virus that affects the liver (from Latin *hepaticus*). This naming convention is imprecise but was the only method available until recently.

Molecular tools now allow viruses to be classified on the basis of genomes and morphology. Thus, related viruses, which may affect different tissues, are grouped into the same families. Differentiating viruses based on their genetic information provides valuable information for treatment and prevention.

363   380   384   385

## Viruses to Treat Cancer

In the early 1900s, physicians observed that tumors regressed in patients with concurrent viral infections. Experimentally induced viral infections in cancer patients during the 1920s suggested that viruses might have antitumor activity. These tumor-destroying viruses, or **oncolytic viruses**, selectively infect and kill tumor cells or cause an immune response against tumor cells. Several viruses are known to selectively infect cancer cells, and these are being genetically modified to remove virulence genes and add colony-stimulating factor genes to promote white blood cells. In 2015, the FDA approved the first oncolytic viral therapy. The new product, Imlygic™, is a herpesvirus to treat melanoma.

> **CHECK YOUR UNDERSTANDING**
>
> ✔ **13-11** What is a provirus?
>
> ✔ **13-12** How can an RNA virus cause cancer if it doesn't have DNA to insert into a cell's genome?

## Latent Viral Infections

### LEARNING OBJECTIVE

**13-13** Provide an example of a latent viral infection.

A virus can remain in equilibrium with the host and not actually produce disease for a long period, often many years. The oncogenic viruses just discussed are examples of such latent infections. All of the human herpesviruses can remain in host cells throughout the life of an individual. When herpesviruses are reactivated by immunosuppression (for example, AIDS), the resulting infection may be fatal. The classic example of such a **latent infection** is the infection of the skin by *Simplexvirus*, which produces cold sores. This virus inhabits the host's nerve cells but causes no damage until it is activated by a stimulus such as fever or sunburn—hence the term *fever blister*.

In some individuals, viruses are produced, but symptoms never appear. Even though a large percentage of the human population carries *Simplexvirus*, only 10–15% of people carrying the virus exhibit the disease.

The chickenpox virus (*Varicellovirus*) can also exist in a latent state. Chickenpox (varicella) is a skin disease that is usually acquired in childhood. The virus gains access to the skin via the blood. From the blood, some viruses may enter nerves, where they remain latent. Later, changes in the immune (T cell) response can activate these latent viruses, causing shingles (zoster). The shingles rash appears on the skin along the nerve in which the virus was latent. Shingles occurs in 10–20% of people who have had chickenpox.

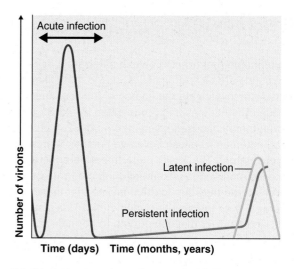

**Figure 13.21 Latent and persistent viral infections.**

**Q** How do latent and persistent infections differ?

## Persistent Viral Infections

### LEARNING OBJECTIVE

**13-14** Differentiate persistent viral infections from latent viral infections.

A **persistent viral infection** (or **chronic viral infection**) occurs gradually over a long period. Typically, persistent viral infections are fatal.

A number of persistent viral infections have in fact been shown to be caused by conventional viruses. For example, several years after causing measles, the measles virus can be responsible for a rare form of encephalitis called subacute sclerosing panencephalitis (SSPE). A persistent viral infection is apparently different from a latent viral infection in that, in most persistent viral infections, detectable virions gradually builds up over a long period, rather than appearing suddenly (**Figure 13.21**).

Several examples of latent and persistent viral infections are listed in Table 13.5.

> **CHECK YOUR UNDERSTANDING**
>
> ✔ **13-13, 13-14** Is shingles a persistent or latent infection?

## Plant Viruses and Viroids

### LEARNING OBJECTIVES

**13-15** Differentiate virus, viroid, and prion.

**13-16** Describe the lytic cycle for a plant virus.

Plant viruses resemble animal viruses in many respects: plant viruses are morphologically similar to animal viruses, and they

**TABLE 13.5 Examples of Latent and Persistent Viral Infections in Humans**

| Disease | Primary Effect | Causative Virus |
|---|---|---|
| **Latent** | **No symptoms during latency; viruses not usually released** | |
| Cold sores | Skin and mucous membrane lesions; genital lesions | HHV-1 and HHV-2 |
| Leukemia | Increased white blood cell growth | HTLV-1 and -2 |
| Shingles | Skin lesions | *Varicellovirus* (Herpesvirus) |
| **Persistent** | **Viruses continuously released** | |
| Cervical cancer | Increased cell growth | Human papillomavirus |
| HIV/AIDS | Decreased CD4$^+$ T cells | HIV-1 and -2 (*Lentivirus*) |
| Liver cancer | Increased cell growth | Hepatitis B virus |
| Persistent enterovirus infection | Mental deterioration associated with AIDS | Echoviruses |
| Progressive encephalitis | Rapid mental deterioration | Rubella virus |
| Subacute sclerosing panencephalitis (SSPE) | Mental deterioration | Measles virus |

have similar types of nucleic acid (Table 13.6). In fact, some plant viruses can multiply inside insect cells. Plant viruses cause many diseases of economically important crops, including beans (bean mosaic virus), corn and sugarcane (wound tumor virus), and potatoes (potato yellow dwarf virus). Viruses can cause color change, deformed growth, wilting, and stunted growth in their plant hosts. Some hosts, however, remain symptomless and only serve as reservoirs of infection.

Plant cells are generally protected from disease by an impermeable cell wall. Viruses must enter through wounds or be assisted by other plant parasites, including nematodes, fungi, and, most often, insects that suck the plant's sap. Once one plant is infected, it can spread infection to other plants in its pollen.

In laboratories, plant viruses are cultured in protoplasts (plant cells with the cell walls removed) and in insect cell cultures.

**TABLE 13.6 Classification of Some Major Plant Viruses**

| Characteristic | Viral Family | Viral Genus or Unclassified Members | Morphology | Method of Transmission |
|---|---|---|---|---|
| Double-stranded DNA, nonenveloped | Caulimoviridae | Cauliflower mosaic virus | | Aphids |
| Single-stranded RNA, + strand, nonenveloped | Bunyaviridae | Watermelon wilt | | Whiteflies |
| | Virgaviridae | *Tobamovirus* | | Wounds |
| Single-stranded RNA, − strand, enveloped | Rhabdoviridae | Potato yellow dwarf virus | | Leafhoppers and aphids |
| Double-stranded RNA, nonenveloped | Reoviridae | Wound tumor virus | | Leafhoppers |

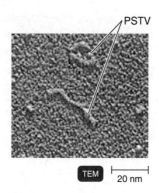

Figure 13.22 **Linear and circular potato spindle tuber viroid (PSTV).**

**Q** How do viroids differ from prions?

Some plant diseases are caused by **viroids**, short pieces of naked RNA, only 300 to 400 nucleotides long, with no protein coat. The nucleotides are often internally paired, so the molecule has a closed, folded, three-dimensional structure that presumably helps protect it from destruction by cellular enzymes. Some viroids, called **virusoids**, are enclosed in a protein coat. Virusoids cause disease only when the cell is infected by a virus. Viroids and virusoids are replicated continuously by host RNA polymerase in the cell nucleus or chloroplasts. When the enzyme reaches the end of the viroid RNA, it goes to the beginning. The viroid RNA is a ribozyme that cuts the continuous RNA into viroid segments. The RNA doesn't code for any proteins and may cause disease by gene silencing (see page 221). Annually, infections by viroids, such as potato spindle tuber viroid, result in losses of millions of dollars from crop damage (**Figure 13.22**).

Current research on viroids has revealed similarities between the base sequences of viroids and introns. Recall from Chapter 8 (page 215) that introns are sequences of genetic material that don't code for polypeptides. This observation has led to the hypothesis that viroids evolved from introns, leading to speculation that researchers may eventually discover animal viroids. Thus far, viroids and virusoids have been conclusively identified as pathogens only of plants, although hepatitis D may be caused by a virusoid. Hepatitis D is RNA enclosed in a protein coat and requires coinfection by Hepatitis B virus. Some researchers call HDV *satellite RNA*.

# Prions

## LEARNING OBJECTIVE

**13-17** Discuss how a protein can be infectious.

A few infectious diseases are caused by prions. In 1982, American neurobiologist Stanley Prusiner proposed that infectious

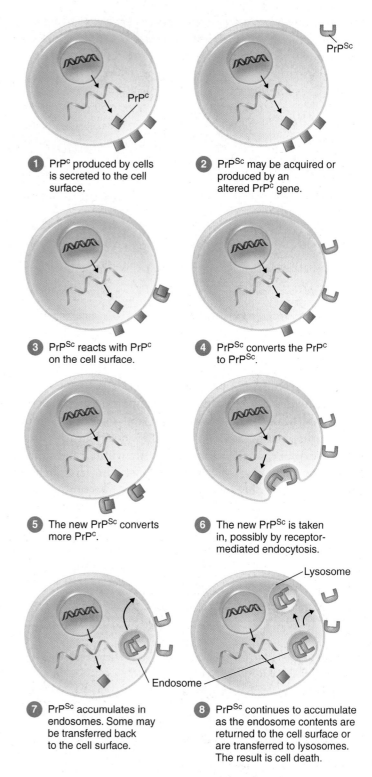

1 PrP$^C$ produced by cells is secreted to the cell surface.

2 PrP$^{Sc}$ may be acquired or produced by an altered PrP$^C$ gene.

3 PrP$^{Sc}$ reacts with PrP$^C$ on the cell surface.

4 PrP$^{Sc}$ converts the PrP$^C$ to PrP$^{Sc}$.

5 The new PrP$^{Sc}$ converts more PrP$^C$.

6 The new PrP$^{Sc}$ is taken in, possibly by receptor-mediated endocytosis.

7 PrP$^{Sc}$ accumulates in endosomes. Some may be transferred back to the cell surface.

8 PrP$^{Sc}$ continues to accumulate as the endosome contents are returned to the cell surface or are transferred to lysosomes. The result is cell death.

Figure 13.23 **How a protein can be infectious.** If an abnormal prion protein (PrP$^{Sc}$) enters a cell, it changes a normal prion protein PrP$^C$ to PrP$^{Sc}$, which now can change another normal PrP$^C$, resulting in an accumulation of the abnormal PrP$^{Sc}$ in the cell and on the cell surface.

**Q** How do prions differ from viruses?

proteins caused a neurological disease in sheep called scrapie. The infectivity of scrapie-infected brain tissue is reduced by treatment with proteases but not by treatment with radiation, suggesting that the infectious agent is pure protein. Prusiner coined the name **prion** for *pro*teinaceous *in*fectious particle.

Nine animal diseases now fall into this category, including the "mad cow disease" that emerged in cattle in Great Britain in 1987. All nine are neurological diseases called *transmissible spongiform encephalopathies* because large vacuoles develop in the brain (Figure 22.18b, page 643). The human diseases are kuru, Creutzfeldt-Jakob disease (CJD), Gerstmann-Sträussler-Scheinker syndrome, and fatal familial insomnia. (Neurological diseases are discussed in Chapter 22.) These diseases often run in families, which indicates a possible genetic cause. However, they aren't only inherited, because mad cow disease arose from feeding scrapie-infected sheep meat to cattle, and the new (bovine) variant was transmitted to humans who ate undercooked beef from infected cattle. Additionally, CJD has been transmitted with transplanted nerve tissue and contaminated surgical instruments.

These diseases are caused by the conversion of a normal host glycoprotein called PrP$^C$ (for cellular prion protein) into an infectious form called PrP$^{Sc}$ (for scrapie protein). The gene for PrP$^C$ is located on chromosome 20 in humans. Recent evidence suggests that PrP$^C$ is involved in preventing cell death. (See the discussion of apoptosis on page 490.) One hypothesis for how an infectious agent that lacks any nucleic acid can reproduce is shown in **Figure 13.23**.

The actual cause of cell damage isn't yet known. Fragments of PrP$^{Sc}$ molecules accumulate in the brain, forming plaques; these plaques are used for postmortem diagnosis, but they don't appear to be the cause of cell damage.

Play Prion Reproduction: *Overview, Characteristics, Diseases* @MasteringMicrobiology

**CHECK YOUR UNDERSTANDING**

✔ **13-15, 13-17** Contrast viroids and prions, and for each name a disease it causes.

✔ **13-16** How do plant viruses enter host cells?

# Study Outline

Go to @MasteringMicrobiology for **Interactive Microbiology**, *In the Clinic* videos, MicroFlix, MicroBoosters, 3D animations, practice quizzes, and more.

## General Characteristics of Viruses (pp. 362–363)

1. Depending on one's viewpoint, viruses may be regarded as exceptionally complex aggregations of nonliving chemicals or as exceptionally simple living microbes.
2. Viruses contain a single type of nucleic acid (DNA or RNA) and a protein coat, sometimes enclosed by an envelope composed of lipids, proteins, and carbohydrates.
3. Viruses are obligatory intracellular parasites. They multiply by using the host cell's synthesizing machinery to cause the synthesis of specialized elements that can transfer the viral nucleic acid to other cells.

### Host Range (pp. 362–363)

4. *Host range* refers to the spectrum of host cells in which a virus can multiply.
5. Most viruses infect only specific types of cells in one host species.
6. Host range is determined by the specific attachment site on the host cell's surface and the availability of host cellular factors.

### Viral Size (p. 363)

7. Viral size is ascertained by electron microscopy.
8. Viruses range from 20 to 1000 nm in length.

## Viral Structure (pp. 363–366)

1. A virion is a complete, fully developed viral particle composed of nucleic acid surrounded by a coat.

### Nucleic Acid (pp. 363–364)

2. Viruses contain either DNA or RNA, never both, and the nucleic acid may be single- or double-stranded, linear or circular, or divided into several separate molecules.
3. The proportion of nucleic acid in relation to protein in viruses ranges from about 1% to about 50%.

### Capsid and Envelope (pp. 364–365)

4. The protein coat surrounding the nucleic acid of a virus is called the capsid.
5. The capsid is composed of subunits, capsomeres, which can be a single type of protein or several types.
6. The capsid of some viruses is enclosed by an envelope consisting of lipids, proteins, and carbohydrates.
7. Some envelopes are covered with carbohydrate-protein complexes called spikes.

### General Morphology (pp. 365–366)

8. Helical viruses are hollow cylinders surrounding the nucleic acid.
9. Polyhedral viruses are many-sided.
10. Enveloped viruses are covered by an envelope and are roughly spherical but highly pleomorphic.
11. Complex viruses have complex structures. For example, many bacteriophages have a polyhedral capsid with a helical tail attached.

## Taxonomy of Viruses (pp. 366–370)

1. Classification of viruses is based on type of nucleic acid and strategy for replication.
2. Virus family names end in *-viridae*; genus names end in *-virus*.
3. A viral species is a group of viruses sharing the same genetic information and ecological niche.

## Isolation, Cultivation, and Identification of Viruses (pp. 370–372)

1. Viruses must be grown in living cells.
2. The easiest viruses to grow are bacteriophages.

### Growing Bacteriophages in the Laboratory (p. 370)

3. The plaque method mixes bacteriophages with host bacteria and nutrient agar.
4. After several viral multiplication cycles, the bacteria in the area surrounding the original virus are destroyed; the area of lysis is called a plaque.
5. Each plaque originates with a single viral particle; the concentration of viruses is expressed as plaque-forming units.

### Growing Animal Viruses in the Laboratory (pp. 370–372)

6. Cultivation of some animal viruses requires whole animals.
7. Simian AIDS and feline AIDS provide models for studying human AIDS.
8. Some animal viruses can be cultivated in embryonated eggs.
9. Cell cultures are animal or plant cells growing in culture media.
10. Viral growth can cause cytopathic effects in the cell culture.

### Viral Identification (p. 372)

11. Serological tests, RFLPs, and PCR are used most often to identify viruses.

## Viral Multiplication (pp. 372–384)

1. Viruses don't contain enzymes for energy production or protein synthesis.
2. For a virus to multiply, it must invade a host cell and direct the host's metabolic machinery to produce viral enzymes and components.

### Multiplication of Bacteriophages (pp. 373–376)

3. During the lytic cycle, a phage causes the lysis and death of a host cell.
4. Some viruses can either cause lysis or have their DNA incorporated as a prophage into the DNA of the host cell. The latter situation is called lysogeny.
5. In penetration, phage lysozyme opens a portion of the bacterial cell wall, the tail sheath contracts to force the tail core through the cell wall, and phage DNA enters the bacterial cell. The capsid remains outside.
6. In biosynthesis, transcription of phage DNA produces mRNA coding for proteins necessary for phage multiplication. Phage DNA is replicated, and capsid proteins are produced. During the eclipse period, separate phage DNA and protein can be found.
7. During maturation, phage DNA and capsids are assembled into complete viruses.

8. During release, phage lysozyme breaks down the bacterial cell wall, and the new phages are released.
9. During the lysogenic cycle, prophage genes are regulated by a repressor coded for by the prophage. The prophage is replicated each time the cell divides.
10. Because of lysogeny, lysogenic cells become immune to reinfection with the same phage and may undergo phage conversion.
11. A lysogenic phage can transfer bacterial genes from one cell to another through transduction. Any genes can be transferred in generalized transduction, and specific genes can be transferred in specialized transduction.

### Multiplication of Animal Viruses (pp. 376–384)

12. Animal viruses attach to the plasma membrane of the host cell.
13. Entry occurs by receptor-mediated endocytosis or fusion.
14. Animal viruses are uncoated by viral or host cell enzymes.
15. The DNA of most DNA viruses is released into the nucleus of the host cell. Transcription of viral DNA and translation produce viral DNA and, later, capsid proteins. Capsid proteins are synthesized in the cytoplasm of the host cell.
16. Multiplication of RNA viruses occurs in the cytoplasm of the host cell. RNA-dependent RNA polymerase synthesizes a double-stranded RNA.
17. After assembly, viruses are released. One method of release (and envelope formation) is budding. Nonenveloped viruses are released through ruptures in the host cell membrane.

## Viruses and Cancer (pp. 384–386)

1. The earliest relationship between cancer and viruses was demonstrated in the early 1900s, when chicken leukemia and chicken sarcoma were transferred to healthy animals by cell-free filtrates.

### The Transformation of Normal Cells into Tumor Cells (pp. 384–385)

2. When activated, oncogenes transform normal cells into cancerous cells.
3. Viruses capable of producing tumors are called oncogenic viruses.
4. Several DNA viruses and retroviruses are oncogenic.
5. The genetic material of oncogenic viruses becomes integrated into the host cell's DNA.
6. Transformed cells contain virus-specific antigens (TSTA), exhibit chromosome abnormalities, and can produce tumors when injected into susceptible animals.

### DNA Oncogenic Viruses (p. 385)

7. Oncogenic viruses are found among the Adenoviridae, Herpesviridae, Poxviridae, Papovaviridae, and Hepadnaviridae.

### RNA Oncogenic Viruses (p. 385)

8. The virus's ability to produce tumors is related to the presence of reverse transcriptase. The DNA synthesized from the viral RNA becomes incorporated as a provirus into the host cell's DNA.

### Viruses to Treat Cancer (p. 386)

9. Oncolytic viruses infect and lyse cancer cells.

### Latent Viral Infections (p. 386)

1. A latent viral infection is one in which the virus remains in the host cell for long periods without producing an infection.
2. Examples are cold sores and shingles.

### Persistent Viral Infections (pp. 386)

1. Persistent viral infections are disease processes that occur over a long period and are generally fatal.
2. Persistent viral infections are caused by conventional viruses; viruses accumulate over a long period.

### Plant Viruses and Viroids (pp. 386–388)

1. Plant viruses must enter plant hosts through wounds or with invasive parasites, such as insects.

2. Some plant viruses also multiply in insect (vector) cells.
3. Viroids are infectious pieces of RNA that cause some plant diseases.
4. Virusoids are viroids enclosed in a protein coat.

### Prions (pp. 388–389)

1. Prions are infectious proteins first discovered in the 1980s.
2. Prion diseases involve the degeneration of brain tissue.
3. Prion diseases are the result of an altered protein; the cause can be a mutation in the normal gene for PrP$^C$ or contact with an altered protein (PrP$^{Sc}$).

## Study Questions

For answers to the Knowledge and Comprehension questions, turn to the Answers tab at the back of the textbook.

## Knowledge and Comprehension

### Review

1. Why do we classify viruses as obligatory intracellular parasites?
2. List the four properties that define a virus. What is a virion?
3. Describe the four morphological classes of viruses, then diagram and give an example of each.
4. DRAW IT Label the principal events of attachment, biosynthesis, entry, and maturation of a + stranded RNA virus. Draw in uncoating.

5. Compare biosynthesis of a + stranded RNA and a – stranded RNA virus.
6. Some antibiotics activate phage genes. MRSA releasing Panton-Valentine leukocidin causes a life-threatening disease. Why can this happen following antibiotic treatment?
7. Recall from Chapter 1 that Koch's postulates are used to determine the etiology of a disease. Why is it difficult to determine the etiology of
   a. a viral infection, such as influenza?
   b. cancer?
8. Persistent viral infections such as (a) _____ might be caused by (b) _____ that are (c) _____.
9. Plant viruses can't penetrate intact plant cells because (a) _____; therefore, they enter cells by (b) _____. Plant viruses can be cultured in (c) _____.

10. NAME IT Identify the viral family that infects skin, mucosa, and nerve cells; causes infections that can recur because of latency, and has polyhedral geometry.

### Multiple Choice

1. Place the following in the most likely order for biosynthesis of a bacteriophage: (1) phage lysozyme; (2) mRNA; (3) DNA; (4) viral proteins; (5) DNA polymerase.
   a. 5, 4, 3, 2, 1
   b. 1, 2, 3, 4, 5
   c. 5, 3, 4, 2, 1
   d. 3, 5, 2, 4, 1
   e. 2, 5, 3, 4, 1
2. The molecule serving as mRNA can be incorporated in the newly synthesized virus capsids of all of the following *except*
   a. + strand RNA picornaviruses.
   b. + strand RNA togaviruses.
   c. − strand RNA rhabdoviruses.
   d. double-stranded RNA reoviruses.
   e. *Rotavirus*.
3. A virus with RNA-dependent RNA polymerase
   a. synthesizes DNA from an RNA template.
   b. synthesizes double-stranded RNA from an RNA template.
   c. synthesizes double-stranded RNA from a DNA template.
   d. transcribes mRNA from DNA.
   e. none of the above
4. Which of the following would be the first step in the biosynthesis of a virus with reverse transcriptase?
   a. A complementary strand of RNA must be synthesized.
   b. Double-stranded RNA must be synthesized.
   c. A complementary strand of DNA must be synthesized from an RNA template.
   d. A complementary strand of DNA must be synthesized from a DNA template.
   e. none of the above
5. An example of lysogeny in animals could be
   a. slow viral infections.
   b. latent viral infections.
   c. T-even bacteriophages.

d. infections resulting in cell death.
e. none of the above

6. The ability of a virus to infect an organism is regulated by
   a. the host species.
   b. the type of cells.
   c. the availability of an attachment site.
   d. cell factors necessary for viral replication.
   e. all of the above

7. Which of the following statements is *false*?
   a. Viruses contain DNA or RNA.
   b. The nucleic acid of a virus is surrounded by a protein coat.
   c. Viruses multiply inside living cells using viral mRNA, tRNA, and ribosomes.
   d. Viruses cause the synthesis of specialized infectious elements.
   e. Viruses multiply inside living cells.

8. Place the following in the order in which they are found in a host cell: (1) capsid proteins; (2) infective phage particles; (3) phage nucleic acid.
   a. 1, 2, 3
   b. 3, 2, 1
   c. 2, 1, 3
   d. 3, 1, 2
   e. 1, 3, 2

9. Which of the following does *not* initiate DNA synthesis?
   a. a double-stranded DNA virus (Poxviridae)
   b. a DNA virus with reverse transcriptase (Hepadnaviridae)
   c. an RNA virus with reverse transcriptase (Retroviridae)
   d. a single-stranded RNA virus (Togaviridae)
   e. none of the above

10. A viral species is *not* defined on the basis of the disease symptoms it causes. The best example of this is
    a. polio.
    b. rabies.
    c. hepatitis.
    d. chickenpox and shingles.
    e. measles.

## Analysis

1. Discuss the arguments for and against the classification of viruses as living organisms.

2. In some viruses, capsomeres function as enzymes as well as structural supports. Of what advantage is this to the virus?

3. Why was the discovery of simian AIDS and feline AIDS important?

4. Prophages and proviruses have been described as being similar to bacterial plasmids. What similar properties do they exhibit? How are they different?

## Clinical Applications and Evaluation

1. A 40-year-old man who was seropositive for HIV experienced abdominal pain, fatigue, and low-grade fever (38°C) for 2 weeks. A chest X-ray examination revealed lung infiltrates. Gram and acid-fast stains were negative. A viral culture revealed the cause of his symptoms: a large, enveloped polyhedral virus with double-stranded DNA. What is the disease? Which virus causes it? Why was a viral culture done after the Gram and acid-fast stain results were obtained?

2. A newborn female developed extensive vesicular and ulcerative lesions over her face and chest. What is the most likely cause of her symptoms? How would you determine the viral cause of this disease without doing a viral culture?

3. By May 14, two people living in the same household had died within 5 days of each other. Their illnesses were characterized by abrupt onset of fever, muscle pain, headache, and cough, followed by the rapid development of respiratory failure. By the end of the year, 36 cases of this disease, with a 50% mortality rate, had been confirmed. A member of the Orthomyxoviridae, Bunyaviridae, or Adenoviridae could cause this disease. Differentiate among these families by method of transmission, morphology, nucleic acid, and type of replication. The reservoir for this disease is mice. Name the disease. (*Hint*: Refer to Chapter 23.)

# Principles of Disease and Epidemiology 14

Now that you have a basic understanding of the structures and functions of microorganisms and some idea of the variety of microorganisms that exist, we can consider how the human body and various microorganisms interact in terms of health and disease.

We all have defenses to keep us healthy. In spite of these, however, we are still susceptible to **pathogens** (disease-causing microorganisms). A rather delicate balance exists between our defenses and the pathogenic mechanisms of microorganisms. When our defenses resist these pathogenic capabilities, we maintain our health—when the pathogen's capabilities overcome our defenses, disease results. After the disease has become established, an infected person may recover completely, suffer temporary or permanent damage, or die.

This chapter discusses the general principles of disease, starting with a discussion of the meaning and scope of pathology. In the last section of this chapter, "Epidemiology," you will learn how these principles are useful in studying and controlling disease. Understanding of these principles is vital to prevent disease transmission to patients in health care settings. Healthcare-associated infection by the *Clostridium difficile* bacteria shown in the photograph is discussed in the Clinical Case. Infections contracted outside the health care setting are called **community-acquired infections**.

◀ *Clostridium difficile* bacteria can cause severe diarrhea.

## In the Clinic

As a county public health nurse, you follow up on reports of communicable diseases. The annual reported incidence of cryptosporidiosis in your state (population of 3.1 million) is 7.5 cases per 100,000 people. This past year you saw 65 cases in your county of 430,000 people. **Calculate the reported county incidence rate. How does this compare to the state rate? Assume that only one in ten people with diarrhea due to cryptosporidiosis seeks medical diagnosis and treatment. Under these circumstances, what is the true incidence in your county?**

Play **In the Clinic** Video
@MasteringMicrobiology

# Pathology, Infection, and Disease

**LEARNING OBJECTIVE**

14-1 Define *pathology, etiology, infection*, and *disease*.

**Pathology** is the scientific study of disease (*pathos* = suffering; *logos* = science). Pathology is first concerned with the cause, or **etiology**, of disease. Second, it deals with **pathogenesis**, the manner in which a disease develops. Third, pathology is concerned with the *structural* and *functional changes* brought about by disease and their effects on the body.

Although the terms *infection* and *disease* are sometimes used interchangeably, they differ somewhat in meaning. **Infection** is the invasion or colonization of the body by pathogenic microorganisms; **disease** occurs when an infection results in any change from a state of health. Disease is an abnormal state in which part or all of the body is incapable of performing its normal functions. An infection may exist in the absence of detectable disease. For example, the body may be infected with the virus that causes AIDS but experience no symptoms of the disease.

The presence of a particular type of microorganism in a part of the body where it is not normally found is also called an infection—and may lead to disease. For example, although large numbers of *E. coli* are normally present in the healthy intestine, their infection of the urinary tract usually results in disease.

Few microorganisms are pathogenic. In fact, the presence of some microorganisms can even benefit the host. Therefore, before we discuss the role of microorganisms in causing disease, let's examine the relationship of microorganisms to the healthy human body.

**CHECK YOUR UNDERSTANDING**

✔ 14-1 What are the objectives of pathology?

# Human Microbiome

**LEARNING OBJECTIVES**

14-2 Describe how the human microbiome is acquired.

14-3 Compare commensalism, mutualism, and parasitism, and give an example of each.

14-4 Contrast normal microbiota and transient microbiota with opportunistic microorganisms.

Recent research indicates that normal and characteristic microbial populations begin to establish

Play Interactive Microbiology
@Mastering**Microbiology**
*See how the microbiome affects a patient's health*

themselves in an individual before birth (in utero). The placental microbiome consists of only few different bacteria, mostly Enterobacteriaceae and *Propionibacterium*. These bacteria are found in the newborn's intestine. Just before a woman gives birth, lactobacilli in her vagina multiply rapidly, and they become the predominant

organisms in the newborn's intestine (see Exploring the Microbiome on page 395). These lactobacilli also colonize the newborn's intestine. More microorganisms are introduced to the newborn's body from the environment when breathing and feeding start. An individual's microbiome changes rapidly during the first three years as the personal microbiome becomes established. After birth, *E. coli* and other bacteria acquired from foods, people, and pets begin to inhabit the large intestine. These microorganisms remain there throughout life and, in response to altered environmental conditions, may increase or decrease in number and contribute to health and disease.

Many other usually harmless microorganisms establish themselves inside other parts of the normal adult body and on its surface. A typical human body contains $3 \times 10^{13}$ body cells, and harbors as many bacteria cells—an estimated $4 \times 10^{13}$ bacterial cells. This gives you an idea of the abundance of microorganisms that normally reside in the human body. The **Human Microbiome Project** began in 2007 to analyze microbial communities called *microbiomes* that live in and on the human body. Its goal is to determine the relationship between changes in the human microbiome and human health and disease. The human microbiome is more diverse than previously thought. Much of the new information coming from studies of the human microbiome is featured in Exploring the Microbiome boxes found in each chapter of this textbook.

Currently, researchers are comparing the microbiomes of healthy volunteers and volunteers with specific diseases. The microorganisms that establish more or less permanent residence (colonize) but that do not produce disease under normal conditions are members of the body's **normal microbiota**. Historically they were referred to as *normal flora** (**Figure 14.1**). Others, called **transient microbiota**, may be present for several days, weeks, or months and then disappear. Microorganisms are not found throughout the entire human body but are localized in certain regions, as shown in Table 14.1 on page 396.

Many factors determine the distribution and composition of the normal microbiota. Among these are nutrients, physical and chemical factors, the host's defenses, and mechanical factors. Microbes vary with respect to the types of nutrients they can use as an energy source. Accordingly, microbes can colonize only those body sites that can supply the appropriate nutrients. These nutrients may be derived from dead cells, food in the gastrointestinal tract, secretory and excretory products of cells, and substances in body fluids.

A number of physical and chemical factors affect the growth of microbes and thus the growth and composition of the normal microbiota. Among these are temperature, pH, available oxygen and carbon dioxide, salinity, and sunlight.

---

*Historically, bacteria once were classified as plants, and thus bacteria on the human body were called *normal flora*.

# Connections between Birth, Microbiome, and Other Health Conditions

Vaginal birth is the point at which people first come into contact with a vast array of microbes. Both the type and timing of our earliest microbial interactions impact development of a healthy immune system, along with the kinds of microbes that ultimately become part of our microbiome. However, in the past 60 years, cesarean section (C-section) births have skyrocketed. In that same time period, incidence of type 1 diabetes, asthma, and obesity have also become major health problems in the developed world. Could these two things be related—and could microbiota be the link between them?

Recent studies indicate asthma, obesity, and type 1 diabetes all appear more commonly in people delivered by C-section than they do in people delivered vaginally. Likewise, there are some microbiota differences between the two birth methods. For instance, *Lactobacillus* and *Bacteroides*

are prevalent in the microbiota of infants delivered vaginally, but the microbiome of infants delivered by C-section more closely resembles that of the human skin, with abundant *Staphylococcus aureus*.

The path from C-section delivery to increased risk of diabetes, obesity, or asthma is complex and not fully understood. Nonetheless, some mothers who deliver by C-section are now asking caregivers to provide a "microbial profile" to their infants that resembles what naturally comes from a vaginal birth. In one study, sterile gauze was placed in the mother's vagina for one hour before C-section delivery. After birth, the gauze was swiped over the newborn's face and body. Over the next month, the microbiota of babies who received these vaginal swabs was tracked. It remained similar to the microbial profiles of infants delivered vaginally in the study (and different from the other C-section babies who did not receive the swab).

Larger, longer-term studies are required to determine whether such treatments might also one day reduce risk of diabetes, asthma, and obesity in people born via C-section.

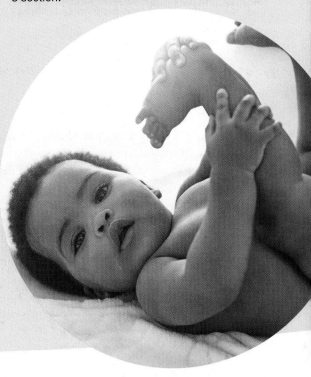

You'll learn in Chapters 16 and 17 that the human body has certain defenses against microbes. These defenses include a variety of molecules and activated cells that kill microbes, inhibit their growth, prevent their adhesion to host cell surfaces, and neutralize toxins that microbes produce. These defenses are extremely important against pathogens. Childhood exposure to microorganisms helps the immune system develop. Indeed, it has been proposed that insufficient exposure to microorganisms in childhood may interfere with the development of the immune system and may play a role in increasing rates of allergies and other immune disorders. This idea, known as the *hygiene hypothesis*, is discussed in Chapter 19 on page 525.

Certain regions of the body are subjected to mechanical forces that may affect colonization by the normal microbiota. For example, the chewing actions of the teeth and tongue movements can dislodge microbes attached to tooth and mucosal surfaces. In the gastrointestinal tract, the flow of saliva and digestive secretions and the various muscular movements of the throat, esophagus, stomach, and intestines can remove unattached microbes. The flushing action of urine also removes unattached microbes. In the respiratory system, mucus traps microbes, which cilia then propel toward the throat for elimination.

The conditions provided by the host at a particular body site vary from one person to another. Among the factors that also affect the normal microbiota are age, nutritional status, diet, health status, disability, hospitalization, stress, climate, geography, personal hygiene, living conditions, occupation, and lifestyle.

## CLINICAL CASE  Bathroom Break

Jamil Carter is in the bathroom—again. Ever since he was hospitalized for a urinary tract infection (UTI) 6 months ago, Jamil has been plagued with fever, chills, and severe diarrhea. He has lost 15 pounds since his hospitalization. Jamil is 75 years old, is retired, and lives with his wife and adult son. He does not smoke and rarely drinks alcohol. While in the hospital, Jamil was treated with the antibiotics ceftriaxone and ciprofloxacin for the UTI. He developed diarrhea 3 days after being discharged from the hospital and has had it ever since.

**What could be causing Jamil's diarrhea and other symptoms? Read on to find out.**

395    408    411    413    418

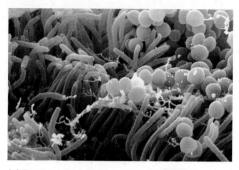

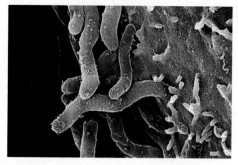

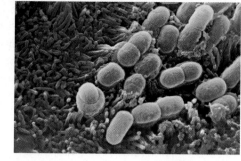

**(a)** Bacteria (orange) on the surface of the nasal epithelium SEM ⊢⊣ 2 μm

**(b)** Bacteria (brown) on the lining of the stomach SEM ⊢⊣ 2.5 μm

**(c)** Bacteria (orange) in the small intestine SEM ⊢⊣ 1 μm

**Figure 14.1 Representative normal microbiota for different regions of the body.**

**Q** Of what value are normal microbiota?

The principal normal microbiota in different regions of the body and some distinctive features of each region are listed in **Table 14.1**. Normal microbiota are also discussed more specifically in Part Four.

Animals with no microbiota whatsoever can be reared in the laboratory. These so-called germ-free mammals used in research are obtained by breeding them in a sterile environment. On the one hand, research with germ-free animals has

| TABLE 14.1 Representative Normal Microbiota by Body Region | | |
|---|---|---|
| **Region** | **Principal Components** | **Comments** |
| Skin | *Propionibacterium, Staphylococcus, Corynebacterium, Micrococcus, Acinetobacter, Brevibacterium; Candida* (fungus), and *Malassezia* (fungus) | • Most of the microbes in direct contact with skin don't become residents because secretions from sweat and oil glands have antimicrobial properties.<br>• Keratin is a resistant barrier, and the low pH of the skin inhibits many microbes.<br>• The skin has a relatively low moisture content. |
| Eyes (Conjunctiva) | *Staphylococcus epidermidis, S. aureus,* diphtheroids, *Propionibacterium, Corynebacterium,* streptococci, and *Micrococcus* | • The conjunctiva, a continuation of the skin or mucous membrane, contains basically the same microbiota found on the skin.<br>• Tears and blinking eliminate some microbes or inhibit others from colonizing. |
| Nose and Throat (Upper Respiratory System) | *Staphylococcus aureus, S. epidermidis,* and aerobic diphtheroids in the nose; *S. epidermidis, S. aureus,* diphtheroids, *Streptococcus pneumoniae, Haemophilus,* and *Neisseria* in the throat | • Although some normal microbiota are potential pathogens, their ability to cause disease is reduced by microbial antagonism.<br>• Nasal secretions kill or inhibit many microbes, and mucus and ciliary action remove many microbes. |
| Mouth | *Streptococcus, Lactobacillus, Actinomyces, Bacteroides, Veillonella, Neisseria, Haemophilus, Fusobacterium, Treponema, Staphylococcus, Corynebacterium,* and *Candida* (fungus) | • Abundant moisture, warmth, and the constant presence of food make the mouth an ideal environment that supports very large and diverse microbial populations on the tongue, cheeks, teeth, and gums.<br>• Biting, chewing, tongue movements, and salivary flow dislodge microbes. Saliva contains several antimicrobial substances. |
| Large Intestine | *Escherichia coli, Bacteroides, Fusobacterium, Lactobacillus, Enterococcus, Bifidobacterium, Enterobacter, Citrobacter, Proteus, Klebsiella,* and *Candida* (fungus) | • The large intestine contains the largest numbers of resident microbiota in the body because of its available moisture and nutrients.<br>• Mucus and periodic shedding of the lining prevent many microbes from attaching to the lining of the gastrointestinal tract, and the mucosa produces several antimicrobial chemicals. |
| Urinary and Reproductive Systems | *Staphylococcus, Micrococcus, Enterococcus, Lactobacillus, Bacteroides,* aerobic diphtheroids, *Pseudomonas, Klebsiella,* and *Proteus* in urethra; lactobacilli, *Streptococcus, Clostridium, Candida albicans* (fungus), and *Trichomonas vaginalis* (protozoan) in vagina | • The lower urethra in both sexes has a resident population; the vagina has an acid-tolerant population of microbes because of the nature of its secretions.<br>• Mucus and periodic shedding of the lining prevent microbes from attaching to the lining; urine flow mechanically removes microbes, and the pH of urine and urea are antimicrobial.<br>• Cilia and mucus expel microbes from the cervix of the uterus into the vagina, and the acidity of the vagina inhibits or kills microbes. |

shown that microbes aren't absolutely essential to animal life. On the other hand, this research has shown that germ-free animals have undeveloped immune systems and are unusually susceptible to infection and serious disease. Germ-free animals also require more calories and vitamins than do normal animals.

## Relationships between the Normal Microbiota and the Host

Once established, the normal microbiota can benefit the host by preventing the overgrowth of harmful microorganisms. This phenomenon is called **microbial antagonism**, or **competitive exclusion**. Microbial antagonism involves competition among microbes. One consequence of this competition is that the normal microbiota protect the host against colonization by potentially pathogenic microbes by competing for nutrients, producing substances harmful to the invading microbes, and affecting conditions such as pH and available oxygen. When this balance between normal microbiota and pathogenic microbes is upset, disease can result. For example, the normal bacterial microbiota of the adult human vagina maintains a local pH of about 4. The presence of normal microbiota inhibits the overgrowth of the yeast *Candida albicans*, which can grow when the pH is altered. If the bacterial population is eliminated by antibiotics, excessive douching, or deodorants, the pH of the vagina reverts to nearly neutral, and *C. albicans* can flourish and become the dominant microorganism there. This condition can lead to a form of vaginitis (vaginal infection).

Another example of microbial antagonism occurs in the large intestine. *E. coli* cells produce *bacteriocins*, proteins that inhibit the growth of other bacteria of the same or closely related species, such as pathogenic *Salmonella* and *Shigella*. A bacterium that makes a particular bacteriocin isn't killed by that bacteriocin but may be killed by other ones. Bacteriocins are being investigated for use in treating infections and preventing food spoilage.

A final example involves another bacterium, *Clostridium difficile*, also in the large intestine. The normal microbiota of the large intestine effectively inhibit *C. difficile*, possibly by making host receptors unavailable, competing for available nutrients, or producing bacteriocins. However, if the normal microbiota are eliminated (for example, by antibiotics), *C. difficile* can become a problem. This microbe is responsible for nearly all gastrointestinal infections that follow antibiotic therapy, from mild diarrhea to severe or even fatal colitis (inflammation of the colon). In 2013, a Canadian infectious disease specialist successfully treated *C. difficile* infections with pills containing normal intestinal microbiota. The normal microbiota were obtained from patients' relatives.

The relationship between the normal microbiota and the host illustrates **symbiosis**, a relationship between two organisms in which at least one organism is dependent on the other (**Figure 14.2**). In the symbiotic relationship called **commensalism**, one of the organisms benefits, and the other is unaffected. Many of the microorganisms that make up our normal microbiota are commensals; these include *Staphylococcus epidermidis* bacteria that inhabit the surface of the skin, the corynebacteria that inhabit

ASM: Microorganisms, cellular and viral, can interact with both human and nonhuman hosts in beneficial, neutral, or detrimental ways.

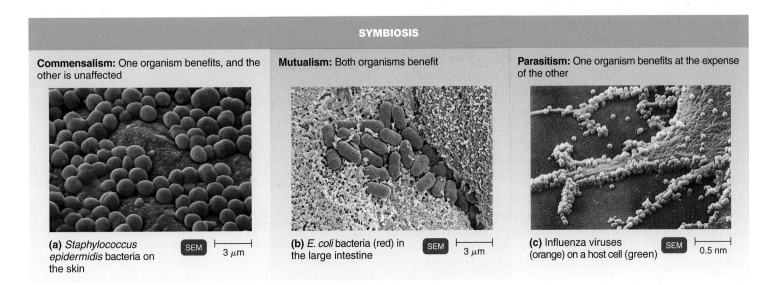

**SYMBIOSIS**

**Commensalism:** One organism benefits, and the other is unaffected

**Mutualism:** Both organisms benefit

**Parasitism:** One organism benefits at the expense of the other

**(a)** *Staphylococcus epidermidis* bacteria on the skin — SEM — 3 μm

**(b)** *E. coli* bacteria (red) in the large intestine — SEM — 3 μm

**(c)** Influenza viruses (orange) on a host cell (green) — SEM — 0.5 nm

**Figure 14.2 Symbiosis.**

**Q** Which type of symbiosis is best represented by the relationship between humans and *E. coli*?

the surface of the eye, and certain saprophytic mycobacteria that inhabit the ear and external genitals. These bacteria live on secretions and sloughed-off cells, and they bring no apparent benefit or harm to the host.

**Mutualism** is a type of symbiosis that benefits both organisms. For example, the large intestine contains bacteria, such as *E. coli*, that synthesize vitamin K and some B vitamins. These vitamins are absorbed into the bloodstream and distributed for use by body cells. In exchange, the large intestine provides nutrients used by the bacteria, allowing them to survive.

Recent genetics studies have found hundreds of antibiotic-resistance genes in the intestinal bacteria. It may seem desirable to have these bacteria survive while a person is taking antibiotics for an infectious disease; however, these beneficial bacteria may be able to transfer antibiotic-resistance genes to pathogens.

In still another kind of symbiosis, one organism benefits by deriving nutrients at the expense of the other; this relationship is called **parasitism**. Many disease-causing bacteria are parasites.

## Opportunistic Microorganisms

Although categorizing symbiotic relationships by type is convenient, keep in mind that the relationship can change under certain conditions. For example, given the proper circumstances, a mutualistic organism, such as *E. coli*, can become harmful. *E. coli* is generally harmless as long as it remains in the large intestine; but if it gains access to other body sites, such as the urinary bladder, lungs, spinal cord, or wounds, it may cause urinary tract infections, pulmonary infections, meningitis, or abscesses, respectively. Microbes such as *E. coli* are called **opportunistic pathogens**. They don't cause disease in their normal habitat in a healthy person but may do so in a different environment. For example, microbes that gain access through broken skin or mucous membranes can cause opportunistic infections. Or, if the host is already weakened or compromised by infection, microbes that are usually harmless can cause disease. AIDS is often accompanied by a common opportunistic infection, *Pneumocystis* pneumonia, caused by the opportunistic organism *Pneumocystis jirovecii* (see Figure 24.19, page 714). This secondary infection can develop in AIDS patients because their immune systems are suppressed. Before the AIDS epidemic, this type of pneumonia was rare. Opportunistic pathogens possess other features that contribute to their ability to cause disease. For example, they're present in or on the body or in the external environment in relatively large numbers. Some opportunistic pathogens may be found in locations in or on the body that are somewhat protected from the body's defenses, and some are resistant to antibiotics.

In addition to the usual symbionts, many people carry other microorganisms that are generally regarded as pathogenic but that may not cause disease in those people. Among the pathogens that are frequently carried in healthy individuals are echoviruses (*echo* comes from *e*nteric *c*ytopathogenic *h*uman *o*rphan), which can cause intestinal diseases, and adenoviruses, which can cause respiratory diseases. *Neisseria meningitidis*, which often resides benignly in the respiratory tract, can cause meningitis, a disease that inflames the coverings of the brain and spinal cord. *Streptococcus pneumoniae*, a normal resident of the nose and throat, can cause a type of pneumonia.

## Cooperation among Microorganisms

It isn't only competition among microbes that can cause disease; cooperation among microbes can also be a factor in causing disease. For example, pathogens that cause periodontal disease and gingivitis have been found to have receptors, not for the teeth, but for the oral streptococci that colonize the teeth.

> **CHECK YOUR UNDERSTANDING**
>
> ✔ **14-2** How do normal microbiota differ from transient microbiota?
>
> ✔ **14-3** Give several examples of microbial antagonism.
>
> ✔ **14-4** How can opportunistic pathogens cause infections?

# The Etiology of Infectious Diseases

**LEARNING OBJECTIVE**

**14-5** List Koch's postulates.

Some diseases—such as polio, Lyme disease, and tuberculosis—have a well-known etiology. Some have an etiology that isn't completely understood, for example, the relationship between certain viruses and cancer. For still others, such as Alzheimer's disease, the etiology is unknown. Of course, not all diseases are caused by microorganisms. For example, the disease hemophilia is an *inherited (genetic) disease*, and osteoarthritis and cirrhosis are *degenerative diseases*. There are several other categories of disease, but here we'll discuss only *infectious diseases*, those caused by microorganisms. To see how microbiologists determine the etiology of an infectious disease, we'll discuss in greater detail the work of Robert Koch, which was introduced in Chapter 1 (page 10).

## Koch's Postulates

In the historical overview of microbiology presented in Chapter 1, we briefly discussed Koch's famous postulates. Recall that Koch was a German physician who played a major role in establishing that microorganisms cause specific diseases. In 1877, he published some early papers on anthrax, a disease of cattle that can also occur in humans. Koch demonstrated that certain bacteria, today known as *Bacillus anthracis*, were always present in the blood of animals that had the

# Koch's Postulates: Understanding Disease

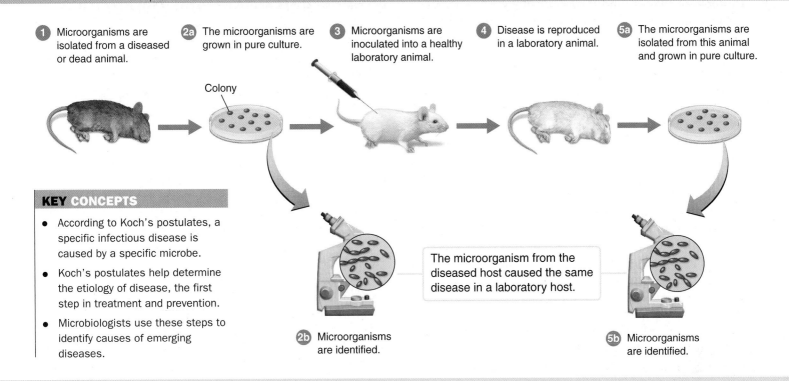

① Microorganisms are isolated from a diseased or dead animal.

②ₐ The microorganisms are grown in pure culture.

③ Microorganisms are inoculated into a healthy laboratory animal.

④ Disease is reproduced in a laboratory animal.

⑤ₐ The microorganisms are isolated from this animal and grown in pure culture.

Colony

The microorganism from the diseased host caused the same disease in a laboratory host.

②ᵦ Microorganisms are identified.

⑤ᵦ Microorganisms are identified.

**KEY CONCEPTS**

- According to Koch's postulates, a specific infectious disease is caused by a specific microbe.

- Koch's postulates help determine the etiology of disease, the first step in treatment and prevention.

- Microbiologists use these steps to identify causes of emerging diseases.

disease and weren't present in healthy animals. He knew that the mere presence of the bacteria did not prove that they had caused the disease; the bacteria could have been there as a result of the disease. Thus, he experimented further.

He took a sample of blood from a sick animal and injected it into a healthy one. The second animal developed the same disease and died. He repeated this procedure many times, always with the same results. (A key criterion in the validity of any scientific proof is that experimental results be repeatable.) Koch also cultivated the microorganism in fluids outside the animal's body, and he demonstrated that the bacterium would cause anthrax even after many culture transfers.

Koch showed that a specific infectious disease (anthrax) is caused by a specific microorganism (*B. anthracis*) that can be isolated and cultured on artificial media. He later used the same methods to show that the bacterium *Mycobacterium tuberculosis* is the causative agent of tuberculosis.

Koch's research provides a framework for the study of the etiology of any infectious disease. Today, we refer to Koch's experimental requirements as **Koch's postulates (Figure 14.3)**. They are summarized as follows:

1. The same pathogen must be present in every case of the disease.

2. The pathogen must be isolated from the diseased host and grown in pure culture.

3. The pathogen from the pure culture must cause the disease when it's inoculated into a healthy, susceptible laboratory animal.

4. The pathogen must be isolated from the inoculated animal and must be shown to be the original organism.

## Exceptions to Koch's Postulates

Although Koch's postulates are useful in determining the causative agent of most bacterial diseases, there are some exceptions. For example, some microbes have unique culture requirements. The bacterium *Treponema pallidum* is known to cause syphilis, but virulent strains have never been cultured on artificial media. The causative agent of leprosy, *Mycobacterium leprae*, has also never been grown on artificial media. Moreover, rickettsial and viral pathogens cannot be cultured on artificial media because they multiply only within cells.

The discovery of microorganisms that can't grow on artificial media has necessitated some modifications of Koch's postulates and the use of alternative methods of culturing and detecting certain microbes. For example, when researchers looking for the microbial cause of legionellosis (Legionnaires' disease) were unable to isolate the microbe directly from a victim, they took the alternative step of inoculating a victim's lung tissue into guinea pigs. These guinea pigs developed the disease's pneumonia-like symptoms, whereas guinea pigs inoculated with tissue from an unafflicted person did not.

Then tissue samples from the diseased guinea pigs were cultured in yolk sacs of chick embryos, a method (see Figure 13.7, page 371) that reveals the growth of extremely small microbes. After the embryos were incubated, electron microscopy revealed rod-shaped bacteria in the chick embryos. Finally, modern immunological techniques (discussed in Chapter 18) were used to show that the bacteria in the chick embryos were the same bacteria as those in the guinea pigs and in afflicted humans.

In a number of situations, a human host exhibits certain signs and symptoms that are associated only with a certain pathogen and its disease. For example, the pathogens responsible for diphtheria and tetanus cause distinguishing signs and symptoms that no other microbe can produce. They are unequivocally the only organisms that produce their respective diseases. But some infectious diseases are not as clear-cut and provide another exception to Koch's postulates. For example, nephritis (inflammation of the kidneys) can involve any of several different pathogens, all of which cause the same signs and symptoms. Thus, it is often difficult to know which particular microorganism is causing a disease. Other infectious diseases that sometimes have poorly defined etiologies are pneumonia, meningitis, and peritonitis (inflammation of the peritoneum, the membrane that lines the abdomen and covers the organs within them).

Some pathogens can cause several disease conditions, resulting in still another exception to Koch's postulates. *Mycobacterium tuberculosis*, for example, is implicated in diseases of the lungs, skin, bones, and internal organs. *Streptococcus pyogenes* (strep'tō-KOK-kus pī-AH-jen-ēz) can cause sore throat, scarlet fever, skin infections (such as erysipelas), and osteomyelitis (inflammation of bone), among other diseases. When clinical signs and symptoms are used together with laboratory methods, these infections can usually be distinguished from infections of the same organs by other pathogens.

Ethical considerations may also impose an exception to Koch's postulates. For example, some agents that cause disease in humans have no other known host. An example is HIV, the cause of AIDS. This poses the ethical question of whether humans can be intentionally inoculated with infectious agents. In 1721, King George I told several condemned prisoners they could be inoculated with smallpox to test a smallpox vaccine (see Chapter 18). He promised their freedom if they lived. Human experiments with untreatable diseases are unacceptable today, although accidental inoculation does occur sometimes. For example, a contaminated red bone marrow transplant satisfied the third Koch's postulate to prove that a herpesvirus caused cancer (see page 378).

**CHECK YOUR UNDERSTANDING**

✔ **14-5** Explain some exceptions to Koch's postulates.

# Classifying Infectious Diseases

## LEARNING OBJECTIVES

**14-6** Differentiate a communicable from a noncommunicable disease.

**14-7** Categorize diseases according to frequency of occurrence.

**14-8** Categorize diseases according to severity.

**14-9** Define *herd immunity*.

Every disease that affects the body alters body structures and functions in particular ways, and these alterations are usually indicated by several kinds of evidence. For example, the patient may experience certain **symptoms**, or changes in body function, such as pain and *malaise* (a vague feeling of body discomfort). These *subjective* changes aren't apparent to an observer. The patient can also exhibit **signs**, which are *objective* changes the physician can observe and measure. Frequently evaluated signs include lesions (changes produced in tissues by disease), edema (swelling), fever, and paralysis. A specific group of symptoms or signs may always accompany a particular disease; such a group is called a **syndrome**. The diagnosis of a disease is made by evaluation of the signs and symptoms, together with the results of laboratory tests.

Diseases are often classified in terms of how they behave within a host and within a given population. A **communicable disease** is a disease in which an infected person transmits an infectious agent, either directly or indirectly, to another person who in turn becomes infected. Chickenpox, measles, influenza, genital herpes, typhoid fever, and tuberculosis are examples. Chickenpox and measles are also examples of **contagious diseases**, that is, diseases that are very communicable and capable of spreading easily and rapidly from one person to another. A **noncommunicable disease** is not spread from one host to another. These diseases are caused by microorganisms that normally inhabit the body and only occasionally produce disease or by microorganisms that reside outside the body and produce disease only when introduced into the body. An example is tetanus: *Clostridium tetani* produces disease only when it is introduced into the body via abrasions or wounds.

## Occurrence of a Disease

To understand the full scope of a disease, we should know something about its occurrence. The **incidence** of a disease is the number of people in a population who develop a disease during a particular time period. It's an indicator of the spread of the disease. The **prevalence** of a disease is the number of people in a population who develop a disease at a specified time, regardless of when it first appeared. Prevalence takes into account both old and new cases. It's an indicator of how seriously and how long a disease affects a population. For example, the incidence of AIDS in the United States in 2014 was 37,600, whereas the prevalence in that same year was estimated to be

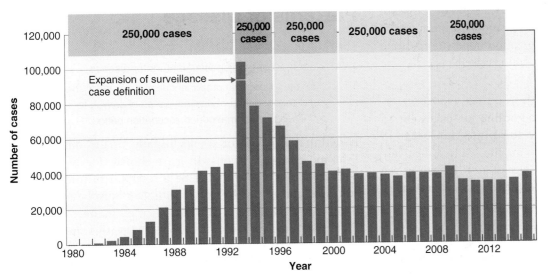

**Figure 14.4 Reported HIV/AIDS cases in the United States.** Notice that the first 250,000 cases occurred over a 12-year period, whereas the second through fifth 250,000 cases in this epidemic occurred in just 3 to 6 years. Much of the increase shown for 1993 is due to an expanded definition of AIDS cases adopted in that year. Reported cases likely rose in 2008 because of another definition change that expanded to include HIV infections in the number of AIDS cases.

*Source:* CDC.

**Q** What was the incidence of AIDS in 2010?

about 1.1 million. Knowing the incidence and the prevalence of a disease in different populations (for example, in populations representing different geographic regions or different ethnic groups) enables scientists to estimate the range of the disease's occurrence and its tendency to affect some groups of people more than others.

Frequency of occurrence is another criterion that is used in the classification of diseases. If a particular disease occurs only occasionally, it is called a **sporadic disease**; typhoid fever in the United States is such a disease. A disease constantly present in a population is called an **endemic disease**; an example of such a disease is the common cold. If many people in a given area acquire a certain disease in a relatively short period, it is called an **epidemic disease**; influenza is an example of a disease that often achieves epidemic status. **Figure 14.4** shows the epidemic incidence of AIDS in the United States. Some authorities consider gonorrhea and certain other sexually transmitted infections to be epidemic at this time as well (see Figure 26.5, page 766). An epidemic disease that occurs worldwide is called a **pandemic disease**. We experience pandemics of influenza from time to time. AIDS is another example of a pandemic disease.

## Severity or Duration of a Disease

Another useful way of defining the scope of a disease is in terms of its severity or duration. An **acute disease** is one that develops rapidly but lasts only a short time; a good example is influenza. A **chronic disease** develops more slowly. The body's reactions may be less severe, but the disease is likely to continue or recur for long periods. Infectious mononucleosis, tuberculosis, and hepatitis B fall into this category. A disease that is intermediate between acute and chronic is described as a **subacute disease**; an example is subacute sclerosing panencephalitis, a rare brain disease characterized by diminished intellectual function and loss of nervous function. A **latent disease** is one in which the causative agent remains inactive for a time but then becomes active to produce symptoms of the disease; an example is shingles, one of the diseases caused by *Varicellovirus*. The virus can enter nerves and remain latent (dormant). Later, changes in the immune response can activate the virus, causing shingles. Another example of a latent disease is cold sores, which are caused by *Simplexvirus*. The virus resides in the nerve cells of the body but causes no damage until it is activated by a stimulus such as sunburn or fever.

The rate at which a disease or an epidemic spreads and the number of individuals involved are determined in part by the immunity of the population. Vaccination can provide long-lasting and sometimes lifelong protection of an individual against certain diseases. People who are immune to an infectious disease may carry the pathogen but not have the disease, thereby reducing the occurrence of the disease. Immune individuals act as a barrier to the spread of infectious agents. Even though a highly communicable disease may cause an epidemic, many nonimmune people will be protected because of the unlikelihood of their coming into contact with an infected person. A great advantage of vaccination is that it protects enough individuals in a population to prevent the disease's rapid spread to those in the population who aren't vaccinated. If most people in a population (herd) are immune to a particular disease, this form of immunity is referred to as **herd immunity**. Where there is herd immunity, outbreaks are limited to sporadic cases because there are not enough susceptible individuals to support the spread of the disease to epidemic proportions. Susceptible individuals include children who are too young to be vaccinated or whose parents refuse to vaccinate them, people with immune disorders, and those who are too ill to be vaccinated (for example, some cancer patients). In addition, improper vaccinations occur globally, and many regions do not have access to vaccinations. An example of a disease that has been eradicated by vaccination and herd immunity is smallpox. The World Health Organization hopes to eradicate other diseases, such as measles and polio, as well.

## Extent of Host Involvement

Infections can also be classified according to the extent to which the host's body is affected. A **local infection** is one in which the invading microorganisms are limited to a relatively small area of the body. Some examples of local infections are boils and abscesses. In a **systemic (generalized) infection**, microorganisms or their products are spread throughout the body by the blood or lymph. Measles is an example of a systemic infection. Very often, agents of a local infection enter a blood or lymphatic vessel and spread to other specific parts of the body, where they are confined to specific areas of the body. This condition is called a **focal infection**. Focal infections can arise from infections in areas such as the teeth, tonsils, or sinuses.

Sepsis is a toxic inflammatory condition arising from the spread of microbes, especially bacteria or their toxins, from a focus of infection. **Septicemia**, also called blood poisoning, is a systemic infection arising from the multiplication of pathogens in the blood. Septicemia is a common example of sepsis. The presence of bacteria in the blood is known as **bacteremia**. **Toxemia** refers to the presence of toxins in the blood (as occurs in tetanus), and **viremia** refers to the presence of viruses in blood.

The state of host resistance also determines the extent of infections. A **primary infection** is an acute infection that causes the initial illness. A **secondary infection** is one caused by an opportunistic pathogen after the primary infection has weakened the body's defenses. Secondary infections of the skin and respiratory tract are common and are sometimes more dangerous than the primary infections. *Pneumocystis* pneumonia as a consequence of AIDS is an example of a secondary infection; streptococcal bronchopneumonia following influenza is an example of a secondary infection that is more serious than the primary infection. A **subclinical infection**, also called *inapparent infection*, is one that doesn't cause any noticeable illness. Poliovirus and hepatitis A virus, for example, can be carried by people who never develop the illness.

### CHECK YOUR UNDERSTANDING

✔ **14-6** Does *Clostridium perfringens* (page 738) cause a communicable disease?

✔ **14-7** Distinguish the incidence from the prevalence of a disease.

✔ **14-8** List two examples of acute and chronic diseases.

✔ **14-9** How does herd immunity develop?

# Patterns of Disease

### LEARNING OBJECTIVES

**14-10** Identify four predisposing factors for disease.

**14-11** Put the following in proper sequence, according to the pattern of disease: period of decline, period of convalescence, period of illness, prodromal period, incubation period.

A definite sequence of events usually occurs during infection and disease. As you will learn shortly, for an infectious disease to occur, there must be a reservoir of infection as a source of pathogens. Next, the pathogen must be transmitted to a susceptible host by direct contact, by indirect contact, or by vectors. Transmission is followed by invasion, in which the microorganism enters the host and multiplies. Following invasion, the microorganism injures the host through a process called pathogenesis (discussed further in the next chapter). The extent of injury depends on the degree to which host cells are damaged, either directly or by toxins. Despite the effects of all these factors, the occurrence of disease ultimately depends on the resistance of the host to the activities of the pathogen.

## Predisposing Factors

Certain predisposing factors also affect the occurrence of disease. A **predisposing factor** makes the body more susceptible to a disease and may alter the course of the disease. Gender is sometimes a predisposing factor; for example, females have a higher incidence of urinary tract infections than males, whereas males have higher rates of pneumonia and meningitis. Other aspects of genetic background may play a role as well. For example, sickle cell disease is a severe, life-threatening form of anemia that occurs when the genes for the disease are inherited from both parents. Individuals who carry only one sickle cell gene have a condition called *sickle cell trait* and appear normal unless specially tested. However, they are relatively resistant to the most serious form of malaria. The potential that individuals in a population might inherit life-threatening sickle cell disease is more than counterbalanced by protection from malaria among carriers of the gene for sickle cell trait. Of course, in countries where malaria isn't present, sickle cell trait is an entirely negative condition.

Climate and weather seem to have some effect on the incidence of infectious diseases. In temperate regions, the incidence of respiratory diseases increases during the winter. This increase may be related to the fact that when people stay indoors, the closer contact with one another facilitates the spread of respiratory pathogens.

Other predisposing factors include vaccination and herd immunity in decreasing the spread of diseases, age (the very young and elderly are more susceptible to infections), antigenic variants and antigenic drift, behavioral and religious practices, inadequate nutrition, environment, habits, lifestyle, fatigue, occupation, preexisting illness, and chemotherapy. It

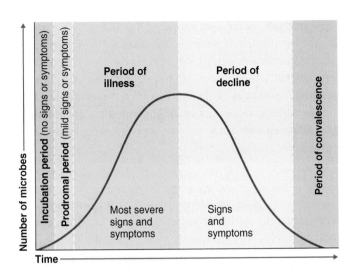

**Figure 14.5  The stages of a disease.**

**Q** During which periods can a disease be transmitted?

is often difficult to know the exact relative importance of the various predisposing factors.

## Development of Disease

Once a microorganism overcomes the defenses of the host, development of the disease follows a certain sequence that tends to be similar whether the disease is acute or chronic (Figure 14.5).

### Incubation Period

The **incubation period** is the interval between the initial infection and the first appearance of any signs or symptoms. In some diseases, the incubation period is always the same; in others, it is quite variable. The time of incubation depends on the specific microorganism involved, its growth rate, the number of infecting microorganisms, and the resistance of the host. Whether a disease can be transmitted during the incubation period depends on the specific pathogen. Infectious virions will not be present in the eclipse phase of their replication. (See Table 15.1, page 425, for the incubation periods of a number of microbial diseases.)

### Prodromal Period

The **prodromal period** is a relatively short period that follows the period of incubation in some diseases. The prodromal period is characterized by early, mild symptoms of disease, such as general aches and malaise. It is often difficult to differentiate the common cold from the prodromal symptoms related to other diseases such as measles, chickenpox, or cytomegalovirus infection.

### Period of Illness

During the **period of illness**, the disease is most severe. The person exhibits overt signs and symptoms of disease, such as

fever, chills, muscle pain (myalgia), sensitivity to light (photophobia), sore throat (pharyngitis), lymph node enlargement (lymphadenopathy), and gastrointestinal disturbances. During the period of illness, the number of white blood cells may increase or decrease. Generally, the patient's immune response and other defense mechanisms overcome the pathogen, and the period of illness ends. If the disease is not successfully overcome (or successfully treated), the patient dies during this period.

### Period of Decline

During the **period of decline**, the signs and symptoms subside. The fever decreases, and the feeling of malaise diminishes. During this phase, which may take from less than 24 hours to several days, the patient is vulnerable to secondary infections.

### Period of Convalescence

During the **period of convalescence**, the person regains strength and the body returns to its prediseased state. Recovery has occurred.

We all know that during the period of illness, people can serve as reservoirs of disease and can easily spread infections to other people. However, you should also know that people can spread infection during incubation and prodrome as well. This is especially true of diseases such as typhoid fever and cholera, in which the convalescing person carries the pathogenic microorganism for months or even years.

---

**CHECK YOUR UNDERSTANDING**

✔ **14-10** What is a predisposing factor?

✔ **14-11** The incubation period for a cold is 3 days, and the period of disease is usually 5 days. If the person next to you has a cold, when will you know whether you contracted it?

---

# The Spread of Infection

**LEARNING OBJECTIVES**

**14-12** Define *reservoir of infection*.

**14-13** Contrast human, animal, and nonliving reservoirs, and give one example of each.

**14-14** Explain three methods of disease transmission.

Now that you have an understanding of normal microbiota, the etiology of infectious diseases, and the types of infectious diseases, we'll examine the sources of pathogens and how diseases are transmitted.

## Reservoirs of Infection

For a disease to perpetuate itself, there must be a continual source of the disease organisms. This source can be either a living organism or an inanimate object that provides a pathogen with adequate conditions for survival and multiplication and an opportunity for transmission. Such a source is called a **reservoir of infection**. These reservoirs may be human, animal, or nonliving.

### Human Reservoirs

*[handwritten: AIDs ... gonorrhea]*

The principal living reservoir of human disease is the human body itself. Many people harbor pathogens and transmit them directly or indirectly to others. People with signs and symptoms of a disease may transmit the disease; in addition, some people can harbor pathogens and transmit them to others without exhibiting any signs of illness. These people, called **carriers**, are important living reservoirs of infection. For example, immune adolescents and adults carry *Bordetella pertussis* and can transmit an infection to a nonvaccinated infant. Some carriers have inapparent infections for which no signs or symptoms are ever exhibited. Other people, such as those with latent diseases, carry a disease during its symptom-free stages— during the incubation period (before symptoms appear) or during the convalescent period (recovery). Typhoid Mary is an example of a carrier (see page 732). Human carriers play an important role in the spread of such diseases as diphtheria, typhoid fever, hepatitis, gonorrhea, amebic dysentery, and streptococcal infections.

*[handwritten left margin: apparent or latent]*

### Animal Reservoirs

*[handwritten: Zoonoses... animals to humans]*

Both wild and domestic animals are living reservoirs of microorganisms that can cause human diseases. Diseases that occur primarily in wild and domestic animals and can be transmitted to humans are called **zoonoses** (ZŌ-ō-nō-sēz) (singular: *zoonosis*). Rabies (found in bats, skunks, foxes, dogs, and coyotes), and Lyme disease (found in field mice) are examples of zoonoses. Other representative zoonoses are presented in Table 14.2.

*[handwritten: → ticks]*

About 150 zoonoses are known. Zoonoses can be transmitted to humans via one of many routes: by direct contact with infected animals; by direct contact with domestic pet waste (such as cleaning a litter box or bird cage); by contamination of food and water; by air from contaminated hides, fur, or feathers; by consuming infected animal products; or by arthropod vectors (insects and ticks that transmit pathogens).

### Nonliving Reservoirs

The two major nonliving reservoirs of infectious disease are soil and water. Soil harbors such pathogens as fungi, which cause mycoses such as ringworm and systemic infections; *Clostridium botulinum*, the bacterium that causes botulism; and *C. tetani*, the bacterium that causes tetanus. Because both species of clostridia are part of the normal intestinal microbiota of horses and cattle, the bacteria are found especially in soil where animal feces are used as fertilizer.

Water that's been contaminated by the feces of humans and other animals is a reservoir for several pathogens, notably those responsible for gastrointestinal diseases. These include *Vibrio cholerae*, which causes cholera; *Cryptosporidium*, one cause of diarrhea; and *Salmonella* Typhi, which causes typhoid fever. Other nonliving reservoirs include foods that are improperly prepared or stored. They may be sources of diseases such as trichinellosis and salmonellosis.

## Transmission of Disease

The causative agents of disease can be transmitted from the reservoir of infection to a susceptible host by three principal routes: contact, vehicles, and vectors.

### Contact Transmission

*[handwritten: living reservoir → host]*

**Contact transmission** is the spread of a disease agent by direct contact, indirect contact, or droplet transmission. **Direct contact transmission**, also known as *person-to-person transmission*, is the direct transmission of an agent by physical contact between its source and a susceptible host; no intermediate object is involved (**Figure 14.6a**). The most common forms of direct contact transmission are touching, kissing, and sexual intercourse. Among the diseases that can be transmitted by direct contact are viral respiratory tract diseases (the common cold and influenza), staphylococcal infections, hepatitis A, measles, scarlet fever, and sexually transmitted infections (syphilis, gonorrhea, and genital herpes). Direct contact is also one way to spread AIDS, syphilis, and infectious mononucleosis. To guard against person-to-person transmission, health care workers use gloves and other protective measures (Figure 14.6b). Potential pathogens can also be transmitted by direct contact from animals (or animal products) to humans. Examples are the pathogens causing rabies (direct contact via bites) and anthrax.

**Congenital transmission** is the transmission of diseases from mother to fetus or newborn at birth. This occurs when a pathogen present in the mother's blood is capable of crossing the placenta or through direct contact with a pathogen in the mother's blood or vaginal secretions during delivery. See the Big Picture in Chapter 22, pages 634–635, for a discussion of congenitally transmitted pathogens.

**Indirect contact transmission** occurs when the agent of disease is transmitted from its reservoir to a susceptible host by means of a nonliving object. The general term for any nonliving object involved in the spread of an infection is a **fomite**. Examples of fomites are stethoscopes, clothing from health care personnel, tissues, handkerchiefs, towels, bedding, diapers, drinking cups, eating utensils, toys, money, and thermometers (Figure 14.6c). Contaminated syringes serve as fomites in transmitting AIDS and hepatitis B. Other fomites may transmit diseases such as tetanus, methicillin-resistant *S. aureus* (MRSA), and impetigo.

## TABLE 14.2 Selected Zoonoses

| Disease | Causative Agent | Reservoir | Transmission Due To | Chapter Reference |
|---|---|---|---|---|
| **VIRAL** | | | | |
| Influenza (some types) | Influenzavirus | Swine, birds | Direct contact | 24 |
| Rabies | Lyssavirus | Bats, skunks, foxes, dogs, raccoons | Direct contact (bite) | 22 |
| West Nile encephalitis | Flavivirus | Horses, birds | Aedes and Culex mosquito bite | 22 |
| Hantavirus pulmonary syndrome | Hantavirus | Rodents (primarily deer mice) | Direct contact with rodent saliva, feces, or urine | 23 |
| **BACTERIAL** | | | | |
| Anthrax | Bacillus anthracis | Domestic livestock | Direct contact with contaminated hides or animals; air; food | 23 |
| Brucellosis | Brucella spp. | Domestic livestock | Direct contact with contaminated milk, meat, or animals | 23 |
| Plague | Yersinia pestis | Rodents | Flea bites | 23 |
| Cat-scratch disease | Bartonella henselae | Domestic cats | Direct contact | 23 |
| Ehrlichiosis | Ehrlichia spp. | Deer, rodents | Tick bites | 23 |
| Leptospirosis | Leptospira spp. | Wild mammals, domestic dogs and cats | Direct contact with urine, soil, water | 26 |
| Lyme disease | Borrelia burgdorferi | Field mice | Tick bites | 23 |
| Psittacosis (ornithosis) | Chlamydophila psittaci | Birds, especially parrots | Direct contact | 24 |
| Rocky Mountain spotted fever | Rickettsia rickettsii | Rodents | Tick bites | 23 |
| Salmonellosis | Salmonella enterica | Poultry, reptiles | Ingestion of contaminated food and water and putting hands in mouth | 25 |
| Endemic typhus | Rickettsia typhi | Rodents | Flea bites | 23 |
| **FUNGAL** | | | | |
| Ringworm | Trichophyton Microsporum Epidermophyton | Domestic mammals | Direct contact; fomites (nonliving objects) | 21 |
| **PROTOZOAN** | | | | |
| Malaria | Plasmodium spp. | Monkeys | Anopheles mosquito bite | 23 |
| Toxoplasmosis | Toxoplasma gondii | Cats and other mammals | Ingestion of contaminated meat or by direct contact with infected tissues or fecal matter | 23 |
| **HELMINTHIC** | | | | |
| Tapeworm (pork) | Taenia solium | Pigs | Ingestion of undercooked contaminated pork | 25 |
| Trichinellosis | Trichinella spiralis | Pigs, bears | Ingestion of undercooked contaminated meat | 25 |

*within 1 meter!*

**Droplet transmission** is a third type of contact transmission in which microbes are spread in *droplet nuclei* (mucus droplets) that travel only short distances (Figure 14.6d). These droplets are discharged into the air by coughing, sneezing, laughing, or talking and travel less than 1 meter from the reservoir to the host. One sneeze may produce 20,000 droplets. Disease agents that travel such short distances are not considered airborne (airborne transmission is discussed shortly). Pathogens can travel at varying distances depending on the size and shape of the particles, initial velocity (coughing, sneezing,

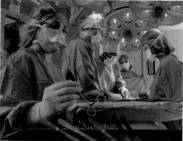

**(a)** Direct contact transmission

**(b)** Preventing direct contact transmission through the use of gloves, masks, and goggles

**(c)** Indirect contact transmission

**(d)** Droplet transmission

**Figure 14.6  Contact transmission.**

**Q** Name a disease transmitted by direct contact, a disease transmitted by indirect contact, and a disease transmitted by droplet transmission.

or normal exhalation), and environmental conditions such as humidity. Examples of diseases spread by droplet transmission are influenza, pneumonia, and pertussis (whooping cough).

### Vehicle Transmission

**Vehicle transmission** is the transmission of disease agents by a medium, such as air, water, or food (**Figure 14.7**). Other media include blood and other body fluids, drugs, and intravenous fluids. An outbreak of *Salmonella* infections caused by vehicle transmission is described in the box in Chapter 25 (page 731). Here we'll discuss air, water, and food as transmission vehicles.

**Airborne transmission** refers to the spread of agents of infection by droplet nuclei in dust that travel more than 1 meter from the reservoir to the host. For example, microbes are spread by droplets, which may be discharged in a fine spray from the mouth and nose during coughing and sneezing (see Figure 14.6d). These droplets are small enough to remain airborne for prolonged periods. The virus that causes measles and the bacterium that causes tuberculosis can be transmitted via

airborne droplets. Dust particles can harbor various pathogens. Staphylococci and streptococci can survive on dust and be transmitted by the airborne route. Spores produced by certain fungi are also transmitted by the airborne route and can cause such diseases as histoplasmosis, coccidioidomycosis, and blastomycosis (see Chapter 24).

In **waterborne transmission**, pathogens are usually spread by water contaminated with untreated or poorly treated sewage. Diseases transmitted via this route include cholera, waterborne shigellosis, and leptospirosis. In **foodborne transmission**, pathogens are generally transmitted in foods that are incompletely cooked, poorly refrigerated, or prepared under unsanitary conditions. Foodborne pathogens cause diseases such as food poisoning and tapeworm infestation. Foodborne transmission frequently occurs because of **cross-contamination**, the transfer of pathogens from one food to another. This can occur when pathogens on hands, gloves, knives, cutting boards, countertops, utensils, and cooking equipment spread to food. This takes place when pathogens on the surface of raw meat,

**(a)** Air

**(b)** Water

**(c)** Food

**Figure 14.7  Vehicle transmission.**

**Q** How does vehicle transmission differ from contact transmission?

**Figure 14.8 Mechanical transmission.**

**Q** How do mechanical transmission and biological transmission by vectors differ?

poultry, seafood, and vegetables and raw eggs are transferred to ready-to-eat foods (foods that do not require cooking or have already been cooked, such as salads and sandwiches). Cross-contamination is responsible for numerous cases of food poisoning.

Both waterborne and foodborne transmission also provide a transfer of microbes by **fecal-oral transmission**. In the cycle, some microbes are ingested contaminated in water or food. The pathogens usually enter the water or food after being shed in the feces of people or animals infected with them. The cycle is interrupted by effective sanitation practices in food handling and production.

### Vectors

Arthropods are the most important group of disease **vectors**—animals that carry pathogens from one host to another. (Insects and other arthropod vectors are discussed in Chapter 12, page 355.) Arthropod vectors transmit disease by two general methods. **Mechanical transmission** is the passive transport of the pathogens on the insect's feet or other body parts (**Figure 14.8**). If the insect makes contact with a host's food, pathogens can be transferred to the food and later swallowed by the host. Houseflies, for instance, can transfer the pathogens of typhoid fever and bacillary dysentery (shigellosis) from the feces of infected people to food.

**Biological transmission** is an active process and is more complex. The arthropod bites an infected person or animal and ingests some of the infected blood (see Figure 12.31, page 355). The pathogens then reproduce in the vector, and the increase in the number of pathogens increases the possibility that they will be transmitted to another host. Some parasites reproduce in the gut of the arthropod; these can be passed with feces. If the arthropod defecates or vomits while biting a potential host, the parasite can enter the wound. Other parasites reproduce in the vector's gut and migrate to the salivary gland; these are directly injected into a bite. Some protozoan and helminthic parasites use the vector as a host for a developmental stage in their life cycle.

Table 14.3 lists a few important arthropod vectors and the diseases they transmit.

| **TABLE 14.3** Representative Arthropod Vectors and the Diseases They Transmit | | | |
|---|---|---|---|
| **Disease** | **Causative Agent** | **Arthropod Vector** | **Chapter Reference** |
| Malaria | *Plasmodium* spp. (protozoan) | *Anopheles* (mosquito) | 23 |
| African trypanosomiasis | *Trypanosoma brucei gambiense* and *T. b. rhodesiense* (protozoan) | *Glossina* (tsetse fly) | 22 |
| Chagas' disease | *T. cruzi* (protozoan) | *Triatoma* (kissing bug) | 23 |
| Yellow fever | *Alphavirus* (yellow fever virus) | *Aedes* (mosquito) | 23 |
| Dengue | *Alphavirus* (dengue virus) | *Ae. aegypti* (mosquito) | 23 |
| Arthropod-borne encephalitis | *Alphavirus* (encephalitis virus) | *Culex* (mosquito) | 22 |
| Ehrlichiosis | *Ehrlichia* spp. | *Ixodes* spp. (tick) | 23 |
| Epidemic typhus | *Rickettsia prowazekii* | *Pediculus humanus* (louse) | 23 |
| Endemic murine typhus | *R. typhi* | *Xenopsylla cheopis* (rat flea) | 23 |
| Rocky Mountain spotted fever | *R. rickettsii* | *Dermacentor andersoni* and other species (tick) | 23 |
| Plague | *Yersinia pestis* | *X. cheopis* (rat flea) | 23 |
| Zika virus disease | Zika virus | *Aedes, Anopheles* (mosquitoes) | 22 |
| Lyme disease | *B. burgdorferi* | *Ixodes* spp. (tick) | 23 |

**CHECK YOUR UNDERSTANDING**

- ✓ 14-12 Why are carriers important reservoirs of infection?
- ✓ 14-13 How are zoonoses transmitted to humans?
- ✓ 14-14 Give an example of contact transmission, vehicle transmission, mechanical transmission, and biological transmission.

# Healthcare-Associated Infections (HAIs)

**LEARNING OBJECTIVES**

14-15 Define *healthcare-associated infections*, and explain their importance.

14-16 Define *compromised host*.

14-17 List several methods of disease transmission in hospitals.

14-18 Explain how healthcare-associated infections can be prevented.

**Healthcare-associated infections (HAIs)** are infections patients acquire while receiving treatment for other conditions at a health care facility, such as a nursing home, hospital, same-day surgery center, outpatient clinic, or in-home health care environment. Traditionally these were called **nosocomial infections** (*nosocomial* is Latin for hospital).

The Centers for Disease Control and Prevention (CDC) estimates that on any given day, about 1 in 25 hospital patients has at least one HAI. The work of pioneers in aseptic techniques such as Lister and Semmelweis (Chapter 1, page 10) decreased the rate of HAIs considerably. However, despite modern advances in sterilization techniques and disposable materials, the rate of HAIs has increased 36% during the last 20 years. In the United States, about 2 million people per year contract HAIs, and over 70,000 die as a result. HAIs represent the eighth leading cause of death in the United States (the top three are heart disease, cancer, and strokes).

HAIs result from the interaction of several factors: (1) microorganisms in the hospital environment, (2) the compromised (or weakened) status of the host, and (3) the chain of transmission in the hospital. **Figure 14.9** illustrates that the presence of any one of these factors alone is generally not enough to cause infection; it is the interaction of all three factors that poses a significant risk of HAI.

1. UTI
2. surg. wound inf.
3. LRI
4. Cutaneous
5. Bacteremias

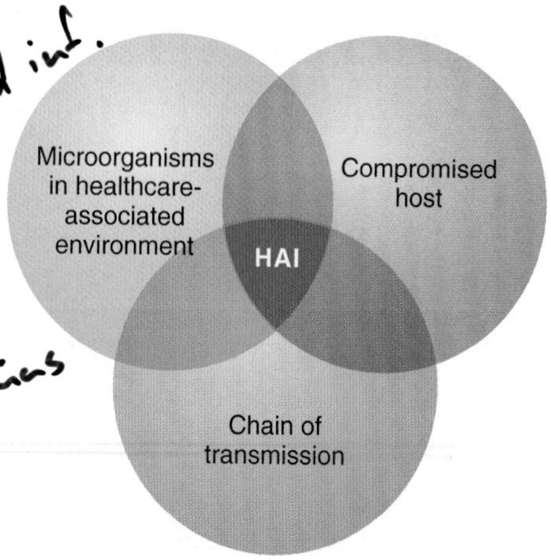


**Figure 14.9 Healthcare-associated infections.**

**Q** What are some settings where HAIs occur?

## Microorganisms in the Hospital

Although every effort is made to kill or check the growth of microorganisms in the hospital, the hospital environment is a major reservoir for a variety of pathogens. One reason is that certain normal microbiota of the human body are opportunistic and present a particularly strong danger to hospital patients. In fact, most of the microbes that cause HAIs don't cause disease in healthy people but are pathogenic only for individuals whose defenses have been weakened by illness or therapy (see the boxes on pages 197 and 417).

In the 1940s and 1950s, most HAIs were caused by gram-positive microbes. *Staphylococcus aureus* was at that time the primary cause of all HAIs. In the 1970s, gram-negative rods, such as *E. coli* and *Pseudomonas aeruginosa*, were the most common causes of HAIs. Then, during the 1980s, antibiotic-resistant gram-positive bacteria—*Staphylococcus aureus*, coagulase-negative staphylococci (see page 428), and *Enterococcus* spp.—emerged as healthcare-associated pathogens. By the 1990s, these gram-positive bacteria accounted for 34% of nosocomial infections, and four gram-negative pathogens accounted for 32%. In the 2000s, antibiotic resistance in HAIs is a major concern. *Clostridium difficile* is now the leading cause of HAIs. In the past few years, improved prevention methods have led to a decrease in the total incidence of HAIs. The principal microorganisms involved in HAIs are summarized in Table 14.4.

**CLINICAL CASE**

Jamil calls his physician to discuss his symptoms and makes an appointment for that afternoon. His wife, Charlene, drives Jamil to the appointment; he is not sure he can make it that far without stopping. Initially, Jamil's physician thinks he has noroviral gastroenteritis, but the symptoms have gone on too long. She orders a stool sample to be sent to the local laboratory for culturing. The results return positive for *C. difficile*.

**Where could Jamil have contracted *C. difficile*?**

395 | 408 | 411 | 413 | 418

**TABLE 14.4** Microorganisms Involved in Healthcare-Associated Infections

| Microorganism | Most Common Infection Type | Percentage of Total Infections | Percentage Resistant to Antibiotics |
|---|---|---|---|
| Coagulase-negative staphylococci | Bloodstream | 11% | Not reported |
| Staphylococcus aureus | Surgical wound | 16% | 55% |
| Clostridium difficile | Diarrhea after abdominal surgery | 15% | Not reported |
| Enterococcus spp. | Bloodstream | 14% | 83% |
| Candida spp. (fungus) | Urinary tract infections | 9% | Not reported |
| Escherichia coli | Urinary tract infections (most common cause) | 12% | 20% |
| Pseudomonas aeruginosa | Urinary tract and pneumonia | 8% | 10% |
| Klebsiella pneumoniae | All sites | 8% | 29% |
| Enterobacter spp. | All sites | 5% | 38% |
| Acinetobacter baumannii | All sites | 2% | 68% |

Source: CDC, Healthcare-Associated Infections.

In addition to being opportunistic, some microorganisms in the hospital become resistant to antimicrobial drugs, which are commonly used there. For example, *P. aeruginosa* and other such gram-negative bacteria tend to be difficult to control with antibiotics because of their R factors, which carry genes that determine resistance to antibiotics (see Chapter 8, page 230). As the R factors recombine, new and multiple resistance factors are produced. These strains become part of the microbiota of patients and hospital personnel and become progressively more resistant to antibiotic therapy. In this way, people become part of the reservoir (and chain of transmission) for antibiotic-resistant strains of bacteria. Usually, if the host's resistance is high, the new strains are not much of a problem. However, if disease, surgery, or trauma has weakened the host's defenses, secondary infections may be difficult to treat.

## Compromised Host

A **compromised host** is one whose resistance to infection is impaired by disease, therapy, or burns. Two principal conditions can compromise the host: broken skin or mucous membranes, and a suppressed immune system.

As long as the skin and mucous membranes remain intact, they provide formidable physical barriers against most pathogens. Burns, surgical wounds, trauma (such as accidental wounds), injections, invasive diagnostic procedures, ventilators, intravenous therapy, and urinary catheters (used to drain urine) can all break the first line of defense and make a person more susceptible to disease in hospitals. Burn patients are especially susceptible to nosocomial infections because their skin is no longer an effective barrier to microorganisms.

The risk of infection is also related to other invasive procedures, such as administering anesthesia, which may alter breathing and contribute to pneumonia, and tracheotomy, in which an incision is made into the trachea to assist breathing. Patients who require invasive procedures usually have a serious underlying disease, which further increases susceptibility to infections. Invasive devices provide a pathway for microorganisms in the environment to enter the body; they also help transfer microbes from one part of the body to another. Pathogens can also proliferate on the devices themselves (see Figure 1.10 on page 17).

In healthy individuals, white blood cells called T cells (T lymphocytes) provide resistance to disease by killing pathogens directly, mobilizing phagocytes and other lymphocytes, and secreting chemicals that kill pathogens. White blood cells called B cells (B lymphocytes), which develop into antibody-producing cells, also protect against infection. Antibodies provide immunity by such actions as neutralizing toxins, inhibiting the attachment of a pathogen to host cells, and helping to lyse pathogens. Drugs, radiation therapy, steroid therapy, burns, diabetes, leukemia, kidney disease, stress, and malnutrition can all adversely affect the actions of T and B cells and compromise the host. In addition, the AIDS virus destroys certain T cells.

A summary of the principal sites of HAIs is presented in **Figure 14.10**.

## Chain of Transmission

Given the variety of pathogens (and potential pathogens) in health care settings and the compromised state of the host, routes of transmission are a constant concern. The principal routes of transmission of HAIs are (1) direct contact

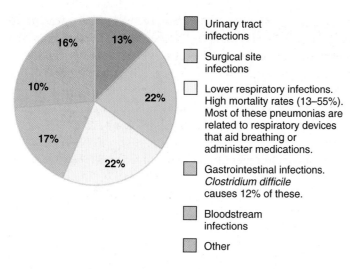

16%  13%
10%
17%
22%
22%

- Urinary tract infections
- Surgical site infections
- Lower respiratory infections. High mortality rates (13–55%). Most of these pneumonias are related to respiratory devices that aid breathing or administer medications.
- Gastrointestinal infections. *Clostridium difficile* causes 12% of these.
- Bloodstream infections
- Other

**Figure 14.10 Principal sites of healthcare-associated infections.**
*Source:* Data from CDC. Healthcare-associated infections, 2016.

**Q** Which type of HAI is most prevalent?

transmission from hospital staff to patient and from patient to patient and (2) indirect contact transmission through fomites and the hospital's ventilation system (airborne transmission).

Because health care personnel are in direct contact with patients, they can often transmit disease. For example, a physician or nurse may transmit microbes to a patient when changing a dressing, or a kitchen worker who carries *Salmonella* can contaminate a food supply.

Certain areas of health care facilities are reserved for specialized care; these include the burn, hemodialysis, recovery, intensive care, and oncology units. Unfortunately, these units also group patients together and provide environments for the epidemic spread of infections from patient to patient.

Many diagnostic and therapeutic hospital procedures provide a fomite route of transmission. The urinary catheter used to drain urine from the urinary bladder is a fomite in many HAIs. Intravenous catheters, which pass through the skin and into a vein to provide fluids, nutrients, or medication, can also transmit HAIs. Respiratory aids can introduce contaminated fluids into the lungs. Needles may introduce pathogens into muscle or blood, and surgical dressings can become contaminated and promote disease (see the Clinical Focus box on page 417).

## Control of Healthcare-Associated Infections

**Universal precautions** (see Appendix C) are employed to reduce the transmission of microbes in health care and residential settings. The precautions are designed to protect patients/residents, staff, and visitors from contact with pathogens. The various precautions can be grouped into two general categories: standard precautions and transmission-based precautions.

**Standard precautions** are basic, minimum practices designed to prevent transmission of pathogens from one person to another and are applied to every person every time. They are employed at all levels of health care, regardless of whether a patient's infection status is confirmed, suspected, or unknown. Among the standard precautions are hand hygiene, use of personal protective equipment (gloves, gowns, facemasks), respiratory hygiene and cough etiquette, disinfection of patient-care equipment and instruments, environmental cleaning and disinfection, safe injection practices, patient placement, and safe resuscitation and lumbar puncture procedures.

**Transmission-based precautions** are procedures designed to supplement standard precautions in individuals with known or suspected infections that are highly transmissible or epidemiologically important pathogens. They are employed when standard precautions do not completely interrupt the transmission route. There are three categories of transmission-based precautions: contact, droplet, and airborne.

- *Contact precautions.* These are used for patients that have infections that can be spread by contact with the patients' feces, urine or other body fluids, skin, vomit, or wounds or by equipment or environmental surfaces contaminated by the patient. *Salmonella, Shigella,* and *Clostridium difficile* are examples of pathogens that require contact precautions.

- *Droplet precautions.* These are used for patients that have infections that can be spread through close contact with droplet nuclei from respiratory secretions that spread only short distances. Examples of diseases that require droplet precautions are influenza, pneumonia, the common cold, whooping cough, and meningitis.

- *Airborne precautions.* These are used for patients that have an infection that can spread by droplet nuclei over long distances. Examples include chickenpox, tuberculosis, and measles.

Control measures aimed at preventing HAIs vary from one institution to another, but universal precautions are always employed to reduce the number of pathogens to which individuals are exposed. Following are some examples of the application of control measures to prevent or reduce HAIs.

According to the CDC, handwashing is the single most important means of preventing the spread of infection. Nevertheless, the CDC reports that health care workers frequently fail to follow recommended handwashing procedures. On average, health care workers wash their hands before interacting with patients only 40% of the time.

In addition to handwashing, tubs used to bathe patients should be disinfected between uses so that bacteria from the previous patient won't contaminate the next one. Respirators and humidifiers provide both a suitable growth environment for some bacteria and a method of airborne transmission. These sources of HAIs must be kept scrupulously clean and disinfected, and materials used for bandages and intubation

*C. difficile* is a bacterium that is involved in 15–25% of all healthcare-associated infections and almost half of all cases of diarrhea. It was first identified in 1935 as part of the normal intestinal microbiota. *C. difficile* was associated with diarrhea in 1977. Infection can range from asymptomatic colonization of patients to diarrhea or colitis. Mortality in older patients is 10–20%. After making sure Jamil is not taking any antibiotics, his physician prescribes the antibiotic metronidazole to treat the *C. difficile*.

**Why does Jamil's physician make sure he is not taking antibiotics before she treats the *C. difficile* infection?** (*Hint:* See page 738.)

395    408    **411**    413    418

(insertion of tubes into organs, such as the trachea) should be single-use disposable or sterilized before use. Packaging used to maintain sterility should be removed aseptically. Physicians can help improve patients' resistance to infection by prescribing antibiotics only when necessary, avoiding invasive procedures if possible, and minimizing the use of immunosuppressive drugs.

Accredited hospitals should have an infection control committee. Most hospitals have at least an infection control nurse or epidemiologist (an individual who studies disease in populations). The role of these staff members is to identify problem sources, such as antibiotic-resistant strains of bacteria and improper sterilization techniques. The infection control officer should make periodic examinations of hospital equipment to determine the extent of microbial contamination. Samples should be taken from tubing, catheters, respirator reservoirs, and other equipment.

> ▶ **Play Nosocomial Infections:** *Overview, Prevention*
> @Mastering**Microbiology**

**CHECK YOUR UNDERSTANDING**

✔ **14-15** What interacting factors result in nosocomial infections?

✔ **14-16** What is a compromised host?

✔ **14-17, 14-18** How are nosocomial infections primarily transmitted, and how can they be prevented?

# Emerging Infectious Diseases

### LEARNING OBJECTIVE

**14-19** List several probable reasons for emerging infectious diseases, and name one example for each reason.

**Emerging infectious diseases (EIDs)** are diseases that are new or changing, are showing an increase in incidence in the recent past, or show a potential to increase in the near future (see Chapter 1). An emerging disease can be caused by a virus, a bacterium, a fungus, a protozoan, or a helminth. About 75% of emerging infectious diseases are zoonotic, mainly of viral origin, and are likely to be vector-borne.

Several criteria are used for identifying an EID. For example, some diseases present symptoms that are clearly distinctive from all other diseases. Some are recognized because improved diagnostic techniques allow the identification of a new pathogen. Others are identified when a local disease becomes widespread, a rare disease becomes common, a mild disease becomes more severe, or an increase in life span permits a slow disease to develop. Examples of emerging infectious diseases are listed in Table 14.5 and described in the boxes in Chapters 8 and 13 (pages 218 and 367).

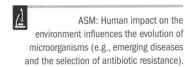

ASM: Human impact on the environment influences the evolution of microorganisms (e.g., emerging diseases and the selection of antibiotic resistance).

A variety of factors contribute to the emergence of new infectious diseases:

- New strains, such as *E. coli* O157:H7 and avian influenza (H5N1), may result from genetic recombination between organisms.

- A new serovar, such as *Vibrio cholerae* O139, may result from changes in or the evolution of existing microorganisms.

- The widespread, and sometimes unwarranted, use of antibiotics and pesticides encourages the growth of more resistant populations of microbes and the vectors (mosquitoes, lice, and ticks) that carry them.

- Global warming and changes in weather patterns may increase the distribution and survival of reservoirs and vectors, resulting in the introduction and dissemination of diseases such as malaria and *Hantavirus* pulmonary syndrome.

- Known diseases, such as Zika virus disease, chikungunya, dengue, and West Nile encephalitis, may spread to new geographic areas by modern transportation. This was less likely 100 years ago, when travel took so long that infected travelers either died or recovered during passage.

- Insect vectors transported to new areas can transmit infections brought by human travelers. The African yellow fever mosquito, *Aedes aegypti*, came to the Americas with the first European explorers. Yellow fever virus was also brought to the Americas with those first explorers, and *A. aegypti* transmitted the disease to native populations and immigrants alike. The Asian tiger mosquito, *A. albopictus*, was inadvertently brought to Texas on a cargo ship from Japan in 1985. Both *Aedes* species are now established throughout the southern and southwestern states. And both are vectors for Zika, chikungunya, dengue, and West Nile viruses.

- Previously unrecognized infections may appear in individuals living or working in regions undergoing ecological changes

**TABLE 14.5** Emerging Infectious Diseases

| Microorganism | Year of Emergence | Disease Caused | Chapter Reference |
|---|---|---|---|
| **BACTERIA** | | | |
| *Elizabethkingia anophelis* | 2013 | Meningitis | 22 |
| *Clostridium difficile* | 2004 | Diarrhea, colitis, and hemorrhagic necrosis | 25 |
| *Bordetella pertussis* | 2000 | Whooping cough | 24 |
| *Mycobacterium ulcerans* | 1998 | Buruli ulcer | 21 |
| Methicillin-resistant *Staphylococcus aureus* | 1968 | Bacteremia, pneumonia | 20 |
| Vancomycin-resistant *Staphylococcus aureus* | 1997 | Bacteremia, pneumonia | 20 |
| *Streptococcus pneumoniae* | 1995 | Antibiotic-resistant pneumonia | 24 |
| *Streptococcus pyogenes* | 1995 | Streptococcal toxic shock syndrome | 21 |
| *Corynebacterium diphtheriae* | 1994 | Diphtheria epidemic, eastern Europe | 24 |
| *Vibrio cholerae* O139 | 1992 | New serovar of cholera, Asia | 25 |
| Vancomycin-resistant enterococci | 1988 | Urinary tract infections, bacteremia, endocarditis | 26, 23 |
| *Bartonella henselae* | 1983 | Cat-scratch disease | 23 |
| *Escherichia coli* O157:H7 | 1982 | Hemorrhagic diarrhea | 25 |
| **FUNGI** | | | |
| *Candida auris* | 2017 | Systemic | 23 |
| *Pneumocystis jirovecii* | 1981 | Pneumonia in immunocompromised individuals | 24 |
| **PROTOZOA** | | | |
| *Trypanosoma cruzi* | 2007 | Chagas' disease in United States | 23 |
| *Cyclospora cayetanensis* | 1993 | Severe diarrhea and wasting syndrome | 25 |
| **HELMINTHS** | | | |
| *Baylisascaris procyonis* | 2001 | Raccoon roundworm encephalitis, in humans | |
| **VIRUSES** | | | |
| Zika virus disease | 2007 | Congenital microcephaly | 22 |
| Chikungunya virus | 2013 | Chikungunya disease, Americas | 23 |
| Middle East respiratory syndrome coronovirus (MERS-CoV) | 2013 | Middle East respiratory syndrome (MERS) | 24 |
| SARS-associated coronavirus | 2002 | Severe acute respiratory syndrome (SARS) | 24 |
| Ebola virus | 1976 | Causes sporadic epidemics | 23 |
| West Nile virus | 1999 | West Nile encephalitis | 22 |
| Nipah virus | 1998 | Encephalitis, Malaysia | 22 |
| Influenza A virus | 1997, 2009 | Avian influenza (H5N1), Swine flu (H1N1) | 24 |
| Hendra virus | 1994 | Encephalitis-like symptoms, Australia | 24 |
| *Hantavirus* | 1993 | *Hantavirus* pulmonary syndrome | 23 |
| Hepatitis C virus | 1989 | Hepatitis | 25 |
| Monkeypox virus | 1985 | Chickenpox-like disease | 21 |
| Dengue virus | 1984 | Dengue | 23 |
| HIV | 1983 | AIDS | 19 |
| **PRIONS** | | | |
| Bovine spongiform encephalitis agent | 1996 | Mad cow disease, Great Britain | 22 |

brought about by natural disaster, construction, wars, and expanding human settlement. In California, the incidence of coccidioidomycosis increased tenfold following the Northridge earthquake of 1994. Workers clearing South American forests are now contracting Venezuelan hemorrhagic fever.

- Even animal control measures may affect the incidence of a disease. The increase in Lyme disease in recent years could be due to rising deer populations resulting from the killing of deer predators.

- Failures in public health measures may be a contributing factor to the emergence of previously controlled infections. For example, the failure of adults to get a diphtheria booster vaccination led to a diphtheria epidemic in the newly independent republics of the former Soviet Union in the 1990s.

- **Bioterrorism**, the use of pathogens or toxins to produce death and disease in humans, animals, or plants as an act of violence and intimidation, is another factor that could affect the occurrence of emerging infectious diseases. The pathogens

## CLINICAL CASE

Antibiotics can kill competing bacteria, thus allowing growth of *C. difficile*. When Jamil's physician learns of the cause of Jamil's diarrhea, she checks with the hospital to see whether any other patients have developed *C. difficile* diarrhea and colitis. It turns out that 20 other patients are also infected with *C. difficile*. The local health department completes an epidemiological study of the outbreak and releases the following information:

### Rate of infection for patients

| | |
|---|---|
| Single room | 7% |
| Double room | 17% |
| Triple room | 26% |

### Rate of environmental isolations of *C. difficile*

| | |
|---|---|
| Bed rail | 10% |
| Commode | 1% |
| Floor | 18% |
| Call button | 6% |
| Toilet | 3% |

### *C. difficile* on hands of hospital personnel after contact with patients who were culture-positive for *C. difficile*

| | |
|---|---|
| Used gloves | 0% |
| Did not use gloves | 59% |
| Had *C. difficile* before patient contact | 3% |
| Washed with nondisinfectant soap | 40% |
| Washed with disinfectant soap | 3% |
| Did not wash hands | 20% |

**What is the most likely mode of transmission, and how can transmission be prevented?**

395   408   411   **413**   418

or toxins can be disseminated through aerosolization, food, human carriers, water, or infected insects. Recent examples of bioterrorism are discussed in the Big Picture in Chapter 24 (pages 696–697).

The CDC, the National Institutes of Health (NIH), and the World Health Organization (WHO) have developed plans to address issues relating to EIDs. Their priorities include the following:

1. To detect, promptly investigate, and monitor emerging infectious pathogens, the diseases they cause, and factors that influence their emergence
2. To expand basic and applied research on ecological and environmental factors, microbial changes and adaptations, and host interactions that influence EIDs
3. To enhance the communication of public health information and the prompt implementation of prevention strategies regarding EIDs
4. To establish plans to monitor and control EIDs worldwide

Because of the importance of emerging infectious diseases to the scientific community, the CDC publishes a monthly journal called *Emerging Infectious Diseases*.

### CHECK YOUR UNDERSTANDING

✔ 14-19 Give several examples of emerging infectious diseases.

# Epidemiology

### LEARNING OBJECTIVES

14-20 Define *epidemiology*, and describe three types of epidemiologic investigations.

14-21 Identify the function of the CDC.

14-22 Define the following terms: *morbidity*, *mortality*, and *notifiable infectious diseases*.

In today's crowded, overpopulated world, in which frequent travel and the mass production and distribution of food and other goods are a way of life, diseases can spread rapidly. A contaminated food or water supply, for example, can affect many thousands of people very quickly. Identifying the causative agent of a disease is desirable so it can be effectively controlled and treated. It is also desirable to understand the mode of transmission and geographical distribution of the disease. The science that studies when and where diseases occur and how they are transmitted in populations is called **epidemiology** (EP-i-dē-mē-ol-ō-jē).

Modern epidemiology began in the mid-1800s with three now-famous investigations. John Snow, a British physician, conducted a series of investigations related to outbreaks of cholera in London. As the cholera epidemic of 1848 to 1849 raged, Snow analyzed the death records attributed to cholera, gathered information about the victims, and interviewed survivors who lived in the neighborhood. Using the information he compiled, Snow made a map showing that most individuals who died of cholera drank or brought water from the Broad Street pump; those who used other pumps (or drank beer, like the workers at a nearby brewery) did not get cholera. He concluded that contaminated water from the Broad Street pump was the source of the epidemic. When the pump's handle was removed and people could no longer get water from this location, the number of cholera cases dropped significantly.

Between 1846 and 1848, Ignaz Semmelweis meticulously recorded the number of births and maternal deaths at Vienna General Hospital. The First Maternity Clinic had become a source of gossip throughout Vienna because the death rate due to puerperal sepsis ranged between 13% and 18%, four times that of the Second Maternity Clinic. Puerperal sepsis (childbirth fever) is a nosocomial infection that begins in the uterus as a result of childbirth or abortion. It is frequently caused by *Streptococcus pyogenes*. The infection progresses to the abdominal cavity (peritonitis) and in many cases to septicemia (proliferation of microbes in the blood). Wealthy women did not go to the clinic, and poor women had learned they had a better chance of surviving childbirth if they gave birth elsewhere before going to the hospital. Looking at his data, Semmelweis identified a common factor among the wealthy women and the poor women who had given birth prior to entering the clinic: they were not examined by the medical students, who had spent their mornings dissecting cadavers. In May 1847, he ordered all medical students to wash their hands with chloride of lime before entering the delivery room, and the mortality rate dropped to under 2%.

Florence Nightingale recorded statistics on epidemic typhus in the English civilian and military populations. In 1858, she published a thousand-page report using statistical comparisons to demonstrate that diseases, poor food, and unsanitary conditions were killing the soldiers. Her work resulted in reforms in the British Army and to her admission to the Statistical Society, their first female member.

These three careful analyses of where and when a disease occurred and how it was transmitted within a population constituted a new approach to medical research and demonstrated the importance of epidemiology. The works of Snow, Semmelweis, and Nightingale resulted in changes that lowered the incidence of diseases even though knowledge of the causes of infectious disease was limited. Most physicians believed that the symptoms they saw were the causes of the disease, not the result of disease. Koch's work on the germ theory of disease was still 30 years in the future.

An epidemiologist not only determines the etiology of a disease but also identifies other possibly important factors and patterns concerning the people affected. An important part of the epidemiologist's work is assembling and analyzing data such as age, sex, occupation, personal habits, socioeconomic status, history of immunization, presence of any other diseases, and the common history of affected individuals (such as eating the same food or visiting the same doctor's office). Also important for the prevention of future outbreaks is knowledge of the site at which a susceptible host came into contact with the agent of infection. In addition, the epidemiologist considers the period during which the disease occurs, either on a seasonal basis (to indicate whether the disease is prevalent during a particular season) or on a yearly basis (to indicate the effects of immunization or an emerging or reemerging disease).

An epidemiologist is also concerned with various methods for controlling a disease. The strategies for controlling diseases include the use of drugs (chemotherapy) and vaccines (immunization). Other methods include the control of human, animal, and nonliving reservoirs of infection, water treatment, proper sewage disposal (enteric diseases), cold storage, pasteurization, food inspection, adequate cooking (foodborne diseases), improved nutrition to bolster host defenses, changes in personal habits, and screening of transfused blood and transplanted organs.

**Figure 14.11** contains graphs indicating the incidence of selected diseases. Such graphs provide information about whether disease outbreaks are sporadic or epidemic and, if epidemic, how the disease might have spread. By establishing the frequency of a disease in a population and identifying the factors responsible for its transmission, an epidemiologist can provide physicians with information that is important in determining the prognosis and treatment of a disease. Epidemiologists also evaluate how effectively a disease is being controlled in a community—by a vaccination program, for example. Finally, epidemiologists can provide data to help in evaluating and planning overall health care for a community.

Epidemiologists use three basic types of investigations when analyzing the occurrence of a disease: descriptive, analytical, and experimental.

## Descriptive Epidemiology

**Descriptive epidemiology** entails collecting all data that describe the occurrence of the disease under study. Usually, descriptive epidemiology is used once a problem has been identified. Relevant information usually includes information about the affected individuals and the place and period in which the disease occurred. Snow's search for the cause of the cholera outbreak in London is an example of descriptive epidemiology.

Such a study is generally *retrospective* (looking backward after the episode has ended). In other words, the epidemiologist backtracks to the cause and source of the disease (see the boxes in Chapters 21 through 26). The search for the cause of an increase in microcephaly (incomplete brain development) is an example of a fairly recent retrospective study. In the initial phase of an epidemiological study, retrospective studies are more common than *prospective* (looking forward) studies, in which an epidemiologist chooses a group of people who are free of a particular

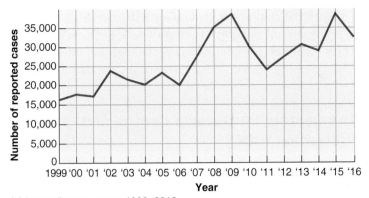

(a) Lyme disease cases, 1999–2016

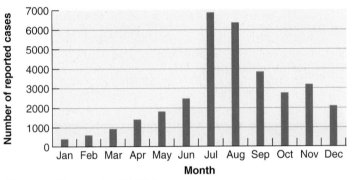

(b) Lyme disease by month, 2016

(c) Reported tuberculosis cases, 1945–2016

**Figure 14.11 Epidemiological graphs. (a)** Lyme disease cases, showing the annual occurrence of the disease during the covered period. **(b)** A different perspective of Lyme disease shows the seasonal pattern that enabled epidemiologists to draw some conclusions about the disease's epidemiology. **(c)** A graph of the incidence of tuberculosis shows a rapid decrease in the rate of infection from 1948 to 1957. This graph records the number of cases per 100,000 people, rather than the total number of cases.

*Source:* Data from CDC.

**Q** **What does graph (b) indicate about transmission of Lyme disease? What can you conclude from graph (c)?**

disease to study. The group's subsequent disease experiences are then recorded for a given period. Prospective studies were used to test the Salk polio vaccine in 1954 and 1955.

## Analytical Epidemiology

**Analytical epidemiology** analyzes a particular disease to determine its probable cause. It can use different variables to discover possible routes and rates of infection. This study can be done in two ways. With the *case control method*, the epidemiologist looks for factors that might have preceded the disease. A group of people who have the disease is compared with another group of people who are free of the disease. For example, one group with meningitis and one without the disease might be matched by age, sex, socioeconomic status, and location. These statistics are compared to determine which of all the possible factors—genetic, environmental, nutritional, and so forth—might be responsible for the meningitis. Nightingale's work was an example of analytical epidemiology, in which she compared disease in soldiers and civilians. With the *cohort method*, the epidemiologist studies two populations: one that has had contact with the agent causing a disease and another that has not (both groups are called *cohort groups*). For example, a comparison of one group composed of people who have received blood transfusions and one composed of people who have not could reveal an association between blood transfusions and the incidence of hepatitis B virus.

## Experimental Epidemiology

**Experimental epidemiology** begins with a hypothesis about a particular disease; experiments to test the hypothesis are then conducted. Semmelweis's use of handwashing is an example of experimental epidemiology. Testing on humans is called a **clinical trial**. A clinical trial can be used to test a hypothesis regarding the effectiveness of a drug. A group of infected individuals is selected and divided randomly so that some receive the drug (test group) and others (control group) receive a *placebo*, a substance that has no effect. In a single blind test, the individuals don't know whether they are in the test or control group. It's called a *double blind test* if the treating physicians also don't know the group identity. If all other factors are kept constant between the two groups, and if those people who received the drug recover more rapidly than those who received the placebo, it can be concluded that the drug was the experimental factor (variable) that made the difference.

## Case Reporting

We noted earlier in this chapter that establishing the chain of transmission for a disease is extremely important. Once known, the chain can be interrupted to slow down or stop the spread of the disease.

An effective way to establish the chain of transmission is *case reporting*, a procedure that requires health care workers to report specified diseases to local, state, and national health officials. **Notifiable infectious diseases**, listed in Table 14.6, are diseases for which physicians are required by law to report cases to the U.S. Public Health Service. These data provide early warning of possible outbreaks. As of 2017, a total of 62 infectious diseases were reported at the national level. Case reporting provides epidemiologists with an approximation of the incidence and prevalence of a disease. This information helps officials decide whether or not to investigate a given disease. The reporting of data also allows epidemiologists to monitor emerging infectious diseases (page 411) in conjunction with comparing the results to previous records of infection. Use of these reports makes it possible to decrease the probability of large-scale infections.

Case reporting provided epidemiologists with valuable leads regarding the origin and spread of AIDS. In fact, one of the first clues about AIDS came from reports of young men with Kaposi's sarcoma, formerly a disease of older men. Using these reports, epidemiologists began various studies of the patients. If an epidemiological study shows that a disease affects a large enough segment of the population, epidemiologists then attempt to isolate and identify its causative agent. Identification is accomplished by a number of different microbiological methods. Identifying the causative agent often provides valuable information regarding the reservoir for the disease.

Once the chain of transmission is discovered, it's possible to apply control measures to stop the disease from spreading. These might include elimination of the source of infection, isolation and segregation of infected people, the development of vaccines, and, as in the case of AIDS, education.

## The Centers for Disease Control and Prevention (CDC)

Epidemiology is a major concern of state and federal public health departments. The **Centers for Disease Control and Prevention (CDC)**, a branch of the U.S. Public Health Service located in Atlanta, Georgia, is a central source of epidemiological information in the United States.

The CDC issues a publication called the *Morbidity and Mortality Weekly Report* (www.cdc.gov). *MMWR*, as it is called, is read by microbiologists, physicians, and other hospital and public health professionals. *MMWR* contains data on **morbidity**, the incidence of specific notifiable diseases, and **mortality**,

---

**TABLE 14.6** Nationally Notifiable Infectious Diseases, 2017

| | | |
|---|---|---|
| Anthrax | Hepatitis A, B, and C | Severe acute respiratory syndrome-associated coronavirus disease |
| Arboviral diseases: neuroinvasive, non-neuroinvasive | HIV infection | Shiga toxin-producing *E. coli* |
| Babesiosis | Influenza-associated pediatric mortality | Shigellosis |
| Botulism | Invasive pneumococcal disease | Smallpox |
| Brucellosis | Legionellosis | Spotted fever rickettsiosis |
| Campylobacteriosis | Leptospirosis | Streptococcal toxic shock syndrome |
| Chancroid | Listeriosis | Syphilis |
| Chlamydia trachomatis infection | Lyme disease | Tetanus |
| Cholera | Malaria | Toxic shock syndrome (nonstreptococcal) |
| Coccidioidomycosis | Measles | Trichinellosis |
| Congenital syphilis | Meningococcal disease | Tuberculosis |
| Cryptosporidiosis | Mumps | Tularemia |
| Cyclosporiasis | Novel influenza A virus infections | Typhoid fever |
| Dengue virus infections | Pertussis (whooping cough) | Vancomycin-intermediate *Staphylococcus aureus* (VISA) and Vancomycin-resistant *S. aureus* (VRSA) |
| Diphtheria | Plague | Varicella |
| Ehrlichiosis and anaplasmosis | Poliovirus infection, paralytic and nonparalytic | Vibriosis |
| Giardiasis | Psittacosis | Viral hemorrhagic fever |
| Gonorrhea | Q fever | Yellow fever |
| Haemophilus influenza invasive disease, | Rabies, animal or human | Zika virus disease and congenital infection |
| Hansen's disease (leprosy) | Rubella and congenital rubella syndrome | |
| Hantavirus pulmonary syndrome and nonpulmonary | Salmonellosis | |
| Hemolytic uremic syndrome, post-diarrheal | | |

# Healthcare-Associated Infections

In this box, you will encounter a series of questions the epidemiologists ask themselves as they try to trace an outbreak to its source. Try to answer each question before going to the next one.

1. Dwayne Jackson, the epidemiologist at a city hospital, would like to find out why during one year, 5287 patients developed bacteremia during their hospital stays. All patients had a fever (>38°C), chills, and low blood pressure; 14% had severe necrotizing fasciitis (see page 597). Dr. Jackson looks at the results of the blood cultures, which were grown on mannitol-salt agar; the bacteria are identified as coagulase-positive, gram-positive cocci (Figure A).

   What organisms are possible agents of infection?

2. The biochemical tests confirm that *Staphylococcus aureus* is the culprit. Antibiotic-sensitivity testing shows that all the isolates are methicillin resistant. Six are vancomycin-intermediate resistant, and one is vancomycin resistant. Methicillin-resistant *S. aureus* (MRSA) can cause a life-threatening, necrotizing illness due to a leukocidin toxin (see page 433).

   What else does Dr. Jackson need to know?

3. Polymerase chain reaction (PCR) is used to determine that strain USA100 caused 80% of the MRSA cases in Dr. Jackson's hospital. USA100 is the cause of 92% of healthcare-acquired strains. The majority (89%) of community-acquired MRSA infections are the USA300 strain. The incidence of MRSA in the community (not hospitalized) is 0.02–0.04%. Dr. Jackson compares the number of patients with MRSA to the procedures performed and cross-references that information with antibiotic use by the patients (Table A).

   Based on the information in the table, which procedure increases the likelihood of infection most?

4. Each year, an estimated 250,000 cases of bloodstream infections occur in hospitals in the United States from inserting needles into veins to deliver intravenous (IV) solutions, and the estimated mortality for these infections is 12–25%. Dr. Jackson sees that people receiving hemodialysis are especially vulnerable to infections because they require access to veins for prolonged periods and undergo frequent punctures of the access site (Figure B).

   How does antimicrobial therapy contribute?

5. The first VRSA (vancomycin-resistant *S. aureus*) infection in the United States occurred in a dialysis patient in 2002. The patient had been treated with vancomycin for a MRSA infection.

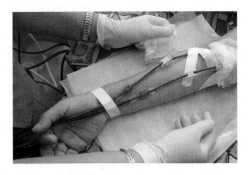

**Figure B Hemodialysis procedure.**

The VRSA isolate contained the *vanA* vancomycin-resistance gene from enterococci. VRSA are always methicillin resistant. Only 14 VRSA cases have been reported in the United States; however, 183 VISA (vancomycin-intermediate *S. aureus*) cases were reported in 2015. Antimicrobial therapy for hemodialysis-associated infections increases the prevalence of antimicrobial resistance. Susceptible bacteria are killed, and bacteria with a mutation that confers resistance are able to grow without competition.

Source: Adapted from *MMWR* 64(37): 1056, September 25, 2015 and *MMWR* 64 (53), August 11, 2017.

**Figure A The gram-positive cocci grown on mannitol-salt agar.**

| TABLE A | | |
| --- | --- | --- |
| Procedure | MRSA-Infected Patients | Total Number of Patients Receiving Procedure |
| Hemodialysis | 813 | 1807 |
| Intravenous (IV) catheter | 1057 | 16,516 |
| Surgery | 945 | 5659 |
| Urinary bladder catheter | 1750 | 7919 |
| Ventilator (invasive airway) | 722 | 7367 |
| Antibiotic Use during the 6 Months Prior to Infection | | |
| Vancomycin | 21 | 41 |
| Fluoroquinolone | 49 | 113 |
| Ceftriaxone | 14 | 41 |

the number of deaths from these diseases. These data are usually organized by state. **Morbidity rate** is the number of people affected by a disease in a given period of time in relation to the total population. **Mortality rate** is the number of deaths resulting from a disease in a population in a given period of time in relation to the total population.

*MMWR* articles include reports of disease outbreaks, case histories of special interest, and summaries of the status of particular diseases during a recent period. These articles often include recommendations for procedures for diagnosis, immunization, and treatment. Several graphs and other data in this textbook are from *MMWR*, and the Clinical Focus boxes are adapted from reports from this publication. See the box on page 417, for example.

 **Play Epidemiology:** *Overview, Occurrence of Disease, Transmission of Disease*
@Mastering**Microbiology**

### CHECK YOUR UNDERSTANDING

✔ **14-20** After learning that 40 hospital employees developed nausea and vomiting, the hospital infection control officer determined that 39 ill people ate green beans in the hospital cafeteria, compared to 34 healthy people who ate in the cafeteria the same day but did not eat green beans in the hospital cafeteria. What type of epidemiology is this?

✔ **14-21** What is the CDC's function?

✔ **14-22** In 2012, the morbidity of West Nile encephalitis was 5674, and the mortality was 286. The morbidity of listeriosis was 121; the mortality was 13. Which disease is more likely to be fatal?

\* \* \*

In the next chapter, we consider the mechanisms of pathogenicity. We'll discuss in more detail the methods by which microorganisms enter the body and cause disease, the effects of disease on the body, and the means by which pathogens leave the body.

### CLINICAL CASE Resolved

Transmission of *C. difficile* can be prevented by wearing gloves for contact with all body substances, using disposable rectal thermometers, and stopping antibiotic overuse. *C. difficile* is acquired by ingesting the bacteria or its endospores from direct contact between people or indirect contact via fomites; it's the most common HAI and is considered an epidemic. Jamil responds well to treatment; he is gaining most of his weight back and no longer spends most of his time in the bathroom.

395 408 411 413 **418**

---

# Study Outline

 Go to @Mastering**Microbiology** for **Interactive Microbiology,** *In the Clinic* videos, *MicroFlix, MicroBoosters, 3D animations, practice quizzes, and more.*

## Introduction (p. 393)

1. Disease-causing microorganisms are called pathogens.
2. Pathogenic microorganisms have special properties that allow them to invade the human body or produce toxins.
3. When a microorganism overcomes the body's defenses, a state of disease results.

## Pathology, Infection, and Disease (p. 394)

1. Pathology is the scientific study of disease.
2. Pathology is concerned with the etiology (cause), pathogenesis (development), and effects of disease.
3. Infection is the invasion and growth of pathogens in the body.
4. A host is an organism that shelters and supports the growth of pathogens.
5. Disease is an abnormal state in which part or all of the body is not properly adjusted or is incapable of performing normal functions.

## Human Microbiome (pp. 394–398)

1. Microorganisms begin colonization in and on the surface of the body soon after birth.

2. Microorganisms that establish permanent colonies inside or on the body without producing disease make up the normal microbiota.

 **Play Interactive Microbiology**
@Mastering**Microbiology**
*See how the microbiome affects a patient's health*

3. Transient microbiota are microbes that are present for various periods and then disappear.

## Relationships between the Normal Microbiota and the Host (pp. 397–398)

4. The normal microbiota can prevent pathogens from causing an infection; this phenomenon is known as microbial antagonism.
5. Normal microbiota and the host exist in symbiosis (living together).
6. The three types of symbiosis are commensalism (one organism benefits, and the other is unaffected), mutualism (both organisms benefit), and parasitism (one organism benefits, and one is harmed).

### Opportunistic Microorganisms (p. 398)

7. Opportunistic pathogens do not cause disease under normal conditions but cause disease under special conditions.

### Cooperation among Microorganisms (p. 398)

8. In some situations, one microorganism makes it possible for another to cause a disease or produce more severe symptoms.

## The Etiology of Infectious Diseases (pp. 398–400)

### Koch's Postulates (pp. 398–399)

1. Koch's postulates are criteria for establishing that specific microbes cause specific diseases.

2. Koch's postulates have the following requirements: (1) the same pathogen must be present in every case of the disease; (2) the pathogen must be isolated in pure culture; (3) the pathogen isolated from pure culture must cause the same disease in a healthy, susceptible laboratory animal; and (4) the pathogen must be reisolated from the inoculated laboratory animal.

### Exceptions to Koch's Postulates (pp. 399–400)

3. Koch's postulates are modified to establish etiologies of diseases caused by viruses and some bacteria, which cannot be grown on artificial media.

4. Some diseases, such as tetanus, have unequivocal signs and symptoms.

5. Some diseases, such as pneumonia and nephritis, may be caused by a variety of microbes.

6. Some pathogens, such as *S. pyogenes*, cause several different diseases.

7. Certain pathogens, such as HIV, cause disease in humans only.

## Classifying Infectious Diseases (pp. 400–402)

1. A patient may exhibit symptoms (subjective changes in body functions) and signs (measurable changes), which a physician uses to make a diagnosis (identification of the disease).

2. A specific group of symptoms or signs that always accompanies a specific disease is called a syndrome.

3. Communicable diseases are transmitted directly or indirectly from one host to another.

4. A contagious disease is a very communicable disease that is capable of spreading easily and rapidly from one person to another.

5. Noncommunicable diseases are caused by microorganisms that normally grow outside the human body and are not transmitted from one host to another.

### Occurrence of a Disease (pp. 400–401)

6. Disease occurrence is reported by incidence (number of people contracting the disease) and prevalence (number of people with the disease) in a defined population, in a specified time.

7. Diseases are classified by frequency of occurrence: sporadic, endemic, epidemic, and pandemic.

### Severity or Duration of a Disease (pp. 401–402)

8. The scope of a disease can be defined as acute, chronic, subacute, or latent.

9. Herd immunity is the presence of immunity to a disease in most of the population.

### Extent of Host Involvement (p. 402)

10. A local infection affects a small area of the body; a systemic infection is spread throughout the body via the circulatory system.

11. A primary infection is an acute infection that causes the initial illness.

12. A secondary infection can occur after the host is weakened from a primary infection.

13. A subclinical, or inapparent, infection does not cause any signs or symptoms of disease in the host.

## Patterns of Disease (pp. 402–403)

### Predisposing Factors (pp. 402–403)

1. A predisposing factor is one that makes the body more susceptible to disease or alters the course of a disease.

2. Examples include gender, climate, age, fatigue, and inadequate nutrition.

### Development of Disease (p. 403)

3. The incubation period is the interval between the initial infection and the first appearance of signs and symptoms.

4. The prodromal period is characterized by the appearance of the first mild signs and symptoms.

5. During the period of illness, the disease is at its height, and all disease signs and symptoms are apparent.

6. During the period of decline, the signs and symptoms subside.

7. During the period of convalescence, the body returns to its prediseased state, and health is restored.

## The Spread of Infection (pp. 403–408)

### Reservoirs of Infection (pp. 403–404)

1. A continual source of infection is called a reservoir of infection.

2. People who have a disease or are carriers of pathogenic microorganisms are human reservoirs of infection.

3. Zoonoses are diseases that affect wild and domestic animals and can be transmitted to humans.

4. Some pathogenic microorganisms grow in nonliving reservoirs, such as soil and water.

### Transmission of Disease (pp. 404–408)

5. Transmission by direct contact involves close physical contact between the source of the disease and a susceptible host. Congenital transmission occurs from mother to fetus.

6. Transmission by fomites (inanimate objects) constitutes indirect contact.

7. Transmission via saliva or mucus in coughing or sneezing is called droplet transmission.

8. Transmission by a medium such as air, water, or food is called vehicle transmission.

9. Airborne transmission refers to pathogens carried on water droplets or dust for a distance greater than 1 meter.

10. Arthropod vectors carry pathogens from one host to another by both mechanical and biological transmission.

## Healthcare-Associated Infections (HAIs) (pp. 408–411)

1. Healthcare-associated infections (HAIs) include those acquired in settings such as hospitals, nursing homes, surgical centers, and health care clinics.

2. About 4% of patients acquire HAIs in the treatment environment.

### Microorganisms in the Hospital (pp. 408–409)

3. Certain normal microbiota are often responsible for HAIs when they are introduced into the body through such medical procedures as surgery and catheterization.

4. Opportunistic bacteria are the most frequent causes of HAIs.

### Compromised Host (p. 409)

5. Patients with burns, surgical wounds, and suppressed immune systems are the most susceptible to HAIs.

### Chain of Transmission (pp. 409–410)

6. HAIs are transmitted by direct contact between staff members and patients and between patients.

7. Fomites such as catheters, syringes, and respiratory devices can transmit HAIs.

### Control of Healthcare-Associated Infections (pp. 410–411)

8. Aseptic techniques can prevent HAIs.

9. Hospital infection control staff members are responsible for overseeing the proper cleaning, storage, and handling of equipment and supplies.

### Emerging Infectious Diseases (pp. 411–413)

1. New diseases and diseases with increasing incidences are called emerging infectious diseases (EIDs).

2. EIDs can result from the use of antibiotics and pesticides, climatic changes, travel, the lack of vaccinations, and improved case reporting.

3. The CDC, NIH, and WHO are responsible for surveillance and responses to emerging infectious diseases.

### Epidemiology (pp. 413–418)

1. The science of epidemiology is the study of the transmission, incidence, and frequency of disease.

2. Modern epidemiology began in the mid-1800s with the works of Snow, Semmelweis, and Nightingale.

3. In descriptive epidemiology, data about infected people are collected and analyzed.

4. In analytical epidemiology, a group of infected people is compared with an uninfected group.

5. In experimental epidemiology, controlled experiments designed to test hypotheses are performed.

6. Case reporting provides data on incidence and prevalence to local, state, and national health officials.

7. The Centers for Disease Control and Prevention (CDC) is the main source of epidemiological information in the United States.

8. The CDC publishes the *Morbidity and Mortality Weekly Report* to provide information on morbidity (incidence) and mortality (deaths).

## Study Questions

For answers to the Knowledge and Comprehension questions, turn to the Answers tab in the back of the textbook.

## Knowledge and Comprehension

### Review

1. Differentiate the terms in each of the following pairs:
   a. etiology and pathogenesis
   b. infection and disease
   c. communicable disease and noncommunicable disease

2. Define *symbiosis*. Differentiate commensalism, mutualism, and parasitism, and give an example of each.

3. Indicate whether each of the following conditions is typical of subacute, chronic, or acute infections.
   a. The patient experiences a rapid onset of malaise; symptoms last 5 days.
   b. The patient experiences cough and breathing difficulty for months.
   c. The patient has no apparent symptoms and is a known carrier.

4. Among hospital patients who have infections, one-third did not enter the hospital with the infection but rather acquired it in the hospital. How do they acquire these infections? What is the method of transmission of these infections? What is the reservoir of infection?

5. Distinguish symptoms from signs as signals of disease.

6. How can a local infection become a systemic infection?

7. Why are some organisms that constitute the normal microbiota described as commensals, whereas others are described as mutualistic?

8. Put the following in the correct order to describe the pattern of disease: period of convalescence, prodromal period, period of decline, incubation period, period of illness.

9. NAME IT This microbe is acquired by humans as infants and is essential for good health. Acquiring a closely related strain causes severe stomach cramps, bloody diarrhea, and vomiting. What is the microbe?

10. DRAW IT Using the data below, draw a graph showing the incidence of influenza during a typical year. Indicate the endemic and epidemic levels.

| Month | Percentage of Physician Visits for Influenza-like Symptoms |
|-------|:---:|
| Jan | 2.33 |
| Feb | 3.21 |
| Mar | 2.68 |
| Apr | 1.47 |
| May | 0.97 |
| Jun | 0.30 |
| Jul | 0.30 |
| Aug | 0.20 |
| Sep | 0.20 |
| Oct | 1.18 |
| Nov | 1.54 |
| Dec | 2.39 |

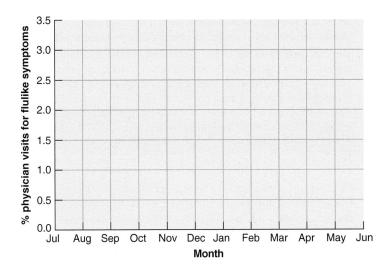

swollen lymph nodes under both arms. On S
been scratched and bitten by a cat. The cat w
and *Yersinia pestis* was isolated from the cat. (
administered to the boy from September 7
from him. On September 17, the boy's te
and on September 22, he was released fr

6. Identify the incubation period for
   a. September 3–5
   b. September 3–6                    d. Septem

7. Identify the prodromal period for this disease.
   a. September 3–5            c. September 6–7
   b. September 3–6            d. September 6–17

Use the following information to answer questions 8–10.

A Maryland woman was hospitalized with dehydration; *Vibrio cholerae* and *Plesiomonas shigelloides* were isolated from the patient. She had neither traveled outside the United States nor eaten raw shellfish during the preceding month. She had attended a party 2 days before her hospitalization. Two other people at the party had acute diarrheal illness and elevated levels of serum antibodies against *Vibrio*. Everyone at the party ate crabs and rice pudding with coconut milk. Crabs left over from this party were served at a second party. One of the 20 people at the second party had onset of mild diarrhea; specimens from 14 of these people were negative for vibriocidal antibodies.

8. This is an example of
   a. vehicle transmission.
   b. airborne transmission.
   c. transmission by fomites.
   d. direct contact transmission.
   e. healthcare-associated transmission.

9. The etiologic agent of the disease is
   a. *Plesiomonas shigelloides*.        d. coconut milk.
   b. crabs.                              e. rice.
   c. *Vibrio cholerae*.

10. The source of the disease was
   a. *Plesiomonas shigelloides*.        d. coconut milk.
   b. crabs.                              e. rice.
   c. *Vibrio cholerae*.

## Analysis

1. Ten years before Robert Koch published his work on anthrax, Anton De Bary showed that potato blight was caused by the alga *Phytophthora infestans*. Why do you suppose we use Koch's postulates instead of something called "De Bary's postulates"?

2. Florence Nightingale gathered the following data in 1855.

| Population Sampled | Deaths from Contagious Diseases |
| --- | --- |
| Englishmen (in general population) | 0.2% |
| English soldiers (in England) | 18.7% |
| English soldiers (in Crimean War) | 42.7% |
| English soldiers (in Crimean War) after Nightingale's sanitary reforms | 2.2% |

Discuss how Nightingale used the three basic types of epidemiological investigation. The contagious diseases were primarily cholera and typhus; how are these diseases transmitted and prevented?

3. Name the method of transmission of each of the following diseases:
   a. malaria              c. salmonellosis
   b. tuberculosis         d. streptococcal pharyngitis

## Multiple Choice

1. The emergence of new infectious diseases is probably due to all of the following *except*
   a. the need of bacteria to cause disease.
   b. the ability of humans to travel by air.
   c. changing environments (e.g., flood, drought, pollution).
   d. a pathogen crossing the species barrier.
   e. the increasing human population.

2. All members of a group of ornithologists studying barn owls in the wild have had salmonellosis (*Salmonella* gastroenteritis). One birder is experiencing her third infection. What is the most likely source of their infections?
   a. The ornithologists are eating the same food.
   b. They are contaminating their hands while handling the owls and nests.
   c. One of the workers is a *Salmonella* carrier.
   d. Their drinking water is contaminated.

3. Which of the following statements is *false*?
   a. *E. coli* never causes disease.
   b. *E. coli* provides vitamin K for its host.
   c. *E. coli* often exists in a mutualistic relationship with humans.
   d. A disease-causing strain of *E. coli* causes bloody diarrhea.

4. Which of the following is *not* one of Koch's postulates?
   a. The same pathogen must be present in every case of the disease.
   b. The pathogen must be isolated and grown in pure culture from the diseased host.
   c. The pathogen from pure culture must cause the disease when inoculated into a healthy, susceptible laboratory animal.
   d. The disease must be transmitted from a diseased animal to a healthy, susceptible animal by direct contact.
   e. The pathogen must be isolated in pure culture from an experimentally infected lab animal.

5. Which one of the following diseases is *not* correctly matched to its reservoir?
   a. influenza— animal
   b. rabies—animal
   c. botulism—nonliving
   d. anthrax—nonliving
   e. toxoplasmosis—cats

Use the following information to answer questions 6–7.

On September 6, a 6-year-old boy experienced fever, chills, and vomiting. On September 7, he was hospitalized with diarrhea and

g. mononucleosis
e. measles
f. hepatitis A

h. tetanus
i. hepatitis B
j. chlamydial urethritis

The following graph shows the incidence of typhoid fever in the United States from 1954 through 2016. Mark the graph to show when this disease occurred sporadically and epidemically. What appears to be the endemic level? What would have to be shown to indicate a pandemic of this disease? How is typhoid fever transmitted?

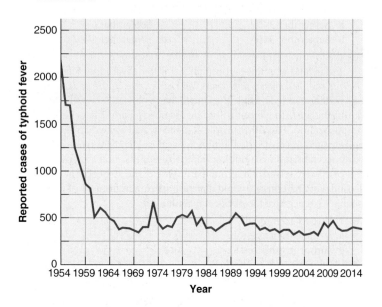

**Year**

## Clinical Applications and Evaluation

1. Three days before a nurse developed meningococcemia, she assisted with intubation of a patient with a *Neisseria meningitidis* infection. Of the 24 medical personnel involved, only this nurse became ill. The nurse recalled that she was exposed to nasopharyngeal secretions and did not receive antibiotic prophylaxis. What two mistakes did the nurse make? How is meningitis transmitted?

2. Three patients in a large hospital acquired infections of *Burkholderia cepacia* during their stay. All three patients received cryoprecipitate, which is prepared from blood that has been frozen in a standard plastic blood transfer pack. The transfer pack is then placed in a water bath to thaw. What is the probable origin of the infections? What characteristics of *Burkholderia* would allow it to be involved in this type of infection?

3. Following is a case history of a 49-year-old man. Identify each period in the pattern of disease that he experienced. On February 7, he handled a parakeet with a respiratory illness. On March 9, he experienced intense pain in his legs, followed by severe chills and headaches. On March 16, he had chest pains, cough, and diarrhea, and his temperature was 40°C. Appropriate antibiotics were administered on March 17, and his fever subsided within 12 hours. He continued taking antibiotics for 14 days. (*Note*: The disease is psittacosis. Can you find the etiology?)

4. *Mycobacterium avium-intracellulare* is prevalent in AIDS patients. In an effort to determine the source of this infection, hospital water systems were sampled. The water contained chlorine.

**Percentage of Samples with *M. avium***

| Hot Water | | Cold Water | |
|---|---|---|---|
| February | 88% | February | 22% |
| June | 50% | June | 11% |

What is the usual method of transmission for *Mycobacterium*? What is a probable source of infection in hospitals? How can such healthcare-associated infections be prevented?

# Microbial Mechanisms of Pathogenicity 15

N ow that you have a basic understanding of how microorganisms cause disease, we will take a look at some of the specific properties of microorganisms that contribute to pathogenicity, the ability to cause disease by overcoming the defenses of a host, and virulence, the degree or extent of pathogenicity. (As discussed throughout the chapter, the term *host* usually refers to humans.)

Microbes don't try to cause disease; the microbial cells are getting food and defending themselves. Sometimes the presence of microbial cells or cell parts can induce symptoms in a host. An example due to *Burkholderia* (in the photograph) is described in the Clinical Case.

To humans, it doesn't make sense for a parasite to kill its host. However, nature does not have a plan for evolution; the genetic variations that give rise to evolution are due to random mutations, not to logic. According to natural selection, organisms best adapted to their environments will reproduce. Coevolution between a parasite and its host seems to occur: the behavior of one influences that of the other. For example, the cholera pathogen, *Vibrio cholerae*, quickly induces diarrhea, threatening the host's life from a loss of fluids and salts but providing a way to transmit the pathogen to another person by contaminating the surrounding environment.

Keep in mind that many of the properties contributing to microbial pathogenicity and virulence are unclear or unknown. We do know, however, that if the microbe overpowers the host defenses, disease results.

◀ *Burkholderia* bacteria like the ones shown here form biofilms that cause infections in hospitalized patients.

## In the Clinic

You are a transplant nurse taking care of a liver transplant patient. He tells you he is concerned that his physician stopped his iron supplements. The patient knows the supplements were used to treat his anemia. **What will you tell the patient?**

*Hint: Read about siderophores on page 430.*

Play **In the Clinic** Video
@Mastering**Microbiology**

# How Microorganisms Enter a Host

## LEARNING OBJECTIVES

**15-1** Identify the principal portals of entry.

**15-2** Define $ID_{50}$ and $LD_{50}$.

**15-3** Using examples, explain how microbes adhere to host cells.

As noted earlier, **pathogenicity** is the ability to cause disease by overcoming host defenses, whereas **virulence** is the degree of pathogenicity. To cause disease, most pathogens must gain access to the host, adhere to host tissues, penetrate or evade host defenses, and damage the host tissues. However, some microbes do not cause disease by directly damaging host tissue. Instead, disease is due to the accumulation of microbial waste products. Some microbes, such as those that cause dental caries and acne, can cause disease without penetrating the body. Pathogens can gain entrance to the human body and other hosts through several avenues, which are called **portals of entry**.

## Portals of Entry

The portals of entry for pathogens are mucous membranes, skin, and direct deposition beneath the skin or membranes (the parenteral route).

### Mucous Membranes

Many bacteria and viruses gain access to the body by penetrating mucous membranes lining the respiratory tract, gastrointestinal tract, genitourinary tract, and conjunctiva, a delicate membrane that covers the eyeballs and lines the eyelids. Most pathogens enter through the mucous membranes of the gastrointestinal and respiratory tracts.

The respiratory tract is the easiest and most frequently traveled portal of entry for infectious microorganisms. Microbes are inhaled into the nose or mouth in drops of moisture and dust particles. Diseases that are commonly contracted via the respiratory tract include the common cold, pneumonia, tuberculosis, influenza, and measles.

Microorganisms can gain access to the gastrointestinal tract in food and water and via contaminated fingers. Most microbes that enter the body in these ways are destroyed by hydrochloric acid and enzymes in the stomach or by bile and enzymes in the small intestine. Those that survive can cause disease. Microbes in the gastrointestinal tract can cause poliomyelitis, hepatitis A, typhoid fever, amebic dysentery, giardiasis, shigellosis (bacillary dysentery), and cholera. These pathogens are then eliminated with feces and can be transmitted to other hosts via contaminated water, food, or fingers.

The genitourinary tract is a portal of entry for pathogens that are contracted sexually. Some microbes that cause sexually transmitted infections (STIs) may penetrate an unbroken mucous membrane. Others require a cut or abrasion of some type. Examples of STIs are HIV infection, genital warts, chlamydia, human herpesvirus-2, syphilis, and gonorrhea.

### Skin

The skin is the largest organ of the body, in terms of surface area and weight, and is an important defense against disease. Unbroken skin is impenetrable by most microorganisms. Some microbes gain access to the body through openings in the skin, such as hair follicles and sweat gland ducts. Larvae of the hookworm actually bore through intact skin, and some fungi grow on the keratin in skin or infect the skin itself.

The conjunctiva is a delicate mucous membrane that lines the eyelids and covers the white of the eyeballs. Although it is a relatively effective barrier against infection, certain diseases such as conjunctivitis, trachoma, and ophthalmia neonatorum are acquired through the conjunctiva.

### The Parenteral Route

Other microorganisms gain access to the body when they are deposited directly into the tissues beneath the skin or into mucous membranes when these barriers are penetrated or injured. This route is called the **parenteral route**. Punctures, injections, bites, cuts, wounds, surgery, and splitting of the skin or mucous membrane due to swelling or drying can all establish parenteral routes. HIV, the hepatitis viruses, and bacteria that cause tetanus and gangrene can be transmitted parenterally.

Even after microorganisms have entered the body, they do not necessarily cause disease. The occurrence of disease depends on several factors, only one of which is the portal of entry.

---

### CLINICAL CASE  The Eyes Have It

Kerry Santos, a board-certified ophthalmologist for 20 years, has had a long day. She performed outpatient cataract surgery on ten patients (**Figure A**) today. As she checks over her patients in the recovery area, she notes that eight of the ten patients have an unusual degree of inflammation and that their pupils are fixed and do not respond to light.

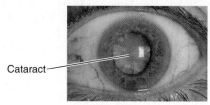

Cataract

**Figure A  A cataract is a clouding of the eye lens that distorts vision.**

What could have caused this complication? Read on to find out.

424   430   435   439   441

## The Preferred Portal of Entry

Many pathogens have a preferred portal of entry that is a prerequisite to their being able to cause disease. If they gain access to the body by another portal, disease might not occur. For example, the bacteria of typhoid fever, *Salmonella* Typhi, produce all the signs and symptoms of the disease when swallowed (preferred route), but if the same bacteria are rubbed on the skin, no reaction (or only a slight inflammation) occurs. *Streptococcus pneumoniae* that is inhaled (preferred route) can cause pneumonia; generally it does not produce signs or symptoms if swallowed. Some pathogens, such as *Yersinia pestis*, the microorganism that causes plague, and *Bacillus anthracis*, the

causative agent of anthrax, can initiate disease from more than one portal of entry. The preferred portals of entry for some common pathogens are listed in Table 15.1.

## Numbers of Invading Microbes

If only a few microbes enter the body, they will probably be overcome by the host's defenses. However, if large numbers of microbes gain entry, the stage is probably set for disease. The likelihood of disease increases as the number of pathogens increases.

The virulence of a microbe is often expressed as the $ID_{50}$ (infectious dose for 50% of a sample population). The 50 is

**TABLE 15.1** Portals of Entry for the Pathogens of Some Common Diseases

| Portal of Entry | Pathogen* | Disease | Incubation Period |
|---|---|---|---|
| **MUCOUS MEMBRANES** | | | |
| Respiratory tract | *Streptococcus pneumoniae* | Pneumococcal pneumonia | 1–3 days |
| | *Mycobacterium tuberculosis*[†] | Tuberculosis | 2–12 weeks |
| | *Bordetella pertussis* | Whooping cough (pertussis) | 12–20 days |
| | Influenza virus *(Influenzavirus)* | Influenza | 18–36 hours |
| | Measles virus *(Morbillivirus)* | Measles (rubeola) | 11–14 days |
| | Rubella virus *(Rubivirus)* | German measles (rubella) | 2–3 weeks |
| | Epstein-Barr virus *(Lymphocryptovirus)* | Infectious mononucleosis | 2–6 weeks |
| | Varicella-zoster virus *(Varicellovirus)* | Chickenpox (varicella) (primary infection) | 14–16 days |
| | *Histoplasma capsulatum* (fungus) | Histoplasmosis | 5–18 days |
| Gastrointestinal tract | *Shigella* spp. | Shigellosis (bacillary dysentery) | 1–2 days |
| | *Brucella* spp. | Brucellosis (undulant fever) | 6–14 days |
| | *Vibrio cholerae* | Cholera | 1–3 days |
| | *Salmonella enterica* | Salmonellosis | 7–22 hours |
| | *Salmonella* Typhi | Typhoid fever | 14 days |
| | Hepatitis A virus *(Hepatovirus)* | Hepatitis A | 15–50 days |
| | Mumps virus *(Rubulavirus)* | Mumps | 2–3 weeks |
| | *Trichinella spiralis* (helminth) | Trichinellosis | 2–28 days |
| Genitourinary tract | *Neisseria gonorrhoeae* | Gonorrhea | 3–8 days |
| | *Treponema pallidum* | Syphilis | 9–90 days |
| | *Chlamydia trachomatis* | Nongonococcal urethritis | 1–3 weeks |
| | Human herpesvirus-2 | Herpes virus infections | 4–10 days |
| | Human immunodeficiency virus (HIV)[‡] | AIDS | 10 years |
| | *Candida albicans* (fungus) | Candidiasis | 2–5 days |
| **SKIN OR PARENTERAL ROUTE** | | | |
| | *Clostridium perfringens* | Gas gangrene | 1–5 days |
| | *Clostridium tetani* | Tetanus | 3–21 days |
| | *Rickettsia rickettsii* | Rocky Mountain spotted fever | 3–12 days |
| | Hepatitis B virus *(Hepadnavirus)*[‡] | Hepatitis B | 6 weeks–6 months |
| | Rabiesvirus *(Lyssavirus)* | Rabies | 10 days–1 year |
| | *Plasmodium* spp. (protozoan) | Malaria | 2 weeks |

*All pathogens are bacteria, unless indicated otherwise. For viruses, the viral species and/or genus name is given.
[†]These pathogens can also cause disease after entering the body via the gastrointestinal tract.
[‡]These pathogens can also cause disease after entering the body via the parenteral route. Hepatitis B virus and HIV can also cause disease after entering the body via the genitourinary tract.

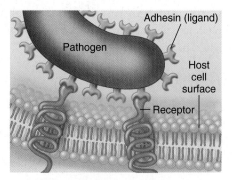

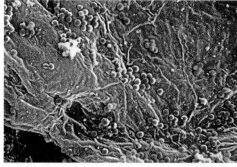

**(a)** Surface molecules on a pathogen, called adhesins or ligands, bind specifically to complementary surface receptors on cells of certain host tissues.

**(b)** *E. coli* bacteria (yellow) on human urinary bladder cells  SEM  5 μm

**(c)** Bacteria (purple) adhering to human skin  SEM  9 μm

**Figure 15.1  Adherence.**

**Q** Of what chemicals are adhesins composed?

not an absolute value; rather, it is used to compare relative virulence under experimental conditions. *Bacillus anthracis* can cause infection via three different portals of entry. The $ID_{50}$ through the skin (cutaneous anthrax) is 10 to 50 endospores; the $ID_{50}$ for inhalation anthrax is inhalation of 10,000 to 20,000 endospores; and the $ID_{50}$ for gastrointestinal anthrax is ingestion of 250,000 to 1,000,000 endospores. These data show that cutaneous anthrax is significantly easier to acquire than either the inhalation or the gastrointestinal forms. A study of *Vibrio cholerae* showed that the $ID_{50}$ is $10^8$ cells; but if stomach acid is neutralized with bicarbonate, the number of cells required to cause an infection decreases significantly.

The potency of a toxin is often expressed as the $LD_{50}$ (lethal dose for 50% of a sample population). For example, the $LD_{50}$ for botulinum toxin in mice is 0.03 ng/kg;[*] for Shiga toxin, 250 ng/kg; and staphylococcal enterotoxin, 1350 ng/kg. In other words, compared to the other two toxins, a much smaller dose of botulinum toxin is needed to cause symptoms.

## Adherence

Almost all pathogens have some means of attaching themselves to host tissues at their portal of entry. For most pathogens, this attachment, called **adherence** (or **adhesion**), is a necessary step in pathogenicity. (Of course, nonpathogens also have structures for attachment.) The attachment between pathogen and host is accomplished by means of surface molecules on the pathogen called **adhesins** or **ligands** that bind specifically to complementary surface **receptors** on the cells of certain host tissues (**Figure 15.1**). Adhesins may be located on a microbe's glycocalyx or on other microbial surface structures, such as pili, fimbriae, and flagella (see Chapter 4).

The majority of adhesins on the microorganisms studied so far are glycoproteins or lipoproteins. The receptors on host cells are typically sugars, such as mannose. Adhesins on different strains of the same species of pathogen can vary in structure. Different cells of the same host can also have different receptors that vary in structure. If adhesins, receptors, or both can be altered to interfere with adherence, infection can often be prevented (or at least controlled).

There is great diversity of adhesins. *Streptococcus mutans*, a bacterium that plays a key role in tooth decay, attaches to the surface of teeth by its glycocalyx. An enzyme produced by *S. mutans*, called glucosyltransferase, converts glucose into a sticky polysaccharide called dextran, which forms the glycocalyx. *Actinomyces* bacterial cells have fimbriae that adhere to the glycocalyx of *S. mutans*. The combination of *S. mutans*, *Actinomyces*, and dextran makes up dental plaque and contributes to dental caries (tooth decay; see Chapter 25, page 724).

Microbes have the ability to come together in masses, cling to surfaces, and take in and share available nutrients in communities called **biofilms** (discussed in more detail in Chapter 6, page 157). Examples of biofilms include the dental plaque on teeth, the algae on the walls of swimming pools, and the scum that accumulates on shower doors. A biofilm forms when microbes adhere to a particular surface that is typically moist and contains organic matter. The first microbes to attach are usually bacteria. Once they adhere to the surface, they multiply and secrete a glycocalyx that further attaches the bacteria to each other and to the surface (see Figure 6.5, page 158). In some cases, biofilms can be several layers thick and may contain several different species of microbes. Biofilms represent another method of adherence and are important because they resist disinfectants and antibiotics. This characteristic is significant, especially when biofilms colonize structures such as teeth, medical catheters, stents, heart

---

[*]Nanogram (ng) equals one-billionth of a gram; kilogram (kg) equals 1000 grams.

# Skin Microbiota Interactions and the Making of MRSA

Thanks to horizontal gene transfer, any bacteria that make contact with one another may end up sharing genes that confer virulence factors. This includes the many species that live as members of the skin microbiome.

Staphylococcus epidermidis is the staphylococcal species most frequently isolated from human skin. By contrast, S. aureus is rarely found on other areas of the skin beyond its preferred site, the nares. Researchers believe that one of the reasons S. epidermidis is so successful at colonizing skin is that it contains arginine catabolic mobile element (ACME), a genetic segment that confers resistance to the antibacterial peptides found on human skin.

However, ACME is a "mobile genetic element"—so named because bacteria easily transfer it to new cells via plasmids, bacteriophages, or transposons. Genetic analysis indicates that S. epidermidis transferred ACME to some S. aureus, making it easier for evolving versions of S. aureus to colonize more of the skin, and to be transferred by direct contact. Likewise, laboratory studies on the bacterial mecA gene led researchers to suspect that resistance to methicillin, oxacillin, and other beta-lactam antibiotics also moved from S. epidermis to S. aureus. These two gene transfers between normally harmless members of the microbiome are some of the key events that helped create methicillin-resistant Staphylococcus aureus (MRSA)—a major health concern in both health care settings and the general community today.

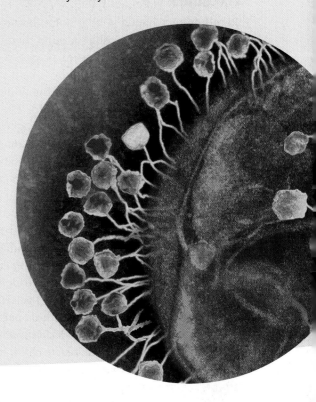

Traits can be transferred to S. aureus by bacteriophages.

valves, hip replacement components, and contact lenses. Dental plaque is actually a biofilm that mineralizes over time, creating what is commonly known as tartar. It is estimated that biofilms are involved in 65% of all human bacterial infections.

Enteropathogenic strains of E. coli (those responsible for gastrointestinal disease) have adhesins on fimbriae that adhere only to specific kinds of cells in certain regions of the small intestine. After adhering, some strains of Shigella and E. coli induce receptor-mediated endocytosis as a vehicle to enter host cells and then multiply within them (see Figure 25.7, page 729). Treponema pallidum, the causative agent of syphilis, uses its tapered end as a hook to attach to host cells. Listeria monocytogenes, which causes meningitis, spontaneous abortions, and stillbirths, produces an adhesin for a specific receptor on host cells. Neisseria gonorrhoeae, the causative agent of gonorrhea, also has fimbriae containing adhesins, which in this case permit attachment to cells with appropriate receptors in the genitourinary tract, eyes, and pharynx. Staphylococcus aureus, which can cause skin infections, produces adhesins that bind to laminin and fibronectin on skin cells (see Exploring the Microbiome, above).

## CHECK YOUR UNDERSTANDING

✔ 15-1 List three portals of entry, and describe how microorganisms gain access through each.

✔ 15-2 The $LD_{50}$ of botulinum toxin is 0.03 ng/kg; the $LD_{50}$ of Salmonella toxin is 12 mg/kg. Which is the more potent toxin?

✔ 15-3 How would a drug that binds mannose on human cells affect a pathogenic bacterium?

# How Bacterial Pathogens Penetrate Host Defenses

## LEARNING OBJECTIVES

15-4 Explain how capsules and cell wall components contribute to pathogenicity.

15-5 Compare the effects of coagulases, kinases, hyaluronidase, and collagenase.

15-6 Define and give an example of antigenic variation.

15-7 Describe how bacteria use the host cell's cytoskeleton to enter the cell.

15-8 Identify six mechanisms of avoiding destruction by phagocytosis.

Although some pathogens can cause damage on the surface of tissues, most must penetrate tissues to cause disease. Here we will consider several factors that contribute to the ability of bacteria to invade a host.

## Capsules

Recall from Chapter 4 that some bacteria make glycocalyx material that forms capsules around their cell walls; this property increases the virulence of pathogenic species. The capsule resists the host's defenses by impairing phagocytosis, the process by which certain cells of the body engulf and destroy microbes (see Chapter 16, page 456). The chemical nature of the capsule appears to prevent the phagocytic cell from adhering to the bacterium. However, the human body can produce antibodies against the capsule, and when these antibodies are present on the capsule surface, the encapsulated bacteria are easily destroyed by phagocytosis.

One bacterium that owes its virulence to the presence of a polysaccharide capsule is *Streptococcus pneumoniae*, the causative agent of pneumococcal pneumonia (see Figure 24.11, page 703). Strains of this bacterium with capsules are virulent, but strains without capsules are avirulent because they are susceptible to phagocytosis. Other bacteria that produce capsules related to virulence are *Klebsiella pneumoniae*, a causative agent of bacterial pneumonia; *Haemophilus influenzae*, a cause of pneumonia and meningitis in children; *Bacillus anthracis*, the cause of anthrax; and *Yersinia pestis*, the causative agent of plague. Keep in mind that capsules are not the only cause of virulence. Many nonpathogenic bacteria produce capsules, and the virulence of some pathogens is not related to the presence of a capsule.

## Cell Wall Components

The cell walls of certain bacteria contain chemical substances that contribute to virulence. For example, *Streptococcus pyogenes* produces a heat-resistant and acid-resistant protein called **M protein** (see Figure 21.6, page 597). This protein is found on both the cell surface and fimbriae. It mediates attachment of the bacterium to epithelial cells of the host and helps the bacterium resist phagocytosis by white blood cells. The M protein thereby increases the virulence of the microorganism. Immunity to *S. pyogenes* depends on the body's production of an antibody specific to M protein. *Neisseria gonorrhoeae* grows inside human epithelial cells and leukocytes. These bacteria use **fimbriae** and an outer membrane protein called **Opa** to attach to host cells. Following attachment by both Opa and fimbriae, the host cells take in the bacteria. (Bacteria that produce Opa form *opa*que colonies on culture media.) The **waxy lipid** (mycolic acid) that makes up the cell wall of *Mycobacterium tuberculosis* also increases virulence by resisting digestion by phagocytes, and the bacteria can even multiply inside phagocytes.

## Enzymes

Microbiologists think that the virulence of some bacteria is aided by the production of extracellular enzymes (*exoenzymes*) and related substances. These chemicals can digest materials between cells and form or digest blood clots, among other functions.

**Coagulases** are bacterial enzymes that coagulate (clot) the fibrinogen in blood. Fibrinogen, a plasma protein produced by the liver, is converted by coagulases into fibrin, the threads that form a blood clot. The fibrin clot may protect the bacterium from phagocytosis and isolate it from other defenses of the host. Coagulases are produced by some members of the genus *Staphylococcus*; they may be involved in the walling-off process in boils produced by staphylococci. However, some staphylococci that do not produce coagulases are still virulent. In these cases, capsules may be more important to their virulence.

Bacterial **kinases** are bacterial enzymes that break down fibrin and thus digest clots formed by the body to isolate the infection. One of the better-known kinases is *fibrinolysin (streptokinase)*, which is produced by such streptococci as *Streptococcus pyogenes*. Streptokinase is used therapeutically to break down blood clots causing heart attacks.

**Hyaluronidase** is another enzyme secreted by certain bacteria, such as streptococci. It hydrolyzes hyaluronic acid, a type of polysaccharide that holds together certain cells of the body, particularly cells in connective tissue. This digesting action is thought to be involved in the tissue blackening of infected wounds and to help the microorganism spread from its initial site of infection. Hyaluronidase is also produced by some clostridia that cause gas gangrene. For therapeutic use, hyaluronidase may be mixed with a drug to promote the spread of the drug through a body tissue.

Another enzyme, **collagenase**, produced by several species of *Clostridium*, facilitates the spread of gas gangrene. Collagenase breaks down the protein collagen, which forms the connective tissue of muscles and other body organs and tissues.

As a defense against adherence of pathogens to mucosal surfaces, the body produces a class of antibodies called IgA antibodies. There are some pathogens with the ability to produce enzymes, called **IgA proteases**, that can destroy these antibodies. *N. gonorrhoeae* has this ability, as does *N. meningitidis*, the causative agent of meningococcal meningitis, and other microbes that infect the central nervous system.

## Antigenic Variation

*Adaptive immunity* refers to a specific defensive response of the body to an infection or to antigens (see Chapter 17). In the presence of antigens, the body produces proteins called antibodies, which bind to the antigens and inactivate them or target them for destruction by phagocytes. However, some pathogens can alter their surface antigens, by a process called **antigenic variation**. Thus, by the time the body mounts an immune response against a pathogen, the pathogen has already altered its antigens and is unaffected by the antibodies. Some microbes can activate alternative genes, resulting in antigenic changes. For example, *N. gonorrhoeae* has several copies of the Opa-encoding gene, resulting in cells with different antigens and in cells that express different antigens over time.

A wide range of microbes is capable of antigenic variation. Examples include *Influenzavirus*, the causative agent of influenza (flu); *Neisseria gonorrhoeae*, the causative agent of gonorrhea; and *Trypanosoma brucei gambiense* (tri-PA-nō-sō-mah BROOS-e-ē GAM-bē-ens), the causative agent of African trypanosomiasis (sleeping sickness). See Figure 22.16, page 640.

## Penetration into the Host

As previously noted, microbes attach to host cells by adhesins. The interaction triggers signals in the host cell that activate factors that can result in the entrance of some bacteria. The actual mechanism is provided by the host cell cytoskeleton. The microfilaments of the eukaryotic cytoskeleton (see page 98) are composed of a protein called actin, which is used by some microbes to penetrate host cells and by others to move through and between host cells.

*Salmonella* strains and *E. coli* make contact with the host cell plasma membrane. This leads to dramatic changes in the membrane at the point of contact. The microbes produce surface proteins called **invasins** that rearrange nearby actin filaments of the cytoskeleton. For example, when S. Typhimurium makes contact with a host cell, invasins of the microbe cause the appearance of the host cell plasma membrane to resemble the splash of a drop of a liquid hitting a solid surface. This effect, called *membrane ruffling*, is the result of disruption in the cytoskeleton of the host cell (**Figure 15.2**). The microbe sinks into the ruffle and is engulfed by the host cell.

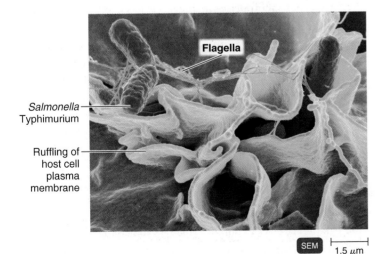

Salmonella Typhimurium

Flagella

Ruffling of host cell plasma membrane

SEM 1.5 μm

**Figure 15.2** *Salmonella* **entering intestinal epithelial cells as a result of ruffling.**

**Q** What are invasins?

Once inside the host cell, certain bacteria such as *Shigella* species and *Listeria* species can actually use actin to propel themselves through the host cell cytoplasm and from one host cell to another. The condensation of actin on one end of the bacteria propels them through the cytoplasm. The bacteria also make contact with membrane junctions that form part of a transport network between host cells. The bacteria use a glycoprotein called *cadherin*, which bridges the junctions, to move from cell to cell.

Still other microbes have the ability to survive inside phagocytes. *Coxiella burnetii*, the causative agent of Q fever, actually requires the low pH inside a phagolysosome to replicate. *L. monocytogenes*, *Shigella* (the causative agent of shigellosis), and *Rickettsia* species (the causative agents of Rocky Mountain spotted fever and typhus) have the ability to escape from a phagosome before it fuses with a lysosome. *Mycobacterium tuberculosis* (the causative agent of tuberculosis), HIV (the causative agent of AIDS), *Chlamydia* (the causative agent of trachoma, nongonococcal urethritis, and lymphogranuloma venereum), *Leishmania* (the causative agent of leishmaniasis), and *Plasmodium* (malarial parasites) can prevent both the fusion of a phagosome with a lysosome and the proper acidification of digestive enzymes. The microbes then multiply within the phagocyte, almost completely filling it. In most cases, the phagocyte dies, and the microbes are released by autolysis to infect other cells. Still other microbes, such as the causative agents of tularemia and brucellosis, can remain dormant within phagocytes for months or years at a time.

## Biofilms

Biofilms also play a role in evading phagocytes. Bacteria that are part of biofilms are much more resistant to  **Play Interactive Microbiology** @Mastering**Microbiology** *See how biofilms affect a patient's health* phagocytosis. Phagocytes do not move through the viscous carbohydrates of the extracellular polymeric substance (EPS) of biofilms. The EPS may shield antigens so they are not recognized by the immune system. The EPS of *Pseudomonas aeruginosa* kills phagocytes.

The study of the numerous interactions between microbes and host cell cytoskeleton is a very intense area of investigation on virulence mechanisms.

### CHECK YOUR UNDERSTANDING

✔ **15-4** What function do capsules and M proteins have in common?

✔ **15-5** Would you expect a bacterium to make coagulase and kinase simultaneously?

✔ **15-6** Many vaccines provide years of protection against a disease. Why doesn't the influenza vaccine offer more than a few months of protection?

✔ **15-7** How does *E. coli* cause membrane ruffling?

✔ **15-8** How does each of these bacteria avoid destruction by phagocytes? *Streptococcus pneumoniae*, *Staphylococcus aureus*, *Listeria monocytogenes*, *Mycobacterium tuberculosis*, *Rickettsia*

# How Bacterial Pathogens Damage Host Cells

**LEARNING OBJECTIVES**

15-9 Describe the function of siderophores.

15-10 Provide an example of direct damage, and compare this to toxin production.

15-11 Contrast the nature and effects of exotoxins and endotoxins.

15-12 Outline the mechanisms of action of A-B toxins, membrane-disrupting toxins, superantigens, and genotoxins.

15-13 Identify the importance of the LAL assay.

15-14 Using examples, describe the roles of plasmids and lysogeny in pathogenicity.

When a microorganism invades a body tissue, it initially encounters phagocytes of the host. If the phagocytes are successful in destroying the invader, no further damage is done to the host. But if the pathogen overcomes the host's defense, then the microorganism can damage host cells in four basic ways:

1. By using the host's nutrients.
2. By causing direct damage in the immediate vicinity of the invasion.
3. By producing toxins, transported by blood and lymph, that damage sites far removed from the original site of invasion.
4. By inducing hypersensitivity reactions.

This fourth mechanism is considered in detail in Chapter 19. For now, we will discuss only the first three mechanisms.

## Using the Host's Nutrients: Siderophores

Iron is required for the growth of most pathogenic bacteria. However, the concentration of free iron in the human body is fairly low because most of the iron is tightly bound to iron-transport proteins, such as lactoferrin, transferrin, and ferritin, as well as hemoglobin. These are discussed in more detail in Chapter 16. To obtain iron, some pathogens secrete proteins called **siderophores** (**Figure 15.3**). Siderophores are released into the medium, where they take the iron away from iron-transport proteins by binding the iron even more tightly. Once the iron-siderophore complex is formed, it is taken up by siderophore receptors on the bacterial surface. Then the iron is brought into the bacterium. In some cases, the iron is released from the complex to enter the bacterium; in other cases, the iron enters as part of the complex.

As an alternative to iron acquisition by siderophores, some pathogens have receptors that bind directly to iron-transport proteins and hemoglobin. Then these are taken into the bacterium directly along with the iron. Also, it is possible that some bacteria produce toxins (described shortly) when iron levels are low. The toxins kill host cells, releasing their iron and thereby making it available to the bacteria.

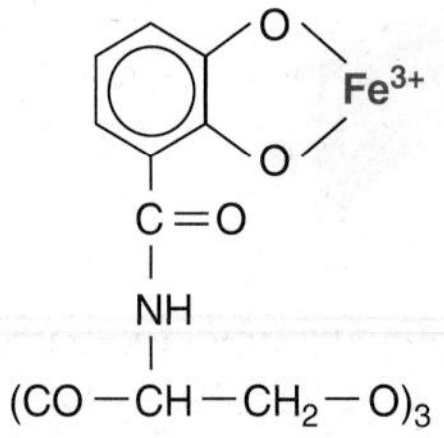


**Figure 15.3 Structure of enterobactin, one type of bacterial siderophore.** The iron ($Fe^{3+}$) is indicated in red.

**Q Of what value are siderophores?**

## Direct Damage

Once pathogens attach to host cells, they can cause direct damage as the pathogens use the host cell for nutrients and produce waste products. As pathogens metabolize and multiply in cells, the cells usually rupture. Many viruses and some intracellular bacteria and protozoa that grow in host cells are released when the host cell ruptures. Following their release, pathogens that rupture cells can spread to other tissues in even greater numbers. Some bacteria, such as *E. coli, Shigella, Salmonella,* and *Neisseria gonorrhoeae,* can induce host epithelial cells to engulf them by a process that resembles phagocytosis. These pathogens can disrupt host cells as they pass through and can then be extruded from the host cells by exocytosis, enabling them to enter other host cells. Some bacteria can also penetrate host cells by excreting enzymes or by their own motility; such penetration can itself damage the host cell. Most damage by bacteria, however, is done by toxins.

> **Play Virulence Factors:** *Penetrating Host Tissues, Hiding from Host Defenses, Enteric Pathogens*; **Phagocytosis:** *Microbes That Evade It* @MasteringMicrobiology

## Production of Toxins

**Toxins** are poisonous substances that are produced by certain microorganisms. They are often the primary factor contributing

**CLINICAL CASE**

Dr. Santos suspects toxic anterior segment syndrome (TASS), which is a reaction to a toxin or other chemical. TASS is caused by (1) chemicals on surgical instruments, resulting from improper or insufficient cleaning; (2) products introduced into the eye during surgery, such as washing solutions or medications; or (3) other substances that enter the eye during or after surgery, such as topical ointments or talc from surgical gloves.

**Why does Dr. Santos suspect an intoxication and not an infection?**

424 430 435 439 441

# Mechanisms of Exotoxins and Endotoxins

## exotoxins

Proteins produced inside pathogenic bacteria, most commonly gram-positive bacteria, as part of their growth and metabolism. The exotoxins are then secreted into the surrounding medium during log phase.

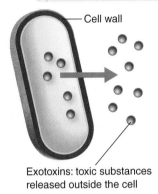

Cell wall

Exotoxins: toxic substances released outside the cell

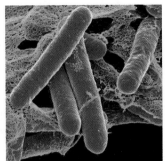

*Clostridium botulinum*, an example of a gram-positive bacterium that produces exotoxins

SEM 1.3 μm

### KEY CONCEPTS

- Toxins are of two general types: exotoxins and endotoxins.

- Bacterial toxins can cause damage to host cells.

- Toxins can elicit an inflammatory response in the host, as well as activate the complement system.

- Some gram-negative bacteria may release minute amounts of endotoxins, which may stimulate natural immunity.

## endotoxins

Lipid portions of lipopolysaccharides (LPS) that are part of the outer membrane of the cell wall of gram-negative bacteria (lipid A). The endotoxins are liberated when the bacteria die and the cell wall lyses, or breaks apart.

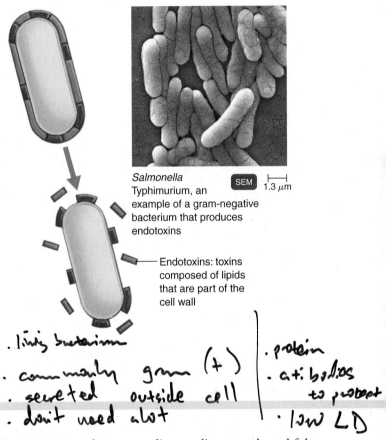

*Salmonella* Typhimurium, an example of a gram-negative bacterium that produces endotoxins

SEM 1.3 μm

Endotoxins: toxins composed of lipids that are part of the cell wall

*[handwritten notes:]*
. lin4 bacterium
. commonly gram (+)
. secreted outside cell
. don't need alot

. protein
. antibodies to protect
. low LD

---

to the pathogenic properties of those microbes. The capacity of microorganisms to produce toxins is called **toxigenicity**. Toxins transported by the blood or lymph can cause serious, and sometimes fatal, effects. Some toxins produce fever, cardiovascular disturbances, diarrhea, and shock. Toxins can also inhibit protein synthesis, destroy blood cells and blood vessels, and disrupt the nervous system by causing spasms. Of the 220 or so known bacterial toxins, nearly 40% cause disease by damaging eukaryotic cell membranes. The term **toxemia** refers to the presence of toxins in the blood. Toxins are of two general types, based on their position relative to the microbial cell: exotoxins and endotoxins. **Intoxications** are caused by the presence of a toxin; not by microbial growth.

### Exotoxins

**Exotoxins** are produced inside some bacteria (mostly gram-negative) as part of their growth and metabolism and are secreted

by the bacterium into the surrounding medium or released following lysis (**Figure 15.4**). *Exo-* refers to "outside," which in this context refers to the fact that exotoxins are secreted to the outside of the bacterial cells that produce them. Exotoxins are proteins, and many are enzymes that catalyze only certain biochemical reactions. Because of the enzymatic nature of most exotoxins, even small amounts are quite harmful because they can act over and over again. Bacteria that produce exotoxins may be gram-positive or gram-negative. The genes for most (perhaps all) exotoxins are carried on bacterial plasmids or phages. Because exotoxins are soluble in body fluids, they can easily diffuse into the blood and are rapidly transported throughout the body.

Exotoxins work by destroying particular parts of the host's cells or by inhibiting certain metabolic functions. They are highly specific in their effects on body tissues. Exotoxins are among the most lethal substances known. Only 1 milligram of the botulinum exotoxin is enough to kill 1 million guinea pigs.

Fortunately, only a few bacterial species produce such potent exotoxins.

Diseases caused by bacteria that produce exotoxins are often caused by minute amounts of exotoxins, not by the bacteria themselves. It is the exotoxins that produce the specific signs and symptoms of the disease. Thus, exotoxins are disease-specific. For example, botulism is usually due to ingestion of the exotoxin, not to a bacterial infection. Likewise, staphylococcal food poisoning is an *intoxication*, not an infection.

The body produces antibodies called **antitoxins** that provide immunity to exotoxins. When exotoxins are inactivated by heat or by formaldehyde, iodine, or other chemicals, they no longer cause the disease but can still stimulate the body to produce antitoxins. Such altered exotoxins are called **toxoids**. When toxoids are injected into the body as a vaccine, they stimulate antitoxin production so that immunity is produced. Diphtheria and tetanus can be prevented by toxoid vaccination.

**Naming Exotoxins** Exotoxins are named on the basis of several characteristics. One is the type of host cell that is attacked. For example, *neurotoxins* attack nerve cells, *cardiotoxins* attack heart cells, *hepatotoxins* attack liver cells, *leukotoxins* attack leukocytes, *enterotoxins* attack the lining of the gastrointestinal tract, and *cytotoxins* attack a wide variety of cells. Some exotoxins are named for the diseases with which they are associated. Examples include *diphtheria toxin* (cause of diphtheria) and *tetanus toxin* (cause of tetanus). Other exotoxins are named for the specific bacterium that produces them, for example, *botulinum toxin* (*Clostridium botulinum*) and *Vibrio enterotoxin* (*Vibrio cholerae*).

**Types of Exotoxins** Exotoxins are divided into three principal types on the basis of their structure and function: (1) A-B toxins, (2) membrane-disrupting toxins, and (3) superantigens.

**A-B Toxins** **A-B toxins** were the first toxins to be studied intensively and are so named because they consist of two parts designated A and B, both of which are polypeptides. Most exotoxins are A-B toxins. The A part is the active (enzyme) component, and the B part is the binding component. An example of an A-B toxin is the diphtheria toxin, which is illustrated in **Figure 15.5**.

Some gram-negative bacteria, including *Haemophilus ducreyi* and *Helicobacter* spp., make **genotoxins**, which cause breaks in eukaryotic DNA. This causes mutations, disrupts cell division, and may lead to cancer. Two of the bacterial genotoxins, *cytolethal distending toxin* and *typhoid toxin*, are A-B toxins. The third, *colibactin*, has not been characterized.

**Membrane-Disrupting Toxins** **Membrane-disrupting toxins** cause lysis of host cells by disrupting their plasma membranes. Some do this by forming protein channels in the plasma membrane; others disrupt the phospholipid portion of the membrane. The cell-lysing exotoxin of *Staphylococcus aureus* is an example of an exotoxin that forms protein channels, whereas that of

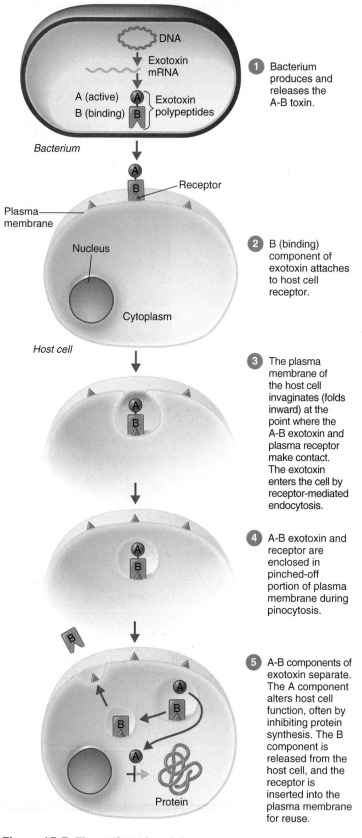

1 Bacterium produces and releases the A-B toxin.

2 B (binding) component of exotoxin attaches to host cell receptor.

3 The plasma membrane of the host cell invaginates (folds inward) at the point where the A-B exotoxin and plasma receptor make contact. The exotoxin enters the cell by receptor-mediated endocytosis.

4 A-B exotoxin and receptor are enclosed in pinched-off portion of plasma membrane during pinocytosis.

5 A-B components of exotoxin separate. The A component alters host cell function, often by inhibiting protein synthesis. The B component is released from the host cell, and the receptor is inserted into the plasma membrane for reuse.

**Figure 15.5 The action of an A-B exotoxin.** A proposed model for the mechanism of action of diphtheria toxin.

**Q** Why is this called an A-B toxin?

*most common ... every gram (−) have this, released when lysed and die*

*fever malaise ↑ lethal dose*

*Clostridium perfringens* is an example of an exotoxin that disrupts the phospholipids. Membrane-disrupting toxins contribute to virulence by killing host cells, especially phagocytes, and by aiding the escape of bacteria from vacuoles within phagocytes (phagosomes) into the host cell's cytoplasm.

Membrane-disrupting toxins that kill phagocytic leukocytes (white blood cells) are called **leukocidins**. They act by forming protein channels. Leukocidins are also active against macrophages, the phagocytes present in tissues. Most leukocidins are produced by staphylococci and streptococci. The damage to phagocytes decreases host resistance. Membrane-disrupting toxins that destroy erythrocytes (red blood cells), also by forming protein channels, are called **hemolysins**. Important producers of hemolysins include staphylococci and streptococci. Hemolysins produced by streptococci are called **streptolysins**. One kind, called *streptolysin O (SLO)*, is so named because it is inactivated by atmospheric oxygen. Another kind of streptolysin is called *streptolysin S (SLS)* because it is stable in an oxygen environment. Both streptolysins can cause lysis not only of red blood cells, but also of white blood cells (whose function is to kill the streptococci) and other body cells. A number of intracellular pathogens secrete pore-forming toxins that lyse phagocyte cell membranes once inside the phagocyte. For example, *Trypanosoma cruzi* (the causative agent of Chagas' disease) and *Listeria monocytogenes* (the causative agent of listeriosis), produce membrane attack complexes that lyse phagolysosome membranes and release the microbes into the cytoplasm of the phagocyte, where they propagate. Later the microbes secrete more membrane attack complexes that lyse the plasma membrane, resulting in release of microbes from the phagocyte and infection of neighboring cells.

**Superantigens** Superantigens are antigens that provoke a very intense immune response. They are bacterial proteins that combine with a protein on macrophages; this nonspecifically stimulates the proliferation of immune cells called T cells. These cells are types of white blood cells (lymphocytes) that act against foreign organisms and tissues (transplants). In response to superantigens, T cells are stimulated to release enormous amounts of chemicals called cytokines. *Cytokines* are small protein molecules produced by various body cells, especially T cells, that regulate immune responses and mediate cell-to-cell communication (see Chapter 17, page 477). The excessively high levels of cytokines released by T cells enter the bloodstream and give rise to a number of symptoms, including fever, nausea, vomiting, diarrhea, and sometimes shock and even death. Bacterial superantigens include the staphylococcal toxins that cause food poisoning and toxic shock syndrome. A summary of diseases produced by exotoxins is shown in Table 15.2.

### Endotoxins

**Endotoxins** differ from exotoxins in several ways. *Endo-* means "within," in this context referring to the fact that the endotoxins are part of bacterial cells, and not a metabolic product. Endotoxins are part of the outer portion of the cell wall of gram-negative bacteria (Figure 15.4). Gram-negative bacteria have an outer membrane surrounding the peptidoglycan layer of the cell wall (see Chapter 4). This outer membrane consists of lipoproteins, phospholipids, and lipopolysaccharides (LPS) (see Figure 4.13c, page 82). The lipid portion of LPS, called **lipid A**, is the endotoxin. Thus, endotoxins are lipids, whereas exotoxins are proteins. ←

Endotoxins are released during bacterial multiplication and when gram-negative bacteria die and their cell walls undergo lysis. Antibiotics used to treat diseases caused by gram-negative bacteria can lyse the bacterial cells; this reaction releases endotoxin and may lead to an immediate worsening of the symptoms, but the condition usually improves as liver lipase breaks down the endotoxin. Endotoxins exert their effects by stimulating macrophages to release cytokines in very high concentrations. At these levels, cytokines are toxic. (See page 477 for a discussion of cytokines.) All endotoxins produce the same signs and symptoms, regardless of the species of microorganism, although not to the same degree. These include chills, fever, weakness, generalized aches, and, in some cases, shock and even death. Endotoxins can also induce miscarriage.

Another consequence of endotoxins is the activation of blood-clotting proteins, causing the formation of small blood clots. These blood clots obstruct capillaries, and the resulting decreased blood supply induces the death of tissues. This condition is referred to as *disseminated intravascular coagulation (DIC)*.

The fever (pyrogenic response) caused by endotoxins is believed to occur as depicted in **Figure 15.6**. Bacterial cell death caused by lysis can also produce fever by this mechanism. Both aspirin and acetaminophen reduce fever by inhibiting the synthesis of prostaglandins. (The function of fever in the body is discussed in Chapter 16, page 462.)

**Shock** refers to any life-threatening decrease in blood pressure. Shock caused by bacteria is called **septic shock**. Gram-negative bacteria cause *endotoxic shock*. Like fever, the shock produced by endotoxins is related to the secretion of a cytokine by macrophages. Phagocytosis of gram-negative bacteria causes the phagocytes to secrete tumor necrosis factor (TNF), sometimes called *cachectin*. TNF binds to many tissues in the body and alters their metabolism in a number of ways. One effect of TNF is damage to blood capillaries; their permeability is increased, and they lose large amounts of fluid. The result is a drop in blood pressure that results in shock. Low blood pressure has serious effects on the kidneys, lungs, and gastrointestinal tract. In addition, the presence of gram-negative bacteria such as *Haemophilus influenzae* type b in cerebrospinal fluid causes the release of TNF and another cytokine called IL-1. These, in turn, cause a weakening of the blood–brain barrier that normally protects the central nervous system from infection. The weakened barrier lets phagocytes in, but this also lets more bacteria enter from the bloodstream. In the United States, about 3 out of every 1000 people develop septic shock each year. One-third of the patients die within a month, and nearly half die within 6 months.

### TABLE 15.2  Diseases Caused by Exotoxins

| Disease | Bacterium | Type of Exotoxin | Mechanism |
|---|---|---|---|
| Botulism | *Clostridium botulinum* | A-B | Neurotoxin prevents transmission of nerve impulses; flaccid paralysis results. |
| Tetanus | *Clostridium tetani* | A-B | Neurotoxin blocks nerve impulses to muscle relaxation pathway; results in uncontrollable muscle contractions. |
| Diphtheria | *Corynebacterium diphtheriae* | A-B | Cytotoxin inhibits protein synthesis, especially in nerve, heart, and kidney cells. |
| Scalded skin syndrome | *Staphylococcus aureus* | A-B | Exotoxin causes skin layers to separate and slough off. |
| Cholera | *Vibrio cholerae* | A-B | Enterotoxin causes secretion of large amounts of fluids and electrolytes that result in diarrhea. |
| Traveler's diarrhea | Enterotoxigenic *Escherichia coli* and *Shigella* spp. | A-B | Enterotoxin causes secretion of large amounts of fluids and electrolytes that result in diarrhea. |
| Anthrax | *Bacillus anthracis* | A-B | Two A components enter the cell via the same B. The A proteins cause shock and reduce the immune response. |
| Gastric (stomach) cancer | *Helicobacter* spp. | A-B toxin | Genotoxin causes breaks in DNA. |
| Skin and soft tissue infection | Methicillin-resistant *S. aureus* | Membrane-disrupting | The Panton-Valentine leukocidin found in the community-acquired strain of MRSA makes pores in WBC membranes. |
| Gas gangrene and food poisoning | *Clostridium perfringens* and other species of *Clostridium* | Membrane-disrupting | One exotoxin (cytotoxin) causes massive red blood cell destruction (hemolysis); another exotoxin (enterotoxin) is related to food poisoning and causes diarrhea. |
| Antibiotic-associated diarrhea | *Clostridium difficile* | Membrane-disrupting | Enterotoxin causes secretion of fluids and electrolytes that results in diarrhea; cytotoxin disrupts host cytoskeleton. |
| Food poisoning | *S. aureus* | Superantigen | Enterotoxin causes secretion of fluids and electrolytes that results in diarrhea. |
| Toxic shock syndrome (TSS) | *S. aureus* | Superantigen | Toxin causes secretion of fluids and electrolytes from capillaries that decreases blood volume and lowers blood pressure. |

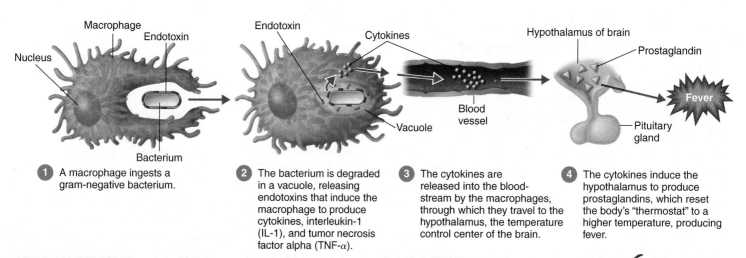

1  A macrophage ingests a gram-negative bacterium.

2  The bacterium is degraded in a vacuole, releasing endotoxins that induce the macrophage to produce cytokines, interleukin-1 (IL-1), and tumor necrosis factor alpha (TNF-α).

3  The cytokines are released into the blood-stream by the macrophages, through which they travel to the hypothalamus, the temperature control center of the brain.

4  The cytokines induce the hypothalamus to produce prostaglandins, which reset the body's "thermostat" to a higher temperature, producing fever.

**Figure 15.6 Endotoxins and the pyrogenic response.** The proposed mechanism by which endotoxins cause fever.

**Q** What is an endotoxin?

*pyrogenic (fever)*

## TABLE 15.3 Exotoxins and Endotoxins

| Property | Exotoxins | Endotoxins |
|---|---|---|
| | | |
| Bacterial Source | Gram-positive and gram-negative bacteria | Gram-negative bacteria |
| Relation to Microorganism | Metabolic product of growing cell | Present in LPS of outer membrane of cell wall and released with destruction of cell or during cell division |
| Chemistry | Proteins, usually with two parts (A-B) | Lipid portion (lipid A) of LPS of outer membrane |
| Pharmacology (Effect on Body) | Specific for a particular cell structure or function in the host (mainly affects cell functions, nerves, and gastrointestinal tract) | General, such as fever, weaknesses, aches, and shock; all produce the same effects |
| Heat Stability | Unstable; can usually be destroyed at 60–80°C (except staphylococcal enterotoxin) | Stable; can withstand autoclaving (121°C for 1 hour) |
| Toxicity (Ability to Cause Disease) | High | Low |
| Fever-Producing | No | Yes |
| Immunology (Relation to Vaccines) | Can be converted to toxoids to immunize against toxin; neutralized by antitoxin | Not easily neutralized by antitoxin; therefore, effective toxoids cannot be made to immunize against toxin |
| Lethal Dose | Small | Considerably larger |
| Representative Diseases | Gas gangrene, tetanus, botulism, diphtheria, scarlet fever, cyanobacterial intoxication | Typhoid fever, urinary tract infections, and meningococcal meningitis |

Endotoxins do not promote the formation of effective antitoxins. Antibodies are produced, but they tend not to counter the effect of the toxin; sometimes, in fact, they actually enhance its effect.

Representative microorganisms that produce endotoxins are *Salmonella* Typhi (the causative agent of typhoid fever), *Proteus* spp. (frequently the causative agents of urinary tract infections), and *Neisseria meningitidis* (the causative agent of meningococcal meningitis).

It is important to have a sensitive test to identify the presence of endotoxins in drugs, medical devices, and body fluids. Materials that have been sterilized may contain endotoxins, even though no bacteria can be cultured from them. One such laboratory test is called the **Limulus amebocyte lysate (LAL) assay**, which can detect even minute amounts of endotoxin. The hemolymph (blood) of the Atlantic coast horseshoe crab, *Limulus polyphemus*, contains white blood cells called amebocytes, which have large amounts of a protein (lysate) that causes clotting. In the presence of endotoxin, amebocytes in the crab hemolymph lyse and liberate their clotting protein. The resulting gel-clot (precipitate) is a positive test for the presence of

**Play Virulence Factors:**
*Exotoxins, Endotoxins*
@MasteringMicrobiology

endotoxin. The degree of the reaction is measured using a spectrophotometer (see Figure 6.21, page 173).

Table 15.3 compares exotoxins and endotoxins.

### CLINICAL CASE

An infection could not have happened this quickly; infections usually take 3 to 4 days to show symptoms. Dr. Santos checks to make sure that the autoclave used to sterilize the ophthalmic equipment is functioning normally, and that single-use topical iodine antiseptic was used. A new sterile tip for the corneal extraction was used for each patient. Both the epinephrine used during surgery and the enzymatic solution for the ultrasonic bath used to clean surgical instruments are sterile, and medications with different lot numbers were used in each surgery. But Dr. Santos knows that the toxin had to have come from somewhere or something connected to the surgeries. Even though the enzymatic solution is sterile, Dr. Santos sends it to a laboratory for an LAL assay.

**Why does Dr. Santos send the enzymatic solution for an LAL assay?**

424   430   **435**   439   441

## Plasmids, Lysogeny, and Pathogenicity

Plasmids are small, circular DNA molecules that are not connected to the main bacterial chromosome and are capable of independent replication. (See Chapter 4, page 90, and Chapter 8, page 229.) One group of plasmids, called R (resistance) factors, is responsible for the resistance of some microorganisms to antibiotics. In addition, a plasmid may carry the information that determines a microbe's pathogenicity. Examples of virulence factors that are encoded by plasmid genes are tetanus neurotoxin, heat-labile enterotoxin, and staphylococcal enterotoxin D. Other examples are dextransucrase, an enzyme produced by *Streptococcus mutans* that is involved in tooth decay; adhesins and coagulase produced by *Staphylococcus aureus*; and a type of fimbria specific to enteropathogenic strains of *E. coli*.

In Chapter 13, we noted that some bacteriophages (viruses that infect bacteria) can incorporate their DNA into the bacterial chromosome, becoming a prophage, and thus remain latent and do not cause lysis of the bacterium. Such a state is called *lysogeny*, and cells containing a prophage are said to be lysogenic. One outcome of lysogeny is that the host bacterial cell and its progeny may exhibit new properties encoded by the bacteriophage DNA. Such a change in the characteristics of a microbe due to a prophage is called **lysogenic conversion**. As a result of lysogenic conversion, the bacterial cell is immune to infection by the same type of phage. In addition, lysogenic cells are of medical importance because some bacterial pathogenesis is caused by the prophages they contain.

Among the bacteriophage genes that contribute to pathogenicity are the genes for diphtheria toxin, erythrogenic toxins, staphylococcal enterotoxin A and pyrogenic toxin, botulinum neurotoxin, and the capsule produced by *Streptococcus pneumoniae*. The gene for Shiga toxin in *E. coli* O157 is encoded by phage genes. Pathogenic strains of *Vibrio cholerae* carry lysogenic phages. These phages can transmit the cholera toxin gene to nonpathogenic *V. cholerae* strains, increasing the number of pathogenic bacteria.

### CHECK YOUR UNDERSTANDING

✔ **15-9** Of what value are siderophores?

✔ **15-10** How does toxigenicity differ from direct damage?

✔ **15-11** Differentiate an exotoxin from an endotoxin.

✔ **15-12** Food poisoning can be divided into two categories: food infection and food intoxication. On the basis of toxin production by bacteria, explain the difference between these two categories.

✔ **15-13** Washwater containing *Pseudomonas* was sterilized and used to wash cardiac catheters. Three patients developed fever, chills, and hypotension following cardiac catheterization. The water and catheters were sterile. Why did the patients show these reactions? How should the water have been tested?

✔ **15-14** How can lysogeny turn the normally harmless *E. coli* into a pathogen?

## Pathogenic Properties of Viruses

**LEARNING OBJECTIVE**

**15-15** List nine cytopathic effects of viral infections.

The pathogenic properties of viruses depend on their gaining access to a host, evading the host's defenses, and then causing damage to or death of the host cell while reproducing themselves.

### Viral Mechanisms for Evading Host Defenses

Viruses have a variety of mechanisms that enable them to evade destruction by the host's immune response (see Chapter 17, page 476). For example, viruses can penetrate and grow inside host cells, where components of the immune system cannot reach them. Viruses gain access to cells because they have attachment sites for receptors on their target cells. When such an attachment site is brought together with an appropriate receptor, the virus can bind to and penetrate the cell. Some viruses gain access to host cells because their attachment sites mimic substances useful to those cells. For example, the attachment sites of rabies virus can mimic the neurotransmitter acetylcholine. As a result, the virus can enter the host cell along with the neurotransmitter.

The AIDS virus (HIV) goes further by hiding its attachment sites from the immune response and by attacking components of the immune system directly. Like most viruses, HIV is cell-specific, in this case attacking only particular body cells that have a surface marker called the CD4 protein. Most of these cells are immune system T cells (T lymphocytes). Binding sites on HIV are complementary to the CD4 protein. The surface of the virus is folded to form ridges and valleys, and the HIV binding sites are located on the floors of the valleys. CD4 proteins are long enough and slender enough to reach these binding sites, whereas antibody molecules made against HIV are too large to make contact with the sites. As a result, it is difficult for these antibodies to destroy HIV.

### Cytopathic Effects of Viruses

Infection of a host cell by an animal virus usually kills the host cell. Death can be caused by the accumulation of large numbers of multiplying viruses, by the effects of viral proteins on the permeability of the host cell's plasma membrane, or by inhibition of host DNA, RNA, or protein synthesis. The visible effects of viral infection are known as **cytopathic effects (CPE)**. Those cytopathic effects that result in cell death are called *cytocidal effects*; those that result in cell damage but not cell death are called *noncytocidal effects*. CPEs are used to diagnose many viral infections.

Cytopathic effects vary by virus. One difference is the point in the viral infection cycle at which the effects occur. Some viral infections result in early changes in the host cell; in other infections, changes are not seen until a much later stage.

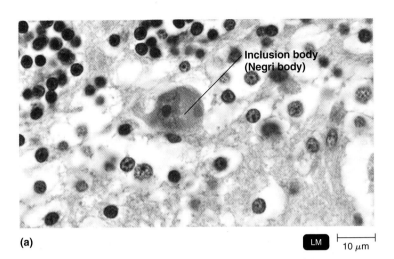

Inclusion body
(Negri body)

(a)

LM | 10 μm

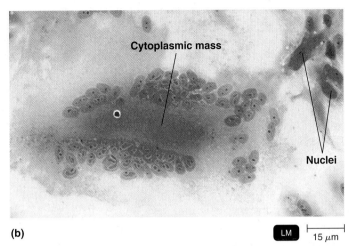

Cytoplasmic mass

Nuclei

(b)

LM | 15 μm

**Figure 15.7 Some cytopathic effects of viruses. (a)** Cytoplasmic inclusion body in brain tissue from a person who died of rabies. **(b)** Portion of a syncytium (giant cell) formed in a cell infected with measles virus. The cytoplasmic mass is probably the Golgi complexes of fused cells.

**Q** What are cytopathic effects?

A virus can produce one or more of the following cytopathic effects:

1. The macromolecular synthesis within the host cell stops. Some viruses, such as *Simplexvirus*, irreversibly stop mitosis.

2. Host cell lysosomes are made to release their enzymes, resulting in destruction of intracellular contents and host cell death.

3. **Inclusion bodies** are found in the cytoplasm or nucleus of some infected cells (**Figure 15.7a**). These granules are sometimes viral parts—nucleic acids or proteins in the process of being assembled into virions. The granules vary in size, shape, and staining properties. They are characterized by their ability to stain with an acidic stain (acidophilic) or with a basic stain (basophilic). Other inclusion bodies arise at sites of earlier viral synthesis but do not contain assembled viruses or their components. Inclusion bodies are important because they can help identify the causative agent. For example, in most cases, rabies virus produces inclusion bodies (Negri bodies) in the cytoplasm of nerve cells, and their presence in animal brain tissue has been used as one diagnostic tool for rabies. Diagnostic inclusion bodies are also associated with measles virus, vaccinia virus, smallpox virus, herpesviruses, and adenoviruses.

4. At times, several adjacent infected cells fuse to form a very large multinucleate cell called a **syncytium** (Figure 15.7b). Such giant cells are produced from infections by viruses that cause measles, mumps, and the common cold.

5. Some viral infections result in changes in the host cell's functions with no visible changes in the infected cells.

For example, when measles virus attaches to its receptor called CD46, the CD46 prompts the cell to reduce production of a cytokine called IL-12, reducing the host's ability to fight the infection.

6. Many viral infections induce antigenic changes on the surface of the infected cells. These antigenic changes are the result of viral-gene encoded proteins, and they elicit a host antibody response against the infected cell, and thus they target the cell for destruction by the host's immune system.

7. Some viruses induce chromosomal changes in the host cell. For example, some viral infections result in chromosomal damage to the host cell, most often chromosomal breakage. Frequently, oncogenes (cancer-causing genes) may be contributed or activated by a virus.

8. Viruses capable of causing cancer *transform* host cells, as discussed in Chapter 13. Transformation results in an abnormal, spindle-shaped cell that does not recognize **contact inhibition**, meaning cells don't stop growing when they come in close contact with other cells (**Figure 15.8**). The loss of contact inhibition results in unregulated cell growth.

Some virus-infected host cells produce substances called alpha and beta **interferons**. Viral infection induces cells to produce these interferons, but the host cell's DNA actually codes for them. Alpha and beta interferons protect neighboring uninfected cells from viral infection in two ways: (1) they inhibit synthesis of viral proteins and host cell proteins; and (2) they kill virus-infected host cells by apoptosis (programmed cell death). However, almost all viruses have mechanisms that evade interferons by partially blocking their synthesis.

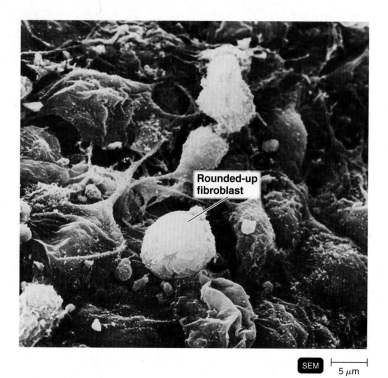

Rounded-up
fibroblast

SEM   ⊢———⊣
5 μm

**Figure 15.8 Human fibroblasts are transformed by Rous sarcoma virus.** Normal fibroblasts grow as flat, spread-out cells.

**Q** What is contact inhibition?

Some representative viruses that cause cytopathic effects are presented in Table 15.4. In Part Four of the book we will discuss the pathological properties of viruses in more detail.

**CHECK YOUR UNDERSTANDING**

✔ 15-15 Define *cytopathic effects*, and give five examples.

**TABLE 15.4 Cytopathic Effects of Selected Viruses**

| Virus (*Genus*) | Cytopathic Effect |
|---|---|
| Poliovirus (*Enterovirus*) | Cytocidal (cell death) |
| Genital warts virus (*Papillomavirus*) | Acidophilic inclusion bodies in nucleus |
| Adenovirus (*Mastadenovirus*) | Basophilic inclusion bodies in nucleus |
| *Lyssavirus* | Acidophilic inclusion bodies in cytoplasm |
| CMV (*Cytomegalovirus*) | Acidophilic inclusion bodies in nucleus and cytoplasm |
| Measles virus (*Morbillivirus*) | Cell fusion |
| *Polyomavirus* | Transformation |
| HIV (*Lentivirus*) | Destruction of T cells |

# Pathogenic Properties of Fungi, Protozoa, Helminths, and Algae

**LEARNING OBJECTIVE**

15-16 Discuss the causes of symptoms in fungal, protozoan, helminthic, and algal diseases.

This section describes some general pathological effects of fungi, protozoa, helminths, and algae that cause human disease. Specific diseases caused by fungi, protozoa, and helminths, along with the pathological properties of these organisms, are discussed in detail in Chapters 21 to 26.

## Fungi

Although fungi cause disease, they do not have a well-defined set of virulence factors. Some fungi have metabolic products that are toxic to human hosts. In such cases, however, the toxin is only an indirect cause of disease, because the fungus is already growing in or on the host. Chronic fungal infections, such as athlete's foot, can also provoke an allergic response in the host.

*Trichothecenes* are fungal toxins that inhibit protein synthesis in eukaryotic cells. Ingestion of these toxins causes headaches, chills, severe nausea, vomiting, and visual disturbances. These toxins are produced by *Fusarium* (fū-SAR-ē-um) and *Stachybotrys* (STA-kē-bah-tris) growing on grains and wallboard in homes.

There is evidence that some fungi do have virulence factors. Two fungi that can cause skin infections, *Candida albicans* and *Trichophyton* (trik-ō-FĪ-ton), secrete proteases. These enzymes may modify host cell membranes to allow attachment of the fungi. *Cryptococcus neoformans* (KRIP-tō-kok-kus nē-ō-FOR-manz) is a fungus that causes a type of meningitis; it produces a capsule that helps it resist phagocytosis. Some fungi have become resistant to antifungal drugs by decreasing their synthesis of receptors for these drugs.

The disease called ergotism, common in Europe during the Middle Ages, is caused by a toxin produced by an ascomycete plant pathogen, *Claviceps purpurea* (KLA-vi-seps pur-pur-Ē-ah), that grows on grains. The toxin is contained in **sclerotia**, highly resistant portions of the mycelia of the fungus that can detach. The toxin itself, **ergot**, is an alkaloid that can cause hallucinations resembling those produced by LSD (lysergic acid diethylamide); in fact, ergot is a natural source of LSD. Ergot also constricts capillaries and can cause gangrene of the limbs by preventing proper blood circulation in the body. Although *C. purpurea* still occasionally occurs on grains, modern milling usually removes the sclerotia.

Several other toxins are produced by fungi that grow on grains or other plants. For example, peanut butter is occasionally recalled because of excessive amounts of **aflatoxin**, which has carcinogenic properties. Aflatoxin is produced by the growth of the mold *Aspergillus flavus*. When ingested, the toxin might be altered in a human body to a mutagenic compound.

Although the sample was sterile, the lab test results show that the solution from the ultrasonic bath is positive for endotoxins. Gram-negative bacteria such as *Burkholderia* found in liquid reservoirs and moist environments can colonize water pipes (see the figure) and, in turn, the laboratory containers used to hold water. In this case, the bacteria from the biofilms were washed into the enzymatic solution.

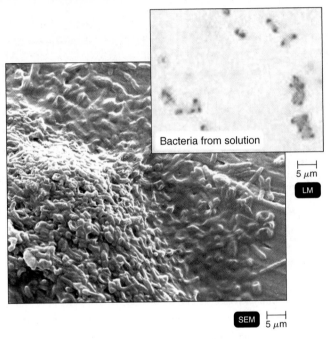

Bacteria from solution

5 μm

LM

SEM  5 μm

**How did endotoxins get in the sterile solutions?**

424   430   435   **439**   441

A few mushrooms produce fungal toxins called **mycotoxins**. Examples are **phalloidin** and **amanitin**, produced by *Amanita phalloides* (A-man-ī-tah fal-LOI-dēz), commonly known as the deathcap. These neurotoxins are so potent that ingestion of the *Amanita* mushroom may result in death.

## Protozoa

The presence of protozoa and their waste products often produces disease symptoms in the host (see Table 12.4, page 347). Some protozoa, such as *Plasmodium*, the causative agent of malaria, invade host cells and reproduce within them, causing their rupture. *Toxoplasma* attaches to macrophages and gains entry by phagocytosis. The parasite prevents normal acidification and digestion; thus, it can grow in the phagocytic vacuole. Other protozoa, such as *Giardia intestinalis*, the causative agent of giardiasis, attach to host cells by a sucking disc (see Figure 12.18a, page 343) and digest the cells and tissue fluids.

Some protozoa can evade host defenses and cause disease for very long periods of time. For example, *Giardia*, which causes diarrhea, and *Trypanosoma*, which causes African trypanosomiasis (sleeping sickness), both use antigenic variation (page 428) to stay one step ahead of the host's immune system. The immune system is alerted to recognize foreign substances called antigens; the presence of antigens causes the immune system to produce antibodies designed to destroy them (see Chapter 17). When *Trypanosoma* is introduced into the bloodstream by a tsetse fly, it produces and displays a specific antigen. In response, the body produces antibodies against that antigen. However, within 2 weeks, the microbe stops displaying the original antigen and instead produces and displays a different one (see Figure 22.16, page 640). Thus, the original antibodies are no longer effective. Because the microbe can make up to 1000 different antigens, such an infection can last for decades.

## Helminths

The presence of helminths also often produces disease symptoms in a host (see Table 12.4, page 347). Some of these organisms actually use host tissues for their own growth or produce large parasitic masses; the resulting cellular damage evokes the symptoms. An example is the roundworm *Wuchereria bancrofti* (woo-kur-ER-ē-ah ban-KROF-tē), the causative agent of elephantiasis. This parasite blocks lymphatic circulation, leading to an accumulation of lymph and eventually causing grotesque swelling of the legs and other body parts. Waste products of the metabolism of these parasites can also contribute to the symptoms of a disease.

## Algae

A few species of algae produce neurotoxins. For example, some genera of dinoflagellates, such as *Alexandrium*, are important medically because they produce a neurotoxin called **saxitoxin**. Although mollusks that feed on the dinoflagellates that produce saxitoxin show no symptoms of disease, people who eat the mollusks develop paralytic shellfish poisoning, with symptoms similar to botulism. Domoic acid is a toxin produced by *Pseudo-nitzschia* diatoms. Crustaceans, fish, and mollusks can accumulate domoic acid. The toxin can cause permanent loss of short-term memory, a condition called *amnesic shellfish poisoning*), coma, or death in humans. Public health agencies frequently prohibit human consumption of mollusks during periods of high algal growth (see Figure 27.10, page 798).

**CHECK YOUR UNDERSTANDING**

✓ **15-16** Identify one virulence factor that contributes to the pathogenicity of each of the following: fungi, protozoa, helminths, and algae.

# Microbial Mechanisms of Pathogenicity

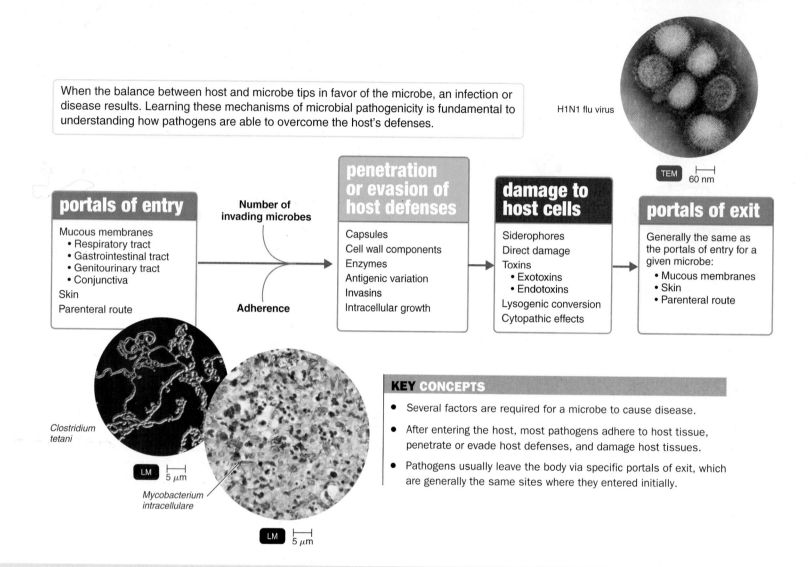

When the balance between host and microbe tips in favor of the microbe, an infection or disease results. Learning these mechanisms of microbial pathogenicity is fundamental to understanding how pathogens are able to overcome the host's defenses.

H1N1 flu virus

TEM    60 nm

### portals of entry

Mucous membranes
- Respiratory tract
- Gastrointestinal tract
- Genitourinary tract
- Conjunctiva

Skin

Parenteral route

Number of invading microbes

Adherence

### penetration or evasion of host defenses

Capsules
Cell wall components
Enzymes
Antigenic variation
Invasins
Intracellular growth

### damage to host cells

Siderophores
Direct damage
Toxins
- Exotoxins
- Endotoxins
Lysogenic conversion
Cytopathic effects

### portals of exit

Generally the same as the portals of entry for a given microbe:
- Mucous membranes
- Skin
- Parenteral route

Clostridium tetani

LM   5 μm

Mycobacterium intracellulare

LM   5 μm

### KEY CONCEPTS

- Several factors are required for a microbe to cause disease.

- After entering the host, most pathogens adhere to host tissue, penetrate or evade host defenses, and damage host tissues.

- Pathogens usually leave the body via specific portals of exit, which are generally the same sites where they entered initially.

# Portals of Exit

### LEARNING OBJECTIVE

15-17 Differentiate portal of entry and portal of exit.

Just as microbes enter the body through a preferred route, they also leave the body via specific routes called **portals of exit** in secretions, excretions, discharges, or tissue that has been shed. In general, portals of exit relate to the infected part of the body, with microbes tending to use the same portal for entry and exit. Portals of exit let pathogens spread through a population by moving from one susceptible host to another. This type of information about the dissemination of a disease is very important to epidemiologists (see Chapter 14).

The most common portals of exit are the respiratory and gastrointestinal tracts. Many pathogens living in the respiratory tract exit in discharges from the mouth and nose, expelled during coughing or sneezing. These microorganisms are found in droplets formed from mucus. Pathogens that cause tuberculosis, whooping cough, pneumonia, scarlet fever, meningococcal meningitis, chickenpox, measles, mumps, smallpox, and influenza are discharged through the respiratory route. Other pathogens exit via the gastrointestinal tract in feces or saliva. Feces may be contaminated with pathogens associated with salmonellosis, cholera, typhoid fever, shigellosis, amebic dysentery, and poliomyelitis. Saliva can also contain pathogens, such as those that cause rabies, mumps, and infectious mononucleosis.

Another important route of exit is the genitourinary tract. Microbes responsible for sexually transmitted infections are found in secretions from the penis and vagina. Urine can also contain the pathogens responsible for typhoid fever and brucellosis, which can exit via the urinary tract. Skin or wound infections are other portals of exit. Infections transmitted from the skin include yaws, impetigo, ringworm, *Simplexvirus*, and warts. Drainage from wounds can spread infections to another person directly or by contact with a contaminated fomite. Infected blood can be removed and reinjected by biting insects and contaminated needles and syringes to spread infection within a population. Examples of diseases transmitted by biting insects are yellow fever, plague, tularemia, and malaria. AIDS and hepatitis B may be transmitted by contaminated needles and syringes.

### CHECK YOUR UNDERSTANDING

✔ 15-17 Which are the most often used portals of exit?

\*\*\*

In the next chapter, we will examine a group of nonspecific defenses of the host against disease. But before proceeding, examine **Figure 15.9** carefully. It summarizes some key concepts of the microbial mechanisms of pathogenicity we have discussed in this chapter.

### CLINICAL CASE  Resolved

Although the bacteria are killed by autoclaving, endotoxins can be released from the dead cells into solutions during autoclaving. Therefore, the solution was the disease source, even though it was sterile.

Dr. Santos treats her patients with prednisone, a topical anti-inflammatory drug, and they all fully recover from the reaction. (She did not prescribe antibiotics, because TASS is not an infection.) She holds a staff meeting to make sure the correct sterilizing procedures are being followed. Dr. Santos also stresses to her employees that preventing TASS depends primarily on using appropriate protocols for cleaning and sterilizing surgical equipment and paying careful attention to all solutions, medications, and ophthalmic devices used during surgery.

424 ▸ 430 ▸ 435 ▸ 439 ▸ **441**

## Study Outline

 Go to @MasteringMicrobiology for **Interactive Microbiology,** *In the Clinic videos, MicroFlix, MicroBoosters, 3D animations, practice quizzes, and more.*

### Introduction (p. 423)

1. Pathogenicity is the ability of a pathogen to produce a disease by overcoming the defenses of the host.
2. Virulence is the degree of pathogenicity.

### How Microorganisms Enter a Host (pp. 424–427)

1. The specific route by which a particular pathogen gains access to the body is called its portal of entry.

### Portals of Entry (p. 424)

2. Many microorganisms can penetrate mucous membranes of the conjunctiva and the respiratory, gastrointestinal, and genitourinary tracts.
3. Most microorganisms cannot penetrate intact skin; they enter hair follicles and sweat ducts.
4. Some microorganisms can gain access to tissues by inoculation through the skin and mucous membranes in bites, injections, and other wounds. This route of penetration is called the parenteral route.

### The Preferred Portal of Entry (p. 425)

5. Many microorganisms can cause infections only when they gain access through their specific portal of entry.

### Numbers of Invading Microbes (pp. 425–426)

6. Virulence can be expressed as $LD_{50}$ (lethal dose for 50% of the inoculated hosts) or $ID_{50}$ (infectious dose for 50% of the inoculated hosts).

### Adherence (pp. 426–427)

7. Surface projections on a pathogen called adhesins (ligands) adhere to complementary receptors on the host cells.
8. Adhesins can be glycoproteins or lipoproteins and are frequently associated with fimbriae.

9. Mannose is the most common receptor.
10. Biofilms provide attachment and resistance to antimicrobial agents.

### How Bacterial Pathogens Penetrate Host Defenses (pp. 427–429)

#### Capsules (p. 428)

1. Some pathogens have capsules that prevent them from being phagocytized.

#### Cell Wall Components (p. 428)

2. Proteins in the cell wall can facilitate adherence or prevent a pathogen from being phagocytized.

#### Enzymes (p. 428)

3. Local infections can be protected in a fibrin clot caused by the bacterial enzyme coagulase.
4. Bacteria can spread from a focal infection by means of kinases (which destroy blood clots), hyaluronidase (which destroys a mucopolysaccharide that holds cells together), and collagenase (which hydrolyzes connective tissue collagen).
5. IgA proteases destroy IgA antibodies.

#### Antigenic Variation (pp. 428–429)

6. Some microbes vary expression of antigens, thus avoiding the host's antibodies.

#### Penetration into the Host (p. 429)

7. Bacteria may produce proteins that alter the actin of the host cell's cytoskeleton, allowing bacteria into the cell.

#### Biofilms (p. 429)

8. Phagocytes are inactivated or killed by the EPS of biofilms.

## How Bacterial Pathogens Damage Host Cells (pp. 430–436)

### Using the Host's Nutrients: Siderophores (p. 430)

1. Bacteria get iron from the host using siderophores.

### Direct Damage (p. 430)

2. Host cells can be destroyed when pathogens metabolize and multiply inside the host cells.

### The Production of Toxins (pp. 430–435)

3. Poisonous substances produced by microorganisms are called toxins; toxemia refers to the presence of toxins in the blood. The ability to produce toxins is called toxigenicity.

4. Exotoxins are produced by bacteria and released into the surrounding medium. Exotoxins, not the bacteria, produce the disease symptoms.

5. Antibodies produced against exotoxins are called antitoxins.

6. A-B toxins consist of an active component that inhibits a cellular process and a binding component that attaches the two portions to the target cell, e.g., diphtheria toxin, *Helicobacter* genotoxin.

7. Membrane-disrupting toxins cause cell lysis, e.g., hemolysins.

8. Superantigens cause release of cytokines, which cause fever, nausea, and other symptoms; e.g., toxic shock syndrome toxin.

9. Endotoxins are the lipid A component of the cell wall of gram-negative bacteria.

10. Bacterial cell death, antibiotics, and antibodies may cause the release of endotoxins.

11. Endotoxins cause fever (by inducing the release of interleukin-1) and shock (because of a TNF-induced decrease in blood pressure).

12. The Limulus amebocyte lysate (LAL) assay is used to detect endotoxins in drugs and on medical devices.

### Plasmids, Lysogeny, and Pathogenicity (p. 436)

13. Plasmids may carry genes for antibiotic resistance, toxins, capsules, and fimbriae.

14. Lysogenic conversion can result in bacteria with virulence factors, such as toxins or capsules.

## Pathogenic Properties of Viruses (pp. 436–438)

1. Viruses avoid the host's immune response by growing inside cells.

2. Viruses gain access to host cells because they have attachment sites for receptors on the host cell.

3. Visible signs of viral infections are called cytopathic effects (CPE).

4. Some viruses cause cytocidal effects (cell death), and others cause noncytocidal effects (damage but not death).

5. Cytopathic effects include stopping mitosis, lysis, formation of inclusion bodies, cell fusion, antigenic changes, chromosomal changes, and transformation.

## Pathogenic Properties of Fungi, Protozoa, Helminths, and Algae (pp. 438–439)

1. Symptoms of fungal infections can be caused by capsules, toxins, and allergic responses.

2. Symptoms of protozoan and helminthic diseases can be caused by damage to host tissue or by the metabolic waste products of the parasite.

3. Some protozoa change their surface antigens while growing in a host, thus avoiding destruction by the host's antibodies.

4. Some algae produce neurotoxins that cause paralysis when ingested by humans.

## Portals of Exit (pp. 440–441)

1. Pathogens leave a host by portals of exit.

2. Three common portals of exit are the respiratory tract via coughing or sneezing, the gastrointestinal tract via saliva or feces, and the genitourinary tract via secretions from the vagina or penis.

3. Arthropods and syringes provide a portal of exit for microbes in blood.

# Study Questions

For answers to the Knowledge and Comprehension questions, turn to the Answers tab at the back of the textbook.

## Knowledge and Comprehension

### Review

1. Compare pathogenicity with virulence.

2. How are capsules and cell wall components related to pathogenicity? Give specific examples.

3. Describe how hemolysins, leukocidins, coagulase, kinases, hyaluronidase, siderophores, and IgA proteases might contribute to pathogenicity.

4. Explain how drugs that bind each of the following would affect pathogenicity:
   a. iron in the host's blood
   b. *Neisseria gonorrhoeae* fimbriae
   c. *Streptococcus pyogenes* M protein

5. Compare and contrast the following aspects of endotoxins and exotoxins: bacterial source, chemistry, toxigenicity, and pharmacology. Give an example of each toxin.

6. DRAW IT  Label this diagram to show how the Shiga toxin enters and inhibits protein synthesis in a human cell.

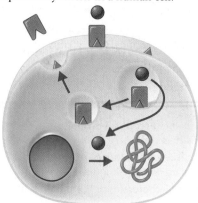

7. Describe the factors contributing to the pathogenicity of fungi, protozoa, and helminths.

8. Which of the following genera is the most infectious?

| Genus | ID$_{50}$ | Genus | ID$_{50}$ |
|---|---|---|---|
| *Legionella* | 1 cell | *Shigella* | 200 cells |
| *Salmonella* | $10^5$ cells | *Treponema* | 52 cells |

9. How can viruses and protozoa avoid being killed by the host's immune response?

10. NAME IT The *Opa* gene is used to identify this endotoxin-producing bacterium that grows well in the high-$CO_2$ conditions inside phagocytes.

## Multiple Choice

1. The removal of plasmids reduces virulence in which of the following organisms?
   a. *Clostridium tetani*
   b. *Escherichia coli*
   c. *Salmonella enterica*
   d. *Streptococcus mutans*
   e. *Clostridium botulinum*

2. What is the LD$_{50}$ for the bacterial toxin tested in the example below?

| Dilution (μg/kg) | No. of Animals Died | No. of Animals Survived |
|---|---|---|
| a. 6 | 0 | 6 |
| b. 12.5 | 0 | 6 |
| c. 25 | 3 | 3 |
| d. 50 | 4 | 2 |
| e. 100 | 6 | 0 |

3. Which of the following is *not* a portal of entry for pathogens?
   a. mucous membranes of the respiratory tract
   b. mucous membranes of the gastrointestinal tract
   c. skin
   d. blood
   e. parenteral route

4. All of the following can occur during bacterial infection. Which would prevent all of the others?
   a. vaccination against fimbriae
   b. phagocytosis
   c. inhibition of phagocytic digestion
   d. destruction of adhesins
   e. alteration of cytoskeleton

5. The ID$_{50}$ for *Campylobacter* sp. is 500 cells; the ID$_{50}$ for *Cryptosporidium* sp. is 100 cells. Which of the following statements is *false*?
   a. Both microbes are pathogens.
   b. Both microbes produce infections in 50% of the inoculated hosts.
   c. *Cryptosporidium* is more virulent than *Campylobacter*.
   d. *Campylobacter* and *Cryptosporidium* are equally virulent; they cause infections in the same number of test animals.
   e. *Cryptosporidium* infections are acquired more easily than *Campylobacter* infections.

6. An encapsulated bacterium can be virulent because the capsule
   a. resists phagocytosis.
   b. is an endotoxin.
   c. destroys host tissues.
   d. kills host cells.
   e. has no effect; because many pathogens do not have capsules, capsules do not contribute to virulence.

7. A drug that binds to mannose on human cells would prevent
   a. the entrance of *Vibrio* enterotoxin.
   b. the attachment of pathogenic *E. coli*.
   c. the action of botulinum toxin.
   d. streptococcal pneumonia.
   e. the action of diphtheria toxin.

8. The earliest smallpox vaccines were infected tissue rubbed into the skin of a healthy person. The recipient of such a vaccine usually developed a mild case of smallpox, recovered, and was immune thereafter. What is the most likely reason this vaccine did not kill more people?
   a. Skin is the wrong portal of entry for smallpox.
   b. The vaccine consisted of a mild form of the virus.
   c. Smallpox is normally transmitted by skin-to-skin contact.
   d. Smallpox is a virus.
   e. The virus mutated.

9. Which of the following does *not* represent the same mechanism for avoiding host defenses as the others?
   a. Rabies virus attaches to the receptor for the neurotransmitter acetylcholine.
   b. *Salmonella* attaches to the receptor for epidermal growth factor.
   c. Epstein-Barr (EB) virus binds to the host receptor for complement.
   d. Surface protein genes in *Neisseria gonorrhoeae* mutate frequently.
   e. none of the above

10. Which of the following statements is true?
    a. The primary goal of a pathogen is to kill its host.
    b. Evolution selects for the most virulent pathogens.
    c. A successful pathogen doesn't kill its host before it is transmitted.
    d. A successful pathogen never kills its host.

## Analysis

1. The graph below shows confirmed cases of enteropathogenic *E. coli*. Why is the incidence seasonal?

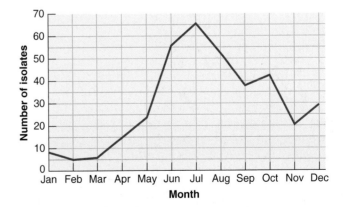

2. The cyanobacterium *Microcystis aeruginosa* produces a peptide that is toxic to humans. According to the graph below, during what season is this bacterium most toxic?

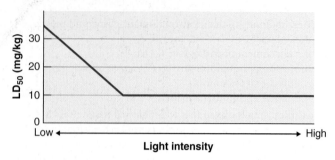

3. When injected into rats, the $ID_{50}$ for *Salmonella* Typhimurium is $10^6$ cells. If sulfonamides are injected with the salmonellae, the $ID_{50}$ is 35 cells. Explain the change in $ID_{50}$ value.

4. How do each of the following strategies contribute to the virulence of the pathogen? What disease does each organism cause?

| Strategy | Pathogen |
|---|---|
| Changes its cell wall after entry into host | *Yersinia pestis* |
| Uses urea to produce ammonia | *Helicobacter pylori* |
| Causes host to make more receptors | *Rhinovirus* |

## Clinical Applications and Evaluation

1. On July 8, a woman was given an antibiotic for presumptive sinusitis. However, her condition worsened, and she was unable to eat for 4 days because of severe pain and tightness of the jaw. On July 12, she was admitted to a hospital with severe facial spasms. She reported that on July 5 she had incurred a puncture wound at the base of her big toe; she cleaned the wound but did not seek medical attention. What caused her symptoms? Was her condition due to an infection or an intoxication? Can she transmit this condition to another person?

2. Explain whether each of the following examples is a food infection or intoxication. What is the probable etiological agent in each case?
   a. Eighty-two people in Louisiana developed diarrhea, nausea, headache, and fever from 4 hours to 2 days after eating shrimp.
   b. Two people in Vermont developed malaise, nausea, blurred vision, breathing difficulty, and numbness 3 to 6 hours after eating barracuda caught in Florida.

3. Cancer patients undergoing chemotherapy are normally *more* susceptible to infections. However, a patient receiving an antitumor drug that affects eukaryotic cytoskeletons was resistant to *Salmonella*. Provide a possible mechanism for the resistance.

# Innate Immunity: Nonspecific Defenses of the Host 16

From our discussion to this point, you can see that pathogenic microorganisms are endowed with special properties that enable them to cause disease if given the right opportunity. If microorganisms never encountered resistance from the host, we would constantly be ill and would die of various diseases after a short life. In most cases, however, our body's defenses prevent this from happening. Some of these defenses are designed to keep out microorganisms altogether, other defenses remove the microorganisms if they do get in, and still others combat them if they remain inside.

In this chapter we discuss the first two lines of defense against pathogens, which we call the *innate immunity defenses*. The first line of defense is our skin and mucous membranes. The second line of defense consists of phagocytes, inflammation, fever, and antimicrobial substances produced by the body. The Clinical Case in this chapter describes one problem that can occur if phagocytes (blue in the photograph) don't function properly.

See pages 446–447 for a **Big Picture** overview of the entire immune system.

◀ A neutrophil (blue) phagocytizing *Aspergillus* spores (red).

## In the Clinic

You are an emergency room nurse caring for Madge, a 30-year-old kidney transplant recipient. She is currently receiving treatment for septic shock—her third episode of this infection in her lifetime. Madge says that since childhood she has always been very susceptible to recurring infections. She is thankful that so far, her transplanted kidney is functioning well and shows no signs of rejection or damage. You run tests that show leukocytosis, normal levels of antibodies, and a C6 deficiency. **What is the cause of Madge's frequent infections, and how does it relate to her transplant tolerance?**

*Hint: Read about leukocyte response to infections on page 455, the complement system on pages 463–467, and testing for complement in the Clinical Focus box on page 470.*

 Play **In the Clinic** Video
@Mastering**Microbiology**

## Every day the human body wages war with microbial pathogens needing a place to live.

### First-Line Defenses

First-line defenses keep pathogens on the outside or neutralize them before infection begins. The skin, mucous membranes, and certain antimicrobial substances are part of these defenses.

### Second-Line Defenses

Second-line defenses slow or contain infections when first-line defenses fail. They include proteins that produce inflammation, fever that enhances cytokine activity, and phagocytes and natural killer (NK) cells, which attack and destroy cancer cells and virus-infected cells.

### Third-Line Defenses

Third-line defenses, shown in the table below, include lymphocytes that target specific pathogens for destruction when the second-line defenses don't contain infections. It includes a memory component that allows the body to more effectively respond to that same pathogen in the future.

First- and second-line defenses are part of the innate immune system, whereas the third-line defenses are referred to as the adaptive immune system. Many leukocytes (white blood cells) coordinate efforts in controlling infections in the second and third lines of immune defense.

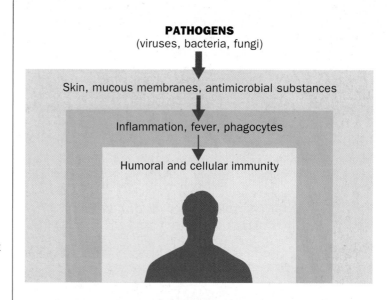

**PATHOGENS**
(viruses, bacteria, fungi)

Skin, mucous membranes, antimicrobial substances

Inflammation, fever, phagocytes

Humoral and cellular immunity

| Innate or Adaptive | Cell Type | Description | Function |
|---|---|---|---|
| INNATE | Basophil | Granulocyte | Releases histamines that cause inflammation. |
| | Eosinophil | Granulocyte | Kills parasites with oxidative burst. |
| | Mast cell | Granulocyte | Antigen-presenting cells; produce antibacterial peptides. |
| BOTH | Neutrophil | Granulocyte | Phagocytizes bacteria and fungi. |
| | Monocyte | Agranulocyte | Precursor to macrophages. Some macrophages are fixed in certain organs while others wander tissues, causing inflammation. All perform phagocytosis. |
| | Dendritic cell | Agranulocyte (many surface projections) | In skin and respiratory and intestinal mucosa, phagocytizes bacteria and presents antigens to T cells. |
| | Natural killer (NK) cell | Agranulocyte (lymphocyte) | Kills cancer cells and virus-infected cells. |
| ADAPTIVE | Plasma cell, B cell | Agranulocyte (lymphocyte) | Recognizes antigens and produces antibodies. |
| | T cells<br><br>T Helper ($T_H$) cell<br><br>Cytotoxic T lymphocyte (CTL)<br><br>T regulatory ($T_{reg}$) cell | Agranulocyte (lymphocyte) | $T_H$ cells secrete cytokines. They are CD4$^+$ cells that bind MHC class II molecules on antigen-presenting cells (APCs). CTLs recognize and kill specific "nonself" cells. They are CD8$^+$ cells that bind to MHC class I molecules. $T_{reg}$ cells are CD4$^+$ cells that destroy cells that do not correctly recognize "self" cells. |

# What Do Blood Cell Counts Tell Us About Patient Health?

## White Blood Cell Counts

White blood cell (WBC) counts measure the number of leukocytes found in the blood. A related test, called *differential white blood cell count*, breaks down the white blood cell count further, identifying the percentages of eosinophils, basophils, neutrophils, monocytes, and lymphocytes. Abnormal blood cell counts give health care providers important clues for diagnosing infections and other conditions.

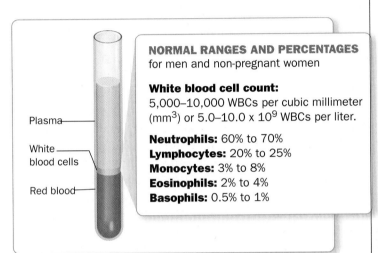

Plasma

White blood cells

Red blood

**NORMAL RANGES AND PERCENTAGES**
for men and non-pregnant women

**White blood cell count:**
5,000–10,000 WBCs per cubic millimeter (mm$^3$) or 5.0–10.0 x 10$^9$ WBCs per liter.

**Neutrophils:** 60% to 70%
**Lymphocytes:** 20% to 25%
**Monocytes:** 3% to 8%
**Eosinophils:** 2% to 4%
**Basophils:** 0.5% to 1%

## High White Blood Cell Counts

A high white blood cell count shows the patient is producing a higher than average number of leukocytes. This typically occurs when the patient battles a bacterial infection. High white blood cell counts may also stem from autoimmune disorders that result in too much inflammatory response, such as rheumatoid arthritis, and from leukemia, a cancer of the blood. Some drugs can cause high white blood cell counts as a side effect; these include certain asthma medications such as albuterol, epinephrine, and corticosteroids.

## Low White Blood Cell Counts

A low white blood cell count shows the patient has fewer leukocytes than expected. A low neutrophil count, in particular, is instructive. Even bacteria that generally live in the mouth and digestive track without being pathogenic may result in disease when the patient's neutrophil count drops below 500 neutrophils per mm$^3$ of blood. Low white blood cell counts may result from viral infections or pneumonia. They may also be caused from autoimmune diseases such as lupus; certain cancers, such as lymphoma; and radiation and other cancer treatments. White blood cell counts may also be low when a patient has an extremely severe bacterial infection, such as septicemia. Finally, numerous drugs may also cause low white blood cell counts, including a variety of antibiotics, diuretics, and anticancer medications.

▲ T cell (blue) infected with HIV. Virus particles (pink) can be seen budding from the infected cell. HIV infects T$_H$ lymphocytes.

SEM | 350 nm

◄ Serial monitoring of the white blood cell count is used to monitor treatment for infants battling the bacterial infection pertussis. Rapidly rising high white blood cell counts have been shown to be associated with higher mortality among infants.

**KEY CONCEPTS**

- Innate immunity involves the first- and second-line defenses.
- Adaptive immunity involves third-line defenses.
- Innate immune actions are fast but nonspecific. Adaptive immune actions are slower, but specific to pathogens, and have a memory component.

# The Concept of Immunity

16-1 Differentiate innate and adaptive immunity.

16-2 Define *Toll-like receptors*.

When microbes attack our bodies, we defend ourselves by utilizing our various mechanisms of immunity. **Immunity**, also called **resistance**, is the ability to ward off disease caused by microbes or their products and to protect against environmental agents such as pollen, chemicals, and animal dander. Lack of immunity is referred to as **susceptibility**. In general, there are two types of immunity: innate and adaptive. **Innate immunity** refers to defenses that are present at birth. They are always available to provide rapid responses to protect us against disease. Innate immunity does not involve recognition of a specific microbe. Further, innate immunity has no memory response, that is, a more rapid and stronger immune reaction to the same microbe at a later date. Innate immunity first-line defenses include skin and mucous membranes, and the second-line defenses include natural killer cells, phagocytes, inflammation, fever, and antimicrobial substances. Innate immune responses represent immunity's early-warning system and are designed to prevent microbes from gaining access into the body and to help eliminate those that do gain access.

**Adaptive immunity** is based on a specific response to a specific microbe once a microbe has breached the innate immunity defenses. It adjusts to handle a particular microbe. Unlike innate immunity, adaptive immunity is slower to respond, but it does have a memory component that allows the body to more effectively target the same pathogens in the future. Adaptive immunity involves lymphocytes (a type of white blood cell) called T cells (T lymphocytes) and B cells (B lymphocytes) and will be discussed in detail in Chapter 17. Here, we concentrate on innate immunity.

Responses of the innate system are activated by protein receptors in the plasma membranes of defensive cells. Among these activators are **Toll-like receptors*** (**TLRs**). These TLRs attach to various components commonly found on pathogens that are called **pathogen-associated molecular patterns (PAMPs)** (see Figure 16.8). Examples include the lipopolysaccharide (LPS) of the outer membrane of gram-negative bacteria, the flagellin in the flagella of motile bacteria, the peptidoglycan in the cell wall of gram-positive bacteria, the DNA of bacteria, and the DNA and RNA of viruses. TLRs also attach to components of fungi and parasites.

You will learn later in this chapter that two of the defensive cells involved in innate immunity are called macrophages and dendritic cells and provide a link between innate immunity and adaptive immunity. When the TLRs on these cells encounter the PAMPs of microbes, such as the LPS of gram-negative bacteria, the TLRs induce the defensive cells to release chemicals called cytokines. **Cytokines** (*cyto-* = cell; *-kinesis* = motion) are proteins that regulate the intensity and duration of immune responses.

One role of cytokines is to recruit other macrophages and dendritic cells, as well as other defensive cells, to isolate and destroy the microbes as part of the inflammatory response. Cytokines can also activate the T cells and B cells involved in adaptive immunity. You will learn more about the different cytokines and their functions in Chapter 17, pages 477–478.

As you learn the individual and unique components of the innate and adaptive immune systems, you will also see that they do not operate independently. In fact, they function as a highly interactive and cooperative "supersystem" that produces a combined response that is more effective than either component can produce separately. Certain molecular and cellular components of the immune system serve important functions in both types of immunity. One example of the cooperation between the two immune systems, which you will learn about later in the chapter, involves macrophages and dendritic cells. These cells provide a link between innate and adaptive immunity.

Play Host Defenses: *The Big Picture* @Mastering**Microbiology**

**CHECK YOUR UNDERSTANDING**

✔ **16-1** Which defense system, innate or adaptive immunity, prevents entry of microbes into the body?

✔ **16-2** What relationship do Toll-like receptors have to pathogen-associated molecular patterns?

# First Line of Defense: Skin and Mucous Membranes

16-3 Describe the role of the skin and mucous membranes in innate immunity.

16-4 Differentiate physical from chemical factors, and list five examples of each.

16-5 Describe the role of normal microbiota in innate immunity.

---

*Fruit flies defend themselves from fungal infections by means of a protein called *Toll*—named for the German word for weird or strange. The term derives from the fact that the Toll protein also is involved in development of the fruit fly embryo and that flies without it have a strange, or weird, appearance.

The skin and mucous membranes are the body's first line of defense against environmental pathogens. This function results from both physical and chemical factors.

## Physical Factors

Physical factors include barriers to entry and processes that remove microbes from the body's surface. The intact **skin** is the human body's largest organ in terms of surface area and

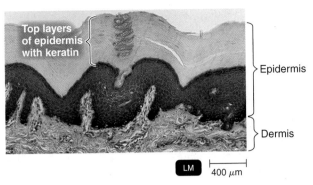

**Figure 16.1  A section through human skin.** The thin layers at the top of this photomicrograph contain keratin. These layers and the darker purple cells beneath them make up the epidermis. The region below the epidermis is the dermis.

**Q** What is the function of keratin in the epidermis?

weight and is an extremely important component of the first line of defense. It consists of the dermis and the epidermis (**Figure 16.1**). The **dermis**, the skin's inner, thicker portion, is composed of connective tissue. The **epidermis**, the outer, thinner portion, is in direct contact with the external environment. The epidermis consists of many layers of continuous sheets of tightly packed epithelial cells with little or no material between the cells. The top layer of epidermal cells is dead and contains a protective protein called **keratin**. The periodic shedding of the top layer helps remove microbes at the surface. In addition, the dryness of the skin is a major factor in inhibiting microbial growth on the skin. When the skin is frequently moist, as in hot, humid climates, skin infections are quite common, especially fungal infections such as athlete's foot. These fungi hydrolyze keratin when water is available.

If we consider the closely packed cells, continuous layering, the presence of keratin, and the dryness and shedding of the skin, we can see why the intact skin provides such a formidable barrier to the entrance of microorganisms. Microorganisms rarely, if ever, penetrate the intact surface of healthy epidermis. However, when the epithelial surface is broken as a result of burns, cuts, puncture wounds, or other conditions, a subcutaneous (below-the-skin) infection often develops. The bacteria most likely to cause infection are the staphylococci that normally inhabit the epidermis, hair follicles, and sweat and oil glands of the skin.

Epithelial cells called *endothelial cells* that line blood and lymphatic vessels are not closely packed like those of the epidermis. This arrangement permits defensive cells to move from blood into tissues during inflammation, but it also permits microbes to move into and out of blood and lymph.

**Mucous membranes** also consist of an epithelial layer and an underlying connective tissue layer. They are an important component of the first line of defense. Mucous membranes line the entire gastrointestinal, respiratory, and genitourinary

tracts. The epithelial layer of a mucous membrane secretes a fluid called **mucus**, a slightly viscous (thick) glycoprotein produced by goblet cells of a mucous membrane. Among other functions, mucus prevents the tracts from drying out. Some pathogens that can thrive on the moist secretions are able to penetrate the membrane if the microorganism is present in sufficient numbers. *Treponema pallidum* is such a pathogen. This penetration may be facilitated by toxic substances produced by the microorganism, prior injury by viral infection, or mucosal irritation.

Besides the physical barrier presented by the skin and mucous membranes, several other physical factors help protect certain epithelial surfaces. One mechanism that protects eyes is the **lacrimal apparatus**, a group of structures that manufactures and drains tears (**Figure 16.2**). The lacrimal glands, located toward the upper, outermost portion of each eye socket, produce the tears and pass them under the upper eyelid. From here, tears pass toward the corner of the eye near the nose and into two small holes that lead through tubes (lacrimal canals) to the nose. The tears are spread over the surface of the eyeball by blinking. Normally, the tears evaporate or pass into the nose as fast as they are produced. This continual washing action helps keep microorganisms from settling on the surface of the eye. If an irritating substance or large numbers of microorganisms come in contact with the eye, the lacrimal glands start to secrete heavily, and the tears accumulate more rapidly than they can be carried away. This excessive production is a protective mechanism because the excess tears dilute and wash away the irritating substance or microorganisms before infection can occur.

In a cleansing action very similar to that of tears, **saliva**, produced by the salivary glands, helps dilute the numbers of microorganisms and wash them from the surface of the teeth and the mucous membrane of the mouth. This helps prevent colonization by microbes.

**CLINICAL CASE** Missing in Action

Two-year-old Jacob is back in the pediatrician's office with another high fever. Jacob has a history of recurrent skin infections, fever, and chronically enlarged lymph nodes.

Jacob's pediatrician notices that his lung sounds are not clear, so he sends Jacob for a chest X-ray exam. Results show a mass in Jacob's right lung. The mass is pneumonia, and Jacob's pediatrician treats him with antibiotics. A few weeks after he finishes his antibiotics, Jacob develops pneumonia again. This time Jacob's pediatrician orders a biopsy of the lung mass, and the culture reveals *Aspergillus* fungus.

**Why isn't Jacob's innate immunity protecting him from infections? Read on to find out.**

449  455  459  463  469  470

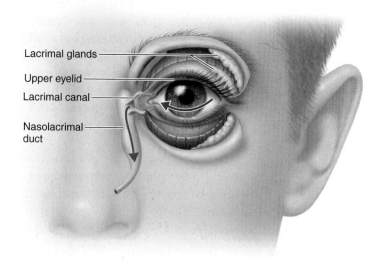

**Figure 16.2 The lacrimal apparatus.** The washing action of tears over the surface of the eyeball is shown by the red arrow. Tears produced by the lacrimal glands pass across the surface of the eyeball into two small holes that convey the tears into the lacrimal canals and the nasolacrimal duct. From here tears pass into the nose, as shown by the gray arrow.

**Q** How does the lacrimal apparatus protect the eyes against infections?

The respiratory and gastrointestinal tracts have many physical forms of defense. **Mucus** traps many of the microorganisms that enter these tracts. The mucous membrane of the nose also has mucus-coated **hairs** that filter inhaled air and trap particles greater than 10 μm. Smaller (to 2 μm) particles will be trapped in mucus of the lower respiratory tract. The cells of the mucous membrane of the lower respiratory tract are covered with **cilia**. By moving synchronously, these cilia propel inhaled dust and microorganisms that have become trapped in mucus upward toward the throat. This so-called

**ciliary escalator** (**Figure 16.3**) keeps the mucus blanket moving toward the throat at a rate of 1 to 3 cm per hour; coughing and sneezing speed up the escalator. Some substances in cigarette smoke are toxic to cilia and can seriously impair the functioning of the ciliary escalator by inhibiting or destroying the cilia. Mechanically ventilated patients are vulnerable to respiratory tract infections because the ciliary escalator mechanism is inhibited. Microorganisms are also prevented from entering the lower respiratory tract by a small lid of cartilage called the **epiglottis**, which covers the larynx (voicebox) during swallowing. The external ear canal contains hairs and **earwax** (*cerumen*), which help prevent microbes, dust, insects, and water from entering the ear.

The cleansing of the urethra by the flow of **urine** is another physical factor that prevents microbial colonization in the genitourinary tract. As you will see later, when urine flow is obstructed—by catheters, for example—urinary tract infections may develop. **Vaginal secretions** likewise move microorganisms out of the female body.

**Peristalsis, defecation, vomiting,** and **diarrhea** also expel microbes. Peristalsis is a series of coordinated contractions that propels food along the gastrointestinal tract. Mass peristalsis of large intestinal contents into the rectum results in defecation. In response to microbial toxins, the muscles of the gastrointestinal tract contract vigorously, resulting in vomiting and/or diarrhea, which may also rid the body of microbes.

## Chemical Factors

Physical factors alone do not account for the high degree of resistance shown by skin and mucous membranes against microbial invasion. Certain chemical factors also play important roles.

Sebaceous (oil) glands of the skin produce an oily substance called **sebum** that prevents hair from drying and becoming brittle. Sebum also forms a protective film over the surface of the skin. One of the components of sebum is unsaturated fatty

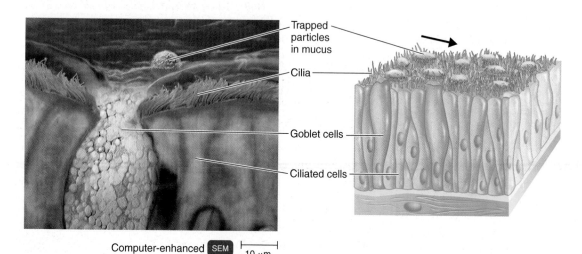

Computer-enhanced SEM |⸺⸺| 10 μm

**Figure 16.3 The ciliary escalator.**

**Q** What can happen if the ciliary escalator is inhibited?

Trapped particles in mucus

Cilia

Goblet cells

Ciliated cells

acids, which inhibit the growth of certain pathogenic bacteria and fungi. The low pH of the skin, between pH 3 and 5, is caused in part by the secretion of fatty acids and lactic acid. The skin's acidity probably discourages the growth of many other microorganisms.

Bacteria that live commensally on the skin decompose sloughed-off skin cells, and the resultant organic molecules and end-products of their metabolism produce body odor. Certain bacteria commonly found on the skin metabolize sebum, and this metabolism forms free fatty acids that cause the inflammatory response associated with acne (as we'll see in Chapter 21). Isotretinoin, a derivative of vitamin A that prevents sebum formation, is a treatment for a very severe type of acne called cystic acne.

The sweat glands of the skin produce **perspiration**, which helps maintain body temperature, eliminates certain wastes, and flushes microorganisms from the surface of the skin. Perspiration also contains **lysozyme**, an enzyme capable of breaking down cell walls of gram-positive bacteria and, to a lesser extent, gram-negative bacteria (see Figure 4.13, page 82). Specifically, lysozyme breaks chemical bonds on peptidoglycan, which destroys the cell walls. Lysozyme is also found in tears, saliva, nasal secretions, tissue fluids, and urine, where it exhibits its antimicrobial activity. Alexander Fleming was studying lysozyme in 1928 when he accidentally discovered the antimicrobial effects of penicillin (see Figure 1.6, page 11).

**Earwax**, besides serving as a physical barrier, also functions as a chemical protectant. It is a mixture of secretions from glands producing earwax as well as from the sebaceous glands, which produce sebum. The secretions are rich in fatty acids, giving the ear canal a low pH, between 3 and 5, which inhibits the growth of many pathogenic microbes. Earwax also contains many dead cells from the lining of the ear canal.

**Saliva** contains not only the enzyme salivary amylase that digests starch, but also a number of substances that inhibit microbial growth. These include lysozyme, urea, and uric acid. The slightly acidic pH of saliva (6.55–6.85) also inhibits some microbes. Saliva additionally contains an antibody (IgA) that prevents attachment of microbes so that they cannot penetrate mucous membranes.

**Gastric juice** is produced by the glands of the stomach. It is a mixture of hydrochloric acid, enzymes, and mucus. The very high acidity of gastric juice (pH 1.2–3.0) is sufficient to destroy bacteria and most bacterial toxins, except those of *Clostridium botulinum* and *Staphylococcus aureus*. However, many enteric pathogens are protected by food particles and can enter the intestines via the gastrointestinal tract. In contrast, the bacterium *Helicobacter pylori* neutralizes stomach acid, thereby allowing the bacterium to grow in the stomach. Its growth initiates an immune response that results in gastritis and ulcers.

**Vaginal secretions** play a role in antibacterial activity in two ways. Glycogen produced by vaginal epithelial cells is broken down into lactic acid by *Lactobacillus* spp. This creates an acidic pH (3–5) that inhibits microbes. Cervical mucus also has some antimicrobial activity.

**Urine**, in addition to containing lysozyme, has an acidic pH (average 6) that inhibits microbes.

Later in the chapter, we will discuss another group of chemicals, the antimicrobial peptides, which play a very important role in innate immunity.

## Normal Microbiota and Innate Immunity

Recall from Chapter 14 (pages 394–398) that certain microbes establish more or less permanent residence (colonization) in and on the body but normally do not produce disease. They constitute the normal microbiota and play an important role preventing the overgrowth of harmful microbes. Technically speaking, the normal microbiota are not normally considered part of the innate immune system, but they are discussed here because of the considerable protection they afford. Basically, the normal microbiota provide resistance to disease in three principal ways.

1. They are well adapted to a limited number of attachment sites on which they live and have a competitive advantage over pathogenic microbes for these colonization sites by competing for available space and nutrients (competitive exclusion). This colonization resistance is especially effective against microbes such as *Clostridium difficile*, *Salmonella*, *Shigella*, and *Candida albican*.

2. Normal microbiota produce substances that inhibit or kill pathogens. For example, *E. coli* bacteria in the large intestines produce bacteriocins that inhibit or kill bacteria of the same or closely related species. Some normal microbiota, such as *Lactobacillus* in the vagina, produce hydrogen peroxide under anaerobic conditions. This has shown to be effective against infections caused by *Chlamydia trachomatis*, *Gardnerella vaginalis*, and *Candida albicans*.

3. Development of the immune system is dependent on the presence of microbiota even before birth. See Exploring the Microbiome (page 452).

In **commensalism**, one organism uses the body of a larger organism as its physical environment and may make use of the body to obtain nutrients. Thus in commensalism, one organism benefits while the other is unaffected. Most microbes that are part of the commensal microbiota are found on the skin and in the gastrointestinal tract. The majority of such microbes are bacteria that have highly specialized attachment mechanisms

# The Microbiome's Shaping of Innate Immunity

Until recently, colonization with microbes at and after birth was assumed to be the main stimulus to the development of our immune system. However, we now know that the development of myeloid cells is influenced by microbiota, and this influence actually begins before birth.

Studies show us that offspring of mice that are treated with antibiotics during pregnancy have lower numbers of blood neutrophils than offspring of mothers who didn't receive antibiotics. Intestinal colonization with microorganisms in a pregnant mouse increases the number of monocytes in newborn mice, too. The continuous presence of microbiota-derived Toll-like receptors also causes neutrophil aging. Aged neutrophils express a receptor that allows their clearance in the bone marrow. Ongoing studies are trying to determine whether microbiota cells directly influence the development of myeloid cells or whether these changes might be due to metabolic products produced by the microbes.

Studies on germ-free mice, which are bred in sterile environments, show even starker immune system changes. Germ-free animals have underdeveloped immune systems compared to mice raised normally, exposed to microbes. The germ-free mice have little lymphoid tissue and very low levels of immune proteins in their body fluids. When germ-free mice do come in contact with microorganisms, the animals are unusually susceptible to infection and serious disease. Normal development of myeloid cells in the bone marrow is reduced, which results in fewer phagocytes and a delayed response to bacterial infections.

Strict asepsis is required to raise germ-free mice.

and precise environmental requirements for survival. Normally, such microbes are harmless, but they may cause disease if their environmental conditions change. These opportunistic pathogens include *E. coli*, *S. aureus*, *S. epidermidis*, *Enterococcus faecalis*, *Pseudomonas aeruginosa*, and oral streptococci.

Recent interest in the importance of bacteria to human health has led to the study of probiotics. **Probiotics** (*pro* = for, *bios* = life) are live microbial cultures applied to or ingested that are intended to exert a beneficial effect. Probiotics may be administered with *prebiotics*, which are chemicals that selectively promote the growth of beneficial bacteria. If lactic acid bacteria (LAB) colonize the large intestine, the lactic acid and bacteriocins they produce can inhibit the growth of certain pathogens. Several studies have shown that ingesting certain LAB can alleviate diarrhea and prevent colonization by *Salmonella enterica* during antibiotic therapy. Researchers are also testing the use of LAB to prevent surgical wound infections caused by *S. aureus* and vaginal infections caused by *E. coli*. In a Stanford University study, HIV infection was reduced in

women treated with LAB that were genetically modified to produce CD4 protein that binds to HIV. Results of several studies suggest that giving probiotics with antibiotics reduces the risk of developing *Clostridium difficile*–associated diarrhea. However, probiotics may not always work. A recent French study found that although use of probiotics reduced incidence of pneumonia acquired in intensive care units (ICUs) and length of ICU stay, probiotics did not significantly reduce hospital mortality rates.

## CHECK YOUR UNDERSTANDING

✔ **16-3** Identify one physical factor and one chemical factor that prevent microbes from entering the body through skin and mucous membranes.

✔ **16-4** Identify one physical factor and one chemical factor that prevent microbes from entering or colonizing the body through the eyes, digestive tract, and respiratory tract.

✔ **16-5** Distinguish microbial antagonism from commensalism.

## Second Line of Defense

When microbes penetrate the first line of defense, they encounter a second line of defense that includes defensive cells, such as phagocytic cells; inflammation; fever; and antimicrobial substances.

Before we look at the phagocytic cells, it will be helpful to first have an understanding of the cellular components of blood.

## Formed Elements in Blood

### LEARNING OBJECTIVES

**16-6**  Classify leukocytes, and describe the roles of granulocytes and monocytes.

**16-7**  Describe the eight different types of white blood cells, and name a function for each type.

Blood consists of fluid, called **plasma**, and **formed elements**—that is, cells and cell fragments suspended in plasma. The formed elements include **erythrocytes**, or **red blood cells (RBCs)**; **leukocytes**, or **white blood cells (WBCs)**; and **platelets**. Formed elements are created in red bone marrow by stem cells in a process called **hematopoiesis** (hēm-a-tō-POY-ē-sis). It begins when a cell called a *pluripotent stem cell* develops into two other types of cells called *myeloid stem cells* and *lymphoid stem cells*. From these two stem cells, all of the formed elements develop. All of these blood cells are shown in **Figure 16.4**. More detailed descriptions of the formed elements that concern us most for innate immunity—the leukocytes—are found in Table 16.1.

Leukocytes are divided into two main categories based on their appearance under a light microscope: granulocytes and agranulocytes. **Granulocytes** owe their name to the presence of large granules in their cytoplasm that can be seen with a light microscope after staining. They are differentiated into three types of cells based on how the granules stain: neutrophils, basophils, and eosinophils. **Neutrophils** stain pale lilac with a mixture of acidic and basic dyes. Neutrophils are also commonly called *polymorphonuclear leukocytes (PMNs)*, or *polymorphs*. (The term *polymorphonuclear* refers to the fact that the nuclei of neutrophils contain two to five lobes.) Neutrophils, which are highly

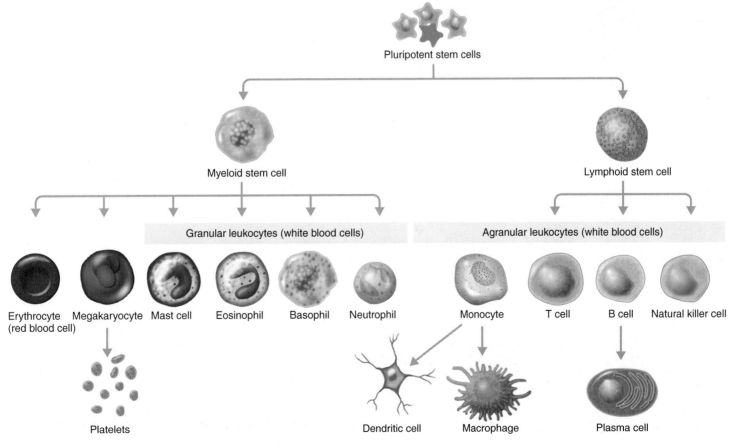

**Figure 16.4  Hematopoiesis.**  The process begins in red bone marrow with a pluripotent stem cell.

**Q** What are the parent cells for granular leukocytes? For agranular leukocytes?

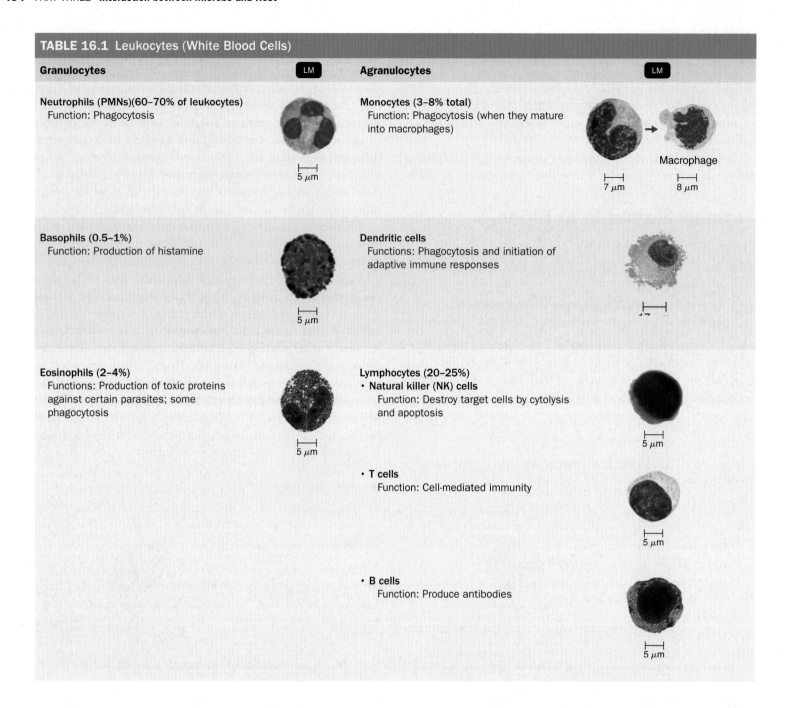

**TABLE 16.1** Leukocytes (White Blood Cells)

| Granulocytes | LM | Agranulocytes | LM |
|---|---|---|---|
| **Neutrophils (PMNs)(60–70% of leukocytes)**<br>Function: Phagocytosis | 5 μm | **Monocytes (3–8% total)**<br>Function: Phagocytosis (when they mature into macrophages) | Macrophage<br>7 μm  8 μm |
| **Basophils (0.5–1%)**<br>Function: Production of histamine | 5 μm | **Dendritic cells**<br>Functions: Phagocytosis and initiation of adaptive immune responses | |
| **Eosinophils (2–4%)**<br>Functions: Production of toxic proteins against certain parasites; some phagocytosis | 5 μm | **Lymphocytes (20–25%)**<br>• **Natural killer (NK) cells**<br>  Function: Destroy target cells by cytolysis and apoptosis | 5 μm |
| | | • **T cells**<br>  Function: Cell-mediated immunity | 5 μm |
| | | • **B cells**<br>  Function: Produce antibodies | 5 μm |

phagocytic and motile, are active in the initial stages of an infection. They have the ability to leave the blood, enter an infected tissue, and destroy microbes and foreign particles.

**Basophils** stain blue-purple with the basic dye methylene blue. Basophils release substances, such as histamine, that are important in inflammation and allergic responses.

**Eosinophils** stain red or orange with the acidic dye eosin. Eosinophils are somewhat phagocytic and also have the ability to leave the blood. Their major function is to kill certain parasites, such as helminths. Although eosinophils are physically too small to ingest and destroy helminths, they can attach to the outer surface of the parasites and discharge peroxide ions that destroy them (see Figure 17.16, page 492). Their number increases significantly during certain parasitic worm infections and hypersensitivity (allergy) reactions.

**Agranulocytes** also have granules in their cytoplasm, but the granules are not visible under the light microscope after staining. There are three different types of agranulocytes: monocytes, dendritic cells, and lymphocytes. **Monocytes** are not actively phagocytic until they leave circulating blood, enter body tissues, and mature into **macrophages**. In fact, the proliferation of lymphocytes is one factor responsible for the swelling of lymph nodes during an infection. As blood and lymph that contain microorganisms pass through organs with macrophages, the microorganisms are removed by phagocytosis. Macrophages also dispose of worn out blood cells.

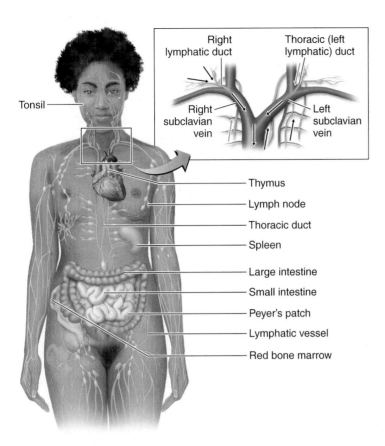

Figure labels:
- Tonsil
- Right lymphatic duct
- Thoracic (left lymphatic) duct
- Right subclavian vein
- Left subclavian vein
- Thymus
- Lymph node
- Thoracic duct
- Spleen
- Large intestine
- Small intestine
- Peyer's patch
- Lymphatic vessel
- Red bone marrow

**Figure 16.5 The lymphatic system is shown in green.** Inset: The arrows indicate the direction of lymph flow.

**Q** Why do lymph nodes swell during an infection?

**Dendritic cells** are also believed to be derived from the same precursor cells as monocytes. They have long extensions that resemble the dendrites of nerve cells, thus their name. Dendritic cells are especially abundant in the epidermis of the skin, mucous membranes, the thymus, and lymph nodes. Dendritic cells destroy microbes by phagocytosis and initiate adaptive immune responses (see Chapter 17, page 476).

**Lymphocytes** include natural killer cells, T cells, and B cells. **Natural killer (NK) cells** are found in blood and in the spleen, lymph nodes, and red bone marrow. NK cells have the ability to kill a wide variety of infected body cells and certain tumor cells. NK cells attack any body cells that display abnormal or unusual plasma membrane proteins. The binding of NK cells to a target cell, such as an infected human cell, causes the release of toxic substances from lytic granules in NK cells. *Lytic granules* are a secretory organelle unique to NK cells. Some granules contain a protein called **perforin**, which inserts into the plasma membrane of the target cell and creates channels (perforations) in the membrane. As a result, extracellular fluid flows into the target cell and the cell bursts, a process called **cytolysis** (sī-TOL-i-sis; *cyto-* = cell; *-lysis* = loosening). Other granules of NK cells release **granzymes**, which are protein-digesting enzymes that induce the target cell to undergo apoptosis, or self-destruction. This type of attack kills infected cells but not the microbes inside the cells; the released microbes, which may or may not be intact, can be destroyed by phagocytes.

**T cells** and **B cells** are not usually phagocytic but play a key role in adaptive immunity (see Chapter 17). They occur in lymphoid tissues of the lymphatic system and also circulate in blood.

During many kinds of infections, especially bacterial infections, the total number of white blood cells increases as a protective response to combat the microbes; this increase is called *leukocytosis*. During the active stage of infection, the leukocyte count might double, triple, or quadruple, depending on the severity of the infection. Diseases that might cause such an elevation in the leukocyte count include meningitis, infectious mononucleosis, appendicitis, pneumococcal pneumonia, and gonorrhea. Other diseases, such as salmonellosis and brucellosis, and some viral and rickettsial infections may cause a *decrease* in the leukocyte count, called *leukopenia*. Leukopenia may be related to either impaired white blood cell production or the effect of increased sensitivity of white blood cell membranes to damage by complement, antimicrobial serum proteins discussed later in the chapter. Leukocyte increase or decrease can be detected by a **differential white blood cell count**, which is a calculation of the percentage of each kind of white cell in a sample of 100 white blood cells. The percentages in a normal differential white blood cell count are shown in parentheses in Table 16.1.

## The Lymphatic System

### LEARNING OBJECTIVE

16-8 Differentiate the lymphatic and blood circulatory systems.

The **lymphatic system** consists of a fluid called *lymph*, vessels called *lymphatic vessels*, a number of structures and organs containing *lymphoid tissue*, and *red bone marrow*, where stem cells develop into blood cells, including lymphocytes (**Figure 16.5**). Lymphoid tissue

---

**CLINICAL CASE**

Jacob's pediatrician also sends Jacob's blood to the lab for a complete blood count (CBC). The results are shown below:

| Red blood cells | 4 million/μl |
| --- | --- |
| Neutrophils | 9700/μl |
| Basophils | 200/μl |
| Eosinophils | 600/μl |
| Monocytes | 1140/μl |

**Which cells should be protecting Jacob from infection? Based on the results of the CBC, how does Jacob's pediatrician know that something is wrong?**

449 **455** 459 463 469 470

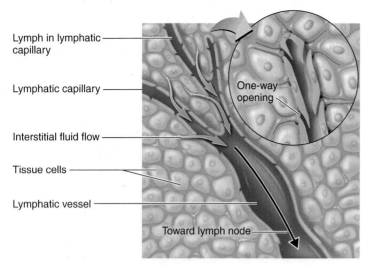

Lymph in lymphatic capillary

Lymphatic capillary

One-way opening

Interstitial fluid flow

Tissue cells

Lymphatic vessel

Toward lymph node

**(a)** Lymphatic capillaries and lymphatic vein

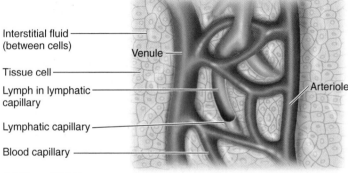

Interstitial fluid (between cells)

Venule

Tissue cell

Lymph in lymphatic capillary

Arteriole

Lymphatic capillary

Blood capillary

**(b)** Flow of fluid between arteriole, blood capillaries, lymphatic capillaries, and venule

**Figure 16.6 Lymphatic capillaries.** Fluid circulating between tissue cells (interstitial fluid) is picked up by lymphatic capillaries.

**Q** Where does lymph fluid go?

contains large numbers of lymphocytes, including T cells, B cells, and phagocytic cells that participate in immune responses. Lymph nodes are the sites of activation of T cells and B cells, which destroy microbes by immune responses (Chapter 17). Also within lymph nodes are reticular fibers, which trap microbes, and macrophages and dendritic cells, which destroy microbes by phagocytosis.

Lymphatic vessels begin as microscopic *lymphatic capillaries* located in spaces between cells (**Figure 16.6**). The lymphatic capillaries permit interstitial fluid derived from blood plasma to flow into them, but not out. Within the lymphatic capillaries, the fluid is called lymph. Lymphatic capillaries converge to form larger lymphatic vessels. These vessels, like veins, have one-way valves to keep lymph flowing in one direction only. At intervals along the lymphatic vessels, lymph flows through bean-shaped *lymph nodes* (Figure 16.5). All lymph eventually passes into the *thoracic (left lymphatic) duct* and *right lymphatic duct* and then into their respective subclavian veins, where the fluid is now called blood plasma. The blood plasma moves through the cardiovascular system and ultimately becomes interstitial fluid between tissue cells, and another cycle begins.

Lymphoid tissues and organs are scattered throughout the mucous membranes that line the gastrointestinal, respiratory, urinary, and reproductive tracts. They protect against microbes that are ingested or inhaled. Multiple large aggregations of lymphoid tissues are located in specific parts of the body. These include the *tonsils* in the throat and *Peyer's patches* in the small intestine. See Figure 17.9, page 486.

The *spleen* contains lymphocytes and macrophages that monitor the blood for microbes and secreted products such as toxins, much like lymph nodes monitor lymph. The *thymus* serves as a site for T cell maturation. It also contains dendritic cells and macrophages.

**Play Host Defenses:** *Overview*
@MasteringMicrobiology

**CHECK YOUR UNDERSTANDING**

✔ **16-6** Compare the structures and function of monocytes and neutrophils.

✔ **16-7** Define *differential white blood cell count.*

✔ **16-8** What is the function of lymph nodes?

# Phagocytes

**LEARNING OBJECTIVES**

**16-9** Define *phagocyte* and *phagocytosis.*

**16-10** Describe the process of phagocytosis, and include the stages of adherence and ingestion.

**Phagocytosis** (from Greek words meaning eat and cell) is the ingestion of a microorganism or other substance by a cell. We have previously mentioned phagocytosis as the method of nutrition of certain protozoa. Phagocytosis is also involved in clearing away debris such as dead body cells and denatured proteins. In this chapter, phagocytosis is discussed as a means by which cells in the human body counter infection as part of the second line of defense.

## Actions of Phagocytic Cells

Cells that perform phagocytosis are collectively called **phagocytes**. All are types of white blood cells or white blood cell derivatives. When an infection occurs, both granulocytes (especially neutrophils, but also eosinophils) and monocytes migrate to the infected area. These leave the blood and migrate into tissues where they enlarge and develop into macrophages. Some macrophages, called **fixed macrophages**, or *histiocytes* are resident in certain tissues and organs of the body. Fixed macrophages are found in the liver (Kupffer's cells), lungs (alveolar macrophages), nervous system (microglial cells), bronchial tubes, spleen (splenic macrophages), lymph nodes, red bone marrow, and the peritoneal cavity surrounding abdominal organs (peritoneal macrophages). Other macrophages are motile and are called **free (wandering) macrophages**, which roam the tissues and gather at sites of infection or

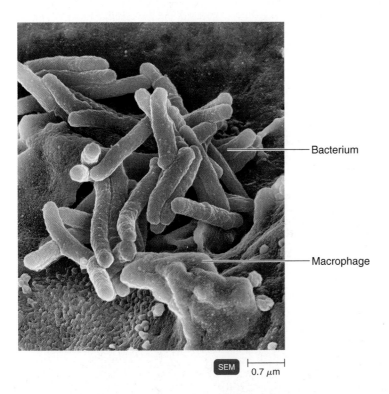

Bacterium

Macrophage

SEM | 0.7 μm

**Figure 16.7 A macrophage engulfing rod-shaped bacteria.**
Macrophages in the mononuclear phagocytic system remove microorganisms after the initial phase of infection.

**Q** What are monocytes?

inflammation. The various macrophages of the body constitute the **mononuclear phagocytic (reticuloendothelial) system**.

During the course of an infection, a shift occurs in the type of white blood cell that predominates in the bloodstream. Granulocytes, especially neutrophils, dominate during the initial phase of bacterial infection, at which time they are actively phagocytic; this dominance is indicated by their increased number in a differential white blood cell count. However, as the infection progresses, the macrophages dominate; they scavenge and phagocytize remaining living bacteria and dead or dying bacteria (**Figure 16.7**). The increased number of monocytes (which develop into macrophages) is also reflected in a differential white blood cell count.

## The Mechanism of Phagocytosis

How does phagocytosis occur? For the convenience of study, we will divide phagocytosis into four main phases: chemotaxis, adherence, ingestion, and digestion (**Figure 16.8**).

### Chemotaxis and Adherence

❶ **Chemotaxis** is the chemical attraction of phagocytes to microorganisms. (The mechanism of chemotaxis is discussed in Chapter 4, page 78.) Among the chemotactic chemicals that attract phagocytes are microbial products, components of white blood cells and damaged tissue cells,

cytokines released by other white blood cells, and, finally, peptides derived from complement—a system of host defense proteins discussed later in the chapter.

As it pertains to phagocytosis, **adherence** is the attachment of the phagocyte's plasma membrane to the surface of the microorganism or other foreign material. Adherence is facilitated by the attachment of pathogen-associated molecular patterns (PAMPs) of microbes to receptors, such as Toll-like receptors (TLRs), on the surface of phagocytes. The binding of PAMPs to TLRs not only initiates phagocytosis, but also induces the phagocyte to release specific cytokines that recruit additional phagocytes.

In some instances, adherence occurs easily, and the microorganism is readily phagocytized. Microorganisms can be more readily phagocytized if they are first coated with certain serum proteins that promote attachment of the microorganisms to the phagocyte. This coating process is called **opsonization**. The proteins that act as *opsonins* include some components of the complement system and antibody molecules (described later in this chapter and in Chapter 17).

### Ingestion

❷ Following adherence, **ingestion** occurs. The plasma membrane of the phagocyte extends projections called **pseudopods** that engulf the microorganism. (See also Figure 16.7.)

❸ Once the microorganism is surrounded, the pseudopods meet and fuse, surrounding the microorganism with a sac called a **phagosome**, or *phagocytic vesicle*. The membrane of a phagosome has enzymes that pump protons ($H^+$) into the phagosome, reducing the pH to about 4. At this pH, hydrolytic enzymes are activated.

### Phagolysosome Formation and Digestion

Next, the phagosome pinches off from the plasma membrane and enters the cytoplasm, where it contacts lysosomes that contain digestive enzymes and bactericidal substances (see Chapter 4, page 100).

❹ On contact, the phagosome and lysosome membranes fuse to form a single, larger structure called a **phagolysosome**.

❺ The contents of the phagolysosome brought in by ingestion are digested in the phagolysosome.

Lysosomal enzymes that attack microbial cells directly include lysozyme, which hydrolyzes peptidoglycan in bacterial cell walls. Lipases, proteases, ribonuclease, and deoxyribonuclease hydrolyze other macromolecular components of microorganisms. Lysosomes also contain enzymes that can produce toxic oxygen products such as superoxide radical ($O_2^-$), hydrogen peroxide ($H_2O_2$), nitric oxide (NO), singlet oxygen ($^1O_2^-$), and hydroxyl radical (OH·) (see Chapter 6, pages 156–157).

# The Phases of Phagocytosis

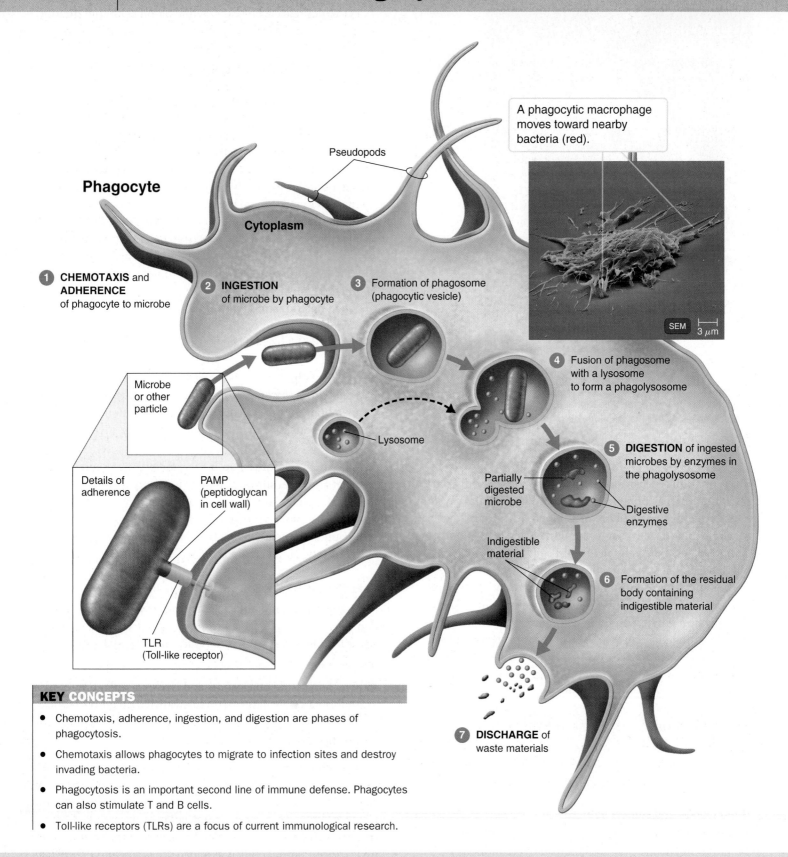

Pseudopods

**Phagocyte**

Cytoplasm

A phagocytic macrophage moves toward nearby bacteria (red).

SEM  3 μm

**1** **CHEMOTAXIS** and **ADHERENCE** of phagocyte to microbe

**2** **INGESTION** of microbe by phagocyte

**3** Formation of phagosome (phagocytic vesicle)

**4** Fusion of phagosome with a lysosome to form a phagolysosome

Microbe or other particle

Lysosome

**5** **DIGESTION** of ingested microbes by enzymes in the phagolysosome

Partially digested microbe

Digestive enzymes

Indigestible material

**6** Formation of the residual body containing indigestible material

Details of adherence

PAMP (peptidoglycan in cell wall)

TLR (Toll-like receptor)

**7** **DISCHARGE** of waste materials

## KEY CONCEPTS

- Chemotaxis, adherence, ingestion, and digestion are phases of phagocytosis.

- Chemotaxis allows phagocytes to migrate to infection sites and destroy invading bacteria.

- Phagocytosis is an important second line of immune defense. Phagocytes can also stimulate T and B cells.

- Toll-like receptors (TLRs) are a focus of current immunological research.

Leukocytes, including neutrophils and macrophages, are the cells listed in the test results that fight infections. According to the lab results, Jacob's white blood cell count is a little high. Leukocytosis, a white blood cell count above normal range, occurs during a fungal infection, but Jacob's pediatrician is concerned that the leukocytes are not doing their job. To check this, he then requests a nitroblue tetrazolium (NBT) test, which is performed on a blood smear on a microscope slide.

Normal neutrophils will reduce the yellow dye, NBT, to an insoluble blue precipitate. Jacob's neutrophils do not produce this result; his neutrophils are not working as they should. Normally, the adherence of a target cell, such as a bacterium, to the neutrophil's plasma membrane stimulates the neutrophil to produce NADPH. This is followed by a lethal oxidative burst of hydrogen peroxide.

**What metabolic pathway produces NADPH for a cell?**

449  455  **459**  463  469  470

Toxic oxygen products are produced by an *oxidative burst*. Other enzymes can make use of these toxic oxygen products in killing ingested microorganisms. For example, the enzyme myeloperoxidase converts chloride ($Cl^-$) ions and hydrogen peroxide into highly toxic hypochlorous acid (HOCl). The acid contains hypochlorite ions, which are found in household bleach and account for its antimicrobial activity (see Chapter 7, pages 191–192).

**6** After enzymes have digested the contents of the phagolysosome brought into the cell by ingestion, the phagolysosome contains indigestible material and is called a *residual body.*

**7** This residual body then moves toward the cell boundary and discharges its wastes outside the cell.

> ▶ **Play Virulence Factors:** *Hiding from Host Defenses, Inactivating Host Defenses;* **Phagocytosis:** *Overview, Mechanism* @MasteringMicrobiology

**CHECK YOUR UNDERSTANDING**

✔ **16-9** What do fixed and wandering macrophages do?

✔ **16-10** What is the role of TLRs in phagocytosis?

\* \* \*

In addition to providing innate resistance for the host, phagocytosis plays a role in adaptive immunity. Macrophages help T and B cells perform vital adaptive immune functions—this will be discussed in Chapter 17.

In the next section, we will see how phagocytosis often occurs as part of another innate mechanism of resistance: inflammation.

# Inflammation

**LEARNING OBJECTIVES**

**16-11** List the stages of inflammation.

**16-12** Describe the roles of vasodilation, kinins, prostaglandins, and leukotrienes in inflammation.

**16-13** Describe phagocyte migration.

Damage to the body's tissues triggers a local defensive response called **inflammation**, another component of the second line of defense. The damage can be caused by microbial infection, physical agents (such as heat, radiant energy, electricity, or sharp objects), or chemical agents (acids, bases, and gases). Certain signs and symptoms are associated with inflammation, which you can remember by thinking of the acronym **PRISH**:

*Pain* due to the release of certain chemicals.

*Redness* because more blood goes to the affected area.

*Immobility* that results from local loss of function in severe inflammations.

*Swelling* caused by an accumulation of fluids.

*Heat,* which is also due to an increase in blood flow to the affected area.

Inflammation has the following functions: (1) to destroy the injurious agent, if possible, and to remove it and its by-products from the body; (2) if destruction is not possible, to limit the effects on the body by confining or walling off the injurious agent and its by-products; and (3) to repair or replace tissue damaged by the injurious agent or its by-products.

Inflammation can be classified as acute or chronic, depending on a number of factors. In **acute inflammation**, the signs and symptoms develop rapidly and usually last for a few days or even a few weeks. It is usually mild and self-limiting, and the principal defensive cells are neutrophils. Examples of acute inflammation are a sore throat, appendicitis, cold or flu, bacterial preumonia, and a scratch on the skin. In **chronic inflammation**, the signs and symptoms develop more slowly and can last for up to several months or years. It is often severe and progressive, and the principal defensive cells are monocytes and macrophages. Examples of chronic inflammation are mononucleosis, peptic ulcers, tuberculosis, rheumatoid arthritis, and ulcerative colitis.

During the early stages of inflammation, microbial structures, such as flagellin, lipopolysaccharides (LPS), and bacterial DNA stimulate the Toll-like receptors of macrophages to produce cytokines, such as *tumor necrosis factor alpha (TNF-α)*. In response to TNF-α in the blood, the liver synthesizes a group of proteins called **acute-phase proteins**; other acute-phase proteins are present in the blood in an inactive form and are converted to an active form during inflammation. Acute-phase proteins induce both local and systemic responses and include proteins such as C-reactive protein, mannose-binding lectin (page 465), and several specialized proteins such as fibrinogen for blood clotting and kinins for vasodilation.

All of the cells involved in inflammation have receptors for TNF-α and are activated by it to produce more of their own TNF-α. This amplifies the inflammatory response. Unfortunately, excessive production of TNF-α may lead to disorders such as rheumatoid arthritis and Crohn's disease. Monoclonal antibodies are used therapeutically to treat such inflammatory disorders (see Chapter 18, pages 508–509).

For purposes of our discussion, we will divide the process of inflammation into three stages: vasodilation and increased permeability of blood vessels, phagocyte migration and phagocytosis, and tissue repair.

## Vasodilation and Increased Permeability of Blood Vessels

Immediately following tissue damage (**Figure 16.9a**), blood vessels dilate (increase in diameter) in the area of damage, and their permeability increases (Figure 16.9b). Dilation of blood vessels, called **vasodilation**, is responsible for the redness (erythema) and heat associated with inflammation.

**Increased permeability** permits defensive substances normally retained in the blood to pass through the walls of the blood vessels and enter the injured area. The increase in permeability, which permits fluid to move from the blood into tissue spaces, is responsible for the **edema** (accumulation of fluid) of inflammation. The pain of inflammation can be caused by nerve damage, irritation by toxins, or the pressure of edema.

❶ Vasodilation and the increase in permeability of blood vessels are caused by a number of chemicals released by damaged cells in response to injury. These chemicals are called **vasoactive mediators**. One such substance is **histamine**, a chemical present in many cells of the body, especially in mast cells in connective tissue, circulating basophils, and blood platelets. Histamine is released in direct response to the injury of cells that contain it; it is also released in response to stimulation by certain components of the complement system (discussed later).

**Figure 16.9 The process of inflammation. (a)** Damage to otherwise healthy tissue—in this case, skin. **(b)** Vasodilation and increased permeability of blood vessels allows phagocyte migration. Phagocytosis by macrophages and neutrophils removes bacteria and cellular debris. Macrophages develop from monocytes. **(c)** The repair of damaged tissue.

**Q** What are the signs and symptoms of inflammation?

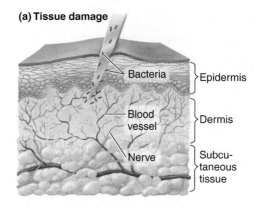

**(a) Tissue damage**

Bacteria — Epidermis
Blood vessel — Dermis
Nerve — Subcutaneous tissue

**(b) Vascular reactions and phagocytosis**

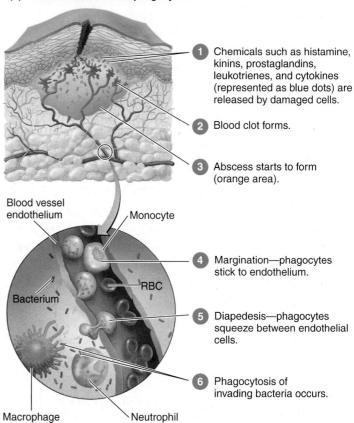

❶ Chemicals such as histamine, kinins, prostaglandins, leukotrienes, and cytokines (represented as blue dots) are released by damaged cells.

❷ Blood clot forms.

❸ Abscess starts to form (orange area).

Blood vessel endothelium — Monocyte

❹ Margination—phagocytes stick to endothelium.

Bacterium — RBC

❺ Diapedesis—phagocytes squeeze between endothelial cells.

Macrophage — Neutrophil

❻ Phagocytosis of invading bacteria occurs.

**(c) Tissue repair**

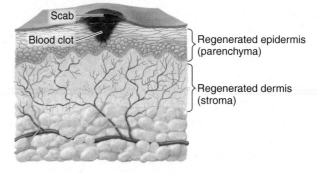

Scab
Blood clot — Regenerated epidermis (parenchyma)
Regenerated dermis (stroma)

Phagocytic granulocytes attracted to the site of injury can also produce chemicals that cause the release of histamine.

**Kinins** are another group of substances that cause vasodilation and increased permeability of blood vessels. Kinins are present in blood plasma, and once activated, they play a role in chemotaxis by attracting phagocytic granulocytes, chiefly neutrophils, to the injured area.

**Prostaglandins**, substances released by damaged cells, intensify the effects of histamine and kinins and help phagocytes move through capillary walls. Despite their positive role in the inflammatory process, prostaglandins are also associated with the pain related to inflammation. **Leukotrienes** are substances produced by mast cells (cells especially numerous in the connective tissue of the skin and respiratory system, and in blood vessels) and basophils. Leukotrienes cause increased permeability of blood vessels and help attach phagocytes to pathogens. Various components of the *complement system* produced by the liver stimulate the release of histamine, attract phagocytes, and promote phagocytosis.

Activated fixed macrophages also secrete **cytokines**, which bring about vasodilation and increased permeability. Vasodilation and the increased permeability of blood vessels also help deliver clotting elements of blood into the injured area.

❷ The blood clots that form around the site of activity prevent the microbe (or its toxins) from spreading to other parts of the body.

❸ As a result, there may be a localized collection of **pus**, a mixture of dead cells and body fluids, in a cavity formed by the breakdown of body tissues. This focus of infection is called an **abscess**. Common abscesses include pustules and boils.

A summary of vasoactive mediators in inflammation is presented in Table 16.2.

The next stage in inflammation involves the migration of phagocytes to the injured area.

## Phagocyte Migration and Phagocytosis

Generally, within an hour after the process of inflammation is initiated, phagocytes appear on the scene. ❹ As the flow of blood gradually decreases, phagocytes (both neutrophils and monocytes) begin to stick to the inner surface of the endothelium (lining) of blood vessels. This sticking process in response to local cytokines is called **margination**. The cytokines alter cellular adhesion molecules on cells lining blood vessels, causing the phagocytes to stick at the site of inflammation. (Margination is also involved in red bone marrow, where cytokines can release phagocytes into circulation when they are needed.) ❺ Then the collected phagocytes begin to squeeze between the endothelial cells of the blood vessel to reach the damaged area. This migration, which resembles ameboid movement, is called **diapedesis**; it can take as little as 2 minutes. ❻ The phagocytes then begin to destroy invading microorganisms by phagocytosis.

As mentioned earlier, certain chemicals attract neutrophils to the site of injury (chemotaxis). These include chemicals produced by microorganisms and even other neutrophils; other chemicals are kinins, leukotrienes, chemokines, and components of the complement system. Chemokines are cytokines that are chemotactic for phagocytes and T cells and thus stimulate both the inflammatory response and an adaptive immune response. The availability of a steady stream of neutrophils is ensured by the production and release of additional granulocytes from red bone marrow.

As the inflammatory response continues, monocytes follow the granulocytes into the infected area. Once the

| **TABLE 16.2** Summary of Vasoactive Mediators of Inflammation | | |
|---|---|---|
| **Vasoactive Mediator** | **Source** | **Effect** |
| Histamine | Mast cells, basophils, and platelets | Vasodilation and increased permeability of blood vessels |
| Kinins | Blood plasma | Chemotaxis by attracting neutrophils |
| Prostaglandins | Damaged cells | Intensify the effects of histamine and kinins and help phagocytes move through capillary walls |
| Leukotrienes | Mast cells and basophils | Increase permeability of blood vessels and help attach phagocytes to pathogens |
| Complement (See Figure 16.12) | Blood plasma | Stimulates release of histamine, attracts phagocytes, and promotes phagocytosis |
| Cytokines | Fixed macrophages | Vasodilation and increased permeability of blood vessels |

monocytes are contained in the tissue, they undergo changes in biological properties and become free macrophages. The granulocytes predominate in the early stages of infection but tend to die off rapidly. Macrophages enter the picture during a later stage of the infection, once granulocytes have accomplished their function. They are several times more phagocytic than granulocytes and are large enough to phagocytize tissue that has been destroyed, granulocytes that have been destroyed, and invading microorganisms.

After granulocytes or macrophages engulf large numbers of microorganisms and damaged tissue, they themselves eventually die. As a result, pus forms, and its formation usually continues until the infection subsides. At times, the pus pushes to the surface of the body or into an internal cavity for dispersal. On other occasions the pus remains even after the infection is terminated. In this case, the pus is gradually destroyed over a period of days and is absorbed by the body.

As effective as phagocytosis is in contributing to innate resistance, there are times when the mechanism becomes less functional in response to certain conditions. For example, with age, there is a progressive decline in the efficiency of phagocytosis. Recipients of heart or kidney transplants have impaired innate defenses as a result of receiving drugs that prevent the rejection of the transplant. Radiation treatments can also depress innate immune responses by damaging red bone marrow. Even certain diseases such as AIDS and cancer can cause defective functioning of innate defenses. Finally, individuals with certain genetic disorders produce fewer or impaired phagocytes.

## Tissue Repair

The final stage of inflammation is tissue repair, the process by which tissues replace dead or damaged cells (Figure 16.9c). Repair begins during the active phase of inflammation, but it cannot be completed until all harmful substances have been removed or neutralized at the site of injury. The ability to regenerate, or repair, depends on the type of tissue. For example, skin has a high capacity for regeneration, whereas cardiac muscle tissue has a low capacity to regenerate.

A tissue is repaired when its stroma or parenchyma produces new cells. The *stroma* is the supporting connective tissue, and the *parenchyma* is the functioning part of the tissue. For example, the capsule around the liver that encloses and protects it is part of the stroma because it is not involved in the functions of the liver; liver cells (hepatocytes) that perform the functions of the liver are part of the parenchyma. If only parenchymal cells are active in repair, a perfect or near-perfect reconstruction of the tissue occurs. A familiar example of perfect reconstruction is a minor skin cut, in which

parenchymal cells are more active in repair. However, if repair cells of the stroma of the skin are more active, scar tissue is formed.

As noted earlier, some microbes have various mechanisms that enable them to evade phagocytosis. Such microbes often induce a chronic inflammatory response, which can result in significant damage to body tissues. The most significant feature of chronic inflammation is the accumulation and activation of macrophages in the infected area. Cytokines released by activated macrophages induce fibroblasts in the tissue stroma to synthesize collagen fibers. These fibers aggregate to form scar tissue, a process called *fibrosis*. Because scar tissue is not specialized to perform the functions of the previously healthy tissue, fibrosis can interfere with the normal function of the tissue.

Play Inflammation:
*Overview, Steps*
@Mastering**Microbiology**

# Fever

**LEARNING OBJECTIVE**

16-14 Describe the cause and effects of fever.

While inflammation is a local response of the body to injury, there are also systemic, or overall, responses. One of the most important is **fever**, an abnormally high body temperature, a third component of the second line of defense. The most frequent cause of fever is infection from bacteria (and their toxins) or viruses.

The brain's hypothalamus is sometimes called the body's thermostat, and it is normally set at 37°C (98.6°F). It is believed that certain substances affect the hypothalamus by setting it at a higher temperature. Recall from Chapter 15 that when phagocytes ingest gram-negative bacteria, the lipopolysaccharides (LPS) of the cell wall are released. LPS causes the phagocytes to release the cytokines interleukin-1 along with TNF-α. These cytokines cause the hypothalamus to release prostaglandins that reset the hypothalamic thermostat at a higher temperature, thereby causing fever (see Figure 15.6, page 434).

The body will continue to maintain the higher temperature until the cytokines are eliminated. The thermostat is then reset to 37°C. As the infection subsides, heat-losing mechanisms such as vasodilation and sweating go into operation. This phase of the fever, called the **crisis**, indicates that body temperature is falling.

Up to a certain point, fever is considered a defense against disease. Interleukin-1 helps step up the production of T cells. High body temperature intensifies the effect of antiviral interferons (page 467) and increases production of transferrins that decrease the iron available to microbes (page 468). Also, because the high temperature speeds up the body's reactions, it may help body tissues repair themselves more quickly. The higher temperature may slow the growth rate of some bacteria.

Among the complications of fever are tachycardia (rapid heart rate), which may compromise older persons with cardiopulmonary disease; increased metabolic rate, which may produce acidosis; dehydration; electrolyte imbalances; seizures in young children; and delirium and coma. As a rule, death results if body temperature rises above 44° to 46°C (112° to 114°F).

---

**CHECK YOUR UNDERSTANDING**

✔ 16-14 How is fever beneficial?

---

**CLINICAL CASE**

The pentose phosphate pathway produces the NADPH. (See figure.) Jacob's pediatrician realizes that Jacob's neutrophils must be missing NADPH oxidase and can't oxidize the NADPH. He diagnoses Jacob with chronic granulomatous disease (CGD), an inherited X-linked recessive disorder in which phagocytes do not function as they should. It is caused by a mutation in the gene for NADPH oxidase.

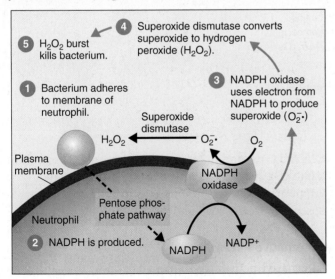

① Bacterium adheres to membrane of neutrophil.
② NADPH is produced.
③ NADPH oxidase uses electron from NADPH to produce superoxide ($O_2^-\cdot$)
④ Superoxide dismutase converts superoxide to hydrogen peroxide ($H_2O_2$).
⑤ $H_2O_2$ burst kills bacterium.

**What is the function of NADPH oxidase? (*Hint:* See page 113.)**

---

# Antimicrobial Substances

## LEARNING OBJECTIVES

**16-15** List the major components of the complement system.

**16-16** Describe three pathways of activating complement.

**16-17** Describe three consequences of complement activation.

**16-18** Define *interferons*.

**16-19** Compare and contrast the actions of IFN-α and IFN-β with IFN-γ.

**16-20** Describe the role of iron-binding proteins in innate immunity.

**16-21** Describe the role of antimicrobial peptides in innate immunity.

The body produces certain antimicrobial substances, a final component of the second line of defense, in addition to the chemical factors mentioned earlier. Among the most important of these are the proteins of the complement system, interferons, iron-binding proteins, and antimicrobial peptides.

## The Complement System

The **complement system** consists of over 30 proteins produced by the liver that circulate in blood serum  **Play Interactive Microbiology** @Mastering**Microbiology** *See how the complement system affects a patient's health* and within tissues throughout the body. (See the box on page 470.) The system is so named because it "completes," or enhances, cells of the immune system in destroying microbes. The complement system is not adaptable, never changing over a person's lifetime. Therefore it is considered part of the innate immune system. However, it can be recruited into action by the adaptive immune system. This is another example of the cooperation between the innate and adaptive immune systems. Together, proteins of the complement system destroy microbes by cytolysis, opsonization, and inflammation (see Figure 16.12), and they also prevent excessive damage to host tissues.

Complement proteins are inactive until split into fragments (products), which activates them. Activated fragments carry out the destructive actions. Complement proteins are usually designated by an uppercase letter C and are numbered C1 through C9, named for the order in which they were discovered. Activated fragments are indicated by lowercase letters *a* and *b*. For example, inactive complement protein C3 is split into activated fragments, C3a and C3b.

Complement proteins act in a *cascade*, where one reaction triggers another, which in turn triggers another. More product is formed with each succeeding reaction in the cascade, amplifying the effects. The cascade of complement proteins that occurs during an infection is called **complement activation**. It may occur in three pathways that end in the activation of C3.

# pathways of complement activation

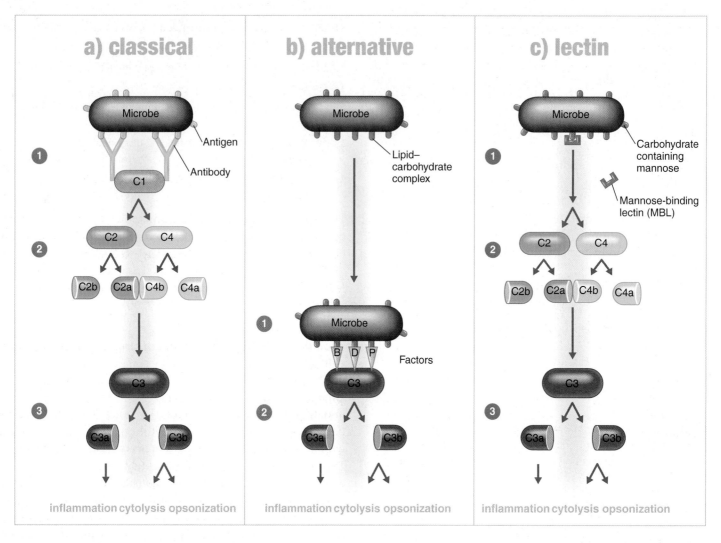

**Figure 16.10 Pathways of complement activation.** The classical pathway begins with an antigen–antibody reaction. The alternative pathway begins by contact between certain complement proteins and a pathogen; it doesn't involve antibodies. In the lectin pathway, lectin binds to mannose on the surface of a microbe.

**Q** How is the alternative pathway similar to the classical pathway, and how are the lectin and alternative pathways different from the classical pathway?

### The Classical Pathway

The **classical pathway** was the first discovered. It is initiated when antibodies bind to antigens, as shown in **Figure 16.10a**:

1. Antibodies attach to antigens (for example, proteins or large polysaccharides on the surface of a bacterium or other cell), forming antigen–antibody complexes. The antigen–antibody complexes bind to and activate C1.

2. Next, activated C1 activates C2 and C4 by splitting each of them. Then C2 splits into fragments C2a and C2b, and C4 is split into C4a and C4b.

3. C2a and C4b combine and together activate C3 by splitting it into C3a and C3b fragments. C3a participates in inflammation, and C3b functions in cytolysis and opsonization.

### The Alternative Pathway

The **alternative pathway** is so named because it was discovered after the classical pathway. Unlike the classical pathway, it does not involve antibodies. The alternative pathway is activated by contact between certain complement proteins and a pathogen, as shown in Figure 16.10b:

❶ C3, constantly present in the blood, combines with complement proteins called factor B, factor D, and factor P (properdin) on the microbe's surface. The complement proteins are attracted to microbial cell surface material (mostly lipid–carbohydrate complexes of certain bacteria and fungi).

❷ Once the complement proteins combine and interact, C3 splits into fragments C3a and C3b. As in the classical pathway, C3a participates in inflammation, and C3b functions in cytolysis and opsonization.

### The Lectin Pathway

The **lectin pathway** is the most recently discovered mechanism for complement activation. When macrophages ingest bacteria, viruses, and other foreign matter by phagocytosis, they release cytokines that stimulate the liver to produce **lectins**, proteins that bind to carbohydrates, as shown in Figure 16.10c:

❶ **Mannose-binding lectin (MBL)** binds to the carbohydrate mannose. MBL binds to many pathogens because MBL molecules recognize a distinctive pattern of carbohydrates that includes mannose, which is found in bacterial cell walls and on some viruses.

❷ As a result of binding, MBL functions as an opsonin to enhance phagocytosis and activates C2 and C4.

❸❹ C2a and C4b activate C3. As with the other two mechanisms, C3 splits into fragments C3a, which participates in inflammation; and C3b, which functions in cytolysis and opsonization.

### Outcomes of Complement Activation

As mentioned previously, the classical, alternative, and lectin pathways result in complement cascades that activate C3. Activation of C3, in turn, can lead to cytolysis, opsonization, and inflammation.

**Cytolysis** *Cytolysis* of microbial cells involves the membrane attack complex (MAC), as shown in Figure 16.12a:

❶ Activated C3 splits into C3a and C3b.

❷ C3b splits C5 into C5a and C5b.

❸ Fragments C5b, C6, C7, and C8 bind together sequentially and insert into the plasma membrane of the invading cell. C5b through C8 act as a receptor that attracts a C9 fragment. Additional C9 fragments are added to form a transmembrane channel.

Together, C5b through C8 and the multiple C9 fragments form the **membrane attack complex (MAC)** (**Figure 16.11**). The MAC creates a hole on a pathogen's cell membrane and makes transmembrane channels, allowing for flow of extracellular

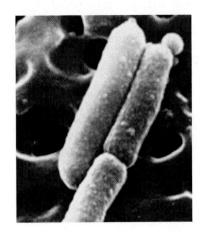
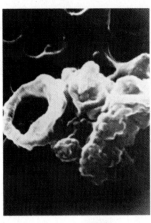

SEM  $\vdash\!\!\dashv$  2 μm

**Figure 16.11 The MAC results in cytolysis.** Micrograph of a rod-shaped bacterium before cytolysis (left) and after cytolysis (right).

*Source:* Reprinted from Schreiber, R.D., et al. "Bacterial Activity of the Alternative Complement Pathway Generated from 11 Isolated Plasma Proteins," *Journal of Experimental Medicine*, 149:870–882, 1979.

**Q** How does complement help fight infections?

fluid into the pathogen. The fluid inflow bursts the microbial cell.

Plasma membranes of the host cell contain proteins that protect against lysis by preventing the MAC proteins from attaching to their surfaces. Also, the MAC forms the basis for the complement fixation test used to diagnose some diseases. This is explained in the Clinical Focus box on page 470 and in Chapter 18 (see Figure 18.10, page 514).

Gram-negative bacteria are more susceptible to cytolysis because they have only one or very few layers of peptidoglycan to protect the plasma membrane from the effects of complement. Gram-positive bacteria have many layers of peptidoglycan, which limit complement's access to the plasma membrane and thus interfere with cytolysis. Bacteria that are not killed by the MAC are said to be *MAC resistant.*

**Opsonization** *Opsonization*, or immune adherence, promotes attachment of a phagocyte to a microbe. This enhances phagocytosis, as **Figure 16.12b** shows:

❶ Activated C3 splits into activated C3a and C3b.

❷ C3b binds to the surface of a microbe, and receptors on phagocytes attach to the C3b.

**Inflammation** This is outlined in Figure 16.12c:

❶ Activated C3 splits into C3a and C3b.

❷ C3a and C5a bind to mast cells and cause them to release histamine and other chemicals that increase blood vessel permeability during *inflammation*. C5a also functions as a very powerful chemotactic factor that attracts phagocytes to the site of an infection.

# Outcomes of Complement Activation

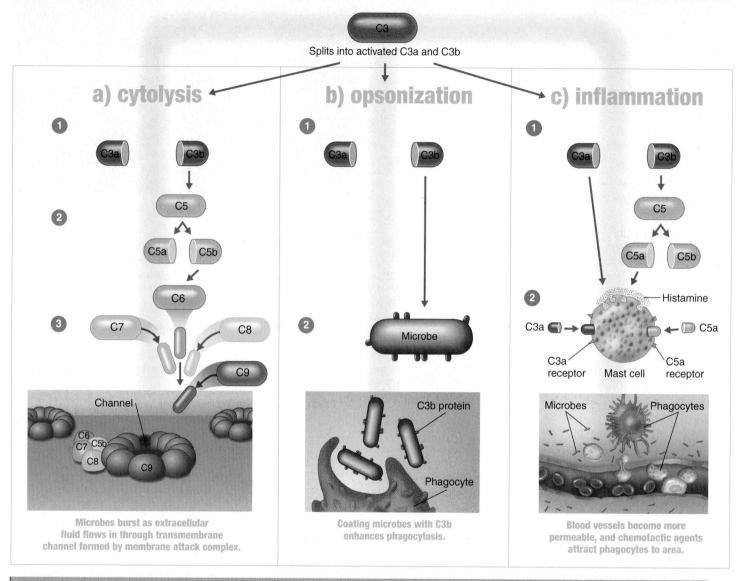

C3

Splits into activated C3a and C3b

## a) cytolysis

**1**  C3a    C3b

**2**  C5

C5a    C5b

C6

**3**  C7    C8

C9

Channel

C6
C7   C5b
C8   C9

Microbes burst as extracellular
fluid flows in through transmembrane
channel formed by membrane attack complex.

## b) opsonization

**1**  C3a    C3b

**2**  Microbe

C3b protein

Phagocyte

Coating microbes with C3b
enhances phagocytosis.

## c) inflammation

**1**  C3a    C3b

C5

C5a    C5b

**2**  Histamine

C3a →    ← C5a

C3a          C5a
receptor  Mast cell  receptor

Microbes          Phagocytes

Blood vessels become more
permeable, and chemotactic agents
attract phagocytes to area.

## KEY CONCEPTS

- The complement system is another way the body fights infection and destroys pathogens. This component of innate immunity "complements," or enhances, other immune reactions.

- Complement is a group of over 30 proteins circulating in serum that are activated in a cascade: one complement protein triggers the next.

- The cascade can be activated by a pathogen directly or by an antibody–antigen reaction.

- Together these proteins destroy microbes by (a) cytolysis, (b) enhanced phagocytosis, and (c) inflammation.

Figure 16.13 shows inflammation stimulated by complement in more detail than Figure 16.12c.

### Regulation of Complement

Once complement is activated, its destructive capabilities usually cease very quickly to minimize the destruction of host cells. This is accomplished by various regulatory proteins in the host's blood and on certain cells, such as blood cells. The regulatory proteins are present at higher concentrations than the complement proteins. The proteins bring about the breakdown or inhibition of activated complement. One example of a regulatory protein is *CD59*, which prevents the assembly of C9 molecules to form the MAC.

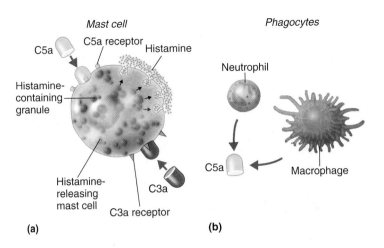

*Mast cell*

*Phagocytes*

(a)          (b)

**Figure 16.13  Inflammation stimulated by complement.**
**(a)** C3a and C5a bound to mast cells, basophils, and platelets trigger release of histamine, which increases blood vessel permeability. **(b)** C5a functions as a chemotactic factor and attracts phagocytes to the site of complement activation.

**Q** How is complement inactivated?

## Complement and Disease

In addition to its importance in defense, the complement system assumes a role in causing disease as a result of inherited deficiencies. C1, C2, or C4 deficiencies cause collagen vascular disorders that result in hypersensitivity (anaphylaxis); deficiency of C3, though rare, results in increased susceptibility to recurrent infections with pyogenic (pus-producing) microbes; and C5 through C9 defects result in increased susceptibility to *Neisseria meningitidis* and *N. gonorrhoeae* infections. Complement may play a role in diseases with an immune component, such as systemic lupus erythematosus, asthma, various forms of arthritis, multiple sclerosis, and inflammatory bowel disease. Complement is also implicated in Alzheimer's disease and other neurodegenerative disorders.

## Evading the Complement System

Some bacteria evade the complement system by means of their capsules, which prevent complement activation. For example, some capsules contain large amounts of a monosaccharide called sialic acid, which discourages opsonization and MAC formation. Other capsules inhibit the formation of C3b and C4b and cover the C3b to prevent it from making contact with the receptor on phagocytes. Some gram-negative bacteria, such as *Salmonella*, can lengthen the O polysaccharide of their LPS (see page 81), which prevents MAC formation. Other gram-negative bacteria, such as *Neisseria gonorrhoeae*, *Bordetella pertussis*, and *influenzae type b*, attach their sialic acid to the sugars in the outer membrane, ultimately inhibiting MAC formation. Gram-positive cocci release an enzyme that breaks down C5a, the fragment that serves as a chemotactic factor that attracts phagocytes.

With respect to viruses, some viruses, such as the Epstein-Barr virus, attach to complement receptors on body cells to initiate their life cycle.

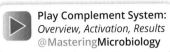

**Play Complement System:**
*Overview, Activation, Results*
@MasteringMicrobiology

**CHECK YOUR UNDERSTANDING**

✓ **16-15** What is complement?

✓ **16-16** List the steps of complement activation via the classical, alternative, and lectin pathways.

✓ **16-17** Summarize outcomes of complement activation.

## Interferons

Because viruses hijack host cells to carry out viral multiplication, it is difficult for the immune system to inhibit viral infections without affecting body cells too. One way an infected host cell counters viral infections is with a family of cytokines called **interferons (IFNs)**. These are a class of proteins produced by certain animal cells, such as lymphocytes and macrophages. Just as different animal species produce different interferons, different types of cells within the same animal also produce different interferons. Interferons produced by people protect human cells, but they produce little antiviral activity for cells of other species, such as mice or chickens. However, the interferons of a species are active against a number of different viruses. They typically play a major role in infections that are acute, such as colds and influenza.

There are three main types of human interferons: **alpha interferon (IFN-α)**, **beta interferon (IFN-β)**, and **gamma interferon (IFN-γ)**. There are also various subtypes of interferons within each of the principal groups. In humans, interferons are produced by fibroblasts in connective tissue and by lymphocytes and other leukocytes. All interferons are small proteins, with molecular masses between 15,000 and 30,000. They are quite stable at low pH and are fairly resistant to heat.

Both IFN-α and IFN-β are produced by virus-infected host cells only in very small quantities that diffuse to uninfected neighboring cells (**Figure 16.14**). Both types are host-cell–specific but not virus-specific. They react with plasma or nuclear membrane receptors, inducing the uninfected neighboring cells to manufacture mRNA for the synthesis of **antiviral proteins (AVPs)**. These proteins are enzymes that disrupt various stages of viral multiplication. For example, one AVP, called *oligoadenylate synthetase*, degrades viral mRNA. Another, called *protein kinase*, inhibits protein synthesis.

Both IFN-α and IFN-β stimulate NK cells, which produce IFN-γ. Gamma interferon is produced by lymphocytes and induces neutrophils and macrophages to kill bacteria. IFN-γ causes macrophages to produce nitric oxide that appears to kill

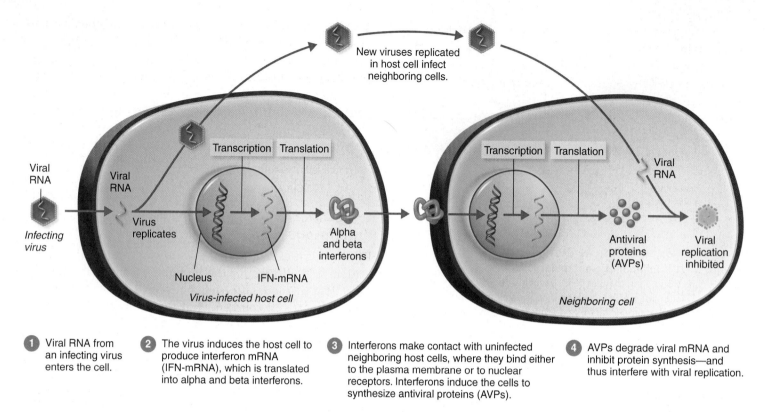

**1** Viral RNA from an infecting virus enters the cell.

**2** The virus induces the host cell to produce interferon mRNA (IFN-mRNA), which is translated into alpha and beta interferons.

**3** Interferons make contact with uninfected neighboring host cells, where they bind either to the plasma membrane or to nuclear receptors. Interferons induce the cells to synthesize antiviral proteins (AVPs).

**4** AVPs degrade viral mRNA and inhibit protein synthesis—and thus interfere with viral replication.

**Figure 16.14 Antiviral action of alpha and beta interferons (IFNs).** Interferons are host-cell–specific but not virus-specific.

**Q** How does interferon stop viruses?

bacteria as well as tumor cells by inhibiting ATP production. IFN-γ increases the expression of MHC molecules and antigen presentation.

Interferons would seem to be ideal antiviral substances, but certain problems do exist. They are stable for short periods of time in the body, so their effect is limited. When injected, interferons have side effects such as nausea, fatigue, headache, vomiting, weight loss, and fever. High concentrations of interferons are toxic to the heart, liver, kidneys, and red bone marrow. Another problem is that interferons have no effect on viral multiplication in cells already infected, and some viruses (such as adenoviruses) have resistance mechanisms that inhibit antiviral proteins.

The importance of interferons in protecting the body against viruses, as well as their potential as anticancer agents, has made their production in large quantities a top health priority. Several groups of scientists have successfully applied recombinant DNA technology in inducing certain species of bacteria to produce interferons. (This technique is described in Chapter 9.) The interferons produced with recombinant DNA techniques, called *recombinant interferons (rIFNs)*, are important for two reasons: they are pure, and they are plentiful.

In clinical trials, IFNs have exhibited no effects against some types of tumors and only limited effects against others. Alpha interferon (Intron® A) is approved in the United States for treating several virus-associated disorders. One is Kaposi's sarcoma, a cancer that often occurs in patients infected with HIV. Other approved uses for IFN-α include treating hepatitis B and C, malignant melanoma, and hairy cell leukemia. A form of IFN-β (Betaseron®) slows the progression of multiple sclerosis (MS) and lessens the frequency and severity of MS attacks. Another form of IFN-β (Actimmune®) is being used to treat osteoporosis.

## Iron-Binding Proteins

Most pathogenic bacteria require iron for their vegetative growth and reproduction (see Chapter 15). Humans use it as a component of cytochromes in the electron transport chain, a cofactor for enzyme systems, and as a part of hemoglobin, which transports oxygen in the body. Many pathogens also require iron to survive. So an infection creates a situation where pathogens and humans compete for available iron.

The concentration of free iron in the human body is low because most of it is bound to **iron-binding proteins**— molecules such as transferrin, lactoferrin, ferritin, and hemoglobin—whose function is to transport and store iron. **Transferrin** is found in blood and tissue fluids. **Lactoferrin** is found in milk, saliva, and mucus. **Ferritin** is located in the liver, spleen, and red bone marrow, and **hemoglobin** is located within red blood cells. The iron-binding proteins not only transport and store iron but also, by doing so, deprive most pathogens of the available iron.

To survive in the human body, many pathogenic bacteria obtain iron by secreting proteins called **siderophores** (see Figure 15.3 on page 430). Recall that siderophores compete to take away iron from iron-binding proteins by binding it more tightly. Once the iron–siderophore complex is formed, it is taken up by siderophore receptors on the bacterial surface and brought into the bacterium; then the iron is split from the siderophore and utilized. (In some cases, the iron enters the bacterium while the siderophore remains outside.)

A few pathogens do not use the siderophore mechanism to obtain iron. For example, *Neisseria meningitidis* produces receptors on its surface that bind directly to human iron-binding proteins. Then the iron-binding protein, along with its iron, is taken into the bacterial cell. Some pathogens, such as *Streptococcus pyogenes*, release hemolysin, a protein that causes the lysis (destruction) of red blood cells. The hemoglobin is then degraded by other bacterial proteins to capture the iron.

## Antimicrobial Peptides

Although fairly recently discovered, **antimicrobial peptides (AMPs)** may be one of the most important components of innate immunity (see also Chapter 20, page 585). Antimicrobial peptides are short peptides that consist of a chain of about 12 to 50 amino acids synthesized on ribosomes. They were first discovered in the skin of frogs, the lymph of insects, and human neutrophils; to date, over 600 AMPs have been discovered in nearly all plants and animals. AMPs have a broad spectrum of antimicrobial activities, including activity against bacteria, viruses, fungi, and eukaryotic parasites. Synthesis of AMPs is triggered by protein and sugar molecules on the surface of microbes. Cells produce AMPs when chemicals in microbes attach to Toll-like receptors (see page 448).

The modes of action of AMPs include inhibiting cell wall synthesis; forming pores in the plasma membrane, resulting in lysis; and destroying DNA and RNA. Among the AMPs produced by humans are *dermcidin*, produced by sweat glands; *defensins* and *cathelicidins*, produced by neutrophils, macrophages, and epithelium; and *thrombocidin*, produced by platelets.

Scientists are especially interested in AMPs for a number of reasons. Besides their broad spectrum of activity, AMPs have shown synergy (working together) with other antimicrobial agents, so that the effect of them working together is greater than that of either working separately. AMPs are also very stable over a wide range of pH. What is particularly significant is that microbes do not appear to develop resistance even though the microbes are exposed to them for long periods of time.

In addition to their killing effect, AMPs also participate in a number of other immune functions. For example, AMPs can sequester the LPS shed from gram-negative bacteria preventing endotoxic shock. AMPs have been found to vigorously attract dendritic cells, which destroy microbes by phagocytosis and initiate an adaptive immune response. AMPs have also been shown to recruit mast cells, which increase blood vessel permeability and vasodilation. This brings about inflammation, which destroys microbes, limits the extent of damage, and initiates tissue repair.

## Other Factors

In addition to the aforementioned factors that provide resistance to infection, several others also play a role. **Genetic resistance** is an inherited trait in a person's genome that provides resistance to a disease. It confers a selective survival advantage. One example is the relationship between sickle cell trait and *Plasmodium falciparum*; individuals who have sickle cell trait are relatively protected against *P. falciparum* malaria. Another example is the relationship between prions and spongiform encephalopathy; a naturally occurring variant of a human prion has been found that completely protects against the disease. With respect to **age**, the very young (whose immune systems are still developing) and the elderly (whose immune systems are less responsive) are more susceptible to disease. Observing healthy protocols such as proper handwashing, controlling sneezing, employing

---

**CLINICAL CASE**

In neutrophils, NADPH is reoxidized to $NADP^+$ by a membrane complex called NADPH oxidase, which uses the electron to produce a superoxide radical ($O_2^-$) from $O_2$. The superoxide will be converted to hydrogen peroxide, and the resulting burst of $H_2O_2$ kills the pathogen. Because Jacob's neutrophils cannot function properly, his pediatrician needs to find a way to stimulate his immune system to kill the fungi that have invaded Jacob's bloodstream.

**What enzyme converts superoxide radicals to hydrogen peroxide? (See the Clinical Case figure on page 463.) What treatment do you think Jacob's pediatrician should suggest for Jacob's CGD?**

449   455   459   463   **469**   470

It is common to draw more than one blood sample for laboratory tests. The blood is collected in tubes with caps of different colors (Figure A). Whole blood may be needed to culture microbes or to type the blood. Serum may be needed to test for enzymes or other chemicals in the blood. Serum is the straw-colored liquid remaining after blood is allow to clot. Blood plasma is the liquid remaining after formed elements are removed from unclotted blood by, for example, centrifugation.

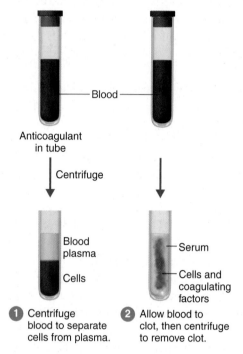

**Figure A Collecting blood cells and serum.**

### Why test for complement?

Complement activity is measured because complement deficiency may be associated with recurrent bacterial infections. Moreover, complement is a key component in immune complex diseases. A decrease in serum complement, which occurs as complement is used in immune complexes, can be used to monitor progress and treatment of immune complex diseases such as systemic lupus erythematosus and rheumatoid arthritis.

Figure B shows how total complement activity is measured. Dilutions of the patient's serum are mixed with sheep red blood cells (RBCs) and antibodies against sheep RBCs (anti-RBCs). Following incubation at 37°C for 20 minutes, the degree of hemolysis, bursting of red blood cells, is determined.

### What is the purpose of the RBCs and anti-RBCs?

The antibodies will react with the antigen (RBCs). This will activate complement in the patient's serum. The degree of lysis is relative to the amount of complement present and is expressed as a percentage of the hemolysis.

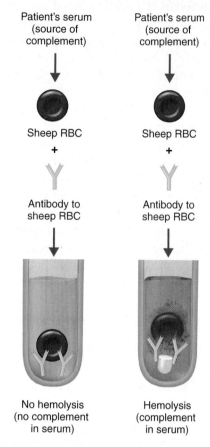

**Figure B Testing for complement.**

---

**CLINICAL CASE** Resolved

Superoxide dismutase (SOD) is an enzyme used to convert superoxide radicals into molecular oxygen and hydrogen peroxide. Jacob's pediatrician suggests treating Jacob with recombinant gamma interferon (IFN-γ), which will incite Jacob's neutrophils and macrophages to kill bacteria. The method of action isn't known, and IFN-γ isn't a cure; Jacob must take IFN-γ for the rest of his life or until he can receive a bone marrow transplant.

449   455   459   463   469   **470**

standard precautions, avoiding cross-contamination and fecal-oral transmission, and safe sex practices also provide resistance to infection.

Table 16.3 contains a summary of innate immunity defenses.

**CHECK YOUR UNDERSTANDING**

✓ **16-18** What is interferon?

✓ **16-19** Why do IFN-α and IFN-β share the same receptor on target cells, yet IFN-γ has a different receptor?

✓ **16-20** What is the role of siderophores in infection?

✓ **16-21** Why are scientists interested in AMPs?

## TABLE 16.3 Summary of Innate Immunity Defenses

| Component | Functions |
| --- | --- |
| **FIRST LINE OF DEFENSE: SKIN AND MUCOUS MEMBRANES** | |
| **PHYSICAL FACTORS** | |
| Epidermis of skin | Intact skin forms a physical barrier to the entrance of microbes; shedding helps remove microbes. |
| Mucous membranes | Inhibit the entrance of many microbes, but not as effectively as intact skin. |
| Mucus | Traps microbes in respiratory and gastrointestinal tracts. |
| Lacrimal apparatus | Provides tears that wash away microbes; tears contain lysozyme, which destroys cell walls, especially of gram-positive bacteria. |
| Saliva | Dilutes and washes microbes from mouth. |
| Hairs | Filter and trap microbes and dust in nose. |
| Cilia | Together with mucus form a ciliary escalator, which traps and removes microbes from lower respiratory tract. |
| Epiglottis | Prevents microbes from entering lower respiratory tract. |
| Earwax | Prevents microbes from entering ear. |
| Urine | Washes microbes from urethra preventing colonization in genitourinary tract. |
| Vaginal secretions | Move microbes out of body. |
| Peristalsis, defecation, vomiting, and diarrhea | Expel microbes from body. |
| **CHEMICAL FACTORS** | |
| Sebum | Forms a protective acidic film over the skin surface that inhibits microbial growth. |
| Earwax | Fatty acids in earwax inhibit the growth of bacteria and fungi. |
| Perspiration | Flushes microbes from skin and contains lysozyme; lysozyme is also present in tears, saliva, nasal secretions, urine, and tissue fluids. |
| Saliva | Contains lysozyme, urea, and uric acid, which inhibit microbes; and immunoglobin A, which prevents attachment of microbes to mucous membranes. Slight acidity discourages microbial growth. |
| Gastric juice | High acidity destroys bacteria and most toxins in stomach. |
| Vaginal secretions | Glycogen breakdown into lactic acid provides slight acidity, which discourages bacterial and fungal growth. |
| Urine | Contains lysozyme. Slight acidity discourages microbial growth. |
| **NORMAL MICROBIOTA** | Compete with pathogens for available space and nutrients and produce substances that inhibit or kill pathogens. |
| **SECOND LINE OF DEFENSE** | |
| **DEFENSIVE CELLS** | |
| Phagocytes | Phagocytosis by cells such as neutrophils, eosinophils, dendritic cells, and macrophages. |
| Natural killer (NK) cells | Kill infected cells by releasing granules of perforin and granzymes. Phagocytes then kill the cells. |
| **INFLAMMATION** | Confines and destroys microbes and initiates tissue repair. |
| **FEVER** | Intensifies the effects of interferons, and speeds up body reactions that aid repair. |
| **ANTIMICROBIAL SUBSTANCES** | |
| Complement system | Causes cytolysis of microbes, promotes phagocytosis, and contributes to inflammation. |
| Interferons | Protect uninfected host cells from viral infection. IFN-$\gamma$ increases phagocytosis. |
| Iron-binding proteins | Inhibit growth of certain bacteria by reducing the amount of available iron. |
| Antimicrobial peptides (AMPs) | Inhibit cell wall synthesis, form pores in the plasma membrane that cause lysis; and destroy DNA and RNA. |
| Other factors | Other factors that provide resistance to infection include genetic resistance, age, and observing healthy protocols. |

# Study Outline

**Go to** @MasteringMicrobiology for **Interactive Microbiology,** *In the Clinic videos, MicroFlix, MicroBoosters, 3D animations, practice quizzes, and more.*

## Introduction (p. 445)

1. The ability to ward off disease through body defenses is called immunity.
2. Lack of immunity is called susceptibility.

## The Concept of Immunity (p. 448)

1. Innate immunity refers to all body defenses that protect the body against any kind of pathogen.
2. Adaptive immunity refers to defenses (antibodies) against specific microorganisms.
3. Toll-like receptors in plasma membranes of macrophages and dendritic cells bind to invading microbes.

## First Line of Defense: Skin and Mucous Membranes (pp. 448–452)

1. The body's first line of defense against infections is a physical barrier and the nonspecific chemicals of the skin and mucous membranes.

### Physical Factors (pp. 448–450)

1. The structure of intact skin and the waterproof protein keratin provide resistance to microbial invasion.
2. The lacrimal apparatus protects the eyes from irritating substances and microorganisms.
3. Saliva washes microorganisms from teeth and gums.
4. Mucus traps many microorganisms that enter the respiratory and gastrointestinal tracts; in the lower respiratory tract, the ciliary escalator moves mucus up and out.
5. The flow of urine moves microorganisms out of the urinary tract, and vaginal secretions move microorganisms out of the vagina.

### Chemical Factors (pp. 450–451)

1. Fatty acids in sebum and earwax inhibit the growth of pathogenic bacteria.
2. Perspiration washes microorganisms off the skin.
3. Lysozyme is found in tears, saliva, nasal secretions, and perspiration.
4. The high acidity (pH 1.2–3.0) of gastric juice prevents microbial growth in the stomach.

### Normal Microbiota and Innate Immunity (pp. 451–452)

1. Normal microbiota change the environment, a process that can prevent the growth of pathogens.

## Second Line of Defense (pp. 453–471)

1. A microbe's penetration of the first line of defense encourages production of phagocytes, inflammation, fever, and antimicrobial substances.

## Formed Elements in Blood (pp. 453–455)

1. Blood consists of plasma (fluid) and formed elements (cells and platelets).
2. Leukocytes (white blood cells) are divided into granulocytes (neutrophils, basophils, eosinophils) and agranulocytes.

## The Lymphatic System (pp. 455–456)

1. The lymphatic system consists of lymph vessels, lymph nodes, and lymphoid tissue.
2. Interstitial fluid is returned to blood plasma via lymph vessels.

## Phagocytes (pp. 456–459)

1. Phagocytosis is the ingestion of microorganisms or particulate matter by a cell.
2. Phagocytosis is performed by phagocytes, certain types of white blood cells or their derivatives.

### Actions of Phagocytic Cells (pp. 456–457)

3. Enlarged monocytes become wandering macrophages and fixed macrophages.
4. Fixed macrophages are located in selected tissues and are part of the mononuclear phagocytic system.
5. Granulocytes, especially neutrophils, predominate during the early stages of infection, whereas monocytes predominate as the infection subsides.

### The Mechanism of Phagocytosis (pp. 457–459)

6. Chemotaxis is the process by which phagocytes are attracted to microorganisms.
7. Toll-like receptors on a phagocyte adhere to the microbial cells. Adherence may be facilitated by opsonization—coating the microbe with serum proteins.
8. Pseudopods of phagocytes engulf the microorganism and enclose it in a phagosome to complete ingestion.
9. Many phagocytized microorganisms are killed by lysosomal enzymes and oxidizing agents.

## Inflammation (pp. 459–462)

1. Inflammation is a bodily response to cell damage; it is characterized by redness, pain, heat, swelling, and sometimes the loss of function.
2. TNF-α stimulates production of acute-phase proteins.

### Vasodilation and Increased Permeability of Blood Vessels (pp. 460–461)

3. The release of histamine, kinins, and prostaglandins causes vasodilation and increased permeability of blood vessels.
4. Blood clots can form around an abscess to prevent dissemination of the infection.

### Phagocyte Migration and Phagocytosis (pp. 461–462)

5. Phagocytes have the ability to stick to the lining of the blood vessels (margination) and also have the ability to squeeze through blood vessels (diapedesis).

6. Pus is the accumulation of damaged tissue and dead microbes, granulocytes, and macrophages.

### Tissue Repair (p. 462)

7. A tissue is repaired when the stroma (supporting tissue) or parenchyma (functioning tissue) produces new cells.
8. Stromal repair by fibroblasts produces scar tissue.

### Fever (pp. 462–463)

1. Fever is an abnormally high body temperature produced in response to a bacterial or viral infection.
2. Bacterial endotoxins, interleukin-1, and TNF-α can induce fever.
3. A chill indicates a rising body temperature; crisis (sweating) indicates that the body's temperature is falling.

### Antimicrobial Substances (pp. 463–470)

#### The Complement System (pp. 463–467)

> **Play Interactive Microbiology** @Mastering**Microbiology** *See how the complement system affects a patient's health*

1. The complement system consists of a group of serum proteins that activate one another to destroy invading microorganisms.
2. Complement proteins are activated in a cascade.
3. C3 activation can result in cell lysis, inflammation, and opsonization.

4. Complement is activated via the classical pathway, the alternative pathway, and the lectin pathway.
5. Complement deficiencies can result in an increased susceptibility to disease.
6. Some bacteria evade destruction by complement by means of capsules, surface lipid–carbohydrate complexes, and enzymatic destruction of C5a.

### Interferons (pp. 467–468)

7. IFN-α and IFN-β induce uninfected cells to produce antiviral proteins (AVPs) that prevent viral replication.
8. IFN-γ activates neutrophils and macrophages to kill bacteria.

### Iron-Binding Proteins (pp. 468–469)

9. Iron-binding proteins transport and store iron, depriving most pathogens of the available iron.

### Antimicrobial Peptides (p. 469)

10. Antimicrobial peptides (AMPs) inhibit cell wall synthesis, form pores in plasma membranes, and destroy DNA and RNA.
11. Antimicrobial peptides are produced by nearly all plants and animals, and bacterial resistance to AMPs has not yet been seen.

### Other Factors (pp. 469–470)

12. Other factors that influence resistance to infection include genetic resistance, age, and observing healthy protocols.

## Study Questions

For answers to the Knowledge and Comprehension questions, turn to the Answers tab at the back of the textbook.

## Knowledge and Comprehension

### Review

1. Identify at least one physical factor and one chemical factor that prevent microbes from entering the body through each of the following:
   a. urinary tract
   b. reproductive tract
2. Define *inflammation*, and list its characteristics.
3. What are interferons? Discuss their roles in innate immunity.
4. How can the complement system cause endotoxic shock?
5. Patients with X-linked chronic granulomatous disease are susceptible to infections because their neutrophils don't generate an oxidative burst. What is the relation of the oxidative burst to infection?
6. Why does hemolysis of red blood cells occur when a person receives a transfusion of the wrong type of blood?
7. Give several examples of how microbes evade the complement system.
8. DRAW IT Label the following processes that result in phagocytosis: margination, diapedesis, adherence, and phagolysosome formation.

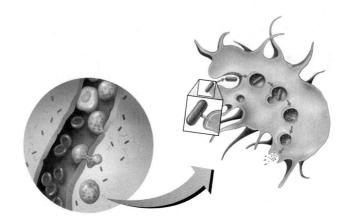

9. Are the following involved in innate or in adaptive immunity? Identify the role of each in immunity:
   a. TLRs                          c. antimicrobial peptides
   b. transferrins
10. NAME IT These agranulocytes are not phagocytic until they wander out of the blood.

## Multiple Choice

1. *Legionella* uses C3b receptors to enter monocytes. This
   a. prevents phagocytosis.
   b. degrades complement.
   c. inactivates complement.
   d. prevents inflammation.
   e. prevents cytolysis.

2. *Chlamydia* can prevent the formation of phagolysosomes and therefore can
   a. avoid being phagocytized.
   b. avoid destruction by complement.
   c. prevent adherence.
   d. avoid being digested.
   e. none of the above

3. If the following are placed in the order of occurrence, which would be the *third* step?
   a. diapedesis
   b. digestion
   c. formation of a phagosome
   d. formation of a phagolysosome
   e. margination

4. If the following are placed in the order of occurrence, which would be the *third* step?
   a. activation of C5 through C9
   b. cell lysis
   c. antigen–antibody reaction
   d. activation of C3
   e. activation of C2 through C4

5. A human host can prevent a pathogen from getting enough iron by all of the following *except*
   a. reducing dietary intake of iron.
   b. binding iron with transferrin.
   c. binding iron with hemoglobin.
   d. binding iron with ferritin.
   e. binding iron with siderophores.

6. A decrease in the production of C3 would result in
   a. increased susceptibility to infection.
   b. increased numbers of white blood cells.
   c. increased phagocytosis.
   d. activation of C5 through C9.
   e. none of the above

7. In 1884, Elie Metchnikoff observed cells collected around a splinter inserted in a sea star embryo. This was the discovery of
   a. blood cells.
   b. sea stars.
   c. phagocytosis.
   d. immunity.
   e. none of the above

8. *Helicobacter pylori* uses the enzyme urease to counteract a chemical defense in the human organ in which it lives. This chemical defense is
   a. lysozyme.
   b. hydrochloric acid.
   c. superoxide radicals.
   d. sebum.
   e. complement.

9. Which of the following statements about IFN-α is *false*?
   a. It interferes with viral replication.
   b. It is host-cell–specific.
   c. It is released by fibroblasts.
   d. It is virus-specific.
   e. It is released by lymphocytes.

10. Which of the following does *not* stimulate phagocytes?
    a. cytokines
    b. IFN-γ
    c. C3b
    d. lipid A
    e. histamine

## Analysis

1. What role does transferrin play in fighting an infection?

2. A variety of drugs with the ability to reduce inflammation are available. Comment on the danger of misuse of these anti-inflammatory drugs.

3. The following list provides examples of complement-evading techniques. For each microbe, identify the disease it causes, and describe how its strategy enables it to avoid destruction by complement.

| Pathogen | Strategy |
|---|---|
| Group A streptococci | C3 does not bind to M protein |
| *Haemophilus influenzae* type b | Has a capsule |
| *Pseudomonas aeruginosa* | Sheds cell wall polysaccharides |
| *Trypanosoma cruzi* | Degrades C1 |

4. The list below identifies a virulence factor for a selected microorganism. Describe the effect of each factor listed. Name a disease caused by each organism.

| Microorganism | Virulence Factor |
|---|---|
| *Influenzavirus* | Causes release of lysosomal enzymes |
| *Mycobacterium tuberculosis* | Inhibits lysosome fusion |
| *Toxoplasma gondii* | Prevents phagosome acidification |
| *Trichophyton* | Secretes keratinase |
| *Trypanosoma cruzi* | Lyses phagosomal membrane |

## Clinical Applications and Evaluation

1. People with *Rhinovirus* infections of the nose and throat have an 80-fold increase in kinins and no increase in histamine. What do you expect for rhinoviral symptoms? What disease is caused by rhinoviruses?

2. A hematologist often performs a differential white blood cell count on a blood sample. Why are these numbers important? What do you think a hematologist would find in a differential white blood cell count of a patient with mononucleosis? With neutropenia? With eosinophilia?

3. Leukocyte adherence deficiency (LAD) is an inherited disease resulting in the inability of neutrophils to recognize C3b-bound microorganisms. What are the most likely consequences of LAD?

4. The neutrophils of individuals with Chédiak-Higashi syndrome (CHS) have fewer than normal chemotactic receptors and lysosomes that spontaneously rupture. What are the consequences of CHS?

5. About 4% of the human population have a mannose-binding lectin deficiency. How might this deficiency affect a person?

Unlike innate immunity, adaptive immunity is designed to recognize self from nonself, and it mounts reactions that are specific to the particular substance or pathogen at hand. The cells and chemical factors involved in adaptive immunity come into play when first- and second-line defenses of the innate system fail. The first time these cells and chemicals encounter a pathogen, responses can take days or longer to develop. However, the adaptive immune system also has a memory component, which engages future defenses against the same pathogen much more quickly.

Adaptive immunity is described as a dual system, with humoral and cellular components. Humoral immunity primarily involves B cells and neutralizes threats outside human cells. Cellular immunity primarily involves T cells and deals with threats inside cells. Both involve specialized immune cell receptors that recognize antigens, followed by activation and production of cells, chemical messengers, and other factors that help destroy the antigen in question or allow the body to remember it later, for speedier future interactions.

This chapter provides a simple introduction to an extremely complex subject. Figure 17.19 (on page 495) summarizes how the main components of the adaptive system work together. For perspective on how both the innate and adaptive immune system combine to provide all our immune defenses, review the Big Picture on pages 446–447.

◀ A lymphocyte (red) attaches to a cancer cell (blue).

## In the Clinic

As a perinatal nurse, you must discuss the results of parvovirus B19 antibody testing with your 22-year-old patient. The patient is pregnant and has a high parvovirus IgM titer. **What does the high IgM titer indicate? What if she had high IgG and low IgM titers?**

*Hint: Read about immunoglobulin classes on pages 480–481. Parvovirus B19 infection is discussed on page 607.*

Play **In the Clinic** Video
@MasteringMicrobiology

# The Adaptive Immune System

### LEARNING OBJECTIVE

**17-1** Compare and contrast adaptive and innate immunity.

For centuries it's been recognized that immunity to certain infectious diseases can be acquired through exposure. For instance, after recovering from measles, a person would almost always be immune to that disease for the rest of his or her life. This protection is due to **adaptive immunity**, which involves a range of microbial defenses that target specific pathogens after exposure.

One of the most important developments in medicine is *vaccination* (*immunization*), a procedure that harnesses the adaptive immune response. A vaccine formulated with a harmless version of a pathogen incites an adaptive response, rendering people immune to the illness without the damage and danger of a full-blown infection that are otherwise needed to obtain these benefits. (See Chapter 18, pages 500–504, for more on vaccination). Vaccination against smallpox, the first disease for which the procedure was developed, actually predated the establishment of the germ theory of disease by nearly a hundred years (see Chapter 14, pages 398–399).

A crucial element of the adaptive immune system is its ability to differentiate between normal "self" cells and "nonself." Without this ability, the immune system might attack components of the body it is designed to protect. In fact, this is exactly what occurs with certain immune disorders, when the body's own tissues are erroneously targeted and damaged by immune cells. (For more on autoimmune disorders, see Chapter 19, pages 536–538.)

The adaptive immune system comes into play only when innate defenses—physical barriers such as skin and mucous membranes, phagocytes, and inflammation—fail to stop a microbe. Innate system responses are always immediate and uniform, regardless of the foreign substance encountered. However, the adaptive system tailors its fight to specific pathogens, toxins, or other substances. The first time the adaptive immune system meets and combats a particular antigen is called the *primary response*, which involves a lag or latent period of 4 to 14 days. Later interactions with that same cell or substance will cause a *secondary response*, which is faster and more effective as a result of the "memory" formed during the primary response. This memory component is also exclusive to the adaptive immune system.

Play Host Defenses:
*The Big Picture*
@Mastering**Microbiology**

---

**CHECK YOUR UNDERSTANDING**

✔ **17-1** Is vaccination an example of innate or of adaptive immunity?

# Dual Nature of the Adaptive Immune System

### LEARNING OBJECTIVE

**17-2** Differentiate humoral from cellular immunity.

Adaptive immunity is considered a **dual system**, with *humoral* and *cellular* components that also contribute to the innate immune system. (See Figure 17.19 for an overview of both components of the adaptive immune system.) Cells of the adaptive immune system originate from pluripotent stem cells in the bone marrow or the fetal liver (see Figure 16.4 on page 453). Although cells of humoral and cellular immunity mature differently, the cells of the adaptive immune system are all found primarily in blood and lymphoid organs.

## Overview of Humoral Immunity

The term **humoral immunity** derives from the word *humors*, an ancient name for body fluids such as blood, phlegm, black bile, and yellow bile. Humoral immunity describes immune actions taking place in these extracellular fluids, brought about by protective molecules called **antibodies**. Another term for *antibody* is **immunoglobulin (Ig)**. Antibodies recognize and combat foreign molecules called **antigens**.

Humoral immunity involves **B lymphocytes**, more commonly known as **B cells**. Immunoglobulins corresponding to specific antigens coat the surfaces of B cells. Activated B cells secrete the same specific immunoglobulin that reacts with a particular antigen component of a virus, bacterium, toxin, or other extracellular material in body tissue fluids and blood. Because humoral immunity fights invaders outside cells, efforts tend to focus on bacteria that live extracellularly (as well as their toxins) and on viruses before they penetrate a target cell.

B cells were named for the *bursa of Fabricius*, the specialized organ of birds where researchers first observed these cells. In humans, lymphocytes are initially produced in the fetal liver. By about the third month of fetal development, the site of B cell creation and maturation (known as *schooling*) becomes the red bone marrow. Once mature, B cells are found primarily in the blood and lymphoid organs.

## Overview of Cellular Immunity

**T lymphocytes**, or **T cells**, are the basis of **cellular immunity**, also called **cell-mediated immunity**. T cells do not directly bind to antigens. Instead, phagocytic cells, such as macrophages or dendritic cells, process and present antigenic peptides to them. T cells have **T cell receptors (TCRs)** that recognize an antigenic peptide attached to a specialized presenting molecule on a cell. When T cells are activated, some

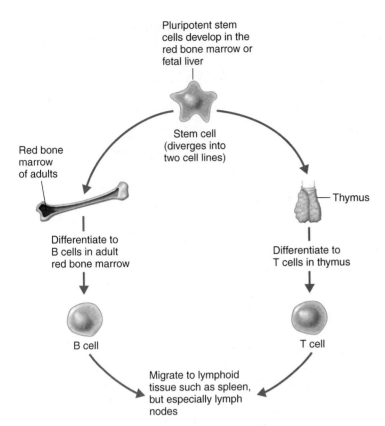

Pluripotent stem cells develop in the red bone marrow or fetal liver

Stem cell (diverges into two cell lines)

Red bone marrow of adults

Thymus

Differentiate to B cells in adult red bone marrow

Differentiate to T cells in thymus

B cell

T cell

Migrate to lymphoid tissue such as spleen, but especially lymph nodes

**Figure 17.1  T cell and B cell development.** Both B cells and T cells originate from stem cells in adult red bone marrow (or in the liver in fetuses). Some cells pass through the thymus and emerge as mature T cells. Others remain in the red bone marrow and become B cells. Both B cells and T cells then migrate to lymphoid tissues, such as the lymph nodes or spleen.

**Q** Which cells, T or B, make antibodies?

destroy target cells that present a particular antigenic peptide. Others can proliferate and secrete chemical messengers, called *cytokines* (discussed next), that induce other cells to perform a function.

T cells owe their name to the thymus, the organ where these particular cells mature (**Figure 17.1**). Once mature, T cells are found in the same places as B cells—primarily in the blood and lymphoid organs.

Cellular immune responses focus on recognizing antigens that have already entered a cell. This immunity is generally best at fighting viruses and some intracellular bacteria such as *Listeria monocytogenes* or *Mycobacterium leprae*.

**CHECK YOUR UNDERSTANDING**

✔ **17-2** What type of cell is most associated with humoral immunity, and what type of cell is the basis of cellular immunity?

# Cytokines: Chemical Messengers of Immune Cells

**LEARNING OBJECTIVE**

**17-3** Identify at least one function of each of the following: cytokines, interleukins, chemokines, interferons, TNF, and hematopoietic cytokines.

Adaptive immunity requires complex interactions between different cells. This communication is mediated by chemical messengers called **cytokines**, which are soluble proteins or glycoproteins. They are produced by nearly all types of immune cells in response to a stimulus. More than 200 cytokines are thought to exist. Most have common names that reflect functions known at the time of the discovery; some are now known to have multiple functions. A cytokine only acts on a cell that has a receptor for it. Cytokines are classified by structure or function. Several important types include interleukins, chemokines, interferons, tumor necrosis factors, and hematopoietic cytokines. Table 17.1 lists some cytokines and their target cells and effects.

**Interleukins (ILs)** serve as communicators primarily between leukocytes. ILs often target immune cells to stimulate cell proliferation, maturation, migration, or activation during an immune response. This type of cytokine may sometimes be useful as a drug treatment designed to stimulate the immune system and treat certain infectious diseases or cancers.

**Chemokines**, a family of small cytokines, induce leukocytes to migrate to areas of infection or tissue damage where they can begin to act against an infection. The name is based on *chemotaxis*, the term for movement of an organism in response

**CLINICAL CASE**  It's Just a Scratch

JoAnna Marsden is a medical examiner at a large city hospital. She receives an autopsy order for Maria Vasquez, a 44-year-old woman. According to the medical file, Vasquez went to the emergency department complaining of not feeling well for several days, with headaches and general body aches. Initial examination by the physician revealed a drop in blood pressure and a scratch on Vasquez's right forearm, but no sign of infection. Vasquez was treated with oxygen but became progressively worse. She died 4 hours after admission. When Dr. Marsden performs the autopsy, she finds internal bleeding and multiple small clots throughout. The abnormal clotting is indicative of disseminated intravascular coagulation triggered by infection or with sepsis.

**What caused Mrs. Vasquez to die from disseminated intravascular coagulation and septic shock? Read on to find out.**

477  478  482  484  488  491

**TABLE 17.1** Key Cytokines and Their Roles

| Cytokine | Source | Target Cell(s) | Effect |
|---|---|---|---|
| IL-4 | $T_H2$ | Naïve $T_H$; B cells | Proliferate $T_H2$ cells; Class switching to IgE |
| IL-12 | Dendritic cells, macrophages, neutrophils | Naïve $T_H$; NK cells | Stimulates growth and function of $T_H1$ cells; Stimulates IFN-$\gamma$ and TNF-$\alpha$ |
| IL-17 | $T_H17$ | Neutrophils | Inflammation |
| IL-22 | $T_H17$ | Epithelial cells | Stimulates epithelial cells to make antimicrobial proteins |
| Gamma interferon (IFN-$\gamma$) | $T_H1$ | CTL, macrophages | Promote phagocytosis; activate macrophages and humoral response |
| Chemokines | Varies; macrophages and epithelial cells | Neutrophils | Chemotaxis |
| TNF-alpha (TFN-$\alpha$) | Macrophages, $T_H$, NK cells | Tumor cells | Inflammation |
| GM-CSF | Macrophages, T cells, NK cells | Myeloid stem cells | Increase macrophages and granulocytes |

to a chemical stimulus. Chemokines are especially important for infections by HIV (see Chapter 19, page 546).

**Interferons (IFNs)** were originally named for one of their functions: the ability to interfere with viral infections in host cells (see Chapter 16). A number of antiviral interferons are available as commercial products in treating disease conditions, including hepatitis and some cancers. Gamma interferon (IFN-$\gamma$) stimulates the immune system.

**Tumor necrosis factor alpha (TNF-$\alpha$)** originally earned its name because tumor cells were observed to be one of its targets. These important cytokines are a strong factor in inflammatory reactions of autoimmune diseases such as rheumatoid arthritis (page 537). Monoclonal antibodies (see pages 508–510) that block the action of TNFs are an available therapy for some of these conditions.

Dr. Marsden performs a Gram stain on a blood smear (see the figure) from the patient. She needs to know what bacteria, if any, caused Mrs. Vasquez's disseminated intravascular coagulation, septic shock, and subsequent death.

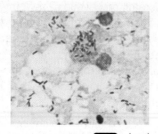

LM ⊢ 10 μm

**Describe the bacteria in the figure.**

477   478   482   484   488   491

**Hematopoietic cytokines** help control the pathways by which stem cells develop into red blood cells or different white blood cells (see Figure 16.4, page 453). Some of these are interleukins (mentioned above). Others are termed *colony-stimulating factors (CSFs)*. An example is granulocyte colony-stimulating factor (G-CSF). This particular CSF stimulates the production of neutrophils from the granulocyte precursors. Another, granulocyte macrophage colony-stimulating factor (GM-CSF), is used therapeutically to increase the numbers of protective macrophages and granulocytes in patients undergoing red bone marrow transplants.

One of the many things cytokines can do is stimulate cells to produce yet more cytokines. This feedback loop occasionally gets out of control, resulting in a **cytokine storm**. This overabundance of cytokines can do significant damage to tissues, which appears to be a factor in the pathology of certain diseases such as influenza, Ebola virus disease, graft-versus-host disease, and sepsis.

**CHECK YOUR UNDERSTANDING**

✔ 17-3 What is the function of cytokines?

# Antigens and Antibodies

**LEARNING OBJECTIVES**

**17-4** Define *antigen*, *epitope*, and *hapten*.

**17-5** Explain antibody function, and describe the structural and chemical characteristics of antibodies.

**17-6** Name one function for each of the five classes of antibodies.

## Antigens

Substances that induce production of antibodies are called **antigens** (short for *antibody gen*erators). Most antigens are either proteins or large polysaccharides. Lipids and nucleic acids are usually antigenic only when combined with proteins and

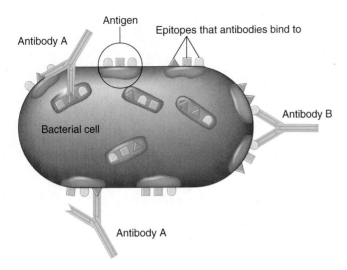

Antigen

Antibody A

Epitopes that antibodies bind to

Bacterial cell

Antibody B

Antibody A

**Figure 17.2 Epitopes (antigenic determinants).** In this illustration, the epitopes are components of the antigenic bacterial cell wall. Each Y-shaped antibody molecule has two binding sites that attach to a specific epitope on an antigen.

**Q** Which would have more epitopes: a protein or a lipid? Why?

polysaccharides. Pathogens can have multiple antigenic sites. Components of invading microbes—such as capsules, cell walls, flagella, fimbriae, bacterial toxins, and viral coats—all tend to be antigenic. However, a compound doesn't have to be part of an invading pathogen to be deemed antigenic by the immune system. Nonmicrobial antigens may include pollen, egg white, blood cell surface molecules, serum proteins from other individuals or species, and surface molecules of transplanted tissues and organs.

Detection of an antigen provokes production of highly specific, corresponding antibodies (described in more detail next). Antigens that cause such a response are often known as *immunogens*. Generally speaking, antibodies react with specific regions on antigens called **epitopes** or **antigenic determinants** (**Figure 17.2**). The nature of the interaction depends on the size, shape, and chemical structure of the binding site on the antibody molecule. Similarly, epitopes can be displayed by antigen presenting cells (such as when macrophages and dendritic cells present them to T cells). A bacterium or virus may have several epitopes that cause the production of different antibodies.

Pathogenic bacteria characteristically possess a number of recognizable antigens called *pathogen-associated molecular patterns* (*PAMPs*, discussed on page 448). PAMPs serve as warning flags of an invading organism that the host can recognize by means of receptors. The best-known of these receptors is the extended family of Toll-like receptors (TLRs).

Most antigens have a molecular mass of 10,000 da or higher. A foreign substance that has a low molecular mass is often not antigenic unless it is attached to a carrier molecule. These low-molecular-mass compounds are called **haptens** (**Figure 17.3**). Once an antibody against the hapten has been formed, the

antibody will react with the hapten independent of the carrier molecule. Penicillin is a good example of a hapten. This drug is not antigenic by itself, but some people develop an allergic reaction to it. (Allergic reactions, discussed on pages 525–530, are a type of hypersensitivity reaction that occurs when the immune system reacts to something that is normally non-pathogenic, such as pollen.) In these people, when penicillin combines with host proteins, the resulting combined molecule initiates an immune response. This concept has therapeutic applications. Conjugated vaccines, which combine an antigen with a protein, work in the same fashion (see page 503).

## Humoral Immunity: Antibodies

Structurally speaking, **antibodies** are compact, relatively soluble proteins. They are designed to recognize and bind to a specific antigen. Antibodies are either secreted by plasma cells or attached to the cell membrane of a B cell.

Each antibody has at least two identical antigen-binding sites that bind to identical epitopes. The number of antigen-binding sites on an antibody is called the **valence** of that antibody. For example, most human antibodies have two binding sites; therefore, they are bivalent. Because a bivalent antibody has the simplest molecular structure, it is called a **monomer**. A typical antibody monomer has four protein chains: two identical *light chains* and two identical *heavy chains*. ("Light" and "heavy" refer to the relative molecular masses.) The chains are joined by disulfide links and other bonds to form a Y-shaped molecule. The Y-shaped molecule is flexible and can assume a T shape (notice the hinge region in **Figure 17.4a**).

The two sections located at the ends of the Y arms are called *variable (V) regions*. These are the sites that bind to the epitopes (Figure 17.4b). The amino acid sequences and, therefore, the three-dimensional structure of these two variable regions are identical on any one antibody. The stem of the antibody monomer and the lower parts of the arms of the Y are called the *constant (C) regions*. They are the same for a particular class of immunoglobulin. There are five major types of C regions,

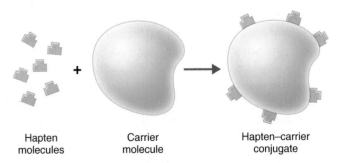

Hapten molecules

+

Carrier molecule

Hapten–carrier conjugate

**Figure 17.3 Haptens.** A hapten is a molecule too small to stimulate antibody formation by itself. By combining with a larger carrier molecule (often a serum protein) the hapten and its carrier form a conjugate that can stimulate an immune response.

**Q** How does a hapten differ from an antigen?

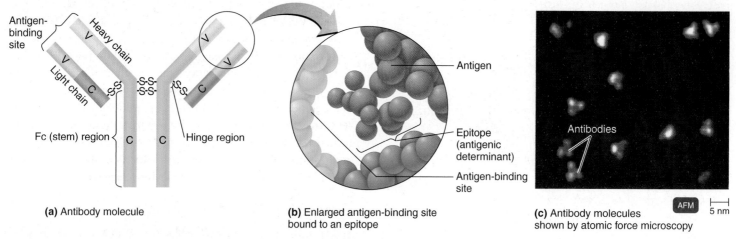

**(a)** Antibody molecule

**(b)** Enlarged antigen-binding site bound to an epitope

**(c)** Antibody molecules shown by atomic force microscopy

**Figure 17.4 Structure of a typical antibody.** The Y-shaped molecule is composed of two light chains and two heavy chains linked by disulfide bridges (S—S). Most of the molecule is made up of constant regions (C), which are the same for all antibodies of the same class. The amino acid sequences of the variable regions (V), which form the two antigen-binding sites, differ for each B cell.

**Q** What is responsible for the specificity of each different antibody?

which account for the five major classes of immunoglobulins (described shortly).

The stem of the Y-shaped antibody monomer is called the *Fc region*, so named because when antibody structure was first being identified, it was a fragment (F) that crystallized (c) in cold storage.

These Fc regions are often important in immunological reactions. If left exposed after both antigen-binding sites attach to an antigen such as a bacterium, the Fc regions of adjacent antibodies can bind complement. This leads to the destruction of the bacterium (see Figure 16.12, page 466). Conversely, the Fc region may bind to a cell, leaving the antigen-binding sites of adjacent antibodies free to react with antigens.

### Immunoglobulin Classes

The simplest and most abundant immunoglobulins are monomers, but they can also assume some different sizes and arrangements. The five classes of Igs are designated IgG, IgM, IgA, IgD, and IgE. Each class has a different role in the immune response. The structures of IgG, IgD, and IgE molecules are Y-shaped monomers. Molecules of IgA and IgM are aggregates of two or five monomers, respectively, that are joined together. The structures and characteristics of the immunoglobulin classes are summarized in Table 17.2.

**IgG** The name **IgG** is derived from the part of blood, called *gamma globulin*, that contains antibodies. IgG accounts for about 80% of all antibodies in serum. In regions of inflammation, these monomer antibodies readily cross the walls of blood vessels and enter tissue fluids. Maternal IgG antibodies,

for example, can cross the placenta and confer passive immunity to a fetus. They protect against circulating bacteria and viruses, neutralize bacterial toxins, trigger the complement system, and, when bound to antigens, enhance the effectiveness of phagocytic cells.

**IgM** Antibodies of the **IgM** class (the M refers to *macro*, reflecting their relatively large size) make up 6% of the antibodies in serum. IgM has a pentamer structure, with five monomers held together by a polypeptide called a *joining (J) chain*. The large size of the molecule prevents IgM from moving about as freely as IgG does, so IgM antibodies generally remain in blood vessels and do not enter surrounding tissues or cross the placenta.

IgM is the predominant type of antibody involved in the response to the ABO blood group antigens on the surface of red blood cells. It is much more effective than IgG in causing the clumping of cells and viruses and in reactions involving the activation of complement. (See Table 19.2 on page 531 for more on the ABO blood group antibodies; pages 511–512 for more on agglutination; and Figure 16.10 on page 464 for more on complement activation.)

The fact that IgM appears first in response to a primary infection and is relatively short-lived makes it uniquely valuable in diagnosing disease. If high concentrations of IgM against a pathogen are detected in a patient, it is likely that the symptoms observed are caused by that pathogen. The detection of IgG, which is relatively long-lived, may indicate only that immunity against a particular pathogen was acquired in the more distant past.

## TABLE 17.2 A Summary of Immunoglobulin Classes

| Class | IgG | IgM | IgA | IgD | IgE |
|---|---|---|---|---|---|
| Structure | Monomer | Pentamer | Dimer (with secretory component) | Monomer | Monomer |
| % Total Serum Antibody | 80% | 6% | 13%* | 0.02% | 0.002% |
| Location | Blood, lymph, intestine | Blood, lymph, B cell surface (as monomer) | Secretions (tears, saliva, mucus, intestine, milk), blood, lymph | B cell surface, blood, lymph | Bound to mast and basophils throughout body, blood |
| Molecular Mass | 150,000 | 970,000 | 405,000 | 175,000 | 190,000 |
| Half-Life in Serum | 23 days | 5 days | 6 days | 3 days | 2 days |
| Complement Fixation | Yes | Yes | No† | No | No |
| Placental Transfer | Yes | No | No | No | No |
| Known Functions | Enhances phagocytosis; neutralizes toxins and viruses; protects fetus and newborn | Especially effective against microorganisms and agglutinating antigens; first antibodies produced in response to initial infection | Localized protection on mucosal surfaces | Serum function not known; presence on B cells functions in initiation of immune response | Allergic reactions; possibly lysis of parasitic worms |

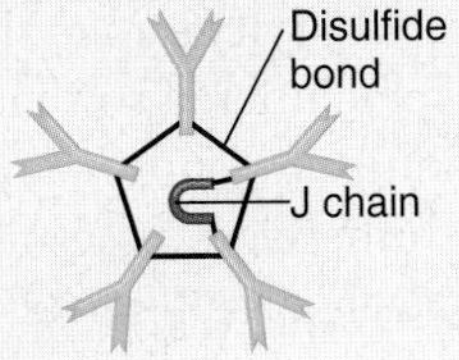

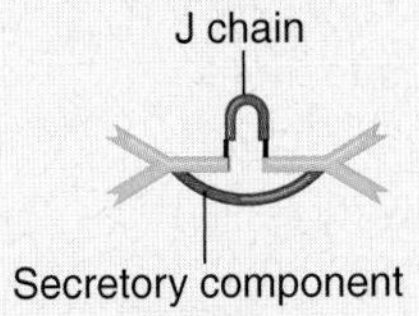

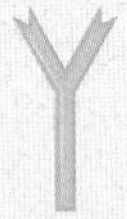

In the illustrations: IgM shows labels "Disulfide bond" and "J chain"; IgA shows labels "J chain" and "Secretory component".

*Percentage in serum only; if mucous membranes and body secretions are included, percentage is much higher.
†May be yes via alternative pathway.

**IgA** IgA accounts for only about 13% of the antibodies in serum, but it is by far the most common form in mucous membranes and in body secretions such as mucus, saliva, tears, and breast milk. If we take this into consideration, IgA is the most abundant immunoglobulin in the body. The form of IgA that circulates in serum, *serum IgA*, is usually in the form of a monomer. The most effective form of IgA, however, consists of two connected monomers that form a *dimer* called *secretory IgA*. It is produced in this form by plasma cells in the mucous membranes—as much as 15 grams per day, mostly by intestinal epithelial cells. Each dimer then enters and passes through a mucosal cell, where it acquires a polypeptide called *secretory component* that protects it from enzymatic degradation. The main function of secretory IgA is probably to prevent the attachment of microbial pathogens to mucosal surfaces. This is especially important in resistance to intestinal and respiratory pathogens. Because IgA immunity is relatively short-lived, the length of immunity to many respiratory infections is correspondingly short. IgA's presence in a mother's milk, especially the colostrum, probably helps protect infants from gastrointestinal infections.

**IgD** IgD antibodies make up only about 0.02% of the total serum antibodies. Their structure resembles that of IgG molecules. IgD antibodies are found in blood, lymph, and particularly on the surfaces of B cells. Serum IgD has no well-defined function; on B cells it assists in the immune response.

**IgE** Antibodies of the **IgE** class where shown to be potent inducers of erythema (superficial reddening or rash of the skin). IgE molecules are slightly larger than IgG molecules; they constitute only 0.002% of the total serum antibodies. IgE molecules bind tightly by their Fc (stem) regions to receptors on mast cells and basophils, specialized cells that participate in allergic reactions (see Chapter 19). When an antigen such as pollen links with the IgE antibodies attached to a mast cell or basophil (see Figure 19.1a, page 526), that cell releases histamine and other chemical mediators. These chemicals provoke a response—for example, an allergic reaction such as hay fever. However, the response can be protective as well, for it attracts complement and phagocytic cells. This is especially useful when the antibodies bind to parasitic worms. The concentration of IgE is greatly increased during some allergic reactions and parasitic infections, which is often diagnostically useful.

Gram-negative rods are visible inside white blood cells from Mrs. Vasquez's sample. Dr. Marsden identifies them as *Capnocytophaga canimorsus* bacteria. This species is found in cats or dogs and can be transmitted to people through bites, licking, scratching, and other contact with an infected animal. Dr. Marsden talks to Mr. Vasquez, who confirms that the scratch on his late wife's arm was from the family dog. However, Dr. Marsden is puzzled, because most people who have contact with dogs do not develop *Capnocytophaga* infections.

**What molecules normally made by B cells combat bacterial infections?**

477  478  **482**  484  488  491

**CHECK YOUR UNDERSTANDING**

✔ **17-4** What part of an antibody reacts with the epitope of an antigen?

✔ **17-5** The original theoretical concepts of an antibody called for a rod with antigenic determinants at each end. What is the primary advantage of the Y-shaped structure that eventually emerged?

✔ **17-6** Which class of antibody is most likely to protect you from a common cold?

# Humoral Immunity Response Process

**LEARNING OBJECTIVES**

**17-7** Compare and contrast T-dependent and T-independent antigens.

**17-8** Differentiate plasma cell from memory cell.

**17-9** Describe clonal selection.

**17-10** Describe how a human can produce different antibodies.

Humoral immune actions take place in the extracellular spaces within the body. So-called *free antigens* found here need to be identified and processed, so that specific antibodies can be created to neutralize them. Likewise, these antibodies need to be remembered so that future interactions with the same type of antigen result in a quicker immune response the next time. Both these processes begin with an inactivated B cell.

Each B cell carries immunoglobulins on its surface. The majority are IgM and IgD, all of which are specific for recognition of the same epitope (portion of an antigen). Some B cells may have other immunoglobulin families on their surface. For example, B cells in the intestinal mucosa are rich in IgA. B cells may carry more than 100,000 identical immunoglobulin molecules, all recognizing the same epitope.

B cells selected to mature can be activated. This allows activated B cells to produce plasma cells that make antibodies as well as memory cells. This process is called **clonal expansion**, or *proliferation*. An antigen that requires a type of T cell called a **T helper ($T_H$) cell** (pages 488–489) to activate a B cell is known as a **T-dependent antigen**. T-dependent antigens are mainly proteins of the type found on viruses, bacteria, red blood cells, and haptens with their carrier molecules. For antibodies to be produced in response to a T-dependent antigen, both B and T cells must recognize and interact with different epitopes on a given antigen. This ensures specificity of the attack and also helps prevent an unintentional autoimmune response. B cells can be activated directly by some antigens, called **T-independent antigens**, without assistance of T cells. First, we'll walk through the steps involved with responding to a T-dependent antigen.

## Activation and Clonal Expansion of Antibody-Producing Cells

When an antigen-presenting cell (APC, see pages 486–487) makes contact with an antigen that can combine with its

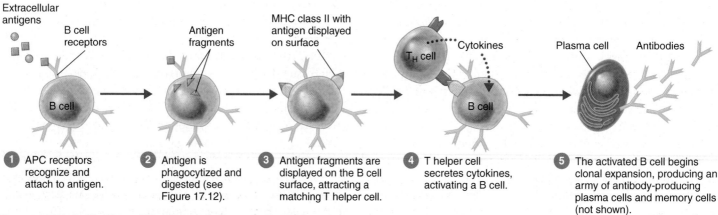

**Figure 17.5 Activation of B cells to produce antibodies.** In this illustration, the B cell is producing antibodies against a T-dependent antigen.

1 APC receptors recognize and attach to antigen.

2 Antigen is phagocytized and digested (see Figure 17.12).

3 Antigen fragments are displayed on the B cell surface, attracting a matching T helper cell.

4 T helper cell secretes cytokines, activating a B cell.

5 The activated B cell begins clonal expansion, producing an army of antibody-producing plasma cells and memory cells (not shown).

**Q** How does activation by T-independent antigens differ from this figure?

particular receptor, the APC and antigen bind. The APC then internalizes the antigen and digests it. Next, the APC displays digested antigen fragments on its surface by combining them with its **major histocompatibility complex (MHC)**, as shown in **Figure 17.5** (steps 1–3).

The MHC is a collection of glycoproteins embedded in the plasma membrane. Class I MHC are found on all mammalian nucleated cells. Their presence identifies a cell as "self," preventing the immune system from making antibodies that would be harmful to our own tissues. Class II MHC are found on APCs (B cells, macrophages, and dendritic cells).

Displaying antigen fragments bound to its MHC class II molecules attracts the appropriate T helper cell to the APC cell. The $T_H$ cell makes contact with the fragment presented on the APC, and then the T helper cell produces cytokines

that activate a B cell, which divides into a large clone of cells. Some of the B cell clones differentiate into antibody-producing plasma cells. Others become long-lived memory cells responsible for the enhanced secondary response to an antigen. Clonal selection and expansion is shown in **Figure 17.6**.

T-independent antigens stimulate B cells directly, without the help of T cells. T-independent antigens tend to be molecules consisting of repeating subunits, such

Play Antigen Processing and Presentation: *Overview;* Humoral Immunity: *Clonal Selection and Expansion* @Mastering**Microbiology**

as polysaccharides or lipopolysaccharides. Bacterial capsules are a common example of T-independent antigens. The repeating subunits, as shown in **Figure 17.7**, can bind to multiple B cell receptors, which is probably why this class of antigens

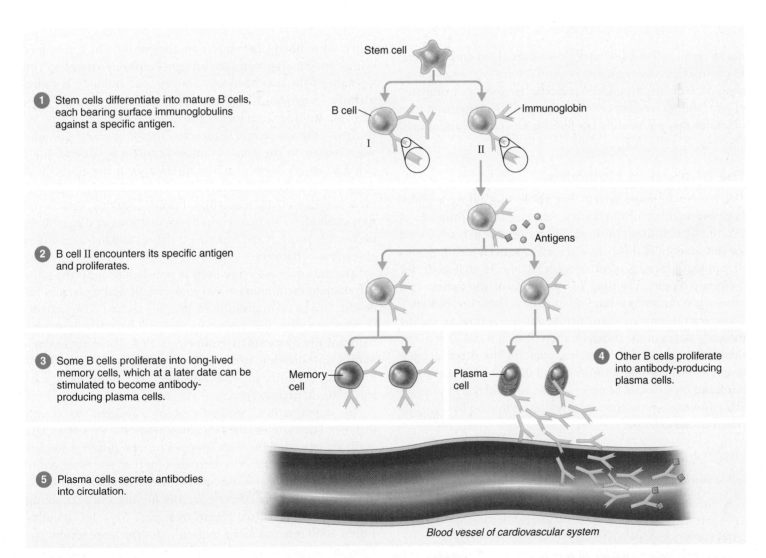

1 Stem cells differentiate into mature B cells, each bearing surface immunoglobulins against a specific antigen.

2 B cell II encounters its specific antigen and proliferates.

3 Some B cells proliferate into long-lived memory cells, which at a later date can be stimulated to become antibody-producing plasma cells.

4 Other B cells proliferate into antibody-producing plasma cells.

5 Plasma cells secrete antibodies into circulation.

Stem cell

B cell

Immunoglobin

I

II

Antigens

Memory cell

Plasma cell

*Blood vessel of cardiovascular system*

**Figure 17.6 Clonal selection and differentiation of B cells.** B cells can recognize an almost infinite number of antigens, but each particular cell recognizes only one type of antigen. An encounter with a particular antigen triggers the proliferation of a cell that is specific for that antigen (here, B cell "II") into a clone of cells with the same specificity, hence the term *clonal selection.* The initial antibodies produced are generally IgM, but later the same cell might produce different classes of antibody, such as IgG or IgE; this is called *class switching.*

**Q** What caused cell "II" to respond?

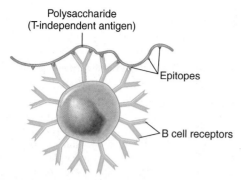

Polysaccharide
(T-independent antigen)

Epitopes

B cell receptors

**Figure 17.7 T-independent antigens.** T-independent antigens have repeating units (epitopes) that can cross-link several antigen receptors on the same B cell. These antigens stimulate the B cell to make antibodies without the aid of T helper cells. The polysaccharides of bacterial capsules are examples of this type of antigen.

**Q How can you differentiate T-dependent from T-independent antigens?**

doesn't require T cell assistance. However, these antigens tend to provoke a weaker immune response than T-dependent antigens do. The T-independent response is composed primarily of IgM, and no memory cells are generated. The immune system of infants may not be stimulated by T-independent antigens until about age 2.

## The Diversity of Antibodies

The human immune system is capable of creating a mind-boggling number of antibodies—an estimated minimum of $1 \times 10^{15}$ (quadrillion) of them. The number of genes required for this amount of diversity is actually relatively small, thanks to random rearrangement of gene segments that code for antigen receptors, resulting in variations of the amino acid sequence at the antigen-binding site (the variable site, or V site). These rearrangements occur before antigen is present, during the early stages of the differentiation of each B cell. Due to the random nature of their creation, some of the B cell antigen receptors made would ultimately lead to antibodies that could harm our own tissues. However, these harmful B cells are usually eliminated at the immature lymphocyte stage by a process called **clonal deletion**.

### CLINICAL CASE

Antibodies, especially IgM antibodies, are made in response to bacterial infections. In her research, Dr. Marsden learns that *Capnocytophaga* possesses T-independent antigens.

**What steps are required for an antibody response to these antigens?**

477  478  482  **484**  488  491

# Results of the Antigen–Antibody Interaction

**LEARNING OBJECTIVE**

**17-11** Describe four outcomes of an antigen–antibody reaction.

When an antibody encounters an antigen for which it is specific, their binding forms an **antigen–antibody complex**. The strength of the bond between antigen and antibody is called **affinity**. In general, the closer the physical fit between antigen and antibody, the higher the affinity. Antibodies tend to recognize the amino acid sequence of the antigen's epitope, giving *specificity* to the antigen–antibody interaction. Antibodies can distinguish between minor differences in the amino acid sequence of a protein and even between two amino acid isomers (see Figure 2.13, page 40). Clinically speaking, this means that antibodies can be used in diagnostic tests to differentiate between hepatitis B and hepatitis C viruses and between different strains of bacteria.

The antibody molecule itself is not damaging to the antigen. Rather, the binding marks foreign cells and molecules for destruction or neutralization by phagocytes and complement. Foreign organisms and toxins are rendered harmless by several mechanisms, as summarized in **Figure 17.8**. These are agglutination, opsonization, neutralization, antibody-dependent cell-mediated cytotoxicity, and activation of complement (see Figure 16.12, page 466).

In **agglutination**, antibodies cause antigens to clump together. For example, the two antigen-binding sites of an IgG antibody can combine with epitopes on two different foreign cells, aggregating the cells into clumps that are more easily ingested by phagocytes. Because of its more numerous binding sites, IgM is more effective at cross-linking and aggregating particulate antigens (see Figure 18.5, page 511). IgG requires 100 to 1000 times as many molecules for the same results. (In Chapter 18, we will see how agglutination is important in the diagnosis of some diseases.)

*Opsonization* is the coating of antigens with antibodies or complement proteins. This enhances ingestion and

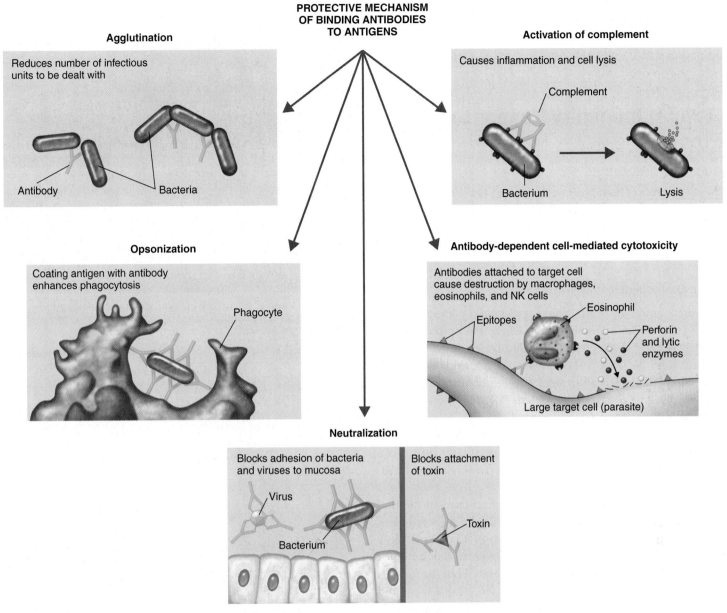

**Figure 17.8  The results of antigen–antibody binding.** The binding of antibodies to antigens forms antigen–antibody complexes that tag foreign cells and molecules for destruction by phagocytes and complement.

**Q** What are some of the possible outcomes of an antigen–antibody reaction?

lysis by phagocytic cells. **Antibody-dependent cell-mediated cytotoxicity** resembles opsonization in that the target organism becomes coated with antibodies; however, in this case the target cell is not engulfed but remains external to the phagocytic cell attacking it (also see Figure 17.16, page 492).

In **neutralization**, IgG antibodies inactivate microbes by blocking their attachment to host cells. IgG can neutralize toxins in a similar manner. By surrounding specific pathogenic components of a microbe, the antibodies can reduce pathogenicity or toxicity.

Finally, either IgG or IgM antibodies may trigger **activation of the complement system**. For example, inflammation

is caused by infection or tissue injury (see Figure 16.9, page 460). One aspect of inflammation is that it will often cause microbes in the inflamed area to become coated with certain proteins. This, in turn, leads to the attachment to the microbe of an antibody–complement complex. This complex lyses the microbe, which then attracts phagocytes and other defensive immune system cells to the area. (For a review of complement, see Figure 16.12, page 466.)

Play Humoral Immunity: *Antibody Function* @MasteringMicrobiology

# Cellular Immunity Response Process

### LEARNING OBJECTIVES

**17-12** Describe at least one function of each of the following: M cells, $T_H$ cells, CTLs, $T_{reg}$ cells, NK cells.

**17-13** Differentiate T helper, T cytotoxic, and T regulatory cells.

**17-14** Differentiate $T_H1$, $T_H2$, and $T_H17$ cells.

**17-15** Define *apoptosis*.

**17-16** Define *antigen-presenting cell*.

Humoral antibodies are effective against pathogens that are circulating freely in the body, where the antibodies can make contact with them. But intracellular antigens, such as a virus, certain bacteria, and some parasites, are not exposed to these circulating antibodies since they enter host cells. T cells likely evolved to combat the problem posed by these intracellular pathogens. They are also the way in which the immune system recognizes other cells that are abnormal—especially cancer cells. Like B cells, each T cell is specific for only a certain antigen. However, T cells will recognize only antigen fragments bound to MHC.

About 98% of immature T cells are eliminated in the thymus, which is akin to clonal deletion in B cells. This reflects a weeding-out process, called **thymic selection**, that allows only those T cells that correctly recognize foreign materials and host cells to continue. Once mature, T cells migrate from the thymus by way of the blood and lymphatic system to various lymphoid tissues, where they are most likely to encounter antigens. (See Figure 16.5, page 455, for a review of the lymphatic system.)

Pathogens destined to live intracellularly most frequently enter the body via the gastrointestinal or respiratory tracts. Each of these tracts is lined with a barrier of epithelial cells. Normally, something can pass this barrier in the gastrointestinal tract only by way of a scattered array of gateway cells called **microfold cells**, or **M cells (Figure 17.9)**. Instead of the myriad of fingerlike microvilli found on the surface of absorptive epithelial cells of the intestinal tract, M cells have microfolds. M cells are located over **Peyer's patches**, which are secondary lymphoid organs located on the intestinal wall. M cells take up antigens from the intestinal tract and allow their transfer to the lymphocytes and antigen-presenting cells of the immune system found throughout the intestinal tract, just under the epithelial-cell layer but especially in the Peyer's patches. It is also here that antibodies, mostly IgA essential for mucosal immunity, are formed and migrate to the mucosal lining.

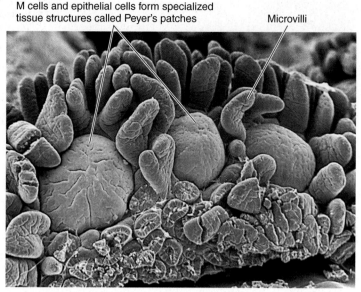

**(a)** M cells are part of the makeup of Peyer's patches (green). Note the tips of the closely packed microvilli on the surrounding epithelial cells. SEM ⊢ 27 μm

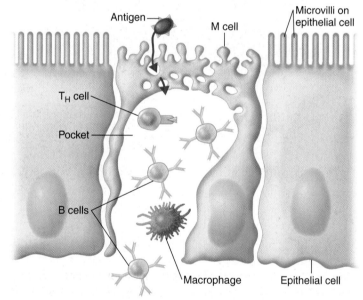

**(b)** M cells facilitate contact between antigens passing through the intestinal tract and cells of the body's immune system.

**Figure 17.9 M cells.** M cells are located within Peyer's patches (see Figure 16.5, page 455), which are located on the intestinal wall. Their function is to transport antigens encountered in the digestive tract to contact lymphocytes and antigen-presenting cells (see this page) of the immune system.

**Q** Why are M cells especially important for immune defenses against diseases affecting the digestive system?

## Antigen-Presenting Cells (APCs)

**Antigen-presenting cells (APCs)** associated with cellular immunity include B cells, dendritic cells, and activated macrophages. All APCs have MHCs on their surfaces that presents potential

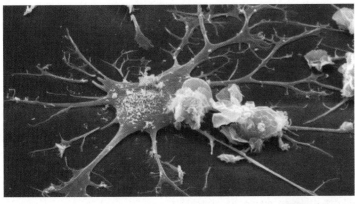

**Figure 17.10  A dendritic cell.** These cells link innate immunity and adaptive immunity by presenting antigens to T cells. The dendritic cell (pink) here is interacting with lymphocytes (yellow) that have been infected by a virus and are producing abnormal endogenous antigens.

**Q** What is the role of dendritic cells in immunity?

antigenic fragments to T cells. T cells interacting with the antigenic epitope and MHC will lead to T cell

Play Antigen Processing and Presentation: *MHC* @MasteringMicrobiology

activation. APCs produce IL-12, which activates $T_H1$ cells.

### Dendritic Cells

**Dendritic cells (DCs)** have long extensions called *dendrites* (**Figure 17.10**) because they resemble nerve cell dendrites. Dendritic cells are the main APCs that induce immune responses by T cells. Dendritic cells in the skin and genital tract are called *Langerhans cells* or *Langerhans DCs*. Other dendritic cell populations are found in the lymph nodes, spleen, thymus, blood, and various tissues—except the brain. Dendritic cells that act as sentinels in these tissues engulf invading microbes, degrade them, and transfer them to lymph nodes for display to T cells located there.

### Macrophages

**Macrophages** (from the Greek for *large eaters*) are usually found in a resting state. We have already discussed the function of these cells related to phagocytosis. They are important for innate immunity and for ridding the body of worn-out blood cells (about 200 billion per day) and other debris, such as cellular remnants from apoptosis. Their motility and phagocytic capabilities are greatly increased when they are stimulated to become **activated macrophages** (**Figure 17.11**). This activation can be initiated by ingestion of antigenic material. Other stimuli, such as cytokines produced by an activated T helper cell, can further enhance the capabilities of macrophages. Once activated, macrophages are more effective as phagocytes and as APCs. Activated macrophages are important factors in

the control of cancer cells, virus-infected cells, and intracellular pathogens such as the tubercle bacillus. Their appearance becomes recognizably different as well—they are larger and become ruffled.

After taking up an antigen anywhere in the body, APCs tend to migrate to lymph nodes or other lymphoid centers on the mucosa, where they present the antigen to T cells located there. T cells carrying receptors that are capable of binding with any specific antigen are present in relatively limited numbers. Migration of APCs increases the opportunity for these particular T cells to encounter the antigen for which they are specific.

## Classes of T Cells

There are classes of T cells that have different functions, rather like the classes of immunoglobulins. As mentioned previously, **T helper cells** (**$T_H$ cells**) cooperate with B cells in the production of antibodies, mainly through cytokine signaling. For their role in cellular immunity, T cells interact more directly with antigens.

T cells are also classified by certain glycoproteins on their cell surface called **clusters of differentiation**, or **CD**. These are membrane molecules that are especially important for adhesion to receptors. The CDs of greatest interest are CD4 and CD8; T cells that carry these molecules are named **CD4$^+$** and **CD8$^+$** cells, respectively. (For the importance of these molecules in HIV infection, see Figure 19.14 on page 545.) T helper cells are classified as CD4$^+$, which bind to MHC class II molecules on B cells and other APCs. CTL cells are classified as CD8$^+$, which bind to MHC class I molecules. T cells that have not encountered a pathogen are called *naïve*. After contact with a pathogen, the T cell is activated and can form effector and memory cells.

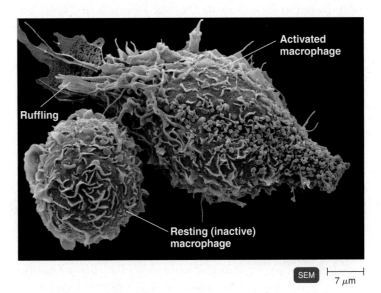

**Figure 17.11  Activated macrophages.** When activated, macrophages become larger and ruffled.

**Q** How do macrophages become activated?

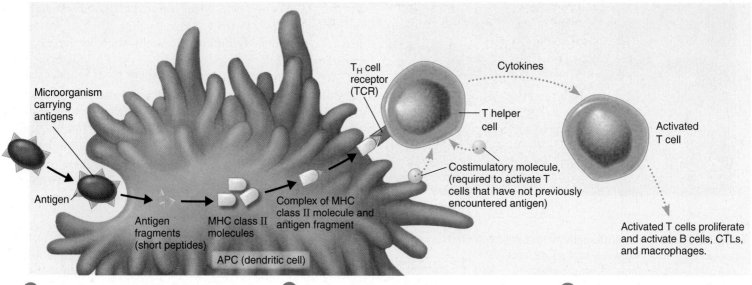

**1** An APC encounters and ingests a microorganism. The antigen is enzymatically processed into short peptides, which combine with MHC class II molecules and are displayed on the surface of the APC.

**2** A receptor (TCR) on the surface of the CD4$^+$ T helper cell ($T_H$ cell) binds to the MHC–antigen complex. The $T_H$ cell or APC is stimulated to secrete a costimulatory molecule. These two signals activate the $T_H$ cell, which produces cytokines.

**3** The cytokines cause the $T_H$ cell (which recognizes a dendritic cell that is producing costimulatory molecules) to become activated.

**Figure 17.12 Activation of CD4$^+$ T helper cells.** To activate a T helper cell, at least two signals are required: the first is the binding of the TCR to the processed antigen, and the second signal requires a costimulating cytokine, such as IL-2 and others. Once activated, the $T_H$ cell secretes cytokines that affect the effector functions of multiple cell types of the immune system.

**Q** Which cells are antigen-presenting cells?

## T Helper Cells (CD4$^+$ T Cells)

T helper cells can recognize an antigen presented on the surface of a macrophage and activate it, making the macrophage more effective in both phagocytosis and in antigen presentation. Dendritic cells are especially important in the activation of T helper cells and in developing their effector functions (**Figure 17.12**).

### CLINICAL CASE

The immune response to antigens takes place primarily in the secondary lymphoid organs, such as the lymph nodes, the mucosa-associated lymphoid tissue, and the spleen. During a primary immune response, pathogens and their constituents are transported to these tissues, where microbial antigens are presented to B cells that constantly enter and leave the secondary lymphoid organs. Dr. Marsden reviews her autopsy notes and sees that Mrs. Vasquez does not have a spleen; it had been removed following an automobile accident some years before.

**Why is the fact that the patient has no spleen important?**

477 478 482 484 **488** 491

For a T helper cell to become activated, its T cell receptor recognizes antigen fragments held in a complex with proteins of MHC class II on the surface of the APC. This is the initial signal for activation; a second, costimulatory signal that comes from either the APC or T helper cell is also required for activation. The activated T helper cell begins to proliferate and secrete cytokines. Secretion of cytokines causes the proliferating T helper cells to differentiate into populations of T helper cell subsets, such as $T_H1$, $T_H2$, and $T_H17$. These subsets act on different cells of the body's defensive systems. They also form a population of long-lived memory cells.

The cytokines produced by **$T_H1$ cells**, especially IFN-$\gamma$, activate cells related to delayed hypersensitivity (see page 535) and are responsible for activation of macrophages (see page 487). They also stimulate the production of antibodies that promote phagocytosis and are especially effective in enhancing the activity of complement, such as opsonization and inflammation (see Figure 16.12, page 466). As shown in **Figure 17.13**, the generation of cytotoxic T lymphocytes requires action by a $T_H1$ cell.

**$T_H2$ cells** produce cytokines, including IL-4. They are associated primarily with the production of antibodies, especially IgE, that are important in allergic reactions (see the discussion of hypersensitivity on page 525). They are also important in

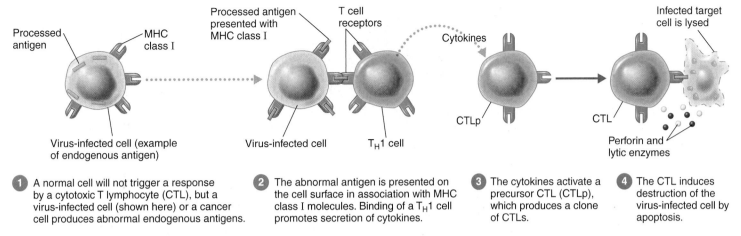

**1** A normal cell will not trigger a response by a cytotoxic T lymphocyte (CTL), but a virus-infected cell (shown here) or a cancer cell produces abnormal endogenous antigens.

**2** The abnormal antigen is presented on the cell surface in association with MHC class I molecules. Binding of a $T_H1$ cell promotes secretion of cytokines.

**3** The cytokines activate a precursor CTL (CTLp), which produces a clone of CTLs.

**4** The CTL induces destruction of the virus-infected cell by apoptosis.

**Figure 17.13  Killing of virus-infected target cell by cytotoxic T lymphocyte.**

**Q** Differentiate a cytotoxic T lymphocyte cell from a T helper cell.

the activation of the eosinophils that defend against infections by extracellular parasites, such as helminths (see Figure 17.16 on page 492).

A third subset is named $T_H17$ **cells** because of their production of large quantities of the cytokine IL-17. The discovery of $T_H17$ cells answered the questions raised by the observation that $T_H1$ and $T_H2$ cells were not effective in dealing with certain infections by extracellular bacteria and fungi. IL-17 acts as a chemokine to recruit neutrophils. Excessive amounts of $T_H17$ cells probably contribute to the inflammation and injury to tissue found in certain autoimmune diseases such as multiple sclerosis, psoriasis, rheumatoid arthritis, and Crohn's disease. They may be associated with the pathologic effects of diseases such as asthma and allergic dermatitis. But they also serve, helpfully, to combat microbial infections of the mucosa by the production of cytokines such as IL-22, which stimulate epithelial cells to produce antimicrobial proteins. Therefore, a severe deficiency of $T_H17$ cells may make one more susceptible to opportunistic infections.

The functions of the three subsets directly involved in the body's defenses against external microbial threats are summarized in **Figure 17.14**, with their primary cytokine referenced in the figure.

▶ Play Antigen Processing and Presentation: *Steps*
@MasteringMicrobiology

### T Regulatory Cells

**T regulatory ($T_{reg}$) cells** make up about 5–10% of the T cell population. They are a subset of the T helper cells and are distinguished by carrying an additional CD25 molecule. Their primary function is to combat autoimmune reactions by suppressing T cells that escape deletion in the thymus without the necessary "education" to avoid reacting against the body's self. They are also useful in protecting the resident microbiota that live in our intestines and aid digestion. Similarly, in pregnancy

they may play a role in protecting the fetus from rejection as nonself. Recently, researchers have discovered evidence of $T_{reg}$ involvement in establishing the skin microbiome (see Exploring the Microbiome).

### Cytotoxic T Lymphocytes (CD8+ T Cells)

A class of T cells called **precursor T cytotoxic cells (CTLp)** can differentiate into an effector cell called a cytotoxic T lymphocyte. **Cytotoxic T lymphocytes,** or **CTLs,** are not capable of attacking any target cell as they emerge from the thymus as CTLp cells. However, they quickly attain this capability. This differentiation requires sequential—and complex—activation of the CTLp by an antigen processed by an antigen-presenting cell and interaction with a T helper cell and costimulatory signals. The resulting CTL is an effector cell that has the ability to recognize and kill target cells that are considered nonself (see Figure 17.13). Primarily, these target cells are self-cells that have been altered by infection with a pathogen, especially viruses. On the infected cell's surface, they carry fragments of **endogenous antigens** that are generally synthesized within the cell and are mostly of viral or parasitic origin. Other important target cells are tumor cells (see Figure 19.12, page 542) and transplanted foreign tissue. Rather than reacting with antigenic fragments presented by an APC in complex with MHC class II molecules, the CTL recognizes endogenous antigens on the target cell's surface that are in combination with an MHC class I molecule. MHC class I molecules are found on nucleated cells; therefore, a CTL can attack almost any cell of the host that has been altered.

In its attack, a CTL attaches to the target cell and releases a pore-forming protein, **perforin.** Pore formation contributes to the subsequent death of the cell and is similar to the action of the complement membrane attack complex described in Chapter 16 (see Figure 16.12a, page 466). **Granzymes,** proteases

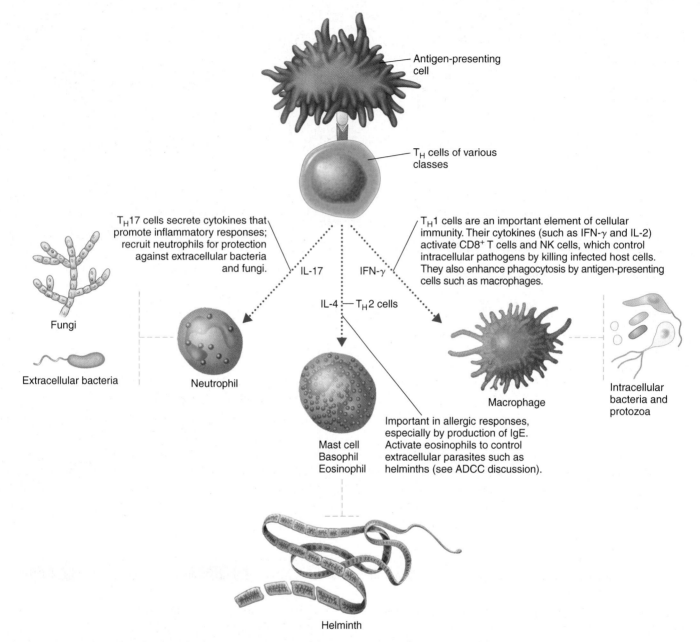

**Figure 17.14 Lineage of effector T helper cell classes and pathogens targeted.**

**Q** Why would IFN-γ be used to treat tuberculosis?

that induce apoptosis, are then able to enter through the pore. **Apoptosis** (a-pah-TO-sis; from the Greek for falling away like leaves) is also called *programmed cell death*. This is a necessary process in multicellular organisms.*

Tracking the death of cells and determining whether the demise is natural or due to trauma or disease are important. If the cell's death is due to trauma or disease, then the body's

defense and repair mechanisms mobilize. Programmed cell death is also an infection-fighting mechanism of last resort: if a cell cannot clear a pathogen any other way, it will die. This helps prevent spread of pathogens, particularly viruses, to nearby healthy cells. Cells that die from apoptosis first cut their genome into fragments, and the external membranes bulge outward in a manner called *blebbing* (**Figure 17.15**). Signals are displayed on the cell's surface that attract circulating phagocytes to digest the remains before any significant leakage of contents occurs.

Play Cell-Mediated Immunity: *Cytotoxic T Lymphocytes* @MasteringMicrobiology

---

* Apoptosis expert Gerry Melino says that without apoptosis, a human body would accumulate 2 tons of bone marrow and lymph nodes and a 16-kilometer intestine by age 80!

**The Relationship between Your Immune Cells and Skin Microbiota**

Over 200 genera of bacteria are permanent or transient members of the skin's microbiome. In fact, over a million microbes populate each square centimeter of skin, and another million T-cells are located near capillaries there. This brings up the question of why certain members of the microbial community are allowed to live whereas others are targeted and destroyed.

Take the example of *Staphylococcus epidermidis* bacteria. Studies show that its ubiquitous presence on the skin actually protects against infection by pathogens. This is partially due to competitive exclusion with other microbes. But it's also because *S. epidermidis* promotes production of the pro-inflammatory cytokines, IL-1 and IL-17, by effector T cells. Yet the T cells do not initiate an inflammatory response against *S. epidermidis* itself.

In normal human skin, 90% of the lymphocytes present are T regulatory cells found near hair follicles. Genetic analysis shows that the majority of Treg cells found in adult skin are memory cells that suppress T helper cells from inciting a fight against nearby microbes (namely, all that *S. epidermidis*). This explains why the immune system is tolerant of its presence.

By contrast, fetal skin contains no memory cells. Babies are also born without normal microbiota on their skin yet—and an immature immune system, too. The skin microbiota and our immune systems develop codependently. Studies in germ-free mice show that contact with *S. epidermidis* during the first few weeks of life is actually necessary to induce the memory T regulatory cells in skin. This time frame corresponds with migration of T regulatory cells to the skin during the first 2 weeks after birth.

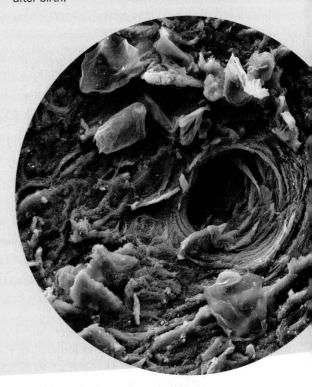

*Staphylococcus* (purple) on human skin.

---

CLINICAL CASE Resolved

T cells, B cells, and dendritic cells are found in the spleen. In fact, about half of all blood lymphocytes circulate through the spleen daily. Normally, phagocytic cells of the spleen clear antibody- and complement-coated microorganisms very rapidly, thereby preventing the dissemination of infectious organisms to important organs. *Capnocytophaga* bacteria cause a range of infections, from self-limiting cellulitis to fatal septicemia. Most fatal infections are contracted from dogs and occur in people without spleens. Unfortunately, Mrs. Vasquez contracted the infectious bacteria from her pet, and because her spleen had been removed some time before, she did not have the immune response necessary to combat the resulting deadly infection.

477 478 482 484 488 491

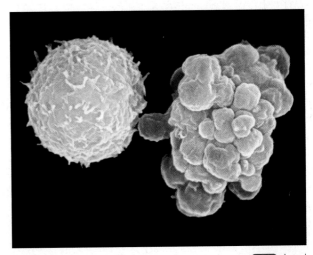

SEM — 4 μm

**Figure 17.15 Apoptosis.** A normal B cell is shown on the left. On the right, a B cell is undergoing apoptosis. Notice the bubble-like blebs.

**Q** What is apoptosis?

# Nonspecific Cells and Extracellular Killing by the Adaptive Immune System

### LEARNING OBJECTIVE

**17-17** Describe the function of natural killer cells.

**17-18** Describe the role of antibodies and natural killer cells in antibody-dependent cell-mediated cytotoxicity.

CTLs are not the only cells that can lead to the destruction of a target cell. **Natural killer (NK) cells** can also destroy certain virus-infected cells and tumor cells and can attack parasites, which are normally much larger than bacteria (**Figure 17.16**). NK cells are large granular leukocytes (10–15% of all circulating lymphocytes). In contrast to CTLs, NK cells are not immunologically specific, meaning that they do not need to be stimulated by a particular antigen. Instead, NK cells distinguish normal cells from transformed cells, or cells infected with intracellular pathogens.

NK cells first contact the target cell and determine whether it expresses MHC class I self-antigens. If it does not—which is often the case in the early stages of viral infection and with some infecting viruses that have developed a system of interfering with the usual presentation of antigens on an APC—they kill the target cell by mechanisms similar to that of a CTL. Tumor cells also have a reduced number of MHC class I molecules on their surfaces. NK cells cause pores to form in the target cell, which leads to either lysis or apoptosis.

The functions of NK cells and the other principal cells involved in cellular immunity are briefly summarized in Table 17.3.

With the help of antibodies produced by the humoral immune system, the cell-mediated immune system can stimulate natural killer cells and cells of the innate defense system,

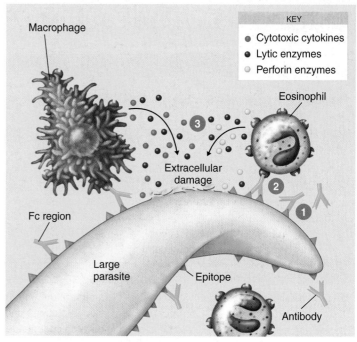

Organisms, such as many parasites, that are too large for ingestion by phagocytic cells must be attacked externally.

**Figure 17.16 Antibody-dependent cell-mediated cytotoxicity (ADCC).** If an organism—for example, a parasitic worm—is too large for ingestion and destruction by phagocytosis, it can be attacked by immune system cells that remain external to it. **1** The target cell is first coated with antibodies. **2** Cells of the immune system, such as eosinophils, macrophages, and NK cells, bind to the Fc regions of the attached antibodies. **3** The target cell is then lysed by substances secreted by the cells of the immune system.

**Q** **Why is ADCC important protection against parasitic protozoa and helminths?**

such as macrophages, to kill targeted cells. In this way, an organism such as a fungus, protozoan, or helminth that is too large to be phagocytized can be attacked by immune system cells. This is referred to as **antibody-dependent cell-mediated cytotoxicity (ADCC)**. As is illustrated in Figure 17.16, the target cell is first coated with antibodies. A variety of cells of the immune system bind to the Fc regions of these antibodies and, thus, to the target cell. The attacking cells secrete substances that then lyse the target cell.

| Principal Cells That Function | |
|---|---|
| **TABLE 17.3** in Cell-Mediated Immunity | |
| **Cell** | **Function** |
| **T Helper (T$_H$1) Cell** | Activates cells related to cell-mediated immunity: macrophages, CTLs, and natural killer cells |
| **T Helper (T$_H$2) Cell** | Stimulates production of eosinophils, IgM, and IgE |
| **T Helper (T$_H$17) Cell** | Recruits neutrophils; stimulates production of antimicrobial proteins |
| **Cytotoxic T Lymphocyte (CTL)** | Destroys target cells on contact by inducing apoptosis |
| **T Regulatory (T$_{reg}$) Cell** | Regulates immune response and helps maintain self-tolerance |
| **Activated Macrophage** | Enhanced phagocytic activity; attacks cancer cells |
| **Natural Killer (NK) Cell** | Attacks and destroys target cells; participates in antibody-dependent cell-mediated cytotoxicity |

# Immunological Memory

## LEARNING OBJECTIVE

**17-19** Distinguish a primary from a secondary immune response.

Antibody-mediated immune responses intensify after the primary response, where a particular antigen is first met and corresponding antibodies are produced. This **secondary response** is also called the **memory response**, or **anamnestic response**. As shown in **Figure 17.17**, this response is comparatively more rapid, reaching a peak in only 2 to 7 days, longer in duration, and considerably greater in magnitude than the primary response. The secondary response is due to the portion of activated B cells that, instead of transforming into antibody-secreting plasma cells, become memory cells. As mentioned earlier, IgM is the first antibody that B cells make during the primary response to an antigen. But an individual B cell is also capable of making different classes of antibody, such as IgG, IgE, or IgA, all with unchanged antigenic specificity. Termed **class switching**, this is observed, especially, in the case of the primary and secondary immune response (compare IgG and IgM in Figure 17.17). Generally speaking, when IgG begins to be produced in the secondary response, the production of IgM will decrease or be sharply curtailed.

Memory cells do not reproduce, but they are long lived. Years or even decades later, if these cells are stimulated by the same antigen, they very rapidly differentiate into antibody-producing plasma cells.

The intensity of the antibody-mediated humoral response can be reflected by the **antibody titer**, the relative amount of antibody found in the blood serum. During a primary immune response, the exposed person's serum contains no detectable antibodies against an antigen for 4 to 7 days. Then there is a slow rise in antibody titer: first, IgM class antibodies are produced, followed by IgG peaking in about 10 to 17 days, after which antibody titer gradually declines. This pattern is characteristic of a primary response to an antigen. A similar response occurs with T cells, which, as we see in Chapter 19, is necessary for establishing the lifelong memory for distinguishing self from nonself.

**Play Humoral Immunity:** *Primary Immune Response, Secondary Immune Response* @Mastering**Microbiology**

**CHECK YOUR UNDERSTANDING**

✔ **17-19** Is the anamnestic response primary or secondary?

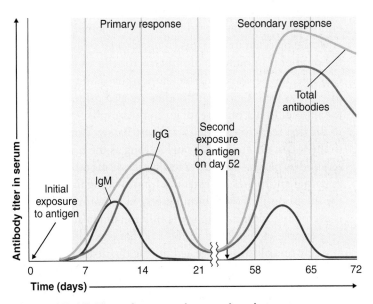

**Figure 17.17 The primary and secondary immune responses to an antigen.** IgM appears first in response to the initial exposure. IgG follows and provides longer-term immunity. The second exposure to the same antigen stimulates the memory cells (formed at the time of initial exposure) to rapidly produce a large amount of antibody. The antibodies produced in response to this second exposure are mostly IgG.

**Q** Why do many diseases, such as measles, occur only once in a person, yet others, such as colds, occur more than once?

# Types of Adaptive Immunity

**17-20**  Contrast the four types of adaptive immunity.

Immunity is acquired *actively* when a person is exposed to microorganisms or foreign substances and the immune system responds. Immunity is acquired *passively* when antibodies are transferred from one person to another. Passive immunity in the recipient lasts only as long as the antibodies are present—in most cases, a few weeks. Both actively acquired immunity and passively acquired immunity can be obtained by natural or artificial means (**Figure 17.18**).

   The four types of adaptive immunity can be summarized as follows:

- **Naturally acquired active immunity** develops from exposure to antigens, illness, and recovery. Once acquired, immunity is lifelong for some diseases such as measles. In other cases, especially for intestinal diseases, immunity may last only a few years. *Subclinical infections,* or *inapparent infections* (those that produce no noticeable symptoms or signs of illness), can also confer immunity.

- **Naturally acquired passive immunity** is the transfer of antibodies from a mother to her infant. Maternal antibodies cross the placenta to the fetus (*transplacental transfer*). If the mother is immune to diphtheria, rubella, or polio, for example, the newborn will be temporarily immune to these diseases as well. Certain antibodies are also passed from the mother to her nursing infant in breast milk, especially in the first secretions, called *colostrum.* Passive immunity in the infant generally lasts only as long as the transmitted antibodies persist—usually a few weeks or months. These maternal antibodies are essential for providing immunity to the infant until its own immune system matures. Colostrum is even more important to some other mammals; calves, for example, do not have antibodies that cross the placenta and rely on colostrum ingested during the first day of life. Researchers often specify fetal calf serum for certain experimental uses because it does not contain maternal antibodies.

- **Artificially acquired active immunity** is the result of vaccination—which will be discussed in Chapter 18. **Vaccines,** also called **immunizations,** introduce antigens to the body. For example, killed or inactivated bacteria can

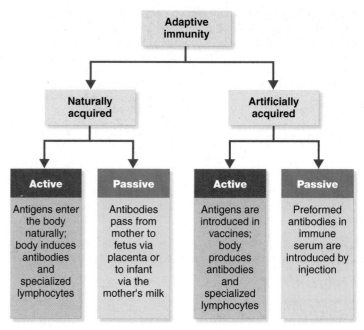

**Figure 17.18  Types of adaptive immunity.**

**Q** Which type of immunity, active or passive, lasts longer?

be injected into the body resulting in an immune response without causing infection.

- **Artificially acquired passive immunity** involves the injection of antibodies (rather than antigens) into the body. These antibodies come from an animal or a human who is already immune to the disease. This is used for postexposure prophylaxis from diseases such as rabies or in immunotherapy (discussed in Chapters 18 and 19).

   When an individual is given artificially acquired passive immunity, it confers an immediate passive protection against the disease. However, although artificially acquired passive immunity is immediate, it is short-lived because antibodies are degraded by the recipient. The half-life of an injected antibody (the time required for half of the antibodies to disappear) is typically about 3 weeks.

   **Figure 17.19** shows how various parts of the innate and adaptive immune system interact with each other.

**CHECK YOUR UNDERSTANDING**

✔ **17-20** What type of adaptive immunity is involved when gamma globulin is injected into a person?

# The Dual Nature of the Adaptive Immune System

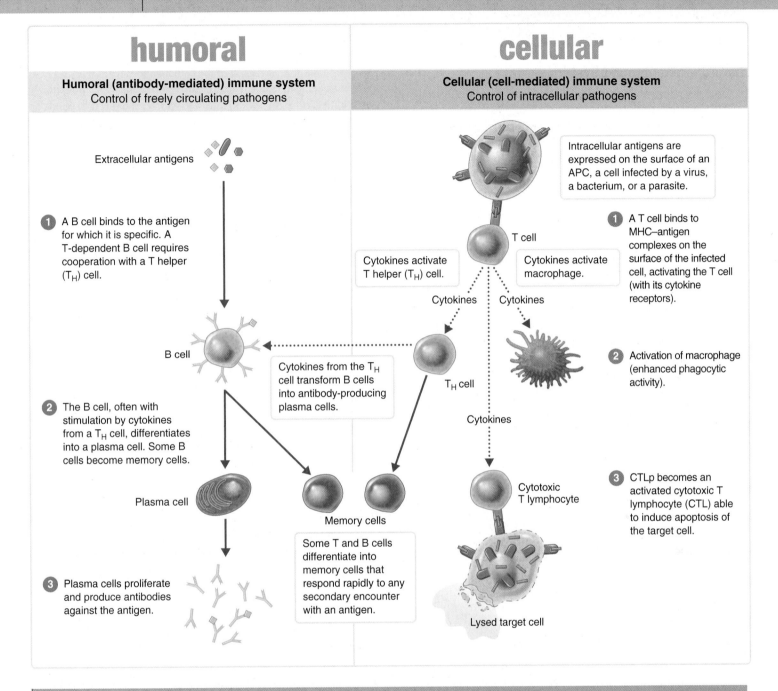

## humoral

**Humoral (antibody-mediated) immune system**
Control of freely circulating pathogens

Extracellular antigens

**1** A B cell binds to the antigen for which it is specific. A T-dependent B cell requires cooperation with a T helper (T$_H$) cell.

B cell

Cytokines from the T$_H$ cell transform B cells into antibody-producing plasma cells.

**2** The B cell, often with stimulation by cytokines from a T$_H$ cell, differentiates into a plasma cell. Some B cells become memory cells.

Plasma cell

Memory cells

Some T and B cells differentiate into memory cells that respond rapidly to any secondary encounter with an antigen.

**3** Plasma cells proliferate and produce antibodies against the antigen.

## cellular

**Cellular (cell-mediated) immune system**
Control of intracellular pathogens

Intracellular antigens are expressed on the surface of an APC, a cell infected by a virus, a bacterium, or a parasite.

T cell

Cytokines activate T helper (T$_H$) cell.

Cytokines activate macrophage.

Cytokines    Cytokines

**1** A T cell binds to MHC–antigen complexes on the surface of the infected cell, activating the T cell (with its cytokine receptors).

T$_H$ cell

Cytokines

**2** Activation of macrophage (enhanced phagocytic activity).

Cytotoxic T lymphocyte

**3** CTLp becomes an activated cytotoxic T lymphocyte (CTL) able to induce apoptosis of the target cell.

Lysed target cell

- The adaptive immune system is divided into two parts, each responsible for dealing with pathogens in different ways. These two systems function interdependently to keep the body free of pathogens.
- **Humoral immunity**, also called antibody-mediated immunity, is directed at freely circulating pathogens and depends on B cells.
- **Cellular immunity**, also called cell-mediated immunity, depends on T cells to eliminate intracellular pathogens, reject foreign tissue recognized as nonself, and destroy tumor cells.
- The adaptive immune sytem provides specificity, clonal expansion, and memory.

# Study Outline

 **Go to** @MasteringMicrobiology **for** Interactive Microbiology, *In the Clinic* *videos, MicroFlix, MicroBoosters, 3D animations, practice quizzes, and more.*

## The Adaptive Immune System (p. 476)

1. Adaptive immunity is the body's ability to react specifically to a microbial infection.

2. The body's response to the first contact with a particular antigen is called the primary response. Specific cells are activated to destroy the antigen.

3. Memory cells respond to subsequent contact with the same antigen.

## Dual Nature of the Adaptive Immune System

(pp. 476–477)

1. Humoral immunity involves antibodies, which are found in serum and lymph and are produced by B cells.

2. Lymphocytes that mature in red bone marrow become B cells.

3. Cellular immunity involves T cells.

4. Lymphocytes that migrate through the thymus become T cells.

5. T cell receptors recognize antigens presented on MHC.

6. Cellular immunity responds to intracellular antigens; humoral immunity responds to antigens in body fluids.

## Cytokines: Chemical Messengers of Immune Cells (pp. 477–478)

1. Cells of the immune system communicate with each other by means of chemicals called cytokines.

2. Interleukins (IL) are cytokines that serve as communicators between leukocytes.

3. Chemokines cause leukocytes to migrate to an infection.

4. Interferon-$\gamma$ stimulates the immune response; other INFs protect cells against viruses.

5. Tumor necrosis factor promotes the inflammatory reaction.

6. Hematopoietic cytokines promote development of white blood cells.

7. Overproduction of cytokines leads to a cytokine storm, which results in tissue damage.

## Antigens and Antibodies (pp. 478–484)

### Antigens (pp. 478–479)

1. An antigen (or immunogen) is a chemical substance that causes the body to produce specific antibodies.

2. As a rule, antigens are proteins or large polysaccharides. Antibodies are formed against specific regions on antigens called epitopes, or antigenic determinants.

3. A hapten is a low-molecular-mass substance that cannot cause the formation of antibodies unless combined with a carrier molecule; haptens react with their antibodies independently of the carrier molecule.

### Humoral Immunity: Antibodies (pp. 479–481)

4. An antibody, or immunoglobulin, is a protein produced by B cells in response to an antigen and is capable of combining specifically with that antigen.

5. Typical monomers consist of four polypeptide chains: two heavy chains and two light chains. They have two antigen-binding sites.

6. Within each chain is a variable (V) region that binds the epitope and a constant (C) region that distinguishes the different classes of antibodies.

7. An antibody monomer is Y-shaped or T-shaped: the V regions form the tips, and the C regions form the base and Fc (stem) region.

8. The Fc region can attach to a host cell or to complement.

9. IgG antibodies are the most prevalent in serum; they provide naturally acquired passive immunity, neutralize bacterial toxins, participate in complement fixation, and enhance phagocytosis.

10. IgM antibodies consist of five monomers held by a joining chain; they are involved in agglutination and complement fixation.

11. Serum IgA antibodies are monomers; secretory IgA antibodies are dimers that protect mucosal surfaces from invasion by pathogens.

12. IgD antibodies are on B cells; they may assist the immune response.

13. IgE antibodies bind to mast cells and basophils and are involved in allergic reactions.

## Humoral Immunity Response Process (pp. 482–484)

1. B cells have antibodies on their surfaces, which recognize specific epitopes.

2. For T-independent antigens: a clone of B cells is selected.

3. For T-dependent antigens: the B cell's immunoglobulins combine with an antigen, and the antigen fragments, combined with MHC class II, activate $T_H$ cells. The $T_H$ cells activate a B cell.

### Activation and Clonal Expansion of Antibody-Producing

Cells (pp. 482–485)

4. Activated B cells differentiate into plasma cells and memory cells.

5. Plasma cells produce IgM antibodies and then produce other classes, usually IgG.

6. B cells that recognize self are eliminated by clonal deletion.

7. Immunoglobulin genes in B cells recombine so that mature B cells each have different genes for the V region of their antibodies.

## Results of the Antigen–Antibody Interaction

(pp. 484–485)

1. An antigen–antibody complex forms when an antibody binds to its specific epitopes on an antigen.

2. Agglutination results when an antibody combines with epitopes on two different cells.

3. Opsonization enhances phagocytosis of the antigen.

4. Antibodies that attach to microbes or toxins and prevent them gaining access to the host or performing their action cause neutralization.

5. Complement activation results in cell lysis.

## Cellular Immunity Response Process (p. 486)

1. T cells mature in the thymus gland. Thymic selection removes T cells that don't recognize MHC molecules of the host and T cells that will attach host cells presenting self proteins in MHC.

2. Helper T cells recognize antigens processed by antigen-presenting cells and presented with MHC II.

3. Cytotoxic T cells recognize antigens processed by all host cells and presented with MHC class I.

### Antigen-Presenting Cells (APCs) (pp. 486–487)

4. APCs include B cells, dendritic cells, and macrophages.

5. Dendritic cells are the primary APCs.

6. Activated macrophages are effective phagocytes and APCs.

7. APCs carry antigens to lymphoid tissues where T cells that recognize the antigen are located.

### Classes of T Cells (pp. 487–490)

8. T cells are classified according to their functions and cell-surface glycoproteins called CDs.

9. T helper ($CD4^+$ T) cells differentiate into $T_H1$ cells, which are involved in cellular immunity; $T_H2$ cells, which are involved in humoral immunity and are associated with allergic reactions and parasitic infections; and $T_H17$ cells, which activate innate immunity.

10. T regulatory cells ($T_{reg}$) suppress T cells against self.

11. Cytotoxic lymphocytes (CTLs), or $CD8^+$ cells, are activated by endogenous antigens and MHC class I on a target cell and are transformed into effector and memory CTLs.

12. CTLs lyse or induce apoptosis in the target cell.

### Nonspecific Cells and Extracellular Killing by the Adaptive Immune System (pp. 492–493)

1. Natural killer (NK) cells lyse virus-infected cells, tumor cells, and parasites. They kill cells that do not express MHC class I antigens.

2. In antibody-dependent cell-mediated cytotoxicity (ADCC), NK cells and macrophages lyse antibody-coated cells.

### Immunological Memory (p. 493)

1. The relative amount of antibody in serum is called the antibody titer.

2. The peak IgG titer in the primary response occurs 10–17 days after exposure to an antigen.

3. The peak titer in the secondary response occurs 2–7 days after exposure.

### Types of Adaptive Immunity (p. 494)

1. Immunity resulting from infection is called naturally acquired active immunity; this type of immunity may be long-lasting.

2. Antibodies transferred from a mother to a fetus (transplacental transfer) or to a newborn in colostrum results in naturally acquired passive immunity in the newborn; this type of immunity can last up to a few months.

3. Immunity resulting from vaccination is called artificially acquired active immunity and can be long-lasting.

4. Artificially acquired passive immunity refers to humoral antibodies acquired by injection; this type of immunity can last for a few weeks.

5. Serum containing antibodies is often called antiserum or gamma globulin.

---

# Study Questions

For answers to the Knowledge and Comprehension questions, turn to the Answers tab at the back of the textbook.

## Knowledge and Comprehension

### Review

1. Contrast the terms in the following pairs:
   a. innate and adaptive immunity
   b. humoral and cellular immunity
   c. active and passive immunity
   d. $T_H1$ and $T_H2$ cells
   e. natural and artificial immunity
   f. T-dependent and T-independent antigens
   g. $CD8^+$ T cell and CTL
   h. immunoglobulin and TCR

2. What does MHC stand for? What is the function of MHC? What types of T cells interact with MHC class I? With MHC class II?

3. **DRAW IT** Label the heavy chains, light chains, and variable and $F_c$ regions of this typical antibody. Indicate where the antibody binds to antigen. Sketch an IgM antibody.

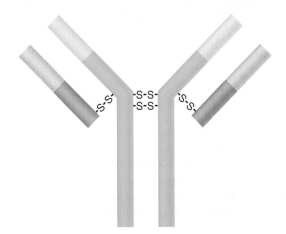

4. Diagram the roles that T cells and B cells play in immunity.

5. Explain a function for the following types of cells: CTL, $T_H$, and $T_{reg}$. What is a cytokine?

6. DRAW IT
   a. In the graph below, at time *A* the host was injected with tetanus toxoid. Show the response to a booster dose at time *B*.
   b. Draw the antibody response of this same individual to exposure to a new antigen at time *B*.

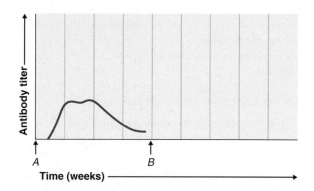

7. How would each of the following prevent infection?
   a. antibodies against *Neisseria gonorrhoeae* fimbriae
   b. antibodies against host cell mannose

8. How can a human make over 10 billion different antibodies with only 25,000 different genes?

9. Explain why a person who recovers from a disease can attend others with the disease without fear of contracting it.

10. NAME IT This cell is found in skin and lymphoid tissue. It is a phagocyte and activates T cells.

## Multiple Choice

Match the following choices to questions 1–4:
   a. innate resistance
   b. naturally acquired active immunity
   c. naturally acquired passive immunity
   d. artificially acquired active immunity
   e. artificially acquired passive immunity

1. The type of protection provided by the injection of diphtheria toxoid.

2. The type of protection provided by the injection of antirabies serum.

3. The type of protection resulting from recovery from an infection.

4. A newborn's immunity to yellow fever.

Match the following choices to the statements in questions 5–7:
   a. IgA          d. IgG
   b. IgD          e. IgM
   c. IgE

5. Antibodies that protect the fetus and newborn.

6. The first antibodies synthesized; especially effective against microorganisms.

7. Antibodies that are bound to mast cells and involved in allergic reactions.

8. Put the following in the correct sequence to elicit an antibody response: (1) $T_H$ cell recognizes B cell; (2) APC contacts antigen; (3) antigen fragment goes to surface of APC; (4) $T_H$ recognizes antigen digest and MHC; (5) B cell proliferates.
   a. 1, 2, 3, 4, 5          d. 2, 3, 4, 1, 5
   b. 5, 4, 3, 2, 1          e. 4, 5, 3, 1, 2
   c. 3, 4, 5, 1, 2

9. A kidney-transplant patient experienced a cytotoxic rejection of his new kidney. Place the following in order for that rejection: (1) apoptosis occurs; (2) $CD8^+$ T cell becomes CTL; (3) granzymes released; (4) MHC class I activates $CD8^+$ T cell; (5) perforin released.
   a. 1, 2, 3, 4, 5          d. 3, 4, 5, 1, 2
   b. 5, 4, 3, 2, 1          e. 2, 3, 4, 1, 5
   c. 4, 2, 5, 3, 1

10. Patients with Chédiak-Higashi syndrome suffer from various types of cancer. These patients are most likely lacking which of the following?
   a. $T_{reg}$ cells          d. NK cells
   b. $T_H1$ cells          e. $T_H2$ cells
   c. B cells

## Analysis

1. Injections of CTLs completely removed all hepatitis B viruses from infected mice, but they killed only 5% of the infected liver cells. Explain how CTLs cured the mice.

2. Why is dietary protein deficiency associated with increased susceptibility to infections?

3. A positive tuberculin skin test shows cellular immunity to *Mycobacterium tuberculosis*. How could a person acquire this immunity?

4. On her vacation to Australia, Janet was bitten by a poisonous sea snake. She survived because the emergency department physician injected her with antivenin to neutralize the toxin. What is antivenin? How is it obtained?

## Clinical Applications and Evaluation

1. A woman had life-threatening salmonellosis that was successfully treated with anti-*Salmonella*. Why did this treatment work, when antibiotics and her own immune system failed?

2. A patient with AIDS has a low $T_H$ cell count. Why does this patient have trouble making antibodies? How does this patient make any antibodies?

3. A patient with chronic diarrhea was found to lack IgA in his secretions, although he had a normal level of serum IgA. What was this patient found to be unable to produce?

4. Newborns (under 1 year) who contract dengue have a higher chance of dying from it if their mothers had dengue prior to pregnancy. Explain why.

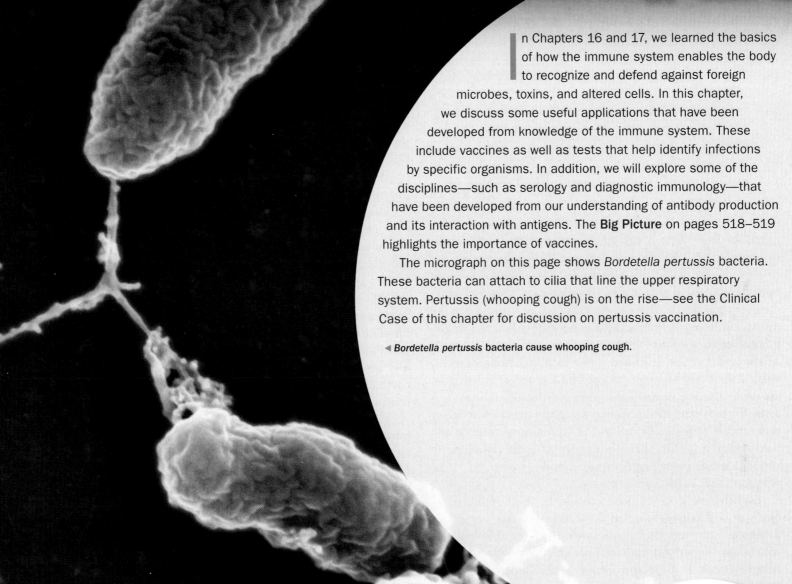

# Practical Applications of Immunology 18

I n Chapters 16 and 17, we learned the basics of how the immune system enables the body to recognize and defend against foreign microbes, toxins, and altered cells. In this chapter, we discuss some useful applications that have been developed from knowledge of the immune system. These include vaccines as well as tests that help identify infections by specific organisms. In addition, we will explore some of the disciplines—such as serology and diagnostic immunology—that have been developed from our understanding of antibody production and its interaction with antigens. The **Big Picture** on pages 518–519 highlights the importance of vaccines.

The micrograph on this page shows *Bordetella pertussis* bacteria. These bacteria can attach to cilia that line the upper respiratory system. Pertussis (whooping cough) is on the rise—see the Clinical Case of this chapter for discussion on pertussis vaccination.

◀ *Bordetella pertussis* bacteria cause whooping cough.

## In the Clinic

As the nurse in a vaccine clinic, you meet Eric, a healthy infant in for his 2-month appointment. His mother asks what vaccinations are due, and you explain that today Eric should receive his second dose of the hepatitis B vaccine, along with first doses of vaccinations designed to protect against rotavirus, diphtheria, tetanus, pertussis, *Haemophilus influenzae* type b, pneumococcus, and polio. **Eric's mother is alarmed that her child must receive "so many shots all at once, when he is so little." How should you respond to her?**

*Hint: Read about vaccines on pages 500–504 and in Tables 18.1 and 18.2.*

Play **In the Clinic** Video
@MasteringMicrobiology

# Vaccines

## LEARNING OBJECTIVES

18-1  Define *vaccine*.

18-2  Explain why vaccination works.

18-3  Differentiate the following and provide an example of each: attenuated, inactivated, subunit, toxoid, VPL, and conjugated vaccines.

18-4  Contrast nucleic acid vaccines and recombinant vector vaccines.

18-5  Compare and contrast the production of attenuated and killed vaccines, recombinant vaccines, and DNA vaccines.

18-6  Define *adjuvant*.

18-7  Explain the value of vaccines, and discuss acceptable risks for vaccines.

Long before vaccines existed, it was known that people who recovered from certain diseases were immune to the same infections thereafter. Chinese physicians were the first to try to prevent disease by exploiting this phenomenon. Records indicate that from at least the 1400s onward, they had children inhale dried smallpox scabs. This usually resulted in mild disease, followed by immunity. Called **variolation**, the smallpox prevention procedure spread through Asia, parts of North Africa, and Eurasia. It gained a foothold in Europe in 1717, when Lady Mary Montagu of England learned of it while in Turkey. In Europe, variolation usually was done by introducing a small amount of the agent into a healthy person through a skin scratch. Montagu herself was severely scarred from smallpox and had lost a brother to the disease. After she championed variolation, the procedure became common in England, parts of Europe, and in the British colonies.

Unfortunately, variolation occasionally resulted in a serious case of smallpox and had a 1% mortality rate, according to eighteenth-century English records. However, this was still well below the 50% mortality rate of smallpox. By 1798, physician Edward Jenner began to deliberately inoculate people with cowpox in an attempt to prevent smallpox. He did so because it had been observed that milkmaids infected by the mild cowpox disease did not contract the more deadly smallpox later. Cowpox inoculation proved much safer than variolation, and it became the main method to prevent smallpox during the nineteenth century. To honor Jenner's work, Louis Pasteur coined the term *vaccination* (from the Latin *vacca*, meaning cow).

Today, a **vaccine** is a suspension of organisms or fractions of organisms used to induce immunity. Thanks to modern vaccines, smallpox has been eradicated worldwide, as has rinderpest, a viral disease of livestock. Measles, polio, and several other infectious viral

**Play Vaccines:** *Function*
@MasteringMicrobiology

diseases of humans are also targeted for elimination through the use of vaccines. See the Clinical Focus box on page 506.

> **CHECK YOUR UNDERSTANDING**
>
> ✔ **18-1** What is the etymology (origin) of the word *vaccine*?

## Principles and Effects of Vaccination

Development of vaccines based on the model of the smallpox vaccine is the single most important application of immunology. An effective vaccine is the most desirable method of disease control. It prevents the targeted disease from ever occurring, thereby ceasing suffering before it ever begins. Disease prevention is generally the most economical public health option as well. Costs of prevention and treatment are especially important in the developing world.

We now know that Jenner's inoculations worked because the cowpox virus, which is not a serious pathogen, is closely related to the smallpox virus. The injection, by skin scratches, provoked a primary immune response that led to formation of antibodies and long-term memory cells. If the recipient encountered the smallpox virus later, memory cells were stimulated and produced a rapid, intense secondary immune response that prevented the disease from progressing (see Figure 17.17, page 493). This response mimics the immunity the person gained by recovering from the disease. The cowpox vaccine was later replaced by a vaccinia virus vaccine, a related poxvirus.

Many communicable diseases can be controlled by behavioral and environmental methods. For example, proper sanitation can prevent the spread of cholera, and the use of latex condoms can slow spread of sexually transmitted infections. If prevention fails, bacterial diseases can often be treated with antibiotics. However, there are few antiviral drugs. Therefore, vaccination is frequently the only feasible method of controlling viral disease. Controlling a disease does not necessarily require that everyone be immune to it. If a high percentage of the population is immune, a phenomenon called **herd immunity**, outbreaks are limited to sporadic cases because there are not enough susceptible individuals to support the spread of epidemics.

It is remarkable that there are still no widespread, useful vaccines against a number of pathogenic microbes, including chlamydias, fungi, protozoa, or helminthic parasites of humans. However, vaccines are under development for AIDS, malaria, and Zika. To create effective vaccines, the developers need to overcome a number of hurdles: understanding the most effective antigens that will cause an immune response, fully understanding the life cycle or stages of a microorganism, finding effective animal models to test efficacy, and funding and coordinated research on a particular vaccine. The principal vaccines used to prevent bacterial and viral diseases in the

**TABLE 18.1 CDC-Recommended Vaccines to Prevent Bacterial Diseases**

| Disease(s) | Vaccine | Recommendation | Booster |
|---|---|---|---|
| *Haemophilus influenzae* type b meningitis | Polysaccharide from *Haemophilus influenzae* type b | Children 2–18 months. | None recommended |
| Meningococcal meningitis | Purified polysaccharide from *Neisseria meningitidis* | For people with substantial risk of infection; recommended for college freshmen, especially if living in dormitories. | Need not established |
| Pneumococcal pneumonia | Purified polysaccharide from 13 or 23 strains of *Streptococcus pneumoniae* | PV23 for adults with certain chronic diseases; people over 65; PV13 for children 2–18 months; years 4–6. | None if first dose administered ≥ 24 months |
| Tetanus, diphtheria, and pertussis | DTaP (children younger than 3), Tdap (older children and adults), Td (booster for tetanus and pertussis) | DTaP (children 2-18 months; 4-6 years); Tdap (similar to Td; single dose for children aged 11–12 years and adults). | Tdap or Td every 10 years |

United States are listed in Tables 18.1 and 18.2. Many of these are recommended within the first few years of childhood. U.S. travelers who might be exposed to cholera, yellow fever, or other diseases not endemic in the United States can obtain other immunization recommendations from the U.S. Public Health Service and local public health agencies.

> **CHECK YOUR UNDERSTANDING**
>
> ✔ **18-2** Why is vaccination often the only feasible way to control most viral diseases?

## Types of Vaccines and Their Characteristics

There are several basic types of vaccines. Some newer vaccines take full advantage of knowledge and technology developed in recent years.

### Live Attenuated Vaccines

Deliberate weakening, called *attenuation*, can lead to the production of live **attenuated vaccines**. Attenuated vaccines are prepared using a living pathogen with reduced virulence. When the practice of producing viruses in the laboratory for study came into use, it was realized that an extended period of maintaining the virus in cell culture, embryonated eggs, or live (nonhuman) animals was, itself, a means to attenuate pathogenic viruses. This finding expanded the number of preventable diseases. Attenuation has also been created by specifically mutating virulence genes in an organism. An example of this is the attenuated bacterial vaccine (Ty21a) to protect against typhoid.

Live vaccines closely mimic actual infection. The pathogen in the vaccine reproduces within the host, and cellular, as well as humoral, immunity usually is induced. Lifelong immunity, especially in the case of viruses, is often achieved without booster immunizations, and an effectiveness rate of 95% is not unusual. This long-term effectiveness probably occurs because the attenuated viruses *replicate* in the body, magnifying the effect of the original dose and thus acting as a series of secondary (booster) immunizations. However, attenuated vaccines carry a risk: it is possible that the replicating attenuated viruses or bacteria might mutate to a more pathogenic form. People with a compromised or weakened immune system should not receive live vaccines, because the attenuated virus or bacteria may cause infections.

### Inactivated Killed Vaccines

**Inactivated killed vaccines** use whole microbes that have been killed, usually by formalin or phenol, after being grown in the laboratory. This keeps the pathogen intact so the immune system can recognize it, but it destroys the pathogen's ability to replicate. Inactivated virus vaccines for humans include rabies and influenza vaccines and the Salk polio vaccine. Inactivated bacterial vaccines include those for pneumococcal pneumonia and cholera. Generally speaking, inactivated killed vaccines are considered safer than live vaccines. However, there may be a risk of incomplete inactivation. Compared to live attenuated vaccines, these inactivated vaccines often require repeated

**CLINICAL CASE** An Ounce of Prevention

Esther Kim, a 3-week-old infant girl, is brought to the emergency department by her parents. She has had a fever and cough for the last 5 days, but now she is coughing so hard that she is vomiting. Esther's 7-year-old brother, Mark, has also been ill with a runny nose and mild cough. The Kims did not think that Esther's illness was serious until she began vomiting. Dr. Roscelli, the resident physician, admits baby Esther to the hospital for tests and observation. She is hospitalized for 5 days.

**What infection does Esther have? Read on to find out.**

501  504  505  510  515  520

**TABLE 18.2** CDC-Recommended Vaccines to Prevent Viral Diseases

| Disease | Vaccine | Recommendation | Booster |
|---|---|---|---|
| Chickenpox | Attenuated virus | For infants aged 12 months. | (Duration of immunity not known) |
| Hepatitis A | Inactivated virus | Children at age 1 year; live in or travel to endemic area; homosexual men; street-drug users; receive blood-clotting factors. | Duration of protection estimated at about 10 years |
| Hepatitis B | Antigenic fragments of virus | For infants and children; for adults, especially health care workers, homosexual men, injecting street-drug users, heterosexual people with multiple partners, and household contacts of hepatitis B carriers. | Duration of protection at least 7 years; need for boosters uncertain |
| Herpes zoster | Attenuated virus | Adults over age 60. | None recommended |
| Human papillomavirus | Antigenic fragments of virus | Boys and girls ages 11–12. | Duration at least 5 years |
| Influenza | Injected vaccine, inactivated virus (A nasally administered vaccine with attenuated virus is not available for the 2016–2017 flu season.) | Everyone over 6 months of age. | Annual |
| Measles | Attenuated virus | For infants aged 15 months. | Adults if exposed during outbreak |
| Mumps | Attenuated virus | For infants aged 15 months. | Adults if exposed during outbreak |
| Poliomyelitis | Killed virus | For children; for adults, as risk to exposure warrants. | (Duration of immunity not known) |
| Rabies | Killed virus | For field biologists in contact with wildlife in endemic areas; for veterinarians; for people exposed to rabies virus by bites. | Every 2 years |
| Rotavirus | Rota Teq®, modified rotaviruses; Rotarix® vaccine, attenuated strain | Oral, for infants up to 8 months | None recommended |
| Rubella | Attenuated virus | For infants aged 15 months; for women of childbearing age who are not pregnant. | Adults if exposed during outbreak |
| Smallpox | Live vaccinia virus | Certain military and health care personnel. | Duration of protection estimated at about 3 to 5 years |

booster doses because there is no replication within the host. As such, they induce a mostly humoral antibody immunity, which makes them less effective than attenuated vaccines that can induce cellular immunity. Several long-used inactivated vaccines are being replaced by newer, more effective subunit vaccines.

**Subunit Vaccines**

**Subunit vaccines** contain only selected antigenic fragments of a microorganism that best stimulate an immune response. This avoids the dangers associated with the use of live or killed pathogenic organisms. Subunit vaccines can be bacteria or virus components. Subunit vaccines can also be produced by genetically modifying other, nonpathogenic microbes to produce the desired antigenic fraction—such vaccines are called **recombinant vaccines**. For example, the hepatitis B vaccine consists of a portion of the viral protein coat that is produced by a genetically modified yeast (see Chapter 9).

This also avoids the need to use viral host cells to grow viruses for vaccines.

**Toxoid Vaccines** contain inactivated toxins produced by a pathogen, eliciting an antibody response against that particular pathogen component. The tetanus, diphtheria, and acellular pertussis toxoids are part of the standard childhood immunization series that requires a series of injections, followed by boosters every 10 years to maintain full immunity. Many older adults have not received boosters and likely have low levels of protection.

**Virus-like Particle (VLP) Vaccines** resemble intact viruses but do not contain any viral genetic material. For example, the human papilloma vaccine consists of viral proteins produced by a genetically modified yeast. The proteins assemble themselves into a VLP.

**Polysaccharide Vaccines** are made from molecules in a pathogen's capsule. Although not very immunogenic, polysaccharide vaccines include those for *N. meningitidis* and pneumococcal pneumonia. Some pathogens, most notably *Streptococcus pneumoniae* (pneumococcus), are virulent primarily because their polysaccharide capsule makes them resistant to phagocytosis. However, polysaccharides are T-independent antigens and are not effective in stimulating children's immune systems until the age of 15 to 24 months (see Figure 17.7, page 484).

**Conjugated Vaccines** have been developed in recent years to deal with children's poor immune response to vaccines based on capsular polysaccharides. The polysaccharides are combined with proteins such as diphtheria or tetanus toxoid. The two separate components chemically linked together create a stronger immune response. This approach has led to the very successful vaccine for *Haemophilus influenzae* type b (Hib), which gives significant protection even at 2 months.

## Nucleic Acid (DNA) Vaccines and Recombinant Vector Vaccines

**Nucleic Acid Vaccines**, often called *DNA vaccines*, are among the newest and most promising vaccines. Naked or encapsulated DNA that encodes specific protein antigens is injected into the patient so that the receiving cell will synthesize the protein. Although the recipient's body synthesizes the protein, the antigen is recognized as foreign, thus triggering an immune response (**Figure 18.1**). The injection can be made by conventional needle or, more efficiently, by a "gene gun," which delivers the vaccine into many cell nuclei. The antigens encoded by the DNA vaccine are expressed on the cell to stimulate both humoral and cellular immunity. DNA vaccines tend to be expressed for extended times, with good immunological memory. A DNA vaccine against Zika virus is currently in clinical trials.

**Recombinant Vector Vaccines** focus on how recombinant DNA is delivered into the target cell. For example, many viruses

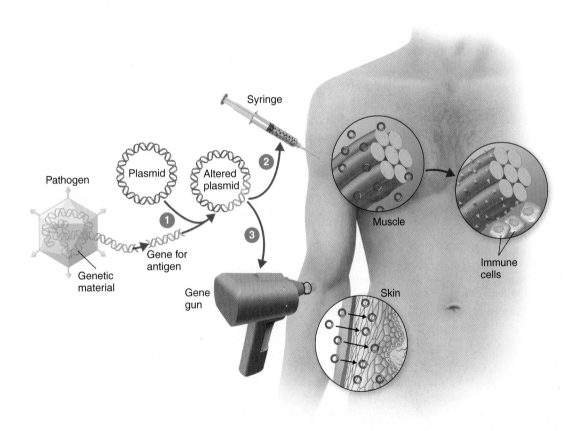

**Figure 18.1  DNA vaccines.** Genes that code for antigens are removed from a pathogen, inserted into a plasmid, and then injected into cell nuclei as part of a vaccine. A small number of host cells at the injection site will take in the vaccine and produce the antigens its DNA encodes. The appearance of the antigens then stimulates an immune response.

**Q** How does a DNA vaccine differ from a recombinant vaccine?

**CLINICAL CASE**

Dr. Roscelli examines Esther every day she is in the hospital. At no time does he wear a face mask. Dr. Roscelli does not suspect pertussis (whooping cough) until the attending physician suggests that he swab Esther's throat and request a PCR test for the bacteria. Sure enough, Esther's throat swab is positive for *Bordetella pertussis* DNA.

**Why do you think Dr. Roscelli did not suspect pertussis when Esther was admitted to the ER?**

501 **504** 505 510 515 520

attach to a cell membrane and inject their DNA into the cell. Vectors take advantage of this natural strategy for delivery. Avirulent viruses or bacteria are used as delivery systems or vectors. The vector is genetically modified to contain genes that create the protein antigen along with promotors to maximize expression (see Figure 9.1 on page 244). The DNA can be fine-tuned through recombinant DNA technology to maximize expression in the target species.

Clinical tests in humans are under way testing DNA vaccines for a number of different diseases, including influenza, HIV, and hepatitis C. A promising Ebola DNA vaccine (rVSV-ZEBOV) expresses a glycoprotein from the Zaire strain of *Ebolavirus*. In its most recent clinical tests, the vaccine was 70–100% effective against Ebola virus disease. Recombinant vector vaccines are being developed for HIV, rabies, and measles. Both DNA vaccines and recombinant vector vaccines are used in animals to immunize against a variety of diseases, including poxviruses, rabies for some species, and West Nile encephalitis.

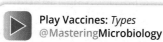
▶ Play Vaccines: *Types*
@Mastering**Microbiology**

**CHECK YOUR UNDERSTANDING**

✔ **18-3** Experience has shown that attenuated vaccines tend to be more effective than inactivated vaccines. Why?

✔ **18-4** What type of vaccine is an adenovirus that expresses the malaria-CS protein?

## Vaccine Production, Delivery Methods, and Formulations

Historically, vaccine production often required growing the pathogen in animals, embryonated eggs, or cell cultures. Recombinant vaccines, DNA vaccines, and recombinant vector vaccines do not need a cell or animal host to grow the vaccine's microbe. This avoids the problems involved in using live attenuated virus, including egg protein in a vaccine, or the difficulty of propagating certain viruses in cell culture. The very successful subunit vaccine for hepatitis B was the first of these recombinant vaccines.

Plants are also a potential production system for doses of antigenic proteins that would be taken orally as pills or as an injection. Tobacco plants are a leading candidate for this use because they are unlikely to contaminate the food chain.

In terms of administration, oral vaccines are favored for many reasons even beyond eliminating a need for injections. For one, they would be especially effective in protecting against the diseases caused by pathogens invading the body through mucous membranes. Examples of current oral vaccines include ones for polio, rotavirus, adenovirus, and cholera. Oral vaccines for tuberculosis, *C. difficile*, and influenza are being developed.

In less-developed regions of the world, minimally trained personnel are called on to vaccinate large numbers of people in less than ideal circumstances. This lack of training and resources poses problems with injected vaccines: single-use doses can be expensive, and sterilization for reusable needles may be uncertain. One alternative delivery method under development is a skin patch (Nanopatch™) that administers a dry formulation of a vaccine. Skin tissue contains a high number of antigen-presenting cells—more than the muscle tissue reached by conventional needles—making skin a good delivery location. Another advantage is that dry vaccines such as the skin patch require no refrigeration. This is especially important; the World Health Organization estimates that half the vaccines used in Africa are ineffective because of faulty refrigeration of vials of injectable vaccines. Nanopatch™ vaccines against influenza and polio are currently being tested. A patch delivery for a vaccine against ticks is being investigated.

Even in the developed world, the sheer number of injections required for infants and children makes creating more multiple-combination vaccines desirable. For example, there are five combination vaccines approved by the U.S. Food and Drug Administration (FDA), including one for pertussis, diphtheria, tetanus, polio, and *Haemophilus influenzae* type b (Hib).

At present, vaccines for many diseases are under development, ranging from those for prominent deadly diseases such as AIDS, malaria, and type I diabetes. Researchers are also investigating vaccines' potential for treating and preventing drug addiction, Alzheimer's disease, cancer, and allergies.

Most current vaccines, especially the inactivated or subunit vaccines, act by causing the production of humoral antibodies. There is a need for vaccines that confer T cell–based immunity, which would be especially useful against tuberculosis, HIV, and cancer. Antigenic variability remains a problem as well; for example, the influenza virus changes antigens each year and requires a new vaccine annually. As a practical matter, if the antigen changes more rapidly than once a year—HIV, for example, changes its antigenic structure daily—it cannot be controlled by conventional vaccines. Computers now allow us to search the structure of

# Microbiome May Enhance Response to Oral Vaccines

Oral vaccines are highly valued because they are so simple and safe to administer. However, for many diseases, creating effective oral vaccines has been challenging. Historically, this difficulty has been attributed to everything from differences in patient nutrition or socioeconomic status to genetics. Now more recent research indicates that patient microbiome composition may be a major determining factor in how strong an immune response is elicited by certain oral vaccines.

Recent research using an orally administered vaccine for typhoid fever showed that the individuals who had better cell-mediated responses to it also had a more diverse intestinal microbiome, with an abundance of Clostridiales. Other studies

of mice show that these same bacteria promote differentiation of T helper cell differentiation in mice.

Administering certain probiotic bacteria, such as *Lactobacillus* species and *Bifidobacterium* species, enhanced antibody responses to oral vaccines against rotavirus, *Salmonella*, polio, and cholera in adult human volunteers. Infants who received an oral rotavirus vaccine along with probiotics also showed higher antibody responses, particularly IgA, which is mostly found in mucous membranes, and IgM, which is one of the first classes of antibodies to appear when an infection is detected.

So far, human studies have been small, but determining the effects of normal microbiota on response to vaccines may

lead to new vaccination methods that include altering the intestinal microbiota.

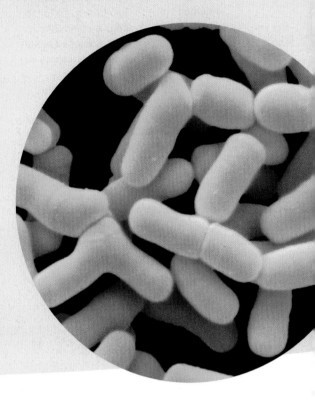

The lactic-acid bacterium, *Bifidobacterium*, is a normal part of the intestinal microbiome.

---

a pathogen's genome for antigens that produce a protective immune response. This "reverse vaccinology" is becoming an essential tool for vaccine development.

## Adjuvants

The early days of commercial vaccine production saw occasional problems with contamination. Unexpectedly, once the contamination was corrected it was found that the purified vaccine was often less effective. This led to experiments designed to determine whether chemical additives could improve effectiveness. A chance discovery showed certain aluminum salts, generally grouped under the term *alum* and called **adjuvants**, allows vaccines to be more effective. At this time, alums and a derivative of lipid A (from LPS) called monophosphoryl lipid A are the only adjuvants approved for use in humans in the United States. See Exploring the Microbiome for current research on bacterial adjuvants. Other adjuvants, such as MF59 (an oil-and-water emulsion) are used in Europe and elsewhere. Some adjuvants are approved only for use in animals. The exact mechanism by which adjuvants work is not known in all detail, but they are known to improve the innate immune response, especially activation of Toll-like receptors.

### CHECK YOUR UNDERSTANDING

✔ 18-5 Which type of vaccine is a live measles virus: inactivated, attenuated, recombinant, or DNA?

✔ 18-6 What is the value of an *adjuvant*?

### CLINICAL CASE

Vaccination has been so successful in reducing childhood infections that many younger physicians have never seen a case of pertussis. Nine days after initial exposure to Esther's illness, Dr. Roscelli exhibits a runny nose and, 4 days later, a cough. Dr. Roscelli assumes he has a cold and declines recommended prophylaxis with erythromycin. Further investigation identifies seven other pertussis cases in health care workers (a respiratory therapist, a radiologic technician, and five student nurses), all of whom work in the emergency department but not in pediatrics.

**How did Dr. Roscelli and the seven health care workers get the infection?**

501 | 504 | **505** | 510 | 515 | 520

# Measles: A World Health Problem

As you read through this box, try to answer each question yourself before moving to the next one.

1. Teenager Maria and her church group returned home to Ohio from a trip to the Philippines. Soon after, she and 383 others from the church developed mouth rashes that looked like tiny red spots with blue-white centers. A few days later, they developed maculopapular skin rashes that spread from face to trunk and extremities, along with fever (≥38°C) and other coldlike symptoms. Testing for IgM measles antibodies confirmed the diagnosis: measles. This highly contagious

in the United States compared to 200,000 cases worldwide (Figure B).
*What would happen if we stopped vaccinating against measles?*

4. Before the vaccine existed, nearly 500,000 measles cases and over 400 deaths occurred every year in the United States. Without vaccinations, the United States would undergo many more measles outbreaks and even epidemics, leading to increased hospitalizations and the grim results that often follow measles infections: blindness, deafness, seizure disorders, and mental retardation.

The Measles Initiative—led by the American Red Cross, the United Nations Foundation, UNICEF, the U.S. Centers for Disease Control and Prevention, and the World Health Organization—has supported the vaccination of nearly 2 billion children in over 80 countries. In 2000, measles caused about 757,000 deaths. By 2015, measles deaths worldwide were reduced to 134,200 people.

Other vaccine-preventable diseases are still quite prevalent in other parts of the world.

*Data source: CDC, 2017.*

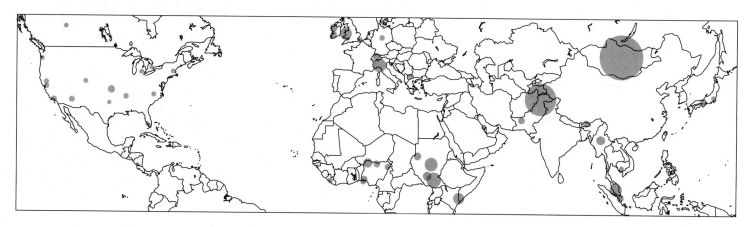

**Figure A  Measles outbreaks (orange circles) in 2016.**
(Council on Foreign Relations)

viral illness can cause pneumonia, diarrhea, encephalitis, and death.
*How did Maria and the others contract measles?*

2. Maria spent 2 weeks in the Philippines with her church group. She and the other infected people had never been vaccinated against measles and so had no defense against it when they encountered it overseas (Figure A).
*If measles is highly contagious, why didn't this outbreak spread to more people?*

3. The measles vaccine provides good protection against the disease. Since the 1960s, when that immunization came on the market, a majority of people in the United States have been vaccinated against measles. In 2016, only 72 cases of measles were reported

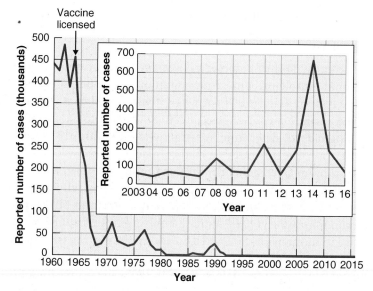

**Figure B  Reported numbers of measles cases in the United States, 1960–2016.**
(CDC, 2017)

## Vaccine Safety

As we have already discussed, variolation, the first attempt to provide immunity to smallpox, sometimes *caused* the disease it was intended to prevent. At the time, the risk was considered very worthwhile, given the mortality rate of the actual disease. Safety issues still arise with various vaccine formulations—there are pluses and minuses when it comes to enduring the immune response and potential side effects. For example, the oral polio vaccine, on rare occasions, may cause the disease. In 1999, a vaccine to prevent infant diarrhea caused by rotaviruses was withdrawn from the market because several recipients developed a life-threatening intestinal obstruction. Minor side effects vary according to vaccine but may include tenderness at the injection site, headache, fever, mild rash, and fatigue.

Public reaction to what is a minuscule individual risk of a bad outcome from vaccination has changed over the decades. People who witness firsthand the damage and deaths associated with serious infections usually feel the risk of vaccine side effects is worthwhile. However, most parents today have never seen a case of polio or measles, and they may view the risk of vaccine side effect as more worrisome than the risk of the disease itself. Moreover, bad science has led people to avoid certain vaccines for their children. In particular, a bogus connection between the measles, mumps, and rubella (MMR) vaccine and the developmental disorder autism led to widespread fears at the beginning of the twenty-first century. The study that sparked both this controversy and a resulting dip in vaccination levels was later debunked and retracted—even so, false information lives on for many years on the Internet. As a result, nearly two decades since the false association between autism and vaccines first arose, preventable diseases such as pertussis, mumps, and measles are reemerging in developed countries.

Extensive scientific surveys have provided no evidence to support a connection between the usual childhood vaccines and autism or any other disease condition. No vaccine will ever be perfectly safe or perfectly effective—nor is any antibiotic or most other drugs, for that matter. Nevertheless, vaccines still remain the safest and most effective means of preventing infectious disease, especially in children.

---

**CHECK YOUR UNDERSTANDING**

✔ **18-7** Why could the oral (Sabin) polio vaccine sometimes cause polio, but the injected (Salk) vaccine does not?

---

# Diagnostic Immunology

### LEARNING OBJECTIVES

**18-8**  Differentiate sensitivity from specificity in a diagnostic test.

**18-9**  Define *monoclonal antibodies*, and identify their advantage over conventional antibody production.

**18-10**  Explain how precipitation reactions and immunodiffusion tests work.

**18-11**  Differentiate direct from indirect agglutination tests.

**18-12**  Differentiate agglutination from precipitation tests.

**18-13**  Define *hemagglutination*.

**18-14**  Explain how a neutralization test works.

**18-15**  Differentiate precipitation from neutralization tests.

**18-16**  Explain the basis for the complement-fixation test.

**18-17**  Compare and contrast direct and indirect fluorescent-antibody tests.

**18-18**  Explain how direct and indirect ELISA tests work.

**18-19**  Explain how Western blotting works.

**18-20**  Explain the importance of monoclonal antibodies.

Throughout most of history, diagnosing a disease was mostly a matter of observing a patient's signs and symptoms—if these were unique, then diagnosis was easy. But when signs and symptoms were more general, accurate diagnosis was often impossible. Today, we have diagnostic tests that can pinpoint various diseases with a high degree of accuracy. Essential elements of diagnostic tests are sensitivity and specificity. **Sensitivity** is the probability that the test is reactive if the specimen is a true positive. **Specificity** is the probability that a test will *not* be reactive if a specimen is a true negative.

More than 100 years ago, Robert Koch was trying to develop a vaccine against tuberculosis and accidentally laid the groundwork for a diagnostic test. He observed that when guinea pigs with tuberculosis were injected with a suspension of *Mycobacterium tuberculosis*, the site of the injection became red and slightly swollen a day or two later. This symptom is now used as a positive result for the widely used tuberculin skin test (see Figure 24.9, page 700)—many colleges and universities require this test as part of admission procedures. Koch, of course, had no idea of the mechanism of cell-mediated immunity that caused this phenomenon, nor did he know of the existence of antibodies.

Immunology gives us many other invaluable diagnostic tools, most of which are based on interactions of humoral antibodies with antigens. A known antibody can be used to identify an *unknown* pathogen (antigen) by its reaction with it. This reaction can be reversed, and a *known* pathogen can be used, for example, to determine the presence of an unknown antibody in a person's blood—which would determine whether he or she had immunity to the pathogen. The main problem that must be overcome in antibody-based diagnostic tests is that antibodies cannot be seen directly. Even at magnifications of well over 100,000×, they appear only as fuzzy, ill-defined particles (see Figure 17.4c on page 480). Therefore, their presence must be established indirectly. We will describe a number of ingenious solutions to this problem.

Other challenges that come with designing good antibody-related diagnostic tests is that specific antibodies produced in animals are made in relatively small quantities and are not purified from other types of antibodies the animal has made.

**CHECK YOUR UNDERSTANDING**

✔ 18-8 What property of the immune system suggested its use as an aid for diagnosing disease: specificity or sensitivity?

## Use of Monoclonal Antibodies

As soon as it was determined that antibodies were produced by B cells, it was understood that if a B cell producing a single type of antibody could be isolated and cultivated, it would be able to produce the desired antibody in nearly unlimited quantities and without contamination by other antibodies. Unfortunately, a B cell reproduces only a few times under the normal cell culture conditions. This problem was largely solved by tapping cancerous plasma B cells for culture. These cancerous plasma B cells, known as *myelomas*, no longer make antibodies but can be isolated and grown indefinitely in cell culture. Combining an "immortal" cancerous B cell with an antibody-producing normal B cell creates a **hybridoma** that, when grown in culture, produces the type of antibody characteristic of the ancestral B cell indefinitely. This allows us to procure immense quantities of identical antibody molecules. Because all of these antibody molecules are produced by a single hybridoma clone, they are called **monoclonal antibodies**, or **Mabs (Figure 18.2)**.

Monoclonal antibodies are uniform, are highly specific, and can be produced in large quantities. Because of these qualities, they are enormously important diagnostic tools. Commercial kits use them to recognize several bacterial pathogens, and nonprescription pregnancy tests use monoclonal antibodies to indicate the presence of a hormone excreted only in the urine of a pregnant woman (see Figure 18.13, page 517).

Monoclonal antibodies have also become a clinically important, frequently used class of drugs. Currently more than 62 have been approved for human therapy. These include treatments for multiple sclerosis, Crohn's disease, psoriasis, cancer, asthma, and arthritis. There are hundreds more drugs of this type currently being developed worldwide for a wide range of diseases and conditions.

The modes of therapeutic action of monoclonal antibodies vary. Certain inflammatory diseases such as rheumatoid arthritis require the action of tumor necrosis factor (TNF; see page 459). Mabs that neutralize TNF block the progression of the disease. One such Mab is infliximab. Other Mabs block a receptor site; an example is omalizumab. This drug treats allergic asthma by preventing the binding of IgE on mast cells and basophils (see Figure 19.1 on page 526). The Mab rituximab is used to treat inflammatory diseases that the TNF-blocking Mabs cannot. This class of Mab binds to antigen-presenting cells, depleting their supply and thereby blocking the progression of the disease.

The therapeutic use of monoclonal antibodies had been limited because they were once produced only by mouse (murine) cells. The immune systems of some patients reacted against the foreign mouse proteins, leading to rashes, swelling, and even occasional kidney failure, plus the destruction of the antibodies. New generations of monoclonal antibodies aim to minimize the murine component, so they are less likely to cause side effects. Essentially, the more human the antibody, the more successful it is likely to be. Researchers are exploring several approaches. **Chimeric monoclonal antibodies** use genetically modified mice to make a human–murine hybrid. (A *chimera* is an animal or tissue made up of elements derived from genetically distinct individuals.) The variable part of the antibody molecule, which includes the antigen-binding sites (see Figure 17.4a on page 480), is murine. The constant region of the antibody molecule is derived from a human source. These monoclonal antibodies are about 66% human. An example is rituximab, which treats leukemias and some autoimmune disorders that are characterized by excessive numbers of B cells. **Humanized antibodies** are constructed so the murine portion is limited to antigen-binding sites, with about 90% of the rest of the molecule being derived from human sources. An example is trastuzumab (Herceptin®), used to treat breast cancer in patients with a certain gene mutation.

Even humanized antibodies can cause unwanted immune responses, which has spurred research and development of **fully human antibodies**. One approach has been to genetically modify mice so they contain human antibody genes. An early fully human example is adalimumab, used to treat rheumatoid arthritis and psoriatic arthritis.

Most types of monoclonal antibodies end in *mab*; the letter(s) directly before that indicate derivation. *Umab* means human-derived, *omab* are from mice, *ximab* are chimera, and *zumab* are humanized. A name may also reference the particular disease or tumor being treated. For instance, biciromab are monoclonal antibodies derived from mice that treat a cardiovascular condition (*cir* being short for *circulatory system*). Mabs are used in the diagnostic tests described throughout the rest of this chapter.

# The Production of Monoclonal Antibodies

1. A mouse is injected with a specific antigen that will induce production of antibodies against that antigen.

2. The spleen of the mouse is removed and homogenized into a cell suspension. The suspension includes B cells that produce antibodies against the injected antigen.

3. The spleen cells are then mixed with myeloma cells that are capable of continuous growth in culture but have lost the ability to produce antibodies. Some of the antibody-producing spleen cells and myeloma cells fuse to form hybrid cells. These hybrid cells are now capable of growing continuously in culture while producing antibodies.

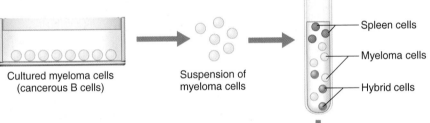

Cultured myeloma cells (cancerous B cells)

Suspension of myeloma cells

Antigen

Spleen

Suspension of spleen cells

Spleen cells

Myeloma cells

Hybrid cells

4. The mixture of cells is placed in a selective medium that allows only hybrid cells to grow.

5. Hybrid cells proliferate into clones called hybridomas. The hybridomas are screened for production of the desired antibody.

6. The selected hybridomas are then cultured to produce large quantities of monoclonal antibodies. Isolated antibodies are used for treating and diagnosing disease.

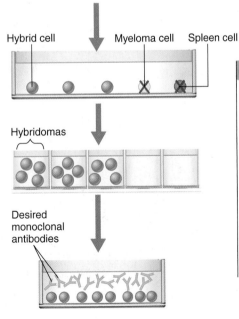

Hybrid cell    Myeloma cell    Spleen cell

Hybridomas

Desired monoclonal antibodies

## KEY CONCEPTS

- The fusion of cultured myeloma cells (cancerous B cells) with antibody-producing spleen cells forms a hybridoma.

- Hybridomas can be cultured to produce large quantities of identical antibodies, called monoclonal antibodies.

- Monoclonal antibody production is an important advancement in medicine and also is integral to common diagnostic and therapeutic tools. A monoclonal antibody can attach to a target cell while carrying a diagnostic marker or an anticellular toxin.

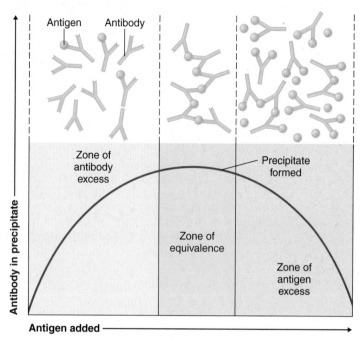

**Figure 18.3 A precipitation curve.** The curve is based on the ratio of antigen to antibody. The maximum amount of precipitate forms in the zone of equivalence, where the ratio is roughly equivalent.

**Q** How does precipitation differ from agglutination?

**CHECK YOUR UNDERSTANDING**

✔ 18-9 The blood of an infected cow would have a considerable amount of antibodies against the infectious pathogen in its blood. How would an equivalent amount of monoclonal antibodies be more useful?

**CLINICAL CASE**

Dr. Roscelli did not wear a mask at any time he examined Esther, and consequently he contracted pertussis. He should have worn a mask to prevent transmission of respiratory infections; he also should have accepted antibiotic treatment for his symptoms. Dr. Roscelli could have transmitted the infection to his colleagues in the emergency department, and, in turn, the infected hospital personnel could have transmitted pertussis to vulnerable patients. When investigating Esther's illness, health care workers discover that neither Esther nor her brother has been vaccinated. The Kims were fearful because they had heard that vaccinations could cause serious side effects and even death.

**Did the Kims make a mistake?**

501    504    505    **510**    515    520

## Precipitation Reactions

**Precipitation reactions** involve the reaction of soluble antigens with IgG or IgM antibodies. Precipitation reactions occur in two stages. Within seconds, the antigens and antibodies rapidly form small antigen–antibody complexes. Next, over the course of minutes or hours, the antigen–antibody complexes form larger, interlocking molecular aggregates called *lattices* that precipitate from the solution. Precipitation reactions normally occur only when the ratio of antigen to antibody is optimal. **Figure 18.3** shows that no visible precipitate forms when either component is in excess. The optimal ratio can be achieved when separate solutions of antigen and antibody are placed adjacent to each other in agar gel or in solution. If they are allowed to diffuse together, an **immunodiffusion test**, or **precipitin ring test** (**Figure 18.4**), will show a cloudy line of precipitation in the area in which the optimal ratio has been reached (the *zone of equivalence*).

Other tests use electrophoresis to speed up the movement of antigen and antibody in a gel, sometimes in less than an hour, with this method. The techniques of immunodiffusion and electrophoresis can be combined in a procedure called **immunoelectrophoresis**. The procedure is used in research to separate proteins in human serum and is the basis of certain diagnostic tests. It is an essential part of the Western blot test used in AIDS testing (see Figure 10.13, page 284).

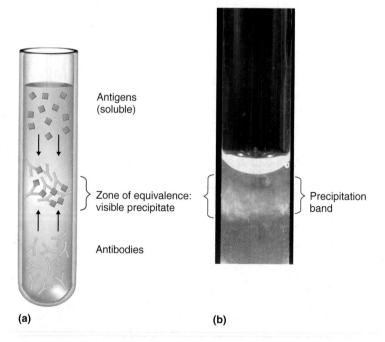

**(a)**    **(b)**

**Figure 18.4 The precipitin ring test. (a)** This drawing shows the diffusion of antigens and antibodies toward each other in a small-diameter test tube. Where they reach equal proportions, in the zone of equivalence, a visible line or ring of precipitate is formed. **(b)** A photograph of a precipitin band.

**Q** What causes the visible line?

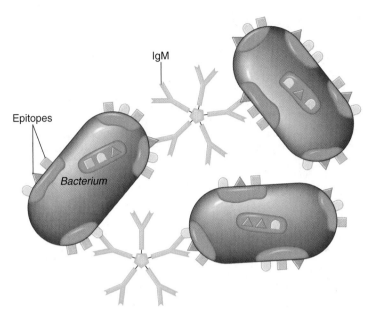

IgM

Epitopes

*Bacterium*

**Figure 18.5 An agglutination reaction.** When antibodies react with epitopes on antigens carried on neighboring cells, such as these bacteria (or red blood cells), the particulate antigens (cells) agglutinate. IgM, the most efficient immunoglobulin for agglutination, is shown here, but IgG also participates in agglutination reactions.

**Q** Draw an agglutination reaction involving IgG.

**CHECK YOUR UNDERSTANDING**

✔ 18-10 Why does the reaction of a precipitation test become visible only in a narrow range?

## Agglutination Reactions

Whereas precipitation reactions involve *soluble* antigens, agglutination reactions involve either *particulate* antigens (particles such as cells that carry antigenic molecules) or soluble antigens adhering to particles. These antigens can be linked together by antibodies to form visible aggregates, a reaction called **agglutination** (Figure 18.5). Agglutination reactions are very sensitive, relatively easy to read (see Figure 10.11, page 283), and available in great variety. Agglutination tests are classified as either direct or indirect.

### Direct Agglutination Tests

**Direct agglutination tests** detect antibodies against relatively large cellular antigens, such as those on red blood cells, bacteria, and fungi. They are usually done in plastic *microtiter plates* that contain many shallow wells. The amount of particulate antigen in each well is the same, but the amount of serum that contains antibodies is diluted, so that each successive well has half the antibodies of the previous well. These tests are used, for example, to test for brucellosis and to separate *Salmonella* isolates into serovars, types defined by serological means.

Clearly, the more antibody we start with, the more dilutions it will take to lower the amount to the point where there is not enough antibody for the antigen to react with. This is the measure of **titer**, or concentration of serum antibody (**Figure 18.6**).

Enlarged photo of wells

Top view of wells

| 1:20 | 1:40 | 1:80 | 1:160 | 1:320 | 1:640 | Control |

**(a)** Each well in this microtiter plate contains, from left to right, only half the concentration of serum that is contained in the preceding well. Each well contains the same concentration of particulate antigens, in this instance red blood cells.

Side view of well

**(b)** Agglutinated: In a positive (agglutinated) reaction, sufficient antibodies are present in the serum to link the antigens together, forming a mat of antigen–antibody complexes on the bottom of the well.

Side view of well

**(c)** Nonagglutinated: In a negative (nonagglutinated) reaction, not enough antibodies are present to cause the linking of antigens. The particulate antigens roll down the sloping sides of the well, forming a pellet at the bottom. In this example, the antibody titer is 160 because the well with a 1:160 concentration is the most dilute concentration that produces a positive reaction.

**Figure 18.6 Measuring antibody titer with the direct agglutination test.**

**Q** What is meant by the term *antibody titer*?

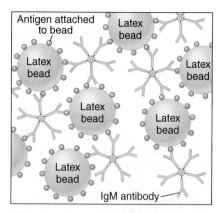

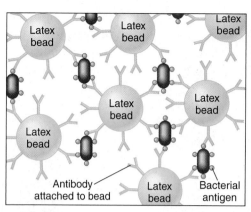

**Figure 18.7 Reactions in indirect agglutination tests.** These tests are performed using antigens or antibodies coated onto particles such as minute latex spheres.

**Q** Differentiate direct from indirect agglutination tests.

**(a)** Reaction in a positive indirect test for antibodies. When particles (latex beads here) are coated with antigens, agglutination indicates the presence of antibodies, such as the IgM shown here.

**(b)** Reaction in a positive indirect test for antigens. When particles are coated with monoclonal antibodies, agglutination indicates the presence of antigens.

For infectious diseases in general, the higher the serum antibody titer, the greater the immunity to the disease. However, the titer alone is of limited use in diagnosing an existing illness. There is no way to know whether the measured antibodies were generated in response to the immediate infection or to an earlier illness. For diagnostic purposes, *a rise in titer* is significant; that is, the titer is higher later in the course of the disease than at its outset. Also, if it can be demonstrated that the person's blood had no antibody titer before the illness but has a significant titer while the disease is progressing, this change, called **seroconversion**, is also diagnostic. This situation is frequently encountered with HIV infections.

Some diagnostic tests specifically identify IgM antibodies. Short-lived IgM is more likely to reflect a response to a current disease condition (see Chapter 17). Specific immunodeficiency diseases, such as the inability to make IgG antibodies, can be identified by notable titer differences from a healthy patient exposed to the same antigen.

### Indirect (Passive) Agglutination Tests

Antibodies against soluble antigens can be detected by agglutination tests if the antigens are attached onto particles such as bentonite clay or, most often, extremely small latex beads. Such tests, known as *latex agglutination tests*, are commonly used for the rapid detection of serum antibodies against bacterial and viral diseases. In such **indirect (passive) agglutination tests**, the antibody reacts with the attached antigen or in reverse by using particles coated with antibodies to detect the antigens against which they are specific (**Figure 18.7**). The particles then agglutinate with one another, much as particles do in the direct agglutination tests. This approach is especially common in tests for the streptococci that cause sore throats. A diagnosis can be completed in about 10 minutes.

### Hemagglutination

When agglutination reactions involve the clumping of red blood cells, the reaction is called **hemagglutination**. These reactions involve red blood cell surface antigens and their complementary antibodies. They are used routinely in blood typing (see Table 19.2, page 531) and in the diagnosis of infectious mononucleosis.

## Neutralization Reactions

Certain viruses, such as those causing mumps, measles, and influenza, can agglutinate red blood cells without an antigen–antibody reaction—a process called **viral hemagglutination** (**Figure 18.8**). **Neutralization** is an antigen–antibody reaction

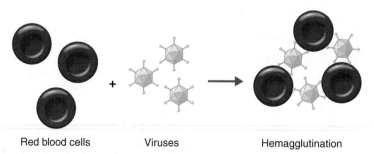

Red blood cells          Viruses          Hemagglutination

**Figure 18.8 Viral hemagglutination.** Viral hemagglutination is not an antigen–antibody reaction.

**Q** What causes agglutination in viral hemagglutination?

in which antibodies block the harmful effects of a bacterial exotoxin or a virus. These reactions were first described in 1890, when investigators observed that immune serum could neutralize the toxic substances produced by the diphtheria pathogen, *Corynebacterium diphtheriae.* Such a neutralizing substance, which is called an *antitoxin,* is a specific antibody produced by a host as it responds to a bacterial exotoxin or its corresponding toxoid (inactivated toxin). The antitoxin combines with the exotoxin to neutralize it (**Figure 18.9a**).

This reaction has led to their use as diagnostic tests. Viruses that exhibit their cytopathic effects in cell culture or embryonated eggs can be used to detect the presence of neutralizing viral antibodies (see pages 436–437). If the serum to be tested contains antibodies against the particular virus, the antibodies will prevent that virus from infecting cells in the cell culture or eggs, and no cytopathic effects will be seen. Such tests, known as *in vitro* neutralization tests, can be used both to identify a virus and to ascertain the viral antibody titer.

A neutralization test used mostly for the serological typing of viruses is the **viral hemagglutination inhibition test**. Certain viruses, such as those causing influenza, mumps, and measles, have surface proteins that will cause the agglutination of red blood cells. This test is most commonly used for subtyping of influenza viruses, although more laboratories are likely to be familiar with ELISA tests for this purpose. If a person's serum contains antibodies against these viruses, these antibodies

will react with the viruses and neutralize them (compare Figure 18.8 and Figure 18.9b). For example, if hemagglutination occurs in a mixture of measles virus and red blood cells but does not occur when the patient's serum is added to the mixture, this result indicates that the serum contains antibodies that have bound to and neutralized the measles virus.

## Complement-Fixation Reactions

In Chapter 16 (pages 463–467), we discussed a group of serum proteins collectively called complement. During most antigen–antibody reactions, a complement serum protein binds to the antigen–antibody complex and is used up, or fixed. This process of **complement fixation** can be used to detect very small amounts of antibody. Complement fixation was once used to diagnose syphilis (Wassermann test) and is still used to diagnose certain viral, fungal, and rickettsial diseases. The complement-fixation test requires great care and good controls. This is one reason why newer, simpler tests such as ELISA and PCR-based tests are increasingly replacing

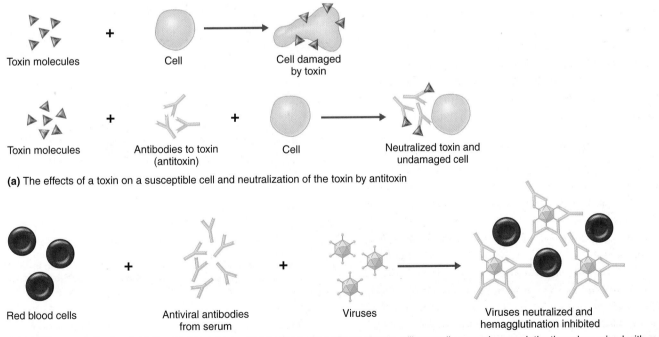

(a) The effects of a toxin on a susceptible cell and neutralization of the toxin by antitoxin

(b) Viral hemagglutination inhibition test to detect antibodies to a virus. These viruses will normally cause hemagglutination when mixed with red blood cells. If antibodies to the virus are present, as shown here, they neutralize the virus and inhibit hemagglutination.

**Figure 18.9 Reactions in neutralization tests.**

**Q** Why does hemagglutination indicate that a patient does not have a specific disease?

it. The test is performed in two stages: complement fixation and indicator (**Figure 18.10**).

## Fluorescent-Antibody Techniques

**Fluorescent-antibody (FA) techniques** can identify microorganisms in clinical specimens and can detect the presence of a specific antibody in serum (**Figure 18.11**). These techniques combine fluorescent dyes such as fluorescein isothiocyanate (FITC) with antibodies that fluoresce when exposed to ultraviolet light (see Figure 3.6, page 57). These procedures are quick, sensitive, and very specific; the FA test for rabies can be performed in a few hours and has an accuracy rate close to 100%.

Fluorescent-antibody tests are of two types. **Direct fluorescent-antibody (FA) tests** are usually used to identify a microorganism in a clinical specimen (Figure 18.11a). During this procedure, the specimen containing the antigen to be identified is fixed onto a slide. Fluorescein-labeled antibodies are then added, incubated briefly, and washed to remove any antibody not bound to antigen. Yellow-green fluorescence under the fluorescence microscope from the bound antibody will be visible even if the antigen, such as a virus, is submicroscopic in size.

**Indirect fluorescent-antibody (FA) tests** are used to detect the presence of a specific antibody in serum following exposure to a microorganism (Figure 18.11b). They are often more sensitive than direct tests. A known antigen is fixed onto a slide, and then the test serum added. If antibody that is specific to that microbe is present, it reacts with the antigen to form a bound complex. To make the antigen–antibody complex visible, fluorescein-labeled **anti-human immune serum globulin (anti-HISG)**, an antibody that reacts specifically with *any* human antibody, is added to the slide. Anti-HISG will be present only if the specific antibody has reacted with its antigen and is therefore present as well. After the slide has been incubated and washed (to remove unbound antibody), it is examined under a fluorescence microscope. If the known antigen fixed to the slide appears fluorescent, the antibody specific to the test antigen is present.

One adaptation of fluorescent antibodies is the **fluorescence-activated cell sorter (FACS)**. In Chapter 17, we learned that T cells carry antigenically specific molecules such as CD4 and CD8 on their surface, and these are characteristic of certain groups of T cells. The depletion of $CD4^+$ T cells is used to follow the progression of AIDS; their populations can be determined with a FACS.

The FACS is a modification of a *flow cytometer*, in which a suspension of cells leaves a nozzle as droplets containing no more than one cell per drop. A laser beam strikes each cell-containing droplet and is then received by a detector that determines certain

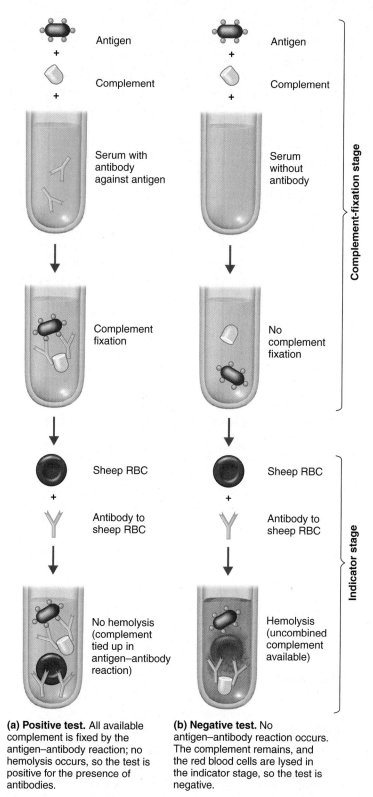

(a) **Positive test.** All available complement is fixed by the antigen–antibody reaction; no hemolysis occurs, so the test is positive for the presence of antibodies.

(b) **Negative test.** No antigen–antibody reaction occurs. The complement remains, and the red blood cells are lysed in the indicator stage, so the test is negative.

**Figure 18.10 The complement-fixation test.** This test indicates the presence of antibodies to a known antigen. Complement will combine (be fixed) with an antibody that is reacting with an antigen. If all the complement is fixed in the complement-fixation stage, then none remains to cause hemolysis of the red blood cells in the indicator stage.

**Q** Why does red blood cell lysis indicate that the patient does not have a specific disease?

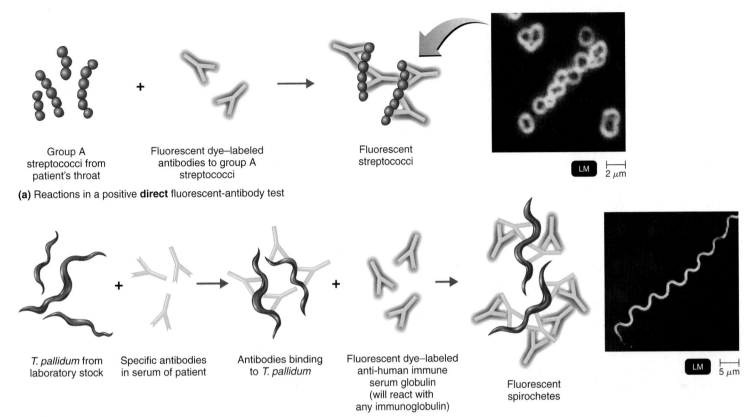

Group A
streptococci from
patient's throat

Fluorescent dye–labeled
antibodies to group A
streptococci

Fluorescent
streptococci

LM   2 μm

**(a)** Reactions in a positive **direct** fluorescent-antibody test

*T. pallidum* from
laboratory stock

Specific antibodies
in serum of patient

Antibodies binding
to *T. pallidum*

Fluorescent dye–labeled
anti-human immune
serum globulin
(will react with
any immunoglobulin)

Fluorescent
spirochetes

LM   5 μm

**(b)** Reactions in a positive **indirect** fluorescent-antibody test

**Figure 18.11  Fluorescent-antibody (FA) techniques.** **(a)** A direct FA test to identify group A streptococci. **(b)** In an indirect FA test such as that used in the diagnosis of syphilis, the fluorescent dye is attached to anti-human immune serum globulin, which reacts with any human immunoglobulin (such as the *Treponema pallidum*–specific antibody) that has previously reacted with the antigen. The reaction is viewed through a fluorescence microscope, and the antigen with which the dye-tagged antibody has reacted fluoresces (glows) in the ultraviolet illumination.

**Q** Differentiate a direct from an indirect FA test.

characteristics, such as size (**Figure 18.12**). For example, if the cells have FA-bound antibody markers to identify them as CD4$^+$ or CD8$^+$ T cells, the detector can measure this fluorescence and cell size. As the laser beam detects a cell of a preselected size or fluorescence, an electrical charge, either positive or negative, can be imparted to it. As the charged droplet falls between electrically charged plates, it is attracted to one receiving tube or another, effectively separating cells of different types. Millions of cells can be separated in an hour with this process, all under sterile conditions, which allows them to be used in experimental work.

An interesting application of the flow cytometer is sorting sperm cells to separate male (Y-carrying) and female (X-carrying) sperm. The female sperm (meaning that it will result in a female embryo when it fertilizes the egg) contains more DNA, 2.8% more in humans, 4% in animals. When the sperm is stained with a fluorescent dye specific for DNA, the female sperm glows more brightly when illuminated by the laser beam and can be separated out. The technique was developed for agricultural purposes. However, it has received medical approval for use in human couples who carry genes for inherited diseases that affect only boys.

**CHECK YOUR UNDERSTANDING**

✔ 18-17 Which test is used to detect antibodies against a pathogen: the direct or indirect fluorescent-antibody test?

**CLINICAL CASE**

Data show that about half of babies with whooping cough get the illness from their parents, and an additional 25% to 35% get the infection from another household member. Neither Mr. nor Mrs. Kim has been ill, but Mark's throat is cultured for *B. pertussis*, which comes back positive. As is often the case in older children and adults, Mark's symptoms are milder than his infant sister's; infants with whooping cough can become severely ill and die.

**Why is vaccination against pertussis so important?**

501   504   505   510   **515**   520

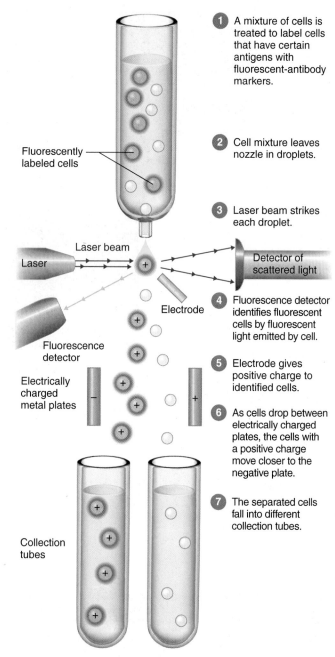

1. A mixture of cells is treated to label cells that have certain antigens with fluorescent-antibody markers.

Fluorescently labeled cells

2. Cell mixture leaves nozzle in droplets.

3. Laser beam strikes each droplet.

Laser beam

Laser

Detector of scattered light

Electrode

4. Fluorescence detector identifies fluorescent cells by fluorescent light emitted by cell.

Fluorescence detector

Electrically charged metal plates

5. Electrode gives positive charge to identified cells.

6. As cells drop between electrically charged plates, the cells with a positive charge move closer to the negative plate.

Collection tubes

7. The separated cells fall into different collection tubes.

**Figure 18.12 The fluorescence-activated cell sorter (FACS).** This technique can be used to separate different classes of T cells. For example, a fluorescence-labeled antibody reacts with the CD4 molecule on a T cell.

**Q** Provide an application of FACS to follow the progress of HIV infection.

## Enzyme-Linked Immunosorbent Assay (ELISA)

The **enzyme-linked immunosorbent assay (ELISA)** is the most widely used of a group of tests known as *enzyme immunoassay (EIA)*. There are two basic methods. The *direct ELISA* detects antigens, and the *indirect ELISA* detects antibodies. A microtiter plate with numerous shallow wells is used in both procedures (see Figure 10.12a, page 283). ELISA procedures are popular primarily

because they are sensitive and require little interpretive skill to read. Because the procedure and results are highly automated, the results tend to be clearly positive or clearly negative.

Many ELISA tests are available for clinical use in the form of commercially prepared kits. Some tests based on this principle are also available for use by the public, including a commonly available home pregnancy test (**Figure 18.13**).

### Direct ELISA

The direct ELISA method is shown in **Figure 18.14a**. A common use of the direct ELISA test is to detect the presence of drugs in urine. For these tests, antibodies specific for the drug are attached to a well of the microtiter plate. (The availability of monoclonal antibodies has been essential to the widespread use of the ELISA test.) When the patient's urine sample is added to the well, any of the drug that it contained would bind to the antibody and is captured. The well is rinsed to remove any unbound drug. To make a visible test, more antibodies specific to the drug are now added (these antibodies have an enzyme attached to them—therefore, the term *enzyme-linked*) and will react with the already-captured drug, forming a "sandwich" of antibody/drug/enzyme-linked antibody. This positive test can be detected by adding a substrate for the linked enzyme; a visible color is produced by the enzyme reacting with its substrate.

### Indirect ELISA

The indirect ELISA test, illustrated in Figure 18.14b, detects antibodies in a patient's sample rather than an antigen such as a drug. Indirect ELISA tests are used, for example, to screen blood for antibodies to HIV (see page 550). For such a purpose, the microtiter wells contain antigens, such as the inactivated virus that causes the disease the test is designed to diagnose. A sample of the patient's serum is added to the well; if it contains antibodies against the virus, they will bind the antigen. The well is rinsed to remove unbound antibodies. If antibodies in the serum and the virus in the well have attached to each other, they will remain in the well—a positive test. To make a positive test visible, some anti-HISG (an immunoglobulin that will attach to *any* antibody, including the one in the patient's serum that has attached to the virus in the well; see page 514) is added. The anti-HISG is linked to an enzyme. A positive test consists of a "sandwich" of a virus/antibody/enzyme-linked-anti-HISG. At this point, the substrate for the enzyme is added, and a positive test is detected by the color change caused by the enzyme linked to the anti-HISG.

## Western Blotting (Immunoblotting)

**Western blotting**, often simply called *immunoblotting*, can identify a specific protein in a mixture (such as proteins extracted from a blood sample). The components of the mixture are separated by electrophoresis in a gel and then transferred to a protein-binding sheet (blot). There the blot is flooded with an enzyme-linked antibody specific for the antigen. The location of the antigen and the enzyme-linked antibody reactant can

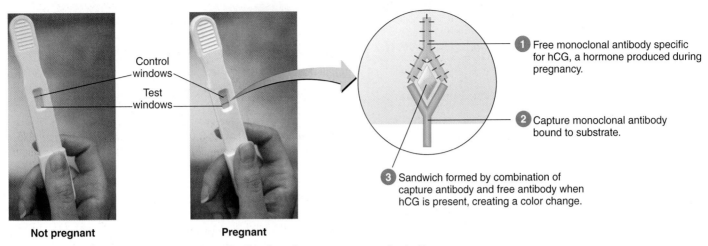

**Figure 18.13 The use of monoclonal antibodies in a home pregnancy test.** Home pregnancy tests detect a hormone called human chorionic gonadotropin (hCG) that is excreted only in the urine of a pregnant woman.

**Q** What is the antigen in the home pregnancy test?

be visualized, usually with a color-reacting label similar to an ELISA test reaction (see Figure 18.14). Figure 10.13, page 284, shows the steps of the Western blot for diagnosing Lyme disease. The most frequent application is in a confirmatory test for HIV infection (see page 550).

**CHECK YOUR UNDERSTANDING**

✔ **18-18** Which test is used to detect antibodies against a pathogen: the direct or the indirect ELISA test?

✔ **18-19** How are antibodies detected in Western blotting?

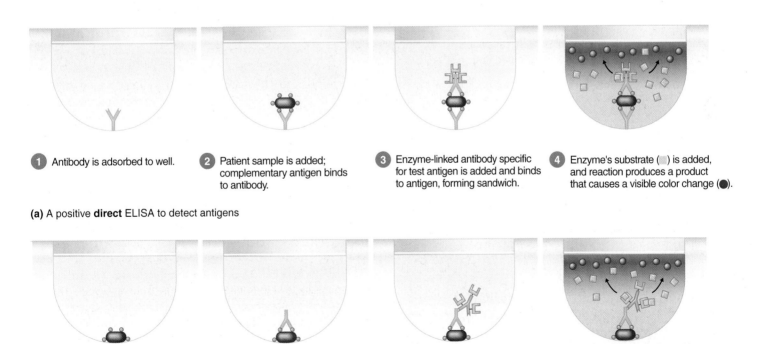

1 Antibody is adsorbed to well.
2 Patient sample is added; complementary antigen binds to antibody.
3 Enzyme-linked antibody specific for test antigen is added and binds to antigen, forming sandwich.
4 Enzyme's substrate (▢) is added, and reaction produces a product that causes a visible color change (●).

**(a)** A positive **direct** ELISA to detect antigens

1 Antigen is adsorbed to well.
2 Patient serum is added; complementary antibody binds to antigen.
3 Enzyme-linked anti-HISG is added and binds to bound antibody.
4 Enzyme's substrate (▢) is added, and reaction produces a product that causes a visible color change (●).

**(b)** A positive **indirect** ELISA to detect antibodies

**Figure 18.14 The ELISA method.** The components are usually contained in small wells of a microtiter plate. For an illustration of a technician carrying out an ELISA test on such a microtiter plate and the use of a computer to read the results, see Figure 10.12 on page 283.

**Q** Differentiate a direct from an indirect ELISA test.

# Vaccines have saved millions. So why do people still contract some vaccine-preventable illnesses?

Life before and after public health immunization campaigns is starkly different. To name a few examples:

- In 1921, before the vaccine, more than 15,000 Americans died from diphtheria. Only a single case of diphtheria has been reported to the Centers for Disease Control and Prevention (CDC) since 2003.

- A rubella epidemic in 1964–1965 infected 12.5 million Americans, killed 2000 babies, and caused 11,000 miscarriages. In 2016, only 5 cases of rubella were reported to CDC.

- It's estimated that in the twentieth century alone, smallpox infections resulted in 300 to 500 million deaths. Smallpox was eliminated worldwide in 1980 thanks to vaccines.

- In 1952, 21,000 cases of paralytic polio occurred in the United States. The nation's last case of wild-type polio virus occurred only 7 years later, in 1959.

- At the beginning of the twentieth century, an average of 48,164 cases of smallpox occurred in the United States every year. The last U.S. smallpox breakout occurred in 1949.

- In the early 2000s, vaccines against several types of human papillomavirus (HPV) were developed. Researchers predict that up to 90% of cervical, penile, and anal cancers seen today can be prevented with widespread use of HPV vaccines.

## Why Are People Still Getting Vaccine-Preventable Diseases?

Despite these public health triumphs, certain vaccine-preventable diseases persist. For instance, it used to be that nearly every child in the United States contracted pertussis, with thousands dying every year. Vaccination reduced incidence greatly, but even so, the United States still averages about 15,000 cases per year, with some deaths. Likewise, measles cases had plummeted after introduction of a vaccine, but in recent years, U.S. outbreaks have occurred. Why?

I won't spread flu to my patients or my family.

Even healthy people can get the flu, and it can be serious.

Everyone 6 months and older should get a flu vaccine. This means you.

This season, protect yourself—and those around you—by getting a flu vaccine.

For more information, visit: http://www.cdc.gov/flu

In some regions of the world, poverty and lack of health infrastructure lead to vaccination rates that are lower than required to fully eliminate incidence. That means that highly infectious diseases such as measles can persist and be transmitted to unvaccinated travelers. In developed nations, it has been discovered that some vaccine formulations need extra boosters to remain effective—such is the case with pertussis.

## Vaccine Concerns

Still other people in the United States have recently avoided vaccination because of concerns over safety. Like any medication, vaccines may cause side effects. The most common ones are mild, such as soreness at the injection site. Vaccines do not cause autism spectrum disorder, but misinformation and "fake science" regarding this issue persist on the Internet. One vaccine ingredient that came under scrutiny is thimerosal, a mercury-based preservative used to prevent contamination of multidose vials of vaccines. However, by 2000, thimerosal was either entirely removed or reduced to trace amounts in all childhood vaccines except for some flu vaccine. This was done as part of a broader national effort to reduce all types of mercury exposure in children.

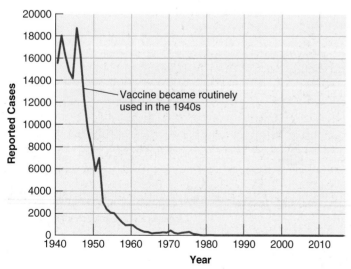

**U.S. diphtheria cases, 1940–2016.** *Source:* Centers for Disease Control and Prevention.

# Working Towards New Disease Elimination and Eradication Goals

In the United States, the incidence of vaccine-preventable diseases has decreased by 99%. Public health groups would like to see the same sort of reduction in diseases throughout the rest of the world, too. Infections for which there is only one reservoir—humans—are particularly good targets for vaccines, since the microbe would have no place to live if all humans were vaccinated.

## Measles and Rubella Initiative

Launched in 2001, the Measles & Rubella (M&R) Initiative is a global partnership focused on disease elimination led by the American Red Cross, United Nations, CDC, and World Health Organization (WHO). The M&R Initiative is committed to ensuring that no child dies from measles or is born with congenital rubella syndrome by 2020.

## Global Polio Eradication Initiative

In many ways, polio is more difficult to eradicate than smallpox was—a significant number of polio cases are asymptomatic, meaning it's not always obvious when the disease is still hiding among a population. Despite that challenge, the Global Polio Eradication Initiative has made great progress, eliminating the illness from nearly all the world. The virus is still being transmitted in three countries with poor health care, poor sanitation, and political instability: Afghanistan, Nigeria, and Pakistan. The hope is that with further efforts, polio will one day be fully eradicated.

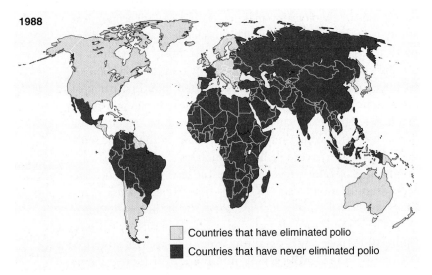

1988

| | Countries that have eliminated polio |
| | Countries that have never eliminated polio |

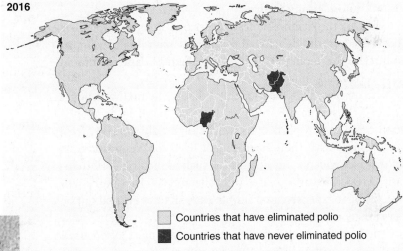

2016

| | Countries that have eliminated polio |
| | Countries that have never eliminated polio |

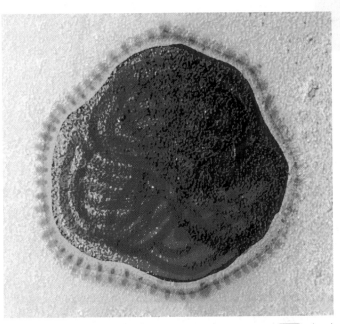

**Measles virus.**

TEM ⊢——⊣ 10 nm

---

**KEY CONCEPTS**

- Disease incidence tracking and other follow-up studies are key to identifying and addressing public health challenges. **(See Chapter 14, "Epidemiology," pages 413–416.)**

- Vaccinated people act as an important barrier to infection for individuals who are not immune **(See Chapter 14, herd immunity, pages 401–402.)**

- Depending, in part, on how they are made, vaccines have different risk factors and effects. **(See Chapter 18, "Vaccines," pages 500–507.)** For pertussis, moving to a vaccine with fewer side effects has meant shorter-term immunity.

## The Future of Diagnostic and Therapeutic Immunology

The introduction of monoclonal antibodies has revolutionized diagnostic immunology by making available large, economical amounts of specific antibodies. This has led to many newer diagnostic tests that are more sensitive, specific, rapid, and simpler to use. For example, tests to diagnose sexually transmitted chlamydial infections and certain protozoan-caused intestinal parasitic diseases are coming into common use. These tests had previously required relatively difficult culture or microscopic methods for diagnosis. At the same time, the use of many of the classic serological tests, such as complement-fixation tests, is declining.

The use of certain *nonimmunological* tests, such as nucleic acid amplification tests (NAATs) that were discussed in Chapter 10 (page 286), is increasing. NAATs are becoming automated to a significant degree. For example, a DNA chip (see Figure 10.18, page 288) containing over 50,000 single-stranded DNA probes can identify genetic information, including that of possible pathogens in a test sample. This chip is scanned and its data automatically analyzed.

Most of the diagnostic tests described in this chapter are those used in the developed world. In many less-developed countries, the money available for diagnosis and treatment alike is tragically small.

The diseases that most of these diagnostic methods target are also those that are more likely to be found in developed countries. In many parts of the world, especially tropical Africa and tropical Asia, there is an urgent need for diagnostic tests for diseases endemic in those areas, such as malaria, leishmaniasis, AIDS, Chagas' disease, and tuberculosis. These tests will need to be inexpensive and simple enough to be carried out by personnel with minimal training.

The tests described in this chapter are most often used to detect existing disease. In the future, diagnostic testing could be directed at *preventing* disease. In the United States we

**CLINICAL CASE** Resolved

A childhood vaccine against pertussis is recommended, yet pertussis has increased in incidence during the past 20 years. Adults and adolescents might be a community reservoir for *B. pertussis* because immunity from childhood vaccination declines beginning 5 to 15 years after the last pertussis vaccine dose.

The public's sense of the importance of childhood vaccinations has also declined; consequently, many children are not fully immunized. The CDC now recommends that adults get revaccinated with a combination vaccine that prevents tetanus, diphtheria, and pertussis.

501 504 505 510 515 **520**

regularly see reports of outbreaks of foodborne disease. Fresh produce does not carry the identifying marks that facilitate the tracking of packaged food products. Thus, sampling methods that would allow complete identification (including specific pathogenic serovars) within a few hours, or even minutes, would save valuable time in tracking outbreaks of infectious disease carried by fruits and vegetables. This saving in time would be translated into immense economic savings for growers and retailers. It would also lead to less human illness.

Not every topic discussed in this chapter is necessarily directed at detecting and preventing disease. As was mentioned on pages 508–509, Mabs have applications in the therapy of disease as well. These are already in use to treat certain cancers such as breast cancer and non-Hodgkin's lymphoma, as well as inflammatory diseases such as rheumatoid arthritis.

**CHECK YOUR UNDERSTANDING**

✔ 18-20 How has the development of monoclonal antibodies revolutionized diagnostic immunology?

## Study Outline

 Go to @MasteringMicrobiology for **Interactive Microbiology,** *In the Clinic* videos, MicroFlix, MicroBoosters, 3D animations, practice quizzes, and more.

### Vaccines (pp. 500–507)

1. Edward Jenner developed the modern practice of vaccination when he inoculated people with cowpox virus to protect them against smallpox.

### Principles and Effects of Vaccination (pp. 500–501)

2. Herd immunity results when most of a population is immune to a disease.

### Types of Vaccines and Their Characteristics (pp. 501–504)

3. Live attenuated vaccines consist of attenuated (weakened) microorganisms; attenuated virus vaccines generally provide lifelong immunity.
4. Inactivated vaccines consist of killed bacteria or viruses.

5. Subunit vaccines consist of antigenic fragments of a microorganism; these include toxoids, virus-like particles, and polysaccharides.
6. Conjugated vaccines combine the desired antigen with a protein that boosts the immune response.
7. Nucleic acid (DNA) vaccines cause the recipient to make the antigenic protein.
8. Recombinant vector vaccines are avirulent viruses or bacteria genetically modified to produce a desired antigen.

### Vaccine Production, Delivery Methods, and Formulations (pp. 504–507)

9. Viruses for vaccines may be grown in animals, cell cultures, or chick embryos.

10. Recombinant vaccines and nucleic acid vaccines do not need to be grown in cells or animals.
11. Genetically modified plants may someday provide edible vaccines.
12. Dry skin patch vaccines don't need refrigeration.
13. Oral administration and combining several vaccines would eliminate the number of injections required for vaccinations.
14. Adjuvants improve the effectiveness of some antigens.
15. Vaccines are the safest and most effective means of controlling infectious diseases.

## Diagnostic Immunology (pp. 507–520)

1. Many tests based on the interactions of antibodies and antigens have been developed to determine the presence of antibodies or antigens in a patient.
2. The sensitivity of a diagnostic test is determined by the percentage of positive samples it correctly detects; and its specificity is determined by the percentage of negative results it gives when the specimens are negative.
3. Direct tests are used to identify specific microorganisms.
4. Indirect tests are used to demonstrate the presence of antibody in serum.
5. Diseases can be diagnosed by a rising titer or seroconversion (from no antibodies to the presence of antibodies).

### Use of Monoclonal Antibodies (pp. 508–509)

6. Hybridomas are produced in the laboratory by fusing a cancerous B-cell with an antibody-secreting plasma cell.
7. A hybridoma cell culture produces large quantities of the plasma cell's antibodies, called monoclonal antibodies.
8. Monoclonal antibodies are used to treat diseases and in diagnostic serological tests.

### Precipitation Reactions (pp. 510–511)

9. The interaction of soluble antigens with IgG or IgM antibodies leads to precipitation reactions.
10. Precipitation reactions depend on the formation of lattices and occur best when antigen and antibody are present in optimal proportions.
11. Immunodiffusion procedures are precipitation reactions carried out in an agar gel medium.
12. Immunoelectrophoresis combines electrophoresis with immunodiffusion for the analysis of serum proteins.

### Agglutination Reactions (pp. 511–512)

13. The interaction of particulate antigens (cells that carry antigens) with antibodies leads to agglutination reactions.

14. Diseases may be diagnosed by combining the patient's serum with a known antigen.
15. Antibodies cause visible agglutination of soluble antigens affixed to latex spheres in indirect or passive agglutination tests.
16. Hemagglutination reactions involve agglutination reactions using red blood cells. Hemagglutination reactions are used in blood typing, the diagnosis of certain diseases, and the identification of viruses.

### Neutralization Reactions (pp. 512–513)

17. In neutralization reactions, the harmful effects of a bacterial exotoxin or virus are eliminated by a specific antibody.
18. An antitoxin is an antibody produced in response to a bacterial exotoxin or a toxoid that neutralizes the exotoxin.
19. In a virus neutralization test, the presence of antibodies against a virus can be detected by the antibodies' ability to prevent cytopathic effects of viruses in cell cultures.
20. In viral hemagglutination inhibition tests, antibodies against certain viruses can be detected by their ability to interfere with viral hemagglutination.

### Complement-Fixation Reactions (pp. 513–514)

21. Complement-fixation reactions are serological tests based on the depletion of a fixed amount of complement in the presence of an antigen–antibody reaction.

### Fluorescent-Antibody Techniques (pp. 514–515)

22. Fluorescent-antibody techniques use antibodies labeled with fluorescent dyes.
23. A fluorescence-activated cell sorter can be used to detect and count cells labeled with fluorescent antibodies.

### Enzyme-Linked Immunosorbent Assay (ELISA) (p. 516)

24. ELISA techniques use antibodies linked to an enzyme.
25. Antigen–antibody reactions are detected by enzyme activity. If the indicator enzyme is present in the test well, an antigen–antibody reaction has occurred.

### Western Blotting (Immunoblotting) (pp. 516–517)

26. Serum antibodies separated by electrophoresis are identified with an enzyme-linked antibody.

### The Future of Diagnostic and Therapeutic Immunology (p. 520)

27. The use of monoclonal antibodies and nucleic acid amplification tests will make new diagnostic tests possible.

# Study Questions

For Answers to the Knowledge and Comprehension questions, turn to the Answers tab at the back of the textbook.

## Knowledge and Comprehension

### Review

1. Classify the following vaccines by type. Which could cause the disease it is supposed to prevent?

a. attenuated measles virus
b. dead *Rickettsia prowazekii*
c. *Vibrio cholerae* toxoid
d. hepatitis B antigen produced in yeast cells
e. purified polysaccharides from *Streptococcus pyogenes*
f. *Haemophilus influenzae* polysaccharide bound to diphtheria toxoid
g. a plasmid containing genes for influenza A protein

2. Define the following terms, and give an example of how each reaction is used diagnostically:
   a. viral hemagglutination
   b. hemagglutination inhibition
   c. passive agglutination

3. DRAW IT Label the components of the direct and indirect FA tests in the following situations. Which test is direct? Which test provides definitive proof of disease?

**(a)** Rabies can be diagnosed postmortem by mixing fluorescent-labeled antibodies with brain tissue.

**(b)** Syphilis can be diagnosed by adding the patient's serum to a slide fixed with *Treponema pallidum*. Anti-human immune serum globulin tagged with a fluorescent dye is added.

4. How are monoclonal antibodies produced?

5. Explain the effects of excess antigen and antibody on the precipitation reaction. How is the precipitin ring test different from an immunodiffusion test?

6. DRAW IT Label components of the direct and indirect ELISA tests in the situations shown below. Which test is direct? Which test provides definitive proof of disease?

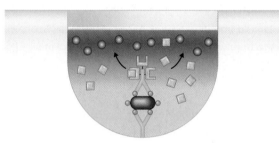

**(a)** Respiratory secretions to detect respiratory syncytial virus

**(b)** Blood to detect human immunodeficiency virus antibodies

7. How does the antigen in an agglutination reaction differ from that in a precipitation reaction?

8. Match the following serological tests in column A to the descriptions in column B.

| Column A | Column B |
| --- | --- |
| _____ **a.** Precipitation | **1.** Occurs with particulate antigens |
| _____ **b.** Western blotting | **2.** Uses an enzyme for the indicator |
| _____ **c.** Agglutination | **3.** Uses red blood cells for the indicator |
| _____ **d.** Complement fixation | **4.** Uses anti-human immune serum globulin |
| _____ **e.** Neutralization | **5.** Occurs with a free soluble antigen |
| _____ **f.** ELISA | **6.** Used to determine the presence of antitoxin |

9. Match each of the following tests in column A to its positive reaction in column B.

| Column A | Column B |
| --- | --- |
| _____ **a.** Agglutination | **1.** Peroxidase activity |
| _____ **b.** Complement fixation | **2.** Harmful effects of agents not seen |
| _____ **c.** ELISA | **3.** No hemolysis |
| _____ **d.** FA test | **4.** Cloudy white line |
| _____ **e.** Neutralization | **5.** Cell clumping |
| _____ **f.** Precipitation | **6.** Fluorescence |

10. NAME IT A purified protein from *Mycobacterium tuberculosis* is injected into a person's skin. A hardened, red area develops around the injection site within 3 days.

## Multiple Choice

Use the following choices to answer questions 1 and 2:
   a. hemolysis
   b. hemagglutination
   c. hemagglutination inhibition
   d. no hemolysis
   e. precipitin ring forms

1. Patient's serum, influenza virus, sheep red blood cells, and anti-sheep red blood cells are mixed in a tube. What happens if the patient has antibodies against influenza?

2. Patient's serum, *Chlamydia*, guinea pig complement, sheep red blood cells, and anti-sheep red blood cells are mixed in a tube. What happens if the patient has antibodies against *Chlamydia*?

3. The examples in questions 1 and 2 are
   a. direct tests.
   b. indirect tests.

Use the following choices to answer questions 4 and 5:
   a. anti-*Brucella*
   b. *Brucella*
   c. substrate for the enzyme

4. Which is the third step in a direct ELISA test?

5. Which item is from the patient in an indirect ELISA test?

6. In an immunodiffusion test, a strip of filter paper containing diphtheria antitoxin is placed on a solid culture medium. Then bacteria are streaked perpendicular to the filter paper. If the bacteria are toxigenic,
   a. the filter paper will turn red.
   b. a line of antigen–antibody precipitate will form.
   c. the cells will lyse.
   d. the cells will fluoresce.
   e. none of the above

Use the following choices to answer questions 7–9.
   a. direct fluorescent antibody
   b. indirect fluorescent antibody
   c. rabies immune globulin
   d. killed rabies virus
   e. none of the above

7. Treatment given to a person bitten by a rabid bat.

8. Test used to identify rabies virus in the brain of a dog.

9. Test used to detect the presence of antibodies in a patient's serum.

10. In an agglutination test, eight serial dilutions to determine antibody titer were set up: Tube 1 contained a 1:2 dilution; tube 2, a 1:4, and so on. If tube 5 is the last tube showing agglutination, what is the antibody titer?
   a. 5
   b. 1:5
   c. 32
   d. 1:32

## Analysis

1. What problems are associated with the use of live attenuated vaccines?

2. Many of the serological tests require a supply of antibodies against pathogens. For example, to test for *Salmonella*, anti-*Salmonella* antibodies are mixed with the unknown bacterium. How are these antibodies obtained?

3. A test for antibodies against *Treponema pallidum* uses the antigen cardiolipin and the patient's serum (suspected of having antibodies). Why do the antibodies react with cardiolipin? What is the disease?

## Clinical Applications and Evaluation

1. Which of the following is proof of a disease state? Why doesn't the other situation confirm a disease state? What is the disease?
   a. *Mycobacterium tuberculosis* is isolated from a patient.
   b. Antibodies against *M. tuberculosis* are found in a patient.

2. Streptococcal erythrogenic toxin is injected into a person's skin in the Dick test. What results are expected if a person has antibodies against this toxin? What type of immunological reaction is this? What is the disease?

3. The following data were obtained from FA tests for anti-*Legionella* in four people. What conclusions can you draw? What is the disease?

| | Antibody Titer | | | |
|---|---|---|---|---|
| | **Day 1** | **Day 7** | **Day 14** | **Day 21** |
| Patient A | 128 | 256 | 512 | 1024 |
| Patient B | 0 | 0 | 0 | 0 |
| Patient C | 256 | 256 | 256 | 256 |
| Patient D | 0 | 0 | 128 | 512 |

4. Alana decided against the relatively new chickenpox vaccine and used her parents' method: she wanted her children to get chickenpox so that they would develop natural immunity. Her two children did get chickenpox. Her son had slight itching and skin vesicles, but her daughter was hospitalized for months with streptococcal cellulitis and underwent several skin grafts before recovering. Alana's housekeeper contracted chickenpox from the children and subsequently died. Almost half of the deaths due to chickenpox occur in adults.
   a. What responsibilities do parents have for their children's health?
   b. What rights do individuals have? Should vaccination be required by law?
   c. What responsibilities do individuals (e.g., parents) have for the health of society?
   d. Vaccines are given to healthy people, so what risks are acceptable?

# 19 Disorders Associated with the Immune System

Normally cells of the immune system remove or neutralize injurious agents, like the two lymphocytes shown attacking a cancer cell in the photograph. However, in this chapter we see that not all immune system responses produce a desirable result. A familiar example is the itchy eyes and runny nose of hay fever, which results from repeated exposure to plant pollen or other environmental antigens. Most of us also know about the importance of blood typing for transfusions or organ transplants to avoid rejection reactions. Another undesirable response occurs when the immune system mistakenly attacks one's own tissue, causing diseases we classify as autoimmune disorders.

Some people are born with a defective immune system (see the Clinical Case in this chapter), and in all of us the effectiveness of our immune system declines with age. Our immune systems can be deliberately crippled (*immunosuppressed*) to prevent the rejection of transplanted organs. Disease can also impair the immune system, especially infection by HIV, a virus that specifically attacks the immune system.

The **Big Picture** on pages 528–529 describes the role of the human microbiome in developing a healthy immune system.

▶ Two lymphocytes attacking a cancer cell (blue).

## In the Clinic

As a nurse who specializes in the treatment of AIDS patients, you discuss the HIV status of a newborn with Jessica, her HIV-positive mother. The infant tested positive by ELISA and Western blot, but the PCR test was negative for HIV. **Does the infant have HIV infection? How do you explain the seemingly conflicting test results to Jessica? And what advice do you give Jessica to prevent transmitting HIV to her baby?**

*Hint: Read about diagnostic methods for HIV on pages 550–551.*

Play **In the Clinic** Video
@Mastering**Microbiology**

# Hypersensitivity

## LEARNING OBJECTIVES

**19-1** Define *hypersensitivity*.

**19-2** Describe the mechanism of anaphylaxis.

**19-3** Compare and contrast systemic and localized anaphylaxis.

**19-4** Explain how allergy skin tests work.

**19-5** Define *desensitization* and *blocking antibody*.

**19-6** Describe the mechanism of cytotoxic reactions and how drugs can induce them.

**19-7** Describe the basis of the ABO and Rh blood group systems.

**19-8** Explain the relationships among blood groups, blood, transfusions, and hemolytic disease of the newborn.

**19-9** Describe the mechanism of immune complex reactions.

**19-10** Describe the mechanism of delayed cell-mediated reactions, and name two examples.

The term **hypersensitivity** refers to an antigenic response that results in undesirable effects. Allergies are a familiar example. Hypersensitivity responses occur in individuals who have been *sensitized* by previous exposure to an antigen, which in this context is often called an **allergen**. When a sensitized individual is exposed to that antigen again, the body's immune system reacts to it in a damaging manner. The study of hypersensitivity reactions is called **immunopathology**. The four principal types of hypersensitivity reactions, summarized in Table 19.1, are anaphylactic, cytotoxic, immune complex, and cell-mediated (or delayed-type) reactions.

## Allergies and the Microbiome

Incidence of food and environmental allergies is increasing in developed nations. The *hygiene hypothesis* suggests that limiting childhood exposure to bacteria and parasites may lower immune tolerance and limit the body's ability to cope with harmless antigens, such as food or pollen. Parasites such as worms are commonly found in developing areas but are mostly absent from places with modern sanitation and good health care infrastructure. Studies show that mammals lacking early exposure to microorganisms are more susceptible to asthma and allergies. The resident microbiota within the human body is also being studied as a factor in certain autoimmune diseases. For more on this, see the **Big Picture** on the human microbiome on pages 528–529.

> **CHECK YOUR UNDERSTANDING**
>
> ✔ **19-1** Are all immune responses beneficial?

## Type I (Anaphylactic) Reactions

**Type I reactions (anaphylactic reactions)** occur 2 to 30 minutes after a sensitized person is reexposed to an antigen. *Anaphylaxis*

means "the opposite of protected," from the prefix *ana-*, meaning against, and the Greek *phylaxis*, meaning protection. **Anaphylaxis** is an inclusive term for the reactions caused when antigens combine with IgE antibodies. Anaphylactic responses can be *systemic reactions*, producing shock and breathing difficulties that are sometimes fatal. However, they may also be *localized reactions*, including common allergic conditions such as hay fever, asthma, and hives (slightly raised, often itchy and reddened areas of the skin).

The IgE antibodies produced in response to an antigen (such as insect venom or plant pollen) bind to the surfaces of mast cells and basophils. Both cell types are similar in morphology and in their contribution to allergic reactions. **Mast cells** are especially prevalent in the mucosal and connective tissue of the skin and respiratory tract and in surrounding blood vessels.* **Basophils** in the bloodstream are recruited to the site of an infection. Mast cells and basophils contain granules of histamine and other chemical mediators (**Figure 19.1a**).

Mast cells and basophils can have as many as 500,000 sites per cell for IgE attachment. The Fc (stem) region of an IgE antibody (see Figure 17.4, page 480) can attach to one of these specific receptor sites on such a cell, leaving two antigen-binding sites free. Of course, the attached IgE monomers will not all be specific for the same antigen. But when an antigen encounters two adjacent antibodies of the same appropriate specificity, it can bind to one antigen-binding site on each antibody, bridging the space between them. This bridge triggers the mast cell or basophil to undergo **degranulation**, a cellular process that releases the granules inside these cells along with the mediators they contain (Figure 19.1b).

These mediators cause the unpleasant and damaging effects of an allergic reaction. The best-known mediator, **histamine**,

---

**CLINICAL CASE** Who Are You?

Malik, a 10-day-old infant, has just been brought home from the neonatal intensive care (NICU) unit after surgery to repair a heart defect. When his mother is changing his diaper, she notices a rash on his buttocks. Assuming diaper rash, Malik's mother treats the area with ointment and thinks nothing else of it. As the day progresses, the rash spreads to Malik's face, giving him the appearance of being slapped. By the time Malik's parents bring him to the emergency department (ED), the rash is lobster-red and has spread to his entire body.

**What is causing Malik's rash? Read on to find out.**

**525** ▸ 530 ▸ 541 ▸ 544 ▸ 554

---

*The name is from the German word *mastzellen*, meaning well fed; mast cells are packed with granules that at one time were mistakenly thought to have been ingested.

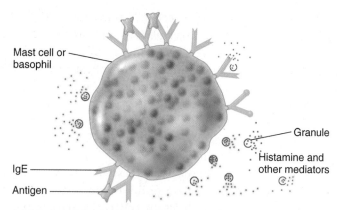

**(a)** IgE antibodies, produced in response to an antigen, coat mast cells and basophils. When an antigen bridges the gap between two adjacent antibody molecules of the same specificity, the cell undergoes degranulation and releases histamine and other mediators.

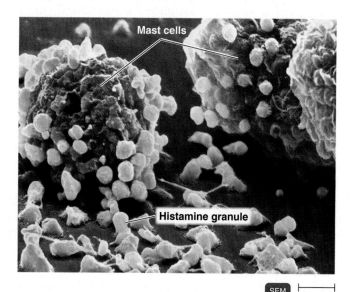

**(b)** A degranulated mast cell that has reacted with an antigen and released granules of histamine and other reactive mediators

**Figure 19.1 The mechanism of anaphylaxis.**

**Q** What type of cells do IgE antibodies bind to?

is stored in the granules. Released histamine increases blood flow and the permeability of blood capillaries, resulting in edema (swelling) and erythema (redness). Other effects include increased mucus secretion (a runny nose, for example) and smooth muscle contraction, which in the respiratory bronchi results in breathing difficulty. Other mediators, such as **leukotrienes** and **prostaglandins**, are not preformed and stored in the granules, but rather are synthesized by the antigen-triggered cell. Because leukotrienes tend to cause prolonged contractions of certain smooth muscles, their action contributes to spasms of the bronchial tubes that occur during asthma attacks. Prostaglandins affect smooth muscles of the respiratory system and increase mucus secretion. Collectively, all these mediators serve as chemotactic agents that, in a few hours, attract neutrophils and eosinophils to the site of the degranulated cell.

### Systemic Anaphylaxis

**Systemic anaphylaxis** (or *anaphylactic shock*) results when release of mediators causes peripheral blood vessels throughout the body to dilate, resulting in a drop in blood pressure (shock). In addition, symptoms including narrowing of airway passages causing respiratory distress, flush or skin rash, tingling sensations, and nausea. This reaction can be fatal within a few minutes. There is very little time to act once someone develops systemic anaphylaxis. Treatment usually involves self-administration with a preloaded syringe of epinephrine, a drug that constricts blood vessels and raises the blood pressure.

Even a small dose of the antigen in question may cause a systemic reaction in someone who is sensitized to it. Injected antigens are more likely to cause a dramatic response than antigens introduced via other portals of entry. People who are allergic to venom of stinging insects such as bees and wasps are at risk for a systemic reaction. The risk of anaphylaxis is one of the reasons why health care providers ask patients about any known drug allergies before treatment occurs.

| Type of Reaction | Time Before Clinical Signs | Characteristics | Examples |
|---|---|---|---|
| **Type I (Anaphylactic)** | <30 min | IgE binds to mast cells or basophils; causes degranulation of mast cell or basophil and release of reactive substances such as histamine. | Anaphylactic shock from drug injections and insect venom; common allergic conditions, such as hay fever, asthma |
| **Type II (Cytotoxic)** | 5–12 hours | Antigen causes formation of IgM and IgG antibodies that bind to target cell; when combined with action of complement, destroys target cell. | Transfusion reactions, Rh incompatibility |
| **Type III (Immune Complex)** | 3–8 hours | Antibodies and antigens form complexes that cause damaging inflammation. | Arthus reactions, serum sickness |
| **Type IV (Delayed Cell-Mediated, or Delayed Hypersensitivity)** | 24–48 hours | Antigens activate CTLs that kills target cell. | Rejection of transplanted tissues; contact dermatitis, such as poison ivy; certain chronic diseases, such as tuberculosis |

**TABLE 19.1 Types of Hypersensitivity**

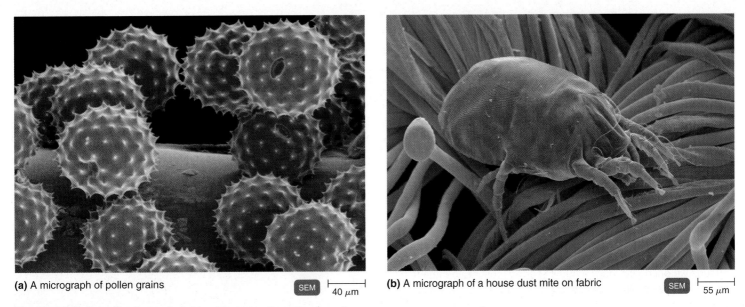

(a) A micrograph of pollen grains    SEM  ├─────┤ 40 μm

(b) A micrograph of a house dust mite on fabric    SEM  ├─────┤ 55 μm

**Figure 19.2  Localized anaphylaxis.** Inhaled antigens such as these are a common cause of localized anaphylaxis.

**Q** Compare localized and systemic anaphylaxis.

You may know someone who is allergic to penicillin. In the United States, this drug allergy occurs in about 3–10% of the population. The common antibiotic is a hapten that combines with a carrier serum protein to cause an immune response in allergic individuals. (See Chapter 17 for a review on haptens.) Skin tests for penicillin sensitivity are available. Patients with a positive skin test can be effectively desensitized for a treatment (see page 530). This is done by orally administering a series of increasing doses of penicillin V over a short time period (completed within 4 hours) immediately prior to the procedure. Allergy to penicillin also includes risk from exposure to some related drugs, such as amoxicillin and carbapenem (page 568).

### Localized Anaphylaxis

**Localized anaphylaxis** is usually immediate, temporary, and less severe than systemic anaphylaxis. Localized anaphylaxis is associated with antigens that are ingested (foods) or inhaled (pollen) (**Figure 19.2a**). The symptoms depend primarily on the route by which the antigen enters the body. In allergies involving the upper respiratory system (such as hay fever), sensitization and production of IgE subsequently involves mast cells that release histamine in the mucous membrane of the upper respiratory tract. The airborne antigen might be a common environmental material such as plant pollen, fungal spores, feces of house dust mites (Figure 19.2b), or animal dander.*

---

*_Dander_ is a general term for very small flakes of old skin cells from the fur, skin, or feathers of animals. It is analogous to dandruff in humans. Pet dander accumulates in upholstery and carpeting in homes. People with allergies to mice, gerbils, and similar small animals are more likely to be allergic to components of urine accumulating in cages.

The typical symptoms are itchy and teary eyes, congested nasal passages, coughing, and sneezing. Antihistamine drugs, which compete for histamine receptor sites, are often used to treat these symptoms.

Asthma is an allergic reaction that mainly affects the lower respiratory system. Symptoms such as wheezing and shortness of breath are caused by the constriction of smooth muscles in the bronchial tubes. For unknown reasons, asthma is becoming a near epidemic, affecting about 8.6% of children in the United States. The _hygiene hypothesis_, described previously, may be a factor in the increased incidence of asthma. Environment, infections, and stress can be a contributing factor in precipitating asthma. Symptoms are usually controlled by aerosol inhalants such as albuterol that relax smooth muscles. Another drug, omalizumab (Xolair®), a humanized monoclonal antibody, blocks IgE for people with moderate to severe allergic asthma or hives not controlled by antihistamines or leukotriene-blockers such as montelukast (Singulair®).

Antigens that enter the body via the gastrointestinal tract can also sensitize an individual. Frequently, so-called food allergies may not be related to hypersensitivity at all and are more accurately described as _food intolerances_. For example, many people are unable to digest the lactose in milk because they lack the enzyme that breaks it down. The lactose enters the intestine, where it osmotically retains fluid, causing diarrhea.

Gastrointestinal upset is a common symptom of food allergies, but it can also result from many other factors. Hives are more characteristic of a true food allergy, and ingestion of the antigen may result in systemic anaphylaxis. Death has even resulted when a person sensitive to fish ate french fries that had

# Is dysbiosis increasing incidence of immune-related disorders?

Incidence of immune-related disorders such as asthma, food allergies, and inflammatory bowel diseases are on the rise throughout the developed world. These disorders involve a variety of cells and modes of action but broadly share origins in a damaging immune response that results in inflammation and tissue damage. Growing research suggests that the increase in autoimmune disorders may be related to increased **dysbiosis**, a harmful imbalance of normal microbiota, among people living modern lifestyles.

The connection between dysbiosis and autoimmune disorders is explained through the *hygiene hypothesis*, which posits that childhood exposure to microorganisms is essential for training the immune system to become tolerant of the harmless microbes and antigens we encounter daily. Without early exposure, the immune system overreacts when it encounters microbes that historically provoked little or no response.

Ironically, the good hygiene, sanitation, and antibiotics that people in developed nations enjoy may have a drawback: helpful microbes are eliminated along with the harmful ones. Studies show that people living in developing nations, with less stringent sanitation and health/hygiene measures systematically in place, tend to have greater diversity of resident microbes. According to the hygiene hypothesis, this lack of childhood exposure to microbes results in less microbiome diversity, and that, in turn, may encourage development of autoimmune disorders.

## Allergies and Asthma

Researchers first observed that the prevalence of allergies was lower among indigenous tribal populations compared to urban populations in the 1970s. More recently, this same difference was seen between children growing up on farms versus children in

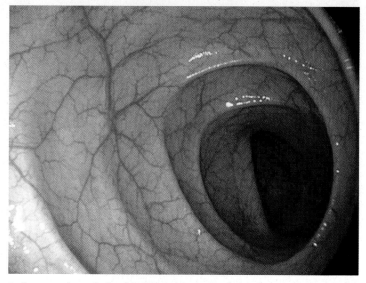

**Endoscope view of a healthy colon.**

urban areas. The farm children encountered a wider range of microbial exposures and had a lower asthma prevalence.

## Inflammatory Bowel Diseases

Dysbiosis is now being closely studied as a possible cause of inflammatory bowel diseases, which include ulcerative colitis and Crohn's disease. Some metabolic products of normal microbiota, such as butyrates, exert an anti-inflammatory effect on the body—so the lack of these normal microbiota could result in more inflammation. Antibiotic usage may also contribute to these diseases as studies demonstrate that the microbiome may not recover its full

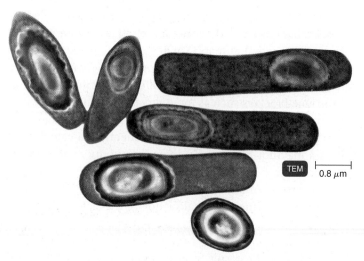

**Clostridium difficile**, a bacterium that can proliferate and cause bowel disease when antibiotics kill other resident microbiota.

TEM — 0.8 μm

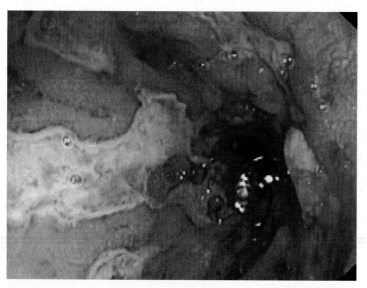

**Endoscope view of an inflamed and ulcerated colon of a patient with Crohn's disease.**

# Diseases with a Dysbiosis Link

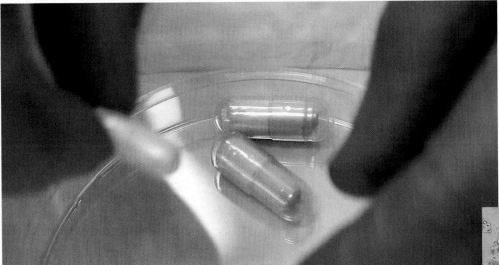

**"Poop pills" developed for fecal transplant treatments.**

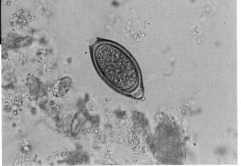

**Egg of *Trichuris suis*, the pig whipworm used to treat Crohn's disease.**

diversity after antibiotic treatment. This may lead to loss of organisms that would typically keep inflammation under control.

## Crohn's Disease

Crohn's disease causes swelling of the gastrointestinal (GI) tract and is often characterized by excessive amounts of the cytokines tumor necrosis factor alpha (TNF-$\alpha$) and interleukin-12 (IL-12). Researchers hypothesize the excess could result from disruption in normal microbiota that usually keep inflammatory cytokines under control. Microbes are a possible treatment for Crohn's disease, too. Helminths such as the whipworm suppress certain T helper cell pathways that are overactive in Crohn's disease. In one study, Crohn's patients ingested whipworm eggs and saw remission rates as high as 73%. Since the worms don't take up residence in humans, the treatment must be repeated periodically to maintain the effect.

## *Clostridium difficile* Infections

*Clostridium difficile* is a major concern in health care settings, where patients undergoing antibiotic therapy can develop debilitating GI infections from it after other, less drug-resistant microbiota are killed. Scientists found success treating *C. difficile* infections with fecal transplants. This entails inoculating the patient with a fecal sample from a healthy individual (usually a family member). Fecal transplants from a donor are more effective than those from self. The microbiota in pills are surrounded by a triple layer of gel to prevent breakdown in the stomach. The healthy microbes found in the donor's sample can restore a healthy microbiome to the patient. Because fecal transplants are much more effective than antibiotic treatments, it has become a common treatment for IBD, and some physicians are trying fecal transplants to treat obesity.

## KEY CONCEPTS

- Normal microbiota are important in maintaining a healthy immune system. **(See Chapter 14, "Relationships Between Normal Microbiota and the Host," pages 397–398.)**

- The Human Microbiome Project is sequencing the genes for 16S ribosomal RNA to help scientists catalogue normal microbiota that are difficult to culture and identify in the laboratory **(See Chapter 9, "Genome Projects," pages 256–257.)**

- *Trichuris suis* is a roundworm related to *T. trichiura*. **(See Chapter 12, "Nematodes," pages 353–355.)**

- Inflammatory diseases are characterized by increased amounts of cytokines produced by T helper cells, including tumor necrosis factor alpha and interleukins. **(See Chapter 16, "Inflammation," pages 459–462.)**

been prepared in oil previously used to fry fish. Skin tests are not reliable indicators for diagnosing food-related allergies, and completely controlled tests for hypersensitivity to ingested foods are very difficult to perform. Only eight foods are responsible for 97% of food-related allergies: eggs, peanuts, tree-grown nuts, milk, soy, fish, wheat, and peas. Most children exhibiting allergies to milk, egg, wheat, and soy develop tolerance as they age, but reactions to peanuts, tree nuts, and seafood tend to persist.

Sulfites are a common cause of allergy. Their use is widespread as a preservative in foods (such as dried fruits and processed meats) and beverages (such as wine); although food labels should indicate their presence, they may be difficult to avoid in practice. A food product may have come into contact with a food allergen through processing machinery or cookware previously used for other foods. In one U.S. Food and Drug Administration report, 25% of bakery, ice cream, and candy products tested positive for peanut allergens even though peanuts were not listed on the required product labels.

### Preventing Anaphylactic Reactions

Some individuals experience an allergic reaction after eating an assortment of foods and may not know exactly what food they are sensitive to. In some cases, skin tests might be of use in diagnosis (**Figure 19.3**). These tests are used for a variety of allergies and involve inoculating small amounts of the suspected antigen just beneath the epidermis. Sensitivity to the antigen is indicated by a rapid inflammatory skin reaction that produces redness, swelling, and itching at the inoculation site. This small affected area is called a *wheal*.

Once the responsible antigen has been identified, the person can either try to avoid contact with it or undergo **desensitization**, or **subcutaneous allergen-specific immunotherapy**. This procedure usually consists of a series of gradually increasing dosages of the antigen carefully injected beneath the skin. The objective is to cause the production of IgG rather than IgE antibodies, in the hope that the circulating IgG antibodies will act as *blocking antibodies* to intercept and neutralize the antigens before they can

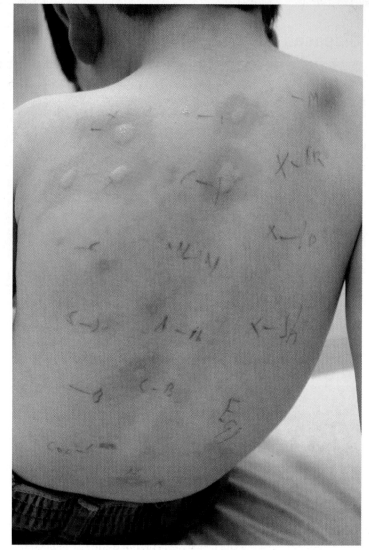

**Figure 19.3 A skin test to identify allergens.** Drops of fluid containing test substances are placed on the skin. A light scratch is made with a needle to allow the substances to penetrate the skin. Reddening and swelling at the site identify the substance as a probable cause of an allergic reaction.

**Q** What is scratched into the skin in a skin test?

react with cell-bound IgE. Desensitization is not a "cure" for allergies and has varying levels of effectiveness, with the most success seen against inhaled and injected (insect venom) allergens.

### CLINICAL CASE

The emergency department physician quickly rules out erythema infectiosum (fifth disease) because the rash did not begin on his face. (Fifth disease is a viral illness; one of the first symptoms is a facial rash.) Malik has no other symptoms, such as difficulty breathing or low blood pressure, that would indicate an anaphylactic reaction. When speaking with the parents, the physician finds out that Malik recently underwent surgery to repair a heart defect and was given a routine blood transfusion. During the surgery it was discovered that Malik doesn't have a thymus gland.

**What is the role of the thymus gland? (Hint: See Chapter 17.)**

525    **530**    541    544    554

### CHECK YOUR UNDERSTANDING

✔ **19-2** In what tissues do we find the mast cells that are major contributors to allergic reactions such as hay fever?

✔ **19-3** Which is the more dangerous to life: systemic or localized anaphylaxis?

✔ **19-4** How can we tell whether a person is sensitive to a particular allergen, such as a tree pollen?

✔ **19-5** Which antibody types need to be blocked to desensitize a person subject to allergies?

**TABLE 19.2** The ABO Blood Group System

| Blood Group | Erythrocyte (Red Blood Cell) Antigens | Illustration | Plasma Antibodies | Cells That Can Be Received | Frequency (% U.S. Population) | | |
|---|---|---|---|---|---|---|---|
| | | | | | White | Black | Asian |
| A | A | | Anti-B | A, O | 41 | 27 | 28 |
| B | B | | Anti-A | B, O | 9 | 20 | 27 |
| AB | A and B | | Neither anti-A nor anti-B antibodies | A, B, AB, O | 3 | 4 | 5 |
| O | Neither A nor B | | Anti-A and Anti-B | O | 47 | 49 | 40 |

## Type II (Cytotoxic) Reactions

**Type II (cytotoxic) reactions** generally involve the activation of complement by the combination of IgG or IgM antibodies with an antigenic cell. This activation stimulates complement to lyse the affected cell, which might be either a foreign cell or a host cell that carries a foreign antigenic determinant (such as a drug) on its surface. Additional cellular damage may be caused within 5 to 8 hours by the action of macrophages and other cells that attack antibody-coated cells.

The most familiar cytotoxic hypersensitivity reactions are *transfusion reactions*, in which red blood cells are destroyed as a result of reacting with circulating antibodies. These reactions involve blood group systems that include the ABO and Rh antigens.

### The ABO Blood Group System

In 1901, Karl Landsteiner discovered that human blood could be grouped into four principal types, which were designated A, B, AB, and O. This method of classification is called the **ABO blood group system**. Since then, other blood group systems have been discovered, but our discussion will be limited to two of the best known, the ABO and the Rh systems. The main features of the ABO blood group system are summarized in Table 19.2.

A person's ABO blood type depends on the presence or absence of carbohydrate antigens located on the cell membranes of red blood cells (RBCs). Blood group antigens are also found on other cells, including cells inside blood vessels. Anti-A and anti-B antibodies are not present in the newborn;

they appear in the first years of life. ABO involves three carbohydrate antigens on the surfaces of cells: A, B, and H. The H gene produces the H antigen, which is modified by enzymes produced by the A and B genes, if they are present. Thus a type O person expresses only the H antigen, and type A individuals convert the H antigen to the A antigen.

The ABO antigens are unique among the antigens on human cells because antibodies called *isoantibodies* are produced in people with different ABO antigens. The anti-A and anti-B antibodies are produced by an individual against the A or B antigens on the cells of other antigen groups. In a person with A antigens, the plasma will contain antibodies against B antigens. Cells of blood type O lack both A and B antigens. As such, type O individuals have antibodies against both A and B antigens. Individuals with type AB cells have plasma with no antibodies to either A or B antigens. Table 19.2 shows that the plasma of individuals with a given blood type, such as A, have antibodies against the alternative blood type, anti-B antibody. The origin of these antibodies is described in Exploring the Microbiome on page 532.

Antigen–antibody reaction causes agglutination and activates complement, which in turn causes lysis of the donor's RBCs as they enter the recipient's system. Therefore, whole blood is not used in transfusions. For example, if whole blood from a type O donor were given to a type B recipient, the anti-B antibodies in the donor would immediately react with the recipient cells (that have B antigens), fixing complement and lysing cells. Or, when type B blood is transfused into a person with type A blood, the antigens on the type B blood cells will react with anti-B antibodies in the recipient's serum.

**The Link between Blood Type and Composition of the Intestinal Microbiome**

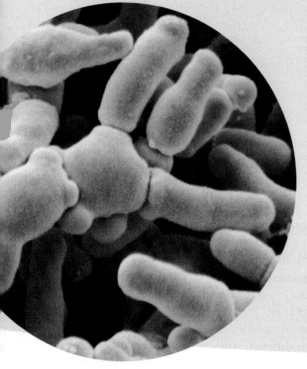

*Bifidobacterium*

During the first year of life, a newborn's B cells start producing antibodies against bacteria in its developing microbiome. Some bacterial antigens are identical to molecules that comprise our personal blood type. If those B cells are allowed to mature, their antibodies would result in widespread tissue damage. Through screening mechanisms, the body destroys the B cells not exhibiting self-tolerance. This scenario is the basis for rejection reactions in blood transfusions. It may also be related to what sorts of bacterial profiles are found in different people's intestines.

In addition to being found on blood cells, blood antigens appear in saliva, mucus, and body fluids of about 80% of the population. This is due to a secreter gene these individuals possess that produces a water-soluble form of blood antigens. More and different bacteria are found in the intestines of nonsecretors. However, *Bifidobacterium* species are rarely present in the intestines of nonsecretors, but they are commonly found in the intestines of those with the secretion gene.

Some anaerobic intestinal bacteria produce exoenzymes that degrade blood antigens. It has been observed that B-degrading bacteria are more abundant in type B individuals. It is likely that the monosaccharides resulting from the exoenzyme effects are used as nutrients by many intestinal bacteria.

From time to time, blood group-degrading bacteria may disappear from the intestines for a week or two. Then the colon is exposed to antigenic structures that normally are degraded. Such antigens may include the polysaccharide O antigens on the cell walls of Enterobacteriaceae bacteria. This B-like antigen on *E. coli* would normally be degraded by bacterial blood-group degrading enzymes. Several studies have shown that immune responses to Enterobacteriaceae play a role in inflammatory bowel disease. Hence, population changes in the intestinal ecosystem may have a significant role in these diseases.

The American physician Charles Drew (**Figure 19.4**) invented the technique for plasma separation that allowed blood to be stored, or "banked." Cells and plasma are separated so that the O cells, lacking A and B antigens, can be transfused to any recipient.

A relationship between blood types and certain diseases has been observed, which may be related to the natural selection of blood types in the populations of certain geographical areas. For example, individuals with type O blood are more susceptible to the incidence and severity of cholera and other diarrheas, whereas individuals with type B are much less affected. This tendency seems to be reflected in the blood types found in the Indian subcontinent, where type B is common and type O less so. More than half of the population of tropical Africa is of type O, which tends to be less severely affected by malaria.

### The Rh Blood Group System

In the 1930s, researchers Karl Landsteiner and Alexander Wiener discovered the presence of a different surface antigen on human red blood cells that also existed on rhesus monkeys. The antigen was named the **Rh factor** (*Rh* for rhesus monkey). The roughly 85% of the population whose cells possess this antigen are called $Rh^+$; those lacking this antigen (about 15%) are $Rh^-$. Antibodies that react with the Rh antigen do not occur naturally in the serum of $Rh^-$ individuals, but exposure to this antigen can sensitize their immune systems to produce anti-Rh antibodies.

**Blood Transfusions and Rh Incompatibility** If blood from an $Rh^+$ donor is given to an $Rh^-$ recipient, the donor's RBCs stimulate the production of anti-Rh antibodies in the recipient. If the recipient then receives $Rh^+$ red blood cells in a subsequent transfusion, a rapid, serious hemolytic reaction will develop.

**Hemolytic Disease of the Newborn** Blood transfusions are not the only way in which an $Rh^-$ person can become sensitized to $Rh^+$ blood. When an $Rh^-$ woman and an $Rh^+$ man produce a child, there is at least a 50% chance that the child will be $Rh^+$ (**Figure 19.5**). If the child is $Rh^+$, the $Rh^-$ mother can become sensitized to this antigen during birth when the placental membranes tear and the fetal $Rh^+$ RBCs enter the maternal

**Figure 19.4 American physician Charles Drew invented the technique for plasma separation that allowed blood to be stored, or "banked."**

circulation, causing the mother's body to produce anti-Rh antibodies of the IgG type. If the fetus in a subsequent pregnancy is Rh$^+$, the mother's anti-Rh antibodies will cross the placenta and destroy the fetal RBCs. The fetal body responds to this immune attack by producing large numbers of immature RBCs called erythroblasts. Thus, the term *erythroblastosis fetalis* was once used to describe what is now called **hemolytic disease of the newborn (HDNB)**. Before the birth of a fetus with this condition, the maternal circulation removes most of the toxic by-products of fetal RBC disintegration. After birth, however, the fetal blood is no longer purified by the mother, and the newborn develops jaundice and severe anemia.

HDNB is usually prevented today by passive immunization of the Rh$^-$ mother at the time of delivery of any Rh$^+$ infant with anti-Rh antibodies, which are available commercially (RhoGAM®). RhoGAM® is administered via intramuscular or intravenous injection to the mother at 28 weeks of pregnancy and soon after delivery. These anti-Rh antibodies combine with any fetal Rh$^+$ RBCs that have entered the mother's circulation, so it is much less likely that she will become sensitized to the Rh antigen. If the disease is not prevented, the newborn's Rh$^+$ blood, contaminated with maternal antibodies, may have to be replaced by transfusion of blood that does not have the mother's anti-Rh antibodies.

### Drug-Induced Cytotoxic Reactions

Blood platelets (thrombocytes) are minute cell-like fragments pinched off from megakaryocytes. These key components

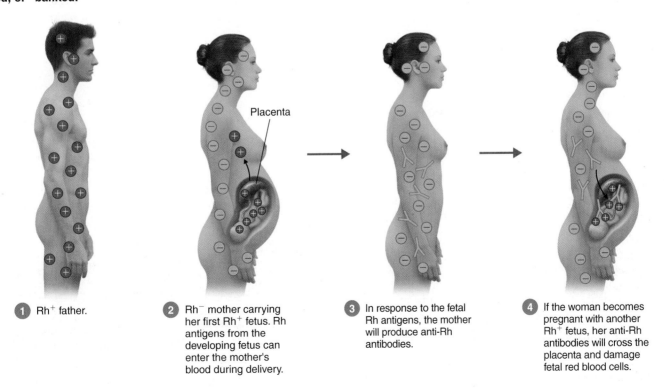

1 Rh$^+$ father.

2 Rh$^-$ mother carrying her first Rh$^+$ fetus. Rh antigens from the developing fetus can enter the mother's blood during delivery.

3 In response to the fetal Rh antigens, the mother will produce anti-Rh antibodies.

4 If the woman becomes pregnant with another Rh$^+$ fetus, her anti-Rh antibodies will cross the placenta and damage fetal red blood cells.

Placenta

**Figure 19.5 Hemolytic disease of the newborn.**

**Q** What type of antibodies cross the placenta?

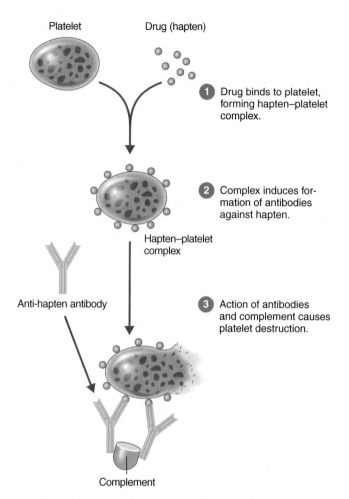

1 Drug binds to platelet, forming hapten–platelet complex.

2 Complex induces formation of antibodies against hapten.

Hapten–platelet complex

Anti-hapten antibody

3 Action of antibodies and complement causes platelet destruction.

Complement

**Figure 19.6 Drug-induced thrombocytopenic purpura.** Molecules of a drug such as quinine accumulate on the surface of a platelet and stimulate an immune response that destroys the platelet.

**Q** What actually destroys the platelets in thrombocytopenic purpura?

in blood clots are destroyed by drug-induced cytotoxic reactions in the disease called **thrombocytopenic purpura**. The drug molecules are usually haptens because they are too small to be antigenic by themselves. In the situation illustrated in **Figure 19.6**, a platelet has become coated with molecules of a drug (quinine is a familiar example), and the combination is antigenic. Both antibody and complement are needed for lysis of the platelets. Loss of platelets results in hemorrhages that appear on the skin as purple spots or *purpura*, which means purple.

Drugs may bind similarly to white or red blood cells, causing local hemorrhaging and yielding symptoms described as "blueberry muffin" skin mottling. When red blood cells are destroyed in this manner, the condition is termed **hemolytic anemia**. Immune-caused destruction of granulocytic white cells is called **agranulocytosis**, and it affects the body's phagocytic defenses.

## Type III (Immune Complex) Reactions

**Type III reactions** involve antibodies against soluble antigens circulating in the serum. In contrast, type II immune reactions are directed against antigens located on cell or tissue surfaces. The antigen–antibody complexes are deposited in organs and cause inflammatory damage.

**Immune complexes** form only when certain ratios of antigen and antibody occur. The antibodies involved are usually IgG. A significant excess of antibody leads to the formation of complement-fixing complexes that are rapidly removed from the body by phagocytosis. When there is a significant excess of antigen, soluble complexes form that do not fix complement and do not cause inflammation. However, when a certain antigen–antibody ratio exists, usually with a slight excess of antigen, the soluble complexes that form are small and escape phagocytosis. These small complexes can fix complement (Chapter 16, pages 463–467).

**Figure 19.7** illustrates the consequences. These complexes circulate in the blood, pass between endothelial cells of the blood vessels, and become trapped in the basement membrane beneath

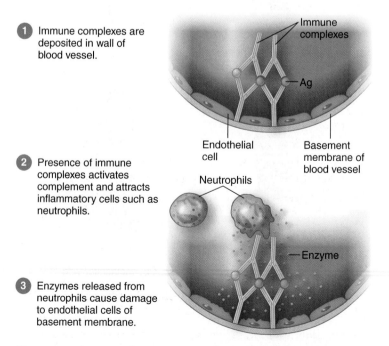

1 Immune complexes are deposited in wall of blood vessel.

Immune complexes

Ag

Endothelial cell

Basement membrane of blood vessel

2 Presence of immune complexes activates complement and attracts inflammatory cells such as neutrophils.

Neutrophils

Enzyme

3 Enzymes released from neutrophils cause damage to endothelial cells of basement membrane.

**Figure 19.7 Immune complex–mediated hypersensitivity.**

**Q** Name one immune complex disease.

the cells. In this location, they may activate complement and cause a transient inflammatory reaction, attracting neutrophils that release enzymes. Repeated introduction of the same antigen can lead to more serious inflammatory reactions, causing damage to the basement membrane's endothelial cells within 2 to 8 hours.

The **Arthus reaction**, a rare side effect of toxoid-containing vaccines, is a type III reaction that can occur in glomeruli and other vessel walls. The reaction occurs because of complement activation in a patient with already circulating IgG antibodies to an injected antigen. This leads to acute local inflammation, edema, and sometimes necrosis. Serum sickness is a less severe type III reaction with swelling and inflammation that occurs from injection of a foreign protein or serum.

## Type IV (Delayed Cell-Mediated) Reactions

Up to this point we have discussed humoral immune responses involving IgE, IgG, or IgM. Type IV reactions involve cell-mediated immune responses and are caused mainly by T cells. **Delayed cell-mediated reactions** (or **delayed hypersensitivity**) are not apparent for a day or more. A major factor in the delay is the time required for the participating T cells and macrophages to migrate to and accumulate near the foreign antigens. Transplant rejection is most commonly mediated by cytotoxic T lymphocytes (CTLs), but other mechanisms are by antibody-dependent cell-mediated cytotoxicity (page 492) or complement-mediated lysis. Another example is described in the Clinical Focus box on page 537.

### Causes of Delayed Cell-Mediated Reactions

Sensitization for delayed hypersensitivity reactions occurs when certain foreign antigens, particularly those that bind to tissue cells, are phagocytized by macrophages and then presented to receptors on the T cell surface. Contact between the antigenic determinant sites and the appropriate T cell causes the T cell to proliferate into mature differentiated T cells and memory cells.

When a sensitized person is reexposed to the same antigen, a delayed hypersensitivity reaction might result. Memory cells from the initial exposure activate T cells, which release destructive cytokines in their interaction with the target antigen. In addition, some cytokines contribute to the inflammatory reaction to the foreign antigen by attracting macrophages to the site and activating them.

### Delayed Cell-Mediated Hypersensitivity Reactions of the Skin

One delayed hypersensitivity reaction that involves the skin is the familiar skin test for tuberculosis. Because *Mycobacterium tuberculosis* is often located within macrophages, this organism can stimulate a delayed cell-mediated immune response. As a screening test, protein components of the bacteria are injected into the skin. If the recipient has (or has had) a previous infection by tuberculosis bacteria, an inflammatory reaction to the injection of these antigens will appear in 1 to 2 days (see Figure 24.9, page 700); this interval is typical of delayed hypersensitivity reactions.

**Allergic contact dermatitis**, another common manifestation of delayed cell-mediated hypersensitivity, is usually caused by haptens that combine with proteins (particularly the amino acid lysine) in the skin of some people to produce an immune response. Reactions to poison ivy (**Figure 19.8**), cosmetics, and the metals in jewelry (especially nickel) are familiar examples of these allergies.

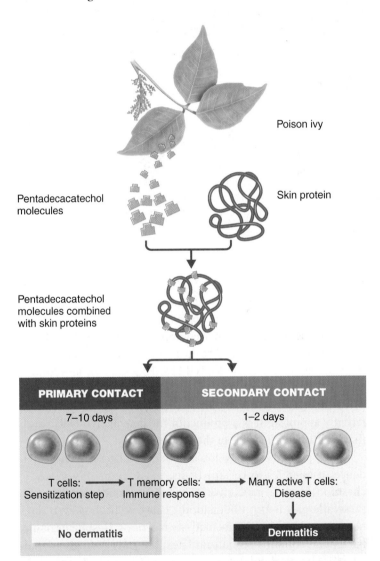

**Figure 19.8 The development of an allergy (allergic contact dermatitis) to catechols from the poison ivy plant.** Pentadecacatechol is a mixture of catechols, oils secreted by the plant that dissolve easily in skin oils and penetrate the skin. In the skin, the catechols function as haptens—they combine with skin proteins to become antigenic and provoke an immune response. The first contact with poison ivy sensitizes the susceptible person, and subsequent exposure results in contact dermatitis.

**Q How does a hapten cause an allergic reaction?**

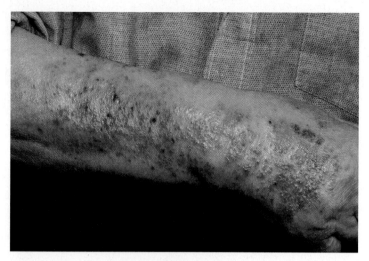

**Figure 19.9 Allergic contact dermatitis.** This person's hand exhibits a severe case of delayed contact dermatitis from wearing latex surgical gloves.

Q What is allergic contact dermatitis?

The increasing exposure to latex in condoms, in certain catheters, and in gloves used by health care workers has led to a greater awareness of hypersensitivity to latex. In many health care environments, latex has been replaced by nonlatex gloves and products. Among health care workers, 8–17% report this type of hypersensitivity to latex surgical gloves (**Figure 19.9**). Synthetic polymers such as vinyl and, especially, nitrile are alternatives to latex, but even nitrile gloves occasionally cause allergic reactions. Most gloves that are made of natural latex, as well as those made of nitrile and neoprene, contain chemical additives called accelerators. Accelerators promote cross-linkage, which adds strength and resiliency, but they have been implicated in allergic reactions. One type of nitrile glove that does not contain accelerators has been developed and has received U.S. Food and Drug Administration (FDA) registration as a medical device that can be labeled as nonallergenic. Another alternative glove, Yulex®, has been recently approved. It is a product of the guayule (wī-Ū-lē) shrub that is native to arid areas of the U.S. Southwest.

Many people who develop an allergy to latex also have evidence of allergies to certain fruits, most commonly avocado, chestnut, banana, and kiwi. (The exact relationship between these allergies is not understood.) Latex paint, however, does not pose a threat of hypersensitivity reactions. Despite its name, latex paint contains no natural latex, but only synthetic nonallergenic chemical polymers.

The identity of the environmental factor causing dermatitis can usually be determined by a *patch test*. Samples of suspected materials are taped to the skin; after 48 hours, the area is examined for inflammation.

**CHECK YOUR UNDERSTANDING**

✓ **19-10** What is the primary reason for the delay in a delayed cell-mediated reaction?

# Autoimmune Diseases

**LEARNING OBJECTIVES**

**19-11** Describe a mechanism for self-tolerance.

**19-12** Give an example of cytotoxic, immune complex, and cell-mediated autoimmune diseases.

When the immune system acts in response to self-antigens and causes damage to one's own organs, the result is an **autoimmune disease**. More than 80 autoimmune diseases have been identified. They affect more than 10% of the population in the developed world.

About 75% of the cases of autoimmune disease selectively affect women. Reasons for this are still being explored, but hormone differences, genetic susceptibility (females have two X chromosomal genes and undergo X chromosome inactivation, whereas males have one), previous infections, and vitamin D deficiency have been explored as possibilities. Treatments for autoimmune diseases are improving as knowledge of the mechanisms controlling immune reactions expands.

Autoimmune diseases occur when there is a loss of **self-tolerance**, the immune system's ability to discriminate self from nonself. In the generally accepted model by which T cells become capable of distinguishing self from nonself, the cells acquire this ability during their passage through the thymus. Any T cells that will target host cells are eliminated by clonal deletion in the thymus (called *thymic selection*; see Chapter 17, page 486). This makes it unlikely that the T cell will attack its own tissue cells. However, when this process goes awry, loss of self-tolerance leads to the production of antibodies against self (*autoantibodies*) or a response by sensitized T cells against a person's own tissue antigens. Autoimmune reactions, and the diseases they cause, can be cytotoxic, immune complex, or cell-mediated in nature.

## Cytotoxic Autoimmune Reactions

**Multiple sclerosis** is one of the more common autoimmune diseases, affecting mostly younger adult women in temperate areas. It is a neurological disease in which autoantibodies, T cells, and macrophages attack the myelin sheath of nerves. This compromises nerve impulse conduction and leads to scarring. Symptoms range from fatigue and weakness to, in some cases, eventual severe paralysis. There is considerable evidence of genetic susceptibility from several genes that interact. The etiology of multiple sclerosis is unknown, but epidemiological evidence indicates that it probably involves some infective agent or agents acquired during early adolescence. The Epstein-Barr virus (page 670) is frequently mentioned as a prime suspect. No cure exists, but treatments with interferons, monoclonal antibodies, and several drugs that interfere with immune processes can significantly slow progression of symptoms.

A Delayed Rash

As you read through this box, you will encounter a series of questions that health care professionals ask as they determine the cause of a patient's symptoms. Try to answer each question before going on to the next one.

1. A 65-year-old woman made a routine dental appointment. Because of her hip and shoulder implants, antibiotics were routinely prescribed for 2 days after any dental work. The woman asked for her usual prescription of cephalothin. The nurse practitioner prescribed penicillin, saying it was less expensive.
   Why are patients with medical implants more susceptible to infection from dental work?

2. Oral bacteria introduced to the bloodstream during dental work can colonize on medical implants such as artificial joints, stents, or catheters. The resulting biofilm may be a source of serious systemic infections. The dental cleaning went well. Seven days later, the woman developed a maculopapular rash over her legs and torso (see the photo).
   What are the most likely causes of a rash, in the absence of fever or other signs of infection?

3. A rash is likely due to an allergic reaction.
   What questions would you ask the patient?

4. The patient had not tried any new foods, cleaning agents, or clothing. She said the only thing different during the past 10 days was her taking the penicillin. The nurse practitioner said penicillin couldn't be the cause because responses to penicillin occur within minutes to hours after exposure.
   Was the nurse practitioner correct?

5. Immediate reactions that occur within minutes or hours suggest an antibody-mediated allergy. Delayed reactions, occurring after days to weeks of an exposure to an allergen in

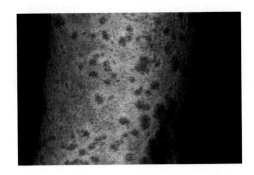

a sensitized person, suggest a type IV cell-mediated reaction. This patient's response fits the profile of a cell-mediated reaction to penicillin.
   What cells are responsible for a type IV hypersensitivity? What antibodies are involved in a type I hypersensitivity?

6. Sensitized T cells are involved in delayed hypersensitivity reactions, including antibiotic-induced rashes. Drug-specific IgE antibodies are responsible for type I immediate hypersensitivity reactions.
   What should the nurse practitioner have asked?

7. The nurse practitioner should have asked whether the patient had any drug allergies before prescribing any antibiotic. However, in this case, the patient had no prior drug-induced allergy.
   Was this the patient's first exposure to penicillin?

8. No. Allergic reactions do not occur on the first exposure to an antigen. The prior exposure could have occurred once during the patient's life. Many immunologists feel that the overuse of penicillin 40 years ago for bacterial infections resulted in an increased frequency of allergic reactions. Most patients who have a history of penicillin allergy, however, will tolerate cephalosporins.

## Immune Complex Autoimmune Reactions

**Graves' disease** is a condition in which the thyroid gland is stimulated to produce increased amounts of thyroid hormones. Normally, the pituitary gland in the brain releases a hormone called thyroid-stimulating hormone (TSH), which induces the thyroid gland to produce its hormones. However, in Graves' disease there is a malfunction of the immune system, and abnormal antibodies are released that mimic TSH. These abnormal antibodies cause the thyroid to produce excessive amounts of hormones, causing pounding of the heart, trembling, and sweating. The most striking external signs of the disease are goiter (a disfiguring swelling of the thyroid gland in the neck) and markedly bulging eyes.

**Myasthenia gravis** is a disease in which muscles become progressively weaker. It is caused by antibodies that coat the acetylcholine receptors at the junctions at which nerve impulses reach the muscles. Eventually, the muscles controlling the

diaphragm and the rib cage may fail to receive the necessary nerve signals, and respiratory arrest and death result.

**Systemic lupus erythematosus** is a systemic autoimmune disease involving immune complex reactions, which mainly affects women. The etiology of the disease is not completely understood, but afflicted individuals produce antibodies directed at components of their own cells, including DNA, which is probably released during the normal breakdown of tissues, especially the skin. The most damaging effects of the disease result from deposits of immune complexes in the kidney glomeruli.

Crippling **rheumatoid arthritis** is a disease in which immune complexes of IgM, IgG, and complement are deposited in the joints. In fact, immune complexes called *rheumatoid factors* may be formed by IgM binding to the Fc region of normal IgG. These factors are found in 70% of individuals suffering from rheumatoid arthritis. The chronic inflammation caused by this deposition eventually leads to severe damage to the cartilage and bone of the joints.

**TABLE 19.3** Selected Autoimmune Disorders

| Autoimmune Disease | Possible Cause |
|---|---|
| Chronic fatigue syndrome | May be due to antibodies attaching to acetylcholine receptors in nerve cells (see Chapter 22, page 645) |
| Rheumatoid arthritis | Cross-reaction with antibodies against streptococcal antigen |
| Systemic lupus erythematosus | Immune complexes involving antibodies against DNA |
| Multiple sclerosis | T cells and macrophages attack the myelin sheath |
| Type I diabetes | T cells destroy insulin-secreting cells |
| Guillain-Barré syndrome | T cells produce cytokines that cause destruction of the myelin sheath of nerve cells (see Chapter 22, pages 632 and 638, and Chapter 25, page 737) |
| Psoriasis | T cells produce cytokines that can induce keratinocyte (skin cells) growth |
| Graves' disease | Antibodies attached to certain receptors in the thyroid gland cause it to enlarge and produce excessive hormones |
| Myasthenia gravis | Antibodies against acetylcholine receptors |
| Crohn's disease | Antibacterial antibodies react with intestinal mucosa |
| Inflammatory bowel disease | Antibacterial antibodies react with intestinal mucosa (see the Big Picture, pages 528–529) |

## Cell-Mediated Autoimmune Reactions

**Insulin-dependent diabetes mellitus** is a common condition caused by immunological destruction of insulin-secreting cells of the pancreas. T cells are clearly implicated in this disease; animals that are genetically likely to develop diabetes fail to do so when their thymus is removed in infancy.

The fairly common skin condition **psoriasis** is an autoimmune disorder characterized by itchy, red patches of thickened skin. As many as 25% of patients develop **psoriatic arthritis**. Several topical and systemic therapies such as corticosteroids and methotrexate are available to help control psoriasis of the skin. Psoriasis is considered to be a $T_H1$ disease and can be treated effectively with immunosuppressants that target T cells and especially the cytokine TNF-α (see page 459), an important factor in inflammation. For psoriatic arthritis, as well as rheumatoid arthritis, the most effective treatments are injections of monoclonal antibodies that inhibit TNF-α. Autoimmune diseases discussed in this book are summarized in Table 19.3.

**CHECK YOUR UNDERSTANDING**

✔ **19-11** What is the importance of clonal deletion in the thymus?

✔ **19-12** What organ is affected in Graves' disease?

# Reactions to Transplantation

**LEARNING OBJECTIVES**

**19-13** Define *HLA complex*, and explain its importance in disease susceptibility and tissue transplants.

**19-14** Explain how a transplant is rejected.

**19-15** Define *privileged site*.

**19-16** Discuss the role of stem cells in transplantation.

**19-17** Define *autograft, isograft, allograft,* and *xenotransplantation products*.

**19-18** Explain how graft-versus-host disease occurs.

**19-19** Explain how rejection of a transplant is prevented.

The inherited genetic characteristics of individuals are expressed not only in the color of their eyes or curl of their hair, but also in the composition of the self molecules on their cell surfaces. Some of these are called **histocompatibility antigens**. The most important of these self molecules are known as the **major histocompatibility complex (MHC)**. In humans, these genes are called the **human leukocyte antigen (HLA) complex**. We encountered these self molecules in Chapter 17 (page 483), where we saw that most antigens can stimulate an immune reaction only if they are associated with an MHC molecule.

A process called *HLA typing* is used to identify and compare HLAs. Certain HLAs are related to an increased susceptibility to specific diseases; one medical application of HLA typing is to identify such susceptibility. A few of these relationships are summarized in Table 19.4.

Another important medical application of HLA typing is in transplant surgery, in which the donor and the recipient must be matched by *tissue typing*. The serological technique shown in **Figure 19.10** is the one most often used. In serological tissue typing, the laboratory uses standardized antisera or monoclonal antibodies that are specific for particular HLAs.

A newer, more accurate technique for analyzing HLA is the use of polymerase chain reaction (PCR) to amplify the cell's DNA (see Figure 9.4, page 248). If this is done for both

**TABLE 19.4** Selected Diseases Related to Specific Human Leukocyte Antigens (HLAs)

| Disease | Increased Risk of Occurrence with Specific HLA* | Description |
|---|---|---|
| **INFLAMMATORY DISEASES** | | |
| Multiple sclerosis | 5 times | Progressive inflammatory disease affecting nervous system |
| Rheumatic fever | 4–5 times | Cross-reaction with antibodies against streptococcal antigen |
| **ENDOCRINE DISEASES** | | |
| Addison's disease | 4–10 times | Deficiency in production of hormones by adrenal gland |
| Graves' disease | 10–12 times | Disorder in which antibodies attached to certain receptors in the thyroid gland cause it to enlarge and produce excessive hormones |
| **MALIGNANT DISEASE** | | |
| Hodgkin's lymphoma | 1.4–1.8 times | Cancer of lymph nodes |

*Compared to the general population.

donor and recipient, a match between donor DNA and recipient DNA can then be made. Having such a DNA match and matching ABO blood type between the donor and the recipient should result in a much higher success rate in transplant surgery.

Other factors may be involved in the success of a transplant, however. According to one hypothesis, the body's reaction to transplanted foreign tissue may be a response to surgery-damaged cells. In other words, tissue rejection may result from a learned reaction to the danger signal posed by damaged cells, rather than a learned reaction to nonself.

Transplants recognized as nonself are rejected—attacked by T cells that directly lyse the grafted cells, by macrophages activated by T cells, and, in certain cases, by antibodies, which activate the complement system and injure blood vessels supplying the transplanted tissue. However, transplants that are not rejected can add many healthy years to a person's life.

**CHECK YOUR UNDERSTANDING**

19-13 What is the relationship between the major histocompatibility complex in humans and the human leukocyte antigen complex?

**Privileged Sites and Privileged Tissue**

Some transplants or grafts do not stimulate an immune response. A transplanted cornea, for example, is rarely rejected, mainly because antibodies usually do not circulate into that portion of the eye. The cornea is therefore considered an immunologically **privileged site**. (However, rejections do occur, especially when the cornea has developed many blood vessels from corneal infections or damage.) The brain is also an immunologically privileged site, probably

because it does not have lymphatic vessels and because the walls of the blood vessels in the brain differ from blood vessel walls elsewhere in the body (the blood–brain barrier is discussed in Chapter 22).

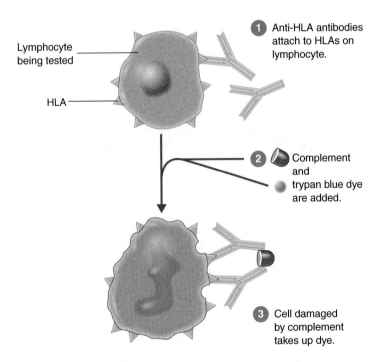

**Figure 19.10 Tissue typing, a serological method.** Lymphocytes from the person being tested are incubated with laboratory test stocks of anti-HLA antibodies specific for a particular HLA. If the antibodies react with the antigens on a lymphocyte, then complement damages the lymphocyte, and dye can enter the cell. Such a positive test result indicates that the person has the particular HLA being tested for.

**Q** Why is tissue typing done?

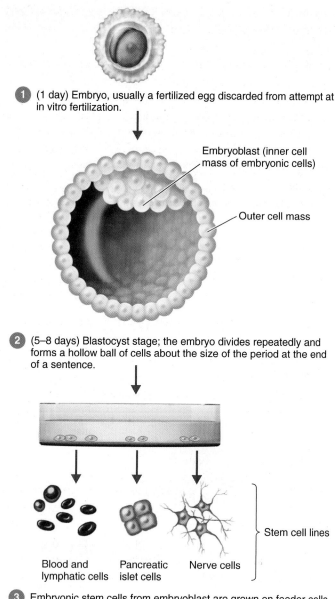

① (1 day) Embryo, usually a fertilized egg discarded from attempt at in vitro fertilization.

Embryoblast (inner cell mass of embryonic cells)

Outer cell mass

② (5–8 days) Blastocyst stage; the embryo divides repeatedly and forms a hollow ball of cells about the size of the period at the end of a sentence.

Stem cell lines

Blood and lymphatic cells

Pancreatic islet cells

Nerve cells

③ Embryonic stem cells from embryoblast are grown on feeder cells in culture medium. Stem cell lines and groups of stem cells form colonies in culture medium. Different conditions, as well as growth factors added to culture medium, direct stem cells to become stem cell lines for various tissues of the body (e.g., blood and lymphatic cells, pancreatic islet cells, nerve cells).

**Figure 19.11 Derivation of embryonic stem cells.**

Q **What does *pluripotent* mean?**

It is possible to transplant **privileged tissue**, which does not stimulate an immune rejection. For example, a person's damaged heart valve can be replaced with a valve from a pig's heart. However, privileged sites and tissues are more the exception than the rule. Transplanted privileged tissue from a pig requires that the pig tissue be modified by decellularization, meaning that the pig tissue's antigenic cellular elements are removed either physically or chemically.

How animals tolerate pregnancy without rejecting the fetus is only partially understood. During pregnancy, the tissues of two genetically different individuals are in direct contact. An important factor seems to be that the MHC classes I and II in the cells that form the outer layer of the placenta—and come into contact with maternal tissue—are not of the specific types that stimulate a cellular immune response. The fetus is also protected by certain proteins it synthesizes, which have immunosuppressive properties. But there is no single, simple explanation.

### Stem Cells

A development that promises to transform transplantation medicine is the use of **stem cells** (see Figure 17.1, page 477), cells that are capable of renewing themselves and can be differentiated into other organ-specific specialized cells. Speaking generally, there are **embryonic stem cells (ESCs)**, which can be harvested from a blastocyst (early stage of embryonic development from days 5 to 8 after fertilization). See **Figure 19.11**. In the medical community, use of embryonic stem cells in therapy is a topic that attracts great interest because they maintain the potential to become any cell type of the body. For example, theoretically these cells could be used to regenerate damaged heart tissue or the failing insulin-producing cells in the pancreas that lead to diabetes. Damaged cartilage in the joints of rheumatoid arthritis patients might be replaced. Advances in **tissue engineering** may make it possible to grow organs in the laboratory. Using patients' cells, which eliminates rejection, researchers have already successfully grown and transplanted urinary bladders.

The nomenclature employed reflects the potential of various stem cells. If a cell can generate all types of cells, it is termed *pluripotent*. These cells naturally occur in mammals only very early in embryo development—hence their other name, *embryonic stem cells*. Later in embryonic development, stem cells specialize and are able to give rise only to specific families of cell types, such as blood, skin, or muscle. These cells are called *multipotent*. After birth, these multipotent stem cells are called **adult stem cells**. They replenish cells, as needed, for various body organs. For instance, skin stem cells continue to grow skin and hair. It is now possible to create **induced pluripotent stem cells (iPSCs)** from adult stem cells in the laboratory. These are made by introducing transcription factor regulatory genes by transduction, or by adding the regulatory proteins to culture media. When these cells are isolated and grown in culture, they give rise to embryonic stem cells. It was originally believed transplants derived from ESCs and IPSCs would not be rejected. Experimentation with xenotransplants in mice suggests that T cells are involved in rejection of ESCs and iPSCs.

### Bone Marrow Transplants

Transplants of bone marrow, known as *hematopoietic stem cell transplants*, are an example of transplantation of adult stem cells. The recipients are usually individuals who lack the capacity to produce B cells and T cells vital for immunity or who are suffering from leukemia. Bone marrow stem cells give rise to red blood cells and immune system lymphocytes (see Chapter 17). The goal of bone marrow transplants is to enable the recipient to produce healthy red blood, or immune system, cells. However, such transplants can result in **graft-versus-host (GVH) disease**. The transplanted bone marrow contains immunocompetent cells that mount primarily a cell-mediated immune response against the tissue into which they have been transplanted. Because the recipients lack effective immunity, GVH disease is a serious complication and can even be fatal.

An extremely promising technique for avoiding this problem is the use of *umbilical cord blood* instead of bone marrow. This blood is harvested from the placenta and umbilical cords of newborns—materials that would otherwise be discarded. It is very rich in the stem cells found in bone marrow. Not only do these cells proliferate into the variety of cells required by the recipient, but also, because stem cells from this source are younger and less mature, the "matching" requirements are also less stringent than with bone marrow. As a result, GVH disease is less likely to occur.

### Grafts

A **graft** is the transfer of a tissue from one part of the body to another, or from one person to another, without the transfer of the blood supply of the grafted tissue. When one's own tissue is grafted to another part of the body, as is done in burn treatment or in plastic surgery, the graft is not rejected. Recent technology has made it possible to use a few cells of a burn patient's uninjured skin to culture extensive sheets of new skin. This new skin is an example of an **autograft**. Identical twins have the same genetic makeup; therefore, skin or organs such as kidneys may be transplanted between them without provoking an immune response. Such a transplant is called an **isograft**. Grafts between people who are not identical twins are called **allografts**. Most transplants fall under this last category, and they will cause an immune response. Matching the HLAs of the donor and recipient as closely as possible reduces the chances of rejection. Because HLAs of close relatives are most likely to match, blood relatives, especially siblings, are the preferred donors.

The need for donor organs far outstrips the current supply. Medical researchers hope to increase the success of **xenotransplantation products**, which are tissues or organs that have been transplanted from animals. However, the body tends to mount an especially severe immune assault on such transplants. To be successful, xenotransplantation products must overcome **hyperacute rejection**, the development of antibodies in early infancy against all distantly related animals such as pigs. With the aid of complement, these antibodies attack the transplanted animal tissue and destroy it within an hour. Hyperacute rejection occurs in human-to-human transplants only when antibodies have been preformed after previous transfusions, transplantations, or pregnancies. Liver transplantation

among humans is unusual in one respect; this organ usually resists hyperacute rejection, and HLA typing is not as important as in other types of tissue.

Unsatisfactory attempts have been made to transplant organs from baboons and other nonhuman primates into people. Research interest is high in genetically modifying pigs, since the animal is plentiful in supply and the right size for its organs to be compatible with humans. (Pigs also generate relatively little public sympathy or protests—another factor that could make them acceptable donors of organs.) Beyond rejection, the primary concern about xenotransplantation products is the possibility of transferring harmful animal viruses along with the donor tissue.

### CHECK YOUR UNDERSTANDING

☑ **19-14** What immune system cells are involved in the rejection of nonself transplants?

☑ **19-15** Why is a transplanted cornea usually not rejected as nonself?

☑ **19-16** Differentiate an embryonic stem cell from an adult stem cell.

☑ **19-17** Which type of transplant is most subject to hyperacute rejection?

☑ **19-18** When red bone marrow is transplanted, many immunocompetent cells are included. How can this be bad?

## Immunosuppression to Prevent Transplant Rejection

To keep the problem of transplant rejection in perspective, it is useful to remember that the immune system is simply doing its job and has no way of recognizing that its attack against the transplant is not helpful. In an attempt to prevent rejection, the recipient of an allograft usually receives treatment to suppress this normal immune response against the graft.

In transplantation surgery, it is generally desirable to suppress cell-mediated immunity, the most important factor in transplant rejection. If humoral (antibody-based) immunity is not suppressed, much of the ability to resist microbial infection will remain. In 1976, the drug *cyclosporine* was isolated from a mold. Cyclosporine suppresses the secretion of interleukin-2

### CLINICAL CASE

Lymphocytes originate from stem cells in the red bone marrow. From the red bone marrow the lymphocytes migrate to the thymus gland, where they mature into T cells. Malik was diagnosed with DiGeorge syndrome: a deletion in chromosome 22 that results in the underdevelopment or complete absence of the thymus gland. Malik, without an effective thymus, cannot develop T cells.

**What is causing Malik's symptoms?**

525    530    **541**    544    554

(IL-2), disrupting cell-mediated immunity by cytotoxic T cells. The successful transplantation of organs such as hearts and livers generally dates from this discovery. Side effects include kidney toxicity and vomiting and diarrhea. Following the success of this drug, other immunosuppressant drugs soon followed. *Tacrolimus* (FK506) has a mechanism similar to that of cyclosporine and is a frequent alternative. Serious side effects include increased risk of certain cancers, diabetes, and weight loss.

Neither cyclosporine nor tacrolimus has much effect on antibody production by the humoral immune system. Both of these drugs remain the mainstay for most regimens to prevent rejection of transplants. Some newer drugs, such as *sirolimus* are among those that inhibit both cell-mediated and humoral immunity. This can be an advantage if chronic or hyperacute rejection by antibodies are a consideration. Sirolimus is best known for its use in stents, cylindrical meshes designed to keep blood vessels open after removal of blockages. Drugs such as *mycophenolate* inhibit the proliferation of T cells and B cells. Biological agents such as the chimeric monoclonal antibody (pages 508–510) *basiliximab* block IL-2 and are frequently prescribed immunosuppressants. Immunosuppressive agents are usually administered in combinations.

Occasionally, a transplant recipient discontinues using immunosuppressant drugs but, surprisingly, does not reject the transplant. Research has provided an insight into a possible procedure to duplicate this deliberately. In these studies, a patient's immune system was treated before kidney transplantation surgery to deplete the body's supply of immune system T cells, which normally patrol for foreign invaders to attack. The transplanted tissue was then surgically implanted, along with bone marrow cells that had been harvested and stored before the patient's T cells were depleted. The subsequent results were unexpected: the immune system was rebuilt as a chimera—a hybrid mixture of the cells of the donated and patient's own cells. As a consequence, the donated organ was accepted as self and not rejected. This retraining of the immune system often allows the patient to stop taking antirejection drugs less than a year after surgery. A puzzling aspect is that the chimeric state is not permanent; the patient's immune system eventually returns to its original state—but still without rejecting the transplanted tissue. Carried to a logical extreme, this suggests the possibility of eventually using nonhuman organs as transplants.

**CHECK YOUR UNDERSTANDING**

✔ 19-19 What cytokine is usually the target of immunosuppressant drugs intended to block transplant rejection?

# The Immune System and Cancer

## LEARNING OBJECTIVES

19-20 Describe how the immune system responds to cancer and how cells evade immune responses.

19-21 Give two examples of immunotherapy.

Like an infectious disease, cancer represents a failure of the body's defenses, including the immune system. Some of the most promising avenues for effective cancer therapy make use of immunological techniques.

It has long been recognized that cancer cells arise frequently in the body and that they are usually eliminated by the immune system much like any other invading cell—the concept of **immune surveillance**. It was postulated that the cell-mediated immune system probably arose to combat cancer cells and that the appearance of a cancerous growth represented a failure of the immune system. This concept has been supported by the observation that cancers occur most often in older adults, whose immune systems are becoming less efficient (called *immunosenescence*), or in the very young, whose immune systems may not have developed fully or properly. Also, individuals who are immunosuppressed by either natural or artificial means are more susceptible to certain cancers.

A cell becomes cancerous when it undergoes transformation and begins to proliferate without control (see Chapter 15, page 437). The surfaces of tumor cells acquire tumor-associated antigens that mark them as nonself to the immune system. **Figure 19.12** illustrates the attack on such a cancer cell by activated $T_C$ cells (cytotoxic T lymphocytes, or CTLs). Activated macrophages can also destroy cancer cells. Although a healthy immune system serves to prevent most cancers, it has limitations. In some cases there is no antigenic epitope for the

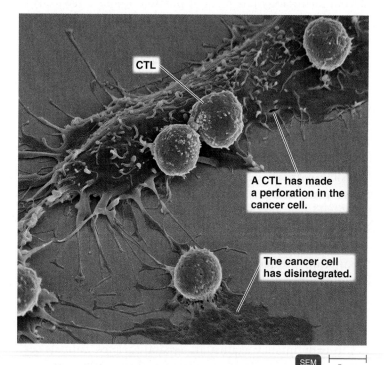

CTL

A CTL has made a perforation in the cancer cell.

The cancer cell has disintegrated.

SEM | 5 μm

**Figure 19.12 The interaction between cytotoxic T lymphocytes (CTL) and cancer cells.** The lymphocytes cause apoptosis (self-destruction) of the cancer cell.

**Q** CTLs can lyse cancer cells. How do they do this? (*Hint:* See Figure 17.13.)

immune system to target. Tumor cells can even reproduce so rapidly that they exceed the capacity of the immune system to deal with them. Finally, some cancerous cells from a metastatic tumor may be transported or reside in a separate location without forming new tumors, in a phenomenon called *latent metastasis* that allows the cancer to become invisible to the immune system.

## Immunotherapy for Cancer

Taking advantage of the immune system to prevent or cure cancer led to **immunotherapy**. At the turn of the twentieth century, William Coley, a physician at a New York City hospital, observed that if cancer patients contracted typhoid fever, their cancers often diminished noticeably. Coley made mixtures of killed gram-positive streptococci and gram-negative *Serratia marcescens* bacteria. These so-called Coley's toxins were injected into cancer patients to simulate a bacterial infection. Some of this work was promising, but results were inconsistent, and advances in surgery and radiation treatment caused it to be nearly forgotten. It is now recognized that endotoxins from such bacteria are powerful stimulants for the production of TNF-α by macrophages. TNF-α interferes with the blood supply of cancers in animals.

Other research determined many years ago that if animals were injected with dead tumor cells, as with a vaccine, they did not develop tumors when injected with live cells from these tumors. Similarly, cancers sometimes undergo spontaneous remission that is probably related to the immune system gaining the advantage.

Cancer vaccines might be either *therapeutic* (used to treat existing cancers) or *prophylactic* (to prevent development of cancers). A vaccine for strains of the human papilloma virus (HPV) that are linked to cervical, anal, and throat cancer is now part of the recommended childhood immunizations in the United States. A vaccine against the cancer-linked hepatitis B virus (HBV) is also available.

An immune therapy for prostate cancer, sipuleucel-T, boosts the patient's immune response. The cancer antigen is attached to the patient's antigen-presenting cells in the lab, and then the cells are injected back into the patient so the antigen will be presented to T cells and spark an immune response.

Monoclonal antibodies are a promising tool for delivering cancer treatment. A humanized monoclonal antibody, *trastuzumab* (Herceptin®) (see Chapter 18, pages 508–510), is currently being used to treat a form of breast cancer. Herceptin® specifically neutralizes a genetically determined growth factor, HER2, that promotes the proliferation of the cancer cells. Monoclonal antibodies may also be used to boost immune response by flagging cancer cells as something to attack. Another approach is to combine a monoclonal antibody with a toxic agent, forming an **immunotoxin**. Theoretically, an immunotoxin might be used to specifically target and kill cells of a tumor with little damage to healthy cells. Adcetris®, a conjugate of the monoclonal antibody *brentuximab* and an attached cytotoxic drug (*vedotin*), treats Hodgkin's lymphoma. There are also radioactively labeled antibodies that attach to cancer cells (called *radioimmunotherapy*). An example is *ibritumomab tiuxetan*, used to treat some types of non-Hodgkin's lymphoma. Radioactive elements are attached to monoclonal antibodies, which bind CD20 on B cells, allowing radiation to kill the cells.

**CHECK YOUR UNDERSTANDING**

✔ **19-20** How do immune-system cells recognize cancerous cells?

✔ **19-21** Give an example of a prophylactic cancer vaccine that is in current use.

# Immunodeficiencies

**LEARNING OBJECTIVE**

**19-22** Compare and contrast congenital and acquired immunodeficiencies.

The absence of a sufficient immune response is called an **immunodeficiency**, which can be either congenital or acquired.

## Congenital Immunodeficiencies

Some people are born with an abnormal immune system. Defects in, or absence of, a number of genes can result in **congenital immunodeficiencies** (also known as **primary immunodeficiencies**). Genetic immune deficiencies can affect complement, phagocytes, B cells, T cells, or a combination of the various immune system actors. For example, individuals with the recessive trait known as DiGeorge syndrome do not have a thymus gland, so they lack cell-mediated immunity. Nude (hairless) mice are a type of animal used in transplantation research that also lack thymus glands (**Figure 19.13**). Hairless because the deleted gene coincidentally controls both thymus and hair development, these mice cannot produce T cells, and so they do not reject transplanted tissue. Even chicken skin, complete with feathers, is readily accepted as a graft by these animals in the lab.

## Acquired Immunodeficiencies

A variety of drugs, cancers, or infectious agents can result in **acquired immunodeficiencies (secondary immunodeficiencies)**. For example, Hodgkin's lymphoma (a type of cancer) lowers the cell-mediated response. Many viruses infect and kill lymphocytes, lowering the immune response. Removal of the spleen decreases humoral immunity. Table 19.5 summarizes several of the better known immune deficiency conditions, including AIDS.

**CHECK YOUR UNDERSTANDING**

✔ **19-22** Is AIDS an acquired or a congenital immunodeficiency?

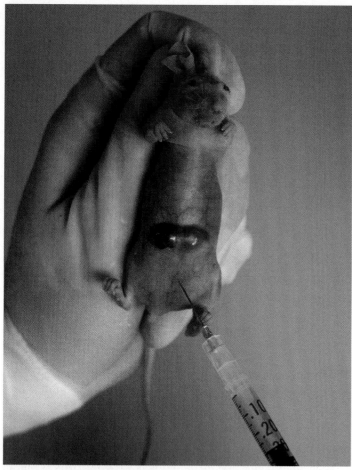

**Figure 19.13 A nude (hairless) mouse that has received a tumor graft.** Immunodeficient mouse models like this have been used to study xenografts. The researcher is anesthetizing the mouse to remove tumor cells.

**Q** What is the role of the thymus gland in immunity?

**CLINICAL CASE**

Without T cells, Malik lacks an effective immune system. The transfused blood contained immunologically competent lymphocytes, including T cells. A normal immune system would have neutralized these cells. In Malik's case, the transfused lymphocytes saw Malik as nonself and attacked his cells, meaning that he had developed graft-versus-host disease (GVHD). In this condition, the transfused T cells recognize and attack nonself cells in the new host. This recognition requires the T cells to attach to a T cell receptor and to a coreceptor such as CD3. When the T cell attaches to the receptor–CD3 complex, it is stimulated to proliferate and attacks the antigen. A monoclonal antibody, muromonab-CD3 (Mab-CD3), is often used to treat immunological tissue rejection.

**What role does the monoclonal antibody play in Malik's recovery? (Hint: See Chapter 18.)**

525   530   541   **544**   554

# Acquired Immunodeficiency Syndrome (AIDS)

**LEARNING OBJECTIVES**

19-23 Give two examples of how infectious diseases emerge.

19-24 Explain the attachment of HIV to a host cell.

19-25 List two ways in which HIV avoids the host's antibodies.

19-26 Describe the stages of HIV infection.

19-27 Describe the effects of HIV infection on the immune system.

19-28 Describe how HIV infection is diagnosed.

| **TABLE 19.5** Selected Immunodeficiencies | | |
|---|---|---|
| **Disease** | **Cells Affected** | **Comments** |
| Acquired immunodeficiency syndrome (AIDS) | Virus destroys CD4$^+$ T cells | Allows cancer and bacterial, viral, fungal, and protozoan diseases; caused by HIV infection |
| Selective IgA immunodeficiency | B, T cells | Affects about 1 in 700, causing frequent mucosal infections; specific cause uncertain |
| Common variable hypogammaglobulinemia | B, T cells (decreased immunoglobulins) | Frequent viral and bacterial infections; second most common immune deficiency, affecting about 1 in 70,000; inherited |
| Reticular dysgenesis | B, T, and stem cells (a combined immunodeficiency; deficiencies in B and T cells and neutrophils) | Usually fatal in early infancy; very rare; inherited; bone marrow transplant a possible treatment |
| Severe combined immunodeficiency (SCID) | B, T, and stem cells (deficiency of both B and T cells) | Affects about 1 in 100,000; allows severe infections; inherited; treated with bone marrow and fetal thymus transplants; gene therapy treatment is promising |
| Thymic aplasia (DiGeorge syndrome) | T cells (defective thymus causes deficiency of T cells) | Absence of cell-mediated immunity; usually fatal in infancy from *Pneumocystis* pneumonia or viral or fungal infections; due to failure of the thymus to develop in embryo |
| Wiskott-Aldrich syndrome | B, T cells (few platelets in blood, abnormal T cells) | Frequent infections by viruses, fungi, protozoa; eczema, defective blood clotting; usually causes death in childhood; inherited on X chromosome |
| X-linked infantile (Bruton's) agammaglobulinemia | B cells (decreased immunoglobulins) | Frequent extracellular bacterial infections; affects about 1 in 200,000; the first immunodeficiency disorder recognized (1952); inherited on X chromosome |

19-29 List the routes of HIV transmission.

19-30 Identify geographic patterns of HIV transmission.

19-31 List the current methods of preventing and treating HIV infection.

In 1981, a cluster of cases of *Pneumocystis* pneumonia appeared in the Los Angeles area. Investigators soon correlated the appearance of this infrequently seen disease with an unusual incidence of a rare form of cancer of the skin and blood vessels called Kaposi's sarcoma. The people affected were all young homosexual men, and all showed a loss of immune function. By 1983, the pathogen causing the loss of immune function had been identified as a virus that selectively infects T helper cells. This virus is now known as *human immunodeficiency virus*, or *HIV* (see Figure 13.20b, page 384).

## The Origin of AIDS

Studies show that HIV-1 (the primary HIV found worldwide in humans) is genetically related to another *Lentivirus*, simian immunodeficiency virus (SIV), which is carried by monkeys, mangabeys, and chimpanzees in central Africa. The most commonly accepted theory is that humans in this part of Africa ate SIV-infected bushmeat (wild game), allowing the virus to cross over into people. The earliest known sample of HIV comes from Kinshasa, Democratic Republic of the Congo, in 1920. Availability of transportation links, high populations of migrants, and the sex trade in Kinshasa might explain the initial spread of HIV. The virus was probably introduced into other areas of the

world several times before it widely spread. Although AIDS was not recognized as a clinical syndrome until 1981, isolated cases have since been documented outside Africa from before then. Frozen tissue samples of a Missouri patient who died in 1969 confirmed HIV infection. In 1974, a Norwegian sailor, his wife, and daughter all died of what was later confirmed as HIV/AIDS. The sailor had traveled to West Africa before returning home and marrying.

### CHECK YOUR UNDERSTANDING

✔ 19-23 On what continent did the HIV-1 virus arise?

## HIV Infection

### The Structure of HIV

HIV, of the genus *Lentivirus*, is a retrovirus (see Figure 13.19, page 383). It has two identical + stranded RNA molecules, the enzymes reverse transcriptase and integrase, and an envelope of phospholipid (**Figure 19.14**). The envelope has glycoprotein spikes termed **gp120** (the notation for a glycoprotein with a molecular mass of 120,000 da) and **gp41**.

### The Infectiveness and Pathogenicity of HIV

There is a strong association between HIV infection and the immune system. HIV is often spread by dendritic cells that reside in mucosal linings, which pick up the virus and carry it to the lymphoid organs. There it contacts cells of the immune

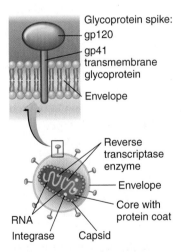

Glycoprotein spike:
- gp120
- gp41 transmembrane glycoprotein
- Envelope

Reverse transcriptase enzyme
Envelope
Core with protein coat
RNA
Integrase      Capsid

**Structure of HIV and infection of a CD4+ T cell.** The gp120 glycoprotein spike on the membrane attaches to a receptor on the CD4+ cell. The gp41 transmembrane glycoprotein probably facilitates fusion by attaching to a fusion receptor on the CD4+ cell.

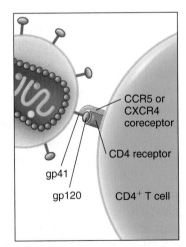

CCR5 or CXCR4 coreceptor
CD4 receptor
gp41
gp120      CD4+ T cell

**1** **Attachment.** The gp120 spike attaches to a receptor and to a CCR5 or CXCR4 coreceptor on the cell. The gp41 participates in fusion of the HIV with the cell.

Viral envelope

**2** **Fusion.** The virus envelope merges with the cell membrane.

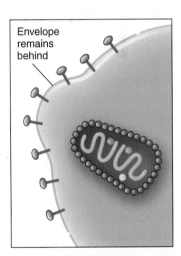

Envelope remains behind

**3** **Entry.** Following fusion with the cell, an entry pore is created. After entry, the viral envelope remains behind and the HIV uncoats, releasing the RNA core for directing synthesis of the new viruses.

**Figure 19.14 HIV structure and attachment to receptors on target T cell.**

**Q** Why does HIV preferentially infect CD4+ cells?

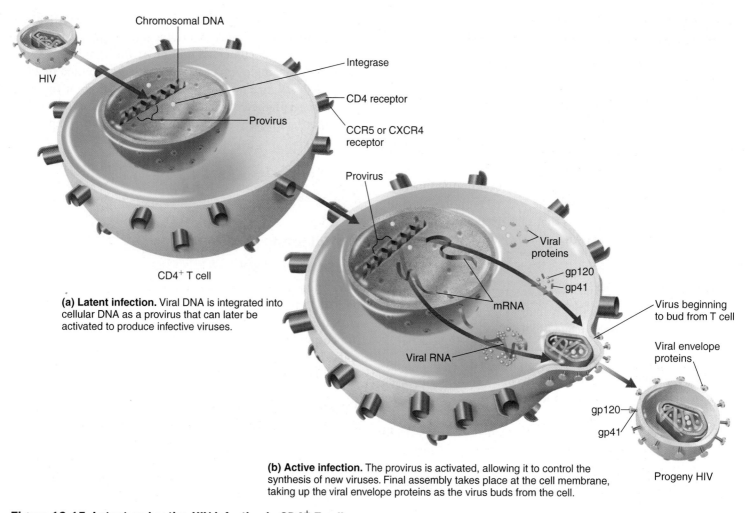

**Chromosomal DNA**

**Integrase**

**HIV**

**CD4 receptor**

**Provirus**

**CCR5 or CXCR4 receptor**

**CD4⁺ T cell**

**(a) Latent infection.** Viral DNA is integrated into cellular DNA as a provirus that can later be activated to produce infective viruses.

**Provirus**

**Viral proteins**

**gp120**

**gp41**

**mRNA**

**Viral RNA**

**Virus beginning to bud from T cell**

**Viral envelope proteins**

**gp120**

**gp41**

**Progeny HIV**

**(b) Active infection.** The provirus is activated, allowing it to control the synthesis of new viruses. Final assembly takes place at the cell membrane, taking up the viral envelope proteins as the virus buds from the cell.

**Figure 19.15 Latent and active HIV infection in CD4⁺ T cells.**

**Q** What is a latent infection?

system, most notably activated T cells, and stimulates an initial strong immune response.

To persist and multiply, HIV must go through the steps of host cell attachment, fusion, and entry in a manner similar to what is shown in Figure 19.14. Attachment depends on the virus's glycoprotein spike (gp120) combining with the CD4 receptor of the virus's preferred target—a T helper cell. Approximately 65,000 of these receptors are found on each T helper cell. Certain coreceptors may also be required for attachment. The two best-known chemokine coreceptors are named CCR5 and CXCR4.* Macrophages and dendritic cells also carry CD4 molecules. Note that many

cells that do not express the CD4 molecule can also become infected, an indication that other receptors can also serve for infection by HIV. Attachment causes a change in gp41, which leads to fusion of the cell membrane and viral envelope.

Once inside the host cell, viral RNA is released and transcribed into DNA by the enzyme reverse transcriptase. The viral DNA then becomes integrated with the host's DNA with the help of viral integrase. The DNA may control the production of an active infection in which new viruses bud from the host cell, as shown in **Figure 19.15**.

Alternatively, this integrated DNA may not produce new HIV, but hide in the host cell's chromosome as a *provirus* (see Figure 19.15a and **Figure 19.16a**). HIV produced by a host cell is not necessarily released from the cell but may remain as *latent virions* in vacuoles within the cell. In fact, a subset of the HIV-infected cells, instead of being killed, become long-lived memory T cells in which the reservoir of latent HIV can persist for decades. This ability of

---

* This nomenclature is based on the beginning amino acid sequence in these proteins. The term CCR5 indicates that the beginning sequence consists of cysteines, thus CC. The letter R is a convention representing the balance of the protein molecule, and the number is for identification. If some other amino acid is located between the first two cysteines, this is shown in the naming—for example, CXCR4.

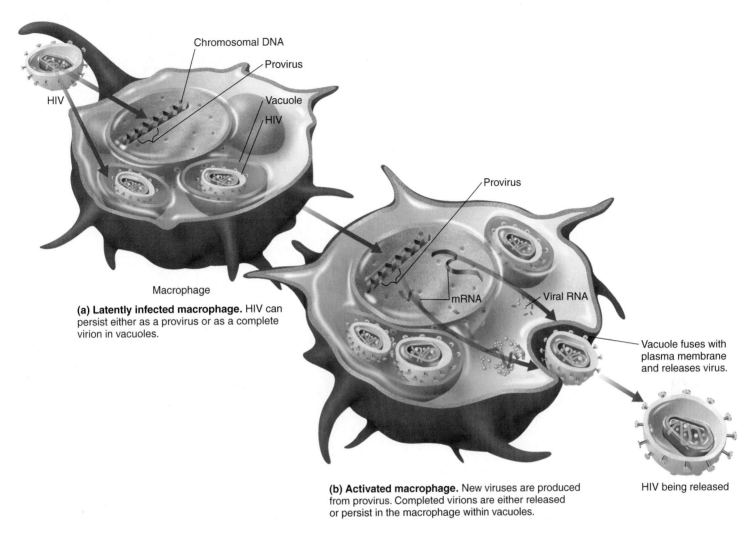

(a) **Latently infected macrophage.** HIV can persist either as a provirus or as a complete virion in vacuoles.

(b) **Activated macrophage.** New viruses are produced from provirus. Completed virions are either released or persist in the macrophage within vacuoles.

HIV being released

**Figure 19.16 Latent and active HIV infection in macrophages.**

**Q** How does an active infection differ from a latent infection?

the virus to remain as a provirus or latent virus within host cells shelters it from the immune system. Another way HIV evades the immune system is *cell–cell fusion*, by which the virus moves from an infected cell to an adjacent uninfected cell.

The virus also evades immune defenses by undergoing rapid antigenic changes. Retroviruses, with their reverse transcriptase enzyme step, have a high mutation rate compared to DNA viruses. They also lack the corrective "proofreading" capacity of DNA viruses. As a result, new variants of HIV abound, complicating the development of vaccines and causing problems related to drug resistance.

**Subtypes of HIV**

There are two major subtypes of HIV: HIV-1 and HIV-2. HIV-1 accounts for about 99% of cases worldwide. HIV-1 viruses are further subdivided into groups that are assigned letter combinations. Group M (for "majority") accounts for about 90% of

cases. Naming conventions become further complicated with lettered subtypes. For example, HIV-1 group M has at least nine subtypes. Subtype B is the dominant HIV found in Europe, Australia, Japan, and the Americas, and almost half of people diagnosed with HIV-1 have subtype C. HIV-2 is endemic in West Africa, but its spread throughout the world has been limited. HIV-2 is generally characterized by having a longer asymptomatic period with lower viral load and mortality rate.

**The Stages of HIV Infection**

The progress of HIV infection in adults can be divided into three clinical phases (**Figure 19.17**).

**Phase 1** The number of viral RNA molecules per milliliter of blood plasma may reach more than 10 million in the first week or so, during the acute infection stage (review Figure 13.21). Billions of $CD4^+$ T cells may be infected within a few weeks, lowering their numbers. Immune responses, and fewer uninfected

# The Progression of HIV Infection

Understanding how the HIV infection progresses in a host is integral to understanding the diagnosis, transmission, and prevention of this pandemic. Although there is no cure, see information below on drug treatments.

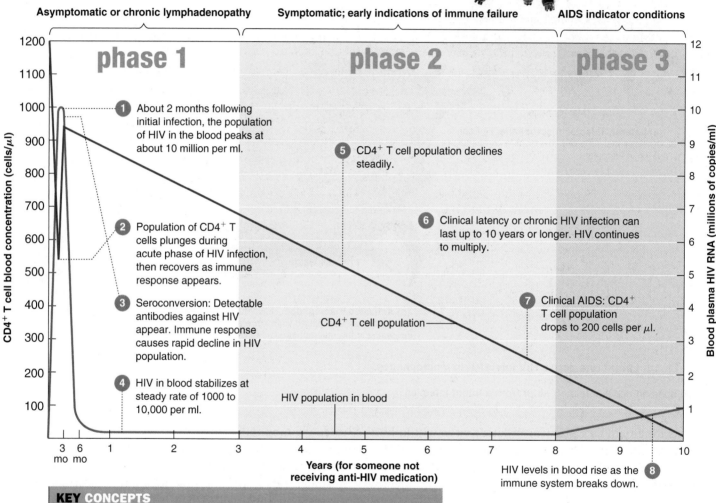

**Asymptomatic or chronic lymphadenopathy**     **Symptomatic; early indications of immune failure**     **AIDS indicator conditions**

phase 1     phase 2     phase 3

**1** About 2 months following initial infection, the population of HIV in the blood peaks at about 10 million per ml.

**2** Population of CD4+ T cells plunges during acute phase of HIV infection, then recovers as immune response appears.

**3** Seroconversion: Detectable antibodies against HIV appear. Immune response causes rapid decline in HIV population.

**4** HIV in blood stabilizes at steady rate of 1000 to 10,000 per ml.

**5** CD4+ T cell population declines steadily.

**6** Clinical latency or chronic HIV infection can last up to 10 years or longer. HIV continues to multiply.

**7** Clinical AIDS: CD4+ T cell population drops to 200 cells per μl.

**8** HIV levels in blood rise as the immune system breaks down.

CD4+ T cell population

HIV population in blood

CD4+ T cell blood concentration (cells/μl)

Blood plasma HIV RNA (millions of copies/ml)

Years (for someone not receiving anti-HIV medication)

## KEY CONCEPTS

- HIV progresses as it destroys the CD4+ T cells essential for the body's defenses against infectious disease and cancer.
- AIDS is the final stage in this progressive infection.

micro tip

Although there is no cure for HIV, combinations of anti-HIV medications (antiretroviral therapy) are the recommended treatment. Under current treatments available, transmission rates are less than 1% if treatment is followed properly. The medications work in a variety of ways: disabling or stopping viral enzymes from making more copies of HIV, disrupting building blocks needed for viral replication, or blocking entry of the virus into cells.

cells to target, deplete the viral numbers in blood plasma sharply within a few weeks. The infection may be asymptomatic or cause *lymphadenopathy* (swollen lymph nodes).

**Phase 2**  The numbers of $CD4^+$ T cells decline steadily. HIV replication continues but at a relatively low level, probably controlled by $CD8^+$ T cells (see page 489 in Chapter 17) and occurs mainly in lymphatic tissue. Only a relatively few infected cells release HIV, although many may contain viruses in latent or proviral form. There are few disease symptoms, but persistent infections by the yeast *Candida albicans*, which can appear in the mouth, throat, or vagina, may signal a decline in immune response. Other conditions may include fever and persistent diarrhea. Oral leukoplakia (whitish patches on oral mucosa), which results from reactivation of latent Epstein-Barr viruses, may appear, as well as other indications of declining immunity, such as shingles.

**Phase 3**  In clinical AIDS, $CD4^+$ T cell counts fall below 200 cells/$\mu$l, and susceptibility to opportunistic infections is high. (The normal population in a healthy individual is 500 to 1500 $CD4^+$ T cells/$\mu$l.) Important AIDS indicator conditions appear, such as *C. albicans* infections of bronchi, trachea, or lungs; cytomegalovirus eye infections; tuberculosis; *Pneumocystis* pneumonia; toxoplasmosis of the brain; and Kaposi's sarcoma. HIV infection devastates the immune system, which is then unable to respond effectively to pathogens. The diseases or conditions most commonly associated with HIV infection and AIDS are summarized in Table 19.6. Success in treating these conditions has extended the lives of many HIV-infected people.

### HIV Positive versus AIDS Disease Status

During phases 1 and 2 of the infection, patients are classified as HIV-positive. During phase 3, patients are classified as having AIDS. The purpose of dividing patients into these categories is primarily to furnish treatment guidelines on when to administer certain drugs. In the United States, antiretroviral therapy is recommended for all HIV-positive individuals. T cell population counts, performed regularly, are indicators of progression of disease.

The progression from initial HIV infection to AIDS usually takes about 10 years in adults without treatment in

**TABLE 19.6  Some Diseases Commonly Associated with AIDS**

| Pathogen or Disease | Disease Description |
|---|---|
| **PROTOZOA** | |
| *Cryptosporidium hominis* | Persistent diarrhea |
| *Toxoplasma gondii* | Encephalitis |
| *Isospora belli* | Gastroenteritis |
| **VIRUSES** | |
| Cytomegalovirus | Fever, encephalitis, blindness |
| Herpes simplex virus (HIV-2) | Vesicles of skin and mucous membranes |
| Varicella-zoster virus | Shingles |
| **BACTERIA** | |
| *Mycobacterium tuberculosis* | Tuberculosis |
| *M. avium-intracellulare* | May infect many organs; gastroenteritis and other highly variable symptoms |
| **FUNGI** | |
| *Pneumocystis jirovecii* | Life-threatening pneumonia |
| *Histoplasma capsulatum* | Disseminated infection |
| *Cryptococcus neoformans* | Disseminated, but especially meningitis |
| *Candida albicans* | Overgrowth on oral and vaginal mucous membranes (phase 2 of HIV infection) |
| *C. albicans* | Overgrowth in esophagus, lungs (phase 3 of HIV infection) |
| **CANCERS OR PRECANCEROUS CONDITIONS** | |
| Kaposi's sarcoma | Cancer of skin and blood vessels (caused by HHV-8) |
| Hairy leukoplakia | Whitish patches on mucous membranes; commonly considered precancerous |
| Cervical dysplasia | Abnormal cervical growth |

industrialized countries; in developing countries, it is often about half this. Cellular warfare on an immense scale occurs during this time. Untreated, at least 100 billion HIVs are generated every day, each with a remarkably short half-life of about 6 hours. These viruses must be cleared by the body's defenses, which include antibodies, cytotoxic T cells, and macrophages. Almost all HIVs are produced by infected CD4$^+$ T cells, which survive for only about 2 days instead of the normal life span of several years. Every day, about 2 billion CD4$^+$ T cells are produced in an attempt to compensate for losses. Over time, however, there is a daily net loss of at least 20 million CD4$^+$ T cells. The most recent studies show that the decrease in CD4$^+$ T cells is not due entirely to direct viral destruction of the cells; rather, it is caused primarily by shortened life of the cells and the body's failure to compensate by increasing production of replacement T cells.

### Variations in Response to HIV Exposure

Several factors affect infection and progression of disease.

**Impact of Age on Survival with HIV Infection** Older adults are less able to replace CD4$^+$ T cell populations. Infants and younger children have an immune system that is not fully developed, leaving them much more susceptible to opportunistic infections.

Infants born to HIV-positive mothers are not always infected—in fact, only about 20% are. Infants with a high viral load survive less than 18 months. HIV-positive mothers who are actively undergoing antiretroviral treatment can reduce the risk of transmitting the virus to their offspring by 99%.

**Exposed, But Not Infected, Population** Certain high-risk people are repeatedly exposed to HIV but remain free of infection. About 1–3% of the populations of the Western world do not have a gene for CCR5, a coreceptor typically found on T helper cells that HIV binds to before entering the cell. Therefore, people without CCR5 receptors are highly resistant to HIV infections.

The role of CCR5 in natural resistance has led to research into drugs that block the receptor. Experiments are under way for the use of gene therapy and personalized gene editing to treat AIDS by replacing the patient's T cells with those that are not susceptible to infection. The initial step is to remove some T cells from patients and modify them by deleting their CCR5. These populations of modified cells would then be multiplied and infused back into the patients. In the small group of patients in which this is being tested, there is encouraging evidence that the numbers of these modified cells are slowly increasing in the patient's bloodstream. Other novel studies are under way to remove the HIV provirus from HIV-infected cells.

**Long-Term Survivors** Occasionally, certain untreated individuals who have been infected with HIV for more than 10 years have not progressed to AIDS. These people are called *long-term nonprogressors*. Other individuals who are repeatedly infected with the virus never progress to AIDS and have little to no virus in their blood. They are called *elite controllers*. These long-term survivors make up less than 5% of the HIV-infected population. Genetics, viral variants, and limited susceptibility to HIV are among the explanations for this population. Some of these individuals have CTLs with unusual powers that are capable of destroying fast-mutating viruses such as HIV. These long-term survivors are of exceptional interest because they might provide insights into treatments for all HIV-infected persons.

## Diagnostic Methods

The CDC recommends routine screening for HIV infections. Recommendations for screening vary based on risk of exposure, from once in a lifetime to every 6 months for people who engage in high-risk behaviors (men who have sex with men, people with multiple sex partners, or those who use intravenous street drugs). The standard procedure for detecting HIV antibodies has been blood tests using an ELISA test (see Figure 18.14b, page 517) to detect HIV antibodies. There are now several relatively inexpensive, rapid tests available for HIV screening that are used in point-of-care clinics, as well as in developing, resource-poor countries. The tests use urine or fingerstick amounts of blood, and the OraQuick® test can even use an oral swab to check for HIV antibodies. They return results in 20 to 30 minutes. Several of these tests are available for home testing. An estimated 13% of HIV-positive Americans do not realize they are infected; this lack of knowledge fuels the spread of the disease. Positive screening tests for antibodies must be confirmed by additional testing, usually by the Western blot test (see Figure 10.13, page 284).

A problem with antibody-type testing is the window of time between infection and the appearance of detectable antibodies, or **seroconversion**. This interval, which can be as long as 3 months, is illustrated in step ❸ of Figure 19.17, where seroconversion follows the peak number of viruses in circulation. Because of this delay, the recipient of an organ transplant or a blood transfusion can become infected with

HIV even though antibody tests did not show the presence of the virus in the donor.

Additions to the Western blot confirmatory test include nucleic acid amplification tests (NAATs). For example, instead of using antibodies, the APTIMA® assay detects the RNA of the HIV-1 virus using real-time PCR, and it is easier to read than the Western blot test. This test can also be used to detect early HIV infections, before appearance of antibodies. Its sensitivity is comparable to tests used to measure **plasma viral load (PVL)** in the blood of patients to monitor the treatment and progression of AIDS. Conventional PVL tests that detect viral RNA use methods such as PCR, are costly, and require 2 or 3 days to complete. Viral RNA can be detected in 7 to 10 days and, less reliably, in 2 to 4 days. To ensure safety of the blood supply as much as possible, the American Red Cross routinely tests for anti-HIV antibody and utilizes NAAT for viral HIV.

A caution to keep in mind in testing for HIV is that current tests may not reliably detect all of the myriad variants of rapidly mutating HIV, especially subtypes that are not normally present in a population. Furthermore, PVL tests sample only the virions circulating in the blood, which is very low compared to the estimated several hundred billion HIV-infected cells.

### CHECK YOUR UNDERSTANDING

✔ **19-28** What form of nucleic acid is detected in a PVL test for HIV?

## HIV Transmission

The transmission of HIV requires the transfer of, or direct contact with, infected body fluids. The two most important fluids in terms of infection risk are blood, which contains 1000 to 100,000 infective viruses per milliliter, and semen, which contains about 10 to 50 viruses per milliliter. HIV is often located within cells in these fluids, especially in macrophages. Saliva generally contains less than 1 virus per milliliter, and kissing is not known to transmit HIV. The virus can survive more than 1.5 days inside a cell but only about 6 hours outside a cell. In developed countries, transmission by transfusion is unlikely because blood is tested for HIV or HIV antibodies.

Routes of HIV transmission include sexual contact, breast milk, transplacental infection of a fetus, blood-contaminated needles, organ transplants, artificial insemination, and blood transfusion. The risk of infection from needlestick injury is 3 out of 1000, or 0.3%. Avoiding exposure is the health care worker's first line of defense against HIV. The CDC has developed the strategy of following *universal precautions* in *all* health care settings.

Probably the most high-risk form of sexual contact for HIV infection is anal-receptive intercourse. These tissues are much more vulnerable to transmission of disease organisms. Vaginal intercourse is much more likely to transmit HIV from man to woman than vice versa, and transmission either way is much greater when genital lesions are present. Although rare, transmission can occur by oral-genital contact.

### CHECK YOUR UNDERSTANDING

✔ **19-29** What is considered to be the most dangerous form of sexual contact for transmission of HIV?

## AIDS Worldwide

Approximately 36 million people are infected with, and living with, HIV today (**Figure 19.18**). An estimated 70% of these are in Africa, including the majority of the world's HIV-infected children. South and southeast Asia, with their dense populations, also have a high number of cases, an estimated 3.5 million. In Western Europe and the United States, the mortality from AIDS has decreased because of the availability of effective antiviral drugs (see Figure 14.4 on page 401).

Worldwide, heterosexual intercourse is the most common mode of HIV transmission. High-risk sexual behavior and street drug use contribute to the spread of HIV/AIDS. Currently, one-third of all HIV infections in eastern Europe and central and southeast Asia are by street drug use. These infections are also important as a bridge leading to sexual transmission.

### CHECK YOUR UNDERSTANDING

✔ **19-30** What is the most common mode, worldwide, by which HIV is transmitted?

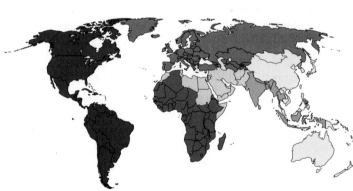

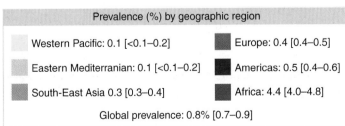

Prevalence (%) by geographic region

| | |
|---|---|
| Western Pacific: 0.1 [<0.1–0.2] | Europe: 0.4 [0.4–0.5] |
| Eastern Mediterranian: 0.1 [<0.1–0.2] | Americas: 0.5 [0.4–0.6] |
| South-East Asia 0.3 [0.3–0.4] | Africa: 4.4 [4.0–4.8] |

Global prevalence: 0.8% [0.7–0.9]

**Figure 19.18 Distribution of HIV infection and AIDS in regions of the world.** Shows percentage of adults (ages 15–49) living with HIV/AIDS.

**Q** Where do you think the most accurate figures would be available?

## Preventing and Treating AIDS

At present, for most of the world the only practical means of controlling AIDS is to minimize transmission. *Biomedical interventions* include condoms, health services, HIV testing, and needle programs. Examples of *behavioral intervention* include sex education, safe infant feeding programs, and counseling. *Structural interventions* focus on making change in social, economic, political, and environmental factors that make individuals or groups vulnerable to HIV.

**Preexposure prophylaxis (PrEP)** and **postexposure prophylaxis (PEP)** for HIV are used to prevent infection after a recent exposure. PrEP and PEP use drug combinations that are also used in HIV treatment. For example, Truvada® is a combination of tenofovir and emtricitabine. The treatment requires strict adherence to daily HIV dosages to lower the chance of getting infected.

### Antiretroviral Therapy (ART)

The rapid reproductive rate and frequent occurrence of drug-resistant mutations dictates that multiple drugs, given simultaneously, must be used. The current treatment is termed **highly active antiretroviral therapy (HAART)**. This therapy consists of administering drug combinations. Patients are often required to take as many as 40 pills a day on a complex schedule. Even so, resistant strains of the virus are likely to emerge. The majority of AIDS patients in the United States receive multiple-drug therapy to minimize survival of resistant strains. The drugs are commonly combined in a single pill to simplify administration. Experience has also shown that eliminating all viruses in latent form in lymphoid tissue is especially difficult. The number of HIV in circulation is often reduced to fewer than can be detected, but this is not the same as eradication.

Because of the increasing number of drugs that at least temporarily control reproduction of the virus, HIV infection went from a swift death sentence to a treatable chronic disease in the developed world. Accessibility and affordability of HIV treatments continue to be a challenge in developing nations. In the United States, antiretroviral therapy (ART) is recommended for all people with HIV. In fact, the term **antiretroviral** has come to imply a drug that is used to treat HIV infections. Major obstacles to treating HIV, as with developing a vaccine, are the high mutation rate that quickly leads to resistant strains and the persistence of latent viral reservoirs. If the effective drugs are interrupted or discontinued, the virus rebounds rapidly. Research into the reproductive mechanisms of HIV has increased the number of potential targets for chemical intervention. **Figure 19.19** shows the major types and some of the drugs available.

**Fusion and Entry Inhibitors** For infection to occur, the virus must attach to the cell's CD4 receptors; an interplay between the gp120 spike on the virus and the coreceptor (such as CCR5) must occur; and finally, there must be a fusion with the cell to allow viral entry. Drugs to block these steps are grouped as *cell entry inhibitors*; some of the drugs of this group target the gp41 region of the viral envelope, which facilitates fusion. An example is *enfuvirtide*, which is expensive and requires daily injections. Another cell entry inhibitor is *maraviroc*, which blocks the chemokine receptor CCR5 to which HIV must bind. Trials are in progress to produce patient immune cells that lack CCR5. These cells would resist HIV attachment and could allow the cells to persist without infection.

**Reverse Transcriptase Inhibitors** After virus fusion with the host cell, reverse transcription from the RNA genome produces a double-stranded cDNA copy of the HIV genome. The first target of anti-HIV drugs was the enzyme reverse transcriptase, an enzyme not present in human cells. There are currently 12 different FDA-approved inhibitors in this category, including *emtricitabine*. In fact, the term **antiretroviral** has come to imply a drug that is used to treat HIV infections. The *nucleoside reverse transcriptase inhibitors (NRTIs)* are analogs of nucleosides and cause the termination of viral DNA by competitive inhibition. There are other drugs that inhibit reverse transcription that are not analogs of nucleic acids; these are the so-called *non-nucleoside reverse transcriptase inhibitors (NNRTIs)*. Atripla® is a combination of two NRTIs plus *efavirenz* (a NNRTI).

One particularly successful application of chemotherapy has been reducing transmission of HIV from an infected mother to her newborn. The administration of even one NRTI drug greatly reduces transmission. Another promising application of chemotherapy is a vaginal gel containing *tenofovir* (an NRTI), which is in current African trials, and has significantly lowered infection rates.

**Integrase Inhibitors** After reverse transcription, the cDNA of the HIV enters the nucleus. There the cDNA must be integrated into the host chromosome to form the HIV provirus. This step requires an enzyme, HIV integrase, which is a target for drugs called *integrase inhibitors*. *Raltegravir, dolutegravir, and elvitegravir* are examples.

**Protease Inhibitors** A third enzyme target is HIV proteases. Proteases perform the essential process of cleaving lengthy viral precursor proteins into smaller, mature structural proteins (such as the capsid proteins) and functional proteins (such as essential enzymes). Most of this occurs as the virus is budding from the cell membrane and shortly thereafter. *Protease inhibitor* drugs such as *atazanavir, indinavir,* and *saquinavir* have proved especially effective when combined with reverse transcriptase inhibitors.

There are several other targets for which drugs are being developed. For example, some **maturation inhibitors** affect conversion of a precursor of capsid protein to mature capsid

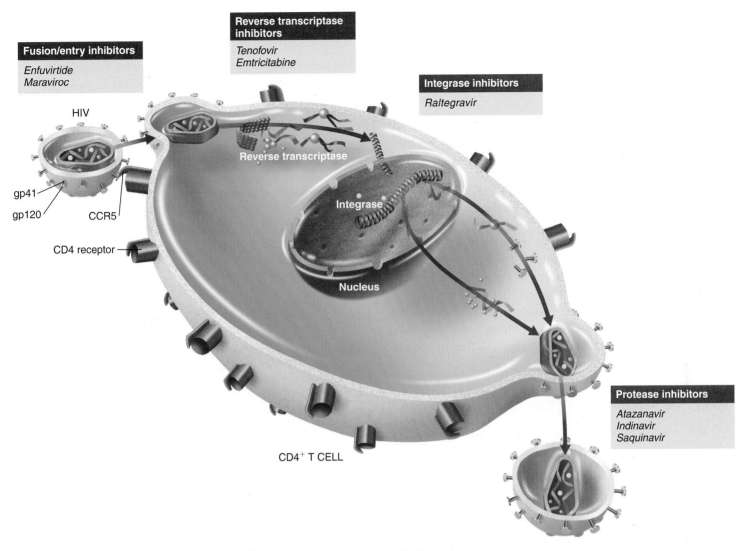

**Fusion/entry inhibitors**

*Enfuvirtide*
*Maraviroc*

**Reverse transcriptase inhibitors**

*Tenofovir*
*Emtricitabine*

**Integrase inhibitors**

*Raltegravir*

HIV

Reverse transcriptase

Integrase

Nucleus

gp41
gp120

CCR5

CD4 receptor

CD4⁺ T CELL

**Protease inhibitors**

*Atazanavir*
*Indinavir*
*Saquinavir*

**Figure 19.19 Drugs that inhibit the HIV life cycle.** The sites of action of antiretroviral drugs are shown.

**Q** Why does a drug that binds to CCR5 on the host cell prevent viral reproduction?

protein, resulting in an aberrant capsid that renders the virus noninfectious. Other potential drugs are **tetherins**, which tether the newly formed virus to the cell, preventing release and spread. Research will probably reveal new targets and chemotherapeutic agents affecting them. Especially important is the need to find drugs that can eradicate the virus in its latent reservoirs. Gene therapy and gene editing are being explored to combat HIV.

### The Challenges of Developing HIV Vaccines

Drugs extended the lives of millions, but they have had little effect on ending the pandemic. Overcoming AIDS likely requires a preventive vaccine—something that so far eludes researchers. Many unsuccessful HIV vaccine trials have come and gone over the decades. One challenge is that there is no model of natural immunity to mimic with a vaccine since, so far, not one of the

millions of people infected by the virus has ever been shown to successfully eradicate HIV via the immune system. The use of attenuated viruses is possible in certain vaccines, but for HIV it is too dangerous an option. Retroviruses also quickly integrate themselves into the DNA of the host cell and then remain latent, meaning that the antigenic qualities of HIV are largely invisible to the immune system most of the time. HIV's high mutation rate complicates matters as well. The virus developed clades that differ substantially from one geographic area to another, making vaccine development that much more difficult.

Ideally, a vaccine would produce antibodies that prevent infection. However, in natural HIV infections, neutralizing antibodies develop rather slowly, appearing 2 months or so after transmission. By the time the immune system produces effective numbers of these antibodies, the target envelope protein of the HIV has mutated and eludes neutralization.

Moreover, the ability of HIV to elude immune defenses by infecting new cells by cell-to-cell fusion contrasts with almost all other viral infections and is a challenge to any vaccine.

In short, any successful future HIV vaccine will likely require a fundamentally different approach to vaccine development. Still, the potential benefits of a vaccine mean that researchers continue to search for a solution to these challenges. An HIV vaccine would have to induce immunity before reservoirs of latent virus are established (see Figure 19.16)—this usually occurs within 5 to 10 days of infection. In fact, a potential target for vaccination might be to prevent or regulate latency. It would also have to stimulate production of cytotoxic T cells that are more effective than those usually produced in response to a natural infection. Finally, an HIV vaccine would have to be affordable in poor regions where infection rates are high. All these factors make the development of an HIV vaccine an extremely difficult task. One creative vaccine being developed contains gp120 with portions of the CD4 receptor. This vaccine would generate antibodies to exposed parts of the virus as it attempts to find its coreceptor, like CCR5.

## CLINICAL CASE Resolved

Once the autoimmune rejection condition is recognized, Malik is successfully treated with Mab-CD3. This drug binds the T cell receptor–CD3 complex on the surface of circulating T cells, preventing the T cells from attacking host cells. Malik also receives cyclosporine, which suppresses IL-2 secretions—an essential chemical messenger in discrimination between self and nonself. If Malik had been diagnosed with DiGeorge syndrome before his transfusion, the blood would have been irradiated to destroy the white blood cells in it, preventing his reaction. Malik recovers from the GVHD, but he will eventually need a thymus transplant.

525 | 530 | 541 | 544 | **554**

## CHECK YOUR UNDERSTANDING

19-31 Why are reverse transcriptase, integrase, and proteases good targets for chemotherapy?

# Study Outline

 Go to @MasteringMicrobiology for **Interactive Microbiology**, *In the Clinic videos, MicroFlix, MicroBoosters, 3D animations, practice quizzes, and more.*

## Introduction (p. 524)

1. Hay fever, transplant rejection, and autoimmunity are examples of harmful immune reactions.
2. Immunosuppression is inhibition of the immune system.

## Hypersensitivity (pp. 525–536)

1. Hypersensitivity reactions occur when a person has been sensitized to an antigen.
2. Hypersensitivity reactions represent immunological responses to an antigen (allergen) that lead to tissue damage rather than immunity.
3. Hypersensitivity reactions can be divided into four classes: types I, II, and III are immediate reactions based on humoral immunity, and type IV is a delayed reaction based on cell-mediated immunity.

### Allergies and the Microbiome (p. 525)

4. Childhood exposure to microbes may decrease development of allergies.

### Type I (Anaphylactic) Reactions (pp. 525–530)

5. Anaphylactic reactions involve the production of IgE antibodies that bind to mast cells and basophils to sensitize the host.
6. The binding of two adjacent IgE antibodies to an antigen causes the target cell to release chemical mediators, such as histamine, leukotrienes, and prostaglandins, which cause the observed allergic reactions.
7. Systemic anaphylaxis may develop in minutes after injection or ingestion of the antigen; this may result in circulatory collapse and death.
8. Localized anaphylaxis is exemplified by hives, hay fever, and asthma.

9. Skin testing is useful in determining sensitivity to an antigen.
10. Desensitization to an antigen can be achieved by repeated injections of the antigen, which leads to the formation of blocking (IgG) antibodies.

### Type II (Cytotoxic) Reactions (pp. 531–534)

11. Type II reactions are mediated by IgG or IgM antibodies and complement.
12. The antibodies are directed toward foreign cells or host cells. Complement fixation may result in cell lysis. Macrophages and other cells may also damage the antibody-coated cells.
13. Human blood may be grouped into four principal types, designated A, B, AB, and O.
14. The presence or absence of two carbohydrate antigens designated A and B on the surface of the red blood cell determines a person's blood type.
15. Naturally occurring antibodies against the opposite AB antigen are present in serum.
16. Incompatible blood transfusions lead to the complement-mediated lysis of the donor red blood cells.
17. The absence of the Rh antigen in certain individuals ($Rh^-$) can lead to sensitization upon exposure to it.
18. When an $Rh^-$ person receives $Rh^+$ blood, that person will produce anti-Rh antibodies. Subsequent exposure to $Rh^+$ cells will result in a rapid, serious hemolytic reaction.
19. An $Rh^-$ mother carrying an $Rh^+$ fetus will produce anti-Rh antibodies. Subsequent pregnancies involving Rh incompatibility may result in hemolytic disease of the newborn (HDNB).

20. HDNB may be prevented by passive immunization of the mother with anti-Rh antibodies.

21. In the disease thrombocytopenic purpura, platelets are destroyed by antibodies and complement.

22. Agranulocytosis and hemolytic anemia result from antibodies against one's own blood cells coated with drug molecules.

### Type III (Immune Complex) Reactions (pp. 534–535)

23. Immune complex diseases occur when IgG antibodies and soluble antigen form small complexes that lodge in the basement membranes of cells.

24. Subsequent complement fixation results in inflammation.

25. Glomerulonephritis is an immune complex disease.

### Type IV (Delayed Cell-Mediated) Reactions (pp. 535–536)

26. Delayed cell-mediated hypersensitivity reactions are due primarily to T cell proliferation.

27. Sensitized T cells secrete cytokines in response to the appropriate antigen.

28. Cytokines attract and activate macrophages and initiate tissue damage.

29. The tuberculin skin test and allergic contact dermatitis are examples of delayed hypersensitivities.

## Autoimmune Diseases (pp. 536–538)

1. Autoimmunity results from a loss of self-tolerance.

2. Self-tolerance occurs during fetal development; T cells that will target host cells are eliminated through thymic selection (clonal deletion).

3. The immune system attacks the myelin sheath of nerves in multiple sclerosis.

4. Graves' disease, myasthenia gravis, and rheumatoid arthritis are immune complex autoimmune diseases.

5. Insulin-dependent diabetes mellitus and psoriasis are cell-mediated autoimmune reactions.

## Reactions Related to Transplantation (pp. 538–542)

1. MHC self molecules located on cell surfaces express genetic differences among individuals; these antigens are called HLAs in humans.

2. To prevent the rejection of transplants, HLA and ABO blood group antigens of the donor and recipient are matched as closely as possible.

3. Transplants recognized as foreign antigens may be lysed by T cells and attacked by macrophages and complement-fixing antibodies.

4. Transplantation to a privileged site (such as the cornea) or of a privileged tissue (such as pig heart valves) does not cause an immune response.

5. Pluripotent stem cells differentiate into a variety of tissues that may provide tissues for transplant.

6. Four types of transplants have been defined on the basis of genetic relationships between the donor and the recipient: autografts, isografts, allografts, and xenotransplantation products.

7. Bone marrow transplants (with immunocompetent cells) can cause graft-versus-host disease.

8. Successful transplant surgery often requires immunosuppressant drugs to prevent an immune response to the transplanted tissue.

## The Immune System and Cancer (pp. 542–543)

1. Cancer cells are normal cells that have undergone transformation, divide uncontrollably, and possess tumor-associated antigens.

2. The response of the immune system to cancer is called immunological surveillance.

3. Cytotoxic T lymphocytes recognize and lyse cancerous cells.

### Immunotherapy for Cancer (p. 543)

4. Prophylactic vaccines against liver and cervical cancer are available.

5. HPV and HBV vaccines protect against some cancers. Trastuzumab (Herceptin®) consists of monoclonal antibodies against a breast cancer growth factor.

6. Immunotoxins are chemical poisons linked to a monoclonal antibody; the antibody selectively locates the cancer cell for release of the poison.

## Immunodeficiencies (pp. 543–544)

1. Immunodeficiencies can be congenital or acquired.

2. Congenital immunodeficiencies are due to defective or absent genes.

3. A variety of drugs, cancers, and infectious diseases can cause acquired immunodeficiencies.

## Acquired Immunodeficiency Syndrome (AIDS) (pp. 544–554)

### The Origin of AIDS (p. 545)

1. HIV is thought to have originated in Africa and was brought to other countries by modern transportation and unsafe sexual practices.

### HIV Infection (pp. 545–550)

2. AIDS is the final stage of HIV infection.

3. HIV is a retrovirus with single-stranded RNA, reverse transcriptase, and a phospholipid envelope with gp120 spikes.

4. HIV spikes attach to CD4 and coreceptors on host cells; the CD4 receptor is found on T helper cells, macrophages, and dendritic cells.

5. Viral RNA is transcribed to DNA by reverse transcriptase. The viral DNA becomes integrated into the host chromosome to direct synthesis of new viruses or to remain latent as a provirus.

6. HIV evades the immune system in latency, in vacuoles, by using cell–cell fusion, and by antigenic change.

7. HIV infection is categorized by clinical phases: phase 1 (asymptomatic), phase 2 (indicator opportunistic infections), and phase 3 ($CD4^+$ cells < 200 cells/$\mu$l).

8. The progression from HIV infection to AIDS takes about 10 years.

9. The life of an AIDS patient can be prolonged by the proper treatment of opportunistic infections.

10. Long-term survivors and elite controllers may hold the key to HIV treatment.

### Diagnostic Methods (pp. 550–551)

11. HIV antibodies are detected by ELISA and Western blotting.

12. Plasma viral load tests detect viral nucleic acid and are used to quantify HIV in blood.

### HIV Transmission (p. 551)

13. HIV is transmitted by sexual contact, breast milk, contaminated needles, transplacental infection, artificial insemination, and blood transfusion.

14. In developed countries, blood transfusions are not a likely source of infection because blood is tested for HIV antibodies.

### AIDS Worldwide (p. 551)

15. Heterosexual intercourse is the primary method of HIV transmission.

### Preventing and Treating AIDS (pp. 552–554)

16. Transmission can be reduced by biomedical, behavioral, and structural interventions.

17. Current chemotherapeutic agents target cell entry and the virus enzymes, including reverse transcriptase, integrase, and protease.

18. Vaccine development is difficult because there are different antigenic clades and the virus remains inside host cells.

# Study Questions

For answers to the Knowledge and Comprehension questions, turn to the Answers tab at the back of the textbook.

## Knowledge and Comprehension

### Review

1. **DRAW IT** Label IgE, antigen, and mast cell, and add an antihistamine to the following figure. What type of cell is this? Singulair® stops inflammation by blocking leukotriene receptors. Add this action to the figure.

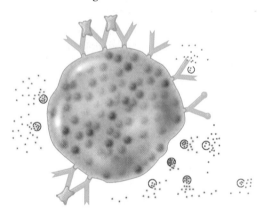

2. In the laboratory, blood is typed by looking for hemagglutination. For example, anti-A antibodies and type A RBCs clump. In a type A person, anti-A antibodies will cause hemolysis. Why?

3. Discuss the roles of antibodies and antigens in an incompatible tissue transplant.

4. Explain what happens when a person develops a contact sensitivity to the poison oak plant.
   a. What causes the observed symptoms?
   b. How did the sensitivity develop?
   c. How might this person be desensitized to poison oak?

5. Why does an ANA (antinuclear antibody) test diagnose lupus?

6. Differentiate the three types of autoimmune diseases. Name an example of each type.

7. Summarize the causes of immunodeficiencies. What is the effect of an immunodeficiency?

8. In what ways do tumor cells differ antigenically from normal cells? Explain how tumor cells may be destroyed by the immune system.

9. If tumor cells can be destroyed by the immune system, how does cancer develop? What does immunotherapy involve?

10. **NAME IT** The Fc region of this protein causes degranulation when it binds to basophils.

## Multiple Choice

1. Desensitization to prevent an allergic response can be accomplished by injecting small, repeated doses of
   a. IgE antibodies.
   b. the antigen (allergen).
   c. histamine.
   d. IgG antibodies.
   e. antihistamine.

2. What does *pluripotent* mean?
   a. ability of a single cell to develop into an embryonic or adult stem cell
   b. ability of a stem cell to develop into many different cell types
   c. a cell without MHC I and MHC II antigens
   d. ability of a single stem cell to heal different types of diseases
   e. ability of an adult cell to become a stem cell

3. Cytotoxic autoimmunity differs from immune complex autoimmunity in that cytotoxic reactions
   a. involve antibodies.
   b. do not involve complement.
   c. are caused by T cells.
   d. do not involve IgE antibodies.
   e. none of the above

4. Antibodies against HIV are ineffective for all of the following reasons *except*
   a. the fact that antibodies aren't made against HIV.
   b. transmission by cell–cell fusion.
   c. antigenic changes.
   d. latency.
   e. persistence of virus particles in vacuoles.

5. Which of the following is *not* the cause of a natural immunodeficiency?
   a. a recessive gene resulting in lack of a thymus gland
   b. a recessive gene resulting in few B cells
   c. HIV infection
   d. immunosuppressant drugs
   e. All of the above are causes of natural immunodeficiency.

6. Which antibodies will be found naturally in the serum of a person with blood type A, Rh$^+$?
   a. anti A, anti B, anti Rh
   b. anti A, anti Rh
   c. anti A
   d. anti B, anti Rh
   e. anti B

Use the following choices to match the type of hypersensitivity to the examples in questions 7 through 10.
   a. type I hypersensitivity
   b. type II hypersensitivity
   c. type III hypersensitivity
   d. type IV hypersensitivity
   e. all of the above

7. Localized anaphylaxis.
8. Allergic contact dermatitis.
9. Due to immune complexes.
10. Reaction to an incompatible blood transfusion.

## Analysis

1. When and how does our immune system discriminate between self and nonself antigens?
2. The first preparations used for artificially acquired passive immunity were antibodies in horse serum. A complication that resulted from the therapeutic use of horse serum was immune complex disease. Why did this occur?

3. Do people with AIDS make antibodies? If so, why are they said to have an immunodeficiency?
4. What are the methods of action of anti-AIDS drugs?

## Clinical Applications and Evaluation

1. Fungal infections such as athlete's foot are chronic. These fungi degrade skin keratin but are not invasive and do not produce toxins. Why do you suppose that many of the symptoms of a fungal infection are due to hypersensitivity to the fungus?
2. After working in a mushroom farm for several months, a worker develops these symptoms: hives, edema, and swelling lymph nodes.
   a. What do these symptoms indicate?
   b. What mediators cause these symptoms?
   c. How may sensitivity to a particular antigen be determined?
   d. Other employees do not appear to have any immunological reactions. What could explain this?

   (*Hint:* The allergen is conidiospores from molds growing in the mushroom farm.)

3. Physicians administering live, attenuated mumps and measles vaccines prepared in chick embryos are instructed to have epinephrine available. Epinephrine will not treat these viral infections. What is the purpose of keeping this drug on hand?
4. A woman with blood type A$^+$ once received a transfusion of AB$^+$ blood. When she carried a type B$^+$ fetus, the fetus developed hemolytic disease of the newborn. Explain why this fetus developed this condition even though another type B$^+$ fetus in a different type A$^+$ mother was normal.

# 20 Antimicrobial Drugs

When the body's normal defenses cannot prevent or overcome a disease, it often can be treated by chemotherapy with antimicrobial drugs. Like the disinfectants discussed in Chapter 7, antimicrobial drugs act by killing or by interfering with the growth of microorganisms. Unlike disinfectants, however, antimicrobial drugs must act *within* the host without damaging the host. This is the important principle of selective toxicity.

Antibiotics were one of the most important discoveries of modern medicine. Not too long ago, little could be done to treat many lethal infectious diseases. The introduction of penicillin, sulfanilamide, and other antimicrobials to treat conditions such as a pneumonia or so-called blood poisoning (sepsis) resulted in cures that seemed almost miraculous.

Today, we are seeing the advances represented by these miracle drugs threatened by the development of antibiotic resistance. For example, there are frequent reports of staphylococcal pathogens that are resistant to practically all the available antibiotics. Certain populations of the pathogens causing tuberculosis are now resistant to essentially all of the available antibiotics that were once effective. The Clinical Case in this chapter describes an infection caused by the antibiotic-resistant *Pseudomonas aeruginosa* shown in the photograph. In some cases, medicine now has only a few more weapons to treat the diseases caused by these pathogens than were available a century ago.

▶ *Pseudomonas aeruginosa* bacteria (blue) are resistant to many antibiotics.

## In the Clinic

Your family knows you are a nurse and always asks you for advice. Your brother asks you about his cough, which started 2 weeks ago. Now he's coughing up mucus and feels sure that it's bronchitis. He asks you whether he should use the amoxicillin he saved from a prescription he received when he had bronchitis last winter. **What should you tell your brother?**

*Hint: Read about preventing microbial resistance on pages 579–584.*

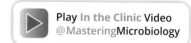

Play In the Clinic Video
@MasteringMicrobiology

# The History of Chemotherapy

## LEARNING OBJECTIVES

**20-1** Identify the contributions of Paul Ehrlich and Alexander Fleming to chemotherapy.

**20-2** Name the microbes that produce most antibiotics.

The birth of modern chemotherapy is credited to the efforts of Paul Ehrlich in Germany during the early part of the twentieth century. While attempting to stain bacteria without staining the surrounding tissue, he speculated about some "magic bullet" that would selectively find and destroy pathogens but not harm the host. This idea provided the basis for both **selective toxicity** and **chemotherapy**, a term he coined.

In 1928, Alexander Fleming observed that the growth of the bacterium *Staphylococcus aureus* was inhibited in the area surrounding the colony of a mold that had contaminated a Petri plate (Figure 1.6, page 11). The mold was identified as *Penicillium notatum*, and its active compound, which was isolated a short time later, was named penicillin. Similar inhibitory reactions between colonies on solid media are commonly observed in microbiology, and the mechanism of inhibition is called *antibiosis* (**Figure 20.1**). From this word comes the term **antibiotic**, a substance produced by microorganisms that in small amounts inhibits another microorganism. Therefore, the wholly synthetic sulfa drugs, for example, are technically **antimicrobial drugs**, not antibiotics, a distinction often ignored in practice. The discovery of sulfa drugs arose from a systematic survey of chemicals by German industrial scientists beginning in 1927. In 1932, a compound termed Prontosil Red, a sulfanilamide-containing dye, was found to control streptococcal infections in mice. During World War II, the Allied armies made wide use of this sulfanilamide compound. The discovery and use of sulfa drugs made it clear that practical antimicrobials could be effective against systemic bacterial infections and resurrected interest in the earlier reports of penicillin.

In 1940 the first clinical trials of penicillin took place. Under wartime conditions in the United Kingdom, research into the development and large-scale production of penicillin was not possible, and this work was transferred to the United States. The original culture of *P. notatum* was not a very efficient producer of the antibiotic. It was soon replaced by a more prolific strain. This valuable organism (a strain of *Penicillium chrysogenum*) was first isolated from a moldy cantaloupe bought at a market in Peoria, Illinois.

## Antibiotic Use and Discovery Today

Antibiotics are actually rather easy to discover, but few are of medical or commercial value. Some are used commercially rather than for treating disease—for example, as a supplement in animal feed (see the Clinical Focus box on page 584). There is an

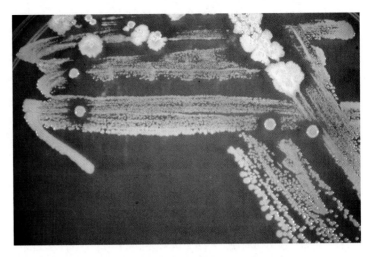

**Figure 20.1  Laboratory observation of antibiosis.** Anyone plating out microbes from natural environments, especially soil, will frequently see examples of bacterial inhibition by antibiotics produced by bacteria, most commonly *Streptomyces* species.

**Q** Would there be any advantage to a soil microbe to produce an antibiotic?

increasing urgency to find new antibiotics to address the growing problem of **antibiotic resistance**, a phenomenon in which formerly effective medications have less and less impact on bacteria.

More than half of our antibiotics are produced by species of *Streptomyces*, filamentous bacteria that commonly inhabit soil. A few antibiotics are produced by endospore-forming

---

**CLINICAL CASE**  Sight Unseen

Ophthalmic surgeon Dr. Vanessa Singh has performed hundreds of corneal transplants without incident. She is understandably worried when a 76-year-old woman on whom she operated yesterday develops a corneal infection. Dr. Singh had given her patient the proper prophylactic subconjunctival injection of gentamicin following transplantation, so she is puzzled by the presence of infection. Postoperative gentamicin is recommended for corneal transplants because *Staphylococcus epidermidis* and *S. aureus* are the most common organisms causing postoperative eye infection.

Dr. Singh takes a culture from the patient's eye and sends it to the lab for analysis. The culture comes back positive for *Pseudomonas aeruginosa*. Dr. Singh checks with the eye bank and discovers that a 30-year-old man who received the other cornea from the donor developed a *P. aeruginosa* infection within 24 hours of surgery. This patient also had received prophylactic gentamicin to prevent infection.

**What does Dr. Singh need to know? Read on to find out.**

559  570  579  582  586

**TABLE 20.1 Representative Sources of Antibiotics**

| Microorganism | Antibiotic |
| --- | --- |
| **GRAM-POSITIVE RODS** | |
| *Bacillus subtilis* | Bacitracin |
| *Paenibacillus polymyxa* | Polymyxin |
| **ACTINOMYCETES** | |
| *Streptomyces nodosus* | Amphotericin B |
| *Streptomyces venezuelae* | Chloramphenicol |
| *Streptomyces aureofaciens* | Chlortetracycline and tetracycline |
| *Saccharopolyspora erythraea* | Erythromycin |
| *Streptomyces fradiae* | Neomycin |
| *Streptomyces griseus* | Streptomycin |
| *Micromonospora purpurea* | Gentamicin |
| **FUNGI** | |
| *Cephalosporium* spp. | Cephalothin |
| *Penicillium griseofulvum* | Griseofulvin |
| *Penicillium chrysogenum* | Penicillin |

bacteria such as *Bacillus*, and others are produced by molds, mostly of the genera *Penicillium* and *Cephalosporium* (SEF-ah-lō-spor'ē-um). See Table 20.1 for the sources of many antibiotics in use today—a surprisingly limited group of organisms. One study screened 400,000 microbial cultures that yielded only three useful drugs. It's especially interesting to note that practically all antibiotic-producing microbes have some sort of sporulation process.

Most antibiotics in use today were discovered by methods that required identifying and growing colonies of antibiotic-producing organisms, mostly by screening soil samples. It's rather easy to identify microbes in such samples that have antimicrobial activity (Figure 20.1); however, many are toxic or not commercially useful. Also, these proved to be examples of "low-hanging fruit," and continued work often resulted in discovery of the same antibiotics. For example, about 1 in every 100 actinomycetes in soil produces streptomycin, and 1 in 250 produces tetracycline. To find an antibiotic that is produced by only one soil or sea microbe in 10 million is a daunting task. Even modern *high-throughput methods*, which rapidly screen very large numbers of microbes in the search for new antibiotics, have failed to yield many new discoveries. In fact, in the past 40 years, research using established methods has led to the clinical use of only a couple of new structural types of microbial inhibitors.

**CHECK YOUR UNDERSTANDING**

✔ **20-1** Who coined the term *chemotherapy*?

✔ **20-2** More than half our antibiotics are produced by a certain genus of bacteria. What is it?

# Spectrum of Antimicrobial Activity

**LEARNING OBJECTIVES**

**20-3** Describe the problems of chemotherapy for viral, fungal, protozoan, and helminthic infections.

**20-4** Define the following terms: *spectrum of activity, broad-spectrum antibiotic, superinfection.*

It's comparatively easy to find or develop drugs that are effective against prokaryotic cells and that do not affect the eukaryotic cells of humans. These two cell types differ substantially in many ways, such as in the presence or absence of cell walls, the structure of their ribosomes, and details of their metabolism. Thus, selective toxicity has numerous targets. The problem is more difficult when the pathogen is a eukaryotic cell, such as a fungus, protozoan, or helminth. At the cellular level, these organisms resemble the human cell much more closely than a bacterial cell does. So, a drug that targets these pathogens usually damages the host, too. Our arsenal against these types of pathogens is much more limited than our arsenal of antibacterial drugs. Viral infections are also particularly difficult to treat because the pathogen is within the human host's cells and because the genetic information of the virus is directing the human cell to make viruses rather than to synthesize normal cellular materials.

Some drugs have a **narrow spectrum of microbial activity**, or range of different microbial types they affect. Penicillin G, for example, affects gram-positive bacteria but very few gram-negative bacteria. Antibiotics that affect a broad range of gram-positive or gram-negative bacteria are therefore called **broad-spectrum antibiotics**.

A primary factor involved in the selective toxicity of antibacterial action lies in the lipopolysaccharide outer layer of gram-negative bacteria and the porins that form water-filled channels across this layer (see Figure 4.13c, page 82). Drugs that pass through the porin channels must be relatively small and preferably hydrophilic. Drugs that are lipophilic (having an affinity for lipids) or especially large do not enter gram-negative bacteria readily.

Table 20.2 summarizes the spectrum of activity of a number of chemotherapeutic drugs. Because the identity of the pathogen is not always immediately known, a broad-spectrum drug would seem to have an advantage in treating a disease by saving valuable time. The disadvantage is that these drugs destroy many normal microbiota of the host. The normal microbiota ordinarily compete with and check the growth of pathogens or other microbes. If the antibiotic does not destroy certain organisms in the normal microbiota but does destroy their competitors, the survivors may flourish and become opportunistic pathogens. An example that sometimes occurs is overgrowth by the yeastlike fungus *Candida albicans*, which is not sensitive to bacterial antibiotics. This overgrowth is called a **superinfection**, a term that is also applied to growth of a target

# Major Action Modes of Antibacterial Drugs

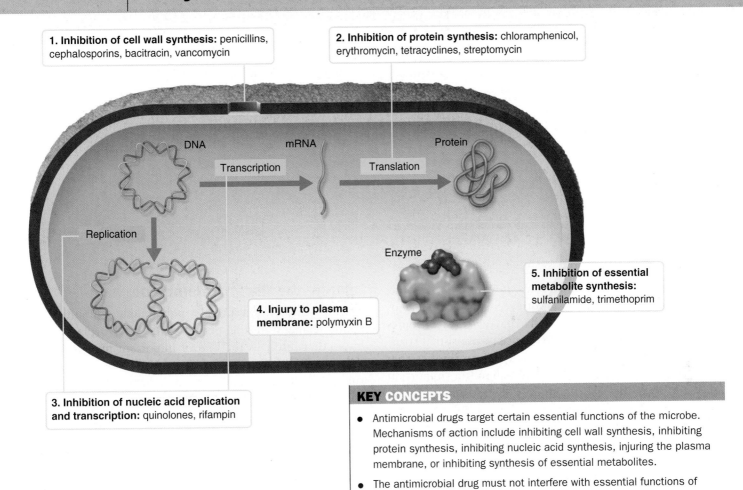

**1. Inhibition of cell wall synthesis:** penicillins, cephalosporins, bacitracin, vancomycin

**2. Inhibition of protein synthesis:** chloramphenicol, erythromycin, tetracyclines, streptomycin

DNA

mRNA

Protein

Transcription

Translation

Replication

Enzyme

**5. Inhibition of essential metabolite synthesis:** sulfanilamide, trimethoprim

**4. Injury to plasma membrane:** polymyxin B

**3. Inhibition of nucleic acid replication and transcription:** quinolones, rifampin

## KEY CONCEPTS

- Antimicrobial drugs target certain essential functions of the microbe. Mechanisms of action include inhibiting cell wall synthesis, inhibiting protein synthesis, inhibiting nucleic acid synthesis, injuring the plasma membrane, or inhibiting synthesis of essential metabolites.

- The antimicrobial drug must not interfere with essential functions of the microbe's host.

---

pathogen that has developed resistance to the antibiotic. In this situation, such an antibiotic-resistant strain replaces the original sensitive strain, and the infection continues.

### CHECK YOUR UNDERSTANDING

- ✔ **20-3** Identify at least one reason why it's so difficult to target a pathogenic virus without damaging the host's cells.

- ✔ **20-4** Why are antibiotics with a very broad spectrum of activity not as useful as one might first think?

## The Action of Antimicrobial Drugs

### LEARNING OBJECTIVE

**20-5** Identify five modes of action of antimicrobial drugs.

Antimicrobial drugs are either **bactericidal** (they kill microbes directly) or **bacteriostatic** (they prevent microbes from growing). In bacteriostasis, the host's own defenses, such as phagocytosis and antibody production, usually destroy the microorganisms. The major modes of action are summarized in **Figure 20.2**.

## Inhibiting Cell Wall Synthesis

Penicillin, the first true antibiotic to be discovered and used if one does not also consider the sulfa drugs, is an example of an inhibitor of cell wall synthesis.

The cell wall of a bacterium consists of a macromolecular network called *peptidoglycan*. Recall from Chapter 4 that peptidoglycan is found only in bacterial cell walls. Penicillin and certain other antibiotics prevent the synthesis of intact peptidoglycan; consequently, the cell wall is greatly weakened, and the cell undergoes lysis (**Figure 20.3**). Because penicillin targets the synthesis process, only actively growing cells are affected by these antibiotics—and, because human cells do not have peptidoglycan cell walls, penicillin has very little toxicity for host cells.

## Inhibiting Protein Synthesis

Because protein synthesis is a common feature of all cells, whether prokaryotic or eukaryotic, it would seem an unlikely target for selective toxicity. One notable difference

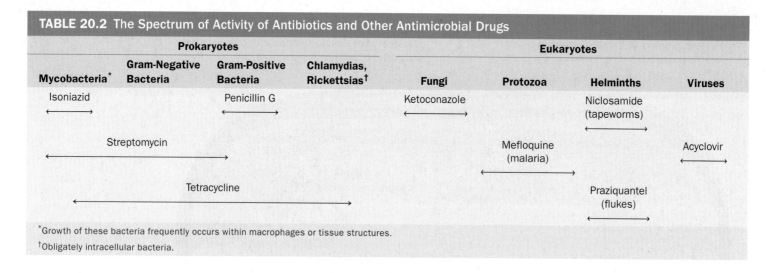

**TABLE 20.2** The Spectrum of Activity of Antibiotics and Other Antimicrobial Drugs

| Prokaryotes | | | | Eukaryotes | | | |
|---|---|---|---|---|---|---|---|
| Mycobacteria[*] | Gram-Negative Bacteria | Gram-Positive Bacteria | Chlamydias, Rickettsias[†] | Fungi | Protozoa | Helminths | Viruses |
| Isoniazid | | Penicillin G | | Ketoconazole | | Niclosamide (tapeworms) | |
| | Streptomycin | | | | Mefloquine (malaria) | | Acyclovir |
| | Tetracycline | | | | | Praziquantel (flukes) | |

[*]Growth of these bacteria frequently occurs within macrophages or tissue structures.
[†]Obligately intracellular bacteria.

between prokaryotes and eukaryotes, however, is the structure of their ribosomes. Eukaryotic cells have 80S ribosomes, whereas prokaryotic cells have 70S ribosomes (Chapter 4, pages 90–91). The difference in ribosomal structure accounts for the selective toxicity of antibiotics that affect protein synthesis. However, mitochondria (important eukaryotic organelles) also contain 70S ribosomes similar to those of bacteria. Antibiotics targeting the 70S ribosomes can therefore have adverse effects on the cells of the host. Among the antibiotics that interfere with protein synthesis are chloramphenicol, erythromycin, streptomycin, and the tetracyclines (**Figure 20.4**).

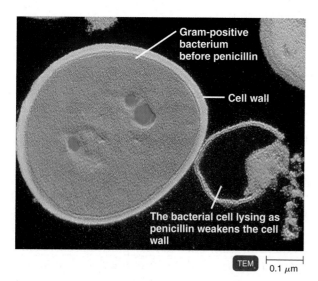

**Figure 20.3 The inhibition of bacterial cell wall synthesis by penicillin.**

Q Why don't penicillins affect human cells?

## Injuring the Plasma Membrane

Certain antibiotics, especially polypeptide antibiotics, bring about changes in the permeability of the plasma membrane that result in the loss of important metabolites from the microbial cell. Some of the polypeptide antibiotics disrupt both the inner and outer membranes of gram-negative bacteria.

Ionophores are antibiotics produced by several soil bacteria and fungi. They allow uncontrolled movement of cations across the plasma membrane. Ionophores are not used in human medicine. They are used in cattle feed because they alter microbiota within the animal's rumen, which improves digestion and promotes cattle growth.

Some antifungal drugs are effective against a considerable range of fungal diseases. These drugs combine with sterols in the fungal plasma membrane to disrupt the membrane (**Figure 20.5**). Because bacterial plasma membranes generally lack sterols, these antibiotics do not act on bacteria.

## Inhibiting Nucleic Acid Synthesis

A number of antibiotics interfere with the processes of DNA replication and transcription in microorganisms. These drugs block bacterial topoisomerase or RNA polymerase (see Table 8.1 on page 210).

## Inhibiting the Synthesis of Essential Metabolites

A particular enzymatic activity of a microorganism can be *competitively inhibited* by a substance (*antimetabolite*) that closely

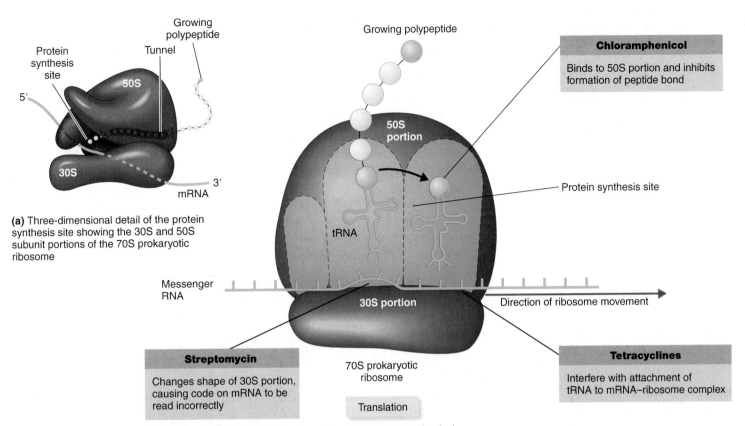

(a) Three-dimensional detail of the protein synthesis site showing the 30S and 50S subunit portions of the 70S prokaryotic ribosome

**Chloramphenicol**

Binds to 50S portion and inhibits formation of peptide bond

Protein synthesis site

**Streptomycin**

Changes shape of 30S portion, causing code on mRNA to be read incorrectly

**Tetracyclines**

Interfere with attachment of tRNA to mRNA–ribosome complex

Translation

(b) Diagram indicating the different points at which chloramphenicol, the tetracyclines, and streptomycin exert their activities

**Figure 20.4 The inhibition of protein synthesis by antibiotics. (a)** The inset shows how the 70S prokaryotic ribosome is assembled from two subunits, 30S and 50S. Note how the growing peptide chain passes through a tunnel in the 50S subunit from the site of protein synthesis. **(b)** The diagram shows the different points at which chloramphenicol, the tetracyclines, and streptomycin exert their activities.

**Q** Why do antibiotics that inhibit protein synthesis affect bacteria and not human cells?

resembles the normal substrate for the enzyme (see Figure 5.7, page 116). An example of competitive inhibition is the relationship between the antimetabolite sulfanilamide (a sulfa drug) and **para-aminobenzoic acid (PABA)**. In many microorganisms, PABA is the substrate for an enzymatic reaction leading to the synthesis of folic acid, a vitamin that functions as a coenzyme for the synthesis of the purine and pyrimidine bases of nucleic acids and many amino acids.

Play Chemotherapeutic Agents: *Modes of Action*
@MasteringMicrobiology

**CHECK YOUR UNDERSTANDING**

✔ **20-5** What cellular function is inhibited by tetracyclines?

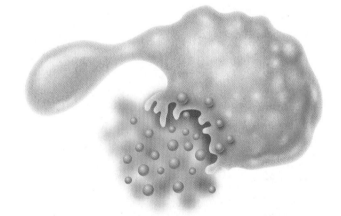

**Figure 20.5 Injury to the plasma membrane of a yeast cell caused by an antifungal drug.** The cell releases its cytoplasmic contents as the plasma membrane is disrupted by the antifungal drug miconazole.

**Q** Many antifungal drugs combine with sterols in the plasma membrane. Why don't they combine with sterols in human cell membranes?

# Common Antimicrobial Drugs

## LEARNING OBJECTIVES

20-6 Explain why drugs in this section are bacteria-specific.

20-7 List the advantages of each of the following over penicillin: semisynthetic penicillins, cephalosporins, and vancomycin.

20-8 Explain why isoniazid and ethambutol are antimycobacterial agents.

20-9 Describe how each of the following inhibits protein synthesis: aminoglycosides, tetracyclines, chloramphenicol, macrolides.

20-10 Compare polymyxin B, bacitracin, and neomycin in their modes of action.

20-11 Describe how rifamycins and quinolones kill bacteria.

20-12 Describe how sulfa drugs inhibit microbial growth.

20-13 Explain modes of action of current antifungal drugs.

20-14 Explain modes of action of current antiviral drugs.

20-15 Explain modes of action of current antiprotozoan and antihelminthic drugs.

Table 20.3 summarizes the commonly used antibacterial drugs. Table 20.4 summarizes the cephalosporins, one group of antibacterial drugs. Table 20.5 summarizes the commonly used drugs effective against fungi, viruses, protozoans, and helminths.

## Antibacterial Antibiotics: Inhibitors of Cell Wall Synthesis

For antibiotics to function as a "magic bullet," they usually must target microbial structures or functions that are not shared with mammalian structures or functions. The eukaryotic mammalian cell usually does not have a cell wall; instead, it has only a plasma membrane (see Chapter 4). For this reason, the microbial cell wall is an attractive target for the action of antibiotics.

### Penicillin

The term **penicillin** refers to a group of over 50 chemically related antibiotics (**Figure 20.6**). All penicillins have a common core structure containing a β-lactam ring called the nucleus. Types are differentiated by the chemical side chains attached to their nuclei. Penicillins prevent the cross-linking of the peptidoglycans, which interferes with the final stages of the synthesis of the cell walls, primarily of gram-positive bacteria (see Figure 4.13a, page 82). Penicillins can be produced either naturally or semisynthetically.

**Natural Penicillins** Penicillins extracted from cultures of *Penicillium* fungi are the so-called **natural penicillins** (Figure 20.6a). The prototype compound of all the penicillins

| Drugs by Mode of Action | Comments |
|---|---|
| **TABLE 20.3 Antibacterial Drugs** | |
| **INHIBITORS OF CELL WALL SYNTHESIS** | |
| **Natural Penicillins** | |
| Penicillin G | Against gram-positive bacteria, requires injection |
| Penicillin V | Against gram-positive bacteria, oral administration |
| **Semisynthetic Penicillins** | |
| Oxacillin | Narrow spectrum, resistant to penicillinase |
| Ampicillin | Broad spectrum |
| Amoxicillin | Broad spectrum; combined with inhibitor of penicillinase |
| **Carbapenems** | |
| Imipenem | Very broad spectrum |
| **Monobactams** | |
| Aztreonam | Effective against gram-negative bacteria, including *Pseudomonas* spp. |
| **Cephalosporins** | |
| Cephalothin | First-generation cephalosporin; activity similar to penicillin; requires injection |
| Ceftaroline | Fifth-generation cephalosporin; activity against MRSA |
| **Polypeptide Antibiotics** | |
| Bacitracin | Against gram-positive bacteria; topical application |
| Vancomycin | A glycopeptide type; penicillinase resistant; against gram-positive bacteria |
| **Antimycobacterial Antibiotics** | |
| Isoniazid | Inhibits synthesis of mycolic acid component of cell wall of *Mycobacterium* spp. |
| Ethambutol | Inhibits incorporation of mycolic acid into cell wall of *Mycobacterium* spp. |

**TABLE 20.3** *(continued)*

| Drugs by Mode of Action | Comments |
|---|---|
| **INHIBITORS OF PROTEIN SYNTHESIS** | |
| Nitrofurantoin | Urinary bladder infections |
| Chloramphenicol | Broad spectrum, potentially toxic |
| **Aminoglycosides** | |
| Streptomycin | Broad spectrum, including mycobacteria |
| Neomycin | Topical use, broad spectrum |
| Gentamicin | Broad spectrum, including *Pseudomonas* spp. |
| **Tetracyclines** | |
| Tetracycline, oxytetracycline, chlortetracycline | Broad spectrum, including chlamydias and rickettsias |
| **Glycylcyclines** | |
| Tigecycline | Broad spectrum, especially MRSA and *Acinetobacter* |
| **Macrolides** | |
| Erythromycin | Alternative to penicillin |
| Azithromycin, clarithromycin | Semisynthetic; broader spectrum and better tissue penetration than erythromycin |
| Telithromycin | New generation of semisynthetic macrolides; used to cope with resistance to other macrolides |
| **Streptogramins** | |
| Quinupristin and dalfopristin (Synercid®) | Alternative for treating vancomycin-resistant gram-positive bacteria |
| **Oxazolidinones** | |
| Linezolid | Useful primarily against penicillin-resistant gram-positive bacteria |
| **Pleuromutilins** | |
| Mutilin, retpamulin | Inhibit gram-positive bacteria |
| **INJURY TO THE PLASMA MEMBRANE** | |
| **Lipopeptides** | |
| Daptomycin | To treat MRSA infections |
| Polymyxin B | Topical use, gram-negative bacteria, including *Pseudomonas* spp. |
| **NUCLEIC ACID SYNTHESIS INHIBITORS** | |
| **Rifamycins** | |
| Rifampin | Inhibits synthesis of mRNA; treatment of tuberculosis |
| **Quinolones and Fluoroquinolones** | |
| Nalidixic acid, ciprofloxacin | Inhibit DNA synthesis; broad spectrum; urinary tract infections |
| **COMPETITIVE INHIBITORS OF THE SYNTHESIS OF ESSENTIAL METABOLITES** | |
| **Sulfonamides** | |
| Sulfamethoxazole-trimethoprim | Broad spectrum; combination is widely used |

**TABLE 20.4** Differential Grouping of Cephalosporins

| Generation | Description | Example |
|---|---|---|
| First | Primarily against gram-positive bacteria | Cephalothin |
| Second | More extended gram-negative spectrum | Cefamandole (IV) Cefaclor (oral) |
| Third | Most active against gram-negative bacteria, including *P. aeruginosa* | Ceftazidime |
| Fourth | Require injections; most extended spectrum of activity | Cefepime |
| Fifth | Effective against gram-negative bacteria and MRSA | Ceftaroline |

**TABLE 20.5** Antifungal, Antiviral, Antiprotozoan, and Antihelminthic Drugs

| | Comments |
|---|---|
| **ANTIFUNGAL DRUGS** | |
| **Agents affecting fungal sterols (plasma membrane)** | |
| **Polyenes** | |
| Nystatin | Thrush |
| Amphotericin B | Systemic fungal infections |
| **Azoles** | |
| Clotrimazole, miconazole | Topical use |
| Ketoconazole | Systemic fungal infections |
| **Allylamines** | |
| Terbinafine, naftifine | Treatment of diseases resistant to azoles |
| **Agents affecting fungal cell walls** | |
| **Echinocandins** | |
| Caspofungin | Systemic fungal infections |
| **Agents inhibiting nucleic acids** | |
| Flucytosine | *Candida, Cryptococcus* |
| **Other antifungal drugs** | |
| Griseofulvin | Inhibits mitotic microtubules; fungal infections of the skin |
| Tolnaftate | Athlete's foot |
| Pentamidine | *Pneumocystis* pneumonia |
| **ANTIVIRAL DRUGS** | |
| **Entry and Fusion Inhibitors** | |
| Maraviroc | Binds CCR5 of HIV |
| **Uncoating, Genome Integration, and Nucleic Acid Synthesis Inhibitors** | |
| Amantadine, zimantadine | Widespread resistance in *Influenzavirus* |
| Zidovudine (AZT), tenofovir, emtricitabine | Inhibit HIV reverse transcriptase |
| Acyclovir, ganciclovir, ribavirin, lamivudine | Used primarily against herpesviruses |
| Adefovir dipivoxil | Treatment of lamivudine-resistant infections of hepatitis B virus |
| Cidofovir | Cytomegalovirus infections |
| **Assembly and Exit Inhibitors** | |
| Saquinavir | HIV protease inhibitors |
| Boceprevir | Hepatitis C virus protease inhibitor |
| Zanamivir, oseltamivir, peramivir | Inhibits neuraminidase of *Influenzavirus* |
| **Interferons** | |
| Alpha interferon | Hepatitis B, D, C |
| **ANTIPROTOZOAN DRUGS** | |
| Chloroquine | Malaria; effective against red blood cell stage only |
| Artemisinin | Malaria |
| Diiodohydroxyquin | Amebic infections; amebicidal |
| Metronidazole, tinidazole | Giardiasis, amebiasis, trichomoniasis |
| Miltefosine | Amebic encephalitis |
| **ANTIHELMINTHIC DRUGS** | |
| Niclosamide | Prevents ATP generation in mitochondria, tapeworm infections |
| Praziquantel | Alters plasma membrane, kills flatworms |
| Pyrantel pamoate | Neuromuscular block; kills roundworms |
| Mebendazole, albendazole | Inhibit absorption of nutrients, intestinal roundworms |
| Ivermectin | Paralyzes intestinal roundworms |

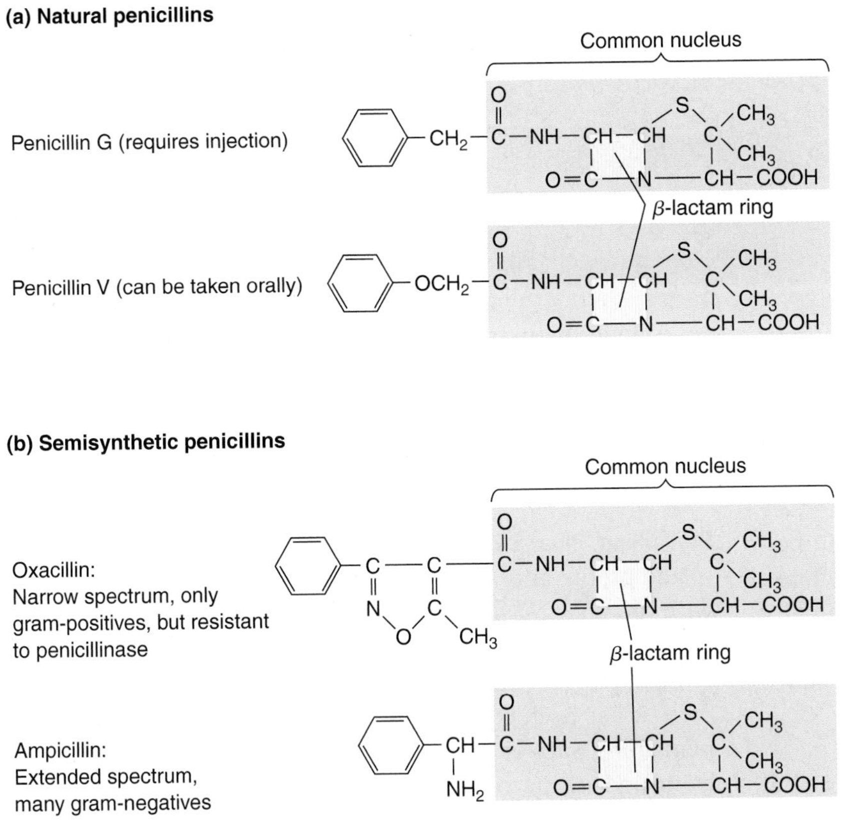

**(a) Natural penicillins**

Common nucleus

Penicillin G (requires injection)

β-lactam ring

Penicillin V (can be taken orally)

**(b) Semisynthetic penicillins**

Common nucleus

Oxacillin:
Narrow spectrum, only
gram-positives, but resistant
to penicillinase

β-lactam ring

Ampicillin:
Extended spectrum,
many gram-negatives

**Figure 20.6 The structure of penicillins, antibacterial antibiotics.** The portion that all penicillins have in common—which contains the (yellow) β-lactam ring—is shaded in purple. The unshaded portions represent the side chains that distinguish one penicillin from another.

**Q** What does *semisynthetic* mean?

is *penicillin G*. It has a narrow but useful spectrum of activity and is often the drug of choice against most staphylococci, streptococci, and several spirochetes. When injected intramuscularly, penicillin G is rapidly excreted from the body in 3 to 6 hours (**Figure 20.7**). When the drug is taken orally, the acidity of the digestive fluids in the stomach diminishes its concentration. *Procaine penicillin*, a combination of the drugs procaine and penicillin G, is retained at detectable concentrations for up to 24 hours; the concentration peaks at about 4 hours. Still longer retention times can be achieved with *benzathine penicillin*, a combination of benzathine and penicillin G. Although retention times of as long as 4 months can be obtained, the concentration of the drug is so low that the organisms must be very sensitive to it. Penicillin V, which is stable in stomach acids and can be taken orally, and penicillin G are the natural penicillins most often used.

Natural penicillins have some disadvantages. Chief among them are their narrow spectrum of activity and their susceptibility to penicillinases. *Penicillinases* are enzymes produced by many bacteria, most notably *Staphylococcus* species, that cleave the β-lactam ring of the penicillin molecule (**Figure 20.8**). Because of this characteristic, penicillinases are sometimes called β-*lactamases*.

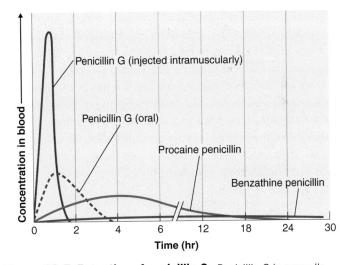

**Figure 20.7 Retention of penicillin G.** Penicillin G is normally injected (solid red line); when administered by this route, the drug is present in high concentrations in the blood but disappears quickly. Taken orally (dotted red line), penicillin G is destroyed by stomach acids and is not very effective. It's possible to improve retention of penicillin G by combining it with compounds such as procaine and benzathine (blue and purple lines). However, the blood concentration reached is low, and the target bacterium must be extremely sensitive to the antibiotic.

**Q** How does a low concentration of penicillin G select for penicillin-resistant bacteria?

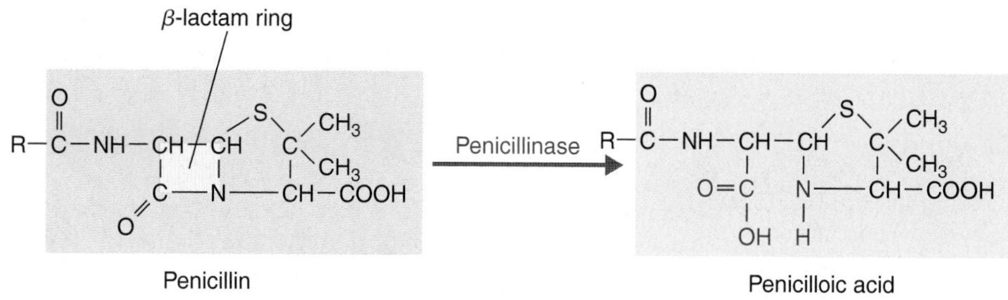

β-lactam ring

Penicillin

Penicilloic acid

**Figure 20.8 The effect of penicillinase on penicillins.** Bacterial production of this enzyme, which is shown breaking the β-lactam ring, is by far the most common form of resistance to penicillins. R is an abbreviation for the chemical side groups that differentiate similar or otherwise-identical compounds.

**Q** What is penicillinase?

**Semisynthetic Penicillins** A large number of **semisynthetic penicillins** have been developed in attempts to overcome the disadvantages of natural penicillins (Figure 20.6b). Scientists develop these penicillins in either of two ways. First, they can interrupt synthesis of the molecule by *Penicillium* and obtain only the common penicillin nucleus for use. Second, they can remove the side chains from the completed natural molecules and then chemically add other side chains that make them more resistant to penicillinase, or the scientists can give them an extended spectrum. Thus the term *semisynthetic*: part of the penicillin is produced by the mold, and part is added synthetically.

**Penicillinase-Resistant Penicillins** Resistance of staphylococcal infections to penicillin soon became a problem because of a plasmid-borne gene for β-lactamase. Antibiotics that were relatively resistant to this enzyme, such as the semisynthetic penicillin *methicillin*, were introduced, but resistance to them also soon appeared; thus the organisms were termed **methicillin-resistant *Staphylococcus aureus* (MRSA)**, usually pronounced *mersa* (see the Clinical Case box in Chapter 1, page 3). Resistance became so prevalent that methicillin has been discontinued in the United States. The term has come to be applied to strains that have developed resistance to a wide range of penicillins and cephalosporins. This includes other penicillinase-resistant antibiotics, such as *oxacillin*, and those combined with β-lactamase inhibitors. See the discussion of antibiotic resistance on pages 579–584.

**Extended-Spectrum Penicillins** To overcome the problem of the narrow spectrum of activity of natural penicillins, broader-spectrum semisynthetic penicillins have been developed. These new penicillins are effective against many gram-negative bacteria as well as gram-positive ones, although they are not resistant to penicillinases.

The first such penicillins were the aminopenicillins, such as *ampicillin* and *amoxicillin*.

When bacterial resistance to these became more common, the carboxypenicillins were developed. Members of this group, such as *carbenicillin* and *ticarcillin*, have even greater activity against gram-negative bacteria and have the special advantage of activity against *Pseudomonas aeruginosa*.

Among the more recent additions to the penicillin family are the ureidopenicillins, such as *mezlocillin* and *azlocillin*. These broader-spectrum penicillins are modifications of the structure of ampicillin. The search for even more effective modifications of penicillin continues.

**Penicillins Plus β-Lactamase Inhibitors** A different approach to the proliferation of penicillinase is to combine penicillins with *potassium clavulanate (clavulanic acid)*, a product of a streptomycete. Potassium clavulanate is a noncompetitive inhibitor of penicillinase with essentially no antimicrobial activity of its own. It has been combined with some new broader-spectrum penicillins, such as *amoxicillin*. The combination is best known by its trade name, Augmentin®.

## Carbapenems

The **carbapenems** are a class of β-lactam antibiotics that substitute a carbon atom for a sulfur atom and add a double bond to the penicillin nucleus. These antibiotics inhibit cell wall synthesis and have an extremely broad spectrum of activity. Representative of this group is a combination of *imipenem* and *cilastatin*. The cilastatin has no antimicrobial activity but prevents the imipenem from being degraded in the kidneys. Tests have demonstrated that imipinem-cilastatin is active against 98% of all organisms isolated from hospital patients. One of the few antibiotics introduced in recent years (2007) is *doripenem*, a carbapenem. It's especially useful against *Pseudomonas aeruginosa* infections.

## Monobactams

Another method of avoiding the effects of penicillinase is demonstrated by *aztreonam*, which is the first member of a new class of antibiotics. It's a synthetic antibiotic that has only a single ring rather than the conventional β-lactam double ring, and it's therefore known as a **monobactam**. Aztreonam's spectrum of activity is remarkable for a penicillin-related compound—this antibiotic, which has unusually low toxicity, affects only certain gram-negative bacteria, including pseudomonads and *E. coli*.

## Cephalosporins

In structure, the nuclei of **cephalosporins** resemble those of penicillin (**Figure 20.9**). Cephalosporins inhibit cell

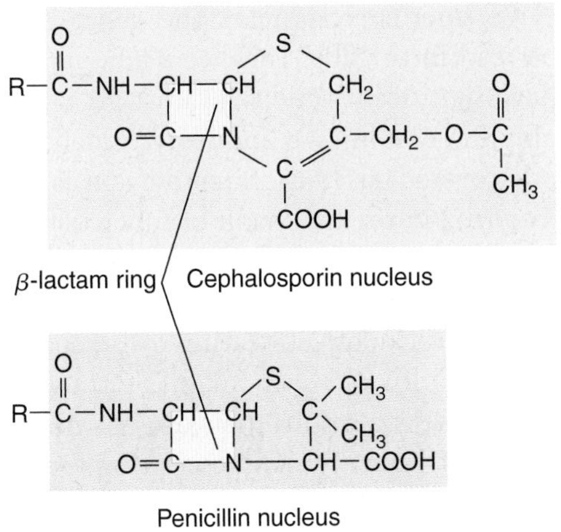

**Figure 20.9 The nuclear structures of cephalosporin and penicillin compared.**

**Q** Would a β-lactamase effective against penicillin G be likely to affect cephalosporins?

wall synthesis in essentially the same way as do penicillins. They are more widely used than any other β-lactam antibiotics. Their β-lactam ring differs slightly from that of penicillin, but bacteria have developed β-lactamases that inactivate them.

Cephalosporins are most commonly grouped according to their generations, reflecting their continued development, as described in Table 20.4.

### Polypeptide Antibiotics

**Bacitracin** *Bacitracin* (the name is derived from its source, a *Bacillus* isolated from a wound on a girl named *Tracy*) is a polypeptide antibiotic effective primarily against gram-positive bacteria, such as staphylococci and streptococci. Bacitracin inhibits the synthesis of cell walls at an earlier stage than do penicillins and cephalosporins. It interferes with the synthesis of the linear strands of the peptidoglycans (see Figure 4.13a, page 82). Its use is restricted to topical application for superficial infections.

**Vancomycin** *Vancomycin* (optimistically named from the word *vanquish*) is one of a small group of glycopeptide antibiotics derived from a species of *Streptomyces* found in the jungles of Borneo. Originally, toxicity of vancomycin was a serious problem, but improved purification procedures in its manufacture have largely corrected this. Although it has a very narrow spectrum of activity, which is based on inhibition of cell wall synthesis, vancomycin has been extremely important in addressing the problem of MRSA (see page 417). Vancomycin has been considered the last line of antibiotic defense for treatment of *Staphylococcus aureus* infections that are resistant to other antibiotics. The widespread use of vancomycin to treat

MRSA has led to the appearance of **vancomycin-resistant enterococci (VRE)**. These are opportunistic, gram-positive pathogens that are particularly troublesome in hospital settings. (See the discussion of healthcare-associated infections in Chapter 14, pages 408–411, and the Clinical Focus box on page 197.) This appearance of vancomycin-resistant pathogens, leaving few effective alternatives, is considered a medical emergency. Telavancin, a semisynthetic derivative of vancomycin, has been introduced and approved for a limited range of uses.

**Teixobactin** *Teixobactin*, discovered in 2015, represents a new class of antibiotics called *acyldepsipeptides*. It inhibits cell wall synthesis in gram-positive bacteria and mycobacteria. It is produced by a soil bacterium, *Eleftheria terrae* (ē-lef-THER-ē-ah TER-rī), that was cultured using a device called an iChip to grow previously unculturable bacteria (see Chapter 28, page 820). Teixobactin is not toxic to mammalian cells, but clinical trials on teixobactin have not yet begun.

### Antimycobacterial Antibiotics

The cell wall of members of the genus *Mycobacterium* differs from the cell wall of most other bacteria. It incorporates mycolic acids that are a factor in their staining properties, causing them to stain as acid-fast (see page 66). The genus includes important pathogens, such as those that cause leprosy and tuberculosis.

**Isoniazid (INH)** is a very effective synthetic antimicrobial drug against *Mycobacterium tuberculosis*. The primary effect of INH is to inhibit the synthesis of mycolic acids, which are components of cell walls only of the mycobacteria. It has little effect on nonmycobacteria. When used to treat tuberculosis, INH is usually administered simultaneously with other drugs, such as rifampin or ethambutol. This minimizes development of drug resistance. Because the tubercle bacillus is usually found only within macrophages or walled off in tissue, any antitubercular drug must be able to penetrate into such sites.

**Ethambutol** is effective only against mycobacteria. The drug inhibits incorporation of mycolic acid into the cell wall, making the cell wall weaker. It's a comparatively weak antitubercular drug; its principal use is as the secondary drug to avoid resistance problems.

#### CHECK YOUR UNDERSTANDING

✔ **20-6** One of the most successful groups of antibiotics targets the synthesis of bacterial cell walls; why does the antibiotic not affect the mammalian cell?

✔ **20-7** What phenomenon prompted the development of the first semisynthetic antibiotics, such as methicillin?

✔ **20-8** What genus of bacteria has mycolic acids in the cell wall?

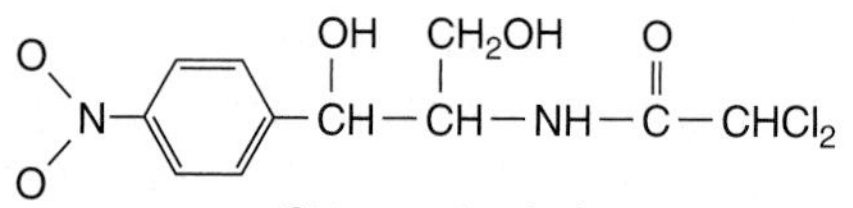

**Figure 20.10 The structure of the antibacterial antibiotic chloramphenicol.** Notice the simple structure, which makes synthesizing this drug less expensive than isolating it from *Streptomyces*.

**Q** How does the binding of chloramphenicol to the 50S portion of the ribosomes affect a cell?

## Inhibitors of Protein Synthesis

### Nitrofurantoin

*Nitrofurantoin* is a synthetic drug introduced in the 1950s. It is used to treat urinary bladder infections because it concentrates in urine as the kidneys remove it from the blood. Nitrofurantoin is converted by bacterial nitrate reductases to intermediates that attack bacterial ribosomal proteins, thus inhibiting protein synthesis and a variety of bacterial enzymes. Resistance to nitrofurantoin is not common because of the variety of target sites of the drug.

### Chloramphenicol

*Chloramphenicol* inhibits the formation of peptide bonds in the growing polypeptide chain by reacting with the 50S portion of the 70S prokaryotic ribosome. Because of its simple structure (**Figure 20.10**), it is less expensive for the pharmaceutical industry to synthesize it chemically than to isolate it from *Streptomyces*. It is relatively inexpensive and has a broad spectrum, so chloramphenicol is often used where low cost is essential. Its small molecular size promotes diffusion into areas of the body that are inaccessible to many other drugs. However, chloramphenicol has serious adverse effects that include suppression of bone marrow activity. This affects the formation of blood cells, resulting in a severe, often fatal condition called *aplastic anemia*. Physicians are advised not to use the drug for trivial conditions or if suitable alternatives are available.

Other antibiotics that inhibit protein synthesis by binding at the same ribosomal site as chloramphenicol are *clindamycin* and *metronidazole* (see page 577). These three drugs are structurally unrelated, but all have potent activity against anaerobes. Clindamycin is often implicated in *Clostridium difficile*–associated diarrhea (see page 738). Its effectiveness against anaerobes has led to its use in the treatment of acne.

### Aminoglycosides

**Aminoglycosides** are a group of antibiotics in which amino sugars are linked by glycoside bonds. Aminoglycoside antibiotics interfere with the initial steps of protein synthesis by changing the shape of the 30S portion of the 70S prokaryotic ribosome. This interference causes the genetic code of the mRNA to be read incorrectly. They were among the first antibiotics to have significant activity against gram-negative bacteria. Probably the best-known aminoglycoside is *streptomycin*, which was discovered in 1944. Streptomycin is still used as an alternative drug in the treatment of tuberculosis, but rapid development of resistance and serious toxic effects have diminished its usefulness.

Aminoglycosides can affect hearing by causing permanent damage to the auditory nerve, and damage to the kidneys has also been reported. As a result, their use has been declining. *Neomycin* is present in many nonprescription topical preparations. *Gentamicin* (spelled with an "i" to reflect its source, the filamentous bacterium *Micromonospora*) is especially useful against *Pseudomonas* infections. Pseudomonads are a major problem for persons suffering from cystic fibrosis. The aminoglycoside *tobramycin* is administered in an aerosol to help control infections that occur in patients with cystic fibrosis.

### Tetracyclines

**Tetracyclines** are a group of closely related broad-spectrum antibiotics produced by *Streptomyces* spp. The tetracyclines interfere with the attachment of the tRNA carrying the amino acids to the ribosome at the 30S portion of the 70S ribosome, preventing the addition of amino acids to the growing polypeptide chain. They do not interfere with mammalian ribosomes, but can affect mitochondrial ribosomes. Tetracyclines are effective against gram-positive and gram-negative bacteria. Tetracyclines not only are effective against gram-positive and gram-negative bacteria but also penetrate body tissues well and are especially valuable against the intracellular rickettsias and chlamydias. Three of the more commonly used tetracyclines are *oxytetracycline*, *chlortetracycline*, and tetracycline itself (**Figure 20.11**).

Some semisynthetic tetracyclines, such as *doxycycline* and *minocycline*, are available. They have the advantage of longer retention in the body.

**CLINICAL CASE**

D r. Singh asks the eye bank for more information about the donor. She finds out that the cornea donor was a previously healthy 30-year-old victim of a motorcycle crash who was on ventilator support for 4 days before his death. The donor's corneas were harvested 3 days before the transplantations took place and were stored in a buffered medium containing 100 μg/mL gentamicin at 4°C.

**How would you determine susceptibility of this *P. aeruginosa* to gentamicin?**

559    **570**    579    582    586

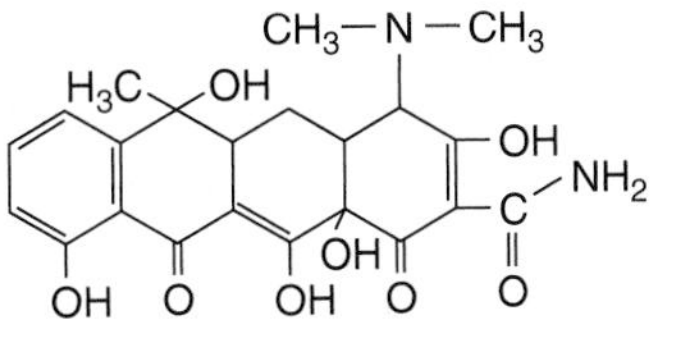

Tetracycline

**Figure 20.11 The structure of the antibacterial antibiotic tetracycline.** Other tetracycline-type antibiotics share the four-cyclicring structure of tetracycline and closely resemble it.

**Q** How do tetracyclines affect bacteria?

Tetracyclines are used to treat many urinary tract infections, mycoplasmal pneumonia, and chlamydial and rickettsial infections. They are also frequently used as alternative drugs for such diseases as syphilis and gonorrhea. Because of their broad spectrum, tetracyclines often suppress the normal intestinal microbiota, causing gastrointestinal upsets and often leading to superinfections, particularly by the fungus *Candida albicans*. They are not advised for children, who might experience a brownish discoloration of the teeth, or for pregnant women, in whom they might cause liver damage.

## Glycylcyclines

The **glycylcyclines** are a newer class of antibiotics discovered in the 1990s. They are structurally similar to the tetracyclines. The best-known example is *tigecycline*. This is a broad-spectrum, bacteriostatic antibiotic that binds to the 30S ribosomal subunit, blocking protein synthesis. An important advantage is that it inhibits the effects of rapid efflux, an important mechanism for bacterial antibiotic resistance (see page 580). Among its disadvantages is that it must be administered by slow intravenous infusion. It is especially useful against MRSA and multidrug-resistant strains of *Acinetobacter baumanii*.

## Macrolides

**Macrolides** are a group of antibiotics named for the presence of a macrocyclic lactone ring. The best-known macrolide in clinical use is *erythromycin* (**Figure 20.12**). Its mode of action is the inhibition of protein synthesis, apparently by blocking the tunnel shown in Figure 20.4a. However, erythromycin is not able to penetrate the cell walls of most gram-negative bacilli. Its spectrum of activity is therefore similar to that of penicillin G, and it is a frequent alternative drug to penicillin. Because it can be administered orally, an orange-flavored preparation of erythromycin is a frequent substitute for penicillin in treating streptococcal and staphylococcal infections in children. Erythromycin is also the drug of choice for the treatment of legionellosis, mycoplasmal pneumonia, and several other infections.

A member of a new class of macrocyclic antibiotics is *fidaxomicin*. It has a rather narrow spectrum of activity and is mainly used to treat *Clostridium difficile* and other clostridia infections. It is a frequent replacement for vancomycin. Other macrolides now available include *azithromycin* and *clarithromycin*. Compared to erythromycin, they have a broader antimicrobial spectrum and penetrate tissues better. This is especially important in the treatment of conditions caused by intracellular bacteria such as *Chlamydia*, a frequent cause of sexually transmitted infection.

A new generation of semisynthetic macrolides, the **ketolides**, is being developed to cope with increasing resistance to other macrolides. The prototype of this generation is *telithromycin*. However, its toxicity to the liver limits its use to treat acute respiratory infections.

## Streptogramins

We mentioned previously that the appearance of vancomycin-resistant pathogens constitutes a serious medical problem. One answer may be a unique group of antibiotics, the **streptogramins**. Synercid® is a combination of two cyclic peptides, *dalfopristin* and *quinupristin*, which are distantly related to the macrolides. They block protein synthesis by attaching to the 50S portion of the ribosome, as do other antibiotics such as chloramphenicol. Synercid®, however, acts at uniquely different points on the ribosome. Dalfopristin blocks an early step in protein synthesis, and quinupristin blocks a later step. The combination causes incomplete peptide chains to be released and is synergistic in its action (see page 583). Synercid® is effective against a broad range of gram-positive bacteria that are resistant to other antibiotics. This makes Synercid® especially valuable, even though it's expensive and has a high incidence of adverse side effects.

## Oxazolidinones

The **oxazolidinones** are another class of antibiotics developed in response to vancomycin resistance. When the FDA approved

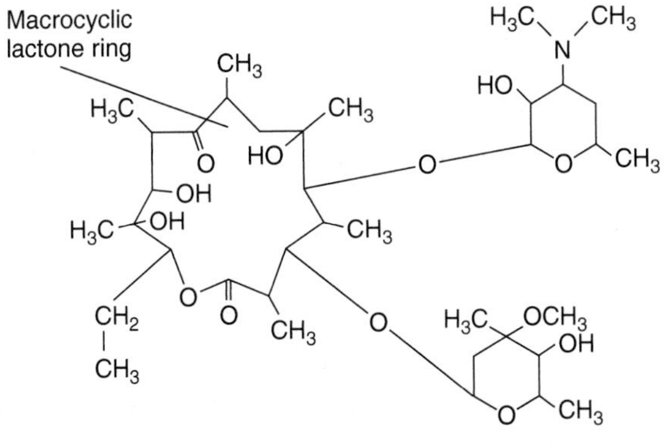

Erythromycin

**Figure 20.12 The structure of the antibacterial antibiotic erythromycin, a representative macrolide.** All macrolides have the macrocyclic lactone ring shown here.

**Q** How do macrolides affect bacteria?

this class of antibiotic in 2001, it represented the first new class of antibiotics to come to market in 25 years. Like several other antibiotics that inhibit protein synthesis, oxazolidinone antibiotics act on the ribosome (see Figure 20.4). However, they are unique in their target, binding to the 50S ribosomal subunit close to the point where it interfaces with the 30S subunit. These drugs are totally synthetic, which may make resistance slower to develop. One member of this antibiotic group is *linezolid*, used mainly to combat MRSA.

### Pleuromutilins

**Pleuromutilin** derivatives and the oxazolidinones represent two of the classes of antibiotics developed since 2000 (also see the following discussion of lipopeptides). They have a unique mode of action that interferes with protein synthesis. The first antibiotic of this class approved for human use is *retapamulin*. It was, however, limited to only topical use. Pleuromutilins are effective against gram-positive bacteria. Originally a product of the *Pleurotis mutilus* mushroom, most are now semisynthetic derivatives.

> **CHECK YOUR UNDERSTANDING**
>
> ✔ 20-9 Why does erythromycin, a macrolide antibiotic, have activity limited largely to gram-positive bacteria even though its mode of action is similar to that of the broad-spectrum tetracyclines?

## Injury to Membranes

The synthesis of bacterial plasma membranes requires the synthesis of fatty acids as building blocks. Researchers in search of an attractive target for new antibiotics have focused their attention on this metabolic step, which is distinct from the fatty acid biosynthesis in humans. A weakness in this approach, however, is that many bacterial pathogens are able to take up preformed fatty acids from serum. In the soil environment from which the antibiotic-producing streptomycetes were isolated, fatty acids are not available. Examples of successful antimicrobials that target fatty acid synthesis are the tuberculosis drug *isoniazid* (page 569) and the antibacterial *triclosan* (page 190).

### Lipopeptides

A **lipopeptide** antibiotic effective only against gram-positive bacteria is *daptomycin*, produced by a streptomycete. Its use is approved for certain skin infections. Daptomycin binds to the bacterial membrane, and resistance is uncommon. This mechanism is distinctive enough that the antibiotic is often used for infections caused by bacteria resistant to numerous antibiotics.

*Polymyxin B* is a bactericidal antibiotic effective against gram-negative bacteria; it acts by binding to the outer membrane of the cell wall. It's primarily used in topical treatment of superficial infections, for which it's available in nonprescription antiseptic ointments.

*Polymyxin E* (colistin) is used to treat antibiotic-resistant ventilator-associated pneumonia caused by gram-negative bacteria. In 2015, colistin resistance was reported in China and Europe. And in 2016, colistin-resistant *E. coli* bacteria were isolated from two unrelated patients in the United States.

Both *bacitracin* and *polymyxin B* are available in nonprescription antiseptic ointments, in which they are usually combined with *neomycin*, a broad-spectrum aminoglycoside. In a rare exception to the rule, these antibiotics do not require a prescription.

Many of the antimicrobial peptides discussed on page 585 target the synthesis of the plasma membrane.

> **CHECK YOUR UNDERSTANDING**
>
> ✔ 20-10 Of the three drugs often found in over-the-counter antiseptic creams—polymyxin B, bacitracin, and neomycin—which has a mode of action most similar to that of penicillin?

## Nucleic Acid Synthesis Inhibitors

### Rifamycins

The best-known derivative of the **rifamycin** family of antibiotics is *rifampin*. These drugs are structurally related to the macrolides and inhibit the synthesis of mRNA. By far the most important use of rifampin is against mycobacteria in the treatment of tuberculosis and leprosy. A valuable characteristic of rifampin is its ability to penetrate tissues and reach therapeutic levels in cerebrospinal fluid and abscesses. This characteristic is probably an important factor in its antitubercular activity, because the tuberculosis pathogen is usually located inside tissues or macrophages. An unusual side effect of rifampin is the appearance of orange-red urine, feces, saliva, sweat, and even tears.

### Quinolones and Fluoroquinolones

In the early 1960s, the synthetic drug *nalidixic acid* was developed—the first of the **quinolone** group of antimicrobials. It exerted a unique bactericidal effect by selectively inhibiting an enzyme (DNA gyrase) needed for the replication of DNA. Although nalidixic acid found only limited use (its only application is for urinary tract infections), it led to the development in the 1980s of a prolific group of synthetic quinolones, the **fluoroquinolones**.

The fluoroquinolones are divided into groups, each of which has a progressively broader spectrum of activity. The earliest generation includes *ciprofloxacin* (Cipro®). A newer group of fluoroquinolones includes *gemifloxacin* and *moxifloxacin*. These antibiotics, with the exception of moxifloxacin, are used to treat urinary tract infections and certain types of pneumonia. As a group, the fluoroquinolones can cause ruptured tendons. Physicians are advised to use these drugs only when there is no other treatment option. Resistance to fluoroquinolones can develop rapidly, even during a course of treatment.

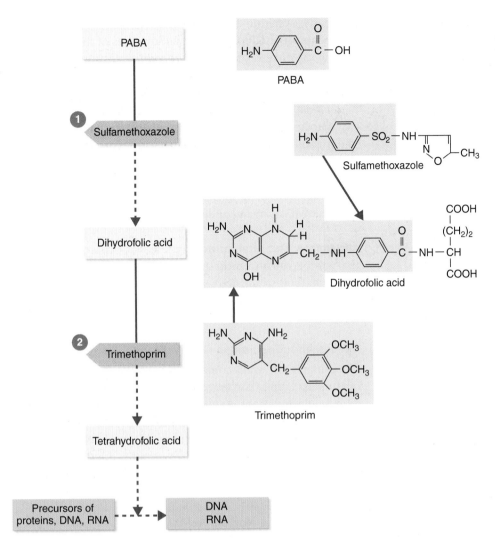

**Figure 20.13 Actions of the antibacterial synthetics sulfamethoxazole and trimethoprim.** SMZ-TMP works by inhibiting different stages in the synthesis of precursors of DNA, RNA, and proteins. Together the drugs are synergistic.

❶ Sulfamethoxazole, a sulfonamide that is a structural analog of PABA, competitively inhibits the synthesis of dihydrofolic acid from PABA.
❷ Trimethoprim, a structural analog of a portion of dihydrofolic acid, competitively inhibits the synthesis of tetrahydrofolic acid.

**Q** Define *synergism*.

---

**CHECK YOUR UNDERSTANDING**

✔ **20-11** What group of antibiotics interferes with the DNA-replicating enzyme DNA topoisomerase?

## Competitive Inhibition of Essential Metabolites

Blocking a cell's ability to synthesize essential metabolites is another mode of drug action.

### Sulfonamides

As noted earlier, **sulfonamides**, or **sulfa drugs**, were some of the first antimicrobial therapies created. Folic acid is an important coenzyme needed for synthesis of proteins, DNA, and RNA. Sulfa drugs are structurally similar to a folic acid precursor called *para-aminobenzoic acid* (PABA), allowing them to competitively bind with the enzyme meant for PABA and thereby block folic acid production. The drugs are bacteriostatic and do not harm human cells because we take up folic acid from our diet rather than synthesize it.

A combination of *sulfamethoxazole* and *trimethoprim* (SMZ-TMP) is widely used today. **Figure 20.13** shows its mode of action. The combined concentration requires only 10% of the amount of the drugs that would be needed if they were used separately—an example of drug synergism (page 583). The combination also has a broader spectrum of action and greatly reduces emergence of resistant strains.

Antibiotics have diminished the importance of sulfa drugs. However, they continue to be effective treatments for certain urinary tract infections, and the drug silver sulfadiazine is also used to control infections in burn patients.

---

**CHECK YOUR UNDERSTANDING**

✔ **20-12** Both humans and bacteria need PABA to make folic acid, so why do sulfa drugs adversely impact only bacterial cells?

## Antifungal Drugs

Eukaryotes, such as fungi, use the same mechanisms to synthesize proteins and nucleic acids as higher animals. Therefore

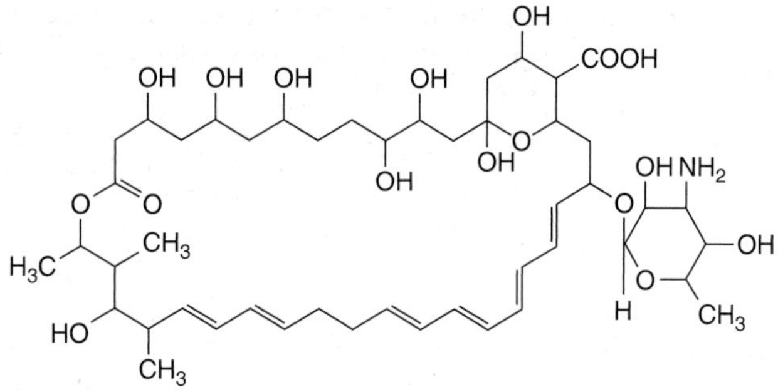

Nystatin

**Figure 20.14 The structure of the antifungal drug nystatin, representative of the polyenes.**

**Q** Why do polyenes injure fungal plasma membranes and not bacterial membranes?

it's more difficult to find a point of selective toxicity in eukaryotes than in prokaryotes. Moreover, fungal infections are becoming more frequent because of their role as opportunistic infections in immunosuppressed individuals, especially people with AIDS.

## Agents Affecting Fungal Sterols

Many antifungal drugs target the sterols in the plasma membrane. In fungal membranes, the principal sterol is ergosterol; in animal membranes, cholesterol. When the biosynthesis of ergosterol in a fungal membrane is interrupted, the membrane becomes excessively permeable, killing the cell. Inhibition of ergosterol biosynthesis is the basis for the selective toxicity of many antifungals, which include members of the polyene, azole, and allylamine groups.

**Polyenes** *Nystatin* is the most commonly used member of the antifungal **polyenes (Figure 20.14)**. It is used to treat thrush (oral *Candida* infections). Amphotericin B is the mainstay of clinical treatment for systemic fungal diseases such as histoplasmosis, coccidioidomycosis, and blastomycosis. The drug's toxicity, particularly to the kidneys, is a strongly limiting factor in these uses.

**Azoles** Some of the most widely used antifungal drugs are represented among the **azole** antibiotics. Before they made their appearance, the only drugs available for systemic fungal infections were amphotericin B and flucytocine (discussed later). The first azoles were **imidazoles**, such as *clotrimazole* and *miconazole* (**Figure 20.15**), which are now sold without a prescription for topical application for treatment of cutaneous mycoses, such as athlete's foot and vaginal yeast infections. An important addition to this group was *ketoconazole*, which has an unusually broad spectrum of activity among fungi. Ketoconazole, taken orally, proved to be an alternative to amphotericin B

for many systemic fungal infections. Ketoconazole topical ointments are used to treat dermatomycoses of the skin.

The use of ketoconazole for systemic infections diminished when the less toxic **triazole** antifungal antibiotics were introduced. The original drugs of this type were *fluconazole* and *itraconazole*. They are much more water soluble, making them easier to use and more effective against systemic infections. The triazole group was expanded with the introduction of *voriconazole*, which has become the new standard for treatment of *Aspergillus* infections in immunocompromised patients. The newest triazole drug to be approved is *posaconazole*, which is used to treat *Aspergillus* and *Candida* systemic fungal infections.

**Allylamines** The **allylamines** represent a class of antifungals that inhibit the biosynthesis of ergosterols in a manner that is functionally distinct. *Terbinafine* and *naftifine*, examples of this group, are frequently used when resistance to azole-type antifungals arises.

## Agents Affecting Fungal Cell Walls

The fungal cell wall contains compounds unique to these organisms. Other than ergosterol, a primary target for selective toxicity among these compounds is β-glucan. A new class of antifungal drugs is the **echinocandins**, which inhibit the biosynthesis of glucans, resulting in an incomplete cell wall and cell lysis. A member of the echinocandin group, *caspofungin*, is used to treat systemic *Aspergillus* infections and *Candida* infections.

## Agents Inhibiting Nucleic Acids

*Flucytosine*, an analog of the pyrimidine cytosine, interferes with the biosynthesis of RNA and therefore protein synthesis. The selective toxicity lies in the ability of the fungal cell to convert flucytocine into 5-fluorouracil, which is incorporated into RNA and eventually disrupts protein synthesis. Mammalian cells lack the enzyme to make this conversion of the drug. Flucytosine has a narrow spectrum of activity and is most effective against *Candida* and *Cryptococcus*. It is usually used with another antifungal because resistance develops quickly.

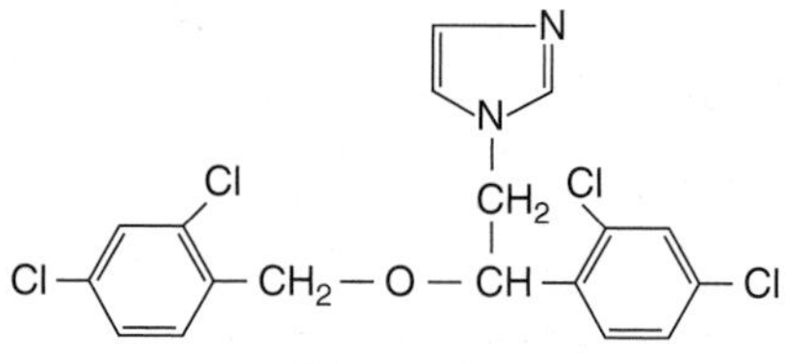

Miconazole

**Figure 20.15 The structure of the antifungal drug miconazole, representative of the imidazoles.**

**Q** How do azoles affect fungi?

### Other Antifungal Drugs

*Griseofulvin* is an antibiotic produced by a species of *Penicillium*. It has the interesting property of being active against superficial dermatophytic fungal infections of the hair (tinea capitis, or ringworm) and nails, even though its route of administration is oral. The drug apparently binds selectively to the keratin found in the skin, hair follicles, and nails. Its mode of action is primarily to block microtubule assembly, which interferes with mitosis and thereby inhibits fungal reproduction.

*Tolnaftate* is a common alternative to miconazole as a topical agent for the treatment of athlete's foot. Its mechanism of action is most likely inhibition of sterol synthesis. *Undecylenic acid* is a fatty acid that has antifungal activity against athlete's foot, although it isn't as effective as tolnaftate or the imidazoles.

*Pentamidine* is used in treating *Pneumocystis* pneumonia, a frequent complication of AIDS. It also is useful in treating several protozoan-caused tropical diseases. The drug's mode of action is unknown, but it appears to bind to DNA.

> **CHECK YOUR UNDERSTANDING**
>
> ✔ **20-13** What sterol in the cell membrane of fungi is the most common target for antifungal action?

## Antiviral Drugs

In developed parts of the world, it's estimated that at least 60% of infectious illnesses are caused by viruses, and about 15% by bacteria. Every year, at least 90% of the U.S. population suffers from a viral disease. Even so, compared to the number of antibiotics available for treating bacterial diseases, there are relatively few antiviral drugs. Many of the recently developed antiviral drugs are directed against HIV, the pathogen responsible for the pandemic of AIDS. Therefore, as a practical matter the discussion of antivirals is often separated into agents that are directed at HIV (see pages 552–553) and those for other infections, such as influenza, herpes, and hepatitis (see Table 20.5).

Because viruses replicate within the host's cells, very often using the genetic and metabolic mechanisms of the host's own cells, it's relatively difficult to target the virus without damaging the host's cellular machinery. Many of the antivirals in use today are analogs of components of viral DNA or RNA. However, as more becomes known about the reproduction of viruses, more targets suggest themselves for antiviral action.

### Entry and Fusion Inhibitors

Drugs that block the initial steps in viral infection—absorption and penetration—are **entry inhibitors**. Several entry inhibitors are being investigated for treating hepatitis B, hepatitis C, and influenza. Entry inhibitors approved to treat HIV target the receptors that HIV uses to bind to the cell before entry, such

as CCR5 (see Figure 19.14, page 545). The first of this class of drugs to target HIV infection step is *maraviroc*. Entry of HIV into the cell can also be blocked by **fusion inhibitors**, such as *enfuvirtide*. This is a synthetic peptide that blocks fusion of the virus and cell by mimicking a region of the gp41 HIV-1 envelope (again, see Figure 19.13).

### Uncoating, Genome Integration, and Nucleic Acid Synthesis Inhibitors

Prior to the start of viral replication, viral nucleic acids are released from their protein coat. Drugs that prevent this uncoating include anti–influenza A drugs *amantadine* and *rimantadine*. However, the CDC no longer recommends their use because of widespread resistance. Once the virus has entered the cell and uncoated, the nucleic acids are free within the host cell. In order for HIV to replicate, the viral DNA must integrate into the genome of the host cell. This requires the enzyme integrase. Drugs such as *raltegravir* and *elvitegravir* are competitive inhibitors of integrase.

Nucleic acid synthesis is an important target for antivirals, especially for the treatment of HIV and herpes infections. Many of these drugs are analogs of nucleic acids (page 44), which inhibit RNA or DNA synthesis once the analog is incorporated. An obvious target for nucleic acid inhibition is the enzyme reverse transcriptase, used by HIV and hepatitis B virus during synthesis of DNA, because this step is not used in human DNA synthesis (page 251 and Figure 9.9). A nucleoside analog, *acyclovir* is used for treating herpes infections, especially genital herpes. It's an especially useful treatment in immunosuppressed individuals. Acyclovir is selectively used by the viral enzyme thymidine kinase (**Figure 20.16**). The drugs *famciclovir* and *ganciclovir* are derivatives of acyclovir and have a similar mode of action. *Ribavirin* resembles the nucleoside guanine and accelerates the already high mutation rate of RNA viruses until the accumulation of errors reaches a crisis point, killing the virus. It is used to treat hepatitis C infections. More recently, a nucleotide analog, *adefovir dipivoxil*, has been introduced for patients whose hepatitis B infections are resistant to *lamivudine*. This drug competitively inhibits the virus's reverse transcriptase. Another nucleoside analog, *cidofovir*, is currently used to treat cytomegalovirus infection of the eye.

Not all drugs that inhibit reverse transcriptase are nucleoside or nucleotide analogs. For example, a few **non-nucleoside inhibitors**, such as *nevirapine*, block RNA synthesis by other mechanisms.

### Assembly and Exit Inhibitors

The production of infectious viral particles requires an important group of enzymes called proteases. Viral replication requires the enzymatic cleavage of protein precursors. Some

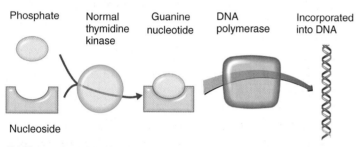

**(a)** Acyclovir structurally resembles the nucleoside deoxyguanosine.

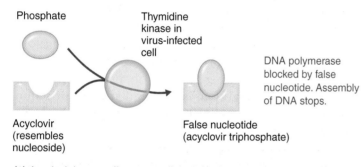

**(b)** The enzyme thymidine kinase combines phosphates with nucleosides to form nucleotides, which are then incorporated into DNA.

Phosphate

Thymidine kinase in virus-infected cell

DNA polymerase blocked by false nucleotide. Assembly of DNA stops.

Acyclovir (resembles nucleoside)

False nucleotide (acyclovir triphosphate)

**(c)** Acyclovir has no effect on a cell not infected by a virus, that is, with normal thymidine kinase. In a virally infected cell, the thymidine kinase is altered and converts the acyclovir (which resembles the nucleoside deoxyguanosine) to a false nucleotide, which blocks DNA synthesis by DNA polymerase.

**Figure 20.16  The structure and function of the antiviral drug acyclovir.**

**Q** Why are viral infections generally difficult to treat with chemotherapeutic agents?

drugs, called **protease inhibitors**, block this step. Examples are *saquinavir* for HIV and *boceprevir* and *telaprevir* for hepatitis C infections.

For a virus to reproduce, it must bud from the host cell. For influenza viruses, this requires the enzyme neuraminidase (page 709). Currently there are three drugs that are competitive inhibitors of this enzyme, allowing them to block release of viral particles. They are *zanamivir* (Relenza®), *oseltamivir* (Tamiflu®), and peramivir (Rapivab®).

### Interferons

Cells infected by a virus often produce alpha and beta interferons, which inhibit further spread of the infection. Interferons are classified as cytokines, discussed in Chapter 17. *Alpha interferon* (see Chapter 16, page 467) is currently a drug of choice for viral hepatitis infections. The production of interferons can be stimulated by a recently introduced antiviral, *imiquimod*. This drug is often prescribed to treat genital warts.

> **CHECK YOUR UNDERSTANDING**
>
> ✔ **20-14** One of the most widely used antivirals, acyclovir, inhibits the synthesis of DNA. Humans also synthesize DNA, so why is the drug still useful in treating viral infections?

### Antiretrovirals for Treating HIV/AIDS

The interest in effective treatments for the pandemic of HIV infections has led to many antiviral drugs developed for this. Several of the drugs discussed previously are used to treat HIV. HIV is an RNA virus, and its reproduction depends on the enzyme reverse transcriptase, which controls the synthesis of DNA from RNA (see page 251). The term **antiretroviral** currently implies a drug that is used to treat HIV infections. Antiretroviral drugs are discussed on pages 552–553.

## Antiprotozoan and Antihelminthic Drugs

For hundreds of years, quinine from the bark of the cinchona tree was the only drug known to be effective for treating a parasitic infection. Peruvian natives had observed that quinine, which is an effective muscle relaxant, controlled the shivering symptomatic of malarial fever. Actually, this characteristic is unrelated to quinine's toxicity to the protozoan that causes malaria. It was first introduced into Europe in the early 1600s. There are now several antiprotozoan and antihelminthic drugs, although many of them are still considered experimental. Qualified physicians can request and receive several of these drugs from the Centers for Disease Control and Prevention (CDC).

### Antiprotozoan Drugs

*Quinine* is still used to control the protozoan disease malaria, but synthetic derivatives, such as *chloroquine*, have largely replaced it. For preventing malaria in areas where the disease has developed resistance to chloroquine, *mefloquine* is often recommended, although serious psychiatric side effects have been reported.

As resistance to the most widely used and cheapest drug, chloroquine, becomes almost universal, the products of a Chinese shrub, *artemisinin* and *artemisinin-based combination therapies (ACTs)*, have become the principal treatment of malaria. Artemisinin was a traditional Chinese medicine long used for controlling fevers: Chinese scientists, following this lead, identified its antimalarial properties in 1971. ACTs kill the asexual and sexual stages of *Plasmodium* spp. in the blood (Figure 12.20 on page 345) by forming free radicals (pages 156–157) that damage proteins. Compared to choroquinine, ACTs are expensive—a problem in malaria-prone areas. This has led to widespread distribution of low-cost, but ineffective, counterfeit ACTs. Some of these contain enough of the genuine drug to evade simple tests, but these low dosages are accelerating development of resistance.

*Quinacrine* is the drug of choice for treating the protozoan disease giardiasis. *Diiodohydroxyquin (iodoquinol)* is an important drug prescribed for several intestinal amebic diseases, but its dosage must be carefully controlled to avoid optic nerve damage.

*Metronidazole* is one of the most widely used antiprotozoan drugs. It's unique in that it acts not only against parasitic protozoa but also against obligately anaerobic bacteria. It's the drug of choice for vaginitis caused by *Trichomonas vaginalis*. It's also used in treating giardiasis and amebic dysentery. Under anaerobic conditions, metronidazole is reduced, and this reduced metronidazole interferes with DNA synthesis.

*Tinidazole*, a drug similar to metronidazole, is effective in treating giardiasis, amebiasis, and trichomoniasis. Another antiprotozoan agent, and the first to be approved for the chemotherapy of diarrhea caused by *Cryptosporidium hominis*, is *nitazoxanide*. It's active in treating giardiasis and amebiasis. Because it interferes with an enzyme used in the anaerobic conversion of pyruvic acid to acetyl-CoA, it is also used to treat some bacterial infections.

*Miltefosine*, first developed as an anticancer drug, is a core drug listed by the World Health Organization to treat leishmaniasis (see pages 679–680). This drug inhibits cytochrome oxidase in mitochondria. It is available from the CDC to treat life-threatening amebic encephalitis (see pages 640–642).

### Antihelminthic Drugs

Tapeworm infections have decreased in developing countries because of improved sewage treatment. However, with the increased popularity of sushi, a Japanese specialty often made with raw fish, the CDC has begun to notice an increased incidence of tapeworm infections in the United States and other developed countries. To estimate the incidence, the CDC documents requests for *niclosamide*, which is the usual first choice in treatment. The drug is effective because it inhibits ATP production under aerobic conditions. *Praziquantel* is about equally effective for treating tapeworms; it kills worms by altering the permeability of their plasma membranes. Praziquantel has a broad spectrum of activity and is highly recommended for treating several fluke-caused diseases, especially schistosomiasis. It causes the helminths to undergo muscular spasms and also makes them susceptible to attack by the immune system. Apparently, its action exposes surface antigens, which antibodies can then reach.

*Mebendazole* and *albendazole* are broad-spectrum antihelminthics that have few side effects and have become the drugs of choice for treating many intestinal helminthic infections. Both drugs inhibit the formation of microtubules in the cytoplasm, which interferes with the absorption of nutrients by the parasite. These drugs are also widely used in the livestock industry; for veterinary applications, they are relatively more effective in ruminant animals.

*Ivermectin* is a drug with a wide range of applications. It's known to be produced by only one species of organism, *Streptomyces avermectinius*, which was isolated from the soil near a Japanese golf course. It's effective against many nematodes (roundworms) and several mites (such as scabies), ticks, and insects (such as head lice). (Some mites and insects happen to share certain similar metabolic channels with affected helminths.) Ivermectin has been used primarily in the livestock industry as a broad-spectrum antihelminthic. Its exact mode of action is uncertain, but the final result is paralysis and death of the helminth without affecting mammalian hosts.

> **CHECK YOUR UNDERSTANDING**
>
> 🖉 **20-15** What was the first drug for parasitic infections?

## Tests to Guide Chemotherapy

**LEARNING OBJECTIVE**

**20-16** Describe two tests for microbial susceptibility to chemotherapeutic agents.

Different microbial species and strains have different degrees of susceptibility to chemotherapeutic agents. Moreover, the susceptibility of a microorganism can change with time, even during therapy with a specific drug. Thus, a physician must know the sensitivities of the pathogen before treatment can be started. However, physicians often can't wait for sensitivity tests and must begin treatment based on their "best guess" estimation of the most likely pathogen causing the illness.

Several tests can be used to indicate which chemotherapeutic agent is most likely to combat a specific pathogen. However, if the organisms have been identified—for example, *Pseudomonas aeruginosa*, beta-hemolytic streptococci, or gonococci—certain drugs can be selected without specific testing for susceptibility. Tests are necessary only when susceptibility isn't predictable or when antibiotic resistance problems develop.

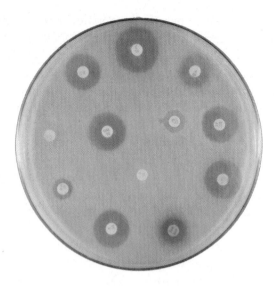

**Figure 20.17 The disk-diffusion method for determining the activity of antimicrobials.** Each disk contains a different chemotherapeutic agent, which diffuses into the surrounding agar. The clear zones indicate inhibition of growth of the microorganism swabbed onto the agar surface.

**Q** Which agent is the most effective against the bacterium being tested?

## The Diffusion Methods

Probably the most widely used, although not necessarily the best, method of testing is the **disk-diffusion method**, also known as the *Kirby-Bauer test* (**Figure 20.17**). A Petri plate containing an agar medium is inoculated ("seeded") uniformly over its entire surface with a standardized amount of a test organism. Next, filter paper disks impregnated with known concentrations of chemotherapeutic agents are placed on the solidified agar surface. During incubation, the chemotherapeutic agents diffuse from the disks into the agar. The farther the agent diffuses from the disk, the lower its concentration. If the chemotherapeutic agent is effective, a **zone of inhibition** forms around the disk after a standardized incubation. The diameter of the zone can be measured; in general, the larger the zone, the more sensitive the microbe is to the antibiotic. For a drug with poor solubility, however, the zone of inhibition indicating that the microbe is sensitive will be smaller than for another drug that is more soluble and has diffused more widely. The zone diameter is compared to a standard table for that drug and concentration, and the organism is reported as *sensitive, intermediate,* or *resistant*. Results obtained by the disk-diffusion method are often inadequate for many clinical purposes. However, the test is simple and inexpensive and is most often used when more sophisticated laboratory facilities aren't available.

A more advanced diffusion method, the **E test**, enables a lab technician to estimate the **minimal inhibitory concentration** (**MIC**), the lowest antibiotic concentration that prevents visible bacterial growth. A plastic-coated strip contains a gradient of antibiotic concentrations, and the MIC can be read from a scale printed on the strip (**Figure 20.18**).

## Broth Dilution Tests

A weakness of the diffusion method is that it doesn't determine whether a drug is bactericidal and not just bacteriostatic. A **broth dilution test** is often useful in determining the MIC and the **minimal bactericidal concentration (MBC)** of an antimicrobial drug. The MIC is determined by making a sequence of decreasing concentrations of the drug in a broth, which is then inoculated with the test bacteria (**Figure 20.19**). The wells that don't show growth (higher concentration than the MIC) can be cultured in broth or on agar plates free of the drug. If growth occurs in this broth, the drug was not bactericidal, and the MBC can be determined. Determining the MIC and MBC is important because it avoids the excessive or erroneous use of expensive antibiotics and minimizes the chance of toxic reactions that larger-than-necessary doses might cause.

Dilution tests are often highly automated. The drugs are purchased already diluted into broth in wells formed in a plastic tray. A suspension of the test organism is prepared and inoculated into all the wells simultaneously by a special inoculating device. After incubation, the turbidity may be read visually, although clinical laboratories with high workloads may read the trays with spectrophotometers that enter the data into a computer that provides a printout of the MIC.

**Figure 20.18 The E test (for epsilometer), a gradient diffusion method that determines antibiotic sensitivity and estimates minimal inhibitory concentration (MIC).** The plastic strip, which is placed on an agar surface inoculated with test bacteria, contains an increasing gradient of the antibiotic. The MIC in µg/ml is clearly shown.

**Q** What is the MIC of this E test?

Concentration of
drug on plates

Highest ◄─────► Lowest

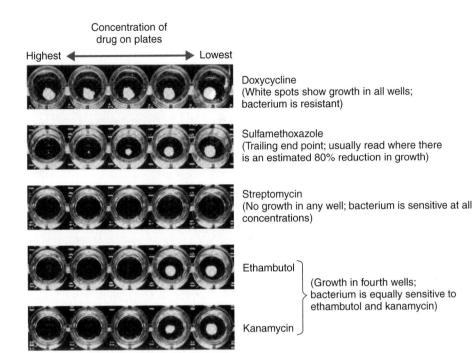

Doxycycline
(White spots show growth in all wells;
bacterium is resistant)

Sulfamethoxazole
(Trailing end point; usually read where there
is an estimated 80% reduction in growth)

Streptomycin
(No growth in any well; bacterium is sensitive at all
concentrations)

Ethambutol ⎫
⎬ (Growth in fourth wells;
⎬ bacterium is equally sensitive to
⎬ ethambutol and kanamycin)
Kanamycin ⎭

**Figure 20.19 A microdilution, or microtiter, plate used for testing for minimal inhibitory concentration (MIC) of antibiotics.** Such plates contain as many as 96 shallow wells that contain measured concentrations of antibiotics. They are usually purchased frozen or freeze dried (page 165). The test microbe is added simultaneously, with a special dispenser, to all the wells in a row of test antibiotics. To ensure that the microbe is capable of growth in the absence of the antibiotic, wells that contain no antibiotic are also inoculated (positive control). To ensure against contamination by unwanted microbes, wells that contain nutrient broth but no antibiotics or inoculum are included (negative control).

**Q** What is *MIC*?

Other tests are also useful for the clinician; a determination of the microbe's ability to produce β-lactamase is one example. One popular, rapid method makes use of a cephalosporin that changes color when its β-lactam ring is opened. In addition, a measurement of the *serum concentration* of an antimicrobial is especially important when toxic drugs are used. These assays tend to vary with the drug and may not always be suitable for smaller laboratories.

The hospital personnel responsible for infection control prepare periodic reports called **antibiograms** that record the susceptibility of organisms encountered clinically. These reports are especially useful for detecting the emergence of strains of pathogens resistant to the antibiotics in use at the institution.

**CHECK YOUR UNDERSTANDING**

✔ 20-16 In the disk-diffusion test, the zone of inhibition indicating sensitivity around the disk varies with the antibiotic. Why?

**CLINICAL CASE**

Dr. Singh sends her sample of *P. aeruginosa* to the CDC for analysis. The ophthalmologist in the other *P. aeruginosa* case also sends a sample. With a broth dilution assay, the MIC against these bacteria is 100 μg/ml. The decimal reduction time (DRT) of gentamicin against this bacterium at 4°C was determined to be 4 days and at 23°C, 20 min.

**How much time would be required to kill 200 cells at each temperature? (Hint: See Chapter 7, page 182.)**

# Resistance to Antimicrobial Drugs

**LEARNING OBJECTIVE**

20-17 Describe the mechanisms of drug resistance.

One of the triumphs of modern medicine has been the development of antibiotics and other antimicrobials. But the development of resistance to them by the target microbes is a worldwide public health problem. When first exposed to a new antibiotic, the susceptibility of microbes tends to be high, and their mortality rate is also high; there may be only a handful of survivors from a population of billions. The surviving microbes usually have some genetic characteristic that accounts for their survival, and natural selection favors their survival. A term adopted for these bacteria is **persister cells**.

Play Interactive
Microbiology
@MasteringMicrobiology
*See how antibiotic-resistance mechanisms affect a patient's health*

Once acquired, however, the mutation is transmitted *vertically* by normal reproduction, and the progeny carry the genetic characteristics of the parent microbe. Because of the rapid reproductive rate of bacteria, only a short time elapses before practically the entire population is resistant to the new antibiotic.

Thousands of antibiotic-resistance genes and proteins against 240 antibiotics are known. Some genetic differences arise from random mutations. These mutational differences can be spread *horizontally* among bacteria by processes such as conjugation (page 234) or transduction (page 235). Drug resistance is often carried by plasmids or by small segments of DNA called transposons, which can jump from one piece of DNA to another (page 231).

# Bacterial Resistance to Antibiotics

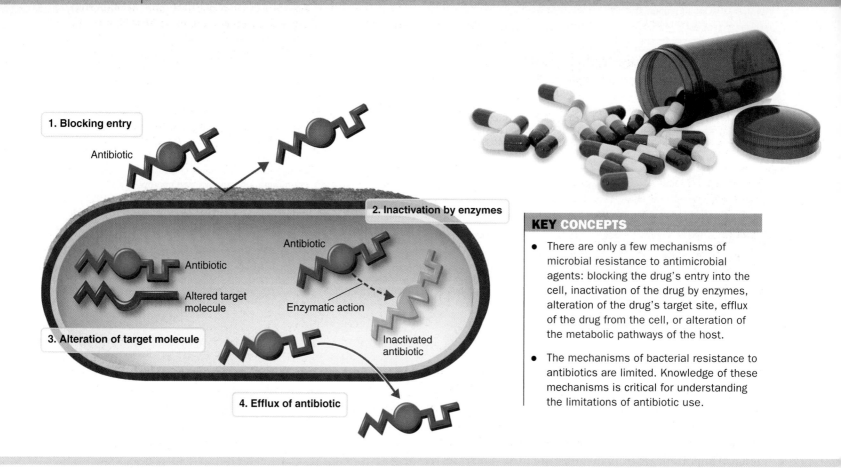

**1. Blocking entry**

Antibiotic

**2. Inactivation by enzymes**

Antibiotic

Antibiotic

Altered target molecule

Enzymatic action

**3. Alteration of target molecule**

Inactivated antibiotic

**4. Efflux of antibiotic**

## KEY CONCEPTS

- There are only a few mechanisms of microbial resistance to antimicrobial agents: blocking the drug's entry into the cell, inactivation of the drug by enzymes, alteration of the drug's target site, efflux of the drug from the cell, or alteration of the metabolic pathways of the host.

- The mechanisms of bacterial resistance to antibiotics are limited. Knowledge of these mechanisms is critical for understanding the limitations of antibiotic use.

Bacteria that are resistant to large numbers of antibiotics are popularly designated as **superbugs**. Although the most publicized of superbugs is MRSA (page 417), superbug status has also been assigned to a range of bacteria, both gram-positive and gram-negative. Faced with infections by such pathogens, medical science has only limited treatment options.

Play Antibiotic Resistance:
*Origins of Resistance*
@Mastering**Microbiology**

## Mechanisms of Resistance

There are only a few major mechanisms by which bacteria become resistant to chemotherapeutic agents. See **Figure 20.20**. At least one clinically troublesome bacterium, *Acinetobacter baumanii*, has developed resistance by means of all of the major target sites illustrated in Figure 20.20.

Play Antibiotic Resistance:
*Forms of Resistance*
@Mastering**Microbiology**

### Enzymatic Destruction or Inactivation of the Drug

Destruction or inactivation by enzymes mainly affects antibiotics that are natural products, such as the penicillins and cephalosporins. Totally synthetic chemical groups of antibiotics such as the fluoroquinolones are less likely to be affected in

this manner, although they can be neutralized in other ways. This may simply reflect the fact that the microbes have had fewer years to adapt to these unfamiliar chemical structures.

The penicillin/cephalosporin antibiotics, and also the carbapenems, share a structure, the β-lactam ring, which is the target for β-lactamase enzymes that selectively hydrolyze it. Nearly 200 variations of these enzymes are now known, each effective against minor variations in the β-lactam ring structure. When this problem first appeared, the basic penicillin molecule was modified. The first of these penicillinase-resistant drugs was methicillin (see page 568), but resistance to methicillin soon appeared. The best-known of these resistant bacteria is the widely publicized pathogen MRSA, which is resistant to practically all antibiotics, not just methicillin (see the Clinical Focus box on page 417). In a recent year, the CDC ascribed 10,000 deaths to this pathogen. In hospital patients, the mortality rate of invasive infections with MRSA can be as high as 20%. Also, *S. aureus* is not the only bacterium of concern; other important pathogens, such as *Streptococcus pneumoniae*, have also developed resistance to β-lactam antibiotics. Furthermore, MRSA has continued to develop resistance against a succession of new drugs such as vancomycin (the "antibiotic of last resort"), even though this antibiotic has a mode of action against cell wall synthesis that is totally different from that of the penicillins. These highly adaptable bacteria have even developed resistance against antibiotic combinations that include *clavulanic acid*, specifically developed as an inhibitor of β-lactamases (see page 568).

At first, MRSA was almost exclusively a problem in hospitals and similar health-related settings, accounting for about 20% of bloodstream infections there. However, it's now the cause of frequent outbreaks in the general community, is more virulent, and affects otherwise healthy individuals. These strains produce a toxin, a leukocidin, that destroys neutrophils, a primary innate defense against infection. Therefore, the descriptive terminology now differentiates *community-associated MRSA* from *healthcare-associated MRSA*. There is an obvious need for rapid tests to detect MRSA bacteria (generally from nasal swabs) so that infections can be isolated and transmission reduced. The most promising of these are based on PCR technology and yield good results within 1 or 2 hours.

### Prevention of Penetration to the Target Site within the Microbe

Gram-negative bacteria are relatively more resistant to antibiotics because of the nature of their cell wall, which restricts absorption of many molecules to movements through openings called porins (see page 81). Some bacterial mutants modify the porin opening so that antibiotics are unable to enter the periplasmic space. Perhaps even more important, when β-lactamases are present in the periplasmic space, the antibiotic that enters is degraded in the periplasmic space, before it can enter the cell.

### Alteration of the Drug's Target Site

The synthesis of proteins involves the movement of a ribosome along a strand of messenger RNA, as shown in Figure 20.4. Several antibiotics, especially those of the aminoglycoside, tetracycline, and macrolide groups, utilize a mode of action that inhibits protein synthesis at this site. Minor modifications at this site can neutralize the effects of antibiotics without significantly affecting cellular function.

Interestingly, the main mechanism by which MRSA gained ascendancy over methicillin was not by a new inactivating enzyme, but by modifying the penicillin-binding protein (PBP) on the cell's membrane. β-Lactam antibiotics act by binding with the PBP, which is required to initiate the cross-linking of peptidoglycan and form the cell wall. MRSA strains become resistant because they have an additional, modified, PBP. The antibiotics continue to inhibit the activity of the normal PBPs, preventing their participation in forming the cell wall. But the additional PBP present on the mutants, although it binds weakly with the antibiotic, still allows synthesis of cell walls that is adequate for survival of MRSA strains.

### Rapid Efflux (Ejection) of the Antibiotic

Certain proteins in the plasma membranes of gram-negative bacteria act as pumps that expel antibiotics, preventing them from reaching an effective concentration. This mechanism was originally observed with tetracycline antibiotics, but it confers resistance among practically all major classes of antibiotics. Bacteria normally have many such efflux pumps to eliminate toxic substances.

### Variations of Mechanisms of Resistance

Variations on these mechanisms also occur. For example, a microbe could become resistant to trimethoprim by synthesizing very large amounts of the enzyme against which the drug is targeted. Conversely, polyene antibiotics can become less effective when resistant organisms produce smaller amounts of the sterols against which the drug is effective. Of particular concern is the possibility that such *resistant mutants* will increasingly replace the susceptible normal populations. **Figure 20.21** shows how rapidly bacterial numbers increase as resistance develops.

## Antibiotic Misuse

Antibiotics have been much misused, nowhere more so than in the less-developed areas of the world. Well-trained personnel are scarce, especially in rural areas, which is perhaps one reason why antibiotics can almost universally be purchased without prescriptions in these countries. A survey in rural Bangladesh, for example, showed that only 8% of antibiotics

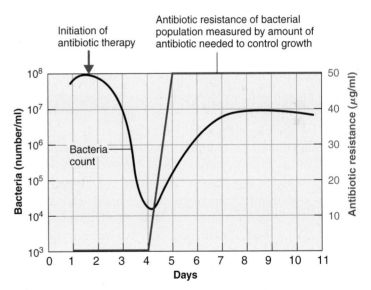

**Figure 20.21 The development of an antibiotic-resistant mutant during antibiotic therapy.** The patient, suffering from a chronic kidney infection caused by a gram-negative bacterium, was treated with streptomycin. The red line records the antibiotic resistance of the bacterial population. Until about the fourth day, essentially all of the bacterial population is sensitive to the antibiotic. Then persister cells (mutants in the population that are resistant to streptomycin) appear. The bacterial population in the patient rises as these mutants replace the sensitive population.

**Q** This test used streptomycin and a gram-negative bacterium. What would the lines have looked like if penicillin G had been the antibiotic?

It would take 12 days to kill 200 cells at 4°C and 60 minutes at 23°C. The gentamicin is more effective at the warmer temperature, but the tissues will deteriorate too quickly at this temperature. Hence, the corneas are stored at 4°C to preserve the tissue even though gentamicin is less effective at 4°C.

**How might storage of the corneas in gentamicin have contributed to these infections?**

559  570  579  **582**  586

had been prescribed by a physician. In much of the world, antibiotics are sold to treat headaches and for other inappropriate uses (**Figure 20.22**). Even when the use of antibiotics is appropriate, dose regimens are usually shorter than needed to eradicate the infection, thereby encouraging the survival of resistant strains of bacteria. Outdated, adulterated (impure), and even counterfeit antibiotics are common.

The developed world is also contributing to the rise of antibiotic resistance. The CDC estimates that in the United States, 30% of the antibiotic prescriptions for ear infections, 100% of the prescriptions for the common cold, and 50% of prescriptions for sore throats were unnecessary or inappropriate to treat the problem pathogen. At least 80% of the antibiotics produced in the United States each year are not used to treat disease but are used in animal feeds to promote growth—a practice that is being discouraged by the CDC, FDA, and consumers (see the Clinical Focus box on page 584). In 2006, the use of antibiotics as a growth enhancer in animals was banned in European Union countries, and in the United States in 2012, the FDA prohibited the use of cephalosporin-class antibiotics in food-producing animals. In 2013, the FDA also created a voluntary plan for the industry to phase out use of some antibiotics. Livestock growers are beginning to substitute essential oils (page 190) to prevent infections in animals. Some chicken farmers are now using carvacrol, an essential oil found in oregano (Figure 7.7e on page 190), to reduce *Campylobacter* in chickens.

## Cost and Prevention of Resistance

Antibiotic resistance is costly in many ways beyond those that are apparent in higher rates of disease and mortality. Developing new drugs to replace those that have lost effectiveness is costly. Almost all of these drugs will be more expensive, sometimes priced in a range that makes them difficult to afford even in highly developed countries. In less-developed parts of the world, the costs are simply unaffordable.

There are many strategies that patients and health care workers can adopt to prevent resistance from developing. Even if they feel they have recovered, patients should always finish the full regimen of their antibiotic prescriptions to discourage the

**Figure 20.22 Antibiotics have been sold without prescriptions for many decades in much of the world.**

**Q** How does this practice lead to development of resistant strains of pathogens?

survival and proliferation of the antibiotic-resistant microbes. Patients should never use leftover antibiotics to treat new illnesses or use antibiotics that were prescribed to someone else. Health care workers should avoid unnecessary prescriptions and ensure that the choice and dosages of antimicrobials are appropriate to the situation. Prescribing the most specific antibiotic possible, instead of broad-spectrum antimicrobials, also decreases the chances that the antibiotic will inadvertently cause resistance among the patient's normal microbiota.

Strains of bacteria that are resistant to antibiotics are particularly common among hospital workers, where antibiotics are in constant use. When antibiotics are injected, as many are, the syringe must first be held vertically and cleared of air bubbles, a practice that causes aerosols of the antibiotic solution to form. When the nurse or physician inhales these aerosols, the microbial inhabitants of the nostrils are exposed to the drug.

Gentamicin is used in commercial storage medium for corneas because it has been reported to be more effective than penicillin or cephalothin in reducing the colony counts of staphylococci and gram-negative rods in a buffered storage medium. Adding gentamicin is intended to preserve the medium before use, not to sterilize corneal tissue. Storage in an antibiotic could select for antibiotic-resistant bacteria.

**What antimicrobial drug would work best to treat *P. aeruginosa*?**

559  570  579  **582**  586

Inserting the needle into sterile cotton to expel air bubbles can prevent aerosols from forming. Many hospitals have special monitoring committees to review the use of antibiotics for effectiveness and cost.

CHECK YOUR UNDERSTANDING

✔ 20-17 What is the most common mechanism that a bacterium uses to resist the effects of penicillin?

## Antibiotic Safety

In our discussions of antibiotics, we have occasionally mentioned side effects. These may be potentially serious, such as liver or kidney damage or hearing impairment. Administering almost any drug involves assessing risks against benefits; this is called the *therapeutic index*. Sometimes, the use of another drug can cause toxic effects that do not occur when the drug is taken alone. One drug may also neutralize the intended effects of the other. For example, rifampin neutralizes the effectiveness of contraceptive pills. Also, some individuals may have hypersensitivity reactions, for example, to penicillins (see the Clinical Focus box on page 537).

A pregnant woman should take only those antibiotics that are classified by the U.S. Food and Drug Administration as presenting no evidence of risk to the fetus.

## Effects of Combinations of Drugs

LEARNING OBJECTIVE

20-18 Compare and contrast synergism and antagonism.

The chemotherapeutic effect of two drugs given simultaneously is sometimes greater than the effect of either given alone (**Figure 20.23**). This phenomenon, called **synergism**, was introduced earlier. For example, in the treatment of bacterial endocarditis, penicillin and streptomycin are much more effective when taken together than when either drug is taken alone. Damage to bacterial cell walls by penicillin makes it easier for streptomycin to enter.

Other combinations of drugs can show **antagonism**. For example, the simultaneous use of penicillin and tetracycline is often less effective than when either drug is used alone. By stopping the growth of the bacteria, the bacteriostatic drug tetracycline interferes with the action of penicillin, which requires bacterial growth.

CHECK YOUR UNDERSTANDING

✔ 20-18 Tetracycline sometimes interferes with the activity of penicillin. How?

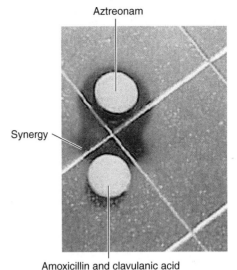

Aztreonam

Synergy

Amoxicillin and clavulanic acid

**Figure 20.23 An example of synergism between two different antibiotics.** The photograph shows the surface of an agar medium inoculated with bacteria. The clear area of no growth around the paper disk at the top shows inhibition by the antibiotic aztreonam. There is a small zone of inhibition around the bottom disk containing amoxicillin plus clavulanic acid. The additional, larger clear area between the two disks illustrates inhibition of bacterial growth through the effects of synergy.

**Q** **What would the plate look like if the two antibiotics had been antagonistic?**

## Future of Chemotherapeutic Agents

LEARNING OBJECTIVE

20-19 Name three areas of research on new chemotherapeutic agents.

As pathogens develop resistance to current chemotherapeutic agents, the need for new ones becomes more pressing. However, developing new antimicrobial agents is not especially easy. There is genuine concern that we may be approaching a post-antibiotic time when infections of minor cuts or scratches might again cause deaths.

Existing antibiotics continue to encounter problems with resistance in large part because their developers have relied on a limited range of targets (see Figure 20.2). A truly new approach to controlling pathogens is to target their virulence factors rather than the microbe producing them. For example, instead of targeting the cholera bacillus, a drug might target the cholera toxin, neutralizing or destroying it. Another potential target is to sequester iron, which pathogens need for growth (see page 430). A drug that sequesters iron would therefore limit proliferation of the pathogens.

The FDA requires that antibiotics be tested against exponentially growing pathogens. This has led to a near-absence of drugs to combat dormant cells, especially the persisters (page 579). Most drugs fail when tested against such cells. Another basic problem awaiting a solution is a lack of drugs

Antibiotics in Animal Feed Linked to Human Disease

As you read through this box, you will encounter a series of questions that microbiologists ask as they combat antibiotic resistance. Try to answer each question before going on to the next one.

1. Livestock growers use antibiotics in the feed of closely penned animals because the drugs reduce the number of bacterial infections and accelerate the animals' growth. Today, more than half the antibiotics used worldwide are given to farm animals.

   Meat and milk that reach the consumer's table are not heavily laden with antibiotics, so what is the risk of using antibiotics in animal feed?

2. The constant presence of antibiotics in these animals is an example of "survival of the fittest." Antibiotics kill some bacteria, but other bacteria have properties that help them survive.

   How do bacteria acquire resistance genes?

3. Resistance to antimicrobial drugs in bacteria results from mutations. These mutations can be transmitted to other bacteria via horizontal gene transfer (Figure A).

What evidence would show that veterinary use of antibiotics promotes resistance?

4. Vancomycin-resistant *Enterococcus* spp. (VRE) were first isolated in France in 1986 and were found in the United States in 1989. Vancomycin and another glycopeptide, avoparcin, were widely used in animal feed in Europe. In 1997, veterinary use of avoparcin was banned in Europe. After the ban, VRE-positive samples decreased from 100% to 25%, and the human carrier rate dropped from 12% to 3%.

   *Campylobacter jejuni* is a commensal in the intestines of poultry. What human disease does *C. jejuni* cause?

5. Annually in the United States, *Campylobacter* causes over 1.3 million foodborne infections. Fluoroquinolone (FQ)-resistant *C. jejuni* in humans emerged in the 1990s (Figure B).

   What FQs are used to treat human infections? (*Hint*: See Table 20.3.)

6. The emergence corresponds with the presence of FQ-resistant *C. jejuni* in chicken meat purchased from grocery stores. FQ-resistant *C. jejuni* could

be selected for in patients who had previously taken an FQ. However, a study of *Campylobacter* isolates from patients between 1997 and 2001 showed that patients infected with FQ-resistant *C. jejuni* had not taken an FQ prior to their illness and had not traveled out of the United States.

Suggest a way to decrease FQ resistance.

7. The use of FQ in chicken feed was banned in 2005 in the United States, in hope of reducing FQ resistance. A variety of approaches may be necessary to reduce the possibility of illness: (1) prevent colonization in the animals at the farm, (2) reduce fecal contamination of meat during processing at the slaughterhouse, and (3) use proper storage and cooking methods.

*Data sources:* CDC and National Microbial Resistance Monitoring System.

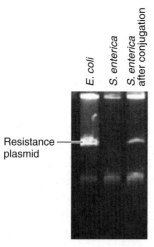

**Figure A  Cephalosporin resistance in *E. coli* transferred by conjugation to *Salmonella enterica* in the intestinal tracts of turkeys.**

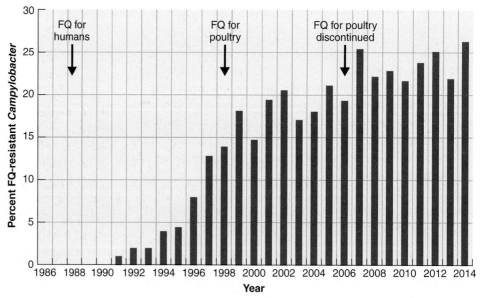

**Figure B  Fluoroquinolone-resistant *Campylobacter jejuni* isolated from humans in the United States, 1986–2014.**

**Looking to the Microbiome for the Next Great Antibiotic**

We know that the human body contains an ecosystem of microorganisms that interact with their environment, the human body, and each other. Normal microbiota keep other bacteria from using resources they need and thereby inhibit growth of some pathogens that would otherwise colonize. In looking for the mechanisms of this sort of competitive exclusion, researchers have found new antibiotics by looking at the molecules produced by many of our body's normal residents.

One source of a potential drug is *Lactobacillus gasseri*, a resident of the vagina. It produces lactocillin, a chemical that experiments show to be a promising agent for inhibiting gram-positive bacteria. Molecularly speaking, lactocillin is a

thiopeptide that works by binding to the 50S ribosomal subunit. Work is under way to create thiopeptide antibiotics based on this same mode of action.

Another potential drug active ingredient is lugdunin, a large molecule produced by the bacterium *Staphylococcus lugdunensis*. Researchers in Germany discovered the compound after noticing people whose noses were colonized by *S. lugdunensis* were less likely to carry pathogenic strains of *S. aureus*. An antibiotic based on lugdunin could be very helpful because the molecule has been shown to inhibit MRSA, *Enterococcus* species, and other bacteria where antibiotic resistance is an ever-growing threat.

These new compounds are not ready for clinical trials, but their discovery is

encouraging researchers to look for more novel antibiotics in the human microbiome.

**Growth of MRSA is inhibited by a medium (pH 7) in which *Lactobacillus* spp. grew (in the center well).**

that treat gram-negative infections. The cell wall of gram-negative bacteria makes them inherently resistant to most antibiotics. In fact, the antibiotics developed over the past few decades have worked only against gram-positive species. Also, over 99% of bacterial species found in nature are incapable of being grown on conventional laboratory media. Attempting to reproduce the cellular environment in the lab to grow and test antibiotic sensitivity of noncultivatable bacteria is complicated and expensive. Multidrug resistance of the bacteria comprising biofilms is another unsolved problem. Finally, drug misuse based on unregulated distribution of antimicrobial drugs speeds up antibiotic resistance in many parts of the world.

New, exotic ecological niches, such as deep-sea sediments, will need to be explored. The most promising new avenue of research to develop new antibiotics will probably be based on knowledge of the basic genetic structure of microbes. A computerized analysis of the genomes of the human microbiome has recently led to the discovery of new antimicrobial compounds.

Microorganisms are not the only organisms that produce antimicrobial substances. Many birds, amphibians, plants, and mammals often produce **antimicrobial peptides**. In fact, such peptides are part of the defense systems of most forms of life, and literally hundreds of such peptides have been identified. Amphibian skin glands are a rich source of antimicrobial peptides that attack

bacterial membranes. The best-known of these are the *magainins* (from the Hebrew for shield). It's especially interesting that this antimicrobial has existed for an indefinite time without significant development of resistance. Another antimicrobial substance, a steroid named *squalamine*, has been isolated from sharks.

Many bacteria produce antimicrobial peptides called *bacteriocins* (see Chapter 14, page 397). Research has shown some of these exhibit a broad spectrum of activity, whereas others have a narrow spectrum. The mechanisms of action differ from those of most antibiotics. Some bacteriocins affect the cell membrane, and others affect protein production. The oral toxicity of bacteriocins is very low. Early trials of a few bacteriocins against *Clostridium difficile* and other gram-positive bacteria gram-positive bacteria show promising results. See Exploring the Microbiome for other examples.

**Phage therapy** has been used in Russia, Georgia, and Poland for more than 50 years. Bacteriophages are viruses that can kill their host bacterial cells (see Chapter 13, page 373). In recent years, the United States and Europe have started looking at bacteriophages to replace some antibiotic use. Phages are specific for their host bacteria and may be useful to treat antibiotic-resistant infections. Environmentally, the soil is full of bacteriophages, and it's said that every couple of days they kill about half the bacteria on Earth.

Serendipity, or accidental discovery, is always a consideration. For example, it's worth mentioning that the first quinolone, nalidixic acid, was discovered as an intermediate in the synthesis of an antimalarial drug, chloroquine, and that the oxazolidinones were originally developed to treat plant diseases.

Finally, there is a special need for new antiviral drugs as well as antifungal and antiparasitic drugs effective against helminths and protozoans, because our arsenal in these categories is very limited.

**CHECK YOUR UNDERSTANDING**

✔ 20-19 What are defensins? (*Hint:* See Chapter 16.)

**CLINICAL CASE** Resolved

Dr. Singh prescribes doripenem for her patient. Doripenem is a carbapenem, which has an extremely broad spectrum of activity and is especially effective against *P. aeruginosa*. The patient recovers from her infection and has no further complications from her surgery.

559  570  579  582  **586**

# Study Outline

 Go to @MasteringMicrobiology for **Interactive Microbiology**, *In the Clinic videos, MicroFlix, MicroBoosters, 3D animations, practice quizzes, and more.*

## Introduction (p. 558)

1. An antimicrobial drug is a chemical substance that destroys pathogenic microorganisms with minimal damage to host tissues.
2. Chemotherapeutic agents include chemicals that combat disease in the body.

## The History of Chemotherapy (pp. 559–560)

1. Paul Ehrlich developed the concept of chemotherapy to treat microbial diseases.
2. Sulfa drugs came into prominence in the 1930s.
3. Alexander Fleming discovered the first antibiotic, penicillin, in 1928; its first clinical trials were done in 1940.
4. Most antibiotics are made by *Streptomyces* bacteria.

## The Spectrum of Antimicrobial Activity (pp. 560–561)

1. Antibacterial drugs affect many targets in a prokaryotic cell.
2. Fungal, protozoan, and helminthic infections are more difficult to treat because these organisms have eukaryotic cells.
3. Narrow-spectrum drugs affect only a select group of microbes (gram-positive cells, for example); broad-spectrum drugs affect a more diverse range of microbes.
4. Small, hydrophilic drugs can affect gram-negative cells.
5. Antimicrobial agents should not cause excessive harm to normal microbiota.
6. Superinfections occur when a pathogen develops resistance to the drug being used or when normally resistant microbiota multiply excessively.

## The Action of Antimicrobial Drugs (pp. 561–563)

1. Antimicrobials generally act either by directly killing microorganisms (bactericidal) or by inhibiting their growth (bacteriostatic).
2. Some agents, such as penicillin, inhibit cell wall synthesis in bacteria.
3. Other agents, such as chloramphenicol, tetracyclines, and streptomycin, inhibit protein synthesis by acting on 70S ribosomes.
4. Ionophore and polypeptide antibiotics damage plasma membranes.

5. Some agents inhibit nucleic acid synthesis.
6. Agents such as sulfanilamide act as antimetabolites by competitively inhibiting enzyme activity.

## Common Antimicrobial Drugs (pp. 564–577)

### Antibacterial Antibiotics: Inhibitors of Cell Wall Synthesis (pp. 564–569)

1. All penicillins contain a β-lactam ring.
2. Natural penicillins produced by *Penicillium* are effective against gram-positive cocci and spirochetes.
3. Penicillinases (β-lactamases) are bacterial enzymes that destroy natural penicillins.
4. Semisynthetic penicillins are resistant to penicillinases and have a broader spectrum of activity than natural penicillins.
5. Carbapenems are broad-spectrum antibiotics that inhibit cell wall synthesis.
6. The monobactam aztreonam affects only gram-negative bacteria.
7. Cephalosporins inhibit cell wall synthesis and are used against penicillin-resistant strains.
8. Polypeptides such as bacitracin inhibit cell wall synthesis primarily in gram-positive bacteria.
9. Vancomycin inhibits cell wall synthesis and may be used to kill penicillinase-producing staphylococci.
10. Isoniazid (INH) and ethambutol inhibit cell wall synthesis in mycobacteria.

### Inhibitors of Protein Synthesis (pp. 570–572)

11. Chloramphenicol, aminoglycosides, tetracyclines, glycylcyclines, macrolides, streptogramins, oxazolidinones, and pleuromutilins inhibit protein synthesis at 70S ribosomes.

### Injury to Membranes (p. 572)

12. Lipopeptides polymyxin B and bacitracin cause damage to plasma membranes.

### Nucleic Acid Synthesis Inhibitors (p. 572)

13. Rifamycin inhibits mRNA synthesis; it's used to treat tuberculosis.
14. Quinolones and fluoroquinolones inhibit DNA gyrase.

**Competitive Inhibition of Essential Metabolites** (p. 573)

15. Sulfonamides competitively inhibit folic acid synthesis.

16. SMZ-TMP competitively inhibits dihydrofolic acid synthesis.

**Antifungal Drugs** (pp. 573–575)

17. Polyenes, such as nystatin and amphotericin B, combine with plasma membrane sterols and are fungicidal.

18. Azoles and allylamines interfere with sterol synthesis and are used to treat cutaneous and systemic mycoses.

19. Echinocandins interfere with fungal cell wall synthesis.

20. The antifungal agent flucytosine is an antimetabolite of cytosine.

21. Griseofulvin interferes with eukaryotic cell division and is used primarily to treat skin infections caused by fungi.

**Antiviral Drugs** (pp. 575–576)

22. Entry inhibitors and fusion inhibitors bind to viral attachment and receptor sites.

23. Nucleoside and nucleotide analogs, such as acyclovir and zidovudine, inhibit DNA or RNA synthesis.

24. Inhibitors of viral enzymes prevents viral assembly and exit.

25. Alpha interferons inhibit the spread of viruses to new cells.

**Antiprotozoan and Antihelminthic Drugs** (pp. 576–577)

26. Chloroquine, artemisinin, quinacrine, diiodohydroxyquin, pentamidine, and metronidazole are used to treat protozoan infections.

27. Antihelminthic drugs include mebendazole, praziquantel, and ivermectin.

## Tests to Guide Chemotherapy (pp. 577–579)

1. Tests are used to determine which chemotherapeutic agent is most likely to combat a specific pathogen.

2. These tests are used when susceptibility cannot be predicted or when drug resistance arises.

**The Diffusion Methods** (p. 578)

3. In the disk-diffusion test, also known as the Kirby-Bauer test, a bacterial culture is inoculated on an agar medium, and filter paper disks impregnated with chemotherapeutic agents are overlaid on the culture.

4. After incubation, the diameter of the zone of inhibition is used to determine whether the organism is sensitive, intermediate, or resistant to the drug.

5. Minimal inhibitory concentration (MIC) is the lowest concentration of drug capable of preventing microbial growth; MIC can be estimated using the E test.

**Broth Dilution Tests** (pp. 578–579)

6. In a broth dilution test, the microorganism is grown in liquid media containing different concentrations of a chemotherapeutic agent.

7. The lowest concentration of a chemotherapeutic agent that kills bacteria is called the minimum bactericidal concentration (MBC).

## Resistance to Antimicrobial Drugs (pp. 579–584)

1. Many bacterial diseases, previously treatable with antibiotics, have become resistant to antibiotics.

2. Superbugs are bacteria that are resistant to several antibiotics.

3. Drug resistance factors are transferred horizontally between bacteria.

4. Resistance may be due to enzymatic destruction of a drug, prevention of penetration of the drug to its target site, cellular or metabolic changes at target sites, alteration of the target site, or rapid efflux of the antibiotic.

> **Play Interactive Microbiology**
> @Mastering**Microbiology**
> *See how antibiotic-resistance mechanisms affect a patient's health*

5. The discriminating use of drugs in appropriate concentrations and dosages can minimize resistance.

## Antibiotic Safety (p. 584)

1. The risk (e.g., side effects) versus the benefit (e.g., curing an infection) must be evaluated before antibiotics are used.

## Effects of Combinations of Drugs (p. 584)

1. Some combinations of drugs are synergistic; they are more effective when taken together.

2. Some combinations of drugs are antagonistic; when taken together, both drugs become less effective than when taken alone.

## Future of Chemotherapeutic Agents (pp. 584–586)

1. New agents include antimicrobial peptides, bacteriocins, and bacteriophages.

2. Virulence factors rather than cell growth factors may provide new targets.

# Study Questions

For answers to the Knowledge and Comprehension questions, turn to the Answers tab at the back of the textbook.

## Knowledge and Comprehension

### Review

1. **DRAW IT** Show where the following antibiotics work: ciprofloxacin, tetracycline, streptomycin, vancomycin, polymyxin B, sulfanilamide, rifampin, erythromycin.

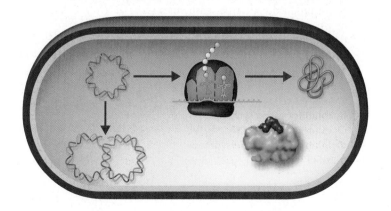

2. List and explain five criteria used to identify an effective antimicrobial agent.

3. What similar problems are encountered with antiviral, antifungal, antiprotozoan, and antihelminthic drugs?

4. Define *drug resistance*. How is it produced? What measures can be taken to minimize drug resistance?

5. List the advantages of using two chemotherapeutic agents simultaneously to treat a disease. What problem can occur when two drugs are used?

6. Why does a cell die from the following antimicrobial actions?
   **a.** Colistimethate binds to phospholipids.
   **b.** Kanamycin binds to 70S ribosomes.

7. How does each of the following inhibit translation?
   **a.** chloramphenicol          **d.** streptomycin
   **b.** erythromycin             **e.** oxazolidinone
   **c.** tetracycline             **f.** streptogramin

8. Dideoxyinosine (ddI) is an antimetabolite of guanine. The –OH is missing from carbon 3′ in ddI. How does ddI inhibit DNA synthesis?

9. Compare the method of action of the following pairs:
   **a.** penicillin and echinocandin
   **b.** imidazole and polymyxin B

10. NAME IT This microorganism is not susceptible to antibiotics or neuromuscular blocks, but it's susceptible to protease inhibitors.

## Multiple Choice

1. Which of the following pairs is *mismatched*?
   **a.** antihelminthic—inhibition of oxidative phosphorylation
   **b.** antihelminthic—inhibition of cell wall synthesis
   **c.** antifungal—injury to plasma membrane
   **d.** antifungal—inhibition of mitosis
   **e.** antiviral—inhibition of DNA synthesis

2. All of the following are modes of action of antiviral drugs *except*
   **a.** inhibition of protein synthesis at 70S ribosomes.
   **b.** inhibition of DNA synthesis.
   **c.** inhibition of RNA synthesis.
   **d.** inhibition of uncoating.
   **e.** All of the above are modes of action of antiviral drugs.

3. Which of the following modes of action would *not* be fungicidal?
   **a.** inhibition of peptidoglycan synthesis
   **b.** inhibition of mitosis
   **c.** injury to the plasma membrane
   **d.** inhibition of nucleic acid synthesis
   **e.** All of the above are fungicidal modes of action.

4. An antimicrobial agent should meet all of the following criteria *except*
   **a.** selective toxicity.
   **b.** the production of hypersensitivities.
   **c.** a narrow spectrum of activity.
   **d.** no production of drug resistance.
   **e.** All of the above are necessary criteria for an antimicrobial.

5. The most selective antimicrobial activity would be exhibited by a drug that
   **a.** inhibits cell wall synthesis.
   **b.** inhibits protein synthesis.
   **c.** injures the plasma membrane.
   **d.** inhibits nucleic acid synthesis.
   **e.** all of the above

6. Antibiotics that inhibit translation have side effects
   **a.** because all cells have proteins.
   **b.** only in the few cells that make proteins.
   **c.** because eukaryotic cells have 80S ribosomes.
   **d.** at the 70S ribosomes in eukaryotic cells.
   **e.** None of the above is correct.

7. Which of the following will *not* affect eukaryotic cells?
   **a.** inhibition of the mitotic spindle
   **b.** binding with sterols
   **c.** binding to 80S ribosomes
   **d.** binding to DNA
   **e.** All of the above will affect them.

8. Cell membrane damage causes death because
   **a.** the cell undergoes osmotic lysis.
   **b.** cell contents leak out.
   **c.** the cell plasmolyzes.
   **d.** the cell lacks a wall.
   **e.** None of the above is correct.

9. A drug that intercalates into DNA has the following effects. Which one leads to the others?
   **a.** It disrupts transcription.
   **b.** It disrupts translation.
   **c.** It interferes with DNA replication.
   **d.** It causes mutations.
   **e.** It alters proteins.

10. Chloramphenicol binds to the 50S portion of a ribosome, which will interfere with
   **a.** transcription in prokaryotic cells.
   **b.** transcription in eukaryotic cells.
   **c.** translation in prokaryotic cells.
   **d.** translation in eukaryotic cells.
   **e.** DNA synthesis.

## Analysis

1. Which of the following can affect human cells? Explain why.
   **a.** penicillin
   **b.** indinavir
   **c.** erythromycin
   **d.** polymyxin

2. Why is idoxuridine effective if host cells also contain DNA?

3. Some bacteria become resistant to tetracycline because they don't make porins. Why can a porin-deficient mutant be detected by its inability to grow on a medium containing a single carbon source such as succinic acid?

4. The following data were obtained from a disk-diffusion test.

| Antibiotic | Zone of Inhibition |
| --- | --- |
| A | 15 mm |
| B | 0 mm |
| C | 7 mm |
| D | 15 mm |

   **a.** Which antibiotic was most effective against the bacteria being tested?
   **b.** Which antibiotic would you recommend for treating a disease caused by this bacterium?
   **c.** Was antibiotic A bactericidal or bacteriostatic? How can you tell?

5. Why do you suppose *Streptomyces griseus* produces an enzyme that inactivates streptomycin? Why is this enzyme produced early in metabolism?

6. The following results were obtained from a broth dilution test for microbial susceptibility.

| Antibiotic Concentration | Growth | Growth in Subculture |
|---|---|---|
| 200 µg/ml | – | – |
| 100 µg/ml | – | + |
| 50 µg/ml | + | + |
| 25 µg/ml | + | + |

a. The MIC of this antibiotic is _____.
b. The MBC of this antibiotic is _____.

## Clinical Applications and Evaluation

1. Vancomycin-resistant *Enterococcus faecalis* was isolated from a foot infection of a 40-year-old man. The patient had a chronic diabetes-related foot ulcer and underwent amputation of a gangrenous toe. He subsequently developed methicillin-resistant *Staphylococcus aureus* bacteremia. The infection was treated with vancomycin. One week later, he developed a vancomycin-resistant *S. aureus* (VRSA) infection. This was the first case of VRSA in the United States. What is the most likely source of the VRSA?

2. A patient with a urinary bladder infection took nalidixic acid, but her condition did not improve. Explain why her infection disappeared when she switched to a sulfonamide.

3. A patient with streptococcal sore throat takes penicillin for 2 days of a prescribed 10-day regimen. Because he feels better, he then saves the remaining penicillin for some other time. After 3 more days, he suffers a relapse of the sore throat. Discuss the probable cause of the relapse.

# 21 Microbial Diseases of the Skin and Eyes

The skin, which covers and protects the body, is the body's first line of defense against pathogens. As a physical barrier, intact skin is almost impossible for pathogens to penetrate. Microbes can, however, enter through skin breaks that are not readily apparent, and the larval forms of a few parasites can penetrate intact skin.

The skin is an inhospitable place for most microorganisms because the secretions of the skin are acidic and most of the skin contains little moisture. Some parts of the body, though, such as the armpit and the area between the legs, have enough moisture to support relatively large bacterial populations. Drier regions, such as the scalp, support rather small numbers of microorganisms. A few microbes that colonize skin can cause disease. One such bacterium is *Pseudomonas aeruginosa*, shown in the photograph. The Clinical Case in this chapter describes how this opportunistic pathogen can cause a skin infection.

▶ *Pseudomonas aeruginosa* bacteria can cause skin infections.

## In the Clinic

As the intake nurse at the clinic, you see a 5-year-old boy with a rash on his hands and feet. His mother explains that he has had the rash for the past few days. There are no other symptoms. All other family members are well. The boy attends kindergarten and day care. His temperature is 37.6°C. You first need to assess the rash. You note that it is maculopapular with a few vesicles. You've seen a few of these rashes this month.
**Is this an exanthem or enanthem? What is the disease?**

*Hint: See the discussion about viral diseases of the skin on pages 602–607.*

**Answers to In the Clinic questions are found online @MasteringMicrobiology.**

# Structure and Function of the Skin

## LEARNING OBJECTIVE

**21-1** Describe the structure of the skin and mucous membranes and the ways pathogens can invade the skin.

The skin of an average adult occupies a surface area of about 1.9 m² and varies in thickness from 0.05 to 3.0 mm. Skin consists of two principal parts, the epidermis and the dermis (**Figure 21.1**). The **epidermis** is the thin outer portion, composed of several layers of epithelial cells. The outermost layer of the epidermis, the *stratum corneum*, consists of many rows of dead cells that contain a waterproofing protein called **keratin**. The epidermis, when unbroken, is an effective physical barrier against microorganisms.

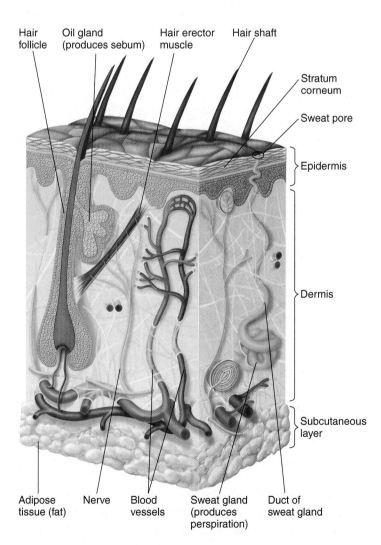

**Figure 21.1 The structure of human skin.** Notice the passageways between the hair follicle and hair shaft, through which microbes can penetrate the deeper tissues. They can also enter the skin through sweat pores.

**Q** What do you perceive from this illustration to be the weak points that would allow microbes to reach the underlying tissue by penetrating intact skin?

The **dermis** is the inner, relatively thick portion of skin, composed mainly of connective tissue. The hair follicles, sweat gland ducts, and oil gland ducts in the dermis provide passageways through which microorganisms can enter the skin and penetrate deeper tissues.

*Perspiration* provides moisture and some nutrients for microbial growth. However, it contains salt, which inhibits many microorganisms; the enzyme lysozyme, which is capable of breaking down the cell walls of certain bacteria; and antimicrobial peptides.

*Sebum*, secreted by oil glands, is a mixture of lipids (unsaturated fatty acids), proteins, and salts that prevents skin and hair from drying out. Although the fatty acids inhibit the growth of certain pathogens, sebum, like perspiration, is also nutritive for many microorganisms.

## Mucous Membranes

In the linings of body cavities that open to the exterior, such as those associated with the gastrointestinal, respiratory, urinary, and genital tracts, the outer protective barrier differs from the skin. It consists of sheets of tightly packed epithelial cells. These cells are attached at their bases to a layer of extracellular material called the *basement membrane*. Many of these cells secrete mucus—hence the name **mucous membrane**, or **mucosa**. Other mucosal cells have cilia; and, in the respiratory system, the mucous layer traps particles, including microorganisms, which the cilia sweep upward out of the body (see Figure 16.3, page 450). Mucous membranes are often acidic, which tends to limit their microbial populations. Also, the membranes of the eyes are mechanically washed by tears, and the lysozyme in tears destroys the cell walls of certain bacteria. Mucous membranes are often folded to maximize surface area; the total surface area in an average human is about 400 m², much more than the surface area of the skin.

> **CHECK YOUR UNDERSTANDING**
>
> ✔ **21-1** Moisture in perspiration encourages microbial growth. What perspiration factors discourage growth?

**CLINICAL CASE** Swimming Lessons

Molly Seidel, a pediatric nurse practitioner, is examining 9-year-old Donald and his 6-year-old sister, Sharon. According to their mother, both children developed rashes around dinner time the evening before. The rashes are similarly distributed over the children's front torsos and thighs. A cloudy fluid discharges when the children scratch the itchy, raised pimples. Molly has already seen several cases of skin rashes in children today. She has diagnosed two children with chickenpox, and she has prescribed penicillin for another child with staphylococcal folliculitis.

**What should Molly do next? Read on to find out.**

**591**  601  608  610  613

# Normal Microbiota of the Skin

## LEARNING OBJECTIVE

**21-2** Provide examples of normal skin microbiota, and state the general locations and ecological roles of its members.

Although the skin is generally inhospitable to most microorganisms, certain microbes are part of the normal microbiota (see Exploring the Microbiome, page 594). On superficial skin surfaces, certain aerobic bacteria produce fatty acids from sebum. These acids inhibit many microbes and allow better-adapted bacteria to flourish.

Microorganisms that find the skin a satisfactory environment are resistant to drying and to relatively high salt concentrations. The skin's normal microbiota contain relatively large numbers of gram-positive cocci, such as staphylococci and micrococci. These bacteria tend to be resistant to dry environments, and to the high osmotic pressures found in concentrated salt or sugar solutions. Scanning electron micrographs show that bacteria on the skin are often grouped into small clumps. Vigorous washing can reduce their numbers but will not eliminate them. Microorganisms remaining in hair follicles and sweat glands after washing will soon reestablish the normal populations. Areas of the body with more moisture, such as the armpits and between the legs, have higher populations of microbes. These metabolize secretions from the sweat glands and are the main contributors to body odor.

Also part of the skin's normal microbiota are gram-positive pleomorphic rods called *diphtheroids*. Some diphtheroids, such as *Cutibacterium (Propionibacterium) acnes*, are typically anaerobic and inhabit hair follicles. Their growth is supported by secretions from the oil glands (sebum), which, as we will see, makes them a factor in acne. These bacteria produce propionic acid, which helps maintain the low pH of skin, generally between 3 and 5. Other diphtheroids, such as *Corynebacterium xerosis* (zer-Ō-sis), are aerobic and occupy the skin surface.

A few gram-negative bacteria, especially *Acinetobacter*, colonize the skin. A yeast, *Malassezia furfur*, is capable of growing on oily skin secretions and is associated with the scaling skin condition known as *dandruff*. Shampoos for treating dandruff contain the antibiotic ketoconazole or zinc pyrithione or selenium sulfide. All are active against this yeast.

> **CHECK YOUR UNDERSTANDING**
>
> ✔ **21-2** Are skin bacteria more likely to be gram-positive or gram-negative?

# Microbial Diseases of the Skin

## LEARNING OBJECTIVES

**21-3** Differentiate staphylococci from streptococci, and name several skin infections caused by each.

**21-4** List the causative agent, mode of transmission, and clinical symptoms of *Pseudomonas* dermatitis, otitis externa, acne, and Buruli ulcer.

**21-5** List causative agents, modes of transmission, and symptoms of warts, smallpox, monkeypox, chickenpox, shingles, cold sores, measles, rubella, fifth disease, hand-foot-and-mouth disease, and roseola.

**21-6** Differentiate cutaneous from subcutaneous mycoses, and provide an example of each.

**21-7** List the causative agent and predisposing factors for candidiasis.

**21-8** List the causative agent, mode of transmission, clinical symptoms, and treatment for scabies and pediculosis.

Rashes and lesions on the skin do not necessarily indicate an infection of the skin; in fact, many skin lesions are actually caused by systemic diseases affecting internal organs. Preliminary diagnosis is often based on the appearance of the rash, so it is important to understand the terms that describe rashes. For example, small, fluid-filled lesions are **vesicles** (**Figure 21.2a**). Vesicles larger than about 1 cm in diameter are termed **bullae** (Figure 21.2b). Flat, reddened lesions are known as **macules** (Figure 21.2c). Raised lesions are called **papules** or, when they contain pus, **pustules** (Figure 21.2d). A skin rash that arises from disease conditions is called an **exanthem**; on mucous membranes, such as the interior of the mouth, such a rash is called an **enanthem**.

## Bacterial Diseases of the Skin

Two genera of bacteria, *Staphylococcus* and *Streptococcus*, are frequent causes of skin-related diseases and merit special discussion. We will also discuss these bacteria in subsequent chapters in relation to other organs and conditions. Superficial staphylococcal and streptococcal infections of the skin are very common. Both genera also may produce invasive enzymes and damaging toxins.

### Staphylococcal Skin Infections

Staphylococci are spherical gram-positive bacteria that form irregular clusters like grapes (see Figure 4.1d, page 74, and Figure 11.22, page 314). For almost all clinical purposes, these bacteria can be divided into those that produce **coagulase**, an enzyme that coagulates (clots) fibrin in blood, and those that do not.

Coagulase-negative strains, such as *Staphylococcus epidermidis*, are very common on the skin, where they may represent 90% of the normal microbiota. They are generally pathogenic only when the skin barrier is broken or is invaded by medical procedures, such as the insertion and removal of catheters into veins. On the surface of the catheter (**Figure 21.3**), the bacteria are surrounded by a slime layer of capsular material that protects them from desiccation and disinfectants (see discussions of biofilms on pages 157–159). This is a primary factor in their importance as a healthcare-associated pathogen.

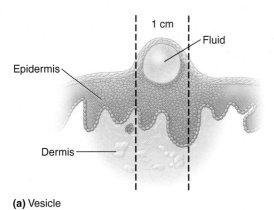

**(a)** Vesicle

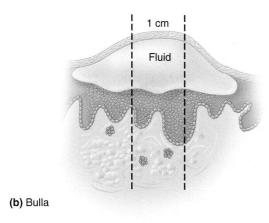

**(b)** Bulla

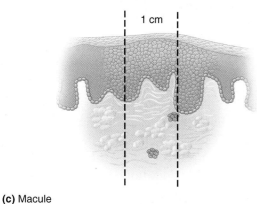

**(c)** Macule

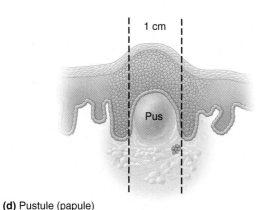

**(d)** Pustule (papule)

**Figure 21.2 Skin lesions. (a)** Vesicles are small, fluid-filled lesions. **(b)** Bullae are larger fluid-filled lesions. **(c)** Macules are flat lesions that are often reddish. **(d)** Papules are raised lesions; when they contain pus, as shown here, they are called pustules.

**Q** Are these skin lesions exanthems or enanthems?

*S. aureus* is the most pathogenic of the staphylococci (also see the discussion of MRSA in Chapter 20 on page 568). It is a permanent resident of the nasal passages of 20% of the population, and an additional 60% carry it there occasionally. It can survive for months on surfaces. Typically, it forms golden-yellow colonies. This pigmentation protects against the antimicrobial effects of sunlight; mutants without it are also more susceptible to killing by neutrophils. Compared to its more innocuous relative *S. epidermidis*, *S. aureus* has about 300,000 more base pairs in its genome—much of it devoted to an impressive array of virulence factors and means of evading host defenses. Almost all pathogenic strains of *S. aureus* are coagulase-positive. This is significant because there is a high correlation between the bacterium's ability to form coagulase and its production of damaging toxins, several of which facilitate the spread of the organism in tissue, damage tissue, or are lethal to host defenses. In addition, some strains can cause life-threatening sepsis (Chapter 23, pages 652–653), and others produce *enterotoxins* that affect the gastrointestinal tract (see Chapter 25, page 728).

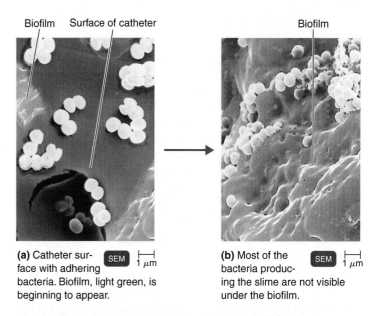

**(a)** Catheter surface with adhering bacteria. Biofilm, light green, is beginning to appear.

**(b)** Most of the bacteria producing the slime are not visible under the biofilm.

**Figure 21.3 Coagulase-negative staphylococci.** These slime-producing bacteria are the most common causative agents of infection by indwelling devices. They adhere to surfaces such as the plastic catheter in the photos. Once they have adhered to the surface **(a)** they begin to divide. Eventually **(b)** the entire surface is coated with a biofilm containing the organisms.

**Q** What is the most likely source of the bacteria that grew on the catheter?

# Normal Skin Microbiota and Our Immune System: Allies in "Skin Wars"

Our skin physically rubs up against countless varieties of microbes daily, forcing the established microbiota to continually defend territory against newcomers who would otherwise displace them on the epidermis. In this battle, healthy microbiota have evolved mechanisms that benefit us as well as them.

For example, *Staphylococcus epidermidis*, a nonpathogenic bacterium that features prominently in healthy skin microbiota, enlists our own immune system to help prevent growth of a competing *Staphylococcus* species that happens to be pathogenic. *Sta. epidermidis* does this by activating a Toll-like receptor on epidermal cells, which induces skin cells to produce beta-defensin. Beta-defensin is an antimicrobial peptide that works as part of our innate immune system. It inhibits growth of *Sta. aureus* and group A *Streptococcus* – but not *Sta. epidermidis*.

*Sta. epidermidis* also ferments glycerol, the molecular basis of fats that is naturally produced by skin cells. During fermentation, these bacteria make succinic acid, which has been shown in several studies to inhibit *Cutibacterium acnes* from growing in mice and in disk-diffusion assays. One day, chronic, nonhealing wounds may be treated using the benign skin microbiome as probiotics. In fact, clinical trials are already under way for a lotion containing microbes that inhibit growth of *Sta. aureus* as a treatment for eczema.

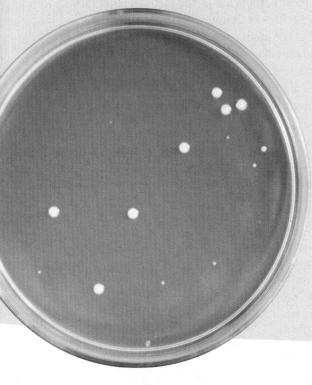

*Staphylococcus aureus* (large colonies) and *Sta. epidermidis* (small colonies).

Once *S. aureus* infects the skin, it stimulates a vigorous inflammatory response, and macrophages and neutrophils are attracted to the site of infection. However, the bacteria have several ways to evade these host defenses. Most strains secrete a protein that blocks chemotaxis of neutrophils to the infection site, and if the bacterium does encounter phagocytic cells, it often produces toxins that kill them. It is resistant to opsonization (see page 451), but, failing this, it can survive well within the phagosome. Other proteins it secretes neutralize the antimicrobial peptide defensins on skin, and its cell wall is lysozyme resistant (see page 83). It sometimes responds to the immune system as a superantigen (see page 433) but often is able to evade the adaptive immune system entirely. All humans possess antibodies against *S. aureus*, but they do not effectively prevent repeated infections. Antibiotic-resistant strains of *S. aureus* have emerged in hospitals and in the community. (See the Clinical Focus box on page 600.)

Because this organism is so commonly present in human nasal passages, it is often transported from there to the skin, where it can enter the body through natural openings such as the hair follicle (see Figure 21.1). Such infections, called **folliculitis**, often occur as pimples. The infected follicle of an eyelash is called a **sty**. A more serious hair follicle infection is the **furuncle (boil)**, which is a type of **abscess**, a localized region of pus surrounded by inflamed tissue. Antibiotics do not penetrate well into abscesses, and the infection is therefore difficult to treat. Draining pus from the abscess is frequently a preliminary step to successful treatment.

When the body fails to wall off a furuncle, neighboring tissue can be progressively invaded. The extensive damage is called a **carbuncle**, a hard, round deep inflammation of tissue under the skin. At this stage of infection, the patient usually exhibits the symptoms of generalized illness with fever.

Staphylococci are the most important causative organism of **impetigo**. This is a highly contagious skin infection mostly affecting children 2 to 5 years of age, among whom it is spread by direct contact. *Streptococcus pyogenes*, a pathogen that we will be discussing shortly, can also cause impetigo, although in fewer cases. Sometimes both *Sta. aureus* and *Str. pyogenes* are involved. *Nonbullous impetigo* (see the bulla in Figure 21.2b) is the more common form. The pathogen usually enters through some minor break in the skin. The infection can also spread to surrounding areas—a process called

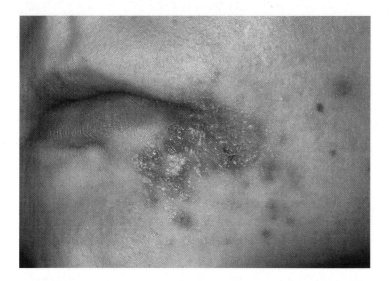

**Figure 21.4 Lesions of impetigo.** This disease is characterized by isolated pustules that become crusted.

Ⓠ **What bacteria most often cause impetigo?**

*autoinoculation*. Symptoms result from the host's response to the infection. The lesions eventually rupture and form light-colored crusts, as shown in **Figure 21.4**. Topical antibiotics are sometimes applied, but the lesions generally heal without treatment and without scarring.

The other type of impetigo, *bullous impetigo*, is caused by a staphylococcal toxin and is a localized form of staphylococcal **scalded skin syndrome**. Actually, there are two exotoxins; exfoliative toxin A, which remains localized, causes bullous impetigo, and exfolioative toxin B, which circulates to distant sites, causes scalded skin syndrome, as shown in **Figure 21.5**. Both toxins cause a separation of the skin layers, *exfoliation*. Outbreaks of bullous impetigo are a frequent problem in hospital nurseries, where the condition is known as **pemphigus neonatorum**, or *impetigo of the newborn*. (See the discussion of hexachlorophene in Chapter 7, page 190.)

Scalded skin syndrome is also characteristic of the late stages of **toxic shock syndrome (TSS)**. In this potentially life-threatening condition, fever, vomiting, and a sunburn-like rash are followed by shock and sometimes organ failure, especially of the kidneys. TSS originally became known as a result of staphylococcal growth associated with the use of a new type of highly absorbent vaginal tampon; the correlation is especially high for cases in which the tampons remain in place too long. A novel staphylococcal toxin called *toxic shock syndrome toxin 1 (TSST-1)* is formed at the growth site and circulates in the bloodstream. The symptoms are thought to be a result of the superantigenic properties of the toxin. (See the discussion of cytokine storms in Chapter 17 on page 478.)

Today a minority of the cases of TSS are associated with menstruation. Nonmenstrual TSS occurs from staphylococcal infections that follow nasal surgery in which absorbent packing is used, after surgical incisions, and in women who have just given birth.

## Streptococcal Skin Infections

Streptococci are gram-positive spherical bacteria. Unlike staphylococci, streptococcal cells usually grow in chains (see Figure 11.23, page 314). Prior to division, the individual cocci elongate on the axis of the chain, and then the cells divide (see Figure 4.1a, page 74). Streptococci cause a wide range of disease conditions beyond those covered in this chapter, including meningitis, pneumonia, sore throat, otitis media, endocarditis, puerperal fever, and even dental caries.

As streptococci grow, they secrete toxins and enzymes, virulence factors that vary with the different streptococcal species. Among these toxins are *hemolysins*, which lyse red blood cells. Depending on the hemolysin they produce, streptococci are categorized as alpha-hemolytic, beta-hemolytic, and

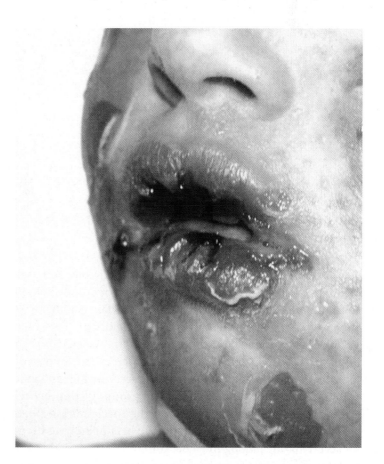

**Figure 21.5 Lesions of scalded skin syndrome.** Some staphylococci produce a toxin that causes the skin to peel off in sheets, as on the face of this child. It is especially likely to occur in children under age 2.

Ⓠ **What is the name of the toxin that produces this syndrome?**

# Macular Rashes

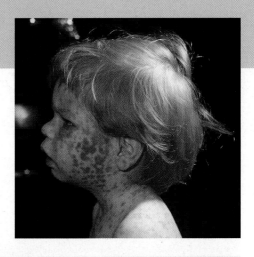

Differential diagnosis is the process of identifying a disease from a list of possible diseases that fit the information derived from examining a patient. A differential diagnosis is important for providing initial treatment and for laboratory testing. For example, a 4-year-old boy with a history of cough, conjunctivitis, and fever (38.3°C) now has a macular rash that started on his face and neck and is spreading to the rest of his body. Use the table below to identify infections that could cause these symptoms. For the solution, go to @MasteringMicrobiology.

| Disease | Pathogen | Portal of Entry | Symptoms | Method of Transmission | Treatment |
|---|---|---|---|---|---|
| **VIRAL DISEASES.** Usually diagnosed by clinical signs and symptoms and may be confirmed by serology or PCR. | | | | | |
| **Measles** (rubeola) | *Morbillivirus* | Respiratory tract | Reddish macules first appearing on face and spreading to trunk and extremities | Aerosol | No treatment; preexposure vaccine |
| **Rubella** (German measles) | *Rubivirus* | Respiratory tract | Mild disease with a macular rash resembling measles, but less extensive and disappearing in 3 days or less | Aerosol | No treatment; preexposure vaccine |
| **Fifth Disease** (erythema infectiosum) | Human parvovirus B19 | Respiratory tract | Mild disease with a macular facial rash | Aerosol | None |
| **Roseola** | *Roseolovirus* (HHV-6, HHV-7) | Respiratory tract | High fever followed by macular body rash | Aerosol | None |
| **Hand-Foot-and-Mouth Disease** | Enteroviruses | Mouth | Flat or raised rash | Aerosol; direct contact | None |
| **FUNGAL DISEASE.** Confirmed by Gram staining of skin scrapings. | | | | | |
| **Candidiasis** | *Candida albicans* | Skin; mucous membranes | Macular rash | Direct contact; endogenous* infection | Miconazole, clotrimazole (topically) |

*Endogenous infections are infections caused by microorganisms already part of the host microbiota.

gamma-hemolytic (actually nonhemolytic) streptococci (see Figure 6.9, page 162). Hemolysins can lyse not only red blood cells, but almost any type of cell. It is uncertain, though, just what part they play in streptococcal pathogenicity.

Beta-hemolytic streptococci are often associated with human disease. This group is further differentiated into serological groups, designated A through T, according to antigenic carbohydrates in their cell walls. The **group A streptococci (GAS)**, which are synonymous with the species *Streptococcus pyogenes*, are the most important of the beta-hemolytic streptococci. They are among the most common human pathogens and are responsible for a number of human diseases—some of them deadly. These streptococci also produce certain enzymes, called *streptolysins*, that lyse red blood cells and are toxic to neutrophils. This group of pathogens is divided into over 80 immunological types according to the antigenic properties of the M protein found in some strains (**Figure 21.6**).

This protein is external to the cell wall on a fuzzy layer of fimbriae. The M protein prevents the activation of complement and allows the microbe to evade phagocytosis and killing by neutrophils (see page 463). It also appears to help the bacteria adhere to and colonize mucous membranes. Another virulence factor of the GAS is their capsule of hyaluronic acid. Exceptionally virulent strains have a mucoid appearance on blood-agar plates from heavy encapsulation and are rich in M protein. Hyaluronic acid is poorly immunogenic (it resembles human connective tissue), and few antibodies against the capsule are produced.

The GAS produce substances that promote the rapid spread of infection through tissue and by liquefying pus. Among these are *streptokinases* (enzymes that dissolve blood clots), *hyaluronidase* (an enzyme that dissolves the hyaluronic acid in the connective tissue, where it serves to cement the cells together), and *deoxyribonucleases* (enzymes that degrade DNA).

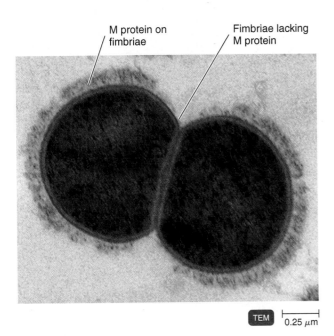

Figure 21.6 The M protein of group A beta-hemolytic
streptococci. Part of each cell shown carries M protein on fimbriae.

M protein on fimbriae

Fimbriae lacking M protein

TEM  0.25 μm

**Q** Is the M protein more likely to be antigenic than a
polysaccharide capsule?

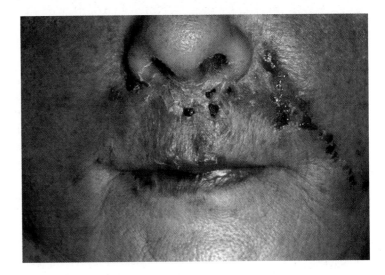

**Figure 21.7 Lesions of erysipelas, caused by group A beta-hemolytic streptococcal toxins.**

**Q** What is the name of the toxin that produces skin reddening?
(Hint: See Chapter 15, page 436.)

Streptococcal skin infections are generally localized, but if the bacteria reach deeper tissue, they can be highly destructive.

When *S. pyogenes* infects the dermal layer of the skin, it causes a serious disease, **erysipelas**. In this disease, the skin erupts into reddish patches with raised margins **(Figure 21.7)**. It can progress to local tissue destruction and even enter the bloodstream, causing sepsis (page 652). The infection usually appears first on the face and often has been preceded by a streptococcal sore throat. High fever is common. Fortunately, *S. pyogenes* has remained sensitive to β-lactam-type antibiotics, especially cephalosporin.

Some 1000 cases of invasive group A streptococcal infection, caused by the "flesh-eating bacteria," occur each year in the United States. The infection may be precipitated by minor breaks in the skin, and early symptoms are often unrecognized, delaying diagnosis and treatment—with serious consequences. Once established, **necrotizing fasciitis (Figure 21.8)** may destroy tissue as rapidly as a surgeon can remove it, and mortality rates from systemic toxicity can exceed 40%. Streptococci are considered the most common causative organism, although other bacteria cause similar conditions. Pyrogenic toxins produced by certain streptococcal M-protein types act as superantigens, causing the immune system to contribute to the damage. Broad-spectrum antibiotics are usually prescribed because of the possibility that multiple bacterial pathogens are present.

Necrotizing fasciitis is often associated with **streptococcal toxic shock syndrome (streptococcal TSS)**, which resembles staphylococcal TSS, described earlier in the chapter. In cases of streptococcal TSS, a rash is less likely to be present, but bacteremia is more likely to occur. M proteins shed from the surfaces of these streptococci form a complex with fibrinogen that binds to neutrophils. This causes activation of neutrophils, precipitating the release of damaging enzymes and consequent shock and organ damage. The mortality rate is much higher than with staphylococcal TSS—up to 80% has been reported.

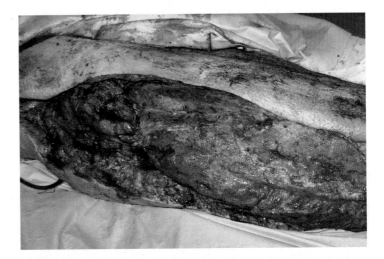

**Figure 21.8 Necrotizing fasciitis of a leg due to group A streptococci.** Extensive damage to the fascia (sheet of connective tissue binding the muscles) may require reconstructive surgery or even amputation of limbs.

**Q** What is the name of the primary toxin that leads to tissue invasion by the pathogen?

Vesicular and Pustular Rashes

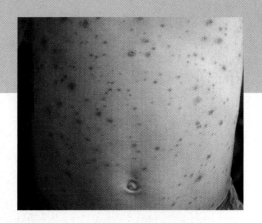

An 8-year-old boy has a rash consisting of vesicular lesions of 5 days' duration on his neck and stomach. Within 5 days, 73 students in his elementary school have an illness matching the case definition for this disease. Use the table below to provide a differential diagnosis and identify infections that could cause these symptoms. For the solution, go to @MasteringMicrobiology.

| Disease | Pathogen | Portal of Entry | Symptoms | Method of Transmission | Treatment |
|---|---|---|---|---|---|
| BACTERIAL DISEASE. Usually diagnosed by culturing the bacteria. | | | | | |
| Impetigo | *Staphylococcus aureus* | Skin | Vesicles on skin | Direct contact; fomites | Topical antibiotics |
| VIRAL DISEASES. Usually diagnosed by clinical signs and symptoms and may be confirmed by serology or PCR. | | | | | |
| Smallpox (variola) | Smallpox (variola) virus | Respiratory tract | Pustules that may be nearly confluent on skin | Aerosol | None |
| Monkeypox | Monkeypox virus | Respiratory tract | Pustules, similar to smallpox | Direct contact with or aerosols from infected small mammals | None |
| Chickenpox (varicella) | *Varicellovirus* (HHV-3) | Respiratory tract | Vesicles in most cases confined to face, throat, and trunk | Aerosol | Acyclovir for adults; preexposure vaccine |
| Shingles (herpes-zoster) | *Varicellovirus* (HHV-3) | Endogenous* infection of peripheral nerves | Vesicles typically on one side of waist, face and scalp, or upper chest | Recurrence of latent chickenpox infection | Acyclovir; preventive vaccine |
| Herpes Simplex | *Simplexvirus* (HHV-1) | Skin; mucous membranes | Vesicles around mouth; can also affect other areas of skin and mucous membranes | Initial infection by direct contact; recurring latent infection | Acyclovir |

*Endogenous infections are infections caused by microorganisms already part of the host microbiota.

### Infections by Pseudomonads

Pseudomonads are aerobic gram-negative rods that are widespread in soil and water. Capable of surviving in any moist environment, they can grow on traces of unusual organic matter, such as soap films or cap liner adhesives used for many product containers. They are resistant to many antibiotics and disinfectants. The most prominent species is *Pseudomonas aeruginosa*, which is considered the model of an opportunistic pathogen.

Pseudomonads frequently cause outbreaks of **Pseudomonas dermatitis**. This is a self-limiting rash of about 2 weeks' duration, often associated with swimming pools and pool-type saunas and hot tubs. When many people use these facilities, the alkalinity rises, and the chlorines become less effective; at the same time, the concentration of nutrients that support the growth of pseudomonads increases. Hot water causes hair follicles to open wider, facilitating the entry of bacteria. Competition swimmers are often troubled with **otitis externa**, or *"swimmer's ear,"* a painful infection of the external ear canal leading to the eardrum that is frequently caused by pseudomonads.

*P. aeruginosa* produces several exotoxins that account for much of its pathogenicity. It also produces an endotoxin. *P. aeruginosa* often grows in dense biofilms that contribute to its frequent identification as a cause of healthcare-associated infections of indwelling medical tubes or devices. This bacterium

# Patchy Redness and Pimple-Like Conditions

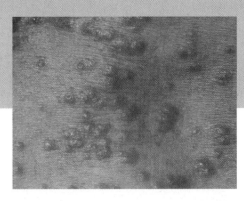

An 11-month-old boy comes to a clinic with a 1-week history of an itchy red rash under his arms. The rash seems to bother him more at night, and he has no fever. Use the table below to provide a differential diagnosis and identify infections that could cause these symptoms. For the solution, go to @MasteringMicrobiology.

| Disease | Pathogen | Portal of Entry | Symptoms | Method of Transmission | Treatment |
|---|---|---|---|---|---|
| **BACTERIAL DISEASES.** Usually diagnosed by culturing the bacteria. | | | | | |
| **Folliculitis** | *Staphylococcus aureus* | Hair follicle | Infection of hair follicle | Direct contact; fomites; endogenous* infection | Draining of pus; topical antibiotics |
| **Toxic Shock Syndrome** | *Staphylococcus aureus* | Surgical incisions | Fever, rash, shock | Endogenous* infection | Antibiotics, depending on sensitivity profile (antibiogram) |
| **Necrotizing Fasciitis** | *Streptococcus pyogenes* | Skin abrasions | Extensive soft-tissue destruction | Direct contact | Surgical tissue removal; broad-spectrum antibiotics |
| **Erysipelas** | *Streptococcus pyogenes* | Skin; mucous membranes | Reddish patches on skin; often with high fever | Endogenous* infection | Cephalosporin |
| ***Pseudomonas* Dermatitis** | *Pseudomonas aeruginosa* | Skin abrasions | Superficial rash | Swimming water; hot tubs | Usually self-limiting |
| **Otitis Externa** | *Pseudomonas aeruginosa* | Ear | Superficial infection of external ear canal | Swimming water | Fluoroquinolones |
| **Acne** | *Cutibacterium (Propionibacterium) acnes* | Sebum channels | Inflammatory lesions originating with accumulations of sebum that rupture a hair follicle | Direct contact | Benzoyl peroxide, isotretinoin, azelaic acid |
| **Buruli Ulcer** | *Mycobacterium ulcerans* | Skin | Localized swelling or hardness progressing to deep ulcer | Contaminated water | Antimycobacterial drugs |
| **VIRAL DISEASE.** Usually diagnosed by clinical signs and symptoms. | | | | | |
| **Warts** | *Papillomavirus* | Skin | A horny projection of the skin formed by proliferation of cells | Direct contact | Removal by liquid nitrogen cryotherapy, electro-desiccation, acids, lasers |
| **FUNGAL DISEASES.** Diagnosis is confirmed by microscopic examination. | | | | | |
| **Ringworm (tinea)** | *Microsporum, Trichophyton, Epidermophyton* | Skin | Skin lesions of highly varied appearance; on scalp may cause local loss of hair | Direct contact; fomites | Griseofulvin (orally); miconazole, clotrimazole (topically) |
| **Sporotrichosis** | *Sporothrix schenckii* | Skin abrasions | Ulcer at site of infection spreading into nearby lymphatic vessels | Soil | Potassium iodide solution (orally) |
| **PARASITIC INFESTATIONS.** Diagnosis is confirmed by microscopic examination of parasite. | | | | | |
| **Scabies** | *Sarcoptes scabiei* (mite) | Skin | Papules, itching | Direct contact | Gamma benzene hexachloride, permethrin (topically) |
| **Pediculosis (lice)** | *Pediculus humanus capitis* | Skin | Itching | Primarily direct contact; possible fomites such as bedding, combs | Topical insecticide preparations |

*Endogenous infections are infections caused by microorganisms already part of the host microbiota.

# Infections in the Gym

As you read through this box, you will encounter a series of questions that epidemiologists ask themselves as they try to trace an outbreak to its source. Try to answer each question before going on to the next one.

1. Jason F., a 21-year-old college wrestler, goes to the college health center with an 11 cm × 5 cm area of redness on his right thigh. It is swollen and warm and tender when touched. His temperature is normal. He is given trimethoprim-sulfamethoxazole.
   What is Jason's probable diagnosis?

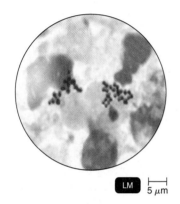

LM | 5 μm

**Figure A**

Negative control     Isolate from patient

**Figure B**

2. Jason probably has some form of bacterial skin infection, for which he is prescribed antibiotics. After 2 days, Jason returns and says the area is worse. Examination reveals a broader area of redness. He is diagnosed with cellulitis. The pustule is opened and drained.
   What do you need to do now?

3. The pus is sent to the lab for a Gram stain and a coagulase test on the culture. The results of the Gram stain and the coagulase test are shown in Figure A and Figure B, respectively.
   What is the cause of the infection?

4. The presence of gram-positive, coagulase-positive cocci indicates *Staphylococcus aureus*. The bacterium is sent for sensitivity testing.
   Why is sensitivity testing necessary?

5. Sensitivity testing is necessary to identify the antibiotic that will be most effective in killing the bacteria. The results are shown in Figure C. (P = penicillin, M = methicillin, E = erythromycin, V = vancomycin, X = trimethoprim-sulfamethoxazole.)
   What treatment is appropriate?

6. Based on the sensitivity testing, the most appropriate treatment is vancomycin. Over a 3-month period, 47 wrestlers who competed in a tournament, including one member from each team, reported skin lesions. Seven are hospitalized; one receives surgical debridement and skin grafts.
   What is the most likely source of the methicillin-resistant *Staphylococcus aureus* (MRSA)?

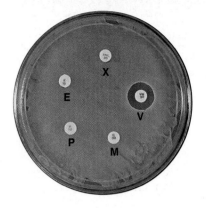

**Figure C**

Three factors might have contributed to transmission in this outbreak: (1) Abrasions and other skin trauma, which can facilitate entry of pathogens, were likely to have occurred. (2) Wrestling involves frequent physical contact among players, which allows for easy person-to-person transmission of *S. aureus* and other skin microbiota. (3) Shared equipment that is not cleaned between users could be a vehicle for *S. aureus* transmission.

Investigation of outbreaks of MRSA among professional athletes showed that all of the infections occurred at the site of a turf burn and rapidly progressed to large abscesses that required surgery to drain. MRSA was recovered from whirlpools and taping gel and from 35 of the 84 nasal swabs from players and staff members.

The CDC recommends cleaning locker rooms and shared equipment with detergent-based cleaners and excluding wrestlers with uncovered skin lesions from competition.

*Source:* Adapted from *MMWR* 64(20):559–560, May 29, 2015.

---

is also a serious opportunistic pathogen for patients with the genetic lung disease cystic fibrosis; biofilm formation plays a prominent part in this.

*P. aeruginosa* is also a very common and serious opportunistic pathogen in burn patients, particularly those with second- and third-degree burns. Infection may produce blue-green pus, whose color is caused by the bacterial pigment **pyocyanin**. Of concern in many hospitals is the ease with which *P. aeruginosa* grows in flower vases, mop water, and even dilute disinfectants.

The relative resistance to antibiotics that characterizes pseudomonads is still a problem. However, in recent years,

several new antibiotics have been developed, and chemotherapy to treat these infections is not as restricted as it once was. The quinolones and the newer, antipseudomonal β-lactam antibiotics are the usual drugs of choice. Silver sulfadiazine is very useful in the treatment of burn infections by *P. aeruginosa*.

## Buruli Ulcer

**Buruli ulcer**, named for a now-renamed region of Uganda in Africa, is an emerging disease found primarily in western and central Africa. Although widespread in tropical Africa, it was first accurately described in Australia in 1948 and since then has been reported in localized tropical and subtropical areas around the globe—including Mexico and areas of South America. The disease is caused by *Mycobacterium ulcerans*, which is similar to the mycobacteria that cause tuberculosis and leprosy. When the pathogen is introduced into the skin, it causes a disease that progresses slowly with few serious early signs or symptoms. Eventually, however, the result is a deep ulcer that often becomes massive and seriously damaging. Untreated, this can be so extensive as to require amputation or plastic surgery. This tissue damage is attributed to the production of a toxin, *mycolactone*. Epidemiologically, the infection is associated with contact with swamps and slow-flowing waters. The pathogen probably enters through a break in the skin from a minor cut or insect bite.

In 2004, the World Health Organization (WHO) identified Buruli ulcer as a global threat to public health. By 2014, WHO's early detection and treatment initiative had decreased the number of new cases by 50%.

Buruli ulcer is diagnosed primarily by the appearance of the ulcer, although awareness is higher in endemic areas, and is treated by antimycobacterial drugs such as streptomycin-rifampicin combinations.

## Acne

**Acne** is probably the most common skin disease in humans, affecting an estimated 17 million people in the United States. More than 85% of all teenagers have the problem to some degree. Acne can be classified by type of lesion into three categories: comedonal acne, inflammatory acne, and nodular cystic acne. They require different treatments.

Normally, skin cells that are shed inside the hair follicle are able to leave, but acne develops when cells are shed in higher than normal numbers, combine with sebum, and the mixture clogs the follicle. As sebum accumulates, whiteheads (comedos) form; if the blockage protrudes through the skin, a blackhead (comedone) forms. The dark color of blackheads is due not to dirt, but to lipid oxidation and other causes. Topical agents do not affect sebum formation, which is a root cause of acne and depends on hormones such as estrogens or androgens. Diet has no known effect on sebum production, but pregnancy, some hormone-based contraceptive methods, and hormonal changes with age do affect sebum formation and influence acne.

**Comedonal (mild) acne** is usually treated with topical agents such as azelaic acid, salicyclic acid preparations, or retinoids (which are derivatives of vitamin A, such as tretinoin, tazarotene, or adapalene). These topical agents do not affect sebum formation.

**Inflammatory (moderate) acne** arises from bacterial action, especially *Cutibacterium acnes*, an anaerobic diphtheroid commonly found on the skin. *P. acnes* has a nutritional requirement for glycerol in sebum; in metabolizing the sebum, it forms free fatty acids that cause an inflammatory response. Neutrophils that secrete enzymes that damage the wall of the hair follicle are attracted to the site. The resulting inflammation leads to the appearance of pustules and papules. At this stage, therapy is usually focused on preventing formation of sebum; topical agents are not effective for this.

Inflammatory acne can also be treated by targeting *P. acnes* with antibiotics. The familiar nonprescription acne treatments containing benzoyl peroxide are effective against some bacteria, especially *P. acnes*, and also cause drying that helps loosen plugged follicles. Benzoyl peroxide is also available as a gel and in products where it is combined with antibiotics such as clindamycin and erythromycin.

Alternatives to chemical treatments have been approved by the U.S. Food and Drug Administration (FDA) for treatment of mild to moderate acne. The CLEARLight® system, which bathes the skin with high-intensity blue light (405–420 nm), and Smoothbeam® treatment, which uses laser light, penetrate the skin surface to speed healing and prevent pimples from forming. Also approved is a handheld device, ThermaClear®, which delivers a brief pulse of heat to the lesions.

Some patients with acne progress to **nodular cystic (severe) acne**. Nodular cystic acne is characterized by nodules or cysts, which are inflamed lesions filled with pus deep within the skin (**Figure 21.9**). These leave prominent scars on the face and upper body, which often leave psychological scars as well. An effective treatment for cystic acne is isotretinoin, which reduces the formation of sebum. Under the trade name of Accutane®, its distribution in the United States has been discontinued by the manufacturer. It is, however, distributed outside this country under the name Roaccutane®. Anyone considering the use of the drug should be warned that it is highly *teratogenic*, meaning it can cause serious damage to the developing fetus in a

**CLINICAL CASE**

The similarity in the siblings' rashes prompts Molly to reexamine her records and obtain more detailed information about the children who came to her office with comparable rashes.

After speaking to the children's parents, Molly learns that all five children have been to the same community swimming pool in the past 72 hours. Molly notifies the health department; they contact the only other general medical practice in this small town and obtain a list of similar cases. In these cases, the patients have rashes on the chest and abdomen (90%), buttocks (67%), arms (71%), legs (86%), and also the hands, feet, and head and neck.

**What pathogens can cause itchy, pimple-like rashes?**

591   601   608   610   613

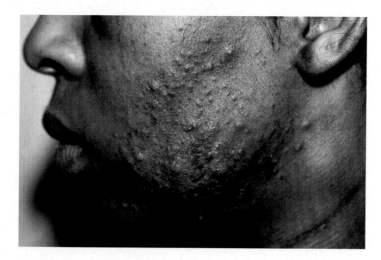

**Figure 21.9 Severe acne.**

**Q** Isotretinoin often leads to dramatic improvement for cases of severe acne, but what precautions must be observed?

pregnant woman. Other side effects may include inflammatory bowel disease and ulcerative colitis.

> **CHECK YOUR UNDERSTANDING**
>
> ✔ **21-3** Which bacterial species features the virulence factor M protein?
>
> ✔ **21-4** What is the common name for otitis externa?

## Viral Diseases of the Skin

Many viral diseases, although systemic and transmitted by respiratory or other routes, are most apparent by their effects on the skin. Although not always very significant in adults, several of these diseases, such as rubella, chickenpox, fifth disease, and herpes simplex, can cause serious damage in a developing fetus.

### Warts

**Warts**, or papillomas, are generally benign skin growths caused by viruses. It was long known that warts can be transmitted from one person to another by contact, even sexually, but it was not until 1949 that viruses were identified in wart tissues. More than 50 types of *papillomavirus* are now known to cause different kinds of warts, often with greatly varying appearances.

After infection, there is an incubation period of several weeks before the warts appear. The most common medical treatments for warts are to apply extremely cold liquid nitrogen (cryotherapy), dry them with an electrical current (electrodesiccation), or burn them with acids. There is evidence that compounds containing salicylic acids are especially effective. Topical application of prescription drugs such as podofilox, which inhibits cell division, or imiquimod, which activates

toll-like receptors (page 448), is often effective. Warts that do not respond to any other treatments can be treated with lasers or injected with bleomycin, an antitumor drug.

Although warts are not a form of cancer, some skin and cervical cancers are associated with certain papillomaviruses. Genital warts is the most common sexually transmitted infection (Chapter 26, page 778).

### Smallpox (Variola)

During the Middle Ages, an estimated 80% of the population of Europe contracted **smallpox\*** at some time during their lives. Those who recovered from the disease retained disfiguring scars. The disease, introduced by American colonists, was even more devastating to Native Americans, who had had no previous exposure and thus little resistance.

Smallpox is caused by an orthopoxvirus known as the smallpox (variola) virus. There are two basic forms of this disease: **variola major**, with a mortality rate of 20–60% and over 80% in children, and **variola minor**, with a mortality rate of less than 1%.

Transmitted by the respiratory route, the viruses infect many internal organs before they eventually move into the bloodstream, infecting the skin and producing more recognizable symptoms. The growth of the virus in the epidermal layers of the skin causes lesions that become pustular after 10 days or so (**Figure 21.10**).

Smallpox was the first disease to which immunity was artificially induced (see pages 10–11) and the first to be eradicated from the human population. The last victim of a natural case of smallpox is believed to be an individual who recovered from variola minor in 1977 in Somalia. (However, 10 months after this case, there was a smallpox fatality in England caused by escape of the virus from a hospital research laboratory.) The eradication of smallpox was possible because an effective vaccine was developed and because there are no animal host reservoirs for the disease. A concerted worldwide vaccination effort was coordinated by the World Health Organization.

Today, only two sites are known to maintain the smallpox virus, one in the United States and one in Russia. Dates for the destruction of these collections have been set and then postponed.

Smallpox would be an especially dangerous agent for bioterrorism. Vaccination in the United States ended in the early 1970s. People who were vaccinated prior to that time have waning immunity; however, they probably have some remaining protection that would at least moderate the disease. Stocks

---

\*The origin of the name *smallpox* reportedly arose in the late fifteenth century in France, where syphilis had just been introduced. These patients exhibited a severe skin rash called *la grosse verole*, or "the great pox." The rash was compared with that of an endemic disease of the time, which was then referred to as *la petite verole*, or "the small pox." In English, the endemic disease became known as smallpox.

in Africa and is endemic there in small animals. There are occasional outbreaks among humans in those areas, and one outbreak of more than 50 cases in the United States in 2003 was attributed to contact with pet prairie dogs. These animals apparently were infected by being housed in pet stores with Gambian giant rats imported from western Africa. Monkeypox closely resembles smallpox in symptoms and, while smallpox was endemic, was probably mistaken for it. The *monkeypox virus*, like smallpox virus, is an orthopoxvirus, and vaccination for smallpox has a protective effect. Monkeypox is known to jump from animals to humans, but fortunately its transmission from human to human has been very limited. The World Health Organization is monitoring recent outbreaks to see whether human-to-human transmission increases.

### Chickenpox (Varicella) and Shingles (Herpes Zoster)

**Chickenpox** (varicella) is a relatively mild disease when contracted, as it usually is, in childhood. The mortality rate from chickenpox is very low and is usually from complications such as encephalitis (infection of the brain) or pneumonia. Almost half of such deaths occur in adults.

Chickenpox (**Figure 21.11a**) is the result of an initial infection with a herpesvirus *Varicellovirus*. The species is varicellazoster or officially, *human herpesvirus* (HHV-3); see Chapter 13, page 378. The disease is acquired when the virus enters the respiratory system, and the infection localizes in skin cells after about 2 weeks. The infected skin is vesicular for 3 to 4 days. During that time, the vesicles fill with pus, rupture, and form a scab before healing. Lesions are mostly confined to the face, throat, and lower back but can also occur on the chest and shoulders. If varicella infection occurs during early pregnancy, serious fetal damage may occur in about 2% of cases. (See also Diseases in Focus 21.2.)

**Reye's syndrome** is an occasional severe complication of chickenpox, influenza, and some other viral diseases. A few days after the initial infection has receded, the patient persistently vomits and exhibits signs of brain dysfunction, such as extreme drowsiness or combative behavior. Coma and death can follow. At one time, the death rate of reported cases approached 90%, but this rate has been declining with improved care and is now 30% or lower when the disease is recognized and treated in time. Survivors may show neurological damage, especially if very young. Reye's syndrome affects children and teenagers almost exclusively. The use of aspirin to lower fevers in chickenpox and influenza increases the chances of acquiring Reye's syndrome.

Like all herpesviruses, a characteristic of HHV-3 is its ability to remain latent within the body. Following a primary infection, the virus enters the peripheral nerves and moves to a central nerve ganglion where it persists as viral DNA. Humoral antibodies cannot penetrate into the nerve cell, and

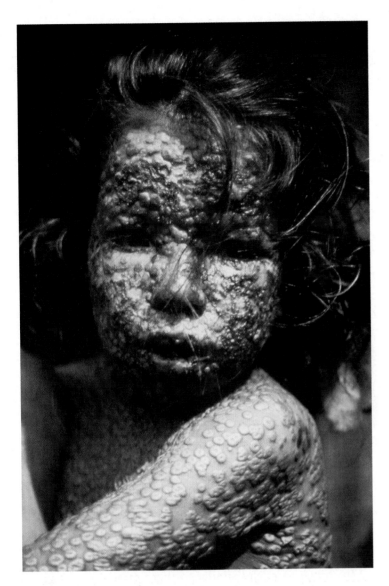

**Figure 21.10 Smallpox lesions.** In some severe cases, the lesions nearly run together (are confluent).

**Q** How do these lesions differ from chickenpox?

of smallpox vaccine are being accumulated as a precaution. No general vaccination program of the entire population is contemplated. However, certain groups, among them military and health care workers, may be an exception. Administered to the general population, the vaccine would cause a significant number of deaths, especially among immunosuppressed individuals.

Complications from the smallpox vaccine can be treated with vaccinia immune globulin, which contains antibodies to the virus. The antiviral drug, cidofovir, can also be administered.

With the disappearance of smallpox, there has been some concern with a similar disease, **monkeypox**. This disease was first identified in 1958 in laboratory monkeys that originated

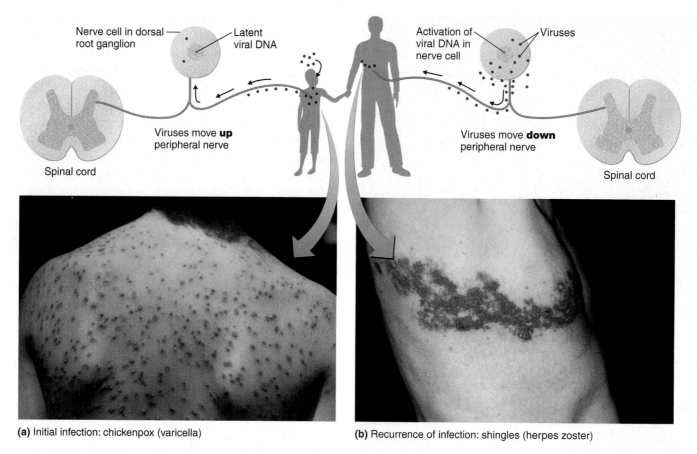

(a) Initial infection: chickenpox (varicella)

(b) Recurrence of infection: shingles (herpes zoster)

**Figure 21.11 Chickenpox (varicella) and shingles (herpes zoster). (a)** Initial infection with the virus, usually during childhood, causes chickenpox. The lesions are vesicles, eventually becoming pustules that rupture and form scabs. The virus then moves to a dorsal root ganglion near the spine, where it remains latent indefinitely. **(b)** Later, usually in late adulthood, the latent virus becomes reactivated, causing shingles. Reactivation can be caused by stress or weakening of the immune system. The skin lesions are vesicles.

**Q** Does the photo in (a) illustrate an early or late stage of chickenpox?

because no viral antigens are expressed on the surface of the nerve cell, cytotoxic T cells are not activated. Therefore, neither arm of the adaptive immune system disturbs the latent virus.

Latent HHV-3 is located in the dorsal root ganglion near the spine. Later, perhaps as long as decades later, the virus may be reactivated (Figure 21.11b). The trigger can be stress or simply the lower immune competence associated with aging. The virions produced by the reactivated DNA move along the peripheral nerves to the cutaneous sensory nerves of the skin, where they cause a new outbreak of the virus in the form of **shingles** (herpes zoster).

Shingles is simply a different expression of the virus that causes chickenpox: different because the patient, having had chickenpox, now has partial immunity to the virus. Exposing unvaccinated children to shingles has led to their contracting chickenpox. Shingles seldom occurs in people under age 20, and by far the highest incidence is among older adults. It is unusual for a patient to develop shingles more than once.

In shingles, vesicles similar to those of chickenpox occur but are localized in distinctive areas. Typically, they are distributed about the waist (the name *shingles* is derived from the Latin *cingulum*, for girdle or belt), although facial shingles and infections of the upper chest and back also occur (see Figure 21.11b). The infection follows the distribution of the affected cutaneous sensory nerves and is usually limited to one side of the body at a time because these nerves are unilateral. Occasionally, such nerve infections can result in nerve damage that impairs vision or even causes paralysis. Severe burning or stinging pain is a frequent symptom; occasionally this persists for months or years, a condition called *postherpetic neuralgia*.

The antiviral drugs acyclovir, valacyclovir, and famciclovir are approved for treatment of shingles. For immunocompromised patients, in which a mortality rate of 17% is reported, and patients with ocular involvement, treatment with antivirals is mandatory.

A live, attenuated varicella vaccine was licensed in 1995. Since then, cases of the disease have declined steadily. There is evidence that the effectiveness of the vaccine, which is about 97% at outset, declines with time. Therefore, varicella in

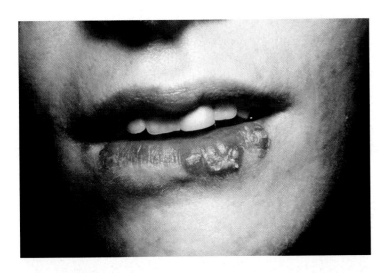

**Figure 21.12  Cold sores, or fever blisters, caused by herpes simplex virus.** Lesions are located mainly at the margin of the red area of the lips.

**Q** Why can cold sores reappear, and why do they recur in the same place?

previously vaccinated persons, called **breakthrough varicella**, is fairly common. Because the vaccine is at least partially effective, it is a relatively mild disease with a rash that does not look much like typical varicella. A booster dose of the vaccine may eventually be needed for complete control of varicella.

Another concern is that the waning effectiveness of the childhood vaccination will lead to a population of susceptible adults, for whom the disease tends to be more severe. Therefore, the current recommendation is that adults 60 years of age or older receive a newly approved *zoster vaccine* even if the subject has had chickenpox or shingles previously.

### Herpes Simplex

Herpes simplex viruses (HSV) can be separated into two identifiable groups, HSV-1 and HSV-2. The name *herpes simplex virus*, used here, is the common or vernacular name. The official names are *human herpesvirus 1* and *2*. HSV-1 is transmitted primarily by oral or respiratory routes, and infection usually occurs in infancy. Serological surveys show that about 90% of the U.S. population has been infected. Frequently, this infection is subclinical, but many cases develop lesions known as **cold sores** or **fever blisters**. These are painful, short-lived vesicles that occur near the outer red margin of the lips (**Figure 21.12**).

Cold sores, caused by herpesvirus infections, are often confused with **canker sores**. The cause of canker sores is unknown, but their occurrence is often related to stress or menstruation. While similar to cold sores in appearance, canker sores usually appear in different areas. They occur as painful sores on movable mucous membranes, such as those on the tongue, cheeks, and inner surface of the lips. They ordinarily heal in a few days but often recur.

HSV-1 usually remains latent in the trigeminal nerve ganglia communicating between the face and the central nervous system (**Figure 21.13**). Recurrences of HSV-1 infection can be triggered by events such as excessive exposure to ultraviolet radiation from the sun, emotional upsets, or the hormonal changes associated with menstruation.

HSV-1 infection can be transmitted by skin contact among wrestlers; this is colorfully termed **herpes gladiatorum**. Incidence as high as 3% has been reported among high school wrestlers. Nurses, physicians, and dentists are occupationally susceptible to **herpetic whitlow**, infections of the finger caused by contact with HSV-1 lesions—as are children with herpetic oral ulcers.

A very similar virus, HSV-2, is transmitted primarily by sexual contact. It is the usual cause of genital herpes (see Chapter 26). HSV-2 is differentiated from HSV-1 by its antigenic makeup and by its effect on cells in cell culture. It is latent in the sacral nerve ganglia found near the base of the spine, a different location from that of HSV-1.

Very rarely, either type of the herpes simplex virus may spread to the brain, causing **herpes encephalitis**. Infections by HSV-2 are more serious, with a fatality rate as high as 70% if untreated. Only about 10% of survivors can expect to lead healthy lives. When administered promptly, acyclovir often cures such encephalitis. Even so, the mortality rate in certain outbreaks is still 28%, and only 38% of the survivors escape serious neurological damage.

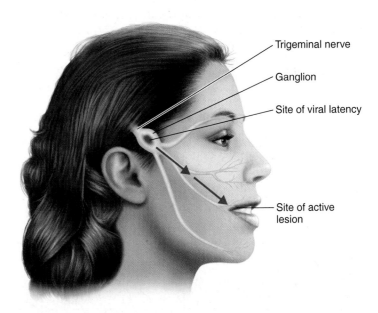

Trigeminal nerve

Ganglion

Site of viral latency

Site of active lesion

**Figure 21.13  Site of latency of herpes simplex type 1 in the trigeminal nerve ganglion.**

**Q** Why is this nerve system called *trigeminal*?

## Measles (Rubeola)

**Measles (rubeola)** is an extremely contagious viral disease spread by the respiratory route. Because a person with measles is infectious before symptoms appear, quarantine is not an effective measure of prevention.

The measles vaccine, now usually administered as the MMR vaccine (measles, mumps, rubella), has almost eliminated measles in the United States. Since the vaccine's introduction measles cases have declined from an estimated 5 million cases a year to virtual disappearance. As with smallpox, there is no animal reservoir for measles, but because the virus is so much more infectious than smallpox, herd immunity is difficult to obtain. Therefore, the present worldwide target is to control measles by vaccination. This approach has met with considerable success; compared to an annual estimated 2.6 million deaths worldwide before 1980, there were 134,200 in 2015. The goal is elimination of measles by 2020. (See the Clinical Focus box in Chapter 18.)

Although the vaccine is about 95% effective, cases continue to occur among people who do not develop or retain good immunity. Some of these infections are caused by contact with infected people from outside the United States.

An unexpected result of the measles vaccine is that many cases of measles today occur in children under age 1. Measles is especially hazardous to infants, who are more likely to have serious complications. In prevaccination days, measles was rare at this age because infants were protected by maternal antibodies derived from their mothers' recovery from the disease. Unfortunately, maternal antibodies made in response to the vaccine are not as effective in providing protection as are antibodies made in response to the disease. Because the vaccine is not effective when administered in early infancy, the child does not receive the initial vaccination before 12 months. Therefore, the child is vulnerable for a significant time.

Similar to smallpox and chickenpox, infection begins in the upper respiratory system. After an incubation period of 10 to 12 days, symptoms develop resembling those of a common cold. Soon, a macular rash appears, beginning on the face and spreading to the trunk and extremities (**Figure 21.14**). Measles lesions of the oral cavity include *Koplik's spots*, small red spots with central blue-white specks, on the oral mucosa opposite the molars. The presence of Koplik's spots is a diagnostic indicator of measles. Serological tests conducted a few days after appearance of the rash can be used to confirm the diagnosis. (See also Diseases in Focus 21.1.)

Measles is an extremely dangerous disease, especially in infants and very old people. It is frequently complicated by middle ear infections or pneumonia caused by the virus itself or by a secondary bacterial infection. Encephalitis strikes approximately 1 in 1000 measles victims; its survivors are often left with permanent brain damage. As many as 1 in 3000 cases is fatal, mostly in infants. A rare complication of measles (about

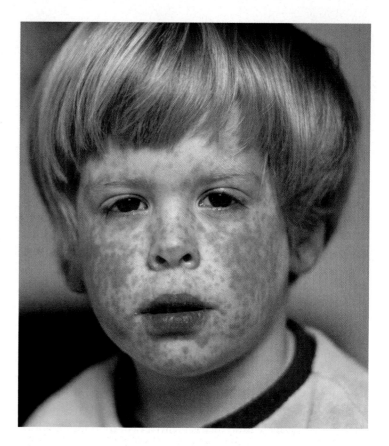

**Figure 21.14  The rash of small raised spots typical of measles (rubeola).** The rash usually begins on the face and spreads to the trunk and extremities.

**Q** Why is it potentially possible to eradicate measles?

1 in 1,000,000 cases) is **subacute sclerosing panencephalitis**. Occurring mostly in men, it appears about 1 to 10 years after recovery from measles. Severe neurological symptoms result in death within a few years.

## Rubella

**Rubella**, or *German measles* (so called because it was first described by German physicians in the eighteenth century), is a much milder viral disease than rubeola (measles) and often goes undetected. A macular rash of small red spots and a light fever are the usual symptoms (**Figure 21.15**). Complications are rare, especially in children, but encephalitis occurs in about 1 case in 6000, mostly in adults. The rubella virus is transmitted by the respiratory route, and an incubation of 2 to 3 weeks is the norm. Recovery from clinical or subclinical cases appears to give a firm immunity.

The seriousness of rubella was not appreciated until 1941, when certain severe birth defects were associated with maternal infection during the first trimester (3 months) of pregnancy, a condition called **congenital rubella syndrome**. If a pregnant woman contracts the disease during this time, there is about a 35% incidence of serious fetal damage, including

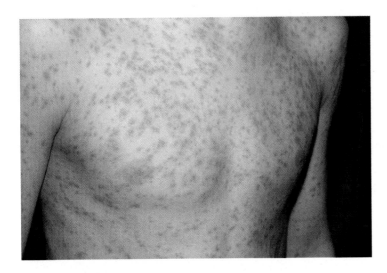

**Figure 21.15 The rash of red spots characteristic of rubella.** The spots are not raised above the surrounding skin.

**Q** What is congenital rubella syndrome?

deafness, eye cataracts, heart defects, mental retardation, autism spectrum disorder, and death. Some 15% of babies with congenital rubella syndrome die during their first year. The last major epidemic of rubella in the United States occurred during 1964 and 1965. At least 20,000 severely impaired children were born during that epidemic.

It is therefore important to identify women of childbearing age who are not immune to rubella. At one time, the blood test required for a marriage license included a test for rubella antibodies; however, only a few states now require premarital blood tests.

A rubella vaccine was introduced in 1969. Follow-up studies indicate that more than 90% of vaccinated individuals are protected for at least 15 years. Because of these preventive measures, fewer than 10 annual cases of congenital rubella syndrome are now reported.

The vaccine is not recommended for pregnant women. However, in hundreds of cases in which women were vaccinated 3 months before or 3 months after their presumed date of conception, no case of congenital rubella syndrome defects has occurred.

### Other Viral Rashes

**Fifth Disease (Erythema Infectiosum)** Parents with young children are often baffled by a diagnosis of fifth disease, which they have never heard of before. The name derives from a 1905 list of skin rash diseases: measles, scarlet fever, rubella, Filatov Dukes' disease (a mild form of scarlet fever), and the fifth disease on the list. This **fifth disease**, or **erythema infectiosum**, produces no symptoms at all in about 20% of individuals infected by the virus (*human parvovirus B19*, first identified in 1989). Symptoms are similar to a mild case of influenza, but there is a distinctive "slapped-cheek" facial rash that slowly fades. In adults who

missed an immunizing infection in childhood, the disease may cause anemia, an episode of arthritis, or, rarely, miscarriage.

**Roseola** Roseola is a mild, very common childhood disease. The child has a high fever for a few days, which is followed by a rash over much of the body lasting for a day or two. Recovery leads to immunity. The pathogens are *human herpesviruses 6 (HHV-6)* and *7 (HHV-7)*—the latter is responsible for 5–10% of roseola cases. Both viruses are present in the saliva of most adults.

**Hand-Foot-and-Mouth Disease** Caused by several enteroviruses, **hand-foot-and-mouth disease** is spread by contact with mucous or saliva of an infected person. It most commonly occurs among children in day care, preschool, and kindergarten. Limited epidemics can occur, especially during summer and fall. The usual incubation is 3–7 days, with initial symptoms of fever followed by sore throat. Soon after, a rash (either flat or raised) appears on areas such as the hands, feet, mouth, tongue, and interior cheeks. It is rare for patients to require hospitalization, but occasionally—when the disease is caused by *Enterovirus* 71—it can be accompanied by neurological conditions such as encephalitis, meningitis, and even a paralysis resembling polio. Adults with normal immune systems are less likely to be infected. There is no treatment.

**CHECK YOUR UNDERSTANDING**

✔ 21-5 How did the odd naming of "fifth disease" arise?

## Fungal Diseases of the Skin and Nails

As mentioned previously the skin is most susceptible to microorganisms that can resist high osmotic pressure and low moisture. It is not surprising, therefore, that fungi cause a number of skin disorders. Any fungal infection of the body is called a **mycosis**.

### Cutaneous Mycoses

Fungi that colonize the hair, nails, and the outer layer (stratum corneum) of the epidermis (see Figure 21.1) are called **dermatophytes**; they grow on the keratin present in those locations. Termed **dermatomycoses**, these fungal infections are more informally known as *tineas* or *ringworm*. **Tinea capitis**, or ringworm of the scalp, is fairly common among elementary school children and can result in bald patches. The infections tend to expand circularly, hence the term *ringworm* (**Figure 21.16a**). The infection is usually transmitted by contact with fomites. Dogs and cats are also frequently infected with fungi that cause ringworm in children. Ringworm of the groin, or jock itch, is known as **tinea cruris**, and ringworm of the feet, or athlete's foot, is known as **tinea pedis** (Figure 21.16b). The moisture in such areas favors fungal infections. Ringworm of the fingernails or toenails is called **tinea unguium**, or *onychomycosis*.

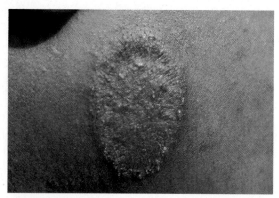

(a) Ringworm. This infection on the cheek is called tinea barbae.

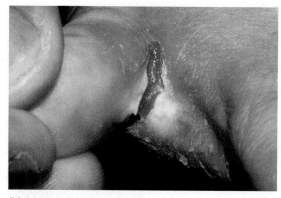

(b) Athlete's foot (Tinea pedis)

**Figure 21.16 Dermatomycoses.**

[Q] Is ringworm caused by a helminth?

Three genera of fungi are involved in cutaneous mycosis. *Trichophyton* (trik-ō-FĪ-ton) can infect hair, skin, or nails; *Microsporum* (mī-krō-SPOR-um) usually involves only the hair or skin; *Epidermophyton* (ep'i-der-mō-FĪ-ton) affects only the skin and nails. The topical drugs available without prescription for tinea infections include miconazole and clotrimazole. Athlete's foot is often difficult to cure. Topical allylamine preparations containing terbinafine or naftifine, as well as another allylamine, butenavine, are recommended and are now available without a prescription. Extended application is usually required. When hair is involved, topical treatment is not very effective. An oral antibiotic, griseofulvin, is often useful in such infections because it can localize in keratinized tissue, such as skin, hair, or nails. When nails are infected, oral itraconazole and terbinafine are the drugs of choice, but treatment may require weeks, and both must be used with caution because of potential severe side effects.

## Subcutaneous Mycoses

**Subcutaneous mycoses** are more serious than cutaneous mycoses. Even when the skin is broken, cutaneous fungi do not seem to be able to penetrate past the stratum corneum, perhaps because they cannot obtain sufficient iron for growth in the epidermis and the dermis. Usually subcutaneous mycoses are caused by fungi that inhabit the soil, especially decaying vegetation, and penetrate the skin through a small wound that allows entry into subcutaneous tissues.

In the United States, the most common disease of this type is **sporotrichosis**, caused by the dimorphic fungus *Sporothrix schenkii*. Most cases occur among gardeners or other people working with soil. The infection frequently forms a small ulcer on the hands. The fungus often enters the lymphatic system in the area and there forms similar lesions. The condition is seldom fatal and is effectively treated with itraconazole, or by ingesting a dilute solution of potassium iodide.

## Candidiasis

The bacterial microbiota of the mucous membranes in the genitourinary tract and mouth usually suppress the growth of such fungi as *Candida albicans*. Several other species of *Candida*, for example *C. tropicalis* or *C. krusei* (KROO-sē-ē), may also be involved. The morphology of these organisms is not always yeastlike but can exhibit the formation of pseudohyphae, long cells that resemble hyphae. In this form, *Candida* is resistant to phagocytosis, which may be a factor in its pathogenicity (**Figure 21.17a**). Because the fungus is not affected by antibacterial drugs, it sometimes

591   601   **608**   610   613

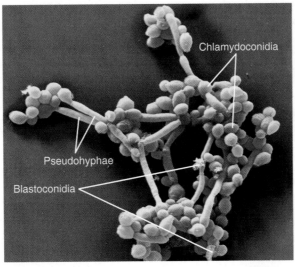

(a) *Candida albicans*  SEM |—| 10 μm  (b) Oral candidiasis, or thrush

**Figure 21.17 Candidiasis. (a)** *Candida albicans*. Notice the spherical chlamydoconidia (resting bodies formed from hyphal cells) and the smaller blastoconidia (asexual spores produced by budding) (see Chapter 12). **(b)** This case of oral candidiasis, or thrush, produced a thick, creamy coating on the tongue.

**Q** How can antibacterial drugs lead to candidiasis?

overgrows mucosal tissue when antibiotics suppress the normal bacterial microbiota. Changes in the normal mucosal pH may have a similar effect. Such overgrowths by *C. albicans* are called **candidiasis**. Newborn infants, whose normal microbiota have not become established, often suffer from a whitish overgrowth of the oral cavity, called **thrush** (Figure 21.17b). *C. albicans* is also a very common cause of vaginitis (see Chapter 26).

Immunosuppressed individuals, including AIDS patients, are unusually prone to *Candida* infections of the skin and mucous membranes. On people who are obese or diabetic, the areas of the skin with more moisture tend to become infected with this fungus. The infected areas become bright red, with lesions on the borders. Skin and mucosal infections by *C. albicans* are usually treated with topical applications of miconazole, clotrimazole, or nystatin. If candidiasis becomes systemic, as can happen in immunosuppressed individuals, *fulminating disease* (one that appears suddenly and severely) and death can result. The usual drug of choice to treat systemic candidiasis is fluconazole. Several new treatments are now also available; for example, some of the new echinocandin class antifungals, such as micafungin and anidulafungin, are now approved for this use.

**CHECK YOUR UNDERSTANDING**

✔ **21-6** How do sporotrichosis and tineas differ? How are they similar?

✔ **21-7** How might penicillin use result in candidiasis?

## Parasitic Infestation of the Skin

Parasitic organisms such as some protozoa, helminths, and microscopic arthropods can infest the skin and cause disease conditions. We will describe two examples of common arthropod infestation, scabies and lice.

### Scabies

Probably the first documented connection between a microscopic organism (330–450 μm) and a disease in humans was **scabies**, which was described by an Italian physician in 1687. The disease involves intense local itching and is caused by the tiny mite *Sarcoptes scabiei* burrowing under the skin to lay its eggs (**Figure 21.18**). The burrows are often visible as slightly elevated, serpentine lines about 1 mm in width. However, scabies may appear as a variety of inflammatory skin lesions, many of them secondary infections from scratching. The mite is transmitted by intimate contact, including sexual contact, and is most often seen in family members, nursing home residents, and teenagers infected by children for whom they baby-sit.

About 500,000 people seek treatment for scabies in the United States each year; in developing countries, it is even more prevalent. The mite lives about 25 days, but by that time eggs have hatched and produced a dozen or so progeny. Scabies is usually diagnosed by microscopic examination of skin scrapings and usually is treated by topical application of permethrin. Difficult cases are sometimes treated with oral ivermectin. (See also Diseases in Focus 21.3.)

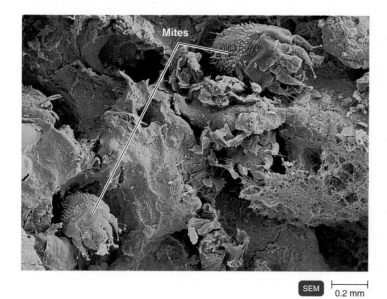

**Figure 21.18 Scabies mites in skin.**

**Q** Would a microscope be required to identify this pathogen?

## Pediculosis (Lice)

Infestations by lice, called **pediculosis**, have afflicted humans for thousands of years. Although usually associated in the public mind with poor sanitation, outbreaks of head lice among middle- and upper-class schoolchildren in the United States are common. Parents are usually appalled, but head lice are easily transferred by head-to-head contact, such as occurs among children who know each other well. The head louse, *Pediculus humanus capitis*, is not the same as the body louse, *Pediculus humanus corporis*. These are subspecies of *Pediculus humanus* that

### CLINICAL CASE

*P. aeruginosa* is isolated from the 26 cases that were tested. The health department obtains samples of pool water and takes environmental swabs from the tile around the pool and from an 18-foot inflatable device from the children's pool. The samples are cultured on nutrient agar. Water chlorination is adequate; the water tests negative for bacteria. *P. aeruginosa* is found on the tile at the shallow end of the pool and on the inflatable. Twenty-five of the patients with rashes and none of the controls had used the inflatable.

The inflatable is not watertight; during use, inflation is maintained with an air pump. The inflatable is used about 1 hour a day, 3 days a week, and stored next to the pool when not in use. Water is visibly seeping from the seams of the inflatable.

**Why is *P. aeruginosa* a likely candidate for this type of infection?**

591  601  608  **610**  613

have adapted to different areas of the body. Only the body louse spreads diseases, such as epidemic typhus.

Lice (see Figure 12.32a, page 356) require blood from the host and feed several times a day. The victim is often unaware of these silent passengers until itching, which is a result of sensitization to louse saliva, develops several weeks later. Scratching can result in secondary bacterial infections. The head louse has legs especially adapted to grasp scalp hairs (**Figure 21.19a**). During a life span of a little over a month, the female louse produces several eggs (nits) a day. The eggs are attached to hair shafts close to the scalp (Figure 21.19b) to benefit from a warmer incubation temperature, and they hatch in about a week. The very young stages of the louse are also called nits. Empty egg cases are whitish and more visible. They do not necessarily indicate the presence of live lice. As the hair grows (at the rate of about 1 cm a month), the attached nit moves away from the scalp.

A point of interest is that in the United States, lice have become adapted to the cylindrical hair shafts found on whites. In Africa, lice have adapted to the noncylindrical hair shafts of blacks.

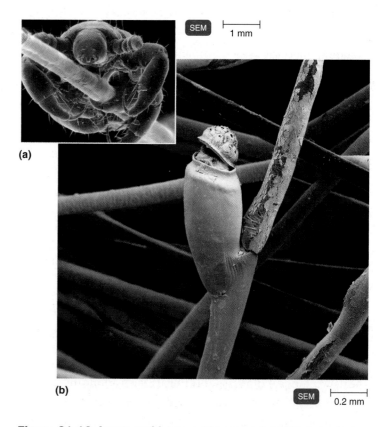

(a)

(b)

**Figure 21.19 Louse and louse egg case.** (a) Adult louse grasping hair. (b) This egg case (nit) contains the nymphal stage of the louse, which is in the process of exiting through the cap (operculum). It does this by gulping air and forcing it out the anus until it pops free, much like a champagne cork.

**Q** How is pediculosis transmitted?

# Microbial Diseases of the Eye

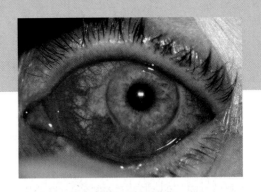

In the morning a 20-year-old man has eye redness with a crust of mucus. The condition resolves with topical antibiotic treatment. Use the table below to provide a differential diagnosis and identify infections that could cause these symptoms. For the solution, go to @MasteringMicrobiology.

| Disease | Pathogen | Portal of Entry | Symptoms | Method of Transmission | Treatment |
|---------|----------|-----------------|----------|------------------------|-----------|
| **BACTERIAL DISEASES** | | | | | |
| **Conjunctivitis** | *Haemophilus influenzae* | Conjunctiva | Redness, itchiness, mucous discharge | Direct contact; fomites | None |
| **Ophthalmia Neonatorum** | *Neisseria gonorrhoeae* | Conjunctiva | Acute infection with much pus formation | Through birth canal | Prevention: tetracycline, erythromycin, or povidone-iodine |
| **Inclusion Conjunctivitis** | *Chlamydia trachomatis* | Conjunctiva | Swelling of eyelid; mucus and pus formation | Through birth canal; swimming pools | Tetracycline |
| **Trachoma** | *Chlamydia trachomatis* | Conjunctiva | Conjunctivitis | Direct contact; fomites; flies | Azithromycin |
| **Ocular Syphilis** | *Treponema pallidum* | Mucous membranes | Redness, blurry vision | Sexually transmitted infection | See Chapter 26, page 774 |
| **VIRAL DISEASES** | | | | | |
| **Conjunctivitis** | Adenoviruses | Conjunctiva | Redness | Direct contact | None |
| **Herpetic Keratitis** | *Simplexvirus* (HHV-1) | Conjunctiva; cornea | Keratitis | Direct contact; recurring latent infection | Trifluridine may be effective |
| **PROTOZOAN DISEASE** | | | | | |
| ***Acanthamoeba* Keratitis** | *Acanthamoeba* spp. | Corneal abrasion; soft contact lenses may prevent removal of ameba by blinking | Keratitis | Contact with fresh water | Topical propamidine isethionate or miconazole; corneal transplant or eye removal surgery may be required |

Treatments of head lice abound, recalling the medical adage that if there are many treatments for a condition, it is probably because none of them are really good. Nonprescription medications such as permethrin insecticide and pyrethrin insecticide are usually the first-choice treatment, but resistance has become common. Other topical preparations containing insecticides such as malathion and the more toxic lindane are also available (lindane is banned in some areas). A single-dose treatment with orally administered ivermectin is occasionally used. A silicone-based product, LiceMD®, is effective and nontoxic. The active principle, *dimethicone*, blocks the breathing tubes of the louse. Combing out the nits with fine-toothed louse combs is another treatment option. This is a difficult, time-consuming procedure that has actually led to the appearance of professional removal services in some cities: expensive, but often worth the price to busy parents.

## CHECK YOUR UNDERSTANDING

✔ **21-8** What diseases, if any, are spread by head lice, such as *Pediculus humanus capitis*?

# Microbial Diseases of the Eye

## LEARNING OBJECTIVES

21-9 Define *conjunctivitis*.

21-10 List the causative agent, mode of transmission, and clinical symptoms of these eye infections: ophthalmia neonatorum, inclusion conjunctivitis, trachoma.

21-11 List the causative agent, mode of transmission, and clinical symptoms of these eye infections: herpetic keratitis, *Acanthamoeba* keratitis.

The epithelial cells covering the eye can be considered a continuation of the skin or mucosa. Many microbes can infect the eye, largely through the *conjunctiva*, the mucous membrane that lines the eyelids and covers the outer white surface of the eyeball. It is a transparent layer of living cells replacing the skin. Diseases of the eye are summarized in Diseases in Focus 21.4.

## Inflammation of the Eye Membranes: Conjunctivitis

**Conjunctivitis** is an inflammation of the conjunctiva, often called by the common name **red eye**, or **pinkeye**. *Haemophilus influenzae* is the most common bacterial cause; viral conjunctivitis is usually caused by adenoviruses. However, a broad group of bacterial and viral pathogens as well as allergies can also cause this condition.

The popularity of contact lenses has been accompanied by an increased incidence of infections of the eye. This is especially true of the soft-lens varieties, which are often worn for extended periods. Among the bacterial pathogens that cause conjunctivitis are pseudomonads, which can cause serious eye damage. To prevent infection, contact lens wearers should not use homemade saline solutions, which are a frequent source of infection, and should scrupulously follow the manufacturer's recommendations for cleaning and disinfecting the lenses. The most effective methods for disinfecting contact lenses involve applying heat; lenses that cannot be heated can be disinfected with hydrogen peroxide, which is then neutralized.

## Bacterial Diseases of the Eye

The bacterial microorganisms most commonly associated with the eye usually originate from the skin and upper respiratory tract.

### Ophthalmia Neonatorum

**Ophthalmia neonatorum** is a serious form of conjunctivitis caused by *Neisseria gonorrhoeae*, the cause of gonorrhea. Large amounts of pus are formed; if treatment is delayed, ulceration of the cornea will usually result. The disease is acquired as the infant passes through the birth canal, and infection carries a high risk of blindness. Early in the twentieth century, legislation required that the eyes of all newborn infants be treated with a 1% solution of silver nitrate, which proved to be a very effective treatment in preventing this eye infection, which previously accounted for nearly one-quarter of all cases of blindness in the United States. Silver nitrate has been almost entirely replaced by antibiotics because of frequent coinfections by gonococci and sexually transmitted chlamydias, which silver nitrate is not effective against. In parts of the world where the cost of antibiotics is prohibitive, a dilute solution of povidone-iodine has proven effective.

### Inclusion Conjunctivitis

Chlamydial conjunctivitis, or **inclusion conjunctivitis**, is quite common today. It is caused by *Chlamydia trachomatis*, a bacterium that grows only as an obligate intracellular parasite. In infants, who acquire it in the birth canal, the condition tends to resolve spontaneously in a few weeks or months, but in rare cases it can lead to scarring of the cornea. Chlamydial conjunctivitis also appears to spread in the unchlorinated waters of swimming pools; in this context, it is called *swimming pool conjunctivitis*. Tetracycline applied as an ophthalmic ointment is an effective treatment.

### Trachoma

A serious eye infection, and probably the greatest single cause of blindness by an infectious disease, is **trachoma**—an ancient name derived from the Greek word for rough. It is caused by certain serotypes of *Chlamydia trachomatis*, but not the same ones that cause genital infections (see pages 770–775). In the arid parts of Africa and Asia, almost all children are infected early in their lives. Worldwide, there are probably 500 million active cases and 7 million blinded victims. Trachoma also occurs occasionally in the southwestern United States, especially among Native Americans.

The disease is a conjunctivitis transmitted largely by hand contact or by sharing such personal objects as towels. Flies may also carry the bacteria. Repeated infections cause inflammation (**Figure 21.20a**), leading to *trichiasis*, an in-turning of the eyelashes (Figure 21.20b). Abrasion of the cornea, especially by the eyelashes, eventually causes scarring of the cornea and blindness. Trichiasis can be corrected surgically, a procedure shown in ancient Egyptian papyri. Secondary infections by other bacterial pathogens are also a factor in the disease. Antibiotics to eliminate chlamydia, especially oral azithromycin, are useful in treatment. The disease can be controlled through sanitary practices and health education.

---

**CHECK YOUR UNDERSTANDING**

✔ 21-9 What is the common name of inclusion conjunctivitis?

✔ 21-10 Why have antibiotics almost entirely replaced the less expensive use of silver nitrate for preventing ophthalmia neonatorum?

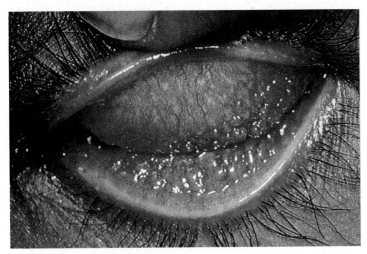

**(a)** Chronic inflammation of the eyelid

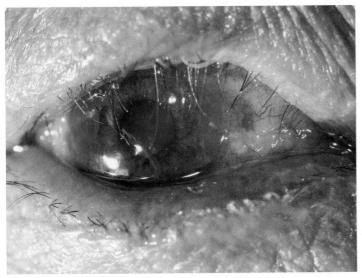

**(b)** Trichiasis, inturned eyelashes, abrading the cornea

**Figure 21.20 Trachoma. (a)** Repeated infection with *Chlamydia trachoma* causes chronic inflammation. The eyelid has been pulled back to show the inflammatory nodules that are in contact with the cornea. The abrasion caused by this damages the cornea and makes it susceptible to secondary infections. **(b)** In later stages of trachoma, the eyelashes turn inward (trichiasis) as shown here, further abrading the cornea.

**Q How is trachoma transmitted?**

## Other Infectious Diseases of the Eye

The diseases discussed here are characterized by inflammation of the cornea, which is called *keratitis*. In the United States, keratitis is mostly bacterial in origin. Fungi and helminths can also cause eye diseases. In Africa and Asia, eye infections are mostly caused by fungi, such as *Candida, Fusarium,* and *Aspergillus.*

See pages 614–615 for the **Big Picture** on Neglected Tropical Diseases for more information.

### Herpetic Keratitis

**Herpetic keratitis** is caused by the same *herpes simplex type 1 virus* (HSV-1) that causes cold sores and is latent in the trigeminal nerves (see Figure 21.13). The disease is an infection of the cornea, often resulting in deep ulcers, and that may be the most common cause of infectious blindness in the United States. The drug trifluridine is often an effective treatment.

### *Acanthamoeba* Keratitis

The first case of ***Acanthamoeba* keratitis** was reported in 1973 in a Texas rancher. Since then, well over 4000 cases have been diagnosed in the United States. This ameba has been found in fresh water, tap water, hot tubs, and soil. Most recent cases have been associated with the wearing of contact lenses, although any cornea damaged by trauma or infection is susceptible. Contributing factors are inadequate, unsanitary, or faulty disinfecting procedures (only heat will reliably kill the cysts), homemade saline solutions, and wearing the contact lenses overnight or while swimming.

In its early stages, the infection consists of only a mild inflammation, but later stages are often accompanied by severe pain. If started early, treatment with 2% chlorhexidine and propamidine isethionate eye drops or topical neomycin has been successful. Damage is often so severe as to require a corneal transplant or even removal of the eye. Diagnosis is confirmed by the presence of trophozoites and cysts in stained scrapings of the cornea.

---

**CHECK YOUR UNDERSTANDING**

✔ **21-11** Of the two eye diseases herpetic keratitis and *Acanthamoeba* keratitis, which is the more likely to be caused by an organism actively reproducing in saline solutions for contact lenses?

---

**CLINICAL CASE** Resolved

*P. aeruginosa* is able to withstand relatively high levels of chlorine, so eradicating it from swimming pools is difficult. Its ability to produce a biofilm may be a factor in its hardiness. Because the inflatable never completely dries, the bacteria probably grow inside while it is in storage. The bacteria leak out the seams and enter the body through minor abrasions, possibly obtained by contact with the inflatable. The rash patterns are consistent with handling the inflatable. The one patient who had a rash on her legs but had not used the inflatable most likely acquired her rash from the tile.

*Pseudomonas* dermatitis outbreaks usually occur as a result of low levels of water disinfectant in pools and hot tubs. In this case, the ability of *Pseudomonas* to grow on organic molecules inside the inflatable contributed to the outbreak. Guidelines for disinfecting pool equipment without damaging the equipment are being developed.

# Neglected Tropical Diseases (NTDs) are a group of 16 diseases contracted by more than a billion people per year.

## A Different Kind of Public Health Campaign

Traditionally, public health campaigns target the largest health threats one disease at a time. An unfortunate consequence of this approach is that certain serious infections with lower incidence rates never meet the criteria for major health campaigns. Thus, awareness, prevention, and treatment efforts may fall through the cracks.

Historically, this was the case for 16 infections now known as Neglected Tropical Diseases (NTDs). They disproportionately infect the poorest people living in the least developed areas. NTDs cause a wide variety of maladies, including blindness (trachoma, onchocerciasis); disfigurement (leprosy, lymphatic filariasis, Buruli ulcer); heart problems (Chagas disease); liver or lung disease (schistosomiasis, fascioliasis, leishmaniasis, echinococcosis); bone, joint, or other movement-related disabilities (yaws, dengue, dracunculiasis); malaise, malnutrition, and cognitive impairment (soil-transmitted helminthiasis); and neurological damage (rabies, cysticercosis, and African trypanosomiasis). Taken as a group, NTDs impact over a billion people annually, with over half a million deaths, according to the Centers for Disease Control and Prevention (CDC).

Despite the wide variance in causes and effects, management strategies for NTDs are frequently similar (see the table). Simultaneous infection with more than one NTD is also common, making a group-management approach a good option for the public health response.

In 2012, the World Health Organization (WHO) issued a declaration that set NTD reduction targets for the year 2020 and outlined ways to attain the goals. Over a dozen major organizations joined forces. Main approaches include early diagnosis and treatment; zoonotic disease management; preventive chemotherapy; vector control and pesticide management; and improvement of sanitation and drinking water safety.

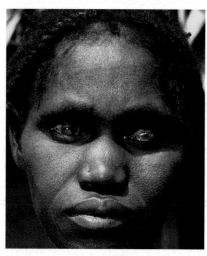

River blindness, caused by the nematode *Onchocerca volvulus*, is the world's second leading cause of blindness. It affects 18 million people in western and central Africa, the Middle East, and Latin America.

| Infection Type | Disease | Management Strategies |
| --- | --- | --- |
| PROTOZOAN | | |
| | African trypanosomiasis | Vector (tsetse fly) control, preventive chemotherapy, intensified disease management, veterinary public health |
| | Chagas disease | Vector (triatoma) control, intensified disease management |
| | Leishmaniasis | Vector (sandfly) control, preventive chemotherapy, intensified disease management |
| HELMINTHIC | | |
| | Cysticercosis | Veterinary public health, improved sanitation and hygiene |
| | Dracunculiasis (Guinea worm disease) | Vector (copepod crustaceans) control, improved hygiene and sanitation |
| | Echinococcosis | Veterinary public health |
| | Fascioliasis (foodborne trematodiases) | Veterinary public health, preventive chemotherapy |
| | Lymphatic filariasis (elephantiasis) | Vector (mosquito) control, preventive chemotherapy, intensified disease management |
| | Onchocerciasis (river blindness) | Vector (black fly) control, preventive chemotherapy |
| | Schistosomiasis (soil-transmitted intestinal worms) | Preventive deworming drugs, improved hygiene and sanitation |
| BACTERIAL | | |
| | Trachoma | Vector (fly) control, improved sanitation and hygiene |
| | Leprosy (Hansen's disease) | Preventive chemotherapy, intensified disease management |
| | Buruli ulcer | Control of aquatic insect reservoirs; rapid diagnosis and treatment, improved sanitation and hygiene |
| | Yaws (endemic treponematosis) | Improved hygiene |
| VIRAL | Dengue | Vector control |
| | Rabies | Veterinary public health |

# A Few Strategies Can Greatly Reduce Incidence of Neglected Tropical Diseases

By 2020, WHO hopes to eradicate dracunculiasis and eliminate lymphatic filariasis, leprosy, trachoma, and African trypanosomiasis. Efforts to reduce NTDs include the following.

## Preventive Chemotherapy

Pharmaceutical firms donate medication and share technology and data to develop new treatments. The World Bank funds initiatives to provide treatments. Preventive medicines for multiple NTDs are packaged and priced at less than $1 per person. Teachers receive training to administer deworming tablets to students. These efforts have confined the transmission of dracunculiasis.

## Innovative, Intensified Disease Management

Most people infected with NTDs live in remote areas, so aid groups host community events where people can receive vaccines, vitamins, and drugs outside the clinical setting. Grants from organizations such as the Bill and Melinda Gates Foundation fund development of portable testing devices, allowing for rapid diagnosis and immediate treatment.

## Veterinary Care

Veterinary care is expensive, and therefore rare, in developing nations. Treating pets, cattle, and pigs for parasites and bacterial and viral diseases helps break transmission to humans of zoonotic diseases such as rabies, cysticercosis, echinococcosis, foodborne trematodiases, and African trypanosomiasis.

## Vector Control

Safe use of pesticides for vector control reduces incidence of African trypanosomiasis, Chagas' disease, leishmaniasis, dengue, dracunculiasis, lymphatic filariasis, and trachoma.

## Improved Sanitation and Hygiene Services

Safe drinking water and improved sanitation systems can reduce prevalence of many diseases, including dracunculiasis, schistosomiasis, trachoma, Buruli ulcer, and yaws.

Access to clean water allows daily hand and face washing, which can reduce the transmission of many diseases.

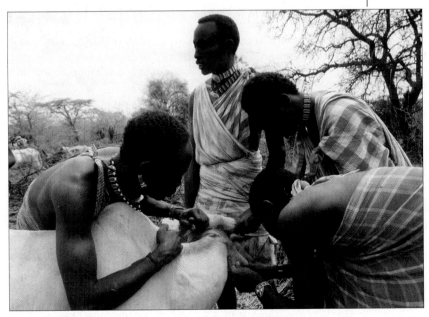

Kenyan Masai take a blood sample from cattle to check for trypanosomes.

### KEY CONCEPTS

- Vectors are animals that carry diseases to humans. **(See Chapter 12, "Arthropods as Vectors," pages 355–357, and Chapter 14, "Transmission of Disease," page 404-407.)**

- Over half the world's population is infected with eukaryotic pathogens, such as protozoans and worms. **(See Chapter 12, "Medically Important Protozoa," pages 342-346, and "Helminths," pages 347-355.)**

- Epidemiologists study patterns in transmission and distribution in order to develop infection control strategies. **(See Chapter 14, "Epidemiology," pages 413-416.)**

- Pet vaccination is prohibitively expensive in most of Africa, Asia, and Latin America, where rabies deaths are much more common. **(See Chapter 22, "Rabies," pages 632-636.)**

# Study Outline

 Go to @MasteringMicrobiology for **Interactive Microbiology**, *In the Clinic videos, MicroFlix, MicroBoosters, 3D animations, practice quizzes, and more.*

## Introduction (p. 590)

1. The skin is a physical barrier against microorganisms.
2. Moist areas of the skin support larger populations of bacteria than dry areas.

## Structure and Function of the Skin (p. 591)

1. The outer portion of the skin (epidermis) contains keratin, a waterproof coating.
2. The inner portion of the skin, the dermis, contains hair follicles, sweat ducts, and oil glands that provide passageways for microorganisms.
3. Sebum and perspiration are secretions of the skin that can inhibit the growth of microorganisms.
4. Sebum and perspiration provide nutrients for some microorganisms.
5. Body cavities are lined with epithelial cells. When these cells secrete mucus, they constitute the mucous membrane.

## Normal Microbiota of the Skin (p. 592)

1. Microorganisms that live on skin are resistant to desiccation and high concentrations of salt.
2. Gram-positive cocci predominate on the skin.
3. Washing does not completely remove the normal skin microbiota.
4. Members of the genus *Cutibacterium* metabolize oil from the oil glands and colonize hair follicles.
5. *Malassezia furfur* yeast grows on oily secretions and may be the cause of dandruff.

## Microbial Diseases of the Skin (pp. 592–602)

1. Vesicles are small fluid-filled lesions; bullae are vesicles larger than 1 cm; macules are flat, reddened lesions; papules are raised lesions; and pustules are raised lesions containing pus.

### Bacterial Diseases of the Skin (pp. 592–611)

2. The majority of skin microbiota consist of coagulase-negative *Staphylococcus epidermidis*.
3. Almost all pathogenic strains of *S. aureus* produce coagulase.
4. Pathogenic *S. aureus* can produce enterotoxins, leukocidins, and exfoliative toxin.
5. Localized infections (sties, pimples, and carbuncles) result from entry of *S. aureus* through openings in the skin.
6. Impetigo is a highly contagious superficial skin infection caused by *S. aureus*.
7. Toxemia occurs when toxins enter the bloodstream; staphylococcal toxemias include scalded skin syndrome and toxic shock syndrome.
8. Streptococci are classified according to their hemolytic enzymes and cell wall antigens.
9. Group A beta-hemolytic streptococci produce a number of virulence factors: M protein, deoxyribonuclease, streptokinases, and hyaluronidase.
10. Invasive group A beta-hemolytic streptococci cause severe and rapid tissue destruction.
11. *Pseudomonas aeruginosa* produces an endotoxin and several exotoxins.
12. Diseases caused by *P. aeruginosa* include otitis externa, respiratory infections, burn infections, and dermatitis.
13. *P. aeruginosa* infections have a characteristic blue-green pus caused by the pigment pyocyanin.
14. *Mycobacterium ulcerans* causes deep-tissue ulceration.
15. Metabolic products (fatty acids) of *Cutibacterium* acnes cause inflammatory acne.

### Viral Diseases of the Skin (pp. 602–607)

16. Papillomaviruses cause skin cells to proliferate and produce a benign growth called a wart or papilloma.
17. Warts are spread by direct contact.
18. Warts may regress spontaneously or be removed chemically or physically.
19. Variola virus causes two types of skin infections: variola major and variola minor.
20. Smallpox is transmitted by the respiratory route, and the virus is moved to the skin via the bloodstream.
21. Smallpox has been eradicated as a result of a vaccination effort by the World Health Organization.
22. HHV-3 is transmitted by the respiratory route and is localized in skin cells, causing a vesicular rash.
23. Complications of chickenpox include encephalitis and Reye's syndrome.
24. After chickenpox, the virus can remain latent in nerve cells and subsequently activate as shingles.
25. Shingles is characterized by a vesicular rash along the affected cutaneous sensory nerves.
26. HHV-3 can be treated with acyclovir. An attenuated live vaccine is available.
27. Herpes simplex infection of mucosal cells results in cold sores and occasionally encephalitis.
28. The virus remains latent in nerve cells, and cold sores can recur when the virus is activated.
29. HSV-1 is transmitted primarily by oral and respiratory routes.
30. Herpes encephalitis occurs when herpes simplex viruses infect the brain.
31. Acyclovir has proven successful in treating herpes encephalitis.
32. Measles is caused by measles virus and is transmitted by the respiratory route.
33. Vaccination against measles provides effective long-term immunity.
34. After the measles virus has incubated in the upper respiratory tract, macular lesions appear on the skin, and Koplik's spots appear on the oral mucosa.
35. Complications of measles include middle ear infections, pneumonia, encephalitis, and secondary bacterial infections.
36. The rubella virus is transmitted by the respiratory route and causes a red rash and light fever.
37. Congenital rubella syndrome can affect a fetus when a woman contracts rubella during the first trimester of her pregnancy.
38. Vaccination with live, attenuated rubella virus provides immunity of unknown duration.
39. Human parvovirus B19 causes fifth disease, and HHV-6 and HHV-7 cause roseola.
40. Hand-foot-and-mouth disease is an infection in young children caused by several enteroviruses.

### Fungal Diseases of the Skin and Nails (pp. 607–609)

41. Fungi that colonize the outer layer of the epidermis cause dermatomycoses.

42. *Microsporum, Trichophyton,* and *Epidermophyton* cause dermatomycoses called ringworm, or tinea.
43. These fungi grow on keratin-containing epidermis, such as hair, skin, and nails.
44. Diagnosis is based on the microscopic examination of skin scrapings or fungal culture.
45. Sporotrichosis results from a soil fungus that penetrates the skin through a wound.
46. The fungi grow and produce subcutaneous nodules along the lymphatic vessels.
47. *Candida albicans* causes infections of mucous membranes and is a common cause of thrush (in oral mucosa) and vaginitis.
48. Topical antifungal chemicals may be used to treat fungal diseases of the skin.

### Parasitic Infestation of the Skin (pp. 609–611)

49. Scabies is caused by a mite burrowing and laying eggs in the skin.
50. Pediculosis is an infestation by *Pediculus humanus.*

## Microbial Diseases of the Eye (pp. 612–613)

1. The mucous membrane lining the eyelid and covering the eyeball is the conjunctiva.

### Inflammation of the Eye Membranes: Conjunctivitis (p. 612)

2. Conjunctivitis is caused by several bacteria and can be transmitted by improperly disinfected contact lenses.

### Bacterial Diseases of the Eye (p. 612)

3. Bacterial microbiota of the eye usually originate from the skin and upper respiratory tract.
4. Ophthalmia neonatorum is caused by the transmission of *Neisseria gonorrhoeae* from an infected mother to an infant during its passage through the birth canal.
5. Inclusion conjunctivitis is an infection of the conjunctiva caused by *Chlamydia trachomatis.* It is transmitted to infants during birth and is transmitted in unchlorinated swimming water.
6. Trachoma is transmitted by hands, fomites, and perhaps flies.

### Other Infectious Diseases of the Eye (p. 613)

7. *Fusarium* and *Aspergillus* fungi can infect the eye.
8. Herpetic keratitis causes corneal ulcers. The etiology is HSV-1 that invades the central nervous system and can recur.
9. *Acanthamoeba* protozoa, transmitted via water, can cause a serious form of keratitis.

## Study Questions

For answers to the Knowledge and Comprehension questions, turn to the Answers tab at the back of the textbook.

## Knowledge and Comprehension

### Review

1. Discuss the usual mode of entry of bacteria into the skin. Compare bacterial skin infections with infections caused by fungi and viruses with respect to mode of entry.
2. What bacteria are identified by a positive coagulase test? What bacteria are characterized as group A beta-hemolytic?
3. DRAW IT On the figure below, show the sites of the following infections: impetigo, folliculitis, acne, warts, shingles, sporotrichosis, pediculosis.

4. Complete the table of epidemiology below.

| Disease | Etiologic Agent | Clinical Symptoms | Mode of Transmission |
|---|---|---|---|
| Acne | | | |
| Pimples | | | |
| Warts | | | |
| Chickenpox | | | |
| Hand-foot-and-mouth disease | | | |
| Measles | | | |
| Rubella | | | |

5. Why do some states require a test for antibodies against rubella for women before issuing a marriage license?
6. Identify the diseases based on the symptoms in the chart below.

| Symptoms | Disease |
|---|---|
| Koplik's spots | |
| Macular rash | |
| Vesicular rash | |
| Small, spotted rash | |
| Recurrent "blisters" on oral mucosa | |
| Corneal ulcer and swelling of lymph nodes | |

7. What complications can occur from HSV-1 infections?

8. What is in the MMR vaccine?

9. A patient exhibits inflammatory skin lesions that itch intensely. Microscopic examination of skin scrapings reveals an eight-legged arthropod. What is your diagnosis? How is the disease treated? What would you conclude if you saw a six-legged arthropod?

10. NAME IT This anaerobic, gram-positive rod is found on the skin. Infections are often treated with retinoids or benzoyl peroxide.

## Multiple Choice

Use the following information to answer questions 1 and 2. A 6-year-old girl was taken to the physician for evaluation of a slowly growing bump on the back of her head. The bump was a raised, scaling lesion 4 cm in diameter. A fungal culture of material from the lesion was positive for a fungus with numerous conidia.

1. The girl's disease was
   a. rubella.
   b. candidiasis.
   c. dermatomycosis.
   d. a cold sore.
   e. none of the above

2. Besides the scalp, this disease can occur on all of the following *except*
   a. feet.
   b. nails.
   c. the groin.
   d. subcutaneous tissue.
   e. The disease can occur on all of these areas.

Use the following information to answer questions 3 and 4. A 12-year-old boy had a fever, rash, headache, sore throat, and cough. He also had a macular rash on his trunk, face, and arms. A throat culture was negative for *Streptococcus pyogenes*.

3. The boy most likely had
   a. streptococcal sore throat.
   b. measles.
   c. rubella.
   d. smallpox.
   e. hand-foot-and-mouth disease.

4. All of the following are complications of this disease *except*
   a. middle ear infections.
   b. pneumonia.
   c. birth defects.
   d. encephalitis.
   e. All are complications of this disease.

5. A patient has conjunctivitis. If you isolated *Pseudomonas* from the patient's mascara, you would most likely conclude all of the following *except* that
   a. the mascara was the source of the infection.
   b. *Pseudomonas* is causing the infection.
   c. *Pseudomonas* has been growing in the mascara.
   d. the mascara was contaminated by the manufacturer.
   e. All of the above are valid conclusions.

6. You microscopically examine scrapings from a case of *Acanthamoeba* keratitis. You expect to see
   a. nothing.
   b. viruses.
   c. gram-positive cocci.
   d. eukaryotic cells.
   e. gram-negative cocci.

Use the following choices to answer questions 7 through 9.
   a. *Pseudomonas*
   b. *S. aureus*
   c. scabies
   d. *Sporothrix*
   e. virus

7. Nothing is seen in microscopic examination of a scraping from the patient's rash.

8. Microscopic examination of the patient's ulcer reveals 10 μm ovoid cells.

9. Microscopic examination of scrapings from the patient's rash shows gram-negative rods.

10. Which of the following pairs is *mismatched*?
   a. leading cause of blindness—*Chlamydia*
   b. chickenpox—shingles
   c. HSV-1—encephalitis
   d. Buruli ulcer—stomach acid
   e. none of the above

## Analysis

1. A laboratory test used to determine the identity of *Staphylococcus aureus* is its growth on mannitol salt agar. The medium contains 7.5% sodium chloride (NaCl). Why is it considered a selective medium for *S. aureus*?

2. Is it necessary to treat a patient for warts? Explain briefly.

3. Analyses of nine conjunctivitis cases provided the data in the table below. How were these infections transmitted? How could they be prevented?

| No. | Etiology | Isolated from Eye Cosmetics or Contact Lenses |
|---|---|---|
| 5 | S. epidermidis | + |
| 1 | Acanthamoeba | + |
| 1 | Candida | + |
| 1 | P. aeruginosa | + |
| 1 | S. aureus | + |

4. What factors made the eradication of smallpox possible? What other diseases meet these criteria?

## Clinical Applications and Evaluation

1. A hospitalized patient recovering from surgery develops an infection that has blue-green pus and a grapelike odor. What is the probable etiology? How might the patient have acquired this infection?

2. A 12-year-old diabetic girl using continuous subcutaneous insulin infusion to manage her diabetes developed a fever (39.4°C), low blood pressure, abdominal pain, and erythroderma. She was supposed to change the needle-insertion site every 3 days after cleaning the skin with an iodine solution. Frequently she did not change the insertion site more often than every 10 days. Blood culture was negative, and abscesses at insertion sites were not cultured. What is the probable cause of her symptoms?

3. A teenaged male with confirmed influenza was hospitalized when he developed respiratory distress. He had a fever, rash, and low blood pressure. *S. aureus* was isolated from his respiratory secretions. Discuss the relationship between his symptoms and the etiological agent.

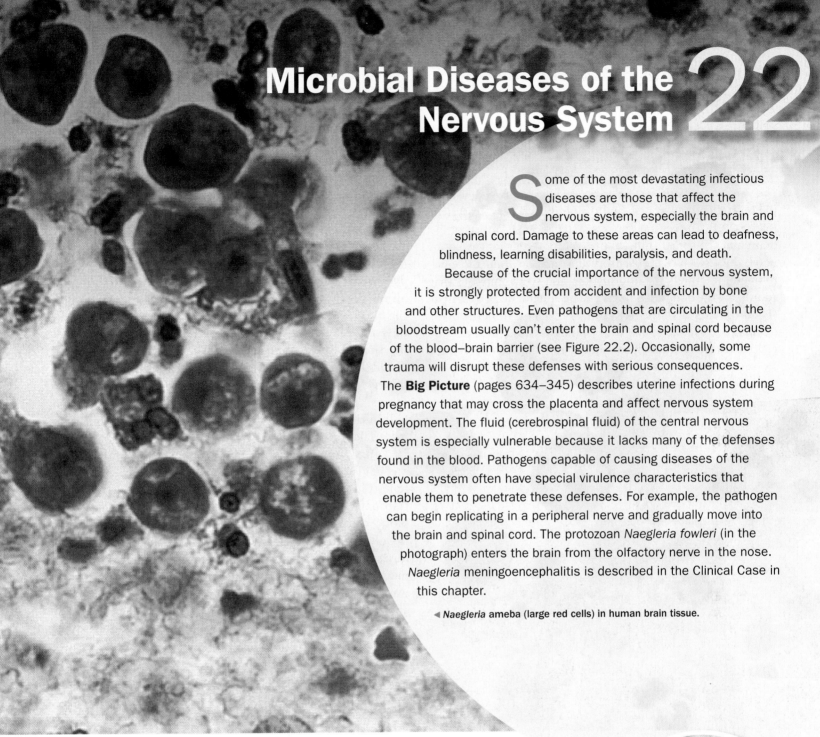

Some of the most devastating infectious diseases are those that affect the nervous system, especially the brain and spinal cord. Damage to these areas can lead to deafness, blindness, learning disabilities, paralysis, and death. Because of the crucial importance of the nervous system, it is strongly protected from accident and infection by bone and other structures. Even pathogens that are circulating in the bloodstream usually can't enter the brain and spinal cord because of the blood–brain barrier (see Figure 22.2). Occasionally, some trauma will disrupt these defenses with serious consequences. The **Big Picture** (pages 634–345) describes uterine infections during pregnancy that may cross the placenta and affect nervous system development. The fluid (cerebrospinal fluid) of the central nervous system is especially vulnerable because it lacks many of the defenses found in the blood. Pathogens capable of causing diseases of the nervous system often have special virulence characteristics that enable them to penetrate these defenses. For example, the pathogen can begin replicating in a peripheral nerve and gradually move into the brain and spinal cord. The protozoan *Naegleria fowleri* (in the photograph) enters the brain from the olfactory nerve in the nose. *Naegleria* meningoencephalitis is described in the Clinical Case in this chapter.

◀ *Naegleria* ameba (large red cells) in human brain tissue.

## In the Clinic

You are a nurse in the neonatal intensive care unit, and your newest patient is a 32-week-old infant whose mother had flulike symptoms prior to delivery. The infant required supplemental oxygen for a few hours of life but soon was weaned off oxygen and tolerated her first feeding without difficulty. At 22 hours, you observe a drop in her heart rate, and despite resuscitation efforts, the infant dies. The next morning you receive a microbiology laboratory report that blood cultures drawn just before the infant's death are growing gram-positive rods. **What is the most likely cause of the infant's infection?**

*Hint: Read about bacterial causes of meningitis in this chapter (pages 622–625).*

Answers to In the Clinic questions are found online @MasteringMicrobiology.

# Structure and Function of the Nervous System

### LEARNING OBJECTIVES

**22-1** Define *central nervous system* and *blood–brain barrier*.

**22-2** Differentiate meningitis from encephalitis.

The human nervous system is organized into two divisions: the central nervous system and the peripheral nervous system (**Figure 22.1**). The **central nervous system (CNS)** consists of the brain and the spinal cord. As the control center for the entire body, the CNS picks up sensory information from the environment, interprets the information, and sends impulses that coordinate the body's activities. The **peripheral nervous system (PNS)** consists of all the nerves that branch off from the brain (cranial nerves) and spinal cord (spinal nerves). These peripheral nerves are the lines of communication between the central nervous system, the various parts of the body, and the external environment.

Both the brain and the spinal cord are covered and protected by three continuous membranes called *meninges* (**Figure 22.2**). These are the outermost *dura mater*, the middle *arachnoid mater*, and the innermost *pia mater*. Between the pia mater and arachnoid membranes is a space called the *subarachnoid space*, in which an adult has 100 to 160 ml of **cerebrospinal fluid (CSF)** circulating. Because CSF has low levels of complement and circulating antibodies and few phagocytic cells, bacteria can multiply in it with few checks.

Late in the nineteenth century, experiments in which dyes were injected into the body resulted in the staining of all the organs of the body—with the important exception of the brain. Conversely, when the CSF was injected with dyes, only the brain was stained. These remarkable results were the first evidence of an important feature of anatomy: the **blood–brain barrier**. Certain capillaries permit some substances to pass from the blood into the brain but restrict others. These capillaries are less permeable than others within the body and are therefore more selective in passing materials.

Drugs cannot cross the blood–brain barrier unless they are lipid-soluble. (Glucose and many amino acids are not lipid-soluble, but they can cross the barrier because special transport systems exist for them.) The lipid-soluble antibiotic chloramphenicol enters the brain readily. Penicillin is only slightly lipid-soluble; but, if it is taken in very large doses, enough may cross the barrier to be effective. Inflammations of the brain tend to alter the blood–brain barrier in a way that allows antibiotics to cross it when they wouldn't be able to otherwise. Probably the most common routes of CNS invasion are the bloodstream and the lymphatic system (see Chapter 23), when inflammation alters permeability of the blood–brain barrier.

An inflammation of the meninges is called **meningitis**. An inflammation of the brain itself is called **encephalitis**. If both the brain and the meninges are affected, the inflammation is called **meningoencephalitis**.

Exploring the Microbiome (page 644) describes how the intestinal microbiome may influence the CNS.

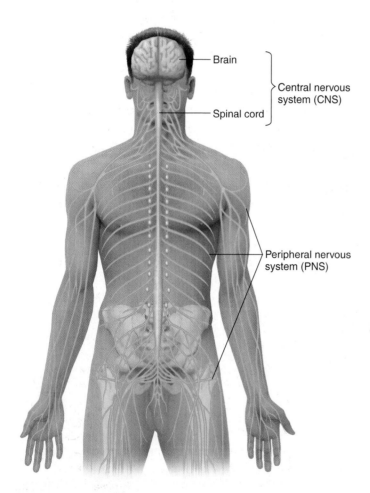

**Figure 22.1 The human nervous system.** This view shows the central and peripheral nervous systems.

**Q** Is meningitis an infection of the CNS or the PNS?

Brain

Central nervous system (CNS)

Spinal cord

Peripheral nervous system (PNS)

### CHECK YOUR UNDERSTANDING

✔ **22-1** Why can the antibiotic chloramphenicol readily cross the blood–brain barrier, whereas most other antibiotics cannot?

✔ **22-2** Encephalitis is an inflammation of what organ or organ structure?

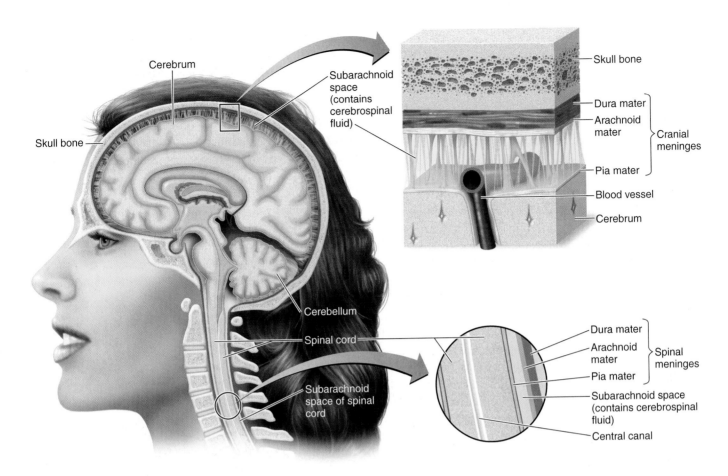

**Figure 22.2 The meninges and cerebrospinal fluid.** The meninges, whether cranial or spinal, consist of three layers: dura mater, arachnoid mater, and pia mater. Between the arachnoid and the pia mater is the subarachnoid space, in which cerebrospinal fluid circulates. The CSF is vulnerable to contamination by microbes carried in the blood that are able to penetrate the blood–brain barrier at the walls of the blood vessels.

**Q** If a patient has meningitis, what barriers would need to be crossed to result in encephalitis?

# Bacterial Diseases of the Nervous System

## LEARNING OBJECTIVES

**22-3** Discuss the epidemiology of meningitis caused by *Haemophilus influenzae*, *Neisseria meningitidis*, *Streptococcus pneumoniae*, and *Listeria monocytogenes*.

**22-4** Explain how bacterial meningitis is diagnosed and treated.

**22-5** Discuss the epidemiology of tetanus, including mode of transmission, etiology, disease symptoms, and preventive measures.

**22-6** State the causative agent, symptoms, suspect foods, and treatment for botulism.

**22-7** Discuss the epidemiology of leprosy, including mode of transmission, etiology, disease symptoms, and preventive measures.

Microbial infections of the CNS are infrequent but often have serious consequences. In preantibiotic times, they were almost always fatal.

**CLINICAL CASE** Keep Your Head above Water

As her parents look on, EMTs load 9-year-old Patricia Scott into the back of the ambulance. Patricia's mother tells one of the EMTs that 3 days earlier Patricia complained of a severe headache. Over the next 3 days she experienced nausea and vomiting. When normally energetic Patricia became increasingly lethargic, and then unresponsive, Patricia's father called 911.

**What could be causing Patricia's sickness? Read on to find out.**

621   625   626   642   643   645

## Bacterial Meningitis

The initial symptoms of meningitis are not especially alarming: a triad of fever, headache, and a stiff neck. Nausea and vomiting often follow. Eventually, meningitis may progress to convulsions and coma. The mortality rate varies with the pathogen but is generally high for an infectious disease today. Many people who survive an attack suffer some degree of neurological damage.

Meningitis can be caused by different types of pathogens, including viruses, bacteria, fungi, and protozoa. **Viral meningitis** (not to be confused with viral encephalitis, page 637) is probably much more common than bacterial meningitis but tends to be a mild disease. Most cases occur in the summer and fall months and are usually caused by a varied group of viruses termed *enteroviruses* (see Table 13.2, page 368). Enteroviruses grow well in the throat and intestinal tract and are responsible for an assortment of mostly minor diseases. Viral meningitis can also be an occasional complication of viral infections such as mumps, chickenpox, and influenza.

Historically, only three bacterial species have been responsible for most of the meningitis cases and their resulting deaths. Meningitis caused by *Haemophilus influenzae* type B, once responsible for a majority of cases, has been nearly eliminated in the United States since introduction of an effective vaccine. In adult patients, that is, older than 16 years, about 80% of meningitis cases are now caused by *Neisseria meningitidis* and *Streptococcus pneumoniae*. A conjugate vaccine against *S. pneumoniae* is coming into widespread use and is expected to lower its incidence as a cause of meningitis, especially among children. This vaccine may also produce a herd immunity that will benefit the adult population. All three of these pathogens possess a capsule that protects them from phagocytosis as they replicate rapidly in the bloodstream, from which they might enter the cerebrospinal fluid. Death from bacterial meningitis often occurs very quickly, probably from shock and inflammation caused by the release of endotoxins of the gram-negative pathogens or the release of cell wall fragments (peptidoglycans and teichoic acids) from gram-positive bacteria.

Nearly 50 other species of bacteria have been reported to be opportunistic pathogens that occasionally cause meningitis. Especially important are *Listeria monocytogenes*, group B streptococci, staphylococci, and certain gram-negative bacteria.

### Haemophilus influenzae Meningitis

*Haemophilus influenzae* is an aerobic, gram-negative bacterium that is a common member of the normal throat microbiota. Occasionally, however, it enters the bloodstream and causes several invasive diseases. In addition to causing meningitis, it is also frequently a cause of pneumonia (page 706), otitis media (page 693), and epiglottitis (page 690). The carbohydrate capsule of the bacterium is important to its pathogenicity, especially those bacteria with capsular antigens of type b.

(Strains that lack a capsule are called *nontypable*.) Medically, the bacterium is often referred to by the acronym *Hib*.

It was given the name *Haemophilus influenzae* because the microorganism was erroneously thought to be the causative agent of the influenza pandemics of 1889 and World War I. *H. influenzae* was probably only a secondary invader during those virus-caused pandemics. *Haemophilus* refers to the need the microorganism has for factors in blood for growth (*hemo* = blood; *philus* = loving).

Hib-caused meningitis occurs mostly in children under age 4, especially at about 6 months, when antibody protection provided by the mother weakens. The incidence is decreasing because of the Hib vaccine, which was introduced in 1988. *H. influenzae* meningitis has accounted for most of the cases of reported bacterial meningitis (45%), with a mortality rate of about 6%.

### Neisseria meningitidis Meningitis (Meningococcal Meningitis)

**Meningococcal meningitis** is caused by *Neisseria meningitidis* (the **meningococcus**). This is an aerobic, gram-negative diplococcus with a polysaccharide capsule that is important to its virulence. Like Hib and the pneumococcus, it is frequently present in the nose and throat of carriers without causing disease symptoms (**Figure 22.3**). These carriers, up to 40% of the population,

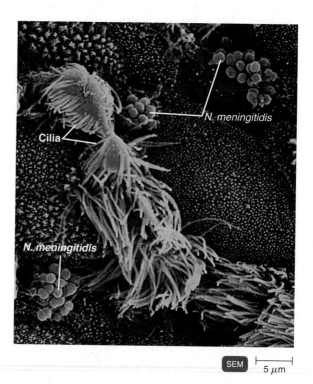

Cilia

*N. meningitidis*

*N. meningitidis*

SEM ⊢——⊣ 5 μm

**Figure 22.3 *Neisseria meningitidis*.** This scanning electron micrograph shows *Neisseria meningitidis* in clusters attached to cells on the mucous membrane in the pharynx.

**Q** What would be the effect if the cilia are inactivated by this infection?

are a reservoir of infection. Transmission is by droplet aerosols or direct contact with secretions.

The symptoms of meningococcal meningitis are mostly caused by an endotoxin that is produced very rapidly and is capable of causing death within just a few hours. The most distinguishing feature is a rash that does not fade when pressed. A case of meningococcal meningitis typically begins with a throat infection, leading to bacteremia and eventually meningitis. It usually occurs in children under 2 years. Significant numbers of these children have residual damage, such as deafness.

Death can occur within a few hours after the onset of fever; however, antibiotic therapy has helped reduce the mortality rate to about 9–12%. Without chemotherapy, mortality rates approach 80%.

The meningococcus occurs in six capsular serotypes associated with invasive disease (A, B, C, W-135, X, and Y). The distribution and frequency of these serotypes varies continuously. Local outbreaks are facilitated by modern ease of travel, which often exposes populations to serotypes that are otherwise uncommon in the area. Meningococcal meningitis is a global problem with the highest incidence of disease found in the so-called meningitis belt of sub-Saharan Africa where epidemics occur every 5 to 12 years.

The incidence in industrialized countries is sporadic and varies by age, occurring most often in infants who have not yet developed protective antibodies. In arid regions of Africa and Asia, dry air causes the nasal mucous membranes to become less resistant to bacterial invasion. This contributes to widespread epidemics, mostly of serotype A and C. However, conjugate serogroups A and C have had encouraging results, leading to hopes that epidemic meningitis might eventually be eliminated from the area.

In the United States, sporadic meningococcal outbreaks occur among college students, presumably as a result of crowding of susceptible populations in dormitories. Before vaccination was introduced in 1982, these outbreaks were a major problem for the U.S. military in recruit barracks. Vaccination is often recommended for students entering college and is required by some institutions.

The three meningococcal serogroups that most commonly circulate and cause disease in the United States are B, C, and Y. Vaccines containing polysaccharide capsular material of serogroups A, C, Y, and W-135 (sometimes simply referred to as W) have been available since the 1980s. The newer serogroup B (MenB) vaccine, licensed in 2014, contains bacterial surface proteins. MenB vaccine is recommended for 16- to 23-year-olds who are at risk of contracting the disease during outbreaks. A previously rare serogroup called X, for which a vaccine is not available, emerged in sub-Saharan Africa during the 2000s.

### *Streptococcus pneumoniae* Meningitis (Pneumococcal Meningitis)

*Streptococcus pneumoniae*, like *H. influenzae*, is a common inhabitant of the nasopharyngeal region. About 70% of the general population are healthy carriers. The pneumococcus, so called because it is best known as a cause of pneumonia (Chapter 24, page 703), is a gram-positive, encapsulated diplococcus. It is the leading cause of bacterial meningitis, now that an effective Hib vaccine is in use. In addition to approximately 6000 cases of meningitis, each year *S. pneumoniae* causes 400,000 cases of pneumonia and millions of cases of painful otitis media (earache). Most of the cases of pneumococcal meningitis occur among children between the ages of 1 month and 4 years. For a bacterial disease, the mortality rate is high: about 8% in children and 22% in the elderly.

A conjugated vaccine, modeled after the Hib vaccine, is recommended for infants under the age of 2 (see Table 18.1, page 501). One useful side effect of this vaccine is that it results in about a 6–7% decrease in cases of otitis media. The large number of serotypes of the pneumococcus will make it difficult to develop vaccines against all of them.

### Diagnosis and Treatment of the Most Common Types of Bacterial Meningitis

A diagnosis of bacterial meningitis requires a sample of cerebrospinal fluid obtained by a spinal tap, or lumbar puncture (**Figure 22.4**). A simple Gram stain is often useful; it will frequently determine the identity of the pathogen with considerable reliability. Cultures are also made from the fluid. For this purpose, prompt and careful handling is required because many of the likely pathogens are very sensitive and won't survive much storage time or even changes in temperature. The most frequently used type of serological tests performed on CSF are latex agglutination tests. Results are available within about 20 minutes. However, a negative result doesn't eliminate the possibility of less common bacterial pathogens or nonbacterial causes.

Bacterial meningitis is life-threatening and develops rapidly. Therefore, prompt treatment of any type of bacterial meningitis is essential, and chemotherapy of suspected cases is usually initiated before identification of the pathogen is complete. Broad-spectrum third-generation cephalosporins are usually the first choice of antibiotics; some experts recommend including vancomycin. As soon as identification is confirmed, or perhaps when antibiotic sensitivity has been determined from cultures, the antibiotic treatment may be changed. Antibiotics are also valuable in protecting patient contacts against the spread of an outbreak.

### Listeriosis

*Listeria monocytogenes* is a gram-positive rod known to cause stillbirth and neurological disease in animals long before it

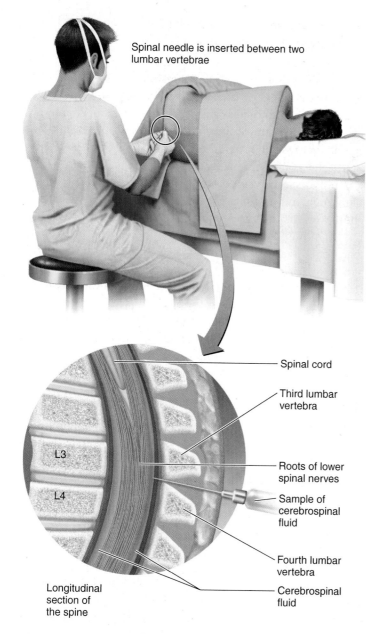

Spinal needle is inserted between two lumbar vertebrae

Longitudinal section of the spine

- Spinal cord
- Third lumbar vertebra
- Roots of lower spinal nerves
- Sample of cerebrospinal fluid
- Fourth lumbar vertebra
- Cerebrospinal fluid

L3

L4

**Figure 22.4 Spinal tap (lumbar puncture).** Diseases affecting the central nervous system, such as meningitis, often require a spinal tap for diagnosis. A needle is inserted between two vertebrae in the lower spine. A sample of cerebrospinal fluid, which is contained in the subarachnoid space (see Figure 22.2), is withdrawn for laboratory examination.

**Q** Microscopically, what would you see in CSF from a healthy person? A person with meningococcal meningitis?

was recognized as causing human disease. Excreted in animal feces, it is widely distributed in soil and water. The name is derived from the proliferation of monocytes (a type of leukocyte) found in some animals infected by it. In recent years, the disease **listeriosis** has changed from a disease of very limited importance to a major concern for the food industry and health authorities. Since the introduction of the Hib

vaccination, listeriosis has become the fourth most common cause of bacterial meningitis.

The disease appears in two basic forms: in infected adults and as an infection of the fetus and newborn. In adult humans, it is usually a mild, often symptomless disease, but the microbe sometimes invades the CNS, causing meningitis. This is most likely to happen to persons whose immune system is compromised, such as persons with cancer, diabetes, or AIDS, or who are taking immunosuppressive medications. Occasionally, *L. monocytogenes* invades the bloodstream and causes a wide range of disease conditions, especially sepsis. Recovering or apparently healthy individuals often shed the pathogen indefinitely in their feces. An important factor in its virulence is that when *L. monocytogenes* is ingested by phagocytic cells, it isn't destroyed; it even proliferates within them, primarily in the liver. It also has the unusual capability of moving directly from one phagocyte to an adjacent one (**Figure 22.5**).

*L. monocytogenes* is especially dangerous when it infects a pregnant woman. She usually suffers no more than mild, flu-like symptoms. The fetus, however, can be infected via the placenta, often resulting in an abortion or stillborn infant. In some cases the disease isn't manifested until a few weeks after birth, usually as meningitis, which can result in significant brain injury or death. The infant mortality rate associated with this type of infection is about 60%.

In human outbreaks, the organism is mostly foodborne. It's frequently isolated from a wide variety of foods; ready-to-eat deli meats and dairy products have been involved in several outbreaks. *L. monocytogenes* is one of the few pathogens capable of growth at refrigerator temperatures, which can

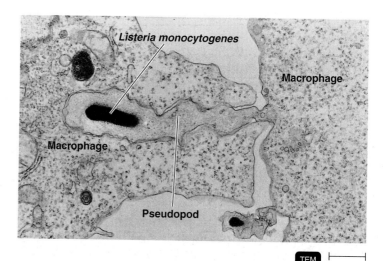

*Listeria monocytogenes*

Macrophage

Macrophage

Pseudopod

TEM | 1 μm

**Figure 22.5 Cell-to-cell spread of *Listeria monocytogenes*, the cause of listeriosis.** Notice that the bacterium has caused the macrophage on the right, in which it resided, to form a pseudopod that is now engulfed by the macrophage on the left. The pseudopod will soon be pinched off and the microbe transferred to the macrophage on the left.

**Q** How is listeriosis contracted?

lead to an increase in its numbers during a food's shelf life. A bacteriophage-containing spray capable of killing at least 170 *L. monocytogenes* strains for treating ready-to-eat meats has been approved by the U.S. Food and Drug Administration (FDA). If it meets with consumer approval, it may be a model for similar sprays to control other foodborne pathogens.

Efforts to improve the methods of detecting *L. monocytogenes* in foods are ongoing. Considerable progress has been made with selective growth media and rapid biochemical tests. However, DNA probes and serological tests using monoclonal antibodies eventually are expected to be the most satisfactory (see Chapter 10, pages 281 and 287). Diagnosis in humans depends on isolating and culturing the pathogen, usually from blood or cerebrospinal fluid. Penicillin G is the antibiotic of choice for treatment.

Microbial causes of meningitis and encephalitis are summarized in Diseases in Focus 22.1.

**Figure 22.6 An advanced case of tetanus.** A drawing of a British soldier during the Napoleonic wars. These spasms, known as opisthotonos, can actually result in a fractured spine. (Drawing by Charles Bell of the Royal College of Surgeons, Edinburgh.)

**Q** What is the name of the toxin that causes opisthotonos?

---

### CHECK YOUR UNDERSTANDING

✔ 22-3 Why is meningitis caused by the pathogen *Listeria monocytogenes* frequently associated with ingestion of refrigerated foods?

✔ 22-4 What body fluid is sampled to diagnose bacterial meningitis?

## Tetanus

The causative agent of **tetanus**, *Clostridium tetani*, is an obligately anaerobic, endospore-forming, gram-positive rod. It's especially common in soil contaminated with animal fecal wastes.

The symptoms of tetanus are caused by an extremely potent neurotoxin, *tetanospasmin*, that is released upon death and lysis of the growing bacteria (see Chapter 15, page 434). It enters the CNS via the peripheral nerves or the blood. The bacteria themselves don't spread from the infection site, and there is no inflammation.

In a muscle's normal operation, a nerve impulse initiates contraction of the muscle. At the same time, an opposing muscle receives a signal to relax so as not to oppose the contraction. The tetanus neurotoxin blocks the relaxation pathway so that both opposing sets of muscles contract, resulting in the characteristic muscle spasms. The muscles of the jaw are affected early in the disease, preventing the mouth from opening, a condition known as *lockjaw*. In extreme cases, spasms of the back muscles cause the head and heels to bow backward, a condition called *opisthotonos* (**Figure 22.6**). Gradually, other skeletal muscles become affected, including those involved in swallowing. Death results from spasms of the respiratory muscles.

Because the microbe is an obligate anaerobe, the wound by which it enters the body must provide anaerobic growth conditions—for example, improperly cleaned deep wounds such as those caused by rusty (and therefore presumably dirt-contaminated) nails. Injecting street-drug users are at high risk: sanitation during injection is not a priority, and the drugs are often contaminated. However, many cases of tetanus arise from trivial injuries, such as sitting on a tack, that the person considers too minor to bring to the attention of a physician.

Effective vaccines for tetanus have been available since the 1940s. But vaccination was not always as common as it is today, where it is part of the standard DTaP (diphtheria, tetanus, and acellular pertussis) childhood vaccine. Currently, about 94% of 6-year-olds in the United States have good immunity. The tetanus vaccine is a *toxoid*, an inactivated toxin that stimulates the formation of antibodies that neutralize the toxin produced by the bacteria. A booster is required every 10 years to maintain good immunity, but many people do not obtain these vaccinations. Serological surveys show that at least 40% of the U.S. adult population does not have adequate protection. Almost all reported cases of tetanus occur in people who either have never received the primary series of tetanus vaccines or who completed a primary series but have not had a booster vaccination in the past 10 years. Deaths are more likely to occur in persons age 60 years and older and in persons who are diabetic.

---

### CLINICAL CASE

Once Patricia reaches the emergency department, the attending physician notes her neurological symptoms and orders a lumbar puncture for a bacterial culture and cell count. While performing the lumbar puncture, the physician notes that the cerebrospinal fluid (CSF), which is normally clear in a healthy person, is bloody and cloudy. The lab reports a high white blood cell count, but the bacterial culture is negative.

**Based on these findings, what differential diagnosis can the attending physician make?**

621  **625**  626  642  643  645

Even so, immunization has made tetanus in the United States a rare disease—typically, fewer than 10 cases a year. In 1903, 406 people died of fireworks-related tetanus injuries alone. (Fireworks explosions drive soil particles deep into human tissue.) Worldwide, an estimated 1 million cases occur annually, at least half in newborns. In many parts of the world, the severed umbilical cords of infants are dressed with materials such as soil, clay, and even cow dung. Estimates are that the mortality rate from tetanus is about 50% in developing areas; in the United States, it is about 25%.

When a wound is severe enough to need a physician's attention, the doctor must decide whether it is necessary to provide protection against tetanus. Usually there isn't enough time to administer the toxoid to produce antibodies and block the progression of the infection, even if given as a booster to a patient who has been immunized. However, temporary immunity can be conferred by *tetanus immune globulin (TIG)*, prepared from the antibody-containing serum of immunized humans. (Prior to World War I, long before tetanus toxoid became available, similar preparations of preformed antibodies called *antisera* were used. Made by inoculating horses, antisera were very effective in lowering the incidence of tetanus in injured people.)

A physician's decision for treatment depends largely on the extent of the deep injuries and the immunization history of the patient, who may not be conscious. People with extensive injuries who have previously had a complete primary vaccination series and up-to-date boosters would be considered protected, requiring no action. For extensive wounds in patients with unknown or low immunity, TIG would be given to provide temporary protection. In addition, the first of a toxoid series would be administered to provide more permanent immunity. When TIG and toxoid are both injected, different sites must be used to prevent the TIG from neutralizing the toxoid. Adults receive a Td (tetanus and diphtheria) vaccine that also boosts immunity to diphtheria. To minimize the production of more toxin, damaged tissue that provides growth conditions for the pathogen should be removed, a procedure called **debridement** (dē-BRĒD-ment) and antibiotics should be administered. However, once the toxin has attached to the nerves, such therapy is of little use.

### CHECK YOUR UNDERSTANDING

✔ **22-5** Is the tetanus vaccine directed at the bacterium or the toxin produced by the bacterium?

## Botulism

**Botulism**, a form of food poisoning, is caused by *Clostridium botulinum*, an obligately anaerobic, endospore-forming gram-positive rod found in soil and many aquatic sediments. Ingesting the endospores usually does no harm, as will be explained shortly. However, in anaerobic environments, such as sealed cans, the microorganism produces an exotoxin.

This neurotoxin is highly specific for the synaptic end of the nerve, where it blocks the release of acetylcholine, a chemical necessary for transmitting nerve impulses across synapses.

Individuals suffering from botulism undergo a progressive *flaccid paralysis* for 1 to 10 days and may die from respiratory and cardiac failure. Nausea, but no fever, may precede the neurological symptoms. The initial neurological symptoms vary, but nearly all sufferers have double or blurred vision. Other symptoms include difficulty swallowing and general weakness. Incubation time varies, but symptoms typically appear within a day or two. As with tetanus, recovery from the disease does not confer immunity because the toxin is usually not present in amounts large enough to be effectively immunogenic.

Botulism was first described as a clinical disease in the early 1800s, when it was known as the sausage disease (*botulus* is the Latin word for sausage). Blood sausage, the type usually involved, was made by filling a pig stomach with blood and ground meats, tying shut all the openings, boiling it for a short time, and smoking it over a wood fire. The sausage was then stored at room temperature. This attempt at food preservation included most of the requirements for an outbreak of botulism. It killed competing bacteria but allowed the more heat-stable *C. botulinum* endospores to survive, and it provided anaerobic conditions and an incubation period for toxin production.

The botulinal toxin will be destroyed by most ordinary cooking methods that bring the food to a boil. Sausage rarely causes botulism today, largely because nitrites are added to it. Nitrites prevent *C. botulinum* from growing after the endospores germinate.

### CLINICAL CASE

An alert laboratory technician's suspicions are aroused when he sees Patricia's negative bacterial culture. He is certain there must be a reason for pus to be present in Patricia's CSF. The technician prepares a wet mount of the CSF (see the figure) to see whether he can detect a microbial presence in Patricia's CSF.

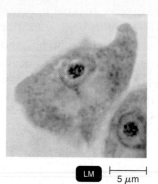

LM    5 µm

**What does the lab technician see in the wet mount of the CSF fluid? How does this affect the physician's diagnosis?**

621    625    **626**    642    643    645

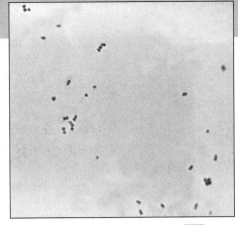

**Gram stain of cerebrospinal fluid.** LM ⊢ 5 µm

Differential diagnosis is the process of identifying a disease from a list of possible diseases that fit the information derived from examining a patient. A differential diagnosis is important for providing initial treatment and for laboratory testing. For example, a worker in a day-care center in eastern North Dakota becomes ill with fever, rash, headache, and abdominal pain. The patient has a precipitous clinical decline and dies on the first day of hospitalization. A Gram stain of the patient's cerebrospinal fluid is shown in the figure. Use the table below to provide a differential diagnosis of the infections that could cause these symptoms. For the solution, go to @MasteringMicrobiology.

| Disease | Pathogen | Portal of Entry | Method of Transmission | Treatment | Prevention |
|---|---|---|---|---|---|
| **BACTERIAL DISEASES** | | | | | |
| *Haemophilus influenzae* Meningitis | *H. influenzae* | Respiratory tract | Endogenous infection; aerosols | Cephalosporin | Capsular Hib vaccine |
| Meningococcal Meningitis | *Neisseria meningitidis* | Respiratory tract | Aerosols | Cephalosporin | Capsular vaccines against serotypes A, B, C, Y, W |
| Pneumococcal Meningitis | *Streptococcus pneumoniae* | Respiratory tract | Aerosols | Cephalosporin | Polysaccharide vaccine |
| Listeriosis | *Listeria monocytogenes* | Mouth | Foodborne infection | Penicillin G | Pasteurizing and cooking food |
| **FUNGAL DISEASE** | | | | | |
| Cryptococcosis | *Cryptococcus neoformans, C. grubii, C. gattii* | Respiratory tract | Inhaling soil contaminated with spores | Amphotericin B, flucytosine | None |
| **PROTOZOAN DISEASES** | | | | | |
| Primary Amebic Meningoencephalitis | *Naegleria fowleri* | Nasal mucosa | Swimming | Amphotericin B, miltefosine | None |
| Granulomatous Amebic Encephalitis | *Acanthamoeba* spp.; *Balamuthia mandrillaris* | Mucous membranes | Swimming | Amphotericin B, miltefosine | None |

Botulinal toxin doesn't form in acidic foods (below pH 4.7). Such acidic foods can therefore be safely preserved without the use of a pressure cooker. There have been cases of botulism from acidic foods that normally would not have supported the growth of the botulism organisms; however, most of these episodes are related to mold growth, which metabolized enough acid to allow *C. botulinum* to begin growing.

### Botulinal Types

There are several serological types of the botulinal toxin produced by different strains of the pathogen. These differ considerably in their virulence and other factors.

*Type A toxin* is probably the most virulent. Deaths have resulted from type A toxin when the food was only tasted but not swallowed. It is even possible to absorb lethal doses through skin breaks while handling laboratory samples. In untreated cases, the mortality rate is 60–70%. The type A endospore is the most heat-resistant of all *C. botulinum* strains. In the United States, it is found mainly in California, Washington, Colorado, Oregon, and New Mexico. The type A organism is usually proteolytic (the breakdown of proteins by clostridia releases amines with unpleasant odors), but obvious spoilage odor is not always apparent in low-protein foods, such as corn and green beans (**Figure 22.7**).

*Type B toxin* is responsible for most European outbreaks of botulism and is the most common type in the eastern United States. The mortality rate in cases without treatment is about 25%. Type B botulism organisms occur in both proteolytic and nonproteolytic strains.

**Figure 22.7 Funeral of an Oregon family wiped out by botulism in 1924.** The outbreak was caused by home-canned string beans. Altogether there were 12 deaths, but two funerals were held at a different church.

**Q** Why would such a drastic outcome be unlikely today?

*Type E toxin* is produced by botulism organisms that are often found in marine or lake sediments. Therefore, outbreaks commonly involve seafood and are especially common in the Pacific Northwest, Alaska, and the Great Lakes area. The endospore of type E botulism is less heat resistant than that of other strains and is usually destroyed by boiling. Type E is nonproteolytic, so the chance of detecting spoilage by odor in high-protein foods such as fish is minimal. The pathogen is also capable of producing toxin at refrigerator temperatures and requires less strictly anaerobic conditions for growth.

### Incidence and Treatment of Botulism

Botulism is not a common disease. Only a few cases are reported each year, but outbreaks from social gatherings or restaurants occasionally involve 20 to 30 cases. About half the cases are type A, and types B and E account about equally for the balance. Alaskan native people probably have the highest rate of botulism in the world, mostly of type E. The problem arises from food preparation methods that reflect a cultural tradition of avoiding the use of scarce fuels for heating or cooking. For example, one food involved in Alaskan outbreaks of botulism is *muktuk*. Muktuk is prepared by slicing the flippers of seals or whales into strips and then drying them for a few days. To tenderize them, they are stored anaerobically in a container of seal oil for several weeks until they approach putrefaction.

Botulism organisms do not seem to be able to compete successfully with the normal intestinal microbiota, so the production of toxin by ingested bacteria almost never causes botulism in adults. However, the intestinal microbiota of infants is not well established, and they may suffer from **infant botulism.** Nearly 100 cases occur in the United States annually, several

times more than any other form of botulism. Although infants have ample opportunity to ingest soil and other materials contaminated with the endospores of the organism, many reported cases have been associated with honey. Endospores of *C. botulinum* are recovered with some frequency from honey, and a lethal dose may be as few as 2000 bacteria. The recommendation is not to feed honey to infants under 1 year of age; there is no problem with older children or adults who have normal intestinal microbiota. For treatment of botulism in infants, a special preparation is available, BabyBIG®. The acronym BIG represents Botulism Immune Globulin. The immune globulin consists of antibodies to botulism toxin that have been derived from human sources.

Botulism is diagnosed by inoculating mice with samples from patient serum, stool, or vomitus specimens (**Figure 22.8**). Different sets of mice are immunized with type A, B, or E antitoxin. All the mice are then inoculated with the test toxin; if, for example, those protected with type A antitoxin are the only survivors, then the toxin is type A. The toxin in food can similarly be identified by mouse inoculation.

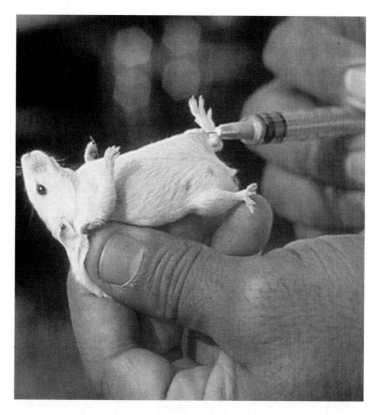

**Figure 22.8 Diagnosing botulism by identifying botulinal toxin type.** To determine whether botulinal toxin is present, mice are injected with the liquid portion of food extracts or cell-free cultures. If the mice die within 72 hours, toxin is present. To determine the specific type of toxin, groups of mice are passively immunized with antisera specific for *C. botulinum* type A, B, or E. If one group of mice receiving a specific antitoxin lives and the other mice die, the type of toxin in the food or culture has been identified.

**Q** What are the symptoms of botulism?

The botulism pathogen can also grow in wounds in a manner similar to that of clostridia causing tetanus or gas gangrene (see Chapter 23, pages 659–660). Such episodes of **wound botulism** occur occasionally.

The treatment of botulism relies heavily on supportive care. Recovery requires that the nerve endings regenerate; it therefore proceeds slowly. Extended respiratory assistance may be needed, and some neurological impairment may persist for months. Antibiotics are of almost no use because the toxin is preformed. Antitoxins aimed at neutralizing A, B, and E toxins are available and are usually administered together. This trivalent antitoxin will not affect the toxin already attached to the nerve endings and is probably more effective on type E than on types A and B. The antitoxin used in adults is derived from horses and has serious side effects, including *serum sickness* (immune complexes formed by reaction with antigens in the antitoxin) and potential anaphylaxis.

The deadly toxin of botulism (Botox®) has therapeutic uses for a number of medical conditions, such as migraines and urinary incontinence. It is also useful for relieving painful muscle contractions in conditions such as cerebral palsy, Parkinson's disease, and multiple sclerosis. Injections in the area of facial wounds prevent muscle movement during healing and result in more presentable scar formation. It has been approved to control involuntary eyelid twitching (blepharospasm), crossed eyes (strabismus), and even excessive sweating (hyperhidrosis). However, the most publicized application has been purely cosmetic: periodic local injections of Botox® to eliminate forehead wrinkles (worry lines).

> **CHECK YOUR UNDERSTANDING**
>
> ✔ **22-6** The very name *botulism* is derived from the fact that sausage was the most common food causing the disease. Why is sausage now rarely a cause of botulism?

## Leprosy

*Mycobacterium leprae* was considered to be the only bacterium that grows in the peripheral nervous system. This distinction, however, is likely also shared with the recently discovered (in 2008) leprosy-causing bacterium *M. lepromatosis*, which is found mostly in Mexico and the Caribbean. *M. leprae* was first isolated and identified around 1870 by Gerhard A. Hansen of Norway; his discovery was one of the first links ever made between a specific bacterium and a disease. **Hansen's disease** is the more formal name for **leprosy**; it is sometimes used to avoid the dreaded name.

These bacteria have an optimum growth temperature of 30°C and show a preference for the outer, cooler portions of the human body. They survive ingestion by macrophages and eventually invade cells of the myelin sheath of the peripheral nervous system, where their presence causes nerve damage from a cell-mediated immune response. It's estimated that *M. leprae* has a very long generation time, about 12 days. *M. leprae*

and *M. lepromatosis* have never been grown on artificial media. Armadillos have been found to be a useful way to culture the leprosy bacillus; they have a body temperature of 30–35°C and are often infected in the wild. The animals are now fairly common in warmer states ranging from Texas to Florida. Several people have actually contracted leprosy from contact with armadillos in Texas. Probably the most efficient way of culturing *M. leprae* now is the inoculation of the footpads of nude mice. The ability to grow the bacteria in an animal is invaluable for evaluating chemotherapeutic drugs.

Leprosy occurs in two main forms (although borderline forms are also recognized) that apparently reflect the effectiveness of the host's cell-mediated immune system. The *tuberculoid (neural) form* is characterized by discolored regions of the skin that have lost sensation and may be surrounded by a border of nodules (**Figure 22.9a**). This disease form is roughly the same as *paucibacillary* in the WHO leprosy classification system. Tuberculoid disease occurs in people with effective immune reactions. Recovery sometimes occurs spontaneously.

In the *lepromatous (progressive) form* of leprosy (which is much the same as *multibacillary* in the WHO system), skin cells are infected, and disfiguring nodules form all over the body. Patients with this type of leprosy have had the least effective cell-mediated immune response, and the disease has progressed from the tuberculoid stage. Mucous membranes of the nose tend to become affected, and a lion-faced appearance is associated with this type of leprosy. Deformation of the hand into a clawed form and considerable necrosis of tissue can also occur (**Figure 22.9b**). The progression of the disease is unpredictable, and remissions may alternate with rapid deterioration.

The exact means of transfer of the leprosy bacillus is uncertain, but patients with lepromatous leprosy shed large numbers in their nasal secretions and in exudates (oozing matter) of their lesions. Most people probably acquire the infection when secretions containing the pathogen contact their nasal mucosa. However, leprosy is not very contagious and usually is transmitted only between people in fairly intimate and prolonged contact. The time from infection to the appearance of symptoms is usually measured in years, although children can have a much shorter incubation period. Death usually results not from the leprosy itself, but from complications, such as tuberculosis.

Much of the public's fear of leprosy can probably be attributed to biblical and historical references to the disease. In the Middle Ages, people with leprosy were rigidly excluded from normal European society and sometimes even wore bells so that people could avoid them. This isolation might have contributed to the near disappearance of the disease in Europe. But patients with leprosy are no longer kept in isolation, because they can be made noncontagious within a few days by the administration of sulfone drugs. The National Leprosy Hospital in Carville, Louisiana, once housed several hundred patients but was closed in 1999. Most patients today are treated on an outpatient basis.

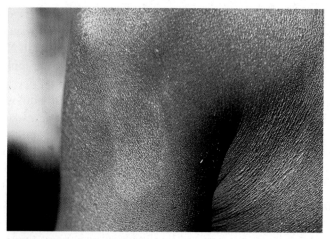

**(a)** Tuberculoid (neural) leprosy

**(b)** Lepromatous (progressive) leprosy

**Figure 22.9 Leprosy lesions. (a)** The depigmented area of skin is typical of tuberculoid (neural) leprosy. **(b)** If the immune system fails to control the disease, the result is lepromatous (progressive) leprosy. This severely deformed hand shows the progressive tissue damage to the cooler parts of the body typical of this later stage.

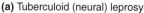 Which form of leprosy is more likely to occur in immunosuppressed individuals? Why?

Currently, about 100 cases are reported each year. Most are imported by infected immigrants from endemic countries; the disease is usually found in tropical climates. Millions of people, most of them in Asia, Africa, and Brazil, suffer from leprosy today, and over half a million new cases are reported each year.

The standard diagnostic test for leprosy is seeing acid-fast bacteria in infected skin or nerves, but permanent nerve damage has probably occurred by this time. An inexpensive blood test for detecting antibodies against *M. leprae* as early as 9 to 12 months, which is in advance of the most damaging clinical symptoms, has been registered for use in Brazil.

Dapsone (a sulfone drug), rifampin, and clofazimine, a fat-soluble dye, are the principal drugs used for treatment, usually in combination. The WHO treatment regimen for paucibacillary leprosy requires 6 months; for the multibacillary form, treatment is extended to 24 months. Clinical trials of a vaccine began in India in 2016. The Bacillus Calmette-Guérin (BCG) vaccine for tuberculosis (also caused by a *Mycobacterium* species) has been found to be somewhat protective against leprosy.

**CHECK YOUR UNDERSTANDING**

✔ **22-7** Why are nude mice and armadillos important in the study of leprosy?

# Viral Diseases of the Nervous System

## LEARNING OBJECTIVES

**22-8** Compare the Salk and Sabin polio vaccines.

**22-9** Compare the preexposure and postexposure treatments for rabies.

**22-10** Discuss the epidemiology of poliomyelitis, rabies, and arboviral encephalitis, including mode of transmission, etiology, and disease symptoms.

**22-11** Explain how arboviral encephalitis can be prevented.

Most viruses affecting the nervous system enter it by circulating in the blood or the lymphatic system. However, some viruses can enter peripheral nerve axons and move along them toward the CNS.

## Poliomyelitis

**Poliomyelitis (polio)** is best known as a cause of paralysis. However, the paralytic form of poliomyelitis probably affects fewer than 1% of people infected with the *poliovirus*. The great majority of cases are asymptomatic or exhibit only mild symptoms, such as headache, sore throat, fever, and nausea.

Polio made its first appearance in the United States in an outbreak in Vermont in the summer of 1894. After that, for decades the country was terrified by summertime epidemics. These annual outbreaks increasingly affected adolescents and young adults, and the number of paralytic cases steadily increased. Many victims died as their respiratory muscles were paralyzed, and thousands of infants and youths were left with their extremities permanently crippled. Later in the twentieth century, development of the iron lung (**Figure 22.10**) kept alive thousands with paralyzed respiratory systems.

Why did this disease seem to appear so suddenly? The answer is paradoxical—probably because of improved sanitation. The primary mode of transmission is ingestion of water contaminated with feces containing the virus. Improved sanitation delayed exposure to polioviruses in feces until after the protection provided by maternal antibodies had waned. At one time, exposure to the

Two different vaccine types are available. In 1955, the *Salk vaccine* (named after Jonas Salk, its developer) was introduced. It consists of viruses of all three types that have been inactivated (killed) by treatment with formalin. Vaccines of this types, called *inactivated polio vaccines (IPV)*, require a series of injections. A version with enhanced potency was introduced in 1988.

The other vaccine type, introduced in 1963, contains living, attenuated (weakened) strains of the virus in a suspension that is ingested. This *Sabin vaccine*, named after its developer (Albert Sabin), is more commonly called *oral polio vaccine (OPV)*. It usually contains the three types of polio virus (trivalent, *tOPV*). It is less expensive to manufacture and is simpler to administer because it does not require the trained personnel and equipment needed for safe, sterile injections. This vaccine mimics an actual infection and induces excellent, and probably life-long, immunity, although its use is precluded in immunodeficient individuals. The live virus is also shed by the recipient and has the effect of immunizing others within the community. However, this shedding can represent a serious disadvantage—the attenuated strains of the disease occasionally revert to virulence and cause the disease. The incidence of this varies by region but is usually about 1 case per 750,000 recipients.

The history of polio vaccination in the United States began with the use of the Salk IPV, which was the first available. When OPV was licensed in 1963, its advantages, especially in administration, led to near-universal adoption. Eventually, however, high rates of vaccination led to the disappearance of polio—with the exception of a few cases every year that were caused by the vaccine-derived virus. In 2000, the United States therefore changed the recommendation from OPV back to IPV. In 2015, vaccine-derived polio outbreaks occurred in Ukraine and Mali. The WHO plans to discontinue OPV by 2018. Read more about vaccine-preventable diseases in the Big Picture on pages 518–519.

### Epidemiology and Eradication Efforts

In the epidemiology of the poliovirus, naturally occurring, *wild-type (WPV)* virus is distinguished from *vaccine-derived virus (VDPV)*. VDPV is an attenuated vaccine virus that has reverted to virulence and is in circulation.

The WHO launched a campaign in 1988 to eradicate polio by 2000. The vaccine used was tOPV. Although the goal of eradication was not met, great gains were made, and by 2000 the number of case reports had fallen by 99%. Also encouraging was the extinction of WPV 2 poliovirus—indicating that eradication might be possible. However, persistent reservoirs of WPV remain in Nigeria, Pakistan, and Afghanistan.

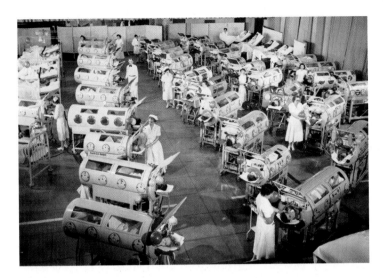

**Figure 22.10 Polio patients in iron lungs.** Many polio patients were able to breathe only with these mechanical aids. Some 6–8 survivors from these polio epidemics still use these machines, at least part of the time. Others are able to use portable respiratory aids.

**Q** What percentage of polio cases resulted in paralysis?

poliovirus was frequent (and is still so today in parts of the world with poor sanitation). Infants were usually exposed to poliovirus while still protected by maternal antibodies. The result was usually an asymptomatic case of the disease and a lifelong immunity. When infection is delayed until adolescence or early adulthood, the paralytic form of the disease appears more frequently.

Because the infection begins when the virus is ingested, its primary areas of multiplication are the throat and small intestine. This accounts for the initial sore throat and nausea. Next, the virus invades the tonsils and the lymph nodes of the neck and ileum (the terminal portion of the small intestine). From the lymph nodes, the virus enters the blood, resulting in *viremia*. In most cases the viremia is only transient, the infection does not progress past the lymphatic system, and clinical disease does not result. If the viremia persists, however, the virus eventually penetrates the capillary walls and enters the CNS. Once in the CNS, the virus displays a high affinity for nerve cells, particularly motor nerve cells in the upper spinal cord. The virus does not infect the peripheral nerves or the muscles. As the virus multiplies within the cytoplasm of the motor nerve cells, the cells die, and paralysis results. Death can result from respiratory failure.

### Diagnosis

Polio is usually diagnosed by isolating the virus from feces and throat secretions. Cell cultures can be inoculated, and cytopathic effects on the cells can be observed (see Table 15.4, page 438).

### Vaccines

There is no cure for polio; it can only be prevented. There are three different serotypes of the poliovirus: types 1, 2, and 3. Immunity must be provided for all three.

## Rabies

**Rabies** (the word is from the Latin for rage or madness) is a disease that almost always results in fatal encephalitis. The causative agent is the *rabies virus*, a member of the genus *Lyssavirus* having a characteristic bullet shape (see Figure 13.18a and discussion on page 380). Lyssaviruses (*lyssa*, from the Greek for frenzy) are single-stranded RNA viruses with no proofreading capability, and mutant strains develop rapidly. Worldwide, humans usually are infected with the rabies virus in saliva from the bite of an infected animal—especially dogs. On rare occasions, the virus can be transmitted through fresh skin abrasions and may cross the mucous membranes of the nose, mouth, and even eyes. The virus proliferates in the PNS and moves, fatally, toward the CNS (**Figure 22.11**). In the United States, the most common cause of rabies is a variant of the virus found in silver-haired bats. (Domestic animals have a high rate of vaccination.) Because deaths from rabies are frequently misdiagnosed, several cases of rabies have been traced to transplanted body tissues, especially corneas.

Rabies is unique in that the incubation period is usually long enough to allow immunity to develop from postexposure vaccination. The natural immune response is ineffective because the viruses are introduced into the wound in numbers too low to provoke it; also, they do not travel through the bloodstream or lymphatic system, where the immune system could best respond. Initially, the virus multiplies in skeletal muscle and connective tissue, where it remains localized for periods ranging from days to months. Then it enters a motor neuron and travels, at the rate of 15 to 100 mm per day, along peripheral nerves to the CNS, where it causes encephalitis. In some extreme cases, incubation periods of as long as 6 years have been reported, but the average is 30 to 50 days. Bites in areas rich in nerve fibers, such as the hands and face, are especially dangerous, and the resulting incubation period tends to be short.

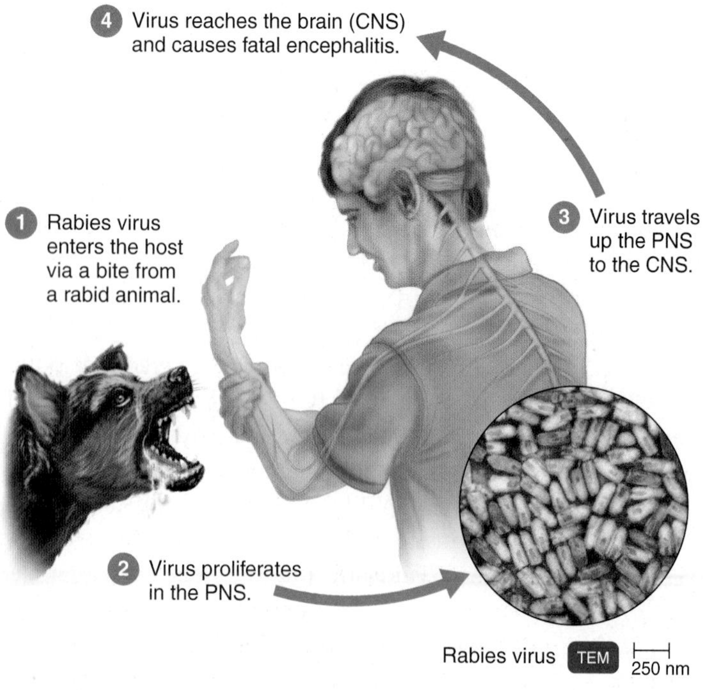

Once the virus enters the peripheral nerves, it isn't accessible to the immune system until cells of the CNS begin to be destroyed, which triggers a belated and ineffective immune response.

Preliminary symptoms are mild and varied, resembling several common infections. When the CNS becomes involved, the patient tends to alternate between periods of agitation and intervals of calm. At this time, a frequent symptom is spasms of the muscles of the mouth and pharynx that occur when the patient feels air drafts or swallows liquids. In fact, even the mere sight or thought of water can set off the spasms—thus the common name *hydrophobia* (fear of water). The final stages of the disease result from extensive damage to the nerve cells of the brain and the spinal cord.

Animals with **furious (classic) rabies** are at first restless, then become highly excitable and snap at anything within reach. The biting behavior is essential to maintaining the virus in the animal population. Humans also exhibit similar symptoms of rabies, even biting others. When paralysis sets in, the flow of saliva increases as swallowing becomes difficult, and nervous control is progressively lost. The disease is almost always fatal within a few days.

Some animals suffer from **paralytic (dumb** or **numb) rabies**, in which there is only minimal excitability. This form is especially common in cats. The animal remains relatively quiet and even unaware of its surroundings, but it might snap irritably if handled. A similar manifestation of rabies occurs in humans and is often misdiagnosed as *Guillain-Barré syndrome*, a form of paralysis that is usually transient but sometimes fatal, or other neurological conditions. There is some speculation that the two forms of the disease may be caused by slightly different forms of the virus.

### Diagnosis

Rabies is usually diagnosed in the laboratory by detection of the viral antigen using the **direct fluorescent-antibody (DFA) test**, which is nearly 100% sensitive and highly specific. These tests can be done on saliva samples, blood, CSF, and skin; postmortem samples are usually taken from the brain. For less-developed parts of the world, the CDC has developed a **rapid immunohistochemical test (RIT)**. It requires only the use of an ordinary light microscope and has a sensitivity and specificity equivalent to the standard DFA test.

### Prevention of Rabies

Only high-risk individuals, such as laboratory workers, animal control professionals, and veterinarians, are routinely vaccinated against rabies before known exposure. If a person is bitten, the wound should be thoroughly washed with soap

④ Virus reaches the brain (CNS) and causes fatal encephalitis.

① Rabies virus enters the host via a bite from a rabid animal.

③ Virus travels up the PNS to the CNS.

② Virus proliferates in the PNS.

Rabies virus   TEM   ⊢—⊣ 250 nm

**Figure 22.11   Pathology of rabies infection.**

**Q** What is the postexposure treatment for rabies?

and water. If the animal is positive for rabies, the person must undergo **postexposure prophylaxis (PEP)**—meaning a series of antirabies vaccine and immune globulin injections. Another indication for antirabies treatment is any unprovoked bite by a skunk, bat, fox, coyote, bobcat, or raccoon not available for examination. Treatment after a dog or cat bite, if the animal cannot be found, is determined by the prevalence of rabies in the area. The bite of a bat may not be perceptible and may be impossible to rule out in cases where the bat had access to sleeping persons or small children. Therefore, the CDC recommends PEP after any significant encounter with a bat—unless the bat can be tested and shown to be negative for rabies.

The original Pasteur treatment, in which the virus was attenuated by drying in the dissected spinal cords of rabies-infected rabbits, has long been replaced by **human diploid cell vaccine (HDCV)**, or chick embryo–grown vaccines. These vaccines are administered in a series of four injections at intervals during a 14-day period. Passive immunization is provided simultaneously by injecting **human rabies immune globulin (RIG)** that has been harvested from people who are immunized against rabies.

### Treatment of Rabies

Once the symptoms of rabies appear, there is very little in the way of effective treatment—only a handful of survivors have been reported. Most of these survivors had received PEP before the appearance of symptoms. There have been only a couple of cases of survival of a patient who had not received PEP. The primary treatment, which succeeds in a minority of cases, is to induce an extended coma to minimize excitability while administering antiviral drugs. This procedure was first used in the case of a Wisconsin girl bitten by a rabid cat and has come to be called the Milwaukee protocol.

### Distribution of Rabies

Rabies occurs all over the world, mostly a result of dog bites. Vaccination of pets is prohibitively expensive in most of Africa, Latin America, and Asia. In these areas, 40,000 to 70,000 deaths by rabies occur annually. In the United States, the vaccination of pets is nearly universal, but rabies is widespread among wildlife, predominantly bats, skunks, foxes, and raccoons, although it is also found in domesticated animals (**Figure 22.12**). As many as 40,000 people are administered postexposure rabies vaccine each year, often as a precaution when the rabies status of the biting animal cannot be determined. Rabies is almost never found in squirrels, rabbits, rats, or mice. The disease has long been endemic in vampire bats of South America. In Europe and North America, there are ongoing efforts to immunize wild animals with live vaccinia virus, genetically modified to produce a rabies virus glycoprotein that is added to food left for the animals to find. This has been highly successful in Europe, and several countries have been declared free of rabies as a result.

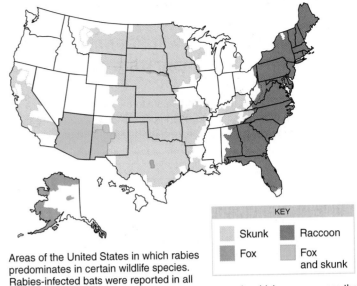

Areas of the United States in which rabies predominates in certain wildlife species. Rabies-infected bats were reported in all 48 contiguous states and Alaska. In eastern states in which raccoons are the predominant rabies-infected animal, many cases were also reported in foxes and skunks.

KEY
Skunk   Raccoon
Fox   Fox and skunk

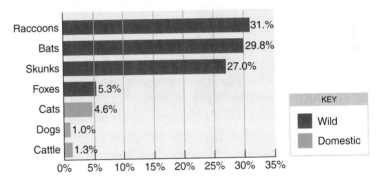

Raccoons 31.%
Bats 29.8%
Skunks 27.0%
Foxes 5.3%
Cats 4.6%
Dogs 1.0%
Cattle 1.3%

0%  5%  10%  15%  20%  25%  30%  35%

KEY
Wild
Domestic

Rabies cases in various wild and domestic animals in the United States. Rabies in domestic animals such as dogs and cats is uncommon because of high vaccination rates. Raccoons, skunks, and bats are the animals most likely to be infected with rabies. Most human cases are caused by bat bites. Worldwide, most human cases are caused by dog bites.

**Figure 22.12 Reported cases of rabies in animals.** Rabies-infected bats were reported in all 48 contiguous states and Alaska. Rabies in foxes include different species in different geographical areas. *Source:* CDC 2017.

**Q** What is the primary reservoir for the rabies virus in your area?

In the United States, 7000 to 8000 cases of rabies are diagnosed in animals each year, but in recent years, only one to six cases have been diagnosed in humans annually (see the Clinical Focus box on page 636). In developing nations, rabies has a much higher incidence.

### Related *Lyssavirus* Encephalitis

In recent years, a few fatal cases of encephalitis that are clinically indistinguishable from classic rabies have occurred in Australia and Scotland—countries considered free of rabies. These cases were found to be caused by genotypes of the

## Infections that cause mild or moderate illness in adults can have catastrophic effects when they pass from mother to child during pregnancy or birth.

In 2016, some athletes did not participate in the Summer Olympics in Brazil because of fears surrounding an outbreak of Zika virus disease there. That same year, the U.S. Centers for Disease Control and Prevention (CDC) issued a notice advising pregnant women not to travel to Florida because of the presence of mosquitoes carrying Zika virus. In 2016, 41 of the 875 babies born to Zika-infected mothers in the United States had birth defects.

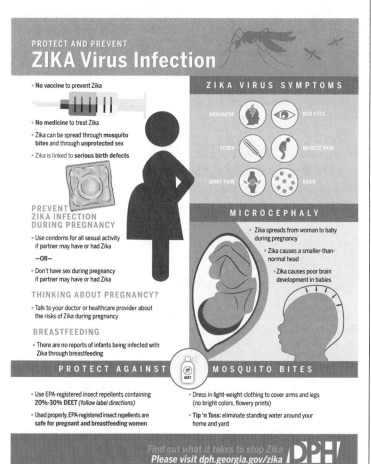

Zika public health notice.

## Most Microbes Don't Cross the Placenta, But Those That Do Can Cause Serious Damage

Congenital (existing before or at birth) infections are generally transmitted vertically, from the mother to child, during pregnancy or delivery. Most common microbes (such as influenza virus or *Candida* yeast) won't cross the placenta to affect a fetus. So the primary treatment concern for most pregnant patients is finding antimicrobial drugs that won't harm the baby as they cure the mother. However, a select group of pathogens possesses the ability to cross the placenta and enter a fetus. These diseases may show minor symptoms in the

mother even as they wreak havoc on the developing baby. Viral diseases in which vertical transmission is a factor include the following.

- **Zika virus disease** crosses the placenta during pregnancy and targets nerve stem cells. It is linked to microcephaly, calcium deposits in the brain, and other brain and eye abnormalities.

- **Neonatal herpes** usually results from transmission of the herpes simplex viruses (HSV-1or HSV-2) from the mother at delivery. It has a mortality rate of 60%, and survivors have central nervous system disorders such as seizures.

- ***Cytomegalovirus* (CMV) infections**, caused by another type of human herpes virus, are the most important cause of congenital infection in developed countries. Transmission to the fetus is highest if the mother acquires the infection during the first half of pregnancy. In the United States, about one out of every 150 babies is born with congenital CMV infection. However, only about 20% of infected babies will be sick. Symptoms of congenital CMV include microcephaly, hearing and vision loss, and seizures.

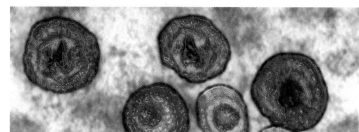

Cytomegalovirus.

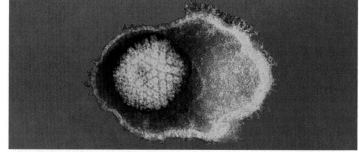

Human *Simplexvirus*.

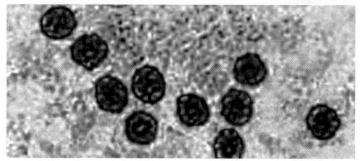

Zika virus, ZikV.

# Bacteria as Well as Viruses Can Cross the Placenta to Cause Disease

- ***Treponema pallidum*** causes congenital syphilis, which can result in miscarriage, stillbirth, and early infant death. The fetus is susceptible to infection after the fourth month of pregnancy. Surviving infants may show altered bone development. In the United States, approximately 429 cases were reported in 2016. Penicillin taken by the mother is 98% effective in preventing the disease.

- Congenital ***Listeria monocytogenes*** infection results in premature delivery, miscarriage, or stillbirth.

- **Group B *Streptococcus*** infection results in deafness or learning disabilities.

- ***Elizabethkingia***, a newly discovered pathogen, can cross the placenta and cause meningitis in newborns.

- Congenital infection by the protozoan ***Toxoplasma gondii*** can result in microcephaly, hydrocephaly (accumulation of CSF in the brain), convulsions, and motor impairment.

## The TORCH Screen Tests Pregnant Women for Diseases That Can Transmit Vertically

TORCH is a panel of tests that was originally developed to screen mothers for antibodies to microbes that can harm the fetus. TORCH stands for Toxoplasmosis, Other, Rubella, Cytomegalovirus, and Herpes simplex virus. The "Other" category includes many diseases, among which are syphilis, chickenpox, HIV, measles, mumps, and hepatitis B.

Since prevention is almost always better than treatment after the fact, women who plan to become pregnant should be up to date on their recommended vaccinations, including those for measles, mumps, and rubella—all of which can cause serious congenital problems.

**Microcephaly is caused by several agents, including Zika virus and *Toxoplasma*.**

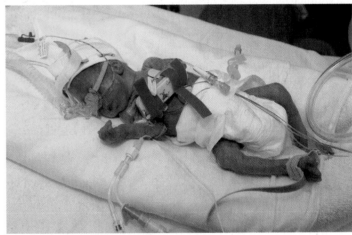

**Congenital *Listeria* infection can cause premature birth.**

## KEY CONCEPTS

- Cytomegalovirus (CMV) is the most common cause of congenital virus infection. **(See Chapter 23, page 670.)**

- Diseases of the reproductive system can be transmitted at birth. **(See Chapter 26, "Gonorrhea," page 766; "Congenital Syphilis," page 774; and "Neonatal Herpes," page 777.)**

- Vaccination before pregnancy can prevent infectious congenital diseases. **(See Chapter 18, "Vaccines," pages 500–507.)**

# A Neurological Disease

As you read through this problem, you'll see questions that clinicians ask themselves as they proceed through a diagnosis and treatment. Try to answer each question before going on to the next one.

1. On September 30, Yolanda, a 10-year-old girl, has pain and stiffness in her right arm and temperature of 38.3°C (101°F). On October 3, she begins vomiting and has increased arm pain and numbness.
   *What could these symptoms indicate?*

2. The high fever could indicate some sort of bacterial or viral infection. Yolanda's pediatrician orders a rapid group A streptococcal antigen test, which is negative. Yolanda is hospitalized on October 7, when she has difficulty swallowing. Her tongue has a whitish coating and is protruding from her mouth.
   *What infections are possible?*

3. The whitish coating on her tongue could indicate mucosal candidiasis. Yolanda is given fluconazole to combat the fungus. On October 8, a lumbar puncture shows elevated numbers of white blood cells.
   *What does this indicate?*

4. An elevated white blood cell count in Yolanda's CSF indicates some type of microbial infection of the CNS. Yolanda is treated with vancomycin for meningoencephalitis. She then experiences hypersalivation and lethargy.
   *What does this suggest? How would you confirm the disease?*

5. Rabies is confirmed by direct fluorescent-antibody staining of a skin biopsy for rabies virus antigens. Yolanda

dies on November 2. An abundance of rabies viral inclusions (Negri bodies) are seen in the brain stem (Figure A).
*How would you treat people who had contact with Yolanda in October and November?*

6. Postexposure prophylaxis (PEP) is administered to 66 people, including 31 people in Yolanda's school.
   *Did the delay in diagnosis affect the outcome of the disease?*

7. Early diagnosis cannot usually save a patient; however, it may help minimize the number of potential exposures and the need for PEP.
   *What else must be determined about this case?*

8. In mid-June, Yolanda had awakened during the night and said a bat had flown into her bedroom window and bitten her. Her mother cleaned a small mark on the girl's arm with an over-the-counter antiseptic but assumed the incident was

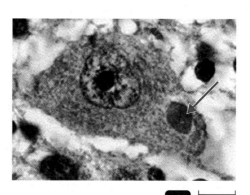

LM  |—————| 15 μm

**Figure A Negri body at the arrow in an infected neuron.**

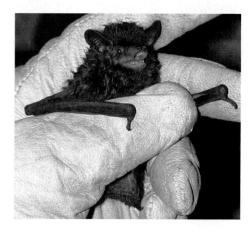

**Figure B Silver-haired bat.**

a nightmare. Two days later, an older sibling took a dead bat away from the yard. The mother did not associate the bat with the previous event and did not seek rabies PEP for the girl.

The nucleotide sequence of the PCR product was used to identify a rabies virus variant associated with silver-haired bats (Figure B).
*Why is rabies surveillance and case reporting important in the United States?*

9. During 2006–2016, 17 of 26 reported human rabies cases acquired in the United States were epidemiologically linked to bats. Human rabies is preventable with proper wound care and timely, appropriate administration of human rabies immune globulin and rabies vaccine before onset of clinical symptoms.

*Data source: CDC, January 2017.*

genus *Lyssavirus* (see page 380) that are closely related to classical rabies virus: the *Australian bat lyssavirus (ABLV)* and the *European bat lyssavirus (EBLV).** 

Classic rabies is caused by one of 11 known genotypes of the genus *Lyssavirus* and is widespread worldwide. Other, non-rabies, lyssaviruses causing encephalitis are indigenous to Europe, Australia, Africa, and the Philippines, most commonly in bats. Different species of bats are infected with distinct variants of the rabies-related lyssaviruses.

---

*A lengthy list of diseases—rabies and similar lyssavirus diseases, as well as the SARS, Ebola, Hendra, and Nipah viruses—are all now known, or strongly suspected, to be transmitted by bats. There are reasons why bats make good disease reservoirs: there are more than a thousand species to occupy various niches; they are long-lived (5 to 50 years), which lends them stability as a reservoir; they tend to roost in close assemblies, which facilitate viral spread; and they fly relatively long distances as they forage for food—some are even migratory. Finally, bats seem to be able to carry viruses for long periods without clearing the infection or becoming ill.

**CHECK YOUR UNDERSTANDING**

✔ **22-9** Why is postexposure vaccination for rabies a practical option?

## Arboviral Encephalitis

Encephalitis caused by mosquito-borne viruses (called arboviruses) is rather common in the United States. (*Arbovirus* is short for *ar*thropod-*bo*rne virus. This terminology represents a functional grouping; it is not a formal taxonomic term.) The incidence of disease increases in the summer months, coinciding with the proliferation of adult mosquitoes. *Sentinel animals,* such as caged chickens, are tested periodically for antibodies to arboviruses. This gives health officials information on the incidence and types of viruses in their area.

A number of clinical types of arboviral encephalitis have been identified; all can cause symptoms ranging from subclinical to severe, including rapid death. Active cases of these diseases are characterized by chills, headache, and fever. As the disease progresses, mental confusion and coma occur. Survivors may suffer from permanent neurological problems.

Horses as well as humans are affected by these viruses; thus, there are strains causing *eastern equine encephalitis (EEE)* and *western equine encephalitis (WEE).* These two viruses are the most likely to cause severe disease in humans. EEE is the more severe; the mortality rate is 30% or more, and survivors experience a high incidence of brain damage, deafness, and other neurological problems. EEE is uncommon (its main mosquito vector prefers to feed on birds); only about 100 cases a year are reported. No cases of WEE have been reported in over 10 years; it has a mortality rate of 5%.

*St. Louis encephalitis (SLE)* acquired its name from the location of an early major outbreak (in which it was originally discovered that mosquitoes are involved in the transmission of these diseases). SLE is distributed from southern Canada to Argentina, but mostly in the central and eastern United States.

Fewer than 1% of people infected exhibit symptoms; it can, however, be a severe disease with a mortality rate in symptomatic patients of about 20%.

*California encephalitis (CE)* was first identified in that state, but most cases occur elsewhere. In fact, there has been only one case of CE in California in 60 years. A relatively mild illness, it is seldom fatal.

A new arbovirus disease, now the most common arbovirus in the United States (**Figure 22.13**), was introduced into the United States in 1999. First reported in the New York City area, it was quickly identified as being caused by *West Nile virus (WNV),* which, like the virus causing SLE, is related to the virus causing Japanese encephalitis (see page 638). The disease is maintained in a bird-mosquito-bird cycle. The primary mosquito is a species of *Culex,* which can overwinter as adults in temperate climates. Birds serve as amplifying hosts; some species, such as house sparrows, can have high levels of viremia without dying. But mortality of infected crows, ravens, and blue jays is high, and public health officials sometimes request reports of dead birds of these species. Most human cases of WNV are subclinical or mild, but the disease can cause a polio-like paralysis or fatal encephalitis, especially in older adults.

Other encephalitis viruses are appearing in the United States. As of 2017, 30 cases of **Heartland virus disease** (a member of the Bunyaviridae) have occurred in the United States is unknown. Infections by **human Powassan virus** (POW, a member of the Flaviviridae) have been recognized in northern latitudes. In the United States, POW virus disease has been reported primarily from northeastern states and the Great Lakes region. See Diseases in Focus 22.2 on page 641 for a summary of the predominant arbovirus-caused diseases of the United States.

**Figure 22.13  West Nile serogroup arbovirus cases: 2016.** This is the most common arbovirus encephalititis in the United States. The majority of cases in this serogroup are of the La Crosse virus.
*Source:* CDC 2017.

**Q** Why do arboviral infections occur during the summer months?

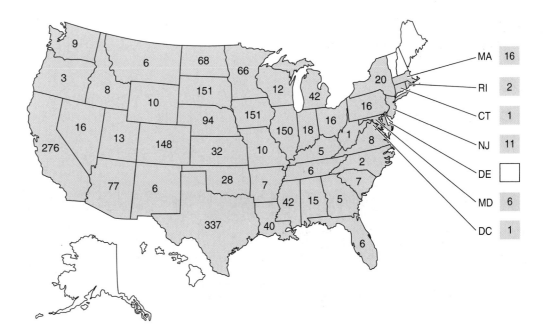

The Far East and South Asia also have endemic arboviral encephalitis. **Japanese encephalitis** is the best known; it is a serious public health problem, especially in Japan, Thailand, Korea, China, and India. Vaccines are used to control the disease in these countries and are often recommended for visitors. Only about 1% of people infected show clinical symptoms, which may involve seizures and paralysis—and a mortality rate of 20–30%.

Arboviral encephalitis is diagnosed by serological tests, usually ELISA tests to identify IgM antibodies. The most effective preventive measure is local control of the mosquitoes.

### Zika Virus Disease

**Zika virus disease (ZVD)** is caused by the *Zika virus*, or *ZIKV*, a member of the Flaviviridae. It is transmitted primarily through the bite of infected *Aedes* spp. mosquitoes, the same mosquitoes that carry the viruses that cause dengue fever, yellow fever, and chikungunya. Zika virus may also be transmitted sexually, from mother to child during pregnancy and delivery, and through blood transfusions. ZIKV was first discovered in 1947 in rhesus monkeys and is named after the Zika forest in Uganda. It was subsequently identified in humans in 1952 in the United Republic of Tanzania. The first outbreak outside Africa occurred in Micronesia, on the island of Yap. The virus moved to French Polynesia and Brazil in 2014. The virus reached the continental United States in 2015 (**Figure 22.14**). Since then, 224 locally contracted cases have been reported.

The most common signs and symptoms of ZVD, which last about a week, affect about 20% of infected individuals. They are usually mild and include fever, headache, muscle and joint pain, malaise, skin rash, and conjunctivitis. Because people usually do not become ill enough to require hospitalization, they may not even realize that they are infected. However, ZIKV

infection during pregnancy greatly increases the risk of fetuses and infants developing a condition called **microcephaly**. In this disorder, the infant's head is much smaller than normal because of abnormal brain development. Infants so affected exhibit developmental delays that range from mild to severe. ZVD is also associated with *Guillain-Barré syndrome*, a temporary muscular weakness and tingling in the upper and lower limbs.

Diagnosis of ZVD is confirmed by reverse transcriptase PCR, and treatment involves rest, intake of sufficient fluids, and pain and fever reduction with common medications. In 2016, the National Institute of Allergy and Infectious Diseases began human testing of a DNA vaccine (see page 503). As with arboviral encephalitis, the best prevention is reducing mosquito-breeding sites and reducing contact between mosquitoes and humans.

---

**CHECK YOUR UNDERSTANDING**

✔ **22-10** Which disease is transmitted by mosquitoes: polio, rabies, or arboviral encephalitis?

✔ **22-11** When there are serious local outbreaks of arboviral encephalitis, what is the usual response to minimize its transmission?

---

# Fungal Disease of the Nervous System

### LEARNING OBJECTIVE

**22-12** Identify the causative agent, reservoir, symptoms, and treatment for cryptococcosis.

The central nervous system is seldom invaded by fungi. However, one pathogenic fungus in the genus *Cryptococcus* is well adapted to growth in CNS fluids.

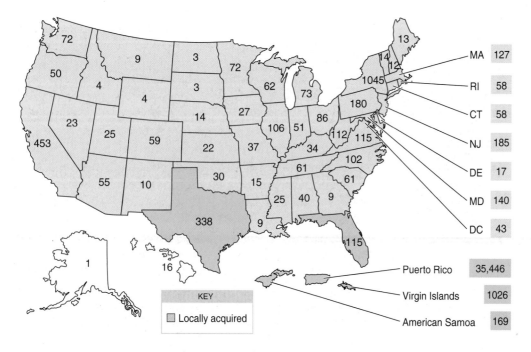

**Figure 22.14 Reported cases of Zika virus disease in the United States, 2015–2017.** Local transmission has occurred around Miami Beach, Florida, and Brownsville, Texas. Data through August 2017.

**Q** What factors determine whether the virus can be locally transmitted?

| | |
|---|---|
| MA | 127 |
| RI | 58 |
| CT | 58 |
| NJ | 185 |
| DE | 17 |
| MD | 140 |
| DC | 43 |

| | |
|---|---|
| Puerto Rico | 35,446 |
| Virgin Islands | 1026 |
| American Samoa | 169 |

KEY

☐ Locally acquired

## *Cryptococcus neoformans* Meningitis (Cryptococcosis)

The disease **cryptococcosis** is caused by fungi of the genus *Cryptococcus*. They form spherical cells resembling yeasts, reproduce by budding, and produce extremely heavy polysaccharide capsules (**Figure 22.15**). The primary species pathogenic for humans are *Cryptococcus neoformans* and *C. grubii*. These organisms are widely distributed, especially in areas contaminated by bird droppings most notably from pigeons, which excrete an estimated 25 pounds a year. The disease is transmitted mainly by the inhalation of dried, contaminated droppings. The inhaled fungi multiply in persons with compromised immune systems, such as AIDS patients, disseminate to the CNS, and cause meningitis that has a high mortality rate. In recent years there have been outbreaks of cryptococcosis in AIDS patients in California caused by *C. gattii* (GAHT-tē-ē), a species that had previously been reported only in tropical regions. However, an association has now been observed with trees native to subtropical and temperate regions, as well; the fungus inhabits an ecological niche in decayed hollows of mature trees. See the Clinical Case on page 325. From there the basidiospores (see page 330) can contaminate surrounding soils or be spread with the distribution of wood products. This species has now been isolated in cases of cryptococcosis, even in otherwise healthy individuals, in several areas of western North America as far north as Vancouver Island in Canada. This disease is likely to continue to spread southward and may eventually affect areas as far as Florida.

The best serological diagnostic test is a latex agglutination test to detect cryptococcal antigens in serum or cerebrospinal fluid. The drugs of choice for treatment are amphotericin B and

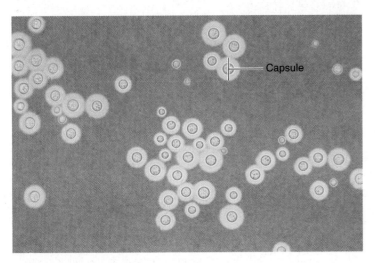

**Figure 22.15  *Cryptococcus neoformans.*** This yeastlike fungus has an unusually thick capsule. In this photomicrograph, the capsule is made visible by suspending the cells in dilute India ink.

**Q** What is the significance of the extremely heavy polysaccharide capsule found in *C. neoformans*?

flucytosine in combination. Even so, the mortality rate may approach 30%.

**CHECK YOUR UNDERSTANDING**

✓ 22-12 What is the most common source of airborne cryptococcal infections?

# Protozoan Diseases of the Nervous System

**LEARNING OBJECTIVE**

22-13 Identify the causative agent, vector, symptoms, and treatment for African trypanosomiasis and amebic meningoencephalitis.

Protozoa capable of invading the CNS are rare. However, those that can reach it cause devastating effects.

## African Trypanosomiasis

**African trypanosomiasis**, or sleeping sickness, is a protozoan disease that affects the nervous system. In 1907, Winston Churchill described Uganda during an epidemic of sleeping sickness as a "beautiful garden of death." Even today, estimates are that as many as half a million Africans are infected, and there are about 100,000 new cases yearly.

The disease is caused by two subspecies of *Trypanosoma brucei* that infect humans: *Trypanosoma brucei gambiense* and *T.b. rhodesiense*. They are morphologically indistinguishable but differ significantly in their epidemiology—that is, in their ability to infect nonhuman hosts. Humans are the only significant reservoir for *T.b. gambiense*, whereas *T.b. rhodesiense* is a parasite of domestic livestock and many wild animals. These protozoans are flagellates (see Figure 23.22 on page 675 for the appearance of a similar organism) that are spread by tsetse fly vectors. *T.b. gambiense* is transmitted by a tsetse fly species that inhabits stream vegetation, where there are also concentrations of human populations. It is distributed throughout west and central Africa and is sometimes termed West African trypanosomiasis. More than 97% of reported cases in humans are of this type. Once a person becomes infected, there are few symptoms for weeks or months. Eventually, a chronic form of disease with fever, headaches, and a variety of other symptoms develops that indicates involvement and deterioration of the CNS. Coma and death are inevitable without effective treatment.

In contrast, infections by *T.b. rhodesiense* are transmitted by species of tsetse flies that inhabit savannahs (grasslands with scattered trees) of eastern and southern Africa. Wild animals inhabiting these areas are well adapted to the parasite and are little affected, but humans and domestic animals become acutely ill. This has had a profound effect on sub-Saharan Africa, an area nearly the size of the United States. Agricultural

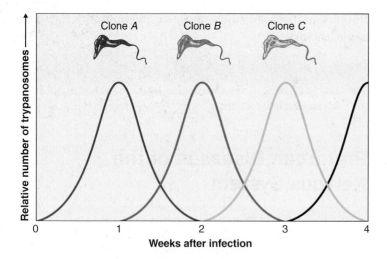

**Figure 22.16 How trypanosomes evade the immune system.** The population of each trypanosome clone drops nearly to zero as the immune system suppresses its members, but a new clone with a different antigenic surface then replaces the previous clone. The black line represents the population of clone D.

**Q** What viral disease that is causing a worldwide pandemic would make for a similar figure?

development has been practically prohibited because domestic food and working animals become infected. Infections of humans follow a more acute course than that caused by *T.b. gambiense*; symptoms of illness are apparent within a few days or so of infection. Death occurs within weeks or a few months, sometimes from cardiac problems even before the CNS is affected.

Typanosomiasis is treated with suramin and pentamidine, but these do not alter the course of the disease once the CNS is affected. The drug that does alter the disease's course, melarsoprol, is very toxic. Eflornithine crosses the blood–brain barrier and blocks an enzyme required for proliferation of the parasite. It requires an extended series of injections, but it is so dramatically effective against even late stages of *T.b. gambiense* that it has been called the resurrection drug. (Its effectiveness against *T.b. rhodesiense* is variable; melarsoprol is still recommended.)

The current primary approach in combating the disease is to attempt elimination of the vector, the tsetse fly. The use of tentlike, insecticide-treated traps that mimic the color and odor of animal hosts of the insect, combined with large-scale releases of sterile males have eliminated the tsetse fly on the offshore island of Zanzibar. (Female tsetse flies mate only once; the release of artificially reared, radiation-sterilized males in vast numbers prevents females that mate with them from producing young.) The insect is a weak flyer, and health care officials hope to repeat this eradication on selected areas of the mainland.

A vaccine is being developed, but a major obstacle is that the trypanosome is able to change protein coats at least 100 times

and can thus evade antibodies aimed at only one or a few of the proteins. Each time the body's immune system is successful in suppressing the trypanosome, a new clone of parasites appears with a different antigenic coat (**Figure 22.16**).

## Amebic Meningoencephalitis

There are three species of free-living protozoa that cause amebic meningoencephalitis, a devastating disease of the nervous system. These protozoa are both found in recreational freshwater. Human exposure to them is apparently widespread; many in the population carry antibodies—fortunately, symptomatic disease is rare. *Naegleria fowleri* is a protozoan (ameba) that causes a neurological disease, **primary amebic meningoencephalitis (PAM)** (**Figure 22.17**). Although scattered cases are reported in most parts of the world, only a few cases are reported in the United States annually. The most common victims are children who swim in warm ponds or streams. The organism initially infects the nasal mucosa and later penetrates to the brain and proliferates, feeding on brain tissue. The fatality rate is nearly 100%, death occurring within a few days after symptoms appear. Because of the rarity of the disease, there is a low "index of suspicion"; also, the symptoms resemble those of encephalitis caused by other, more common, pathogens. Diagnosis is typically made at autopsy. There have been only a very

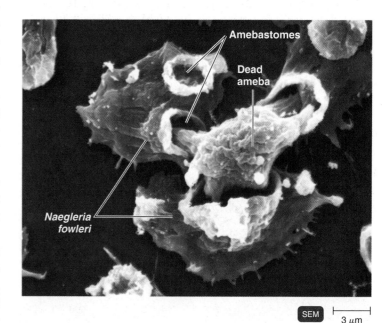

**Figure 22.17 *Naegleria fowleri.*** This photo shows two vegetative stages of *N. fowleri* beginning to devour a presumably dead ameba. The suckerlike structures (called amebastomes) function in phagocytic feeding—usually on bacteria or assorted debris that may include host tissue. This protozoan also has a spherical cyst stage and an ovoid flagellated stage (which is most likely to be the infective form) that allows it to swim rapidly in its aquatic habitat.

**Q** How is amebic meningoencephalitis transmitted?

# Types of Arboviral Encephalitis

Arboviral encephalitis is usually characterized by fever, headache, and altered mental status ranging from confusion to coma. Vector control to decrease contacts between humans and mosquitoes is the best prevention. Mosquito control includes removing standing water and using insect repellent while outdoors. An 8-year-old girl in rural Wisconsin has chills, headache, and fever and reports having been bitten by mosquitoes. Use the table below to determine which types of encephalitis are most likely. How would you confirm your diagnosis? For the solution, go to @MasteringMicrobiology.

*Culex* mosquito engorged with human blood.

| Disease | Pathogen | Mosquito Vector | Reservoir | U.S. Distribution | Epidemiology | Mortality |
|---|---|---|---|---|---|---|
| California Encephalitis (La Crosse and Jamestown Canyon Strains) | CE virus (*Bunyavirus*) | *Aedes* | Small mammals | | Affects mostly 4- to 18-year age groups in rural or suburban areas. Rarely fatal; about 10% have neurological damage | 1% of those hospitalized |
| Eastern Equine Encephalitis | EEE virus (*Togavirus*) | *Aedes, Culiseta* | Birds, horses | | Affects mostly young children and younger adults; relatively uncommon in humans | >30% |
| Heartland Encephalitis | Heartland virus (*Phlebovirus*) | *Amblyomma americanum* tick | May be deer or raccoons | | Neurological problems, thrombocytopenia, leukopenia | 20% |
| Powassan Encephalitis (light gold) | POW virus (*Flavivirus*) | *Ixodes* spp. ticks | May be white-footed mice | | Long-term neurological problems may occur | 10–15% |
| St. Louis Encephalitis | SLE virus (*Flavivirus*) | *Culex* | Birds | | Mostly urban outbreaks; affects mainly adults over 40 | 20% |
| West Nile Encephalitis | WN virus (*Flavivirus*) | Primarily *Culex* | Primarily birds, assorted rodents, and large mammals | | Most cases asymptomatic—otherwise symptoms vary from mild to severe; likelihood of severe neurological symptoms and fatality increases with age | 4–18% of those hospitalized |

The CSF fluid contains slow-moving ameboid cells. The technician performs an indirect immunofluoresence test to determine what specific microorganism is in Patricia's CSF. The test shows antibodies against *Naegleria fowleri* at a dilution of 1:4096. The news is serious: Patricia has primary amebic meningoencephalitis, usually a rapidly fatal disease. *N. fowleri* is a euglenozoa that lives as an ameba in warm freshwater. In low nutrients, the trophozoite forms a rapidly motile cell with two flagella. The trophozoite encysts during cold or dry conditions and reemerges when conditions improve (see the figure).

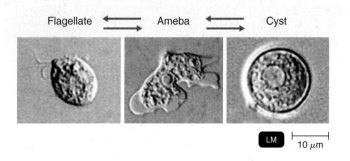

Flagellate ⟷ Ameba ⟷ Cyst

LM | 10 μm

**How is *N. fowleri* transmitted?**

621 | 625 | 626 | **642** | 644 | 645

few survivors of PAM. They were treated with a combination of several antibiotics.

A similar neurological disease is **granulomatous amebic encephalitis (GAE)**. GAE is caused by *Acanthamoeba* spp. It is chronic, slowly progressive, and fatal in a matter of weeks or months. GAE has an unknown incubation period, and months may elapse before symptoms appear. Granulomas (see Figure 23.28, page 683) form around the organism in response to an immune reaction. The portal of entry is not known but is probably mucous membranes. Multiple lesions are formed in the brain and other organs, especially the lungs.

*Balamuthia mandrillaris* is a free-living ameba that causes GAE in mammals. The ameba, present in soil, is probably transmitted by inhalation or through skin lesions. Approximately 150 cases of balamuthiasis have been reported worldwide since the disease was recognized in 1990; 10 of those cases were in the United States. Of the 150 people affected, only 7 survived.

Miltefosine is available from the CDC to treat infections caused by *N. fowleri*, *Acanthamoeba* spp., and *B. mandrillaris*.

**CHECK YOUR UNDERSTANDING**

✔ **22-13** What insect is the vector for African trypanosomiasis?

# Nervous System Diseases Caused by Prions

**LEARNING OBJECTIVE**

**22-14** List the characteristics of diseases caused by prions.

Several fatal diseases affecting the human central nervous system are caused by prions. To explain the term **prion**, we need to recall from the discussion of enzymes in Chapter 5 that the shape of an enzyme's protein component is essential for its operation. A certain protein is normally found on the surface of brain cell neurons and is even found on the surface of certain stem cells in red bone marrow and cells that become neurons; called normal prion protein, PrP$^C$. Its function is uncertain, but there is evidence that it may guide maturation of nerve cells. Certainly, the protein's shape causes no damage. But this protein can assume two folded shapes, one normal and the other abnormal (there is no change in the amino acid sequence). If the PrP$^C$ encounters an *abnormally folded protein*, the normal protein changes its shape and also becomes abnormally folded—this is called PrP$^{Sc}$. In fact, a chain reaction of protein misfolding occurs. Therefore, a single infective prion may lead to a cascade of new prions, which then clump together to form the fibril aggregations of misfolded proteins that are found in diseased brains. See **Figure 22.18a**. Autopsies of this infected brain tissue also show that it exhibits a characteristic spongiform degeneration (it is porous, like a sponge), as shown in Figure 22.18b. (Also see the discussion of prions in Chapter 13, pages 388–389, and Figure 13.23.) In recent years, the study of these diseases, called **transmissible spongiform encephalopathies (TSE)**, has been one of the most interesting areas of medical microbiology.

A typical prion disease in animals is **sheep scrapie**, which has been long known in Great Britain and made its first appearance in the United States in 1947. The infected animal scrapes itself against fences and walls until areas of its body are raw. During a period of several weeks or months, the animal gradually loses motor control and dies. The infection can be experimentally passed to other animals by injecting brain tissue from one animal to the next. Similar conditions are seen in mink, possibly resulting from the animals' being fed mutton. A prion disease, **chronic wasting disease**, affects wild deer and elk in the western United States and Canada. It is invariably fatal, and there are concerns that it might infect humans who eat venison and might eventually infect domestic livestock.

Humans suffer from TSE diseases similar to scrapie; **Creutzfeldt-Jakob disease (CJD)** is an example. CJD is rare (about 300 cases per year in the United States). It often occurs in families, an indication of a genetic component. This form of CJD is sometimes referred to as classic CJD to differentiate it from similar variants that have appeared. There is no doubt that an infective agent is involved because transmission via corneal transplants and accidental scalpel nicks of a surgeon during

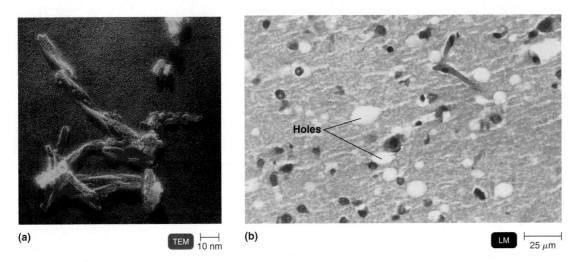

(a)  TEM ⊢—⊣ 10 nm     (b)  LM ⊢—⊣ 25 µm

**Figure 22.18 Spongiform encephalopathies.** These diseases, caused by prions, include bovine spongiform encephalopathy, scrapie in sheep, and Creutzfeldt-Jakob disease in humans. All are similar in their pathology. **(a)** Brain tissue showing characteristic fibrils produced by prion diseases. These fibrils are insoluble aggregates of abnormally folded proteins (prions). Individual prions are not visible by any known technology. **(b)** Brain tissue showing the clear holes that give it a spongiform appearance.

**Q** What are prions?

autopsy have been reported. Several cases have been traced to the injection of a growth hormone derived from human tissue. Boiling and irradiation have no effect, and even routine autoclaving is not reliable. This has led to suggestions that surgeons use disposable instruments where there is a risk of exposure to CJD. To sterilize reusable instruments, the WHO currently recommends a strong solution of sodium hydroxide combined with extended autoclaving at 134°C. However, there are reports that applications of a simple cleaning detergent combined with protease enzymes to disrupt the prions may prove an effective solution to the problem.

Some tribes in New Guinea have suffered from a TSE disease called **kuru** (a native word for shaking or trembling). Transmission of kuru is apparently related to the practice of cannibalistic rituals. Carleton Gajdusek received the Nobel Prize for Physiology and Medicine in 1976 for his investigations of kuru.

## Bovine Spongiform Encephalopathy and Variant Creutzfeldt-Jakob Disease

A TSE that is much in the news is **bovine spongiform encephalopathy (BSE)**. The disease is better known as *mad cow disease* because of the behavior of the animals. The outbreak that began in 1986 in Great Britain was eventually controlled by drastic culling of herds. The origin of the disease is usually ascribed to feed supplements containing meat and bone meal contaminated with prions from sheep infected with scrapie, a long-endemic neurological disease. Another hypothesis proposes that BSE resulted from a spontaneous mutation in a cow and that there is no connection with scrapie.

There are no tests to detect PrP$^{Sc}$ in live animals. A Western blot test is used to identify PrP$^{Sc}$ in postmortem brain tissue. In attempts to prevent introduction of BSE into the United States, there are rules prohibiting the use of meat from "downer" animals (fallen and unable to rise and walk) for any purpose and the use of animal protein as a feed supplement. The FDA has banned for human and pet consumption certain portions of the cattle carcass that are most likely to contain a neurological pathogen. Only a small percentage of animal carcasses in the United States are tested for BSE—in Europe and Japan, practically all slaughtered animals are tested.

CLINICAL CASE

Cysts can be inhaled with dust, and amebae can be forced into the nose when a swimmer dives underwater. The ameba crosses the nasal mucosa to enter the central nervous system. The ameba secretes hydrolytic enzymes that digest the nasal mucosa and nerve cells, allowing it to enter the subarachnoid space. The ameba then feeds on the digested nerve cells. A week earlier, Patricia and her family had been swimming in Deep Creek Hot Springs. The little girl did not heed the sign warning swimmers to keep their head above water. The attending physician also tests Patricia's parents for their antibody titers. Patricia's father has a low (1:16) antibody titer against *N. fowleri* but is not ill; her mother's serum is negative for antibody.

**What is the treatment for amebic meningoencephalitis?**

# Microbes Impacting the CNS

We know intestinal microbiota help us by preventing colonization of pathogens in the GI tract. They also facilitate nutrient absorption by metabolizing indigestible dietary compounds, such as amylose in starch. Now mounting evidence suggests that intestinal microbes also send chemical signals that affect the central nervous system. This means microbiota may play a role in human moods or emotions and related disorders.

Seventy percent of all neurons in the peripheral nervous system are in the digestive tract, and these nerves are directly connected to the central nervous system through the vagus nerves. The idea that the intestinal microbiome affects the nervous system and depression or anxiety disorders first surfaced in 1910, when humans given live lactic acid bacteria showed improvement in depression symptoms. Not much happened for the next 100 years. Then in 2004, an increased stress reaction was seen in germ-free mice compared to normal mice—sparking new interest in the link between the intestinal microbiome and the central nervous system.

In mice studies, researchers found the amount of brain neurotransmitters produced was related to the presence of certain intestinal bacteria. This is thought to be caused by short-chain fatty acids that the bacteria make. Genera that seem to reduce anxiety- and depression-related behavior in mice include *Bacteroides*, *Propionibacterium*, *Lactobacillus*, and *Prevotella*. Another study found that human brain activity changed when the test subjects consumed a fermented dairy product containing a variety of bacteria, including *Bifidobacterium animalis* subsp *lactis*, *Streptococcus thermophilus*, *Lactobacillus bulgaricus*, and *Lactococcus lactis*.

Altering mood and mental health are not the only ways that microbes may affect the nervous system. Recently researchers found that patients with Parkinson's disease had fewer *Prevotella* bacteria and more Enterobacteriaceae than healthy people. Parkinson's disease is a nervous system disorder characterized by tremors and rigid muscles. People suffering with the disease have reduced levels of the neurotransmitter dopamine, which affects nerve transmission to muscles. Probiotic bacteria may be useful to increase dopamine. These early data suggest that probiotics could potentially be used as a first-line treatment for some neurological disorders.

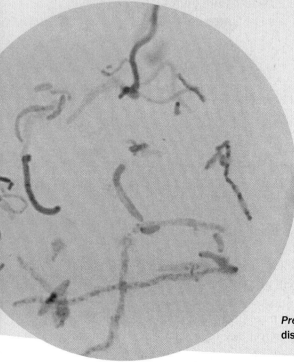

*Prevotella* bacteria are one of the genera being investigated for a link to Parkinson's disease.

If this disease were to establish itself in domestic cattle in the United States, it would be economically devastating. However, there is another aspect—that the disease could be passed on to humans. In Great Britain and a few other locales around the world, a few cases of apparent classic CJD appeared in relatively young humans. CJD rarely occurs in this age group, and a connection with BSE was feared. Investigation also showed that this variant of CJD (vCJD) differed in significant ways from classic CJD (Table 22.1). A few hundred cases have been identified so far. Considering the long incubation times

**TABLE 22.1** Comparative Characteristics of Classic and Variant Creutzfeldt-Jakob Disease

| Characteristic | Classic CJD | Variant CJD |
|---|---|---|
| Median Age at Death (yr) | 68 (range 23–97) | 28 (range 14–74) |
| Median Duration of Illness (mo) | 4 to 5 | 13 to 14 |
| Clinical Presentation | Dementia; early neurological signs | Prominent psychiatric and behavioral symptoms; delayed neurological signs |
| Genotype* | Other amino acid combinations | Methionine/methionine |

*Victims are homozygous at codon 129, that is, both of their PrP genes (one from each parent) have methionine coded at this position. This is characteristic of only about 37% of Caucasians. Other members of this population have different amino acid combinations at this position—and, although a handful of cases that have appeared were exceptions (homozygous valine/valine, or heterozygous methionine/valine), no one with these genotypes has contracted vCJD to date.

of prion diseases and that an estimated 1 million cattle had been infected with BSE, it was feared that large numbers of vCJD cases might eventually appear. However, this concern has subsided, especially since the number of cases declined from a small peak in 2000 and after it was shown that the affected patients shared a certain limited genetic profile.

**CHECK YOUR UNDERSTANDING**

✔ 22-14 What are the recommendations for sterilizing reusable surgical instruments when prion contamination might be a factor?

# Diseases Caused by Unidentified Agents

## LEARNING OBJECTIVE

22-15 List some possible causes of Bell's palsy, acute flaccid myelitis, and chronic fatigue syndrome.

In August 2014 and November 2016, the CDC received increased reports of people with **acute flaccid myelitis (AFM)**. Symptoms include limb weakness and one or more of the following: facial droop/weakness, drooping eyelids, and difficulty with swallowing or slurred speech. The 2014 increase coincided with a national outbreak of severe respiratory illness caused by a nonpolio enterovirus called EV-D68. Preliminary studies suggest EV-D68 may be the cause of AFM.

**Bell's palsy** occurs when a nerve that controls facial muscles is inflamed and can't communicate with muscles. This results in symptoms of drooping eyelid or mouth on one side of the face. This inflammation may be caused by one of the herpes viruses: HHV-1, HHV-3, HHV-4, and HHV-5 have all been suggested. Acyclovir may shorten the course of the disease. However, most people recover within 6 months with or without treatment.

There is no diagnostic test for **chronic fatigue syndrome (CFS)**. It is diagnosed by a persistent, unexplained fatigue that lasts at least 6 months with at least four of these symptoms: sore throat, tender lymph nodes, muscle pain, pain in multiple joints, headaches, unrefreshing sleep, malaise after exercise, and impaired short-term memory or concentration. The condition affects an estimated 800,000 to 2.5 million people in the United States. A number of viruses have been suggested as the cause, but current studies suggest that no one pathogen causes CFS and that the illness may be triggered by a variety of infections.

See Exploring the Microbiome (on the facing page) for the role of the human microbiome in other nervous system disorders.

**CHECK YOUR UNDERSTANDING**

✔ 22-15 Name one common disease that may be associated with Bell's palsy.

\*\*\*

Diseases in Focus 22.3 summarizes the main causes of microbial disease involving neurological symptoms and paralysis.

**CLINICAL CASE**  Resolved

Patricia is treated with the antibiotics amphothericin B and rifampin. *N. fowleri* amebae are widespread, but infection is rare. As many as 100 amebae per liter of water may be necessary for infection. Inapparent infection is not uncommon, and the father's low antibody titer suggests he does have an infection. Patricia is one of fewer than 10 patients reported to survive primary amebic meningoencephalitis. Patricia survives because of the laboratory technician's quick thinking; her infection is diagnosed early, and she is placed on antiamebic therapy immediately.

621  625  626  642  643  **645**

# Microbial Diseases with Neurological Symptoms or Paralysis

After eating canned chili, two children experience cranial nerve paralysis followed by descending paralysis. The children are on mechanical ventilation. Leftover canned chili is tested by mouse bioassay. Use the table below to make a differential diagnosis and identify infections that could cause these symptoms. For the solution, go to @MasteringMicrobiology.

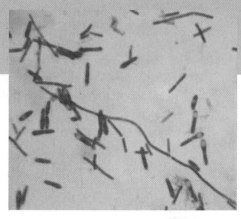

Gram stain from canned chili.

| Disease | Pathogen | Symptoms | Method of Transmission | Treatment | Prevention |
|---|---|---|---|---|---|
| **BACTERIAL DISEASES** | | | | | |
| **Tetanus** | *Clostridium tetani* | Lockjaw; muscle spasms | Puncture wound | Tetanus immune globulin; antibiotics | Toxoid vaccine (DTaP, Td) |
| **Botulism** | *Clostridium botulinum* | Flaccid paralysis | Foodborne intoxication | Antitoxin | Proper canning of foods; infants should not eat honey |
| **Leprosy** | *Mycobacterium leprae, M. lepromatosis* | Loss of sensation in skin; disfiguring nodules | Prolonged contact with contaminated secretions | Dapsone, rifampin, clofaximine | Possibly BCG vaccine |
| **VIRAL DISEASES** | | | | | |
| **Poliomyelitis** | Poliovirus | Headache, sore throat, stiff neck; paralysis if motor nerves infected | Ingesting contaminated water (fecal–oral route) | Mechanical breathing aid | Inactivated polio vaccine (IPV) |
| **Rabies** | *Lyssavirus* | Fatal infection; agitation, muscle spasms, difficulty swallowing | Animal bite | Postexposure treatment: rabies immunoglobulin plus vaccine | Human diploid cell vaccine for high-risk individuals; vaccination of domestic animals |
| **PROTOZOAN DISEASE** | | | | | |
| **African Trypanosomiasis** | *Trypanosoma brucei rhodesiense, T. b. gambiense* | Fatal infection; early symptoms (headache, fever) progress to coma | Tsetse fly | Suramin; pentamidine, melarsoprol, eflornithine | Vector control |
| **PRION DISEASES** | | | | | |
| **Creutzfeldt-Jakob Disease** | Prion | Fatal infection; neurological symptoms include trembling | Inherited; ingested; transplants | None | None |
| **Kuru** | Prion | Same as Creutzfeldt-Jakob disease | Contact or ingestion | None | None |

## Study Outline

 **Go to** @MasteringMicrobiology for **Interactive Microbiology**, *In the Clinic videos, MicroFlix, MicroBoosters, 3D animations, practice quizzes, and more.*

### Structure and Function of the Nervous System

(pp. 620–621)

1. The central nervous system (CNS) consists of the brain, which is protected by the skull bones, and the spinal cord, which is protected by the backbone.

2. The peripheral nervous system (PNS) consists of the nerves that branch from the CNS.

3. The CNS is covered by three layers of membranes called meninges: the dura mater, arachnoid mater, and pia mater. Cerebrospinal fluid (CSF) circulates between the arachnoid mater and the pia mater in the subarachnoid space.

4. The blood–brain barrier normally prevents many substances, including antibiotics, from entering the brain.

5. Microorganisms can enter the CNS through trauma, along peripheral nerves, and through the bloodstream and lymphatic system.

6. An infection of the meninges is called meningitis. An infection of the brain is called encephalitis.

### Bacterial Diseases of the Nervous System (pp. 621–630)

#### Bacterial Meningitis (pp. 622–625)

1. The three major causes of bacterial meningitis are *Haemophilus influenzae, Streptococcus pneumoniae,* and *Neisseria meningitidis.*

2. Nearly 50 other species of opportunistic bacteria can cause meningitis.

3. *H. influenzae* is part of the normal throat microbiota. It requires blood factors for growth; serotypes are based on capsules.

4. *H. influenzae* type b is the most common cause of meningitis in children under 4 years old.

5. An Hib conjugated vaccine directed against the capsular polysaccharide antigen is available.

6. *N. meningitidis* causes meningococcal meningitis. This bacterium is found in the throats of healthy carriers and is transmitted by droplet aerosols or direct contact with secretions.

7. Meningococci probably gain access to the meninges through the bloodstream. The bacteria may be found in leukocytes in CSF.

8. A purified capsular polysaccharide vaccine against serotypes A, C, Y, and W-135 is available.

9. *S. pneumoniae* is commonly found in the nasopharynx.

10. Young children are most susceptible to *S. pneumoniae* meningitis. Untreated, it has a high mortality rate.

11. A *S. pneumoniae* conjugated vaccine is available.

12. *Listeria monocytogenes* causes meningitis in newborns, the immunosuppressed, pregnant women, and cancer patients.

13. Acquired by ingestion of contaminated food, listeriosis may be asymptomatic in healthy adults.

14. *L. monocytogenes* can cross the placenta and cause spontaneous abortion and stillbirth.

#### Tetanus (pp. 625–626)

15. Tetanus is caused by an exotoxin produced by *Clostridium tetani.*

16. *C. tetani* produces the neurotoxin tetanospasmin, which causes the symptoms of tetanus: spasms, contraction of muscles controlling the jaw, and death resulting from spasms of respiratory muscles.

17. Acquired immunity results from DTaP immunization.

18. Following an injury, an immunized person may receive a booster of tetanus toxoid. An unimmunized person may receive (human) tetanus immune globulin.

19. Debridement (removal of tissue) and antibiotics may be used to control the infection.

#### Botulism (pp. 626–629)

20. Botulism is caused by an exotoxin produced by *C. botulinum* growing in foods.

21. Serological types of botulinum toxin vary in virulence; type A is the most virulent.

22. The toxin is a neurotoxin that inhibits the transmission of nerve impulses.

23. Blurred vision occurs in 1 to 2 days; progressive flaccid paralysis follows for 1 to 10 days, possibly resulting in death from respiratory and cardiac failure.

24. *C. botulinum* will not grow in acidic foods or in an aerobic environment. Endospores are killed by proper canning. Adding nitrites to foods inhibits growth of *C. botulinum.*

25. The toxin is heat labile and is destroyed by boiling (100°C) for 5 minutes.

26. Infant botulism results from the growth of *C. botulinum* in an infant's intestines.

27. Wound botulism occurs when *C. botulinum* grows in anaerobic wounds.

28. For diagnosis, mice protected with antitoxin are inoculated with toxin from the patient or foods.

#### Leprosy (pp. 629–630)

29. Leprosy, or Hansen's disease, is caused by *Mycobacterium leprae* or *M. lepromatosis.*

30. These bacteria have never been cultured on artificial media. They can be cultured in armadillos and mouse footpads.

31. The tuberculoid form of the disease is characterized by loss of sensation in the skin surrounded by nodules.

32. In the lepromatous form, disseminated nodules and tissue necrosis occur.

33. Leprosy is not highly contagious and is spread by prolonged contact with exudates.

34. Untreated individuals often die of secondary bacterial complications, such as tuberculosis.

35. Laboratory diagnosis is based on observations of acid-fast rods in a skin biopsy.

### Viral Diseases of the Nervous System (pp. 630–638)

#### Poliomyelitis (pp. 630–631)

1. The symptoms of poliomyelitis are usually sore throat and nausea, and occasionally paralysis (fewer than 1% of cases).

2. Poliovirus is transmitted by the ingestion of water contaminated with feces.

3. Poliovirus first invades lymph nodes of the neck and small intestine. Viremia and spinal cord involvement may follow.

4. Diagnosis is based on isolation of the virus from feces and throat secretions.

5. The Salk vaccine (an inactivated polio vaccine [IPV]) involves the injection of formalin-inactivated viruses and boosters every few years. The Sabin vaccine (an oral polio vaccine [OPV]) contains three live, attenuated strains of poliovirus and is administered orally.

6. Polio is a good candidate for elimination through vaccination.

### Rabies (pp. 632–633, 636)

7. Rabies virus (*Lyssavirus*) causes an acute, usually fatal, encephalitis called rabies.

8. Rabies may be contracted through the bite of a rabid animal or invasion through skin. The virus multiplies in skeletal muscle and connective tissue.

9. Encephalitis occurs when the virus moves along peripheral nerves to the CNS.

10. Symptoms of rabies include spasms of mouth and throat muscles followed by extensive brain and spinal cord damage and death.

11. Laboratory diagnosis may be made by DFA tests of saliva, serum, and CSF or brain smears.

12. Reservoirs for rabies in the United States include skunks, bats, foxes, and raccoons. Domestic cattle, dogs, and cats may get rabies. Rodents and rabbits seldom get rabies.

13. Postexposure treatment includes administration of human rabies immune globulin (RIG) along with multiple intramuscular injections of vaccine.

14. Preexposure treatment consists of vaccination.

15. Other genotypes of *Lyssavirus* cause rabies-like diseases.

### Arboviral Encephalitis (pp. 637–638)

16. Symptoms of encephalitis are chills, headache, fever, and eventually coma.

17. Many types of viruses transmitted by mosquitoes (called arbo-viruses) cause encephalitis.

18. The incidence of arboviral encephalitis increases in the summer months, when mosquitoes are most numerous.

19. Zika virus disease is often mild in adults, but the virus can cause CNS birth defects if it infects a fetus.

20. Control of the mosquito vector is the most effective way to control arboviral infections.

## Fungal Disease of the Nervous System (pp. 638–639)

### *Cryptococcus neoformans* Meningitis (Cryptococcosis) (p. 639)

1. *Cryptococcus* spp. are encapsulated yeastlike fungi that cause cryptococcosis.

2. The disease may be contracted by inhaling dried infected pigeon or chicken droppings.

3. The disease begins as a lung infection and may spread to the brain and meninges.

4. Immunosuppressed individuals are most susceptible to cryptococcosis.

5. Diagnosis is based on latex agglutination tests for cryptococcal antigens in serum or CSF.

## Protozoan Diseases of the Nervous System (pp. 639–642)

### African Trypanosomiasis (pp. 639–640)

1. African trypanosomiasis is caused by the protozoa *Trypanosoma brucei gambiense* and *T.b. rhodesiense* and is transmitted by the bite of the tsetse fly.

2. The disease affects the nervous system of the human host, causing lethargy and eventually coma. It is commonly called sleeping sickness.

3. Vaccine development is hindered by the protozoan's ability to change its surface antigens.

### Amebic Meningoencephalitis (pp. 640–642)

4. Encephalitis caused by the protozoan *Naegleria fowleri* is almost always fatal.

5. Granulomatous amebic encephalitis, caused by *Acanthamoeba* spp. and *Balamuthia mandrillaris*, is a chronic disease.

## Nervous System Diseases Caused by Prions (pp. 642–645)

1. Prions are self-replicating proteins with no detectable nucleic acid.

2. Diseases of the CNS that progress slowly and cause spongiform degeneration are caused by prions.

3. Transmissible spongiform encephalopathies are caused by prions that are transferable from one animal to another.

4. Creutzfeldt-Jakob disease and kuru are human diseases similar to scrapie. They are transmitted between humans.

## Diseases Caused by Unidentified Agents (p. 645)

1. The causes of acute flaccid myelitis, Bell's palsy, and chronic fatigue syndrome have not been definitively established, but viruses have been implicated.

# Study Questions

For answers to the Knowledge and Comprehension questions, turn to the Answers tab at the back of the textbook.

## Knowledge and Comprehension

### Review

1. If *Clostridium tetani* is relatively sensitive to penicillin, why doesn't penicillin cure tetanus?

2. What treatment is used against tetanus under the following conditions?
   a. before a person suffers a deep puncture wound
   b. after a person suffers a deep puncture wound

3. Why is the following description used for wounds that are susceptible to *C. tetani* infection: ". . . Improperly cleaned deep puncture wounds . . . ones with little or no bleeding . . ."?

4. Provide the following information on poliomyelitis: etiology, method of transmission, symptoms, prevention. Why aren't the Salk and Sabin vaccines considered treatments for poliomyelitis?

5. Fill in the following table:

| Causative Agent of Meningitis | Susceptible Population | Transmission | Treatment |
|---|---|---|---|
| N. meningitidis | | | |
| H. influenzae | | | |
| S. pneumoniae | | | |
| L. monocytogenes | | | |
| C. neoformans | | | |

6. Fill in the following table.

| Disease | Etiology | Transmission | Symptoms | Treatment |
|---|---|---|---|---|
| Arboviral encephalitis | | | | |
| African trypanosomiasis | | | | |
| Botulism | | | | |
| Leprosy | | | | |

7. ▓DRAW IT▓ On the figure below, identify the portal of entry of *H. influenzae*, *C. tetani*, botulinum toxin, *M. leprae*, poliovirus, *Lyssavirus*, arboviruses, and *Acanthamoeba*.

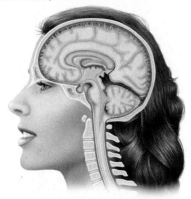

8. Outline the procedures for treating rabies after exposure. Outline the procedures for preventing rabies prior to exposure. What is the reason for the differences in the procedures?

9. Provide evidence that Creutzfeldt-Jakob disease is caused by a transmissible agent.

10. ▓NAME IT▓ This organism causes meningitis and is transmitted mainly by the inhalation of dried, contaminated bird droppings. Infections are treated with amphotericin B and flucytosine.

## Multiple Choice

1. Which of the following is *false*?
   a. Only puncture wounds by rusty nails result in tetanus.
   b. Rabies is seldom found in rodents (e.g., rats, mice).
   c. Polio is transmitted by the fecal–oral route.
   d. Arboviral encephalitis is rather common in the United States.
   e. All of the above are true.

2. Which of the following does *not* have an animal reservoir or vector?
   a. listeriosis
   b. cryptococcosis
   c. amebic meningoencephalitis
   d. rabies
   e. African trypanosomiasis

3. A 12-year-old girl hospitalized for Guillain-Barré syndrome had a 4-day history of headache, dizziness, fever, sore throat, and weakness of legs. Seizures began 2 weeks later. Bacterial cultures

were negative. She died 3 weeks after hospitalization. An autopsy revealed inclusions in brain cells that tested positive in an immunofluoresence test. She probably had
   a. rabies.
   b. Creutzfeldt-Jakob disease.
   c. botulism.
   d. tetanus.
   e. leprosy.

4. After receiving a corneal transplant, a woman developed dementia and loss of motor function; she then became comatose and died. Cultures were negative. Serological tests were negative. Autopsy revealed spongiform degeneration of her brain. She most likely had
   a. rabies.
   b. Creutzfeldt-Jakob disease.
   c. botulism.
   d. tetanus.
   e. leprosy.

5. Endotoxin is responsible for symptoms caused by which of the following organisms?
   a. *N. meningitidis*
   b. *S. pyogenes*
   c. *L. monocytogenes*
   d. *C. tetani*
   e. *C. botulinum*

6. The increased incidence of encephalitis in the summer months is due to
   a. maturation of the viruses.
   b. increased temperature.
   c. the presence of adult mosquitoes.
   d. an increased population of birds.
   e. an increased population of horses.

Match the following choices to the statements in questions 7 and 8:
   a. antirabies antibodies
   b. HDVC

7. Produces longest lasting protection.

8. Used for passive immunization.

Use the following choices to answer questions 9 and 10:
   a. *Cryptococcus*
   b. *Haemophilus*
   c. *Listeria*
   d. *Naegleria*
   e. *Neisseria*

9. Microscopic examination of cerebrospinal fluid reveals gram-positive rods.

10. Microscopic examination of cerebrospinal fluid from a person who washes windows on a building in a large city reveals ovoid cells.

## Analysis

1. Most of us have been told that a rusty nail causes tetanus. What do you suppose is the origin of this adage?

2. OPV is no longer used for routine vaccination. Provide the rationale for this policy.

## Clinical Applications and Evaluation

1. A 1-year-old infant was lethargic and had a fever. When admitted to the hospital, he had multiple brain abscesses with gram-negative coccobacilli. Identify the disease, etiology, and treatment.

2. A 40-year-old bird handler was admitted to the hospital with soreness over his upper jaw, progressive vision loss, and bladder dysfunction. He had been well 2 months earlier. Within weeks he lost reflexes in his lower extremities and subsequently died. Examination of CSF showed lymphocytes. What etiology do you suspect? What further information do you need?

3. A normal baby gained weight appropriately for 12 weeks. Then she stopped feeding. Her right eardrum was inflamed, she had a stiff neck, and her temperature was 40°C. Examination of CSF revealed Gram-negative coccobacilli. Identify the disease and treatment.

# 23

# Microbial Diseases of the Cardiovascular and Lymphatic Systems

The cardiovascular system consists of the heart, blood, and blood vessels. The lymphatic system consists of the lymph, lymph vessels, lymph nodes, and the lymphoid organs, which include the tonsils, appendix, spleen, and thymus. Fluids in both systems circulate throughout the body, intimately contacting many tissues and organs. The blood and lymph distribute nutrients and oxygen to body tissues, carrying away wastes. However, these same qualities make the cardiovascular and lymphatic systems vehicles for the spread of pathogens that enter their circulation when an insect bite, needle, or wound penetrates the skin. Because of this, many of the body's defensive systems are found in the blood and lymph. Circulating phagocytic cells are especially important; these are also in fixed locations such as the lymph nodes and spleen. The blood is an important part of our adaptive immune system; antibodies and specialized cells circulate to intercept pathogens introduced into the blood. Occasionally the defensive systems found in the blood are overwhelmed, and pathogens proliferate explosively with disastrous results. Dengue virus (shown in the photograph) is one such pathogen, and it grows in the immune system's macrophages. Dengue is described in the Clinical Case in this chapter.

▶ Dengue virus (blue) is transmitted by mosquitoes.

## In the Clinic

As an obstetrics nurse, you find out that one of your patients developed mononucleosis-like symptoms during pregnancy. **What testing is now necessary? Why? What advice should pregnant women receive to prevent this infection?**

*Hint: Read about congenital infections on pages 670 and 676.*

Answers to In the Clinic questions are found online @MasteringMicrobiology.

# Structure and Function of the Cardiovascular and Lymphatic Systems

## LEARNING OBJECTIVE

**23-1** Identify the role of the cardiovascular and lymphatic systems in spreading and eliminating infections.

The center of the **cardiovascular system** is the heart (**Figure 23.1**). The function of the cardiovascular system is to circulate blood through the body's tissues so it can deliver certain substances to cells and remove other substances from them.

*Blood* is a mixture of formed elements (see Table 16.1, page 454) and a liquid called blood plasma. The **lymphatic system** is an essential part of the circulation of blood (**Figure 23.2**). As the blood circulates, some blood plasma filters out of the

blood capillaries into spaces between tissue cells, called *interstitial spaces*. The circulating fluid in the interstitial spaces is called *interstitial fluid*. Microscopic lymphatic vessels that surround tissue cells are called *lymph capillaries*. As the interstitial fluid moves around the tissue cells, it's picked up by the lymph capillaries; the fluid is then called *lymph*.

Because lymph capillaries are very permeable, they readily pick up microorganisms or their products. From lymph capillaries, lymph is transported into larger lymph vessels called *lymphatics*, which contain valves that keep the lymph moving toward the heart. Eventually, all the lymph is returned to the blood just before the blood enters the heart. As a result of this circulation, proteins and fluid that have filtered from the plasma are returned to the blood.

At various points in the lymphatic system are oval structures called *lymph nodes* (bean-shaped bodies ranging in size from a few millimeters to as much as 2 cm), through which lymph flows. (Also, see Figure 16.5, page 455.) Within the lymph nodes are fixed macrophages that help clear the lymph of infectious microorganisms. At times the lymph nodes themselves get infected and become visibly swollen and tender; swollen lymph nodes are called **buboes**.

Lymph nodes are also an important component of the body's immune system. Foreign microbes entering lymph nodes encounter two types of lymphocytes: B cells, which are stimulated to become plasma cells that produce humoral antibodies; and T cells, which then differentiate into effector T cells that are essential to the cell-mediated immune system.

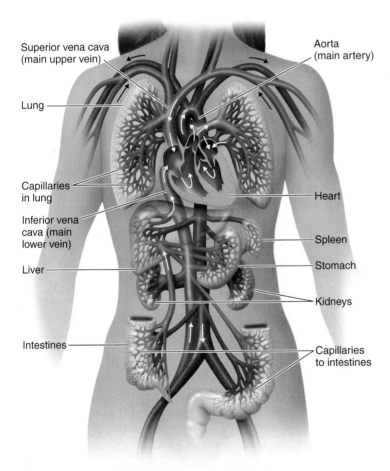

**Figure 23.1 The human cardiovascular system and related structures.** Details of circulation to the head and extremities are not shown in this simplified diagram. The blood circulates from the heart through the arterial system (red) to the capillaries (purple) in the lungs and other parts of the body. From these capillaries, the blood returns through the venous system (blue) to the heart.

Labels: Superior vena cava (main upper vein); Aorta (main artery); Lung; Capillaries in lung; Inferior vena cava (main lower vein); Liver; Intestines; Heart; Spleen; Stomach; Kidneys; Capillaries to intestines

**Q** How can a focal infection become systemic?

---

**CLINICAL CASE** Mosquito Mishap

Katie Tanaka, a normally healthy 34-year-old, has just returned to Rochester, New York, from a 1-week trip to Key West, Florida. Although Katie knows she should be a little tired after the long trip, she is surprised to feel so run down one day after she gets home. Katie makes an appointment with her primary care physician that afternoon, when she develops a fever, headache, and chills. Her doctor orders a urinalysis; the results reveal the presence of bacteria and red blood cells in her urine. Katie's doctor diagnoses her with a urinary tract infection and prescribes antibiotics.

Two days later, Katie returns to her primary care physician with a worsening headache, pain in the back of the eye made worse by eye movement, and complaints of feeling lightheaded, although her fever has resolved. Katie is alert and oriented but has substantial discomfort from her headache. When she is asked to close her eyes and stand with her feet together (touching each other), Katie begins swaying, which is a possible indicator of a brain lesion.

**What infections are possible? Read on to find out.**

651  668  671  677  683

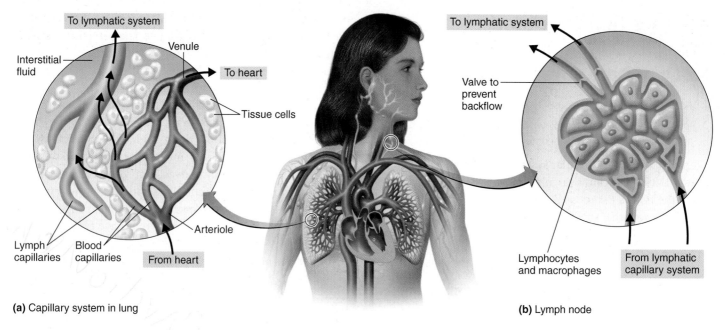

**(a)** Capillary system in lung

**(b)** Lymph node

**Figure 23.2 The relationship between the cardiovascular and lymphatic systems. (a)** From the blood capillaries, some blood plasma filters into the surrounding tissue, where it is called interstitial fluid, and enters the lymph capillaries. This fluid, now called lymph, returns to the heart through the lymphatic circulatory system (green), which channels the lymph to a vein. **(b)** All lymph returning to the heart must pass through at least one lymph node. (See also Figure 16.5, page 455.)

**Q** **What is the role of the lymphatic system in defense against infection?**

**CHECK YOUR UNDERSTANDING**

✔ **23-1** Why is the lymphatic system so valuable for the working of the immune system?

# Bacterial Diseases of the Cardiovascular and Lymphatic Systems

### LEARNING OBJECTIVES

**23-2** List the signs and symptoms of sepsis, and explain the importance of infections that develop into septic shock.

**23-3** Differentiate gram-negative sepsis, gram-positive sepsis, and puerperal sepsis.

**23-4** Describe the epidemiologies of endocarditis and rheumatic fever.

**23-5** Describe the epidemiology of tularemia.

**23-6** Describe the epidemiology of brucellosis.

**23-7** Describe the epidemiology of anthrax.

**23-8** Describe the epidemiology of gas gangrene.

**23-9** List three pathogens that are transmitted by animal bites and scratches.

**23-10** Compare and contrast the causative agents, vectors, reservoirs, symptoms, treatments, and preventive measures for plague, Lyme disease, and Rocky Mountain spotted fever.

**23-11** Identify the vector, etiology, and symptoms of five diseases transmitted by ticks.

**23-12** Describe the epidemiologies of epidemic typhus, endemic murine typhus, and spotted fevers.

Once bacteria gain access to the bloodstream, they become widely disseminated. In some cases, they are also able to reproduce rapidly. Those that don't reproduce may comprise the blood microbiome (see Exploring the Microbiome on page 653).

## Sepsis and Septic Shock

Moderate numbers of microorganisms can enter the bloodstream without causing harm. In hospital conditions, the blood frequently is contaminated as a result of invasive procedures, such as insertion of catheters and intravenous feeding tubes. Blood and lymph contain numerous defensive phagocytic cells. Also, blood is low in available iron, which is a requirement for bacterial growth. However, if the defenses of the cardiovascular and lymphatic systems fail, microbes can proliferate in the blood. An acute illness that is associated with the presence and persistence of pathogenic microorganisms or their toxins in the blood is termed **septicemia**. A similar term that is not equated medically with septicemia is *sepsis*, although there is a tendency to use them interchangeably. **Sepsis** is defined as a

**Is Blood Sterile?**

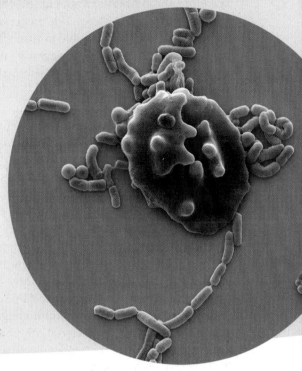

We've always assumed that blood from a healthy donor is sterile. Donated blood is routinely cultured to look for pathogens, and only 0.1% of cultures from blood packs have bacterial growth—those that do are assumed to have been contaminated during the blood draw itself. However, more recent studies indicate that healthy human blood may not be so sterile after all.

Analyses of blood from healthy donors found bacterial 16S DNA. This genetic material is found in all prokaryotes, and because it is evolutionarily conserved, it is also used to track relatedness of various types of bacteria. Most of the blood bacterial DNA is located in the white cells and platelets, about 6% is in red cells, and a tiny amount (0.03%) is in the plasma.

These data suggest that blood does have a microbiome.

Most of the DNA is from the Proteobacteria phylum. A comparison of patients with cardiovascular disease (CVD) showed that CVD patients have more DNA overall in their plasma than healthy people do. *Propionibacterium* predominated in CVD patients, whereas healthy people had more *Pseudomonas* DNA.

The source of bacteria in blood is most likely bacteria from the mouth or intestines that crossed mucous membranes into the blood. Additional research is needed to determine whether this DNA is from live bacteria and whether they may cause CVD.

**Red blood cell and bacteria.**

*systemic inflammatory response syndrome (SIRS)* caused by a focus of infection that releases mediators of inflammation into the bloodstream. The site of the infection itself is not necessarily the bloodstream, and in about half of the cases no microbes can be found in the blood. Sepsis and septicemia are often accompanied by the appearance of **lymphangitis**, inflamed lymph vessels visible as red streaks under the skin, running along the arm or leg from the site of the infection (**Figure 23.3**).

If the body's defenses do not quickly control the infection and the resulting SIRS, the results are progressive and frequently fatal. The first stage of this progression is sepsis. The most obvious signs and symptoms are fever, chills, and accelerated breathing and heart rate. When sepsis results in a drop in blood pressure (*shock*) and dysfunction of at least one organ, it is considered to be **severe sepsis**. Once organs begin to fail, the mortality rate becomes very high. A final stage, when low blood pressure can no longer be controlled by addition of fluids, is **septic shock**. More than one million cases occur each year in the United States with a fatality rate of 28–50%.

### Gram-Negative Sepsis

Septic shock is most likely to be caused by gram-negative bacteria. Recall that the cell walls of many gram-negative bacteria contain endotoxins (toxic lipopolysaccharides [LPS]; see pages 81–83) that are released upon lysis of the cell. These endotoxins can cause a severe drop in blood pressure with its associated signs and symptoms. Septic shock is often called by the alternative names *gram-negative sepsis* or *endotoxic shock*. Less than one-millionth

of a milligram of endotoxin is enough to cause the symptoms. *Klebsiella spp.*, *E. coli*, and *Pseudomonas aeruginosa* are most frequently involved. Recent outbreaks of an emerging pathogen, *Elizabethkingia spp.*, have occurred in the United States. The CDC recently warned healthcare facilities about worldwide emergence of an invasive multidrug-resistant fungus, *Candida auris*.

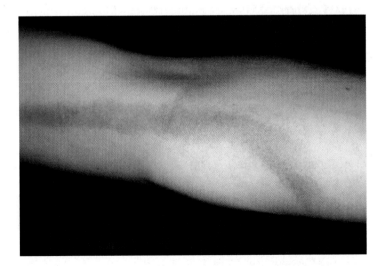

**Figure 23.3 Lymphangitis, one sign of sepsis.** As the infection spreads from its original site along the lymph vessels, the inflamed walls of the vessels become visible as red streaks.

**Q** Why does the red streak sometimes end at a certain point?

653

An effective treatment for severe sepsis and septic shock has been a medical priority for many years. The early symptoms of sepsis are relatively nonspecific and not especially alarming. Therefore, the antibiotic treatments that might arrest it then are frequently not administered. The progression to lethal stages is rapid and generally impossible to treat effectively. Administering antibiotics then may even aggravate the condition by causing the lysis of large numbers of bacteria that then release more endotoxins.

In addition to antibiotics, treatment of septic shock involves attempts to neutralize the LPS components and inflammation-causing cytokines. Attempts to develop an effective drug to do this have been unsuccessful so far.

### Gram-Positive Sepsis

Gram-positive bacteria are now the most common cause of sepsis. Both staphylococci and streptococci produce potent exotoxins that cause toxic shock syndrome, a toxemia discussed in Chapter 21 (page 595). The frequent use of invasive procedures in hospitals allows gram-positive bacteria to enter the bloodstream. Such healthcare-associated infections (HAIs) are a particular risk for patients who undergo regular dialysis for kidney dysfunction. The bacterial components that lead to septic shock in gram-positive sepsis are not known with certainty. Possible sources are various fractions of the gram-positive cell wall or even bacterial DNA.

An especially important group of gram-positive bacteria are the enterococci, which are responsible for many HAIs. The enterococci are inhabitants of the human colon and frequently contaminate skin. Once considered relatively harmless, two species in particular, *Enterococcus faecium* and *Enterococcus faecalis*, are now recognized as leading causes of HAIs of wounds and the urinary tract. Enterococci have a natural resistance to penicillin and have rapidly acquired resistance to other antibiotics. What has made them something of a medical emergency is the appearance of vancomycin-resistant strains. Vancomycin (see page 569) was often the only remaining antibiotic to which these bacteria, especially *E. faecium*, were still sensitive. Among isolates of *E. faecium* from HAIs of the bloodstream, almost 90% are now resistant.

Up to this point our discussion of the streptococci has been focused on serologic group A. There is an emerging awareness of **group B streptococci (GBS)**. *S. agalactiae* (ā-gal-AK-tē-ī) is the only GBS and is the most common cause of life-threatening *neonatal sepsis*. The CDC recommends that pregnant women be tested for vaginal GBS and that women with GBS be offered antibiotics during labor.

### Puerperal Sepsis

**Puerperal sepsis**, also called **puerperal fever** and **childbirth fever**, is an HAI. It begins as an infection of the uterus as a result of childbirth or abortion. *Streptococcus pyogenes*, a group A beta-hemolytic streptococcus, is the most frequent cause, although other organisms may cause infections of this type.

Puerperal sepsis progresses from an infection of the uterus to an infection of the abdominal cavity (*peritonitis*) and in many cases to sepsis. At a Paris hospital between 1861 and 1864, of the 9886 women who gave birth, 1226 (12%) died of such infections. These deaths were largely unnecessary. Some 20 years before, Oliver Wendell Holmes in the United States and Ignaz Semmelweis in Austria had clearly demonstrated that the disease was transmitted by the hands and instruments of the attending midwives or physicians and that disinfecting the hands and instruments could prevent such transmission. Antibiotics, especially penicillin, and modern hygienic practices have now made *S. pyogenes* puerperal sepsis an uncommon complication of childbirth.

### Therapy for Sepsis

An effective therapy for sepsis is a medical priority and will probably require entirely new approaches. For one thing, the symptoms of sepsis are largely caused by the body's response to the infection, a response that has been described as being "unhelpfully exuberant." Any agent that would suppress this response would be independent of the source of the infection. Even in the absence of such therapies, the care of patients with sepsis has improved, and the mortality rate in recent years has declined but is still around 28%.

> **CHECK YOUR UNDERSTANDING**
>
> ✔ **23-2** What are two of the conditions that define the systemic inflammatory response syndrome of sepsis?
>
> ✔ **23-3** Are the endotoxins that cause sepsis from gram-positive or gram-negative bacteria?

## Bacterial Infections of the Heart

The wall of the heart consists of three layers. The inner layer, called the *endocardium*, lines the heart muscle itself and covers the valves. An inflammation of the endocardium is called **endocarditis**.

One type of bacterial endocarditis, **subacute bacterial endocarditis** (so named because it develops slowly; **Figure 23.4**), is characterized by fever, general weakness, and a heart murmur. It is usually caused by alpha-hemolytic streptococci (most often, *Streptococcus viridans*), which are common in the oral cavity, although enterococci or staphylococci may also be involved. The condition probably arises from a focus of infection elsewhere in the body, such as in the teeth or tonsils. Microorganisms are released by tooth extractions or tonsillectomies, enter the blood, and find their way to the heart. Normally, such bacteria would be quickly cleared from the blood by the body's defensive mechanisms. However, in people whose heart valves are abnormal, because of either congenital heart defects or such diseases as rheumatic fever and syphilis, the bacteria lodge in the preexisting lesions. Within the lesions, the bacteria multiply and become entrapped in blood clots that protect them from phagocytes and antibodies. As multiplication progresses and the clot gets

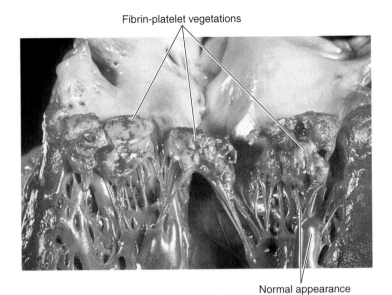

Fibrin-platelet vegetations

Normal appearance

**Figure 23.4 Bacterial endocarditis.** This is a case of subacute endocarditis, meaning that it developed over a period of weeks or months. The heart has been dissected to expose the mitral valve. The cordlike structures connect the heart valve to the operating muscles.

**Q** **How can a tongue piercing lead to subacute bacterial endocarditis?**

larger, pieces of the clot break off and can block blood vessels or lodge in the kidneys. In time, the function of the heart valves is impaired. Left untreated by appropriate antibiotics, subacute bacterial endocarditis is fatal within a few months.

A more rapidly progressive type of bacterial endocarditis is **acute bacterial endocarditis**, which is usually caused by *Staphylococcus aureus*. The organisms find their way from the initial site of infection to normal or abnormal heart valves; the rapid destruction of the heart valves is frequently fatal within a few days or weeks if untreated. Streptococci can also cause **pericarditis**, inflammation of the sac around the heart (the *pericardium*).

**CHECK YOUR UNDERSTANDING**

✔ 23-4 What medical procedures are usually the cause of endocarditis?

## Rheumatic Fever

Streptococcal infections, such as those caused by *Streptococcus pyogenes*, sometimes lead to **rheumatic fever**, which is generally considered an autoimmune complication. It occurs primarily in people aged 4 to 18 and often follows an episode of streptococcal sore throat. The disease is usually first expressed as a short period of arthritis and fever. Subcutaneous nodules at joints often accompany this stage (**Figure 23.5**). In about half of persons affected, an inflammation of the heart, probably from a misdirected immune reaction against streptococcal M protein, damages the valves. Reinfection with streptococci renews the

immune attack. Damage to heart valves may be serious enough to result in eventual failure and death. People who have had an episode of rheumatic fever are at risk of renewed immunological damage with repeated streptococcal sore throats. The bacteria have remained sensitive to penicillin, and patients at particular risk, such as these, often receive a monthly preventive injection of long-acting penicillin G benzathine.

As many as 10% of people with rheumatic fever develop **Sydenham's chorea**, an unusual complication known in the Middle Ages as Saint Vitus' dance. Several months following an episode of rheumatic fever, the patient (much more likely to be a girl than a boy) exhibits purposeless, involuntary movements during waking hours. Occasionally, sedation is required to prevent self-injury from flailing arms and legs. The condition disappears after a few months.

Sepsis and infections of the heart are summarized in Diseases in Focus 23.1.

## Tularemia

**Tularemia** is an example of a *zoonotic* disease, that is, a disease transmitted by contact with infected animals, in this case most commonly rabbits and rodents. The name derives from Tulare County, California, where the disease was originally observed in ground squirrels in 1911. The pathogen is *Francisella tularensis*, a small gram-negative bacillus. It can enter humans by several routes. The most common is penetration of the skin at a minor abrasion, where it creates an ulcer at the site. About a week after infection, the regional lymph nodes enlarge; many will contain pockets of pus. (See the Clinical Focus box on page 659.) The bacterium can multiply in macrophages—as much as a thousand-fold. Mortality is normally less than 3%.

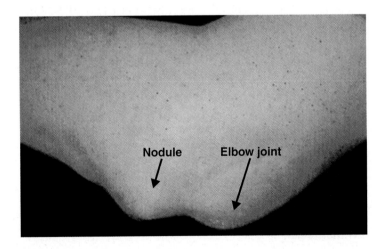

Nodule    Elbow joint

**Figure 23.5 A nodule caused by rheumatic fever.** Rheumatic fever was named, in part, because of the characteristic subcutaneous nodules that appear at the joints, as shown in this patient's elbow. Infection with group A beta-hemolytic streptococci sometimes leads to this autoimmune complication.

**Q** **Is rheumatic fever a bacterial infection?**

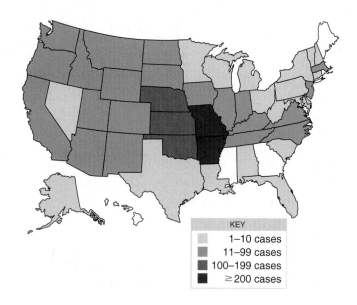

**Figure 23.6 Tularemia cases in the United States (2005–2015).**
*Source:* CDC, January 2017.

**Q** What area reporting tularemia is closest to you?

KEY
1–10 cases
11–99 cases
100–199 cases
≥ 200 cases

If left untreated, the proliferation of *F. tularensis* can lead to sepsis and infection of multiple organs.

Almost 90% of cases in the United States are related to contact with rabbits, and the disease is often known locally as *rabbit fever*. Tularemia is also transmitted in some areas by ticks and deer flies and is known there as *deer fly fever*. Respiratory infection, usually by dust contaminated by urine or feces of infected animals, can cause an acute pneumonia with a mortality rate exceeding 30%. The infective dose is very small, and handling this organism requires biosafety level 3 procedures (see page 162).

At one time, so few cases of tularemia (fewer than 200) were recorded annually in the United States that it was removed from the list of nationally notifiable diseases. However, concern that it might be used as a biological weapon has recently led to its reinstatement on the list. **Figure 23.6** illustrates the geographic distribution of tularemia within the United States.

The intracellular location of the bacterium is a problem in chemotherapy. Antibiotics such as streptomycin, administered for 10 to 15 days, are an effective treatment.

> **CHECK YOUR UNDERSTANDING**
>
> ✔ **23-5** What animals are the most common reservoir for tularemia?

## Brucellosis (Undulant Fever)

With over 500,000 new human cases annually, **brucellosis** is the world's most common bacterial zoonosis. It is endemic in the Mediterranean Basin, South and Central America, Eastern Europe, Asia, Africa, and the Middle East. It is also economically important as a disease of animals in the developing world. Human cases of brucellosis are usually not fatal, but the disease tends to persist in the reticuloendothelial system (see page 457), where the bacteria evade the host's defenses; they are especially adept at evading phagocytic cells. This ability allows long-term survival and replication. The disease often becomes chronic and is capable of affecting any organ system.

*Brucella* bacteria are small, aerobic, gram-negative coccobacilli. During laboratory handling, they easily become airborne and are considered dangerous to handle. In fact, they are considered a potential agent of bioterrorism. There are three species of *Brucella* bacteria that are of greatest interest. *Brucella abortus* (broo-SEL-lah ah-BOR-tus) is found primarily in cattle but also infects camels, bison, and several other animals. *Brucella suis* (SOO-is) is a species mostly infecting swine but is known to infect cattle when they are kept in contact with swine herds. Abattoir (slaughterhouse) workers who come in contact with swine carcasses are at risk of brucellosis from this species. The most serious pathogen, and the cause of most human cases, is *Brucella melitensis* (mel-i-TEN-sis). This species is most commonly found today in goats and sheep.

At the present time, most U.S. cases of brucellosis are caused by *B. melitensis*, predominantly among Hispanics. The disease is endemic in Mexico and is often imported into the United States in unpasteurized food products, such as Mexican soft cheese made from goat milk.

The incubation period is usually 1 to 3 weeks but might be much longer. Symptoms of brucellosis have a wide spectrum, depending on the stage of the disease and the organs affected. Typically they include fever (often rising and falling, which has given the disease an alternative name of *undulant fever*), malaise, night sweats, and muscle aches. Although several serological tests are available, there is still a need for a definitive diagnostic test. The ultimate diagnostic proof is isolation of *Brucella* from the patient's blood or tissue. Because the disease is not common, diagnosis often must start with patient interviews that suggest a contact in endemic areas of the disease.

Antibiotic therapy is possible, and the bacteria have not shown development of resistance. However, treatment must be very long term, usually at least 6 weeks, and involves a combination of at least two antibiotics.

> **CHECK YOUR UNDERSTANDING**
>
> ✔ **23-6** What ethnic group in the United States is most commonly affected by brucellosis, and why?

## Anthrax

In 1877, Robert Koch isolated *Bacillus anthracis*, the bacterium that causes **anthrax** in animals. The endospore-forming bacillus is a large, aerobic, gram-positive microorganism that is

# Human-Reservoir Infections

*D*ifferential diagnosis is the process of identifying a disease by examining a patient and comparing the results to a list of possible diseases. A differential diagnosis is important for providing initial treatment and for laboratory testing. Microorganisms circulating in the blood may reflect a serious, uncontrolled infection. For example, a 27-year-old woman has a fever and cough for 5 days. She is hospitalized when her blood pressure drops. Despite aggressive treatment with fluids and massive doses of antibiotics, she dies 5 hours after hospitalization. Catalase-negative, gram-positive cocci are isolated from her blood. Use the table below to provide a differential diagnosis and identify infections that could cause these symptoms. For the solution, go to @MasteringMicrobiology.

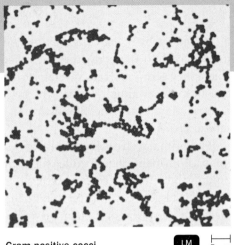

Gram-positive cocci.

LM | 5 μm

| Disease | Pathogen | Symptoms | Reservoir | Method of Transmission | Treatment |
|---|---|---|---|---|---|
| **BACTERIAL DISEASES** | | | | | |
| **Septic Shock** | Gram-negative bacteria, enterococci, group B streptococci | Fever, chills, increased heart rate; lymphangitis | Human body | Injection; catheterization | Antibiotics |
| **Puerperal Sepsis** | *Streptococcus pyogenes* | Peritonitis; sepsis | Human nasopharynx | Nosocomial | Penicillin |
| **Endocarditis Subacute Bacterial Acute Bacterial** | Mostly alpha-hemolytic streptococci; *Staphylococcus aureus* | Fever, general weakness, heart murmur; damage to heart valves | Human nasopharynx | From focal infection | Antibiotics |
| **Pericarditis** | *Streptococcus pyogenes* | Fever; general weakness; heart murmur | Human nasopharynx | From focal infection | Antibiotics |
| **Rheumatic Fever** | Group A beta-hemolytic streptococci | Arthritis, fever; damage to heart valves | Immune reactions to streptococcal infections | Not transmissible | Supportive. Prevention: penicillin to treat streptococcal sore throats |
| **VIRAL DISEASES** | | | | | |
| **Burkitt's Lymphoma** | HHV-4 | Tumor | Unknown | Unknown | Surgery |
| **Infectious Mononucleosis** | HHV-4 | Fever, general weakness | Humans | Saliva | None |
| **Cytomegalovirus** | *Cytomegalovirus* | Mostly asymptomatic; initial infection acquired during pregnancy can be damaging to fetus | Humans | Body fluids | Ganciclovir, fomivirsen |
| **UNKNOWN ETIOLOGY** | | | | | |
| **Kawasaki Syndrome** | Unknown | Fever, rash, coronary artery abnormalities | Unknown | Unknown | None |

apparently able to grow slowly in soil types that meet specific moisture conditions. The endospores have survived in soil tests for up to 60 years. The disease strikes primarily grazing mammals, such as cattle and sheep. The *B. anthracis* endospores are ingested along with grasses, causing a fulminating, fatal sepsis.

The incidence of human anthrax is now rare in the United States. People at risk are those who handle animals, hides, wool, and other animal products from certain foreign countries. (See the Clinical Case, Chapter 2, page 25.)

Infections by *B. anthracis* are initiated by endospores. Once introduced into the body, they are taken up by macrophages, where they germinate into vegetative cells. These are not killed, but multiply, eventually killing the macrophage. The released bacteria then enter the bloodstream, replicate rapidly, and secrete toxins.

The primary virulence factors of *B. anthracis* are two exotoxins. Both toxins share a third toxic component, a cell receptor–binding protein called the *protective antigen*, that binds the toxins to target cells and permits their entry. One toxin, the *edema toxin*, causes local edema (swelling) and interferes with phagocytosis by macrophages. The other toxin, *lethal toxin*, specifically targets and kills macrophages, which disables an essential defense of the host. Furthermore, the capsule of *B. anthracis* is very unusual. It is not a polysaccharide but rather is composed of amino acid residues, which for some reason do not stimulate a protective response by the immune system. Therefore, once the anthrax bacteria enter the bloodstream, they proliferate without any effective inhibition until there are tens of millions per milliliter. These immense populations of toxin-secreting bacteria ultimately kill the host.

Anthrax affects humans in three forms: cutaneous anthrax, gastrointestinal anthrax, and inhalational (pulmonary) anthrax.

**Cutaneous anthrax** results from contact with material containing anthrax endospores. Over 90% of naturally occurring cases of anthrax in humans are cutaneous; the endospore enters at some minor skin lesion. A papule appears and then eventually vesicles, which rupture and form a depressed, ulcerated area that is covered by a black eschar (scab) as shown in **Figure 23.7**. (The name *anthrax* is derived from the Greek word for coal.) In most cases the pathogen does not enter the bloodstream, and other symptoms are limited to a low-grade fever and malaise. However, if the bacteria enter the bloodstream, mortality without antibiotic treatment can reach 20%; with antibiotic therapy, mortality is usually less than 1%.

A relatively rare form of anthrax is **gastrointestinal anthrax** caused by ingestion of undercooked food containing anthrax endospores. Symptoms are nausea, abdominal pain, and bloody diarrhea. Ulcerative lesions occur in the gastrointestinal tract ranging from the mouth and throat to, mainly, the intestines. Mortality is usually more than 50%.

The most dangerous form of anthrax in humans is **inhalational (pulmonary) anthrax**. Endospores inhaled into the lungs have a high probability of entering the bloodstream. Symptoms of the first few days of the infection are not especially alarming: mild fever, coughing, and some chest pain. Antibiotics can arrest the disease at this stage, but unless suspicion of anthrax is high, they are unlikely to be administered. As the bacteria enter the bloodstream and proliferate, the illness progresses in 2 or 3 days into septic shock that usually kills the patient within 24 to 36 hours. The mortality rate is exceptionally high, approaching 100%.

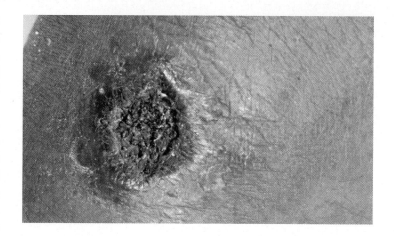

**Figure 23.7 Anthrax lesion.** The swelling and formation of a black scab that forms around the point of infection is a characteristic of cutaneous anthrax.

**Q** What are the other types of anthrax?

Antibiotics are effective in treating anthrax if they are administered in time. Currently recommended drugs are doxycycline or ciprofloxacin plus one or two additional agents that are known to be active against the pathogen. A recent development in the treatment of symptomatic inhalational anthrax is the use of raxibacumab, which inhibits the formation of toxin. This monoclonal antibody has proven to be effective in animal studies. People who have been exposed to anthrax endospores can be given preventive doses of antibiotics for a time as a precaution. This time period is usually quite long because experience has shown that up to 60 days can elapse before the inhaled endospores germinate and initiate active disease.

Vaccination of livestock against anthrax is a standard procedure in endemic areas. A single dose of an effective live, attenuated vaccine is used, which is considered unsafe for use in humans. The only vaccine currently approved for use in humans contains an inactivated form of the protective antigen toxin and is designed to prevent entry of the other two toxins into the host's cells. This vaccine requires a series of five injections over a period of 18 months, followed by annual boosters. Three doses of the vaccine over 4 weeks, along with antibiotic treatment, are recommended for people who have been exposed to *B. anthracis*.

Diagnosis of anthrax has usually consisted of isolating and identifying *B. anthracis* from a clinical specimen—which is too slow for detecting bioterrorism outbreaks. A blood test can detect both inhalational and cutaneous cases of anthrax within an hour. Furthermore, locations such as a few mail-sorting facilities are being equipped with automated electronic sensors that can immediately detect anthrax spores.

**CHECK YOUR UNDERSTANDING**

✔ **23-7** How do animals such as cattle become victims of anthrax?

As you read through this problem, you will encounter questions that primary health care providers ask themselves as they solve a clinical problem. Try to answer each question as a health care provider.

1. On February 15, Maggie, a 12-year-old girl, is seen by her pediatrician for fever, malaise, painful left underarm lymph node, and skin sloughing off her left ring finger. Amoxicillin is prescribed.
   *What diseases are possible?*

2. Intermittent fever and enlarged lymph node persist for 49 days. Maggie then undergoes excisional biopsy of the left axillary lymph node. The excised tissue is cultured; a Gram stain of the bacteria that grew is shown in the figure.
   *What additional tests would you do?*

3. Serological tests revealed the following results:

| Pathogen | Antibody Titer |
| --- | --- |
| Bartonella | 0 |
| Ehrlichia | 0 |
| Francisella | 4,096 |
| Cytomegalovirus | 0 |
| Toxoplasma gondii | 0 |

Maggie improves after treatment with streptomycin.
*What is the cause of the infection? What do you need to know?*

4. PCR is used to confirm identification of *Francisella tularensis*. Maggie liked to play with her cat, often cuddling with the animal. Between January 2 and February 8, the cat bit Maggie on the lip as she kissed the cat.
   *Where will you look for the source of the infection?*

5. Maggie's cat is an indoor/outdoor cat. The cat frequently leaves dead mice at Maggie's front door. Cats can acquire tularemia through contact with an infected animal. Tularemia in cats can range from nonclinical infection to mild illness with fever to anorexia and death. However, even cats with no clinical illness can transmit tularemia.
   *What is the most likely source of infection?*

6. Domestic cats are very susceptible to tularemia and have been known to transmit the bacteria to humans. In one survey, 12% of the domestic cats examined had antibodies to *F. tularensis*. In a Nebraska survey, nearly half of the *F. tularensis* isolates were from cats.

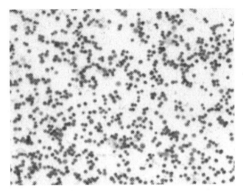

Gram-stained bacteria cultured from lymph node.  LM | 2 μm

The significance of feline *F. tularensis* transmission to humans is unknown and may be underrecognized. Care should be taken when handling any sick or dead animal.

Identifying the organism is important because it is often resistant to antibiotics commonly used for skin and systemic infections and because it is a potential agent of biological terrorism.

*Source:* For recent data on the increase in tularemia: *MMWR* 64(47): 1317–1318, December 4, 2015.

## Gangrene

If a wound causes the blood supply to be interrupted, a condition known as **ischemia**, the wound becomes anaerobic. Ischemia leads to **necrosis**, or tissue death. The death of soft tissue resulting from the loss of blood supply is called **gangrene** (Figure 23.8). These conditions can also occur as a complication of diabetes.

Substances released from dying and dead cells provide nutrients for many bacteria. Various species of the genus *Clostridium*, which are gram-positive, endospore-forming anaerobes widely found in soil and in the intestinal tracts of humans and domesticated animals, grow readily in such conditions. *C. perfringens* is the species most commonly involved in gangrene, but other clostridia and several other bacteria can also grow in such wounds.

Once ischemia and the subsequent necrosis caused by impaired blood supply have developed, **gas gangrene** can develop, especially in muscle tissue. As the *C. perfringens* microorganisms grow, they ferment carbohydrates in the tissue and produce gases (carbon dioxide and hydrogen) that swell the tissue. The bacteria produce toxins that move along

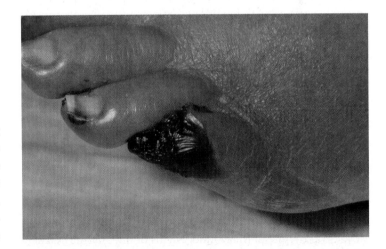

**Figure 23.8  The toes of a patient with gangrene.** This disease is caused by *Clostridium perfringens* and other clostridia. The black, necrotic tissue, resulting from poor circulation or injury, furnishes anaerobic growth conditions for the bacteria, which then progressively destroy adjoining tissue.

**Q** How can gangrene be prevented?

muscle bundles, killing cells and producing necrotic tissue that is favorable for further bacterial growth. Eventually, these toxins and bacteria enter the bloodstream and cause systemic illness. Enzymes produced by the bacteria degrade collagen and proteinaceous tissue, facilitating the spread of the disease. Without treatment, the condition is fatal.

Gas gangrene can also result from improperly performed abortions. *C. perfringens*, which resides in the genital tract of about 5% of all women, can infect the uterine wall and lead to gas gangrene, resulting in a life-threatening bloodstream infection.

The surgical removal of necrotic tissue and amputation are the most common medical treatments for gas gangrene. When gas gangrene occurs in such regions as the abdominal cavity or the reproductive tract, the patient can be treated in a **hyperbaric chamber**, which contains a pressurized oxygen-rich atmosphere. The oxygen saturates the infected tissues and thereby prevents the growth of the obligately anaerobic clostridia. Small chambers are available that can accommodate a gangrenous limb. The prompt cleaning of serious wounds and precautionary treatment with penicillin are the most effective steps in preventing gas gangrene.

> **CHECK YOUR UNDERSTANDING**
>
> ✔ 23-8 Why are hyperbaric chambers effective in treating gas gangrene?

## Systemic Diseases Caused by Bites and Scratches

Animal bites can result in serious infections. About 4.4 million animal bites occur in the United States annually, accounting for about 1% of visits to the emergency rooms in hospitals.

Dog bites make up at least 80% of reported bite incidents; cat bites, only about 10%. Cat bites are, however, more penetrating, resulting in a higher infection rate (30–50%) than the bites of dogs (15–20%). Domestic animals often harbor *Pasteurella multocida* (pas-tyer-EL-lah mul-tō-SID-ah), a gram-negative rod similar to the *Yersinia* bacterium that causes plague (page 661). *P. multocida* is primarily a pathogen of animals, and it causes sepsis (hence the name *multocida*, meaning many-killing).

Humans infected with *P. multocida* have varied responses. For example, local infections with severe swelling and pain can develop at the site of the wound. Forms of pneumonia and sepsis may develop and are life-threatening. Penicillin and tetracycline are usually effective in treating these infections.

In addition to *P. multocida*, an assortment of anaerobic bacterial species are often found in infected animal bites, as well as species of *Staphylococcus*, *Streptococcus*, and *Corynebacterium*. Bites by humans, mostly as a result of fighting, are also prone to serious infections. In fact, before antibiotic therapy became available,

nearly 20% of victims of human bites on extremities required amputation—currently only about 5% of cases require it.

### Cat-Scratch Disease

**Cat-scratch disease**, although it receives little attention, is surprisingly common. An estimated 22,000 or more cases occur annually in the United States, many more than the well-known Lyme disease. People who own or are closely exposed to cats are at risk. The pathogen is an aerobic, gram-negative bacterium, *Bartonella henselae* (bar-ton-EL-lah HEN-sel-ī). Microscopy shows that the bacterium can inhabit the interior of some cat red blood cells. It is connected to the exterior of the cell and to the surrounding extracellular fluid by a pore (**Figure 23.9**). Resident there, the bacteria cause a persistent bacteremia in cats; it is estimated that as many as 40% of domestic and feral (wild) cats carry these bacteria in their blood. The primary mode of transmission is by the scratch of a cat; it is uncertain whether bites of cats or of cat fleas transmit the disease to humans. But the presence of cat fleas is definitely a requirement for the infection to be maintained among cats. *B. henselae* multiplies in the digestive system of the cat flea and survives for several days in flea feces. Cat claws then become contaminated from flea feces.

The initial sign is a papule at the infection site, which appears 3 to 14 days after exposure. Swelling of the lymph nodes and usually malaise and fever follow in a couple of weeks. Cat-scratch disease is ordinarily self-limiting, with a duration of a few weeks, but in severe cases antibiotic therapy may be effective.

### Rat-Bite Fever

In the United States, about 20,000 rat bites occur annually—and a bite may cause the disease **rat-bite fever**. At one time,

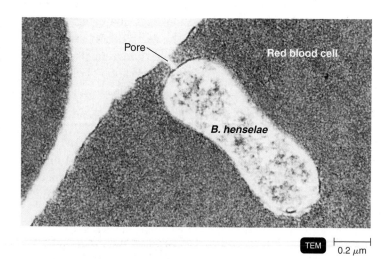

**Figure 23.9 Electron micrograph showing the location of *Bartonella henselae* within a red blood cell.** Only a pore connects the bacterium with the extracellular fluid.

**Q** Why can *B. henselae* infection persist in cats?

the victims of rat bites were small children in substandard housing. Today, rats are popular as laboratory study animals and even as pets; potential patients are now often laboratory technicians who handle rats as well as pet owners and pet store workers. Although about half of both wild and laboratory rats are known to carry the bacterial pathogens, only a minority of rat bites (about 10%) result in disease.

There are two similar but distinct diseases. In North America the more common disease, called *streptobacillary rat-bite fever*, is caused by *Streptobacillus moniliformis* (when the pathogen is ingested, the disease is termed *Haverhill fever*). This is a filamentous, gram-negative, highly pleomorphic, fastidious bacterium that is difficult to culture, although isolation in culture is the best diagnostic method. The symptoms are initially fever, chills, and muscle and joint pain, followed in a few days by a rash on the extremities. Occasionally there are more serious complications; if untreated, mortality is around 10%.

The other bacterial pathogen causing rat-bite fever is *Spirillum minus*. In this case, the disease is called *spirillar fever*; in Asia, where most cases occur, it is known as *sodoku*. It is more likely to occur in bites by wild rodents. The symptoms are similar to those of streptobacillary rat-bite fever. Because the pathogen cannot be cultured, diagnosis is made by microscopic observation of the gram-negative, spiral-shaped bacterium. Treatment by penicillin is usually effective for both forms of rat-bite fever.

Cardiovascular infections transmitted to humans by contact with other animals are summarized in Diseases in Focus 23.2.

> **CHECK YOUR UNDERSTANDING**
>
> ✔ **23-9** *Bartonella henselae*, the pathogen of cat-scratch disease, is capable of growth in what insect?

## Vector-Transmitted Diseases

Vector-borne diseases of the cardiovascular system are summarized in Diseases in Focus 23.3. See pages 672–673 for the **Big Picture** box on how climate change is impacting the vector for chikungunya, a disease discussed later in the chapter.

### Plague

Few diseases have affected human history more dramatically than **plague**, known in the Middle Ages as the Black Death. This term comes from one of its characteristics, the dark blue areas of skin caused by hemorrhages.

The disease is caused by a gram-negative, rod-shaped bacterium, *Yersinia pestis*. Normally a disease of rats, plague is transmitted from one rat to another by the rat flea, *Xenopsylla cheopis* (zē-NOP-sil-lah chē-Ō-pis) (see Figure 12.32b, page 356). In the far West and Southwest, the disease is endemic in wild rodents, especially ground squirrels and prairie dogs.

If its host dies, the flea seeks a replacement host, which may be another rodent or a human. It can jump about 9 cm. A plague-infected flea is hungry for a meal because the growth of the bacteria forms a biofilm that blocks the flea's digestive tract, and the blood the flea ingests is quickly regurgitated. An arthropod vector is not always necessary for plague transmission. Contact from the skinning of infected animals; scratches, bites, and licks by domestic cats; and similar incidents have been reported to cause infection.

In the United States, exposure to plague is increasing, as residential areas encroach on areas with infected animals. In parts of the world where human proximity to rats is common, infection from this source still prevails.

From the flea bite, bacteria enter the human's bloodstream and proliferate in the lymph and blood. One factor in the virulence of the plague bacterium is its ability to survive and proliferate inside phagocytic cells rather than being destroyed by them. An increased number of highly virulent organisms eventually emerges, and an overwhelming infection results. The lymph nodes in the groin and armpit become enlarged, and fever develops as the body's defenses react to the infection. Such swellings, called *buboes*, account for the name **bubonic plague**. This is the most common form, comprising 80–95% of cases today. The mortality rate of untreated bubonic plague is 50–75%. Death, if it occurs, is usually within less than a week after the appearance of symptoms.

A particularly dangerous condition called **septicemic plague** arises when the bacteria enter the blood and proliferate, and cause bleeding into the skin and other organs (**Figure 23.10**) and septic shock. Eventually, the blood carries the

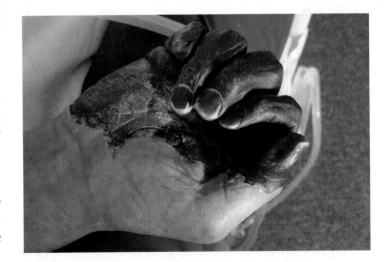

**Figure 23.10  A case of bubonic plague.** Bubonic plague is caused by infection with *Yersinia pestis*. This photograph shows the characteristic black fingers caused by hemorrhaging under the skin. This led to the name "Black Death" in the 14th century.

**Q** In what two ways is plague transmitted?

Infections from Animal Reservoirs Transmitted by Direct Contact

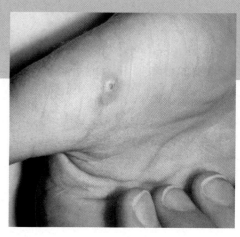

The patient's infected scratch.

The following diseases should be included in the differential diagnosis of patients with exposure to animals. A 10-year-old girl is admitted to a local hospital after having a fever (40°C) for 12 days and back pain for 8 days. Bacteria cannot be cultured from tissues. She has a recent history of dog and cat scratches. She recovers without treatment. Use the table below to make a differential diagnosis and identify infections that could cause these symptoms. For the solution, go to @MasteringMicrobiology.

| Disease | Pathogen | Symptoms | Reservoir | Method of Transmission | Treatment |
|---|---|---|---|---|---|
| **BACTERIAL DISEASES** | | | | | |
| Tularemia | *Francisella tularensis* | Local infection; pneumonia | Rabbits; rodents | Direct contact with infected animals, deer fly bite; inhalation | Streptomycin, doxycycline |
| Brucellosis | *Brucella* spp. | Local abscess; undulating fever | Grazing mammals | Direct contact | Tetracycline, streptomycin |
| Anthrax | *Bacillus anthracis* | Papule (cutaneous); bloody diarrhea (gastrointestinal); septic shock (inhalational) | Soil; large grazing mammals | Direct contact; ingestion; inhalation | Doxycycline, ciprofloxacin |
| Animal Bites | *Pasteurella multocida* | Local infection; sepsis | Animal mouths | Dog/cat bites | Penicillin |
| Rat-Bite Fever | *Streptobacillus moniliformis, Spirillum minus* | Sepsis | Rats | Rat bites | Penicillin |
| Cat-Scratch Disease | *Bartonella henselae* | Prolonged fever | Domestic cats | Cat bites or scratches, fleas | Antibiotics |
| **PROTOZOAN DISEASE** | | | | | |
| Toxoplasmosis | *Toxoplasma gondii* | Mild disease; initial infection acquired during pregnancy can be damaging to fetus; serious illness in AIDS patients | Domestic cats | Ingestion | Pyrimethamine, sulfadiazine, and folinic acid |

bacteria to the lungs, and a form of the disease called **pneumonic plague** results. The mortality rate for this type of plague is nearly 100%. Even today, this disease can rarely be controlled if it isn't recognized within 12 to 15 hours of the onset of fever. People can become infected from inhaling respiratory droplets after close contact with domestic cats and humans with pneumonic plague.

Europe was ravaged by repeated pandemics of plague; from the years 542 to 767, outbreaks occurred repeatedly in cycles of a few years. After a lapse of centuries, the disease reappeared in devastating form in the fourteenth and fifteenth centuries. It is estimated to have killed more than 25% of the population, resulting in lasting effects on the social and economic structure of Europe. A nineteenth-century pandemic primarily affected Asiatic countries; 12 million are estimated to have died in India. The last major rat-associated urban outbreak in the United States occurred in Los Angeles in 1924 and 1925. Plague is established in the ground squirrel and prairie dog

# Infections Transmitted by Vectors

The following diseases should be considered in the differential diagnosis of patients with a history of tick and insect bites or who have traveled to endemic countries. These diseases are all prevented by controlling exposure to insect and tick bites. A 22-year-old soldier returning from a tour of duty in Iraq has three painless skin ulcers.

She reports being bitten by insects every night. Ovoid, protozoa-like bodies are observed within her macrophages by examination with a light microscope. Use the table below to make a differential diagnosis and identify infections that could cause these symptoms. For the solution, go to @MasteringMicrobiology.

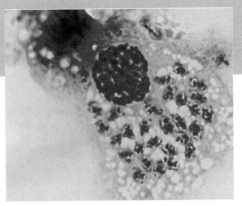

**A macrophage almost entirely filled with ovoid cells.**  LM ⊢ 5 μm

| Disease | Pathogen | Symptoms | Reservoir | Method of Transmission | Treatment |
|---|---|---|---|---|---|
| **BACTERIAL DISEASES** | | | | | |
| Plague | *Yersinia pestis* | Enlarged lymph nodes; septic shock | Rodents | Fleas; inhalation | Gentamicin, fluoroquinolones |
| Relapsing Fever | *Borrelia* spp. | Series of fever peaks | Rodents | Soft ticks | Tetracycline |
| Lyme Disease | *Borrelia burgdorferi; B. mayonii* | Bull's-eye rash; neurologic symptoms | Field mice | *Ixodes* ticks | Doxycycline, amoxicillin |
| Ehrlichiosis and Anaplasmosis | *Ehrlichia* spp. *Anaplasma* spp. | Flulike | Deer | *Ixodes* ticks | Tetracycline |
| Typhus Fever | *Rickettsia prowazekii* | High fever, stupor, rash | Squirrels | *Pediculus humanus corporis* louse | Tetracycline, chloramphenicol |
| Endemic Murine Typhus | *Rickettsia typhi* | Fever; rash | Rodents | *Xenopsylla cheopis* flea | Tetracycline, chloramphenicol |
| Rocky Mountain Spotted Fever | *Rickettsia rickettsii* | Macular rash; fever; headache | Ticks; small mammals | *Dermacentor* ticks | Tetracycline, chloramphenicol |
| **VIRAL DISEASE** | | | | | |
| Chikungunya | Chikungunya virus | Fever; joint pain | Humans | *Aedes* mosquito | Supportive |
| **PROTOZOAN DISEASES** | | | | | |
| Chagas Disease (American Trypanosomiasis) | *Trypanosoma cruzi* | Damage to heart muscle or peristaltic movement of gastrointestinal tract | Rodents, opossums | Reduviid bug | Nifurtimox |
| Malaria | *Plasmodium* spp. | Fever and chills at intervals | Humans | *Anopheles* mosquito | Malarone, artemisinin |
| Leishmaniasis | *Leishmania* spp. | *L. donovani:* systemic disease; *L. tropica:* skin sores; *L. braziliensis:* disfiguring damage to mucous membranes | Small mammals | Sandfly | Amphotericin B, paromomycin, meglumine antimoniate |
| Babesiosis | *Babesia microti* | Fever and chills at intervals | Rodents | *Ixodes* ticks | Atovaquone and azithromycin |

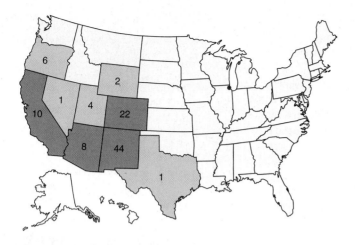

**Figure 23.11 The U.S. geographic distribution of human plague, 2000–2016.** The 2009 Illinois case was acquired from the laboratory.

*Source:* CDC, 2017.

**Q** What area reporting plague is closest to you?

communities in the western states (**Figure 23.11**). About seven cases are reported annually, most resulting from flea bites. In 2014, at least four cases of pneumonic plague were acquired from a dog with the disease.

Plague has been most commonly diagnosed by isolating the bacterium and then sending it to a laboratory for identification. A rapid diagnostic test, however, can reliably detect the presence of the capsular antigen of *Y. pestis* in blood and other fluids of patients within 15 minutes even under remote field conditions. People exposed to infection can be given prophylactic antibiotic protection. A number of antibiotics, including gentamicin and fluoroquinolones, are effective. Recovery from the disease confers reliable immunity. A vaccine is available for people likely to come into contact with infected fleas during field operations or for laboratory workers exposed to the pathogen.

### Relapsing Fever

Except for the species that causes Lyme disease (discussed below), all members of the spirochete genus *Borrelia* cause **relapsing fever**. In the United States, the disease is transmitted by soft ticks that feed on rodents. The disease occurs mostly in the western states. The incidence of relapsing fever increases during the summer months, when the activity of rodents and arthropods increases.

The disease is characterized by fever, sometimes in excess of 40.5°C, jaundice, and rose-colored skin spots. After 3 to 5 days, the fever subsides. Three or four relapses may occur, each shorter and less severe than the initial fever. Each recurrence is caused by a different antigenic type of the spirochete, which evades existing immunity. Diagnosis is made by observing the bacteria in the patient's blood, which is unusual for a spirochete disease. Tetracycline is effective for treatment.

### Lyme Disease (Lyme Borreliosis)

In 1975, a cluster of disease cases in young people that was first diagnosed as rheumatoid arthritis was reported near the city of Lyme, Connecticut. **Lyme disease** is now the most common vector-borne disease in the United States. The seasonal occurrence (summer months), lack of contagiousness among family members, and descriptions of an unusual skin rash that appeared several weeks before the first symptoms suggested a tickborne disease. In 1983, a spirochete that was later named *Borrelia burgdorferi* was identified as the cause. In 2016, another causative bacterium, *B. mayonii* (mā-YŌ-nē-ē), was discovered. In 2013, *B. miyamotoi* (mē-yah-MŌ-toy) was discovered in the northeastern United States. It causes a similar disease without the rash. In Europe and Asia, Lyme disease is usually known as **Lyme borreliosis**. Often in these locales, the tick and *Borrelia* species differ from those in the United States. Tens of thousands of cases are reported annually. In the United States, Lyme disease is most prevalent on the Atlantic coast (**Figure 23.12**).

Field mice are the most important animal reservoir. The nymphal stage of the tick feeds on infected mice and is the most likely to infect humans, even though adult ticks are about twice as likely to carry the bacterial pathogen. This is because nymphal ticks are small and less likely to be noticed before the infection is transmitted. Deer are important in maintaining the disease because the ticks feed and mate on them. They are a dead-end host and do not become infected. Although their blood may contain a few of the pathogens, they are much less likely than mice to carry nymphs or to infect the nymphs.

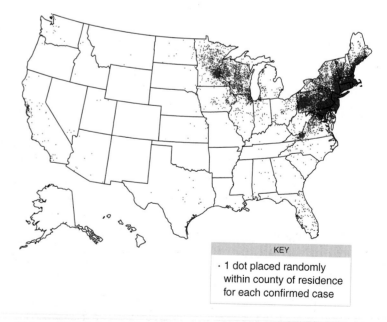

**KEY**

· 1 dot placed randomly within county of residence for each confirmed case

**Figure 23.12 Lyme disease in the United States, reported cases by county, 2015.**

*Source:* CDC.

**Q** What factors are responsible for the geographic distribution of Lyme disease?

The tick (one of two *Ixodes* species) feeds three times during its life cycle (**Figure 23.13a**). The first and second feedings, as a larva and then as a nymph, are usually on a field mouse. The third feeding, as an adult, is usually on a deer. These feedings are separated by several months, and the ability of the spirochete to remain viable in the disease-tolerant field mice is crucial to maintaining the disease in the wild.

On humans, the ticks usually attach from a perch on shrubs or grass. They do not feed for about 24 hours, and it usually requires 2 or 3 days of attachment before transfer of bacteria and infection occur. Probably only about 1% of tick bites result in Lyme disease.

On the Pacific coast, the tick that transmits Lyme disease is the western black-legged tick *Ixodes pacificus* (IKS-ō-des pah-SI-fi-kus) (see also Figure 12.31b, page 355). In the rest of the country, *Ixodes scapularis* (skap-ū-LAR-is) is most often responsible. This latter tick is so small that it is often missed. On the Atlantic coast, almost all *Ixodes* ticks carry the spirochete (Figure 23.13b); on the Pacific coast, few are infected because that tick feeds on lizards that do not carry the spirochete effectively.

The first symptom of Lyme disease is usually a rash that appears at the bite site. It is a red area that clears in the center as it expands to a final diameter of about 15 cm (**Figure 23.14**).

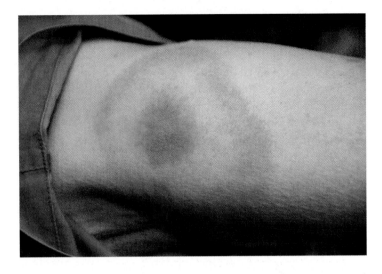

**Figure 23.14  The bull's-eye rash of Lyme disease.**
The rash is not always this obvious.

**Q** What symptoms occur once the rash fades?

This distinctive rash occurs in about 75% of cases. Flulike symptoms appear in a couple of weeks as the rash fades.

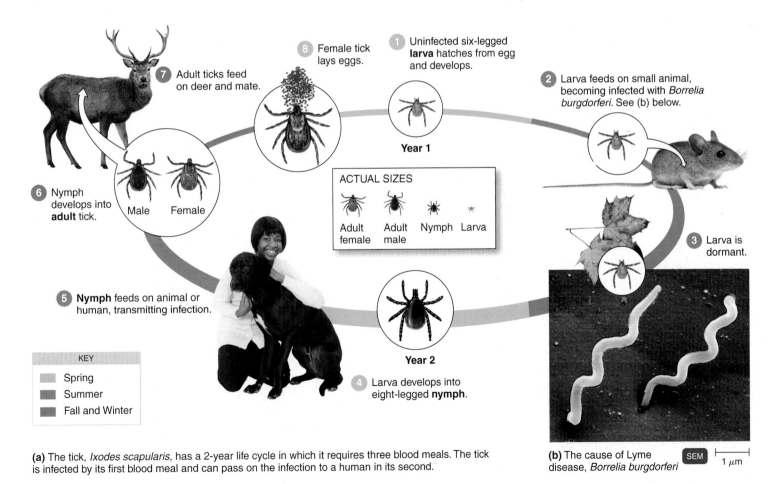

① Uninfected six-legged **larva** hatches from egg and develops.

② Larva feeds on small animal, becoming infected with *Borrelia burgdorferi*. See (b) below.

**Year 1**

③ Larva is dormant.

④ Larva develops into eight-legged **nymph**.

**Year 2**

⑤ **Nymph** feeds on animal or human, transmitting infection.

⑥ Nymph develops into **adult** tick.

⑦ Adult ticks feed on deer and mate.

⑧ Female tick lays eggs.

Male    Female

ACTUAL SIZES
Adult female    Adult male    Nymph    Larva

KEY
Spring
Summer
Fall and Winter

(a) The tick, *Ixodes scapularis*, has a 2-year life cycle in which it requires three blood meals. The tick is infected by its first blood meal and can pass on the infection to a human in its second.

(b) The cause of Lyme disease, *Borrelia burgdorferi*    SEM    ⊢——⊣ 1 μm

**Figure 23.13  The life cycle of the tick vector of Lyme disease.**

**Q** What other diseases are transmitted by ticks?

Antibiotics taken during this interval are very effective in limiting the disease.

During a second phase, in the absence of effective treatment, there is often evidence that the heart is affected. The heartbeat may become so irregular that a pacemaker is required. Incapacitating, chronic neurological symptoms, such as facial paralysis, oppressive fatigue, and memory loss may be present. Some cases result in meningitis and encephalitis. In a third phase, months or years later, some patients develop arthritis that may affect them for years. Immune responses to the presence of the bacteria are probably the cause of this joint damage. A vaccine is available for veterinary use in dogs. Natural immunity to reinfection appears to be variable. For example, patients who have progressed to the stage of Lyme arthritis appear to have considerable immunity against reinfection, whereas patients at the earlier stages of the disease do not.

Diagnosis of Lyme disease depends partly on the symptoms and an index of suspicion based on the prevalence in the geographic area. Physicians are cautioned that serological tests must be interpreted in conjunction with clinical symptoms and the likelihood of exposure to infection. Serological tests are challenging to interpret, and following a positive initial ELISA (page 516) or indirect fluorescent-antibody (FA) test (page 514), confirmation should be attempted with a Western blot test (page 516). Also, after effective antibiotic treatment eliminates the bacteria, antibodies—even IgM antibodies—often persist for years and may confuse later attempts at diagnosis.

Several antibiotics are effective in treating the disease, although in the later stages, large amounts and very extended administration times may be needed.

### Ehrlichiosis and Anaplasmosis

**Human monocytotropic ehrlichiosis (HME)** is caused by *Ehrlichia chafeensis* (er-LIK-ē-ah chaf-FĒ-en-sis). This is a gram-negative, rickettsia-like, obligately intracellular bacterium. Aggregates of bacteria—called *morulae*, the Latin word for mulberry—form within the cytoplasm of monocytes. *E. chafeensis* was first observed in a human case in 1986; previously it had been considered a solely veterinary pathogen. HME is a tick-borne disease; the common name for the usual vector is the Lone Star tick. Cases occasionally occur where this tick is not found, so there may be other vectors. The white-tailed deer is the main animal reservoir, but it does not show signs of illness.

A similar tickborne disease, **human granulocytic anaplasmosis (HGA)**, was formerly called *human granulocytic ehrlichiosis*. The change occurred when the causative organism, an obligate intracellular bacterium formerly grouped with the ehrlichia, was renamed *Anaplasma phagocytophilum* (an-ah-PLAZ-mah fag'ō-si-TAH-fi-lum). The tick vector is *Ixodes scapularis*, the same species of vector as Lyme disease (page 664) and babesiosis (page 680).

The symptoms of these diseases are identical, and HGA was identified only when a case occurred in Wisconsin, where the Lone Star tick was unknown. Patients suffer from a flulike disease with high fever and headache; the fatality rate is less than 5%. The diseases probably occur with a frequency much higher than reported. Cases of HME and HGA are both widespread and sometimes geographically overlap. Once either disease is suspected (often from detection of morulae in blood smears), diagnosis is usually by the indirect FA test for HME and a polymerase chain reaction (PCR) test (page 247) for HGA. Therapy with antibiotics such as doxycycline is usually effective.

### Typhus

The various typhus diseases are caused by rickettsias, bacteria that are obligate intracellular parasites of eukaryotes. Rickettsias, which are spread by arthropod vectors, infect mostly the endothelial cells of the vascular system and multiply within them. The resulting inflammation causes local blockage and rupture of the small blood vessels.

**Typhus Fever (Epidemic Louseborne Typhus)** Typhus fever is caused by *Rickettsia prowazekii* and is carried by the human body louse *Pediculus humanus corporis* (pe-DIK-ū-lus hū-MAN-us KOR-por-is) (see Figure 12.32a, page 356). The pathogen grows in the gastrointestinal tract of the louse and is excreted by it. It is transmitted when the feces of the louse are rubbed into the wound when the bitten host scratches the bite. The disease flourishes in crowded and unsanitary surroundings, when lice can transfer readily from an infected host to a new host. Although a rare disease in the United States, several cases have occurred in

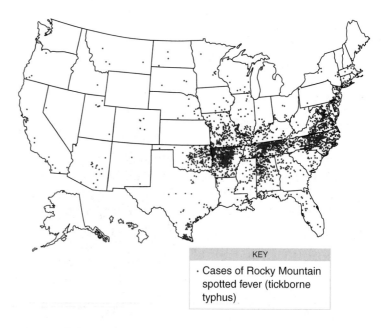

KEY

· Cases of Rocky Mountain spotted fever (tickborne typhus)

**Figure 23.15 Cases of Rocky Mountain spotted fever, reported by county, 2015.**

*Source:* CDC.

**Q** Geographically, is this a disease of rural or urban areas?

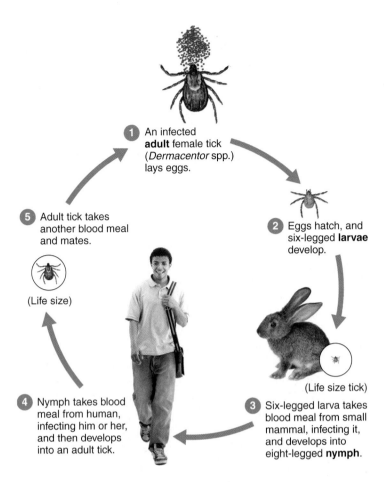

**Figure 23.16 The life cycle of the tick vector (*Dermacentor* spp.) of Rocky Mountain spotted fever.** Mammals are not essential to survival of the pathogen, *Rickettsia rickettsii*, in the tick population; the bacteria may be passed by transovarian passage, so new ticks are infected upon hatching. A blood meal is required for ticks to advance to the next stage in the life cycle.

**Q** What is meant by *transovarian passage*?

the eastern states from contact with flying squirrels or their nests. Anne Frank, the teenaged writer of the famed World War II diary, died of typhus contracted in concentration camp conditions.

Typhus fever produces a high and prolonged fever that lasts at least 2 weeks. Stupor and a rash of small red spots caused by subcutaneous hemorrhaging are characteristic, as the rickettsias invade blood vessel linings. Mortality rates are very high when the disease is untreated.

Tetracycline and chloramphenicol are usually effective against typhus fever, but eliminating conditions in which the disease flourishes is more important. The microbe is considered especially hazardous, and attempts to culture it require extreme care. Vaccines are available for military populations, which historically have been highly susceptible to the disease.

**Endemic Murine Typhus** Transmitted by the rat flea *Xenopsylla cheopis* (see Figure 12.32b, page 356), **endemic murine typhus** occurs sporadically rather than in epidemics. The term *murine*

(derived from Latin for mouse) refers to the fact that rodents, such as rats and squirrels, are the common hosts for this type of typhus. The pathogen responsible for the disease is *Rickettsia typhi*, a common inhabitant of rats. With a mortality rate of less than 5%, the disease is considerably less severe than typhus fever. Except for the reduced severity of the disease, endemic murine typhus is clinically indistinguishable from typhus fever. Tetracycline and chloramphenicol are effective treatments for endemic murine typhus, and rat control is the best preventive measure.

**Spotted Fevers** Tickborne typhus, or **Rocky Mountain spotted fever**, is probably the best-known rickettsial disease in the United States. It is caused by *Rickettsia rickettsii*. Despite its name (it was first recognized in the Rocky Mountain area), it is most common in the southeastern states and Appalachia (**Figure 23.15**). This rickettsia is a parasite of ticks and is usually passed from one generation of ticks to another through their eggs, a mechanism called *transovarian passage* (**Figure 23.16**). Surveys show that in endemic areas, perhaps 1 out of every 1000 ticks is infected. In different parts of the United States, different ticks are involved—in the west, the wood tick *Dermacentor andersoni* (der-mah-SEN-ter AN-der-son-ē); in the east, the dog tick *Dermacentor variabilis* (VAR-ē-ah-bi-lis).

About a week after the tick bites, a macular rash develops that is sometimes mistaken for measles (**Figure 23.17**); however, it often appears on palms and soles, where viral rashes do not occur. The rash is accompanied by fever and headache.

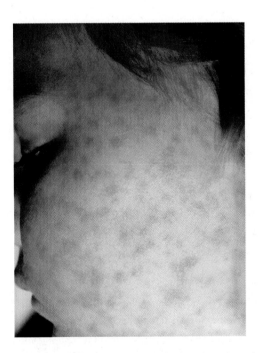

**Figure 23.17 The rash characteristic of Rocky Mountain spotted fever.** This rash is often mistaken for measles. People with dark skin have a higher mortality rate because the rash is often not recognized early enough for effective treatment.

**Q** How can Rocky Mountain spotted fever be prevented?

Katie is referred to a local emergency department (ED) for further evaluation and management. At the ED, Katie has a normal temperature of 37.1°C. A complete blood cell (CBC) count reveals a white blood cell count of 3900/µl and a platelet count of 115,000/µl. Her evaluation includes a computed tomography (CT) scan of her head and lumbar puncture. The CT scan does not reveal any brain trauma or injury, and her cerebrospinal fluid (CSF) does not show the presence of bacteria. Katie's light-headedness resolves later that evening, and she is sent home after spending half the day in the ED.

**What do her CBC results indicate?** (*Hint:* Refer to Chapter 16.)

651   **668**   671   677   683

Death, which occurs in about 3% of the approximately 4000 cases reported each year, is usually caused by kidney and heart failure.

Serological tests don't become positive until late in the illness. Diagnosis before the typical rash appears is difficult because symptoms vary widely. Also, on dark-skinned individuals, the rash is difficult to see. A misdiagnosis can be costly; if treatment is not prompt and correct, the mortality rate is about 20%.

Antibiotics such as tetracycline and chloramphenicol are very effective if administered early enough. No vaccine is available.

**CHECK YOUR UNDERSTANDING**

✔ **23-10** Why is the plague-infected flea so eager to feed on a mammal?

✔ **23-11** What animal does the infecting tick feed on just before it transmits Lyme disease to a human?

✔ **23-12** Which disease is tickborne: epidemic typhus, endemic murine typhus, or Rocky Mountain spotted fever?

# Viral Diseases of the Cardiovascular and Lymphatic Systems

**LEARNING OBJECTIVES**

**23-13** Describe the epidemiologies of Burkitt's lymphoma, infectious mononucleosis, and CMV inclusion disease.

**23-14** Compare and contrast the causative agents, vectors, reservoirs, and symptoms of yellow fever, dengue, severe dengue, and chikungunya fever.

**23-15** Compare and contrast the causative agents, reservoirs, and symptoms of Ebola and *Hantavirus* pulmonary syndrome.

Viruses cause a number of cardiovascular and lymphatic diseases, prevalent mostly in tropical areas. However, one viral disease of this type, infectious mononucleosis, is an especially familiar infectious disease among American college-aged individuals.

## Burkitt's Lymphoma

In the 1950s, Denis Burkitt, an Irish physician working in eastern Africa, noticed the frequent occurrence in children of a fast-growing tumor of the jaw (**Figure 23.18**). Known as **Burkitt's lymphoma**, this is the most common childhood cancer in Africa.

Burkitt suspected a viral cause of the tumor and a mosquito vector. At that time, there was no known virus that caused human cancer, although several viruses were clearly associated with animal cancers. Intrigued by this possibility, in 1964 British virologist Tony Epstein and his student, Yvonne Barr, performed biopsies on the tumors. A virus was cultured from this material, and the electron microscope showed a herpeslike virus in the culture cells; it was named the *Epstein-Barr virus (EB virus)*. The official name of this virus is *human herpesvirus 4 (HHV-4)*.

EB virus is clearly associated with Burkitt's lymphoma, but the mechanism by which it causes the tumor isn't understood. Research eventually showed, however, that mosquitoes don't transmit the virus or the disease. Instead, mosquito-borne malarial infections apparently foster the development of Burkitt's lymphoma by impairing the immune response to

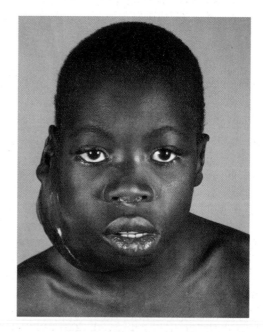

**Figure 23.18  A child with Burkitt's lymphoma.** Cancerous tumors of the jaw caused by Epstein-Barr virus (EB virus) are seen mainly in children. This child was successfully treated.

**Q** **What is the relationship between malarial areas and areas with Burkitt's lymphoma?**

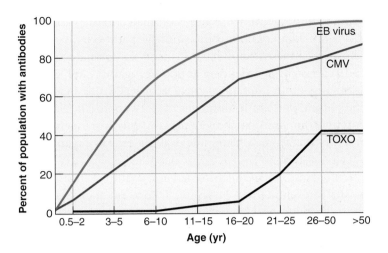

**Figure 23.19 The typical U.S. prevalence of antibodies against Epstein-Barr virus (EB virus), cytomegalovirus (CMV), and *Toxoplasma gondii* (TOXO) by age.**

*Source:* Laboratory Management, June 1987, pp. 23ff.

**Q** To judge from this graph, which of these diseases is more likely to result from early-childhood infections?

EB virus, which is almost universally present in human adults worldwide. Burkitt's lymphoma is rare in the United States but does occur in children and in AIDS patients. The virus has, in fact, become so adapted to humans that it is one of our most effective parasites. It establishes a lifelong infection in most people (**Figure 23.19**) that is harmless and rarely causes disease. In the United States, early treatment with anticancer drugs has a high success rate.

**CHECK YOUR UNDERSTANDING**

✔ **23-13** Although not a disease with an insect vector, why is Burkitt's lymphoma most commonly a disease found in malarial areas?

## Infectious Mononucleosis

The identification of EB virus as the cause of **infectious mononucleosis**, or *mono*, resulted from one of the accidental discoveries that often advance science. A technician in a laboratory investigating EB virus served as a negative control for the virus. While on vacation, she contracted an infection characterized by fever, sore throat, swollen lymph nodes in the neck, and general weakness. The most interesting aspect of the technician's disease was that she now tested serologically positive for EB virus. It was soon confirmed that the same virus that is associated with Burkitt's lymphoma also causes almost all cases of infectious mononucleosis.

In developing parts of the world, infection with EB virus occurs in early childhood, and over 95% of adults have acquired antibodies. Nearly 20% of adults in the United States

carry EB virus in oral secretions. Childhood EB virus infections are usually asymptomatic, but if infection is delayed until young adulthood, as is often the case in the United States, the result is more symptomatic probably because of an intense immunological response. The peak U.S. incidence of the disease occurs at about age 15 to 25. A principal cause of the rare deaths is rupture of the enlarged spleen (a common response to a systemic infection) during vigorous activity. Recovery is usually complete in a few weeks, and immunity is permanent.

The usual route of infection is by the transfer of saliva by kissing or, for example, by sharing drinking vessels. It doesn't spread among casual household contacts, so aerosol transmission is unlikely. The incubation period before appearance of symptoms is 4 to 7 weeks.

EB virus maintains a persistent infection in the mouth and throat, which accounts for its presence in saliva. It is probable that resting memory B cells (see Figure 17.6, page 483) located in lymphoid tissue are the primary site of replication and persistence. Most of the symptoms are attributed to responses of T cells to the infection.

The disease name *mononucleosis* refers to lymphocytes with unusual lobed nuclei that proliferate in the blood during the acute infection (**Figure 23.20**). The infected B cells produce heterophile antibodies, so called from the Greek *hetero* (different) and *phile* (affinity). These are weak antibodies with multispecific activities; their significance is that they are used in the diagnosis of mono. If this test is negative, the symptoms may be caused by cytomegalovirus (see page 670) or several other disease conditions. A fluorescent-antibody test that detects IgM antibodies against EB virus is the most specific diagnostic method. There is no recommended therapy for most patients.

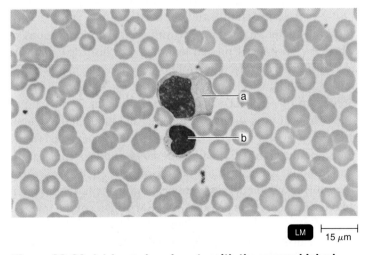

**Figure 23.20 (a) Large lymphocyte with the unusual lobed nucleus that is characteristic of mononucleosis. (b) Normal lymphocyte**

**Q** What antibodies indicate a patient has mono?

## Other Diseases and Epstein-Barr Virus

We have just discussed two diseases, Burkitt's lymphoma and infectious mononucleosis, for which there is a clear association with EB virus. There is a lengthy list of diseases for which there is a suspected, but not proven, relationship with EB virus. Some of the more familiar of these include **multiple sclerosis** (autoimmune attack on the nervous system), **Hodgkin's lymphoma** (tumors of the spleen, lymph nodes, or liver), and **nasopharyngeal** (nose and pharynx) **cancer** among certain ethnic groups in southeast Asia and Inuits.

## *Cytomegalovirus* Infections

Almost all of us will become infected with *Cytomegalovirus (CMV)* during our lifetime. The CMV is a very large herpesvirus that, much like the Epstein-Barr virus, probably remains latent in white blood cells, such as monocytes, neutrophils, and T cells. It is not much affected by the immune system, replicating very slowly and escaping antibody action by moving between cells that are in contact. Carriers of the virus may shed it in body secretions such as saliva, semen, and breast milk. When CMV infects a cell, it causes the formation of distinctive inclusion bodies that are visible by microscopy. When these bodies occur in pairs, they are known as "owl's eyes" and are useful in diagnosis. These inclusion bodies were first reported in 1905 in certain cells of newborn infants affected with congenital abnormalities. The cells were also enlarged, a condition known as *cytomegaly*, from which the virus eventually received its name. This disease of the newborns was given the name of **cytomegalic inclusion disease (CID)**. The inclusion bodies were originally thought to be stages in the life cycle of a protozoan, and a viral cause of the disease was not proposed until 1925. The *Cytomegalovirus* wasn't isolated until some 30 years after that. The official name is *human herpesvirus 5*.

In the United States, about 8000 infants each year are born suffering symptomatic damage from CID, the most serious of which includes severe mental retardation or hearing loss. If the mother is already infected before conception, the rate of transmission to a fetus is less than 2%, but if the primary infection occurs during pregnancy, the rate of transmission is in the range of 40–50%. Tests to determine the immune status of the mother are available, and it's recommended that physicians determine the immune status of female patients of childbearing age. All non-immune women should be informed of the risks of infection during pregnancy. A complicating factor is that women who are CMV positive before conception might still be infected with new strains of CMV and transmit it to the fetus.

In healthy adults, acquiring a CMV infection causes either no symptoms or those resembling a mild case of infectious mononucleosis. It has been said that if CMV were accompanied by a skin rash, it would be one of the better-known childhood diseases. It is therefore not surprising, given that 80% of the population of the United States is estimated to carry the virus, that CMV is a common opportunistic pathogen in persons whose immune systems have become compromised. Figure 23.19 shows the prevalence of antibodies against CMV, Epstein-Barr virus, and *Toxoplasma gondii* (page 676). In developing parts of the world, infection rates of CMV approach 100%. For immunocompromised individuals, CMV is a frequent cause of a life-threatening pneumonia, but almost any organ can be affected. About 85% of AIDS patients exhibit a CMV-caused eye infection, *Cytomegalovirus retinitis*. Without treatment, it results in eventual loss of vision. To prevent transmission of CMV during transplantation procedures, an immunoglobulin preparation containing a standardized amount of antibody is recommended. For treatment of CMV illness, ganciclovir has been the mainstay. A suitable alternative, if resistance to this antiviral develops, is foscarnet.

CMV is transmitted mostly by activities that result in contact with body fluids that contain the virus, such as kissing, and it is very common among children in day-care settings. It can also be transmitted sexually, by transfused blood, and by transplanted tissue. Transmission by transfused blood can be eliminated by filtering out the white cells from the blood or by serological testing of the donor for the virus. Transplanted tissue is usually tested for the virus, and products are now available that contain antibodies to neutralize CMV present in donated tissue. Vaccines are under development, but none is currently available.

## Chikungunya

Another tropical disease now causing concern is **chikungunya** (the name sounds like "chicken-gun-yah," but the disease is often referred to simply as *chik*). The name comes from an African language and means "that which bends up." The symptoms are a high fever and severe, crippling joint pains—especially in the wrists, fingers, and ankles—that can persist for weeks or months. There is often a rash and even massive blisters. The death rate is very low. The virus (CHIKV) is transmitted by the *Aedes* mosquito, primarily *Aedes aegypti* (Ā-e-dēz ē-JIP-tē), which spreads the disease widely in Asia and Africa. Recent outbreaks have also been caused by *A. albopictus* (al-BŌ-pik-tus). A mutation in CHIKV, which is related to the virus causing western equine encephalitis (WEE) and eastern equine encephalitis (EEE) (page 637), has adapted the virus to multiply in this insect. It is uncertain whether there is an animal reservoir. More than 1.7 million cases have occurred throughout the Caribbean since the first case was reported in the Western Hemisphere in 2013. An outbreak has already occurred in Italy, and locally acquired cases have occurred in Florida and Texas. Chik is the most common disease acquired by travelers, and its local transmission demonstrates how modern air travel contributes to disease emergence.

*A. albopictus* is also known as the Asian tiger mosquito because of its bright white stripes. Well adapted to urban settlements, it also survives cold climates and will probably become established eventually even in the northern parts of the United States and the coastal areas of Scandinavia. Because it is an extremely aggressive daytime biter, it is a serious nuisance for outdoor activities. Of greater concern to health officials is that *A. albopictus* is known, so far, to transmit both chikungunya and dengue, a disease that will be discussed shortly.

## Classic Viral Hemorrhagic Fevers

### Yellow Fever

Most hemorrhagic fevers are zoonotic diseases; they appear in humans only from infectious contact with their normal animal hosts. Some of them have been medically familiar for so long that they are considered "classic" hemorrhagic fevers. First among these is **yellow fever**. The yellow fever virus is injected into the skin by a mosquito, *A. aegypti*.

In the early stages of severe cases of the disease, the person experiences fever, chills, and headache, followed by nausea and vomiting. This stage is followed by jaundice, a yellowing of the skin that gave the disease its name. This coloration reflects liver damage, which results in the deposit of bile pigments in the skin and mucous membranes. The mortality rate for yellow fever is high, about 20%.

Yellow fever is still endemic in tropical areas of South America and Africa. At one time, the disease was endemic in the United States and occurred as far north as Philadelphia. The last U.S. case of yellow fever occurred in Louisiana in 1905 during an outbreak that resulted in about 1000 deaths. Mosquito eradication campaigns initiated by the U.S. Army surgeon Walter Reed were effective in eliminating yellow fever in the United States.

Monkeys are a natural reservoir for the virus, but human-to-human transmission can maintain the disease. Local control of mosquitoes and immunization of the exposed population are effective controls in urban areas.

Diagnosis is usually by clinical signs, but it can be confirmed by a rise in antibody titer or isolation of the virus from the blood. There is no specific treatment for yellow fever. The vaccine is an attenuated live viral strain and yields a very effective immunity.

### Dengue and Severe Dengue

Compared to yellow fever, dengue (DEN-gē) is a similar but milder disease also transmitted by *A. aegypti* mosquitoes. This disease is endemic in the Caribbean and other tropical environments. Globally, an estimated 400 million cases in at least 100 countries occur annually. The countries surrounding the Caribbean are reporting an increasing number of cases of dengue. In most years, more than 100 cases are imported into the United States, mostly by travelers from the Caribbean and South America.

Most infections by the dengue virus (DENV) causing dengue are asymptomatic, and the disease itself may vary from a mild case of fever to a severe, fatal disease. Patients who recover without major incident are classified as having **dengue**. If the patient suffers from severe bleeding and organ impairment, the case is classified as **severe dengue**. Severe dengue occurs mostly in patients with their second dengue infection and in infants who have passive immunity from their mother. It appears that antibodies to the virus enhance its ability to enter cells, a development called *antibody enhancement*.

The disease is a leading cause of death among children in endemic countries. It does not appear to have an animal reservoir. The primary mosquito vector for dengue is common in the Gulf states. Several cases of dengue acquired in Florida and Hawaii since 2010 have raised concern about the potential for emergence of dengue in the U.S. An efficient secondary vector, *Aedes albopictus*, has also expanded widely in range in recent years. Attempts to control dengue by vector control have not met with success. Several vaccines are in clinical trials. However, the risk of antibody enhancement and severe dengue is proving difficult to overcome. Effective antiviral drugs have yet to be developed.

---

**CHECK YOUR UNDERSTANDING**

✔ 23-14 Why is the mosquito *Aedes albopictus* a special concern to the populations of temperate climates?

---

**CLINICAL CASE**

Katie's lowered white blood cell count (leukopenia) may indicate a viral infection. Four days later, Katie returns to her primary care provider: her gums are bleeding, and she "just doesn't feel right." On examination, Katie has a temperature of 37.1°C, but now she has a rash on her legs. When asked, Katie explains that the rash is from scratching the numerous mosquito bites she incurred while in Key West. Katie's doctor doesn't think the rash looks like it's from mosquito bites; he sends a serum sample to a private lab for testing. IgM antibodies for dengue are reported in her serum. After Katie's physician notifies the public health department of the test result, Katie's earlier serum specimen, CSF specimen, and a repeat serum specimen are sent to CDC for confirmatory testing.

**What does the presence of IgM antibodies mean?**

651  668  **671**  677  683

*A disease normally found only in Africa and Asia spreads to the Americas, and public health officials warn that climate change may bring a rising tide of vector-borne diseases to the United States.*

Chikungunya is a viral disease with symptoms similar to dengue. Until recently, cases outside Africa and Asia were found only among people who had traveled to endemic regions. However, in 2013 the first indigenous cases appeared in the Western Hemisphere. By 2017, 1.7 million cases had been reported in the Americas. In the United States, there have been over 400 locally acquired cases.

## The Challenge of Controlling Mosquitoes That Spread Chikungunya

There is no vaccine, so the best way to prevent chikungunya's spread is vector control. The vectors are *Aedes aegypti*, known as the yellow fever mosquito, and *A. albopictus*, known as the Asian tiger mosquito. Yellow fever mosquitoes tend to live in tropical or subtropical regions and are also the main vector of dengue. The Asian tiger mosquito can also transmit dengue and West Nile virus. It is an invasive species introduced to the Americas through sea cargo containers.

Unlike many other mosquitoes, both species that transmit chikungunya feed all day long rather than just at dusk. The Asian tiger mosquito prefers feeding on humans over other animals and frequently lives inside buildings or very near them.

The Asian tiger mosquito has been moving north and east since its introduction in 1987. At present, the northernmost boundary is New Jersey and southern New York. If temperatures and rainfall rise as predicted over the next century, then the range of the Asian tiger mosquito will also grow. One study suggests about 50% of the land in the northeastern United States, where about 30 million people reside, could become habitat for the Asian tiger mosquito by the year 2080. This will likely facilitate the spread of chikungunya and other diseases now considered tropical throughout most of the U.S. East Coast as well.

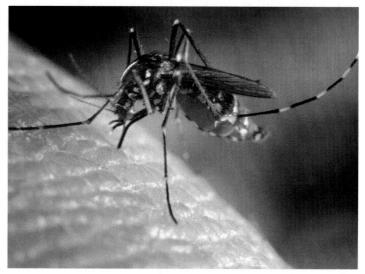

**Asian tiger mosquito, *A. albopictus*, a chikungunya vector.**

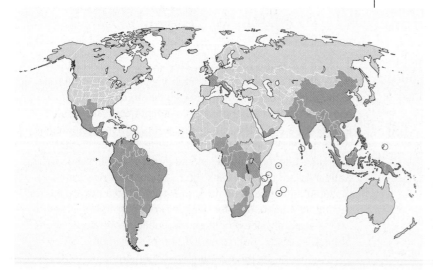

**Local transmission of chikungunya (as of July 2017).**
*Source:* CDC.

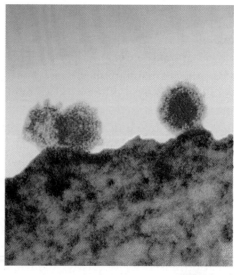

**TEM of chikungunya virus.**  TEM ⊢ 25 nm

# Searching for New Methods of Vector Control

Traditional mosquito control efforts focus on eliminating standing water sources used by mosquitoes to breed, along with spraying larvicides or insecticides in mosquito-prone areas. But total elimination of standing water is nearly impossible, and effective reduction requires full community support. Additionally, insecticides don't work as well for the indoor-dwelling Asian tiger mosquitoes as they do for species that live outside. Bed nets can provide some protection within a home, but because these mosquitoes feed all day long, this method isn't very effective in controlling this species.

## Some Methods of Control

- **Water storage covers** are inexpensive wooden covers on concrete water storage containers. Their use in one community in India ended the main breeding location for Asian tiger mosquitoes.

- **Ovitraps** are cylindrical containers designed to be attractive egg-laying sites. They contain mesh that prevents mature mosquitoes escaping from the containers later. Some ovitraps also have paddles coated with sticky material that trap the egg-laying females.

- **Biological controls** include release of larvae-eating copepods (a crustacean), mosquito fish, dragonflies, and beetle larvae in breeding areas. "Mosquito dunks," some of which contain *Bacillus thuringiensis israelensis* bacteria, are slow-dissolving blocks that can be added to ponds or fountains to kill larvae. There have also been some sites where genetically modified sterile male mosquitoes were released to cut down population growth.

Biological controls include "mosquito dunks" like the one shown above that are placed in ponds or fountains and kill larvae.

## KEY CONCEPTS

- When climate changes, certain insects that vector diseases may spread into new areas, causing new outbreaks as they go. **(See Chapter 12, "Arthropods as Vectors," pages 355–357, and Chapter 14, "Transmission of Disease," pages 404–407.)**

- *Bacillus thuringiensis* is used in mosquito dunks, one method of mosquito control. **(See Chapter 11, "Bacillales," pages 313–314.)**

**Common mosquito habitats around your home**

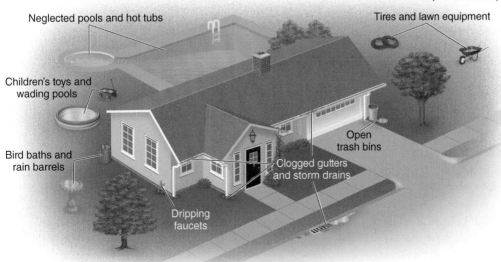

Neglected pools and hot tubs

Tires and lawn equipment

Children's toys and wading pools

Bird baths and rain barrels

Open trash bins

Clogged gutters and storm drains

Dripping faucets

**Eliminating sources of standing water is a main line of attack in controlling mosquito populations. The illustration above shows common sources of standing water in homes.**

## Emerging Viral Hemorrhagic Fevers

Certain other hemorrhagic diseases are considered new or "emerging" hemorrhagic fevers. In 1967, 31 people became ill and 7 died after contact with some African monkeys that were imported into Europe. The virus was strangely shaped (in the form of a filament [filovirus]) and was named for the site of the laboratory outbreak in Germany, the **Marburg virus**, or **green monkey virus**. The symptoms of infection by hemorrhagic viruses are mild at first; headache and muscle pain. But after a few days the victim suffers from high fever and begins vomiting blood and bleeding profusely, both internally and from external openings such as the nose and eyes. Death comes in a few days from organ failure and shock.

A similar hemorrhagic fever, **Lassa fever**, appeared in west Africa in 1969 and was traced to a rodent reservoir. The *Lassa virus*, an arenavirus, is present in the rodent's urine and is the source of human infections. Outbreaks of Lassa fever have killed thousands.

Seven years later, outbreaks in Africa of another highly lethal hemorrhagic fever were caused by *Ebolavirus*, a filovirus similar to the Marburg virus (**Figure 23.21**). The walls of the blood vessels are damaged, the virus interferes with coagulation, and blood leaks into surrounding tissue. Named **Ebola virus disease (EVD)** or **Ebola**, for a regional river, this is now a well-publicized disease, with mortality approaching 90%. The natural host reservoir for *Ebolavirus* is probably a cave-dwelling fruit bat, which is used as food and is not acutely affected by the virus it carries. Once a human is infected and shedding blood, the infection is spread by contact with the blood and body fluids and in many cases by the reuse of needles used on patients. The local custom of washing the body before burial often triggers new infections.

In 2016, a recombinant vaccine composed of live vesicular stomatitis virus carrying *Ebolavirus* glycoprotein was shown to be 100% effective in phase 3 clinical trials involving 10,000 people. The vaccine will likely be used in a ring-containment approach. Ring containment consists of identifying people with the infection, vaccinating everyone who has had contact with them, and then vaccinating people in surrounding areas.

South America has several hemorrhagic fevers caused by Lassa-like viruses (arenaviruses) that are maintained in the rodent population. **Argentine** and **Bolivian hemorrhagic fevers** are transmitted in rural areas by contact with rodent excretions. A handful of deaths in California have been attributed to the **Whitewater Arroyo virus**, an arenavirus with a reservoir in wood rats in southwestern states. These are the first reports of arenavirus-caused hemorrhagic disease in the Northern Hemisphere.

***Hantavirus* pulmonary syndrome**, caused by the *Sin Nombre virus*,* a bunyavirus, has become well known in the United States because of several outbreaks, mostly in the western states. It manifests itself as a frequently fatal pulmonary infection, in which the lungs fill with fluids. The main treatment is mechanical respiration; the antiviral ribavirin is recommended, but its

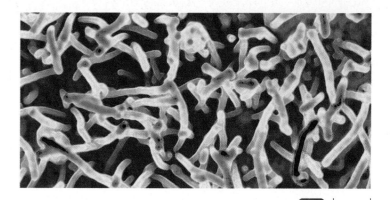

Figure 23.21 *Ebolavirus.* The viruses disrupt the blood clotting system.

**Q** **Can you see why the *Ebolavirus* is called a filovirus?**

value is uncertain. Actually, *Hantavirus*-caused diseases have a long history, especially in Asia and Europe. It is best known there as **hemorrhagic fever with renal syndrome** and primarily affects renal (kidney) function. All these related diseases are transmitted by the inhalation of viruses in dried urine and feces from infected small rodents. Worldwide, there are at least 14 known disease-causing hantaviruses.

Diseases in Focus 23.4 describes the various viral hemorrhagic fevers.

### CHECK YOUR UNDERSTANDING

✔ 23-15 Which disease does Ebola more closely resemble, Lassa fever or *Hantavirus* pulmonary syndrome?

# Protozoan Diseases of the Cardiovascular and Lymphatic Systems

### LEARNING OBJECTIVES

23-16 Compare and contrast the causative agents, modes of transmission, reservoirs, symptoms, and treatments for Chagas disease, toxoplasmosis, malaria, leishmaniasis, and babesiosis.

23-17 Discuss the worldwide effects of these diseases on human health.

Protozoa that cause diseases of the cardiovascular and lymphatic systems often have complex life cycles, and their presence may affect human hosts seriously.

---

*The virus causing the pulmonary *Hantavirus* outbreak in 1993 in the Four Corners area of the southwestern United States (Arizona, Utah, Colorado, and New Mexico) was originally called the Four Corners virus. Local authorities were concerned about the effect of this name on tourism in the area and complained. The name Sin Nombre, Spanish for no name, was then adopted.

# Viral Hemorrhagic Fevers

Viral hemorrhagic fevers are endemic in tropical countries where, except for dengue, they are found in small mammals. However, increasing international travel has resulted in importation of these viruses into the United States. There is no treatment.

The CDC's Special Pathogens Branch has specialized containment facilities to confirm diagnosis of viral hemorrhagic fevers by serology, nucleic acids, and virus culture. Use the table below to make a differential diagnosis and identify the cause of a rash and severe joint pain in a 20-year-old woman. For the solution, go to @MasteringMicrobiology.

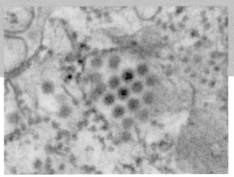

Small viruses seen by electron microscopy in a patient's tissues. Upon isolation, they were identified as single-stranded RNA viruses in the family Flaviviridae.

| Disease | Pathogen | Portal of Entry | Symptoms | Reservoir | Method of Transmission | Prevention |
|---|---|---|---|---|---|---|
| Yellow Fever | Flavivirus (yellow fever virus) | Skin | Fever, chills, headache; jaundice | Monkeys | *Aedes aegypti* | Vaccination; mosquito control |
| Dengue | Flavivirus (dengue virus) | Skin | Fever, muscle and joint pain, rash | Humans | *Aedes aegypti; A. albopictus* | Mosquito control |
| Emerging Viral Hemorrhagic Fevers (Marburg, Ebola, Lassa) | Filovirus, arenavirus | Mucous membranes | Profuse bleeding | Possibly fruit bats and other small mammals | Contact with blood | None; EVD vaccine |
| *Hantavirus* Pulmonary Syndrome | Bunyavirus (Sin Nombre hantavirus) | Respiratory tract | Pneumonia | Field mice | Inhalation | None |

## Chagas Disease (American Trypanosomiasis)

**Chagas disease**, also known as **American trypanosomiasis**, is a protozoan disease of the cardiovascular system. The causative agent is *Trypanosoma cruzi* (tri′pa-nō-SŌ-mah KROOZ-ē), a flagellated protozoan (**Figure 23.22**). The protozoan was discovered in its insect vector by the Brazilian microbiologist Carlos Chagas in 1910. The disease is endemic in Central America and parts of South America, where it chronically infects an estimated 8 million and kills as many as 50,000 annually. It has been introduced into the United States by population migration. In 2006, blood banks began screening for the disease.

The reservoir for *T. cruzi* is a wide variety of wild animals, including rodents, opossums, and armadillos. The arthropod vector is the reduviid bug, called the "kissing bug" because it often bites people near the lips (see Figure 12.32d, page 356). The insects live in the cracks and crevices of mud or stone huts with thatched roofs. One study of reduviid bugs in the state of Arizona showed that 40% of these insects in the Tucson area harbored the parasite. The range of this insect can extend as far north as Illinois. The trypanosomes, which grow in the gut of the bug, are passed on if the bug defecates while feeding. The bitten human or animal

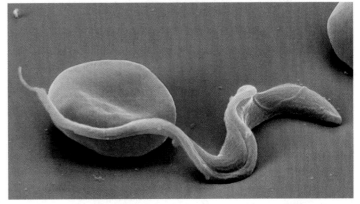

**Figure 23.22** *Trypanosoma cruzi*, **the cause of Chagas disease (American trypanosomiasis).** The trypanosome has an undulating membrane; the flagellum follows the outer margin of the membrane and then projects beyond the body of the trypanosome as a free flagellum. Note the red blood cells in the photo.

**Q** Name a common trypanosomal disease that occurs in another part of the world. (*Hint:* It was discussed in Chapter 22 on page 639.)

often rubs the feces into the bite wound or other skin abrasions by scratching or into the eye by rubbing. The infection progresses in stages. The acute stage, characterized by fever and swollen glands lasting for a few weeks, may not cause alarm. However, 20–30% of people infected will develop a chronic form of the disease—in some cases, 20 years later. Damage to the nerves controlling the peristaltic contractions of the esophagus or colon can prevent them from transporting food. This causes them to become grossly enlarged, conditions known as *megaesophagus* and *megacolon*. Most deaths are caused by damage to the heart, which occurs in about 40% of chronic cases. Pregnancy during the chronic stage can result in congenital infections.

Diagnosis in endemic areas is usually based on symptoms. In the acute phase, the trypanosomes can sometimes be detected in blood samples. During the chronic phase these are undetectable—although patients can transmit the infection by transfusions, transplants, and congenitally. Diagnosis of chronic disease depends on serological tests, which are not very sensitive or specific. Two, or even three, repeated samplings may be required.

Treating Chagas disease is very difficult when chronic, progressive stages have been reached. The trypanosome multiplies intracellularly and is difficult to reach chemotherapeutically. The only drugs currently available are nifurtimox and benznidazole, which are triazole derivatives (see page 574). Benznidazole therapy was found to eliminate infection in about 60% of infected children and is less toxic than nifurtimox. These drugs are effective only during the early, acute phase, when few people realize they are infected, and must be administered for an extended time, 30 to 60 days. Neither drug is effective during the chronic stage; both also have serious side effects.

## Toxoplasmosis

**Toxoplasmosis**, a disease of blood and lymphatic vessels, is caused by the protozoan *Toxoplasma gondii*. *T. gondii* is a spore-forming protozoan, as is the malarial parasite.

Cats are an essential part of the life cycle of *T. gondii* (**Figure 23.23**). Random tests on urban cats have shown that a large number of them are infected with the organism, which causes no

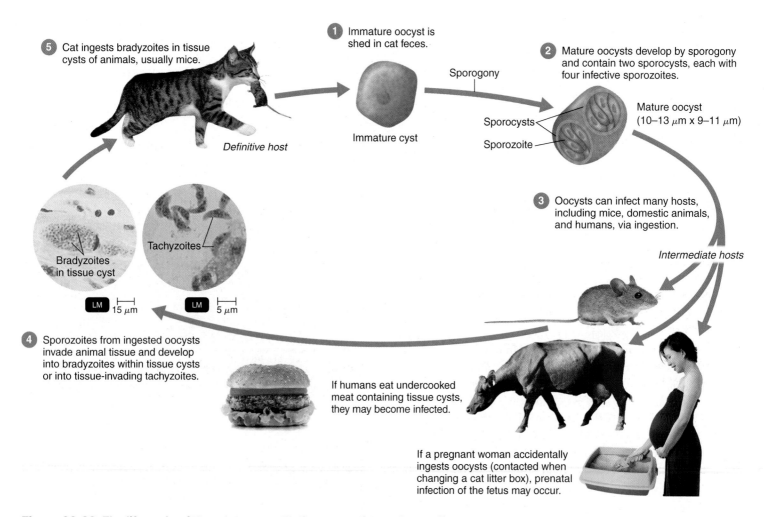

**Figure 23.23 The life cycle of *Toxoplasma gondii*, the cause of toxoplasmosis.** The domestic cat is the definitive host, in which the protozoa reproduce sexually.

**Q** How do humans contract toxoplasmosis?

apparent illness in the cat. (A curiosity of the infection in rodents is that it apparently causes them to lose their normal avoidance behavior toward cats, making them more likely to be caught and thus to infect the cat.) The microbe undergoes its only sexual phase in the intestinal tract of the cat. Millions of oocysts are then shed in the cat's feces for 7 to 21 days and contaminate food or water that can be ingested by other animals. The *oocysts* contain *sporozoites* that invade host cells and form trophozoites called *tachyzoites* (about the size of large bacteria, $2 \times 7$ μm). The intracellular parasite reproduces rapidly (*tachys* is Greek for rapid). The increased numbers cause the rupture of the host cell and the release of more tachyzoites, resulting in a strong inflammatory response.

As the immune system becomes increasingly effective, the disease enters a chronic phase in animals and humans; the infected host cell develops a wall to form a *tissue cyst*. The numerous parasites within such a cyst (in this stage called *bradyzoites; bradys* being Greek for slow) reproduce very slowly, if at all, and persist for years, especially in the brain. These cysts are infective when ingested by intermediate or definitive hosts.

In people with a healthy immune system, toxoplasmosis infection results in only very mild symptoms or none at all. Approximately 22.5% of the U.S. population, without even being aware of it, have been infected with *T. gondii* (see Figure 23.19). Humans generally acquire the infection by ingesting undercooked meats containing tachyzoites or tissue cysts, although there is a possibility of contracting the disease more directly by contact with cat feces. The primary danger is congenital infection of a fetus, resulting in stillbirth or a child with severe brain damage or vision problems. This fetal damage occurs only when the initial infection is acquired during pregnancy. The problem also affects wildlife. Off the California coast, a fatal encephalitis of sea otters and sea lions has appeared, caused by *T. gondii*—apparently, they are being infected by oocysts in wastewater contaminated from the flushed contents of cat litter boxes. Loss of immune function, AIDS being the best example, allows the inapparent infection to be reactivated from tissue cysts. It often causes severe neurological impairment and may damage vision from the reactivation of tissue cysts in the eye.

Toxoplasmosis can be detected by serological tests. Toxoplasmosis is usually diagnosed by detecting *Toxoplasma*-specific IgG and IgM antibodies. IgM antibodies and PCR for *Toxoplasma* DNA are used to determine a current infection, which is particularly important during pregnancy. Toxoplasmosis can be treated with pyrimethamine in combination with sulfadiazine and folinic acid. This does not, however, affect the chronic bradyzoite stage and is quite toxic.

## Malaria

**Malaria** is characterized by chills and fever and often by vomiting and severe headache. These symptoms typically appear at intervals of 2 to 3 days, alternating with asymptomatic periods.

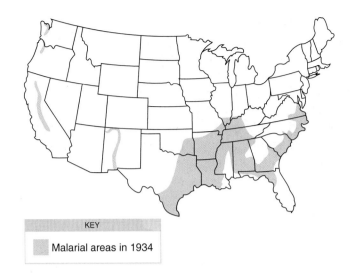

KEY

☐ Malarial areas in 1934

**Figure 23.24  Malaria in the United States.**

**Q** If malaria was eliminated in the United States in 1949, why do 1700 cases occur annually?

Malaria occurs wherever the *Anopheles* mosquito vector is found and there are human hosts for the *Plasmodium* protozoan.

The disease was once widespread in the United States (**Figure 23.24**), but effective mosquito control and a reduction in the number of human carriers resulted in elimination of malaria in 1949. Since then, approximately 1,700 cases of malaria have been reported every year in the United States, almost all in recent travelers. Reported malaria cases reached a 40-year high of 1925 in 2011. There have been 63 outbreaks of locally transmitted mosquito-borne malaria in several states. Occasionally, malaria has been transmitted by unsterilized syringes used by drug addicts. Blood transfusions from people who have been in an endemic area are also a potential risk. Malaria is endemic in tropical Asia, Africa, and Latin America, where it is still a serious problem. It is estimated that malaria

**CLINICAL CASE**

gM antibodies are the first antibodies made in response to an infection, and they are relatively short lived. Therefore, their presence indicates a current infection. The technicians at the CDC discover that both of Katie's serum specimens are positive for dengue IgM antibodies. Dengue virus serotype 1 (DENV-1) is detected by reverse transcription polymerase chain reaction from the CSF specimen. By the time the health department interviews Katie, it is 2 weeks after she reported her original symptoms. Katie has been steadily improving since then and is now almost completely recovered.

**How is dengue transmitted?**

651   668   671   **677**   683

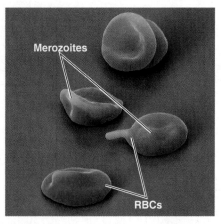

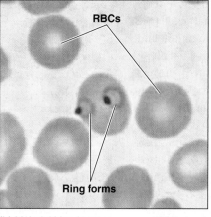

(a) Merozoites (greenish) inside RBCs | SEM | 2 μm

(b) Malarial blood smear; note the ring forms. | LM | 2 μm

**Figure 23.25 Malaria.** **(a)** Some of the red blood cells (RBCs) are lysing and releasing merozoites that will infect new RBCs. **(b)** Blood smears are used in diagnosing malaria; the protozoa can be detected growing in the RBCs. In the early stages, the feeding protozoan resembles a ring within the RBC. The light central area within the circular ring is the food vacuole of the protozoan, and the dark spot on the ring is the nucleus.

**Q** Look at the life cycle of the malarial parasite in Figure 12.20. Which of these stages, (a) or (b), actually occurs first?

affects 300 million people worldwide and causes 500,000 deaths, half of these among children, annually. Africa, where 90% of the mortality from the disease occurs, suffers the most from malaria.

Four species of *Plasmodium* cause malaria in humans. *Plasmodium vivax* is widely distributed because it can develop in mosquitoes at a lower temperature and is the cause of the most prevalent form of malaria. Sometimes referred to as "benign" malaria, the cycle of paroxysms (recurrent intensifications of symptoms) occurs every 2 days, and the patients generally survive even without treatment. An important factor in the life cycle of *P. vivax* is that it can remain dormant in the liver of the patient for months and even years. This bridges the cold-weather gap in the life cycle of mosquitoes in temperate countries—providing them with a continuing supply of infected subjects. *P. ovale* and *P. malariae* also cause a relatively benign malaria, but even so, the victims lack energy. These latter two malarial types are lower in incidence and rather restricted geographically.

The most dangerous malaria is that caused by *P. falciparum*. Perhaps one reason for the virulence of this type of malaria is that humans and the parasite have had less time to become adapted to each other. It is believed that humans have been exposed to this parasite (through contact with birds) only in relatively recent history. Referred to as "malignant" malaria, untreated it eventually kills about half of people infected. The highest mortality rates occur in young children. More red blood cells (RBCs) are infected and destroyed than in other forms of malaria. The resulting anemia severely weakens the victim. Furthermore, the RBCs develop surface knobs **(Figure 23.25a)** that cause them to stick to the walls of the capillaries, which become clogged. This clogging prevents the infected RBCs from reaching the spleen, where phagocytic cells would eliminate them. The blocked capillaries and subsequent loss of blood supply leads to death of the tissues. Kidney and liver damage is caused in this fashion. The brain is frequently affected, and *P. falciparum* is the usual cause of cerebral malaria.

Malaria and its symptoms are intimately related to the protozoan's complex reproductive cycle (see Figure 12.20, page 345). Infection is initiated by the bite of a mosquito, which carries the *sporozoite* stage of the *Plasmodium* protozoan in its saliva. About 300 to 500 sporozoites enter the bloodstream of the bitten human and within about 30 minutes enter the liver cells. The sporozoites in the liver cells undergo reproductive *schizogony* by a series of steps that finally results in the release of about 30,000 *merozoite* forms into the bloodstream.

The merozoites infect RBCs. Within the RBCs they again undergo schizogony, and after about 48 hours, the RBCs rupture and each releases about 20 new merozoites. Laboratory diagnosis of malaria is usually made by examining a blood smear (Figure 23.25b) for infected RBCs. With the release of the merozoites there is also a simultaneous release of toxic compounds, which is the cause of the paroxysms of chills and fever that are characteristic of malaria. The fever reaches 40°C, and a sweating stage begins as the fever subsides. Between paroxysms, the patient feels normal.

Many of the released merozoites infect other RBCs within a few seconds to renew the cycle in the bloodstream. If only 1% of the RBCs contain parasites, an estimated 100 billion parasites will be in circulation at one time in a typical malaria patient. Some of the merozoites develop into male or female *gametocytes*. When these enter the digestive tract of a feeding mosquito, they pass through a sexual cycle that produces new infective sporozoites. It took the combined labors of several generations of scientists to unravel this complex life cycle of the malaria parasite.

People who survive malaria acquire a limited immunity. Although they can be reinfected, they tend to have a less severe form of the disease. This relative immunity almost disappears if the person leaves an endemic area with its periodic reinfections. Malaria is especially dangerous during pregnancy because adaptive immunity is suppressed.

### Malaria Vaccines

The malarial parasite reproduces in a series of stages. The number of parasites that would serve as targets for a vaccine vary greatly in these stages. The sporozoite stage involves few pathogens and was an early target for experimental vaccines. In the liver stage, the vaccine would need to deal with hundreds of pathogens. Once the parasite begins to proliferate in the blood, the numbers quickly reach into the trillions. Vaccines targeting this stage are likely only to moderate symptoms. An intriguing concept is the *transmission-blocking vaccine*. The idea is to use the human host to generate antibodies and deliver them to the biting mosquito. There, instead of dealing with trillions of parasites, the vaccine needs to deal only with the relative handful in the mosquito. Obviously, the disadvantage is that the recipients of the vaccine still get sick but would have the questionable satisfaction of knowing they are not likely to be passing it on to someone else.

In any event, large amounts of money and resources are now directed at developing a vaccine or vaccines. A truly global malarial vaccine would have to control not only *P. falciparum* but also the widespread, although milder, *P. vivax*. There are special problems in developing a malarial vaccine. For example, in its various stages, the pathogen has as many as 7000 genes that can mutate. The result is that the parasite is very efficient at evading the human immune response. The current goal is to have a vaccine for young children that is at least 50% effective and lasts longer than a year. One such vaccine, made from a sporozoite protein produced in yeast, will be tested in three countries in 2017 to determine whether the clinical trial results can be replicated in real life.

### Diagnosis of Malaria

The most common diagnostic test for malaria is the blood smear, which requires a microscope. It is also time-consuming and requires skill in interpretation. It is still the "gold standard" for diagnosis when a well-trained staff is available. Rapid, antigen-detecting diagnostic tests that can be performed by staff with minimal training have been developed but are relatively expensive. High-quality rapid diagnostic tests that are affordable and perform reliably under field conditions are urgently needed. In endemic areas malaria is commonly diagnosed by simply observing symptoms, mainly fever, but this frequently leads to misdiagnosis. It has been found that only about half of such patients given prescriptions for antimalarial drugs actually had the disease.

### Prophylaxis and Therapy for Malaria

There are two considerations for antimalarial drugs: for prophylaxis (prevention) or for treatment.

**Prophylaxis** For travel to the few areas in which the *Plasmodium* is still sensitive to it, chloroquine is the drug of choice. In chloroquine-resistant areas, a combination of atovaquone and proguanil is the best tolerated. Travelers to malarial areas are often prescribed mefloquine. It requires only a weekly dosage, but users must be cautioned about serious side effects. These include dizziness and loss of balance, which may become permanent. Some of the psychiatric symptoms, such as depression and hallucinations, may persist for years, even after the drug is no longer being used.

**Therapy** There is a lengthy list of antimalarial drugs available; recommendations and requirements vary with cost, likelihood of developing resistance, and other factors. In the United States, if the species cannot be identified it should be assumed that the patient is infected with *P. falciparum*. If the patient is from an area still sensitive to chloroquine, it is the drug of choice; for patients coming from chloroquine-resistant zones, there are several options. The two currently preferred are malarone or oral quinine plus an antibiotic such as doxycycline.

The WHO recommends artemisinin combination therapies (ACT) for treatment of malaria worldwide. They are not used for prophylaxis. Examples of artemisinin derivatives are artesunate (not licensed in the United States) and artemether. The short-lived artemisinin component of ACT is intended to remove most of the parasites; the partner drug, with an extended period of activity, is intended to eliminate the remainder.

As with other tropical diseases, the availability of medications is limited by the very low income of the people affected, which makes their development unprofitable. The most profitable application of antimalarials will probably continue to be prophylaxis of travelers to malarial areas.

### Prevention

Effective control of malaria is not in sight. It will probably require a combination of vector control and chemotherapeutic and immunological approaches. Currently, the most promising preventative method is the use of insecticide-treated bed nets, because the *Anopheles* mosquito is a night feeder. In malarial areas, a sleeping room often will contain hundreds of mosquitoes, 1–5% of which are infectious. The expense of these efforts and the need for an effective political organization in malarial areas are probably going to be as important in controlling the disease as are advances in medical research.

## Leishmaniasis

**Leishmaniasis** is a widespread and complex disease that exhibits several clinical forms. The protozoan pathogens are of about 20 *Leishmania* species, often categorized into three groups described below. Leishmaniasis is transmitted by the bite of female sandflies, about 30 species of which are found in much of the tropical world and around the Mediterranean. These insects are smaller than mosquitoes and often penetrate the mesh of standard netting. Small mammals are an unaffected reservoir of the protozoans. The infective form, the *promastigote*, is in the saliva of the insect. It loses its flagellum when it penetrates the skin of the mammalian victim, becoming an *amastigote* that proliferates in phagocytic cells, mostly in fixed locations in tissue. These amastigotes are then ingested by feeding sandflies, renewing the cycle. Contact with contaminated blood from transfusions or shared needles can also lead to infection.

Visceral leishmaniasis has emerged as an important opportunistic infection associated with HIV. In southern Europe, up to 70% of cases of visceral leishmaniasis in adults are associated with HIV infection.

### *Leishmania donovani* Infection (Visceral Leishmaniasis)

*Leishmania donovani* (lĭsh-MAN-ē-ah DAH-nō-van-ē) infection occurs in much of the tropical world, although 90% of the cases occur in India, Ethiopia, Somalia, South Sudan, Sudan, and Brazil. Estimates are that about half a million cases occur per year. Visceral leishmaniasis is often fatal. Early symptoms, following infection by as long as a year, resemble the chills and sweating of malaria. As the protozoa proliferate in the liver and spleen, these organs enlarge greatly. Eventually, kidney function is also lost as these organs are invaded. This is a debilitating disease that, if untreated, will lead to death within a year or two.

Several inexpensive serological tests that are easy to use have been developed to diagnose visceral leishmaniasis. These have generally replaced microscopic examination of blood and tissues to demonstrate the parasite. PCR tests are very good to confirm diagnosis but usually require a central laboratory.

The first-line treatment in Europe and United States is liposomal amphotericin B, but it's relatively expensive for endemic countries. In many of these areas, conventional formulations of amphotericin B or meglumine antimoniate, which contains the toxic metal antimony, are in use. The first effective oral drug is miltefosine. It has demonstrated a cure rate as high as 82%, but it is teratogenic, resistance develops rapidly, and it is toxic to a significant number of recipients. An inexpensive injectable aminoglycoside antibiotic, paromomycin, has shown good effectiveness.

### *Leishmania tropica* Infection (Cutaneous Leishmaniasis)

*Leishmania tropica* (TROP-i-kah) and *L. major* infection causes a cutaneous form of leishmaniasis sometimes called *oriental sore*. A papule appears at the bite site after a few weeks of incubation (**Figure 23.26**). The papule ulcerates and, after healing, leaves a prominent scar. This form of the disease is the most common and is found in much of Asia, Africa, and the Mediterranean region. It has been reported in Latin America; since 2000, locally acquired cases have occurred in Texas and Oklahoma.

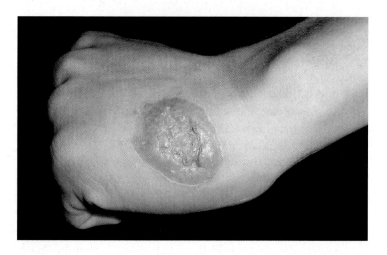

**Figure 23.26 Cutaneous leishmaniasis.** Lesion on the back of the hand of a patient.

**Q** Is this case likely to progress to visceral leishmaniasis?

### *Leishmania braziliensis* Infection (Mucocutaneous Leishmaniasis)

*Leishmania braziliensis* (brah-ZIL-ē-en-sis) infection is known as mucocutaneous leishmaniasis because it affects mucous membranes as well as skin. It causes disfiguring destruction of the tissues of the nose, mouth, and upper throat. This form of leishmaniasis is most commonly found in the Yucatán Peninsula of Mexico and in the rain forest areas of Central and South America; it often affects workers harvesting the chicle sap used for making chewing gum. This disease is often referred to as *American leishmaniasis*.

Diagnosis of cutaneous and mucocutaneous leishmaniasis in the areas where they are endemic usually depends on clinical appearance and microscopic examination of the lesion scrapings.

Mild cases of cutaneous and mucocutaneous disease will often eventually heal without treatment.

## Babesiosis

There have been increased reports of **babesiosis**, a tickborne disease once thought to be restricted to animals. It is now a nationally notifiable disease. Rodents are the reservoir in the wild; the tick vectors are most commonly *Ixodes* species. The field of medical entomology largely arose from investigations in the nineteenth century by the American microbiologist Theobald Smith into bovine babesiosis, or tick fever, in Texas cattle. The human disease in the United States is caused by a protozoan, usually *Babesia microti*. The disease resembles malaria in some respects and has been mistaken for it; the parasites replicate in the RBCs and cause a prolonged illness of fever, chills, and night sweats. It can be much more serious, sometimes fatal, in immunocompromised patients. For example, the first human cases were observed in persons who had undergone splenectomy (removal of the spleen). Simultaneous treatment with the drugs atovaquone and azithromycin has been effective.

# DISEASES IN FOCUS 23.5 Infections Transmitted by Soil and Water

A minority of systemic infections are acquired by contact with soil and water. The pathogens usually enter through a break in the skin. For example, a 65-year-old man with poor circulation in his legs develops an infection following injury to a toe. Dead tissue further reduces circulation, requiring amputation of two toes. Use the table below to make a differential diagnosis and identify infections that could cause these symptoms. For the solution, go to @MasteringMicrobiology.

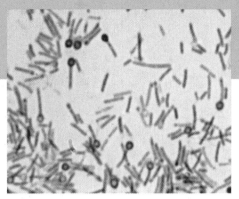

Gram-stained bacteria from the patient's toe.   2.5 μm

| Disease | Pathogen | Symptoms | Reservoir | Method of Transmission | Treatment |
|---------|----------|----------|-----------|------------------------|-----------|
| **BACTERIAL DISEASE** | | | | | |
| Gangrene | *Clostridium perfringens* | Tissue death at infection site | Soil | Puncture wound | Surgical removal of necrotic tissue |
| **HELMINTHIC DISEASE** | | | | | |
| Schistosomiasis | *Schistosoma* spp. | Inflammation and tissue damage at site of granulomas (e.g., liver, lungs, bladder) | Definitive host; humans | Cercariae penetrate skin | Praziquantel Prevention: sanitation; elimination of host snail |

## CHECK YOUR UNDERSTANDING

✔ **23-16** What tickborne disease in the United States is sometimes mistaken for malaria when blood smears are inspected?

✔ **23-17** Eliminating which of these diseases, malaria or Chagas disease, would have the greater effect on the well-being of the population of Africa?

# Helminthic Disease of the Cardiovascular and Lymphatic Systems

## LEARNING OBJECTIVE

**23-18** Diagram the life cycle of *Schistosoma*, and show where the cycle can be interrupted to prevent human disease.

Many helminths use the cardiovascular system for part of their life cycle. Schistosomes find a home there, shedding eggs that are distributed in the bloodstream. See Diseases in Focus 23.5.

## Schistosomiasis

**Schistosomiasis** is a debilitating disease caused by a small fluke. It is probably second only to malaria in the number of people it kills or disables. The symptoms of the disease result from eggs shed by adult schistosomes in the human host. These adult helminths are 15 to 20 mm long, and the slender female lives permanently in a groove in the body of the male, from which is derived the name: *schistosome*, or split-body (**Figure 23.27a**). The union between the male and female produces a continuing supply of new eggs. Some of these eggs lodge in tissues. Defensive reactions of the human host to these foreign bodies cause local tissue damage called **granulomas** (**Figure 23.28**). Other eggs are excreted and enter the water to continue the cycle.

The life cycle of *Schistosoma* is depicted in Figure 23.27b. The disease is spread by human feces or urine carrying eggs of the schistosome that enter water supplies with which humans come into contact. In the developed world, sewage and water treatment minimizes the contamination of the water supply. Also, snails of certain species are essential for one stage of the life cycle of the schistosomes. They produce the cercariae that penetrate the skin of a human entering contaminated water. In most areas of the United States, a suitable host snail isn't present. Therefore, even though it's estimated that schistosome eggs are being shed by many immigrants, the disease is not being propagated.

There are two primary types of schistosomiasis. The disease caused by *Schistosoma haematobium* (shis-tō-SŌ-mah hē-mah-TŌ-bē-um), sometimes called urinary schistosomiasis,

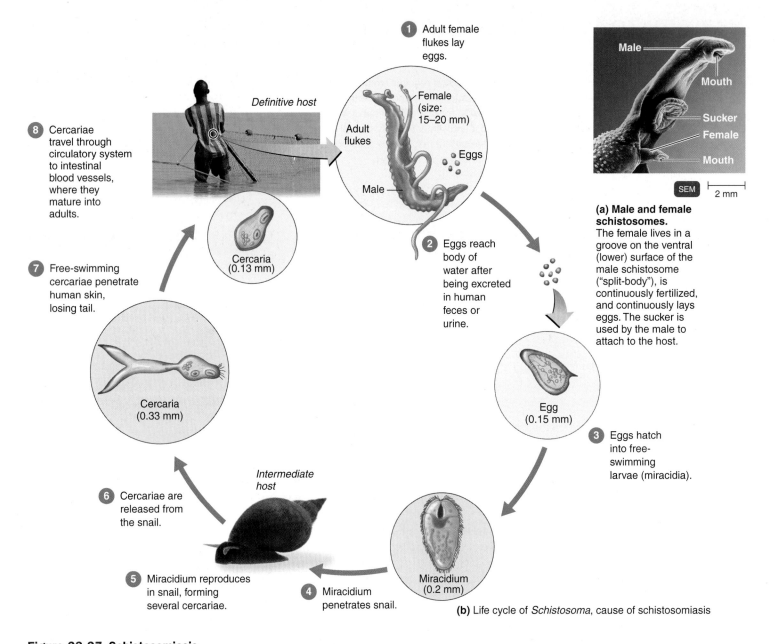

(a) **Male and female schistosomes.** The female lives in a groove on the ventral (lower) surface of the male schistosome ("split-body"), is continuously fertilized, and continuously lays eggs. The sucker is used by the male to attach to the host.

1 Adult female flukes lay eggs.

2 Eggs reach body of water after being excreted in human feces or urine.

3 Eggs hatch into free-swimming larvae (miracidia).

4 Miracidium penetrates snail.

5 Miracidium reproduces in snail, forming several cercariae.

6 Cercariae are released from the snail.

7 Free-swimming cercariae penetrate human skin, losing tail.

8 Cercariae travel through circulatory system to intestinal blood vessels, where they mature into adults.

Definitive host

Intermediate host

Adult flukes
Female (size: 15–20 mm)
Male
Eggs

Cercaria (0.13 mm)

Cercaria (0.33 mm)

Egg (0.15 mm)

Miracidium (0.2 mm)

(b) Life cycle of *Schistosoma*, cause of schistosomiasis

**Figure 23.27 Schistosomiasis.**

**Q** What is the role of sanitation and snails in maintaining schistosomiasis in a population?

results in inflammation of the urinary bladder wall. Similarly, *S. haemotobium*, *S. japonicum*, and *S. mansoni* cause intestinal inflammation. Depending on the species, schistosomiasis can cause damage to many different organs when eggs migrate in the bloodstream to different areas—for example, damage to the liver or lungs, urinary bladder cancer, or, when eggs lodge in the brain, neurological symptoms. Geographically, *S. japonicum* is found in east Asia. *S. haematobium* infects many people throughout Africa and the Middle East, most particularly Egypt. *S. mansoni* has a similar distribution but also is endemic in South America and the Caribbean, including Puerto Rico.

It's estimated that more than 250 million of the world's population are affected.

The adult worms appear to be unaffected by the host's immune system. Apparently, they quickly coat themselves with a layer that mimics the host's tissues.

Laboratory diagnosis consists of microscopic identification of the flukes or their eggs in fecal and urine specimens, intradermal tests, and serological tests such as complement-fixation and precipitin tests.

Schistomiasis is treated with praziquantel. Sanitation and elimination of the host snail are also useful forms of control.

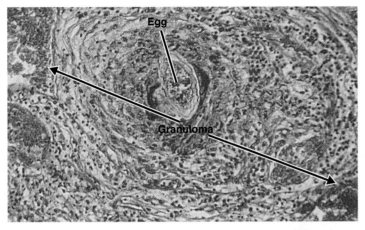

Figure 23.28 A granuloma from a patient with schistosomes. Some of the eggs laid by the adult schistosomes lodge in the tissue, and the body responds to the irritant by surrounding it with scarlike tissue, forming a granuloma.

 Why is the immune system ineffective against adult schistosomes?

**CHECK YOUR UNDERSTANDING**

✔ 23-18 What freshwater organism is essential to the life cycle of the pathogen causing schistosomiasis?

# Disease of Unknown Etiology

**LEARNING OBJECTIVE**

23-19  Recognize the clinical features of Kawasaki syndrome.

## Kawasaki Syndrome

Probably the most common cause of acquired heart disease in the United States (replacing rheumatic fever) is an acute febrile illness of unknown etiology, **Kawasaki syndrome (KS)**. As many as 5000 cases are diagnosed in the United States annually. It most often affects younger children, especially boys under the age of 5. Patients with the disease suffer a high and persistent fever, widespread skin rash, and swelling of the hands and feet and of the lymph glands in the neck. Without treatment, mortality may be about 1% but is much lower with effective treatment, involving aspirin (which affects blood clotting) and an intravenously administered immunoglobulin. KS is diagnosed primarily from its clinical signs and symptoms; there is no laboratory test available. KS may be triggered by an infection, although no specific pathogen is known.

**CHECK YOUR UNDERSTANDING**

✔ 23-19 What diseases of the cardiovascular and lymphatic systems need to be ruled out before a clinician can conclude that a patient has Kawasaki syndrome?

**CLINICAL CASE** Resolved

Dengue is transmitted by mosquitoes. Worldwide there are 100 million cases of dengue each year. Cases of dengue in returning U.S. travelers have increased steadily during the past 20 years. Dengue is now the leading cause of acute febrile illness in U.S. travelers returning from the Caribbean, South America, and Asia. Many of these travelers are still viremic upon return to the United States and are potentially capable of introducing dengue virus into a community with competent mosquito vectors. Katie's illness, which occurred in 2009, represents the first dengue case acquired in the continental United States outside the Texas–Mexico border since 1945 and the first locally acquired case in Florida since 1934. Concern about the potential for emergence of dengue in the continental United States has increased in recent years. The most efficient mosquito vector, *Aedes aegypti*, is found in the southern and southeastern United States. A secondary vector, *A. albopictus*, has spread throughout the southeastern United States since its introduction in 1985 and was responsible for a dengue outbreak in Hawaii in 2001.

651  668  671  677  **683**

---

# Study Outline

▶ **Go to** @MasteringMicrobiology for **Interactive Microbiology**, *In the Clinic videos, MicroFlix, MicroBoosters, 3D animations, practice quizzes, and more.*

## Structure and Function of the Cardiovascular and Lymphatic Systems (p 651)

1. The heart, blood, and blood vessels make up the cardiovascular system.
2. Lymph, lymph vessels, lymph nodes, and lymphoid organs constitute the lymphatic system.
3. Plasma transports dissolved substances. Red blood cells carry oxygen. White blood cells are involved in the body's defense against infection.
4. Fluid that filters out of capillaries into spaces between tissue cells is called interstitial fluid.
5. Interstitial fluid enters lymph capillaries and is called lymph; vessels called lymphatics return lymph to the blood.
6. Lymph nodes contain fixed macrophages, B cells, and T cells.

## Bacterial Diseases of the Cardiovascular and Lymphatic Systems (pp. 652–658)

### Sepsis and Septic Shock (pp. 652–654)

1. Sepsis is an inflammatory response caused by the spread of bacteria or their toxin from a focus of infection. Septicemia is sepsis that involves proliferation of pathogens in the blood.

2. Gram-negative sepsis can lead to septic shock, characterized by decreased blood pressure. Endotoxin causes the symptoms.

3. Antibiotic-resistant enterococci and group B streptococci cause gram-positive sepsis.

4. Puerperal sepsis begins as an infection of the uterus following childbirth or abortion; it can progress to peritonitis or septicemia.

5. *Streptococcus pyogenes* is the most frequent cause of puerperal sepsis.

6. Oliver Wendell Holmes and Ignaz Semmelweiss demonstrated that puerperal sepsis was transmitted by the hands and instruments of midwives and physicians.

### Bacterial Infections of the Heart (pp. 654–655)

7. The inner layer of the heart is the endocardium.

8. Subacute bacterial endocarditis is usually caused by alpha-hemolytic streptococci, staphylococci, or enterococci.

9. The infection arises from a focus of infection, such as a tooth extraction.

10. Preexisting heart abnormalities are predisposing factors.

11. Signs include fever, weakness, and heart murmur.

12. Acute bacterial endocarditis is usually caused by *Staphylococcus aureus*.

13. The bacteria cause rapid destruction of heart valves.

### Rheumatic Fever (p. 655)

14. Rheumatic fever is an autoimmune complication of streptococcal infections.

15. Rheumatic fever is expressed as arthritis or inflammation of the heart. It can result in permanent heart damage.

16. Antibodies against group A beta-hemolytic streptococci react with streptococcal antigens deposited in joints or heart valves or cross-react with the heart muscle.

17. Rheumatic fever can follow a streptococcal infection, such as streptococcal sore throat. Streptococci might not be present at the time of rheumatic fever.

18. Prompt treatment of streptococcal infections can reduce the incidence of rheumatic fever.

19. Penicillin is administered as a preventive measure against subsequent streptococcal infections.

### Tularemia (pp. 655–656)

20. Tularemia is caused by *Francisella tularensis*. The reservoir is small wild mammals, especially rabbits.

21. Signs include ulceration at the site of entry, followed by septicemia and pneumonia.

### Brucellosis (Undulant Fever) (p. 656)

22. Brucellosis can be caused by *Brucella abortus*, *B. melitensis*, and *B. suis*.

23. The bacteria enter through minute breaks in the mucosa or skin, reproduce in macrophages, and spread via lymphatics to liver, spleen, or bone marrow.

24. Signs include malaise and fever that spikes each evening (undulant fever).

25. Diagnosis is based on serological tests.

### Anthrax (pp. 656–658)

26. *Bacillus anthracis* causes anthrax. In soil, endospores can survive for up to 60 years.

27. Grazing animals acquire an infection after ingesting the endospores.

28. Humans contract anthrax by handling hides from infected animals. The endospores enter through cuts in the skin, respiratory tract, or mouth.

29. Entry through the skin results in a papule that can progress to sepsis. Entry through the respiratory tract can result in septic shock.

30. Diagnosis is based on isolating and identifying the bacteria.

### Gangrene (pp. 659–660)

31. Soft tissue death from ischemia (loss of blood supply) is called gangrene.

32. Microorganisms grow on nutrients released from gangrenous cells.

33. Gangrene is especially susceptible to the growth of anaerobic bacteria such as *Clostridium perfringens*, the causative agent of gas gangrene.

34. *C. perfringens* can invade the wall of the uterus during improperly performed abortions.

35. Surgical removal of necrotic tissue, hyperbaric chambers, and amputation are used to treat gas gangrene.

### Systemic Diseases Caused by Bites and Scratches (pp. 660–661)

36. *Pasteurella multocida*, introduced by the bite of a dog or cat, can cause septicemia.

37. Anaerobic bacteria infect deep animal bites.

38. Cat-scratch disease is caused by *Bartonella henselae*.

39. Rat-bite fever is caused by *Streptobacillus moniliformis* and *Spirillum minus*.

### Vector-Transmitted Diseases (pp. 661–668)

40. Plague is caused by *Yersinia pestis*. The vector is usually the rat flea (*Xenopsylla cheopis*).

41. Relapsing fever is caused by *Borrelia* spp. and transmitted by soft ticks.

42. Lyme disease is caused by *Borrelia burgdorferi* and is transmitted by a tick (*Ixodes*).

43. Human ehrlichiosis and anaplasmosis are caused by *Ehrlichia* and *Anaplasma* and are transmitted by *Ixodes* ticks.

44. Typhus is caused by rickettsias, obligate intracellular parasites of eukaryotic cells.

## Viral Diseases of the Cardiovascular and Lymphatic Systems (pp. 668–674)

### Burkitt's Lymphoma (pp. 668–669)

1. Epstein-Barr virus (EB virus, HHV-4) causes Burkitt's lymphoma.

2. Burkitt's lymphoma tends to occur in patients whose immune system has been weakened; for example, by malaria or AIDS.

**Infectious Mononucleosis** (p. 669)

3. Infectious mononucleosis is caused by EB virus.

4. The virus multiplies in the parotid glands and is present in saliva. It causes the proliferation of atypical lymphocytes.

5. The disease is transmitted by the ingestion of saliva from infected individuals.

6. Diagnosis is made by an indirect fluorescent-antibody technique.

7. EB virus may cause other diseases, including cancers and multiple sclerosis.

**Other Diseases and Epstein-Barr Virus** (p. 670)

8. EB virus is associated with certain cancers and autoimmune diseases.

**Cytomegalovirus Infections** (p. 670)

9. CMV (HHV-5) causes intranuclear inclusion bodies and cytomegaly of host cells.

10. CMV is transmitted by saliva and other body fluids.

11. CMV inclusion disease can be asymptomatic, a mild disease, or progressive and fatal. Immunosuppressed patients may develop pneumonia.

12. If the virus crosses the placenta, it can cause congenital infection of the fetus, resulting in impaired mental development, neurological damage, and stillbirth.

**Chikungunya Fever** (pp. 670–671)

13. The chikungunya virus, which causes fever and severe joint pain, is transmitted by *Aedes* mosquitoes.

**Classic Viral Hemorrhagic Fevers** (p. 671)

14. Yellow fever is caused by the yellow fever virus. The vector is the *Aedes aegypti* mosquito.

15. Signs and symptoms include fever, chills, headache, nausea, and jaundice.

16. Diagnosis is based on the presence of virus-neutralizing antibodies in the host.

17. No treatment is available, but there is an attenuated, live viral vaccine.

18. Dengue is caused by the dengue virus and is transmitted by the *Aedes* mosquito.

19. Signs are fever, muscle and joint pain, and rash.

20. Severe dengue is characterized by bleeding and organ failure.

21. Mosquito abatement is necessary to control the disease.

**Emerging Viral Hemorrhagic Fevers** (p. 674)

22. Human diseases caused by Marburg, Ebola, and Lassa fever viruses were first noticed in the late 1960s.

23. *Ebolavirus* is found in fruit bats; Lassa fever viruses are found in rodents. Rodents are the reservoirs for Argentine and Bolivian hemorrhagic fevers.

24. *Hantavirus* pulmonary syndrome and hemorrhagic fever with renal syndrome are caused by *Hantavirus*. The virus is contracted by inhalation of dried rodent urine and feces.

## Protozoan Diseases of the Cardiovascular and Lymphatic Systems (pp. 674–681)

**Chagas Disease (American Trypanosomiasis)** (pp. 675–676)

1. *Trypanosoma cruzi* causes Chagas disease. The reservoir includes many wild animals. The vector is a reduviid, the "kissing bug."

**Toxoplasmosis** (pp. 676–677)

2. Toxoplasmosis is caused by *Toxoplasma gondii*.

3. *T. gondii* undergoes sexual reproduction in the intestinal tract of domestic cats, and oocysts are eliminated in cat feces.

4. In the host cell, sporozoites reproduce to form either tissue-invading tachyzoites or bradyzoites.

5. Humans contract the infection by ingesting tachyzoites or tissue cysts in undercooked meat from an infected animal or contact with cat feces.

6. Congenital infections can occur. Signs and symptoms include severe brain damage or vision problems.

**Malaria** (pp. 677–679)

7. The signs and symptoms of malaria are chills, fever, vomiting, and headache, which occur at intervals of 2 to 3 days.

8. Malaria is transmitted by *Anopheles* mosquitoes. The causative agent is any one of four species of *Plasmodium*.

9. Sporozoites reproduce in the liver and release merozoites into the bloodstream, where they infect red blood cells and produce more merozoites.

**Leishmaniasis** (pp. 679–680)

10. *Leishmania* spp., which are transmitted by sandflies, cause leishmaniasis.

11. The protozoa reproduce in the liver, spleen, and kidneys.

12. Leishmaniasis is treated with liposomal amphotericin B.

**Babesiosis** (p. 680)

13. Babesiosis is caused by the protozoan *Babesia microti* and is transmitted to humans by ticks.

## Helminthic Disease of the Cardiovascular and Lymphatic Systems (pp. 681–683)

**Schistosomiasis** (pp. 681–683)

1. Species of the blood fluke *Schistosoma* cause schistosomiasis.

2. Eggs eliminated with feces hatch into larvae that infect the intermediate host, a snail. Free-swimming cercariae are released from the snail and penetrate the skin of a human.

3. The adult flukes live in the veins of the liver or urinary bladder in humans.

4. Granulomas are from the host's defense to eggs that remain in the body.

5. Observation of eggs or flukes in feces, skin tests, or indirect serological tests may be used for diagnosis.

6. Chemotherapy is used to treat the disease; sanitation and snail eradication are used to prevent it.

## Disease of Unknown Etiology (p. 683)

**Kawasaki Syndrome** (p. 683)

1. Kawasaki syndrome is characterized by fever, rash, and swollen lymph nodes in the neck. The cause is unknown.

# Study Questions

For answers to the Knowledge and Comprehension questions, turn to the Answers tab at the back of the textbook.

## Knowledge and Comprehension

### Review

1. **DRAW IT** Show the path of *Streptococcus* from a focal infection to the pericardium. Identify the portals of entry for *Trypanosoma cruzi*, *Hantavirus*, and cytomegalovirus.

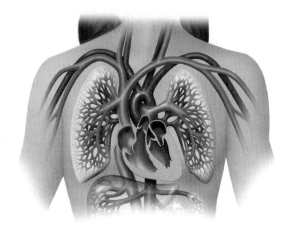

2. Complete the following table.

| Disease | Frequent Causative Agent | Predisposing Condition(s) |
|---|---|---|
| Puerperal sepsis | | |
| Subacute bacterial endocarditis | | |
| Acute bacterial endocarditis | | |
| Rheumatic fever | | |

3. Compare and contrast epidemic typhus, endemic murine typhus, and tickborne typhus.

4. Complete the following table.

| Disease | Causative Agent | Vector | Treatment |
|---|---|---|---|
| Malaria | | | |
| Yellow fever | | | |
| Dengue | | | |
| Relapsing fever | | | |
| Leishmaniasis | | | |

5. Complete the following table.

| Disease | Causative Agent | Transmission | Reservoir |
|---|---|---|---|
| Tularemia | | | |
| Brucellosis | | | |
| Anthrax | | | |
| Lyme disease | | | |
| Ehrlichiosis | | | |
| Cytomegalic inclusion disease | | | |
| Plague | | | |

6. List the causative agent, method of transmission, and reservoir for schistosomiasis, toxoplasmosis, and Chagas disease. Which disease are you most likely to get in the United States? Where are the other diseases endemic?

7. Compare and contrast cat-scratch disease and toxoplasmosis.

8. Why is *Clostridium perfringens* likely to grow in gangrenous wounds?

9. List the causative agent and method of transmission of infectious mononucleosis.

10. **NAME IT** Most people have been infected with this microorganism, often without symptoms. Infection during pregnancy can result in deafness or mental retardation in the newborn.

## Multiple Choice

Use the following choices to answer questions 1 through 4:
- a. ehrlichiosis
- b. Lyme disease
- c. septic shock
- d. toxoplasmosis
- e. viral hemorrhagic fever

1. A patient presents with vomiting, diarrhea, and a history of fever and headache. Bacterial cultures of blood, CSF, and stool are negative. What is your diagnosis?

2. A patient was hospitalized because of continuing fever and progression of symptoms including headache, fatigue, and back pain. Tests for antibodies to *Borrelia burgdorferi* were negative. What is your diagnosis?

3. A patient complained of headache. A CT (computed tomography) scan revealed cysts of varying size in her brain. What is your diagnosis?

4. A patient presents with mental confusion, rapid breathing and heart rate, and low blood pressure. What is your diagnosis?

5. A patient has a red circular rash on his arm and fever, malaise, and joint pain. The most appropriate treatment is
- a. antibiotics.
- b. chloroquine.
- c. anti-inflammatory drugs.
- d. antimony.
- e. no treatment.

6. Which of the following is not a tickborne disease?
- a. babesiosis
- b. ehrlichiosis
- c. Lyme disease
- d. relapsing fever
- e. tularemia

Use the following choices to answer questions 7 and 8:

a. brucellosis
b. malaria
c. relapsing fever
d. Rocky Mountain spotted fever
e. Ebola

7. The patient's fever spikes each evening. Oxidase-positive, gram-negative coccobacilli were isolated from a lesion on his arm. What is your diagnosis?

8. The patient was hospitalized with fever and headache. Spirochetes were observed in her blood. What is your diagnosis?

9. Which of these diseases has the highest incidence in the U.S.?

a. brucellosis
b. Ebola
c. malaria
d. plague
e. Rocky Mountain spotted fever

10. Nineteen workers in a slaughterhouse developed fever and chills, with the fever spiking to 40°C each evening. The most likely method of transmission of this disease is

a. a vector.
b. the respiratory route.
c. a puncture wound.
d. an animal bite.
e. water.

## Analysis

1. Indirect fluorescent-antibody (FA) tests on the serum of three 25-year-old women, each of whom is considering pregnancy, provided the information below. Which of these women may have toxoplasmosis? What advice might be given to each woman with regard to toxoplasmosis?

| Patient | Antibody Titer | | |
|---------|------|------|------|
| | **Day 1** | **Day 5** | **Day 12** |
| Patient A | 1024 | 1024 | 1024 |
| Patient B | 1024 | 2048 | 3072 |
| Patient C | 0 | 0 | 0 |

2. What is the most effective way to control malaria and dengue?

3. In 2016, a researcher noted "We have a vaccine that enhances dengue. Vaccine recipients less than 5 years old had five to seven times more rates of hospitalizations for severe dengue than placebo controls." Offer an explanation for this.

## Clinical Applications and Evaluation

1. A 19-year-old man went deer hunting. While on the trail, he found a partially dismembered dead rabbit. The hunter picked up the front paws for good luck charms and gave them to another hunter in the party. The rabbit had been handled with bare hands that were bruised and scratched from the hunter's work as an automobile mechanic. Festering sores on his hands, legs, and knees were noted 2 days later. What infectious disease do you suspect the hunter has? How would you proceed to prove it?

2. On March 30, a 35-year-old veterinarian experienced fever, chills, and vomiting. On March 31, he was hospitalized with diarrhea, left armpit bubo, and secondary bilateral pneumonia. On March 27, he had treated a cat that had labored respiration; an X-ray image revealed pulmonary infiltrates. The cat died on March 28 and was disposed of. Chloramphenicol was administered to the veterinarian. On April 10, his temperature returned to normal, and on April 20, he was released from the hospital. Sixty human contacts were given tetracycline. Identify the incubation and prodromal periods for this case. Explain why the 60 contacts were treated. What was the etiologic agent? How would you identify the agent?

3. Three of five patients who underwent heart valve replacement surgery developed bacteremia. The causative agent was *Enterobacter cloacae*. What were the patients' signs and symptoms? How would you identify this bacterium? A manometer used in the operations was culture-positive for *E. cloacae*. What is the most likely source of this contaminant? Suggest a way of preventing such occurrences.

4. In August and September, six people who each at different times spent a night in the same cabin developed fever, as shown in the graph below. Three recovered after tetracycline (TET) therapy, two recovered without therapy, and one was hospitalized with septic shock. What is the disease? What is the incubation period of this disease? How do you account for the periodic temperature changes? What caused septic shock in the sixth patient?

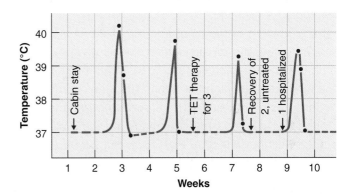

5. A 67-year-old man worked in a textile mill that processed imported goat hair into fabrics. He noticed a painless, slightly swollen pimple on his chin. Two days later he developed a 1-cm ulcer at the pimple site and a temperature of 37.6°C. He was treated with tetracycline. What is the etiology of this disease? Suggest ways to prevent it.

# 24 Microbial Diseases of the Respiratory System

With every breath, we inhale several microorganisms; therefore, the upper respiratory system is a major portal of entry for pathogens. In fact, respiratory system infections are the most common type of infection—and among the most damaging. Some pathogens that enter via the respiratory route can infect other parts of the body, causing such diseases as measles, mumps, and rubella.

The upper respiratory system has several anatomical defenses against airborne pathogens. Coarse hairs in the nose filter large dust particles from the air. The nose is lined with a mucous membrane that contains numerous mucus-secreting cells and cilia. The upper portion of the throat also contains a ciliated mucous membrane. The mucus moistens inhaled air and traps dust and microorganisms. The cilia help remove these particles by moving them toward the mouth for elimination.

At the junction of the nose and throat are masses of lymphoid tissue, the tonsils, which contribute immunity to certain infections. Because the nose and throat are connected to the sinuses, nasolacrimal apparatus, and middle ear, infections commonly spread from one region to another. Microbes that escape these defenses may be able to cause infection. Such an infection caused by the *Chlamydophila psittaci* shown in the photograph is described in the Clinical Case in this chapter.

▶ When infecting a host cell, chlamydial bacteria produce infectious elementary bodies (small, brown) and noninfectious-reticulate bodies (red).

## In the Clinic

As a nurse educator in a large urban hospital, you are asked to provide flu classes for patients and staff. Differentiate influenza from the common cold. **What prevention measures will you recommend to your students?**

*Hint: Read about the common cold on page 693 and influenza on page 709.*

Answers to In the Clinic questions are found online @MasteringMicrobiology.

**Figure 24.1 Structures of the upper respiratory system.**

**Q** Name the upper respiratory system's defenses against disease.

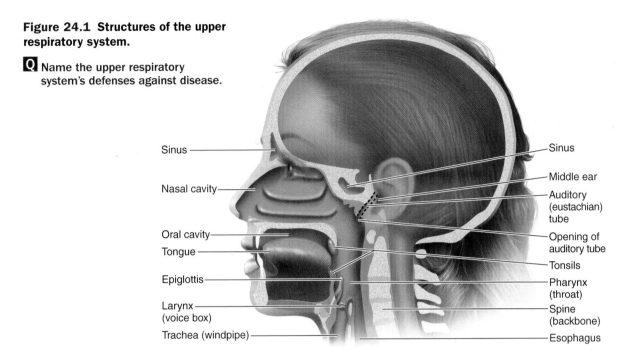

Sinus

Nasal cavity

Oral cavity

Tongue

Epiglottis

Larynx (voice box)

Trachea (windpipe)

Sinus

Middle ear

Auditory (eustachian) tube

Opening of auditory tube

Tonsils

Pharynx (throat)

Spine (backbone)

Esophagus

# Structure and Function of the Respiratory System

### LEARNING OBJECTIVE

24-1 Describe how microorganisms are prevented from entering the respiratory system.

It's convenient to think of the respiratory system as being composed of two divisions: the upper respiratory system and the lower respiratory system. The **upper respiratory system** consists of the nose, the pharynx (throat), larynx (voice box), and the structures associated with them, including the middle ear and the auditory (eustachian) tubes (**Figure 24.1**). Ducts from the sinuses (air-filled spaces in certain skull bones) and the nasolacrimal ducts from the lacrimal (tear-forming) apparatus empty into the nasal cavity (see Figure 16.2, page 450). The auditory tubes from the middle ear empty into the upper portion of the throat.

The **lower respiratory system** consists of the trachea (windpipe), bronchial tubes, and *alveoli* (**Figure 24.2**). Alveoli are air sacs that make up the lung tissue; within them, oxygen and carbon dioxide are exchanged between the lungs and blood. Our lungs contain more than 300 million alveoli, with an area for gas exchange of 70 or more square meters in an average adult. The double-layered membrane enclosing the lungs is the *pleura*, or pleural membranes. A ciliated mucous membrane lines the lower respiratory system down to the smaller bronchial tubes and helps prevent microorganisms from reaching the alveoli.

Particles trapped in the larynx, trachea, and larger bronchial tubes are moved up toward the throat by a ciliary action called the *ciliary escalator* (see Figure 16.3, page 450). If microorganisms actually reach the lungs, phagocytic cells called *alveolar macrophages* usually locate, ingest, and destroy most of them. IgA antibodies in such secretions as respiratory mucus, saliva, and tears also help protect mucosal surfaces of the respiratory system from many pathogens. Thus, the body has several mechanisms for removing the pathogens that cause airborne infections.

### CHECK YOUR UNDERSTANDING

✔ 24-1 What is the function of hairs in the nasal passages?

---

**CLINICAL CASE** It's for the Birds

For the past 2 days, Caille Nguyen has had a fever and feels out of sorts. In fact, her entire family is sick. Her three children, Gabbie, Steven, and Tre, also have fevers. Caille's husband, Art, and Gabbie and Steven have no appetite and are beginning to lose weight. Everyone has a dry cough. At first Caille thinks the children are just sad because of the loss of Bitsy, their beloved cockatiel. The family had purchased the cockatiel from a local pet store 2 months earlier. Unfortunately, Bitsy had an increasingly difficult time breathing and could not stand erect; she had to be euthanized by a local veterinarian the week before.

**What could be causing the Nguyen family's symptoms? Read on to find out.**

689   706   707   709   711   715

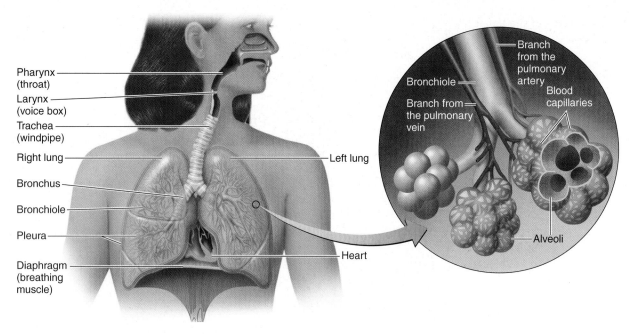

**Figure 24.2 Structures of the lower respiratory system.**

🅠 Name the lower respiratory system's defenses against disease.

## Normal Microbiota of the Respiratory System

**LEARNING OBJECTIVE**

24-2 Characterize the normal microbiota of the upper and lower respiratory systems.

A number of potentially pathogenic microorganisms are part of the normal microbiota in the respiratory system. However, they usually do not cause illness because the predominant microorganisms of the normal microbiota suppress their growth by competing with them for nutrients and producing inhibitory substances.

The lung microbiome is described in the Exploring the Microbiome box.

**CHECK YOUR UNDERSTANDING**

✔ 24-2 Normally, the lower respiratory tract is nearly sterile. What is the primary mechanism responsible?

## Microbial Diseases of the Upper Respiratory System

**LEARNING OBJECTIVE**

24-3 Differentiate pharyngitis, laryngitis, tonsillitis, sinusitis, and epiglottitis.

As most of us know from personal experience, the respiratory system is the site of many common infections. We will soon discuss **pharyngitis**, inflammation of the mucous membranes of the throat, or sore throat. When the larynx is the site of infection, we suffer from **laryngitis**, which affects our ability to speak. The microbes that cause pharyngitis also can cause inflamed tonsils, or **tonsillitis**.

The nasal sinuses are cavities in certain cranial bones that open into the nasal cavity. They have a mucous membrane lining that is continuous with that of the nasal cavity. Infection of a sinus involving heavy nasal discharge of mucus is called **sinusitis**. If the opening by which the mucus leaves the sinus becomes blocked, internal pressure can cause pain or a sinus headache. These diseases are almost always *self-limiting*, meaning that recovery will usually occur even without medical intervention.

Probably the most threatening infectious disease of the upper respiratory system is **epiglottitis**, inflammation of the epiglottis. The epiglottis is a flaplike structure of cartilage that prevents ingested material from entering the larynx (see Figure 24.1). Epiglottitis is a rapidly developing disease that can result in death within a few hours. It is caused by opportunistic pathogens, usually *Haemophilus influenzae* type b. The Hib vaccine, although directed primarily at meningitis (page 622), has significantly reduced the incidence of epiglottitis in the vaccinated population.

**Discovering the Microbiome of the Lungs**

Until as recently as 2009, many scientists and medical professionals believed that healthy lungs are sterile, based on findings from studies using traditional culture techniques. However, genetic studies related to the Human Microbiome Project show this isn't the case. It would actually be remarkable if the alveoli didn't routinely come into contact with at least some microbes, considering that the average adult inhales over 10,000 liters of microbe-filled air every day.

The most common bacterial genera present in healthy lungs are *Prevotella*, *Veillonella*, *Streptococcus*, and *Pseudomonas*. These bacteria are also common members of the oral microbiome. This suggests that, surprisingly, the primary source of the lung microbiota is likely the mouth rather than inhaled air. *Prevotella*, *Veillonella*, and *Streptococcus* are all anaerobic bacteria. It may seem illogical for species that don't use oxygen to live in the lower respiratory system, the site of gas exchange. However, although these bacteria do not use oxygen, they all share the ability to avoid oxygen's toxic effects by producing antioxidant enzymes, including peroxidase and superoxide dismutase.

Patients with chronic pulmonary diseases such as asthma, cystic fibrosis, or chronic obstructive pulmonary disease (COPD) have a predominance of gram-negative bacteria compared to individuals without chronic lung disease. One study also found that nitric-oxide-metabolizing bacteria (*Nitrosomonas* spp.) were present in the lungs of asthma patients. This is notable in that airway inflammation in asthma patients is often diagnosed by measuring exhaled nitric oxide. Higher-than-normal levels indicate that the airways may be inflamed. It could be that the *Nitrosomonas* bacteria in lungs are responsible for this phenomenon that clinicians use as a diagnostic tool.

Lung microbiome research is new, so many unanswered questions remain. For instance, researchers don't yet know whether core species change with age or environment or whether the microbiota are evenly dispersed throughout the lungs or concentrated in specific areas. Depending on how discoveries proceed, we may one day have lung disease therapies that use probiotics.

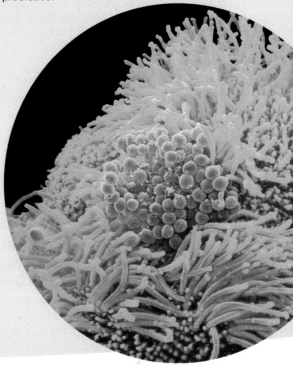

***Staphylococcus* sp. in the trachea.**

---

**CHECK YOUR UNDERSTANDING**

✔ 24-3 Which one of the following is most likely to be associated with a headache: pharyngitis, laryngitis, sinusitis, or epiglottitis?

## Bacterial Diseases of the Upper Respiratory System

**LEARNING OBJECTIVE**

24-4 List the causative agent, symptoms, prevention, preferred treatment, and laboratory identification tests for streptococcal pharyngitis, scarlet fever, diphtheria, cutaneous diphtheria, and otitis media.

Airborne pathogens make their first contact with the body's mucous membranes as they enter the upper respiratory system. Many respiratory or systemic diseases initiate infections here.

### Streptococcal Pharyngitis (Strep Throat)

**Streptococcal pharyngitis (strep throat)** is an upper respiratory infection caused by *group A streptococci (GAS)*. This gram-positive bacterial group consists solely of *Streptococcus pyogenes*, the same bacterium responsible for many skin and soft tissue infections, such as impetigo, erysipelas, and acute bacterial endocarditis.

Pharyngitis is characterized by local inflammation and a fever (**Figure 24.3**). Frequently, tonsillitis occurs, and the lymph nodes in the neck become enlarged and tender. Another frequent complication is otitis media (see page 693).

The pathogenicity of GAS is enhanced by their resistance to phagocytosis. They are also able to produce special enzymes, called *streptokinases*, which lyse fibrin clots, and *streptolysins*, which are cytotoxic to tissue cells, red blood cells, and protective leukocytes.

At one time, the diagnosis of pharyngitis was based on culturing bacteria from a throat swab. Results took overnight or longer, but, beginning in the early 1980s, rapid antigen detection tests that were capable of detecting GAS directly on throat swabs became available. A physician can perform a rapid test in the office. These rapid tests have high specificity. However, negative samples should be cultured because of the varying sensitivity of these tests. (Specificity and sensitivity are

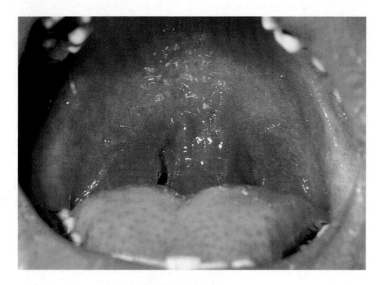

**Figure 24.3 Streptococcal pharyngitis.** Note the inflammation.

**Q** How is strep throat diagnosed?

discussed in Chapter 18, page 507.) Actually, the majority of patients seen for sore throats do not have a streptococcal infection. Some cases are caused by other bacteria, but most are caused by viruses—for which antibiotic therapy is ineffective. GAS should be confirmed and treated in children older than 3 years to prevent development of rheumatic fever. Fortunately, GAS have remained sensitive to penicillin.

Pharyngitis is now most commonly transmitted by respiratory secretions, but epidemics of streptococcal pharyngitis spread by unpasteurized milk were once frequent.

## Scarlet Fever

When the *Streptococcus pyogenes* strain causing streptococcal pharyngitis produces an *erythrogenic* (reddening) *toxin*, the resulting infection is called **scarlet fever**. When the strain produces this toxin, it has been lysogenized by a bacteriophage (see Figure 13.12, page 375). Recall that this means the genetic information of a bacteriophage (bacterial virus) has been incorporated into the chromosome of the bacterium, so the characteristics of the bacterium have been altered. The toxin causes a pinkish red skin rash, which is probably the skin's hypersensitivity reaction to the circulating toxin, and a high fever. The tongue has a spotted, strawberry-like appearance and then, as it loses its upper membrane, becomes very red and enlarged. Classically, scarlet fever has been considered to be associated with streptococcal pharyngitis, but it might accompany a streptococcal skin infection.

It is usually a mild illness, but scarlet fever needs antibiotic treatment to prevent later development of rheumatic fever.

## Diphtheria

Another bacterial infection of the upper respiratory system is **diphtheria**. Until 1935, it was the leading infectious killer of children in the United States. The disease begins with a sore throat and fever, followed by general malaise and swelling of the neck. The organism responsible is *Corynebacterium diphtheriae*, a gram-positive, non–endospore-forming rod. Its morphology is pleomorphic, frequently club-shaped, and it stains unevenly (**Figure 24.4**).

Part of the normal immunization program for children in the United States is the **DTaP vaccine**, which protects against diphtheria, tetanus, and pertussis. In this name, the D represents diphtheria toxoid, an inactivated toxin that causes the body to produce antibodies against the diphtheria toxin.

*C. diphtheriae* has adapted to a generally immunized population, and relatively nonvirulent strains are found in the throats of many symptomless carriers. The bacterium is well suited to droplet transmission and is very resistant to drying.

A tough grayish membrane that forms in the throat in response to the infection is characteristic of diphtheria (from the Greek word for leather). It contains fibrin, dead tissue, and bacterial cells and can totally block the passage of air to the lungs.

Although the bacteria don't invade tissues, those that have been lysogenized by a phage can produce a powerful exotoxin, which then circulates in the bloodstream and interferes with protein synthesis. Historically, diphtheria was the first disease for which a toxic cause was identified. Only 0.01 mg of this highly virulent toxin can be fatal. Thus, if antitoxin therapy is to be effective, it must be administered before the toxin enters the tissue cells. When such organs as the heart and kidneys are affected by the toxin, the disease can rapidly be fatal. In other cases the nerves can be involved, and partial paralysis results.

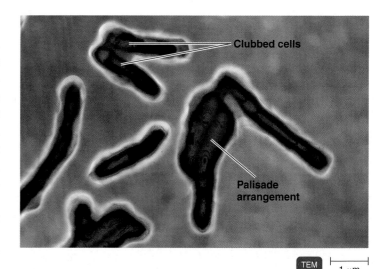

Clubbed cells

Palisade arrangement

TEM ⊢———⊣ 1 μm

**Figure 24.4 *Corynebacterium diphtheriae*, the cause of diphtheria.** This image shows the club-shaped morphology; the dividing cells are often observed folding together to form V- and Y-shaped figures. Also notice the side-by-side palisade arrangement.

**Q** Are corynebacteria gram-positive or gram-negative?

The number of diphtheria cases reported in the United States each year is currently five or fewer. The disease occurs mainly in unvaccinated children and travelers to developing countries. When diphtheria was more common, repeated contacts with toxigenic strains reinforced the immunity, which otherwise weakens with time. Many adults lack immunity because immunity wanes over time. A booster dose of vaccine should be administered every 10 years to maintain protective antibody levels. Some surveys indicate effective immune levels in as few as 20% of the adult population. In the United States, when any trauma in adults requires tetanus toxoid, it is usually combined with diphtheria toxoid (Td vaccine).

Diphtheria is also expressed as **cutaneous diphtheria**. In this form of the disease, *C. diphtheriae* infects the skin, usually at a wound or similar skin lesion, and there is minimal systemic circulation of the toxin. In cutaneous infections, the bacteria cause slow-healing ulcerations that are covered by a gray membrane. Cutaneous diphtheria is fairly common in tropical countries.

In the past, diphtheria was spread mainly to healthy carriers by droplet infection. Respiratory cases have been known to arise from contact with cutaneous diphtheria.

Laboratory diagnosis by bacterial identification is difficult, requiring several selective and differential media. Identification is complicated by the need to differentiate toxin-forming isolates from strains that are not toxigenic; both may be found in the same patient.

Even though antibiotics such as erythromycin and penicillin control the growth of the bacteria, they do not neutralize the diphtheria toxin. Thus antibiotics should be used only in conjunction with antitoxin.

**CHECK YOUR UNDERSTANDING**

✔ 24-4 Among streptococcal pharyngitis, scarlet fever, or diphtheria, which two diseases are usually caused by the same genus of bacteria?

## Otitis Media

One of the more uncomfortable complications of the common cold, or of any infection of the nose or throat, is infection of the middle ear, **otitis media**, or earache. The pathogens cause the formation of pus, which builds up pressure against the eardrum and causes it to become inflamed and painful (**Figure 24.5**). The condition is most frequent in early childhood because the auditory tube connecting the middle ear to the throat is small and more horizontal than in adults and so is more easily blocked by infection (see Figure 24.1).

A number of bacteria can cause otitis media. Until this century, *S. pneumoniae* was the most common cause of otitis media. The conjugate vaccine to prevent *S. pneumoniae* pneumonia has reduced the incidence of otitis media caused by the strains in the vaccine. The bacteria most frequently involved

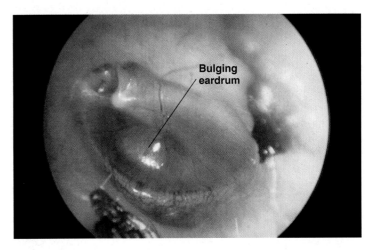

**Figure 24.5 Acute otitis media, with bulging eardrum.**

**Q** What is the most common bacterium causing middle ear infections?

are other *S. pneumoniae*, nonencapsulated *H. influenzae*, *Moraxella catarrhalis* (more-ax-EL-lah ka-tar-RA-lis) and *S. pyogenes*. In about 30% of cases, no bacteria can be detected. Viral infections may be responsible in these instances; respiratory syncytial viruses (see page 708) are the most common isolate.

Otitis media affects 85% of children before the age of 3 and accounts for nearly half of office visits to pediatricians—an estimated 8 million cases each year in the United States. It is estimated that ear infections account for about one-fourth of the prescriptions for antibiotics; however, antibiotics should be prescribed only if the infection is caused by a bacterium. Broad-spectrum penicillins, such as amoxicillin, are usually the first choice for children.

# Viral Disease of the Upper Respiratory System

**LEARNING OBJECTIVE**

24-5 List the causative agents and treatments for the common cold.

Probably the most prevalent disease of humans, at least those living in the temperate zones, is a viral disease affecting the upper respiratory system—the common cold.

## The Common Cold

More than one virus is involved in the etiology of the **common cold**. In fact, there are hundreds—more than 200 different viruses that are members of several different families of viruses are known to cause colds. Identification procedures that require isolation and culturing often fail to identify the cause of a cold. However, new techniques that use PCR to look for the viral DNA or RNA make culturing unnecessary and frequently turn up previously unknown cold viruses. Most cold viruses are **rhinoviruses** (30–50%); **coronaviruses** (10–15%)

Microbial Diseases of the Upper Respiratory System

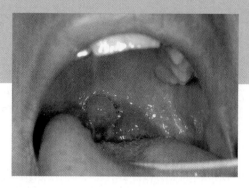

Grayish membrane in the throat is a characteristic of this disease.

The differential diagnosis for the following diseases is usually based on clinical symptoms, and throat swabs may be used to culture bacteria. For example, a patient presents with fever and a red, sore throat. Later a grayish membrane appears in the throat. Gram-positive rods were cultured from the membrane. Use the table below to make a differential diagnosis and identify infections that could cause these symptoms. For the solution, go to @MasteringMicrobiology

| Disease | Pathogen | Symptoms | Treatment |
|---|---|---|---|
| **BACTERIAL DISEASES** | | | |
| Epiglottitis | *Haemophilus influenzae* | Inflammation of the epiglottis | Antibiotics; maintain airway Prevention: Hib vaccine |
| Streptococcal Pharyngitis (strep throat) | Streptococci, especially *Streptococcus pyogenes* | Inflamed mucous membranes of the throat | Penicillin |
| Scarlet Fever | Erythrogenic toxin-producing strains of *Streptococcus pyogenes* | Streptococcal exotoxin causes reddening of skin and tongue and peeling of affected skin | Penicillin |
| Diphtheria | *Corynebacterium diphtheriae* | Grayish membrane forms in throat; cutaneous form also occurs | Erythromycin and antitoxin Prevention: DTaP vaccine |
| Otitis Media | Several agents, especially *Streptococcus pneumoniae*, *Haemophilus influenzae*, and *S. pyogenes* | Accumulations of pus in middle ear cause painful pressure on eardrum | Broad-spectrum antibiotics if bacterial Prevention: pneumococcal vaccine |
| **VIRAL DISEASE** | | | |
| Common Cold | Rhinoviruses, coronaviruses; *Enterovirus* | Familiar symptoms of coughing, sneezing, runny nose | Supportive |

are also important. However, 20–30% of viruses that cause colds are classified by researchers as previously unknown.

We tend to accumulate immunities against cold viruses during our lifetime, which may be a reason why older people usually have fewer colds. Immunity is based on the ratio of IgA antibodies to single serotypes and has a good short-term effectiveness. Isolated populations may develop a group immunity, and colds disappear among them until a new set of viruses is introduced.

The symptoms of the common cold are familiar to all of us. They include sneezing, excessive nasal secretion, and congestion. The infection can easily spread from the throat to the sinuses, the lower respiratory system, and the middle ear, leading to complications of laryngitis and otitis media. The uncomplicated cold usually is not accompanied by fever. It is generally in the interest of a cold-causing virus not to make the cold sufferer too ill—the host needs to move around, shedding the virus to others, especially in mucus.

Rhinoviruses thrive at a temperature slightly below that of normal body temperature, such as might be found in the upper respiratory system, which is open to the outside environment. No one knows exactly why the number of colds seems to increase with colder weather in temperate zones. It is not known whether closer indoor contact promotes epidemic-type transmission or whether physiological changes increase susceptibility.

A single rhinovirus deposited on the nasal mucosa is often sufficient to cause a cold. However, there is surprisingly little agreement on how the cold virus is transmitted to a site in the nose. Experiments with guinea pigs and the influenza virus show that viruses tend to be carried on airborne droplets of water vapor. In the dry air (low humidity) typical of low temperatures, the droplets are smaller and remain airborne longer, facilitating person-to-person transmission. At the same time, the cooler air causes the cilia of the ciliary escalator to work more slowly, allowing inhaled viruses to spread in the upper respiratory system.

Research has shown that during the first 3 days of a cold, nasal mucus contains a high concentration of cold viruses that multiply in nasal cells. (If mucus is green, the reason is that it contains many white blood cells with iron-containing components directed at destroying pathogens.) The viruses in mucus remain viable on surfaces touched by contaminated fingers for at least several hours. The conventional wisdom is that the virus is most likely transmitted through finger contact with the nostrils and eyes (tear ducts communicate with the nose). Transmission also occurs when the cold viruses in airborne droplets from coughing and sneezing land on suitable tissues in the nose and eyes.

*Enterovirus* D68 (EV-D68) causes coldlike symptoms. In 2014, the United States experienced a nationwide outbreak of EV-D68 associated with severe respiratory illness. Small numbers of EV-D68 have been reported regularly since 1987. Children and teenagers are most likely to get infected with EV-D68 and become ill because they do not yet have immunity from previous exposures to these viruses. Some children may have difficulty breathing, although most people recover within a few days.

Because colds are caused by viruses, antibiotics are of no use in treatment. Symptoms can be relieved by cough suppressants and antihistamines, but these medications do not speed recovery. There is still considerable truth in the medical adage that an untreated cold will run its normal course to recovery in a week, whereas with treatment it will take 7 days.

The diseases affecting the upper respiratory system are summarized in Diseases in Focus 24.1.

**CHECK YOUR UNDERSTANDING**

✔ 24-5 Which viruses, rhinoviruses or coronaviruses, cause about half of the cases of the common cold?

# Microbial Diseases of the Lower Respiratory System

Many of the same bacteria and viruses that infect the upper respiratory system can also infect the lower respiratory system. As the bronchi become involved, **bronchitis** or **bronchiolitis** develops (see Figure 24.2). A severe complication of bronchitis is **pneumonia**, in which the pulmonary alveoli become involved.

## Bacterial Diseases of the Lower Respiratory System

### LEARNING OBJECTIVES

24-6 List the causative agent, symptoms, prevention, preferred treatment, and laboratory identification tests for pertussis and tuberculosis.

24-7 Compare and contrast the seven bacterial pneumonias discussed in this chapter.

24-8 List the etiology, method of transmission, and symptoms of melioidosis.

Bacterial diseases of the lower respiratory system include tuberculosis and the many types of pneumonia caused by bacteria. Lesser-known diseases such as psittacosis and Q fever also fall into this category.

### Pertussis (Whooping Cough)

Infection by the bacterium *Bordetella pertussis* results in **pertussis**, or **whooping cough**. *B. pertussis* is a small, obligately aerobic, gram-negative coccobacillus. The virulent strains possess a capsule. The bacteria attach specifically to ciliated cells in the trachea, first impeding their ciliary action and then progressively destroying the cells (**Figure 24.6**). This prevents the ciliary escalator system from moving mucus. *B. pertussis* produces several toxins. *Tracheal cytotoxin*, a fixed cell wall fraction of the

bacterium, is responsible for damage to the ciliated cells, and *pertussis toxin* enters the bloodstream and is associated with systemic symptoms of the disease.

Primarily a childhood disease, pertussis can be quite severe. The initial stage, called the *catarrhal stage*, resembles a common cold. Prolonged sieges of coughing characterize the *paroxysmal stage*, or second stage. (The name *pertussis* is derived from the Latin *per*, meaning thoroughly, and *tussis*, meaning cough.) When ciliary action is compromised, mucus accumulates, and the infected person desperately attempts to cough up these mucus accumulations. The violence of the coughing

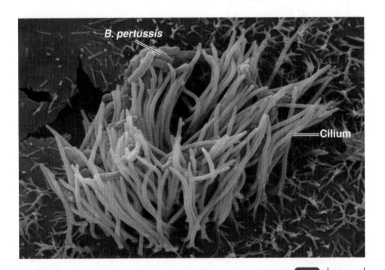

**Figure 24.6 Ciliated cells of the respiratory system infected with *Bordetella pertussis*.** Cells of *B. pertussis* (orange) can be seen growing on the cilia; they will eventually cause the loss of the ciliated cells.

**Q** What is the name of the toxin produced by *Bordetella pertussis* that causes the loss of cilia?

*Biological agents were first tapped by armies, and now by terrorists. Today, technology and ease of travel increase the potential damage.*

## History of Bioweapons

Biological weapons (bioweapons)—pathogens intentionally used for hostile purposes—are not new. The "ideal" bioweapon is one that disseminates by aerosol, spreads efficiently from human to human, causes debilitating disease, and has no readily available treatment.

The earliest recorded use of a bioweapon occurred in 1346 during the Siege of Kaffa, in what is now known as Feodosia, Ukraine. There the Tartar army catapulted their own dead soldiers' plague-ridden bodies over city walls to infect opposing troops. Survivors from that attack went on to introduce the "Black Death" to the rest of Europe, sparking the plague pandemic of 1348–1350.

In the eighteenth century, blankets contaminated with smallpox were intentionally introduced into Native American populations by the British during the French and Indian War. And during the Sino-Japanese War (1937–1945), Japanese planes dropped canisters of fleas carrying *Yersinia pestis* bacteria, the causative agent of plague, on China. In 1975, *Bacillus anthracis* endospores were accidentally released from a bioweapon production facility in Sverdlovsk.

A citadel in Ukraine, location of the first known biowarfare attack in history.

## Biological Weapons Banned in the Twentieth Century

The Geneva Conventions are internationally agreed upon standards for conducting war. Written in the 1920s, they prohibited deploying bioweapons—but did not specify that possessing or creating them was illegal. As such, most powerful nations in the twentieth century continued to create bioweapons, and the growing stockpiles posed an ever-growing threat. In 1975, the Biological Weapons Convention banned both possession and development of biological weapons. The majority of the world's nations ratified the treaty, which stipulated that any existing bioweapons be destroyed and related research halted.

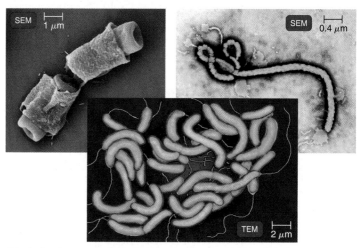

*(Clockwise from top left): Bacillus anthracis, Ebolavirus,* and *Vibrio cholerae* are just a few microbes identified as potential bioterrorism agents.

## Emergence of Bioterrorism

Unfortunately, the history of biowarfare doesn't end with the ratification of the Biological Weapons Convention. Since then, the main actors engaging in biowarfare have not been nations but rather radical groups and individuals. One of the most publicized bioterrorism incidents occurred in 2001, when five people died from, and many more were infected with, anthrax that an army researcher sent through the mail in letters.

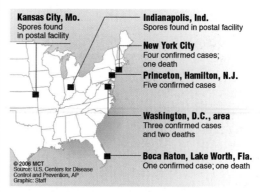

Map showing location of 2001 bioterrorism anthrax attacks.

| Selected Diseases Identified as Potential Bioweapons | |
|---|---|
| **Bacterial** | **Viral** |
| Anthrax (*Bacillus anthracis*) | Nonbacterial meningitis (Arenaviruses) |
| Psittacosis (*Chlamydophila psittaci*) | *Hantavirus* disease |
| Botulism (*Clostridium botulinum* toxin) | Hemorrhagic fevers (Ebola, Marburg, Lassa) |
| Tularemia (*Francisella tularensis*) | Monkeypox |
| Cholera (*Vibrio cholerae*) | Nipah virus infection |
| Plague (*Yersinia pestis*) | Smallpox |

# Public Health Authorities Try to Meet the Threat of Bioterrorism

One of the problems with bioweapons is that they contain living organisms, so their impact is difficult to control or even predict. However, public health authorities have created some protocols to deal with potential bioterrorism incidents.

**CAUTION**

**BIOHAZARD**

Biological hazard symbol.

## New Technologies and Techniques to Identify Bioweapons

Monitoring public health, and reporting incidence of diseases of note, is the first step in any bioterrorism defense plan. The faster a potential incident is uncovered, the greater the chance for containment. Rapid tests are being investigated to detect genetic changes in hosts due to bioweapons even before symptoms develop. Early-warning systems, such as DNA chips or recombinant cells that fluoresce in the presence of a bioweapon, are also being developed.

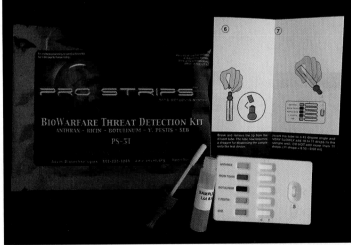

Pro Strips Rapid Screening System, developed by ADVNT Biotechnologies LLC, is the first advanced multi-agent biowarfare detection kit that tests for anthrax, ricin toxin, botulinum toxin, plague, and SEB (staphylococcal enterotoxin B).

## Vaccination: A Key Defense

When the use of biological agents is considered a possibility, military personnel and first -responders (health care personnel and others) are vaccinated—if a vaccine for the suspected agent exists. New vaccines are being developed, and existing vaccines are being stockpiled for use where needed.

The current plan to protect civilians in the event of an attack with a microbe is illustrated by the smallpox preparedness plan. This killer disease has been eradicated from the population, but unfortunately, a cache of the virus remains preserved in research facilities, meaning that it might one day be weaponized. It's not practical to vaccinate all people against the disease. Instead, the U.S. government's strategy following a confirmed smallpox outbreak includes "ring containment and voluntary vaccination." A "ring" of vaccinated/protected individuals is built around the bioterrorism infection case and their contacts to prevent further transmission.

Examining mail for *B. anthracis*.

### KEY CONCEPTS

- Vaccination is critical to preventing spread of infectious diseases, especially those that can be weaponized. **(See Chapter 18, "Principles and Effects of Vaccinations," pages 500–501.)**

- Many organisms that could be used for weapons require BSL-3 facilities. **(See Chapter 6, "Special Culture Techniques," pages 161–162.)**

- Tracking pathogen genomics provides information on its source. **(See Chapter 9, "Forensic Microbiology," pages 258–260.)**

in small children can actually result in broken ribs. Gasping for air between coughs causes a whooping sound, hence the informal name of the disease. Coughing episodes occur several times a day for 1 to 6 weeks. The *convalescence stage*, the third stage, may last for months. Because infants are less capable of coping with the effort of coughing to maintain an airway, irreversible damage to the brain occasionally occurs.

Diagnosis of pertussis is primarily based on clinical signs and symptoms. The pathogen can be cultured from a throat swab inserted through the nose on a thin wire and held in the throat while the patient coughs. Culture of the fastidious pathogen requires care. As alternatives to culture, PCR methods can also be used to test the swabs for presence of the pathogen, a procedure that is required to diagnose the disease in infants.

Treatment of pertussis with antibiotics, most commonly erythromycin or other macrolides, is not effective after onset of the paroxysmal coughing stage but may reduce transmission.

> **CHECK YOUR UNDERSTANDING**
>
> ✔ 24-6 Another name for pertussis is whooping cough. This symptom is caused by the pathogens' attack on which cells?

## Tuberculosis

In Europe during the seventeenth through the nineteenth centuries, **tuberculosis (TB)** was responsible for an estimated 20–30% of all deaths. This probably exerted a strong selection pressure for genes that protected against TB in this population. However, in recent decades co-infection with HIV has been a prominent cause of increasing susceptibility to infection and also of rapid progression from infection to active disease. Other factors are the increasing populations of susceptible individuals in prisons and other crowded facilities, as well as the elderly or undernourished.

Tuberculosis is an infectious disease caused by the bacterium *Mycobacterium tuberculosis*, a slender rod and an obligate aerobe. The rods grow slowly (20-hour or longer generation time), sometimes form filaments, and tend to grow in clumps (**Figure 24.7**). On the surface of liquid media, their growth appears moldlike, which suggested the genus name *Mycobacterium* (*myco* means fungus).

Another mycobacterial species, *Mycobacterium bovis* (BŌ-vis), is a pathogen mainly of cattle. *M. bovis* is the cause of **bovine tuberculosis**, which is transmitted to humans via contaminated milk or food. Bovine tuberculosis accounts for fewer than 1% of TB cases in the United States. It seldom spreads from human to human, but before the days of pasteurized milk and the development of control methods such as tuberculin testing of cattle herds, this disease was a frequent form of tuberculosis in humans. *M. bovis* infections cause TB that primarily affects the bones or lymphatic system. At one time, a common manifestation of this type of TB was hunchbacked deformation of the spine.

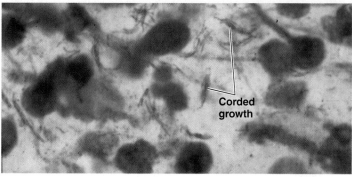

LM ⊢—⊣ 2.5 μm

**Figure 24.7 *Mycobacterium tuberculosis*.** The filamentous, red-stained funguslike growth shown here in a smear from lung tissue is responsible for the organism's name. Under other conditions, it grows as slender, individual bacilli. A waxy component of the cell, cord factor, is responsible for this ropelike arrangement. An injection of cord factor causes pathogenic effects exactly like those caused by tubercle bacilli.

**Q** **What characteristic of this bacterium suggests use of the prefix *myco-*?**

Other mycobacterial diseases also affect people in the late stages of HIV infection. A majority of the isolates are of a related group of organisms known as the *M. avium-intracellulare* (ā-vē-um in-trah-SEI-ū-lar) complex. In the general population, infections by these pathogens are uncommon.

Mycobacteria stained with carbolfuchsin dye cannot be decolorized with acid-alcohol and are therefore classified as *acid-fast* (see page 66). This characteristic reflects the unusual composition of the cell wall, which contains large amounts of lipids. These lipids might also be responsible for the resistance of mycobacteria to environmental stresses, such as drying. In fact, these bacteria can survive for weeks in dried sputum and are very resistant to chemical antimicrobials used as antiseptics and disinfectants (see Table 7.7, page 198).

Tuberculosis is a particularly good illustration of the ecological balance between host and parasite in infectious disease. A host is not usually aware of tuberculosis pathogens that invade the body and are defeated, which occurs 90% of the time. If immune defenses fail, however, the host becomes very much aware of the resulting disease.

A tragic demonstration of individual variation in resistance was the Lübeck disaster in Germany in 1926. By error, 249 babies were inoculated with virulent tuberculosis bacteria instead of the attenuated vaccine strain. Even though all received the same inoculum, there were only 76 deaths, and the remainder did not become seriously ill.

Tuberculosis is most commonly acquired by inhaling the bacillus. Only very fine particles containing one to three bacilli reach the lungs, where they are usually phagocytized by a macrophage in the alveoli (see Figure 24.2). The macrophages of a healthy individual become activated by the presence of the bacilli and usually destroy them. About three-fourths of TB cases affect the lungs, but other organs can also become infected.

## Pathogenesis of Tuberculosis

The pathogenesis of TB is shown in **Figure 24.8**. An important factor in the pathogenicity of the mycobacteria probably is that the mycolic acids of the cell wall strongly stimulate an inflammatory response in the host. The figure depicts the situation in which the body's defenses fail and the disease progresses to a fatal conclusion. However, most healthy people will defeat a potential infection with activated macrophages, especially if the infecting dose is low.

**①–②** If the infection progresses, the host isolates the pathogens in a walled-off lesion called a *tubercle* (meaning lump or knob), a characteristic that gives the disease its name.

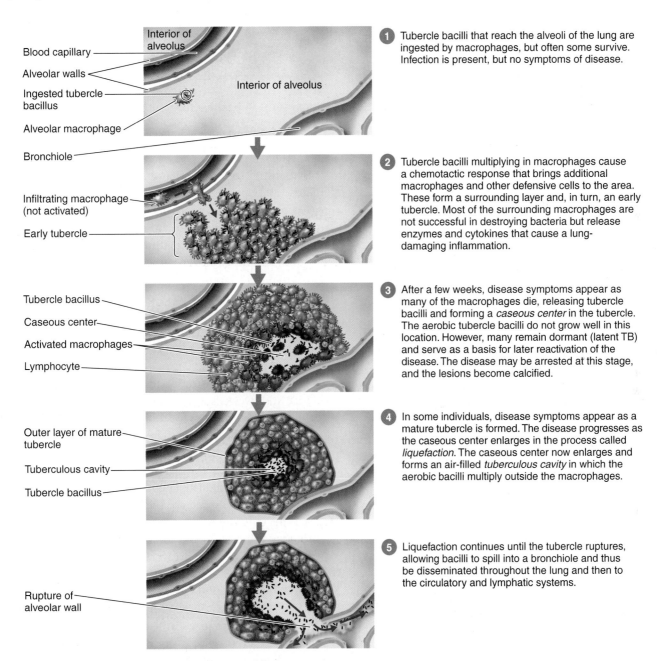

❶ Tubercle bacilli that reach the alveoli of the lung are ingested by macrophages, but often some survive. Infection is present, but no symptoms of disease.

❷ Tubercle bacilli multiplying in macrophages cause a chemotactic response that brings additional macrophages and other defensive cells to the area. These form a surrounding layer and, in turn, an early tubercle. Most of the surrounding macrophages are not successful in destroying bacteria but release enzymes and cytokines that cause a lung-damaging inflammation.

❸ After a few weeks, disease symptoms appear as many of the macrophages die, releasing tubercle bacilli and forming a *caseous center* in the tubercle. The aerobic tubercle bacilli do not grow well in this location. However, many remain dormant (latent TB) and serve as a basis for later reactivation of the disease. The disease may be arrested at this stage, and the lesions become calcified.

❹ In some individuals, disease symptoms appear as a mature tubercle is formed. The disease progresses as the caseous center enlarges in the process called *liquefaction*. The caseous center now enlarges and forms an air-filled *tuberculous cavity* in which the aerobic bacilli multiply outside the macrophages.

❺ Liquefaction continues until the tubercle ruptures, allowing bacilli to spill into a bronchiole and thus be disseminated throughout the lung and then to the circulatory and lymphatic systems.

**Figure 24.8 The pathogenesis of tuberculosis.** This figure represents the progression of the disease when the defenses of the body fail. In most otherwise healthy individuals, the infection is arrested, and fatal tuberculosis does not develop.

**Q** Almost a third of Earth's population is infected with *Mycobacterium tuberculosis*—does a study of this figure show why this is not the same as a third of Earth's population *having* tuberculosis?

❸ When the disease is arrested at this point, the lesions slowly heal, becoming calcified. These show up clearly on X-ray films and are called *Ghon's complexes*. (Computed tomography [CT] is more sensitive than X rays in detecting lesions of TB.) The bacteria may remain viable for years, in which case the disease is called **latent TB**. Such individuals are infected with *M. tuberculosis* but do not have TB. The only sign of TB infection is a positive reaction to the tuberculin skin test or TB blood test. Persons with latent TB infection are not infectious and cannot spread TB infection to others.

❹ Macrophages ingest and surround the tubercle bacilli, forming a barrier outer layer.

❺ If the body's defenses fail at this stage, the tubercle breaks down and releases virulent bacilli into the airways of the lung and then the cardiovascular and lymphatic systems.

Coughing, the more obvious symptom of the lung infection, also spreads the infection by bacterial aerosols. Sputum may become bloodstained as tissues are damaged, and eventually blood vessels may become so eroded that they rupture, resulting in fatal hemorrhaging. The disseminated infection is called *miliary tuberculosis* (the name is derived from the numerous millet seed–sized tubercles formed in the infected tissues). The body's remaining defenses are overwhelmed, and the patient suffers weight loss and a general loss of vigor. At one time, TB was also known as *consumption*.

### Diagnosis of Tuberculosis

People infected with tuberculosis respond with cell-mediated immunity against the bacterium. This form of immune response, rather than humoral immunity, develops because the pathogen is located mostly within macrophages. This immunity, involving sensitized T cells, is the basis for the **tuberculin skin test (Figure 24.9)**, a screening test for infection. A positive test does not necessarily indicate active disease. In this test, a purified protein derivative of the tuberculosis bacterium, derived by precipitation from broth cultures, is injected cutaneously. If the injected person has been infected with TB in the past, sensitized T cells react with these proteins, and a delayed hypersensitivity reaction occurs in about 48 hours. This reaction appears as an induration (hardening) and reddening of the area around the injection site. In this test, known as the *Mantoux test*, dilutions of 0.1 ml of antigen are injected and the reacting area of the skin is measured.

A positive tuberculin test in the very young is a probable indication of an active case of TB. In older individuals, it might indicate only hypersensitivity resulting from a previous infection or vaccination, not a current active case. Nonetheless, it is an indication that further examination is needed, such as a chest X-ray or CT examination to detect lung lesions and attempts to isolate the bacterium.

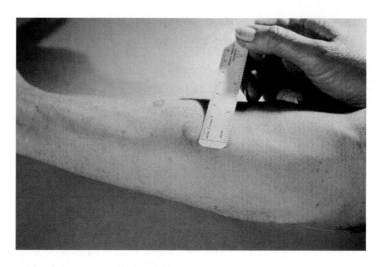

**Figure 24.9  A positive tuberculin skin test on an arm.**

🇶 What does a positive tuberculin skin test indicate?

The initial step in laboratory diagnosis of active cases is a microscopic examination of smears, such as sputum. According to recent medical opinion, the commonly used 125-year-old microscopic exam routinely misses half of all cases. Confirming a diagnosis of TB by isolating the bacterium poses difficulties because the pathogen grows very slowly. A colony might take 3 to 6 weeks to form, and completing a reliable identification series may add another 3 to 6 weeks.

Blood tests measure release of IFN-γ from white blood cells after exposure to mycobacterial antigen in a test tube. They are the preferred tests for a person who has received BCG vaccinations.

Nucleic acid amplification tests (NAATs) can detect *M. tuberculosis* 1 to 2 weeks sooner than cultures and at the same time can determine resistance to a major TB antibiotic, rifampin.

Evidence indicates that, compared to the skin test, these rapid tests have higher specificity and less cross-reactivity with BCG vaccination (see the discussion of TB vaccines, following). They do not distinguish latent from active infection. These assays seem likely to replace the tuberculin skin test for many uses, especially where cross-reactivity with BCG vaccination is a problem. If they could be adopted worldwide at centers for TB treatment, they would help avert millions of TB-related deaths.

### Treatment of Tuberculosis

The first effective antibiotic for TB treatment was streptomycin, which was introduced in 1944. Streptomycin is still in use, and all of the currently used drugs were developed decades ago. Even the *short course* of treatment for TB (there are variations in the regimen, depending on sensitivity of the organism and other factors) requires the patient to adhere to a minimum of 6 months of therapy. Multiple-drug therapy is needed to minimize the emergence of resistant strains. This typically includes four drugs, isoniazid, rifampin, ethambutol, and pyrazinamide, which are considered **first-line drugs**. If the strain of *M. tuberculosis* is susceptible to the drugs, this regimen can lead to a cure. The likelihood that resistance may develop is increased because many patients fail to faithfully follow such a prolonged regimen, which can involve 130 doses of the drugs.

In addition to the first-line drugs, there are a number of **second-line drugs** that can be used, mainly if resistance develops to alternatives. These include several aminoglycosides, fluoroquinolones, streptomycin, and para-aminosalicylic acid (PAS). These drugs are either less effective than first-line drugs, have toxic side effects, or may be unavailable in some countries.

The prolonged treatment is necessary because the tubercle bacillus grows very slowly or is only dormant (the only drug effective against the dormant bacillus is pyrazinamide), and many antibiotics are effective only against growing cells. Also, the bacillus may be hidden for long periods in macrophages or other locations that are difficult to reach with antibiotics.

Not surprisingly, problems have arisen with cases of TB that are caused by **multi-drug-resistant (MDR)** strains. These are defined as being resistant to the two most effective first-line drugs, isoniazid and rifampin. In addition, strains have arisen that are also resistant to the most effective second-line drugs, such as any fluoroquinolone, and to at least one of three injectable second-line drugs, such as the aminoglycosides amikacin or kanamycin, as well as the polypeptide capreomycin. These cases, defined as **extensively drug-resistant (XDR)**, are virtually untreatable and are emerging globally. An additional consideration is that anywhere from 30–90% of persons with TB are also HIV positive—with the accompanying damage to the immune system. In one study, all patients testing positive for both HIV and XDR tuberculosis died within 3 months of diagnosis.

Obviously, there is a pressing need for new, effective drugs to treat TB, especially XDR cases. In 2012, bedaquiline was approved to treat MDR TB.

### Testing for Drug Susceptibility

Solid-media culture-based methods for drug susceptibility testing can take as long as 4 to 8 weeks for finalized results. However, *M. tuberculosis* grows faster in liquid media. These assays are simultaneously useful for both diagnosis and determination of drug susceptibility. The Microscopic-Observation Drug-Susceptibility Assay (MODS) is based on direct observation of the typical cording growth (see Figure 24.7) of *M. tuberculosis* in liquid cultures, requires only 6 to 8 days, and is relatively inexpensive. The determination of susceptibility for rifampin can be considered a marker for potential resistance to other drugs. Recall that NAATs also rapidly test for rifampin resistance. NAATs are needed to determine resistance to the other first-line and second-line anti-TB drugs. The biggest problem is that most drug resistance, aside from resistance to

rifampin, is not due to a single gene. One new test, MTBDRsl, does identify resistance to fluoroquinolones.

### Tuberculosis Vaccines

The **BCG vaccine** is a live culture of *M. bovis* that has been made avirulent by long cultivation on artificial media. (BCG stands for bacillus of Calmette and Guérin, the French scientists who originally isolated the strain.) The BCG vaccine has been available since the 1920s and is one of the most widely used vaccines in the world. In 1990, it was estimated that 70% of the world's schoolchildren received it. In the United States, however, the vaccine is currently recommended only for certain children at high risk who have negative skin tests. People who have received the vaccine show a positive reaction to tuberculin skin tests. This has always been one argument against its widespread use in the United States. Another argument against the universal administration of BCG vaccine is its very uneven effectiveness. Experience has shown that it is fairly effective when given to young children, but for adolescents and adults it sometimes has an effectiveness approaching zero. Worse, it has been found that HIV-infected children, who need it most, frequently will develop a fatal infection from the BCG vaccine. Recent work indicates that exposure to members of the *M. avium-intracellulare* complex that is often encountered in the environment may interfere with the effectiveness of the BCG vaccine—which might explain why the vaccine is more effective early in life, before much exposure to such environmental mycobacteria. A number of new vaccines are in the experimental pipeline, but they will require large numbers of human samples and several years of follow-up to evaluate.

### Worldwide Incidence of Tuberculosis

Tuberculosis has emerged as a global pandemic (**Figure 24.10a**). Estimates are that more than 10 million people develop active tuberculosis every year and that infections result in nearly 2 million deaths annually. (Worldwide, the incidence of TB per capita is falling at about 1% a year. However, the world population is growing at about 2% a year—therefore, the total number of new TB cases is still rising.) Probably a third of the world's population has latent TB. Also, HIV and tuberculosis are almost inseparable, and tuberculosis is the leading direct cause of death in much of the world affected by HIV.

TB incidence in the United States steadily decreased for decades (see Figure 14.11c, page 415). The incidence has been approximately three cases per 100,000 people for the past few years, with two-thirds of the cases occurring among foreign-born people (Figure 24.10b).

## Bacterial Pneumonias

The term *pneumonia* is applied to many pulmonary infections, most of which are caused by bacteria. Pneumonia caused by *Streptococcus pneumoniae* is the most common, about two-thirds

**(a)** Estimated tuberculosis incidence worldwide in 2016, per 100,000 population

KEY

- ☐ 0–24.9
- ☐ 25–99
- ☐ 100–199
- ☐ 200–299
- ■ ≥300
- ☐ No data
- ☐ Not applicable

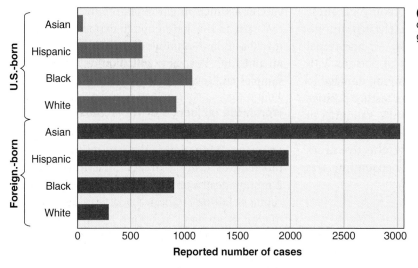

**(b)** Reported tuberculosis cases among American ethnic groups in 2016

**Figure 24.10 Distribution of tuberculosis. (a)** Tuberculosis worldwide. **(b)** Tuberculosis in the United States. Rates among American ethnic groups.
*Source:* World Health Organization (WHO), 2017; *MMWR* 65(11): 273–278, March 24, 2017.

**Q** How can tuberculosis be eliminated?

of cases, and is therefore referred to as *typical pneumonia*. Pneumonias caused by other microorganisms, which can include fungi, protozoa, viruses, and other bacteria, especially mycoplasma, are termed *atypical pneumonias*. This distinction is becoming increasingly blurred in practice.

Pneumonias also are named after the portions of the lower respiratory tract they affect. For example, if the lobes of the lungs are infected, it is called *lobar pneumonia*; pneumonias caused by S. *pneumoniae* are usually of this type. *Bronchopneumonia* indicates that the alveoli of the lungs adjacent to the bronchi are infected. *Pleurisy* is often a complication of various pneumonias, in which the pleural membranes become painfully inflamed. (See Diseases in Focus 24.2.)

## Pneumococcal Pneumonia

Pneumonia caused by *S. pneumoniae* is called **pneumococcal pneumonia**. *S. pneumoniae* is a gram-positive, ovoid bacterium (**Figure 24.11**). This microbe is also a common cause of otitis media, meningitis, and sepsis. The cell pairs are surrounded by a dense capsule that makes the pathogen resistant to phagocytosis. These capsules are also the basis of serological differentiation of pneumococci into at least 90 serotypes. Most human infections are caused by only 23 variants, and these are the basis of current vaccines.

Pneumococcal pneumonia involves both the bronchi and the alveoli (see Figure 24.2). Symptoms include high fever, breathing difficulty, and chest pain. (Atypical pneumonias usually have a slower onset and less fever and chest pain.) The lungs have a reddish appearance because blood vessels are dilated. In response to the infection, alveoli fill with some red blood cells, neutrophils (see Table 16.1, page 454), and fluid from surrounding tissues. The sputum is often rust colored from blood coughed up from the lungs. Pneumococci can invade the bloodstream, the pleural cavity surrounding the lung, and occasionally the meninges. No bacterial toxin has been clearly related to pathogenicity.

A presumptive diagnosis can be made by isolating the pneumococci from the throat, sputum, and other fluids. Pneumococci can be distinguished from other alpha-hemolytic streptococci by observing the inhibition of growth next to a disk of optochin (ethylhydrocupreine hydrochloride) or by performing a bile solubility test. A latex indirect agglutination test (see Figure 18.7, page 512) that detects a capsule antigen of *S. pneumoniae* in the urine can be performed in a physician's office and, with 93% accuracy, can make a diagnosis in 15 minutes.

There are many healthy carriers of the pneumococcus. Virulence of the bacteria seems to be based mainly on the carrier's resistance, which can be lowered by stress. Many illnesses of older adults terminate in pneumococcal pneumonia.

A recurrence of pneumococcal pneumonia is not uncommon, but the serological types are usually different. Before chemotherapy was available, the mortality rate was as high as 25%. This has now been lowered to 5–7%. About 90% of the 900,000 infections that occur annually are in adults.

Antibiotic resistance is an increasing problem. Treatment usually begins with a broad-spectrum cephalosporin until antibiotic-sensitivity testing is done (see Figure 20.17, page 578). Possible drugs include a β-lactam, macrolide, or fluoroquinoline.

A conjugated pneumococcal vaccine has been effective in preventing infection by the 13 serotypes included in it. It has also had an indirect herd effect shown by reduction in other diseases, such as otitis media, attributable to the pneumococcus.

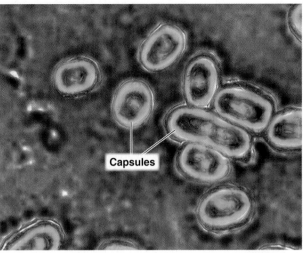

**Figure 24.11** *Streptococcus pneumoniae,* **the cause of pneumococcal pneumonia.** Some of the cocci in the photo are undergoing division and appear as extended ovals. The prominent capsule appears as a bright outline.

**Q** What component of the cell is the primary antigen?

The pneumococcal polysaccharide vaccine, recommended for people with asthma and for older adults, protects against 23 strains of the bacteria.

### *Haemophilus influenzae* Pneumonia

*Haemophilus influenzae* is a gram-negative coccobacillus, and a Gram stain of sputum will differentiate this type of pneumonia from pneumococcal pneumonia. Children under 5 and adults over 65 are most at risk for infection. The Hib vaccine has reduced the incidence in children by 99%. Diagnostic identification of the pathogen uses special media that determine requirements for X and V factors (see page 305). Third-generation cephalosporins are resistant to the β-lactamases produced by many *H. influenzae* strains and are therefore usually the drugs of choice.

### Mycoplasmal Pneumonia

The mycoplasmas, which do not have cell walls, do not grow under the conditions normally used to recover most bacterial pathogens. Because of this characteristic, pneumonias caused by mycoplasmas are often confused with viral pneumonias.

The bacterium *Mycoplasma pneumoniae* is the causative agent of **mycoplasmal pneumonia**. This type of pneumonia was first discovered when such atypical infections responded to tetracyclines, indicating that the pathogen was nonviral. Mycoplasmal pneumonia is a common type of pneumonia in young adults and children. It may account for as much as 20% of pneumonias, although it is not a reportable disease. The symptoms, which persist for 3 weeks or longer, are low-grade fever, cough,

# Common Bacterial Pneumonias

Pneumonia is a leading cause of illness and death among children worldwide and the seventh leading case of death in the United States. Pneumonia can be caused by a variety of viruses, bacteria, and fungi. To prove that a bacterium is causing the pneumonia, the bacterium is isolated from cultures of blood or, in some cases, lung aspirates.

A 27-year-old man with a history of asthma is hospitalized with a 4-day history of progressive cough and 2 days of spiking fevers. Gram-positive cocci in pairs are cultured from a blood sample. Use the table below to identify infections that could cause these symptoms. For the solution, go to @MasteringMicrobiology.

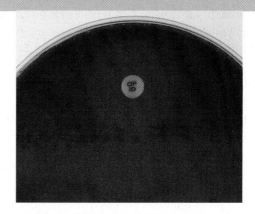

An optochin-inhibition test of the cultured bacteria on blood agar.

| Disease | Pathogen | Symptoms | Reservoir | Diagnosis | Treatment |
| --- | --- | --- | --- | --- | --- |
| Pneumococcal Pneumonia | Streptococcus pneumoniae | Infected alveoli of lung fill with fluids; interferes with oxygen uptake | Humans | Positive optochin inhibition test or bile solubility test; presence of capsular antigen | Macrolides Prevention: pneumococcal vaccine |
| Haemophilus influenzae Pneumonia | Haemophilus influenzae | Symptoms resemble pneumococcal pneumonia | Humans | Isolation; special media for nutritional requirements | Cephalosporins Prevention: Hib vaccine |
| Mycoplasmal Pneumonia | Mycoplasma pneumoniae | Mild but persistent respiratory symptoms; low fever, cough, headache | Humans | Isolation of bacteria | Tetracyclines |
| Legionellosis | Legionella pneumophila | Potentially fatal pneumonia | Water | Culture on selective media | Azithromycin |
| Psittacosis (Ornithosis) | Chlamydophila psittaci | Symptoms, if any, are fever, headache, chills | Birds | Bacterial culture or PCR | Tetracyclines |
| Chlamydial Pneumonia | Chlamydophila pneumoniae | Mild respiratory illness; resembles mycoplasmal pneumonia | Humans | PCR | Azithromycin |
| Q Fever | Coxiella burnetii | Mild respiratory disease lasting 1–2 weeks; occasional complications such as endocarditis occur | Large mammals; can be transmitted via unpasteurized milk | Increasing antibody titer | Doxycycline and chloroquine |

and headache. Occasionally, they are severe enough to lead to hospitalization. Other terms for the disease are *primary atypical* (that is, the most common pneumonia not caused by the pneumococcus) and *walking pneumonia*.

When isolates from throat swabs and sputum grow on a medium containing horse serum and yeast extract, some form distinctive colonies with a "fried-egg" appearance (**Figure 24.12**). The colonies are so small that they must be observed with magnification. The mycoplasmas are highly varied in appearance because they lack cell walls (see Figure 11.24, page 316).

Diagnosis based on recovering the pathogens might not be useful in treatment because as long as 3 or more weeks may be required for the slow-growing organisms to develop. Rapid PCR tests are becoming available; however, they are expensive and have not been clinically validated.

Treatment with antibiotics such as tetracycline usually hastens the disappearance of symptoms but does not eliminate the bacteria, which the patient continues to carry for several weeks.

## Legionellosis

**Legionellosis**, or **Legionnaires' disease**, first received public attention in 1976, when a series of deaths occurred among members of the American Legion who had attended a meeting in Philadelphia. Because no obvious bacterial cause could be found, the deaths were attributed to viral pneumonia.

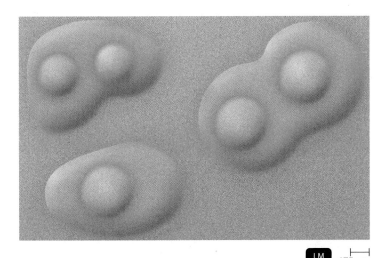

LM | 175 μm

**Figure 24.12   Colonies of *Mycoplasma pneumoniae*, the cause of mycoplasmal pneumonia.**

**Q** **Could you see these colonies without magnification?**

Close investigation, mostly with techniques directed at locating a suspected rickettsial agent, eventually identified a previously unknown bacterium, an aerobic gram-negative rod now known as *Legionella pneumophila*, which is capable of replication within macrophages. Over 44 species of *Legionella* have now been identified; not all of them cause disease.

The disease is characterized by a high fever of 40.5°C, cough, and general symptoms of pneumonia. No person-to-person transmission seems to be involved. Recent studies have shown that the bacterium can be readily isolated from natural waters. In addition, the microbes can grow in the water of air-conditioning cooling towers, perhaps indicating that some epidemics in hotels, urban business districts, and hospitals were caused by airborne transmission. Recent outbreaks have been traced to whirlpool spas, humidifiers, showers, decorative fountains, and even potting soil.

The organism has also been found to inhabit the water lines of many hospitals. Most hospitals keep the temperature of hot water lines relatively low (43–55°C) as a safety measure, and in cooler parts of the system this inadvertently maintains a good growth temperature for *Legionella*. This bacterium is considerably more resistant to chlorine than most other bacteria and can survive for long periods in water with a low level of chlorine. Evidence indicates *Legionella* exist primarily in biofilms that are highly protective. The bacteria are often ingested by waterborne amebae when these are present but continue to proliferate and may even survive within encysted amebae. The most successful method for water disinfection in hospitals with a need to control *Legionella* contamination has been installation of copper-silver ionization systems.

The disease appears to have always been fairly common, if unrecognized. More than 5000 cases are reported each year, but the actual incidence is estimated at over 25,000 annually.

Men over 50 are the most likely to contract legionellosis, especially smokers or the chronically ill. (See the Clinical Focus box on page 708.)

*L. pneumophila* is also responsible for **Pontiac fever**, which is essentially another form of legionellosis. Its symptoms include fever, muscular aches, and usually a cough. The condition is mild and self-limiting. During outbreaks of legionellosis, both forms may occur.

The best diagnostic method is culture on a selective charcoal–yeast extract medium. Serological tests to detect O antigen in urine are available. However, these tests detect only one serogroup. Azithromycin and other macrolide antibiotics are the drugs of choice for treatment.

### Psittacosis (Ornithosis)

The term **psittacosis** is derived from the disease's association with psittacine birds, such as parakeets and other parrots. It was later found that the disease can also be contracted from many other birds, such as pigeons, chickens, ducks, and turkeys. Therefore, the more general term **ornithosis** has come into use.

The causative agent is *Chlamydophila psittaci* (SIT-tah-sē), a gram-negative, obligate intracellular bacterium. The taxonomy of this organism has recently been revised. The genus name has been changed from *Chlamydia* to *Chlamydophila*. This taxonomic change has also been made with *C. pneumoniae* (see the discussion of chlamydial pneumonia that follows). We will continue to use the generic terms *chlamydial* and *chlamydiae*. One way chlamydiae differ from rickettsias, which are also obligate intracellular bacteria, is that chlamydiae form tiny **elementary bodies** as one part of their life cycle (see Figure 11.15, page 310). Unlike most rickettsias, elementary bodies are resistant to environmental stress; therefore, they can be transmitted through air and do not require a bite to transfer the infective agent directly from one host to another.

Psittacosis is a form of pneumonia that usually causes fever, coughing, headache, and chills. Subclinical infections are very common, and stress appears to enhance susceptibility to the disease. Disorientation, or even delirium in some cases, indicates that the nervous system can be involved.

The disease is seldom transmitted from one human to another but is usually spread by contact with the droppings and other exudates of fowl. One of the most common modes of transmission is inhalation of dried particles from droppings. The birds themselves usually have diarrhea, ruffled feathers, respiratory illness, and a generally droopy appearance. Parakeets and other parrots sold commercially are usually (but not always) free of the disease. Many birds carry the pathogen in their spleen without symptoms, becoming ill only when stressed. Pet store employees and people involved in raising turkeys are at greatest risk of contracting the disease.

Psittacosis is diagnosed by isolating the bacterium in embryonated eggs or by cell culture. PCR can be used to identify the *Chlamydia* species. No vaccine is available, but tetracyclines are

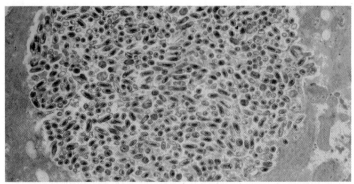

(a) Masses of *Coxiella burnetii* growing in a placental cell  TEM 1.5 μm

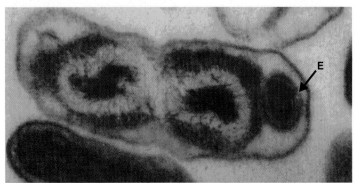

(b) This cell has just divided; notice the endospore-like body (E), which is probably responsible for the relative resistance of the organism.  TEM 0.17 μm

**Figure 24.13** *Coxiella burnetii*, **the cause of Q fever.**

**Q** **By what two methods is Q fever transmitted?**

effective antibiotics in treating humans and animals. Effective immunity does not result from recovery, even when high titers of antibody are present in the serum.

Most years, fewer than 100 cases and very few deaths are reported in the United States. The main danger is late diagnosis. Before antibiotic therapy was available, the mortality rate was about 15–20%. *C. psittasci* is listed by the CDC as a potential bioterrorism weapon (see the **Big Picture**, pages 696–697).

### Chlamydial Pneumonia

Outbreaks of a respiratory illness in populations of college students were found to be caused by a chlamydial organism. Originally the pathogen was considered a strain of *C. psittaci*, but it has been assigned the species name *Chlamydophila pneumoniae*, and the disease is known as **chlamydial pneumonia**. Clinically, it resembles mycoplasmal pneumonia. (There is also strong evidence of association between *C. pneumoniae* and atherosclerosis, the deposition of fatty deposits that block arteries.)

The disease is apparently transmitted from person to person, probably by the respiratory route. Nearly half the U.S. population has antibodies against the organism, an indication

that this is a common illness. PCR is the preferred diagnostic method because culturing the bacteria is slow and serological tests don't distinguish species. The most effective antibiotic is azithromycin.

### Q Fever

In 2015, six Americans were diagnosed with Q fever after being injected with fetal sheep cells in Germany. This xenotransplant has unsubstantiated claims of improving vitality (anti-aging). Q fever was first described in Australia during the mid-1930s, when a previously unreported flulike pneumonia made an appearance. In the absence of an obvious cause, the affliction was labeled **Q fever** (for *query*), much as one might say "X fever." The causative agent was subsequently identified as the obligately parasitic, intracellular bacterium *Coxiella burnetii* (KOKS-ē-el-lah ber-NE-tē-ē) (**Figure 24.13a**). Currently, it is classified as a member of the gammaproteobacteria. It has the ability to multiply intracellularly. Most intracellular bacteria, such as rickettsia, are not resistant enough to survive airborne transmission, but this microorganism is an exception.

Q fever has a wide range of clinical symptoms, and systematic testing shows that about 60% of cases are not even symptomatic. Cases of *acute Q fever* usually feature symptoms of high fever, headaches, muscle aches, and coughing. A feeling of malaise may persist for months. The heart becomes involved in about 2% of acutely ill patients and is responsible for the rare fatalities. In cases of *chronic Q fever*, the best known manifestation is endocarditis (see page 654). Some 5 to 10 years might elapse between the initial infection and the appearance of endocarditis; and, because these patients show few signs of acute disease, the association with Q fever is often missed. Antibiotic therapy and earlier diagnosis have lowered the mortality rate from chronic Q fever to under 5%.

*C. burnetii* is a parasite of several arthropods, especially cattle ticks, and it is transmitted among animals by tick bites. Infected animals include cattle, goats, and sheep, as well as

**CLINICAL CASE**

When the family's symptoms worsen, Caille makes appointments with Dr. Cantwell, the family physician. Because of the family's respiratory symptoms, Dr. Cantwell orders chest X-ray exams, which confirm lobar pneumonia in Caille, Art, and Steven. While in the doctor's office, the children tell Dr. Cantwell about Bitsy and how much they miss their pet cockatiel. Dr. Cantwell, recognizing that cockatiels are psittacine birds, takes a blood sample for antibody testing, prescribes tetracycline, and asks everyone to return in 1 month for convalescent serum samples.

**Why does Dr. Cantwell want the sera tested?**

689 **706** 707 709 711 715

most domestic mammalian pets. In animals the infection is usually subclinical. Cattle ticks spread the disease among dairy herds, and the microbes are shed in the feces, milk, and urine of infected cattle. Once the disease is established in a herd, it is maintained by aerosol transmission. The disease is spread to humans by ingesting unpasteurized milk and by inhaling aerosols of microbes generated in dairy barns.

Inhaling a single pathogen is enough to cause infection, and many dairy workers have acquired at least subclinical infections. Workers in meat- and hide-processing plants are also at risk. The pasteurization temperature of milk, which was originally aimed at eliminating tuberculosis bacilli, was raised slightly in 1956 to ensure the killing of *C. burnetii*. In 1981, an endospore-like body was discovered, which may account for this heat resistance (Figure 24.13b). This resistant body resembles the elementary body of chlamydiae more than typical bacterial endospores.

Diagnosis is based on an increasing titer of antibodies against *Coxiella*. The pathogen can be identified by isolation and growth in chick embryos in eggs or in cell culture. Laboratory workers testing for *Coxiella*-specific antibodies in a patient's serum can use serological tests.

A disease found worldwide, most cases of Q fever in the United States occur in the western states. The disease is endemic to California, Arizona, Oregon, and Washington. A vaccine for laboratory workers and other high-risk personnel is available. Doxycycline has been recommended for treatment. When growth within macrophages in chronic infections renders *C. burnetii* resistant, the killing activity can be restored by combining doxycycline with chloroquine, an antimalarial. The chloroquine raises the pH of the phagosome, increasing doxycycline's efficiency.

## Melioidosis

In 1911, a new disease was reported among drug addicts in Rangoon, Burma (now Myanmar). The bacterial pathogen, *Burkholderia pseudomallei*, is a gram-negative rod formerly placed in the genus *Pseudomonas*. It closely resembled the bacterium causing glanders, a disease of horses. Therefore, the disease was named **melioidosis** (mel-ē-oi-DŌ-sis) from the Greek *melis* (distemper of asses) and *eidos* (resemblance). It is now recognized as a major infectious disease in tropical regions of the world where the pathogen is widely distributed in moist soils. Over 150,000 cases occur annually.

Clinically, melioidosis is most commonly seen as pneumonia. Mortality arises from dissemination, manifesting itself as septic shock. The mortality rate in southeast Asia is about 50% and in Australia approaches 20%. However, it can also appear as abscesses in various body tissues that resemble necrotizing fasciitis (see Figure 21.8, page 597), as severe sepsis, and even as encephalitis. Transmission is primarily by inhalation, but alternative infective routes are by inoculation through puncture wounds and ingestion. Incubation periods can be very long, and occasional delayed-onset cases still surface in this population.

Dr. Cantwell suspects psittacosis because of the evidence of respiratory disease and recent exposure to a cockatiel. The Nguyens are all feeling better when they return to provide convalescent serum the following month. Indirect FA test results of their sera are shown below.

| Nguyen Family Member | Titer against *Chlamydophila psittaci* | |
|---|---|---|
| | Acute Serum | Convalescent Serum |
| Caille | 0 | 0 |
| Art | 32 | 16 |
| Gabbie | 64 | 32 |
| Steven | 64 | 32 |
| Tre | 128 | 64 |

**What do these data indicate?**

689  706  **707**  709  711  715

There have been outbreaks of melioidosis in Asia in recent years, causing concern that the disease will spread to other countries. A cluster of cases in a remote village in western Australia occurred following an earthquake, and in Taiwan an outbreak was reported in the aftermath of the damaging typhoon of 2010.

Diagnosis is usually made by isolating the pathogen from body fluids. Serological tests in endemic areas are problematic because of widespread exposure to a similar, nonpathogenic bacterium. A rapid PCR test is used by public health laboratories. Treatment by antibiotic is uncertain in effectiveness; the most commonly used is ceftazidime, a β-lactam antibiotic, but months of treatment may be required.

**CHECK YOUR UNDERSTANDING**

✔ **24-7** What group of bacterial pathogens causes what is informally called "walking pneumonia"?

✔ **24-8** The bacterium causing melioidosis in humans also causes a disease of horses known as what?

# Viral Diseases of the Lower Respiratory System

**LEARNING OBJECTIVE**

**24-9** List the causative agent, symptoms, prevention, and preferred treatment for viral pneumonia, RSV, and influenza.

For a virus to reach the lower respiratory system and initiate disease, it must pass numerous host defenses designed to trap and destroy it. Nonetheless, numerous respiratory ailments are caused by viruses. Rapid PCR kits that can identify several respiratory viruses are available.

As you read through this problem, you will see questions that epidemiologists ask themselves and each other as they solve a clinical problem. Try to answer each question as though you were an epidemiologist.

1. Jerry Roberts, a 64-year-old man, sees his primary care physician, complaining of fever, malaise, and a cough. His vaccinations are up-to-date, including DTaP. His condition worsens over several days; he has difficulty breathing, and his temperature rises to 40.4°C. He is hospitalized, and his lungs show signs of mild inflammation with thin, watery secretion. A Gram stain of bacteria isolated from the patient is shown in the photo.
   **What diseases are possible?**

2. The same day, Antonio Viviano, a 57-year-old man, goes to the emergency department because he has shortness of breath, fatigue, and cough. The day before he had fever and chills, with a maximum body temperature of 38.6°C.
   **What additional tests would you do on both patients?**

3. PCR, lab cultures, and serological tests should be run on both patients. Both patients have an antibody titer >1024 against *Legionella pneumophila* serogroup 1. The local health department is contacted because two patients are hospitalized with legionellosis.
   **What do you need to know now?**

4. Both patients should be asked whether they have traveled recently, and if so, where. One week before hospitalization, both men stayed in the same hotel within one day of each other. Seven additional cases of legionellosis were identified at other hospitals. A follow-up questionnaire is given to all nine patients to ascertain travel that preceded the illness, including location, accommodations, dates, and information about exposures to common sources for infection (see the table).
   **What are likely sources of infection?**

5. Epidemic legionellosis usually results from exposure of susceptible individuals to an aerosol generated by an environmental source of water contaminated with *Legionella*.
   **Why is it important to identify the source?**

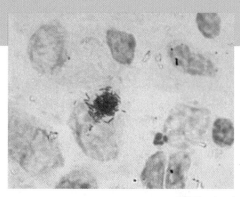

**Gram stain showing bacteria within a tissue sample**.  LM ⊢⊣ 4 μm

6. Retrospective identification of cases allows control and remediation efforts. *L. pneumophila* of the same monoclonal antibody type was recovered from the hot water storage tanks, cooling tower, and showers and faucets in rooms occupied by patients and well guests.
   **Why didn't other hotel guests get sick?**

7. During outbreaks, attack rates tend to be highest in specific high-risk groups, including older adults, smokers, and immunocompromised persons.
   **What are your recommendations for remediation?**

   Shower necks and faucets are disinfected with bleach. The spa filter was cleaned, and the potable water system is hyperchlorinated.

   Hotels have been common locations for legionellosis outbreaks since the disease was first recognized among hotel guests in Philadelphia in 1976.

   *Source:* Adapted from CDC data, 2016.

| Patient's Travel History | |
|---|---|
| Age | ≥50 |
| Gender | 9 male |
| Number of Nights at Hotel | 1–4 (average: 3) |
| Diabetes, coronary artery disease | 6 |
| Smoker | 4 |
| Showered in Hotel | 9 |
| Used Hotel Whirlpool Spa | 1 |
| Used Beach Shower | 4 |

## Viral Pneumonia

**Viral pneumonia** can occur as a complication of influenza, measles, or even chickenpox. A number of Enteroviruses and other viruses have been shown to cause viral pneumonia, but viruses are isolated and identified in fewer than 1% of pneumonia-type infections because few laboratories are equipped to test clinical samples properly for viruses. In those cases of pneumonia for which no cause is determined, viral etiology is often assumed if mycoplasmal pneumonia has been ruled out.

In recent years, coronaviruses have emerged as causative agents of pneumonia. In 2003, **SARS-associated coronavirus (SARS)** emerged in Asia and spread to several countries, killing over 900 people. Since 2004, no known cases of SARS have been reported anywhere in the world. Isolation and quarantine appear to have successfully eliminated that virus, although related viruses are found in bats in the eastern hemisphere. In 2012, **Middle East respiratory syndrome coronavirus (MERS-CoV)** was first reported in Saudi Arabia and then spread to several other countries. PCR is used to confirm SARS and MERS-CoV. In 2014, two travel-related cases were reported in the United States. Additional cases of MERS-CoV occurred in Saudi Arabia in 2017.

## Respiratory Syncytial Virus (RSV)

**Respiratory syncytial virus (RSV)** is probably the most common cause of viral respiratory disease in infants. It can also cause a life-threatening pneumonia in older adults, where it is easily misdiagnosed as influenza. There are about 14,000 deaths from RSV each year in the United States, mostly in older adults. Epidemics occur during the winter and early spring. Virtually all children become infected by age 2—of whom about 1% require hospitalization. We have previously mentioned that RSV is sometimes implicated in cases of otitis media. The name of the virus is derived from its characteristic of causing cell fusion (*syncytium* formation, Figure 15.7b, page 437) when grown in cell culture. The symptoms are coughing and wheezing that last for more than a week. Fever occurs only when there are bacterial complications. Several rapid serological tests are now available that use samples of respiratory secretions to detect both the virus and its antibodies.

Naturally acquired immunity is very poor. The humanized monoclonal antibody, palivizumab, is recommended for prophylaxis in immunocompromised and other high-risk patients. A recombinant vaccine for pregnant women to provide passive immunity for the infant is in clinical trials.

## Influenza (Flu)

The developed countries of the world are probably more aware of **influenza (flu)** than any other disease except the common cold. The flu is characterized by chills, fever, headache, and muscular aches. Recovery normally occurs in a few days, and coldlike symptoms appear as the fever subsides. Still, an estimated 3000 to 50,000 Americans die annually of flu-related complications, even in nonepidemic years. Diarrhea is not a normal symptom of the disease, and the intestinal discomfort attributed to "stomach flu" is probably viral gastroenteritis (see page 745).

### The Influenza Virus

Viruses in the genus *Influenzavirus* consist of eight separate RNA segments of differing lengths enclosed by an inner layer of protein and an outer lipid bilayer (Figure 13.3b, page 365, and **Figure 24.14**). Embedded in the lipid bilayer are numerous projections that characterize the virus. There are two types of projections: *hemagglutinin (HA) spikes* and *neuraminidase (NA) spikes*.

The HA spikes, of which there are about 500 on each virus, allow the virus to recognize and attach to body cells before infecting them. Antibodies against the influenza virus are directed mainly at these spikes. The term *hemagglutinin* refers to the agglutination of red blood cells (hemagglutination) that occurs when the viruses are mixed with them. This reaction is important in serological tests, such as the hemagglutination inhibition test often used to identify the influenza virus and some other viruses.

The NA spikes, of which there are about 100 per virus, differ from the HA spikes in appearance and function. They

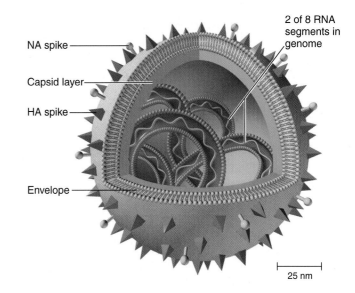

**Figure 24.14 Detailed structure of the influenza virus.** The virus is composed of a protein coat (capsid) that is covered by a lipid bilayer (envelope) and two types of spikes. The genome is composed of eight segments of RNA: six are devoted to internal proteins, and two are devoted to HA and NA spike proteins. Morphologically, under certain environmental conditions, the influenza virus assumes a filamentous form.

**Q** **What is the primary antigenic structure on the influenza virus?**

enzymatically help the virus separate from the infected cell as the virus exits after intracellular reproduction. NA spikes also stimulate the formation of antibodies, but these are less important in the body's resistance to the disease than those produced in response to the HA spikes.

---

**CLINICAL CASE**

The titers confirm Dr. Cantwell's suspicions that the Nguyen family has had psittacosis. The decreasing titers show they are recovering. Fewer than 50 cases of human psittacosis are reported each year. Infections may occur more often than reflected by reported cases for several reasons: (1) *C. psittaci* infections may be only mildly symptomatic; (2) physicians may not elicit a history of bird exposure when evaluating patients because they may not suspect the diagnosis or because patients may not recall transient bird exposure; (3) convalescent-phase serum samples may not be obtained on patients who show clinical improvement on therapy; and (4) prompt initiation of appropriate antibiotic therapy may blunt the antibody response to *C. psittaci*, making convalescent serologies unreliable.

Dr. Cantwell calls the Nguyen's veterinarian to obtain more information about Bitsy's death.

**What does Dr. Cantwell need to know about Bitsy?**

689   706   707   **709**   711   715

There are three serotypes of human influenza viruses. Influenza A and B viruses cause seasonal epidemics of disease almost every winter in the United States. The emergence of new influenza A viruses can cause an influenza pandemic. Influenza C infections generally cause a mild respiratory illness and are not thought to cause epidemics. Influenza A viruses are identified by variation in the HA and NA antigens. The different forms of the antigens are assigned numbers—for example HI, H2, H3, Nl, and N2. There are 16 subtypes of HA and 9 of NA. Each number change represents a substantial alteration in the protein makeup of the spike. These variations are determined by two processes, antigenic drift and antigenic shift. High mutation rates are a characteristic of RNA viruses, which lack the "proofreading" ability of DNA viruses. Accumulations of these mutations, **antigenic drift**, eventually allow the virus to elude much host immunity. The virus might still be designated as H2N2, for example, but viral strains arise that reflect minor antigenic changes. So far, the only truly human-adapted viruses are H1N1, H2N2, and H3N2. In an evolutionary sense, from the point of view of the virus it is desirable to accumulate mutations that favor transmission with minimal pathogenicity. (If the virus quickly kills or makes the host bedridden, it is less likely to be transmitted.)

**Antigenic shifts** mark changes great enough to evade most of the immunity developed in the human population. This is responsible for the outbreaks, including the pandemics of 1918, 1957, and 1968. Antigenic shifts involve a major genetic recombination, called *reassortment*, involving the eight segments of viral RNA (see Figure 24.14). To visualize reassortment, think of the symbols on the wheels of a slot machine.

The virus occurs in avian and mammalian strains; humans generally are not infected by avian strains. However, swine and many wild birds can be infected with both avian and mammalian strains of the influenza virus. Swine are, therefore, good "mixing vessels" in which reassortment occurs. (See the Clinical Focus box in Chapter 13 on page 367).

### Epidemiology of Influenza

Almost every year, epidemics of the flu spread rapidly through large populations, although not always as a worldwide pandemic. The mortality rate from the disease is usually not high, less than 1%, and these deaths are mainly among the very young and the very old. However, so many people are infected in a major epidemic that the total number of deaths is often large.

The most recent pandemic, in 2009, involved an H1N1 virus. This strain is always of special interest because the lethal 1918 pandemic (see the following discussion) was caused by an H1N1 virus. This strain had apparently been circulating indefinitely in pigs in Mexico and Central America and had not been detected because there was little surveillance in that area. Mutations of the influenza virus are more likely to occur in humans, who have a long life span. The virus must continue to mutate in order to evade accumulating immunological resistance. Swine and poultry, in contrast, have short life spans, especially if farm-reared, and the viruses infecting them are less likely to accumulate mutations. An H1N1 influenza virus, which has little pressure to mutate in farm-reared swine, tends to remain little changed in successive generations of the animal.

### Influenza Vaccines

Thus far, it has not been possible to make a vaccine for influenza that gives long-term immunity to the general population. Although it isn't difficult to make a vaccine for a particular antigenic strain of virus, each new strain of circulating virus must be identified in time, usually about February, for the useful development and distribution of a new vaccine later that year. Strains of the influenza virus are collected in about 100 centers worldwide, then analyzed in central laboratories. This information is then used to decide on the composition of the vaccines to be offered for the next flu season. The vaccines are usually *multivalent*—directed at the three or four most important strains in circulation at the time.

A major problem is that production methods for vaccines require growing the virus in egg embryos (see Figure 13.7 on page 371). This process is labor intensive and also requires lead times of 6 to 9 months.

The use of eggs to create vaccines can be avoided by *cell-culture techniques*, in which the virus is grown in vats of cells. A recombinant influenza vaccine has been produced from HA protein expressed by a baculovirus (an insect pathogen) and grown in insect cells. Cell-based vaccines can be produced faster because the cells can be kept frozen. Additionally, such vaccines are not a problem for people with allergies to eggs.

The ultimate goal is a flu vaccine that will protect against all flu strains. An example would be to use a *conserved protein* as a target antigen. Such proteins are identical, or nearly so, in all flu viruses and are essential to the virus. Such a target might even not be in the virus itself, but might appear, for example, in the membrane of the infected cells. Therefore, targeting a conserved protein would lead to the destruction of the infected cells as well as the virus itself. Another example is in the hemagluttinin stalk. The globular head consists of proteins that change rapidly, whereas the stalk proteins, which are also necessary for infection, are conserved. However, the conserved proteins are not strongly antigenic, but molecules could be attached to them to provoke a stronger response.

## The 1918–1919 Pandemic

In any discussion of influenza, the great pandemic of 1918–1919 must be mentioned.* Worldwide, 20 to 50 million people died, including an estimated 675,000 in the United States. No one is sure why it was so unusually lethal. Today, the very young and very old are the principal victims, but in 1918–1919, young adults had the highest mortality rate, often dying within a few hours, probably from a "cytokine storm." The infection is usually restricted to the upper respiratory system, but some change in virulence allowed the virus to invade the lungs and cause lethal hemorrhaging.

Evidence also suggests that the virus was able to infect cells in many organs of the body. In 2005, analysis of material preserved from the lungs of U.S. soldiers killed by the flu and from the exhumed body of a victim buried in permanently frozen soil in Alaska led to the complete genetic sequencing of the 1918 virus. The process of reverse genetics was then used to recreate the virus and grow it in chicken embryos and mice.

Bacterial complications also frequently accompanied the infection and, in those preantibiotic days, were often fatal. The 1918 viral strain apparently became endemic in the U.S. swine population and may have originated there. (See the Clinical Focus box in Chapter 13 on page 367.) Occasionally, influenza is still spread to humans from this reservoir, but the disease has not propagated as the virulent disease of 1918 did.

## Diagnosis of Influenza

Influenza is difficult to diagnose reliably from clinical symptoms, which numerous respiratory diseases share. However, there are now several commercially available techniques that can diagnose influenza A and B within 20 minutes from a sample taken in a physician's office (from nasal washes or nasal swabs). These rapid tests have varying sensitivity and are most useful during influenza season. PCR is used to document strains that are circulating.

## Treatment of Influenza

The antiviral drugs zanamivir and oseltamivir significantly reduce the symptoms of influenza A if administered promptly. They are inhibitors of neuraminidase. If taken within 24 hours of onset of influenza, these drugs slow replication. This action allows the immune system to be more effective, which shortens

*There will always be uncertainty concerning the origin of this most famous pandemic. The best reliable reports place the first well-documented cases among U.S. Army recruits at Camp Funston, Kansas, in March of 1918. The initial wave of influenza was a relatively mild illness that spread rapidly among the crowded troops and reached France as they were dispatched overseas. There the virus underwent a lethal mutation, seriously incapacitating troops on both sides of the front. Military censorship concealed this, and the first newspaper descriptions were published when the outbreak reached the population of neutral Spain, hence the name assigned to the pandemic: the **Spanish flu**. This second wave of influenza, with its high mortality, soon spread throughout the world and reentered the United States in the autumn and winter of 1918.

**CLINICAL CASE**

Dr. Cantwell asks the Nguyens' veterinarian what, if any, symptoms Bitsy had before the decision was made to euthanize her and whether any tests were run on the bird after she was euthanized. The veterinarian consults his notes and tells Dr. Cantwell that chlamydial antigen was detected by ELISA from both cloacal (intestinal) and throat swabs of the euthanized cockatiel, but cultures for *C. psittaci* were not obtained.

**Based on these results, what is the most likely mode of transmission, and how can transmission be prevented?**

689 706 707 709 **711** 715

durations of symptoms and lowers the mortality rate. The bacterial complications of influenza are amenable to treatment with antibiotics.

**CHECK YOUR UNDERSTANDING**

✔ 24-9 Is reassortment of the RNA segments of the influenza virus the cause of antigenic shift or antigenic drift?

# Fungal Diseases of the Lower Respiratory System

**LEARNING OBJECTIVE**

24-10 List the causative agent, mode of transmission, preferred treatment, and laboratory identification tests for four fungal diseases of the respiratory system.

Fungi often produce spores that are disseminated through the air. It is therefore not surprising that several serious fungal diseases affect the lower respiratory system. The rate of fungal infections has been increasing in recent years. Opportunistic fungi are able to grow in immunosuppressed patients, and AIDS, transplant drugs, and anticancer drugs have created more immunosuppressed people than ever before.

## Histoplasmosis

**Histoplasmosis** superficially resembles tuberculosis. In fact, it was first recognized as a widespread disease in the United States when X-ray surveys showed lung lesions in many people who were tuberculin-test-negative. Although the lungs are most likely to be initially infected, the pathogens may spread in the blood and lymph, causing lesions in almost all organs of the body.

Symptoms are usually poorly defined and mostly subclinical, and the disease passes for a minor respiratory infection. In a

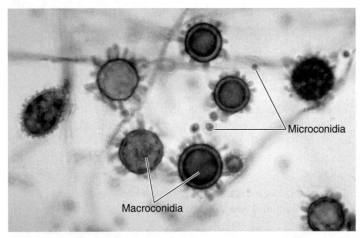

The macroconidia of *Histoplasma capsulatum* are especially useful for diagnostic purposes. Microconidia bud off from hyphae and are the infectious form. At 37°C in tissues, the organism converts to a yeast phase composed of oval, budding yeasts.

LM | 12 μm

**Figure 24.15** *Histoplasma capsulatum*, a dimorphic fungus that causes histoplasmosis.

**Q** What does the term *dimorphic* mean?

few cases, perhaps fewer than 0.1%, histoplasmosis progresses, and it becomes a severe, generalized disease. This occurs with an unusually heavy inoculum or upon reactivation, when the infected person's immune system is compromised.

The causative organism, *Histoplasma capsulatum* (his-TŌ-plaz-mah kap-sū-LAH-tum), is a dimorphic fungus; that is, it has a yeastlike morphology in tissue growth, and, in soil or artificial media, it forms a filamentous mycelium carrying reproductive conidia (**Figure 24.15**). In the body, the yeastlike form is found intracellularly in macrophages, where it survives and multiplies.

Although histoplasmosis is rather widespread throughout the world, it has a limited geographic range in the United States (**Figure 24.16**). In general, the disease is found in the states adjoining the Mississippi and Ohio rivers. More than 75% of the population in some of these states have antibodies against the infection. In other states—Maine, for example—a positive test is a rare event. Approximately 50 deaths are reported in the United States each year from histoplasmosis.

Humans acquire the disease from airborne conidia produced under conditions of appropriate moisture and pH levels. These conditions occur especially where droppings from birds and bats have accumulated. Birds themselves, because of their high body temperature, do not carry the disease, but their droppings provide nutrients, particularly a source of nitrogen, for the fungus. Bats, which have a lower body temperature than birds, carry the fungus, shed it in their feces, and infect new soil sites.

Clinical signs and history, serological tests for *Histoplasma* antigen and, most important, either isolating the

pathogen or identifying it in tissue specimens are necessary for proper diagnosis. Currently, the most effective chemotherapy is itraconazole.

## Coccidioidomycosis

Another fungal pulmonary disease, also rather restricted geographically, is **coccidioidomycosis**. The causative agent is *Coccidioides immitis*, a dimorphic fungus. The arthroconidia are found in dry, alkaline soils of the American Southwest and in similar soils of South America and northern Mexico. Because of its frequent occurrence in the San Joaquin Valley of California, it is sometimes known as *Valley fever* or *San Joaquin fever*. In tissues, the organism forms a thick-walled body called a *spherule* filled with endospores (**Figure 24.17**). In soil, it forms filaments that reproduce by the formation of arthroconidia. The wind carries the arthroconidia to transmit the infection. Arthroconidia are often so abundant that simply driving through an endemic area can result in infection, especially during a dust storm. An estimated 150,000 infections occur each year.

Most infections are not apparent, and almost all patients recover in a few weeks, even without treatment. The symptoms of coccidioidomycosis include chest pain and perhaps fever, coughing, and weight loss. In less than 1% of cases, a progressive disease resembling tuberculosis disseminates throughout the body. A substantial proportion of adults who are long-time residents of areas where the disease is endemic have evidence of prior infection with *C. immitis* by the skin test.

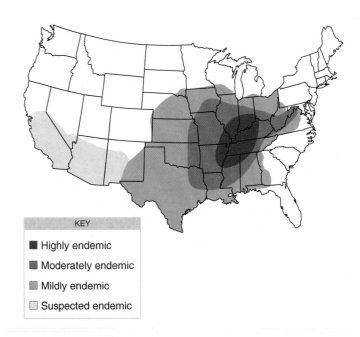

KEY
- ■ Highly endemic
- ■ Moderately endemic
- ■ Mildly endemic
- ☐ Suspected endemic

**Figure 24.16 Histoplasmosis distribution.**
*Source:* CDC.

**Q** Compared with the disease distribution shown in the map in Figure 24.18, what can you determine about the moisture requirements in the soil for the two fungi involved?

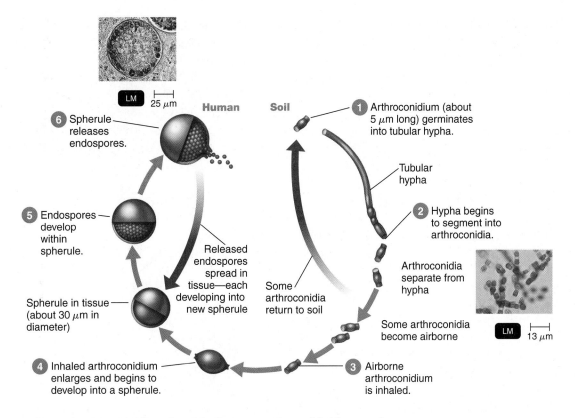

**Human**          **Soil**

① Arthroconidium (about 5 μm long) germinates into tubular hypha.

⑥ Spherule releases endospores.

Tubular hypha

② Hypha begins to segment into arthroconidia.

⑤ Endospores develop within spherule.

Arthroconidia separate from hypha

Released endospores spread in tissue—each developing into new spherule

Some arthroconidia return to soil

Some arthroconidia become airborne

Spherule in tissue (about 30 μm in diameter)

④ Inhaled arthroconidium enlarges and begins to develop into a spherule.

③ Airborne arthroconidium is inhaled.

**Figure 24.17 The life cycle of *Coccidioides immitis*, the cause of coccidioidomycosis.**

**Q** What is the natural habitat of *Coccidioides*?

The incidence of coccidiodomycosis has been increasing recently in California and Arizona (**Figure 24.18**). Contributing factors include a prolonged drought, an increased number of older residents, and an increased prevalence of HIV/AIDS. Outbreaks may occur after an earthquake or other event that disturbs large amounts of soil. About 50 to 100 deaths occur annually from this disease in the United States.

Diagnosis is most reliably made by identifying the spherules in tissue or fluids. The organism can be cultured from fluids or lesions, but laboratory workers must use great care because of the possibility of infectious aerosols. Several serological tests and DNA probes are available for identifying isolates. A tuberculin-like skin test is used in screening.

Fluconazole or itraconazole is used to treat coccidioidomycosis.

## *Pneumocystis* Pneumonia

***Pneumocystis* pneumonia (PCP)** is caused by *Pneumocystis jirovecii* (ye-rō-VET-zē-ē), formerly *P. carinii* (**Figure 24.19**). The taxonomic position of this microbe has been uncertain ever since its discovery in 1909, when it was thought to be a developmental stage of a trypanosome. Since that time, there has been no universal agreement about whether it is a protozoan or a fungus. It has some characteristics of both groups.

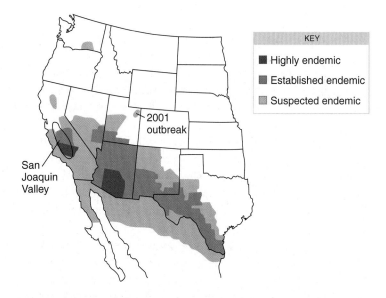

KEY

■ Highly endemic
■ Established endemic
▨ Suspected endemic

2001 outbreak

San Joaquin Valley

**Figure 24.18 The U.S. endemic area for coccidioidomycosis.**
The outlined area in California is the San Joaquin Valley. Because of the very high incidence there, the disease is also sometimes called Valley Fever. The small area on the map in northeastern Utah indicates an outbreak in 2001 in which ten archeologists working on excavations at the Dinosaur National Monument were infected.
*Source:* CDC.

**Q** Why does the incidence of coccidioidomycosis increase after ecological disturbances, such as earthquakes and construction?

**Figure 24.19 The life cycle of *Pneumocystis jirovecii*, the cause of *Pneumocystis* pneumonia.** Long classified as a protozoan, this organism is now usually considered to be a fungus, but it has characteristics of both groups.

**Q** Of what value is proper classification of this organism?

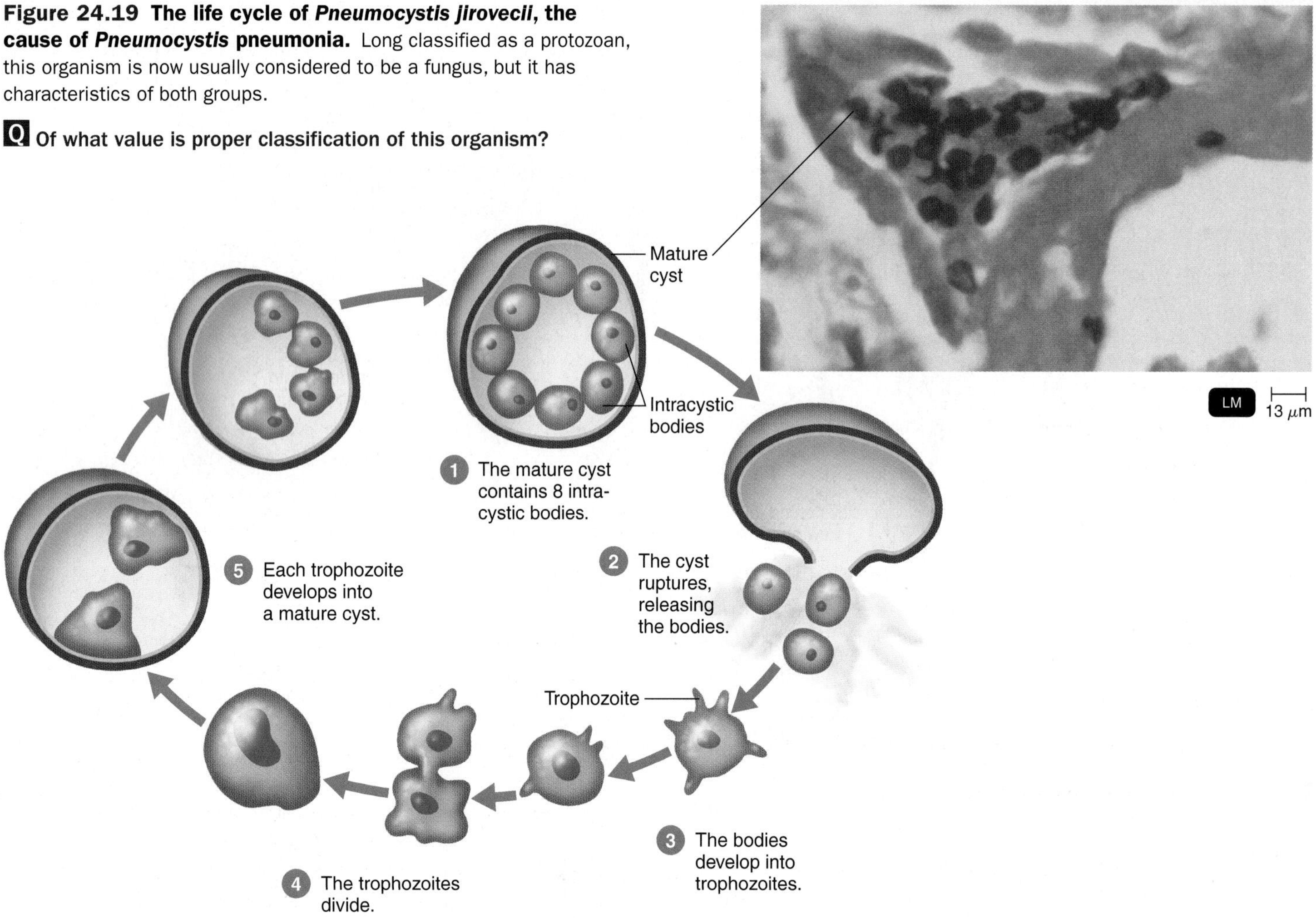

Mature cyst

Intracystic bodies

① The mature cyst contains 8 intracystic bodies.

⑤ Each trophozoite develops into a mature cyst.

② The cyst ruptures, releasing the bodies.

Trophozoite

③ The bodies develop into trophozoites.

④ The trophozoites divide.

LM 13 μm

Analysis of RNA and certain other structural characteristics indicate that it's closely related to certain yeasts, and it's usually reported as a fungus.

The pathogen is sometimes found in healthy human lungs. Immunocompetent adults have few or no symptoms, but newly infected infants occasionally show symptoms of a lung infection. This population may also be the reservoir of the organism, which is not found in the environment, animals, or very often in healthy humans. Persons with compromised immunity are the most susceptible to symptomatic PCP. This portion of the population has also expanded greatly in recent decades. For example, before the AIDS epidemic, PCP was an uncommon disease; perhaps 100 cases occurred each year. By 1993, it had become a primary indicator of AIDS, with more than 20,000 annual reported cases. Presumably, the loss of an effective immune defense allows the activation of a latent infection. Other groups that are very susceptible to this disease are people whose immunity is depressed because of cancer or who are receiving immunosuppressive drugs to minimize rejection of transplanted tissue.

In the human lung, the microbes are found mostly in the lining of the alveoli. Diagnosis is usually made from sputum samples in which cysts are detected. There, they form a thick-walled cyst in which spherical intracystic bodies successively divide as part of a sexual cycle. The mature cyst contains eight such bodies (see Figure 24.19). Eventually the cyst ruptures and releases them, and each body develops into a trophozoite. The trophozoite cells can reproduce asexually by fission, but they may also enter the encysted sexual stage.

The fatality rate is 100% without treatment. The drug of choice for treatment is currently trimethoprim-sulfamethoxazole.

## Blastomycosis (North American Blastomycosis)

**Blastomycosis** is usually called **North American blastomycosis** to differentiate it from a similar South American blastomycosis. It is caused by the fungus *Blastomyces dermatitidis* (blas′tō-MĪ-sēz der-mah-TI-ti-dis), a dimorphic fungus endemic in soil around the Great Lakes and Mississippi River Valley (**Figure 24.20**). Approximately 100 cases are reported each year, although most infections are asymptomatic.

**Figure 24.20 Blastomycosis endemic area.**

*Source:* CDC.

**Q** What other respiratory mycosis has a similar geographic distribution? How would you differentiate the two diseases?

The infection begins in the lungs after inhalation of conidiospores. It resembles bacterial pneumonia and can spread rapidly. Cutaneous ulcers commonly appear when the yeast are disseminated in circulating monocytes. Abscesses may form, with extensive tissue destruction. The pathogen can be isolated from pus and biopsy specimens. Itraconazole or amphotericin B is usually an effective treatment.

## Other Fungi Involved in Respiratory Disease

Many other opportunistic fungi may cause respiratory disease, particularly in hosts who are immunosuppressed or have been exposed to massive numbers of spores. **Aspergillosis** is an important example; it is airborne by the conidia of *Aspergillus fumigatus* (FŪ-mi-gah-tus) and other species of *Aspergillus*, which are widespread in decaying vegetation. Compost piles are ideal sites for growth, and farmers and gardeners are most often exposed to infective amounts of these conidia.

Similar pulmonary infections sometimes result when individuals are exposed to spores of other mold genera, such as *Rhizopus* and *Mucor*. Such diseases can be very dangerous, particularly invasive infections of pulmonary aspergillosis. Predisposing factors include an impaired immune system, cancer, and diabetes. As with most systemic fungal infections, there is only a limited arsenal of antifungal agents available; itraconazole and amphotericin B have proved the most useful.

---

**CHECK YOUR UNDERSTANDING**

✔ **24-10** The droppings of both blackbirds and bats support the growth of *Histoplasma capsulatum*; which of these two animal reservoirs is normally actually infected by the fungus?

\* \* \*

Diseases in Focus 24.3 summarizes the microbial respiratory diseases affecting the lower respiratory system discussed in this chapter.

---

**CLINICAL CASE** Resolved

Domestic and imported pet birds, as well as humans, are at risk for infection with and transmission of *C. psittaci* because shipping, crowding, and breeding promote shedding of the organism. Avian infection, which has a prevalence of less than 5%, may increase to 100% under such circumstances. The U.S. Department of Agriculture (USDA) requires a 30-day quarantine period for all imported birds to prevent the introduction of Newcastle disease (a viral disease that affects fowl); during this period, psittacine birds receive medicated chlortetracycline (CT) feed to prevent transmission of *C. psittaci* to USDA staff. Unless treatment is continued for 45 days, infected birds arriving to distributors from breeders and from quarantine may shed *C. psittaci* and continue to do so after purchase by consumers. Therefore, breeders and importers should ensure that all domestic nestlings and imported birds receive prophylactic CT for 45 continuous days to prevent future outbreaks of human psittacosis.

689  706  707  709  711  **715**

Microbial Diseases
of the Lower Respiratory System

Three weeks after working on the demolition of an abandoned building in Kentucky, a worker is hospitalized for acute respiratory illness. At the time of demolition, a colony of bats inhabited the building. An X-ray examination reveals a lung mass. A purified protein derivative test is negative; a cytological examination for cancer is also negative. The mass is surgically removed. Microscopic examination of the mass reveals ovoid yeast cells. Use the table below to make a differential diagnosis and identify infections that could cause these symptoms. For the solution, go to @MasteringMicrobiology.

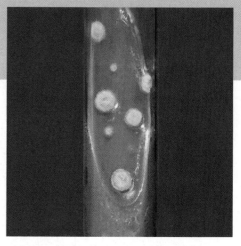

**Mycelial culture grown from the patient's lung mass.**

| Disease | Pathogen | Symptoms | Reservoir | Diagnosis | Treatment |
|---|---|---|---|---|---|
| **BACTERIAL DISEASES** | | | | | |
| **Bacterial Pneumonia** (see Diseases in Focus 24.2, page 704) | | | | | |
| **Pertussis** (whooping cough) | *Bordetella pertussis* | Spasms of intense coughing to clear mucus | Humans | Bacterial culture | Erythromycin Prevention: DTaP vaccine |
| **Tuberculosis** | *Mycobacterium tuberculosis* *M. bovis* *M. avium-intracellulare* | Cough, blood in mucus | Humans, cows: can be transmitted via unpasteurized milk | X-ray imaging; presence of acid-fast bacilli in sputum; tests for IFN-γ; PCR test for *M. tuberculosis* | Multiple-antimycobacterial drugs Prevention: pasteurization of milk; BCG vaccine |
| **Melioidosis** | *Burkholderia pseudomallei* | Pneumonia, or as tissue abscesses and severe sepsis | Moist soil | Bacterial culture, PCR | Ceftazidime |
| **VIRAL DISEASES** | | | | | |
| **Respiratory Syncytial Virus (RSV) Disease** | Respiratory syncytial virus | Pneumonia in infants | Humans | Serological tests | Prophylaxis: Palivizumab |
| **Influenza** | *Influenzavirus*; several serotypes | Chills, fever, headache, and muscular aches | Humans, pigs, birds | Serological tests, PCR | Zanamivir, oseltamivir |
| **FUNGAL DISEASES** | | | | | |
| **Histoplasmosis** | *Histoplasma capsulatum* | Resembles tuberculosis | Soil; widespread in Ohio and Mississippi River valleys | Serological tests | Itraconazole |
| **Coccidioidomycosis** | *Coccidioides immitis* | Fever, coughing, weight loss | Desert soils of U.S. Southwest | Serological tests | Fluconazole or itraconazole |
| ***Pneumocystis* Pneumonia** | *Pneumocystis jirovecii* | Pneumonia | Most likely humans | Microscopy | Trimethoprim-sulfamethoxazole |
| **Blastomycosis** | *Blastomyces dermatitidis* | Resembles bacterial pneumonia; extensive tissue damage | Soil in Great Lakes and Mississippi River valleys | Isolation of pathogen | Itraconazole or amphotericin B |

# Study Outline

 Go to @MasteringMicrobiology for **Interactive Microbiology,** *In the Clinic* *videos, MicroFlix, MicroBoosters, 3D animations, practice quizzes, and more.*

## Introduction (p. 688)

1. Infections of the upper respiratory system are the most common type of infection.
2. Pathogens that enter the respiratory system can infect other parts of the body.

## Structure and Function of the Respiratory System (p. 689)

1. The upper respiratory system consists of the nose, pharynx, and associated structures, such as the middle ear and auditory tubes.
2. Coarse hairs in the nose filter large particles from air entering the respiratory tract.
3. The ciliated mucous membranes of the nose and throat trap airborne particles and remove them from the body.
4. Lymphoid tissue, tonsils, and adenoids provide immunity to certain infections.
5. The lower respiratory system consists of the larynx, trachea, bronchial tubes, and alveoli.
6. The ciliary escalator of the lower respiratory system helps prevent microorganisms from reaching the lungs.
7. Microbes in the lungs can be phagocytized by alveolar macrophages.
8. Respiratory mucus contains IgA antibodies.

## Normal Microbiota of the Respiratory System (p. 690)

1. The normal microbiota of the nasal cavity and throat can include pathogenic microorganisms.

## Microbial Diseases of the Upper Respiratory System (pp. 690–695)

1. Specific areas of the upper respiratory system can become infected to produce pharyngitis, laryngitis, tonsillitis, sinusitis, and epiglottitis.
2. These infections may be caused by several bacteria and viruses, often in combination.
3. Most respiratory tract infections are self-limiting.
4. *H. influenzae* type b can cause epiglottitis.

## Bacterial Diseases of the Upper Respiratory System (pp. 691–693)

### Streptococcal Pharyngitis (Strep Throat) (pp. 691–692)

1. This infection is caused by group A beta-hemolytic streptococci, the group that consists of *Streptococcus pyogenes*.
2. Symptoms of this infection are inflammation of the mucous membrane and fever; tonsillitis and otitis media may also occur.
3. Rapid diagnosis is made by enzyme immunoassays.
4. Immunity to streptococcal infections is type-specific.

### Scarlet Fever (p. 692)

5. Strep throat, caused by an erythrogenic toxin-producing *S. pyogenes*, results in scarlet fever.

6. *S. pyogenes* produces erythrogenic toxin when lysogenized by a phage.
7. Symptoms include a red rash, high fever, and a red, enlarged tongue.

### Diphtheria (pp. 692–693)

8. Diphtheria is caused by exotoxin-producing *Corynebacterium diphtheriae*.
9. Exotoxin is produced when the bacteria are lysogenized by a phage.
10. A membrane, containing fibrin and dead human and bacterial cells, forms in the throat and can block the passage of air.
11. The exotoxin inhibits protein synthesis, and heart, kidney, or nerve damage may result.
12. Laboratory diagnosis is based on isolation of the bacteria and the appearance of growth on differential media.
13. Routine immunization in the United States includes diphtheria toxoid in the DTaP vaccine.
14. Slow-healing skin ulcerations are characteristic of cutaneous diphtheria.
15. There is minimal dissemination of the exotoxin in the bloodstream.

### Otitis Media (p. 693)

16. Earache, or otitis media, can occur as a complication of nose and throat infections.
17. Pus accumulation causes pressure on the eardrum.
18. Bacterial causes include *Streptococcus pneumoniae*, non-encapsulated *Haemophilus influenzae*, *Moraxella catarrhalis*, and *Streptococcus pyogenes*.

## Viral Disease of the Upper Respiratory System (pp. 693–695)

### The Common Cold (pp. 693–695)

1. Any one of approximately 200 different viruses, including rhinoviruses, coronaviruses, and EV-D68, can cause the common cold.
2. The incidence of colds increases during cold weather, possibly because of increased interpersonal indoor contact or physiological changes.

## Microbial Diseases of the Lower Respiratory System (pp. 695–715)

1. Many of the same microorganisms that infect the upper respiratory system also infect the lower respiratory system.
2. Diseases of the lower respiratory system include bronchitis and pneumonia.

## Bacterial Diseases of the Lower Respiratory System (pp. 695–707)

### Pertussis (Whooping Cough) (pp. 695–698)

1. Pertussis is caused by *Bordetella pertussis*.
2. The initial stage of pertussis resembles a cold and is called the catarrhal stage.

3. The accumulation of mucus in the trachea and bronchi causes deep coughs characteristic of the paroxysmal (second) stage.

4. The convalescence (third) stage can last for months.

5. Regular immunization for children has decreased the incidence of pertussis.

### Tuberculosis (pp. 698–701)

6. Tuberculosis is caused by *Mycobacterium tuberculosis*.

7. *Mycobacterium bovis* causes bovine tuberculosis and can be transmitted to humans by unpasteurized milk.

8. *M. avium-intracellulare* complex infects patients in the late stages of HIV infection.

9. *M. tuberculosis* may be ingested by alveolar macrophages; if not killed, the bacteria reproduce in the macrophages.

10. Lesions formed by *M. tuberculosis* are called tubercles; macrophages and bacteria form the caseous lesion that might calcify and appear in an X-ray image as a Ghon's complex.

11. Liquefaction of the caseous lesion results in a tuberculous cavity in which *M. tuberculosis* can grow.

12. New foci of infection can develop when a caseous lesion ruptures and releases bacteria into blood or lymph vessels; this is called miliary tuberculosis.

13. A positive tuberculin skin test can indicate either an active case of TB, prior infection, or vaccination and immunity to the disease.

14. Active infections can be diagnosed by detection of IFN-γ or rapid PCR test for *M. tuberculosis*.

15. Chemotherapy usually involves three or four drugs taken for at least 6 months; multidrug-resistant *M. tuberculosis* is becoming prevalent.

16. BCG vaccine for tuberculosis consists of a live, avirulent culture of *M. bovis*.

### Bacterial Pneumonias (pp. 701–707)

17. Typical pneumonia is caused by *S. pneumoniae*.

18. Atypical pneumonias are caused by other microorganisms.

19. Pneumococcal pneumonia is caused by encapsulated *Streptococcus pneumoniae*.

20. Children under 5 and adults over 65 are most susceptible to *H. influenzae* pneumonia.

21. *Mycoplasma pneumoniae* causes mycoplasmal pneumonia; it is an endemic disease.

22. Legionellosis is caused by the aerobic gram-negative rod *Legionella pneumophila*.

23. *Chlamydophila psittaci*, the bacterium that causes psittacosis (ornithosis), is transmitted by contact with contaminated droppings and exudates of fowl.

24. *Chlamydophila pneumoniae* causes pneumonia; it is transmitted from person to person.

25. Obligately parasitic, intracellular *Coxiella burnetii* causes Q fever.

### Melioidosis (p. 707)

26. Melioidosis, caused by *Burkholderia pseudomallei*, is transmitted by inhalation, ingestion, or through puncture wounds. Symptoms include pneumonia, sepsis, and encephalitis.

## Viral Diseases of the Lower Respiratory System (pp. 707–711)

### Viral Pneumonia (p. 708)

1. A number of viruses can cause pneumonia as a complication of infections such as influenza.

2. The etiologies are not usually identified in a clinical laboratory because of the difficulty in isolating and identifying viruses.

### Respiratory Syncytial Virus (RSV) (pp. 708–709)

3. RSV is the most common cause of pneumonia in infants.

### Influenza (Flu) (pp. 709–711)

4. Influenza is caused by *Influenzavirus* and is characterized by chills, fever, headache, and general muscular aches.

5. Hemagglutinin (HA) and neuraminidase (NA) spikes project from the outer lipid bilayer of the virus.

6. Viral strains are identified by antigenic differences in their protein coats (A, B, and C); influenza A is further subdivided by differences in the HA and NA spikes.

7. Antigenic shifts and antigenic drift enable the virus to evade natural immunity.

8. Multivalent vaccines are available.

9. Zanamivir and oseltamivir are effective drugs against influenza A virus.

## Fungal Diseases of the Lower Respiratory System (pp. 711–715)

1. Fungal spores are easily inhaled; they may germinate in the lower respiratory tract.

2. The incidence of fungal diseases has been increasing in recent years.

3. The mycoses in the following sections can be treated with itraconazole.

### Histoplasmosis (pp. 711–712)

4. *Histoplasma capsulatum* causes a subclinical respiratory infection that only occasionally progresses to a severe, generalized disease.

5. The disease is acquired by inhaling airborne conidia.

6. Isolating or identifying the fungus in tissue samples is necessary for diagnosis.

### Coccidioidomycosis (pp. 712–713)

7. Inhaling the airborne arthroconidia of *Coccidioides immitis* can result in coccidioidomycosis.

### *Pneumocystis* Pneumonia (pp. 713–714)

8. *Pneumocystis jirovecii* is found in healthy human lungs.

9. *P. jirovecii* causes disease in immunosuppressed patients.

### Blastomycosis (North American Blastomycosis) (pp. 714–715)

10. *Blastomyces dermatitidis* is the causative agent of blastomycosis.

11. The infection begins in the lungs and can spread to cause extensive abscesses.

### Other Fungi Involved in Respiratory Disease (p. 715)

12. Opportunistic fungi can cause respiratory disease in immuno-suppressed hosts, especially when large numbers of spores are inhaled.

13. Among these fungi are *Aspergillus*, *Rhizopus*, and *Mucor*.

# Study Questions

For answers to the Knowledge and Comprehension questions, turn to the Answers tab at the back of the textbook.

## Knowledge and Comprehension

### Review

1. DRAW IT  Show the location of the following diseases: common cold, diphtheria, coccidioidomycosis, influenza, pneumonia, scarlet fever, tuberculosis, whooping cough

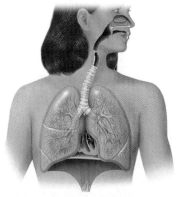

2. Compare and contrast mycoplasmal pneumonia and viral pneumonia.

3. List the causative agent, symptoms, and treatment for four viral diseases of the respiratory system. Separate the diseases according to whether they infect the upper or lower respiratory system.

4. Complete the following table.

| Disease | Causative Agent | Symptoms | Treatment |
|---|---|---|---|
| Streptococcal pharyngitis | | | |
| Scarlet fever | | | |
| Diphtheria | | | |
| Whooping cough | | | |
| Tuberculosis | | | |
| Pneumococcal pneumonia | | | |
| *H. influenzae* pneumonia | | | |
| Chlamydial pneumonia | | | |
| Otitis media | | | |
| Legionellosis | | | |
| Psittacosis | | | |
| Q fever | | | |
| Epiglottitis | | | |
| Melioidosis | | | |

5. Under what conditions can the saprophytes *Aspergillus* and *Rhizopus* cause infections?

6. A patient has been diagnosed as having pneumonia. Is this sufficient information to begin treatment with antimicrobial agents? Briefly discuss why or why not.

7. List the causative agent, mode of transmission, and endemic area for the diseases histoplasmosis, coccidioidomycosis, blastomycosis, and *Pneumocystis* pneumonia.

8. Briefly describe the procedures and positive results of the tuberculin test and what is indicated by a positive test.

9. Identify the bacteria involved in respiratory infections using the following laboratory test results:

   Gram-positive cocci
      Catalase-positive: a. _____
      Catalase-negative
         Beta-hemolytic, bacitracin inhibition: b. _____
         Alpha-hemolytic, optochin inhibition: c. _____
   Gram-positive rods
      Non–acid-fast: d. _____
      Acid-fast: e. _____
   Gram-negative cocci: f. _____
   Gram-negative rods
      Aerobes
         Coccobacilli: g. _____
         Rods
            Grow on nutrient agar: h. _____
            Require special media: i. _____
      Facultative anaerobes
         Coccobacilli: j. _____
   Intracellular parasites
      Form elementary bodies: k. _____
      Do not form elementary bodies: l. _____
   Wall-less: m. _____

10. NAME IT  These aerobic, gram-negative bacteria produce tracheal cytotoxin that kills ciliated cells of the trachea.

## Multiple Choice

1. A patient has fever, difficulty breathing, chest pains, fluid in the alveoli, and a positive tuberculin skin test. Gram-positive cocci are isolated from the sputum. The recommended treatment is
   a. a macrolide.
   b. antitoxin.
   c. isoniazid.
   d. tetracyclines.
   e. none of the above.

2. No bacterial pathogen can be isolated from the sputum of a patient with pneumonia. Antibiotic therapy has not been successful. The next step should be
   a. culturing for *Mycobacterium tuberculosis*.
   b. culturing for *Mycoplasma pneumoniae*.
   c. culturing for fungi.
   d. a change in antibiotics.
   e. none; nothing more can be done.

Match the following choices to the culture descriptions in questions 3 through 6:
   a. *Chlamydophila*
   b. *Coccidioides*
   c. *Histoplasma*
   d. *Mycobacterium*
   e. *Mycoplasma*

3. Your culture from a pneumonia patient appears not to have grown. You do see colonies, however, when the plate is viewed at 100X.

4. This pneumonia etiology requires cell culture.

5. Microscopic examination of a lung biopsy shows ovoid cells in macrophages. You suspect these are the cause of the patient's symptoms, but your culture grows a filamentous organism.

6. Microscopic examination of a lung biopsy shows spherules.

7. In San Francisco, ten animal health care technicians developed pneumonia 2 weeks after 130 goats were moved to the animal shelter where they worked. Which of the following is *false*?
   a. Diagnosis is made by a blood agar culture of sputum.
   b. The cause is *Coxiella burnetii*.
   c. The bacteria produce endospores.
   d. The disease was transmitted by aerosols.
   e. Diagnosis is made by complement-fixation tests for antibodies.

8. Which of the following leads to all the rest?
   a. catarrhal stage
   b. cough
   c. loss of cilia
   d. mucus accumulation
   e. tracheal cytotoxin

Match the following choices to the statements in questions 9 and 10:
   a. *Bordetella pertussis*
   b. *Corynebacterium diphtheriae*
   c. *Legionella pneumophila*
   d. *Mycobacterium tuberculosis*
   e. none of the above

9. Causes the formation of a membrane across the throat.

10. Resistant to destruction by phagocytes.

## Analysis

1. Differentiate *S. pyogenes* causing strep throat from *S. pyogenes* causing scarlet fever.

2. Why might the influenza vaccine be less effective than other vaccines?

3. Explain why it would be impractical to include cold and influenza vaccinations in the required childhood vaccinations.

## Clinical Applications and Evaluation

1. In August, a 24-year-old man from Virginia developed difficulty breathing and bilateral lobe infiltrates 2 months after driving through California. During initial evaluation, typical pneumonia was suspected, and he was treated with antibiotics. Efforts to diagnose the pneumonia were unsuccessful. In October, a laryngeal mass was detected, and laryngeal cancer was suspected; treatment with steroids and bronchodilators did not result in improvement. Lung biopsy and laryngoscopy detected diffuse granular tissue. He was treated with amphotericin B and discharged after 5 days. What was the disease? What might have been done differently to decrease the patient's recovery time from 3 months to 1 week?

2. During a 6-month period, 72 clinic staff members became tuberculin-positive. A case-control study was undertaken to determine the most likely source of *M. tuberculosis* infection among the staff. A total of 16 cases and 34 tuberculin-negative controls were compared.

Pentamidine isethionate is not used for TB treatment. What disease was probably being treated with this drug? What is the most likely source of infection?

| | Cases | Control |
|---|---|---|
| Works ≥40 hr/week | 100% | 62% |
| In room during aerosolized pentamidine isethionate therapy in TB patients | 31 | 3 |
| Patient contact | 94 | 94 |
| Lunch eaten in staff lounge | 38 | 35 |
| Resident of western Palm Beach | 75 | 65 |
| Female | 81 | 77 |
| Cigarette smoker | 6 | 15 |
| Contact with nurse diagnosed with TB | 15 | 12 |
| In unventilated room during collection of TB-positive sputum samples | 13 | 8 |

3. In a 2-week period, eight infants in an intensive care nursery (ICN) developed pneumonia caused by RSV. Complement-fixation (CF) screening and ELISA for viral antigens were performed to diagnose infections. RSV-positive patients were placed in a separate room. A 2-week-old girl from the newborn nursery, adjacent to the ICN, also developed an RSV infection. Toward the end of this outbreak, CF tests and direct ELISA tests were made of ten ICN staff members. ELISA–viral antigen tests were negative; RSV titers as determined by CF are shown below.

| Staff | RSV Titer |
|---|---|
| A | 0 |
| B | 64 |
| C | 32 |
| D | 128 |
| E | 256 |
| F | 0 |
| G | 0 |
| H | 32 |
| I | 32 |
| J | 16 |

Comment on the probable source of this outbreak. Explain the apparent discrepancy between the CF test and ELISA results. How can RSV infections in nurseries be prevented?

# Microbial Diseases of the Digestive System 25

icrobial diseases of the digestive system are second only to respiratory diseases as causes of illness in the United States. Most such diseases result from ingesting food or water contaminated with pathogenic microorganisms or their toxins. These pathogens usually enter the food or water supply after being shed in the feces of people or animals infected with them. Therefore, microbial diseases of the digestive system are typically transmitted by a **fecal–oral cycle**. This cycle is interrupted by effective sanitation practices in food production and handling. Modern methods of sewage treatment and disinfection of water are essential. There is also an increasing awareness of the need for new tests that will rapidly and reliably detect pathogens in foods (a perishable commodity).

The Centers for Disease Control and Prevention (CDC) estimates that about 76 million cases of foodborne disease resulting in about 5000 deaths occur annually in the United States. As more of our food products—especially fruits and vegetables—are grown in countries with poor sanitation standards, outbreaks of foodborne disease from imported pathogens are expected to increase. Some *Escherichia coli* cause disease by making a toxin called Shiga toxin. The bacteria (shown in the photo) that make these toxins are called Shiga toxin-producing *E. coli* (STEC). A STEC infection is described in the Clinical Case in this chapter.

◀ *Escherichia coli* bacteria are essential members of the human microbiome unless they produce a toxin, such as the Shiga toxin produced by *E. coli* O157:H7.

## In the Clinic

As the county public health nurse, you were contacted regarding a woman who had acute gastroenteritis after dining at a local restaurant with friends. You interviewed the diners and confirmed that three of the seven in the party had consumed New England clam chowder. Within 1 to 4 hours after consumption, the three had onset of nausea and vomiting lasting 24 to 48 hours. The four diners who had not eaten the clam chowder did not become ill. The restaurant had kept the clam chowder at 39°C for the lunch and dinner service. **What microbial etiology is most likely?**

*Hint: Make a list of the foodborne diseases covered in this chapter to match to this case.*

**Answers to In the Clinic questions are found online** @MasteringMicrobiology.

# Structure and Function of the Digestive System

**LEARNING OBJECTIVE**

25-1 Name the structures of the digestive system that contact food.

The **digestive system** is essentially a tubelike structure, the *gastrointestinal (GI) tract*, or *alimentary canal*—mainly the mouth, pharynx (throat), esophagus (food tube leading to the stomach), stomach, and the small and large intestines. It also includes *accessory structures* such as the teeth and tongue. Certain other accessory structures such as the salivary glands, liver, gallbladder, and pancreas lie outside the GI tract and produce secretions that are conveyed by ducts into it (**Figure 25.1**).

The purpose of the digestive system is to digest foods—that is, to break them down into small molecules that can be taken up and used by body cells. In a process called *absorption*, these end-products of digestion pass from the small intestine into the blood or lymph for distribution to body cells. Then the food moves through the large intestine, where water and any remaining nutrients are absorbed from it. Over the course of an average life span, about 25 tons of food pass through the GI tract. The resulting undigested solids, called *feces*, are eliminated from the body through the anus. Intestinal gas, or *flatus*, is a mixture of nitrogen from swallowed air and microbially produced carbon dioxide, hydrogen, and methane. On average, we produce 0.5 to 2.0 liters of flatus every day.

There is also a relationship between the body's digestive system and immune system. This is described in Exploring the Microbiome on the next page. Throughout life, the intestinal mucosa continues to be challenged by the antigens of the intestinal microbiota and ingested. As a consequence, an estimated 70% of the immune system is located in the intestinal tract, especially the small intestine. This loosely organized lymphoid tissue and structures such as lymph nodes and Peyer's patches are collectively called *gut-associated lymphoid tissue (GALT)*.

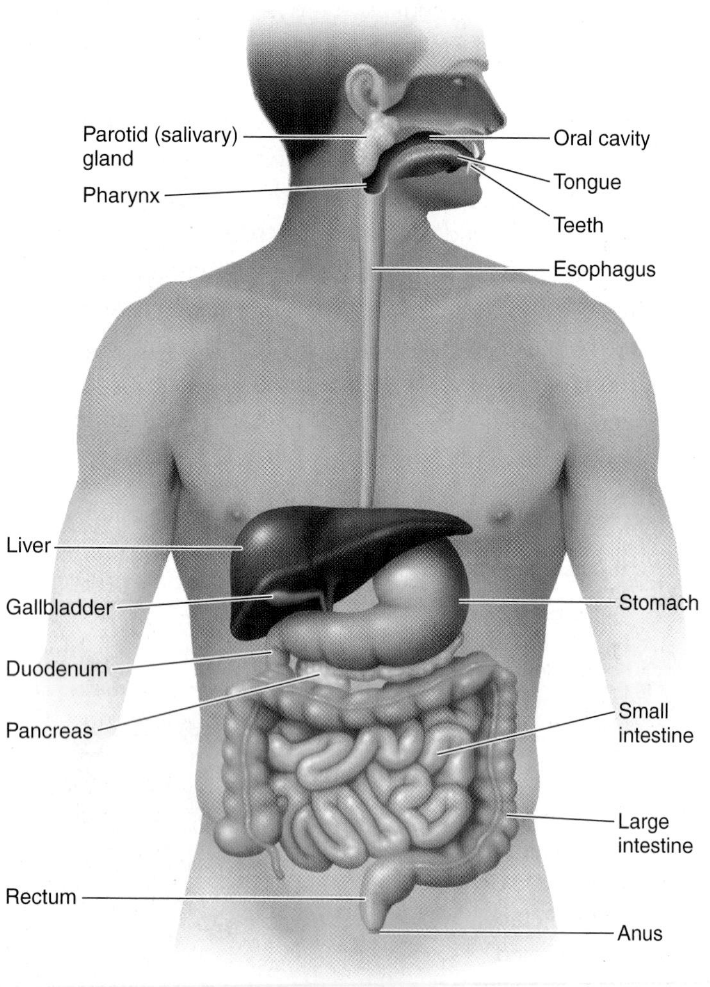
Parotid (salivary) gland
Pharynx
Oral cavity
Tongue
Teeth
Esophagus
Liver
Gallbladder
Duodenum
Pancreas
Rectum
Stomach
Small intestine
Large intestine
Anus

**Figure 25.1 The human digestive system.**

**Q** Where are microorganisms normally found in the digestive system?

**CHECK YOUR UNDERSTANDING**

✓ 25-1 Small explosions have occurred when a surgeon used spark-producing instruments to remove intestinal polyps. What ignited?

# Normal Microbiota of the Digestive System

**LEARNING OBJECTIVE**

25-2 Identify parts of the gastrointestinal tract that normally have microbiota.

Bacteria heavily populate most of the digestive system. In the mouth, each milliliter of saliva can contain millions of bacteria. Because of the hydrochloric acid produced by the stomach and the rapid movement of food through the small intestine, these organs house relatively few microorganisms. By contrast, the large intestine has enormous microbial populations, exceeding 100 billion bacteria per gram of feces. (Up to 40% of fecal mass is microbial cell material.) The population of the large intestine is composed mostly of anaerobes and facultative anaerobes. Most of these bacteria assist in the enzymatic breakdown of foods, especially many polysaccharides that would otherwise be indigestible. Some of them synthesize useful vitamins.

It's important to understand that food passing through the tubelike GI tract, although it's in contact with the body, remains outside the body. Unlike the body's exterior, such as the skin, the GI tract is adapted to absorbing nutrients passing through it. However, at the same time that nutrients

**Sorting Out Good Neighbors from Bad in the GI Tract**

The deeper you go into the gastrointestinal (GI) tract, the more crowded the bacterial scene. About 10 bacterial species per gram of contents are found in the esophagus, while over 1000 bacterial species per gram reside in the large intestine. This brings up a puzzling question: With 70% of the body's immune response originating in the GI tract, how do commensal bacteria survive there while pathogens are simultaneously sought out and destroyed? Current research indicates our resident microbes may have as much to do with this sorting out of good bacteria from the bad as our own immune system does.

We are born with an underdeveloped immune system and very few microbes populating our bodies. The first bacteria that colonize an infant's GI tract seem to teach the immune system to tolerate them over other types of species. This process is enhanced by the fact that the beneficial bacteria then make compounds that hurt pathogens while leaving their and our cells intact.

Gram-negative bacteria such as *Bacteroides* can activate intestinal dendritic cells in the intestinal mucosa, kicking off a process that leads to secretion of immunoglobulin A (IgA) in the GI tract. This antibody class exerts anti-inflammatory effects and creates a barrier in the mucosa that prevents many pathogens from attaching to the epithelial cells there.

Resident microbes also produce antimicrobial peptides (AMPs). These compounds bind to the more negatively charged membranes of bacteria and lyse them, leaving the membranes of our own cells intact. Some bacteria induce our own AMPs. For example, *Bacteroides thetaiotaomicron* induces intestinal cells to make an enzyme that synthesizes the AMP defensin. The short-chain fatty acids produced by *Lactobacillus* and *Bacteroides* cause host cells to make another antimicrobial peptide, cathelicidin. Interestingly, both gram-positive and gram-negative normal microbiota are

resistant to the effects of these AMPs, whereas pathogenic species are sensitive to them.

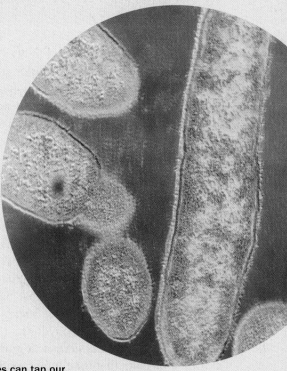

**Commonly found in the GI tract, *Bacteriodes* species can tap our immune system and also make their own defensive chemicals to aid in the fight against pathogens.**

are absorbed from the GI tract, harmful microbes ingested in food and water must be kept from invading the body. An important factor in this defense is the high acid content of the stomach, which eliminates many potentially harmful ingested microbes.

The small intestine also contains important antimicrobial defenses, most significantly, millions of specialized, granule-filled cells called *Paneth cells*. These are capable of phagocytizing bacteria, and they also produce antibacterial proteins called *defensins* (see antimicrobial peptides, page 469) and the antibacterial enzyme *lysozyme*.

### CHECK YOUR UNDERSTANDING

✔ 25-2 How are normal microbiota confined to the mouth and large intestine?

**CLINICAL CASE** A Birthday Surprise

Nadia Abramovic is worried about her 5-year-old daughter, Anna. The week started out like any other; as a matter of fact, Anna was still talking about her birthday party the weekend before. But for the past two days, Anna has been pale and listless and complaining that her stomach hurts. When she sees that Anna has slimy, bloody diarrhea, Mrs. Abramovic immediately calls the pediatrician's office to make an appointment. Anna's pediatrician sends a stool sample to the local laboratory for a bacterial culture.

**How will the laboratory test for the etiology of Anna's illness? Read on to find out.**

723    733    739    745    755

# Bacterial Diseases of the Mouth

## LEARNING OBJECTIVE

**25-3** Describe the events that lead to dental caries and periodontal disease.

The mouth, which is the entrance to the digestive system, provides an environment that supports a large and varied microbial population.

## Dental Caries (Tooth Decay)

The teeth, unlike any other exterior surface of the body, are hard and do not shed surface cells (**Figure 25.2**). This allows masses of microorganisms and their products to accumulate. These accumulations, called **dental plaque**, are a type of biofilm (see page 157 in Chapter 6) and are intimately involved in the formation of **dental caries**, or tooth decay.

Oral bacteria convert sucrose and other carbohydrates into lactic acid, which in turn attacks the tooth enamel. The microbial population on and around the teeth is very complex. Based on ribosomal identification methods (see the discussion of FISH on page 288 in Chapter 10), over 700 species of bacteria have been identified in the oral cavity. Probably the most important *cariogenic* (caries-causing) bacterium is *Streptococcus mutans*, a gram-positive coccus that has important virulence characteristics (**Figure 25.3a**). *S. mutans* is capable of metabolizing a wide range of carbohydrates, tolerates a high level of acidity, and synthesizes *dextran*, a gummy polysaccharide of glucose molecules that is an important factor in the formation of dental plaque (Figure 25.3b). Some other species of streptococci are also cariogenic but play a lesser role in initiating caries.

The initiation of caries depends on the attachment of *S. mutans* or other streptococci to the tooth. These bacteria don't

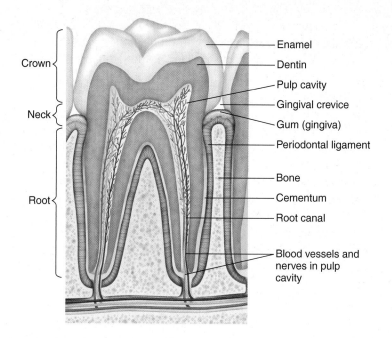

**Figure 25.2 A healthy human tooth.**

**Q** Why can a biofilm accumulate on teeth?

adhere to a clean tooth, but a freshly brushed tooth will become coated with a pellicle (thin film) of proteins from saliva within minutes. Within a couple of hours, cariogenic bacteria become established on this pellicle and begin to produce dextran (see Figure 25.3b). In the production of dextran, the bacteria first hydrolyze sucrose into its component monosaccharides, fructose and glucose. The enzyme glucosyltransferase then assembles the glucose molecules into dextran. The residual fructose is the primary sugar fermented into lactic acid. Accumulations of bacteria and dextran adhering to the teeth make up dental plaque.

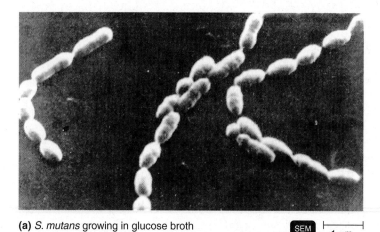

**(a)** *S. mutans* growing in glucose broth

**(b)** *S. mutans* growing in sucrose broth; note the accumulations of dextran. Arrows point to *S. mutans* cells.

**Figure 25.3 *Streptococcus mutans.*** The dextran allows *S. mutans* to adhere to teeth.

**Q** What makes dental plaque a type of biofilm?

The bacterial population of plaque may harbor over 400 bacterial species but is predominantly streptococci and filamentous members of the genus *Actinomyces*. (Older, calcified deposits of plaque are called *dental calculus*, or *tartar*.) *S. mutans* especially favors crevices or other sites on the teeth protected from the shearing action of chewing or the flushing action of the liter or so of saliva produced in the mouth each day. On protected areas of the teeth, plaque accumulations can be several hundred cells thick. Because plaque is not very permeable to saliva, the lactic acid produced by bacteria isn't diluted or neutralized, and it breaks down the enamel of the teeth to which the plaque adheres.

Although saliva contains nutrients that encourage the growth of bacteria, it also contains antimicrobial substances, such as *lysozyme*, that help protect exposed tooth surfaces. Some protection is also provided by *crevicular fluid*, a tissue exudate that flows into the gingival crevice (see Figure 25.2) and is closer in composition to serum than saliva. It protects teeth by virtue of its flushing action, its phagocytic cells, and immunoglobulin content.

Localized acid production within deposits of dental plaque results in a gradual softening of the external *enamel*. Enamel low in fluoride is more susceptible to the effects of the acid. This is the reason for fluoridation of water and toothpastes, a significant factor in the decline in tooth decay in the United States.

The stages of tooth decay are shown in **Figure 25.4**. If the initial penetration of the enamel by caries remains untreated, bacteria can penetrate into the interior of the tooth. The composition of the bacterial population involved in spreading the decayed area from the enamel into the *dentin* is entirely different from that of the population initiating the decay. The dominant microorganisms are gram-positive rods and filamentous bacteria; *S. mutans* is present in small numbers. Although once

considered the cause of dental caries, *Lactobacillus* spp. actually play no role in initiating the process. However, these very prolific lactic acid producers are important in advancing the front of the decay once it has become established.

The decayed area eventually advances to the *pulp* (see Figure 25.4), which connects with the tissues of the jaw and contains the blood supply and the nerve cells. Almost any member of the normal microbiota of the mouth can be isolated from the infected pulp and roots. Once this stage is reached, root canal therapy is required to remove the infected and dead tissue and to provide access for antimicrobial drugs that suppress renewed infection. If untreated, the infection may advance from the tooth to the soft tissues, producing dental abscesses caused by mixed bacterial populations that contain many anaerobes.

Although dental caries is probably one of the more common infectious diseases in humans today, it was scarce in the Western world until about the seventeenth century. In human remains from older times, only about 10% of the teeth contain caries. The introduction of table sugar, or sucrose, into the diet is highly correlated with our present level of caries in the Western world. Studies have shown that sucrose, a disaccharide composed of glucose and fructose, is much more cariogenic than either glucose or fructose individually (see Figure 25.3). People living on high-starch diets (starch is a polysaccharide of glucose) have a low incidence of tooth decay unless sucrose is also a significant part of their diet. The contribution of bacteria to tooth decay has been shown by experiments with germ-free animals. Such animals don't develop caries even when fed a sucrose-rich diet designed to encourage their formation.

Sucrose is pervasive in the modern Western diet. However, if sucrose is ingested only at regular mealtimes, the protective and repair mechanisms of the body are usually not overwhelmed. It is the sucrose that is ingested between meals that

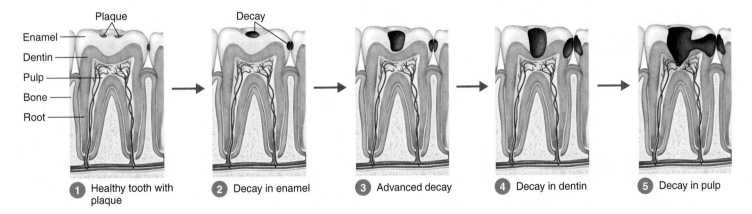

| Enamel — | Dentin — | Pulp — | Bone — | Root — |

1 Healthy tooth with plaque    2 Decay in enamel    3 Advanced decay    4 Decay in dentin    5 Decay in pulp

**Figure 25.4 The stages of tooth decay.** 1 A tooth with plaque accumulation in difficult-to-clean areas. 2 Decay begins as enamel is attacked by acids formed by bacteria. 3 Decay advances through the enamel. 4 Decay advances into the dentin. 5 Decay enters the pulp and may form abscesses in the tissues surrounding the root.

**Q** How does the formation of plaque contribute to tooth decay?

is most damaging to teeth. Sugar alcohols, such as mannitol, sorbitol, and xylitol, are not cariogenic; xylitol even appears to inhibit carbohydrate metabolism in *S. mutans*. This is why these sugar alcohols are used to sweeten "sugarless" candies and chewing gum.

The best strategies for preventing dental caries are minimal ingestion of sucrose; brushing, flossing, and professional cleaning to remove plaque; and the use of fluoride. Professional removal of plaque and tartar at regular intervals lessens the progression to periodontal disease.

## Periodontal Disease

Even people who avoid tooth decay might, in later years, lose their teeth to **periodontal disease**, a term for a number of conditions characterized by inflammation and degeneration of structures that support the teeth (**Figure 25.5**). The roots of the tooth are protected by a covering of specialized connective tissue called *cementum*. As the gums recede with age or with overly aggressive brushing, the formation of caries on the cementum becomes more common.

### Gingivitis

In many cases of periodontal disease, the infection is restricted to the gums, or *gingivae*. This resulting inflammation, called **gingivitis**, is characterized by bleeding of the gums while the teeth are being brushed (see Figure 25.5). This is a condition experienced by at least half of the adult population. It has been shown experimentally that gingivitis will appear in a few weeks if brushing is discontinued and plaque is allowed to accumulate. An assortment of streptococci, actinomycetes, and anaerobic gram-negative bacteria predominate in these infections.

### Periodontitis

Gingivitis can progress to a chronic condition called **periodontitis**, an insidious condition that generally causes little discomfort. About 35% of adults suffer from periodontitis, which is increasing in incidence as more people retain their teeth into old age. The gums are inflamed and bleed easily. Sometimes pus forms in *periodontal pockets* surrounding the teeth (see Figure 25.5). As the infection continues, it progresses toward the root tips. The bone and tissue that support the teeth are destroyed, eventually leading to loosening and loss of the teeth. Numerous bacteria of many different types, primarily *Porphyromonas* (POR-fi-rō-mō-nas) species, are found in these infections; the damage to tissue is done by an inflammatory response to the presence of these bacteria. Periodontitis can be treated surgically by eliminating the periodontal pockets.

**Acute necrotizing ulcerative gingivitis**, also termed **Vincent's disease** or **trench mouth**, is one of the more common serious mouth infections. The disease causes enough pain to make normal chewing difficult. Foul breath (halitosis) also accompanies the infection. Among the bacteria usually associated with this condition is *Prevotella intermedia* (prev-ō-TEL-lah in-ter-MĒ-dē-ah), averaging up to 24% of the isolates. Because these pathogens are usually anaerobic, treatment includes hydrogen peroxide after debridement. Antibiotics may be effective. Bacterial diseases of the mouth are summarized in Diseases in Focus 25.1.

> **CHECK YOUR UNDERSTANDING**
>
> ✔ 25-3 Why are "sugarless" candies and gum, which actually contain sugar alcohols, not considered cariogenic?

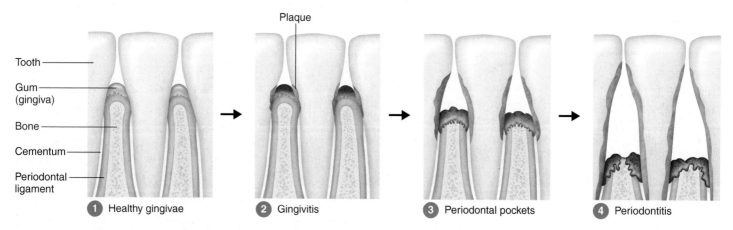

**Figure 25.5 The stages of periodontal disease.** ❶ Teeth firmly anchored by healthy bone and gum tissue (gingiva). ❷ Toxins in plaque irritate gums, causing gingivitis. ❸ Periodontal pockets form as the tooth separates from the gingiva. ❹ Gingivitis progresses to periodontitis. Toxins destroy the gingiva and bone that support the tooth and the cementum that protects the root.

**Q** What is the cause of "pink tooth brush"?

# Bacterial Diseases of the Mouth

Most adults show signs of gum disease, and about 14% of U.S. adults aged 45 to 54 have a severe case. Use the table below to identify infections that could cause persistent sore, swollen, red, or bleeding gums, as well as tooth pain or sensitivity and bad breath. For the solution, go to @MasteringMicrobiology.

This gram-negative rod grown on blood agar accounts for nearly one-quarter of cases.

| Disease | Pathogen | Symptoms | Treatment | Prevention |
|---|---|---|---|---|
| Dental Caries | Primarily *Streptococcus mutans* | Discoloration or hole in tooth enamel | Removal of decayed area | Brushing, flossing, reducing dietary sucrose |
| Periodontal Disease | Various, primarily *Porphyromonas* spp. | Bleeding gums, pus pockets | Removal of damaged area; antibiotics | Plaque removal |
| Acute Necrotizing Ulcerative Gingivitis | *Prevotella intermedia* | Pain with chewing, halitosis | Removal of damaged area; antibiotics | Brushing, flossing |

# Bacterial Diseases of the Lower Digestive System

## LEARNING OBJECTIVE

**25-4** List the causative agents, suspect foods, signs and symptoms, and treatments for staphylococcal food poisoning, shigellosis, salmonellosis, typhoid fever, cholera, gastroenteritis, and peptic ulcer disease.

Diseases of the digestive system are essentially of two types: infections and intoxications.

An **infection** occurs when a pathogen enters the GI tract and multiplies. Microorganisms can penetrate into the intestinal mucosa and grow there, or they can pass through to other systemic organs. **M (microfold) cells** translocate antigens and microorganisms to the other side of the epithelium where they can contact lymphoid tissues (Peyer's patches) to initiate an immune response (see page 486, Figure 17.9, and Figure 25.7). Infections of the GI tract are characterized by a delay in the appearance of gastrointestinal disturbance while the pathogen increases in numbers or affects invaded tissue. There is also usually a fever, one of the body's general responses to an infective organism.

Some pathogens cause disease by forming toxins that affect the GI tract. An **intoxication** is caused by the ingestion of such a preformed toxin. Most intoxications, such as that caused by *Staphylococcus aureus*, are characterized by a very sudden appearance (usually in only a few hours) of symptoms of a GI disturbance. Fever is less often one of the symptoms.

Both infections and intoxications often cause *diarrhea*, which most of us have experienced. Severe diarrhea accompanied by blood or mucus is called **dysentery**. Both types of digestive system diseases are also frequently accompanied by *abdominal cramps, nausea,* and *vomiting*. Diarrhea and vomiting are both defensive mechanisms designed to rid the body of harmful material.

The general term **gastroenteritis** is applied to diseases causing inflammation of the stomach and intestinal mucosa. Botulism is a special case of intoxication because the ingestion of the preformed toxin affects the nervous system rather than the GI tract (see Chapter 22, page 626).

In developing countries, diarrhea is a major factor in infant mortality. Approximately one child in every four dies of it before the age of 5. It is estimated that mortality from childhood diarrhea could be halved by *oral rehydration therapy* (replacement of lost fluids and electrolytes). This is usually a solution of sodium chloride, potassium chloride, glucose, and sodium bicarbonate to replace lost fluids and electrolytes. These solutions are sold in the infant supply department of many stores. Recently, the World Health Organization (WHO) has issued a recommendation to replenish losses of zinc during diarrheal episodes with a course of zinc tablets. This has been found to reduce the duration and severity of diarrheal episodes and even help prevent future episodes for 2 or 3 months. Public health departments often determine the incidence of diarrhea in the population by receiving weekly reports on the sales of oral rehydration preparations.

Diseases of the digestive system are often related to food ingestion.

## Staphylococcal Food Poisoning (Staphylococcal Enterotoxicosis)

A leading cause of gastroenteritis is **staphylococcal food poisoning**, an intoxication caused by ingesting an enterotoxin produced by *S. aureus*. Staphylococci are comparatively resistant to environmental stresses, as discussed on page 314. They also have a fairly high resistance to heat; vegetative cells can tolerate 60°C for half an hour. Their resistance to drying and radiation helps them survive on skin surfaces. Resistance to high osmotic pressures helps them grow in foods, such as cured ham, in which the high osmotic pressure of salts inhibits the growth of competitors.

*S. aureus* is often an inhabitant of the nasal passages, from which it contaminates the hands. It's also a frequent cause of skin lesions on the hands. From these sources, it can readily enter food. If the microbes are allowed to incubate in the food, a situation called **temperature abuse**, they reproduce and release enterotoxin into the food. These events, which lead to outbreaks of staphylococcal intoxication, are illustrated in **Figure 25.6**.

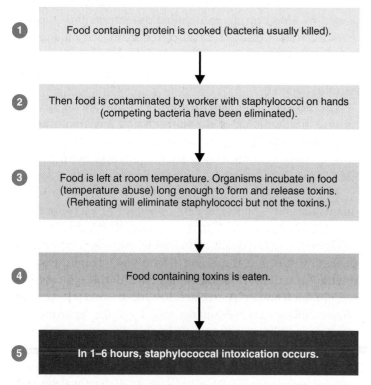

**1** Food containing protein is cooked (bacteria usually killed).

**2** Then food is contaminated by worker with staphylococci on hands (competing bacteria have been eliminated).

**3** Food is left at room temperature. Organisms incubate in food (temperature abuse) long enough to form and release toxins. (Reheating will eliminate staphylococci but not the toxins.)

**4** Food containing toxins is eaten.

**5** In 1–6 hours, staphylococcal intoxication occurs.

**Figure 25.6 The sequence of events in a typical outbreak of staphylococcal food poisoning.**

**Q** How does this differ from foodborne illness caused by a virus?

*S. aureus* produces several toxins that damage tissues or increase the microorganism's virulence. The production of the toxin of serological type A (which is responsible for most cases) is often correlated with the production of an enzyme that coagulates blood plasma. Such bacteria are described as *coagulasepositive*. No direct pathogenic effect can be attributed to the enzyme, but it is useful in the tentative identification of types that are likely to be virulent.

Generally, a population of about 1 million bacteria per gram of food will produce enough enterotoxin to cause illness. The growth of the microbe is facilitated if the competing microorganisms in the food have been eliminated—by cooking, for example. It is also more likely to grow if competing bacteria are inhibited by a higher-than-normal osmotic pressure or by a relatively low moisture level. *S. aureus* tends to outgrow most competing bacteria under these conditions.

Custards, cream pies, and ham are examples of high-risk foods. Competing microbes are minimized in custards by the high osmotic pressure of sugar and by cooking. In ham they are inhibited by curing agents, such as salts and preservatives. Because food contamination by human handlers can't be avoided completely, the most reliable method of preventing staphylococcal food poisoning is adequate refrigeration during storage to prevent toxin formation. The toxin itself is heat stable and can survive up to 30 minutes of boiling. Therefore, once the toxin is formed, it is not destroyed when the food is reheated, although the bacteria will be killed.

The mortality rate of staphylococcal food poisoning is almost zero among otherwise healthy people, but it can be significant in weakened individuals, such as residents of nursing homes.

The diagnosis of staphylococcal food poisoning is usually based on the symptoms, particularly the short incubation time characteristic of intoxication. If the food hasn't been reheated so the bacteria are still alive, the pathogen can be recovered and grown. *S. aureus* isolates can be tested by *phage typing*, a method used in tracing the source of the contamination (see Figure 10.14, page 285). These bacteria grow well in 7.5% sodium chloride, so this concentration is often used in media for their selective isolation. Pathogenic staphylococci usually ferment mannitol, produce hemolysins and coagulase, and form golden-yellow colonies. They cause no obvious spoilage when growing in foods. Detecting the toxin in food samples has always been a problem; there may be only 1 to 2 nanograms in 100 g of food. Reliable serological methods have become commercially available only recently.

## Shigellosis (Bacillary Dysentery)

Bacterial infections, such as salmonellosis and shigellosis, usually have longer incubation periods (12 hours to 2 weeks) than bacterial intoxications, reflecting the time needed for the microorganism to grow in the host. Bacterial infections are often characterized by some fever, indicating the host's response to the infection.

**Shigellosis**, also known as **bacillary dysentery** to differentiate it from amebic dysentery (page 750), is a severe form of diarrhea caused by a group of facultatively anaerobic gramnegative rods of the genus *Shigella*. The genus was named for the Japanese microbiologist Kiyoshi Shiga. The bacteria don't have any natural reservoir in animals and spread only from person to person. Outbreaks are most often seen in families, day-care facilities, and similar settings.

There are four species of pathogenic *Shigella*: *S. sonnei* (SŌN-nē-ē), *S. dysenteriae* (dis-en-TE-rē-ī), *S. flexneri* (FLEKS-ner-ē), and *S. boydii* (BOY-dē-ē). These bacteria are residents only of the intestinal tract of humans, apes, and monkeys. They are closely related to the pathogenic *E. coli*.

The most common species in the United States is *S. sonnei*; it causes a relatively mild dysentery. Many cases of so-called traveler's diarrhea might be mild forms of shigellosis. At the other extreme, infection with *S. dysenteriae* often results in a severe dysentery and prostration. The toxin responsible is unusually virulent and is known as the **Shiga toxin** (see enterohemorrhagic *E. coli*, page 736). *S. dysenteriae*, fortunately, is the least common species of pathogenic *Shigella* in the United States.

The infectious dose required to cause disease is small; the bacteria are not much affected by stomach acidity. They proliferate to immense numbers in the small intestine, but the primary site of disease is the large intestine. There, the bacteria attach to epithelial M cells (page 486). Membranous cellular ruffles surrounding the cell take the bacterium into the cell, similar to invasion by *Salmonella* (shown in Figure 15.2, page 429). The bacteria multiply in the cell and soon spread to neighboring cells, producing Shiga toxin that destroys tissue (**Figure 25.7**). Dysentery is the result of damage to the intestinal wall.

Shigellosis can cause as many as 20 bowel movements in one day. Additional symptoms of infection are abdominal cramps and fever. *Shigella* bacteria rarely invade the bloodstream. Macrophages not only fail to kill *Shigella* bacteria that they phagocytize, but also are killed by them. Diagnosis is usually based on recovery of the microbes from rectal swabs.

The CDC estimates that about 500,000 cases of shigellosis occur annually; most are caused by *S. sonnei* and chiefly affect children under the age of 5. *S. dysenteriae* has a significant mortality rate, however, and the death rate in developing countries where it's prevalent can be as high as 20%. Some immunity seems to result from recovery, but a satisfactory vaccine has not yet been developed.

In severe cases of shigellosis, antibiotic-sensitivity testing is necessary to determine appropriate antibiotic treatment.

## Salmonellosis (*Salmonella* Gastroenteritis)

The *Salmonella* bacteria (named for their discoverer, Daniel Salmon) are gram-negative, facultatively anaerobic

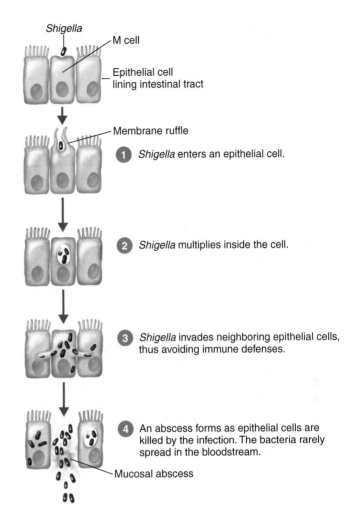

1  *Shigella* enters an epithelial cell.

2  *Shigella* multiplies inside the cell.

3  *Shigella* invades neighboring epithelial cells, thus avoiding immune defenses.

4  An abscess forms as epithelial cells are killed by the infection. The bacteria rarely spread in the bloodstream.

*Shigella*
M cell
Epithelial cell lining intestinal tract
Membrane ruffle
Mucosal abscess

**Figure 25.7  Shigellosis.** This shows the sequence of infection of the intestinal wall.

**Q** Why do *Shigella* rarely spread to the bloodstream?

rods. Their normal habitat is the intestinal tracts of humans and many animals. All salmonellae are considered pathogenic to some degree, causing **salmonellosis**, or *Salmonella* **gastroenteritis**. Pathogenically, salmonellae are separated into *typhoidal salmonellae* (see typhoid fever, page 732) and the *nontyphoidal salmonellae*, which cause the milder disease of salmonellosis.

The nomenclature of the *Salmonella* microbes differs from the norm. There are only two species: *S. enterica* and *S. bongori*. Infections are most often caused by *S. enterica*. There are more than 2000 serotypes (or serovars), of *S. enterica*, only about 50 of which are isolated with any frequency in the United States. (For a discussion of the nomenclature of the salmonellae, see page 304.) Strains are referred to as, for example, *S. enterica* serotype Typhimurium or *S.* Typhimurium.

The salmonellae first invade the intestinal mucosa and multiply there (see Figure 15.2, page 429). Occasionally they manage to pass through the intestinal mucosa at M cells to enter the lymphatic and cardiovascular systems, and from there they

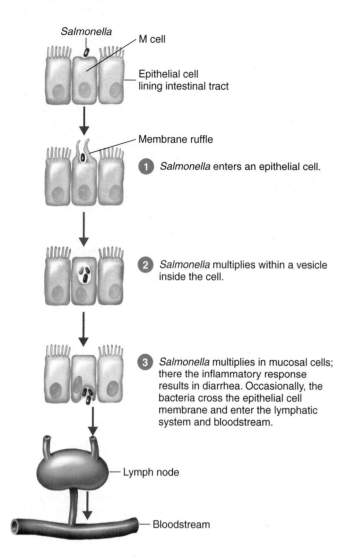

Figure 25.8 Salmonellosis. This shows the sequence of infection of the intestinal wall. Compare with Figure 25.7 showing infection with *Shigella*. Note that invasion of the bloodstream, which happens infrequently, can result in septic shock.

**Q** Why does salmonellosis have a longer incubation period than a bacterial intoxication?

may spread and eventually affect many organs (**Figure 25.8**). They replicate readily within macrophages. Salmonellosis has an incubation time of about 12 to 36 hours. There is usually a moderate fever accompanied by nausea, abdominal pain and cramps, and diarrhea. As many as 1 billion salmonellae per gram can be found in an infected person's feces during the acute phase of the illness.

The mortality rate is overall very low, probably less than 1%. However, the death rate is higher in infants and among the very old; death is usually from septic shock. The infectious dose is usually large, 1000 cells or more. However, the severity and incubation time can depend on the number of *Salmonella* ingested. Normally, recovery will be complete in a

few days, but many patients will continue to shed the organisms in their feces for up to 6 months. Antibiotic therapy isn't useful in treating salmonellosis or, indeed, many diarrheal diseases; treatment consists of oral rehydration therapy.

Salmonellosis is probably greatly underreported. It is estimated that 1 million cases and 380 deaths occur annually (**Figure 25.9**). Meat products are particularly susceptible to contamination by *Salmonella*. The sources of the bacteria are the intestinal tract of many animals. Pet reptiles, such as turtles and iguanas, are also a source; their carriage rate is as high as 90%. In fact, the sale of small turtles (<10 cm) as pets is now prohibited by the FDA because of the risk that children may put them in the mouth. *S.* Enteritidis and *S.* Typhimurium are especially well adapted to chickens and turkeys. Hens are highly susceptible to infection, and the bacteria contaminate the eggs. The bacteria have developed the ability to survive in the albumin, which contains natural preservatives such as *lysozyme* (see page 83) and *lactoferrin* (which binds iron the bacteria require). Estimates are that 1 in 20,000 eggs in this country is contaminated by *Salmonella*. Health authorities caution the public to eat only well-cooked eggs. An often unsuspected factor is the presence of inadequately cooked or raw eggs in foods such as hollandaise sauce, cookie batter, and Caesar salad. Fruits have been frequent sources of foodborne illness from ingestion of *Salmonella* (see the Clinical Focus box on the facing page).

Prevention also depends on good sanitation practices to deter contamination and on proper refrigeration to prevent increases in bacterial numbers. The microbes are generally

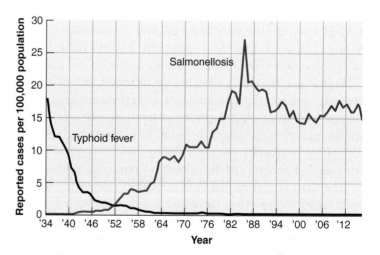

Figure 25.9 The incidence of salmonellosis and typhoid fever. An important factor in comparing the two diseases is that typhoid transmission is almost entirely human-to-human, and salmonellosis is transmitted primarily between animal products and humans.

Source: CDC. *MMWR* 65(52), January 6, 2017.

**Q** Can you suggest reasons for the change in prevalence of those two diseases?

# A Foodborne Infection

As you read through this problem, you'll see questions that epidemiologists ask themselves as they solve a clinical problem. Try to answer each question as an epidemiologist.

1. On August 29, Joanie, a 36-year-old woman in Ohio, is hospitalized with a 3-day history of nausea, vomiting, and diarrhea. She has a temperature of 39.5°C, and she is dehydrated.

   What sample is needed from Joanie to determine the cause of her signs and symptoms?

2. A stool culture grows gram-negative, non–lactose-fermenting bacteria.

   Can you identify the bacteria? (See the photograph.)

3. Joanie is one of 907 culture-confirmed cases in a 40-state outbreak of salmonellosis.

   What information would you try to obtain from these patients?

4. No single restaurant or restaurant chain is associated with the outbreak.

   How will you determine the source of the infection?

5. Epidemiologists conduct a case-control study to compare 53 case-patients with 53 healthy controls from the same geographic locations. All 106 people are asked to complete a questionnaire about foods eaten. (See the table below.)

   Odds ratio (OR) is a measure of the probability (risk) that an event will result in disease. The OR must be calculated for each source of exposure.

   Using the 2 × 2 table as a guide, complete the remaining calculations to determine the probable source of the infection.

6. There is a strong association between illness and consumption of cucumbers. The implicated cucumbers were of one

Some *Salmonella* colonies form red colonies with black centers, which allows them to be differentiated from red colonies of *Shigella*.

variety (Persian) and distributed to markets by one company.

   What would you do now?

7. *Salmonella* Poona was isolated from surfaces in the packing shed and from cardboard boxes.

   What factors might contribute to cucumbers as a vehicle of transmission?

The U.S. Food and Drug Administration (FDA) investigated the grower but could not identify the source of contamination. They noted concerns about wastewater management and sanitizing of the prewash and storage areas.

Consumers should follow safe handling guidelines: cucumbers, like most produce, should be washed thoroughly, scrubbed with a clean produce brush before peeling or cutting, and refrigerated as soon as possible to prevent multiplication of bacteria such as *Salmonella*.

*Source:* Adapted from *MMWR* 65(5051): 1430–1433, December 30, 2016.

| Exposure | Exposed | | Not Exposed | | Odds Ratio |
|---|---|---|---|---|---|
| | (a) Ill | (b) Not Ill | (c) Ill | (d) Not Ill | |
| Eggs | 47 | 40 | 6 | 13 | 2.55 |
| Chicken | 32 | 20 | 21 | 33 | |
| Lettuce | 34 | 30 | 19 | 23 | |
| Milk | 42 | 39 | 11 | 14 | |
| Cucumbers | 47 | 24 | 6 | 29 | |

**Calculation using a statistical 2 × 2 contingency table**

| | Ill | Not Ill | Odds Ratio |
|---|---|---|---|
| Ate _____ | (a) | (b) | $(e) = \dfrac{a}{a + b}$ |
| Did not eat _____ | (c) | (d) | $(f) = \dfrac{c}{c + d}$ |
| Odds ratio = | $\dfrac{a \times d}{b \times c} = $ _____ | | = Times more likely to become ill by eating that item |

destroyed by normal cooking. Chicken, for example, should be cooked to an internal temperature of 74°C and ground beef to 71°C. However, contaminated food can contaminate a surface, such as a cutting board. Then another food subsequently prepared on the board might not be cooked.

Diagnosis usually depends on isolating the pathogen from the patient's stool or from leftover food. Serotyping is used in outbreaks to identify strains. State public health labs get a DNA fingerprint (see Chapter 9, pages 258–260) of isolates. These fingerprints can be used to track an outbreak to its source.

## Typhoid Fever

The most virulent serotype of *Salmonella, S.* Typhi, causes the bacterial disease **typhoid fever**. Unlike the salmonellae that cause salmonellosis, this pathogen is not found in animals; it's spread only in the feces of other humans. Before the days of proper sewage disposal, water treatment, and food sanitation, typhoid was an extremely common disease. Its incidence has been declining in the United States, whereas that of salmonellosis has been increasing (see Figure 25.9). Typhoid fever is still a frequent cause of death in parts of the world with poor sanitation. Globally, an estimated 21 million cases occur annually, causing tens of thousands of deaths.

Instead of being destroyed by phagocytic cells, *S.* Typhi multiply within them and are disseminated into multiple organs, especially the spleen and liver. Eventually, the phagocytic cells lyse and release *S.* Typhi into the bloodstream. The time required for this explains why the incubation period of typhoid fever (2 or 3 weeks) is much longer than for salmonellosis (12 to 36 hours). The patient with typhoid fever suffers from a high fever of about 40°C and continual headache. Diarrhea appears only during the second or third week, and the fever then tends to decline. In severe cases, which can be fatal, ulceration and perforation of the intestinal wall can occur. Before antibiotic therapy was available, a mortality rate of 20% was common; with the treatments available today, it is less than 1%.

Substantial numbers of recovered patients, about 1–3%, become *chronic carriers*. They harbor the pathogen in the gallbladder and continue to shed bacteria for several months. A number of such carriers continue to shed the organism indefinitely. The classic example of a typhoid carrier was Mary Mallon, also known as Typhoid Mary. She worked as a cook in New York state in the early part of the twentieth century and was responsible for several outbreaks of typhoid and three deaths. She became well known through the attempts of the state to restrain her from working at her chosen trade.

In recent years there have been about 300 to 350 annual cases of typhoid fever in the United States, of which 70% were acquired during foreign travel. Normally, there are fewer than three deaths each year.

When the antibiotic chloramphenicol was introduced in 1948, typhoid became a treatable disease. However, chloramphenicol-resistant *Salmonella* emerged in the 1970s, and ceftriaxone or azithromycin is now largely used instead. Treatment of the chronic carrier might require weeks of antibiotic therapy.

Recovery from typhoid confers lifelong immunity. Vaccines are seldom used in developed countries except for high-risk laboratory or military personnel. Vaccination is recommended before travel to endemic countries. The vaccine that has long been in use is a killed-organism type, which must be injected.

A live attenuated vaccine that can be taken orally in three or four doses protects well for as long as 7 years.

**CHECK YOUR UNDERSTANDING**

✔ 25-4 Why was typhoid fever almost entirely eliminated in developed countries by modern sewage treatment whereas salmonellosis has not been?

## Cholera

The causative agent of **cholera**, one of the most serious gastrointestinal diseases, is *Vibrio cholerae*, a slightly curved, gram-negative rod with a single polar flagellum (**Figure 25.10**). Cholera bacilli grow in the small intestine and produce an exotoxin, *cholera toxin* (see Chapter 15, pages 431 and 436), that causes host cells to secrete water and electrolytes, especially potassium. The result is watery stools containing masses of intestinal mucus and epithelial cells—called "rice water stools" from their appearance. As much as 12 to 20 liters (3 to 5 gallons) of fluids can be lost in a day, and the sudden loss of these fluids and electrolytes causes shock, collapse, and often death. The blood, lacking fluids, may become so viscous that vital organs are unable to function properly. Violent vomiting generally also occurs. The microbes are not invasive, and a fever is usually not present. The severity of cholera varies greatly, and the number of subclinical cases might be several times the number reported. Untreated cases of cholera may have a mortality rate of 50%, although with proper supportive care it is usually less than 1%. The diagnosis is based upon symptoms and culturing of *V. cholerae* from feces.

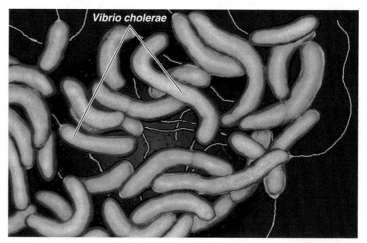

**Figure 25.10 *Vibrio cholerae*, the cause of cholera.** Notice the slightly curved morphology.

**Q** What are the effects of the sudden loss of fluid and electrolytes during infection with *V. cholerae*?

Cholera bacteria, and other members of the genus *Vibrio* in general, are strongly associated with brackish (salty) waters characteristic of estuaries, although they are also readily spread in contaminated freshwater. They form biofilms and colonize copepods (tiny crustaceans), algae, and other aquatic plants and plankton, which aids their survival. It has even been reported that because of this growth habit, straining contaminated water through folded layers of finely woven cloth (such as saris worn by Indian women) often removes these attached bacteria and makes the water safe to drink. Under unfavorable conditions, *V. cholerae* may become dormant; the cell shrinks into a nonculturable, spherical state. A favorable change in the environment causes them to revert rapidly to the culturable form. Both forms are infectious.

Although they survive well in their aquatic environment, cholera bacteria are exceptionally sensitive to stomach acids. Persons with impaired stomach acid secretion or who are taking antacids are at higher risk of infection. Normal individuals may require infectious doses on the order of 100 million bacteria to cause severe cholera. Recovery from cholera results in an effective immunity, but only to bacterial strains of the same antigenic characteristics. The serogroup O:1 (see the footnote in Chapter 11, page 304), which caused a pandemic in the 1880s, is known as the *classical* strain. A later pandemic was caused by a biotype of O:1 named *El Tor* (for the El Tor quarantine camp for pilgrims to Mecca, where it was first isolated). Until the 1990s it was thought that only *V. cholerae* O:1 caused epidemic cholera, but a widespread epidemic in India and Bangladesh by a new serogroup, O:139, changed this view. There are also nonepidemic strains of *V. cholerae*, non-O:1/O:139, that are only infrequently associated with large-scale outbreaks of cholera. They occasionally cause wound infections or sepsis, especially in people with liver disease or who are immunosuppressed.

In the United States there have been occasional cases of cholera caused by the O:1 serogroup. These have all occurred in the Gulf Coast area, and the pathogen may be endemic in these coastal waters. Outbreaks of cholera in this country are limited by high standards of sanitation. This represents the primary means of control and is important because stools may contain 100 million *V. cholerae* per gram. An example of how this can change quickly was illustrated in 2010, when the Caribbean nation of Haiti experienced an earthquake that seriously damaged much of the water supply and other systems. (See the **Big Picture** on cholera after natural disasters on pages 734–735.) An outbreak of cholera caused several hundred deaths when cholera bacteria of a strain usually found in Asia were introduced from an outside source. Available oral vaccines provide immunity of relatively short duration and only moderate effectiveness.

## CLINICAL CASE

The fecal sample is cultured on sorbitol–MacConkey agar. A sorbitol-negative colony is tested for lactose fermentation. The bacteria produce acid from lactose and do not use citrate as their sole carbon source.

**Identify the bacteria by using the identification key in Figure 10.8 on page 281. What does Anna's pediatrician need to know about her history?**

723   **733**   739   745   755

The most effective therapy is intravenous replacement of the lost fluids and electrolytes. As much as 10% of the patient's body weight within a few hours may be required. Rehydration therapy is so effective that in Bangladesh, for example, where cholera is common, deaths are considered "unusual." Severe cholera may be treated with doxycycline.

### Noncholera Vibrios

At least 11 species of *Vibrio*, in addition to *V. cholerae*, can cause human illness. Most are adapted to life in salty coastal waters. *Vibrio parahaemolyticus* (par'ah-hē-mō-LI-ti-kus) is found in saltwater estuaries in many parts of the world. It is morphologically similar to *V. cholerae* and the most common cause of gastroenteritis by *Vibrio* spp. in humans. The noncholera *Vibrio* bacteria cause about 80,000 illnesses and 100 deaths in the United States every year. Raw oysters and crustaceans, such as shrimp and crabs, have been associated with several outbreaks of gastroenteritis in the United States in recent years.

These infections are life threatening and require early antibiotic therapy for successful treatment.

### *Escherichia coli* Gastroenteritis

One of the most prolific microorganisms in the human intestinal tract is *Escherichia coli*. Because it's so common and so easily cultivated, microbiologists often regard it as something of a laboratory pet. *E. coli* are normally harmless, but certain strains can be pathogenic. Mobile genetic elements can turn *E. coli* into a highly adapted pathogen causing a range of diseases. Some toxin-secreting pathogenic strains are well adapted to invasion of intestinal epithelial cells, causing *E. coli gastroenteritis*. Other locations, such as the urinary tract, bloodstream, and central nervous system, can also be affected. Five pathogenic varieties (pathotypes) of *E. coli* have been well characterized.

**Enteropathogenic *E. coli* (EPEC)** is a major cause of diarrhea in developing countries and is potentially fatal in infants. As the bacteria attach to the intestinal wall, they eliminate surrounding microvilli and stimulate host-cell actin to form

**Cholera is one of the most feared diseases that occurs after natural disasters. Accurate tracking of the true causes of these epidemics can lead to better treatment and prevention.**

Many people assume that the greatest source of disease that breaks out after natural disasters is dead bodies. However, studies show that displacement of the survivors and disruption of access to safe water are the biggest contributors. Cholera is a diarrheal disease that can increase when sanitation and modern sewage disposal systems are compromised. In 1991, an outbreak in Peru caused over 1 million cases and 10,000 deaths. An epidemic of over 16,000 cases occurred after flooding in West Bengal in 1998.

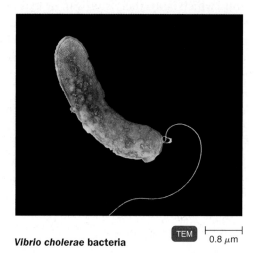

***Vibrio cholerae* bacteria**   TEM   ⊢—⊣ 0.8 μm

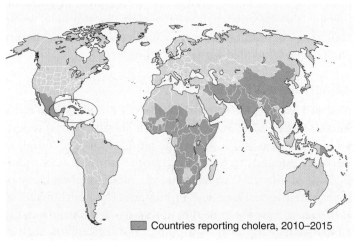

☐ Countries reporting cholera, 2010–2015

*V. cholerae* nearly identical to a strain circulating in Nepal infected and killed thousands with cholera in the Caribbean (circled area on map) 2010–2015.

*Source:* World Health Organization.

**People in Haiti struggle to find clean drinking water after the 2016 hurricane.**

In 2004, 17,000 cases of diarrheal disease, including cholera, struck Bangladesh after severe floods. In 2010, a cholera epidemic affected over 600,000 people in earthquake-ravaged Haiti, resulting in over 7000 deaths. And after a hurricane in 2016, cholera cases again surged in Haiti, with an estimated 770 new cases per week.

## What Caused Cholera to Return to the Caribbean after the 2010 Earthquake?

Cholera's appearance in Haiti after its major earthquake was something of a mystery. No outbreaks had occurred in the region for about a century. The epidemic also occurred nearly 10 months after the earthquake, well into relief and rebuilding efforts. Using gel electrophoresis and whole-genome sequence typing, epidemiologists determined that the *Vibrio cholerae* responsible for most cases was an Asian strain nearly identical to one circulating in Nepal the same year Haiti's outbreak began. Infected Nepalese soldiers who were part of a United Nations (U.N.) peacekeeping force brought the pathogen to Haiti, where a deficient septic system at the Nepalese base allowed a stream to be contaminated with base sewage. The stream flows into a river many Haitians use for drinking water. The initial outbreak developed among people who drank water downstream from the base.

Cholera remains on the islands of Hispaniola and Cuba, with over 400,000 cases reported since the 2010 earthquake. In 2014, the same bacterial strain caused a cholera outbreak in Mexico. This outbreak began with flooding caused by a tropical storm and hurricane.

# Disaster and Disease—Searching for Solutions

## Strategies for Disaster Preparedness

- **Oral Rehydration Solutions** Cholera's mortality rate is significantly reduced by treating victims with oral rehydration solutions made from salt, sugar, and water. Since the 1970s, this therapy is estimated to have saved over 40 million lives. Teaching global citizens how to prepare this life-saving solution can prevent many deaths from diarrheal diseases after disasters. At health care facilities, cholera cots, specially designed beds, are also used to collect and measure feces lost during infection so the same amount of fluid can be replaced in the patient.

- **Stockpiling Vaccines** Disaster preparedness experts have learned from the Haiti earthquake and subsequent cholera outbreak that stockpiling vaccines, when possible, can help to prevent future outbreaks like this one. One study on oral cholera vaccine in the aftermath of the Haiti earthquake demonstrated that vaccinating even half of an area's population made outbreaks less likely by providing herd immunity to the community at large.

Oral vaccine stockpiles, when quickly dispensed, can help curb outbreaks before they become widespread. According to the World Health Organization, about 3–5 million cholera cases occur annually, with 100,000–120,000 deaths due to fluid loss.

## The Ultimate Solution

While oral rehydration therapy and vaccines can be helpful once a cholera outbreak begins, the ultimate solution to a cholera epidemic is to engineer proper sanitation, a goal that is more long-range and expensive. The World Health Organization estimates that over 760 million people lack access to safe water and 2.5 billion people lack proper sanitation—and that's within their daily lives, not as a result of disruptions in normal sanitation after a disaster event. Many public and private agencies are developing programs to tackle this large goal. For example, the CDC's WASH (Water, Sanitation and Hygiene) Program promotes safe household water storage techniques, handwashing interventions, and training of community health workers. Their efforts have resulted in a 25% decrease in childhood diarrheal infections within four Central American countries, and 50% fewer diarrheal infections in children receiving weekly handwashing lessons.

## KEY CONCEPTS

- Although natural disasters do not automatically cause disease outbreaks, damage to water and sanitation infrastructure can increase the risk of diarrheal diseases such as cholera. **(See Chapter 27, "Water Treatment" page 799 and "Sewage (Wastewater) Treatment" page 800.)**

- Vaccination of a majority of the population can lead to herd immunity that protects the unvaccinated within that community. **(See Chapter 18, "Principles and Effects of Vaccination" pages 500–501.)**

- Tracking the genomics of pathogens has become a mainstay of monitoring, preventing, and controlling infectious disease outbreaks. **(See Chapter 9, "Forensic Microbioligy" pages 258–260.)**

**A health educator demonstrating how to make an oral rehydration solution from water, salt, and sugar in Thiruvananthapuram, India. Treating cholera victims with this solution can significantly reduce mortality.**

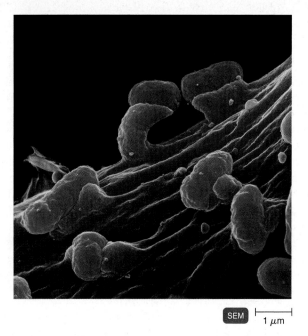

**Figure 25.11 Pedestal formation by Enterohemorrhagic *E. coli* (EHEC) O157:H7.** As EHEC bacteria (orange) adhere to the epithelial wall, they destroy the surface microvilli and cause the formation of a pedestal-like projection (green) on which they rest. The function of these actin-rich structures is unclear, but they may facilitate the bacteria's spread to adjacent cells.

**Q Is adhesion a factor in the pathogenicity of a microbe?**

pedestals beneath their site of attachment (**Figure 25.11**). EPEC bacteria secrete a number of effector proteins that are translocated into host cells, some contributing to diarrhea.

**Enteroinvasive *E. coli* (EIEC)** is generally agreed to be almost synonymous with *Shigella*—it has the same pathogenic mechanisms. EIEC gain access to the submucosa of the intestinal tract through M cells (see Figure 25.7) in the same manner as *Shigella*. This invasion results in inflammation, fever, and a *Shigella*-like dysentery.

**Enteroaggregative *E. coli* (EAEC)** is a group of coliforms found only in humans. They are named for their growth habit, in which the bacteria cause a "stacked-brick" configuration when grown with epithelial cells. EAEC are not invasive but produce an enterotoxin causing a watery diarrhea. Some studies suggest that another pathotype, *diffusely adherent E. coli*, is also associated with diarrheal illness.

In recent years, strains of **enterohemorrhagic *E. coli* (EHEC)** have caused several outbreaks of serious disease in the United States. The primary virulence factor in these bacteria is a Shiga-like toxin. Shiga toxins are a family of toxins that are closely related. Some *E. coli* strains that produce Shiga-like toxins are termed **Shiga-toxin-producing *E. coli* (STEC)**. True Shiga toxin is produced only by *Shigella dysenteriae*. Most outbreaks are due to EHEC serotype O157:H7. Other lesser known strains include O121 and O104:H21. (Refer to page 304 for an explanation of this numerical nomenclature.) Because the toxin is released

upon the cell's lysis, antibiotic therapy can worsen the attack by causing the release of more toxin.

Cattle, which are not affected by the pathogen, are the main reservoir; infections are spread by contaminated food or water. Currently, 10–30% of domestic cattle carry STEC, which contaminate the carcass at slaughter. There are requirements for testing ground meats for the presence of this strain of *E. coli*, especially if intended for export. Leafy vegetables may also be contaminated, sometimes by runoff from feedlots. Ingested food is not the only infection source; some cases have been associated with children's visits to farms or petting zoos. The infectious dose is estimated to be very small, probably fewer than 100 bacteria.

STEC bacteria are responsible for more than 265,000 illnesses each year. About 6% of infected people develop inflammation of the colon (the last part of the large intestine, ending just above the rectum) involving profuse bleeding, called *hemorrhagic colitis*. Unlike *Shigella*, these *E. coli* bacteria don't invade the intestinal wall (see Figure 25.7), but rather release the toxin into the intestinal lumen (space).

Another dangerous complication is *hemolytic uremic syndrome (HUS)*. Characterized by blood in the urine, often leading to kidney failure, HUS occurs when the kidneys are affected by the toxin. Some 5–10% of young children who have been infected progress to this stage, which has a mortality rate of about 5%. Management of these patients primarily involves intravenous rehydration and careful monitoring of serum electrolytes. Some survivors of HUS may require kidney dialysis or even transplants.

It's recommended that public health laboratories test routinely for STEC. A standard method is to use media that differentiate *E. coli* O157 bacteria by their inability to ferment sorbitol (Figure 10.8, page 281). In 2016, an ELISA test was developed by the USDA to identify Shiga-toxin-producing strains.

Vaccines that greatly lower the numbers of O157: H7 bacteria in cattle are available, but it is uncertain whether they will find widespread use.

A pathogenic group of *E. coli* called **enterotoxigenic *E. coli* (ETEC)** secretes enterotoxins that cause diarrhea. The illness is frequently fatal for children under 5. One of the enterotoxins ETEC produces resembles the cholera toxin in function. ETEC bacteria are not invasive and remain in the intestinal lumen.

**Traveler's Diarrhea**

It has long been observed that travel broadens the mind and loosens the bowels, leading to the common name of **traveler's diarrhea**. The most common bacterial cause is ETEC; the second most frequent isolate is EAEC. Traveler's diarrhea can also be caused by other gastrointestinal pathogens, such as *Salmonella*, *Shigella*, and *Campylobacter*—as well as by various unidentified bacterial pathogens, viruses, and protozoan parasites. In fact, in most cases the causative agent is never identified, and chemotherapy is not attempted. Once contracted, the best treatment is the usual oral rehydration recommended for all diarrhea. In severe cases, antimicrobial drugs may be necessary. Prescribed antibiotics may provide some protection; another option is to take an over-the-counter drug such as Lomotil® to treat the symptoms, but the best advice in risky areas is to prevent infection.

## Campylobacteriosis (*Campylobacter* Gastroenteritis)

*Campylobacter* are gram-negative, microaerophilic, spirally curved bacteria that have emerged as the leading cause of diarrhea in the United States. They adapt well to the intestinal environment of animal hosts, especially poultry. Culturing *Campylobacter* requires conditions of low oxygen and high carbon dioxide developed in special apparatus. The bacteria's optimum growth temperature of about 42°C approximates that of their animal hosts, but the bacteria do not replicate in food. Almost all retail chicken is contaminated with *Campylobacter*. Nearly 60% of cattle excrete the organism in feces and milk, but retail red meats are less likely to be contaminated.

There are more than an estimated 1 million cases of **campylobacteriosis** in the United States annually, usually caused by *C. jejuni*. The infectious dose is fewer than 1000 bacteria. Clinically, it is characterized by fever, cramping abdominal pain, and diarrhea or dysentery. Normally, recovery follows within a week.

An unusual complication of campylobacterial infection is that it is linked, in about 1 in 1000 cases, to the neurological disease *Guillain-Barré syndrome*, a temporary paralysis. Apparently, a surface molecule of the bacteria resembles a lipid component of nervous tissue and provokes an autoimmune attack.

Campylobacteriosis is prevented by thoroughly cooking chicken and pasteuring milk. Infections may be treated with azithromycin.

## *Helicobacter* Peptic Ulcer Disease

In 1982, a physician in Australia cultured a spiral-shaped, microaerophilic bacterium observed in the biopsied tissue of stomach ulcer patients. Now named *Helicobacter pylori*, it is accepted that this microbe is responsible for most cases of **peptic ulcer disease**. This syndrome includes gastric and duodenal ulcers. (The duodenum is the first few centimeters of the small intestine.) About 30–50% of the population in the developed world become infected; the infection rate is higher elsewhere. Only about 15% of people infected develop ulcers, so certain host factors are probably involved. For example, people with type O blood are more susceptible, which is also true of cholera. *H. pylori* is also designated as a carcinogenic bacterium. Gastric cancer develops in about 3% of people infected with these bacteria.

The stomach mucosa contains cells that secrete gastric juice containing proteolytic enzymes and hydrochloric acid that activates these enzymes. Other specialized cells produce a layer of mucus that protects the stomach itself from digestion.

If this defense is disrupted, an inflammation of the stomach (gastritis) results. This inflammation can then progress to an ulcerated area (**Figure 25.12**). Through an interesting adaptation, *H. pylori* can grow in the highly acidic environment of the stomach, which is lethal for most microorganisms. *H. pylori* produces large amounts of an especially efficient urease, an enzyme that converts urea to the alkaline compound ammonia, resulting in a locally high pH in the area of growth.

The eradication of *H. pylori* with antimicrobial drugs usually leads to the disappearance of peptic ulcers. Several antibiotics, usually administered in combination, have proven effective. Bismuth subsalicylate (Pepto-Bismol®) is also effective and is often part of the drug regimen. When the bacteria are successfully eliminated, the recurrence rate of the ulcer is only about 2–4% a year. Reinfection can result from many environmental sources but is less likely in areas with high standards of sanitation; in fact, there is some evidence that infection by *H. pylori* is slowly disappearing in developed countries.

The most reliable diagnostic test requires a biopsy of tissue and culture of the organism. An interesting diagnostic approach is the urea breath test. The patient swallows radioactively labeled urea; if the test is positive, $CO_2$ labeled with radioactivity can be detected in the breath within about 30 minutes. This test is most useful for determining the effectiveness of chemotherapy because a positive test is an indication of live *H. pylori*. Diagnostic tests of stools to detect antigens (not antibodies) for *H. pylori* are suitable for follow-up tests following therapy. They are the noninvasive test of choice, especially for children. Serological tests to detect antibodies are inexpensive but not useful in determining eradication.

## *Yersinia* Gastroenteritis

Other enteric pathogens being identified with increasing frequency are *Yersinia enterocolitica* (en'ter-ō-kōl-IT-ik-ah) and *Y. pseudotuberculosis* (soo'dō-too'ber-kū-LŌ-sis). These gram-negative bacteria are intestinal inhabitants of many domestic animals and are most often transmitted by eating raw or undercooked pork. Both microbes are distinctive in their ability to grow at refrigerator temperatures of 4°C. This ability increases their numbers in stored refrigerated blood, to the extent that their endotoxins can result in shock to the blood recipient. *Yersinia* has occasionally been the cause of severe reactions when it contaminates transfused blood.

About 100,000 cases of **Yersinia gastroenteritis**, or **yersiniosis**, caused by these pathogens occur annually in the United States. The symptoms are diarrhea, fever, headache, and abdominal pain. The pain is often severe enough to cause a misdiagnosis of appendicitis. Diagnosis requires culturing the organism, which can then be identified using biochemical or molecular tests. Adults suffering from yersiniosis usually recover in 1 or 2 weeks; children may take longer. Treatment with antibiotics and oral rehydration may be helpful.

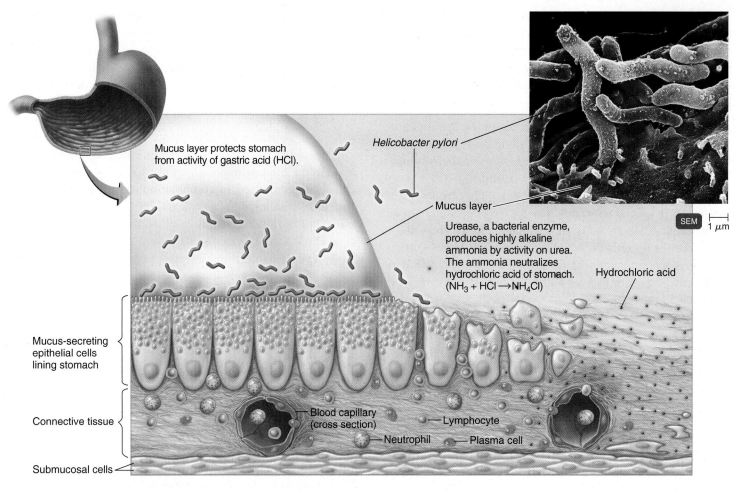

**Figure 25.12 *Helicobacter pylori* infection, leading to ulceration of the stomach wall.**
To survive in the acidic environment of the stomach, *H. pylori* bacteria must neutralize the gastric acid, hydrochloric acid (HCl). They do this by producing large amounts of the enzyme urease. Urea, which is normally secreted into the stomach, is converted into carbon dioxide and ammonia [$(NH_2)_2CO + H_2O \rightarrow CO_2 + 2NH_3$]. The ammonia neutralizes the gastric HCl ($NH_3 + HCl \rightarrow NH_4Cl$).

**Q** How can ammonia be used to diagnose *Helicobacter* infection?

## *Clostridium perfringens* Gastroenteritis

One of the more common, if underrecognized, causes of food poisoning in the United States is caused by *Clostridium perfringens*, a large, gram-positive, endospore-forming, obligately anaerobic rod. This bacterium is also responsible for human gas gangrene (see Chapter 23, page 659).

Most outbreaks of *Clostridium perfringens* **gastroenteritis** are associated with meats or meat stews contaminated with intestinal contents of the animal during slaughter. Such foods meet the pathogen's nutritional requirement for amino acids, and when the meats are cooked, the oxygen level is lowered enough for clostridial growth. The endospores survive most routine heatings, and the generation time of the vegetative bacterium is less than 20 minutes under ideal conditions. Large populations can therefore build up rapidly when foods are being held for serving or when inadequate refrigeration leads to slow cooling.

The microbe grows in the intestinal tract and produces an exotoxin that causes the typical symptoms of abdominal pain and diarrhea. Most cases are mild and self-limiting and probably are never clinically diagnosed. If treatment is required, oral rehydration is recommended. The symptoms usually appear 8 to 12 hours after ingestion. Diagnosis is based on finding at least $10^6$ *C. perfringens* endospores per gram of stool within 48 hours of the onset of illness.

## *Clostridium difficile*–Associated Diarrhea

*Clostridium difficile*–**associated diarrhea** is a disease condition that has appeared in recent decades and has been described as being responsible for more deaths than all other intestinal infections combined. *C. difficile* is a gram-positive, endospore-forming anaerobe found in the stool of many healthy adults. The exotoxins it produces cause a disease that manifests itself in symptoms ranging from a mild case of diarrhea to life-threatening colitis (inflammation of the colon). The colitis can result in ulceration, and possible perforation, of the intestinal wall. The disease is usually precipitated by the extended use of antibiotics. The elimination of most competing intestinal bacteria permits rapid proliferation of the toxin-producing *C. difficile*. It occurs mostly in health care settings such as hospitals and nursing homes. Outbreaks have occurred in daycare centers, and caregivers have been known to acquire it from patients. The mortality rate is highest in elderly patients. *C. difficile* causes a half-million infections and as many as 29,000 deaths annually. Infections are treated with vancomycin or fidaxomicin, although recurrence is common.

### *Bacillus cereus* Gastroenteritis

*Bacillus cereus* (SEER-ē-us) is a large, gram-positive, endospore-forming bacterium that is very common in soil and vegetation and is generally considered harmless. It has, however, been identified as the cause of outbreaks of foodborne illness. Heating the food does not always kill the spores, which germinate as the food cools. Because competing microbes have been eliminated in the cooked food, *B. cereus* grows rapidly and produces toxins. Rice dishes served in Asian restaurants seem especially susceptible.

There are two different clinical syndromes associated with *B. cereus* gastroenteritis, which correspond to two different toxins elaborated by the bacteria. Some cases of ***Bacillus cereus gastroenteritis*** resemble *C. perfringens* intoxications and are almost entirely diarrheal in nature (usually appearing 8 to 16 hours after ingestion). Other episodes resemble staphylococcal food poisoning, with nausea and vomiting 2 to 5 hours after ingestion. It's suspected that different toxins are involved in producing the differing symptoms. Both forms of the disease are self-limiting. The diseases can be differentiated by isolating at least $10^5$ *B. cereus* per gram of suspected food.

Bacterial diseases of the GI tract are summarized in Diseases in Focus 25.2.

## Viral Diseases of the Digestive System

### LEARNING OBJECTIVES

25-5 List the causative agents, modes of transmission, sites of infection, and symptoms for mumps.

25-6 Differentiate hepatitis A, hepatitis B, hepatitis C, hepatitis D, and hepatitis E.

25-7 List the causative agents, mode of transmission, and symptoms of viral gastroenteritis.

Although viruses don't reproduce within the contents of the digestive system like bacteria, they invade many organs associated with the system.

### Mumps

The targets of the mumps virus, the parotid glands, are located just below and in front of the ears (see Figure 25.1). Because the parotids are one of the three pairs of salivary glands of the digestive system, it is appropriate to include a discussion of mumps in this chapter.

**Mumps** typically begins with painful swelling of one or both parotid glands 16 to 18 days after exposure to the virus (**Figure 25.13**). The virus is transmitted in saliva and respiratory secretions, and its portal of entry is the respiratory tract.

**Figure 25.13 A case of mumps.** This patient shows the typical swelling of mumps.

**Q** How is the mumps virus transmitted?

# Bacterial Diseases of the Lower Digestive System

An 8-year-old boy has diarrhea, chills, fever (39.3°C), abdominal cramps, and vomiting for 3 days. The next month, his 12-year-old brother experiences the same symptoms. Two weeks before the first patient became ill, the family had purchased a small (<10 cm) red-eared slider turtle at a flea market. Use the table below and the information on pages 727–739 to identify infections that could cause these symptoms. For the solution, go to @MasteringMicrobiology.

**Red-eared slider pets should be >10 cm (4 in.), large enough that children can't put the turtles in their mouths.**

| Disease | Pathogen | Symptoms | Intoxication/Infection | Diagnostic Test | Treatment |
|---------|----------|----------|------------------------|-----------------|-----------|
| Staphylococcal Food Poisoning | Staphylococcus aureus | Nausea, vomiting, and diarrhea | Intoxication (enterotoxin) | Phage typing | None |
| Shigellosis (bacillary dysentery) | Shigella spp. | Tissue damage and dysentery | Infection (endotoxin and Shiga toxin, exotoxin) | Isolation of bacteria on selective media | Usually none needed |
| Salmonellosis | Salmonella enterica | Nausea and diarrhea | Infection (endotoxin) | Isolation of bacteria on selective media, serotyping | Oral rehydration |
| Typhoid Fever | Salmonella Typhi | High fever, significant mortality | Infection (endotoxin) | Isolation of bacteria on selective media, serotyping | Requires antibiotic-susceptibility testing. Preventive: vaccine |
| Cholera | Vibrio cholerae O:1 and O:139 | Diarrhea with large water loss | Infection (exotoxin) | Isolation of bacteria on selective media | Rehydration; doxycycline |
| Vibrio parahaemolyticus Gastroenteritis | V. parahaemolyticus | Cholera-like diarrhea, but generally milder | Infection (enterotoxin) | Isolation of bacteria on 2–4% NaCl | Rehydration |
| Escherichia coli Gastroenteritis | EPEC, EIEC, EAEC, ETEC | Watery diarrhea | Infection (exotoxins) | Isolation on selective media, DNA fingerprinting | Oral rehydration |
| Shiga Toxin–Producing Enterohemorrhagic E. coli | E. coli O157:H7 | Shigella-like dysentery; hemorrhagic colitis, HUS | Infection, Shiga toxin (exotoxin) | Isolation, sorbitol fermentation test, DNA fingerprinting | Intravenous rehydration, serum electrolyte monitoring |
| Campylobacteriosis (Campylobacter Gastroenteritis) | Campylobacter jejuni | Fever, abdominal pain, diarrhea | Infection | Isolation in low $O_2$, high $CO_2$ | Azithromycin |
| Helicobacter Peptic Ulcer Disease | Helicobacter pylori | Peptic ulcers | Infection | Urea breath test, bacterial culture | Antibiotics |
| Yersinia Gastroenteritis | Yersinia enterocolitica | Abdominal pain and diarrhea, usually mild; may be confused with appendicitis | Infection (endotoxin) | Culture, biochemical or molecular tests | Oral rehydration |
| Clostridium perfringens Gastroenteritis | Clostridium perfringens | Usually limited to diarrhea | Infection (exotoxin) | Isolation of $10^6$ endospores/g feces | Oral rehydration |
| C. difficile–Associated Diarrhea | Clostridium difficile | Mild diarrhea to colitis; 1–2.5% mortality | Infection (exotoxin) | Cytotoxin assay | Vancomycin, fidaxomicin |
| Bacillus cereus Gastroenteritis | B. cereus | May take form of diarrhea, nausea, vomiting | Intoxication | Isolation of $\geq 10^5$ B. cereus/g food | None |

An infected person is most infective to others during the first 48 hours before clinical symptoms appear. Once the viruses have begun to multiply in the respiratory tract and local lymph nodes in the neck, they reach the salivary glands via the blood. Viremia (the presence of virus in the blood) begins several days before the onset of mumps symptoms and before the virus appears in saliva. The virus is present in the blood and saliva for 3 to 5 days after the onset of the disease and in the urine after about 10 days.

Mumps is characterized by inflammation and swelling of the parotid glands, fever, and pain during swallowing. About 4 to 7 days after the onset of symptoms, the testes can become inflamed, a condition called *orchitis*. This happens in about 20–40% of men past puberty; sterility is a possible but rare consequence. Other possible complications include meningitis, inflammation of the ovaries, and pancreatitis.

An effective attenuated live vaccine is available and is often administered as part of the trivalent measles, mumps, rubella (MMR) vaccine. Second attacks are rare, and cases involving only one parotid gland or subclinical cases (about 15–20% of those infected), are as effective as bilateral mumps in conferring immunity.

If confirmation of the diagnosis (which usually is based only on symptoms) is desired, the virus can be isolated by embryonated egg or cell culture techniques and identified by ELISA tests.

---

**CHECK YOUR UNDERSTANDING**

✔ 25-5 Why is mumps included with the diseases of the digestive system?

---

## Hepatitis

**Hepatitis** is an inflammation of the liver. At least five different viruses cause hepatitis, and probably more remain to be discovered or become better known. Hepatitis is an occasional result of infections by other viruses such as Epstein-Barr virus (EBV) or cytomegalovirus (CMV). Drug and chemical toxicity can also cause acute hepatitis that is clinically identical to viral hepatitis. The characteristics of the various forms of viral hepatitis are summarized in Diseases in Focus 25.3 on page 743.

### Hepatitis A

The *hepatitis A virus (HAV)* is the causative agent of **hepatitis A**. The virus contains single-stranded RNA and lacks an envelope. It can be grown in cell culture.

After a typical entrance via the oral route, HAV multiplies in the epithelial lining of the intestinal tract. Viremia eventually occurs, and the virus spreads to the liver, kidneys, and spleen. The virus is shed in the feces and can also be detected in the blood and urine. The amount of virus excreted is greatest before symptoms appear and then declines rapidly. Therefore,

a food handler responsible for spreading the virus might not appear to be ill at the time. The virus can probably survive for several days on such surfaces as cutting boards. HAV is resistant to chlorine disinfectants at concentrations ordinarily used in water, a characteristic that enhances fecal contamination of food or drink. Mollusks, such as oysters, that live in contaminated waters are also a source of infection.

At least 50% of infections with HAV are subclinical, especially in children. In clinical cases, the initial symptoms are anorexia (loss of appetite), malaise, nausea, diarrhea, abdominal discomfort, fever, and chills. These symptoms are more likely to appear in adults; they last 2 to 21 days, and the mortality rate is low. Nationwide epidemics occur about every 10 years, mostly in people under 14. In some cases, there is also jaundice (signs are yellowing of the skin and the whites of the eyes) and the dark urine typical of liver infections. In these cases, the liver becomes tender and enlarged.

There is no chronic form of hepatitis A, and the virus is usually shed only during the acute stage of disease. The incubation time averages 4 weeks and ranges from 2 to 6 weeks, making epidemiological studies for the source of infections difficult. There are no animal reservoirs.

Acute disease is diagnosed by the detection of IgM anti-HAV because these antibodies appear about 4 weeks after infection and disappear about 3 to 4 months after infection. Recovery results in lifelong immunity.

No specific treatment for the disease exists, but people at risk of exposure or who have been exposed to hepatitis A can be given immune globulin, which provides protection for several months. Inactivated vaccines are recommended for travelers to areas of endemic disease and for high-risk groups, such as homosexual men and injecting street-drug users (IDUs). HAV vaccination is now part of the recommended childhood vaccination schedule. The annual number of reported cases has declined from 30,000, before the vaccine, to 1800 in 2006.

### Hepatitis B

**Hepatitis B** is caused by the *hepatitis B virus (HBV)*. HBV and HAV are completely different viruses: HBV is larger, its genome is double-stranded DNA, and it is enveloped. HBV is a unique DNA virus; instead of replicating its DNA directly, it passes through an intermediate RNA stage resembling a retrovirus.

The serum from patients with hepatitis B contains three distinct particles (**Figure 25.14**). The largest is the complete virion; it's infectious and capable of replicating. It's often referred to as a *Dane particle*, after the virologist who first observed it. There are also smaller *spherical particles*, about half the size of a complete virion, and *filamentous particles*, which are tubular particles similar in diameter to the spherical particles but about ten times as long. The spherical and filamentous particles are unassembled components of the virion without nucleic acids; assembly is evidently not very efficient, and large numbers

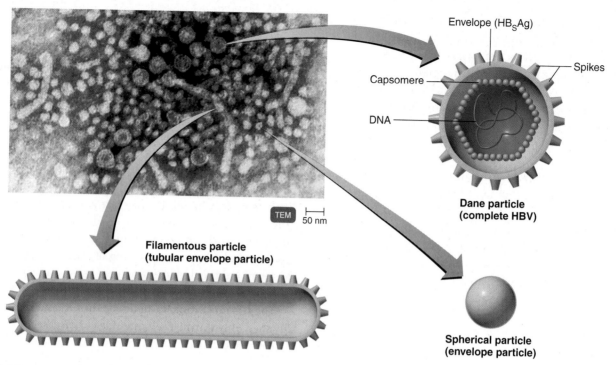

**Figure 25.14 Hepatitis B virus (HBV).** The micrograph and illustrations depict the distinct types of HBV particles discussed in the text.

**Q** What are other causes of viral hepatitis?

of these unassembled components accumulate. Fortunately, these numerous unassembled particles contain *hepatitis B surface antigen (HBsAg)*, which can be detected with antibodies to them. Such antibody tests make convenient screening of blood for HBV possible.

A third of the world's population shows serological evidence of past infection, but most people have cleared the virus. More than 350 million people have become chronic carriers of the virus. Most of these carriers are Asians and Africans, with a considerable proportion from Mediterranean countries. They acquired the infection at birth or in the first couple of years after birth. Many chronic carriers eventually die of liver cancer or cirrhosis (hardening and degeneration; see photo of the liver in Diseases in Focus 25.3).

Infection by HBV can take a varied path. There is a marked difference between acute and chronic HBV infection. If an individual is infected with HBV, acute hepatitis can result. Most such cases will resolve spontaneously as the patient clears the virus. About 5% of cases of acute hepatitis B will progress to chronic hepatitis B.

**Acute Hepatitis B** Many cases of acute hepatitis B are subclinical; the infected person is often entirely unaware. In about a third of the cases, the patient exhibits symptoms of disease—the person feels unwell and often suffers from low-grade fever,

nausea, and abdominal pain. Eventually, jaundice, dark urine, and other evidence of liver damage appears. A long period of gradual recovery marked by fatigue and malaise follows as the damaged liver recovers. However, in a few cases (less than 1%), the patient develops *fulminant hepatitis*, causing sudden, massive liver damage; survival without a liver transplant is uncommon. If a case of hepatitis persists for more than 6 months, the condition is considered to have become chronic.

**Chronic Hepatitis B** Most individuals suffering from acute hepatitis B clear the virus successfully, but some fail to do so and develop chronic hepatitis B. People infected when very young are the most likely to become chronic carriers. The risk for infants is about 90%; in children of 1 to 5 years, about 25–50%. Adolescents and young adults have a much lower risk, only 6–10%. Overall, up to 10% of infected patients become chronic carriers of the virus. For some, the condition is essentially asymptomatic: they are considered inactive carriers and have a low risk of progressing to clinical disease. Many others suffer from malaise, loss of appetite, and general fatigue—but usually without evidence of jaundice. In cases in which the chronic infection results in liver cirrhosis, the patient becomes seriously ill. Tests of liver function usually follow, leading to a diagnosis. Without treatment, the prognosis is poor but this varies. Liver cancer develops in some cases. In fact, liver cancer

Healthy liver.

Liver damaged by hepatitis C.

Hepatitis is an inflammation of the liver. Chronic hepatitis may be asymptomatic, or there may be evidence of liver disease (including cirrhosis or liver cancer). Hepatitis can be caused by a variety of viruses, alcohol, or drugs; however, it is most often caused by one of the following viruses. Use the table below to determine which virus is the most likely cause of this infection: After eating at one restaurant, 355 people are diagnosed with the same hepatitis virus. For the solution, go to @MasteringMicrobiology.

| Hepatitis Type | Pathogen | Symptoms | Incubation Period | Method of Transmission | Diagnostic Test | Treatment | Vaccine |
|---|---|---|---|---|---|---|---|
| A | Hepatitis A virus, Picornaviridae | Mostly subclinical; fever, headache; malaise, jaundice in severe cases; no chronic disease | 2–6 weeks | Ingestion | IgM antibodies | Immunoglobulin | Inactivated virus. Postexposure immune globulin |
| B | Hepatitis B virus, Hepadnaviridae | Frequently subclinical; similar to HAV, but no headache; more likely to progress to severe liver damage; chronic disease occurs | 4–26 weeks | Parenteral; sexual contact | IgM antibodies | Interferon and nucleoside analogues | Genetically modified vaccine produced in yeast |
| C | Hepatitis C virus, Flaviviridae | Similar to HBV, more likely to become chronic | 2–22 weeks | Parenteral | PCR for viral RNA | Enzyme inhibitors, interferon, and ribavirin | None |
| D | Hepatitis D virus, Deltaviridae | Severe liver damage; high mortality rate; chronic disease may occur | 6–26 weeks | Parenteral; requires coinfection with hepatitis B | Antibodies and PCR for viral RNA | None | HBV vaccine is protective |
| E | Hepatitis E virus, Caliciviridae | Similar to HAV, but pregnant women may have high mortality; no chronic disease | 2–6 weeks | Ingestion | Antibodies against HEV or HEV RNA | None | None |

is the most prevalent form of cancer in sub-Saharan Africa and East Asia, areas where hepatitis B is extremely common.

Hepatitis is a worldwide disease, but there is a significant difference in the clinical expression of hepatitis between areas of high prevalence and those of low prevalence.

In high-prevalence (Asian) countries, HBV infection tends to be acquired around the time of birth (perinatal) from infected mothers. As a consequence, the immune system doesn't recognize a difference between the virus and the host, and a high level of immunologic tolerance ensues. Because of this tolerance, the infection isn't accompanied by acute hepatitis; instead, a chronic, usually lifelong infection is established. This is the case in about 90% of infected persons. In spite of the immunologic tolerance to HBV, some liver injury occurs, and there is a high risk of death from liver disease, especially among men.

By contrast, in low-prevalence (Western) countries, most acute infections by HBV occur from exposure to infected blood or other body fluids. It is often a disease of young adults participating in risky behaviors—injecting street drugs or sexual promiscuity, for example. Long-term intimate nonsexual contact with an infected individual can also transmit HBV. Infected people who are immunocompetent develop a strong immune response, and the virus is cleared in all but about 1% of those infected. These patients have a much lower incidence of chronic disease and of liver cancer.

The diagnosis of HBV is usually based on symptoms, followed by tests of liver function. Serological tests can detect HBV antigens and antibodies. The presence of hepatitis B surface antigen (HBsAg) indicates the presence of the virus in the blood. After the virus is cleared, anti-HBsAg appears, and the patient is considered immune. Detection of the hepatitis B "e" antigen (HBeAg), a marker for the core of the virus, usually means that the virus is replicating vigorously. If this antigen disappears and is replaced by antibodies against it, this usually means that liver disease associated with viral reproduction has diminished. It also means that the patient is less infectious to others.

HBV is transmitted through sexual contact; needles, syringes, or other drug-injection equipment; and from mother to baby at birth. Health care workers and others who are in daily contact with blood have a considerably higher incidence of hepatitis B than members of the general population. Vaccination is recommended for all health care and public safety workers. All donated blood is tested for HBV and hepatitis C virus.

Preventing HBV infection involves several strategies, including using disposable needles and syringes and barrier-type contraception. Transmission from mother to infant can be prevented by administering hepatitis B immune globulin (HBIG) to the newborn immediately after birth. These babies should also be vaccinated. The HBV vaccine has become widespread worldwide and is now part of the childhood immunization schedule in the United States. It hasn't been possible to cultivate HBV in cell culture, a step that was necessary for the development of vaccines for polio, mumps, measles, and rubella. The available HBV vaccines use HBsAg produced by a genetically modified yeast. The annual incidence has declined from 30,000 cases before the vaccine to under 3000 in 2016, and eventual elimination of the disease is conceivable.

There is no specific treatment for acute HBV. For chronic HBV infection, there are currently seven approved treatments. However, none of these is reliably curative, largely because the DNA of the virus becomes integrated into the genome of the host. The aim of treatment for chronic HBV infections is to diminish the DNA of the virus to levels that are undetectable with a PCR assay.

Treatment decisions are made on the basis of several factors, such as patient age and the stage of the disease. Coinfections with HIV often occur and complicate treatment. Available antivirals include alpha interferon (see page 467 for a discussion of interferons) as well as several nucleoside analogues, such as lamivudine, adefovir, entecavir, telbivudine, and tenofovir DF. The course of treatment typically extends over several months. Combinations of at least two drugs are recommended to minimize development of resistance. Liver transplantation is often a final option in treatment.

### Hepatitis C

In the 1960s, a previously unsuspected form of transfusion-transmitted hepatitis, now called **hepatitis C**, appeared. This new form of hepatitis soon constituted almost all transfusion-transmitted hepatitis—as testing eliminated HBV in the blood supply. Eventually, serological tests to detect hepatitis C virus (HCV) antibodies were developed that similarly reduced the transmission of HCV to very low levels. However, there is a delay of about 70 to 80 days between infection and the appearance of detectable HCV antibodies. The presence of HCV in contaminated blood cannot be detected during this interval, and about 1 in 100,000 transfusions can still result in infection. Blood-collecting facilities in the United States can now detect HCV-contaminated blood within 25 days of infection. A PCR test can detect viral RNA within 1 to 2 weeks after infection.

HCV has a single strand of RNA and is enveloped. The virus doesn't kill the infected cell, but it triggers an immune inflammatory response that either clears the infection or slowly destroys the liver. (See the photo on page 743.) The virus is capable of rapid genetic variation to evade the immune system. This characteristic, along with the fact that currently HCV is cultured very inefficiently, complicates the search for an effective vaccine.

Hepatitis C has been described as a silent epidemic, killing more people than AIDS in the United States. It is often clinically inapparent—few people have recognizable symptoms until about 20 years have elapsed. Probably as many as a third of individuals infected with HCV clear the virus spontaneously. Even today, probably only a minority of infections have been diagnosed. Often, hepatitis C is detected only during some routine testing, such as for insurance or blood donation. A majority of cases, perhaps as high as 85%, progress to chronic hepatitis, a much higher rate than with HBV. Surveys indicated an estimated 3.2 million of the U.S. population are chronically infected. About 25% of chronically infected patients develop liver cirrhosis or liver cancer. Hepatitis C is probably the major reason for liver transplantation. Persons infected with HCV should be immunized against both HAV and HBV (a combination vaccine is now available) because they cannot afford the risk of further liver damage.

Preventing HCV is limited to minimizing exposure—even sharing of items such as razors, toothbrushes, or nail clippers is dangerous. A common source of infection is the sharing of injection equipment among IDUs. At least 80% of this group is infected with HCV. In one exceptional case, the disease was transmitted by means of a straw shared for inhaling cocaine. Interestingly, in more than one-third of the cases, a mode of transmission—by contaminated blood, sexual contact, or other means—cannot be identified.

Treatment includes a combination of HCV protease and polymerase inhibitors, interferon, and ribavirin. Complete eradication of HCV is attained in many cases.

### Other Hepatitis Viruses

**Hepatitis D** can occur as either acute (*coinfection form*) or chronic (*superinfection form*) hepatitis. In people with a case of self-limiting acute hepatitis B, coinfection with HDV disappears as the HBV is cleared from the system, and the condition resembles a typical case of acute hepatitis B. However, if the HBV infection progressed to the chronic stage, superinfection with HDV is often accompanied by progressive liver damage and a fatality rate several times that of people infected with HBV alone.

Hepatitis D is linked to the epidemiology of hepatitis B. In the United States and northern Europe, the disease occurs predominantly in high-risk groups, such as IDUs.

**Hepatitis E (HEV)** is spread by fecal–oral transmission, much like hepatitis A, which it clinically and structurally resembles. It is endemic in areas with poor sanitation, especially India and Southeast Asia. HEV does not cause chronic liver disease, but for some unexplained reason it is responsible for a mortality rate in excess of 20% in pregnant women.

### CHECK YOUR UNDERSTANDING

✔ 25-6 Of the several hepatitis diseases, HAV, HBV, HCV, HDV, and HEV, which two now have effective vaccines to prevent them?

## Viral Gastroenteritis

Acute gastroenteritis is one of the most common diseases of humans. About 90% of cases of acute viral gastroenteritis are caused by either the *Rotavirus* or *Norovirus*.

### Rotavirus

*Rotavirus* (**Figure 25.15**) is probably the most common cause of viral gastroenteritis, especially in children. It's estimated to cause about 3 million cases, but fewer than 100 deaths, every year in the United States. Mortality is much higher in less developed countries because rehydration therapy isn't as available. Acquired immunity then makes *rotavirus* infections, except for certain strains, much less common in adults. In most cases, following an incubation period of 2 to 3 days, the patient suffers from low-grade fever, diarrhea, and vomiting, which persists for about a week.

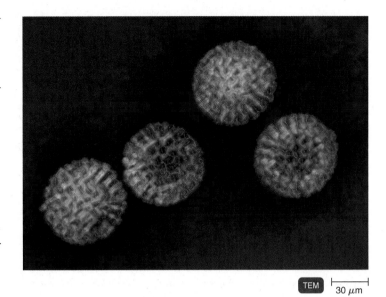

TEM ⊢ 30 μm

**Figure 25.15 *Rotavirus.*** This negatively stained electron micrograph shows the morphology of the *rotavirus* (*rota* = wheel), which gives the virus its name.

**Q** What disease does *rotavirus* cause?

*Rotavirus* cases usually peak during the cooler winter months. An infectious dose is estimated to be fewer than 100 viruses, and patients shed billions in every gram of stool.

### CLINICAL CASE

The health department makes a visit to the petting zoo to investigate the animals with which Anna had contact. Rectal swabs or fecal samples are cultured from the various animals at the zoo (see the table.)

**Isolation of STEC O157 from Fecal Sample/Rectal Swabs Collected at the Petting Zoo**

| Animal(s) | No. of Animals | No. of Animals with Isolates Identical to the DNA Pattern of Anna's STEC O157 |
|---|---|---|
| Deer | 8 | 1 |
| Donkey | 1 | 0 |
| Goats | 8 | 2 |
| Guinea fowl | 5 | 0 |
| Llama | 1 | 0 |
| Peacock | 1 | 0 |
| Pig | 1 | 0 |
| Rabbits | 10 | 0 |
| Sheep | 4 | 3 |

**Based on these results, what is the most likely mode of transmission, and how can transmission be prevented?**

723 733 739 **745** 755

In 2006 a live, orally administered vaccine was licensed. Before the vaccine, more than 90% of children in the United States were infected by the age of 3. In some cases, parents also become infected. The vaccine has decreased incidence by 98%.

*Rotavirus* infections are routinely diagnosed by several types of commercially available tests, such as enzyme immunoassays. Treatment is usually limited to oral rehydration therapy.

### Norovirus

**Noroviruses** were first identified following an outbreak of gastroenteritis in Norwalk, Ohio, in 1968. The responsible agent was identified in 1972 and called the *Norwalk virus*. Several similar viruses were later identified, and this group was termed *Norwalk-like viruses*. All were determined to be members of the caliciviruses (named for the Latin *calyx*, meaning cup—cup-shaped depressions are visible on the viruses) and are now termed *noroviruses*. It is not practical to culture them, and they do not infect the usual laboratory animals. Humans become infected by fecal–oral transmission from food and water and even aerosols from vomiting. The infectious dose may be as low as 10 viruses. The viruses continue to be shed for several days after the patient is asymptomatic. More than 20 million cases of *norovirus* gastroenteritis occur annually in the United States, but only about 300 deaths. About half of adult Americans show serological evidence that they have been infected. (See the Clinical Focus box in Chapter 9, page 264.) The currently dominant strain of noroviruses made its appearance around 2002, which is attributed to several possible factors. This strain may be more virulent, or more environmentally stable; also, fewer people may have had resistance to it from previous exposure. Natural resistance to a particular strain may last only a few months—at most about 3 years.

Cleanup and prevention of transmission following an outbreak on a cruise ship or restaurant, for example, has proved to be a challenging problem. The viruses are unusually persistent on environmental surfaces, including door handles or elevator buttons. The CDC recommends washing hands with soap and water, especially after using the toilet and changing diapers, and always before eating, preparing, or handling food. Sanitizing hand gels may be used after washing but are not a substitute for washing. The noroviruses don't have a lipid envelope and are, therefore, not reliably inactivated by ethanol. Most of the effectiveness of such measures is probably related to mechanical removal, as with hand soap. To decontaminate hard, nonporous surfaces requires solutions containing 1000 to 5000 ppm of hypochlorite (a 1:50 or 1:10 solution of household [5.26%] bleach, respectively).

To detect noroviruses in stool samples, laboratories use sensitive PCR and enzyme immunoassay (EIA) tests. The availability of such new and sensitive assays has led to recognition of noroviruses as the most common cause (at least half of recent foodborne outbreaks in the United States) of nonbacterial gastroenteritis.

Following an incubation period of 18 to 48 hours, the patient suffers from vomiting and/or diarrhea for 2 or 3 days. Diarrhea is the most prevalent symptom in children; most adults experience diarrhea, although many adult patients experience only vomiting. The severity of symptoms often depends upon the size of the infectious dose.

The only treatment for viral gastroenteritis is oral rehydration or, in exceptional cases, intravenous rehydration.

Viral diseases of the GI tract are summarized in Diseases in Focus 25.4.

### CHECK YOUR UNDERSTANDING

✔ **25-7** Two very common causes of viral gastroenteritis are rotaviruses and noroviruses. Which of these now can be prevented by a vaccine?

# Fungal Diseases of the Digestive System

### LEARNING OBJECTIVE

**25-8** Identify the causes of ergot poisoning and aflatoxin poisoning.

Some fungi produce toxins called *mycotoxins* that cause blood diseases, nervous system disorders, kidney damage, liver damage, and even cancer. Mycotoxin intoxication is considered when multiple patients have similar clinical signs and symptoms. Diagnosis is usually based on finding the fungi or mycotoxins in food (Diseases in Focus 25.5, page 748).

### Ergot and Aflatoxin Poisoning

Mycotoxins produced by *Claviceps purpurea* (KLA-vi-seps pur-pur-Ē-ah), a fungus causing smut infections on grain crops, cause **ergot poisoning** when rye or other cereal grains contaminated with the fungus are ingested. The toxin can restrict blood flow in the limbs, with resulting gangrene. It may also cause hallucinogenic symptoms, producing bizarre behavior similar to that caused by LSD.

*Aflatoxin* is a mycotoxin produced by the fungus *Aspergillus flavus*, a common mold. It has been found in many foods but is particularly likely to be found on peanuts. **Aflatoxin poisoning** can cause serious damage to livestock when their feed is contaminated with *A. flavus*. Although risk to humans is unknown, there is strong evidence aflatoxin contributes to cirrhosis of the liver and cancer of the liver in parts of the world, such as India and Africa, where food is subject to aflatoxin contamination.

# Viral Diseases of the Digestive System

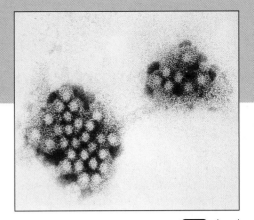

An outbreak of diarrhea begins in mid-June, peaks in mid-August, and tapers off in September. The case definition for this outbreak is defined as diarrhea (three loose stools during a 24-hour period) in a member of a swim club. The virus shown at right is isolated from one patient. Use the table below to identify infections that could cause these symptoms. For the solution, go to @MasteringMicrobiology.

**TEM** ⊢ 50 nm

Virus cultured from the patient's stool.

| Disease | Pathogen | Symptoms | Incubation Period | Diagnostic Test | Prevention |
|---|---|---|---|---|---|
| **Mumps** | Mumps virus, Paramyxoviridae | Painful swelling of parotid glands | 16–18 days | Symptoms; virus culture | Attenuated vaccine |
| **Viral Gastroenteritis** | *Rotavirus* | Vomiting, diarrhea for 1 week | 1–3 days | Enzyme immunoassay for viral antigens in feces | Attenuated vaccine |
| | *Norovirus* | Vomiting, diarrhea for 2–3 days | 18–48 hr | PCR | Thorough handwashing |

**Hepatitis** (See Diseases in Focus 25.3 on page 743.)

---

## CHECK YOUR UNDERSTANDING

✔ **25-8** What is the connection between the occasional hallucinogenic symptoms produced by ergot poisoning and a modern illicit drug?

# Protozoan Diseases of the Digestive System

## LEARNING OBJECTIVE

**25-9** List the causative agents, modes of transmission, symptoms, and treatments for giardiasis, cryptosporidiosis, cyclosporiasis, and amebic dysentery.

Several pathogenic protozoa complete their life cycles in the human digestive system (Diseases in Focus 25.5, page 748). Usually they are ingested as resistant, infective cysts and are shed in greatly increased numbers as newly produced cysts.

## Giardiasis

*Giardia intestinalis* (also known as *G. lamblia* and occasionally as *G. duodenalis*) is a flagellated protozoan that is able to attach firmly to a human's intestinal wall (**Figure 25.16**). In 1681, van Leeuwenhoek described them as having "bodies . . . somewhat longer than broad and their belly, which was flatlike, furnished with sundry little paws."

G. *intestinalis* is the cause of **giardiasis**, a prolonged diarrheal disease. Sometimes persisting for weeks, giardiasis is characterized by malaise, nausea, flatulence (intestinal gas), weakness, weight loss, and abdominal cramps. The distinctive odor of hydrogen sulfide can often be detected in the breath or stools. The protozoa sometimes occupy so much of the intestinal wall that they interfere with food absorption.

Outbreaks of giardiasis in the United States occur often, especially during camping and swimming seasons. About 7%

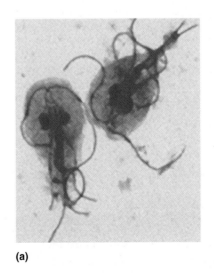

(a)

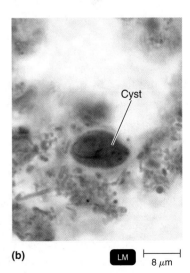

Cyst

(b)

**LM** ⊢ 8 μm

**Figure 25.16 *Giardia intestinalis*, the flagellated protozoan that causes giardiasis.** (a) The trophozoite attaches to the intestinal wall. (b) The cyst provides protection from the environment before it is ingested by a new host.

**Q** How are *Giardia* cysts eliminated from drinking water?

# DISEASES IN FOCUS 25.5 Fungal, Protozoan, and Helminthic Diseases of the Lower Digestive System

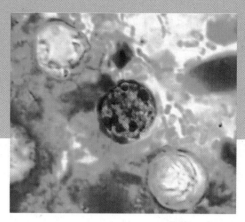

3 μm

**Acid-fast stain from the patient's feces.**

Public health officials in Pennsylvania are notified of cases of watery diarrhea, with frequent, sometimes explosive, bowel movements among persons associated with a residential facility (e.g., residents, staff, and volunteers). The disease is associated with eating snow peas. Use the table below to identify possible causes of these symptoms. For the solution, go to @MasteringMicrobiology.

| Disease | Pathogen | Symptoms | Reservoir or Host | Diagnostic Test | Treatment |
|---|---|---|---|---|---|
| **FUNGAL DISEASES** | | | | | |
| **Ergot Poisoning** | *Claviceps purpurea* | Restricted blood flow to limbs; hallucinogenic. | Mycotoxin produced by fungus growing on grains | Finding fungal sclerotia in food | None |
| **Aflatoxin Poisoning** | *Aspergillus flavus* | Liver cirrhosis; liver cancer. | Mycotoxin produced by fungus growing on food | Immunoassay for toxin in food | None |
| **PROTOZOAN DISEASES** | | | | | |
| **Giardiasis** | *Giardia intestinalis* | Protozoan adheres to intestinal wall, may inhibit nutritional absorption; diarrhea. | Water; mammals | FA | Metronidazole; nitazoxanide |
| **Cryptosporidiosis** | *Cryptosporidium hominis, C. parvum* | Self-limiting diarrhea; may be life-threatening in immunosuppressed patients. | Cattle; water | Acid-fast stain | Nitazoxanide |
| **Cyclosporiasis** | *Cyclospora cayetanensis* | Watery diarrhea. | Humans; birds; usually ingested with fruits and vegetables | Acid-fast stain | Trimethoprim and sulfamethoxazole |
| **Amebic Dysentery** (amebiasis) | *Entamoeba histolytica* | Ameba lyses epithelial cells of intestine, causes abscesses; significant mortality rate. | Humans | Microscopy; EIA | Metronidazole |
| **HELMINTHIC DISEASES** | | | | | |
| **Tapeworms** | *Taenia saginata, T. solium, Diphyllobothrium latum* | Adults cause few symptoms; pork tapeworm larvae may encyst in many organs (neurocysticercosis) and cause damage. | Intermediate host: cattle, pigs, fish; definitive host: humans | Microscopic exam of feces | Praziquantel; niclosamide |
| **Hydatid Disease** | *Echinococcus granulosus* | Larvae form in body; may be very large and cause damage. | Intermediate host: sheep, humans; definitive host: dogs | Serology; X-ray exam | Surgical removal; albendazole |
| **Pinworms** | *Enterobius vermicularis* | Itching around anus. | Intermediate and definitive hosts: humans | Microscopic exam | Pyrantel pamoate |
| **Hookworms** | *Necator americanus, Ancyclostoma duodenale* | Large infections may result in anemia. | Larvae enter skin from soil; definitive hosts: humans | Microscopic exam | Mebendazole |
| **Ascariasis** | *Ascaris lumbricoides* | Helminths live off undigested intestinal contents, causing few symptoms. | Intermediate and definitive hosts: humans | Microscopic exam | Mebendazole |
| **Whipworm** | *Trichuris trichiura* | Diarrhea, malnutrition. | Intermediate and definitive hosts: humans | Microscopic exam of feces | Albendazole, mebendazole |
| **Trichinellosis** | *Trichinella spiralis, T. nativa* | Larvae encyst in striated muscle; usually few symptoms, but large infections may be fatal. | Intermediate and definitive hosts: mammals (including humans) | Biopsy; ELISA | Mebendazole; corticosteroids |

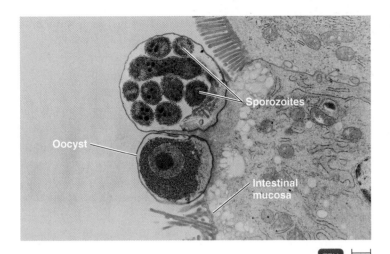

**Figure 25.17 Cryptosporidiosis.** Oocysts of *Cryptosporidium hominis* are shown here embedded in the intestinal mucosa.

**Q** How is cryptosporidiosis transmitted?

of the population are healthy carriers and shed the cysts in their feces. The pathogen is also shed by a number of wild mammals, especially beavers, and the disease occurs in backpackers who drink from untreated wilderness waters.

Most outbreaks are transmitted by contaminated water supplies. In a recent national survey of surface waters serving as sources for U.S. municipalities, the protozoan was detected in 18% of the samples. Because the cyst stage is relatively insensitive to chlorine, filtration or boiling of water supplies is usually necessary to eliminate the cysts from water.

Microscopic examination is often used for diagnosis. Because *G. intestinalis* isn't reliably excreted, stool samples collected on 3 successive days may be needed. The CDC currently recommends serological tests that use antibodies to test for the presence of trophozoites and cysts in feces. These tests are especially useful for epidemiological screening. Testing of drinking water for *Giardia* is difficult but often necessary to prevent or trace disease outbreaks. These tests are frequently combined with tests for *Cryptosporidium* protozoa, discussed in the next section.

Treatment with metronidazole or quinacrine hydrochloride is usually effective within a week. Nitazoxanide is used to treat both cryptosporidiosis (see Figure 25.17) and giardiasis. Like metronidazole, it affects anaerobic metabolic pathways, but it requires a shorter treatment regimen.

**CHECK YOUR UNDERSTANDING**

✔ **25-9** Is giardiasis caused by ingestion of a cyst or an oocyst?

## Cryptosporidiosis

**Cryptosporidiosis** is caused by the protozoan *Cryptosporidium*. The most prevalent species affecting humans are *C. parvum* and *C. hominis.* The term *cryptosporidiosis* (medical personnel often

refer to it more simply as *crypto*) describes infections by either organism. Infection occurs when humans ingest the cryptosporidian oocysts (**Figure 25.17**). The oocysts eventually release sporozoites into the small intestine. The motile sporozoites invade the epithelial cells of the intestine and undergo a cycle that eventually releases oocysts to be excreted in the feces. (Compare with the similar life cycle of *Toxoplasma gondii* in Figure 23.23, page 676). The disease is a cholera-like diarrhea lasting 10 to 14 days. In immunodeficient individuals, including AIDS patients, the diarrhea becomes progressively worse and is life-threatening.

The infection is transmitted to humans largely through recreational and drinking water systems contaminated with oocysts of *Cryptosporidium*, mostly from animal wastes, especially cattle. Studies in the United States show that many, if not most, lakes, streams, and even wells are contaminated. The oocysts, like the cysts of *G. intestinalis*, are resistant to chlorination and must be removed from water by filtration. Even filtration sometimes fails. This is especially true of swimming pools, where both chlorination and filtration systems are ineffective in removing oocysts. Alternatives to routine chlorination are ultraviolet radiation, ozonation, and chlorine dioxide. See the Clinical Focus box in Chapter 12, page 351. An infectious dose may be as low as ten oocysts. Fecal–oral transmission resulting from poor sanitation also occurs; many outbreaks have occurred in day-care settings.

Testing of water is important, but methods have been described as being cumbersome, time-consuming, and inefficient. Most widely used is an FA test that can simultaneously detect both *G. intestinalis* cysts and *Cryptosporidium* oocysts. Municipal drinking water is filtered at the water treatment plant to remove *Giardia* cysts and *Cryptosporidium* oocysts.

The recommended drug for treatment is nitazoxanide, which is also effective in treating giardiasis.

Cryptosporidiosis is most reliably diagnosed in the laboratory by detecting oocysts in fecal samples by microscopic examination of acid-fast stains (**Figure 25.18a**).

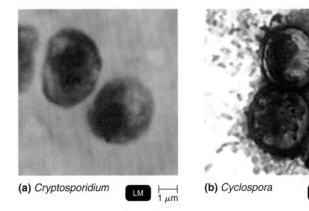

**(a)** *Cryptosporidium*  LM  1 µm       **(b)** *Cyclospora*  LM  2.5 µm

**Figure 25.18 Oocysts of two parasitic protozoa.** Acid-fast stains of fecal samples. These oocysts are stained with a cold acid-fast stain that uses more concentrated dyes.

**Q** How would you differentiate these two parasites?

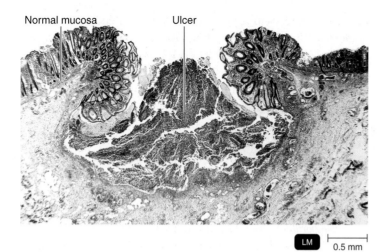

Normal mucosa     Ulcer

LM     0.5 mm

**Figure 25.19 Section of intestinal wall showing a typical flask-shaped ulcer caused by *Entamoeba histolytica*.**

**Q** If this lesion progressed far enough, could it be life-threatening?

## Cyclosporiasis

A protozoan discovered in 1993 is responsible for a series of recent diarrheal disease outbreaks. The pathogen has since been named *Cyclospora cayetanensis*.

The symptoms of **cyclosporiasis** are a few days of watery diarrhea, but in some cases it may persist for weeks. The disease is especially debilitating for immunosuppressed people, such as AIDS patients. It is uncertain whether humans are the only host for the protozoan. Most outbreaks have been associated with the ingestion of oocysts in water, on contaminated berries, or similar uncooked foods. The foods are presumed to have been contaminated by oocysts shed in human feces or possibly from birds in the field.

Diagnosis is by acid-fast staining of feces (Figure 25.18b). Oocysts can also be detected using fluoresence microscopy because the oocysts are naturally fluorescent. There is really no satisfactory test to detect contamination of foods. The antibiotic combination of trimethoprim and sulfamethoxazole is used for treatment.

## Amebic Dysentery (Amebiasis)

**Amebic dysentery**, or **amebiasis**, is spread mostly by food or water contaminated by cysts of the protozoan ameba *Entamoeba histolytica* (see Figure 12.19b, page 344). Although stomach acid can destroy trophozoites, it does not affect the cysts. In the intestinal tract, the cyst wall is digested away, and the trophozoites are released. They then multiply in the epithelial cells of the wall of the large intestine. A severe dysentery results; the feces characteristically contain blood and mucus. The trophozoites feed on tissue in the gastrointestinal tract (**Figure 25.19**).

Severe bacterial infections result if the intestinal wall is perforated. Abscesses might have to be treated surgically, and the

invasion of other organs, particularly the liver, is not uncommon. Perhaps 5% of the U.S. population are asymptomatic carriers of *E. histolytica*. Worldwide, one person in ten is estimated to be infected, mostly asymptomatically, and about 10% of these infections progress to the more serious stages.

Diagnosis largely depends on recovering and identifying the pathogens in feces. (Red blood cells, ingested as the parasite feeds on intestinal tissue and observed within the trophozoite stage of an ameba, help identify *E. histolytica*.) Several EIA serological tests are available. Such tests are especially useful when the affected areas are outside the intestinal tract and the patient is not passing amebae.

Metronidazole is the drug of choice in treatment.

# Helminthic Diseases of the Digestive System

### LEARNING OBJECTIVE

25-10 List the causative agents, modes of transmission, symptoms, and treatments for tapeworms, hydatid disease, pinworms, hookworms, whipworms, ascariasis, and trichinellosis.

Helminthic parasites are very common in the human intestinal tract, especially in warm, moist regions where sanitation is poor. **Figure 25.20** shows the worldwide estimated incidence of infection with some intestinal helminths. These diseases are called *Neglected Tropical Diseases (NTDs)* (see the Big Picture in Chapter 21, pages 614–615) because they infect 1.5 billion people in the poorest countries and are not yet controlled. In spite of the parasites' size and formidable appearance, light infections often produce few symptoms. They have become so well adapted to their human hosts, and vice versa, that when their presence is revealed, it is often a surprise. Heavy infections can cause a range of symptoms described below.

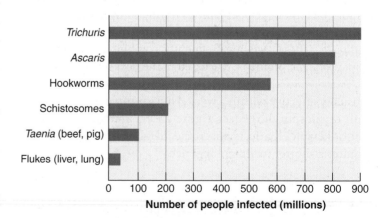

**Figure 25.20 The worldwide prevalence of human infections with the most common helminths, 2015.** Two billion people are infected.

*Source:* World Health Organization.

**Q** How is each of these diseases transmitted?

## Tapeworms

The life cycle of a typical **tapeworm** extends through three stages. The adult worm lives in the intestine of a human host, where it produces eggs that are excreted in the feces (see Figure 12.28, page 353). The eggs are ingested by animals such as grazing cattle, where the egg hatches into a larval form called a *cysticercus* (plural: *cysticerci*) that lodges in the animal's muscles. Human infections by tapeworms begin with the consumption of undercooked beef, pork, or fish containing cysticerci. The cysticerci develop into adult tapeworms that attach to the intestinal wall by suckers on the scolex (see Figure 12.27, page 352).

The adult beef tapeworm, *Taenia saginata* (TE-nē-ah sa-jin-AH-tah), seldom causes significant symptoms beyond a vague abdominal discomfort. However, psychological distress can result when a meter or more of detached segments (proglottids) occasionally break loose and unexpectedly slip out of the anus.

*Taenia solium* (SŌ-lē-um), the pork tapeworm, has a life cycle similar to that of the beef tapeworm. An important difference is that *T. solium* may produce the larval stage in the human host. **Taeniasis** develops when the adult tapeworm infects the human intestine. This is a generally benign, asymptomatic condition, but the host continuously expels eggs of *T. solium*, which contaminate hands and food under poor sanitary conditions. **Cysticercosis**, infection with the larval stage, can develop when humans or swine ingest *T. solium* eggs. These eggs can leave the digestive tract and develop into larvae that lodge in tissue (usually brain or muscles). Cysticerci in muscle tissue are relatively benign and cause few serious symptoms, but the larvae occasionally lodge in an eye, causing **ophthalmic cysticercosis** and affecting vision (**Figure 25.21**). The most serious, and much more common, disease is **neurocysticercosis**, which arises when the larvae develop in areas of the central nervous system, such as the brain. Neurocysticercosis, which is endemic in Mexico and Central America, has become a fairly common condition in parts of the United States with large Mexican and Central American immigrant populations.

The symptoms often mimic those of epilepsy or a brain tumor. The number of cases reported reflects, in part, the use of computed tomography (CT) scanning or magnetic resonance imaging (MRI) in diagnosis. In endemic areas, neurological patients can be screened with serological tests for antibodies to *T. solium*.

The fish tapeworm *Diphyllobothrium latum* (dī'fil-lō-BAH-thrē-um LĀ-tum) is found in pike, trout, perch, and salmon. The CDC has issued warnings about the risks of fish tapeworm infection from sashimi and sushi (Japanese dishes prepared from raw fish), foods that have become increasingly popular. To relate a vivid example, about 10 days after eating, one person developed symptoms of abdominal distention, flatulence, belching, intermittent abdominal cramping, and diarrhea. Eight days later, the patient passed a tapeworm 1.2 m (4 ft) long, identified as a species of *Diphyllobothrium*.

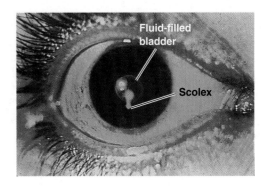

**Figure 25.21 Ophthalmic cysticercosis.** Some cases of cysticercosis affect the eye.

**Q** What organ is most likely to be affected by neurocysticercosis?

Laboratory diagnosis of tapeworms consists of identifying the tapeworm eggs (**Figure 25.22a** and **b**) or segments in feces. Adult tapeworms in the intestinal stage can be eliminated with antiparasitic drugs such as praziquantel and niclosamide. Cases of neurocysticercosis can sometimes be treated with drugs, but these often worsen the situation, and surgery may be required to remove cysticerci.

**CHECK YOUR UNDERSTANDING**

✔ 25-10 What species of tapeworm is the cause of cysticercosis?

## Hydatid Disease

One of the most dangerous tapeworms is *Echinococcus granulosus* (ē-KĪ-nō-kok'kus gran-ū-LŌ-sus), which is only a few millimeters in length (see Figure 12.28, page 353). Dogs shed the tapeworm eggs in their feces, which are ingested by grazing mammals. Once ingested, the eggs hatch and develop into cysts in the internal organs. The disease is most commonly found in people involved in raising sheep.

Once ingested by a human, the eggs of *E. granulosus* hatch, and larvae may migrate to various tissues of the body. The liver and lungs are the most common sites, but the brain and numerous other sites also may be infected. Once in place, the egg develops into a **hydatid cyst** that can grow to a diameter of 1 cm in a few months (**Figure 25.23**). In some locations, cysts may not be apparent for many years. Some, where they are free to expand, become enormous, containing up to 15 liters (4 gallons) of fluid.

Damage may arise from the size of the cyst in such areas as the brain or the interior of bones. If the cyst ruptures in the host, it can lead to the development of a great many daughter cysts. Another factor in the pathogenicity of such cysts is that the fluid contains proteinaceous material to which the host becomes sensitized. If the cyst suddenly ruptures, the result can be life-threatening anaphylactic shock.

For diagnosis, several serological tests that detect circulating antibodies are useful in screening. If available, physical imaging methods such as X rays, CT, and MRI are best.

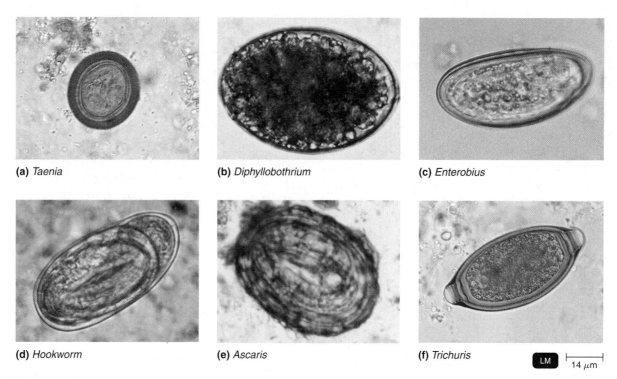

(a) *Taenia*    (b) *Diphyllobothrium*    (c) *Enterobius*

(d) *Hookworm*    (e) *Ascaris*    (f) *Trichuris*    LM ⊢——⊣ 14 μm

**Figure 25.22 Helminth eggs.** Helminth infections are diagnosed by microscopic examination of feces for helminth eggs.

**Q** Make a key to identify these eggs.

Treatment is usually surgical removal, but care must be taken to avoid release of the fluid and the potential spread of infection or anaphylactic shock. If removal isn't feasible, the drug albendazole can kill the cysts.

## Nematodes

### Pinworms

Most of us are familiar with the **pinworm**, *Enterobius vermicularis* (see Figure 12.29, page 354). This tiny worm (females are 8–13 mm in length, males 2–5 mm) migrates out of the anus of the human host to lay its eggs, causing local itching. Whole households may become infected. Diagnosis is usually based on finding eggs (Figure 25.22c) around the anus. These can be viewed by pressing transparent cellulose tape, sticky side down, against the skin, transferring the tape to a microscope slide and viewing the slide under a microscope. Such drugs as pyrantel pamoate (often available without a prescription) and mebendazole are usually effective in treatment.

### Hookworms

**Hookworm** infections were once a very common parasitic disease in the southeastern states. In the United States, the species most often seen is *Necator americanus*. Another species, *Ancylostoma duodenale*, is widely distributed around the world.

The hookworm attaches to the intestinal wall and feeds on blood and tissue rather than on partially digested food

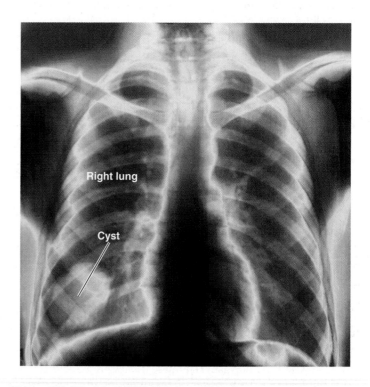

Right lung

Cyst

**Figure 25.23 A hydatid cyst formed by *Echinococcus granulosus*.** A large cyst can be seen in this X-ray image of the lung of an infected individual.

**Q** How do hydatid cysts affect the body?

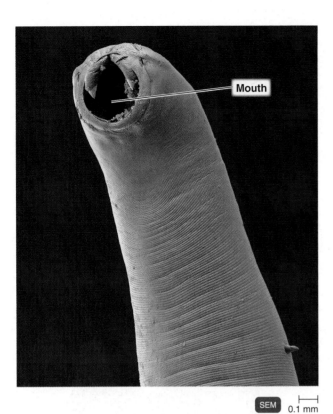

**Figure 25.24** *Ancylostoma* **hookworm.** The mouth of the hookworm is adapted to attachment and feeding on tissue. Males are typically 8–11 mm in length, females 10–13 mm.

**Q** Why can a hookworm infection lead to anemia?

(Figure 25.24), so the presence of large numbers of worms can lead to anemia and lethargic behavior. Heavy infections can also lead to a bizarre symptom known as *pica*, a craving for peculiar foods, such as laundry starch or soil containing a certain type of clay. Pica is a symptom of iron deficiency anemia.

Because the life cycle of the hookworm requires human feces to enter the soil and bare skin to contact contaminated soil, the incidence of the disease has declined greatly with improved sanitation and the practice of wearing shoes. Hookworm infections are diagnosed by finding parasite eggs in feces (Figure 25.22d) and can be treated effectively with mebendazole.

### Ascariasis

One of the most widespread helminthic infections is **ascariasis**, caused by *Ascaris lumbricoides*. This condition is familiar to many American physicians. In the southeastern United States it is quite common, with a reported incidence of 20–60% in the childhood population. Worldwide, perhaps 30% of the population is infected. Diagnosis is often made when an adult worm emerges from the anus, mouth, or nose (see Chapter 12, page 354). These worms can be quite large, up to 30 cm (about 1 ft) in length (Figure 25.25). In the intestinal tract, they live on partially digested food and cause few symptoms.

The worm's life cycle begins when eggs (upwards of 200,000 per day) are shed in a person's feces and, under poor sanitary conditions, are ingested by another person. In the upper intestine, the eggs hatch into small wormlike larvae that pass into the bloodstream and then into the lungs. There they migrate into the throat and are swallowed. The larvae develop into egg-laying adults in the intestines.

In the lungs, the tiny larvae may cause some pulmonary symptoms. Extremely large numbers may block the intestine, bile duct, or pancreatic duct. The worms do not usually cause severe symptoms, but their presence can be manifested in distressing ways. The most dramatic consequences of infection with *A. lumbricoides* are from the migrations of adult worms. Worms have been known to leave the body of small children through the umbilicus (navel) and to escape through the nostrils of a sleeping person. Microscopic examination of feces for eggs (Figure 25.22e) is used for diagnosis. Once ascariasis is diagnosed, it can be effectively treated with mebendazole or albendazole.

### Whipworm (*Trichuris trichiura*)

**Whipworm** infestations, known as *trichuriasis*, are widespread in tropical areas of the world, especially Asia. The nematode's name, *Trichuris trichiura* (Greek *trichos* = a hair, and *oura* = tail), is derived from its morphology. The worms are 30 to 50 mm in length. The main body is thin and hairlike, but the posterior end abruptly becomes thick, resembling a coiled whip with a handle—therefore, the common name of *whipworm*. In the United States, its distribution and incidence are similar to those of *A. lumbricoides*. Medical technicians microscopically inspecting fecal samples will occasionally encounter the whipworm's distinctive egg (Figure 25.22f). Nationally, eggs are present in a little over 1% of the population. In southeastern states, children pick up infective eggs from contaminated soil; in that area of the United States, the incidence of whipworm in children is about 20%.

**Figure 25.25** *Ascaris lumbricoides*, **the cause of ascariasis.** These intestinal worms are large, the female can be 30 cm in length.

**Q** What are the principal features of the life cycle of *A. lumbricoides*?

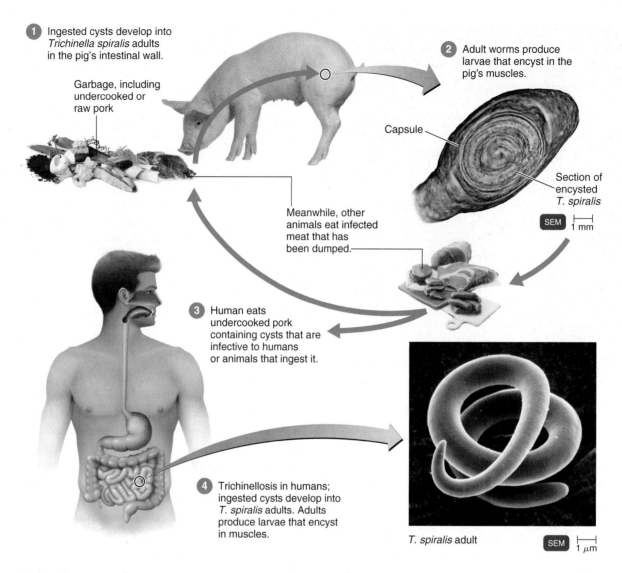

1 Ingested cysts develop into *Trichinella spiralis* adults in the pig's intestinal wall.

Garbage, including undercooked or raw pork

2 Adult worms produce larvae that encyst in the pig's muscles.

Capsule

Section of encysted *T. spiralis*

SEM | 1 mm

Meanwhile, other animals eat infected meat that has been dumped.

3 Human eats undercooked pork containing cysts that are infective to humans or animals that ingest it.

4 Trichinellosis in humans; ingested cysts develop into *T. spiralis* adults. Adults produce larvae that encyst in muscles.

*T. spiralis* adult

SEM | 1 μm

**Figure 25.26 The life cycle of *Trichinella spiralis*, the causative agent of trichinellosis.**

**Q** What is the most common vehicle of infection of *T. spiralis*?

When an embryonated egg is ingested, it hatches and enters the intestinal glands (crypts of Lieberkühn, deep crevices lined with cells that secrete intestinal juice). There the worm grows and slowly begins to tunnel back to the interior intestinal surface. Eventually, the worm positions itself such that the posterior end extends into the intestinal lumen and the hairlike anterior remains buried in the mucosa. The worm lives there for several years as a tissue parasite, feeding on cell contents and blood. Light infections of fewer than 100 worms usually pass unnoticed, but very heavy infestations may cause abdominal pain and diarrhea. Trichuriasis can also lead to anemia and malnutrition, resulting in significant weight loss and retarded growth. Treatment is with mebendazole or albendazole, although most cases do not require medical attention.

### Trichinellosis

Most infections by the small roundworm *Trichinella spiralis*, called **trichinellosis** (formerly called *trichinosis*), are insignificant. The larvae, in encysted form, are located in muscles of the host. In 1970, routine autopsies of human diaphragm muscles showed that about 4% of cadavers tested carried this parasite.

The severity of the disease is generally proportional to the number of larvae ingested. Ingesting raw or undercooked meat (especially bear, pig, cougar, or dog) puts a person at risk of infection (**Figure 25.26**).

Any ground meat can be contaminated from machinery previously used to grind contaminated meats. Freezing pork for prolonged periods (for example, −23°C for 10 days) kills *T. spiralis*. However, freezing doesn't kill some species found in wild game, such as *Trichinella nativa*.

In the muscles of intermediate hosts such as pigs, the *T. spiralis* larvae are encysted in the form of short worms about 1 mm in length. When a human ingests the flesh of an infected animal, digestive action in the intestine removes the cyst wall. The organism then matures into the adult form. The adult worms spend only about a week in the intestinal mucosa and produce larvae that invade tissue. Eventually, the larvae encyst in muscle (common sites include the diaphragm and eye muscles), where they are barely visible in biopsied specimens.

Symptoms of trichinellosis include fever, swelling around the eyes, and gastrointestinal upset. Small hemorrhages under the fingernails are often observed. Biopsy specimens, as well as a number of serological tests, can be used in diagnosis. A serological ELISA test that detects the parasite in meats has been developed. Treatment consists of administering albendazole or mebendazole to kill intestinal worms and corticosteroids to reduce inflammation.

The overall number of reported cases has decreased because of improved pig-raising and commercial and home freezing. Outbreaks can occur when multiple people share the same wild game. In the past 10 years, the number of cases reported annually in the United States has varied from 13 to 30. Death is extremely rare.

## CLINICAL CASE  Resolved

Because Anna was in contact with three different animals that have an identical strain of *E. coli*, it is more than likely she became infected from the animals at the petting zoo. Petting-zoo–associated STEC O157 has been linked to direct animal contact (i.e., touching or feeding), indirect contact (e.g., touching sawdust or shavings), and exposure from contaminated clothes, shoes, strollers, or other fomites.

Petting zoo visits are popular leisure activities and also have become an important feature of education for children. Visitors of petting zoos appear to face only a small risk of becoming infected with STEC O157 from the livestock or the farm environment, given the relatively small number of human cases each year in proportion to the large number of visitors. Cattle and other ruminants, such as sheep and goats, are important natural reservoirs of STEC O157. It isn't practical to try to exclude animals carrying STEC O157, because they usually don't show clinical symptoms, and shedding appears to be intermittent and transient. Colonization of cattle with STEC O157 typically lasts 2 months or less. The CDC recommends that animal parks provide adequate hand washing stations and post guidelines telling visitors to wash their hands after leaving animal areas. Anna drinks plenty of fluids and recovers in 5 days.

723  733  739  745  **755**

---

## Study Outline

 **Go to** @MasteringMicrobiology for **Interactive Microbiology**, *In the Clinic videos, MicroFlix, MicroBoosters, 3D animations, practice quizzes, and more.*

### Introduction (p. 721)

1. Diseases of the digestive system are the second most common illnesses in the United States.
2. Diseases of the digestive system usually result from ingesting microorganisms or their toxins in food and water.
3. The fecal–oral cycle of transmission can be broken by the proper disposal of sewage, the disinfection of drinking water, and proper food preparation and storage.

### Structure and Function of the Digestive System (p. 722)

1. The gastrointestinal (GI) tract, or alimentary canal, consists of the mouth, pharynx, esophagus, stomach, small intestine, and large intestine.
2. In the GI tract, with mechanical and chemical help from the accessory structures, large food molecules are broken down into smaller molecules that can be transported by blood or lymph to cells.
3. Feces, the solids resulting from digestion, are eliminated through the anus.
4. GALT is part of the immune system.

### Normal Microbiota of the Digestive System (pp. 722–723)

1. Large numbers of bacteria colonize the mouth.
2. The stomach and small intestine have few resident microorganisms.
3. Bacteria in the large intestine assist in degrading food and synthesizing vitamins.
4. Up to 40% of fecal mass is microbial cells.

### Bacterial Diseases of the Mouth (pp. 724–726)

#### Dental Caries (Tooth Decay) (pp. 724–726)

1. Dental caries begin when tooth enamel and dentin are eroded and the pulp is exposed to bacterial infection.
2. *Streptococcus mutans*, found in the mouth, uses sucrose to form dextran from glucose and lactic acid from fructose.
3. Bacteria adhere to teeth by the sticky dextran, forming dental plaque.
4. Acid produced during carbohydrate fermentation destroys tooth enamel at the site of the plaque.
5. Gram-positive rods and filamentous bacteria can penetrate into dentin and pulp.

6. Carbohydrates such as starch, mannitol, sorbitol, and xylitol are not used by cariogenic bacteria to produce dextran and do not promote tooth decay.

### Periodontal Disease (p. 726)

7. Caries of the cementum and gingivitis are caused by streptococci, actinomycetes, and anaerobic gram-negative bacteria.

8. Chronic gum disease (periodontitis) can cause bone destruction and tooth loss; periodontitis is due to an inflammatory response to a variety of bacteria growing on the gums.

9. Acute necrotizing ulcerative gingivitis is often caused by *Prevotella intermedia*.

## Bacterial Diseases of the Lower Digestive System (pp. 727–739)

1. A gastrointestinal infection is caused by the growth of a pathogen in the intestines.

2. Incubation times range from 12 hours to 2 weeks. Symptoms of infection generally include a fever.

3. A bacterial intoxication results from ingesting preformed bacterial toxins.

4. Symptoms appear 1 to 48 hours after ingestion of the toxin. Fever is not usually a symptom of intoxication.

5. Infections and intoxications cause diarrhea, dysentery, or gastroenteritis.

6. These conditions are usually treated with fluid and electrolyte replacement.

### Staphylococcal Food Poisoning (Staphylococcal Enterotoxicosis) (p. 728)

7. Staphylococcal food poisoning is caused by the ingestion of an enterotoxin produced in improperly stored foods.

8. *S. aureus* is inoculated into foods during preparation. The bacteria grow and produce enterotoxin in food stored at room temperature.

9. Boiling for 30 minutes is not sufficient to denature the exotoxin.

10. Foods with high osmotic pressure and those not cooked immediately before consumption are most often the source of staphylococcal enterotoxicosis.

11. Laboratory identification of *S. aureus* isolated from foods is used to trace the source of contamination.

### Shigellosis (Bacillary Dysentery) (pp. 728–729)

12. Shigellosis is caused by any of four species of *Shigella*.

13. Symptoms include blood and mucus in stools, abdominal cramps, and fever. Infections by *S. dysenteriae* result in ulceration of the intestinal mucosa.

### Salmonellosis (*Salmonella* Gastroenteritis) (pp. 729–731)

14. Salmonellosis, or *Salmonella* gastroenteritis, is caused by many *Salmonella enterica* serovars.

15. Symptoms include nausea, abdominal pain, and diarrhea and begin 12 to 36 hours after eating large numbers of *Salmonella*. Septic shock can occur in infants and in the elderly.

16. Mortality is lower than 1%, and recovery can result in a carrier state.

### Typhoid Fever (p. 732)

17. *Salmonella* Typhi causes typhoid fever; the bacteria are transmitted by contact with human feces.

18. Fever and malaise occur after a 2-week incubation period. Symptoms last 2 to 3 weeks.

19. *S.* Typhi is harbored in the gallbladder of carriers.

20. Vaccines are available for high-risk people and travelers.

### Cholera (pp. 732–733)

21. *Vibrio cholerae* O:1 and O:139 produce an exotoxin that alters the membrane permeability of the intestinal mucosa; the resulting vomiting and diarrhea cause loss of body fluids.

22. The symptoms last for a few days. Untreated cholera has a 50% mortality rate.

### Noncholera Vibrios (p. 733)

23. Ingestion of other *V. cholerae* serotypes can result in mild diarrhea.

24. *Vibrio* gastroenteritis can be caused by *V. parahaemolyticus*.

25. These diseases are contracted by eating contaminated crustaceans or contaminated mollusks.

### *Escherichia coli* Gastroenteritis (pp. 733, 736–737)

26. Enterotoxigenic, enteroinvasive, and enteroaggregative strains of *E. coli* cause diarrhea.

27. Enterohemorrhagic *E. coli*, such as *E. coli* O157:H7, produces Shiga toxins that cause inflammation and bleeding of the colon, including hemorrhagic colitis and hemolytic uremic syndrome.

28. The most common causes of traveler's diarrhea are enterotoxigenic and enteroaggregative *E. coli*.

### Campylobacteriosis (*Campylobacter* Gastroenteritis) (p. 737)

29. *Campylobacter* is the most common cause of diarrhea in the United States.

30. *Campylobacter* is transmitted in chicken and unpasteurized milk.

### *Helicobacter* Peptic Ulcer Disease (p. 737)

31. *Helicobacter pylori* produces ammonia, which neutralizes stomach acid; the bacteria colonize the stomach mucosa and cause peptic ulcer disease.

32. Bismuth and several antibiotics may be useful in treating peptic ulcer disease.

### *Yersinia* Gastroenteritis (p. 737)

33. *Y. enterocolitica* and *Y. pseudotuberculosis* are transmitted in undercooked pork.

34. *Yersinia* can grow at refrigeration temperatures.

### *Clostridium perfringens* Gastroenteritis (p. 738)

35. *C. perfringens* causes a self-limiting gastroenteritis.

36. Endospores survive heating and germinate when foods (usually meats) are stored at room temperature.

37. Exotoxin produced when the bacteria grow in the intestines is responsible for the symptoms.

### *Clostridium difficile*–Associated Diarrhea (p. 738–739)

38. Growth of *C. difficile* following antibiotic therapy can result in mild diarrhea or colitis.

39. The condition is usually associated with health care environments and day-care centers.

### *Bacillus cereus* Gastroenteritis (p. 739)

40. Ingesting food contaminated with the soil saprophyte *Bacillus cereus* can result in diarrhea, nausea, and vomiting.

## Viral Diseases of the Digestive System (pp. 739–746)

**Mumps** (pp. 739–741)

1. Mumps virus enters and exits the body through the respiratory tract.
2. About 16 to 18 days after exposure, the virus causes inflammation of the parotid glands, fever, and pain during swallowing. About 4 to 7 days later, orchitis may occur.
3. After onset of the symptoms, the virus is found in the blood, saliva, and urine.
4. A measles, mumps, rubella (MMR) vaccine is available.

**Hepatitis** (pp. 741–745)

5. Inflammation of the liver is called hepatitis. Symptoms include loss of appetite, malaise, fever, and jaundice.
6. Viral causes of hepatitis include hepatitis viruses, Epstein-Barr virus (EBV), and cytomegalovirus (CMV).
7. Hepatitis A virus (HAV) is transmitted via the fecal–oral route.
8. Hepatitis B virus (HBV) is transmitted via blood and semen.
9. Hepatitis C virus (HCV) is transmitted via blood.
10. Hepatitis D virus (HDV) occurs as a superinfection or coinfection with hepatitis B virus.
11. Hepatitis E virus (HEV) is spread by the fecal–oral route.

**Viral Gastroenteritis** (pp. 745–746)

12. Viral gastroenteritis is most often caused by a *Rotavirus* or *Norovirus*.
13. The incubation period is 2 to 3 days; diarrhea lasts up to 1 week.

## Fungal Diseases of the Digestive System (pp. 746–747)

1. Mycotoxins are toxins produced by some fungi.
2. Mycotoxins affect the blood, nervous system, kidneys, or liver.
3. Cereal grains are the crop most often contaminated with ergot, the *Claviceps* mycotoxin.
4. Peanuts are the crop most often contaminated with aflatoxin-producing *Aspergillus flavus*.

## Protozoan Diseases of the Digestive System

(pp. 747–750)

**Giardiasis** (pp. 747, 749)

1. *Giardia intestinalis* grows in the intestines of humans and wild animals and is transmitted in contaminated water.
2. Symptoms of giardiasis are malaise, nausea, flatulence, weakness, and abdominal cramps that persist for weeks.

**Cryptosporidiosis** (p. 749)

3. *Crytosporidium* spp. cause diarrhea; in immunosuppressed patients, the disease is prolonged for months.
4. The pathogen is transmitted in contaminated water.

**Cyclosporiasis** (p. 750)

5. *Cyclospora cayetanensis* causes diarrhea; the protozoan was first identified in 1993.
6. It is transmitted in contaminated produce.

**Amebic Dysentery (Amebiasis)** (p. 750)

7. Amebic dysentery is caused by *Entamoeba histolytica* growing in the large intestine.
8. The ameba feeds on red blood cells and GI tract tissues. Severe infections result in abscesses.

## Helminthic Diseases of the Digestive System

(pp. 750–755)

**Tapeworms** (p. 751)

1. Tapeworms are contracted by the consumption of undercooked beef, pork, or fish containing encysted larvae (cysticerci).
2. The scolex attaches to the intestinal mucosa of humans (the definitive host) and matures into an adult tapeworm.
3. Eggs are shed in the feces and must be ingested by an intermediate host.
4. Neurocysticercosis in humans occurs when the pork tapeworm larvae encyst in humans.

**Hydatid Disease** (pp. 751–752)

5. Humans infected with the tapeworm *Echinococcus granulosus* might have hydatid cysts in their lungs or other organs.
6. Dogs are usually the definitive hosts, and sheep are the intermediate hosts for *E. granulosus*.

**Nematodes** (pp. 752–755)

7. Humans are the definitive host for pinworms, *Enterobius vermicularis*.
8. Hookworm larvae bore through skin and migrate to the intestine to mature into adults.
9. *Ascaris lumbricoides* adults live in human intestines.
10. Ingested *Trichuris trichiura* eggs hatch in the large intestine. Larvae live attached to the intestinal lining.
11. *Trichinella spiralis* larvae encyst in muscles of humans and other mammals to cause trichinellosis.

## Study Questions

For answers to the Knowledge and Comprehension questions, turn to the Answers tab at the back of the textbook.

## Knowledge and Comprehension

### Review

1. Complete the following table:

| Disease | Causative Agent | Method of Transmission | Symptoms | Treatment |
|---|---|---|---|---|
| Aflatoxin poisoning | | | | |
| Cryptosporidiosis | | | | |
| Pinworms | | | | |
| Whipworms | | | | |

2. Complete the following table:

| Causative Agent | Suspect Foods | Treatment | Prevention |
|---|---|---|---|
| *Vibrio parahaemolyticus* | | | |
| *V. cholerae* | | | |
| *E. coli* O157 | | | |
| *Campylobacter jejuni* | | | |
| *Yersinia enterocolitica* | | | |
| *Clostridium perfringens* | | | |
| *Bacillus cereus* | | | |
| *Staphylococcus aureus* | | | |
| *Salmonella enterica* | | | |
| *Shigella* spp. | | | |

3. **DRAW IT** Identify the site colonized by the following organisms: *Echinococcus granulosus, Enterobius vermicularis, Giardia, Helicobacter pylori,* hepatitis B virus, mumps virus, rotavirus, *Salmonella, Shigella, Streptococcus mutans, Trichinella spiralis, Trichuris.*

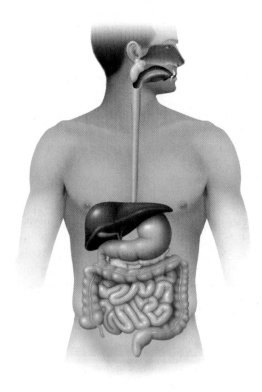

4. *E. coli* bacteria are part of the normal microbiota of the intestines and can cause gastroenteritis. Explain why this one species is both beneficial and harmful.

5. Define *mycotoxin*. Give an example of a mycotoxin.

6. Explain how the following diseases differ and how they are similar: giardiasis, amebic dysentery, cyclosporiasis, and cryptosporidiosis.

7. Differentiate among the following factors of bacterial intoxication and bacterial infection: prerequisite conditions, causative agents, onset, duration of symptoms, and treatment.

8. Complete the following table:

| Disease | Causative Agent | Mode of Transmission | Site of Infection | Symptoms | Prevention |
|---|---|---|---|---|---|
| Mumps | | | | | |
| Hepatitis A | | | | | |
| Hepatitis B | | | | | |
| Viral gastro-enteritis | | | | | |

9. Look at life cycle diagrams for human tapeworm and trichinellosis. Indicate stages in the life cycles that could be easily broken to prevent these diseases.

10. **NAME IT** Cysts of this flagellated organism survive in water; when ingested, the trophozoite grows in the intestine, causing diarrhea.

## Multiple Choice

1. All of the following can be transmitted by recreational (i.e., swimming) water sources *except*
   a. amebic dysentery.
   b. cholera.
   c. giardiasis.
   d. hepatitis B.
   e. salmonellosis.

2. A patient with nausea, vomiting, and diarrhea within 5 hours after eating most likely has
   a. shigellosis.
   b. cholera.
   c. *E. coli* gastroenteritis.
   d. salmonellosis.
   e. staphylococcal food poisoning.

3. Isolation of *E. coli* from a stool sample is diagnostic proof that the patient has
   a. cholera.
   b. *E. coli* gastroenteritis.
   c. salmonellosis.
   d. typhoid fever.
   e. none of the above

4. Gastric ulcers are caused by
   a. stomach acid.
   b. *Helicobacter pylori.*
   c. spicy food.
   d. acidic food.
   e. stress.

5. Microscopic examination of a patient's fecal culture shows comma-shaped bacteria. These bacteria require 2–4% NaCl to grow. The bacteria probably belong to the genus
   a. *Campylobacter.*
   b. *Escherichia.*
   c. *Salmonella.*
   d. *Shigella.*
   e. *Vibrio.*

6. A cholera epidemic in Peru had all of the following characteristics. Which one *led* to the others?
   a. eating raw fish
   b. sewage contamination of water
   c. catching fish in contaminated water

**d.** *Vibrio* in fish intestine

**e.** including fish intestines with edibles

Use the following choices to answer questions 7–10:

**a.** *Campylobacter*

**b.** *Cryptosporidium*

**c.** *Escherichia*

**d.** *Salmonella*

**e.** *Trichinella*

**7.** Identification is based on the observation of oocysts in feces.

**8.** A characteristic disease symptom caused by this microorganism is swelling around the eyes.

**9.** Microscopic observation of a stool sample reveals gram-negative helical cells.

**10.** This microbe is frequently transmitted to humans via raw eggs.

## Analysis

**1.** Why is a human infection of *Trichinella* considered a dead-end for the parasite?

**2.** Complete the following table:

| Disease | Conditions Necessary for Microbial Growth | Basis for Diagnosis | Prevention |
|---|---|---|---|
| Staphylococcal food poisoning | | | |
| Salmonellosis | | | |
| *C. difficile* diarrhea | | | |

**3.** Match the foods in column A with the microorganism (column B) most likely to contaminate each:

| Column A | Column B |
|---|---|
| _____ a. Beef | 1. *Vibrio* |
| _____ b. Delicatessen meats | 2. *Campylobacter* |
| _____ c. Chicken | 3. *E. coli* O157:H7 |
| _____ d. Milk | 4. *Listeria* |
| _____ e. Oysters | 5. *Salmonella* |
| _____ f. Pork | 6. *Trichinella* |

What disease does each microbe cause? How can these diseases be prevented?

**4.** Which diseases of the gastrointestinal tract can be acquired by swimming in a pool or lake? Why are these diseases not likely to be acquired while swimming in the ocean?

## Clinical Applications and Evaluation

**1.** In New York on April 26, patient A was hospitalized with a 2-day history of diarrhea. An investigation revealed that patient B had onset of watery diarrhea on April 22. On April 24, three other people (patients C, D, and E) had onset of diarrhea. All three had vibriocidal antibody titers ≥ 640. In Ecuador on April 20, B bought crabs that were boiled and shelled. He shared crabmeat with two people (F and G), then froze the remaining crab in a bag. Patient A returned to New York on April 21 with the bag of crabmeat in his suitcase. The bag was placed in a freezer overnight and thawed on April 22 in a double-boiler for 20 minutes. The crab was served 2 hours later in a crab salad. The crab was consumed during a 6-hour period by A, C, D, and E. Individuals F and G did not become ill. What is the etiology of this disease? How was it transmitted, and how could it have been prevented?

**2.** The 2130 students and employees of a public school system developed diarrheal illness on April 2. The cafeteria served chicken that day. On April 1, part of the chicken was placed in water-filled pans and cooked in an oven for 2 hours at a dial setting of 177°C. The oven was turned off, and the chicken was left overnight in the warm oven. The remainder of the chicken was cooked for 2 hours in a steam cooker and then left in the device overnight at the lowest possible setting (43°C). Two serotypes of gram-negative, cytochrome oxidase-negative, lactose-negative rods were isolated from 32 patients. What is the pathogen? How could this outbreak have been prevented?

**3.** A 31-year-old man became feverish 4 days after arriving at a vacation resort in Idaho. During his stay, he ate at two restaurants that were not associated with the resort. At the resort, he drank soft drinks with ice, used the hot tub, and went fishing. The resort is supplied by a well that was dug 3 years ago. He went to the hospital when he developed vomiting and bloody diarrhea. Gram-negative, lactose-negative bacteria were cultured from his stool. The patient recovered after receiving intravenous fluids. What microorganism most likely caused his symptoms? How is this disease transmitted? What is the most likely source of his infection, and how would you verify the source?

**4.** Three to 5 days after eating Thanksgiving dinner at a restaurant, 112 people developed fever and gastroenteritis. All the food had been consumed except for five "doggie" bags. Bacterial analysis of the mixed contents of the bags (containing roast turkey, giblet gravy, and mashed potatoes) showed the same bacterium that was isolated from the patients. The gravy had been prepared from giblets of 43 turkeys that had been refrigerated for 3 days prior to preparation. The uncooked giblets were ground in a blender and added to a thickened hot stock mixture. The gravy was not reboiled and was stored at room temperature throughout Thanksgiving Day. What was the source of the illness? What was the most likely etiologic agent? Was this an infection or an intoxication?

# 26 Microbial Diseases of the Urinary and Reproductive Systems

The **urinary system** is composed of organs that regulate the chemical composition and volume of the blood and as a result excrete mostly nitrogenous waste products and water. Because it provides an opening to the outside environment, the urinary system is prone to infections from external contacts. The mucosal membranes that line the urinary system are moist and, compared to skin, more supportive of bacterial growth. The *Leptospira interrogans* bacteria shown in the photo infect the kidneys (leptospirosis) but enter through a cut or mucous membranes of the nose or mouth. Leptospirosis is the subject of the Clinical Case in this chapter.

The **reproductive system** shares several of the organs with the urinary system. Its function is to produce gametes that propagate the species and, in the female, to support and nourish the developing embryo and fetus. In the same fashion as the urinary system, it provides openings to the external environment and is therefore prone to infections. This is especially true because intimate sexual contact can promote exchange of microbial pathogens between individuals. It isn't surprising, therefore, that certain pathogens have adapted to this environment and a sexual mode of transmission. As a result, they are often no longer able to survive in more rigorous environments.

▶ *Leptospira interrogans* bacteria are transmitted in urine of infected animals.

## In the Clinic

You are a nurse in a community sexual health clinic. Your first patient today is Kylin, a 20-year-old college student who dropped in for her first pelvic exam. She has had two sexual partners in the past year. She has not noticed vaginal discharge, sores, or painful urination. During her pelvic examination, you observe that her cervix appears inflamed and that a watery discharge is present. **What are the most likely microbial infections? What laboratory tests would confirm a diagnosis?**

*Hint: Read about bacterial diseases of the reproductive systems on pages 766–776.*

**Answers to In the Clinic questions are found online at @MasteringMicrobiology.**

# Structure and Function of the Urinary System

## LEARNING OBJECTIVE

**26-1** List the antimicrobial features of the urinary system.

The **urinary system** consists of two *kidneys*, two *ureters*, a single *urinary bladder*, and a single *urethra* (**Figure 26.1**). Certain wastes, collectively called *urine*, are removed from the blood as it circulates through the kidneys. The urine passes through the ureters into the urinary bladder, where it is stored prior to elimination from the body through the urethra. In the female, the urethra conveys only urine to the exterior. In the male, the urethra is a common tube for both urine and seminal fluid.

Where the ureters enter the urinary bladder, physiological valves prevent the backflow of urine to the kidneys. This mechanism helps shield the kidneys from lower urinary tract infections. In addition, the urea in urine has some antimicrobial properties. The flushing action of urine during urination also tends to remove potentially infectious microbes.

> **CHECK YOUR UNDERSTANDING**
>
> ✔ **26-1** How can the flow of urine prevent infection?

# Structure and Function of the Reproductive Systems

## LEARNING OBJECTIVE

**26-2** Identify the portals of entry for microbes into the female and male reproductive systems.

The **female reproductive system** consists of two *ovaries*, two *uterine (fallopian) tubes*, the *uterus*, including the *cervix*, the *vagina*, and *external genitals* (**Figure 26.2**). The ovaries produce female sex hormones and ova (eggs). When an ovum is released during the process of ovulation, it enters a uterine tube, where fertilization may occur if viable sperm are present. The fertilized ovum (zygote) descends the tube and enters the uterus. It implants in the inner wall of the uterus and remains there while it develops into an embryo and, later, a fetus. The external genitals (*vulva*) include the clitoris, labia, and glands that produce a lubricating secretion during copulation.

The **male reproductive system** consists of two *testes*, a system of *ducts*, *accessory glands*, and the *penis* (**Figure 26.3**). The testes produce male sex hormones and sperm. To exit the body, sperm cells pass through a series of ducts: the epididymis, ductus (vas) deferens, ejaculatory duct, and urethra.

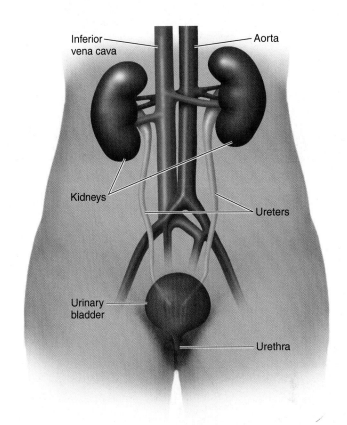

**Figure 26.1 Organs of the human urinary system, shown here in the female.**

**Q** What features of the urinary system help prevent colonization by microbes?

---

**CLINICAL CASE** Swimming Upstream

Maricel Quimuyog, a 25-year-old professional freestyle kayaker, is having difficulties training for her next kayaking event. Although Maricel normally enjoys her outdoor activities and is in good physical shape, she hasn't been feeling well. Thinking at first her headache, fever, and muscle pain are simply the flu, Maricel tries to take it easy. But when her skin and the whites of her eyes start to look jaundiced and she has trouble catching her breath, Maricel becomes concerned and goes to her physician. On physical examination, Maricel is alert and her lungs are clear. Her physician sends blood and urine samples to a local laboratory for blood counts and cultures; the blood count shows 9500 leukocytes/mm$^3$ (88% neutrophils, 10% lymphocytes, and 2% monocytes). Maricel's 24-hour urine output, however, is nearly double the normal amount. Maricel's physician is concerned because of the dehydration and loss of sodium and magnesium in the urine.

**What is causing Maricel's symptoms? Read on to find out.**

761    765    770    776

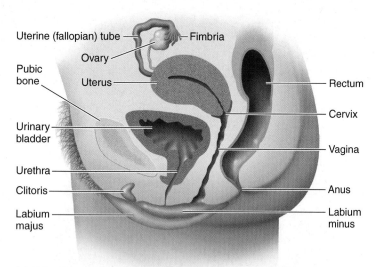

Uterine (fallopian) tube — Fimbria
Ovary
Pubic bone
Uterus
Rectum
Cervix
Urinary bladder
Vagina
Urethra
Clitoris
Anus
Labium majus
Labium minus

**(a)** Side view section of female pelvis showing reproductive organs

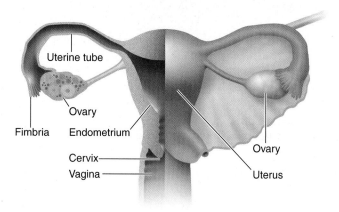

Uterine tube
Ovary
Fimbria
Endometrium
Cervix
Ovary
Vagina
Uterus

**(b)** Front view of female reproductive organs, with the uterine tube and ovary to the left in the sectioned drawing. Movement of the fimbriae causes fluid to flow, propelling the egg into the uterine tube.

**Figure 26.2 Female reproductive organs.**

**Q** Where are normal microbiota found in the female reproductive system?

**CHECK YOUR UNDERSTANDING**

✔ 26-2 Look at Figure 26.2. If a microbe enters the female reproductive system (the uterus, etc.), does it also necessarily enter the urinary bladder, causing cystitis?

# Normal Microbiota of the Urinary and Reproductive Systems

**LEARNING OBJECTIVE**

26-3 Describe the normal microbiota of the upper urinary tract, the male urethra, and the female urethra and vagina.

Normal urine is not sterile. The normal microbiota of the urinary tract is currently being studied, see the Exploring the

Microbiome box (page 763). Urine may become contaminated with microbiota of the skin near the end of its passage through the urethra. Therefore, urine collected directly from the urinary bladder has fewer microbial contaminants than voided urine.

The predominant bacteria in the vagina are the lactobacilli. These bacteria produce lactic acid, which maintains the acidic pH (3.8 to 4.5) of the vagina, inhibiting the growth of most other microbes. Most vaginal lactobacilli produce hydrogen peroxide, which also inhibits growth of other bacteria. Estrogens (sex hormones) promote the growth of lactobacilli by enhancing the production of glycogen by vaginal epithelial cells. The glycogen quickly breaks down into glucose, which the lactobacilli metabolize into lactic acid.

Other bacteria, such as streptococci, various anaerobes, and some gram-negatives, are also found in the vagina. The yeastlike fungus *Candida albicans* (see page 779) is part of the normal microbiota of 10–25% of women, even when they are asymptomatic.

Pregnancy and menopause are often associated with higher rates of urinary tract infections. The reason is that estrogen levels are lower, resulting in lower populations of lactobacilli and therefore less vaginal acidity.

Semen picks up bacteria from the urethra; however, the recently discovered microbiome of the seminal vesicles, including *Proprionibacterium*, *Corynebacterium*, and *Pseudomonas*, may affect sperm production.

**CHECK YOUR UNDERSTANDING**

✔ 26-3 What is the association between estrogens and the microbiota of the vagina?

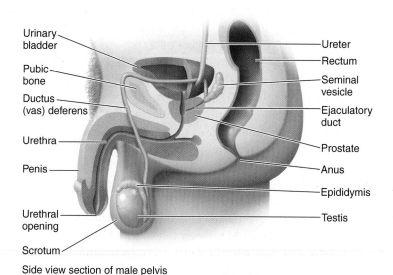

Urinary bladder
Pubic bone
Ductus (vas) deferens
Urethra
Penis
Urethral opening
Scrotum
Ureter
Rectum
Seminal vesicle
Ejaculatory duct
Prostate
Anus
Epididymis
Testis

Side view section of male pelvis

**Figure 26.3 Male reproductive and urinary organs.**

**Q** What factors protect the male urinary and reproductive systems from infection?

# Resident Microbes of the Urinary System

Contrary to common belief, urine is not sterile. This idea probably arose before clean water and antibiotics were widely available, when urine was used to clean wounds. The notion was then reinforced by the fact that clinical cultures of urine used to identify pathogens favor finding fast-growing bacteria such as *E. coli*. Standard clinical techniques are generally not designed to grow slow-growing or fastidious bacteria.

Using anaerobic culture techniques and genetic analysis, researchers are now studying the urinary microbiome in earnest. Urine samples are collected midstream to reduce contamination by skin microbiota. So far, 35 different genera have been identified in the urinary tract. These bacteria are different from those normally found on the skin and in the intestine.

The most prevalent genera in urine are *Lactobacillus, Corynebacterium, Streptococcus, Actinomyces*, and *Staphylococcus*. Urine from females contains greater species diversity, with more Actinobacteria and Bacteroidetes than found in male urine.

Urine from patients with urgency incontinence (overactive bladder) had lower bacterial diversity in urine than healthy volunteers. One study found an increase in *Gardnerella* and decrease in *Lactobacillus* species in patients with overactive bladder versus people without the disorder. The urinary bladder microbiota may be an important link between the coordination of neurotransmitters, nerves, and urinary bladder muscles. While it's too early to know whether this loss in bacterial diversity is a cause or a result of overactive bladder,

further research may one day lead to new treatments.

**Study of urine microbiota is still in its early days.**

## Diseases of the Urinary System

The urinary system normally contains few microbes, but it's subject to opportunistic infections that can be quite troublesome. Almost all such infections are bacterial, although occasional infections by schistosome parasites, protozoa, and fungi occur. In addition, as we will see in this chapter, sexually transmitted infections often affect the urinary system as well as the reproductive system.

## Bacterial Diseases of the Urinary System

### LEARNING OBJECTIVES

26-4 Describe the modes of transmission for urinary and reproductive system infections.

26-5 List the microorganisms that cause cystitis, pyelonephritis, and leptospirosis, and name the predisposing factors for these diseases.

Urinary system infections are most frequently initiated by an inflammation of the urethra, or *urethritis*. Infection of the urinary bladder is called *cystitis*, and infection of the ureters is *ureteritis*. The most significant danger from lower urinary tract infections is that they may move up the ureters and affect the kidneys, causing *pyelonephritis*. Occasionally the kidneys are affected by systemic bacterial diseases, such as *leptospirosis*. The pathogens causing these diseases are found in excreted urine.

Bacterial infections of the urinary system are usually caused by microbes that enter the system from external sources. In the United States, over 8 million urinary tract infections occur each year. About 500,000 cases are healthcare-associated, and 75% of these are associated with urinary catheters. Because of the proximity of the anus to the urinary opening, intestinal bacteria predominate in urinary tract infections. Most infections of the urinary tract are caused by *Escherichia coli*. Infections by *Pseudomonas*, because of their natural resistance to antibiotics, are especially troublesome.

Diseases of the urinary system are summarized in Diseases in Focus 26.1.

# Bacterial Diseases of the Urinary System

A 20-year-old woman feels a stinging sensation when urinating and feels an urgent need to urinate, even if very little urine is excreted. Lactose-fermenting, gram-negative rods are cultured from her urine (see the photo). Use the table below to identify infections that could cause these symptoms. For the solution, go to @MasteringMicrobiology.

MacConkey agar culture from the patient's urine. This medium is designed to selectively grow gram-negative bacteria and differentiate them by their ability to ferment lactose.

| Disease | Pathogen | Symptoms | Diagnosis | Treatment |
|---|---|---|---|---|
| **Cystitis** (urinary bladder infection) | *Escherichia coli*, *Staphylococcus saprophyticus* | Difficulty or pain in urination | $>10^3$ CFU/ml of one species and + LE test | Nitrofurantoin |
| **Pyelonephritis** (kidney infection) | Primarily *E. coli* | Fever; back or flank pain | $>10^5$ CFU/ml of one species and + LE test | Cephalosporin |
| **Leptospirosis** (kidney infection) | *Leptospira interrogans* | Headaches, muscular aches, fever; kidney failure a possible complication | Serological test | Doxycycline |

## Cystitis

**Cystitis** is a common inflammation of the urinary bladder in females. Symptoms often include *dysuria* (difficult, painful, urgent urination) and pyuria.

The female urethra is less than 2 inches long, and microorganisms traverse it readily. It's also closer than the male urethra to the anal opening and its contaminating intestinal bacteria. These considerations are reflected in the fact that the rate of urinary tract infections in women is eight times that of men. In either gender, most cases are due to infection by *E. coli*, which can be identified by cultivation on differential media such as MacConkey's agar. Another frequent bacterial cause is the coagulase-negative *Staphylococcus saprophyticus* (STAF-i-lō-kok'kus sa-prō-FI-ti-kus).

As a general rule, a urine sample with more than 1,000 colony forming units (CFU) per milliliter of a single species indicates cystitis. The diagnosis should also include a positive urine test for *leukocyte esterase* (LE), an enzyme produced by neutrophils—which indicates an active infection. Nitrofurantoin usually clears cases of cystitis quickly. A fluoroquinolone is often successful if drug resistance is encountered.

## Pyelonephritis

In 25% of untreated cases, cystitis may progress to **pyelonephritis**, an inflammation of one or both kidneys. Symptoms are fever and flank or back pain. In females, it's often a complication of lower urinary tract infections. The causative agent in about 75% of the cases is *E. coli*. Pyelonephritis generally results in bacteremia; blood cultures and a Gram stain of the urine for bacteria are useful for diagnosis. A urine sample of more than 100,000 CFUs/ml of a single species indicates pyelonephritis. If pyelonephritis becomes chronic, scar tissue forms in the kidneys and severely impairs their function. Because pyelonephritis is a potentially life-threatening condition, treatment usually begins with intravenous, extended-term administration of a broad-spectrum antibiotic, such as a second- or third-generation cephalosporin.

## Leptospirosis

**Leptospirosis** is primarily a disease of domestic or wild animals, but it can be passed to humans and sometimes causes severe kidney or liver disease. The causative agent is the spirochete *Leptospira interrogans* (in-TER-ah-ganz'), shown in **Figure 26.4**. *Leptospira* has a characteristic shape: an exceedingly fine spiral, only about 0.1 μm in diameter, wound so tightly that it is barely discernible under a darkfield microscope. Like other spirochetes, *L. interrogans* (so named because the hooked ends suggest a question mark) stains poorly and is difficult to see under a normal light microscope. It's an obligate aerobe

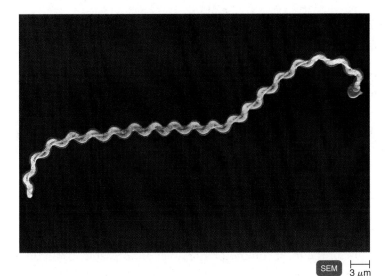

SEM  |——| 3 μm

**Figure 26.4** *Leptospira interrogans*, **the cause of leptospirosis.**

**Q** On what basis is *L. interrogans* named?

that can be grown in a variety of artificial media supplemented with rabbit serum.

Animals infected with the spirochete shed the bacteria in their urine for extended periods. In rats, the bacteria inhabit renal tubules, an immunologically privileged site, where they continue to reproduce and are shed, copiously, in urine for months. Worldwide, leptospirosis is probably the most common zoonosis; it's endemic in temperate and tropical environments. Humans become infected by contact with urine-contaminated water from freshwater lakes or streams, soil, or sometimes with animal tissue. People whose occupations expose them to animals or animal products are most at risk. Infections have been associated with recreational water sports. Usually the pathogen enters through minor abrasions in the skin or mucous membranes. When ingested, it enters through the mucosa of the upper digestive system. In the United States, dogs and rats are the most common sources. Domestic dogs have a sizable rate of infection; even when immunized, they may continue to shed leptospira.

After an incubation period of 1 to 2 weeks, headaches, muscular aches, chills, and fever abruptly appear. Several days later, the acute symptoms disappear, and the temperature returns to normal. A few days later, however, a second episode of fever may occur. Leptospires are observed within nonphagocytic cells of infected patients. It's uncertain how the pathogens enter host cells, but they use this as a mechanism to spread to target organs and evade the immune system. Because of this, the immune response is delayed long enough (1 or 2 weeks) for the population in the blood and tissues to reach enormous numbers. In a small number of cases the kidneys and liver become seriously infected (*Weil's disease*); kidney failure is the most common cause of death. An emerging form of leptospirosis, *pulmonary hemorrhagic syndrome*, has appeared globally. Affecting the lungs with massive bleeding, it has a fatality rate of more than 50%. Recovery results in a solid immunity, but only to the particular serovar involved. There are usually about 50 cases of Weil's disease reported each year in the United States, but because the clinical symptoms are not distinctive, many cases are probably never diagnosed. Fifty percent of the 100–200 annual cases in the United States occur in Hawaii.

Most cases of leptospirosis are diagnosed by a serological test that is complicated and usually done by central laboratories. However, a number of rapid serological tests are available for a preliminary diagnosis. Also, a diagnosis can be made by sampling blood, urine, or other fluids for the organism or its DNA. Doxycycline (a tetracycline) is the recommended antibiotic for treatment; however, administration of antibiotics in later stages is often unsatisfactory. That immune reactions are responsible for pathogenesis in later stages may be an explanation.

**CHECK YOUR UNDERSTANDING**

✓ **26-4** Why is urethritis, an infection of the urethra, frequently preliminary to further infections of the urinary tract?

✓ **26-5** Why is *E. coli* the most common cause of cystitis, especially in females?

**CLINICAL CASE**

Maricel's physician receives the results of her blood and urine cultures. The serological tests for STIs and HIV are negative. In response to her physician's questions about possible travel exposure, Maricel reports that she had been on a 2-week kayaking trip in Costa Rica the previous month. Maricel had really enjoyed the excursion because it included kayaking in streams near isolated rural villages, so she was able to get a feel for the "real" Costa Rica.

**What should Maricel's physician test for next?**

761  **765**  770  776

# Diseases of the Reproductive Systems

Microbes causing infections of the reproductive systems are usually very sensitive to environmental stresses and require intimate contact for transmission.

Most diseases of the reproductive systems transmitted by sexual activity have been called **sexually transmitted diseases (STDs)**. In recent years, this term has been replaced with **sexually transmitted infections (STIs)**. The reason is that the concept of "disease" implies obvious signs and symptoms. Because many persons infected by the more common sexually transmitted pathogens don't have apparent signs or symptoms, the term *STI* often seems more appropriate and is used in this book. More than 30 bacterial, viral, or parasitic infections have been identified as sexually transmitted. In the United States, 20 million new cases of STIs are estimated to occur annually, half of which occur in 15- to 24-year-olds. Most bacterial STIs can be prevented by the use of condoms. For information on new diagnostic options, see the **Big Picture** on STI home kits on pages 768–769.

## Bacterial Diseases of the Reproductive Systems

### LEARNING OBJECTIVE

26-6 List the causative agents, symptoms, methods of diagnosis, and treatments for gonorrhea, nongonococcal urethritis

(NGU), pelvic inflammatory disease (PID), syphilis, lymphogranuloma venereum (LGV), chancroid, and bacterial vaginosis.

Bacteria are the causative agents in about 20% of all STIs. Most bacterial STIs will not cause harm; however, some have the potential to cause serious health problems if not diagnosed and treated early. Bacterial STIs can be successfully treated with antibiotics.

### Gonorrhea

One of the most common reportable, or notifiable, communicable diseases in the United States is **gonorrhea**, an STI caused by the gram-negative diplococcus *Neisseria gonorrhoeae*. An ancient disease, gonorrhea was described and given its present name by the Greek physician Galen in 150 C.E. (*gon* = semen + *rhea* = flow; a flow of semen—apparently, he confused pus with semen). The incidence of gonorrhea in the United States has increased from a low of 98.1 cases per 100,000 people in 2009 to 130 cases per 100,000 in 2016 (**Figure 26.5a**). The true number of cases is probably much larger, probably two or three times those reported (Figure 26.5b). More than 60% of patients with gonorrhea are aged 15 to 24.

To infect, the gonococcus must attach to the mucosal cells of the epithelial wall by means of fimbriae. The pathogen

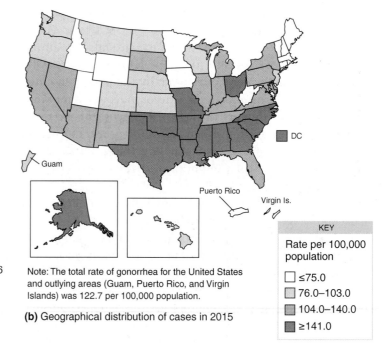

(a) Incidence of gonorrhea in the United States, 1941–2016

Note: The total rate of gonorrhea for the United States and outlying areas (Guam, Puerto Rico, and Virgin Islands) was 122.7 per 100,000 population.

(b) Geographical distribution of cases in 2015

KEY
Rate per 100,000 population
☐ ≤75.0
☐ 76.0–103.0
☐ 104.0–140.0
☐ ≥141.0

**Figure 26.5 The U.S. incidence and distribution of gonorrhea.**
*Source:* CDC, 2017.

**Q** How do gonococci attach to mucosal epithelial cells?

invades the spaces separating columnar epithelial cells, which are found in the oral-pharyngeal area, the eyes, rectum, urethra, opening of the cervix, and the external genitals of prepubertal females. The invasion sets up an inflammation and, when leukocytes move into the inflamed area, the characteristic pus forms. In men, a single unprotected exposure results in infection with gonorrhea 20–35% of the time. Women become infected 60–90% of the time from a single exposure.

Men become aware of a gonorrheal infection by painful urination and a discharge of pus-containing material from the urethra (**Figure 26.6**). About 80% of infected men show these obvious symptoms after an incubation period of only a few days; most others show symptoms in less than a week. In the days before antibiotic therapy, symptoms persisted for weeks. A common complication is urethritis, although this is more likely to be the result of coinfection with *Chlamydia*, which will be discussed shortly. An uncommon complication is *epididymitis*, an infection of the epididymis. Usually only unilateral, this is a painful condition resulting from the infection ascending along the urethra and the ductus deferens (see Figure 26.3).

In women, the disease is more insidious. Only the cervix, which contains columnar epithelial cells, is infected. The vaginal walls are composed of stratified squamous epithelial cells, which are not colonized. Very few women are aware of the infection. Later in the course of the disease, there might be abdominal pain from complications such as pelvic inflammatory disease (discussed on pages 771–772).

In both men and women, untreated gonorrhea can disseminate and become a serious, systemic infection. Complications of gonorrhea can involve the joints, heart (*gonorrheal endocarditis*), meninges (*gonorrheal meningitis*), eyes, pharynx, or other parts of the body. *Gonorrheal arthritis*, which is caused by the growth of the gonococcus in fluids in joints, occurs in about 1% of gonorrhea cases. Joints commonly affected include the wrist, knee, and ankle.

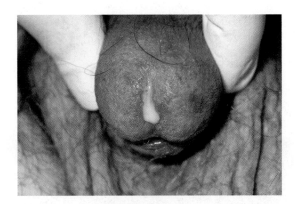

**Figure 26.6 Pus-containing discharge from the urethra of a man with an acute case of gonorrhea.**

**Q** What causes pus formation in gonorrhea?

If a mother is infected with gonorrhea, the eyes of her infant can become infected as it passes through the birth canal. This condition, **ophthalmia neonatorum**, can result in blindness. Because of the seriousness of this condition and the difficulty of being sure the mother is free of gonorrhea, antibiotics are placed in the eyes of all newborn infants. If the mother is known to be infected, an intramuscular injection of antibiotic is also administered to the infant. Some sort of prophylaxis is required by law in most states. Gonorrheal infections can also be transferred by hand contact from infected sites to the eyes of adults.

Gonorrheal infections can be acquired at any point of sexual contact; pharyngeal and anal gonorrhea are not uncommon. The symptoms of **pharyngeal gonorrhea** often resemble those of the usual septic sore throat. **Anal gonorrhea** can be painful and accompanied by discharges of pus. In most cases, however, the symptoms are limited to itching.

Increased sexual activity with a series of partners, and the fact that in women the disease may go unrecognized, contributed considerably to the increased incidence of gonorrhea and other STIs during the 1960s and 1970s. The widespread use of oral contraceptives also contributed to the increase. Oral contraceptives often replaced condoms and spermicides, which help prevent disease transmission.

There is no effective adaptive immunity to gonorrhea. The conventional explanation is that the gonococcus exhibits extraordinary antigenic variability—which is true. Lately, though, an alternative theory has appeared that provides an additional mechanism. The gonococcus has certain proteins, Opa proteins (see Chapter 15, page 428), that are essential for it to bind to the cells lining the host's urinary and reproductive tracts. Recent research has shown that an Opa protein variant binds to a certain receptor (CD66) on CD4$^+$ T cells that is needed for activation and proliferation of the cells. This inhibits the development of an immunological memory response against the gonococcus. Almost all clinical isolates of the gonococcus have been found to carry this Opa protein variant. This suppression of immunity may also explain why people with gonorrhea are more susceptible to other STIs, including HIV.

### Diagnosis of Gonorrhea

Gonorrhea in men is diagnosed by finding gonococci in a stained smear of pus from the urethra. The typical gram-negative diplococci within the phagocytic leukocytes are readily identified (**Figure 26.7**). It's uncertain whether these intracellular bacteria are in the process of being killed or whether they survive indefinitely. Probably at least a fraction of the bacterial population remains viable. Gram staining of exudates isn't as reliable with women. Usually, a culture is taken from within the cervix and grown on special media. Cultivation of the nutritionally fastidious bacterium requires an atmosphere enriched in carbon dioxide. The gonococcus is very sensitive

**Millions of STI cases go undiagnosed every year. Home test kits can speed diagnosis and treatment, allowing individuals who might otherwise avoid the health clinic to privately begin the screening process.**

### At-Home Test for Sexually Transmitted Infections

Sexually transmitted infections (STIs) are a major public health concern throughout the world and are at record-high numbers in the United States. Over 2 million new cases of STIs were reported in 2016, but the Centers for Disease Control and Prevention (CDC) estimates that many more undiagnosed infections occur annually.

In the hope of discovering and treating more STIs, researchers at Johns Hopkins School of Medicine created "I Want the Kit," a self-administered screening currently available to residents of Maryland, Washington, D.C., and Alaska.

Users collect samples at home and mail them to a lab where nucleic acid amplification testing (NAAT) screens for chlamydial infection, gonorrhea, and trichomoniasis. The user can obtain results in 1–2 weeks using a password over the phone or online. People with positive tests receive referrals to nearby clinics for counseling and treatment options. There is also an option on the website allowing the user to anonymously notify sexual partners that they may also be infected. Planned Parenthood has similar kits to test for *Chlamydia* and gonorrhea available to residents of California, Minnesota, Idaho, and Washington.

A recent study indicated that home or pharmacy testing is feasible. At present, the FDA has approved at-home test kits only for hepatitis C, HIV, and UTIs.

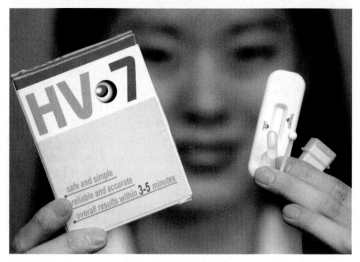

**Female test kit contents for "I Want the Kit"**

### Other Home Test Options for HIV and Urinary Tract Infections

In 2013, the FDA approved OraQuick®, an HIV oral testing kit. The OraQuick strip is similar to an indirect ELISA test. Priced at about $40 per test, it uses an HIV antigen and enzyme indicator to test the oral mucosa for antibodies against HIV. Clinical studies showed this test produces about one false positive in every 5000 uninfected individuals, and one false negative in every 12 HIV-infected individuals.

While urinary tract infections (UTIs) are not technically STIs, sometimes women misattribute symptoms of an STI as being from a UTI. At-home testing for UTIs is also available. A dipstick is held in the urine stream, and the test strip then indicates presence of nitrites, which are usually produced by bacteria that cause UTIs. These tests also look for presence of leukocytes, which indicate an immune response to an infection.

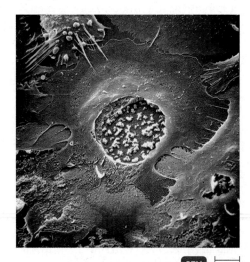

SEM | 3 μm

***Chlamydia* (blue) replicating inside vesicle within a cell**

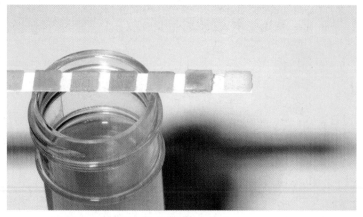

**Uritest, a home test for urinary tract infection**

# Are At-Home Test Kits a Good Public Health Strategy?

## Pros of At-Home Testing

- **More Cases Diagnosed** In a 5-year period, the "I Want the Kit" initiative detected more chlamydial infections than traditional clinics did in the areas where it was available. Its creators estimate that this method of testing would save about $41,000 in direct medical costs per 10,000 women, when compared with clinic-based screening.

- **Better Access for Patients** Home testing kits can also be very helpful for people with limited transportation or for residents of rural areas who live far from medical facilities. Most home test kits are available at drugstores or can be ordered online or over the phone.

- **Quicker Treatment** Providing a screening method that works for people who are reluctant or unable to visit a clinic means more diagnoses and quicker treatment. It can also lead to fewer complications and better prognosis. For example, a positive result from a UTI home test kit may prompt health care providers to prescribe antibiotics without requiring patients to first provide a urine sample for culturing. In this case, faster treatment results in less discomfort and downtime for patients and could prevent a UTI from progressing to a kidney infection.

## Some Cons of Home-Testing

- **Cost** Although home test kits reduce public health costs, they increase consumer costs because they are typically not covered by health insurance plans.

- **Privacy Concerns** Allowing access to test information through a hotline or website raises concerns that results may fall into the hands of someone other than the patient. People using home kits should be mindful of keeping PINs or other test-related paperwork away from prying eyes.

- **Concerns about Accuracy** Unfortunately, not every home test kit sold online today is necessarily effective. Users should seek kits that are FDA approved. And regardless of test results, a person with persistent or worsening symptoms should always see a health care provider.

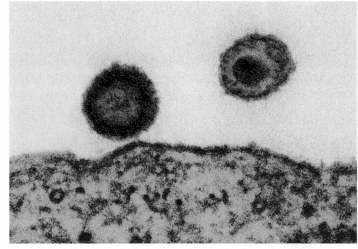

TEM | 50 nm

**HIVs (human immunodeficiency viruses) infecting a cell**

Home screening tests provide an alternative for people without easy access to clinics.

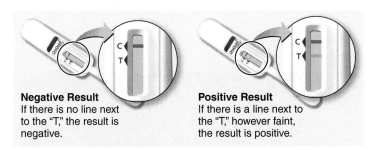

**Negative Result**
If there is no line next to the "T," the result is negative.

**Positive Result**
If there is a line next to the "T," however faint, the result is positive.

A positive OraQuick® HIV test contains a synthetic HIV gp-41 protein. If the sample contains antibodies against gp-41, an enzyme reaction causes the T strip to change color.

## KEY CONCEPTS

- Certain gram-negative organisms convert nitrates to nitrites, so the presence of nitrites in urine can indicate urinary tract infection. **(See Chapter 5, "Anaerobic Respiration," page 128.)**

- Nucleic acid amplification is used to screen for chlamydiasis, gonorrhea, and trichomoniasis infections. **(See Chapter 9, "Polymerase Chain Reaction," pages 247–248.)**

- Some home tests for HIV, such as OraQuick, are similar to indirect ELISA tests. **(See Chapter 18, "Enzyme-Linked Immunosorbent Assay," page 516, and Figures 18.13, 18.14.)**

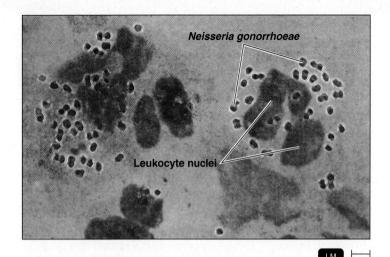

**Figure 26.7 A smear of pus from a patient with gonorrhea.**
The *Neisseria gonorrhoeae* bacteria are contained within phagocytic leukocytes. These gram-negative bacteria are visible here as pairs of cocci. The large stained bodies are the nuclei of the leukocytes.

**Q** How is gonorrhea diagnosed?

to adverse environmental influences (desiccation and temperature) and survives poorly outside the body. It even requires special transporting media to keep it viable for short intervals before the cultivation is under way. Cultivation has the advantage of allowing determination of antibiotic sensitivity.

Diagnosis of gonorrhea has been aided by the development of an ELISA test that detects *N. gonorrhoeae* in urethral pus or on cervical swabs within about 3 hours with high accuracy. This and other rapid tests use monoclonal antibodies against antigens on the surface of the gonococcus. Nucleic acid amplification tests (NAATs) are very accurate for identifying clinical isolates from suspected cases.

### Treatment of Gonorrhea

The guidelines for treating gonorrhea require constant revision as resistance appears (see the Clinical Focus box). Because of increasing multi-drug resistance, the only recommended treatment for all forms of gonorrhea (cervical, urethral, rectal, pharyngeal) is 250 mg of ceftrixone and 1 g azithromycin. Unless a coinfection by *Chlamydia trachomatis* (see the discussion of non-gonococcal urethritis, following) can be ruled out, the patient should also be treated for this organism. It's also standard practice to treat sex partners of patients to decrease the risk of reinfection and to decrease the incidence of STIs in general.

### Nongonococcal Urethritis (NGU)

**Nongonococcal urethritis (NGU)**, also known as **nonspecific urethritis (NSU)**, refers to any inflammation of the urethra not caused by *Neisseria gonorrhoeae*. Symptoms include painful urination and a watery discharge.

### *Chlamydia trachomatis*

The most common pathogen associated with NGU is *Chlamydia trachomatis*. More than 1.5 million cases are reported annually in the United States. Many people suffering from gonorrhea are coinfected with *C. trachomatis*, which infects the same columnar epithelial cells as the gonococcus. *C. trachomatis* is also responsible for the STI lymphogranuloma venereum (discussed on page 775) and trachoma (see page 612). Of special importance is the fact that twice as many cases are reported in women as in men. In women, it is responsible for many cases of pelvic inflammatory disease (discussed on pages 771–772), plus eye infections and pneumonia in infants born to infected mothers. Untreated genital chlamydial infections are also associated with an increased risk of cervical cancer. *Chlamydia*-infected cells may be more susceptible to infection with human papillomavirus (page 778).

Because the symptoms are often mild in men and because women are usually asymptomatic, many cases of NGU go untreated. Although complications are not common, they can be serious. Men may develop inflammation of the epididymis. In women, inflammation of the uterine tubes may cause scarring, leading to sterility. As many as 60% of such cases may be from chlamydial rather than gonococcal infection. It's estimated that about 50% of men and 70% of women are unaware of their chlamydial infection.

NAATs are the most reliable methods of diagnosis, and they can be done quickly. Urine samples can be used, but the sensitivity is lower than with swabs. Swab specimens (urethral or vaginal, as the case might be) collected by the patients themselves are often used.

In view of the serious complications often associated with infections by *C. trachomatis*, it's recommended that physicians routinely screen sexually active women 25 years of age and younger for infection. Screening is also recommended for other higher-risk groups, such as persons who are unmarried, had a prior risk of STIs, and have multiple sexual partners.

Bacteria other than *C. trachomatis* can also be implicated in NGU. *Mycoplasma genitalium* (mī-kō-PLAZ-mah jen-i-TAL-ē-um) causes up to 30% of urethritis cases in males. This

The physician requests an anti-leptospira antibody test on Maricel's blood. The result is a titer of 1:100, which indicates that Maricel is or has been infected with *Leptospira interrogans* bacteria. Now, at day 15 of Maricel's illness, the physician draws another blood sample for a second microscopic agglutination test. The titer is now 1:800.

**Why is a second serological test necessary?**

761    765    **770**    776

# Survival of the Fittest

As you read through this problem, you will see questions that health care providers ask themselves and each other as they solve a clinical problem. Try to answer each question as you read through the problem.

1. On May 24, Jason, a 35-year-old man, goes to the Denver STI Clinic with a history of painful urination and urethral discharge of approximately 1 month's duration.
   *What other information do you need about Jason's history?*

2. On March 11, Jason returned from a "dating tour" in Thailand, during which he had sexual contact with seven or eight female prostitutes; he denies having had any sexual contact since returning to the United States.
   *What sample should be taken, and how should it be tested?*

3. *Neisseria gonorrhoeae* is identified by polymerase chain reaction (PCR) of urethral discharge. Jason is treated with a single 250-mg dose of ceftriaxone orally.
   *What is the advantage of PCR or enzyme immunoassay (EIA) over cultures for diagnosis?*

4. PCR and EIA provide results within a few hours, eliminating the need for the patient to return for treatment. Jason returns to the clinic on June 7 with continuing symptoms. *N. gonorrhoeae* is again detected in urethral discharge. Jason denies having had any sexual contact since the previous visit.

The attending physician requests antimicrobial susceptibility testing of the *N. gonorrhoeae* isolates.
*Why would the physician be interested in antimicrobial susceptibility test results on this patient's specimen?*

5. One reason for Jason's failure to respond to ceftriaxone may be due to infection with a resistant *N. gonorrhoeae*. Susceptibility testing would be helpful to explore this possibility.
   The treatment and control of gonorrhea has been complicated by the ability of *N. gonorrhoeae* to develop resistance to antimicrobial agents (see the graph).
   *How does antibiotic resistance emerge?*

6. In an environment filled with antibiotics, bacteria that have mutations for antibiotic resistance will have a selective advantage and be the "fittest" to survive.
   *How is antibiotic susceptibility determined?*

7. *N. gonorrhoeae* must be grown in culture for disk-diffusion or broth dilution tests for antimicrobial susceptibility. The increasing use of nonculture methods for gonorrhea diagnosis such as PCR and EIA is a major challenge to monitoring antimicrobial resistance in *N. gonorrhoeae*.

Source: Data from CDC. *Sexually Transmitted Disease Surveillance 2016.*

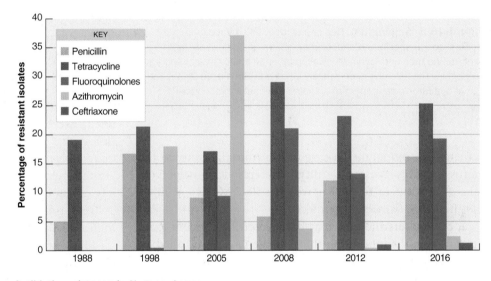

Antibiotic resistance in *N. gonorrhoeae*.

---

bacterium may cause cervicitis in women. Another cause of urethritis is *Ureaplasma urealyticum* (u'rē-ah-PLAZ-mah u'rē-ah-LIT-i-kum). This pathogen is a member of the mycoplasma (bacteria without a cell wall).

Both chlamydia and mycoplasma are sensitive to azithromycin and doxycyline.

## Pelvic Inflammatory Disease (PID)

**Pelvic inflammatory disease (PID)** is a collective term for any extensive bacterial infection of the female pelvic organs, particularly the uterus, cervix, uterine tubes, or ovaries. During their reproductive years, one in ten women suffers from PID, and one in four of these will have serious complications, such as infertility or chronic pain.

Pelvic inflammatory disease is considered to be a *polymicrobial infection*—that is, a number of different pathogens might be the cause, including coinfections. The two most common microbes are *N. gonorrhoeae* and *C. trachomatis*. The onset of chlamydial PID is relatively more insidious, with fewer initial inflammatory symptoms than when caused by *N. gonorrhoeae*. However, the damage to the uterine tube may be greater with chlamydia, especially with repeated infections.

The bacteria may attach to sperm cells and be transported by them from the cervical region to the uterine tubes. Women who use barrier contraceptives, especially with spermicides, have a significantly lower rate of PID.

Infection of the uterine tubes, or **salpingitis**, is the most serious form of PID (**Figure 26.8**). Salpingitis can result in

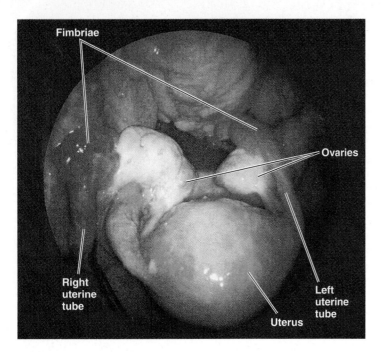

**Figure 26.8 Salpingitis.** This photograph, taken through a laparoscope (a specialized endoscope), shows an acutely inflamed right uterine tube and inflamed, swollen fimbriae and ovary, caused by salpingitis. The left tube is only mildly inflamed. (See Figure 26.2.) The use of a laparoscope is the most reliable diagnostic method for PID.

**Q** What is PID?

scarring that blocks the passage of ova from the ovary to the uterus, possibly causing sterility. One episode of salpingitis causes infertility in 10–15% of women; 50–75% become infertile after three or more such infections.

A blocked uterine tube may cause a fertilized ovum to be implanted in the tube rather than the uterus. This is called an *ectopic* (or *tubal*) *pregnancy*, and it is life-threatening because of the possibility of rupture of the tube and resulting hemorrhage. The reported cases of ectopic pregnancies have been increasing steadily, corresponding to the increasing occurrence of PID.

A diagnosis of PID depends strongly on signs and symptoms, in combination with laboratory indications of a gonorrheal or chlamydial infection of the cervix. The recommended treatment for PID is the simultaneous administration of doxycycline and cefotetan (a β-lactam). This combination is active against both the gonococcus and chlamydia. Such recommendations are constantly being reviewed.

### Syphilis

The causative agent of **syphilis** is a gram-negative spirochete, *Treponema pallidum* (**Figure 26.9**). Thin and tightly coiled, *T. pallidum* stains poorly with the usual bacterial stains. (The bacterial name is derived from the Greek words for twisted thread and pale.) *T. pallidum* lacks the enzymes necessary to build many complex molecules; therefore, it relies on the host for

many of the compounds necessary for life. Outside the mammalian host, the organism loses infectiveness within a short time. For research purposes the bacteria are usually propagated in rabbits, but they grow slowly, with a generation time of 30 hours or more. They can be grown in cell culture, at low oxygen concentrations, but only for a few generations.

*T. pallidum* has no obvious virulence factors such as toxins, but it produces several lipoproteins that induce an inflammatory immune response. This is what apparently causes the tissue destruction of the disease. Almost immediately on infection, the organisms rapidly enter the bloodstream and invade deeper tissues, easily crossing the junctions between cells. They have a corkscrew-like motility that allows them to swim readily in gel-like tissue fluids.

The earliest reports of syphilis date back to the end of the fifteenth century in Europe, when the return of Columbus from the New World gave rise to a hypothesis that syphilis was introduced to Europe by his men. One English description of the "Morbus Gallicus" (French disease) seems to clearly describe syphilis as early as 1547 and ascribes its transmission in these terms: ". . . It is taken when one pocky person doth synne in lechery one with another."

Separate strains of *T. pallidum* (subspecies *T.p. pertenue*) are responsible for certain tropically endemic skin diseases such as **yaws**. These cause skin lesions but are not sexually transmitted. However, there is evidence of a probable historical association with syphilis. Current research based on genetic analysis of *Treponema* spp. indicates that a pathogen of yaws found in South America adjacent to the Caribbean mutated into a sexually transmitted disease on contact with European explorers.

The number of new syphilis cases in the United States has increased almost every year since 2001 (**Figure 26.10**).

Many states discontinued the requirements for premarital syphilis tests because so few cases were detected. At present, the population most at risk is men having sex with men,

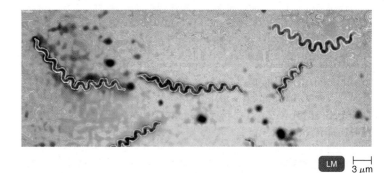

**Figure 26.9 *Treponema pallidum*, the cause of syphilis.** The microbes are made more visible in this brightfield micrograph by merging multiple photos.

**Q** A diagnostic method for syphilis is the darkfield microscope. Why not use a brightfield microscope?

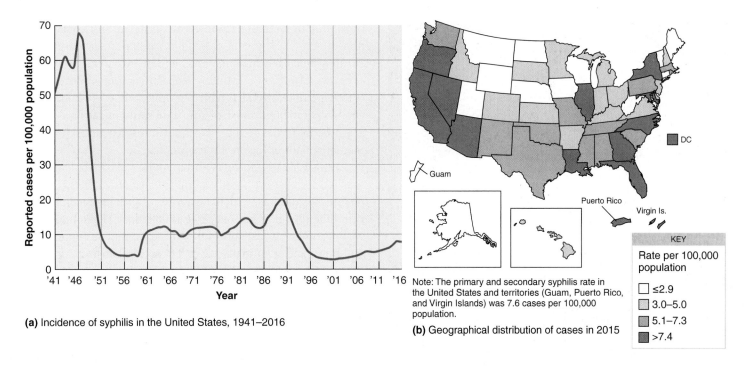

(a) Incidence of syphilis in the United States, 1941–2016

Note: The primary and secondary syphilis rate in the United States and territories (Guam, Puerto Rico, and Virgin Islands) was 7.6 cases per 100,000 population.

(b) Geographical distribution of cases in 2015

KEY
Rate per 100,000 population
☐ ≤2.9
☐ 3.0–5.0
☐ 5.1–7.3
☐ >7.4

**Figure 26.10 The U.S. incidence and distribution of primary and secondary syphilis.**
*Source:* CDC, 2017.

**Q** How is syphilis diagnosed?

although in 2015 the incidence increased in both males and females in every racial/ethnic group.

Syphilis is transmitted by sexual contact of all kinds via syphilitic infections of the genitals or other body parts. The incubation period averages 3 weeks but can range from 2 weeks to several months. The disease progresses through several recognized stages.

### Primary Stage Syphilis

In the *primary stage* of the disease, the initial sign is a small, hard-based **chancre**, or sore, which appears at the site of infection 10 to 90 days following exposure—on average, about 3 weeks (**Figure 26.11a**). The chancre is painless, and an exudate of serum forms in the center. This fluid is highly infectious, and examination with a darkfield microscope shows many spirochetes. In a couple of weeks, this lesion disappears. None of these symptoms causes any distress. In fact, many women are entirely unaware of the chancre, which is often on the cervix. In men, the chancre sometimes forms in the urethra and isn't visible. During this stage, bacteria enter the bloodstream and lymphatic system, which distribute them widely in the body.

### Secondary Stage Syphilis

Several weeks after the primary stage (the exact length of time varies and may overlap), the disease enters the *secondary stage*, characterized mainly by oral sores and skin rashes of varying appearance. The rash is widely distributed on the skin and mucous membranes and is especially visible on the palms and on the soles (Figure 26.11b). The damage done to tissues at this stage and the later tertiary stage is caused by an inflammatory response to circulating immune complexes that lodge at various body sites. Other symptoms often observed are the loss of patches of hair, malaise, and mild fever. A few people may exhibit neurological symptoms.

At this stage, the lesions of the rash contain many spirochetes and are very infectious. Transmission from sexual contact occurs during the primary and secondary stages. Dentists and other health care workers coming into contact with fluid from these lesions can become infected by the spirochete entering through minute breaks in the skin. Such nonsexual transmission is possible, but the microbes don't survive long on environmental surfaces and are very unlikely to be transmitted via such objects as toilet seats. Secondary syphilis is a subtle disease; at least half of the patients diagnosed with this stage can recall no lesions at all. Symptoms usually resolve within 3 months.

### Latent Period

The symptoms of secondary syphilis will go away with or without treatment, and the disease enters a *latent period*. During this period, there are no symptoms. After 2 to 4 years of latency, the disease is not normally infectious, except for transmission from mother to fetus. The majority of cases don't progress beyond the latent stage, even without treatment.

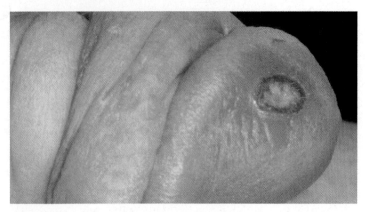

**(a)** Chancre of primary stage on a male in genital area

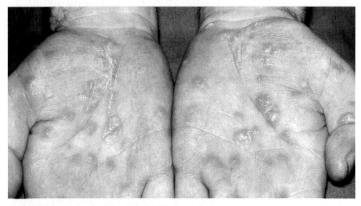

**(b)** Lesions of secondary syphilis rash on hands; any surface of the body may be afflicted with such lesions.

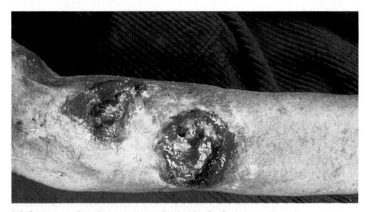

**(c)** Gummas of tertiary stage on the back of a forearm; gummas such as these are rarely seen today in the era of antibiotics.

**Figure 26.11 Characteristic lesions associated with various stages of syphilis.**

**Q** How are the primary, secondary, and tertiary stages of syphilis distinguished?

## Tertiary Stage Syphilis

Because the symptoms of primary and secondary syphilis aren't disabling, people may enter the latent period without having received medical attention. In up to 25% of untreated cases, the disease reappears in a *tertiary stage*. This stage occurs only after an interval of many years from the onset of the latent phase.

*T. pallidum* has an outer layer of lipids that stimulates little effective immune response, especially from cell-destroying complement reactions. It has been described as a "Teflon pathogen." Nonetheless, most of the symptoms of tertiary syphilis are probably due to the body's immune reactions (of a cell-mediated nature) to surviving spirochetes.

Tertiary, or late-stage, syphilis can be classified generally by affected tissues or type of lesion. *Gummatous syphilis* is characterized by **gummas**, which are a form of progressive inflammation that appear as rubbery masses of tissue (Figure 26.11c) in various organs (most commonly the skin, mucous membranes, and bones) after about 15 years. There they cause local destruction of these tissues but usually not incapacitation or death.

*Cardiovascular syphilis* results most seriously in a weakening of the aorta. In preantibiotic days, it was one of the more common symptoms of syphilis; it is now rare. Few, if any, pathogens are found in the lesions of the tertiary stage, and they are not considered very infectious. Today, it is rare for cases of syphilis to be allowed to progress to this stage.

Complications of untreated syphilis involving the eyes or central nervous system can occur during any stage of the disease. *Ocular syphilis* causes blurred vision and permanent blindness. Symptoms of *neurosyphilis* can vary widely. The patient can suffer from personality changes and other signs of dementia (*paresis*), seizures, loss of coordination of voluntary movement (*tabes dorsalis*), partial paralysis, loss of ability to use or comprehend speech, loss of sight or hearing, or loss of bowel and bladder control.

## Congenital Syphilis

One of the most distressing and dangerous forms of syphilis, called **congenital syphilis**, is transmitted across the placenta to the unborn fetus. Damage to mental development and other neurological symptoms are among the more serious consequences. This type of infection is most common when pregnancy occurs during the latent period of the disease. A pregnancy during the primary or secondary stage is likely to produce a stillbirth. Treating the mother with antibiotics during the first two trimesters (6 months) will usually prevent congenital transmission.

## Diagnosis of Syphilis

Diagnosis of syphilis is complex because each stage of the disease has unique requirements. Tests fall into three general groups: visual microscopic inspection, nontreponemal serological tests, and treponemal serological tests. For preliminary screening, laboratories use either nontreponemal serological tests or microscopic examination of exudates from lesions when these are present. If a screening test is positive, the results are confirmed by treponemal serological tests.

*Microscopic tests* are important for screening for primary syphilis because serological tests for this stage are not reliable; antibodies take 1 to 4 weeks to form. The spirochetes can be detected in exudates of lesions by microscopic examination with a darkfield microscope (see Figure 3.4b, page 56). A darkfield microscope is necessary because the bacteria stain poorly and are only about 0.2 µm in diameter, near the lower limit of resolution for a brightfield microscope. Similarly, a **direct fluorescent-antibody test (DFA-TP)** using monoclonal antibodies (see Figure 18.11a, page 515) will both show and identify the spirochete. Figure 26.9 shows *T. pallidum* under brightfield illumination made possible by a computer enhancing technique.

At the secondary stage, when the spirochete has invaded almost all body organs, serological tests are reactive. *Nontreponemal serological tests* are so called because they are nonspecific; they detect *reagin-type antibodies*, not the antibodies produced against the spirochete itself. Generally, these tests are used for screening. Reagin-type antibodies are apparently a response to lipid materials the body forms as an indirect reaction to infection by the spirochete. The antigen used in such tests is thus not the syphilis spirochete but an extract of beef heart (cardiolipin) that seems to contain lipids similar to those that stimulated the reagin-type antibody production. These tests will detect only about 70–80% of primary syphilis cases, but they will detect 99% of secondary syphilis cases. An example of nontreponemal tests is the slide agglutination **VDRL test** (for Venereal Disease Research Laboratory). Also used are modifications of the **rapid plasma reagin (RPR) test**, which is similar. The newest nontreponemal test is an ELISA test that uses the VDRL antigen.

There are also *treponemal-type serological tests* that react directly with the spirochete. Certain **enzyme immunoassay (EIA)** treponemal tests can be done in many laboratories and offer high-throughput screening. There are also simple **rapid diagnostic tests (RDTs)** of this type that can be done from a finger-prick blood sample in a physician's office. Neither of these groups of tests will distinguish a prior from an active infection; confirmatory tests, which usually must be done at a central reference laboratory, are required.

Only treponemal-type tests are used for confirmatory testing. An example is the **fluorescent treponemal antibody absorption test**, or **FTA-ABS test**, an indirect fluorescent-antibody test (see Figure 18.11b, page 515). Treponemal tests are not used for screening because about 1% of the results will be false positives, but a positive test with both treponemal and nontreponemal types is highly specific.

### Treatment of Syphilis

Benzathine penicillin, a long-acting formulation that remains effective in the body for about 2 weeks, is the usual antibiotic treatment of syphilis. The serum concentrations achieved by this formulation are low, but the spirochete has remained very sensitive to this antibiotic.

For penicillin-sensitive people, several other antibiotics, such as azithromycin, doxycycline, and tetracycline, have also proven effective.

## Lymphogranuloma Venereum (LGV)

Several STIs that are uncommon in the United States occur frequently in the tropical areas of the world. For example, *Chlamydia trachomatis*, the cause of the eye infection trachoma and a major cause of NGU, is also responsible for **lymphogranuloma venereum (LGV)**, a disease found in tropical and subtropical regions. It is apparently caused by serovars of *C. trachomatis* that are invasive and tend to infect lymphoid tissue. In the United States, there are usually 200 to 400 cases each year, mostly in homosexual men, many of whom are also HIV-positive.

The microorganisms invade the lymphatic system, and the regional lymph nodes become enlarged and tender. Suppuration (a discharge of pus) may also occur. Inflammation of the lymph nodes results in scarring, which occasionally obstructs the lymph vessels. This blockage sometimes leads to massive enlargement of the external genitals in men. In women, involvement of the lymph nodes in the rectal region leads to narrowing of the rectum. These conditions may eventually require surgery.

For diagnosis, blood tests for antibodies to the serovars of *C. trachomatis* causing the disease are satisfactory. Direct tests for chlamydia antigens, such as ELISA or NAAT, are available through public health laboratories. The drug of choice for treatment is doxycycline.

## Chancroid (Soft Chancre)

The STI known as **chancroid (soft chancre)** occurs most frequently in tropical areas, where it is seen more often than syphilis. The number of reported cases in the United States has been declining from a peak of 5000 cases in 1988. Like syphilis, chancroid is a risk factor in the transmission and acquisition of HIV infection. Because chancroid is so seldom seen by some physicians and is difficult to diagnose, it's probably underreported. It's very common in Africa, Asia, and the Caribbean.

In chancroid, a swollen, painful ulcer that forms on the genitals involves an infection of the adjacent lymph nodes. Infected lymph nodes in the groin area sometimes even break through and discharge pus to the surface. Such lesions are an important factor in the sexual transmission of HIV, especially in Africa. Lesions might also occur on such diverse areas as the tongue and lips. The causative agent is *Haemophilus ducreyi* (hē-MAH-fil-us doo-KRĀ-ē), a small gram-negative rod that can be isolated from exudates of lesions. Symptoms and the culture of these bacteria are the primary means of diagnosis. The recommended antibiotics include single doses of azithromycin or ceftriaxone.

## Bacterial Vaginosis

Inflammation of the vagina due to infection, or **vaginitis**, is most commonly caused by one of several organisms: mainly the fungus *Candida albicans* (KAN-dē-dah AL-bi-kanz), the protozoan *Trichomonas vaginalis* (TRIK-ō-mō-nas va-ji-NA-lis), or the bacterium *Gardnerella vaginalis*, a small, pleomorphic gram-variable rod (see Diseases in Focus 26.2 on page 779). Most of these cases are attributed to the presence of *G. vaginalis* and are termed **bacterial vaginosis**. (Because there is no sign of inflammation, the term *vaginosis* is preferred to *vaginitis*).

The condition is something of an ecological mystery. It's believed that bacterial vaginosis is precipitated by some event that decreases the number of vaginal *Lactobacillus* bacteria that normally produce hydrogen peroxide. This competitive change allows bacteria, especially *G. vaginalis*, to proliferate, producing amines that contribute to a further rise in pH. These various bacteria, most of which are commonly found in the vaginas of asymptomatic women, are assumed to be metabolically interdependent. This situation does not lend itself to the application of Koch's postulates to determine a specific cause. There is no corresponding disease condition in men, but *G. vaginalis* is often present in their urethras. Therefore, the condition may be sexually transmitted, but it also occurs occasionally in women who have never been sexually active. The prevalence of vaginosis among women between 14 and 49 years of age is about 30%.

Bacterial vaginosis is characterized by a vaginal pH above 4.5 and a copious, frothy vaginal discharge. When tested with a potassium hydroxide solution, these vaginal secretions emit a fishy odor from amines produced by *G. vaginalis*. Diagnosis is based on the vaginal pH, fishy odor (the *whiff test*), and microscopic observation of *clue cells* in the discharge. These clue cells

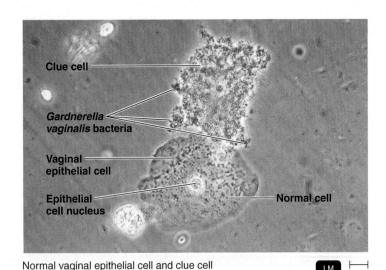

Clue cell

*Gardnerella vaginalis* bacteria

Vaginal epithelial cell

Epithelial cell nucleus

Normal cell

Normal vaginal epithelial cell and clue cell

LM  ⊢────┤ 9 μm

**Figure 26.12  Clue cell.** *Gardnerella* bacteria coat the surface of a vaginal epithelial cell.

**Q** What symptoms would cause you to look for clue cells?

IgG antibodies against *L. interrogans* can persist for years, and their presence may only indicate a previous exposure. The rise in titer on the second test indicates a current infection. Maricel is treated with doxycycline for leptospirosis and is soon able to resume training for her next kayaking event.

Occupational exposure most likely accounts for 30–50% of human cases of leptospirosis. The main occupational groups at risk include farm workers, veterinarians, pet shop owners, plumbers and sewer workers, meat handlers and slaughterhouse workers, and military troops. However, since 1970 leptospirosis has been increasingly associated with recreational activities. Prolonged water exposure from swimming or kayaking in freshwater lakes or streams, for example, has been associated with *L. interrogans* infection.

761  765  770  **776**

are sloughed-off vaginal epithelial cells covered with a biofilm of bacteria, mostly *G. vaginalis* (**Figure 26.12**). The disease has been considered more of a nuisance than a serious infection, but it's now seen as a factor in many premature births and low-birth-weight infants.

Treatment is primarily by metronidazole, a drug that eradicates the anaerobes essential to continuation of the disease but allows the normal lactobacilli to repopulate the vagina. Treatments designed to restore the normal population of lactobacilli, such as application of acetic acid gels and even yogurt, have not been shown conclusively to be effective.

**CHECK YOUR UNDERSTANDING**

✔ 26-6 Why is the disease condition of the female reproductive system principally featuring growth of *Gardnerella vaginalis* termed *vaginosis* rather than *vaginitis*?

# Viral Diseases of the Reproductive Systems

**LEARNING OBJECTIVE**

26-7 Discuss the epidemiology of genital herpes and genital warts.

Viral diseases of the reproductive system are difficult to treat, and so they represent an increasing health problem.

## Genital Herpes

A much publicized STI is **genital herpes**, usually caused by *herpes simplex virus type 2* (HSV-2). Herpes simplex virus type 1 (HSV-1) is primarily responsible for oral cold sores (see page 605), but it

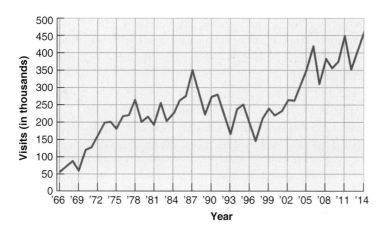

**Figure 26.13 Genital herpes: initial visits to physicians' offices, United States, 1966 to 2014.**

*Source:* CDC, 2015.

**Q** What are possible causes for changes in the incidence of a disease, such as shown on this graph?

can also cause genital herpes. The official names are human herpesvirus 1 and 2.

In the United States, one in four persons over the age of 30 is infected with HSV-2—but most are unaware they are infected. There has been a marked increase in genital HSV-1 infections, which is usually acquired by oral–genital contact, and this now constitutes about half of cases of genital herpes in this country (**Figure 26.13**).

Genital herpes lesions appear after an incubation period of up to 1 week and cause a burning sensation. After this, vesicles appear (**Figure 26.14**). In both men and women, urination can be painful, and walking is quite uncomfortable; the patient is even irritated by clothing. Usually, the vesicles heal in a couple of weeks.

The vesicles contain infectious fluid, but many times the disease is transmitted when no lesions or symptoms are apparent. Semen may contain the virus. Condoms may not provide protection because in women the vesicles are usually on the external genitals (seldom on the cervix or within the vagina), and in men the vesicles may be on the base of the penis.

One of the most distressing characteristics of genital herpes is the possibility of recurrences. There is an element of truth in the medical adage that, unlike love, herpes is forever. As in other herpes infections, such as cold sores or chickenpox-shingles, the virus enters a lifelong latent state in nerve cells. Some people have several recurrences a year; for others, recurrence is rare. Men are more likely to experience recurrences than women. Reactivation appears to be triggered by factors that lower the immune system, including menstruation, emotional stress or illness (a factor that is also involved in the appearance of cold sores), and being "run down." About 90% of patients with HSV-2 and about 50% of those with HSV-1 will

have recurrences. Recurrence rates decrease over time, regardless of treatment. Low numbers of viruses may be produced at any time during latency; thus, the virus is transmitted even in the absence of visible symptoms.

Diagnosis of genital herpes can be done by culture of the virus taken from a vesicle; however, PCR testing of such samples has proven more sensitive and is potentially faster. If there are no lesions to be sampled, serological testing can identify HSV infections or confirm clinical diagnosis by symptoms.

There is no cure for genital herpes, although research on its prevention and treatment is intensive. Discussions of chemotherapy use terms such as *suppression* or *management* rather than *cure*. Currently, the antiviral drugs acyclovir, famciclovir, and valacyclovir are recommended for treatment. They are fairly effective in alleviating the symptoms of a primary outbreak; there is some relief of pain and slightly faster healing. Taken over several months, they lower the chances of recurrence during that time.

**Neonatal Herpes**

*Neonatal herpes* is a serious consideration for women of childbearing age. The virus can cross the placental barrier and affect the fetus, causing spontaneous abortion or serious fetal damage. If untreated, a survival rate of only about 40% can be expected, and even treated survivors will have considerable disability. Herpes infection of the newborn is most likely to have serious consequences when the mother acquires the initial herpes infection during the pregnancy. If tests show a pregnant woman has no antibodies against herpes viruses, she requires special counseling to avoid an initial infection. Exposure to recurrent or asymptomatic herpes is much less likely to damage the fetus, most likely because of protective maternal antibodies.

Most infections of the newborn occur from exposure to HSV during delivery. HSV-2 infections are likely to be more serious than HSV-1 infections. If genital sores that might be caused by herpes infection are present at the time of delivery, a sample can be taken and isolates tested to determine whether they are HSV-1 or HSV-2. If the culture is negative but a herpes infection is still suspected, a PCR test for viral

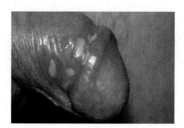

**Figure 26.14 Vesicles of genital herpes on a penis.**

**Q** What microbe causes genital herpes?

DNA can be done. It is rather common for pregnant women to be shedding HSV-2 even though they show no evidence of an infection. Even so, fewer than 1% of newborns develop neonatal herpes, which is also probably because of protective antibodies.

Some newborns have infections that are confined to the skin, mucous membranes, and eyes. With proper treatment, the outcome of these cases is usually good. However, about 30% of the cases are associated with damage to the central nervous system that may include developmental delays, blindness, hearing loss, or epilepsy. Disseminated viral infections can result in death of the newborn.

Culture and identification of the virus take a few days, but fluorescent antibody tests can quickly detect viral protein, or PCR tests can detect the presence of viral DNA. Treatment usually involves intravenous administration of acyvlovir. No vaccine is currently available.

## Genital Warts

Warts are an infectious disease caused by viruses known as papillomaviruses. (See Chapter 21, page 602, for more-familiar, skin-associated, warts.) Many papillomaviruses have a predilection for growth not on the skin, but on the mucous membranes that line organs such as the respiratory tract, mouth, anus, and genitalia. Such **genital warts** (or condyloma acuminata) are usually transmitted sexually and are an increasing problem. Nearly 5 million new cases are estimated to occur in the United States each year; the incidence in women has decreased since introduction of the HPV vaccine. Worldwide, genital warts may well be the most common STI.

There are more than 60 types of human papillomaviruses (HPV), and certain serotypes tend to be linked with certain

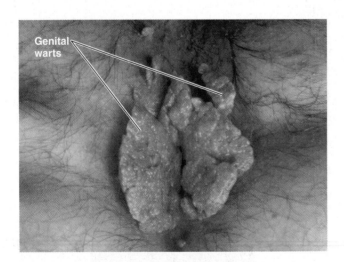

**Figure 26.15 Genital warts on a vulva.**

**Q** What is the relationship between genital warts and cervical cancer?

forms of genital warts. For example, some genital warts are extremely large, resembling a cauliflower with multiple finger-like projections, whereas others are relatively smooth or flat (**Figure 26.15**).

Penile lesions are often flat and quite inapparent, an important factor in male to female transmission. The incubation period is usually a matter of a few weeks or months. Visible genital warts are most often caused by serotypes 6 and 11. These serotypes rarely cause cancer, which is the most serious concern with these infections. The types most likely to cause cancer are types 16 and 18, but these have a relatively low prevalence. Even so, cervical cancer caused by HPV kills at least 4000 women annually in the United States. Oral, anal, and penile cancers are also attributed to HPV infections.

The quadrivalent HPV vaccine protects against infection by HPV types 6, 11, 16, and 18. The more expensive nine-valent vaccine protects against the four HPVs in the quadrivalent vaccine plus serotypes 31, 33, 45, 52, and 58. The vaccines are recommended for adolescents aged 11 through 12 years and are even required in some areas. The immune response to the vaccines is much more effective than that resulting from a natural infection, which is relatively weak.

Warts can be treated but not cured (see the discussion on page 602), but approximately 90% of cases clear spontaneously within 2 years. The available methods used for warts, such as surgery or cryotherapy, are not as effective against genital warts. Two patient-applied gels, podofilox and imiquimod, are often useful treatments. Imiquimod stimulates the body to produce interferon (pages 467–468), which appears to account for its antiviral activity.

## AIDS

**AIDS**, or HIV infection, is a viral disease that is frequently transmitted by sexual contact. However, its pathogenicity is based on damage to the immune system, so it is discussed on pages 544–554. It's important to remember that the lesions resulting from many of the diseases of bacterial and viral origin facilitate the transmission of HIV.

*****

There are other sexually transmitted viruses that do not infect the genitourinary system. Like HIV, these viruses carry out their pathogenicity in other organ systems. These viruses are listed in Diseases in Focus 26.3 (page 781).

> **CHECK YOUR UNDERSTANDING**
>
> ✔ 26-7 Both genital herpes and genital warts are caused by viruses; which one is the greater danger to a pregnancy?

# Characteristics of the Most Common Types of Vaginitis and Vaginosis

Vaginitis, or inflammation of the vagina, often accompanies vaginal infections. Vaginitis may be caused by microbial infections. The cause of vaginitis cannot be determined on the basis of symptoms or physical examination alone. Usually, diagnosis involves examining a specimen of vaginal fluid under a microscope (see the photo). Use the table below to identify the infection caused by the organism in the figure. For the solution, go to @MasteringMicrobiology.

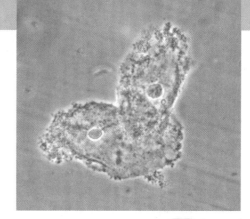

LM ⊢ 25 μm

**Epithelial cells covered with rod-shaped bacteria from a vaginal swab.**

| Disease | Pathogen | Symptoms | | | | Diagnosis | Treatment |
|---------|----------|----------|---|---|---|-----------|-----------|
| | | Odor, Color, and Consistency of Discharge | Amount of Discharge | Appearance of Vaginal Mucosa | pH (normal pH is 3.8–4.2) | | |
| **Bacterial Vaginosis** | Bacterium, *Gardnerella vaginalis* | Fishy; gray-white, thin, frothy | Copious | Pink | >4.5 | Presence of clue cells | Metronidazole |
| **Candidiasis** | Fungus, *Candida albicans* | Yeasty or none, white, curdy | Varies | Dry, red | <4 | Microscopic exam | Clotrimazole, fluconazole |
| **Trichomoniasis** | Protozoan, *Trichomonas vaginalis* | Foul, greenish-yellow; frothy | Copious | Tender, red | 5–6 | Microscopic exam; DNA probes; monoclonal antibody | Metronidazole |

# Fungal Disease of the Reproductive Systems

### LEARNING OBJECTIVE

26-8 Discuss the epidemiology of candidiasis.

The fungal disease described here is the well-known *yeast infection* for which nonprescription treatments are advertised.

## Candidiasis

Vaginal infections by yeastlike fungi of the genus *Candida* are responsible for millions of physician office visits every year. By the time they reach the age of 25, an estimated half of college women will have had at least one physician-diagnosed episode. Nonprescription antifungal therapies to treat these infections are among the best-selling over-the-counter products in the United States. *Candida albicans* is the most common species, causing 85–90% of cases. Infections by other species, such as *C. glabrata,* are more likely to be resistant to antifungals and to be chronic or recurrent.

*C. albicans* often grows on mucous membranes of the mouth, intestinal tract, and genitourinary tract (see Diseases in Focus 26.2; see also Figure 21.17, page 609). Infections are usually a result of opportunistic overgrowth when the competing microbiota are suppressed by antibiotics or other factors. *C. albicans* is the cause of **oral candidiasis**, or thrush (see Chapter 21, page 609). It's also responsible for occasional cases of NGU in men and for **vulvovaginal candidiasis**, which is the most common cause of vaginitis. About 75% of all women experience at least one episode.

The lesions of vulvovaginal candidiasis resemble those of thrush but produce more irritation: severe itching; a thick, yellow, cheesy discharge; and yeasty or no odor. *C. albicans*, the *Candida* species responsible for most cases, is an opportunistic pathogen. Predisposing conditions include the use of oral contraceptives and pregnancy, which cause an increase of glycogen in the vagina (see the discussion of the normal vaginal microbiota on page 762). Hormones are probably a factor; candidiasis is much less common in girls before puberty or in women after menopause. Yeast infections are a frequent symptom in women suffering from uncontrolled diabetes. Thus, diabetes and antibiotic therapy are predisposing factors to *C. albicans* vaginitis.

A yeast infection is diagnosed by microscopic identification of the fungus in scrapings of lesions and by isolation of the fungus in culture. Treatment usually consists of topical application of nonprescription antifungal drugs such as clotrimazole and miconazole. An alternative treatment is a single dose of oral fluconazole or other azole-type antifungal.

---

**CHECK YOUR UNDERSTANDING**

☑ **26-8** What changes in the vaginal bacterial microbiota tend also to favor the growth of the yeast *Candida albicans*?

---

# Protozoan Disease of the Reproductive Systems

### LEARNING OBJECTIVES

26-9 Discuss the epidemiology of trichomoniasis.

The only STI caused by a protozoan affects mostly young, sexually active women. It may be the most common STI—nearly 8 million cases per year are reported in the United States—but it isn't widely known. Its prevalence in certain STI clinics is 25% or higher.

### Trichomoniasis

The anaerobic protozoan *Trichomonas vaginalis* is frequently a normal inhabitant of the vagina in women and of the urethra in many men (**Figure 26.16**). It's usually sexually transmitted. If the normal acidity of the vagina is disturbed, the protozoan may overgrow the normal microbial population of the genital mucosa and cause **trichomoniasis**. (Men rarely have any symptoms resulting from the presence of the protozoan.) The infection is often accompanied by a coinfection with gonorrhea. In response to the protozoan infection, the body accumulates leukocytes at the infection site. The resulting discharge is profuse, greenish yellow, and characterized by a foul odor. This discharge is accompanied by irritation and itching. Up to half the cases, however, are asymptomatic.

The incidence of trichomoniasis is higher than that of gonorrhea or chlamydia, but it is considered relatively benign

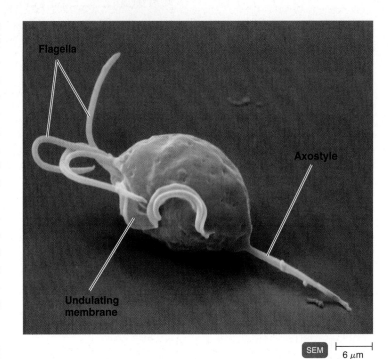

Flagella

Axostyle

Undulating membrane

SEM ⊢———⊣ 6 μm

**Figure 26.16** *Trichomonas vaginalis.* The flagella and undulating membrane provide motility. The axostyle attaches the protozoan to vaginal cells.

**Q** Are there any harmful effects from infection by this protozoan?

and is not a reportable disease. It's known, however, to cause preterm delivery and associated problems, such as low birth weight.

Diagnosis is usually made by microscopic examination and identification of the organisms in the discharge. They can also be isolated and grown on laboratory media. The pathogen can be found in semen or urine of male carriers. New rapid tests making use of DNA probes and monoclonal antibodies are now available. Treatment is by oral metronidazole, administered to both sex partners, which readily clears the infection. The major microbial diseases of the urinary and reproductive systems are summarized in Diseases in Focus 26.3 on the next page.

We have seen in this and previous chapters that a number of diseases can cause serious infections or birth defects in newborns when they infect a pregnant woman. See the Big Picture in Chapter 22 on pages 634–635 for a discussion of other infections that may be transmitted from mother to fetus.

---

**CHECK YOUR UNDERSTANDING**

☑ **26-9** What are the symptoms of the presence of *Trichomonas vaginalis* in the male reproductive system?

---

# Microbial Diseases of the Reproductive Systems

A 26-year-old woman has abdominal pain, painful urination, and fever. Cultures grown in a high-$CO_2$ environment reveal gram-negative diplococci. Use the table below to identify infections that could cause these symptoms. For the solution, go to @MasteringMicrobiology.

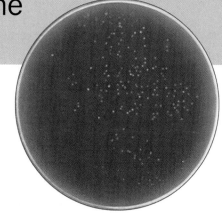

Gram-negative diplococci on Thayer-Martin agar, which is a blood-containing medium with antibiotics to encourage the growth of this pathogen and inhibit the growth of unwanted microbes.

| Disease | Pathogen | Symptoms | Treatment |
|---|---|---|---|
| **BACTERIAL DISEASES** | | | |
| Gonorrhea | Neisseria gonorrhoeae | Men: painful urination and discharge of pus. Women: few symptoms but possible complications, such as PID. | Ceftriaxone and azithromycin |
| Nongonococcal Urethritis (NGU) | Chlamydia trachomatis, Mycoplasma genitalium | Painful urination and watery discharge. In women, possible complications, such as PID. | Azithromycin, doxycycline |
| Pelvic Inflammatory Disease (PID) | N. gonorrhoeae, C. trachomatis | Chronic abdominal pain; possible infertility | Doxycycline and cefotetan |
| Syphilis | Treponema pallidum | Initial sore at site of infection, later skin rashes and mild fever; final stages may be severe lesions, damage to cardiovascular and nervous systems. | Benzathine penicillin |
| Lymphogranuloma Venereum (LGV) | C. trachomatis | Swelling in lymph nodes in groin | Doxycycline |
| Chancroid (Soft Chancre) | Haemophilus ducreyi | Painful ulcers of genitals; swollen lymph nodes in groin | Erythromycin; ceftriaxone |
| Bacterial Vaginosis | See Diseases in Focus 26.2, page 779 | | |
| **VIRAL DISEASES** | | | |
| Genital Herpes | Herpes simplex virus type 2; HSV type 1 | Painful vesicles in genital area | Acyclovir |
| Genital Warts | Human papillomaviruses | Warts in genital area | Podofilox, imiquimod vaccine |
| **VIRAL STIs WITHOUT GENITOURINARY PATHOGENICITY** | | | |
| AIDS | See Chapter 19, pages 544–554 | | Prevention: HPV vaccine |
| Hepatitis B | See Chapter 25, pages 741–744 | | |
| Hepatitis C | See Chapter 25, pages 744–745 | | |
| Zika virus disease | See Chapter 22, page 638 | | |
| **FUNGAL DISEASE** | | | |
| Candidiasis | See Diseases in Focus 26.2, page 779 | | |
| **PROTOZOAN DISEASE** | | | |
| Trichomoniasis | See Diseases in Focus 26.2, page 779 | | |

## Study Outline

 Go to @MasteringMicrobiology for **Interactive Microbiology**, *In the Clinic* videos, MicroFlix, MicroBoosters, 3D animations, practice quizzes, and more.

### Introduction (p. 760)

1. The urinary system regulates the chemical composition and volume of the blood and excretes nitrogenous waste and water.
2. The reproductive system produces gametes for reproduction and, in the female, supports the growing embryo.
3. Microbial diseases of these systems can result from infection from an outside source or from opportunistic infection by members of the normal microbiota.

### Structure and Function of the Urinary System (p. 761)

1. Urine is transported from the kidneys through ureters to the urinary bladder and is eliminated through the urethra.
2. Valves prevent urine from flowing back to the urinary bladder and kidneys.
3. The flushing action of urine and normal urine itself have some antimicrobial value.

### Structure and Function of the Reproductive Systems (pp. 761–762)

1. The female reproductive system consists of two ovaries, two uterine tubes, the uterus, the cervix, the vagina, and the external genitals.
2. The male reproductive system consists of two testes, ducts, accessory glands, and the penis; seminal fluid leaves the male body through the urethra.

### Normal Microbiota of the Urinary and Reproductive Systems (p. 762)

1. Gram-positive bacteria predominate in the urinary tract.
2. Lactobacilli dominate the vaginal microbiota; actinobacteria dominate the microbiome of the seminal vesicle.

## Diseases of the Urinary System (pp. 763–765)

### Bacterial Diseases of the Urinary System (pp. 763–765)

1. Urethritis, cystitis, and ureteritis are inflammations of tissues of the lower urinary tract.
2. Pyelonephritis can result from lower urinary tract infections or from systemic bacterial infections.
3. Opportunistic gram-negative bacteria from the intestines often cause urinary tract infections.
4. Heathcare-associated infections of the urinary system can occur following catheterization.
5. Treatment of urinary tract infections depends on isolation and antibiotic sensitivity testing of the causative agents.

### Cystitis (p. 764)

6. Inflammation of the urinary bladder, or cystitis, is common in females.
7. The most common etiologies are *E. coli* and *Staphylococcus saprophyticus.*

### Pyelonephritis (p. 764)

8. Inflammation of the kidneys, or pyelonephritis, is usually a complication of lower urinary tract infections.
9. About 75% of pyelonephritis cases are caused by *E. coli.*

### Leptospirosis (pp. 764–765)

10. The spirochete *Leptospira interrogans* is the cause of leptospirosis.
11. The disease is transmitted to humans by urine-contaminated water.
12. Leptospirosis is characterized by chills, fever, headache, and muscle aches.

## Diseases of the Reproductive Systems (pp. 766–780)

1. Most infections of the reproductive system are sexually transmitted infections (STIs). The incidence of STIs is at an all-time high.
2. Most STIs can be prevented by the use of condoms.

### Bacterial Diseases of the Reproductive Systems (pp. 765–776)

#### Gonorrhea (pp. 766–770)

1. *Neisseria gonorrhoeae* causes gonorrhea.
2. Gonorrhea is a common reportable communicable disease in the United States.
3. *N. gonorrhoeae* attaches to mucosal cells of the oral-pharyngeal area, genitals, eyes, and rectum by means of fimbriae.
4. Symptoms in men are painful urination and pus discharge. Blockage of the urethra and sterility are complications of untreated cases.
5. Women might be asymptomatic unless the infection spreads to the uterus and uterine tubes (see pelvic inflammatory disease).
6. Gonorrheal endocarditis, gonorrheal meningitis, and gonorrheal arthritis are complications that can affect both sexes if gonorrheal infections are untreated.
7. Ophthalmia neonatorum is an eye infection acquired by infants during passage through the birth canal of an infected mother.
8. Gonorrhea is diagnosed by ELISA or NAATS.

#### Nongonococcal Urethritis (NGU) (pp. 770–771)

9. Most cases of nongonococcal urethritis (NGU), or nonspecific urethritis (NSU), are caused by *Chlamydia trachomatis.*
10. *C. trachomatis* infection is the most common STI.
11. Symptoms of NGU are often mild or lacking, although uterine tube inflammation and sterility may occur.
12. *C. trachomatis* can be transmitted to infants' eyes at birth.
13. Diagnosis is based on detection of chlamydial DNA in urine.
14. *Ureaplasma urealyticum* and *Mycoplasma genitalium* also cause NGU.

### Pelvic Inflammatory Disease (PID) (pp. 771–712)

15. Extensive bacterial infection of the female pelvic organs, especially of the reproductive system, is called pelvic inflammatory disease (PID).

16. PID is caused by *N. gonorrhoeae, C. trachomatis*, and other bacteria. Infection of the uterine tubes is called salpingitis.

### Syphilis (pp. 772–775)

17. Syphilis is caused by *Treponema pallidum*, a spirochete that has not been cultured in vitro. Laboratory cultures are grown in rabbits or cell cultures.

18. The primary lesion is a small, hard-based chancre at the site of infection. The bacteria then invade the blood and lymphatic system, and the chancre spontaneously heals.

19. The appearance of a widely disseminated rash on the skin and mucous membranes marks the secondary stage. Spirochetes are present in the lesions of the rash.

20. The patient enters a latent period after the secondary lesions spontaneously heal.

21. At least 10 years after the secondary lesion, tertiary lesions called gummas can appear on many organs.

22. Congenital syphilis, resulting from *T. pallidum* crossing the placenta during the latent period, can cause neurological damage in the newborn.

23. *T. pallidum* is identifiable through darkfield microscopy of fluid from primary and secondary lesions.

24. Many serological tests, such as VDRL, RPR, and FTA-ABS, can be used to detect the presence of antibodies against *T. pallidum* during any stage of the disease.

### Lymphogranuloma Venereum (LGV) (p. 775)

25. *C. trachomatis* causes lymphogranuloma venereum (LGV), which is primarily a disease of tropical and subtropical regions.

26. The bacteria are spread in the lymph system and cause enlargement of the lymph nodes, obstruction of lymph vessels, and swelling of the external genitals.

27. Diagnosis is by ELISA or NAAT.

### Chancroid (Soft Chancre) (p. 775)

28. Chancroid, a swollen, painful ulcer on the mucous membranes of the genitals or mouth, is caused by *Haemophilus ducreyi*.

### Bacterial Vaginosis (p. 776)

29. Bacterial vaginosis is an infection without inflammation caused by *Gardnerella vaginalis*.

30. Diagnosis of *G. vaginalis* is based on the presence of clue cells.

## Viral Diseases of the Reproductive Systems (pp. 776–778)

### Genital Herpes (pp. 776–778)

1. Herpes simplex viruses (HSV-1 and HSV-2) cause genital herpes.

2. Symptoms of the infection are painful urination, genital irritation, and fluid-filled vesicles.

3. The virus might enter a latent stage in nerve cells. Vesicles reappear following trauma and hormonal changes.

4. Neonatal herpes is contracted during fetal development or birth. It can result in neurological damage or infant fatalities.

### Genital Warts (p. 778)

5. Human papillomaviruses cause warts.

6. Some human papillomaviruses that cause genital warts cause cancer.

### AIDS (p. 778)

7. AIDS is a sexually transmitted disease of the immune system (see Chapter 19, pages 544–554).

8. Other viral STIs that do not infect the genitourinary system include hepatitis B, hepatitis C, and Zika virus disease (see Chapter 22, page 638, and Chapter 25, pages 741–745).

## Fungal Disease of the Reproductive Systems (pp. 779–780)

### Candidiasis (pp. 779–780)

1. *Candida albicans* causes NGU in men and vulvovaginal candidiasis, or yeast infection, in women.

2. Vulvovaginal candidiasis is characterized by lesions that produce itching and irritation.

3. Predisposing factors are pregnancy, diabetes, and broad-spectrum antibacterial chemotherapy.

4. Diagnosis is based on observation of the fungus and its isolation from lesions.

## Protozoan Disease of the Reproductive Systems (p. 780)

### Trichomoniasis (p. 780)

1. *Trichomonas vaginalis* causes trichomoniasis when the pH of the vagina increases.

2. Diagnosis is based on observation of the protozoa in purulent discharges from the site of infection.

# Study Questions

For answers to the Knowledge and Comprehension questions, turn to the Answers tab at the back of the textbook.

## Knowledge and Comprehension

### Review

1. **DRAW IT** Diagram the pathway taken by *E. coli* to cause cystitis. Do the same for pyelonephritis. Diagram the pathway taken by *Neisseria gonorrhoeae* to cause PID.

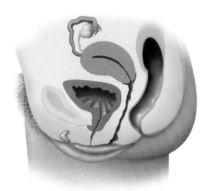

2. How are urinary tract infections acquired?

3. Explain why *E. coli* is frequently implicated in cystitis in females.

4. Name one organism that causes pyelonephritis. What are the portals of entry for microbes that cause pyelonephritis?

5. Complete the following table:

| Disease | Causative Agent | Symptoms | Method of Diagnosis | Treatment |
|---|---|---|---|---|
| Bacterial vaginosis | | | | |
| Gonorrhea | | | | |
| Syphilis | | | | |
| PID | | | | |
| NGU | | | | |
| LGV | | | | |
| Chancroid | | | | |

6. Describe the symptoms of genital herpes. What is the causative agent? When is this infection least likely to be transmitted?

7. Name one fungus and one protozoan that can cause reproductive system infections. What symptoms would lead you to suspect these infections?

8. List the genital infections that cause congenital and neonatal infections. How can transmission to a fetus or newborn be prevented?

9. **NAME IT** Intracellular reticulate bodies of this gram-negative bacterium convert to elementary bodies that can infect a new host cell.

## Multiple Choice

1. Which of the following is usually transmitted by contaminated water?
   a. *Chlamydia*
   b. leptospirosis
   c. syphilis
   d. trichomoniasis
   e. none of the above

Use the following choices to answer questions 2–5:
   a. *Candida*
   b. *Chlamydia*
   c. *Gardnerella*
   d. *Neisseria*
   e. *Trichomonas*

2. Microscopic examination of vaginal smear shows flagellated eukaryotes.

3. Microscopic examination of vaginal smear shows ovoid eukaryotic cell.

4. Microscopic examination of vaginal smear shows epithelial cells covered with bacteria.

5. Microscopic examination of vaginal smear shows gram-negative cocci in phagocytes.

Use the following choices to answer questions 6–8:
   a. candidiasis
   b. bacterial vaginosis
   c. genital herpes
   d. lymphogranuloma venereum
   e. trichomoniasis

6. Difficult to treat with chemotherapy

7. Fluid-filled vesicles

8. Frothy, fishy discharge

Use the following choices to answer questions 9 and 10:
   a. *Chlamydia trachomatis*
   b. *Escherichia coli*
   c. *Mycobacterium hominis*
   d. *Staphylococcus saprophyticus*

9. The most common cause of cystitis

10. In cases of NGU, diagnosis is made using PCR to detect microbial DNA.

## Analysis

1. The tropical skin disease called yaws is transmitted by direct contact. Its causative agent, *Treponema pallidum pertenue*, is indistinguishable from *T. pallidum*. Syphilis epidemics in Europe coincided with the return of Columbus from the New World. How might *T. pallidum pertenue* have evolved into *T. pallidum* in the temperate climate of Europe?

2. Why can frequent douching be a predisposing factor to bacterial vaginosis, vulvovaginal candidiasis, or trichomoniasis?

3. *Neisseria* is cultured on Thayer-Martin media, consisting of chocolate agar and nystatin, incubated in a 5% $CO_2$ environment. How is this selective for *Neisseria*?

4. The list below is a key to selected microorganisms that cause genitourinary infections. Complete this key by listing genera discussed in this chapter in the blanks that correspond to their respective characteristics.

Gram-negative bacteria

  Spirochete

    Aerobic: a. _____

    Anaerobic: b. _____

  Coccus

    Oxidase-positive: c. _____

  Bacillus, nonmotile

    Requires X factor: d. _____

  Gram-positive wall: e. _____

  Obligate intracellular parasite: f. _____

  Lacking cell wall

    Urease-positive: g. _____

    Urease-negative: h. _____

  Fungus

    Pseudohyphae: i. _____

  Protozoa

    Flagella: j. _____

No organism observed/cultured from patient: k. _____

# Clinical Applications and Evaluation

1. A previously healthy 19-year-old woman was admitted to a hospital after 2 days of nausea, vomiting, headache, and neck stiffness. Cerebrospinal fluid and cervical cultures showed gram-negative diplococci in leukocytes; a blood culture was negative. What disease did she have? How was it probably acquired?

2. A 28-year-old woman was admitted to a Wisconsin hospital with a 1-week history of arthritis of the left knee. Four days later, a 32-year-old man was examined for a 2-week history of urethritis and a swollen, painful left wrist. A 20-year-old woman seen in a Philadelphia hospital had pain in the right knee, left ankle, and left wrist for 3 days. Pathogens cultured from synovial fluid or urethral culture were gram-negative diplococci that required proline to grow. Antibiotic sensitivity tests gave the following results:

| Antibiotic | MIC Tested (µg/ml) | Susceptible MIC (µg/ml) |
|---|---|---|
| Cefoxitin | 0.5 | ≤2 |
| Penicillin | 8 | ≤0.06 |
| Spectinomycin | 64 | ≤32 |
| Tetracycline | 4 | ≤0.25 |

What is the pathogen, and how is this disease transmitted? Which of the antibiotics should be used for treatment? What is the evidence that these cases are related?

3. Using the following information, determine what the disease is and how the infant's illness might have been prevented:

May 11: A 23-year-old woman has her first prenatal examination. She is $4\frac{1}{2}$ months pregnant. Her VDRL results are negative.

June 6: The woman returns to her physician complaining of a labial lesion of a few days' duration. A biopsy is negative for malignancy, and herpes test results are negative.

July 1: The woman returns to her physician because the labial lesion continues to cause some discomfort.

Sept. 15: The baby's father has multiple penile lesions and a generalized body rash.

Sept. 25: The woman delivers her baby. Her RPR is 32, and the infant's is 128.

Oct. 1: The woman takes her infant to a pediatrician because the baby is lethargic. She is told the infant is healthy and not to worry.

Oct. 2: The baby's father has a persistent body rash and plantar and palmar rashes.

Nov. 8: The infant becomes acutely ill with pneumonia and is hospitalized. The admitting physician finds signs of osteochondritis.

# 27 Environmental Microbiology

Microbes, especially those that belong to the Domains Bacteria and Archaea, make up the **Earth microbiome**. Microbes live on plant leaves and roots, in insects, and in the most widely varied habitats on Earth. They are found in boiling hot springs, and as many as 5000 bacteria have been isolated from each milliliter of snow at the South Pole. Microbes have been recovered from minute openings in rocks a kilometer (0.62 mile) or more below the planet surface. Explorations of the deepest ocean have revealed large numbers of microbes living there, in eternal darkness and subject to incredible pressures. Microbes are also found in clear mountain streams flowing from a melting glacier and in waters nearly saturated with salts, such as those of the Dead Sea.

In previous chapters, we focused primarily on the disease-causing capabilities of microorganisms. Controlling infectious disease is one aspect of environmental microbiology; for example, environmental health microbiologists test drinking water regularly to ensure that it is free of pathogens. One such pathogen, *Vibrio cholerae*, shown in the photo, is the topic of the Clinical Case in this chapter. In this chapter you will also learn about many of the positive functions microbes perform in the environment. The ecological services provided by the Earth microbiome, such as recycling nitrogen and removing pollutants, are essential for maintaining life on Earth.

▶ *Vibrio cholerae* bacteria are curved cells with a single flagellum.

## In the Clinic

As an environmental health nurse, you are investigating ways to reduce nitrous oxide ($N_2O$) levels in the atmosphere. Global warming is a public health concern, and $N_2O$ is a greenhouse gas that can absorb 300 times more radiant energy than carbon dioxide. You know that nitrate fertilizers are essential for crops but that $N_2O$ emissions are linked to soil nitrite levels. In your experiments, fertilizer with microbial inhibitors was associated with lower $N_2O$ levels than was fertilizer without inhibitors. **Explain why the microbial inhibitors reduced $N_2O$ production.**

*Hint: Read about the nitrogen cycle on pages 789–791.*

Answers to In the Clinic questions are found online at @MasteringMicrobiology.

# Microbial Diversity and Habitats

**27-1** Define *extremophile*, and identify two "extreme" habitats.

**27-2** Define *symbiosis*.

**27-3** Define *mycorrhiza*, differentiate endomycorrhizae from ectomycorrhizae, and give an example of each.

The diversity of microbial populations indicates that they take advantage of any niches found in their environment. Different amounts of oxygen, light, or nutrients may exist within a few millimeters in the soil. As a population of aerobic organisms uses up the available oxygen, anaerobes are able to grow. If the soil is disturbed by plowing, earthworms, or other activity, the aerobes will again be able to grow to repeat this succession.

> ASM: Because the true diversity of microbial life is largely unknown, its effects and potential benefits have not been fully explored.

Microbes that live in extreme conditions of temperature, acidity, alkalinity, or salinity are called **extremophiles**. Most are members of the Archaea. The enzymes (**extremozymes**) that make growth possible under these conditions have been of great interest to industries because they can tolerate extremes of temperature, salinity, and pH that would inactivate other enzymes.

Microorganisms live in an intensely competitive environment and must exploit any advantage they can. They may metabolize common nutrients more rapidly or use nutrients that competing organisms cannot metabolize. Some, such as the lactic acid bacteria that are so useful in making dairy products, are able to make an environmental niche inhospitable to competing organisms. The lactic acid bacteria are unable to use oxygen as an electron acceptor and are able to ferment sugars only to lactic acid, leaving most of the energy unused. However, the acidity inhibits the growth of more efficient, competing microbes.

## Symbiosis

**Symbiosis** is a close association between two unlike organisms that is beneficial to one or both of them (recall Chapter 14). Economically, the most important example of an animal-microbe symbiosis is that of the ruminants, animals that have a tanklike digestive organ called a *rumen*. Ruminants, such as cattle and sheep, graze on cellulose-rich plants. Bacteria in the rumen ferment the cellulose into compounds that are absorbed into the animal's blood and subsequently are used for carbon and energy. Rumen protozoa keep the bacterial population under control by eating bacteria. Similarly, wood-eating insects such as termites harbor cellulose-degrading bacteria in their digestive tracts.

Another important example of symbiosis is the relationship between plant roots and certain fungi, called **mycorrhizae**, or mycorrhizal symbionts (*myco* = fungus; *rhiza* = root).

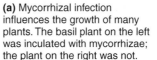

**(a)** Mycorrhizal infection influences the growth of many plants. The basil plant on the left was inoculated with mycorrhizae; the plant on the right was not.

**(b)** Truffles. An ectomycorrhiza, usually of oak trees.

**Figure 27.1 Mycorrhizae and their considerable commercial value.**

**Q** Why are mycorrhizae valuable for the uptake of phosphorus?

There are two primary types of these fungi: *endomycorrhizae*, also known as *arbuscular mycorrhizae*; and *ectomycorrhizae*. Both types function like root hairs on plants; that is, they extend the surface area through which the plant can absorb nutrients, especially phosphorus, which is not very mobile in soil.

Most grasses and other plants are surprisingly dependent on these fungi for proper growth, and their presence is nearly universal in the plant kingdom. Managers of commercial pine tree farms, for example, must ensure that seedlings are inoculated with soil containing effective mycorrhizae (**Figure 27.1a**).

Truffles, known as a food delicacy, are ectomycorrhizae, usually of oak trees (Figure 27.1b). These "underground mushrooms" have developed a different, nonaerial method of distributing their spores. This distribution depends upon the truffles' ability to attract the attention of animals that will ingest them and then deposit undigested spores into new locations.

> CHECK YOUR UNDERSTANDING
>
> ✔ **27-1** Identify two habitats for extremophile organisms.
> ✔ **27-2** What is the definition of *symbiosis*?
> ✔ **27-3** Is a truffle an endomycorrhiza or an ectomycorrhiza?

# Soil Microbiology and Biogeochemical Cycles

**27-4** Define *biogeochemical cycle*.

**27-5** Outline the carbon cycle, and explain the roles of microorganisms in this cycle.

**27-6** Outline the nitrogen cycle, and explain the roles of microorganisms in this cycle.

**27-7**  Define *ammonification, nitrification, denitrification,* and *nitrogen fixation.*

**27-8**  Outline the sulfur cycle, and explain the roles of microorganisms in this cycle.

**27-9**  Describe how an ecological community can exist without light.

**27-10**  Compare and contrast the carbon cycle and the phosphorus cycle.

**27-11**  Give two examples of the use of bacteria to remove pollutants.

**27-12**  Define *bioremediation.*

The soil microbiome consists of billions of microbes. Typical soil has millions of bacteria in each gram. The microbial population of soil is largest in the top few centimeters and declines rapidly with depth. The most numerous organisms in soil are bacteria. Although actinomycetes are bacteria, they are usually considered separately.

 ASM: Microorganisms and their environment interact with and modify each other.

Bacterial soil populations are usually estimated using plate counts on nutrient media, and the actual numbers are probably greatly underestimated by this method. No single nutrient medium or growth condition can possibly meet all the nutritional and other requirements of soil microorganisms. Metagenomics (see Chapter 9, page 257) is currently being used to look at soil microbiomes. This technique looks for rRNA genes in soil samples. Using metagenomics, microbiologists have discovered more new microorganisms in the twenty-first century than at any time since van Leeuwenhoek first saw microorganisms. However, these methods provide only limited information on the metabolic activities of the microorganisms in soil and can't differentiate between DNA in live and dead bacteria, so there is much left to discover.

We can think of soil as a "biological fire." A leaf falling from a tree is consumed by this fire as microbes in the soil metabolize its organic matter. Elements in the leaf enter the **biogeochemical cycles** for carbon, nitrogen, and sulfur that we will discuss in this chapter. In biogeochemical cycles, elements are oxidized and reduced by microorganisms to meet their metabolic needs. (See the discussion of oxidation-reduction in Chapter 5, pages 117–118.) Without biogeochemical cycles, life on Earth would cease to exist.

## The Carbon Cycle

The primary biogeochemical cycle is the **carbon cycle** (**Figure 27.2**). All organisms, including plants, microbes, and animals, contain large amounts of carbon in the form of organic compounds such as cellulose, starches, fats, and proteins. Let's take a closer look at how these organic compounds are formed.

Recall from Chapter 5 that autotrophs perform an essential role for all life on Earth by reducing carbon dioxide to form organic matter. When you look at a tree, you might think that its mass has come from the soil where it grows. In fact, its great mass of cellulose is derived from the 0.04% of carbon dioxide in the atmosphere. This occurs as a result of photosynthesis, the first step of the carbon cycle in which photoautotrophs *fix* (incorporate) carbon dioxide into organic matter using energy from sunlight.

In the next step of the cycle, chemoheterotrophs such as animals and protozoa eat autotrophs and may in turn be eaten by other animals. Thus, as the organic compounds of the autotrophs are digested and resynthesized, the carbon atoms of carbon dioxide are transferred from organism to organism up the food chain.

Chemoheterotrophs, including animals, use some of the organic molecules to satisfy their energy requirements. When this energy is released through respiration, carbon dioxide immediately becomes available to start the cycle over again. Much of the carbon remains within the organisms until they excrete it as wastes or die. When plants and animals die, these organic compounds are decomposed by bacteria and fungi. During decomposition, the organic compounds are oxidized, and $CO_2$ is returned to the cycle.

Carbon is stored in rocks, such as limestone ($CaCO_3$), and is dissolved as carbonate ions ($CO_3^{2-}$) in oceans. Vast deposits of fossil organic matter exist in the form of fossil fuels, such as coal and petroleum. Burning these fossil fuels releases $CO_2$, increasing the amount of $CO_2$ in the atmosphere. This increased atmospheric carbon dioxide is causing a **global warming** of the Earth.

An interesting aspect of the carbon cycle is methane ($CH_4$) gas. Sediments on the ocean floor contain an estimated 10 trillion tons of methane, about twice as much as the Earth's deposits of fossil fuels such as coal and petroleum. Furthermore, methanogenic bacteria in the ocean's depths are constantly producing more (see page 796). Methane is much more

**CLINICAL CASE**  Clean Water—A Matter of Life and Death

Two days ago, Charity, a 48-year-old journalist from Miami, returned to the United States from a 6-week trip to several countries for a story she is writing on recovery progress after major earthquakes. When she first came home, Charity began to experience diarrhea, which became worse as the day progressed. After the second day of severe diarrhea and the beginning of leg pain, Charity seeks care at a local outpatient health care facility. She reports no vomiting or fever but has 10 watery stools per day without visible blood or mucus.

**What immediate treatment does Charity need? Read on to find out.**

788   797   799   803   804   805

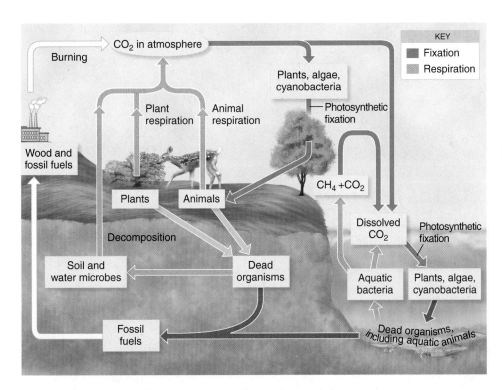

KEY
- ■ Fixation
- ▨ Respiration

**Figure 27.2 The carbon cycle.** On a global scale, the return of $CO_2$ to the atmosphere by respiration closely balances its removal by fixation. However, the burning of wood and fossil fuels adds more $CO_2$ to the atmosphere. The destruction of forests and wetlands removes $CO_2$-fixing organisms; as a result, the amount of atmospheric $CO_2$ is steadily increasing.

**Q** How does the accumulation of carbon dioxide in the atmosphere affect Earth's climate?

potent as a greenhouse gas than is carbon dioxide, and the Earth's environment would be dangerously altered if all this gas escaped to the atmosphere.

**CHECK YOUR UNDERSTANDING**

✔ 27-4 What biogeochemical cycle is much publicized as contributing to global warming?

✔ 27-5 What is the main source of the carbon in the cellulose forming the mass of a forest?

## The Nitrogen Cycle

The **nitrogen cycle** is shown in **Figure 27.3**. All organisms need nitrogen to synthesize protein, nucleic acids, and other nitrogen-containing compounds. Molecular nitrogen ($N_2$) makes up almost 80% of the Earth's atmosphere. For plants to assimilate and use nitrogen, it must be fixed, that is, taken up and combined into organic compounds. The activities of specific microorganisms are important to the conversion of nitrogen to usable forms.

### Ammonification

Almost all the nitrogen in the soil exists in organic molecules, primarily in proteins. When an organism dies, the process of microbial decomposition results in the hydrolytic breakdown of proteins into amino acids. In a process called **deamination**, the amino groups of amino acids are removed and converted into ammonia ($NH_3$). This release of ammonia is called **ammonification** (see Figure 27.3). Ammonification, brought about by numerous bacteria and fungi, can be represented as follows:

$$\text{Proteins from dead cells and waste products} \xrightarrow[\text{decomposition}]{\text{Microbial}} \text{Amino acids}$$

$$\text{Amino acids} \xrightarrow[\text{decomposition}]{\text{Microbial}} \text{Ammonia (NH}_3\text{)}$$

Microbial growth releases extracellular proteolytic enzymes that decompose proteins. The resulting amino acids are transported into the microbial cells, where ammonification occurs. The fate of the ammonia produced by ammonification depends on soil conditions (see the discussion of denitrification, which follows). Because ammonia is a gas, it rapidly disappears from dry soil, but in moist soil it becomes solubilized in water, and ammonium ions ($NH_4^+$) are formed:

$$NH_3 + H_2O \longrightarrow NH_4OH \longrightarrow NH_4^+OH^-$$

Ammonium ions from this sequence of reactions are used by bacteria and plants for amino acid synthesis.

### Nitrification

The next sequence of reactions in the nitrogen cycle involves the oxidation of the nitrogen in the ammonium ion to produce nitrate, a process called **nitrification**. Living in the soil are autotrophic nitrifying bacteria, such as those of the genera *Nitrosomonas* and *Nitrobacter*. These microbes obtain energy by oxidizing ammonia or nitrite. In the first stage, *Nitrosomonas* oxidizes ammonium to nitrites:

$$\underset{\text{Ammonium ion}}{NH_4^+} \xrightarrow{\textit{Nitrosomonas}} \underset{\text{Nitrite ion}}{NO_2^-}$$

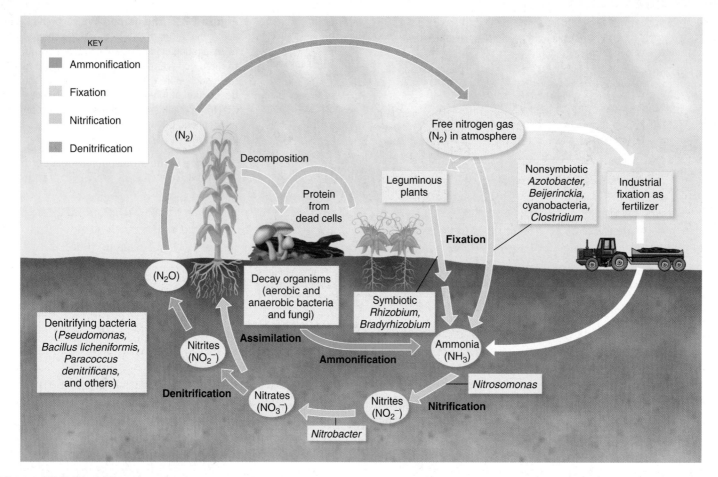

**Figure 27.3 The nitrogen cycle.** In general, nitrogen in the atmosphere goes through fixation, nitrification, and denitrification. Nitrates assimilated into plants and animals after nitrification go through decomposition, ammonification, and then nitrification again.

**Q** Which processes are performed exclusively by bacteria?

In the second stage, such organisms as *Nitrobacter* oxidize nitrites to nitrates:

$$NO_2^- \xrightarrow{\textit{Nitrobacter}} NO_3^-$$
$$\text{Nitrite ion} \qquad\qquad \text{Nitrate ion}$$

Plants tend to use nitrate as their source of nitrogen for protein synthesis because nitrate is highly mobile in soil and is more likely than ammonium to encounter a plant root. Ammonium ions would actually make a more efficient source of nitrogen because they require less energy to incorporate into protein, but these positively charged ions are usually bound to negatively charged clays in the soil, whereas the negatively charged nitrate ions are not bound.

### Denitrification

The form of nitrogen resulting from nitrification is fully oxidized and no longer contains any biologically usable energy. However, it can be used as an electron acceptor by microbes metabolizing other organic energy sources in the absence of atmospheric oxygen (see the discussion of anaerobic respiration

on page 128). This process, called **denitrification**, can lead to a loss of nitrogen to the atmosphere, especially as nitrogen gas. Denitrification can be represented as follows:

$$NO_3^- \longrightarrow NO_2^- \longrightarrow N_2O \longrightarrow N_2$$
$$\text{Nitrate ion} \quad\; \text{Nitrate ion} \quad\; \begin{array}{c}\text{Nitrous}\\ \text{oxide}\end{array} \quad \begin{array}{c}\text{Nitrogen}\\ \text{gas}\end{array}$$

Denitrification occurs in waterlogged soils, where little oxygen is available. In the absence of oxygen as an electron acceptor, denitrifying bacteria substitute the nitrates of agricultural fertilizer. This converts much of the valuable nitrate into gaseous nitrogen that enters the atmosphere and represents a considerable economic loss.

### Nitrogen Fixation

We live at the bottom of an ocean of nitrogen gas. The air we breathe is about 79% nitrogen, and above every acre of soil (the area of an American football field from the goal line to the opposite 10-yard line, or $50.6 \times 80$ meters) stands a column of nitrogen weighing about 32,000 tons. But only a few species of bacteria, including cyanobacteria, can use it directly as a nitrogen source. The process by which they convert nitrogen gas to ammonia is known as **nitrogen fixation**.

Bacteria that are responsible for nitrogen fixation all rely on the same enzyme, *nitrogenase*. It is estimated that Earth's entire supply of this essential enzyme could fit into a single large bucket. Nitrogenase is inactivated by oxygen. Therefore, it probably evolved early in the history of the planet, before the atmosphere contained much molecular oxygen and before nitrogen-containing compounds were available from decaying organic matter. Nitrogen fixation is brought about by two types of microorganisms: free-living and symbiotic. (Agricultural fertilizers are made up of nitrogen that has been fixed by industrial physical–chemical processes.)

**Free-Living Nitrogen-Fixing Bacteria** Free-living nitrogen-fixing bacteria are found in particularly high concentrations in the *rhizosphere*, a region roughly 2 millimeters from the plant root. The rhizosphere represents something of a nutritional oasis in the soil, especially in grasslands. Among the free-living bacteria that can fix nitrogen are aerobic species such as *Azotobacter*. These aerobic organisms apparently shield the anaerobic nitrogenase enzyme from oxygen by, among other things, having a very high rate of oxygen use that minimizes the diffusion of oxygen into the interior of the cell, where the enzyme is located.

Another free-living obligate aerobe that fixes nitrogen is *Beijerinckia* (bī-yer-INK-ē-ah). Some anaerobic bacteria, such as certain species of *Clostridium* (klos-TRID-ē-um), also fix nitrogen. The bacterium *C. pasteurianum* (PAS-tyer-ē-ā-num), an obligately anaerobic, nitrogen-fixing microorganism, is a prominent example.

There are many species of aerobic, photosynthesizing cyanobacteria that fix nitrogen. Because their energy supply is independent of carbohydrates in soil or water, they are especially useful suppliers of nitrogen to the environment. Cyanobacteria usually carry their nitrogenase enzymes in specialized structures called **heterocysts** that provide anaerobic conditions for fixation (see Figure 11.13a, page 308).

Most of the free-living nitrogen-fixing bacteria are capable of fixing large amounts of nitrogen under laboratory conditions. However, in the soil there is usually a shortage of usable carbohydrates to supply the energy needed to reduce nitrogen to ammonia, which is then incorporated into protein. Nevertheless, these nitrogen-fixing bacteria make important contributions to the nitrogen economy of such areas as grasslands, forests, and the arctic tundra.

**Symbiotic Nitrogen-Fixing Bacteria** Symbiotic nitrogen-fixing bacteria play an even more important role in plant growth for crop production. Members of the genera *Rhizobium*, *Bradyrhizobium*, and others infect the roots of leguminous plants, such as soybeans, beans, peas, peanuts, alfalfa, and clover. (These agriculturally important plants are only a few of the thousands of known leguminous species, many of which are bushy plants or small trees found in poor soils in many parts of the world.) Rhizobia, as these bacteria are commonly known, are specially adapted to particular leguminous plant species, on which they form **root nodules** (**Figure 27.4**). Nitrogen is then fixed by a symbiotic process of the plant and the bacteria. The plant furnishes anaerobic conditions and growth nutrients for the bacteria, and the bacteria fix nitrogen that can be incorporated into plant protein.

There are similar examples of symbiotic nitrogen fixing *Frankia* in nonleguminous plants, such as alder trees. The growth of 1 acre of alder trees can fix about 50 kg of nitrogen each year; the trees thus make a valuable addition to the forest economy.

Another important contribution to the nitrogen economy of forests is made by **lichens**, which are a combination of fungus and an alga or a cyanobacterium in a mutualistic relationship (see Figure 12.11, page 336). When one symbiont is a nitrogen-fixing cyanobacterium, the product is fixed nitrogen that eventually enriches the forest soil. Free-living cyanobacteria can fix significant amounts of nitrogen in desert soils after rains and on the surface of arctic tundra soils. Rice paddies can accumulate heavy growths of such nitrogen-fixing organisms. The cyanobacteria also form a symbiosis with a small floating fern, *Azolla*, which grows thickly in rice paddy waters (**Figure 27.5**).

> **CHECK YOUR UNDERSTANDING**
>
> ✔ **27-6** What is the common name for the group of microbes that oxidize soil nitrogen into a form that is mobile in soil and likely to be used by plants for nutrition?
>
> ✔ **27-7** Bacteria of the genus *Pseudomonas*, in the absence of oxygen, will use fully oxidized nitrogen as an electron acceptor, a process in the nitrogen cycle that is given what name?

## The Sulfur Cycle

The **sulfur cycle** (**Figure 27.6**) and nitrogen cycle resemble each other in the sense that they represent numerous oxidation states of these elements. The most reduced forms of sulfur are the sulfides, such as the odorous gas hydrogen sulfide ($H_2S$). Like the ammonium ion of the nitrogen cycle, this is a reduced compound that generally forms under anaerobic conditions. In turn it represents a source of energy for autotrophic bacteria. These bacteria convert the reduced sulfur in $H_2S$ into elemental sulfur granules and fully oxidized sulfates ($SO_4^{2-}$).

Several phototrophic bacteria, such as the green and purple sulfur bacteria, also oxidize $H_2S$, forming internal sulfur granules (see Figure 11.14, page 309). Like *Thiomargarita*, they can further oxidize the sulfur to sulfate ions (see Figure 11.28, page 320).

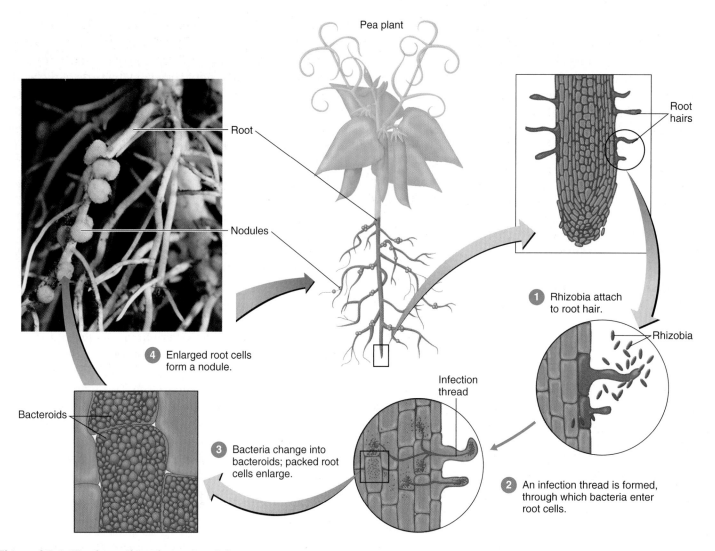

Pea plant

Root

Nodules

① Rhizobia attach to root hair.

Root hairs

Rhizobia

④ Enlarged root cells form a nodule.

Bacteroids

③ Bacteria change into bacteroids; packed root cells enlarge.

Infection thread

② An infection thread is formed, through which bacteria enter root cells.

**Figure 27.4 The formation of a root nodule.** Members of the nitrogen-fixing genera *Rhizobium* and *Bradyrhizobium* form these nodules on legumes. This mutualistic association is beneficial to both the plant and the bacteria.

**Q** In nature, are leguminous plants more likely to be valuable in rich agricultural soils or in poor desert soils?

It is important to recognize that these organisms are using light for energy; the hydrogen sulfide is used to reduce $CO_2$ (see Chapter 5, page 136).

Plants and bacteria incorporate sulfates to become part of sulfur-containing amino acids for humans and other animals. There, they form disulfide links that give structure to proteins. As proteins are decomposed, in a process called **dissimilation**, the sulfur is released as hydrogen sulfide to reenter the cycle.

## Life without Sunshine

Interestingly, it is possible for entire biological communities to exist without photosynthesis by exploiting the energy in $H_2S$. Chemoautotrophs oxidize $H_2S$ to produce NADH, which goes to the electron transport chain. Such communities occur, for example, around deep-sea vents. See Exploring the Microbiome,

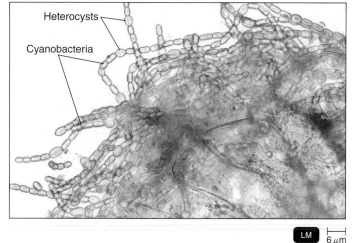

Heterocysts

Cyanobacteria

LM | 6 μm

**Figure 27.5 The *Azolla*–cyanobacteria symbiosis.** A section through the leaf of an *Azolla* freshwater fern. The cyanobacterium *Anabaena azollae* is visible as chains of cells within the leaf cavity.

**Q** What is the major contribution of cyanobacteria as symbionts?

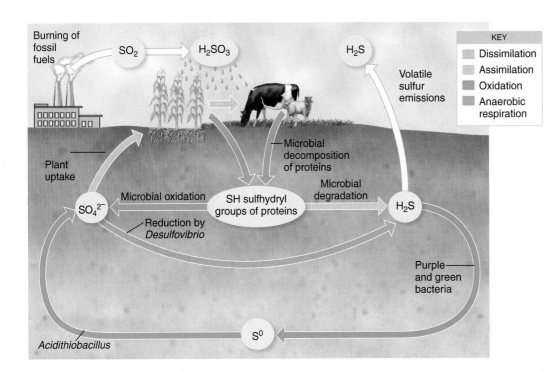

**Figure 27.6 The sulfur cycle.**
Reduced forms of sulfur such as $H_2S$ and elemental sulfur ($S°$) are energy sources for some microbes under aerobic or anaerobic conditions. Under anaerobic conditions, $H_2S$ can be used as a substitute for $H_2O$ in photosynthesis by purple and green bacteria (see page 308) to produce $S°$. Oxidized forms of sulfur, such as sulfates ($SO_4^{2-}$), are used as electron acceptors, as a substitute for oxygen, under anaerobic conditions by certain bacteria. Many organisms assimilate sulfates to make the —SH groups of proteins.

**Q** Why is a source of sulfur necessary for all organisms?

page 794. Deep caves, totally isolated from sunlight, have been discovered that also support entire biological communities. The **primary producers** in these systems are chemoautotrophic bacteria rather than photoautotrophic plants or microbes.

Another microbial ecosystem operating far from sunlight has been discovered over 1 km deep within rocks, including shales, granites, and basalts. Such bacteria are called **endoliths** (inside rocks), which must grow in the near absence of oxygen and with minimal nutrient supplies. Carbon dioxide dissolved in the water serves as a carbon source, and cellular organic matter is produced. Some is excreted, or is released upon the death and lysis of the microbe, and becomes available for the growth of other microbes. Nutrient inputs, especially nitrogen, are very small in this environment, and generation times may be measured in many years.

## The Phosphorus Cycle

Another important nutritional element that is part of a biogeochemical cycle is phosphorus. The availability of phosphorus may determine whether plants and other organisms can grow in an area. The problems associated with excess phosphorus (eutrophication) are described later in the chapter.

Phosphorus exists primarily as phosphate ions ($PO_4^{3-}$) and undergoes very little change in its oxidation state. The **phosphorus cycle** instead involves changes from soluble to insoluble forms and from organic to inorganic phosphate, often in relation to pH. For example, phosphate in rocks can be solubilized by the acid produced by bacteria such as *Acidithiobacillus*. Unlike the other cycles, there is no volatile phosphorus-containing product to return phosphorus to the atmosphere in the way carbon dioxide, nitrogen gas, and sulfur dioxide are returned.

Therefore, phosphorus tends to accumulate in the seas. It can be retrieved by mining the above-ground sediments of ancient seas, mostly as deposits of calcium phosphate. Seabirds also mine phosphorus from the sea by eating phosphorus-containing fish and depositing it as guano (bird droppings). Certain small islands inhabited by such birds have long been mined for these deposits as a source of phosphorus for fertilizers.

### CHECK YOUR UNDERSTANDING

✔ **27-8** Certain nonphotosynthetic bacteria accumulate granules of sulfur within the cell; are the bacteria using hydrogen sulfide or sulfates as an energy source?

✔ **27-9** What chemical usually serves as an energy source for organisms that survive in darkness?

✔ **27-10** Why does phosphorus tend to accumulate in the seas?

## The Degradation of Synthetic Chemicals in Soil and Water

We take for granted that soil microorganisms will degrade materials entering the soil. Natural organic matter, such as falling leaves or animal residues, is in fact readily degraded. However, in this industrial age many chemicals that do not occur in nature (**xenobiotics**), such as plastics, enter the soil in large amounts. In fact, plastics comprise about a fourth of all municipal wastes. A proposed solution to the problem is to develop biodegradable plastics made from polylactide (PLA) produced by lactic acid fermentation. When composted (see Figure 27.8), PLA plastic degrades in a few weeks. PLA-based plastics are appearing in a number of commercial products, such as disposable water bottles and drinking cups. Another version of

**Resident Microbes of Earth's Most Extreme Environments**

Until humans explored the deep-ocean floor, scientists believed that only a few forms of life could survive in that high-pressure, completely dark, oxygen-poor environment. Then, in 1977, *Alvin*, the deep-sea submersible, carried two scientists 2600 meters below the surface at the Galápagos Rift (about 350 km northeast of the Galápagos

Islands). There, amid the vast expanse of barren basalt rocks, the scientists found unexpectedly rich oases of life. Superheated water from beneath the seafloor was rising through fractures in the Earth's crust called vents. They discovered that mats of bacteria were growing along the sides of the vents, where temperatures exceeded 100°C (see the figure).

## Ecosystem of the Hydrothermal Vents

Life at the surface of the world's oceans depends on photosynthetic organisms, such as plants and algae, which harness the sun's energy to fix carbon dioxide ($CO_2$) to make carbohydrates. At the deep-ocean floor, where no light penetrates, photosynthesis is not possible. The scientists found that the primary producers at the ocean floor are chemoautotrophic bacteria. Using chemical energy from hydrogen sulfide ($H_2S$) as a source of energy to fix $CO_2$, the chemoautotrophs create an environment that supports higher life forms. Hydrothermal vents in the seafloor supply the $H_2S$ and $CO_2$.

A microbial biofilm is visible on this deep-sea hydrothermal vent. Water with black sulfide particles is being emitted through the ocean floor at temperatures above 100°C.

## New Products from Hydrothermal Vents

Terrestrial fungi and bacteria have had a major impact on the development of biotechnology. Hydrothermal vents are the next frontier in the hunt for new products. In 2010 a peptide produced by *Thermovibrio ammonificans* was shown to induce apoptosis (cell death) and thus potential anticancer activity. Researchers are growing *Pyrococcus furiosus* because it produces alternative fuels, hydrogen gas and butanol. DNA polymerases (enzymes that synthesize DNA) isolated from two archaea living near deep-sea vents are being used in the polymerase chain reaction (PCR), a technique for making many copies of DNA. In PCR, single-stranded DNA is made by heating a chromosome fragment to 98°C and cooling it so that DNA polymerase can copy each strand. DNA polymerases from *Thermococcus litoralis*, called Vent®, and from *Pyrococcus*, called Deep Vent®, are not denatured at 98°C. These enzymes can be used in automatic thermalcyclers to repeat the heating and cooling cycles, allowing many copies of DNA to be made easily and quickly.

biodegradable plastic, also made by bacterial fermentation, is called polyhydroxyalkanoate, or PHA. Products made of PHA degrade more easily and can stand higher temperatures in use, but they are more expensive than PLA.

Many synthetic chemicals, such as pesticides, are highly resistant to degradation by microbial attack. A well-known example is the insecticide DDT, which proved so resistant that it accumulated to damaging levels in the environment.

Small differences in chemical structure can make large differences in biodegradability. The classic example is that of two herbicides: 2,4-D (the common chemical used to kill lawn weeds) and 2,4,5-T (used to kill shrubs); both were components of Agent Orange, which was used to defoliate jungles during the Vietnam war. The addition of a single chlorine atom to the structure of 2,4-D extends its life in soil from a few days to an indefinite period.

A growing problem is the leaching into groundwaters of toxic materials that are not biodegradable or that degrade very slowly. The sources of these materials may include landfills, illegal industrial dumps, or pesticides applied to agricultural crops.

### Bioremediation

The use of microbes to detoxify or degrade pollutants is called **bioremediaton**. Oil spills from wrecked tankers and drilling accidents represent some of the most dramatic examples of chemical pollution. Bioremediation occurs naturally as microbes attack the petroleum if conditions are aerobic. However, microbes usually obtain their nutrients in aqueous solution, and oil-based products are relatively nonsoluble. Also, petroleum hydrocarbons are deficient in essential elements, such as nitrogen and phosphorus. Bioremediation of oil spills

**(a)** Oily rocks after the Exxon Valdez oil spill

**(b)** One month after addition of nitrogen-phosphorous fertilizer

**Figure 27.7 Bioremediation of oil.**

**Q** All living organisms require the chemical elements represented in the acronym CHONPS. Which elements can a bacterium obtain from petroleum?

is greatly enhanced if the resident bacteria are provided with "fertilizer" containing nitrogen and phosphorus (**Figure 27.7**).

One of the most promising successes for bioremediation occurred on an Alaskan beach following the *Exxon Valdez* oil spill in 1989. Scientists hit on a very simple way to speed up the process: they simply dumped ordinary nitrogen and phosphorus plant fertilizers (**bioenhancers**) onto a test beach. The number of oil-degrading bacteria increased compared with that on unfertilized control beaches, and oil was quickly cleared from the test beach. This technique works on land, but keeping fertilizer near oil in open water oil spills, such as the 2010 well blowout in the Gulf of Mexico, is proving difficult.

Bioremediaton may also make use of microbes that have been selected for growth on a certain pollutant or of genetically modified bacteria that are specially adapted to metabolize petroleum products. The addition of such specialized microbes is called **bioaugmentation**.

### Solid Municipal Waste

Solid municipal waste (garbage) is most frequently placed into large compacted landfills. Conditions are largely anaerobic, and even presumably biodegradable materials such as paper are not very effectively attacked by microorganisms. In fact, recovering a 20-year-old newspaper in readable condition is not at all unusual. But such anaerobic conditions promote activity of the same methanogens used in the operation of anaerobic sludge digesters to treat sewage (see page 803). The methane produced can be tapped with drill holes and burned to generate electricity or purified and introduced into natural gas pipeline systems (see Figure 28.13, page 823). Such systems are part of the design of many large landfills in the U.S., some of which provide energy for industrial plants and homes.

**Composting** is a process gardeners use to convert plant remains into the equivalent of natural humus. A pile of leaves or grass clippings will undergo microbial degradation.

Under favorable conditions, thermophilic bacteria will raise the temperature of the compost to 55–60°C in a couple of days. After the temperature declines, the pile can be turned to renew the oxygen supply, and a second temperature rise will occur. Over time, the thermophilic microbial populations are replaced by mesophilic populations that slowly continue the conversion to a stable material similar to humus. Municipal wastes are composted in windrows (long, low piles). Municipal waste disposal now also makes increasing use of composting methods (**Figure 27.8**). Farmers then use the compost in their fields.

### CHECK YOUR UNDERSTANDING

✔ **27-11** Why are petroleum products naturally resistant to metabolism by most bacteria?

✔ **27-12** What is the definition of the term *bioremediation*?

## Aquatic Microbiology and Sewage Treatment

### LEARNING OBJECTIVES

**27-13** Describe the freshwater and seawater habitats of microorganisms.

**27-14** Explain how wastewater pollution is a public health problem and an ecological problem.

**27-15** Discuss the causes and effects of eutrophication.

**27-16** Explain how water is tested for bacteriological purity.

**27-17** Describe how pathogens are removed from drinking water.

**27-18** Compare primary, secondary, and tertiary sewage treatment.

**27-19** List some of the biochemical activities that take place in an anaerobic sludge digester.

**27-20** Define *biochemical oxygen demand (BOD)*, *activated sludge system*, *trickling filter*, *septic tank*, and *oxidation pond*.

Solid municipal wastes being turned by a specially designed machine

**Figure 27.8 Composting municipal wastes.**

**Q** A compost pile of grass and leaves is very high in carbon; does it have much nitrogen?

Aquatic microbiology refers to the study of microorganisms and their activities in natural waters, such as lakes, ponds, streams, rivers, estuaries, and oceans.

## Aquatic Microorganisms

Large numbers of microorganisms in a body of water generally indicate high nutrient levels in the water. Water contaminated by inflows from sewage systems or from biodegradable industrial organic wastes is relatively high in bacterial numbers. Similarly, ocean estuaries (fed by rivers) have higher nutrient levels and therefore larger microbial populations than other shoreline waters.

In water, particularly with low nutrient concentrations, microorganisms tend to grow on stationary surfaces and on particulate matter. In this way, a microorganism has contact with more nutrients than if it were randomly suspended and floating freely with the current. Many bacteria whose main habitat is water often have appendages and holdfasts that attach to various surfaces. One example is *Caulobacter* (see Figure 11.2, page 299).

### Freshwater Microbiota

A typical lake or pond serves as an example of the various zones and the kinds of microbiota found in a body of fresh water. The **littoral zone** along the shore has considerable rooted vegetation, and light penetrates throughout it. The **limnetic zone** consists of the surface of the open water area away from the shore. The **profundal zone** is the deeper water under the limnetic zone. The **benthic zone** contains the sediment at the bottom.

Microbial populations of freshwater bodies tend to be affected mainly by the availability of oxygen and light. In many ways, light is the more important resource because photosynthetic algae are the main source of organic matter, and hence of energy, for the lake. These organisms are the primary producers of a lake that support a population of bacteria, protozoa, fish, and other aquatic life. Photosynthetic algae are located in the limnetic zone.

Oxygen does not diffuse into water very well, as any aquarium owner knows. Microorganisms growing on nutrients in stagnant water quickly use up the dissolved oxygen in the water. In the oxygenless water, fish die, and anaerobic activity produces odors. Wave action in shallow layers, or water movement in rivers, tends to increase the amount of oxygen throughout the water and aid in the growth of aerobic populations of bacteria. Movement thus improves the quality of water and aids in the degradation of polluting nutrients.

Deeper waters of the profundal and benthic zones have low oxygen concentrations and less light. As light passes through the water column, longer wavelengths are absorbed by the water. It is not unusual for photosynthetic microbes in deeper zones to use different wavelengths of light from those used by surface-layer photosynthesizers (see Figure 12.12a, page 338).

Purple and green sulfur bacteria are found in the benthic zone. These bacteria are anaerobic photosynthetic organisms that metabolize $H_2S$ to sulfur and sulfate in the bottom sediments of the benthic zone.

The sediment in the benthic zone includes bacteria such as *Desulfovibrio* that use sulfate ($SO_4^{2-}$) as an electron acceptor and reduce it to $H_2S$. Methane-producing bacteria are also part of these anaerobic benthic populations. In swamps, marshes, or bottom sediments, they produce methane gas. *Clostridium* species are common in shallow-bottom sediments and may include botulism organisms, particularly those causing outbreaks of botulism in waterfowl. The clostridia grow on animal carcasses, and their toxin is concentrated in maggots feeding on the carcasses. Birds eating the maggots then ingest the toxin. This carcass–maggot cycle can kill millions of birds in a season.

### Seawater Microbiota

As knowledge of the microbial life of the oceans expands, largely identified by ribosomal RNA methods (see the discussion of FISH on page 288 in Chapter 10), biologists are becoming more conscious of the importance of oceanic microbes. Seafloor sediments have been found to have large populations of bacteria. These organisms are mostly Archaea, which adapt well to environmental stresses and have low energy requirements. One conclusion, so far, has been that nearly a third of all life on the planet consists of microbes that live, not in ocean waters, but under the seafloor. These microbes make immense amounts of methane gas that could be environmentally damaging if it were to be released into the atmosphere.

In the upper, relatively sunlit waters of the ocean, photosynthetic cyanobacteria of the genera *Synechococcus* (si-nek-ō-KOK-kus) and *Prochlorococcus* (prō-KLŌR-ō-kok-kus) are abundant. Populations of different strains vary at different depths according to their adaptations to available sunlight. A drop of seawater might contain 20,000 cells of *Prochlorococcus*, a tiny sphere less than 0.7 µm in diameter. This unseen population of microscopic organisms fills the upper 100 meters of ocean and exerts a profound influence on life on Earth. The support of oceanic life depends largely on such photosynthetic microscopic life, the marine **phytoplankton** (a term derived from the Greek for wandering plants).

Photosynthetic bacteria such as these form the basis of the oceanic food chain. Billions of these in every liter of seawater double in number every few days and are consumed at about the same rate by microscopic predators. They fix carbon dioxide to form organic matter that is eventually released as dissolved organic matter and is used by the ocean's heterotrophic bacteria. A cyanobacterium, *Trichodesmium* (trik-ō-DES-mē-um), fixes nitrogen and helps replenish the nitrogen that is lost as organisms sink to oceanic depths. Immense populations of another bacterium, *Pelagibacter ubique*, metabolize the waste products of these photosynthetic populations (see Chapter 11, page 297).

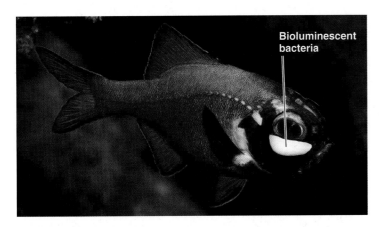

Bioluminescent bacteria

**Figure 27.9 Bioluminescent bacteria as light organs in fish.** This is a deep-sea flashlight fish (*Photoblepharon palpebratus*). *Aliivibrio fischeri* bacteria in the organ under the eye produce light.

**Q** What enzyme is responsible for bioluminescence?

In waters below about 100 meters, members of the Archaea begin to dominate microbial life. Planktonic members of this group of the genus *Crenarchaeota* (kren-ar-KĒ-ō-tah) account for much of the microbial biomass of the oceans. These organisms are well adapted to the cool temperatures and low oxygen levels of oceanic depths. Their carbon is primarily derived from dissolved $CO_2$.

Microbial **bioluminescence**, or light emission, is an interesting aspect of deep-sea life. Many bacteria are luminescent, and some have established symbiotic relationships with benthic-dwelling fish. These fish sometimes use the glow of their resident bacteria as an aid in attracting and capturing prey in the complete darkness of the ocean depths (**Figure 27.9**). These bioluminescent organisms have an enzyme called *luciferase* that picks up electrons from flavoproteins in the electron transport chain and then emits some of the electron's energy as a photon of light.

> **CHECK YOUR UNDERSTANDING**
>
> ✔ **27-13** Purple and green sulfur bacteria are photosynthetic organisms, but they are generally found deep in freshwater rather than at the surface. Why?

## The Role of Microorganisms in Water Quality

Water in nature is seldom totally pure. Even rainfall is contaminated as it falls to Earth.

### Water Pollution

The form of water pollution that is our primary interest is microbial pollution, especially by pathogenic organisms.

**The Transmission of Infectious Diseases** Water that moves below the ground's surface undergoes a filtering that removes most microorganisms. For this reason, water from springs and deep wells is generally of good quality. The most dangerous

form of water pollution occurs when feces enter the water supply. Many diseases are perpetuated by the fecal–oral route of transmission, in which a pathogen is shed in human or animal feces, contaminates water, and is ingested (see Chapter 25). Globally, it is estimated that waterborne diseases are responsible for over 2 million deaths each year, mostly among children under the age of 5.

Examples of such diseases are typhoid fever and cholera, caused by bacteria that are shed only in human feces. Improvements in sanitation, including the use of sand filter beds, in developed nations have greatly reduced incidence.

**Chemical Pollution** Preventing chemical contamination of water is a difficult problem. Industrial and agricultural chemicals leached from the land enter water in great amounts and in forms that are resistant to biodegradation. Rural waters often have excessive amounts of nitrate from agricultural fertilizers. When ingested, the nitrate is converted to nitrite by bacteria in the gastrointestinal tract. Nitrite competes for oxygen in the blood and is especially likely to harm infants.

An example of chemical pollution was brought about by the synthetic detergents developed immediately after World War II. These rapidly replaced many of the soaps then in use. Because these new detergents were not biodegradable, they rapidly accumulated in the waterways. In some rivers, large rafts of detergent suds could be seen traveling downstream. These detergents were replaced by biodegradable synthetic formulations.

Biodegradable detergents, however, still present a major environmental problem because they often contain phosphates. Unfortunately, phosphates pass almost unchanged through sewage systems and can lead to **eutrophication**, which is caused by an overabundance of nutrients in lakes and streams.

To understand the concept of eutrophication, recall that algae and cyanobacteria get their energy from sunlight and

**CLINICAL CASE**

The physician prescribes a single dose of doxycycline and tells Charity to drink plenty of liquids. He also asks what countries Charity had visited. She tells him that she had been to China, the Philippines, Haiti, Chile, and Indonesia. Before she became ill, Charity had been in good health. Just before leaving Haiti to go home, Charity had eaten fried shrimp and prawns purchased from a local market and prepared by a local family. She also recalled drinking a half glass of water with her dinner; she did not know whether it was bottled water.

**What should the physician suspect to be the cause of Charity's severe diarrhea?**

788  **797**  799  803  804  805

**Figure 27.10 A red tide.** These blooms of aquatic growth are caused by excess nutrients in water. The color is from the pigmentation of the dinoflagellates.

**Q** What is the primary energy source of the dinoflagellates that cause such aquatic blooms?

their carbon from carbon dioxide dissolved in water. In most waters, only nitrogen and phosphorus supplies, therefore, remain inadequate for algal growth. Both of these nutrients can enter water from domestic, farm, and industrial wastes when waste treatment is absent or inefficient. These additional nutrients cause dense aquatic growths called **algal blooms**. Because many cyanobacteria can fix nitrogen from the atmosphere, these photosynthesizing organisms require only traces of phosphorus to initiate blooms. Once eutrophication results in blooms of algae or cyanobacteria, the eventual effect is the same as adding biodegradable organic matter. In the short run, these algae and cyanobacteria produce oxygen. However, they eventually die and are degraded by bacteria. During the degradation process, the oxygen in the water is used up, killing the fish. Undegraded remnants of organic matter settle to the bottom and hasten the filling of the lake.

Blooms of toxin-producing phytoplankton (**Figure 27.10**), which were mentioned in Chapter 12 (page 340), are probably caused by excessive nutrients from oceanic upwellings or terrestrial wastes. In addition to eutrophication effects, this type of biological bloom can affect human health. Seafood, especially clams or similar mollusks, that ingest these plankton become toxic to humans.

Municipal waste containing detergents is likely to be the main source of phosphates in lakes and streams. As a result, phosphate-containing detergents and lawn fertilizers are banned in many places.

## Water Purity Tests

Historically, most of our concern about water purity has been related to the transmission of disease. Therefore, tests have been developed to determine the safety of water; many of these tests are also applicable to foods.

It is not practical, however, to look only for pathogens in water supplies. For one thing, if we were to find the pathogens causing typhoid or cholera in the water system, the discovery would already be too late to prevent an outbreak of the disease. Moreover, such pathogens would probably be present only in small numbers and might not be included in tested samples.

The tests for water purity in use today are aimed instead at detecting particular **indicator organisms**. There are several criteria for an indicator organism. The most important criterion is that the microbe be consistently present in human feces in substantial numbers so that its detection is a good indication that human wastes are entering the water. The indicator organisms should also survive in the water at least as well as the pathogens would. The indicator organisms must be detectable by simple tests that can be carried out by people with relatively little training in microbiology.

In the United States, the usual indicator organisms in freshwater are the *coliform bacteria*.* **Coliforms** are defined as aerobic or facultatively anaerobic, gram-negative, non–endospore-forming, rod-shaped bacteria that ferment lactose to form gas within 48 hours of being placed in lactose broth at 35°C. Because some coliforms are not solely enteric bacteria but are more commonly found in plant and soil samples, many standards for food and water specify the identification of *fecal coliforms*. The predominant fecal coliform is *E. coli*, which constitutes a large proportion of the human intestinal population. There are specialized tests to distinguish fecal coliforms from nonfecal coliforms. Note that coliforms are not themselves pathogenic under normal conditions, although certain strains can cause diarrhea (see Chapter 25, page 736) and opportunistic urinary tract infections (see Chapter 26, page 764).

The methods for determining the presence of coliforms in water are based largely on the lactose-fermenting ability of coliform bacteria. The multiple-tube method can be used to estimate coliform numbers by the most probable number (MPN) method (see Figure 6.19, page 172). The membrane filtration method is a more direct method of determining the presence and numbers of coliforms. This is possibly the most widely used method in North America and Europe. It makes use of a filtration apparatus similar to that shown in Figure 7.4 (page 185). In this application, though, the bacteria collected on the surface of a removable membrane filter are placed on an appropriate medium and incubated. Coliform colonies have a distinctive appearance and are counted. This method is suitable for low-turbidity waters that do not clog the filter and have relatively few noncoliform bacteria that would mask the results.

---

*The U.S. Environmental Protection Agency (EPA) recommends the use of *Enterococcus* bacteria as a safety indicator for waters in oceans and bays. Populations of the enterococci decrease more uniformly than coliforms in both freshwater and seawater.

A newer and more convenient method of detecting coliforms, specifically the fecal coliform *E. coli*, makes use of media containing the two substrates *o*-nitrophenyl-β-D-galactopyranoside (ONPG) and 4-methylumbelliferyl-β-D-glucuronide (MUG). Coliforms produce the enzyme β-galactosidase, which acts on ONPG and forms a yellow color, indicating their presence in the sample. *E. coli* is unique among coliforms in almost always producing the enzyme β-glucuronidase, which acts on MUG to form a fluorescent compound that glows blue when illuminated by long-wave UV light. It can also be applied to solid media, such as in the membrane filtration method. The colonies fluoresce under UV light. These simple tests, or variants of them, can detect the presence or absence of coliforms or *E. coli* and can be combined with the multiple-tube method to enumerate them.

Coliforms have been very useful as indicator organisms in water sanitation, but they have limitations. One problem is the growth of coliform bacteria embedded in biofilms on the inner surfaces of water pipes. These coliforms do not, then, represent external fecal contamination of the water, and they are not considered a threat to public health. Standards governing the presence of coliforms in drinking water require that any positive water sample be reported, and occasionally these indigenous coliforms have been detected. This has led to unnecessary community orders to boil water.

A more serious problem is that some pathogens, especially viruses and protozoan cysts and oocysts, are more resistant than coliforms to chemical disinfection. Through the use of sophisticated methods of detecting viruses, it has been found that chemically disinfected water samples that are free of coliforms are often still contaminated with enteric viruses. The cysts of *Giardia intestinalis* and oocysts of *Cryptosporidium* are so resistant to chlorination that completely eliminating them by this method is probably impractical; mechanical methods such as filtration are necessary. A general rule for chlorination is that viruses are more resistant to treatment than is *E. coli* and

that the cysts of *Cryptosporidium* and *Giardia* are 100 times more resistant than viruses.

**CHECK YOUR UNDERSTANDING**

✔ **27-14** Which disease is more likely to be transmitted by polluted water, cholera or influenza?

✔ **27-15** Name a microorganism that will grow in water even if there is no source of organic matter for energy or a nitrogen source—but does require small inputs of phosphorus.

✔ **27-16** Coliforms are the most common bacterial indicator of health-threatening water pollution in the United States. Why is it usually necessary to specify *fecal* coliforms?

## Water Treatment

When water is obtained from uncontaminated reservoirs fed by clear mountain streams or from deep wells, it requires minimal treatment to make it safe to drink. Many cities, however, obtain their water from badly polluted sources, such as rivers that have received municipal and industrial wastes upstream. The steps used to purify this water are shown in **Figure 27.11**. Water treatment is not intended to produce sterile water, but rather water that is free of disease-causing microbes.

### Coagulation and Filtration

Very turbid (cloudy) water is allowed to stand in a holding reservoir for a time to allow as much particulate suspended matter as possible to settle out. The water then undergoes **flocculation**, the removal of colloidal materials such as clay, which is so small (smaller than 10 μm) that it would otherwise remain in suspension indefinitely. A flocculant chemical, such as aluminum potassium sulfate (alum), forms aggregations of fine suspended particles called *floc*. As these aggregations slowly settle out, they entrap colloidal material and carry it to the bottom. Large numbers of viruses and bacteria are also removed this way. Alum was used to clear muddy river water during the first half of the nineteenth century in the military forts of the American West, long before the germ theory of disease was developed.

After flocculation, the water is treated by **filtration**—that is, passing it through beds of 2 to 4 feet of fine sand or crushed anthracite coal. As mentioned previously, some protozoan cysts and oocysts are removed from water only by such filtration treatment. The microorganisms are trapped mostly by surface adsorption onto the sand particles. They do not penetrate the tortuous routing between the particles, even though the openings might be larger than the microbes that are filtered out. These filters are periodically backflushed to clear them of accumulations. Water systems of cities that have an exceptional concern for toxic chemicals supplement sand filtration with filters of activated charcoal (carbon). Charcoal removes not only particulate matter but also most dissolved organic chemical pollutants. A properly operated water treatment plant

**CLINICAL CASE**

The physician suspects cholera, and he sends a stool sample to a local laboratory. The stool culture yields colonies suspected of being *Vibrio cholerae*. This result is confirmed by the county public health laboratory. Latex agglutination tests in the state public health laboratory confirm that the colonies are producing cholera toxin. Further testing at the CDC identifies the isolate as the El Tor biotype of *V. cholerae* O:1. DNA fingerprinting shows that this is the same strain of *V. cholerae* that is causing an epidemic in Haiti.

**How is cholera transmitted? How does an earthquake promote transmission?**

788    797    **799**    803    804    805

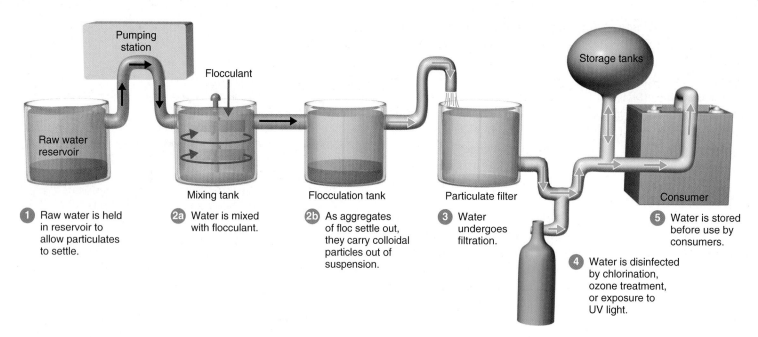

**Figure 27.11 The steps involved in water treatment in a typical municipal water purification plant.**

Q Does removal of "colloidal particles" by flocculation involve living organisms?

will remove viruses (which are harder to remove than bacteria and protozoa) with an efficiency of about 99.5%. Low-pressure *membrane filtration systems* are now coming into use. These systems have pore openings as small as 0.2 μm and are more reliable for removal of *Giardia* and *Cryptosporidium*.

### Disinfection

Before entering the municipal distribution system, the filtered water is chlorinated. Because organic matter neutralizes chlorine, the plant operators must pay constant attention to maintaining effective levels of chlorine.

As noted in Chapter 7 (page 197), another disinfectant for water is ozone treatment. Ozone ($O_3$) is a highly reactive form of oxygen that is formed by electrical spark discharges and UV light. (The fresh odor of air following an electrical storm or around a UV light bulb is from ozone.) Ozone for water treatment is generated electrically at the site of treatment. Ozone treatment is also valued because it leaves no taste or odor. Because it has little residual effect, ozone is usually used as a primary disinfectant treatment and is followed by chlorination. The use of UV light is also a supplement or alternative to chemical disinfection. Ultraviolet tube lamps are arranged so that water flows close to them. This is necessary because of the low penetrating power of UV radiation.

#### CHECK YOUR UNDERSTANDING

✔ **27-17** How do flocculants such as alum remove colloidal impurities, including microorganisms, from water?

## Sewage (Wastewater) Treatment

Sewage, or wastewater, includes all the water from a household that is used for washing and toilet wastes. Rainwater flowing into street drains and some industrial wastes enter the sewage system in many cities. Sewage is mostly water and contains little particulate matter, perhaps only 0.03%. Even so, in large cities the solid portion of sewage can total more than 1000 tons of solid material per day.

Until environmental awareness intensified, a surprising number of large American cities had only a rudimentary sewage treatment system or no system at all. Raw sewage, untreated or nearly so, was simply discharged into rivers or oceans. A flowing, well-aerated stream is capable of considerable self-purification. Therefore, until expanding populations and their wastes exceeded this capability, this casual treatment of municipal wastes did not cause problems. In the United States, most cases of simple discharge have been improved. But this is not true in much of the world. Many of the communities bordering the Mediterranean dump their unprocessed sewage into the sea.

### Primary Sewage Treatment

The usual first step in sewage treatment is called **primary sewage treatment** (Figure 27.12a). In this process, large floating materials in incoming wastewater are screened out, the sewage is allowed to flow through settling chambers to remove sand and similar gritty material, skimmers remove floating oil and grease, and floating debris is shredded and ground. After this step, the sewage passes through sedimentation tanks, where

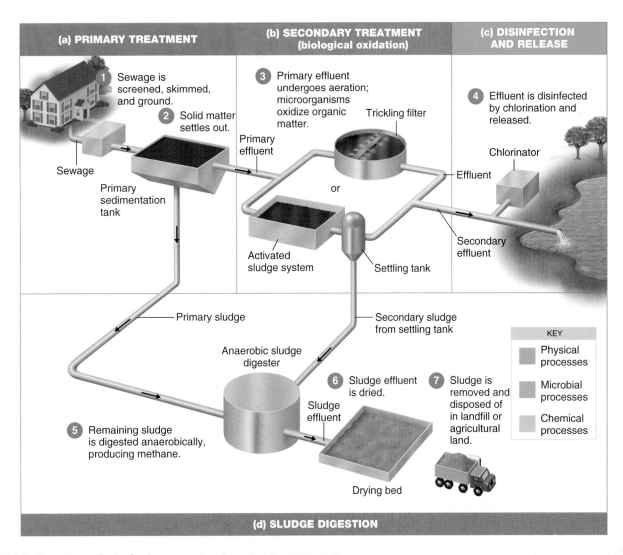

**Figure 27.12 The stages in typical sewage treatment.** Microbial activity occurs aerobically in trickling filters or activated sludge aeration tanks and anaerobically in the anaerobic sludge digester. A particular system would use either activated sludge aeration tanks or trickling filters, not both, as shown in this figure. Methane produced by sludge digestion is burned off or used to power heaters or pump motors.

**Q** Which processes require oxygen?

more solid matter settles out. Sewage solids collecting on the bottom are called **sludge**—at this stage, *primary sludge.* About 40–60% of suspended solids are removed from sewage by this settling treatment, and flocculating chemicals that increase the removal of solids are sometimes added at this stage. Biological activity is not particularly important in primary treatment, although some digestion of sludge and dissolved organic matter can occur during long holding times. The sludge is removed on either a continuous or an intermittent basis, and the effluent (the liquid flowing out) then undergoes secondary treatment.

### Biochemical Oxygen Demand

An important concept in sewage treatment and in the general ecology of waste management is **biochemical oxygen demand**

**(BOD)**, a measure of the biologically degradable organic matter in water. Primary treatment removes about 25–35% of the BOD of sewage.

BOD is determined by the amount of oxygen required by bacteria to metabolize the organic matter. The classic method of measurement is the use of special bottles with airtight stoppers. Each bottle is first filled with test water or dilutions. The water is initially aerated to provide a relatively high level of dissolved oxygen and is seeded with bacteria if necessary. The filled bottles are incubated in the dark for 5 days at 20°C, and the decrease in dissolved oxygen is determined by a chemical or electronic testing method. The more oxygen that is used up as the bacteria degrade the organic matter in the sample, the greater the BOD, which is usually expressed in milligrams

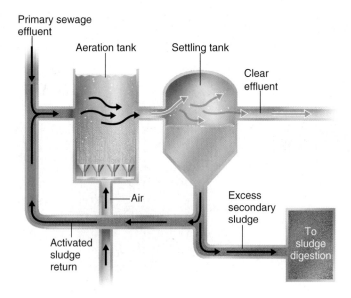

**Figure 27.13 An activated sludge system of secondary sewage treatment.**

**Q** What are the similarities between winemaking and activated sludge sewage treatment?

of oxygen per liter of water. The amount of oxygen that normally can be dissolved in water is only about 10 mg/liter; typical BOD values of wastewater may be 20 times this amount. If this wastewater enters a lake, for example, bacteria in the lake begin to consume the organic matter responsible for the high BOD, rapidly depleting the oxygen in the lake water. (See the discussion of eutrophication on pages 797–798.)

### Secondary Sewage Treatment

After primary treatment, the greater part of the BOD remaining in the sewage is in the form of dissolved organic matter. **Secondary sewage treatment**, which is predominantly biological, is designed to remove most of this organic matter and reduce the BOD (Figure 27.12b). In this process, the sewage undergoes strong aeration to encourage the growth of aerobic bacteria and other microorganisms that oxidize the dissolved organic matter to carbon dioxide and water. Two commonly used methods of secondary treatment are activated sludge systems and trickling filters.

In the aeration tanks of an **activated sludge system**, air or pure oxygen is passed through the effluent from primary treatment (**Figure 27.13**). The name is derived from the practice of adding some of the sludge from a previous batch to the incoming sewage. This inoculum is termed *activated sludge* because it contains large numbers of sewage-metabolizing microbes. The activity of these aerobic microorganisms oxidizes much of the sewage organic matter into carbon dioxide and water. Especially important members of this microbial community are species of *Zoogloea* bacteria, which form bacteria-containing

masses in the aeration tanks called floc, or *sludge granules*. Soluble organic matter in the sewage is incorporated into the floc and its microorganisms. Aeration is discontinued after 4 to 8 hours, and the contents of the tank are transferred to a settling tank, where the floc settles out, removing much of the organic matter. These solids are subsequently treated in an anaerobic sludge digester, which will be described shortly. Probably more organic matter is removed by this settling-out process than by the relatively short-term aerobic oxidation by microbes. The clear effluent is disinfected and discharged.

Occasionally, the sludge will float rather than settle out; this phenomenon is called **bulking**. When this happens, the organic matter in the floc flows out with the discharge effluent, resulting in local pollution. Bulking is caused by the growth of filamentous bacteria of various types; *Sphaerotilus natans* and *Nocardia* species are frequent offenders. Activated sludge systems are quite efficient: they remove 75–95% of the BOD from sewage.

**Trickling filters** are the other commonly used method of secondary treatment. In this method, the sewage is sprayed over a bed of rocks or molded plastic (**Figure 27.14**). The components of the bed must be large enough so that air penetrates to the bottom but small enough to maximize the surface area available for microbial activity. A biofilm (see page 157) of aerobic microbes grows on the rock or plastic surfaces. Because air circulates throughout the rock bed, these aerobic microorganisms in the slime layer can oxidize much of the organic matter trickling over the surfaces into carbon dioxide and water. Trickling filters remove 80–85% of the BOD, so they are generally less efficient than activated sludge systems. However, they are

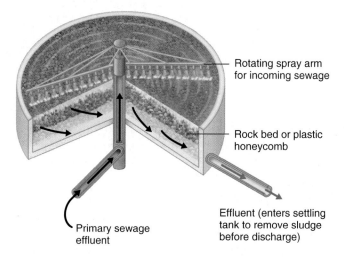

**Figure 27.14 A trickling filter of secondary sewage treatment.** The sewage is sprayed from the system of rotating pipes onto a bed of rocks or plastic honeycomb designed to have a maximum surface area and to allow oxygen to penetrate deeply into the bed.

**Q** Which would make the more efficient bed for a trickling filter system, fine sand or golf balls?

usually less troublesome to operate and have fewer problems from overloads or toxic sewage. Note that sludge is also a product of trickling filter systems.

Another biofilm-based design for secondary sewage treatment is the **rotating biological contactor** system. This is a series of disks several feet in diameter, mounted on a shaft. The disks rotate slowly, with their lower 40% submerged in wastewater. Rotation provides aeration and contact between the biofilm on the disks and the wastewater. The rotation also tends to cause the accumulated biofilm to slough off when it becomes too thick. This is about the equivalent of floc accumulation in activated sludge systems.

### Disinfection and Release

Treated sewage is disinfected, usually by chlorination, before being discharged (Figure 27.12c). The discharge is usually into an ocean or into flowing streams. The treated wastewater can be used to irrigate orchards, vineyards, and nonfood crops to avoid phosphorous contamination of waterways and to conserve freshwater. The soil to which this water is applied acts as a trickling filter to remove chemicals and microorganisms before the water reaches groundwater and surface water supplies.

### Sludge Digestion

Primary sludge accumulates in primary sedimentation tanks; sludge also accumulates in activated sludge and in trickling filter secondary treatments. For further treatment, these sludges are often pumped to **anaerobic sludge digesters** (Figure 27.12d and **Figure 27.15**). The process of sludge digestion is carried out in large tanks from which oxygen is almost completely excluded.

In secondary treatment, emphasis is placed on maintaining aerobic conditions so that organic matter is converted to carbon dioxide, water, and solids that can settle out. An anaerobic sludge digester, however, is designed to encourage the growth of anaerobic bacteria, especially methane-producing bacteria

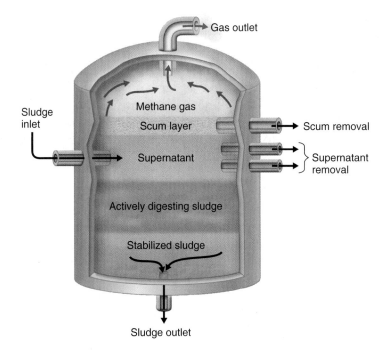

**Figure 27.15 Sludge digestion.**

**Q** **What might be some uses for the stabilized sludge?**

that decrease these organic solids by degrading them to soluble substances and gases, mostly methane (60–70%) and carbon dioxide (20–30%). Methane and carbon dioxide are relatively innocuous end-products, comparable to the carbon dioxide and water from aerobic treatment. The methane is routinely used as a fuel for heating the digester and is also frequently used to run power equipment in the plant.

There are essentially three stages in the activity of an anaerobic sludge digester. The first stage is the production of carbon dioxide and organic acids from anaerobic fermentation of the sludge by various anaerobic and facultatively anaerobic microorganisms. In the second stage, the organic acids are metabolized to form hydrogen and carbon dioxide, as well as organic acids such as acetic acid. These products are the raw materials for a third stage, in which the methane-producing bacteria produce methane ($CH_4$). Most of the methane is derived from the energy-yielding reduction of carbon dioxide by hydrogen gas:

$$CO_2 + 4H_2 \longrightarrow CH_4 + 2H_2O$$

Other methane-producing microbes split acetic acid ($CH_3COOH$) to yield methane and carbon dioxide:

$$CH_3COOH \longrightarrow CH_4 + CO_2$$

After anaerobic digestion is completed, large amounts of undigested sludge still remain, although it is relatively stable and inert. To reduce its volume, this sludge is pumped to shallow drying beds or water-extracting filters. Following this step, the sludge can be used for landfill or as a soil conditioner,

*V. cholerae* is transmitted by the fecal–oral route. Prior to the earthquake, only 63% of Haiti's population had access to an improved drinking water source (a closed well, chlorine, or filtration and safe storage containers), and only 17% had access to adequate sanitation. Many people used springs for their drinking water. Cholera spread rapidly across Haiti 9 months after the earthquake because of the lack of safe water and sanitation and the large numbers of displaced people. The fatality rate from cholera in Haiti is 3.3%.

**Charity recovers uneventfully; why is the case fatality rate so high in Haiti?**

sometimes under the name *biosolids*. Sludge is assigned to two classes: class A sludge contains no detectable pathogens, and class B sludge is treated only to reduce numbers of pathogens below certain levels. Most sludge is class B, and public access to application sites is limited. Sludge has about one-fifth the growth-enhancing value of normal commercial lawn fertilizers but has desirable soil-conditioning qualities, much as do humus and mulch. A potential problem is contamination with heavy metals that are toxic to plants.

## Septic Tanks

Homes and businesses in areas of low population density that are not connected to municipal sewage systems often use a **septic tank**, a device whose operation is similar in principle to primary treatment (**Figure 27.16**). Sewage enters a holding tank, and suspended solids settle out. The sludge in the tank must be pumped out periodically and disposed of. The effluent flows through a system of perforated piping into a leaching (soil drainage) field. The effluent entering the soil is decomposed by soil microorganisms. The microbial action necessary for proper

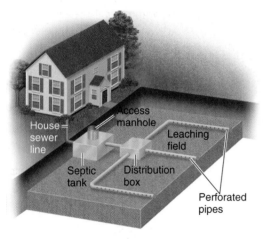

(a) Overall plan. Most soluble organic matter is disposed of by percolation into the soil.

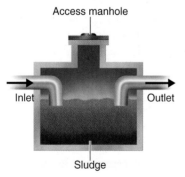

(b) A section of a septic tank

### Figure 27.16  A septic tank system.

**Q** Which type of soil would require the larger drainage area, clay or sandy?

788    797    799    803    **804**    805

functioning of a septic tank can be impaired by excessive amounts of products such as antibacterial soaps, drain cleaners, medications, "every flush" toilet bowl cleaners, and bleach.

These systems work well when not overloaded and when the drainage system is properly sized to the load and soil type. Heavy clay soils require extensive drainage systems because of the soil's poor permeability. The high porosity of sandy soils can result in chemical or bacterial pollution of nearby water supplies.

## Oxidation Ponds

Many industries and small communities use **oxidation ponds**, also called *lagoons* or *stabilization ponds*, for water treatment. These are inexpensive to build and operate but require large areas of land. Designs vary, but most incorporate two stages. The first stage is analogous to primary treatment; the sewage pond is deep enough that conditions are almost entirely anaerobic. Sludge settles out in this stage. In the second stage, which roughly corresponds to secondary treatment, effluent is pumped into an adjoining pond or system of ponds shallow enough to be aerated by wave action. Because it is difficult to maintain aerobic conditions for bacterial growth in ponds with so much organic matter, the growth of algae is encouraged to produce oxygen. Bacterial action in decomposing the organic matter in the wastes generates carbon dioxide. Algae, which use carbon dioxide in their photosynthetic metabolism, grow and produce oxygen, which in turn encourages the activity of aerobic microbes in the sewage. Large amounts of organic matter in the form of algae accumulate, but this is not a problem because the oxidation pond, unlike a lake, already has a large nutrient load.

Some small sewage-producing operations, such as isolated campgrounds and highway rest stop areas, use an *oxidation ditch* for sewage treatment. In this method, a small oval channel in the shape of a racetrack is filled with sewage water. A paddle wheel propels the water in a self-contained flowing stream aerated enough to oxidize the wastes.

### Tertiary Sewage Treatment

The effluent from secondary treatment plants contains some residual BOD. It also contains about 50% of the original nitrogen and 70% of the original phosphorus, which can greatly affect an aquatic ecosystem when discharged into small streams or recreational lakes. Sewage can be treated to a level of purity that allows its use as drinking water—winningly termed "toilet to tap." This **tertiary sewage treatment** is the practice now in some arid-area cities in the United States and may be expanded. Lake Tahoe in the Sierra Nevada, surrounded by extensive development, is the site of one of the best-known tertiary sewage treatment systems. Similar systems are used to treat wastes entering the southern portion of San Francisco Bay and in eastern Australia.

Tertiary treatment is designed to remove essentially all the BOD, nitrogen, and phosphorus. Tertiary treatment depends less on biological treatment than on physical and chemical treatments. Phosphorus is precipitated out by combining with such chemicals as lime, alum, and ferric chloride. Filters of fine sands and activated charcoal remove small particulate matter and dissolved chemicals. Nitrogen is converted to ammonia

and discharged into the air in stripping towers. Some systems encourage denitrifying bacteria to form volatile nitrogen gas. Finally, the purified water is disinfected.

Tertiary treatment provides water that is suitable for drinking. The water is discharged into waterways, used to recharge groundwater, or used on crops.

### CHECK YOUR UNDERSTANDING

✔ **27-18** Which type of sewage treatment is designed to remove almost all phosphorus from sewage?

✔ **27-19** What metabolic group of anaerobic bacteria is especially encouraged by operation of a sludge digestion system?

✔ **27-20** What is the relationship between BOD and the welfare of fish?

### CLINICAL CASE Resolved

Improvements in water quality and sanitation are necessary to reduce transmission of cholera. Because cholera can progress quickly to severe dehydration, shock, and death, rapid rehydration is the mainstay of cholera treatment. However, rehydration therapy requires clean water, and water treatment must be inexpensive.

788   797   799   803   804   **805**

---

# Study Outline

 **Go to** @Mastering**Microbiology** for **Interactive Microbiology**, *In the Clinic videos, MicroFlix, MicroBoosters, 3D animations, practice quizzes, and more.*

## Microbial Diversity and Habitats (p. 787)

1. Microorganisms make up the Earth microbiome. They live in a wide variety of habitats because of their ability to use a variety of carbon and energy sources and to grow under different physical conditions.
2. Extremophiles live in extreme conditions of temperature, acidity, alkalinity, or salinity.

## Symbiosis (p. 787)

3. Symbiosis is a relationship between two different organisms or populations.
4. Symbiotic fungi called mycorrhizae live in and on plant roots; they increase the surface area and nutrient absorption of the plant.

## Soil Microbiology and Biogeochemical Cycles (pp. 787–795)

1. In biogeochemical cycles, certain chemical elements are recycled.
2. Microorganisms in the soil decompose organic matter and transform carbon-, nitrogen-, and sulfur-containing compounds into usable forms.

3. Microbes are essential to the continuation of biogeochemical cycles.
4. Elements are oxidized and reduced by microorganisms during these cycles.

## The Carbon Cycle (pp. 788–789)

5. Carbon dioxide is incorporated, or fixed, into organic compounds by photoautotrophs and chemoautotrophs.
6. These organic compounds provide nutrients for chemoheterotrophs.
7. Chemoheterotrophs release $CO_2$ that is then used by photoautotrophs.
8. Carbon is removed from the cycle when it is in $CaCO_3$ and fossil fuels.

## The Nitrogen Cycle (pp. 789–791)

9. Microorganisms decompose proteins from dead cells and release amino acids.
10. Ammonia is liberated by microbial ammonification of amino acids.
11. The nitrogen in ammonia is oxidized to produce nitrates for energy by nitrifying bacteria.

12. Denitrifying bacteria reduce the nitrogen in nitrates to molecular nitrogen ($N_2$).

13. $N_2$ is converted into ammonia by nitrogen-fixing bacteria including free-living genera such as *Azotobacter* and cyanobacteria, and the symbiotic bacteria *Rhizobium* and *Frankia*.

14. Ammonium and nitrate are used by bacteria and plants to synthesize amino acids that are assembled into proteins.

### The Sulfur Cycle (pp. 791–792)

15. Hydrogen sulfide ($H_2S$) is used by autotrophic bacteria; the sulfur is oxidized to form $S^0$ or $SO_4^{2-}$.

16. Plants, algae, and bacteria can reduce $SO_4^{2-}$ to make certain amino acids. These amino acids are in turn used by animals.

17. $H_2S$ is released by decay or dissimilation of these amino acids.

### Life without Sunshine (pp. 792–793)

18. Chemoautotrophs are the primary producers in deep-sea vents and within deep rocks.

### The Phosphorus Cycle (p. 793)

19. Phosphorus (as $PO_4^{3-}$) is found in rocks and bird guano.

20. When solubilized by microbial acids, the $PO_4^{3-}$ is available for plants and microorganisms.

### The Degradation of Synthetic Chemicals in Soil and Water (pp. 793–795)

21. Many synthetic chemicals, such as pesticides, are resistant to degradation by microbes.

22. The use of microorganisms to remove pollutants is called bioremediation.

23. Municipal landfills prevent decomposition of solid wastes because they are dry and anaerobic.

24. In some landfills, methane produced by methanogens can be recovered for an energy source.

## Aquatic Microbiology and Sewage Treatment (pp. 795–805)

### Aquatic Microorganisms (pp. 796–797)

1. The study of microorganisms and their activities in natural waters is called aquatic microbiology.

2. Natural waters include lakes, ponds, streams, rivers, estuaries, and the oceans.

3. The concentration of bacteria in water is proportional to the amount of organic material in the water.

4. Most aquatic bacteria tend to grow on surfaces rather than in a free-floating state.

5. The number and location of freshwater microbiota depend on the availability of oxygen and light.

6. Photosynthetic algae are the primary producers of a lake; they are found in the limnetic zone.

7. Microbes in stagnant water use available oxygen and can cause odors and the death of fish.

8. Purple and green sulfur bacteria are found in the profundal zone, which contains light and $H_2S$ but no oxygen.

9. *Desulfovibrio* reduces $SO_4^{2-}$ to $H_2S$ in benthic mud.

10. Methane-producing bacteria are also found in the benthic zone.

11. Phytoplankton are the primary producers of the open ocean.

12. *Pelagibacter ubique* is a decomposer in ocean waters.

13. Archaea predominate below 100 m.

14. Some algae and bacteria are bioluminescent. They possess the enzyme luciferase, which can emit light.

### The Role of Microorganisms in Water Quality (pp. 797–799)

15. Microorganisms are filtered from water that percolates into groundwater supplies.

16. Some pathogenic microorganisms are transmitted to humans in drinking and recreational waters.

17. Resistant chemical pollutants may be concentrated in animals in an aquatic food chain.

18. Nutrients such as phosphates cause algal blooms, which can lead to eutrophication of aquatic ecosystems.

19. Tests for the bacteriological quality of water are based on the presence of indicator organisms, the most common of which are coliforms.

20. Coliforms are aerobic or facultatively anaerobic, gram-negative, non–endospore-forming rods that ferment lactose with the production of gas within 48 hours of being placed in a medium at 35°C.

21. Fecal coliforms, predominantly *E. coli*, are used to indicate the presence of human feces.

### Water Treatment (pp. 799–800)

22. Drinking water is held in a holding reservoir long enough that suspended matter settles.

23. Flocculation treatment uses a chemical such as alum to coalesce and then settle colloidal material.

24. Filtration removes protozoan cysts and other microorganisms.

25. Drinking water is disinfected with chlorine to kill remaining pathogenic bacteria.

### Sewage (Wastewater) Treatment (pp. 800–805)

26. Domestic wastewater is called sewage; it includes household water, toilet wastes, and rainwater.

27. Primary sewage treatment is the removal of solid matter called sludge.

28. Biological activity is not very important in primary treatment.

29. Biochemical oxygen demand (BOD) is a measure of the biologically degradable organic matter in water.

30. Primary treatment removes about 25–35% of the BOD of sewage.

31. BOD is determined by measuring the amount of oxygen bacteria require to degrade the organic matter.

32. Secondary sewage treatment is the biological degradation of organic matter after primary treatment.

33. Activated sludge systems, trickling filters, and rotating biological contactors are methods of secondary treatment.

34. Microorganisms degrade the organic matter aerobically.

35. Secondary treatment removes up to 95% of the BOD.

36. Treated sewage is disinfected, usually by chlorination, before discharge onto land or into water.

37. Sludge is placed in an anaerobic sludge digester; bacteria degrade organic matter and produce simpler organic compounds, methane, and $CO_2$.

38. The methane produced in the digester is used to heat the digester and operate other equipment.

39. Excess sludge is periodically removed from the digester, dried, and disposed of (as landfill or soil conditioner) or incinerated.

40. Septic tanks can be used in rural areas to provide primary treatment of sewage.
41. Small communities can use oxidation ponds for secondary treatment.
42. These require a large area in which to build an artificial lake.
43. Tertiary sewage treatment uses physical filtration and chemical precipitation to remove all the BOD, nitrogen, and phosphorus from water.
44. Tertiary treatment provides drinkable water, whereas secondary treatment provides water usable only for irrigation.

# Study Questions

For answers to the Knowledge and Comprehension questions, turn to the Answers tab at the back of the textbook.

## Knowledge and Comprehension

### Review

1. The koala is a leaf-eating animal. What can you infer about its digestive system?
2. Give one possible explanation of why *Penicillium* would make penicillin, given that the fungus does not get bacterial infections.
3. In the sulfur cycle, microbes degrade organic sulfur compounds, such as (a)_____, to release $H_2S$, which can be oxidized by *Acidithiobacillus* to (b)_____. This ion can be assimilated into amino acids by (c)_____ or reduced by *Desulfovibrio* to (d)_____. $H_2S$ is used by photoautotrophic bacteria as an electron donor to synthesize (e)_____. The sulfur-containing by-product of this metabolism is (f)_____.
4. Why is the phosphorus cycle important?
5. DRAW IT Identify where the following processes occur: ammonification, decomposition, denitrification, nitrification, nitrogen fixation. Name at least one organism responsible for each process.

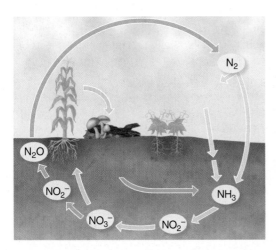

6. The following organisms have important roles as symbionts with plants and fungi; describe the symbiotic relationship of each organism with its host: cyanobacteria, mycorrhizae, *Rhizobium*, *Frankia*.
7. Outline the treatment process for drinking water.
8. The following processes are used in wastewater treatment. Match the stage of treatment with the processes. Each choice can be used once, more than once, or not at all.

| Processes | Treatment Stage |
|---|---|
| _____ a. Leaching field | 1. Primary |
| _____ b. Removal of solids | 2. Secondary |
| _____ c. Biological degradation | 3. Tertiary |
| _____ d. Activated sludge system | |
| _____ e. Chemical precipitation of phosphorus | |
| _____ f. Trickling filter | |
| _____ g. Results in drinking water | |

9. *Bioremediation* refers to the use of living organisms to remove pollutants. Describe three examples of bioremediation.
10. NAME IT These nitrogen-fixing prokaryotes provide nitrogen fertilizer in rice paddies; they live symbiotically in the cells of the freshwater plant *Azolla*.

## Multiple Choice

For questions 1–4, answer whether
  a. the process takes place under aerobic conditions.
  b. the process takes place under anaerobic conditions.
  c. the amount of oxygen doesn't make any difference.

1. Activated sludge system
2. Denitrification
3. Nitrogen fixation
4. Methane production
5. The water used to prepare intravenous solutions in a hospital contained endotoxins. Infection control personnel performed plate counts to find the source of the bacteria. Their results:

| | Bacteria/100 ml |
|---|---|
| Municipal water pipes | 0 |
| Boiler | 0 |
| Hot water line | 300 |

All of the following conclusions about the bacteria can be drawn *except* which one?
  a. They were present as a biofilm in the pipes.
  b. They are gram-negative.
  c. They come from fecal contamination.
  d. They come from the city water supply.
  e. none of the above

Use the following choices to answer questions 6–8:
  a. aerobic respiration
  b. anaerobic respiration
  c. anoxygenic photosynthesis
  d. oxygenic photosynthesis

6. $CO_2 + H_2S \xrightarrow{\text{Light}} C_6H_{12}O_6 + S^0$

7. $SO_4^{2-} + 10H^+ + 10e^- \longrightarrow H_2S + 4H_2O$

8. $CO_2 + 8H^+ + 8e^- \longrightarrow CH_4 + 2H_2O$

9. All of the following are effects of water pollution *except*
   a. the spread of infectious diseases.
   b. increased eutrophication.
   c. increased BOD.
   d. increased growth of algae.
   e. none of the above; all of these are effects of water pollution.

10. Coliforms are used as indicator organisms of sewage pollution because
   a. they are pathogens.
   b. they ferment lactose.
   c. they are abundant in human intestines.
   d. they grow within 48 hours.
   e. all of the above

## Analysis

1. Here are the formulas of two detergents that have been manufactured:

$$C—C—C—C—C—C—C—C—C—C\dots$$

$$
\begin{array}{c}
\quad\quad\quad\quad C \\
\quad\quad\quad\quad | \\
\quad\quad\quad\quad C \quad\quad C \\
\quad\quad\quad\quad | \quad\quad\quad | \\
C—C—C—C—C—C\dots \\
\quad\quad\quad | \\
\quad\quad\quad C
\end{array}
$$

   Which of these would be resistant, and which would be readily degraded by microorganisms? (*Hint:* Refer to the degradation of fatty acids in Chapter 5.)

2. Explain the effect of dumping each of the following into a pond on the eutrophication of the pond.
   a. Untreated sewage
   b. Sewage that has primary treatment
   c. Sewage that has secondary treatment
   Contrast your previous answers with the effect of each type of sewage on a fast-moving river.

## Clinical Applications and Evaluation

1. Flooding after two weeks of heavy rainfall in Tooele, Utah, preceded a high rate of diarrheal illness. *Giardia intestinalis* was isolated from 25% of the patients. A comparison study of a town 65 miles away revealed diarrheal illness in 2.9% of the 103 people interviewed. Tooele has a municipal water system and a municipal sewage treatment plant. Explain the probable cause of this epidemic and method(s) of stopping it. What would a fecal coliform test have shown?

2. The bioremediation process shown in the photograph is used to remove benzene and other hydrocarbons from soil contaminated by petroleum. The pipes are used to add nitrates, phosphates, oxygen, or water. Why are each of these added? Why is it not always necessary to add bacteria?

# Applied and Industrial Microbiology 28

In Chapter 27, we saw that microbes are an essential factor in many natural phenomena that make life possible on Earth. In this chapter we will look at how microorganisms are harnessed in such useful applications as the making of food and industrial products. Many of these processes—especially baking, winemaking, brewing, and cheesemaking—have origins long lost in history.

Modern civilization, with its large urban populations, could not be supported without methods of preserving food. In fact, civilization arose only after agriculture produced a year-round, stable food supply so that people were able to give up a nomadic hunting-and-gathering way of life.

In Chapter 9, we discussed industrial applications of genetically modified microorganisms that are at the cutting edge of our knowledge of molecular biology. Many of these applications are now essential to modern industry. With an increasing need for new antibiotics and environmentally safe production of chemicals, researchers are exploring the Earth microbiome for bacteria and fungi that produce useful products without genetic modification. In this chapter, we will explore the microbial production of foods, drugs, and chemicals. The Clinical Case shows the role of microbiologists in ensuring that pathogens such as *Salmonella* (in the photo) are not in foods.

◀ *Salmonella enterica* serovars are frequent causes of foodborne illness.

## In the Clinic

As a nurse with the CDC's Epidemic Intelligence Service, you are notified of two Minnesota patients with listeriosis. You query the database and find three more cases in Illinois, Indiana, and Ohio. You confirm that all five had eaten a soft-ripened Brie cheese from the same manufacturer. Pasteurized milk was used to make the cheese, and a wine-vinegar solution was brushed on the cheese daily during fermentation. **What causes listeriosis? How can contamination occur after pasteurization?**

*Hint: Read about food microbiology on pages 810–817.*

Answers to In the Clinic questions are found online at @MasteringMicrobiology.

# Food Microbiology

## LEARNING OBJECTIVES

**28-1** Describe thermophilic anaerobic spoilage and flat sour spoilage by mesophilic bacteria.

**28-2** Compare and contrast food preservation by industrial food canning, aseptic packaging, radiation, and high pressure.

**28-3** Name four beneficial activities of microorganisms.

Many of the methods of food preservation used today were probably discovered by chance in centuries past. People in early cultures observed that dried meat and salted fish resisted decay. Farmers learned that if grains were kept dry, they did not become moldy. Nomads must have noticed that soured animal milk resisted further decomposition and was still palatable. Moreover, if the curd of the soured milk was pressed to remove moisture and allowed to ripen (in effect, cheesemaking), it was even more effectively preserved and tasted better. And some plants, such as olives and cocoa, were found to be more palatable after "spoilage" (fermentation).

The foods listed in Table 28.1 will give you an idea of the worldwide use of fermentation to preserve foods or make them edible.

## Foods and Disease

As more food products are being prepared at central facilities and widely distributed, it is becoming more likely that food, like municipal water supplies, might be a source of widespread disease outbreaks. To minimize the potential for disease outbreaks, communities have established local agencies whose role is to inspect dairies and restaurants. The United States Food and Drug Administration (FDA) and Department of Agriculture (USDA) also maintain a system of inspectors at ports and central processing locations. The **Hazard Analysis and Critical Control Point (HACCP)** system is designed to prevent contamination by identifying points at which foods are most likely to be contaminated with harmful microbes. Monitoring of these control points can prevent such microbes from being introduced or, if they are present, arrest their proliferation. For

**TABLE 28.1  Some Fermented Foods**

| | | |
|---|---|---|
| **Dairy Products** | | |
| Cheeses (ripened) | Milk curd | *Streptococcus* spp., *Leuconostoc* spp., *Propionibacterium* spp. |
| Kefir | Milk | *Streptococcus lactis, Lactobacillus bulgaricus, Candida* spp. |
| Kumiss | Mare's milk | *Lactobacillus bulgaricus, L. leichmannii, Candida* spp. |
| Yogurt | Milk, milk solids | *Streptococcus thermophilus, L. bulgaricus* |
| **Meat and Fish Products** | | |
| Country-cured hams | Pork hams | *Aspergillus, Penicillium* spp. |
| Dry sausages | Pork, beef | *Pediococcus cerevisiae* |
| Fish sauces | Small fish | Halophilic *Bacillus* spp. |
| **Nonbeverage Plant Products** | | |
| Cocoa beans (chocolate) | Cacao fruits (pods) | *Candida krusei, Geotrichum* spp. |
| Coffee beans | Coffee cherries | *Erwinia dissolvens, Saccharomyces* spp. |
| Kimchi | Cabbage and other vegetables | Lactic acid bacteria |
| Miso | Soybeans | *Aspergillus oryzae, Zygosaccharomyces rouxii* |
| Olives | Green olives | *Leuconostoc mesenteroides, Lactobacillus plantarum* |
| Poi | Taro roots | Lactic acid bacteria |
| Sauerkraut | Cabbage | *Leuconostoc mesenteroides, Lactobacillus plantarum* |
| Soy sauce | Soybeans | *A. oryzae* or *A. sojae, Z. rouxii, Lactobacillus delbrueckii* |
| **Breads** | | |
| Bread | Wheat flours | *Saccharomyces cerevisiae* |
| San Francisco sourdough bread | Wheat flour | *Saccharomyces exiguus, Lactobacillus sanfranciscensis* |

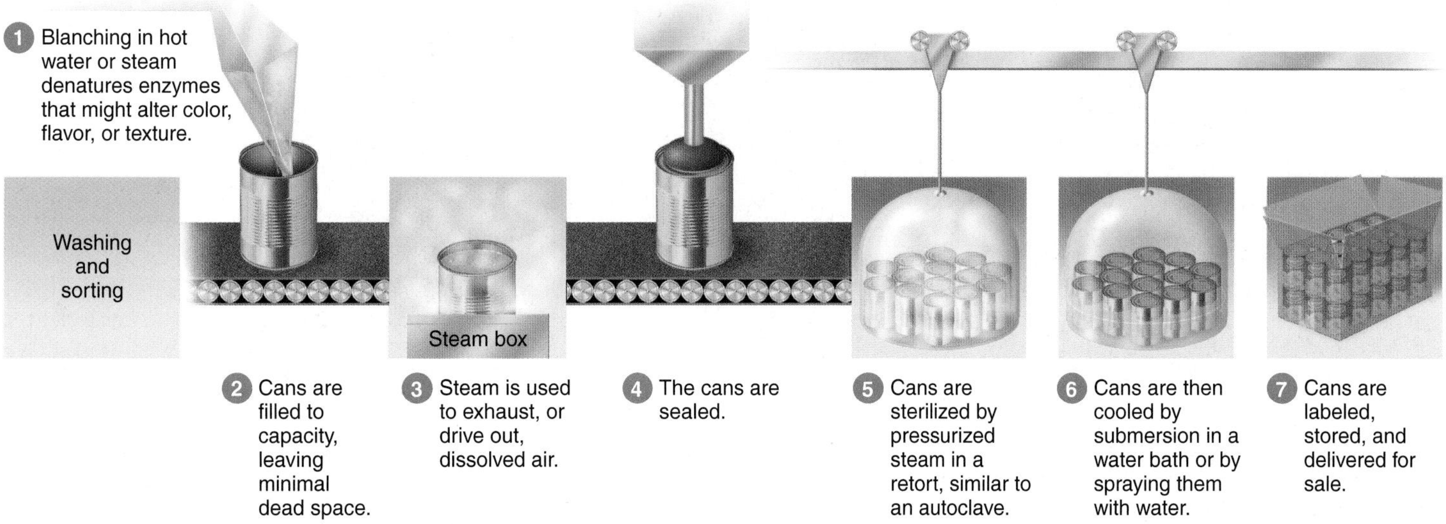

① Blanching in hot water or steam denatures enzymes that might alter color, flavor, or texture.

Washing and sorting

Steam box

② Cans are filled to capacity, leaving minimal dead space.

③ Steam is used to exhaust, or drive out, dissolved air.

④ The cans are sealed.

⑤ Cans are sterilized by pressurized steam in a retort, similar to an autoclave.

⑥ Cans are then cooled by submersion in a water bath or by spraying them with water.

⑦ Cans are labeled, stored, and delivered for sale.

**Figure 28.1  The commercial sterilization process in industrial canning.**

**Q** How does commercial sterilization differ from complete sterilization?

example, the HACCP system can identify steps during processing at which meats or fresh produce are likely to become contaminated. The HACCP system also requires monitoring of adequate temperatures to kill pathogens during processing and adequate storage temperatures to prevent their reproduction.

After harvest, foods must be preserved to prevent spoilage and growth of pathogens. One bacteriocin produced by *Lactococcus lactis*, nisin, is widely used as a food preservative. Canning (discussed below), drying, and refrigeration (Chapter 7, page 185) are other common methods of preserving food.

## Industrial Food Canning

In Chapter 7, you learned that preserving foods by heating a properly sealed container, as in home canning, is not difficult. The challenge in commercial canning is to use the right amount of heat necessary to kill spoilage organisms and dangerous microbes, such as the endospore-forming *Clostridium botulinum*, without degrading the appearance and palatability of food. Thus, much research is applied to determining the exact minimum heat treatment that will accomplish both these goals.

Industrial food canning is much more technically sophisticated than home canning (**Figure 28.1**). Industrially canned goods undergo **commercial sterilization** by steam under pressure in a large **retort**, which operates on the same principle as an autoclave (see Figure 7.2, page 183). Commercial sterilization is intended to destroy *C. botulinum* endospores and is not as rigorous as complete sterilization. The reasoning is that if *C. botulinum* endospores are destroyed, then any other significant spoilage or pathogenic bacteria will also be destroyed.

To ensure commercial sterilization, enough heat is applied for the **12D treatment** (12-decimal reductions, or *botulinal cook*),

by which a theoretical population of *C. botulinum* endospores would be decreased by 12 logarithmic cycles. (See Figure 7.1, page 181, and Table 7.2, page 180.) What this means is that if there were $10^{12}$ (1,000,000,000,000) endospores in a can, after treatment there would be only one survivor. Because $10^{12}$ is an improbably large population, this treatment is considered quite safe. Certain thermophilic endospore-forming bacteria have endospores that are more resistant to heat treatment than those of *C. botulinum*. However, these bacteria are obligate thermophiles and generally remain dormant at temperatures lower than about 45°C. Therefore, they are not a spoilage problem at normal storage temperatures.

### Spoilage of Canned Food

If canned foods are incubated at high temperatures, such as in a truck in the hot sun or next to a steam radiator, the thermophilic bacteria that often survive commercial sterilization

---

**CLINICAL CASE**   Dr. Chang and the Chocolate Factory

Dr. Derrick Chang of the CDC is alerted by PulseNet, the national molecular subtyping network for foodborne disease surveillance. PulseNet has identified an increase of genetically identical *Salmonella* Typhimurium in the United States. This increase shows 120 isolates from 23 states in the last 60 days.

**What is causing this outbreak? Read on to find out.**

**811**  813  816  820  822  824

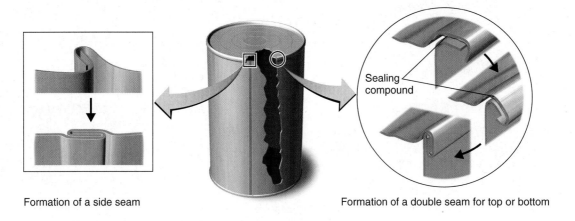

Formation of a side seam

Formation of a double seam for top or bottom

**Figure 28.2 The construction of a metal can.** Notice the seam construction, which was introduced about 1904. During cooling after sterilization (see Figure 28.1, step 6), the vacuum formed in the can may actually force contaminating organisms into the can along with water.

**Q** Why isn't the can sealed before it is placed in the steam box?

can germinate and grow. **Thermophilic anaerobic spoilage** is therefore a fairly common cause of spoilage in low-acid canned foods. The can usually swells from gas, and the contents have a lowered pH and a sour odor. A number of thermophilic species of *Clostridium* can cause this type of spoilage. When thermophilic spoilage occurs but the can is not swollen by gas production, the spoilage is termed **flat sour spoilage**. This type of spoilage is caused by thermophilic organisms such as *Geobacillus stearothermophilus* (jĕ'ō-bah-SIL-lus ste-rō'ther-MAH-fil-us), which is found in the starch and sugars used in food preparation. Many industries have standards for the numbers of such thermophilic bacteria permitted in raw materials. Both types of spoilage occur only when the cans are stored at higher-than-normal temperatures, which permits the growth of bacteria whose endospores are not destroyed by normal processing.

Mesophilic bacteria can spoil canned foods if the food is underprocessed or if the can leaks. Underprocessing is more likely to result in spoilage by endospore formers; the presence of non–endospore-forming bacteria strongly suggests that the can leaks. Leaking cans are often contaminated during the cooling of cans after processing by heat. The hot cans are sprayed with cooling water or passed through a trough filled with water. As the can cools, a vacuum is formed inside, and external water can be sucked through a leak past the heat-softened sealant in the crimped lid (**Figure 28.2**). Contaminating bacteria in the cooling water are drawn into the can with the water. Spoilage from underprocessing or can leakage is likely to produce odors of putrefaction, at least in high-protein foods, and occurs at normal storage temperatures. In such types of spoilage, there is always the potential that botulinal bacteria will be present.

Some acidic foods, such as tomatoes or preserved fruits, are preserved by processing temperatures of 100°C or lower. The reasoning is that the only spoilage organisms that will grow in such acidic foods are easily killed by even 100°C temperatures. Primarily, these would be molds, yeasts, and certain vegetative bacteria.

Occasional problems in acidic foods develop from a few microorganisms that are both heat resistant and acid tolerant. Examples of heat-resistant fungi are the mold *Byssochlamys fulva* (bis-sō-KLA-mis FUL-vah), which produces a *heat-resistant ascospore*, and a few molds, especially species of *Aspergillus* (as-per-JIL-lus), that sometimes produce specialized resistant bodies called *sclerotia*. A spore-forming bacterium, *Bacillus coagulans* (bah-SIL-lus kō-AG-ū-lanz), is unusual in that it is capable of growth at a pH of almost 4.0.

## Aseptic Packaging

The use of **aseptic packaging** to preserve food has been increasing. Packages are usually made of some material that cannot tolerate conventional heat treatment, such as laminated paper or plastic. The packaging materials come in continuous rolls that are fed into a machine that sterilizes the material with a hot hydrogen peroxide solution, sometimes aided by ultraviolet (UV) light. Metal containers can be sterilized with superheated steam or other high-temperature methods. High-energy electron beams can also be used to sterilize the packaging materials. While still in the sterile environment, the material is formed into packages, which are then filled with liquid foods that have been conventionally sterilized by heat. The filled package is not sterilized after it is sealed.

## Radiation and Industrial Food Preservation

It has long been recognized that irradiation is lethal to microorganisms; in fact, a patent was issued in Great Britain in 1905 for the use of ionizing radiation to improve the condition of foodstuffs. X rays were specifically suggested in 1921 as a way

**TABLE 28.2 Approximate Doses of Radiation Needed to Kill Various Organisms (Prions Are Not Affected)**

| Organisms | Dose (kGy)* |
|---|---|
| Higher animals (whole body) | 0.005–0.1 |
| Insects | 0.01–1 |
| Non–endospore-forming bacteria | 0.5–10 |
| Bacterial endospores | 10–50 |
| Viruses | 10–200 |

*Gray is a measure of ionizing irradiation; kGy is 1000 Grays.

*Source:* J. Farkas, "Physical Methods of Food Preservation," in *Food Microbiology: Fundamentals and Frontiers*, 2d ed., M.P. Doyle et al. (Eds) (Washington, DC: ASM Press, 2001).

to inactivate the larvae in pork that are the cause of trichinellosis. Ionizing irradiation inhibits DNA synthesis and effectively prevents microorganisms, insects, and plants from reproducing. The ionizing irradiation is usually X rays or the gamma rays produced by radioactive cobalt-60. Up to certain energy levels, high-energy electrons produced by electron accelerators are also used. The main practical difference is in penetration capabilities. These sources inactivate the target organisms and do *not* induce radioactivity in the food or packaging material. The relative doses of radiation needed to kill various organisms are presented in Table 28.2. Radiation is measured in *Grays*, named for an early radiologist—often in terms of thousands of Grays, abbreviated as kGy.

- *Low doses of irradiation (less than 1 kGy)* are used to kill insects (disinfestation) and inhibit sprouting, as in stored potatoes. Similarly, it can delay ripening of fruits during storage.

- *Pasteurizing doses (1 to 10 kGy)* can be used on meats and poultry to eliminate or critically reduce the numbers of specific bacterial pathogens.

- *High doses (more than 10 kGy)* are used to sterilize, or at least greatly lower, the bacterial populations in many spices. Spices are often contaminated with 1 million or more bacteria per gram, although these are not considered to be normally hazardous to health.

A specialized use of irradiation has been to sterilize meats eaten by American astronauts, and a few health facilities have selectively used irradiation to sterilize foods ingested by immunocompromised patients. Millions of implanted medical devices, such as pacemakers, have been irradiated. Irradiated food is marked in the United States with a radura symbol (⊛) and a printed notice. Unfortunately, this symbol has often been interpreted as a warning rather than the description of an approved processing treatment or preservative. In fact, irradiated foods are not radioactive; consider that the X-ray table in a hospital does not become radioactive from repeated daily exposure to ionizing radiation. Recently, the FDA has

Dr. Chang initiates a case-control study with representatives of the state health departments that had reported S. Typhimurium infections. Fifteen items, suspected as possible vehicles of infection on the basis of the individual case investigations, are listed. State officials determine whether each suspect item was used or consumed by the infected person within the 3 days before onset of illness. The family of each patient identifies two neighborhood controls of the same age and gender as the patient. Controls were asked the same questions as patients, except that they were questioned about the use or consumption of the 15 suspect items during the previous month. Some of the data collected are shown in the table.

| Foil-Wrapped Chocolate Balls | Cases | Controls |
|---|---|---|
| Ate | 38 | 12 |
| Did not eat | 7 | 79 |

**Calculate the odds ratio for this food item. (*Hint:* See page 731.)**

811　**813**　816　820　822　824

allowed, upon special approval, substitution of language such as "pasteurization" rather than "irradiation."

When deep penetration is a requirement, the preferred method for irradiation is gamma rays produced by cobalt-60. However, this type of treatment requires several hours of exposure in isolation behind protective walls (**Figure 28.3**).

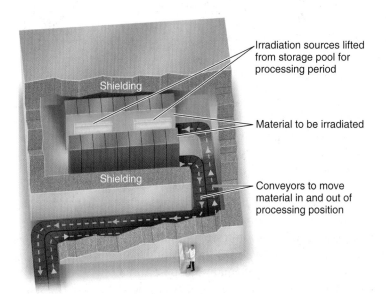

**Figure 28.3 A gamma-ray irradiation facility.** Shown is the path of the material to be irradiated.

Irradiation sources lifted from storage pool for processing period

Material to be irradiated

Conveyors to move material in and out of processing position

Shielding

Shielding

**Q** Can microwaves be used to sterilize foods?

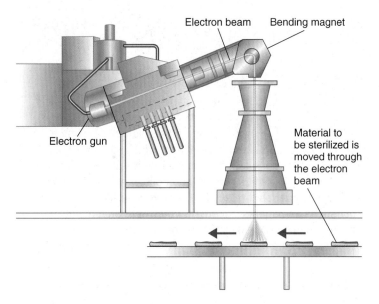

Electron beam    Bending magnet

Electron gun

Material to
be sterilized is
moved through
the electron
beam

**Figure 28.4 Electron-beam accelerator.** These machines
generate an electron stream that is accelerated down a long tube by
electromagnets of the opposite charge. In the drawing, the electron
beam is bent by a "bending magnet." This serves to filter out electrons
of unwanted energy levels, providing a beam of uniform energy. The
vertical beam is swept back and forth over the target as it is moved
past the beam. The penetrating power of the beam is limited: if the
target substance is expressed as an equivalent thickness of water, the
maximum is about 3.9 cm (1.5 in). In contrast, X rays will penetrate
about 23 cm (9 in).

**Q** Are high-energy electrons ionizing radiation?

High-energy electron accelerators (**Figure 28.4**) are much
faster and sterilize in a few seconds, but this treatment has low
penetrating power and is suitable only for sliced meats, bacon,
or similar thin products. Also, plasticware used in microbiology is usually sterilized in this way. Another recent application
is to irradiate mail to kill possible bioterrorism agents that it
might contain, such as anthrax endospores.

## High-Pressure Food Preservation

A recent development in food preservation has been the use
of a high-pressure processing technique (*pascalization*). Pre-
wrapped foods such as fruits, deli meats, and precooked
chicken strips are submerged into tanks of pressurized water.
The pressure can reach 87,000 pounds per square inch (psi)—
which has been compared to the equivalent of about three
elephants standing on a dime. This process kills many bacteria, such as *Salmonella*, *Listeria*, and pathogenic strains of
*E. coli*, by disrupting many cellular functions. It also kills
nonpathogenic microorganisms that tend to shorten the shelf
life of such products.

Because the process does not require additives, it does not
require regulatory approval. It has the advantage of preserving
colors and tastes of foods better than many other methods and
does not provoke the concerns of irradiation.

## The Role of Microorganisms in Food Production

In the latter part of the nineteenth century, microbes used in
food production were grown in pure culture for the first time.
This development quickly led to an improved understanding
of the relationships between specific microbes and their products and activities. This period can be considered the beginning of industrial food microbiology. For example, once it was
understood that a certain yeast grown under certain conditions produced beer and that certain bacteria could spoil the
beer, brewers were better able to control the quality of their
products. Specific industries became active in microbiological
research and selected certain microbes for their special qualities. The brewing industry extensively investigated the isolation and identification of yeasts and selected those that could
produce more alcohol. In this section, we will discuss the role
of microorganisms in the production of several common foods.

### Cheese

The United States leads the world in the making of cheese, producing millions of tons every year. Although there are many
types of cheeses, all require the formation of a **curd**, which
can be separated from the main liquid fraction, or **whey**
(**Figure 28.5**). The curd is made up of a protein, **casein**, and is
usually formed by the action of an enzyme, **rennin** (or chy-
mosin), which is aided by acidic conditions provided by certain lactic acid–producing bacteria. These inoculated lactic
acid bacteria also provide the characteristic flavors and aromas
of fermented dairy products during the ripening process. The
curd undergoes a microbial ripening process, except in the case
of a few unripened cheeses, such as ricotta and cottage cheese.

The milk has been coagulated by the action of rennin (forming curd) and is
inoculated with bacteria for flavor and acidity. The curd is being cut to allow
whey to drain out.

**Figure 28.5  Making cheddar cheese.**

**Q** Are there living bacteria in the final cheese product?

Cheeses are generally classified by their hardness, which is produced in the ripening process. The more moisture lost from the curd and the more the curd is compressed, the harder the cheese.

The hard cheddar and Swiss cheeses are ripened by lactic acid bacteria growing anaerobically in the interior. Such hard, interior-ripened cheeses can be quite large. The longer the incubation time, the higher the acidity and the sharper the taste of the cheese. A *Propionibacterium* (prō-pē′on-ē-bak-TI-rē-um) species produces carbon dioxide, which forms the holes in Swiss cheese. Semisoft cheeses, such as Limburger, are ripened by bacteria and other contaminating organisms growing on the surface. Blue and Roquefort cheeses are ripened by *Penicillium* molds inoculated into the cheese. The texture of the cheese is loose enough that adequate oxygen can reach the aerobic molds. The growth of the *Penicillium* molds is visible as blue-green clumps in the cheese. Camembert, a soft cheese, is ripened in small packets so that the enzymes of *Penicillium* mold growing aerobically on the surface will diffuse into the cheese for ripening.

### Other Dairy Products

*Butter* is made by churning cream until the fatty globules of butter separate from the liquid *buttermilk*. The typical flavor and aroma of butter and buttermilk are from *diacetyls*, a combination of two acetic acid molecules that is a metabolic end-product of fermentation by some lactic acid bacteria. Today, commercially sold buttermilk is usually not a by-product of buttermaking but is made by inoculating skim milk with bacteria that form lactic acid and the diacetyls. *Cultured sour cream* is made from cream inoculated with microorganisms similar to those used to make buttermilk.

Yogurt, a slightly acidic dairy product, is found around the world and is popular in the United States. Commercial yogurt is made from milk, from which at least one-fourth of the water has been evaporated in a vacuum pan. The resulting thickened milk is inoculated with a mixed culture of *Streptococcus thermophilus*, primarily for acid production, and *Lactobacillus delbrueckii bulgaricus* (bul-GAR-i-kus), to contribute flavor and aroma. The temperature of the fermentation is about 45°C for several hours, during which time *S. thermophilus* outgrows *L. d. bulgaricus*. Maintaining the proper balance between the flavor-producing and the acid-producing microbes is the secret of making yogurt.

*Kefir* and *kumiss* are fermented milk beverages that are popular in eastern Europe. The usual lactic acid–producing bacteria are supplemented with a lactose-fermenting yeast to give these drinks an alcohol content of 1–2%.

### Nondairy Fermentations

Historically, milk fermentation allowed dairy products to be stored and then consumed much later. Other microbial fermentations were used to make certain plants edible. For example, pre-Columbian people in Central and South America learned to ferment cacao beans before consumption. The microbial products released during fermentation produce the chocolate flavor.

Microorganisms are also used in baking, especially for bread. The sugars in bread dough are fermented by yeasts. The species of yeast used in baking is *Saccharomyces cerevisiae*. This same species of yeast is also used in the brewing of beer from grains and the fermentation of wines from grapes. (At one time *S. cerevisiae* was classified as multiple species, such as *S. pastorianus*, *S. uvarum*, and *S. c. ellipsoideus*; these and a few other species names are often encountered in older literature.) *S. cerevisiae* will grow readily under both aerobic and anaerobic conditions, although, unlike facultatively anaerobic bacteria such as *E. coli*, it cannot grow anaerobically indefinitely. Various strains of *S. cerevisiae* have been developed over the centuries and are highly adapted to certain fermentation uses.

Anaerobic conditions for producing ethanol by the yeasts are mandatory for producing alcoholic beverages. In baking, carbon dioxide forms the typical bubbles of leavened bread. Aerobic conditions favor carbon dioxide production and are encouraged as much as possible. This is the reason bread dough is kneaded repeatedly. Whatever ethanol is produced evaporates during baking. In some breads, such as rye or sourdough, the growth of lactic acid bacteria produce the typical tart flavor.

Fermentation is also used in the production of such foods as *sauerkraut*, *pickles*, *olives*, and even salami and coffee, in which the beans undergo a fermentation step.

### Alcoholic Beverages and Vinegar

Microorganisms are involved in the production of almost all alcoholic beverages. Beer is the product of grain starches fermented by yeast. Lager is fermented slowly with yeast strains that remain on the bottom (*bottom yeasts*). Ale is fermented relatively rapidly, at a higher temperature, with yeast strains that usually form clumps that are buoyed to the top by $CO_2$ (*top yeasts*). Because yeasts are unable to use starch directly, the starch from grain must be converted to glucose and maltose, which the yeasts can ferment into ethanol and carbon dioxide. In this conversion, called **malting**, starch-containing grains, such as malting barley, are allowed to sprout and then are dried and ground. This product, called **malt**, contains starch-degrading enzymes (amylases) that convert cereal starches into carbohydrates that can be fermented by yeasts. Light beers use amylases or selected strains of yeast to convert more of the starch to fermentable glucose and maltose, resulting in fewer carbohydrates and more alcohol. The beer is then diluted to arrive at an alcohol percentage in the usual range. **Sake**, the Japanese rice wine (which is actually a beer), is made from rice without malting because the mold

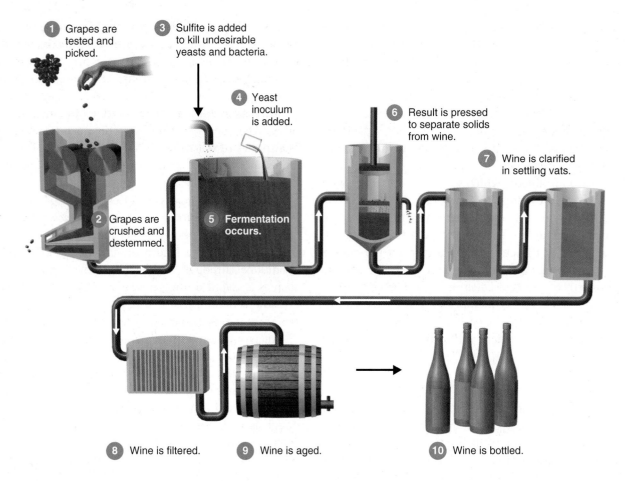

① Grapes are tested and picked.

② Grapes are crushed and destemmed.

③ Sulfite is added to kill undesirable yeasts and bacteria.

④ Yeast inoculum is added.

⑤ Fermentation occurs.

⑥ Result is pressed to separate solids from wine.

⑦ Wine is clarified in settling vats.

⑧ Wine is filtered.

⑨ Wine is aged.

⑩ Wine is bottled.

**Figure 28.6 The basic steps in making red wine.** For white wines, the pressing precedes fermentation so that the color is not extracted from the solid matter.

**Q** What happens if air enters at step 5? At step 10?

*Aspergillus* is first used to convert the rice's starch to sugars that can be fermented. (See the discussion of koji, page 819.) For *distilled spirits*, such as *whiskey*, *vodka*, and *rum*, carbohydrates from cereal grains, potatoes, and molasses are fermented to alcohol. The alcohol is then distilled to make a concentrated alcoholic beverage.

*Wines* are made from fruits, typically grapes, that contain sugars that yeasts can use directly for fermentation; malting is unnecessary in winemaking. Grapes usually need no additional sugars, but other fruits might be supplemented with sugars to ensure enough alcohol production. The steps of winemaking are shown in **Figure 28.6**. Lactic acid bacteria are important when wine is made from grapes that are especially acidic from high concentrations of malic acid. These bacteria convert the malic acid to the weaker lactic acid in a process called **malolactic fermentation**. The result is a less acidic, better-tasting wine than would otherwise be produced.

Wine producers who allowed wine to be exposed to air found that it soured from the growth of aerobic bacteria that converted the ethanol in the wine to acetic acid. The result was *vinegar* (*vin* = wine; *aigre* = sour). The process is now used deliberately to make vinegar. Ethanol is first produced by anaerobic fermentation of carbohydrates by yeasts. The ethanol is then aerobically oxidized to acetic acid by acetic acid–producing bacteria of the genera *Acetobacter* and *Gluconobacter*.

### CLINICAL CASE

Illness due to S. Typhimurium infection has a high association with consumption of foil-wrapped chocolate balls (odds ratio = 35.7). Dr. Chang initiates environmental testing and traceback to locate the source of contamination. From questioning families and examining store invoices, investigators identify the specific chocolate item (identified by the manufacturer's code number). State health department laboratories find that at least 22 of these chocolate samples contain S. Typhimurium.

**How will Dr. Chang find the source of contamination?**

811  813  **816**  820  822  824

# Industrial Microbiology and Biotechnology

## LEARNING OBJECTIVES

**28-4** Define *industrial fermentation* and *bioreactor*.

**28-5** Differentiate primary from secondary metabolites.

**28-6** Describe the role of microorganisms in the production of industrial chemicals and pharmaceuticals.

**28-7** Define *bioconversion*, and list its advantages.

**28-8** List biofuels that can be made by microorganisms.

The word **biotechnology** was first used in 1918 to describe the use of living organisms to produce products—in reference to combining agriculture and technology. The industrial uses of microorganisms had their beginnings in large-scale food fermentations that produced lactic acid from dairy products and ethanol from brewing. These two chemicals also proved to have many industrial uses unrelated to foods. During World Wars I and II, microbial fermentation and similar technologies were used in the production of armament-related chemical compounds, such as butanol and acetone. Present industrial microbiology dates largely from the technology developed to produce antibiotics following World War II. There is renewed interest in some of these classic microbial fermentations, especially if they can be used as feedstocks, can be used for products that are renewable, or, ideally, use products that would otherwise be wasted.

In recent years, biotechnology has been revolutionized by the application of genetically modified organisms. In Chapter 9, we discussed the methods for making these modified organisms using recombinant DNA technology and described some of the products derived from them.

## Fermentation Technology

The industrial production of microbial products usually involves fermentation. *Industrial fermentation* is the large-scale cultivation of microbes or other single cells to produce a commercially valuable substance. We have just discussed the most familiar examples: the anaerobic food fermentations used in the dairy, brewing, and winemaking industries. Much of the same technology, with the frequent addition of aeration, has been adapted to make other industrial products, such as insulin and human growth hormone, from genetically modified microorganisms. Industrial fermentation is also used in biotechnology to obtain

useful products from genetically modified plant and animal cells (see Table 9.2, page 256). For example, animal cells are used to make monoclonal antibodies (see Chapter 18, pages 508–509).

Vessels for industrial fermentation are called **bioreactors** (**Figure 28.7a**); they are designed with close attention to aeration, pH control, and temperature control. There are many different designs, but the most widely used bioreactors are of the continuously stirred type (Figure 28.7b). The air is introduced through a

**(a)** Bioreactors like this are used to produce antibiotics.

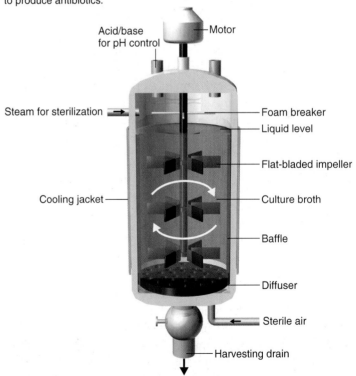

**(b)** Section of a continuously stirred bioreactor for industrial fermentations.

**Figure 28.7 Bioreactors are used for large-scale growth of microorganisms.** Environmental conditions such as pH and aeration can be adjusted to maximize the desired product made by the microbe.

**Q** Identify one essential difference between the bioreactor illustrated and a vat for making wine.

diffuser at the bottom (which breaks up the incoming airstream to maximize aeration), and a series of impeller paddles and stationary wall baffles keep the microbial suspension agitated. Oxygen is not very soluble in water, and keeping the heavy microbial suspension well aerated is difficult. Highly sophisticated designs have been developed to achieve maximum efficiency in aeration and other growth requirements, including medium formulation. The high value of the products of genetically modified microorganisms and eukaryotic cells has stimulated the development of newer types of bioreactors and computerized controls for them.

Bioreactors are sometimes very large, holding as much as 500,000 liters. When the product is harvested at the completion of the fermentation, this is known as *batch production*. There are other designs of fermentors. For *continuous flow production*, in which the substrates (usually a carbon source) are fed continuously past immobilized enzymes or into a culture of growing cells, spent medium and desired product are continuously removed.

Generally speaking, the microbes in industrial fermentation produce either primary metabolites, such as ethanol, or secondary metabolites, such as penicillin. A **primary metabolite** is formed essentially at the same time as the new cells, and the production curve follows the cell population curve almost in parallel, with only minimal lag. **Secondary metabolites** are not produced until the microbe has largely completed its logarithmic growth phase, known as the **trophophase**, and has entered the stationary phase of the growth cycle. The following period, during which most of the secondary metabolite is produced, is known as the **idiophase**. The secondary metabolite may be a microbial conversion of a primary metabolite. Alternatively, it may be a metabolic product of the original growth medium that the microbe makes only after considerable numbers of cells and a primary metabolite have accumulated. Cellular metabolism leaves behind small-molecule chemical fingerprints of the cellular processes: a metabolic profile. The use of these chemical fingerprints to study cellular processes involving metabolites is termed **metabolomics**.

Strain improvement is also an ongoing activity in industrial microbiology. (A microbial **strain** differs physiologically in some significant way. For example, it has an enzyme to carry out some additional activity or lacks such an ability, but this difference is not enough to change its species identity.) A well-known example is that of the mold used for penicillin production. The original culture of *Penicillium* did not produce penicillin in large enough quantities for commercial use. A more efficient culture was isolated from a moldy cantaloupe from a Peoria, Illinois, supermarket. This strain was treated variously with UV light, X rays, and nitrogen mustard (a chemical mutagen). Selections of mutants, including some that arose spontaneously, quickly increased the production rates by a factor of more than 100. Today, the original penicillin-producing molds produce, not the original 5 mg/L, but 60,000 mg/L. Improvements in fermentation techniques have nearly tripled even this yield.

## Immobilized Enzymes and Microorganisms

In many ways, microbes are packages of enzymes. Industries are increasing their use of free enzymes isolated from microbes to manufacture many products, such as high-fructose corn syrups, paper, and textiles. The demand for such enzymes is high because they are specific and do not produce costly or toxic waste products. And, unlike traditional chemical processes that require heat or acids, enzymes work under moderate conditions and are safe and biodegradable. For most industrial purposes, the enzyme must be immobilized on the surface of some solid support or otherwise manipulated so that it can convert a continuous flow of substrate to product without being lost.

Continuous flow techniques have also been adapted to live whole cells and sometimes even to dead cells (**Figure 28.8**). Whole-cell systems are difficult to aerate, and they lack the single-enzyme specificity of immobilized enzymes. However, whole cells are advantageous if the process requires a series of steps that can be carried out by one microbe's enzymes. They also have the advantage of allowing continuous flow processes with large cell populations operating at high reaction rates. Immobilized cells, which are usually anchored to microscopically small spheres or fibers, are currently used to make high-fructose corn syrup, aspartic acid, and numerous other products of biotechnology.

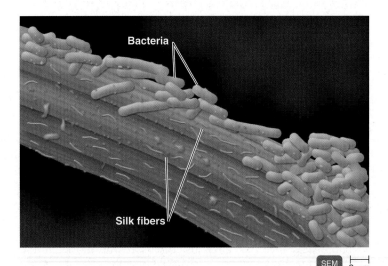

SEM   2 µm

**Figure 28.8 Immobilized cells.** In some industrial processes, the cells are immobilized on surfaces such as the silk fibers shown here. The substrate flows past the immobilized cells.

**Q** How does this process resemble the action of a trickling filter in sewage treatment?

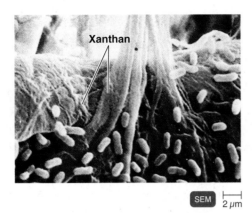

SEM | 2 μm

**Figure 28.9 *Xanthomonas campestris*, producing gooey xanthan.**

**Q** What product can you find at home or in the supermarket that contains xanthan?

## Industrial Products

As mentioned earlier, cheesemaking produces an organic waste called whey. The whey must be disposed of as sewage or dried and burned as solid waste. Both of these processes are costly and ecologically problematic. However, microbiologists have discovered an alternative use for whey. A research team working with the USDA used an enrichment technique to isolate a plant pathogen, *Xanthomonas campestris* (zan'thō-MŌ-nas kam-PES-tris) bacterium, that uses the lactose in whey and produces xanthan. First, they inoculated a whey medium with *X. campestris* and incubated it for 24 hours. Then they transferred an inoculum of this culture to a flask of lactose broth, to select a lactose-utilizing cell. The strain did not have to make xanthan from this broth; it only had to grow and use lactose. A lactose-utilizing strain was isolated through serial transfers, selecting for the strain with the best ability to grow. After incubation for 10 days, an inoculum was transferred to another flask of lactose broth, and the procedure was repeated two more times. When transferred to a flask of whey medium, the final lactose-utilizing bacteria grew in the whey and produced xanthan. The bacterium can convert 40 g/L of whey powder into 30 g/L of xanthan gum (**Figure 28.9**). The xanthan gum is used to thicken a wide variety of products from salad dressing to shampoo.

In this way, microbiologists are devising uses for wastes and creating new products. In this section, we will discuss some of the more important commercial microbial products and the growing alternative energy industry.

### Amino Acids

Amino acids have become a major industrial product from microorganisms. For example, over 1 million tons of *glutamic acid* (L-glutamate), used to make the flavor enhancer monosodium glutamate, are produced every year. Certain amino acids,

such as *lysine* and *methionine*, cannot be synthesized by animals and are present only at low levels in the normal diet. Therefore, the commercial synthesis of lysine and some of the other essential amino acids as cereal food supplements is an important industry. More than 250,000 tons each of lysine and methionine are produced every year.

Two microbially synthesized amino acids, *phenylalanine* and *aspartic acid* (L-aspartate), have become important as ingredients in the sugar-free sweetener aspartame (NutraSweet®). Some 7000 to 8000 tons of each of these amino acids are produced annually in the United States. Because only the L-isomer of an amino acid is desired, microbial production, which forms only the L-isomer, has an advantage over chemical production, which forms both the D-isomer and the L-isomer (see Figure 2.13, page 40).

In nature, microbes rarely produce amino acids in excess of their own needs because feedback inhibition prevents wasteful production of primary metabolites (see Chapter 5, page 116). Commercial microbial production of amino acids depends on specially selected mutants and sometimes on ingenious manipulations of metabolic pathways.

### Citric Acid

*Citric acid* is a constituent of citrus fruits, such as oranges and lemons, and at one time these were its only industrial source. However, over 100 years ago, citric acid was identified as a product of mold metabolism. This discovery was first used as an industrial process when World War I interfered with the picking of the Italian lemon crop. Citric acid has an extraordinary range of uses beyond the obvious ones of giving tartness and flavor to foods. It is an antioxidant and pH adjuster in many foods, and in dairy products it often serves as an emulsifier. Over 1.6 million tons of citric acid are produced every year worldwide. Much of it is produced by a mold, *Aspergillus niger* (NĪ-jer), with molasses used as a substrate.

### Enzymes

Enzymes are widely used in different industries. For example, fabric, especially denim, is treated with fungal cellulase to achieve the worn or distressed look. *Amylases* are used in the production of syrups from corn starch, in the production of fabric and paper sizing (a coating for smoothness, as on this page), and in the production of glucose from starch. The microbiological production of amylase is considered to be the first biotechnology patent issued in the United States, which was given to the Japanese scientist Jokichi Takamine. The basic process by which molds were used to make an enzyme preparation known as **koji** had been used for centuries in Japan to make fermented soy products. Koji is an abbreviation of a Japanese word meaning bloom of mold, reflecting the infiltration of a cereal substrate, either rice or a wheat-soybean mixture, with a filamentous fungus (*Aspergillus*). Primarily, the amylases in koji change starch into sugars, but koji preparations also contain proteolytic enzymes that convert the protein in soybeans into a more digestible and flavorful form. It is the basis of soybean fermentations that are staples of the Japanese diet, such as *soy sauce* and *miso* (a fermented paste of soybeans with a meaty flavor). *Sake*, the well-known Japanese rice wine, makes use of amylases of koji to change the carbohydrates of rice into a form that yeasts can use to produce alcohol. This is roughly the equivalent of the barley malt (page 815) used in beer brewing.

*Glucose isomerase* is an important enzyme; it converts the glucose that amylases form from starches into fructose, which is used in place of sucrose as a sweetener in many foods. Probably half of the bread baked in this country is made with the aid of *proteases*, which adjust the amount of glutens (protein) in wheat so that baked goods are improved or made uniform. Other proteolytic enzymes are used as meat tenderizers or in detergents as an additive to remove proteinaceous stains. About a third of all industrial enzyme production is for this purpose. *Rennin*, an enzyme used to form curds in milk, is usually produced commercially by fungi but more recently by genetically modified bacteria.

## Vitamins

Vitamins are sold in large quantities combined in tablet, chewable, and liquid form and are used as individual food supplements. Microbes can provide an inexpensive source of some vitamins. *Vitamin B$_{12}$* is produced by *Pseudomonas* and *Propionibacterium* species. *Riboflavin (B$_2$)* is another vitamin produced by fermentation, mostly by fungi such as *Ashbya gossypii* (ASH-bē-ah gos-SIP-ē-ē). *Vitamin C* (ascorbic acid) is produced at the rate of 60,000 tons per year by a complicated modification of glucose by *Acetobacter* species.

## Pharmaceuticals

Modern pharmaceutical microbiology developed after World War II, when production of antibiotics was introduced.

All antibiotics were originally the products of microbial metabolism. Many are still produced by microbial fermentations, and work continues on the selection of more productive mutants by nutritional and genetic manipulations. Since Fleming's discovery of penicillin, researchers have focused on culturing soil bacteria to identify new antibiotics. However, most of the Earth microbiome is unculturable. In 2015, researchers at Northeastern University developed the **isolation chip (iChip)** to grow bacteria in their natural environment (**Figure 28.10**). Bacteria are placed in a chamber with pores that let nutrients in and wastes out but are small enough to contain bacteria. The chamber is then put in a soil slurry, seawater, or whatever the bacterium's natural environment is. Bacteria that grow in the chamber and their metabolic products can be studied. The new antibiotic teixobactin was discovered using the iChip.

The following ingredients are combined in the manufacturing of milk chocolate: cacao beans, cocoa butter (fat pressed out of the cacao bean), sugar, lecithin, vanillin, and salt. Cacao beans from Ghana, Nigeria, Brazil, and Ecuador are blended and roasted for 30 minutes at 125°C. The beans are then air cooled and ground. In the mixing room, the dry ingredients (salt, sugar, and ground beans) are blended and then mixed with the Brazilian cocoa butter in 3-ton batches for molding.

The factory microbiologist is responsible for ensuring that the raw ingredients are free of pathogens when they enter the factory. In the past, she has rejected coconut milk and eggs that tested positive for *Salmonella*. She recently rejected a peanut shipment that tested positive for mycotoxins. Dr. Chang asks the factory microbiologist to culture selected items in the production line. Her results are shown in the table.

| | Number of Samples | Number Positive for S. Typhimurium |
|---|---|---|
| Raw material storage area | 56 | 0 |
| Bean roasting room | 16 | 2 |
| Beans | 14 | 0 |
| Cocoa butter | 9 | 0 |
| Lecithin | 7 | 0 |
| Vanillin | 1 | 0 |
| Raw bean room | 11 | 2 |
| Mixing room | 14 | 0 |
| Trash room | 7 | 0 |
| Janitorial equipment | 10 | 0 |
| Chocolate molds | 62 | 2 |
| Tap water | 5 | 0 |
| Production line chocolate samples | 25 | 0 |

**Now where will Dr. Chang look?**

811  813  816  **820**  822  824

At least 6000 antibiotics have been described. One organism, *Streptomyces hygroscopicus*, has different strains that make almost 200 different antibiotics. Antibiotics are typically made industrially in large bioreactors (see Figure 28.7b).

Vaccines are a product of industrial microbiology. Many antiviral vaccines are mass-produced in chicken eggs or cell cultures. The production of vaccines against bacterial diseases usually requires the growth of large amounts of the bacteria. Recombinant DNA technology is increasingly important in the development and production of subunit vaccines (see Chapter 18, page 502).

*Steroids* are a very important group of chemicals that include *cortisone*, which is used as an anti-inflammatory drug,

**Figure 28.10 iChip.** The isolation chip (iChip) allows researchers to grow bacteria in their natural environment. There are about 400 chambers on this iChip, and a different bacterium can be grown in each.

**Q** What is the advantage of an iChip over a flask containing nutrient broth?

and *estrogens* and *progesterone*, which are used in oral contraceptives. Recovering steroids from animal sources or chemically synthesizing them is difficult, but microorganisms can synthesize steroids from sterols or from related, easily obtained compounds. For example, **Figure 28.11** illustrates the conversion of a sterol into a valuable steroid.

## Other Chemicals

Traditional chemical companies are turning to microbiology to develop environmentally sound production methods that minimize toxic wastes and the associated costs. Chemical synthesis of indigo, for example, requires a high pH and produces waste that explodes on contact with air. However, *Pseudomonas putida* (poo-TĒ-dah) produces an enzyme that converts the bacterial by-product indole to indigo.

Microbes can even make plastic. Over 25 bacteria make polyhydroxyalkanoate (PHA) inclusion granules as a food reserve. PHAs are similar to common plastics, and because they are made by bacteria, they are also readily degraded by many bacteria. PHAs could provide a biodegradable alternative to conventional plastic, which is made from petroleum.

## Copper Extraction by Leaching

*Acidithiobacillus ferrooxidans* is used in recovering otherwise unprofitable grades of copper ore, which sometimes contain as little as 0.1% copper. At least 25% of the world's copper is produced this way. *Acidithiobacillus* bacteria get their energy from the oxidation of a reduced form of iron ($Fe^{2+}$) in ferrous sulfide to an oxidized form ($Fe^{3+}$) in ferric sulfate. Sulfuric acid ($H_2SO_4$) is also a product of the reaction. The acidic solution of $Fe^{3+}$-containing water is applied by sprinklers and allowed to percolate downslope through the ore body (**Figure 28.12**). The ferrous iron, $Fe^{2+}$, and *A. ferrooxidans* are normally present in the ore and continue to contribute to the reactions. The $Fe^{3+}$ in the sprinkling water reacts with insoluble copper ($Cu^+$) in *copper sulfides* in the ore to form soluble copper ($Cu^{2+}$), which takes the form of *copper sulfates* ($CuSO_4$). The soluble copper sulfate moves downslope to collection tanks, where it contacts metallic scrap iron. The copper sulfates react chemically with the iron and precipitate out as metallic copper ($Cu^0$). In this reaction, the metallic iron ($Fe^0$) is converted into ferrous iron ($Fe^{2+}$) that is recycled to an aerated oxidation pond, where *Acidithiobacillus* bacteria use it for energy to renew the cycle.

## Microorganisms as Industrial Products

Microorganisms themselves sometimes constitute an industrial product. *Baker's yeast (S. cerevisiae)* is produced in large aerated fermentation tanks. At the end of the fermentation, the contents of the tank are about 4% yeast solids. The cells are harvested by continuous centrifuges and are pressed into the familiar yeast cakes or packets sold for home baking. Wholesale bakers purchase yeast in 50-lb boxes.

Other important microbes that are sold industrially are the symbiotic nitrogen-fixing bacteria *Rhizobium* and *Bradyrhizobium*. These organisms are usually mixed with peat moss to preserve moisture; the farmer mixes the peat moss and bacterial inoculum with the seeds of legumes to ensure infection of the plants with efficient nitrogen-fixing strains (see Chapter 27, page 791). For many years, gardeners have used the insect pathogen *Bacillus thuringiensis* to control leaf-eating insect larvae. This bacterium produces a toxin (Bt-toxin) that kills certain moths, beetles, and

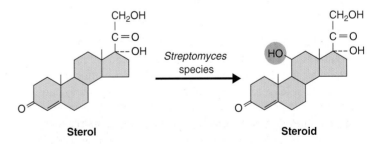

**Figure 28.11 The production of steroids.** Shown here is the conversion of a precursor compound such as a sterol into a steroid by *Streptomyces*. The addition of a hydroxyl group (highlighted in purple on the steroid) to carbon number 11 might require more than 30 steps by chemical means, but the microorganism can add it in only one step.

**Q** Name a commercial product that is a steroid.

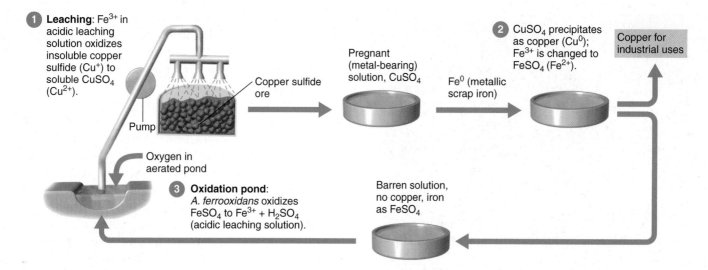

**Figure 28.12 Biological leaching of copper ores.** The chemistry of the process is much more complicated than shown here. Essentially, *Acidithiobacillus ferrooxidans* bacteria are used in a biological/chemical process that changes insoluble copper in the ore into soluble copper that leaches out and is precipitated as metallic copper. The solutions are continuously recirculated.

**Q** Name another metal that is recovered by a similar process.

flies when ingested by their larvae. *B. thuringiensis* subspecies *israelensis* produces Bt-toxin that is especially active against mosquito larvae and is widely used in municipal control programs. Commercial preparations containing Bt-toxin and endospores of *B. thuringiensis* are available at almost any gardening supply store. A new approach to insect control using the insect microbiome is described in the Exploring the Microbiome box (page 823).

> **CHECK YOUR UNDERSTANDING**
>
> ✔ 28-6 At one time, citric acid was extracted on an industrial scale from lemons and other citrus fruits. What organism is used to produce it today?

**CLINICAL CASE**

The beans seem to be the most likely source of the bacteria. Dr. Chang asks how the beans are harvested and stored. He is told that after harvest, cacao beans are fermented at the farm in wooden boxes that are often covered with banana leaves. He is also told there has been only one recorded incident of *Salmonella* contamination of the raw beans. Hearing this, Dr. Chang suspects that contamination must have occurred in the room at the factory where the raw beans are kept. Looking at the room, Dr. Chang spots a discolored area on an overhead pipe in the bean room. No one has noticed the leak. The quality control microbiologist swabs the discolored area, which grows *Salmonella*.

**What characteristics of chocolate prevent microbial growth?**

## Alternative Energy Sources Using Microorganisms

As our supplies of fossil fuels diminish or become more expensive, interest in the use of renewable energy resources will increase. Prominent among these is **biomass**, the collective organic matter produced by living organisms, including crops, trees, and municipal wastes. Microbes can be used for **bioconversion**, the process of converting biomass into alternative energy sources. Bioconversion can also decrease the amount of waste materials requiring disposal.

**Methane** is one of the most convenient energy sources produced from bioconversion. Many communities produce useful amounts of methane from wastes in landfill sites (**Figure 28.13**).

> **CHECK YOUR UNDERSTANDING**
>
> ✔ 28-7 Landfills are the site of a major form of bioconversion—what is the product?

## Biofuels

**Biofuels** are energy sources produced from living organisms, rather than from fossils of organisms that lived over 300 million years ago. Interest in renewable biofuels is increasing because they can provide clean, sustainable energy sources. The initial interest has focused on **ethanol**, which is already widely used as a supplement to gasoline (90% gasoline + 10% ethanol), and the technology is well established. Brazil, for example, produces large amounts of ethanol from sugarcane, about a third of its transportation fuel. Ethanol has, however, a number of

## EXPLORING THE MICROBIOME  Using Bacteria to Stop the Spread of Zika Virus

*Wolbachia* is quite possibly the most common bacterial genus on Earth. Members of this gram-negative bacterial genus commonly live in arthropods, most notably in insects, including mosquitoes. In some cases they behave as commensal microbes in insects, and in other cases their behavior classifies them as parasites.

In some insects, *Wolbachia* destroys males of its host species. *Wolbachia* can turn males into females by interfering with the male hormone. In mosquitoes, if only the male is infected, the mating pair create eggs that fail to hatch. If the female (or both partners) are infected, the next generation of mosquitoes is born infected with the bacteria. The infected new generation mount an immune response against the bacteria that makes the mosquitoes less-than-optimum hosts for arthropod-borne viruses that can transmit to humans too.

Now areas of Brazil, Colombia, southeast Asia, and California are starting to tap this bacteria as a tool in the fight against Zika infections. *Wolbachia*-infected mosquitoes raised in a lab, called symbiotically modified organisms (SMOs), are released into areas where people live among *Aedes aegypti* mosquitoes. This mosquito is associated with Zika transmission to humans. When wild, uninfected females mate with male SMOs, the nonviable eggs that result help reduce the overall mosquito population. The female SMOs mate and pass their *Wolbachia* infections on to a new generation in the wild—but the young mosquitoes will not transmit the Zika virus. Additionally, this procedure reduces transmission of viruses that cause chikungunya and dengue by 67% and 37%, respectively.

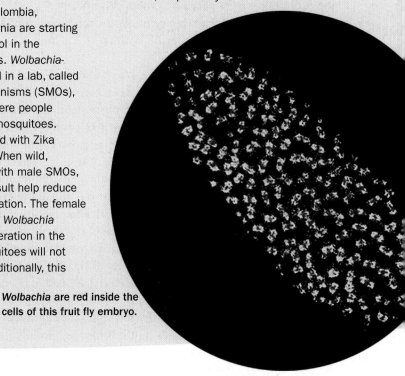

**Wolbachia** are red inside the cells of this fruit fly embryo.

---

**Figure 28.13 Methane production from solid wastes in landfills.** Methane accumulates in landfills and can be used for energy. This installation near Los Angeles has 50 microturbines that produce electricity from methane produced by the landfill. Immediately behind the microturbines are five gas flaring stacks that mask the flames from excess flared methane—a requirement so that aircraft will not confuse it with airport lighting.

Gas flaring stacks

Microturbines produce electricity from methane

**Q** How is methane produced in a landfill?

deficiencies: it cannot be transported by conventional pipelines (because it absorbs water so avidly), and it has 30% less energy content than gasoline. Also, to produce ethanol from corn creates pressures on the supply and price of a valuable foodstuff.

These drawbacks have increased interest in biofuels derived from cellulosic materials, such as cornstalks, wood, and wastepaper, and from exotic nonfood plants such as jatropha, camelina, and miscanthus. In the United States there is special interest in switchgrass—which once carpeted the prairies of the Midwest. Such grasses are perennials and require little more attention than harvesting. The technology for producing ethanol from cellulose is less well known and more expensive than that from corn or sugarcane. The sugar molecules that make up cellulose may be broken apart by enzymes—in fact, genes to synthesize these enzymes have been genetically introduced into *E. coli. Cellulose* sources also contain significant amounts of a similar component, *hemicellulose*, which will require organisms capable of digesting it—probably genetically modified microbes. The digestibly resistant cellulosic component *lignin* could be burned to heat early steps in fermentation processing.

"Higher" alcohols such as butanol, which have longer carbon chains, and especially branched alcohols such as isobutanol and isobutyraldehyde, have advantages over conventional ethanol. They have less capacity to absorb water and have a higher energy content. Bacteria have been genetically modified to produce several forms of higher alcohols from glucose. A basic problem in microbial production of biofuels is that we want the microbe to excrete fuel so that we eliminate the expensive step of harvesting them periodically.

A theoretically attractive organism for producing biofuels is algae. Algae offer a number of advantages; for one, they do not take up valuable farmland needed for food production. Also, algae produce 40 times the energy per acre that corn produces—and the land the algae grow on can be agriculturally nonproductive as long as it has abundant sunlight. Experimental algal production sites have even used the carbon dioxide emissions from power plants to accelerate growth. The algae can be harvested on an almost daily basis. Oils squeezed from them can be turned into biodiesel fuel and possibly jet fuel: typical algae yield 20% of their weight in oil, and some even more. After oil extraction, the remainder, rich in carbohydrates and proteins, can be used to produce ethanol or as animal feed.

Hydrogen is an attractive candidate as a replacement for fossil fuels, especially if it can be produced by splitting water. It can be used in fuel cells to generate electricity and, if burned to generate energy, does not produce harmful residues. Most research into the production of hydrogen has concentrated on physical and chemical methods, but it is also potentially possible to use bacteria or algae to produce hydrogen from the fermentation of various waste products or by modifications of photosynthesis.

Electron transfer in microorganisms is also being explored as a source of electricity. Bacteria or algae that can generate an electric current are called *exoelectrogens*. In **microbial fuel cells**, exoelectrogens are grown in a nutrient medium such as soil or wastewater. Electrons generated in the electron transport chain are transferred to an electrode and then to a wire.

The technologies outlined above will require time to reach their potential. Currently, science is in the early phases of the learning curves that all new technologies face.

**CHECK YOUR UNDERSTANDING**

✔ 28-8 How can microbes provide fuels for cars and electricity?

## Industrial Microbiology and the Future

Microbes have always been exceedingly useful to humankind, even when their existence was unknown. They will remain an essential part of many basic food-processing technologies. The development of recombinant DNA technology has further intensified interest in industrial microbiology by expanding the potential for new products and applications. As the supplies of fossil energy become more scarce, interest in renewable energy sources, such as hydrogen and ethanol, will increase. The use of specialized microbes to produce such products on an industrial scale will probably become more important. As new biotechnology applications and products enter the marketplace, they will affect our lives and well-being in ways that we can only speculate about today.

**CLINICAL CASE** Resolved

The low moisture, high fat, and high sugar content of chocolate does not favor bacterial growth, but it does significantly increase the heat resistance of bacteria. Consequently, bacteria may survive roasting.

To address the risk posed by *Salmonella*, all food safety agencies have pursued an ongoing strategy to reduce the prevalence of the pathogen in the food chain. However, despite all the efforts, the number of salmonellosis cases remains high.

811   813   816   820   822   **824**

# Study Outline

 Go to @MasteringMicrobiology for **Interactive Microbiology**, *In the Clinic* videos, *MicroFlix, MicroBoosters, 3D animations, practice quizzes, and more.*

### Food Microbiology (pp. 810–816)

1. The earliest methods of preserving foods were drying, the addition of salt or sugar, and fermentation.

### Foods and Disease (pp. 810–811)

2. Food safety is monitored by the FDA and USDA and also by use of the HACCP system.

### Industrial Food Canning (pp. 811–812)

3. Commercial sterilization of food is accomplished by steam under pressure in a retort.

4. Commercial sterilization heats canned foods to the minimum temperature necessary to destroy *Clostridium botulinum* endospores while minimizing alteration of the food.

5. The commercial sterilization process uses sufficient heat to reduce a population of *C. botulinum* by 12 logarithmic cycles (12D treatment).

6. Endospores of thermophiles can survive commercial sterilization.

7. Canned foods stored above 45°C can be spoiled by thermophilic anaerobes.

8. Thermophilic anaerobic spoilage is sometimes accompanied by gas production; if no gas is formed, the spoilage is called flat sour spoilage.

9. Spoilage by mesophilic bacteria is usually from improper heating procedures or leakage.

10. Acidic foods can be preserved by heat of 100°C because micro-organisms that survive are not capable of growth in a low pH.

11. *Byssochlamys, Aspergillus,* and *Bacillus coagulans* are acid-tolerant and heat-resistant microbes that can spoil acidic foods.

### Aseptic Packaging (p. 812)

12. Presterilized materials are assembled into packages and aseptically filled with heat-sterilized liquid foods.

### Radiation and Industrial Food Preservation (pp. 812–814)

13. Gamma and X-ray radiation can be used to sterilize food, kill insects and parasitic worms, and prevent the sprouting of fruits and vegetables.

### High-Pressure Food Preservation (p. 814)

14. Pressurized water (pascalization) is used to kill bacteria in fruit and meat.

### The Role of Microorganisms in Food Production (pp. 814–816)

15. The milk protein casein curdles because of the action by lactic acid bacteria or the enzyme rennin.

16. Old-fashioned buttermilk was produced by lactic acid bacteria growing during the butter-making process.

17. Sugars in bread dough are fermented by yeast to ethanol and $CO_2$; the $CO_2$ causes the bread to rise.

18. Carbohydrates obtained from grains, potatoes, or molasses are fermented by yeasts to produce ethanol in the production of beer, wine, and distilled spirits.

## Industrial Microbiology and Biotechnology (pp. 817–824)

1. Microorganisms produce alcohols and acetone that are used in industrial processes.

2. Industrial microbiology has been revolutionized by the ability of genetically modified cells to make many new products.

3. Biotechnology is a way of making commercial products by using living organisms.

### Fermentation Technology (pp. 817–818)

4. The growth of cells on a large scale is called industrial fermentation.

5. Industrial fermentation is carried on in bioreactors, which control aeration, pH, and temperature.

6. Primary metabolites such as ethanol are formed as the cells grow (during the trophophase).

7. Secondary metabolites such as penicillin are produced during the stationary phase (idiophase).

8. Mutant strains that produce a desired product can be selected.

9. Enzymes or whole cells can be bound to solid spheres or fibers. When substrate passes over the surface, enzymatic reactions change the substrate to the desired product.

### Industrial Products (pp. 819–822)

10. Microbes produce xanthan, amino acids, vitamins, and citric acid used in foods and medicine.

11. Enzymes used in manufacturing foods, medicines, and other goods are produced by microbes.

12. Vaccines, antibiotics, and steroids are products of microbial growth.

13. The metabolic activities of *Acidithiobacillus ferrooxidans* can be used to recover uranium and copper ores.

14. Yeasts are grown for wine- and breadmaking; other microbes (*Rhizobium, Wolbachia,* and *Bacillus thuringiensis*) are grown for agricultural use.

### Alternative Energy Sources Using Microorganisms (p. 822)

15. Organic waste, called biomass, can be converted by microorganisms into the alternative fuel methane, a process called bioconversion.

16. Fuels produced by microbial fermentation are methane, ethanol, and hydrogen.

### Biofuels (pp. 822–824)

17. Biofuels include alcohols and hydrogen (from microbial fermentation) and oils (from algae).

### Industrial Microbiology and the Future (p. 824)

18. Recombinant DNA technology will continue to enhance the ability of industrial microbiology to produce medicines and other useful products.

# Study Questions

For answers to the Knowledge and Comprehension questions, turn to the Answers tab at the back of the textbook.

## Knowledge and Comprehension

### Review

1. What is industrial microbiology? Why is it important?

2. How does commercial sterilization differ from sterilization procedures used in a hospital or laboratory?

3. Why is a can of blackberries preserved by commercial sterilization typically heated to 100°C instead of at least 116°C?

4. Outline the steps in the production of cheese, and compare the production of hard and soft cheeses.

5. Beer is made with water, malt, and yeast; hops are added for flavor. What is the purpose of the water, malt, and yeast? What is malt?

6. Why is a bioreactor better than a large flask for industrial production of an antibiotic?

7. The manufacture of paper includes the use of bleach and formaldehyde-based glue. The microbial enzyme xylanase whitens paper by digesting dark lignins. Oxidase causes the fibers to stick together, and cellulase will remove ink. List three advantages of using these microbial enzymes over traditional chemical methods for making paper.

8. Describe an example of bioconversion. What metabolic processes can result in fuels?

9. DRAW IT Label the trophophase and idiophase in this graph. Indicate when primary and secondary metabolites are formed.

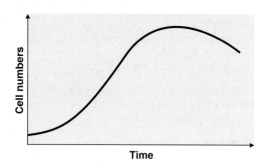

10. NAME IT Van Leeuwenhoek was the first to see this budding microbe with a nucleus and cell wall; although humans have used it since before the beginning of recorded history, Louis Pasteur was the first to figure out what it does.

## Multiple Choice

1. Foods packed in plastic for microwaving are
   a. dehydrated.
   b. freeze-dried.
   c. packaged aseptically.
   d. commercially sterilized.
   e. autoclaved.

2. *Acetobacter* is necessary for only one of the steps of vitamin C manufacture. The easiest way to accomplish this step would be to
   a. add substrate and *Acetobacter* to a test tube.
   b. affix *Acetobacter* to a surface and run substrate over it.
   c. add substrate and *Acetobacter* to a bioreactor.
   d. find an alternative to this step.
   e. none of the above

Use the following choices to answer questions 3–5:
   a. *Bacillus coagulans*
   b. *Byssochlamys*
   c. flat sour spoilage
   d. *Lactobacillus*
   e. thermophilic anaerobic spoilage

3. The spoilage of canned foods due to inadequate processing, accompanied by gas production.

4. The spoilage of canned foods caused by *Geobacillus stearothermophilus*.

5. A heat-resistant fungus that causes spoilage in acidic foods.

6. The term *12D treatment* refers to
   a. heat treatment sufficient to kill 12 bacteria.
   b. the use of 12 different treatments to preserve food.
   c. a $10^{12}$ reduction in *C. botulinum* endospores.
   d. any process that destroys thermophilic bacteria.

7. Which one of the following is *not* a fuel produced by microorganisms?
   a. algal oil
   b. ethanol
   c. hydrogen
   d. methane
   e. uranium

8. Which type of radiation is used to preserve foods?
   a. ionizing
   b. nonionizing
   c. radiowaves
   d. microwaves
   e. all of the above

9. Which of the following reactions is undesirable in winemaking?
   a. Sucrose → ethanol
   b. Ethanol → acetic acid
   c. Malic acid → lactic acid
   d. Glucose → pyruvic acid

10. Which of the following reactions is an oxidation carried out by *Acidithiobacillus ferrooxidans*?
   a. $Fe^{2+} \rightarrow Fe^{3+}$
   b. $Fe^{3+} \rightarrow Fe^{2+}$
   c. $CuS \rightarrow CuSO_4$
   d. $Fe^0 \rightarrow Cu^0$
   e. none of the above

## Analysis

1. Which bacteria seem to be most frequently used in the production of food? Propose an explanation for this.

2. *Methylophilus methylotrophus* can convert methane ($CH_4$) into proteins. Amino acids are represented by this structure:

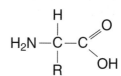

   Diagram a pathway illustrating the production of at least one amino acid.

3. Faded, worn-look denim is produced with cellulase. How does cellulase accomplish the look and feel of scores of washings? What is the source of the cellulase?

## Clinical Applications and Evaluation

1. Suppose you are culturing a microorganism that produces enough lactic acid to kill itself in a few days. The graph below shows conditions in the bioreactor:

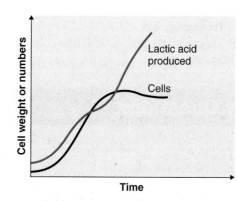

   a. How can the use of a bioreactor help you maintain the culture for weeks or months?
   b. If your desired product is a secondary metabolite, when can you begin collecting it?
   c. If your desired product is the cells themselves and you want to maintain a continuous culture, when can you begin harvesting?

2. Researchers at the CDC inoculated apple cider with $10^5$ *E. coli* O157:H7 cells/ml to determine the fate of the bacteria in apple cider (pH 3.7). They obtained the following results:

| | *E. coli* **O157:H7 CFU/ml after 25 days** |
|---|---|
| Apple cider at 25°C | $10^4$ (mold growth evident by 10 days) |
| Apple cider with potassium sorbate at 25°C | $10^3$ |
| Apple cider at 8°C | $10^2$ |

What conclusions can you reach from these data? What disease is caused by *E. coli* O157:H7? (*Hint:* See Chapter 25.)

3. The antibiotic efrotomycin is produced by *Streptomyces lactamdurans*. *S. lactamdurans* was grown in 40,000 liters of medium. The medium consisted of glucose, maltose, soybean oil, $(NH_4)_2SO_4$, NaCl, $KH_2PO_4$, and $Na_2HPO_4$. The culture was aerated and maintained at 28°C. The following results were obtained from analyses of the culture medium during cell growth:

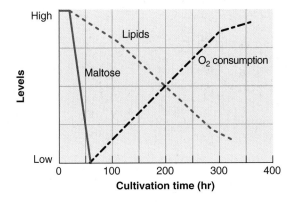

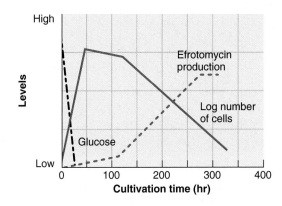

a. Under what conditions is the most efrotomycin produced? Is it a primary or secondary metabolite?
b. Which is used first, maltose or glucose? Suggest a reason for this.
c. What is the purpose of each ingredient in the growth medium? (*Hint:* See Chapter 6.)
d. What is *Streptomyces*? (*Hint:* See Chapter 11.)

# Answers to Knowledge and Comprehension Study Questions

## Chapter 1

### Review

1. People came to believe that living organisms arise from nonliving matter because they would see flies coming out of manure, maggots coming out of dead animals, and microorganisms appearing in liquids after a day or two.

2. **a.** Certain microorganisms cause diseases in insects. Microorganisms that kill insects can be effective biological control agents because they are specific for the pest and do not persist in the environment.
   **b.** Carbon, oxygen, nitrogen, sulfur, and phosphorus are required for all living organisms. Microorganisms convert these elements into forms that are useful for other organisms. Many bacteria decompose material and release carbon dioxide into the atmosphere, which plants use. Some bacteria can take nitrogen from the atmosphere and convert it into a form that plants and other microorganisms can use.
   **c.** Normal microbiota are microorganisms that are found in and on the human body. They do not usually cause disease and can be beneficial.
   **d.** Organic matter in sewage is decomposed by bacteria into carbon dioxide, nitrates, phosphates, sulfate, and other inorganic compounds in a wastewater treatment plant.
   **e.** Recombinant DNA techniques have resulted in insertion of the gene for insulin production into bacteria. These bacteria can produce human insulin inexpensively.
   **f.** Microorganisms can be used as vaccines. Some microbes can be genetically modified to produce components of vaccines.
   **g.** Biofilms are aggregated bacteria adhering to each other and to a solid surface.

3. **a.** 1, 3   **c.** 1, 4, 5   **e.** 5   **g.** 6
   **b.** 8   **d.** 2   **f.** 3   **h.** 7

4. **a.** 7   **c.** 3   **e.** 6   **g.** 1
   **b.** 4   **d.** 2   **f.** 5

5. **a.** 11   **e.** 3   **i.** 1   **m.** 7   **q.** 13
   **b.** 14   **f.** 9   **j.** 12   **n.** 5   **r.** 16
   **c.** 15   **g.** 10   **k.** 18   **o.** 6
   **d.** 17   **h.** 2   **l.** 4   **p.** 8

6. **a.** *B. thuringiensis* is sold as a biological insecticide.
   **b.** *Saccharomyces* is the yeast sold for making bread, wine, and beer.

7. Bacterium

8.

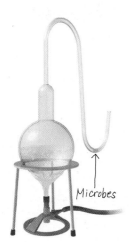

Microbes

## Multiple Choice

1. a   6. e
2. c   7. c
3. d   8. a
4. c   9. c
5. b   10. a

## Chapter 2

### Review

1. Atoms with the same atomic number and chemical behavior are classified as chemical elements.

2. 

3. **a.** Ionic
   **b.** Single covalent bond
   **c.** Double covalent bonds
   **d.** Hydrogen bond

4. **a.** Synthesis reaction, condensation, or dehydration
   **b.** Decomposition reaction, digestion, or hydrolysis
   **c.** Exchange reaction
   **d.** Reversible reaction

5. The enzyme speeds up this decomposition reaction.

6. **a.** Lipid
   **b.** Protein
   **c.** Carbohydrate
   **d.** Nucleic acid

7. **a.** Amino acids
   **b.** Right to left
   **c.** Left to right

8. The entire protein shows tertiary structure, held by disulfide bonds. No quaternary structure.

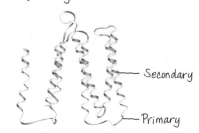

— Secondary

— Primary

9.

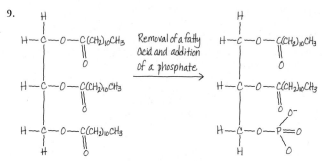

Removal of a fatty acid and addition of a phosphate →

10. Fungus

## Multiple Choice

| | | | | |
|---|---|---|---|---|
| 1. c | 3. b | 5. b | 7. a | 9. b |
| 2. b | 4. e | 6. c | 8. a | 10. c |

## Chapter 3

### Review

1. **a.** $10^{26}$ m        **b.** 1 nm        **c.** $10^3$ nm

2. **a.** Compound light microscope
   **b.** Darkfield microscope
   **c.** Phase-contrast microscope
   **d.** Fluorescence microscope
   **e.** Electron microscope
   **f.** Differential interference contrast microscope

3.

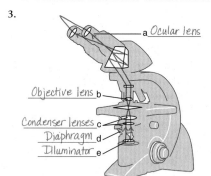

a *Ocular lens*
*Objective lens* b
*Condenser lenses* c
*Diaphragm* d
*Illuminator* e

4.
| Ocular Lens Magnification | × | Oil Immersion Lens Magnification | = | Total Magnification of Specimen |
|---|---|---|---|---|
| 10× | | 100× | | 1000× |

5. **a.** 1500×
   **b.** 10,000,000×
   **c.** 0.2 μm
   **d.** 10 pm
   **e.** Seeing three-dimensional detail.

6. In a Gram stain, the mordant combines with the basic dye to form a complex that will not wash out of gram-positive cells. In a flagella stain, the mordant accumulates on the flagella so that they can be seen with a light microscope.

7. A counterstain stains the colorless non–acid-fast cells so that they are easily seen through a microscope.

8. In the Gram stain, the decolorizer removes the color from gram-negative cells. In the acid-fast stain, the decolorizer removes the color from non–acid-fast cells.

9. **a.** Purple        **e.** Purple
   **b.** Purple        **f.** Purple
   **c.** Purple        **g.** Colorless
   **d.** Purple        **h.** Red

10. An acid-fast bacterium (*Mycobacterium*)

## Multiple Choice

| | | | | |
|---|---|---|---|---|
| 1. c | 3. b | 5. a | 7. d | 9. a |
| 2. d | 4. a | 6. e | 8. b | 10. c |

## Chapter 4

### Review

1.  *a. and e.*        *b. and e.*

   *c.*

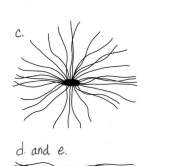

   *d. and e.*

2. **a.** sporogenesis
   **b.** certain adverse environmental conditions
   **c.** germination
   **d.** favorable growth conditions

3.

a.    d.
b.    e.
c.    f.

4. **a.** 4
   **b.** 6
   **c.** 1
   **d.** 3
   **e.** 1, 5
   **f.** 3, 9
   **g.** 2, 8
   **h.** 7

5. An endospore is called a resting structure because it provides a method for one cell to "rest," or survive, as opposed to grow and reproduce. The protective endospore wall allows a bacterium to withstand adverse conditions in the environment.

6. **a.** Both allow materials to cross the plasma membrane from a high concentration to a low concentration without expending energy. Facilitated diffusion requires carrier proteins.
   **b.** Both require enzymes to move materials across the plasma membrane. In active transport, energy is expended.
   **c.** Both move materials across the plasma membrane with an expenditure of energy. In group translocation, the substrate is changed after it crosses the membrane.

7. **a.** Diagram (a) refers to a gram-positive bacterium because the lipopolysaccharide–phospholipid–lipoprotein layer is absent.
   **b.** The gram-negative bacterium initially retains the violet stain, but it is released when the outer membrane is dissolved by the decolorizing agent. After the dye–iodine complex enters, it becomes trapped by the peptidoglycan of gram-positive cells.
   **c.** The outer layer of the gram-negative cells prevents penicillin from entering the cells.
   **d.** Essential molecules diffuse through the gram-positive wall. Porins and specific channel proteins in the gram-negative outer membrane allow passage of small water-soluble molecules.
   **e.** Gram-negative.

8. An extracellular enzyme (amylase) hydrolyzes starch into disaccharides (maltose) and monosaccharides (glucose). A carrier enzyme (maltase) hydrolyzes maltose and moves one glucose into the cell. Glucose can be transported by group translocation as glucose-6-phosphate.

9. **a.** 3
   **b.** 4
   **c.** 7
   **d.** 1
   **e.** 6
   **f.** 2
   **g.** 5

10. Actinomycetes

## Multiple Choice

| | | | | |
|---|---|---|---|---|
| 1. e | 3. b | 5. d | 7. b | 9. a |
| 2. d | 4. a | 6. e | 8. e | 10. b |

## Chapter 5

### Review

1. **(a)** is the Calvin-Benson cycle, **(b)** is glycolysis, and **(c)** is the Krebs cycle.
   **a.** Glycerol is catabolized by pathway (b) as dihydroxyacetone phosphate. Fatty acids by pathway (c) as acetyl groups.
   **b.** In pathway (c) at α-ketoglutaric acid.
   **c.** Glyceraldehyde-3-phosphate from the Calvin-Benson cycle enters glycolysis. Pyruvic acid from glycolysis is decarboxylated to produce acetyl for the Krebs cycle.
   **d.** In (a) between 3-phosphoglyceric acid and 1,3-diphosphoglyceric acid and between glyceraldehyde 3-phosphate and ribulose diphosphate. In (b), between glucose and glyceraldehyde-3-phosphate.
   **e.** The conversion of pyruvic acid to acetyl, isocitric acid to α-ketoglutaric acid, and α-ketoglutaric acid to succinyl CoA.
   **f.** By pathway (c) as acetyl groups.
   **g.**

| | Uses | Produces |
|---|---|---|
| Calvin-Benson cycle | 6 NADPH | |
| Glycolysis | | 2 NADH |
| Pyruvic acid → acetyl | | 1 NADH |
| Isocitric acid → α-ketoglutaric acid | | 1 NADH |
| α-ketoglutaric acid → Succinyl CoA | | 1 NADH |
| Succinic acid → Fumaric acid | | 1 FADH$_2$ |
| Malic acid → Oxaloacetic acid | | 1 NADH |

   **h.** Dihydroxyacetone phosphate; acetyl; oxaloacetic acid; α-ketoglutaric acid.

2.

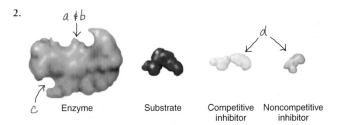

e. When the enzyme and substrate combine, the substrate molecule will be transformed.

When the competitive inhibitor binds to the enzyme, the enzyme will not be able to bind with the substrate.

When the noncompetitive inhibitor binds to the enzyme, the active site of the enzyme will be changed so the enzyme cannot bind with the substrate.

3.

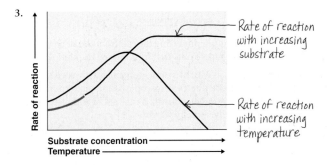

4. Oxidation-reduction: A coupled reaction in which one substance loses electrons and another gains electrons.
   **a.** The final electron acceptor in aerobic respiration is molecular oxygen; in anaerobic respiration, it is another inorganic molecule.
   **b.** An electron transport chain is used in respiration but not in fermentation. The final electron acceptor in respiration is usually inorganic; in fermentation it is usually organic.
   **c.** In cyclic photophosphorylation, electrons are returned to chlorophyll. In noncyclic photophosphorylation, chlorophyll receives electrons from hydrogen atoms.

5. **a.** Photophosphorylation
   **b.** Oxidative phosphorylation
   **c.** Substrate-level phosphorylation

6. oxidation

7. **a.** CO$_2$          **e.** CO$_2$
   **b.** Light           **f.** Inorganic molecules
   **c.** Organic molecules  **g.** Organic molecules
   **d.** Light           **h.** Organic molecules

8. Protons are pumped from one side of the membrane to the other; transfer of protons back across the membrane generates ATP. **a** and **b.** Outer portion is acidic and has a positive electrical charge. **c.** Energy-conserving sites are the three loci where protons are pumped out. **d.** Kinetic energy is realized at ATP synthase.

9. NAD$^+$ is needed to pick up more electrons. NADH is usually reoxidized in respiration. NADH can be reoxidized in fermentation.

10. Chemoautotroph

### Multiple Choice

| | | | | |
|---|---|---|---|---|
| 1. a | 3. b | 5. c | 7. b | 9. c |
| 2. d | 4. c | 6. b | 8. a | 10. b |

## Chapter 6

### Review

1. In binary fission, the cell elongates, and the chromosome replicates. Next, the nuclear material is evenly divided. The plasma membrane invaginates toward the center of the cell. The cell wall thickens and grows inward between the membrane invaginations; two new cells results.

2. **Carbon:** synthesis of molecules that make up a living cell. **Hydrogen:** source of electrons and component of organic molecules. **Oxygen:** component of organic molecules; electron acceptor in aerobes. **Nitrogen:** component of amino acids. **Phosphorus:** in phospholipids and nucleic acids. **Sulfur:** In some amino acids.

3. **a.** Catalyzes the breakdown of $H_2O_2$ to $O_2$ and $H_2O$.
   **b.** $H_2O_2$; peroxide ion is $O_2^{2-}$.
   **c.** Catalyzes the breakdown of $H_2O_2$;
   $$2H^+ + H_2O_2 \xrightarrow{\text{Peroxidase}} 2H_2O$$
   **d.** $O_2^-$; this anion has one unpaired electron.
   **e.** Converts superoxide to $O_2$ and $H_2O_2$;
   $$2O_2^- + 2H^+ \xrightarrow{\text{Superoxide dismutase}} O_2 + H_2O_2$$
   The enzymes are important in protecting the cell from the strong oxidizing agents, peroxide and superoxide, that form during respiration.

4. Direct methods are those in which the microorganisms are seen and counted. Direct methods are direct microscopic count, plate count, filtration, and most probable number. Growth is inferred by indirect methods: turbidity, metabolic activity, and dry weight.

5. The growth rate of bacteria slows down with decreasing temperatures. Mesophilic bacteria will grow slowly at refrigeration temperatures and will remain dormant in a freezer. Bacteria will not spoil food quickly in a refrigerator.

6. Number of cells $\times 2^{n \text{ generations}} = $ Total number of cells
   $$6 \quad \times \quad 2^7 \quad = \quad 768$$

7. Petroleum can meet the carbon and energy requirements for an oil-degrading bacterium; however, nitrogen and phosphate are usually not available in large quantities. Nitrogen and phosphate are essential for making proteins, phospholipids, nucleic acids, and ATP.

8. A chemically defined medium is one in which the exact chemical composition is known. A complex medium is one in which the exact chemical composition is not known.

9.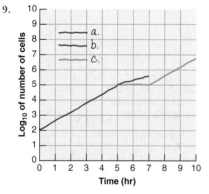

10. Cold, salty, aerobic

### Multiple Choice

| | | | | |
|---|---|---|---|---|
| 1. c | 3. c | 5. c | 7. e | 9. b |
| 2. a | 4. a | 6. d | 8. c | 10. b |

## Chapter 7

### Review

1. Autoclave. Because of the high specific heat of water, moist heat is readily transferred to cells.

2. Pasteurization destroys most organisms that cause disease or rapid spoilage of food.

3. Variables that affect determination of the thermal death point are
   - The innate heat resistance of the strain of bacteria
   - The past history of the culture, whether it was freeze-dried, wetted, etc.
   - The clumping of the cells during the test
   - The amount of water present
   - The organic matter present
   - Media and incubation temperature used to determine viability of the culture after heating

4. **a.** the ability of ionizing radiation to break DNA directly. However, because of the high water content of cells, free radicals ($H \cdot$ and $OH \cdot$) that break DNA strands are likely to form.
   **b.** formation of thymine dimers.

5. All cells do not die at once.

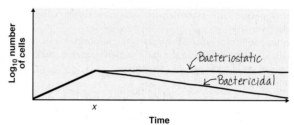

6. All three processes kill microorganisms; however, as moisture and/or temperatures are increased, less time is required to achieve the same result.

7. Salts and sugars create a hypertonic environment. Salts and sugars (as preservatives) do not directly affect cell structures or metabolism; instead, they alter the osmotic pressure. Jams and jellies are preserved with sugar; meats are usually preserved with salt. Molds are more capable of growth in high osmotic pressure than are bacteria.

8. Disinfectant B is preferable because it can be diluted more and still be effective.

9. Quaternary ammonium compounds are most effective against gram-positive bacteria. Gram-negative bacteria that were stuck in cracks or around the drain of the tub would not have been washed away when the tub was cleaned. These gram-negative bacteria could survive the washing procedure. Some pseudomonads can grow on quats that have accumulated.

10. Pseudomonads (*Pseudomonas* and *Burkholderia*)

### Multiple Choice

| | | | | |
|---|---|---|---|---|
| 1. d | 3. d | 5. a | 7. b | 9. b |
| 2. b | 4. d | 6. A | 8. b | 10. b |

## Chapter 8

### Review

1. DNA consists of a strand of alternating sugars (deoxyribose) and phosphate groups with a nitrogenous base attached to each sugar. The bases are adenine, thymine, cytosine, and guanine. DNA exists in a cell as two strands twisted together to form a double helix. The two strands are held together by hydrogen bonds between their nitrogenous bases. The bases are paired in a specific, complementary way: A-T and C-G. The information held in the sequence of nucleotides in DNA is the basis for synthesis of RNA and proteins in a cell.

2.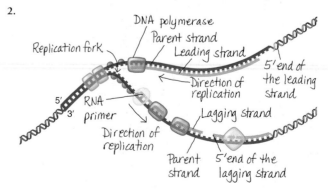

3. **a.** 2     **d.** 1
   **b.** 4     **e.** 5
   **c.** 3

4. **a.** ATATTACTTTGCATGGACT
   **b.** met-lys-arg-thr-(end)
   **c.** TATAATGAAACGTTCCTGA
   **d.** No change
   **e.** Cysteine substituted for arginine
   **f.** Proline substituted for threonine (missense mutation)
   **g.** Frameshift mutation
   **h.** Adjacent thymines might polymerize
   **i.** ACT

5. Iron deficiency could stimulate miRNA that is complementary to RNA encoding iron-requiring proteins.

6. **a.** After translation          **c.** Before transcription
   **b.** After transcription        **d.** Before transcription

7. CTTTGA. Endospores and pigments offer protection against UV radiation. Additionally, repair mechanisms can remove and replace thymine polymers.

8. **a.** Culture 1 will remain the same. Culture 2 will convert to $F^+$ but will have its original genotype.
   **b.** The donor and recipient cells' DNA can recombine to form combinations of $A^+B^+C^+$ and $A^-B^-C^-$. If the F plasmid also is transferred, the recipient cell may become $F^+$.

9. Mutation and recombination provide genetic diversity. Environmental factors select for the survival of organisms through natural selection. Genetic diversity is necessary for the survival of some organisms through the processes of natural selection. Organisms that survive may undergo further genetic change, resulting in the evolution of the species.

10. *Escherichia coli*

### Multiple Choice

| | | | | |
|---|---|---|---|---|
| 1. c | 3. c | 5. c | 7. a | 9. d |
| 2. d | 4. d | 6. b | 8. c | 10. a |

## Chapter 9

### Review

1. **a.** Both are DNA. cDNA is a segment of DNA made by RNA-dependent DNA polymerase. It is not necessarily a gene; a gene is a transcribable unit of DNA that codes for protein or RNA.
   **b.** Both are DNA. A RFLP is a segment of DNA produced when a restriction endonuclease hydrolyzes DNA. It is not usually a gene; a gene is a transcribable unit of DNA that codes for protein or RNA.
   **c.** Both are DNA. A DNA probe is a short, single-stranded piece of DNA. It is not a gene; a gene is a transcribable unit of DNA that codes for protein or RNA.
   **d.** Both are enzymes. DNA polymerase synthesizes DNA one nucleotide at a time using a DNA template; DNA ligase joins pieces (strands of nucleotides) together.
   **e.** Both are DNA. Recombinant DNA results from joining DNA from two different sources; cDNA results from copying a strand of RNA.
   **f.** The proteome is the expression of the genome. An organism's genome is one complete copy of its genetic information. The proteins encoded by this genetic material comprise the proteome.

2. In protoplast fusion, two wall-less cells fuse together to combine their DNA. A variety of genotypes can result from this process. In b, c, and d, specific genes are inserted directly into the cell.

3. **a.** *Bam*HI, *Eco*RI, and *Hind*III make sticky ends.
   **b.** Fragments of DNA produced with the same restriction enzyme will spontaneously anneal to each other at their sticky ends.

4. The gene can be spliced into a plasmid and inserted into a bacterial cell. As the cell grows, the number of plasmids will increase.
   The polymerase chain reaction can make copies of a gene using DNA polymerase and a primer for the gene.

5.

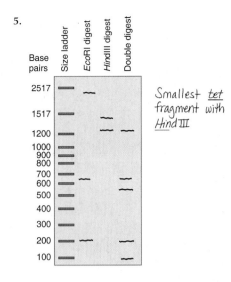

Smallest *tet* fragment with *Hind* III

6. In a eukaryotic cell, RNA polymerase copies DNA; RNA processing removes the introns, leaving the exons in the mRNA. cDNA can be made from the mRNA by reverse transcriptase.

7. See Tables 9.2 and 9.3.

8. You probably used a few plant cells in a Petri plate for your experiment. You can grow these cells on plant-cell culture media with tetracycline. Only the cells with the new plasmid will grow.

9. In RNAi, siRNA binds mRNA, creating double-stranded RNA, which is enzymatically destroyed.

10. Retroviridae

### Multiple Choice

| | | | | |
|---|---|---|---|---|
| 1. b | 3. b | 5. c | 7. c | 9. e |
| 2. b | 4. b | 6. d | 8. b | 10. a |

## Chapter 10

### Review

1. A and D appear to be most closely related because they have similar GC moles %. No two are the same species.

2. A and D are most closely related.

3. The purpose of a cladogram is to show the degree of relatedness between organisms. A dichotomous key can be used for identification but doesn't show relatedness like the cladogram. *Mycoplasma* and *Escherichia* are on one branch in the key, but the cladogram indicates *Mycoplasma* is more closely related to *Clostridium*.

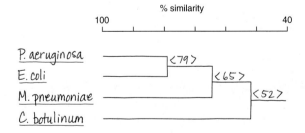

4. One possible key is shown below. Alternative keys could be made starting with morphology or glucose fermentation.

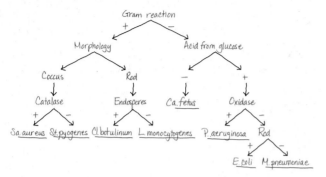

5. *Bordetella bronchiseptica*

## Multiple Choice

| | | | | |
|---|---|---|---|---|
| 1. b | 3. d | 5. e | 7. a | 9. a |
| 2. e | 4. b | 6. a | 8. e | 10. b |

## Chapter 11

### Review

1. **a.** *Clostridium*      **h.** *Spirillum*
   **b.** *Bacillus*        **i.** *Pseudomonas*
   **c.** *Streptomyces*     **j.** *Escherichia*
   **d.** *Mycobacterium*    **k.** *Mycoplasma*
   **e.** *Streptococcus*    **l.** *Rickettsia*
   **f.** *Staphylococcus*   **m.** *Chlamydia*
   **g.** *Treponema*

2. **a.** Both are oxygenic photoautotrophs. Cyanobacteria are prokaryotes; algae are eukaryotes.
   **b.** Both are chemoheterotrophs capable of forming mycelia; some form conidia. Actinomycetes are prokaryotes; fungi are eukaryotes.
   **c.** Both are large rod-shaped bacteria. *Bacillus* forms endospores; *Lactobacillus* is a fermentative non–endospore-forming rod.
   **d.** Both are small rod-shaped bacteria. *Pseudomonas* has an oxidative metabolism; *Escherichia* is fermentative. *Pseudomonas* has polar flagella; *Escherichia* has peritrichous flagella.
   **e.** Both are helical bacteria. *Leptospira* (a spirochete) has an axial filament. *Spirillum* has flagella.
   **f.** Both are gram-negative, rod-shaped bacteria. *Escherichia* bacteria are facultative anaerobes, and *Bacteroides* bacteria are anaerobes.
   **g.** Both are obligatory intracellular parasites. *Rickettsia* are transmitted by ticks; *Chlamydia* have a unique developmental cycle.
   **h.** Both are atypical gram-positive bacteria. *Mycobacterium* is a high G+C, acid-fast genus. *Mycoplasma* is a low G+C genus that lacks cell walls.

3. There are many ways to draw a key. Here is one example.

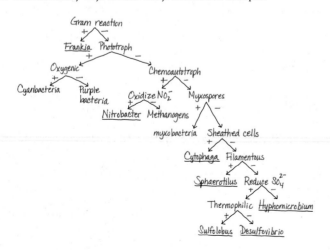

4. Methanogens

## Multiple Choice

| | | | | |
|---|---|---|---|---|
| 1. b | 3. e | 5. b | 7. e | 9. b |
| 2. b | 4. a | 6. c | 8. b | 10. a |

## Chapter 12

### Review

1. **a.** Systemic
   **b.** Subcutaneous
   **c.** Cutaneous
   **d.** Superficial
   **e.** Systemic

2. **a.** *E. coli*
   **b.** *P. chrysogenum*

3. Arthroconidia (*Trichophyton*)

4. As the first colonizers on newly exposed rock or soil, lichens are responsible for the chemical weathering of large inorganic particles and the consequent accumulation of soil. Algae are primary producers in aquatic food chains and are important oxygen-producers.

5. Cellular slime molds exist as individual ameboid cells. Plasmodial slime molds are multinucleate masses of protoplasm. Both survive adverse environmental conditions by forming spores.

6. **a.** Flagella
   **b.** *Giardia*
   **c.** None
   **d.** *Nosema*
   **e.** Pseudopods
   **f.** *Entamoeba*
   **g.** None
   **h.** *Plasmodium*
   **i.** Cilia
   **j.** *Balantidium*
   **k.** Flagella
   **l.** *Trypanosoma*
   **m.** Flagella
   **n.** *Trichomonas*

7. *Trichomonas* cannot survive for long outside a host because it does not form a protective cyst. *Trichomonas* must be transferred from host to host quickly.

8. Ingestion

9. The male reproductive organs are in one individual, and the female reproductive organs are in another. Roundworms belong to the Phylum Nematoda.

10.

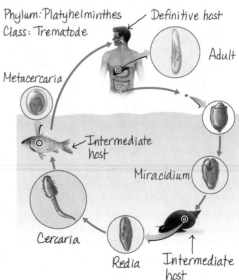

## Multiple Choice

| | | | | |
|---|---|---|---|---|
| 1. d | 3. b | 5. a | 7. b | 9. a |
| 2. b | 4. a | 6. d | 8. d | 10. c |

## Chapter 13

### Review

1. Viruses absolutely require living host cells to multiply.

2. A virus
   - contains DNA or RNA;
   - has a protein coat surrounding the nucleic acid;
   - multiplies inside a living cell using the synthetic machinery of the cell; and
   - causes the synthesis of virions.
   A virion is a fully developed virus particle that transfers the viral nucleic acid to another cell and initiates multiplication.

3. Polyhedral (Figure 13.2); helical (Figure 13.4); enveloped (Figure 13.3); complex (Figure 13.5).

4.

5. Both produce double-stranded RNA, with the − strand being the template for more + strands. + strands act as mRNA in both virus groups.

6. Antibiotic treatment of *S. aureus* can activate phage genes that encode P-V leukocidin.

7. **a.** Viruses cannot easily be observed in host tissues. Viruses cannot easily be cultured in order to be inoculated into a new host. Additionally, viruses are specific for their hosts and cells, making it difficult to substitute a laboratory animal for the third step of Koch's postulates.
   **b.** Some viruses can infect cells without inducing cancer. Cancer may not develop until long after infection. Cancers do not seem to be contagious.

8. **a.** subacute sclerosing panencephalitis
   **b.** common viruses
   **c.** Answers will vary. One example of a possible mechanism is latent, in an abnormal tissue.

9. **a.** of the rigid cell walls
   **b.** vectors such as sap-sucking insects
   **c.** plant protoplasts and insect cell cultures

10. Herpesviridae

### Multiple Choice

| | | | | |
|---|---|---|---|---|
| 1. e | 3. b | 5. b | 7. c | 9. d |
| 2. c | 4. a | 6. e | 8. d | 10. c |

## Chapter 14

### Review

1. **a.** Etiology is the study of the cause of a disease, whereas pathogenesis is the manner in which the disease develops.
   **b.** *Infection* refers to the colonization of the body by a microorganism. Disease is any change from a state of health. A disease may, but does not always, result from infection.
   **c.** A communicable disease is a disease that is spread from one host to another, whereas a noncommunicable disease is not transmitted from one host to another.

2. *Symbiosis* refers to unlike organisms living together. Commensalism—one of the organisms benefits and the other is unaffected; e.g., corynebacteria living on the surface of the eye. Mutualism—both organisms benefit; e.g., *E. coli* receives nutrients and a constant temperature in the large intestine and produces vitamin K and certain B vitamins that are useful for the human host. Parasitism—one organism benefits while the other is harmed; e.g., *Salmonella enterica* receives nutrients and warmth in the large intestine, and the human host experiences gastroenteritis or typhoid fever.

3. **a.** Acute
   **b.** Chronic
   **c.** Subacute

4. Hospital patients may be in a weakened condition and therefore predisposed to infection. Pathogenic microorganisms are generally transmitted to patients by contact and airborne transmission. The reservoirs of infection are the hospital staff, visitors, and other patients.

5. Changes in body function that the patient feels are called *symptoms*. Symptoms such as weakness or pain are not measurable by a physician. Objective changes that the physician can observe and measure are called *signs*.

6. When microorganisms causing a local infection enter a blood or lymph vessel and are spread throughout the body, a systemic infection can result.

7. Mutualistic microorganisms are providing a chemical or environment that is essential for the host. Commensal organisms are not essential; another microorganism might serve the function as well.

8. Incubation period, prodromal period, period of illness, period of decline (may be crisis), period of convalescence.

9. *Escherichia coli*

10.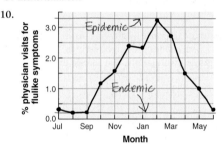

### Multiple Choice

| | | | | |
|---|---|---|---|---|
| 1. a | 3. a | 5. d | 7. c | 9. c |
| 2. b | 4. d | 6. a | 8. a | 10. b |

## Chapter 15

### Review

1. The ability of a microorganism to produce a disease is called *pathogenicity*. The degree of pathogenicity is *virulence*.

2. Encapsulated bacteria can resist phagocytosis and continue growing. *Streptococcus pneumoniae* and *Klebsiella pneumoniae* produce capsules that are related to their virulence. M protein found in the cell walls of *Streptococcus pyogenes* and mycolic acid in the cell walls of *Mycobacterium* help these bacteria resist phagocytosis.

3. Hemolysins lyse red blood cells; hemolysis might supply nutrients for bacterial growth. Leukocidins destroy neutrophils and macrophages that are active in phagocytosis; this decreases host resistance to infection. Coagulase causes fibrinogen in blood to clot; the clot may protect the bacterium from phagocytosis and other host defenses. Bacterial kinases break down fibrin; kinases can destroy a clot that was made to isolate the bacteria, thus allowing the bacteria to spread. Hyaluronidase hydrolyzes the hyaluronic acid that binds cells together; this could allow the bacteria to spread through tissues. Siderophores take iron from host iron-transport proteins, thus allowing bacteria to get iron for growth. IgA proteases destroy IgA antibodies; IgA antibodies protect mucosal surfaces.

4.  **a.** Would inhibit bacteria.
    **b.** Would prevent adherence of *N. gonorrhoeae*.
    **c.** *S. pyogenes* would not be able to attach to host cells and would be more susceptible to phagocytosis.

5.

|  | Exotoxin | Endotoxin |
|---|---|---|
| Bacterial source | Gram + | Gram − |
| Chemistry | Proteins | Lipid A |
| Toxigenicity | High | Low |
| Pharmacology | Destroy certain cell parts or physiological functions | Systemic, fever, weakness, aches and shock |
| Example | Botulinum toxin | Salmonellosis |

6.
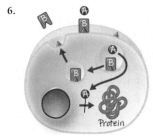

7.  Pathogenic fungi do not have specific virulence factors; capsules, metabolic products, toxins, and allergic responses contribute to the virulence of pathogenic fungi. Some fungi produce toxins that, when ingested, produce disease. Protozoa and helminths elicit symptoms by destroying host tissues and producing toxic metabolic wastes.

8.  *Legionella*

9.  Viruses avoid the host's immune response by growing inside host cells; some can remain latent in a host cell for prolonged periods. Some protozoa avoid the immune response by mutations that change their antigens.

10. *Neisseria gonorrhoeae*

### Multiple Choice

| | | | | |
|---|---|---|---|---|
| 1. d | 3. d | 5. c | 7. b | 9. d |
| 2. c | 4. a | 6. a | 8. a | 10. c |

## Chapter 16

### Review

1.  Physical: movement out; Chemical: lysozome; acids
    Physical: movement out; Chemical: acidic environment in female

2.  Inflammation is the body's response to tissue damage. The characteristic symptoms of inflammation are pain, redness, immobility, swelling, and heat.

3.  Interferons are defensive proteins. Alpha interferon and beta interferon induce uninfected cells to produce antiviral proteins. Gamma interferon is produced by lymphocytes and activates neutrophils to kill bacteria.

4.  Endotoxin binds C3b, which activates C5–C9 to cause cell lysis. This can result in free cell wall fragments, which bind more C3b, resulting in C5–C9 damage to host-cell membranes.

5.  Toxic oxygen products can kill pathogens.

6.  The recipient's antibodies combine with donor antigens and fix complement; the activated complement causes hemolysis.

7.  Inhibit formation of C3b; prevent MAC formation; hydrolyze C5a.

8.
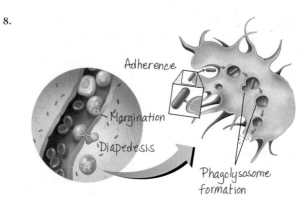

9.  **a.** Innate. Facilitate adherence of phagocyte and pathogen.
    **b.** Innate. Bind iron.
    **c.** Innate. Kill or inhibit bacteria.

10. Monocyte (macrophage)

### Multiple Choice

| | | | | |
|---|---|---|---|---|
| 1. a | 3. c | 5. e | 7. c | 9. d |
| 2. d | 4. d | 6. a | 8. b | 10. e |

## Chapter 17

### Review

1.  **a.** Adaptive immunity is the resistance to infection obtained during the life of the individual; it results from the production of antibodies and T cells. Innate immunity refers to the resistance of species or individuals to certain diseases that is not dependent on antigen-specific immunity.
    **b.** Humoral immunity is due to antibodies (and B cells). Cellular immunity is due to T cells.
    **c.** Active immunity refers to antibodies produced by the individual who carries them. Passive immunity refers to antibodies produced by another source and then transferred to the individual who needs the antibodies.
    **d.** $T_H1$ cells produce cytokines that activate T cells. Cytokines produced by $T_H2$ cells activate B cells.
    **e.** Natural immunity is acquired naturally, i.e., from mother to newborn or following an infection. Artificial immunity is acquired from medical treatment, i.e., by injection of antibodies or by vaccination.
    **f.** T-dependent antigens: Certain antigens must combine with self-antigens to be recognized by $T_H$ cells and then by B cells. T-independent antigens can elicit an antibody response without T cells.
    **g.** CTLp cells cannot kill other cells until they are activated to CTLs by an antigen-presenting cell.
    **h.** Immunoglobins = antibodies; TCRs = antigen receptors on T cells.

2.  The major histocompatability complex (MHC) are self-antigens. CTL cells react with MHC I; $T_H$ cells react with MHC II.

3.
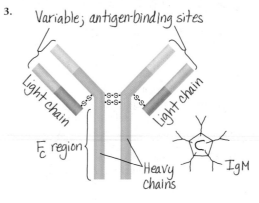

4. See Figure 17.19.

5. CTLs destroy target cells on contact. $T_H$ cells interact with an antigen to "present" it to a B cell for antibody formation. $T_R$ cells suppress the immune response. Cytokines are chemicals released by cells that initiate a response by other cells.

6.
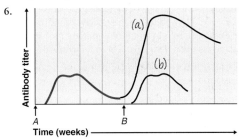

7. Both would prevent attachment of the pathogen; (a) interfere with the attachment site on the pathogen and (b) interfere with the pathogen's receptor site.

8. Rearrangement of the V region genes during embryonic development produces B cells with different antibody genes.

9. The person recovered because he or she produced antibodies against the pathogen. The memory response will continue to protect the person against that pathogen.

10. Dendritic cell

**Multiple Choice**

| | | | | |
|---|---|---|---|---|
| 1. d | 3. b | 5. d | 7. c | 9. c |
| 2. e | 4. c | 6. e | 8. d | 10. d |

## Chapter 18

### Review

1. **a.** Whole-agent. Live, avirulent virus that can cause the disease if it mutates back to its virulent state.
   **b.** Whole-agent; (heat-) killed bacteria.
   **c.** Subunit; (heat- or formalin-) inactivated toxin.
   **d.** Subunit
   **e.** Subunit
   **f.** Conjugated
   **g.** Nucleic acid

2. **a.** Some viruses are able to agglutinate red blood cells. This reaction is used to detect the presence of large numbers of virions capable of causing hemagglutination (e.g., *Influenzavirus*).
   **b.** Antibodies produced against viruses that are capable of agglutinating red blood cells will inhibit the agglutination. Hemagglutination inhibition can be used to detect the presence of antibodies against these viruses.
   **c.** This is a procedure to detect antibodies that react with soluble antigens by first attaching the antigens to insoluble latex spheres. This procedure may be used to detect the presence of antibodies that develop during certain mycotic or helminthic infections.

3.
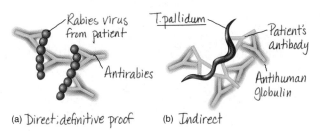
(a) Direct; definitive proof   (b) Indirect

4. See Figure 18.2.

5. If excess antibody is present, an antigen will combine with several antibody molecules. If excess antigen is present, an antibody will combine with several antigens. Refer to Figure 18.3.

6.
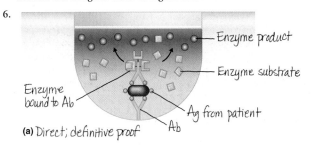
(a) Direct; definitive proof
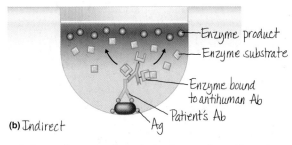
(b) Indirect

7. Particulate antigens react in agglutination reactions. The antigens can be cells or soluble antigens bound to synthetic particles. Soluble antigens take part in precipitation reactions.

8. **a.** 5   **d.** 3
   **b.** 4, 6   **e.** 6
   **c.** 1   **f.** 2, 4

9. **a.** 5   **d.** 6
   **b.** 3   **e.** 2
   **c.** 1   **f.** 4

10. Positive tuberculin skin test; the person has antibodies against *M. tuberculosis*.

**Multiple Choice**

| | | | | |
|---|---|---|---|---|
| 1. c | 3. b | 5. a | 7. c | 9. b |
| 2. d | 4. c | 6. b | 8. a | 10. c |

## Chapter 19

### Review

1.
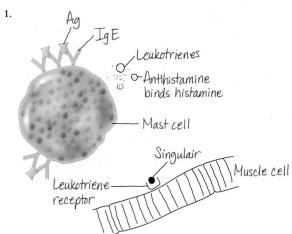

2. Recipient's serum contains complement; activated complement causes hemolysis.

3. Recipient's antibodies will react with donor's tissues.

4. Refer to Figure 19.8.
   a. The observed symptoms are due to lymphokines.
   b. When a person contacts poison oak initially, the antigen (catechols on the leaves) binds to tissue cells, is phagocytized by macrophages, and is presented to receptors on the surface of T cells. Contact between the antigen and the appropriate T cell stimulates the T cell to proliferate and become sensitized. With subsequent exposure to the antigen, sensitized T cells release lymphokines, and a delayed hypersensitivity occurs.
   c. Small repeated doses of the antigen are believed to cause the production of IgG (blocking) antibodies.

5. Lupus patients have antibodies directed at their own DNA.

6. Cytotoxic: Antibodies react with cell-surface antigens.
   Immune complex: Antibody–complement complexes deposit in tissues.
   Cell-mediated: T cells destroy self cells. See Table 19.1.

7. Natural
      Inherited
      Viral infections, most notably HIV
   Artificial
      Induced by immunosupression drugs
   Result: Increased susceptibility to various infections depending on the type of immune deficiency.

8. Tumor cells have tumor-specific antigens. CTL cells may react with tumor-specific antigens, initiating lysis of the tumor cells.

9. Some malignant cells can escape the immune system by antigen modulation or immunological enhancement. Immunotherapy might trigger immunological enhancement. The body's defense against cancer is cell-mediated and not humoral. Transfer of lymphocytes could cause graft-versus-host disease.

10. IgE antibody

## Multiple Choice

| | | | | |
|---|---|---|---|---|
| 1. b | 3. b | 5. d | 7. a | 9. c |
| 2. b | 4. a | 6. e | 8. d | 10. b |

## Chapter 20

### Review

1.

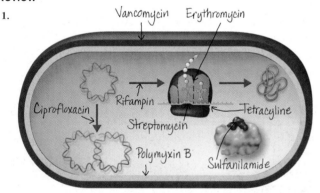

2. The drug (1) should exhibit selective toxicity; (2) should have a broad spectrum; (3) should not produce hypersensitivity in the host; (4) should not produce drug resistance; and (5) should not harm normal microbiota.

3. Because a virus uses the host cell's metabolic machinery, it is difficult to damage the virus without damaging the host. Fungi, protozoa, and helminths possess eukaryotic cells. Therefore, antiviral, antifungal, antiprotozoan, and antihelminthic drugs must also affect eukaryotic cells.

4. Drug resistance is the lack of susceptibility of a microorganism to a chemotherapeutic agent. Drug resistance may develop when microorganisms

are constantly exposed to an antimicrobial agent. Ways to minimize the development of drug-resistant microorganisms include judicious use of antimicrobial agents; following directions on the prescription; or administering two or more drugs simultaneously.

5. Simultaneous use of two agents can prevent the development of resistant strains of microorganisms, take advantage of the synergistic effect, provide therapy until a diagnosis is made, and lessen the toxicity of individual drugs by reducing the dosage of each in combination. One problem that can result from simultaneous use of two agents is an antagonistic effect.

6. a. Like polymyxin B, causes leaks in the plasma membrane.
   b. Interferes with translation.

7. a. Inhibits formation of peptide bond.
   b. Prevents translocation of ribosome along mRNA.
   c. Interferes with attachment of tRNA to mRNA-ribosome complex.
   d. Changes shape of 30S portion of ribosome, resulting in misreading mRNA.
   e. Prevents 70S ribosomal subunits from forming.
   f. Prevents release of peptide from ribosome.

8. DNA polymerase adds bases to the 3′–OH.

9. a. Penicillin inhibits bacterial cell wall synthesis. Echinocandin inhibits fungal cell wall synthesis.
   b. Imidazole interferes with fungal plasma membrane synthesis. Polymyxin B disrupts any plasma membrane.

10. Human immunodeficiency virus

### Multiple Choice

| | | | | |
|---|---|---|---|---|
| 1. b | 3. a | 5. a | 7. e | 9. c |
| 2. a | 4. b | 6. d | 8. b | 10. c |

## Chapter 21

### Review

1. Bacteria usually enter through inapparent openings in the skin. Fungal pathogens (except subcutaneous) often grow on the skin itself. Viral infections of the skin (except warts and herpes simplex) most often gain access to the body through the respiratory tract.

2. *Staphylococcus aureus; Streptococcus pyogenes*

3.

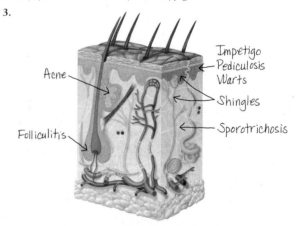

4.

| Etiologic Agtent | Clinical Symptoms | Mode of Transmission |
|---|---|---|
| *P. acnes* | Infected oil glands | Direct contact |
| *S. aureus* | Infected hair follicles | Direct contact |
| Papovavirus | Benign tumor | Direct contact |
| Herpesvirus | Vesicular rash | Respiratory route |
| Enteroviruses | Flat or raised rash | Direct contact |
| Paramyxovirus | Papular rash, Koplik's spots | Respiratory route |
| Togavirus | Macular rash | Respiratory route |

5. The test determines the woman's susceptibility to rubella. If the test is negative, she is susceptible to the disease. If she acquires the disease during pregnancy, the fetus could become infected. A susceptible woman should be vaccinated.

6.

| Symptoms | Disease |
|---|---|
| Koplik's spots | Measles or rubeola |
| Macular rash | Measles |
| Vesicular rash | Chickenpox |
| Small, spotted rash | German measles |
| "Blisters" | HHV-1 or HHV-2 infection |
| Corneal ulcer | Herpetic keratitis |

7. The central nervous system can be invaded following herpetic keratitis; this results in encephalitis.

8. Attenuated measles, mumps, and rubella viruses.

9. The patient has scabies, an infestation of mites in the skin. It is treated with permethrin insecticide or gamma benzene hexachloride. The presence of a six-legged arthropod (insect) indicates pediculosis (lice).

10. *Propionibacterium acnes*

**Multiple Choice**

| | | | | |
|---|---|---|---|---|
| 1. c | 3. b | 5. d | 7. e | 9. a |
| 2. d | 4. c | 6. d | 8. d | 10. d |

---

## Chapter 22

### Review

1. The symptoms of tetanus are due to neurotoxin, not to bacterial growth (infection and inflammation).

2. **a.** Vaccination with tetanus toxoid.
   **b.** Immunization with antitetanus toxin antibodies.

3. "Improperly cleaned" because *C. tetani* is found in soil that might contaminate a wound. "Deep puncture" because it is likely to be anaerobic. "No bleeding" because a flow of blood ensures an aerobic environment and some cleansing.

4. Etiology—Picornavirus (poliovirus).
   Transmission—Ingestion of contaminated water.
   Symptoms—Headache, sore throat, fever, nausea; rarely, paralysis.
   Prevention—Sewage treatment.

These vaccinations provide artificially acquired active immunity because they cause the production of antibodies, but they do not prevent or reverse damage to nerves.

5.

| Causative Agent | Susceptible Population | Transmission | Treatment |
|---|---|---|---|
| *N. meningitidis* | Children; military recruits | Respiratory | Penicillin |
| *H. influenzae* | Children | Respiratory | Rifampin |
| *S. pneumoniae* | Children; elderly | Respiratory | Penicillin |
| *L. monocytogenes* | Anyone | Foodborne | Penicillin |
| *C. neoformans* | Immunosuppressed individuals | Respiratory | Amphotericin B |

6.

| Disease | Etiology | Transmission | Symptoms | Treatment |
|---|---|---|---|---|
| Arboviral encephalitis | Togaviruses, Arboviruses | Mosquitoes (*Culex*) | Headache, fever, coma | Immune serum |
| African trypanosomiasis | *T. b. gambiense, T. b. rhodesiense* | Tsetse fly | Decreased physical activity and mental acuity | Suramin; melarsoprol |
| Botulism | *C. botulinum* | Ingestion | Flaccid paralysis | Antitoxin |
| Leprosy | *M. leprae* | Direct contact | Areas of sensation loss in skin | Dapsone |

7.

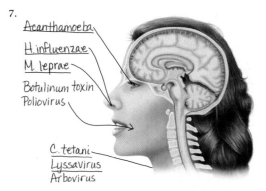

Acanthamoeba
H. influenzae
M. leprae
Botulinum toxin
Poliovirus
C. tetani
Lyssavirus
Arbovirus

8. Postexposure treatment—Passive immunization with antibodies followed by active immunization with HDCV. Preexposure treatment—Active immunization with HDCV.

Following exposure to rabies, antibodies are needed immediately to inactivate the virus. Passive immunization provides these antibodies. Active immunization will provide antibodies over a longer period of time, but they are not formed immediately.

9. The causative agent of Creutzfeldt-Jakob disease (CJD) is transmissible. Although there is some evidence for an inherited form of the disease, it has been transmitted by transplants. Similarities with viruses are (1) the prion cannot be cultured by conventional bacteriological techniques and (2) the prion is not readily seen in patients with CJD.

10. *Cryptococcus neoformans*

**Multiple Choice**

| | | | | |
|---|---|---|---|---|
| 1. a | 3. a | 5. a | 7. b | 9. c |
| 2. c | 4. b | 6. e | 8. a | 10. a |

# Chapter 23

## Review

1.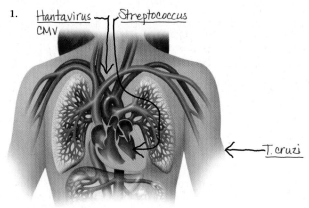

Hantavirus — Streptococcus
CMV

← T. cruzi

2.

| Disease | Causative Agent | Predisposing Conditions |
|---|---|---|
| p.s. | *Str. pyogenes* | Abortion or childbirth |
| s.b.e. | alpha-hemolytic strep. | Preexisting lesions |
| a.b.e. | *Sta. aureus* | Abnormal heart valves |
| r.f. | *Str. pyogenes* | Autoimmune |

3. All are vectorborne rickettsial diseases. They differ from each other in (1) etiologic agent, (2) vector, (3) severity and mortality, and (4) incidence (e.g., epidemic, sporadic).

4.

| Causative Agent | Vector | Treatment |
|---|---|---|
| *Plasmodium* | *Anopheles* | Quinine derivative |
| *Flavivirus* | *Aedes aegypti* | None |
| *Flavivirus* | *Aedes aegypti* | None |
| *Borrelia* | Soft ticks | Tetracycline |
| *Leishmania* | Sandflies | Amphotericin B, paromomycin, meglumine antimoniate |

5.

| Disease | Causative Agent | Transmission | Reservoir |
|---|---|---|---|
| Tularemia | *Francisella tularensis* | Skin abrasions, ingestion, inhalation, bites | Rabbits |
| Brucellosis | *Brucella* spp. | Ingestion of milk, direct contact | Cattle |
| Anthrax | *Bacillus anthracis* | Skin abrasions, inhalation, ingestion | Soil, cattle |
| Lyme disease | *Borrelia burgdorferi* | Tick bites | Deer, mice |
| Ehrlichiosis | *Ehrlichia* spp. | Tick bites | Deer |
| Cytomegalic inclusion disease | HHV-5 | Saliva, blood | Humans |
| Plague | *Yersinia pestis* | Flea bites, inhalation | Rodents |

6.

| Disease | Causative Agent | Transmission | Reservoir | Endemic Area |
|---|---|---|---|---|
| Schistosomiasis | *Schistosoma* spp. | Penetrate skin | Aquatic snail | Asia, South America |
| Toxoplasmosis | *Toxoplasma gondii* | Ingestion, inhalation | Cats | United States |
| Chagas disease | *Trypanosoma cruzi* | "Kissing bug" | Rodents | Central America |

7.

| | Reservoir | Etiology | Transmission | Symptoms |
|---|---|---|---|---|
| Cat-scratch disease | Cats | *Bartonella henselae* | Scratch; touching eyes, fleas | Swollen lymph nodes, fever, malaise |
| Toxoplasmosis | Cats | *Toxoplasma gondii* | Ingestion | None, congenital infections, neurologic damage |

8. Gangrenous tissue is anaerobic and has suitable nutrients for *C. perfringens*.

9. Infectious mononucleosis is caused by EB virus and is transmitted in oral secretions.

10. Rubella virus

## Multiple Choice

| | | | | |
|---|---|---|---|---|
| 1. a | 3. d | 5. a | 7. a | 9. c |
| 2. e | 4. c | 6. e | 8. c | 10. c |

# Chapter 24

## Review

1.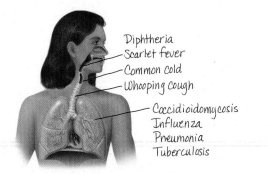

Diphtheria
Scarlet fever
Common cold
Whooping cough
Coccidioidomycosis
Influenza
Pneumonia
Tuberculosis

2. Mycoplasmal pneumonia is caused by *Mycoplasma pneumoniae* bacteria. Viral pneumonia can be caused by several different viruses. Mycoplasmal pneumonia can be treated with tetracyclines, whereas viral pneumonia cannot.

3.

| Disease | Causative Agent | Symptoms |
|---|---|---|
| *Upper Respiratory System* | | |
| Common cold | Coronaviruses, rhinoviruses, EV-D68 | Sneezing, excessive nasal secretions, congestion |
| *Lower Respiratory System* | | |
| Viral pneumonia | Several viruses | Fever, shortness of breath, chest pains |
| Influenza | *Influenzavirus* | Chills, fever, headache, muscular pains |
| RSV | Respiratory syncytial virus | Coughing, wheezing |

Zanamivir and oseltamivir are used to treat influenza; palivizumab, for life-threatening RSV.

4.

| Disease | Symptoms |
|---|---|
| Streptococcal pharyngitis | Pharyngitis and tonsillitis |
| Scarlet fever | Rash and fever |
| Diphtheria | Membrane across throat |
| Whooping cough | Paroxysmal coughing |
| Tuberculosis | Tubercles, coughing |
| Pneumococcal pneumonia | Reddish lungs, fever |
| *H. influenzae* pneumonia | Similar to pneumococcal pneumonia |
| Chamydial pneumonia | Low fever, cough, and headache |
| Otitis media | Earache |
| Legionellosis | Fever and cough |
| Psittacosis | Fever and headache |
| Q fever | Chills and chest pain |
| Epiglottitis | Inflamed, abscessed epiglottis |
| Melioidosis | Pneumonia |

Refer to Diseases in Focus 24.1, 24.2, and 24.3 to complete the table.

5. Inhalation of large numbers of spores from *Aspergillus* or *Rhizopus* can cause infections in individuals with impaired immune systems, cancer, and diabetes.

6. No. Many different organisms (gram-positive bacteria, gram-negative bacteria, and viruses) can cause pneumonia. Each of these organisms is susceptible to different antimicrobial agents.

7.

| Disease | Endemic Areas in the United States |
|---|---|
| Histoplasmosis | States adjoining the Mississippi and Ohio rivers |
| Coccidioidomycosis | American Southwest |
| Blastomycosis | Mississippi river valleys and Great Lakes area |
| *Pneumocystis* pneumonia | Ubiquitous |

Refer to Diseases in Focus 24.3 to complete the table.

8. In the tuberculin test, purified protein derivative (PPD) from *M. tuberculosis* is injected into the skin. Induration and reddening of the area around the injection site indicate an active infection or immunity to tuberculosis.

9. a. *Staphylococcus aureus*
   b. *Streptococcus pyogenes*
   c. *S. pneumoniae*
   d. *Corynebacterium diphtheriae*
   e. *Mycobacterium tuberculosis*
   f. *Moraxella catarrhalis*
   g. *Bordetella pertussis*
   h. *Burkholderia pseudomallei*
   i. *Legionella pneumophila*
   j. *Haemophilus influenzae*
   k. *Chlamydophila psittaci*
   l. *Coxiella burnetii*
   m. *Mycoplasma pneumoniae*

10. *Bordetella pertussis*

## Multiple Choice

| | | | | |
|---|---|---|---|---|
| 1. a | 3. e | 5. c | 7. a | 9. b |
| 2. c | 4. a | 6. b | 8. e | 10. d |

## Chapter 25
## Review

1.

| | Causative Agent | Method of Transmission |
|---|---|---|
| Aflatoxin poisoning | *Aspergillus flavus* | Ingestion of toxin |
| Cryptosporidiosis | *Cryptosporidium hominis* | Ingestion |
| Pinworms | *Enterobius vermicularis* | Ingestion |
| Whipworms | *Trichuris trichiura* | Ingestion |

Refer to Diseases in Focus 25.5 to complete the table.

2.

| Causative Agent | Suspect Foods | Prevention |
|---|---|---|
| *V. parahaemolyticus* | Oysters, shrimp | Cooking |
| *V. cholerae* | Water | |
| *E. coli* O157 | Water, vegetables, ground beef | Cooking |
| *C. jejuni* | Chicken | Cooking |
| *Y. enterocolitica* | Meat, milk | Cooking |
| *C. perfringens* | Meat | Refrigeration after cooking |
| *B. cereus* | Rice dishes | Refrigeration after cooking |
| *S. aureus* | Creamy, salty | Refrigerating food |
| *S. enterica* | Eggs, poultry, vegetables | Cooking |
| *Shigella* spp. | Water, environmental fecal contamination | Disinfection |

Refer to Diseases in Focus 25.2 to complete the table.

3.

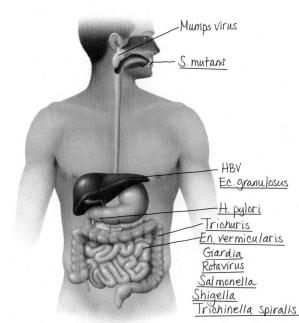

4. Certain strains of *E. coli* may produce an enterotoxin or invade the epithelium of the large intestine. *E. coli* bacteria produce vitamin K and folic acid that can be used by humans and produce bacteriocins that inhibit pathogens.

5. Toxin produced by fungi; see p. 746.

6. All four are caused by protozoa. The infections are acquired by ingesting protozoa in contaminated water. Giardiasis is a prolonged diarrhea. Amebic dysentery is the most severe dysentery, with blood and mucus in the stools. *Cryptosporidium* and *Cyclospora* produce severe diseases in persons with immunodeficiencies.

7. **Bacterial intoxication:** Microorganisms must be allowed to grow in food from the time of preparation to the time of ingestion. This usually occurs when foods are stored unrefrigerated or improperly canned. The etiologic agents (*Staphylococcus aureus* or *Clostridium botulinum*) must produce an exotoxin. Onset: 1 to 48 hours. Duration: A few days. Treatment: Antimicrobial agents are ineffective. The patient's symptoms may be treated.

   **Bacterial infection:** Viable microorganisms must be ingested with food or water. The organisms could be present during preparation and survive cooking or be inoculated during later handling. The etiologic agents are usually gram-negative organisms (*Salmonella, Shigella, Vibrio,* and

*Escherichia*) that produce endotoxins. *Clostridium perfringens* is a gram-positive bacterium that causes food infection. Onset: 12 hours to 2 weeks. Duration: Longer than intoxication because the microorganisms are growing in the patient. Treatment: Rehydration.

8. 

| Disease | Site | Symptoms |
|---------|------|----------|
| Mumps | Parotid glands | Inflammation of the parotid glands and fever |
| Hepatitis A | Liver | Anorexia, fever, diarrhea |
| Hepatitis B | Liver | Anorexia, fever, joint pains, jaundice |
| Viral gastroenteritis | Lower GI tract | Nausea, diarrhea, vomiting |

Refer to Diseases in Focus 25.3 and 25.4 to complete this question.

9. Cook meat thoroughly. Eliminate the source of contamination to cattle and pigs.

10. *Giardia*

### Multiple Choice

| | | | | |
|---|---|---|---|---|
| 1. d | 3. e | 5. e | 7. b | 9. a |
| 2. e | 4. b | 6. b | 8. e | 10. d |

## Chapter 26

### Review

1.

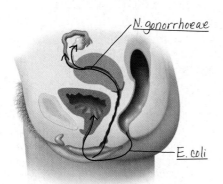

2. Urinary tract infections may be acquired by improper personal hygiene and contamination during medical procedures. They are often caused by opportunistic pathogens.

3. The proximity of the anus to the urethra and the relatively short length of the urethra can allow contamination of the urinary bladder in females. Gastrointestinal infections are also a predisposing factor for cystitis in females.

4. *Escherichia coli* causes about 75% of the cases. Portals of entry are from the lower urinary tract or systemic infections.

5. 

| Disease | Symptoms | Diagnosis |
|---------|----------|-----------|
| Bacterial vaginosis | Fishy odor | Odor, pH, clue cells |
| Gonorrhea | Painful urination | Isolation of *Neisseria* |
| Syphilis | Chancre | FTA–ABS |
| PID | Abdominal pain | Culture of pathogen |
| NGU | Urethritis | Absence of *Neisseria* |
| LGV | Lesion, lymph node enlargement | Observation of *Chlamydia* in cells |
| Chancroid | Swollen ulcer | Isolation of *Haemophilus* |

Refer to Diseases in Focus 26.2 and 26.3 to complete table.

6. Symptoms—Burning sensation, vesicles, painful urination.
Etiology—Herpes simplex virus type 2 (sometimes type 1). When the lesions are not present, the virus is latent and transmissible.

7. *Candida albicans*—Severe itching; thick, yellow, cheesy discharge.
*Trichomonas vaginalis*—Profuse yellow discharge with disagreeable odor.

8. 

| Disease | Prevention of Congenital Disease |
|---------|----------------------------------|
| Gonorrhea | Treatment of newborn's eyes |
| Syphilis | Prevention and treatment of mother's disease |
| NGU | Treatment of newborn's eyes |
| Genital herpes | Cesarean delivery during active infection |

9. *Chlamydia trachomatis*

### Multiple Choice

| | | | | |
|---|---|---|---|---|
| 1. b | 3. a | 5. d | 7. c | 9. b |
| 2. e | 4. c | 6. c | 8. b | 10. a |

## Chapter 27

### Review

1. The koala should have an organ housing a large population of cellulose-degrading microorganisms.

2. *Penicillium* might make penicillin to reduce competition from faster-growing bacteria.

3. **a.** amino acids
   **b.** $SO_4^{2-}$
   **c.** plants and bacteria
   **d.** $H_2S$
   **e.** carbohydrates
   **f.** $S^0$

4. Phosphorus must be available for all organisms.

5.

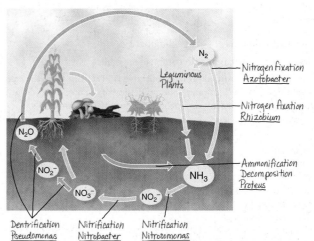

6. Cyanobacteria: With fungi, cyanobacteria act as the photoautotrophic partner in a lichen; they may also fix nitrogen in the lichen. With *Azolla*, they fix nitrogen.
Mycorrhizae: Fungi that grow in and on the roots of higher plants; increase absorption of nutrients.
*Rhizobium*: In root nodules of legumes; fix nitrogen.
*Frankia*: In root nodules of alders, roses, and other plants; fix nitrogen.

7. Settling
Flocculation treatment
Sand filtration (or activated charcoal filtration)
Chlorination
The amount of treatment prior to chlorination depends on the amount of inorganic and organic matter in the water.

8. 

| | |
|---|---|
| **a.** 2 | **e.** 3 |
| **b.** 1 | **f.** 2 |
| **c.** 2 | **g.** 3 |
| **d.** 2 | |

9. Biodegradation of sewage, herbicides, oil, or PCBs.

10. Cyanobacteria (*Anabaena*)

## Multiple Choice

| | | | | |
|---|---|---|---|---|
| 1. a | 3. b | 5. c | 7. b | 9. e |
| 2. b | 4. b | 6. c | 8. b | 10. c |

## Chapter 28

### Review

1. Industrial microbiology is the science of using microorganisms to produce products or accomplish a process. Industrial microbiology provides (1) chemicals such as antibiotics that would not otherwise be available, (2) processes to remove or detoxify pollutants, (3) fermented foods that have desirable flavors or enhanced shelf life, and (4) enzymes for manufacturing a variety of goods.

2. The goal of commercial sterilization is to eliminate spoilage and disease-causing organisms. The goal of hospital sterilization is complete sterilization.

3. The acid in the berries will prevent the growth of some microbes.

4. Milk $\xrightarrow{\text{Lactic Acid Bacteria}}$ Curd + Whey
   $\qquad\qquad\qquad\qquad\quad\downarrow\qquad\downarrow$
   $\qquad\qquad\qquad\qquad$ Cheese Waste

   Hard cheese is ripened by lactic acid bacteria growing anaerobically in the interior of the curd. Soft cheese is ripened by molds growing aerobically on the outside of the curd.

5. Nutrients must be dissolved in water; water is also needed for hydrolysis. Malt is the carbon and energy source that the yeast will ferment to make alcohol. Malt contains glucose and maltose from the action of amylases on starch in seeds (e.g., barley).

6. A bioreactor provides the following advantages over simple flask containers:
   - Larger culture volumes can be grown.
   - Process instrumentation for monitoring and controlling critical environmental conditions such as pH, temperature, dissolved oxygen, and aeration can be used.
   - Sterilization and cleaning systems are designed in place.
   - It offers aseptic sampling and harvest systems for in-process sampling.
   - Improved aeration and mixing characteristics result in improved cell growth and high final cell densities.
   - A high degree of automation is possible.
   - Process reproducibility is improved.

7. (1) Enzymes don't produce hazardous wastes. (2) Enzymes work under reasonable conditions; for example, they don't require high temperatures or acidity. (3) Use of enzymes eliminates the need to use petroleum in chemical syntheses of solvents such as alcohol and acetone. (4) Enzymes are biodegradable. (5) Enzymes are not toxic.

8. The production of ethyl alcohol from corn; or methane from sewage. Alcohols and hydrogen are produced by fermentation; methane is produced by anaerobic respiration.

9. 

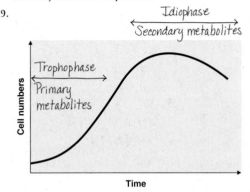

10. *Saccharomyces cerevisiae*

## Multiple Choice

| | | | | |
|---|---|---|---|---|
| 1. c | 3. e | 5. b | 7. e | 9. b |
| 2. b | 4. c | 6. c | 8. a | 10. a |

# Appendix A

## Metabolic Pathways

**Figure A.1 The Calvin-Benson cycle for photosynthetic carbon metabolism.**

**❶–❸** The initial fixation and reduction of carbon occur, generating the three-carbon compounds glyceraldehyde 3-phosphate and dihydroxyacetone phosphate, **❹** which are interconvertible. **Ⓐ–Ⓓ** On average, 2 of every 12 three-carbon molecules are used in the synthesis of glucose. **❺** Ten of every 12 three-carbon molecules are used to generate ribulose 5-phosphate by a complex series of reactions. **❻** The ribulose 5-phosphate is then phosphorylated at the expense of ATP, forming ribulose 1,5-diphosphate, the acceptor molecule with which the sequence began. (See Figure 5.26, page 137, for a simplified version of the Calvin-Benson cycle.)

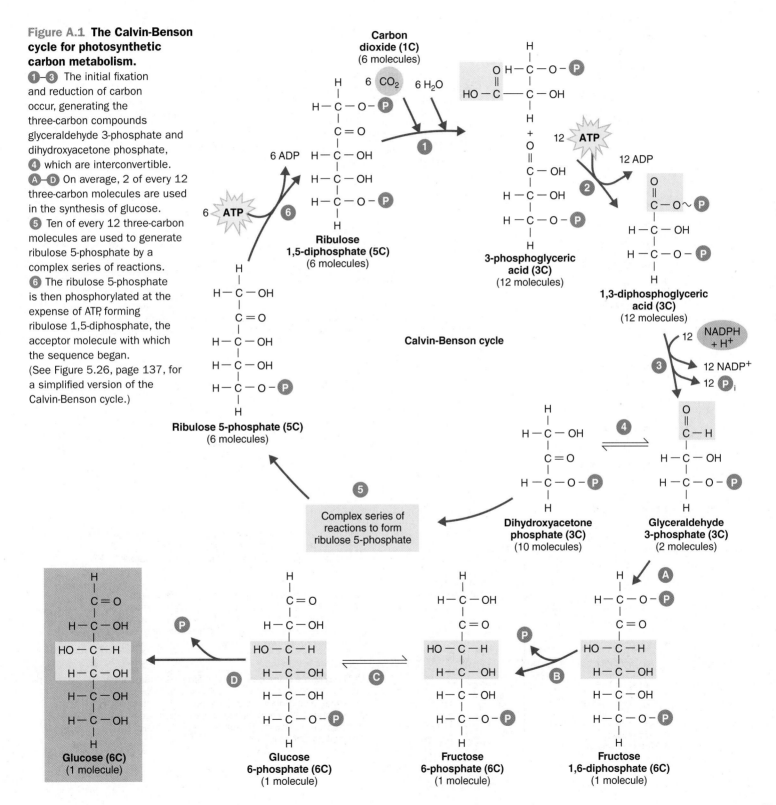

Appendices

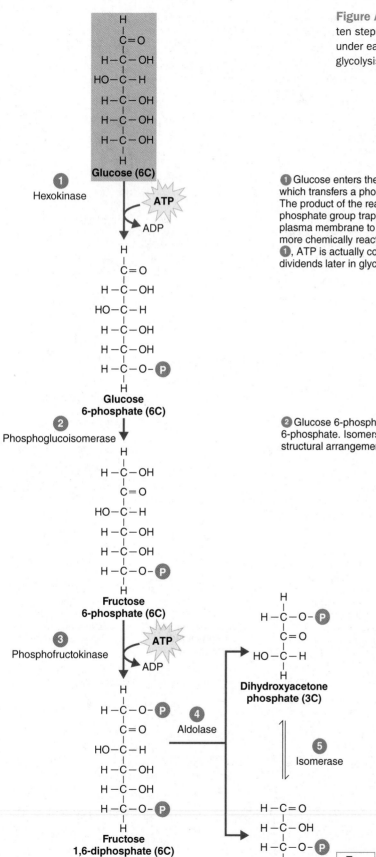

**Figure A.2  Glycolysis (Embden-Meyerhof pathway).** Each of the ten steps of glycolysis is catalyzed by a specific enzyme, which is named under each step number. (See Figure 5.12, page 122, for a simplified version of glycolysis.)

❶ Glucose enters the cell and is phosphorylated by the enzyme hexokinase, which transfers a phosphate group from ATP to the number 6 carbon of the sugar. The product of the reaction is glucose 6-phosphate. The electrical charge of the phosphate group traps the sugar in the cell because of the impermeability of the plasma membrane to ions. Phosphorylation of glucose also makes the molecule more chemically reactive. Although glycolysis is supposed to *produce* ATP, in step ❶, ATP is actually consumed—an energy investment that will be repaid with dividends later in glycolysis.

❷ Glucose 6-phosphate is rearranged to convert it to its isomer, fructose 6-phosphate. Isomers have the same number and types of atoms but in different structural arrangements.

❸ In this step, still another molecule of ATP is invested in glycolysis. An enzyme transfers a phosphate group from ATP to the sugar, producing fructose 1,6-diphosphate.

❹ This is the reaction from which glycolysis gets its name ("sugar splitting"). An enzyme cleaves fructose 1,6-diphosphate into two different three-carbon sugars: glyceraldehyde 3-phosphate and dihydroxyacetone phosphate. These two sugars are isomers.

❺ The enzyme isomerase interconverts the three-carbon sugars. The next enzyme in glycolysis uses only glyceraldehyde 3-phosphate as its substrate. This pulls the equilibrium between the two three-carbon sugars in the direction of glyceraldehyde 3-phosphate, which is removed as fast as it forms.

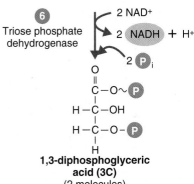

**6**
Triose phosphate
dehydrogenase

2 NAD⁺
2 NADH + H⁺
2 Pᵢ

O
‖
C—O~P
H—C—OH
H—C—O-P
H

**1,3-diphosphoglyceric
acid (3C)**
(2 molecules)

**7**
Phospho-
glycerokinase

2 ADP
2 ATP

O
‖
C—OH
H—C—OH
H—C—O-P
H

**3-phosphoglyceric
acid (3C)**
(2 molecules)

**8**
Phospho-
glyceromutase

O
‖
C—OH
H—C—O-P
H—C—OH
H

**2-phosphoglyceric
acid (3C)**
(2 molecules)

**9**
Enolase

H₂O

O
‖
C—OH
C—O~P
‖
H—C
H

**Phosphoenolpyruvic acid (PEP) (3C)**
(2 molecules)

**10**
Pyruvate kinase

2 ADP
2 ATP

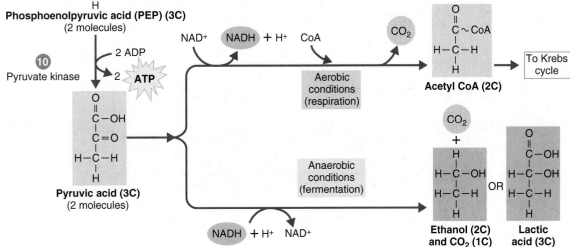

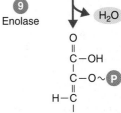

**6** An enzyme now catalyzes two sequential reactions while it holds glyceraldehyde 3-phosphate in its active site. First, the sugar is oxidized at the number 1 carbon and NAD⁺ is reduced, resulting in the formation of NADH + H⁺. Second, the enzyme couples this reaction to the creation of a high-energy phosphate bond at the number 1 carbon of the oxidized substrate. The source of the phosphate is inorganic phosphate, which is always present in the cell. As products, the enzyme releases NADH + H⁺ and 1,3-diphosphoglyceric acid. Notice in the figure that the new phosphate bond is symbolized with a high-energy bond (~), which indicates that the bond is at least as energetic as the phosphate bonds of ATP.

**7** At this step, glycolysis produces ATP. The phosphate group, with its high-energy bond, is transferred from 1,3-diphosphoglyceric acid to ADP. For each glucose molecule that began glycolysis, step **7** produces two molecules of ATP, because every product after the sugar-splitting step (step **4**) is doubled. Of course, two ATPs were invested to get sugar ready for splitting. The ATP ledger now stands at zero. By the end of step **7**, glucose has been converted to two molecules of 3-phosphoglyceric acid.

**8** Next, an enzyme relocates the remaining phosphate group of 3-phosphoglyceric acid to form 2-phosphoglyceric acid. This prepares the substrate for the next reaction.

**9** An enzyme forms a double bond in the substrate by extracting a water molecule from 2-phosphoglyceric acid to form phosphoenolpyruvic acid. This results in the electrons of the substrate being arranged in such a way that the remaining phosphate bond becomes very unstable.

**10** The last reaction of glycolysis produces another molecule of ATP by transferring the phosphate group from phosphoenolpyruvic acid to ADP. Because this step occurs twice for each glucose molecule, the ATP ledger now shows a net gain of two ATPs. Thus, the glycolysis of one molecule of glucose results in two molecules of pyruvic acid, two molecules of NADH + H⁺, and two molecules of ATP. Each molecule of pyruvic acid can now undergo respiration or fermentation.

**Figure A.3 The pentose phosphate pathway.**
This pathway, which operates simultaneously with glycolysis, provides an alternative route for the oxidation of glucose and plays a role in the synthesis of biological molecules, depending on the needs of the cell. Possible fates of the various intermediates are shown in the boxes on the right. (See Chapter 5, page 121.)

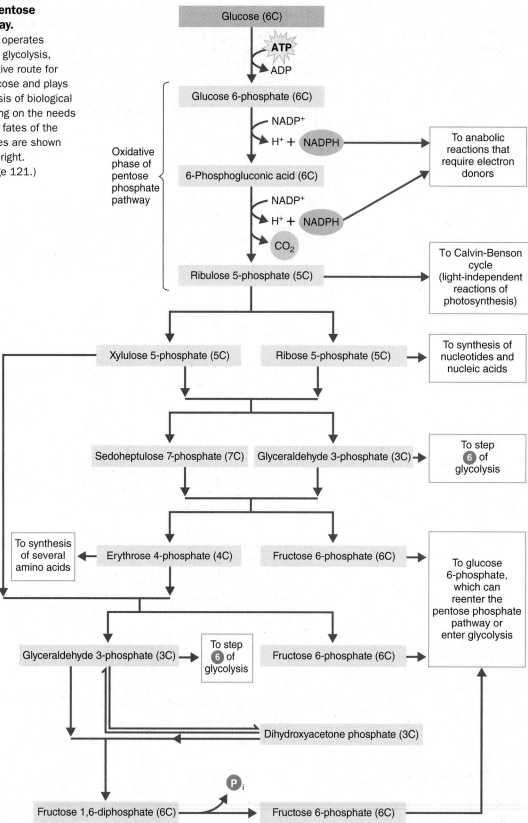

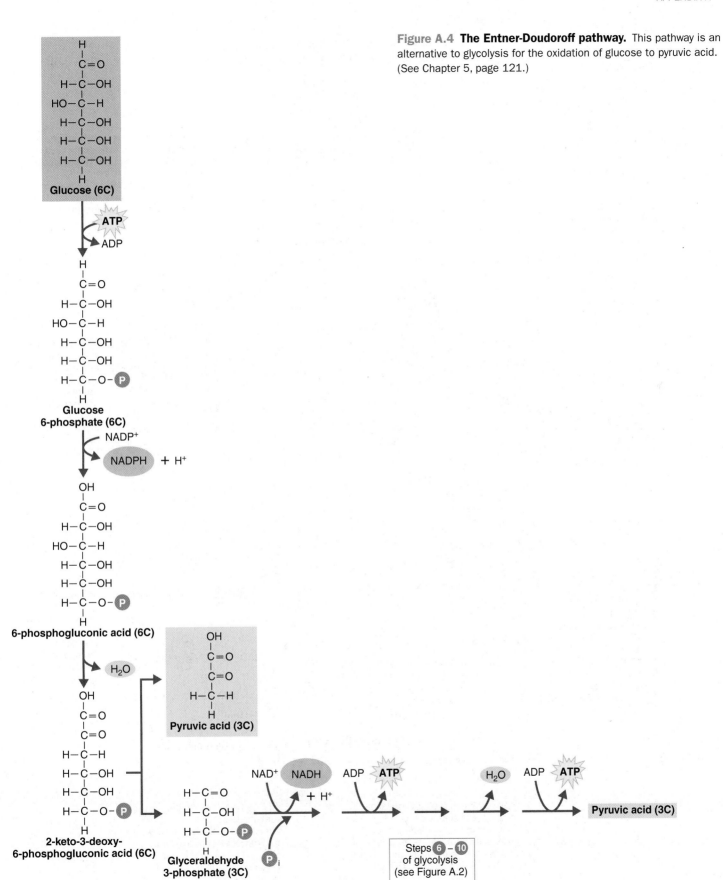

**Figure A.4 The Entner-Doudoroff pathway.** This pathway is an alternative to glycolysis for the oxidation of glucose to pyruvic acid. (See Chapter 5, page 121.)

**Figure A.5 The Krebs cycle.**
See Figure 5.13, page 124, for a
simplified version.

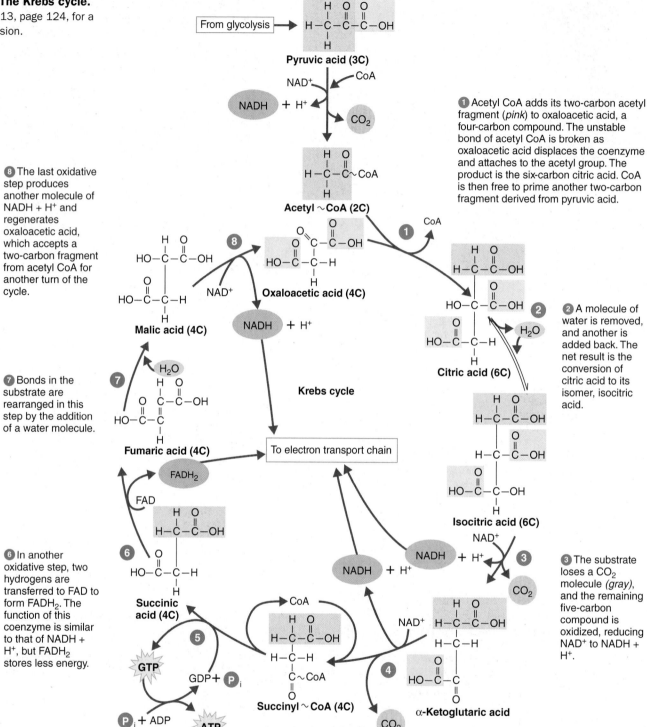

**From glycolysis** → Pyruvic acid (3C)

NAD⁺ ＋ CoA
NADH + H⁺ → $CO_2$

Acetyl ～CoA (2C)

**①** Acetyl CoA adds its two-carbon acetyl fragment (*pink*) to oxaloacetic acid, a four-carbon compound. The unstable bond of acetyl CoA is broken as oxaloacetic acid displaces the coenzyme and attaches to the acetyl group. The product is the six-carbon citric acid. CoA is then free to prime another two-carbon fragment derived from pyruvic acid.

**⑧** The last oxidative step produces another molecule of NADH + H⁺ and regenerates oxaloacetic acid, which accepts a two-carbon fragment from acetyl CoA for another turn of the cycle.

Malic acid (4C)

NAD⁺   Oxaloacetic acid (4C)

⑧   NADH + H⁺

CoA   ①

Citric acid (6C)

**②** A molecule of water is removed, and another is added back. The net result is the conversion of citric acid to its isomer, isocitric acid.

$H_2O$   ②

**⑦** Bonds in the substrate are rearranged in this step by the addition of a water molecule.

⑦   $H_2O$

Fumaric acid (4C)

**Krebs cycle**

**To electron transport chain**

Isocitric acid (6C)

NAD⁺

FADH₂

FAD

NADH + H⁺   ③

NADH + H⁺

$CO_2$

**③** The substrate loses a $CO_2$ molecule (*gray*), and the remaining five-carbon compound is oxidized, reducing NAD⁺ to NADH + H⁺.

**⑥** In another oxidative step, two hydrogens are transferred to FAD to form FADH₂. The function of this coenzyme is similar to that of NADH + H⁺, but FADH₂ stores less energy.

⑥

Succinic acid (4C)

GTP

⑤

GDP + Ⓟᵢ

Ⓟᵢ + ADP

**ATP**

CoA

Succinyl ～CoA (4C)

④

NAD⁺

α-Ketoglutaric acid

$CO_2$

**④** $CO_2$ (*gray*) is lost; the remaining four-carbon compound is oxidized by the transfer of electrons to NAD⁺ to form NADH + H⁺ and is then attached to CoA by an unstable bond.

**⑤** Substrate-level phosphorylation occurs in this step. CoA is displaced by a phosphate group, which is then transferred to GDP to form guanosine triphosphate (GTP). GTP is similar to ATP, which is formed when GTP donates a phosphate group to ADP.

# Appendix B

## Exponents, Exponential Notation, Logarithms, and Generation Time

### Exponents and Exponential Notation

Very large and very small numbers, such as 4,650,000,000 and 0.00000032, are cumbersome to work with. It is more convenient to express such numbers in exponential notation—that is, as a power of 10. For example, $4.65 \times 10^9$ is in standard exponential notation, or **scientific notation**: 4.65 is the *coefficient*, and 9 is the power or *exponent*. In standard exponential notation, the coefficient is always a number between 1 and 10, and the exponent can be positive or negative.

To change a number into exponential notation, follow two steps. First, determine the coefficient by moving the decimal point so there is only one nonzero digit to the left of it. For example,

The coefficient is 3.2. Second, determine the exponent by counting the number of places you moved the decimal point. If you moved it to the left, the exponent is positive. If you moved it to the right, the exponent is negative. In the example, you moved the decimal point seven places to the right, so the exponent is −7. Thus,

$$0.00000032 = 3.2 \times 10^{-7}$$

Now suppose you are working with a larger number instead of a very small number. The same rules apply, but the exponential value will be positive rather than negative. For example,

$$4,650,000,000 = 4.65 \times 10^9$$

To multiply numbers written in exponential notation, multiply the coefficients and *add* the exponents. For example,

$$(3 \times 10^4) \times (2 \times 10^3) =$$
$$(3 \times 2) \times (10^{4+3}) = 6 \times 10^7$$

To divide, divide the coefficient and *subtract* the exponents. For example,

$$\frac{3 \times 10^4}{2 \times 10^3} = \frac{3}{2} \times 10^{4-3} = 1.5 \times 10^1$$

Microbiologists use exponential notation in many situations. For instance, exponential notation is used to describe the number of microorganisms in a population. Such numbers are often very large (see Chapter 6). Another application of exponential notation is to express concentrations of chemicals in a solution—chemicals such as media components (Chapter 6), disinfectants (Chapter 7), or antibiotics (Chapter 20). Such numbers are often very small. Converting from one unit of measurement to another in the metric system requires multiplying or dividing by a power of 10, which is easiest to carry out in exponential notation.

### Logarithms

A **logarithm (log)** is the power to which a base number is raised to produce a given number. Usually we work with logarithms to the base 10, abbreviated $\log_{10}$. The first step in finding the $\log_{10}$ of a number is to write the number in standard exponential notation. If the coefficient is exactly 1, the $\log_{10}$ is simply equal to the exponent. For example,

$$\log_{10} 0.00001 = \log_{10} (1 \times 10^{-5})$$
$$= -5$$

If the coefficient is not 1, as is often the case, the logarithm function on a calculator must be used to determine the logarithm.

Microbiologists use logs for calculating pH levels and for graphing the growth of microbial populations in culture (see Chapter 6).

### Calculating Generation Time

As a cell divides, the population increases exponentially. Numerically this is equal to 2 (because one cell divides into two) raised to the number of times the cell divided (generations);

$$2^{\text{number of generations}}$$

To calculate the final concentration of cells:

$$\text{Initial number of cells} \times 2^{\text{number of generations}} = \text{Number of cells}$$

For example, if 5 cells were allowed to divide 9 times, this would result in

$$5 \times 2^9 = 2560 \text{ cells}$$

To calculate the number of generations a culture has undergone, cell numbers must be converted to logarithms. Standard logarithm values are based on 10. The log of 2 (0.301) is used because one cell divides into two.

$$\text{Number of generations} = \frac{\substack{\text{log number} \\ \text{of cells (end)}} - \substack{\text{log number of} \\ \text{cells (beginning)}}}{0.301}$$

To calculate the generation of time for a population:

$$\frac{60 \text{ min/hr} \times \text{hours}}{\text{number of generations}} = \textit{minutes/generation}$$

As an example, we will calculate the generation time if 100 bacterial cells growing for 5 hours produced 1,720,320 cells:

$$\frac{\log 1,720,320 - \log 100}{0.301} = 14 \text{ generations}$$

$$\frac{60 \text{ min/hr} \times 5 \text{ hours}}{14 \text{ generations}} = 21 \text{ minutes/generation}$$

A practical application of the calculation is determining the effect of a newly developed food preservative on the culture. Suppose 900 of the same species were grown under the same conditions as the previous example, except that the preservative was added to the culture medium. After 15 hours, there were 3,276,800 cells. Calculate the generation time, and decide whether the preservative inhibited growth.

*Answer:* 75 minutes/generation. The preservative did inhibit growth.

# Appendix C

## Methods for Taking Clinical Samples

To diagnose a disease, it is often necessary to obtain a sample of material that may contain the pathogenic microorganism. Samples must be taken aseptically. The sample container should be labeled with the patient's name, room number (if hospitalized), date, time, and medications being taken. Samples must be transported to the laboratory immediately for culture. Delay in transport may result in the growth of some organisms, and their toxic products may kill other organisms. Pathogens tend to be fastidious and die if not kept in optimum environmental conditions.

In the laboratory, samples from infected tissues are cultured on differential and selective media in an attempt to isolate and identify any pathogens or organisms that are not normally found in association with that tissue.

### Universal Precautions[*]

The following procedures should be used by all health care workers, including students, whose activities involve contact with patients or with blood or other body fluids. These procedures were developed to minimize the risk of transmitting HIV or AIDS in a health care environment, but adherence to these guidelines will minimize the transmission of *all* healthcare-associated infections.

1. Wear gloves when touching blood and body fluids, mucous membranes, and nonintact skin and when handling items or surfaces soiled with blood or body fluids. Change gloves after contact with each patient.
2. Wash hands and other skin surfaces immediately and thoroughly if contaminated with blood or other body fluids. Wash hands immediately after gloves are removed.
3. Wear masks and protective eyewear or face shields during procedures that are likely to generate droplets of blood or other body fluids.
4. Wear gowns or aprons during procedures that are likely to generate splashes of blood or other body fluids.
5. To prevent needlestick injuries, needles should not be recapped, purposely bent or broken, or otherwise manipulated by hand. After disposable syringes and needles, scalpel blades, and other sharp items are used, place them in puncture-resistant containers for disposal.
6. Although saliva has not been implicated in HIV transmission, mouthpieces, resuscitation bags, and other ventilation devices should be available for use in areas in which the need for resuscitation is predictable. Emergency mouth-to-mouth resuscitation should be minimized.
7. Health care workers who have exudative lesions or weeping dermatitis should refrain from all direct patient care and from handling patient-care equipment.
8. Pregnant health care workers are not known to have a greater risk of contracting HIV infection than health care workers who are not pregnant; however, if a health care worker develops HIV infection during pregnancy, the infant is at risk of infection. Because of this risk, pregnant health care workers should be especially familiar with, and strictly adhere to, precautions to minimize the risk of HIV transmission.

### Instructions for Specific Sampling Procedures

#### Wound or Abscess Culture

1. Cleanse the area with a sterile swab moistened in sterile saline.
2. Disinfect the area with 70% ethanol or iodine solution.
3. If the abscess has not ruptured spontaneously, a physician will open it with a sterile scalpel.

4. Wipe the first pus away.
5. Touch a sterile swab to the pus, taking care not to contaminate the surrounding tissue.
6. Replace the swab in its container, and properly label the container.

#### Ear Culture

1. Clean the skin and auditory canal with 1% tincture of iodine.
2. Touch the infected area with a sterile cotton swab.
3. Replace the swab in its container.

#### Eye Culture

This procedure is often performed by an ophthalmologist.
1. Anesthetize the eye with topical application of a sterile anesthetic solution.
2. Wash the eye with sterile saline solution.
3. Collect material from the infected area with a sterile cotton swab. Return the swab to its container.

#### Blood Culture

1. Close the room's windows to avoid contamination.
2. Clean the skin around the selected vein with 2% tincture of iodine on a cotton swab.
3. Remove dried iodine with gauze moistened with 80% isopropyl alcohol.
4. Draw a few milliliters of venous blood.
5. Aseptically bandage the puncture.

#### Urine Culture

1. Provide the patient with a sterile container.
2. Instruct the patient to first void a small volume from the urinary bladder before collection (to wash away extraneous bacteria of the skin microbiota) and then to collect a midstream sample.
3. A urine sample may be stored under refrigeration (4–6°C) for up to 24 hours.

#### Fecal Culture

For bacteriological examination, only a small sample is needed. This may be obtained by inserting a sterile swab into the rectum or feces. The swab is then placed in a tube of sterile enrichment broth for transport to the laboratory. For examination for parasites, a small sample may be taken from a morning stool. The sample is placed in a preservative (polyvinyl alcohol, buffered glycerol, saline, or formalin) for microscopic examination for eggs and adult parasites.

#### Sputum Culture

1. A morning sample is best because microorganisms will have accumulated while the patient is sleeping.
2. The patient should rinse his or her mouth thoroughly to remove food and normal microbiota.
3. The patient should cough deeply from the lungs and expectorate into a sterile glass wide-mouth jar.
4. Care should be taken to avoid contaminating health care workers.
5. If there is little sputum (for example, in a patient with tuberculosis), stomach aspiration may be necessary.
6. Infants and children tend to swallow sputum. A fecal sample may be of some value in these cases.

[*]*Source:* Centers for Disease Control and Prevention and National Institutes of Health. *Biosafety in Microbiological and Biomedical Laboratories.*

# Appendix D

## Pronunciation Rules and Word Roots

### Rules for Pronouncing Scientific Names

The easiest way to learn new material is to talk about it, and that requires saying scientific names. Scientific names may look difficult at first glance, but keep in mind that generally every *syllable* is pronounced. The primary requirement in saying a scientific name is to communicate it.

The rules for the pronunciation of scientific names depend, in part, on the derivation of the root word and its vowel sounds. We have provided some general guidelines here. Pronunciations frequently do not follow the rules because a common usage has become "accepted" or the derivation of the name cannot be determined. For many scientific names there are alternative correct pronunciations.

### Vowels

Pronounce all the vowels in scientific names. Vowels marked with a line above the letter are pronounced with a long sound, as in *rate (rāt)*. Vowels not marked by a line above the letter are pronounced as a short sound, as in *rat*. Two vowels written together and pronounced as one sound are called a *diphthong* (for example, the *ou* in *sound*). A special comment is needed about the pronunciation of the vowel endings *-i* and *-ae:* There are two alternative ways to pronounce each of these. In this book, we usually give the pronunciation of a long *e (ē)* to the *-i* ending, and a long *i (ī)* to the *-ae* ending. However, the reverse pronunciations are also correct and in some cases are preferred. For example, *coli* is usually pronounced KŌ-lī. Other vowel sounds are *oy* as in *oil* and *oo* as in *boot*.

### Consonants

When *c* or *g* is followed by *ae, e, oe, i,* or *y,* it has a soft sound. Examples of words with soft *c* and *g* include *circus* and *giraffe*. When *c* or *g* is followed by *a, o, oi,* or *u,* it has a hard sound. When a double *c* is followed by *e, i,* or *y,* it is pronounced as *ks* (e.g., *cocci*).

### Accent

The accented syllable is usually the next-to-last or third-to-last syllable. The most strongly accented syllable appears in capital letters. A secondary accent is noted with a prime ('), as in *Staphylococcus* (STAF-i-lō-kok'kus).

1. The accent is on the next-to-last syllable:
   a. When the name contains only two syllables. Example: PES-tis.
   b. When the next-to-last syllable is a diphthong. Example: ah-kan-thah-MÊ-bah.
   c. When the vowel of the next-to-last syllable is long. Example: trep-ō-NÊ-mah. The vowel in the next-to-last syllable is long in words ending in the following suffixes:

| Suffix | Example |
|---|---|
| -ales | Orders such as Eubacteriales |
| -ina | Sarcina |
| -anus, -anum | pasteurianum |
| -uta | diminuta |

   d. When the word ends in one of the following suffixes:

| Suffix | Example |
|---|---|
| -atus, -atum | caudatum |
| -ella | Salmonella |

2. The accent is on the third-to-last syllable in family names. Families end in *-aceae,* which is always pronounced -Ā-sē-ē. The most strongly accented syllable appears in capital letters. A secondary accent is noted with a prime ('). For example *Staphylococcus* (STAF-i-lō-kok'kus).

### Pronunciation Key for This Text

| | | | |
|---|---|---|---|
| a hat | g go | ō go | u cup |
| ā age | i sit | or order | û use |
| ah father | ī ice | oo boot | x zero |
| ch child | ks tax | oy oil | z zero |
| e let | kw quiz | ou out | zh seizure |
| ē see | ng long | sh she | |
| er term | o hot | th thin | |

### Word Roots Used in Microbiology

The Latin rules of grammar pertain to singular and plural forms of scientific names.

| | Gender | | |
|---|---|---|---|
| | Feminine | Masculine | Neuter |
| Singular | -a | -us | -um |
| Plural | -ae | -i | -a |
| Examples | alga, algae | fungus, fungi | bacterium, bacteria |

**a-, an-** absence, lack. Examples: abiotic, in the absence of life; anaerobic, in the absence of air.

**-able** able to, capable of. Example: viable, having the ability to live or exist.

**actino-** ray. Example: actinomycetes, bacteria that form star-shaped (with rays) colonies.

**aer-** air. Examples: aerobic, in the presence of air; aerate, to add air.

**albo-** white. Example: *Streptomyces albus* produces white colonies.

**ameb-** change. Example: ameboid, movement involving changing shapes.

**amphi-** around. Example: amphitrichous, tufts of flagella at both ends of a cell.

**amyl-** starch. Example: amylase, an enzyme that degrades starch.

**ana-** up. Example: anabolism, building up.

**ant-, anti-** opposed to, preventing. Example: antimicrobial, a substance that prevents microbial growth.

**archae-** ancient. Example: archaeobacteria, "ancient" bacteria, thought to be like the first form of life.

**asco-** bag. Example: ascus, a baglike structure holding spores.

**aur-** gold. Example: *Staphylococcus aureus,* gold-pigmented colonies.

**aut-, auto-** self. Example: autotroph, self-feeder.

**bacillo-** a little stick. Example: bacillus, rod-shaped.

**basid-** base, pedestal. Example: basidium, a cell that bears spores.

**bdell-** leech. Example: *Bdellovibrio,* a predatory bacterium.

**bio-** life. Example: biology, the study of life and living organisms.

**blast-** bud. Example: blastospore, spores formed by budding.

**bovi-** cattle. Example: *Mycobacterium bovis,* a bacterium found in cattle.

**brevi-** short. Example: *Lactobacillus brevis,* a bacterium with short cells.

**butyr-** butter. Example: butyric acid, formed in butter, responsible for rancid odor.

**campylo-** curved. Example: *Campylobacter*, curved rod.

**carcin-** cancer. Example: carcinogen, a cancer-causing agent.

**caseo-** cheese. Example: caseous, cheeselike.

**caul-** a stalk. Example: *Caulobacter*, appendaged or stalked bacteria.

**cerato-** horn. Example: keratin, the horny substance making up skin and nails.

**chlamydo-** covering. Example: chlamydoconidia, conidia formed inside hypha.

**chloro-** green. Example: chlorophyll, green-pigmented molecule.

**chrom-** color. Examples: chromosome, readily stained structure; metachromatic, intracellular colored granules.

**chryso-** golden. Example: *Streptomyces chryseus*, golden colonies.

**-cide** killing. Example: bactericide, an agent that kills bacteria.

**cili-** eyelash. Example: cilia, a hairlike organelle.

**cleisto-** closed. Example: cleistothecium, completely closed ascus.

**co-, con-** together. Example: concentric, having a common center, together in the center.

**cocci-** a berry. Example: coccus, a spherical cell.

**coeno-** shared. Example: coenocyte, a cell with many nuclei not separated by septa.

**col-, colo-** colon. Examples: colon, large intestine; *Escherichia coli*, a bacterium found in the large intestine.

**conidio-** dust. Example: conidia, spores developed at the end of aerial hypha, never enclosed.

**coryne-** club. Example: *Corynebacterium*, club-shaped cells.

**-cul** small form. Example: particle, a small part.

**-cut** the skin. Example: Firmicutes, bacteria with a firm cell wall, gram-positive.

**cyano-** blue. Example: cyanobacteria, blue-green pigmented organisms.

**cyst-** bladder. Example: cystitis, inflammation of the urinary bladder.

**cyt-** cell. Example: cytology, the study of cells.

**de-** undoing, reversal, loss, removal. Example: deactivation, becoming inactive.

**di-, diplo-** twice, double. Example: diplococci, pairs of cocci.

**dia-** through, between. Example: diaphragm, the wall through or between two areas.

**dys-** difficult, faulty, painful. Example: dysfunction, disturbed function.

**ec-, ex-, ecto** out, outside, away from. Example: excrete, to remove materials from the body.

**en-, em-** in, inside. Example: encysted, enclosed in a cyst.

**entero-** intestine. Example: *Enterobacter*, a bacterium found in the intestine.

**eo-** dawn, early. Example: *Eobacterium*, a 3.4-billion-year-old fossilized bacterium.

**epi-** upon, over. Example: epidemic, number of cases of a disease over the normally expected number.

**erythro-** red. Example: erythema, redness of the skin.

**eu-** well, proper. Example: eukaryote, a proper cell.

**exo-** outside, outer layer. Example: exogenous, from outside the body.

**extra-** outside, beyond. Example: extracellular, outside the cells of an organism.

**firmi-** strong. Example: *Bacillus firmus* forms resistant endospores.

**flagell-** a whip. Example: flagellum, a projection from a cell; in eukaryotic cells, it pulls cells in a whiplike fashion.

**flav-** yellow. Example: *Flavobacterium* cells produce yellow pigment.

**fruct-** fruit. Example: fructose, fruit sugar.

**-fy** to make. Example: magnify, to make larger.

**galacto-** milk. Example: galactose, monosaccharide from milk sugar.

**gamet-** to marry. Example: gamete, a reproductive cell.

**gastr-** stomach. Example: gastritis, inflammation of the stomach.

**gel-** to stiffen. Example: gel, a solidified colloid.

**-gen** an agent that initiates. Example: pathogen, any agent that produces disease.

**-genesis** formation. Example: pathogenesis, production of disease.

**germ, germin-** bud. Example: germ, part of an organism capable of developing.

**-gony** reproduction. Example: schizogony, multiple fission producing many new cells.

**gracili-** thin. Example: *Aquaspirillum gracile*, a thin cell.

**halo-** salt. Example: halophile, an organism that can live in high salt concentrations.

**haplo-** one, single. Example: haploid, half the number of chromosomes or one set.

**hema-, hemato-, hemo-** blood. Example: *Haemophilus*, a bacterium that requires nutrients from red blood cells.

**hepat-** liver. Example: hepatitis, inflammation of the liver.

**herpes** creeping. Example: herpes, or shingles, lesions appear to creep along the skin.

**hetero-** different, other. Example: heterotroph, obtains organic nutrients from other organisms; other feeder.

**hist-** tissue. Example: histology, the study of tissues.

**hom-, homo-** same. Example: homofermenter, an organism that produces only lactic acid from fermentation of a carbohydrate.

**hydr-, hydro-** water. Example: dehydration, loss of body water.

**hyper-** excess. Example: hypertonic, having a greater osmotic pressure in comparison with another.

**hypo-** below, deficient. Example: hypotonic, having a lesser osmotic pressure in comparison with another.

**im-** not, in. Example: impermeable, not permitting passage.

**inter-** between. Example: intercellular, between the cells.

**intra-** within, inside. Example: intracellular, inside the cell.

**io-** violet. Example: iodine, a chemical element that produces a violet vapor.

**iso-** equal, same. Example: isotonic, having the same osmotic pressure when compared with another.

**-itis** inflammation of. Example: colitis, inflammation of the large intestine.

**-karyo, -caryo** a nut. Example: eukaryote, a cell with a membrane-enclosed nucleus.

**kin-** movement. Example: streptokinase, an enzyme that lyses or moves fibrin.

**lacti-** milk. Example: lactose, the sugar in milk.

**lepis-** scaly. Example: leprosy, disease characterized by skin lesions.

**lepto-** thin. Example: *Leptospira*, thin spirochete.

**leuko-** whiteness. Example: leukocyte, a white blood cell.

**lip-, lipo-** fat, lipid. Example: lipase, an enzyme that breaks down fats.

**-logy** the study of. Example: pathology, the study of changes in structure and function brought on by disease.

**lopho-** tuft. Example: lophotrichous, having a group of flagella on one side of a cell.

**luc-, luci-** light. Example: luciferin, a substance in certain organisms that emits light when acted upon by the enzyme luciferase.

**lute-, luteo-** yellow. Example: *Micrococcus luteus*, yellow colonies.

**-lysis** loosening, to break down. Example: hydrolysis, chemical decomposition of a compound into other compounds as a result of taking up water.

**macro-** largeness. Example: macromolecules, large molecules.

**mendosi-** faulty. Example: mendosicutes, archaeobacteria lacking peptidoglycan.

**meningo-** membrane. Example: meningitis, inflammation of the membranes of the brain.

**meso-** middle. Example: mesophile, an organism whose optimum temperature is in the middle range.

**meta-** beyond, between, transition. Example: metabolism, chemical changes occurring within a living organism.

**micro-** smallness. Example: microscope, an instrument used to make small objects appear larger.

**-mnesia** memory. Examples: amnesia, loss of memory; anamnesia, return of memory.

**molli-** soft. Example: Mollicutes, a class of wall-less eubacteria.

**-monas** a unit. Example: *Methylomonas*, a unit (bacterium) that utilizes methane as its carbon source.

**mono-** singleness. Example: monotrichous, having one flagellum.

**morpho-** form. Example: morphology, the study of the form and structure of organisms.

**multi-** many. Example: multinuclear, having several nuclei.

**mur-** wall. Example: murein, a component of bacterial cell walls.

**mus-, muri-** mouse. Example: murine typhus, a form of typhus endemic in mice.

**mut-** to change. Example: mutation, a sudden change in characteristics.

**myco-, -mycetoma, -myces** a fungus. Example: *Saccharomyces*, sugar fungus, a genus of yeast.

**myxo-** slime, mucus. Example: Myxobacteriales, an order of slime-producing bacteria.

**necro-** a corpse. Example: necrosis, cell death or death of a portion of tissue.

**-nema** a thread. Example: *Treponema* has long, threadlike cells.

**nigr-** black. Example: *Aspergillus niger,* a fungus that produces black conidia.

**ob-** before, against. Example: obstruction, impeding or blocking up.

**oculo-** eye. Example: monocular, pertaining to one eye.

**-oecium, -ecium** a house. Examples: perithecium, an ascus with an opening that encloses spores; ecology, the study of the relationships among organisms and between an organism and its environment (household).

**-oid** like, resembling. Example: coccoid, resembling a coccus.

**oligo-** small, few. Example: oligiosaccharide, a carbohydrate composed of a few (7–10) monosaccharides.

**-oma** tumor. Example: lymphoma, a tumor of the lymphatic tissues.

**-ont** being, existing. Example: schizont, a cell existing as a result of schizogony.

**ortho-** straight, direct. Example: orthomyxovirus, a virus with a straight, tubular capsid.

**-osis, -sis** condition of. Examples: lysis, the condition of loosening; symbiosis, the condition of living together.

**pan-** all, universal. Example: pandemic, an epidemic affecting a large region.

**para-** beside, near. Example: parasite, an organism that "feeds beside" another.

**peri-** around. Example: peritrichous, projections from all sides.

**phaeo-** brown. Example: Phaeophyta, brown algae.

**phago-** eat. Example: phagocyte, a cell that engulfs and digests particles or cells.

**philo-, -phil** liking, preferring. Example: thermophile, an organism that prefers high temperatures.

**-phore** bears, carries. Example: conidiophore, a hypha that bears conidia.

**-phyll** leaf. Example: chlorophyll, the green pigment in leaves.

**-phyte** plant. Example: saprophyte, a plant that obtains nutrients from decomposing organic matter.

**pil-** a hair. Example: pilus, a hairlike projection from a cell.

**plankto-** wandering, roaming. Example: plankton, organisms drifting or wandering in water.

**plast-** formed. Example: plastid, a formed body within a cell.

**-pnoea, -pnea** breathing. Example: dyspnea, difficulty in breathing.

**pod-** foot. Example: pseudopod, a footlike structure.

**poly-** many. Example: polymorphism, many forms.

**post-** after, behind. Example: posterior, a place behind a (specific) part.

**pre-, pro-** before, ahead of. Examples: prokaryote, a cell with the first nucleus; pregnant, before birth.

**pseudo-** false. Example: pseudopod, false foot.

**psychro-** cold. Example: psychrophile, an organism that grows best at low temperatures.

**-ptera** wing. Example: Diptera, the order of true flies, insects with two wings.

**pyo-** pus. Example: pyogenic, pus-forming.

**rhabdo-** stick, rod. Example: rhabdovirus, an elongated, bullet-shaped virus.

**rhin-** nose. Example: rhinitis, inflammation of mucous membranes in the nose.

**rhizo-** root. Examples: *Rhizobium,* a bacterium that grows in plant roots; mycorrhiza, a fungus that grows in or on plant roots.

**rhodo-** red. Example: *Rhodospirillum,* a red-pigmented, spiral-shaped bacterium.

**rod-** gnaws. Example: rodents, the class of mammals with gnawing teeth.

**rubri-** red. Example: *Clostridiium rubrum,* red-pigmented colonies.

**rumin-** throat. Example: *Ruminococcus,* a bacterium associated with a rumen (modified esophagus).

**saccharo-** sugar. Example: disaccharide, a sugar consisting of two simple sugars.

**sapr-** rotten. Example: *Saprolegnia,* a fungus that lives on dead animals.

**sarco-** flesh. Example: sarcoma, a tumor of muscle or connective tissues.

**schizo-** split. Example: schizomycetes, organisms that reproduce by splitting and an early name for bacteria.

**scolec-** worm. Example: scolex, the head of a tapeworm.

**-scope, -scopic** watcher. Example: microscope, an instrument used to watch small things.

**semi-** half. Example: semicircular, having the form of half a circle.

**sept-** rotting. Example: septic, presence of bacteria that could cause decomposition.

**septo-** partition. Example: septum, a cross-wall in a fungal hypha.

**serr-** notched. Example: serrate, with a notched edge.

**sidero-** iron. Example: *Siderococcus,* a bacterium capable of oxidizing iron.

**siphon-** tube. Example: Siphonaptera, the order of fleas, insects with tubular mouths.

**soma-** body. Example: somatic cells, cells of the body other than gametes.

**speci-** particular things. Examples: species, the smallest group of organisms with similar properties; specify, to indicate exactly.

**spiro-** coil. Example: spirochete, a bacterium with a coiled cell.

**sporo-** spore. Example: sporangium, a structure that holds spores.

**staphylo-** grapelike cluster. Example: *Staphylococcus,* a bacterium that forms clusters of cells.

**-stasis** arrest, fixation. Example: bacteriostasis, cessation of bacterial growth.

**strepto-** twisted. Example: *Streptococcus,* a bacterium that forms twisted chains of cells.

**sub-** beneath, under. Example: subcutaneous, just under the skin.

**super-** above, upon. Example: superior, the quality or state of being above others.

**sym-, syn-** together, with. Examples: synapse, the region of communication between two neurons; synthesis, putting together.

**-taxi** to touch. Example: chemotaxis, response to the presence (touch) of chemicals.

**taxis-** orderly arrangement. Example: taxonomy, the science dealing with arranging organisms into groups.

**tener-** tender. Example: Tenericutes, the phylum containing wall-less eubacteria.

**thallo-** plant body. Example: thallus, an entire macroscopic fungus.

**therm-** heat. Example: *Thermus,* a bacterium that grows in hot springs (to 75°C).

**thio-** sulfur. Example: *Thiobacillus,* a bacterium capable of oxidizing sulfur-containing compounds.

**-thrix** See trich-.

**-tome, -tomy** to cut. Example: appendectomy, surgical removal of the appendix.

**-tone, -tonic** strength. Example: hypotonic, having less strength (osmotic pressure).

**tox-** poison. Example: antitoxin, effective against poison.

**trans-** across, through. Example: transport, movement of substances.

**tri-** three. Example: trimester, three-month period.

**trich-** a hair. Example: peritrichous, hairlike projections from cells.

**-trope** turning. Example: geotropic, turning toward the Earth (pull of gravity).

**-troph** food, nourishment. Example: trophic, pertaining to nutrition.

**-ty** condition of, state. Example: immunity, the condition of being resistant to disease or infection.

**undul-** wavy. Example: undulating, rising and falling, presenting a wavy appearance.

**uni-** one. Example: unicellular, pertaining to one cell.

**vaccin-** cow. Example: vaccination, injection of a vaccine (originally pertained to cows).

**vacu-** empty. Example: vacuoles, an intracellular space that appears empty.

**vesic-** bladder. Example: vesicle, a bubble.

**vitr-** glass. Example: in vitro, in culture media in a glass (or plastic) container.

**-vorous** eat. Example: carnivore, an animal that eats other animals.

**xantho-** yellow. Example: *Xanthomonas,* produces yellow colonies.

**xeno-** strange. Example: axenic, sterile, free of strange organisms.

**xero-** dry. Example: xerophyte, any plant that tolerates dry conditions.

**xylo-** wood. Example: xylose, a sugar obtained from wood.

**zoo-** animal. Example: zoology, the study of animals.

**zygo-** yoke, joining. Example: zygospore, a spore formed from the fusion of two cells.

**-zyme** ferment. Example: enzyme, any protein in living cells that catalyzes chemical reactions.

Appendices

# Appendix E

## Classification of Prokaryotes According to *Bergey's Manual**

**Domain: Archaea**
  **Phylum Crenarchaeota**
    Class: Thermoprotei
      Order: Desulfurococcales
        Family: Pyrodictiaceae
          *Pyrodictium*
      Order: Sulfolobales
        Family: Sulfolobaceae
          *Sulfolobus*
  **Phylum Euryarchaeota**
    Class: Methanobacteria
      Order: Methanobacteriales
        Family: Methanobacteriaceae
          *Methanobacterium*
        Family: Methanosarcinaceae
          *Methanosarcina*
    Class: Methanococci
      Order: Methanococcales
        Family: Methaococcaceae
          *Methanococcus*
          *Methanothermococcus*
    Class: Halobacteria
      Order: Halobacteriales
        Family: Halobacteriaceae
          *Haloarcula*
          *Halobacterium*
    Class: Thermococci
      Order: Thermococcales
        Family: Thermococcaceae
          *Pyrococcus*
          *Thermococcus*
**Domain: Bacteria**
  Unclassified
    *Thermovibrio*
  **Phylum Thermotogae**
    Class: Thermotogae
      Order: Thermotogales
        Family: Thermotogaceae
          *Thermotoga*
  **Phylum Deinococcus-Thermus**
    Class: Deinococci
      Order: Deinococcales
        Family: Deinococcaceae
          *Deinococcus*
      Order: Thermales
        *Thermus*
  **Phylum Chloroflexi**
    Class: Chloroflexi
      Order: Chloroflexales
        Family: Chloroflexaceae
          *Chloroflexus*
  **Phylum Cyanobacteria**
    Class: Cyanobacteria
      *Gloeocapsa*
      *Prochlorococcus*
      *Synechococcus*
      *Spirulina*
      *Anabaena*

**Phylum Chlorobi**
  Class: Chlorobia
    Order: Chlorobiales
      Family: Chlorobiaceae
        *Chlorobium*
**Phylum Proteobacteria**
  Class: Alphaproteobacteria
    Order: Rhodospirillales
      Family: Rhodospirillaceae
        *Azospirillum*
        *Magnetospirillum*
        *Rhodospirillum*
      Family: Acetobacteraceae
        *Acetobacter*
        *Gluconacetobacter*
        *Gluconobacter*
        *Stella*
    Order: Rickettsiales
      Family: Rickettsiaceae
        *Rickettsia*
      Family: Anaplasmataceae
        *Anaplasma*
        *Ehrlichia*
        *Wolbachia*
      Unclassified
        *Pelagibacter*
    Order: Rhodobacterales
      Family: Rhodobacteraceae
        *Paracoccus*
    Order: Caulobacterales
      Family: Caulobacteraceae
        *Caulobacter*
    Order: Rhizobiales
      Family: Rhizobiaceae
        *Agrobacterium*
        *Rhizobium*
      Family: Bartonellaceae
        *Bartonella*
      Family: Brucellaceae
        *Brucella*
      Family: Beijerinckiaceae
        *Beijerinckia*
      Family: Bradyrhizobiaceae
        *Bradyrhizobium*
        *Nitrobacter*
        *Rhodopseudomonas*
      Family: Hyphomicrobiaceae
        *Hyphomicrobium*
  Class: Betaproteobacteria
    Order: Burkholderiales
      Family: Burkholderiaceae
        *Burkholderia*
        *Cupriavidus*
        *Ralstonia*
      Family: Alcaligenaceae
        *Bordetella*
    Unclassified
      *Sphaerotilus*

    Order: Hydrogenophilales
      Family: Hydrogenophilaceae
        *Acidithiobacillus*
    Order: Methylophilales
      Family: Methylophilaceae
        *Methylophilus*
    Order: Neisseriales
      Family: Neisseriaceae
        *Neisseria*
    Order: Nitrosomonadales
      Family: Nitrosomonadaceae
        *Nitrosomonas*
      Family: Spirillaceae
        *Spirillum*
    Order: Rhodocyclales
      Family: Rhodocyclaceae
        *Zoogloea*
  Class: Gammaproteobacteria
    Order: Chromatiales
      Family: Chromatiaceae
        *Chromatium*
      Family: Ectothiorhodospiraceae
        *Ectothiorhodospira*
    Order: Xanthomonadales
      Family: Xanthomonadaceae
        *Xanthomonas*
    Order: Thiotrichales
      Family: Thiotrichaceae
        *Beggiatoa*
        *Thiomargarita*
      Family: Francisellaceae
        *Francisella*
    Order: Legionellales
      Family: Legionellaceae
        *Legionella*
      Family: Coxiellaceae
        *Coxiella*
    Order: Pseudomonadales
      Family: Pseudomonadaceae
        *Azomonas*
        *Azotobacter*
        *Pseudomonas*
      Family: Moraxellaceae
        *Acinetobacter*
        *Moraxella*
    Order: Vibrionales
      Family: Vibrionaceae
        *Aliivibrio*
        *Photobacterium*
        *Vibrio*
    Order: Aeromonadales
      Family: Aeromonadaceae
        *Aeromonas*
    Order: Enterobacteriales
      Family: Enterobacteriaceae
        *Citrobacter*
        *Cronobacter*
        *Enterobacter*

---

*Bergey's Manual of Systematic Bacteriology*, 2nd ed., 5 vols. (2001–2012), is the reference for classification. *Bergey's Manual of Determinative Bacteriology*, 9th ed. (1994), should be used for identifying culturable bacteria and archaea.

*Erwinia*
*Escherichia*
*Klebsiella*
*Pantoea*
*Plesiomonas*
*Proteus*
*Salmonella*
*Serratia*
*Shigella*
*Yersinia*
Order: Pasteurellales
  Family: Pasteurellaceae
    *Haemophilus*
    *Pasteurella*
    *Mannheimia*
Unclassified
  *Carsonella*
Class: Deltaproteobacteria
  Order: Desulfovibrionales
    Family: Desulfovibrionaceae
      *Desulfovibrio*
  Order: Bdellovibrionales
    Family: Bdellovibrionaceae
      *Bdellovibrio*
  Order: Myxococcales
    Family: Myxococcaceae
      *Myxococcus*
Class: Epsilonproteobacteria
  Order: Campylobacterales
    Family: Campylobacteraceae
      *Campylobacter*
    Family: Helicobacteraceae
      *Helicobacter*
**Phylum Firmicutes**
Class: Bacilli
  Order: Bacillales
    Family: Bacillaceae
      *Bacillus*
      *Geobacillus*
    Family: Listeriaceae
      *Listeria*
    Family: Paenibacillaceae
      *Paenibacillus*
    Family: Staphylococcaceae
      *Staphylococcus*
    Family: Thermoactinomycetaceae
      *Thermoactinomyces*
  Order: Lactobacillales
    Family: Lactobacillaceae
      *Lactobacillus*
      *Pediococcus*
    Family: Enterococcaceae
      *Enterococcus*

Family: Leuconostocaceae
  *Leuconostoc*
Family: Streptococcaceae
  *Lactococcus*
  *Streptococcus*
Class: Clostridia
  Order: Clostridiales
    Family: Clostridiaceae
      *Clostridium*
    Family: Veillonellaceae
      *Veillonella*
    Unclassified
      *Epulopiscium*
  Order: Thermoanaerobacteriales
    Family: Thermoanaerobacteriaceae
      *Thermoanaerobacterium*
**Phylum Tenericutes**
Order: Mycoplasmatales
  Family: Mycoplasmataceae
    *Mycoplasma*
    *Ureaplasma*
Order: Entomoplasmatales
  Family: Spiroplasmataceae
    *Spiroplasma*
Order: Anaeroplasmatales
  Family: Erysipelotrichidae
    *Erysipelothrix*
**Phylum Actinobacteria**
Class: Actinobacteria
  Order: Actinomycetales
    Family: Actinomycetaceae
      *Actinomyces*
    Suborder: Micrococcineae
      Family: Micrococcaceae
        *Micrococcus*
      Family: Brevibacteriaceae
        *Brevibacterium*
      Family: Cellulomonadaceae
        *Tropheryma*
      Family: Corynebacteriaceae
        *Corynebacterium*
      Family: Mycobacteriaceae
        *Mycobacterium*
      Family: Nocardiaceae
        *Nocardia*
        *Rhodococcus*
      Family: Micromonosporaceae
        *Micromonospora*
      Family: Propionibacteriaceae
        *Propionibacterium*

Family: Streptomycetaceae
  *Streptomyces*
Family: Frankiaceae
  *Frankia*
Order: Bifidobacteriales
  Family: Bifidobacteriaceae
    *Bifidobacterium*
    *Gardnerella*
**Phylum Planctomycetes**
Order: Planctomycetales
  Family: Planctomycetaceae
    *Gemmata*
    *Planctomyces*
**Phylum Chlamydiae**
Order: Chlamydiales
  Family: Chlamydiaceae
    *Chlamydia*
    *Chlamydophila*
**Phylum Spirochaetes**
Class: Spirochaetes
  Order: Spirochaetales
    Family: Spirochaetaceae
      *Borrelia*
      *Treponema*
    Family: Leptospiraceae
      *Leptospira*
**Phylum Bacteroidetes**
Class: Bacteroidetes
  Order: Bacteroidales
    Family: Bacteroidaceae
      *Bacteroides*
    Family: Porphyromonadaceae
      *Porphyromonas*
    Family: Prevotellaceae
      *Prevotella*
Class: Flavobacteria
  Family: Flavobacteriaceae
    *Capnocytophaga*
    *Elizabethkingia*
Class: Sphingobacteria
  Order: Sphingobacteriales
    Family: Flexibacteraceae
      *Cytophaga*
**Phylum Fusobacteria**
Class: Fusobacteria
  Order: Fusobacteriales
    Family: Fusobacteriaceae
      *Fusobacterium*
      *Streptobacillus*

Appendices

# Glossary

**9 + 2 array**  Attachment of microtubules in eukaryotic flagella and cilia; 9 pairs of microtubules plus two microtubules.

**12D treatment**  A sterilization process that would result in a decrease of the number of *Clostridium botulinum* endospores by 12 logarithmic cycles.

**ABO blood group system**  The classification of red blood cells based on the presence or absence of A and B carbohydrate antigens.

**abscess**  A localized accumulation of pus.

**A-B toxin**  Bacterial exotoxins consisting of two polypeptides.

**acetyl group**

$$H_3C - \overset{\overset{\displaystyle O}{\|}}{C} -$$

**acid**  A substance that dissociates into one or more hydrogen ions ($H^+$) and one or more negative ions.

**acid-fast stain**  A differential stain used to identify bacteria that are not decolorized by acid-alcohol.

**acidic dye**  A salt in which the color is in the negative ion; used for negative staining.

**acidophile**  A bacterium that grows below pH 4.

**acquired immunodeficiency (secondary immunodeficiency)**  The inability, obtained during the life of an individual, to produce specific antibodies or T cells, due to drugs or disease.

**Actinobacteria**  A phylum of gram-positive, chemoheterotrophic bacteria with a high G + C ratio and a signature rRNA sequence.

**activated macrophage**  A macrophage that has increased phagocytic ability and other functions after exposure to mediators released by T cells after stimulation by antigens.

**activated sludge system**  A process used in secondary sewage treatment in which batches of sewage are held in highly aerated tanks; to ensure the presence of microbes efficient in degrading sewage, each batch is inoculated with portions of sludge from a precious batch.

**activation energy**  The minimum collision energy required for a chemical reaction to occur.

**active site**  A region on an enzyme that interacts with the substrate.

**active transport**  Net movement of a substance across a membrane against a concentration gradient; requires the cell to expend energy.

**acute disease**  A disease in which symptoms develop rapidly but last for only a short time.

**acute-phase proteins**  Serum proteins whose concentration changes by at least 25% during inflammation.

**adaptive immunity**  The ability, obtained during the life of the individual, to produce specific antibodies and T cells.

**adenosine diphosphate (ADP)**  The substance formed when ATP is hydrolyzed and energy is released.

**adenosine triphosphate (ATP)**  An important intracellular energy source.

**adherence**  Attachment of a microbe or phagocyte to another's plasma membrane or other surface.

**adhesin**  A carbohydrate-specific binding protein that projects from prokaryotic cells; used for adherence, also called a ligand.

**adjuvant**  A substance added to a vaccine to increase its effectiveness.

**aerobe**  An organism requiring molecular oxygen ($O_2$) for growth.

**aerobic respiration**  Respiration in which the final electron acceptor in the electron transport chain is molecular oxygen ($O_2$).

**aerotolerant anaerobe**  An organism that does not use molecular oxygen ($O_2$) but is not affected by its presence.

**aflatoxin**  A carcinogenic toxin produced by *Aspergillus flavus*.

**agar**  A complex polysaccharide derived from a marine alga and used as a solidifying agent in culture media.

**agglutination**  A joining together or clumping of cells.

**agranulocyte**  A leukocyte without visible granules in the cytoplasm when viewed through a light microscope; includes monocytes and lymphocytes.

**alarmone**  A chemical signal that promotes a cell's response to environmental stress.

**alcohol**  An organic molecule with the functional group —OH.

**alcohol fermentation**  A catabolic process, beginning with glycolysis, that produces ethyl alcohol to reoxidize NADH.

**aldehyde**  An organic molecule with the functional group

**alga** (plural: **algae**)  A photosynthetic eukaryote; may be unicellular, filamentous, or multicellular but lack the tissues found in plants.

**algal bloom**  An abundant growth of microscopic algae producing visible colonies in nature.

**algin**  A sodium salt of mannuronic acid ($C_6H_8O_6$); found in brown algae.

**allergen**  An antigen that evokes a hypersensitivity response.

**allergy**  *See* hypersensitivity.

**allograft**  A tissue graft that is not from a genetically identical donor (i.e., not from self or an identical twin).

**allosteric inhibition**  The process in which an enzyme's activity is changed because of binding to the allosteric site.

**allosteric site**  The site on an enzyme at which a noncompetitive inhibitor binds.

**allylamines**  Antifungal agents that interfere with sterol synthesis.

**amanitin**  A polypeptide toxin produced by *Amanita* spp., inhibits RNA polymerase.

**Ames test**  A procedure using bacteria to identify potential carcinogens.

**amination**  The addition of an amino group.

**amino acid**  An organic acid containing an amino group and a carboxyl group. In alpha-amino acids the amino and carboxyl groups are attached to the same carbon atom called the alpha-carbon.

**aminoglycoside**  An antibiotic consisting of amino sugars and an aminocyclitol ring; for example, streptomycin.

**amino group**  —$NH_2$.

**ammonification**  The release of ammonia from nitrogen-containing organic matter by the action of microorganisms.

**Glossary**

**amphibolic pathway** A pathway that is both anabolic and catabolic.

**amphitrichous** Having flagella at both ends of a cell.

**anabolism** All synthesis reactions in a living organism; the building of complex organic molecules from simpler ones.

**anaerobe** An organism that does not require molecular oxygen ($O_2$) for growth.

**anaerobic respiration** Respiration in which the final electron acceptor in the electron transport chain is an inorganic molecule other than molecular oxygen ($O_2$); for example, a nitrate ion or $CO_2$.

**anaerobic sludge digester** Anaerobic digestion used in secondary sewage treatment.

**anal pore** A site in certain protozoa for elimination of waste.

**analytical epidemiology** Comparison of a diseased group and a healthy group to determine the cause of the disease.

**anamnestic response** *See* memory response.

**anamorph** Ascomycete fungi that have lost the ability to reproduce sexually; the asexual stage of a fungus.

**anaphylaxis** A hypersensitivity reaction involving IgE antibodies, mast cells, and basophils.

**Angstrom (Å)** A unit of measurement equal to $10^{-10}$ m, or 0.1 nm.

**Animalia** The kingdom composed of multicellular eukaryotes lacking cell walls.

**anion** An ion with a negative charge.

**anoxygenic** Not producing molecular oxygen; typical of cyclic photophosphorylation.

**antagonism** Active opposition; (1) When two drugs are less effective than either one alone. (2) Competition among microbes.

**antibiogram** Report of antibiotic susceptibility of a bacterium.

**antibiotic** An antimicrobial agent, usually produced naturally by a bacterium or fungus.

**antibody** A protein produced by the body in response to an antigen, and capable of combining specifically with that antigen.

**antibody-dependent cell-mediated cytotoxicity (ADCC)** The killing of antibody-coated cells by natural killer cells and leukocytes.

**antibody titer** The amount of antibody in serum.

**anticodon** The three nucleotides by which a tRNA recognizes an mRNA codon.

**antigen** Any substance that causes antibody formation; also called an immunogen.

**antigen–antibody complex** The combination of an antigen with the antibody that is specific for it; the basis of immune protection and many diagnostic tests.

**antigen-binding sites** A site on an antibody that binds to an antigenic determinant.

**antigenic determinant** A specific region on the surface of an antigen against which antibodies are formed; also called epitope.

**antigenic drift** A minor variation in the antigenic makeup of influenza viruses that occurs with time.

**antigenic shift** A major genetic change in influenza viruses causing changes in H and N antigens.

**antigenic variation** Changes in surface antigens that occur in a microbial population.

**antigen-presenting cell (APC)** A macrophage, dendritic cell, or B cell that engulfs an antigen and presents fragments to T cells.

**anti-human immune serum globulin (anti-HISG)** An antibody that reacts specifically with human antibodies.

**antimetabolite** A competitive inhibitor.

**antimicrobial peptide (AMP)** An antibiotic that is bactericidal and has a broad spectrum of activity; *see* bacteriocin.

**antiretroviral** A drug used to treat HIV infection.

**antisense DNA** DNA that is complementary to the DNA encoding a protein; the antisense RNA transcript will hybridize with the mRNA encoding the protein and inhibit synthesis of the protein.

**antisense strand (– strand)** Viral RNA that cannot act as mRNA.

**antisepsis** A chemical method for disinfection of the skin or mucous membranes; the chemical is called an antiseptic.

**antiserum** A blood-derived fluid containing antibodies.

**antitoxin** A specific antibody produced by the body in response to a bacterial exotoxin or its toxoid.

**antiviral protein (AVP)** A protein made in response to interferon that blocks viral multiplication.

**apoenzyme** The protein portion of an enzyme, which requires activation by a coenzyme.

**apoptosis** The natural programmed death of a cell; the residual fragments are disposed of by phagocytosis.

**aquatic microbiology** The study of microorganisms and their activities in natural waters.

**archaea** Domain of prokaryotic cells lacking peptidoglycan; one of the three domains.

**arthroconidia** An asexual fungal spore formed by fragmentation of a septate hypha.

**Arthus reaction** Inflammation and necrosis at the site of injection of foreign serum, due to immune complex formation.

**artificially acquired active immunity** The production of antibodies by the body in response to a vaccination.

**artificially acquired passive immunity** The transfer of humoral antibodies formed by one individual to a susceptible individual, accomplished by the injection of antiserum.

**artificial selection** Choosing one organism from a population to grow because of its desirable traits.

**ascospore** A sexual fungal spore produced in an ascus, formed by the ascomycetes.

**ascus** A saclike structure containing ascospores; found in the ascomycetes.

**asepsis** The absence of contamination by unwanted organisms.

**aseptic packaging** Commercial food preservation by filling sterile containers with sterile food.

**aseptic surgery** Techniques used in surgery to prevent microbial contamination of the patient.

**aseptic techniques** Laboratory techniques used to minimize contamination.

**asexual spore** A reproductive cell produced by mitosis and cell division (eukaryotes) or binary fission (actinomycetes).

**atom** The smallest unit of matter that can enter into a chemical reaction.

**atomic force microscopy (AFM)** *See* scanned-probe microscopy.

**atomic number** The number of protons in the nucleus of an atom.

**atomic mass** The total number of protons and neutrons in the nucleus of an atom.

**atrichous** Bacteria that lack flagella.

**attenuated vaccine** A vaccine containing live, attenuated (weakened) microorganisms.

**autoclave**   Equipment for sterilization by steam under pressure, usually operated at 15 psi and 121°C.

**autograft**   A tissue graft from one's self.

**autoimmune disease**   Damage to one's own organs due to action of the immune system.

**autotroph**   An organism that uses carbon dioxide ($CO_2$) as its principal carbon source; chemoautotroph, photoautotroph.

**auxotroph**   A mutant microorganism with a nutritional requirement that is absent in the parent.

**axial filament**   The structure for motility found in spirochetes; also called endoflagellum.

**azole**   Antifungal agents that interfere with sterol synthesis.

**bacillus** (plural: **bacilli**)   (1) Any rod-shaped bacterium. (2) When written as a genus (*Bacillus*) refers to rod-shaped, endospore-forming, facultatively anaerobic, gram-positive bacteria.

**bacteremia**   A condition in which there are bacteria in the blood.

**bacteria**   Domain of prokaryotic organisms, characterized by peptidoglycan cell walls; **bacterium** (singular) when referring to a single organism.

**bacterial growth curve**   A graph indicating the growth of a bacterial population over time.

**bactericide**   A substance capable of killing bacteria.

**bacteriocin**   An antimicrobial peptide produced by bacteria that kills other bacteria.

**bacteriochlorophyll**   A photosynthetic pigment that transfers electrons for photophosphorylation; found in anoxygenic photosynthetic bacteria.

**bacteriology**   The scientific study of prokaryotes, including bacteria and archaea.

**bacteriophage (phage)**   A virus that infects bacterial cells.

**bacteriostasis**   A treatment capable of inhibiting bacterial growth.

**base**   A substance that dissociates into one or more hydroxide ions ($OH^2$) and one or more positive ions.

**base pairs**   The arrangement of nitrogenous bases in nucleic acids based on hydrogen bonding; in DNA, base pairs are A-T and G-C; in RNA, base pairs are A-U and G-C.

**base substitution**   The replacement of a single base in DNA by another base, causing a mutation; also called point mutation.

**basic dye**   A salt in which the color is in the positive ion; used for bacterial stains.

**basidiospore**   A sexual fungal spore produced in a basidium, characteristic of the basidiomycetes.

**basidium**   A pedestal that produces basidiospores; found in the basidiomycetes.

**basophil**   A granulocyte (leukocyte) that readily takes up basic dye and is not phagocytic; has receptors for IgE Fc regions.

**batch production**   An industrial process in which cells are grown for a period of time after which the product is collected.

**B lymocyte (B cell)**   A type of lymphocyte; differentiates into antibody-secreting plasma cells and memory cells.

**BCG vaccine**   A live, attenuated strain of *Mycobacterium bovis* used to provide immunity to tuberculosis.

**beer**   Alcoholic beverage produced by fermentation of starch.

**benthic zone**   The sediment at the bottom of a body of water.

*Bergey's Manual*   *Bergey's Manual of Systematic Bacteriology*, the standard taxonomic reference on bacteria; also refers to *Bergey's Manual of Determinative Bacteriology*, the standard laboratory identification reference on bacteria.

**β-lactam**   Core structure of penicillins.

**beta oxidation**   The removal of two carbon units from a fatty acid to form acetyl CoA.

**binary fission**   Prokaryotic cell reproduction by division into two daughter cells.

**binomial nomenclature**   The system of having two names (genus and specific epithet) for each organism; also called scientific nomenclature.

**bioaugmentation**   The use of pollutant-acclimated microbes or genetically engineered microbes for bioremediation.

**biochemical oxygen demand (BOD)**   A measure of the biologically degradable organic matter in water.

**biocide**   A substance capable of killing microorganisms.

**bioconversion**   Changes in organic matter brought about by the growth of microorganisms.

**bioenhancer**   Nutrients such as nitrate and phosphate that promote microbial growth.

**biofilm**   A microbial community that usually forms as a slimy layer on a surface.

**biofuels**   Energy resources made by living organisms, usually from biomass, e.g., ethanol, methane.

**biogenesis**   The theory that living cells arise only from preexisting cells.

**biogeochemical cycle**   The recycling of chemical elements by microorganisms for use by other organisms.

**bioinformatics**   The science of determining the function of genes through computer-assisted analysis.

**biological transmission**   The transmission of a pathogen from one host to another when the pathogen reproduces in the vector.

**bioluminescence**   The emission of light from the electron transport chain; requires the enzyme luciferase.

**biomass**   Organic matter produced by living organisms and measured by weight.

**bioreactor**   A fermentation vessel with controls for environmental conditions, e.g., temperature and pH.

**bioremediation**   The use of microbes to remove an environmental pollutant.

**Biosafety Level (BSL)**   Safety guidelines for working with live microorganisms in a laboratory, four levels called BSL-1 through BSL-4.

**biotechnology**   The industrial application of microorganisms, cells, or cell components to make a useful product.

**bioterrorism**   Use of a living organism or its product to intimidate.

**biotype**   *See* biovar.

**biovar**   A subgroup of a serovar based on biochemical or physiological properties; also called biotype.

**bisphenol**   Phenolic that contains two phenol groups connected by a bridge.

**blade**   A flat leaflike structure of multicellular algae.

**blastoconidium**   An asexual fungal spore produced by budding from the parent cell.

**blebbing**   Bulging of plasma membrane as a cell dies.

**blood–brain barrier**   Cell membranes that allow some substances to pass from the blood to the brain but restrict others.

**brightfield microscope** A microscope that uses visible light for illumination; the specimens are viewed against a white background.

**broad-spectrum antibiotic** An antibiotic that is effective against a wide range of both gram-positive and gram-negative bacteria.

**broth dilution test** A method of determining the minimal inhibitory concentration by using serial dilutions of an antimicrobial drug.

**bubo** An enlarged lymph node caused by inflammation.

**budding** (1) Asexual reproduction beginning as a protuberance from the parent cell that grows to become a daughter cell. (2) Release of an enveloped virus through the plasma membrane of an animal cell.

**budding yeast** Following mitosis, a yeast cell that divides unevenly to produce a small cell (bud) from the parent cell.

**buffer** A substance that tends to stabilize the pH of a solution.

**bulking** A condition arising when sludge floats rather than settles in secondary sewage treatment.

**bullae** (singular: **bulla**) Large serum-filled vesicles in the skin.

**bursa of Fabricius** An organ in chickens responsible for maturation of the immune system.

**Calvin-Benson cycle** The fixation of $CO_2$ into reduced organic compounds; used by autotrophs.

**capnophile** A microorganism that grows best at relatively high $CO_2$ concentrations.

**capsid** The protein coat of a virus that surrounds the nucleic acid.

**capsomere** A protein subunit of a viral capsid.

**capsule** An outer, viscous covering on some bacteria composed of a polysaccharide or polypeptide.

**carbapenems** Antibiotics that contain a β-lactam antibiotic and cilastatin.

**carbohydrate** An organic compound composed of carbon, hydrogen, and oxygen, with the hydrogen and oxygen present in a 2:1 ratio; carbohydrates include starches, sugars, and cellulose.

**carbon cycle** The series of processes that converts $CO_2$ to organic substances and back to $CO_2$ in nature.

**carbon fixation** The synthesis of sugars by using carbons from $CO_2$. *See also* Calvin-Benson cycle.

**carbon skeleton** The basic chain or ring of carbon atoms in a molecule; for example,

**carboxyl group**

**carboxysome** A prokaryotic inclusion containing ribulose 1,5-diphosphate carboxylase.

**carcinogen** Any cancer-causing substance.

**carrier** Organism (usually refers to humans) that harbors pathogens and transmits them to others.

**carrying capacity** The number of organisms that an environment can support.

**casein** Milk protein.

**catabolism** All decomposition reactions in a living organism; the breakdown of complex organic compounds into simpler ones.

**catabolite repression** Inhibition of the metabolism of alternative carbon sources by glucose.

**catalase** An enzyme that breaks down hydrogen peroxide: $2H_2O_2 \rightarrow 2H_2O + O_2$

**catalyst** A substance that increases the rate of a chemical reaction but is not altered itself.

**cation** A positively charged ion.

**CD (cluster of differentiation)** Number assigned to an epitope on a single antigen, for example, CD4 protein, which is found on T helper cells.

**cDNA (complementary DNA)** DNA made in vitro from an mRNA template.

**cell culture** Eukaryotic cells grown in culture media; also called tissue culture.

**cell theory** All living organisms are composed of cells and arise from preexisting cells.

**cellular immunity (cell-mediated immunity)** An immune response that involves T cells binding to antigens presented on antigen-presenting cells; T cells then differentiate into several types of effector T cells.

**cellular respiration** *See* respiration.

**cell wall** The outer covering of most bacterial, fungal, algal, and plant cells; in bacteria, it consists of peptidoglycan.

**Centers for Disease Control and Prevention (CDC)** A branch of the U.S. Public Health Service that serves a central source of epidemiological information.

**central nervous system (CNS)** The brain and the spinal cord. *See also* peripheral nervous system.

**centriole** A structure consisting of nine microtubule triplets, found in eukaryotic cells.

**centrosome** Region in a eukaryotic cell consisting of a pericentriolar area (protein fibers) and a pair of centrioles; involved in formation of the mitotic spindle.

**cercaria** A free-swimming larva of trematodes.

**CFU (colony-forming unit)** Visible bacterial colonies on solid media.

**chancre** A hard sore, the center of which ulcerates.

**chemical bond** An attractive force between atoms forming a molecule.

**chemical element** A fundamental substance composed of atoms that have the same atomic number and behave the same way chemically.

**chemical reaction** The process of making or breaking bonds between atoms.

**chemically defined medium** A culture medium in which the exact chemical composition is known.

**chemiosmosis** A mechanism that uses a proton gradient across a cytoplasmic membrane to generate ATP.

**chemistry** The science of the interactions between atoms and molecules.

**chemoautotroph** An organism that uses an inorganic chemical as an energy source and $CO_2$ as a carbon source.

**chemoheterotroph** An organism that uses organic molecules as a source of carbon and energy.

**chemokine** A cytokine that induces, by chemotaxis, the migration of leukocytes into infected areas.

**chemotaxis** Movement in response to the presence of a chemical.

**chemotherapy** Treatment of disease with chemical substances.

**chemotroph** An organism that uses oxidation-reduction reactions as its primary energy source.

**chimeric monoclonal antibody** A genetically engineered antibody made of human constant regions and mouse variable regions.

**chlamydoconidium**  An asexual fungal spore formed within a hypha.

**chlorophyll *a***  A photosynthetic pigment that transfers electrons for photophosphorylation; found in plant, algae, and cyanobacteria.

**chloroplast**  The organelle that performs photosynthesis in photoautotrophic eukaryotes.

**chlorosome**  Plasma membrane folds in green sulfur bacteria containing bacteriochlorophylls.

**chromatin**  Threadlike, uncondensed DNA in an interphase eukaryotic cell.

**chromatophore**  An infolding in the plasma membrane where bacterio-chlorophyll is located in photoautotrophic bacteria; also known as thylakoids.

**chromosome**  The structure that carries hereditary information, chromosomes contain genes.

**chronic infection**  An illness that develops slowly and is likely to continue or recur for long periods.

**ciliary escalator**  Ciliated mucosal cells of the lower respiratory tract that move inhaled particulates away from the lungs.

**cilium (plural: cilia)**  A relatively short cellular projection from some eukaryotic cells, composed of nine pairs plus two microtubules. *See* flagellum.

***cis***  Hydrogen atoms on the same side across a double bond in a fatty acid. *See* trans.

**cistern**  A flattened membranous sac in endoplasmic reticulum and the Golgi complex.

**clade**  A group of organisms that share a particular common ancestor; a branch on a cladogram.

**cladogram**  A dichotomous phylogenetic tree that branches repeatedly, suggesting the classification of organisms based on the time sequence in which evolutionary branches arose.

**class**  A taxonomic group between phylum and order.

**class switching**  Ability of a B cell to produce a different class of antibody against one antigen.

**clinical trial**  Research to determine whether a treatment is effective and safe for humans.

**clonal deletion**  The elimination of B and T cells that react with self.

**clonal selection**  The development of clones of B and T cells against a specific antigen.

**clone**  A population of cells arising from a single parent cell.

**clue cells**  Sloughed-off vaginal cells covered with *Gardnerella vaginalis*.

**coagulase**  A bacterial enzyme that causes blood plasma to clot.

**coccobacillus (plural: coccobacilli)**  A bacterium that is an oval rod.

**coccus (plural: cocci)**  A spherical or ovoid bacterium.

**codon**  A sequence of three nucleotides in mRNA that specifies the insertion of an amino acid into a polypeptide.

**coenocytic hypha**  A fungal filament that is not divided into uninucleate cell-like units because it lacks septa.

**coenzyme**  A nonprotein substance that is associated with and that activates an enzyme.

**coenzyme A (CoA)**  A coenzyme that functions in decarboxylation.

**coenzyme Q**  *See* ubiquinone.

**cofactor**  (1) The nonprotein component of an enzyme. (2) A microorganism or molecule that acts with others to synergistically enhance or cause disease.

**coliforms**  Aerobic or facultatively anaerobic, gram-negative, non–endospore-forming, rod-shaped bacteria that ferment lactose with acid and gas formation within 48 hours at 35°C.

**collagenase**  An enzyme that hydrolyzes collagen.

**collision theory**  The principle that chemical reactions occur because energy is gained as particles collide.

**colony**  A visible mass of microbial cells arising from one cell or from a group of the same microbes.

**colony hybridization**  The identification of a colony containing a desired gene by using a DNA probe that is complementary to that gene.

**commensalism**  A symbiotic relationship in which two organisms live in association and one is benefited while the other is neither benefited nor harmed.

**commercial sterilization**  A process of treating canned goods aimed at destroying the endospores of *Clostridium botulinum*.

**communicable disease**  Any disease that can be spread from one host to another.

**community-acquired infection**  Infection contracted outside the health care setting.

**competence**  The physiological state in which a recipient cell can take and incorporate a large piece of donor DNA.

**competitive exclusion**  Growth of some microbes prevents the growth of other microbes.

**competitive inhibitor**  A chemical that competes with the normal substrate for the active site of an enzyme. *See also* noncompetitive inhibitor.

**complement**  A group of serum proteins involved in phagocytosis and lysis of bacteria.

**complementary DNA (cDNA)**  DNA made in vitro from an mRNA template.

**complement fixation**  The process in which complement combines with an antigen–antibody complex.

**complex medium**  A culture medium in which the exact chemical composition is not known.

**complex virus**  A virus with a complicated structure, such as a bacteriophage.

**composting**  A method of solid waste disposal, usually plant material, by encouraging its decomposition by microbes.

**compound**  A substance composed of two or more different chemical elements.

**compound light microscope (LM)**  An instrument with two sets of lenses that uses visible light as the source of illumination.

**compromised host**  A host whose resistance to infection is impaired by another condition.

**condensation reaction**  A chemical reaction in which a molecule of water is released; also called dehydration synthesis.

**condenser**  A lens system located below the microscope stage that directs light rays through the specimen.

**confocal microscopy**  A light microscope that uses fluorescent stains and laser to make two- and three-dimensional images.

**congenital**  Refers to a condition existing at birth; may be inherited or acquired in utero.

**congenital immunodeficiency (primary immunodeficiency)**  The inability, due to an individual's genotype, to produce specific antibodies or T cells.

**conidiophore**  An aerial hypha bearing conidiospores.

**conidiospore** *See* conidium.

**conidium** An asexual spore produced in a chain from a conidiophore.

**conjugated vaccine** A vaccine consisting of the desired antigen and other proteins.

**conjugation** The transfer of genetic material from one cell to another involving cell-to-cell contact.

**conjugative plasmid** A prokaryotic plasmid that carries genes for sex pili and for transfer of the plasmid to another cell.

**constitutive gene** A gene that is produced continuously.

**contact inhibition** The cessation of animal cell movement and division as a result of contact with other cells.

**contact transmission** The spread of disease by direct or indirect contact or via droplets.

**contagious disease** A disease that is easily spread from one person to another.

**continuous cell line** Animal cells that can be maintained through an indefinite number of generations in vitro.

**continuous flow** An industrial fermentation in which cells are grown indefinitely with continual addition of nutrients and removal of waste and products.

**corepressor** A molecule that binds to a repressor protein, enabling the repressor to bind to an operator.

**cortex** The protective fungal covering of a lichen.

**counterstain** A second stain applied to a smear, provides contrast to the primary stain.

**covalent bond** A chemical bond in which the electrons of one atom are shared with another atom.

**crisis** The phase of a fever characterized by vasodilation and sweating.

**crista** (plural: **cristae**) Folding of the inner membrane of a mitochondrion.

**cross-contamination** Transfer of pathogens from one fomite to another.

**crossing over** The process by which a portion of one chromosome is exchanged with a portion of another chromosome.

**CTL (cytotoxic T lymphocyte)** An activated CD8$^+$ T cell; kills cells presenting endogenous antigens.

**culture** Microorganisms that grow and multiply in a container of culture medium.

**culture medium** The nutrient material prepared for growth of microorganisms in a laboratory.

**curd** The solid part of milk that separates from the liquid (whey) in the making of cheese, for example.

**cutaneous mycosis** A fungal infection of the epidermis, nails, or hair.

**cuticle** The outer covering of helminths.

**cyanobacteria** Oxygen-producing photoautotrophic prokaryotes.

**cyclic AMP (cAMP)** A molecule derived from ATP, in which the phosphate group has a cyclic structure; acts as a cellular messenger.

**cyclic photophosphorylation** The movement of an electron from chlorophyll through a series of electron acceptors and back to chlorophyll; anoxygenic; purple and green bacterial photophosphorylation.

**cyst** A sac with a distinct wall containing fluid or other material; also, a protective capsule of some protozoa.

**cysticercus** An encysted tapeworm larva.

**cytochrome** A protein that functions as an electron carrier in cellular respiration and photosynthesis.

**cytokine** A small protein released from human cells that regulates the immune response; directly or indirectly may induce fever, pain, or T cell proliferation.

**cytokine storm** Overproduction of cytokines; can cause damage to the human body.

**cytolysis** The destruction of cells, resulting from damage to their cell membrane, that causes cellular contents to leak out.

**cytopathic effect (CPE)** A visible effect on a host cell, caused by a virus, that may result in host cell damage or death.

**cytoplasm** In a prokaryotic cell, everything inside the plasma membrane; in a eukaryotic cell, everything inside the plasma membrane and external to the nucleus.

**cytoplasmic streaming** The movement of cytoplasm in a eukaryotic cell.

**cytoskeleton** Microfilaments, intermediate filaments, and microtubules that provide support and movement for eukaryotic cytoplasm.

**cytosol** The fluid portion of cytoplasm.

**cytostome** The mouthlike opening in some protozoa.

**cytotoxin** A bacterial toxin that kills host cells or alters their functions.

**dalton (da)** The measure of molecular mass; equals 1g/mole.

**darkfield microscope** A microscope that has a device to scatter light from the illuminator so that the specimen appears white against a black background.

**deamination** The removal of an amino group from an amino acid to form ammonia. *See also* ammonification.

**death phase** The period of logarithmic decrease in a bacterial population; also called logarithmic decline phase.

**debridement** Surgical removal of necrotic tissue.

**decarboxylation** The removal of $CO_2$ from an amino acid.

**decimal reduction time (DRT)** The time (in minutes) required to kill 90% of a bacterial population at a given temperature; also called D value.

**decolorizing agent** A solution used in the process of removing a stain.

**decomposition reaction** A chemical reaction in which bonds are broken to produce smaller parts from a large molecule.

**deep-freezing** Preservation of bacterial cultures at $-50°C$ to $-95°C$.

**defensins** Small peptide antibiotics made by human cells.

**definitive host** An organism that harbors the adult, sexually mature form of a parasite.

**degeneracy** Redundancy of the genetic code; that is, most amino acids are encoded by several codons.

**degerming** The removal of microorganisms in an area; also called degermation.

**degranulation** The release of contents of secretory granules from mast cells or basophils during anaphylaxis.

**dehydration synthesis** *See* condensation reaction.

**dehydrogenation** The loss of hydrogen atoms from a substrate.

**delayed hypersensitivity** Cell-mediated hypersensitivity.

**denaturation** A change in the molecular structure of a protein, usually making it nonfunctional.

**dendritic cell (DC)** A type of antigen-presenting cell characterized by long fingerlike extensions; found in lymphatic tissue and skin.

**denitrification** The reduction of nitrogen in nitrate to nitrite or nitrogen gas.

**dental plaque**    A combination of bacterial cells, dextran, and debris adhering to the teeth.

**deoxyribonucleic acid (DNA)**    The nucleic acid of genetic material in all cells and some viruses.

**deoxyribose**    A five-carbon sugar contained in DNA nucleotides.

**dermatomycosis**    A fungal infection of the skin; also known as tinea or ringworm.

**dermatophyte**    A fungus that causes a cutaneous mycosis.

**dermis**    The inner portion of the skin.

**descriptive epidemiology**    The collection and analysis of all data regarding the occurrence of a disease to determine its cause.

**desensitization**    The prevention of allergic inflammatory responses.

**desulfurization**    Removal of sulfur from an organic compound.

**desiccation**    The removal of water.

**diapedesis**    The process by which phagocytes move out of blood vessels.

**dichotomous key**    An identification scheme based on successive paired questions; answering one question leads to another pair of questions, until an organism is identified.

**differential interference contrast (DIC) microscope**    An instrument that provides a three-dimensional, magnified image.

**differential medium**    A solid culture medium that makes it easier to distinguish colonies of the desired organism.

**differential stain**    A stain that distinguishes objects on the basis of reactions to the staining procedure.

**differential white blood cell count**    The number of each kind of leukocyte in a sample of 100 leukocytes.

**diffusion**    The net movement of molecules or ions from an area of higher concentration to an area of lower concentration.

**dimorphism**    The property of having two forms of growth. *See also* sexual dimorphism.

**dioecious**    Referring to organisms in which organs of different sexes are located in different individuals.

**diplobacilli** (singular: **diplobacillus**)    Rods that divide and remain attached in pairs.

**diplococci** (singular: **diplococcus**)    Cocci that divide and remain attached in pairs.

**diploid cell**    A cell having two sets of chromosomes; diploid is the normal state of a eukaryotic cell.

**diploid cell line**    Eukaryotic cells grown in vitro.

**direct agglutination test**    The use of known antibodies to identify an unknown cell-bound antigen.

**direct contact transmission**    A method of spreading infection from one host to another through some kind of close association between the hosts.

**direct fluorescent-antibody (DFA) test**    A fluorescent-antibody test to detect the presence of an antigen.

**direct microscopic count**    Enumeration of cells by observation through a microscope.

**disaccharide**    A sugar consisting of two simple sugars, or monosaccharides.

**disease**    An abnormal state in which part or all of the body is not properly adjusted or is incapable of performing normal functions; any change from a state of health.

**disinfection**    Any treatment used on inanimate objects to kill or inhibit the growth of microorganisms; a chemical used is called a disinfectant.

**disk-diffusion method**    An agar-diffusion test to determine microbial susceptibility to chemotherapeutic agents; also called Kirby-Bauer test.

**D-isomer**    Arrangement of four different atoms or groups around a carbon atom. *See* L-isomer.

**dissimilation**    A metabolic process in which nutrients are not assimilated but are excreted as ammonia, hydrogen sulfide, and so on.

**dissimilation plasmid**    A plasmid containing genes encoding production of enzymes that trigger the catabolism of certain unusual sugars and hydrocarbons.

**dissociation**    The separation of a compound into positive and negative ions in solution. *See also* ionization.

**disulfide bond**    A covalent bond that holds together two atoms of sulfur.

**DNA base composition**    The moles-percentage of guanine plus cytosine in an organism's DNA.

**DNA chip ("microarray")**    A silica wafer that holds DNA probes; used to recognize DNA in samples being tested.

**DNA fingerprinting**    Analysis of DNA by electrophoresis of restriction enzyme fragments of the DNA.

**DNA polymerase**    Enzyme that synthesizes DNA by copying a DNA template.

**DNA probe**    A short, labeled, single strand of DNA or RNA used to locate its complementary strand in a quantity of DNA.

**DNA vaccine**    Injection of DNA into animal cells so that the cells produce the antigen that will stimulate the immune system.

**domain**    A taxonomic classification based on rRNA sequences; above the kingdom level.

**donor cell**    A cell that gives DNA to a recipient cell during genetic recombination.

**droplet transmission**    The transmission of infection by small liquid droplets carrying microorganisms.

**DTaP vaccine**    A combined vaccine used to provide active immunity, containing diphtheria and tetanus toxoids and *Bordetella pertussis* cell fragments.

**D value**    *See* decimal reduction time.

**dysbiosis**    An imbalance of microbiome in the human body (such as the gastrointestinal tract), leading to adverse health conditions.

**dysentery**    A disease characterized by frequent, watery stools containing blood and mucus.

**eclipse period**    The time during viral multiplication when complete, infective virions are not present.

**ecology**    The study of the interrelationships between organisms and their environment.

**edema**    An abnormal accumulation of interstitial fluid in tissues, causing swelling.

**electron**    A negatively charged particle in motion around the nucleus of an atom.

**electronic configuration**    The arrangement of electrons in shells or energy levels in an atom.

**electron microscope**    A microscope that uses electrons instead of light to produce an image.

**electron shell**    A region of an atom where electrons orbit the nucleus, corresponding to an energy level.

**electron transport chain, electron transport system**    A series of compounds that transfer electrons from one compound to another, generating ATP by oxidative phosphorylation.

**electroporation**    A technique by which DNA is inserted into a cell using an electrical current.

**elementary body**    The infectious form of chlamydiae.

**ELISA (enzyme-linked immunosorbent assay)**    A group of serological tests that use enzyme reactions as indicators.

**embryonic stem cell (ESC)**    A cell from an embryo that has the potential to become a wide variety of specialized cell types.

**emerging infectious disease (EID)**    A new or changing disease that is increasing or has the potential to increase in incidence in the near future.

**Embden-Meyerhof pathway**    *See* glycolysis.

**enanthem**    Rash on mucous membranes. *See also* exanthem.

**encephalitis**    Infection of the brain.

**endemic disease**    A disease that is constantly present in a certain population.

**endergonic reaction**    A chemical reaction that requires energy.

**endocarditis**    Infection of the lining of the heart (endocardium).

**endocytosis**    The process by which material is moved into a eukaryotic cell.

**endoflagellum**    *See* axial filament.

**endogenous**    (1) Infection caused by an opportunistic pathogen from an individual's own normal microbiota. (2) Surface antigens on human cells produced as a result of infection.

**endolith**    An organism that lives inside rock.

**endoplasmic reticulum (ER)**    A membranous network in eukaryotic cells connecting the plasma membrane with the nuclear membrane.

**endospore**    A resting structure formed inside some bacteria.

**endosymbiotic theory**    A model for the evolution of eukaryotes which states that organelles arose from prokaryotic cells living inside a host prokaryote.

**endotoxic shock**    *See* gram-negative sepsis.

**endotoxin**    Part of the outer portion of the cell wall (lipid A) of most gram- negative bacteria; released on destruction of the cell.

**end-product inhibition**    *See* feedback inhibition.

**energy level**    Potential energy of an electron in an atom. *See also* electron shell.

**enrichment culture**    A culture medium used for preliminary isolation that favors the growth of a particular microorganism.

**enteric**    The common name for a bacterium in the family Enterobacteriaceae.

**enterotoxin**    An exotoxin that causes gastroenteritis, such as those produced by *Staphylococcus, Vibrio,* and *Escherichia.*

**Entner-Doudoroff pathway**    An alternative pathway for the oxidation of glucose to pyruvic acid.

**envelope**    An outer covering surrounding the capsid of some viruses.

**enzyme**    A molecule that catalyzes biochemical reactions in a living organism, usually a protein. *See also* ribozyme.

**enzyme immunoassay (EIA)**    *See* ELISA.

**enzyme-linked immunosorbent assay**    *See* ELISA.

**enzyme–substrate complex**    A temporary union of an enzyme and its substrate.

**eosinophil**    A granulocyte whose granules take up the stain eosin.

**epidemic disease**    A disease acquired by many hosts in a given area in a short time.

**epidemiology**    The science that studies when and where diseases occur and how they are transmitted.

**epidermis**    The outer portion of the skin.

**epitope**    *See* antigenic determinant.

**equivalent treatments**    Different methods that have the same effect on controlling microbial growth.

**ergot**    A toxin produced in sclerotia by the fungus *Claviceps purpurea* that causes ergotism.

**essential oils (EOs)**    Volatile oils extracted from plants; smell like the plant.

**ester linkage**    Bonding between fatty acids and glycerol in bacterial and eukaryotic phospholipids:

$$----C—O—\overset{\overset{\displaystyle O}{\|}}{C}----$$

**E test**    An agar diffusion test to determine antibiotic sensitivity using a plastic strip impregnated with varying concentrations of an antibiotic.

**ethambutol**    A synthetic antimicrobial agent that interferes with the synthesis of RNA.

**ethanol**

$$H—\overset{\overset{\displaystyle H}{|}}{\underset{\underset{\displaystyle H}{|}}{C}}—\overset{\overset{\displaystyle H}{|}}{\underset{\underset{\displaystyle H}{|}}{C}}—OH$$

**ether linkage**    Bonding between fatty acids and glycerol in archaeal phospholipids: ----C—O—C-----

**etiology**    The study of the cause of a disease.

**eukarya**    All eukaryotes (animals, plants, fungi, and protists); members of the Domain Eukarya.

**eukaryote**    A cell having DNA inside a distinct membrane-enclosed nucleus.

**eukaryotic species**    A group of closely related organisms that can interbreed.

**eutrophication**    The addition of organic matter and subsequent removal of oxygen from a body of water.

**exanthem**    Skin rash. *See also* enanthem.

**exchange reaction**    A chemical reaction that has both synthesis and decomposition components.

**exergonic reaction**    A chemical reaction that releases energy.

**exon**    A region of a eukaryotic chromosome that encodes a protein.

**exotoxin**    A protein toxin released from living, mostly gram-positive bacterial cells.

**experimental epidemiology**    The study of a disease using controlled experiments.

**exponential growth phase**    *See* log phase.

**extensively drug-resistant (XDR)**    *M. tuberculosis* strains resistant to isoniazid, rifampin, fluoroquinolones, and at least one more drug.

**extracellular polymeric substance (EPS)**    A glycocalyx that permits bacteria to attach to various surfaces.

**extreme thermophile**    *See* hyperthermophile.

**extremophile**    A microorganism that lives in environmental extremes of temperature, acidity, alkalinity, salinity, or pressure.

**extremozymes**    Enzymes produced by extremophiles.

**facilitated diffusion**    The movement of a substance across a plasma membrane from an area of higher concentration to an area of lower concentration, mediated by transporter proteins.

**facultative anaerobe**    An organism that can grow with or without molecular oxygen ($O_2$).

**facultative halophile**   An organism capable of growth in, but not requiring, 1–2% salt.

**FAD**   Flavin adenine dinucleotide; a coenzyme that functions in the removal and transfer of hydrogen ions ($H^1$) and electrons from substrate molecules.

**FAME**   Fatty acid methyl ester; identification of microbes by the presence of specific fatty acids.

**family**   A taxonomic group between order and genus.

**feedback inhibition**   Inhibition of an enzyme in a particular pathway by the accumulation of the end-product of the pathway; also called end-product inhibition.

**fermentation**   The enzymatic degradation of carbohydrates in which the final electron acceptor is an organic molecule, ATP is synthesized by substrate-level phosphorylation, and $O_2$ is not required.

**fermentation test**   Method used to determine whether a bacterium or yeast ferments a specific carbohydrate; usually performed in a peptone broth containing the carbohydrate, a pH indicator, and an inverted tube to trap gas.

**ferritin**   One of several human iron-binding proteins that reduce iron available to a pathogen.

**fever**   An abnormally high body temperature.

**F factor (fertility factor)**   A plasmid found in the donor cell in bacterial conjugation.

**fibrinolysin**   A kinase produced by streptococci.

**filtration**   The passage of a liquid or gas through a screenlike material; a 0.45-μm filter removes most bacteria.

**fimbria (plural: fimbriae)**   An appendage on a bacterial cell used for attachment.

**Firmicutes**   A phylum of gram-positive bacteria with a low G + C ratio that possess a signature rRNA sequence.

**FISH**   Fluorescent in situ hybridization; use of rRNA probes to identify microbes without culturing.

**fission yeast**   Following mitosis, a yeast cell that divides evenly to produce two new cells.

**fixed macrophage**   A macrophage that is located in a certain organ or tissue (e.g., liver, lungs, spleen, or lymph nodes); also called a histiocyte.

**fixing**   (1) In slide preparation, the process of attaching a specimen to a slide. (2) Regarding chemical elements, combining elements so that a critical element can enter the food chain. *See also* Calvin-Benson cycle; nitrogen fixation.

**flagellum (plural: flagella)**   A thin appendage from the surface of a cell; used for cellular locomotion; composed of flagellin in prokaryotic cells, composed of 9 + 2 microtubules in eukaryotic cells.

**flaming**   The process of sterilizing an inoculating loop by holding it in an open flame.

**flat sour spoilage**   Thermophilic spoilage of canned goods not accompanied by gas production.

**flatworm**   An animal belonging to the phylum Platyhelminthes.

**flavoprotein**   A protein containing the coenzyme flavin; functions as an electron carrier in electron transport chains.

**flocculation**   The removal of colloidal material during water purification by adding a chemical that causes colloidal particles to coalesce.

**flow cytometry**   A method of counting cells using a flow cytometer, which detects cells by the presence of a fluorescent tag on the cell surface.

**fluid mosaic model**   A way of describing the dynamic arrangement of phospholipids and proteins comprising the plasma membrane.

**fluke**   A flatworm belonging to the class Trematoda.

**fluorescence**   The ability of a substance to give off light of one color when exposed to light of another color.

**fluorescence-activated cell sorter (FACS)**   A modification of a flow cytometer that counts and sorts cells labeled with fluorescent antibodies.

**fluorescence microscope**   A microscope that uses an ultraviolet light source to illuminate specimens that will fluoresce.

**fluorescent-antibody (FA) technique**   A diagnostic tool using antibodies labeled with fluorochromes and viewed through a fluorescence microscope; also called immunofluorescence.

**FMN**   Flavin mononucleotide; a coenzyme that functions in the transfer of electrons in the electron transport chain.

**focal infection**   A systemic infection that began as an infection in one place.

**folliculitis**   An infection of hair follicles, often occurring as pimples.

**fomite**   A nonliving object that can spread infection.

**frameshift mutation**   A mutation caused by the addition or deletion of one or more bases in DNA.

**free (wandering) macrophage**   A macrophage that leaves the blood and migrates to infected tissue.

**freeze-drying**   *See* lyophilization.

**functional group**   An arrangement of atoms in an organic molecule that is responsible for most of the chemical properties of that molecule.

**fungus**   (plural: **fungi**) An organism that belongs to the Kingdom Fungi; a eukaryotic absorptive chemoheterotroph.

**furuncle**   An infection of a hair follicle.

**fusion**   The merging of plasma membranes of two different cells, resulting in one cell containing cytoplasm from both original cells.

**gamete**   A male or female reproductive cell.

**gametocyte**   A male or female protozoan cell.

**gamma globulin**   The serum fraction containing immunoglobulins (antibodies); also called immune serum globulin.

**gastroenteritis**   Inflammation of the stomach and intestine.

**gas vacuole**   A prokaryotic inclusion for buoyancy compensation.

**gel electrophoresis**   The separation of substances (such as serum proteins or DNA) by their rate of movement through an electrical field.

**gene**   A segment of DNA (a sequence of nucleotides in DNA) encoding a functional product.

**gene editing**   A technique to add, delete, or insert DNA in a chromosome using bacterial enzymes.

**gene silencing**   A mechanism to inhibit gene expression. *See* RNAi.

**gene therapy**   Treating a disease by replacing abnormal genes.

**generalized transduction**   The transfer of bacterial chromosome fragments from one cell to another by a bacteriophage.

**generation time**   The time required for a cell or population to double in number.

**genetic code**   The mRNA codons and the amino acids they encode.

**genetic recombination**   The process of joining pieces of DNA from different sources.

**genetics**   The science of heredity and gene function.

**genetic testing**   Techniques for determining which genes are in a cell's genome.

**genome**   One complete copy of the genetic information in a cell.

**genomic library** A collection of cloned DNA fragments created by inserting restriction enzyme fragments in a bacterium, yeast, or phage.

**genomics** The study of genes and their function.

**genotoxin** A chemical that damages DNA or RNA.

**genotype** The genetic makeup of an organism.

**genus** (plural: **genera**) The first name of the scientific name (binomial); the taxon between family and species.

**germicide** *See* biocide.

**germination** The process of starting to grow from a spore or endospore.

**germ theory of disease** The principle that microorganisms cause disease.

**global warming** Retention of solar heat by gases in the atmosphere.

**glycocalyx** A gelatinous polymer surrounding a cell.

**glycolysis** The main pathway for the oxidation of glucose to pyruvic acid; also called Embden-Meyerhof pathway.

**Golgi complex** An organelle involved in the secretion of certain proteins.

**graft-versus-host (GVH) disease** A condition that occurs when a transplanted tissue has an immune response to the tissue recipient.

**gram-negative bacteria** Bacteria that lose the crystal violet color after decolorizing by alcohol; they stain red after treatment with safranin.

**gram-positive bacteria** Bacteria that retain the crystal violet color after decolorizing by alcohol; they stain dark purple.

**Gram stain** A differential stain that classifies bacteria into two groups, gram-positive and gram-negative.

**granulocyte** A leukocyte with visible granules in the cytoplasm when viewed through a light microscope; includes neutrophils, basophils, and eosinophils.

**granuloma** A lump of inflamed tissue containing macrophages.

**granum** (plural: **grana**) Stack of thylakoid membrane.

**granzymes** Proteases that induce apoptosis.

**green nonsulfur bacteria** Gram-negative, nonproteobacteria; anaerobic and phototrophic; use reduced organic compounds as electron donors for $CO_2$ fixation.

**green sulfur bacteria** Gram-negative, nonproteobacteria; strictly anaerobic and phototrophic; no growth in dark; use reduced sulfur compounds as electron donors for $CO_2$ fixation.

**group translocation** In prokaryotes, active transport in which a substance is chemically altered during transport across the plasma membrane.

**gumma** A rubbery mass of tissue characteristic of tertiary syphilis.

**HAART (highly active antiretroviral therapy)** A combination of drugs used to treat HIV infection.

**halogen** One of the following elements: fluorine, chlorine, bromine, iodine, or astatine.

**halophile** An organism that requires a high salt concentration for growth.

**H antigen** Flagella antigens of enterics, identified by serological testing.

**hapten** A substance of low molecular mass that does not cause the formation of antibodies by itself but does so when combined with a carrier molecule.

**HA (hemagglutinin) spike** Antigenic projections from the outer lipid bilayer of *Influenzavirus*.

**Hazard Analysis and Critical Control Point (HACCP)** System of prevention of hazards, for food safety.

**healthcare-associated infection (HAI)** An infection that develops during a stay at a health care facility and was not present at the time the patient was admitted.

**helminth** A parasitic roundworm or flatworm.

**hemagglutination** The clumping of red blood cells.

**hematopoiesis** The formation of blood cells.

**hemoflagellate** A parasitic flagellate found in the circulatory system of its host.

**hemolysin** An enzyme that lyses red blood cells.

**herd immunity** The presence of immunity in most of a population.

**hermaphroditic** Having both male and female reproductive capacities.

**heterocyst** A large cell in certain cyanobacteria; the site of nitrogen fixation.

**heterolactic** Describing an organism that produces lactic acid and other acids or alcohols as end-products of fermentation; e.g., *Escherichia*.

**heterotroph** An organism that requires an organic carbon source; also called organotroph.

**Hfr cell** A bacterial cell in which the F factor has become integrated into the chromosome; Hfr stands for high frequency of recombination.

**high-efficiency particulate air (HEPA) filter** A screenlike material that removes particles larger than 0.3 μm from air.

**high-temperature short-time (HTST) pasteurization** Pasteurizing at 72°C for 15 seconds.

**histamine** A substance released by tissue cells that causes vasodilation, capillary permeability, and smooth muscle contraction.

**histocompatibility antigen** An antigen on the surface of human cells.

**histone** A protein associated with DNA in eukaryotic chromosomes.

**holdfast** The branched base of an algal stipe.

**holoenzyme** An enzyme consisting of an apoenzyme and a cofactor.

**homolactic** Describing an organism that produces only lactic acid from fermentation; e.g., *Streptococcus*.

**horizontal gene transfer** Transfer of genes between two organisms in the same generation. *See also* vertical gene transfer.

**host** An organism infected by a pathogen. *See also* definitive host; intermediate host.

**host range** The spectrum of species, strains, or cell types that a pathogen can infect.

**hot-air sterilization** Sterilization by the use of an oven at 170°C for approximately 2 hours.

**human leukocyte antigen (HLA) complex** Human cell surface antigens. *See also* major histocompatibility complex.

**Human Microbiome Project** A project to characterize the microbial communities found on the human body.

**humanized antibody** Human antibodies produced by genetically modified mice.

**humoral immunity** Immunity produced by antibodies dissolved in body fluids, mediated by B cells; also called antibody-mediated immunity.

**hyaluronidase** An enzyme secreted by certain bacteria that hydrolyzes hyaluronic acid and helps spread microorganisms from their initial site of infection.

**hybridoma** A cell made by fusing an antibody-producing B cell with a cancer cell.

**hydrogen bond**   A bond between a hydrogen atom covalently bonded to oxygen or nitrogen and another covalently bonded oxygen or nitrogen atom.

**hydrolysis**   A decomposition reaction in which chemicals react with the $H^+$ and $OH^-$ of a water molecule.

**hydroxide**   $OH^-$; the anion that forms a base.

**hydroxyl**   —OH; covalently bonded to a molecule forms an alcohol.

**hydroxyl radical**   A toxic form of oxygen (OH•) formed in cytoplasm by ionizing radiation and aerobic respiration.

**hyperacute rejection**   Very rapid rejection of transplanted tissue, usually in the case of tissue from nonhuman sources.

**hyperbaric chamber**   An apparatus to hold materials at pressures greater than 1 atmosphere.

**hypersensitivity**   An altered, enhanced immune reaction leading to pathological changes; also called allergy.

**hyperthermophile**   An organism whose optimum growth temperature is at least 80°C; also called extreme thermophile.

**hypertonic solution**   A solution that has a higher concentration of solutes than an isotonic solution.

**hypha**   A long filament of cells in fungi or actinomycetes.

**hypotonic solution**   A solution that has a lower concentration of solutes than an isotonic solution.

**ID$_{50}$**   The number of microorganisms required to produce a demonstrable infection in 50% of the test host population.

**idiophase**   The period in the production curve of an industrial cell population in which secondary metabolites are produced; a period of stationary growth following the phase of rapid growth. *See also* trophophase.

**IgA**   The class of antibodies found in secretions.

**IgD**   The class of antibodies found on B cells.

**IgE**   The class of antibodies involved in hypersensitivities.

**IgG**   The most abundant class of antibodies in serum.

**IgM**   The first class of antibodies to appear after exposure to an antigen.

**immune complex**   A circulating antigen-antibody aggregate capable of fixing complement.

**immune serum globulin**   *See* gamma globulin.

**immune surveillance**   The body's immune response to cancer.

**immunity**   *See* adaptive immunity, innate immunity.

**immunization**   *See* vaccination.

**immunodeficiency**   The absence of an adequate immune response; may be congenital or acquired.

**immunodiffusion test**   A test consisting of precipitation reactions carried out in an agar gel medium.

**immunoelectrophoresis**   The identification of proteins by electrophoretic separation followed by serological testing.

**immunofluorescence**   *See* fluorescent-antibody technique.

**immunogen**   *See* antigen.

**immunoglobulin (Ig)**   A protein (antibody) formed in response to an antigen and can react with that antigen. *See also* globulin.

**immunology**   The study of a host's defenses to a pathogen.

**immunotherapy**   Making use of the immune system to attack tumor cells, either by enhancing the normal immune response or by using toxin-bearing specific antibodies. *See also* immunotoxin.

**immunotoxin**   An immunotherapeutic agent consisting of a poison bound to a monoclonal antibody.

**incidence**   The fraction of the population that contracts a disease during a particular period of time.

**inclusion**   Material held inside a cell, often consisting of reserve deposits.

**inclusion body**   A granule or viral particle in the cytoplasm or nucleus of some infected cells; important in the identification of viruses that cause infection.

**incubation period**   The time interval between the actual infection and first appearance of any signs or symptoms of disease.

**indicator organism**   A microorganism, such as a coliform, whose presence indicates conditions such as fecal contamination of food or water.

**indirect (passive) agglutination test**   An agglutination test using soluble antigens attached to latex or other small particles.

**indirect contact transmission**   The spread of pathogens by fomites (nonliving objects).

**indirect FA test**   A fluorescent-antibody test to detect the presence of specific antibodies.

**infection**   The growth of microorganisms in the body.

**infectious disease**   A disease in which pathogens invade a susceptible host and carry out at least part of their life cycle in the host.

**inflammation**   A host response to tissue damage characterized by redness, pain, heat, and swelling; and sometimes loss of function.

**innate immunity**   Host defenses that afford protection against any kind of pathogen. *See also* adaptive immunity.

**inoculum**   Microbes introduced into a culture medium to initiate growth.

**inorganic compound**   A small molecule that does not contain carbon and hydrogen.

**insertion sequence (IS)**   The simplest kind of transposon.

**interferon (IFN)**   A specific group of cytokines. Alpha- and beta-IFNs are antiviral proteins produced by certain animal cells in response to a viral infection. Gamma-IFN stimulates macrophage activity.

**interleukin (IL)**   A chemical that causes T-cell proliferation. *See also* cytokine.

**intermediate host**   An organism that harbors the larval or asexual stage of a helminth or protozoan.

**intoxication**   A condition resulting from the ingestion of a microbially produced toxin.

**intron**   A region in a eukaryotic gene that does not code for a protein or mRNA.

**invasin**   A surface protein produced by *Salmonella* Typhimurium and *Escherichia coli* that rearranges nearby actin filaments in the cytoskeleton of a host cell.

**iodophor**   A complex of iodine and a detergent.

**ion**   A negatively or positively charged atom or group of atoms.

**ionic bond**   A chemical bond formed when atoms gain or lose electrons in the outer energy levels.

**ionization**   The separation (dissociation) of a molecule into ions.

**ionizing radiation**   High-energy radiation with a wavelength less than 1 nm; causes ionization. X rays and gamma rays are examples.

**ischemia**   Localized decreased blood flow.

**isograft**   A tissue graft from a genetically identical source (i.e., from an identical twin).

**isolation chip (iChip)** A chamber composed of a plastic membrane that allows molecules to pass through; used to grow bacteria in their natural environment.

**isomer** One or two molecules with the same chemical formula but different structures.

**isotonic solution** A solution in which, after immersion of a cell, osmotic pressure is equal across the cell's membrane.

**isotope** A form of a chemical element in which the number of neutrons in the nucleus is different from the other forms of that element.

**karyogamy** Fusion of the nuclei of two cells; occurs in the sexual stage of a fungal life cycle.

**keratin** A protein found in epidermis, hair, and nails.

**ketolide** Semisynthetic macrolide antibiotic; effective against macrolide-resistant bacteria

**kinase** (1) An enzyme that removes a ⓟ from ATP and attaches it to another molecule. (2) A bacterial enzyme that breaks down fibrin (blood clots).

**kingdom** A taxonomic classification between domain and phylum.

**kinin** A substance released from tissue cells that causes vasodilation.

**Kirby-Bauer test** *See* disk-diffusion method.

**Koch's postulates** Criteria used to determine the causative agent of infectious diseases.

**koji** A microbial fermentation on rice; usually *Aspergillus oryzae*; used to produce amylase.

**Krebs cycle** A pathway that converts two-carbon compounds to $CO_2$, transferring electrons to $NAD^+$ and other carriers; also called tricarboxylic acid (TCA) cycle or critic acid cycle.

**lactic acid fermentation** A catabolic process, beginning with glycolysis, that produces lactic acid to reoxidize NADH.

**lactoferrin** One of several human iron-binding proteins that reduce iron available to a pathogen.

**lag phase** The time interval in a bacterial growth curve during which there is no growth.

**larva** The sexually immature stage of a helminth or arthropod.

**latent disease** A disease characterized by a period of no symptoms when the pathogen is inactive.

**latent infection** A condition in which a pathogen remains in the host for long periods without producing disease.

**$LD_{50}$** The lethal dose for 50% of the inoculated hosts within a given period.

**lectin** Carbohydrate-binding proteins on a cell, not an antibody.

**leukocidins** Substances produced by some bacteria that can destroy neutrophils and macrophages.

**leukocyte** A white blood cell.

**leukotriene** A substance produced by mast cells and basophils that causes increased permeability of blood vessels and helps phagocytes attach to pathogens.

**L form** Prokaryotic cells that lack a cell wall; can return to walled state.

**lichen** A mutualistic relationship between a fungus and an alga or a cyanobacterium.

**ligand** *See* adhesin.

**light-dependent (light) reaction** The process by which light energy is used to convert ADP and phosphate to ATP. *See also* photophosphorylation.

**light-independent (dark) reactions** The process by which electrons and energy from ATP are used to reduce $CO_2$ to sugar. *See also* Calvin-Benson cycle.

**limnetic zone** The surface zone of an inland body of water away from the shore.

**Limulus amebocyte lysate (LAL) assay** A test to detect the presence of bacterial endotoxins.

**lipase** An enzyme that breaks down triglycerides into their component glycerol and fatty acids.

**lipid** A non–water-soluble organic molecule, including triglycerides, phospholipids, and sterols.

**lipid A** A component of the gram-negative outer membrane; endotoxin.

**lipid inclusion** *See* inclusion.

**lipopolysaccharide (LPS)** A molecule consisting of a lipid and a polysaccharide, forming the outer membrane of gram-negative cell walls.

**ʟ-isomer** Arrangement of four different atoms or groups around a carbon atom. *See* ᴅ-isomer.

**littoral zone** The region along the shore of the ocean or a large lake where there is considerable vegetation and where light penetrates to the bottom.

**local infection** An infection in which pathogens are limited to a small area of the body.

**localized anaphylaxis** An immediate hypersensitivity reaction that is restricted to a limited area of skin or mucous membrane; for example, hayfever, a skin rash, or asthma. *See also* systemic anaphylaxis.

**logarithmic decline phase** *See* death phase.

**log phase** The period of bacterial growth or logarithmic increase in cell numbers; also called exponential growth phase.

**lophotrichous** Having two or more flagella at one end of a cell.

**luciferase** An enzyme that accepts electrons from flavoproteins and emits a photon of light in bioluminescence.

**lymphangitis** Inflammation of lymph vessels.

**lymphocyte** A leukocyte involved in specific immune responses.

**lyophilization** Freezing a substance and sublimating the ice in a vacuum; also called freeze-drying.

**lysis** (1) Destruction of a cell by the rupture of the plasma membrane, resulting in a loss of cytoplasm. (2) In disease, a gradual period of decline.

**lysogenic conversion** The acquisition of new properties by a host cell infected by a lysogenic phage.

**lysogenic cycle** Stages in viral development that result in the incorporation of viral DNA into host DNA.

**lysogeny** A state in which phage DNA is incorporated into the host cell without lysis.

**lysosome** An organelle containing digestive enzymes.

**lysozyme** An enzyme capable of hydrolyzing bacterial cell walls.

**lytic cycle** A mechanism of phage multiplication that results in host cell lysis.

**macrolide** An antibiotic that inhibits protein synthesis; for example, erythromycin.

**macromolecule** A large organic molecule.

**macrophage** A phagocytic cell; a mature monocyte. *See* fixed macrophage, free wandering macrophage.

**macule** A flat, reddened skin lesion.

**magnetosome**   An iron oxide inclusion, produced by some gram-negative bacteria, that acts like a magnet.

**major histocompatibility complex (MHC)**   The genes that code for histocompatibility antigens; also known as human leukocyte antigen (HLA) complex.

**malolactic fermentation**   The conversion of malic acid to lactic acid by lactic acid bacteria.

**malt**   Germinated barley grains containing maltose, glucose, and amylase.

**malting**   The germination of starchy grains resulting in glucose and maltose production.

**margination**   The process by which phagocytes stick to the lining of blood vessels.

**mast cell**   A type of cell found throughout the body that contains histamine and other substances that stimulate vasodilation.

**matrix**   Fluid in mitochondria.

**maximum growth temperature**   The highest temperature at which a species can grow.

**M (microfold) cell**   Cells that take up and transfer antigens to lymphocytes, on Peyer's patches.

**mechanical transmission**   The process by which arthropods transmit infections by carrying pathogens on their feet and other body parts.

**medulla**   A lichen body consisting of algae (or cyanobacteria) and fungi.

**meiosis**   A eukaryotic cell replication process that results in cells with half the chromosome number of the original cell.

**membrane attack complex (MAC)**   Complement proteins C5–C9, which together make lesions in cell membranes that lead to cell death.

**membrane filter**   A screenlike material with pores small enough to retain microorganisms; a 0.45-$\mu$m filter retains most bacteria.

**memory cell**   A long-lived B or T cell responsible for the memory, or secondary, response.

**memory response**   A rapid rise in antibody titer following exposure to an antigen after the primary response to that antigen; also called anamestic response or secondary response.

**meningitis**   Inflammation of the meninges, the three membranes covering the brain and spinal cord.

**merozoite**   A trophozoite of *Plasmodium* found in red blood cells or liver cells.

**mesophile**   An organism that grows between about 10°C and 50°C; a moderate-temperature–loving microbe.

**mesosome**   An irregular fold in the plasma membrane of a prokaryotic cell that is an artifact of preparation for microscopy.

**messenger RNA (mRNA)**   The type of RNA molecule that directs the incorporation of amino acids into proteins.

**metabolic pathway**   A sequence of enzymatically catalyzed reactions occurring in a cell.

**metabolism**   The sum of all the chemical reactions that occur in a living cell.

**metabolomics**   The study of small molecules in and around growing cells.

**metacercaria**   The encysted stage of a fluke in its final intermediate host.

**metachromatic granule**   A granule that stores inorganic phosphate and stains red with certain blue dyes; characteristic of *Corynebacterium diphtheriae*. Collectively known as volutin.

**metagenomics**   The study of the genomes of uncultured organisms by the collection and sequencing of DNA from environmental samples.

**methane**   The hydrocarbon $CH_4$, a flammable gas formed by the microbial decomposition of organic matter; natural gas.

**methylase**   An enzyme that attaches methyl groups ($—CH_3$) to a molecule; methylated cytosine is protected from digestion by restriction enzymes.

**methylate**   Addition of a methyl ($—CH_3$) group to a compound.

**microaerophile**   An organism that grows best in an environment with less molecular oxygen ($O_2$) than is normally found in air.

**microarray**   DNA probes attached to a glass surface, used to identify nucleotide sequences in a sample of DNA.

**microbial fuel cell**   A system used to grow bacteria and transfer electrons from their electron transport systems to a wire (electricity).

**microbiome**   All the microorganisms in an environment.

**micrometer ($\mu$m)**   A unit of measurement equal to $10^{-6}$ m.

**microorganism**   A living organism too small to be seen with the naked eye; includes bacteria, fungi, protozoa, and microscopic algae; also includes viruses.

**microRNA (miRNA)**   Small, single-stranded RNA that prevent translation of a complementary mRNA.

**microtubule**   A hollow tube made of the protein tubulin; the structural unit of eukaryotic flagella and centrioles.

**microwave**   Electromagnetic radiation with wavelength between $10^{21}$ and $10^{-3}$ m.

**minimal bactericidal concentration (MBC)**   The lowest concentration of chemotherapeutic agent that will kill test microorganisms.

**minimal inhibitory concentration (MIC)**   The lowest concentration of a chemotherapeutic agent that will prevent growth of the test microorganisms.

**minimum growth temperature**   The lowest temperature at which a species will grow.

**miracidium**   The free-swimming, ciliated larva of a fluke that hatches from the egg.

**missense mutation**   A mutation that results in the substitution of an amino acid in a protein.

**mitochondrion** (plural: **mitochondria**)   An organelle containing Krebs cycle enzymes and the electron transport chain.

**mitosis**   A eukaryotic cell replication process in which the chromosomes are duplicated; usually followed by division of the cytoplasm of the cell.

**MMWR**   *Morbidity and Mortality Weekly Report;* a CDC publication containing data on notifiable diseases and topics of special interest.

**mobile genetic elements**   Segments of DNA (e.g., plasmids) that can move between chromosomes or between cells.

**mole**   An amount of a chemical equal to the atomic masses of all the atoms in a molecule of the chemical.

**molecular biology**   The science dealing with DNA and protein synthesis of living organisms.

**molecular clock**   An evolution timeline based on nucleotide sequences in organisms.

**molecular mass**   The sum of the atomic masses of all atoms making up a molecule.

**molecule**   A combination of atoms forming a specific chemical compound.

**monoclonal antibody (Mab)**   A specific antibody produced in vitro by a clone of B cells hybridized with cancerous cells.

**monocyte**   A leukocyte that is the precursor of a macrophage.

**monoecious**   Having both male and female reproductive capacities.

**monomer** A small molecule that collectively combines to form polymers.

**monomorphic** Having a single shape; most bacteria always present with a genetically determined shape. *See also* pleomorphic.

**mononuclear phagocytic system** A system of fixed macrophages located in the spleen, liver, lymph nodes, and red bone marrow.

**monosaccharide** A simple sugar consisting of 3–7 carbon atoms.

**monotrichous** Having a single flagellum.

**morbidity** (1) The incidence of a specific disease. (2) The condition of being diseased.

**morbidity rate** The number of people affected by a disease in a given period of time in relation to the total population.

**mordant** A substance added to a staining solution to make it stain more intensely.

**mortality** The number of deaths from a specific notifiable disease.

**mortality rate** The number of deaths resulting from a disease in a given period of time in relation to the total population.

**most probable number (MPN) method** A statistical determination of the number of coliforms per 100 ml of water or 100 g of food.

**motility** The ability of an organism to move by itself.

**M protein** A heat- and acid-resistant protein of streptococcal cell walls and fibrils.

**mucous membranes** Membranes that line body openings, including the intestinal tract, open to the exterior; also called mucosa.

**multi-drug-resistant (MDR)** *M. tuberculosis* strains resistant to isoniazid and rifampin.

**mutagen** An agent in the environment that brings about mutations.

**mutation** Any change in the nitrogenous base sequence of DNA.

**mutation rate** The probability that a gene will mutate each time a cell divides.

**mutualism** A type of symbiosis in which both organisms or populations are benefited.

**mycelium** A mass of long filaments of cells that branch and intertwine, typically found in molds.

**mycolic acid** Long-chained, branched fatty acids characteristic of members of the genus *Mycobacterium*.

**mycology** The scientific study of fungi.

**mycorrhiza** A fungus growing in symbiosis with plant roots.

**mycosis** A fungal infection.

**mycotoxin** A toxin produced by a fungus.

**NAD$^+$** A coenzyme that functions in the removal and transfer of hydrogen ion (H$^-$) and electrons from substrate molecules.

**NADP$^+$** A coenzyme similar to NAD$^+$.

**nanometer (nm)** A unit of measurement equal to $10^{-9}$ m, $10^{-3}$ μm.

**nanotechnology** Making molecular- or atomic-sized products.

**NA (neuraminidase) spikes** Antigenic projections from the outer lipid bilayer of *Influenzavirus*.

**natural killer (NK) cell** A lymphoid cell that destroys tumor cells and virus-infected cells.

**naturally acquired active immunity** Antibody production in response to an infectious disease.

**naturally acquired passive immunity** The natural transfer of humoral antibodies, for example, transplacental transfer.

**natural selection** Process by which organisms with certain inherited characteristics are more likely to survive and reproduce than organisms with other characteristics.

**necrosis** Tissue death.

**negative (indirect) selection** The process of identifying mutations by selecting cells that do not grow using replica plating.

**negative staining** A procedure that results in colorless bacteria against a stained background.

**neutralization** An antigen–antibody reaction that inactivates a bacterial exotoxin or virus.

**neutron** An uncharged particle in the nucleus of an atom.

**neutrophil** A highly phagocytic granulocyte; also called polymorphonuclear leukocyte (PMN) or polymorph.

**nitrification** The oxidation of nitrogen in ammonia to produce nitrate.

**nitrogen cycle** The series of processes that converts nitrogen ($N_2$) to organic substances and back to nitrogen in nature.

**nitrogen fixation** The conversion of nitrogen ($N_2$) into ammonia.

**nitrosamine** A carcinogen formed by the combination of nitrite and amino acids.

**noncommunicable disease** A disease that is not transmitted from one person to another.

**noncompetitive inhibitor** An inhibitory chemical that does not compete with the substrate for an enzyme's active site. *See also* allosteric inhibition; competitive inhibitor.

**noncyclic photophosphorylation** The movement of an electron from chlorophyll to NAD$^+$; plant and cyanobacterial photophosphorylation.

**nonionizing radiation** Short-wavelength radiation that does not cause ionization; ultraviolet (UV) radiation is an example.

**non-nucleoside inhibitor** A drug that binds with and inhibits the action of the HIV reverse transcriptase enzyme.

**nonsense codon** A codon that does not encode any amino acid.

**nonsense mutation** A base substitution in DNA that results in a nonsense codon.

**normal microbiota** The microorganisms that colonize a host without causing disease; also called normal flora.

**nosocomial infection** *See* healthcare-associated infection.

**notifiable infectious disease** A disease that physicians must report to the U.S. Public Health Service; also called reportable disease.

**nuclear envelope** The double membrane that separates the nucleus from the cytoplasm in a eukaryotic cell.

**nuclear pore** An opening in the nuclear envelope through which materials enter and exit the nucleus.

**nucleic acid** A macromolecule consisting of nucleotides; DNA and RNA are nucleic acids.

**nucleic acid amplification test (NAAT)** Test to identify an organism without culturing by making copies (amplifying) nucleic acid sequences that are specific for the organism being detected.

**nucleic acid hybridization** The process of combining single complementary strands of DNA.

**nucleic acid vaccine** A vaccine made up of DNA, usually in the form of a plasmid.

**nucleoid** The region in a bacterial cell containing the chromosome.

**nucleolus (plural: nucleoli)** An area in a eukaryotic nucleus where rRNA is synthesized.

**nucleoside** A compound consisting of a purine or pyrimidine base and a pentose sugar.

**nucleotide**    A compound consisting of a purine or pyrimidine base, a five-carbon sugar, and a phosphate.

**nucleotide (or nucleoside) analog**    A chemical that is structurally similar to the normal nucleotide or nucleoside in nucleic acids but with altered base-pairing properties.

**nucleotide excision repair**    The repair of DNA involving removal of defective nucleotides and replacement with functional ones.

**nucleus**    (1) The part of an atom consisting of the protons and neutrons.
(2) The part of a eukaryotic cell that contains the genetic material.

**numerical identification**    Bacterial identification schemes in which test values are assigned a number.

**nutrient agar**    Nutrient broth containing agar.

**nutrient broth**    A complex medium made of beef extract and peptone.

**O antigen**    Polysaccharide antigens in the outer membrane of gram-negative bacteria, identified by serological testing.

**objective lenses**    In a compound light microscope, the lenses closest to the specimen.

**obligate aerobe**    An organism that requires molecular oxygen ($O_2$) to live.

**obligate anaerobe**    An organism that does not use molecular oxygen ($O_2$) and is killed in the presence of $O_2$.

**obligate halophile**    An organism that requires high osmotic pressures such as high concentrations of NaCl.

**ocular lens**    In a compound light microscope, the lens closest to the viewer; also called the eyepiece.

**Okazaki fragments**    Short strands of DNA made by copying the lagging strand during DNA synthesis.

**oligodynamic action**    The ability of small amounts of a heavy metal compound to exert antimicrobial activity.

**oligosaccharide**    A carbohydrate consisting of 2 to approximately 20 monosaccharides.

**oncogene**    A gene that can bring about malignant transformation.

**oncogenic virus**    A virus that is capable of producing tumors; also called oncovirus.

**oncolytic virus**    Virus that infects and kills tumor cells or causes an immune response against tumor cells.

**oocyst**    An encysted apicomplexan zygote in which cell division occurs to form the next infectious stage.

**Opa**    A bacterial outer membrane protein; cells with Opa form opaque colonies.

**operator**    The region of DNA adjacent to structural genes that controls their transcription.

**operon**    The operator and promoter sites and structural genes they control.

**opportunistic pathogen**    A microorganism that does not ordinarily cause a disease but can become pathogenic under certain circumstances.

**opsonization**    The enhancement of phagocytosis by coating microorganisms with certain serum proteins (opsonins); also called immune adherence.

**optimum growth temperature**    The temperature at which a species grows best.

**order**    A taxonomic classification between class and family.

**organelle**    A membrane-enclosed structure within eukaryotic cells.

**organic compound**    A molecule that contains carbon and hydrogen.

**organic growth factor**    An essential organic compound that an organism is unable to synthesize.

**osmosis**    The net movement of solvent molecules across a selectively permeable membrane from an area of lower solute concentration to an area of higher solute concentration.

**osmotic lysis**    Rupture of the plasma membrane resulting from movement of water into the cell.

**osmotic pressure**    The force with which a solvent moves from a solution of lower solute concentration to a solution of higher solute concentration.

**oxidase test**    Test for an enzyme that oxidizes cytochrome *c*.

**oxidation**    The removal of electrons from a molecule.

**oxidation pond**    A method of secondary sewage treatment by microbial activity in a shallow standing pond of water.

**oxidation-reduction**    A coupled reaction in which one substance is oxidized and one is reduced; also called redox reaction.

**oxidative phosphorylation**    The synthesis of ATP coupled with electron transport.

**oxygenic**    Producing oxygen, as in plant and cyanobacterial photosynthesis.

**ozone**    $O_3$.

**PAMP (pathogen-associated molecular patterns)**    Molecules present on pathogens and not self.

**pandemic disease**    An epidemic that occurs worldwide.

**papule**    Small, solid elevation of the skin.

**parasite**    An organism that derives nutrients from a living host.

**parasitism**    A symbiotic relationship in which one organism (the parasite) exploits another (the host) without providing any benefit in return.

**parasitology**    The scientific study of parasitic protozoa and worms.

**parenteral route**    A portal of entry for pathogens by deposition directly into tissues beneath the skin and mucous membranes.

**pasteurization**    The process of mild heating to kill particular spoilage microorganisms or pathogens.

**pathogen**    A disease-causing organism.

**pathogenesis**    The manner in which a disease develops.

**pathogenicity**    The ability of a microorganism to cause disease by overcoming the defenses of a host.

**pathology**    The scientific study of disease.

**pellicle**    (1) The flexible covering of some protozoa. (2) Scum on the surface of a liquid medium.

**penicillins**    A group of antibiotics produced either by *Penicillium* (natural penicillins) or by adding side chains to the β-lactam ring (semisynthetic penicillins).

**pentose phosphate pathway**    A metabolic pathway that can occur simultaneously with glycolysis to produce pentoses and NADH without ATP production; also called hexose monophosphate shunt.

**peptide bond**    A bond joining the amino group of one amino acid to the carboxyl group of a second amino acid with the loss of a water molecule.

**peptidoglycan**    The structural molecule of bacterial cell walls consisting of the molecules N-acetylglucosamine, N-acetylmuramic acid, tetrapeptide side chain, and peptide side chain.

**perforin**    Protein that makes a pore in a target cell membrane, released by cytotoxic T lymphocytes.

**pericarditis** Inflammation of the pericardium, the sac around the heart.

**period of convalescence** The recovery period, when the body returns to its predisease state.

**peripheral nervous system (PNS)** The nerves that connect the outlying parts of the body with the central nervous system.

**periplasm** The region of a gram-negative cell wall between the outer membrane and the cytoplasmic membrane.

**peritrichous** Having flagella distributed over the entire cell.

**peroxidase** An enzyme that destroys hydrogen peroxide: $H_2O_2 + 2\,H^+ \rightarrow 2\,H_2O$

**peroxide anion** An oxygen anion consisting of two atoms of oxygen $(O_2^{-2})$.

**peroxisome** Organelle that oxidizes amino acids, fatty acids, and alcohol.

**peroxygen** A class of oxidizing-type sterilizing disinfectants.

**persistent viral infection** A disease process that occurs gradually over a long period.

**persister cells** Bacterial cells in a population that avoid being killed by antibiotics because they are dormant, not because they are mutants.

**Peyer's patches** Lymphoid organs on the intestinal wall.

**PFU (plaque-forming units)** Visible clearing in a bacterial culture caused by lysis of bacterial cells by bacteriophages.

**pH** The symbol for hydrogen ion ($H^1$) concentration; a measure of the relative acidity or alkalinity of a solution.

**phage** *See* bacteriophage.

**phage conversion** Genetic change in the host cell resulting from infection by a bacteriophage.

**phage typing** A method of identifying bacteria using specific strains of bacteriophages.

**phagocyte** A cell capable of engulfing and digesting particles that are harmful to the body.

**phagocytosis** The ingestion of particles by eukaryotic cells.

**phagolysosome** A digestive vacuole.

**phagosome** A food vacuole of a phagocyte; also called a phagocytic vesicle.

**phalloidin** A peptide toxin produced by *Amanita phalloides*, affects plasma membrane function.

**phase-contrast microscope** A compound light microscope that allows examination of structures inside cells through the use of a special condenser.

**phenol** OH Also called carbolic acid.

**phenolic** A derivative of phenol used as a disinfectant.

**phenotype** The external manifestations of an organism's genotype, or genetic makeup.

**phosphate group** A portion of a phosphoric acid molecule attached to some other molecule, (P),

$$PO_4^{3-},\ {}^-O-\overset{\overset{\displaystyle O}{\|}}{\underset{\underset{\displaystyle O^-}{|}}{P}}-O^-$$

**phospholipid** A complex lipid composed of glycerol, two fatty acids, and a phosphate group.

**phosphorus cycle** The various solubility stages of phosphorus in the environment.

**phosphorylation** The addition of a phosphate group to an organic molecule.

**photoautotroph** An organism that uses light as its energy source and carbon dioxide ($CO_2$) as its carbon source.

**photoheterotroph** An organism that uses light as its energy source and an organic carbon source.

**photolyase** An enzyme that splits thymine dimers in the presence of visible light.

**photophosphorylation** The production of ATP in a series of redox reactions; electrons from chlorophyll initiate the reactions.

**photosynthesis** The conversion of light energy from the sun into chemical energy; the light-fueled synthesis of carbohydrate from carbon dioxide ($CO_2$).

**phototaxis** Movement in response to the presence of light.

**phototroph** An organism that uses light at its primary energy source.

**phylogeny** The evolutionary history of a group of organisms; phylogenetic relationships are evolutionary relationships.

**phylum** A taxonomic classification between kingdom and class.

**phytoplankton** Free-floating photoautotrophs.

**pilus (plural: pili)** An appendage on a bacterial cell used for conjugation and gliding motility.

**pinocytosis** Taking in molecules by infolding of the plasma membrane, in eukaryotes.

**plankton** Free-floating aquatic organisms.

**Plantae** The kingdom composed of multicellular eukaryotes with cellulose cell walls.

**plaque** A clearing in a bacterial lawn resulting from lysis by phages. *See also* dental plaque.

**plasma** (1) The liquid portion of blood in which the formed elements are suspended. (2) Excited gases used for sterilizing.

**plasma cell** A cell that an activated B cell differentiates into; plasma cells manufacture specific antibodies.

**plasma (cytoplasmic) membrane** The selectively permeable membrane enclosing the cytoplasm of a cell; the outer layer in animal cells, internal to the cell wall in other organisms.

**plasmid** A small circular DNA molecule that replicates independently of the chromosome.

**plasmodium** (1) A multinucleated mass of protoplasm, as in plasmodial slime molds. (2) When written as a genus, refers to the causative agent of malaria.

**plasmogamy** Fusion of the cytoplasm of two cells; occurs in the sexual stage of a fungal life cycle.

**plasmolysis** Loss of water from a cell in a hypertonic environment.

**plate count** A method of determining the number of bacteria in a sample by counting the number of colony-forming units on a solid culture medium.

**pleomorphic** Having many shapes, characteristic of certain bacteria.

**pluripotent** A cell that can differentiate into many different types of tissue cells.

**pneumonia** Inflammation of the lungs.

**point mutation** *See* base substitution.

**polar flagella** Having flagella at one or both ends of a cell.

**polar molecule** A molecule with an unequal distribution of charges.

**polymer**   A molecule consisting of a sequence of similar molecules, or monomers.

**polymerase chain reaction (PCR)**   A technique using DNA polymerase to make multiple copies of a DNA template in vitro. *See also* cDNA.

**polymorphonuclear leukocyte (PMN)**   *See* neutrophil.

**polypeptide**   (1) A chain of amino acids. (2) A group of antibiotics.

**polysaccharide**   A carbohydrate consisting of 8 or more monosaccharides joined through dehydration synthesis.

**porins**   A type of protein in the outer membrane of gram-negative cell walls that permits the passage of small molecules.

**portal of entry**   The avenue by which a pathogen gains access to the body.

**portal of exit**   The route by which a pathogen leaves the body.

**positive (direct) selection**   A procedure for picking out mutant cells by growing them.

**postexposure prophylaxis (PEP)**   Medication or antibodies given after potential exposure to a pathogen or toxin.

**pour plate method**   A method of inoculating a solid nutrient medium by mixing bacteria in the melted medium and pouring the medium into a Petri dish to solidify.

**prebiotics**   Chemicals that promote growth of beneficial bacteria in the body.

**precipitation reaction**   A reaction between soluble antigens and multivalent antibodies to form visible aggregates.

**precipitin ring test**   A precipitation test performed in a capillary tube.

**predisposing factor**   Anything that makes the body more susceptible to a disease or alters the course of a disease.

**prevalence**   The fraction of a population having a specific disease at a given time.

**primary cell line**   Human tissue cells that grow for only a few generations in vitro.

**primary infection**   An acute infection that causes the initial illness.

**primary metabolite**   A product of an industrial cell population produced during the time of rapid logarithmic growth. *See also* secondary metabolite.

**primary producer**   An autotrophic organism, either chemotroph or phototroph, that converts carbon dioxide into organic compounds.

**primary response**   Antibody production in response to the first contact with an antigen. *See also* memory response.

**primary sewage treatment**   The removal of solids from sewage by allowing them to settle out and be held temporarily in tanks or ponds.

**prion**   An infectious agent consisting of a self-replicating protein, with no detectable nucleic acids.

**privileged site (tissue)**   An area of the body (or a tissue) that does not elicit an immune response.

**probiotics**   Microbes inoculated into a host to occupy a niche and prevent growth of pathogens.

**prodromal period**   The time following the incubation period when the first symptoms of illness appear.

**profundal zone**   The deeper water under the limnetic zone in an inland body of water.

**proglottid**   A body segment of a tapeworm containing both male and female organs.

**prokaryote**   A cell whose genetic material is not enclosed in a nuclear envelope.

**prokaryotic species**   A population of cells that share certain rRNA sequences; in conventional biochemical testing, it is a population of cells with similar characteristics.

**promoter**   The starting site on a DNA strand for transcription of RNA by RNA polymerase.

**prophage**   Phage DNA inserted into the host cell's DNA.

**prophylactic**   Anything used to prevent disease.

**prostaglandin**   A hormonelike substance that is released by damaged cells, intensifies inflammation.

**prostheca**   A stalk or bud protruding from a prokaryotic cell.

**protease**   An enzyme that digests protein (proteolytic enzymes).

**protein**   A large molecule containing carbon, hydrogen, oxygen, and nitrogen (and sulfur); some proteins have a helical structure and others are pleated sheets.

**proteobacteria**   Gram-negative, chemoheterotrophic bacteria that possess a signature rRNA sequence.

**proteomics**   The science of determining all of the proteins expressed in a cell.

**protist**   Term used for unicellular and simple multicellular eukaryotes; usually protozoa and algae.

**proton**   A positively charged particle in the nucleus of an atom.

**protoplast**   A gram-positive bacterium or plant cell treated to remove the cell wall.

**protoplast fusion**   A method of joining two cells by first removing their cell walls; used in genetic engineering.

**protozoan** (plural: **protozoa**)   Unicellular eukaryotic organisms; usually chemoheterotrophic.

**provirus**   Viral DNA that is integrated into the host cell's DNA.

**pseudohypha**   A short chain of fungal cells that results from the lack of separation of daughter cells after budding.

**pseudopod**   An extension of a eukaryotic cell that aids in locomotion and feeding.

**psychrophile**   An organism that grows best at about 15°C and does not grow above 20°C; a cold-loving microbe.

**pscyhrotroph**   An organism that is capable of growth between about 0°C and 30°C.

**purines**   The class of nucleic acid bases that includes adenine and guanine.

**purple nonsulfur bacteria**   Alphaproteobacteria; strictly anaerobic and phototrophic; grow on yeast extract in dark; use reduced organic compounds as electron donors for $CO_2$ fixation.

**purple sulfur bacteria**   Gammaproteobacteria; strictly anaerobic and phototrophic; use reduced sulfur compounds as electron donors for $CO_2$ fixation.

**pus**   An accumulation of dead phagocytes, dead bacterial cells, and fluid.

**pustule**   A small pus-filled elevation of skin.

**pyocyanin**   A blue-green pigment produced by *Pseudomonas aeruginosa*.

**pyrimidines**   The class of nucleic acid bases that includes uracil, thymine, and cytosine.

**quaternary ammonium compound (quat)**   A cationic detergent with four organic groups attached to a central nitrogen atom; used as a disinfectant.

**quorum sensing**   The ability of bacteria to communicate and coordinate behavior via signaling molecules.

**R** Used to represent nonfunctional groups of a molecule. *See also* resistance factor.

**rapid diagnostic test (RDT)** A test that allows diagnosis of a disease within a few minutes.

**rapid identification methods** Bacterial identification tools that perform several biochemical tests simultaneously.

**rapid plasma reagin (RPR) test** A serological test for syphilis.

**r-determinant** A group of genes for antibiotic resistance carried on R factors.

**receptor** An attachment for a pathogen on a host cell.

**receptor-mediated endocytosis** A type of pinocytosis in which molecules bound to proteins on the plasma membrane are taken in by infolding of the membrane.

**recipient cell** A cell that receives DNA from a donor cell during genetic recombination.

**recombinant DNA (rDNA)** A DNA molecule produced by combining DNA from two different sources.

**recombinant DNA (rDNA) technology** Manufacturing and manipulating genetic material in vitro; also called genetic engineering.

**recombinant vaccine** A vaccine made by recombinant DNA techniques.

**redia** A trematode larval stage that reproduces asexually to produce cercariae.

**redox reaction** *See* oxidation-reduction.

**red tide** A bloom of planktonic dinoflagellates.

**reducing medium** A culture medium containing ingredients that will remove dissolved oxygen from the medium to allow the growth of anaerobes.

**reduction** The addition of electrons to a molecule.

**refractive index** The relative velocity with which light passes through a substance.

**rennin** An enzyme that forms curds as part of any dairy fermentation product; originally from calves' stomachs, now produced by molds and bacteria.

**replica plating** A method of inoculating a number of solid minimal culture media from an original plate to produce the same pattern of colonies on each plate.

**reservoir of infection** A continual source of infection.

**resistance** The ability to ward off diseases through innate and adaptive immunity.

**resistance (R) factor** A bacterial plasmid carrying genes that determine resistance to antibiotics.

**resistance transfer factor (RTF)** A group of genes for replication and conjugation on the R factor.

**resolution** The ability to distinguish fine detail with a magnifying instrument; also called resolving power.

**respiration** A series of redox reactions in a membrane that generates ATP; the final electron acceptor is usually an inorganic molecule.

**restriction enzyme** An enzyme that cuts double-stranded DNA at specific sites between nucleotides.

**reticulate body** The intracellular growing stage of chlamydiae.

**reticuloendothelial system** *See* mononuclear phagocytic system.

**retort** A device for commercially sterilizing canned food by using steam under pressure; operates on the same principle as an autoclave but is much larger.

**reverse genetics** Genetic analysis that begins with a piece of DNA and proceeds to find out what it does.

**reverse transcriptase** An RNA-dependent DNA polymerase; an enzyme that synthesizes a complementary DNA from an RNA template.

**reversible reaction** A chemical reaction in which the end-products can readily revert to the original molecules.

**RFLP** Restriction fragment length polymorphism; a fragment resulting from restriction-enzyme digestion of DNA.

**Rh factor** An antigen on red blood cells of rhesus monkeys and most humans; possession makes the cells $Rh^+$.

**rhizine** A rootlike hypha that anchors a fungus to a surface.

**ribonucleic acid (RNA)** The class of nucleic acids that comprises messenger RNA, ribosomal RNA, and transfer RNA.

**ribose** A five-carbon sugar that is part of ribonucleotide molecules and RNA.

**ribosomal RNA (rRNA)** The type of RNA molecule that forms ribosomes.

**ribosomal RNA (rRNA) sequencing** Determination of the order of nucleotide bases in rRNA.

**ribosome** The site of protein synthesis in a cell, composed of RNA and protein.

**riboswitch** Part of an mRNA molecule that binds to a substrate; can change the mRNA structure and regulates synthesis of the mRNA.

**ribotyping** Classification or identification of bacteria based on rRNA genes.

**ribozyme** An enzyme consisting of RNA that specifically acts on strands of RNA to remove introns and splice together the remaining exons.

**ring stage** A young *Plasmodium* trophozoite that looks like a ring in a red blood cell.

**RNAi** RNA interference; stops gene expression at transcription by using a short interfering RNA to make double-stranded RNA.

**RNA-induced silencing complex (RISC)** A complex consisting of a protein and siRNA or miRNA that binds complementary mRNA, preventing transcription of the mRNA.

**RNA primer** A short strand of RNA used to start synthesis of the lagging strand of DNA, and to start the polymerase chain reaction.

**root nodule** A tumorlike growth on the roots of certain plants containing symbiotic nitrogen-fixing bacteria.

**rotating biological contactor** A method of secondary sewage treatment in which large disks are rotated while partially submerged in a sewage tank exposing sewage to microorganisms and aerobic conditions.

**rough ER** Endoplasmic reticulum with ribosomes on its surface.

**roundworm** An animal belonging to the phylum Nematoda.

**S (Svedberg unit)** Notes the relative rate of sedimentation during ultra-high speed centrifugation.

**salt** A substance that dissolves in water to cations and anions, neither of which is $H^+$ or $OH^-$.

**sanitization** The removal of microbes from eating utensils and food preparation areas.

**saprophyte** An organism that obtains its nutrients from dead organic matter.

**sarcina** (plural: **sarcinae**) (1) A group of eight bacteria that remain in a packet after dividing. (2) When written as a genus, refers to gram-positive, anaerobic cocci.

**saturation** (1) The condition in which the active site on an enzyme is occupied by the substrate or product at all times. (2) In a fatty acid, having no double bonds.

**saxitoxin**   A neurotoxin produced by some dinoflagellates.

**scanned-probe microscopy**   Microscopic technique used to obtain images of molecular shapes, to characterize chemical properties, and to determine temperature variations within a specimen.

**scanning acoustic microscope (SAM)**   A microscope that uses high-frequency ultrasound waves to penetrate surfaces.

**scanning electron microscope (SEM)**   An electron microscope that provides three-dimensional views of the specimen magnified 1000–10,000×.

**scanning tunneling microscopy (STM)**   *See* scanned-probe microscopy.

**schizogony**   The process of multiple fission, in which one organism divides to produce many daughter cells.

**scientific nomenclature**   *See* binomial nomenclature.

**sclerotia**   The compact mass of hardened mycelia of the fungus *Claviceps purpurea* that fills infected rye flowers; produces the toxin ergot.

**scolex**   The head of a tapeworm, containing suckers and possibly hooks.

**secondary infection**   An infection caused by an opportunistic microbe after a primary infection has weakened the host's defenses.

**secondary metabolite**   A product of an industrial cell population produced after the microorganism has largely completed its period of rapid growth and is in a stationary phase of the growth cycle. *See also* primary metabolite.

**secondary response**   *See* memory response.

**secondary sewage treatment**   Biological degradation of the organic matter in wastewater following primary treatment.

**secretory vesicle**   A membrane-enclosed sac produced by the endoplasmic reticulum (ER); transports synthesized material into cytoplasm.

**selective medium**   A culture medium designed to suppress the growth of unwanted microorganisms and encourage the growth of desired ones.

**selective permeability**   The property of a plasma membrane to allow certain molecules and ions to move through the membrane while restricting others.

**selective toxicity**   The property of some antimicrobial agents to be toxic for a microorganism and nontoxic for the host.

**semiconservative replication**   The process of DNA replication in which each double-stranded DNA molecule contains one original strand and one new strand.

**sense codon**   A codon that codes for an amino acid.

**sense strand (+ strand)**   Viral RNA that can act as mRNA.

**sensitivity**   Percentage of positive samples correctly detected by a diagnostic test.

**sepsis**   The presence of a toxin or pathogenic organism in blood and tissue.

**septate hypha**   A hypha consisting of uninucleate cell-like units.

**septicemia**   The proliferation of pathogens in the blood, accompanied by fever; sometimes causes organ damage.

**septic shock**   A sudden drop in blood pressure induced by bacterial toxins.

**septum**   A cross-wall in a fungal hypha.

**serial dilution**   The process of diluting a sample several times.

**seroconversion**   A change in a person's response to an antigen in a serological test.

**serological testing**   Techniques for identifying a microorganism based on its reaction with antibodies.

**serology**   The branch of immunology that studies blood serum and antigen–antibody reactions in vitro.

**serotype**   *See* serovar.

**serovar**   A variation within a species; also called serotype.

**serum**   The liquid remaining after blood plasma is clotted; contains antibodies (immunoglobulins).

**sexual dimorphism**   The distinctly different appearance of adult male and female organisms.

**sexual spore**   A spore formed by sexual reproduction.

**Shiga toxin**   An exotoxin produced by *Shigella dysenteriae* and entero-hemorrhagic *E. coli*.

**shock**   Any life-threatening loss of blood pressure. *See also* septic shock.

**short tandem repeats (STRs)**   Repeating sequences of two to five nucleotides.

**shotgun sequencing**   A technique for determining the nucleotide sequence in an organism's genome.

**shuttle vector**   A plasmid that can exist in several different species; used in genetic engineering.

**siderophore**   Bacterial iron-binding proteins.

**sign**   A change due to a disease that a person can observe and measure.

**simple stain**   A method of staining microorganisms with a single basic dye.

**singlet oxygen**   Highly reactive molecular oxygen ($O_2^-$).

**siRNA**   Small interfering RNA; An intermediate in the RNAi process in which the long double-stranded RNA has been cut up into short (~21 nucleotides) double-stranded RNA.

**site-directed mutagenesis**   Techniques used to modify a gene in a specific location to produce the desired polypeptide.

**slide agglutination test**   A method of identifying an antigen by combining it with a specific antibody on a slide.

**slime layer**   A glycocalyx that is unorganized and loosely attached to the cell wall.

**sludge**   Solid matter obtained from sewage.

**smear**   A thin film of material containing microorganisms, spread over the surface of a slide.

**smooth ER**   Endoplasmic reticulum without ribosomes.

**snRNP**   Small nuclear ribonucleoprotein (pronounced "snurp"). Short RNA transcript plus protein that combines with pre-mRNA to remove introns and join exons together.

**solute**   A substance dissolved in another substance.

**solvent**   A dissolving medium.

**Southern blotting**   A technique that uses DNA probes to detect the presence of specific DNA in restriction fragments separated by electrophoresis.

**specialized transduction**   The process of transferring a piece of cell DNA adjacent to a prophage to another cell.

**species**   The most specific level in the taxonomic hierarchy. *See also* bacterial species; eukaryotic species; viral species.

**specific epithet**   The second or species name in a scientific binomial. *See also* species.

**specificity**   Percentage of false positive results given by a diagnostic test.

**spectrum of microbial activity**   The range of distinctly different types of microorganisms affected by an antimicrobial drug; a wide range is referred to as a broad spectrum of activity.

**spheroplast**   A gram-negative bacterium treated to damage the cell wall, resulting in a spherical cell.

**spicule**   One of two external structures on the male roundworm used to guide sperm.

**spike** A carbohydrate-protein complex that projects from the surface of certain viruses.

**spiral** *See* spirillum and spirochete.

**spirillum** (plural: **spirilla**) (1) A helical or corkscrew-shaped bacterium. (2) When written as a genus, refers to aerobic, helical bacteria with clumps of polar flagella.

**spirochete** A corkscrew-shaped bacterium with axial filaments.

**spontaneous generation** The idea that life could arise spontaneously from nonliving matter.

**spontaneous mutation** A mutation that occurs without a mutagen.

**sporadic disease** A disease that occurs occasionally in a population.

**sporangiophore** An aerial hypha supporting a sporangium.

**sporangiospore** An asexual fungal spore formed within a sporangium.

**sporangium** A sac containing one or more spores.

**spore** A reproductive structure formed by fungi and actinomycetes. *See also* endospore.

**sporogenesis** *See* sporulation.

**sporozoite** A trophozoite of *Plasmodium* found in mosquitoes, infective for humans.

**sporulation** The process of spore and endospore formation; also called sporogenesis.

**spread plate method** A plate count method in which inoculum is spread over the surface of a solid culture medium.

**staining** Colorizing a sample with a dye to view through a microscope or to visualize specific structures.

**standard precautions** Minimum precautions, such as handwashing, to prevent transmission of infection.

**staphylococci** (singular: **staphylococcus**) Cocci in a grapelike cluster or broad sheet.

**stationary phase** The period in a bacterial growth curve when the number of cells dividing equals the number dying.

**stem cell** An undifferentiated cell that gives rise to a variety of specialized cells.

**stereoisomers** Two molecules consisting of the same atoms, arranged in the same manner but differing in their relative positions; mirror images; also called D-isomer and L-isomer.

**sterile** Free of microorganisms.

**sterilization** The removal of all microorganisms, including endospores.

**steroid** A specific group of lipids, including cholesterol and hormones.

**stipe** A stemlike supporting structure of multicellular algae and basidiomycetes.

**storage vesicle** Organelles that form from the Golgi complex; contain proteins made in the rough ER and processed in the Golgi complex.

**strain** Genetically different cells within a clone. *See* serovar.

**streak plate method** A method of isolating a culture by spreading microorganisms over the surface of a solid culture medium.

**streptobacilli** (singular: **streptobacillus**) Rods that remain attached in chains after cell division.

**streptococci** (singular: **streptococcus**) (1) Cocci that remain attached in chains after cell division. (2) When written as a genus, refers to gram-positive, catalase-negative bacteria.

**streptolysin** A hemolytic enzyme, produced by streptococci.

**subacute disease** A disease with symptoms that are intermediate between acute and chronic.

**subclinical infection** An infection that does not cause a noticeable illness; also called inapparent infection.

**subcutaneous mycosis** A fungal infection of tissue beneath the skin.

**substrate** Any compound with which an enzyme reacts.

**substrate-level phosphorylation** The synthesis of ATP by direct transfer of a high-energy phosphate group from an intermediate metabolic compound to ADP.

**subunit vaccine** A vaccine consisting of an antigenetic fragment.

**sulfhydryl group** —SH.

**sulfur cycle** The various oxidation and reduction stages of sulfur in the environment, mostly due to the action of microorganisms.

**sulfur granule** *See* inclusion.

**superantigen** An antigen that activates many different T cells, thereby eliciting a large immune response.

**superbug** Bacterium resistant to a large number of antibiotics.

**superficial mycosis** A fungal infection localized in surface epidermal cells and along hair shafts.

**superinfection** The growth of a pathogen that has developed resistance to an antimicrobial drug being used; the growth of an opportunistic pathogen.

**superoxide dismutase (SOD)** An enzyme that destroys superoxide:

$$O_2^- + O_2^- + 2H^+ \rightarrow H_2O_2 + O_2$$

**superoxide radical** A toxic anion ($O_2^-$) with an unpaired electron.

**surface-active agent** Any compound that decreases the tension between molecules lying on the surface of a liquid; also called surfactant.

**surfactant** *See* surface-active agent.

**susceptibility** The lack of resistance to a disease.

**symbiosis** The living together of two different organisms or populations.

**symptom** A change in body function that is felt by a patient as a result of a disease.

**syncytium** A multinucleated giant cell resulting from certain viral infections.

**syndrome** A specific group of signs or symptoms that accompany a disease.

**synergism** The principle whereby the effectiveness of two drugs used simultaneously is greater than that of either drug used alone.

**synthesis reaction** A chemical reaction in which two or more atoms combine to form a new, larger molecule.

**synthetic drug** A chemotherapeutic agent that is prepared from chemicals in a laboratory.

**systematics** The science organizing groups of organisms into a hierarchy.

**systemic anaphylaxis** A hypersensitivity reaction causing vasodilation and resulting in shock; also called anaphylactic shock.

**systemic (generalized) infection** An infection throughout the body.

**systemic mycosis** A fungal infection in deep tissues.

**tachyzoite** A rapidly growing trophozoite form of a protozoan.

**T antigen** An antigen in the nucleus of a tumor cell.

**tapeworm** A flatworm belonging to the class Cestoda.

**taxa** Subdivisions used to classify organisms, e.g., domain, kingdom, phylum.

**taxis** Movement in response to an environmental stimulus.

**taxonomy**   The science of the classification of organisms.

**T lymphocyte (T cell)**   A type of lymphocyte, which develops from a stem cell processed in the thymus gland, that is responsible for cell-mediated immunity. *See also* CTL (cytotoxic T lymphocyte), T helper cells, T regulatory cells.

**TCRs (T cell receptors)**   Molecules on T cells that recognize antigens.

**T helper ($T_H$) cell**   A specialized T cell that often interacts with an antigen before B cells interact with the antigen.

**T regulatory ($T_{reg}$) cells**   Lymphocytes that appear to suppress other T cells.

**T-dependent antigen**   An antigen that will stimulate the formation of antibodies only with the assistance of T helper cells. *See also* T-independent antigen.

**teichoic acid**   A polysaccharide found in gram-positive cell walls.

**telomere**   Noncoding regions of DNA at the ends of eukaryotic chromosomes.

**teleomorph**   The sexual stage in the life cycle of a fungus; also refers to a fungus that produces both sexual and asexual spores.

**temperate phage**   A phage capable of lysogeny.

**temperature abuse**   Improper food storage at a temperature that allows bacteria to grow.

**terminator**   The site on a DNA strand at which transcription ends.

**tertiary sewage treatment**   A method of waste treatment that follows conventional secondary sewage treatment; nonbiodegradable pollutants and mineral nutrients are removed, usually by chemical or physical means.

**tetrad**   A group of four cocci.

**thallus**   The entire vegetative structure or body of a fungus, lichen, or alga.

**thermal death point (TDP)**   The temperature required to kill all the bacteria in a liquid culture in 10 minutes.

**thermal death time (TDT)**   The length of time required to kill all bacteria in a liquid culture at a given temperature.

**thermoduric**   Heat resistant.

**thermophile**   An organism whose optimum growth temperature is between 50°C and 60°C; a heat-loving microbe.

**thermophilic anaerobic spoilage**   Spoilage of canned foods due to the growth of thermophilic bacteria.

**thylakoid**   A chlorophyll-containing membrane in a chloroplast. A bacterial thylakoid is also known as a chromatophore.

**thymic selection**   Elimination of T cells that don't recognize self antigens (major histocompatibility complex).

**tincture**   A solution in aqueous alcohol.

**T-independent antigen**   An antigen that will stimulate the formation of antibodies without the assistance of T helper cells. *See also* T-dependent antigen.

**Ti plasmid**   A tumor-inducing plasmid that can be incorporated into a host plant chromosome; found in *Agrobacterium*.

**titer**   An estimate of the amount of antibodies or viruses in a solution; determined by serial dilution and expressed as the reciprocal of the dilution.

**TLR (Toll-like rceptor)**   Transmembrane protein of immune cells that recognizes pathogens and activates an immune response directed against those pathogens.

**total magnification**   The magnification of a microscopic specimen, determined by multiplying the ocular lens magnification by the objective lens magnification.

**toxemia**   The presence of toxins in the blood.

**toxigenicity**   The capacity of a microorganism to produce a toxin.

**toxin**   Any poisonous substance produced by a microorganism.

**toxoid**   An inactivated toxin.

**T plasmid**   An *Agrobacterium* plasmid carrying genes for tumor induction in plants.

**trace element**   A chemical element required in small amounts for growth.

**trans**   Hydrogen atoms on opposite side across a double bond in a fatty acid. *See* cis.

**transamination**   The transfer of an amino group from an amino acid to another organic acid.

**transcription**   The process of synthesizing RNA from a DNA template.

**transduction**   The transfer of DNA from one cell to another by a bacteriophage. *See also* generalized transduction; specialized transduction.

**transferrin**   One of several human iron-binding proteins that reduce iron available to a pathogen.

**transfer RNA (tRNA)**   The type of RNA molecule that brings amino acids to the ribosomal site where they are incorporated into proteins.

**transfer vesicle**   Membrane-bound sacs that move proteins from the Golgi complex to specific areas in the cell.

**transformation**   (1) The process in which genes are transferred from one bacterium to another as "naked" DNA in solution. (2) The changing of a normal cell into a cancerous cell.

**transient microbiota**   The microorganisms that are present in an animal for a short time without causing a disease.

**translation**   The use of mRNA as a template in the synthesis of protein.

**transmission-based precautions**   Precautions used to prevent transmission of infection from individuals with a known or suspected infection.

**transmission electron microscope (TEM)**   An electron microscope that provides high magnifications (10,000–100,000×) of thin sections of a specimen.

**transport media**   Media used to keep microorganisms alive between sample collection and laboratory testing; usually used for clinical samples.

**transport vesicle**   Membrane-bound sacs that move proteins from the rough ER to the Golgi complex.

**transporter protein**   A carrier protein in the plasma membrane.

**transposon**   A small piece of DNA that can move from one DNA molecule to another.

**trickling filter**   A method of secondary sewage treatment in which sewage is sprayed out of rotating arms onto a bed of rocks or similar materials, exposing the sewage to highly aerobic conditions and microorganisms.

**trophophase**   The period in the production curve of an industrial cell population in which the primary metabolites are formed; a period of rapid, logarithmic growth. *See also* idiophase.

**trophozoite**   The vegetative form of a protozoan.

**tuberculin skin test**   A skin test used to detect the presence of antibodies to *Mycobacterium tuberculosis*.

**tumor necrosis factor alpha (TNF-α)**   A polypeptide released by phagocytes in response to bacterial endotoxins.

**tumor-specific transplantation antigen (TSTA)**   A viral antigen on the surface of a transformed cell.

**turbidity**    The cloudiness of a suspension.

**turnover number**    The number of substrate molecules acted on per enzyme molecule per second.

**two-photon microscope**    A light microscope that uses fluorescent stains and long wavelength light.

**ubiquinone**    A low–molecular mass, nonprotein carrier in an electron transport chain; also called coenzyme Q.

**ultra-high-temperature (UHT) treatment**    A method of treating food with high temperatures (140–150°C) for very short times to make the food sterile so that it can be stored at room temperature.

**uncoating**    The separation of viral nucleic acid from its protein coat.

**undulating membrane**    A highly modified flagellum on some protozoa.

**universal precautions**    Procedures used to reduce transmission of microbes in health care settings and residential settings.

**unsaturated**    A fatty acid with one or more double bonds.

**use-dilution test**    A method of determining the effectiveness of a disinfectant using serial dilutions.

**vaccination**    The process of conferring immunity by administering a vaccine; also called immunization.

**vacuole**    An intracellular inclusion, in eukaryotic cells, surrounded by a plasma membrane; in prokaryotic cells, surrounded by a proteinaceous membrane.

**valence**    The combining capacity of an atom or a molecule.

**variolation**    An early method of vaccination using infected material from a patient.

**vasodilation**    Dilation or enlargement of blood vessels.

**VDRL test**    A rapid screening test to detect the presence of antibodies against *Treponema pallidum*. (VDRL stands for Venereal Disease Research Laboratory.)

**vector**    (1) A plasmid or virus used in genetic engineering to insert genes into a cell. (2) An arthropod that carries disease-causing organisms from one host to another.

**vegetative**    Referring to cells involved with obtaining nutrients, as opposed to reproduction.

**vehicle transmission**    The transmission of a pathogen by an inanimate reservoir.

**vertical gene transfer**    Transfer of genes from an organism or cell to its offspring.

**vesicle**    (1) A small serum-filled elevation of the skin. (2) Smooth oval bodies formed in plant roots by mycorrhizae.

**V factor**    NAD$^+$ or NADP$^+$.

**vibrio**    (1) A curved or comma-shaped bacterium. (2) When written as a genus (*Vibrio*), a gram-negative, motile, facultatively anaerobic curved rod.

**viral hemagglutination**    The ability of certain viruses to cause the clumping of red blood cells in vitro.

**viral hemagglutination inhibition test**    A neutralization test in which antibodies against particular viruses prevent the viruses from clumping red blood cells in vitro.

**viral species**    A group of viruses sharing the same genetic information and ecological niche.

**viremia**    The presence of viruses in the blood.

**virion**    A complete, fully developed viral particle.

**viroid**    Infectious RNA.

**virology**    The scientific study of viruses.

**virulence**    The degree of pathogenicity of a microorganism.

**virus**    A submicroscopic, parasitic, filterable agent consisting of a nucleic acid surrounded by a protein coat.

**volutin**    Stored inorganic phosphate in a prokaryotic cell. *See also* metachromatic granule.

**Western blotting**    A technique that uses antibodies to detect the presence of specific proteins separated by electrophoresis.

**whey**    The fluid portion of milk that separates from curd.

**xenobiotics**    Synthetic chemicals that are not readily degraded by microorganisms.

**xenodiagnosis**    A method of diagnosis based on exposing a parasite-free normal host to the parasite and then examining the host for parasites.

**xenograft (xenotransplantation product)**    A tissue graft from another species.

**X factor**    Substances from the heme fraction of blood hemoglobin.

**yeast**    Nonfilamentous, unicellular fungi.

**yeast infection**    Disease caused by growth of certain yeasts in a susceptible host.

**zone of inhibition**    The area of no bacterial growth around an antimicrobial agent in the disk-diffusion method.

**zoonosis**    A disease that occurs primarily in wild and domestic animals but can be transmitted to humans.

**zoospore**    An asexual algal spore; has two flagella.

**zygospore**    A sexual fungal spore characteristic of the zygomycetes.

**zygote**    A diploid cell produced by the fusion of two haploid gametes.

# Credits

**Chapter 10. Opener:** Steve Gschmeissner/Brand X Pictures/Science Photo Library/Getty Images. **ITC:** Budimir Jevtic/Shutterstock. **T10.1.1:** Eye of Science/Science Source. **T10.1.2:** Scimat/Science Source. **T10.1.3:** Dick Briggs, Center for Microscopy and Imaging, Smith College, Northampton, MA. **10.3:** Jerome Pickett Heaps/Science Source. **10.4a:** Georgette Douwma/Nature Picture Library. **10.4b:** Christine Case, Skyline College/Pearson. **10.4c:** Philippe Plailly/Science Source. **10.9:** Christine Case, Skyline College/Pearson. **10.10:** Bsip/Science Source. **10.11:** Christine Case, Skyline College/Pearson. **10.12a:** Sinclair Stammers/Science Source. **10.12b:** BSIP/Newscom. **10.13:** Mcs/Spl/Science Source. **10.14:** Chris Jones/University of Leeds/John Heritage. **10.15:** State Laboratories Division, Hawaii Department of Health. **10.18a:** Arno Massee/Science Source. **10.18d:** Jonathan Frye, USDA. **10.19:** Republished with permission of American Society for Microbiology, from Fluorescent In Situ Hybridization Allows Rapid Identification of Microorganisms in Blood Cultures, V. A. Kempf, K. Trebesius, I. B. Autenrieth, *Journal of Clinical Microbiology,* 38(2); pp. 830–838, © 2000. **CC:** Canada Communicable Disease Report (CCDR) 2005: Volume 31. Public Health Agency of Canada. **EM:** Christine Case, Skyline College/Pearson.

**Chapter 11. Opener:** Eye of Science/Science Source. **ITC:** ChaNaWiT/Shutterstock. **11.1:** USDA/APHIS Animal And Plant Health Inspection Service. **11.2b:** Biology Pics/Science Source. **11.3:** J. R. Factor/Science Source. **11.4:** Nobeastsofierce Science/Alamy. **11.6:** Science Source. **11.7:** Dr. Linda M. Stannard/University of Cape Town/Science Source. **11.8:** James Cavallini/Science Source. **11.9a:** A. Barry Dowsett/Science Source. **11.9b:** Biophoto Associates/Science Source. **11.10:** Alfred Pasieka/Photolibrary/Getty Images. **11.11b:** Heinrich Lünsdorf/Helmholtz Center for Infection Research, Germany. **11.12:** Eye of Science/Science Source. **11.13a:** L. Brent Selinger/Pearson Education. **11.13b:** Claire Ting/Science Source. **11.14:** M. I. Walker/Science Source. **11.15b:** Biomedical Imaging Unit, Southampton General Hospital/Science Source. **11.16:** Jenny Wang, Cheryl Jenkins, Richard I. Webb, and John A. Fuerst. Isolation of Gemmata-Like and Isosphaera-Like Planctomycete Bacteria from Soil and Freshwater. *Applied and Environmental Microbiology,* January 2002, p. 417–422, Vol. 68, No. 1, Fig. 1c. **11.17:** Chris Bjornberg/Science Source. **11.18a:** © 2002 National Academy of Sciences, U.S.A. **11.19:** Scimat/Science Source. **11.20:** Esther Angert, Cornell University. **11.21:** Neil Crickmore. **11.22:** Eye of Science/Science Source. **11.23:** Scimat/Science Source. **CC1:** Kate Curran/The Clinical Chemistry and Hematology Laboratory. **CC2:** Michael Abbey/Science Source. **11.24:** Bsip Sa/Alamy. **CC2:** Kate Curran/The Clinical Chemistry and Hematology Laboratory. **11.25b:** Dr. Jeremy Burgess/Science Source. **11.26:** David M. Phillips/Science Source. **11.27:** Courtesy of Karl O. Stetter. **EM:** Juergen Berger/Science Source. **11.28:** National Library of Medicine.

**Chapter 12. Opener:** E. Gueho/CNRI/Science Source. **ITC:** Avatar_023/Shutterstock. **12.1.1:** Steve Gschmeissner/Science Source. **12.1.2:** James Gathany/CDC. **12.1.3:** Eye of Science/Science Source. **12.1.4:** Sarah Spaulding/USGS. **12.1.5:** Biophoto Associates/Science Source. **12.3:** Christine Case, Skyline College/Pearson. **12.4:** Spl/Science Source. **12.5:** Christine Case, Skyline College/Pearson. **12.6a:** Bsip/Science Source. **12.6b:** Meredith Carlson/Biophoto Associates/Science Source. **12.6c:** David M. Phillips/Science Source. **12.6d:** David M. Phillips/Science Source. **12.6e:** Dr. Jeremy Burgess/Science Source. **12.7:** Christine Case, Skyline College/Pearson. **12.8:** Biophoto Associates/Science Source. **12.9.1:** Republished with permission of Taylor and Francis Group LLC Books, from Food Mycology: A Multifaceted Approach to Fungi and Food, Jan Dijksterhuis, Robert A. Samson, pp.101–117 © 2007. **12.10.1:** Biophoto Associates/Science Source. **12.10.2:** Christine Case, Skyline College/Pearson. **CC1.1:** Lucille K. Georg/CDC. **CC1.2:** Connie Nichols, Duke University Medical Center. **EM:** Steve Gschmeissner/Science Source. **12.11a:** Christine Case, Skyline College/Pearson. **12.12:** Christine Case, Skyline College/Pearson. **12.13a:** Andrew J. Martinez/Science Source. **12.14a:** Steve Gschmeissner/Science Source. **12.16.1:** Chad Bischoff. **12.16.2:** Noble Proctor/Science Source. **12.16.3:** Biophoto Associates/Science Source. **12.17:** M. I. Walker/Science Source. **12.18a:** Dr. Tony Brain/Science Source. **12.18b:** Wim Van Egmond/Science Source. **12.18c:** David M. Phillips/Science Source. **12.19b:** From *Archives of Medical Research,* Vol 17 Supplementary 1 by D. Mirelman, et al, Alteration of isoenzyme patterns of a cloned culture of non pathogenic Entamoeba histolytica upon changes in growth conditions, pp. 187–193, © Elsevier 1986. **12.21b:** Eric V. Grave/Science Source. **12.22:** Dave Pressland/Flpa/Science Source. **12.23:** Christine Case, Skyline College/Pearson. **12.24:** D. O. Rosenberry, 2001, *Malformed Frogs in Minnesota: An Update: USGS Water Fact Sheet* FS-043-01. Photo by David Hoppe. **12.25b:** Biophoto Associates/Science Source. **12.26.1:** Drbimages/Getty Images. **12.26.2:** Masalski Maksim/Shutterstock. **12.26.3:** Eric Isselee/Shutterstock. **CF12.A:** Centers for Disease Control and Prevention. **12.27:** Steve Gschmeissner/Science Source. **12.28.1:** M. B. Hildreth, M. D. Johnson, K. R. Kazacos, Echinococcus *Multilocularis: A Zoonosis of Increasing Concern in the*

*United States, Compendium on Continuing Education for the Practicing Veterinarian.* **12.28.2:** Science Stock Photography/Science Source. **12.28.3:** Maxim Kulko/Shutterstock. **12.28.4:** James Pierce/Shutterstock. **12.28.5:** Michaeljung/Shutterstock. **12.29a:** Steve Gschmeissner/Science Source. **12.30:** Nnehring/E+/Getty Images. **12.31a:** James Gathany/CDC. **12.31b:** Himagine/E+/Getty Images. **SQ1:** Lucille K. Georg/CDC. **SQ2.1:** Sedmi/Shutterstock. **SQ2.2:** Eric Isselee/Shutterstock. **SQ3:** Pascal Goetgheluck/Science Source.

**Chapter 13. Opener:** Hazel Appleton/Health Protection Agency Centre for Infections/Science Source. **ITC:** Li Chaoshu/Shutterstock. **EM:** Science Source. **13.2b:** James Cavallini/Science Source. **13.3b:** Hazel Appleton/Health Protection Agency for Infections/Science Source. **13.4b:** Nixxphotography/iStock / Getty Images. **13.5a:** Science Source. **13.5b:** Eye of Science/Science Source. **13.6:** Christine Case, Skyline College/Pearson. **13.9a:** Science Source. **13.9b:** Dr. E. Walker/Science Source. **13.14a:** Dr. Klaus Boller/Science Source. **13.14b:** Chris Bjornberg/Science Source. **13.16a:** C. Garon and J. Rose/CDC. **13.16b:** Dr. Linda M. Stannard/University of Cape Town/Science Source. **13.18a:** Ami Images/Science Source. **13.18b:** Dr. Linda M. Stannard/University of Cape Town/Science Source. **13.20b:** NIBSC/Science Source. **13.23:** Agricultural Research Service Honey Bree/USDA.

**Chapter 14. Opener:** Biomedical Imaging Unit, Southampton General Hospital/Science Source. **ITC:** Aijiro/Shutterstock. **EM:** PeopleImages/E+/Getty Images. **14.1a:** Juergen Berger/Science Source. **14.1b:** Spl/Science Source. **14.1c:** Stephanie Schuller/Science Source. **14.2a:** David Spears Frps Frms/Corbis Documentary/Getty Images. **14.2b:** Stephanie Schuller/Steven Lewis/Science Source. **14.2c:** Nibsc/Science Source. **14.6a:** Monkey Business Images/Shutterstock. **14.6b:** David Joel/Photographer's Choice/Getty Images. **14.6c:** George Doyle/Stockbyte/Getty Images. **14.6d:** Andrew Davidhazy/Photo Arts and Sciences at Rochester Institute of Technology. **14.7a:** Marlas111/iStock/Getty Images. **14.7b:** Alfaguarilla/Shutterstock. **14.7c:** Slawomir Fajer/Shutterstock. **14.8:** Harald Theissen/ImageBroker/Alamy. **CF14.A:** Christine Case, Skyline College/Pearson. **CF14.B:** Bork/Shutterstock.

**Chapter 15. Opener:** Eye of Science/Science Source. **ITC:** Fotoluminate LLC/Shutterstock. **CC1:** National Eye Institute. **15.1b:** Professor Pietro M. Motta et al/Science Source. **15.1c:** Science Source. **EM:** Ami Images/Science Source. **15.2:** Science Source. **15.4.1:** Dr. Gary Gaugler/Science Source. **15.4.2:** Janice Haney Carr/CDC. **15.7a:** Frederick A. Murphy, Veterinary Medicine Dean's Office, University of California, Davis. **15.7b:** Diana Hardie, University of Cape Town Medical School, South Africa. **15.8:** Biophoto Associates/Science Source. **CC2.1:** Janice Carr/CDC. **CC2.2:** Jason D. Pimentel, MD, FRCPA. **15.9.1:** C. S. Goldsmith and A. Balish/CDC. **15.9.2:** Biophoto Associates/Science Source. **15.9.3:** Dr. Cecil H. Fox/Science Source.

**Chapter 16. Opener:** SPL/Science Source. **ITC:** Vilevi/Shutterstock. **BP1:** PeopleImages/iStock/Getty Images. **BP2:** NIBSC/Science Source. **16.1:** Ed Reschke/Photolibrary/Getty Images. **16.3:** Anatomical Travelogue/Science Source. **EM:** Fotografixx/iStock/Getty Images. **T16.1.1:** Jarun Ontakrai/Shutterstock. **T16.1.2:** Biophoto Associates/Science Source. **T16.1.3:** Alvin Telser/Science Source. **T16.1.4.1:** Biophoto Associates/Science Source. **T16.1.4.2:** Jarun011/iStock/Getty Images. **T16.1.5:** Republished with permission of American Society of Hematology, from *Manipulating Dendritic Cell Biology for the Active Immunotherapy of Cancer,* David W. O'Neill, Sylvia Adams, Nina Bhardwaj, 104(8), 2004. **T16.1.6:** Biophoto Associates/Science Source. **T16.1.7:** Alvin Telser/Science Source. **T16.1.8:** Dr. Mae Melvin/CDC. **16.7:** Spl/Science Source. **16.8:** Eye of Science/Science Source. **16.11:** Republished with permission of Rockefeller University Press, from Cytolysis caused by complement via the classical pathway, RD Schreiber, DC Morrison, ER Podack and HJ Muller-Eberhard, *Journal of Experimental Medicine.* 1979; 149:870–82.

**Chapter 17. Opener:** Steve Gschmeissner/Science Source. **ITC:** Monkey Business Images/Shutterstock. **CC1:** Billie Ruth Bird/CDC. **17.4c:** Courtesy of Dr. Zhifeng Shao, Shanghai Jiao Tong University. **17.9a:** Spl/Science Source. **17.10:** Dr. Marion Schneider, Sektion Experimentelle Anästhesiologie. **17.11:** A. Barry Dowsett/Camr/Science Source. **EM:** Eye of Science/Science Source. **17.15:** Dr. Gopal Murti/Science Source.

**Chapter 18. Opener:** Gary D. Gaugler/Science Source. **ITC:** Creativa Images/Shutterstock. **EM:** Scimat/Science Source. **18.4b:** Christine Case, Skyline College/Pearson. **18.6a:** Amit Roy, University of Kentucky. **18.11a:** Michael Abbey/Science Source. **18.11b:** Scott Camazine/Alamy. **18.13:** Science Source. **BP1:** Centers for Disease Control and Prevention. **BP2:** VEM/BSIP SA/Alamy Stock Photo.

**Chapter 19. Opener:** Steve Gschmeissner/Science Source. **ITC:** Alice-photo/Shutterstock. **19.1b:** Lennart Nilsson/Albert Bonniers Forlag AB/TT

# Trademark Attributions

**Chapter 2** NutraSweet®

**Chapter 4** MacroBID®

**Chapter 6** OxyPlate™

**Chapter 7** Lysol®; pHisoHex®; Betadine®; Clorox®; Chlor-Floc®; Purell®; Germ-X®; Zephiran®; Surfacine®; Xgel; Cēpacol®; Cidex®; Bioquell

**Chapter 9** Glybera®; Flavr Savr™; GloFish®

**Chapter 13** Imylgic™

**Chapter 16** Intron® A; Betaseron®; Actimmune®

**Chapter 18** Nanopatch™; Herceptin®; Rotarix®; Rota Teq®

**Chapter 19** Xolair®; Singulair®; RhoGAM®; Yulex®; Herceptin®; Adcetris®; OraQuick®; APTIMA®; Truvada®; Atripla®

**Chapter 20** Augmentin®; Primaxin®; MacroBID®; Synercid®; Cipro®; Relenza®; Tamiflu®; Truvada®

**Chapter 21** CLEARLight®; Smoothbeam®; ThermaClear®; Accutane®; Roaccutane®; LiceMD®

**Chapter 22** BabyBIG®; Botox®

**Chapter 25** Lomotil®; Pepto-Bismol®

**Chapter 26** OraQuick®; Uritest

**Chapter 27** Vent$_R$®; Deep Vent$_R$™

**Chapter 28** NutraSweet®

# Index

# Taxonomic Guide to Diseases

## Bacteria and the Diseases They Cause

### Alphaproteobacteria

| | | |
|---|---|---|
| Anaplasmosis | *Anaplasma phagocytophilum* | p. 666 |
| Brucellosis | *Brucella* spp. | p. 656 |
| Cat-scratch disease | *Bartonella henselae* | p. 660 |
| Ehrlichiosis | *Ehrlichia* spp. | p. 666 |
| Endemic (murine) typhus | *Rickettsia typhi* | p. 667 |
| Rocky Mountain spotted fever | *R. rickettsii* | pp. 667–668 |
| Typhus fever | *R. prowazekii* | pp. 666–667 |

### Betaproteobacteria

| | | |
|---|---|---|
| Gonorrhea | *Neisseria gonorrhoeae* | pp. 766–767, 770 |
| Melioidosis | *Burkholderia pseudomallei* | p. 707 |
| Meningitis | *N. meningitidis* | pp. 622–623 |
| Healthcare-associated infections | *Burkholderia* spp. | pp. 300–301, 439 |
| Ophthalmia neonatorum | *N. gonorrhoeae* | p. 612 |
| Pelvic inflammatory disease | *N. gonorrhoeae* | pp. 771–772 |
| Rat-bite fever | *Spirillum minor* | pp. 660–661 |
| Pertussis (Whooping cough) | *Bordetella pertussis* | pp. 695, 698 |

### Gammaproteobacteria

| | | |
|---|---|---|
| Animal bites | *Pasteurella multocida* | p. 660 |
| Chancroid | *Haemophilus ducreyi* | p. 775 |
| Cholera | *Vibrio cholerae* | pp. 732–733 |
| Conjunctivitis | *H. influenzae* | p. 612 |
| Cystitis | *Escherichia coli* | p. 764 |
| Dermatitis | *P. aeruginosa* | pp. 598, 600 |
| Epiglottitis | *H. influenzae* | p. 690 |
| Gastroenteritis | *E. coli* | pp. 733, 736 |
| Gastroenteritis | *V. parahaemolyticus* | p. 733 |
| Gastroenteritis | *Yersinia enterocolitica* | p. 737 |
| Gastroenteritis | *Y. pseudotuberculosis* | p. 737 |
| Legionellosis | *Legionella pneumophila* | pp. 704–705 |
| Meningitis | *H. influenzae* | p. 622 |
| Otitis externa | *P. aeruginosa* | pp. 598, 600 |
| Otitis media | *H. influenzae* | p. 693 |
| Otitis media | *Moraxella catarrhalis* | p. 693 |
| Plague | *Y. pestis* | pp. 661–664 |
| Pneumonia | *H. influenzae* | p. 703 |
| Pneumonia | *Klebsiella pneumoniae* | p. 304 |
| Pyelonephritis | *E. coli* | p. 764 |
| Q fever | *Coxiella burnetii* | pp. 706–707 |
| Salmonellosis | *Salmonella enterica* | pp. 729–731, 811 |
| Septicemia | Gram-negative bacteria | pp. 653–654 |
| Shigellosis (Bacillary dystentery) | *Shigella* spp. | pp. 728–729 |
| Tularemia | *Francisella tularensis* | pp. 655–656 |
| Typhoid fever | *S. enterica* Typhi | p. 732 |

### Epsilonproteobacteria

| | | |
|---|---|---|
| Campylobacteriosis | *Campylobacter jejuni* | p. 737 |
| Gastritis, peptic ulcers | *Helicobacter pylori* | p. 737 |

# Taxonomic Guide to Diseases (continued)

## Clostridia

| | | |
|---|---|---|
| Botulism | *Clostridium botulinum* | pp. 626–629 |
| Gangrene | *C. perfringens* | pp. 659–660 |
| Gastroenteritis | *C. difficile* | pp. 738–739 |
| Gastroenteritis | *C. perfringens* | p. 738 |
| Tetanus | *C. tetani* | pp. 625–626 |

## Mollicutes

| | | |
|---|---|---|
| Pneumonia | *Mycoplasma pneumoniae* | pp. 703–704 |
| Urethritis | *Mycoplasma, Ureaplasma* | pp. 770–771 |

## Bacilli

| | | |
|---|---|---|
| Anthrax | *Bacillus anthracis* | pp. 656–658 |
| Bacterial endocarditis | *Staphylococcus aureus* | pp. 654–655 |
| Cystitis | *S. saprophyticus* | p. 764 |
| Dental caries | *Streptococcus mutans* | pp. 724–726 |
| Endocarditis | Alpha-hemolytic streptococci | pp. 654–655 |
| Erysipelas | *Streptococcus pyogenes* | p. 597 |
| Folliculitis | *Staphylococcus aureus* | p. 594 |
| Food poisoning | *Staphylococcus aureus* | p. 728 |
| Gastroenteritis | *B. cereus* | p. 739 |
| Impetigo | *S. aureus* | p. 594 |
| Listeriosis | *Listeria monocytogenes* | pp. 623–625 |
| Meningitis | *Streptococcus agalactiae* | p. 623 |
| Meningitis | *Streptococcus pneumoniae* | p. 623 |
| MRSA infections | *Staphylococcus aureus* | pp. 18, 417, 600 |
| Necrotizing fasciitis | *Streptococcus pyogenes* | p. 597 |
| Otitis media | *S. pneumoniae* | p. 693 |
| Pneumonia | *S. pneumoniae* | p. 703 |
| Puerperal sepsis | *S. pyogenes* | p. 654 |
| Rheumatic fever | *S. pyogenes* | p. 655 |
| Scalded skin syndrome | *Staphylococcus aureus* | p. 595 |
| Scarlet fever | *Streptococcus pyogenes* | p. 692 |
| Sepsis | *Enterococcus* spp. | p. 654 |
| Sepsis | *Streptococcus agalactiae* | p. 654 |
| Strep throat | *S. pyogenes* | pp. 691–692 |
| Toxic shock syndrome | *Staphylococcus aureus* | p. 595 |
| Toxic shock syndrome | *Streptococcus pyogenes* | p. 597 |

## Actinobacteria

| | | |
|---|---|---|
| Abscess | *Mycobacterium abscessus* | p. 197 |
| Acne | *Cutibacterium (Propionibacterium) acnes* | pp. 601–602 |
| Buruli ulcer | *M. ulcerans* | pp. 600–601 |
| Diphtheria | *Corynebacterium diphtheriae* | pp. 692–693 |
| Leprosy | *M. leprae* | pp. 629–630 |
| Mycetoma | *Nocardia asteroides* | p. 317 |
| Tuberculosis | *M. tuberculosis* | pp. 698–701 |
| Tuberculosis | *M. bovis* | pp. 141, 698 |
| Vaginosis | *Gardnerella vaginalis* | p. 776 |

## Chlamydiae

| | | |
|---|---|---|
| Inclusion conjunctivitis | *Chlamydia trachomatis* | p. 612 |
| Lymphogranuloma venereum | *C. trachomatis* | p. 775 |
| Pelvic inflammatory disease | *C. trachomatis* | pp. 771–772 |
| Pneumonia | *Chlamydophila pneumoniae* | p. 706 |
| Psittacosis | *C. psittaci* | pp. 705–706 |
| Trachoma | *Chlamydia trachomatis* | p. 612 |
| Urethritis | *C. trachomatis* | p. 770 |

# Taxonomic Guide to Diseases (continued)

## Spirochetes

| | | |
|---|---|---|
| Leptospirosis | *Leptospira interrogans* | pp. 764–765 |
| Lyme disease | *Borrelia burgdorferi* | pp. 664–666 |
| Relapsing fever | *Borrelia* spp. | p. 664 |
| Syphilis | *Treponema pallidum* | pp. 772–775 |

## Bacteroidetes

| | | |
|---|---|---|
| Acute necrotizing gingivitis | *Prevotella intermedia* | p. 726 |
| Periodontal disease | *Porphyromonas* spp. | p. 726 |
| Septic shock | *Capnocytophaga canimorsus* | p. 482 |

## Fusobacteria

| | | |
|---|---|---|
| Rat-bite fever | *Streptobacillus moniliformis* | pp. 660–661 |

# Fungi and the Diseases They Cause

## Zygomycetes

| | | |
|---|---|---|
| Opportunistic infections | *Mucor, Rhizopus* | p. 715 |

## Microsporidia

| | | |
|---|---|---|
| Opportunistic infections | *Encephalitozoon intestinalis* | p. 330 |

## Ascomycetes

| | | |
|---|---|---|
| Aspergillosis | *Aspergillus fumigatus* | pp. 715, 746 |
| Blastomycosis | *Blastomyces dermatitidis* | pp. 714–715 |
| Candidiasis | *Candida albicans* | pp. 608–609, 779–780 |
| Coccidioidomycosis | *Coccidioides immitis* | pp. 712–713 |
| Ergot poisoning | *Claviceps purpurea* | p. 746 |
| Histoplasmosis | *Histoplasma capsulatum* | pp. 711–712 |
| *Pneumocystis* pneumonia | *Pneumocystis jirovecii* | pp. 713–714 |
| Ringworm, athlete's foot | *Microsporum, Trichophyton* | pp. 607–608 |
| Sporotrichosis | *Sporothrix schenckii* | p. 608 |

## Basidiomycetes

| | | |
|---|---|---|
| Dandruff | *Malassezia furfur* | p. 592 |
| Meningitis | *Cryptococcus* spp. | p. 638 |
| Mycotoxins | | pp. 438–439 |

# Protozoa and the Diseases They Cause

## Diplomonads

| | | |
|---|---|---|
| Giardiasis | *G. intestinalis* | pp. 747, 749 |

## Parabasilids

| | | |
|---|---|---|
| Trichomoniasis | *Trichomonas vaginalis* | p. 780 |

## Euglenozoa

| | | |
|---|---|---|
| African trypanosomiasis | *Trypanosoma brucei* | pp. 639–640 |
| Chagas disease | *T. cruzi* | pp. 675–676 |
| Leishmaniasis | *Leishmania* spp. | pp. 679–680 |
| Meningoencephalitis | *Naegleria fowleri* | pp. 640, 642 |

## Apicomplexa

| | | |
|---|---|---|
| Babesiosis | *Babesia microti* | p. 680 |
| Cryptosporidiosis | *Cryptosporidium* spp. | p. 749 |
| Cyclosporiasis | *Cyclospora cayetanensis* | p. 750 |
| Malaria | *Plasmodium* spp. | pp. 677–679 |
| Toxoplasmosis | *Toxoplasma gondii* | pp. 676–677 |

# Taxonomic Guide to Diseases (continued)

### Amoebozoa

| | | |
|---|---|---|
| Amebic dysentery | *Entamoeba histolytica* | p. 750 |
| Encephalitis | *Acanthamoeba* spp., *Balamuthia mandrillaris* | p. 642 |
| Keratitis | *Acanthamoeba* spp. | p. 613 |

### Ciliates

| | | |
|---|---|---|
| Dysentery | *Balantidium coli* | p. 346 |

## Helminths and the Diseases They Cause

### Platyhelminths

| | | |
|---|---|---|
| Hydatid disease | *Echinococcus granulosus* | pp. 751–752 |
| Lung fluke | *Paragonimus* spp. | pp. 349–350 |
| Schistosomiasis | *Schistosoma* spp. | pp. 681–683 |
| Tapeworm infections | *Taenia* spp. | p. 751 |

### Nematodes

| | | |
|---|---|---|
| Ascariasis | *Ascaris lumbricoides* | p. 753 |
| Hookworms | *Necator americanus*, *Ancylostoma* | pp. 752–753 |
| Pinworms | *Enterobius vermicularis* | p. 752 |
| Trichinellosis | *Trichinella spiralis* | pp. 754–755 |
| Whipworm | *Trichuris trichiura* | pp. 753–754 |

## Algae and the Diseases They Cause

### Diatoms and Dinoflagellates

| | | |
|---|---|---|
| Paralytic shellfish poisoning | *Alexandrium, Pfiesteria* | p. 340, Figure 27.10 |
| Domoic acid toxicosis | *Diatoms* | p. 339 |

### Oomycota

| | | |
|---|---|---|
| Potato blight | *Phytophthora* | p. 340 |

## Arthropods and the Diseases They Cause

| | | |
|---|---|---|
| Pediculosis | *Pediculus humanus* | p. 610 |
| Scabies | *Sarcoptes scabiei* | p. 609 |

## Viruses and the Diseases They Cause

### DNA Viruses

| | | |
|---|---|---|
| Burkitt's lymphoma | Herpesvirus | pp. 668–669 |
| Chickenpox | Herpesvirus | pp. 603–605 |
| Cold sores | Herpesvirus | p. 605 |
| Cytomegalovirus infections | Herpesvirus | p. 670 |
| Fifth disease | Parvovirus | p. 607 |
| Genital herpes | Herpesvirus | pp. 776–777 |
| Genital warts | Papovavirus | p. 778 |
| Hepatitis B | Hepadnavirus | pp. 741–744 |
| Infectious mononucleosis | Herpesvirus | p. 669 |
| Keratitis | Herpesvirus | p. 613 |
| Monkeypox | Poxvirus | p. 603 |
| Neonatal herpes | Herpesvirus | pp. 777–778 |
| Roseola | Herpesvirus | p. 607 |
| Shingles | Herpesvirus | pp. 603–605 |
| Smallpox | Poxvirus | pp. 602–603 |
| Warts | Papovavirus | p. 602 |